Encyclopedia of Caves and Karst Science

Board of Advisers

Andrew Chamberlain, Department of Archaeology and Prehistory, University of Sheffield
Emily Davis, Speleobooks, New York
Derek Ford, School of Geography and Geology, McMaster University
David Gillieson, School of Tropical Environment Studies & Geography, James Cook University
William Halliday, Nashville, Tennessee
Elery Hamilton-Smith, Chair IUCN/WCPA Working Group on Cave and Karst Protection
Alexander Klimchouk, Institute of Geological Sciences, National Academy of Science, Ukraine
David Lowe, British Geological Survey
Art Palmer, Earth Science Department, State University of New York
Trevor Shaw, Karst Research Institute, Postojna, Slovenia
Boris Sket, Department of Biology, University of Ljubljana, Slovenia
Tony Waltham, Department of Civil Engineering, Nottingham Trent University
Paul Williams, Department of Geography, University of Auckland
Paul Wood, Department of Geography, Loughborough University

Encyclopedia of Caves and Karst Science

John Gunn, Editor

Fitzroy Dearborn
An imprint of the Taylor and Francis Group
New York London

Published in 2004 by
Fitzroy Dearborn
An Imprint of the Taylor and Francis Group
29 West 35 Street
New York, NY 10001–2299
www.routledge-ny.com

Published in Great Britain by
Fitzroy Dearborn
An Imprint of the Taylor and Francis Group
11 New Fetter Lane
London EC4P 4EE
www.routledge.co.uk

Copyright © 2004 by Taylor & Francis Books, Inc.

All rights reserved. No part of this book may be reprinted or reproduced or utilized in any form or by any electronic, mechanical, or other means, now known or hereafter invented, including any photocopying and recording, or in any information storage or retrieval system, without permission in writing from the publisher.

10 9 8 7 6 5 4 3 2 1

Printed on acid-free, 250-year-life paper
Manufactured in the United States of America

Front cover illustrations:
Looking out of the 100 m deep "Lost World" entrance to Mangapu Cave, Waitomo, New Zealand. (Photo by John Gunn)
Stalactites and stalagmites in "Castle Grotto", Hollow Hill Cave, Waitomo, New Zealand. (Photo by John Gunn)
A fine polygonal karst of dolines each about 100 m across in the gypsum karst of Sivas, Turkey. (Photo by Tony Waltham)
Camel cricket (*Ceuthophilus cunicularis*), a trogloxene in central Texas caves. (Photo by Steve Taylor)

Back cover illustrations:
Solution pocket developed along a joint in the ceiling of Gruta de Torrinha, Brazil. (Photo by John Gunn)
Classic hand silhouettes in the cave of Ujung Bulo in Sulawesi, Indonesia. (Photo by Tony Waltham)
The troglomorphic crab *Cancrocaeca xenomorpha*. (Photo by Didier Rigal)
Tower karst near Guilin, China. (Photo by Tony Waltham)

Copyright in these and individual photographs in the text is with the photographer, as noted in the caption.

Library of Congress Cataloging-in-Publication Data

Encyclopedia of caves and karst science / edited by John Gunn.
        p. cm.
Includes bibliographical references and index.
   ISBN 1-57958-399-7   (hardback : alk. paper)
 1. Caves—Encyclopedias. 2. Karst—Encyclopedias.   I. Gunn, John. II. Title.
  GB601.E535 2003
  551.44′7′03—dc21

2003006469

Contents

Board of Advisers   ii

Editor's Introduction   vii

Alphabetical List of Entries   xi

Thematic List of Entries   xv

Encyclopedia of Caves and Karst Science   1

Notes on Contributors   789

Index   809

Editor's Introduction

This is the first encyclopedia of Caves and Karst Science and provides a unique, comprehensive, and authoritative reference source that can be used both by subject-specialists who wish to obtain information from outside of their immediate area of knowledge and by non-specialists who wish to gain an understanding of the diverse and multi-disciplinary nature of caves and karst science. It will also be useful to cavers who wish to learn more about the environments in which they undertake their sport and to conservationists, engineers, planners, and others who are charged with developing and managing in a sustainable manner complex karst environments. The 351 entries were selected by a multi-disciplinary Advisory Board of leading scholars, all of whom are cavers. The entries cover a wide range of topics and each entry also includes both references and further reading to enable deeper study. While not intended as an atlas, there is a wide geographical coverage of all scientifically important karst areas, the level of detail (continent, country, region, or individual site or cave) reflecting the Advisory Board's opinions as to the importance of the locality. It is the first encyclopedia to cover all the disciplines involved in cave and karst science—archaeology, biology, chemistry, ecology, geology, geomorphology, history, hydrology, paleontology, and physics as well as exploration, survey, photography, literature, and art. The resources found in caves and in karst areas are outlined, including the underground water that supplies around a quarter of the world's population. Caves and karst environments are fragile and special places so there is appropriate consideration of conservation and management, including protected areas. Contributors are all leading authorities in their area and all entries were subject to review by the Editor, members of the Advisory Board, or other subject specialists.

The term "cave" is commonly applied to natural openings, usually in rocks, that are large enough to permit entry by humans. The term is also sometimes applied to openings constructed by humans but this encyclopedia is confined to natural caves. The majority of these form parts of a wider landscape known as karst, defined by Derek Ford in this volume as "terrain with distinctive hydrology and landforms arising from the combination of high rock solubility and well-developed solution channel (secondary) porosity underground". As well as distinctive landforms and drainage, karst areas have distinctive biological attributes and provide a wide range of resources, together with particular problems for humans who wish to exploit them. Caves are natural museums, preserving important archaeological evidence as well as sediments and deposits that provide clues to past environments. They also present many challenges for explorers. Although several thousand kilometres of cave have been explored, it is certain that many passages have yet to be entered and these represent the only truly unknown parts of the Earth outside of the oceans. Each year several well-equipped teams undertake major expeditions to discover new cave but it remains the case that individuals and groups working steadily on smaller projects also succeed in becoming the first humans ever to set foot in a particular passage. While the thrill of discovery drives many, caving has also become a popular recreational activity in many countries and many thousands more enjoy visits to commercially developed tourist caves.

To discover new cave it is usually necessary to have an understanding of the scientific contexts in which caves are formed while the discovery of new caves provides opportunities for new scientific discoveries. This synergy between science and exploration is one of the factors that makes the study of caves and karst so exciting. Caves are widely distributed, from within the Arctic Circle to the Antarctic and from below sea level to altitudes of several thousand metres. Karst is also widespread and it has been estimated that about a quarter of the world's population draw their water from karst rocks. Cave and karst science is multi-disciplinary, being undertaken by archaeologists, biologists, chemists, ecologists, geologists, geographers, historians, hydrologists, and physicists. It is also applied as conservationists, consultants, engineers, environmental managers, and planners require an understanding of the special characteristics of karst in order to manage and sustainably develop the many resources of karst areas.

Choosing the entries and authors

Given the above it is perhaps surprising that this is the first encyclopedia devoted to Caves and Karst Science. As such it seeks to describe all of the world's important karst areas and the most important caves. However, it is not an atlas and coverage is not, and could not have been, exhaustive. Instead, an initial list of sites was drawn up with the assistance of an appointed board of advisers who have very wide subject expertise and caving experience (see page ii and Notes on Contributors). The list was then widely publicized and suggestions for additional / alternative sites sought from the cave and karst community before a final list was agreed. This comprises 100 entries on the world's most significant cave and karst sites, and regional discussions of the world's largest areas of caves and karst. In addition to assisting in the choice of important sites and areas the Advisory Board helped to draw up a list of topical entries considered to be of primary importance to their particular branch of science. Again there was wide discussion of the lists both between the Advisory Board and also via the encyclopedia web site.

Following the consultative process a final list was produced broken down by themes, although some entries fall into more than one field:

Archaeology, Art in Caves and Paleontology: 29 entries, from dating methods to major sites, and regional discussions of art in caves

Biospeleology: 78 entries on ecology and ecological processes, habitats, cave ecology & man, subterranean fauna, subterranean biodiversity, regional faunas, the world's richest cave faunas, evolution of subterranean fauna, and important subterranean taxa

Caves and Caving: 23 entries on cave media, caving, and the uses of caves

Cave and Karst Regions: 75 entries on the world's largest and most important areas of caves and karst together with 28 entries on individual caves and cave systems

Conservation and Management: 19 entries on topics such as environmental impact assessment, groundwater pollution, and tourist caves

Geoscience: 78 entries on caves and speleogenesis, climate of caves, deposits in caves, karst settings and landforms, processes and techniques, and pseudokarst

History: 27 entries on key events and personalities in the history of cave and karst science together with the history of cave exploration in particular regions of the world

Resources and Development: 22 entries on the resources of karst regions and some of the key problems in their exploitation

The next stage was to allocate a word limit for each entry and to seek suitable authors. Most entries were planned at 1000 words, although it was recognized that some would need to be longer, up to a maximum of 4000 words. The Advisory Board suggested some possible authors who were invited individually. Other authors were approached in more general terms and invited to offer to write entries that remained unassigned. The objective was to include the leading authorities on each topic from a range of different countries. All entries were read by the Editor and by selected members of the Advisory Board or other reviewers.

The finished work has a total of 351 entries on a wide range of topics. A thematic list, an extensive index, and cross-references are provided to help readers explore themes systematically. All the entries have bibliographies and / or suggestions for further reading, pointing the reader toward original research and textbooks that augment the overview approach to which an encyclopedia is inevitably limited. Most entries have either line diagrams expanding on explanations given in the text, or photographs illustrating type examples. There is also a section of colour pages illustrating the range of themes.

Entries and structure

The 351 entries appear in alphabetical order and are of several kinds (for the complete list of entries see page xi). Although each entry is self-contained, the links between entries can be explored in a number of ways. The Thematic List on page xv groups the entries within broad and more specific categories and provides a useful summary of related entries. Almost all of the entries have "See Also" links, both within the text and at the end of the entry, so the reader is encouraged to browse outwards from a starting node. Finally the Index provides a detailed listing of topics, organisms, and sites or countries that do not have their own entry but are discussed within the context of broader entries.

One aspect which came as a surprise to the Editor was the difficulty of deciding the exact titles of several entries and hence the point at which they would appear in the alphabetic listing. Following much discussion it was decided that all organisms would appear alphabetically by scientific name (hence "Chiroptera" not "Bats", and "Pisces" not "Fish") and that where the common name for a site commences with a local word for a cave or geographical feature then the entry should be under the location (hence "Draenen, Ogof Draenen, Wales" and "Encantado, Sistema del Rio, Puerto Rico"). We recognize that this may initially cause some confusion but if you cannot find an entry on a topic you expected to find you should be able to use the Thematic List or Index to locate the title of the entry that contains the topic you are looking for. Thus, for the examples above the Index will contain cross references for Bats, Fish, Ogof Draenen, Rio Encantado, and Sistema del Rio Encantado.

Acknowledgements

My thanks are due to all the contributors who provided excellent material, adhered to length guidelines and submitted their work on time. I also thank those who caused me angst by submitting overlong scripts that needed serious pruning and those from whom the script was finally extricated months after the deadline. You know who you are . . .

Particular thanks are due to fellow cavers who supplied photographs for no reward other than seeing their work in print. Commitment to the wider good of caving is one of the marks of our sport.

The Editorial Advisory Board played a vitally important role in helping to decide what entries should be included and who should write them, as well as in reviewing the submitted essays. Special thanks are due to Paul Wood who reviewed all of the biology entries, many of which were written by scientists who do not have English as their mother tongue. Also to Tony Waltham who reviewed all of the entries on individual cave systems and on the world's largest and most important areas of caves and karst. Tony also edited many of the maps and photographs to ensure consistency and provided new maps and many photographs for entries where the author was unable to do so. As we approached the production phase Tony also provided a great deal of constructive advice on the choice of illustrative material, particularly in the colour section.

From start to finish the project has been overseen in an incredibly efficient manner by Gillian Lindsey, first at Fitzroy Dearborn and then at Routledge in New York. Without her commitment and drive the work may never have come to fruition and on behalf of myself and the members of the Advisory Board I offer her our very grateful thanks.

Special thanks for personal support during the editorial process are due to Ernest, Doreen, Cathy, Eilíse, and Owain Gunn, to Paul Somers and Lian, caver of the future, and to Sarah Davies.

Alphabetical List of Entries

A
Accidents and Rescue
Adaptation: Behavioural
Adaptation: Eyes
Adaptation: Genetics
Adaptation: Morphological (External)
Adaptation: Morphological (Internal)
Adaptation: Physiological
Africa, North
Africa, South: Archaeological Caves
Africa, Sub-Saharan
Africa: Biospeleology
Aggtelek and Slovak Karst, Hungary-Slovakia
Aggtelek Caves, Hungary-Slovakia: Archaeology
Akiyoshi-dai Karst and Caves, Japan
Alpine Karst
Altamira Cave, Spain: Archaeology
America, Central
America, Central and Caribbean Islands: Biospeleology
America, Central: Archaeological Caves
America, North: Archaeological Caves
America, North: Biospeleology
America, North: History
America, South
America, South: Biospeleology
America, South: History
Amphibia
Amphibia: *Proteus*
Anchialine Habitats
Annelida
Antarctica
Appalachian Mountains, United States
Arachnida
Arachnida: Aranae (Spiders)
Arachnida: Acari (Mites and Ticks)
Arachnida: Minor Groups
Archaeologists
Archaeology of Caves: History
Ardèche Caves, France: Archaeology

Art in Caves
Art in Caves: History
Art Showing Caves
Art: Cave Art in Australasia
Art: Cave Art in Europe
Art: Cave Art in the Americas
Asia, Central
Asia, Northeast
Asia, Northeast: History
Asia, Southeast
Asia, Southeast Islands
Asia, Southeast: Archaeological Caves
Asia, Southeast: Biospeleology
Asia, Southeast: History
Asia, Southwest
Asiago Plateau, Italy
Atapuerca Caves, Spain: Archaeology
Australia
Australia: Archaeological and Paleontological Caves
Australia: Biospeleology
Australia: History
Aves (Birds)

B
Bambuí Karst, Brazil
Bauxite Deposits in Karst
Bear Rock Karst, Canada
Belgium: Archaeological Caves
Belize River Caves
Biodiversity in Hypogean Waters
Biodiversity in Terrestrial Cave Habitats
Biofilms
Biokarstification
Biology of Caves
Biospeleologists
Blue Holes of the Bahamas
Boa Vista, Toca, Brazil
Books on Caving
Britain and Ireland: Archaeological and Paleontological Caves

Britain and Ireland: Biospeleology
Britain and Ireland: History
Burials in Caves
Burren Glaciokarst, Ireland

C
Calcareous Alps, Austria
Canada
Canary Islands: Biospeleology
Cape Range, Australia: Biospeleology
Carbon Dioxide-enriched Cave Air
Carbonate Karst
Carbonate Minerals: Precipitation
Caribbean Islands
Caribbean Islands: History
Carlsbad Cavern and Lechuguilla Cave, United States
Carmel Caves, Israel: Archaeology
Castleguard Cave, Canada
Caucasus, Georgia
Caves
Caves in Fiction
Caves in History: The Eastern Mediterranean
Cerknica Polje, Slovenia: History
Cheju-do Lava Caves, South Korea
Chemistry of Natural Karst Waters
Chillagoe and Mitchell-Palmer Karsts, Australia
China
China: Archaeological Caves
Chiroptera (Bats)
Climate of Caves
Coastal Karst
Cockpit Country Cone Karst, Jamaica
Colonization
Communications in Caves
Condensation Corrosion
Cone Karst
Conservation: Cave Biota
Conservation: Protected Areas
Construction on Karst
Crevice Caves
Crimea, Ukraine
Crustacea
Crustacea: Amphipoda
Crustacea: Copepoda
Crustacea: Decapoda (Shrimps, Crayfish, Crabs)
Crustacea: Isopoda (Aquatic)
Crustacea: Isopoda: Oniscidea (Woodlice)
Crustacea: Ostracoda
Crustacea: Syncarida
Cuba
Cupp-Coutunn Cave, Turkmenistan

D
Dams and Reservoirs on Karst
Dating Methods: Archaeological
Dating of Karst Landforms
Dent de Crolles Cave System, France
Di Feng Dong, China
Dinaric Karst

Dinaric Karst: Biospeleology
Dinaride Poljes
Disease
Dissolution: Carbonate Rocks
Dissolution: Evaporite Rocks
Dissolution: Silicate Rocks
Diving in Caves
Dolines
Draenen, Ogof Draenen, Wales

E
Edwards Aquifer, United States
Edwards Aquifer, United States: Biospeleology
Encantado, Sistema del Rio, Puerto Rico
Entrance Habitats
Environmental Impacts Assessment
Erosion Rates: Field Measurements
Erosion Rates: Theoretical Models
Europe, Alpine
Europe, Balkans and Carpathians
Europe, Central
Europe, Central: Archaeological Caves
Europe, Central: History
Europe, Mediterranean
Europe, Mediterranean: Archaeological and Paleontological Caves
Europe, North
Evaporite Karst
Evolution of Hypogean Fauna
Exploration Societies
Exploring Caves
Extraterrestrial Caves

F
Films in Caves
Floral Resources
Fluviokarst
Folklore and Mythology
Food Resources
Forests on Karst
France: Biospeleology
France: History
France, Southern Massif Central
Frasassi Caves, Italy

G
Geophysical Detection of Caves and Karstic Voids
Geoscientists
Gibraltar Caves: Archaeology
Glacier Caves and Glacier Pseudokarst
Glacierized and Glaciated Karst
Golondrinas and the Giant Shafts of Mexico
Grand Canyon, United States
Groundwater in Karst
Groundwater in Karst: Borehole Hydrology
Groundwater in Karst: Conceptual Models
Groundwater in Karst: Mathematical Models
Groundwater Pollution: Dispersed
Groundwater Pollution: Point Sources
Groundwater Pollution: Remediation

Groundwater Protection
Guano
Gunpowder

H
Ha Long Bay, Vietnam
Hawaii Lava Tube Caves, United States
Hawaiian Islands: Biospeleology
Highways on Karst
Hölloch, Switzerland
Hongshui River Fengcong Karst, China
Huanglong and Jiuzhaigou, China
Huautla Cave System, Mexico
Human Occupation of Caves
Hydraulics of Caves
Hydrocarbons in Karst

I
Ice in Caves
Inception of Caves
Indian Subcontinent
Insecta: Apterygota
Insecta: Coleoptera (Beetles)
Insecta: Pterygota
Interstitial Habitats (Aquatic)
Interstitial Habitats (Terrestrial)
Invertebrates: Minor Groups
Iran

J
Jeita Cave, Lebanon
Journals on Caves

K
Kaijende Arête and Pinnacle Karst, Papua New Guinea
Kanin Massif, Slovenia-Italy
Karren
Karst
Karst Evolution
Karst Hydrology: History
Karst Resources and Values
Karst Water Resources
Khammouan, Laos-Vietnam
Kras, Slovenia
Krubera Cave, Georgia

L
Limestone as a Mineral Resource
Littoral Caves

M
Madagascar
Mammoth Cave Region, United States
Mammoth Cave, United States: Biospeleology
Marine Cave Habitats
Mendip Hills, England
Microbial Processes in Caves
Microorganisms in Caves

Military Uses of Caves
Minerals in Caves
Mineral Deposits in Karst
Mollusca
Mona, Puerto Rico
Monitoring
Morphology of Caves
Morphometry of Caves
Morphometry of Karst
Movile Cave, Romania
Mulu, Sarawak
Music in and about Caves
Myriapoda (Centipedes and Millipedes)

N
Nahanni Karst, Canada
Nakanai Caves, Papua New Guinea
New Zealand
Nullarbor Plain, Australia

O
Organic Resources in Caves
Organisms: Classification
Ornamental Use of Limestone

P
Paleoenvironments: Clastic Cave Sediments
Paleoenvironments: Speleothems
Paleokarst
Paleontology: Animal Remains in Caves
Paleotectonics from Speleothems
Palynology
Pamukkale, Turkey
Paragenesis
Patagonia Marble Karst, Chile
Patterns of Caves
Peak District, England
Photographing Caves
Phytokarst
Picos de Europa, Spain
Pierre Saint-Martin, France-Spain
Pinega Gypsum Caves, Russia
Piping Caves and Badlands Pseudokarst
Pisces (Fishes)
Pisces (Fishes): Amblyopsidae
Plitvice Lakes, Croatia
Poljes
Ponors
Postojna-Planina Cave System, Slovenia
Postojna Planina Cave System: Biospeleology
Pseudokarst

Q
Quarrying of Limestone
Quartzite Caves of South America

R
Radiolocation
Radon in Caves

Ramsar Sites—Wetlands of International Importance
Recreational Caving
Religious Sites
Restoration of Caves and Speleothem Repair
Russia and Ukraine

S
Salukkan Kallang, Indonesia: Biospeleology
Sediments: Allochthonous Clastic
Sediments: Autochthonous Clastic
Sediments: Biogenic
Sedom Salt Karst, Israel
Selma Plateau Caves, Oman
Sewu Cone Karst, Java
Shanidar Cave, Iraq: Archaeology
Shilin Stone Forest, China
Siberia, Russia
Siebenhengste, Switzerland
Silicate Karst
Škocjanske Jama, Slovenia
Sof Omar Cave, Ethiopia
Soils on Carbonate Karst
Soil Erosion and Sedimentation
Solution Breccias
Soviet Union: Speleological History
Spannagel Cave, Austria
Speciation
Speleogenesis
Speleogenesis Theories: Early
Speleogenesis Theories: Post–1890
Speleogenesis: Coastal and Oceanic Settings
Speleogenesis: Computer Models
Speleogenesis: Deep-Seated and Confined Settings
Speleogenesis: Unconfined Settings
Speleologists
Speleothem Studies: History
Speleothems: Carbonate
Speleothems: Evaporite
Speleothems: Luminescence
Speleotherapy
Springs
Stamps and Postcards
Stripe Karst
Subterranean Ecology

Subterranean Habitats
Sulfide Minerals in Karst
Surveying Caves
Syngenetic Karst

T
Talus Caves
Thermal Water Habitats
Tourism and Caves: History
Tourist Caves
Tourist Caves: Air Quality
Tourist Caves: Airborne Debris
Tourist Caves: Algae and Lampenflora
Tower Karst
Travertine
Tunnelling and Underground Dams in Karst
Turkey

U
Ukraine Gypsum Caves and Karst
United States of America

V
Valleys in Karst
Vercors, France
Vézère Archaeological Caves, France
Villa Luz, Cueva de, Mexico
Vjetrenica, Bosnia-Herzegovina: Biospeleology
Volcanic Caves
Vulcanospeleology: History

W
Walsingham Caves, Bermuda: Biospeleology
Water Tracing
Water Tracing: History
Wilderness
Wind and Jewel Caves, United States
World Heritage Sites

Y
Yangshuo Karst, China
Yorkshire Dales, England
Yucatán Phreas, Mexico

Thematic List of Entries

Archaeology, Paleontology, and Cave Art
Africa, South: Archaeological Caves
Aggtelek Caves, Hungary-Slovakia: Archaeology
Altamira Cave, Spain: Archaeology
America, Central: Archaeological Caves
America, North: Archaeological Caves
Ardèche Caves, France: Archaeology
Archaeologists
Archaeology of Caves: History
Art in Caves
Art in Caves: History
Art: Cave Art in Australasia
Art: Cave Art in Europe
Art: Cave Art in the Americas
Asia, Southeast: Archaeological Caves
Atapuerca Caves, Spain: Archaeology
Australia: Archaeological and Paleontological Caves
Belgium: Archaeological Caves
Britain and Ireland: Archaeological and Paleontological Caves
Burials in Caves
Carmel Caves, Israel: Archaeology
China: Archaeological Caves
Dating Methods: Archaeological
Europe, Central: Archaeological Caves
Europe, Mediterranean: Archaeological and Paleontological Caves
Gibraltar Caves: Archaeology
Human Occupation of Caves
Paleontology: Animal Remains in Caves
Shanidar Cave, Iraq: Archaeology
Vézère Archaeological Caves, France

Biospeleology (Speleobiology)

Ecology and Ecological Processes
Biodiversity in Hypogean Waters
Biodiversity in Terrestrial Cave Habitats
Biofilms
Biology of Caves
Biokarstification
Food Resources
Guano
Microorganisms in Caves
Paleontology: Animal Remains in Caves
Sediments: Biogenic
Subterranean Ecology

Habitats
Anchialine Habitats
Entrance Habitats
Interstitial Habitats (Aquatic)
Interstitial Habitats (Terrestrial)
Marine Cave Habitats
Subterranean Habitats
Thermal Water Habitats

Cave Ecology and Man
Biospeleologists
Conservation: Cave Biota
Folklore and Mythology
Groundwater Pollution: Dispersed
Groundwater Pollution: Point Source
Tourist Caves: Algae and Lampenflora

Regional Subterranean Faunas
Africa: Biospeleology
America, Central and Caribbean Islands: Biospelology
America, North: Biospeleology
America, South: Biospeleology
Asia, Southeast: Biospeleology
Australia: Biospeleology
Britain and Ireland: Biospeleology
Canary Islands: Biospeleology
Dinaric Karst: Biospelology
France: Biospeleology

Richest Cave Faunas
Cape Range, Australia: Biospeleology
Edwards Aquifer, United States: Biospeleology
Hawaiian Islands: Biospeleology

Mammoth Cave, United States: Biospeleology
Movile Cave, Romania
Postojna-Planina Cave System, Slovenia: Biospeleology
Salukkan Kallang, Indonesia: Biospeleology
Vjetrenica, Bosnia-Herzegovina: Biospeleology
Walsingham Caves, Bermuda: Biospeleology

Evolution of Subterranean Fauna
Adaptation: Behavioural
Adaptation: Eyes
Adaptation: Genetics
Adaptation: Morphological (External)
Adaptation: Morphological (Internal)
Adaptation: Physiological
Colonization
Evolution of Hypogean Fauna
Organisms: Classification
Speciation

Important Subterranean Taxa
Amphibia
Amphibia: *Proteus*
Annelida
Arachnida
Arachnida: Acari (Mites)
Arachnida: Araneae (Spiders)
Arachnida: Minor Groups
Aves (Birds)
Chiroptera (Bats)
Crustacea
Crustacea: Amphipoda
Crustacea: Copepoda
Crustacea: Decapoda (Shrimps, Crayfish, Crabs)
Crustacea: Isopoda (Aquatic)
Crustacea: Isopoda-Oniscidea (Woodlice)
Crustacea: Ostracoda
Crustacea: Syncarida
Insecta: Apterygota
Insecta: Coleoptera (Beetles)
Insecta: Pterygota
Invertebrates: Minor Groups
Mollusca
Myriapoda (Centipedes and Millipedes)
Pisces (Fishes)
Pisces (Fishes): Amblyopsidae

Caves and Caving

Cave Media
Art Showing Caves
Books on Caves
Caves in Fiction
Films in Caves
Folklore and Mythology
Journals on Caves
Music about and in Caves
Stamps and Postcards

Caving
Accidents and Rescue
Communications in Caves
Diving in Caves

Exploring Caves
Photographing Caves
Radiolocation
Recreational Caving
Surveying Caves

Uses of caves
Burials in Caves
Human Occupation of Caves
Military Uses of Caves
Music about and in Caves
Religious Sites
Speleotherapy
Tourist Caves

Caves and Karst Regions

Caves and Cave Systems
Belize River Caves
Blue Holes of Bahamas
Boa Vista, Brazil
Carlsbad Cavern and Lechuguilla Cave, United States
Castleguard Cave, Canada
Cheju-do Lava Caves, South Korea
Cupp-Coutunn Cave, Turkmenistan
Dent de Crolles Cave System, France
Di Feng Dong, China
Draenen, Ogof Draenen, Wales
Encantado, Sistema del Rio, Puerto Rico
Frasassi Caves, Italy
Golondrinas and the Giant Shafts of Mexico
Hawaii Lava Tube Caves, United States
Hölloch, Switzerland
Huautla Cave System, Mexico
Jeita Cave, Lebanon
Krubera Cave, Georgia
Mammoth Cave Region, United States
Movile Cave, Romania
Nakanai Caves, Papua New Guinea
Postojna-Planina Cave System, Slovenia
Siebenhengste, Switzerland
Škocjanske Jama, Slovenia
Sof Omar Cave, Ethiopia
Spannagel Cave, Austria
Villa Luz, Cueva de, Mexico
Wind and Jewel Caves, United States

Karst Regions
Africa, North
Africa, Sub-Saharan
Aggtelek and Slovak Karst, Hungary-Slovakia
Akiyoshi-dai, Japan
America, Central
America, South
Antarctica
Appalachian Mountains, United States
Asia, Central
Asia, Northeast
Asia, Southeast
Asia, Southeast Islands
Asia, Southwest

Asiago Plateau, Italy
Australia
Bambuí Karst, Brazil
Bear Rock Karst, Canada
Burren Glaciokarst, Eire
Calcareous Alps, Austria
Canada
Caribbean Islands
Caucasus, Georgia
Chillagoe and Mitchell-Palmer Karsts, Australia
China
Cockpit Country Cone Karst, Jamaica
Crimea, Ukraine
Cuba
Dinaric Karst
Dinaride Poljes
Edwards Aquifer and the Texas Karst, United States
Europe, Alpine
Europe, Balkans and Carpathians
Europe, Central
Europe, Mediterranean
Europe, North
France, Southern Massif Central
Grand Canyon, United States
Ha Long Bay, Vietnam
Hongshui River Fengcong Karst, China
Huanglong and Jiuzhaigou, China
Indian Subcontinent
Iran
Kaijende Arete and Pinnacle Karst, Papua New Guinea
Kanin Massif, Slovenia-Italy
Khammouan, Laos-Vietnam
Kras, Slovenia
Madagascar
Mendip Hills, England
Mona, Puerto Rico
Mulu, Sarawak
Nahanni Karst, Canada
New Zealand
Nullarbor Plain, Australia
Pamir and Tien Shan Karst, Asia
Pamukkale, Turkey
Patagonia Marble Karst, Chile
Peak District, England
Picos de Europa, Spain
Pierre Saint-Martin, France-Spain
Pinega Gypsum Caves, Russia
Plitvice Lakes, Croatia
Quartzite Caves of South America
Russia and Ukraine
Sedom Salt Karst, Israel
Selma Plateau Caves, Oman
Sewu Cone Karst, Java
Shilin Stone Forest, China
Siberia, Russia
Turkey
Ukraine Gypsum Caves and Karst
United States of America
Vercors, France
Yangshuo Karst, China

Yorkshire Dales, England
Yucatán Phreas, Mexico

Conservation and Management

Carbon Dioxide-enriched Cave Air
Conservation: Cave Biota
Conservation: Protected Areas
Environmental Impact Assessment
Groundwater Pollution: Dispersed
Groundwater Pollution: Point Source
Groundwater Pollution: Remediation
Groundwater Protection
Karst Resources and Values
Monitoring
Ramsar Sites—Wetlands of International Importance
Religious Sites
Restoration of Caves and Speleothem Repair
Tourist Caves
Tourist Caves: Air Quality
Tourist Caves: Airborne Debris
Tourist Caves: Algae and Lampenflora
Wilderness
World Heritage Sites

Geoscience

Caves and Speleogenesis
Caves
Extraterrestrial Caves
Hydraulics of Caves
Inception of Caves
Morphology of Caves
Morphometry of Caves
Paragenesis
Patterns of Caves
Speleogenesis
Speleogenesis: Coastal and Oceanic Settings
Speleogenesis: Computer Models
Speleogenesis: Deep-seated and Confined Settings
Speleogenesis: Unconfined Settings
Speleogenesis Theories: Early
Speleogenesis Theories: Post–1890

Deposits in Caves
Carbonate Minerals: Precipitation
Ice in Caves
Minerals in Caves
Palaeoenvironments: Clastic Sediments
Palaeoenvironments: Speleothems
Palynology
Sediments: Allochthonous Clastic
Sediments: Autochthonous Clastic
Sediments: Biogenic
Speleothems: Carbonate
Speleothems: Evaporite
Speleothems: Luminescence

Karst Geomorphology
Alpine Karst
Carbonate Karst
Coastal Karst
Cone Karst

Dolines
Evaporite Karst
Fluviokarst
Glaciated and Glacierized Karst
Karren
Karst
Karst Evolution
Morphometry of Karst
Paleokarst
Phytokarst
Poljes
Ponors
Silicate Karst
Solution Breccias
Springs
Stripe Karst
Syngenetic Karst
Tower Karst
Travertine
Valleys in Karst

Processes and Techniques
Biokarstification
Carbon Dioxide-enriched Cave Air
Chemistry of Natural Karst Waters
Climate of Caves
Condensation Corrosion
Dating of Karst Landforms
Dissolution: Carbonate Rocks
Dissolution: Evaporite Rocks
Dissolution: Silicate Rocks
Erosion Rates: Field Measurements
Erosion Rates: Theoretical Models
Geophysical Detection of Caves and Karstic Voids
Groundwater in Karst
Groundwater in Karst: Borehole Hydrology
Groundwater in Karst: Conceptual Models
Groundwater in Karst: Mathematical Models
Microbial Processes in Caves
Palaeotectonics from Speleothems
Radon in Caves
Water Tracing

Pseudokarst
Crevice Caves
Glacier Caves and Glacier Pseudokarst
Littoral caves
Piping Caves and Badlands Pseudokarst
Pseudokarst
Talus Caves
Volcanic Caves

History
America, North: History
America, South: History
Archaeologists
Archaeology of Caves: History
Art in Caves: History
Asia, Northeast: History
Asia, Southeast: History
Australia: History
Biospeleologists
Britain and Ireland: History
Caribbean Islands: History
Caves in History: The Eastern Mediterranean
Cerknica Polje, Slovenia: History
Europe, Central: History
Exploration Societies
France: History
Geoscientists
Gunpowder
Karst Hydrology: History
Soviet Union: Speleological History
Speleogenesis Theories: Early
Speleogenesis Theories: Post–1890
Speleologists
Speleothem Studies: History
Tourism and Caves: History
Vulcanospeleology: History
Water Tracing: History

Resources and Development
Bauxite Deposits in Karst
Construction on Karst
Dams and Reservoirs on Karst
Disease
Floral Resources
Forests on Karst
Guano
Gunpowder
Highways on Karst
Hydrocarbons in Karst
Karst Resources and Values
Karst Water Resources
Limestone as a Mineral Resource
Mineral Deposits in Karst
Organic Resources in Caves
Ornamental Use of Limestone
Quarrying of Limestone
Religious Sites
Soils on Carbonate Karst
Soil Erosion and Sedimentation
Sulfide Minerals in Karst
Tunnelling and Underground Dams in Karst

A

ACCIDENTS AND RESCUE

Cave exploration is, by its nature, a potentially hazardous activity, although the relative risk of an accident, and its severity, clearly vary from cave to cave. In all but the simplest of cases, rescue of an injured person from a cave poses problems that are different from those encountered on the surface. A cave is a relatively hostile environment in which to perform a rescue: it is completely dark beyond the entrance zone and there may be other potential hazards, including water, low temperatures, constricted passages, and vertical shafts. Hence, there is a need for specialist rescue teams who are aware of the latest rescue techniques.

The French Cave Rescue Organization has records that have been kept from the early 1900s but it is believed that the world's first formally organized cave rescue team was founded in Yorkshire, England, in 1934 (Eyre & Frankland, 1988). At present, over 20 nations have active cave rescue organizations. The International Cave Rescue Commission within the International Union of Speleology provides a forum for cave rescue organizations to communicate with each other at their congresses and through the internet at http://netdial.caribe.net/~emercado/uis.htm.

Rescue organizations have training seminars and extensive practice rescue from caves. At the training seminars, often international events, experience gathered from real and practice rescues is exchanged and rescue equipment and techniques are demonstrated. Rescue records are published in national journals, often with full details and analysis. In addition, many organizations keep detailed records of their rescues and hence it is possible to generate statistics on the diversity of caving accidents that require rescue. Statistics from the United States show that the majority of incidents (c.30%) involve vertical caving and rope work. This is followed by individuals who have become lost or have experienced light failure (18%); water problems including being trapped by flooding (14%); entrapment (10%); and explosion, rock fall, bad air, and medical problems, each making 5% or less (Hempel & Fregeau-Conover, 2001). Frankland (1991) gives statistics for the causes of injury or death to British cavers: falls (61%), rock fall (15%), drowning (12%), bad air (21%), and others (10%). A more detailed analysis of cave rescue statistics for the Yorkshire Dales area (Forder, 2001) showed that half of the incidents between 1935 and 2000 involved floods or falls and that falls accounted for a quarter of all fatalities, and floods 13%. Standing (1976) showed that 80% of British rescue incidents involved novices and in the United States, 90% of call-outs were for inexperienced or poorly equipped cavers (Hempel & Fregeau-Conover, 2001).

The above statistics illustrate that inexperienced and poorly equipped cavers are most at risk and that falls, usually off short, unprotected climbs or while vertical caving with ropes and ladders, are the most common accidents. In such falls lower limb fractures are the most common injury, followed closely and perhaps surprisingly by spinal fractures, almost entirely without spinal cord damage. Spinal injuries pose particular difficulties for the rescue team.

Rescue of the lost caver requires that the rescue teams have substantial local knowledge and the resources for rapid systematic searches of complex caves or several caves in an area. Speed in any rescue is critical because even if the cavers are not injured they are in danger from hypothermia. Lloyd (1964) found that death from hypothermia could occur within half an hour of the onset of symptoms. The prevention of hypothermia in the lost, trapped, or accident victim is a major consideration of the rescue team, who typically carry equipment such as thick neoprene exposure bags, heat packs, insulating systems, and warm-air rebreathing apparatus.

When a person is injured a factor critical to successful rescue is the time taken to alert the rescue organization. Nevertheless, the difficulty of the cave and the depth of the accident site mean that there may be hours or even days between an injury and the arrival of the rescue team. Hence, the availability of basic survival equipment and knowledge of first aid among the victim's group enhances the probability of a successful rescue.

Many cave systems flood and often flash flood. Special care is required in the lowest levels of caves and close to sumps as in these regions back-flooding can cause water levels to rise exceptionally rapidly. When floodwaters prevent safe exit the only option is to seek a position above the maximum water level, which can sometimes be deduced from foam or mud-coated walls or vegetation lodged in cracks and crevices. Some caves that experience frequent flooding of their entrance passages have designated points where trapped cavers may safely await rescue (e.g. Peak Cavern, England). It is usually advisable to wait until the rescue team arrives unless absolutely certain that the water levels have dropped sufficiently for a safe exit. It is not wise to move with or against a flood, and exceptionally hazardous to attempt to ascend or descend ropes or ladders when the cave is flooding. The worst caving disaster in Britain was in Mossdale Caverns in 1967 when six cavers lost their lives as a result of an unexpected flash flood. Guide books often contain information as to which caves are known to flood and a wise caver will check the weather before entering such caves. However, intense precipitation is not always predictable and in the Mossdale area on 3 May 1986, another flash flood resulted in six separate rescue call-outs. This time there was only a single fatality.

Rock fall or movement can trap a caver, especially when they are negotiating breakdown. Rescue of the trapped can be a frightening experience for the rescue team if the area remains unstable. Rescue of a person from bad air requires both a fast response and great care as accidents in bad air could result in the deaths of the rescuer as well as the victim. Rescue and recovery teams use breathing equipment or wait until the hazardous atmosphere has cleared. Bad air can arise from a number of sources: partial combustion of fuel in fires, stoves, or generators can result in accumulations of carbon monoxide as well as carbon dioxide, and explosives release a variety of exceptionally toxic gases. Less obviously, respiration in a confined space results in hazardous increases of carbon dioxide and reduced oxygen. For example, in England in 1976 six cavers were free diving a $c.8$ m sump in Langstroth Pot with a small air bell where a breath could be taken halfway. Unfortunately they used up the oxygen in the air bell, replacing it with expired carbon dioxide, and there were three fatalities.

Every rescue situation is different and there is no formula for exactly what must be done. On expeditions in remote areas and in most vertical caves emphasis is placed on self-rescue, which in this context means that members of the victim's party organize the rescue. The cavers performing self-rescue must assess the degree of injury and the availability of outside rescue before attempting to remove the victim from the cave. Some rescues cannot wait for an outside team; for example, if the victim is trapped on rope hanging in a harness they should be released as fast as possible as death occurs rapidly to both the conscious and unconscious. If the cavers are trained in first aid and techniques of rescue, especially those needed for rescue in vertical caves (Warild, 1994), then the probability of a successful self-rescue is increased. However, an untrained group would probably fail to rescue a severely injured victim from deep in a cave.

It should be emphasized that the death of a victim can be a harrowing experience for all. The most experienced rescuer should adequately package the body such that the object being handled by the recovery team is minimally visibly human.

Flooded passages (sumps) pose particular problems. Only fellow divers can rescue cave divers who have equipment failure or an accident; divers have also been used to bring out uninjured cavers trapped beyond flooded cave passages. Probably the earliest such rescue was in 1969 when a trapped and unconscious cavers in Meregill Hole, England, was brought through a sump with a diving mask held onto his face. He survived the dive, but died of hypothermia shortly afterwards. Subsequently there has been considerable research into the development of equipment for transporting injured or trapped cavers safely through flooded passages. Unfortunately very few cave diving incidents result in a successful rescue, and body recovery poses its own problems. In the United States the International Underwater Cave Rescue and Recovery team assists local law enforcement agencies and other rescue organizations by supplying specially trained divers to aid in rescues or recoveries.

Modern rescue teams have the benefit of state-of-the-art equipment; the ability to call out cave, mine, and other specialist rescue teams, and to transport experts and victims thousands of kilometres if necessary. No expense is spared if it is believed the cavers are still alive. For example, probably the world's longest, largest, and most expensive cave rescue took place in France in November 1999 when seven cavers were trapped by floods. Shafts were drilled to locate air spaces in which the cavers were likely to be waiting (a first) and the successful rescue took ten days.

JOHN FRANKLAND, EMILY DAVIS, AND JULIA JAMES

Works Cited

Eyre, J. & Frankland, J. 1988. *Race Against Time: The History of The Cave Rescue Organisation's First Fifty Years*, Dent, Yorkshire: Lyon Books

Forder, J. 2001. An analysis of cave rescue statistics, Dales area, UK, 1935–2000. *Cave and Karst Science*, 28(3): 131–34

Frankland, J.C. 1991. Accidents to cavers. In *Caving Practice and Equipment*, revised edition, edited by D. Judson, London: British Cave Research Association

Hempel, J.C. & Fregeau-Conover, A. (editors) 2001. *On Call: A Complete Reference for Cave Rescue*, Huntsville, Alabama: National Speleological Society

Lloyd, O.C. 1964. Cavers dying of cold. *Bristol Medico-Chirurgical Journal*, 79: 261

Standing, I.J. 1976. Cave rescue incidents in Britain 1935–1972. *Cave Science*, 50: 13–20

Warild, A. 1994. *Vertical: A Technical Manual for Cavers*, 3rd edition, Sydney: The Sydney Research Council (updated 2001, on CD only)

Further Reading

Putnam, W.O. (editor) 2000. American caving accidents 1996–1998. *NSS News*, April

Spéléo Secours Français. 2000. *The Cave Rescuer's Manual*, Federation Française de Spéléologie (also published in French as *Manuel Technique du Spéléo-Secours Français*)

ADAPTATION: BEHAVIOURAL

Of all phenotypic traits, behaviour plays the most immediate role in determining the ability of a group of individuals to survive during colonization of the subterranean (hypogean) environment. Changes in behaviour as a response to environmental change are thus crucial to understanding the phenomenon of adaptation to hypogean environments. In the following overview of current understanding of behavioural adaptations in caves, most of the examples presented come from studies of fish, the taxon on which most behavioural experiments have been performed.

Numerous animals colonize the hypogean environment by developing entirely new behaviours such as echo-location (e.g. bats and the nocturnal oilbird *Steatornis caripensis*), bioluminescence (e.g. the New Zealand glow-worm—actually a fly larva—*Arachnocampa luminosa*), or hibernation (e.g. a carp from China, *Varicorhinus* [*Scaphestes*] *macrolepis*). However, these types of behavioural modifications are the exception, not the rule. Most can be grouped into one of the following categories: feeding, reproduction, social behaviour (including aggregation, responses to alarm substances, and antagonistic behaviour), photoresponses, and circadian rhythms.

Feeding
Many cave animals have enlarged sensory systems that allow them to sense the presence of food. To improve their ability to find food, many cave animals move continuously in their environment; for example the troglomorphic form of the fish *Astyanax fasciatus* swims continuously. In studies, these fish also decrease the angle of their body relative to the bottom of the aquarium in order to increase the area of contact between food on the bottom and areas of their skin where there are larger numbers of chemoreceptors (Schemmel, 1980). Similar adaptations have been reported for the toothless blindcat *Trogloglanis pattersoni*, the Somalian cave fish *Phreatichthys andruzzii*, and the catfish *Trichomycterus itacarambiensis*. Coprophagy (feeding on excrement) is not uncommon in caves, and some animals specifically choose such environments. For example, larvae of the pyralid moth *Aglossa pinguinalis* are found inside caves in densities 700 times greater than at the surface. This is apparently due to the fact that in more exposed excrement they cannot compete with other coprophagous specialists, owing to a lack of parental care and slow growth rates for this species (Piñero & López, 1998).

Reproductive Behaviour
The use of chemicals to attract mates has been reported in a number of cave organisms, ranging from crickets to fish. This behaviour has yet to be studied thoroughly in natural conditions. Aggressive behaviour associated with reproduction has been observed to decrease, probably due to a lack of visual information (Parzefall, 2000) (see below).

Aggregation/Schooling
In general, all hypogean animals show a tendency to reduce organized forms of aggregation, from insects (Christiansen, 1970) to fish (Romero, 2001). Romero (1984, 1985a), for example, observed that individuals of *A. fasciatus* morphologically identical to the epigean populations of the same species actively entered a cave in Costa Rica for both feeding and to escape from predators (the fishing bat *Noctilio leporinus*). However, unlike the typical epigean *A. fasciatus*, the population did not form schools.

Responses to Alarm Substances
Many bony fishes contain an "alarm substance" in their skin that is released into the water when the skin is damaged, for example by biting. It is believed that this substance is sensed by conspecifics (members of the same species) of the individual that has been harmed, and thus they use the information to escape or seek refuge. Although the substance is produced by many hypogean fishes, the response to its release is either highly reduced or lost (Fricke, 1988).

Aggression and Antagonistic Behaviour
Aggression has been reported for many cave fishes, and some amphibia and crustaceans (Parzefall, 2000). Aggression is usually reduced, but antagonistic (confrontational behaviour without fighting) behaviour persists in many species, including amblyopsids (where the rituals decrease in complexity in parallel with the degree of cave adaptation, Bechler, 1983), the Ceguinho catfish *Pimelodella kronei*, and the Cueva del Guácharo blind catfish *Tricomycterus conradi*.

Responses to Light
Many blind, depigmented hypogean animals do show behavioural responses to light. Photoresponses are mediated by the pineal organ and—to a certain extent—by extrapineal organs (Langecker, 1992; see Adaptation: Morphological-Internal). They are termed scotophilia (the tendency to stay in the dark side of an aquarium under experimental conditions) and have been reported for many species of hypogean fishes (Romero, 2001). The behaviour of staying away from light is usually more common among recent invaders of the hypogean environment than among more troglomorphic species (Green & Romero, 1997). Also, the degree of scotophilia (in which the fish moves away from light) increases with development, as it does for epigean forms (Romero, 1985b). Although earlier authors made much of this behaviour as an indication that these species maintain such responses in order to stay in caves, all available evidence suggests that such behaviour is an inherited relic from their epigean ancestors (Romero, 1985b).

Circadian Rhythms
This is a system also known as a "biological clock", which controls a series of physiological and behavioural responses in an organism. The rhythmicity generated by these clocks is usually triggered by light and temperature changes in the environment. Many cave animals show a reduction or total loss in their ability to generate biological rhythms even when exposed to light under experimental conditions (Lamprecht & Weber, 1992). As with the photoresponses mentioned above, circadian cycles are more reduced (or totally absent) in the more troglomorphic species (see also Adaptation: Eyes).

Acoustic Behaviour
Hoch (2000) found that the cave planthopper, *Oliarus polyphemus*, utilizes communication systems similar to those of its epigean ancestor, by communicating using substrate-borne vi-

brations, and also found that the cave environment seems to be especially suited to low-frequency sound transmissions. No acoustic communication has been demonstrated for hypogean fish, despite the obvious advantage of such behaviour. More research is needed in this area.

In summary, with the exception of a few very specialized types of behaviour, most cave animals tend to reduce or eliminate many of the typical responses of their epigean ancestors. To understand the role played by behaviour during hypogean colonization, ethological studies on hypogean organisms that have yet to achieve a troglomorphic state are probably the most important avenues to explore.

ALDEMARO ROMERO

See also **Colonization**

Works Cited

Bechler, D.L. 1983. The evaluation of agonistic behaviour in amblyopsid fishes. *Behavioral Ecology and Sociobiology*, 12: 35–42

Christiansen, K. 1970. Experimental studies in aggregation and dispersion of Collembola. *Pedobiologia*, 10: 180–90

Fricke, D. 1988. Reaction to alarm substance in cave populations of *Astyanax mexicanus* (Characidae, Pisces). *Ethology*, 76: 305–08

Green, S. & Romero, A. 1997. Responses to light in two blind cave fishes (*Amblyopsis spelaea* and *Typhlichthys subterraneus*) (Pisces: Amblyopsidae). *Environmental Biology of Fishes*, 50: 167–74

Hoch, H. 2000. Acoustic communication in darkness. In *Subterranean Ecosystems*, edited by H. Wilkens, D.C. Culver & W.F. Humphries, Amsterdam and New York: Elsevier

Lamprecht, G. & Weber, F. 1992. Spontaneous locomotion behaviour in cavernicolous animals—the regression of the endogenous circadian system. In *The Natural History of Biospeleology*, edited by A.I. Camacho, Madrid: Museo de Ciencias Naturales

Langecker, T.G. 1992. Light sensitivity of cave vertebrates—behavioral and morphological aspects. In *The Natural History of Biospeleology*, edited by A.I. Camacho, Madrid: Museo de Ciencias Naturales

Parzefall, J. 2000. Ecological role of aggressiveness in the dark. In *Subterranean Ecosystems*, edited by H. Wilkens, D.C. Culver & W.F. Humphries, Amsterdam and New York: Elsevier

Piñero, F.S. & López, F.J.P. 1998. Coprophagy in Lepidoptera: Observational and experimental evidence in the pyralid moth *Aglossa pinguinalis*. *Journal of Zoology*, 244: 357–62

Romero, A. 1984. Behavior in an "intermediate" population of the subterranean-dwelling characid *Astyanax fasciatus*. *Environmental Biology of Fishes*, 10: 203–07

Romero, A. 1985a. Cave colonization by fish: role of bat predation. *American Midland Naturalist*, 113: 7–12

Romero, A. 1985b. Ontogenetic change in phototactic responses of surface and cave populations of *Astyanax fasciatus* (Pisces: Characidae). *Copeia*, 1985: 1004–11

Romero, A. 2001. It's a wonderful hypogean life: a guide to the troglomorphic fishes of the world. *Environmental Biology of Fishes*, 62: 13–41

Schemmel, C. 1980. Studies on the genetics of feeding behaviour in the cave fish *Astyanax mexicanus* f. *Anoptichthys*, an example of apparent monofactorial inheritance by polygenes. *Zeitschrift für Tierpsychologie*, 53: 9–22

Further Reading

Berti, R. & Masciarelli, L. 1993. Comparative performances of non-visual food search in the hypogean cyprinid *Phreatichthys andruzzii* and in the epigean relative *Barbus filamentosus*. *International Journal of Speleology*, 22: 121–30

Langecker, T.G. & Longley, G. 1994. Morphological adaptations of the Texas blind catfishes *Trogloglanis pattersoni* and *Satan eurystomus* (Siluriformes, Ictaluridae) to their underground environment. *Copeia*, 1993: 976–87

Trajano, E. 1997. Food and reproduction of *Trichomycterus itacarambiensis*, cave catfish from south-eastern Brazil. *Journal of Fish Biology*, 51: 53–63

ADAPTATION: EYES

The lightless regions of caves, referred to as "the deep zone" by Howarth (1993), can harbour aquatic, terrestrial, and even flying organisms. Although troglophiles—and even some animals that occasionally visit caves—often show adaptations involving photoreceptors and sight, few generalizations can be made regarding their eyes. Cave animals such as bats generally possess poor eyesight, but are never blind; other animals that simply visit caves may have grossly enlarged eyes with highly sensitive photoreceptors (Jamaican cave frogs, for instance) or, as in the case of some millipedes, may lack eyes altogether. On the other hand, obligate cave-dwellers frequently have eyes that are reduced in size, and some have lost their eyes entirely. Thus, eyelessness has traditionally been regarded as a sure sign of a "troglobitic" lifestyle. However, with the discovery in New Zealand of cave harvestmen (*Hendea myersi cavernicola*) that possess large, functional eyes but display other typically troglobitic features such as a pale, nonpigmented body, it has become clear that troglobites need not necessarily be blind. The New Zealand cave harvestmen prey upon the luminescent larvae of the so-called glow-worm fly, *Arachnocampa luminosa*, which is a troglophile. For a predatory animal like the cave harvestman, to be able to see its luminescent prey is an obvious advantage. Therefore, as with the occurrence of bioluminescence in mesopelagic organisms, an adaptation which has prevented the widespread evolution of eyelessness in the deep sea—only a few bathypelagic fishes completely lack eyes—provided that light sources still exist and sight still has a role to play, then even a troglobite need not be eyeless.

Notwithstanding the exceptions discussed above, it is clear that lack of light—especially sunlight—is a powerful feature of most cave environments and, undoubtedly, has led to the reduction and even disappearance of eyes and their constituent elements in many cavernicolous vertebrate as well as invertebrate species. In the dark it is energetically demanding to maintain a functioning eye with its photopigments, optical structures, dark / light adaptational mechanisms, nerve connections to the brain, and the ability to integrate and interpret visual signals. Moreover, it is a luxury that can become a handicap, because eyes can become injured, infected, or become a hindrance in locomotion. Consequently, eye degenerations in cave vertebrates are common among troglobitic fishes (see Table and Figure in Pisces: Amblyopsidae) as well as cave newts and salamanders, but nevertheless, tiny eye rudiments under the skin—often invisible from the outside—are usually present.

Embryologically, the eyes of cave fishes and amphibia at first develop quite normally, but then begin to exhibit developmental retardation. Arrested growth and differentiation follows, and finally programmed cell deaths occur. The grafting experiments of the French biospeleologist Jacques Durand showed elegantly and convincingly that this scenario does not depend on the presence or absence of light. He exchanged the larval eyes of the hypogean, aquatic cave salamander *Proteus anguinus* with those of the epigean *Euproctes asper* and observed that in both cases the donor eye behaved in the way that it would have done in the specimen of origin. The degeneration of the eye appears to be controlled by a cascade of regulatory genes; eight to ten genes are now thought to be involved specifically in the eye regression of the cave fish *Astyanax fasciatus*. Eye regression in cave vertebrates is mostly centripetal, which means that it commences peripherally with the cornea and lens, then affects retinal and perceptive structures, and finally engulfs nerves and the optic centres of the brain. Interestingly, cave fish are known to compensate for the loss of their visual system with enhancements to their lateral line organ and taste buds. Eyeless cave arthropods frequently exhibit elongated limbs, antennae, and body hairs, presumably to facilitate the detection of tactile stimuli, water or air currents. In arthropods with compound eyes, reduced eye size and a smaller number of facets are usually the first signs of eye regression: approximately 30 ommatidia are present in the eye of the cave opposum shrimp *Heteromysoides cotti*, three in the cave shrimp *Typhlatya garciai*, and none in *Troglocaris anophthalmus*.

Apparently the more rudimentary the eye, the more variable its appearance: the percentage of misshapen or malformed ommatidia increases, pigmentation becomes irregular, and cellular outlines as well as photoreceptive elements become increasingly less uniform. Such multifariousness in a rudimentary organ has been interpreted as support for Darwin's theory of natural selection, since organs deemed unnecessary—and which have therefore undergone regression—ought to be less and less affected by selection. Functionally, loss of colour-, shape-, and form-vision precede changes in the ability to detect flickering lights, to adjust the visual system to changes in ambient light intensity, and to protect the eye against photic damage. However, whether the latter occurs as a consequence of the cave environment or has prompted a species to establish itself in the cave, needs to be examined from case to case. An entire book has been devoted to one species of cave crustacean, the amphipod *Gammarus minus*. It is interesting that numerous totally eyeless species of terrestrial cave insects and aquatic crustaceans exist, but that almost no cave crab is completely eyeless, either exhibiting only reductions in eye size or lack of pigment and facetation around the eye region as well. Eyeless cave arthropods need not necessarily be completely insensitive to light: extra-ocular photoreceptors (for example, the so-called tail photoreceptor in the sixth abdominal ganglion of the crayfish), may be able to convey information on ambient light levels.

Without daylight, the most important time-keeper to coordinate circadian or seasonal rhythmicities is absent in caves, and many troglobitic animals, whether with or without rudimentary eyes, are indeed non-rhythmic. However, strictly clock-controlled locomotion rhythms are known from some troglobites. Since eyes and photoreception cannot be involved, secondary or even tertiary signals, such as daily movements of cave bats; periods of increased production of faeces; or activity periods of bat guano flies probably are. The question of how long it would take an animal species to lose its eye in total darkness is hard to answer. Fruitflies kept in total darkness for more than 600 generations showed little change in eye size. However, selective pressures were absent in that particular experiment, and undoubtedly the presence of predators would have had some impact. It was estimated that the total loss of the compound eye in a troglobitic arthropod could require some 100 000–1 000 000 generations. However, in view of the fact that deepwater hydrothermal vent shrimps apparently underwent a near-total loss of the dioptric (lens and corneal) elements (but not the photoreceptive cells) of their eyes in only 10 000 generations, estimates for the time-scale of morphological changes in the eyes of cave organisms may need to be revised.

VICTOR BENNO MEYER-ROCHOW

Further Reading

Culver, D.C., Kane, T.C. & Fong, D.W. 1995. *Adaptation and Natural Selection in Caves: The Evolution of* Gammarus minus, Cambridge, Massachusetts: Harvard University Press

Durand, J.P. 1976. Ocular development and involution in the European cave salamander *Proteus anguinus* Laurenti. *Biological Bulletin*, 151: 450–66

Howarth, F.G. 1993. High-stress subterranean habitats and evolutionary change in cave-inhabiting arthropods. *American Naturalist*, 142: 65–77

Jeffrey, W.R., Strickler, A.G., Guiney, S., Heyser, D.G. & Tomarev, S.I. 2000. Prox1 in eye degeneration and sensory organ compensation during development and evolution of the cavefish *Astyanax. Development, Genes, and Evolution*, 210: 223–30

Lamprecht, G. & Weber, F. 1985. Time-keeping mechanisms and their ecological significance in cavernicolous animals. *National Speleological Society Bulletin*, 47: 147–62

Meyer-Rochow, V.B. & Liddle, A. 1987. Structure and function of the eyes of two species of opilionids (*Megalopsalis tumida*: Palpatores and *Hendea myersi cavernicola*: Laniatores) from the New Zealand Glowworm Caves. *Proceedings of the Royal Society of London*, B233: 293–319

Meyer-Rochow, V.B. & Nilsson, H.L. 1999. Compound eyes in polar regions, caves, and the deep-sea. In *Atlas of Arthropod Sensory Receptors: Dynamic Morphology in Relation to Function*, edited by E. Eguchi & Y. Tominaga, Tokyo and New York: Springer

Stringer, I.A. & Meyer-Rochow, V.B. 1997. Flight activity of insects within a Jamaican cave: In search of the Zeitgeber. *Invertebrate Biology*, 116: 348–54

Wilkens, H. 1988. Evolution and genetics of epigean and cave *Astyanax fasciatus* (Characidae, Pisces). *Evolutionary Biology*, 23: 271–367

ADAPTATION: GENETICS

Any particular feature of an organism results from the interplay of genetics and the environment, but it is the genetic component that is passed on to succeeding generations, forming the raw material of evolution, which is, by one definition, simply the change in gene frequency through time. For a trait—whether behavioural, demographic, or morphological—to be adaptive in the cave environment, it must satisfy several requirements (Brandon, 1990), the most important of which are: (1) that some types are better adapted than others in the cave environment; (2) that selection has occurred; and (3) that trait differences are genetically determined. For many of the traits shared by obligate cave animals, the selective advantage is obvious. Thus, any increase in size or complexity of extra-optic sensory structures confers an apparent advantage in the darkness of caves, and any decrease in number of eggs produced confers an apparent advantage in the food-poor environment of caves. The third requirement—that trait differences are genetically determined—is more difficult to document. Purely environmental effects can mimic some of these adaptive changes. For example, colour differences between cave-dwelling and surface-dwelling crayfish (*Procambarus simulans*) are related to the availability of carotenoid pigment precursors, not genetic differences (Maguire, 1960). Starvation can also mimic some effects. Starving animals become emaciated, sometimes giving the appearance of a true cave animal. Even egg production is affected by starvation (Charlesworth, 1980).

There have been two studies that document the genetic basis of adaptation—Fong's study of the heritability of relative antennal length in cave and spring populations of the amphipod *Gammarus minus* (Fong, 1989; Jernigan, Culver & Fong, 1994) and Schemmel's study of taste-bud density and feeding behaviour in river and cave populations of the Mexican characin fish *Astyanax fasciatus* (Schemmel, 1974a, 1974b).

Schemmel found that density of taste buds was high in cave populations, low in surface river populations, and intermediate in phylogenetically young cave populations. In river populations the only area of dense taste buds is at the tip of the mouth, and this is enlarged in all cave populations, especially to the ventral side of the head (Wilkens, 1992). For the most part these differences were shown to be genetic because he measured the differences on long-term captive populations raised under identical conditions. Taking advantage of the captive populations at the Zoological Institute in Hamburg, Schemmel crossed several cave populations in order to learn more about the genetics of taste-bud density differences. By considering the variability of the offspring of these crosses, Schemmel showed that the differences in taste-bud density between river fish and cave fish were due to at least two or three taste-bud genes. The polygenes show an additive manner of expression, but there was a threshold effect, where the density increases abruptly, indicating gene interaction.

Fong studied a series of antennal characters in two spring and two cave populations of *Gammarus minus*. He measured both relative lengths and number of segments in 60-day-old individuals raised under constant darkness and constant temperature. As expected, all relative lengths and number of segments were larger in cave populations than in spring populations. Because he compared groups of siblings, Fong was able to measure heritability, the fraction of the variation that is attributable to genetic as opposed to environmental variation. Fong measured what is called broad-sense heritability (because it does not eliminate the variance due to the common environment of sibs, e.g. the maternal effect). Nevertheless, his results were striking. For the six antennal traits measured in four populations, heritabilities averaged 0.6, and in only one case in 24 was there no statistically significant heritability (Fong, 1989).

If these results are typical of cave organisms, then almost all cases of morphological change associated with isolation in caves have a strong genetic component. The kinds of morphological changes associated with starvation and the absence of pigment precursors listed above only mimic adaptation at a superficial level, and are not a general explanation for the adaptation of animals to the cave environment.

Fong and his colleagues (Jones, Culver & Kane, 1992; Culver *et al.*, 1994; Culver, Kane & Fong, 1995), were able to push the analysis of the genetic basis of adaptation much further because they were able to demonstrate that animals with larger antennae or more antennal segments had larger numbers of eggs and / or were more likely to mate than animals with smaller antennae or fewer antennal segments. Both mating propensity and egg number are components of fitness. This reconnects with the criteria for showing that a trait is adaptive—the differences must be genetic and selected for. In this case, the agent of selection is clearly life in darkness.

Eye reduction has a strong genetic basis in those species that have been carefully studied, including *Astyanax fasciatus* and *Gammarus minus* (Culver & Wilkens, 2000). However, the adaptive nature of these losses is not at all certain. At least in part, eye loss occurs due to the relaxation of selection in favour of maintaining eyes and the accumulation of selectively neutral, structurally reducing mutations. The work of Fong and his colleagues on *Gammarus minus* cited above also indicates that there is a selective advantage to eye loss, due to either energy conservation or simplification of neurological connections.

David C. Culver

See also **Evolution of Hypogean Fauna; Speciation**

Works Cited

Brandon, R.N. 1990. *Adaptation and Environment*, Princeton, New Jersey: Princeton University Press

Charlesworth, B. 1980. *Evolution in Age-structured Populations*, Cambridge and New York: Cambridge University Press

Culver, D.C. & Wilkens, H. 2000. Critical review of the relevant theories of the evolution of subterranean animals. In *Subterranean Ecosystems*, edited by H. Wilkens, W.F. Humphreys & D.C. Culver, Amsterdam and New York: Elsevier

Culver, D.C., Kane, T.C. & Fong, D.W. 1995. *Adaptation and Natural Selection in Caves: The Case of* Gammarus minus, Cambridge, Massachusetts: Harvard University Press

Culver, D.C., Jernigan, R.W., O'Connell, J. & Kane, T.C. 1994. The geometry of natural selection in cave and spring populations of the amphipod *Gammarus minus* (Crustacea: Amphipoda). *Biological Journal of the Linnean Society*, 52: 49–67

Fong, D.W. 1989. Morphological evolution of the amphipod *Gammarus minus* in caves: Quantitative genetic analysis. *American Midland Naturalist*, 121: 361–78

Jernigan, R.W., Culver, D.C. & Fong, D.W. 1994. The dual role of selection and evolutionary history as reflected in genetic correlations. *Evolution*, 48: 587–96

Jones, R., Culver, D.C. & Kane, T.C. 1992. Are parallel morphologies of cave organisms the result of similar selective pressures? *Evolution*, 46: 353–65

Maguire, B. 1960. Regressive evolution in cave animals and its mechanism. *Texas Journal of Science*, 13: 363–70

Schemmel, C. 1974a. Genetische Untersuchungen zur Evolution de Geschmacksapparates bei cavernicolen Fischen. [Genetic investigations on the evolution of the taste apparatus in cave fish]. *Zeitschrift für Zoologische und Systematik Evolution-forschung*, 12: 196–215

Schemmel, C. 1974b. Ist die cavernicole Micos-Population von *Astyanax mexicanus* (Characidae, Pesces) hybriden Ursprungs? [Is the cavernicolous Micos population of *Astyanax mexicanus* (Characidae, Pisces) of hybrid origin?] *Mitteilungen Hamburg Zoologisches Museum und Institut*, 71: 193–201

Wilkens, H. 1992. Neutral mutations and evolutionary progress. In *The Natural History of Biospeleology*, edited by A.I. Camacho, Madrid: Museo Nacional de Ciencias Naturales

Further Reading

Culver, D.C., Kane, T.C. & Fong, D.W. 1995. *Adaptation and Natural Selection in Caves: The Case of Gammarus minus*, Cambridge, Massachusetts: Harvard University Press
The most thorough study to date of the genetics of adaptation in a cave organism.

Kane, T.C. & Richardson, R.C. 1990. The phenotype as the level of selection: Cave organisms as model systems. *Journal of the Philosophy of Science Association*, 1: 151–64
A discussion of adaptation and its demonstration in a cave organism by a philosopher of science and a biologist.

Sket, B. 1985. Why all cave animals do not look alike—discussion on adaptive value of reduction processes. *NSS Bulletin*, 47(2): 78–85
A convincing argument for the adaptive nature of eye reduction.

ADAPTATION: MORPHOLOGICAL (EXTERNAL)

Morphological adaptation is used to mean evolutionary modifications of the external morphology of lineages of organisms, which are associated with their existence in caves. While some of these modifications may be non-adaptive, most putatively increase their chances of survival or competitiveness in caves. Selected works describing such adaptations in specific groups and reviews of general cave adaptation are listed in the references and further reading. A good example of morphological changes can be seen in cave Collembola of the family Entomobryidae where members of the genera *Pseudosinella* and *Sinella* go through very similar evolutionary changes in the caves of Japan, North America, Central America, and Europe (Christiansen, 1961; see Figure 1). These involve increase of adult size, expansion of mesothorax and elongation of furcula, and associated muscle changes producing an enlarged mesothorax, all associated with increased jumping ability. They also involve elongation of legs (also probably associated with increased predation escape), elongation of antennae, and enlargement of antennal sensory organs (both probably associated with increased olfactory and tactile sensitivity). The most studied changes are major changes in the foot complex. These have been shown to be associated first with adhesion on smooth wet rock surfaces and penetration into wet clay, and then with movement over water surfaces. In addition, there is the much-debated loss of eyes and pigment.

Other excellent descriptions of morphological changes involved in lineages of cave-adapted organisms have been given for a number of groups, all of which include eye and pigment reduction or loss. In each case there are more specialized changes. In Amblyopsid fishes (see Pisces: Amblyopsidae) these include hypertrophy of free motion sensing neuromasts and flattening of the head (see Figure 2D in that entry). Specialized changes in beetles of the family Cholevidae and the subfamily Trechinae include wing loss, cuticle thinning, elongation of appendages, narrowing of the prothorax, physogastry or (in some Cholevidae) narrowing of abdomen (see Insecta: Coleoptera). In Crustacea of the family Crangoncytidae (Amphipoda) the special changes include elongation of appendages (see Crustacea: Amphipoda).

It is notable that while earlier works suggested non-neoDarwinian explanations for adaptation in cave animals (Vandel, 1965) almost all recent work has been undertaken with the clear assumption of the applicability of the Darwinian core tenets. A possible recurrence of a view similar to that held by Vandel can be seen in the emphasis on phenoplastic adaptation put forward by Romero (see Evolution of Hypogean Fauna).

Troglomorphy

The term "troglomorph" (Christiansen, 1962) was created to designate phenotypic features that are characteristic of cave ani-

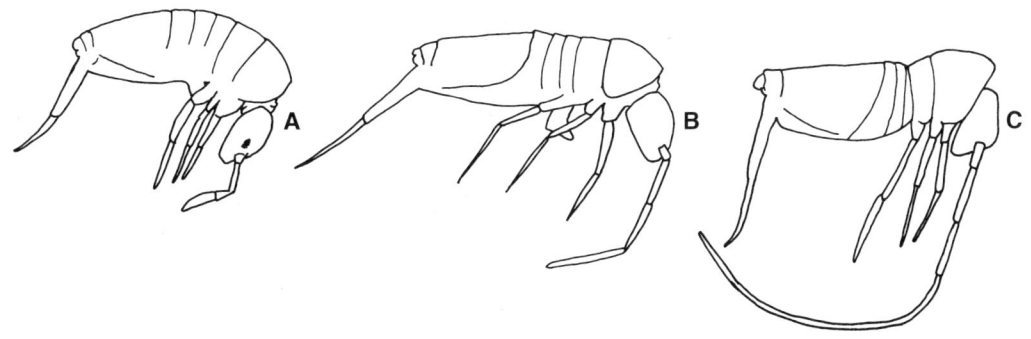

Adaptation: Morphological (External): Figure 1. Stages of increasing troglomorphism in Collembola Entomobryinae. A. *Pseudosinella octopunctata*; B. *P. hirsuta*; C. *P. christianseni*.

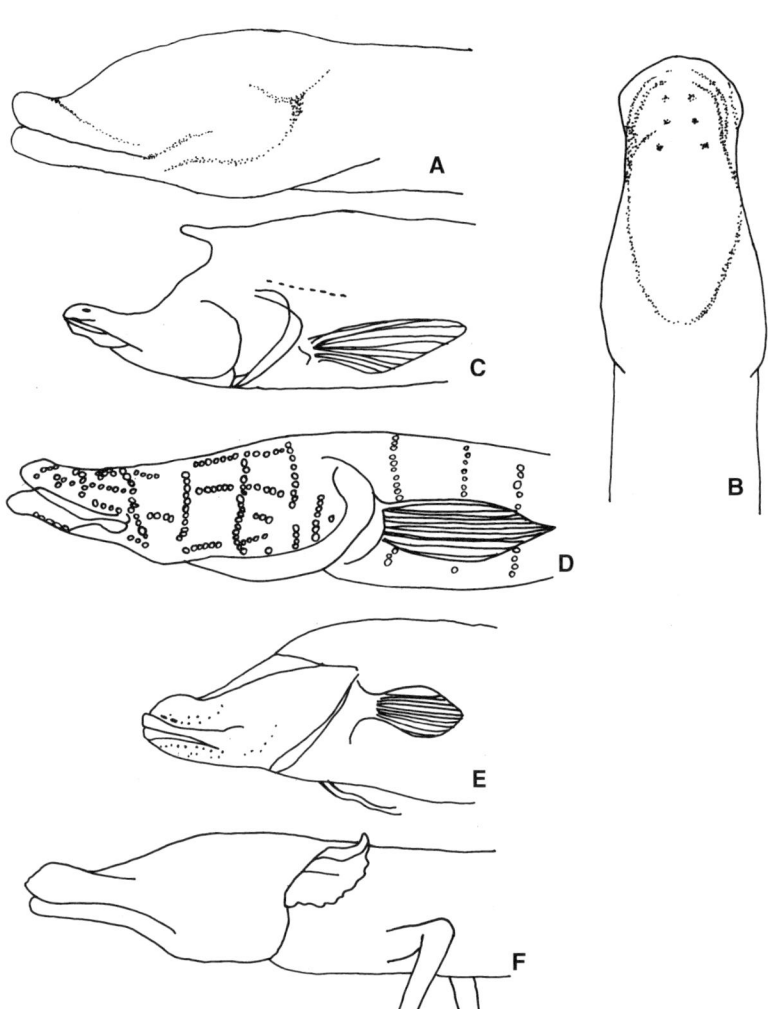

Adaptation: Morphological (External): Figure 2. Various troglomorphic vertebrates showing head flattening. A. lateral and B. dorsal view of head of *Ophisternon inferniale* (Synbranchidae), after Vandel (1965); C. *Synocyclocheilus hyalinus* (Cyprinidae); D. *Speoplatyrhinus poulsoni* (Amblyopsidae), C. and D. modified after Weber in Wilkins *et al.* (2000), with permission from Elsevier Science; E. *Lucifuga subterraneus* (Bythitidae); F. *Typhlomolge rathbuni* (Plethodontidae), E. and F. after Eigenmann (1909).

mal evolution, and served to identify cave-adapted organisms without the unprovable and often erroneous assumption that they live only in caves implied by terms such as troglobite. While the term was originally used for morphological features, subsequent work has shown that it applies equally well to behavioural and physiological features. The Table summarizes some of the major morphological troglomorphic features.

Questions as to the adaptive nature of troglomorphic features are putatively answered by their convergent nature but many recent works have furnished more direct evidence (e.g. Christiansen, 1965; Culver, Kane & Fong, 1995 and references within).

Troglomorphy is not universal among cave organisms. In order for it to occur two factors have to be present: 1) a strong selection pressure for the development of a particular characteristic, and 2) the genetic and physiological or behavioural ability of the organism to respond to this pressure. Many cave organisms lack one or both of these factors. The phreatobite entomostracan crustacea and the extremely edaphic Collembola, such as the members of the family Onychiuridae, rarely show morphological troglomorphy. In many other groups, troglomorphy is questionable or inconsistent (Culver, 1982). It is also absent in cave environments such as guano piles or large masses of organic debris which are extremely energy rich. Peck has pointed out that troglomorphy should not be expected in these cases since troglomorphy only occurs when organisms are exploiting large volume spaces such as the surface of cave walls or floors (Peck, 1973), or large bodies of water. Indeed it has been long noted that the troglomorphic features of cave Entomobryidae are most closely approached in two very different surface habitats. The foot structure is closest to that seen in aquatic Collembola but the body shape features are most similar to those seen in forms (largely tropical) which live above the litter or soil, in trees. It is interesting that something similar to troglomorphy has been found in phreatobites where small size and elongate body form are developed convergently (Boutin & Coineau, 2000).

Where troglomorphy does occur, it allows for a separate analysis of those features clearly affected by the cave environment and those which are unaffected by it. It also gives clear polarity for phylogenetic analysis and permits a measure, or least an indication, of the degree of cave adaptation of different groups of organisms (Christiansen, 1961; Poulson, 1986).

Troglomorphy also shows varying degrees of taxonomic localization. In all cases there is some degree of convergent or parallel evolution in different regions and taxonomic groups. This is

Adaptation: Morphological (External): Common troglomorphic characteristics.

Morphological	Reference
Specialization of sensory organs (touch, chemoreceptor, hygroreceptor, thermoreceptor, pressure receptors)	Vandel, 1964; Weber, 2000
Elongation of appendages	Weber, 2000; Christiansen, 1961; Harvey, Shear & Hoch, 2000; Vandel, 1964
Pseudophysogastry	Vandel, 1964
Reduction of eyes, pigment, wings	Vandel, 1964; Weber, 2000
Compressed or depressed body form (hexapods)	Harvey, Shear & Hoch, 2000
Increased egg volume	Vandel, 1964
Increased size (Collembola, Arachnida)	Christiansen, 1961; Harvey, Shear & Hoch, 2000
Unguis elongation (Collembola)	Christiansen, 1961
Foot modification (Collembola, planthoppers)	Christiansen, 1961, Howarth et al., 1990
Scale reduction or loss (Fish)	Weber, 2000
Loss of pigment cells and deposits	Numerous
Cuticle thinning (terrestrial arthropods)	Numerous
Elongate body form (Teleost fishes, Arachnids)	Weber, 2000
Depressed, shovel-like heads (Teleost fishes and salamanders)	Weber, 2000
Reduction or loss of swim bladder	Romero & Paulson, 2001

usually transgeneric or even transfamilial, as in the case of the foot structure of European, Japanese, and American members of the genera *Pseudosinella* (Figure 1) and *Sinella* of the family Entomobryidae. In this case, as well as in the case of the similar changes in the family Orchesellidae, the adaptive significance of these changes is well known (Christiansen, 1965). In many cases, as with the flattening of the heads of troglomorphic fish and Amphibia (Figure 2), this is not so. Some troglomorphic features are so general (elongation of appendages, loss of pigmentation) as to be considered virtually universal.

Studies of troglomorphic evolution have produced a number of interesting ideas. Most exciting of all are the works of Culver, Kane, and their collaborators on *Gammarus minus*, which unites genetic, population genetic, population dynamic, and ecological analysis to answer the question of whether troglomorphic changes are in fact adaptive (e.g. Kane & Culver, 1992; Richardson & Kane, 1988). These works indicate that the troglomorphic features which have been studied are subject to selection and are in fact adaptive.

KENNETH CHRISTIANSEN

See also **Evolution of Hypogean Fauna**

Works Cited

Boutin, C. & Coineau, N. 2000. Evolutionary rates and phylogenetic age in some stygiobiontic species. In *Subterranean Ecosystems*, edited by H. Wilkens, D.C. Culver & W. Humphreys, Amsterdam and New York: Elsevier

Christiansen, K. 1961. Convergence and parallelism in Cave Entomobryinae. *Evolution*, 15(3): 288–301

Christiansen, K. 1962. Proposition pour la classification des animaux cavernicoles. *Spelunca*, 2: 76–78

Christiansen, K. 1965. Behavior and form in the evolution of cave Collembola. *Evolution*, 19(4): 529–37

Culver, D.C., Kane, T.C. & Fong, D.W. 1995. *Adaptation and Natural Selection in Caves: The Evolution of* Gammarus minus, Cambridge, Massachusetts: Harvard University Press

Culver, D.C. 1982. *Cave Life: Evolution and Ecology*, Cambridge, Massachusetts: Harvard University Press

Harvey, M.S., Shear, W.A. & Hoch, H. 2000. Onycophora, Arachnida, myriapods and Insecta. In *Subterranean Ecosystems*, edited by H. Wilkens, D.C. Culver & W. Humphreys, Amsterdam and New York: Elsevier

Kane, T.C. & Culver, D.C. 1992. Biological processes in space and time: Analysis of adaptation. In *The Natural History of Biospeleology*, edited by A.I. Camacho, Madrid: Museo Nacional de Ciencias Naturales

Peck, S.B. 1973. A systematic revision and the evolutionary biology of the *Ptomaphagus (Adelops)* beetles of North America (Coleoptera, Leiodidae, Catopinae), with emphasis on cave-inhabiting species. *Bulletin of the Museum of Comparative Zoology*, 45(2): 29–162

Poulson, T.L. 1986. Evolutionary reduction by neutral mutations: plausibility arguments and data from Amblyopsid fishes and Linyphiid spiders. *National Speleological Society Bulletin*, 47(2): 109–17

Richardson, R. & Kane, T. 1988. Orthogenesis and evolution in the 19th century, the idea of progress in American neo-Lamarckism. In *Evolutionary Progress*, edited by M. Nitecki, Chicago: University of Chicago Press

Romero, A. & Paulson, K.M. 2001. It's a wonderful hypogean life: A guide to the troglomorphic fishes of the world. *Environmental Biology of Fishes*, 62:13–41

Vandel, A. 1965. *Biospeleology: The Biology of Cavernicolous Animals*, Oxford and New York: Pergamon Press (originally published in French, 1964)

Weber, A. 2000. Subterranean organisms—fish and amphibia. In *Subterranean Ecosystems*, edited by H. Wilkens, D.C. Culver & W. Humphreys, Amsterdam and New York: Elsevier

Further Reading

Barr, T.C. 1968. Cave ecology and the evolution of troglobites. *Evolutionary Biology*, 2: 35–102

Barr, T. 1985. Pattern and process in speciation of Trechine beetles in Eastern North America (Coleoptera: Carabidae: Trechinae). In *Taxonomy, Phylogeny and Zoogeography of Beetles and Ants*, edited by G.E. Ball, Boston and Dordrecht: Junk

Christiansen, K. 1992. Biological processes in space and time: Cave life in the light of modern evolutionary theory. In *The Natural History of Biospeleology*, edited by A.I. Camacho, Madrid: Museo Nacional de Ciencias Naturales

Culver, D. & Wilkens, H. 2000. Critical review of relevant theories of the evolution in subterranean animals. In *Subterranean Ecosystems*, edited by H. Wilkens, D.C. Culver & W. Humphreys, Amsterdam and New York: Elsevier

Culver, D.C. 2001. The dark zone. *The Sciences*, 41(3): 30–35

Eigenmann, C.H. 1909. *Cave Vertebrates of America: A Study in Degenerative Evolution*, Washington, DC: Carnegie Institution of Washington

Hobbs, H.H. 2000. Subterranean organisms—Crustacea. In *Subterranean Ecosystems*, edited by H. Wilkens, D.C. Culver & W. Humphreys, Amsterdam and New York: Elsevier

Holsinger, J.R. 1988. Troglobites: The evolution of cave-dwelling organisms. *American Scientist*, 88: 146–53

ADAPTATION: MORPHOLOGICAL (INTERNAL)

Anatomical adaptations are the internal reflections of the morphological and behavioural modifications of subterranean animals. Total darkness, almost unique to the subterranean environment, favours those organisms that have developed or are in the process of developing non-visual sensory structures. These are represented by any structure employed in any kind of sensory reception: mechanical (air currents, touch), olfactory (smell), gustatory (taste), or auditory (sounds). These sensory structures tend to be more highly developed in cave animals compared to epigean organisms.

In the brains of cave fish the nervous centres responsible for non-visual sensations, memory, learning, and spatial orientation are hypertrophied and can be used to estimate the degree of eye regression, and of sensorial compensation in cave vertebrates. Telencephalon hypertrophy is the most homogenous feature in the cerebral development of hypogean fishes. The cerebellum of many subterranean fishes is larger and its dimension is related to the time since cave colonization. This part of the brain receives information from the lateral line, free neuromasts, and other nerve pathways. The degree of development reflects the locomotory activity of cave species. Sometimes, related stygobitic and stygophilic fishes co-exist in the same pool within a cave. American researchers who compared the brain of the highly adapted *Typhlichthys* sp. and the less adapted *Chologaster* sp. from the River Styx in Mammoth Cave (Kentucky) found obvious differences (Mohr & Poulson, 1966) (see Figure 1). The inner ear structures of hypogean fishes are also modified, generally being larger, and are used to maintain equilibrium by perception of body motion and position.

The pineal organ has a photoreceptive function in fishes and amphibians. It contains many photosensitive cells, similar to receptors within the eye, but also has neuroendocrine cells that synthesize and release the hormone melatonin in the presence of light. Thus, the absence of light in caves will induce the degeneration of the exterior parts of pineal photoreceptors of stygobitic animals. Some cavernicolous species with completely reduced eyes can still have functional pineal cells. This has been clearly demonstrated in the fish *Astayanax fasciatus* (Langecker, 2000).

The development of sensory structures in the dark is relatively well documented for some subterranean species. In contrast, the development of other organs not directly linked to sight has been very poorly studied, for example exocrine glands, producing pheromones. Cave beetles from the sub-family Leptodirinae that possess several types of glands with varying degrees of development or involution (depending on the phyletic line) have been described. The sternal gland is a multicellular gland present in the males of several Leptodirinae species. The *Speonomus* spp. are characterized by enhanced development of this gland but in marked contrast, *Batysciola* spp. are characterized by reduction of the sternal gland (Juberthie-Jupeau & Cazals, 1984; Moldovan & Juberthie, 1998) (see Figure 2). In the latter case, there is also a change from a secretion with less volatile particles to a very volatile secretion. However, this gland is absent in beetles highly adapted to life in subterranean habitats.

In subterranean environments, food input is not uniformly distributed in space or time and the primary adaptation of animals living in hypogean habitats is the improved ability to accumulate and store large energy reserves. Their lipids contain twice the energy per unit weight compared to proteins and carbohydrates. The adipose tissue can increase through excessive feeding, increased feeding efficiency, and improved metabolic pathways favouring lipid deposition during food-rich periods (Hüppop, 2000). This has also been observed in epigean animals, for example hibernating birds and mammals. In several terrestrial and aquatic cave animals an increase in fat content has been observed. Exceptional development of the adipose tissue has been recorded

Adaptation: Morphological (Internal): Figure 1. Differences between the brains of the stygobitic fish *Typhlichthys* sp. and the stygophilic *Chologaster* sp. (after Mohr & Poulson, 1966, modified).

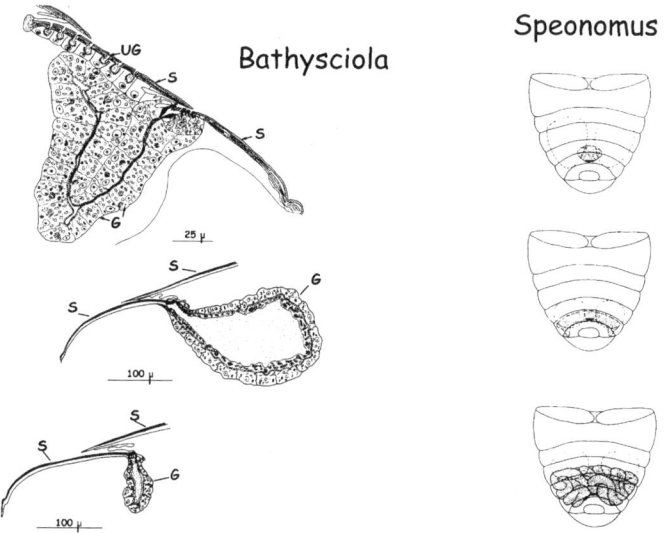

Adaptation: Morphological (Internal): Figure 2. Evolution of the sternal gland in two subterranean genera: *Bathysciola*—a soil species (top) and a cavernicolous species (bottom), and *Speonomus*—from different cavernicolous species: *S. delarouzei*: (top), *S. pyrenaus* (middle), and *S. hydrophilus* (bottom). UG = unicelluar gland, S = sternite, G = sternal gland.

in springtails, Leptodirinae beetles, remipedes, and shrimps. For example, cave beetles can survive for several months without feeding. In cave decapod crustaceans, the hepatopancreas, a major lipid storage organ, is larger than that of other decapods and lipids are also deposited in oleospheres, specialized compartments within the hepatopancreas. An interesting example is that of the larvae of the most highly adapted cave beetles from the Leptodirinae family that hatch from large, yolk-rich eggs, do not feed at all, have an atrophied and non-functional digestive apparatus, and break down the lipid reserves to protein (Deleurance-Glaçon, 1963). The adult digestive apparatus of soil and cave species only display differences linked to the presence of the chitinous spines in proventricular and mesenteric caecums. Cave animals accumulate large volumes of fat reserves in the internal spaces occupied by other structures in epigean species, such as the wing muscles and the sternal gland, and which are completely absent in subterranean species.

The storage of lipids in fish has been extensively studied. The primary organs used for storage are the subcutis, muscles, viscera, orbit of the reduced eyeball, and cranium. An interesting example is provided by two Texas cave fishes *Trogloglanis pattersoni* and *Satan eurystomus* (Siluriforme), whose swim bladders are totally reduced and their place filled by fat deposits (Langecker & Longley, 1993). This allows the maintenance of a small body, which is advantageous for animals living in subterranean crevices.

Food scarcity and stability of the environment also help explain the tendency for paedomorphosis and neoteny in some subterranean vertebrates. Besides the morphological features that characterize this retardation of somatic development, some anatomical changes have been observed in cave fish and amphibians, including reduced muscles and ossification. Together with a reduced metabolic rate, oxygen consumption and accumulation of lipids are all characteristic of hypothyroidism. In some subterranean fish underdevelopment of the thyroid has been recorded. However, in *Proteus anguinus* the thyroid is highly developed and in this instance we probably observe the insensitivity of the tissues to thyroxin, the hormone secreted by the thyroid (Langecker, 2000).

There have been relatively few detailed studies of anatomical adaptations to subterranean life and much remains to be learnt from the organisms that inhabit hypogean habitats.

OANA MOLDOVAN

Works Cited

Deleurance-Glaçon, S. 1963. Recherches sur les coléoptères troglobies de la sous-famille des Bathysciinae. *Annales des Sciences Naturelles, Zoologie*, 5(1): 1–172

Hüppop, K. 2000. Reduced, patchy or periodical food supply. In *Subterranean Ecosystems*, edited by H. Wilkens, D.C. Culver & W.F. Humphreys, Amsterdam and New York: Elsevier

Juberthie-Jupeau, L. & Cazals, M. 1984. Les différents types de glande sternale tubuleuse propre aux mâles de certains Coléoptères Bathysciinae souterrains. *Compte Rendu de l'Academie de Science de Paris*, 298(14): 393–96

Langecker, T.G. 2000. Lack of light. In *Subterranean Ecosystems*, edited by H. Wilkens, D.C. Culver & W.F. Humphreys, Amsterdam and New York: Elsevier

Langecker, T.G. & Longley, G. 1993. Morphological adaptations of the Texas blind catfishes *Trogloglanis pattersoni* and *Satan eurystomus* (Siluriformes: Ictaluridae) to their underground environment. *Copeia*, 93(4): 976–86

Mohr, C.E. & Poulson, T.L. 1966. *The Life of the Cave*, New York: McGraw-Hill

Moldovan, O.T. & Juberthie, C. 1998. Sternal gland in the species of *Bathysciola* (Coleoptera: Cholevidae: Leptodirinae). *Mémoires de Biospéologie*, 25: 107–10

ADAPTATION: PHYSIOLOGICAL

In subterranean ecosystems, the main environmental factors influencing physiological adaptations in cave organisms are darkness, lack of food, and also hypoxic conditions. These factors have a high selective value in regard to adaptive features such as a drastic reduction in eye size and a reduced metabolic rate compared with that of surface-dwelling relatives. To study the physiological adaptations of cave organisms to the subterranean environment, it is necessary to compare troglobites with troglophilous and epigean species. Studies have been carried out on aquatic and terrestrial invertebrates and vertebrates by investigating oxygen consumption and metabolic responses.

Oxygen Consumption

The most meaningful information on metabolic rate is given by measuring the oxygen consumption of the organism's whole body rather than just part of it. In most comparisons, hypogean species show generally lower metabolic rates than their epigean relatives. For example, the hypogean crustaceans *Stenasellus virei*, *Niphargus virei*, and *N. rhenorhodanensis* have a reduced metabolic rate because they show O_2 consumption rates in normal oxygenation 1.6–4.5 times lower than the epigean *Asellus aquaticus* and *Gammarus fossarum* (Hervant et al., 1998). The same was found with salamanders, among which the obligate cave-dweller *Proteus anguinus* has an oxygen consumption as low as one-eighth that of the facultative cave-dweller *Euproctus asper* (Hervant et al., 2000).

Reduced activity and decreased metabolic rates are a response to a selection pressure to economize on energy use because food and oxygen are in short supply (Culver, 1982; Malard & Hervant, 1999). Usually, cave animals have a higher basal metabolic rate than interstitial fauna. Culver & Poulson (1971) showed that individuals of *Stygobromus emarginatus* inhabiting more open water have higher activity rates than those of *S. spinatus* living deeper in the gravels of the stream bed. In the same way, Mathieu (1980, 1983) found interstitial *N. rhenorhodanensis* to be less active than cave-adapted ones.

A low metabolic rate is also an adaptation to a dysoxic environment. During periods of hypoxic stress, hypogean crustacean species possess a lower critical pO_2 (oxygen partial pressure) than epigean ones, which may indicate that these organisms are better adapted to low O_2 tensions (Hervant et al., 1998), and probably maintain a more efficient aerobic metabolism for a longer time in declining pO_2 pressure instead of partly switching to anaerobic metabolism. Post-hypoxic recovery results in a high oxygen debt (the excess O_2 consumed during post-hypoxia). The main expla-

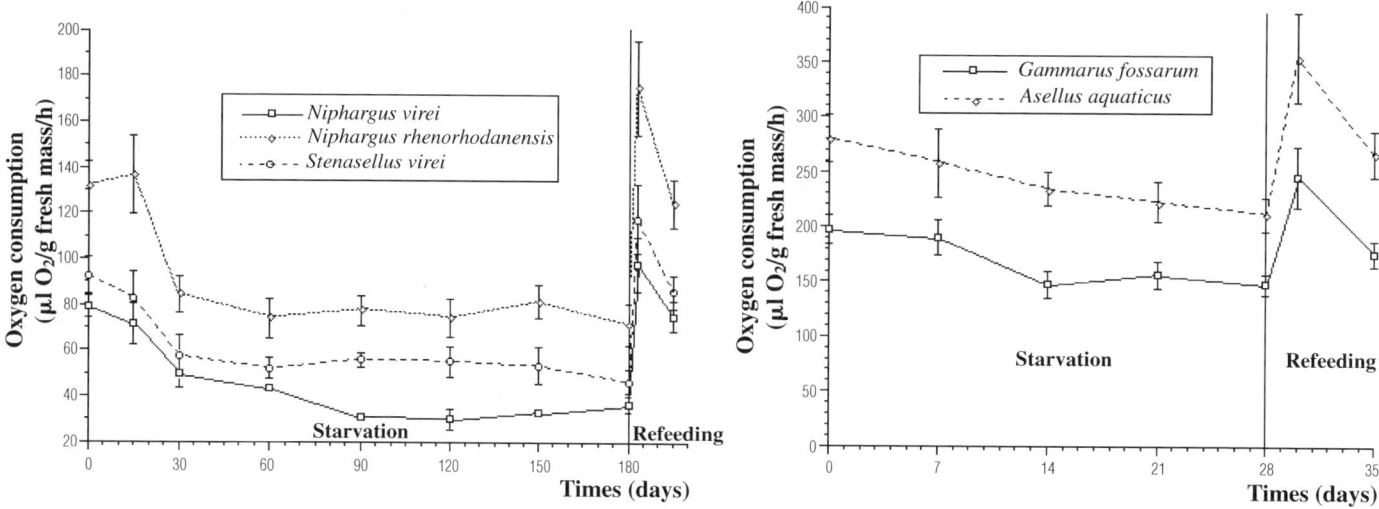

Adaptation: Physiological: Oxygen consumption of the epigean *Gammarus fossarum* and *Asellus aquaticus*, and of the hypogean *Niphargus virei*, *N. rhenorhodanensis*, and *Stenasellus virei* during long-term starvation and subsequent feeding at 11°C and in darkness. Values are means ± standard deviation for $n = 8–10$ animals (modified from Hervant et al., 1997).

nation for the smaller O_2 debt shown by hypogean species is the lower energetic expenditure during hypoxia, due to a decrease in movement and ventilation rates.

During long-term starvation, the respiratory—but also locomotory and ventilation—rates are drastically reduced in hypogean crustacean species, whereas epigean species show a smaller decrease in these rates and respond by a marked and transitory hyperactivity (Hervant et al., 1997) (see Figure). During such a long-term starvation, the troglobite vertebrate *Proteus anguinus* shows a slight increase in oxygen consumption, after which its respiratory rate decreases. By contrast, the epigean *Euproctes asper* shows an immediate and high increase in its rate of respiration, which decreases later to a value similar to that for well-fed individuals (Hervant et al., 2001).

Metabolic Adaptations

The physiological adaptations of cave animals may also be understood by examining intermediary and energy metabolism. Both epigean and hypogean animals respond to long-term severe hypoxia with classic anaerobic metabolism, characterized by a decrease in ATP and phosphagen, the use of glycogen, and the accumulation of lactate (Hervant et al., 1995; 1996). Compared with epigean crustaceans, the hypogean amphipod *N. virei*, which uses coupled fermentation of glycogen and amino acids, stores large amounts of glycogen and arginine phosphate. After severe hypoxia, *N. virei* appears to implement a strategy of lactate removal quite different from that observed in epigean crustaceans, favouring lactate-supported gluco- and glyconeogenesis and rapid glycogen replenishment, instead of rapid lactate removal via oxidative pathways.

The metabolic response to prolonged food deprivation in epigean amphipods appears to be monophasic, showing an immediate and large decline in all energy reserves. In contrast, hypogean amphipods display successive periods of glucidic, proteo-glucidic, and then lipidic-dominant catabolism during food deprivation. Lipids and proteins are the most metabolized substrates, whereas glycogen contributes little energy (Hervant et al., 2001).

This high degree of storage, the rapid restoration of fermentable metabolic fuel without feeding, and the ability to reduce glycolytic flux and energy expenditure due to a decrease in locomotion and ventilation activity, allow hypogean species to survive harsh conditions, and consequently are very good adaptations to the subterranean environment.

Sensory Functions

It is generally assumed that the distinctive morphology of cave animals results from selection for increased sensory organs on appendages, which in turn results in lengthened appendages to compensate for eye and pigment cell degeneration (Culver, 1982). A good example of such adaptations is given by the amphipod *Gammarus minus*, which can be found in large caves and in springs in Virginia. Cave-dwelling *G. minus* populations have smaller eyes and longer antennae than surface populations, and they also have smaller optic ganglia and larger olfactory lobes in their central nervous system. In some other species of arthropods, the antennae of hypogean species are relatively short compared to those of their epigean relatives (Culver, 1982). It has also been demonstrated that the blind crayfish *Orconectes australis packardi* learns about its environment using both tactile and chemosensory cues, mainly located on the antennae but also elsewhere on the body (Li & Cooper, 2001).

A classic example of sensory compensation and cave adaptation in general is also provided by primitive vertebrates, one obvious feature being the hypertrophy of the lateral line system observed in cave fishes or amphibians. This is clear in the amblyopsid cave fishes, where, because of differences in neuromast morphology, some species are more sensitive to general water movement, and others are more sensitive to the direction of the current (Poulson, 1963). Hypertrophy and differential sensitivity certainly confer an increased ability to find food and detect predators in an environment where food is scarce.

JACQUES MATHIEU AND FRÉDÉRIC HERVANT

Works Cited

Culver, D.C. 1982. *Cave Life: Evolution and Ecology*, Cambridge, Massachusetts: Harvard University Press

Culver, D.C. & Poulson, T.L. 1971. Oxygen consumption and activity in closely related amphipod populations from cave and surface habitats. *American Midland Naturalist*, 85(1): 74–84

Hervant, F., Mathieu, J. & Durand, J.P. 2000. Metabolism and circadian rhythms of the European blind salamander *Proteus anguinus* and a facultative cave dweller, the Pyrenean newt (*Euproctes asper*). *Canadian Journal of Zoology*, 78: 1427–32

Hervant, F., Mathieu, J. & Durand, J.P. 2001. Behavioural, physiological and metabolic responses to long-term starvation and refeeding in a blind cave-dwelling (*Proteus anguinus*) and a surface-dwelling (*Euproctes asper*) salamander. *Journal of Experimental Biology*, 204: 269–81

Hervant, F., Mathieu, J. & Messana, G. 1998. Oxygen consumption and ventilation in declining oxygen tension and posthypoxic recovery in epigean and hypogean aquatic crustaceans. *Journal of Crustacean Biology*, 18: 717–27

Hervant, F., Mathieu, J., Garin, D. & Freminet, A. 1995. Behavioural, ventilatory, and metabolic responses to severe hypoxia and subsequent recovery of the hypogean *Niphargus rhenorhodanensis* and the epigean *Gammarus fossarum* (Crustacea: Amphipoda). *Physiological Zoology*, 68(2): 223–44

Hervant, F., Mathieu, J., Garin, D. & Freminet, A. 1996. Behavioural, ventilatory, and metabolic responses of the hypogean amphipod *Niphargus virei* and the epigean isopod *Asellus aquaticus* to severe hypoxia and subsequent recovery. *Physiological Zoology*, 69(6): 1277–1300

Hervant, F., Mathieu, J., Barré, H., Simon, K. & Pinon, C. 1997. Comparative study on the behavioural, ventilatory, and respiratory responses of hypogean crustaceans to long-term starvation and subsequent refeeding. *Comparative Biochemical Physiology*, 118A(4): 1277–83

Li, H. & Cooper, R.L. 2001. Spatial familiarity in the blind cave crayfish *Orconectes australis packardi*. *Crustaceana*, 74(5): 417–33

Malard, F. & Hervant, F. 1999. Oxygen supply and the adaptations of animals in groundwater. *Freshwater Biology*, 41: 1–30

Mathieu, J. 1980. Activité locomotrice et métabolisme respiratoire de deux populations de *Niphargus rhenorhodanensis* mesurés à une température de 11°C. *Crustaceana*, suppl. 6: 160–69

Mathieu, J. 1983. Activité locomotrice de *Niphargus rhenorhodanensis* en fonction de différentes conditions expérimentales. *Mémoires de Biospéologie*, 10: 401–05

Poulson, T.L. 1963. Cave adaptation in amblyopsid fishes. *American Midland Naturalist*, 70: 257–90

Further Reading

Camacho, A.I. (editor) 1992. *The Natural History of Biospeleology*, Madrid: Museo Nacional de Ciencias Naturales

Culver, D.C., Kane, T.C. & Fong, D.W. 1995. *Adaptation and Natural Selection in Caves: The Evolution of* Gammarus minus, Cambridge, Massachusetts: Harvard University Press

Ginet, R. & Decou, V. 1977. *Initiation à la biologie et à l'écologie souterraines*, Paris: Delarge

Vandel, A. 1965. *Biospeleology: The Biology of Cavernicolous Animals*, Oxford and New York: Pergamon Press (originally published in French, 1964)

AFRICA, NORTH

The African continent is poor in limestone outcrops in general, but the northern zone has abundant karst and Algeria has the longest and deepest cave systems in Africa (Boll Maza, 18.6 km, Anou Ifflis, −1170 m, Anou Boussouil, −805 m). For a long time Morocco had the continent's deepest cave (Kef Toghobeit, −722 m) and one of the longest cave networks (Wit Tamdoun, 17.5 km). There is very good potential for further discoveries in both countries. Tunisia has only a few karstified massifs, which are not very well known. Libya possesses a few areas of interesting gypsum karst, and Egypt has very few caves. Climates of North Africa vary with annual rainfalls ranging from >1000 mm in high mountains (Rif and Atlas ranges) to <150 mm in the Sahara. Karst morphology is equally variable.

Morocco

In Morocco, less than 1000 caves have been discovered, but the country has more than 100 000 km² of karst terrain (see Figure 1). There are three geological domains north of the Saharan border, which correspond to three orogenic phases: (1) the Anti-Atlas domain near the Saharan border has folded Precambrian rocks forming mountains ranging from 2000 to 3000 m in elevation; (2) the Atlas domain is a second range with Caledonian, Hercynian, and Alpine structures; it is the largest of the three domains and has very different stratigraphies and lithologies; and (3) the Rifin domain, the smallest and youngest domain from the late Alpine. These three domains are separated by major fault zones. Karst landscapes, especially on limestone, are well represented in these three domains. Paleozoic limestone occurs in the Anti-Atlas domain (18 000 km²) where it is more than 300 m thick in the western part, becoming thinner in the east. Superficial karst forms are developed, but caves are rare.

Lias limestone and dolostone form the largest karst terrain in Morocco, 30 000 km² in extent, mainly in the Middle and High Atlas ranges. These formations are large reservoirs that sustain the base flow of Morocco's major rivers. They measure up to 500 m thick in the calcareous High Atlas. Alpine karst is well developed with karren, dolines, poljes, and shafts in the higher altitudes of the High Atlas with good rainfall (600 to 1000 mm annual rainfall) and snow cover during winter. One of Africa's deepest caves, Kef Toghobeit (−722 m), is also developed in this formation, in the Rif (Northern Morocco). In the Middle Atlas, lower than the High Atlas, the karst landforms are also dense and various, but the limestones facies are not very

Africa, North: Table 1. Major Caves of Morocco.

Name	Region	Length/Depth
Wit Tamdoun	Tazroukht, Agadir	17 500 m
Rhar Chara	Middle Atlas, Taza	7650 m
Kef Aziza	Central Atlas, Bou Denib	3950 m
Kef Toghobeit	Rif, Bab Taza, Chaouene	3918 m
Rhar Chiker	Middle Atlas, Taza	3865 m
Kef Toghobeit	Rif, Bab Taza, Chaouene	−722 m
Kef Tikhoubaï	Middle Atlas, Taza	−310 m

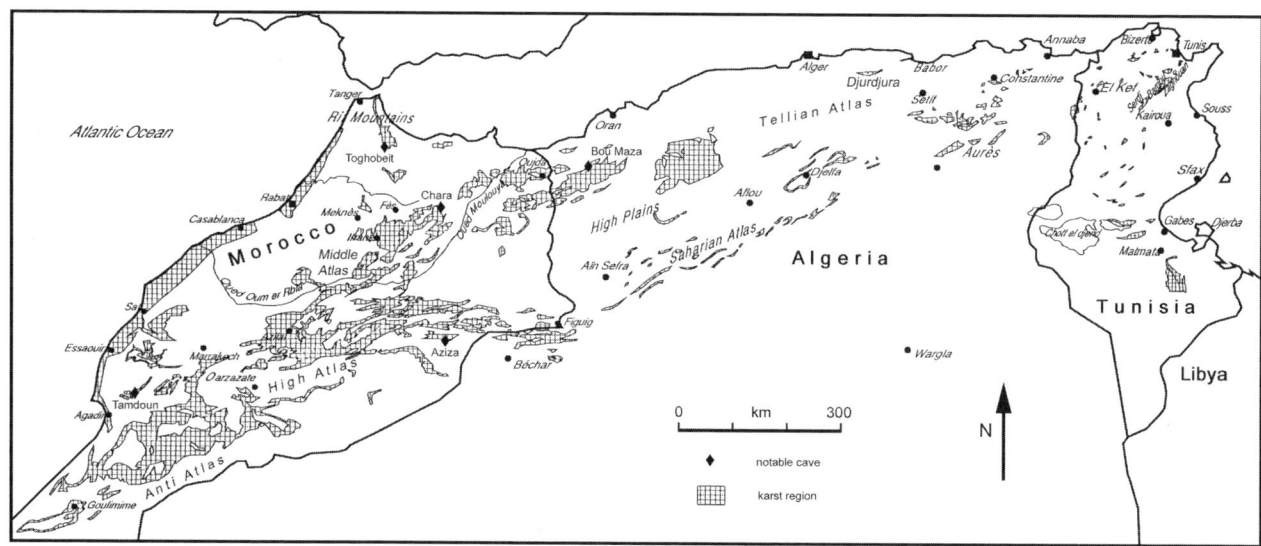

Africa, North: Figure 1. The main karst regions and locations of the major caves in the Moroccan and Algerian sectors of the Atlas Mountains in North Africa.

good for cave development. The more frequent landforms are poljes developed upon the limestone plateaus. Among the other types of closed depressions are "issianes", shallow basins due to superficial corrosion following the truncation of paleovalleys, and large dolines whose origin lies in the dissolution of underlying Triassic salt beds. In the fresh volcanic landscapes, rugged holes are scattered across the outcrops of basaltic sheets. Some may be related to the collapse of lava tunnels, but most are

Africa, North: Figure 2. The Ighi bou Ilaghamane is a clean shaft in the floor of a dry valley in the Moroccan karst. (Photo by Pete Hart)

interstratal karstic collapses into the underlying limestones (Martin, 1981, and see photo in Dolines entry).

Middle and Upper Jurassic limestones are also very common in the High Atlas, Middle Atlas, and Eastern Morocco ($c.20\ 000\ km^2$). They are thinner and are less homogenous. Some interesting karst landforms have been described on these limestones including "wave karst", a succession of small parallel dry valleys with asymmetrical dolines on the Ait Abdi plateau (Azilal), which are aligned with the main wind direction due to control by snow accumulations (Couvreur, 1974; Perritaz & Monbaron, 1998). On this plateau, an ancient cave network with vertical shafts suggests that climates were wetter in the past. Radiometric dating of speleothems yielded ages between 3200 and 220 000 years, and some beyond the range of the U–Th method ($> 400\ 000$ years). The lateral flow is conducted by an interstratal network, inactive and dry in the upper part, but active at the base near the regional aquiclude, attesting three karstification phases. Morocco's longest cave, the Wit Tamdoun underground river, is developed in Malm limestone, in the Western High Atlas, near Agadir.

Nearly everywhere in Morocco, karst springs are associated with calcareous tufas. Some deposits are very large and form attractive landforms, especially where they obstruct valleys. The Ouzoud cascades and the Imi n'Ifri natural bridge (Azilal) are well-known examples.

Algeria

In Algeria people have used caves over a long period of time as shelters, sanctuaries, and sheepfolds. Cave wall paintings have been discovered in the shelters in Tassili N'Ajjer, in the southern part of Algeria. Many caves are still used today by shepherds (Collignon, 1997). The first modern cave explorations were by French scientists and speleologists at the beginning of the 20th century. In 1948, Anou Boussouil, then 505 m deep, was briefly the world's second deepest known cave, and the Tafna underground river was Africa's longest known cave in 1933. During the war for independence (1954–62), explorations could not be made, but caves were widely used as refuges, arsenals, or small hospitals. Systematic explorations by French and Belgian immigrants began again after 1973. Expeditions were organized by European cavers and geologists, and Algerian caving groups were founded, but the civil war that began in 1991 again stopped exploration. There are many small limestone ranges scattered throughout northern Algeria, and four terrains have been found to contain significant karst—the Oran Meseta (plateau), the Tell mountain ranges, the Constantine tabular massifs, and salt and gypsum diapirs.

The Oran Meseta are large karstic plateaus in western Algeria comprising some 6000 km^2 of mostly Jurassic limestone and dolostone between 800 m and 1500 m in elevation. Large underground rivers, siphons, and the main Algerian poljes are found there. In the Tlemcen and Saïda mountains, about ten stream-sinks and springs have been dived, from which the country's largest underground rivers can be reached (including Tafna, 18.6 km of large galleries). These rivers are generally of low gradient and represent some of the region's main water resources, especially during the dry summer season. Some caves are used as pumping stations and speleologists have significantly contributed to their exploitation at Rhar El Khal and Hassi Dermam.

The Tellian Atlas contains many small karst areas, of which Djurdjura and Babors, composed of Jurassic and Eocene limestones folded into one another, are the most spectacular. The highest parts (c.2300 m) are all composed of limestone. Classical high mountain range karst landforms (karren, snow pits, dolines, perennial underground ice) are well developed, in areas with the highest rainfalls in the country (more than 1000 mm a^{-1}). In 1950, Anou Boussouil, a large pit in the Djurdjura, was explored to a depth of 505 m (the world's second deepest pit at that time). In 1980 it became Africa's deepest cave (-805 m). New groups of cavers came to Djurdjura and discovered Anou Ifflis (-1170 m, Africa's only cave >1000 m deep) and about 20 other pits, of which three are over 200 m deep.

In the Constantine region, Cretaceous neritic limestone with many dolines and shafts forms small anticlines that overlook the surrounding eastern Algerian plains in a terrain of 18 000 km^2 where it is more than 300 m thick in the western part, becoming thinner in the east. Superficial karst forms are developed, but caves are rare.

In most of northern Algeria, Jurassic limestones overlie Triassic sequences with abundant salt and gypsum layers in anticlines,

Africa, North: Figure 3. The AA4 shaft on the Ait Abdi plateau in Morocco's Central High Atlas, that descends into a cave is rich in speleothems that are evidence of more humid climates in the past. (Photo by Luc Perritaz)

diapirs, or salt domes. The Djebel Nador, in the east of the country, possesses one of the largest diapiric structures in Africa. In such zones there are abundant dolines and pits. The rock is weak and roofs are often collapsed in many places, dividing long galleries into many smaller caves. In Algeria some of the world's deepest gypsum caves (e.g. Dahredj Ghar Kef in Djebel Nador, 2450 m long, 212 m deep) and salt caves (El Outaya, -55 m) have been discovered. Salt outcrops are quickly dissolved by rainwater, and caves exist only in very dry regions with active diapiric uplift.

Algeria is tectonically active and has many thermal springs. Some of these hydrothermal flows cross the limestone massifs, where they have produced caves with cupolas and calcite and gypsum encrustations (especially when water contains carbon dioxide and hydrogen sulfide). The small Biban massif in the Tellian Atlas has some wonderful caves (Rhar Es Skhoun, Rhar Amalou, Rhar Medjraba) with very rare speleothems, but exploration is hindered by the high temperatures.

Tunisia

In spite of the fact that a large portion of the Tunisian landscape is composed of limestone (or dolomite), there are only a few karstified ranges (for example djebels Zaghouan Oust, Ben Saïdane, and Fkirine, see Figure 1). In addition to this, the carbonates are relatively thin, except in the djebels Serdj, Bargou, Taboursouk, and Cap Bon. The Djebel Serdj (1357 m high) has been extensively explored and contains the country's largest cave networks: the Rhar Ain et Tsab cave (2700 m long, 160 m deep) and the Mine cave (also called Rhar Djebel Serdj cave, 1700 m

Africa, North: Table 2. Major caves of Algeria.

Name	Region	Length/Depth
Rhar Bou Maza (Tafna)	Tlemcen	18 600 m
Kef El Kaous	Traras	4160 m
Anou Boussouil	Djurdjura	3200 m
Anou Ifflis	Djurdjura	-1170 m
Anou Boussouil	Djurdjura	-805 m
Anou Achra Lemoun	Djurdjura	-323 m

long, 267 m deep). Lhopiteau (1980) grouped Tunisian caves into oasis caves (mainly short and dry); many types of southern djebel caves (in climatic transition zones); northern Tunisia caves (very similar to European ones); gypsum caves (small and not deep, with good mineral concretions); and thermal springs and caves.

Libya

In spite of the desert climate, which is unfavourable for karstification, some remarkable caves have nonetheless developed. A gypsum karst is associated with the Upper Jurassic Bir al Ghanam formation, extending 100 km southeast of Tripoli to, and beyond, the border with Tunisia. The formation, which is 400 m thick and horizontal, consists of two gypsum beds separated by an essentially dolomitic bed. Both the upper and lower gypsum beds are karstified and contain numerous caves. Some 7 km of passage has been surveyed, including Umm al Masabih cave (3593 m long). The caves are mainly linear, carrying ephemeral streams (active several hours a year during rain-generated floods), and they display vadose morphology. Bedding planes and joints both play a role in passage development. In the karst, the upper gypsum is removed by erosion and the plateau surface lies on the more resistant dolomite. Underlying caves cause collapse features, and many of these contain cave entrances and swallets (Klimchouk *et al.*, 1996).

Egypt

About half the area of Egypt is on outcrops of limestone and chalk, but there is minimal karst development in the harsh desert climates. A low plateau of Mesozoic and Tertiary limestones separates the Nile Valley from the sand seas of the Western Desert, and the same limestones extend eastwards to the north end of the Gulf of Suez. Across most of these areas, karst landforms are barely seen in terrains of bare rock escarpments and aeolian dune fields (Stringfield, LaMoreaux & LeGrand, 1974). The few small springs and caves are known because they provide water and shelter in the desert; typically, the location of St Antony's Monastery, east of Cairo, is due to its karstic spring and caves (only a few metres long). In the Western Desert, the Farafra oasis is famed for its spectacular "White Desert", a basin 100 km across distinguished by thousands of pinnacles of snow-white chalk that stand clear of the orange desert sand (Waltham, 2001). These are relict karst features, mostly pinnacles 2–4 m tall, but there are small areas of towers 10–15 m tall. They were formed in wetter climates of the late Tertiary, and are now being degraded by thermal shattering and aeolian sand abrasion. Travertines in the nearby Kharga oasis, which are now dry and largely predate 750 000 years ago, are probably contemporary with the pinnacle and tower karsts. In the Eastern Desert, about 70 km southeast of Beni-Suef city, an unusual cave in alabaster was intersected by blasting. Sannur Cave is a crescent-shaped chamber which at 275 m is the largest cave in Egypt (Günay *et al.*, 1997).

LUC PERRITAZ

Works Cited

Collignon, B. 1997. La spéléologie française en Afrique: Algérie. *Spelunca Mémoires*, 23: 141–43

Couvreur, G. 1974. Le Rôle de la neige dans l'évolution des formes karstiques de haute montagne du Haut Atlas central (Maroc). *Phénomènes karstiques*, vol. 2, Paris: CNRS (Mémoires et documents du CNRS, new series 15)

Günay, G., Ekmekçi, M., Bayari, C.S. & Kurttas, T. 1997. Sannur Cave: A crescent-shaped cave in Egypt. In *Karst Waters and Environmental Impacts*, edited by G. Günay & A.I. Johnson, Rotterdam: Balkema

Klimchouk, A., Lowe, D., Cooper, A. & Sauro, U. (editors) 1996. Gypsum karst of the world, *International Journal of Speleology*, Theme Issue 25(3–4)

Lhopiteau, J.-J. 1980. La Tunisie spéléologique. *Spéléo-Drack*, 14

Martin, J. 1981. Le Moyen Atlas central: Etude géomorphologique. *Notes et Mémoires du Service Géologique du Maroc*, 258

Perritaz, L. & Monbaron, M. 1998. Geomorphological approach to the Aït Abdi Karst Plateau (Central High Atlas, Morocco). *Zeitschrift für Geomorphologie* N.F., Suppl.-Bd. 109: 83–104

Quinif, Y. 1976. Les Karsts du Constantinois: Aspects spéléologiques. *Recueil d'articles parus dans Subterra*, 64–65 (1975), 66–68 (1976)

Stringfield, V.T., LaMoreaux, P.E. & LeGrand, H.E. 1974. Western Desert of Egypt. *Bulletin of Geological Survey of Alabama*, 105: 62–106

Waltham, T. 2001. Pinnacles and barchans in the Egyptian desert. *Geology Today*, 17: 101–104

Further Reading

Baritaud, T. 1997. La Spéléologie française en Afrique: Tunisie. *Spelunca Mémoires*, 23: 151–52

Collignon, B. 1992. La Spéléologie en Algérie. *Spelunca*, 48: 14–24

Halliday, W.R. 1998. Caves of Egypt. *D.C. Speleograph*, May 1998

Lips, B. 1997. La Spéléologie française en Afrique: Maroc. *Spelunca Mémoires*, 23: 148–49

AFRICA, SOUTH: ARCHAEOLOGICAL CAVES

The archaeological caves of South Africa are remarkable in that together they contain a long history of evolution of hominids (hominins, by some authorities), beginning with the Australopithecines at over three million years ago, through to the first tool users in the early Stone Age (ESA; 2.5 million to 250 000 years ago), the middle Stone Age (MSA; 250 000 to 22 000 years ago), *Homo sapiens* in the MSA (at least 100 000 years ago) and the late Stone Age (LSA; 22 000 to 2000 years ago) and up to the historical period (see Figure 1). The Iron Age in southern Africa is marked by the arrival of the taller Bantu-type peoples—farmers—into an area that had been predominantly bushman territory. The bushmen, mainly hunter-gatherers, are also noted for their rock art, and they currently live in the drier areas of southern West Africa.

The first Australopithecine to be uncovered was the Taung child skull found at the Buxton Lime Quarry, in the Kalahari Desert, near the Namib border in 1924. This site is not a cave but a very large set of tufa (travertine) deposits, and the skull was unearthed by blasting operations, along with fossil baboons and other fauna. Raymond Dart at Witwatersrand University, Johannesburg, assigned the skull to a new genus, *Australopithecus africanus* (meaning "southern ape" for the genus and "of Africa"

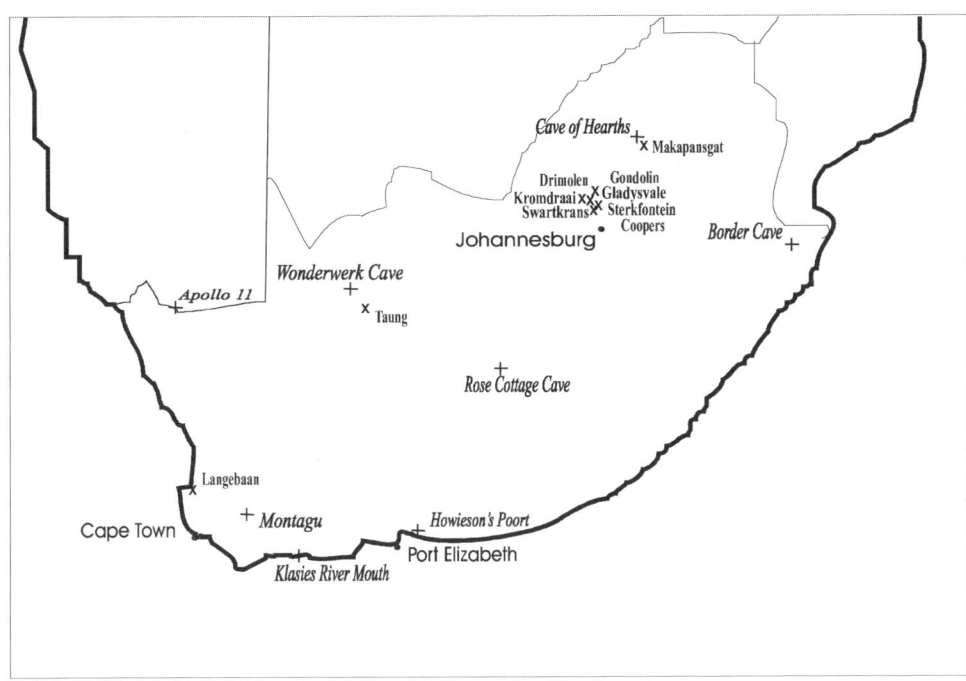

Africa, South: Archaeological Caves: Figure 1. Southern Africa locations of caves mentioned in the text. "X" marks the Australopithecine sites north of Johannesburg. "+" signs represent *Homo* and later cave sites. Langebaan is an open-air site that has provided a rich source of mammalian fauna for the late Miocene. It has been used to aid faunal dating of Makapansgat and other Australopithecine sites.

for the species name) claiming a "missing-link" status for it. It took more than 30 years for the skull and similar finds to be accepted by British authorities, such as Sir Arthur Keith (of Piltdown fame), as being germane to the evolution of humans. By the association of the faunal remains at the site to similar dated fauna of East Africa, the Taung skull is currently reckoned to be about 2.4 to 2.6 million years old. Berger & Clarke (1995) have reconstructed an interesting story of how the skull and other fauna may have resulted from eagle predation.

The Australopithecine caves of South Africa formed in Proterozoic dolomites and, in addition to the usual clastic infill, originally contained massive amounts of pure speleothems in the form of stalagmites and flowstones. In the early part of the 1900s these speleothems were mined for gold extraction in the Witwatersrand goldfields around Johannesburg and Pretoria. The Australopithecine fossils turned up with the speleothems and with the unwanted breccias. Among the hominid finds was the skull, "Mrs Ples", at Sterkfontein, blasted out of a flowstone in two pieces. Robert Broom, who came to be in charge of the acquired site, assigned it to *Plesianthropus transvaalensis*. Later, Robinson reassigned it to *Australopithecus*. Sterkfontein has turned out to be a rich site: the latest find, by Ron Clarke and Nkwane Molefe, being an almost complete skeleton known as "Little Foot" from the Silberberg Grotto near the bottom of a shaft inside the cave (Clarke, 1998). By magnetostratigraphy, the skeleton is currently assigned an age of between 3.30 and 3.33 Ma BP (Partridge *et al.*, 1999).

Across the shallow valley from Sterkfontein lies the mined site of Swartkrans. The fossil hominid sites of Sterkfontein, Swartkrans, and Kromdraai were declared a World Heritage Site in 1999. Swartkrans too contained *Australopithecus* remains, although the skulls of many of the finds are more robust than the early finds at Sterkfontein. The Swartkrans skulls possess a sagittal crest, more robust jaws and more massive molar teeth, indicating a more powerful masticatory action than the gracile Australopithecines of Sterkfontein and Makapansgat. The robust species are referred to either as *Australopithecus robustus* or to a new genus, *Paranthropus robustus.* There are active debates as to how distinct these genera are from each other and which of these or the East African Australopithecines is a direct ancestor to the *Homo* lineage. The later, long, excavation at Swartkrans was largely the work of C.K. "Bob" Brain, of the Transvaal Museum, Pretoria, who went about reconstruction of the stratigraphy in a painstaking enterprise (see Figure 2). Brain also tested Dart's ideas about the nature of the remains at another site, the Makapansgat Limeworks Cave. This huge cave is in the Northern Province (old Transvaal) about 300 km north of the Johannesburg group. Dart had been shown shattered animal long bones that seemed to have been burned artificially. From these, he pictured a savage Australopithecine culture using bones, teeth, and horns as weapons and tools and using fire inside the cave, and he coined the term "osteodontokeratic" (bone–teeth–horn) culture. It was later shown that the burnt appearance was due to infiltration by manganese dioxide, which is black. Brain carried out a series of heating experiments on bone to show how different stages of burning might appear on site. He and others also investigated hyena dens to show, firstly, that the thick bone accumulations at the Limeworks Cave and elsewhere were due to animal denning and, secondly, that the marks on the bones (when properly prepared from the matrix in the laboratory) were due to hyena tooth marks and the gnawing of other animals such as porcupines.

The study of how buried bones and artefacts result in their present state is known as taphonomy, and Dart and Brain are thus two of the earliest researchers to promote cave taphonomy. As an interesting example of a story in taphonomy, Brain showed fairly convincingly that the puncture marks on one of the juvenile Australopithecine skulls resulted from being dragged by a leopard to the cave site. The upper canines of a leopard closely

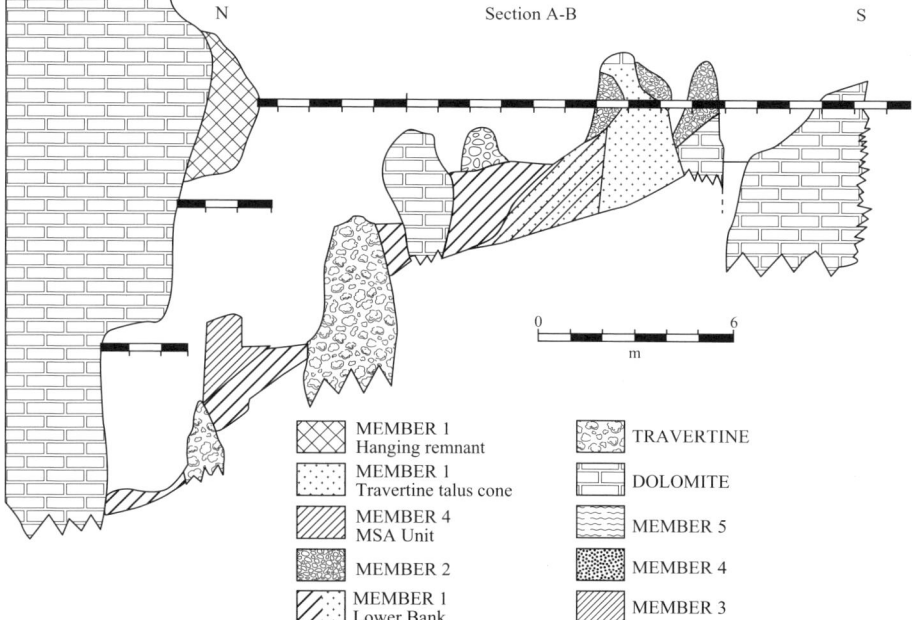

Africa, South: Archaeological Caves: Figure 2. Plan and section of Swartkrans Cave (from Brain, 1993, with permission). The section shows the cave in its present state of partial excavation.

fit the puncture marks in the top of the skull cap (calvaria). This in turn has implications on whether the cave functioned as an Australopithecines' "home" site, in the way that baboons often use caves, or was a leopard (or felid) lair (Brain, 1981).

The later deposits of Sterkfontein, Swartkrans, and other caves have yielded remains of tool-making *Homo*, variously suggested to belong to either *H. habilis*, *H. erectus*, or *H. ergaster*. The remains occur together with some of the earliest ESA stone tools in the form of simple choppers of Oldowan type (Clark, 1993).

How are the remains and artefacts dated? The two main problems with all these sites is that they are very difficult to unravel stratigraphically and the deposits in which they lie are not readily amenable to direct, radiometric, dating. This is in contrast to their East African Rift, and Ethiopian, counterparts where there is often simple "layer-cake" stratigraphy and where there is often recognizable inter-site stratigraphic correlation. Riverine and lake sediments can be dated by magnetostratigraphy, and ash horizons are datable by potassium–argon (K–Ar) dating (or Ar–Ar dating).

The Johannesburg group of Swartkrans, Sterkfontein, Drimolen, Kromdraai, Gondolin, Coopers, and Gladysvale are mainly eroded rift systems to the surface (avens from below), with complex collapse phases of breccias and flowstones with uncertain contacts and rapid lateral changes of facies. Sterkfontein and other caves still possess streams flowing sluggishly under a low gradient. In most cases, the miners got there before the archaeologists, and their excavations reveal some relationships but obscure and confuse others. Magnetostratigraphy of the sediment and flowstone successions is also an uncertain game, since the various polarities that are recorded in the deposits are stretched and compressed due to different depositional rates and can therefore be fitted to almost any part of the global polarity time-scale (GPTS). As for radiometric methods, there are no ash deposits (no nearby volcanoes) and so Ar–Ar cannot be used. The hominid-bearing speleothem deposits are not less than 500 000 years old, and so they are beyond the range of the U–Th method. Electron spin resonance (ESR) techniques have been tried at Sterkfontein, but the ages are thought to be tentative, mainly because of the difficulty of estimating reliably the long-term dose rates (see Dating Methods: Archaeological) (Schwarcz et al., 1994).

The Makapansgat Limeworks Cave is different in that it appears to have operated as a huge resurgence to a mountain river system, probably in the late Miocene (ending at about 5.1 million years ago). Toward the end of its life about 3–3.5 million years ago, it contained an estimated 60 000 tonnes of speleothems and thick layers of sediments and breccias. The present 2 ha cavern is an eroded remnant estimated to be about half its original height, revealing speleothems at the surface. Nearly all the speleothems have been mined out, although, interestingly, some distinctive subaqueous, mammillary, deposits are still in place. A number of Australopithecine remains were retrieved from miner's breccia dumps, and a 1 m thick bone breccia band, known as the "Grey Breccia" contained a dense assemblage consisting of over 80% bovid bones, the rest being baboons, equids, large felids, hyenas, Australopithecus, and smaller animals.

At about the same time as the Limeworks dumps were being sorted, from about 1945 to the 1960s, some of the Witwatersrand researchers (Van Riet Lowe, Kitching, Tobias, Hughes, Malan, and others) were working at another mine site about 1 km up the valley—the Cave of Hearths. Under the direction of Revil Mason, a long sequence of early *Homo* occupation layers were uncovered, containing stone tools from the ESA, MSA, LSA, and up to historic times. The Cave of Hearths, together with Wonderwerk Cave, Northern Cape, and Montagu Cave in the Western Cape, is one of the few South African caves to contain artefacts of the Acheulian culture (MSA) (Deacon & Deacon, 1999). As Mason blasted his way down through the calcite-hardened breccias he uncovered hearth-like structures firstly in the upper layers and then toward the base. Tentative claims were made for the earliest known use of fire although this still remains to be substantiated. This site has undergone recent reappraisal, chiefly by a team from Liverpool University (England). The lowermost archaeological layers are normally magnetized, showing that they are less than 780 000 years old. The sediments contain no associated speleothems in the unexcavated sediment left as a witness section that might lend themselves to U–Th dating, and they are probably too old for thermoluminescence or optically stimulated luminescence dating. In comparison to Wonderwerk Cave, the Cave of Hearths has a much lower density of faunal material, and this suggests that it probably did not operate as a home base in the way that was once thought likely.

In 1925, a poorly excavated skull, with a brain capacity of 1280 ml and with some Neanderthal-like features, was unearthed from a fissure in a dolomite cave at Kabwe (the Broken Hill lead–zinc quarry), in Zambia. The maxilla shows dental caries and abscessing, and Bartsiokas & Day (1993) have suggested that the skull displays evidence of lead poisoning.

Those sites of the Middle Stone Age that contain not only artefacts but also human remains are relevant to the question about the transition of *Homo erectus*, associated with the Acheulian hand-axe technology, to *Homo sapiens* associated with the Upper Paleolithic (in Europe) and more sophisticated stone tools of the preformed core type (in Europe, the Levallois technique). Migration of *Homo erectus* out of Africa (the migration known popularly as "Out of Africa I") is thought to have occurred between 1.5 and 2 million years ago. In Europe, these hominids were succeeded, probably ancestrally, by the Neanderthals. When and where the first anatomically modern humans appeared also finds its answers in Africa. Several DNA studies suggest that humans appeared in Africa at about 200 000 years ago. In South Africa, there are several caves containing modern-looking skeletal remains, including Klasies River (Mouth) Cave and Border Cave. Klasies River Cave, near Port Elizabeth, is a set of openings, 6 to 18 m above sea level. It was excavated by Wymer and Singer from 1967 to 1968 (Singer & Wymer, 1982) and then again by H. Deacon in the 1990s. The human remains uncovered included jaws, teeth, and fragments of crania (Rightmire & Deacon, 1991) that have been dated faunally and by ESR to between about 115 000 to 60 000 years ago (Klein, 1999). In Europe, by contrast, the first anatomically modern humans (AMH; old term, Cro-Magnon) succeeded the Neanderthals from about 40 000 to 30 000 years ago (Eastern Europe to Iberia). It should be realized that although this discussion restricts itself to evidence from cave sites, there is a broader picture of hominid evolution and culture from all African sites. Although many archaeologists and paleoanthropologists are persuaded by the origin of modern humans by replacement out of Africa ("Out-of-Africa II") there is a sizeable minority that thinks that modern humans evolved more or less in every region of the Old World (including, possibly, Australia which appears to have witnessed the first entry of humans about 40 000 to 60 000 years ago). Milford Wolpoff is the chief dissenting voice on the Out-of-Africa II migration, but see also, for example, Smith (1992) and, for the argument itself, see Delson et al. (2000).

Near the Cape, around 1910, John Hewitt excavated a rock shelter at Howiesons Poort and uncovered a fine, small, and distinctive stone tool technology of the MSA that was not to be approximated till the late Stone Age (22 000 to 2000 years ago; Deacon & Deacon, 1999). The fine blades appear to have supported hafted tools, and the Howiesons Poort technology has been found at many other sites in southern Africa, including Klasies River Mouth Cave and in Nelson's Bay Cave, between the Cape and Port Elizabeth. At Rose Cottage Cave, excavated currently by Lyn Wadley of Witwatersrand Archaeology Department, there was occupation before and after Howiesons Poort times. Dated to between 70 000 and 60 000 years ago (Vogel, 2001), the Howiesons Poort culture is used as a marker horizon in southern African sites as far north as Lake Eyasi in central Tanzania and west to the Apollo 11 Cave in Namibia (Deacon & Deacon, 1999).

Many archaeologically important cave sites are situated on the southern coast, and they are important as recorders of the use of seafood. Sea levels were high around 125 000 years ago during interglacial isotope stage 5e, and this appears to have resulted in the scouring of the Klasies River Mouth Cave openings. After that, as sea levels fell, the evidence suggests that the caves were occupied sporadically by hunter-gatherers who scavenged and hunted seals and penguins and also lived off shellfish.

The fact that the MSA human skeletal remains are thought to be modern has suggested to some anthropologists that fully modern humans came out of Africa at least by 120 000 years ago (the "Out-of-Africa II" event—see Carmel Caves, Israel: Archaeology). The aforementioned Howieson's Poort MSA technology has also featured in this issue (Parkington, 1990). A reconstruction of ash layers, tools, various adornments, and other cultural evidence on site, has prompted Deacon & Deacon (1999) also to suggest that, "They were probably the direct ancestors of the Later Stone Age peoples"—and that must include the Bushmen of today. In effect, the Bushmen of southern Africa may well be the direct descendants of the earliest humans. This adds impetus to the great interest in old and new Bushman wall paintings.

ALF LATHAM

Works Cited

Bartsiokas, A. & Day, M.H. 1993. Lead poisoning and dental caries in the Broken Hill hominid. *Journal of Human Evolution*, 24: 243–49

Berger, L.R. & Clarke, R.J. 1995. Eagle involvement in the accumulation of the Taung child fauna. *Journal of Human Evolution*, 29: 275–99

Brain, C.K. 1981. *The Hunters or the Hunted? An Introduction to African Cave Taphonomy*, Chicago: University of Chicago Press

Brain, C.K. (editor) 1993. *Swartkrans: A Cave's Chronicle of Early Man*, Pretoria: Transvaal Museum

Clark, J.D. 1993. Stone artefact assemblages from Members 1–3, Swartkrans Cave, In *Swartkrans: A Cave's Chronicle of Early Man*, edited by C.K. Brain, Pretoria: Transvaal Museum

Clarke, R.J. 1998. The first ever discovery of a well-preserved skull and associated skeleton of Australopithecus. *South African Journal of Science*, 94: 460–63

Deacon, H.J. & Deacon, J. 1999. *Human Beginnings in South Africa: Uncovering the Secrets of the Stone Age*, Walnut Creek, California: Altamira Press

Delson, E., Tattersall, I., Van Couvering, C. & Brooks, A. (editors) 2000. *Encyclopedia of Human Evolution and Prehistory*, 2nd edition, New York: Garland

Klein, R.G. 1999. *The Human Career: Human Biological and Cultural Origins*, 2nd edition, Chicago: University of Chicago Press

Parkington, J. 1990. A critique of the consensus view on the age of Howieson's Poort assemblages in South Africa. In *The Emergence of Modern Humans: An Archaeological Perspective*, edited by P. Mellars, Ithaca, New York: Cornell University Press and Edinburgh: Edinburgh University Press

Partridge, T.C., Shaw, J., Heslop, D. & Clarke, R.J. 1999. The new hominid skeleton from Sterkfontein, South Africa: age and preliminary assessment. *Journal of Quaternary Science*, 14: 293–98

Rightmire, G.P. & Deacon, H.J. 1991. Comparative studies of Late Pleistocene humans from Klasies River Mouth, South Africa. *Journal of Human Evolution*, 20: 131–56

Schwarcz, H.P., Grün, R. & Tobias, P.V. 1994. ESR dating studies of the Australopithecine site of Sterkfontein, South Africa. *Journal of Human Evolution*, 26: 175–81

Singer, R. & Wymer, J.J. 1982. *The Middle Stone Age at Klasies River Mouth in South Africa*, Chicago: Chicago University Press

Smith, F.H. 1992. Models and realities in modern human origins: the African fossil evidence. In *The Origin of Modern Humans and the Impact of Chronometric Dating*, edited by M.J. Aitken, C.B. Stringer & P. Mellars, London: Royal Society and Princeton, New Jersey: Princeton University Press

Vogel, J.C. 2001. Radiometric dates for the Middle Stone Age in South Africa. In *Humanity from African Naissance to Coming Millennia*, edited by P.V. Tobias, M.A. Raath, J. Moggi-Cecchi & G.A. Doyle, Firenze: Firenze University Press

Further Reading

Johanson, D. & Edgar, B. 1996. *From Lucy to Language*, New York: Simon and Schuster and London: Weidenfeld and Nicolson

Jones, S., Martin, R.D. & Pilbeam, D.R. (editors) 1992. *The Cambridge Encylopedia of Human Evolution*, Cambridge and New York: Cambridge University Press

Partridge, T.C. & Maud, R.R. (editors) 2000. *The Cenozoic of Southern Africa*, Oxford and New York: Oxford University Press

AFRICA, SUB-SAHARAN

Sub-Saharan Africa has 42 countries, climates ranging from humid tropical to arid, and a predominance of ancient Precambrian rocks. Although only a small proportion of the area is karst, and many of the karst areas have never been fully explored, the subcontinent has many significant sites: caves with early hominid remains dating to 4.5–5.0 Ma, the largest underground lake in the world and the fourth deepest siphon, a 305 m deep cave in quartzite, the sixth and ninth longest lava tube caves in the world, caves in pyroclastics that are frequented by elephants and buffalo to lick salt from their walls, and a cave with 5 m speleothems in the floor of an active volcanic crater. It also has an example of the rare labyrinth karst landform, human-induced subsidence into karst cavities as a result of gold-mining activities, and a sea cave in quartzite with large carbonate speleothems and the remains of the first modern humans.

Caves: The two longest caves in sub-Saharan Africa are Sof Omar in Ethiopia and Apocalypse Pothole in South Africa, both of which are maze caves. Sof Omar (see separate entry) is 15.1 km long but only 1.2 km of this is river passage, the remainder being in a maze formed near the resurgence, giving the cave a total of 42 entrances. Apocalypse Pothole in dolomite has 12.1 km of mapped passages. The deepest caves are Mwenga Mwena (Zimbabwe) in quartzite (−305 m), and Bushmangat (South Africa) in dolomite (−295 m). Cango Cave (5.3 km long) in the Western Cape of South Africa is the most important tourist cave and is possibly the best decorated.

The sub-continent has spectacular cenotes, as well as caves with large underground lakes. In the Otavi Mountains of Namibia are the cenotes of Aikab, Aigamas, Guinas Lake, and Otji-

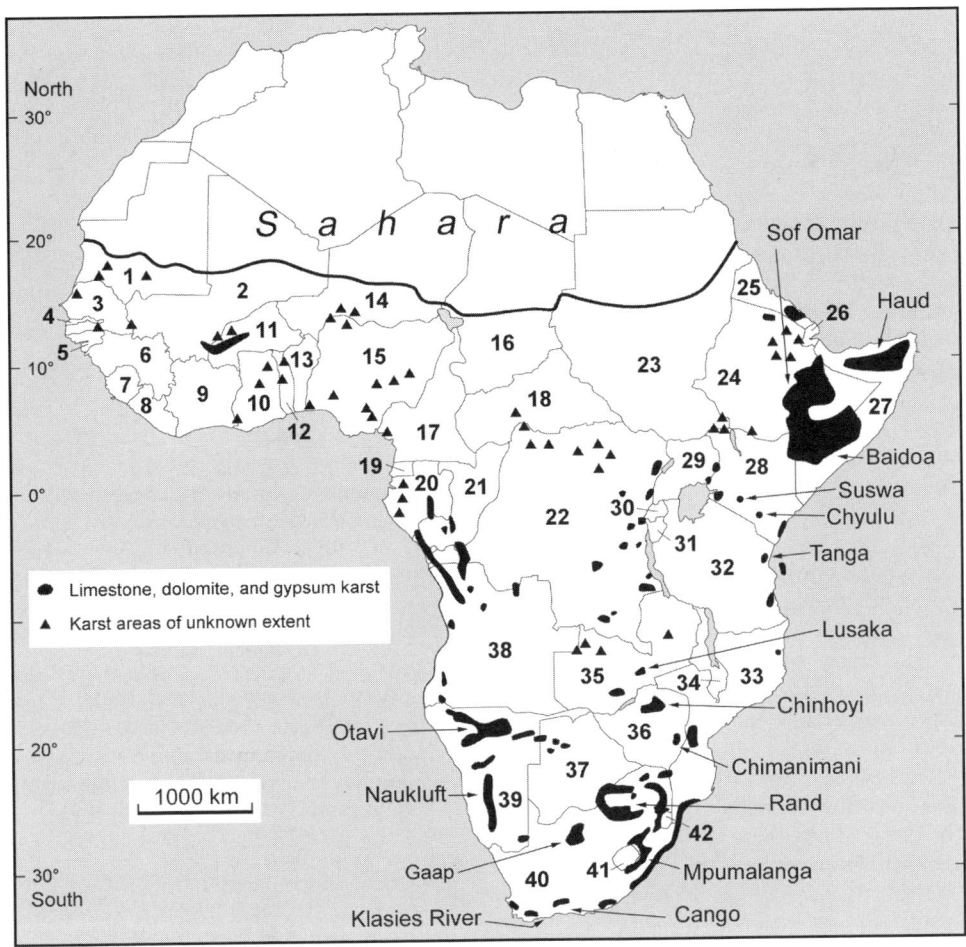

Africa: Sub-Saharan: Figure 1. The main karst areas of sub-Saharan Africa. Nations are numbered:
1 = Mauritania.
2 = Mali. 3 = Senegal.
4 = Gambia. 5 = Guinea Bissau.
6 = Guinea. 7 = Sierra Leone.
8 = Liberia. 9 = Ivory Coast.
10 = Ghana. 11 = Burkino Fasso.
12 = Togo. 13 = Benin. 14 = Niger.
15 = Nigeria. 16 = Chad.
17 = Cameroon.
18 = Central African Republic.
19 = Equatorial Guinea. 20 = Gabon.
21 = Congo. 22 = Zaire. 23 = Sudan.
24 = Ethiopia. 25 = Eritrea.
26 = Djibouti. 27 = Somalia.
28 = Kenya. 29 = Uganda.
30 = Rwanda.
31 = Burundi. 32 = Tanzania.
33 = Mozambique. 34 = Malawi.
35 = Zambia. 36 = Zimbabwe.
37 = Botswana. 38 = Angola.
39 = Namibia. 40 = South Africa.
41 = Lesotho. 42 = Swaziland.

koto Lake (see Figure 2), and also underground lakes at Harasib and Dragon's Breath. The latter, discovered in 1986, may have the largest underground lake in the world, 260 × 180 m, more than 90 m deep, and 60 m below ground. Harasib Cave is the fourth deepest cave on the subcontinent (−210 m) with a lake 135 × 35 m and 90 m deep. Lakes in Aikab, Guinas, and Aigamas are 120 × 100 × 63 m deep, 145 × 70 × 120 m deep, and 115 × 20 × 40 m deep. Guinas is one of two sites with the multicoloured indigenous fish, *Tilapia guinasana*, while Aigamas is the only site with the catfish, *Clarias cavernicola*. Otjikoto Lake, 100 m in diameter and 75 m deep, is a National Monument because during World War I German troops dumped Krupp ammunition wagons and field guns into the lake before surrendering to British troops. Today this military equipment is an underwater museum at depths of 48–52 m. Divers have recovered speleothems up to 130 ka old from depths up to 40 m in the Otavi lakes, indicating much lower ground water levels in the past. The Wonderhole at Chinhoyi Caves, northwest of Harare, Zimbabwe, is a circular cenote 46 m in diameter and 190 m deep, with a 144 m deep lake—the Sleeping Pool. Bushmangat in the Gaap Plateau dolomites of the Northern Cape, South Africa, contains a lake 30 × 12 m at the surface and 264 m deep, making it the fifth deepest siphon in the world. The lake widens to 247 × 75 m in plan extent at a depth of 100 m.

Three of the five deepest caves in sub-Saharan Africa are in the Chimanimani Mountains of eastern Zimbabwe in quartzite: Mawenga Mwena (−305 m), Jungle Pot (−221 m), and Bounding Pot (−194 m). Big End Chasm has the world's largest chamber in quartzite: 90 m high and 70 × 15 m in plan. The famous Klasies River Mouth sea caves on the South African Tsitsikama coast are also in quartzite. They contain the oldest

Africa: Sub-Saharan: Figure 2. Otjikoto Lake, a cenote in the Otavi Mountains of Namibia. (Photo by Tony Waltham)

remains of anatomically modern humans, dating to more than 100 ka (see entry Africa, South: Archaeological Caves), and also house carbonate speleothems up to 1 m in diameter and 4 m high, deposited by dripwaters draining from shell-rich aeolian sands on the surface above the cave.

The largest and most important caves in East Africa are in volcanic rocks. Kenya has the sixth longest lava tube cave in the world, Leviathan Cave, 12.5 km long and 3–10 m in diameter; the ninth longest, Mount Suswa, at 11.0 km; and six of the top 60. Almost 100 lava tube caves are known in the Chyulu Range between Nairobi and Mombasa, while Mt Suswa has 40 caves in an area of 3 km^2 that are segments of a complex braided tube system with up to 3 levels of passages, and 67 entrances. Passages are 6–10 m in diameter increasing to 20–30 m where collapse has occurred. The 19th longest lava tube cave in the world is Ubuvomo bwa Musanze on Mt Ruhengueri in Rwanda at 4.6 km and 210 m deep, while nearby Ubuvomo bwa Nyrabadogo is 1.5 km long. All are developed in pahoehoe lava flows and various caves have lavafalls 10 m high, lava stalactites and flowstones, rare silica dripstones and flowstones, calcium phosphate deposits beneath bat guano, and bones of rhino, giraffe, leopard, baboon, and antelope.

In 1990, Hades Cave was discovered in the crater floor of Oldoinyo Lengai in northern Tanzania, the world's only active carbonatite volcano. The cave, 20–30 m in diameter and 5–8 m deep, was formed by the collapse of the 0.5–1.0 m thick roof of an old spatter cone. Pale yellow stalactites 1–4 cm in diameter and up to 5 m long stretched from the ceiling, probably formed from lava spatters that stuck to the roof of the void. As there was considerable volcanic activity within 100 m of the cave, it probably did not survive for long.

There are also numerous small caves in tuffs and agglomerates; these are up to 300 m long but have breakdown chambers up to 100 m wide, the best known being Kitum and Makingen caves on Mt Elgon in Kenya. The entrances to these caves are behind waterfalls and are frequented by elephant and buffalo for the salt they contain, as seen by tusk marks on the walls.

Many dolomite caves in southern and eastern Africa are water-table caves formed by solution in the upper part of the phreatic zone. Lowering of the water table focused solution deeper in the dolomite and left upper passages aerated and open to the surface via vertical shafts allowing the influx of soils and making them death traps for animals falling into them. In Botswana, Angola, Namibia, and South Africa caves of this kind contain fossiliferous, cemented breccias of middle Miocene to Holocene age that include baboon, hominoid, and hominid remains. At Berg Aukas, Namibia, remains of two hominoids, 13 and 9–10 Ma old, were recovered from sediments in dolomite fissures. Several dolomite caves northwest of Johannesburg, South Africa, including Sterkfontein, and Makapansgat Limeworks near Potgietersrus, contain important fossils. Basal breccias at Sterkfontein in the upper, older part of the cave are more than 4 Ma old, and an almost complete hominid skeleton recovered from the cave is 3.30–3.33 Ma old.

Surface karst varies considerably across the region in response to both the climate and the rock lithology. In the humid tropics, rare labyrinth karst has formed in the Tanga limestone of Tanzania, with fissures up to 15 m deep and 1–2 m wide isolating sharp pinnacles and ridges and in places coalescing to form internal closed depressions. Cone and tower karst with depressions occurs in an area including parts of south Congo, western Zaire, and Angola. Elsewhere on the subcontinent doline karst is more typical with occasional uvalas and poljes. Coastal limestones in Western Cape, South Africa, have closed depressions up to 20 km long, dry valleys, large springs, and extensive dry phreatic cave systems. Hundreds of shallow dolines, dry valleys, and depressions up to 8 km long dot the covered karst near Alexandria, Eastern Cape. The Gaap Plateau, Northern Cape Province is pockmarked by closed depressions with ephemeral lakes typically 3–5 m deep and 100–500 m long, while strings of dolines 5–10 m deep mark old stream drainage lines. There is also doline karst in calcrete 5–10 m thick in the upper catchments of the Tsondab, Gaub, and Kuiseb rivers of the central Namib Desert where annual rainfall is only 150–250 mm. This karst extends from the escarpment westwards for 40–50 km with dolines up to 25 m across and 2–3 m deep.

Lusaka, the capital of Zambia, sits atop a marble plateau with dolines 10–100 m across and up to 10 m deep. Almost everywhere the marble is buried beneath a shallow layer of lateritic gravel. As there is no surface drainage, the marble surface is susceptible to flooding after very heavy rains when infiltration capacity is exceeded. The worst flooding was in 1926 when the whole township was flooded and boats were rowed down Cairo Road in the city centre. Despite construction of flood drains the city was flooded again in 1978–79. In areas where the lateritic gravels have been excavated, marble pinnacles 3–10 m high have been exposed (see Figure 3). The pinnacles are disappearing rapidly as they are being broken into pieces for sale on the street as aggregate.

The Rand gold mining area of South Africa, west of Johannesburg, is perhaps the best-known area of karst in the subcontinent. Gold is found in beds of quartzite and conglomerate overlain by 1220 m of dolomite largely hidden beneath 9–150 m of unconsolidated sediment. The dolomites are divided into water compartments by a series of vertical syenite dikes. To facilitate mining, groundwater was pumped from the Venterspost compartment in 1955, and from the Oberholzer and Bank compart-

Africa: Sub-Saharan: Figure 3. Marble pinnacles exposed from beneath lateritic gravel in Lusaka, Zambia. (Photo by George Brook)

ments in 1960 and 1972, the watertight nature of the dikes making it possible for each compartment to be de-watered independently. By 1957 sinkholes appeared in the Venterspost compartment by compaction of surficial sediments (up to several km long) and by their collapse into subsurface cavities. In 1962 a vertical-sided sinkhole, 55 m across and 30 m deep, formed in the Oberholzer compartment, engulfing the crushing plant of the West Driefontein mine and causing 29 deaths. As sinkholes continued to develop, some small towns had to be evacuated. In 1964 the West Driefontein mine was extended into the Bank compartment which had not been de-watered. On 26 October, 1968 a fissure opened up in the roof of the workings beneath the Bank compartment and a torrent of water flowed in. Fortunately, 13 500 men were evacuated safely but it took 26 days to stem the flow of water. The source of the inrush was the saturated, cavernous dolomites of the Bank compartment with 850 m head. Following the disaster the Bank compartment was de-watered.

Evaporite karst is unusual in sub-Saharan Africa but is well developed in gypsum and anhydrite on the northern Haud Plateau in Somalia. Shallow solution dolines up to 100 m in diameter are common, while many collapse dolines up to 50 m in diameter and 15 m deep lead into caves 0.5–20 m below the surface, such as Ail Afwein (1275 m long) and Las Anod (1455 m long). Many caves contain water that is used for camels and goats. Gypsum speleothems and crystals are common, and hyenas occupy many entrances. On the limestone Baidoa Plateau in southern Somalia there are several 1–3 km long structural/erosional basins with lakes including those at Ted, Moragavi, and Burdo. In the east, the plateau has a thick soil cover and some of the numerous shallow depressions contain perched water bodies.

Quartzite karst has developed in eastern Mpumalanga Province, South Africa (see Silicate Karst) where dolines with pinnacled slopes occur along abandoned streamlines in 20–25 m of quartzite resting on granite. Seventeen caves up to several hundred metres long are known, some with vertical shafts to 20 m deep and chambers 60 × 25 m and 15 m high. Most have water flowing through them, disappearing and reappearing several times. Small limonite stalactites and flowstones are present. Quartzite karst has also developed in the Cape Peninsula of South Africa including numerous small caves, with Ronan's Well the longest at over 700 m.

Spring and waterfall tufas occur in semiarid and arid parts of sub-Saharan Africa. Notable are the massive, relict, perched spring and waterfall tufas in South Africa, Namibia, and Somalia, and there are smaller examples in southeastern Zimbabwe. Since the Miocene, thick masses of tufa have accumulated along the dolomite Gaap Escarpment of South Africa, particularly in gorges cut back into the scarp. Individual tufa masses such as Gorrokop are 1.8 km long and average about 30 m thick. At Buxton-Norlim, in 1924, the famous type specimen of the species *Australopithecus africanus* was discovered in the Thabaseek tufa carapace, which was being quarried for lime. Smaller waterfall tufas about 45 m across and 10 m high, but up to 1200 m across and 30 m high, occur as plugs in incised valleys of dolomite areas in Northern and Mpumalanga Province. Numerous massive waterfall tufas occur at different elevations in the Naukluft Mountains of Namibia, a dolomitic nappe complex. The Blässkranz tufa, 500 m long and 150 m high, becomes a huge waterfall after uncommon heavy rains. Large waterfall tufas also occur in Kaokoland near Warmquelle, in the Kuiseb River Valley where average annual rainfall is less than 50 mm, and in the Swartkloof Mountains of southwest Namibia. The largest tufas in Somalia are around the southeastern edge of the Baidoa plateau at Deileb Wanei and Isha Baidoa. At Isha Baidoa the tufa is more than 20 m thick and extends for hundreds of metres along the escarpment. In northern Somalia relict spring and waterfall tufas up to 300 m long and 40 m high are found at several locations in the Golis Mountains and on the Gulf of Aden coast.

GEORGE A. BROOK

Further Reading

Alvarez, P. 1997. Morphologies karstique et implications minières en République Centrafricaine. *Journal of African Earth Sciences*, 293(2): 293–305

Brink, A.B.A. & Partridge, T.C. 1965. Transvaal karst: Some considerations of development and morphology. *The South African Geographical Journal*, 47: 11–34

Brook, G.A. & Ford, D.C. 1978. The origin of labyrinth and tower karst and the climatic conditions necessary for their development. *Nature*, 275(5680): 493–96

Brook, G.A., Marais, E. & Cowart, J.B. 1999. Evidence of wetter and drier conditions in Namibia from tufas and submerged speleothems. *Cimbebasia*, 15: 29–39

Butzer, K.W., Stuckenrath, R., Bruzewicz, A.J. & Helgren, D.M. 1978. Late Cenozoic paleoclimates of the Gaap Escarpment, Kalahari margin, South Africa. *Quaternary Research*, 10: 310–39

Davies, G.J. 1998. "Hades"—a remarkable cave on Oldoinyo Lengai in the East African Rift Valley. *International Journal of Speleology*, 27B(1/4): 57–67

Deacon, H.J. & Geleijnse, V.B. 1988. The stratigraphy and sedimentology of the main site sequence, Klasies River, South Africa. *South African Archaeological Bulletin*, 43: 5–14

Marker, M.E. 1971. Waterfall tufas: A facet of karst geomorphology in South Africa. *Zeitschrift für Geomorphologie* N.F., 12: 138–52

Marker, M.E. 1988. Karst. In *The Geomorphology of Southern Africa*, edited by B.P. Moon & G.F. Dardis, Johannesburg: Southern Book Publishers: 175–97

Partridge, T.C. 2000. Hominid-bearing cave and tufa deposits. In *The Cenozoic of Southern Africa*, edited by T.C. Partridge & R.R. Maud, Oxford and New York: Oxford University Press: 100–30

Pickford, M., Mein, P. & Senut, B. 1994. Fossiliferous Neogene karst fillings in Angola, Botswana and Namibia. *Suid-Afrikaanse Tydskrif vir Wetenskap*, 90: 227–30

Simons, J.W.E. 1998. Volcanic caves of East Africa. *International Journal of Speleology*, 27B(1/4): 11–20

AFRICA: BIOSPELEOLOGY

Africa spans temperate to tropical latitudes, with a great variety of different ecosystems, making it an extremely varied continent with high levels of biodiversity. Few zones have been explored from a biospeleological perspective, nonetheless the results obtained so far show considerable subterranean diversity. Biospeleological research began in the first decades of the 20th century and was concentrated in eastern (Ethiopia, Kenya, and Somalia), northern (especially Morocco), and southern (Namibia and South Africa) Africa, and on Madagascar. Several researchers explored these territories and collected a large amount of zoological material.

The French zoologists C. Alluaud and R. Jeannel were among the first to research in detail the subterranean fauna of Africa. In 1911–12 they visited several African countries where they collected a large amount of troglobitic material, and the first African stygobites. The results of their research were published in *Biospeologica* (Jeannel & Racovitza, 1914), where the first review of several African caves and their fauna can be found. In the 1950s, French scholars led extensive research in Madagascar, while more recently activity has been re-established in northern Africa by French and Moroccan researchers and, in southern Africa, by Namibian and South African researchers. At the beginning of the 20th century and later, in the 1970s, Italian researchers collected and presented data on biospeleological aspects of Ethiopia and Somalia. In the rest of the continent biospeleological research has been sporadic and scattered.

African caves host almost all animal taxa, with many troglophiles and trogloxenes, but only a few troglobites. Stygobites, stygophiles, and stygoxenes on the contrary are numerous and are primarily restricted to karstic areas developed in the limestone deposited during Cretaceous ($c.$ 100 million years ago) marine invasions. In some cases, such as Amphipoda and Isopoda of eastern and southern Africa, these are apparently the only representatives of the taxa in the region.

Although the continent has not been extensively explored from a biospeleological perspective, the list of African subterranean-dwelling animals is long. Excluding Protozoa, flatworms, worms, and all hyporheic (interstitial spaces within the sediments of a stream bed) and marine interstitial (living in the spaces between sand grains) taxa, the subterranean African diversity is presented below by taxa.

Mollusca
A number of terrestrial and aquatic snails and clams are reported from African caves, although no true troglobitic species are known.

Crustacea
The presence of Copepoda in subterranean waters in Africa has been reported from almost every explored area; however the reports are scattered and fragmentary. Ostracoda is a widely distributed taxon in subterranean waters, although few species are strictly stygophilic or stygobitic. A single stygobitic species has been found in Zaire and several others in the wells of Somalia. Of the Malacostraca, several African forms of Bathynellacea have been described, mostly interstitial. Several species of Thermosbaenacea are reported from African subterranean waters (Morocco, Somalia, South Africa, and Tunisia). The first to be described was *Thermosbaena mirabilis* from the El Hamma hot springs in Tunisia. No stygobitic Mysid (Mysidacea, or Opossum shrimps) is present in Africa, though stygophilic species are reported from Kenya, Madagascar, Aldabra, and Zanzibar Islands. A single species of Spelaeogriphacea, *Spelaeogriphus lepidops*, has been described from South Africa.

There are many stygobitic Isopoda species distributed in eastern, western, northern, and southern Africa belonging especially to the Cirolanidae and Stenasellidae families. Some of them show a striking aspect such as the spiny *Acanthastenasellus forficuloides*, known from a single female specimen collected in Somalian subterranean waters. Subterranean species of African terrestrial Isopoda are mostly troglophilic. Only termitophilous taxa show true troglobitic adaptations.

Amphipoda has many stygobitic species, belonging to several families (Hadziidae, Salentinellidae, and Gammaridae being the most represented), distributed in eastern, western, northern, and southern Africa.

Of the Order Decapoda (shrimps), several stygobitic and stygophilic species of the Athyidae family are found in Kenya, Madagascar, Somalia, Tanzania, and Zaire. One species of the Palaemonidae, *Typhlocaris lethaea* from Libya and several stygoxene decapods of the Penaeidae and Athyidae families are also present in African caves.

Onychophora
Only two true troglobitic species of velvet worms exist in the world, of which one, *Peripatopsis alba*, lives in a South African cave.

Arachnida
The best-represented arachnid orders in African caves are the Scorpiones (scorpions), Pseudoscorpiones (pseudoscorpions), Solifugae (sun-spiders), Araneae (spiders), Opiliones (harvestmen), Amblypigi (whip spiders), and Acari (ticks and water mites). It is impossible to enumerate all the Arachnida that can be found waiting for their prey in African caves. Numerous parasitic ticks live on bats, and terrestrial taxa that are associated with guano, predate on eggs or insect larvae, feed on fungi, or scavenge are common in caves. Several water mites (Halacarids and Hydracarids) can be found in subterranean waters.

Myriapoda
Centipedes (Chilopoda) are often present in caves, although no troglobites are reported so far from Africa. Millipedes (Diplopoda) are present with a few troglophilic and trogloxene species in caves. In Somalia specimens of *Graphidostreptus gigas* were found in Showli Berdi cave.

Hexapoda
Although numerous species of springtails (Collembola) have been found in African caves, only two species from Madagascar are described as true troglobites: *Cyphoderopsis madagascarensis* and *Troglobius coprophagus*. Only one Diplura genus, *Anysocampa*, is reported from the Ethiopian region. A single species is truly troglobitic in Zaire, *Austrjapyx leleupi*.

Insecta
Only one species of silver fish (Thysanura) is reported from South Africa as troglobitic, *Lepidospora makapaan*. Cavernico-

lous or troglophilic Orthoptera (crickets) are present in most African caves, they are mostly scavenger and detritus feeders. They belong to only two families, the Phalangopsidae and Rhaphidophoridae. The fleas (Siphonaptera) are well known to any speleologist and to those working in African caves in particular. They cannot be considered true cavernicoles because they are all parasites of vertebrates such as bats, and belong to the subterranean world only because of this specialization. No troglobitic species of wasps or bees (Hymenoptera) are known but many nest at the cool and dark entrance of many caves in Africa. They are quite aggressive, especially the bees, thus making some caves a risky place to enter. Beetles (Coleoptera) are the most diversified insect taxon to have colonized the subterranean environment. They are present in African caves especially with many carabids (ground beetle), cholevids, curculionids (snout beetle), pselaphids, staphilinids (rove beetle), and even with aquatic forms of the Hydrophilidae and Dytiscidae families (diving beetle). Only one earwig (Dermaptera) species is known from Africa. This is the first ever described species of a troglobitic earwig, *Dyplatis milloti*. Of cockroaches (Blattodea), only the family Nocticolidae has troglobitic species in Africa, where several troglophilic species are present. Cicadas and aphids (Homoptera) have very few cave representatives in Africa: one species of Hypochthonellidae in Zimbabwe and only one troglobitic species of Cixidae in Madagascar. Only a few species of moths (Lepidoptera) of the families Oecophoridae, Plutellinae, Alucitidae, and Noctuidae have been so far reported from African caves, but they cannot be considered troglobitic species. The four reported African species of troglobitic flies (Diptera) have guanophagous (feeding on bat guano) larvae and haematophagous (blood-sucking) adults.

Fish
Eight species of cave-dwelling fish were described in Africa between 1921 and 1959: *Caecobarbus geertsii* (former Zaire); *Uegitglanis zammaranoi* (Somalia); *Phreatichthys andruzzii* (Somalia); *Barbopsis devecchii* (Somalia); *Typhleotris madagascariensis* (Madagascar); *Clarias cavernicola* (Namibia); and *Typhleotris pauliani* (Madagascar). The first cave-dwelling fish, a small cyprinid discovered in 1921 in the Belgian Congo, was the first subterranean fish in the world to be officially protected (by the Belgian colonial authorities). Somalia has the highest cave fish diversity in Africa, with three species. Italian interest has remained strong and these species have been extensively studied in recent years.

Amphibia and reptiles
Several toads and frogs searching for cool, humid places rich in insects, can be easily found in African caves. No troglobitic snake or lizard has been recorded in any African cave. Most of the reptiles enter the caves to feed on bats, bird eggs, or mice as is the case with *Naja* sp. (cobra), *Bitis* sp. (puff adder), and *Varanus* sp.

Birds
Several species of swallows (*Apus* sp.) inhabit the entrances of African caves, but so far none is reported to have adaptations like those observed for Asian species, which use echolocation deep in the cave. Only one other bird has been reported to nest inside caves in Africa, *Picathartes orea* (bald picathartes) of Gabon.

Mammals
Several species of mammals spend part of their time in caves. Bats are certainly the best-represented group: several insectivorous and frugivorous bats roost in African caves, often forming huge colonies. Other mammals, especially the Insectivora and Rodentia such as the Hystricidae (Old World porcupines), may be found in caves, even in the deeper zones. The signalled presences of Carnivora especially felids such as the African wildcat (*Felis lybica*) or the leopard (*Panthera pardus*) are numerous. The African elephant *Loxodonta africana* has been documented to enter the Kitum Cave (Mt Elgon, Kenya) and to dig with its tusks the salt that can be found there.

GIUSEPPE MESSANA

Further Reading

Botosaneanu, L. (editor) 1986. *Stygofauna Mundi, A Faunistic, Distributional, and Ecological Synthesis of the World Fauna Inhabiting Subterranean Waters (Including the Marine Interstitial)*, Leiden: Brill

Jeannel, R. & Racovitza, E. 1914. Biospeologica. XXXIII. Enumeration des grottes visitées. 1911–1913. *Archives de Zoologie Expérimentale et Générale*, 53: 325–558

Juberthie, C. & Decu V. (editors) 1994. *Encyclopaedia Biospeologica*, vol 1; vol. 2 (1998); vol. 3 (2001), Moulis and Bucharest: Société de Biospéologie (vols 1 and 2 are by taxa; pp. 1461–1740 in vol. 3 refer to Africa)

Wilkens, H., Culver, D.C. & Humphreys, W.F. (editors) 2000. *Subterranean Ecosystems*, Amsterdam and New York: Elsevier

AGGTELEK AND SLOVAK KARST, HUNGARY–SLOVAKIA

The Aggtelek and Slovak area comprises a major karst region, in which the caves have been designated as a World Heritage Site, approved in 1995. It straddles the southern foothills of the Carpathians, and consists of the Aggtelek National Park and Biosphere Reserve in Hungary (19 708 ha) and the Slovak Karst Protected Landscape Area and Biosphere Reserve in Slovakia (36 165 ha), together with the nearby Dobšina Ice Cave. Topographically, the main block comprises a series of rolling plateaux in thick limestones and dolostones, dissected by a few deep river valleys. Relief ranges from altitudes of 200 to 900 m. Geologically, it is a part of the Szilice Overthrust in the Carpathians, displaying complex, highly deformed structure. More than 700 solution caves are known. The climate is humid continental, with mean annual temperatures between 5 and 12°C and 600–1000 mm precipitation, depending on altitude. Natural vegetation is an oak, hornbeam, and beech deciduous forest.

Aggtelek-Domica Region
The principal karst area is shown in Figure 1. The limestone (Wetterstein Formation) and dolostone (Guttenstein Formation) are platform deposits of Triassic age, uplifted and deformed during the Cretaceous. Denudation commenced in the late Cretaceous under tropical or subtropical conditions, forming *terra rossa* soils, with some bauxites in karst depressions. In the mid-

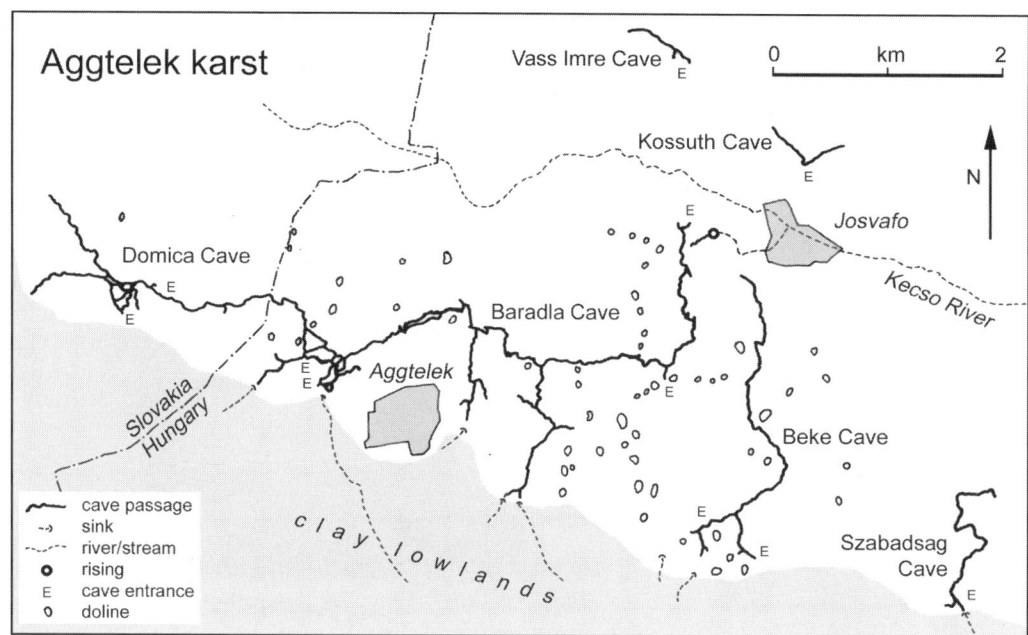

Aggtelek and Slovak Karst: Figure 1. Outline map of the cave systems within the Aggtelek karst beneath the border between Hungary and Slovakia.

Tertiary (Miocene), the region was partly buried by deposits of the Pannonian Inland Sea, chiefly sands and clays from the mountains further north. Erosion has since exhumed the higher ground, which now forms a rounded massif to altitudes of 450–500 m, above remaining clay lowlands at $c.300$ m. Kecso River (Figure 1) entrenched the massif, permitting early springs to drain to it, at an altitude of $c.270$ m. A knickpoint has receded through the Kecso valley, lowering modern spring levels to 220–230 m. The region experienced severe cold at times during the Quaternary, with solifluction and other periglacial features indicating periodic loss of natural forest cover, although there was no growth of glaciers (Zambo, 1993).

It is important to appreciate that the modern (post-Pannonian) karst has developed upon an older, well-fissured subtropical karst that remains partly buried. The principal surface karst landforms are shapely solution dolines, 10–100 m in diameter and up to 30 m in depth. Larger examples retain Pliocene sediments at their bases, covered by a few metres of colder climate brown or black earths, with up to one metre of man-induced erosional detritus on top. A few have become blocked, creating ponds. Modern dissolutional activity is studied intensely in Beké Doline, a major karst monitoring site (Zambo & Ford, 1997). There are also extensive subsoil karren, frequently exposed by erosion.

The principal caves are excellent examples of dendritic river cave development of the water table-shallow phreatic type (Ford, 2000). Baradla–Domica Cave (24 km long) is the lengthiest, an underground linking of six streams that collect on the clays and sink at the limestone contact (Figure 1), plus rivulets from the overlying karst. The entrance galleries at Aggtelek and Domica have a complex, multi-level structure, apparently due to Quaternary episodes of clastic filling, with paragenesis and incomplete re-excavation (Bolner-Takacs et al., 1989). The main trunk passage is slightly sinuous and $c.7$ km long. It descends through the highly deformed strata, at a regular gradient of $4–8$ m km^{-1}. Its cross-sectional area *decreases* downstream, despite the addition of tributary streams, due to progressive undercapture to modern springs downstream of the Kecso River knickpoint. Flood overflow now reaches a sinkpoint in a boulder pile beneath the Oriasokterem (Giants' Hall) breakdown chamber, on average once every two years. The spring cave is constricted; flood discharge there is $c.20$ m^3 s^{-1}.

The cave is profusely decorated with calcite speleothems, including stalagmites and bosses up to 10 m or more in height or girth. U-series dating has established that most growth is no older than the last interglacial ($c.125\,000$ years BP); the cave was partly filled with alluvial debris during the succeeding Würm cold phases and some larger stalagmites were toppled by extreme flooding. There is also significant condensation corrosion damage to speleothems in tall chambers, which can function as cave climate chimneys.

Beké Cave (6.4 km long), Szabadsag (3.4 km long), Kossuth (1.3 km long), and Vass Imre (1 km long) are similar, low gradient river caves but with lesser catchments and smaller passages. Beké Cave is noted for the erosional effects of aggradation that it displays, such as multiple, undulating corrosional-corrasional notching in the rock walls. Beké, Kossuth, and Vass Imre are sites of long-term hydrologic, hydrochemical, and cave climatic studies.

Higher on the plateaux, 162 steep swallet caves or vertical shafts beneath dolines are known, reaching depths of 100–250 m. Sixty collapse caves are also recorded. Miners discovered relict hydrothermal caves in some isolated limestone blocks at the extremities of the karst, including the spectacular Ochtina Aragonite Cave (in Slovakia) and Rákóczi Cave, which are open to tourists.

Baradla Cave was one of the earliest centres of biospeleological research. More than 500 troglobite and troglophile species and sub-species are now known from here and other caves of the region, some of them unique. There are 21 species of bats, many of which winter in the caves.

History

The earliest surviving published reference on Baradla Cave is from 1549 AD. Guided visits took place quite frequently during the 18th century. The first accurate map, 1794, showed 2170 m of passages from the Aggtelek entrance. In 1825, the county engineer, Imre Vass, passed a downstream obstacle to extend the known cave, publishing a striking map, with 7.9 km of galleries, in 1829. Tourism increased vigorously in the following years. A partly artificial entrance was engineered alongside Voros-to (Red Lake, a pond doline) in 1890, permitting 6 km through-trips from Aggtelek. A second tunnel, in 1930, allowed entrance or exit at the downstream end of the cave, at Jósvafő. The connection between Baradla and Domica Cave was discovered in 1932.

Dobšina Ice Cave

Dobšina (German: Dobschau) Ice Cave is first known to have been entered in 1870, via a small collapse doline at an altitude of 969 m in a north-facing slope. This gave access to a series of breakdown chambers containing large, perennial ice masses and descending to a depth of $c.110$ m (Figure 2). The discovery attracted great attention, was opened for tourists the following year, and fitted with electric lighting as early as 1884. In 2000, it was added to the World Heritage List, as an adjunct of the Slovak Karst. The cave is the truncated, relict, upstream end of the Stratenska Cave system, an 18.3 km, multi-level river cave in the Wetterstein limestone (Tulis, 1986). Breakdown and sedimentation have separated the Dobšina end (1232 m of passages) from the remainder with a seal that is nearly airtight and watertight, although there is limited through drainage.

The main ice body is a "textbook" example of a glacière. It largely fills two chambers, has a volume of $c.110\,000$ m^3 and a maximum thickness of 26.5 m, making it the largest compact

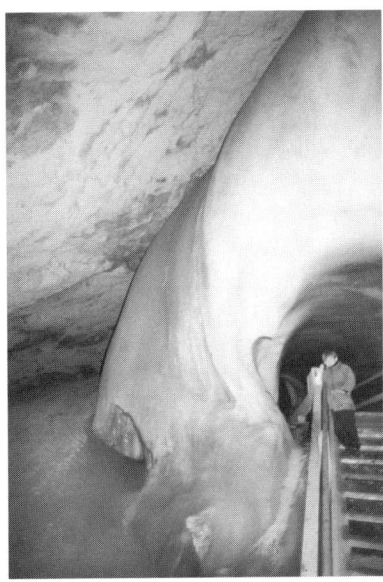

Aggtelek and Slovak Karst: Figure 3. Tunnel cut through the ice in Dobšina Ice Cave. (Photo by John Gunn)

glacière in a show cave. It has an extensive flat floor, sometimes used for skating parties, and large ice stalagmites, columns, and draperies. The glacière accreted from films of flowing water in regular wet season layers because the sealed cavity is a cold trap, receiving dense cool air in winter that cannot be displaced by the warmer air of other seasons. Measurements in the 1880s found that the annual range of air temperature was -0.15 to $-0.86°C$ around the glacière, in contrast to a mean annual temperature of approximately $+6°C$ outside the cave.

There are 515 m of tourist paths to display the ice, including a tunnel cut through it that reveals the layered structure (Figure 3). Modern ice behaviour is closely monitored to prevent degradation due to visitors. Between 1981 and 1990, floor ice was observed to increase by 3–115 mm a^{-1} at different locations, much of the water deriving from the melt of nearby wall ice at 1–27 mm a^{-1}. The mass was also deforming plastically (flowing) at rates of ~ 10 mm a^{-1} horizontally and up to 27 mm a^{-1} vertically downwards (Lalkovic, 1995).

DEREK FORD

See also **Aggtelek Caves: Archaeology; Ice in Caves**

Works Cited

Bolner-Takacs, K., Eszterhás, I., Juhasz, M. & Kraus, S. 1989. The caves of Hungary. *Karszt-es Barlang*, Special Issue, 17–30

Ford, D.C. 2000. Caves Branch, Belize, and the Baradla-Domica System, Hungary and Slovakia. In *Speleogenesis; Evolution of Karst Aquifers*, edited by A.B. Klimchouk, D.C. Ford, A.N. Palmer & W. Dreybrodt, Huntsville, Alabama: National Speleological Society

Lalkovič, M. 1995. On the problems of the ice filling in the Dobšina Ice Cave. *Acta Carsologica*, 24: 314–22

Pelech, J.E. 1879. *The valley of Stracena and the Dobschau Icecavern (Hungary)*, London: Trübner

Tulis, J. 1986. Survey of the cave, Stratenska Jaskyna. *Bulletin of the Slovak Speleological Society*, 1–2: 11–15

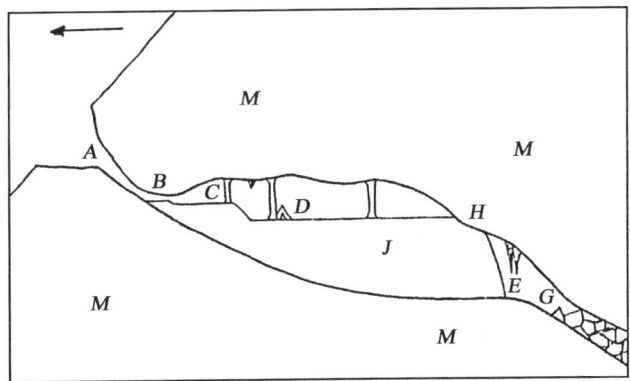

Aggtelek and Slovak Karst: Figure 2. A sketch long section of Dobšina Ice Cave drawn by Dr. Johann E. Pelech (1879). It indicates the principles of the *glacière* or cold trap very well. The direction of the arrow indicates north. (A) External visible hollow in the limestone. (B) Actual entrance. (C, D) Upper portion of the Cavern with the three Ice Pillars, the Ice Saloon, Ice Tent, and Ice Hillock. (E) Lower portion of the Cavern, the Corridor with the Ice Wall. Near (G) the Cavern contracts, and is filled with fragments of rock; through these the water flows away. (J). The great Ice Mass. (M) The Limestone Rock. At (H) the top of the ice is in contact with the roof of the Cavern, thus dividing it into two parts, the Upper and Lower Stages. (Pelech, 1879)

Zambo, L. 1993. Physical geographical characteristics of Aggtelek Karst. *Annales, Universitatis Rolando Eotvos Nominatae, Sectio Geographica*, 22–23: 280–88
Zambo, L. & Ford, D.C. 1997. Limestone dissolution processes in Beke Doline, Aggtelek National Park, Hungary. *Earth Surface Processes and Landforms*, 2: 531–43

Further Reading

Bolner-Takacs, K. 1998. *The Caves of the Aggtelek Karst*, Jósvafő: Aggtelek National Park

Hazslinszky, T. 1992. Visitor's books of the Baradla Cave from the last century. *Karszt-es Barlang*, Special Issue; 41–46
Izapy, G. & Maucha, L. 1989. Subsurface water chemical matter-transportation values of karstic areas in Hungary. *Proceedings of the 10th International Congress of Speleology*, edited by T. Hazslinsky & K. Bolner-Takacs, Budapest: Hungarian Speleological Society
Jakucs, L. 1977. *Morphogenetics of Karst Regions: Variants of Karst Evolution*, New York: Wiley and Bristol: Hilger (originally published in Hungarian, 1977)

AGGTELEK CAVES, HUNGARY–SLOVAKIA: ARCHAEOLOGY

The Baradla–Domica Cave System, one of the most important archaeological sites in Central Europe, where artefacts of the Neolithic Bükk culture have been found, straddles the boundary between northeastern Hungary and southern Slovakia. The cave system is almost 25 km long and is the most important landform of both the Aggtelek (Hungary) and the Slovak karst. It was excavated by the erosive activity of the underground Styx River, mainly in Middle Triassic (Ladinian) Wetterstein limestone. However, there are substantial cave deposits consisting of gravel and cave loam. The Baradla–Domica caves probably served as temporary or permanent dwellings for paleolithic hunters. In the Middle Ages they also provided sanctuary from the invading Tartars and Turks. But the people of the Bükk culture inhabited the caves for the longest period—during the Middle Neolithic, c.6000–4500 BC.

The Bükk culture derives its name from the Bükk Mountains in northeastern Hungary, which was the main centre from which it spread. The people subsisted by farming, animal husbandry (especially sheep and goats), the obsidian trade, and textile and ceramic production. Their ceramics include some of the finest examples of Central European Neolithic pottery. Apart from their settlements on the river terraces and hillsides, the people also favoured caves as dwelling-places. From this point of view, the Baradla–Domica Cave System (especially the Domica Cave in Slovakia) represents the most important archaeological site for the Bükk culture.

The first archaeological discoveries in the Domica Cave sediments were made in 1926. However, the first archaeological research only began in 1932–33, about 50 years later than in the Hungarian Baradla Cave, where Jenő Nyári had already begun work in 1876–77. Jaroslav Böhm led the research at Domica Cave in the 1930s, in cooperation with American archaeologists (V.J. Fewkes and his team). The Hungarian scientists Hubert Kessler and Maria Mottl also excavated in this cave during World War II (in 1940). New research into the cave settlement was initiated by the Archaeological Department of the Slovak Academy of Sciences in 1956 (by Juraj Bárta) and continued in 1963 (Bárta, 1965; Lichardus, 1969).

A large quantity of new data on the occupancy of the Domica Cave have been collected during this research. The finding of paleolithic leaf-shaped stone "spike" (tool) of the Szeletian culture in allochthonous sediments was especially noteworthy. This is evidence of the presence of prehistoric hunters in the cave or nearby. However, a substantial proportion of the cave artefacts date back to the Neolithic Period. In the cave, six cultural horizons (Domica Ia, Ib, IIa, IIb, III, and IV) have been found (Lichardus, 1969). These horizons cover the time span from the end of the Early Neolithic to the Middle–Late Neolithic boundary. The painted ceramics of the older Neolithic cultures (a sphere from the Starčevo–Karanovo–Kriš culture and pottery shards from the Gemer painted ceramics culture) have been found in the Domica Ia to IIa horizons, whereas artefacts of the Bükk culture have been excavated only from younger horizons (Domica IIb to IV) (Bárta, 1965; Lichardus, 1969).

Unlike the painted ceramics, the basic motif of the Bükk culture line pottery is a spiral, sculpted decoration, often encrusted with white, yellow, and red pigment. Round vessels predominate, with amphora-like vessels being less common. Bone tools (awls, smoothers, comb-like tools, grinders, rods, a bone ring, pendants, needles, and daggers), stone tools (polished axes and splintered blades), and earthen spindles (evidence of the textile production) have also been found here. Post-holes, connected with the Bükk culture and also known from the Baradla Cave, suggest that dwellings were constructed inside the caves. Several fireplaces were also found, some of which served as sources of charcoal for radiometric dating. On the basis of radiocarbon analysis, the Bükk culture charcoal layer was dated to 6122 ± 75 BP (4120 ± 75 BC). This is in good agreement with the age (6492 ± 100 BP, 4490 ± 100 BC) obtained by radiocarbon dating of samples from the Bükk culture settlement of Arkatul, near Korlát in Hungary (Bárta, 1965).

The unique discovery of three geometrical Neolithic charcoal drawings in 1931 was the most important find both in the Domica Cave (Benický, 1945) and the whole cave system. These drawings (see Figure)—the only prehistoric murals found in Central Europe—may represent symbols of fertility, and a stylized female figure. They are situated in the back of the cave section known as the "Sacred Corridor", which has a shape of a female lap. The way in which the drawings are located in this part of the cave is presumably related to a sacred site in which religious ceremonies could take place. The corridor was separated from the other cave spaces in two places, probably by some kind of curtains fastened to wooden beams. Traces of post-holes on the cave floor tend to confirm this theory (Bárta, 1965).

From the anthropological point of view, only two human jawbones (one of which belonged to an elderly individual) have been found in the Domica Cave, whereas a Neolithic burial-ground with 13 human skeletons has been discovered in Baradla Cave. The individuals were buried in a crouched position, face down, and were covered by large flat stones. Along with these

Aggtelek Caves: Archaeology: One of three charcoal drawings in the "Sacred Corridor" of the Domica Cave. (Reproduced by the permission of the Slovak Museum of Nature Protection and Speleology in Liptovský Mikuláš, Slovakia.)

human bones, footprints of Neolithic people were preserved on the floors of both caves. Unfortunately, these footprints were inadvertently destroyed during the first exploration of the cave system.

It is known that the Domica Cave was settled by the Bükk people, especially during the winter, when it was warmer in the cave than outside (the average temperature in the cave is 10°C). The cave could have served as a religious site and/ or a ceramics production centre. Artefacts of the Bükk culture have also been discovered in other caves of the Baradla–Domica system (e.g. Baradla Cave, Čertová pec Cave, etc.), but their importance is not as great as those in Domica Cave. However, because of the presence of unique karst phenomena and rare archaeological finds, the Baradla–Domica Cave System was registered as a UNESCO World Heritage Site in 1995.

Martin Sabol

See also **Aggtelek and Slovak Karst; Art: Cave Art in Europe**

Works Cited

Bárta, J., 1965. Príspevok k pravekému osídleniu jaskýň Domickej sústavy [Contribution to prehistoric settlement of the Domica Cave System]. *Slovenský kras—Acta Carsologica Slovaca*, 5: 58–73

Benický, V., 1945. Ako boly objavené kresby neolitického človeka v Domici [How drawings of Neolithic Man have been discovered in the Domica Cave]. *Krásy Slovenska*, 22(6–8): 147–48

Lichardus, J., 1969. Neolitické osídlenie Slovenského krasu vo svetle nových archeologických výskumov [Neolithic settlement of the Slovak Karst in the light of new archaeological research]. *Slovenský kras—Acta Carsologica Slovaca*, 7(1967–1968): 63–79

Further Reading

Droppa, A., 1961. *Domica–Baradla jaskyne predhistorického človeka* [Domica–Baradla Caves of Prehistoric Man], Bratislava: Šport

AKIYOSHI-DAI KARST AND CAVES, JAPAN

The Akiyoshi-dai (plateau) karst is located in Yamaguchi Prefecture, in the western part of mainland Honshu, the largest of the Japanese islands, and is the most famous and largest karst plateau in Japan. Permo-Carboniferous limestones crop out over an area of some 130 km² in one of the main caving regions in Japan. The limestone plateau is shaped like a parallelogram, measuring 17 km in an eastnortheast–westsouthwest direction and 8 km in a northnorthwest–southsoutheast direction, and is surrounded by mainly clastic facies of Permo-Carboniferous age. The western part is covered by Triassic and the northern part by Cretaceous rocks.

The limestone plateau is at altitude of 100–400 m with a gently undulating surface and is mostly surrounded by steep marginal slopes. The valley of the Koto River cuts through the central part of the area from north to south, and divides the Akiyoshi-dai (plateau) into two areas: Higashino-dai in the east and Nishino-dai to the west. About 2200 dolines have been located, and the mean doline density for the total area is *c.*20 per square kilometre on the plateau. The highest doline density is 140–160 per square kilometre in the central part of the Higashino-dai (Miura, 1991). There are also 81 karst springs and 429 caves distributed all over the plateau. About 45 km² of the Akiyoshi-dai Plateau was designated a Quasi-National Park of Japan in 1955, and 13.8 km² was designated a Special National Monument of Japan in 1964, based on its great significance to the geological sciences.

The Akiyoshi Limestone Group, which forms the Akiyoshi-dai Plateau, has a total thickness of *c.*770 m, ranging in age from early Carboniferous (late Tournasian) to late Permian (Guadalupian), and yields abundant fusulinids, bryozoans, crinoids, brachiopods, corals, and other fossils. There is no terrigenous clastic material in any part of the succession. The limestones were deposited in five main environments: talus slope, true reef, beach, marginal lagoon, and central lagoon. The limestone rests conformably on a greenstone succession of alkaline volcaniclastic rocks. The reef complex was formed above a ridge as the result of submarine volcanic activity (Ota, 1977). The northwestern part of the Akiyoshi limestone complex is inverted (recumbent folded and subsequently overturned structures) and the southwestern part displays the normal succession (Schwan & Ota, 1977). In the Carboniferous limestone belonging to the upper part of the *Millerella yowarensis* zone, emergence surfaces with evidence of paleokarst are known to occur (Nagai, 1993).

The caves developed in the Akiyoshi-dai limestone plateau can be divided into three groups on the basis of elevation: high-altitude (higher than 250 m above sea level), middle altitude (250–130 m above sea level) and low-altitude (lower than 130 m above sea level) caves. The high-altitude caves are mostly pits at altitudes of 320–280 m on the Choujagamori Plain. The middle-altitude caves range from an altitude of 160 to 180 m, and occur mostly on the Wakatakebara Plain. The lower-altitude caves mostly open at the foot of the plateau (Kawano, 1983).

Akiyoshi-dai Karst and Caves, Japan: The terraced gour pools of Hyakumai-zara in the main stream cave of Akiyoshi-dai. (Photo by Akiyoshi-dai Museum)

Akiyoshi-do Cave is the longest cave in the plateau and is designated a Special Natural Monument of Japan. It is one of the lower-altitude caves, with a natural entrance at the foot of a 50 m high limestone cliff; the entrance is 20 m high and 8 m wide. Before the 1960s, the known passageway of Akiyoshi-do Cave was c.2 km long including the tourist route, but underwater explorations by various teams have since revealed that it is more than 7.5 km long. A connection was also found between Akiyoshi-do Cave and Kuzuga-ana Cave, and the combined cave system is currently over 8.5 km long. The tourist route, 1 km long, in Akiyoshi-do Cave runs parallel to an underground stream, and passes a complex of 500 terraced rimstone pools (see Figure), large stalagmites, and large columns.

Taisho-da Cave lies in the northern part of Higashi-dai Plateau and opens at an altitude of 159 m of the northern foot of Mt Mana. Taisho-do Cave is designated a Natural Monument of Japan. The cave is complex, developed with three passage levels and is 1000 m in total length. The upper-level passage (at an altitude of 180–175 m) includes a large chamber containing many speleothems and notches at an altitude of 178 m on the cave walls. The middle level (at an altitude of 175–160 m) is a labyrinthine passage. The lower-level passages (at an altitude of 160–120 m) reach to the water table. Within these three levels, many phreatic solution features (bore passage, natural bridges, pillars, pockets, and anastomoses) are developed. Taisho-do Cave consists of a phreatic cave system linked with Sano-ana Cave, Inugamori-no-ana Cave, and Current-mark-no-ana Cave. The total length of these caves is c.2200 m. (Nakagawa et al., 1979).

Takaga-ana Cave is located on the northern part of Nishino-dai Plateau. The vertical entrance shaft of 42 m in a collapse doline is located at an altitude of 241.6 m. Takaga-ana Cave lies in the middle-altitude group and is the second-longest cave in the Akiyoshi-dai Plateau. It is a complex cave with shafts and passages on four levels at heights of 180 m, 170–160 m, 140–130 m and 90 m levels. The main cave passages are more than 1500 m long, developed at the 170–160 m level. The total surveyed length including the 41 branch caves is 4525 m, and the volume is c.120 000 m^3, developed along many fissures. The cave is decorated with numerous speleothems of various types. *Macaca* (macaques), *Lutra* (river otters), *Paleoloxodon naumani* (an extinct elephant), and other vertebrate remains have been obtained by surface collection of the cave deposits (Shuho-cho et al., 1981).

Recently, polluted waste waters have sometimes been detected in the groundwaters of the Akiyoshi-do Cave. It is obvious that the pollution is derived from domestic waste water from the tourist hotels and installations built on the surface of the Akiyoshi-dai Plateau. It is essential that the pollution is prevented by dealing with the waste-water problem and periodically monitoring the groundwater in the caves.

NARUHIKO KASHIMA

See also map in **Asia, Northeast**

Works Cited

Kawano, M. 1983. Considerations on the history of development of the limestone caves at the Akiyoshi Plateau (in Japanese with English abstract). *Bulletin of the Akiyoshi-dai Museum of Natural History*, 18: 1–20

Miura, H. 1991. Surface features and their relations to the caves in the Akiyoshi Plateau in Japan. In *Geography of Akiyoshi Karst*, edited by H. Miura, Shimonoseki, Yamaguchi: Shunhou-sha (paper in English, in mixed Japanese and English book)

Nagai, K. 1993. Discontinuity structures and paleokarst in the Akiyoshi limestone group (in Japanese with English abstract). *Journal of the Speleological Society of Japan*, 18: 42–55

Nakagawa, K., Imamura, O. & Hiramoto, T. 1979. Taisho-do drainage cave system at the northern part of the Akiyoshi-dai Plateau: a model of development of phreatic cave system (in Japanese with English abstract). *Journal of the Speleological Society of Japan*, 4: 32–41

Ota, M. 1977. Geological studies of Akiyoshi. Part I. General geology of the Akiyoshi limestone group. *Bulletin of the Akiyoshi-dai Science Museum*, 12: 1–33

Ota, M., Sugimura, A. & Haikawa, T. 1980. The Akiyoshi limestone group and geologic structures (in Japanese). In *Limestone Caves in Akiyoshi-dai: Sciences of Limestone Caves*, edited by M. Kawano, Shimonoseki, Yamaguchi: Shunhou-sha

Schwan, W. & Ota, M. 1977. Geological studies of Akiyoshi. Part II. Structural tectonics of the Akiyoshi limestone group and its surroundings (southwest Japan). *Bulletin of the Akiyoshi-dai Science Museum*, 12: 35–100

Shuho-cho & The Board of Education, Shuho-cho. 1981. *Takaga-ana Limestone Cave, Akiyoshi Plateau, Western Japan* (in Japanese with English abstract), Shimonoseki, Yamaguchi: Shunhou-sha

Further Reading

Kanmera, K., Sano, H. & Isozaki, Y. 1990. Akiyoshi terrane. In *Pre-Cretaceous Terranes of Japan*, edited by K. Ichikawa, S. Mizutani, I. Hara, S. Hada & A. Yao, Osaka: Osaka City University (Publication of IGCP Project No. 224)

Kawano, M. (editor). 1980. *Limestone Caves in Akiyoshi-dai: Sciences of Limestone Caves*, Shimonoseki, Yamaguchi: Shunhou-sha (in Japanese)

Tanaka, K. & Nozawa, T. (editors) 1977. *Geology and Mineral Resources of Japan*, 3rd edition, Kawasaki-shi, Japan: Geological Survey of Japan

ALPINE KARST

Alpine karst is the solutional landscape occurring at high altitudes throughout the mountains of the world. It is distinctive primarily by virtue of localized colder conditions resulting from the altitude, often resulting in restricted vegetation, seasonal snow and snow melt, and past and present glaciers. Exceptionally steep hydraulic gradients develop between upland recharge areas and adjacent valleys. Alpine conditions do not particularly favour karst development: runoff is seasonal, dissolution subdued by reduced biological activity, and past and present geomorphic processes compete and hinder evolution of karst.

Hydrology

Alpine environments are characterized by strong climatic gradients; average temperatures decline by about 6°C per km elevation and upland areas capture additional precipitation, much of which may be in the form of snow. Many alpine karsts therefore experience a strong spring-summer freshet, supplemented only so far as snow patches, glacier ice, and summer rainfall sustain runoff. Meltwater from snow and ice initially contains little dissolved gas or minerals and hence does not have much inherent capacity to dissolve limestone. Carbon dioxide solubility is higher under cold conditions, but partial pressures of gases are reduced under lower atmospheric pressures. Furthermore, poorly developed or skeletal soils combined with a limited growing season result in little enhancement of solutional capacity from soil CO_2 during infiltration.

Groundwater hydraulics under mountains are strongly influenced by relief. Saw-tooth (dissected cuesta) terrain characterized by continuous slopes between peak and valley tends to develop recharge in the upper fraction and discharge throughout the lower fraction with respective linkage of the highest and lowest components. More classical alpine terrain of cirques, dissected plateaus, and entrenched valleys exhibits much deeper circulation and strong groundwater convergence in valleys below.

Processes

Alpine karst processes reflect the seasonal hydrology and subdued chemistry. For much of the year cold conditions may limit runoff, and lack of soil water storage may result in intermittent rather than sustained summer flows. Alpine snow banks are important in sustaining infiltration and stream flow. Closed depressions preferentially accumulate and sustain snow into the summer, and the resulting meltwater may enhance their development in otherwise dry summer conditions.

Much solutional potential may be expended on preferential dissolution of fine carbonates generated by glacial abrasion, and present in glacially derived soils. Evapotranspiration losses are low in comparison with many other environments and hence the runoff percentage tends to be high, although this is more than compensated for by low pCO_2 in the absence of biogenic sources. Hence, karst denudation rates are not high, and alpine karst generally develops quite slowly, unless there is substantial runoff or a supplementary source of acidity. Oxidation of crushed pyrite to produce sulfuric acid is one such process. Two factors, currently poorly understood, are the impact of initial freshet water that is chemically enriched (especially where acidic air pollution is high), and the very rapid kinetics of dissolution arising from water containing no dissolved minerals.

Cold conditions, seasonally abundant water, rapid runoff, and steep gradients strongly sustain other geomorphic processes, most of which can be viewed as competing with karst development. Frozen ground may allow accumulation of massive ice and perennial snow in surface shafts. This effectively cuts off the water supply and may prevent exploration, although in the Julian Alps explorers have been able to penetrate over 200 m down ice-filled shafts by following narrow routes at the contact between ice and the rock walls. Frost shattering of susceptible materials can devastate surface karst forms and choke closed depressions. Mass movement by solifluction, creep, rockfall, and debris flows may also obscure karst features, or restrict groundwater recharge.

Glaciers play an ambiguous role in alpine karst where many closed depression are both karstic and glacial in development. Closed depressions tend to be preferred sites for snow accumulation and glacier development, and glacier erosion at valley heads tends to produce overdeepened (closed) cirques that may function in a similar way to dolines following ice-melt. The bed temperature of most alpine glaciers is fixed at melting point. As a result, glaciers protect the ground from freezing and so permit groundwater recharge at altitudes otherwise sustaining permanently frozen ground. They may also prevent frost shattering, and scour accumulated debris from surfaces. As a result the finest alpine karst terrain is largely found in areas recently exposed by glacier retreat. Small cirque glaciers may coexist with active karst, as they produce little debris. However, more active, extensive glaciers generate large quantities of till that accumulates in lower reaches as moraines. Much alpine karst is obscured by a mantle of such glacial deposits that not only neutralize solutional erosion, but may also preclude groundwater recharge, and support lakes which act as sustained traps for fine sediments. Glacial action is particularly enhanced by the steep topographic and climatic gradients of alpine regions. At higher altitudes, erosion dominates, generally disrupting surface catchments (especially closed depressions) and truncating shallower karst systems. At lower altitudes, glaciers tend to deposition and the resulting glacial, fluvioglacial, and glaciolacustrine sediments occupy valley floors, raising local base levels and often burying karst springs.

Karst Landforms

The composition of carbonate mountain ranges reflects the regional geological history. As most mountains are developed at plate margins, the carbonates of mountain ranges may be considerably thicker than is found in plate interiors. Intense folding, faulting, and alteration of rocks are common in mountain ranges, resulting in varied juxtaposition of various grades of carbonate with insoluble silicates, and sometimes containing exotic minerals arising from hydrothermal and volcanic activity. Many of the features of alpine karst may be linked to the active tectonics associated with mountain development. Valleys and poljes may occupy down-faulted blocks or zones of weakness; horizontally bedded blocks form plateaus. Tilted blocks tend to generate saw-tooth ridges that may be separated by strike-aligned valleys developed in weaker strata; regional karst drainage is often along the strike of such blocks discharging into larger, cross-cutting glacial valleys. Carbonate tectonic units in the European Alps

Alpine Karst: View across the dipping limestone pavements on the Slovene side of Monte Kanin. (Photo by John Gunn)

may contain independent karst systems isolated from adjacent rocks by the impermeable fault surface.

The paucity of surface vegetation due to glacial erosion, soil erosion, and the harsh climate renders alpine karst features more easily seen and appreciated than in other environments. Solutional features like karren are often well developed in suitably massive, pure carbonates, and the resulting lapiaz or limestone pavements can cover many square kilometres (see figure). Such surfaces are composed of numerous, structurally controlled closed depressions, indicating highly fragmented recharge. Larger closed depressions may occupy tectonic structures, or may occupy cirque floors or valleys previously overdeepened by ice. However, many alpine closed depressions act as agents of their own burial. They are filled with clastic deposits generated by weathering, mass movement, glaciation, and runoff, and the sink points are unable to carry the coarse materials involved. In time, if the terrain stabilizes, suffosion dolines may develop in the fill, eventually leading to reopening of the ponor. For example, Medicine Lake, Jasper National Park, Canada occupies a karst depression blocked by a rock slide, glacial deposits, and river sediments. It has failed to develop a significant ponor due to continuous infilling of nascent openings.

Alpine Caves

The fundamental features of alpine caves are those of any cave system, although there may be considerably greater vertical development and longer preservation of ancient passages. Vertical shafts are common underground, and are often exposed at the surface, truncated by glaciation and filled with shattered rock and sediment. They are the most common alpine cave entrance, but relatively few provide access to a cave system. Vadose channels may downcut rapidly, rather than widening, creating narrow meandering canyons linking sequences of shafts, as at Krubera Cave (see separate entry). Phreatic sections may also develop considerable vertical range, including dramatic upwards flowing segments to phreatic loops, reflecting the fracture frequency, and the tendency to deep groundwater circulation in alpine terrain.

Mountains are dynamic landscapes at the time scale of karst development. Tectonic, climatic, and geomorphic patterns may change many times during evolution of cave systems. Changes to the alpine terrain and base level will alter patterns of recharge and discharge, disrupting the fundamental trajectories of cave system development. Recharge shafts may entirely lose their surface catchment and become relict. Springs developed at the valley floor may become perched above base level as a consequence of later entrenchment, and the associated passages become progressively abandoned. Passage infill and overflows may develop in response to alluvial infilling of the valley floor, for example the more than 60 separate springs recognized in and below Maligne Canyon, Jasper National Park, Canada. Within the cave system, phreatic conduits may be abandoned and reactivated as a response to cycles of valley deepening and infilling, or glaciation of the recharge surface and valleys. Reorganized surface recharge may form new vertical shafts, graded to a new local base level and cutting directly through former passage networks. These are type examples of invasion vadose caves (see Speleogenesis: Unconfined Settings). Backfilling in valleys may result in these shafts becoming flooded, and overflows may reactivate ancient phreatic or choked conduits. Such rejuvenation and reactivation are normal in many karst regions. However, in alpine karst, thick carbonates, the great vertical range of cave development, and the marked transitions in controls over time may result in more radical changes and permit greater preservation of cave systems than in lower relief regions. Multilevel passages, shaft complexes, and a wide suite of clastic and mineral deposits provide a tantalising record of the past environment.

Exploration

The great relief of alpine regions permits development of the great deep caves of the world, and the complex history of many alpine caves makes them extraordinarily extensive. In many cases, active streams occupy relatively few of the known passages. The possibility of ever greater depths of exploration has attracted considerable interest from speleologists. However, apart from relief, there is no common form for deep alpine caves. Some are simple shafts (e.g. Epos Chasm, Astraka Plateau, Greece), or are developed along simple contacts or fractures (e.g. Krubera). Some are developed into complex horizontal networks, linked by shafts (e.g. Gouffre Berger, see Vercors entry). Others are trunk systems gathering water from a number of distinct sub-catchments (e.g. Pierre Saint-Martin, see separate entry). The challenge is often to gain entry to the system, for the recharge surface may be disrupted. In fact, many great caves are not entered by hydrologically prominent depressions, for these are often infilled. Rather, inconspicuous shafts, lacking any surface catchment often survive to permit explorers to enter (e.g. Yorkshire Pot, Alberta, Canada). A number of alpine caves have been explored from the lower entrance. In some (e.g. Castleguard Cave, see separate entry) exploration has required only persistence, as all the large shafts are descents, despite penetrating upstream. In others, extraordinary technical virtuosity has been required to push exploration up vertical shafts (for examples see entries on Dent de Crolles and on Siebenhengste). Fortunately, upward exploration, often pursuing the cave wind, may reveal upper entrances, permitting more ready access to the inner reaches of the system (e.g. Hölloch, see separate entry, and Nettlebed Cave, New Zealand).

Alpine cave exploration is almost always highly technical, requiring skill and equipment to tackle the vertical sections. Not only is there a considerable quantity of equipment to transport, but the length of many trips demands establishment of underground camps. In many systems, water also poses a challenge, in some cases by the intensity of flow, in others by closure of passages by floods. Exploration may be restricted to wintertime when there is much less runoff. Low temperatures mean that ice may form, preventing access, or possibly converting a difficult wet section to a straightforward frozen pool.

Resources

Alpine karst aquifers may be used in water supply and power generation. For example Vienna (Austria) obtains at least 60% of its water from alpine karst. However, the strong seasonality of flow, lack of quality protection, and poor moderation of recharge may render the water resource of limited value. Quarrying poses a threat in some regions as many alpine caves are developed in carbonates attractive to the aggregate and cement industries. Much alpine karst lacks surface expression, but retains its hydrological function. Reservoirs may prove ineffective, if built on such terrain. Surface karst is often extremely fragile, although nature may be the primary vandal. However, economic development of alpine terrain often results in wholesale destruction of karst surfaces for development of roads and alpine resorts, and significant compromise of water quality.

CHRIS SMART

See also **Asiago Plateau, Italy; Caucasus, Georgia; Europe, Alpine; Glacierized and Glaciated Karst; Kanin Massif, Slovenia/Italy; Picos de Europa, Spain**

Further Reading

Audra, P. 1995. Alpine karst speleogenesis: Case studies from France (Vercors, Chartreuse, Ile de Cremieu) and Austria (Tennengebirge). *Cave and Karst Science*, 21: 75–80

Bögli, A. 1964. Le Schichttreppenkarst. *Revue Belge de Géographie*, 88: 64–82

Ford, D.C. 1979. A review of alpine karst in the southern Rocky Mountains of Canada. *National Speleological Society Bulletin*, 41: 53–65

Ford, D.C. (editor) 1983. Castleguard Cave and Karst, Columbia Icefields area, Rocky Mountains of Canada: A symposium. *Arctic and Alpine Research*, 15(4): 425–544

Ford, D.C. & Williams, P.W. 1989. *Karst Geomorphology and Hydrology*, London and Boston: Unwin Hyman

Maire, R. 1977. Les cavités de haute montagne. *Revue Spelunca*, 1: 3–8

Maire, R. 1990. La haute montagne calcaire. *Karstologia Mémoires*, 3

Smart, C.C. 1988. Quantitative tracing of the Maligne Karst Aquifer, Alberta, Canada. *Journal of Hydrology*, 98: 185–204

ALTAMIRA CAVE, SPAIN: ARCHAEOLOGY

Located 2 km south of the village of Santillana del Mar (Santander), near the north coast of Spain, the cave of Altamira, nicknamed the Sistine Chapel of Cave Art, was decorated at various times between *c.*16 000 and 14 000 years ago. First discovered by a hunter in 1868, it was visited in 1876 by a local landowner, Don Marcelino Sanz de Sautuola, who noticed some black painted signs on a wall at the back, but thought little of them. In 1879 he returned to do some excavating and, while he was digging in the cave floor, searching for prehistoric tools and portable art of the kind he had recently seen displayed at a Paris exhibition, his 8-year-old daughter Maria was playing in the cavern. Suddenly she spotted the cluster of great polychrome bison paintings on the ceiling.

Her father, at first incredulous, became more interested when he found that the figures seemed to be done with a fatty paste, and noticed a close similarity in style between these huge figures and the small portable depictions from the Ice Age which he had seen at the Paris exhibition; he therefore deduced that the cave art was of similar age, but his attempts to present his views and his discovery to the academic establishment met with widespread rejection and accusations of naivety or fraud. Sanz de Sautuola died prematurely in 1888, a sad and disillusioned man (see Art in Caves: History).

Altamira Cave is 296 m long, comprises a series of chambers and passages, and ends in a very long, narrow section known as the "Horse's Tail". Although the site is best known for its magnificent decorated ceiling, its galleries contain an abundance of engravings, including some particularly fine deer heads identical to some engraved on deer shoulder-blades in the cave's occupation layers. There are also some meandering finger tracings, some of which form a bovine head. One remarkable feature is a series of masks, where natural rock-shapes were turned into humanoid faces by the addition of eyes and other details; most of these masks can only be noticed when one is leaving, rather than entering, the "Horse's Tail".

The great hall, with its high vault, has engravings and also some red compartmentalized quadrilateral signs, similar to those of the cave of El Castillo in the same region. The cave also has black paintings (black figures often occur in different zones from red figures), some stencilled hands, and some (far rarer) positive painted hand prints.

To the left, as one enters Altamira, is the great hall of paintings, measuring about 20 m by 10 m. The floor has been lowered in order to allow visitors easier access and viewing of the very low ceiling on which a score of large painted animals are spread: there are 18 bison, a horse and a hind, the latter being 2.5 m in length—the biggest figure in the cave. They are polychromes, done in ochre, manganese, and charcoal. Most animals are standing, but a few natural bosses in the ceiling are occupied by curled-up bison, which thus appear three-dimensional. Two or three painted figures on the ceiling have often been described as boars (a very rare animal in Ice Age art) but they are now seen as streamlined bison (especially since one has horns).

The curled-up bison on the bosses have been described as sleeping, wounded or dying, falling down a cliff, or as clear pictures of females giving birth. Currently the dominant view is that they are males, rolling in dust impregnated with their urine, in order to rub their scent on territorial markers, even though one of them has udders! In fact, they may simply be bison drawn to fit the bosses—they have the same volume, form, and dorsal line as those standing around them, but their legs are bent and their heads are down. Some researchers see this chamber as a symbolic pound, with a bison-drive depicted on the ceiling (the curled-up animals at the centre are dead, while those around them stand and face the hunters—there are male humans engraved at the edge); another interpretation is that the ceiling is a depiction of a bison herd in rutting season.

Although the Altamira ceiling has sometimes been taken as a single accumulated composition, it actually comprises a series of superimpositions: researchers have distinguished five separate phases of decoration, beginning with some continuous-line engravings, followed by figures in red flat-wash, then some multiple-line engravings, some black figures, and finally the famous polychromes. The multiple-line figures are identical to some portable specimens from the cave, dated to 14 480 years ago, so it is clear that the two earlier phases pre-date them, while the black figures and polychromes are younger. Charcoal used in some polychrome bison on the painted ceiling has produced radiocarbon dates from 14 820 to 13 130 years ago. The cave was probably blocked shortly after this period.

Sanz de Sautuola saw the ceiling as a unified work, and many subsequent researchers have declared that one artist of genius could have done all the cave polychromes. Recent detailed observations have confirmed these intuitions that one expert artist was probably responsible for at least all the polychrome bison on the ceiling. The different radiocarbon dates, if accurate, may indicate subsequent retouching.

The occupation layers at Altamira have yielded material from the Mousterian; the Solutrean, including classic shouldered-points and the engraved deer shoulder-blades; and the early Magdalenian, with perforated antler batons, and antler spear-points with complex decorative motifs. The Magdalenian fauna is dominated by red deer, and has abundant seashells reflecting the cave's proximity to the coast; seal bones occurred in the Solutrean layers. Between the Solutrean and Magdalenian periods, the cave was used by cave bears for hibernation. Altamira's wealth of occupation material and of portable and parietal art (literally art on walls) suggests strongly that it was an important regional focus at times, perhaps the scene of seasonal or periodic aggregations when people from a wide area might meet for ritual, economic, and social activities.

Altamira became a World Heritage Site in 1985. In July 2001, a magnificent facsimile of Altamira Cave, constructed a few hundred metres from the original, was opened to the public; it not only contains a perfect replica of the decorated ceiling, with every crack, engraving and painting on it, but places it in its original context by reconstructing the whole of the cave entrance chamber and its gaping mouth, which has not been seen since it collapsed towards the end of the Ice Age.

PAUL G. BAHN

See also **Art: Cave Art in Europe**

Further Reading

Apellániz, J.-M. 1983. El autor de los bisontes tumbados del techo de los polícromos de Altamira [The author of the bison on the polychrome ceiling of Altamira]. In *Homenaje al Prof. Martin Almagro Basch*, vol. 1, Madrid: Ministerio de Cultura

Bahn, P.G. 2001. Cloning Altamira. *Archaeology*, 54(2): 72–75

Bahn, P.G. & Vertut, J. 1997. *Journey Through the Ice Age*, London: Weidenfeld and Nicolson and Berkeley: University of California Press

Beltrán, A. (editor) 1999. *The Cave of Altamira*, New York: Abrams (original Spanish edition, 1998)

Breuil, H. & Obermaier, H. 1935. *The Cave of Altamira at Santillana del Mar, Spain*, Madrid: Tipografia de Archivos

Cartailhac, E. & Breuil, H. 1906. *La Caverne d'Altamira à Santillane, près Santander (Espagne)*, Monaco: Imprimerie de Monaco

Freeman, L. & González Echegaray, J. 2001. *La Grotte d'Altamira*, Paris: La Maison des Roches

Freeman, L.G., González Echegaray, J., Bernaldo de Quirós, F. & Ogden, J. (editors) 1987. *Altamira Revisited and Other Essays on Early Art*, Chicago: Institute for Prehistoric Investigations and Santander: Centro de Investigación y Museo de Altamira

García Guinea, M.A. 1979. *Altamira y otros cuevas de Cantabria* [Altamira and other caves of Cantabria], Madrid: Silex

Jordá Cerdá, F. 1972. Las superposiciones en el gran techo de Altamira [Superimpositions on the great ceiling of Altamira]. In *Santander Symposium*, International Symposium on Parietal Art, 1970, Santander, Spain, Santander: Patronato de las Cuevas Prehistóricas

Jordá Cerdá, F. 1981. El gran techo de Altamira y sus santuarios superpuestos [The great ceiling of Altamira and its superimposed sanctuaries]. In *Altamira Symposium*, Madrid: Ministerio de Cultura

Madariaga de la Campa, B. 2001. *Sanz de Sautuola and the Discovery of the Caves of Altamira*, Santander: Fundación Marcelino Botín

AMERICA, CENTRAL

Central America, here defined as the isthmus between the United States and South America including the Yucatán Peninsula, contains many significant carbonate karst landscapes, with a regional karst area totalling about 431 300 km², or 17% of the total land area (Figure 1). Over 90% of Central America's karst is in Mexico, particularly in the southern and eastern states of Oaxaca, Guerrero, Chiapas, Pueblo, and Tamaulipas, and on the Yucatán Peninsula. Significant karst also occurs in Guatemala, Belize, and Honduras.

Geologically, the bulk of Mexico, excluding the Yucatán, is related structurally to North America and not to the remainder of Central America. The carbonate rocks of Central America range in age from Quaternary to Jurassic, representing discontinuous carbonate deposition over more than 200 million years. Considerable geologic, topographic, and environmental heterogeneity characterizes the region, but Central America contains a number of dramatic karst landscapes. These include cockpits, towers, dry valleys, dolines of various types and sizes, cenotes, and extensive cave systems, together with an impressive marine karst landscape (Table 1). The world's second longest barrier reef is located off the Caribbean coast of the Yucatán, Belize, and Honduras.

Karst landscapes in Central America have been, and still are, influenced by tectonic, eustatic, and climatic changes (Gardner, 1987) and they have also undergone significant alterations as the result of human activity (Day, 1993).

The most extensive Central American carbonate karst areas are in Mexico, in the Sierra Madre Oriental, in the Yucatán Peninsula, and in the Southern Mountains. The Sierra Madre Oriental includes some 130 000 km² of rugged, mountainous

America, Central: Figure 1. The main karst regions in Central America, and the location of significant caves referred to in the text.

karst with extensive dolines and deep shafts punctuating an elevated plateau, bounded by steep escarpments and dissected by deep canyons. The Southern Mountains are a structurally complex range that extends into Guatemala. They include the Huautla Plateau and the Chiapas Highlands, in a karst area extending over 60 000 km², and a similarly sized area further west in northern Guerrero. In the Yucatán, over 115 000 km² of Tertiary and Quaternary carbonates give rise to a subdued karst landscape, characterized by shallow dolines, low residual hills, cenotes, and flooded cave systems.

Another extensive karst area, including cockpit and tower karst developed in Cretaceous and Tertiary carbonates, extends through the Peten of east central Guatemala into western Belize, covering nearly 15 000 km², and there are also significant elevated karst areas in the Alta Verapaz and Huehuetenango Departments of Guatemala. Honduras has three major karst areas, covering approximately 10 000 km²: the Montana Santa Barbara in the northwest, the Cordillera Agalta in central Honduras, and the Sierra de Colon and Cordillera Entre Rios in the southeast bordering Nicaragua. The karst in Montana Santa Barbara and Cordillera Agalta has received only moderate scientific attention, and the karst on the Sierra de Colon has received little attention, due to its remoteness and dense vegetation. The folded limestone mountains of the Sierra de Colon extend across the border into neighbouring Nicaragua. The Cretaceous carbonates of the Atima Limestone of the Yojoa Group are several hundred metres thick and are heavily karstified. There are also significant carbonate karst areas elsewhere in Nicaragua and throughout Costa Rica (Mora, 1992; Peacock & Hempel, 1993; Troester *et al.*, 1987). In Panama, there is karst in the northwest along the border with Costa Rica, in the Archipielego de Bocas del Toro, in central Panama in the Maje Mountains near the Rio Chepo o Bayano, and in the eastern Darien Department (Reeves, 2000). El Salvador has less than 300 km² of karst, located along the border with Honduras and Guatemala, south of Anguiatu.

Considerable topographic variation characterizes the karst of Central America as a whole, although three distinct karst terrain styles—doline, polygonal (cockpit/cone), and tower—are recog-

America, Central: The areas of karst and significant landforms in Central America.

Country / Area	Karst Area (km²)	Towers	Cockpits	Dolines	Fluviokarst	Marine
Belize	5000	✓	✓	✓	✓	✓
Costa Rica	2000			✓	✓	✓
El Salvador	300			✓		
Guatemala	15 000	✓	✓	✓	✓	
Honduras	10 000		✓	✓		✓
Yucatán Peninsula	115 000			✓		✓
Rest of Mexico	276 700	✓		✓	✓	
Nicaragua	5000			✓		
Panama	2000			✓		✓

America, Central: Figure 2. A corner of the extensive calcite dripstone deposits in the Gruta del Palmito in the Bustamente karst of northern Mexico. (Photo by Andy Eavis)

nized. Dry or underdrained valleys and subdued depressions or dolines occur throughout the region but have received little scientific attention. Cockpit, cone, and tower karst occurs in southern Mexico, Belize, Guatemala, Honduras, and Nicaragua. Variations of these types occur throughout Central America and are not restricted to the areas described above. Flooded shafts, cenotes, are the characteristic surface landform across the northern Yucatán Peninsula.

The karst rocks range from pure, dense, hard, fractured, crystalline limestones, some much altered from their original state, to impure, powdery, soft, porous, amorphous carbonates. Their depositional environments were highly variable, but the low-lying limestones of the Yucatán Peninsula were formed as extensive carbonate platforms and reefs analogous to those off the Caribbean shore today. Some interior karstlands, especially in mountainous areas adjacent to volcanoes, are mantled by volcanic ash. Some carbonates, particularly in Belize and Guatemala, are brecciated as a result of extreme tectonic disruption or meteoric impact. Others have been extensively folded and faulted as a consequence of orogenic mountain building. Karst landscape elevations range from sea level to an altitude of 3000 m; some are mountainous and restricted in area, others planar and extensive; some are hydrologically isolated, while others receive allogenic surface drainage from higher, adjacent non-karst terrains.

There has been considerable but disparate cave exploration and scientific karst research in Central America. Some of the longer and deeper cave systems have been explored and studied extensively, but other potentially significant karst areas have received little attention. The history of cave exploration dates back to the pre-Hispanic Maya, who penetrated some caves for considerable distances as part of ceremonial and ritual practices, and made extensive and varied uses of caves. Modern cave research started in the 19th century with, for example, the exploration of the Gruta de Palmito in Nuevo Leon, Mexico, between 1835 and 1875. Exploration of the Guerrero caves began in earnest in the 1930s, with the traverses of the underground rivers of Chontalcoatlan and San Jeronimo. More recently, the Grupo Espeleologico Mexicano, which was formed around 1960, and the Association for Mexican Cave Studies, formed in 1962, have promoted cave and karst research throughout Mexico, increasingly with international cooperation. The Yucatán karst (see separate entry) has received considerable attention in the context of differential dissolution in groundwater mixing zones and has become famous for its very long underwater cave systems.

Widely published scientific research has been conducted in some other Central American karst areas, notably in Belize (see Veni, 1996) and Costa Rica (Peacock & Hempel, 1993; Troester et al., 1987), but overall the regional karst offers considerable scope for future research. The karst areas of Honduras and Nicaragua warrant further study, although this may be hindered by problems of access.

Extensive cave systems occur through Central America, although the best known are those in the Sierra Madre Oriental, the Southern Highlands, the Yucatán Peninsula, Guatemala, and Belize. The Sierra Madre Oriental includes many famous deep and long cave systems, including Sistema Purificacion, 94 km long. Sistema Cuetzalan is an extensive, dendritic river cave system with over 35 km of passage and as much again in adjacent caves not yet connected. Sótano de Las Golondrinas (see separate entry, Golondrinas and the Giant Shafts of Mexico) is the most famous of many large bell-shaped daylight shafts. In Puebla, at the southern tip of the Sierra Madre Oriental, there is another group of significant deep cave systems including Akemati (1130 m deep) and Ocotempa (1063 m deep).

The Southern Mountains, including the Huautla Plateau, the Chiapas Highlands and northern Guerrero, constitute another of the world's premier cave areas. Sistema Huautla (see separate entry) is 56 km long and at 1475 m deep the deepest cave in the western hemisphere, while, on the opposite side of the Santa Domingo Canyon, Sistema Cheve is 1386 m deep and drains to a canyon-floor resurgence 2540 m below the Cheve entrance (Hose, 2000). Spectacular river caves in Chiapas include the resurgence system of Veshtucoc (5 km long and 380 m deep), while the Sumidero Yochib is a challenging system over 3 km long and over 200 m deep with polished canyons containing waterfalls and deep plunge pools. The Guerrero karst includes well-known caves, such as hoyo de San Miguel (455 m deep), the very well decorated show cave grutus de Juxtlahuaca (5 km), and the river caves of Rio Chontalcoatlan (6 km), and gruta del Rio San Jeronimo (6 km). In Quintana Roo, the extensive water-filled conduit caves of the coastal Yucatán have been a focus of research into carbonate dissolution in freshwater-seawater mixing zones (Back et al., 1986).

In Belize the caves of the Chiquibul River System are particularly well known (see separate entry, Belize River Caves). The Chiquibul Cave System contains four segments, three in Belize (Cebada Cave, Actun Tun Kul, and the Kabal Group), and the fourth (Xibalba) in Guatemala, with a total length exceeding 50 km. There are also extensive river caves in Guatemala, including the Sistema del Rio Candaleria in Alta Verapaz, with 12 km of passages in seven segments and at least 55 entrances (Bordier, 1976).

Central America's natural karstland vegetation varies from xerophytic scrub to wet tropical broadleaf forest, including both deciduous and evergreen trees, although much of the original forest has been cleared, with only fragments remaining in remote karst areas. The region also supports one of the world's most diverse wildlife assemblages, and specifics of the regional karstland ecology warrant additional studies. Regionally, levels of endemism and species diversity are high, particularly in terms of plant species and terrestrial vertebrates.

Human impacts on Central American karst landscapes and caves have been long-term and severe, in particular through forest clearance, species introduction, agriculture, degradation of water resources, and industrial activities, including mining, quarrying, and construction (Day, 1993). Important archaeological sites, both surface and subterranean, are significant facets of karstlands throughout Central America. Mayan archaeological sites within karst areas include Tikal (in Guatemala), Caracol (in Belize), and Chichen-Itza (in Yucatán). The Mayan cave art in Naj Tunich, Guatemala is less well known but no less significant. Pre-Hispanic influence on regional karst areas was extensive, particularly in terms of forest clearance and agricultural activities.

Contemporary threats to the Yucatán karst include hotel expansion along the coast, illegal quarrying activities, and the potential impacts of inappropriate attempts to restore quarried areas. Groundwater contamination, as a result of inadequate sewage disposal practices, is a perennial problem in the Yucatán (Back, 1999). In Belize, adverse impacts range from agricultural expansion to increasing tourism. Forest reserves continue to be logged, despite local opposition, and quarrying of limestone for construction projects is ongoing. In Guatemala and Honduras, major threats include the exploitation of floral and faunal resources, the establishment of settlements within protected areas, and unclear or ineffective legislation. Overall, about 10% of the Central American karst has been designated as some form of protected area, including 68% of the karst in Belize. However, no karstlands are yet designated as protected areas in Nicaragua or El Salvador, and much of the most significant Mexican karst is not protected.

MICK DAY AND JEFF KUENY

See also **Caribbean Islands; Cone Karst; Tower Karst; Villa Luz, Cueva de**

Works Cited

Back, W. 1999. The Yucatán Peninsula, Mexico. In *Karst Hydrology and Human Activities: Impacts, Consequences and Implications*, edited by D. Drew & H. Hötzl, Rotterdam: Balkema

Back, W., Hanshaw, B.B., Herman, J.S. & Van Driel, J.N. 1996. Differential dissolution of a Pleistocene reef in the ground-water mixing zone of coastal Yucatan, Mexico. *Geology*, 14: 137–40

Back, W., Hanshaw, B.B. & Van Driel, J.N. 1984. Role of groundwater in shaping the eastern coastline of the Yucatan Peninsula, Mexico. In *Groundwater as a Geomorphic Agent*, edited by R.G. LaFleur, Boston: Allen and Unwin

Bordier, B. (editor) 1976. Guatemala. *Spelunca*, supplement 3

Day, M.J. 1993. Human impacts on Caribbean and Central American karst. In *Karst Terrains: Environmental Changes and Human Impact*, edited by P.W. Williams. Cremlingen-Destedt: Catena

Gardner, T.W. 1987. Overview of Caribbean geomorphology. In *Geomorphic Systems of North America*, edited by W.L. Graf, Boulder, Colorado: Geological Society of America

Hose, L.D. 2000. Speleogenesis of Sistema Cheve, Oaxaca, Mexico. In *Speleogenesis: Evolution of Karst Aquifers*, edited by A.B. Klimchouk, D.C. Ford, A.N. Palmer & W. Dreybrodt, Huntsville, Alabama: National Speleological Society

Mora, S. 1992. Controls on karst in Costa Rica. In *Hydrology of Selected Karst Regions*, edited by W. Back, J.S. Herman & H. Paloc, Hannover: Heinz Heise

Peacock, N. & Hempel, J.C. 1993. Studies in the Rio Corredor Basin. *NSS Bulletin*, 55(1/2)

Troester, J.W., Back, W. & Mora, S.C. 1987. Karst of the Caribbean. In *Geomorphic Systems of North America*, edited by W.L. Graf, Boulder, Colorado: Geological Society of America

Veni, G. (editor) 1996. Special theme issue on Belize. *Journal of Cave and Karst Studies*, 58(2)

AMERICA, CENTRAL AND THE CARIBBEAN ISLANDS: BIOSPELEOLOGY

Central America (including Mexico) and the Caribbean Islands encompass a vast area ranging from the temperate zone in the north to the tropics in the south. The hypogean fauna is highly diverse and contains species derived from northern and southern elements and from freshwater and marine ancestors. The subterranean fauna of much of this area is poorly known with large areas still essentially unexplored. This is especially true of southern Central America where remoteness and a long history of political instability have prevented any comprehensive study of the fauna.

Volcanic deposits cover much of southern Central America but extensive limestone outcrops occur in Belize, Guatemala, Honduras, Nicaragua, Costa Rica, and Panama. Lava tubes occur in all of the countries of this area except for Belize. No biospeleological studies have been conducted in Nicaragua and only a few guano-associated species have been reported from El Salvador. The only cave system in Panama that has been extensively surveyed is the Chilibrillo Caves. This large bat cave contains no troglobitic fauna. Several studies have been conducted in Costa Rica, but only one troglobite (a phalangodid harvestman) and one stygobite (a pseudothelphusid crab) have been described from caves. Studies in the caves of Honduras have produced a few troglobites but only one collembolan has been described. The interstitial habitat has produced stygobitic mites in Costa Rica and copepods in El Salvador and Honduras.

American, Canadian, French, and Italian biospeleologists have studied areas of Guatemala. A few caves have been investigated in the provinces of Izabal and Petén but no troglobites or stygobites are known here. However, the highland provinces of Alta Verapaz and Huehuetenango have a rich fauna of troglobites and stygobites. The troglobitic fauna includes pseudoscorpions, spiders, millipedes, collembolans, entotrophs, crickets, and carabid and leiodid beetles. All have close affinities with the fauna of southern Mexico. The stygobitic fauna includes triclad planarians, amphipods, isopods, and crabs. The triclad and amphipods are of marine origin, whereas the isopods and crabs are derived from freshwater ancestors.

The troglobites of Belize are closely related to the fauna of the Yucatán Peninsula, but include an unusual endemic millipede and two genera of Opiliones not otherwise represented by troglobites in Mexico. Other troglobites include schizomids, spiders, pseudoscorpions, and crickets. The stygobitic fauna includes two species (a pseudothelphusid crab and a pimelodid catfish) of freshwater origin. These genera also contain species in southern Mexico. The remaining stygobites have been obtained from anchialine habitats off the coast of Belize and include Remipedia, copepods, and cirolanid isopods. These species have affinities with other anchialine species from the coast of Yucatán and the Caribbean Islands.

Reddell (1981) detailed the history of biospeleological studies in Mexico. The Association for Mexican Cave Studies continues study in various parts of Mexico. Recent work under the direction of José Palacios-Vargas from Mexico City has resulted in major faunal discoveries. Approximately 300 troglobites and 140 stygobites have been described from Mexico but many more will probably be recorded. Few areas of the country have been adequately sampled and large areas remain unexplored. In particular, many of the numerous isolated mountain ranges in northern Mexico have yet to be visited. One lava tube in Veracruz contains a prolific fauna and other volcanic caves probably contain fauna of interest. Recent discoveries of endemic stygobites in caves and springs in northern Mexico suggest that many additional species await discovery and description.

The distribution and composition of the cavernicole fauna of Mexico is a result of a combination of past climatic and geological history and the complex physiography of the country. The majority of the terrestrial fauna is composed of species closely related to taxa still inhabiting the surface. A few species in the extreme north of Mexico are identical to or congeneric with species found in Texas to the north. Numerous species in northern Mexico occupy caves in isolated mountain ranges surrounded by desert. Of special interest are two families of millipede, the Cambalidae and Trichopetalidae, which are common elements of the southeastern United States fauna but occur only in caves in Mexico. These appear to be relicts derived from leaf litter inhabitants of forests that once extended along the Gulf Coast south into Veracruz. These forests are now restricted to high elevations in Mexico, but some species of these families are now restricted to caves at both high and low elevations, with the more highly cave-adapted species found in lowland caves.

The aquatic fauna in Mexico is derived both from freshwater and marine ancestors. The freshwater groups include some dugesiid flatworms, hydrobiid snails, asellid and stenasellid isopods, crayfish, palaemonid shrimps, pseudothelphusid crabs, dytiscid beetles, and fishes. Two species of earthworm and one trichonisicd isopod appear to have become secondarily adapted to freshwater in caves. The marine derivatives include species that are apparently relicts of the vast Cretaceous marine embayment that covered much of Mexico. These species presumably became adapted to freshwater as the seas retreated. This fauna includes amphipods, isopods, mysids, and alpheid shrimp. A large number of species have been described from the caves of the Yucatán Peninsula (see separate entry), largely as the result of diving. Although some of the species inhabiting groundwater in this area are fully adapted to freshwater, others have been found only in anchialine habitats. These latter species have their closest affinities with the fauna of the islands of the Caribbean. Among notable groups inhabiting these waters are remipedes, copepods, ostracods, cirolanid isopods, mysids, amphipods, and fishes.

The nutrient-rich caves of the lowlands in the eastern foothills of the Sierra Madre Oriental, Oaxaca, Veracruz, and Tabasco contain a preponderance of troglophiles, many associated with bat guano. The troglobitic fauna include species apparently derived from taxa inhabiting leaf litter and include large isopods, arachnids, centipedes, millipedes, thysanurans, and histerid beetles. The spiders (particularly the Pholcidae) are well represented, but troglobitic schizomids, pseudoscorpions, and ricinuleids are also known. One tarantula from the family Theraphosidae is a troglobite in the lowland caves of Oaxaca. The stygobitic fauna includes species of both freshwater and marine origin. The freshwater derivatives include snails, palaemonid shrimp, crayfish, and fish. The marine derivates include dimarcusid flatworms, cirolanid isopods, mysids, and alpheid and atyid shrimps. Of special interest are several species of fish. Two species of the ictalurid genus *Prietella* have been found in caves along the eastern slopes of the Sierra Madre Oriental and one in the extreme north of Mexico near the border with Texas. One pimelodid catfish of the genus *Rhamdia* has been found in caves in Oaxaca. Two species that possess troglomorphic populations are of special interest. The poeciliid *Poecilia mexicana* displays varying degrees of eye and pigment reduction within Cueva de Villa de Luz, a remarkable sulfur-based chemoautotrophic cave ecosystem (see separate entry). The characid *Astyanax fasciatus* has apparently colonized caves many times in the states of San Luis Potosí and Tamaulipas and has become the best-studied cave fish in the world. This reflects the ability of subterranean populations to fully interbreed with surface ancestral populations and the ease with which it can be raised in captivity.

High elevation caves throughout Mexico are typically more nutrient-poor and contain a much more diverse troglobitic fauna than caves at lower elevations. The fauna of the Sierra Madre Oriental includes trichoniscid isopods, scorpions, schizomids, amblypygids, pseudoscorpions, opilionids, numerous species of spider, centipedes, millipedes, collembolans, campodeid entotrophs, silverfish, phalangopsid crickets, and beetles. Troglobitic beetles, which are absent from lowland caves, are especially speciose with numerous species of carabid beetles belonging to several genera. The family Leiodidae also contains troglobites in this area. Many genera reach their northern limits in the isolated mountain ranges of Durango, Nuevo León, and Coahuila. Several schizomids, ricinuleids, spiders, pseudoscorpions, opilionids, centipedes, millipedes, thysanurans, and carabid beetles are now relicts at higher elevations on mountains surrounded by desert. Stygobites found at higher elevations include dugesiid flatworms, earthworms, amphipods, isopods, and dytiscid beetles. The troglobitic fauna of the highlands of southern Mexico typically contains the same groups as that of the Sierra Madre Oriental but includes several families of spider and millipede that reach their northern limits of distribution in extreme southern Mexico. Pseudothelphusid crabs have also been found in Chiapas. The stygobitic fauna includes a number of unusual species, including a polychaete worm from a cave in Guerrero.

The troglobitic fauna of the Yucatán Peninsula includes trichoniscid isopods, scorpions, amblypygids, pseudoscorpions, spiders, millipedes, collembolans, and phalangopsid crickets. The stygobitic fauna is particularly rich and includes remipedes, copepods, ostracods, thermosbaenaceans, mysids, amphipods, cirolanid isopods, and fish. The fish *Ogilbia pearsei* is a marine relict, whereas the eel *Ophisternon infernale* is presumably of freshwater origin.

Numerous species of bat inhabit Mexican caves. These include carnivorous, piscivorous, insectivorous, and haematophagous species. Attempts to eradicate vampire bats in southern Mexico have led to the destruction of many colonies of all species. Other vertebrates occurring as trogloxenes in Mexican caves include frogs, salamanders, lizards, snakes, and rodents.

The cavernicole fauna of the Caribbean islands is generally poorly studied although the fauna of Cuba has been fairly well documented and research in Jamaica and Puerto Rico has considerably extended our knowledge of the troglobitic fauna of these islands (Peck, 1999; Peck *et al.*, 1998). Very little is known about the terrestrial cave fauna of the island of Hispaniola. The aquatic fauna of many of the islands is now well understood due to the extensive studies by a number of researchers studying the interstitial and subterranean fauna. Extensive diving in many caves has greatly expanded our knowledge of the anchialine fauna (Illiffe, 2000; see also Walsingham Caves, Bermuda: Biospeleology). A troglobitic onychophoran from Jamaica is only one of two in the world, the other being in South Africa. The only troglobitic isopod recorded from the Caribbean Islands is an undescribed species from Jamaica. With few exceptions, the troglobitic arachnids from the Caribbean islands are represented by endemic genera. Troglobitic schizomids are known from Cuba and Jamaica, amblypygids from Cuba, pseudoscorpions from the Dominican Republic, Cuba, and Jamaica, opilionids from Jamaica, and spiders from Cuba, Jamaica, and Mona Island. Of special interest are troglobitic mygalomorph spiders from Cuba (Barychelidae) and Jamaica (Diplura). Two species of troglobitic centipede are known from Cuba. The absence of troglobitic millipedes from the Caribbean islands is puzzling. The troglobitic insect fauna of the Caribbean islands is far more limited than on the mainland. Collembola are known from Cuba, Guadeloupe, Haiti, and Jamaica, thysanurans on Cuba, pentacentrid and phalangopsid crickets on Cuba, roaches on Cuba, Jamaica, and Puerto Rico, fulgoroid homopterans on Jamaica and possibly Mona Island, and carabid beetles on Jamaica.

The stygobitic fauna in the Caribbean is almost entirely derived from marine ancestors. The only exceptions appear to be some copepods and crayfish on Cuba and elmid beetles on Haiti. Many mites have been sampled from the interstitial zone below rivers in Cuba and Haiti. There is a remarkable diversity of species inhabiting both freshwater and anchialine habitats in the Caribbean islands. With the exception of triclad planarians on Jamaica and in the Venezuelan Islands, polychaetes in the Bahama Islands and Netherlands Antilles, and a chaetognath in the Bahama Islands, the invertebrates all belong to the Crustacea. Remipedes have been found in the Bahama Islands and Turks and Caicos Islands; themosbaenaceans in Cuba, República Dominicana, Guadeloupe, Jamaica, Puerto Rico, Venezuelan Islands, and Virgin Islands; ostracods in the Bahama Islands, Cuba, Haiti, Jamaica, and Turks and Caicos; copepods in the Bahama Islands, Barbados, Cuba, Guadeloupe, Haiti, Jamaica, Netherlands Antilles, and Venezuelan Islands; cumaceans in the Bahama Islands; isopods in the Bahama Islands, Cayman Islands, Cuba, Dominican Republic, Guadeloups, Haiti, Jamaica, and Netherlands Antilles; amphipods in the Bahama Islands, Barbados, Cayman Islands, Cuba, Dominican Republic, Guadeloupe, Haiti, Jamaica, Leeward Islands, Netherlands Antilles, Puerto Rico, Turks and Caicos, Venezuelan Islands, and Virgin Islands; mysids in the Bahama Islands, Cuba, Dominican Republic, Jamaica, and Puerto Rico; grapsid crabs in the Bahama Islands and Jamaica; and shrimps in the Bahama Islands, Cayman Islands, Cuba, Dominican Republic, Guadeloupe, Jamaica, Leeward Islands, Mona Island, Netherlands Antilles, Puerto Rico, and Turks and Caicos. Stygobitic fish, all belonging to the genus *Lucifuga*, have been found only in the Bahama Islands, Cuba, and Jamaica.

The stygobitic fauna of the Caribbean islands exhibits many puzzling distributional patterns, with some genera endemic to a particular island whereas others are widespread throughout the Caribbean, including Belize and the Yucatán Peninsula. Some genera also occur in the Galapagos Islands, some on both sides of the Atlantic Ocean, and a few have recently also been found in Australia. The distribution of many of these taxa clearly indicates a great age for some lineages.

The subterranean fauna of Central America and the Caribbean islands is severely threatened by a variety of pressures. The greatest threat in many areas is deforestation with resultant erosion, loss of habitat for bats, and reduction in sources of nutrient input into caves. The construction of extensive resorts along the Caribbean coast of Mexico and on many of the islands of the Caribbean has destroyed some caves and led to severe problems with pollution from inadequate sewage disposal facilities. Many caves in Mexico lie in or near small villages that routinely dispose of trash and raw sewage directly into entrances. A proposal to dispose of sewage from cities into wells in northern Yucatán threatens the entire water supply of the Peninsula. Overpumping, especially along the Caribbean coast of the Yucátan Peninsula and some Caribbean islands, threatens to diminish groundwater supplies but also increase the salinity of caves that open onto the coast. Although national parks and other nature reserves exist in some areas, these form only a limited part of the area needed to protect the natural subterranean biodiversity of this region.

JAMES REDDELL

Works Cited

Iliffe, T.M. 2000. Anchialine cave ecology. In *Subterranean Ecosystems*, edited by H. Wilkens, D.C. Culver, & W.F. Humphreys, New York: Elsevier

Kuny, J.A. & Day, M.J. 2002. Designation of protected karstlands in Central America: A regional assessment. *Journal of Cave and Karst Studies*, 64: 165–74

Peck, S.B. 1999. Synopsis of diversity of subterranean invertebrate faunas of the West Indian Island of Hispaniola. *Novitates Caribaea*, 1999(1): 14–32

Peck, S.B., Ruiz-Baliú, A.E., & Garcés González, G.F. 1998. The cave-inhabiting beetles of Cuba (Insecta: Coleoptera): Diversity, distribution and ecology. *Journal of Cave and Karst Studies*, 60: 156–66

Reddell, J.R. 1981. A review of the cavernicole fauna of Mexico, Guatemala, and Belize. *Texas Memorial Museum Bulletin*, 27: 327 pp.

Further Reading

Juberthie, C. & Decu, V. (editors) 1994. *Encyclopaedia Biospeologica*, vol. 1, Moulis and Bucharest: Société de Biospéologie

Romero, A. & Paulson, K.M. 2001. It's a wonderful hypogean life: A guide to the troglomorphic fishes of the world. *Environmental Biology of Fishes*, 62: 13–41

Silva-Taboada, G. 1988. *Sinopsis de la espeleofauna Cubana*, Ciudad de la Habana: Editorial Científico-Técnica

AMERICA, CENTRAL: ARCHAEOLOGICAL CAVES

The geographical area discussed here corresponds with the region known to archaeologists as Mesoamerica. It includes the area of pre-Columbian high cultures in central and southern Mexico, all of Guatemala, Belize, and the western portions of Honduras and El Salvador.

For Mesoamerican cultures, caves were perhaps the most sacred features in the natural landscape. Their importance is a reflection of a basic Amerindian religious focus on the Earth as a sacred and animate entity. Growing out of that focus was a fundamental concern with place. In central Mexico, the Nahuatl word for community, *altépetl*, literally means "water-filled mountain". The glyphic rendering is a mountain with a cave at its base. A people were identified with their place, their sacred mountain, and its cave. In many Maya languages the name for the principal indigenous deity translates as "hill-valley", and the deity is considered to be the owner of that "hill-valley". In a very real sense, the landscape is personified and deified. The word for cave in many indigenous languages translates as "stone house" because caves were seen as the residence of the deities. This may be the reason that 16th-century Yucatec Maya used the term *actun* for both caves and stone buildings like temples. Neither is this the only case. In Maya inscriptions, pyramids were called "hills" and represented sacred mountains. Thus, place and its surrounding landscape were so important that the dominant human constructions, pyramids and temples, were models of mountains and caves.

Historical Development of Cave Archaeology

The archaeological investigation of caves in Mexico and Central America has a long history, dating back to the travels and descriptions of John Lloyd Stevens and Frederick Catherwood in the 1840s. Formal archaeological investigations that met or exceeded the standards of the day were initiated before the end of the 19th century at a number of sites in the Maya area. Henry Mercer published *Hill-Caves of Yucatan* in 1896, Edward Thompson wrote *Cave of Loltun* (1897), the *Caverns of Copan* was published by George Gordon (1898), and Eduard Seler reported on Quen Santo (1901) in the Guatemalan Highlands.

During the period between the world wars, cave studies languished, with the only noteworthy studies being the British Museum Expedition to Pusilhá in the Maya area (Gruning, 1930; Joyce *et al.*, 1928; Joyce, 1929) and the Cueva Encantada, Morelos (Arellano & Müller, 1948; Müller, 1948) in Central Mexico. During this period, cave studies disappeared as an important element of the archaeological literature and never again approached the discipline's highest standards of field methodology.

After World War II, the situation slowly improved, with sound scientific studies being produced by the Carnegie Institution of Washington's Mayapan Project and by David Pendergast in Belize. At the same time, a number of high-profile discoveries focused attention on the often-spectacular cave remains. The most impressive was the opening of a blocked passage at Balankanche near Chichen Itza, which revealed dozens of elaborate incense burners and other artefacts, all in their original context. The study of the cave was noteworthy for the recording of a modern Maya cave ritual that coincided with the beginning of the study (Andrews, 1970). Two caves containing Olmec paintings were discovered in Guerrero, but unfortunately no archaeological study accompanied the recording of the artwork. Finally, the discovery of a cave beneath the Pyramid of the Sun at Teotihuacan was probably the single most important cave discovery in Central Mexico, because it engendered a great deal of rethinking of the importance of caves. In the Maya area, the discovery of Naj Tunich served to launch cave archaeology as a recognized subdiscipline (Brady, 2000).

While cave investigation has a 150-year history, no attempt was made to synthesize the data and produce a coherent statement on the function of caves until the publication of J. Eric Thompson's *The Role of Caves in Maya Culture* in 1959. Thompson's views, however, were not widely circulated until the appearance of a revised version of his original article (Thompson, 1975). Thompson's synthesis is noteworthy because, with the exception of the use of cenotes as sources of drinking water, all of his major functions are religious. Furthermore, habitation, the function most frequently proposed by field archaeologists, was discarded completely. That same year, Doris Heyden (1975) published the first synthesis in English on Central Mexican cave use. Heyden's discussion is noteworthy because she is the first to propose that a cave was of such importance that it determined the placement, size, and orientation of one of the largest structures in all of Mesoamerica. Interestingly, neither Thompson nor Heyden ever did archaeological work in caves. By the late 1980s a new trend was notable as archaeologists specializing in caves began to dominate theoretical discussions, using data that they had collected in the field (Bonor Villarejo, 1989; Brady, 2000; Stone, 1995)

The Role of Caves in Legitimizing Settlement

The founding of a new settlement in Mesoamerica was a matter of great cosmological importance and was always accompanied by ritual. After their conquest, the Spanish found that the easiest way to map a community's boundaries was to have the foundation ritual reperformed, because this ceremony formally established those boundaries. As a result, there are many descriptions of such rituals. It is clear from these descriptions that indigenous people searched for a particular landscape configuration that had cosmological significance and in which caves were central. Angel García-Zambrano (1994: 218) notes that these caves, "when ritually dedicated to the divinities, became the pulsing heart of the new town, providing the cosmogonic referents that legitimized the settlers' rights for occupying that space and for the ruler's authority over that site."

From 1990 to 1993 a cave survey was carried out at the Maya site of Dos Pilas in the Peten area of northern Guatemala, in order to determine if a relationship between surface architecture and caves could be detected. The study found a far more pervasive pattern of locating site architecture in relation to caves than reported for any other site. Relationships are present on three levels. Two of the site's three large public architectural complexes are built directly over caves. Hieroglyphic inscriptions at the largest complex appear to refer to the cave in the toponym, or place name. The third complex may be aligned with a cave feature (springs), and there is the possibility that a cave also runs below it. At a nearby site, a cave runs beneath the central plaza, and the architecture was laid out so that a skylight entrance could be utilized as an important religious feature. A number of important, but secondary, complexes also appear to be formed around large caves. Finally, small residential complexes and even individual house structures are associated with small caves generally less than 10 m in length (Brady, 1997; Brady *et al.*, 1997). Thus, it appears that the caves at Dos Pilas structured the site layout from the largest public architectural complexes down to individual house mounds.

While the Dos Pilas data are the most systematic and compelling, a number of examples of cave/architecture relationships have been recorded. The earliest example was Edward Thompson's discovery in 1896 of a cave beneath the centre of the Ossario at Chichen Itza (Thompson, 1938). Other early examples were reported at Tulum (Lothrop, 1924: 109), Cozumel (Mason, 1927: 278), and Polol (Lundell, 1934: 177). Because the importance of the caves was not appreciated, no attempt was made to interpret what appears to be a clear and widespread pattern.

Artificial Caves

As noted above, within Mesoamerica there is a widely shared conceptual landscape in which caves are a major feature. Interestingly, most of the volcanic areas of Mesoamerica have few if any natural caves. In these areas the conceptual landscape is imposed on and reproduced in local landscapes by the excavation of artificial caves. In recent years, large numbers of these have been documented in the Maya Highlands (Brady & Veni, 1992) and in Central Mexico (Manzanilla, López & Freter, 1996; Medina Jaen, 2000).

Artificial caves offer clear proof that the relationship between architecture and natural caves noted at Dos Pilas and other sites was deliberate, because artificial caves replicate the same patterns. These features are also important because their form reflects the decisions of their makers. Several of the elaborate artificial caves appear to be models of the seven-chambered cave of origin, suggesting that this may be what all the caves in site centres are meant to represent. The association of other artificial caves with sacred locations and pilgrimage sites adds another dimension to our understanding of the role of caves. The presence of the cave would appear to be one of those markers that alerts visitors to the fact that this a place of supernatural power.

Artificial caves appear to be constructed along fairly regular plans, so they are clearly an architectural form in their own right. Some of the caves reflect construction on a monumental scale, requiring as much or more labour input as pyramidal structures. The truly massive scale of some constructions is of central importance, because the lavish expenditure of resources should alert archaeologists to the fact that the focus of such attention is somehow central to the concerns of the society.

Caves and Ritual

The preceding sections have dealt with caves that were located in political core areas and so were the focus of large-scale public rituals. To this could be added the many pilgrimage caves. The vast majority of all caves, however, are located in peripheral areas, but these almost invariably show some type of ancient utilization. This suggests that these features, no matter how small or how remote, were never overlooked.

What type of rituals were carried out in these caves? The rural Mesoamerica religious tradition, which is still very much alive, suggests that the rituals revolved around the agricultural cycle. Villages often make processions to ask the permission of the Earth Lord before clearing and burning fields, and perhaps again at the time of planting. The discovery of small corn cobs, four to five centimetres long, at many caves indicates that ceremonies for the offering of the first young ears may have been widely practised. The Day of the Cross (3 May) is celebrated throughout Mesoamerica. Occurring just before the onset of the rainy season, the celebrations traditionally include petitions for rain. The ceremonies are often made in a cave, because rain is considered to be a terrestrial rather than a celestial phenomenon. Rain, clouds, and lightning are formed within caves and then sent by the rain deities into the sky. Special rituals are undertaken in cases of drought, which are usually interpreted as a supernatural punishment for some oversight or infraction. Thus, rituals involving caves appear to occur at every significant point in the agricultural cycle.

Besides providing rain and fertility, the earth is also the source of disease and pestilence. In Yucatán, rituals are performed to keep the evil airs from escaping from caves/cenotes. Curing ceremonies often involve cave rituals. In many cases, the disease-causing agent may be removed by a shaman passing an egg over the person's body or sucking out a poisonous intrusion. The disease-bearing object will then be returned to its source in the cave. Not surprisingly then, caves are often thought to be places of witchcraft where disease can be given to an enemy.

Caves appear to have been used by all segments of society in pre-Columbian Mesoamerica. These features carried a variety of meanings. As the place of human origin they were connected to the miracle of creation and were the focus of group identity. The association with the earth made them sacred places where earth deities could be propitiated and petitioned for the basic necessities of agrarian life. They were frightening places as well. From caves, the deities could send drought and disease to punish humans for misconduct. The huge artefact assemblages recovered from caves suggest that the visitation for all these purposes was heavy and prolonged.

Show Caves

There are a number of show caves open to the public in the region but only a few of the better developed can be mentioned here, and many of the large caves, such as the Chiquibul System, have not been excavated professionally. In northern Mexico near Ciudad Madera, Chihuahua, the Cuartena Casas (Forty Houses) Archaeological Zone is formed around a series of cliff dwellings related to Southwestern rather than Mesoamerican culture. The sites were occupied from about 950 AD until perhaps the 14th century and appear to be related to Casas Grandes (Paquimé).

In Central Mexico, the most famous cave is the man-made tunnel under the Pyramid of the Sun at Teotihuacan, which has been discussed in detail by Heyden (1975). Special permission is needed, however, to get access. Another man-made cave at the site of Xochicalco, Morelos, is open to the public. The main cave appears to have been used as a solar observatory. The position of the sun was charted by sunlight entering through a hole in the ceiling.

In the Maya area, Balankanche Cave near the World Heritage Site of Chichen Itza has regular tours in an assortment of languages. Both the cave and the original context of many of the artefacts were heavily modified in opening the cave to tourism. Loltun Cave, near Oxkutzcab is one of the most impressive caves with large passages. The site is noteworthy for its rock art (Thompson, 1897) as well as a sequence of use dating back to the end of the Pleistocene. Finally, the Cave of Bolanchen, Campeche, made famous by Frederick Catherwood's painting of the use of a huge ladder in the 1840s, has been the subject of a good deal of archaeological investigation (Zapata Peraza, Benevides Castillo & Peña Castillo, 1991).

In Guatemala, the Cueva de las Pinturas, 20 km south of Flores, Peten, contains a large polychrome inscription and substantial architectural modifications dating to the Preclassic. Just across from Flores in Santa Elena, parts of Actun Kan are open to the public. The cave, formerly known as Jobitzina, is quite extensive with another entrance on the opposite side of the hill in which the cave is located. In Alta Verapaz, Lanquin Cave has long been a tourist destination but is also sacred to the Q'eqchi' Maya. Finally, the man-made caves at Utatlan can be visited but are often in use by the K'iche' Maya.

In Honduras, Tauleve, a show cave near Lake Yojoa, is sacred to the Lenca who conduct ceremonies at the cave on April 24. The Talgua Cave (Cave of the Glowing Skulls) near Catacamas is now open to tourism but visitors are not allowed to enter the chambers containing archaeological material. In Belize, visitors can see many intact vessels in their original context at Actun Chichem Ha, near San Ignacio.

JAMES E. BRADY

See also **Art: Cave Art in the Americas**

Works Cited

Andrews, E.W. 1970. *Balankanche, Throne of the Tiger Priest*, New Orleans: Middle American Research Institute, Tulane University

Arellano, A.R.V. & Müller, F. 1948. La cueva encantada de Chimalacatlan, Morelos. *Sociedad Mexicana de Geografía y Estadística*, 66: 481–91

Bonor Villarejo, J.L. 1989. *Las Cuevas Mayas: Simbolismo y Ritual*, Madrid: Universidad Complutense de Madrid

Brady, J.E. 1997. Settlement configuration and cosmology: the role of caves at Dos Pilas. *American Anthropologist*, 99(3): 602–18

Brady, J.E. 2000. ¿Un Chicomostoc en Teotihuacan? The Contribution of the Heyden Hypothesis to Mesoamerican Cave Studies. In *In Chalchihuitl in Quetzalli, Precious Greenstone, Precious Quetzal Feather: Mesoamerican Studies in Honor of Doris Heyden*, edited by E. Quiñones Keber, Lancaster, California: Labyrinthos Press

Brady, J.E. & Veni, G. 1992. Man-made and pseudo-karst caves: the implications of sub-surface geologic features within Maya centers. *Geoarchaeology*, 7(2): 149–67

Brady, J.E., Scott, A., Cobb, A., Rodas, I., Fogarty, J. & Urquizú, M. 1997. Glimpses of the dark side of the Petexbatun Regional Archaeological Project: the Petexbatun Regional Cave Survey. *Ancient Mesoamerica*, 8(2): 353–64

García-Zambrano, A.J. 1994. Early colonial evidence of Pre-Columbian rituals of foundation. In *Seventh Palenque Round Table, 1989*, edited by M.G. Robertson & V. Field, San Francisco: Pre-Columbian Art Research Institute

Gordon, G.B. 1898. *Caverns of Copan, Honduras: Report on Explorations by the Museum, 1896–97*, Cambridge, Massachusetts: Peabody Museum, Harvard University; reprinted New York: Kraus, 1970

Gruning, E.L. 1930. Report on the British Museum expedition to British Honduras, 1930. *Journal of the Royal Anthropological Institute*, 60: 477–83

Heyden, D. 1975. An interpretation of the cave underneath the Pyramid of the Sun in Teotihuacan, Mexico. *American Antiquity*, 40: 131–47

Joyce, T.A. 1929. Report on the British Museum expedition to British Honduras, 1929. *Journal of the Royal Anthropological Institute*, 59: 439–59

Joyce, T.A., Gann, T., Gruning, E.L. & Long, R.C.E. 1928. Report on the British Museum expedition to British Honduras, 1928. *Journal of the Royal Anthropological Society*, 58: 323–49

Lothrop, S.K. 1924. *Tulum: an Archaeological Study of the East Coast of Yucatan*, Washington, DC: Carnegie Institution of Washington

Lundell, C.L. 1934. *Ruins of Polol and other Archaeological Discoveries in the Department of Peten, Guatemala*, Washington, DC: Carnegie Institution of Washington

Manzanilla, L., López, C. & Freter, A. 1996. Dating results from excavations in quarry tunnels behind the Pyramid of the Sun at Teotihuacan. *Ancient Mesoamerica*, 7(2): 245–66

Mason, G. 1927. *Silver Cities of Yucatan*, New York and London: G.P. Putnam

Medina Jaen, M. 2000. *Las Cuevas de Tepeaca—Acatzingo, Puebla: Estudio Arqueológico, Etnohistórico, Etnográfico*, Tesis de Licencido, Mexico: Escuela Nacional de Antropología e Historia

Mercer, H.C. 1896. *The Hill-Caves of Yucatan: A Search for Evidence of Man's Antiquity in the Caverns of Central America*, Philadelphia: Lippincott; reprinted Norman: University of Oklahoma Press, 1975

Müller, F. 1948. Chimalacatlan. *Acta Anthropologica*, 3(1)

Seler, E. 1901. *Die Alten Ansiedlungen von Chaculá, im Distrikte Nenton des Departments Huehuetenango der Republik Guatemala*, Berlin: Dietrich Reiner

Stone, A.J. 1995. *Images from the Underworld: Naj Tunich and the Tradition of Maya Cave Painting*, Austin: University of Texas Press
Thompson, E.H. 1897. *Cave of Loltun, Yucatan: Report of Explorations by the Museum, 1888–89 and 1890–91*, Cambridge, Massachusetts: Peabody Museum, Harvard University
Thompson, E.H. 1938. *The High Priest's Grave, Chichen Itza, Yucatan, Mexico, a Manuscript*, Chicago: Field Museum of Natural History; reprinted New York: Kraus, 1968
Thompson, J.E. 1975. Introduction to the reprint edition. In *The Hill-Caves of Yucatan*, by H.C. Mercer, Norman: University of Oklahoma Press
Zapata Peraza, R.L., Benevides Castillo, A. & Peña Castillo, A. 1991. *La Gruta de Xtacumbilxunaan, Campeche*, Instituto Nacional de Antropología e Historia, Mexico

Further Reading

Brady, J.E. 1999. *Sources for the Study of Mesoamerican Ritual Cave Use*, 2nd edition, Los Angeles: California State University
An extensive bibliography on Mesoamerican ritual cave use.

Brady, J.E. & Ashmore, W. 1999. Mountains, caves, water: Ideational landscapes of the ancient Maya. In *Archaeologies of Landscapes: Contemporary Perspectives*, edited by W. Ashmore & A.B. Knapp, Oxford: Blackwell
Discusses the use of caves in the context of ritual landscapes.
Heyden, D. 1987. Caves. In *Encyclopedia of Religion*, edited by M. Eliade, vol. 3, New York: Macmillan
A good overview of cave use in all parts of the world.
MacLeod, B. & Puleston, D.E. 1978. Pathways into darkness: The search for the road to Xibalbá. *Tercera Mesa Redonda de Palenque*, vol. 4, edited by M.G. Robertson & D.C. Jeffers, Monterey, California: Herald
A wide-ranging discussion of the Maya's use of Petroglyph Cave in Belize.
Vogt, E.Z. 1981. Some aspects of the sacred geography of Highland Chiapas. In *Mesoamerican Sites and World Views*, edited by E.P. Benson, Washington, DC: Dumbarton Oaks Research Library and Collection
An excellent discussion of modern Maya beliefs concerning sacred landscapes.

AMERICA, NORTH: ARCHAEOLOGICAL CAVES

Archaeological cave sites in North America occur from Alaska to the Yucatán Peninsula. The only significant area lacking archaeological caves is the Canadian Shield. The two most significant areas of prehistoric deep cave use occur in the karst regions of the Central Lowlands of the United States and the Mayan cave region of southern Mexico and Central America (see America, Central: Archaeological Caves). Most archaeological caves are prehistoric habitation sites, where people sought shelter under cliff overhangs, or rock shelters and cave entrances. A significant number of cave sites, however, contain evidence of human activity far beyond the reach of daylight, that includes the mining of minerals and stone resources, drawing pictographs and petroglyphs, placement or burial of the dead, and performance of rituals.

In North America, as in all parts of the world, humans have used the natural shelter afforded by cave entrances for everyday activities: processing food, cooking, making tools and clothing, and, undoubtedly, other social activities. Many cave entrances contain deeply stratified deposits, some dating to the earliest, well-accepted occupation of the Americas, $c.10 000–12 000$ years BP. Deeply stratified cave deposits were used to define regional cultural chronologies in many regions of North America, thus making them important sites in the history of archaeology.

Two important cave sites, with late Pleistocene Paleoindian occupations, are Wasden/Owl Cave, Idaho, and Bluefish Caves, Yukon. Owl Cave is a collapsed lava tube with a large bison bone bed, stone tools, and worked bone, ^{14}C dated between 8200 and 7800 BP. Underlying this strata, is a bone bed of broken and modified mammoth, bison, and camelid remains, with a fluted, Paleoindian point association (Miller, 1982). Bluefish Cave I contains an early deposit of stone tools and other artefacts, with the remains of mammoth, bison, horse, and other mammals. Bone collagen ^{14}C dates range from 25 000–12 000 BP, and the tool assemblage is similar to late Pleistocene tool traditions known from Siberia (Cinq-Mars, 1978). The association between the human artefacts and Pleistocene animal bone, however, is still problematic.

Three important stratified cave sites in the western United States are Ventana Cave, Arizona, Danger Cave, Utah, and Bat Cave, New Mexico. Ventana Cave is a deep rock shelter formed by a seep spring in volcanic agglomerate. Completely excavated in the 1940s (Haury, 1950), the cave yielded stratified deposits from the late Pleistocene to historic Tohono O'odham (Papago) Indians. The earliest deposit contained bones of extinct Pleistocene fauna and a few human artefacts, suggesting a possible Paleoindian occupation. Later ^{14}C dating of this layer established that it was early Holocene ($c.10 600–8800$ years BP) in age, and that the early tool industry was more closely related to later Archaic occupations than Paleoindian. Danger Cave is a dry cave that has a long history of excavation, beginning in the 1930s (Jennings, 1957). A stratified sequence of occupations dating from 10 500 BP up to historic Indian occupations, was used to define the Desert Archaic tradition, a successful hunting and gathering way of life adapted to the desert and mountain environments of western North America. Bat Cave also has a long sequence of occupation ($c.10 500$ BP to historic times); however, it is best known because it contains the earliest association of maize with human occupation outside of Mesoamerica (Dick, 1965). Re-analysis of the deposits, and direct ^{14}C dating of the maize remains, confirmed its spread into North America by 3500–3000 BP.

In central and eastern United States, three important stratified cave sites are Russell Cave, Alabama, Dust Cave, Alabama, and Graham Cave, Missouri. Russell Cave is a true limestone cave that is now a US National Monument. Excavated in the 1950s and 1960s, archaeological deposits in the entrance were more than 9 m deep, dating from 9000 to 400 BP (Griffin, 1974). The deposits yielded a wealth of chronological information on tool and artefact manufacture, especially during the Archaic Period ($c.9000–3000$ BP). Ongoing excavations at nearby Dust Cave, with deposits dating from 10 500 to 3250 BP, are yielding new information on the depositional context, paleoenvironment and human subsistence during this early time period, as well as more traditional information on prehistoric tool manufacture and use. Graham Cave is an unusual cave, formed in sandstone, which also contains early archaeological deposits, primarily dating from 9700 to 7000 BP (Logan, 1952). Prehistoric occupation spans the transition from the late Paleoindian to Archaic Period hunters and gatherers in an important ecotone between the western prairies and eastern hardwood forests.

In the eastern United States, the first evidence of deep cave exploration by prehistoric Indians is dated $c.4600$ BP. In a remote passage of a cave in Tennessee, approximately 1.5 km from the entrance, were found some 275 foot impressions in a soft mud floor. A thin scatter of cane torch charcoal (*Arundinaria* sp.), and charcoal smudges on the walls and ceiling of the passage, were used to date this prehistoric exploration. It appears from the footprints that nine individuals (possibly in two separate trips) travelled to the end of this passage and then returned. A number of other cave sites in Tennessee, Kentucky, and Indiana, contain evidence of prehistoric exploration dating prior to 3000 BP. Based on the limited remains found in these caves, the objective of these early cave trips appears to be simply exploring. However, these early prehistoric cave explorers were apparently adept at penetrating remote and demanding cave passage and successfully returning to the entrance.

A number of new activities begin to appear $c.3000$ BP, in the record of prehistoric cave use in the eastern United States. Third Unnamed Cave, Tennessee, contains evidence for quarrying chert cobbles from a remote sandy-floored passage, testing and reducing the cobbles by knapping, and the engraving of petroglyphs onto the limestone ceiling and walls (Simek, Franklin & Sherwood, 1998). The petroglyphs are primarily geometric: rayed circles, semi-circles, checkerboards, chevron patterns, and numerous other enigmatic groups of lines. Wyandotte Cave, Indiana, was explored in prehistoric times and variously mined for chert and aragonite. The exploration and possibly chert mining occurred earlier, $c.4150-2200$ BP, but the mining of a large aragonite stalagmite appears to date $c.2200-1150$ BP (Munson & Munson, 1990). Artefacts made of Wyandotte Cave aragonite have been identified at several sites in the Midwestern United States, suggesting that this cave resource was widely valued.

The most extensive prehistoric mining activity in the caves of North America is found in Mammoth Cave and Salts Cave, Mammoth Cave National Park, Kentucky. Dating $c.3000-2200$ BP, several kilometres of both caves were systematically mined for gypsum ($CaSO_4 \cdot 2H_2O$) in the form of crusts and flowers battered from the walls and ceiling, and selenite needles dug from the floor sediments. In more limited quantities, mirabilite ($Na_2SO_4 \cdot 10H_2O$) is also present in the caves and appears to have been collected in prehistoric times. It is not clear why so much effort was spent mining sulfate minerals, although gypsum may be made into a white paint or paste and the crystals themselves can be quite spectacular. Mirabilite, if ingested in sufficient quantities, is a medicinal cathartic.

The archaeology of Mammoth and Salts caves is well known for the perishable remains preserved in the dry passages: unburned torch material, cordage, woven foot wear, twined textile fragments, gourd containers, wooden bowls, digging sticks, mussel shell scrapers, and climbing poles are commonly found in the mined passages (Watson, 1969; 1974). Historically, at least two desiccated prehistoric bodies were found in the cave, both appearing to be mining accident victims. One is a $c.45$-year old man killed by rock fall, and the other a $c.9$-year old boy who appears to have died from a fall. Also among the dry remains preserved in the caves, are hundreds of human paleofaeces (hence the strong suggestion that mirabilite was consumed in the cave for its cathartic effect), which is an unparalleled source of material for the study of prehistoric diet and parasitic infection during this time period. There are also a small number of enigmatic charcoal pictographs and petroglyphs found in Mammoth and Salts caves, associated with the mining, although the rendering of glyphs appears to have been a minor activity.

During later prehistoric periods in North America ($c.2000$ BP to historic times), native groups appear to have used caves increasingly for ceremonial or ritualistic settings, including use as burial sites and decoration of the walls and ceilings with glyphs. The Copena culture of northern Alabama and Georgia used caves to inter individuals with elaborate burial items, including artefacts made of imported copper, galena, mica, marine shell, and steatite (Beck, 1995). Vertical shafts or pit caves, with surface openings, were used to deposit bodies in many areas of North America. This is not a very well-studied phenomenon, but the greatest concentration of burial pit caves is reported from southwest Virginia and east Tennessee. Human remains, incorporated into talus cones at the base of pit openings, are complex depositional environments awaiting more detailed, systematic study.

So-called "cave art", in the dark zone of caves, is turning out to be widespread in the eastern United States. Pictographs, petroglyphs, and mud-glyphs, found in numerous caves, contain an array of geometric shapes, but increasingly zoomorphic, anthropomorphic, and iconographic elements dominate the art. Some of the art appears related to motifs found on artefacts widespread in the southeastern United States, and mythological figures of historic Indian groups, such as Cherokee underworld creatures (Faulkner, Deane & Earnest, 1984). The performance of rituals or shamanistic acts, associated with the drawing of cave art, has been inferred for many of these late prehistoric deep cave sites.

One of the best archaeological examples of the use of an inner cave setting for ritual behaviour, is Feather Cave, New Mexico. An inner room of the cave, rediscovered in 1964, contained some 400 perishable artefacts in their original position, thought to be at least 600 years old. The room contained a large assortment of miniature bows, reed arrows, crook pahos (prayer sticks) and other artefacts, along with pictographs adorning some walls (especially hand prints outlined in white clay-like pigment). The site was interpreted, with the help of elders from the nearby Pueblo Indians, as a Mogollon sun and earth shrine visited during biannual solar ceremonies, similar to Pueblo ceremonial caves still in use today (Ellis & Hammack, 1968). Parallels may also be drawn between Pueblo ceremonial caves and concepts regarding caves and supernatural beings in Mesoamerican cultures. It is debatable whether parallel beliefs between cultures are due to diffusion of religious ideas, or similar geological features (in this case caves) influence similar concepts among different cultures.

The ritual use of caves is also a renewed topic of interest in Mesoamerican archaeology. Classic Maya sites ($c.1700-1050$ BP) appear to be associated with important caves, often incorporating local caves into the settlement layout. From these new cave discoveries, a more complete picture of ancient Maya ritual practice is emerging. As in the case of the historic Cherokee in the eastern United States, and Pueblo groups in the southwestern United States, belief systems of modern Maya Indians provide a link between sacred cave use and ritual knowledge in the present to the beginning of prehistoric concepts of caves a millennium or more in the past.

GEORGE M. CROTHERS

See also **Art: Cave Art in the Americas**

Works Cited

Beck, L.A. 1995. Regional cults and ethnic boundaries in "Southern Hopewell". In *Regional Approaches to Mortuary Analysis*, edited by L.A. Beck. New York: Plenum Press

Cinq-Mars, J. 1978. Bluefish Cave I: A late Pleistocene Eastern Beringian cave deposit in the Northern Yukon. *Canadian Journal of Archaeology*, 3(1): 1–32

Dick, H.W. 1965. *Bat Cave*, Santa Fe, New Mexico: School of American Research

Ellis, F.H. & Hammack, L. 1968. The inner sanctum of Feather Cave, A Mogollon sun and earth shrine linking Mexico and the Southwest. *American Antiquity*, 33(1): 25–44

Faulkner, C.H., Deane, B. & Earnest, Jr, H.H. 1984. A Mississippian Period ritual cave in Tennessee. *American Antiquity*, 49(2): 350–361

Griffin, J.W. 1974. *Investigations in Russell Cave*, Washington DC: National Park Service

Haury, E.W. 1950. *The Stratigraphy and Archaeology of Ventana Cave*, Tucson: University of Arizona Press

Jennings, J.D. 1957. *Danger Cave*, Salt Lake City: University of Utah Press

Logan, W.D. 1952. *Graham Cave: An Archaic Site in Montgomery County, Missouri*, Columbia: Missouri Archaeological Society

Miller, S.J. 1982. The archaeology and geology of an extinct megafauna/fluted point association at Owl Cave, the Wasden Site, Southeastern Idaho: A preliminary report. In *Peopling of the New World*, edited by J.E. Ericson, R.E. Taylor & R. Berger, Los Altos, California: Bellena Press

Munson, P.J. & Munson, C.A. 1990. *The Prehistoric and Early Historic Archaeology of Wyandotte Cave and Other Caves in Southern Indiana*, Indianapolis: Indiana Historical Society

Simek, J.F., Franklin, J.D. & Sherwood, S.C. 1998. The context of early Southeastern prehistoric cave art: A report on the archaeology of 3rd Unnamed Cave. *American Antiquity*, 63(4): 663–77

Watson, P.J. 1969. *The Prehistory of Salts Cave, Kentucky*, Reports of Investigations, No. 16, Springfield: Illinois State Museum

Watson, P.J. (editor) 1974. *Archaeology of the Mammoth Cave Area*, New York: Academic Press

Further Reading

Crothers, G.M., Faulkner, C.H., Simek, J.F., Watson, P.J. & Willey, P. 2002. Woodland Cave archaeology in Eastern North America. In *The Woodland Southeast*, edited by D.G. Anderson & R.C. Mainfort, Jr, Tuscaloosa: University of Alabama Press

Faulkner, C.H. (editor) 1986. *The Prehistoric Native American Art of Mud Glyph Cave*, Knoxville: University of Tennessee Press

Goldman-Finn, N.S. & Driskell, B.N. (editors) 1994. Preliminary archaeological papers on Dust Cave, Northwest Alabama. *Journal of Alabama Archaeology*, 40(1–2): 1–255

AMERICA, NORTH: BIOSPELEOLOGY

Faunal Distributions: a Result of History

North America—Canada and the continental United States (northern Mexico is discussed in the entry on America, Central & the Caribbean Islands: Biospeleology) contains a diverse and widely distributed cave fauna. The current distribution of cavernicoles is the result of interwoven historical parameters such as glaciation, shifting of vegetative communities, moisture (or aridity), and the rise and fall of shallow inland seas. Continental drift (elements of the North American cave fauna have holarctic affinities) and the availability of suitable caves—typically in limestone, but also lava tubes and other types of caves—play a role in explaining the current biogeographic setting.

Pleistocene glaciation is a factor limiting the northern extent of the range of many North American troglobites and stygobites, especially terrestrial taxa. Most terrestrial cavernicoles see their greatest diversity south of the maximum reach of Pleistocene glaciation (see biodiversity map of United States in colour plate section). Notable exceptions occur, such as Castleguard Cave in western Canada where a troglobitic mite (*Robustocheles occulta*) is thought to have survived the Pleistocene in this cave beneath the Columbia Ice Field. Distributions of aquatic faunas, too, are sometimes limited by the glacial history—for example, the troglobitic flatworms of the genus *Sphalloplana* are found almost exclusively to the south of the maximum extent of Pleistocene glaciation. Among the aquatic taxa there are a number of lineages that seem to have survived the episodic advance and retreat of glaciers in the groundwater. Distributions of various amphipod crustaceans (e.g. species of *Stygobromus* and *Bactrurus*) are especially indicative of such a history.

The Pleistocene climatic fluctuations have influenced not only through the direct impacts of glaciers, but also through shifts in vegetative zones. Much of the North American terrestrial troglobite fauna has affinities with epigean leaf-litter/soil communities. Faunal elements associated with moist leaf litter typical of cooler montane forests may have entered caves during cooler periods when such forests extended further south and to lower elevations. As the forests retreated north and to higher elevations some taxa—especially those with features that pre-adapt them to life in caves—were isolated from their epigean ancestors.

The relatively depauperate fauna of western North America is explained only in part by somewhat lower densities of caves. Here, the more extreme aridity of the climate may have had much the same influence on terrestrial troglobites as glaciation had on the cave fauna of the northern parts of eastern North America. Arid western caves may have historically lacked the capacity to sustain populations of terrestrial troglobites because of both insufficient moisture and a lack of suitable energy input in the form of organic debris entering the caves from the more sparse vegetative communities in this arid climate. Finally, forest leaf litter communities in the southwestern United States are thought to have historically lacked a diverse community of cool/moist adapted invertebrates through most of the Pleistocene. Such ancestral, pre-adapted, taxa are thought to have been precursors to many of the terrestrial troglobitic arthropods of the eastern United States.

The aquatic fauna of North American caves is also influenced by ancient seas. Stygobitic freshwater isopods of the largely marine family Cirolanidae occur in southern Texas caves (and southward into Mexico) and an endemic cirolanid (*Antrolana lira*) is also known from a small area in Virginia. These taxa may be relicts that were able to adapt to the freshwater cave environment as ancient shallow seas receded. Other aquatic cavernicoles have clear links to extant epigean aquatic taxa which co-occur in the same region. Stygobites in these groups may represent more recent invasions from the surface.

Richness and diversity of the fauna

There are approximately 1300 to 1500 described species of troglobites and stygobites in North America (north of Mexico). Approximately 70% of these are terrestrial, the remainder aquatic. Recent estimates as high as 6000 species have been made when undescribed taxa are included. Such estimates generally ignore microbes, protozoa, and fungi. The richness of the North American stygobite and troglobite faunas is not equally distributed among major taxonomic groups.

Among the vertebrates, troglobites and stygobites are found only among the fish (six or seven species) and salamanders (at least ten taxa). Several species of stygobitic fish occur in the family Amblyopsidae with several species across the eastern United States (*Amblyopsis rosae* in the Ozark Plateau and *Amblyopsis spelaea* and *Typhlichthys subterraneus* in the Interior Low Plateaux). The most extremely troglomorphic species among them is *Speoplatyrhinus poulsoni*, endemic to a very small area in northern Alabama. Two small catfish species are known from the large aquifer of the Edward Plateau. The grotto sculpin, *Cottus* sp. (Cottidae), of southeastern Missouri is likely to be described as a species distinct from a closely related surface relative.

The salamander fauna of North American caves includes a number of troglophiles and troglobites, dominated by members of the family Plethodontidae. Among these are *Haideotriton wallacei* (Florida and Georgia), *Typhlotriton spelaeus* (Ozark Plateau), two species and several subspecies of *Gyrinophilus* (Interior Low Plateaux and Appalachians), the Cave Salamander, *Eurycea lucifuga*, and several plethodontid salamanders in the Edwards Plateau (see Edwards Aquifer: Biospeleology, and photo of Texas Blind Cave Salamander in Amphibia).

Many North American bats are trogloxenes, and utilize caves as roost sites for over-wintering, for migration stopovers, as maternity sites, and as bachelor roosts. Those species that roost in large colonies produce large deposits of guano that serve as an energy-rich food source for a guanophillic community that includes a variety of flies, beetles, mites, millipedes, and other invertebrates. Among the more significant large-colony bat species are two eastern North American species, the Indiana Bat and the Gray Bat, and, further south, the Mexican Free-Tailed bat. All three species can congregate in large colonies which sometimes number many thousands of individuals. A variety of other bat species, especially in the genera *Myotis*, *Pipistrellus*, and *Eptesicus*, are frequent inhabitants of North American caves.

Other trogloxenic vertebrates regularly utilizing caves include packrats (*Neotoma* spp.), mice, Turkey Vultures (*Cathartes aura*), and Eastern Phoebes (*Sayornis phoebe*), all of which commonly nest in or near the entrances or twilight zones of caves. Raccoons (*Procyon lotor*), who commonly venture deeper into caves, may leave behind significant deposits of faeces which can comprise an important energy input into some caves.

North America's invertebrate cave fauna includes a variety of insect, myriopod, arachnid, and crustacean species. Other groups of cavernicolous organisms—generally less well studied than the macroinvertebrates and vertebrates—include aquatic oligochetes (Annelida), branchiobdellidans, horsehair worms, entocytherid ostracods, copepods, and a variety of bacteria, fungi, and protozoans. Among the flatworms (Turbellaria), stygobites are known in several genera, including *Macrocotyla* and *Sphalloplana*.

Few molluscs are adapted for life in caves. Several groups contain stygobites, and the hydrobiid genus *Fontigens* is notable among these, as it includes several stygobites found in caves of the eastern and southeastern United States.

The arachnid fauna of North America is diverse and fascinating. Several genera of opilionids are troglobites, including *Phalangodes* and *Bishopella* in the eastern United States, *Texella* species in the Edwards Plateau and Guadelupe Mountains, and *Cryptobunus* in the west. Some of these species are troglobites and others are troglophiles. A number of other opilionids may be found in caves, such as the Holarctic genus *Sabacon*. The common genus *Lieobunum* includes trogloxenes that occur in caves from Canada to Texas (and into Mexico). A southwestern species, *Lieobunum townsendii*, frequently occurs in huge aggregations of many hundreds of individuals in or near cave entrances. This species forages above ground at night, and thus may function as an important energy source for some caves.

Mites of caves are not well studied. The family Rhagidiidae occurs in caves across much of North America and includes several troglobites. Pseudoscorpionida is one of the more diverse cavernicolous arachnid groups. There are more than 20 troglobitic pseudoscorpions in the Appalachians alone. Some cave pseudoscorpions are notably larger than their surface relatives. Troglobites occur in several families. *Apochthonius* (Chthoniidae) is one of the more diverse genera, with troglobites found in caves across the United States. As with many of the arachnid groups, there are a number of species that are endemic to one or few sites.

Across much of the United States the spider families Linyphiidae, Leptonetidae, and Nesticidae contain troglobites. Other families (e.g., Telemidae and Dictynidae) contain numerous cave-adapted species in certain regions. In the eastern United States there are several cave-limited *Islandia* and *Nesticus* species. In Texas, the genera *Eidmanella*, *Neoleptoneta*, and *Cicurina* contain a variety of troglobites, and the genus *Cicurina* (Dictynidae) includes epigean species, troglophiles, and narrowly endemic troglobites. In the western United States (California), the genera *Telema*, *Blabomma*, and *Cybaeozyga* each contain several troglobites. The most commonly encountered troglobitic/troglophilic spider in the eastern United States is the widespread linyphiid, *Phanetta subterranea*. Several pholcids are common troglophilic spiders. The common troglophilic spider *Meta ovalis* is one of the more obvious arachnid inhabitants of caves in eastern North America, and is closely related to a European species.

The stygobitic crustaceans include more than 30 decapods—primarily crayfish species (especially *Orconectes* and *Procambarus*) but also three narrowly endemic species of freshwater paleomonid shrimps. Troglobitic crayfish are known from the Ozark Plateau and the Interior Low Plateaus, but the group is most diverse in the aquifers of Florida, where up to three species coexist. One notable decapod community outside of Florida is the faunal assemblage of Shelta Cave (Huntsville, Alabama) where three species of crayfish and a shrimp (one of the three stygobitic *Palaemonias* species) are known.

The amphipod fauna of caves and other groundwater in North America is diverse and dominated by crangonyctid and gammarid taxa. Among the genera frequently encountered are *Crangonyx, Stygobromus, Bactrurus,* and *Gammarus*. Detailed studies of *Gammarus minus* and other Appalachian cave fauna are the basis for some of the theoretical framework of current studies of aquatic cavernicoles in North America. Several other amphipod families occur in North American caves and groundwater, most notably in the Edwards Aquifer of Texas.

In addition to the cirolanid isopods mentioned above, there are many stygobitic or phreatobitic species of the asellid isopod genus *Caecidotea*. *Salmasellus* occurs in a few caves of western North America. Cave-adapted terrestrial isopods of the family Trichoniscidae may also be found in caves of the southeastern and western United States.

Of the myriopods, only the millipedes contain an abundance of troglobites. Troglobitic millipedes are frequently encountered in caves throughout the eastern United States and occur in caves across the country. Genera with troglobites include *Pseudotremia, Cambala, Tingupa,* and *Scoterpes*, but there are a number of others. The centipedes commonly encountered in North American caves are troglophiles.

Among the hexapods, there are a number of troglobites among the springtails (*Arrhapolites, Tomocera, Sinella, Pseudosinella,* and *Oncopodura* are examples of genera containing troglobites) and diplurans (e.g. *Litocampa, Haplocampa, Eumesocampa*).

Troglophiles and trogloxenes are found in several insect orders. The colourful moth *Scolioptrix libatrix* over-winters in caves, and other moths are associated with guano deposits. Flies of the families Heleomyzidae, Phoridae, Sphaeroceridae, and Mycetophilidae are frequent in caves. In the eastern United States the larval stage of the fungus gnat *Macrocera* builds webs to entrap prey. Mosquitoes (Culicidae) of several genera over-winter in caves across much of North America.

Cave crickets (e.g. *Haedenoecus* spp.) and camel crickets (*Ceuthophilus* spp.; Figure 1) are important components of many North American cave communities, and several species may co-occur in a single cave. Trogloxenic *Ceuthophilus* species forage outside of caves at night and rest in the caves during the daytime. Faeces, bodies, and eggs of these animals comprise an important energy source for terrestrial cave communities. In the western United States, some lava tube caves contain another interesting orthopteran, *Grylloblatta* (Grylloblattodea).

Hymenoptera are generally not important components of North American cave communities, with the exception of the Red Imported Fire Ant (*Solenopsis invicta*). This introduced species forages in caves of the Edwards Plateau (and perhaps elsewhere in the southeastern United States) and is thought to have a negative impact on cave communities.

America, North: Biospeleology: Figure 1. *Ceuthophilus cunicularis*, a trogloxene in central Texas caves. Cave crickets are important components of many North American cave communities. (Photo by Steve Taylor, courtesy of Natural Resources Branch, Fort Hood, Texas)

Insect troglobite diversity is greatest in the beetles. Several staphylinids occur in caves, with the troglophile genus *Quedius* being the most frequently encountered. Spider beetles (Ptinidae) include some troglobitic species (e.g. *Niptus*), and the mold beetle family Pselaphidae is rich in troglobites, especially in the widespread genus *Batrisodes*, but also several other genera. Mold beetles occur in many caves of the Appalachians, Interior Low

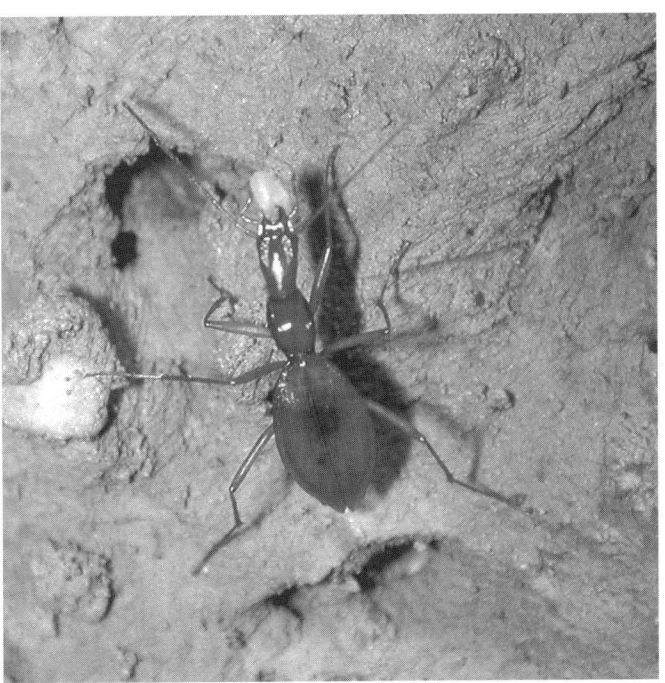

America, North: Biospeleology: Figure 2. *Rhadine reyesi*, a predatory troglobitic beetle that feeds on the eggs of cave crickets. (Photo by Jean Krejca and Steve Taylor, courtesy of Natural Resources Branch, Fort Hood, Texas)

Plateaux, and Edwards Plateau. Aquatic beetles occur in the Edwards Aquifer. The Leiodidae is one of the important cave beetle families, with more than 18 troglobites in the genus *Ptomaphagus* in the southeastern United States. Much of the cavernicole diversity in the Coleoptera occurs in the Carabidae, with many troglophilic and troglobitic species especially in the genera *Pseudanophthalmus* which is especially rich in species in the Appalachians and Interior Low Plateaus and *Rhadine* (Figure 2) which is most diverse in the Edwards Plateau and occurs in caves throughout much of the southwestern United States. Several species of beetles in each of these two genera are closely associated with cave or camel crickets, and prey upon the eggs of those crickets.

North America's cave fauna is threatened by a variety of environmental problems associated with human activities. These problems, discussed in greater detail in the entry Conservation: Cave Biota, include 1) microbial contamination and organic enrichment of karst aquifers through inadequate treatment of human, livestock, and poultry waste; 2) increased sedimentation of subterranean streams resulting from deforestation, agricultural activities, and development; 3) spills of toxic materials (e.g. petroleum and acids) from pipelines, tanker trucks, and freight

America, North: Biospeleology. Endangered species utilizing caves in the continental United States (excludes spring and groundwater taxa). Modified after Elliott (2000).

Common Name	Scientific Name	Year listed as Federally Endangered	Occurrence
Snail:			
Tumbling Creek Cavesnail	*Antrobia culveri*	2001	Missouri
Arachnids:			
Bee Creek Cave Harvestman	*Texella reddelli*	1988	Texas
Bone Cave Harvestman	*Texella reyesi*	1988	Texas
Robber Baron Cave Harvestman	*Texella cokendolpheri*	2000	Texas
Tooth Cave Pseudoscorpion	*Tartarocreagris texana*	1988	Texas
Tooth Cave Spider	*Neoleptoneta myopica*	1988	Texas
Madla's Cave Spider	*Cicurina madla*	2000	Texas
Robber Baron Cave Spider	*Cicurina baronia*	2000	Texas
cave spider*	*Cicurina venii*	2000	Texas
Vesper Cave Spider	*Cicurina vespera*	2000	Texas
Government Canyon Cave Spider	*Neoleptoneta microps*	2000	Texas
Insects:			
Coffin Cave Mold Beetle	*Batrisodes texanus*	1988	Texas
Helotes Mold Beetle	*Batrisodes venyivi*	2000	Texas
Kretschmarr Cave Mold Beetle	*Texamaurops reddelli*	1988	Texas
Tooth Cave Ground Beetle	*Rhadine persephone*	1988	Texas
ground beetle*	*Rhadine exilis*	2000	Texas
ground beetle*	*Rhadine infernalis*	2000	Texas
Crustaceans:			
cave crayfish*	*Cambarus aculabrum*	1993	Arkansas
cave crayfish*	*Cambarus zophonastes*	1987	Arkansas
Lee County Cave Isopod	*Lirceus usdagalun*	1992	Virginia
Alabama Cave Shrimp	*Palaemonias alabamae*	1988	Alabama
Kentucky Cave Shrimp	*Palaemonias ganteri*	1983	Kentucky
Illinois Cave Amphipod	*Gammarus acherondytes*	1998	Illinois
Fish and Amphibia:			
Alabama Cavefish	*Speoplatyrhinus poulsoni*	1977	Alabama
Salamander			
Texas Blind Salamander	*Typhlomolge rathbuni*	1967	Texas
Bats:			
Gray Bat	*Myotis grisescens*	1976	southeastern US
Indiana Bat	*Myotis sodalis*	1967	eastern US
Lesser Long-Nosed Bat	*Leptonycteris curasoae yerbabuenae*	1988	Arizona and New Mexico, to Central America
Mexican Long-Nosed Bat	*Leptonycteris nivalis*	1988	New Mexico, Texas, to Central America
Ozark Big-Eared Bat	*Corynorhinus townsendii ingens*	1979	Arkansas, Missouri, Oklahoma
Virginia Big-Eared Bat	*Corynorhinus townsendii virginianus*	1979	Kentucky, North Carolina, West Virginia, Virginia

*no common name

trains; and 4) reductions in recharge, resulting from increased impervious cover (buildings, roads, parking lots). Continued human population growth also increases impact visitation to caves (habitat degradation), as well as increased rates of cave destruction through quarrying and filling of caves. In the continental United States, 23 troglobitic invertebrates, one fish, one salamander and six bats—all taxa that utilize caves—are currently listed as federally endangered (see Table). In addition, a variety of other cave-dependent species have been proposed as candidates for federal recognition.

STEVEN J. TAYLOR

See also **Edwards Aquifer: Biospeleology; Mammoth Cave: Biospeleology**

Further Reading

Culver, D.C., Christman, M.C., Elliott, W.R., Hobbs, III, H.H. & Redell, J.R. 2003. The North American obligate cave fauna: regional patterns. *Biodiversity and Conservation*, 12(2): 441–468

Culver, D.C., Hobbs, III, H.H., Christman, M.C. & Master, L.L. 1999. Distribution map of caves and cave animals in the United States. *Journal of Cave and Karst Studies*, 61(3): 139–40

Culver, D.C. & Holsinger, J.R. 1992. How many species of troglobites are there? *NSS Bulletin*, 54: 79–80

Culver, D.C. & Sket, B. 2000. Hotspots of subterranean biodiversity in caves and wells. *Journal of Cave and Karst Studies*, 62(1): 11–17

Elliott, W.R. 2000. Conservation of the North American cave and karst biota. In *Subterranean Ecosystems*, edited by H. Wilkens, D.C. Culver & W.F. Humphreys, Amsterdam and New York: Elsevier

Gertsch, W.J. 1992. Distribution patterns and speciation in North American cave spiders with a list of the troglobites and revision of the cicurinas of the subgenus *Cicurella*. In *Studies on the Cave and Endogean Fauna of North America II*, edited by J.R. Reddell, Austin: University of Texas (Texas Memorial Museum, Speleological Monographs, 3)

Hobbs, III, H.H. 1992. Caves and springs. In *Biodiversity of the Southeastern United States. Aquatic Communities*, edited by C.T. Hackney, S.M. Adams & W.H. Martin, New York: Wiley

Peck, S.B. 1973. A review of the invertebrate fauna of volcanic caves in western North America. *NSS Bulletin*, 35(4): 99–107

Peck, S.B. 1988. A review of the cave fauna of Canada, and the composition and ecology of the invertebrate fauna of caves and mines in Ontario. *Canadian Journal of Zoology*, 66: 1197–13

Peck, S.B. 1997. Origin and diversity of the North American cave fauna. In *Conservation and Protection of the Biota of Karst: Symposium at Nashville, Tennessee, February 13–16, 1997*, edited by I.D. Sasowsky, D.W. Fong & E.L. White, Charles Town, West Virginia: Karst Waters Institute (Karst Waters Institute Special Publication, 3)

Peck, S.B. 1998. A summary of diversity and distribution of the obligate cave-inhabiting faunas of the United States and Canada. *Journal of Cave and Karst Studies*, 60(1): 18–26

Reddell, J.R. (editor) 1986. *Studies on the Cave and Endogean Fauna of North America*, Austin: University of Texas (Texas Memorial Museum, Speleological Monographs, 1)

Reddell, J.R. (editor) 1992. *Studies on the Cave and Endogean Fauna of North America II*, Austin: University of Texas (Texas Memorial Museum, Speleological Monographs, 3)

Reddell, J.R & Cokendolpher, J.C. (editors) 2001. *Studies on the Cave and Endogean Fauna of North America III*, Austin: University of Texas (Texas Memorial Museum, Speleological Monographs 5)

Ricketts, T. H., E. Dinerstein, D. M. Olsen, C. J. Leucks, *et al.*, (editors) 1999. *Terrestrial Ecoregions of North America: A Conservation Assessment*, Washington DC: Island Press.

AMERICA, NORTH: HISTORY

Canada and the United States have karsts and pseudokarsts in many differing terrains, with differing landforms, and with differing spelean histories. In broad terms, the United States has about a dozen major calcareous karst areas and one in gypsum. Canada has almost as many in carbonate rocks, and several in gypsum (see entries on Canada and United States of America). The density and types of caves and other features vary markedly between the areas, and many of their subordinate units have their own history of exploration and study, and of significant events.

The study of cave exploration in the United States and Canada began in the 18th century with a simple mention of caves as curiosities. Longer accounts followed, largely in locally oriented tourist guides and early travel books and articles. Gazetteers pulled together scattered accounts, but these tended to be oriented toward specific localities. Eventually, caves were studied as interesting features of natural history; the study of caves for their own sake came relatively late. When a distinctly American speleology began to arise in the 19th century, it too was oriented toward local caves and their specific patterns and features. Thus it was largely independent of evolving European concepts, and perhaps somewhat suspicious of them.

In 1832 the erratic genius Constantine Rafinesque chose an inopportune time to utilize some of these European concepts. In the first systematic effort in American speleology (a classification of the caves of Kentucky) he began with his own observations on swallet and resurgence caves, rock shelters and crevice caves, and saltpetre caves with huge passages. To these he added Buckland's theology-based antediluvial concepts of bone caves, and a brand-new concept of limestone as a lava-like rock extruded from caves through crater-like dolines. At least in part, Buckland soon abandoned the antediluvial theory, and hardly anyone else ever thought of limestone as an extruded rock.

At the north end of the Appalachian Highlands karst area, Kastning has shown how a host of little-remembered figures contributed significantly to American speleology through local observations and reports (Kastning, 1981). To some degree, similar unsung participants contributed to equivalent progress in other karstic areas. Benchmark papers (e.g. those of Nathaniel Shaler, William M. Davis, and other well-known later speleologists) thus evolved naturally from a broadly based continuum, rather than as isolated studies.

Shaw (1992) mentions several writers who, in addition to Constantine Rafinesque, merit special note in this period: W.S. Blatchley, H.C. Hovey and his son E.O. Hovey, Edmund Lee, David Owen, Louisa Owen, and Charles Wright. These notables were a small fraction of the potential list. Not all were obscure. Not all were academics. But all were fluent in English. Early French and Spanish reports (for example, on what now is Florida

Caverns, in 1693), were almost entirely unknown to pioneer American speleologists.

In the United States, the first wave of interest in caves was based largely on the pioneers' need for nitrates to manufacture gunpowder. The second wave began in Kentucky, Indiana, Virginia, and New York in the 1880s, peaked around the turn of the century, and ebbed at the time of World War I. The third began to gain momentum in the 1930s and 1940s and achieved unprecedented success after World War II.

The Europeans founded their initial settlements along the Atlantic coast and its rivers, with littoral and riverside caves being the first to be found. Important riverside caves included Cave-in-Rock on the Ohio River, Indian Cave on the Tennessee River and the largely forgotten Carver's Cave in what is now St Paul, Minnesota. Away from the rivers, small caves in Pennsylvania and in the Virginia highlands were recorded in the 1740s and 1750s.

The names of well-educated "founding fathers" of the United States are prominent in this early period. George Washington left his signature in a small cave near what is now Harpers Ferry, West Virginia. Thomas Jefferson made the first map of a US cave (Madison's Cave, Virginia) and a later edition of his celebrated book (1853) *Notes on the State of Virginia* reproduced the third and fourth American cave maps—of Madison's Cave and what now is Grand Caverns, Virginia, and of Mammoth Cave, Kentucky (Figure 1). Jefferson consistently used scientific

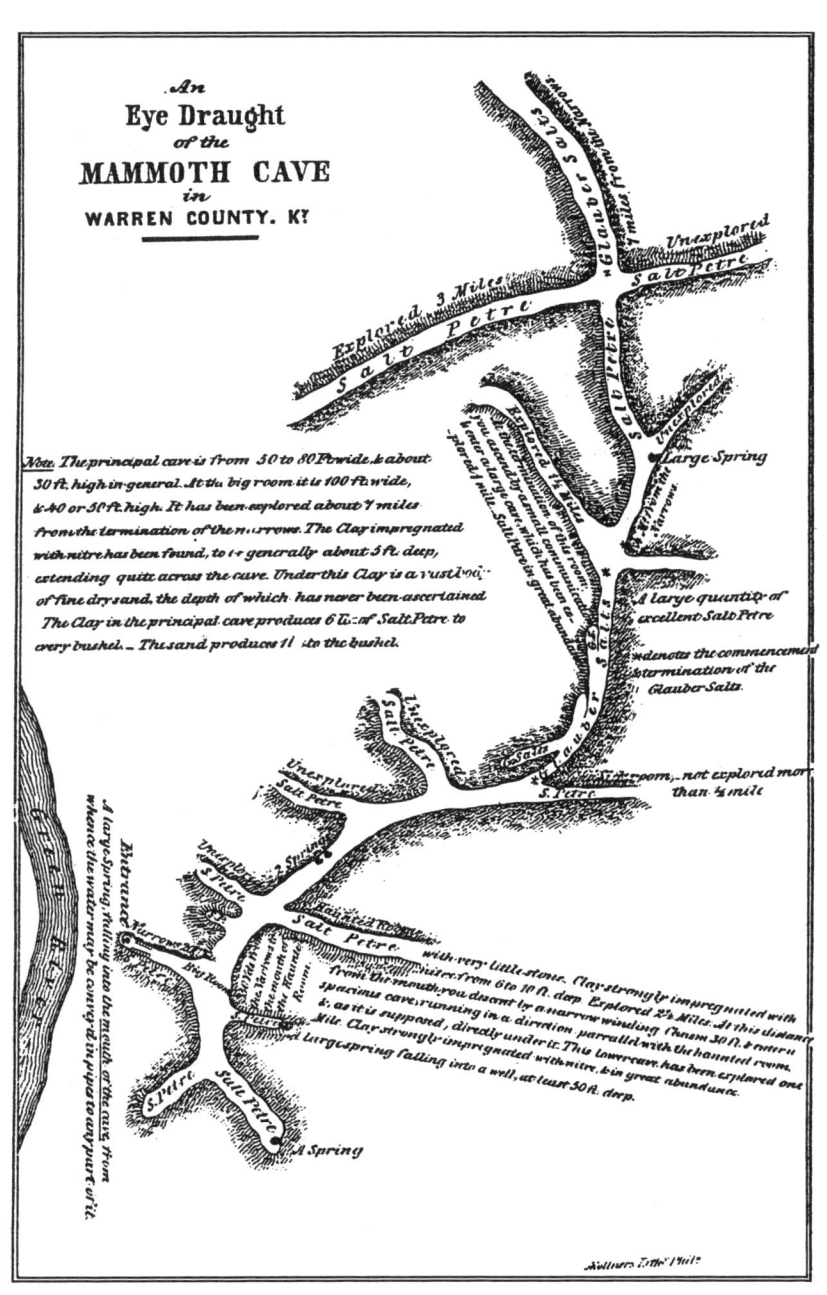

America, North, History: Figure 1. 1811 sketch map of Mammoth Cave, Kentucky, as reproduced in the 1853 edition of *Notes on the State of Virginia* by Thomas Jefferson. William R. Halliday Collection.

America, North, History: Figure 2. Horace C. Hovey, father of American speleology. From an 1871 portrait by a photographer named Thompson. William R. Halliday Collection.

principles to evaluate aspects of caves. His studies of remains of a ground sloth from Organ Cave, West Virginia—the first ever discovered—led to its scientific name: *Megalonyx jeffersoni*.

In the mid-18th century, "long hunters" (hunters from pioneer settlements who stayed away from home for up to several months at a time) and a few hardy settlers began making gunpowder from nitrates in caves in what was then western Virginia (see Gunpowder). During the American Revolution, several of these caves were briefly pressed into full production. Around the turn of the century, Mammoth Cave and dozens of others were explored for nitrates, with production eventually reaching a very large scale until the Kentucky producers were devastated by the New Madrid earthquakes. Then the entire nitrate market collapsed at the end of the War of 1812.

American cave explorers followed in the footsteps of early cavers of the French and Spanish eras, but apparently without learning anything from them. Constructive plagiarism by fledgling American literary and scientific journals and travel books kept literate Americans somewhat informed about European caves and concepts. But 20 years after the collapse of the nitrate industry, "cave hunting was little else than a bizarre and venturesome underground diversion" (Clarke, 1914).

Canadian caves were even less celebrated. One cave near Lake Ontario was mentioned (fancifully and apparently second hand) in Winterbotham's 1796 *View of the United States of America*. Two others were reported in Quebec, in 1822 and 1823, with a new discovery in Ontario in 1824 and a flowery account of a cave in Nova Scotia in 1836. Gibb's landmark: *On Canadian Caverns*, however, did not appear until 1861. The now-famous Steam Cave at Banff was not seen by a European until 1875 and most of the other long-famous Canadian Rockies caves were not encountered until after 1900. The hell-roaring Nakimu Caves were discovered in 1904; within a few years they were developed in an early version of ecotourism. Alpinist A.O. Wheeler led North America's first real alpine cave exploration in 1911, down 76 m of pitches in Arctomys Cave.

In the rebel southern states of the United States, the American Civil War (1861–65) precipitated a renewed wave of exploitation of saltpetre caves, and interest heightened throughout the reunited nation. Notable academic geologists, archaeologists, and biologists like Nathaniel Shaler and A.S. Packard, published landmark reports in 1876 and 1888.

However, it was a middle-aged clergyman, Horace C. Hovey, who triggered the second major wave of American speleology. Beginning in 1878, his writings had an impact unmatched in American spelean history. Hovey began caving as a youth but pursued it only intermittently. Nevertheless he dominated American speleology for 35 years. To the public he brought genial, unexaggerated accounts with which each reader could identify. His timing was fortunate: cave diggers broke through into Virginia's magnificent Luray Caverns shortly after publication of his 1878 article. When the news broke, *Scientific American* magazine rushed him to Luray for a scoop, and suddenly Hovey had enough material for his famous 1882 book *Celebrated American Caverns*. It became the most influential American cave book of all time. Almost overnight he became "final in experience and judgment in that phase of science now designated by the unlovely term speleology" (Clarke, 1914).

Hovey never travelled west of Arizona's Grand Canyon, but his wave of enthusiasm swept across the entire country. Show caves sprang into existence as far away as California. Back home in Indiana, W.S. Blatchley published the first state-wide report on caves. In 1900, E.S. Balch followed with *Glacières, or Freezing Caverns*, the first exhaustive American work on cave meteorology.

The Hovey wave ebbed during World War I, but a new wave of interest began in 1925 when Floyd Collins died in a detached part of Mammoth Cave. Later that year, members of the Explorers Club light-heartedly sought the end of Virginia's Endless Cavern, with considerable publicity. Publicity also was building at Carlsbad Cavern, where the ill-fated Nicholson Expedition of 1930 provided another prolonged media event—and triggered the creation of Carlsbad Caverns National Park. Raconteur Russell Trall Neville lectured widely in the 1930s, and Clay Perry's books and articles about American caves brought still further cave consciousness to a wide audience.

In the scientific world, notable papers by Weller (1927) and William M. Davis (1930) strongly stimulated speleogenetic thought in the United States. Within a few years, Swinnerton, Gardner, and Malott all provided the beginnings of a new, ongoing flood of papers expanding Davis's concepts in specific localities. That of J Harlen Bretz in 1942 was especially significant. From it, ordinary American cavers learned to read the silent language of speleogenetic sequences. More recently, contributions by William B. White, Arthur Palmer, and Ernst Kastning are among many meriting special mention.

State-wide cave reports began to appear. Ralph Stone's *Caves of Pennsylvania* was published in 1930. Soon it was followed by McGill's *Caves of Virginia* and W.E. Davies's *Caves of West Virginia* and *Caves of Maryland*. Bretz's *Caves of Missouri* and many others followed.

The word "speleogenesis" broke into print in 1960 at a symposium where the old concept of speleogenetic cycles was laid to rest (Halliday, 1960). (Kyrle had used its German equivalent by 1923.) Also advanced at that symposium was a then-heretical concept: not all karstic caves have precisely the same origin.

Extensive US participation in international speleological congresses accelerated in 1981 when the 8th was convened in Kentucky. The purely American brand of speleology became more and more internationalized. With an extensive publication programme culminating in the appearance of the internationally edited book *Speleogenesis* in 2000, the National Speleological Society took on a leadership role in international speleology.

The National Speleological Society was founded in 1941. After World War II, the third wave of American speleology rose to new heights. The number of recorded caves skyrocketed; 8000 are now listed in the state of Tennessee alone. New US depth records were repeatedly broken in the Rocky Mountains, then surpassed by Lechuguilla Cave at 478 m (the vertical ranges of long lava tube caves don't count; see Volcanic Caves). That depth wasn't much by world standards, but the Cave Research Foundation of especially dedicated explorers (an offshoot of the National Speleological Society) pushed the mapped length of Mammoth Cave to 587 km by 1996, by far the greatest in the world. And a barely separate segment of its system—Fisher Ridge Cave—reached 161 km in 2001.

Opened in 1986 by two years' digging, the crystalline magnificence of Lechuguilla Cave somewhat distracted attention from its statistics. With 176 km currently mapped, it became the third longest in the United States.

The explored length of "little" Jewel Cave in South Dakota increased rapidly to 204 km, surpassing neighbouring Wind Cave (just over 160 km). Even California, not renowned for its caves, was found to have a long cave: Lilburn Cave at 27 km and another which is 367 m deep (Bigfoot Cave).

Progress in conservation has been no less important. Most Americans think in terms of national parks and monuments rather than World Heritage Sites, but three caves or groups of caves have been designated as national parks (Mammoth Cave, Carlsbad Caverns, and Wind Cave). Four others are national monuments: Jewel Cave, beautiful little Timpanogos Cave in Utah, Oregon Cave, and Russell Cave in Alabama (primarily an archaeological reserve, but also with notable karstic features). When Mount St Helens National Volcanic Monument was created, the Caves Basalt Lava Flow comprised one of its three units, and many other American and Canadian national parks and monuments are karstic or pseudokarstic.

In Canada, cave studies in the 20th century followed a somewhat different route. The arrival of Derek Ford and other keen British cavers at McMaster University provided an international scientific flavour long lacking in the United States. They also brought in their special skills and determination, and experience in alpine karsts. Soon a secondary Canadian centre of speleology was spun off to Calgary, and the famous Nakimu Caves soon lost their status on the list of deep Canadian caves. In the 1960s and 1970s, three of the five deepest caves in Canada were discovered, explored and/or studied in the Canadian Rockies: Arctomys Cave, Castleguard Cave, Yorkshire Pot, and later, Close to the Edge Cave. Castleguard Cave—Ford's speciality and Canada's longest—was mapped for 20.1 km. A member of his team found that it ends with ice plugs from a major glacier overhead. Even in the remote Northwest Territory, Quebec cavers and members of Ford's group found several caves over 1 km long in the isolated Nahanni National Park. Here, Grotte Valerie is notable for its exceptionally diversified underground climate.

Meanwhile in Quebéc, the vigorous and well-organized Société Québecois de Spéléologie was formed in the early 1970s, with a distinct French flavour (Beaupré & Caron, 1986). It soon became noted for excellent scientific and educational programmes, but found itself frustrated by the lack of large caves in Quebéc. Its activities soon became thoroughly international.

Another caving centre developed in and near Victoria, British Colombia. With an initial impetus in the early 1960s from Clarence Hronek in Vancouver and Seattle's Cascade Grotto of the National Speleological Society, several vigorous groups emerged in the 1960s, 1970s, and 1980s. Vancouver Island now has the greatest number and density of caves in any Canadian region, including five each from the dozen longest and dozen deepest caves in Canada.

Cave diving in the continental United States began to come of age in 1953, in enormous underwater cave passages in Florida. Pleistocene mastodon and other paleontological remains were found at the start of a huge underwater borehole at Wakulla Spring. Here, current explorations are more than 5 km from the entrance, with an average depth of nearly 100 m *en route*. Garry Salsman and Wally Jenkins are acclaimed for early achievements here. The nearby Woodville Karst Plain system may be far larger, with more than 20 interconnected spring entrances and a claimed underwater length of 32.9 km in 2001. The longest recorded underwater traverse in Florida appears to be 4268 m, between Big Dismal Sink and Cheryl Sink. Elsewhere, cave diving in the remarkable karstic springs of the Ozark Plateau has shown that several are outlets of other huge karstic conduits at depth, reached by underwater corridors sloping uniformly downward from the swirling springs. In Nevada, however, the National Park Service forbids diving in the amazing Devil's Hole, so its remarkable depth remains unmeasured. The late Sheck Exley is widely credited with the primary role in advancing American cave diving, with Bill Stone noted for technological advances.

Biologists initiated systematic study of gypsum caves in Kansas in the mid-20th century. In the great gypsum karst of Texas, David Belski subsequently began an extensive geomorphological project, still ongoing. Interest in crevice caves and other limestone landforms resulting from deeply buried gypsum karst in northeastern Arizona is largely recent. In Canada, gypsum caves are widespread but interest in them has been desultory. An apparently world-class example in Wood Buffalo National Park is full of very cold water, and the others all seem to be less than 300 m long (D. Ford, personal communication, 2001).

Interest in American and Canadian pseudokarst and pseudokarstic caves has been sporadic. William R. Halliday, Donald Davis, Louise Hose, and the Colorado, San Diego and Southern California Grottoes (Chapters) of the National Speleological

Society have provided notable leadership. The smooth, water-polished surfaces of Greenhorn Cave and other granite stream caves of California have been appreciated by local cavers, but the talus caves in the Pinnacles National Monument have received more publicity. Regions especially noted for piping caves include Oregon's John Day country, Texas's Palo Duro Canyon, and the Anza-Borrego area of southern California.

More than a century ago, Israel Russell found lengthy glacier caves in Alaska, with a subglacial river about 50 km long coursing beneath the entire width of the Malaspina Glacier's piedmont. Direct scientific observation in Alaskan glacier caves shows a lack of progress since the 1970s. The Paradise Ice Caves—large glacier caves on Mount Rainier—were commonly visited by 1908, and its summit geothermal firn caves saved the lives of the first party to ascend it. The geothermal summit caves of Mount Rainier and Mount Baker were primarily investigated by Eugene Kiver in the 1970s. The glacier containing the Paradise Ice Caves shrivelled, re-expanded, and then disappeared completely during the last decades of the 20th century, though in 2001, the caves were reported to be re-forming in firn.

WILLIAM R. HALLIDAY

Works Cited

Beaupré, M. & Caron, D. 1986. *Découvrez le Québec souterrain*, Sillery, Québec, Presses de l'Université du Québec

Bretz, J Harlen 1942. Vadose and phreatic features of limestone caverns. *Journal of Geology*, 50: 675–809

Clarke, H. 1914. Memoir of Horace Carter Hovey. *Geological Society of America Bulletin*, 26: 21–27

Davis, W. 1930. Origin of limestone caverns. *Geological Society of America Bulletin*, 30: 475–628

Halliday, W. 1960. Changing concepts of speleogenesis. *National Speleological Society Bulletin*, 22: 23–29

Kastning, E. 1981. Pioneers of American cave and karst science prior to 1930. *Journal of Spelean History*, 15: 36–37

Rafinesque, C. 1832. The caves of Kentucky. *Atlantic Journal*, 1: 27–30

Shaler, N. 1876. *Antiquity of the Caverns and Cavern Life of the Ohio Valley*, Frankfort, Kentucky: Kentucky Geological Survey

Shaw, T.R. 1992. *History of Cave Science: The Exploration and Study of Limestone Caves, to 1900*, 2nd edition, Broadway, New South Wales: Sydney Speleological Society

Weller, J.M. 1927. *The Geology of Edmonson County*, Frankfort, Kentucky: Kentucky Geology Survey

Further Reading

Journal of Spelean History, vol. 1 (1966) to present. Published by the American Spelean History Association—the history section of the National Speleological Society, Huntsville, Alabama

Exley, S. 1994. *Caverns Measureless to Man*, St Louis, Missouri: Cave Books

George, A.I. 2001. *The Saltpeter Empires of Great Saltpeter Cave and Mammoth Cave*, Louisville, Kentucky: HMI Press

Greene, M.T. 1982. *Geology in the Nineteenth Century*, Ithaca, New York: Cornell University Press

Halliday, W.R. 1959. *Adventure is Underground: The Story of the Great Caves of the West and the Men who Explore Them*, New York: Harper

Halliday, W.R. 1966, 1976. *Depths of the Earth: Caves and Cavers of the United States*, revised edition, New York: Harper and Row

Hovey, H.C. 1882. *Celebrated American Caverns*, Cincinnati: Clarke; reprinted 1970 with a new introduction by W.R. Halliday, New York: Johnson Reprint company

Nymeyer, R. & Halliday, W.R. 1991. *Carlsbad Cavern: The Early Years*, Carlsbad, New Mexico: Carlsbad Caverns–Guadalupe Mountains Association

Perry, C. 1948. *Underground Empire: Wonders and Tales of New York Caves*, New York: Stephan Daye Press

Thompson, P. (editor) 1976. *Cave Exploration in Canada*, Edmonton, Alberta: Canadian Caver Magazine

Vineyard, J. & Feder, G.L. 1974. *Springs of Missouri*, Rolla, Missouri: Missouri Geological Survey and Water Resources

AMERICA, SOUTH

Compared with other continents, South America contains relatively small areas of carbonate, probably less than 2% of its total surface. Carbonate karst is concentrated within the ancient plateaux of central Brazil and in a few scattered areas of the Andean countries, although some small bodies of carbonate rock are dispersed throughout the continent (Figure 1). Nevertheless, there are some important South American karst areas with significant caves (see Table), including the world's best-developed quartzite karst (see Quartzite Caves of South America).

Venezuela is second only to Brazil in the area covered by karst scenery. The Roraima quartzite occupies a major part of the country's southern sector, hosting impressive vertical caves on the top of quartzite towers (known locally as tepuis), which rise abruptly from the rainforest. Many deep fissure-like caves and massive shafts occur in this area. Carbonate karst is concentrated in the northern coastal mountains of the Andean zone, the location of c.95% of all caves recorded in Venezuela. In this area, karst is developed in carbonate rocks of Jurassic–Tertiary age, mostly as isolated outcrops usually of local extent, surrounded by noncarbonate rocks. The most significant karst is developed in five areas: Oriental, Central, Falcón, Andes, and Perija. In the Oriental area, the best-developed karst occurs in the Caripe–Mata de Mango area in the state of Monagas, where about 30 caves are developed in Cretaceous limestone. Venezuela's most celebrated cave, the 10 200 m long Cueva del Guácharo is located in this area, protected since 1949 by the Alejandro de Humboldt Natural Monument. The central part of this area hosts significant caves in the metamorphosed limestones of the Cordillera de la Costa (Coastal Range), for example, the 4292 m long Cueva Alfredo Jahn.

In the Falcón area, karst occurs in the San Luis and Serranía Churuguara mountains. The former area shows well-developed karst topography on Oligocene reefal limestones. Many deep caves occur within this zone, such as the 305 m deep Haitón del Guarataro. The Andes area shows Cretaceous reefal limestone at Humocaro, La Azulita, and Guaraque. The karst region of the Sierra Perijá, close to the Colombian border, has recently yielded major caves. In this area, lower Cretaceous limestones outcrop in a very wet rainforest environment. Many active caves exist in the area, including Venezuela's longest known cave, the 18 200 m long Cueva del Samán and many other important caves such as Cueva Sumidero La Retirada (6080 m long), Cueva los Encantos (4600 m), and Cueva los Laureles (4300 m).

America, South: Figure 1. Main areas of carbonate and quartzite karst in South America.

Little is known about karst in Surinam, Guyana, and French Guyana. Carbonate rocks appear to be absent, although the Roraima quartzite does extend towards Guyana. Short caves in granite and under laterite (ferricrete) layers have been noted in French Guyana.

Over a hundred caves have been recorded in Colombia. Carbonate karst occurs mainly within the Cordillera Oriental, where isolated karst areas contain a number of significant caves. The best-known area lies in the Santander Department, where some major vertical caves have been recorded, among them the renowned Hoyo del Aire, a 120 m deep, 100+ m wide pit which gives access to a massive passage terminating at a depth of 270 m, and the 4926 m long Sistema Hermosura, Colombia's longest cave. In southern Colombia, a small karst area hosts the Cueva del Indio (3507 m long) and the important Cueva de los Guácharos (1000 m long), both of which have been protected since 1960 by the Cueva de los Guácharos National Park. One significant sandstone cave has been explored in the Tolima area—the Cueva del Cunday, which is 850 m long and 160 m deep.

Carbonate karst in Ecuador is concentrated mostly in the Oriental sector of the Andes. The Cretaceous limestones of the Napo, Pastaza, and Morona Santiago provinces host most of

America, South: The longest and deepest caves of South America, as at January 2003.

Longest caves			Deepest caves		
Cave	Country	Length (m)	Cave	Country	Depth (m)
1. Toca da Boa Vista	Brazil	97 000	1. Sima Pumacocha	Peru	638
2. Toca da Barriguda	Brazil	28 700	2. Gruta do Centenário	Brazil	481
3. Cueva del Samán	Venezuela	18 200	3. Millpu de Kaukiran	Peru	407
4. Gruta do Padre	Brazil	16 400	4. Gruta da Bocaina	Brazil	404
5. Boqueirão	Brazil	15 170	5. Sima Aonda	Venezuela	383
6. Lapa do Angélica	Brazil	14 100	6. Perte du Futur	Chile	376
7. Gruna da Água Clara	Brazil	13 880	7. Sima Auyán-tepui Noroeste	Venezuela	370
8. Lapa São Mateus III	Brazil	10 610	8. Sima Aonda 3	Venezuela	335
9. Cueva del Guácharo	Venezuela	10 200	9. El Tragadero	Peru	334
10. Lapa São Vicente I	Brazil	10 130	10. Sima Aonda 2	Venezuela	325

the c.400 caves known so far. The Napo Province hosts many caves, including the 2460 m long Cueva de San Bernardo. The Pastaza Province is home to several caves, but difficulties of access have precluded systematic exploration. The deepest and longest caves in Ecuador are located in the Morona Santiago area. The Cueva de los Tayos de Coangos is 4800 m long and 201 m deep. The same region contains the 2305 m long Cueva de Shimpiz. Important lava caves occur in the volcanic Galápagos Islands. Most of the 67 caves that were recorded up to 1988 are concentrated in Santa Cruz Island, including the 3010 m long Cueva del Cascajo and the partially commercialized 2331 m long Cueva de Gallardo, the most important lava tube cave in South America.

Carbonate crops out in many areas of Peru, although some of the best-developed karst lies in the northern and central region. The country is noted for its high-altitude karst, some of which occurs as high as 4800 m in altitude. In the northern portion of the country, significant caves occur in the Cajamarca area, one of the most important being the hydrological system known as Uchkopisjo, with a combined length of 2350 m. The San Andrés de Cutervo area has been a National Park since 1961 and contains, among other caves, the Gruta de los Guácharos (1134 m long and 180 m deep) and the 334 m deep El Tragadero (Figure 2). Southeast of the town of Cajamarca there is a high-altitude karst area (4000–4200 m in altitude) with an extraordinary collection of deep pits, many with depths around 200 m. The area near the town of Tarma (Central Peru) hosts the Millpu de Kaukiran, the second deepest carbonate cave in South America (407 m deep and 2141 m long). Its resurgence is the Gruta Huagapo, the longest Peruvian cave at 2396 m in length. Ongoing exploration in an area in south-central Peru has yielded South America's deepest carbonate cave, Sima Pumacocha, which is 638 m deep.

There are only a few karst areas in Bolivia, but the most well-known area is the Torotoro Karst near Cochabamba, which comprises Cretaceous limestones hosting the Gruta Umajalanta, the deepest and longest cave in the country at 4.6 km long and 164 m deep. Other nearby caves include the resurgence of Chiflonkkakka and the 154 m deep Caverna Huayllas. The area belongs to the Torotoro National Park, which was established in 1989. There are some areas of gypsum and sandstone karst in the country, which host respectively the 413 m long Gruta de San Pedro and the 350 m long Gruta San Miserato. Precambrian carbonates also occur near the border with Brazil, at Puerto Suarez. These outcrops have received little attention to date and no significant caves have yet been noted.

The most important karst zone in Paraguay is located near the border with Brazil, where Precambrian dolomites form a

America, South: Figure 2. The large open sink of the Rio Charcay at the Tragadero de San Andes in the high Andean karst of Peru. (Photo by Chris Birkhead)

cone karst with isolated hills above the alluvial plain of the Paraguay River and its tributaries. A number of caves have been explored in this area, none of which is of a significant length. Uruguay is of little speleological importance due to the near absence of carbonates. The best-known cave in the country is the 41 m long Cueva de Arequita, developed in granite and located near the town of Minas.

The Chilean karst is characterized by extreme climatic conditions. Limestone, gypsum, and salt are known to outcrop in the hyper-arid Atacama Desert (precipitation of around 1 mm a^{-1}) in northern Chile. Surface karst features have been noted in these rocks, and the Cueva del Valle de la Luna, in salt, comprise two separate caves respectively 500 and 250 m in length. At the southern end of Chile, the Patagonian region hosts some spectacular marble karst, especially on the islands of Diego de Almagro and Madre de Dios (see Patagonia Marble Karst, Chile), under high precipitation ($c.$7000 mm a^{-1}) and extreme conditions of wind and temperature. Unique wind-controlled karren have been observed, and recent exploration in the Madre de Dios has yielded the 376 m deep Perte du Futur. Near Puerto Natales, the celebrated Cueva Mylodon is probably South America's best-known paleontological site. The cave is a massive passage 200 m long and 100 m wide, developed in conglomerate, where well-preserved ground sloth and human fossil remains have been studied since the end of the 19th century.

Only 1000 km^2 of Argentina is made up of karst, which is a fairly small area for such a large country. The most important carbonate and gypsum areas occur in the provinces of Mendoza and Neuquén. The two most important carbonate areas are the Las Brujas region, near Malargüe, Mendoza, with the 1343 m long Cueva de las Brujas, a cave of supposedly hypogene origin (Forti, Benedetto & Costa, 1993), and the Cuchillo Cura area in Las Lajas, Neuquén, home of the country's longest cave, the 3433 m long Sistema Cuchillo Cura. Gypsum karst occurs in La Yesera del Tiromen and Curymil, both in the Neuquén Province, and at Sierra de Reyes and Sierra de Cara Cura in the Mendoza Province. The most significant gypsum cave in Argentina is the 853 m long Cueva del León in Las Lajas, Neuquén. Some lava caves also exist, such as the Cueva de Doña Otilia in Malargüe, Mendoza, at 838 m in length.

Brazil contains the largest area of karst in South America. Carbonate karst occupies $c.$200 000 km^2, mostly distributed in the southern half of the country. The majority of the karst develops over Precambrian carbonates, under a seasonal climate (with a rainfall of between 500 and 2000 mm a^{-1}). The Bambuí karst (see Bambuí Karst, Brazil) is the largest carbonate area in the country, hosting about half of the $c.$3500 known caves, including some of the country's most significant caves. The Una Karst, located mostly within a semi-arid area, in northern Bahia State, is home to the longest cave in South America, the 97 km long Toca da Boa Vista (see entry on Boa Vista, Toca, Brazil) besides significant caves in the vicinity of the town of Iraquara. In southern Brazil, the outcrops of the Açungui Group contains many major active caves in tropical rain forest, including the world's tallest cave entrance at Gruta Casa de Pedra (215 m, Figure 3). Caves in this area are generally very well decorated (Figure 4), and most of this area is protected by the Upper Ribeira and Jacupiranga State Parks (Karmann & Ferrari, 2000). In western Brazil, many caves have been identified in the isolated carbonate hills of the Corumbá and Araras Groups, including important

America, South: Figure 3. Part of the entrance to Casa de Pedra cave, Ribeira karst, Brazil. The entrance is thought to be the highest in the world at 215 m. The stream-sink at the bottom left leads to a fine river passage. (Photo by John Gunn)

flooded caves near the town of Bonito. Other important carbonate areas include Cretaceous limestones of the Apodi Group in the northeasternmost part of Brazil, and karst developed over freshwater carbonates of the Caatinga Formation, in northern Bahia State.

Brazilian quartzite caves occur mostly in the eastern portion of the country (see Quartzite Caves of South America), and comprise the deepest and longest caves known in this type of rock. Sandstone karst is widespread throughout Brazil, from the

America, South: Figure 4. Fallen stalactite block in Casa de Pedra cave, Ribeira Karst, Brazil. The block appears to have detached slowly and rotated through 180° landing upside down on the floor. (Photo by John Gunn)

southern provinces to Amazonia. The best-developed areas lie in rocks of the Furnas and Botucatu formations, including the 1400 m long Caverna Aroe Jari at Chapada dos Guimarães, Mato Grosso State. In Amazonia, the Prainha area has some long sandstone caves, such as the 1297 km long Gruta Planaltina. The longest sandstone cave in Brazil is currently the 1633 m long Toca das Confusões in the semi-arid region of the state of Piauí. Another cave of note is the 1600 m long Caverna dos Ecos, developed in mica schists (Karmann, Sánchez & Fairchild, 2001). Caves in granite, talus, bauxite, and under laterite have also been recorded.

AUGUSTO AULER

Works Cited

Forti, P., Benedetto, C. & Costa, G. 1993. Las Brujas Cave (Malargüe, Argentina): An example of the oil pools control on the speleogenesis. *Theoretical and Applied Karstology*, 6: 87–93

Karmann, I. & Ferrari, J.A. 2000. Karst and Caves of the Upper Ribeira State Park (PETAR), Southern São Paulo State. http://www.unb.br/ig/sigep/sitio043/sitio043english.htm

Karmann, I., Sánchez, L.E. & Fairchild, T.R. 2001. Caverna dos Ecos (Central Brazil): Genesis and geomorphologic context of a cave developed in schist, quartzite and marble. *Journal of Cave and Karst Studies*, 63: 41–47

Further Reading

Auler, A., Rubbioli, E. & Brandi, R. 2001. *As Grandes Cavernas do Brasil* [The Great Caves of Brazil], Belo Horizonte: Grupo Bambuí de Pesquisas Espeleológicas

Decu, V., Urbani, F. & Bordon, C. 1994. Venezuela. In *Encyclopaedia Biospeologica*, vol. 1, edited by C. Juberthie and V. Decu, Moulis: Société de Biospéologie

Gilbert, A. 1988. Les cavités dans la lave de l'île Santa Cruz, Iles Galapagos, Equateur [Caves in lava in Santa Cruz Island, Galápagos Islands, Ecuador]. In *Proceedings of the 1st Congresso de Espeleologia da América Latina e do Caribe*, Belo Horizonte, Sociedade Brasileira de Espeleologia: 56–64

Karmann, I. & Sánchez, L.E. 1980. Distribuição das rochas carbonáticas e províncias espeleológicas do Brasil. [Distribution of carbonate rocks and speleological provinces in Brazil.] *EspeleoTema*, 13: 105–167

Salomon, J.N. 1995. Le Chili: Pays de karsts extrêmes [Chile. Country of extreme karst]. *Karstologia*, 24: 52–56

AMERICA, SOUTH: BIOSPELEOLOGY

South America, with an area of 17 800 000 km² mostly within the tropical zone and a high diversity of physiography from lowland forests to semi-desert high mountains, encompasses several regions of biological megadiversity. The subterranean fauna was therefore predicted to be accordingly rich and diversified due to the diversity of potential colonizers. Intensive, systematic biospeleological studies started late in comparison with Europe and North America. However, rapid progress has been made in the last 30 years.

In 1799, the naturalist Alexander von Humboldt made the first written reference to a South American cave animal, the oilbird or "guácharo" (*Steatornis caripensis*), from caves in Venezuela. Other naturalists travelling throughout the continent in the 19th century occasionally mentioned cave animals, especially bats and birds. By the end of that century, the first South American troglobite, the blind catfish—*Pimelodella kronei*, was described from caves in southeastern Brazil. During the first half of the 20th century, biospeleology in South America was mainly restricted to occasional descriptions of taxa from caves scattered across the continent. Extensive studies started in the 1950s and 1960s, with foreign expeditions to several countries in addition to efforts by some native biologists, especially in Venezuela, and later in Brazil. Brazil currently has the best-studied cave fauna in South America, following intensive studies starting in the 1980s that ranged from faunistic surveys to community studies and to investigation of the biology of different taxa.

Based on such studies, ecological and evolutionary patterns began to emerge, revealing that differences with respect to temperate cave ecosystems were less than previously assumed. Although the taxonomic composition of subterranean communities differ as a consequence of biogeographic factors, ecological and evolutionary mechanisms and processes appear to be similar. Caves with huge amounts of guano produced by very large colonies of bats and echo-locating birds, frequently cited as characteristic of tropical regions, are not as common in South America as generally assumed. As observed in temperate areas, many caves studied in South America are oligotrophic, with food resources for cavernicoles represented by vegetal debris, relatively small guano piles, and dead and living epigean organisms accidentally entering caves or carried in by water (streams or percolating water).

The most frequent subterranean taxa found in continental South America are presented here, organized by ecological affinities. Some taxa are distributed throughout the continent, whereas others are restricted either to fully tropical, warm caves (>22°C) to the north of Tropic of Capricorn, or to subtropical to temperate caves in southeast Brazil, Uruguay, and Chile.

South America is remarkable for its chiropteran (bat) diversity, only equalled by that in Southeast Asia. Many bat species frequently use caves, especially phyllostomids that include frugivorous, nectarivorous, insectivorous, carnivorous, and haematophagous species. In inhabited karst areas with livestock, vampire bats may be particularly numerous in some caves. Different kinds of bat guano support distinct communities of invertebrates; it is noteworthy that vampire bats, with their distinctive black guano due to their specialized diet based on blood, are exclusively neotropical. Other trogloxene mammals include didelphids (e.g., the four-eyed opossum, *Philander opossum*, the river otter (*Lontra longicaudis*), and rodents such as the water rat (*Nectomys squamipes*), spiny rats (*Proechmys*), and agouti (*Agouti paca*). In northwest South America, oilbirds form large, noisy colonies nesting in some caves. Their "guano", actually a mixture of faeces and regurgitated, partially digested fruits, supports rich communities of detritivorous invertebrates. Barn owls and some other birds are observed in caves throughout South America, but never far from entrances. Among invertebrates, it has been shown that large opilionids (harvestmen) of the genus *Goniosoma* form trogloxene populations in southeast Brazil, reproduce near the entrances, and leave caves regularly to feed.

Most terrestrial invertebrates regularly found in South American caves are troglophiles. Conspicuous omnivorous/detritivorous macroinvertebrates include phalangopsid crickets, mainly *Endecous*, cockroaches (in warm caves), harvestmen, and millipedes; rhaphidophorid crickets, typical of North temperate caves, have been found in Chile. Their predators are mainly spiders, which present a high diversity in South American caves—33 out of 54 families occurring in the neotropical region have been recorded in Brazilian caves. The most frequent are large wandering ctenids (mainly *Ctenus*), sicariids (*Loxosceles*), and web-building pholcids, theridiosomatids (mainly *Plato*), and theridiids. Additional predators are centipedes, reduviid bugs and, in fully tropical caves, amblypigids (especially the large *Heterophrynus*) and scorpions. Among detritivores, cave millipedes are also diversified in South America, including Polydesmida (e.g., chelodesmids), pseudonannolenids, and spirostreptids. Most cave harvestmen are gonyleptids, especially pachylines; cosmetids, agoristenids, and cranaids are found in fully tropical caves.

Soil animals are typically found in sediment banks, concentrating in plant debris and animal matter deposited on these banks or on rocky surfaces. These include earthworms (e.g., enchytraeids), the ubiquitous springtails (isotomids, paronellids, entomobryids, arrhopalitids) and mites, campodeid diplurans, oniscidean isopods (mainly phylosciids, platyarthrids, and styloniscids), millipedes, larvae of tineid moths, larvae and adults of beetles such as cholevids (usually on animal matter), and ptilodactylids. Their predators include predaceous mites, beetles (e.g., carabids, pselaphids, histerids, staphylinids), pseudoscorpions (mainly chernetids), small spiders (e.g., ochyroceratids, hahniids), centipedes, and, in tropical caves, schizomids. Some species are mainly found in guano, certain taxa being restricted to or showing strong preference for particular kinds of guano (e.g., *Acherontides* springtails, larvae of *Drosophila*, *Fannia*, and *Psilochaeta* flies in vampire bat guano; cydnid and lygaeid homopterans, and lithobiomorph centipedes in frugivorous bat guano).

As expected in view of their overall diversity, beetles and flies are the most diversified cave insects in South America. Several families of small to medium-sized predaceous and detritivorous beetles have been recorded. As in temperate karst areas, the predaceous carabids and the detritivorous cholevids are among the commonest cave beetles; however, the subgroups that contribute with the highest diversity in most temperate caves—trechines within carabids, bathysciines within cholevids, including many troglobitic (exclusively subterranean) species—are rare in South American caves. Several famlies of dipterans have been recorded in South American caves, including psychodids, chironomids, keroplatids, sphaerocerids, phorids, milichiids and muscids.

Nymphs and larvae of flying insects such as chironomid flies, mayflies, and caddisflies live in cave streams. Their adults, which may attain very high population densities at certain times of the year, are captured in the webs of spiders and of larvae of keroplatid flies. In addition to immature insects, the aquatic fauna includes adult heteropterans (e.g., veliids) and beetles (e.g., elminthids, dytiscids), molluscs such as hydrobiid gastropods, and crustaceans. Among crustaceans, amphipods, mainly bogidiellids and hyalellids, are frequent in both streams and phreatic water bodies throughout South America. Other crustacean taxa have more localized distributions: spelaogriphaceans in phreatic water bodies from southwest Brazil; Calabozoa isopods in phreatic waters from Venezuela and northeast Brazil; the anomuran *Aegla* in cave streams of southeast Brazil; and crabs mainly in northwest South America. Shrimps are rare in South American caves, usually as accidental or trogloxene occurrences.

The total diversity of terrestrial troglobites in South America is moderate, much lower than the observed in Northern hemisphere temperate caves especially for some groups such as spiders and beetles. The great majority of South American terrestrial troglobites evolved from soil animals—several springtails, isopods, a few campodeid diplurans, rare thysanurans, chthoniid pseudoscorpions, a few, generally small spiders, beetles (mainly pselaphids and carabids), and polydesmida diplopods. Other terrestrial troglobites are harvestmen (gonyleptids in Brazil, agoristenids in Venezuela, a triaenonychid in Argentina), a few cockroaches (Venezuela and Brazil) and a cixiid homopteran (Argentina).

On the other hand, South America is relatively rich in aquatic stygobites, including both stream-dwelling and phreatobic crustaceans (syncarids, amphipods, isopods, spelaeogriphaceans, decapods such as *Aegla* anomurans, and pseudothelphusid crabs), hydrobiid gastropods, a few planarians, and fishes. South America is distinguished by its richness of stygomorphic fishes. At least 19 exclusively subterranean species are known, mostly Siluriformes (catfishes). The most frequent are catfishes of the genus *Trichomycterus* from Brazil, Venezuela, Colombia, and Bolivia; armoured catfishes of the genus *Ancistrus* from Brazil and Venezuela; and heptapterine catfishes from Brazil. In addition, several epigean stygomorphic species (with reduction of eyes and pigmentation) have been found in turbid rivers of the Amazon Basin.

While the cave fauna is better known in Brazil, the interstitial aquatic fauna has been better surveyed in countries such as Venezuela, Argentina, and Chile. A diversified stygobitic crustacean fauna including several species of copepods (mainly harpacticoids), ostracods, syncarids, isopods (e.g., protojanirids in Argentina) and amphipods (bogidiellids, ingolfiellids), in addition to hydracarians, have been found in interstitial habitats.

As proposed for temperate karst areas, paleoclimatic fluctuations resulting in periodic elimination of epigean populations and, consequently, isolation of troglophilic populations in subterranean habitats (which they colonized during more favourable periods) and subsequent allopatric speciation, have been evoked as a major cause of diversification of the South American troglobitic fauna, at least for terrestrial taxa. In the case of aquatic troglobites, there is evidence that other isolation mechanisms such as stream capture, or even parapatric or sympatric speciation, may also be important.

ELEONORA TRAJANO

Further Reading

Decu, V., Orghidan, T., Dancau, D., Bordon, C. Linares, O., Urbani, F., Tronchoni, J. & Bosque, C. (editors) 1987. *Fauna hipogea y hemiedáfica de Venezuela y de otros países de América del Sur*, Bucharest: Academiei Republicii Socialialiste România

Gnaspini, P. & Trajano, E. 2000. Guano communities in tropical caves. In *Subterranean Ecosystems*, edited by H. Wilkens, D.C. Culver & W.F. Humphreys, Amsterdam and New York: Elsevier

Juberthie, C. & Decu, V. (editors) 1994. *Encyclopaedia Biospeologica*, vol. 1, Moulis and Bucharest: Société de Biospéologie

Mahnert, V. 2001. Cave-dwelling pseudoscorpions (Arachnida, Pseudoscorpiones) from Brazil. *Revue suisse de Zoologie*, 108(1): 95–148

Pinto-da-Rocha, R. 1995. Sinopse da fauna cavernícola do Brasil (1907–1994) [Synopsis of Brazilian cave fauna (1907–1994]. *Papéis avulsos de Zoologia*, 39(6): 61–173

Trajano, E. 1995. Evolution of tropical troglobites: Applicability of the model of Quaternary climatic fluctuations. *Mémoires de Biospéologie*, 22: 203–09

Trajano, E. 2000. Cave faunas in the Atlantic tropical rainforest: Composition, ecology, and conservation. *Biotropica*, 32(4b): 882–93

Trajano, E., Golovatch, S.I., Geoffroy, J.-J., Pinto-da-Rocha, R. & Fontanetti, C. 2000. Synopsis of Brazilian cave-dwelling millipedes (Diplopoda). *Papéis avulsos de Zoologia*, 41(15): 213–41

Trajano, E. 2001. Ecology of subterranean fishes: An overview. *Environmental Biology of Fishes*, 62(1–3): 133–60

AMERICA, SOUTH: HISTORY

The relationship between man and caves in South America precedes the arrival of the first Europeans at the end of the 15th century. South American inhabitants did not appear to have strong cultural ties with caves and karst features in the same way as the Maya of Central America and Mexico (see America, Central: Archaeological Caves), but nevertheless there is numerous evidence of use of caves by prehistoric civilizations. Throughout the Brazilian karst, cave entrances were used as shelters from the end of the Pleistocene/beginning of the Holocene, and many of the country's most important archaeological sites are located in karst areas. Some caves were apparently used as burial sites, as large numbers of human remains have been located in selected sites within caves. In Venezuela, the indians of Caripe have hunted the Guácharo or oil bird (*Steatornis caripensis*) inside caves since at least 3500 BP (Perera, 1976). The vocabulary of some indian cultures in eastern Brazil also records a few words that designate karst features, such as *itararé* ("hollow rock") and *anhonhecanhuva* ("water that sinks").

Following European colonization there were some sparse references to caves, mostly in Venezuela, during the period 1500–1700. The majority of the descriptions refer to caves with some kind of indian usage. The first recorded reference to a Venezuelan cave dates from 1548 when Diego de Vallejo and others entered a cave in the state of Trujillo (Urbani, 1989). In 1579, in the same area, Alonso Pacheco and others gave a description of Indian rituals in a cave (Urbani, 1993). The Cueva del Guácharo (Venezuela) has the longest and richest history of any cave in South America. It was probably first visited in 1659, and the first published reference dates from 1666 (Urbani, 1996). Since then, numerous visitors of varied scientific relevance have described the cave, among them Tauste (first description of the cave in 1678), Jiménez Perez (1773), Ibarra (1795), Humboldt and Bonpland (1799), Grisel (1834), L'Herminier (1834), Codazzi (1835), Bellerman (first paintings of the cave in 1843), and Lisboa (first sketch of the cave in 1866) (Urbani, 1999).

In 1691, Francisco de Mendonça Mar, a peasant of strong religious inclination established a shrine at a cave by the São Francisco River (presently known as Bom Jesus da Lapa), eastern Brazil, which has since become very popular as a pilgrimage site. Other caves in Brazil became centres of pilgrimage from the 18th century onwards. The Portuguese colonial government in Brazil financed the excavation of saltpetre in hundreds of caves throughout the country for the manufacture of gunpowder. Saltpetre exploitation in Brazil reached its climax during the 18th and 19th centuries, not ending until early in the 20th century.

Numerous naturalists and travellers, mostly Europeans, described caves in South America during the late 18th and 19th centuries, among them Ferreira, Eschwege, Pohl, Spix and Martius, Saint Hilaire, Lund, Liais, Fonseca, and Krone in Brazil; Goudot and Boulin and Cuervo in Colombia; Darwin at Galápagos Islands (Ecuador); Humboldt, Marcoy, Castelnau, and Raimondi in Peru; Berg in Uruguay; and Humboldt, Codazzi, Lisboa, Funk, Goering, Sievers, Scharfernorth, and Marcano in Venezuela. Among them, at least four deserve special mention. Alexander von Humboldt was the forerunner of cave science in Venezuela, following his visit (together with Bonpland) to the Cueva del Guácharo in 1799. Humboldt explored the first 422 m of this famous cave and described scientifically the unique Guácharo in his *Personal Narratives of Travels to the Equinoctial Regions of America during the years 1799–1804*. Humboldt also made a pioneering visit in 1802 to the Uchkupisjo caves in Peru.

In Brazil, the Danish naturalist Peter Wilhelm Lund laid the foundations of Brazilian cave science during intensive research in the caves of central Minas Gerais state from 1834 to 1844. Lund explored over 1000 caves in search of fossil bones. He hired the Norwegian Peter Andreas Brandt, a merchant turned cartographer and painter, who was responsible for surveying the caves and drawing caving scenes. At least 30 cave maps and cave and karst paintings have been preserved in Danish museums.

Besides his more immediate interest in vertebrate paleontology, Lund was the first to suggest that cave saltpetre has its origins in nitrates in the soil above the cave, an explanation not adopted again until the 20th century. Lund also published valuable data on karst landscape, cave origin, cave sedimentation, and fossil emplacement in his series of "memories", published in Danish and translated partially in other languages (Lund, 1840).

The intrepid Father Romualdo Cuervo may well be considered the pioneer of vertical caving in South America after descending suspended in a basket, in 1851, the 120 m deep entrance shaft of Hoyo del Aire in Colombia. The Venezuelan chemist Vicente Marcano was the main explorer of Venezuelan caves in the 19th century, studying over 30 caves between 1883 and 1890. His immediate interest was the study of guano deposits for use as a fertilizer.

The 20th century saw the beginning of organized caving in South America. Several people made important contributions in both exploration and scientific studies of caves, including the first description of a troglobitic fish (in Brazil) by Miranda Ribeiro in 1907. In 1937 the first caving club of South America, the still active Sociedade Excursionista e Espeleológica (SEE), was founded in the town of Ouro Preto, Brazil. Other clubs followed in the 1950s and 1960s, such as the Speleology Section of the Sociedad Venezolana de Ciencias Naturales, formed in 1952, later evolving to the Sociedad Venezolana de Espeleologia in 1967. In Brazil, many of these earlier groups were emulated by expatriate cavers of European origin, mainly French. The 1960s also saw the foundation of national caving bodies, the Sociedad Peruana de Espeleologia (now defunct) in 1965 and the Sociedade Brasileira de Espeleologia in 1969.

Cave exploration had a great impetus from the 1960s to the present. Most of the major caves of the continent were explored during this period by enthusiastic local caving clubs and foreign expeditions. The contribution of European expeditions was particularly significant in countries such as Colombia, Peru, Bolivia, and Ecuador, and they are responsible for the majority of what is known about karst and caves in these countries. Argentina also experienced a rapid growth of its speleology from the early 1970s, when many active clubs were formed. In 2000 the Federación Argentina de Espeleología was founded.

The present status of cave and karst research and exploration in South America varies between countries. Well-organized (but small) local caving communities exist in Argentina, Brazil, and Venezuela, with incipient or non-existent local speleology in the remaining countries. Cave and karst research is performed in a few universities and established postgraduate programs allow the regular production of theses on karst related topics. Cave exploration remains restricted mostly to the upper classes of financially debilitated countries, aided by frequent foreign expeditions. Nevertheless, the technical level of both exploration and science is equal with what is done in more developed countries, although performed on a much more restricted scale. Cave preservation is increasingly a topic of concern in South America. Several conservation areas have recently been designated to protect karst resources. In 1997 the Brazilian government created a centre for the protection and management of caves (CECAV), the first of its kind in South America.

AUGUSTO AULER

See also **Caribbean Islands: History**

Works Cited

Lund, P.W. 1840. View of the fauna of Brazil previous to the last geological revolution. *Magazine of Natural History*, 4: 1–8, 49–57, 105–12, 153–61, 207–13, 251–59, 307–17, 373–89
This has good information on the paleontological contents of caves.

Perera, M.A. 1976. Notas sobre una excavación en la Cueva del Guácharo (Mo.1), estado Monaguas, Venezuela [Notes on an excavation in the Cueva del Guácharo (Mo.1), Monaguas State, Venezuela]. *Boletín de la Sociedad Venezolana de Espeleología*, 7(14): 249–65

Urbani, F. 1989. Cuevas venezolanas conocidas en los siglos XV al XVIII [Venezuelan caves known in the 15th to 18th centuries]. *Boletín de Historia de las Geociencias en Venezuela*, 37: 1–78

Urbani, F. 1993. Vida y obra de los iniciadores de la espeleología en Venezuela. Parte 5: Siglos XV y XVI [Life and works of the pioneers of speleology in Venezuela. Part 5: 15th and 16th centuries]. *Boletín de la Sociedad Venezolana de Espeleología*, 27: 7–13

Urbani, F. 1996. Vida y obra de los iniciadores de la espeleología en Venezuela. Parte 7. Siglos XVI al XVIII [Life and works of the pioneers of speleology in Venezuela. Part 7. 16th to 18th centuries]. *Boletín de la Sociedad Venezolana de Espeleología*, 30: 38–55

Urbani, F. 1999. Historia espeleologica venezolana. Parte 10. Una cronologia de la Cueva del Guácharo [History of speleology in Venezuela. Part 10. A chronology of Cueva del Guácharo]. *Boletín de la Sociedad Venezolana de Espeleología*, 33: 51–69

Further Reading

Juberthie, C. & Decu, V. (editors) 1994. *Encyclopaedia Biospeologica*, vol. 1, Moulis: Societé de Biospéologie (chapters on South American countries include good regional histories as well as biological details)

Urbani, F. 1982–97. Vida y obra de los iniciadores de la espeleologia venezolana. Rojas [Life and works of the pioneers of speleology in Venezuela]

1. (1982) Jean-Baptiste Boussingault, Agustín Codazzi y Arístides. *Boletín de la Sociedad Venezolana de Espeleología*, 10(18): 17–47
2. (1982) François Depons, Jean J. Dauxion Lavaysse, James Mudie Spence, Ramón Bolet, Herman F.C. Ten Kate y Leonard V. Dalton. *Boletín de la Sociedad Venezolana de Espeleología*, 10(19): 143–173
3. (1984) John Princep, José Maria Del Real, Alexander Walker, Francisco Zea, Pál Rosti, Simón Ugarte, Achille Müntz y Bonifacio Marcano. *Boletín de la Sociedad Venezolana de Espeleologia*, 21: 33–50
4. (1986). Autores diversos 1855–1881. G.A. Gardiner, M.M. Lisboa (1809–1881). *Boletín de la Sociedad Venezolana de Espeleologia*, 22: 29–44
5. (1997) Gaspar Marcano (1850–1910), Vicente Marcano (1848–1891), exploraciones del Ing. Juan de Dios Monserrate en 1894. *Boletín de la Sociedad Venezolana de Espeleología*, 31: 37–52

AMPHIBIA

The characteristic features of amphibia (frogs, salamanders, and gymnophiona) include a slimy skin that is not well suited to prevent desiccation, and a body temperature corresponding directly to ambient temperature. It is not surprising therefore that amphibia are commonly encountered in the entrance or twilight zone of caves, which are characterized by high air humidity and buffered temperature changes. Many amphibia visit caves occasionally as temporary shelter from unfavourable environmental conditions outside, such as summer drought or winter frost. Caves are also used as breeding sites by some species. However, the impact of those visitors on the cave ecosystem is low since they only make a small contribution to the subterranean food web. More important for biospeleology are the troglophilic salamander species which are rarely found far from caves, and can also feed on subterranean invertebrates. Well-known examples are the Pyrenean salamander *Euproctus asper*, members of the genus *Hydromantes* in southern France, northern to central Italy, Sardinia, and California, the long-tailed salamander *Plethodon longicauda*, the common cave salamander *P. lucifuga*, the slimy salamander *P. glutinosus*, the purple salamander *Gyrinophilus porphyriticus*, all from the eastern United States, and *Eurycea neotenes* from Texas. The high number of amphibian species which can be observed occasionally in caves is in marked contrast to the low number of obligate cave dwellers or troglobites, which cannot survive outside caves. Only ten amphibian troglobitic species or subspecies are known and all are salamanders (see Table 1). Troglobitic members are restricted to two families, the Proteidae and the Plethodontidae. The Proteidae comprise six species, but only one is troglobitic. This is the famous blind cave salamander *Proteus anguinus* (see Amphibia: *Proteus*) from the Dinaric karst. The family Plethodontidae comprises about 246 species and is distributed mainly from North America down to northern South America, with a few members in Europe. However, the nine troglobitic taxa are restricted to the United States. Cave salamanders share some morphological characteristics with the other group of troglobitic vertebrates, the cave fishes (see Pisces entry). Advanced cave-adapted forms have reduced eyes and body pigmentation, a slender body, and a comparatively large, anteriorly depressed, shovel-like head with a large mouth. Cave plethodontids show a trend to markedly elongated thin legs, giving them a fragile appearance. With the exception of *Proteus* which may reach up to 40 cm in length, all other cave salamanders are small, rarely exceeding 15 cm.

Amphibians are generally characterized by a life cycle with two stages: an aquatic larvae which metamorphoses into a terrestrial adult after some time. With the exception of only two species, *Gyrinophilus subterraneus* and the cave salamander *Typhlotriton spelaeus*, all other troglobitic salamanders usually attain sexual maturity without metamorphosis, i.e. they remain aquatic for the whole life and retain some larval characteristics, such as the feathery external gills and a fin-shaped tail. Because of the blood in the capillaries the external gills appear brightly red and are in sharp contrast to the whitish body. The persistence of a larval body form in the adult is called neoteny. It seems that this life strategy is advantageous in subterranean habitats since approximately half of all known neotenic salamander taxa are troglobitic. However, the reasons for this adaptation are still not fully understood and may even differ between the species.

Food scarcity in the caves may play an important role. Salamanders tend to eat any prey they can handle. In caves these are usually crustaceans, insect larvae, snails, and worms, but occasionally also other salamanders—even cannibalism has been observed. Although troglobitic vertebrates are usually small they are nevertheless the largest predators in the subterranean community and therefore at the top of the food chain or web. Hence, the energetic costs of metamorphosis might be too high and selection would act towards prolonging the larval phase. Another argument might be that food scarcity in the terrestrial habitat is usually stronger than in the aquatic habitat. Accordingly, neotenic salamanders that remain aquatic can exploit the more reliable food source. Climate may also have an influence on the development of neoteny. For example, the neotenic troglobite *Eurycea tridentifera* lives in caves of the Edwards Plateau in central Texas.

Amphibia: Troglobitic salamanders (Urodela, Amphibia) and their distribution.

Proteidae	
Proteus anguinus	Dinaric karst from Trieste, Italy in the northwest through Slovenia and Croatia to the Trebisnica River in Bosnia-Herzegovina in the southeast
Plethodontidae	
Eurycea tridentifera	Caves in the Cibolo sinkhole plain and adjacent regions, Comal and Bexar Counties, Edwards Plateau, Texas, United States
Gyrinophilus gulolineatus	Caves of the east bank of the Tennessee River Valley (Appalachian Valley), Tennessee, United States
Gyrinophilus palleucus necturoides	Big Mouth Cave, Elk River drainage, Grundy County, southern Cumberland Plateau, southern Tennessee, United States
Gyrinophilus palleucus palleucus	Caves of the southern Cumberland Plateau, southern Tennessee and northern Alabama, United States
Gyrinophilus subterraneus	General Davis Cave near Alderson, Greenbrier County, (Greenbrier Valley), West Virginia, United States
Haideotriton wallacei	Subterranean waters of the Dougherty Plain of southwestern Georgia and adjacent northwestern Florida, United States
Typhlomolge rathbuni	San Marcos pool of the Balcones Aquifer, Hays County, Texas, United States
Typhlomolge robusta	Balcones Aquifer to the north and east of the Blanco River, Hays County, Texas, United States
Typhlotriton spelaeus	Southwestern Ozark Plateaus (Springfield and Salem Plateaus) of southwest Missouri, southeast Kansas, northeast Oklahoma, and northwest Arkansas, United States

This area is not very suitable for amphibians because it is very dry. Therefore, a surface-living close relative, *Eurycea neotenes*, can only survive in damp headwater canyons, springs, and cave entrances. This troglophilic species exhibits a remarkable plasticity in its life cycle because some populations are completely neotenic while others have a variable proportion of metamorphosed animals. It seems that the reliability of spring discharge is the decisive factor. Only under conditions of a permanently sufficient discharge is metamorphosis an advantageous strategy. Springs with strong discharge fluctuations only support neotenic populations and if the discharge becomes too low the animals retreat into the underground spring. The evolution of the permanent cave dweller *Eurycea tridentifera* potentially started this way. A very remarkable life cycle can be observed in the cave salamander or Ozark blind salamander *Typhlotriton spelaeus*. This species has the largest distribution area of all troglobitic plethodontids and can be found in food-rich caves of the southwestern Ozark Plateau. *T. spelaeus* still exhibits metamorphosis as a life history characteristic. The larvae possess well-developed functional eyes and a marked body colouration. They usually live in cave entrances or even outside where they feed on small invertebrates. With the onset of metamorphosis after 2–3 years the animals retreat deep into the cave. Colouration becomes pale and the eyes are overgrown by the eyelids. In addition, the photoreceptors in the retina degenerate, i.e. the animal becomes blind. Also as adults the animals stay close to the water. They feed on terrestrial insects, isopods, and snails. It is thought that all troglobitic salamanders lay eggs instead of giving birth to free-living larvae. The internal insemination probably happens via a spermatophore which has been produced and deposited by the male and is taken up by the female. Not all troglobitic salamanders live in "real" caves, i.e. in caverns accessible to man. Some, like the Georgia blind salamander *Haideotriton wallacei* or the Texas blind salamander *Typhlomolge rathbuni* inhabit groundwater and are usually found only in wells after pumping activities. *T. rathbuni* is regarded as the most cave-adapted and phylogenetically oldest troglobitic plethodontid with a supposed Miocene origin. It shows the highest degree of eye reduction and the longest limbs. The animals move very slowly, stalking around on their thin legs with the large head held high to sense every possible prey organism in the surrounding. Like all aquatic salamanders, *T. rathbuni* possesses well-developed sensory organs for smell, taste, mechanoreception, and even electroreception. If disturbed the animal swims away and tries to hide.

Most salamanders encountered in caves are legally protected and should not be collected. They are potentially at risk due to restricted distribution and low population sizes. Water pollution and excessive groundwater pumping are serious threats for these fascinating animals.

AXEL WEBER

See also **Edwards Aquifer: Biospeleology**

Further Reading

Durand, J.P. 1998. Amphibia. In *Encyclopaedia Biospeologica*, vol. 2, edited by C. Juberthie & V. Decu, Moulis and Bucarest: Societé de Biospéologie (in French)

Sweet, S.S. 1986. Caudata. In *Stygofauna Mundi, A Faunistic, Distributional, and Ecological Synthesis of the World Fauna Inhabiting Subterranean Waters (Including the Marine Interstitial)*, edited by L. Botosaneanu, Leiden: Brill

Weber, A. 2000. Fish and Amphibia. In *Subterranean Ecosystems*, edited by H. Wilkens, D.C. Culver & W.F. Humphreys, Amsterdam and New York: Elsevier

AMPHIBIA: *PROTEUS*

Proteus anguinus anguinus, popularly called the blind cave salamander or "human fish" due to its pink skin, is the sole species of the genus *Proteus*, and is only found in Europe. It is the largest, most remarkable, and world-famous model for troglobites. The distribution of *Proteus* is entirely within the Dinaric karst; nearly 200 localities, from the lower reaches of the Isonzo–Soča River in Italy in the northwest through to the southern part of Slovenia to the Trebišnjica River in Herzegovina in the southeast are known (Sket, 1997). As noted in Folklore and Mythology, *Proteus* were once considered to be the offspring of dragons.

Proteus is an obligate neotene and preserves some larval or immature characters in adulthood; the major reason for cessation of its metamorphosis, as in other obligate neotenes, probably lies in the insensitivity of its tissues to thyroid hormones. (Neoteny is retention of some larval or immature traits in adulthood; metamorphosis is a series of postembryonic changes involving structural, physiological, biochemical, and behavioural transformations.) *Proteus* retains three pairs of outer gills, two pairs of gill slits, an integument with many larval characteristics, and typical visceral skeletal elements. It has an extended, narrow body and a reduced number of digits, that is three on the anterior legs and two on the posterior legs. *Proteus* shows some general troglomorphic characteristics: specialization of the sensory organs, elongation of individual body parts, especially the disproportionate growth of the head in length, reduced eyes, skin depigmentation, slow metabolism, resistance to starvation, increased lifespan, and probably the reduction of intraspecific aggressive behaviour. Proteiids characteristically possess no maxillae (i.e. the upper jawbones in vertebrates). Absence of maxillae is clearly paedomorphic (the phenomenon in which larval or immature features of ancestors become adult characteristics of descendants).

Proteus anguinus parkelj ssp. nov. from southeastern Slovenia differs from *Proteus anguinus anguinus* in its dark pigmentation, better-developed eyes, and many other morphological characteristics. This black, non-troglomorphic form was discovered in 1986 and has been attributed to a subspecies (Sket & Arntzen, 1994). Further research on the distribution and the racial polymorphism within *Proteus* is in progress. However, allozyme analysis also revealed the probability of speciation within this genus.

The embryonic development of *Proteus* takes about 130 days at 11°C, and larval development takes about 100 days at the same temperature (Durand & Delay, 1981). The water temperature in its natural habitat varies from 8°C during the winter to 11°C in the summer and autumn. They reach sexual maturity in about their 16th year. Owing to their slow metabolism, their lifespan may be as long as 70 years. Nutritional research on *Proteus* in its natural environment has proved that it is carnivorous, and has determined the main organisms on which it feeds, i.e. Crustacea, Insecta larvae, and Gastropoda. Its extraordinarily low metabolic rate, typical of troglobites, reflects its complete adaptation to conditions in cave habitats (Istenič, 1986).

Those sensory organs that have not undergone regression and for which proper stimuli are expected to be found in the underground waters of the Dinaric karst include the inner ear, the lateral line system, and sensory receptors that detect variations in the Earth's magnetic field, all of which are worthy of study (Bulog, 1989a, 1989b; Istenič & Bulog, 1984; Schlegel & Bulog, 1997; Bulog & Schlegel, 2000). In *Proteus*, these sensory organs, including the chemoreceptors, obviously play an important role in orientation, searching for prey, intraspecies communication, and mutual recognition.

Little is known about the hearing of *Proteus*, but occasionally observed reactions to sounds have suggested a hearing capacity, especially underwater. This would be of adaptive value in dark caves—in order to recognize particular sounds and to locate prey. Recently performed behavioural studies on the hearing ability of *Proteus* (Bulog & Schlegel, 2000) indicate that unpigmented specimens showed the greatest sensitivity to sounds at *c*.1.5 kHz, while the black specimens were most sensitive to sounds at about 2 kHz. The fact that *Proteus* reacts spontaneously and consistently well to sounds demonstrates the biological significance and adaptive value of its underwater hearing.

The lateral-line sensory system is an aggregation of epidermal sensory organs scattered over the head and along the body in fish and aquatic amphibians. It includes ciliary mechanoreceptive neuromasts and electroreceptive ampullary organs. The discovery of ampullary organs in *Proteus* indicated that it is a candidate for passive electroreceptivity, which may play an important role in animals in underground water habitats (Istenič & Bulog, 1984). Behavioural experiments have confirmed that *Proteus* is sensitive to weak electric fields (Schlegel & Bulog, 1997).

Recently, some behavioural evidence of sensitivity to the orientation of the Earth's magnetic field has also been found. It is not yet known whether the magnetic field is detected by electroreceptors, or if the magnetoreceptivity is based on a different sensory mechanism.

The visual system undergoes some degree of regression in most cave animals. Among different changes in the eye structure, such as degeneration of the lens and optic nerve, it often includes changes in the structure of photoreceptor cells. Continuous darkness is known to result in the degeneration of outer segments in the retinal as well as in pineal photoreceptors. *Proteus anguinus anguinus* has small eye rudiments, seen as black dots, which lie buried deeply under the skin. The retina, lens, and vitreous body are strongly reduced in size and complexity in comparison to non cave-dwelling salamander species. Although eye development begins normally in the embryo, in the larval stage it is retarded and the degeneration becomes soon evident (Durand, 1971). Light and electron-microscopic studies of the retina of dark-coloured *Proteus* specimens led to the initial comparative analysis of the differentiation of these sensory organs (Bulog, 1992).

Proteus anguinus anguinus has a zone of less-pigmented skin on the anterior-dorsal part of its head. This zone is even more obvious on the head of the black subspecies *Proteus anguinus parkelj*. These pale areas are very sensitive to directional point-source illumination and are located above or in front of the pineal gland. The pineal organ of the lower vertebrates is a photosensitive organ with many functions. Mediated by the pineal hormone melatonin, it controls circadian rhythms, gonadotropic activity, and body colour changes. It contains photoreceptor cells, which in their general features resemble retinal photoreceptors. In cave animals, which live under conditions of permanent darkness and constant temperature, the role of the pineal organ as a circadian pacemaker would be expected to become biologically useless.

Proteus has played a significant role in biospeleological study, both in the wild and in captive populations. The comparison of *Proteus anguinus anguinus* with its non-troglomorphic relative, the sighted and black-pigmented *Proteus anguinus parkelj*, is expected to provide new evidence about the development of photoreceptor structures in cave animals. The species is unfortunately threatened by water pollution and unscrupulous international dealers, despite being listed in the CITES convention, which prohibits trade in rare wild animals. The species is also protected under Slovenian law, and listed as vulnerable in the World Conservation Union (IUCN) Red List of Threatened Species.

BORIS BULOG

See also **Dinaric Karst: Biospeleology; Postojna–Planina Cave System, Slovenia: Biospeleology**

Works Cited

Bulog, B. 1989a. Tectorial structures of the inner ear sensory epithelia of *Proteus anguinus* (Amphibia, Caudata). *Journal of Morphology*, 201: 59–68

Bulog, B. 1989b. Differentiation of the inner ear sensory epithelia of *Proteus anguinus* (Urodela, Amphibia). *Journal of Morphology*, 202: 325–38

Bulog, B. 1992. Ultrastructural analysis of the retina of *Proteus sp.*— Dark pigmented specimens (Urodela, Amphibia). In *Proceedings of the 10th European Congress on Electron Microscopy, Granada, Spain*, vol. 3, edited by A. Rios *et al.*, Granada: Universidad de Granada

Bulog, B. & Schlegel, P. 2000. Functional morphology of the inner ear and underwater audiograms of *Proteus anguinus* (Amphibia, Urodela). *European Journal of Physiology*, 439 (Suppl): 165–67

Durand, J.P. 1971. Recherches sur l'appareil visuel du Protée, *Proteus anguinus* Laurenti, Urodele hypogé. *Annales de Spéléologie*, 26: 497–824

Durand, J.P. & Delay, B. 1981. Influence of temperature on the development of *Proteus anguinus* (Caudata: Proteidae) and relation with its habitat in the subterranean world. *Journal of Thermal Biology*, 6: 53–57

Istenič, L. 1986. Evidence of hypoxic conditions in the habitat of the cave salamander *Proteus anguinus* in the Planinska jama. In *Proceedings of the 9th International Congress of Speleology*, Barcelona

Istenič, L. & Bulog, B. 1984. Some evidence for the ampullary organs in the European cave salamander *Proteus anguinus* (Urodela, Amphibia). *Cell Tissue Research*, 235: 394–402

Schlegel, P. & Bulog, B. 1997. Population-specific behavioral electrosensitivity of the European blind cave salamander (*Proteus anguinus*, Amphibia, Caudata). *Journal de Physiologie (Paris)*, 91: 75–79

Sket, B. & Arntzen, J.W. 1994. A black, non-troglomorphic amphibian from the karst of Slovenia: *Proteus anguinus parkelj* n. ssp. (Urodela: Proteidae). *Bijdragen tot de Dierkunde*, 64(1): 33–53

Sket, B. 1997. Distribution of *Proteus* (Amphibia: Urodela: Proteidae) and its possible explanation. *Journal of Biogeography*, 24: 263–80

Further Reading

Aljančič, M., Bulog, B., Kranjc, A., Josipovič, D., Sket, B. & Skoberne, P. 1993. *Proteus: the Mysterious Ruler of Karst Darkness*, Ljubljana: Vitrum

Atema, J., Fay, R.R., Popper, A.N. & Tavolga, W.N. (editors) 1988. *Sensory Biology of Aquatic Animals*, New York: Springer

Briegleb, W. 1962. Zur Biologie und Ökologie des Grottenolmes (*Proteus anguinus* Laur. 1768). *Zeitschrift für Morphologie und Ökologie der Tiere*, 51: 271–334

Duellman, W.E. & Trueb, L. 1986. *Biology of Amphibians*, New York: McGraw Hill

Kos, M., Bulog, B., Szél, Á. & Röhlich, P. 2001. Immunocytochemical demonstration of visual pigments in the degenerate retinal and pineal photoreceptors of the blind cave salamander (*Proteus anguinus*). *Cell Tissue Research*, 303: 15–25

Schlegel, P. 1996. Behavioral evidence and possible physical and physiological mechanisms for earth-magnetic orientation in the European blind cave salamander *Proteus anguinus*. *Mémoires de Biospéologie*, 23: 5–16

Schlegel, P. 1997. Behavioral sensitivity of the European blind cave salamander, *Proteus anguinus*, and Pyrenean newt *Euproctus asper*, to electrical fields in water. *Brain, Behavior and Evolution*, 49: 121–31

Yun-Bo Shi 2000. *Amphibian Metamorphosis: From Morphology to Molecular Biology*, New York: Wiley-Liss

Useful Websites

Amphibia web http://elib.cs.berkeley.edu/aw/index.html

ANCHIALINE HABITATS

Anchialine (anchihaline) cave habitats are located in water-filled voids near the coast, where there is a demonstrable marine influence shown either by mixohalinity (different brackish salinities) or by the presence of marine-derived animal species (Sket, 1996). The hydrological connections to the sea—as well as to the land—are underground, and there is restricted contact between the water body and the open air. Anchialine cave waters may function as corridors to open anchialine pools (Holthuis, 1973). The distinction between anchialine and marine caves is not always clear. A cave with euhaline (purely marine) water should only be considered as an anchialine habitat if it contains troglomorphic species. Riedl's (1966) concept of the marginal cave also included a theory about its origin, which differs from the archialine cave concept.

Most bodies of anchialine cave water have been observed in tropical to warm-temperate regions. The caves mostly developed during sea-level lowstands in the Pleistocene (see Speleogenesis: Coastal and Oceanic Settings); they retained the freshwater influx from the land, while the subsequent sea-water influx changed their ecology. Anchialine habitats in coral reefs and coastal lava fields have a different origin, but a similar morphology. Blue holes, anchialine pools, metahaline pools and some other habitats may exhibit some characteristics of anchialine cave waters. In particular, some stygobites may be present in these habitats, due to a diminished number of competitors in the generally unfavourable conditions of anchialine habitats.

Ecological Conditions

Besides the mixohalinity itself, one of the most striking abiotic characteristics of anchialine waters is their stratification (Figure 1). The primary stratification is in salinity; the extreme upper and lower layers may be limnic and euhaline (marine) respectively. This depends on the influx intensities of fresh and marine water, and therefore will change throughout the year or even within one day.

Great differences in the densities of waters of variable salinities prevent any mixing of layers, which might otherwise be expected during changes in temperature. Stratification is usually gradual, with a thin layer of rapid transition (the halocline) between the denser and the less-dense layers. This stable salinity and density stratification also allows stratification in other characteristics. Consumption of oxygen by animals and bacteria, particularly on the bottom, causes oxygen depletion. This deficit may be neutralized by exchange with sea water at the bottom and by diffusion from the surface in the uppermost layers, and therefore remains most extreme in the halocline layer, where there is least mixing. Accumulation of hydrogen sulfide may coincide with the oxygen deficit. If the water body is exposed to temperature changes associated with the atmosphere at the

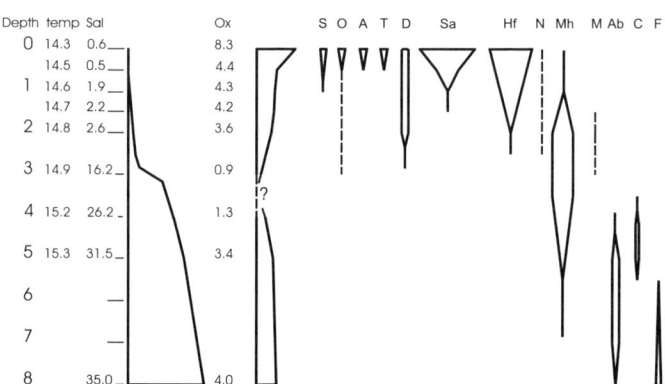

Anchialine Cave Habitats: Ecological stratification in the anchialine lake of the cave Šipun near Cavtat, Croatia, in September 1975. Depth in metres; temp, temperature in °C; Sal, salinity in ppt.; Ox, oxygen content in mg l^{-1}; S, *Saxurinator sketi* (Gastropoda); O, Oligochaeta; A, *Acanthocyclops venustus*, T, *Thermocyclops dybowskii*, and D, *Diacyclops antrincola* (Copepoda); Sa, *Salentinella angelieri*, Hf, *Hadzia fragilis*, and N, *Niphargus salonitanus* (Amphipoda); Mh, *Monodella halophila* (Thermosbaenacea); M, *Metacyclops trisetosus* (Copepoda); Ab, *Ammonia beccarii* (Foraminifera); C, *Caecum glabrum* (Gastropoda); F, *Filogranula annulata*. The population density is measured in estimated relative values.

land's surface, a temperature inversion may occur in the winter, with the upper layers being the coolest (Sket, 1986).

Fauna

The main biological characteristic of this habitat is the presence of stygobitic species together with species of marine provenance. The unfavourably diverse and variable degree of mixohalinity further reduces the number of species that can colonize these habitats. On the other hand, contact with the faunistically rich marine ecosystem potentially increases diversity.

The number of species known from anchialine cave waters increased from less than 30 in the 1960s (Riedl, 1966) to c.150 in the 1980s, and close to 400 species in the 1990s. The Table shows species numbers by groups; numbers of known taxa are still increasing rapidly, especially for copepods, but also for amphipods and other groups. Similarly to hypogean waters in general, crustaceans are by far the most prevalent group in anchialine waters. This is evident in the list of known species as well as within any local fauna. Since some crustaceans are larger (except for vertebrates) and more numerous than most other cave inhabitants, Crustacea are also the most important group in relation to biomass. In the Šipun Cave (Cavtat, Croatia; Figure 1) which exhibits a complete salinity range from limnic to polyhaline, nine crustacean species and only six other species have been found. The ratios of individuals and biomass between crustacea and other species is even more striking—at least 10:1. In Cave C-28, on Cape Range Peninsula, Australia, only five species have been found below the pycnocline (the thin layer of a rapid density change), with all but one (a fish) being crustaceans. In the anchialine waters of Quintana Roo, Mexico, 12–14 species have been identified, all of which are crustaceans.

It is difficult to assess exactly the number of anchialine species. While the distinction between anchialine cavernicole and interstitial fauna is seldom ambiguous, it is often questionable if a non- or slightly troglomorphic animal is anchialine or only a marine-benthic generalist. As in freshwater cave habitats, some (here necessarily euryhaline, or able to live in waters of a wide range of salinities) stygoxenes and stygophiles occur in anchialine waters. Another facultative presence—again euryhaline—are stygobitic species from freshwater caves. A very common non-specialized marine species in anchialine caves in the Caribbean is the fish *Eleotris pisonis*, and freshwater species include the copepod *Diacyclops bicuspidatus odessanus* and amphipod *Niphargus hebereri*, both found in anchialine caves on the Dinaric coast. It may be fortuitous that some of these animals may be known only from anchialine caves, without necessarily being confined to them.

Most stygobitic (and usually troglomorphic) species are related to and probably derived from the extant shallow-marine fauna—most amphipods (e.g. Hadziidae) and shrimps (e.g. *Typhlatya* spp.). Representatives of some groups show close phylogenetic relationships to deep-sea species. These include the cavernicole Tantulocarida, Mictacea (*Mictocaris halope* in Bermuda), Decapoda Anomoura (*Munidopsis polymorpha* in the Canary Islands) and some other decapods, and some Ostracoda (*Danielopolina* spp.). The only higher taxonomic group limited to anchialine caves and therefore of a non-identifiable origin seems to be the Remipedia. Some specialized inhabitants of anchialine caves which are unquestionably invaders from fresh waters are known mainly from the Dinaric coast; for example, the amphipods *Niphargus salonitanus* and *N. pectencoronatae*.

General Comments

Some characteristics of anchialine fauna make them very important for understanding the colonization of less favourable habitats, such as non-marine and hypogean waters. The largely anchialine shrimps *Typhlatya* spp. are among the few members of the mainly circumtropical freshwater family Atyidae, which are still marginally marine. Amphipoda in anchialine waters are globally represented primarily by the family Hadziidae and only a few other species, although a huge variety of amphipod groups come into contact with these habitats. A number of groups are represented circumglobally (resembling the extension of the ancient Tethys ocean), including Remipedia, Thermosbaenacea, genera *Typhlatya* (Decapoda), and *Spelaeomysis* (Mysidacea). The strictly stygobitic species—such as *Niphargus hebereri* from the Dinaric coasts—may build dense populations in sunlit areas if no epigean competitors are present. Some delicate elements—such as Thermosbaenacea—may show different apparent salinity preferences, depending on the presence or absence of competitors.

Boris Sket

See also **Cape Range, Australia: Biospeleology; Walsingham Caves, Bermuda: Biospeleology**

Works Cited

Holthuis, L.B. 1973. Caridean shrimps found in land-locked saltwater pools at four Indo-west Pacific localities (Sinai Peninsula, Funafuti Atoll, Maui and Hawaii Islands), with the description of the new genus and four new species. *Zoologische Verhandelingen*, 128: 1–48
Riedl, R. 1966. *Biologie der Meereshoehlen* [Biology of Sea-Caves], Hamburg: Parey
Sket, B. 1986. Ecology of the mixohaline hypogean fauna along the Yugoslav coast. *Stygologia*, 2(4): 317–38
Sket, B. 1996. The ecology of the anchihaline caves. *Trends in Ecology and Evolution*, 11(5): 221–25

Further Reading

Danielopol, D.L. 1990. The origin of the anchialine fauna—the "deep sea" versus the "shallow water" hypothesis tested against the empirical evidence of the Thaumatocyprididae (Ostracoda). *Bijdragen tot de Dierkunde*, 60(3–4): 137–43

Anchialine Habitats: Approximate numbers of essentially anchialine species known in the 1990s (according to Sket, 1996).

Crustacea		Other groups	
Remipedia	11	Porifera	3
Ostracoda	35–40	Turbellaria	1
Copepoda	40–45	Gastropoda	5
Tantulocarida	1	Annelida	10
Leptostraca	1	Chaetognatha	4
Decapoda	45–50		
Thermosbaenacea	30–35	Pisces	10
Mysidacea	30–35		
Mictacea	1		
Tanaidacea	2		
Isopoda	30–35		
Amphipoda	80–95		

Humphreys, W.F. 1993. The significance of the subterranean fauna in biogeographical reconstruction: examples from Cape Range peninsula, Western Australia. In *The Biogeography of Cape Range, Western Australia*, edited by W.F. Humphreys, Perth: Western Australian Museum (Records of the Western Australian Museum, Supplement 45)

Iliffe, T.M. 2000. Anchialine caves. In *Subterranean Ecosystems*, edited by H. Wilkens, W.F. Humphreys & D.C. Culver, Amsterdam and New York: Elsevier

Por, F.D. 1985. Anchialine pools—comparative hydrobiology. In *Hypersaline Ecosystems: The Gavish Sabkha*, edited by G.M. Friedman & W.E. Krumbein, Berlin and New York: Springer

Riedl, R. & Ozretic, B. 1969. Hydrobiology of marginal caves. Part I. General problems and introduction. *Internationale Revue der Gesamten Hydrobiologie*, 54(5): 661–83

Stock, J.H. 1994. Biogeographic synthesis of the insular groundwater faunas of the (sub)tropical Atlantic. *Hydrobiologia*, 287: 105–17

Stock, J.H., Iliffe, T.M. & Williams, D. 1986. The concept "anchialine" reconsidered. *Stygologia*, 2(1–2): 90–92

ANNELIDA

Annelida or annelids are segmented, usually worm-like animals without articulated legs. The body consists of a chain of body segments, or somites; in front of them is the head lobe or prostomium, behind is an anal lobe or pygidium. A paired ventral nerve cord extends along the body with a pair of ganglia in each somite. The mouth and the anus are close to both ends of the body, connected by a straight intestine.

The main groupings of annelids are Polychaeta and Clitellata. The predominantly marine Polychaeta are characterized by pairs of bifurcated limbs or parapodia on each somite, with appendages called rami on the parapodia, each ending with a bundle of chaetae (bristles). Polychaetes are gonochoristic, i.e. they have separate sexes. Clitellata lack parapodia, and are hermaphroditic. The basal subgroup of clitellates is Oligochaeta, where bundles of, or single chaetae, are still present in the wall of each somite and the coelomic cavity is normally segmented. Leeches (Hirudinea) developed from within oligochaetes. They are devoid of chaetae and their coelomic cavity is restructured into a system of tubes that replaces the primary circulatory system. Leeches possess an oral and a caudal sucker, used for locomotion.

Polychaeta or polychaetes are principally a marine group with few freshwater representatives, the latter being mostly hypogean. The free-swimming and worm-like "Errantia" (names within quotation marks are not real taxonomic units) are represented in marine, anchialine, and fresh cave waters, mostly in the tropics, by representatives of the family Nereididae. Nereids have short antennae and palpi on their prostomia, their evertible mouth cavity has strong, deeply forked, cuticular jaws. *Namanereis hummelincki* is a stygobitic nereid from marine or anchihaline habitats in the Caribbean while *N. beroni* has been recorded from a cave at an altitude of 1600 m in Papua New Guinea. Representatives of the "Archiannelida" family Nerillidae are common in interstitial marine waters and anchihaline caves. They may be a few millimetres long, usually with two palpi on the prostomium and with up to ten somites. Some *Nerilla* species are benthic, others are stygobitic or in the marine interstitial

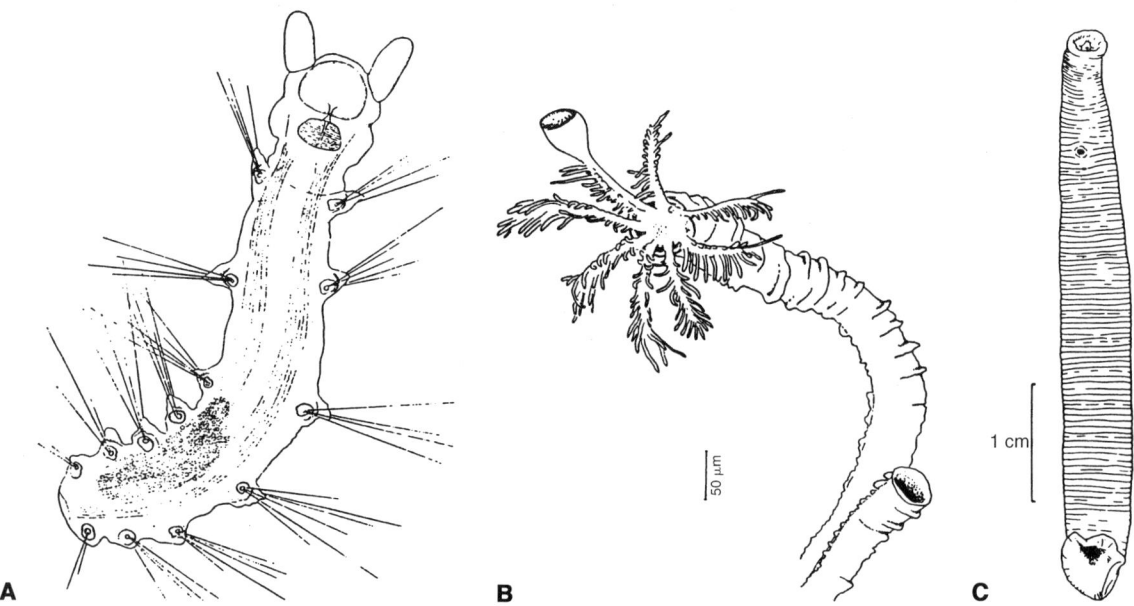

Annelida: Figure 1. (A) Polychaeta: *Troglochaetus beranecki*; (B) Polychaeta: *Marifugia cavatica*; (C) Hirudinea: *Dina absoloni*. Reproduced with permission from Botosaneanu (1986).

habitat. *Troglochaetus beranecki*, 0.5 mm long, is widely distributed in fresh interstitial or cave waters in Europe, from Finland to northern Italy; it has been known since 1921. Similar species have been found in Japan and recently in North America.

The only known stygobitic representative of the substrate-bound "Sedentaria" group is the cave tube worm, Marifugia (*Marifugia cavatica*), the only freshwater species of the family Serpulidae. Marifugia secrete a calcareous tube of less than 1 mm external diameter; the first few centimetres are attached to the substrate, while the youngest parts are erect. Apically, feathery ciliated tentacles function as gills for respiration and filter feeding. The ontogeny starts with a ciliated, free-living larva which soon attaches to the substratum and develops a tiny tube. Marifugia are present patchily throughout the whole Dinaric karst. While some authors consider it to be an ancient marine relict in caves, it has recently been suggested to have invaded cave waters via extensive Pliocene-Pleistocene freshwater lakes in the region.

Oligochaetes are a group found in the soil, in freshwater, and in marine water. They possess chaetae, whose shape and position are taken into consideration when carrying out a systematic study, as are the position and morphology of the genital organs. Some aquatic (Haplotaxidae, Lumbriculidae, Naididae, Tubificidae, and Parvidrilidae), semiaquatic and terrestrial families (Enchytraeidae, Lumbricidae, and Ocnerodrilidae) are found in caves. The Lumbricidae are found in humid soil, wet guano, and in the sediment of underground watercourses and gours. They are generally considered to be tubloxenes. The terrestrial Ocnerodrilidae are found only in South American caves. The Enchytraeidae are common in cave waters with genera such as *Enchytraeus*, *Cernosvitoviella*, and *Marionina*. The Haplotaxidae are also present with the three genera *Delaya*, *Haplotaxis*, and *Villiersia*, *D. bureschi* being largely distributed in Europe. The Naididae are present in caves with ubiquitous species; the *Pristina* and *Pristinella* genera comprise stygophilic species and some stygobitic species, found only in the Antilles. Lumbriculidae and Tubificidae exhibit a high level of diversity and comprise many stygobitic taxa. The presence of freshwater species belonging to the marine genera of Tubificidae (*Tubificoides*, *Spiridion*, *Aktedrilus*, *Abyssidrilus*, *Gianius*, and *Phallodriloides*) highlights the links between some groundwater taxa and their relatives in marine habitats.

The recently discovered holarctic family of Parvidrilidae, which has only two species (the European stygobitic *Parvidrilus spelaeus* and the North American interstitial *P. strayeri*) highlights the relationship between this family and the Gondwanian Phreodrilidae.

The biodiversity of underground oligochaetes is rather high: up to 100 stygobitic species have been identified worldwide. Some subterranean Oligochaeta show characters probably related to their adaptation to a subterranean way of life, such as a reduction in body size (for example the stygobitic species of Phallodrilinae and Parvidrilidae), or in body diameter on the gonadal region, due to the shifting or the asymmetrical bending of the spermathecae.

Hirudinea (leeches) are represented in caves only by few blood-sucking species. These include troglophilic, mollusc-sucking representatives of the family Glossiphoniidae, such as *Glossiphonia complanata* in the Postojna–Planina Cave System, Slovenia. There are also some land leeches, Haemadipsidae, which suck bats, for example *Leiobdella jawarerensis* in Papua New Guinea and a *Haemadipsa* sp. in Yunnan, China; both are without skin pigmentation but retain pigmented eyes.

All eyeless cave leeches are predators of small invertebrates, for example *Haemopis caeca* of Haemopidae in Movile Cave, Romania. Most, however, belong to the family Erpobdellidae, moreover to the genus *Dina* or its phylogenetic vicinity. The epigean and troglophilic species *D. krasensis* is endemic to southwestern Slovenia and Croatian Istra, with a partially depigmented population in the Postojna–Planina Cave System. The oldest known (by Johansson, 1913) stygobitic leech *D. absoloni*, is eyeless and widespread in southeastern parts of the Dinaric karst. Other stygobitic *Dina* species and probably *Trocheta* species have developed from southeastern France—northern Italy—Dinaric karst—Bulgaria—Georgia. The *Dina*-related *Croatobranchus mestrovi* from cold (5°C) shafts in the Croatian Mt Velebit is extremely modified. It exhibits unique extendible tentacles around its mouth and *c.*10 pairs of finger-shaped gills

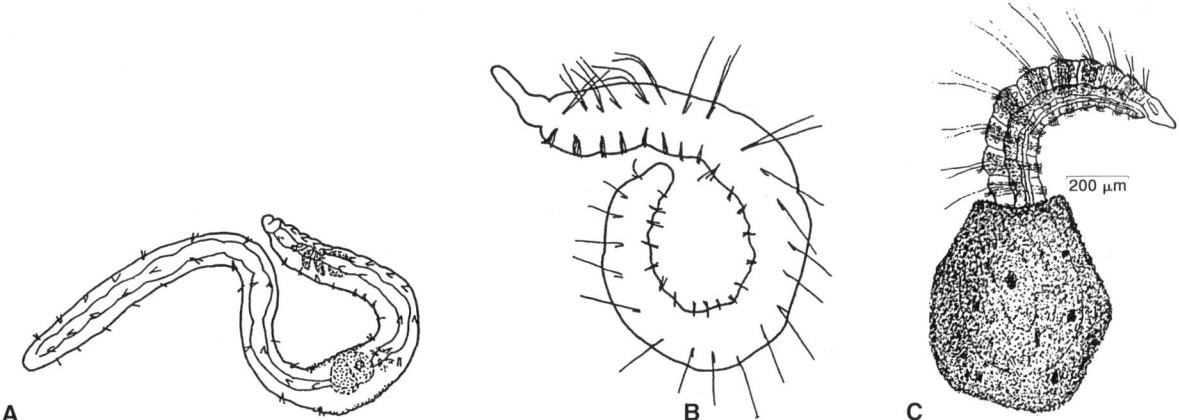

Annelida: Figure 2. Oligochaeta. (A) the enchitreid *Marionina* argentea; (B) the naidid *Pristina aequiseta*; (C) the tubificid *Rhyacodrilus amphigenus*. Reproduced with permission from Timm (1999).

along the trunk. Undescribed stygobitic erpobdellid leeches have been found in Texas and Japan.

BEATRICE SAMBUGAR AND BORIS SKET

Further Reading

Botosaneanu, L. (editor) 1986. *Stygofauna Mundi: A Faunistic, Distributional, and Ecological Synthesis of the World Fauna Inhabiting Subterranean Waters*, Leiden: Brill

Giani, N., Sambugar B., Rodriguez P. & Martinez-Ansemil, E. 2001. Oligochaetes in southern European groundwater: New records and an overview. *Hydrobiologia*, 463: 65–74

Hartmann-Schroeder, G. 1986. Polychaeta (incl. Archiannelida). In *Stygofauna Mundi, A Faunistic, Distributional, and Ecological Synthesis of the World Fauna Inhabiting Subterranean Waters (Including the Marine Interstitial)*, edited by L. Botosaneanu, Leiden: Brill

Juberthie, C. & Decu, V. 1998. Annelida Polychaeta. In *Encyclopaedia Biospeologica*, vol. 2, edited by C. Juberthie & V. Decu, Moulis and Bucharest: Société de Biospéologie

Juget, J. & Dumnicka, E. 1986. Oligochaeta (incl. Aphanoneura) des eaux souterraines continentales. In *Stygofauna Mundi, A Faunistic, Distributional, and Ecological Synthesis of the World Fauna Inhabiting Subterranean Waters (Including the Marine Interstitial)*, edited by L. Botosaneanu, Leiden: Brill

Martinez-Ansemil, E., Sambugar, B. & Giani, N. 2002. First record of Parvidrilidae (Annelida, Oligochaeta) in Europe with the description of a new species (*Parvidrilus spelaeus* sp. nov.) and comments on the family and its phyletic relationships. *Journal of Zoology*, 256(4): 495–504

Sambugar, B., Giani, N. & Martinez-Ansemil, E. 1999. Groundwater Oligochaetes from southern-Europe. Tubificidae with marine phyletic affinities: new data with description of a new species, review and consideration on their origin. *Mémoires de Biospéologie*, 26: 107–16

Sket, B. 1986. Hirudinea. In *Stygofauna Mundi, A Faunistic, Distributional, and Ecological Synthesis of the World Fauna Inhabiting Subterranean Waters (Including the Marine Interstitial)*, edited by L. Botosaneanu, Leiden: Brill

Sket, B., Dovč, P., Jalžić, B., Kerovec, M., Kučinić, M. & Trontelj, P. 2001. A cave leech (Hirudinea, Erpobdellidae) from Croatia with unique morphological features. *Zoologica Scripta*, 30(3): 223–29

Timm, T. 1999. *A Guide to the Estonian Annelida*, Tallinn: Estonian Academy Publishers

Turquin, M.-J. 1994. Hirudinea. In *Encyclopaedia Biospeologica*, vol. 1, edited by C. Juberthie & V. Decu, Moulis and Bucharest: Société de Biospéologie

ANTARCTICA

The first caver on Antarctica was probably Emile Racoviţă, who arrived in the Peninsula in January 1898. An ice cave entrance appears in a photo taken during the Scott expedition in 1911, but it wasn't until the mid-1970s that the first cave exploration took place. During the Tazieff expedition to Mount Erebus, some fumarole caves were explored in the volcano, near the US MacMurdo base. In the 1980s, some classic glacier caves, created by water flow, were explored in the Schirmacher Oasis, near the Russian base of Novolazarevskaya (Queen Maud Land). The caves, in the glacier front, consist of some outflow conduits draining internal glacier waters. Later, in 1985, other fumarole caves were explored on Mt Melbourne, by Italian glaciologists. These early explorations were made during the course of research with other objectives and the first caving expedition to Antarctica was in March 2000, when small ice caves (moulins) were explored in the Collins Glacier on King George Island, South Shetlands.

Karst phenomena in glaciers develop where the average yearly temperature is around 0°C, either at high altitudes or at high latitudes (see Glacier Caves and Glacier Pseudokarst). The whole Antarctic continent has an average yearly temperature well below zero; only the edge of the Peninsula and the surrounding islands are not far from the isotherm at 0°C. Hence there are only minor examples of typical glacial karst phenomena, such as moulins and bedières. However, there are also glacier caves in other parts of the continent, that are not formed by simple waterstreams, but by water flow, volcanic heat, local thermal imbalance, and by ice fractures.

Water Flow

Most glacier caves are formed by runoff, where melt water enters the glacier through a "moulin". These caves are endoglacial because they are excavated inside the glacier core. Antarctica is too cold to allow the formation of large, meltwater rivers. For example, the ablation zone in Collins Glacier (King George Island) has a very limited extension (up to 100 m above sea level) but in spite of this, small (<50 m deep) moulins are formed on the sides of the main ice stream falling into the sea. Also the streams are very small, with a typical discharge of 1 l s^{-1}. In some regions of the main continent, yet to be explored, the seasonal runoff may be stronger and the interaction of meltwater streams with very cold ice may create morphologies different from the typical temperate glacier karst, in which small water streams can flow away almost everywhere.

Volcanic Heat

Some glacier caves form by volcanic heat released at fumaroles. These caves are subglacial and are carved on the contact of ice with rock. This cave type is well known in Iceland, where "warm" (0°C) ice tolerates the presence of meltwater. In contrast, the caves seen in Antarctica, on Mt Erebus and Mt Melbourne, formed in ice at −30 to −40°C, and so the water was immediately refrozen. However, strong airflow permits active sublimation, resulting in vapourization of ice in warmer areas and crystallization on cooler zones near the cave entrances. The main interest of these caves, explored for about 100 m, is in their biological material. They are temperate oases surrounded by an extremely adverse environment (the maximum yearly temperature is −20°C) and they host some endemic bacteria. All the areas around these fumarole caves are protected to avoid contamination: visitors are forbidden and special care is necessary to explore the caves.

Thermal Imbalance

Some very large caves have formed at the contact between cold glaciers flowing from the Plateau (−20°C) and warmer sea ice (at −2°C). The contrast in ice temperature results in strong differences in equilibrium vapour pressure in the adjacent atmosphere (~100 Pa at −20°C, and ~500 Pa at −2°C). In a closed cavity, separating two such contrasting ice bodies, the resulting vapour pressure gradient will result in significant sublimation and redistribution of ice. Marine salt probably plays a role, but further studies are necessary to understand these caves that can play a role in the breaking of floating ice tongues.

Ice Fractures

Crevasses in ice are largely tectonic in origin and are not normally classed as glacier caves. Crevasse depth depends upon temperature. For soft, temperate ice (0°C), the ice plasticity is high and the crevasse depths range from 25–30 m. However, when the ice temperature is lower, crevasses may be much deeper: at −50°C (ice temperature of the Antarctic Plateau) the crevasses reach 3–400 m. If these crevasses act as cold traps and are filled with the coldest winter air (−80°C), the depth may increase up to 700–1000 m. However, to date this has not been confirmed by exploration.

Although most of Antarctica is covered by ice, about 3000 m of horizontally bedded sedimentary rocks, ranging in age from Devonian to Jurassic, are exposed in the central Transantarctic Mountains. On the western flank of the Queen Elizabeth Range, beneath Mount Counts, the Lower to Middle Cambrian Shackleton Limestone crops out and is unconformably overlain by the Pagoda Formation (made up of tillite, sandstone, and shale). Lindsay (1970) describes a large depression, thought to be a doline, and a cave up to 6.2 m wide and 5.5 m high, that is completely filled with sediment of pre-Permian age. The site is remote and difficult to access but it is likely that similar paleokarstic deposits, and even open caves, may be discovered in the future.

GIOVANNI BADINO

See also **Glacier Caves and Glacier Pseudokarst**

Further Reading

Badino, G. & Meneghel, M. 2001. Le grotte nei ghiacci dell'Antartide, *Speleologia*, 43: 52–58

Badino, G. & Meneghel M. 2001. Caves in the glaciers of Terra Nova Bay, Victoria Land, *Proceedings of the 13th International Congress of Speleology*, edited by F. Lino *et al.*, Brasilia: Sociedade Brasiliera de Espeleologia (CD-ROM)

Lindsay, J.F. 1970. Paleozoic cave deposit in the Central Transantarctic Mountains. *New Zealand Journal of Geology and Geophysics*, 13: 1018–49

APPALACHIAN MOUNTAINS, UNITED STATES

The Appalachian mountains extend for 3000 km from Alabama (United States) to Newfoundland (Canada). They contain significant limestone and dolostone formations, especially in western and southern areas, and commonly these are less than 300 m in thickness. There are many sinking streams because of the juxtaposition of carbonates with other rocks, and a large number of extensive cave systems have been found in strata that vary from flat-lying to steeply dipping.

The northern half of the Appalachians, north of New York, consist of relatively flat uplands with isolated mountain ranges. The Proterozoic and Paleozoic strata contain few carbonates and no long caves. The southern half of the Appalachians consist of a series of parallel mountain ranges with local relief of several hundred metres, and is divided into four zones. The eastern two zones, the Piedmont and the Blue Ridge, consist largely of Proterozoic and Lower Paleozoic metamorphosed sedimentary and volcanic rocks, with few carbonates and no long caves. The two western zones, the Valley and Ridge and the Appalachian Plateau, have many extensive caves, and more than 10 000 caves have been documented in these two zones. The locations of the caves at least 5 km in length are shown in Figure 1, and are clearly clustered in certain areas.

The Valley and Ridge zone consists of long parallel ridges and valleys. Lower Paleozoic limestones and sandstones form the ridges and shales form the valleys. Stratal dips are often steep, with the result that many caves have linear patterns, being elongated along the strike. The caves of Burnsville Cove in Virginia are typical. Over 80 km of caves have been mapped within a small area in a 230 m sequence of Silurian and Devonian limestones. The principal caves are the Butler–Sinking Creek Cave System (28 km), which is largely formed along a synclinal axis, and the Chestnut Ridge Cave System (22 km), which is partly formed along the adjacent, parallel, anticlinal axis.

The Appalachian Plateau is an extensive area of mostly flat-lying Carboniferous strata. The plateau is known in the north as the Allegheny Plateau and in the south as the Cumberland Plateau. Upper Carboniferous shales and sandstones outcrop across most of these plateaus, and are underlain by Lower Car-

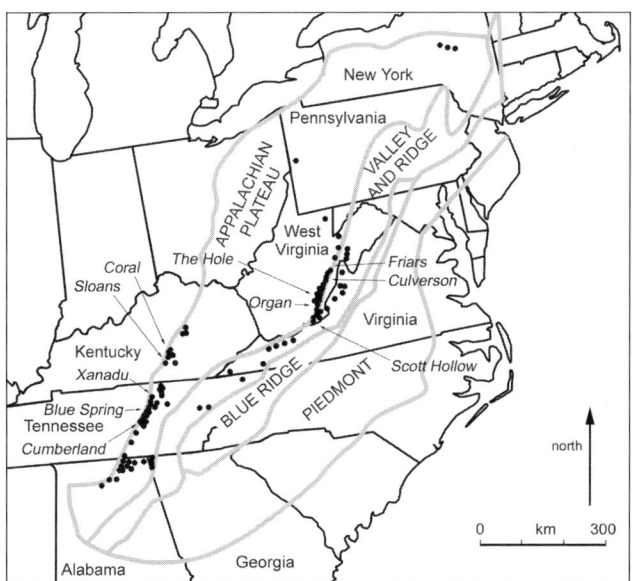

Appalachian Mountains: Figure 1. The four structural zones of the Appalachian karst in the United States, with all caves containing more than 5 km of mapped passage marked by spots, and the ten longest caves named.

Appalachian Mountains: The longest caves in the Appalachians.

Rank	Cave	State	Length (km)
1	Friars Hole Cave System	West Virginia	72
2	Organ Cave System	West Virginia	64
3	Blue Spring Cave	Tennessee	53
4	Cumberland Caverns	Tennessee	44
5	Scott Hollow Cave	West Virginia	43
6	Sloans Valley System	Kentucky	40
7	Xanadu Cave System	Tennessee	38
8	The Hole	West Virginia	37
9	Coral Cave System	Kentucky	36
10	Culverson Creek Cave	West Virginia	33

boniferous strata which include limestones. The most extensive caves have been found in the 40–90 m thick Monteagle Limestone (from Alabama to Kentucky) and the 100–400 m thick Greenbrier Group (in West Virginia). Most of the long caves in the Monteagle Limestone are close to the western edge of the Cumberland Plateau. The outcrop of the Greenbrier Group varies in width from less than 100 m in northern West Virginia to more than 10 km in the south. In the latter area the strata dip gently to the west. Streams that sink along the eastern edge of the limestone have formed distinctive caves at the contact between the limestone and the underlying shale, with vadose passages being incised into the shales. These "contact caves" include Organ Cave (64 km), Scott Hollow Cave (43 km), and The Hole (37 km). Nearby are caves that have been formed by streams sinking at the top of the limestone, and these include Friars Hole Cave System (72 km) and Culverson Creek Cave (33 km).

Friars Hole Cave System is the longest known cave in the Appalachians (Figure 2). It is located close to the boundary of the Valley and Ridge and Appalachian Plateau zones, and bears features typical of caves in both zones. The entrances of the cave are found in a series of small limestone inliers along Friars Hole, a 200 m deep valley. A series of streams flow off predominantly shales, with minor sandstones and limestones, and drain an area of 86 km^2. These streams sink where they reach the Union Limestone, a 50 m thick member of the Greenbrier Group, which has a structural dip of 2°. The Union Limestone is composed principally of sparites and micrites, but has four impure beds which are 1–3 m in thickness and have a clay content of about 50%. The 30 m thick Pickaway Limestone underlies the Union Limestone. The sinking streams descend through the vadose zone in a stairstep fashion, descending joints and then flowing downdip to the northwest, perched on the impure beds for distances up to 100 m before descending on another joint. Much of the complexity of the cave is due to the large number of such inlets, and more than 250 have been mapped, of which about 100 currently carry water.

The recharge to the cave throughout its history has been dominated by flow from the sinking streams of Hills Creek and Bruffy Creek, which together provide more than half the recharge of the cave. These streams sink several kilometres to the northeast of the cave, and the water is seen in the cave as Rocky River, a large strike-oriented passage. A second drainage, in the central part of the cave, includes the streams sinking at the Snedegars, Toothpick, Rubber Chicken, and Crookshank entrances. The cave stream descends to the lower Pickaway where it is 80 m stratigraphically below the top of the limestone, but the cave is 188 m deep because it extends almost 2 km downdip from the cave entrances. There is a third major stream in the southern part of the cave, which includes the stream sinking at Friars Hole Cave. All three cave streams flow to a spring which lies 10 km to the southwest on Spring Creek.

The water table has dropped 130 m since the earliest passages in the cave were formed. Uranium series and paleomagnetic dating of speleothems puts the age of the earliest passages at about four million years, and since that time a series of strike-oriented trunk passages has formed in succession. Most of these major passages are roughly rectangular in shape, with heights and widths of 5–10 m. Many were formed at the intersection between low-angle thrust faults and either the Union-Pickaway contact or at the impure bed which is 5 m above it. As the water table dropped, old trunk passages were abandoned and new ones were formed, usually at the same stratigraphic horizon, but several hundred metres downdip and thus at a lower elevation. The exact sequences and interrelations of these major relict passages are difficult to determine because only a small fraction of the passages in the drainage basin have been found. Many major passages in the cave terminate at sediment blockages or breakdown. The complexity in Friars Hole Cave System is caused by multiple sinking streams with their associated loads of clastic rocks, a lithology of interbedded pure and impure limestones, and a long history of karstification, and is typical for major caves in the Appalachians.

STEPHEN R.H. WORTHINGTON

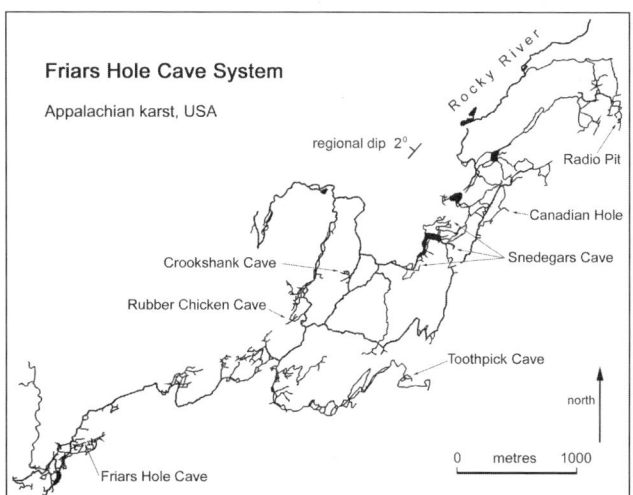

Appalachian Mountains: Figure 2. Outline map of the Friars Hole Cave System in the Appalachian karst.

Further Reading

Jones, W.K. 1997. *Karst Hydrology Atlas of West Virginia*, Charles Town, West Virginia: Karst Waters Institute (Special Publication 4)

Kastning, E.H. & Kastning, K.M. (editors) 1991. *Appalachian Karst: Proceedings of the Appalachian Karst Symposium, Radford, Virginia*, Huntsville, Alabama: National Speleological Society

Medville, D.M. 1981. Geography of the Friars Hole Cave System, U.S.A. In *Proceedings of the 8th International Congress of Speleology*, edited by B.F. Beck, Huntsville, Alabama: National Speleological Society: 412–13

Sasowsky, I. & White, W.B. 1994. The role of stress-relief fracturing in the development of cavernous porosity in carbonate aquifers. *Water Resources Research*, 30: 3523–30

Saunders, J.W., Medville, D.M. & Koerschner, W.F. 1977. Karst drainage patterns in the long mountains of the eastern United States. In *Proceeding of the 7th International Speleological Congress, Sheffield, England*, edited by T.D. Ford, British Cave Research Association: 375–76

White, W.B. 2000. Cave development in Burnsville Cove, Virginia, U.S.A. In *Speleogenesis: Evolution of Karst Aquifers*, edited by A. Klimchouk, D.C. Ford, A.N. Palmer & W. Dreybrodt, Huntsville, Alabama: National Speleological Society: 362–66

White, W.B. & Schmidt, V.A. 1966. Hydrology of a karst area in east-central West Virginia. *Water Resources Research*, 2: 549–60

ARACHNIDA

The Arachnida are one of the most important groups inhabiting subterranean environments. Not only are they extremely abundant and widespread, but also include a large number of species, several of which show troglomorphic adaptations. Most arachnids are active predators and are involved in the ecological equilibrium of the subterranean food web.

The class Arachnida consists of 11 extant orders, nine of which include cavernicolous species. The Orders Araneae and Acari are the main groups and each has a separate entry. They have colonized subterranean habitats worldwide and their adaptation to the hypogean world has yielded a large number of troglomorphic species. An entry on Minor Groups includes the Orders Opilionida and Pseudoescorpionida, which are abundant in most subterranean habitats, but include fewer troglomorphic species, and the Orders Scorpionida, Amblypygi, Schizomida, Palpigradida, and Ricinulei, which are scarce in subterranean environments, although no less interesting. The Orders Uropigia and Solifuga are the only groups that do not have cavernicolous representatives.

Several factors have influenced the adaptation of Arachnida to subterranean environments: the particular physical characteristics of this environment, the scarce energy available, and the fact that Arachnida are primarily predators. This adaptation has involved morphological changes as well as changes related to their life cycles. From a morphological perspective, the main adaptive changes are the reduction and loss of ocular structures, a marked or total loss of pigmentation and the lengthening of appendages. When considering life cycles, adaptation to the subterranean environment is reflected by a reduced metabolic rate, which results in the loss of the nocturnal rhythm, a lengthening of their life cycle, and a low reproductive rate.

CARLES RIBECA

ARACHNIDA: ARANEAE (SPIDERS)

Spiders are one of the most common groups inhabiting subterranean environments. They are predators, and are usually abundant inside caves, from the entrance to the deepest galleries. Not all species occurring in caves are exclusive to the subterranean environment: around 25–30% are accidentals (trogloxenes) and appear in or around the entrance zone; around 50% of the species are regularly found in caves but also occur in the epigean environment (troglophiles); and between 20–25% are strictly cavernicolous (troglobitic). These troglobitic species are the most interesting from a taxonomical and evolutionary perspective since they are, in many cases, the living representatives of ancient evolutionary lineages that are now extinct in the epigean environment.

The main morphological adaptations of troglobitic Araneae species are the reduction and loss of eyes, a remarkable or total depigmentation, and the lengthening of appendages. Their adaptation to the subterranean environment also affects their life cycle. Generally, they show a lengthening of their embryonic and post-embryonic development, as well as their adult life. Several interesting studies on this subject have been performed on troglomorphic species belonging to the genera *Phaneta* and *Anthrobia* in the United States (Poulson, 1981) and *Telema, Leptoneta*, and *Troglohyphantes* in Europe (Juberthie, 1985; Deeleman-Reinhold, 1978). In general, when compared to their epigean relatives, cavernicolous species show a low metabolic rate and decreased activity. Their reproductive effort is diminished, the number of eggs per brood is lower, and the eggs are considerably larger. As a result, cavernicolous species have adapted their biological parameters to the environment they inhabit: they regulate their energetic expense according to the amount of energy available in the subterranean environment.

Distribution

Spiders are very abundant in all subterranean habitats worldwide. Nevertheless, the number of species in each cave varies according to the latitude and biogeographical region. Caves in temperate climates rich in organic matter can host up to ten different species. In contrast, caves in subpolar regions are much poorer in energetic resources as well as in diversity. The areas richest in troglomorphic species are those of temperate climates, mainly Mediterranean: the Mediterranean region, central and southern United States, Mexico, Japan, and Korea. In the southern hemisphere, they are abundant in South Africa, Australia, and in several regions in continental Asia, although these are less known. Tropical areas show a high biodiversity but host a low number of troglomorphic species. Possibly, when the large karst areas in Asia, Africa, and South America are explored, the number of troglomorphic species will increase considerably.

Main Troglomorphic Groups and Species

The Order Araneae includes three Suborders: Liphistiomorpha, Mygalomorpha, and Arachnomorpha. Currently, there are 40 000 spider species known, which are grouped into more than

Arachnida: Araneae: Figure 1. *Spermophorides justoi* (Araneae: Pholcidae), a troglobitic spider from caves of El Hierro, Canary Islands. The *Spermophorides genus* has other cave-dwelling species occurring in Tenerife and La Palma. (Photo by Pedro Oromi)

120 different families. Half of the families have troglomorphic species, giving just under 1000 troglomorphic species in total.

The Suborder Liphistiomorpha occurs in Southeast Asia, from Japan to Vietnam, and comprises the spiders that are considered the most primitive: they have a segmented opisthosoma (rear part of body). This suborder does not include a large number of species, and only two are troglomorphic: *Liphistius batuensis*, from Batu Cave, Malaysia, and *Liphistius langkawi* from Thailand.

The Suborder Mygalomorpha is common in tropical areas. Some species colonize warm regions in temperate climates, although it is absent from cold climate zones. Species collected inside caves are generally nocturnal, living under rocks found in the entrance zone of caves. Although they may be abundant, most are trogloxenes, some are troglophilic, and a few are troglomorphic—all of which occur in tropical regions. Examples from the American continent and the Caribbean area include *Speleoctenizia ashmolei* from Ecuador, *Masteria pecki* from Jamaica, three species of *Euagrus* and six species of *Speleopelma* from Mexico, and *Troglotele caeca* from Cuba. Only one species has been described in Africa: *Aporoptychus stercoricola* (Guinea). From Australia and New Zealand there is *Troglodiplura lowryi* and *Hexatella cavernicola* respectively, and *Masteria caeca* from Asia (Philippines).

The Suborder Araneomorpha comprises most of the known spider species and is well represented in the subterranean environment. Many of its families are cosmopolitan, or worldwide in distribution, such as the Pholcidae, which are abundant in most caves of tropical and temperate regions in both hemispheres. Most Pholcidae species are troglophilic, but some are troglomorphic. The most common genera that include troglobitic species are *Pholcus* and *Spermophorides* (see Figure 1) in Europe; *Physocyclus, Modisimus, Anopsicus, Metagonia, Pholcophora*, and *Psilochorus* in Mexico, Jamaica, and Belize; *Priscula* in Venezuela; *Coryssocnemis* in Ecuador; and *Artema* in New Guinea.

The Telemidae are an interesting family since their troglomorphic species are distributed across temperate as well as across tropical regions. This fact suggests that the Telemidae are an ancient family, and some of their representatives can be considered as living fossils. *Telema tenella*, one of the most studied troglomorphic species, occurs in the eastern Pyrenees (France and Spain) and is the only Telemidae found in the European continent. Some other troglobitic species are *T. niponica* from Japan and *T. mayana* from Mexico, *Cangoderces lewisi* and *Apneumonella oculata* from the African continent, and several species belonging to the genus *Usofila* from the United States and New Caledonia.

The Leptonetidae are small spiders inhabiting small rock piles on the floor of caves as well as cracks and rough areas on the walls. Most species are troglomorphic. The most important genera in Europe are *Leptoneta, Teloleptoneta, Paraleptoneta, Protoleptoneta, Sulcia, Barusia*, and *Leptonetella*, which are distributed across the northern side of the Mediterranean basin, from the Iberian Peninsula to Turkey. In Asia, the main genera are *Leptoneta, Saturana, Falcileptoneta*, and *Masirana* (occurring in China, Japan, and Korea). American species belong to the genera *Leptoneta, Archoleptoneta, Neoleptoneta, Appaleptoneta*, and *Callileptoneta* (United States, Mexico, and Brazil)

Species belonging to the family Ochyroceratidae are spread across the southern hemisphere and they occupy the same niche as the Leptonetidae in the northern hemisphere. Ochyroceratidae are abundant among fallen leaves and vegetal litter in forest undergrowth, and are also common in caves, where several troglomorphic species have been described. Most of the troglomorphic species belong to the genera *Ochyrocera, Theotima, Fageicera*, and *Speocera* in South and Central America, the West Indies, and Hawaii; *Speleoderces* from Africa and *Simonicera, Psiloderces, Althepus, Theotima*, and *Speocera* from Asia.

The Dysderidae are a family with a Palaearctic distribution. They comprise several troglobitic species belonging to the genera *Dysdera* (see Figure 2), *Speleoharpactea, Harpactea, Stalita, Folkia, Stalagtia, Minotauria, Rhode*, and *Rhodera*. Troglomorphic species occur on the northern side of the Mediterranean basin, in North Africa, and in the Canary Islands.

The Linyphiidae are one of the most widespread spider families in the epigean environment. A large number of troglomorphic species are found in the northern hemisphere, most of them belonging to the genera *Troglohyphantes, Lepthyphantes, Porrhomma, Centromerus, Caviphantes, Thypholyniphia, Icariella*, and *Troglodytia* (Europe), *Walkenaeria* (Canary Islands), *Phanetta*

Arachnida: Araneae: Figure 2. *Dysdera unguimmanis* (Araneae: Dysderidae) is by far the most troglomorphic among the eight cave-dwelling species of the genus occurring in Canary Island caves. (Photo by Pedro Oromi)

and *Anthrobia*, (United States), *Meioneta* (Hawaii), *Allomengea* and *Jacsonella* (Korea), and *Dunedinia* (Australia).

The Nesticidae are spread across the northern hemisphere, as the Leptonetidae, but have also been recorded in Brazil. The main genera that include troglomorphic species are *Nesticus* (Europe, United States, China, Japan, Korea, and Ceylon), *Canarionesticus* (Canary Islands), *Typhlonesticus* and *Carpatonesticus* (Europe), *Gaucelmus* and *Eidmanella* (United States and Mexico), and *Nesticiella*, which is distributed across Japan, Russia, China, Vietnam, Central Africa, Hawaii, and Fiji.

The Agelenidae are abundant in tropical regions. Several troglomorphic species have been described in temperate areas: they belong to the genera *Tegenaria*, from the Western Mediterranean region, and *Cicurina* and *Blabomma* from Mexico, United States, Japan, and Korea.

The families Clubionidae, Liocranidae, Gnaphosidae, Lycosidae, and Prodidomidae are poorly represented in subterranean environments, and most species are epigean. Nevertheless, some extremely troglomorphic exceptions exist, such as *Agraecina cristiani* from Movile Cave in Romania, *Brachyanillus liocraninus* from Spain, *Berlandia tenebricola* from Tanganyika (Clubionidae); two *Lygromma* species from the Galapagos Islands and *Herpsillus suavis* from Cuba (Gnaphosidae); and *Lycosa howarthi* and *Adelocosa anops* from Hawaii (Lycosidae).

The Theridiidae are a set of species closely related to vegetation and ground fauna. They are common in tropical caves. Most species are troglophilic, but there are some troglomorphic species such as *Achaearanea mundula*, and a few more belonging to the genera *Coscinidia* and *Stemops*, from New Guinea, as well as to *Steatoda* and *Pholcomma* from Australia. Some troglomorphic species from the genus *Theridion* have been described in oceanic islands (Azores and Galapagos).

Hickmania troglodytes, from Tasmania, deserves special attention: it belongs to the Austrochilidae family, considered the most primitive within the Suborder Araneomorpha.

The genera *Meta* and *Metellina* of the family Metidae are troglophilic, inhabiting caves in temperate regions worldwide. Several species belonging to *Loxosceles* (Loxoscelidae) and the families Theridiosomatidae, Ctenidae, Oonopidae, Tetrablemidae, and Anapidae, are also troglomorphic, inhabiting tropical caves.

CARLES RIBERA

Works Cited

Juberthie, C. 1985. Cycle vital de *Telema tenella* dans la grotte-laboratoire de Moulis et strategies de reproduction chez les araignées cavernicoles. *Mémoires de Biospéologie*, 12: 77–89

Poulson, T.L. 1981. Variation in life history of Liniphiid cave spiders. *Proceedings of the 8th International Speleological Congress*, edited by B.F. Beck, Huntsville, Alabama: National Speleological Society

Deeleman-Reinhold, C.L. 1978. Revision of the cave-dwelling and related spiders of the genus *Troglohyphantes* (Linyphiidae) with special reference to the Yugoslav species. *Slovenska akademija znanosti in umetnosti, Classis IV: Historia Naturalis*, 23(6): 1–221

Further Reading

Ribera, C. & Juberthie, C. 1994. Araneae. In *Encyclopaedia Biospeologica*, vol. 1, edited by C. Juberthie & V. Decu, Moulis and Bucharest: Société de Biospéologie

ARACHNIDA: ACARI (MITES AND TICKS)

Acari (mites and ticks) and spiders are the most diverse and abundant groups of Arachnida in the subterranean environment. Over 45 000 Acari species are recorded, and more than 1000 have been reported from caves. Most Acari found in caves are troglophilic. Considering their environment and their lifestyle, Acari occurring in subterranean habitats can be divided into three groups.

Acari Terrestria

This group includes those Acari that are totally or optionally free living and they are the most common and abundant group in subterranean environments and occur in most caves. Several species live in close relation to guano deposits, common in many caves. Bat guano may support densities of between tens of thousands and millions of Acari individuals per square metre. These are generally guanobitic, although some species are bat parasites during some periods of their lives, tending to be ecologically and evolutionarily related to the caves they inhabit. Some authors consider them as recent cavernicolous species (eucavernicolous). Another set of species comprises the typical ground species associated with other organic detritus. They are generally predators that consume Collembola and other microarthropods, eggs, and insect larvae, but can also feed on fungi or may be detritophagous or necrophagous. This second group includes many species for which only a few individuals are known and therefore neither their biology nor their ecological requirements have been reported. These forms are considered troglophilic or troglobitic depending on their external morphological characteristics. A lengthening of appendages and a lack of pigmentation are the most conspicuous adaptations to the subterranean environment.

Among the Acari terrestria the Notostigmata (Opilioacarida) are considered the most primitive group, showing abdominal segmentation. They are cosmopolitan and comprise 20 species, four of which have only been reported from caves (*Opilioacarus orghidani* and *O. vanderhammeni* from Cuba and *Siamacarus dalgeri* and *S. withi* from Thailand).

The Mesostigmata (Gamasida) are common and abundant in most biotopes. All are eyeless and the most common groups in caves are predators, but some are saprophagous or guanophagous. Some species are parasites of small rodents and insectivores. Most species can be considered as troglophilic, but there are several troglomorphic taxa, such as *Eugamasus cavernicola* (Parasitidae), a European neotroglobitic species, and 31 of the 50 species belonging to the family Uropodidae that have been reported from caves.

The Prostigmata (Actinedida) include seven different families. Most species are troglophilic, but there are several troglobitic taxa such as *Bonzia brownei* from England and *Ischnothrombium diploctenum* from Cuba. *Proterorhagia oztotloica* (Proterorhagiidae), from Mexican caves, shows an extraordinary adaptation to the subterranean environment. It has lost all pigmentation, its lateral eyes are absent, and its legs and chelicerae are lengthened. It displays highly developed sensorial setae.

The family Rhagidiidae includes 22 troglomorphic species belonging to the genera *Rhagidia* and *Poecilophysis*; they are cosmopolitan and predators. The North American genus *Foveacheles* includes several troglobitic species, as does the genus Rubostocheles. *Flabellorhagidia* and *Troglocheles* also comprise troglomorphic species, and *Troglocheles vandeli*, from France, is one of the most highly evolved troglomorphic species.

The Astigmata (Acaridida) also include some cave species. They tend to be saprophagous and/or fungivorous, or associated with bat guano. Most species can be considered troglophilic, although some are troglobitic. The Cryptostigmata (Oribatida) is the most common group recorded in the subterranean environment.

Acari Parasiti

Parasitic Acari that live in association with bats and other little mammals are common in caves; their evolution is related to that of their hosts. The Acari parasiti can be subdivided into three groups: (i) temporary parasitic species that are only known from caves and are adapted to the subterranean environment; (ii) temporary or periodically parasitic species that do not display any adaptation to the subterranean environment; and (iii) permanent parasites evolutionarily related to their hosts but not to the subterranean environment.

The first group interests us the most and includes ticks, Trombiculids, and some guanobites. They spend most of their adult life in cracks on the walls or in guano and they look for a host only for feeding. Generally, their parasitism is not extremely specific and they parasitize almost any bat species. Most species are well adapted to parasitic life but they do not display adaptations to the subterranean environment. Nevertheless, some species show an extraordinary morphological and physiological adaptation to the subterranean environment. Some Ixodid species, which are parasites of Chiroptera, show a noticeable lengthening of their appendages and develop at a relatively low temperature (15°C). About 30 cavernicolous species are known within the family Argasidae (Metastigmata). The families Spelaeorhychidae (specific Chiroptera parasites), Trombiculidae, and Wenhoekiidae (Prostigmata) include 210 species, most of them troglophilic.

The second group, comprising temporary or occasional parasitic species that do not display adaptations to the subterranean environment, parasitize rodents and insectivores (Family Laelapidae) or they may be specific Chiroptera parasites. They are exclusively haematophagous. Almost 100 species have been reported, and all of them are considered as troglophilic.

The third group's occurrence in caves is accidental and they are considered trogloxenes. They parasitize hairs, exocuticles, endocuticles, the cornea, nose, mouth, stomach, and intestines. Several authors think their coexistence is very ancient.

Hydracarina

Subterranean Hydracarina (or water mites) are primarily interstitial, although some may also inhabit the water table (Limnohalacaridae and some hydrachnellas). Most species found in the subterranean interstitial environment come into caves from epigean watercourses. The true subterranean Hydracarina (stygobites) originate from epigean watercourses but have adapted to the subterranean environment (hyporheic interstitial environment). They are closely associated with the grains of sand that are more or less in contact with the water flow. All subterranean Hydracarina are predators, feeding on insect larvae (mainly Chironomidae), copepods, ostracoda, and other aquatic micro-invertebrates.

Adaptive characteristics shown include absence of visual structures (cornea and retina); a marked diminishment of pigmentation, although in some species the ocular pigment remains; and reduced size but a body elongation—some highly evolved species show a worm-like aspect that helps them move between the grains of sand. Unlike terrestrial cavernicolous species, subterranean Hydracarina show a shortening of the first three pairs of legs; the first pair of legs tend to be used as feelers. The fourth pair, far more robust, are used for moving. The stygobites have also reduced their reproductive effort: they produce fewer eggs per brood.

The Hydracarina (Suborder Actinedida) comprises ten superfamilies. The superfamily Hydrachnellae is the most abundant in the subterranean environment (several hundred species). The superfamily Halacaroidea includes 20 stygobitic species. The superfamily Stygotrombioidea includes 19 species, all of which are stygobitic. Several other superfamilies have subterranean representatives including Trombidioidea, Eylaioidea, and Hydryphantoidea.

CARLES RIBERA

Further Reading

Dusbábek, F. 1999. Acari Parasiti. In *Encyclopaedia Biospeologica*, vol. 2, Moulis and Bucharest: Société de Biospéologie

Palacios Vargas, J.G. *et al.* 1999. Acari Terrestria. In *Encyclopaedia Biospeologica*, vol. 2

Schwarz, A.E. *et al.* 1999. Hydracarina. In *Encyclopaedia Biospeologica*, vol. 2

ARACHNIDA: MINOR GROUPS

The following section provides a brief overview of Arachnida that occur relatively infrequently or are poorly represented in subterranean environments.

Scorpionida

The Order Scorpionida includes about 1300 species grouped into nine families, five of which (Chactidae, Vaejovidae, Diplocemtridae, Chaerilidae, and Ischnuridae) have troglomorphic representatives. The first records of cave-dwelling scorpions were from Sumatra (*Chaerilus cavernicola*) and Malacca (*Chaerilus agilis*). Three species have been recorded in Europe, although none of these can be considered as troglobitic. *Belisarius xambeui*, endemic to the eastern Pyrenees (France and Spain), shows a slight loss of pigmentation as well as a lack of its central eyes and a marked reduction of laterals, but it inhabits both hyogean and epigean habitats. The highest number of troglomorphic species is found in the Mexican karst, where ten species have been recorded. In addition, troglobitic species have been recorded from Ecuador, Sarawak, and Australia (the recently described Ischnuridae: *Liocheles polisorum*).

Pseudoscorpionida

The Order Pseudoscorpionida is composed of about 3000 species distributed worldwide, with the exception of polar areas. Pseudoscorpions are abundant in most subterranean habitats, and a large number of troglophilic species have been described. Troglomorphic species are not numerous, but they cannot be considered as scarce. The most common families in the subterranean environment are Neobisiidae, Chthoniidae (see Figure), Bochicidae, Vachoniidae, and Syarinidae. A spectacular lengthening of appendages, the absence of eyes, and a complete loss of pigmentation are common adaptations to the subterranean environment. In some cases, due to the fact that life cycles are unknown, it is difficult to distinguish troglophilic from troglobitic species.

Troglomorphic species are abundant in the Mediterranean climate zone of the northern hemisphere (Mediterranean basin, United States, Mexico, Japan, and Korea). There are also records from South American caves, the West Indies, Australia, and Tasmania, although they seem to be less abundant. The presence or absence of cavernicolous pseudoscorpionida remains unknown in many regions of the world, and therefore the current distribution map will probably be modified in the future.

Opilionida

Harvestmen (Order Opilionida) are common and abundant in many subterranean habitats. About 5000 species are currently known, and 125 of these can be considered as cave dwellers, including 80 troglobitic species. The Order Opilionida is divided into three Suborders: Cyphophthalmi, Laniatores, and Palpatores.

The Cyphophthalmi are small and have similarities with Acari, possessing short legs and hard integuments. Just over 100 species have been described, of which 12 are troglobitic. The main family with cavernicolous representatives is the Sironidae. They inhabit temperate areas in the Northern hemisphere and include 22 species, all of them eyeless, of which ten troglobitic species occur in Europe. The Pettalidae include 41 eyeless species of Gondwanic distribution, with only one troglomorphic species, *Speleosiro argasiformis* from South Africa. Other families with troglomorphic representatives are Troglosironidae (*Troglosiro aelleni*) from New Caledonia and *Neogovea mexasca* (Neogoveidae) from Mexico.

The Suborder Laniatores is the most abundant, occurring in most regions of the globe, but appears to be more abundant in tropical zones. Its species are humicolous and lapidicolous, with large, spinous, and prehensile pedipalpi. The main groups including troglobitic species are the Phalangodinae (50 troglobitic species), which inhabit temperate and tropical areas in America, Europe, and Asia, and the Assamiidae, distributed across tropical regions of Africa and Southeast Asia. The Triaenonychidae, which includes 20 troglobitic species, have an austral distribution (Australia, New Zealand, Africa, and South America) although they also have a representative in the United States. The Trabunidae, which occur in Europe, United States, and Japan, include some troglomorphic species that are relicts from the ancient fauna that colonized the European continent. Finally, the Gonyleptidae and the Agoristeridae, with a Neotropical distribution, are abundant in South American caves, but they have very few troglomorphic representatives.

The Superfamily Palpatores is cosmopolitan, although it is notably more abundant in the Palaearctic region. The most important groups belong to the families Nemastomatidae, which has nine troglomorphic species in Europe plus two in the United States, and Ischyropsalidae, which has more than ten troglobitic species in Europe.

Finally, the Suborder Phalangodidea is constituted by the typical long-legged harvestmen, with a cosmopolitan distribution. Within this group, there are no troglomorphic representatives, although they are common and abundant in the threshold zone of caves.

Amblypygi

Amblypygi (whip spiders) are largely nocturnal and colonize caves readily. They are distributed across the tropical regions of America, Africa, and Asia. Their main features are a flat body and large, powerful, and spinous pedipalpi. The first pair of legs (antenniform legs) are very long, with sub-segmentation of both the tarsi and the tibia, which are used as feelers. Most genera include epigean and cavernicolous species, some inhabiting both hypogean and epigean environments. Some genera, such as *Stygophrynus*, appear to be restricted to caves. About 40 species, among over 100 Amblypygi species known, have been reported from caves. Nevertheless, only a few can be considered as troglobitic.

There are no real differences between epigean and cave species with the exception of a few cave species that have reduced eyes. In some cavernicolous species, the pedipalpi may be extremely elongated. In others, medial eyes are reduced in size (*Paraphrynus baeops*) or even absent (*P. velame*). Many species of *Paraphrynus* inhabiting Mexican caves show reduced medial eyes. A few *Charinus* species also have reduced eyes. *C. (Speleophrynus) tronchonii* has no median eyes and the lateral eyes partially reduced, and *C. (Speleophrynus) bordoni* is completely blind, both taxa being from Venezuela. A similar eye reduction has occurred in the genus *Tricharinus* from Surinam and Jamaica. The cave-dwelling African *Paracharon caecus* lacks eyes, but also inhabits termite nests. There

Arachnida: Minor Groups. *Tyrannochthonius superstes* (Pseudoscorpiones: Chthoniidae), can be found in the lava tubes of Tenerife, together with other pseudoscorpions well adapated to cave life. (Photo by Pedro Oromí)

are also troglobitic species (*Charinides cubensis* and *C. wanlessi*) in Cuba. Other troglobitic species are known from Venezuela, Guinea, Zanzibar, Tanzania, and Bardera in Somalia.

Schizomida

The Order Schizomida includes more than 100 species with lucifugous (light-avoiding) and hygrophilic behaviour that are spread across the intertropical region. They are small, between 2 and 8 mm long. Their dorsal scutum is divided into three parts: pro-, meso- and metapeltidium. The opistosome has a long, multiarticulated flagellum that varies between species. They show a marked sexual dimorphism. The first pair of legs is long and thin, and used as feelers. Most Schizomida are endogean (inhabiting the soil zone immediately below the surface) and it is very difficult to tell troglobitic species from endogean ones.

A total of 32 species have been reported from caves, of which 25 can be considered as troglobitic. Most cavernicolous species belong to the genus *Schizomus* and troglomorphic species have been described from Mexico (Chiapas, Tamaulipas, Veracruz, and Yucatan), Guatemala, California, Cuba, Jamaica, Puerto Rico, Ecuador, Venezuela, Tanzania, Zanzibar, Rodriguez Island, and Angola. *Schizomus* is the most diverse and widespread genus. The genus *Agastoschizomus* also includes troglobitic species: *A. lucifer*, from Sierra de El Abra, and *A. huitzomolotitlensis* from El Sotano de Huitzomolotitlam, San Luis de Potosí, both from Mexico. Cave species have also been reported from Africa and Asia: *Trithyreus parvus* from Gabon and *T. pileti* from Batu Caves. Only one species is known from India, and two from Japan.

Ricinulei

The Order Ricinulei includes only about 30 species that are distributed in tropical and subtropical areas in the American continent and Western Africa. They are small (10 mm), flat, slow, eyeless, and cryptic (coloured to camouflage) and have a preference for humid environments. The only characteristic identifying troglobitic species is a slight lengthening of their appendages, which makes it very difficult to assign them to a clear ecological category.

Known species are grouped into three genera: *Ricinoides* (seven species from Western Africa), *Cryptocellus* (21 species from tropical and subtropical America), and *Heteroricinoides*, (two species from Venezuela and Colombia). There are ten cave-dwelling species, of which nine belong to the genus *Cryptocellus* (eight from Mexico and one from Cuba); the tenth cave-dwelling species is *Heteroricinoides bordoni* from Venezuela. This group has not been reported from African caves.

Palpigradida

The Order Palpigradida comprises the smallest and most difficult to find Arachnida: they rarely reach 2 mm in length. They are agile and extremely fragile. Their most conspicuous characteristic is a long flagellum used as a feeler. Their pedipalpi are ambulatory and the first pair of legs is antenniform. All palpigradida are blind and totally depigmented. They are thought to be of intertropical origin.

The Palpigradida includes about 80 species, 27 of which have been found only in caves (21 from Europe, one from Cuba and five from tropical Asia). These 27 species can be considered troglobitic: since all palpigradida are blind and lack pigmentation, it is difficult to distinguish between endogean and cavernicolous species. Generally, cavernicolous species tend to be bigger than endogean species, and they also show notably longer appendages (particularly the chelicera).

From a taxonomical perspective, the Palpigradida are divided into two families: the Protokoeneniidae, with two genera (*Protokoenenia* and *Triadokoenenia*), and the Eukoeneniidae, with four genera (*Eukoenenia, Allokoenenia, Koenenioides*, and *Leptokoenenia*). With the exception of Triadokoenenia and Leptokoenenia), the remaining genera are found in caves. The largest and most modified species belong to the genus *Eukoenenia*, which comprises the species that are most adapted to subterranean environments.

Carles Ribera

Further Reading

Condi, B., 1999. Palpigradida. In *Encyclopaedia Biospeologica*, vol. 2, edited by C. Juberthie & V. Decu, Moulis and Bucharest: Société de Biospéologie

Georgescu, M. 1994. Schizomida. In *Encyclopaedia Biospeologica*, vol. 1, edited by C. Juberthie & V. Decu, Moulis and Bucharest: Société de Biospéologie

Heurtault, J. 1994. Pseudoscorpions. In *Encyclopaedia Biospeologica*, vol. 1

Juberthie, C. 1994. Ricinulei. In *Encyclopaedia Biospeologica*, vol. 1

Lourengo, W. 1994. Scorpiones. In *Encyclopaedia Biospeologica*, vol. 1

Rambla, M. & Juberthie, C. 1994. Opiliones. In *Encyclopaedia Biospeologica*, vol. 1

Weygold, P. 1994. Amblypygi. In *Encyclopaedia Biospeologica*, vol. 1

ARCHAEOLOGISTS

An archaeologist is defined as one who makes a "study of human antiquities, especially of the prehistoric period and usually by excavation" (*Concise Oxford Dictionary*). This entry reviews the life and work of eight of those who had the greatest influence on understanding the development of the human race before historical records began, using evidence from caves. No aspect could be more crucial than the antiquity of humankind. Until the 1850s, it was normal to believe that humans were a recent addition to the world. There was too little time since the Creation, then thought to have been some 6000 years ago, for humans to have existed at the same time as extinct animals, those destroyed in the biblical Flood. Georges Cuvier, with all his authority, had in 1812 expressly denied the existence of such early humans. So, included here is a succession of people who gradually came to realize that deposits in caves contradicted this belief. None of these researchers were what would now be called archaeologists. They were either geologists, priests, or doctors, taking an interest in this branch of natural history. Nevertheless, the work was archaeological and some of the most important ever to have taken place. Three more conventional archaeologists are also described.

Johann Friedrich Esper (1732–81)

Esper (Figure 1) was the parish priest of Uttenreuth, near Erlangen in Germany. Excavating during 1771 in the Bavarian cave of Burggaillenreuth, he found human bones together with those of cave bear and other extinct animals (Esper, 1774, p.26). He considered very carefully whether or not they were contemporary with each other, but concluded that he had insufficient evidence for such a momentous interpretation.

> . . . I do not, however, suppose without adequate reason that these human remains are of the same age as the other animal petrifactions. They must have come together with the others by chance.

This extract is translated from the original German version of Esper's book; in the French version published in the same year, the sense was reversed by mistake to say that the human and animal remains were of the same age.

John McEnery (1796–1841)

McEnery was the first person to convince himself that early humans were contemporary with extinct animals. His excavations in Kent's Cavern (Devon, United Kingdom) took place between 1825 and 1829, but none of his findings was printed until many years later and even now some of them remain unpublished. What he did and found, however, is known, as are his conclusions and also the opinions of Buckland with whom he was in touch.

McEnery was born in Ireland. Although his name is often spelled as MacEnery, he himself wrote it as McEnery. He was ordained as a priest in 1819 and in 1822 became chaplain to the family living at Torre Abbey in Torquay, where he remained until his early death. The existence of bones in the nearby Kent's Cavern was already known but they had not been seriously studied before McEnery (1825–28?; 1869). He left lengthy notes

Archaeologists: Figure 1. J. F. Esper.

Archaeologists: Figure 2. William Buckland.

Archaeologists: Figure 3. P.C. Schmerling.

Archaeologists: Figure 4. William Pengelly.

Archaeologists: Figure 5. William Boyd Dawkins.

Archaeologists: Figure 6. The Abbé Breuil in 1919 or before.

on his work, which are in the library of the Torquay Natural History Society. Most of them were published in 1869, showing that he had found human bones and teeth, flint tools, pottery, and charcoal, as well as bones and teeth of *Homotherium latidens* (sabre-toothed tiger), rhinoceros, hyaena, and bear.

What is not immediately obvious is McEnery's own opinion. Did he or did he not believe that the human remains were of the same age as the animal bones found with them? What he wrote in his manuscript notes, intended as a basis for publication, conflicts with the opinion he expressed in private letters to Buckland.

It must be remembered that this young (30 years old in 1826) and inexperienced excavator was exposed to two of the world's authorities of the day, both of whom were opposed to the idea. Cuvier, already mentioned, identified some of his specimens and McEnery visited him in Paris. Buckland, with whom McEnery was in close contact, was deliberately cautious, believing that the apparent association at the same level in the deposits could be explained by supposing that humans had dug pits in which their bones and artefacts were found. These two must have been powerful influences on the young man, whose theological training may also have had an effect. This is what he wrote in his manuscript (McEnery, 1869, p.226):

> I am persuaded that if due attention is paid to the place in which these remains occur—and the manner that they are intermingled with the soil and bones reputed fossil, it will be in every case found ... that they are not coeval neither with one or the other but that they had been added subsequently to the deposition of the former and commingled with them into a common heap by causes such as operated here I mean the visits of man, or according to their position by the disturbing action of running waters.
>
> Had I not devoted so long a period to personal examination of all the circumstances attending this delicate question, in common with others I should have fallen into the error of supposing human remains to be contemporaneous because conjoined with the deposit of mud and bones.

However, what were presumably his true views, forthright and unconstrained by any intention of publication, are contained in two letters written to Buckland in 1828 or 1829 and never previously published (McEnery, 1825–28?):

> Nothing could give me greater pleasure than to concur with you in this and in every other point, but it would not be just to truth to give up the impressions of three years observation without the fullest conviction to the contrary.
>
> Hitherto I had rather bowed to your high authority than to evidence. But the matter is now clear as day.

William Buckland (1784–1856)

Buckland (Figure 2) was one of the most influential geologists of the 19th century, but only a small part of his work related to humankind. At Oxford, his lectures were so highly regarded that a new chair of geology was created for him. His sense of fun drew many listeners and it also prompted humorous drawings relating to his cave interests. Having seen the Burggaillenreuth cave where Esper had worked, he was keen to investigate the newly found cave of Kirkdale in Yorkshire, but no human remains were involved. The fact that the cave contained bones of animals too large to get through the entrance and that they were nearly all broken and chewed, led him to conclude that hyaenas, the other occupants, had taken them there as food. This unconventional idea (Buckland, 1822) earned him the Royal Society's Copley Gold Medal. Buckland's only involvement with human remains in caves was in Torquay, as already described. He himself had expected that humankind would be found coeval with extinct animals but he did not accept that the Kent's Cavern evidence was convincing enough for such a contentious conclusion.

Philippe Charles Schmerling (1791–1836)

Although McEnery had been excavating in the 1820s, it was Schmerling who, in 1833, first published a definite statement that the human remains he had found were contemporary with those of extinct animals. Schmerling (Figure 3) trained as a doctor in Belgium and in 1825 began to practise in Liege. Only four years later he gave up medicine as a result of a chance meeting with some fossil animal bones that some of his patients' children were playing with. They came from a cave at Chokier, opened by quarrying and not previously disturbed. Stimulated by these, he spent the next four years exploring more than 40 similar caves. Besides 60 species of animals, he found fossil human bones, together with chipped stone tools and carved bones. They lay beneath flowstone floors, scattered and abraded just like the animal bones that were with them, so they could not have been buried later into an earlier deposit, as Buckland had supposed at Kent's Cavern. At Engis, a child's skull, known now to be a Neanderthal, was next to a mammoth tooth, and a Cro-Magnon skull was with rhinoceros teeth.

Schmerling published his findings in great detail and concluded (1833, p.66): "There can be no doubt that the human bones were buried at the same time and by the same cause as the other extinct species." Charles Lyell visited Schmerling and referred to his discoveries but it was only later that he accepted their significance. At first he had insisted that coexistence could only be proven if the remains were found together in stratified deposits, but he afterwards acknowledged that Schmerling's specimens had been found "under circumstances far more difficult to get over than any I have previously heard of" (Lyell, 1881, 1: 401–02).

William Pengelly (1812–94)

Pengelly's great contribution to archaeology was his careful and precisely recorded excavation, carried out in a Brixham cave in Devon in 1858–59, which was sufficient to convince the doubters that humankind and the extinct animals there really were contemporary. The lesson was reinforced by the much more extensive excavations he directed in Kent's Cavern from 1864 to 1879. There is some analogy here with Darwin's achievement, in making evolution generally accepted by means of massive evidence. Darwin, however, had to argue his case; with Pengelly the evidence largely did it for him.

Pengelly (Figure 4) had only an elementary school education and was largely self-taught, yet his researches brought him Fellowship of the Royal Society. He started a school in Torquay and later took private pupils, who included members of European royal families. He studied geology and also did an immense amount of detailed work on such diverse subjects as dialect and rainfall, and he reprinted all he could find written about Devon caves, amounting to many hundred pages. His publication of the McEnery (1869) manuscript is the most accurate available.

When the Brixham cave was found in 1858, it had been sealed since prehistoric times and its flowstone floor, beneath which the deposits were excavated, was undisturbed. It was this, coupled with his system of excavation, close supervision and accurate recording, which made his findings accepted. A grid system fixed the location of every object and the dig followed the stratification layers (Prestwich, 1874). The Kent's Cavern excavation was less important in that it came second, and previous disturbance of the deposits made it more complex, but the facts and their publication year by year to the wide audience of the British Association for the Advancement of Science members, helped the antiquity of humankind to become generally recognized. McEnery's conclusions of 40 to 50 years before were vindicated and Lyell, who had been sceptical of Schmerling's opinion, was converted.

William Boyd Dawkins (1837–1929)

At last here is someone who was known as an archaeologist who worked in caves, in contrast to the geologists and others, whose crucial evidence for the antiquity of humankind had been archaeological. Boyd Dawkins (Figure 5) was not associated with the same vital breakthrough but he was, on the other hand, better known than many of his predecessors. His work, his writing and his long career, meant that he was regarded as *the* cave archaeologist for much of the 19th and 20th centuries, though more of his work was on Pleistocene animals rather than on humans.

Boyd Dawkins became interested in geology while at Oxford and in 1861 was appointed to the Geological Survey of Great Britain. In 1866, he was elected FRS and from 1874 to 1909 he was Professor of Geology and Anthropology in Owen's College, Manchester, the predecessor of Manchester University. Inspired by Pengelly's results at Brixham, he excavated in the Wookey Hole Hyaena Den, finding Mousterian tools together with extinct animals (Dawkins, 1862–63). His classic book *Cave Hunting* (1874), brought together facts on the physical aspects of caves, as well as their contents in many European countries. He himself carried out many cave excavations until 1903, when he was 66 years old, but still he was more concerned with mammal remains than human.

Henri Édouard Prosper Breuil (1877–1961)

Finally we come to a man who was widely acclaimed throughout his long life, for work in a special field of cave archaeology. It was he who founded, early in the 20th century, the new study of prehistoric cave art (see separate entry, Art in Caves: History). Ordained priest in 1900, the Abbé Breuil (Figure 6) was lecturer in Prehistory and Ethnography in the University of Fribourg 1905–10, Professor of Prehistoric Ethnography at the Institut de Paléontologie Humaine in Paris from 1910, and Professor of Prehistory at the Collège de France 1929–47. He was elected a member of the Institute de France in 1938, received gold medals from the Society of Antiquaries in London and the National Academy of Sciences in Washington, and held honorary degrees from several foreign universities.

While still a young man, he often accompanied Émile Cartailhac (1845–1921), the doyen of prehistory, who had previously denied the antiquity of cave art. Breuil was at the cave of La Mouthe in 1902, when Cartailhec accepted it with his famous "Mea culpa d'un sceptique". He had seen the painted caves of Les Combarelles and Font-de-Gaume, when they were discovered in 1901, and he went to Altamira in 1902. He was also involved in the discovery of Tuc d'Audoubert in 1912 and Les Trois-Frères in 1916. His own main contributions were his skilled and laborious copying of the art, its stylistic analysis and interpretation, and its publication in numerous magnificent books. In 1940, it was he who first assessed the Lascaux paintings, but towards the end of his life, he denied the authenticity of those found at Rouffignac in 1956. His bibliography exceeds 793 items (Lantier, 1957).

Pei Wenzhong (1904–1982)

After graduating in geology from Peking University in 1927, Pei had a long and distinguished career as an archaeologist and paleo-anthropologist, including cave excavations of paleolithic remains in many parts of China. Best known and most significant, was his work at the cave of Zhoukoudian (then spelled Choukoutien), 47 km southwest of Beijing. In 1927, a single tooth from there had been recognized by the Canadian, Davidson Black of Beijing, as coming from a previously unknown hominid called Peking Man, *Sinanthropus pekinensis* at that time and now *Homo erectus pekinensis*. Pei joined the excavation there in 1928, and in the very next year he discovered the first complete skull of Peking Man. It was a predecessor of both Neanderthal and modern man, dated to the middle Pleistocene period. It postdates Java Man and is considered to be more advanced than the other early hominids, with a range of cranial capacity overlapping that of modern man. The discovery of the skull was followed, under Pei's direction, by stone tools, charred wood, and animal bones, some of which were burnt. The Abbé Breuil visited the site in 1931 and from 1935 to 1937, Pei studied under him in Paris, where he gained his doctorate.

TREVOR SHAW

See also **Archaeology of Caves: History**

Works Cited

Buckland, W. 1822. Account of an assemblage of fossil teeth and bones of elephant, rhinoceros, hippopotamus, bear, tiger, and hyaena, and sixteen other animals; discovered in a cave at Kirkdale, Yorkshire, in the year 1821. *Philosophical Transactions of the Royal Society*, 122(1): 171–236

Dawkins, W.B. 1862–63. On a hyaena-den at Wookey-Hole, near Wells. *Quarterly Journal of the Geological Society of London*, 18: 115–25 and 19: 260–74

Dawkins, W.B. 1874. *Cave Hunting: Researches on the Evidence of Caves Respecting the Early Inhabitants of Europe*, London: Macmillan

Esper, J.F. 1774. *Ausführliche Nachricht von neuentdeckten Zoolithen unbekannter vierfüsiger Tiere* [Description of Newly Discovered Fossils of Unknown Quadrupeds], Nuremberg: Knorrs

Lantier, R. 1957. *Hommage à M. l'abbé Henri Breuil (pour son quatre-vingtième anniversaire)* [A Tribute to the Abbé Breuil on his 80th Birthday], Paris: Henri-Martin (with a bibliography of his work)

Lyell, K.M. 1881. *Life, Letters and Journals of Sir Charles Lyell*, 2 vols, London: Murray

McEnery, J. 1825–1828? Six letters addressed to William Buckland. Manuscript in the library of the Karst Research Institute, Postojna

McEnery, J. 1869. Origin of cavern researches. In *The Literature of Kent Cavern*, edited by W. Pengelly, part 2: 203–482. *Report and Transactions of the Devonshire Association*, 3(1): 191–482

Prestwich, J. 1874. Report on the exploration of Brixham Cave, conducted ... under the superintendence of Wm. Pengelly ... *Philosophical Transactions of the Royal Society*, 163(2): 471–572

Schmerling, P.C. 1833. *Recherches sur les ossemens fossiles découverts dans les cavernes de la province de Liege*, vol. 1 [Researches on the Fossil Bones Found in the Caves of Liège Province], Liege: Collardin

Further Reading

Alexander, E.M.M. 1964. Father John MacEnery: Scientist or charlatan? *Report and Transactions of the Devonshire Association*, 96: 113–46

Boylam, P.J. 1967. Dean William Buckland, 1784–1856: A pioneer in cave science, *Studies in Speleology*, 1(5): 237–53

Brodrick, A.H. 1963. *The Abbé Breuil, Prehistorian: A Biography*, London: Hutchinson

Heller, F. (editor) 1972. *Die Zoolithenhöhle bei Burggaillenreuth ... 200 Jahre wissenschaftliche Forschung 1771–1971* [The Fossil Animal Cave near Burggaillenreuth: 200 Years of Scientific Research, 1771–1971]. Erlangen: Üniversitätsbund (on Esper and others)

Jackson, J.W. 1967. Sir William Boyd Dawkins (1837–1929): A biographical sketch. *Cave Science*, 5(39) for 1966: 397–412

Pengelly, H. 1897. *A Memoir of William Pengelly of Torquay*, London: Murray (with a list of his published papers)

Shapiro, H.L. 1975. *Peking Man*, New York: Simon and Schuster; London: Allen and Unwin, 1976

Ubaghs, G. 1975. Schmerling, Philippe-Charles. In *Dictionary of Scientific Biography*, edited by C.C. Gillespie, 16 vols, New York: Scribner

ARCHAEOLOGY OF CAVES: HISTORY

Archaeology was born during the first half of the 19th century through the systematic excavation of deposits found in the caves of Europe. Those excavations were undertaken by inquisitive intellectuals, geologists, and antiquarians, who were delving into the history of the world and especially of humankind. They established a past for humanity by providing great time depth and previously undocumented variation over human history. In the New World, archaeology in caves helped resolve questions about the tenure of Native Americans, establishing the complexity of their cultures and contributing to our knowledge about agriculture in the prehistoric Americas. Cave archaeology has thus been a fundamental part of scientific understandings of human origins and development, and dramatically changed our view of our world and ourselves.

The birth of archaeology was from the mouths of caves in Western Europe, wrapped up in one of the most momentous intellectual changes in human history: the establishment of human antiquity. Beginning in the late 18th-century Enlightenment, European scholars began to question the biblical account of divine Creation, which according to Bishop Ussher placed the creation of all things, including people, at 6000 years ago. Excavations inside limestone caves and dolines, however, were showing the presence of extinct animals in what we now know were deposits from the last Ice Age. Frenchman Georges Cuvier argued persuasively in 1795 that extinctions had occurred, resulting from catastrophic floods that killed off some species and produced the stony deposits, "diluvium", in which their bones came to rest. He believed that all such events occurred during the 6000-year biblical span, and that humans had nothing to do with extinct animals. Contradictions to Cuvier's view came from caves, in Britain, France, and Belgium.

In 1822, bones were discovered in Paviland Cave in Wales, bringing eminent geologist William Buckland, advocate of the biblical chronology, to the cave. Buckland confirmed the presence of extinct animals and uncovered a partial human skeleton surrounded by ivory ornaments and buried in red ochre. The "Red Lady of Paviland" (actually a male, see Britain and Ireland: Archaeological Caves) was of great interest, but Buckland dismissed it as intrusive, denying any association between human and extinct animal remains. A modern radiocarbon age indicates the Paviland hominid is more than 18 000 years old. Buckland denied all other associations of humans with antediluvian deposits, because for him, the human species was recent. But in this, he also abandoned equating the biblical Deluge to the diluvium of the geologists. The door was open for events not described in the biblical account.

Excavations in Kent's Cavern (in Torquay, England) by the Rev. John McEnery began in 1825. Over the next four years, McEnery dug beneath a thick travertine floor to find extinct animal bones and stone tools of clear human manufacture. Here was evidence for antediluvian humanity. McEnery recognized this during his excavations but later denied antediluvian status for the Kent's Cavern finds under pressure from Buckland (see Archaeologists entry). In 1842, however, further work at the cave by R.A.C. Austen reaffirmed the association between stone tools and extinct animals, and Austen argued strongly for an antediluvian origin for the deposits and thus for humans in Britain. But with Buckland's great authority, Austen's claims had little effect.

Continental caves became the focus of interest for both human antiquity and cave archaeology. In the mid-1820s Marcel de Serres, Paul Tournal, and Jules de Christol began to develop the science of cave excavation in French sites such as Lunel-Viel, Pondres, Souvignargues, and in caves near Bize. Tournal began to publish on Bize in 1828, when he argued that there was a stratigraphic sequence in the caves including upper levels with human remains, modern animals, and pottery, and lower layers with human remains and extinct animals. Tournal cited

Christol's excavations in the Gard, where antediluvian animals such as rhinoceros were found with artefacts, as evidence for a very long sequence of human occupation beginning long before the biblical chronology. For him, the biblical Deluge was a recent historic event in this sequence and unrelated to diluvial evidence. In short, Tournal defined "prehistoric times", although he did not use the term. De Serres concurred based on his work at Fauzan Cave. The French intellectual establishment, led by Cuvier, tried to reject the southern French cave data, but the list of relevant sites was growing too long to ignore.

In Belgium, physician Philippe-Charles Schmerling had begun to explore caves around Liege, finding extinct animal bones, stone tools, and human remains. At Engis in 1832, two human skulls were discovered with the bones of rhinoceros and elephants; one cranium was badly eroded, but the other had a strange shape and bony prominences at the back and over the eyes. It was a Neanderthal, although the definition of that type would have to wait two decades until the type specimen was found at Felderhof Cave in the Neander Valley of Germany. Nevertheless, Schmerling made strong arguments as to the ancient origins of his collections (see Archaeologists entry).

So compelling were the Belgian data that Buckland and Charles Lyell, the eminent geologist who would ultimately synthesize geological and archaeological proofs for the antiquity of the world, were forced to admit the confusion that complex deposition processes in cave sites brought to their position on human antiquity. They pushed archaeologists outside caves to find data they would accept, leading to a series of open-air excavations around Amiens and Abbeville in northern France. These digs would finally lead to acceptance of antediluvian human prehistory, but the beginnings of the debate were carried out in caves.

Cave archaeology in the New World has less history than Europe. The first Europeans to penetrate North American woodlands in the late 18th century frequently entered deep caves seeking saltpetre (nitrates) for their gunpowder, leaving their names as incised wall graffiti, and they could not have missed evidence for prehistoric occupation littering the cave floors. But the invaders' opinion of Native American capabilities was low, and they did not believe that the materials they saw were of native origin.

Yet, American scholars began to investigate the contents of deep caves. At first, as in contemporary Europe, extinct Pleistocene animal bones were of interest. A giant ground sloth (*Megalonyx* sp.) was discovered in a West Virginia cave in the 1790s, causing great interest among the scientists of the day, including Thomas Jefferson. Finds like this showed that extinct animals had populated the New World just as in the Old. In the first part of the 19th century, the brilliant jurist John Haywood catalogued a number of archaeological cave sites in his *Natural and Aboriginal History of Tennessee* (1825), including caves with human effigies sculpted in stone. For Haywood, the aborigines could neither have produced these nor the mounds and earthworks he observed across Tennessee, so he attempted to prove that these must be evidence for lost Hebrew tribes in North America. His view was typical of early American archaeologists generally, despite contradictions from the likes of Thomas Jefferson, whose careful excavations in a mound at Monticello led him to argue a native origin for the varied archaeological features seen on the American landscape.

By the last half of the 19th century, American archaeologists had come to accept the antiquity of Native Americans and their role in producing the archaeological record. Yet this acceptance did not lead to the recognition of the importance of cave sites. In Mesoamerica, however, archaeologists in the 1880s and 1890s began to notice a relation between large Maya settlements such as Copan, Palenque, and Tikal, and caves located nearby or beneath the towns. Caves were clearly used by Maya religious leaders as ceremonial precincts; paintings, burials, and shrines were discovered, reflecting the Maya belief that the underground world was an important source of power and balance for the outer and upper worlds frequented by humans (see America, Central: Archaeological Caves).

There were a few archaeological incursions into North American dark caves (see America, North: Archaeological Caves). In 1875, F.W. Putnam examined the archaeological record of Kentucky's caves, noting their use for human burials and listing an impressive array of artefacts that he recovered in Salts and Mammoth caves, including torch remnants, moccasins, and containers. Also in 1875, local men found the desiccated body of a young boy (misidentified as a female and dubbed "Little Alice") in Salts Cave, igniting the public's imagination and interest in prehistoric cave use. In 1896, W.S. Blatchley excavated a trench in the dark zone of Wyandotte Cave (Indiana) looking for tools used by prehistoric mineral miners; his work was based on archaeological observations made since 1877 by various scholars in that cave, who noted the removal of gypsum and mirabilite from the walls and hammerstones, torches, and containers associated with the mineral removals. Back in Kentucky, Colonel Bennett Young performed the first systematic excavations in Salts Cave, publishing his results in 1910 and detailing a remarkably preserved record including numerous perishable artefacts. Young showed that prehistoric cavers had traversed many miles of dark zone passage in Salts Cave with their simple caving technologies. In 1916, Nels Nelson trenched the "Historic Entrance" to Mammoth Cave, recording a long sequence of occupations there. Another prehistoric cadaver was discovered in Mammoth Cave in 1935. "Lost John" was an adult male who had died when a boulder he was undercutting, presumably in search of cave minerals, collapsed on him.

The largest and most influential cave archaeology project in North America began in Salts and Mammoth caves in 1963 under the direction of Patty Jo Watson. That project entailed large-scale excavations in the vestibule of Salts Cave and the Historic Entrance of Mammoth Cave, archaeological surveys of the inner passageways of both caves, and specialized scientific analysis of various products of fieldwork including numerous human coprolites found scattered about the caves. These studies documented two seminal findings. First, as early as 3000 years ago, Native Americans undertook extensive and dangerous forays into the deep recesses of large caves seeking minerals that had little or no subsistence utility, i.e., their activities were more likely religious or social than purely economic. Second, analysis of human coprolites showed that these people were undergoing a transition to agriculture from a simple hunting / gathering way of life; surprisingly, they were domesticating local plants like sunflower, chenopods, and marsh elder rather than importing plants from Middle America, at that time seen as the source of American agriculture. These discoveries revolutionized American prehistory and have formed the basis of most subsequent work in the Eastern Woodlands.

New discoveries continue to this day. In 1979, a cave was discovered in Tennessee containing prehistoric art in the dark zone. For Charles Faulkner who studied it, the art comprises religious symbols traced into soft mud, giving the cave its name, Mud Glyph Cave. Over the past two decades, nearly 50 such caves have been found in southeastern North America, comprising a prehistoric Native American cave art tradition spanning 4000 years (see Art: Cave Art in the Americas). Caves clearly played a central role in the economic and ceremonial lives of ancient Americans.

Finally, mention must be made of caves in Africa, which, while excavated more recently than elsewhere, have yielded information on the very origins of humanity (see Africa, South: Archaeological Caves). In 1924, Raymond Dart discovered the remains of a young animal that was obviously similar to humans; he named it *Australopithecus africanus* and argued that it was the earliest known human antecedent, well over 1 million years old. Over the next decades, many similar fossils were discovered in South African caves and dolines, such as Sterkfontein, Swartkrans, and Makapansgat. *Australopithecus* was shown to have great time depth (more than 3 million years) and to comprise several different species. African caves like Klasies River Mouth have also yielded evidence for the origins of modern humans. At Klasies, a sea cave on the southern Cape, and at inland sites like Border Cave, human fossils show that anatomically modern humans evolved in Africa more than 200 000 years ago and probably spread from there to colonize much of the Old World, replacing or blending with existing human populations like the Neanderthals in Europe.

Caves around the world have been central in the history of archaeology. Deposits in caves have helped to establish human antiquity, yielded data on the very origins of humanity, and provided information about evolving human cultures across the globe. New information continues to come to light as archaeologists and cavers join forces to examine the deeper recesses of the Earth for evidence of ancient human activities.

JAN F. SIMEK

Further Reading

Bassie-Sweet, K. 1996. *At the Edge of the World: Caves and Late Classic Maya World View*, Norman, Oklahoma: University of Oklahoma Press

Daniel, G. 1963. *The Idea of Prehistory*, Cleveland, Ohio: World Publishing

Faulkner, C. (editor) 1986. *The Prehistoric Native American Art of Mud Glyph Cave*, Knoxville, Tennessee: University of Tennessee Press

Grayson, D. 1983. *The Establishment of Human Antiquity*, London and New York: Academic Press

Sherwood, S. & Simek, J. (editors) 2001. *Cave Archaeology in the Eastern Woodlands*, special issue of *Midcontinental Journal of Archaeology*, 26(2)

Watson, P.J. 1969. *The Prehistory of Salts Cave, Kentucky*, Springfield, Illinois: Illinois State Museum

Watson, P.J. (editor) 1974. *Archaeology of the Mammoth Cave Area*, New York: Academic Press

ARDÈCHE CAVES, FRANCE: ARCHAEOLOGY

The Ardèche, a region of southeast France around the Rhône valley, contains numerous caves in its limestone massifs, and many of them were decorated during the last Ice Age, often during the Solutrean Period, around 20 000 years ago. They continue to be discovered today, usually by speleologists and enthusiasts rather than archaeologists, and a whole series was found by Jean-Marie Chauvet and his companions, culminating in 1994 with the biggest and grandest of all, the cave which now rightly bears his name.

The Ardèche region was among the first in the world where Ice Age cave art was discovered—but unfortunately its age and its significance were not realized for a long time. In 1870, local archaeologist Jules Ollier de Marichard wrote in his excavation notebook that he had seen "signs of the zodiac" engraved in a deep gallery of the cave of Ebbou, and in 1873 he added that he had seen animal silhouettes sketched on the walls. But Ebbou had to wait until 1946 before its art was rediscovered. In 1879, de Marichard wrote a letter to the eminent French prehistorian Emile Cartailhac, stating that some caves in Ardèche had red paintings of "fantastic animals", but he put nothing in print. In 1878, however, a local schoolteacher, Léopold Chiron, did print something: he had noticed the deep engravings in the cave of Chabot (Gard), and not only took photographs and imprints, but also published a note about them, although he could not know their date. He mistakenly thought he could see birds and people among the lines, because the Chabot engravings are difficult to decipher, and its figures, including mammoths, are far from clear. In May 1879, Chiron wrote to the prehistorian Gabriel de Mortillet to tell him of the discovery of a cave with Ice Age flint tools and with engravings on the walls—Chiron had no doubt the drawings were ancient, since they were covered in calcite. But his pioneering observation was ignored. He again called attention to the Chabot engravings at conferences in 1889 and 1893, but still found no support, while in 1890 he also mentioned engravings in the nearby Cave of Le Figuier (Ardèche).

Some of these sites—such as Chabot—proved to contain occupation deposits from only the Solutrean Period. But one site in the region—La Tête-du-Lion—became one of the first well-dated examples of cave art. Its decorated panel features red-ochre drawings of an aurochs, deer, and ibex, as well as dots. Immediately below it are the remains of a fireplace, clearly used for lighting by the artist at work, with spots of red ochre next to it on the ground. Analysis of these spots showed that they were of exactly the same composition as the drawings on the wall, and so a radiocarbon date of 19 700 BC from charcoal in the fireplace provided a solid age for this cave's decoration in the Early Solutrean Period, like much of the other art in this region.

At the end of 1994, Jean-Marie Chauvet, Eliette Brunel Deschamps, and Christian Hillaire discovered the Chauvet Cave, easily the biggest cavern of the entire region, and far bigger than the other decorated cavities known in the Ardèche. It has proved to contain over 420 painted and engraved figures. It also has numerous and varied traces of human and animal visitors—including footprints, and abundant bones of cave bear. One cave bear skull has been purposely placed on a natural, isolated rock, and has inevitably led to speculations about a cave bear cult, although it could easily have been placed there by a child. Bear clawmarks also occur underneath, or in other places, on top of some of the engraved figures on the walls.

The painted figures include some hand prints, as well as panels of large red dots made with paint on the palm of the hand, and a wide range of "signs". Red and black figures tend to be located in different parts of the cave. A number of different animal species—at least 15—are depicted, including not only the usual range (40 horses, 31 bison, 10 aurochs, 20 ibex, 25 deer), but also many rarer species: 66 mammoths, 15 bears, 72 big cats, and 65 rhinos. Many of the rhino figures show a puzzling broad black stripe across the middle of the body. There are also a spotted panther, an owl, a possible human female, and other figures. Overall, mammoths, rhinos, bears, and big cats constitute up to two-thirds of the cave's animal figures.

Some of the animal figures feature a great deal of infill and very subtle shading, where the paint was spread by hand or with a tool to depict relief or the nuances of the animal's coat. Some of the black drawings in the cave are surrounded by lighter areas of wall which seem to have been scraped to produce increased contrast.

Initial assessments of the content and style of Chauvet's art assigned it to around 20 000 years ago, so it was a shock when radiocarbon dates from charcoal in several black figures came out at more than 30 000 years ago, placing the cave in the Aurignacian Period. Nevertheless, these dates were initially accepted by most specialists as valid, and they may well prove to be correct. However, serious doubts still remain as to whether they are indeed correct, especially since the cave's art contains so many features which have hitherto been associated with much later phases of the Ice Age—bison heads seen from the front; animals in movement, such as an aurochs with splayed legs and a pair of "fighting" rhinos; the "M" mark and shoulder stripes on some horses; a bison drawn with seven or eight legs and multiple outlines, perhaps to indicate movement; many different ways of showing perspective; and so forth. These features, together with the sheer technical sophistication of the depictions, are certainly causing many specialists to wonder if these radiocarbon dates are indeed correct, and further results by different laboratories as well as analyses of possible natural contamination will be needed to establish once and for all whether Chauvet Cave's art is the earliest known cave art in the world or, alternatively, a more conventional—albeit outstanding—set of images.

PAUL G. BAHN

See also **Art: Cave Art in Europe**

Further Reading

1984. *L'Art des cavernes: atlas des grottes ornées paléolithiques françaises* [Cave art: Atlas of the French Paleolithic decorated caves], Paris: Ministère de la Culture

Bahn, P.G. & Vertut, J. 1997. *Journey Through the Ice Age*, London: Weidenfeld and Nicolson and Berkeley: University of California Press

Chabredier, L. 1975. Les gravures paléolithiques de la grotte d'Ebbou (Ardèche) [The palaeolithic engravings of the cave of Ebbou]. *Archéocivilisation*, nouvelle série, numéro spécial, 14/15

Chauvet, J.-M., Brunel Deschamps, E. & Hillaire, C. 1995. *Chauvet Cave: The Discovery of the World's Oldest Paintings*, London: Thames and Hudson; as *Dawn of Art, The Chauvet Cave: The Oldest Known Paintings in the World*, New York: Abrams (original French edition, 1995)

Clottes, J. (editor) 2001. *La Grotte Chauvet. L'Art des origines* [Chauvet Cave: The Art of Our Origins], Paris: Le Seuil

Clottes, J. et al. 1995. Les peintures paléolithiques de la Grotte Chauvet-Pont d'Arc, à Vallon-Pont d'Arc (Ardèche, France): datations directes et indirectes par la méthode du radiocarbone. [The palaeolithic paintings of the Chauvet Cave: direct and indirect radiocarbon dates.] *Comptes Rendus Académie des Sciences de Paris*, 320, série IIa, 1133–40

Combier, J. 1984. Grottes ornées de l'Ardèche. [Decorated caves of the Ardèche.] In *Les Premiers Artistes, Dossier de l'Archéologie*, 87: 80–86

Combier, J., Drouot, E. & Huchard, P. 1958. Les grottes solutréennes à gravures pariétales du canyon inférieur de l'Ardèche. [The Solutrean caves with parietal engravings in the lower canyon of the Ardèche.] *Mémoires de la Société Préhistorique Française*, 5: 61–117

Gély, B., 2000. *Les Grottes Ornées de l'Ardèche* [The Decorated Caves of the Ardèche], Veurey (Isère): Editions du Dauphiné Libéré

Roudil, J.-L. 1995. *Préhistoire de l'Ardèche: Le temps des chasseurs et leur art* [Prehistory of the Ardèche. The time of the hunters and their art]. Privas: Conseil Départemental de la Culture de l'Ardèche

Züchner, C. 1999. La Cueva Chauvet, datada arqueológicamente. [The Chauvet Cave, dated archaeologically.] *Edades, Revista de Historia*, 6: 167–85

ART IN CAVES

Although the vast majority of prehistoric rock art occurs in the open air or in shallow rock shelters—for example in South Africa, North and South America, and Australia—there is nevertheless an important body of art (mostly paintings and engravings) lying in total darkness within deep caves in a few parts of the world. The best-known examples, of course, are some of the hundreds of decorated caves from the Ice Age of Eurasia (see Art: Cave Art in Europe), though art from more recent periods of prehistory is also found deep inside European caves—for example, in several caves in northern Norway, from about 3000 years ago—as well as sealed inside the chambers of megalithic monuments.

The phenomenon is by no means uniquely European. A whole series of deep caves in Australia have been found to contain a profusion of finger markings, such as Koonalda Cave, 15 000–24 000 years old (see Art: Cave Art in Australasia) and engraved motifs, while paintings—mostly hand stencils—also occur in caves in Tasmania, New Guinea, Sulawesi (see figure), Borneo, China, and elsewhere. Easter Island has abundant decoration inside many small caves, in the form of both paintings and rock carvings and engravings. Increasing numbers of decorated deep caves are also being discovered in various parts of the United States (see Art: Cave Art in the Americas), especially in Tennessee and Alabama, the best known being the Mud Glyph

Art in Caves: Classic hand silhouettes in the cave of Ujung Bulo in Sulawesi, Indonesia. (Photo by John Middleton)

Cave of Tennessee which recently had to be blocked up to prevent further damage to its fragile finger-drawings in soft clay. There is a wealth of decorated caves in Cuba and the Dominican Republic, but alas, the decimation of the local people after the arrival of the Europeans has ensured that we know nothing of the meaning of their cave art.

Although some cave art was always known to exist by local people, its discovery by the world of scholarship can be traced back to the second half of the 19th century in Europe, when a few examples of Ice Age cave art were spotted, culminating in the realisation that Altamira Cave (see separate entry) contained paintings from the Ice Age. The first published account of Maya cave painting in the Yucatán (Mexico) dates to 1897, when Edward Thompson, the American consul in that part of the world, found some paintings in the cave of Loltun, as well as some rock carvings; however he accorded them scant importance, merely describing them as curious symbols and grinning skulls.

Because caves appear mysterious and menacing places to us, there has long been a tendency to associate their art with secret, esoteric, exclusive rites redolent of fear and awe. Rock art in the daylight and the open-air seems far less "private". But it would be simplistic to interpret art of the past in relation to these modern impressions. As we know from Australia, for example, open-air sites can be just as imbued with power and taboo as anything underground. Indeed, some of the art in deep caves appears to be "public", being easily visible in large, readily accessible chambers. However, a great deal of it is undeniably private, in small niches, or chambers only accessible through a long journey or after negotiating difficult physical obstacles necessitating climbs, crawls, or tight squeezes. There are cases—like the famous Ice Age clay bison of Le Tuc d'Audoubert in the French Pyrenees—where the very act of making the journey and of producing the images seems to have been what mattered; the artist(s) never returned to visit their work.

Why was art placed in such inaccessible locations? Deep caves are strange environments, bereft not only of light but also of sounds—except perhaps for dripping water or, at times, bats. One experiences utter blackness, total silence, a loss of sense of direction, a change of temperature, and a frequent sense of claustrophobia. To enter a deep cave is to leave the everyday world and cross a boundary into the unknown—a supernatural world. It is easy to imagine that caves therefore symbolized transitions in human life and could be used for rituals linked with those transitions, especially puberty rites. Or perhaps it was felt that by entering this world one could better commune with or summon up the supernatural forces which dwelt there, and hence the images were made to reach and compel those forces. Cave decoration certainly requires strong motivation, since it involves negotiating such obstacles and taking both equipment and lighting into the site.

As in the Caribbean, we have no ethnographic information from Australia on the use of deep caves—indeed they seem to have been avoided by Australian Aborigines in historical times. But we can gain some insights into potential motivations by studying the deep caves which were decorated by the Maya of Central America (see Art: Cave Art in the Americas). Since theirs was not a prehistoric culture, and since some of their cave decoration consisted of their glyphs (which were their writing), we can both read the texts and learn from Maya ethnohistory what they were doing in caves. It is certainly clear that, for the Maya, it was the act of production that counted, not the durability of the art. "Hidden" or private images are sometimes found in especially awkward or even dangerous locations. Their use of caves involved altars and "chapels", and water was a focus of the rituals here. Caves were considered dangerous and chaotic places, contrasting starkly with the domestic community, and they were visited to make contact with gods and spirits in some way. Sanctity was proportional to spatial remoteness and this factor would certainly help to explain the decorated "sanctuaries" in remote corners of many prehistoric caves: for example, the Tuc d'Audoubert's clay bison are located at the very furthest point of a 900 m long cave, in a chamber reached only after an often uncomfortable and difficult journey. In Maya terms, therefore, these bison were left in a sacred location, unsullied through their utter remoteness from daily human life. The realm of total darkness and silence was the greatest contrast imaginable to the everyday human world.

PAUL G. BAHN

See also **Photos 18 & 19 in colour section**

Further Reading

Bahn, P.G. 1998. *The Cambridge Illustrated History of Prehistoric Art*, Cambridge and New York: Cambridge University Press

Bahn, P.G. & Vertut, J. 1997. *Journey Through the Ice Age*, London: Weidenfeld and Nicolson and Berkeley: University of California Press

Faulkner, C.H. (editor) 1986. *The Prehistoric Native American Art of Mud Glyph Cave*, Knoxville: University of Tennessee Press

Lewis-Williams, D. 2002. *The Mind in the Cave: Consciousness and the Origins of Art*, London: Thames and Hudson

Núñez Jiménez, A. 1975. *Cuba: Dibujos Rupestres*, Havana: Editorial de Ciencias Sociales

Stone, A.J. 1995. *Images from the Underworld: Naj Tunich and the Tradition of Maya Cave Painting*, Austin: University of Texas Press

Stone, A.J. & Bahn, P.G. 1993. A comparison of Franco-Cantabrian and Maya art in deep caves: Spatial strategies and cultural considerations. In *Time and Space: Dating and Spatial Considerations in Rock Art Research*, edited by J. Steinbring *et al.*, Melbourne: Australian Rock Art Research Association

ART IN CAVES: HISTORY

The history of the study of rock art in deep caves is widely regarded as having commenced with the discovery of the paleolithic art in Altamira, Spain, in 1879 (see Altamira entry). However, ever since the earliest artists created their works others will have viewed them and undoubtedly drawn inspiration for their own subsequent works. For instance, Neolithic art, Roman art, and later inscriptions in the vicinity of paleolithic cave art all suggest that the art was viewed at these various times. Even much of the famous cave art of Lascaux (in the Périgord, Dordogne see colour plate section) is probably not from the Pleistocene, but may have been created in the Holocene, in response to earlier art. In 1458, Pope Calixtus III decreed that the religious ceremonies held in "the Spanish cave with the horse pictures" had to cease. Although it is unknown which site he referred to, it was almost certainly a site of paleolithic art. This decree also implies the use of the ancient rock art in religious practices in late medieval times. By the 19th century, however, all knowledge of this rock art seems to have been lost, much to the detriment of its rediscoverer.

The life of Don Marcelino Santiago Tomás Sanz de Sautuola (1831–88) was destroyed through his discovery of paleolithic art in Altamira. The archaeological establishment judged the cave art to be a crude joke or a hoax, and considered its discoverer to be either a charlatan or a dupe. De Sautuola produced immaculate publications in 1880 and 1882, trying in vain to secure acceptance of his find, but most of his opponents refused to even inspect the site (de Sautuola, 1880). He died prematurely six years later, a broken and bitter man, in the full knowledge that he had made one of the greatest discoveries in the history of archaeology.

Léopold Chiron had already found engravings deep in the French cave of Chabot in 1878, and in 1890 found more in another site, Figuier. In 1883, Francois Daleau excavated engravings on a wall in Pair-non-Pair that had been covered by Ice Age sediments. However, de Sautuola's treatment by the discipline deterred others from publicizing such new finds. In 1895, a bison engraving was discovered in the French cave of La Mouthe. Emile Rivière, who had seen the Altamira paintings, then found more rock art in La Mouthe, and four years later a paleolithic lamp. Thus the evidence in favour of paleolithic rock art mounted, but full acceptance by the archaeological establishment did not occur until the end of the century.

At that time, a young Catholic priest had begun to develop a great fascination for the subject of European cave art. Abbé Henri Breuil was to dominate the field for the next six decades, and a great deal of our knowledge of the paleolithic rock art traditions is attributable to his unparalleled life work (Breuil, 1952, and see Archaeologists entry). His reign was followed by that of André Leroi-Gourhan (1968), after whose death Jean Clottes became the key scholar of paleolithic cave art. Throughout the 20th century, a stylistic sequence for the art was refined and honed by successive scholars. Its bases were the stylistic genres perceived by the leading researchers, which were often constructs of a very tenuous nature (see also Art in Caves). Although significant changes were made to this stylistic sequence from time to time, it remained unchanged in its essential evolutionary basis. A distinctive development from the most simple and primitive to the most complex and ornate remained its most fundamental tenet until 1995, when it was refuted by Bednarik (1995). This was the result of new discoveries, most especially that of Chauvet Cave in France, whose dating by Clottes *et al.* (1995) demonstrated that the most sophisticated paleolithic cave art was also the earliest. During the 1990s, the introduction of direct dating of European cave art and the demise of stylistic dating, instances of fakes, and rejections of scientific dating results prompted various controversies, culminating in 1995 in what Michel Lorblanchet later described in *La Recherche* as an earthquake in paleolithic rock art research.

Cave art is not, however, limited to Europe; it is found in all continents except Antarctica. A second tradition of Ice Age cave art occurs along the southern coast of Australia. The first site discovered was Koonalda Cave, presented by Alexander Gallus (1968). The scientific investigation and the recognition as a specific tradition of Australian rock art only began with the discovery of the Mount Gambier corpus in 1980 (Bednarik, 1990). The first sites located there, Malangine and Koongine Caves, were subjected to direct dating of the rock art by Robert Bednarik in 1980. This was in fact the introduction of scientific dating of any form of rock art, whereas it took another ten years for direct dating techniques to be adopted by French cave art specialists. In contrast to the Franco-Cantabrian cave art chronology, the Australian cultural sequence has not given rise to controversy. This is because it has not been developed through stylistic constructs of individual archaeologists, but through scientific data obtained from substances physically related to the art, and through the identification of specific behavioural traces. In Australia, the Parietal Markings Project is responsible for the discovery of about 90% of all known sites, including all 40 cave sites at Mount Gambier (see also Art: Cave Art in Australia).

Another region noted for its cave art includes parts of Central America and the Caribbean islands (see Art: Cave Art in the Americas). Specific clusters of sites occur in Cuba, Hispaniola (two specific concentrations) and in the general area of Belize, Guatemala, and the Yucatán Peninsula. The cave art of the last-mentioned region is attributed to the Maya and some 22 sites are currently known. This region has been studied especially by Andrea Stone (1995). The dozen or so sites in the Dominican Republic have been presented by Fernando Morban Laucer (1978). There are also minor numbers of cave art sites located on several other Caribbean islands. All cave art of this region is assumed to be substantially less than 2000 years old.

Finally, a remarkable series of rock art sites has been located in several caves in the Kentucky–Alabama region of North America. They are popularly known as "mud glyph caves", but the rock art, which is thought to be fairly recent, seems to occur on moonmilk rather than mud. This series, discovered since 1980, has been presented by Charles Faulkner (1986).

ROBERT BEDNARIK

Works Cited

Bednarik, R.G. 1990. The cave petroglyphs of Australia. *Australian Aboriginal Studies*, 2: 64–68

Bednarik, R.G. 1995. Refutation of stylistic constructs in Palaeolithic rock art. *Comptes Rendus de l'Académie de Sciences Paris*, 321(série IIa, No. 9): 817–21

Breuil, H. 1952. *Four Hundred Centuries of Cave Art*, Montignac: Centre d'Études et de Documentation Préhistoriques

Clottes, J., Chauvet, J.-M., Brunel-Deschamps, E., Hillaire, C., Daugas, J.-P., Arnold, M., Cachier, H., Evin, J., Fortin, P., Oberlin, C., Tisnerat, N. & Valladas, H. 1995. Les peintures paléolithiques de la Grotte Chauvet-Pont d'Arc, Vallon-Pont-d'Arc (Ardèche, France): datations directes et indirectes par la méthode du radiocarbone. *Comptes Rendus de l'Académie des Sciences de Paris*, 320: 1133–40

Faulkner, C.H. (editor) 1986. *The Prehistoric Native American Art of Mud Glyph Cave*, Knoxville: University of Tennessee Press

Gallus, A. 1968. Parietal art in Koonalda Cave, Nullarbor Plain. *Helictite*, 6: 43–49

Leroi-Gourhan, A. 1968. *The Art of Prehistoric Man in Western Europe*, London: Thames and Hudson (original French edition 1965)

Morban Laucer, F. 1978. *El arte rupestre de la Republica Dominicana, petroglifos de la Provincia de Azua*, Santo Domingo: Fundación García Arévalo

de Sautuola, M.S. 1880. *Breves apuntes sobre algunos objetos prehistóricos de la provincia de Santander*, Santander: self-published

Stone, A.J. 1995. *Images of the Underworld: Naj Tunich and the Tradition of Maya Cave Painting*, Austin: University of Texas Press

Further Reading

Bahn, P.G. 1998. *The Cambridge Illustrated History of Prehistoric Art*, Cambridge and New York: Cambridge University Press

ART SHOWING CAVES

From the earliest of times, man's imagination has been captured by the mysterious underground. It was much later that artists began to create images of caves, and preserve them for us to view today. The early artists largely ignored the question of science as they concentrated on the mystery, fear, and magic of caves. This entry will briefly review the chronology of artwork inspired by caves, from the ancient Greeks to the present day. It will review some of the individual artists who have made an impact on the development of technique and show how people have seen caves through time.

The first caves in art were crude, pictorial descriptions of caves. They would generally and often inaccurately show a passage or cavern as one room and there would be small icons showing bones, mystical characters, and demons. Caves were fearful places. The oldest surviving remnant of a cave illustration is a map of a decorated cave made in the time of Assyrian King Salmanassar III *c.*852 BC. Its survival over such a long period is because it was made in bronze relief (see Figure 1 in Caves in History: The Eastern Mediterranean).

The ancient Greeks regarded springs as sacred and viewed caves as homes to nymphs and deities. It was here that philosophers would retreat for contemplation and inspiration. They, in turn, suggested to architects that they build grottos or artificial caves around springs, which would be adorned with shells and crystals to emulate the natural structure of decorated passages in the deeper parts of caves. Basins, like gour pools, would catch and hold the sacred springs and water would be directed though lion-headed waterspouts carved out of tufa.

Later, the Romans used grotto or fountain constructions to decorate theatres or stage sets, to give a more rustic appearance emphasizing the atmosphere of a landscape or scene. These structures, called "nymphaea", were often copied and used as fountained courtyards in large public buildings. Smaller grotto structures kept their more traditional form in private gardens as the sites for small shrines.

The excavations in the 15th and 16th centuries of ancient Roman sites in Italy, such as the Domus Aurea (Golden House) of Nero in Rome, led to a revival of interest in using cave themes in architectural artwork. The High Renaissance Classicism of Raphael and his contemporaries (early 16th century), which was the prevailing style at the time of the initial excavations, was eventually (*c.*1530–90) succeeded by Mannerism, a style that emphasized the tension between nature and art. Cave imagery was a frequent Mannerist device. They distinguished the styles of man-made grotto between the "architectural grotto" and the "natural grotto". The former retained its old manner of shell, tufa, stone, and moss, finished with small pools and sculptures. The "natural grotto" became more elaborate and took the natural form to extremes of fantasy. Stalactites would hang in profusion and the walls would be covered in a mosaic of coloured pebbles, shells, and pieces of coral. Then came the Mannerist "ruined grotto". These gave the impression of threatening to collapse upon entering. The innards were a concoction of three-dimensional images with seas of strange mythical creatures, with swaths and garlands of exotic blossoms in arbors and wrapped around columns. Mirrors often represented water. Examples of these extravagant temples of the underground can be seen in Grotta Grande designed by Bernardo Buontalenti in Florence *c.*1583–85, Grotta Pavese in Genoa *c.*1594, and Schloss Hellbrunn in Salzberg *c.*1615 by Santino Solari. Fredrich Sutris designed a typical grotto wall, in 1581–86, at the Grottenhof Residenz in Munich, which was very much influenced by the Italian Mannerists.

Art Showing Caves: Blue Grotto in Capri. Painting by Y. Gianni. With permission from the owner, Carolina Shrewsbury. (reproduced in colour section as Figure 20)

Later came the introduction of the "automata", a mechanical device that artificially pumped water around a course. The automata was used about 2000 years ago as a prop in Dionysian plays, and was reinvented as a representation of the subterranean forces of nature and man's power to control them. The fact that the automata were powered by water made the artificial grotto the ideal place for them. The only fully functional automata that has been preserved is at the Hellbrunn water garden in Salzberg, Austria. The religious and mystical approach to the decorated grotto was largely confined to central Europe. German royalty, in particular, held a deep fascination for the mystery and magic of the subterranean world.

Progression of the inspiration behind caves in art throughout history is superbly witnessed by representations of the Blue Grotto (Grotta Azzura) on the island of Capri. It has been used within ancient Roman mosaics and mimicked in the architecture of the man-made grottos of famous Italian, French, and German palaces. The cave itself was once adorned with statues and items from Roman worship, which marked it as probably a Nymphaeum. It is indeed a mysterious and wonderful place made famous by the blue and aquamarine light that is reflected from the crystal-clear waters. For centuries after the Romans had been there, the cave seemed forgotten. In 1826 the German painter, August Kopisch, was taken in a small fishing boat to the cave after locals had enthused over the "marvelously blue grotto". From then on, the rediscovered secret of Capri became a popular haunt for painters, poets, and writers. It became fashionable in the 19th century for many wealthy Europeans to do a "Grande Tour", and to acquire a memento or souvenir at each port of call. This would often be a small painting, usually a watercolour of a known landmark of the area—in the case of Capri it would be the Blue Grotto. King Ludwig II of Bavaria was very much inspired and influenced by the Blue Grotto in Capri. Twice, he sent one of his architects to Capri to study the cave. He had the Venus Grotto built for him at the Schloss Linderhof. It is a kitschy, extravagant fantasia of something between the Blue Grotto and the Venushöhle in Horselberg from Richard Wagner's *Tannhauser*. A large painting depicting a scene adorns an artificial stage, fronted by a large well-lit lake with a boat fashioned into a shell upon which the eccentric King would recline, as a live orchestra played his favourite pieces of music from a hidden pit. Built in 1876–67, it had the first coloured spotlighting used in stage sets. Its grandiosity brought the fashion for artificial grottoes to a close.

During the 1670s and 1680s, Baron Johann Valvasor (see Speleologists) visited Adelsberg Cave, the first European show cave now known as Postojnska Jama in Slovenia. He documented the whole cave with maps and sketches. His formations are shown as the typical view of the cave, full of terrifying creatures from the underworld. His cave dimensions were quite unreliable. Almost a century later, Joseph Nagel (see Speleologists) invited several artists to help illustrate his own volumes on the cave. Sebastian Rosenstingel and Carlo Beduzzi were regular "expedition artists" just as we have photographers and film makers today, depicting the explorer showing royalty and scholars the great chasms.

In the 1790s, a hunter found what was later to become known as Mammoth Cave in Kentucky, United States. This was to become as famous as the Adelsberg cave in Europe. Stephen Bishop, a slave who worked as a guide at the cave, explored most of the system at a time when deep cave exploration was in its infancy. Bishop's work can be seen as a detailed and accurate map of the cave, which he drew freehand by pencil as a document for a guidebook produced by his owner, Doctor John Crogan, who owned the cave at that time.

In the 18th century, in the wake of the garden grottoes and the beginnings of world travel, art inspired by caves was adopted by the classical artist of the day. Walking in the footsteps of August Kopisch, they branched out to incorporate new wonders of earthly magnitude and nature's still power, often portrayed in the magnificent painting of light and towering mountains within the Rococo movement. One of these artists was a German, Caspar David Friedrich (1774–1840), who painted caves in religious scenes. Though rarely seen, as this type of art has had difficulty as an understood and recognized subject compared to his landscapes, they convey a powerful imagery of reflected light within a dark, mysterious world.

One of the early artists of caves in this Romantic era was Caspar Wolf, a Swiss artist and draughtsman born in 1735 who moved into the French landscape movement and took up a keen interest in mountains and the forces of nature. Many of his works feature the movement of water and it was probably curiosity about the sources (springs) that led him to the many cave areas in Germany and Switzerland. Several of his cave works are on permanent view in the Museum in Aargau, Switzerland, his most famous being of Beatushöhle, a major cave system of the middle alpine regions of Switzerland. He is one of the first known artists to have taken his materials into the cave to produce his own oil and pencil sketches from nature. He was also the first documented artist to understand the geological nature of his subject that, besides the retaining of the Rococo instability, is portrayed clearly in his work.

William Westall arrived in North Craven in 1817 when Weathercote and Jordas (now Yordas) caves were the most frequented show caves. Seven of twelve of his engravings of the area were of caves. The opinion is that Westall made the finest drawings of caves in Britain at that time. He wrote of the entrance of Jordas, "The peculiar character of this view is derived from the great masses of rock that hang from the roof, apparently loose which, combined with the deep gloom of the interior, usually occasion some degree of terror on the first visit to the cave." A few days after working there, a torrent of water from a burst underground stream caused a rock fall and the cave entrance was made considerably smaller.

Over the rest of the 19th century, various artists started to go further into caves to penetrate a deeper realm of huge darkness. Mostly they would retreat to the confines of their warm studios, leaving the cave explorer to regale terrifying experiences of his treacherous expeditions. Alois Schaffenrath, in the mid-19th century, spent much of his time working in Adelsberg Cave as it was being opened up for public show-cave access. His work is significant because he clearly shows how he lit the cave for his sketches, by strategically placing torches in much the same way as artists still do today. It was amongst the explorers of caves at this time that mapping the underground accurately and writing in journals with illustration became more of a priority than impressing a general public with the skills of artistic technique. From the second half of the 19th century into the first decades of the 20th century, there are countless records of descents and well-lit tours underground. Unfortunately, there is little record of who these artists were, but their beautiful, detailed studies can be found in journals, newspapers, and books of this period all over the world. The artwork of the French artist, Lucien Rudaux, who accompanied the travels of Edouard Alfred Martel was however well documented, and can be seen on the web. His magnificent artwork includes *Bramabiau*, 1888, *Padrirac*, 1889, and *Aven Armand*, 1897, all painted in France.

From the end of the 19th century, art played a massive role in bringing awareness of the natural wonder of the underground to general public notice. In the United States, there was significant growth in the discovery of show caves. As speleologists strove for new income during the depression of the 1930s, chasms opened in abundance for any wandering traveller's curiosity. Posters and huge billboards dotted Central America with wonderful displays of an exaggerated, colourful underground. Bowling Green in Kentucky, the home of Mammoth Cave, in particular, has a very long history of artists up to the present day who have been inspired by the variety of formation and passage within Mammoth Cave.

In 1839, photography was born, though it was not for many years that the camera was successfully used underground (see Photographing Caves). This was a new form of art and very much the preference when recording scientific discoveries. Painting found a challenger who could outdo the need for realism in many fields of scientific research. As photography took over the mantle as the leading medium, the oil painters and watercolourists of the day lost their vogue, as far as the needs of reproducing underground images for books, papers, and other media went. Painting caves and taking a sketch book on speleological expeditions and surveys became a private domain. Within the caving ranks, artists worked independently of one another, creating countless cartoons, picture cards, magazine covers, and, if found out, were asked to reproduce photographs.

The American artist John Agnew and the Canadian artist Linda Heslop are among the most well-known and most-reproduced artists of speleology in the world today, with work in a variety of caving books and journals. But they are not the only ones by far. The speleoartist is beginning to attract larger audiences and there are specialized groups that cave specifically to light caves and sketch, often unconsciously working out the nature of geological faults before they know much about speleology! Exhibitions are popular and artist institutions include the St Lukas Gilde (Austria), the International Society of Speleological Art, and the international group SpeleoArt. Although photography has not lost its position as the challenging medium for the engineer and scientist, it cannot overcome the challenge of human talent held in the hands of an artist. That is: to put the metaphor into an image of a cave.

A comment about the future for an art that is firmly entrenched in the speleological world concludes this brief history of caves in art. As well as the great scientific and environmental importance of the karstic environment, there are strong arguments that it should be preserved because of its aesthetic qualities. One of the roles of the artist is to make such natural beauty visible to those who either cannot see it or can see it only poorly. Caves are storybooks of time that nature has created, and another task for artists is to educate a public who are better developed to understand a metaphorical approach rather than a purely scientific one.

CAROLINA SHREWSBURY

See also **Photographing Caves; Stamps and Postcards**

Further Reading

Agnew, J. 2001. *Painting the Secret World of Nature*, Cincinnati, Ohio: North Light
Brook, C. & Hogbin, B. 1993. *Caves and Caving*, 61: 18–19 and cover
Danielopol, D. 1998. Conservation and protection of the biota of karst: assimilation of scientific ideas through artistic perception. *Journal of Cave and Karst Studies*, 60(2): 67
Elderkin, G.W. 1941. The natural and artificial grotto. *Hesperia*, x/2: 125–37
Heslop, L. 1996. *The Art of Caving*, Saint Louis: Cave Books
Jackson, D.D. 1982. *Underground Worlds*, Alexandria, Virginia: Time Life Books
Miller, N. 1982. *Heavenly Caves: Reflections on the Garden Grotto*, London and Boston: Allen and Unwin
Mitchell, W.R. 1989. *Yorkshire's Hollow Mountains*, Settle: Mitchell; previously published as *The Hollow Mountains*, 1961
National Speleological Society, 1982. Cave lights. *NSS News*, June 1982: 163–77 and cover
National Speleological Society, 2001. Euphoria! *NSS News*, July 2001: 205
Shaw, T.R. 1967. *Cave Illustrations before 1900: A Catalogue of Non-Photographic Illustrations of Caves*, Settle; British Speleological Association
St Lukas Gilde. 1994. *Gildenpost*, 3
St Lukas Gilde. 1996. *Gildenpost*, 4
Time-Life Books, 1970. *The Camera*, New York: Time-Life Books

Useful Websites

Martel Gallery, Paintings from Lucien Rudaux, http://members.aol.com/bkliebhan/spelhist/martel-g.htm
SpeleoArt http://www.speleoart.net/

ART: CAVE ART IN AUSTRALASIA

Since cave rock art is restricted almost exclusively to limestone karsts, its distribution in Australia is determined by the occurrence of such landforms. While research on Australasian islands has yielded only limited results, four karst areas have been successfully surveyed for cave art on the Australian mainland. These are the Perth region, the Nullarbor, Buchan, and Mt Gambier, the only one to have been surveyed comprehensively, where about 40 sites are currently known. (The map in the Australia entry shows the location of these four regions.) Mt Gambier is also the only Australasian region that has yielded evidence of a lengthy cultural sequence of rock art traditions occurring in darkness, as well as dating information from both archaeological floor deposits and the rock art itself. With the exception of a single site on the Nullarbor Plain, Koonalda Cave, it is also the only cave art region in this part of the world that has had a significant impact on archaeological theories. Australian cave art comprises a major Pleistocene component, which is entirely nonfigurative and whose motif range matches that of middle paleolithic art traditions of the Old World.

On the Nullarbor Plain the occurrence of cave art has been confirmed in five sites, although further discoveries are still likely. In most cases this art occurs in the form of rock paintings, especially red-ochre hand stencils. Koonalda Cave represents that area's only major known instance of engravings and finger flutings (Gallus, 1968), which form the rock art corpus in three other cave art regions: near Perth, where three sites are presently known (as well as one cave painting site); at Buchan, with only one site; and the large body of cave petroglyphs between Portland and Mt Gambier (Bednarik, 1990). It needs to be appreciated that within the massive rock art corpus of Australia—the largest in the world (several hundred thousand sites)—the continent's cave art still represents only a minuscule fraction of one per cent.

The Mt Gambier region shows a sequence of several distinctive marking traditions. These are often separated chronologically by evidence of geomorphological events, such as tectonic adjustments, roof falls, subsidences, inundations, speleo-weathering or animal scratch marks (including megafaunal marks), and sometimes physically by layers of calcite speleothem (Bednarik, 1998). Finger flutings, made by drawing fingers over a soft, moonmilk-covered surface, seem to represent the earliest tradition (Bednarik, 1986). The traces of another distinctive behaviour occurring early in the sequence are large concentrations of impact marks, often in the form of deeply pounded holes forming discrete wall panels. These deep pits may be as much as 20 cm deep, and individual tool marks remain frequently recognizable in them. They are thought to be the most extreme form of the ubiquitous cup marks (cupules) occurring widely above ground and on much harder rock, including quartzite and granite, in similar large groups. Such cup marks occur in their hundreds of thousands in the continent's north, where they are generally recognized as representing the earliest surviving rock art.

A very different subsequent tradition has been named the Karake style, after the Mt Gambier cave where it was first recognized. It consists of a variety of geometrical motifs, usually deeply pounded and abraded into cave walls, with groove depths of up to 40 mm (Aslin, Bednarik & Bednarik, 1985). Circles are particularly prominent, sometimes forming groups of a few dozen, occasionally with internal vertical parallel lines, lozenge patterns or central pits. Other motifs include simple linear designs resembling bird or kangaroo tracks, but it is not at all certain that this is what they were intended to depict. Such designs grade into arrangements that are defined as CLMs (convergent lines motifs, in which several lines are bunched together, converging at one end but not necessarily touching). Occasional other motif types are zigzag or wavy lines, arcuate designs, and parallel lines. Again, some characteristic features of the tradition can be recognized in the pre-figurative petroglyphs above ground. A distinctive arrangement of several circles at an open granite site has been dated to about 27 000 years (Bednarik, 2001), and the art of one of the Karake sites, Malangine Cave near Mt Gambier, has been shown to be in excess of 28 000 years old, through uranium – thorium dating of a covering layer of reprecipitated calcite (Bednarik, 1999). Thus, dates obtained from this cave art tradition match dating of similar rock art at open sites.

Art: Cave Art in Australasia: Cave petroglyphs in Paroong Cave, Mt Gambier. (Photo by Robert Bednarik)

The more recent cave petroglyphs at Mt Gambier consist of shallow abraded engravings of Holocene age, but it must be emphasized that some of the finger flutings, too, appear to be of that period. Another form of cave markings, frequent near Mt Gambier and Perth, consists of apparently unstructured assemblages of tool marks made either with limestone clasts or with chert tools. These are sometimes exceptionally well preserved. Only one of the Mt Gambier sites presents minor pigmented rock art, fully exposed to daylight. Similarly, there is much painted rock art in shelters and cave entrances in a further limestone area, the Chillagoe region of north Queensland, but this is not cave art in the sense that it does not occur in darkness.

Five of the sites of cave petroglyphs along the southern coast have been subjected to archaeological excavations: Orchestra Shell, Koonalda, Koongine, Malangine, and New Guinea 2 Caves. Although some of these yielded Pleistocene occupation evidence, for none of these cases has a satisfactory nexus been established between sedimentary dates and the rock art. At six of the cave petroglyph sites, subterranean chert or chalcedony mining traces are quite prominent, and it is possible that the mining activities coincided chronologically with some of the rock art production. This is the major corpus of Pleistocene underground mining evidence in the world (Bednarik, 1995).

In Tasmania, three caves with minor rock painting are known, but the direct dating evidence obtained from two motifs, reportedly from blood residues, remains controversial. Cave art also occurs in Papua New Guinea, including small examples of finger fluting in two deep caves, Kalate Egeanda and an unpublished site (Ballard, 1992). Although rock art occurs widely on the islands of Australasia, there are no further cases known of true cave art, even though much of this art does occur in limestone regions. Examples are the shelter paintings of Vanuatu, or the numerous limestone painting sites of New Zealand, none of which occurs more than four metres from the dripline.

ROBERT BEDNARIK

Works Cited

Aslin, G.D., Bednarik, E.K. & Bednarik, R.G. 1985. The "Parietal Markings Project": a progress report. *Rock Art Research*, 2: 71–74

Ballard, C. 1992. First report of digital fluting from Melanesia: the cave art site of Kalate Egenada, Southern Highlands Province, Papua New Guinea. *Rock Art Research*, 9: 119–21

Bednarik, R.G. 1986. Parietal finger markings in Europe and Australia. *Rock Art Research*, 3: 30–61, 159–70

Bednarik, R.G. 1990. The cave petroglyphs of Australia. *Australian Aboriginal Studies*, 1990(2): 64–68

Bednarik, R.G. 1995. Untertag-Bergbau im Pleistozän. *Quartär*, 45/46: 161–75

Bednarik, R.G. 1998. Direct dating results from Australian cave petroglyphs. *Geoarchaeology*, 13: 411–18

Bednarik, R.G. 1999. The speleothem medium of finger flutings and its isotopic geochemistry. *The Artefact*, 22: 49–64

Bednarik, R.G. 2001. Pilbara petroglyphs dated. *Rock Art Research*, 18: 55–57

Gallus, A. 1968. Parietal art in Koonalda Cave, Nullarbor Plain. *Helictite*, 6: 43–49

Further Reading

Flood, J. 1997. *Rock Art of the Dreamtime*, chapters 2–4, Sydney: Angus and Robertson

ART: CAVE ART IN EUROPE

Many caves in Europe, as elsewhere, have been frequented since the Ice Age and throughout history by new occupants, by shepherds, by the curious or adventurous. In some areas, such as the Basque Country, superstitions and religious traditions associated with caves are probably extremely ancient, perhaps even extending back to the Ice Age. Although art of the Neolithic and Bronze Age is found on the walls of many caves in Europe, from France and Italy to Norway, and some markings from Roman and even medieval times occur in a few, the vast majority of known cave art in Europe is attributable to the last Ice Age, dating from about 35 000 to 10 000 years ago.

While one or two scholars noticed art on cave walls in the 1860s and 1870s, nobody realised its significance or its possible age until the discovery of the paintings of Altamira Cave (see separate entry) in northern Spain. Because of the initial doubts about that cave's authenticity, it was more than 20 years later, in 1902, that the discovery of art in France's Vézère caves (see separate entry) led to the existence and authenticity of Ice Age cave art being finally accepted.

Numerous examples were then found during the first years of the 20th century, but major sites continued to be discovered (e.g. Lascaux in 1940), and indeed one or two are still found every year in France or Spain. Currently about 300 decorated caves and rock shelters of the period are known in Eurasia—unfortunately we do not know if these comprise the majority of the original total, or just a tiny sample.

The known sites vary enormously in size and in numbers of images—some contain a single image, others have many hundreds. The art is found primarily on cave walls, although some ceilings were decorated—most famously that of Altamira—and in a very few caves in the French Pyrenees, drawings have been found in the clay floor. Since most caves have had their floors destroyed by subsequent visitors, we have probably lost untold quantities of floor-art, while abundant images in caves have been covered over by calcite deposits or lost to erosion through being located close to or inside the entrances.

Ice Age cave art comprises a very wide range of techniques. The simplest was finger-markings on clay or the soft mondmilch (moonmilk) layer on walls. By far the commonest technique in parietal art is engraving, which was done with any sharp flint tool. Bas-relief sculpture was only done in rock shelters or cave mouths, but work in clay—on present evidence limited to the French Pyrenees—ranges from bas-relief to haut-relief (the two famous clay bison of le Tuc d'Audoubert), and even fully three-dimensional figures, such as the clay bear in the cave of Montespan. Where pigments were concerned, two colours were used overwhelmingly—red (which is always iron oxide, red ochre, haematite), and black (which is usually charcoal, though manganese dioxide sometimes occurs). The paints were applied with the finger, or with some kind of tool—though none has survived. Some lumps of pigments are in the form of "crayons" with use-wear, but most painting was certainly done with liquid paints, applied with a brush or a pad. In certain cases, particularly hand stencils, the paint was spat from the mouth, or sprayed by blowing through a tube. A great deal of cave art seems to have been done in intense episodes of artistic activity by a small number of highly skilled people—indeed, in some cases such as Covalanas in Spain, it is virtually certain that a single person was responsible for every figure in the cave.

The subject matter of cave art comprises primarily animal figures—almost all adults drawn in profile, with no groundlines or landscapes, and very few recognisable scenes. Only a limited array of species was depicted—this is not an Ice Age bestiary. They are dominated by horse and bovids (bison and aurochs); deer, ibex, and mammoths are secondary, while carnivores and rhinos are rare, except in Chauvet Cave. Occasionally one species is overwhelmingly dominant, such as the mammoth in Rouffignac Cave.

Humans or "humanoids" (anthropomorphs) occur far more rarely, which constitutes one of the enigmas of cave art—since the artists were clearly capable of drawing them when they wished (there are numerous examples in portable art objects), then their scarcity in cave art may be explained by their depiction being taboo for some reason or, perhaps more probably, simply irrelevant to what the artists were trying to do. The final category of cave art is the range of apparently "abstract" or "non-figurative" motifs, known as "signs". While the simpler signs, such as dots and lines, are inevitably ubiquitous, the more complex ones are highly localized in space and time and have been interpreted as some kind of ethnic marker.

Some cave art was "public", on open view, but much is hidden or inaccessible, which points to a religious motivation for at least some of it. Much appears linked to a complex mythology, and no single explanation can possibly suffice for the phenomenon. Interpretations have varied over the years: at first, the art was assumed to have no particular meaning, it was "art for art's sake" produced by hunters with time on their hands. Subsequently, "sympathetic magic" became the dominant theory—it was assumed that the pictures were symbolically killed (hunting magic) or made to encourage the animals' reproduction (fertility magic). Unfortunately, there are no hunting scenes or copulation scenes among the many thousands of images, very few animals are marked with possible missiles or wounds, and on the whole the artists were not drawing the species which formed the basis of their diet.

In the mid-20th century, attention shifted to developing more complex (and probably more accurate) interpretations, and it was decided to focus on the art itself—by quantifying each category of image, and treating cave art not as a random accumulation of individual pictures, but as a carefully laid-out composition within each cave (an idea developed primarily by the French prehistorian André Leroi-Gourhan).

A more recent approach has been the study of the acoustic properties of decorated caves and rock shelters, an intriguing and long-neglected factor which tries to revive something that one might imagine gone for ever: the sound dimension which accompanied whatever rituals may have been carried out in such sites. Echoes are common in and around decorated rock shelters all over the world, as is the equally mysterious and impressive phenomenon of cavities which focus and project sound, while several decorated caves contain "lithophones"—stalactites which bear percussion marks and sometimes even painted marks.

Overall, since Ice Age art spans such a huge area of Europe—from Portugal to the Urals—and such a huge timespan, and such a tremendous range of techniques and styles, it is obvious that no single explanation can ever suffice for the phenomenon.

Each of the many interpretations already put forward probably contains some truth, while there are doubtless many more which have not yet been imagined.

PAUL G. BAHN

See also **Altamira Cave, Spain; Ardèche Caves, France; Vézère Caves, France**

Further Reading

Bahn, P.G. 1998. *The Cambridge Illustrated History of Prehistoric Art*, Cambridge and New York: Cambridge University Press
Bahn, P.G. & Vertut, J. 1997. *Journey Through the Ice Age*, London: Weidenfeld and Nicolson and Berkeley: University of California Press
Breuil, H. 1952. *Four Hundred Centuries of Cave Art*, Montignac: Centre d'Etudes et de Documentation Préhistorique
Graziosi, P. 1960. *Palaeolithic Art*, London: Faber and New York: McGraw Hill
Leroi-Gourhan, A. 1995. *Préhistoire de l'Art Occidental*, 3rd edition, Paris: Mazenod; as *Treasures of Prehistoric Art*, New York: Abrams, 1967, as *The Art of Prehistoric Man in Western Europe*, London: Thames and Hudson, 1968 (original French edition, 1965)
Lewis-Williams, D. 2002. *The Mind in the Cave: Consciousness and the Origins of Art*, London: Thames and Hudson
Lorblanchet, M. 1995. *Les Grottes Ornées de la Préhistoire: Nouveaux Regards*, Paris: Errance
Sieveking, A. 1979. *The Cave Artists*, London: Thames and Hudson
Ucko, P.J. & Rosenfeld, A. 1967. *Palaeolithic Cave Art*, New York: McGraw Hill and London: Weidenfeld and Nicolson
Zervos, C. 1959. *L'Art de l'Epoque du Renne en France*, Paris: Cahiers d'Art

ART: CAVE ART IN THE AMERICAS

Some 300–400 cave art sites are known in the Western Hemisphere, although few of them have been adequately investigated. Numerically, this corpus rivals the European Paleolithic; however, New World cave art presents a more fragmented picture, occurring in a wider range of cultural and geological settings, including sandstone and basalt fissures in addition to deep limestone caves. The present overview encompasses all of these environments if researchers consider them to be caves. Cave art is found in North, Central, and South America, and the Caribbean Islands, which possess the largest concentration in the hemisphere. Cave art sites utilized by hunter-gatherers over 3000 years ago are known in the Americas; however, many sites are associated with sedentary agriculturalists, such as the protohistoric Aztecs of Mexico, the Pueblos of the American Southwest, the Mississippian cultures of the Southeast, and the Taínos of the Caribbean. The most deeply sequestered art is linked with the pilgrimage activities and elaborate ceremonialism of village-dwelling, farming populations who believed that caves were sacred places where spirits could be contacted.

Issues surrounding cave art studies in the Western Hemisphere parallel those of the parent field of rock art research. One of its primary concerns is the relationship between rock art and the earliest human populations—research aided by new techniques for directly dating pictographs and coatings on petroglyphs, the most reliable being ^{14}C AMS analysis of pigments. Cave art in the New World does not approach the antiquity of European or Australian examples. The oldest New World examples occur in South America, in modest sandstone caves, typically 10–20 m deep (see the South American chapters in Strecker & Bahn, 1999, and Whitley, 2001). Several caves with Late Pleistocene paintings have been discovered in Brazil, where the Caverna da Pedra Pintada revealed datable pigment splashes with a minimum age of 9000 BC. In Argentina, the oldest cave art, assigned to the Toldense culture (9000 BC), features negative hand prints and paintings of guanacos. Hand prints in the Cueva de las Manos in Patagonia have been dated to 7300 BC. Small caves are also associated with the art of Early Holocene hunter-gatherers in the high Andes in southern Peru and Chile. However, none of the early sites reported in South America have dark zone art. The earliest New World dark zone art is found in the southeastern United States (Tennessee and Kentucky) and dates between 1600 and 1000 BC. These Archaic sites contain both "mud-glyph" art (see below) and incised petroglyphs (Simek *et al.*, 1998).

In the western United States, cave art typically occurs in basalt formations. Some 517 painted motifs, mostly circles and zigzags dating between 1000 AD and the historic period, have been documented in lava tubes in the Modoc Plateau of California (Whitley, Loubser & Hann, 2003). Pictographs have also been discovered in one lava tube in Oregon. Owl Cave in Washington houses paintings in an 18 m diagonal basalt fissure (Keyser et al., 1998). In the western United States, cave art is generally given a shamanic interpretation. Other cave art sites are widely dispersed in the west (Montana, Colorado, and Wyoming).

Caves with pictographs, including four with dark zone art, are scattered across the American Southwest (Greer & Greer, 1999). The Mogollon site of Feather Cave, New Mexico (1000 AD) is unique in the Americas in having dark zone art associated with undisturbed perishable materials (Ellis & Hammack, 1968). An isolated chamber (Arrow Grotto) has preserved miniature bows and arrows, wooden bowls, and prayer sticks. Arrows were also found lying under a stencilled hand, and feathers were set into wall crevices surrounding groups of pictographs.

A few cave art sites are found in the American Midwest. Five have been reported in Missouri (Greer & Greer, 1999). One has dark zone pictographs dating to the Mississippian period. The newly discovered Tainter Cave in southwestern Wisconsin has the only dark zone cave art known in the Upper Midwest. Charcoal pictographs dating to 900 AD occur at 70 m in a sandstone tunnel. One portrays a hunter shooting a pregnant deer with a bow and arrow.

Mud-glyph cave art—designs incised into mud-covered embankments and walls—has been known in the southeastern United States since 1979. Five of seven investigated sites date to the Mississippian period (1000–1600 AD) and have Southern Cult imagery, such as the trilobed arrow and weeping eye; however, Archaic sites with related art suggest a more ancient origin for the mud-glyph tradition (Simek et al., 1998). Other caves in the region have incised petroglyphs and mud-drawn pictographs, while Mammoth Cave and Salts Cave have charcoal pictographs drawn with burnt cane torches. With roughly 20 decorated limestone caves, the southeastern United States has the largest number of dark zone cave art sites in North America.

Mesoamerica is another important repository of New World cave art. Greer & Greer (1998) note sites in San Luis Potosí, Mexico. Closer to Mexico City, a 12.5 m lava tube houses an Aztec-style painting. Two cave art sites are associated with the Middle Formative (900–400 BC) Olmec culture (see summary in Stone, 1995). Their large naturalistic paintings are the oldest stylistically datable cave art in Mesoamerica. They are also the most deeply sequestered: at Juxtlahuaca, Guerrero paintings lie 1000 m from the entrance. The greatest concentration of Mesoamerican cave art occurs in the heavily karstified Maya Lowlands, where some 50 sites are known. Strecker did extensive work on cave petroglyphs in Yucatán. Painted sites are summarized in Stone (1995), and new cave art continues to be discovered. Classic period cave painting (300–900 AD) produced by Maya elites has shed light on the role of caves in archaic states (Figure 1). In Naj Tunich, Guatemala, hieroglyphic inscriptions reveal the politicized nature of elite cave shrines. A more rustic form of Maya cave art is anthropomorphically carved speleothems, also found in Caribbean caves. Evidence suggests that the Maya viewed these "stalagmite idols" as gods (Brady, 1999). Central America south of the Maya area has revealed few cave art sites, although two with pictographs were recently discovered in Honduras (Brady et al., 2000).

Art: Cave Art in the Americas: Figure 1. Black line painting of a deity ball-player next to the steps of a ball-court. Painting 22 cm high, date c.700 AD. Drawing 21, Naj Tunich, Guatemala. (Photo by Chip and Jennifer Clark)

Ironically, the region with the greatest concentration of cave art, the Caribbean Islands of Cuba, Puerto Rico, Hispaniola (Dominican Republic), and Jamaica, has the least accessible literature for an English-speaking readership. All indications are that hundreds of cave art sites have been located in the Greater Antilles, although they have not been scientifically investigated. They are associated with the protohistoric Taíno and earlier groups. The Cuban archipelago boasts 130 sites (Nuñez, 1985) with cave art spanning several thousand years of Pre-Columbian occupation. Although petroglyphs do occur, the art is dominated by thickly drawn pictographs of geometrics, animals, and frontal figures and faces (Figure 2). Dubelaar et al. (1999) have published a guide to Puerto Rican rock art, mentioning dozens of cave art sites found largely in the island's mountainous interior. While dark zone art in limestone caves has major expressions in most regions of the Americas, little has been found in South America. Decorated limestone caves are known in the Chapada Diamantina, Bahia, and the state of Minas Gerais, Brazil; however, pictographs and petroglyphs occur principally in lighted areas near entrances, for instance, in the Lapa da Sol near Iraquara, Bahia (see also Figure 3).

The wealth of cave art found in the Americas offers unique insights into the ritual utilization of the landscape and symbolic systems of populations running the entire gamut of social complexity. Increasing archaeological and public interest in rock art

Art: Cave Art in the Americas: Figure 2. Red paintings of large, thickly drawn symbols; grouping at right, roughly 1.2 m high. Cueva Pichardo, Cuba. (Photo by Reinaldo Morales Jr.)

Art: Cave Art in the Americas: Figure 3. A wall of paintings 10 m high in an alcove at the upstream end of the Gruta do Janelão in the Peracu National Park, Minas Gerais state, in Brazil. (Photo by Tony Waltham)

promises a bright future for our understanding of this rich expression of indigenous American art.

ANDREA STONE

See also **Art: Cave Art in Europe; Art in Caves: History**

Works Cited

Brady, J. 1999. The Gruta de Jobonche: An analysis of speleothem rock art. In *Land of the Turkey and the Deer*, edited by R. Gubler, Lancaster, California: Labyrinthos

Brady, J., Begley, C., Fogarty, J., Stierman, D., Luke, B. & Scott, A. 2000. Talgua Archaeological Project: A preliminary assessment. *Mexicon*, 22(5): 111–18

Dubelaar, C., Hayward, M. & Cinquino, M. 1999. *Puerto Rican Rock Art: A Resource Guide*, New York: Panamanian Consultants

Ellis, F. & Hammack, L. 1968. The inner sanctum of Feather Cave, a Mogollon sun and earth shrine linking Mexico and the Southwest. *American Antiquity*, 33(1): 25–44

Greer, J. & Greer, M. 1999. Dark zone and twilight zone pictographs in U-Bar Cave, southwestern New Mexico. *Rock Art Papers*, vol. 14, edited by K. Hedges, San Diego: San Diego Museum of Man

Keyser, J., Pederson, C., Bettis, G., Poetschat, G. & Hiczun, H. 1998. Owl Cave. In *Columbia Plateau Rock Art*, Portland: Oregon Archaeological Society

Nuñez Jiménez, A. 1985. *Arte rupestre de Cuba—Rock Art of Cuba—L'arte rupestre di Cuba*, Milan: Jaca Books
Includes an English text with a non-technical overview, accompanied by colour illustrations, of some of Cuba's most important cave art sites.

Simek, J., Franklin, J. & Sherwood, S. 1998. The context of early southeast prehistoric cave art: a report on the archaeology of 3rd unnamed cave. *American Antiquity*, 63(4): 663–67

Strecker, M. & Bahn, P. (editors) 1999. *Dating and the Earliest Known Rock Art*, Oxford: Oxbow Books
The chapters in this book detail evidence for early cave art in South America.

Whitley, D. (editor) 2001. *Handbook of Rock Art Research*, Walnut Creek, California: AltaMira Press
A scholarly overview of the rock art field, including technical analysis, interpretation, and world rock art, with essays by leading experts and information on cave art sites.

Whitley, D., Loubser, J. & Hann, D. 2003. Friends in low places: rock-art and landscape on the Modoc Plateau. In *Pictures in Place*, edited by C. Chippendale & G. Nash, Cambridge and New York: Cambridge University Press

Further Reading

Dacal Moure, R. & Rivero de la Calle, M. 1997. *Art and Archaeology of Pre-Columbian Cuba*, Pittsburgh: University of Pittsburgh Press
Provides black-and-white and colour illustrations of cave pictographs and petroglyphs and the most up-to-date discussion in English of Cuban prehistory.

DiBlasi, P. 1996. Prehistoric expressions from the central Kentucky karst. In *Of Caves and Shell Mounds*, edited by K. Carstens & P.J. Watson, Tuscaloosa: University of Alabama Press
A fairly recent overview of the mud-glyph cave art tradition in Kentucky.

Faulkner, C. (editor) 1986. *The Prehistoric Native American Art of Mud Glyph Cave*, Knoxville: University of Tennessee Press
A collection of essays on the best-known southeastern United States Mississippian cave art site.

Greer, J. & Greer, M. 1998. *Dark Zone Rock Art in North America. Rock Art Papers*, vol. 13, edited by K. Hedges, San Diego: San Diego Museum of Man
The only regional synthesis available of dark zone cave art in the

ASIA, CENTRAL

This region covers deep interior parts of Eurasia within Kazakhstan, Turkmenistan, Uzbekistan, Tajikistan, and Kyrgyzstan (see map in Russia entry). Most of the region lies within the internal catchment basins of the Caspian and Aral Seas and Lake Balkhash. The relief ranges from lowlands of the eastern Caspian (-132 m) to the high uplands of the Pamir, and some of the highest mountain peaks of the Tien Shan. Because of its interior continental position and isolation from the summer monsoons, arid or ultra-arid climates are characteristic for most of the region. Annual precipitation is commonly within 100–300 mm, but is below 100 mm in many places in lowlands and mountains. Stony, clayey, and sandy deserts on lowlands and uplands are the most common landscapes in Central Asia.

Although soluble rocks crop out through large parts of the region, in both plains and mountains, the dry climate results in generally poor contemporary development of limestone karst. Typical doline-dominated karst landscapes on limestones are found in only a few places in the mountains where local conditions create higher precipitation, as on the Kyrktau Plateau in the Zeravshansky range (see Figure 1). However, caves develop in some mountain areas even below seemingly non-karstic, barren, and fissured surfaces (e.g. in the Bajsuntau region of Tien Shan). It is likely that this cave development is maintained mainly by condensation recharge from the epikarstic zone. A specific "Central Asiatic" arid type of epikarst can be distinguished, characterized by scarcity of dolines and by the presence of the densely fissured near-surface zone, mantles of limestone clasts (the result of physical weathering), and collapse shaft entrances. Another feature of the region is the many hypogenic caves produced by hydrothermal and sulfuric acid speleogenesis. Opened by surface denudation, small relict caves of this type are abundant throughout the mountains of Tien Shan, Pamir, and Kopetdag. Some large caves are also known, both relict and active (e.g. Syjkyrdu Cave in Pamir, Karljuksky and the Fergansky caves in Tien Shan, Bakharden Cave in Kopetdag). Low precipitation is enough to develop pronounced karst in gypsum and salt wherever these lithologies are at or close to the surface. There are many spectacular evaporite karsts in the region, and it is the aridity that allows karst landscapes to survive on evaporites.

Based on geological structure, Central Asia can be subdivided into several large regions (Grozdetsky, 1981): (1) the Turansky Plain; (2) Caledonian and Hercynian mountains of the Tien Shan; (3) the plains and low mountains of Central Kazakhstan; and (4) the alpine mountains of the Kopetdag range and Pamir.

The Turansky Plain has a Paleozoic folded basement mantled with Mesozoic and Cenozoic cover containing carbonates and evaporites. It is a vast desert region that extends between the Caspian Sea, the Tien Shan and Central Kazakhstan, the Ural Mountains, and the Kopetdag range. The Hercynian carbonate structures of the basement locally crop out in the cores of anticlines (e.g. Mangyshlak) or as low mountain ranges. Most of the Turansky Plain is occupied by the large sand deserts of Karakum, Kysylkum, and Mujunkum. Prominent is the Ustjurt tableland that rises above lowlands between the Caspian and the Aral Seas. The plateau is composed of Neogene limestones and underlying gypsum, and is bordered with distinct denudational cliffs. Karst features include large closed depressions, solution dolines, collapse dolines with vertical walls, and many small caves. The longest cave is Sarykamyshskaya (200 m long) and the deepest shaft is Bolojuk (120 m). Adjacent to Ustjurt on the southwest is the Karabogazgol Gulf, where a contemporary deposition of evaporites occurs; it is connected to the Caspian Sea by a channel 4 km long. In the eastern part of the Turansky plain gypsum and limestone karst is reported in the Betpak-Dala desert and the Chujsky region (in Carboniferous gypsum).

The Tien Shan is a block folded system that originated during the Caledonian and Hercynian orogenies and regenerated during the Alpine cycle. The ranges of Tien Shan stretch east to west for almost 3000 km through Uzbekistan, Tajikistan, Kyrgyzstan, Kazakhstan, and northern China, and are up to 400 km wide. The highest peaks exceed 7000 m. The Caledonian northern Tien Shan includes ranges north and east of the Fergansky range and the Talaso-Fergansky fault. The Hercynian southern Tien Shan has two series of ranges separated by the Fergansky depression; the Karatau, Ugamsky, Pskemsky, Chatkalsky, and Kuraminsky ranges lie to the north of the depression, while the

Asia, Central: Figure 1. Doline on the Kyrktau Plateau in the Zeravshansky range. (Photo by John Gunn)

Asia, Central: Figure 2. The splendid limestone escarpment of the Bajsuntau ridge, with a number of cave systems truncated in its scarp face. (Photo by Andy Eavis)

Alajsky, Turkestansky, Zeravshansky, Gissarsky, Bajsuntau, and Kugitangtau ranges lie to the south. Thick and extensive Paleozoic carbonates are widespread in the Tien Shan. Mesozoic and Cenozoic sediments mainly fill depressions but locally form ridges folded and uplifted during the Alpine regeneration (e.g. in the Bajsuntau and Kugitangtau ranges). Ice caves are found in mountain glaciers of the northern Tien Shan (Mikhalev, 1989).

Karst is better studied in the southern Tien Shan. In the ranges to the north of the Fergansky depression, notable karst landscapes and caves are reported from the Ugamsky and Pskemsky ranges. Among many vertical shafts, the Zajdmana Cave (length 830 m/depth 506 m) and Uluchurskaja Cave (1500 m/ 280 m) are the largest. Many limestone massifs lie within the ranges south and southwest of the Fergansky depression, but doline landscapes and active caves are found only in a few localities. Instead, there are a great many small hydrothermal caves. Some larger caves of this type are found along the southern edge of the Fergansky depression, including Kun-Ee-Gout (a 3000 m long, 3-D system of integrated karst cavities and ancient lead-silver mine workings), the 1480 m long Pobednaja Cave, and the 220 m deep Fersman's Cave in Tjuja-Mujun. A spectacular karst is associated with the Neogene salt and gypsum caprocks in the Kysyl Dzhar Mountains; it is densely pitted with dolines, and swallow holes, and numerous caves are reported but not documented.

The most renowned karst of the southern Tien Shan is the Kyrktau plateau, at the western end of the Zeravshansky range at 2100–2400 m. Its classical karst landscapes (Figure 1) include great karren fields, numerous dolines and many vadose shafts in Silurian limestones and dolostones (Klimchouk *et al.*, 1981). During the 1970s, over 100 shafts were explored there including the Kievskaja Cave, 990 m deep and then the deepest cave in Asia.

The Gissarsky includes the Bajsuntau, Chul' Bair, and Kugitangtau ranges, with extensive carbonate and gypsum formations of Jurassic, Cretaceous, and Paleogene ages folded during the Alpine orogeny. There are several remarkable areas of gypsum karst with densely packed dolines, blind valleys, and many caves.

These include the Yakkabag Mountains (on Gissarsky), Mingchukur and Kukhisurkh (on the Bajsuntau) with the Kjaptarkhana Cave over 1000 m long, and Khodzharustam (on the Kugitangtau) with many 20–60 m deep shafts (Gvozdetsky, 1981).

The limestone escarpments of Bajsuntau and Chul' Bair reach more than 3000 m high. Their many important caves include Bojbulok (14 270 m/1415 m, currently the deepest cave in Asia) and the Festival'naja-Ledopadnaja system (13 000 m/625 m) (Bernabei & De Vivo, 1992). On the southwestern edge of the Kugitangtau range, in Turkmenistan, the Karljuksky area is renowned for its highly mineralized caves ramifying in three dimensions, which originated by sulfuric acid dissolution. These include the Cupp-Coutunn Cave System (see separate entry), Khashim-Ojuk Cave (6400 m/170 m), and Geofizicheskaya Cave (4300 m/100 m; see photograph in Speleogenesis: Deep-Seated and Confined Settings). In the nearby lower Gaurdaksky range a large tectonically controlled cave of the same name is another example of sulfuric acid speleogenesis. It begins in Jurassic gypsum but extends horizontally for 11 010 m in underlying limestones.

Central Kazakhstan lies between the Western Siberian platform, the Ural Mountains, the Altai-Sayan folded belt (see Siberia entry), and the northern Tien Shan. Low hills and plains areas are broken by wide valleys and extensive depressions. Most of the region is occupied by dry steppes with salt lakes in the northern part, and by deserts to the west and south of Lake Balkhash.

The area is composed of Caledonian and Hercynian folded Precambrian and Paleozoic rocks, with carbonates, some heavily metamorphosed in the latter. Surface karst features are rare, though features of karst hydrology include intermittent streams, springs, and small karst lakes in the eastern parts. There are many paleokarstic features, attributed to the Cretaceous and late Oligocene, including dolines and large depressions filled with clays and bauxites. Large deposits of bauxites and phosphorites, associated with the upper Devonian and Carboniferous limestones, are of karst origin. These beds are often heavily karstified at depth, with porosities reaching 26–28% and high yields of karst water. Deep-seated cavities are encountered in mines in

Asia, Central: Figure 3. An abandoned trunk passage in the Festival'naya cave system in the Bajsuntau escarpment. (Photo by Andy Eavis)

many bauxite and iron ore deposits. In sulfide ore deposits in the eastern parts, some large cavities were formed due to sulfuric acid dissolution.

The Pamir Mountains, situated mainly within Tajikistan, are essentially of Alpine origin. The central and southern parts of the Pamir belong to the same orogenic belt as the adjacent mega-ranges of the Hindu Kush, Karakorum, and Kun-Lun. The eastern Pamir is a high upland with a very arid climate, with the floors of valleys and depressions at 3600–4000 m and mountains rising to 6000 m. The western Pamir is a series of jagged ranges that rise for 1500–3500 m above the bottoms of deep valleys at altitudes of 1500–2500 m. The highest peaks of the Pamir, in the central and northern ranges, rise to over 7000 m.

Current knowledge of karst and caves is limited to a few areas within the Pamir Upland, particularly in its northern part. Limestones of Silurian, Devonian, Carboniferous, and Triassic ages have extensive outcrops, but contemporary karst development is negligible due to the very arid climate. Cave-riddled, mogote-like, rocky hills near Rangkul' Lake are regarded as paleokarst (Gvozdetsky, 1981). The most notable karst site is the Syjkyrdu (Rangkulskaya) Cave, in Triassic limestones at 4600 m above sea level to the south of the lake. The cave, of presumably hydrothermal origin, is a complex 3-D system with 2050 m of passages mapped so far within a vertical range of 268 m. It contains remnants of a "fossil" underground glacier, "cave loess" sediments, and archaeological artefacts.

Gypsum karst with closely packed doline fields is known at many high altitude sites in the Petra Pervogo and Zaalajsky ranges. Gypsum caves in the Petra Pervogo range include one 400 m long, climbing 126 m above the entrance level.

Spectacular salt karst is formed on the Khodja-Mumyn and Khodja-Sartis salt domes, near Kuljab in the Tadhiksky depression. The domes rise up to 900 m above the surrounding plain and support a great variety of karstic features including dolines, deep collapses and shafts, various karren forms, pinnacles, and giant salt "mushrooms" capped with residual boulders of gypsum (Dzens-Litovsky, 1966). The many caves in the Khodja-Mumyn dome include Dnepropetrovskaya (2500 m/100 m), Komsomol'skaya (1800 m/100 m), Soljenoye Chjudo (870 m/60 m), and Vershinnaja (338 m/120 m).

ALEXANDER KLIMCHOUK

See also **Russia and Ukraine**

Works Cited

Bernabei, T. & De Vivo, A. 1992. *Grotte e storie dell'Asia Centrale*, Padova: Cento Editorale Veneto (in Italian and English)

Gvozdetsky, N.A. 1981. *Karst*, Moscow: Mysl (in Russian)

Dzens-Litovsky, A.I. 1966. *Salt Karst of the USSR*, Leningrad: Nedra (in Russian)

Klimchouk, A.B., Rogozhnikov, V.Ja. & Lomaev, A.A. 1981. *Karst of the Kyrktau Massif, Zeravshan Ridge, Tien Shan*, Kiev: Institute of Geological Science (in Russian)

Mikhalev, V.N. 1989. *Karst of Kirgizstan*, Frunze: Ilim (in Russian, extended abstract and figure captions in German)

ASIA, NORTHEAST

Eastern Asia has the world's most extensive karst landscapes, largely in the provinces of Guangxi and Guizhou in southern China (see separate China entry). There are also many other significant karst regions within China, but the adjacent lands of Tibet, Mongolia, Korea, and Taiwan all have only minimal karst, while there is little more in Japan. Similarly there is little significant karst in Siberia, though there is a profusion of karst in adjacent countries to the south (see Asia, Southeast entry). Climates in the region vary from wet sub-tropical in southern China to cold, dry continental in Mongolia. This does have a major influence on karst morphologies, but the areas with less favourable climates broadly coincide with those with less limestone, thus enhancing the regional differences.

Mongolia has scattered outcrops of limestone beneath its grassland steppes and barren mountains. Many fragments of relict cave have been found, none longer than 500 m, but no active caves are known. Modern karst processes above and below ground appear to be on minimal scales in the cold, dry climate.

Korea has two belts of broken outcrops of strongly folded Lower Paleozoic limestones across its mountainous terrain. The smaller belt, in South Korea, contains a number of short caves, of which a high proportion are very well decorated with calcite and have been developed as tourist sites. Its deepest cave, Namgamduk, is a shaft system reaching 181 m. There are some areas of doline karst, but most of the outcrop appears as steep wooded slopes with incomplete underground drainage. Little is recorded about North Korea and there are no known regions of conspicuous karst landforms, though a number of decorated caves have been developed for tourist access. All the longest caves in Korea are lava tubes in Cheju Island (see separate entry), off the coast of South Korea.

Japan has over 1500 km^2 of karst on small, disconnected outcrops of limestone scattered across this plate boundary terrain of fold mountains and active volcanoes. Karst occurs on all four main islands, but the most significant sites are on Honshu. In the north, the Iwate karst has a number of long caves including Akka-Do with over 8 km of mapped passages. On the west coast, the Niigata karst contains the country's deepest caves with Byakuren-Do (422 m deep) and three others over 300 m deep. The single most important karst is on the Permo-Carboniferous limestone of the Akiyoshi Plateau (see separate entry), near the southwestern tip of Honshu. Uplifted reef limestones are widespread on Okinawa and the Ryukyu Islands, where there are areas of doline karst, numerous caves including Gyokusan-Do with over 5 km of passages, and marine platforms of fretted limestone. There are lava tubes in various of the basaltic volcanoes, with caves over 2 km long on the island of Fukue.

Taiwan has little limestone and little karst. In the eastern flank of the core mountains, the Taroko Gorge has its most deeply incised section through a belt of marbles, but it is a fluvial feature characterized by surface flow on steep slopes. Springs in its floor carry only a small proportion of drainage from the marble outcrops.

TONY WALTHAM

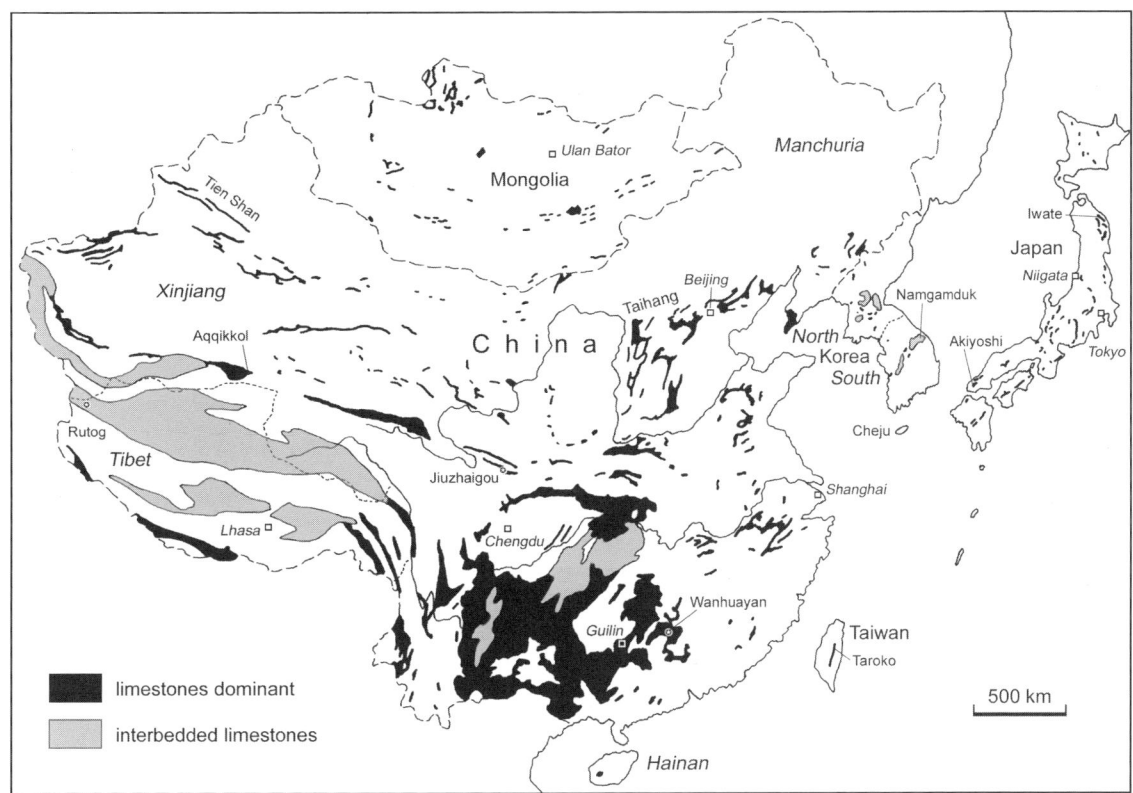

Asia, Northeast: Generalized map of the carbonate outcrops of northeast Asia. These areas are much larger than the areas with significant karst and include areas of dolomitic limestone. Locations of many key karst sites in southern China are marked on the enlarged map in the China entry.

See also **Akiyoshi-dai Karst and Caves; Asia, Southeast; Cheju-do Lava Caves; China; Di Feng Dong; Hongshui Karst Plain; Huanglong and Jiuzhaigou; Russia and Ukraine; Shilin Stone Forest, China; Yangshuo Karst**

Further Reading

Urushibara-Yoshino, K. & Kashima, N. 1998. Karst of Japan. In *Global Karst Correlation*, edited by Yuan D. & Liu Z., New York: Science Press

ASIA, NORTHEAST: HISTORY

The countries covered in this entry are China, Tibet, Japan, Korea, Mongolia, and Taiwan. The Chinese and Tibetans have been exploring and using caves for shelter and religion for thousands of years. Reference to karst in China is found frequently in early Chinese literature and is depicted in their art, particularly of the towers of Guilin, while caves featuring Buddhist art have been tourist attractions since ancient times. Studies of Japanese karst have a substantial history in the 20th century and those in Korea less so, while in Mongolia and Taiwan caves seem to have remained unnoticed until comparatively recently.

China

As befits the country of such an ancient civilization, much was known about Chinese karst and caves from very early times up to the mid-17th century. Then there was a long interval before the next phase of karst research in China began after World War II with international links and exchange programmes. The archaeological excavation in the 1920s of "Peking Man" from the cave at Zhoukoudian is described in the entry China: Archaeological Caves.

It is said (Sweeting, 1995, p.33) that caves in north China were described in a book written before 221 BC. More definite, because the texts are known today, are descriptions of speleothems and their use in powdered form as medicine from the 4th century BC to the 16th century AD. Their efficacy in this respect was very real, for people with a calcium-deficient diet benefit enormously from them, just as modern Chinese benefit from drinking crushed Pleistocene "dragon's teeth" to alleviate heart conditions. The pharmaceutical use of speleothems led to their being studied themselves, in the same way that European medicine led to physic gardens and the study of botany (Shaw, 1992).

Ko Hung, in the *Pao Phu Tzu* of about 300 AD, described two objects which are clearly stalagmites. In the Shao-Shih

Mountain, "there is a stone pillar standing more than a hundred feet from the entrance to the caves, having on top of it a stone looking like an upturned lid, ten feet high" (quoted from *Science and Civilization in China* by Needham, 1959; vol.3, p.606). Drips, he says, fall into this from the cave roof above. Again: "The shih-kuei chih (stone cassia mushroom) grows in the caves of holy mountains. It resembles the kuei tree (cinnamon), but it is really a stone. It is more than one foot in height, and about one foot in diameter. It is bright and shining. Its taste is sharp, it has branches and twigs" (quoted from translation by Eugene Feifel, in *Monumenta Serica*, 11:7). In 1175 AD Fang Chengda wrote that "Milky water drips continuously in the cave, with the process of condensation taking place simultaneously. [Each stalactite] hollow in its centre is just like a goose feather . . . and the milky bed as white as snow is formed by the condensation of stone solution" (quoted from *A Brief Introduction to China's Research in Karst*, Yuan Daoxian, 1981, p.3).

A Chinese drawing (Figure 1) of speleothems was included in the *Pên Tshao Kang Mu* (The Great Pharmacopoeia) of Lin Shih-Chen in 1596. The earliest known reference to speleothems being used medicinally is their inclusion among the *materia medica* in the early Chinese pharmacopoeias. These include the *Chi Ni Tzu* which dates from the 4th century BC and the *Shen Nung Pên Tshao Ching* (Pharmacopeia of the Heavenly Husbandman) about the 1st century BC. A silver box containing fragments of medicinal stalactite dating from the 8th century AD was unearthed in 1970 at a Tang site in Shensi province. In about 300 AD, Ko Hung's stone mushrooms in caves were thought to provide a recipe for long life, either by drinking the water from them or from the material itself, pounded and eaten. The continuing importance of these substances is shown by their being

Asia, Northeast: History: Figure 2. Xu Xiake.

discussed, with their various uses, in the later mineralogical books such as Tu Wan's *Yün Lin Shih Phu* (Cloud Forest Lapidary) of 1130 and Chou Chhü-Fei's *Lin Wai Tai Ta* of 1178. In 1596 Li Shih Chen's Great Pharmacopoeia listed under different names seven kinds of stalactites and stalagmites, separating them not only by coarseness and shape, but also by their being different parts of the speleothems, such as the "root part". At about this time a printed reference of 1541 records a cave fish, now called *Sinocyclocheilus hyalinus*, being found in the Alu Caves of Yunnan (see Biospeleologists entry).

All the above are but interesting details compared with the comprehensive work of Xu Xiake, the assumed name of Xu Hongzu (1587–1641) (Figure 2). He is rightly regarded as the father of Chinese speleology and would perhaps have been the father of world speleology, coming as he did even before Kircher and Valvasor, if his writings had been more widely known to his successors. His book, which was written over a period up to 1640, was the first to describe karst landforms systematically. The original manuscript is in the National Library of China in Beijing, and it was first printed (in Chinese) in 1776. The book is a diary recording his travels around China, undertaken on foot and usually alone over a period of more than 30 years. Between 1637 and 1640 he was in the karst of south and southwest China, spending 976 days there in all. The karst features described include most of the karst landforms that we recognize today, such as karren, dolines, uvalas, poljes, natural bridges, dry valleys, disappearing rivers, tower karst, blind valleys, water swallets, cave deposits, and caves themselves. His descriptions of these karst features are detailed and basically accurate. Some modern special terms in karst studies in China, such as fenglin or peak forest, originated from Xu Xiake's descriptions.

In Guangxi province alone, Xu Xiake explored more than 160 named caves and tens of caves without names. Many of these caves had never previously been explored. When he explored them, he measured their size, depth, and other dimensions by pacing or by visual estimation. He also collected samples

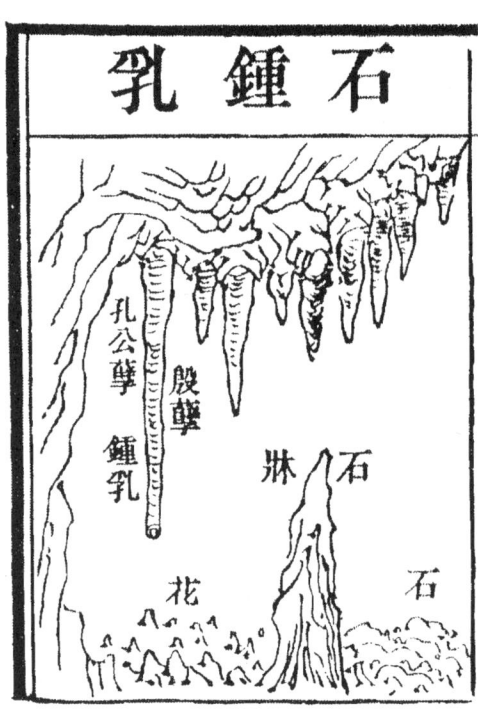

Asia, Northeast: History: Figure 1. Speleothems in Li Shih-Chen's book of 1596.

of cave deposits. His descriptions included the distribution, structure, and relationships of the caves; the shape, size, and the direction of cave passages; direction, size, and shape of cave entrances; the shape, size, and colour of cave deposits; flow of water; and explanations of the formations of deposits. Thus, of a big column hanging from the roof: "this is the result of solution and precipitation of rock". Ting (1921) said that Xu Xiake's diary "read more like a 20th-century surveyor's field book than the diary of a 17th-century traveller". A modern assessment of Xu Xiake's work (Hu, 1991), from which much of this summary is drawn, gives a route map of his travels in the karst. A statue of him now stands outside the Institute of Karst Geology in Guilin.

There followed several centuries in which no significant work on karst took place in China. Some studies began again in the 1930s and 1940s and then, after the People's Republic of China was formed in 1949, Europeans from the Eastern bloc were encouraged to visit the tower karst of Guilin where the Karst Institute was to be founded in 1976. In the 1970s and 1980s other Europeans were also welcomed and Sweeting (1995) describes her own and other collaborations. There was Paul Williams's work with the Chinese Academy of Sciences in 1976, followed by a Franco-Chinese hydrological project, and Anglo-Chinese and Chinese-American projects. Of particular note is the China Caves Project, an informal collaboration between British cavers and scientists from the Institute of Karst Geology in Guilin. The project commenced in 1982 (Waltham, 1986), and is ongoing. There have been over 15 expeditions to China and many exchange visits. More recently a most productive liaison has developed since 1995 between the governments of China and Slovenia with exchange visits and joint work by the Karst Research Institute in Postojna, the Yunnan Institute of Geography, and the Institute of Geology at the Chinese Academy of Sciences in Beijing (Chen Xiaoping *et al.*, 1998). Work by the Chinese themselves has developed considerably and all these collaborative projects ensure that they are fully aware of the techniques and traditions of research elsewhere.

Japan

The earliest description of a Japanese cave seems to be the report on Akiyoshi Cave published in 1904 in Japanese (Torii, 1957–59). More widely known in the West is the slightly later account by an English resident, the schoolmaster George Edward Gauntlett (1868–1956) who visited it in 1907. The cave had already been known for a long time but it was difficult to access. Gauntlett (1909) published a plan and offered to pay for some of the passages to be enlarged and a boat to be provided to encourage visitors. A set of picture postcards was issued at the same time. Gauntlett also reported seven other caves nearby, including a vertical one which he planned to explore. A stream sinking 10 km to the northeast was said to have been traced to the main cave by throwing young rice stalks into the water.

Systematic exploration of Japanese caves was first undertaken by the geologist Denzo Sato from 1920–25 (Torii, 1957–59). Professor Ohhashi had investigated a cave at Akita in 1915 and some in Okinawa were explored by Awatsu in 1920. Biospeleological research was started in 1927 by Professor Masuzo Uémo, later the President of the Speleological Society of Japan. There was more work in the 1930s: Torii collected at 62 sites between 1936 and 1939 and Dr Verhoeff published on the subject from 1939–43. Relatively little was done on prehistoric cave remains but Ice Age humans were found in several caves in 1938, and the discovery of human bones of Pleistocene age was published in 1950. Japanese biospeleologists also worked in Korea with their colleagues there, notably in 1966 (see below). Several other expeditions abroad took place in later years including to Korea in 1976, to the Philippines in 1982 and 1983, and to Taiwan from 1996.

The Speleological Society of Japan was founded in 1975 and started to publish its *Journal*, which still continues, in the next year. The Japan Caving Association was formed in 1959 and its *Japan Caving* commenced in 1969.

Korea

The 1500 m long Dongryong-gul cave in North Korea is said to have been known since the 7th or 8th century (Shikama, 1948) and to have been used as a refuge in the Russo-Japanese and Japanese-Chinese wars. The same cave was surveyed in 1929. Cave fauna was collected there and elsewhere around 1939 and biospeleology has driven much of the subsequent cave research in Korea. In 1966 a joint Korean-Japanese expedition in South Korea explored 21 limestone caves and two lava caves, surveying some of them, and collecting fauna from them and from 49 wells. More biological work followed.

The Korean Caving Association was founded in 1966 and by 1975 had become the Korean Association for Conservation of Caverns which afterwards surveyed those caves designated as national monuments. A cave survey team in the National Science Museum listed 92 caves in 1974, with location, and published references, and the cave Exploration Club of Dongguk University, formed in 1970, reported 66 species from more than 150 caves. The Speleological Society of Korea was formed in 1973 and published a journal from 1975. Its members worked closely with their Japanese counterparts who visited Korea again in 1976 (Lee, 1989).

Mongolia

The karst of Mongolia was almost unrecognized until the 1990s, though the Czech ethnographer Pavel Poucha had noticed a cave near the Russian border in 1956. A Czech-led international expedition in 1994 published a map showing widespread patches of karst throughout the country, especially in the extreme north near Chövsgöl nuur. They surveyed several caves up to 300 m long, remarked on the presence of permafrost and noted a cave in pseudokarst in western Mongolia (Holúbek, 1995). Three British visits from 1994 to 1996 investigated caves in the west of the country (including one with paleolithic-style wall paintings of animals), and the Gobi desert.

Tibet

In 1715 two Jesuit priests, Desideri and Freyre, visited the Padmasambhava cave at Chiu Gompa monastery in Ngari province (Gebauer *et al.*, 1995, p.224). The Tibetan karst has also been studied by the Chinese since 1951 (Sweeting, 1995, p.39, pp.212–19), and Loubière (1987) reported a 100 m long cave 15 km north of Lhasa that contained water as well as religious paintings. In the 1980s and 1990s Sweeting and Zhang Dian recorded caves between 4000 m and 5000 m above sea level in northern Tibet, in the Dingri area on the slopes of the Himalayas, and near Lhasa. All were 100 m or less in length and all were dry, having been formed when the water table was much higher. Evidence from the caves suggests a tropical Tertiary period, a warm and wet early and middle Pleistocene phase from which most of the dated speleothems came, followed by the present cold and arid climate from the late Pleistocene.

Taiwan

Less is known about caves and karst in Taiwan than in any other part of northeast Asia. There is reef limestone in the south of the island and Hsu (1975) mentioned three caves there, indicating that they had been little studied. The Japanese started serious investigations in that Taroko Gorge limestone area in 1996.

TREVOR SHAW

Works Cited

Chen X. *et al.* (editors) 1998. *South China Karst I*, Ljubljana: Založba ZRC

Gauntlett, [G].E.[L]. 1909. The caves of Yamaguchi. *The Yorkshire Ramblers' Club Journal*, 3(9): 41–44

Gebauer, H.D., Mansfield, R., Chabert, C. & Kusch, H. 1995. *Speleological bibliography of South Asia 1995*, Kathmandu: Armchair Adventure Press

Holúbek, P. 1995. International expedition AGUJ 1994. In the depths of Mongolia. *International Caver*, 13: 11–16

Hsu, T.L. 1975. A note on caves of Taiwan. *The Cascade Caver*, 14(2 & 3): 28

Hu B. 1991. Xu Xiake, A Chinese traveller of the seventeenth century and his contribution to karst studies. *Cave Science*, 18(3): 153–57

Lee B.H. 1989. Speleology in Korea with special reference to biological surveys. *Proceedings of the 10th International Congress of Speleology*, edited by T. Hazslinszky & K. Bolner-Takacs, Budapest: Hungarian Speleological Society: 762–65

Loubière, J.F. 1987. Tibet. *Spelunca*, 28: 12–13

Shaw, T.R. 1992. *History of Cave Science: The Exploration and Study of Limestone Caves, to 1900*, 2nd edition, Broadway, New South Wales: Sydney Speleological Society

Shikama, T. 1948. Toryukutu cavern, North Korea. *Cave Science*, 1(4): 121–22

Sweeting, M.M. 1995. *Karst in China: Its Geomorphology and Environment*, Berlin: Springer

Ting, V.K. 1921. On Hsü Hsia-K'o . . . explorer and geographer. *The New China Review*, 3(5): 325–37

Torii, H.S. 1957–59. Kurze Geschichte der Höhlen- und Karstforschung in Japan [A short history of cave and karst research in Japan], *Die Höhle*, 8(4): 104–107; 9(2): 37–40; 10(1): 11; 10(4): 99–102

Waltham, A.C. (editor) 1986. *China Caves '85: The First Anglo-Chinese Project in the Caves of South China*, London: Royal Geographical Society

ASIA, SOUTHEAST

Karst is a widespread and important phenomenon in mainland Southeast Asia. The countries of Myanmar, Thailand, Laos, Vietnam, Cambodia, and mainland Malaysia all contain significant limestone karst areas, which cover around 215 000 km², or 10% of the region. Large areas of sandstone and buried evaporite karst are also present. The karst occurs mostly on carbonate rocks older than Jurassic, with the Permian and Carboniferous being the most important, located primarily on structural highs in association with Paleozoic and Mesozoic formations.

Cambodia has limited karst areas. Tower karst dominates the south near Kampot and Tuk Meas. More extensive areas are in the northwest near Pailin, Battambang, and Sisophon, with some around Stung Treng. Laos has the larger part of the extensive Khammouan karst (see separate entry, Khammouan). Other extensive karst is found at Vangvieng, Kasi, Luang Prabang, the Nam Ou valley, and Viengsai. Mainland Malaysia has scattered small karst areas in the central part in Perak, Kelantan, and Pahang States, and near Kuala Lumpur. In the northwest, Perlis State and Langkawi Islands have larger areas of karst. Myanmar has large karst areas on the 500 km wide Shan Plateau, with more at Mogok, near Hpa-an / Mawlamyine and in the Myeik archipelago. In Thailand, the 12 000 km² "Western Karst Complex" lies near the Myanmar border. Other large karst areas are at Phangnga, Krabi, Saraburi, Mae Hong Son, Chiang Dao and the Loei / Chum Phae area. In Vietnam, the largest karst areas include the continuation of the Khammouan from Laos and many areas in the north at Lang Son / Cao Bang, Ha Giang, Song Da / Son La and Ha Long Bay (see separate entry, Ha Long Bay).

Climatic Conditions and Vegetation

The region has a tropical to sub-tropical monsoon climate, with strong seasonality. Rainfall usually arrives from May to October (southwest monsoon) and more locally from November to January (northeast monsoon). Total rainfall is around 2500 mm a^{-1} at low to moderate elevations, increasing to more than 5000 mm on the mountain sides exposed towards the southwest. Average annual temperature at low elevations varies from 28°C in the south to 25°C in the north, and to 21°C at altitudes of 1000 m. High temperatures, strongly seasonal rainfall, and high evapotranspiration lead to long periods when there is autogenic recharge to the karst aquifer. Southeast Asia has been subject to moderate fluctuations in climate during Quaternary glacial episodes. The effect of such periods was reduced rainfall and temperature, with only limited impact on karst processes. During some interglacials, it was warmer and wetter than the present.

Vegetation type and density is very variable on the karst. In equatorial areas, rain forest may be present and quite dense. In unspoilt areas of Laos, forest has a lower density, and it is better developed on shaly limestones. In more arid areas, such as Kanchanaburi, Thailand, the vegetation is more xerophytic and deciduous. The native vegetation has been heavily destroyed in many areas by human activities.

Tectonic History

Morphogenetic evolution of the karst is the result of a complex tectonic history. Notwithstanding many uncertainties, three main periods may be considered. The oldest carbonates, with a real importance, are Ordovician. They are partly metamorphosed and outcrop discontinuously in the mountains from northwestern Thailand to mainland Malaysia. Silurian carbonates are found in Vietnam. Devonian carbonates are found mainly on the Shan Plateau, Myanmar, and in northern Vietnam. Carboniferous carbonates are well developed in the Khammouan karst of Laos and Vietnam. Permian carbonates occur

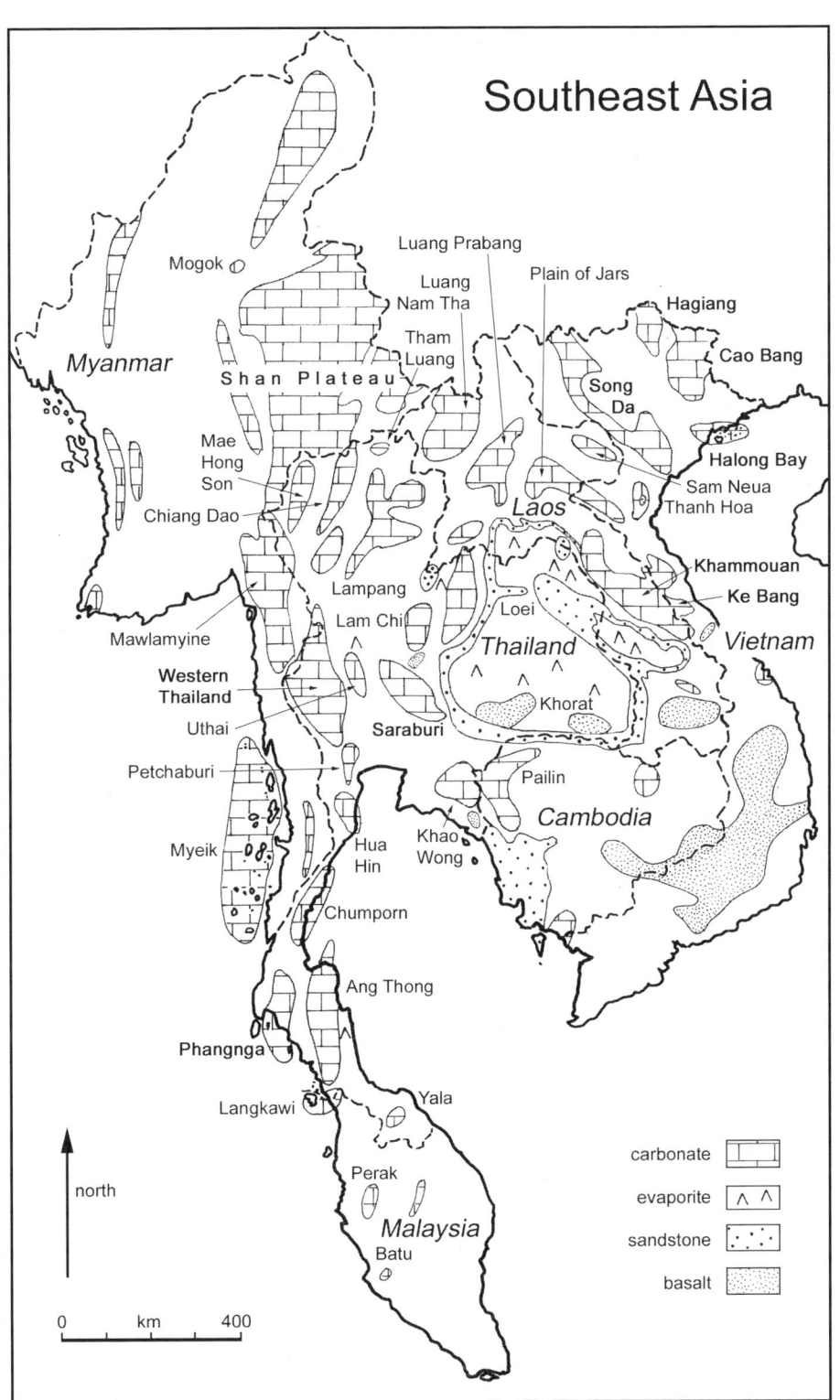

Asia, Southeast: Figure 1. The main karst regions of continental Southeast Asia.

widely throughout Southeast Asia. In Thailand, they are composed of massive limestone. On the Shan Plateau, Myanmar, they are in the thick Devonian to Triassic sequence that is pervasively dolomitized carbonate.

The Indosinian orogeny occurred in the middle Triassic. This significant compression, uplift, and erosion episode affected much of Southeast Asia, bringing many carbonates into subaerial positions, resulting in a prolonged karstification episode. Upper Triassic half-grabens, formed via tectonic relaxation, were filled in part by lacustrine limestone. Occasional Liassic limestones, up to 50 m thick, are interbedded with thick continental formations.

Regional uplift began around 65 million years ago. Combined with the effects of the young Himalayan orogeny and the opening of the South China Sea, it led to large-scale erosion. For example, some 3250 m has been eroded from the central Khorat Plateau in northeast Thailand. As carbonates of various ages became exposed, karst evolution began in earnest. Indosinian karst landforms have often been exhumed and rejuvenated and play a significant role in the evolution of the Cenozoic karst. This long-lasting uplift generated a relative deepening in base level, leading to extensive planation surfaces, large and deep poljes, tower karst, and fengcong karst.

Paleokarsts and Cave Fills

Paleokarsts have formed mainly during the Indosinian and the Cenozoic erosion episodes. The stratigraphic gap of the Indosinian period reaches a maximum of 60 million years in Central Laos, though effective erosion time may only have lasted 20–35 million years. Paleokarst landforms of the Permo-Carboniferous limestones include karstified surfaces (southern Thailand), valleys and low rolling hills (north central Thailand and central Laos), and tower karst (same regions plus northwest Cambodia and Ha Long Bay). Karst breccias, with a reddish, calcareous silty matrix, are common and have been dissected by later caves and quarries. Paleokarst fills were subsequently buried and underwent diagenesis to the same degree as the host carbonate.

The offshore Nang Nuan oil field, in the Gulf of Thailand, appears to produce from caves in Permian to Jurassic limestones buried beneath thick Cenozoic sediments. In Myanmar, the Yadana gas field, also offshore, is located in the karstified, though highly porous, upper part of an upper Oligocene to lower Miocene isolated carbonate platform.

The current erosion episode has generated more paleokarsts that are moderately lithified, as in Cambodia. Pleistocene paleokarst fills and associated faunas have been described from Thailand and Myanmar (vertebrates at Mogok). Partial cave fills, many with prehistoric remains, have been found throughout the region, notably at Tham Hang and Tham Pha Loi, in northern Laos, and in Cambodia.

Surface Morphology

Southeast Asia contains some of the most spectacular surface karsts in the world, with a wide variety of tropical karst landforms. Lithological differences determine variations in morphology. Thinly bedded, argillaceous, and partly metamorphosed Ordovician limestones in Thailand and Malaysia, and Liassic limestones in Thailand and Laos, show a massive morphology with moderately steep slopes and few holokarst features. Massively bedded Permo-Carboniferous limestones, though sometimes partly dolomitic, evolve into holokarsts. The advanced diagenesis, fracturing, and purity of these limestones produces sharp-edged and spectacular landforms. Their salient feature is the existence of karst massifs, with very steep slopes or cliffs rising from karst plains, which merge with the subdued topography of other rocks. Karst plains result from planation at the rainy season water-table level. Lateral dissolution forms notches, resulting in cliff collapse and increased slope angles. In Ratchaburi, Hpa-an, and Kampot, a raised paleo-sea level probably generated some planation through wave erosion and mixing corrosion.

Positive landforms consist of fengcong, towers, and pinnacles, some of which are highly developed (e.g. the tsingy morphology of the Khammouan karst). Negative features include poljes, uvalas, dolines, and large shafts, giving access to underground streams (e.g. near Kanchanaburi). Major canyons often cut through karst terrains, for example, those of northwest Vietnam and the parallel Salween and Mae Khlong Rivers of Myanmar and Thailand, respectively. Karren are very well developed and excellent examples of most forms can be seen on virtually any bare rock surface. Travertine waterfalls are common features and the huge falls near Pyin-Oo-Lwin, Myanmar, are a good example.

Caves

In holokarst areas, caves are extremely numerous and often large. The Phong Nha system in Vietnam, is to date 44.5 km long, but it is made up of separate caves. The longest known caves of the Southeast Asian countries are: Laos—Nam Non Cave, 22.1 km; Vietnam—Hang Khe Ry, 18.9 km; Thailand—Tham Phra Wang Daeng, 13.1 km; mainland Malaysia—Tempurung Cave, 4.8 km; Myanmar—Mundewa Cave, 1.8 km; and Cambodia—Roung Dei Ho / Roung Thom Ken system, 1.8 km. The largest cave entrance is Tham En in Khammouan, 215 m wide and 30 m high. Large chambers include: 260×240 m in Tham En, and 210×155 m in Tham Koun Dôn, both in Laos. Tham Toutche, near Kanchanaburi, has a chamber $200 \times 180 \times 50$ m, entered via a passage 100×35 m and 300 m long. The largest subterranean river flows through the 6.5 km long, 40×40 m main passage of the Xé Bang Fai Cave, at an average 68 $m^3 s^{-1}$. The deepest cave is Cong Nuoc in Vietnam (600 m deep). Tham Phi Seau in Laos gently rises up to $+315$ m.

Caves are predominantly subhorizontal. Multilevel systems, containing fine examples of phreatic and vadose morphologies, have been found. Caves in mountainous areas often show complex evolutionary histories, notably Tham Chiang Dao in northern Thailand. Vertical systems are not so common, though some deep pits do exist. Speleothems of all kinds are frequent and sometimes spectacular, such as the 60 m high column in Tham Sao Hin, Kanchanaburi, a 2 m wide shield in Tham Nam Thieng, Khammouan, and cave pearls to more than 20 cm in size in the latter cave.

High atmospheric CO_2 levels are encountered in some caves, with up to 8% being measured in Tham Kubio, northeast Thailand. High CO_2 is particularly prevalent in caves of western and northern Thailand and near Taunggyi, Myanmar. The possibility of a deep-seated CO_2 source is likely, as there are numerous thermal springs in the region and other geological indications. A biogenic source is more likely in Phangnga area, southern Thailand.

Partly Drowned Karst

Several coastal karst areas have been inundated by the sea, due to tectonic subsidence (from mainly extensional deformation) and / or rising eustatic sea level. The best examples are the islands found along the coast of the Andaman Sea from the Myeik Archipelago to Phangnga Bay, the Langkawi Islands, the Ang Thong Islands in the Gulf of Thailand, and Ha Long Bay, Vietnam. Karst morphology, prior to inundation, was mature to very mature. Ang Thong consists of relatively massive hills, though tower karst is most common in other locations. Only towers and hills protrude from the sea with the karst plain, the area between them being fully submerged. The towers may form groups that start inland and extend out into the sea.

A prominent solution notch often marks present-day sea level. Locally, another notch occurs at 4–6 m above, representing a former high sea-level stand. Caves may pierce the towers as inland stream caves, as tidal range caves (some opening in the notches) and completely submerged caves, with the best examples in the Andaman Sea and Ha Long Bay. Another interesting feature is the "hong", located within many of the islands. It is a partly drowned chamber, with the top often partly opened to the sky, usually accessible via boats through a cave at low tide. "Hong lakes" are large dolines, with marine to brackish lakes, surrounded by sub-vertical cliffs, and can reach more than 300 m in diameter. Good examples are found in Phangnga Bay, Langkawi, Ang Thong, and Ha Long Bay.

Most of the caves and "hongs" undoubtedly formed prior to invasion by the sea. However, karstification, due to mixing corrosion at mainly rainy-season haloclines, is a powerful process. It modifies inherited morphologies and creates new ones, such as the many tidal range caves and notches seen in the area. Terrigenous sediments, supplied by rivers, form overall prograding tide-dominated coastal deposits that progressively cover the karst until eventual burial. The tower karst of Mawlamyine, Myanmar, is undergoing this process. In such areas, a sheet of freshwater may cover the area over long periods during the rainy season.

Dolomite Karst

Dolomite fluviokarst covers large parts of the Shan Plateau, Myanmar. The low competence of the rock produces smooth, rounded landforms, including cupola-shaped hills with low angle slopes and rounded valleys. Cliffs are found mostly along the main valleys that progressively deepen into canyons and reach the plateau edge. Travertine dams and flowstones are common along the streams. The deep gorge at Gogteik is the most famous and has travertine caves. Cave formation processes are depressed in dolomite. Explored caves have relatively small cross sections. The most significant are Mundewa Cave near Taunggyi (1.8 km) and Peik Kyin Myaung Cave near Pyin-Oo-Lwin, both with streams.

Mineral-Bearing Karst

Limestones and interbedded silici-clastic rocks at Mogok, Myanmar, have undergone high regional metamorphism and intrusion by acid igneous dykes. Rubies, sapphires, spinels and other high-quality gemstones have crystallized in the limestone. Karstification produced a characteristic mountainous landscape with barren lands and crypto-lapies. A large proportion of the area has a thick soil cover. Some dolines are present, sometimes very impressive. Many caves are found, mostly with small passages. Gemstones are concentrated in cave fills, in colluvium trapped between the rockhead pinnacles and in valleys. Hydrothermal ore-bearing veins are present in some karst areas. Cassiterite (tin ore), eroded from nearby granites, occurs as karst placer deposits in peninsula Malaysia. The tin-bearing alluvium is exploited from depressions in the karst surface and from cave fills. Saltpetre is extracted from certain caves in Myanmar, Laos, and Vietnam.

Sandstone Karst

Karst features and caves are known in the thick Mesozoic sandstones of the Khorat Plateau, northeast Thailand. Surface karst is often relatively bare and the landforms are somewhat different to those of the carbonates. The sandstones tend to suffer more disintegration than solution, with sharp-edged features being rare. In planar areas, with low to moderate dip, the surface drainage converges into dip-directed, incised channels. A few of these end in blind valleys, where the stream sinks underground as at Tham Patiharn. Runiform landscapes and pinnacles occur where a duricrust or a conglomeratic bed has selectively protected from erosion. Dolines are rare. Caves rarely exceed 100 m in length. The exception is Tham Patiharn at 784 m long, with a boulder choke at the end. The upper and lower level passages in this seasonal stream cave are similar to limestone caves and show large sections. A few siliceous speleothems are present. Many of the small caves are outlets whose formation was comparable to that of soil pipes.

Human Use of Caves and Karst

Caves and karst in Southeast Asia have been widely used by Man since prehistoric times (see separate entries, Asia, Southeast: Archaeological Caves; Burials in Caves). Buddhism is the dominant religion in the region and makes use of many caves as hermit retreats, underground temples, and places of worship. It has been estimated that Buddhism utilizes several hundred caves in Thailand alone (see Religious Sites). Some very famous Buddhist caves are Pindaya Cave, Shan Plateau, containing more than 8000 Buddha images, the Pak Ou Caves near Luang Prabang, Laos, Tham Khao Luang in Thailand and the Marble Caves in Vietnam. The Batu Caves in Malaysia (Figure 2) are used for worship by followers of Hinduism. Animism related to caves still exists in Laos, Myanmar, and northern Thailand.

Caves have long been used as hiding places for people and valuables. Villagers once hid in caves in Kanchanaburi to evade Burmese warriors. Caves in Laos were used by soldiers as moving

Asia, Southeast: Figure 2. The main chamber of the Temple Cave in the Batu cave system in Malaya, with the concrete floor in readiness for frequent religious festivals. (Photo by Paul Wycherley)

bases during World War II. Caves were used as shelters during the Vietnam War in Vietnam and Laos, especially along the Ho Chi Minh trail. In the Viengsai area, they were converted into factories, schools, administrative centres, hospitals and dormitories, etc. Phatet Lao political leaders lived in natural and reshaped caves in Viengsai for around ten years. Caves on the seashore, which have an open entrance at low tide, have provided pirates with hideouts in the Andaman Sea. The 7.5 km long Nam Hin Boun underground river in Khammouan has been used as a throughfare for more than a century and is traversed by boats ferrying both passengers and goods. Caves such as the 1.2 km long main way in Tham Heup, Khammouan, are used by people walking to reach their remote villages in Laos and Vietnam.

Natural resources gathered from caves include gemstones, tin ore, guano, saltpetre, water for drinking and agriculture, and food in the form of fish, bats, swifts and swift nests. Karst quarrying is a massive industry. Offshore Nang Nuan oil field, in the Gulf of Thailand, and Yadana gas field, in the Andaman Sea, are in karstified limestone.

CLAUDE MOURET

See also **Khammouan, Laos-Vietnam**

Further Reading

Allen, T. 1993. Caves in the province of Lang Son, Vietnam. *International Caver*, 9: 3–10

Bates, J. & Nwe, D.T. 2002. Myanmar: An atlas of karst conservation. *International Caves 2001*, 36–39

Dunkley, J., Sefton, R., Nichterlein, D. & Taylor, J. 1989. Karsts and caves of Burma (Myanmar). *Cave Science*, 16(3): 123–31

Dunkley, J. 1995. *The Caves of Thailand*, Sydney: Speleological Research Council

Khang, P. 1991. Présentation des régions karstiques du Vietnam. [Presentation of karst regions of Vietnam]. *Karstologia*, 18: 1–12

Lagrou, D., Coessens, V. & Masschelein, J. 1997. Belgian-Vietnamese speleological expedition in Son La 1995–1996. In *Proceedings of the 12th International Congress of Speleology*, vol. 1, La-Chaux-de-Fonds, Switzerland: Swiss Speleological Society

Laumanns, M. 1996. Cambodia 1996. *International Caver*, 17: 34–39

Mouret, C. 1994. Paleokarsts at the Permian-Triassic boundary in Southeast Asia. An introduction. *Supplement to Proceedings of the 11th International Congress of Speleology, Beijing, 1993*, Beijing: Chinese Academy of Science

Mouret, C. 1997. Human use of caves in Myanmar. *Proceedings of the 12th International Congress of Speleology*, vol. 1, La-Chaux-de-Fonds, Switzerland: Swiss Speleological Society

Mouret, C. 1997. Human use of caves in Laos. *Proceedings of the 12th International Congress of Speleology*, vol. 1, La-Chaux-de-Fonds, Switzerland: Swiss Speleological Society

Price, L. 1998. *Malaysian Cave Bibliography (up to 1997)*, Kuala Lumpur: Gua Publications

ASIA, SOUTHEAST ISLANDS

With a total area of about 230 000 km^2, the islands of Southeast Asia contain some of the most varied karst regions in the world (Gillieson & Spate, 1997). Many are of high relief, with spectacular arrays of tower and cone karst, several of which have now been added to the World Heritage List in recognition of their special geomorphology and biology. They are scattered throughout the islands of the Malaysian archipelago as well as the adjoining fringe of the Asian mainland. Karst is found in Borneo, the Philippines, Indonesia, and the island of New Guinea (Figure 1). The tectonic setting of the Southeast Asian region is complex, due to the interactions between the Philippine, Pacific, Australian, and Eurasian plates. An older, stable region comprising the Asian mainland, peninsular Indo-China and Borneo abuts a younger unstable region associated with widespread neotectonism coupled with island-arc volcanism. The most extensive limestones in Borneo are the massive shallow-water Eocene to early Miocene rocks in northern Sarawak, including the karst of Gunung Mulu. Smaller carbonate outcrops of Cretaceous age are found in northern Kalimantan and Sabah, with younger Miocene uplifted reef limestones along the northwest coast. The Sunda and Banda volcanic arcs form the underlying structure for the karst of Indonesia. The karst is discontinuous, and is formed on Mesozoic and Tertiary limestones, with the largest continuous area in Gunung Sewu. In contrast, the karst of Sulawesi is formed on Tertiary limestone in the central and southwestern parts of the island. The Philippines contain a complex array of karst landscapes formed on rocks ranging from Cretaceous to Tertiary in age, and also widespread Quaternary raised reefs.

Three genetic types of tower karst can be identified in the region:

1. Residual hills protruding from a peneplaned carbonate surface covered by a veneer of alluvium. This is a very common karst style throughout Southeast Asia, where alluvial deposition has occurred adjacent to the limestone. Examples include the tower karst of south Sulawesi, Indonesia (McDonald, 1976).
2. Residual hills emerging from carbonate inliers in a planed surface cut mainly across non-carbonate rocks. This style is more common where older Paleozoic limestones occur as strike belts in volcanic or other sedimentary rock types. Examples include the karst of the Emia and Mendi valleys, Southern Highlands, Papua New Guinea.
3. Isolated carbonate towers rising from steeply sloping pedestal bases of varying lithologies, e.g. the steep mountain karst of East Timor.

Cone karst is widespread throughout insular Southeast Asia and finds its best expression in the Gunung Sewu karst of Java and the Bohol karst of the Philippines. Here polygonal depressions with residual hills entirely cover the limestone plateau surfaces. The depressions are frequently drained by bedrock shafts, which are partially or totally blocked by soils and debris from vegetation. Many show alignment of the karst depressions along major lineaments, and there may be cone asymmetry due to the underlying geological structure. In many parts of southeast Asia these depressions are partially infilled with soils derived from volcanic ash deposits.

Pinnacle karst is a spectacular small-scale landform found principally in the humid tropics and subtropics, with the best-known examples coming from Gunung Api, Mulu, and Mount

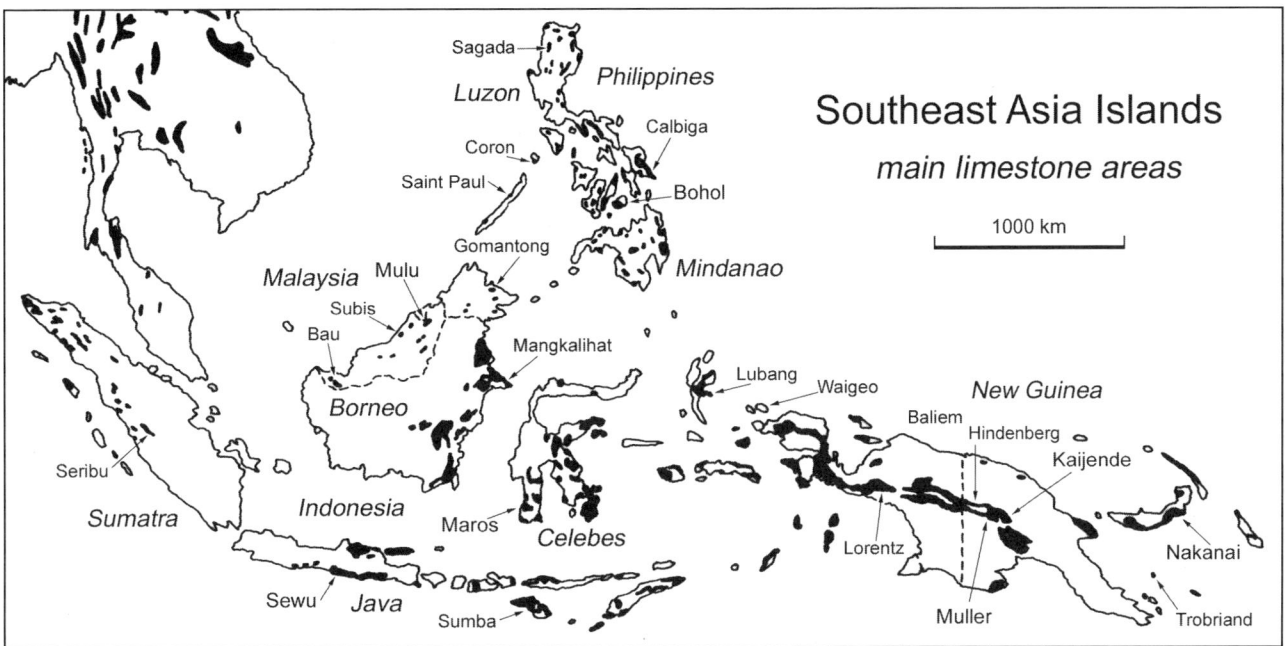

Asia, Southeast Islands: Figure 1. The main karst areas of Island Southeast Asia (details are not shown on the mainland).

Kaijende, New Guinea. Typically the pinnacles are razor-edged limestone blades, closely packed, which can attain heights of thirty to fifty metres. The intervening depressions frequently contain thin organic soils, dense root mats, and deep fissures. Hundreds of pinnacles may be found in an area of a few hectares.

The Maros karst area of southern Sulawesi, Indonesia, has attracted a number of French caving expeditions. Here spectacular tower karst of dominantly Tertiary age occurs in a wide belt, with gorges, springs, and some natural arches. The karst has been dissected by marginal rivers, and the tower bases have a large number of cliff-foot caves and through caves. In addition, relict passages and overhangs occur at varying levels on the towers and are used as burial sites by the Torajan people.

Gunung Mulu National Park in Sarawak, Borneo (see Mulu Sarawak), is world famous with large caves that have been systematically explored since the first expedition in 1978. The karst of the Gunung Sewu area of southern Java has been studied extensively since the 1930s and presents the classic cone karst developed on Miocene limestones.

The Puerto Princesa Subterranean River National Park is situated on the northwestern coast of Palawan, Philippines, and formed part of the land bridge between Borneo and Luzon during glacial epochs. The limestone is of middle Miocene age and is quarried locally as marble. The karst landscape includes extensive limestone towers and polygonal karst, as well as a large polje at Culiatan. The towers attain altitudes of 600 to 1000 m, where their tops have extensive spitzkarren and rough sharp pinnacles termed "assegai karst" (Longman & Brownlee, 1980), while the flanks show deep solution runnels. The limestone massif is drained by the St Paul's Underground River (Figure 2). This strike-oriented main passage is essentially horizontal and navigable by boat for more than 6 km making it probably the world's longest tidally influenced river cave (Hayllar, 1981). The present limit of exploration is a sump that is relatively close to a major stream-sink. Given that the entire catchment is within the reserve, this site offers significant potential for ongoing studies of tropical karst landform processes, as well as interactions between terrestrial and marine processes on a limestone coast. In 1999 the National Park was inscribed on the World Heritage List.

The island of Bohol in the south-central Philippines is world famous for a distinctive karst known as the "chocolate hills". The limestone covers an area of about 600 km^2. This landscape is characterized by smooth, cone-shaped isolated hills of limestone, which have elongated interfluvial ridges. Three sets of summit elevations have been discerned, while the individual form of the mogotes reflects the structural geology of the area. The karst depressions are usually elongated, and are sometimes closed. These narrow valleys open into flat valleys which resemble poljes. Individual karst landforms include extensive caves, stream-sinks and springs, and tufa dams on the watercourses. Over 50% of the karst landscape is flat and supports irrigated agriculture, principally rice. Steeper slopes are terraced and may be planted with rice or support plantations. There has been a profound decline of about 40% in the outflow of springs in the karst over the last 20 years. This is not explained by climatic change and instead shifts in land use and resource allocation seem to be implicated.

The large island of New Guinea contains karst areas occupying around 15% of the land area and extending from sea level to nearly 4000 m. There is a great diversity of karst types in this tectonically active landscape, where present uplift rates are 2–3 mm annually. The major rivers of the central cordillera of New Guinea cut through the karsts, and there are extensive alluviated karst plateaux, one of which has an area of 15 000 km^2 in the Kikori–Kutubu area of the Southern Highlands. On the large islands of New Britain and New Ireland there are extensive poly-

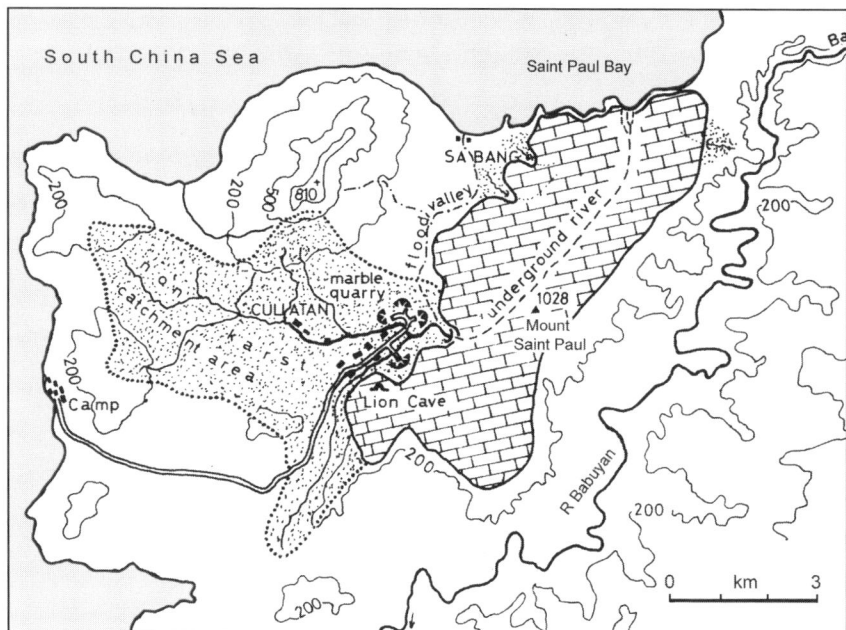

Asia, Southeast Islands: Figure 2. Outline map of the karst at Saint Paul, on Palawan Island, the Philippines, where the river cave reaches from the sink east of Cullatan out to Saint Paul Bay (after map by D. Balazs).

gonal and doline karsts—the Nakanai area contains New Guinea's deepest cave, Muruk, which is over 1100 m deep. This karst system drains to large springs at sea level on the nearby coast. Karst areas are of local importance for water supplies and for their soil resources. More recently, considerable gas and some oil reserves have been found in the Southern Highlands karst belt, and there are plans to pipe the gas across the Torres Strait to Australia.

The combination of rugged relief, karst terrain, and high rainfall makes the limestone ranges of western Papua New Guinea some of the least populated and least-visited areas on Earth. These karsts lie on the southern flank of the central ranges and are bounded by huge scarps up to 700 m high. The maximum recorded annual rainfall is 11 000 mm at the Ok Tedi mine site on Mt Fubilan. With a topographic relief of around 700–800 m, a sequence of pure Miocene limestones up to 1500 m thick, and rainfall in excess of 5000 mm over much of the area, high rates of karst denudation and cave development can be expected. The related Mamo Kananda and Atea Kananda caves (Gillieson, 1985) underly the Muller Range and achieve a total length of 50 km and a depth of 525 m. The caves contain large river passages, flood-prone phreatic mazes, and abandoned upper canyons. Further west in the Victor Emanuel Range the huge passages of Selminum Tem extend for 25 km and drain to springs below the escarpment (Figure 3).

The western half of the island of New Guinea contains extensive karst areas ranging from sea level to 5000 m at Puntjak Jaya (Carstenz Pyramid). The escarpment to the south of the glaciated alpine karst descends nearly 4000 m to the coastal plains, and is now part of the Lorenz National Park. This area has some of most spectacular alpine karren fields, giant dolines, and cliffs in the world, and its caves are still relatively unexplored. The huge Baliem Valley near Wamena drains into a very large cave, partially explored, which flash floods with overnight rises of 30 to 5 m (Figure 4). Other caves in the Baliem are used as burial sites by the Dani people. To the east the Star Mountains massif on the border of West Papua and Papua New Guinea contains glaciated karst with large dolines and some caves. Extensive Quaternary raised reefs in West Papua, Ambon, and Waigeo are unstudied by karst scientists and unexplored by speleologists. On Waigeo there are areas of karst with large karst springs and

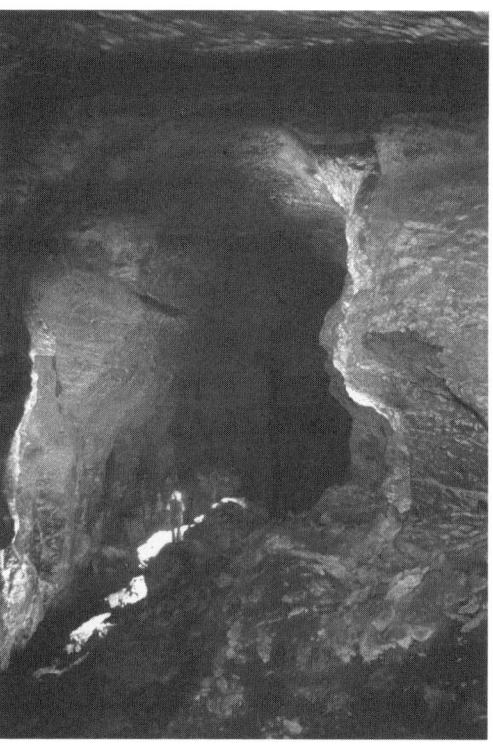

Asia, Southeast Islands: Figure 3. Warp Drive, the major abandoned passage in the Selminum Tem Cave in Papua New Guinea. (Photo by David Gillieson)

Asia, Southeast Islands: Figure 4. The main stream passage in Kutiuleruk Cave in the Wamena karst of Irian Jaya. (Photo by Andy Eavis)

cave entrances visible from the air, extending into the sea to form an archipelago of circular islands, drowned cones with sea-level notches.

DAVID GILLIESON

See also **Kaijende Arête and Pinnacle Karst; Mulu, Sarawak; Nakanai Caves; Sewu Cone Karst**

Works Cited

Day, M. & Urich, P. 2000. An assessment of protected karst landscapes in Southeast Asia. *Cave and Karst Science*, 27: 61–70

Gillieson, D. 1985. Geomorphic development of limestone caves in the highlands of Papua New Guinea. *Zeitschrift für Geomorphologie*, 29: 51–70

Gillieson, D. & Spate, A. 1997. Karst and caves in Australia and New Guinea. In *Global Karst Correlation*, edited by Yuan Daoxian & Liu Zaihua, Beijing and New York: Science Press

Hayllar, T. 1981. Caving in Palawan. *Journal of the Sydney Speleological Society*, 25(12): 215–31

Longman, M.W. & Brownlee, D.N. 1980. Characteristics of karst topography, Palawan, Philippines. *Zeitschrift für Geomorphologie*, 24(3): 299–317

McDonald, R.C. 1976. Limestone morphology in South Sulawesi, Indonesia. *Zeitschrift für Geomorphologie, Supplement* 26: 79–91

Further Reading

Brook, D. 1977. Caves and karst of the Hindenberg ranges. *Geographical Journal*, 143: 27–41

James, J. M. & Dyson, H. J. 1980. Caves and karst of the Muller Range. *Speleological Research Council: Sydney*, 150 pp.

ASIA, SOUTHEAST: ARCHAEOLOGICAL CAVES

Archaeological research in Southeast Asian caves was started in the late 19th century, by colonial officials in Indonesia, Malaysia, and Vietnam. Their investigations and subsequent research shows that caves have been used as temporary shelters and burial places by humans over the past 40 000 years and serve as post-depositional containers for their stone, bone and shell tools, food remains and other material residues. The remains of these small bands of hunters and gatherers, probably of Melanesoid, or at least non-Mongoloid type, are usually called the Hoabinhian Culture, from the region in northern Vietnam, where they were first recognized early in the 20th century. Fossilized remains of extinct hominid species have also been found in a few limestone fissure deposits in Vietnam and Thailand. Many caves also contain evidence for the non-anthropogenic climatic and environmental changes from the late Pleistocene into the mid-Recent period.

During the Iron Age (c.500 BC–AD 200) and later, many caves in western Thailand and East Malaysia were used to place hollowed log coffins, or large pottery jars for secondary burial, as in coastal regions of Indonesia and the Philippines. Others contain prehistoric rock paintings and engravings and from the mid-first millennium AD. Buddhist images, shrines, and inscriptions were created in caves in Burma, Thailand, and central Vietnam (see Asia, Southeast, and Religious Sites).

Thailand

Cave research started later in Thailand than in neighbouring Malaysia, but over 100 caves of archaeological significance have been found. However, only one flexed Hoabinhian burial is so far known from a cave (Sai Yok) in Thailand, in contrast to the many found in Malaysia, Vietnam, and Indonesia. The best-known archaeological cave in the country is Spirit Cave (Tham Pi) in Mae Hong Son province in the northwest. Excavated in the 1960s and 1970s, Spirit Cave has produced evidence of a range of plant foods eaten by the cave occupants, from about 12 000 to 7500 years BP. In the late 2nd millennium BC, later peoples left their pottery and stone adzes as evidence of their brief visits—Spirit Cave, as well as other caves in the area such as Tham Lod, contains wooden coffins dated to about 2200–1200 years BP.

Caves at Ban Kao and Tham Khao Thalu in Kanchanaburi Province, have also yielded a large amount of prehistoric finds, including log coffins. Limestone caves and scarps along the Thai–Burmese border, in the Petchabuan Range of Central Thailand, and overlooking the Mekong River in Nakorn Sawan Province, all contain galleries of rock paintings, making Thailand an important area for prehistoric art.

Human occupation at Lang Rong Rien cave, in Krabi Province of Peninsular Thailand, has been dated from c.40 000 BP, making this one of the oldest inhabited sites in Southeast Asia and an important location for studying the long-term human occupation of the region. Further south, in Satun Province, recent exploration by Thai archaeologists shows that this also has many caves with significant archaeological deposits. In 1999, a discovery of possible *Homo erectus* skull fragments, at a fissure deposit in Lampang province, northern Thailand, may extend human use of caves in Southeast Asia to over 500 000 BP. If confirmed, the find would be the first evidence for the existence of *H. erectus* in Asia outside China, Vietnam, and Indonesia.

Vietnam

The numerous caves and rock shelters in the limestones of Hoa Binh, Than Hoa, and Bacson Provinces, long ago attracted the attention of French geologists, such as Madeleine Colani and Henri Mansuy, and their research, and that of later Vietnamese researchers, has produced the earliest and richest sequences of human occupation in Southeast Asia. At Lang Cuom Cave, Colani uncovered a deposit of about 100 skulls in 1927, with very few post-cranial bones—perhaps the remains of headhunting or ancestor cults. Other caves contain flexed inhumations sprinkled with red or yellow ochre. Vietnamese archaeologists date the Hoabinhian and Bacson Cultures to between *c*.12 000 and 7000 years ago and refer to them as "Early Neolithic" but it seems that no evidence for food production has been identified. Some *H. erectus* teeth, found in Tham Khuyen and Lang Trang fissure deposits, were dated between 480 ka and 148 ka.

The drowned karst towers of Ha Long Bay (see separate entry) contain caves that preserve the remains of human occupation from over 15 000 years ago, when the sea level was much lower and the towers stood in a forested plain drained by fast flowing rivers. Some islands (e.g. Hon Vung Ba Cua) also contain red ochre rock paintings—the first so far known in Vietnam.

Malaysia

In the 1880s, cave excavations on Peninsular Malaysia were pioneered by scientists with interests in geology. There are many caves with archaeologically important remains, especially Gua Cha in Kelantan, which has been relatively preserved from commercial destruction through its isolation. Gua Cha contains a Hoabinhian occupation and burial with clear evidence of the replacement of Hoabinhian culture by intrusive Neolithic, probably Mongoloid settlers, who brought a more settled agricultural way of life, made pottery, polished stone tools, and buried their dead laid out in individual burial pits with many grave goods.

The Lenggong Valley in Perak contains Peninsular Malaysia's only Palaeolithic site, a karst area with some 40 small caves, many of which have archaeological remains. An open-air 75 000 year-old stone tool workshop was found at Kota Tampan, which is thought to have been abandoned, following heavy ash falls from a volcanic eruption in Sumatra. At Gua Gunung Runtuh, excavated in 1990, the oldest complete skeleton in Malaysia, dated to *c*.11 000 years ago, was found together with stone tools and food remains; this is popularly known as "Perak Man". Other caves in the area contained stone tools, bones, pottery, and jewellery, dating over the last 10 000 years. The oldest rock paintings in Malaysia are about 2000 years old at Gua Tambun, on an exposed cliff in Perak.

In Sarawak, East Malaysia, the Niah Cave complex is internationally known from the pioneering research of Tom Harrisson and is now a major tourist attraction, both on account of the enormous caverns with their daring birds' nest collectors, and

Asia Southeast: Archaeological Caves: Wall art in the Painted Cave of the Niah cave group in the Subis Hills of Sarawak. (Photo by Tony Waltham)

the 40 000 year old sequence of human occupation. The earliest evidence for human presence in Malaysia was a skull dated to *c*.38 000 years, found in the Great Cave at Niah. Harrisson's excavations revealed late Pleistocene chopping tools and flakes, and Neolithic edge-ground axes, bone points, adzes, pottery, shell jewellery, mats, nets, boats, then iron tools, ceramics, and glass beads dating to the Iron Age. Tom and Barbara Harrisson excavated 166 burials—39 of them Mesolithic, the rest Neolithic. The site is being re-excavated by a joint Malaysian-British team to determine the accuracy of Harrisson's work, to obtain a more refined chronology and recover evidence for environmental changes made possible by recent archaeological scientific methods. The Painted Cave at Niah has paintings roughly 1200 years old, depicting boats and people practising their funerary rites (see photo). A number of carved wooden coffin "death ships" were found in 1958 on the floor below the paintings.

In Sabah, caves at Baturong were inhabited from 17 000 years ago, when the site became accessible after a lake, which had been formed by volcanic deposits blocking the nearby river, dried up. Later human occupation dated to about 3000–2500 years BP and later still from the first millennium AD, has also been excavated by Bellwood and others in the Madai Caves near the eastern coast of Sabah. Wooden coffins, dating to the 14th century, have also been found in many Sabah caves.

Indonesia

Limestone caves with archaeological remains, ancient burials, and rock paintings are numerous in Indonesia, from Sumatra to Moluku and West Papua, but relatively few have been well excavated and reported, except in South Sulawesi and Timor. The site of Guwa Lawa, in east-central Java, was recognized to be rich in sub-fossil fauna, and human burials and occupational debris, following work there by van Es and van Stein Callenfels in the 1920–30s, but only sketchy reports were published. Recently a joint Indonesian-French project has restarted cave research in central Java, with very promising results. Flexed burials of Hoabinhian affinity are known from caves in Java, including Guwa Lawa and, as in Vietnam and Malaysia, on the basis of large molars, rugged skulls and lack of mongoloid characters, have been labelled "Melanesoid".

In the Maros karst region of South Sulawesi, some 40 km inland from the provincial capital of Makassar, deposits at a rock shelter named Leang Burung 2, have been dated by Glover to between $c.31\,000-20\,000$ BP, and similar dates have been obtained from another such site near Pangkajene. At Ulu Leang 1, a cliff foot cave in the Lealeang Valley, with a rich sequence of occupation with freshwater and marine molluscs, stone and bone tools, and plant remains, was also excavated by Glover and dated from $c.10\,000-2500$ BP. Many other caves in the region contain similar materials, as well as red ochre hand stencils and paintings of animals, such as the *Babirusa*—an endemic wild suid. Higher level caves, such as Ulu Lleang 2, contain jar burials of a late prehistoric Iron Age. Caves, sometimes artificially enlarged as rock-cut tombs, in the Toraja Highlands of Central Sulawesi, have long been used for burial and are now popular tourist attractions.

The remote island of Timor in eastern Indonesia, is also rich in limestone caves with human occupation deposits. Research started there with the work of Bühler and Willems in the 1930s, was continued from Australia by Glover in the 1960s and very recently archaeologists from Australia, led by Spriggs and O'Oonnor, have resumed work in East Timor, finding evidence for occupation at Lene Hara Cave at the eastern tip of the island dated to $c.31\,000$ BP. Other important archaeological cave deposits have been excavated at Uai Bobo 1 and 2 (from 13 000 BP), at Lie Siri (from 9000 BP) and at Nikiniki in western Timor.

The Philippines

Limestone caves with archaeological remains are found in many parts of the Philippines, especially in the Cagayan Valley of northern Luzon and in Palawan Island, where a National Museum team, headed by the late Robert Fox, excavated a skull, teeth, and jaw at Tabon Cave in 1962, originally dated to $c.22\,000$ BP, but now re-dated to $c.16\,000$ BP. The cave, at an altitude of 33 m, is one of more than 30 in the outcrop, once 30 km from the sea. After people abandoned Tabon, Guri Cave became an important shelter between $c.5000$ and 2000 years ago. Flake stone tools, bones of wild pig and deer, were found and marine shells show the sea had reached its present level by that time. Other stone age sites include Laurente and Rebel Caves, near Penablanca in the Cagayan Valley of northern Luzon, inhabited about 16 000 years BP, and at Ayub Cave at Maitum, Cotabatu Province, Mindanao Island, containing spectacular anthropomorphic burial jars representing individual persons buried there nearly two thousand years ago.

IAN GLOVER AND LIZ PRICE

See also **Burials in Caves**

Further Reading

Anderson, D. 1997. Cave archaeology in Southeast Asia. *Geoarchaeology*, 12(6): 607–38

Barker, G. *et al.* 2000. The Niah Caves: Preliminary report on the first (2000) season. *Sarawak Museum Journal*, 55 (76): 111–49

Bartstra, G.-J. 1998. Short history of the archaeological exploration of the Maros caves in South Sulawesi. *Modern Quaternary Research in Southeast Asia*, 15: 193–22

Bellwood, P. 1997. *Prehistory of the Indo-Malaysian Archipelago*, revised edition, Honolulu: University of Hawaii Press

Dizon, E. & Santiago, R. 1996. *Faces from Maitum: The Archaeological Excavations of Ayub Cave*, Manila: National Museum of the Philippines

Fox, R. 1970. *The Tabon Caves: Archaeological Explorations and Excavations on Palawan Island, Philippines*, Manila: National Museum

Glover, I.C. 1977. The Late Stone Age in Eastern Indonesia. *World Archaeology*, 9(1): 42–61

Glover, I.C. 1986. *Archaeology in Eastern Timor, 1966–67*, Canberra: Department of Prehistory, Research School of Pacific Studies, Australian National University

Ha, Van Tan 1980. Nouvelles recherches préhistoriques et protohistoriques au Vietnam. *Bulletin de l'École française d'Extrême-Orient*, 68: 113–54

Harrisson, T. 1954. The caves of Niah: A history of prehistory. *The Sarawak Museums Journal*, 8(12): 549–95

Higham, C. 1989. *The Archaeology of Mainland Southeast Asia: From 10,000 BC to the Fall of Angkor*, Cambridge and New York: Cambridge University Press

Higham, C. & Thosarat, R. 1998. *Prehistoric Thailand: From Early Settlement to Sukhothai*, Bangkok: River Books and London: Thames and Hudson

Price, L. 2001. *Caves and Karst of Peninsular Malaysia*, Kuala Lumpur: Gua Publications

ASIA, SOUTHEAST: BIOSPELEOLOGY

Southeast Asia is not a homogeneous biogeographical area. Situated at the contact between several tectonic plates, its paleogeographical history is probably the most complex and the least known among the large land masses of the world (Whitmore, 1987). Limestone rocks of various geological ages and extent are present up to the highest altitudes (Sulawesi: Mt Mekongga, 2800 m), representing all the classical karst types, from coral limestone to immense plateaux and cone karsts. An important feature controlling the diversification of cave fauna is habitat fragmentation. Southeast Asian karsts are extremely fragmented. Thousands of limestone islands are scattered from the Philippines to Indonesia, and a huge number of limestone outcrops exist inland in Java, Sumatra, Sulawesi, and Peninsular Malaysia. Climate is another influential factor for subterranean biodiversity. The region is mostly under an hyperhumid tropical climate, which is thought to hamper the evolutionary shift towards obligate subterranean life, but some parts of the region (in Laos, central and eastern Thailand, and parts of Java, Sulawesi, and Eastern Maluku) experience a relatively dry or seasonal climate.

The biospeleology of Southeast Asia has a long history. As long ago as 1888, the Italian zoologist Leonardo Fea made important zoological collections in Burma (present-day Myanmar) caves near Moulmein (present-day Mawlamyine). Eugene Simon, the French arachnologist, collected and described the first cave species of the Philippines in 1892. In 1899, the Skeat expedition collected cave animals in southern Thailand. The Batu Caves in Peninsular Malaysia were sampled and studied in great detail by several biologists, from Ridley in 1898 to McClure in 1967. Very few cave-adapted species were found in these caves and it is probably the main reason why tropical caves were thought to have a low troglobitic richness, an idea which lingered until the work of Frank Howarth (summarized in Howarth, 1983). It was only in the 1980s that troglobitic and troglomorphic cave species began to be discovered in large numbers, first in Mulu caves of Sarawak (Chapman, 1984), then in Thailand and southern Sulawesi (Deharveng & Leclerc, 1989). There is now evidence that limestone caves in all Southeast Asia regions host rich troglobitic and stygobitic fauna (Deharveng & Bedos, 2000). However, overall knowledge of the subterranean fauna of Southeast Asia remains poor and uneven, and there are few well-documented sites (Figure 1). No biological information is available for the large limestone ranges of Kalimantan or Timor, and even in Java, only 3 karstic regions out of a total of 60 have been reasonably sampled for their subterranean fauna. There are also many taxonomic gaps, especially for microarthropods. For example, fewer stygobitic copepods are known from subterranean habitats in the whole of Southeast Asia (Juberthie & Decu, 2001: five species), than from Afghanistan alone (seven species). Furthermore, habitats have been unevenly sampled. Cave troglobites and stygobites are the focus of collecting, while anchialine and interstitial habitats, and to a lesser degree guano, are rarely sampled. Finally, virtually nothing is known about the biology and ecology of cave species in the region, making the ecological status of many of them disputable. In spite of these considerable gaps, some general traits of the subterranean fauna of the region are taking form following the important discoveries of the last two decades.

Despite its complex paleogeographical history, the subterranean fauna of Southeast Asia is surprisingly homogeneous at the level of the highest taxa. Compared to temperate regions, and as in other tropical areas, Arachnida tend to outnumber insects in diversity within terrestrial habitats, while decapods and isopods tend to outnumber amphipods in aquatic habitats. Differences with other tropical cave fauna are marked by the presence of several characteristic zoological groups, all of which are functionally important in the ecosystem. Some of these taxa may extend into the Indian subcontinent or Australia, and some may be lacking in Southeast Asia margins (Maluku islands in Indonesia, or northern Vietnam), but most are present in all the documented areas of the region. They are: (1) the swiftlets of the genus *Collocalia*, guano producers at the same level as bats; (2) the bat-eating snake *Elaphe taeniura* and its related forms; (3) the large rhaphidophorid crickets of the Rhaphidophorini tribe, an oriental group, which form the basis of invertebrate food webs in all caves of the region; (4) the giant Arthropods (crickets, sparassid spiders, amblypygids, centipedes), which haunt energy-rich habitats of lowland caves; (5) the cambalopsid millipedes, which live in dense populations at the surface of guano piles; and (6) the Nocticolidae, a cockroach family rich in troglomorphic species and present in many caves of the region.

The characteristic faunal elements listed above represent a small part of the subterranean fauna of Southeast Asia in terms of biodiversity. According to the most recent works (Juberthie & Decu, 2001), about 150 obligate subterranean species have been described in Southeast Asia, and many more are listed as undescribed morphospecies, or as species of uncertain ecological status. The number of non-obligate but regular subterranean species (mainly guanobites) is probably also very high. Most taxa of this cave fauna are narrowly endemic, many being known only from a single karstic unit or a single cave. Among the troglobitic species, a number of species belong to monospecific genera, and several are highly troglomorphic. Compared to epigean fauna, these rare taxa and their remarkable adaptive features give an exceptional biological value to the subterranean fauna of the region, in spite of its low species richness.

The origin of subterranean fauna may be approached at different levels. At a broad scale, it has been possible to relate in a few cases the distribution of taxa to major paleo- or biogeographical features. The eastern part of Southeast Asia is the transition zone between Australian and Asian fauna, indicated by Wallace's and related lines (Whitmore, 1981). This major biogeographical limit is reflected in part by the regional subterranean fauna. Several generic or suprageneric taxa are not known east of Java, for example *Cyphoderopsis* (Collembola) or carinate cambalopsid millipedes. However, no subterranean taxon of the Australian region has its western limit near Wallace's line. At Batu Lubang in Halmahera, the easternmost documented cave site of Southeast Asia, the known troglobites do not exhibit any particular relationship with Australian fauna. From Thailand to Maluku, the bulk of the subterranean fauna is clearly constituted of lineages from the Oriental Region, eventually extending to Australia such as Nocticolidae cockroaches. The Kra Isthmus in southern Thailand is a second important chorological limit, clearly marked in the flora. It is also significant for some cave fauna: within their Southeast Asian range, the Collembola *Cyphoderopsis* (troglobitic paronellids) and *Willemia nadchatrami* (a guanobitic hypogastrurid) are only present south of the isthmus, while *Troglopedetes* is only known north of the isthmus in Asia. Another important faunal change occurs in continental Southeast Asia from west to east, best illustrated by the distribution of cave entomobryoid Collembola, where *Troglopedetes* of western Thailand gives way to a troglomorphic line of *Coecobrya* in eastern Thailand, then to *Lepidonella* in Vietnam. This pattern is in agreement with the existence of an Indochina block independent

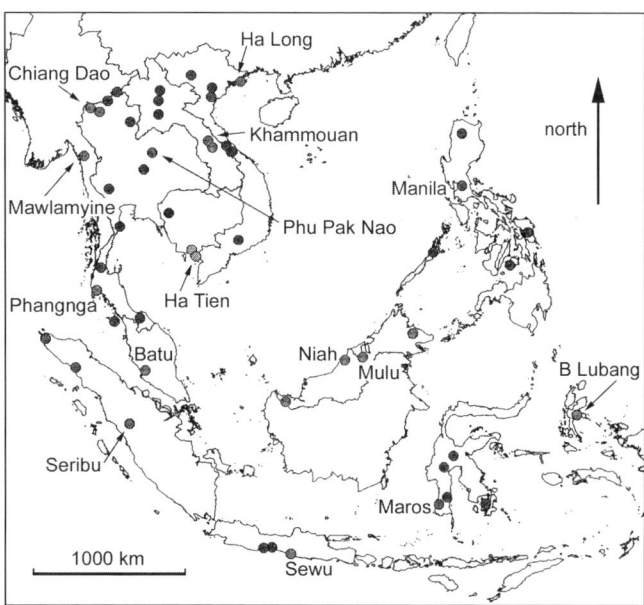

Asia, Southeast: Biospeleology: Figure 1. Locations of the main caves and karst areas in Southeast Asia that have been surveyed for subterranean fauna.

from other parts of Southeast Asia until about 30 Ma years ago, based on recent plate tectonic reconstitutions (Whitmore, 1987). However, the relevance of these limits and the broad scale paleogeographical reconstructions to the whole cave fauna is not supported by several taxa, and chorological data are too loose to derive any interpretation for most groups.

The majority of the obligate subterranean species and guanobites belong to groups which are well represented in epigean habitats of the region, such as Collembola, spiders, or crabs. A few taxa do not have epigean relatives anywhere in the world, but are diversified in subterranean habitats of Southeast Asia, such as *Stenasellus* isopods or bogidiellid amphipods among stygobites. Relictual subterranean taxa, i.e. without known relatives in the regional epigean fauna, but with epigean relatives elsewhere in the world, were thought until recently to be absent from tropical caves (Chapman, 1986). They might be actually present in Southeast Asia, as suggested by recent discoveries of isolated mono- or oligo-specific genera, like the Opilioacari *Siamacarus* of Thailand, or the Coleoptera *Mateuellus* of Sulawesi and *Laosapliaenops* of Laos.

Beside these remarkable taxa, endemic species swarms have been discovered in increasing number among the subterranean fauna of Southeast Asia. They are composed of closely related and geographically isolated species, which fit more or less the fragmentation pattern of subterranean habitats. For instance, at least four different species of the stygobitic genus *Stenasellus* (Isopoda) have been recognized in the fragmented limestone habitats of Gunung Seribu (West Sumatra). Swarms of this kind are to be expected among many obligate subterranean lineages of Southeast Asia, as suggested by taxonomic studies on ochyroceratid spiders (Deeleman-Reinhold, 1995), or paronellid Collembola (Deharveng & Gers, 1993).

As a rule, troglobitic and freshwater stygobitic assemblages are exclusively constituted of narrow endemic species. Patterns are different in guano and anchialine subterranean communities, where widespread forms (such as tineid moths for guano, or the shrimp *Parhippolite uveae* for anchialine habitats) co-occur with species of more limited distribution.

The high level of endemism and the high taxonomic differentiation of Southeast Asia subterranean fauna can be explained by the high fragmentation of the habitat and by the selective constraints imposed by the underground ecological conditions (see Subterranean Habitats), which makes the adaptive shift from epigean to subterranean life particularly demanding (Gibert & Deharveng, 2002). To cope with these conditions, obligate cave species have developed remarkable morphological and biological adaptations in temperate regions, which were long considered to be rare in the tropics. It is not actually the case. Biological adaptations have never been investigated, but highly troglomorphic species have been discovered since the 1980s in Southeast Asia. Troglomorphic taxa exhibit a regression of pigment and eyes, a loss of wings (in Pterygota), and an elongation of appendages (in Arthropoda) compared to their epigean relatives. They are found in most subterranean lineages and in all areas of Southeast Asia, but they seem to be more frequent and often more modified in seasonal tropics like Laos or southern Sulawesi (see Salukkan Kallang, Indonesia: Biospeleology).

Guanobites do not usually differ morphologically from epigean species. This morphological contrast with highly modified troglobites living in the same ecosystem illustrates how various

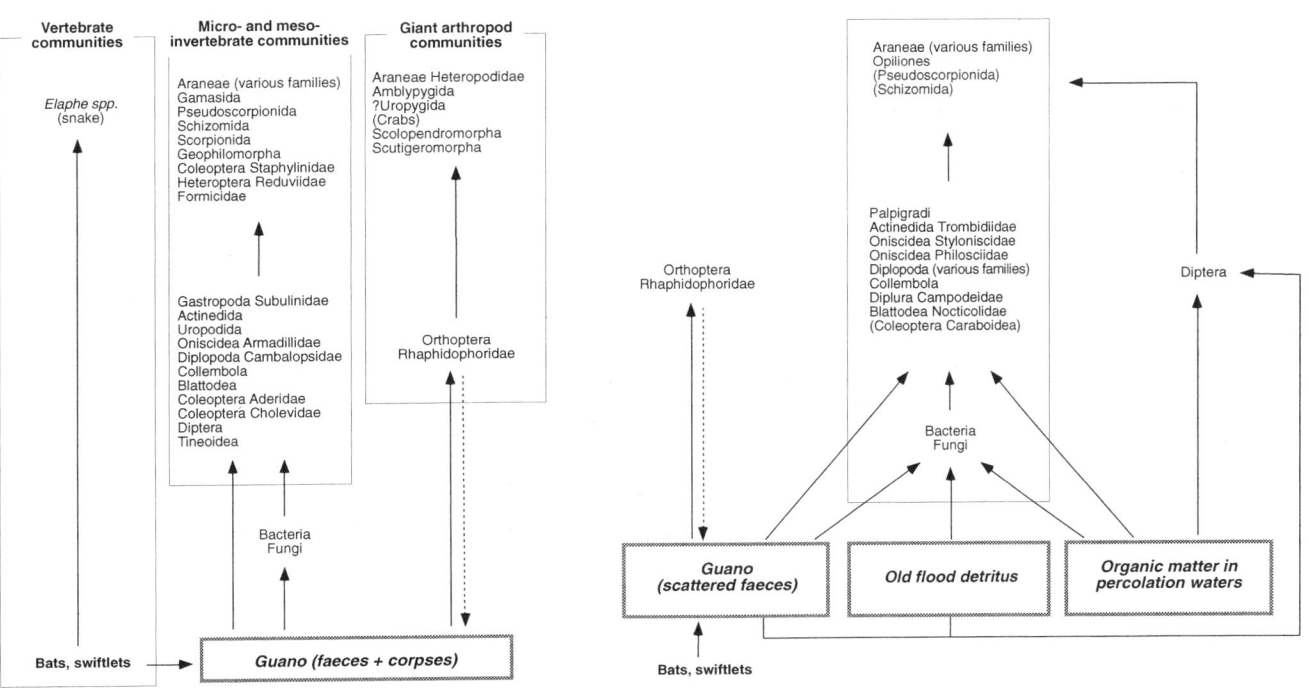

Asia, Southeast: Biospeleology: Figure 2. Generalized terrestrial food webs in Southeast Asia caves, excluding parasitic taxa (after Deharveng & Bedos, 2000, modified).

the subterranean habitats and their associated evolutionary constraints can be. Almost all subterranean habitats have been encountered in Southeast Asia: freshwater interstitial, oligotrophic freshwater, anchialine habitats, and energy-poor to extremely energy-rich terrestrial habitats. However, the MSS (Mesovoid Shallow Substratum) or Superficial Underground Compartment has not been recorded, though a variant may exist associated with volcanic boulders. Each of these habitats has its own faunal assemblage. Guano accumulations, produced by bats or swiftlets, are conspicuous and pungently smelling features in many Southeast Asia caves. They host giant arthropod assemblages in the region, and very dense populations of meso- and micro-arthropods, according to the generalized food web summarized on Figure 2 (Deharveng & Bedos, 2000). In contrast to assemblages of guano and other energy-rich habitats (flood debris, root mats), troglobitic and stygobitic communities are restricted to oligotrophic habitats. Troglobites are mainly found foraging around micropatches of food resources scattered in an overall azoic environment, mostly disseminated faeces of swiftlets, bats, or rhaphidophorid crickets (Figure 2). They never colonize guano accumulations. Stygobitic species are found in small endogeneous inlets, phreatic waters, percolating waters, and more rarely in exogenous streams which have been filtered through fine boulders. They are absent in large sink-resurgence systems, where epigean fauna are often abundant. In principle, Amphipods feed on particulate or dissolved organic matter, and Isopods are predaceous, but some species are able to shift from predaceous to saprophagous, others are omnivorous, and the feeding habits of most Southeast Asia stygobites (from fish to aquatic beetles) are so poorly documented that it is impossible to summarize their trophic relationships.

With its high proportion of endemics, phylogenetically isolated lineages and highly evolved taxa, the subterranean fauna of Southeast Asia represents a unique and fragile biological heritage. Quarrying with its many collateral and irreversible damages, water pollution, deforestation, and over-frequentation of caves are currently taking a heavy toll on karstic ecosystems of the region. Guano caves appear particularly vulnerable, because of the sensitivity of bats and swiftlets to disturbance. Traditional swallow nest exploitation has so far permitted preservation of a significant part of this biodiversity, but the prospects are more uncertain for other kinds of subterranean habitats. In spite of the increasing concern of public opinion in Southeast Asia, this biological heritage is at risk in many karstic areas (Vermeulen & Whitten, 1999).

LOUIS DEHARVENG

See also **Salukkan Kallang, Indonesia: Biospeleology**

Works Cited

Chapman, P. 1984. The invertebrate fauna of the caves of Gunung Mulu National Park, Sarawak. *Sarawak Museum Journal*, 30(51), Part II: 1–18

Chapman, P. 1986. Non-relictual cavernicolous invertebrates in tropical Asian and Australasian caves. *Proceedings of the 9th International Congress of Speleology, Barcelona (Spain)*, vol. 2: 16–63

Deeleman-Reinhold, C.L. 1995. The Ochyroceratidae of the Indo-Pacific region (Araneae). *The Raffles Bulletin of Zoology*, suppl. 2: 1–103

Deharveng, L. & Bedos, A. 2000. The cave fauna of Southeast Asia: ecology, origin, evolution. In *Subterranean Ecosystems*, edited by H. Wilkens, D.C. Culver & W.F. Humphreys, Amsterdam and New York: Elsevier

Deharveng, L. & Gers, C. 1993. Ten new species of *Troglopedetes* Absolon, 1907 from caves of Thailand (Collembola, Paronellidae). *Bijdragen tot de Dierkunde*, 63(2): 103–13

Deharveng, L. & Leclerc, P. 1989. Recherches sur les faunes cavernicoles d'Asie du Sud-Est [Researches on the cave fauna of Southeast Asia]. *Mémoires de Biospéologie*, 16: 91–110

Gibert, J. & Deharveng, L. 2002. Subterranean ecosystems: A truncated functional biodiversity. *Bioscience*, 52(6): 473–81

Howarth, F.G. 1983. Ecology of cave arthropods. *Annual Revue of Entomology*, 28: 365–89

Juberthie, C. & Decu, V. (editors) 2001. *Encyclopaedia Biospeologica*, vol. 3, Moulis and Bucharest: Société Internationale de Biospéologie (entries on Burma, Cambodia, Indonesia, Laos, Malaysia, Philippines, Thailand, Vietnam)

Vermeulen, J. & Whitten, T. (editors) 1999. *Biodiversity and Cultural Property in the Management of Limestone Resources: Lessons from East Asia*, Washington DC: World Bank

Whitmore, T.C. (editor) 1981. *Wallace's Line and Plate Tectonics*, Oxford: Clarendon Press and New York: Oxford University Press

Whitmore, T.C. (editor) 1987. *Biogeographical Evolution of the Malay Archipelago*, Oxford: Clarendon Press and New York: Oxford University Press

ASIA, SOUTHEAST: HISTORY

Although there are few indigenous speleologists and no regional journal or serial publications, Southeast Asian caves witness great cultural, religious, and natural heritage, their human use dating back perhaps 50 000 years. This review documents use and exploration of caves. About 10–15% of the region is karst (less in Malaysia and Cambodia), often in remote locations. Nevertheless karst has an immense significance for sustainable agriculture, water quality, tourism, and biodiversity conservation in the region.

Accessible, published knowledge of caves must be separated from the indisputable antiquity and continuity of their cultural importance. Although caves preserve the evidence better than open spaces, early hominids clearly occupied and exploited the region's karstlands (see entry on Asia, Southeast: Archaeological Caves). Protohistoric cave artefacts evidence ethnic and cultural flux and well-developed trading networks extending to India and beyond; some peninsular sites suggest early links with Indian Brahmano-Buddhist culture. Across the north, many caves contain wooden coffins (Kiernan, Spies & Dunkley, 1988), inadequately studied but tentatively dated at 0–1000 AD. Tin has been mined from karst areas in Perak (Malaysia) for 1500 years. The Chinese visited caves in Malaya, Borneo, and elsewhere during the Ming Dynasty (1368–1634 AD), trading birds' nests, an industry still flourishing in southern Thailand and forming a significant tourist attraction in Phangnga Bay.

Many caves have been renowned as religious sites for more than a millennium. About 1345 AD Li Thai, King of Sukhothai described the veneration of caves and their integration into the

Buddhist community. Thailand alone hosts perhaps 200 Buddhist caves, mostly in tower karst, many preserving examples of stylistic changes in iconography extending perhaps 1300 years. The darker recesses are frequently meditation sites and occasionally, as in Tham Tab Tao near Chiang Mai, icons nearly a kilometre inside evidence early exploration by monks. Pindaya Cave in Burma contains perhaps 10 000 images, while the best-known cave in Laos, Tham Thing near Luang Prabang, may once have housed 37 000.

Speleological Investigations

Early European accounts described littoral caves and others close to the coast, or along routes of travellers and explorers such as Henri Mouhot. Rugged topography, civil unrest, and limited support infrastructure inhibited exploration and indigenous research until relatively recently, with most early work undertaken by colonial officers and expatriates, especially in Malaysia, Indonesia, Vietnam, and Burma (see, for example, Annandale, Brown & Gravely, 1913). In 1924 Sawicki wrote the first modern account of karst in Thailand. Later, occasional peripatetic scientists and speleologists such as Denes Balazs (1968; 1976) in Indonesia and Philippines and Heinrich Kusch in Thailand greatly advanced understanding of the region's potential. However, the era of organized expeditions awaited relatively cheap travel in the 1970s, and Mansfield's (1985) valuable historical survey, though timely, was published when half the region's countries were still inaccessible to travellers.

Much karst lies far from urban areas, in rugged, remote and sparsely populated pockets of some of the world's poorest countries. Many expeditions seeking grand adventure instead encountered unpredictable logistics and capricious officialdom, experiencing the truism that it is better to travel hopefully than to arrive. Multiskilled interdisciplinary teams cooperating with local groups usually achieved most, with British, Australian, Italian, and French expeditions predominating. Some British expeditions (especially those studying Mulu) attracted significant sponsorship, while Australians have been self-funded. Continuity of study has been rare, notable exceptions being Limbert's Vietnam expeditions and the Australian and French (Association Pyrénéenne de Spéléologie) projects in Thailand and Indonesia. Small, self-funded foreign expeditions visit Peninsular Malaysia annually.

Malaysia and Thailand are well documented; local groups explore caves, commercial caving adventure trips are available, and Dean Smart spent several productive years as Thailand's cave specialist. Successive, focused Australian and French expeditions surveyed some of the longest caves in the region and awakened indigenous awareness of karst heritage. The Siam Society's Journal occasionally publishes research on Thailand's caves, especially archaeological sites. However a reasonably continuous tradition of indigenous study has emerged only in Malaysia, led until recently by expatriates, culminating in major compendiums (Bullock, 1965; Price, 2001). The entry on Mulu, Sarawak describes the remarkable outcomes of work there.

Elsewhere exploration generally remains at reconnaissance level, confounded by political, military, insurgent, and logistical difficulties. Laos and Vietnam were inaccessible until 1990. In Laos, several groups explored Vang Vieng, while Claude Mouret's 11 expeditions prospected the vast Khammouane Karst and other regions, including the impressive Nam Hin Boun, providing a solid foundation for further studies. Burma is severely restricted, graphically highlighted in the only significant speleological account (Dunkley et al., 1989). Small French expeditions have since returned but vast areas remain completely unvisited. Cambodia is still speleological *terra incognita*.

Vietnam encourages foreign expeditions on a joint venture, multiskilled basis, primarily where a local benefit exists, especially tourist potential. Among others, Howard Limbert's six expeditions since 1990, mostly with Hanoi University, concentrated on underground river systems in the rugged central and northern border ranges contiguous with karst in Laos and China, such as Hang Khe Rhy, Hang Over, Hang Phong Nha, Nguom Sap, and Nguom Lung Sam.

After Balazs, little was published about Indonesia and Philippines, and comprehensive reviews are needed to facilitate further work. In Indonesia, Dr Robbie Ko coordinates cave studies. Large river caves in the Gunung Sewu karst of southern Java proved irresistible to several foreign groups, and others roamed Timor, Sumba, Seram, Sumatra, Sulawesi, and Halmahera. A dozen frustrating forays into Irian Jaya barely sampled its tantalising prospects despite massive limestone in high relief, karst rising to nearly 5000 m on Jaya Peak (Carstensz Pyramid), and possibly the world's largest stream sink (Baliem River).

The Philippines Mountaineering Federation encompasses indigenous cave exploration, hosting joint ventures in Luzon, the Visayas, Mindanao, and Samar. The magnificent, 21 km long St Paul's Cave on Palawan was mapped by several groups, and the extraordinary karst of Coron Island was explored (Piccini & Rossi, 1994). On Bohol, caves were investigated for their hydrological significance for agriculture.

Greater political stability, development pressures, and increasing local interest throughout Southeast Asia combine to improve the prospect both of further spectacular discoveries and of more intensive, focused research.

JOHN ROBERT DUNKLEY

Works Cited

Annandale, N., Brown, J. & Gravely, F. 1913. The limestone caves of Burma and the Malay Peninsula. *Journal of the Asiatic Society of Bengal*, 9(10): 391–423

Balazs, D. 1968. Karst regions in Indonesia. *Karszt-es Barlangkutatas*, 5: 3–61

Balazs, D. 1976. Karst types in the Philippines. *Proceedings of the 6th International Congress of Speleology*, vol. 2, Prague: Czechoslovak Academy of Sciences

Bullock, J.A. (editor) 1965. *Malayan Nature Journal*, 19(1): 1–112 (special Malaysian caves issue)

Dunkley, J.R., Sefton, M., Nichterlein, D. & Taylor, J. 1989. Karst and caves of Burma (Myanmar). *Cave Science*, 16(3): 123–31

Kiernan, K., Spies, J. & Dunkley, J. 1988. Prehistoric occupation and burial sites in the mountains of the Nam Khong area, northwestern Thailand. *Australian Archaeology*, 27: 24–44

Mansfield, R. 1985. Speleology in South-east Asia: An historic background. *Shepton Mallet Caving Club Journal*, 7(9): 23–33

Piccini, L. & Rossi, G. 1994. Italian caving exploration in the island of Palawan, Philippines. *Speleologia*, 31: 5–61

Price, L. 2001. *Caves and Karst of Peninsular Malaysia*, Kuala Lumpur: Gua Publications

Sawicki, L. 1924. A karst in Siam—Koh Si Chang. In *En Recueil de travaux, offert a M. Jovan Cvijic* (Collection of works, presented to M. Jovan Cvijic), Belgrade

Further Reading

Association Pyrenéenne de Spéléologie 1986, 1987, 1988. *Expedition Thai Maros 85; Expedition Thai-Maros 85; Expedition Thai 87–Thai 88*, Toulouse: Association Pyrenéenne de Spéléologie (in French)
  Investigations of karst and caves and biospeleological studies in Thailand, Sulawesi, and Maluku.
Dunkley, J.R. 1994. *The Caves of Thailand*, Sydney: Speleological Research Council
  Describes 2000 caves, comprehensively referenced, with articles on physical environment and cultural significance of caves.
Kusch, H. 1975–1988. (series of articles in) *Die Hohle*, 26(4); 27(3); 28(3); 31(3); 32(2); 33(3); 33(4); 35(1); 36(3); 38(3); 39(4)
  Descriptions of cave explorations in Malaysia, Thailand, Burma, and Indonesia (in German)
Meredith, M. & Wooldridge, J. 1992. *Giant Caves of Borneo*, Kuala Lumpur: Tropical Press
  Illustrated popular account of British expeditions to Mulu Caves, Sarawak
Mouret, C. 2001. Scientific and human activities during cave and karst exploration in Laos. *Proceedings of the 13th International Congress of Speleology*, Brasilia: Brazilian Speleological Society (CD-ROM)
Munier, C. 1998. *Sacred Rocks and Buddhist Caves in Thailand*, Bangkok: White Lotus
  Catalogue of Buddhist cave sites with discussion of their historical and cultural significance.
Price, L. 1998. *Malaysian Cave Bibliography*, Kuala Lumpur: Gua Publications
  Reference manual listing over 1600 publications and nearly 1000 newspaper articles (unannotated).
Valli, E. & Summers, D. 1990. *The Nest Gatherers of Tiger Cave*, London: Thames and Hudson; as *Shadow Hunters: The Nest Gatherers of Tiger Cave*, Charlottesville, Virginia: Eastman Kodak
  Coffee-table book celebrating the traditional birds' nest industry.

ASIA, SOUTHWEST

The dry lands of southwest Asia, colloquially known as the Near and Middle East, include a belt of eroded fold mountains, reaching from Turkey to Afghanistan, along with the Arabian Peninsula to their immediate south. They are distinguished by desert and semi-desert terrains, where summer temperatures normally exceed 40°C, while winters frequently drop below freezing, except on the coastal strips. Annual rainfalls in the interior are typically 200 mm or less, rising to around 1000 mm on the coastal mountains of the Mediterranean, notably in Lebanon and much of Turkey, and to about 500 mm on the mountains of Yemen and southern Oman, which are caught by the edge of the Indian Ocean monsoons.

Southwest Asia has large tracts of Mesozoic and Tertiary limestones, together with chalk, dolomite, and gypsum, all inherited from Tethyan deposition. Many were folded within the Alpine-Himalayan orogenic belt, while others have been left little deformed on rigid basement blocks. Karst development is greatly restrained by the modern dry climate, so that potentially cavernous rocks extend across large areas and over high ranges of altitude, with few or no known caves of any significant size. Throughout the region, the Pleistocene was marked by pluvial stages, when fluvial and karstic processes were active. Some relict karst features and scattered remnants of cave passages and sediments survive from those times.

The caves and karst of the region are best reviewed by country, in sequence roughly from west to east.

Turkey (see separate entry) has more known caves than the rest of the region put together. Much of the Taurus Mountains are limestone karst and the region also contains the great travertine deposits of Pamukkale (see separate entry), while the gypsum karst of central Turkey is of major importance.

Syria provides a complete contrast, with minimal visible signs of karst. However, beneath its largely empty deserts there are four major karstic aquifers. The Jurassic limestones that form its western mountains are a minor extension of the karst in Lebanon, but there has been little investigation of any caves on the high land. The vauclusian rising of Fijeh has a mean outflow of 7.5 $m^3 s^{-1}$, and supplies much of the water for Damascus. It lies below an anticlinal ridge of Cretaceous limestone, and has been dived to a depth of 77 m. Further east, the same limestones have minimal signs of karst on their low desert ridges, where they yielded the excellent building stone for Palmyra. Tertiary limestones in the southeast form the large dry Chami Plateau, with its cap of thick calcrete. They also contain the Ras-el-Ain springs, at the head of the Khabour River, which have a total mean flow of 38 $m^3 s^{-1}$; they lie close to the Turkish border, with most of their catchment in Turkey. The water rises from the buried aquifer along faults through overlying Miocene clay, and emerges from collapse depressions in covering alluvium (Burdon & Safadi, 1963). Karstic conduits within Miocene gypsum of the Fars Series account for underground drainage of some large basins in the northeastern desert, where Cater Magara has 7300 m of passages.

Lebanon has a major karst region on the Jurassic Kesrouane Limestone of the Jebel Liban, west of the Beqa'a Valley. The spectacular mountains have many deep caves. Faouar Dara has a large, clean streamway descending 22 shafts to a depth of 622 m. Qattine Azar, 548 m deep, has 2 km of large stream passage, reached by a shaft of 182 m. The water from both these caves resurges close to the sea, at Fawar Antelias, 18 km from and 1540 m below Faouar Dara. The coastal foothills contain a number of large river caves, including the spectacular Jeita (see separate entry). The Mediterranean coast of Lebanon is notable for its major submarine resurgences in Mesozoic and Tertiary karstic aquifers (Ghannam, Ayoub & Acra, 1998). Springs beneath 45 m of water, in Chekka Bay, may yield 60 $m^3 s^{-1}$ from fissures in the Cretaceous Sannine Limestone, fed in part from the deeper Jurassic limestone.

Israel has the rolling karst hills of soft Cretaceous and Tertiary limestone and chalk in the Judean desert. The Dead Sea scrolls were found in small fissure caves in the scarp face at Qumran, and the Mount Carmel caves (see separate entry) are of archaeological importance but there are few significant known caves of length or depth. Much more significant for earth scientists is the salt karst of Mount Sedom (see separate entry).

Jordan has even less karst east of the Dead Sea. Cretaceous limestones have some collapse dolines and calcrete crusts on the hills north and south of Kerak, while the spectacular tufa cascades of Hammamat Ma'in are fed by radon-rich, geothermal waters that circulate through limestones at depth.

Iraq has no karst in its central Mesopotamian lowlands. In the northwest, the Sulevani Plain has dolines, some short caves and underground drainage in the Miocene Fars gypsum that forms a continuation of the karst plain in Syria. The northeast encompasses a small sector of the Zagros limestone belt of Iran, where stream-sinks, caves and springs have been reported in the Mosul region and Shanidar Cave (see separate entry) is an important archaeological site.

Saudi Arabia has a basement of granite and Paleozoic sandstone, exposed in the west, that has subsided beneath the Tethyan carbonates and mudrocks forming the petroleum-rich belt along the Persian Gulf. Karstic dissolution has contributed to the permeability of some of the buried reservoir rocks. Fossil groundwater is also being "mined" from non-renewable limestone aquifers, for irrigation in parts of the eastern interior. The main outcrops of the Jurassic to Eocene carbonates and sulfates are on the escarpments northwest and south of Riyadh, and also those nearer the Gulf coast. Of the many large collapse dolines, Ain Hit is the best known, 88 m deep to its water surface and dived into a series of flooded chambers; these lie in limestone, though the entrance descends through Jurassic anhydrite. Further north, the Cave of the Falling Star is a shaft, 100 m deep, that bells out into a large chamber. Over the border, in Kuwait, dissolution of the Eocene Dammam Limestone has caused some deep collapses in thick cover sediments.

Yemen has the same limestones, better exposed in parts of its mountainous terrain. The Jurassic Amran Limestone around San'a has various hot springs, and also caves reported to yield steam clouds from hydrothermal waters, but these await investigation. Tertiary limestones form the plateaus either side of Wadi Hadramaut (in the east of the country), but details of the karst are little known. The island of Socotra has limestone mountains, where Hoq Cave has 3000 m of mapped passages and Giniba Cave has a chamber 200 m long and 150 m wide beneath the Deksam Plateau.

Oman has its Dhofar karst in the far south on the same Tertiary limestones as those over the border in Yemen, but this area has a wetter climate that frequently shrouds the Qara karst plateau in fog. It is distinguished by some large collapse dolines, of which Tawi Atayr is 130 m across and 210 m deep. Quanaf Cave has a large passage 1500 m long, descending 223 m to a long lake, unexplored because of bad air. Hatat Lohum is a smaller shaft system, reaching 207 m in depth. Banks of travertine hundreds of metres high cascade into Wadi Durbat, but their tree cover indicates a level of activity that has declined in the modern climate. In northern Oman, the Selma Plateau (see separate entry), southeast of Muscat, has the deepest and finest caves of the Middle East, also in Tertiary limestone. Southwest of Muscat, Jabal Akhdar is the higher escarpment on a massive breached anticline of Mesozoic Wasia Limestone. Its surface is bare limestone fretted by microkarren and dry wadis, while its caves include Kahf Hoti (see figure), a fine through system 5 km long and 262 m deep (Waltham, Brown & Middleton, 1985). Percolation water entering Hoti is at 32°C, while its many lakes are cooled to 21°C, by evaporation into the through wind. Many of Oman's northern caves (and some in Abu Dhabi) are distinguished by large masses of banded flowstone that are cut through

Asia, Southwest: A tall canyon cut along joints creates the feeling of an underground wadi prone to flash floods in Kahf Hoti in the Jebel Akhdar of Oman. (Photo by Tony Waltham).

by narrow stream trenches. The flowstone is a relict of karstification in pluvial stages of the Pleistocene—which were synchronous with interglacial stages of the higher latitudes (Burns *et al.*, 2001), and not with the glacial stages, as had previously been assumed.

Abu Dhabi is the only one of the United Arab Emirates that has significant known karst. Jebel Hafeet is an isolated anticline of Eocene limestone that crosses the border into Oman. It has no surface karst features larger than microkarren, but it has a handful of caves (none with more than 500 m of passages). Bulbous phreatic tunnels perched 800 m above the modern water table may have been formed by hydrothermal dissolution and may predate part of the late Tertiary tectonic uplift. Kahf Hamam is a singularly unpleasant cave, with a temperature of 33°C and 100% humidity. There are also mature dissolutional foot caves carved into both the limestone and the veneer of old, cemented, bahada debris. Further north, hills of Mesozoic limestone in Ras al Khaimah contain some small relict caves, but little is known of any karst on the limestone's furthest outcrops on the Musandam peninsula.

Iran (see separate entry) has extensive tracts of limestone in the Zagros Mountains. There are few caves, but Ghar Parau reaches 751 m deep, and Ghar Alisadr has over 11 km of mapped passages.

TONY WALTHAM

Works Cited

Burdon, D.J. & Safadi, C. 1963. Ras-el-Ain: the great karst spring of Mesopotamia. *Journal of Hydrology*, 1: 58–95

Burns, S.J., Fleitmann, D., Matter, A., Neff, U. & Mangini, A. 2001. Speleothem evidence from Oman for continental pluvial events during interglacial periods. *Geology*, 29: 623–26

Ghannam, J., Ayoub, G.M. & Acra, A. 1998. A profile of the submarine springs in Lebanon as a potential water resource. *Water International*, 23: 278–86

Waltham, A.C., Brown, R.D. & Middleton, T.C. 1985. Karst and caves in the Jabal Akhdar, Oman. *Cave Science*, 12: 69–79

Further Reading

Burdon, D.J. & Safadi, C. 1964. The karst groundwaters of Syria. *Journal of Hydrology*, 2: 324–47

Jeannin, P-Y. 1992. Expedition suisse aux Emirats Arabes Unis. *Stalactite*, 42: 47–55

Voigt, S. & Schnadwinkel, M. 1995. Caving beneath the desert: Cater Magara, Syria. *International Caver*, 14: 15–26

Waltham, T. & Fogg, T. 1998. Limestone caves in Jebel Hafeet, United Arab Emirates. *Cave & Karst Science*, 25: 15–22

ASIAGO PLATEAU, ITALY

The Asiago Plateau, also called the plateau of 7 'Comuni', is the largest mountain group in the Venetian Pre-Alps and one of the main caving regions in Italy. The plateau area covers c.600 km² and is c.25 km long (north–south) and up to 30 km wide, being delimited by large and steep scarps (Figure 1). The range in elevation is from a few hundred metres to an altitude of 2300 m.

The dolomite of the late Triassic (Dolomia principale), which constitutes the basement of the plateau, was deposited in a platform environment and ranges over a thickness of 1000 m. The

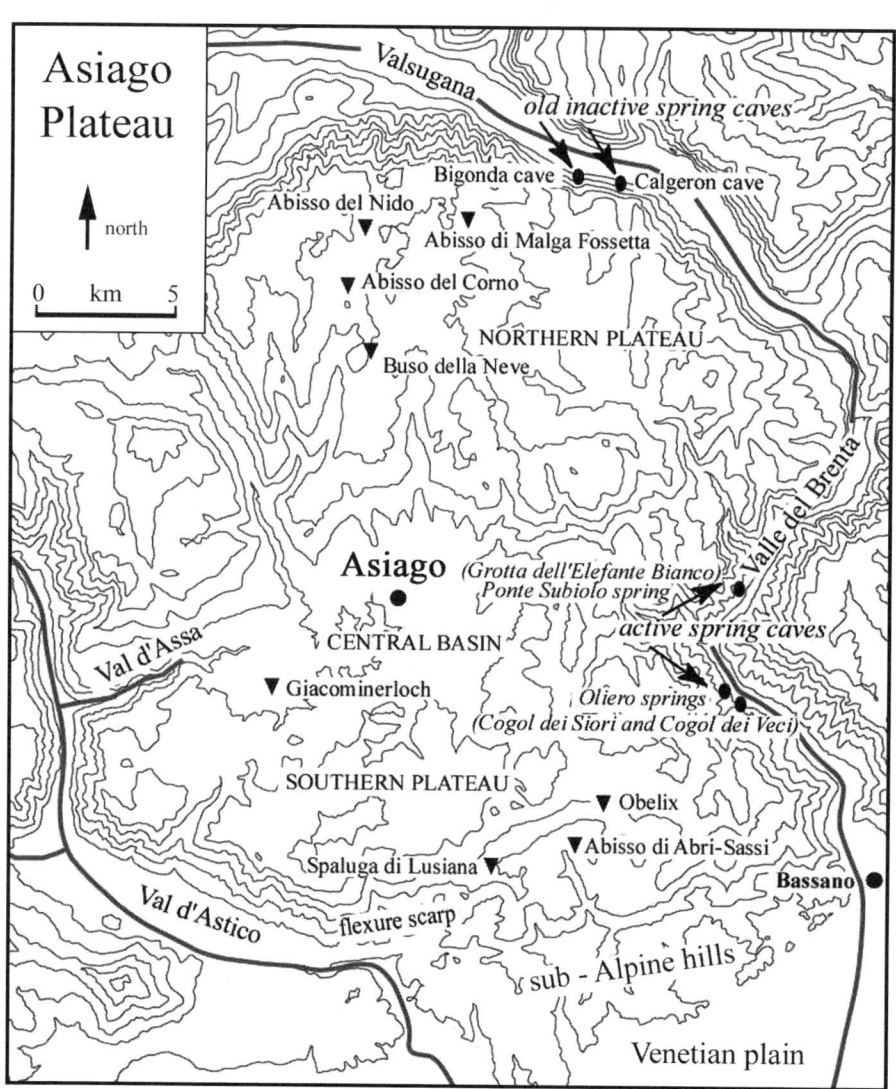

Asiago Plateau, Italy: Figure 1. Map of the karst geosystem of the Asiago plateau.

overlying limestones, mostly of Jurassic and Cretaceous age and ranging over slightly less than 1000 m in thickness, were deposited in three main environments: lagoon, platform, and basin. The Calcari grigi formation of the lower Jurassic is an expression of a lagoon environment and is characterized by lignite and clay bands. Rosso Ammonitico, of the middle and upper Jurassic, is a condensed unit rich in iron oxides and ammonites. Biancone, of the lower and middle Cretaceous, is a chalk-type limestone, rich in clay, especially in its middle and upper parts. Scaglia Rossa, of the upper Cretaceous, is found outcropping only in small areas of the plateau. Paleogene basaltic lavas and tuffs are locally interbedded with the limestones. An early speleogenesis inside the carbonate sequence is probably linked with Paleogene vulcanism.

The mountain group constitutes the upper part of a tectonic wedge produced in the framework of the alpine orogenesis. The wedge is delimited on the southern margin by the thrust of Bassano–Valdobbiadene and on the northern margin by the backthrust of Val di Sella. The southern part of the extruded wedge is bent to form an asymmetrical anticline, the central part a syncline, and the northern part another asymmetrical anticline.

In accordance with these tectonic structures, three main geomorphological subunits are recognizable: the southern plateau, corresponding to the top of the southern anticline; a central basin corresponding to the central syncline; and a northern and upper plateau corresponding to the northern anticline (Sauro, 1995; Mietto & Sauro, 2000).

The southern plateau is characterized by a network of dry valleys, with wide floors and steep slopes. This fluviokarstic network evolved within the Jurassic limestones through the fluvial network, which developed in the overlying rock units during uplift. The central basin presents a dense network of V-shaped dry valleys, which mostly developed under periglacial conditions. The ridges between the valleys are covered by thick eluvial and colluvial mantles, mostly composed of soil and loess-like sediments, clay, and weathered chert fragments. The northern plateau is characterized by both karst and glacial forms, such as glacio-karstic closed depressions and limestone pavements.

The evolution of the morphostructure was marked by the gradual arrangement of the tectonic structure and by the accompanying erosional events. The old karst drainage system fed large karst springs at the foot of the northern scarp of the plateau. Two large cave systems (Bigonda and G.B. Trener caves, at an altitude of $c.350$ m), that developed in epiphreatic conditions just below the old water table and now hang 150 m above the Valsugana (upper Brenta Valley) valley floor, have been mapped for over 30 km. During exceptional flood conditions, such as the November 1966 flood, these caves behave like overflow outlets.

The present-day spring caves are in the Valle del Brenta at the foot of the eastern scarp of the plateau (Oliero and Ponte Subiolo spring-caves, at an altitude of $c.150$ m). The average discharge of the three main springs is estimated to be $c.15$ m s^{-1} (Meneghel et al., 1986). Divers have explored more than 5 km of the underwater caves. The underground piracy, with the development of new springs, and of relative underground network, is a consequence of the deepening of the lower Brenta Valley, through fluvial and glacial erosion during the middle and upper Pleistocene.

The more than 2300 explored caves are mostly vertical in development (Mietto & Sauro, 2000). The best known are listed in the Table. It is possible to distinguish some main types of cave, and in particular:

1. alpine-type shaft systems, consisting of successions of vertical shafts separated by narrow passages, especially frequent in the northern plateau;
2. old swallowing systems, consisting in large underground meandering gorges connected by vertical shafts, especially frequent in the southern plateau;
3. cave systems with intermediate characters between the two previous types;
4. simple shafts and chambers, relics of old systems cut by the erosional surfaces;
5. simple small shafts resulting from inception-type speleogenesis, mostly in the epikarstic zone at the transition between rock layers with different hydraulic conductivity, in particular inside or near the Rosso Ammonitico rock unit (Klimchouk et al., 1997);

Asiago Plateau, Italy: Some of the best-known karst caves in the Sette Comuni Plateau (for types, see text).

Name	Hydrogeological zone	Type	Development in metres	Vertical development in metres
Bigonda cave	vadose	7	26 720	−100, + 350
Calgeron cave (also called G B Trener Cave)	vadose	7	5220	−130, + 250
Abisso di Malga Fossetta	vadose	1	4200	−974
Giacominerloch	vadose	2	3100	−507
Buso della Neve di Zingarella	vadose	2	2900	−245
Cogol dei Veci	epiphreatic, submerse	6	2350	−59 (below spring level)
Cogol dei Siori	epiphreatic, submerse	6	3000	−63 (below spring level)
Abri Sassi system	vadose	2	1865	−353
Abisso del Nido	vadose	1	1504	−466
Abisso del Corno	vadose	1	945	−472
Obelix	vadose	3	1500	−700
Spaluga di Lusiana	vadose	3	630	−270
Grotta dell'Elefante Bianco	epiphreatic, submerse	6	365	−139 (below spring level)

6. epiphreatic spring caves developed mostly just below the water table;
7. old epiphreatic caves, now active only during exceptional floods.

The different types of caves are an expression of the evolution of the various geomorphological environments of the karst mountain group. In particular, the large swallowing systems probably began to develop as swallow holes inside the old network of fluvial valleys, afterwards being dried by the development of the karst network. Some of these systems have been reactivated during the cold phases of the Pleistocene, swallowing the melting water of the local glaciers, as documented by the filling deposits (e.g., Giacominerloch). The alpine-type shaft systems have developed gradually during the denudation and the uplifting of the plateau. The bottom of the Abisso di Malga Fossetta is not far from a branch of the old epiphreatic cave of Bigonda. In many caves of the upper plateau, there are ice deposits, also with considerable volumes (Busellato & Gruppo Grotte Schio, 1991).

Perched aquifers are stored, both in till deposits, and in the densely fractured chalk-type limestones. The water circulates very fast through the main cave passages of the karstic aquifer.

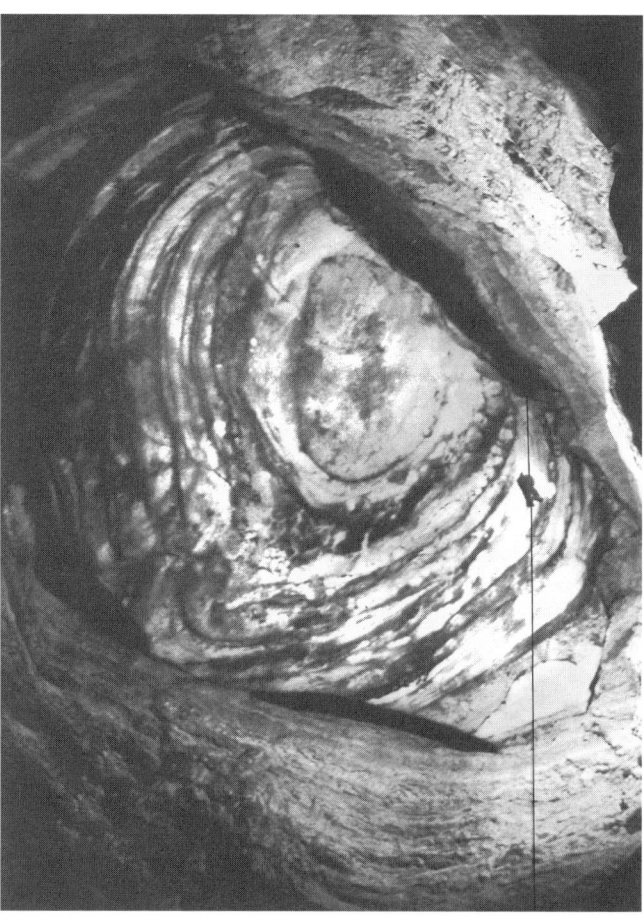

Asiago Plateau, Italy: Figure 2. Domed ice roof of a large chamber in Buso della Neve di Zingarella. The ice block, seen from below, shows its layered structure. A speleologist is descending on a rope (from Busellato et al., 1991, with the permission of the authors).

A tracing experiment between an active swallow hole and the main springs revealed an average speed of about 400 m h^{-1} over a distance of about 10 km and a difference in elevation of 1000 m (Gennari et al., 1989). The response of the springs to the input of heavy precipitation may be only a few hours (Celi & Sauro, 1995).

During the cold phases of the Pleistocene, a plateau glacier developed in the northern part of the mountain group, and at the last glacial maximum, the extent of the local glacier was $c.270$ km^2. Some glacial tongues reached the central basin, which was an important glacial outlet. After deglaciation, the plateau was rich in creeks, lakes, and springs, but during the Holocene, the reopening of karst swallow holes caused a progressive "desertification" of the plateau (Sauro, 2001). The Gelpack creek, which crosses the central basin and which discharges its waters in many shallow holes, is a relict of the glacial outlet.

During the holiday periods, the central basin becomes like a small town, attaining a population of 100 000. In connection with the urbanization, a flourishing quarrying industry has developed. There is also an integrated system of industries producing modern cattle-breeding farms and related installations for dairy production. Over the last 30 years the average quarried limestone exceeds the limestone eroded naturally by karst solution by about three times (about 1000 tons/day as opposed to 300 tons/day) (Meneghel et al., 1986).

The increase in water consumption has required the recycling of deep karstic water from the main springs which, after use, is returned as wastewater that is polluted with organic substances, chemical compounds, etc. This recycling accounts for a quantity that is of the order of 0.5% of total precipitation. Although the various forms of impact are severe, the quality of the karst water is still relatively good. Some specimens of *Proteus anguinus* (cave salamander), introduced into the local aquifer from the Carso di Trieste during the mid-19th century, have reproduced and have continued to survive. The impact of urbanization has certainly been mitigated by the presence, in the central basin, of the chalk-type limestone and by the thick alluvial cover. The mostly deserted upper plateau is a much more vulnerable environment, but here human impact is now relatively scarce.

UGO SAURO

Works Cited

Busellato, L. & Gruppo Grotte Schio 1991. *Dimensione buio*, Vicenza: Safigraf

Celi, M. & Sauro, U. 1995. Primi risultati del monitoraggio delle acque della Sorgente del Cogol dei Siori (Fiume Oliero, Valle del Brenta, Prealpi Venete). *Studi Tridentini di Scienze Naturali. Acta Geologica*, 70: 53–67

Gennari, G. Landi, M. & Sauro, U. 1989. Un'esperienza con traccianti sull'altopiano dei Sette Comuni (Prealpi Venete). In *Atti del XV Congresso Nazionale di Speleologia*, Castellana Grotte (Bari): Istituto Italiano di Speleologia: 369–80

Klimchouk, A. Sauro, U. & Lazzarotto, M. 1997. "Hidden" shafts at the base of the epikarstic zone: A case study from Sette Comuni plateau, Venetian Pre-Alps, Italy. *Cave and Karst Science*, 23(3): 101–08

Meneghel, M., Sauro, U., Baciga, M.L., Fileccia, A., Frigo, G., Toniello, V. & Zampieri, D. 1986. Sorgenti carsiche ed erosione chimica nelle Prealpi Venete. *Studi Trentini di Scienze Naturali. Acta Geologica*, 62: 145–72

Mietto, P. & Sauro, U. 2000. *Le Grotte del Veneto: paesaggi carsici e grotte del Veneto*, Verona: Regione del Veneto and La Grafica

Regione Veneto 2001. *Hydrogeological map of Sette Comuni Plateau*, Regione Veneto Direzione Geologia e Ciclo dell'Acqua-Kater

Sauro, U. & Lanzingher, M. 1995. The study of the morphokarstic unit of Sette Comuni Plateau (Venetian Fore-Alps): State-of-the-art. *Studi Tridentini di Scienze Naturali. Acta Geologica*, 70: 7–8

Sauro, U. 2001. Ambienti geomorfologici e paesaggi dell'altopiano dei Sette Comuni fra il Pleistocene finale e l'Olocene. In *Atti del Convegno: Le incisioni rupestri della Val d'Assa: ipotesi a confronto*, Vicenza: Comune di Gallio: 13–26

Further Reading

Belloni, S., Martinis, B. & Orombelli, G. 1972. Karst of Italy. In *Karst: Important Karst Regions in the Northern Hemisphere*, edited by M. Herak & V.T. Stringfield, Amsterdam and New York: Elsevier

Cucchi, F. & Forti, P. 1986. Map of karst areas in Italy. In *Caving in Italy*, edited by A. Cigna, Bologna: Sociéta speleologica italiana

Sauro, U. 1993. Human impact on the karst of the Venetian Fore-Alps (Southern Alps, Northern Italy). *Environmental Geology*, 21(3): 115–21

Sauro, U. 1995. Dinamica geomorfologica e vulnerabilità della risorsa acqua nell'Altopiano dei Sette Comuni (Prealpi Venete). *Studi Tridentini di Scienze Naturali. Acta Geologica*, 70: 43–51

Verico, P. & Zorzin, R. (editors) 1996. *Alpine Caves: Alpine Karst Systems and their Environmental Context, Proceedings of the International Congress, Asiago, 1992*, Vicenza: Federazione Speleologica Veneta

ATAPUERCA CAVES, SPAIN: ARCHAEOLOGY

The Sierra de Atapuerca, located 15 km east of Burgos in northern Spain, is the world's largest known repository of fossil humans from the Middle Pleistocene period, between one million and 127 000 years ago. Archaeologists have been working here since 1976, but have barely scratched the surface of this extraordinary set of sites.

At the turn of the 20th century, a stone-quarrying company cut a great railway trench, or "trinchera", through the southwestern part of the sierra, providing a first glimpse of a host of sediment-filled limestone caves which, it is now known, were occupied by early humans as well as by carnivores and bats. Most of them had filled up with sediments and the detritus of collapsed roofs and entrances. At one site known as the Galeria, archaeologists have been excavating since the 1980s in a hard breccia containing pebble tools and flakes of quartz, quartzite, and flint, as well as Acheulian hand axes and cleavers. Uranium / thorium dating has attributed stalagmitic deposits at the top of the Galeria to about 125 000 years ago. From estimates of the rate at which sediments accumulated in the cave, it has been calculated that the first human presence in the levels below occurred around 350 000 years ago, with sporadic occupation up to 200 000 or 150 000 years ago. A fragment of a human parietal (skull) bone, about 250 000 years old, was found here in close proximity to a hand axe.

The nearby Gran Dolina, another collapsed limestone cave, contains sediments some 20 m deep spanning a period from more than one million to about 120 000 years ago, the bottom half belonging to the Lower Pleistocene (before 780 000 BP) and the top half to the Middle Pleistocene. Primitive stone tools were first found in layer TD4, known to be more than 700 000 years old. Since 1994, primitive, blue-stained human teeth as well as bone fragments from at least six individuals have been recovered from a higher level, TD6, together with numerous stone tools including hammer-stones, flakes and cores made from quartzite river pebbles and flints, but no hand axes or cleavers. This layer is 0.6 m below a level dated by paleomagnetism to 780 000 years ago, so the human bones must be between 800 000 and a million years old, perhaps the oldest in Europe. They have been ascribed to a new species, *Homo antecessor*, thought to be ancestral to modern humans. Cutmarks on some of the human bones have been interpreted as evidence for cannibalism. The tools from TD4, located below TD6, are clearly at least a million years old.

A few hundred metres from the Trinchera is an enormous cave, the Cueva Mayor, with a red painting of a horsehead near the entrance which was long supposed to be from the Ice Age, though it is now thought to be modern. At the back of the cave, an arduous 500 m from the entrance, is the famous Sima de los Huesos (Pit of the Bones) where investigation was triggered in 1976 by the discovery of a human jaw more than 200 000 years old. It was found in sediments which, for centuries, had been yielding cave bear remains. A side chamber contains cave bear dens and claw marks gouged in pockets of clay, looking as fresh as if they had been done yesterday. The whole cave is marked with graffiti, the oldest from 1561. Local people have been coming here for hundreds of years to obtain bear bones and teeth—according to a tradition, every young man of the region must present his fiancée with a bear tooth from the Cueva Mayor as a sign of his prowess.

In 1976 Spanish paleontologist Trinidad Torres came here to collect sediment containing bear bones. It was this sediment that yielded the human jaw that eventually inspired the current investigations. Excavations began in 1983, and showed that beneath the thousands of cave bear bones there were rich deposits of human remains. Some time after the pit was no longer used by humans, hibernating bears began to fall into it. So far, the pit has yielded more human bones of the period—more than 200 000 years ago—than all other sites combined. Even small phalanges and other hand and foot bones are unearthed, while sieving of the sediments recovers the tiny bones of the inner ear.

The pit has revolutionized knowledge of *Homo heidelbergensis*, a transitional species between *Homo erectus* and Neanderthals. Although only 2–3% of the pit's deposits have so far been examined by the slow, painstaking excavation work, almost 2000 bones have been found from at least 33 individuals, and perhaps as many as 50. All parts of the skeleton are present, and males and females are equally represented, with most individuals being adolescents or young adults. It seems that, over several generations, bodies were carried into the cave and tossed into the shaft in a form of crude mortuary ritual—essentially the earliest known burials in the world. The absence of herbivore bones and stone tools indicates that this was not an occupation site, and the lack of carnivore damage suggests that the bones were not left there by animals.

These bones represent about 90% of all pre-Neanderthal bones ever found in Europe. The people seem to have been robust; their teeth are very worn from chewing plants, but there are no cavities from dental disease, and overall the bones show few signs of illness or trauma. The remains include three remarkably well-preserved skulls found in 1992, with large Neanderthal-like brow-ridges and projecting faces, and a complete pelvis has also been recovered.

Another site in the Trinchera, the Sima del Elefante (Pit of the Elephant), where work began recently, has yielded a few tools of c.200 000 years old, and it seems to be connected to the Cueva Mayor hundreds of metres away, which means that in this site alone there remain millions of cubic metres of unexplored Pleistocene sediments. This fact, together with the other caves and shelters in the sierra which have not yet been excavated, ensures that Atapuerca will remain a site of world importance for decades to come.

PAUL G. BAHN

Further Reading

Arsuaga, J.L. et al. 1993. Three new human skulls from the Sima de los Huesos Middle Pleistocene site in Sierra de Atapuerca, Spain. *Nature*, 362: 534–47

Arsuaga, J.L., Bermúdez de Castro, J.M. & Carbonell, E. (editors) 1997. The Sima de los Huesos hominid site. *Journal of Human Evolution*, 33: 105–421 (special issue)

Bermúdez de Castro, J.M. et al. 1997. A hominid from the Lower Pleistocene of Atapuerca, Spain: Possible ancestor to Neandertals and modern humans. *Science*, 276: 1392–95

Bermúdez de Castro, J.M., Arsuaga, J.L., Carbonell, E. & Rodriguez, J. (editors) 1999. *Atapuerca: Nuestros Antecesores*, Salamanca: Junta de Castillón y León

Bermúdez de Castro, J.M., Carbonell, E. & Arsuaga, J.L. (editors) 1999. Gran Dolina Site: TD6 Aurora Stratum (Burgos, Spain). *Journal of Human Evolution*, 37: 309–700 (special issue)

Carbonell, E. et al. 1995. Lower Pleistocene hominids and artifacts from Atapuerca-TD6 (Spain). *Science*, 269: 826–30

Carbonell, E. et al. 1998. La revolución de Atapuerca. *Revista de Arqueología*, 210(October): 14–24

Cervera, J., Arsuaga, J.L., Bermúdez de Castro, J.M. & Carbonell, E. 1998. *Atapuerca: un Millón de Años de Historia*, Madrid: Plot Ediciones

AUSTRALIA

A relative paucity of karst in Australia is partly attributable to the great age of the ancient shield of which much of the continent is comprised, which predates most of the life forms from which biogenic carbonates are derived. The shield is also sufficiently old that ample time has elapsed for such limestone as was deposited to have been eroded away (Jennings, 1983). Nevertheless, many aspects of Australian karst make it of great scientific interest. Karst and caves have developed in carbonate formations that range in age from Pleistocene to Precambrian, and the 200 000 km^2 Nullarbor Plain area is one of the larger karst areas on Earth (see Nullarbor Karst). Australia is also the only continent to lack the Tertiary orogenic belts that are elsewhere rich in caves, but the resulting history of tectonic stability has enabled the survival of some very ancient karst phenomena. The Wombeyan karst in New South Wales and Eugenana karst, Tasmania, have been exposed since the Devonian, and karst features older than the Permian have been reported from several other karsts.

Great climatic variation across the continent contributes to Australia's considerable karst geodiversity (Figure 1). Australia is the driest continent apart from Antarctica and this, coupled with the low relative relief and hence lower potential hydraulic head, has inhibited karst evolution. Thus, in the Tertiary limestones of the vast Nullarbor, only 300 caves and 1400 dolines have been recorded while in contrast, highly evolved karst occurs in similar rocks in wetter southeastern South Australia. Karsts in the humid Eastern Highlands are the most highly evolved despite sedimentary environments in the Tasman mobile zone having been only marginally suitable for carbonate deposition. Most of these karsts are small and surrounded by non-carbonate rocks, aggressive runoff from which rapidly attacks and erodes the limestone.

Climate change has also been of great consequence to karst evolution. Although the evidence is ambiguous, the principal caves beneath the Nullarbor may have formed under previously wetter conditions. There is less ambiguity in the Eastern Highlands and Tasmania, where many karsts bear a legacy of previously cooler conditions. Continental Australia and Tasmania are now entirely deglaciated, and while late Cainozoic glaciation of mainland Australia involved only a few small cirque glaciers distant from any karst, ice caps and valley and cirque glaciers previously covered c.7000 km^2 of Tasmania. Oligocene glacial sediments in the Forth Valley, immediately upstream from the Lorinna karst, imply that the present broad topography was already established, again highlighting the relative antiquity of even these most alpine of Australian karsts.

A reduction in forest biomass during cooler and drier episodes in the late Cainozoic allowed periglacial slope instability to extend to low altitudes in Tasmania, mantling many karsts with allogenic regolith. At Junee-Florentine, cave streams were displaced by colluvial and alluvial sediments causing temporary reversion to surface drainage (Goede, 1973). Periglacial effects have also been recorded from Yarrangobilly, New South Wales and elsewhere in the southeastern highlands of mainland Australia.

Some Australian karsts

The youngest carbonates in which karst has formed are dune limestones (<700 ka old) in coastal southwestern Western Australia, on the Eyre Peninsula and Kangaroo Island in South Australia, southern Victoria and scattered localities on the Bass Strait islands and western Tasmania. Syngenetic karst (see separate entry) has developed in these poorly consolidated aeolianites. Solution pipes typically connect the surface to caves at the water table. Easter Cave, southwestern Western Australia, contains >7.5 km of passages, and nearby Strongs Cave contains straw stalactites to 9.8 m in length.

The arid–semi arid Nullarbor karst contains some very large caverns such as those of Koonalda, Abrakurrie, and Webubbie

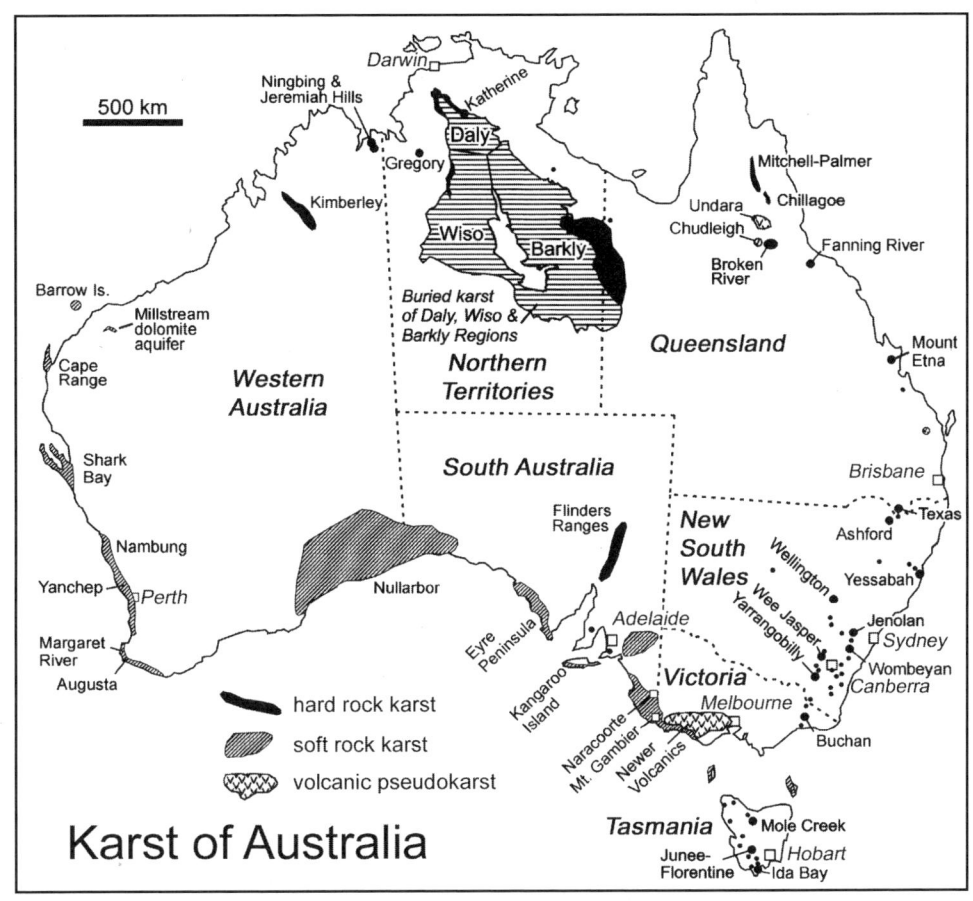

Australia: Figure 1. Location of principal karst areas and caves in Australia.

caves but the longest, Old Homestead Cave (~28 km), is a maze of relatively small passages. The karst province of southeastern South Australia, formed in Oligocene limestones, includes a World Heritage-listed Tertiary mammalian fossil site at Victoria Fossil Cave, Naracoorte (see Australia: Archaeological and Paleontological Caves). Karst formed in Miocene limestone also occurs in the Murray Valley, South Australia.

Notwithstanding the general absence of Tertiary orogenic influences in Australia, karst in the arid Cape Range, Western Australia, has formed in four emerged post-Miocene marine terraces, the caves containing a well-developed invertebrate fauna with strong Gondwanan affinities (see entry Cape Range, Australia: Biospeleology). Karst also occurs on Australia's external Territory of tropical Christmas Island, on the boundary between the Indo-Australian and Caroline plates. The carbonates here comprise raised coralline shore platforms that fringe a core of Oligocene–Miocene reef limestones overlying volcanic rocks. To date, only minor karren have been recorded from poorly known Miocene–Oligocene limestone in the remote sub-Antarctic Australian Territory of Heard Island in the southern Indian Ocean.

Limestones of Permian age host karsts at Kempsey in New South Wales. Relatively widespread but impure Permian limestones in Tasmania are known to contain small caves only at Gray and on the coast at Maria Island. Caves formed in Carboniferous limestone at Texas, Queensland, were inundated by an artificial reservoir constructed to provide water for the local tobacco industry.

The tropical, semi-arid Limestone Ranges of the Kimberley region, Western Australia comprise an exhumed Devonian barrier reef system with a relief of $c.80$ m. The limestone has been dissected to form an alluviated labyrinth karst. Other important karsts formed in Devonian limestone occur in Queensland at Mt Etna, previously the scene of a long-running conflict over limestone quarrying, and at Broken River. Karst in rocks of this age also occurs in New South Wales at Wee Jasper, Timor, and Wellington, the latter an important megafaunal fossil site, and in Victoria at Buchan where incision probably resulted from tectonic uplift (Webb et al., 1992). Other small karsts formed in Devonian limestone include the Lake Sydney glaciokarst in Tasmania, where stream-sinks buried by glacial sediment during the most recent glaciation are presently being exhumed as underground drainage is re-established.

The Chillagoe tower karst, northern Queensland, is formed in Silurian limestone (see Chillagoe and Mitchell-Palmer Karsts). More than 150 karst towers up to 100 m high also occur further north in the seasonally arid Mitchell-Palmer karst, characterized by very rugged karren and containing caves of considerable biological interest. Silurian limestones also host some of the most significant karsts in New South Wales, at Jenolan, Yarrangobilly, Cooleman Plain, Wombeyan, Bungonia, Colong, Bendithera, and Wyanbene. About 45 km of passages have been surveyed within the Jenolan tourist caves complex, which lies within the Greater Blue Mountains World Heritage Area (Kiernan, 1988).

A greater proportion of carbonate rocks accumulated in those parts of the sedimentary basins that were to become Tasmania. Rainfall totals and effectiveness are greater in Tasmania than over most of Australia and it is here that Australia's deepest caves are located, near the retreating margins of caprocks that have protected the carbonates, thus permitting retention of relatively high local relief. Current information suggests the deepest is Niggly Cave (375 m) in the Junee-Florentine area, but Anne-A-Kananda (373 m) at Mt Anne, and the 12 km long Ice Tube–Growling Swallet system (360 m), also at Junee-Florentine, are also within the range of potential survey error.

The most highly evolved of the c.300 Tasmanian karsts are formed in Ordovician limestones. The most intensively studied are at Mole Creek, Junee-Florentine, and at Ida Bay where >23 km of locally massive passages have been mapped to date in Exit Cave, one of numerous karsts within the Western Tasmanian Wilderness World Heritage Area (Figures 2–4) and the scene of another conflict over limestone quarrying. Recognition of a rich Pleistocene human occupation site in Kutikina Cave (formerly Fraser Cave) in the Franklin River Valley played a major role in establishment of this World Heritage Area and the defeat of proposals for hydro-electric reservoirs that would have inundated many riverine karsts. A number of additional cave occupation sites have since been discovered in southwestern Tasmania, including Pleistocene art at Ballawinnie Cave in the Maxwell Valley and Wargata Mina in the Cracroft Valley. The altitude of some caves at Precipitous Bluff on Tasmania's south coast is broadly coincident with emerged coast-facing terraces of probable Tertiary age, making these caves potentially significant for

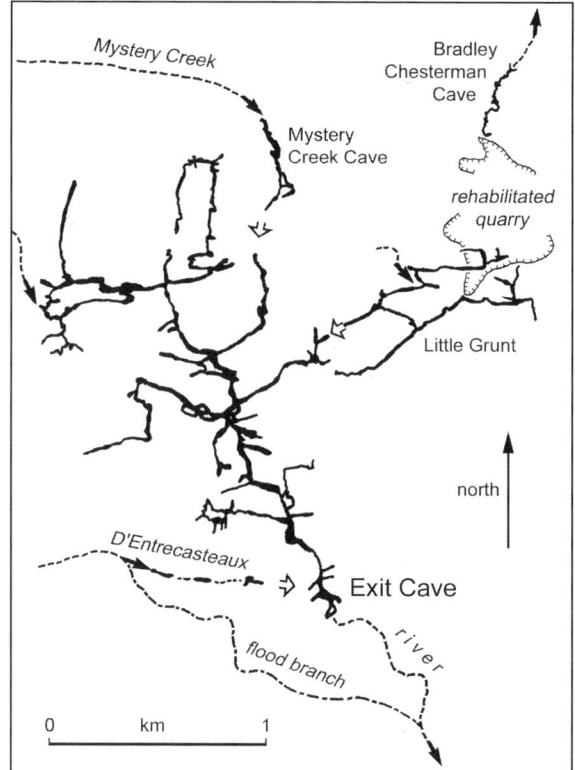

Australia: Figure 3. Map of Exit Cave area, Ida Bay karst, Tasmania.

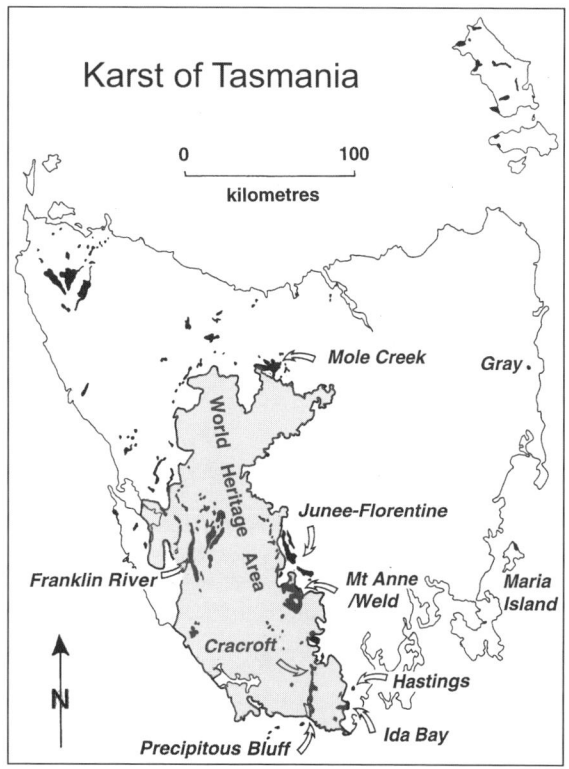

Australia: Figure 2. Tasmanian karst areas and location of the Tasmanian Wilderness World Heritage Area.

Australia: Figure 4. Massive blockfall breakdown in the main canyon of Exit Cave, Tasmania. (Photo by Andy Eavis)

studies of neotectonic uplift around the Tasmanian coast. Several Tasmanian karsts are of biological significance. Karsts at Abercrombie and Walli, New South Wales, are also formed in Ordovician carbonates.

Extensive karstification has occurred in Cambrian carbonates in the Barkly Tableland of the Northern Territory and northern Queensland. Water table caves in low karst towers of possible Cretaceous age occur in the Katherine area. The Bullita Cave system, Gregory National Park, is Australia's longest surveyed cave, comprising a 65 km long maze with many daylight holes. A Miocene vertebrate fossil locality in the Riversleigh karst has World Heritage status. Hydrothermal waters are present in some of these karsts and particularly extensive tufa accumulations occur at Lawn Hill. Caves in the low relief Camooweal karst descend 60–70 m to a water table that may rise 20 m after storms. In the Flinders and Mt Lofty Ranges, South Australia, cave development in Cambrian carbonates has been inhibited by aridity.

A complex of caves formed in Precambrian dolomite at Hastings includes one of Tasmania's more important tourist attractions (Newdegate Cave). The greatest topographic relief attained by carbonate rocks anywhere in Australia is ~600 m, at the important Mt Anne glaciokarst in southwestern Tasmania (Figure 2). Surface karst landforms, warm springs, and small caves in the Tarkine area, northwestern Tasmania, are formed in extensive magnesite deposits and hence are rare on a world scale. Karst has also formed in Precambrian carbonates at Coorow, Western Australia and in the Flinders Ranges, South Australia.

Some karst features have also formed in non-carbonate rocks. Dissolution of silica in relatively insoluble rocks has produced some large caves, including stream caves, in Upper Proterozoic quartz sandstones. They include Whalemouth Cave, North Kimberley, Western Australia, and others in Arnhem Land, Northern Territory, and adjacent parts of northern Queensland. Runiform relief is well developed in the Ruined City, Arnhem Land. Notwithstanding a lack of subsurface drainage, the sandstone ridges and towers of the Bungle Bungle Range, Western Australia, have also been interpreted as karstic because dissolution of the quartz cement in the sandstone has been fundamental to tower evolution. Other caves have developed in the Triassic Hawkesbury sandstone of the Sydney Basin and rockshelters in sandstone are widespread. Formation of a cave by dissolution of autometasomatic alteration products in dolerite has been reported from Wayatinah, Tasmania and caves in basalt at Coolah, New South Wales, also appear to have formed by dissolution of a zeolite zone. Little evaporite karst is known apart from a few small gypsum caves in the Wolf Creek meteorite crater, Western Australia.

Turning to non-karst caves, some impressive pits appear to be the result of subjacent karst collapse, including the 100 m deep Big Hole, formed in Devonian sandstones and shales near Braidwood, New South Wales. A system of lava tubes extends for $c.100$ km from the Undara Volcano, North Queensland, which erupted 23 km^3 of lava $c.0.19$ million years ago. The longest single cave has been explored for about 1 km (Atkinson, 1990). Numerous lava tube caves are also present in the Pliocene to Holocene lava flows of western Victoria, on sub-Antarctic Heard Island, and elsewhere. The stability and climate of the Australian continent has favoured development of extensive areas of duricrust formed over weak, decomposed material that is susceptible to erosion by piping. Despite considerable potential for caves there has been negligible investigation of these duricrusts. The most extensive caves may have formed beneath ferricretes. The greater hydraulic gradient in the weathering profile at breakaways and jump-ups facilitates mass removal, but percolines more distant from scarps suggest prior preparation may be involved. Some raised sea caves are of considerable scientific interest, including important archaeological sites at Rocky Cape, northwestern Tasmania, and on Hunter Island, Bass Strait. Iron Monarch Cave, King Island, is a raised sea cave formed in non-carbonate rocks but contains many speleothems formed of carbonate leached from overlying dune sands.

KEVIN KIERNAN

See also **Australia: Archaeological and Paleontological Caves; Cape Range, Australia: Biospeleology; Chillagoe and Mitchell-Palmer Karsts; Nullarbor Karst**

Works Cited

Atkinson, A. 1990. The Undara lava tube system and its caves. *Helictite*, 28: 3–14

Goede, A. 1973. Hydrological observations at the Junee resurgence and a brief regional description of the Junee area, Tasmania. *Helictite*, 11: 3–24

Jennings, J.N. 1983. A map of karst cave areas in Australia. *Australian Geographical Studies*, 21(2): 183–96

Kiernan, K. 1988.The geomorphology of the Jenolan Caves area. *Helictite*, 26(2): 6–21

Webb, J.A., Fabel, D., Finlayson, B.L., Ellaway, M., Shu, Li & Spiertx, H-P. 1992. Denudation chronology from cave and river terrace levels: the case of Buchan Karst, southeastern Australia. *Geological Magazine*, 129(3): 307–17

Further Reading

Gillieson, D. & Spate, A. 1998. Karst and caves in Australia and New Guinea. In *Global Karst Correlation*, edited by Yuan D. & Liu Z., New York: Science Press

Jennings, J.N. 1967. Some karst areas of Australia. In *Landform Studies from Australia and New Guinea*, edited by J.N. Jennings & J.A. Mabbutt, Canberra: Australian National University Press

Kiernan, K. 1988. *The Management of Soluble Rock Landscapes: An Australian Perspective*, Sydney: Speleological Research Council

Kiernan, K. 1995. *An Atlas of Tasmanian Karst*, 2 vols, Hobart: Tasmanian Forest Research Council

Matthews, P.G. 1985. *Australian Karst Index*, Melbourne: Australian Speleological Federation

Osborne, A. & Branagan, D.F. 1988. Karst landscapes of New South Wales, Australia. *Earth Science Reviews*, 25: 467–80

AUSTRALIA: ARCHAEOLOGICAL AND PALEONTOLOGICAL CAVES

In 1831, a small party of Europeans, less than 50 years in Australia, climbed into a doline (Breccia Cave) in the Wellington Valley, on the western slopes of the ranges northwest of Sydney. George Ranken fixed his rope to a projection that gave way, revealing that it was in fact a gigantic bird femur. From the red cave earths, Ranken and explorer Major Thomas Mitchell collected specimens of extinct marsupial and avian megafauna, and dispatched them to scientists in Edinburgh and London (Rich, van Tets & Knight, 1985). So was born vertebrate paleontology in Australia.

The oldest and most diverse cave fossils in Australia are at Riversleigh, northern Queensland (Archer, Hand & Godthelp, 1991). From the late Miocene on, caves developed in the Oligo-Miocene and Cambrian limestones. Some fossils are contemporary with the deposition of the younger limestones, but others accumulated in pit-fall caves and fissures or as the prey remains of carnivorous bats. In the mid-Tertiary Riversleigh was rich lowland rainforest, with the most diverse faunal assemblages of any known from the Australian continent, containing at least 68 mammal species in 27 families as well as birds, reptiles, amphibians, fish, molluscs, and insects. As the continent drifted north, the climate became drier and the rich rainforest faunas all but disappeared. Today, the ancient cave deposits are found as surface outcrops on low, dry, and dusty hills. Riversleigh, protected within Lawn Hill National Park, is part of the Australian Fossil Mammal Sites World Heritage Area, recognized in 1994.

Fossils have been found in Australia in all kinds of caves and everywhere caves occur. In deep, wet caves in older limestones in the Great Dividing Range, along the east coast of Australia, they tend to be disarticulated jumbles of broken bones in red cave earths, sometimes consolidated into hard breccias. Much of the fossil deposits at Wellington Caves fall into this category, although the remaining caves here are old and inactive.

Caves in the younger Tertiary and Quaternary limestones and calcarenites along the southern coasts of Australia tend to be large, horizontal tunnels, where sediment, containing the remains of passing wildlife captured by dolines, has built up huge talus piles. At Naracoorte Caves, South Australia, the tunnel floors are strewn with skulls and skeletons of Late Quaternary giant extinct marsupials (Wells, Moriarty & Williams, 1984). The Naracoorte Caves Conservation Area (300 ha) is also part of the Australian Mammal Fossil Sites World Heritage Area.

Fossil deposits derived from predators occur in all types of caves. Owls roosted in entrances to deep caves, or crevices and fissures on limestone cliff faces, and marsupial carnivores, especially the Tasmanian devil, *Sarcophilus*, once widespread on mainland Australia, made their dens not only in limestone caves, but also in lava tunnels, common in the volcanic plains of western Victoria. Most known predator deposits are late Pleistocene or Holocene in age, but Riversleigh preserves much older accumulations made by carnivorous ghost bats (Megadermatidae), whose prey included fish, frogs, lizards, birds, and small mammals including other bats. These caves with younger predator-derived faunal assemblages provide much information about local faunas before the impact of Europeans. They also commonly contain archaeological material. Karst areas with important archaeological sites include the southwest of Australia, including the Nullarbor Plain, the southeast highlands, Tasmania, and also New Guinea, which like Tasmania, lies on the Australian continental plate and was connected by dry land at times of glacial low sea levels. The term "cave" tends to be used indiscriminately by Australian archaeologists for both karst and non-karst caves and rock shelters. Many famous archaeological "caves", in Arnhem Land in Northern Territory, the Carnarvon Ranges in Queensland (Kenniff Cave), the Grampians in Victoria, and on the New South Wales coast, around Sydney, occur in sandstone ranges or sea caves. Flood (1999) is a useful introduction to the wide range of archaeological caves and shelters in Australia.

There are ethnographic accounts of Aboriginal fear of dark caves, but there is much archaeological evidence of dark zone penetration. About 24 000 BP, just before the height of the last glaciation, Aboriginal people climbed down 20 m cliffs into the huge Koonalda doline on the Nullarbor Plain, the karst plateau edging the Great Australian Bight. They moved into the high dark tunnels of the cave, and by torchlight, quarried flint and marked abstract patterns in the soft cave walls.

In southwestern Tasmania, caves such as Kutikina, with rich faunal assemblages, preserve evidence of human settlement of periglacial uplands between at least 30 000 and 12 000 BP. After an early phase, most were unoccupied during the height of the Last Glacial Maximum, then intensively used as hunting camps during the time of ameliorating climate, 17 000–13 000 BP. The deposits are dominated by the bones of wallabies (*Macropus rufogriseus*) and wombats, but are also noted for artefacts made from "Darwin glass", from the Mt Darwin meteorite crater. Most sites were no longer used after 13 000 BP, probably due to the reduction of the preferred open habitat of the game species as forests expanded (Porch & Allen, 1995).

Many of these important cave sites lie within the Tasmanian Wilderness, inscribed on the World Heritage List in 1989. In 1995 Kutikina, Ballawinne, and Wargata Mina were returned to Tasmanian Aboriginal community ownership, the first time Aborigines had been given both title and sole management of land in a World Heritage Area. Wargata Mina contains the most southern cave art in Australia; hand stencils there, with human blood protein as a constituent of the red pigment, date to 11 000 BP (Loy *et al.*, 1990).

A review of Aboriginal use of karst caves in southeastern Australia by Spate (1997) identified about 30 localities used as occupation or art sites, and for the disposition of the dead, dating back as far as 23 000 years. The area with the richest record, Cooleman Caves, in the alpine Kosciuszko National Park, has perhaps the harshest climate.

Many archaeological cave sites have extant faunal remains of mixed provenance, and also extinct megafauna, though doubts remain whether any demonstrate Aboriginal predation on the latter. In Cloggs Cave, in southeast Victoria, stone artefacts occur in an owl pellet deposit, with rare bones of the extinct kangaroo *Sthenurus*; Seton Rockshelter, Kangaroo Island, South Australia, has quartz artefacts, rare *Sthenurus* teeth, but mainly bone fragments typical of *Sarcophilus*. These sites have similar dates, 17–16 000 BP, associated with the *Sthenurus* levels. Aboriginal people regularly, if sporadically, lived in Devils Lair, in far southwest Western Australia, from 33 000 to 6000 BP. Bone points occur at 22 000 BP, and bone beads at 13 000 BP. The remains of wallabies and other animals in Devils Lair were hunted by both humans and *Sarcophilus*, and lower layers contain extinct marsupial megafauna. Nombe, a karst rock shelter in the New Guinea highlands, has a similar antiquity and contents: wallabies and possums hunted by the occupants, as well as the occasional bones of extinct megafauna (Mountain, 1993). *Sarcophilus* apparently never reached New Guinea, so the highly distinctive chewed fragments of its prey are missing from this site. A pattern emerges: extinct Pleistocene species occur in lower units, overlaid by rich, and often mixed, archaeological and predator accumulations of bone, but no firm evidence of human predation of the megafauna.

Finally, New Guinea also has the highest and most unusual karst archaeological site, Mapala Rockshelter, at 4000 m, set beneath a limestone block perched by retreating ice across lateral moraines near the Carstenz Glacier in the Indonesian west of the island. Again, this is a hunting camp, containing wallaby and possum bones, dating back more than 4000 BP, but used in living memory by the Damal people (Hope & Hope, 1976).

JEANNETTE HOPE

See also **Art: Cave Art in Australasia**

Works Cited

Archer, M., Hand, S.J. & Godthelp, H. 1991. *Riversleigh: The Story of Animals in Ancient Rainforest of Inland Australia*, Sydney: Reed Books

Flood, J. 1999. *The Riches of Ancient Australia: A Journey into Prehistory*, 3rd edition, St Lucia: University of Queensland Press

Hope, G.S. & Hope, J.H. 1976. Man on Mt Jaya. In *The Equatorial Glaciers of New Guinea*, edited by G.S. Hope, J.A. Peterson, U. Radok & I. Allison, Rotterdam: Balkema

Loy, T.H., Jones, R., Nelson, D.E., Meehan, B., Vogel, J., Southon, J. & Cosgrove, R. 1990. Accelerator radiocarbon dating of human blood proteins in pigments from Late Pleistocene art sites in Australia. *Antiquity*, 64: 110–16

Mountain, M.J. 1993. Bones, hunting and predation in the Pleistocene of northern Sahul. In *Sahul in Review: Pleistocene Archaeology in Australia, New Guinea and Island Melanesia*, edited by A. Smith, M. Spriggs & B. Fankhauser, Canberra: Australian National University, Department of Prehistory

Porch, N. & Allen, J. 1995. Tasmania: Archaeological and palaeoecological perspectives. In special issue: Transitions Pleistocene to Holocene in Australia and New Guinea. *Antiquity*, 69: 714–32

Rich, P.V., van Tets, G.F. & Knight, F. (editors) 1985. *Kadimakara Extinct Vertebrates of Australia*, Melbourne: Pioneer Design Studio; 2nd edition, Princeton, New Jersey: Princeton University Press, 1990

Spate, A. 1997. Karsting around for bones: Aborigines and karst caves in southeastern Australia. *Australian Archaeology*, 45: 35–44

Wells, R., Moriarty, K. & Williams, D.L.G. 1984. The fossil vertebrate deposits of Victoria Fossil Cave Naracoorte: An introduction to the geology and fauna. *Australian Zoologist*, 21(4): 305–33

AUSTRALIA: BIOSPELEOLOGY

In the late 19th century scientists recognized that Australian caves contained terrestrial and aquatic species that belonged to ancient and relict groups of value to studies in evolution. Widespread studies of Australian cave life did not commence until the 1970s and there have still been few systematic surveys of cave or groundwater (aquifer) faunas. While a few detailed studies have been conducted of hypogean ecology, environment, and evolution, much of the known fauna remains undescribed. The composition of Australian cave fauna is influenced by past connections with the supercontinents Pangaea and Gondwana, and by it having formed part of the eastern seaboard of the Tethys Ocean during the Cretaceous. Above all, it is shaped by the onset of aridity, following the separation of Australia from Antarctica about 45 million years ago, which displaced well-watered, cool temperate, and subtropical forests, as Australia drifted northwards. Aridity appears to have played a prominent role in isolating cave species now far from the humid ecosystems of their ancestors.

Australia, which covers 7.7 million km^2 and is of generally low relief, contains extensive shield regions that have remained above sea level since the Paleozoic and so support numerous relictual lineages of freshwater stygobites. Over 500 discrete areas of carbonate rocks occur in Australia (see map in Australia entry). The caves and karst support terrestrial, fresh and saltwater, and anchialine ecosystems.

Hamilton-Smith and Eberhard (2000) presented a regional classification of the subterranean fauna based on recognized bioclimatic zones. These include tropical climates in the north, a large subtropical dry province occupying the centre and the western seaboard, a transitional zone with winter rain along southwestern coasts, and a warm temperate/tropical transition zone largely covering the Eastern Highlands.

Until recently, the subterranean fauna of Australia—with the exception of that in Tasmania which has had a more humid climatic history than the mainland with periods of Cainozoic glaciation—was perceived as being much less diverse than those from the well-studied caves of Europe and North America. This coincidence in Tasmania of glaciation with speciose troglobite communities was consistent with the Northern Hemisphere model in which glaciation was perceived as the driving factor in troglobite evolution. However, recently this perception has been overturned, especially by discoveries within the tropics and the arid parts of the continent. Notably diverse cave faunas occur in Tasmania, Jenolan Caves (New South Wales), Far North Queensland (Chillagoe karst and Undara lava tube), and Cape Range and Barrow Island (see Cape Range, Australia: Biospeleology). Areas where strongly cave-adapted species are sparse include the aeolian limestones of the southwestern coasts and the extensive Nullarbor karst (especially stygofauna).

The island state of Tasmania, with 300 discrete karst areas, exhibits a higher proportion of troglobites than found on the mainland in New South Wales, with 200 discrete karst areas, having 34 and 17 troglobitic genera respectively, distributed among a comparable number of families. However, by comparison, some tropical caves are species rich (Malipatil & Howarth, 1990) and the arid tropical Cape Range and Barrow Island karst in Western Australia supports 51 obligate subterranean genera in 42 families (32 troglobite and 19 stygobite genera). In the tropics, numerous cave-adapted planthoppers (Homoptera: Fulgoroidea) occur in which the adults retain the underground root-feeding behaviour common amongst the nymphs. These co-occur with epigean species and provide a graded series of morphological change associated with adaptation to cave life. Different species display all grades of cave adaptation from almost no modification from surface relatives to the loss of eyes and pigment and strongly reduced wings (Hoch & Howarth, 1989).

Amphipods, a major component of world cave faunas, were long considered sparse in Australian inland waters, and thought especially to abhor the tropics. Recent work has uncovered a diversity of amphipods in the arid zone and in the tropics, represented by a number of families considered to belong to both marine and freshwater lineages. The distribution of the latter, along with other ancient freshwater crustacea, coincides closely with areas not covered by marine incursions during the Cretaceous (Bradbury, 1999).

In southwestern Australia, root mats forming in shallow streams of caves in aeolian limestone support a diverse aquatic community, largely lacking overt modification to cave life (Jasinska et al., 1996). Whether these mats also support a terrestrial component, such as seen in the tropical planthoppers, is unrecorded.

Many species are restricted to single karst areas, especially in the impounded karsts of eastern Australia (Thurgate et al., 2001), but even within a given karst area there is often no interbreeding between populations of troglobites in adjacent caves (Humphreys & Adams, 2001). Unexpected small-scale endemicity also occurs in the stygofauna inhabiting groundwater calcretes of arid Australia. These are limestone masses deposited from groundwater flow into the alluvium filling ancient paleochannels. They form where the baselevel of the groundwater approaches the surface and are found upstream of the saltlakes (playas) that occur at intervals along the length of the paleochannels. For example, most of the 49 known species of stygal diving beetles (Dytiscidae) are restricted to a single calcrete body (Watts & Humphreys, 2000). Molecular data suggest that separate dytiscid lineages invaded groundwater in the middle Miocene with the onset of general aridity (Cooper et al. 2002), but other calcretes appear to support much more ancient faunas with Gondwanan affinities (Poore & Humphreys, 1998), or contain rich assemblages of amphipods.

Owing to sparse research in Australia, meaningful comparison with the diversity of subterranean faunas of other continents is premature. Certain areas, such as Chillagoe, Undara, Jenolan, and Cape Range, contain rich troglobite faunas by world standards. But there are extensive karst areas, such as the Barkley Tableland, the Nullarbor and the syngenetic karst in aeolian dunes of the southwestern coasts, that apparently contain low to very low diversity of strongly cave-adapted species. These factors, combined with the generally low proportion of carbonate landscape in the fragments of Gondwana, suggest that the overall diversity in Australia may not be high by world standards, but the case remains open.

Biospeleology has resulted in the discovery in Australia of a number of higher taxonomic groups (e.g. Class Remipedia; Orders Thermosbaenacea, Misophrioida and Spelaeogriphacea), and of important evolutionary linkages between phreatoicidean isopods (see figure) from Africa, Australia, and New Zealand (Wilson & Keable, 1999). A number of troglobitic and stygobitic species and several cave communities are specifically protected under fauna legislation, part of an Australian ethos of cave and cave fauna protection that is projected strongly internationally (Watson et al., 1997).

WILLIAM FRANK HUMPHREYS

See also **Cape Range, Australia: Biospeleology**

Works Cited

Bradbury, J.H. 1999. The systematics and distribution of Australian freshwater amphipods: A review. In *Crustaceans and the Biodiversity Crisis: Proceedings of the Fourth International Crustacean Congress, Amsterdam, The Netherlands, July 20–24, 1998,* edited by F.R. Schram & J.C. von Vaupel Klein, Leiden: Brill

Cooper, S.J.B., Hinze, S., Leys, R., Watts, C.H.S. & Humphreys, W.F. 2002. Islands under the desert: Molecular systematics and evolutionary origins of stygobitic water beetles (Coleoptera: Dytiscidae) from central Western Australia. *Invertebrate Systematics,* 16(4): 589–90

Hamilton-Smith, E. & Eberhard, S.M. 2000. Conservation of cave communities in Australia. In *Subterranean Ecosystems,* edited by H. Wilkens, D.C. Culver & W.F. Humphreys, Amsterdam and New York: Elsevier

Hoch, H. & Howarth, F.G. 1989. The evolution of cave-adapted cixiid planthoppers in volcanic and limestone caves in North Queensland, Australia (Homoptera: Fulgoroidea). *Mémoires de Biospéologie,* 16: 17–24

Humphreys, W.F. & Adams, M. 2001. Allozyme variation in the troglobitic millipede *Stygiochiropus communis* (Diplopoda: Paradoxosomatidae) from arid tropical Cape Range, northwestern Australia: population structure and implications for the management of the region. *Records of the Western Australian Museum,* Supplement 64: 15–36

Jasinska, E.J., Knott, B. & McComb, A.J. 1996. Root mats in ground water: A fauna-rich cave habitat. *Journal of the North American Benthological Society,* 15: 508–19

Malipatil, M.B. & Howarth, F.G. 1990. Two new species of *Micropolytoxus* Elkins from Northern Australia (Hemiptera:

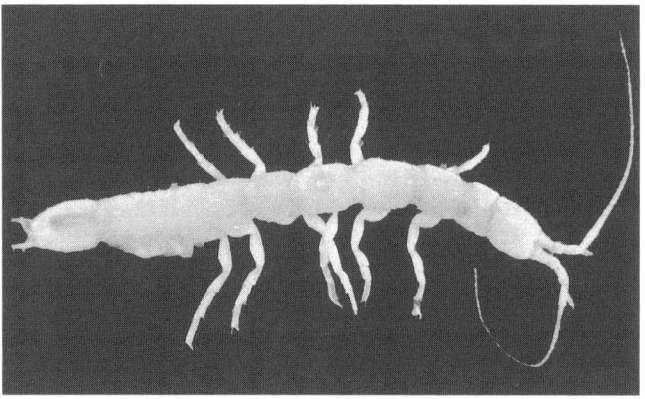

Australia: Biospeleology: *Phreatoicoides gracilis,* a groundwater phreatoicidean isopod from Australia, showing lack of eyes and pigment characteristic of subterranean animals. (Photo by G.D.F. Wilson)

Reduviidae: Saicinae). *Journal of the Australian Entomological Society*, 29: 37–40
Poore, G.C.B. & Humphreys, W.F. 1998. First record of Spelaeogriphacea from Australasia: A new genus and species from an aquifer in the arid Pilbara of Western Australia. *Crustaceana*, 71: 721–42
Thurgate, M.E, Gough, J.S., Clarke, A.K., Serov, P. & Spate, A. 2001. Stygofauna diversity and distribution in Eastern Australian cave and karst areas. *Records of the Western Australian Museum*, Supplement 64: 49–62
Watson, J., Hamilton-Smith, E., Gillieson, D. & Kiernan, K. (editors) 1997. *Guidelines for Cave and Karst Protection*, Gland, Switzerland and Cambridge: International Union for Conservation of Nature and Natural Resources
Watts, C.H.S. & Humphreys, W.F. 2000. Six new species of *Nirridessus* and *Tjirtudessus* (Dytiscidae; Coleoptera) from underground waters in Australia. *Records of the South Australian Museum*, 33: 127–44
Wilson, G.D.F. & Keable, S.J. 1999. A new genus of phreatoicidean isopod (Crustacea) from the North Kimberley Region, Western Australia. *Zoological Journal of the Linnean Society, London*, 126: 51–79

Further Reading

Eberhard, S.M. & Humphreys, W.F. 2003. The crawling, creeping and swimming life of caves. In *Australia Underground: A Tribute to Joe Jennings*, edited by B. Finlayson & E. Hamilton-Smith, Sydney: University of New South Wales Press
Humphreys, W.F. & Harvey, M.S. (editors) 2001. Subterranean biology in Australia 2000. *Records of the Western Australian Museum*, Supplement 64: 1–242

AUSTRALIA: HISTORY

The first verified report of limestone in Australia was of abundant quantities of stratified limestone in the vicinity of the Tamar Estuary in northern Tasmania, in a report to Governor King by Colonel William Patterson in October 1804 (Lane, 1975). The first such report on mainland Australia was made in Governor Macquarie's dispatch to Earl Bathurst of 30 June 1815, which recorded the discovery of limestone by G.W. Evans at Limestone Creek, a tributary of the Belubula River, west of what was to become the town of Bathurst in New South Wales (Carne & Jones, 1919: pp.6–8). The Bathurst district also provided the first record of a European entering an Australian cave, when, on 8 November 1821, William Lawson, one of the trio credited with the crossing of the Blue Mountains, explored Limekilns Cave. Lawson recorded in his journal (Lawson, 1821):

> Camped at the Limestone Hills, here government has a kiln built for burning of lime for the use of the settlement which proves to be the very best quality. Here is a curious cave through a solid rock of limestone. Its entrance is very narrow. At nine o'clock at night I took four men with three candles and proceeded into it about one hundred yards. At the end is a fine pool of clear water. In many places for several yards together I was obliged to creep on my hands and knees. The inside of the cave is very curious and well worth seeing. I got some fine specimens. Came out at one o'clock in the morning.

While it is likely that caves were discovered at Bungonia in New South Wales by early settlers around 1820, the earliest written report was by botanist and explorer Alan Cunningham, who noted in his diary on 27 April 1824 that, passing through limestone country, "we found the land exceedingly cavernous. Orifices four feet in diameter connected with capacious subterraneous Excavations, appeared in every part of the Forestland, of whom some presented yawning fissures of apparently great depth . . ." (Cunningham, 1824). The report went on to describe a visit to what is now known as Drum Cave (B13), though they were stopped by a pit of "unfathomed depth"—later surveyed at 43 m.

The earliest-known illustrations of Australian caves were three watercolours executed in 1826 by a professional painter, Augustus Earle, of "Mosman's Cave, Wellington Valley, NS Wales". These pre-dated by two years the publication of the first written report on the Wellington Caves by Hamilton Hume, a member of Sturt's 1828 expedition (Hamilton-Smith, 1997). The paleontological value of the caves at Wellington was realized about 1830 and became well known following the publication of Major Mitchell's journals in 1838 (Mitchell, 1838). These include a plan of the caves, which was the first accurate cave plan to be published in Australia (Figure 1).

Australia's earliest-known cave map was prepared by surveyor Henry Hellyer in August 1827, but this was not to scale and was not published until 1990 (Figure 2) The cave, at Rocky Cape on the north coast of Tasmania, is in quartzite and was noted by Hellyer as being 10 feet (3 m) wide at the entrance and 80 feet (24 m) deep. The cave became known as Rocky Cape North Cave and yielded important information on the Aboriginal inhabitants of the area over the previous 8000 years (Middleton, 1990).

The first published plans of Australian caves accompanied Capt. John Henderson's record of his observations in New South Wales (Figure 3) (Henderson 1832). Although relatively crude and not to scale, the two sketches are identifiably of Tunnel Cave (BN25–28) at Borenore and Breccia Cave (or Bone Cave No. 3) at Wellington.

Caves were easier to find in Western Australia. In March 1827, two years prior to European settlement, Capt. James Stirling noted that the "heads" at the mouth of the Swan River were composed of limestone which, "subject to the action of the surge, is worn into Caverns". Exploring inland in the vicinity of Cape Naturaliste, Stirling noted "compact limestone, sections of which were seen 200 feet in depth" and "under the limestone cliffs, many magnificent Caverns; some of these are remarkable for their extent and some for the beautiful stalactites and incrustations which they contain" (Stirling, 1827).

While the Jenolan Caves (formerly Binda or Fish River Caves) in the Blue Mountains of eastern New South Wales would have been known to the Gundungura people for millennia, their discovery by Europeans is shrouded in mystery. A widely accepted story is that they were found by James Whalan in 1838 while searching for a cattle thief named McKeown. Unfortunately there is no contemporary record of this event, or any reliable

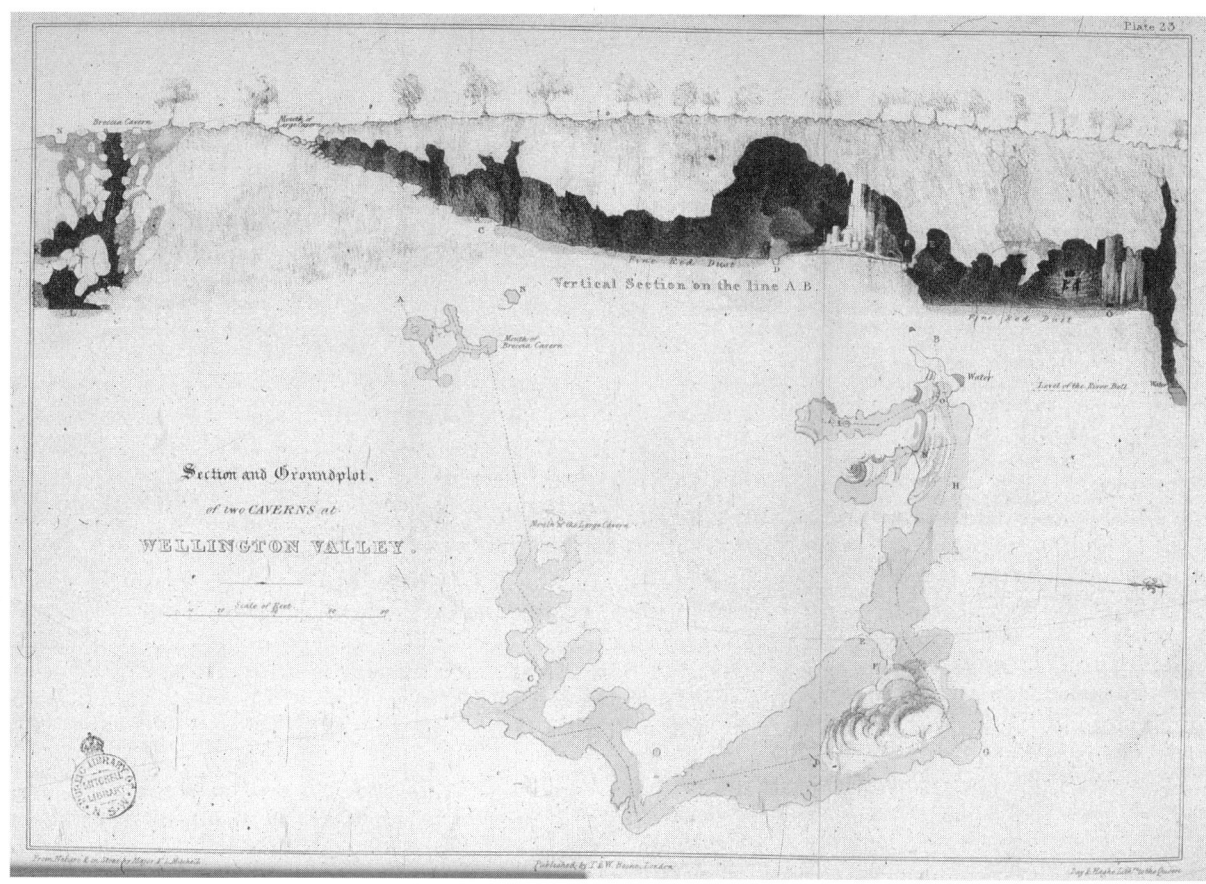

Australia: History: Figure 1. Section and ground plot of two caverns at Wellington Valley. (Reproduced with permission from a copy held in the Mitchell Library, State Library of New South Wales.)

record of McKeown, although the first published name of the caves, in 1856, was "McKeon's Caves" (Havard, 1934). Certainly, James's brother, Charles Whalan, who had the closest farm to the caves, conducted the first tourist parties to them in the 1840s and 1850s. Jeremiah Wilson, an immigrant from Enniskillen, Ireland, settled in the district in 1843 and began exploring the caves in 1855. John Lucas, the member of the Legislative Assembly for the area, visited the caves in 1861 and, as a result of his representations, a reserve was declared over the "Fish River Caves" on 2 October 1866—the first Australian cave reserve. Wilson was appointed the first caretaker on 12 January 1867. Originally, visitors to the caves camped in the Grand

Australia: History: Figure 2. Hellyer's 1827 plan of Rocky Cape North Cave. (Reproduced with permission of the Roberts family and the University of Tasmania Library)

Arch, but Wilson constructed the first accommodation house in 1880 (Dunkley, 1986).

The outstanding cave surveyor of the late 19th and early 20th centuries in Australia was Oliver Trickett (1847–1934). As an employee of the New South Wales Department of Mines, he began surveying caves in 1896 and continued till his retirement in 1919. Although originally employed as a survey draughtsman, he was given responsibility for the management of all of the colony's caves and he supervised much of the work which went into opening caves to the public at Jenolan, Wombeyan, Wellington, Yarrangobilly, and Abercrombie. He produced numerous maps, photographs, detailed reports, models, and informative guidebooks (Middleton, 1991).

In South Australia the Narracoorte Caves (formerly Mosquito Plains Caves), on the traditional lands of the Boandik people, were first entered by Europeans in 1845 when Benjamin Sanders recovered some of his sheep from what became known as Blanche Cave, where they had been driven by the Aborigines. A reserve was placed over the cave in the late 1870s, but the first ranger was not appointed until 1885. William Reddan was in charge of the cave from 1886 and discovered others; he resigned in 1916 (Lewis, 1977).

The speleological significance of Australia's largest area of outcropping limestone, the Nullarbor Plain, appears not to have

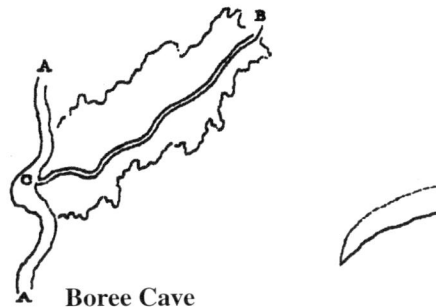
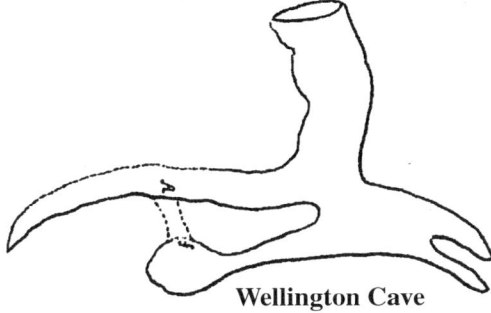

Australia: History: Figure 3. Henderson's sketch plan of "Boree Cave" (Tunnel Cave), Borenore, and a section of Breccia Cave, Wellington. (Reproduced with permission from a copy held in the Mitchell Library, State Library of New South Wales.)

been recognized until Professor Tate went looking for artesian water for grazing cattle (Tate, 1879). The first expeditions specifically to study the caves of the region were conducted by Capt. J. Maitland Thomson between 1930 and 1960 (Dunkley, 1967).

The first recording of karst caves in Tasmania resulted from an excursion by the Lieutenant-Governor, George Arthur, and his party, in January 1829 to inspect land being applied for by the Van Diemens Land Company in the island's northwest. The Assistant Government Surveyor, Thomas Scott, recorded in his diary for 15 January 1829 that, beyond "Moleside" (now Mole Creek) the party "visited a remarkable cavern in the limestone rock". A more detailed account of the cave visit was given in the *Hobart Town Courier* on 28 February 1829 (Clarke, 1999); the cave had obviously been found earlier by local settlers. Concern about vandalism in caves near Mole Creek dates to at least 1873 and arose over the alleged removal of "a cartload of cave formations". By April 1879 the Tasmanian Government had moved to reserve 300 acres around the "new caves" near Sassafras Creek and a further 100 acres around the "old caves" (Oakdens, Wet, or Caveside Caves). In 1881 Henry Judd and a party of four discovered a cave in the Cracroft Valley in southern Tasmania and reported on the 400 yards of passage explored (Kiernan, 1978). Judds Cavern was found to contain Aboriginal hand stencils in 1987, and subsequently became known as Wargata Mina; with Kutikina and Ballawinie caves it was returned to Aboriginal ownership by an Act of the Tasmanian Parliament on 6 December 1995.

Stewart Ryrie recorded his visit to a cave at "Buckan" (Buchan), Victoria, on 7 April 1840; he "descended into one of them a considerable depth but saw nothing remarkable" (Ryrie, 1840). In 1889, James Stirling recorded and mapped caves at Buchan (Stirling, 1889; Hamilton-Smith, 1986). The government took control of 279 acres of land in 1900 and this subsequently became part of the Buchan Caves Reserve. Buchan Caves were opened to the public in 1907 (Kahrau, 1972).

In 1882 caves were first recorded in the Rockhampton area, central Queensland, by John Olsen, a Norwegian immigrant. These were opened to the public in 1884 and remain a local tourist attraction (Caffyn, 1970).

GREGORY MIDDLETON
(assisted by ELERY HAMILTON-SMITH and STEPHEN McCABE)

Works Cited

Caffyn, P.C. 1970. Historical outline of mining, recreation and conservation activities at Mt Etna and Limestone Ridge. In *Mount Etna Caves: A Collection of Papers Covering Several Aspects of the Mt. Etna and Limestone Ridge Caves Areas of Central Queensland*, edited by J.K. Sprent, St Lucia, Queensland: University of Queensland Speleological Society

Carne, J.E. & Jones, L.J. 1919. *The Limestone Deposits of New South Wales*, Sydney: Government Printing Offices

Clarke, A. 1999. Baldocks Cave: the site locality for the Tasmanian cave spider and selected historical references relating the early discovery of limestone in northern Tasmania and the caves of the Chudleigh (Mole Creek) area. In *Cave Management in Australasia 13*, edited by K. Henderson, Melbourne: Australasian Cave and Karst Management Association

Cunningham, A. 1824. Journal of Allan Cunningham, vol. 3, pp.121–24 (original in the Archives Office of NSW), transcribed by J.M. Whaite. In *Bungonia Caves*, edited by R. Ellis et al., Sydney: Sydney Speleological Society, 1972 (Sydney Speleological Society Occasional Paper, 4)

Dunkley, J.R. 1967. The geographical and historical background. In *Caves of the Nullarbor*, edited by J.R. Dunkley & T.M.L. Wigley, Sydney: Speleological Research Council

Dunkley, J.R. 1986. *Jenolan Caves as they were in the Nineteenth Century*, Sydney: Speleological Research Council

Hamilton-Smith, E. 1986. An early report of caves at Buchan. *Journal of the Sydney Speleological Society*, 30(8): 159

Hamilton-Smith, E. 1997. Nineteenth century paintings, drawings and engravings of Australian caves. *Helictite*, 35(1&2):12–38

Havard, W.L. 1934. The romance of Jenolan Caves. *Journal of the Royal Australian Historical Society*, 20(1): 18–65

Henderson, John. 1832. *Observations on Natural History, & a Collected During a Visit to the Colonies of New South Wales and Van Diemen's Land*, Calcutta: Baptist Mission Press

Kahrau, W. 1972. *Australian Caves and Caving*, Melbourne: Periwinkle Books

Kiernan, K. 1978. Caving. In *The South West Book: A Tasmanian Wilderness*, edited by H. Gee & J. Fenton, Melbourne: Australian Conservation Foundation

Lane, E.A. 1975. Lime, limestone and the first caves. *Australian Natural History*, 18(6): 191–95

Lawson, W. 1821. *Journal of a Tour into the Country North of Bathurst Nov. 8–24, 1821*. Unpublished manuscript in Mitchell Library, Sydney

Lewis, I. 1977. *Discover Naracoorte Caves*, Adelaide: Subterranean Foundation

Middleton, G. 1990. Rocky Cape North Cave—an early cave map from Tasmania. *Journal of the Sydney Speleological Society*, 34(11): 213–15

Middleton, G.J. 1991. *Oliver Trickett: Doyen of Australia's cave surveyors 1847–1934*, Sydney: Sydney Speleological Society and Jenolan Caves Historical and Preservation Society (Sydney Speleological Society Occasional Paper, 10)

Mitchell, T.L. 1838. *Three Expeditions into the Interior of Eastern Australia*, London: Boone, 2 vols; 2nd edition, 1839

Ryrie, S. 1840. Journal of a certain tour [original in the Dixon Library, State Library of New South Wales], published in 1991 in *Gippsland Heritage Journal*, 11: 12–17

Stirling, J., 1827. Captain Stirling to Governor Darling (18 April 1827): Narrative of operations. In *Historical Records of Australia*, Series III, 6: 551–84

Stirling, J. 1889. Preliminary report on the Buchan Caves, Goldfields of Victoria. *Reports of the Mining Registrars for the Quarter Ending 31st December 1889*, pp.65–68

Tate, R. 1879. The natural history of the country around the head of the Great Australian Bight. *Transactions of the Philosophical Society of Adelaide*, 1879: 94–128

Further Reading

Hamilton-Smith, E. 1997. Perceptions of Australian caves in the 19th century: the visual record. *Helictite*, 35(1&2): 5–11

Lane, E.A. & Richards, A.M. 1963. The discovery, exploration and scientific investigation of the Wellington Caves, New South Wales. *Helictite*, 2(1): 1–53

AVES (BIRDS)

Birds in karst landscapes can be grouped into three rather indistinct categories: casuals, birds which visit and utilize both karst and non-karst areas; usuals, which regularly favour karstlands as feeding or nesting sites; and habituals, which are restricted or closely tied to karst habitats. Studies dedicated to contemporary birds in karstlands are in their infancy, so this account, which focuses particularly on tropical species, should be regarded as preliminary. For a brief account of fossilized bird remains in cave sediments and what they reveal about paleoenvironments, see Bramwell (1976).

The casuals are the most numerous species in karstlands but their utilization of specifically karstic resources has not been studied in detail. Vegetative structure, food sources, and nesting sites are no less important to the casuals than the other categories and in some respects they may be related to the karst substrate. Cliff-nesting species, for example, will utilize karst and non-karst cliffs without necessary regard to lithology but with due regard to cliff morphology. Barn Owls (*Tyto alba*), Cliff Swallows (*Petrochelidon pyrrhonata*), and flycatchers such as the Eastern Phoebe (*Sayornis phoebe*) of North America sometimes nest or roost in cave entrances, but certainly not exclusively.

On the basis of available studies it is difficult to differentiate between the casuals and the usuals, which include species that forage or nest regularly in karst landscapes, although they may also spend time in adjacent non-karst areas. A study of birds recorded in the karst terrain of Belize (Day, 1989) indicates that over 300 of Belize's approximately 550 bird species have been recorded in the karst, with at least 250 of these occurring on a regular basis. In this context, karstlands may be more significant avian habitats than has been realized previously.

The habituals have received the greatest attention, although they are the least numerous of the three categories. They may be subdivided into two types: those that inhabit exclusively terrestrial karst landscapes, and those that are associated with caves.

The surface dwellers include more than a dozen species that occur almost exclusively in limestone vegetation or in intimate association with limestone rocks and cliffs. In southern Mexico, for example, Sumichrast's Wren (*Hylorchilus sumichrasti*) and Nava's Wren (*Hylorchilus navai*) inhabit karst forests especially near limestone outcrops (BirdLife International, 2000; Stattersfield *et al.*, 1998). Likewise, the Cuban Solitaire (*Myadestes elisabeth*) nests exclusively on steep limestone cliffs in moist montane forest (BirdLife International, 2000; Garrido & Kirkconnell, 2000). In Southeast Asia three species of babbler, the Limestone Wren Babbler (*Napothera crispifrons*), the Streaked Wren Babbler (*Napothera brevicaudata*), and the Sooty Babbler (*Stachyris herberti*) similarly are found in evergreen forest with limestone outcrops (MacKinnon & Phillipps, 2000; Robson, 2000; BirdLife International, 2000; Stattersfield *et al.*, 1998). The Greater Melampitta (*Melampitta gigantean*) roosts in forested dolines in Irian Jaya (Stattersfield *et al.*, 1998) and the Giant White-eye (*Megazosterops palauensis*) occurs in limestone forests and thickets on Palau (BirdLife International, 2000; Stattersfield *et al.*, 1998).

Birds habitually associated with caves may be regarded as habitual trogloxenes, using caves temporarily for shelter from weather, safe nesting, and raising young but completing their life cycles at the surface. Most use the entrance habitat of caves as nesting sites, feeding in adjacent surface landscapes, although often venturing deep into caves beyond the threshold or twilight zone. The vast majority of cave-dwelling birds are insectivorous, with the notable exception of the South American guácharo or oilbird (*Steatornis caripensis*) which feeds on fruit (de Schauensee and Phelps, 1978).

The most numerous cave-nesting birds are the tropical swiftlets (*Collocalia*, and perhaps *Hydrochous* and *Aerodramus* spp.), whose colonies are particularly associated with Southeast Asia, New Guinea, and the South Pacific islands. (Swiflet taxonomy is a vexed issue; the reader is referred to the discussion by Chantler & Driessens, 1995, who consider all swiftlets as belonging to the genus *Collocalia*.) There are some 25 species of cave-nesting swiftlets including the Edible-nest Swiftlet (*Collocalia fuciphaga*; Chantler & Driessens, 1995), whose saliva-based nests provide the essential ingredient of bird's-nest soup. At least five species of swifts (*Chaetura* and *Apus* spp) also nest in caves, although not exclusively, as do the Red-rumped and Cave Swallow (*Hirundo daurica* and *Petrochelidon fulva*). The Red-rumped swallow is found in Europe, Africa, temperate and southeast Asia, in the Greater and Lesser Sundas, and in the Philippines (Jeyarajasingam & Pearson, 1999). The Cave Swallow is found from southern North America to southern Mexico and the West Indies (Terres, 1980). The Grey-necked Rockfowl (*Picathartes oreas*) of West Africa also nests colonially in caves (BirdLife International, 2000; Stattersfield *et al.*, 1998), and the Angolan Cave-chat (*Xenocopsychus ansorgei*) can also be regarded as a trogloxene because it forages and sometimes nests in cave entrances (BirdLife International, 2000).

Perhaps the most fascinating cave-nesting bird is the neotropical guácharo or oilbird (*Steatornis caripensis*). Colonies of this cave-nesting nocturnal frugivore, whose Spanish name may be translated as "one who cries and laments" (Jackson, 1982), occur in South America northward from Bolivia to Venezuela, Guyana, and Trinidad (Thomas, 1999). The colony in Venezuela's Guácharo Cave has been estimated at 19 000 birds (Jackson, 1982). Introduced to European science by Alexander von Humboldt after a visit to Venezuela in 1799, the oilbird is so-called in English because native Americans used oil from the bodies of chicks for cooking and illumination. Like the swiftlets, the guácharo navigates in caves by echolocation, emitting high-pitched tongue clicks. Echolocation, the use in navigation of reflected sound as part of vocalizations, is known among birds only in cave-nesting swiftlets and the oilbird (Gill, 1995). Among swiftlets, some species in the genus *Collocalia* are particularly well adapted to nesting in the interior of caves. Those species that do so are possessed of the ability to echolocate, which they use as an aid to navigation within the darkness of cave nesting sites. The echolocation abilities of swiftlets are not "ultrasound" sonar, as is typical of bats. Instead these are low-frequency sounds, brief "clicks" at 2–10 kHz frequencies, and about 1 millisecond in duration (Medway & Pye, 1977). The oilbird (*Steatornis caripensis*) uses clicks of a longer duration (15–20 milliseconds) and of a wider frequency range (1–15 kHz) (Konishi & Knudsen, 1979).

MICK DAY AND BILL MUELLER

See also **Asia, Southeast: Biospeleology; Guano**

Works Cited

Bramwell, D. 1976. Birds as cave fossils. In *The Science of Speleology*, edited by T.D. Ford & C.H.D. Cullingford, London and New York: Academic Press

BirdLife International, 2000. *Threatened Birds of the World: The Official Source for Birds on the IUCN Red List*, Cambridge: BirdLife International and Barcelona: Lynx

Chantler, P. & Driessens, G. 1995. *A Guide to the Swifts and Treeswifts of the World*, East Sussex: Pica Press

Day, M.J. 1989. Birds of the Hummingbird karst. In *Ecology and Environment in Belize*, edited by D.M. Munro, Edinburgh: Department of Geography, University of Edinburgh

de Schauensee, R.M. and Phelps, W.H. 1978. *A Guide to the Birds of Venezuela*, Princeton, New Jersey: Princeton University Press

Garrido, O.H. & Kirkconnell, A. 2000. *Field Guide to the Birds of Cuba*, Ithaca, New York: Comstock

Gill, F.B. 1995. *Ornithology*, 2nd edition, New York: Freeman

Jackson, D.D. 1982. *Underground Worlds*, Alexandria, Virginia: Time-Life Books

Jeyarajasingam, A. & Pearson, A. 1999. *The Birds of West Malaysia and Singapore*, Oxford and New York: Oxford University Press

Konishi, M. & Knudsen, E.I. 1979. The oilbird: hearing and echolocation. *Science*, 204: 425–27

MacKinnon, J. & Phillipps, K. 2000. *A Field Guide to the Birds of China*, Oxford and New York: Oxford University Press

Medway, Lord & Pye, J.D. 1977. Echolocation and the systematics of swiftlets. In *Evolutionary Ecology*, edited by B. Stonehouse & C. Perrins, Baltimore: University Park Press and London: Macmillan

Robson, C. 2000. *A Guide to the Birds of Southeast Asia*, Princeton, New Jersey: Princeton University Press

Stattersfield, A.J., Crosby, M.J., Long, A.J. & Wege, D.C. 1998. *Endemic Bird Areas of the World: Priorities for Biodiversity Conservation*, Cambridge: BirdLife International

Terres, J.K. 1980. *The Audubon Society Encyclopedia of North American Birds*, New York: Knopf

Thomas, B.T. 1999. Oilbird. In *Handbook of Birds of the World*, edited by J. del Hoyo, A. Elliot & J. Sargatal, vol. 5, Barcelona: Lynx

Further Reading

Chapman, P. 1993. *Caves and Cave Life*, London: HarperCollins

Jefferson, G.T. 1976. Cave faunas. In *The Science of Speleology*, edited by T.D. Ford & C.H.D. Cullingford, London and New York: Academic Press

Sasowsky, I.D., Fong, D.W. & White, E.L. (editors) 1997. *Conservation and Protection of the Biota of Karst*, Charles Town, West Virginia: Karst Waters Institute

Wilkens, H., Culver, D.C. & Humphreys, W.F. (editors) 2000. *Subterranean Ecosystems*, Amsterdam and New York: Elsevier

B

BAMBUÍ KARST, BRAZIL

The carbonate rocks of the Bambuí Group represent by far the largest area (c.105 000 km²) of carbonate karst in Brazil, corresponding to over half of all karst areas in the country (see map in America, South). The Bambuí Group comprises a sequence of pelitic and carbonate marine sediments of Precambrian age. Of the six formations that make up the group, two form karst scenery. The Sete Lagoas Formation lies over the basal glaciogenic Jequitaí Formation, and is characterized by light-grey limestone and dolomite with some clay / silt intercalations. It is overlain by the pelitic Serra de Santa Helena Formation, followed by the Lagoa do Jacaré Formation, another carbonate unit that comprises darker oolitic and pisolitic limestone, again showing some interspersed pelitic lenses. Their thickness can reach a few hundred metres, although rarely more than 200 m lies above the water table. Most of the Bambuí karst lies over the São Francisco Craton, a tectonic unit stabilized in the Precambrian. Thus, the carbonate sequences are mostly undeformed, except where they lie close to the border of the craton.

Well-developed karst terrains occur both in the Sete Lagoas and Lagoa do Jacaré Formations, both with highly seasonal climate (dry winter, wet summer) with precipitation levels ranging between 1000 and 2000 mm a^{-1}. The landscape is usually subdued, with low-gradient rivers and savannah-like vegetation. Much of the Bambuí karst is a covered karst, marked by soil generated by the weathering of pelitic upper units. The karst scenery shows most of the typical karst landforms (Figure 1), and due to the large area of carbonate there is significant variability between sites. Over 2000 caves are known in the Bambuí karst, which is a very small number considering the still great speleological potential of the region.

Within the Bambuí karst, many sites present major karst development and significant caves, such as the Arcos / Pains, Lagoa Santa, Cordisburgo, Montes Claros, Peruaçu (all in Minas Gerais State), São Desidério, Serra do Ramalho (Bahia State), Mambaí and São Domingos (Goiás State). Among these, Lagoa Santa, Peruaçu, Serra do Ramalho and São Domingos deserve special mention due to the outstanding nature of their karst and caves. The Lagoa Santa area is the birthplace of Brazilian speleology and arguably the best-studied karst area in the country, with over 500 known caves. The known caves are mostly short (less than 10 are over 1 km in length) and display extensive sedimentation and paragenetic features, including very rich fossiliferous deposits studied since the 19th century. A complex underground hydrology, including lakes at the base of limestone

Bambuí Karst, Brazil: Figure 1. Razor-sharp ridges of rillenkarren score the bare surfaces between clumps of cactus and thorn bush in typical terrain on the Bambuí limestones. (Photo by Tony Waltham)

cliffs, characterizes this karst. Urbanization is rapidly encroaching, and groundwater pollution, induced doline development, soil erosion, limestone quarrying, and cave vandalism are among the serious environmental problems facing the Lagoa Santa karst.

The Peruaçu area has been ranked among the most scenic karst areas in the world. Its highlight is the outstanding Gruta do Janelão, a 4.7-km long cave that contains a major river passage up to 100 m wide and high (Figure 2). Collapse of the roof has created a series of scenic skylights that allow the entry of natural light and growth of tropical vegetation within the cave, creating an exceptionally beautiful underground landscape (Figure 3). Massive speleothems occur throughout the passage. The Gruta do Janelão is the last cave excavated by the allogenic Peruaçu River. The other three caves upstream are also of large dimensions, although shorter and devoid of the beautiful light effects. The 150 m deep collapse, Peruaçu Canyon, between the caves, and several dry and very well-decorated caves now truncated from the underground drainage (many representing major archaeological sites) complete the landscape, a magnificent fluvio-

Bambuí Karst, Brazil: Figure 3. The magnificent main passage of Gruta do Janelão, lit by the skylights (janelão means big window) that break out into the giant dolines. Note the person silhouetted on the sandbank for scale. (Photo by Tony Waltham)

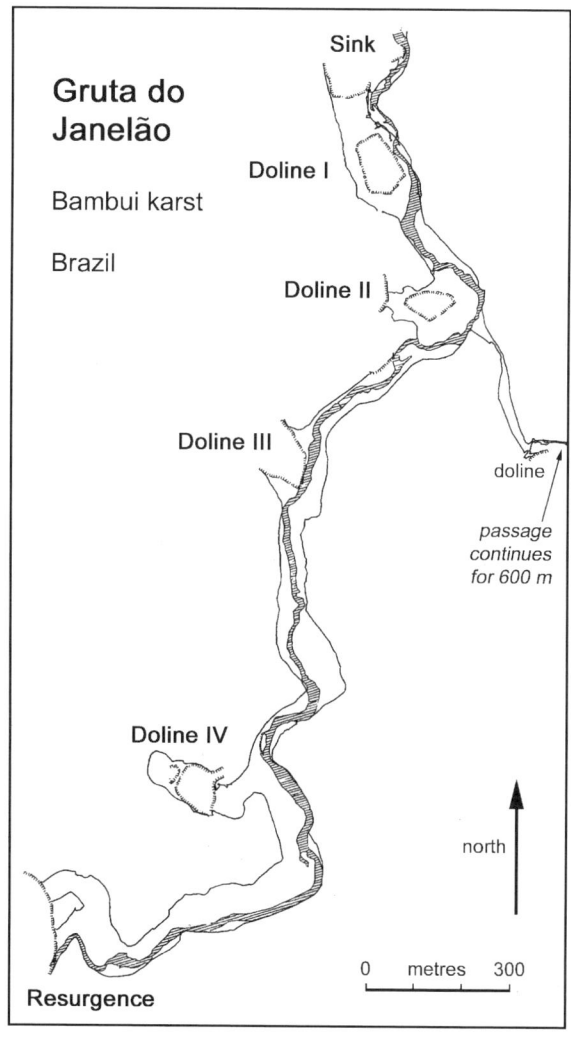

Bambuí Karst, Brazil: Figure 2. Outline map of the large main passage in the Gruta do Janelão, in the Bambuí karst of Brazil.

karst scenery that deservedly has recently been granted National Park status.

The elevated limestone plateau known as Serra do Ramalho is another notable karst area. Recent exploration has revealed a series of major caves with long meandering passages. Among the caves, there are seven over 3 km long, although ongoing exploration will undoubtedly reveal many more. Most caves experience very high flood events. On the edges of the karst many significant caves occur, such as the 16 km long Gruta do Padre, where paleomagnetic dating of clastic sediments has allowed determination of fluvial incision rates of $25–34$ m Ma^{-1}.

The limestone in the São Domingos area outcrops as an elongated ridge parallel to the major scarps of Cretaceous sandstone. Some large rivers (with discharges up to 5 m^3 s^{-1}) pour off the sandstone and sink into the ridge, creating a series of outstanding cave systems, with six caves presently over 5 km in length. The caves are characterized by large river passages with equally large and very well-decorated upper levels. Underground confluences are common in the area, and thus few large springs occur in the distal side of the ridge. The core of the São Domingos karst is now protected by the Terra Ronca State Park.

As a whole the Bambuí karst presents a varied karst scenery that spans 10 degrees of latitude. It has significant scientific potential, since several paleontological and archaeological sites are commonly associated with the caves. Furthermore, a number of medium-sized cities depend on groundwater from the Bambuí karst aquifer. Environmental pressure is increasing in the area, although recently a few conservation units have been created to protect the karst resources.

AUGUSTO AULER

See also **America, South** for location map

Further Reading

Auler, A.S. & Farrant, A.R. 1996. A brief introduction to karst and caves in Brazil. *Proceedings of the University of Bristol Speleological Society*, 20: 187–200

Auler, A.S., Rubbioli, E.L. & Brandi, R. 2001. *As Grandes Cavernas do Brasil* [The Great Caves of Brazil], Belo Horizonte: Grupo Bambuí de Pesquisas Espeleológicas
A summary of the longest and deepest caves in Brazil, including surveys and information on karst areas (in Portuguese)

BAUXITE DEPOSITS IN KARST

Karst bauxites are formed as terrestrial sediments accumulated on carbonate rocks in karstic depressions and weathered under humid tropical and subtropical conditions. In the past they were thought to be derived from residual deposits, but they are now believed to originate from volcanic ash and tephra that is transported into karstic depressions by various agents: water, wind, and gravity. Karstic depressions tend to trap material from the surrounding areas (Bárdossy, 1989), and if there is extensive deep weathering of feldspar-bearing ash, for example, the feldspars will be desilicified, leaving behind deposits—bauxites—rich in aluminium. At the same time, karstification will affect the underlying limestones, and weathering residues from these will be added to the bauxite.

Karst bauxites are clay-like, earthy, non-crystalline, usually reddish deposits found in karstic depressions (karren, dolines, and poljes). The dominant mineral is boehmite (α-$Al_2O_3 \cdot H_2O$) with a low percentage of gibbsite (α-$Al_2O_3 \cdot H_2O$) and diaspore (β-$Al_2O_3 \cdot H_2O$). The fabric of bauxite deposits may be pelletal, pisoidal, ooidal, conglomeratic, arenitic, spheroidal, or microdebris. Karst bauxites from both vadose (oxidizing) and phreatic (reducing) diagenetic environments have been identified (D'Argenio & Mindszenty, 1995). There are seven main karst bauxite belts worldwide, in North America, the Caribbean, the Mediterranean, from Iran to the Himalayas, from the Urals to Siberia, East Asia, and on the margins of the Pacific Ocean (see Figure). In 1998 the world production of bauxite was 125 million tonnes, including both laterite and karst bauxites. Of this, nearly 25% was made up of karst bauxites. The main karst bauxite producing countries are Jamaica, China, Kazakhstan, Hungary, and Greece (see Table). In 2001, bauxite output was expected to increase in Jamaica (with an output of 15 300 thousand tonnes per year) and China (with an output of 2709 thousand tonnes per year).

The largest reserves of karst bauxite are found in Jamaica, where 9.3% of the world's economic reserves are located. The bauxites are found in Cockpit Country infilling 10–30 m deep dolines. Lenticular surface deposits 5–15 m thick cover the limestone plateau. The basement rock is a Middle Eocene–Lower Miocene shelf / slope limestone (Comer, 1974). These bauxites contain *c.*46–47% of Al_2O_3. Total Jamaican reserves amount to some 2000 million tonnes. Mining began in 1952, with the production of 0.5 million tonnes. In the mid-1980s output increased to more than 12 million tonnes annually. At the end of the 1980s output fell below 10 million tonnes, but in the 1990s increased to more 10 million tonnes. The economy of Jamaica depends heavily on exporting alumina and bauxite. The alumina output is 3 million tonnes per year, with five of the mines being open-cast. Mines are currently being operated in Clarendon, Dry Harbour, Ewarton, Kirkvine, and St Elizabeth. The following mines are not producing at the moment: Lydford, St Catherine, St Elizabeth–JBI, and Trelawny (with a combined total of 470 million tonnes of reserves).

The karst bauxites of Hungary occur in the form of lenticular, stratiform, graben, and deep doline deposits in the karst regions of the Trans-Danubian mountain range (the dolines are 10–30 m in diameter and 30–60 m deep). The bauxites contain *c.*48–55% of Al_2O_3. Two Cretaceous horizons have been identified within dolines formed on karstified Triassic limestone and dolomite surfaces, and an Eocene horizon on an Upper Cretaceous karst surface. The bauxites occur mostly below the karst water table. The extraction of subsurface lenticular bauxite beds is carried out by strip-mining. Bauxite production commenced in the Bakony Mountains in 1926. In the 1980s the annual bauxite production reached 3 million tonnes. In order to prevent a fall in the karst water table, the mines of Gánt and Iszkaszentgyörgy were closed down during the bauxite crisis of the 1980s. Economic reserves were estimated at 31.3 million tonnes in 1995, while the production in Hungary was 0.935 million tonnes in 1999, and increased to 1.046 million tonnes in 2000. Out of the total economic reserves, some 4.6 million tonnes occur in operating mines and 17.4 million tonnes are located in explored

Bauxite Deposits in Karst: Karst bauxite production (in thousands of tonnes) 1990–98 (source: World Mineral Statistics 1994; 1994–1998. British Geological Survey, World Metal Statistics Yearbook, 2000).

Country	1990	1995	1996	1997	1998
China	3655	8255	8879	9000	9000
France	490	131	165	164	80
Greece	2511	2200	2452	1877	1714
Hungary	2559	1015	1056	743	1909
Jamaica	10 965	10 857	11 829	11 987	12 646
Kazakhstan	92	3318	3346	3416	3437
Romania	247	175	175	127	135
Turkey	773	232	544	369	458
USA	494	397	258	233	153
Yugoslavia	–	60	323	470	226

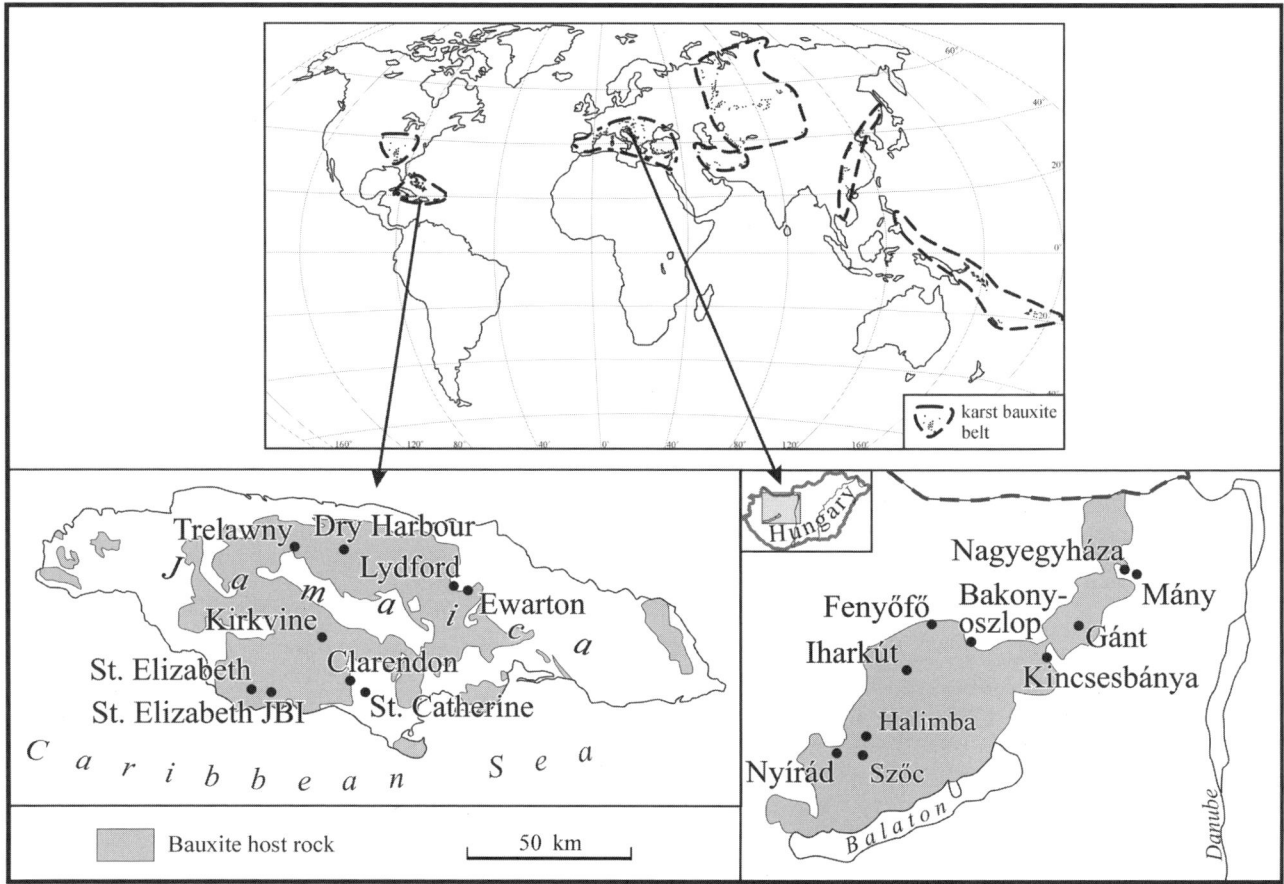

Bauxite Deposits in Karst: Karst bauxite belts of the world, and location of mines in Jamaica and Hungary.

but unexploited areas. The annual output of Hungarian bauxite mines is 1.5 million tonnes. The operating mines are at Halimba, Fenyőfő, Óbarok, Iharkút, and Némétbánya.

Karst bauxites are known to exist in New Guinea, but there is no specific information available about them. Bauxite formation began in the Pleistocene and continues up to the present day. The deposits accumulated in dolines on polygonal karsts. Bauxites and bauxite-like deposits have been described from New Ireland, Sorgeri, Bereina, and Cape Nelson (Hughes, 1990). In Oceania, further occurrences are known from the Philippines, the Solomon Islands, and the Loyalty Islands.

More than 95% of the bauxite extracted worldwide is used in alumina production. Three tonnes of bauxite are needed to produce one tonne of alumina, and two tonnes of alumina are required to produce one tonne of aluminium. The remainder is used by the abrasive, refractory, and chemical industries. During the refining of one tonne of aluminium, 1.2 to 4 tonnes of highly alkaline red mud residue are generated. There are various techniques available to process this mud. One involves mixing it with sandy soil to increase its acidity and increase nutrient retention capacity. If it is mixed with desiccated compost it has a beneficial impact on wetland vegetation and can contribute to the conservation of wetlands. Red mud mixed with gypsum can be used as a filter in sewage treatment systems. After intensive drying, cajunite, an excellent adsorbent, is produced, which is used to remove wet pollutants. Other important uses of red mud are as a substitute for clays in brick and tile manufacturing, in the ceramic industry, for foundations in road construction, or as a filler in PVC and rubber production.

Reclaiming land damaged by bauxite mining is a major environmental task. On former arable lands, revegetation is very slow, so it is more feasible to plant trees on the site or use the land for grazing. Air pollution from the dust produced during the extraction and transport of bauxite can also be a major problem. In areas that have been strip-mined, hazardous waste is often disposed of thoughtlessly, and harmful substances leach very rapidly into the karst system. Legal protection has to be ensured for these areas. During mining, karst water is pumped out and this can lead to the lowering of the karst water table and the drying out of karst caves, lakes, and springs. The first bauxite mine reclamation survey was launched in 1991, by the International Primary Aluminium Institute. The latest report of the survey was published in 2000. It shows that some 70% of former mining areas have been returned to native forest, 17% to agriculture and pasture, and 10% is being used for industrial purposes, urban development, or commercial forest plantations.

Useful and educational geological trails have also been created in abandoned bauxite mines.

ILONA BÁRÁNY-KEVEI

Works Cited

Bárdossy, G. 1989. Bauxites in paleokarst: A systematic and regional review. In *Paleokarst: A Systematic and Regional Review*, edited by P. Bosák, D. Ford, J. Glazek & I. Horácek, Amsterdam: Elsevier

Comer, J.B. 1974. Genesis of Jamaican bauxite. *Economic Geology*, 69: 1251–64

D'Argenio, B. Mindszenty, A. 1995. Bauxites and related paleokarst: Tectonic and climatic event markers at regional unconformities. *Eclogae Geologicae Helvetiae*, 88(3): 453–99

Hughes, F.E. (editor) 1990. *Geology of Mineral Deposits of Australia and Papua New Guinea*, vol. 2, Melbourne: Australian Institute of Mining and Metallurgy

Further Reading

Bárdossy, G. 1982. *Karst Bauxites: Bauxite Deposits on Carbonate Rocks*, Amsterdam and New York: Elsevier (original Hungarian edition 1977)

Sinclair, I.G.L. 1967. Bauxite genesis in Jamaica: New evidence from trace element distribution. *Economic Geology*, 62: 482–86

http://www.world-aluminium.org/production/mining/index.html Useful data, history, and reports from the International Aluminium Institute

http://www.american.edu/ted/BAUXITE.HTM Case study on the impacts of trade on the environment for bauxite production in Jamaica

http://www.unctad.org/en/docs/poitedcomd13.pdf Recent and planned changes in production capacity for bauxite, alumina, and aluminium

http://www.bauxit.hu/bxgb.htm Bakony Bauxite Mines Ltd. An overview of the company

BEAR ROCK KARST, CANADA

Interstratal solution breccias are common and widespread in the geological record (see Solution Breccias). However it is rare to find them widely exposed at the surface, resistant and displaying large secondary karst landforms developed upon them. One example is the karst in the Bear Rock Formation, which is of Lower and Middle Devonian age and extends $>50\,000$ km^2 in subcrop under the Mackenzie Valley and Mackenzie and Franklin Mountains in the Northwest Territories of Canada. In its central area (64–66°N, 125–130°W) it rests unconformably on 500 m or more of thick to massively bedded dolostones overlying >100 m of salt and redbeds, and is overlain by resistant limestones (Hume Formation) succeeded by thick siltstone, shale, and clay deposits of Mesozoic age.

Where it has been cored during oil exploration deep beneath the valley, the intact Formation consists of a regular sequence of dolomite and anhydrite beds (a typical supratidal or "sabkha" facies) totalling 250–300 m in thickness and becoming disturbed (brecciated) towards the top (Meijer Drees, 1993). In outcrop on the flanks of the mountains east and west of the valley, it is always found as a solution mega-breccia, reduced to ~200 m or less in thickness beneath the Hume Formation and thinning to zero upslope (Figure 1). Brecciation and cementation are complete to the base of the Formation, which is now composed of resistant cliff-forming pack breccias and regressive float breccias, both of dolomite and dedolomite clasts cemented by secondary calcite and occasional gypsum. A particularly resistant pack breccia of several metres thickness, that has developed along the surface (the Landry Member), is attributed to near-surface evaporative effects, i.e. exceptionally deep case-hardening. Evidently, both brecciation and cementation have been caused by circulation of meteoric groundwaters, commencing in the mid-Tertiary and continuing to the present.

The outcrops of the Bear Rock Formation range in altitude from 100 to 1000 m above sea level. Mean annual temperatures are -7 to -10°C (January mean about -28°C), supporting a zone of continuous permafrost calculated to be approximately 50 m in depth. Mean annual precipitation ranges 350–450 mm: most falls as snow but heavy rainstorms may occur in summer. The region was covered by the margin of the Laurentide Icesheet during the last (Wisconsinan) glaciation; the ice was comparatively thin and sluggish, depositing very little till in the mountains.

Bear Rock Karst: A dolomite and dedolomite solution breccia created by preferential dissolution of gypsum in the Bear Rock Formation at Bear Rock near Fort Norman, Northwest Territories, Canada. The section is ~15 m in height. (Photo by Derek Ford)

All outcrops display karst landforms. The principal types are:

1. dolines (sinkholes) and compound dolines of solutional and/or collapse origin, tens to hundreds of metres in length or diameter;
2. blind valleys where allogenic streams flow into the karst and sink;
3. greater solutional troughs along the edge of the Bear Rock Formation or following faults within it, many containing seasonal or permanent lakes;
4. small poljes of tectonic origin or created by glacial moraine dams; polje floors are up to 5 km^2 in area and have catchments as great as 90 km^2.

There are also small caves formed by seepage and frost action in cliffs ("frost pockets") and stream caves blocked by detritus or ice.

Periglacial processes attack all features, partially filling them with talus or solifluction lobes. Where the karst development is most intensive, however, (in the Canyon Ranges of the Mackenzie Mountains), the dominant appearance is of a "dissolution drape landscape" that is unlike any the writer has seen elsewhere. It is suspected that the Landry Member thickened and was stable under pre-Ice Age conditions. Either glacier margin meltwaters or groundwater leaks through the permafrost via unfrozen taliks then broke it up into the slabs, that are now seen tipping down into new karstic depressions in the weaker breccia beneath.

Field studies covering three melt seasons established that, despite the presence of nominally continuous permafrost, there is effective groundwater circulation through the outcrops and to depths exceeding 500 m. Recharge is chiefly via allogenic waters able to maintain conduit inflow along the upstream margins, or through ponds of meltwater that build up in the sinkholes until pressure ruptures the ground ice and permits rapid drainage. These latter taliks are of distinctive karstic origin (Ford & Williams, 1989, Figure 10.13). There is also some supra-permafrost flow to shallow springs and seepages in cliff faces. Discharge is via seasonal and some perennial springs: most are located at the lowest topographic positions, indicating that the groundwater systems are mature despite effects of the recent glaciation and modern permafrost.

Early in the melt season the spring waters have a $CaSO_4$ ionic composition representing the dominance of shallow (permafrost active zone) meltwaters. In mid-season the larger springs change to an $MgSO_4$ composition with water temperatures as high as 4°C, indicating the discharge of deeper waters that are accomplishing dedolomitization. The few perennial springs (which create winter "icings") have a significant NaCl component. Modelling suggests that there are three-component groundwater circulation systems involving the Bear Rock Formation, the underlying dolostones and the salt, and which are extending the brecciation process today. Analysis of oxygen isotope patterns in precipitation, sinking streams and springs indicates that mean residence times of the $MgSO_4$ waters are 40–90 days, suggesting flow rates of ~ 20–60 m d^{-1} through the breccia.

DEREK FORD

See also **Canada; Nahanni Karst; Solution Breccias**

Works Cited

Ford, D.C. & Williams, P.W. 1989. *Karst Geomorphology and Hydrology*, London and Boston: Unwin Hyman

Hamilton, J.P. 1995. *Karst Geomorphology of the Northeastern Mackenzie Mountains, N.W.T., Canada*, unpublished PhD thesis, McMaster University

Meijer Drees, N.C. 1993. *The Devonian Succession in the Subsurface of the Great Slave and Great Bear Plains, Northwest Territories*, Ottawa: Geological Survey of Canada (Geological Survey of Canada, Bulletin 393)

BELGIUM: ARCHAEOLOGICAL CAVES

The valleys of the Meuse River and its tributaries are mainly eroded in limestone. Many natural shelters, in the form of sometimes quite deep caves, can be found within this karstic system. The entrances to the caves were occupied continuously by humans, from the Paleolithic to modern times. In contrast, the deeper parts of the caves and galleries only served as occasional refuges during troubled periods (e.g. protohistory, the Roman era). There is abundant archaeological evidence of occupation, with the principal sites including La Naulette, Chaleux, Trou Da Somme, Furfooz, Trou Magrite, Bois Laiterie (Ardenne Meuse Valley), Grottes de Goyet (Samson River), Grotte du Docteur, Trou Sandron (Mehaigne River), Trou Al'Wesse (Hoyoux River), Fond-de-Forêt, and Grotte Walou (Vesdre River) (see map). Here, three of the more important sites (Spy, Engis, and Scladina) are presented in more detail.

Numerous Belgian caves have been excavated since the early 19th century, when the discipline of prehistory was first developed. The first Neanderthal was discovered near Liège at the site of Engis, in 1830, by Philippe-Charles Schmerling (see Archaeologists). The general palaeolithic sequence, established by Edouard Dupont on the basis of key sites in the province of Namur (Trou Magrite, Grottes des Goyet, among others), remains valid for Western Europe. Subsequently, the famed Neanderthal burials of Spy Cave were discovered in 1886 by Marcel De Puydt and Max Lohest, and were studied by Julien Fraipont. This pioneering research utilized methods from several disciplines: geology, stratigraphy, and paleontology. Nowadays the methods used are more strictly archaeological.

Spy Cave

The rocky overhang above the entrance of Spy Cave has provided the original name of the site—"Betche al'Rotche" ("beaked rock" in Walloon). The interior of the cave was initially excavated without any proper methodology, but the terrace was subsequently excavated by a multidisciplinary team from Liège (1885–86). A substantial and important stratigraphic sequence was recognized. Near the middle of the terrace, two Neanderthal skeletons were discovered. Because the skeletons were nearly complete, they were probably buried intentionally. They are associated stratigraphically with a late phase of the Middle Paleo-

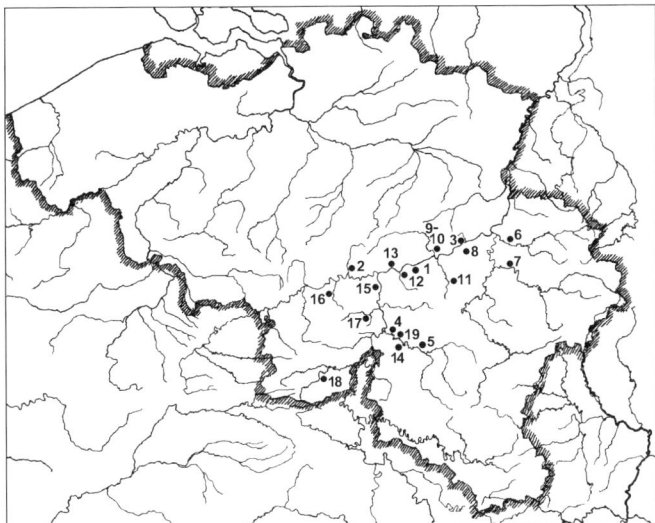

Belgium: Archaeological Caves: Locations of principal Belgian archaeological cave sites. 1. Scladina, 2. Spy, 3. Engis, 4. Trou Magrite, 5. La Naulette, 6. Fonds-de-Forêt and Grotte Walou, 7. Remouchamps, 8. Engihoul, 9. Grotte du Docteur, 10. Trou Sandron, 11. Trou al'Wesse (Modave), 12. Goyet, 13. Marche-les-Dames, 14. Trou Da Somme, 15. Bois Laiterie, 16. Presle, 17. Montaigle, 18. Couvin.

lithic (Charentian or Eastern European Micoquian). Other Mousterian levels, as well as a probable Acheulean level containing bifaces (shaped tools used as hand axes), form the lower part of the sequence. Upper Paleolithic strata are found above the burials. A thick layer called "red" by the excavators (resulting from iron oxide) contained a large number of Aurignacian artefacts, including numerous ivory pendants. Several dispersed artefacts, probably intermediate between the Mousterian and the Aurignacian, include foliate points corresponding with the beginning of the Upper Paleolithic in Northern Europe. A Gravettian level with tanged points and unifacial points ("points à face plane") overlies the Aurignacian level, and this industry is also found across the north European plain (comparable with open-air sites such as the Maisières-Canal, near Mons). The sequence continues with a sterile loess corresponding to the cold phase of the Second Pleniglacial, during which there is no evidence of human occupation in Northern Europe. Finally, traces of human occupation during the Late Upper Paleolithic are found above the loess layer.

Engis Caves

This is a series of four or five shelters and small caves opening at the summit of a limestone cliff near Engis in the province of Liège. Some of them have been destroyed as a result of the extension of alum-extraction quarries, but others have yielded archaeological material, primarily of Mousterian and Gravettian age. The research here was begun in the 1820s by Philippe-Charles Schmerling, a professor at the University of Liège, before the discipline of prehistory was established. Schmerling discovered the remains of two humans, of which only the skulls remain today. One is a Neanderthal child—the first Neanderthal discovered (although not recognized as such until after the eponymous discovery in 1866 in the Neander valley in Germany). The second is now known to be of Mesolithic age, based on a date recently obtained. Schmerling thus established the antiquity of humankind, based on the association of human remains and paleolithic tools with extinct fauna, long before the discovery of Neanderthals and before Darwin's theory of evolution was published.

Scladina Cave

Discovered in the 1970s in the village of Sclayn (province of Namur), Scladina has been the focus of systematic research by a team directed by the University of Liège. It is the main and upper cave of a long network of galleries, opening at different points in the valley. The cave was completely filled in at the time of discovery and the sediments form a long sequence several metres thick. At the base, as yet not fully investigated, the deposits correspond to oxygen isotope stage 6. The deposits excavated to date belong principally to the last interglacial (stage 5) and to the beginning of the following glacial (stage 4). The main human occupation occurred during the first temperate oscillation, around 130 000 years ago (stage 5e?), characterized by a Levallois Mousterian lithic industry and a subsistence economy oriented toward the exploitation of the chamois. Another interglacial stratum, more recent (around 117 000 years ago) yielded the dispersed remains of a Neanderthal child, including the maxilla, the mandible, isolated teeth, and other postcranial remains. Excavation is ongoing in these levels. Near the middle of the glacial deposits (around 38 000 years old), another Mousterian industry—the Charentian—is present. Finally, late paleolithic traces, as well as disturbed Neolithic burials, were found at the top of the deposits.

MARCEL OTTE AND REBECCA MILLER

Further Reading

De Puydt, M., Lohest, M. & Fraipont, J. 1886. L'homme contemporain du mammouth à Spy, province de Namur. Compte rendu des travaux du Congrès tenu à Namur les 17–19 août 1886. *Annales de la Fédération Archéologique et Historique de Belgique*, 2: 205–35

Dewez, M. 1987. *Le Paléolithique Supérieur Récent dans les Grottes de Belgique*. Louvain-la-Neuve: Publications d'Histoire de l'Art et d'Archéologie de l'Université Catholique de Louvain LVII

Dupont, E. 1872. *L'homme pendant les âges de la pierre dans les environs de Dinant-Sur-Meuse*, 2nd edition, Brussels: Deltombe

Otte, M. 1979. *Le Paléolithique supérieur ancien en Belgique*, Brussels: Musées Royaux d'Art et d'Histoire

Schmerling, P.-C. 1833–34. *Recherches sur les ossemens fossiles découverts dans les cavernes de la province de Liège*, Liège: Collardin

Ulrix-Closset, M. 1975. *Le Paléolithique Moyen dans le massin mosan en Belgique*, Wetteren: Universa

BELIZE RIVER CAVES

A small, little-known, and somewhat eccentric country, Belize is a pocket of Creole, Mayan, and British heritage, surrounded by countries of predominantly Spanish origin. Located along the northeast coast of Central America, it measures only 300 km from its southern rainforests to the savannahs of its northern border with Mexico. From the nation's eastern Caribbean coast, site of the world's second longest barrier reef, Belizeans look 100 km westward to their border with Guatemala, which in 1991 relinquished its claim that Belize was a Guatemalan state. Despite its small size, Belize has world-class caves, including some of the world's largest chambers and passages, perhaps one of the greatest concentrations of archaeologically significant caves (see America, Central: Archaeological Caves), and a diverse subterranean fauna that is still largely unstudied (see America, Central and the Carribbean Islands: Biospeleology). The country's rich speleological potential has been explored with increasing frequency since 1970, with much of the work focused on its amazing river caves.

Five main karst areas occur in the southern half of Belize, formed mainly in Cretaceous limestones (Veni, 1996). Southern Belize is dominated by the Maya Mountains, igneous and metamorphic rocks, reaching a maximum elevation of 1122 m at Victoria Peak and receiving over 3000 mm of annual rainfall. The size and locations of Belize's river caves are directly related to the Maya Mountains. Most of the country's large caves have been formed by large volumes of water flowing off the highlands onto the limestones of four of the five main karst regions. Equally important is that the water is chemically undersaturated with calcite. As a result, the water rapidly dissolves the limestone to form vast underground passages and chambers. In general, the caves are best preserved close to, but not next to, the highlands. Both at their contact with the non-karstic highland rocks and furthest away from the highlands, the limestones tend to become thinner, allowing less space for caves. Many caves are truncated by collapse.

The largest caves in Belize tend to be long stream passages, or former stream passages, with relatively little depth. Much of the vertical extent occurs in steep doline entrances, within the larger spaces of the passages, and down the gradient of the long passages. In the Boundary Fault Karst Area of central Belize, the development of multiple levels of passages by underground streams, carving new passages at lower elevations, is especially important in creating deeper caves. The Table lists some of the country's longest caves.

Among the country's most notable caves are the Chiquibul and Caves Branch cave systems. The Chiquibul is located in the Vaca Plateau karst area in west-central Belize. It was formed by the downcutting of the Chiquibul River flowing west off the Maya Mountains and includes a series of large, perennially active, seasonally active, and hydrologically inactive galleries. The main passages are more than 50 m wide by 30 m high and connect to the Belize Chamber. This unusually wide section of passage is one of the largest chambers in the world, at about 250 m long by 200 m wide and at least 50 m high. Three main caves make up the cave network, the longest of which is the linked Cebada Cave and Actun Tunkul. Prior to their connection they had been the two longest caves in the Chiquibul Cave System, and they are now the longest cave in Central America (Miller, 2000). The entire hydrologically linked network is 60 km long and crosses into Guatemala, where the underground Chiquibul River reappears at the base of a 200 m wide by 80 m high cave entrance.

Belize River Caves: Longest caves of Belize.

Cave	Length (km)
Chiquibul System (Cebada Cave and Actun Tunkul)	39.0
Petroglyph–St. Herman's Cave	17.0
Blue Creek Cave	9.0
Actun Chek	8.1
Actun Tunichil Muknal	5.3

Like the Chiquibul, the Caves Branch is a system of hydrologically connected caves some of which have not yet been physically connected by exploration. Petroglyph–St. Herman's Cave System is the longest single cave of this system with a surveyed length of 17 km. It is fed by Actun Chek and Actun Lubul Ha, which capture water from a 64 km² non-karst drainage area that flows north off the Maya Mountains. Caves Branch Cave is the downstream segment of the system and is truncated by several large dolines that have collapsed into its large passages.

Eco-tourism and adventure tours dramatically increased in Belize during the 1990s. Trips are regularly scheduled for Belize's river caves like Caves Branch, Blue Creek Cave, Barton Creek Cave, and a few even into the remote Chiquibul. The large size of the passages makes access relatively easy for non-cavers, and floating in inflatable automobile tire tubes in pleasant 25°C temperature streams is a welcome respite from the tropical heat and humidity on the surface. Archaeological material within the caves lends to the mystery and appeal of the tours but raises concerns about the possibility of inadvertently spurring some to seek out and loot artefacts from these and other caves. However, positive impacts that occur through education include tourist money that supports conservation and research, and proactive management of a few caves by some tour operators. The long-term impacts of these tours on Belize's caves and cultural heritage remain to be determined.

GEORGE VENI

See also **America, Central**

Works Cited

Miller, T. 2000. Inside Chiquibul: exploring Central America's longest cave. *National Geographic*, 197(4): 54–71

Veni, G. (editor) 1996. Special theme issue on Belize. *Journal of Cave and Karst Studies*, 58(2)
  This issue is the only scientific compendium on Belize speleology and includes eight papers on cave exploration, archeology, geology, biology, and conservation, plus a bibliography of caving in Belize.

BIODIVERSITY IN HYPOGEAN WATERS

Biodiversity in hypogean waters (both cave and interstitial) may be considered as the absolute number of all species inhabiting hypogean waters; the number of stygobitic (i.e. obligate hypogean aquatic) species; or the relationship of either of these numbers to surface fauna richness. At a global scale, biodiversity can be associated with specific biogeographical, hydrographical, or political units, or an area with an arbitrarily defined size and shape. It may also be measured as the number of higher taxa (e.g. orders or classes), rather than species. Biodiversity will be considered here according to a combination of these criteria. Due to the wide variety of methods available for computing species richness, and also because in different areas either the epigean or the hypogean fauna may be inadequately studied, comparisons are difficult.

Overall Biodiversity

The number of stygobitic species was estimated to be over 1000, or c.8% of all freshwater metazoan species in Europe, according to data in *Limnofauna europaea* (Illies, 1978), but their true number is probably close to 3000 (Sket, 1999b). This is because proportionately more karst is present in Europe, and its interstitial waters are comparatively well investigated (Marmonier et al., 1993). Since stygophile and stygoxene species have often not been reported as such, their numbers cannot be easily estimated. While over 50% of species in surface fresh waters are insects (either as larval or all stages), insects are poorly represented in hypogean waters. By contrast, 38% of European freshwater Crustacea species are stygobitic. Some of the subgroups are highly stygobitic (Isopoda and Amphipoda at 50% and Bathynellacea at 100%). Approximately 60% of all described stygobitic freshwater species in Europe are Crustacea, and half of these are Malacostraca. In North America, where the Entomostraca have been poorly studied and interstitial waters are not well explored, the Malacostraca represent 76% of the Crustacea (according to data in Peck, 1998) and worldwide in anchialine habitats they represent c.60% of the entire stygobitic fauna. Therefore, Crustacea, and the Malacostraca for which more exact and more recent data are available, are a particularly good model group for demonstrating the diversity of regional stygobitic faunas.

Biodiversity of the Malacostraca

Marine, continental, and hypogean waters are each inhabited by 10 or 11 of the 15 orders of Crustacea. There are approximately the same number of families containing stygobitic species as freshwater species. Three orders are limited to hypogean waters, and a fourth occurs in the deep sea as well as underground. Recently, Sket (1999a) estimated the number of described stygobitic Malacostraca species as 2130, while the estimated number of all described malacostracan taxa was close to 20 000. Thus, if the estimates are correct, some 10% of all malacostracan species are stygobites.

Hot Spots

The diversity of hypogean fauna is spatially very variable (see Table). Among regions of roughly the same area, the wider Dinaric (or west Balkan) region has nearly 400 species and subspecies (according to data in Botosaneanu, 1986), which is by far the richest biodiversity worldwide (see Dinaric Karst: Biospeleology). The biodiversity of the hypogean waters of Slovenia, which forms a small part of the Dinaric region, is as high as in some areas ten times larger. The Dinaric region is followed by the east Balkan, Pyrenean–Aquitanian, and Rhodano-Lotharingian regions, all of which are positioned along the belt where most European centres of high overall biodiversity occur. Some taxonomic groups may be even more concentrated—around 40% of all known stygobitic species of Gastropoda (Mollusca) inhabit the Dinaric region.

Regions outside Europe with similar total numbers of stygobites are much larger in size, and mainly occur within the northern warm temperate belt. Stygobitic faunas in tropical regions are generally less diverse, although their paucity is not as striking as the relative scarcity of terrestrial troglobitic faunas in the tropics. In all comparisons, it should be borne in mind that only in Europe have interstitial fresh waters been moderately well explored; this is not the case even in North America, let alone in tropical countries.

High numbers of regional stygobitic species are also mirrored in rich local faunas. Of the 20 richest caves—in which 20 or more specialized hypogean species have been recorded (Culver & Sket, 2000)—there are 11 with 20 or more stygobites. Among the richest are the Postojna–Planina Cave System (49 stygobites), the Vjetrenica Cave in Bosnia–Herzegovina (40 stygobites), and the anchialine Walsingham Cave System in Bermuda (up to 37 stygobites). These three caves, each with their own encyclopedia entry, differ in interesting ways. Postojna–Planina Cave System has a rich assortment of non-stygobites in addition to the usual stygobites, as do the Walsingham Caves. By contrast, the aquatic fauna of Vjetrenica Cave is almost exclusively stygobitic.

Why are there so many crustacea? Although the difficult environmental conditions, restricted accessibility, and relative ecological homogeneity of the hypogean realm severely limit the potential numbers of stygobitic species in general and also intrinsically exclude some groups (like the Insecta), some other groups may locally be over-represented when compared with epigean habitats. This is particularly the case with malacostracan groups.

Since the under-representation of Insecta underground could possibly be explained by their higher energetic demands, the successful invasion and diversification of the Crustacea may be explained by a number of circumstances. First, the absence of insects has left a number of potential niches. Second, high speciation rates can be achieved by spatial partitioning within a region that has a high degree of spatial isolation. Third, the geographical co-occurrence of a number of species is possible if there is specialization to different ecological niches. Fourth, temperatures in hypogean habitats are stable, and the absence of low winter temperatures allows the survival of species that entered the area in warmer times. Finally, a turbulent geological history may have influenced high speciation rates locally.

High numbers of hypogean species for some groups are compensated for by their restriction to smaller areas. As many as 78% of the stygobitic taxa in the Dinaric region are endemic to that region (153 400 km^2) and 37–67% of the taxa may be endemic to one of its subregions, measuring only 3500–51 000 km^2 in area. According to data from Illies (1978), 50 stygobitic and only six epigean species of Copepoda: Cyclopoida were limited to one of his regions in Europe (which are 150 000–

Biodiversity in Hypogean Waters: Numbers of stygobitic taxa (species and subspecies) in the regions with the highest diversity, according to data in Botosaneanu, 1986 (from Sket, 1999b). Marine interstitial taxa (labelled as "Q" or "Q, P") excluded. A: regions of highest diversity in Europe; B: richest regions outside Europe; C: subregions of the wider Dinaric region (first line in category A).

	Name or region	Number of stygobitic taxa	Approximate area (km^2)
A	West Balkan Province (=Dinaric Region; incl. Slovenia)	396	153 400
	East Balkan Province	245	205 000
	Pyrenean–Aquitanian Province	200	165 000
	Slovenia (south and west)	169	14 900
	Rhodano-Lotharingian Province	168	141 000
B	Japan	212	369 661
	Caribbean Islands	144	
	Appalachian Highlands	136	900 000
	Mexico	100	1 969 367
	New Zealand	76	264 820
C	Slovenia (south and west)	169 (113)	14 900
	Istra	24 (9)	3600
	Croatia (southwest and south)	102 (60)	25 500
	Bosnia–Herzegovina	99 (55)	51 100
	Serbia (west)	15 (6)	33 800
	Montenegro and Kosovo	55 (36)	24 500

350 000 km^2 in area), while only two stygobitic and 45 epigean species were distributed throughout six or more such regions. A number of stygobitic species were known from one locality only.

Species may also be heterogeneously distributed within a cave system, according to their ecological needs, a fact neatly illustrated by the example of Vjetrenica (Sket, 1999a). It has been shown that the number of species in a cave system depends to a great degree on the region, but is positively correlated with the richness of food resources and with the size of the caves (Culver and Sket, 2000). The existence of hundreds of mostly stygobitic *Niphargus* species and subspecies in Europe is possible through their extreme diversity of size and shape of all body parts (indicating a variety of ecological niches) as well as their highly restricted ranges.

BORIS SKET

Works Cited

Botosaneanu, L. (editor) 1986. *Stygofauna Mundi: A Faunistic, Distributional, and Ecological Synthesis of the World Fauna Inhabiting Subterranean Waters (Including the Marine Interstitial)*, Leiden: Brill

Culver, D.C. & Sket, B. 2000. Hotspots of subterranean biodiversity in caves and wells. *Journal of Cave and Karst Studies*, 62(1): 11–17

Illies, J. (editor) 1978. *Limnofauna europaea*. 2nd edition, Stuttgart: Fischer

Marmonier, P., Vervier, P., Gibert, J. & Dole-Olivier, M.-J. 1993. Biodiversity in ground waters. *Trends in Ecology and Evolution*, 8(11): 392–95

Peck, S.B. 1998. A summary of diversity and distribution of the obligate cave-inhabiting faunas of the United States and Canada. *Journal of Cave and Karst Studies*, 60(1): 18–26

Sket, B. 1999a. The nature of biodiversity in hypogean waters and how it is endangered. *Biodiversity and Conservation*, 8(10): 1319–38

Sket, B. 1999b. High biodiversity in hypogean waters and its endangerment—the situation in Slovenia, Dinaric karst, and Europe. *Crustaceana*, 72(8): 767–79

Further Reading

Bowman, T.E. & Abele, L.G. 1982. Classification of the recent Crustacea. In *The Biology of Crustacea*, vol. 1, edited by L.G. Abele, New York: Academic Press

Christman, M.C. & Culver, D.C. 2001. The relationship between cave biodiversity and available habitat. *Journal of Biogeography*, 28: 367–80

Danielopol, D., Pospisil, P. & Rouch, R. 2000. Biodiversity in groundwater: a large scale view. *Trends in Ecology and Evolution*, 15(6): 223–24

Rouch, R. & Danielopol, D.L. 1987. L'origine de la faune aquatique souterraine, entre le paradigme du refuge et le modèle de la colonisation active. *Stygologia*, 3(4): 345–72

Rouch, R. & Danielopol, D.L. 1997. Species richness of microcrustacea in subterranean freshwater habitats. Comparative analysis and approximate evaluation. *Internationale Revue der gesammten Hydrobiologie*, 82(2): 121–45

Sket, B. 1997. The anchihaline habitats, a dispersed "center" of biotic diversity. In *Conservation and Protection of the Biota of Karst*, edited by I.D. Sasowsky, D.W. Fong & E.L. White, Charles Town, West Virginia: Karst Water Institute (Special Publication 3)

Stoch, F. 1995. The ecological and historical determinants of crustacean diversity in groundwaters, or: why are there so many species? *Mémoires de Biospéologie*, 22: 139–60

Vandel, A. 1965. *Biospeleology: The Biology of Cavernicolous Animals*, Oxford and New York: Pergamon Press (originally published in French, 1964)

Ward, J.V. & Voelz, N.J. 1998. Altitudinal distribution patterns of surface water and groundwater faunas. In *Advances in River Bottom Ecology*, edited by G. Bretschko and J. Helešic, Leiden: Backhuys

BIODIVERSITY IN TERRESTRIAL CAVE HABITATS

The subterranean fauna is a mixture of forms with different origins and life histories, characterized by convergent features—adaptations to life in permanent darkness and high humidity. For geographical, geological, ecological, or historical reasons there are regions or even individual caves in the world that shelter an unusually high number of troglobitic species. These biodiversity hot spots include the Lanquedoc aquifers in France and the Edwards Aquifer in Texas, and in terrestrial cave habitats the Dinaric karst in Slovenia, Movile Cave in Romania, and Mammoth Cave in Kentucky (the latter four have separate encyclopedia entries). Most cave animals are invertebrates, dominated by arthropod groups such as Arachnida and Insecta (see Table) but there are also some troglophilic vertebrates. This entry discusses biodiversity in caves by taxonomic group.

The Isopoda are mostly aquatic but include terrestrial cave species belonging to the Suborder Oniscidea (see Crustacea: Isopoda Oniscidea). Sometimes these species do not differ greatly from their epigean relatives, but some are characterized by classic adaptations to life in darkness.

The Arachnida (see separate entries) are very diverse, and have successfully colonized many ecological zones, including caves. Cave scorpions have been recorded only from Mexico, Ecuador, and Sarawak. The smaller Pseudoscorpiones lack the poisonous tail spine of the scorpions. Subterranean species are

Biodiversity in Terrestrial Cave Habitats: The number of some subterranean arthropod species and their global distribution. (Data from *Encyclopaedia Biospeologica*, 1994, 1998.)

Taxonomic group	No. of epigean taxa	No. of hypogean taxa	Geographical distribution
Isopoda terrestria	3500 species 34 families	270 species 14 families 37 genera	Mostly Mediterranean
Scorpiones	1300 species 9 families	13 species 4 families	11 species from Mexico
Pseudoscorpiones	3000 species 23 families	c.64 genera with cave representatives	Worldwide
Araneae	40 000 species c.100 families	c.1000 species 29 cave genera c.50 families	Worldwide
Opiliones	5000 species	115 species (82 troglobitic)	Mostly Mediterranean, North and Central America
Ricinulei	30 species 3 genera	10 species 2 genera	8 species from Mexico
Schizomida	100 species 6 genera	32 species 2 genera	Tropics
Amblypygi	100 species 3 families	40 species	Tropics
Palpigradi	78 species	27 species	Mostly in Europe
Acari	50 000 species	250 species (troglophilic, very few troglobitic)	Worldwide
Chilopoda	2500 species	55 species troglobitic	North America, Cuba, Mediterranean region
Diplopoda	10 000 species	c.80 genera with cave representatives	Mostly from the Mediterranean region, North America, Japan
Collembola	2000 species	240 troglobitic and 200 troglophilic species and dozens of species	Palearctic and Mediterranean regions; Worldwide
Diplura	100 species	92 species	80% Palearctic
Thysanura	700 species	6 species and 4 subspecies	Worldwide
Dermaptera	1100 species	2 slightly adapted and 2 strongly adapted species	Canary Islands and Hawaii
Blattodea	3700 species	29 species (different degrees of adaption)	Tropics, Canary Islands and Australia
Psocoptera	3800 species	12 species	Mostly Mediterranean
Homoptera	33 000 species	59 species	Mostly tropical
Lepidoptera	120 000 species	100 cave species (3 troglobitic)	Worldwide
Diptera	150 000 species	22 troglobitic species	Worldwide
Notoptera	25 species	7 species	North America, Japan, South Korea
Siphonaptera	2500 species	137 parasitic species	Mostly tropical
Hymenoptera	125 000 species	220 species	Worlwide
Coleoptera	300 000 species	2000 troglobitic species	69% Atlantic–Mediterranean regions

present in most families, with some genera being completely restricted to caves. The Araneae (spiders), one of the most abundant groups of land animals, are also common in caves. Some of their morphological adaptations are less reported from other groups: troglobitic spiders have more mechanical receptors than the troglophiles, and relative length of the appendages is directly correlated with the degree of adaptation. The Opiliones, or harvestmen, possess a pair of defensive glands that produce noxious secretions. They are present on every continent and in every terrestrial environment. Some representatives are good biogeographical indicators. The Ricinulei is a small order with species living in caves and tropical leaf litter. Members of this order have been described from American caves, with some populations being relatively well represented (more than 10 000 individuals). Twenty-five troglobitic Schizomida species live on guano or plant debris, and rarely under rocks. Their distribution is limited to the tropics. The Amblypygi (whip spiders) contains some 40 troglophilic and troglobitic species, found in Central and South America, Africa, and southern Asia. Of the 20 species of Palpigradi (miniature whip scorpions), several cave species are known from the Mediterranean region. The Acari are frequent in caves: mites are associated with guano or live on soils, organic matter or in rock cracks, while ticks are parasitic on bats.

The four classes of Myriapoda (see separate entry) are abundant mostly in Europe and temperate regions of North America. Chilopoda (centipedes) have soft bodies with many unfused trunk segments, each one with a pair of legs; the first pair is modified to form large poisonous fangs. Some of the epigean species have troglomorphic features, but data about their adaptations are generally scarce. Diplopoda (millipedes) have two pair of legs on each segment, a calcified cuticle, and no fangs. All three suborders of this class have species living in caves, some of them with more body segments. The only troglobitic Pauropoda have been reported from Japan. The Symphyla, a very old group, include small, eyeless, and depigmented representatives. It has two cave species, both of which are found in Postojna Cave (Slovenia).

The insects (see separate entries) include all arthropods with three pairs of legs and three body segments. Collembola, the oldest group of insects, have occupied all the terrestrial biotopes, especially soils. For example, in the Palearctic and Mediterranean zones alone there are 240 troglobitic species. These species are K-strategists (i.e. their reproduction is regulated to maintain small numbers of offspring, in view of the population capacity limit, K, of the habitat) and many studies have been made of their physiology and behaviour. A pair of posterior cerci or feeler-like appendages characterizes the Diplura, which are pre-adapted for life in darkness. Two families have cave species distributed mostly in the Palearctic region. Pre-adapted representatives of Thysanura have two caudal filaments. Ten cave species are known with representatives on each continent. In the Dermaptera (earwigs), the cerci form heavily sclerotized posterior forceps. Species of two Dermaptera families are found in caves, but only two species, from lava tubes, display strong adaptations. Cave cockroaches belonging to the Blattodea are largely distributed throughout the tropics, both in karst and in lava tube caves. The species have various adaptive features. The Psocoptera (psocids, booklice), a small group with few subterranean species, is distributed mostly in the Mediterranean region. The representatives of the Homoptera (plant bugs) have piercing / sucking mouthparts forming a beak. All cavernicole species belong to groups where the immature stages live close to the soil or in the soil, and they are found in tropical areas and some temperate regions. Hypogean individuals communicate by low-frequency signals transmitted as vibrations through the substratum. Trichoptera (caddisflies), together with Lepidoptera (butterflies and moths) and Diptera (flies), are components of the cave-wall community—most species are subtroglophilic, troglophilic, and guanophilic. Some flies are troglobitic, with more or less obvious adaptations. Out of the Orthoptera group, the crickets and some tettigonids will colonize caves. Some crickets are true troglobites, but others use caves only as shelters. Cave tettigonids are troglophilic, ubiquitous components of the cave-wall community, which display troglomorphic features. Hypogean species of Notoptera are described from the western United States, South Korea, and west Japan. The ectoparasitic fleas from the Order Siphonaptera are associated exclusively with bats or other cave mammals, and have no notable adaptations. Another common component of the cave-wall community is the Hymenoptera (wasps, bees), including troglophilic, subtroglophilic, and trogloxenic species. The Coleoptera (beetles) is the best-represented underground group, with almost 2000 species distributed worldwide; they are frequent in the temperate regions.

Troglobitic vertebrate species are not nearly as numerous as the invertebrates, although the faeces and guano of these vertebrates may, in some caves, represent the only source of food for underground communities. Amphibians (frogs, toads, and salamanders) are usually trogloxenes in entrance habitats, however eight truly troglobitic species are known, seven of them from North America. (*Proteus*, the only troglobitic vertebrate in Europe, is described elsewhere in this Encyclopedia.) Aves (birds; see separate entry) include six troglophilic species. Mammalia (mammals) include troglophilic or trogloxenic species from different groups (marsupials, rats and other rodents, insectivores, pigs, okapi, felines, etc.) and Chiroptera (bats; see separate entry). The diversity of cave vertebrates may have been greater in cave refugia during climatic fluctuations; for example, the recently extinct cave bear during the Pleistocene. At least 93 fossil vertebrate species have been identified in Pleistocene sediments at Victoria Cave, Naracoorte, Australia (see Australia: Archaeological and Paleontological Caves).

OANA MOLDOVAN

Further Reading

Brusca, R.C. & Brusca, G.J. 1990. *Invertebrates*, Sunderland, Massachusetts: Sinauer Associates

Ginet, R. & Decou, V. 1977. *Initiation à la biologie et à l'écologie souterraine*, Paris: Delarge

Harvey, M.S., Shear, W.A. & Hoch, H. 2000. Onychophora, Arachnida, Myriapods and Insecta. In *Subterranean Ecosystems*, edited by H. Wilkens, D.C. Culver & W.F. Humphreys, Amsterdam and New York: Elsevier

Juberthie, C. & Decu, V. (editors) 1994. *Encyclopaedia Biospeologica*, 2 vols, Moulis and Bucharest: Société de Biospéologie

BIOFILMS

Microorganisms are individually very tiny, but early in the history of life on Earth they invented the biofilm: a wonderful method of protecting themselves and extending their influence on the environment. A biofilm is a coating on rock or other surfaces composed of microorganisms, extracellular slime and other materials that the organisms produce, and particles (sediments, organics, or minerals) that are trapped by or precipitated within the film. Caves often contain biofilms on many of their surfaces, formed both subaqueously and subaerially (under both wet and relatively dry conditions). The variety of biofilms in caves can be quite spectacular even to the naked eye (Figure 1). Closer inspection reveals the intimate associations between cells, slime, and particles (Figure 2).

Biofilms are ubiquitous in nature. Examples include the plaque that grows on teeth and algae that grow on the inside of fish tanks. Films can be very thin and invisible to the naked eye or very thick and even geologically significant. Biofilms are probably common because they provide a large number of advantages for the bacteria, fungi, cyanobacteria, and algae that live in them. Protection from desiccation and ultraviolet radiation are major advantages for biofilm organisms that live in surface habitats, but even in the dark and usually humid environments of caves, the biofilm serves many functions. It can provide protection from chemicals and ions that might be harmful to the organisms and serves as a kind of "pantry" for storing potential nutrients and growth factors, allowing organisms to concentrate materials that may be sparse in the larger environment (e.g. Costerton et al., 1994). Protection from extremes of pH (relative degrees of acidity or alkalinity) can allow biofilm organisms to live in conditions that they could not tolerate without their protecting blanket of film. In some cases, the biofilm can retain concentrations of antimicrobial or inhibitory substances that organisms produce to help protect themselves from other microbial competition for resources and from predation by higher organisms such as protists and small invertebrates.

Biofilms in caves are found in many different guises. At least anecdotally, they seem to be unusually diverse compared to those found on the surface (e.g. Boston et al., 2001; Melim et al., 2001). Strings, loops, thick mats, filamentous fabrics, thin slippery coatings, dense encrustations, highly mineralized Fe / Mn deposits, chalky crusts, gelatinous rinds, iron-rich slimes, manganese fluff, and some moonmilks found in caves may all qualify as biofilms by the definition cited above. The specific chemistry, geology, ecology, and microclimatology of a cave helps to dictate the exact details of a biofilm structure (Stoodley et al., 1999), however, the tendency to form these films is widespread.

The nutrient base available to a biofilm is a dominant factor controlling its overall abundance. While most caves probably receive small amounts of organic materials that form the basis of the ecology of a film or mat, the role of organisms that rely for their life energies on inorganic substances (known as chemolithoautotrophs) remains unclear. However, in caves with significant inorganic sources of energy (e.g. reduced sulfur, iron, or manganese) the chemolithoautotrophic contribution to the system may predominate (Kinkle & Kane, 2000; Boston et al., 2001).

Besides trapping particles within the film or mat, biofilms can also strongly affect the nature and rate of dissolution and mineralization processes in the underlying rocky substrate. Biofilms can concentrate substances that chemically attack their substrata, increasing the rate of dissolution compared to analogous abiotic processes. They can also promote the formation of elec-

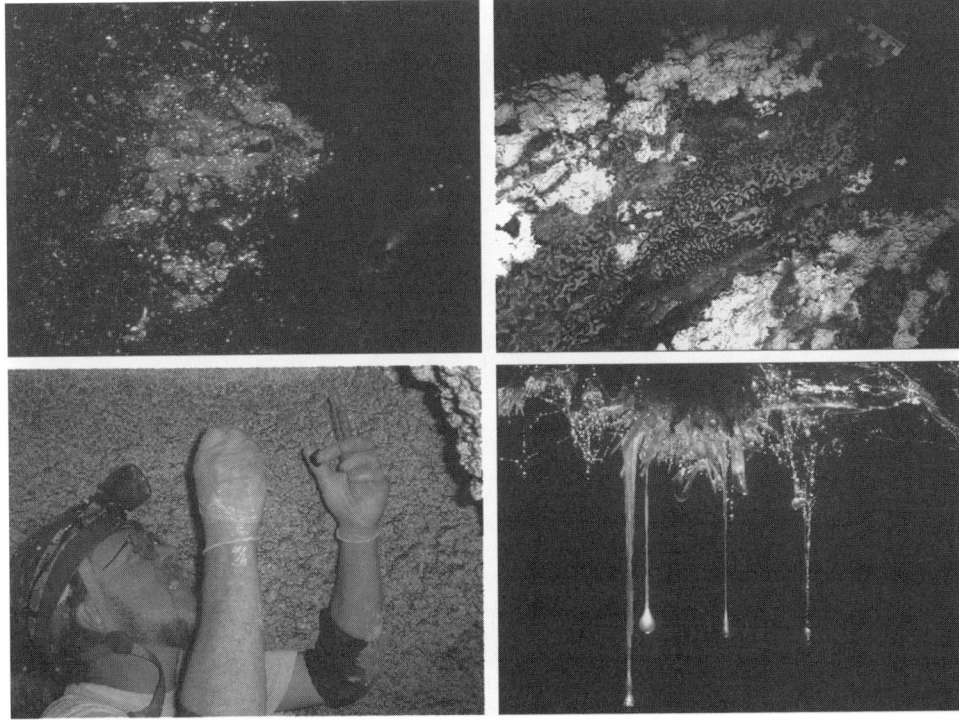

Biofilms: Figure 1. Biofilms occur in many forms from thin, invisible layers to thick, colourful spongy mats. Upper left: actinomycete bacteria form a beautiful pink biofilm in a Hawaiian lava tube cave. Upper right: black swirled patterns are microbial mat "biovermiculations" in Cueva de Villa Luz, a hydrogen sulfide-dominated cave in Tabasco, Mexico. Lower left: mineralogist Mike Spilde samples a mud-like, calcite rich biomat covering all surfaces in passage within Spider Cave, New Mexico, United States. Lower right: stringy biofilms, termed "snottites", hang pendulously from the ceilings and wall projections in the sulfuric acid environment of Cueva de Villa Luz. (Photos by Kenneth Ingham)

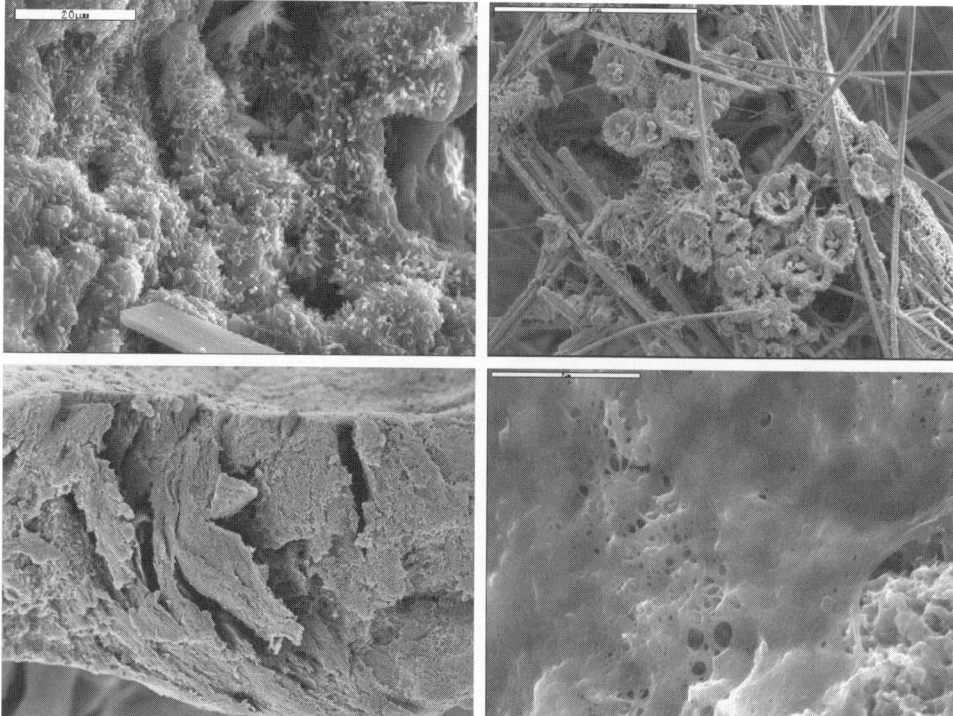

Biofilms: Figure 2. Scanning electron micrographs of cave biofilms: Upper left: *Thiobacilli* (hydrogen sulfide-producing extreme acid-loving bacteria) encrust the sticky slimes that make up the snottites shown in Figure 1 (image by M. Spilde and D. Northup). Gypsum and elemental sulfur crystals are produced in situ by the organisms and incorporated into the biofilm. Upper right: "nests" of bacteria and microfungal hyphae are visible on the calcite fibers found in "Crisco" moonmilk from Spider Cave shown in Figure 1 (image by M. Spilde and P. Boston). Lower left: cut edge through a thick (1.5 cm) biofilm mat from Cueva de Villa Luz showing the dense packing of organisms, filaments, minerals, particles, and slime (image by D. Soroka). Lower right: fossil biofilm from paleo pool fingers, Hidden Cave, New Mexico, United States. Acid etched preparation, image by L. Melim, M. Spilde, and D. Northup.

trochemical corrosion cells (known as tubercles) created by acid-producing sulfide-oxidizing organisms. Of similar importance is the process of *in situ* precipitation of minerals within mats and films. Calcite precipitation of unique crystal morphologies have been documented (Folk & Chafetz, 2000) and manganese mineral precipitation of unique crystals has also been reported (Spilde *et al.*, 2002). The resulting mineral particles, if they prove to be solely biogenic, may provide an important tool for investigation of fossil biofilms and preserved microbial mats as well as elucidating living biofilm formation mechanisms.

Biofilms can form in water under rapidly running conditions and on the humid surfaces in caves with no running water (Northup *et al.*, 2000). This plasticity in the ability to exploit many different habitat types, fundamental cave geochemistries, and hydrological conditions make biofilms and mats a highly significant feature of subterranean ecosystems. Furthermore, because calcium ions seem to play an important role in adhesion of cells to surfaces, carbonate caves are in a unique position to chemically encourage biofilm formation. Unusual properties of calcium ions encourage interactions with protein and polysaccharide adhesin molecules at the cell surface. Calcium helps to neutralize the electrically active double layer between cell and substrate, thus promoting adhesion.

The term "microbial mat" simply refers to a biofilm that has achieved a thickness and size easily visible to the human eye. A mat type common in daily experience is a thick layer of algae, mud, and associated slimey substances that grow around the margins of water bodies. A more elaborate microbial mat type is known as a "stromatolite", that is, a layered structure created by microbial growth at the surface covered by episodic deposits of sediments followed by further growth of the organisms to entomb those particles. In contrast to their vast abundance on early Earth as shown in the geological record, stromatolites are now rare on the Earth's surface and restricted to a few protected lagoonal environments. Without weather and often without a significant population of higher organisms, caves can provide thick microbial mats with a protected environment in which to grow, precipitate minerals within their structure, and potentially fossilize into cave stromatolites that show the banded structure in cross section of their surface counterparts (Melim *et al.*, 2001).

Biofilms are such an effective means of promoting the survival and growth of microorganisms that it has been suggested that they are actually primary units of evolutionary selection rather than the organisms that they contain (Caldwell & Costerton, 1996). The co-evolution and metabolic dependence of biofilm organisms upon one another is quite striking. This may explain why it is often difficult or impossible to grow biofilm organisms in isolation, especially those from caves. The diversity and importance of biofilms and mats in caves makes their study one of the cutting edge research topics of modern microbial biospeleology.

PENNY BOSTON

See also **Microbial Processes in Caves; Micro-organisms in Caves; Sediments: Biogenic**

Works Cited

Boston, P.J., Spilde, M.N., Northup, D.E., Melim, L.A., Soroka, D.S., Kleina, L.G., Lavoie, K.H., Hose, L.D., Mallory, L.M., Dahm, C.N., Crossey, L.J. & Schelble, R.T. 2001. Cave biosignature suites: Microbes, minerals and Mars. *Astrobiology Journal*, 1(1): 25–55

Caldwell, D.E. & Costerton, J.W. 1996. Are bacterial biofilms constrained to Darwin's concept of evolution through natural selection? *Microbiologia SEM*, 12(3): 347–58

Costerton, J.W., Lewandowski, Z., DeBeer, D., Caldwell, D., Korber, D. & James, G. 1994. Minireview: Biofilms, the customized microniche. *Journal of Bacteriology*, 176(8): 2137–42

Folk, R.L. & Chafetz, H.S. 2000. Bacterially induced microscale and nanoscale carbonate precipitates. In *Microbial Sediments*, edited by R. Riding & S.M. Awramik, Berlin and New York: Springer

Kinkle, B.K. & Kane, T.C. 2000. Chemolithotrophic microorganisms and their potential role in subsurface environments. In *Subterranean Ecosystems*, edited by H. Wilkens, D. Culver & W. Humphreys, Amsterdam and New York: Elsevier

Melim, L.A., Shinglman, K.M., Boston, P.J., Northup, D.E., Spilde, M.N. & Queen, J.M. 2001. Evidence of microbial involvement in pool finger precipitation, Hidden Cave, New Mexico. *Geomicrobiology Journal*, 18:11–29

Northup, D.E., Dahm, C.A., Melim, L.A., Spilde, M.N., Crossey, L.J., Lavoie, K.H., Mallory, L.M., Boston, P.J., Cunningham, K.I. & Barns, S.M. 2000. Evidence for geomicrobiological interactions in Guadalupe caves. *Journal of Cave and Karst Studies*, 62(2): 80–90

Spilde, M.N., Boston, P.J., Schelble, R.T. & Papike, J.J. 2002. Mineral precipitation by Mn-oxidizing microbes: Comparing natural and cultured Mn-minerals. In *Lunar and Planetary Science XXXIII: Papers Presented to the 33rd Lunar and Planetary Science Conference*, Houston: Lunar and Planetary Institute (CD-ROM Abstract #1090)

Stoodley, P., Dodds, I., Boyle, J.D. & Lappin-Scott, H.M. 1999. Influence of hydrodynamics and nutrients on biofilm structure. *Journal of Applied Microbiology*, 85: 19S–28S

Further Reading

Bryers, J.D. 1993. Bacterial biofilms. In *Current Opinion in Biotechnology: Biochemical Engineering*, 4: 197–204

Evans, L.V. 2000. *Biofilms: Recent Advances in Their Study and Control*, Amsterdam: Harwood Academic

Stolz, J.F. 2000. Structure of microbial mats and biofilms. In *Microbial Sediments*, edited by R. Riding & S.M. Awramik, Berlin and New York: Springer

BIOKARSTIFICATION

Biokarst refers to karst landforms, usually small in scale, produced mainly by organic action. Biokarstification is an umbrella term for the organic processes involved in the creation of such landforms. These processes can be divided into two main types, organic contributions to weathering and erosion, and biological aids to reprecipitation, or deposition of calcium carbonate. Biokarstifying processes have been suggested as being important in landform production in a whole range of environments, including karren on Mediterranean limestone surfaces (Fiol, Fornos & Gines, 1996), caves (as found, for example, in Lechuguilla Cave, New Mexico by Cunningham *et al.*, 1995) and coastal limestones (e.g. around many Mediterranean coasts as reported by Schneider & Torunski, 1983 and Schneider & Le Campion-Alsumard, 1999). Most biokarstification activity appears to result from the action of a suite of micro-organisms and lower plants, including cyanobacteria, fungi, algae, and lichens, often in mixed biofilm communities. Higher plants may also be involved, largely through root activity, as may animals, for example through the biocorrosive action of urine, or the boring and grazing activity of a suite of coastal molluscs.

In terms of erosional manifestations of biokarstification, the major processes involved are biochemical contributions to carbonate dissolution, biochemical and biophysical processes of boring and grazing, and the biophysical weathering activity of lichens. Many organisms are known to secrete substances which are capable of enhancing limestone dissolution, including organic acids and carbon dioxide. In many cases, it has proved difficult to determine the exact mechanism by which organisms (especially micro-organisms) encourage dissolution, but the pitting of underlying limestones associated with many micro-organism communities (as shown, for example, in the Negev Desert environment, by Danin & Garty, 1983) illustrates that it does occur. Roots can also bore by chemical and physical means, but their role in biokarstification has been much less well studied in recent years.

A whole host of grazing and boring molluscs and other organisms commonly inhabit the intertidal zone of limestone coasts and cause considerable biochemical and biophysical erosion, through the action of chemical secretions and the rasping of teeth and claws. For example, a recent study has illustrated the role played by limpets in weathering and eroding chalk shore platforms in east Sussex, southern England, through both excavation of hollows (homescars) and also ingestion of chalk during grazing on algae on the surface (Andrews & Williams, 2000). Experimental observations here suggest that limpets are responsible for about 12% of the downwearing in areas they inhabit. In most places, combinations of organisms act together, with complex spatial and temporal dynamics, and it can be hard to quantify the overall effect. Some quantitative studies of individual processes and their net action have been made for a range of tropical, Mediterranean and temperate coastal environments (see, for example, the work of Trudgill, 1976, from Aldabra Atoll, and Trudgill, 1987, from the Burren coast, Co. Clare) as reviewed by Spencer (1988).

An important biophysical weathering role has been identified for several lichen species as a response to wetting and drying. Lichens can absorb a huge amount of water, and swell to many times their original size. Drying produces a concomitant shrinkage, and repeated wetting and drying can exert enough force to pluck away fragments of limestone. Moses and Smith (1993) carried out an experimental study on this lichen role, and suggested that it may play an important role in the enlargement of kamenitzas on temperate limestone pavements. Although many studies have isolated individual biokarstic processes, and even quantified the role of individual organisms, it has so far proved difficult to quantify the overall biological contribution in relation to other earth surface processes, such as dissolution.

Depositional forms of biokarst, such as tufas and some speleothem and travertines, also exist. In these cases, organisms play a range of roles in aiding deposition of calcium carbonate and other minerals. Organisms may directly precipitate minerals as part of their life processes, and they can also play indirect roles through altering the chemical micro-environment to favour precipitation of minerals out of solution, and finally they can provide a site (or nucleus) for precipitation. Pentecost (1981) has discussed the various roles that cyanobacteria and algae play in tufa deposition. In essence, organisms can be either active or passive contributors to tufa deposition. Active roles include the direct precipitation of calcium carbonate or alterations of the nearby chemical environment, for example through removal of CO_2. Passive contributions to tufa deposition include the provision of suitable substrates for deposition and nucleation of calcite. Several studies have been made of the balance of processes influencing tufa deposition, for example by Viles and Pentecost (1999), who illustrate the roles of a range of cyanobacteria, algae, and higher plants in the fast build-up of tufa barrages in a small karst stream in South Wales. In most cases, a mixed community of organisms contributes to the formation of depositional biokarst landforms in a whole suite of ways, and it can be very difficult to untangle individual roles.

Although much work on biokarstification has focused on small-scale organisms and their role in the production of small-scale landforms, most karst landscapes have an important, and neglected, biokarstic component, in the form of plant and soil effects. Plants and soils, including associated micro-organisms, play a huge role in dissolution in many karst environments, with plants taking up calcium and magnesium directly and contributing to the production of acidic soils.

Heather Viles

See also **Biofilms; Coastal Karst; Microbial Processes in Caves; Phytokarst**

Works Cited

Andrews, C. & Williams, R.B.G. 2000. Limpet erosion of chalk shore platforms in Southeast England. *Earth Surface Processes and Landforms*, 25: 1371–81

Cunningham, K.I., Northup, D.E., Pollastro, R.M., Wright, W.G. & LaRock, E.J. 1995. Bacteria, fungi and biokarst in Lechuguilla Cave, Carlsbad Caverns National Park, New Mexico. *Environmental Geology*, 25: 2–8

Danin, A. & Garty, J. 1983. Distribution of cyanobacteria and lichens on hillsides in the Negev Highlands and their impact on biogenic weathering. *Zeitschrift für Geomorphologie*, 27: 423–44

Fiol, L., Fornos, J.J. & Gines, A. 1996. Effects of biokarstic processes on the development of solutional rillenkarren in limestone rocks. *Earth Surface Processes and Landforms*, 21: 447–52

Moses, C.A. & Smith, B.J. 1993. A note on the role of *Collema auriforma* in solution basin development on a Carboniferous limestone substrate. *Earth Surface Processes and Landforms*, 18: 363–68

Pentecost, A. 1981. The tufa deposits of the Malham district of North Yorkshire. *Field Studies*, 5: 365–87

Schneider, J. & Torunski, H. 1983. Biokarst on limestone coasts, morphogenesis and sediment production. *Marine Ecology*, 4: 45–63

Schneider, J. & Le Campion-Alsumard, T. 1999. Construction and destruction of carbonates by marine and freshwater cyanobacteria. *European Journal of Phycology*, 34: 417–26

Spencer, T. 1988. Limestone coastal morphology: The biological contribution. *Progress in Physical Geography*, 12: 66–101

Trudgill, S.T. 1976. The marine erosion of limestones on Aldabra Atoll, Indian Ocean. *Zeitschrift für Geomorphologie*, 26: 164–200

Trudgill, S.T. 1987. Bioerosion of intertidal limestone, Co. Clare, Eire. 3: Zonation, processes and form. *Marine Geology*, 74: 111–21

Viles, H.A. & Pentecost, A. 1999. Geomorphological controls on tufa deposition at Nash Brook, South Wales, United Kingdom. *Cave and Karst Science*, 26: 61–68

Further Reading

Viles, H.A. 1984. Biokarst: Review and prospect. *Progress in Physical Geography*, 8: 523–43

Viles, H.A. 1988. Organisms and karst geomorphology. In *Biogeomorphology*, edited by H.A. Viles, Oxford and New York: Blackwell

BIOLOGY OF CAVES

The biology of subterranean habitats (known as speleobiology, biospeleology, cave biology, or subterranean biology) is a research field incorporating all aspects of biological study of the biota from subterranean habitats and / or the study of subterranean habitats themselves. Speleobiology is not a coherent scientific discipline with its own methodology, rather it is a research field that relies upon the techniques and concepts of different biological disciplines (hence the term speleobiology), as well as from physical speleology (hence the traditional name biospeleology). Caving techniques have always been vitally important in providing biological samples during field investigations. However, caving or speleological knowledge is not a prerequisite for all speleobiological studies.

History

The first published account of specialized cave species was in 1541 when a Chinese author mentioned "hyaline fish living in the caves" in Yunnan (see Biospeleologists). These were described as *Synocyclocheilus diaphanus* in 1994. In 1689, Valvasor documented that a spring near Vrhnika in today's Slovenia brought animals from underground, which the locals proclaimed as dragon's young. The same creature was described as *Proteus anguinus* by J. Laurenti in 1768, also from an epigean or surface habitat. It was first described from caves by Jeršinovič von Loewengreif in 1797. In 1772, G.A. Scopoli wrote about fungi in old mine galleries, the first study of underground plants.

The serious search for cave animals began in the period 1830–45 with the first description of a cave beetle, drobnovratnik or slenderneck, *Leptodirus hochenwartii*, by L. Čeč in Postojna cave, Slovenia. After describing the species in 1832, F. Schmidt issued a large reward of 25 forints for the next specimen. Discoveries in the Slovenian karst continued with additional beetles and cave snails. In 1840, Močulski (Motschoulsky) reported cave beetles in the Caucasus, and soon after this the cave fish *Amblyopsis spelaeus* and a number of invertebrates were recorded in the United States by T. Tellkampf and colleagues.

The period 1845–1950 was characterized by the discovery of rich faunistic communities and surveys in several European countries, North and Central America, and New Zealand. Beside taxonomy and faunistic studies, some ecological data and the details of the structure of reduced organs in subterranean organisms was documented, and important theories regarding cave animals began to appear. One of the most important early biospeleologists was the Danish zoologist J.C. Schiödte, who investigated and published records regarding Slovenian cave fauna in the 1840s and 1850s and also made the first attempt at a general ecological classification of cave fauna. J.R. Schiner added a faunistics chapter to Schmidl's (1854) early speleological monograph concerning the same area. He criticized Schiödte's classification system and introduced the nomenclature ("Troglophilen, Troglobien"), which is essentially still in use today.

Additional important early biological contributions were made by Delarouzee in 1857 on cave beetles in Pyrenean caves; Bilimek in 1867 on cave fauna from Mexico, Chilton in 1882 on hypogean animals from New Zealand, and Apfelbeck who initiated systematic cave faunistic investigations in Bosnia and Herzegovina during the late 1880s. In 1871 Packard, a prominent neoLamarckian, wrote a monograph regarding the cave fauna of Mammoth Cave, Kentucky. In 1888, Jurinac published a doctoral thesis devoted to a local (Croatian) cave fauna, while D. Šoštarić investigated the fauna of local wells. The great monograph by O. Hamann, *Europaeische Hoehlenfauna* (1896) cited 400 references on speleobiological subjects. The first century of speleobiology was concluded with the foundation of the first underground biological laboratory, in the Paris catacombs, by A. Vire in 1897.

During the 1920s and 1930s, A. Remane discovered and recorded the existence of interstitial fauna in marine sands of the Nordsee and Ostsee and S. Karaman discovered and recorded the existence of interstitial fauna in fresh waters of Macedonia. Since this pioneering work, interstitial habitats and some other non-karstic hypogean habitats have been included in the research field of speleobiology.

The first half of the 20th century was marked by a number of great monographs, by Vire (1900), Eigenmann (1909), Spandl (1926), Stammer (1932), and Dudich (1932). The Romanian biologist E.G. Racovita (Racovitza, 1907) wrote his "Essai sur les problèmes biospéologiques", thus promoting the name for this research field, which had initially been proposed by Vire in 1904. He also initiated, accompanied mainly by his French colleagues R. Jeannel and P.A. Chappuis, systematic faunistic and taxonomic investigations named "Biospeologica". This resulted in the publication of over 80 monographs and papers between 1907 and 1962. One of the most important results of this period was a great synthetic, but not analytical, list of the world cave fauna (Wolf, 1934–38).

Following World War II an underground laboratory was established at Moulis, France, in 1954. The investigations in Moulis were enormously influential and initiated research in several biological areas and regions. Subsequently major advances have been made in experimental ethology, experimental genetics, population genetics, ecological genetics, physiology, molecular phylogenetics, and biogeography of subterranean organisms. Several comprehensive monographs on cave fauna and subterranean biology in general have been published, including those by Vandel (1965), Barr (1968), Ginet & Decou (1977), Camacho (1992) and Wilkens, Culver & Humphreys (2000). Monographs on specific themes include Gibert, Danielopol & Stanford (1994, mainly interstitial water); Riedl (1966, marine caves); Chapelle (1993, microbial life); Thines (1969, hypogean fishes); and Culver (1982, evolution and ecological genetics). In *Stygofauna Mundi* (Botosaneanu, 1986) there is a list of all stygobitic (i.e. aquatic) species with basic data on their distribution and ecology. The *Encyclopaedia Biospeologica* (Juberthie & Decou, 1994–2001) presents comprehensive reviews of taxonomic groups represented in subterranean habitats as well as the hypogean faunas of a number of countries.

The remainder of this entry reviews the contributions of some branches of biology and speleology, commencing with cladistics and cladistic techniques. These have remained controversial since their inception, but they have had some impact on studies of systematics and evolutionary histories of subterranean taxa. Among the first successful uses of cladistics to delineate the evolutionary history of subterranean species was that of Notenboom (1988) on the amphipod genus *Pseudoniphargus*. Notenboom used morphological characters to construct the cladogram, and molecular characters (especially mtDNA sequences). These have subsequently been extensively used on the spider genus *Nesticus* (e.g., Hedin 1997) and on the amphipod genus *Bactrurus* (Koenemann & Holsinger, 2001).

Subterranean fauna are distributed in all tropical and temperate regions, even close to northern glaciers. The diverity of fauna is richest in Mediterranean Europe, particularly in the Dinaric karst. In the tropics, this fauna is comparatively poor, particularly in terrestrial habitats. From the European Balkan Peninsula, more than 660 aquatic and more than 970 terrestrial obligate subterranean species have been recorded. In North America, whose interstitial fauna is poorly studied, 425 aquatic and 930 obligate terrestrial cave species (of which 350 are Coleoptera) are known; the very optimistic prediction is that over 6000 hypogean taxa exist (see America, North: Biospeology). In Europe around 2000 stygobitic taxa have been recorded, about 8% of the aquatic fauna, and this may rise to around 3000 in total. The richest aquatic group is Crustacea, and beetles (Coleoptera) are the richest terrestrial group. Except for beetles, insects are poorly represented relative to their abundance in epigean habitats. Recent advances in data acquisition and in digital mapping capabilities are leading to rapid increases in geographic analysis of distribution patterns.

Historical biogeographic investigations have been mainly devoted to the mechanisms and causes of subterranean colonization (see Colonization entry). Historically, caves were mainly presented as refugia of passively imported ancient elements, but it is increasingly stressed that "modern" invasive species actively penetrate and colonize hypogean habitats. Active colonization appears to be the most reasonable explanation, since any species will colonize subterranean habitats it is in contact with if it is able to survive and reproduce there. Some species may be isolated underground, but no species will successfully survive and reproduce when forced underground if its gene pool does not contain suitable genes (alleles). There have been attempts to estimate the age of cave populations by degree of eye reduction (see Adaptation: Eyes), as well as by the degree of heterozygosity and by molecular clock hypotheses for DNA sequence data. However, the former is invalid if the reducing mutations are not neutral and the latter may be wrong if a cave population is not the result of a single invasion wave. The results have been widely disparate estimates of the age of different groups and it is not possible at present to distinguish between real differences in age and the errors in the estimates themselves.

The process of cave-related evolution was historically often explained by "Lamarckian" or "neoLamarckian" processes. However, it has been shown that even the slightest "troglomorphisms", like partial skin depigmentation, are often genetically fixed, although discussion about the reasons for regressive evolution (e.g. development of depigmentation and eye reduction) still continues (see Evolution of Hypogean Fauna). The number of authors advocating selection (by economizing on resource use) is similar to the number advocating mutation pressure (reductive mutations being selectively neutral underground). A consensus exists about the selective—and adaptive—advantage of constructive troglomorphisms. The genetic and morphogenetic mechanisms of some reductions are gradually being clarified (e.g. Yamamoto & Jeffery, 2000). This demonstrates the possibility of some slight phenotypic influences of darkness on the visual / neural systems of subterranean organisms (Cooper *et al.*, 2001). Some authors even predict reduction of some genetic mechanisms (of mismatch repair: Langecker, 2000). The degree of adaptation and specialization in relation to the subterranean environment enables us to classify the species (or races) occurring in subterranean habitats. Some very elaborate systems have been developed but they are mostly useless since (1) for a great number of species we can only speculate about their presence / absence on the surface and (2) the latter may depend on local abiotic and biotic conditions (Sket, 1986). Schiner's system as modified by Holsinger & Culver (1988) and combined with the data about troglomorphism appears to be the best employable solution (see also Organisms: Classification entry).

The genetic structure of cave populations has been a topic of interest for decades, particularly because of the potential role of genetic drift in cave organisms. Among the earliest work on this problem was that of Kosswig on genetic polymorphism in *Asellus aquaticus*. Extensive genetic analysis of polymorphisms of soluble enzymes revealed that levels of genetic variation in invertebrate populations were not generally reduced related to surface populations. However, in vertebrates, genetic variability was often low, indicating small population size and an enhanced role for genetic drift. Sbordoni *et al.* (2000) also reported that populations could be differentiated on the scale of a few hundreds of metres.

Physiological investigations have been mainly directed to the measurement of metabolic rate (see also Adaptation: Physiology). It has been found that subterranean species generally have lower metabolic rates than epigean taxa, although there are exceptions. Low metabolic rate allows some cave animals to withstand extremely disoxic or poisonous conditions (Sket, 1986). However, we do not currently know if subterranean organisms can reproduce successfully in these environments. A physiological connection between restricted feeding and longevity has been proven (e.g. Tatar & Rand, 2002) and further studies will probably show whether there is any evolutionary connection between "starvation physiology" and the long lives of cave animals.

Regarding the sensory physiology and ethology of subterranean organisms, reactions to light have been thoroughly investigated. The sensitivity of some eyeless animals to light has been known for some time, but an interesting recent discovery suggests the "blindness" of some animals with well-developed eyes (Cooper *et al.*, 2001). Sensitivity and reactions to other stimuli has not been given enough attention. Some ethological mechanisms, like circadian rhythms and agressive behaviour are the subject of reduction in troglobites while a seasonal periodicity may be observed more often (Lamprecht & Weber, 1992). There have also been changes observed in some feeding behaviour patterns of cave fish while differences in the sexual behaviour unrelated to the cave environment (Juberthie-Jupeau, 1988) may have played a role in speciation.

Most research on the population ecology of cave animals has been concerned with modifications of life history traits that occurred as a result of adaptation to the cave environment. While some r-strategic traits, like progenesis, occur in some interstitial habitats (Coineau, 2000), the stable cave environment facilitated inhabitants a K life strategy with slower ontogeny, neoteny, low offspring numbers, and a long life span (Culver, 1982). This makes troglobites particularly weak competitors with epigean species in relatively food-rich habitats (e.g. if the subterranean habitat is slightly polluted; Sket, 1977), even if a cave species is not physiologically bound to the climatic characters of the subterranean environment. In contrast, cave species are good competitors in the food-poor environment of caves, and competition often involves predation because competitors are also potential food.

Karst hydrogeology is important to speleobiology because of the constraints it imposes on the distribution and movement of cave organisms. A more explicit connection was made by the French biologist Rouch (1986) who pioneered the concept of the karst basin as an ecosystem. Hydrogeology also plays a key role in anchialine habitats (see separate entry).

The search for chemoautotrophic organisms in caves has been the focal point of cave microbiological studies. Early workers, especially Caumartin (1963), were able to isolate chemoautotrophic bacteria from caves. More recent work has emphasized their importance in ecosystem energetics. Relying on a battery of stable isotope and biochemical tests, Sarbu, Kane & Kinkle (1996) were able to demonstrate convincingly that Movile Cave in Romania was a rare case of a chemoautotrophically based subterranean ecosystem (see Movile Cave and Microorganisms in Caves entries). Microbes also figure prominently in the many processes of deposition of minerals in caves (see Microbial Processes in Caves).

Finally, the need to conserve and protect subterranean animals has grown increasingly important. Protection of bats has attracted the most attention (Stebbings, 1988) because of the vulnerability of hibernating and maternity colonies to destruction. Protection of other species is of increasing concern because of the growth of problems of groundwater contamination and development (Sket, 1999; see also Conservation: Cave Biota).

BORIS SKET AND DAVID C. CULVER

See also **Subterranean Ecology; Subterranean Habitats**

Works Cited

Barr, T.C., Jr. 1968. Cave ecology and the evolution of troglobites. *Evolutionary Biology*, 2: 35–102

Botosaneanu, L. (editor) 1986. *Stygofauna Mundi*, Leiden: Brill

Camacho, A.I. (editor) 1992. *The Natural History of Biospeleology*, Madrid: Museo Nacional de Ciencias Naturales

Caumartin, V. 1963. Review of the microbiology of underground environment. *Bulletin of the National Speleological Society*, 25(1): 1–14

Chapelle, F.H. 1993. *Ground-water Microbiology and Geochemistry*, New York: Wiley

Coineau, N. 2000. Adaptations to interstitial groundwater life. In *Subterranean Ecosystems*, edited by H. Wilkens, D.C. Culver & W. Humphreys, Amsterdam and New York: Elsevier

Cooper, R.L., Li H., Long L.Y., Cole, J.L. & Hopper, H.L. 2001. Anatomical comparisons of neural systems in sighted epigean and

troglobitic crayfish species. *Journal of Crustacean Biology*, 21(2): 360–74
Culver, D.C. 1982. *Cave Life: Evolution and Ecology*, Cambridge, Massachusetts: Harvard University Press
Dudich, E. 1932. *Biologie der Aggteleker Tropfsteinhoehle "Baradla" in Ungarn* [Biology of the Aggtelek cave "Baradla" in Hungary]. Wien: Verlag Spelaeologisches Institut
Eigenmann, C.H. 1909. Cave vertebrates of America: A study in degenerative evolution. *Memoirs of Carnegie Institution Washington*, 104: 1–241
Gibert, J., Danielopol, D.L. & Stanford, J.A. (editors) 1994. *Groundwater Ecology*. London: Academic Press
Ginet, R. & Decou, V. 1977. *Initiation a la biologie at a l'ecologie souterraine* [Introduction to the subterranean biology and ecology], Paris: Delarge
Hamann, O. 1896. *Europaeische Hoehlenfauna. Eine Darstellung der in den Hoehlen Europas lebenden Tierwelt mit besonderer Beruecksichtigung der Hoehlenfauna Krains* [European cave fauna. A presentation of the fauna inhabiting caves of Europe, with a particular consideration of cave fauna of Carniola], Jena: Costenoble
Hedin, M.C. 1997. Molecular phylogenetics at the population/ species interface in cave spiders of the southern Appalachians. *Molecular Biology and Evolution*, 14: 309–24
Holsinger, J.R. & Culver, D.C. 1988. The invertebrate cave fauna of Virginia and a part of Eastern Tennessee: Zoogeography and ecology. *Brimleyana*, 14: 1–162
Juberthie, C. & Decou, V. (editors) 1994–2001. *Encyclopaedia Biospeologica* 3 vols., Moulis and Bucarest: Société de Biospeologie
Juberthie-Jupeau, L. 1988. Mating behaviour and barriers to hybridization in the cave beetle of the *Speonomus delarouzeei* complex (Coleoptera, Catopidae, Bathysciinae). *International Journal of Speleology*, 17: 51–63
Koenemann, S. & Holsinger, J.R. 2001. Systematics of the North American Subterranean amphipod genus *Bactrurus* (Crangonyctidae). *Beaufortia*, 51:1–56
Lamprecht, G. & Weber, F. 1992. Spontaneous locomotion behaviour in cavernicolous animals: The regression of the endogenous circadian system. In *The Natural History of Biospeleology*, edited by A.I. Camacho, Madrid: Museo Nacional de Ciencias Naturales
Langecker, T.G. 2000. The effects of continuous darkness on cave ecology and cavernicolous evolution. In *Subterranean Ecosystems*, edited by H. Wilkens, D.C. Culver & W.F. Humphreys, Amsterdam: Elsevier
Notenboom, J. 1988. Phylogenetic relationships and biogeography of the groundwater-dwelling amphipod genus Pseudoniphargus (Crustacea), with emphasis on the Iberian species. *Bijdragen to de Dierkunde*, 58(2): 159–204
Racovitza, E.G. 1907. Les problèmes biospéologiques. *Biospeologica I. Archives de Zoologie Experiméntale et Générale*, 4e série, 6(7): 371–488
Riedl, R. 1966. *Biologie der Meereshoehlen* [Biology of Sea-Caves]. Hamburg/Berlin: Parey
Rouch, R. 1986. Sur l'ecologie des eaux souterraines dans le karst. *Stygologia*, 2(4): 352–98
Rouch, R. & Danielopol, D.L. 1987. L'origine de la faune aquatique souterraine, entre le paradigme du refuge et le modèle de la colonisation active. *Stygologia*, 3(4): 345–72
Sarbu, S., Kane, T.C. & Kinkle, B.K. 1996. A chemoautotrophically based groundwater ecosystem. *Science*, 272: 1953–55
Sbordoni, V., Allegrucci, G. & Cesaroni, D. 2000. Population genetic structure, specation and evolutionary rates in cave-dwelling organisms. In *Subterranean Ecosystems*, edited by H.Wilkens, D.C. Culver & W. Humphreys, Amsterdam: and New York: Elsevier
Schmidl, A. 1854. *Zur Hoehlenkunde des Karstes. Die Grotten und Hoehlen von Adelsberg, Lueg, Planina und Laas* [On the cave science of Karst. Caves of Postojna, Predjama, Planina and Lož], with contributions from Dr. Alois Pokorny, Dr. J. Rud. Schiner and Wilhelm Zippe, Wien: Braumueller
Sket, B. 1977. Gegenseitige Beeinflussung der Wasserpollution und des Höhlenmilieus. *Proceedings of the 6th International Congress of Speleology, Olomouc 1973*, 5: 253–62
Sket, B. 1986. Ecology of the mixohaline hypogean fauna along the Yugoslav coast. *Stygologia*, 2(4): 317–38
Sket, B. 1999. The nature of biodiversity in hypogean waters and how it is endangered. *Biodiversity & Conservation*, 8(10): 1319–38
Spandl, H. 1926. *Die Tierwelt der unterirdiscehn Gewaesser* [The fauna of subterranean waters]. Wien: Verlag Spelaeologisches Institut
Stammer, H.-J. 1932. Die Fauna des Timavo. *Zoologische Jahrbuecher. Abteilung fuer Systematik*, 63: 521–656
Stebbings, R.E. 1988. *Conservation of European Bats*, Bromley, England: Christopher Helm
Tatar, M. & Rand, D.M. 2002. Dietary advice on Q. *Science*, 295(5552): 54–55
Thines, G. 1969. *L'evolution regressive des poissons cavernicoles et abyssaux* [The regressive evolution of cave- and deep sea fishes], Paris: Masson et Cie
Vandel, A. 1965. *Biospeleology: The Biology of Cavernicolous Animals*, Oxford: Pergamon Press
Vire, A. 1900. *La faune souterraine de France* [The subterranean fauna of France], Paris: Bailliere
Wilkens, H., Culver, D.C. & Humphreys, W.F. 2000. *Subterranean Ecosystems*, Amsterdam and New York: Elsevier
Wolf, B. 1934–1938. *Animalium cavernarum catalogus*. 's-Gravenhage: Junk
Yamamoto, Y. & Jeffery, W.R. 2000. Central role of the lens in cave fish eye degeneration. *Science*, 289: 631–33

BIOSPELEOLOGISTS

This entry provides short biographical sketches of researchers, now deceased, who were known for either "firsts" in biospeleological discoveries or for their influence on biospeleological ideas. The order is chronological, according to the time in which they made their major contribution to speleology.

In 1537, the Italian poet and scholar Giovanni Giorgio (GianGiorgio) Trissino (1478–1550) recorded what must have been a form of cave amphipod (probably *Niphargus*) from a cave in northern Italy. He also wrote detailed accounts of Italian caves. However, it is thought that the first printed report of a cave organism was by Yi-Jing Jie, who described an eyeless hyaline fish from the Alu caves of Yunnan in China in 1541. This species, *Sinocyclocheilus hyalinus*, was not scientifically described until 1994. The French engineer and mathematician Jacques Besson (c.1530–73) reported little eels (*petites anguilles*) in a cave stream somewhere in Europe in 1569. However, he did not give a locality, and neither did he describe the fish as blind or depigmented (Romero & Lomax, 2000).

Athanasius Kircher (1602–80), a German-born Jesuit polymath, wrote *Mundus Subterraneus* in 1665, probably the first printed work on earth science to include speleology. There he enumerated numerous cave organisms, including mythical beasts such as dragons, although none of the animals that he mentioned displayed the typical features associated with troglomorphic animals: blindness and depigmentation. (Romero, 2000). In 1674, the English naturalist Martin Lister (1639–1712) published the earliest reference to underground fungi.

The Slovenian historian and naturalist Janez Vajkard [Johann Weichard] Valvasor (1641–93) published the first systematic study of caves in a particular area, and explained the mystery of disappearing karst lakes by a system of underground rivers and reservoirs (see also Speleologists). In his 1689 book *The Glory of the Duchy of Carniola*, Baron Valvasor described the discovery of the blind cave salamander *Proteus anguinus* at a karst spring near Vrhnika. This was the first troglobite to be scientifically described and named, by the Viennese zoologist Josephi Nicolai Laurenti in 1768.

In 1748, Marc-René Marquis de Montalembert (1714–1800), a French aristocrat, reported a blind, subterranean fish in a spring at Gabard, Angoumois, near one of his estates in southwestern France in 1748. However, no drawings or specimens were preserved and his description remains unconfirmed (Romero, 1999a).

The first study of underground plants was published in 1772 by the Italian naturalist Giovanni Antonio Scopoli (1723–88) (Scopoli also sent specimens of *Proteus anguinus* to other researchers in Europe, and it was one of these specimens that Laurenti examined). Green plants in the mines of Freiburg that had not fully developed due to lack of light were described in 1793 by the German naturalist and geographer Alexander von Humboldt (1769–1859). In 1799 he visited Guácharo Cave in Venezuela, and described the nocturnal oilbird, *Steatornis caripensis*, flocks of which emerge from the cave at night. In 1805 he published a description of a freshwater species of catfish ejected from an underground volcano in Quito, Ecuador, although this discovery remains unsubstantiated (Romero & Paulson, 2001).

Karl Franz Anton von Schreibers (1775–1852) performed the first detailed anatomical studies of *Proteus anguinus*, on one of Scopoli's specimens stored in the Vienna Natural History Museum. He also bred *Proteus* in artificial underground caves.

In 1842, August Otto Theodor Tellkampf (1812–83), a German physician who emigrated to America, visited Mammoth Cave, and later described several species of invertebrates from the cave fauna. He also contributed detailed descriptions of the northern cavefish *Amblyopsis spelaea* and concluded that its eyes and those of blind cave crayfishes had become rudimentary from lack of use (Romero, 2002)

Jean Louis Rodolphe Agassiz (1807–73), a Swiss anatomist who emigrated to the United States and served as Director of the Harvard Museum of Comparative Zoology from 1859 to 1873, in 1847 proposed a plan to raise individuals of *A. spelaea* under different light conditions, to study the influence of illumination on its eyes and pigmentation. A creationist, he thought only of the effects of the environment on development, not of the environment influencing evolution. He never carried out the experiments, but several of his students also showed a great deal of interest in cave fauna, and later developed his ideas (Romero, 2002).

The Danish naturalist Jørgen Matthias Christian Schiödte (1815–84) was particularly interested in the correlations between anatomical characteristics and the biological conditions under which organisms live. Schiödte's 1849 work: *Specimen faunæ subterraneae* provided the first classification of cave animals: shade animals, twilight animals, animals in the dark zone, and animals living on stalactites.

Jeffries Wyman (1814–74), the first curator of the Peabody Museum of Yale University, studied *A. spelaea* in great detail. Although it lacked eyes, Wyman found that it had well-developed optic lobes. He proposed in 1854 that this imperfection of the eyes "might be owing to a want of stimulus through a series of generations" and that the organ of vision, however imperfect, "is more like the eyes of other vertebrates". For him, *A. spelaea* was an excellent subject for the study of evolution (Romero, 2001).

In 1854, Ignaz Rudolph Schiner (1813–73) classified cave organisms according to their degree of dependence on the underground environment: troglobites, troglophiles, and "occasional cavernicoles" (trogloxenes). This classification was later modified by Emil Racovitza, and is commonly used today.

Charles Darwin (1809–82) believed that cave fauna supported his theory of evolution. He noted that cave fauna were more closely related to the fauna of the surrounding regions than elsewhere. He first considered the mechanisms of both natural selection and disuse to explain troglomorphic features, i.e. enlargement of some sensory systems and appendages in the former and blindness and depigmentation in the latter. To him this suggested a "contest ... between selection enlarging and disuse alone reducing these organs" (Darwin, *On the Origin of the Species by Means of Natural Selection*, 1859). However, by the third edition of his book in 1861 he relegated the importance of natural selection, eliminating his speculation about a "contest" between selection and disuse.

Alpheus Hyatt (1838–1902), a student of Agassiz, visited Mammoth Cave in 1859, and collected cave fauna. He used the specimens to demonstrate that here were animals whose phylogenetic lineage had become "old" (Romero, 2001).

In 1864, the American paleontologist and evolutionist Edward Drinker Cope (1840–97) described what he thought to be a new species and genus of troglomorphic fish, *Gronias nigrilabris*, from Pennsylvania, without presenting any evidence that such a fish had been captured in the hypogean environment (Romero, 1999b; Romero & Romero, 1999). Cope introduced the idea that evolution was directed by trends, and that when an organism travelled too far down an adaptively specialized path, as in adaptation to living in caves, it could not reverse this process. He also described the life cycles of the cave fauna of Wyandotte Cave, Indiana.

In 1871, Alpheus Spring Packard Jr (1839–1905), another former student of Agassiz, examined Mammoth Cave specimens,

Biospeleologists: Athanasius Kircher (top left); Emil G. Racovitza (top right); René Gabriel Jeannel (bottom left); Albert Vandel (bottom right).

and the fauna, particularly the fish, convinced him of their usefulness as a demonstration of Lamarckian evolution. He thought that cave fauna had a very recent origin and that the loss of certain organs was compensated for by the hypertrophy of others.

Frederic Ward Putnam (1839–1915), another of Agassiz's students, visited Mammoth Cave in 1871, and in 1874 collected a large number of specimens. He criticized Cope's interpretation that *A. spelaea* was able to survive in those waters because of its "projecting under jaw and upward direction of the mouth [which] renders it easy for the fish to feed at the surface of the water", and presented numerous examples that contradicted Cope's assertions. He viewed the amblyopsids as former marine and salt-water estuary fishes that were slowly trapped in that geographical area. He also described a new species of amblyopsid, *Chologaster agassizi*.

Edward Ray Lankester (1847–1929), a British neo-Darwinist, proposed that blindness among cave animals was due to a special type of natural selection. In an 1893 article in *Nature*, he suggested that some animals are, by chance, born with defective eyes. Occasionally a few animals—some with normal eyes and some with defective eyes—fall or are swept into caves. In each generation, those that have good eyes are able to see the light and escape, and eventually only those that are blind remain in the cave. He also believed that one can find organisms degenerating ontogenetically and phylogenetically. In his 1880 book *Degeneration: A Chapter in Darwinism* he defined "degeneration" as "a loss of organization making the descendant far simpler or lower in structure than its ancestor".

Carl H. Eigenmann (1863–1927) was a German-born ichthyologist who studied and worked in the United States. Much of his work between 1887 and 1909 was devoted to understanding how visual structures were lost in cave vertebrates. He described two new species of cave fish: *Amblyopsis rosae* from Missouri, and *Trogloglanis pattersoni* from the artesian waters of San Antonio, Texas. He published more than 40 papers, abstracts, and books on cave fauna—his *Cave Vertebrates of North America* (1909) being the most important. Originally a neo-Lamarckian, Eigenmann thought that the reduction or disappearance of organs among cave animals was a case of convergent evolution. He pointed out that a lack of pigmentation had to be understood as the combination of genetically fixed and epigenetically (environmentally influenced) determined characters (Romero, 1986b). He believed that cave faunas were not the result of "accidents", but rather the product of active colonization.

In 1896, Armand Viré (1869–1951), a French scientist and cave explorer attached to the Muséum National d'Histoire Naturelle, set up the world's first underground laboratory, which operated until 1914 in the catacombs of the Botanical Garden in Paris. He published several articles on European cave fauna, based on his visits to caves.

Emil G. Racovitza (Racoviță) (1868–1947) was a Romanian zoologist who worked in France until returning home in 1920. His *Essai sur les problèmes biospéléologiques* (1907) is considered to mark the birth of biospeleology as an independent science. He initiated an extensive international research programme called "Biospeologica" (primarily intending to document and collect cave fauna). In 1920 he founded the world's first Speleological Institute in Cluj, Romania. He explored 1200 caves in Europe and Africa, collected about 50 000 specimens of cave animals, and published 66 papers on subterranean fauna. For Racovitza, an anti-neo-Darwinist, all cave organisms were "preadapted" to the cave environment where "the function creates the organ." (Motas, 1962).

The French entomologist René Gabriel Jeannel (1879–1965) studied subterranean beetles mostly from Europe and Africa. With Racovitza he founded the journal *Biospéologica* in 1907, became deputy manager of the Speleological Institute in Cluj, and in 1926 published *Faune cavernicole de la France*.

Charles Marcus Breder Jr (1897–1983), the director of the New York Aquarium, led the renaissance of the study of cave fishes by using *Astyanax fasciatus* as his prime subject of research. He emphasized behavioural, physiological, genetic, and ecological approaches (Romero, 1984; Romero, 1986a; Romero, 2001). Many of Breder's contributions are still cited, and several of his associates and students have studied this cave fish. Breder was the dominant figure in hypogean fish research in the 1940s and 1950s.

Curt Kosswig (1903–82), a German scientist who spent many years in exile in Turkey, used the results of his own Mendelian studies to explain the "rudimentation" or loss of structures among cave animals. He founded the cave fish team at the Zoological Museum and Zoological Institute at the University of Hamburg.

Albert Vandel (1894–1980) founded the Laboratoire souterrain de Moulis in 1948. In his highly influential book on cave biology (*Biospéologie: La Biologie des Animaux Cavernicoles*, 1964) he proposed that the evolution of cavernicoles was neither neo-Lamarckian nor neo-Darwinism, but which he called "organicist". According to him, all phyletic lines pass through several successive stages: creation; expansion and diversification; and finally specialization and senescence. The last stage of this cycle was "regressive or gerontocratic" evolution. He considered cavernicoles to be good examples of regressive evolution. However, his ideas lack empirical support.

ALDEMARO ROMERO

Works Cited

Motas, C. 1962. Emil G. Racovitza: Founder of biospeleology. *National Speleological Society Bulletin*, 24: 3–8

Racovitza, E.G. 1907. Éssai sur les problèmes biospéléologiques. *Archives du Zoologie Expérimentale et Générale*, 6: 371–488

Romero, A. 1984. Charles Marcus Breder, Jr. 1897–1983. *National Speleological Society News*, 42: 8

Romero, A. 1986a. Charles Breder and the Mexican blind characid. *National Speleological Society News*, 44: 16–18

Romero, A. 1986b. He wanted to know them all: Eigenmann and his blind vertebrates. *National Speleological Society News*, 44: 379–81

Romero, A. 1999a. The blind cave fish that never was. *National Speleological Society News*, 57: 180–81

Romero, A. 1999b. Myth and reality of the alleged blind cave fish from Pennsylvania. *Journal of Spelean History*, 33: 67–75

Romero, A. 2000. The speleologist who wrote too much. *National Speleological Society News*, 58: 4–5

Romero, A. 2001. Scientists prefer them blind: the history of hypogean fish research. *Environmental Biology of Fishes*, 62: 43–71

Romero, A. 2002. The life and work of a little known biospeleologist: Theodor Tellkampf. *Journal of Spelean History*, 36(2): 68–76

Romero, A. & Lomax, Z. 2000. Jacques Besson, cave eels and other alleged European Fishes. *Journal of Spelean History*, 34: 72–77

Romero A. & Paulson, K.M. 2002. Humboldt's alleged subterranean fish from Ecuador. *Journal of Spelean History*, 35(2): 56–59

Romero, A. & Romero, A. 1999. On Cope, caves, and skeletons in the closet. *National Speleological Society News*, 57: 341–43

Further Reading

Barr, T.C. 1966. Evolution of cave biology in the United States, 1822–1965. *National Speleological Society Bulletin*, 28: 15–21

Shaw, T.R. 1992. *History of Cave Science: The Exploration and Study of Limestone Caves, to 1900*, 2nd edition, Broadway, New South Wales: Sydney Speleological Society

Useful Websites

Biographical information on active biospeleologists can be found in web pages such as: http://members.xoom.it/bioscience/isbios/XXX.htm and http://www.cancaver.ca/bio/people.htm.

BLUE HOLES OF THE BAHAMAS

Blue holes are karst features from the Bahama islands which have been documented for over 100 years. The Bahamas comprise an archipelago consisting of 29 large islands and 661 small islands (called "cays"), extending from near the Florida peninsula southeastwards for 1400 km to Hispaniola. Blue holes are water-filled vertical openings in the carbonate rock of the islands, which exhibit complex morphologies, ecologies, and water chemistries (Figure 1). Their deep-blue colour is a result of a water depth greater than 10 m. Blue holes are the composite result of carbonate deposition and dissolution cycles, which have been controlled by glacial sea-level fluctuations. As a result, blue holes are polygenetic in origin, forming by three possible mechanisms:

1. Glacial melting and sea-level rise can form blue holes by drowning dolines and shafts developed in the vadose zone during sea-level lowstands (Figure 2A);
2. Deep dissolution voids are produced in the phreatic zone of carbonate platforms (Figure 2B). Exploratory drilling has encountered large voids at depths of 21 to 4082 m; the deepest of these voids was large enough to accept 2430 m of broken drill pipe. These voids can collapse towards the surface to create a blue hole—a process called progradational collapse (Mylroie et al., 1995) or aston collapse (Palmer, 1986);
3. The Bahama islands are steep-walled platforms, and there is abundant evidence of fracture along the margins of these platforms. These fractures form parallel to the coast, commonly in an "en echelon" fashion, to create graben-like features (Figure 2C). Where these fractures are open to the surface, it is possible to enter the extensive and deep blue-hole systems (Palmer, 1986).

Blue holes have been widely studied by cave divers over the last 30 years, initially by George Benjamin (1970). They have provided information regarding karst processes, global climatic change, marine ecology, and carbonate geochemistry. The term "blue hole" has been used in a variety of ways, but they are best defined as: "subsurface voids that are developed in carbonate banks and islands; are open to the Earth's surface; contain tidally-influenced waters of fresh, marine, or mixed chemistry; extend below sea-level for a majority of their depth; and may provide access to submerged cave passages." (Mylroie et al. 1995, p.225).

Blue holes are found in two settings: ocean holes open directly into the present marine environment and contain marine water, usually with tidal flow; inland blue holes (Figure 1) are isolated by the present-day topography from marine conditions, and open directly on to the land surface or into an isolated pond or lake, and contain tidally influenced water of a variety of chemistries, from fresh to marine. The most common alternative use of the term "blue hole" is to describe large and deep karst springs.

Blue holes are widely distributed in the Bahamas. While the largest and best-known blue holes are found on North Andros,

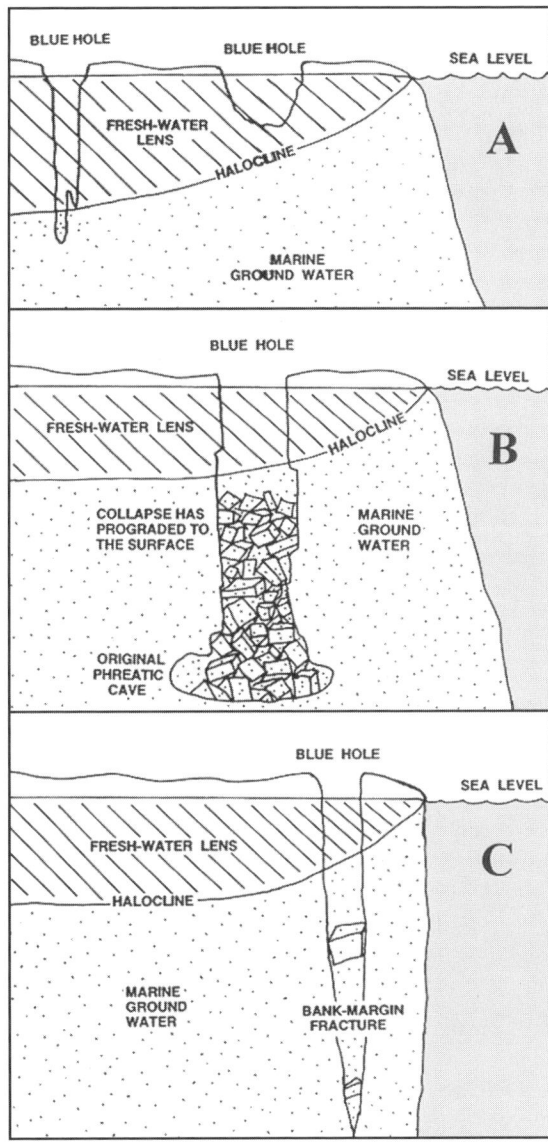

Blue Holes of Bahamas: Figure 1. Watling's Blue Hole, San Salvador Island, Bahamas. (Photo by John Mylroie)

Blue Holes of Bahamas: Figure 2. Dolines and vadose shafts produced during sea-level lowstands are drowned by sea-level rise to produce blue holes (A). Phreatic voids produced deep in the platform, possibly within the freshwater lens during sea-level lowstands, form blue holes when the collapse reaches the surface, and sea-level rises (B). Fracture of the steep bank margins of the Bahamas creates voids that become blue holes (C).

South Andros, and Grand Bahama islands in the northwestern Bahamas, blue holes are found on islands all the way to Great Inagua Island at the southeastern end of the archipelago. In the northwestern Bahamas, blue holes with depths in the 100–125 m range are common, and it was thought that their depth was limited by the position of the lowest glacial sea-level lowstand, which was 125 m below present sea-level. However, exploratory wells commonly intersect voids below that depth, and Dean's Blue Hole on Long Island in the southeastern Bahamas was found to be over 200 m deep, ending in a large chamber. Blue holes commonly lead into major horizontal cave systems, such as Lucayan Caverns (9184 m long) on Grand Bahama Island, and Conch Blue Hole (1153 m long) on North Andros Island.

Blue holes are most often thought of in terms of the Bahamas, but similar features are found worldwide. The cenotes of the Yucatán region of Mexico are very similar (see Yucatán Phreas, Mexico), as they are also water-filled shafts that commonly lead to extensive cave systems underwater. Cenotes with many features similar to blue holes are also found in the Mount Gambier region of southern Australia, well away from the coast. Florida's extensive karst contains many sinks and holes that lead into water in a manner analogous to blue holes, and southern Africa has deep, flooded shafts (Farr, 1991). However, the Bahamian blue holes appear to be unique in that they have strong currents that oscillate with the tide. The currents can be so strong that even the fittest scuba diver can be overcome. The reversing flow of blue holes has been viewed as an aspect of subplatform water flow in the carbonate banks of the Bahamas; flow in some blue holes does not reverse, but is a continuous discharge (Whitaker & Smart, 1997). The reversing flow of blue holes does not appear to be a factor in their formation, but rather an effect of their forming under other conditions and being inherited as pathways for tidal and groundwater flow patterns.

The reversing flow of ocean blue holes creates an important biological environment. The ebb and flow brings nutrients and prey into the blue holes, which support a variety of organic life, and the rock walls of the blue holes are an important hardground for organisms that need to attach themselves to a solid substrate in order to survive. Deeper in the blue holes, there are a number of cave-adapted organisms, such as *Lucifuga speleotes*, the blind Bahamian cave fish. More significantly, a new class of Crustacea, Remipedia, was discovered in the Bahamian blue holes in the late 1970s (Yager, 1981). Inland blue holes commonly have fresh water overlying marine water, to produce a halocline. The collection of organic material at this horizon creates a food source exploited by a variety of organisms. Excess accumulation of organics can lead to anoxic conditions, and the development of an ecology based on chemosynthesis by reducing bacteria.

JOHN MYLROIE

See also **Caribbean Islands; Speleogenesis: Coastal and Oceanic Settings; Yucatán Phreas, Mexico**

Works Cited

Benjamin, G.J. 1970. Diving into the blue holes of the Bahamas. *National Geographic Magazine*, 138: 347–63
Farr, M. 1991. *The Darkness Beckons*, London: Diadem Books
Mylroie, J.E., Carew, J.L. & Moore, A.I. 1995. Blue holes: Definition and genesis. *Carbonates and Evaporites*, 10(2): 225–33
Palmer, R. 1986. The blue holes of South Andros, Bahamas. *Cave Science*, 13(1): 3–6
Whitaker, F.F. & Smart, P.L. 1997. Hydrogeology of the Bahamian Archipelago. In *Geology and Hyrdogeology of Carbonate Islands*, edited by H.L. Vacher & T.M. Quinn, Amsterdam and Oxford: Elsevier Science
Yager, J. 1981. Ramipedia, a new class of Crustacea from a marine cave in the Bahamas. *Journal of Crustacean Biology*, 1(3): 328–33

Further Reading

Cave Science 11(1), 1984, is a special issue devoted to Bahamas blue holes.
Cave Science 12(1), 1985, *Cave Science*, 13(1), 1986, and *Cave Science*, 13(2), 1986 each contain three Bahama blue hole articles.
Cave and Karst Science, 25(2), 1998 is a memorial issue on Bahama blue holes, published to honour Rob Palmer, who died in a scuba accident in the Red Sea in 1997.
Exley, S. 1994. *Caverns Measureless to Man*, St Louis, Missouri: Cave Books.
Sheck Exley finished this book a short time before dying in a cave diving accident in Mexico in 1994. He undertook much early blue-hole diving work in the Bahamas; his death emphasizes the dangers of cave diving.
Palmer, R. 1985. *The Blue Holes of the Bahamas*, London: Jonathan Cape
Palmer, R. 1989. *Deep into Blue Holes*, London: Unwin Hyman; revised and updated 1997, Nassau, Bahamas: Media Publishing

BOA VISTA, TOCA, BRAZIL

Toca da Boa Vista is the longest cave in the Southern Hemisphere—a maze of dry passages over 100 km long that seems to extend interminably beneath the semi-arid terrain of northeastern Brazil. The cave appears to represent a rare (and by far the longest) example of a cave generated by sulfuric acid produced within dolomite through the oxidation of sulfide. Toca da Boa Vista contains also significant fossil remains and calcite deposits that have yielded important paleoenvironmental information.

Toca da Boa Vista is located close to the village of Laje dos Negros (municipality of Campo Formoso) in the dry lands of northern Bahia State (located on Figure 1 of the America, South entry). Mean rainfall is about 480 mm a^{-1}, with considerable interannual variability. The cave is contained within dolomites of the Salitre Formation, Una Group. This Precambrian carbonate sequence (see America, South) extends over much of the northern half of the state of Bahia, generating major karst features around Andaraí, Iraquara, and in the Jacaré River valley. However, the better-developed underground karst occurs near Laje dos Negros, where the two longest caves in Brazil are located close by. Toca da Boa Vista has 102 km of mapped passages (as at January 2003), while Toca da Barriguda is 28.7 km long (see figure). They are separated by a mere 700 m, and undoubtedly share a common genetic history, although some morphological differences exist between them, such as the large rooms of Toca

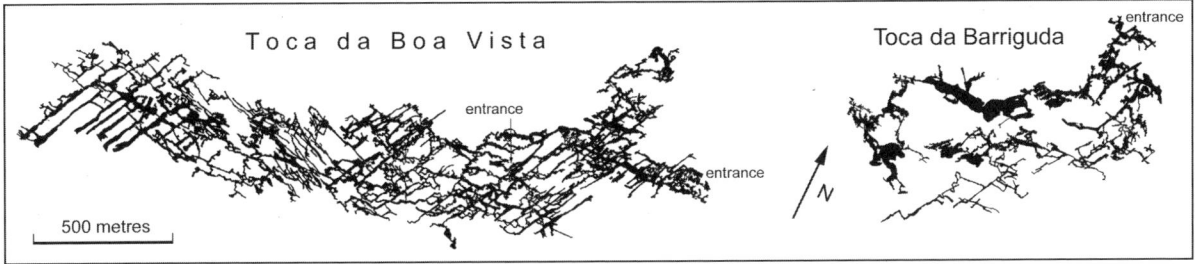

Boa Vista, Toca, Brazil: Outline map of Toca da Boa Vista and Toca da Barriguda, drawn in true relationship to each other. Survey by Grupo Bambuí de Pesquisas Espeleológicas.

da Barriguda. A connection between these two major caves is likely in the future, and would result in a mega-system among the top 10 longest caves of the world.

Toca da Boa Vista is a two-dimensional and roughly horizontal maze that undulates in profile, reaching a maximum depth of 60 m below the surface. Although the cave plan is strongly suggestive of joint control, most passages do not appear to follow any joint at ceiling level, and neither can an initiation horizon or conspicuous bedding plane be observed in the walls. However, the cave floor is almost always obscured by breakdown or sedimentation. The cave shows abrupt variations of cross-section along its passages and lacks vadose features and common fluvial forms such as scallops or meanders. Cupolas, sharp bedrock projections, and cuspate marks abound, and the walls (and speleothems) are heavily corroded by condensation corrosion (see Condensation Corrosion), yielding dolomitic sand. The caves show no relationship to the surface, which is basically flat ground without noticeable karst features. Wall-rock dissolutional phreatic features can be followed up to the cave entrances, showing that the cave passages used to extend above the present surface elevation, having been truncated by denudational lowering.

Toca da Boa Vista shares many of the features of deep-seated hypogenic caves of sulfuric acid origin (see entries on Carlsbad and Lechuguilla Caverns; Speleogenesis: Deep Seated and Confined Settings), although it lacks massive gypsum deposition and a three-dimensional pattern. The local geology does not favour a deep-seated origin because the carbonates were deposited on a craton stabilized in the Precambrian, and no deep sources of acidity (such as hydrocarbon deposits or volcanic emissions) are known to occur. However, the carbonate sequence of the Una Group is known to host massive deposits of sulfide. The oxidation of pyrite beds within the dolomite is a possible source of acidity. Because the pyrite would be oxidized by groundwater, it should not be directly observed in cave walls. Gypsum deposits, due to their high solubility, would be easily removed by later phreatic water flow. Hydrochemical analysis of the Una Group aquifer demonstrates that there is a correlation between groundwater sulfate and bedrock Ca and Mg, indicating that sulfate is being produced by bedrock–water reactions. The lack of sulfate beds within the carbonate suggests that sulfide oxidation is the most likely mechanism for sulfate generation in the area.

Large-scale speleogenesis through metal sulfide oxidation has not been recognized elsewhere in the world. The two-dimensional, mostly horizontal nature of the cave conforms with stratabound deposits of sulfide described in other areas of the Una Group. Such "shallow" hypogenic speleogenesis exhibits a stronger lithological control than either deep-seated or epigenic caves, because the source of the acidity lies within specific carbonate units. Sharp morphological changes are likely to reflect heterogeneities in the distribution of the sulfide.

While the ages of Toca da Boa Vista and Toca da Barriguda exceed the limit of the U-series method, minimum ages for the caves have been obtained through two magnetically reversed fine-grained sediment deposits. It is highly likely that these two sediments represent distinct reversal episodes, yielding a minimum age of 1.07 million years, although it is very probable that the caves started forming earlier in the Tertiary.

Subaqueous speleothem deposition is widespread in certain areas, representing a well-marked former lake level about 13 m above the present water table. Uranium-series ages of shelfstone and subaqueous coralloids, together with dating of surface travertines, have allowed the recognition of a wetter period from the last glacial maximum to the beginning of the Holocene (Auler & Smart, 2001). Remarkably complete and very well-preserved fossil remains have been located in the areas close to the entrances. Some of these fossils represent the first complete skeletons of certain Pleistocene extinct species, such as an extinct cave dog (*Protocyon troglodytes*), a small bear (*Arctotherium brasiliense*), and a large monkey (*Protopithecus brasiliensis*) (Hartwig & Cartelle, 1996). A complete skeleton of a new species of large monkey (*Caipora bambuiorum*) has also been found among the subaqueous deposits of the cave (Cartelle & Hartwig, 1996). The age of the fossils has been constrained through ^{230}Th / ^{234}U dating of calcite layers associated with the bones. The interval of deposition spans much of the last glacial period, but also includes a few older deposits.

The cave does not face any immediate threat, due to its isolated location and difficult access. Its high internal temperature (around 28°C) and the intricate nature of the passages, do not favour commercialization. No special measures have been taken to preserve this outstanding speleological site.

AUGUSTO AULER

See also **America South; Sulfide Minerals in Karst**

Works Cited

Auler, A.S. & Smart, P.L. 2001. Late Quaternary paleoclimate in semiarid northeastern Brazil from U-series dating of travertine and water-table speleothems. *Quaternary Research*, 55: 159–67

Cartelle, C. & Hartwig, W.C. 1996. A new extinct primate among the Pleistocene megafauna of Bahia, Brazil. *Proceedings of the National Academy of Sciences USA*, 93: 6405–09

Hartwig, W.C. & Cartelle, C. 1996. A complete skeleton of the giant South American primate Protopithecus. *Nature*, 381: 307–11

BOOKS ON CAVING

Historical introduction

It is difficult to state the exact number of caving books printed worldwide each year but it ranges between 100 and 200, representing less than one tenth of the annual page output of caving journals (see separate entry). However, their history is far longer, beginning over 2000 years ago. Most of the early references to caves are very short sentences contained in books dedicated to completely different topics (from medicine to poetry, from philosophy to science). Thus, the first known publication completely dedicated to caves was a four-page pamphlet printed in 1535 by Berthold Buchner concerning the exploration of Breitenwinner Höhle in Germany: the single copy still existing is conserved in the State Library of Ulm. In 1665 A. Kircher printed the first edition of *Mundus Subterraneous*, which is generally accepted as the first scientific book on caves and karst (Figure 1). A few years later J.W. Valvasor published the four volumes of his *Die Ehre des Herzogthums Crain* (1689), in which his explorations of several caves in the Karst region are described and the blind cave salamander *Proteus anguinus* is first recorded as a "baby dragon".

Up to the first half of the 19th century only a few hundred speleological books were printed. Many were just reports of caving explorations but some dealt with hydrogeology and speleogenesis (Vallisnieri, 1715), cave morphology (Cappelleri, 1757), cave mineralogy (Zimmermann, 1788), or biology (Configliachi, 1819).

At the end of the 18th and at the beginning of the 19th century tourist guides to famous caves, often enriched by beautiful engravings, became rather popular books: the first general book on the show caves of the world (Lang, 1801) reports on famous caves from Germany, Belgium, England, Portugal, and Greece.

The birth of modern speleology should date from the publication of *Die Grotten und Höhlen von Adelsberg, Lueg, Planina und Laas* (Schmidl, 1854), the first speleological treatise dealing with different caving topics from biology to mineralogy, and from cave exploration to mapping. Later, in 1882, *Celebrated American Caverns* (Hovey, 1882) became the first important caving book printed outside Europe. The caver who most improved caving literature was E.A. Martel, who almost single-handedly brought speleology to the attention of the wider public. In his lifetime he printed over 700 articles and over 20 books on caving topics (see Chabert & Courval, 1971), the most famous being *Les Abîmes* (1894) (Figure 2) and *Nouveau traité des eaux souterraines* (1921). Thanks to Martel's activities, from the end of the 19th century several books on caves and speleology began to be printed worldwide. The Austrian G. Kyrle also contributed

Books on Caving: Figure 1. Title page of *Mundus Subterraneous* (Kircher, 3rd edition, 1678).

Books on Caving: Figure 2. Cover page of *Les Abîmes* (Martel, 1894).

with outstanding papers: in particular his *Grundriss der Theoretischen Speläologie* (1923) should be mentioned here.

Present situation

Nowadays caving books are printed in all five continents. Their subjects include the different scientific fields involved in the underground world, novels and humorous books, technical manuals, show cave guides, cave inventories, and general or local histories of caving activity. Over 60% of published caving books describe explorations, either in the home area or overseas. Of the former, one of the first was the account by Lawrence & Brucker (1955) of the exploration of Floyd Collins Crystal Cave and this was notable as it was produced by a major US publisher. Similarly, *Ghar Parau* (Judson, 1973), one of the first books to tell the story of an overseas expedition, was produced by a major publisher and printed in both Britain and the United States. In contrast, most exploration reports are printed by a single caving club or association. Despite the high quality and general interest of many of these publications, they are not easily accessible outside the country or even the region in which they were printed. Exceptions are represented by a few books such as *Lechuguilla: Jewel of the Underground* (Taylor, 1991) and *Río La Venta: Treasure of Chiapas* (Badino *et al.*, 1999).

Although a few details on the techniques of cave exploration were printed in the first half of the 20th century (Boegan & Bertarelli, 1926), the extremely fast development in exploration techniques experienced from 1960 to 1980 provided an incentive for the production of several technical manuals, the most well known being *Techniques de la spéléologie alpine* (Marbach *et al.*, 1973), which was printed in four different editions from 1973 to 2000 and *On Rope* (Padgett & Smith, 1996).

A popular category of caving books amongst both cavers and non-caving scientists is represented by the general treatises on speleology, which give an overview on the different scientific fields related to caving. The first such book was *La Spéléologie* (Martel, 1900) and practically every country in the world, where speleology exists, has its own manual. Milestones in this field are still *British Caving* (Cullingford, 1953) and *The Science of Speleology* (Ford & Cullingford, 1976).

Some 3–5% of cave books are inventories, although most of these only have a regional distribution. Exceptions are those that give a general worldwide overview such as the *Atlas of the Great Caves of the World* (Courbon *et al.*, 1989). Show cave guides, which represent about 20% of the caving books, normally have high print runs (up to 20 000) but never have a really wide circulation among cavers, being always confined to the tourist audience.

In the last 20 years or so, a growing category of caving book has been the monograph dealing with research in a specific field of scientific speleology. These books normally have a diffuse circulation owing to their being of interest to both specialized researchers and cavers. It is impossible to cite even a small fraction of these monographs, though those surely worthy of mention include *Paleokarst* (James & Choquette, 1988), *Speleogenesis* (Klimchouk *et al.*, 2000), and the *Encyclopaedia Biospeologica* (Juberthie & Decu, 1994). Probably the most successful monograph with worldwide distribution was *Cave Minerals* (Hill, 1976) and the later *Cave Minerals of the World* (Hill & Forti, 1997).

In contrast to research monographs, there have been a small number of books by experienced cavers that are aimed at the general public and which try to give an impression of caves and cavers. Early examples were Casteret's *Dix ans sous terre* and *Au fond des gouffres* published in English as *Ten Years Under the Earth* (Casteret, 1938). *Depths of the Earth* (Halliday, 1966) sold widely in the United States and elsewhere while *Radiant Darkness* (Bögli & Franke, 1965) and *The World of Caves* (Waltham, 1976) were notable for extensive use of large, spectacular, colour photographs. Closer to the Casteret tradition and introducing the general public to the sport of cave diving, was *The Great Caving Adventure* (Farr, 1984). Finally, brief mention must be made of the small group of publishers that specialize in caving books. These include Speleo Projects in Switzerland, Zephyrus Press (United States) which reprinted, or printed English editions of, many famous caving books, and Cave Books (United States), a division of the Cave Research Foundation which publishes a wide range of material from expedition reports to fiction. Many national caving bodies such as the NSS and BCRA also have their own publishing branches.

How to get information

Cave books published prior to 1900 are extremely rare and often expensive, so they are hard to find. One of the largest collections of old caving books is in the F. Anelli Cave Documentation Centre, Bologna (Italy), where over 1250 publications from 1590 to the first years of 1900 are available for direct consultation. Relevant bibliographic information may be extracted from *History of Cave Science* (Shaw, 1992) and *A Guide to Speleological Literature of the English Language 1794–1996* (Northup *et al.*, 1998).

It remains the case that only a few caving books have a true international audience, most being of national or local interest. Thus most books are printed in runs of under 2000. Moreover only a few professional booksellers in Europe and North America specialize in cave and karst books. Hence many new caving books and particularly those of mainly local interest, are distributed through caving clubs and societies, which advertise them through their bulletins.

The best source of information for both caving books and caving journals is the Bibliographic Bulletin of the International Union of Speleology: *Speleological Abstracts*, printed annually by the Swiss Speleological Society and available also on the internet at www.isska.cm/bbs. In the case of old and / or out of print books it is possible to refer to the local UIS *Documentation Centres* (see Caving Journals entry for the list of their addresses) to obtain the required xerox.

PAOLO FORTI

See also **Caves in Fiction; Journals on Caves**

Works Cited

Badino, G. *et al.* (editors) 1999. *Río La Venta: Treasure of Chiapas*, Treviso: Associazione Esplorazioni Geografiche Rio La Venta

Boegan, E. & Bertarelli, L.V. 1926. *Duemila Grotte*, Milan: Touring Club Italiano

Bögli, A. & Franke, H. 1965. *Radiant Darkness: The Wonderful World of Caves*, London: Harrap

Cappelleri, M.A. 1757. *Pilatis Montis Historia*, Basel: Holt

Casteret, N. 1938. *Ten Years under the Earth*, New York: Greystone Press and London: Dent (translation of *Dix ans sous terre* and *Au fond des gouffres*)

Chabert, C. & Courval, M. 1971. *E.-A. Martel, 1859–1938: bibliographie*, Parios: Club Alpin Français

Configliachi, P. 1819. *Del proteo anguino di Laurenti*, Pavia: Fusi

Courbon, P., Chabert, C., Bosted, P. & Lindsley, K. 1989. *Atlas of the Great Caves of the World*, St Louis, Missouri: Cave Books

Cullingford, C.H.D. (editor) 1953. *British Caving: An Introduction to Speleology*, London: Routledge and Kegan Paul; 2nd edition, 1962

Farr, M. 1984. *The Great Caving Adventure*, Yeovil, Somerset: Oxford Illustrated Press

Ford, T.D. & Cullingford, C.H.C. 1976. *The Science of Speleology*, London and New York: Academic Press

Halliday, W.R. 1966. *Depths of the Earth: Caves and Cavers of the United States*, New York: Harper & Row; revised edition, 1974

Hill, C.A. 1976. *Cave Minerals*, Huntsville, Alabama: National Speleological Society

Hill, C.A. & Forti, P. 1997. *Cave Minerals of the World*, 2nd edition, Huntsville, Alabama: National Speleological Society

Hovey, H.C. (editor) 1882. *Celebrated American Caverns, Especially Mammoth, Wyandot, and Luray*, Cincinnati: Clarke

James, N.P. & Choquette, P. 1988 *Paleokarst*, New York: Springer

Juberthie, C. & Decu, V. (editors) 1994. *Encyclopaedia Biospeologica*, 2 vols, Moulis and Bucharest: Société de Biospéologie

Judson, D. 1973. *Ghar Parau*, New York: Macmillan, and London: Cassell

Kircher, A. 1665. *Mundus Subterraneus*, Amsterdam: Janssonium P. Weyerstraten

Klimchouk, A.B., Ford, D.C., Palmer, A.N. & Dreybrodt, W. (editors) 2000. *Speleogenesis: Evolution of Karst Aquifers*, Huntsville, Alabama: National Speleological Society

Kyrle, G. 1923. *Grundriss der Theoretischen Speläologie*, Vienna: Österreichische Staatsdruckerei

Lang, C. 1801. *Gallerie der unterirdischen schöpfungs-wunder un des menschlichen kunstfleisses unter der erde*, Leipzig, Tauchnitz

Lawrence, J.P. & Brucker, R. 1955. *The Caves Beyond: The Story of Floyd Collins' Crystal Cave Exploration*, New York: Funk & Wagnalls

Marbach, G. & Dobrilla, J.P. 1973. *Techniques de la spéléologie Alpine*, France: Marbach

Martel, E.A. 1894. *Les abîmes* [The Depths], Paris: Delagrave

Martel, E.A. 1921. *Nouveau traité des eaux souterraines*, Paris: Doin

Martel, E.A. 1900. *La Spéléologie; ou, Science des Caverns*, Paris: Carré & Naud

Northup, D. et al. 1998. *A Guide to Speleological Literature of the English Language 1794–1996*, Huntsville, Alabama: National Speleological Society

Padgett, A.P. & Smith, B. 1996. *On Rope*, Huntsville, Alabama: National Speleological Society

Schmidl, A. 1854. *Zur Höhlenkunde des Karstes: die Grotten und Höhlen von Adelsberg, Lueg, Planina und Laas*, Vienna: Braumüller

Shaw, T.R. 1992. *History of Cave Science: The Exploration and Study of Limestone Caves, to 1900*, 2nd edition, Broadway, New South Wales: Sydney Speleological Society

Taylor, M. (editor) 1991. *Lechuguilla: Jewel of the Underground*, Basel: Speleo Projects

Vallisnieri, A. 1715. *Lezione accademica intorno all'origine delle fontane*, Venice: Bortoli; 2nd edition, 1726

Valvasor, J.W. 1689. *Die Ehre des Herzogthums Crain*, 4 vols, Nurnberg: Endter

Waltham, A.C. 1976. *The World of Caves*, London: Orbis

Zimmermann, E. 1788. *Voyage à la Nitrière naturelle qui se trouve à Molfetta, dans la terre de Bari en Pouille*, Paris

BRITAIN AND IRELAND: ARCHAEOLOGICAL AND PALEONTOLOGICAL CAVES

The British Isles contain several regions of karst landscape with abundant caves (Figures 1 and 2; see also Figure 1 in Europe, North), while sea caves are also widespread, particularly on the western and northern seaboards of Britain. Archaeological and paleontological remains are frequently found in these caves, but in most instances the finds are fragmentary and they often lack a secure stratigraphic provenance. The remains of Pleistocene animals are fairly common, but there are very few human skeletons that date to earlier than the Holocene, and there are no known examples of paleolithic art on the walls of caves in the British Isles. The archaeological record in British and Irish caves is therefore impoverished compared with comparable sites in continental Europe; nevertheless the discoveries made in British caves in the first half of the 19th century were highly significant to the development of knowledge and ideas concerning human antiquity and in demonstrating that early humans were present in Britain at the same time as extinct Ice Age animals.

Scotland

The Creag nan Uamh caves are formerly phreatic systems developed in Cambro-Ordovician dolomitic limestone in the Allt nan Uamh Valley, 5 km south of Inchnadamph in Sutherland, northwest Scotland. The caves were covered by the Late Devensian ice sheet and the cave sediments—gravels and cave earths—are predominantly derived from glacial materials. Archaeological excavations at the caves in the 1920s revealed many naturally shed reindeer antlers, which were presumed to have been cached in the caves by paleolithic hunter–gatherers (Lawson, 1981). Subsequent radiocarbon dating has shown that the antlers were deposited in at least three separate episodes during the last (Devensian) glaciation, and, together with the absence of evidence for the contemporary presence of humans, the faunal remains are now thought to have accumulated by natural agencies. After the end of the last ice age the caves were used as burial chambers by neolithic and bronze age people.

A series of caves and rock shelters containing archaeological material are located at a present-day altitude of 10 m around the former shoreline of Oban Bay, in Argyll on the northwest coast of Scotland. These former sea caves were uplifted during isostatic rebound of the Scottish coastline after the last glacial maximum, and the cave entrances were then utilized by Mesolithic hunter–gatherers between about 8000 and 4000 BC. Some of the caves were later used as burial sites during the Neolithic, Bronze Age, and Iron Age periods (Saville & Hallen, 1994).

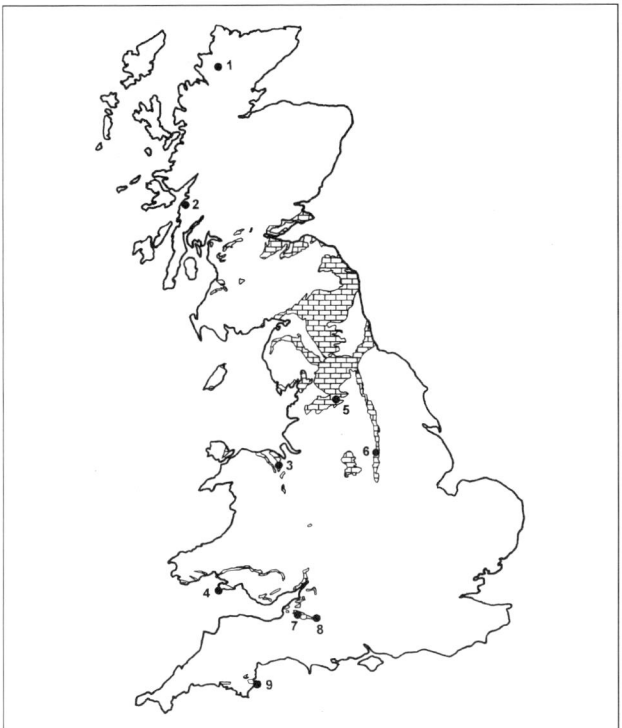

Britain and Ireland: Archaeological and Paleontological Caves: Figure 1. Archaeological caves in Britain mentioned in the text. The principal outcrops of Devonian, Carboniferous and Permian limestones are also shown.
1. Creag nan Uamh Caves. 2. Oban Bay Caves. 3. Pontnewydd Cave.
4. Paviland Cave. 5. Victoria Cave. 6. Creswell Caves. 7. Cheddar Gorge Caves. 8. Westbury Quarry Cave.
9. Kent's Cavern and Brixham Cave.

Wales

Comprehensive reviews of the cave archaeology of Wales can be found in Ford (1989), and only two of the most important sites are mentioned here. Pontnewydd Cave is located near Denbigh, about 10 km from the north coast of Wales. The cave was excavated by Stephen Aldhouse-Green in the 1970s and 1980s (Green, 1984). The site is distinguished archaeologically by its abundance of stone tools of Lower Paleolithic type, together with waste flakes from their manufacture. Accompanying fauna show that the cave was occupied by early humans during an interglacial period about 230 000 years ago (Oxygen Isotope Stage 7). Fragmentary bones and teeth of several human individuals have been excavated from debris-flow deposits in the cave, but radiocarbon dating has shown that at least some of these human remains are derived from disturbed Holocene burials.

Paviland Cave, also known as Goat's Hole Cave, is located on the Gower Peninsula on the south coast of Wales. The cave is the richest early Upper Paleolithic site in the British Isles, and has produced the region's only ceremonial burial from the Ice Age, radiocarbon dated to 26 ka. The burial was discovered in 1823 by William Buckland, who thought that the ochre-covered skeleton was that of a Roman prostitute, hence it is colloquially known as the "Red Lady of Paviland". Subsequent study has shown the skeleton to be that of a young adult male whose grave had been furnished with carved ivory bracelets and wands and perforated seashells. Further excavations in the cave during the 20th century recovered thousands of stone tools and more bone and ivory artefacts, and a recent programme of radiocarbon dating has shown that the cave was occupied episodically throughout the Upper Paleolithic, from about 30 000 to 12 000 years ago (Aldhouse-Green, 2000).

England

In Britain, the caves of South Devon, developed mainly in coastal outcrops of the Devonian Limestone, were among the first such sites to be subject to scientific investigation. Kent's Cavern in Torquay (known earlier as Kent's Hole) was first excavated in the 1820s by the Reverend John McEnery (see Archaeologists). McEnery's controversial claim of finding stone tools that were contemporaneous with the bones of extinct animals was rejected by the eminent geologist William Buckland, who regarded the stone artefacts as intrusive. However, careful excavations by the Torquay Natural History Society, directed by William Pengelly in 1846 and continued in the 1860s, corroborated the work of McEnery through meticulous recording of the stratigraphic levels where the bones and artefacts were found. Brixham Cave, on the Devon coast 8 km south of Kent's Cavern, was also excavated by Pengelly in 1846 and provided additional proof of the antiquity of human activity in Devon.

Most of the faunal remains and artefacts from Kent's Cavern are of Last Glacial and Holocene age, but elements of a much earlier fauna, including cave bear, sabretoothed cat and the pine vole *Pitimys gregaloides* indicate that some deposits are of Hoxnian or earlier age (Oxygen Isotope Stage 10 or earlier). Lower Paleolithic hand axes have also been recovered from the lowest stratigraphic unit in the cave, confirming the great antiquity of the sequence of deposits in this cave. Cave-infill deposits of similar age have been investigated at Westbury Quarry in Somerset. The cave was exposed by quarrying of the Carboniferous Limestone in 1969, revealing a layered sequence of sands, gravels and limestone breccias, and conglomerates. Most of the faunal remains were found in the upper calcareous deposits, and the bones are predominantly of large carnivores and rodents, with the latter deriving from owl pellets (Andrews, 1990). Through comparison with other dated paleontological sites in Britain, the Westbury fauna is thought to date to about 500 000 years ago. The presence of flint artefacts and cut-marked animal bones indicate the presence of early humans at Westbury at this time.

Several other caves on the Carboniferous Limestone outcrop of the Mendip Hills have produced human remains and artefacts, although these mostly date from the late Upper Paleolithic and Holocene. In Cheddar Gorge, Gough's Cave and Sun Hole both contained Upper Paleolithic burials dating to about 14 000 years ago. The human bones from Gough's Cave exhibited cut-marks, showing that the bodies had been defleshed, using stone tools, before they were buried. The nearby caves of Wookey Hole, 10 km to the southeast of Cheddar, include the paleontological sites of Hyaena Den (explored by William Boyd Dawkins in the 1850s), Badger Hole, and Rhinoceros Hole.

Creswell Crags, located on the county boundary between Nottinghamshire and Derbyshire, comprises a natural gorge and cave system within the narrow linear outcrop of Permian age

Magnesian Limestone. Archaeological interest in the caves at Creswell Crags began in the 1870s, when Pin Hole, Church Hole, Robin Hood's Cave and Mother Grundy's Parlour were excavated by the Reverend J. Magins Mello (a local antiquary) and Thomas Heath (the curator of Derby Museum), who were provided with specialist advice by William Boyd Dawkins and Professor George Busk of the Royal College of Surgeons. Excavations continued at the caves in the 20th century, and the caves are now managed as an educational resource by the Creswell Heritage Trust.

The Creswell Caves have yielded important cultural and environmental remains dating from the Middle and Upper Paleolithic (approximately 100 000 to 10 000 years ago). Mousterian (Middle Paleolithic) and early Upper Paleolithic flint tools have been found in the caves, and Creswell Crags is the type site for the Creswellian Industry, a distinctive suite of late Upper Paleolithic ("late glacial") stone tool types that is found throughout northwest Europe. Robin Hood's Cave produced one of the very few genuine examples of ice age art found in Britain, an animal rib engraved with a drawing of a horse's head. This artefact, together with an engraved drawing of a human figure on another rib fragment from Pin Hole, have their closest affinities with portable engraved art found at late Upper Paleolithic sites in northern France and Belgium.

Victoria Cave, near Settle in North Yorkshire, is another site that was systematically explored during the second half of the 19th century. The cave was discovered by Joseph Jackson, an amateur collector, who brought the site to the attention of the Society of Antiquaries of London. Subsequent investigations were conducted in 1870 and 1878 by the Settle Caves Exploration Committee, working under the auspices of the British Association for the Advancement of Science and advised by a scientific panel that included some of the most distinguished geologists and archaeologists of the day. The excavation and recording methods employed at Victoria Cave were based on the innovative principles pioneered by William Pengelly at Kent's Cavern, and included the early use of photography to monitor the progress of the work. The excavations revealed a preglacial bone-bearing deposit sealed by nearly five metres of later sediments, and Victoria Cave is best known today as a paleontological site dating from the Ipswichian Period of the last interglacial, approximately 130 000 years ago.

Ireland

Ireland's caves (Figure 2), like those in Britain, were intensively investigated in the late 19th and early 20th centuries, and much of the evidence for the colonization of Ireland by animal species from continental Europe has come from faunal remains recovered from caves. A large programme of radiocarbon dating of animal and human bones from Irish caves has recently been completed, clarifying the depositional history of many of the cave sites (Woodman *et al.*, 1997). In County Waterford the caves of Ballynamintra, Kilgreany, and Shandon have all produced Upper Pleistocene fauna, together with younger Holocene deposits, which at Kilgreany and Ballynamintra contained evidence of prehistoric human activity, including burials. In County Cork, Castlepook Cave and Foley Cave both contained

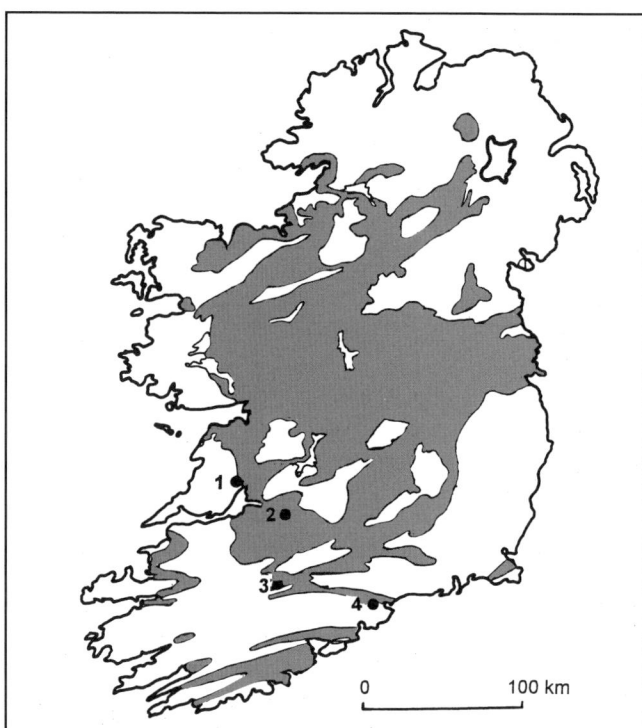

Britain and Ireland: Archaeological and Paleontological Caves: Figure 2. Archaeological caves in Ireland mentioned in the text. The principal outcrops of Carboniferous limestones are also shown. 1. Edenvale and Newhall / Barntick Caves. 2. Annah, Caherguillamore, and Killuragh Caves. 3. Castlepook and Foley Caves. 4. Ballynamintra, Kilgreany, and Shandon Cave.

Pleistocene fauna, with more recent Holocene fauna in the uppermost deposits. The Edenvale and Newhall / Barntick group of caves in County Clare has produced vast amounts of animal bone that was deposited in the late glacial and Holocene periods, and archaeological material is also present in these cave systems.

Natural caves and fissures in Ireland were used as burial chambers during the Neolithic Period—a similar practice to that observed in Britain (Chamberlain, 1996). Three separate examples of such burials have been discovered in County Limerick in caves and rock fissures at Annagh, Cahirguillamore, and Killuragh. Unlike the situation in Britain, caves continued to be used as burial sites in the early historical period (even after Ireland was Christianized) and there are also examples of medieval occupation debris, including charcoal, pottery, metalwork, and bone artefacts at several cave sites in Ireland.

ANDREW T. CHAMBERLAIN

Works Cited

Aldhouse-Green, S. 2000. *Paviland Cave and the "Red Lady": A Definitive Report*, Bristol: Western Academic and Specialist Press

Andrews, P. 1990. *Owls, Caves and Fossils: Predation, Preservation and Accumulation of Small Mammal Bones in Caves, With an*

Analysis of the Pleistocene Cave Faunas from Westbury-sub-Mendip, Somerset, UK, London: Natural History Museum Publications

Chamberlain, A.T. 1996. More dating evidence for human remains in British caves. *Antiquity*, 70: 950–53

Ford, T.D. (editor) 1989. *Limestones and Caves of Wales*, Cambridge and New York: Cambridge University Press

Green, H.S. 1984. *Pontnewydd Cave: A Lower Palaeolithic Hominid Site in Wales: the First Report*, Cardiff: National Museum of Wales

Lawson, T.J. 1981. The 1926–27 excavations of the Creag nan Uamh bone caves, near Inchnadamph, Sutherland. *Proceedings of the Society of Antiquaries of Scotland*, 111: 7–20

Saville, A. & Hallen, Y. 1994. The "Obanian Iron Age": human remains from the Oban cave sites, Argyll, Scotland. *Antiquity*, 68: 715–23

Woodman, P., McCarthy, M. & Monaghan, N. 1997. The Irish Quaternary fauna project. *Quaternary Science Reviews*, 16: 129–59

Further Reading

Buckland, W. 1823. *Reliquiae Diluvianae, or, Observations on the Organic Remains Contained in Caves, Fissures, and Diluvial Gravel, and on Other Geological Phenomena, Attesting the Action of an Universal Deluge*, London: Murray

Campbell, J.B. 1977. *The Upper Palaeolithic of Britain: A Study of Man and Nature in the Late Ice Age*, 2 vols, Oxford: Clarendon Press

Dawkins, W.B. 1874. *Cave Hunting*, London: Macmillan; reprinted Wakefield, EP Publishing, 1973

BRITAIN AND IRELAND: BIOSPELEOLOGY

A relatively small number of hypogean taxa have been recorded in Britain and Ireland, compared to mainland Europe. The low number of troglobitic and stygobitic fauna recorded in subterranean habitats largely reflects the glacial history and post-glacial colonization of taxa. Despite some enthusiastic collection and recording of organisms from caves during the early and mid-20th century, little systematic biospeleological research has been undertaken in Britain and Ireland. Recent research has indicated that new discoveries are still to be made and that this largely neglected area of research may provide valuable information regarding the colonization of troglobitic/stygobitic taxa. However, a number of sites and populations of subterranean organisms are under threat from human disturbance, and the need for active management and conservation of this important biological resource has only recently been recognized (Wood & Gunn, 2000).

A total of 46 troglobitic / stygobitic taxa have been recorded from Britain and Ireland (12 terrestrial and 15 aquatic, plus a further 19 interstitial forms). Collection of fauna from caves in Britain and Ireland began in the mid-19th century, although the primary period of active collection and recording was under the direction of the Biological Recorder of the Cave Research Group, Mary Hazelton, between 1930 and the late 1970s. Records for Great Britain were published in 15 parts, between 1938 and 1978, with a compilation of records for Ireland being published in a separate volume. Additional cavernicolous data has been summarized for specific limestone regions, e.g., northern England, Mendip, Peak District, and Wales. There are no stygobitic fish or amphibians in Britain and Ireland, although some populations of stygophilic fish may exist (Proudlove, 1982), and bat colonies have been recorded within a number of caves.

While the cave fauna of Britain and Ireland is relatively well known, the groundwater fauna of the large chalk aquifers, and hyporheic fauna associated with riverine gravels, has been poorly studied. It is likely that important discoveries will be made in these areas, particularly with regard to microcrustacea such as copepods and cladocera. The majority of freshwater organisms found in British and Irish caves are accidental occupants (e.g., larvae of Ephemeroptera, Plecoptera, and Trichoptera washed into caves), although some are stygophilic. The freshwater amphipod *Gammarus pulex* is often abundant in caves, and some populations are certainly stygophilic, displaying incipient troglomorphic characters such as reduced body pigmentation. Other taxa, such as the diving beetle, *Hydroporus ferrugineus* (Dytiscidae), are clearly stygophilic as adults, although the larvae have only been recorded in subterranean waters and may be stygobitic.

The majority of the freshwater stygophilic taxa in Britain and Ireland are Crustacea, although relatively little is known regarding subterranean Ostracoda, Cladocera, Copepoda, and other meiofaunal groups. The distribution of freshwater stygobitic taxa is strongly associated with the occurrence of carbonate rocks (see map), with numerous records from springs, wells, riverine gravels, and other groundwater dominated habitats. The most diverse hypogean group is the Amphipoda with six taxa in the genera *Niphargus* (five in Britain and two in Ireland) and *Crangonyx subterraneus* being recorded to date, with the most recent discovery (*Niphargus wexfordensis*) being made in 1980.

The biogeographic distribution of most cave-dwelling organisms generally reflects that of carbonate rocks and Pleistocene glacial episodes. However, two species, *Niphargus aquilex* and *Antrobathynella stammeri*, have been recorded some distance north of the last glacial maxima in England, as well as *Antrobathynella stammeri* from Scotland. Other taxa have also colonized areas that have been glaciated, such as *Proasellus cavaticus* and *Niphargus fontanus* (South Wales), *N. kochianus kochianus* (Norfolk and Cambridgeshire), and *N. kochianus irlandicus* (central Ireland).

The terrestrial fauna of British and Irish caves, and other subterranean habitats such as mines, is dominated by accidental and troglophilic organisms. In the threshold region, spiders of the genus *Meta* are common, as are two species of trogloxene moths (*Scolyopteryx libatrix* and *Triphosa dubitata*), which hibernate in the winter. The most important members of the terrestrial communities in the dark zone are Collembola, which act as consuming detritivores, and mites which are predators. More than 70 species of Collembola have been recorded but only 15 are common, with the most frequently recorded being *Tomocerus minor*. Over 40 species of mites have been recorded, of which six are probably troglobites. Other common inhabitants are the isopod *Androniscus dentiger*, the mycetphilid fly *Speolepta leptogaster*, and the beetle *Trechus micros*.

The lack of research on British and Irish cave fauna, and the perceived paucity of aquatic and terrestrial fauna in caves, is reflected in the fact that none of the species currently have any

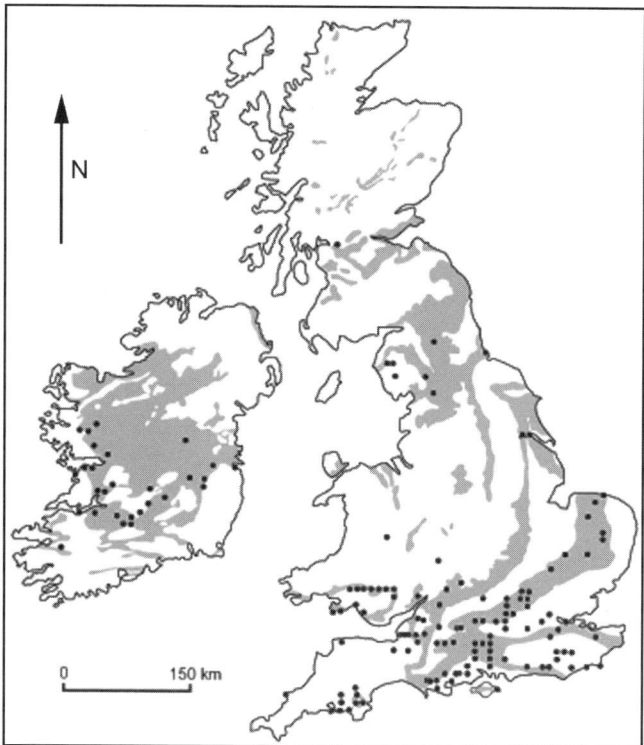

Britain and Ireland: Biospeleology: Map of Britain and Ireland indicating the distribution of selected freshwater hypogean taxa (*Antrobathynella stammeri*, *Crangonyx subterraneus*, *Niphargus aquilex*, *Niphargus fontanus*, *Niphargus glenniei*, *Niphargus kochianus kochianus*, *Niphargus kochianus irlandicus* and *Proasellus cavaticus*). The grey areas are outcrops of rock sequences that contain some limestone units and are therefore more extensive than the actual limestone outcrops or the areas of karst landforms.

form of statutory protection. Around 75% of cave passage in Great Britain lie within Sites of Special Scientific Interest, and 11 caves are recognized for their importance for hibernating bats, but none of the citations mention any cavernicolous fauna.

There is a long history of anthropogenic disturbance of subterranean ecosystems in Britain and Ireland, with clear evidence of bacterial contamination at some sites. Quarrying, changes in agricultural practices, waste disposal and groundwater abstractions all pose serious threats to the fauna of subterranean ecosystems. Many of the potential impacts to the communities that occupy British and Irish caves have been identified, including those associated with show-cave development and recreational caving (Gunn *et al.*, 2000).

Britain and Ireland have few obligate subterranean taxa and are biodiversity "cold-spots" compared to many other karst areas. However, there are numerous troglophilic and stygophilic taxa and much has yet to be learned from the detailed study of the ecology and physiology of subterranean organisms. All obligate subterranean taxa were troglophilic in their evolutionary past, and the paths to obligation and troglomorphy may be revealed through the study of these organisms.

PAUL J. WOOD AND GRAHAM PROUDLOVE

Works Cited

Gunn, J., Hardwick, P. & Wood, P.J. 2000. The invertebrate community of the Peak-Speedwell cave system, Derbyshire, England—Pressures and considerations for conservation management. *Aquatic Conservation: Marine and Freshwater Ecosystems*, 10: 353–69

Proudlove, G.S. 1982. Cave life. Part 2 cave fishes. *Caves and Caving*, 15: 6–7

Wood, P.J. & Gunn, J. 2000. The aquatic invertebrate fauna within a cave system in Derbyshire, England. *Internationale Vereinigung für Theoretische und Angewandte Limnologie*, 27: 901–05

Further Reading

Chapman, P. 1993. *Caves and Cave Life*, London: HarperCollins

Cubbon, B.D. 1976. Cave flora. In *The Science of Speleology*, edited by T.D. Ford & C.H.D. Cullingford, London and New York: Academic Press

The most thorough treatment of the flora of UK caves

Hazelton, M. 1955 *et seq.* Biological Records of the Cave Research Group of Great Britain. *Biological Supplement of the Cave Research Group, Great Britain*, Parts 1–5, *Biological Records Cave Research Group, Great Britain*, Parts 6–8, *Transactions of the Cave Research Group of Great Britain*, 7: 10–19; 9: 162–241; 10: 143–65; 12: 3–26; 13: 167–97; 14: 205–30; 15: 191–254, *Transactions of the British Cave Research Association*, 5: 164–98

Hooper, J.H.D. 1976. Bats in caves. In *The Science of Speleology*, edited by T.D. Ford & C.H.D. Cullingford, London and New York: Academic Press

Jefferson, G.T. 1976. Cave faunas. In *The Science of Speleology*, edited by T.D. Ford & C.H.D. Cullingford, London and New York: Academic Press

BRITAIN AND IRELAND: HISTORY

The earliest explorers of British caves in historical times were undoubtedly lead miners, who encountered natural "opens" in the course of their search for ore. A notable example is the exploration of Pen Park Hole in Gloucestershire, leading to the first survey of a natural cave ever to be published (Shaw, 1992, pp.14–15). Another early scientific reference to British caves is the account of a descent of Elden (*sic*) Hole, Derbyshire, by John Lloyd in the 1772 *Philosophical Transactions of the Royal Society*. Subsequently, early geologists, such as Adam Sedgwick (1785–1873) and Sir Roderick Murchison (1792–1871) came along. They were trying to form the big picture from the complex structures and strata that constitute the British Islands. They found caves with bones buried in silt, and thus the diluvial theory of a universal flood, which supported biblical teachings, persisted for a while. William Buckland (1784–1856) was its principal proponent, with findings in Kirkdale Cave in Northern England, after which he wrote *Reliquiae Diluvianiae* (Relics of the Flood) in 1823. He soon realized the errors of his thinking and came to appreciate the reality that caves simply sat there in the landscape "catching" things that went past and were in fact "museums"

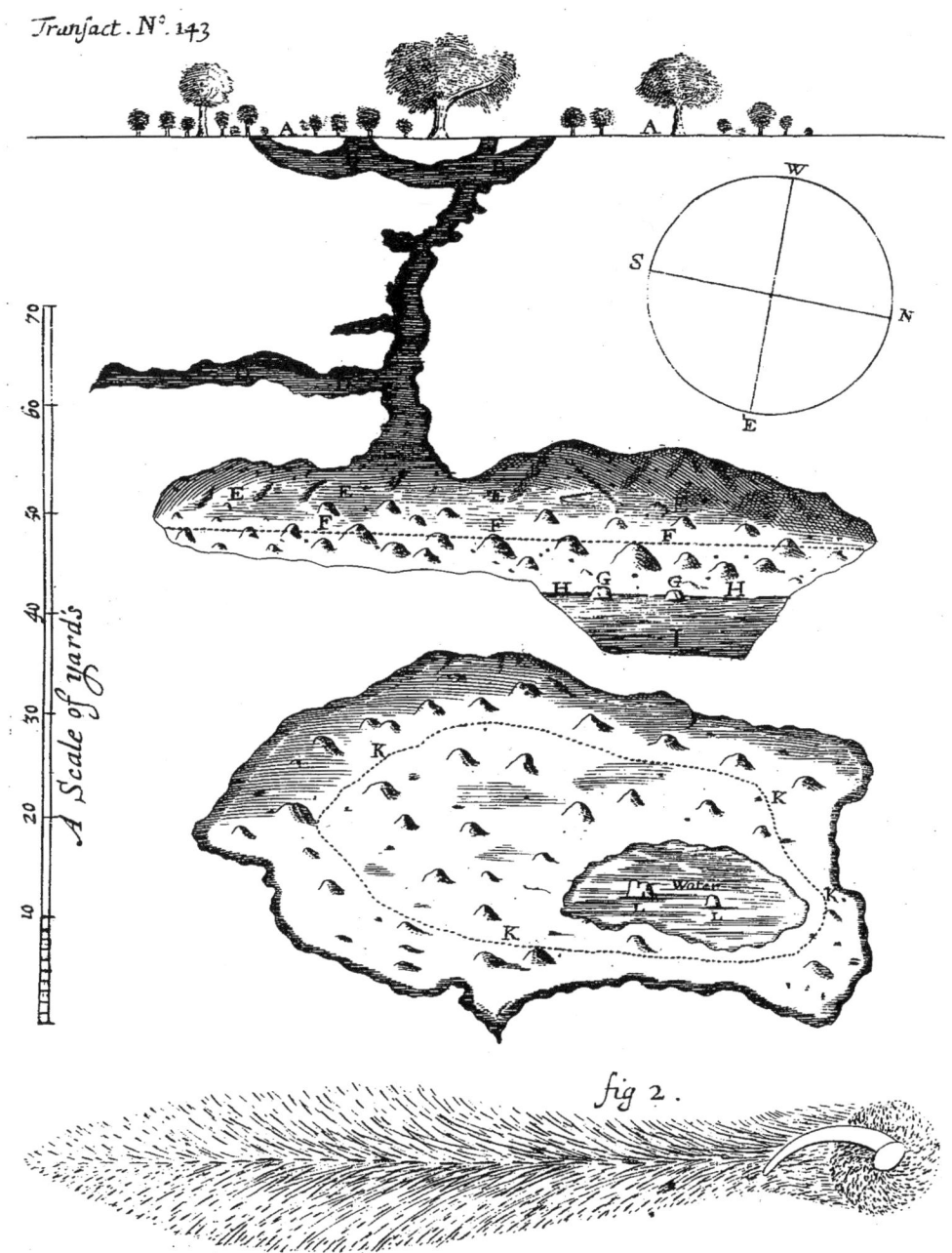

Britain and Ireland: History: Survey of Pen Park Hole Gloucestershire, by Captain Grenville Collins, September 1682. First published in *Philosophical Transactions of the Royal Society*, 143, 1683.

of a changing surface topography. Although there was some exploration for its own sake and for tourism, cave sedimentology developed as the main interest in caves throughout the 19th century, with paleontology and archaeology being the principal themes.

Édouard Martel (1859–1938) visited, and extended several of the best-known British and Irish caves in 1895, and this provided impetus for exploration, first in England and then in Ireland. The Yorkshire Ramblers' Club (YRC), founded in 1892, was thus activated by Martel's successful descent of Gaping Gill Main Shaft (110 m) and made a second descent the following year. It continued with the exploration of most of the open shafts of the Three Peaks area of the Northern Pennines each year, until war broke out in 1914. Harold Brodrick (1875–1946) led the YRC to follow in Martel's footsteps with explorations in County Fermanagh, from 1907. They started with the Boho block, Noon's Hole, Arch Cave (Ooboraghan) and Pollanafrin, and then moved south to the Marble Arch area in 1908. E.A. Baker, and members of the YRC, started to examine the County Clare limestone area in 1911, but it was the University of Bristol

Speleological Society that made major discoveries from 1948 to 1970 (Tratman, 1969). Jack Coleman (1914–71) was at the forefront of Irish cave exploration for many years and wrote what is still the only text on the caves of the whole of Ireland (Coleman, 1965).

Limestone hydrology attracted attention on Mendip, Somerset, with the Wookey Hole resurgence and its sink at Swildons Hole in 1901 (Baker & Balch, 1907). Herbert Balch (1869–1958) formed the Mendip Nature Research Committee in 1908 and they pursued the exploration of Swildons Hole. St Cuthbert's Swallet and Eastwater Cavern also drain to Wookey Hole but connections are not in prospect, there being an intervening deep and tortuous phreas. Wookey was, however, the birthplace of cave diving, with the pioneering activities of Graham Balcombe (1907–2000), Penelope Powell and Jack Sheppard from 1935 (see Diving in Caves entry).

The first comprehensive attempt to coordinate cave science and exploration in Britain came in 1935 when Eli Simpson (1884–1962), Leo Palmer (1891–1962) and other leading academics formed the British Speleological Association (BSA). They had observed the formation of similar bodies across Europe and strongly felt the need for this in Britain. During the war years, the BSA was held together by Eli Simpson (Craven, 2001), but unfortunately there were many personality clashes and although it struggled on as a "national" caving organization with regional groups it never regained its scientific or national credibility. From 1946, the position of a scientific caving society was filled by the Cave Research Group (CRG), which substantially coordinated British cave research through the publication of its Transactions and other scientific publications, together with *British Caving*. It was founders Aubrey Glennie and Mary Hazelton who jointly did much to foster the study of hypogean fauna in Britain. The CRG was blessed with some extremely competent cave surveyors. Lewis Railton (1907–71) set the benchmark for cave survey and presentation with the Ogof Ffynnon Ddu survey publication (Railton, 1953), and with Arthur Butcher set out a clear and well-defined basis for a range of techniques and presentation standards (see Surveying Caves entry). There was much duplication of effort and the two organizations merged to form the British Cave Research Association (BCRA) in 1973. Trevor Ford edited *Transactions* of the CRG and then BCRA for 30 years from 1965, for which he was awarded a Certificate of Merit by the National Speleological Society of the United States in 1995 and an OBE (Order of the British Empire) in 1997.

Exploration of the longest and/or deepest cave systems did not come until quite late in the day. Although the Gaping Gill–Clapham Cave System saw considerable attention from the YRC, the Yorkshire Speleological Association (1906–14) and many others from the early days right through to the present, it has still only been connected by cave divers, with a current total length of 18 km. In the Northern Pennines, it was the Lancaster–Ease Gill System that came to be the most extensive, at 75 km. Exploration did not start until 1946, with the chance discovery of Lancaster Pot by BSA members, who attempted to establish an underground laboratory that was gated and locked. It did not function, but rather led to the fragmentation of the BSA and the formation of many new caving clubs (Eyre, 1989).

The biggest system in the Peak District is now Peak–Speedwell, which came together incrementally from the early efforts of the BSA in the late 1930s up to the present (see Peak District entry). Following the formation of the BCRA in 1973, the one remaining active group of the former BSA became the Technical Projects Group (later Technical Speleo Group, TSG) and it has continued to work on this complex system. With the discovery of "Titan" in 1999, it now boasts the tallest vertical feature in British caving (vertical range 219 m).

There is a scattered area of Devonian reef limestones in south Devon. The caves of Higher Kiln Quarry, Buckfastleigh, Reed's Cave–Baker's Pit, have the distinction of being the most extensive in Britain, developed in pre-Carboniferous limestones—over 3 km. Many were rich in mammalian bone deposits; early excavations being carried out by Buckland and Pengelly. The latter was the inspiration for the establishment at Higher Kiln Quarry in 1962, of what is now the William Pengelly Cave Studies Centre (and Trust).

The geographical layout of the South Wales limestones as one layer in a huge depositional basin, where the cave bearing strata have been exposed by transverse fluvial activity at a number of isolated sites, has provided a complicated explorational picture. In the Swansea Valley, the South Wales Caving Club and its precursors worked first at Dan-yr-Ogof (1936–39) and then Ogof Ffynnon Ddu in the 1946–53 period. Major extensions to both of these systems were made in the 1960–80 period. Exploration of the Mynydd Llangattock area further east began in 1957, with the Agen Allwedd breakthrough. A complex system of as yet unconnected caves gradually came together with discoveries in Ogof Daren Cilau, Eglwys Faen, and Ogof Craig ar Ffynnon. The most recent phase in this region commenced in 1994, with the discovery of Ogof Draenen at the southeastern extremity of the limestone above Blaenafon. Exploration here by many caving clubs is still progressing and the system now exceeds 70 km (see Draenen entry).

In North Wales, there is a substantially concealed area of limestone where most caves are accessible only as a result of dewatering of the aquifers by lead mining activities. The Alyn River Caves were dewatered firstly by the Halkyn Deep Level and then by the Sea Level or Milwr Tunnel, which ceased driving in 1957. Further south, the three cave systems of Dydd Byraf, Llyn Du, and Llyn Parc are fragments of one extensive system first entered by lead miners. The North Wales Caving Club rediscovered these through persistent detective and exploration work throughout the 1970s and 1980s.

There are a number of scattered limestone outcrops across Scotland. Many of these contain relatively small caves, those of Assynt, Sutherland, being the most numerous, extensive and most thoroughly explored, with the Grampian Speleological Group being the chief activist since its formation in 1961.

David Judson

Works Cited

Baker, E.A & Balch, H.E. 1907. *The Netherworld of Mendip: Explorations in the Great Caverns of Somerset, Yorkshire, Derbyshire, and Elsewhere*, Clifton: J. Baker

Coleman, J.C. 1965. *The Caves of Ireland*, Tralee, Ireland: Anvil Books

Craven, S.A. 1999. A history of cave exploration in the Northern Pennines, United Kingdom, up to 1838. *Cave & Karst Science*, 26(2): 53–59

Craven, S.A. 2001. The British Speleological Association (1935–1973) and its founder, Eli Simpson: With particular reference to activities in the northern Pennines of England. *Cave and Karst Science*, 28(3): 99–112

Eyre, J. 1989. *The Ease Gill System: Forty Years of Exploration*, BCRA (Speleo History Series 1)

Railton, C.L. 1953. *The Ogof Ffynnon Ddu System*, Cave Research Group: (CRG Publication No. 6)

Shaw, T.R. 1992. *History of Cave Science: The Exploration and Study of Limestone Caves, to 1900*, 2nd edition, Broadway, New South Wales: Sydney Speleological Society

Tratman, E.K. (editor) 1969. *The Caves of North-west Clare, Ireland*, Newton Abbot, Devon: David & Charles

Further Reading

Cullingford, C.H.D. (editor) 1953. *British Caving: An Introduction to Speleology*, London: Routledge and Kegan Paul

Judson, D. (editor) 1995. *Caving Practice and Equipment*, Leicester and Birmingham, Alabama: Cordee, BCRA and Menasha Ridge Press

BURIALS IN CAVES

Caves accessible to humans have been used from the Paleolithic to the present. They provide a variety of burial conditions, such as graves in the ground, body exposure on the floor, coffins exposed in selected places, storage of bones or pieces of bone in jars, and cremation urns. Habits have varied through time, and also according to the particular people, social status of the deceased person, position in the family, age, and cause of death.

In Mousterian times, and perhaps before, *Homo sapiens neandertalensis* buried their dead in caves and rock shelters such as Tabun Cave, Israel, dated to about 180 000 years ago. The first discovery of a Neanderthal grave was in 1908 at La Chapelle-aux-Saints Cave, Corrèze, France. In La Ferrassie, Dordogne, France, eight bodies were placed into man-made graves, given specific positions and covered with protective flagstones, artefacts, and earth mounds. Clearly, Neanderthal Man was conscious of death and life, as shown by his rituals. In Shanidar Cave, Iraq, the covering was pine branches and flowers (see Shanidar Cave, Iraq: Archaeology), and in Teshik-Tash Cave, Uzbekistan, a boy's body was circled with ibex horns. Archaic *Homo sapiens sapiens* were also given offerings, as in Qafzeh, Palestine, that indicate belief in a life after death. The reality of Mousterian burials has been questioned by a number of authors, such as Gargett (1999). Though their comments bring an interesting insight into the necessary fossilization processes, deliberate burial could be confirmed in most of the cases.

Dordogne caves have provided evidence of burials for nearly every prehistoric period. Some caves were repeatedly used, as for instance, Rouffignac (Neolithic, Bronze, and Iron Ages), or Déroc Caves (Late Neolithic and Chalcolithic).

Burial habits evolved, from mainly collective exposure on the floor during the Neolithic, with up to a hundred bodies in a single cave; partial (accidental, sometimes ritual) burning of bones; graves in the cave floor (Mousterian, Bronze Age); ash deposits near gallery walls after cremation (appearing during Late Bronze II Age); bone storage after flesh removal (Late Bronze Age); body exposure on wooden platforms, as in La Fontanguillère (Bronze Age) and Foissac Caves (Lot, France, Chalcolithic Age); or apparently little use (during Iron Age before Tene III, perhaps because of widespread cremation). In some caves, body remains lie on flowstones or clayey ground subsequently covered with calcite, for instance Early Iron Age remains at Palabres Cave (Lot, France). Some people occasionally hid themselves in caves, but were smoked in with wood fires at the entrance and so died.

The cave environment and noticeable cave features were taken into consideration, probably in relation to belief or specific interest. Late Bronze Age ashes from cremation in Rouffignac Cave entrance were placed in galleries with Magdalenian wall paintings. Wooden platforms in La Fontanguillère and Foissac Caves were located above a permanent stream, perhaps in relation with a water cult or to signify a route to the realm of the dead. At different periods variable artefacts, pottery, and offerings were placed besides bodies.

Comparable habits existed elsewhere in the world. In Vietnam, for instance, a few caves and rock shelters were possibly used for burial during Lower Neolithic Hoabinhian II–III and three caves, Pho Binh Gia, Khéo-Phay, and Lang Cuom, revealed skulls of people of the Bacsonian culture. In the latter, skulls were rubbed with red ochre; 80 to 100 individuals were present—probably a secondary burial. Children's skulls were used as funeral urns or trophies. Many caves with Upper Neolithic remains are very rich in burial-related attributes.

In Laos, Mesolithic—and perhaps older—human remains were found in Tham Pong. A Lower Neolithic burial was uncovered in that cave (an adult on its back) and in Tham Hang South, where six bodies, including three lying on their back on an ash blanket, were placed parallel to the cave wall. Upper Neolithic burial possibly existed in Tham Hang. Burial in karst fissures of Khammouan karst, sometimes associated with pottery, lasted up to 400 to 300 BC. These burials are similar to some modern burials as in Sagada (see below) and suggest a possible use of coffins. Undated, relatively young coffins are also found in the area.

Hardwood coffins are common in nearly 100 caves of the Mae Hong Son and Kanchanaburi areas of Thailand (200 BC–800 AD) (see also Asia, Southeast: Archaeological Caves). Many of these caves have a difficult access and the coffins themselves are stored in inaccessible locations, such as on ridges or ledges, some even being placed horizontally on pole systems. The coffin shapes vary and their length may exceed 2 m. Associated items include glass beads, ceramics, and metal objects.

Caves in the Southeast Asian islands, such as Borneo (Malaysia, Indonesia) and Palawan (Philippines), offer similarities with the mainland and make the link with modern times. Borneo's Niah Cave delivered a 38 000 years old *Homo sapiens* skull, tree trunk Neolithic coffins dated 2460 ± 70 years BP, 1000-year-old boat-shaped coffins, together with many other burials. More coffins are found in several areas of Borneo, as in Sabah State, the upper reaches of Mahakam River, and recent ones in Mulu karst (e.g. Cave of the Winds).

Palawan caves in the Tabon area have demonstrated human presence for at least 45 000 years, including the Tabon Man (22–24 000 years BP) and long time burial habits. In the Early Neolithic, a flexed male body in Duyong Cave was placed face down in a grave, with associated earrings and tools (2680 ± 250 BC). From *c.* 1500–1000 BC in the Late Neolithic, to the 14th century AD, jars offered primary and secondary burial, though skulls were sometimes buried separately.

Jars, covers, vessels, and other items are sometimes very numerous: there being up to 200 in the single-chambered Tabon Cave (200–500 BC)—with glass beads, as in Thailand, and lime containers, probably for betel chewing; more than 500 such items in Tadyaw Cave, with iron tools of the advanced Metal Age (100 BC to 300 AD); more than 78 in Manunggul Cave, many of Late Neolithic (890–710 BC) and Early Metal Age (190 BC), including pottery covers showing a ship of the dead and one trunk-shaped pottery coffin 73 × 34 cm large for secondary burial. Bone painting with haematite occurred during the Neolithic and Early Metal Age (from c. 1500 to 1000 BC, to c. 190 BC) and gradually disappeared before the 13th century AD.

In rare narrow caves, bodies were put into maceration on the ground, probably wrapped in a mat. Skull burial, supine position (primary burial), and bone bundles (up to seven persons) were also observed.

All sites were chosen for their beauty or magnificent scenery. Jars were often placed in the daylight, or sometimes in dark chambers decorated with beautiful stalagmites. Nearly all burial caves in the Tabon area face the sea at various elevations (115 m in a sheer cliff at Manunggul), as do burial caves in the Palawan El Nido area and in southern Luzon. Jars are usually placed along walls, occasionally protected by stones, leaving space for new jars or ritual gatherings (cult-of-the-dead), or in the centre and at the rear of the cave.

Also in Palawan, two bodies and funeral attributes were placed one century ago in a cave of Balagbag Cliffs. In Lungun Cave, a 1.15 m long hardwood log coffin (700–1000 AD) closed with pegs, was placed high above a small stream; iron tools and some pottery were found inside. A beautifully carved, 2.24 m long, wooden boat-coffin was found on a cliff ledge (13th or 14th century).

Wooden coffins are found in several mountain areas of Southeast Asia, especially in Luzon, Philippines, and Sulawesi, Indonesia. In Kabayan, northern Luzon, more than 20 caves are packed with hollowed tree trunks containing mummies in a foetal position, that are up to 400–500 years old. Bodies were rubbed with a medicinal plant decoction and smoked during long funeral rites. Carved coffins, such as a water buffalo-shaped coffin made of fine wood, were reserved for elite people. Skulls and bones are stored in some caves.

In Sagada, northern Luzon, cave and cliff burial is still occasionally performed. More than 50 sites exist, mainly on the karst fringe close to villages. Coffins are exposed on the cave floor, ledges (high above a cave stream in Matangkib Cave), horizontal poles, or placed in fissures, because locals believe that the souls of deceased adults are able to move and must be left free to do so. Selecting a future coffin location takes into account kinship, deceased person's quality and social role, and the availability of suitable locations. Infants, children, and women who died in childbirth are buried outside.

Coffins are traditionally made up of hollowed pine logs with a lid, and closed by vine ties or pegs. Occasionally, lids were carved (lizard, a symbol of felicity), pegs as well: they were reserved for highly appreciated persons. Normally no offering is placed in coffins. Bodies are commonly given a foetal position. Recent coffins are often nailed wooden planks and bear bodies laying on their back. A few burial jars have been discovered.

There can be no long time between death and body transfer to the burial site. The rituals to bring the corpses to the caves are by tradition rather strict. No sharp noise must occur during the body transfer to the coffin already placed in the cave, which explains why selected caves or cliffs are not distant from the village. Once all required sacrifices are completed, the deceased person is considered as definitely departed to another world.

In Tanah Torajah, Sulawesi, log coffins are placed in caves (or man-dug underground chambers) after long ceremonies, with numerous sacrifices. Selected caves are located at ground level or at various elevations on karst slopes. Some are accessible by long bamboo ladders used for one burial only. Rituals during coffin carriage to the burial site are less strict than in Sagada and loud noise is common. Coffin shaking and irregular loops are made, so the soul cannot remember the way back to the village. Then the coffin is quickly carried to its final place. The dead has now attained the status of an ancestor and a statue of him is placed near the burial place, facing the realm of the alive. Children cannot access this status and so are usually buried in hollowed trees. Recent to present burial caves are also known in Papua New Guinea, in Lifou Island, and nearby New Caledonia, in Tahiti and Marquesas Islands, Easter Island, and Patagonia (Alakaluf Indians).

In the United States, cave burials have been found for example in western Kentucky. In Salts Cave, the flexed mummy of a 9 year old boy (dated at 10 BC to 30 AD ± 160 years) was found high upon a ledge, desiccated by the dry, cool atmosphere. In Short Cave, mummies of a baby, an adult sitting in a stone grave with offerings, and a flexed or sitting adolescent, were all dressed with deerskins. In Mammoth Cave, a flexed female body was buried in an oval grave lined with burned grass or fibre matting. Burnt bones were uncovered separately. Short and Salts Caves revealed primary burials of people defleshed for this purpose. Ashes mixed with animal and human bones were found in crevices near the entrance (and also inside Salts Cave). Bones from foetuses to adult individuals were very abundant, with marks of defleshing (questionably burial rituals or cannibalism), and many were partly burnt—probably accidentally. Some ornaments and tools, and many fabrics were present. The remains are likely to belong to several Indian or close populations.

In eastern Kentucky, Ash Cave was used for primary interments of female adults and children; it also contains layers of ashes and vegetal remains with many fabrics. In Tennessee and Missouri, primary and secondary burials have been uncovered in caves. Burnt bones, stone slab graves, and pottery were found in Miller's Cave, Missouri.

Burial caves are known in South and Central America, for instance in Maya areas of Central America. In Brazil, 8–10 000 year old remains of the Lagoa Santa Man were found in Sumidouro and other caves in Minas Gerais, and 40 000 year old human remains in rock shelters of Piaui State.

Africa also has burial caves, e.g. of Dogon people in Bandiagara Cliffs of southern Mali.

Cave burials exist because favourable natural sites, adequate beliefs, and necessity are present at the same time. It is remarkable that this habit has lasted since Paleolithic times and that similarities are found in different locations and cultures. Although a variety of ritual uses of caves do still occur, burials are gradually disappearing under the pressure of Christianity and westernization of indigenous cultures.

CLAUDE MOURET

See also **Asia, Southeast: Archaeological Caves; Human Occupation of Caves; Religious Sites**

Further Reading

Chevillot, C. 1990. L'occupation du milieu souterrain en Périgord durant la Protohistoire [Occupation of underground realm in Périgord during Protohistory]. In *Les Cahiers de Commarque, 2è colloque sur le Patrimoine troglodytique*, Les Eyzies-de-Tayac

De Leon, L. 1976. The mummies of Benguet. *Philippine Panorama*, August 8: 52–54

Fox, R.B. 1970. *Tabon Caves: Archaeological Explorations and Excavations on Palawan Island, Philippines*, Manila: National Museum

Gargett, R.H. 1999. Middle Paleolithic burial is not a dead issue: The view from Qafzeh, Saint-Césaire, Kebara, Amud and Dederiyeh. *Journal of Human Evolution*, 37: 27–90

Mouret, C. 2000. Grottes et falaises sépulcrales de Luzon et Sulawesi [Burial caves and cliffs of Luzon and Sulawesi]. In *Actes de la 10ème Rencontre d'Octobre*, Paris: Speleo-Club de Paris

Robbins, L.M. 1974. Prehistoric people of the Mammoth Cave area. In *Archeology of the Mammoth Cave Area*, edited by P.J. Watson, New York: Academic Press

BURREN GLACIOKARST, IRELAND

The Burren plateau of northwest County Clare is the finest example of a karstic terrain in Ireland. The landforms of the Burren have been strongly influenced by glacial and periglacial processes, as well as by purely solutional, karstic processes, and in this respect the landscape resembles that of the Yorkshire Dales karst in England, with extensive areas of limestone pavement.

The Burren is bounded to the west and north by the Atlantic Ocean and Galway Bay, respectively. The southern boundary is where the limestone passes beneath younger shale and sandstone of Namurian age. This corresponds to a line from Corofin to Kilfenora to Lisdoonvarna to the coast at Doolin. The eastern limit of the Burren is the foot of the scarp at an altitude of approximately 60 m, which extends from Corranroo Bay in the north to Kilnaboy in the southeast (Figure 1). The area defined above is some 360 km² in extent and forms a gently inclined plateau at 200–300 m above sea level in the north and 100–150 m above sea level in the south, bounded by steep scarps on all but the southern flank. Only isolated summit areas exceed an altitude of 300 m and the highest point is the shale-capped Slieve Elva at 344 m. To the west, the three Aran islands are, in many respects, an extension of the main Burren.

The Burren karst is developed in Carboniferous limestones, with some 500 m of limestone succession exposed. Differing characteristics of the limestones are reflected in landforms and in hydrology throughout the Burren. For example, the Maumcaha Unit, 70 m in thickness, with few joints or bedding planes, is overlain by a 140 m thick strongly bedded sequence of limestones (the Aillwee unit), and underlain by a similar sequence, some 150 m thick. The massive unit forms steep, uniform slopes or cliffs, whilst the bedded units form tiers of stepped terraces, particularly on the northern flank of the Burren overlooking Galway Bay

The uppermost stage of the limestone is the Brigantian Slievenaglasha Formation, characterized by more coarse-grained rocks than the main Burren Limestone, and incorporating numerous nodules or sheets of chert between the limestone beds. The chert seems to function in a similar manner to the shale

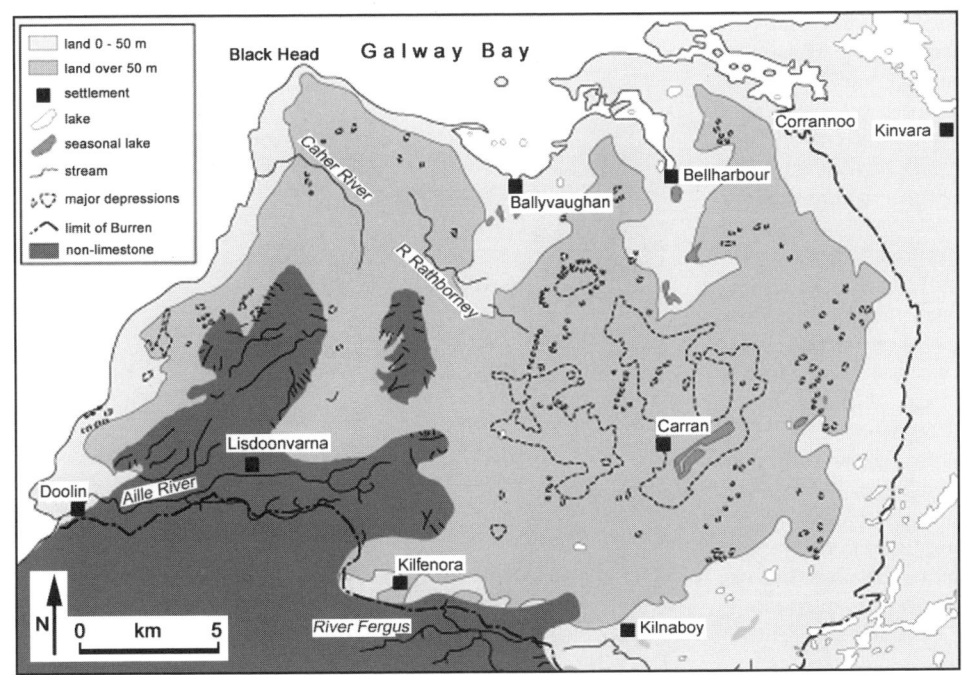

Burren Glaciokarst: Figure 1. Major depressions and seasonal lakes within the Burren Karst, Ireland.

layers, as an important control on the groundwater movement, and to an extent on landform development. Resting unconformably on the Brigantian limestones, are the Namurian sequence of calcareous shales, phosphatic shales, siltstones, and sandstones. In the western Burren, the Namurian rocks reach northwards to form the uppermost part of the hills of Slieve Elva and Knockauns, while the adjacent hill of Poulacapple is capped with a thin outlier of Namurian shales.

The rocks of the Burren are tilted to the southsouthwest with dips of 1–5 degrees over much of the area, with gentle asymmetric folding along northnortheast–southsouthwestern axes, only becoming apparent in the eastern Burren. Two major joint sets, oriented approximately north–south and east–west, occur over the whole area, their relative dominance and spacing varying locally. The slight inclination of the rocks over most of the Burren plateau forms an extensive dip-slope, mainly of Slievenaglasha Limestone, declining gradually in altitude from north to south and contrasting with the steep north-facing scarp slope.

The karst of the Burren has developed as the overlying Namurian strata have been stripped away by fluvial and glacial action, exposing the limestone to solutional erosion. In the past, as now, the contact between the acidic, impermeable Namurian rocks with surface rivers and the limestone, was a zone of intense karstification. Removal of the Namurian strata has extended westwards and southwards and so the oldest karsts are those of the northern and eastern Burren, whilst the youngest are those at the present shale–limestone contact in the Lisdoonvarna–Kilfenora area.

Annual precipitation on the Burren averages approximately 1500 mm, of which 1000–1100 mm becomes recharge. Point recharge via sinking streams is largely confined to the contact between the overlying impermeable Namurian rocks and the limestone in the west and southwest of the Burren. Elsewhere, recharge is diffuse and hence caves are uncommon. Underground drainage is strongly influenced by the geology. Numerous layers of clay, shale, and chert interbedded with the limestones, act as barriers to the vertical percolation of rainwater and encourage sub-lateral flow of water, particularly along bedding partings. The final outlet for the Burren groundwater is via peripheral springs located at hydrological lows, such as the River Fergus valley and the littoral or submarine springs on the Atlantic coast and Galway Bay, or at the contact with the overlying Namurian rocks to the south of the Burren. The springs on the west (Atlantic) coast occur up to 500 m offshore and at depths of up to 15 m below sea level.

Enclosed depressions, the diagnostic landforms of karst regions, are abundant: more than 1500 such features, exceeding 100 m^2 in floor area, occur on the Burren plateau, together with several hundred smaller hollows. One-third of the depressions exceed 1000 m^2 in area. The greatest concentration of enclosed depressions is in the central and eastern parts of the plateau, with a smaller clustering between Slieve Elva-Knockauns Hill and the coast. Only one-quarter of the enclosed depressions are circular in form, the remainder being elongate to some extent, with some 15% resembling sections of gorge or trench, rather than conventional dolines. The dominant orientation of the enclosed depressions is parallel or sub-parallel to the direction of ice movement during the last glacial advance (north–south) and glacial erosion may have enlarged many depressions and accentuated their elongation.

Burren Glaciokarst: Figure 2. Terraced limestone pavements on the nearly horizontal limestones of Slieve Carran. (Photo by Tony Waltham)

In the central Burren are two very large enclosed basins, sufficiently large to be described as poljes or large uvalas, with a combined area of c.16 km^2, enclosed by the 125 m contour. The Carran depression has an area of 9 km^2, extends to an average depth of 30 m below the surrounding plateau surface, and contains a stream fed by springs, which sinks at the southern end of the depression. In wet conditions, the swallow hole is unable to cope with the inflow of water and a lake forms in the depression, even though the floor of the basin is more than 70 m above the water table. To the west of the Carran depression, is the slightly smaller enclosed basin of Kilcorney-Meggagh, with a total floor area of 7 km^2.

Limestone pavements, or decayed pavements with surfaces of loose limestone slabs, cover 20% of the Burren plateau and northern flank (Figures 2 and 3). A further 30% of the plateau surface consists of a mosaic of bare rock and patches of thin rendzina soils. The best pavements develop where the dip of the rocks is low and where this dip corresponds to the slope of the

Burren Glaciokarst. Figure 3. Limestone pavement at Sheshy More. (Photo by John Gunn)

ground surface. Pavements are found on all the limestones of the Burren, but are particularly well developed on the highest beds of the Brigantian, near the contact with the Namurian strata.

Along much of the Burren coastline, and especially on the western Atlantic facing shore, limestone bedrock is exposed on the foreshore. Between high and low tide levels, a distinctive suite of karren, mainly kamenitza, has developed in response to erosion of the limestone, by a combination of biologically driven dissolution and direct erosion by marine organisms (see Coastal Karst).

The great majority of extensive cave systems on the Burren are associated with streams that sink at, or close to, the contact between the limestone and the overlying Namurian shales and sandstones, and are therefore active stream caves. Many of these caves have developed in the north–south joint systems and are oriented north–south down the dip of the strata. Such caves are commonly developed within a single bedding plane for long distances, typically perched above a chert-rich horizon, and occasionally stepping down one bed or more via a vertical drop. The caves are linear or dendritic in plan view. The passage form of these caves typically comprises a wide, low roof section of enlarged bedding plane incised by a narrow trench or canyon that contains the stream. The trench is typically 0.3–3 m wide but may be up to 30 m deep. Passage cross sections are therefore typically T-shaped.

Some of these caves (e.g. Polldubh) are simple, single stream conduits, in which the morphology and dimensions of the cave passage are obviously related to the present day stream and may be of Holocene age. Others have a more complex morphology, with abandoned sections of passage, oxbows, and offset meander belts at different heights above the active stream route, together with calcite and sediment accumulations, for example, the Pollnagollum and Doolin systems. The longest cave known is Pollnagollum, with more than 14 km of passages surveyed. Some three other caves exceed 5 km in length and a further 17 exceed 1 km in length. Vertical development in the caves is limited, with the exception of Poll na gCeim (130 m deep) and Faunarooska Cave (95 m deep)

There are also caves of presumed great age, often in locations wholly disjuncted from the present-day hydrology. Examples of such caves are Vigo Cave and Aillwee Cave. Sediment infills are widespread in these caves, ranging in type from complex bedded sequences of fine sediments (e.g. Pol-an-Ionain and Aillwee Cave) to extensive sand deposits of unknown origin, as in Glencurran Cave. Pol-an-Ionain also contains the largest stalactite in the British Isles (7 m; Self, 1981).

The age of the Burren caves has not been thoroughly investigated, but limited evidence from U-series dating of calcite specimens suggests that the simple stream passage caves may be of Holocene age, whilst the more complex stream caves may originate from at least last interglacial times (120–132 000 years ago). Dates from Aillwee Cave suggest an age considerably greater than half a million years.

DAVID DREW

See also **Europe, North (for location map); Karren**

Works Cited

Self, C.A. 1981. *Caves of County Clare*, Bristol: University of Bristol Spelaeological Society

Further Reading

Drew, D.P. 1990. The hydrology of the Burren. *Irish Geography*, 23(2): 69–89

Drew, D.P. 2001. *Classic Landforms of the Burren Karst*, Sheffield: Geographical Association

Simms, M. 2001. *Exploring the Limestone Landscapes of the Burren and Gort Lowlands*, Belfast: burrenkarst.com

Tratman, E.K. (editor) 1968. *The Caves of North-West Clare, Ireland*, Newton Abbot: David and Charles

Williams, P.W. 1966. Limestone pavements with special reference to western Ireland. *Transactions of the Institute of British Geographers*, 40: 155–72

C

CALCAREOUS ALPS, AUSTRIA

About one sixth of the surface of Austria is comprised of karstified rocks (see map in Europe, Alpine entry), and some 14 000 caves had been surveyed as of 2001. More than 90% of these caves occur within the Northern Calcareous Alps, but there are also notable areas of karst and caves in the Central Alps (e.g. see entry on Spannagel Cave) and in the Southern Calcareous Alps. The great economic significance of these karst terrains is illustrated by the fact that many major cities in Austria are supplied by high-quality karst-derived drinking water. The geological history and evolution of large cave systems in the Eastern Alps of Austria has traditionally been attributed to successive phases of uplift since the Tertiary, subsequently modified by the pervasive Pleistocene glaciations. Recently, Frisch *et al.* (2001) elaborated this model further by combining structural and thermochronological data. According to their model, major cave formation commenced during the Late Miocene under climatic conditions significantly different from the present, and continued throughout the Pliocene and the Quaternary.

The western karst plateaus of the Northern Calcareous Alps, west of the Salzach River, include the massifs of the Leoganger Steinberge, the Steinernes Meer, and the Hagengebirge. In the Leoganger karst, Lamprechtsofen was until 2001 the world's deepest explored cave (-1632 m). It extends from the high plateau down to the Saalach River with a surveyed length of nearly 50 km, and emerges as a water cave where the final part is developed as a show cave. East of the Saalach River, the more extensive plateau of the Steinernes Meer (which translates into English as "rocky sea") contains the Kolkbläser-Monsterhöhlen system with a measured length of 44 km and at least four genetically different passage series. In the Hagengebirge, the major caves are Tantalhöhle (34 km long) and Jägerbrunntroghöhle (30 km long, 1078 m deep). East of the Hagengebirge, the deeply entrenched Salzach River has created a local relief of nearly 1500 m, which is crucial for the latest evolution of the subsurface karst.

Tennengebirge is a major karst plateau, some 20 km southeast of Salzburg. It has more than 700 known caves, including some of the longest in Austria; Eisriesenwelt is one of the largest ice caves in the world, with a surveyed length of about 40 km. The Tennengebirge, like most of the other plateaus of the Calcareous Alps, is composed of thick Upper Triassic Dachstein Limestone, locally with steep dips. As a result of this thickness, in conjunction with successive uplifts and periods of stability, the massif has recorded all of the karstification phases, now laid out as recognizable cave horizons. The evolution of the Tennengebirge is clearly shown in Figure 2 by the morphology of the Cosa-Nostra-Bergerhöhle system (Audra, 1994). Karstification began during the Oligocene, beneath a cover of gravel deposits, known as Augensteine (1), which were progressively stripped from the Lower Miocene onward. During the Miocene, deep karst conduits developed, thanks to the input of major drainage from the non-karstic alpine ranges, which fed into the tectonically controlled networks (2). These horizontal levels, developed during periods of tectonic stability, have been layered by successive phases of uplift (3). The oldest and highest (~ 2100 m), still to be found among old Tertiary cones (Kuppen), corresponds to the Ruinenhöhlen (4), cavities which are currently being destroyed by erosion (including Hornhöhle). At altitudes of 1500–1700 m, large labyrinthic networks occur (5), known as the Riesenhöhlen (the giant-cave level, e.g., Eisriesenwelt). Following a major uplift during the Pliocene, torrential runoff from the Central Alps could no longer reach the karst systems of the Northern Calcareous Alps. Subsequently, vertical passages developed as a result of the increase in hydraulic head (6), and their formation accelerated during glaciations under the influence of local ice caps and the Salzach glacier.

The Cosa-Nostra-Bergerhöhle system includes several generations of cave networks. The entrance of the Cosa Nostra Cave is a tubular gallery of Miocene origin containing reworked weathered sediments (7). The remaining parts of the system are of Pliocene and Quaternary origin. The shaft zone, as much as 750 m deep in places (6), is then linked to the Bergerhöhle-Bierloch tubular horizontal network (8). During the Pliocene,

Calcareous Alps, Austria: Figure 1. Ice-scoured limestone surfaces with massive rinnenkarren on the Rofangebirge, one of the western massifs of the northern Calcareous Alps. (Photo by Tony Waltham)

the valley floor was at an altitude of about 1000 m, and the karst system was fed by sinking streams, transporting Jurassic pebbles into the caves (9). Following the lowering of the water table, the caves became vadose, and the clastic sediments were then overlain by a first generation of large flowstones bearing a reversed paleomagnetic polarity (i.e., older than 780 000 years). During the glacial periods, the deeper parts of the system were episodically flooded up to over 600 m high, and deposition of carbonate-rich varves occurred, dated as younger than 780 000 years by their normal paleomagnetism. These varves cover older sediments. The bottom of the network is linked to the 700 m high phreatic zone (10), which corresponds to a structural barrier. During high stage, water pours into the Brunnecker Cave, where the discharge finally emerges at the present 500 m high base level of the Salzach River (11). Brunnecker Cave contains granite pebbles originating from old Salzach glacial deposits, and these are responsible for the canyon and pothole morphology. At present, a second generation of less abundant flowstones (one has a Th/U date of 5200 years BP) is developing in the Bierloch and Brunnecker Caves.

The eastern karst plateaus include the Totes Gebirge (with the largest area of high alpine karst in Austria) and the Dachstein massif (with the highest peak of the Northern Calcareous Alps at 2994 m). The three longest caves in Austria are Hirlatzhöhle (87 km) and Dachstein-Mammuthöhle (58 km) in the Dachstein, and Raucherkarhöhle (79 km) in the Totes Gebirge, and there are many other significant caves such as DÖF-Sonnenleiterschacht (16 km, −1054 m) and Schwarzmooskogeleishöhle Gebirge (54 km, −1030 m). It is notable that the vadose and epiphreatic zones of some of these caves are already reaching the modern base levels and resurgence levels. Nevertheless, the base flow even of the large karst springs in this area is dominated by small joints rather than by cave passages (Pavuza, 1998). Together with the Eisriesenwelt (in the Tennengebirge) the Dachstein-Rieseneishöhle, and to some extent the Dachstein-Mammuthöhle, represent the rare type of perennial ice caves that contain large underground glaciers. The ice of the Dachstein Caves is surprisingly young, as ^{14}C dates on organic debris from the ice base yielded an ice age of not more than 800 years. This implies that these caves became ice-free at least once during the Holocene. Currently a distinct decline of the ice bodies can be observed (Mais & Pavuza, 2000). On the plateau surface, the extensive karren landscapes emerged after the retreat of the last Pleistocene glaciers some 10 000 years ago. Ongoing measurements (using limestone tablets) indicate that surface corrosion rates are about 100 mm/10 ka, which correspond well with the heights of the uneroded basements of the limestone pedestals.

PHILIPPE AUDRA AND RUDOLF PAVUZA

See also **Alpine Karst; Europe, Alpine**

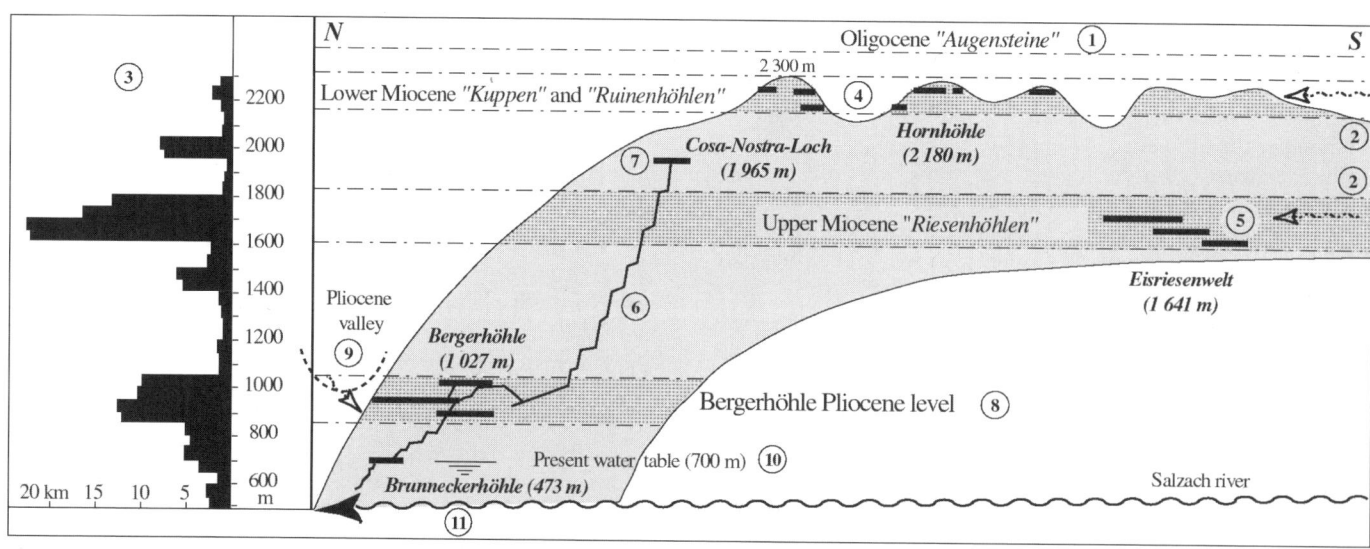

Calcareous Alps, Austria: Figure 2. The Cosa Nostra-Bergerhöhle system, Tennengebirge. The altitudinal distribution of known cave passages is shown on the left.

Works Cited

Audra, P. 1994. Alpine karst speleogenesis: Case studies from France (Vercors, Chartreuse, Ile de Crémieu) and Austria (Tennengebirge). *Cave and Karst Science*, 21(3): 75–80

Frisch, W., Kuhlemann, J., Dunkl, I. & Székely, B. 2001. The Dachstein paleosurface and the Augenstein Formation in the Northern Calcareous Alps—a mosaic stone in the geomorphological evolution of the Eastern Alps. *International Journal of Earth Sciences*, 90(3): 500–18

Mais, K. & Pavuza, R. 2000. Hinweise zu Höhlenklima und Höhleneis in der Dachstein-Mammuthöhle (Oberösterreich) [Remarks on cave climate and cave ice in the Dachstein-Mammoth Cave], *Die Höhle (Wien)*, 51(4): 121–25

Pavuza, R. 1998. Zur Hydrochemie des Hirlatzgebietes [Hydrochemistry of the Hirlatz, Dachstein-Massiv], In *Die Hirlatzhöhle im Dachstein*, edited by Buchegger, G. & Greger, W. *Wissenschaftliche Beihefte zur Zeitschrift Die Höhle*, 52: 214–20

Further Reading

Klappacher, W. et al. (editors). *Salzburger Höhlenbuch* [Cave inventory of Salzburg], Salzburg: Landesverein für Höhlenkunde, 6 vols

CANADA

Karst rocks are widely distributed around the central Precambrian Shield in Canada (Figure 1). In outcrop, there are ~500 000 km² of limestone and marble, 600 000 km² of dolostone, and 80 000 km² of evaporite rocks, a total similar to China, where the world's most extensive karstlands are found. Along the southern and western fringes of the shield, these rocks occur as thin, little disturbed sedimentary veneers of Paleozoic age, dominated by shallow marine, lagoonal, and reefal facies, and passing beneath younger sandstones, shales, and clays downdip. To the east and north, deposits are thicker, more varied, generally folded and faulted. The Western Cordillera contains thick sequences of shallow marine, reefal, and lagoonal facies, plus deeper marine rocks on the west coast. These mountain ranges are Cretaceous–Tertiary in age, with some uplift and deformation continuing today.

Modern ecoclimatic conditions on the karst range from coastal rainforest to high alpine and polar deserts. The southern interior has a continental climate, with moderate precipitation and cold winters. Permafrost is widespread in central regions and continuous in the north.

Almost all of Canada has experienced repeated glaciations, including general submergence under Laurentide Continental and Cordilleran ice sheets during the last (Wisconsinan / Weichselian) glaciation. Effects of glaciation have been chiefly destructive or inhibitive—erasing, dissecting, and burying karst landforms, and clogging aquifers and caves with injected detritus. "Outcrop" is something of a misnomer in much of the lowlands, because the karst bedrocks are buried under many metres of glacial till, outwash, etc. and are often hydrologically inert. However, in suitable conditions, glacial deposits may also preserve karst aquifers and enhance their productivity, while the most rapid rates of cave development anywhere probably occur around glacier margins during the melting stages (Ford, 1987).

To the north, permafrost becomes progressively thicker and more continuous, restricting karst by limiting groundwater circulation to the seasonally thawed zone or to leakages (taliks) into sub-permafrost groundwater. There is little modern development along the Arctic mainland coast or in the islands, where "felsenmeere" (tracts of frost-shattered rubble) are predominant. Some limestone karren and epikarst groundwater flow are reported where hydraulic gradients are steep (e.g. Akpatok Island) and a few shallow dolines and sinking streams lie in gypsum and salt exposed in diapirs.

The Canadian Shield

The Shield is chiefly composed of igneous or metamorphic rocks unsuited to karstification, although dolines are reported in a carbonatite (calcium carbonate igneous rock) in Ontario. Thick marbles in the southern shield in Ontario and Quebec contain some postglacial "shortcut" caves through narrow ridges and a few older phreatic relicts. In the Belcher Islands, Hudson Bay, pavements and small dolines occur on some younger Proterozoic limestones that escaped metamorphism.

The St Lawrence and Western Interior Platforms, Hudson Bay Lowlands

These lowlands are remnants of the general cover of Paleozoic carbonate and evaporite rocks deposited over the worn-down shield. Strata dip gently away from the shield, forming wide plains edged by low but steep escarpments facing inwards. The combination of low relief and widespread burial by glacial deposits limits modern karst to narrow zones of higher hydraulic gradient, such as crests of escarpments or river gorges.

Anticosti Island is wholly carbonate, with relief up to an altitude of 300 m, and an exceptionally strong master joint set. Most formations are thin-bedded and shaly, however, restricting development to shallow epikarst. Thicker limestones, in the Salmon River basin, support a holokarst with deep kluftkarren, desiccated string bogs, and spectacular stream-sinks.

There are many local features between Quebec City and the Ottawa Valley; Boischatel, St Casimir, and Bonnechere are notable postglacial stream caves (Beaupre & Caron, 1986). Beneath and alongside the Ottawa River are ~10 km of "textbook" anastomosing bedding plane passages forming today, with fierce currents and accessible only to divers. Limestone terrain between Lake Ontario and Georgian Bay in Ontario is notable for extensive tracts of shallow pavement, with distinctive alvar ecology.

Niagaran dolostones form a ~50 m escarpment extending 900 km from New York State, through Niagara Falls, the Bruce Peninsula, and Manitoulin Island in Ontario, to the Door Peninsula of Wisconsin. Karren pavements are common along the scarp crest, broadening into peninsulas of holokarst between glacial breaches locally. Attractive sea caves and coastal karren are found along the Great Lakes shores and there are short mazes and stream caves elsewhere.

The Hudson Bay Lowlands are a muskeg (peat bog) terrain on glaciomarine sediments burying carbonate rocks. The land

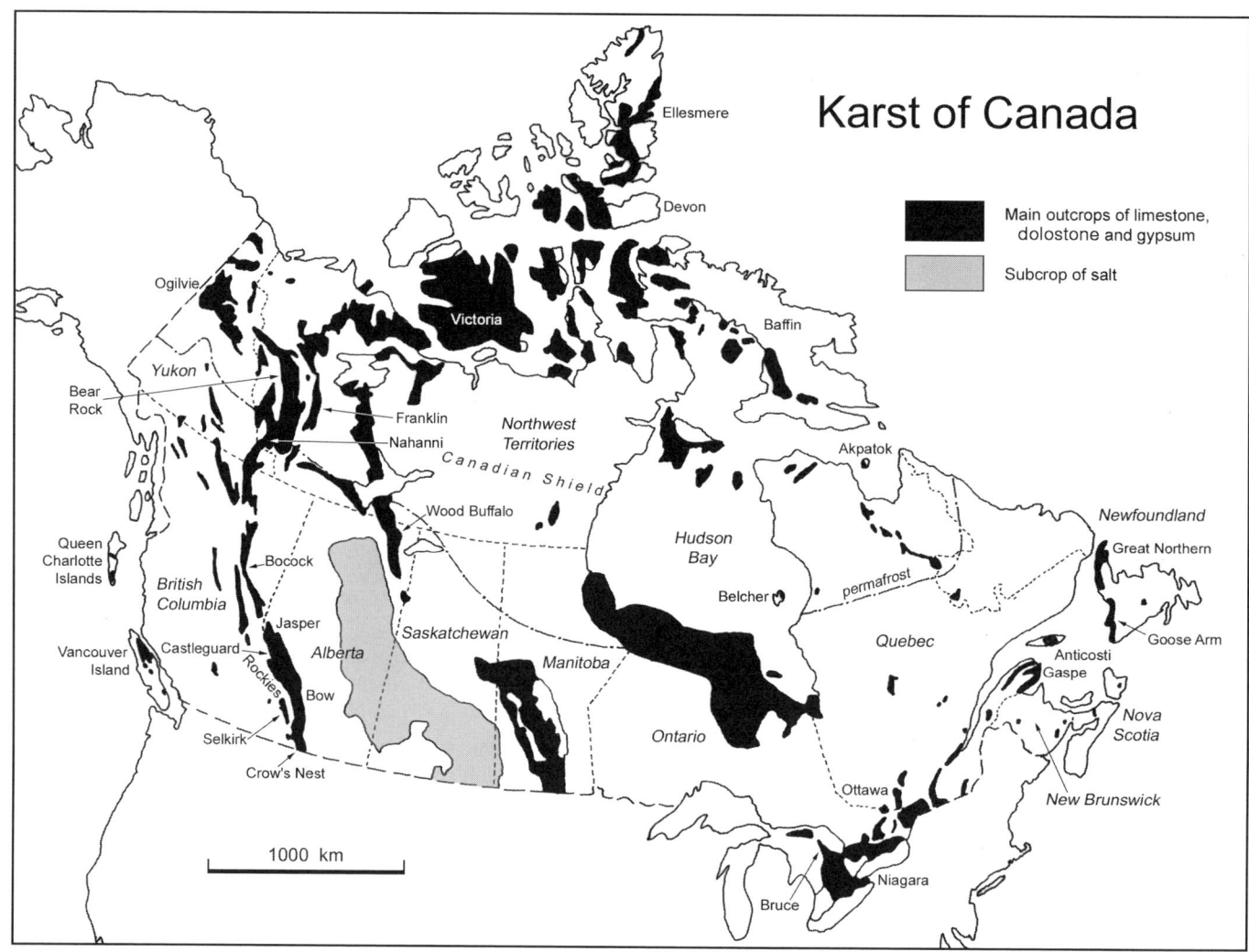

Canada: Figure 1. The outcrop of limestone, dolostone, and sulfate rocks in Canada and the subcrop of the Prairie salt.

is still rebounding vigorously from glacial loading. Karst is seen in small gypsum outcrops and where rivers entrench the rising land. Resistant bioherms (organically derived rock mounds), surrounded by moats of dolines, rise like castles above the muskeg, draining it to the nearest gorge.

In Manitoba, the platform consists of ~400 m of dolostone, minor limestones, and gypsum. The city of Winnipeg rests on a glaciolacustrine clay plain that buries and preserves >400 km² of dolostone pavement, converting it into a productive aquifer. Between Lake Winnipeg and Lake Winnipegosis are tracts of doline karst, some serving as snake hibernacula. Quite exotic is the Lake St Martin impact crater (Jurassic) filled with lagoonal gypsum, that now hosts many short caves (McRitchie, 1991). There are extensive, shallow pavements north of the lakes.

South of the Great Slave Lake there is an area of 50 000 km² underlain by >850 m of carbonates and evaporites. In the south, the Athabasca Tar Sands (the largest known oil deposits) are partly accumulated in paleokarst traps. Further north, Wood Buffalo National Park contains the most extensive gypsum karst in Canada, chiefly suffosion dolines where there is only glacial till on the gypsum, and larger collapse dolines with a thin cover of dolostone. Known caves are few, becoming water-filled or blocked by ice. There are large regional springs (blueholes) with high salt contents (Ford, 1997). West of the park is Pine Point, a remarkable Devonian doline and cavernous paleokarst that trapped important lead/zinc deposits: Quaternary glacial loading rejuvenated some of the dolines.

Beneath much of Alberta and Saskatchewan, the Devonian Prairie Salt is >200 m thick where not thinned by dissolution, >500 000 km² in extent, and buried 200–2500 m beneath later carbonate and clastic rocks. The eastern (up-dip, shallow) edge has receded by an average of 130 km westwards, as a consequence of interstratal solution. More than 100 breccia pipes, and larger solution troughs up to 360 km² in area, are known in Saskatchewan and Alberta. Many of these are linked to effects of glacial loading and recession, although the surface expression is very muted (Anderson & Hinds, 1997).

Atlantic Region

The Great Northern Peninsula of Newfoundland has tracts of limestone pavement and well-developed coastal karst despite the fact that sea ice is fast to the shore for four or more months each winter. The Goose Arm karst displays >250 m of relief, conical hills, and large closed depressions in steeply dipping Ordovician limestones and dolostones. It is interpreted as a cone karst, highly modified by glacial erosion and disruption of groundwater flow.

Further south in Newfoundland, Nova Scotia, and New Brunswick, the karst is chiefly of collapse and suffosion dolines in thick, locally deformed, gypsum and anhydrite, buried by glacial deposits. Densities are up to 100 km^2 and the largest individuals are >500 m in diameter. Where the overburden is thin there are patches of "schlottenkarren" (solutional funnels 0.5–1.5 m in diameter, tightly packed together). Caves are short, with much breakdown. In the Gaspe Peninsula, Quebec, narrow bands of steeply dipping limestone display some modern karst drainage and three ancient caves modified by glacial action.

The Western Cordillera

This area comprises the Rocky Mountains of Alberta–British Columbia, coastal and interior ranges, Vancouver Island and the Queen Charlotte Islands in British Columbia, the Mackenzie Mountains and Franklin Mountains in the Northwest Territories, and outlying ranges in Yukon Territory. The majority of Canada's known caves, including all of the longest and deepest, are found in this large region, as well as the most accentuated surface karst features. The southern Rockies and Vancouver Island are the most intensively explored areas.

The southern Rockies extend 950 km from Crows Nest Pass, to Mt Bocock, near Fort St John. Eastern (Front) ranges are composed chiefly of thick Paleozoic carbonates overthrust on Mesozoic clastic rocks, creating steep narrow cuestas. "Main" ranges are broader, with thick carbonates overlying massive quartzites that contain thin limestones. The mountains are sculpted into classic alpine glacial forms. Some hundreds of cirques and U-shaped valleys display karst drainage. There are major karst depressions overdeepened by glacier scour. The best development occurs on plateaus or benches above deep entrenchments. Pavements, shafts, all types of sinkholes, and relict caves are found, draining to major springs in the valleys (Ford, 1979).

The greatest density of known caves is around Crows Nest Pass, a deep entrenchment that attracts strike-oriented karst drainage from north and south. The topographic Continental Divide (between drainage to the Atlantic and Pacific) is breached by Yorkshire Pot (14 km of passages, 390 m deep) and Gargantua (Figure 2; 6 km long, 290 m deep), two excellent examples of alpine multiphase caves with phreatic loops. Between Banff and Jasper, Castleguard Cave (see separate entry) extends beneath the Columbia Icefield, and is therefore a classic feature of glaciokarst.

The Bow Valley is noted for Banff Hotsprings, a major thermokarst discharge via small caves, and Ratsnest Cave (4 km long, 245 m deep), another multiphase complex with evidence

Canada: Figure 2. The ancient phreatic tubes of Gargantua Cave are truncated in the limestone cliffs beside the Ptolemy Meadows in the high Canadian Rockies; the main entrance to the cave is reached by a traverse across the lower snowfield. (Photo by Tony Waltham)

of many episodes of glacial flooding. Near Jasper is the Maligne River, the largest-known sinking river in Canada. Its catchment of 2300 km^2 drains into a polje, Medicine Lake, where the waters pass underground for 16 km, descending ~400 m to springs aggraded by glacial deposition. Mean discharge is ~13 m^3 s^{-1}, maximum >45 m^3 s^{-1}. Underground flowtime is as short as 11 hours during floods (Smart, 1988). In the Main Ranges west of Jasper, the Mural Formation is a thinner limestone between quartzites. Where it dips steeply, the deepest Canadian cave, Arctomys, has developed as a staircase of vadose potholes descending to a siphon at a depth of 536 m (Rollins, 2002). Where it is more level, at Small River, there is a model sub-glacier karst.

North of Jasper there is active exploration but few studies. Plateaus near Bastille Mountain, at Moon River, and in the Dezaiko Range, are reported to contain hundreds of dolines, many caves, and resurgences. "Close to the Edge" is a window into a spectacular shaft, 254 m deep. The most northerly discoveries are short caves in a 45 m thick Triassic limestone near Mt Bocock (Rollins, 2002).

The best karst development in the interior of British Columbia occurs in narrow, steeply dipping bands of marble, i.e. "stripe karst". Nakimu Caves (4.5 km long, 270 m deep) and the Mt Tupper system in the Selkirk Mountains are the most spectacular, channelling alpine meltwaters in roaring cascades. Further north, near Prince George, there is Fang Cave, another dynamic river system.

The Franklin Mountains and Mackenzie Mountains lie, respectively, east and west of the Mackenzie River, between the latitudes of Great Slave and Great Bear Lakes. They are very extensive and scarcely inhabited. The Franklins and eastern half of the Mackenzies (the "Canyon Ranges") are composed of thick Paleozoic carbonate formations with much gypsum, anhydrite, and salt. The Franklins contain many large dolines, where dolostone and other cover rocks have collapsed into gypsum. Around the west end of Great Bear Lake, van Everdingen (1981) mapped ~1400 dolines, 27 larger compound features, and 63 major springs. This is at the discontinuous–continuous permafrost transition. Dolines diminish in frequency further north.

South Nahanni & Bear Rock Breccia

South Nahanni (see Nahanni entry) is a spectacular karst in the southern Mackenzie Mountains and the Bear Rock Breccia (see Bear Rock Karst entry), 450 km further north, supports another significant karst. Between them are small patches of Nahanni-type corridor karst with springs and caves, a moraine-dammed polje, and some large karst depressions and pavements on dolostones.

Yukon Territory

In Yukon Territory relict caves, a sinking river, and several large perennial springs (icings) are known in limestones in the remote Ogilvie Mountains, a region that escaped glaciation (too dry) but has continuous permafrost. The caves are important as refugia and because of massive, relict speleothems attributed to late Tertiary (pre-permafrost) conditions (Lauriol et al., 2001).

Vancouver Island and the Queen Charlottes

This area has the most dynamic modern karst settings in Canada. The principal karst rock is a fine-grained marine limestone up to 800 m in thickness, underlain and overlain by volcanics and partly metamorphosed rocks. It crops out from sea level to an altitude of 2000 m, mostly in steeply dipping, strike-oriented mountain ranges. Intense glacial erosion has suppressed development of large surface karst features, although there are a few big, ice-modified, closed depressions. Some peaks reach the alpine tundra but most limestone slopes are covered by coniferous forests. Precipitation ranges from 1500 mm a^{-1} at sea level up to a possible 5000 mm or more at the summits.

The principal surface karst landforms are abundant karren and small dolines. Hundreds of caves are known, the majority being explored only for short distances because of constrictions, sediment blockages, or siphons. However, five of the ten longest and deepest Canadian caves are on Vancouver Island, including Arch-Treasure Cave (~8 km), and Thanksgiving Cave, 400 m deep. Speleothems are abundant, chiefly of calcite and postglacial in age. There are mammal and bird remains in entrances and a few instances of low-lying caves being used for interments. Clear-cutting forest practices have substantial impacts in some karst areas, chiefly soil erosion and cave sedimentation (Harding & Ford, 1993).

DEREK FORD

Works Cited

Anderson, N.L. & Hinds, R.C. 1997. Glacial loading and unloading: A possible cause of rock-salt dissolution in the Western Canadian Basin. *Carbonates and Evaporites*, 12(1): 43–52

Beaupre, M. & Caron, D. 1986. *Découvrez le Québec souterrain*, Québec: Presses de l'Université du Québec

Ford, D.C. 1979. A review of alpine karst in the southern Rocky Mountains of Canada, *National Speleological Society of America Bulletin*, 41(3): 53–65

Ford, D.C. 1987. Effects of glaciations and permafrost upon the development of karst in Canada. *Earth Surface Processes and Landforms*, 12: 507–21

Ford, D.C. 1997. Principal features of evaporite karst in Canada. *Carbonates and Evaporites*, 12(1): 15–23

Harding, K.A. & Ford, D.C. 1993. Impacts of primary deforestation upon limestone slopes in northern Vancouver Island, British Columbia. *Environmental Geology*, 21: 137–43

Lauriol, B., Prevost, C., Deschamps, E., Cinq-Mars, J. & Labrecque, S. 2001. Faunal and archeological remains as evidence of climate change in freezing caverns, Yukon Territory, Canada. *Arctic*, 54(2): 135–41

McRitchie, W.D. (editor) 1991. *Caves in Manitoba's Interlake Region*, Winnipeg: Speleological Society of Manitoba

Rollins, J. 2002. *Caves of Canada,* Calgary: Rocky Mountain Books

Smart, C.C. 1988. Quantitative tracing of the Maligne Karst Aquifer, Alberta, Canada. *Journal of Hydrology*, 98: 185–204

Van Everdingen, R.O. 1981. Morphology, hydrology and hydrochemistry of karst in permafrost near Great Bear Lake, Northwest Territories. National Hydrological Research Institute of Canada (Paper 11)

Further Reading

Gadd, B. 1995. *Handbook of the Canadian Rockies*, 2nd edition, Jasper, Alberta: Corax Press

Yonge, C.J. 2001. *Under Grotto Mountain: Rat's Nest Cave*, Calgary, Alberta: Rocky Mountain Books

CANARY ISLANDS: BIOSPELEOLOGY

The Canary Islands support an interesting, highly endemic invertebrate fauna, and are an excellent model for understanding island evolution. The subterranean environment includes a rich fauna that has evolved independently of that occurring on the adjacent African mainland; all species are locally endemic and some belong to unexpected animal groups.

The archipelago is located in the northeast Atlantic between 27°37' and 29°35'N, and 13°20' and 18°10' W. There are seven main islands aligned from east to west: Lanzarote, Fuerteventura (the nearest to Africa, 110 km), Gran Canaria, Tenerife, La Gomera, La Palma (the most distant, 460 km), and El Hierro. They are all of volcanic origin and were built up from the sea floor. In consequence they have never had subaerial contact either to each other (except Lanzarote and Fuerteventura) or to the mainland. The eastern islands, Lanzarote and Fuerteventura, are the oldest (15–21 Ma) and the age diminishes towards the western islands, El Hierro being the youngest (<1 Ma). The most probable geological origin is explained by the hotspot theory, where successive volcanic islands form as tectonic plates move over a stationary hotspot in the mantle, though volcanic activity has started again in some of the older, eastern islands, and stopped in some western (La Gomera) each one following independent cycles according to its stage of erosion, which alternately activates and stops volcanic periods (Carracedo et al., 1998).

The climate is subtropical, strongly influenced by the humid northeast trade winds, which combined with the altitude and the northwest dry winds, produce an inversion layer and marked vegetational zones. The eastern islands are lower and much drier, but the central (Tenerife 3714 m) and western islands are high and more humid, especially on their north slopes.

There are volcanic terrains of acidic and basaltic nature, the latter being more extended and suitable for the development of subterranean habitats. The most abundant caves are lava tubes, formed only in pahoehoe basaltic flows, but there are also volcanic pits and a network of voids and cracks, which together with the mesovoid shallow substratum (MSS, or terrestrial interstitial habitat) also form habitats supporting hypogean fauna. Lava tubes usually collapse or become silted within a few hundred thousand years, thus only persisting in relatively young parts of the islands and being almost absent on Gran Canaria and La Gomera where there has hardly been any recent vulcanism. The richest islands in such caves are Tenerife (131 recorded tubes and pits), Lanzarote (30), La Palma (70), and El Hierro (35). The longest cave systems are Cueva del Viento (17.5 km) and Cueva de Felipe Reventón (1.9 km) in Tenerife, Cueva de Don Justo (6.5 km) in El Hierro, and Cueva de los Verdes (6.5 km) in Lanzarote. However, richness of adapted cave fauna does not depend on the extent of the caves, but on their environmental conditions; thus lengthy but dry caves such as Cueva de los Verdes have no terrestrial troglobites at all, while the small Cueva de los Roques (Tenerife) holds 24 troglobitic species, 4 of them endemic to the cave.

Terrestrial troglobites have evolved locally and are exclusive to each island, or even to a part of an island if underground barriers prevent such animals dispersing. The Anaga peninsula of Tenerife developed as an independent island that joined to Tenerife when this grew about 2 Ma ago; many genera are represented by different, vicariant troglobitic species in Anaga and in the rest of the island, but never overlap their distribution (Oromí & Martín, 1992). When such barriers do not exist, the same species can be present in distant, apparently isolated caves on the same island, because the network of cracks, and especially the MSS, allow dispersion even in terrains lacking tubes. As a result of this phenomenon, very recent lava tubes like Cueva de Todoque (formed in 1949) in La Palma have already been colonized by troglobites. Even on islands without suitable caves (Gran Canaria and La Gomera) there are fauna adapted to the MSS (Medina & Oromí, 1990).

Lanzarote has abundant lava tubes, but has a dry climate with scarce soil cover making most caves almost sterile, with no adapted terrestrial fauna known so far (Martín, 1989). The situation in Fuerteventura is much the same except for Cueva del Llano, a humid cave where two troglobites occur: the long-legged spider *Spermophorides fuertecavensis*, and the harvestman *Maiorerus randoi*, a unique species of the suborder Laniatores in the Canaries (see Figure 1). Dry lava tubes are occupied mainly by lavicolous species such as the cricket *Hymenoptila lanzarotensis*. The aquatic fauna, especially that of anchialine habitats which are widespread in Lanzarote, is much richer. Jameo del Agua and Túnel de la Atlántida are remarkable caves for both their large dimensions and their fauna of marine origin, with at least 19 stygobitic species. These include the abundant polychaete worm *Gesiella jameensis* and the anomuran crab *Munidopsis polymorpha*, a tourist attraction in the show-cave of Jameo del Agua. The anchialine waters of Lanzarote harbour the richest assemblage of misophrioid copepods in the world, with at least six endemic species belonging to five different genera, two of them monotypic and one, *Palpophria aestheta*, being the single representative of the family Palpophriidae. Other crustaceans of special interest are *Stygotantulus stocki* (Tantulocarida), *Halosbaena fortunata* (Thermosbaenacea), *Heteromysoides cotti* (Mysidacea), *Spelaeonicippe buchi* (Amphipoda), and the only

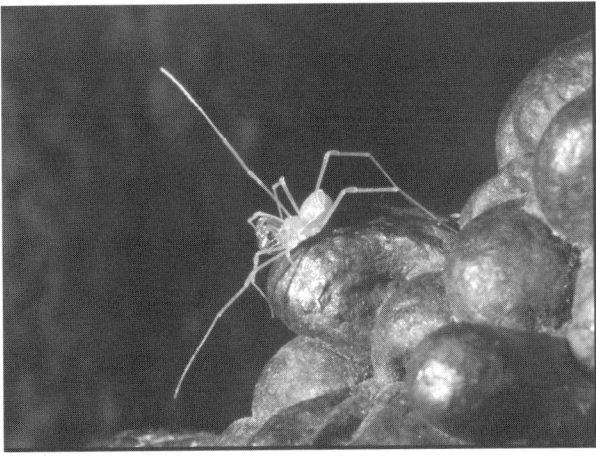

Canary Islands: Figure 1. *Maiorerus randoi* (Opiliones: Phalangodidlae), the only species of a relictual, endemic genus with no relatives in the Canary Islands. It is unique to a cave in Fuerteventura island. (Photo by Pedro Oromí)

remipede species known so far from the eastern Atlantic: *Speleonectes ondinae*.

Tenerife is by far the richest island both in lava tubes and hypogean terrestrial fauna, its species generally being more troglomorphic than in other islands. As many as 64 troglobites are known to date, some with their ancestral epigean equivalent species still living on the surface; this is the case for the millipedes *Glomeris* sp. and *Dolichoiulus* spp., the centipedes *Lithobius speleovulcanus* and *Cryptops vulcanicus*, the wood louse *Venezillo tenerifensis*, and the planthopper *Tachycixius lavatubus*. Others are relict species without extant surface ancestors on the archipelago. For example, the spider genus *Canarionesticus* (1 sp.); the ground beetles *Wolltinerfia* (3 spp.), *Canarobius* (2 spp.), and *Spelaeovulcania* (1 sp.); and the weevils *Oromia* (2 spp.) are all endemic genera including only troglobitic species. The range of underground niches together with long timescales and geological features such as underground barriers and other isolating mechanisms have promoted several cases of adaptive radiation: the spider genus *Dysdera* has seven troglobitic species on the island (most of them being sympatric), and the *Loboptera* cockroaches include 11 species (mostly allopatric with different degrees of adaptation to the underground; see Figure 2). Tenerife and other western islands are remarkable for the variety and abundance of rove-beetles occurring in subterranean habitats, many of them highly troglomorphic: *Alevonota* (6 spp.), *Ocypus* (2 spp.), *Domene* (see Figure 3; 5 spp.), and *Medon* (2 spp.). Since lava tubes in the Canary Islands have no permanent water streams or pools, stygobites occupy only phreatic waters and constitute a poor fauna compared to that of terrestrial habitats. In Tenerife 15 hypogean crustaceans have been recorded; among them the copepods *Parastenocaris* (2 spp.) and the amphipods *Pseudoniphargus associatus* and *Rhipidogammarus mivariae* are found in freshwater habitats, and the remaining seven crustaceans occur in brackish underground waters.

The westernmost La Palma and El Hierro are the other islands with abundant hypogean fauna. Possibly due to the islands' younger age their troglobites are less evolved, and very often the genera colonizing the underground are much the same in these two islands but different to those of Tenerife: *Cixius* and *Mee-*

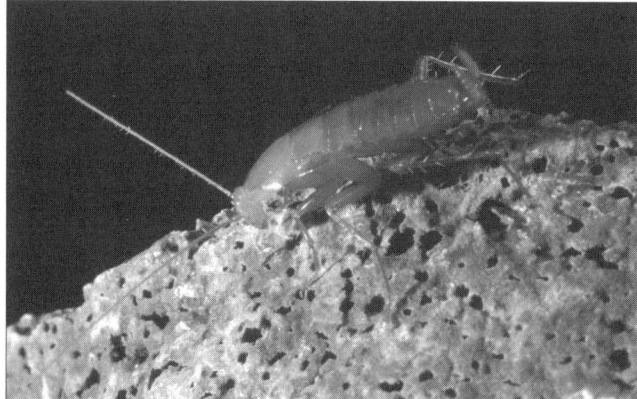

Canary Islands: Biospeleology: Figure 2. *Loboptera troglobia* (Blattaria: Blattellidae), the best-adapted cockroach to cave life among 11 cave-dwelling species of this genus in the Canary Islands. (Photo by Pedro Oromí)

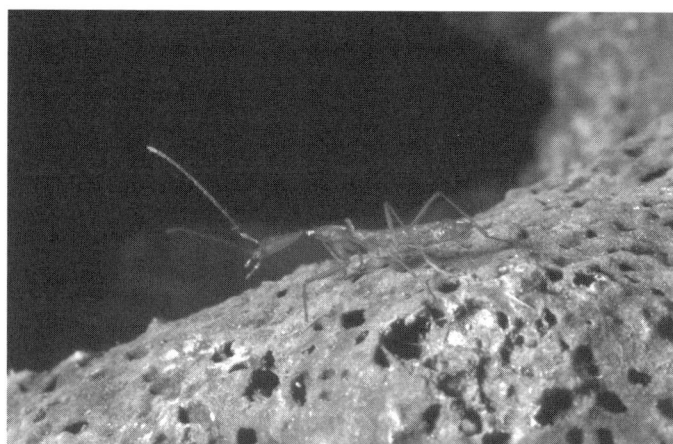

Canary Islands: Figure 3. *Domene vulcanica* (Coleoptera: Staphylinidae), from lava tubes in Tenerife, is probably the most troglobitic insect of the Canary fauna. (Photo by Pedro Oromí)

noplus (planthoppers), *Trechus, Thalassophilus, Licinopsis*, and *Parazuphium* (ground-beetles), *Laparocerus* and *Baezia* (weevils). Furthermore, some species belong to higher groups never found in other caves of the Canaries and even of the Palaearctic, such as terrestrial sandhoppers (Amphipoda), earwigs (Dermaptera), or thread-legged bugs (Hemiptera Reduviidae). The terrains are geologically young and the prevailing MSS is of the volcanic type (Oromí, Medina & Tejedor, 1986), which holds an abundant fauna made up of local ambimorph species of *Loboptera, Licinopsis, Trechus*, and *Alevonota*. In La Palma most of the 31 recorded troglobites are spread all over the island, except two cockroaches (*Loboptera* spp.) and four planthoppers (*Cixius* spp.) which are clearly allopatric, a feature also observed in other islands for these groups of insects. They can be considered as local relicts together with the long-legged bug *Collartida tamausu*, the planthopper *Meenoplus claustrophilus*, and the rove-beetles *Domene benahoarensis* and *Alevonota* (3 spp.), since no epigean relatives are known on the island. However, other animals like *Dysdera ratonensis, Licinopsis angustula, Trechus benahoaritus*, and *Laparocerus zarazagai* have ancestral species on the surface of La Palma. Other non-relictual species are included in groups which have very few troglobites worldwide: the earwig *Anataelia troglobia* which sometimes occurs in the same cave as its epigean relative *A. lavicola*, and the sandhopper *Palmorchestia hypogaea* belonging to a local genus with another representative living in the forest. Only two aquatic species are known from La Palma, both found in brackish interstitial waters: *Pseudoniphargus multidens* and *P. cupicola*.

The hypogean fauna from El Hierro includes 19 troglobites, many of them belonging to genera also represented in La Palma by vicariant species. The cockroach *Loboptera ombriosa* is the most widespread troglobite in the island, showing a cline of morphological adaptations from big, red-brown individuals in the north to smaller yellowish ones in the extreme south. Distributional patterns of other species also indicate differences between the El Golfo area at the northwest covered by humid forests, and the rest of the island with pine forest and drier bushland. Three ambimorph ground beetles of the genus *Licinopsis* are found: *L. picescens* in caves and MSS of El Golfo, and

L. obliterata and the more adapted *L. schurmanni* in the rest of El Hierro. The most troglomorphic animal on the island is the thread-legged bug *Collartida anophthalma*, a relict belonging to a genus from Central and Eastern Africa with the only troglobites on the Canary Islands. The aquatic fauna is scarce and limited to brackish interstitial waters, where two amphipod crustaceans have been found: *Pseudoniphargus salinus* and *Ingolfiella canariensis*.

The caves on El Hierro are generally well preserved but on La Palma and Tenerife some caves have open, easy access and in these there has been damage to speleothems, graffitti on the walls, and accumulation of rubbish. But probably the worst problem is pollution due to domestic waters poured into some caves, in which the adapted fauna has disappeared and alien opportunistic species invade the subterranean environment.

The only inhabited lava tube on Fuerteventura endures continuous alterations and pressure from politicians to become a show cave, which would be the epitaph for the unique *Maiorerus randoi*. An EU Legal Instrument For the Environment (LIFE) project on conservation of the cave habitat was carried out from 1999–2001, but it only concerned the caves located in public protected areas of three islands (Fuerteventura excluded). A study on the current hypogean fauna compared to that recorded 20 years ago, an inventory of the bat species and populations and their underground roosts, and an accurate analysis of the increasing factors of environmental damage were performed for 50 caves in Tenerife, La Palma, and El Hierro. The final report also included a proposal for a sewage network for the area of Cueva del Viento, as well as every necessary action for better conservation of other deteriorated caves. In spite of the initial interest by the Canary Government on this subject, no signs of continuity are evident and none of the recommended actions have been completed; in the meanwhile, this unique fauna remains threatened.

PEDRO OROMÍ

See also **Hawaiian Islands: Biospeleology**

Works Cited

Carracedo, J.C., Day, D., Guillou, H., Rodríguez, E., Canas, J.A. & Pérez, F.J. 1998. Hotspot volcanism close to a passive continental margin: The Canary Islands. *Geological Magazine*, 135(5): 591–604

Martín, J.L. 1989. *Fauna invertebrada del Parque Nacional de Timanfaya*, Santa Cruz de Tenerife: Servicio de Publicaciones de la Caja General de Ahorros de Canarias

Medina, A.L. & Oromí, P. 1990. First data on the superficial underground compartment on La Gomera (Canary Islands). *Mémoires de Biospéologie*, 17: 87–91

Oromí, P. & Martín, J.L. 1992. The Canary Islands. Subterranean fauna, characterization and composition. In *The Natural History of Biospeleology*, edited by A.I. Camacho, Madrid: Museo Nacional de Ciencias Naturales

Oromí, P., Medina, A.L. & Tejedor, M.L. 1986. On the existence of a superficial underground compartment in the Canary Islands. *Proceedings of the 9th International Speleological Congress*, vol. 2: 147–51

Further Reading

Hernández, J.J., Medina, A.L. & Izquierdo, I. 1992. Volcanic caves in El Hierro Island, Canary Islands, Spain. In *Proceedings of the 6th International Symposium on Vulcanospeleology*, edited by G.T. Rea, Huntsville, Alabama: National Speleological Society: 185–96

Juan, C., Emerson, B.C., Oromí, P. & Hewitt, G.M. 2000. Colonization and diversification: Towards a phylogeographic synthesis of the Canary Islands. *Trends in Ecology and Evolution*, 15(3): 104–09

Oromí, P. (editor) 1995. *La Cueva del Viento*, Santa Cruz de Tenerife: Consejería de Política Territorial

Oromí, P., Martín, J.L., Medina, A.L. & Izquierdo, I. 1991. The evolution of the hypogean fauna in the Canary Islands. In *The Unity of Evolutionary Biology*, edited by E.C. Dudley, vol. 2: 380–95, Portland: Dioscorides Press

CAPE RANGE, AUSTRALIA: BIOSPELEOLOGY

Cape Range is the only Tertiary orogenic karst in Australia and supports one of the world's richer subterranean faunas. The fauna occurs in three parts, separate terrestrial faunas in Cape Range proper, on its coastal plain, and a relict fauna, largely of Tethyan origin, occurring in the anchialine groundwater of the coastal plain (see map in Australia entry).

Cape Range is an anticline of Eocene/Miocene marine limestones and forms a peninsula projecting northward into the Indian Ocean on the mid-west coast of Australia. To the west the peninsula is the closest point in Australia to the continental shelf, while to the east it abuts the vast North West Shelf. The peninsula is fringed by a coastal plain and exhibits a series of raised wave-cut terraces to the west. The range rises to an altitude of 330 m and is dissected by gorges which cut through the highly karstic Tulki Limestone into the underlying and marly Mandu Limestone. The gorges thus form a discontinuity in the karst which serve as a barrier to dispersal of the subterranean fauna. In addition the marl retains a groundwater layer at which the caves develop laterally. More than 600 caves are known, mostly vertical solution pipes, with or without collapse, to 100 m depth and similar lateral extent. Five caves within the range reach water and one has extensive development (> 6 km) near the water table.

Cape Range lies just within the tropics and experiences an arid climate (mean annual rainfall 284 mm, pan evaporation 3219 mm) in which the episodic rainfall is largely torrential. Consequently, caves may dry continuously for 1–2 years before being rewetted. The vegetation is largely tussock grass (*Triodia*) with areas of *Acacia, Eucalyptus,* and *Banksia*.

Although Cape Range is now an arid area, far removed from rainforest, the area has a relict rainforest fauna with both temperate and tropical elements. The diverse subterranean fauna encompasses both terrestrial and aquatic ecosystems and is entirely endemic to the range. It includes at least 30 highly troglomorphic species, among them the most troglomorphic cockroach in the world (*Nocticola flabella*) in which only the mouthparts and genitalia are heavily sclerotized. Characteristic taxa include myriapods (centipedes, millipedes), arachnids (micro-whipscorpions, pseudoscorpions, scorpions, spiders, mites), insects (beetles, cockroaches, crickets, planthoppers, japygids), and crustaceans (slaters). A subfamily of pseudoscorpions (Indohyinae) confined to the circum-Indian Ocean region is represented in the cave fauna (*Hyella*), and there are at least six endemic genera, the micro-whipscorpion *Draculoides*, nemobiine cricket *Ngamarlanguia*, pseudoscorpion *Hyella*, paradoxosomatid millipede *Stygiochiropus*, a japygid dipluran, and a ctenid spider. The only aquatic taxon known from the range is an endemic genus of melitid amphipod (*Norcapensis*).

Allozyme analysis has shown that the southern two-thirds of the range contains three distinct genetic provinces, which are largely separated by areas of deep gorges. Even within these provinces *Stygiochiropus* millipedes mostly do not interbreed even between adjacent caves. Within the northern part of the range a suite of distinct species is recognized.

On the coastal plain and flanks of the range a related but quite distinct troglobitic fauna is present. Access is very limited due to sparse surface openings and much information has been derived from bores drilled for limestone resource assessment. A notable attribute of this fauna is the presence of six troglobitic species in three genera of micro-whipscorpions (Schizomida), a seventh, *Draculoides vinei*, being confined to the caves in range proper.

The coastal plain is fringed with the only continental anchialine system known in the southern hemisphere. These near coastal groundwaters, affected by marine tides, typically exhibit highly stratified physicochemical properties (see Anchialine Habitats entry). The characteristic fauna occurs below a thermohalocline and hydrogen sulfide layers, in conditions low in oxygen and supporting chemoautotrophic energy fixation. It is mostly accessed through traditional and later pastoral wells and recent bores, although several caves access this habitat including Bundera Sinkhole, a site globally notable for its fauna (Humphreys, 1999). The anchialine ecosystem contains the only blind cave fish in Australia, a gudgeon (*Milyeringa*), and a swamp eel (*Ophisternon*). Most of the remaining taxa are crustaceans and include halocyprid ostracods, harpacticoid, cyclopoid, calanoid, and misophrioid copepods, atyid shrimps (two species of *Stygiocaris*, sympatric in places), cirolanid isopods, remipeds, thermosbaenaceans, melitid, bogidiellid, and hadziid amphipods, and bathynellid syncarids. A number of other higher taxa are represented including turbellaria and polychaetes (Humphreys, 2000a,b).

Cape Range, Australia: Biospeleology: Some subterranean animals from Cape Range. *In rows from top left:* 1. The millipede, *Stygiochiropus communis*; 2. *Nocticola flabella*, the world's most cave-adapted cockroach; 3. *Lasionectes exleyi*, the only member of the Class Remipedia known from the southern hemisphere; 4. Unnamed blind philosciid isopodp; 5. *Bamazomus vespertinus*, one of seven species of micro-whipscorpions known from Cape Range; 6. *Stygiocaris stylifera*, an atyid shrimp inhabiting groundwater of the coastal plain; 7. *Glennhuntia glennhunti*, a harvestmen endemic to the coastal plain; 8. *Halosbaena tulki*, the only member of the Order Thermosbaenacea known from the southern hemisphere; 9. *Milyeringa veritas* (Eleotridae), one of two cave fishes sympatric on Cape Range. (Photos by Douglas Elford)

This fauna is notable because its affinities principally lie with the fauna from anchialine caves on either side of the North Atlantic. This relationship is supported by numerous congeneric (*Halosbaena, Haptolana, Lasionectes, Danielopolina, Liagoceradocus, Stygocyclopia, Speleophria*) or closely related species (endemic genus *Bunderia*). These lineages occur on the full "Tethyan track" (Stock, 1993), that is, they occupy that part of the ocean that contained congruent epicontinental seas in the Mesozoic (between 200 and 60 million years ago). These "Tethyan" lineages probably have a long stygal history, i.e. restricted to subterranean realms, following vicariant events resulting from the movement of the tectonic plates during the Mesozoic. This fauna includes the only southern hemisphere representatives of the Class Remipedia, the Orders Thermosbaenacea and Misophrioida, and the genera *Danielopolina* and *Haptolana* (Humphreys, 2000b).

About a quarter of the Cape Range karst province lies within Cape Range National Park. However, many of the most significant sites lie in areas of active resource exploitation or with resource potential under active pursuit. Water is abstracted from karst aquifers to supply Exmouth town and military facilities (airforce and naval communications bases) and the rapid growth of tourism attracted largely by the long Ningaloo fringing reef on the west coast. Nearly 100 km^2 are under limestone mining or exploration tenements and there is active petroleum exploration. A number of the terrestrial (e.g. *Hyella* sp., *Stygiochiropus* spp.) and aquatic subterranean species (e.g. *Milyeringa veritas, Ophisternon candidum, Lasionectes exleyi*) are listed under threatened species legislation or are recognized as belonging to endangered communities. Amongst the latter are Camerons Cave, within the Exmouth townsite, and Bundera Sinkhole which lies in a bombing range reserve. The World Heritage potential of Cape Range has been noted independently for its karst features and its subterranean fauna.

WILLIAM FRANK HUMPHREYS

Works Cited

Humphreys, W.F. 1999. Physico-chemical profile and energy fixation in Bundera Sinkhole, an anchialine remiped habitat in north-western Australia. *Journal of the Royal Society of Western Australia*, 82: 89–98

Humphreys, W.F. 2000a. Karst wetlands biodiversity and continuity through major climatic change—an example from arid tropical Western Australia. In *Biodiversity in Wetlands: Assessment, Function and Conservation*, vol. 1, edited by B. Gopal, W.J. Junk & J.A. Davis, Leiden: Backhuys

Humphreys, W.F. 2000b. The hypogean fauna of the Cape Range peninsula and Barrow Island, northwestern Australia. In *Subterranean Ecosystems*, edited by H. Wilkens, D.C. Culver & W.F. Humphreys, Amsterdam and New York: Elsevier

Stock, J.H. 1993. Some remarkable distribution patterns in stygobiont Amphipoda. *Journal of Natural History*, 27: 807–19

Further Reading

Humphreys, W.F. (editor) 1993. The biogeography of Cape Range, Western Australia. *Records of the Western Australian Museum*, Supplement 45: 1–248

Humphreys, W.F. 2000. Relict faunas and their derivation. In *Subterranean Ecosystems*, edited by H. Wilkens, D.C. Culver & W.F. Humphreys, Amsterdam and New York: Elsevier

Humphreys, W.F. & Adams, M. 2001. Allozyme variation in the troglobitic millipede *Stygiochiropus communis* (Diplopoda: Paradoxosomatidae) from arid tropical Cape Range, northwestern Australia: population structure and implications for the management of the region. *Records of the Western Australian Museum*, Supplement 64: 15–36

CARBON DIOXIDE-ENRICHED CAVE AIR

Carbon dioxide (CO_2) is a colourless, odourless gas, which tastes acidic. Most caves have air enriched with carbon dioxide, although in the majority of cases this is only slightly above the average surface atmospheric concentration of 0.04 volume/volume (v/v)%. Carbon dioxide in aqueous solution is a critical component in both the solution and precipitation of speleothems and the solution of limestone. In addition, an elevated level of CO_2 in cave air can be an indicator of pollution in the cave or in its catchment. Thus, it is important to understand how this gas is generated and distributed in caves.

To elucidate the sources of CO_2 and its distribution and movement, both spatially and temporally, within caves, it is necessary to conduct experimental studies. Identification of the CO_2 sources requires measurement of both CO_2 and oxygen (O_2) levels, together with measurement of humidity, temperature, and pressure. Analysis of trace gases, for example hydrogen sulfide and methane, can give additional information as to a CO_2 source. Early methods for the analysis of cave air are reviewed in James (1977). For CO_2 studies in wild caves, Dräger or Gastec gas analysers are recommended, as these instruments are portable, robust, and versatile, enabling O_2 and other gases to be measured in addition to CO_2. Modern methods of measuring air quality are described in the entry on Tourist Caves: Air Quality. Smith (1996) lists the simple CO_2 tests available to the exploration caver.

There are three distinct types of CO_2 enrichment in cave air (James, 1977):

Type I: The addition of CO_2 to cave air with dilution of other air components

This type of CO_2 enrichment of cave air is common, particularly in caves containing active speleothems. As cave waters precipitate calcite, CO_2 is liberated and can degas from solution into the cave atmosphere. In general, the elevated level of carbon dioxide produced in this manner is low and the associated oxygen depletion is modest. Hence, the atmosphere in well-decorated areas is not hazardous. For example, in Tommy Grahams Cave, Nullarbor Plain, Australia, a chamber between two siphons has a measured CO_2 level of 4.0 v/v%. The CO_2 source is degassing of carbonate water, hence the depletion of O_2 (0.8 v/v%) is barely discernible. In contrast, Hanimec Cave, Slovakia, sometimes has a CO_2 level of 36 v/v%, mainly due to volcanic emissions, and calculations show that the O_2 depletion would be in the order of 17.5 v/v%.

Type II: The replacement of O_2 by CO_2

This type of CO_2 enrichment is due to respiration by the flora and fauna in caves. A respiratory quotient of one implies that one molecule of O_2 is replaced with one molecule of CO_2. However, the respiratory quotient varies with the organic material being utilized by the organism and can be as low as 0.7. Type II processes can occur within the cave or in soil above the cave; in the latter instance, CO_2 enters the cave atmosphere by barometric pumping. It is not possible to distinguish these sources from each other.

Type III: Cave air in which there has been depletion of O_2 without an equivalent increase in CO_2

There are many mechanisms leading to Type III cave air, including addition of nitrogen, methane, or hydrogen sulfide to the residual fraction or the removal of O_2 by non-respiratory processes. The table lists caves with Type III CO_2 and possible causes of the Type III atmosphere (James & Dyson, 1981).

Carbon Dioxide-Enriched Cave Air: Caves with Type III CO_2 and possible causes of the Type III atmosphere.

Cave	Cause
Tlamaya Cave, Mexico	Anoxic mud (natural)
Odyssey Cave, Australia	Iron bacteria (natural)
Lower Kane Cave, United States	Oxidation of sulfides (natural)
Kacna Jama, Slovenia	Waste from a paper mill
Sinkhole Plain caves, Kentucky, United States	Domestic and industrial waste

Halbert (1982) recommended the use of Gibbs triangle to categorize CO_2-enrichment in cave air. The triangle can be generated for dry air or, if temperature and humidity are measured, modified for moist air. It is particularly useful when the CO_2-enriched cave air is composed of a mixture of types. For example, both natural and anthropogenic Type III processes are often associated with micro-organisms, which at the same time generate Type II CO_2-enriched cave air. Carbon stable isotope studies can assist in the identification of CO_2 sources and have been used with great success by Klimchouk and Jablokova (1989), who were able to distinguish between aerobic microbial CO_2 production (Type II) and anaerobic microbial CO_2 production from methane (Type III).

The concentration of carbon dioxide is lowered in the cave atmosphere by three major mechanisms: (1) by diffusion of the gas into air with a lower CO_2 concentration; (2) by advection, that is, air currents carrying and venting it from its source; and (3) by solution in water; CO_2 can be dissolved in dripping or condensation waters until they are saturated with it. In caves with streams, floods or periods of steady rainfall can completely clear the gas.

Carbon dioxide, if sufficiently concentrated, can be a hazard to explorers in caves regardless of rock type. The physiological consequences of the combined effects of increased CO_2 and associated depleted oxygen (O_2) are discussed in James, Pavey & Rogers (1975). It is critical for cavers to be able to detect the presence of hazardous CO_2 levels. A number of papers (Renault, 1972 & 1979; James & Dyson, 1981; Ek & Gewelt, 1985) are available to exploration cavers that assist in predicting in which caves, where in them, and under what conditions hazardous levels of CO_2 may occur. It is pertinent here to dismiss the myth that this heavier-than-air gas sinks to the cave floor and ponds. The mixing of CO_2 with air occurs spontaneously by diffusion. The separation of CO_2 from a gas mixture via the reverse process is not possible under cave conditions. As a general rule, the greatest enrichment of CO_2 will be near to its source.

JULIA JAMES

See also **Tourist Caves: Air Quality**

Works Cited

Ek, C. & Gewelt, M. 1985. Carbon dioxide in cave atmospheres. New results in Belgium and comparison with some other countries. *Earth Surface Processes and Landforms*, 10: 173–87

Halbert, E.J.M. 1982. Evaluation of carbon dioxide and oxygen data in cave atmospheres using the Gibbs triangle and the cave air index. *Helictite*, 20(2): 60–68

James, J.M. 1977. Carbon dioxide in the cave atmosphere. *Transactions of the British Cave Research Association*, 4: 417–29

James, J.M. & Dyson H.J. 1981. Carbon dioxide in caves. *Caving International*, 13: 54–59

James, J.M., Pavey, A.J. & Rogers, A.F. 1975. Foul air and the resulting hazards to cavers. *Transactions of the British Cave Research Association*, 2: 79–88

Klimchouk, A.B. & Jablokova, N. 1989. Genesis of carbon dioxide of air in Ukrainian caves. *Proceedings of the 10th International Congress of Speleology*, vol. 3, edited by T. Hazslinszky & K. Bolner-Takacs, Budapest: Hungarian Speleological Society

Renault, P. 1972. Le gaz des cavernes. *Science Progres Decouverte*, 2443: 12–18

Renault, P. 1979. Mesures periodiques de la pCO_2 dans les grottes françaises du cours de ces dix dernieres annees. *Actes du Symposium International sur l'erosion karstique*, Nimes: Union International de Speleologie

Smith G.K. 1996. Naked flame tests for caves' foul air and human tolerance. *Helictite*, 34(2): 39–47

CARBONATE KARST

The carbonate rocks, limestone and dolostone, are the principal hosts of dissolutional caves and karst landforms. Although significantly less soluble than sulfates and salt, they are much more widespread at the Earth's surface and underground, and usually have greater thickness. Limestone is more important than dolostone because of its greater solubility. On every scale, it hosts greater varieties of karst landforms, cave patterns, and morphology than are found in other rocks.

Limestone is composed of calcium carbonate ($CaCO_3$), deposited as the minerals aragonite and calcite. Aragonite has an orthorhombic atomic structure, forming crystals with acicular, prismatic, or tabular shapes, with a density of 2.95. Sr^{2+} and

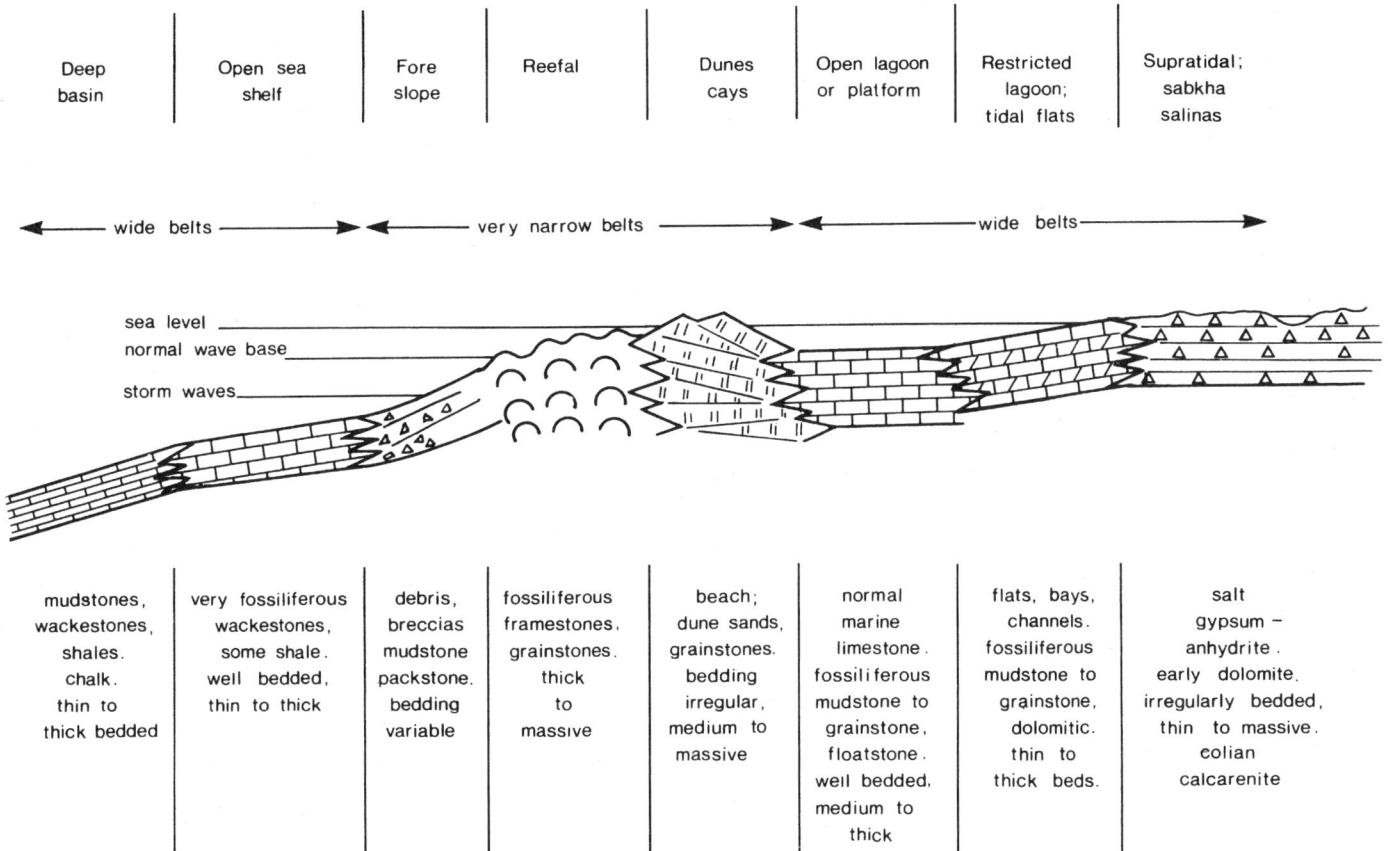

Carbonate Karst: Figure 1. Composite facies model to illustrate deposition of limestone, early dolostone, and evaporite rocks. This is a generalized, simplified picture. Not all facies will be present in any given transect. Narrow belts range from a few metres to a few kilometres in width, wide belts from hundreds of metres to more than 100 km. Modified from Wilson, 1974.

Ba^{2+} are common substitutes for Ca^{2+} atoms in the lattice. Calcite is trigonal, with massive, scalenohedral, or rhombohedral crystals, with a density of 2.71. Mn^{2+}, Fe^{2+}, and Mg^{2+} may substitute for Ca^{2+}. Over geological time spans, aragonite is unstable, with most of it inverting to calcite.

The common environments for limestone deposition range from the deep sea to supratidal mudflats or marshes overwashed by storm waters (Figure 1). Shallow platforms, sheltered lagoons, and open shelves are most common (Scholle, Bebout & Moore, 1983). Limestone forming in such environments is composed of aragonite or calcite marine debris such as shell and skeletal fragments, faecal pellets, and precipitates in algal mats, with interstices partly filled with carbonate mud ("micrite"). Popular petrological classifications by Folk (1962) and Dunham (1962) give textural details. The deposits accumulate in regular layers ("beds") separated by minor discontinuities marking disturbances, such as an influx of clay during a storm. Bedded limestones can range from single beds within sequences of clastic rocks such as sandstones or shales, to continuous carbonate deposits up to five kilometres in thickness. Most karst is developed on and in limestones 10–1000 m in thickness.

Coral reefs and other mounds, such as carbonate sand dunes, make a small but distinctive volumetric contribution to the world's limestones. Reefs range from frameworks tens to hundreds of metres in height composed of successive coral generations, to carbonate sand piles with scattered, isolated corals and algal mats. The earliest mats ("stromatolites") are >3.4 billion years in age; there have been successors in the record ever since.

Most limestones contain some proportion of insoluble impurities, chiefly silicate sands and clays, which increase the closer deposition is to river deltas or other sources of terrestrial detritus. There is a continuous gradation in nature from (nearly) pure carbonates to sandstones, etc. with carbonate cements, as shown in Figure 2. Most karst develops where there is >90% limestone or dolostone but it is occasionally found on impure carbonates or even on calcareous sandstones. More than a few percent of clay inhibits most development because it clogs the initial dissolution channels.

Diagenesis is a term describing processes of compaction and/or cementation of the initially unconsolidated carbonate mud and debris (Moore, 2001). Very young rocks may harden quickly where raised above sea level and exposed to rainfall that dissolves surficial material and reprecipitates it in the interstices below: for example, in the Bahamas, carbonate sand dunes of the last interglacial (125 000 years ago) are now hard rocks with good karst and caves. However, most limestones have experienced

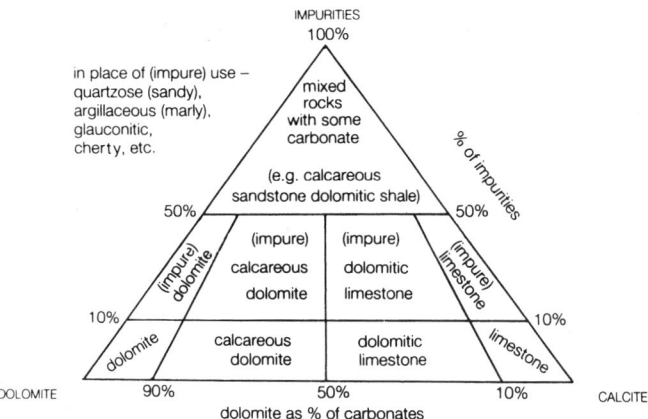

Carbonate Karst: Figure 2. Bulk classification of carbonate rocks (from Leighton & Pendexter, 1962).

deep burial and compaction. Part of the dissolved load is precipitated from sea water as it is expelled, cementing the debris with coarse calcite crystals, "sparite". A rock may be fully recrystallized during the process, destroying much or all of its original fabric and fossils. At depths of several kilometres much of the limestone itself may be dissolved, leaving insoluble residues concentrated along highly irregular solution surfaces termed "stylolites". Marine silica skeletal debris is dissolved and reprecipitates as nodules or layers of chert ("flintstone") in the limestone. These processes reduce the bulk porosity of the rock from initial values greater than 40% to 5%, or less in most instances, at which the beds become essentially impermeable. Rocks with lesser diagenetic alteration and consequent higher porosity such as chalk (15–40%) tend to develop poor karst and few caves because of the dispersal of the dissolutional attack.

Dolostone is a major product of diagenesis. It is composed of the double carbonate mineral, dolomite ($CaMg \cdot 2CO_3$) and has a hexagonal structure, forming rhombohedral, massive, or sugary (saccharoidal) crystals, with a density of 2.85. Some dolomite is precipitated directly, usually in evaporitic lagoonal or supratidal settings. Most, however, is created when Mg-rich sea water passes through earlier aragonite and calcite deposits, dissolving and reprecipitating them, with Mg^{2+} substituting for up to half of the Ca^{2+} ions. Dolomitization may be complete, altering an entire formation, or affect only particular beds within it, or merely patches within a bed. Reefs are particularly prone, due to their greater porosity. Presence of dolostone is most important in cave and karst landform development because the mineral is much less soluble than calcite. The majority of dolostones may develop karstic (dissolutional channel) drainage given sufficient time but caves large enough for human entry are quite rare. Dolostone beds often function as aquitards. Surface landforms are more limited in type and scale.

Metacarbonate (Marble) is produced when limestone or dolostone are metamorphosed by heat. With mild heat (low-grade metamorphism) smaller grains and outer surfaces of larger ones become annealed. With higher grade there is complete melting followed by recrystallization upon cooling, producing a dense, homogeneous rock of low porosity that is ideal for sculpture. Small natural sculptural features such as dissolutional karren and scallops attain their sharpest form on fine marble.

Carbonatites are intrusive igneous rocks composed of 60–90% carbonate minerals, chiefly calcite. They are very rare but a few instances with dolines and caves are known, for example in Precambrian Shield rocks in northern Ontario, Canada (Bell, 1989).

Fissure frequency: In most limestones and dolostones the effective permeability is controlled by penetrable bedding planes, contacts, and joints (products of deposition and diagenesis), and / or by later faults (Klimchouk & Ford, 2000). Density of joints is inversely proportional to thickness of the host beds. Bedding thickness may vary considerably within or between carbonate formations. Those with only thin (<10 cm) or medium (<30 cm) bedding tend to develop poor karst because (as with chalk) the dissolutional attack is widely dispersed. Thick (>30 cm) to massive (>100 cm) bedding with joint separations of 100–300 cm are the most suitable for full cave and karst development, although "sandwich" sequences of a few thin beds between massive ones are also effective. Metamorphism may seal earlier bedding planes and joints, which inhibits karst development in many marble formations. A notable exception is the Nordland marble stripe karst of Norway (see Stripe Karst).

DEREK FORD

Works Cited

Bell, K.L. (editor) 1989. *Carbonatites: Genesis and Evolution*, London and Boston: Unwin Hyman

Dunham, R.J. 1962. Classification of carbonate rocks. In *Classification of Carbonate Rocks*, edited by W.E. Ham, Tulsa, Oklahoma: American Association of Petroleum Geologists

Folk, R.L. 1962. Spectral subdivision of limestone types. In *Classification of Carbonate Rocks*, edited by W.E. Ham, Tulsa, Oklahoma: American Association of Petroleum Geologists

Klimchouk, A. & Ford, D.C. 2000. Lithological and structural controls of dissolutional cave development. In *Speleogenesis: Evolution of Karst Aquifers*, edited by A.B. Klimchouk, D.C. Ford, A.N. Palmer & W. Dreybrodt, Huntsville, Alabama: National Speleological Society

Leighton, M.W. & Pendexter, C. 1962. Carbonate rock types. In *Classification of Carbonate Rocks*, edited by W.E. Ham, Tulsa, Oklahoma: American Association of Petroleum Geologists

Moore, C.H. 2001. *Carbonate Reservoirs: Porosity Evolution and Diagenesis in a Sequence Stratigraphic Framework*, Amsterdam and New York: Elsevier

Scholle, P.A., Bebout, D.G. & Moore, C.H. (editors) 1983. *Carbonate Depositional Environments*, Tulsa, Oklahoma: American Association of Petroleum Geologists

Wilson, J.L. 1974. Characteristics of carbonate platform margins. *American Association of Petroleum Geologists Bulletin*, 58: 810–24

Further Reading

Budd, D.A., Saller, A.H. & Harris, P.M. (editors) 1995. *Unconformities and Porosity in Carbonate Strata*, Tulsa, Oklahoma: American Association of Petroleum Geologists

Ford, D.C. & Williams, P.W. 1989. *Karst Geomorphology and Hydrology*, London and Boston: Unwin Hyman, pp.10–41

Scholle, P.A. 1978. *A Color Illustrated Guide to Carbonate Rock Constituents, Cements and Porosities*, Tulsa, Oklahoma: American Association of Petroleum Geologists

CARBONATE MINERALS: PRECIPITATION

The primary reason for precipitation of calcite is loss of carbon dioxide leading to supersaturation. This entry explains the chemical processes involved. The precipitation of calcite from supersaturated $H_2O–CO_2–CaCO_3$ solutions can create spectacular tufa deposits, such as the sinter terraces of Pamukkale, Turkey, or the Huanglong Ravine in China (see entries on Pamukkale and Huanglong and Jiuzhaogou). The reaction is also responsible for the production of speleothems in cave environments. To understand how these processes operate over geological timescales, it is necessary to know the rates of calcite deposition. Precipitation from a supersaturated solution can be described by the rate law of Plummer, Wigley & Parkhurst (1978) (known as the PWP equation):

$$R = k_4(HCO_3^-)_s(Ca^{2+})_s - k_1(H^+)_s$$
$$- k_2(H_2CO_3^*)_s - k_3 \quad (1)$$

where the parentheses denote the chemical activities of the ions at the surface of the mineral (see Dissolution: Carbonate Rocks). This implies that precipitation is driven by the same three elementary chemical reactions as is dissolution. However, since the solution is supersaturated with respect to calcite, the reverse reactions (precipitation) within the first term of equation 1 predominate over the forward reactions (dissolution) contained in the negative terms. Therefore all chemical and physical processes that affect dissolution simply have to be reversed. As an example: during dissolution, 1 mole of CO_2 has to be converted into H^+ and HCO_3^- to release 1 mole of $CaCO_3$ into the solution. To deposit 1 mole of $CaCO_3$ this must be reversed, and consequently 1 mole of HCO_3^- and H^+ must react to give CO_2, which can outgas from the solution. Precipitation rates can be written as:

$$R = \alpha(c - c_{eq}) \quad (2)$$

where c is the actual calcium concentration in the solution, and c_{eq} is the equilibrium concentration with respect to calcite. The units are mol cm^{-3} = mmol l^{-1}. Owing to the complex processes involved, the kinetic constant α depends on pCO_2, temperature, the ratio V/A of the volume of the solution, and the surface area on which the calcite is deposited, and also on the hydrodynamic flow conditions. For details of the processes that determine the reaction rates, refer to Dissolution: Carbonate Rocks. Values of α for various geological situations are listed by Liu & Dreybrodt (1997) and Baker et al. (1998). The theoretical predictions have been verified experimentally in batch experiments and by observing precipitation from solutions contained in porous media of limestone (Dreybrodt et al., 1997). Close to equilibrium, inhibition occurs, which reduces the reverse reaction, such that an apparent equilibrium is reached at about 1.1 c_{eq}. Field observations also show that precipitation from natural streams is inhibited for saturation indexes (SI) less than 0.8.

Speleothems grow where thin layers ($c.0.1$ mm thick) of supersaturated water flow in either a laminar or turbulent way. When the flow of these films is laminar, the precipitation rate can be obtained from equation 2 by using Figure 1. The figure shows values of α for various temperatures and thicknesses δ of laminar-flowing films open to the atmosphere, with $pCO_2 = 3 \times 10^{-4}$ atm, as being common in caves. The values of α are almost independent of pCO_2 for values between 3×10^{-4} and 5×10^{-3} atm. For cave atmospheres, c_{eq} is about 0.6 mmol l^{-1}. Figure 2 shows growth rates for various film depths, δ, in relation to the calcium concentration (Baker et al., 1998). For waters dripping from cave roofs to form speleothems ($c = 2.5$ mmol l^{-1}), the maximum growth rates are about 0.2 mm a^{-1}.

When water is supplied as drops at time intervals of T, the average growth rate R_{av} is reduced to:

$$R_{av} = (c - c_{eq})[1 - \exp(-T\alpha/\delta)]\delta/T \quad (3)$$

(see Dreybrodt, 1999). Baker et al. (1998) have observed drip rates, and have recorded the chemistry of drip waters feeding stalagmites in caves in England and France. They determined the growth rates of stalagmites by counting annual laminae and found that these agreed well with those predicted by the model. Growth rates of stalagmites were also measured in the laboratory, by dripping a defined supersaturated solution on a $CaCO_3$ sample cut from a stalagmite and observing its weight increase (Buhmann & Dreybrodt, 1985a, b; Dreybrodt, 1988). Again good agreement with the theoretical predictions was obtained.

When supersaturated water flows turbulently, equation 2 remains valid. Since mass transport is now effected through a laminar boundary layer, it becomes dependent on its thickness ϵ (see Dissolution: Carbonate Rocks). Furthermore, due to the slow conversion of HCO_3^- into CO_2, α is also dependent on the depth of the water layer. For a water with $pCO_2 = 1 \times 10^{-3}$ atm, as is common in streams depositing calcite and $\epsilon = 0.01$ cm at 10°C, $\alpha = 2.19 \times 10^{-5}$ cm s^{-1} for $\delta = 0.1$ cm, it rises to 3.04×10^{-5} cm s^{-1} for $\delta = 2$ cm and to 3.28×10^{-5} cm s^{-1} for $\delta = 10$ cm. For higher values of $\delta > 10$ cm its ϵ value stays constant. Values of α dependent on $\delta = V/A$, ϵ, pCO_2, and temperature are listed by Liu & Dreybrodt (1997). Figure 3 illustrates the precipitation rates for various values of ϵ for $\delta = 2$ cm. Note that at realistic conditions ($\epsilon = 0.01$ cm) the rates drop by about a factor of 5 compared to the hypothetically full turbulent flow ($\epsilon = 0$) shown by the uppermost curve.

Field observations of deposition of calcite on to marble tablets in calcite-precipitating streams (Dreybrodt et al., 1992; Liu et al., 1995; Bono et al., 2001) revealed deposition rates between 0.5 mm a^{-1} and 3 mm a^{-1} in close agreement with theoretical predictions. One observation is of particular significance. Rates of deposition on to marble tablets located in fast-flowing water across a rimstone dam were higher, by a factor of about 4, than those measured from tablets placed in the still water of the pool, some metres away from the dam, although the chemistry of the water was identical. This stresses the importance of the boundary layer. Its thickness ϵ becomes smaller with increasing flow velocity.

In summary, precipitation rates of calcite cannot be calculated by use of the PWP equation (equation 1) from the bulk concentrations of the solution as obtained by chemical analysis. Due to slow conversion of $HCO_3^- + H^+ \rightarrow CO_2 + H_2O$ and diffusional mass transport involved during precipitation, the

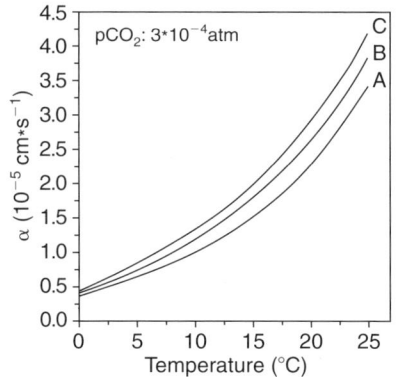

Carbonate Minerals: Precipitation: Figure 1. Values of α as a function of temperature for film thickness δ = 0.005 cm (A), δ = 0.0075 cm (B), and δ = 0.01 cm (C). Flow is laminar. From Baker *et al.* (1998).

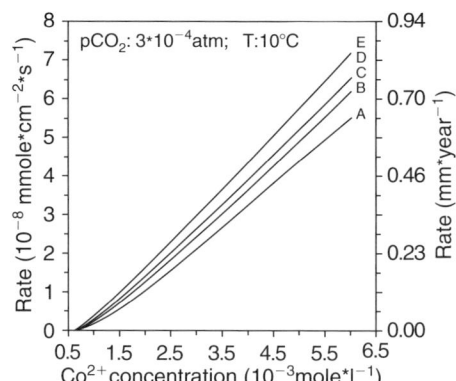

Carbonate Minerals: Precipitation: Figure 2. Precipitation rates at various values of film thickness A (0.005 cm), B (0.0075 cm), C (0.01 cm), D (0.02 cm), E (0.04 cm) as a function of the calcium concentration. Flow is laminar. From Baker *et al.* (1998).

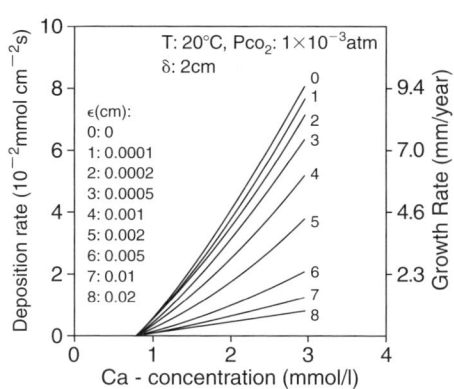

Carbonate Minerals: Precipitation: Figure 3. Deposition rates from solutions in turbulent flow with a depth δ of 2 cm, for various values of boundary-layer thickness.

rates depend not only on the chemical composition of the solution but also critically on the V/A ratio of the solution, and also on the hydrodynamic flow conditions. Each geological situation must be analysed to determine these parameters. Uncritical use of the PWP equation leads to overestimations of the rates by up to a factor of 10. Only when the V/A ratio is high and when the thickness of $\epsilon < 10\mu m$, such that neither CO_2 conversion nor mass transport are significant, does equation 1 yield the correct results.

WOLFGANG DREYBRODT

See also **Minerals in Caves; Speleothems: Carbonate**

Works Cited

Baker, A., Genty, D., Dreybrodt, W., Barnes, W.L., Mockler, N.J. & Grapes, J. 1998. Testing theoretically predicted stalagmite growth rate with recent annually laminated samples: implications for past stalagmite deposition. *Geochimica et Cosmochimica Acta*, 62(3): 393–404

Bono, P., Percopo, C., Vosbeck, K. & Dreybrodt, W. 2001. Inorganic calcite precipitation in Tartare karstic spring (Lazio, central Italy): Field measurements and theoretical prediction of depositional rates. *Environmental Geology*, 41: 305–13

Buhmann, D. & Dreybrodt, W. 1985a. The kinetics of calcite dissolution and precipitation in geologically relevant situations of karst areas: 1. Open system. *Chemical Geology*, 48: 189–211

Buhmann, D. & Dreybrodt, W. 1985b. The kinetics of calcite dissolution and precipitation in geologically relevant situations of karst areas: 2. Closed system. *Chemical Geology*, 53: 109–24

Dreybrodt, W. 1988. *Processes in Karst Systems: Physics, Chemistry, and Geology*, Berlin and New York: Springer

Dreybrodt, W. 1999. Chemical kinetics, speleothem growth and climate. *Boreas*, 28: 347–56

Dreybrodt, W., Buhmann, D., Michaelis, J. & Usdowski, E. 1992. Geochemically controlled calcite precipitation by CO_2-outgassing: field measurements of precipitation rates in comparison to theoretical predictions. *Chemical Geology*, 97: 107–22

Dreybrodt, W., Eisenlohr, L., Madry, B. & Ringer, S. 1997. Precipitation kinetics of calcite in the system $CaCO_3$–H_2O–CO_2: the conversion to CO_2 by the slow process $H^+ + HCO_3^- \rightarrow CO_2 + H_2O$ as a rate limiting step. *Geochimica et Cosmochimica Acta*, 61(18): 3897–904

Liu, Z. & Dreybrodt, W. 1997. Dissolution kinetics of calcium carbonate minerals in H_2O–CO_2 solutions in turbulent flow: the role of the diffusion boundary layer and the slow reaction $H_2O + CO_2 \Leftrightarrow H^+ + HCO_3^-$. *Geochimica et Cosmochimica Acta*, 61: 2879–89

Liu, Z., Svensson, U., Dreybrodt, W., Yuan, D. & Buhmann, D. 1995. Hydrodynamic control of inorganic calcite precipitation in Huanglong Ravine, China: field measurements and theoretical prediction of deposition rates. *Geochimica et Cosmochimica Acta*, 59(15): 3087–97

Plummer, L.N., Wigley, T.M.L. & Parkhurst, D.L. 1978. The kinetics of calcite dissolution in CO_2–water systems at 5°C to 60°C and 0.0 to 1.0 atm CO_2. *American Journal of Science*, 278: 537–73

Further Reading

Dreybrodt, W., Buhmann, D., Michaelis, J. & Usdowski, E. 1992. Geochemically controlled calcite precipitation by CO_2-outgassing: field measurements of precipitation rates in comparison to theoretical predictions. *Chemical Geology*, 97: 107–22

CARIBBEAN ISLANDS

The Caribbean region, also known as the West Indies, consists of hundreds of islands scattered roughly between latitudes 10°N and 28°N and longitudes 59°W and 86°W. The islands can be grouped into three main geological categories: the Bahamas (including the Turks and Caicos Islands), the Greater Antilles, and the Lesser Antilles. Limestones range in age from Holocene to Jurassic (0–200 million years). More than half the total land area of the region consists of karst, i.e. 130 000 km^2, of which 90% is located on the Greater Antilles (Troester *et al.*, 1987).

The Caribbean region has a very unusual geologic history. It did not exist until North and South America began to drift apart during the Triassic, about 200 million years ago. As a result, the oldest rocks in the region date back to only the Middle Mesozoic (about 190 million years ago). They are volcanic in origin and form the nucleus of most of the islands.

The climate is classified as tropical marine with no distinctive summer or winter, but with a dry and wet season. Mean annual temperatures, and temperature difference between the warmest and the coolest months, exhibit a distinctive south to north gradient, while rainfall does not conform to such a systematic pattern because it depends on latitude as well as the geometry and topography of the islands. Temperatures average about 27°C in the summer and about 24°C in the winter. Rainfall ranges from 500 mm a^{-1} in some places to as much as 5000 mm a^{-1} in mountainous areas.

Caves may be found everywhere in the Caribbean, but especially on the Greater Antilles, the Bahamas, the limestone islands of the outer arc of the Lesser Antilles, and the Netherlands Antilles. Submerged caves containing speleothems exist on many of these islands, indicating that they were above water at some time in the past.

The Bahamas archipelago consists of about 700 low-lying carbonate islands and cays located on several carbonate banks or platforms that have a water depth of < 10 m. Most of the islands are long narrow strips because they are old dune ridges that were formed during glacial periods in the Quaternary when reduced sea levels exposed the entire banks. Most caves are formed in these dunes. Soil cover is either very thin or absent, and the exposed rocks show signs of extensive dissolution resulting in razor-sharp edges on rillenkarren that is often referred to as phytokarst because of its biologically controlled genesis.

Several different types of caves can be found on these islands. On San Salvador Island, flank-margin caves formed along the margins of the freshwater lenses just beneath the flanks of the dune ridges. This formation appears to be controlled primarily by sea-level position (Mylroie & Carew, 1991). Flank-margin caves are found all over the Bahamas, and have since been described on many other islands in the Caribbean and elsewhere in the world. Other types of caves commonly found are pit caves, banana holes, and blue holes. Pit caves are vertical shafts, 1–2 m in diameter and up to 15 m deep. Unlike flank-margin caves, pit caves do not appear to be related to past sea-level positions, but are believed to have formed in the epikarst by localized vadose flow. Banana holes are wider than they are deep and are only found at elevations of 6 m or less. They occur in lowland plains, away from the margin of the ridges. Banana holes are of phreatic origin and formed along previous water-table positions where fresh groundwater mixed with infiltrating rainwaters and where oxidation / reduction reactions from decaying organic material enhanced dissolution as well. The last category, blue holes, is described in detail in the entry on Blue Holes of Bahamas.

The Greater Antilles lie astride the Cayman Trough, which is the northern boundary of the Caribbean Plate. They comprise Cuba and the Cayman Islands on the northern side of the trough, with Jamaica, Puerto Rico, the Virgin Islands, and Hispaniola (Haiti and Dominican Republic) on the southern side. The islands have a complicated surface geology with volcanic, metamorphic, and sedimentary rocks. Only the Cayman Islands and Isla de Mona consist entirely of carbonate rocks. Raised limestone marine terraces along the coasts suggest that there was a dome-like uplift during the Quaternary centred on the passage between Cuba and Haiti. The reef terraces were formed during sea-level high stands and were gradually uplifted out of the reach of following high sea-level stands, and were thus preserved. These terraces have been extensively karstified and contain many caves.

A larger variety and some of the most spectacular examples of both surface and subsurface tropical karst occur in the Greater Antilles due to the large size of the islands, the thickness of the limestone deposits (up to 1500 m), and the presence of non-carbonate rocks. The original Indian populations knew of, and used, a large number of caves as is evident from the cave art discovered on all the large islands. The caves contain thick deposits of bat guano that have been mined for fertilizer and gunpowder. Cuba (see separate entry) has a variety of extensive karst areas, which includes the famous cone karst or mogotes and the marine terrace karst. Hispaniola is the most mountainous of the islands and much of its interior consists of uninhabited cone karst in Cretaceous and Tertiary limestones; the name Haiti is actually the local term for cone karst. Haiti is 36% karstified whereas the Dominican Republic is 52% karstified and has five major karst regions. The marine terraces are most abundant on the west coast of Haiti and reduce in number and maximum elevation towards the east coast of the Dominican Republic, from 28 terraces with a maximum height of 640 m in the west to 8 terraces attaining a maximum height of 100 m in the east. Most explored caves are within these Pleistocene reef limestones because of their easy accessibility, unlike the rugged mountainous interior. The caves on Hispaniola reach a maximum of 170 m deep (Bim Sejouene, Haiti) and a maximum length of 1.7 km (Trouin Sene) in Haiti, and 10 km long (Cueva Fun Fun or Boca del Infierno) in the Dominican Republic.

Jamaica has outcrops of Tertiary limestones across two-thirds of its area. The mountainous interior is dominated by karst hills except across the Central Inlier and Blue Mountains which are formed of older non-carbonates. The Cockpit Country (see separate entry) has the finest landscapes of cone karst, with endless tracts of forest-covered hemispherical hills and intervening dolines, but cone karst is the dominant landform across large areas away from the Cockpit Country. The largest of many poljes lie around the edge of the Cockpit Country, and other areas have a rolling terrain with more subdued hills on chalky facies of the limestones. Streams from the Central Inlier drain through the karst, with the longest underground routes beneath the Dry Harbour Mountains, just east of the Cockpit Country. Quashies

Caribbean Islands: The Sleeping Pools, a long lake in an old, horizontal phreatic tube in the Golding River Cave on the southern side of the Central Inlier of Jamaica. (Photo by Tony Waltham)

River cave is the finest of many very active stream-sink caves (Waltham & Smart, 1975), but all routes drop quickly to sumps, where the long phreatic passages beyond remain unexplored. Scattered through the karst are a number of collapse dolines and impressive open shafts 100–200 m deep, but none leads into long cave systems. Fragments of abandoned phreatic trunk caves survive within the karst (Figure 1); many passages are more than 20 m high and wide, and are liberally decorated with secondary calcite deposits. Also notable among the many caves on the island (Fincham, 1997) are Jackson's Bay Caves, with 3300 m of large and beautiful water passages beneath a dry region of low-lying, cactus-covered karst near the south coast.

Puerto Rico is 28% limestone. Within the northern limestone zone, 35% of the area is covered by blanket sands and alluvial soil. The remaining 65% is known as the Karst Belt and shows extensive surficial karst features (Monroe, 1976). Mogotes are isolated karst towers that rise from the blanket sands in the northern section. Towards the south, mogote karst changes into cockpit or cone karst and doline karst. The Arecibo Observatory, the largest radio-telescope antenna in the world, is located in one of the dolines. There are numerous caves. River caves are formed by rivers that originate on non-carbonate rocks disappearing underground at the contact with carbonate rocks. They are enlarged by solution as well as abrasion and are by far the largest on the island. The longest caves on the island are Sistema del Rio Encantado (17 km long and 250 m deep, see separate entry) and the 9 km of passages in unconnected sections along the Rio Camuy. There is also significant karst on the Isla de Mona, located between Hispaniola and Puerto Rico, described in the Mona entry. The karst of the Cayman Islands closely resembles that of the Bahamas and Isla de Mona.

The Lesser Antilles are multiple chains of islands, most with volcanic cores. The Windward Islands (or the Volcanic Caribbees), from Guadeloupe to Grenada, are entirely formed of Tertiary volcanics, except where active volcanism continues today, and they have no carbonate caves. The Leeward Islands (or the Limestone Caribbees) were planed down by the sea and buried beneath thick limestones, all within the Tertiary. Some of the islands, notably Anguilla, Barbuda, and Antigua, are largely karst terrains, but the only known caves are less than 200 m long. The southern islands are more varied, and include the Netherlands Antilles, Trinidad and Tobago, and the limestone island of Barbados, outside the main Antilles arc.

Raised marine terraces occur on Barbados and the Netherlands Antilles and are also the result of a combination of continuous tectonic uplift and Quaternary sea-level fluctuations. On the Netherlands Antilles, there are four terrace levels of which the highest reaches 170 m. They show a limited extent of surface karst, and contain many caves less than 500 m long. On Curaçao as well as Aruba, the highest caves are found in the Higher Terrace at 60 m, whereas the highest cave on Bonaire is in the Middle Terrace at 30 m above sea level. The caves on Aruba are, in general, longer and deeper than those on the other islands, with Lago Colony Cave as the longest at 480 m and Huliba Cave as the deepest at 35 m.

Barbados consists entirely of limestone and is highly karstified. It has over 20 caves of which Harrison's Cave is a tourist cave 780 m long. The longest cave is Bowmanston Cave, 1700 m long and 90 m deep. Trinidad and Tobago are only 15% karstified and little is known about the caves of these islands. On Trinidad, Dunstan Cave is known for its colony of oilbirds (or guácharo); Aripo No. 1 Cave houses a large bat colony, and the longest cave is the 220 m long Oropouche Cumaca.

ROOSMARIJN TARHULE-LIPS

See also **Speleogenesis: Coastal and Oceanic Settings**

Works Cited

Fincham, A. 1997. *Jamaica Underground: The Caves, Sinkholes and Underground Rivers of the Island*, 2nd edition, Kingston: Press University of the West Indies

Monroe, W.H. 1976. *The Karst Landforms of Puerto Rico*, Washington: US Government Printing Office (Geological Survey Professional Paper 899)

Mylroie, J.E. & Carew, J.L. 1991. The flank margin model for dissolution cave development in carbonate platforms. *Earth Surface Processes and Landforms*, 15: 413–24

Troester, J.W., W. Back and S.C. Mora, 1987. Karst of the Caribbean. In *Geomorphic Systems of North America*, edited by W.L. Graf, Boulder, Colorado: Geological Society of America

Waltham, A.C. & Smart, P.L. 1975. Caves of Jamaica. *British Cave Research Association Bulletin*, 19: 25–31

Further Reading

Kueny, J.A. & Day, M.J. 1998. An assessment of protected karst landscapes in the Caribbean. *Caribbean Geography*, 9(2): 87–100

Mylroie, J.E., Carew, J.L. & Vacher, H.L. 1995. Karst development in the Bahamas and Bermuda. In *Terrestrial and Shallow Marine Geology of the Bahamas and Bermuda*, edited by H.A. Curran & B. White, Boulder, Colorado: Geological Society of America

Versey, H.R. 1972. Karst of Jamaica. In *Important Karst Regions in the Northern Hemisphere*, edited by M. Herak & V.T. Stringfield, Amsterdam and New York: Elsevier

CARIBBEAN ISLANDS: HISTORY

On 18 May 1494 Columbus saw two powerful karst springs on the south coast of Cuba, probably in Bahía de Cochinas. This was reported by his contemporary Andres Bernáldez (1930) who heard it from Columbus himself:

> In the ground beneath it [the sea shore], there sprang forth two springs of water, so abundant that the outlet [of] each was the size of a very large orange, and this water spouted up with force. When the tide was coming in, it was so cold and such and so sweet, that no better could be found in the world

Also in Cuba, the fact that native fishermen lived in caves in the western peninsula of Guanahacibibes was recorded by Bartholomé de Las Casas about 1512 (Núñez Jiménez, 1980). Then in the neighbouring island of Hispaniola the seaward end of a limestone cave (probably Boca del Infierno) in Bahía de Samaná was entered by Andreas Morales in the summer of 1513 (Martyr, 1555). So it seems that the earliest written records of karst and caves in the Americas all refer to the Caribbean, within 11 years of Columbus's first landing.

Although the Caribbean area is a geographical entity there has been no common thread in the study of its caves and karst, due largely to the different languages and cultures derived from the European countries who developed the islands. Investigation progressed only by a series of isolated explorations in the various islands. Also relevant were the cave-related "industries" that grew up—water supply, guano, edible birds, paleontology, and early tourism.

In Cuba, of which some 70% is karst, a submarine spring off the coast was noticed in 1734 but it was the European geologist Humboldt whose visit in 1800 made the existence of submarine springs widely known (Humboldt, 1856). He also mentioned sinking rivers and caves in the Matanzas region. It was there that the cave of Bellamar was discovered in 1861 and shown to visitors for $1 "including guides and lights" (see Figure). Cave fish from Cuba were described by Poey in 1856, only 14 years after those in Mammoth Cave (Kentucky) were recorded. Commercial exploitation of bat guano in the early 20th century led to more cave discoveries, some with fossil bones. After 1945 scientific study of the karst developed rapidly, with a geographical institute in Havana and a national cave society.

Jamaica, also with 70% limestone, is another island where caves were recorded quite early. Hans Sloane (1707) noted his 1688 observations on underground rivers, stalactite caves, and guano caves. Long (1774) published a six-page description of Runaway Bay Caves, near Dry Harbour where Columbus had landed, and also Riverhead Cave at the source of the Rio Pedro and another cave no longer known. The Geological Survey Memoir of 1869 contains a vast amount of detailed information on the island's hydrology and caves, including those in the more remote areas. It made the Jamaican karst well known worldwide and was cited by Cvijić and by Martel. Besides the usual bat guano exploitation the bats themselves were studied about 1860. Extensive cave exploration and underground water tracing has taken place since the 1940s, inspired by the University of the West Indies and aided by visits from Europe.

Caves in the Bahamas were noted in 1725 and studied in the 19th and 20th centuries, mainly by geologists and those concerned in guano extraction. Thus McKinnnon (1804) gave an account of a cave in Crooked Island. The submerged cave shafts or Blue Holes in the area, first noted by De Kerasoret (1753), were recorded on charts about 1840 and attracted much attention in the 1890s when it was appreciated that they were evidence of land subsidence. It was in 1853 that the fresh groundwater was shown to exist as a lens floating on top of salt groundwater, with its implication for the depth to which wells should be sunk.

The caves of Trinidad attracted great interest in the 19th century, primarily because of the presence within them of the cave-dwelling bird, the guácharo. Continuing study of the bird and its haunts was supplemented from the 1890s by work on bats, which later culminated in research to control the rabies epidemic of the 1930s. Geological aspects of the caves tended to take second place despite the Geological Survey Memoir of 1860. The guácharo was a desirable food, so finding it had commercial as well as scientific value. Although already well known in Venezuela, the earliest reference to a Trinidad cave (Latham, 1823) is also the first example of the bird living there:

> . . . it is at the time of the new and full moon, in April and May, that the people . . . resort thither; when finding the young ones not sufficiently fledged to be able to fly, they speedily fill their boats, not, however, despising the old ones, many of which are knocked down with sticks.

Descriptions of Trinidad caves themselves occur in the Geological Survey Memoir of 1860 and in accounts of travellers' visits, which were more usually limited to guácharo caves. By 1903 such visits were common enough to cause photographs to be issued on picture postcards. Cave exploration for its own sake began in the 1940s, with the increased number of foreigners working there, and cave fauna studies have also become important.

The cave now known as Bat's Cave in Antigua was shown on a map of 1749 and described in detail by Mrs Lanaghan

Caribbean Islands: History: Bellamar Cave, Cuba, in 1870.

(1844). In the 1890s a small-scale attempt was made to sell bat guano from it. In Barbuda, three inland caves were marked on charts of about 1814. Caves in the Caicos Islands were mentioned in 1773 and guano caves there were described in 1884. In the Cayman Islands, too, there are caves but they were not recorded until the 20th century.

The extensive karst of Hispaniola is relatively little known. After the 1513 Samaná visit already mentioned, caves near the city of Santo Domingo attracted attention from 1849 when Cueva de Borbon was discovered. Caves near Santanna were already much visited by 1873, and 20th-century guano extraction led to others being recorded.

The caves of Curaçao and the other Dutch islands have been well investigated in the 20th century and the cave of Hato was described by Teenstra and others from 1828.

Caves in Puerto Rico have mainly been studied since the 1950s, but guano was being taken from some in the early 20th century and at least one cave was commonly visited by travellers in the 1890s.

TREVOR SHAW

Works Cited

Bernáldez, A. 1930. *Historia de los Reyes Católicos Don Fernando y Doña Isabel* [History of the Catholic Monarchs Fernando and Isabel], extract translated into English on pp.130–33 of *Select Documents Illustrating the Four Voyages of Christopher Columbus*, vol. 1, edited by C. Jane, London: Hakluyt Society. The location of the springs as at Cochinos is deduced by S.E. Morrison, 1974, *The European Discovery of America*, vol. 2: *The Southern Voyages AD 1492–1616*, New York: Oxford University Press: 127. He uses an erroneous translation of the Spanish text, which wrongly indicates that the springs were beneath the sea.

De Kerasoret. 1753. British Library Add. MS. 21384, ff. 15 v, 16 v

Humboldt, A. von 1856. *The Island of Cuba*, London: Sampson Low: 133–34 (original French edition 1826, vol. 1: 55–56)

Lanaghan, Mrs. 1844. *Antigua and the Antiguans*, London: Saunders and Otley, vol. 1: 281–82

Latham, J. 1823. *A General History of Birds*, Winchester, vol. 7: 366

Long, E. 1774. *The History of Jamaica*, London: Lowndes, vol. 2: 95–100

McKinnon, D. 1804. *A Tour through the British West Indies, in the Years 1802 and 1803*, London: White: 161–63

Martyr, P. 1555. *The Decades of the Newe Worlde or West India*, London: Powell (original Latin edition, 1516). The description is reprinted, together with a deduction of the date of the visit and the identity of the cave, in T.R. Shaw, 1995. Earlier knowledge of the American caves was reported in J.W. Valvasor's *Die Ehre des Herzogthums Crain* (1689)—from 1513, *Acta Carsologica*, 24: 545–73

Núñez Jiménez, A. (editor) 1980. *40 Años explorando a Cuba: historia documentada de la Sociedad Espeleológica de Cuba* [40 Years Exploring in Cuba: History of the Speleological Society of Cuba], Havana: Editorial Científico Técnica: has Las Casas's original manuscript

Sloane, H. 1707. *A Voyage to the Islands Madera, Barbados, Nieves, S. Christophers and Jamaica*, London, vol. 1: ix–xii

Further Reading

Contribution to the history of cave studies in West Indies (to commemorate 500 years of Columbus's discovery of America), *Acta Carsologica*, 22, 1993: 7–135

This is a special edition containing papers by J.P.E.C. Darlington, A.G. Fincham, T.M. Iliffe and T.R. Shaw

Fincham, A.G. 1997. *Jamaica Underground: The Caves, Sinkholes and Underground Rivers of the Island*, Kingston: University of the West Indies

Wagenaar Hummelinck, P. 1979. *Caves of the Netherlands Antilles*, Utrecht: Foundation for Scientific Research in Surinam and the Netherlands Antilles (in Dutch and English)

CARLSBAD CAVERN AND LECHUGUILLA CAVE, UNITED STATES

Located in the Chihuahuan Desert of New Mexico and Texas, the Guadalupe Mountains contain little surface karst but are known for spectacular caves, including Carlsbad Cavern and Lechuguilla Cave (DuChene & Hill, 2000; Figure 1). Carlsbad Caverns National Park, which contains both caves plus at least 90 others, was established in 1930 and is administered by the National Park Service. The Park was designated a World Heritage Site in 1995 because of its geological significance and the beauty of its cave features. Carlsbad Cavern is open to guided and self-guided tours year round and receives up to 650 000 visitors each year. Off-trail activities, under permit, are limited to research and mapping. Several other caves in the Park are open to "wild" caving under permit. Access to other caves, including Lechuguilla, is limited to researchers and surveyors. Of the Park's 190 km^2, 70% (including Lechuguilla Cave) has been designated a wilderness area.

The Guadalupes consist of an uplifted Permian reef complex (Figure 2). The abrupt Guadalupe Escarpment, forming the southeastern border, consists of massive Capitan Limestone, a bryozoan-sponge-algal reef with a fine-grained matrix. To the northwest are prominently bedded back-reef dolomites and limestones, which form a high dissected plateau. To the southeast, the reef is bordered by an apron of reef talus, which merges diagonally downward with deeper-water strata of the Delaware Basin. These basinal rocks are overlain by Permian evaporites, mainly gypsum. The Guadalupes owe their height to late Permian, Mesozoic, and especially late Cenozoic uplift, accompanied by faulting, folding, and southeastward tilting. The present crest of the upland is 1300–2300 m above sea level, with the highest point at 2667 m at Guadalupe Peak. The surrounding plains lie at about 1000 m. The average annual rainfall in the Guadalupes is only about 360 mm, and surface karst is limited mainly to scattered karren. In this dry climate, hypogenic cave development is favoured.

The Guadalupe Mountains contain perhaps the world's foremost examples of cave origin by sulfuric acid (Egemeier, 1987; Hill, 1987). Hydrogen sulfide escaped in aqueous solution from oil and gas fields in neighbouring and underlying strata, migrated toward zones of lower head in the permeable reef complex, and oxidized to sulfuric acid at and near the water table. This acid rapidly formed caves in the carbonate rocks, and the dis-

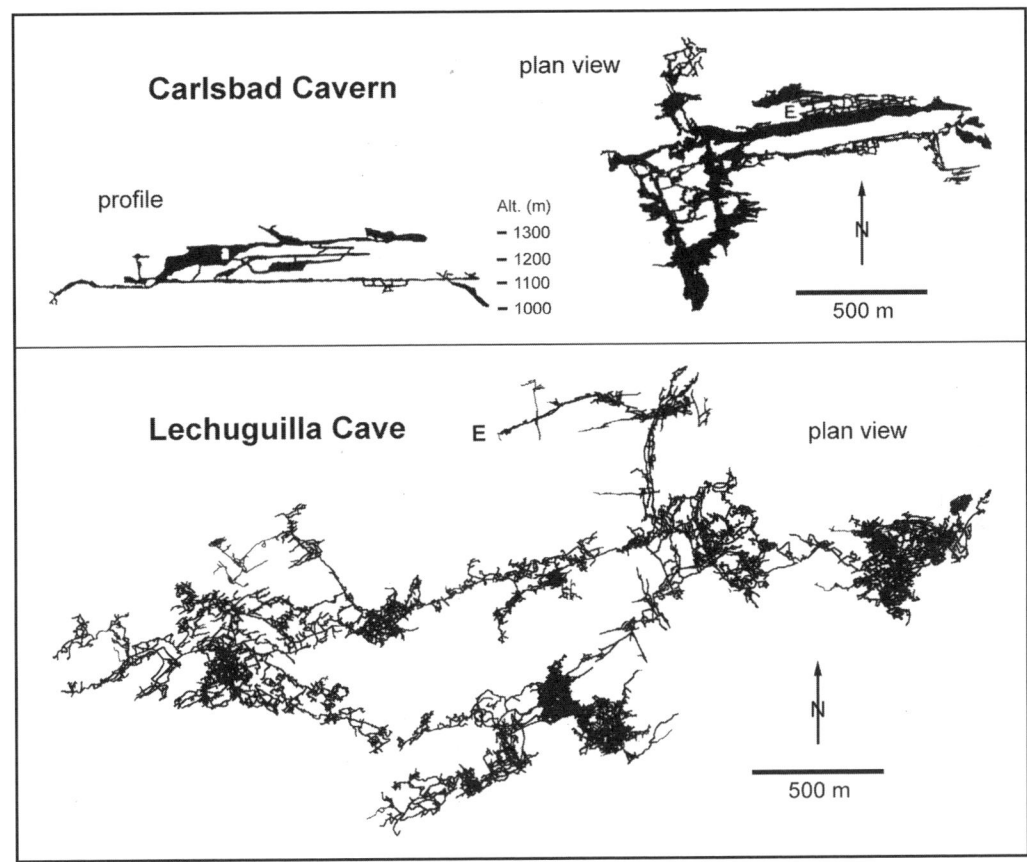

Carlsbad Cavern and Lechuguilla Cave: Figure 1. Outline surveys of Carlsbad Cavern and Lechuguilla Cave, at the same scale.

solved load was carried to springs. This process is episodic and presently dormant.

Guadalupe caves typically have ramifying patterns consisting of irregular rooms and mazes with passages branching outward from them at various levels (Figure 1; also see Patterns of Caves entry). Maps resemble inkblots, with many overlapping tiers. Branches do not converge as tributaries, but instead are distributary outlets at successively lower elevations. Some caves, or parts of caves, have network or spongework patterns. Some involve simple widening of one or more fractures.

Carlsbad Cavern is one of the world's most spectacular caves. Its single natural entrance, located at the highest point in the cave, was originally an outlet for ascending water. It drops into the steeply descending Main Corridor, which leads downward to the main levels, which include the Big Room, the largest known chamber in any US cave. An elevator gives direct access to

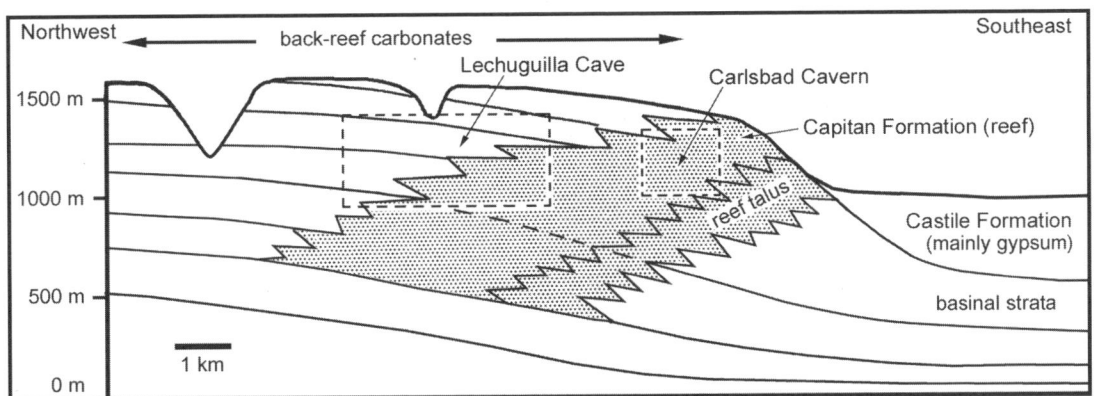

Carlsbad Cavern and Lechuguilla Cave: Figure 2. Diagrammatic profile through the Capitan reef, showing the structural locations of the Carlsbad and Lechuguilla caves.

the Big Room level. In the opposite direction, the Main Corridor contains one of the largest bat colonies in the United States and was the site of early 20th-century guano mining. The dominant speleothems in the cave are massive dripstone. The cave's surveyed length is 49.7 km and its depth is 316 m.

Lechuguilla Cave is the world's fifth longest cave (180 km) and the deepest limestone cave in the United States (478 m). It was discovered in 1986 through digging at the bottom of a 25 m entrance shaft, which, like the Carlsbad entrance, once served as a water outlet. Lechuguilla Cave's passages and rooms are not as large as those of Carlsbad, but they are far more complex, so that many parts of the two-dimensional map are virtually incomprehensible (Figure 1). The steeply descending entrance passage leads to the main levels, which branch in three directions. Each section of the cave contains a variety of steeply inclined passages interconnected with complex spongework mazes. The cave contains some of the world's most spectacular gypsum speleothems, including crystalline "chandeliers" up to 6–8 m long.

Many rooms and galleries in these and other Guadalupe caves form nearly horizontal levels discordant to lithologic boundaries (Figure 3). Deep fissure-like rifts descend from the floors of major levels and pinch downward in narrow constrictions. Steeply sloping passages represent former paths of rising H_2S-rich water below the water table. This water converged with, and mixed with, oxygen-rich water fed by infiltration at higher elevations. Depths of mixing and dissolution extended over vertical ranges as much as 200 m below the water table (Palmer & Palmer, 2000).

The reaction of sulfuric acid with carbonate bedrock produced extensive gypsum deposits. These include rinds on bedrock up to half a metre thick, as well as massive deposits on cave floors that precipitated from saturated solutions (Figure 3). Inherited textures show that some of the gypsum rinds replaced carbonate bedrock on a roughly volumetric basis. Light isotopic ratios of sulfur show that the gypsum is not primary, but instead a by-product of microbially mediated redox reactions (Hill, 1987).

Besides the spectacular calcite and gypsum deposits described above, Guadalupe caves contain a variety of unusual speleothems. Perhaps unique in the world, subaqueous helictites are produced by the common-ion effect, where calcite-saturated water reacts with gypsum-rich water in pools (Davis, Palmer & Palmer, 1990). Some minerals are by-products of sulfuric acid speleogenesis, including sulfur, alunite, hydrated halloysite, and dickite. Evaporation produces aragonite needles in bushes up to 3 m high, as well as lesser deposits of magnesium-rich minerals such as huntite, hydromagnesite, magnesite, and dolomite. Condensation corrosion takes place on upper-level surfaces where moist air ascends from warmer lower levels. Dissolution of bedrock and speleothems by condensation moisture is common, and the dissolved minerals are drawn to zones of evaporation by capillary potential, where they precipitate as speleothems and crusts. Intense wall weathering in some areas has produced a thick corrosion residue of clay and metal oxides harbouring a rich microbial community (Cunningham *et al.*, 1995). In places microbial filaments were coated with later calcite to produce finger-like speleothems of either subaqueous or subaerial origin.

Dissolution by flowing vadose water is virtually absent. However, some bedrock floors and walls contain deep rills apparently

Carlsbad Cavern and Lechuguilla Cave: Figure 3. Lebarge Borehole in Lechuguilla Cave, a horizontal level at 1190 m above sea level, 240 m above the local water table. Note massive gypsum on floor and gypsum wall crust from the reaction of sulfuric acid with the local limestone. The massive reef limestone is clearly shown here. The pool is a perched water body fed by local infiltration. (Photo by Art Palmer)

formed by highly acidic water, but with no corresponding ceiling features. The following process can be observed in still-active H_2S caves: Infiltration or condensation forms droplets and films of fresh water on cave walls. They absorb H_2S that escapes from the water into the cave air, in addition to oxygen from the air, and they react to form sulfuric acid. This reaction depletes the content of the two gases, allowing more of each to be absorbed and to react. The carbonate bedrock is soon replaced by a gypsum rind, which prevents the acid from being neutralized. The pH drops to measured values as low as zero. Rills form where these droplets fall onto carbonate bedrock.

Argon dating of alunite in the Guadalupe caves by Polyak *et al.* (1998) shows consistently increasing age with elevation. Since the alunite is a by-product of the speleogenetic process, these dates indicate the true ages of the caves. The highest levels are 12 million years old, indicating that most tectonic uplift took place since then. Ages decrease downward at a decelerating rate, indicating a decrease in uplift rate and water-table decline with time. The lowest dated cave level is the Big Room of Carlsbad Cavern, at about 4 million years old.

Despite the essentially horizontal water-table control of many passages and rooms, these levels do not correlate well between neighboring caves, or even within individual complex caves (Palmer & Palmer, 2000). Clearly the cave-forming process involved local episodes of H_2S release along different paths at different times. As the water table dropped, bursts of cave enlargement occurred at those times and places where H_2S was rising to the water table in significant quantity, probably during uplift and deformation. When these releases coincided with rather static water tables, distinct horizontal levels resulted.

ARTHUR N. PALMER

See also **Condensation Corrosion; Hydrocarbons in Karst; Photographing Caves (Photo); Speleogenesis: Deep-seated and Confined Settings; United States of America**

Works Cited

Cunningham, K.I., Northup, D.E., Pollastro, R.M., Wright, W.G. & LaRock, E.J. 1995. Bacteria, fungi, and biokarst in Lechuguilla Cave, Carlsbad Caverns National Park, New Mexico. *Environmental Geology*, 25(1): 2–8

Davis, D.G., Palmer, M.V. & Palmer, A.N. 1990, Extraordinary subaqueous speleothems in Lechuguilla Cave, New Mexico. *National Speleological Society Bulletin*, 52: 70–86

DuChene, H.R. & Hill, C.A. (editors) 2000. The caves of the Guadalupe Mountains. *Journal of Cave and Karst Studies*, 62(2)

Egemeier, S.J. 1987. A theory for the origin of Carlsbad Caverns. *National Speleological Society Bulletin*, 49(2): 73–76

Hill, C.A. 1987. *Geology of Carlsbad Cavern and Other Caves in the Guadalupe Mountains, New Mexico and Texas*, Soccorro, New Mexico: New Mexico Bureau of Mines and Mineral Resources, Bulletin 117

Palmer, A.N. & Palmer, M.V. 2000. Hydrochemical interpretation of cave patterns in the Guadalupe Mountains, New Mexico. *Journal of Cave and Karst Studies*, 62(2): 91–108

Polyak, V.J., McIntosh, W.C., Güven, N. & Provencio, P. 1998. Age and origin of Carlsbad Cavern and related caves from $^{40}Ar/^{39}Ar$ of alunite. *Science*, 279: 1919–22

Further Reading

Hill, C.A. 1996. *Geology of the Delaware Basin, Guadalupe, Apache, and Glass Mountains, New Mexico and West Texas*, Midland, Texas: Society for Sedimentary Geology, Permian Basin Section

Jagnow, D.H. 1989. The geology of Lechuguilla Cave, New Mexico. In *Subsurface and Outcrop Examination of the Capitan Shelf Margin, Northern Delaware Basin*, edited by P.M. Harris & G.A. Grover, Tulsa, Oklahoma: Society of Economic Paleontologists and Mineralogists

Sasowsky, I.D. & Palmer, M.V. (editors) 1994. *Breakthroughs in Karst Geomicrobiology and Redox Geochemistry*, Charles Town, West Virginia: Karst Waters Institute (KWI Special Publication 1)

CARMEL CAVES, ISRAEL: ARCHAEOLOGY

The caves of the Mount Carmel area, Israel, along with inland caves and open-air sites, have contributed to the resolution of two major issues in human evolution. The first is the question of when and how anatomically modern humans (*Homo sapiens sapiens*) succeeded the Neanderthals (*Homo sapiens neanderthalensis*). The second concerns the first evidence for the domestication of plants and animals in mesolithic or epipaleolithic times.

The archaeological caves of the Levant occur as recesses in the reef front of Cretaceous – Eocene limestones (Figure 1). The reef front runs parallel to the Levantine coast and culminates in Mount Carmel at Haifa, at an elevation of about 200 m. There are valleys running from the interior that cut through the reef front, and about 20 km south of Haifa and at the mouth of one of these—Wadi el-Mughara (Wadi of the Caves)—are found, in close proximity, the caves of Tabun, Jamal, El-Wad, and Es-Skhul. Further south is the important Neanderthal site, Kebara. The caves lie inland of a coastal strip about 3.5 km wide and from 40 to 60 m above sea level (Jelinek, 1982). It is not certain how these caves formed initially. Although some of the interior sediments are derived beach sands, suggesting a possible origin as former sea caves, their position at the mouth of the wadi and the phreatic form of their chambers suggests that they once functioned as resurgences, possibly below a former sea level. The age of cave formation is not known. El-Wad, Tabun, and Kebara possess curious funnel-shaped pits in the floor, now filled with sediment, that suggest a possible link to lower chambers and it is possible that water flowed up these pits in the early, active phase.

The sediments in the pit at Tabun (as at Kebara) slipped down even while the cave was being occupied in Paleolithic times, and it is curious that Acheulian stone tools and flakes are now aligned at steep angles within the steeply dipping sediment, parallel to the walls of the pit. There are no speleothems that might be dated by U-series methods. Further inland, Hayonim Cave, in which remains from the Mousterian and Natufian cultures have been found, contains fallen, partially rotted, stalagmites and flowstones of poor quality that date from over 250 000 to about 90 000 years ago.

The caves of the Levant were first excavated systematically by Dorothy Garrod, head of the Archaeology Department at the University of Cambridge, mainly from 1929 to 1934. Her work culminated in the two volumes of Garrod & Bate (1937), *The Stone Age of Mount Carmel*. Her analysis and divisions of the Levantine paleolithic to neolithic stone cultures, guided by her knowledge of French lithic typology, have largely stood the test of time (Belfer-Cohen & Bar-Yosef, 1999).

Figure 2 reproduces Garrod's section at Tabun. The skeleton at X, known as Tabun I or C1, has a skull that possesses characteristics of the Western European Neanderthals (e.g. prominent brow ridges and receding chin). Unfortunately, it was not clear to Garrod, or to anyone since, whether the remains truly belonged to layer C, or were interred into layer C but belong in time to layer B. From unpublished Garrod notes, the current

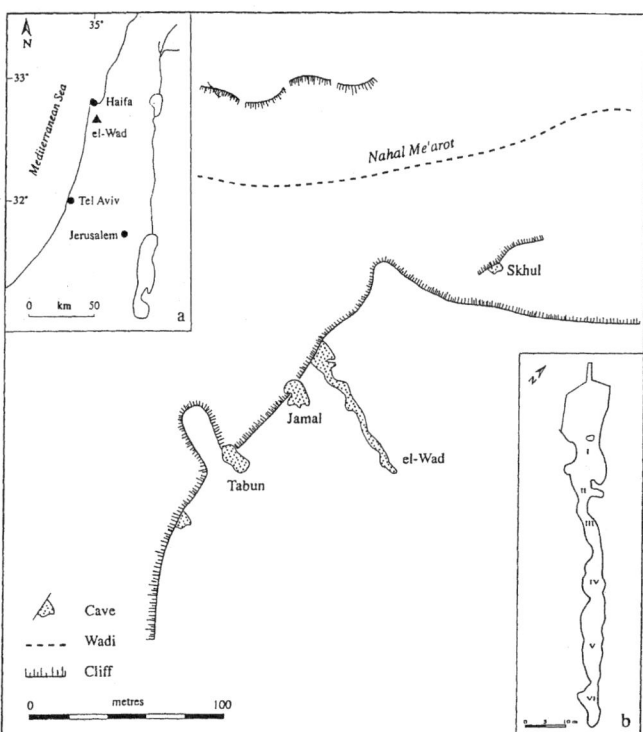

Carmel Caves, Israel: Figure 1. Archaeological caves of Wadi Mughara, Mount Carmel area, Israel (from Garrod & Bate, 1937). Inset (a) is the Levantine coastal strip. Inset (b) is a plan of El-Wad.

estimate is that it belonged to layer B (Bar-Yosef & Callander, 1999). There is a distinct faunal break in both large and small fauna between the two layers, yet the change in lithic assemblage is not that great. In addition to the faunal and cultural associations of Tabun 1, there is the question of whether the skeleton (probably female) is an actual Neanderthal burial. Even more important is the question of how Tabun I is to be dated, since the layer to which it belongs determines the dose rate estimates for TL (thermoluminescence) or ESR (electron spin resonance; see Dating Methods: Archaeological). An attempt to date Tabun I by nondestructive U-series gamma-ray spectrometry gave a tentative and controversial late age of 34 ka (Schwarcz et al., 1998; Millard & Pike, 1999; Bar-Yosef & Callander, 1999). Tabun II is a jawbone found at the base of layer C. What is intriguing is that this jaw looks like it belongs to a more anatomically modern human (AMH; it has a recognizable chin), which implies that, for a few tens of thousands of years, AMHs preceded the Neanderthal types or overlapped with them.

The controversy about Tabun I is part of the issue of how and when the AMHs succeeded the Neanderthals. Up until the modern phases of excavation in the 1970s–1980s, it was widely believed that the Levantine Neanderthal—AMH successions followed in similar fashion to those of Western Europe, namely that the AMH types followed the Neanderthals, the latter dying out at about 35 000 to 30 000 years ago. In addition to Tabun I, Neanderthals were excavated from Kebara Cave, further down the coast, and from Amud Cave near to the Sea of Galilee (see also Shanidar Cave, Iraq: Archaeology). Then AMH remains were found round the corner from Tabun at Es-Skhul and in Qafzeh Cave, near Nazareth. By studying the faunal successions of fossil small rodents from these caves, Einat Tchernov and Ofer Bar-Yosef suggested that the AMHs actually preceded the Neanderthals in the Near East. Unfortunately the remains were outside the range for the ^{14}C dating method. So this gave the impetus to apply other radiometric dating studies (and in no small measure to their development) using U–Th (of teeth), TL (of burnt flint), and ESR (of teeth). These studies showed that the Qafzeh AMH remains date to 110 ka and possibly 120 ka and those at Es-Skhul to 100 ka, whereas the Amud Neanderthals date to 50 ka and those from Kebara to 60–64 ka. There are certainly problems with some of the age assignments, the chief of these being the mismatch in ESR ages for the Tabun layers with the older TL ages estimated from burnt flint in these same layers. But whatever may be the present inconsistencies, the chronologies are good enough to confirm that the AMHs preceded the Neanderthals in the region. The Levant AMHs thus precede those of Europe. The dating of these various remains helped to promote the idea that the AMH peoples originally evolved in Africa (sub-Saharan Africa) and that some of them migrated out of Africa, over 100 000 years ago, and into Europe and Asia, with the Near East acting as a corridor. The earlier migration of *Homo erectus* out of Africa between about 1.5 to 2 Ma is known as the "Out-of-Africa I" migration, so the subsequent AMH migration is known as the "Out-of-Africa II" migration. Bar-Yosef speculated that it was the onset of the last glaciation that induced a subset of the Neanderthal population to colonize the Near East from the north following changes or migrations in the animals on which they depended. As far as the emergence of modern humans is concerned, the Out-of-Africa II migration is set off against the alternative hypothesis of a more general multiregional development (see discussion in Delson et al., 2000).

Dorothy Garrod also excavated the terrace at El-Wad and confirmed the presence of the Natufian culture which she assigned to the Mesolithic (Boyd, 1999). This is now thought to represent the earliest transitional culture to adopt sedentism through the cultivation of cereal. The Natufian is actually named after Wadi en-Natuf in the Judean hills where Garrod excavated a cave in 1928. The El-Wad terrace contained several burials with shells and other ornaments. In addition, excavation turned up a large number of backed stone blades displaying "a high degree of polish along the working edge", together with a number of "sickle hafts". In 1932 a complete grooved bone sickle haft was excavated from Kebara Cave, and Garrod suggested that it was used in a primitive form of agriculture. Since Garrod's time, further work on Natufian sites has added to the idea of cereal cultivation, but—with the possible exception of dogs—not to the domestication of animals. Whereas Garrod had guessed the Natufian to be about 4000–5000 BP, ^{14}C dates from early Jericho place it at 8000–10 000 BP. A major review of the Natufian by various contributors is given in an edited volume by Bar-Yosef and Valla (1991).

ALF G. LATHAM

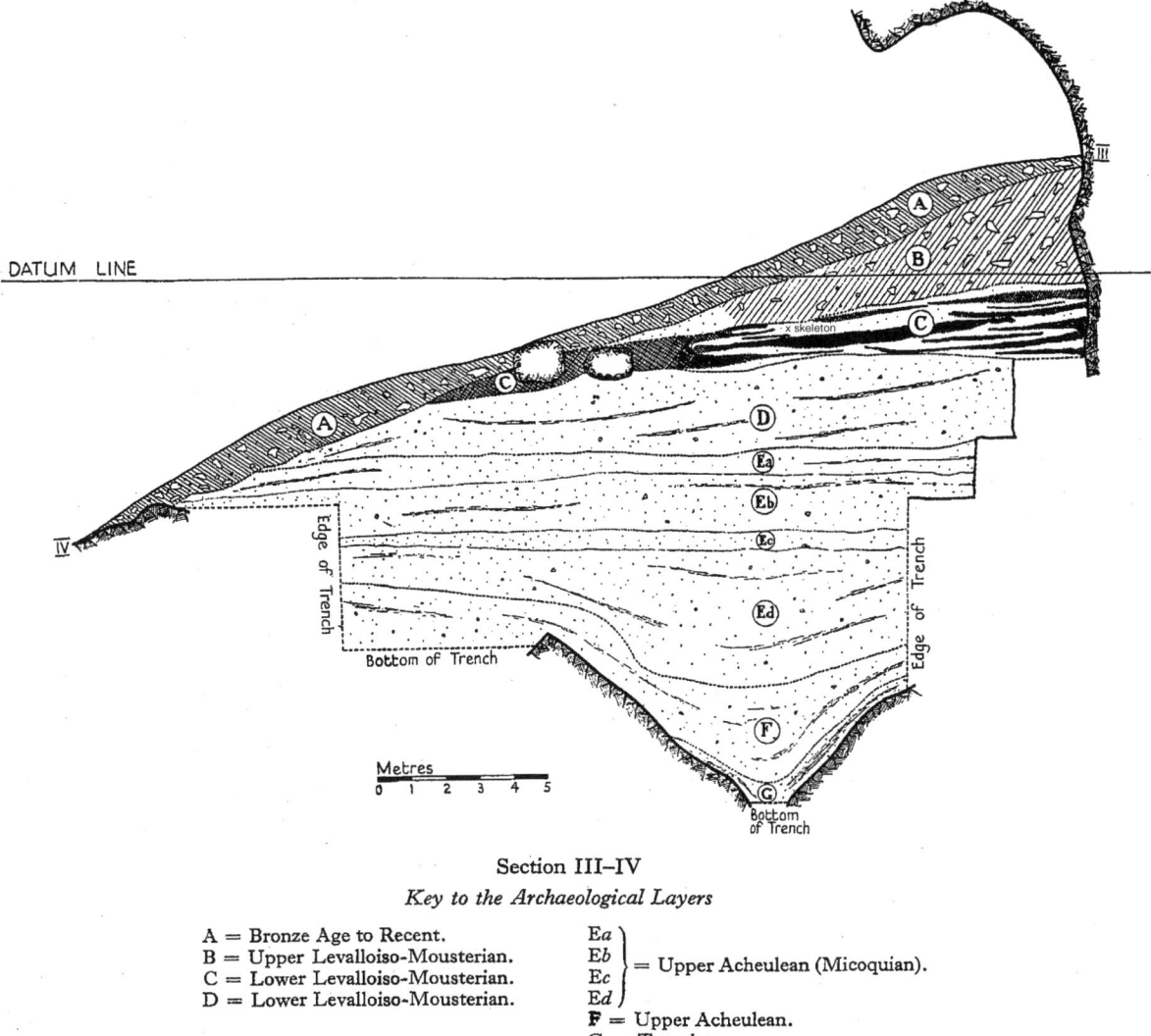

Carmel Caves, Israel: Figure 2. Et-Tabun Cave and Garrod's section from an alcove near the west side out to the hill slope. The position of skeleton Tabun 1 is marked by a cross in layer C immediately below layer B. For the terminology of the Paleolithic, see for example, Delson *et al.* (2000) or Jones *et al.* (1992).

Works Cited

Bar-Yosef, O. & Callander, J. 1999. The woman from Tabun: Garrod's doubts in historical perspective. *Journal of Human Evolution*, 37: 879–85

Bar-Yosef, O. & Valla, F.R. (editors) 1991. *The Natufian Culture in the Levant*, Ann Arbor, Michigan: International Monographs in Prehistory

Belfer-Cohen, A. & Bar-Yosef, O. 1999. The Levantine Aurignacian: 60 years of research. In *Dorothy Garrod and the Progress of the Palaeolithic: Studies in the Prehistoric Archaeology of the Near East and Europe*, edited by W. Davies & R. Charles, Oxford: Oxbow Books

Boyd, B. 1999. "Twisting the Kaleidoscope": Dorothy Garrod and the 'Natufian Culture'. In *Dorothy Garrod and the Progress of the Palaeolithic: Studies in the Prehistoric Archaeology of the Near East and Europe*, edited by W. Davies & R. Charles, Oxford: Oxbow Books

Delson, E., Tattersall, I., Van Couvering, C. & Brooks, A. (editors) 2000. *Encyclopedia of Human Evolution and Prehistory*, 2nd edition, New York: Garland

Garrod, D.A.E. & Bate, D.M.A. 1937. *The Stone Age of Mount Carmel*, 2 vols, Oxford: Clarendon Press

Jelinek, A.J. 1982. The Tabun cave and Palaeolithic man in the Levant. *Science*, 216: 1369–75

Jones, S., Martin, R.D. & Pilbeam, D.R. (editors) 1992. *The Cambridge Encylopedia of Human Evolution*, Cambridge and New York: Cambridge University Press

Millard, A.R. & Pike, A.W.G. 1999. Uranium-series dating of the Tabun Neanderthal: A cautionary note. *Journal of Human Evolution*, 36: 581–86

Schwarcz, H.P., Simpson, J.J. & Stringer, C.B. 1998. Neanderthal skeleton from Tabun: U-series data by gamma-ray spectrometry. *Journal of Human Evolution*, 35: 635–45

Further Reading

Aitken, M., Stringer, C.B. & Mellars, P. (editors) 1992. *The Origin of Modern Humans and the Impact of Chronometric Dating: A Discussion*, London: Royal Society and Princeton, New Jersey: Princeton University Press, 1993

Davies, W. & Charles, R. (editors) 1999. *Dorothy Garrod and the Progress of the Palaeolithic: Studies in the Prehistoric Archaeology of the Near East and Europe*, Oxford: Oxbow Books

Stringer, C.B. & McKie, R. 1996. *African Exodus: The Origins of Modern Humanity*, London: Cape and New York: Holt
A layperson's, non-specialist account.

CASTLEGUARD CAVE, CANADA

Castleguard Cave is the longest currently known cavern in Canada (20 km) and the foremost example anywhere of a cavern extending underneath a modern glacier (Figure 1). It displays many striking features of interaction between glaciers and karst aquifers, a complex modern climate, rich mineralization, and a stygobitic fauna that has possibly survived one or more ice ages beneath deep ice cover in the heart of the Rocky Mountains (Ford, 1983).

Relief in the region rises from an altitude of 1500 m (trunk valley floors) to summits at 3500 m. Mean annual temperatures are 0 to $-14°C$ across this height range. The tree line is at ~2100 m. The Columbia Icefield is 320 km² in area and 200–300 m thick with valley glaciers radiating out for up to 10 km. Ice thickness and extent were much greater during the major glaciations, when only the mountain peaks protruded as nunataks. The karst rocks are resistant carbonates of Cambrian age. The Cathedral Formation (>560 m thick) is massively bedded, crystalline limestone that contains the cave. Above it, the Stephen Formation (80 m) is a leaky calcareous shale aquitard overlain by further thick-bedded limestones and dolostones (Pika and Eldon Formations). The summit strata are weaker shales, sandstones, and dolostones (Figure 1). Stratal dip is S–SE at 4–6°.

On the surface, the Cathedral limestones host a suite of small but typical alpine karst forms such as karren, dolines, and shafts, many of which were overridden and lightly eroded by glaciers during the Little Ice Age of the last 500 years. The glaciers are now receding. In some places streams can be heard sinking into shafts beneath them.

Cave Morphology and Genesis

Explorers can only enter the cave at its downstream end, at an altitude of 2010 m in the north wall of Castleguard River valley, more than 300 m above the valley floor. From there the cave ascends 386 m to terminations underneath the icefield (Figure 2). There are three distinct morphologic sections:

1. a Headward Complex of inlet passages created by repeated glacial blockage and re-routing of sinking waters in the past. The passages are plugged by ice or debris today. Younger vadose shafts pass through and become blocked by constrictions or debris below;
2. a Downstream Complex of low tunnels in one major bedding plane, created by flooding and obstructions by glacial ice in Castleguard Valley in the past;
3. the Central Cave, a sequence of remarkably long, straight conduits created where one master bedding plane is intersected by a pair of vertical joints that are linked by a sedimentary dyke crossing them (Figure 3a).

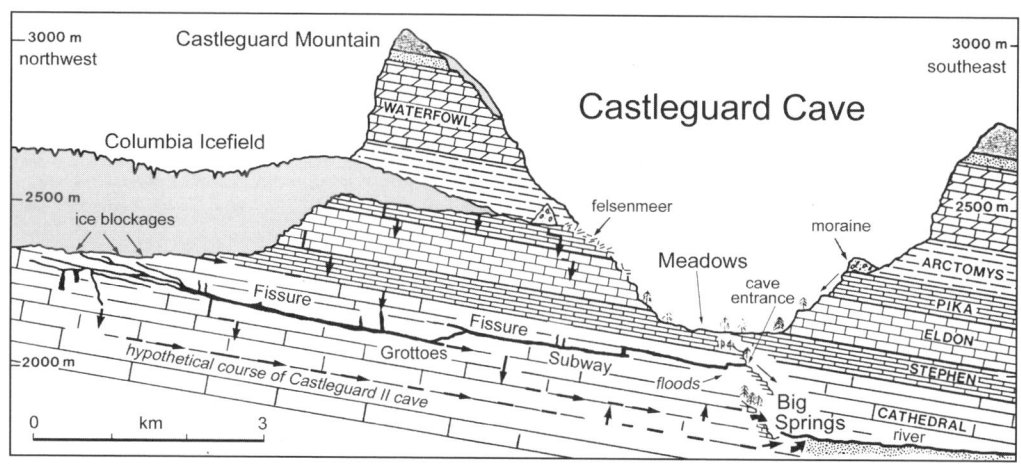

Castleguard Cave: Figure 1. Schematic section through Castleguard Mountain, Alberta, Canada, showing the geology and locations of Castleguard Cave and Castleguard II.

Castleguard Cave: Figure 2. The ice plug at the end of Castleguard's main tube. The ice is squeezed down from the floor of the Columbia Icefield where it sits across the truncated end of the old phreatic tube, but exactly how far the ice has flowed along the tube is unknown. (Photo by Tony Waltham)

The cave possibly originated beneath the Stephen impermeable cover rocks as a single phreatic loop of >370 m depth (Ford et al., 2000). More certainly, as Castleguard Valley was entrenched below the Stephen Formation, the cave became enlarged to its modern dimensions as shown in Figure 3b, two shallow, principal loops consisting of vadose entrenchments up to 15 m deep draining into phreatic tubes of beautiful circularity (Figure 4). Following this main phase, the cave drained through undercapture passages in the bottoms of the loops (Figure 3c). The undercaptures channel local invasion waters today and the Downstream Complex can be flooded from another lifting shaft, Boon's Blunder, and discharged through the explorers' entrance, closing it.

Hydrology

Modern waters drain from the glacier soles, alpine karst, and meadows to a set of springs extending 3 km downstream from Big Springs, dramatic overflows 15–40 m above the valley floor (Figure 1). The waters flow through the putative series of inaccessible caves, Castleguard II (Smart, 1983). Artesian Spring, lowest in elevation, is perennial. As the melt season progresses, upstream springs such as Gravel and Watchman begin to flow. The Big Springs, 100 m higher than Artesian Spring, have a maximum discharge >7 m^3 s^{-1} and handle mean summer

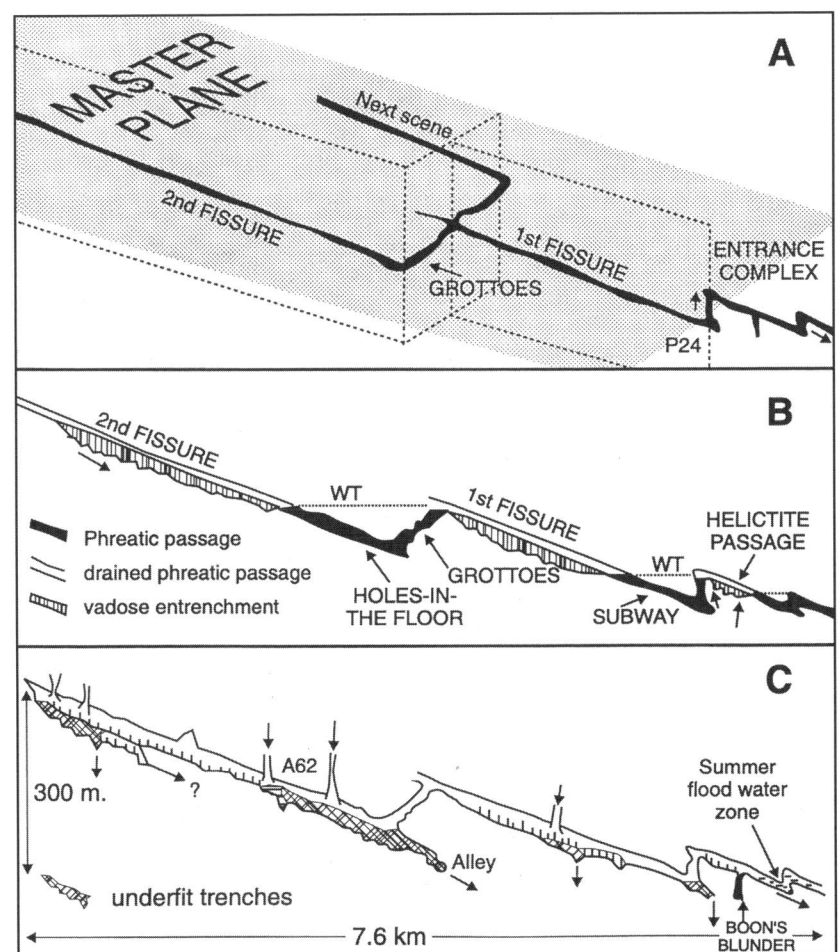

Castleguard Cave: Figure 3. (a) The initial phreatic passages in the Central Cave and Downstream Complex, showing the master bedding plane and intersecting vertical fractures that guided them. (b) The cave at the close of the principal enlargement phase. Drawdown vadose canyons supplied water to a succession of shallow phreatic loops. (c) The modern cave; small invasion vadose streams have cut shafts and underfit trenches in the drained galleries, and are lost into impenetrably small, undercapture passages continuing on down into Castleguard II.

Castleguard Cave: Figure 4. "The Subway" at the lower end of Castleguard Cave, a "textbook" example of a phreatic solution conduit. It is straight and maintains the form and dimensions shown here for a distance of 500 m. (Photo by Tony Waltham)

floods. Their capacity is exceeded when there is very strong melting on the icefield. Groundwater then backs up in the aquifer until it floods the Downstream Complex of the cave, 270 m above Big Springs: cave discharge can be >5 m^3 s^{-1}. Tracer dye placed in glacier edge stream-sinks such as Frost Pot reaches the Big Springs, 4 km distant and 750 m lower, in as little as 3.5 hours. This is a "textbook" example of a dynamic alpine karst aquifer. The large number and great height range of its springs are attributed to repeated disruptions such as debris pluggings, during the glaciations.

Cave sediments, speleothems, and dating

Subglacial boulders, gravel, and sand were swept or bulldozed into the head of the cave. Many were partially cemented by calcite and later eroded, indicating a long history of filling and removal. Throughout the Central Cave there are remnants of three partial fillings with varved silts and clays, separated by phases of erosion or calcite deposition. They are deposits of glacier flour, settled out of suspension when the cave became backflooded with subglacial waters.

Despite its location beneath glaciers or alpine desert, the cave is well decorated with calcite speleothems. There is one double layer of cubic cave pearls, a unique deposit. In the warmer central sector (temperature 2–3°C, relative humidity 95%+) there are small evaporative aureoles of aragonite, huntite, hydromagnesite, gypsum, mirabilite, and epsomite; evaporation in these extreme conditions is due to a strong draft blowing through the cave when not in flood.

U-series dating and paleomagnetic studies of speleothems show that the cave became relict (i.e. Castleguard II was well developed) more than 780 000 years ago. The varved clays are younger. This antiquity is typical of multi-level alpine caves.

Stygobitic fauna

Colonies of two stygobitic crustaceans, blind and unpigmented, live in pools in the Central Cave. An isopod, *Salmasellus steganothrix*, is known elsewhere in the Canadian Rockies. An amphipod, *Stygobromus canadensis*, is known only in Castleguard. It is possible that the cave served as a subglacial refugium for these species during the last glaciation or even longer (Holsinger *et al.*, 1983).

DEREK FORD

See also **Alpine Karst; Glacierized and Glaciated Karst**

Works Cited

Ford, D.C. (editor) 1983. Castleguard Cave and Karst, Columbia Icefields area, Rocky Mountains of Canada: A symposium. *Arctic and Alpine Research*, 15(4): 425–554

Ford, D.C, Lauritzen, S.-E. & Worthington, S. 2000. Speleogenesis of Castleguard Cave, Rocky Mountains, Alberta, Canada. In *Speleogenesis: Evolution of Karst Aquifers*, edited by A.B. Klimchouk, D.C. Ford, A.N. Palmer & W. Dreybrodt, Huntsville, Alabama: National Speleological Society

Holsinger, J.R., Mort, J.S. & Recklies, A.D. 1983. The subterranean crustacean fauna of Castleguard Cave and its zoogeographic significance. *Arctic and Alpine Research*, 15(4): 543–49

Smart, C.C. 1983. The hydrology of the Castleguard Karst, Columbia Icefields, Alberta, Canada. *Arctic and Alpine Research*, 15(4): 471–86

CAUCASUS, GEORGIA

The Caucasus is a large range of fold mountains, structurally continuous with the Crimean Mountains and the Mediterranean Alpine fold belt. It occupies the isthmus between the Black Sea and the Caspian Sea (for location, see map in Russia and Ukraine entry). By both its structure and topography, the Caucasus is divided into three main regions: the Great Caucasus (or the Principal Caucasus Range), the Minor Caucasus, and the Kura-Rionsky depression that separates them. The formation of the large anticlinoriums of the Great Caucasus and Minor Caucasus, as well as of the intervening synclines and depressions, is attributed mainly to late Alpine folding. Since the end of the Pliocene, through the whole Pleistocene, differential uplift continued and caused the formation of the complex tectonic structure and mountainous relief.

The Great Caucasus stretches for about 1500 km, from the Kerchensky channel that connects the Azov and Black seas, to the Caspian Sea near the city of Baku. The Principal Caucasus Range is divided into the Western Caucasus, Central Caucasus (with the highest peak, Mount El'brus at 5633 m), and Eastern Caucasus. Two large monoclinal ranges located to the north of the Principal Caucasus Range, the Peredovoy and Skalisty ranges, are termed the North Caucasus. Most of the Great Caucasus lies within Georgia and its autonomous republic Abkhasia. The Northern Caucasus and the extreme western flank of the Principal Caucasus Range lie within Russia, while the Eastern Caucasus stretches through Azerbaijan. The Minor Caucasus lies mainly within Armenia and Azerbaijan.

The high mountain alpine relief with jagged ridges, large glaciers, and summits exceeding 4000–5000 m is characteristic of the Great Caucasus, while altitudes in the Minor Caucasus are generally lower. The Great Caucasus shows some similarity to the Alps in the regional distribution of karst, with fore-ridges composed of karstified Upper Jurassic and Lower Cretaceous limestones girdling the glaciated core zone of igneous rocks. While the Caucasus holds a diversity of karst types on various lithologies, it is high mountain alpine karst on carbonates that is the most remarkable in the region because of its spectacular glaciokarstic landscapes, tremendous vertical relief, and very deep caves.

On the northern slope of the Great Caucasus, karst is developed in monoclinal and folded structures in Mesozoic limestones, dolomites, and gypsum. Alpine karst massifs and deep caves are known in the western sector of the Northern Caucasus and at the western flank of the core zone of the Principal Range. In the former, the Zagedan Massif, part of the Abishyra-Akhuba range, holds Gorlo Barloga Cave, the deepest and the highest cave in Russia (depth 900 m, length >3000 m) developed in a Devonian and Carboniferous sequence of metamorphosed limestones and schists. Other significant caves in this massif include Zagedanskaya (570 m deep, 5300 m long), Rostovskaya (500 m deep, 1000 m long), and Alekseeva (465 m deep, 4410 m long). The Dzhentu Massif is notable for Majskaya Cave (500 m deep, 4410 m long) with its abundant crystal formations of gypsum and mirabilite. The most spectacular alpine karst is situated at the western flank of the core zone, in the Fisht Massif of Upper Jurassic limestones and above Lower Jurassic volcanic and sedimentary rocks. With its high glaciokarstic terrains (up to 2876 m altitude), the massif holds the westernmost glacier in the Caucasus. Deep caves include Krestik-Turist (633 m deep, 14 000 m long), Parjashchaya Ptitsa (565 m deep, 4500 m long), and Ol'ga (520 m deep, >3500 m long).

On the southern slope of the Great Caucasus, a chain of limestone massifs stretches for about 400 km, parallel to the Principal Range and mainly along the Black Sea shore. The belt of Mesozoic limestones starts near the town of Sochi in Russia and extends through Abkhasia to Western Georgia, where it goes inland. In the Sochi area the Alek, Akhtsy, Vorontsovsky, Dzukhra, Akhshtyr, and Akhun massifs constitute a marginal recharge area of the Sochinsky artesian basin (Dublyansky et al., 1985). With altitudes ranging between 600 and 1000 m and patches of Paleogene insoluble cover in the limestones, these massifs are not alpine karsts, but they have many allogenically fed caves up to 500 m deep.

Further to the southeast, in Abkhasia, the limestone belt continues throughout the Gagrinsky and Bzybsky ranges, where the large Arabika (490 km^2) and Bzybsky (550 km^2) massifs represent outstanding examples of high mountain alpine karst (see Figure 1). The massifs have strongly pronounced glaciokarstic landscapes (Figure 2) and caves at elevations ranging predominantly between 1900 and 2400 m, while the highest peaks rise to 2705 m in Arabika and to 2634 m in the Bzybsky Massif. Both massifs were glaciated during the late Pleistocene, when recharge from glaciers, focused along fissured zones in glacier bodies, contributed considerably to the development of large caves (Klimchouk & Rogozhnikov, 1984; Vakhrushev, Dublyansky & Amelichev, 2001). The most remarkable characteristic of these massifs is the great vertical magnitude of karst groundwater drainage. This has been proved to be up to 2300 m by groundwater tracing from sinking streams to springs in both massifs. However, it might be even greater in reality. In Arabika, karst groundwater is also discharged by submarine springs, some of which rise from the floor of the Black Sea at depths of up to 400 m (Buachidze & Meliva, 1967). Moreover, boreholes near the town of Gagra, at the foot of Arabika are reported to encounter open karst cavities at depths as great as 2300 m (Maruashvili, 1972). It is therefore not surprising that the Arabika and Bzybsky massifs have yielded many deep caves, seven of which are now known to exceed 1000 m and three of which are over 1500 m deep, including the world's second deepest, Krubera (see Krubera Cave, Georgia). These features place the Arabika and Bzybsky massifs among the world's major karst and cave regions.

The Arabika Massif is bordered by the Black Sea shoreline to the southwest, and by the deeply incised Bzyb River valley to the east, separating Arabika from the adjacent Bzybsky Massif. Tectonically, Arabika is on a large anticline in the sub-Caucasian northwest–southeast trend, with a gently dipping southwestern flank and a steeply dipping northeastern flank. Several low-order folds and large faults complicate the major anticline. The core of the massif is Upper Jurassic limestone resting on Lower Jurassic volcanics, while Lower Cretaceous limestones are present on ridges and peaks in the central part and in the fringing foothills. The limestones dip continuously to the Black Sea shore and dip below the modern sea level.

The several hundred caves known in Arabika include Krubera (1710 m deep), the nearby Arabikskaja system (Kujbyshevskaja-Genrikhova Bezdna, 1110 m deep) in the southwest, the Iljukhina system (1240 m deep) in the centre, Dzou Cave (1080 m deep) and Moskovskaja Cave (970 m deep) in the northeast, and Sarma Cave (1530 m deep) in the southeast. Development of the large caves is strongly controlled by the block fault structures. Explorations and groundwater tracing tests have identified at least three large karst hydrological systems (Klimchouk & Kasjan, 2001). The central system has Krubera, Arabikskaja, and Iljukhina caves connected to the large springs on the coast about 2300 m below; in the northern system the Dzou and Moskovskaja caves drain to the Gegsky Vodopad spring in the Gega Valley about 1700 m below. Sarma Cave is presumed to drain to the Goluboe Ozero spring 2100 m below.

The Bzybsky Massif is the structural continuation of Arabika to the southwest, between the Bzyb and Aapsta Rivers. Its southern border is the Kaldakharsky fault which brings Mesozoic limestones into contact with Paleogene non-karstic sediments. The central part is a plateau that is less entrenched by glaciation than the Arabika Massif, at altitudes of 1900–2100 m, with summits rising above 2500 m. Like Arabika, the overall tectonic structure of the Bzybsky Massif is a large asymmetric anticline complicated with several low-order folds. The large Chipshirinsky longitudinal fault divides the massif into the northern and southern blocks, the latter being uplifted about 700 m relative to the former. Many lesser faults further divide the folded structure into a complex system of blocks that exert strong control over

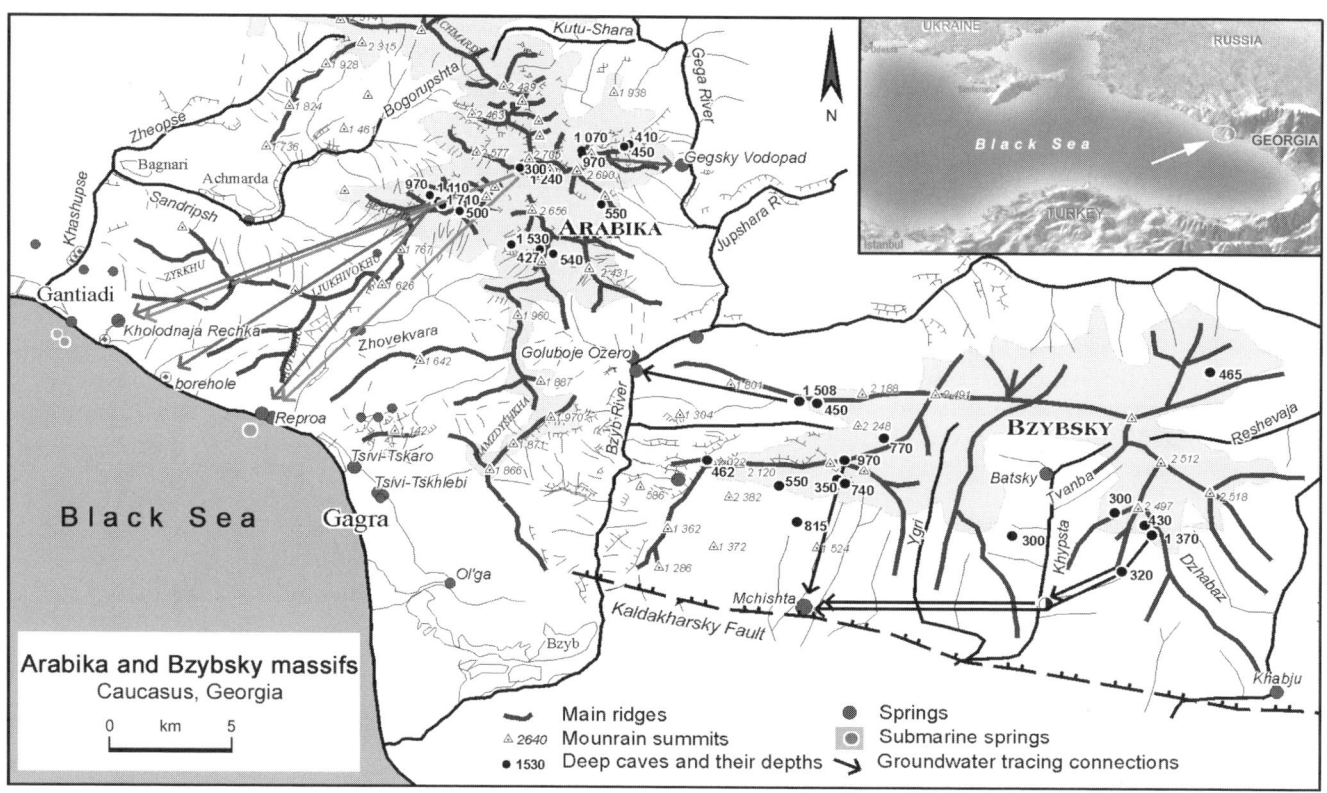

Caucasus, Georgia: Figure 1. Karst features of the Arabika and Bzybsky massifs in the western Caucasus of Abkhasia, Georgia.

Caucasus, Georgia: Figure 2. Glacio Karst of the Arabika Massif with Arabika peak in the background. (Photo by Alexander Klimchouk)

cave development (Vakhrushev, Dublyansky & Amelichev, 2001).

More than 400 caves have been explored in the Bzybsky Massif since 1971, with 55 >100 m deep. The deepest caves are Pantjukhina Cave (1508 m deep, 5530 m long), Snezhnaja-Mezhennogo System (1370 m deep, 19 000 m long), Napra (970 m deep, 3170 m long), Pionerskaja (815 m deep, 1700 m long), Grafsky Proval (770 m deep, 1750 m long), and Forel'naja (740 m deep, 1500 m long). The most remarkable is the Snezhnaja-Mezhennogo System, with its branchwork pattern, an underground river reaches at depth of about 700 m below the Snezhnaja entrance, the large and long river passage, and many large chambers modified by breakdown. The total volume of the cave is estimated to be about 1.7×10^6 m^3. Most of the known deep caves, located within the southern block drain to the Mchishta Spring, the largest spring in the Caucasus with its average discharge of 9.3 m^3 s^{-1} and maximum discharge of 197 m^3 s^{-1}. The spring is on the Kaldakhvarsky fault, and its underground catchment is estimated at 160 km^2. Tracing tests have shown a direct connection to Mchishta from the Snezhnaja-Mezhennogo System (1930 m above) and Napra (2300 m above). Divers in Mchishta have discovered a large cave of total length 4200 m (of which 780 m are underwater) and a volume of 380 000 m^3. The cave starts with a siphon 340 m long and 45 m deep with an underwater shaft in which a depth of 75 m has been reached. The entrance siphon leads to the eastern and western series, which have quite different morphologies. The eastern series is a 3-D maze about 1 km long. The western series is a 1400 m long, linear gallery, 5–20 m high and wide, leading to another siphon 320 m long and 48 m deep. It is hoped that further exploration in Mchishta Cave may give access to the main collector of southern Bzybsky. The northern block of the massif drains to springs in the Bzyb River valley. A tracer injected in Boguminskaya Cave appeared in Pantjukhina Cave at −600 m, and came out at the Bzyb Spring, located just in front of the Goluboje Ozero Spring, which is across the river. The depth of the karstic drainage in this case is 1920 m.

To the north of the Bzybsky Massif, the high Achibakh Massif also demonstrates features of alpine karst, although deep caves are not yet known. Further to the southeast, along the southern slope of the Great Caucasus, the limestone belt continues through Western Georgia, where many important karsts and caves are known at low altitudes (Tintilozov, 1976). Several massifs rise above 2000 m, including Ohachkue, Kvira, Mingaria, Askhi, and Rachinsky, but they have neither signs of glaciation nor deep caves.

ALEXANDER KLIMCHOUK

See also **Alpine Karst; Glacierized and Glaciated Karst**

Works Cited

Buachidze, I.M. & Meliva, A.M. 1967. On the issue of discharge of groundwaters to the Black Sea in the vicinity of Gagra. In *Trudy Laboratorii gidrogeologicheskikh i inzhenerno-geologicheskikh problem*, 3, Tbilisi (in Russian)

Dublyansky, V.N. et al. 1985. *Karst and Groundwaters of Mountain Massifs of the Western Caucasus*, Moscow: Nauka (in Russian)

Klimchouk, A. & Kasjan, J. 2001. Krubera (Voronja): The deepest cave in the world. *Caves & Caving*, 91: 34–41 (The same article is also published in *Spelunca*, 82: 15–24 and *NSS News*, 59(9): 252–57)

Klimchouk, A. & Rogozhnikov, V. 1984. On the impact of late Pleistocene glaciation on the karst of the Arabika massif. *Izvestiya VGO*, 116(2): 165–70 (in Russian)

Maruashvili, L.I. 1972. Chronological and spatial regularities of the cave formation in limestones. In *Karst in Carbonate Rocks*, Moscow: Moscow University (in Russian); Transactions of the Moscow Society of Naturalists, vol. 47

Tintilozov, Z.K. 1976. *Karst Caves of Georgia*, Tbilisi: Metzniereba (in Russian)

Vakhrushev, B.A., Dublyansky, V.N. & Amelichev, G.N. 2001. *Karst of the Bzubsky Range*. Moscow: Russian University of People's Friendship (in Russian)

Further Reading

Klimchouk, A. 1990. Karst circulation systems of the Arabika massif. *Peschery (Caves): Inter-university Scientific Transactions*, Perm: Perm University: 6–16 (in Russian)

Klimchouk, A. 1991. Le grotte del massiccio di Arabika. *La Rivista del CAI*, 112(1): 37–47

CAVES

Various kinds of cavities occur widely beneath the Earth's surface, both natural and artificial (created by humans). The term "cave" is commonly applied to natural openings, usually in rocks, that are large enough for human entry. This definition is clearly anthropocentric, relies on the ambiguous criterion of accessibility by man, has no genetic meaning, and is therefore non-scientific. It also implies that a cavity is connected to the surface through entrances. Caves can be distinguished from surface landforms by morphometric criteria: caves are forms in which the long dimension (length or depth) is greater than the cross-sectional dimensions at the entrance. The anthropocentrism of the above definition of a cave implies that it is largely air-filled, but advances in underwater cave exploration during the second half of the 20th century have dramatically relaxed this limitation. The concept of a cave is, rather, an exploration notion.

Caves are formed by different processes in many rock types and unconsolidated sediments. They can be classified according to their origin and the lithology of host rock or the type of sediment. However, speleogenetic processes differ greatly and many of them do not depend strongly on host rock composition. Numerous attempts have been made to develop genetic classifications that would encompass all caves and relate them to host rocks (e.g. White, 1988; Dubljansky & Andrejchuk, 1989) but none of them appear to be quite harmonious in grouping the processes and genetic types of caves. Instead, some kind of heuristic approach, only partly genetic, is commonly used in distinguishing the most significant classes of caves. The great majority of natural caves, including the largest ones (see Figure), are solution (or karst) caves, that is, caves that have been created principally by the dissolution of bedrock by water circulating through initial openings such as fissures and pores. Solution caves are the most important to cave and karst scientists and so are the principal focus of this Encyclopedia; they are also most important in terms of their interference with human activities. Caves produced by processes other than solution occur in many types

Caves: An example of a large abandoned trunk passage, Beimo Dong, in the Bama karst of Guangxi, China. (Photo by Jerry Wooldridge)

of rocks and are sometimes referred to as pseudokarst or karst-like features (see Pseudokarst entry).

Solution caves occur widely in limestone and gypsum, and, less commonly, in salt, quartzite, and clastic rocks cemented by soluble material (e.g. some sandstones and conglomerates). Although dissolution is the dominant process, other processes such as erosion and gravitational breakdown may take part in their development, particularly at later stages. The lower limit of the size of a conduit (the precursor to a cave) is accepted to be about 5–15 mm. This range encompasses the important thresholds which permit turbulent flow through initial openings under common hydraulic gradients and temperatures, switch calcite dissolution kinetics to boost up the growth rate, and trigger effective sediment transport (Ford, 1988; Palmer, 1991; White, 1977; 1988; see also entry on Speleogenesis). It therefore allows conduits to be distinguished from proto-conduits, and from pores that form during diagenesis of sediments. Speleogenetic consideration implies that conduits can be partly or entirely filled by water, air, and sediments during various stages of their development.

Volcanic caves are formed in several ways (see Volcanic Caves entry), of which the most important is the evacuation of fluid lava from a cooling lava flow. The process operates mainly in pahoehoe basalt flows and is responsible for some very large cave systems, such as the 65.5 km long and 1101 m deep Kazumura Cave in Hawaii. Caves of this type are also termed lava tube caves.

Glacier caves are formed by melting of the ice of glaciers (see Glacier Caves and Glacier Pseudokarst). They form along crevices in the ice and along its contact with bedrock, through the action of invading water usually derived from ice melting on the surface of the glacier. Glacier caves tend to be ephemeral, but some form quite large integrated systems, e.g. Paradise Ice Caves (Washington State, United States) which extend for over 24 km.

Crevice caves may form in any massive rock by disjunction due to various forces (see Crevice Caves). This group includes true tectonic caves, whose openings were formed by tectonic tension forces, and dilatancy / gravity caves usually formed along cliffs and steep slopes due to the combined effect of unloading and gravitational mass movement.

Littoral caves (or sea caves) may form in many types of rock at sea cliffs, due to wave action that involves abrasion (see Littoral Caves). They are abundant in many coastal areas around the world; the most famous cave of this type is Fingal's Cave in basalts of the Hebrides Islands, Scotland.

Piping caves (or suffosion caves) develop in fine-grained, poorly consolidated sediments such as peat, loams, clays, and particularly loess, by removal of tiny clasts suspended in water (see Piping Caves). The process may involve dissolution to facilitate decomposition of the bedrock material but most of the mass transport is mechanical. Piping caves rarely attain a length of more than 100 m.

Erosion caves (or stream-cut caves) form in soft rocks such as shales by mechanical action of water streams. They are numerous in cliffs, being commonly shelters rather than caves. Erosion caves develop best where some resistant bed overlies a weaker bed, though large high-gradient streams may also produce erosion caves in solid rock. Illustrative of the conditional character of the process-based classification of caves is the fact that mechanical erosion plays a considerable part in the formation of many solution caves and it may become the dominant process in mature stages of their development if the caves carry large streams. Some other processes, such as aeolian deflation, hydration, crystallization and contraction, and underground pyrolysis, can also form non-solutional caves, although they are small and rare.

Caves are characterized by their shape and size, the latter being measured by various dimensions considered in the entry on Morphometry of Caves. While most caves do not exceed 1 km in length and 100 m in depth, hundreds are much larger. The longest cave is the 563 km long Mammoth Cave System in Kentucky, United States, the deepest is Gouffre Mirolda (-1733 m) in the Chamonix Alps, France, and the largest underground chamber, by both area and volume, is Sarawak Chamber in Malaysia (respectively 160 000 m^2 and 12 million m^3). The shape of caves, or their morphology, is extremely variable and can be described in several scales. Micro-morphology characterizes minor forms that constitute relief of the host rock surface within passages. Meso-morphology expresses the geometry of the main elements of a cave, that is, of passages (elongated segments in which the length is considerably greater than width or height) and chambers (cave elements in which the width or height is larger compared with adjacent passages). Passages can be linear, angulate, or sinuous, and be horizontal, inclined, or vertical. Vertical passages are termed pits or shafts. (For more details of micro- and meso-morphology features, see Morphology of Caves.) The simplest caves consist of single elements but most are assemblages of passages and chambers with various relationships to each other. Macro-morphology of caves is expressed by cave patterns, overall orientation relative to the gravity field, and tier structure. Several distinctive types of cave patterns are recognized: single-conduit caves, single-void caves, branchwork caves, and maze caves (the latter are further subdivided into network, anastomotic, ramiform, and spongework mazes). Patterns of solution caves are controlled mainly by hydrogeologic factors, by the recharge mode and type of flow in particular (Palmer, 1991, 2000; see also Patterns of Caves). These and other important controls change in regular ways throughout the evolution of karst aquifers, giving rise to the changes of karst / speleogenetic settings and respective karst types (Klimchouk & Ford, 2000). Several settings are recognized that generate distinct styles of solution cave development (see entries on Speleogenesis). The above evolution may result in the presence of different cave patterns in the same region or in their combination in the same cave system.

Caves are sometimes classified according to their orientation relative to the gravity field into horizontal, inclined, and vertical caves, although large systems commonly have elements of various orientations. Vertical caves are dominant in mountain regions where thickness of karstified rocks and the vadose zone is great. Although there are many truly vertical caves that consist of single-drop shafts (up to 640 m deep, as is the Vrtiglavica shaft in the Kanin massif in the Julian Alps, Slovenia), most vertical caves are actually composed of shafts interspersed with inclined passages and have a stair-step profile. In many caves passages are stacked in tiers or levels, often superimposed and interconnected to form complex 3-D cave systems. A number of other approaches have been suggested to further classify solution caves as the most diverse, abundant, and scientifically and practically important category. They are overviewed in Ford & Williams (1989, pp.243–48). Some genetically based classifications are outlined in the entry on Speleogenesis.

Tens of thousands of caves have been explored so far but this is still a small proportion of the underground world represented by explorable cave passages. Caves constitute a particular environment with their specific 3-D morphology, hydrology (see Hydraulics of Caves), sediments, and minerals (see entries on Sediments, Minerals in Caves, and Speleothems), climate (see Climate of Caves) and biota (see entries on Biology of Caves and on Micro-organisms in Caves), and so they can be regarded as special complex natural systems. During long time spans in some distant past many caves functioned as traps for animals that lived at that time (see Paleontology: Animal Remains in Caves) while others served as shelters and dwellings for ancient man who left there many marks of his activity. Being well protected from destructive processes operating on the surface, caves are unique repositories of various forms of information about the past, imprinted in their morphology and contents. This justifies the exceptional scientific value of caves, recognition of which has dramatically increased during recent decades. For the above reasons, and due to the fact that caves offer spectacular scenery with striking natural marvels, beautiful mineral formations, and other unique features, a number of them have been recognized under the UNESCO World Heritage Convention and instituted as protected sites or areas on the national level (see World Heritage Sites).

ALEXANDER KLIMCHOUK

See also **Pseudokarst**

Works Cited

Dubljansky, V.N. & Andrejchuk, V.N. 1989. *Speleologia (terminologia, svyazi s drugimi naukami, klassificatziya polostey)* [Speleology (terminology, relations with other sciences, classification of cavities)], Kungur, Russia: Gorny Institute of the Ural Division of the Academy of Sciences of the USSR

Ford, D.C. 1988. Characteristics of dissolutional cave systems in carbonate rocks. In *Paleokarst*, edited by N.P. James and P.W.Choquette, New York: Springer

Ford, D.C. & Williams, P.W. 1989. *Karst Geomorphology and Hydrology*, London and Boston: Unwin Hyman

Klimchouk, A.B. & Ford, D.C. 2000. Types of karst and evolution of hydrogeological settings. In *Speleogenesis: Evolution of Karst Aquifers*, edited by A.B. Klimchouk, D.C. Ford, A.N. Palmer & W. Dreybrodt, Huntsville, Alabama: National Speleological Society

Palmer, A.N. 1991. Origin and morphology of limestone caves. *Geological Society of America Bulletin*, 103: 1–21

Palmer, A.N. 2000. Hydrologic control of cave patterns. In *Speleogenesis: Evolution of Karst Aquifers*, edited by A.B. Klimchouk, D.C. Ford, A.N. Palmer & W. Dreybrodt, Huntsville, Alabama: National Speleological Society

White, W.B. 1977. Role of solution kinetics in the development of karst aquifers. In *Proceedings of the Twelfth International Congress: Karst Hydrogeology: Huntsville, Alabama [September 21–27, 1975]*, edited by J.S. Tolson and F.L. Doyle, Huntsville, Alabama: UAH Press

White, W.B. 1988. *Geomorphology and Hydrology of Karst Terraines*, Oxford and New York: Oxford University Press

Further Reading

Bögli, A. 1980. *Karst Hydrology and Physical Speleology*, Berlin and New York: Springer (original German edition, 1978)

Gillieson, D. 1996. *Caves: Processes, Development and Management*, Oxford and Cambridge, Massachusetts: Blackwells

Klimchouk, A.B., Ford, D.C., Palmer, A.N. & Dreybrodt, W. (editors) 2000. *Speleogenesis: Evolution of Karst Aquifers*, Huntsville, Alabama: National Speleological Society

Moore, G.W. & Sullivan, N. 1997. *Speleology: Caves and the Cave Environment*, 3rd edition, St Louis, Missouri: Cave Books

CAVES IN FICTION

For as long as humans have walked the Earth we have wondered at and feared what might lie within the dark reaches of caves. For some, the mere mention of caves brings on claustrophobic nightmares, and reading about them is as close as they want to get. Thankfully, there are hundreds of books that take readers to the farthest depths of the Earth from the comfort of their favourite chair. Many of the early stories associated with caves and the underworld invoke images of a dark and mysterious place, inhabited by gods and demons, and of course the final resting-place for the souls of the dead. The stories in Greek and Roman mythology are filled with journeys to the underworld. Many of these underground travels are exemplified in *The Iliad* and *The Odyssey* written by Homer, and in Vergil's, *The Aeneid*.

One of the first writers to combine mythology with the science and religion of the day was Danté Alighieri in *The Divine Comedy*. Written in the early 1300s, this epic poem elaborately describes Danté's spiritual journey through Hell, Purgatory, and Heaven. The most popular of these three books is *Part I: The Inferno*. Here Danté, led by his trusted guide, Virgil, takes the reader down into the centre of the Earth and gives us a guided tour of Satan's kingdom and all its horrors. As they pass into Hell, above its gate they read the inscription "Abandon all Hope, ye who enter here." Deep in the underworld they encounter bizarre scenes, and a cast of characters and demons that would rival any modern-day science fiction film, culminating in an encounter with Satan at the Earth's core. This work is considered a literary masterpiece, and the foundation from which all underground fiction would emerge.

In one of his most famous works, British poet Samuel Taylor Coleridge writes about the underground world in *Kubla Khan*. This elegant poem describes a kingdom where the Alph River flows "through caverns measureless to man." In his writing he describes the cave as "a savage place" that is both "holy and enchanted." It is said that Coleridge found inspiration to write this poem after reading about Florida's springs.

Other writers envision the Earth as hollow, brimming with lost worlds and civilizations. With so much life teeming above ground, why not below too? In 1864 Jules Verne wrote in *A Journey to the Center of the Earth*: "The words which make up the human language are inadequate for those who venture into the depths of the Earth." Although this may well be true, with this book Verne wrote the definitive story against which all other hollow-earth fiction would be judged. In the grandest of styles, Verne mixes scientific fact with fiction and gives us a view of the interior of our planet, filled with vast oceans, prehistoric monsters, mountain landscapes, and man-like creatures. In one of his last books, Verne once again returned to the subterranean, in *Underground City*, also called *Child of the Cavern*. H.G. Wells is considered by many to be the father of modern science fiction. In his 1901 book, *The First Men in the Moon*, he combined the thrill of space travel with the underground exploration of Earth's moon.

Using the hollow-earth theme, Edgar Rice Burroughs produced a seven-novel series that began in 1914 with *At the Earth's Core*. Whereas Verne simply entered the Earth through an extinct volcano, Burroughs uses a rocket-powered burrowing machine to reach his inner-earth world, complete with a miniature Sun, Moon, extinct dinosaurs and mammals, and an intelligent ape-like race. Other cave-related books of his include *Cave Girl* and *Tarzan and the Cave City*.

Even Dorothy and the Wizard of Oz find themselves at the centre of the Earth in L. Frank Baum's *Dorothy and the Wizard in Oz*. After falling into the Earth following an earthquake, the two are reunited and they make a perilous journey back to the surface. In 1865, Lewis Carroll wrote *Alice's Adventures in Wonderland*. Originally titled *Alice's Adventures Underground*, this is the story of a young girl's accidental fall into a rabbit hole and her discovery of an underground world filled with strange people and unusual places.

Many fiction books are modelled on true events and real places. In Robert Penn Warren's novel, *The Cave*, set in a Tennessee cave, he writes of a young man trapped alone in a cave as the world above watches the tragedy unfold. This story and all its characters bears striking resemblance to—and was based on—the plight of Floyd Collins as he lay trapped and ultimately died in Sand Cave in Kentucky in 1925.

One of the best-known writers of the macabre was H.P. Lovecraft. His well-known short story, *The Beast in the Cave*, tells of a man lost from his tour group and stalked by a wild beast in the bowels of Mammoth Cave.

Authors also use the dark, constricting, and unknown passages found in caves as effective means for what they symbolize. Plato, the great Greek philosopher, uses a cave in one of his most famous works, *The Allegory of the Cave*. In it we find helpless prisoners chained in a cave since childhood and unable to move or see each other. They are led to believe that what they are shown by their captors is reality. The cave is a metaphor for understanding the level of human knowledge and its relationship to its surroundings.

In E.M. Forster's 1924 novel, *A Passage to India*, caves play a central role in the mystery and understanding of the story, and they serve as a physical manifestation for the events that take place. Written at a time when Britain ruled over India, we see the caves as complex, confusing, and ungovernable. Forster skilfully uses the caves as a powerful symbol of what happens when two cultures collide.

Caves also make great places to hide from one's enemies. In Ernest Hemingway's *For Whom the Bell Tolls*, loyalists fighting in the Spanish Civil War use a cave as a strategic position as they plan their offensive against the fascists.

The use of caves is prevalent throughout many of the novels written by Mary Shelley. In *Frankenstein*, the monster finds refuge in a high mountain ice cave, and in *The Last Man*, the story focuses on the famed Sibyl's Cave, located in Naples, which contained the "Sibylline Leaves", used to predict the future.

For young readers, nothing beats the danger and adventure that caves can hold. Their dark and mysterious world makes the perfect setting for finding lost gold and solving crime. In Mark Twain's 1876 classic, *The Adventures of Tom Sawyer*, much of the story centres around McDougal's Cave and the trouble that Tom and his friends find in it. Today the cave is called Mark Twain Cave and it is one of Missouri's most popular show caves. Victor Appleton's Tom Swift novels (*Tom Swift in the Caves of Ice* and *Tom Swift in the Caves of Nuclear Fire*) also continue the adventure theme. In Franklin W. Dixon's popular *Hardy Boys* series, *The Secret of the Caves* is just one of the many books where we find our young cavers searching for clues to their latest mystery. And where would we be without the magic words, "Open Sesame!" spoken by Ali Baba in the classic book, *The Arabian Nights*. Speaking these words opened the mouth of a hidden cave containing the treasure of the Forty Thieves.

Caves are used throughout the writings of J.R.R. Tolkien. His popular books *The Hobbit* and *The Lord of the Rings* tell of adventures with orcs, goblins, dwarves, dragons, and a cave troll, all found under the Earth. The strange creature Gollum lived by a subterranean lake beneath the Misty Mountains and his eyes gradually grew big to be able to see in the dim light.

Delving into the Earth is still as popular as ever. Many of Clive Cussler's adventure novels find action underground. His *Inca Gold* takes readers on a wild ride on an underground river in search of treasure. In Jeff Long's *The Descent*, Danté meets Verne as a scientific expedition is launched into the underworld in search of Hell and its ruler. Nevada Barr takes murder underground in *Blind Descent*, a gripping novel, set in New Mexico's Lechuguilla Cave, that pits man against man in the dark wilderness below the Earth's surface. In *Cavern*, by Jake Page, cavers discover that not every prehistoric beast died during the last Ice Age. David Poyer stirs up a mystery in the chilly underwater caves of Florida with his book, *Down to a Sunless Sea*, and Anne Mclean Matthews turns a cave into a blood-soaked killing ground in her disturbing psychological thriller, *The Cave*. James Rollins finds a lost civilization and the answer to one of mankind's biggest mysteries in *Subterranean*, and Paul Preuss follows in the footsteps of Verne on his way to the centre of the Earth in *Core*.

It seems that the lure of caves will forever be ingrained in works of fiction. We have only just begun to unravel a few of the mysteries that lie beneath the surface of the Earth. Every year we plunge deeper and deeper into its dark voids, discovering new life-forms and charting new worlds. Perhaps some of this fiction might one day become fact.

PAUL JAY STEWARD

See also **Films in Caves, Folklore and Mythology**

Works Cited

Anon. 1994. *The Arabian Nights*, retold by N. Philip, New York: Orchard Books

Appleton, V. 1911. *Tom Swift in the Caves of Ice*, New York: Grosset and Dunlap

Appleton, V. II 1956. *Tom Swift in the Caves of Nuclear Fire*, New York: Grosset and Dunlap
Barr, N. 1998. *Blind Descent*, New York: Putnam
Baum, L.F. 1908. *Dorothy and the Wizard in Oz*, Chicago: Reilly and Lee
Carroll, L. 1988. *Alice in Wonderland*, New York: Abrams (originally published 1865)
Cussler, Clive, 1994. *Inca Gold*, New York: Simon and Schuster
Dante Alighieri, 1977. *The Divine Comedy*, translated by John Ciardi, New York: Norton (originally written in Italian, c.1307–21)
Dixon, F.W. 1964. *The Hardy Boys: The Secret of the Caves*, New York: Grosset and Dunlap
Forster, E.M. 1924. *A Passage to India*, New York: Harcourt Brace and London: Arnold
Hemingway, E. 1940. *For Whom the Bell Tolls*, New York: Scribner and London: Cape, 1941
Homer, 1974. *The Iliad*, translated by R. Fitzgerald, Garden City, New York: Anchor (originally composed c.850–800 BC)
Homer, 1996. *The Odyssey*, translated by R. Fagles, New York: Viking (originally composed c.850–800 BC)
Long, J. 1999. *The Descent*, New York: Crown
Lovecraft, H.P. 1996. The Beast in the Cave. In *The Transition of H.P. Lovecraft: The Road to Madness*, edited by H. Phillips, New York: Ballantine (essay originally written 1905 and published 1918)
Matthews, A.Mc. 1997. *The Cave*, New York: Warner Books
Page, J. 2000. *Cavern*, Albuquerque, New Mexico: University of New Mexico Press
Penn Warren, R. 1959. *The Cave*, New York: Random House and London: Eyre and Spottiswoode
Poyer, D. 1996. *Down to a Sunless Sea*, New York: St Martin's Press
Plato, 1970. The Allegory of the Cave. In *The Republic of Plato*, translated by F. MacD. Cornford, Oxford and New York: Oxford University Press (originally composed c.387–347 BC)
Preuss, P., 1993. *Core*, New York: Morrow
Rice Burroughs, E. 1914. *At the Earth's Core*, New York: Ace Books
Rollins, J. 1999. *Subterranean*, New York: Avon Books
Shelley, M. 1985. *Frankenstein*, Harmondsworth and New York: Penguin (originally published 1818)
Shelley, M. *The Last Man*, Oxford and New York: Oxford University Press (originally published 1826)
Tolkien, J.R.R. 1937. *The Hobbit, or There and Back Again*, London: Allen and Unwin, and Boston: Houghton Mifflin, 1938
Tolkien, J.R.R. 1954–55. *The Lord of the Rings*, London: Allen and Unwin and Boston: Houghton Mifflin, 3 vols
Twain, M., 1986. *The Adventures of Tom Sawyer*, New York: Penguin (first published 1876)
Vergil, 1990. *The Aeneid*, translated by R. Fitzgerald, New York: Vintage Books (written 19 BC)
Verne, J. 1877. *Underground City; or, Child of the Cavern*, Philadelphia: Porter and Coates (originally published in French, 1877)
Verne, J. 1992. *A Journey to the Center of the Earth*, Oxford and New York: Oxford University Press (originally published in French, 1864; first English edition 1874)
Wells, H.G. 1968. *The First Men in the Moon*, New York: Lance Books (originally published 1901)

CAVES IN HISTORY: THE EASTERN MEDITERRANEAN

The caves of the eastern Mediterranean region are perhaps the most historically important caves in the western hemisphere. The first historically documented caves are those of Iskender-i-Birkilin, at the head of the Tigris River in Kurdistan. By 1100 BC they were already a well-known attraction on an important trade route. In that year, King Tiglath Pileser of Assyria left a dated cuneiform inscription commemorating his visit. Some 250 years later, King Shalmaneser III also left commemorative inscriptions at the caves and on various monuments in his capital city. Here, the decorations on a bronze gate band included the earliest known depiction of a cave (Figure 1 and Halliday & Shaw, 1995). This and a black obelisk with an incised inscription about the cave are now exhibited in the British Museum. Alexander the Great, known throughout the Middle East as Iskender, is said to have bivouacked his army in the Birkilin Great Cave some five centuries later. The Iskender-i Birkilin caves have borne his name ever since.

Throughout the ancient world, Greek and Roman folklore was full of references to caves, not all of which actually existed. Two different show caves in Crete (and another on the island of Naxos) are advertised as the mythological birthplace of the great god Zeus. Hermes Cave near Corinth was named for the gods' messenger, supposedly fathered by Zeus on the nymph Maia in this cave. Ancient stories related that the orphaned in-

Caves in History: Figure 1: Section of the hinge from the Bronze Gates of Shalmaneser III, with engraving of the Birkilin stream cave (lower panel). © The British Museum.

fant Dionysos (another son of Zeus, by Semele, who had a jealous husband) was fostered in what now is called the Cave of Dionysos, in Arcadia.

Homer's fictional hero Odysseus was noted for amazing spelean adventures. In far northeastern Greece, the town of Maroneia is one of several which claim various caves where he had to blind the one-eyed giant Polyphemus in order to escape captivity (Mallorca's Cave of Arta also has a strong case). Near his mythical home on the island of Ithaca, the Cave of Marmarospilia is listed as the Cave of the Naiads, where Odysseus briefly stashed his treasure before reconnoitring his home, in disguise. The best claimant to the legendary River Styx or River Lethe appears to be in the Cyrenaica karst east of Benghazi, Libya (Preble, 1945). However, the supposed Labyrinth, in Crete, is merely an underground limestone quarry.

In western Turkey, the builders of the long-ruined city of Hieropolis must have known of "Plutonium", "Pit of the Jinns", or "Dog Cave", as a travertine cave in the world-famous terraces of Pamukkale has been variously termed. This cave opens downward within the complex ruins of the Temple of Apollo. Its shape traps hypoxic concentrations of CO_2 released by the thermal springs. Old tourist leaflets announced—probably without much exaggeration—that the classical writers Strabo and Dio Cassius discussed its lethal effects.

The Bible and the Koran contain numerous references to caves, although most biblical caves are mere rock shelters or grottoes. However, in his analysis of caves mentioned in the Bible, Stebbins (1947) confirmed that others were important, both historically and scientifically. One was discussed in Buckland's 1824 *Reliquae Diluvianae*. Bethlehem's famous Grotto of the Nativity and St Paul's Grotto on Malta are not mentioned in the New Testament, and there is doubt that either is a natural cave. Historic Ghar Dalam and many smaller caves in Malta are natural, and some authorities have suggested that a tiny part of St Paul's Grotto was also originally natural.

By the time of the Prophet Muhammad, the fable of the Cave of the Seven Sleepers—a Rip van Winkle precursor—had become a beloved folk tale. The Prophet used it effectively as a parable. Today, at least two cities claim this cave: Ephesus and Tarsus. The better-publicized site is about 3 km east of the principal ruins at Ephesus, near Turkey's Aegean coast. It has been commemorated on a Turkish postage stamp that actually depicts the ruin of an ancient religious complex built against a hillside containing at least two phreatic caverns. Like Scotland's King Robert the Bruce, the Prophet Muhammad was saved by a spider web spun across the entrance of a cave where he was hiding on Mount Thar south of Mecca (the Prophet's enemies were certain that since the web was intact, no one had approached the cave for days). His people were so familiar with caves that this account was accepted universally. This cave also has been commemorated on postage stamps.

In the karsts of Lebanon, a Roman temple was built on top of one of the resurgences at Afka Cave. The site for another Roman temple was a magnificent field of spitzkarren. On the other hand, most of the "cave monasteries", cave temples, and refuges of Lebanon are merely beneath overhangs or in grottoes, rather than in true caves. Although the splendidly decorated inner portions of Jeita Cave are 20th-century discoveries, the great resurgence from its entrance was utilized in antiquity. Israel's well-known Qumran Caves are noted for their caches of religious manuscripts—the famous Dead Sea Scrolls. Israel's largest and most important caves are in salt karst known since biblical days (see Sedom Salt Karst). No hieroglyphic inscription is known in or about any true cave in Egypt, but the history of the caves and karstic spring at St Anthony's Monastery (in the presently arid Eastern Desert) can be traced to about 305 AD.

Throughout Greece, caves and karst have been an integral part of ordinary life since prehistoric times. The Acropolis and the heights of Delphi are formed of well-developed karst. At Delphi, the 100 m Corycian Cave was supposedly the seat of the famous Delphian oracle; the centre of the universe. At a less lofty level of life, well into the 20th century pregnant women in Crete rubbed their bulging bellies ("just for luck") against a pregnant-looking stalagmite in the cave shrine of Eileithya, once worshipped as the goddess of childbirth and motherhood, as well as the mother of Eros. The 260 m Cave of the Dragon, located within the town of Kastoria, has long been a favourite topic among Grecian storytellers. Large littoral caves served as convenient boat houses, refuges, and pirate dens. The vanishing current of sea water disappearing into the rocky shore at Argostoli on the island of Cephalonia must have perplexed early man, but this did not stop him from constructing mills to harness its power. Inland in the Peloponnesus, the famous *katavothres* of the plain of Mantineia—gaping stream-sinks—must also have bemused him.

Pre- and post-Christian shrines and memorials to victims of wars and persecutions are almost innumerable here. Graffiti in the Cave of Antiparos have been dated back to Archilochus (728–650 BC). Near Athens, in the fourth century BC, a skilled sculptor transformed the entire cave of Nympholeptos into a marvellous cave shrine complete with carved statues of nymphs, the Graces, Apollo, Pan—and himself. The shrine at Ton Limnon Cave in the northern Peloponnesus is credited with curing insanity, at least in mythological times, when some unfortunate women were reputedly cured of the delusion that the gods had changed them into cows.

Fine 14th- and 15th-century Byzantine paintings can be admired on the walls of Panagia Eleousa Cave in northern Macedonia.

Greece's Cave of Antiparos (on the island of that name) was the first cave to become celebrated in the mainstream of modern Western thought, but at a terrible cost. In 1673, M. de Nointel, the French ambassador to Istanbul, explored this steeply sloping cave and celebrated Mass in its largest chamber. Subsequently J.P. Tournefort described the events in Jules Verne style. His account was translated into several languages (Tournefort, 1718), accompanied by splendid, highly imaginative engravings which made the cave seem approximately ten times life size. Fanciful illustrations of some later accounts depicted even grander scenes (Figure 2; see also Figure in Speleothems: History). Despite their exaggerations, these wonderful engravings of the Cave of Antiparos had a salutary effect on the dominant establishment of northern and western Europe. Even in enlightened intellectual circles, predominant conventional wisdom heretofore had populated caves with evil trolls and other loathsome imaginary beings; obviously caves were thus places to shun and abhor. Suddenly they now became places of beauty, interest, and curiosity. Isolated Antiparos became a frequent port of call and its cave became more and more celebrated. Unfortunately, huge smoked inscriptions soon defaced the cave, and its glories

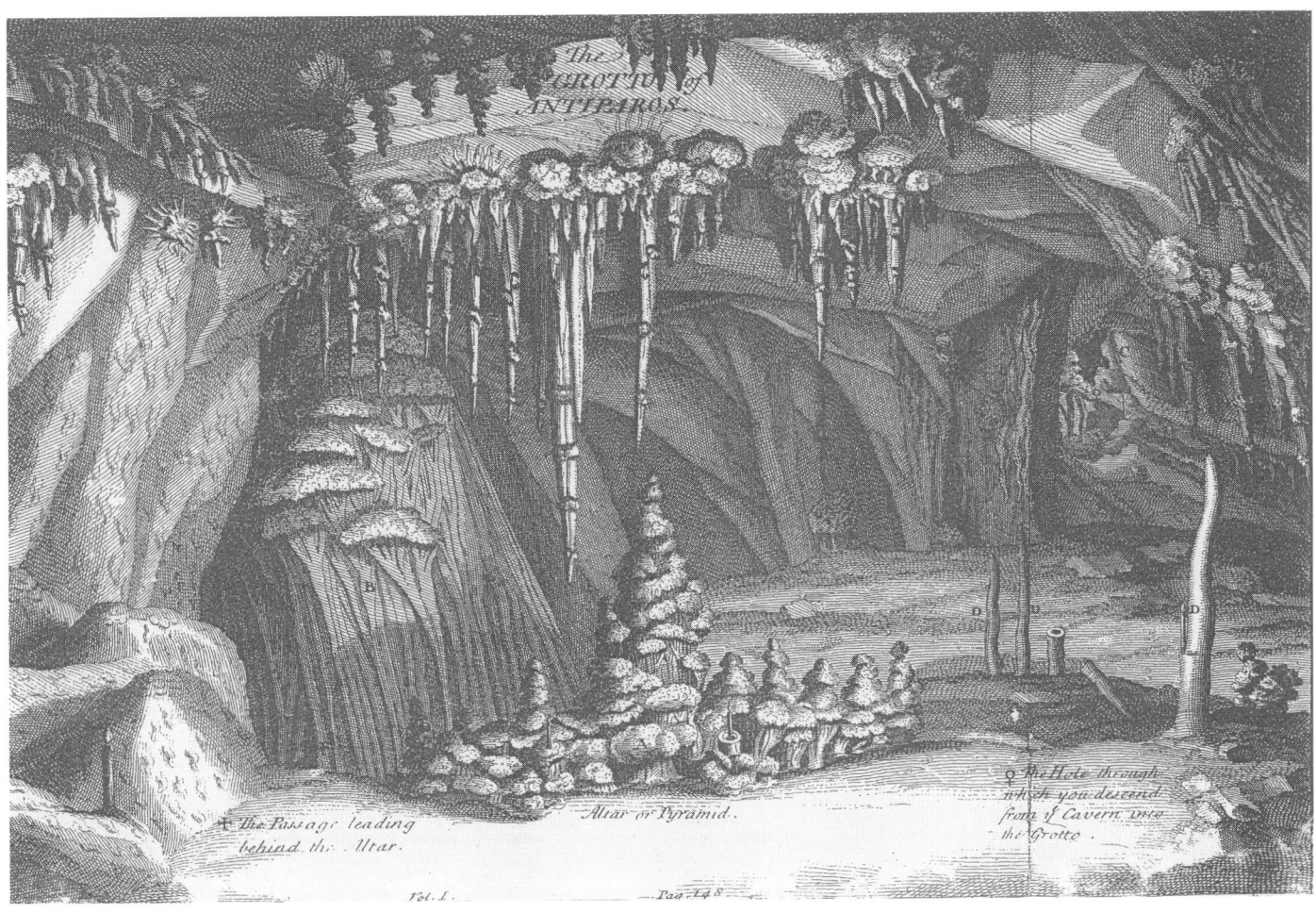

Caves in History: Figure 2: Artist's conception of the Grotte d'Antiparos, drawn 1776 by J.B. Hilair. From *Voyage pittoresque de la Grèce*, vol. 1, 1782, by M.G.A.F. de Choiseul-Gouffier. The artistic merit of this and other early engravings of scenes in this cave profoundly influenced European views about caves but enormously exaggerated the size of rooms and passages. William R. Halliday collection, courtesy of Trevor Shaw.

lived on only in the imaginative engravings. Today, the Cave of Antiparos preserves only traces of its former beauty. This once-great cave remains a sad monument to the impact of fame on unprotected caves everywhere. Hopefully it will remain a truly unforgettable lesson of history.

WILLIAM R. HALLIDAY

See also **Folklore and Mythology; Religious Sites**

Works Cited

Halliday, W. & Shaw, T. 1995. The Iskender-i-Birkilin Caves in the 9th and 12th Centuries, B.C. *Bulletin of the National Speleological Society*, 57: 108–10

Preble, J. 1945. Cave of the River Styx. *Bulletin of the National Speleological Society*, 7: 13–15

Stebbins, C. 1947. Cave references in the Bible. *Bulletin of the National Speleological Society*, 9: 35–37

Tournefort, J.P. 1718. *A Voyage into the Levant*. vol. 1, London: Browne (original French edition, 1717; a reprint of the section on the Cave of Antiparos, without illustrations, appears in *Journal of Spelean History*, 29: 48–51, 1995)

Further Reading

Aygen, T. 1984. *Turkish Caves*, Istanbul: Turkiye Turing vs Otomobil Kurumu Publishing

Fileccia, A. 1983. Carsismo della Cirenaica. *Speleologia*, 9: 17–19

Petrocheilou, A. 1984. *The Greek Caves*, Athens: Ekdotike Athenon

Shaw, T. 1952. The caves of Malta. *Bulletin of the National Speleological Society*, 14: 34–39

Siderides, N. 1911. Les katavothres de Grece. *Spelunca: Bulletin et Mémoires de la Societé de Spéléologie*, 5(3): 63–64

CERKNICA POLJE, SLOVENIA: HISTORY

The seasonal lake of Cerknica occupies a level depression immediately south of the town of Cerknica and 10 km east of Postojna. The floor area of this depression is 38 km², of which two-thirds is flooded every winter to form the lake. Its regular water supply is two-fold: underground from the Obrh stream which sinks nearby in Loško Polje, and also from a few surface streams such as the Cerkniščica. In addition, when the local water table rises as a result of large amounts of rainfall, water pours out of holes (estavelles) in the level bottom of the "lake" and from caves around its edge. When the water table drops again, the lake flows back into many of these holes. It also drains into several overflow caves such as Velika and Mala Karlovica at its northwest corner, with most of the water going underground to feed karst springs in Planinsko Polje.

With such a large, almost flat surface as the "lake" bed, the appearance and disappearance of the water can seem remarkably rapid, and this is what drew attention to it long ago. So strange did this seem, that the place has a long recorded history. A little before 7 BC, Strabo referred to it as the "marsh called Lugeum". In 1551 Wernher published a map of the lake (Figure 1) and stated that: "where not long ago men were fishing you may now bring in a harvest, and even sow for the next one, and trade in the produce; and at the turn of the year go fishing again". When Lazius published his map of the region in 1561, the lake was shown four times too large in comparison with its surroundings. Clearly it was regarded as important. No doubt this early popularity was due in part to the fact that the place was close to the ancient salt road from the Adriatic into central Europe. Once written about, it remained a "tourist attraction" well into the 19th century.

The early fame of the polje inspired writers who sought to explain the mechanism of the disappearing lake. In 1537, a Latin poem by Leonberger left it to the water nymphs to explain. More than a century later, the Jesuit polymath, A. Kircher, in 1665, made the remarkably modern suggestion that the fluctuating level of the lake was due to changes of water level in a natural reservoir beneath the mountains—a kind of water-table hypothesis. This suggestion was disregarded when investigators started to give the lake's behaviour more serious consideration. Their explanations included:

1. Simple flow in single channels such as can easily be simulated in a laboratory experiment;
2. Complex flow through many channels and siphons;
3. Diffuse flow beneath what amounted to a water table.

J.W. Valvasor, well known as a topographer in Slovenia, was also a cave explorer (see Speleologists entry). In 1687 he put forward a simple explanation for the seasonal flooding of the

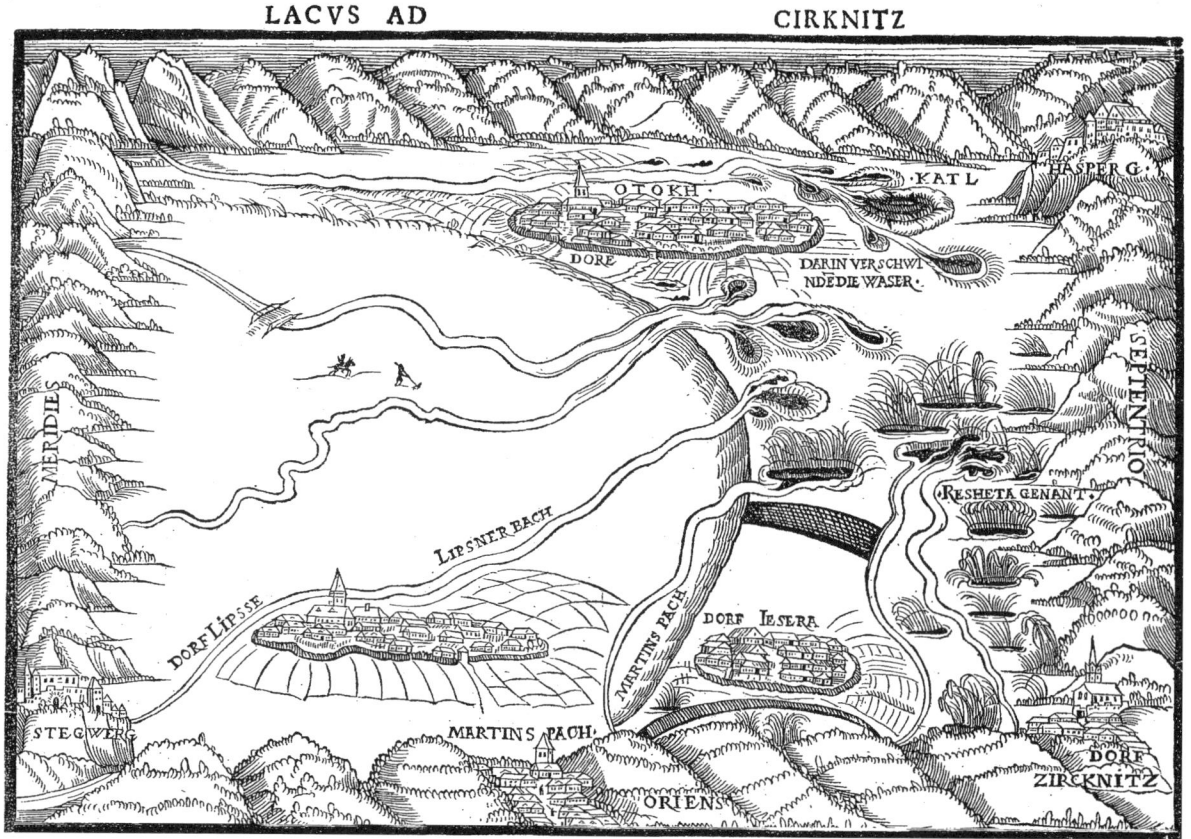

Cerknica Polje, Slovenia: History: Figure 1. Wernher's 1551 map of the Cerknica Lake, showing the main water sinks. South is at the top of the map.

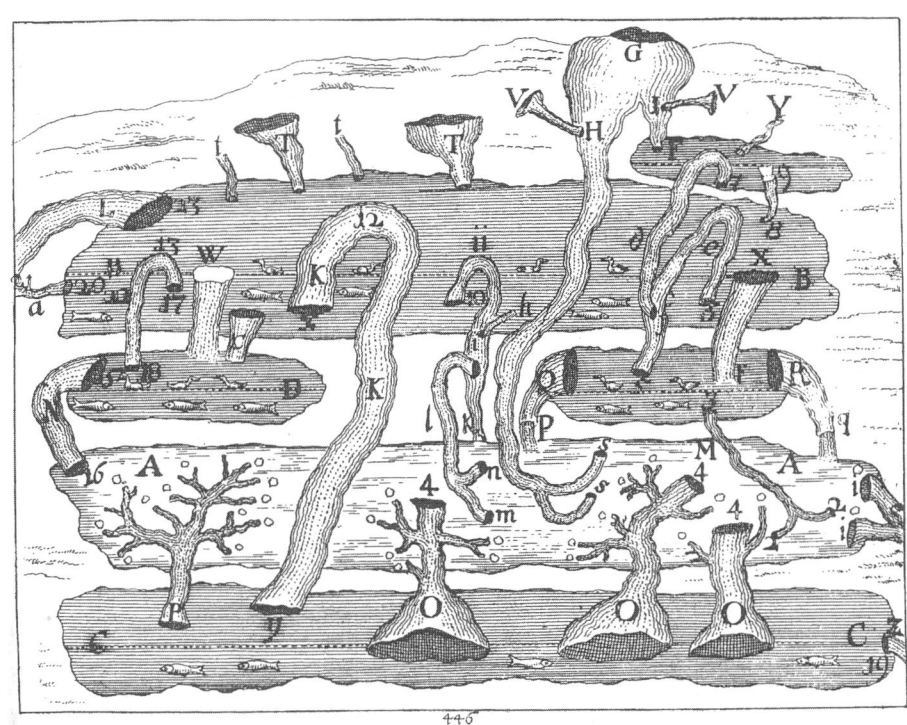

Cerknica Polje, Slovenia: History: Figure 2. Valvasor's diagram of 1689, showing supposed subterranean lakes associated with the Cerknica Lake, and the channels and siphons connecting them.

Cerknica Lake. The lake, he said, was an overflow from underground lakes beneath the surrounding hills, which normally drained elsewhere. This idea was in fact partially correct. Unfortunately, two years later, Valvasor evidently felt that he needed to explain the sudden onset of floodwater flow into the lake. To do this he modified his simple explanation unrecognizably to a system consisting of the lake itself, an underground lake beneath it, more underground lakes in the hills, channels connecting them all, and no fewer than six siphons (Figure 2).

Some 30 years later, F.A. von Steinberg produced a theory that was different but just as complex. It was written about 1720 and published in 1758. He postulated several levels of caves within the hills, interconnected and with siphons. To explain the sudden start of flooding into the lake he supposed that the air entrained in the underground waterfalls caused pressure to build up in some of the caves.

So much for the two really complicated theories. J.A. Nagel produced a simple and elegant theory, not unlike Valvasor's earlier one. Water, he said, will appear in the surface lake only if the rate of supply to an underground system is greater than the off-take. Thus Nagel was basically correct and would have been entirely so if he had not confined his water flow to single channels. His manuscript account remained unpublished in Vienna, so it is not known to what extent he influenced the ideas of his successors.

With the next writer began a trend towards the concepts of groundwater and a water table in limestone. F. Hacquet, in 1778, referred to Steinberg's theory as unnecessarily complicated and he himself produced a refined version of Valvasor's first explanation. The mountains round the lake, he said, are filled with caves and act like a sponge. Only when they are full does the Cerknica Lake begin to fill.

Finally, Tobias Gruber, in 1781, postulated what amounted to groundwater in the rock beneath the lake. This was in effect a multichannel version of Nagel's ideas, which he may or may not have been aware of. He expressed his explanation more clearly than Hacquet, who in any case was concerned with water-filled caves around the lake rather than beneath it. Thus Gruber wrote in his *Briefe hydrographischen und physikalischen Imhalts aus Krain* (pp.111–12):

> Beneath the Zirknitz [Cerknica] lake and its surrounding mountains there flows continuous water . . . All the caves and passages taken together make an underground river bed which stretches away invisibly for so many miles. Only here and there does it break into daylight a little as in the river of Rakov Škocjan and, intermittently, in the Cerknica Lake.

Trevor Shaw

Further Reading

Shaw, T.R. 1992. *History of Cave Science: The Exploration and Study of Limestone Caves, to 1900*, 2nd edition, Broadway, New South Wales: Sydney Speleological Society
Gives much more detail, with original sources.
Shaw, T.R. 2000. *Foreign Travellers in the Slovene Karst 1537–1900*, Ljubljana: Založba ZRC
Reprints of early descriptions of karst features.

CHEJU-DO LAVA CAVES, SOUTH KOREA

Cheju-do, the largest island (1792 km² in area) of Korea, is located in the Korean Strait, c.90 km south of the Korean Peninsula. The island is a long ellipse, 74 km east–west and 39 km north–south. Mt Halla, the highest peak (1950 m) in South Korea, is located near the centre of the island. Geologically the island is a complex shield volcano, which extends with a gentle slope to the coastal region; it is notable for its numerous parasitic cones. Cheju Island is almost wholly composed of volcanic rocks, mainly trachyte, trachy-andesite, andesite, and basalt, with some sedimentary formations (Won, 1975; Lee, 1982). The pre-volcanic basement of the island is not exposed, but the results of a regional gravity survey together with the presence of granitic xenoliths found in the effusive rocks and in a basaltic breccia near the base of the exposed succession suggest that the basement is composed of pre-Tertiary (probably Bulgugsa) granite. According to Won (1975), the sequences of the volcanic eruptions and some sedimentary rocks in Cheju Island are grouped from oldest to youngest as follows: the Upper Pliocene Sogwipo Formation; the Lower Pleistocene Pyosonri basalt, Sogwipo and Jungmun trachytes; the Lower–Upper Pleistocene Seongsan, Hwasun, and Sinyangri Formations; the Upper Pleistocene Jeju and Hahyori basalts and Beopjeongri (including Sanbansan) trachyte; Siheungri, Seongpanag basalts; Hanlasan basalt and trachy-andesite; Baegrogdam basalt; Lower Holocene Volcanic cone I and Volcanic cone II; and the Upper Holocene AD 1007 and 1002 episodes of volcanic activity; Shell-sand Formation.

The basalt is the most widely distributed rock, followed by the trachyte and trachy-andesite, and the sedimentary formations are the least abundant. These volcanic flows erupted more than 79 times during the history of volcanic activity. There are a couple of old manuscripts which date several eruptions to 1002 and 1007 AD.

Cheju Island is a major vulcanospeleological area and one of the main show cave regions in South Korea. The lava caves of Cheju Island have been systematically surveyed since 1977 as a joint project between the Japanese Vulcanospeleological Society and the Speleological Society of Korea. There are 57 lava caves recorded, but some of them have not yet been scientifically surveyed. About 80% of the lava caves are in Pyosonri basalt, and the rest are distributed among the Jeju, Hahyori, Siheungri, and Halasan basalts (Speleological Society of Korea, 1988; Sameshima et al., 1988).

Man-jang Cave is located in the northernmost part of the island. The name of the cave derives from "Man" (ten thousand)-"jang" (c.3 m long), meaning "so long that it reaches 30 000 m".

The total length of the Man-jang Cave is actually 8928 m, with the lower major cave measuring 5164 m, and the upper two caves 2031 m and 1733 m respectively, and the cross-sectional area of the main passage is among the largest in the world (see Photo). At the upper end of an upper cave, there are four floor shafts, which connect to the main cave at the lower level. There are also large-scale lava bridges in the lower reaches of the upper cave. The K–Ar age determination on a lava bridge on the third floor dates it as 190 000 years old (Sawa & Inoue, 1999). Twenty-one lava balls were found in the main cave. The lava balls were probably left on the cave floor because of the slow movement of the lava flow or due to the thinness of the lava. "Turtle Rock" is a large lava ball on a terrace, which has a prominent rim on its side, indicating the lowering of the lava-flow level since the ball was carried down. The other typical features of this cave are "tube in tube" structures, lava pillars, silica pillar formations (lavatites), ripple wave marks, and long ribs / ledges at multiple levels on the walls, left by past stages of lava flowing through the cave. The K–Ar age of the two lava column samples (taken from the area of the ceiling) 1060 m from the entrance of the main show cave, have been dated at 30 000–50 000 and 320 000–420 000 years ago, respectively (Sawa et al., 1990). Gypsum formations and coralloidal diatom speleothems also occur (Kashima et al., 1989). Man-jang Cave (with the associated Kim-nyong Cave) was designated as a Korean Natural Monument in 1962.

Bil-le-mot Cave is situated in the southern area, in Aewol district and was discovered in 1971. This cave is developed in the Siheungri lava, where the lava was ponded by a ridge of the Pyosonri lava. Systematic investigations have been carried out since 1989 as a joint project between Korean, Chinese, and Japanese speleologists. The narrow entrance barely allows a person to go through but leads to grand halls and an extremely complex cave system with passages branching and crossing both horizontally and vertically. There is even a spiral passage between upper and lower parallel caves. The main cave is 2917 m long, but the length of the complex branch caves is 8832 m, making the total length 11 749 m. There are lavatites 28 cm long in this cave, and the largest lava ball is 7 m long and 2.5 m wide and high. It also contains calcite, opal, quartz, and trona (Kashima et al., 1989). This cave was designated as a Korean Natural Monument in 1984. However, it is not open to the public.

The Hyop-jae cave system is in the Hallim district near the northwest coast of the island, in Pyosonri basalt. Ok-san Cave, So-cheon Cave, Han-dul Cave, Ssang-ryong Cave with Hyop-jae Cave were together designated as Korean Natural Monument

Cheju-do Lava Caves, South Korea: List of lava caves longer than 2 km on Cheju Island (compiled by T. Ogawa and I. Sawa in July 2001). Location abbreviations: PJ=Puc-Jeju-Gun; NJ=Nam-Jeju-Gun.

Name of cave	Location	Length (m)	Depth (m)
Bil-le-mot Cave	O-um-Ri, Aewol-Up, PJ	11 749	255
Man-jang Cave	Kim-nyong-Ri, Kujwa-Up, PJ	8924	125
Su-san Cave I	Su-san-Ri, Songsan-Up, NJ	4676	140
So-cheon Cave	Hyo-ghae-Ri, Hallim-Up, PJ	2980	130
Wa-hul Cave	Wa-hul-Ri, Chochon-Up, PJ	2066	130

Cheju-do Lava Caves, South Korea: The main passage in Man-jang Gul is unusually large for a lava tube, and has a kilometre of its length lit for tourist access over a smooth lava floor. (Photo by Tony Waltham)

No. 236 in 1971. A total of 17 km of passages has been mapped in the many caves in this area. This cave system is well known for its abundant carbonate speleothems. Microcoquina sand dunes are widely distributed in the Hallim district, overlying the cavernous Pyosonri basalt. Carbonate-bearing groundwaters percolated into the Hyop-jae cave system and deposited carbonate speleothems (dripstone, flowstone, and other speleothems) which cover the bare lava walls and coexist with lava speleothems (Kashima et al., 1989).

NARUHIKO KASHIMA

Works Cited

Kashima, N., Ogawa, T. & Hong, S.H. 1989. Volcanogenic speleominerals in Cheju Island, Korea. *Journal of the Speleological Society of Japan*, 14: 32–39

Lee, M.W. 1982. Petrology of Cheju Island, Korea—Part 1. Petrography and bulk chemical composition (in Japanese with English abstract). *Journal of Japanese Association of Mineralogists, Petrologists and Economic Geologists*, 77: 203–14

Sameshima, T., Ogawa, T. & Kashima, N. 1988. *5th International Symposium on Vulcanospeleology Excursion Guide Book*, Tokyo: 5th International Symposium on Vulcanospeleology Secretariat

Sawa, I. & Inoue, H. 1999. X-ray fluorescence analysis and K–Ar age determination of a lava bridge in Manjang-gul Cave, Korea. *Journal of the Speleological Society of Japan*, 24: 57–63

Sawa, I., Murata, M., Hong, S.H. & Kashima, N. 1990. Analysis of the twin lava column sample from Manjang-gul Cave, Korea (in Japanese with English abstract). *Journal of the Speleological Society of Japan*, 15: 42–46

Speleological Society of Korea, 1988. The caves of natural monument in Korea (in Korean). *Cave (Journal of Speleological Society of Korea)*, 17: 13–105

Won, C.K. 1975. Study of geologic development and the volcanic activity of the Jeju Island. *Scientific Report of Kon-Kuk University*, 1: 1–42

CHEMISTRY OF NATURAL KARST WATERS

Karst water represents a subset of groundwater whose chemical composition is controlled fundamentally by climate, the type of source materials being dissolved, and residence time. In the case of climate, temperature and rainfall control weathering rates of system solids, with both parameters constraining seasonal and short-term variations in chemical composition. Soil temperature and saturation govern microbial production of CO_2, which dictates the capacity of soil solutions to dissolve carbonate, and are therefore parameters of particular importance in limestone and dolomite karst areas, although not relevant to the formation of gypsum karst.

Source rocks yield solutes at highly varying rates (in the order of fast to slow): ion exchange on clays, dissolution of gypsum, metastable carbonate (aragonite, Mg-calcite, crushed calcite), limestone (calcite), pyrite, dolomite, and finally silicate phases. The atmospheric zone is also a source of karst water solutes, delivering atmospheric dust, pollutants, and marine aerosols in incident rainfall, and organics and particulate matter washed from plant foliage to the ground surface via rain throughfall.

In the dual-porosity soil and aquifer zones (Figure 1, Tooth & Fairchild, 2003), residence time is controlled by hydrology, which determines the nature and degree of water / solid interaction. Primarily, residence time is a function of storage capacity and the distribution of matrix and preferential flow routes in the soil (Figure 1A), and of matrix, fracture, and conduit routes (Figure 1B) within the aquifer. Secondarily, temporal variations occur in soil and aquifer zone flow pathways, in response to rainfall inputs of differing duration and intensity. Resultant karst water chemistries are also constrained by the relative abundance and degree of reactivity of soil and aquifer zone materials. A

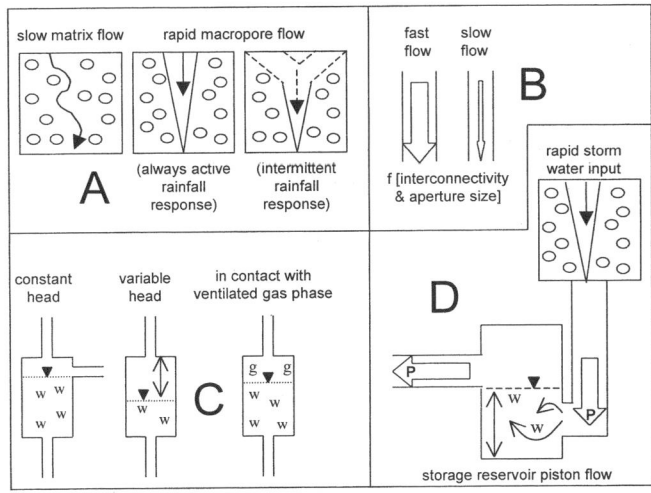

Chemistry of Natural Karst Waters: Figure 1. Potential hydrological controls on karst water geochemistry; **A.** soil zone flow routes; **B.** aquifer zone flow routes; **C.** aquifer reservoir head space and nature of the gas phase; **D.** mixing in piston flow reservoirs.

further control on geochemistry is the relative dominance of the processes of carbonate dissolution and precipitation (Figure 1C) during aquifer throughflow to the cave exit point or spring.

Many geochemical parameters have been measured in karst waters, depending on the purpose of the study. Water hardness represents the amount of dissolution of carbonates (dolomite and calcite) and can be approximated in several ways. Controls on water hardness (technically equivalent to alkalinity) in karstic regions have been extensively investigated by use of electroconductivity (EC), a surrogate for total ionic content, and hence total amount of carbonate rock dissolution in carbonate terrains. Whereas EC can be readily logged automatically, individual species normally have to be analysed on individual samples. Determination of pH close to the time of sampling is also needed to determine the relative saturation of solutions for calcite and dolomite. Alkalinity (bicarbonate and carbonate) is determined by titration with a strong acid, although it can also be approximated by calculating the ionic charge balance, particularly if EC data is available, as this provides an independent check (Rossum, 1975). Determination of the major ions (Ca^{2+}, Mg^{2+}, K^+, Na^+, HCO_3^-, SO_4^{2-}, Cl^-, and NO_3^-) is essential for studies of solute sources, supplemented as required by trace species. Measurement of the intensity of UV-stimulated fluorescence of cave waters is a powerful tool in reconstructing the input of soil-derived dissolved organic material, particularly humic and fulvic acids (e.g. Baker, Barnes & Smart, 1997). Another indication of the introduction of short residence time waters is the dilution of species known to be generated by slow dissolution (e.g. pyrite-derived sulfate). In recent years there has been a surge of interest in the chemistry of cave waters in relation to that of precipitated speleothems, in order to provide modern calibrations for paleoclimate studies of the speleothems. Much work has focused on isotopic parameters, particularly $\delta^{18}O$, because of its relationship to environmental temperature, but also $\delta^{13}C$ because of its relation to the carbonate system and soil processes (e.g. Bar-Matthews et al., 1996). Strontium isotope ratios ($^{87}Sr/^{86}Sr$) reflect cation sources which, in favourable cases, can be correlated with the recharge regime and hence used as a proxy tool for estimating past rainfall (Banner et al., 1996). Isotopes belonging to the decay series of U isotopes, as well as ^{14}C, have also been studied in relation to modern processes that affect our understanding of radiometric dating of speleothems.

An extensive study of spatial and temporal variation in the sinking creeks and springs of south-central Kentucky in the Mammoth Cave area (Hess & White, 1983), clearly demonstrated the importance of seasonal temperature variations and dissolution along the flowpath in controlling water hardness. Rainfall events caused dilution of spring water because flow was concentrated in conduits with limited opportunities for carbonate dissolution. However, in other cases with differing plumbing (Figures 1A, B), piston flow effects (Figure 1D) can lead to an initial increase in ion content. In some cases, the karst above a cave system can be precipitating calcium carbonate, particularly if falling water levels in a conduit or fissure lead to degassing of

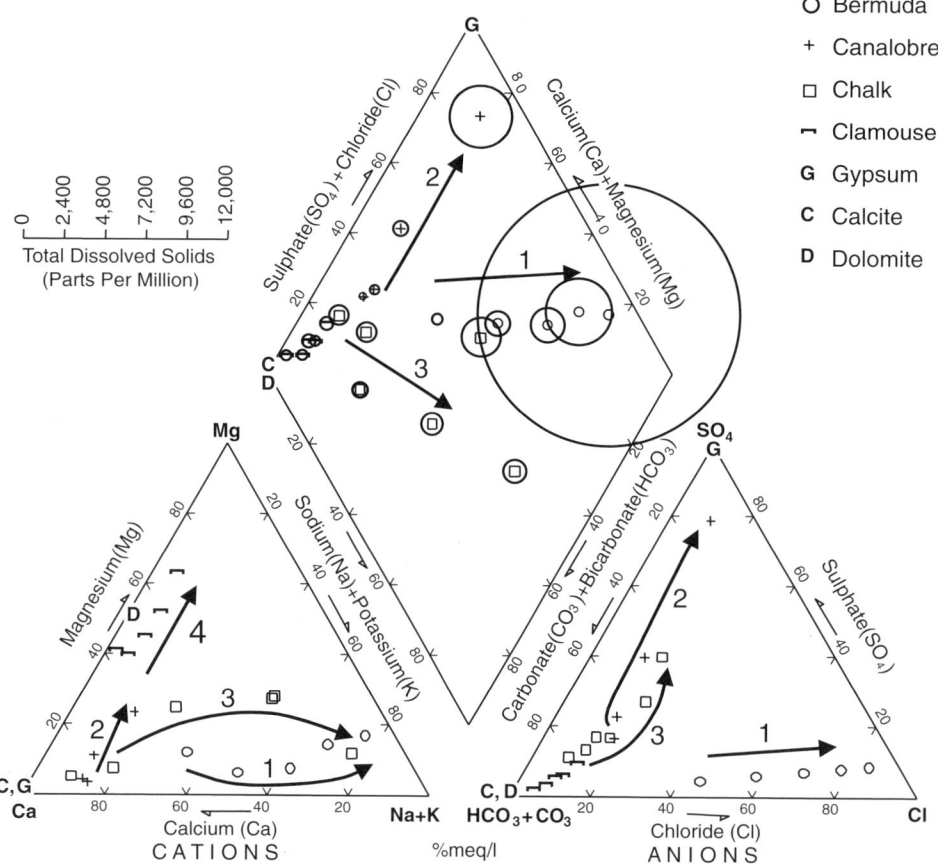

Chemistry of Natural Karst Waters: Figure 2. Piper diagram illustrating some process trends in karst water chemistry (see text for explanation of the numbered trends and the literature sources of data). Points have been selected as representatives from larger data sets in each case. The total dissolved solids in the upper (diamond-shaped) diagram are shown by the size of the circles around the data points (scale in upper left).

carbon dioxide into an air space subject to ventilation (Figure 1C). This process leads to characteristic increases in $\delta^{13}C$ (Bar-Matthews *et al.*, 1996) and also in Mg / Ca ratio (Fairchild *et al.*, 2000), as the $CaCO_3$ precipitates have much lower Mg / Ca than that of the precipitating solution. Thus, in many cases, the Mg / Ca and other ratios in cave water solutions (and speleothems) reflect the water saturation state of the cave and hence could be used in paleoclimate studies. Nevertheless, adjacent drips may have different chemistries, reflecting their differing flowpaths, and long-term evolution of the regolith will also lead to changing elemental compositions (Fairchild *et al.*, 2000).

Figure 2 illustrates some major trends in the composition of karst waters. Coastal karst lies above a marine–freshwater mixing zone, such that samples at progressively greater depths will show increased proportions of the major ions present in seawater (Na, Cl > SO_4 > Mg > Ca, K > HCO_3 > CO_3). This explains the trends shown in the Bermuda data (trend 1 in Figure 2; Plummer *et al.*, 1976), whereas in the freshwater part of the aquifer, the dominant changes are due to the dissolution of aragonite and Mg-calcite and the precipitation of calcite. Data from the Canalobre Cave, southeast Spain (trend 2, Figure 2; Andreu *et al.*, 1999) show a clear evolution from the unsaturated zone in Cretaceous limestones to the saturated zone, in response to dissolution of subsurface calcium sulfate (gypsum), although a soluble salt source of Mg is also required. Complex longer-term processes are responsible for the evolution of groundwaters in the London Chalk Basin, United Kingdom, away from the main recharge zone (trend 3, Figure 2; Elliott, Andrews & Edmunds, 1999) including carbonate dissolution–reprecipitation reactions, infiltration of sulfate-rich waters from oxidation of pyrite, and mixing with older, more saline waters. Finally, the trend of waters from the Clamouse cave in southern France (trend 4, cation diagram, Figure 2; Fairchild *et al.*, 2000) reflects varying amounts of prior calcite precipitation.

IAN J. FAIRCHILD AND ANNA F. TOOTH

See also **Dissolution: Carbonate Rocks; Dissolution: Evaporite Rocks; Dissolution: Silicate Rocks; Erosion Rates (Field Measurements, Theoretical Models)**

Works Cited

Andreu, J.M., Cerón, J.C., Pulido-Bosch, A. & Estévez, A. 1999. Geochemistry of waters in the Canalobre Cave and aquifer of Cabeçó D'Or (south-east Spain). *Carbonates and Evaporites*, 14: 182–90

Baker, A., Barnes, W.L. & Smart, P.L. 1997. Variations in the discharge and organic matter content of stalagmite drip waters in Lower Cave, Bristol. *Hydrological Processes*, 11: 1541–55

Banner, J.L., Musgrove, M., Asmerom, Y., Edwards, R.L. & Hoff, J.A. 1996. High-resolution temporal record of Holocene groundwater chemistry: Tracing links between climate and hydrology. *Geology*, 24: 1049–53

Bar-Matthews, M., Ayalon, A., Matthews, A., Sass, E. & Halicz, L. 1996. Carbon and oxygen isotope study of the active water-carbonate system in a karstic Mediterranean cave: Implications for paleoclimate research in semiarid regions. *Geochimica Cosmochimica Acta*, 60: 337–47

Elliott, T., Andrews, J.N. & Edmunds, W.M. 1999. Hydrochemical trends, palaeorecharge and groundwater ages in the fissured Chalk aquifer of the London and Berkshire Basins, UK. *Applied Geochemistry*, 14: 333–63

Fairchild, I.J., Borsato, A., Tooth, A.F., Frisia, S., Hawkesworth, C.J., Huang, Y., McDermott, F. & Spiro, B. 2000. Controls on trace element (Sr-Mg) compositions of carbonate cave waters: Implications for speleothem climatic records. *Chemical Geology*, 166: 255–69

Hess, J.W. & White, W.B. 1993. Groundwater geochemistry of the carbonate karst aquifer, southcentral Kentucky, U.S.A. *Applied Geochemistry*, 8: 189–204

Plummer, L.N., Vacher, H.L., Mackenzie, F.T., Bricker, O.P. & Land, L.S. 1976. Hydrogeochemistry of Bermuda: A case history of ground-water diagenesis of biocalcarenites. *Geological Society of America Bulletin*, 87: 1301–16

Rossum, J.R. 1975. Checking the accuracy of water analyses through the use of conductivity. *Journal of the American Water Works Association*, 67: 204–05

Tooth, A.F. & Fairchild, I.J. 2003. Soil and karst aquifer hydrological controls on the geochemical evolution of speleothem-forming drip waters, Crag Cave, southwest Ireland. *Journal of Hydrology*, 273: 51–68

CHILLAGOE AND MITCHELL-PALMER KARSTS, AUSTRALIA

Chillagoe is a small town, about 150 km west of Cairns, in tropical north Queensland (see map in Australia entry). West of the Great Dividing Range, the perennial Walsh, Mitchell, and Palmer rivers drain the area to the Gulf of Carpentaria. The Mitchell-Palmer karst extends for 100 km between those two rivers, in remote terrain accessible only by minor tracks. The region has a tropical monsoonal climate, with an annual average rainfall of 830 mm, most of which falls in the short wet season from December to March. Daily maximum temperatures frequently exceed 38°C in the wet season and over 25°C in the dry season. The dominant vegetation of the area is savannah woodland, with Eucalyptus and Corymbia trees. The main land uses are extensive cattle grazing, and minor quarrying of the limestone for lime and building stone. Aboriginal occupation of the area dates back at least 20 000 years, and many caves and overhangs have rock art and prehistoric occupation sites. Europeans settled the area in 1872, when alluvial and reef gold was discovered on the Palmer River. Today marble is quarried at Chillagoe. In spite of the remoteness of the area and the small local population, there is an active caving club that has documented the area and its caves. Tourist caves are well established, receiving between 10 000 and 14 000 visitors each year, and they have developed some specialized ecotourism programmes.

The Silurian to Early Devonian Chillagoe Limestone outcrops for 300 km along the Palmerville Fault as a belt several kilometres wide. The limestones accumulated in shallow marine conditions on a gently sloping ramp. To the west of the Palmerville Fault, are Precambrian metamorphics, while to the east, the Ordovician to Late Devonian Hodgkinson Province is composed of sandstones and siltstones. Compression of the region, during the late Devonian to early Carboniferous, resulted in intense deformation of the limestones, producing near-vertical bedding, coupled with widespread brecciation. Later, broad downwarping of northwestern Queensland in the early Mesozoic

Chillagoe and Mitchell-Palmer Karsts, Australia: Royal Arch tower, located 6 km south of Chillagoe. A broad pediment abruptly meets the steep cliffs of the tower. (Photo by David Gillieson)

resulted in deposition of Jurassic sandstones which unconformably overly the limestones. Since marine retreat in the mid-Cretaceous, the region has been continuously exposed to weathering.

The Silurian limestones form prominent and spectacular towers up to 70 m high near Chillagoe, and are separated by lower undulating terrain formed on chert, sandstone, and volcanic ridges. The ridges and towers lie approximately parallel to the regional strike, which is northwest–southeast near Chillagoe and more northerly in the Mitchell-Palmer karst. The towers are typically up to 1 km long and 300–400 m wide (see Photo). Surrounding the Chillagoe towers are low-angled pediments, several hundred metres wide with rock pavements, thin veneers of rubble, and spalled blocks. Close to the major rivers, the karsts are alluviated at their margins and swamp slots are present. In the Mitchell-Palmer karst, the towers are flanked by steeper bedrock ramps with skeletal soils and some scree. Most of the towers have very well-developed rillenkarren (solution fluting) which drain the upper surfaces of the pinnacles and feed larger rinnenkarren (solution runnels) which may be up to 50 m long and 30 cm deep. There are extensive kamenitza (solution pans) on the upper surfaces of the towers, along with deep grikes (vertical slots) and solution corridors. On the larger towers, large depressions and amphitheatres divide the major limestone blocks; their origin is not clear, but some may relate to cave collapse.

Over a thousand caves have been recorded and mapped in the Chillagoe and Mitchell-Palmer karsts (Chillagoe Caving Club, 1982). The longest is the Queenslander Cave, a joint-controlled maze with 6 km of passages. Few cave entrances are known from the pediments, and most caves are entered from the upper surfaces of the towers or from the angle between the cliffs and the pediments. The dominant passage shapes are rifts that narrow upwards and may intersect the surface to produce daylight holes. Larger chambers 30–50 m wide are located at joint intersections. Solution pendants and spongework suggest that early cave development involved phreatic solution. Speleothem deposits are common and may completely block passages. Cave coral is widespread and phytokarst is common near entrances. False floors of flowstone are to be found up to 20 m above the present cave floors. Cave floors are composed of a mix of breakdown blocks and flat silt or clay; many caves also have deep guano deposits. Deep shafts extend down from the upper surfaces of the towers and one (Christmas Pot) reaches the water table at a depth of 15 m below the pediment.

Palaeokarst features are found at Chillagoe and lend support to a long geomorphic history. There have been at least three phases of sinter copper mineralization in pipes, associated with regional volcanism and faulting. Fissure deposits of chalcopyrite, cuprite, and minor azurite in several caves, suggest that at least some cave passages had formed by the most recent phase 280 Ma ago. To the south of the Mitchell River, the unconformity between the Chillagoe Limestone and the Jurassic sandstone has a local relief of at least 60 m, with steep contacts. In addition, *in situ* sandstone is found at the base of some towers and also has been reported from their summits. This suggests that at least some of the towers had formed in the Mesozoic.

The major rivers of the area, such as the Walsh, Mitchell, and Palmer, are dominantly fed by catchments on non-carbonate rocks. Some of their tributary streams rise from karst springs but these may cease to flow during the six month-long dry season. There are extensive deposits of porous calcite tufas on most streams draining the karst, either as small tufa cascades or as stream bed and bank terraces. The terrace deposits are typically dense and horizontally laminated; limited uranium series dating indicates ages between 260 ka and 80 ka. In the dry season, the water table lies between 10 to 15 m below the plains, and water can be seen in a few caves. During the wet season, deep pools form in many caves, but drain over a period of a few weeks. This water flows directly into the caves from the bare tower surfaces via fissures or down many open shafts. In many caves, horizontal flood lines of organic matter indicate maximum depths, while minor solution notches can be seen in some caves, for example Carpentaria Cave. Small water-filled cavities have been encountered during drilling at depths up to 90 m, but it appears that the karst aquifer does not have a well-integrated subsurface conduit system This is supported by lack of drawdown in drill holes adjacent to a 300 m deep open cut gold mine.

DAVID GILLIESON

See also **Australia; Tower Karst**

Works Cited

Chillagoe Caving Club. 1982. *Chillagoe Karst*, Cairns: Chillagoe Caving Club

Further Reading

Ford, T.D. 1978. Chillagoe—a tower karst in decay. *Transactions of the British Cave Research Association*, 5: 61–84

Williams, P.W. 1978. Interpretations of Australasian karsts. In *Landform Evolution in Australasia*, edited by J.L. Davies & M.A.J. Williams, Canberra: Australian National University Press

Willmot, W.F.& Tresize, D.L. 1989. *Rocks and Landscapes of the Chillagoe District*, Brisbane: Queensland Department of Mines

CHINA

China has an enormous variety of karst, extending over more than 500 000 km² (see map in Asia, Northeast entry), including the world's most extensive and most important karst terrain, largely in the provinces of Guangxi and Guizhou (Figure 1). The karst occurs on limestones that dominate the stratigraphy of southern China from Devonian to Triassic, and most notably on the massive Qixia and Maokou facies of Permian limestones. Climates vary with mean temperatures and annual rainfalls ranging from 20°C and >1200 mm in southern Guangxi to −5°C and <250 mm on the Qinghai-Xizang Plateau. Karst morphology is equally variable, but is dominated by the extensive fengcong and fenglin karsts of southern China. Fengcong (peak cluster) is essentially a mature cone karst with steep hill profiles, which develops over time from doline karst. Fenglin (peak forest) is better known as tower karst with isolated steep-sided towers rising from an alluviated plain. Fenglin is an extreme form of karst that only develops where dissolution rates, tectonic uplift, and alluviation maintain a critical balance (Zhang, 1980). More people live on karst in China than in any other country. Environmental impacts are therefore significant; man has modified the karst, especially within the last 50 years, but the karst has also influenced mankind's development in the region. There is vast exploitation of karst groundwater, while there are numerous cave dams, underground diversions, and even hydro-electric power stations inside caves. China has more subsiding dolines than any other country, with houses, villages, roads, railways, fields, and reservoirs lost in collapse events. Karst research in China is major, but has little contact with Western research; valuable insights are provided by Yuan (1991) and Lu (1986). It should be noted that statistics from China can be misleading—areas of karst generally include large zones where limestones are buried beneath other rocks with non-karstic surface morphology, and cave lengths generally refer to groups of related caves not yet physically connected.

Guangxi is the world's most important karst with respect to its surface morphology. It has extensive outcrops of folded Paleozoic limestones, broken by anticlinal hills of Devonian sandstone that provide allogenic drainage and sediment. Landscape evolution throughout the Tertiary has created many areas of fenglin—the famous and spectacular tower karst on flat alluvi-

China: Figure 1. Simplified map of the karst regions of southern China. Small areas of non-karst occur within the regions marked as karst, and there are small outcrops of limestone within the non-karst regions. Located sites are referred to in the text, and sites outside this area are located on the map in the Asia, Northeast entry.

ated plains. These are interspersed with wide zones of fengcong that create inhospitable terrains of steep hills and deep isolated dolines. Some of the finest examples of both fenglin and fengcong are traversed by the Li Jiang (River) at Yangshuo (see Yangshuo Karst entry), south of Guilin. West of Guilin, the karst plains and fenglin extend across the Hongshui plain around Liuzhou and onward into Vietnam, while fengcong dominates in an adjacent belt to the north that fringes the rise onto the Guizhou Plateau. The most deeply dissected, and some of the most dramatic, karst landscapes occur along this intermediate zone, including that in the Hongshui Basin (see Hongshui River Fengcong karst entry).

The fenglin towers have active foot caves and also relicts at high levels that preserve sediments proving their origins over long timescales. Towards the western end of the fengcong karst there are many major caves distinguished by very large passages. Just west of the Hongshui, the Panyang caves of Bama County contain over 40 km of very large passages, including those of Mawang Dong (10 km long). Each cave lies under a fengcong cluster and is truncated in the marginal dolines (Gill, Lyon & Fowler, 1990). Just to their south, Solue Dong has 7 km of large river passage, but has not yet been followed to the Fulon Dong resurgence. Further northwest, the caves of Leye County lie between giant dolines (tiankengs) and have a total of 30 km of very large passages, chambers, and deep shaft entrances in five disconnected segments (Figures 2 and 3).

Guizhou is a huge, complex, dissected, rolling karst plateau on a folded pile of Paleozoic carbonates up to 3000 m thick. It is surfaced with fengcong hills of all sizes, between isolated basins, perched karst plains, and inliers of non-carbonates (Zhang, Yang & He, 1992). A broad sequence of landscape evolution, from doline karst to fengcong with depressions (Figure 4) to fenglin plains, and then to rejuvenated fengcong with canyons, has been recognized in the Longgong karst, near central Anshun (Smart *et al.*, 1986) and in the southern Dushan area (Song, 1986). Cones with lower slope angles are formed on dolomites, and are known as qiufeng karst. Large rivers cross much of the plateau, and some are deeply entrenched in canyons that cut clean through mature fengcong cones. Many rivers pass through short caves and under high natural bridges; the Ganhe natural bridge north of Shuicheng stands 136 m above the river. Travertine barriers and cascades are a feature of the plateau; they include the Huangguoshu Falls, 70 m high over a knick-point on the Dabang River.

China: Figure 3. The main river cave out of the Dashiwei tiankeng in Leye County. (Photo by Andy Eavis)

Hundreds of caves have been mapped in Guizhou. Many contain relict passages and chambers with massive stalagmites, and dated material shows a long spread of ages in the Middle and Late Pleistocene, with peaks of deposition at 80–115 ka and 31–49 ka (Zhang & Barbary, 1988). The show cave of Daji Dong, near Zhijin, is notable for its giant stalagmites and columns, and the nearby Santang Dong is 7.2 km long. South of Anshun, Gebihe Dong has 11.9 km of very large river passage (broken by an undived sump). The sink entrance is an arch 116 m high, there is a skylight shaft 370 m deep above the river passage, and a chamber near the rising is one of the world's largest, 700 m long and 200 m wide (Collignon, 1992). In the southwestern county of Anlong, Ban Dong is 17.8 km long and contains a chamber 300 m by 150 m and a shaft of 225 m as one of its entrances. Near Shuicheng, Saguo Dong has a streamway with a succession of 17 shafts dropping into a segment of the underground Fala River, and the nearby Wujia Dong is 430 m deep. In northern Guizhou, the karst around Houping has over 23 km of mapped caves.

Yunnan contains a western extension of the Guizhou Plateau, but its fencong karst is generally not as well developed. There are some spectacular river sinks and long underground drainage routes, but few major caves have yet been mapped. The province is best known for its many stone forests of pinnacle karst, including the type site at Shilin (see Shilin Stone Forest entry). Baishuitai is the most spectacular of a number of large flights of travertine terraces. Yanzi Dong at Jianshui is China's largest tourist cave with a huge river passage upstream from its resurgence into some very large and well-decorated chambers.

Sichuan has areas of karst distributed around the sandstone core of its Red Basin. The western part of the province rises towards the Tibet Plateau, and has outcrops of limestone at altitudes up to 5000 m. In the northwest, the Minshan area is

China: Figure 2. A hot-air balloon over the giant doline (tiankeng) of Dashiwei in the Leye karst. (Photo by Zhu Xuewen)

China: Figure 4. Conical hills of the fengcong karst rise above the Shuicheng basin in Guizhou. (Photo by Tony Waltham)

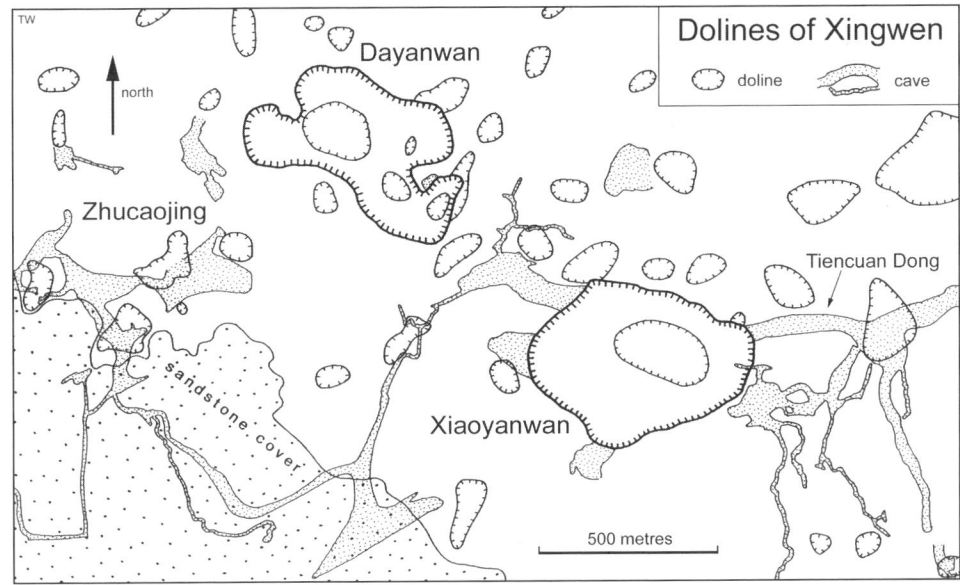

China: Figure 5. Outline map of the giant dolines at Xingwen, Sichuan, in relation to the known cave passages beneath and beside them.

too dry and cold to support mature alpine karst; karren are small, and there are no known caves larger than natural bridges and frost pockets in the spectacular crags. The valley floors contain travertines including those of Huanglong and Jiuzhaigou (see separate entry). South of the Red Basin, the dip slope of the Xingwen escarpment of Permian limestone is carved into a spectacular karst of dolines and small pinnacles. It contains extensive large caves, both active and relict, largely orientated along the strike, with 30 km of passages mapped to date (Waltham, Brook & Bottrell, 1993). They open into the sides of two giant collapse dolines, of which Xiaoyanwan is 650 m long, 490 m wide, and ringed by vertical limestone walls over 100 m high. Three stages of doline evolution are seen, and Xiaoyanwan is the middle, mature stage. Dayanwan is an older degraded doline higher up the dipslope, and Zhucaojing Cave has a series of large chambers that will eventually collapse to form the next of Xingwen's giant dolines (Figures 5 and 6).

Chongqing is now its own municipality, so has taken some of the finest karst out of Sichuan. In the east, the Chang Jiang (Yangtze River) drains out through its famous gorges, with high vertical cliffs that cut through massive escarpments of impure

China: Figure 6. The large chamber beneath the entrance shaft of Zhucaojing Cave which will eventually collapse to form part of the next giant doline in the Xingwen karst (see map, Figure 5). (Photo by Tony Baker)

carbonates. Just south of the gorges, deeply dissected karst in pure limestone contains Di Feng Dong (see separate entry) and also many other caves that contain huge, old passages in parts completely filled with clastic sediments, as well as some very large chambers. The karst of the Wulong region, astride the Wu Jiang, is notable for some deeper caves developed in Cambrian limestones. Furong Dong contains an ancient tunnel richly decorated with aragonite crystals and also a shaft 200 m deep, both beyond large passages developed as a show cave. Higher in the mountains, Qikeng Dong has a shaft system that meets a major streamway 700 m down, and continues to a depth of 920 m. San Qiao has three giant natural bridges surviving as roof remnants across a spectacular collapse gorge, and also a tributary cave 4 km long through from a separate tiankeng (giant doline).

Hubei and **Hunan** have the eastern part of the main region of mature fengcong karst in their western sectors, though most of its conical hills have lower profiles than those in Guangxi. The downstream half of the Yangtze Gorges is in Hubei, and the continuing limestone belt south of them contains the great caves of Tenglong. There are 10 km of active caves containing the Qing River, and a high-level dry cave 8700 m long is an earlier route of the same river. Hunan has a northern continuation of the Guilin karst with large areas of fengcong and less of fenglin. Wanhuayan is a cave notable for its large river passages nearly 6 km long, partly developed as a show cave.

Northern China lacks any continuation of the Upper Paleozoic limestones that support the extensive mature karst further south. Cambrian and Ordovician limestones have large outcrops in Shanxi, notably forming the Taihang Mountains. These are well-fractured rocks, so they act as diffuse-flow aquifers with only a small proportion of conduit flow. These feed many large springs, of which the largest is Nyangziguan with a flow of 10–16 $m^3 s^{-1}$ from a catchment of 3500 km^2. At the north end of the Taihang, in Hebei, the fissure caves of Zhoukoudian are famous for their fossils of "Peking Man" (see China: Archaeological Caves) in a sediment sequence over 40 m deep, where the 780 ka magnetic reversal occurs at a depth of 34 m. There are also zones of gypsum karst along the flanks of the Taihang in Shanxi. Shandong has extensive fengcong karst with low relief on outcrops of thinly bedded Paleozoic limestones and dolomites. Some fluviokarst here is a relict of colder conditions during the Pleistocene, and there are some very large risings but no known caves of significant extent. The arid northwest of China has very little recorded karst. A belt of alpine karst with observed cave entrances lies just south of Aqqikkol Lake in the Kunlun Ranges of southern Xinjiang. Both this and the limestone outcrops of the Tien Shan await exploration and investigation.

Tibet (now the Chinese province of Xijang) may have over 100 000 km^2 of carbonate outcrop, but most are of impure limestones interbedded with shales, all of which have been structurally deformed and metamophosed to some extent. In the cold, dry, periglacial environment limestone is resistant to weathering; many slopes and crags have pinnacle features, but these were formed by frost shattering and are not karstic. The only karstic landforms are widespread microkarren and isolated rinnenkarren downslope of snow patches. Tufa is widespread, but is largely related to geothermal sources; most is inactive and some has been dated to Pleistocene interglacial stages. Modern karst processes are minimal, with very low measured rates of dissolution (Zhang, 1999). A cave at Rutog in the far west has 100 m of passage, but all other known caves are less than 20 m long. Most of these caves are only short fissures; some stalagmites have been dated to >350 ka, but few caves are recognizable as fragments of old systems that were more extensive. Bygone concepts of Tibet having remnants of Tertiary karst features have not been validated by field data. Any karst formed in wetter environments prior to uplift of the Himalayas appears to have been destroyed by subsequent periglacial processes.

TONY WALTHAM

See also **Asia, Northeast**

Works Cited

Collignon, B. 1992. The underground Gebihe River. *International Caver*, 5: 12–18

Gill, D., Lyon, B. & Fowler, S. 1990. The caves of Bama County, Guangxi, China. *Cave Science*, 17: 55–66

Lu Y. 1986. *Karst in China: Landscapes, Types, Rules*, Beijing: Geological Publishing House

Smart, P., Waltham, T., Yang M. & Zhang Y. 1986. Karst geomorphology of western Guizhou, China. *Cave Science*, 13: 89–103

Song L. 1986. Karst geomorphology and subterranean drainage in south Dushan, Guizhou Province, China. *Cave Science*, 13: 49–63

Waltham, T., Brook, D. & Bottrell, S. 1993. The caves and karst of Xingwen, China. *Cave Science*, 20: 75–86

Yuan D. (editor) 1991. *Karst of China*, Beijing: Geological Publishing House

Zhang D. 1999. Field examination of limestone dissolution rates and the formation of active karren on the Tibetan Plateau. *Cave and Karst Science*, 26: 81–86

Zhang S. & Barbary, J. (editors) 1988. *Guizhou Expe 86*, Paris: Fédération française de spéléologie (Spelunca Mémoire, 16)

Zhang Y., Yang M. & He C. 1992. Karst geomorphology and environmental implications in Guizhou, China. *Cave Science*, 19: 13–20

Zhang Z. 1980. Karst types in China. *GeoJournal*, 4: 541–70

Further Reading

Dunton, B. & Laverty, M. 1993. The caves of Doshan, Guizhou Province, China. *Cave Science*, 20: 65–71

Gebauer, D. (editor) 1992. Le reseau souterrain du synclinal de Santang (Guizhou, Chine). *Stalactite*, 42(1+2): 56–66

Senior, K. (editor) 1995. The Yangtze Gorges Expedition: China Caves Project 1994. *Cave and Karst Science*, 22: 53–90

Sweeting, M.M. 1995. *Karst in China: Its Geomorphology and Environment*, Berlin and New York: Springer

Sweeting, M.M., Bao H. & Zhang D. 1991. The problem of paleokarst in Tibet. *Geographical Journal*, 157: 316–25

Waltham, A.C. 1996. Limestone karst morphology in the Himalayas of Nepal and Tibet. *Zeitschrift für Geomorphologie*, 40: 1–22

Waltham, A.C. & Willis, R.G. (editors) 1993. *Xingwen*, British Cave Research Association

Zhang D. 1996. A morphological analysis of Tibetan limestone pinnacles. *Geomorphology*, 15: 79–91

Zhu X. 1988. *Guilin Karst*, Shanghai: Shanghai and Scientific Technical Publishers

CHINA: ARCHAEOLOGICAL CAVES

China is abundant in geological and cave resources, attracting hundreds of foreign cavers and organized expeditions each year. The Karst Dynamics Laboratory (KDL) in Guilin, a karst-rich city in southern China, has led karst studies in China. The major research goals of this lab include basic theory in karstology, rehabilitation of rocky desertification, development and harnessing of karst water resources, and geological hazards prevention. Paleoenvironmental reconstructions using karst records have been especially useful, with KDL establishing continuous records for southern China for the past 36 ka (www.karst.edu.cn, 1995–2002).

The Chinese people have long known of the fossil riches that lay within their hills and caves (see entry Asia, Northeast: History). Called "dragon bones" (*long gu*) or "dragon teeth" (*long ya*), these fossils were systematically harvested for centuries, with the occupation of dragon bone-digger passing from father to son (Boaz & Ciochon, 2001). These "dragon" products were sold in turn for medicinal purposes in apothecaries, since at least the Lungshan Period (and late Xia Dynasty), 5–4 ka. It was some time, though, before the formal disciplines of archaeology and human paleontology gained recognition in China (the early 1920s), and they did not rise to prominence until sometime later. In the years before China was liberated (prior to 1949), these disciplines mainly fell to foreigners, since Chinese citizens were forbidden to research the fossils they discovered (Jia & Huang, 1990). The situation has changed, however, and archaeology has returned to the everyday people of China, who have, with renewed enthusiasm, begun to deluge the Chinese Academy of Sciences with reports of new discoveries (Jia, 1975). China is rich with fossiliferous caves, the most noteworthy being the Middle Pleistocene deposits of Zhoukoudian, where the famous "Peking Man" was discovered (Jia & Huang, 1990). In addition to Zhoukoudian, four other cave sites figure prominently (Figure 1).

Zhoukoudian World Heritage Site: "Peking Man Site at Zhoukoudian" was added to UNESCO's World Heritage List in 1987. The Middle to Late Pleistocene cave and fissure deposits are located 45 km or so southwest of Beijing. The site consists of many fossiliferous deposits, the best known of which is Locality 1, preserving a column of more than 40 m of stratified infilling (Pope & Olsen, 2000), and the source of the largest collection of *Homo erectus* fossils from any single site (Figure 2). KDL notes that "Peking Man Cave at Zhoukoudian ... and the polje at Zhoukoudian are examples of paleokarsts that formed under the warm and humid paleoclimate in Neogene" and are "mainly composed of high peaks, stone pillars and deep gorges ..." (www.karst.edu.cn, 1995–2002).

Excavation at the site was initiated in 1921, and still continues in some localities (though not at Locality 1). So far, work at Longgushan (Chinese for "Dragon Bone Hill"), as the site is called, has uncovered the remains of *Homo erectus*, along with a large sample of stone implements and tool-making debris. These remains have recently been re-dated to between 470 and 670 ka (Grün *et al.*, 1997). Additionally, the remains of *Homo sapiens sapiens* that date to 8–11 ka were found in the Upper Cave above Locality 1. Thousands of faunal remains have also been found within the 40 m thick deposits that once filled the cave.

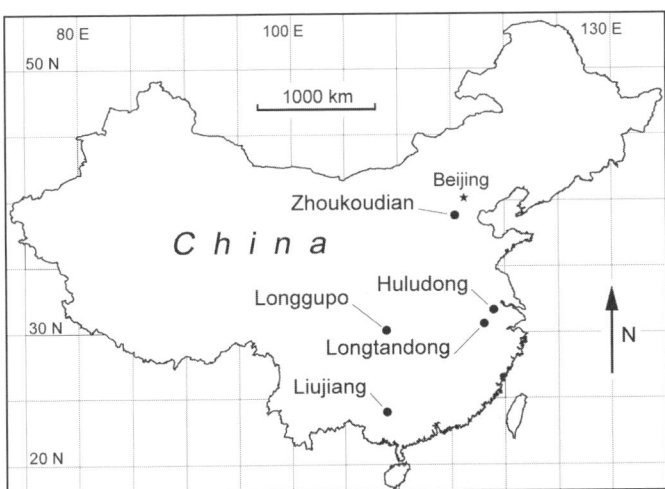

China: Archaeological Caves: Figure 1. Map of China with the location of five significant archaeological cave sites depicted.

China: Archaeological Caves: Figure 2. View of Locality 1, Zhoukoudian from the perspective of Pigeon Hall.

During the years from 1978 to 1980, a massive multidisciplinary campaign was carried out at Zhoukoudian with the leadership of the Institute of Vertebrate Paleontology and Paleoanthropology (IVPP) in Beijing. Furthermore, with the founding of the Zhoukoudian International Paleoanthropological Research Center at Beijing's Institute of Vertebrate Paleontology and Paleoanthropology in 1993, a new era of multidisciplinary work began (Boaz & Ciochon, 2001).

With this reinvestigation of Zhoukoudian, the site has yielded new estimates of its age and number of hominid specimens represented (now over 50 individuals), igniting new debate. Early interpretations by Henri Breuil and Franz Weidenreich portrayed the site as the cave home of "Man, the Great Hunter", where fire was utilized and even cannibalization may have been present. These ideas have since been overturned, beginning in the late 1920s and 1930s. In later years, paleoanthropologists Noel Boaz and Russell Ciochon have individually re-examined the original casts (regrettably, the Longgushan fossil assemblage mysteriously disappeared on December 7 1941, during the aftermath of the World War II Japanese invasion of China, after being packed for shipment to the United States; only their casts remain) and all new specimens collected since the mid-1960s. Boaz and Ciochon found that giant hyaenas, *Pachycrocuta brevirostris*, were the most common and complete fossils represented, suggesting that they, and not humans, were the principal cave tenants. Though alternate interpretations abound, the preponderance of the evidence, along with the condition of the human remains, have led specialists to speculate that the hominid fossils may be the remnants of an ancient giant hyaena cache and not the result of a prolonged cave occupation after all (Boaz & Ciochon, 2001). The results of this most recent research have shown that two-thirds of Longgushan's *Homo erectus* fossils show carnivore damage. This pattern of damage on the skulls is consistent with hyaenid chewing of the delicate facial area in an effort to dine on the fat-rich brains (Boaz & Ciochon, 2001) (Figure 3). In addition, recent analysis by geochemist Steve Weiner has uncovered no evidence of controlled hearths (Weiner et al., 2000). Still, despite this lack of evidence (and apparent evidence to the contrary), proof of human occupancy and of controlled hearths may simply not have been preserved. Only further fieldwork will reveal the truth.

In conclusion, UNESCO boasts, the site of Zhoukoudian "is not only an exceptional reminder of the prehistoric human societies of the Asian continent, but also illustrates the process of evolution" (www.worldheritagesite.org, 2002).

Longtandong: "Dragon Pool Cave", the English translation for the site, is located in Hexian County, Anhui Province (Figure 4). The cave is formed from Cambrian dolomitic limestone, and the in-filling can be divided into five sedimentary layers: (1) brown-red clay, (2) red clay, (3) yellow-green silty and sandy clay, (4) brown-yellow fossiliferous (human and non-human) sandy clay, and (5) yellow-gray sandy clay (Wu & Dong, 1982).

In 1980, Huang Wanbo and his IVPP colleagues excavated a heavily fossilized *Homo erectus* skullcap, mandibular fragments, and four teeth, all presumed to belong to a young male. Further excavation in 1981 yielded additional *Homo erectus* fossils: parts of the frontal and parietal bones, and five teeth (Wu & Dong, 1985). More recently, the left side of a fairly robust mandible and seven teeth, all bearing some resemblance to similar elements from Zhoukoudian, have been found (Wu & Poirier, 1995).

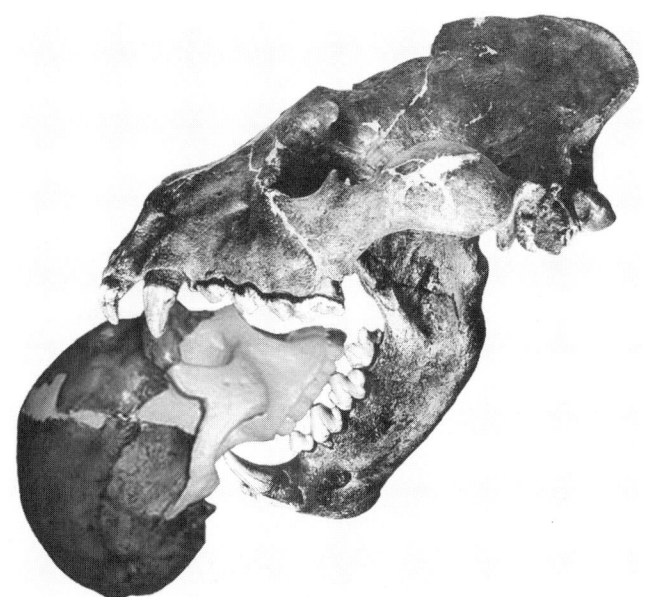

China: Archaeological Caves: Figure 3. Composite image of *Pachycrocuta brevirostris*, the giant hyaena, and *Homo erectus* skulls illustrating how hyaena ingestion of the human face may have proceeded. *Pachycrocuta* was likely responsible for the bulk of the accumulation of human and other animal remains found in Locality 1 at Zhoukoudian.

In 1982, Wu Rukang and Dong Xingren postulated a close relationship between the *Homo erectus* specimens from the Hexian find (as it is referred to) and from Zhoukoudian Locality 1 (Wu & Dong, 1985). Still, "certain progressive features are noted in the Hexian individual. For example, the postorbital constriction is not as pronounced as in 'Peking Man' and is, in fact, less marked than in the Dali hominid from Shaanxi Province, which is thought to be archaic *Homo sapiens*" (Wu & Dong, 1985:87). After further investigation, Wu and colleague made the stronger statement, "the Hexian remains document the

China: Archaeological Caves: Figure 4. View of Longtan Cave, Hexian County, Anhui Province, eastern China. This site yielded the latest dated occurrence of *Homo erectus* in China.

youngest-dated *Homo erectus* finds in China and are perhaps the last known occurrence of this taxon anywhere in the world" (Wu & Poirier, 1995:7). More recently, however, Swisher *et al.* (1996) have dated *Homo erectus* remains from the site of Ngandong in central Java to 27–54 ka. Nevertheless, the remains of *Homo erectus* from Hexian still rank as one of the latest occurrences in the world, and their importance should not be underestimated.

Huludong: This limestone cave is situated on Tangshan (Chinese for "Warm Spring Hill") near Tangshan town, 26 km east of Nanjing, the capital of Jiangsu Province, in eastern China. Huludong (in English, "Calabash Cave") is separated from the Hexian site by 100 km and also by the Changjiang River (Yangtze River). The Hulu karst cave actually is made up of two parts, a major and a minor cave. The minor cave, as the name suggests, is smaller and connects to the larger major cave by a channel. However, the minor cave is the more archaeologically significant because it is where the human remains were found. The Hulu Cave (as the major / minor cave complex is called) developed in Ordovician limestone and has four different layers of infilled sediments. From top to bottom, these layers are: (1) travertine, (2) brown-red sand and silty clay layer, (3) brown-red fossiliferous (human and non-human) clay layer, and (4) brown-red clay (Xu, 1993).

In 1990, local farmers discovered mammalian fossils during quarrying, which they dutifully reported to the Nanjing Institute of Paleontology and Geology (NIPG), Academia Sinica. Excavation was undertaken for a short time in January 1993 by a joint IVPP / NIPG team. However, it was in March 1993 that local workers came upon a hominid skull while clearing the cave in preparation for making it a tourist attraction (Wu & Poirier, 1995). The faunal remains allowed for biostratigraphic age estimates that all of the fauna (including the human skull) belonged to the later Middle Pleistocene. "The specimen is an almost complete anterior portion of a human skull, including an almost complete frontal bone, the left half of the face, a portion of the parietal bones, a part of the occipital, and a small part of the right upper face" (Wu & Poirier, 1995:91). The frontal and brow ridges are robust, with the frontal being very constricted behind the orbit. This skull is attributed to *Homo erectus*.

Longgupo: Known in English as "Dragon Bone Slope", the early Pleistocene site of Longgupo is located 20 km south of the Three Gorges area of the Yangtze River in eastern Sichuan Province (Huang *et al.*, 1995). The strata have yielded *Gigantopithecus blacki* and a hominid, which is for the time being classified as early *Homo*, as well as stone artefacts (Figure 5). Paleomagnetic and biostratigraphic determinations, as well as aminoacid racemization and electron spin resonance, have dated the strata where the hominid remains were discovered at 1.8–2.0 Ma. These same deposits have also been interpreted as being younger (Olsen, 2000).

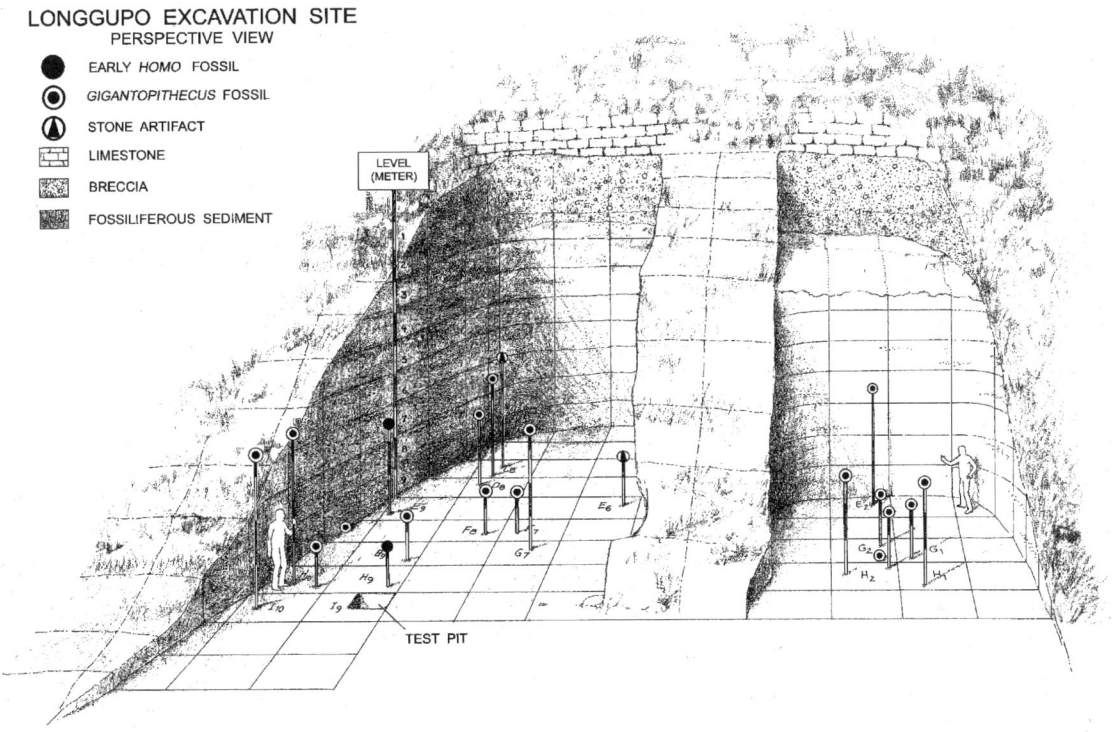

China: Archaeological Caves: Figure 5. Excavation diagram of Longgupo, Sichuan Province, with the locations of early *Homo*, stone artefacts, and *Gigantopithecus* remains indicated. Some researchers view this site as containing the earliest evidence of human occupation in Asia. New excavations in 2003 by a French–Chinese team will test this assertion. Computer-generated image produced by Autumn Noble from a colour drawing by Will Thompson.

Scientists have not reached consensus on the identity of the hominid remains, with some citing similarities with *Homo habilis* and *Homo ergaster*, and others believing them to be either pongine in origin or indeterminate (Olsen, 2000). What is incredible is that, if the hominid interpretation is true (which some prominent researchers do not support; Schwartz & Tattersall, 1996), the 2 Ma age suggests an earlier human arrival from East Africa to Asia than was previously imagined (Larick & Ciochon, 1996a). "The Longgupo finds suggest that soon after evolving in Africa a primitive hominid migrated to Asia with the aid of an elementary technology. They also confirm scattered and ambiguous evidence that *Homo erectus*, once believed to have been the first hominid to colonize Asia, may instead have developed there from an earlier form" (Larick & Ciochon, 1996b: 51). New archaeological excavations by a joint French–Chinese team will begin in 2003 at Longgupo.

Liujiang (Liuchiang or Liukiang) Cave: The "Liujiang Man" remains were discovered in a small cavern near Tongtianyan, in Liujiang County (for which the cave is named), Guangxi Province. Liujiang township has long been a tourist attraction because of its magnificent karst formations. The cave mouth is about 5 m above the ground, at the bottom of a limestone hill (Jia, 1980).

In 1958, local farmers digging up some of the Tongtianyan cave's deposits for fertilizer came upon a human skull and other fossils. The head of the farm reported the matter to the government who sent an IVPP field team (Wu, 2003). Unfortunately, the farm workers could not recall where exactly in the cave the fossils had come from, so provenance could not be ascertained. This has led to many controversies regarding the dating (Wu, 2003).

The human fossils discovered here include: a complete female skull (Figure 6), four thoracic vertebrae with four ribs of varying lengths still articulated, a complete set of lumbar vertebrae, a sacrum, a right pelvic bone, all belonging to a mature male; right and left female femoral fragments were also found. Accompanying these human fossils were miscellaneous faunal remains, which served to date the cave finds as belonging to the Late Pleistocene (Compiling Group of the Atlas, 1980).

It was noted at the time that the hominid-bearing deposits were loose and damp, which was very different from the hard and yellowish mammalian deposits found elsewhere in Guangxi. Researchers Pei Wenzhong and Jia (Jia) Lanpo concluded that the deposits must be of different ages (Jia, 1980). The date of "Liujiang Man" has been projected to be Late Pleistocene by looking at the morphology, itself, and the degree of fossilization present. Calcified plate (stalagmite) dating of the cave provides an age estimate of 67 ka ($\pm$ 6 ka), whereas the U-series dating puts the dates at 101 to 227 ka (Wu, 2003). However, these dates are problematic since the original location of the fossils cannot be precisely determined. Therefore, "nobody can be sure of the association between the human fossils and the samples used for doing [sic] dating. Never!" (Wu, 2004). Still "Liujiang Man" may represent an early type of Mongoloid and even "the earliest Neoanthropic man ever found in China" (Compiling Group of the Atlas, 1980:139). In fact, some studies suggest that South China may be the birthplace of Mongoloids, and even of modern *Homo sapiens* itself!

In summary, human origins have long been the primary concern of paleoanthropology. Africa has yielded many answers to

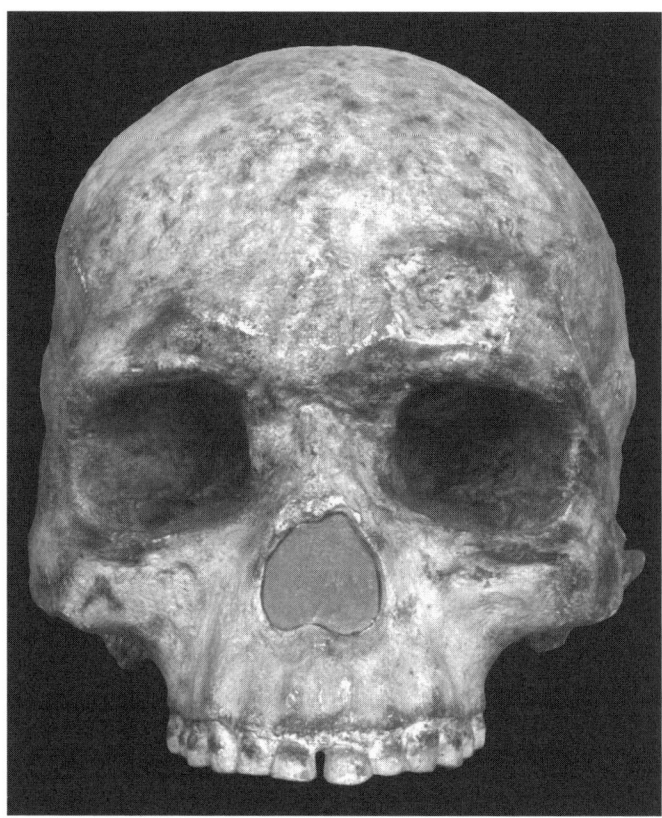

China: Archaeological Caves: Figure 6. Female skull of modern *Homo sapiens* found in Liujiang Cave, Guangxi Province, southern China. Some scholars view this specimen as evidence for the earliest occurrence of anatomically modern humans in China.

the ageless question of where humans came from, but Asia in general, and China in particular, have also provided some intriguing explanations. Despite much work, many questions remain, such as: How long has the genus *Homo* been present in Asia? What is the role of *Homo erectus* and Asia in the process of human evolution? What is the earliest evidence of anatomically modern *Homo sapiens* in Asia? These questions and others drive the scientists today who continue to unearth the answers, just as did the dragon bonediggers of old.

RUSSELL L. CIOCHON AND K. LINDSAY EAVES-JOHNSON

Works Cited

Compiling Group of the Atlas, Institute of Vertebrate Paleontology and Paleoanthropology, Chinese Academy of Sciences (editors) 1980. *Atlas of Primitive Man in China*, Beijing: Science Press

Boaz, N. & Ciochon, R.L. 2001. The scavenging of "Peking Man". *Natural History*, 110: 46–51

Grün, R., Huang, P-H., Wu, X., Stringer, C.B., Thorne, A.G. & McCulloch, M. 1997. ESR analysis of teeth from the paleoanthropological site of Zhoukoudian, China. *Journal of Human Evolution*, 32: 83–91

Huang, W., Ciochon, R., Gu, Y., Larick, R., Fang, Q., Schwarcz, H., Yonge, C., de Vos, C. & Rink, W. 1995. Early *Homo* and associated artefacts from Asia. *Nature*, 378: 275–78

Jia, L. 1975. *The Cave Home of Peking Man*, Peking: Foreign Languages Press

Jia, L. 1980. *Early Man in China*, Beijing: Foreign Languages Press

Jia, L. & Huang, W. 1990. *The Story of Peking Man: From Archaeology to Mystery*, Beijing: Foreign Languages Press and New York: Oxford University Press

Larick, R. & Ciochon, R.L. 1996a. The African emergence and early Asian dispersals of the genus *Homo*. *American Scientist*, 84: 538–51

Larick, R. & Ciochon, R.L. 1996b. The first Asians. *Archaeology*, 49(1): 51–53

Olsen, J.W. 2000. Longgupo. In *Encyclopedia of Human Evolution and Prehistory*, 2nd edition, edited by Eric Delson *et al.*, New York: Garland

Pope, G.G. & Olsen, J.W. 2000. Zhoukoudian. In *Encyclopedia of Human Evolution and Prehistory*, 2nd edition, edited by Eric Delson *et al.*, New York: Garland

Schwartz, J.H. & Tattersall, I. 1996. Whose teeth? *Nature*, 381: 201–02

Swisher, C.C., Rink, W.J., Antón, S.C., Schwarcz, H.P., Curtis, G.H., Suprijo, A. & Widiasmoro. 1996. Latest *Homo erectus* of Java: Potential contemporaneity with *Homo sapiens* in Southeast Asia. *Science*, 274: 1870–74

Weiner, S., Xu, Q., Goldberg, P., Liu, J. & Bar-Yosef, O. 2000. Evidence for the use of fire at Zhoukoudian. *Science*, 281: 251–53

Wu, X. 2004. Notes on the recovery and dating of "Liujiang Man". In *The Human Evolution Source Book*, 2nd edition, edited by R.L. Ciochon & J.G. Fleagle, Englewood Cliffs, New Jersey: Prentice Hall

Wu, R. & Dong, X. 1982. Preliminary study of *Homo erectus* remains from Hexian, Anihui. *Acta Anthropologica Sinica*, 1(1): 2–13

Wu, R. & Dong, X. 1985. *Homo erectus* in China. In *Palaeoanthropology and Palaeolithic Archaeology in the People's Republic of China*, edited by Wu Rukang & J.W. Olsen, Orlando, Florida: Academic Press

Wu, X. & Poirier, F.E. 1995. *Human Evolution in China: A Metric Description of the Fossils and a Review of the Sites*, Oxford and New York: Oxford University Press

Xu, Q. 1993. The discovery of the Middle Pleistocene mammals from Tangshan, Nanjing and its implication. *Chinese Science Bulletin*, 38: 1403–06

http://www.karst.edu.cn, The Karst Dynamics Laboratory (KDL) Website has a summary of the group's findings from 1995–2002

http://www.worldheritagesite.org/sites/site449.html, World Heritage: Peking Man Site at Zhoukoudian Website, 2002

Further Reading

Boaz, N.T., Ciochon, R.L., Xu Q. & Liu, J. 2000. Large mammalian carnivores as a taphonomic factor in bone accumulation at Zhoukoudian. *Acta Anthropologica Sinica*, 19 (supplement): 224–234

Ciochon, R.L. 1995. The earliest Asians yet. *Natural History*, 104(12): 50–54

Etler, D. 2002. Implications of new fossil material attributed to Plio-Pleistocene Asian Hominidae. http://www.chineseprehistory.org//art1.htm#POSITION

Institute of Hydrogeology and Engineering Geology 1976. *Karst in China*, Shanghai: Shanghai People's Publishing House

Sutcliffe, A.J. 1976. Cave palaeontology and archaeology, Part I: Cave palaeontology. In *The Science of Speleology*, edited by T.D. Ford & C.H.D. Cullingford, London and New York: Academic Press

Zhu, X. 1989. *Karst Landscapes in China and Guilin Karst*, #S619. 73 slides available from the NSS AV Library, http://www.caves.org/service/avlibrary/slide_catalog_frame_set.html

CHIROPTERA (BATS)

The Chiroptera form an easily recognizable mammalian order, the only one capable of true flight; colugos (Dermoptera), a few rodents (Rodentia), and some marsupial ringtailed possums (Diprotodontia) are gliders. Approximately 925 species of bat are currently recognized (Koopman, 1993), which makes them one of the largest mammalian orders, second only to rodents. Bats are divided into two sharply differentiated suborders: Megachiroptera (including only the Old World fruit bat family Pteropodidae with 166 species) and Microchiroptera (the remaining 16 families). The vast majority of microchiropterans are grouped in three families: horseshoe bats (Rhinolophidae; 130 species), American leaf-nosed bats (Phyllostomidae; 143 species), and vespertilionid bats (Vespertilionidae; 318 species); two families (Kitti's hog-nosed bats—Craseonycteridae, and New Zealand short-tailed bats—Mystacinidae) are monospecific. New species are still being discovered; for example, 28 new names were proposed between 1983 and 1993, and several new species were recently recognized even within Europe (e.g. Helversen *et al.*, 2001).

Molecular evidence and adaptations to flight favour a monophyletic origin for bats—the view receiving most support today (Simons, 1995). However, the morphology of the nerve paths between the retina and the mid-brain places megachiropterans closer to primates (Primates) and colugos than to microchiropterans, thus suggesting that chiropterans are diphyletic. There is a gap in the fossil record connecting bats with the Upper Cretaceous eutherian ancestors, and the oldest well-preserved true bat (*Icaronycteris index* from the early Eocene) already shows the key characteristics of modern chiropterans. Modern families with the oldest fossil record (horseshoe bats, vespertilionid bats, funnel-eared bats or Natalidae), all of which are microchiropterans, had already appeared by the early Eocene. The remains of fossil bats, which are rare, are mostly found in limnic sediments or in fissure and cave sediments (Storch, 1999). Bats have a worldwide distribution, being absent only from the polar regions. The greatest number of species live in the tropics, with those in tropical America outnumbering those of central Africa and Southeast Asia (Hutson *et al.*, 2001).

Bats range in weight between 2 g and 1600 g, with the vast majority of species weighing 10–40 g; the largest sizes are attained by the megachiropterans. The bones of the hand support the elastic and thin wing membrane (patagium) which stretches between the body, the highly modified bones of the hand (i.e. the much elongated forearm, metacarpals, and fingers) and the hind legs, occasionally also enclosing the tail vertebrae. Because of the demands of flight, movements of the wrist, elbow, and knee joints are restricted to a single plane (Feldhamer *et al.*, 1999). Skeletal modifications for flight also involve fusion of the vertebrae (the seventh cervical and first two thoracic ones), formation of a pectoral ring by the first two ribs and modification of the sternum, the scapula, and the proximal end of the humerus. The hind limbs are thin and rotated 180°, so that the knees point backwards. The work of extending and flexing the wings is done by the large proximal muscles (the pectoral muscle alone forms c.12% of the lean body mass) and transmitted by the nearly inelastic forearm muscles (Vaughan et al., 2000). All muscles contain large amounts of myoglobin, and the special fast-twitch muscle fibres have high to medium densities of mitochondria (Ransome, 1990). The lungs, although not structurally specialized, are relatively large, and the heart is nearly twice the weight of that of a terrestrial mammal of similar size.

All aspects of bat biology and natural history reflect their ability to fly. Unlike birds, bats use a slow but highly manoeuvrable flight, with a consequently low wing loading (the ratio of the body mass to the wing area). During flight, bats emit high-frequency high-intensity sound pulses (with frequencies >20 kHz) from the larynx. They use these sound waves to provide information about the environment from the returning echoes (echo-location). The pulses are either frequency-modulated (FM; as in vespertilionid bats) or constant-frequency (CF; as in horseshoe bats); with the latter, moving objects are detected by their Doppler shift. Calls are emitted either through the nose or through the mouth, which involves a basic dichotomy in skull shape. Complex skin folds ("nose leaves") surrounding the nostrils are characteristic of nasal emitters (horseshoe bats, American leaf-nosed bats). Megachiropterans rely on vision, and only the genus *Rousettus* is capable of echo-location using low-frequency tongue-clicks.

Flight has allowed bats to fill many ecological niches, and the adaptive radiation within the Phyllostomidae is particularly spectacular. Bats can feed on insects, fish, small terrestrial vertebrates, blood, pollen, and fruits. Species with narrow, pointed-tipped wings and low-energy flight are insectivorous and forage in open habitats. Bats living in closed habitats have broad wings with rounded wing-tips and exploit various food sources. In temperate regions with seasonal food supply and cold winters, bats have developed torpor (hypothermia), a strategy which allows them to decrease their normal body temperature to a level closer to the ambient temperature. Daily and winter torpor (hibernation) are widespread in temperate bats; the latter involves a preparation phase (the accumulation of body fat in late summer and early autumn) as well as energy-saving physiological changes during the cold period (low heart and breathing rates, suppression of shivering, vasoconstriction, etc.). At ambient temperatures near 5°C, hibernating bats maintain a body temperature about 1°C above the ambient temperature. As the body temperature drops to a minimum critical level (which is close to freezing in some *Myotis*) bats are in danger of freezing to death. If ambient temperatures drop below 5°C, bats spontaneously raise their metabolic rate and become fully alert, or regulate their body temperature and remain in hypothermia. The period of uninterrupted hibernation is usually up to 30 days, but may reach over 80 days in the little brown bat *Myotis lucifugus* (Ransome, 1990).

Bats will survive torpor only in places providing suitable microclimatic conditions (constant low ambient temperatures and high humidity to compensate for water losses through evaporation) and safety from predation and disturbances. Frequent arousal from hibernation is energetically costly and may exhaust bats' energetic reserves; several studies show that mortality rates are highest at the end of hibernation (Raesly & Gates, 1987).

Consequently, bats are sensitive to disturbance in their roosting places. There are two common extremes of roost type used by bats. Tree holes are exposed to considerable variations in temperature and humidity, while caves provide a more stable environment and are thus important hibernation sites for many bat species. As a consequence, temperate tree bats more frequently show long-distance migration to cope with seasonal changes, while cave-dwelling bats rely on hibernation or combine short-distance migrations with hibernation.

Bats show considerable variability and plasticity regarding roost-type selection. Sites may change with the time of the year and even the time of the day, but will also depend on the local geography. This makes it difficult to classify species simply by roosting site (Ransome, 1990). At least 42% of bat genera ($N = 177$; Koopman, 1993) seek shelter in caves, and 16% are mainly or exclusively associated with this type of roost (data from Nowak, 1999); the roost types of many genera are still little known. Hill and Smith (1984) suggest that caves shelter more bats than most other roost sites combined. Depending on the species and the region, caves may serve as year-round (particularly so in the tropics) or transitional shelters, hibernation sites (in temperate regions), places for maternity colonies (mainly in the tropics), and for copulation. They provide bats with suitable microclimatic conditions (larger roosts may encounter sufficient microclimatic variability, thus reducing the need to change sites) and reduce the risk of predation. If possible, bats will tend to stick to those roosts with less chance of disturbance. Bats are able to become familiar with high-quality roosts where they maintain beneficial social relationships (e.g. stable territorial borders result in decreased aggression between territory holders). The habitat value of a cave can increase as a result of occupancy, simply because guano and urine accumulations stabilize temperature and humidity (Lewis, 1995). The disadvantages of roosting in caves include the risk of bats falling prey to predators that have learned the location of a roost, exposure to external parasites (many insect and mite parasites lay eggs on the walls of the roost or attach to the walls as pupae) and pathogens associated with guano accumulations. Bats occupying spatially scarce and permanent roosts are likely to show higher roost fidelity (Lewis, 1995), and some cave bats are known to return to the same cave for a number of years. As a rule, bats select sites within a roost according to their microclimate preferences, and can change their position (e.g. Ransome, 1990; Betts, 1997). However, selection of hibernation sites is not only influenced by microclimate (temperature and humidity in particular), but also by the structural properties of the site and the landscape surrounding it (e.g. the presence of foraging areas; Raesly & Gates, 1987). In caves with more than a single species, nonrandom associations are common (Rodríguez-Durán, 1998).

A large number of bats may use a single roost; up to 20 million Brazilian free-tailed bats *Tadarida brasiliensis* are found in some Mexican caves, reaching densities up to 3000 bats per square metre (Hutson et al., 2001). A significant amount of organic matter is transported into the caves by bats in the form of guano (annual production in certain north Mexican caves is 18 700 tons) and as a result of mortality in the roosts (for example, at six weeks after parturition, non-volant Brazilian free-tailed bats fell to the cave floor and died there at rates of five to twenty per hour in Davis Blowout Cave, Texas; Wilkins, 1989).

Bats are K-selected mammals with a long lifespan (over 30 years in some species) and a slow reproduction rate (usually a single young per female per year; four is the maximum). All this makes them vulnerable to disturbances, and as many as 778 microchiropteran species (out of 834) are included in one of the six IUCN "red list" categories of threatened species: four are already extinct, 15 are critically endangered, and 30 are endangered (Hutson *et al.*, 2001). Cave dwellers are particularly sensitive, especially in areas where caves are scarce and serve as a key habitat. Kitti's hog-nosed bat *Craseonycteris thonglongyai*, the sole member of its family, is restricted to 21 limestone caves in a small area of Thailand. The stronghold of the southeastern bat *Myotis austroriparius* has always been in Florida, where an estimated 380 000 individuals lived in 15 nursery colonies in caves; only five of those caves were still found to contain bats during a 1991–92 survey (Whitaker and Hamilton, 1998). Reported declines in bat numbers caused by human disturbance are 90% in the greater horseshoe bat *Rhinolophus ferrumequinum* in British caves, 99% in the Brazilian free-tailed bat in the caves of Texas, 50% in the southeastern bat in Florida, 76–89% in maternity colonies of the grey bat *Myotis grisescens* and up to 73% in hibernating Indiana myotis *Myotis sodalis*; in eight out of ten caves in northern Mexico cave-dwelling bats declined 67–100% due to burning, tourism, and mining (Hutson *et al.*, 2001). In addition to habitat destruction and modification, other threats to cave bats include roost disturbances due to mining and quarrying activities, guano mining, bird (cave swiftlets) nest collecting, caving, the expansion of tourism, deliberate disturbances, and sealing of caves and mines (Hutson *et al.*, 2001). Eradication measures against some pest species (the Egyptian fruit-bat *Rousettus aegyptiacus* and the vampire bat *Desmodus rotundus*) in their cave roosts resulted in the destruction of other non-target, mainly insectivorous bats (Hutson *et al.*, 2001). In Israel, several cave bats (Blasius's horseshoe bat, *Rhinolophus blasii*, the lesser and greater mouse-eared bats *Myotis blythi* and *M. myotis*, and Schreibers' bat, *Miniopterus schreibersii*) which were still numerous several decades ago, are now extremely rare or even locally extinct (Makin, 1988), following fumigation of caves to control the Egyptian fruit-bat. On the other hand, proper conservation measures have already had a short-term positive effect on populations of roosting bats. For example, counts of hibernating bats in the Moravian Karst in 1983–87 showed an increase in the populations of several species, while populations of the lesser horseshoe bat *Rhinolophus hipposideros* remained stable (Bauerová *et al.*, 1988). Periodic catastrophic events can also have a major impact on cave bats. For example, the entire population of 85 000 southeastern bats in Sneads Cave, Florida, was destroyed by flooding in 1994; this was 27% of the total population of the species (Whitaker & Hamilton, 1998).

Some cave bats which used to be primarily strict cave-dwellers have found suitable conditions in man-made structures (mines, tunnels, and buildings). The cave-dwelling big brown bat *Eptesicus fuscus* has begun to hibernate in heated buildings, but not in unheated ones (Whitaker & Gummer, 1992). The fortification tunnels of Nietopierek in western Poland, which are used by c.30 000 bats of 11 species, are one of the most important hibernation sites in Northern Europe. This is the largest concentration of bats anywhere in Europe.

Boris Kryštufek

Works Cited

Bauerová, Z., Gaisler, J., Kolařik, M. & Zima, J. 1988. Variation in numbers of hibernating bats in the Moravian karst: results of visual census in 1983–1987. In *European Bat Research 1987*, edited by V. Hanák, I. Horáček & J. Gaisler, Prague: Charles University Press

Betts, B.J. 1997. Microclimate in Hell's canyon mines used by maternity colonies of *Myotis yumanensis*. *Journal of Mammalogy*, 78: 1240–50

Feldhamer, G.A., Drickhamer, L.C., Vessey, S.H. & Merritt, J.F. 1999. *Mammalogy: Adaptation, Diversity, and Ecology*, Boston: McGraw Hill

Helversen, O. von, Heller, K.-G., Meyer, F., Nemeth, A., Vollethe, M. & Gombkötö, P. 2001. Cryptic mammalian species of whiskered bat (*Myotis alcathoe* n.sp.) in Europe. *Naturwissenshaften*, 88: 217–23

Hill, J.E. & Smith, J.D. 1984. *Bats: A Natural History*, London: British Museum of Natural History and Austin: University of Texas Press

Hutson, A.M., Micklenburgh, S.P. & Racey, P.A. 2001. *Microchiropteran Bats: Global Status Survey and Conservation Action Plan*, Gland, Switzerland: IUCN/SSC Chiroptera Specialist Group

Koopman, K.F. 1993. Order Chiroptera. In *Mammal Species of the World: A Taxonomic and Geographic Reference*, 2nd edition, edited by D.E. Wilson & D.A.M. Reeder, 2nd edition, Washington: Smithsonian Institution Press

Lewis, S.E. 1995. Roost fidelity of bats: a review. *Journal of Mammalogy*, 76: 481–96

Makin, D. 1988. The status of bats in Israel. In *European Bat Research 1987*, edited by V. Hanák, I. Horáček & J. Gaisler, Prague: Charles University Press

Nowak, R.M. 1999. *Walker's Mammals of the World*, vol. 1, Baltimore: Johns Hopkins University Press

Ransome, R.D. 1990. *The Natural History of Hibernating Bats*, London: Christopher Helm

Raesley, R.L. & Gates, J.E. 1987. Winter habitat selection by north temperate cave bats. *American Midland Naturalist*, 118: 15–31

Rodríguez-Durán, A. 1998. Nonrandom aggregations and distribution of cave-dwelling bats in Puerto Rico. *Journal of Mammalogy*, 79: 141–46

Simons, N.B. 1995. Bat relationships and the origin of flight. In *Ecology, Evolution and Behaviour of Bats*, edited by P.A. Racey & S.M. Swift, Oxford: Clarendon Press and New York: Oxford University Press

Storch, G. 1999. Order Chiroptera. In *The Miocene Land Mammals of Europe*, edited by E.K. Rössner, München: Friedrich Pfeil

Vaughan, T.A., Ryan, J.M. & Czaplewski, N.J. 2000. *Mammalogy*, 5th edition, Philadelphia: Saunders College Press

Whitaker, J.O. Jr & Gummer, S.L., 1992. Hibernation of the big brown bat, *Eptesicus fuscus*, in buildings. *Journal of Mammalogy*, 73: 312–16

Whitaker, J.O., Jr & Hamilton, W.J. Jr 1998. *Mammals of the Eastern United States*, 3rd edition, Ithaca, New York: Cornell University Press

Wilkins, K.T. 1989. *Tadarida brasiliensis*. *Mammalian Species*, 331: 1–10

CLIMATE OF CAVES

Caves are usually closed environments, because the energy exchanges with outside are generally small and only become important when a watercourse flows through a cave. In any case, the climate of caves tends to be rather constant, though it is influenced both by the outside seasonal variation (in the vicinity of the entrance) and by heat exchange from inner parts. Roughly speaking, the average air temperature of a cave is very close to the outside average temperature. Water temperature depends on the average temperature of the hydrological basin supplying the water, corrected for the transformation of work into heat (a theoretical increase of 0.234°C per 100 m of fall). A cave's climate is characterized mainly by temperature, relative humidity, and airflow but air quality is also important, for example carbon dioxide and radon.

The oldest known reference to cave climate is in Titus Lucretius Carus' poem *De rerum natura* (1st century BC) but it concerns hydrology and radon only (Cigna, 1993). These same arguments were considered later by Leonardo da Vinci in his book on the nature, weight, and movement of waters written from 1506 to 1510 where he described the hydrology of a cave in the North of Italy (probably the Grotta di Fiumelatte close to Lake Como). In 1678 M.J. Herbinio added information on cave airflow to a description of underground hydrology. In particular, he observed that airflow was not constant and changed direction and strength according to the seasons and other factors. In 1712 the French scientist Billerez explained that caves seem to be warm in winter and cold in summer simply because their range of variation of temperature is smaller than the outside range. Therefore, even though on an absolute scale caves are colder in winter and warmer in summer, the sensation of a visitor is the opposite because cave air is colder than outside air in summer and warmer in winter.

Probably the first published paper on cave climate was published by G.-A. De Levis in 1795. He described a small artificial cave excavated in sandstone, containing a small pond of water during summertime, decreasing in wet periods and increasing in dry periods. He found also that the mean air temperature was about 1.2°C higher than the outside air temperature and the water temperature about 0.6°C lower. In the 19th century cave climate started to be studied for scientific purposes, particularly in Germany and Austria (Crammer, 1899), although only temperature was measured and airflow was recorded qualitatively.

During the first half of the 20th century, studies continued to be mainly devoted to recording data without any further development of theory or an explanation of the mechanisms involved that would enable forecasting of climate. In 1913 H. Bock published an outstanding paper in which he developed a mathematical treatment of data from ice caves, both dynamic and static, but unfortunately this paper was totally ignored by the scientists of that time. Only in the second half of the 20th century was mathematics again used in cave climatology and the description of phenomena upgraded from qualitative to quantitative (e.g. Cigna, 1961; Wigley, 1967; Wigley & Brown, 1969). Later, such studies developed to cover most aspects of cave climatology and physics. Andrieux (1970–72) and Badino (1995) are the most complete papers in this field. Since the 1970s technological developments (in particular, inexpensive data loggers) have facilitated more detailed research. The provision of financial support by some show caves for environmental research and evaluation of visitor impact has also been instrumental in the development of cave climatology.

Temperature

Air temperature in the inner part of a cave is mainly in equilibrium with the rock and water temperature. However, near the entrance, changes in the outside temperature propagate into the cave mainly by convection, i.e. mechanical transport of air masses. Such changes of temperature, when considered over a time interval of at least one year, may be treated as "thermal waves". Only in the case of a shaft in thermal equilibrium conditions (i.e. when air density increases with depth) will the heat exchange occur mainly by conduction. In this case the length along which the temperature variation between outside and inside occurs is generally rather short (a few metres). When the mechanism of convection prevails, such a length is of the order of tens of metres. The transmission of heat by conduction through the rock is negligible in most cases because the attenuation of the thermal wave is very large: e.g. a reduction factor of 200 with a delay of about 6 months is found through a 15 m layer of limestone. Therefore, if a thermal wave is recorded inside a cave at distances of hundreds of metres, the propagation from outside must have occurred by means of a displacement of air masses. Calculation of the delay (displacement of a maximum of the sinusoid in days or months) and the attenuation gives important information on the influence of the outside climate on the cave through the propagation of the thermal wave (Figure 1). In the past, spot measurements were made by using a thermometer (alcohol or mercury) but now electronic instruments often coupled to data loggers are used with resistance thermometers. Such probes may easily detect a temperature variation as low as 0.01 or 0.001°C.

Relative Humidity

In caves the relative humidity (expressed as a percentage of the saturation value) is generally very close to 100% other than in passages close to an entrance. The simplest form of forced ventilated hygrometer is the sling or whirling hygrometer, while a more professional instrument is the Assmann hygrometer (sometimes referred to as a "psychrometer") which is still used for calibration purposes or for spot measurements. However, to a large extent these devices have been replaced by humidity sensors which can be classified into two categories: capacitive sensors and dewpoint probes. Capacitive sensors have a serious problem because, when the relative humidity is close to 100% (as is common in caves), they give incorrect results on account of condensation occurring over the sensor. Dewpoint probes are not affected by such condensation but their cost is about one order of magnitude greater than the cost of the capacitive devices. The high cost can be reduced by using an Assmann hygrometer with the thermometers replaced by a multiway differential thermocouple or by a ventilated capacitive sensor. Thus, the sensor would be more easily restored to a dry condition in case of condensation. It is also possible to avoid condensation around 100% humidity by heating the sensor and correcting the result obtained. Nevertheless it must be stressed that the solutions reported above are acceptable only when the relative humidity

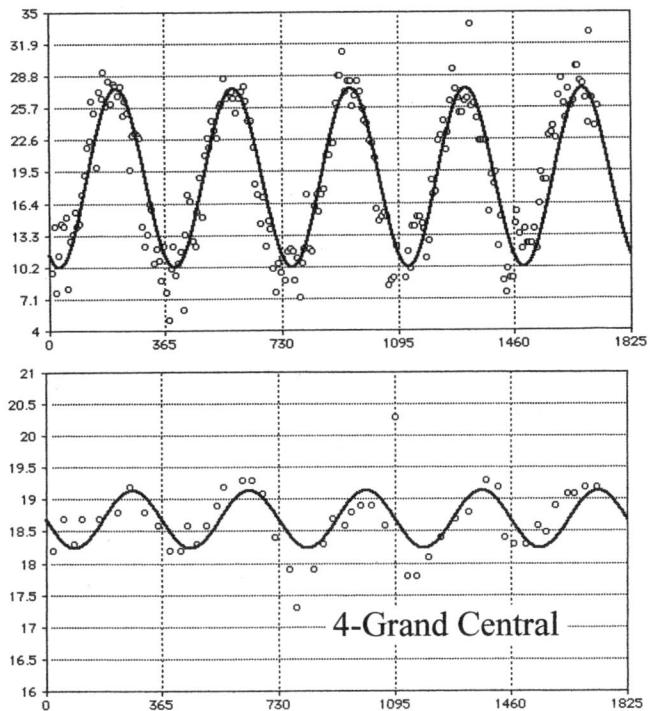

Climate of Caves: Figure 1. Air temperature in the Kartchner Caverns (Arizona, USA). In the upper part, the thermal wave outside (first maximum: day 154); in the lower part, the thermal wave of Station 4—"Grand Central" inside the cave (first maximum: day 214). The delay is of 60 days and the attenuation is about 1/20. Dots: experimental values; line: sinusoidal best fit; horizontal axis: days (Jan 1, 1996 = 0); vertical axis: °C.

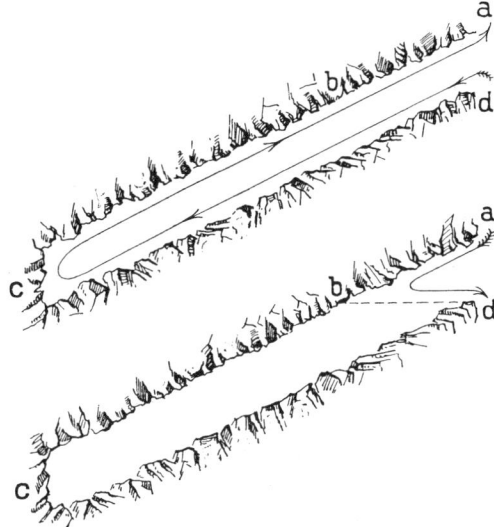

Climate of Caves: Figure 2. Cold trap. Upper figure: in winter the cave air, which is warmer than the outside atmosphere, flows upward. Lower figure: in summer the cave air, which is colder than the outside atmosphere, remains in equilibrium (below the level b-d) and there is only a limited circulation at the entrance (from: Crammer, 1899).

The airflow in caves may also be due to other causes such as the drag of a waterfall or changes of the external atmospheric pressure. A decrease of atmospheric pressure will lead to cave air flowing to the outside and the reverse occurs when atmospheric pressure increases. However, atmospheric pressure does not only

is not very close to 100%. When it is necessary to distinguish, in the vicinity of 100% humidity, between condensing or evaporating conditions, the error affecting the measurements is too large and no longer acceptable. In this case evaporimeters could be used but this technique has still to be developed and tested in the cave environment.

Airflow

Air circulation in a cave depends on the number of entrances and the general shape of the cave. Near the end of the 19th century Crammer (1899) showed that a cave with a single entrance may act either as a warm trap (if the entrance is in the lower part of the cave) or as a cold trap (if the entrance is in the upper part) where the outside air is "trapped", respectively, in the summer or in the winter season (Figures 2 and 3). When there are two entrances at different altitudes, the difference in air density between the cave atmosphere and the corresponding outside air column results in airflow. The flow is directed from the lower to the higher entrance in winter and the reverse in summer (Figure 4). In general the driving force depends on temperature, warm air being less dense than cold air but when the outside temperature is close to the cave temperature relative humidity may play a role, since humid air is less dense than dry air. Where there are more than two entrances, the airflow pattern may be more complicated, being influenced also by the shape and the size of the passages involved.

Climate of Caves: Figure 3. Warm trap. Upper figure: in summer the cave air, which is colder than the outside atmosphere, flows downward. Lower figure: in winter the cave air, which is warmer than the outside atmosphere, remains in equilibrium (above the level a-b) and there is only a limited circulation at the entrance (from: Crammer, 1899).

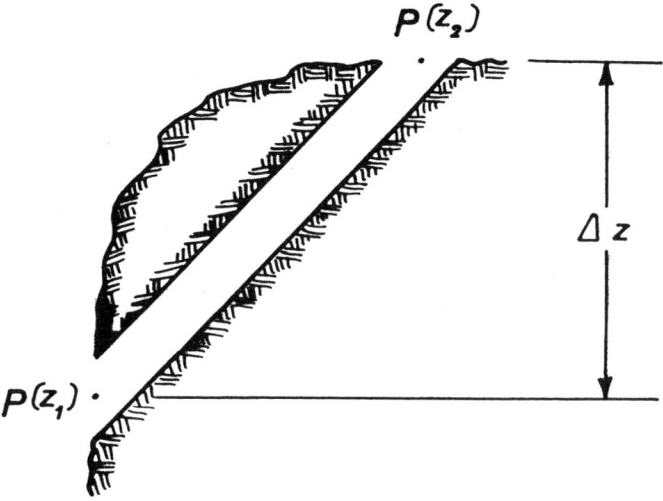

Climate of Caves: Figure 4. Wind cave. When two entrances (at different altitudes) are present, the difference of air density between the cave atmosphere and the corresponding outside air column results in airflow. The flow is directed from the lower to the higher entrance in winter and the reverse in summer. In general the driving force depends on temperature, warm air being lighter than cold air; when the outside temperature is rather close to the cave temperature, then the relative humidity may play a role, since humid air is lighter than dry air.

vary steadily, but also has a number of oscillations with many frequencies. Hence a cave may behave as a very large Helmholtz resonator, being excited by the oscillations with the frequency of the cave itself. This is the same phenomenon as occurs when someone blows across the mouth of a bottle and a note is produced, but, in the case of a cave, the frequency is so low that no sound can be heard, the most common values being around a fraction of 1 Hz. An interesting application of this phenomenon is in the evaluation of the volume of a cave when it consists of a large chamber connected to the outside by a small passage (in comparison to the chamber size). If an entrance to an unexplored passage is found to "breathe" (i.e. with an airflow directed alternatively in and out every some seconds or minutes) then, by applying a formula, it is possible to obtain a rough evaluation of the size of the cave without entering it (Cigna, 1967).

The vane anemometers usually employed in meteorology are not used in caves because the slow air speed (often less than 0.1 m s^{-1}) is outside their range. Hot wire anemometers are more suitable but they detect only the velocity and not the direction of the airflow that, in caves, may have two ways (in or out, with reference to a given environment). The direction may be identified by using two probes, one of which is shielded. By swapping the position of the shields, it is possible to determine the direction of air flow since the probe that is shielded from the prevailing air flow will record a lower air velocity.

Cave Climate Monitoring Networks

Most scientists concerned about the conservation and protection of cave environments recommend that the atmosphere in all show caves should be monitored to determine whether operation of the cave is having any adverse impacts. While spot measurements of air quality are quite common (see entries on Tourist Caves: Air Quality, Radon in Caves) only a few show caves are equipped with climate monitoring stations, although the numbers are increasing. Monitoring networks generally consist of a number of sensors in different parts of the cave, each measuring a separate parameter. Each site, or even each sensor, may have a separate data logger but increasingly sensors are linked to multichannel loggers, or even directly to computers inside or outside the cave. Threshold levels may be established for each parameter being monitored so that an immediate warning can be given if something moves out of the normal range.

ARRIGO A. CIGNA

See also **Carbon Dioxide-Enriched Cave Air; Radon in Caves**

Works Cited

Andrieux, C. 1970–72. Contribution à l'étude du climat des cavités naturelles des massifs karstiques. *Annales de Spéléologie*, 25(2): 444–90; *Annales de Spéléologie*, 25(3): 491–559; *Annales de Spéléologie*, 26(1): 5–30; *Annales de Spéléologie*, 26(2): 367–86; *Annales de Spéléologie*, 27(1): 5–77

Badino, G. 1995. Fisica del clima sotterraneo. *Memorie dell'Istituto Italiano di Speleologia*, series 2, 7: 1–137

Cigna, A.A. 1961. Air temperature distributions near the entrance of caves. In Atti Symposium Internazionale di Speleogenesi 1960. *Memoria* V, vol. 2, *Rassegna Speleologica Italiana*: 259–67

Cigna, A.A. 1967. An analytical study of air circulation in caves. *International Journal of Speleology*, 1–2(3): 41–54

Cigna, A.A. 1993. Speleology by Titus Lucretius Carus. *Atti del Simposio Internazionale sulla Protostoria della Speleologia*, Città di Castello: Nuova Prhomos: 17–24

Crammer, H. 1899. *Eishöhlen- und Windröhren Studien*, Vienna: Ab. K.K. Geographischen Gesellschaft, I: 19–76

Wigley, T.M.L. 1967. Non-steady flow through a porous medium and cave breathing. *Journal of Geophysical Research*, 72: 3199–3205

Wigley, T.M.L. & Brown, M.C. 1969. Geohydrological implications of cave breathing. In *Proceedings of the 5th International Congress of Speleology*, edited by Verband der Deutschen Höhlen- und Karstforscher e.V.: S 23/1–7

Further Reading

de Freitas, C.R. & Littlejohn, R.N. 1987. Cave climate: Assessment of heat and moisture exchange. *Journal of Climatology*, 7: 553–69

de Freitas, C.R., Littlejohn, R.N., Clarkson, T.S. & Kristament, I.S. 1982. Cave climate: Assessment of airflow and ventilation. *Journal of Climatology*, 2: 383–97

These two papers describe research on the climate of the Waitomo Cave, New Zealand.

Michie, N. A. 1997. *An Investigation of the Climate, Carbon Dioxide and Dust in Jenolan Caves, NSW*, unpublished PhD thesis, Department of Physical Geography, Macquarie University

While it is not the policy of the Editor and Publishers to reference unpublished theses an exception is made for this work which contains an up to date and comprehensive review of cave climate processes.

Wigley, T.M.L. & Brown, M.C. 1976. The physics of caves. In *The Science of Speleology*, edited by T.D. Ford & C.H.D. Cullingford, London and New York: Academic Press

A good English language review of the physical principles governing cave climates.

COASTAL KARST

Coastal processes on karst rocks often produce markedly different small-scale forms than those on non-soluble rocks. These small-scale dissolutional and bioerosional forms, called "coastal karren" or "littoral karren", range in size from submillimetre to some tens of metres and develop in the subtidal, intertidal, and supratidal zones. They are often delicately etched yet spectacularly jagged, the forms varying with wetting regime, energy levels, biological activity, and vegetation or sediment cover. The most important requirement is that abrasive tools are rare: close to beaches the rock surface is often smoothly abraded like any other rock type. The second requirement is that the rate of cliff recession is relatively slow: where retreat is fast a steep fracture surface results, as with any other rock type. The most striking coastal karren develops in warm climates, usually behind the protection of an offshore reef.

Coastal processes include abrasion, hydraulic action, wave action/quarrying, wetting and drying, corrosion, dissolution, bioerosion, bioconstruction and, in cold regions, frost action. The balance of processes varies with position above sea level; for example, wetting and drying are maximum at mid-tide. In karst rocks, bioerosion and bioconstruction may be dominant, with dissolution playing a small role. Bioerosion results from: (1) colonization of surfaces by epilithic algae and cyanophytes; (2) boring into rock surfaces by endolithic algae and cyanophytes, fungi, lichen, invertebrates such as gastropods, echinoderms, clionid sponges; and (3) grazing of epi- and endoliths by, e.g. chitons, gastropods, echinoderms, crabs, and parrot fish. Bioconstruction results from the build-up of carbonate encrustations of calcareous algae, vermitid gastropods, and serpulid worms. Protector organisms, such as mussels, are neither bioeroders nor bioconstructors: they simply protect the rock surface from erosion. Coastal karren erosion rates in sheltered areas are typically ~ 1 mm a^{-1}; they may exceed 2.5 mm a^{-1} in exposed regions, but where bioconstruction dominates over bioerosion, they may reach only 0.17 mm a^{-1}.

Littoral zone processes do not preferentially etch joints and bedding planes. The most common form of coastal karren is subcircular basins, often in a hierarchy of sizes up to several metres in scale, with intervening pinnacles. The overall form varies according to organisms present, exposure levels, tidal range

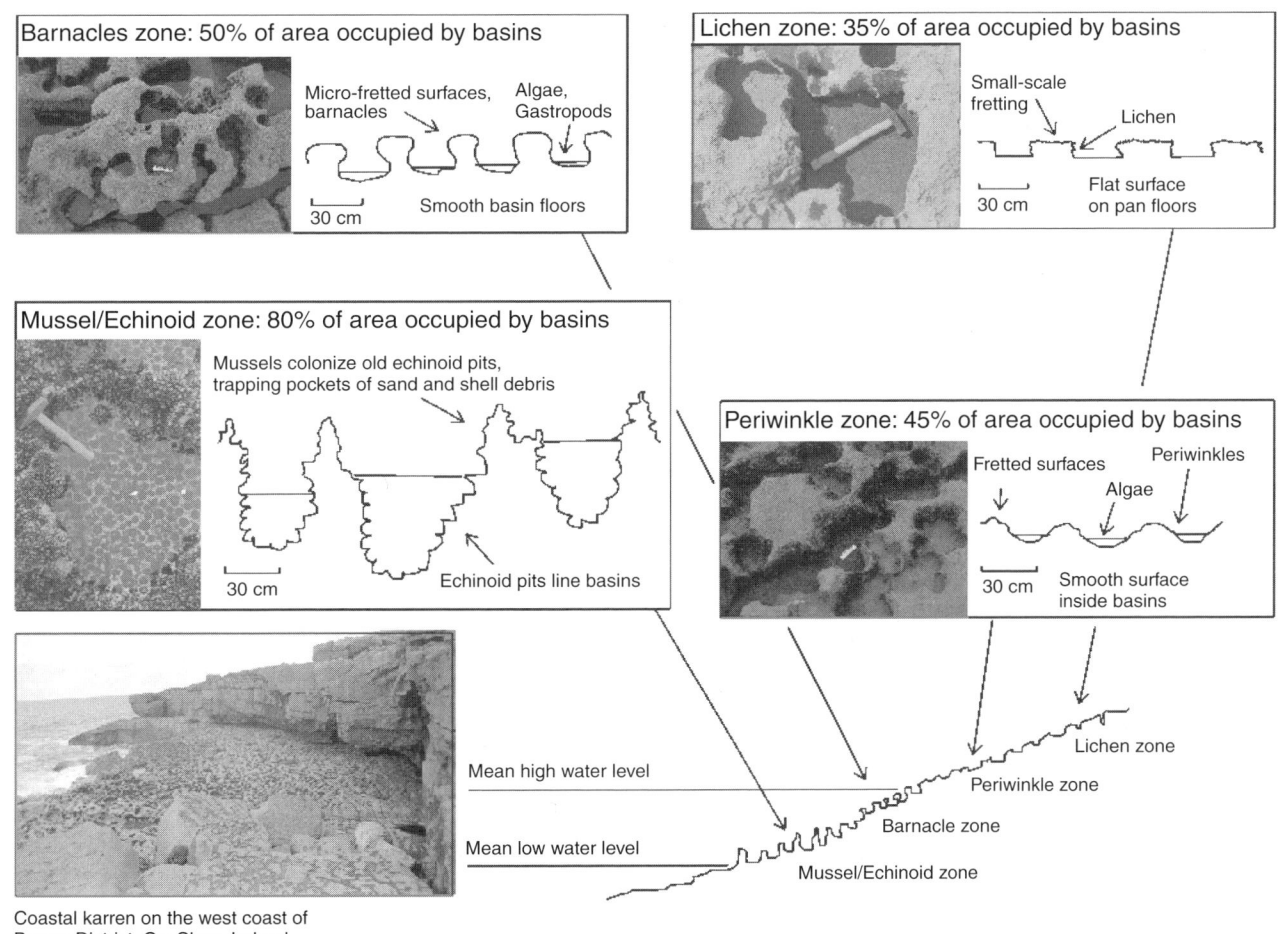

Coastal Karst: Figure 1. Coastal karren of a temperate region, Co. Clare, Ireland.

and regime, input of fresh rainwater, and to some extent, lithological and structural variations. Tropical karren, because of the effect of encrusters, show some unique intertidal features. Otherwise the erosional ramp with pinnacles and basins is the typical form. Most show a morphological zonation corresponding to biological zonation in relation to tidal levels.

In cool and temperate regions, the pinnacle-and-basin ramp is well developed. The example in Figure 1 is from the west coast of Co. Clare, Ireland. Here the spray zone (lichen zone) is characterized by flat-bottomed, steep-sided, shallow, isolated pans cut into the rock "plateau". The pans grade downshore to the splash zone (periwinkle zone), with deeper, round-sided, somewhat round-bottomed, and less isolated basins. The upper intertidal zone (barnacle zone), shows relatively flat-bottomed basins with overhanging rounded edges and barnacle-encrusted rounded inter-basin plateau remnants. The lower intertidal zone (mussel/echinoid zone), shows sharp, jagged, narrow pinnacles encrusted with mussels, separating deep, rounded to conical basins that are lined with echinoderm pits. The subtidal zone shows subdued relief, with no basin or pit karren; in some regions there is a shallow cliff at low tide with a slight subtidal notch.

In warmer regions, the erosional ramp with pinnacles and basins is supplemented by unique intertidal and subtidal features that relate to organisms present, exposure levels, and wave energy. Four units may be present, although not all coasts will have all of these units (Figure 2).

The splash and spray zone is usually a ramp with basins and pinnacles showing sharp, jagged projections between relatively flat-bottomed basins, often with overhanging rims. Direct bioerosion is minor. Jaggedness increases with increasing splash. Where splash is minor, basins dominate, ~3–4 m wide and ~1 m deep. In the intertidal zone, biogenic activity, abrasion, hydraulic action, and solutional activity focused at the water level produces a notch, incised typically from 1 to 5 m. The roof and overhanging rock forms the visor, the topmost surface of which is often pitted and fretted.

In sheltered and/or microtidal environments, processes are focused on a small vertical expanse of cliff, where the notch of small height digs deeply into the rock at mid-tide level. As tidal range or exposure increase, processes are spread out over a larger expanse of cliff face; the notch will be less incised but taller. In higher energy regions, where water is warm and turbulent, carbonate-encrusting organisms abound and a tidal platform develops. Calcareous algal coatings, vermitid gastropods, and serpulid worm encrustations protect the rock from erosion. The encrusters, intolerant of emersion, concentrate below mean tide level. The rock above, without this protection, is open to normal notching processes. Thus, over time, the upper notched part retreats while the encrusted zone remains intact. The remnant form is the tidal platform. The outer edge may have a thick accumulation; in some localities a tidal platform is built out from a steep rock surface by encrustation alone. In higher energy

Coastal karren on the south coast of Dominican Republic. Pinnacles and basins of splash zone in foreground, Surf platform in midground, visor and notch in background

Miocene limestones, Isle de Mona, Inter-tidal notching forms a "mushroom rock".

Puerto Rico, eolianites: Surf platform exposed at very low tide with vasques (wide, shallow, rimmed pools). Flat-bottomed basins and jagged pinnacles in foreground

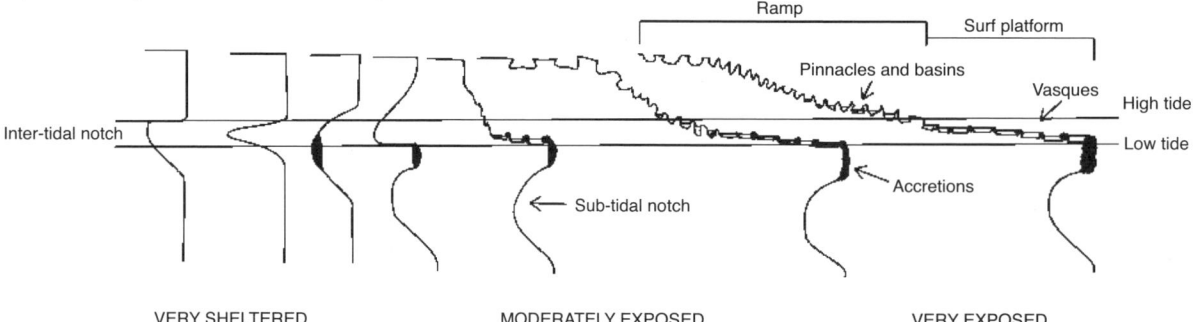

Model of tropical coastal karst development with increasing energy levels

Coastal Karst: Figure 2. Coastal karren of tropical regions.

environments, the tidal platform often has a ramped surface, made up of a series of steps separating wide, shallow pools or "vasques". Each riser is a narrow, sinuous, lobed ridge, protected by encrusters that thrive on the rims over which water continually drains. Beneath the tidal platform, where encrustations are less important and bioerosion can continue, a low-tidal or subtidal notch develops.

All four units may not be present at one location; the coastal profile that develops relates to exposure level. A dynamic equilibrium develops as the cliff edge, notches, and platform retreat together. In the splash zone, downcutting is important in addition to lateral retreat: as the splash zone basins and pinnacles are eroded down they are transformed into the tidal platform with vasques.

JOYCE LUNDBERG

See also **Biokarstification; Speleogenesis: Coastal and Oceanic Settings**

Further Reading

Focke, J.W. 1978. Limestone cliff morphology on Curaçao (Netherlands Antilles) with special attention to the origin of notches and vermetid/coralline algal surf benches ("cornices", "trottoirs"). *Zeitschrift für Geomorphologie*, 22: 329–49

Ford, D.C. & Williams, P. 1989. *Karst Geomorphology and Hydrology*, London and Boston: Unwin Hyman (especially pp.393–96)

Guilcher, A. 1953. Essai sur la zonation et la distribution des formes littorales de dissolution du calcaire. *Annales Geographie*, 62: 161–79

Jennings, J.N. 1985. *Karst Geomorphology*, 2nd edition, Oxford and New York: Blackwell (especially pp.215–21)

Ley, R.G. 1979. The development of marine karren along the Bristol Channel coastline. *Zeitschrift für Geomorphologie*. Suppl. Band 32: 75–89

Schneider, J. 1976. Biological and inorganic factors in the destruction of limestone coasts. *Contributions to Sedimentology*, 6: 1–112

Spencer, T. 1988. Limestone coastal morphology: The biological contribution. *Progress in Physical Geography*, 66–101

Spencer, T. & Viles, H. 2002. Bioconstruction, bioerosion and disturbance on tropical coasts: Coral reefs and rocky limestone shores. *Geomorpholoy*, 48: 23–50.

Trenhaile, A.S. 1987. *The Geomorphology of Rock Coasts*, Oxford: Clarendon Press and New York: Oxford University Press (especially Chapter 10, pp.240–65)

Trudgill, S.T. 1985. *Limestone Geomorphology*, London and New York: Longman (especially Chapter 10, pp.156–73)

COCKPIT COUNTRY CONE KARST, JAMAICA

Jamaica's Cockpit Country is the spectacular humid tropical "type example" of what is sometimes described as the "egg-box" style of polygonal karst. Centred on Trelawny Parish, in northwestern Jamaica, the Cockpit Country covers about 600 km². The cockpits, which dominate the landscape (Figure 1), are steep-sided, more-or-less enclosed lobate depressions, some over 100 m deep and 1 km in diameter, surrounded by residual hills or ridges, and are so named because they resemble the arenas formerly used for cock fighting. The residual hills and ridges are notched by elevated saddles, and many cockpits are connected to one or more of their neighbours by a lower corridor. Some cockpits are elongated, reflecting structural influences or inheritance from abandoned surface drainage courses. The term "cone karst" is in some ways inappropriate, since the residual hills are rarely conical or isolated, rather being linked at their bases as irregular ridge remnants, and it is the enclosed depressions, rather than the hills, which are the focus of geomorphic activity.

Cockpit slopes and the surrounding hilltops and ridges are highly irregular, with patchy clay soil cover. Slopes consist of combinations of vertical cliffs, inclined bedrock surfaces, "staircases", or talus accumulations. By contrast, the depression bases often have a deep regolith cover, and some contain relict, debris-choked vertical shafts. Internal drainage is centripetal, although dominantly vertical (Day, 1979). Regional drainage is largely autogenic and northward, although there are some allogenic inputs on the southern boundary. On the northern periphery, underground drainage emerges at a series of springs, which feed rivers draining to the north coast.

Most of the Cockpit Country is developed in Eocene carbonates of the White Limestone Formation, although some of the southern area is developed on older rocks of the Yellow Limestone group. The White Limestones are generally extremely pure, mechanically competent, and well-bedded, with blocks of strata dipping towards the north–northwest and separated by northwest–southeast and northeast–southwest-trending faults. The oldest formation, the Troy-Claremont, is approximately 300 m in thickness. The Troy Limestone is generally unfossiliferous, recrystallized, and dolomitized; the Claremont is a fossiliferous, biomicritic limestone. The younger Swanswick Formation, which outcrops in the northern part of the Cockpit Country, is a rubbly, fossiliferous biosparite up to 100 m in thickness. There continues to be uncertainty about the origin of the bauxite

Cockpit Country Cone Karst: Figure 1. Air view of a small fraction of the tree-covered cones and cockpits that constitute Jamaica's distinctive Cockpit Country. (Photo by Tony Waltham)

deposits that occur in the vicinity of the Cockpit Country. Zans (1959) and others considered them to have been derived from the non-carbonate Central Inlier and deposited within the karst as alluvial clays, but others, including Smith *et al.* (1972) were convinced that the material is a weathered residue derived from the insoluble component of the White Limestone itself. A third possibility, perhaps the most plausible, is that the bauxite is derived from Miocene bentonitic clays of volcanic origin (Comer, 1974) (see also Bauxite Deposits in Karst).

Cave systems, associated with the Cockpit Country, include some active river caves flanking the Central Inlier to the south and some large abandoned phreatic conduit systems around the periphery. Although the most important type of cave in Jamaica is the very gently graded, large river passage, and these presumably exist at depth beneath the surface, most known caves within the Cockpit Country proper are either debris-choked pits within cockpits (Smith *et al.*, 1972) or fragmentary dry passages within the residual hills and ridges (Brown & Ford, 1973). Few caves show any evidence of being developed along fault planes, although bedding and joint-plane control is important in some

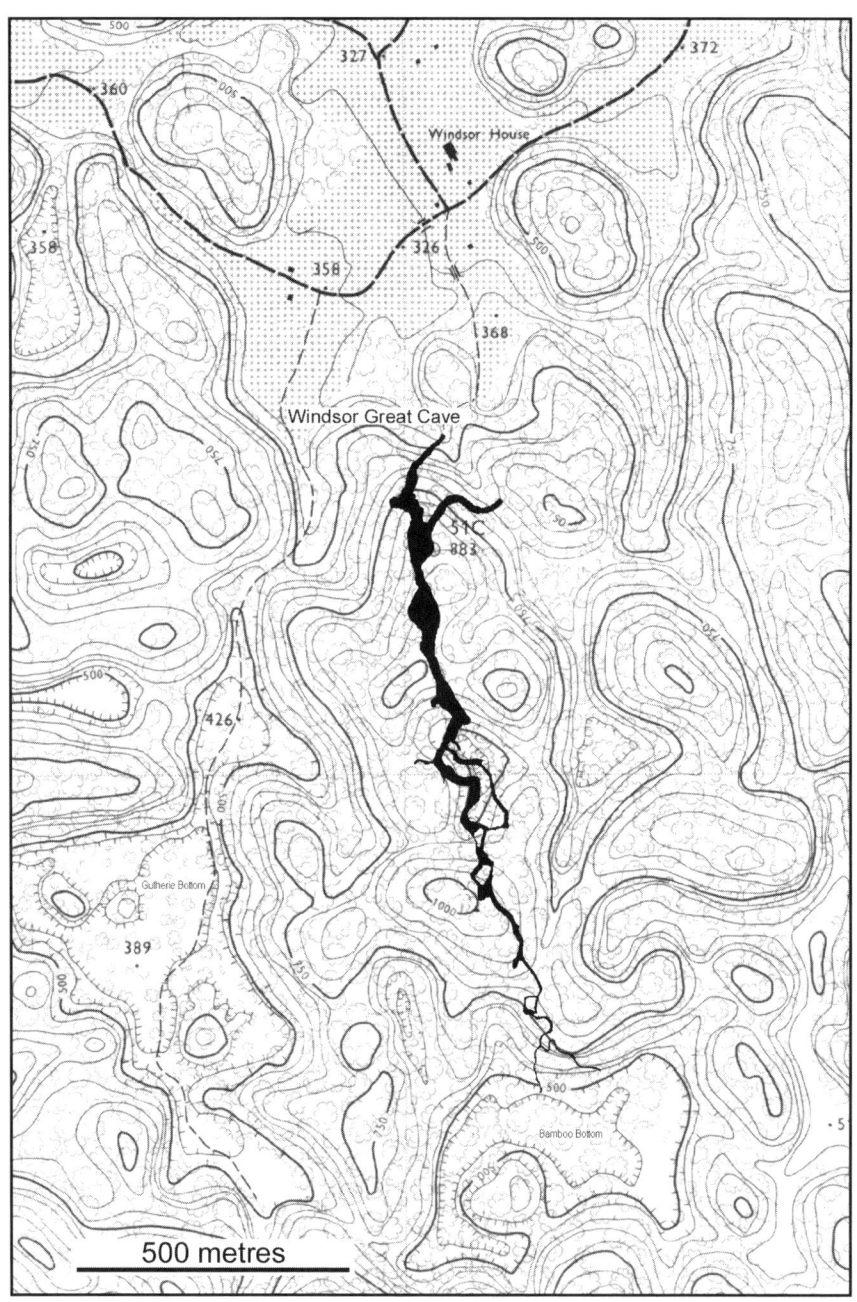

Cockpit Country Cone Karst: Figure 2. Topographic map of a small part of Jamaica's Cockpit Country with contours at intervals of 50 feet (15 m) to show morphology of the conical hills and depressions. The outline of Windsor Great Cave is superimposed; it has a large old passage, and a smaller streamway in from Bamboo Bottom.

areas, and lithology plays an important role in influencing cave morphology (Wadge & Draper, 1977a,b). Many of the larger peripheral caves, such as the spacious, old phreatic tunnel of Windsor Great Cave (Figure 2), have significant bat populations and have been commercially exploited for guano (Fincham, 1997).

The dramatic Cockpit Country landscape attracted the attention of 19th-century European geologists, such as J. Sawkins in 1869, but the first systematic geomorphological studies were undertaken in the 1950s by the Jamaica Geological Survey under the direction of V.A. Zans (Zans, 1951) and by visiting Europeans such as Herbert Lehmann, Marjorie Sweeting, and Harold Versey. In 1955, Zans and Sweeting traversed the Cockpit Country via the Troy-Windsor trail, an ambulation that convinced Sweeting (1958) that the terrain was the result of surface dissolution, rather than collapse or groundwater upwelling. Versey (1972) nevertheless continued to stress the mechanical significance of groundwater upwelling during heavy rainfall events, particularly in peripheral areas where large, complex depressions, such as Bamboo Bottom, have deeply alluviated floors with incised drainage channels and estavelles.

The surficial dissolutional origin is clearly complex, and it has been noted by Conrad Aub in particular that the vegetation canopy obscures the true irregularity of the cockpit slope surfaces, and that "rafting" of rain by the canopy results in approximately 14% more water reaching cockpit floors than the surrounding summits (Aub, 1969).

The Cockpit Country vegetation includes a range from wet to dry limestone forest, in which there is considerable floristic diversity and an extraordinary number of endemic species. The Cockpit Country fauna is also significant, including threatened species such as bats, snakes, frogs, and all but one of Jamaica's 28 endemic bird species.

Human influences have, to date, been limited by inaccessibility and lack of surface water, although there exists an extensive trail system, and many peripheral cockpits have been used for agriculture. In the 1700s, escaped slaves known as the Maroons used the Cockpit Country as a base for guerrilla activities against the British army. Treaties gave the Maroons a degree of autonomy that they have maintained ever since.

Since 1950, much of the Cockpit Country has been designated as a Forest Reserve, although there has been little enforcement of conservation directives. The immediate vicinity has a population of some 10 000 people, and is exploited for bauxite mining and agriculture. In the Forest Reserve, illegal logging, farming, hunting, and trapping for the pet trade are particular problems. More recently, the Cockpit Country has been proposed as a UN World Heritage Site, and there are currently plans to inscribe it as a national park.

MICK DAY AND SEAN CHENOWETH

See also **Caribbean Islands; Cone Karst**

Works Cited

Aub, C.F.T. 1969. The nature of cockpits and other depressions in the karst of Jamaica. *Proceedings of the 5th International Speleological Congress*, edited by Verband der Deutschen Höhlen- und karstforscher e.V.

Brown, M.C. & Ford, D.C. 1973. Caves and groundwater patterns in a tropical karst environment: Jamaica, West Indies. *American Journal of Science*, 273: 622–33

Comer, J.B. 1974. The genesis of Jamaican bauxite. *Economic Geology*, 69: 1251–64

Day, M.J. 1979. The hydrology of polygonal karst depressions in northern Jamaica. *Zeitschrift für Geomorphologie N.F., Supplementbande*, 32: 25–34

Fincham, A.G. 1997. *Jamaica Underground: The Caves, Sinkholes and Underground Rivers of the Island*, 2nd edition, Kingston: The Press, University of the West Indies

Smith, D.I., Drew, D.P. & Atkinson, T.C. 1972. Hypotheses of karst landform development in Jamaica. *Transactions of the Cave Research Group of Great Britain*, 14: 159–73

Sweeting, M.M. 1958. The karstlands of Jamaica. *The Geographical Journal*, 124(2): 184–99

Versey, H.R. 1972. Karst in Jamaica. In *Karst: Important Karst Regions of the Northern Hemisphere*, edited by M. Herak & V.T. Springfield, Amsterdam and New York: Elsevier

Wadge, G. & Draper, G. 1977a. Tectonic control on speleogenesis in Jamaica. *Proceedings of the 7th International Speleological Congress*, edited by T. Ford, Sheffield: British Cave Research Association

Wadge, G. & Draper, G. 1977b. The influence of lithology on Jamaican cave morphology. *Proceedings of the 7th International Speleological Congress*, edited by T. Ford, Sheffield: British Cave Research Association

Zans, V.A. 1951. On karst hydrology in Jamaica. *International Geodesic and Geophysical Union, Scientific Hydrology Assembly, Brussels*, vol. 2

Zans, V.A. 1959. Recent views on the origin of bauxite. *Geonotes*, 1(5): 123–32

Further Reading

Day, M.J. 1993. Human impacts on Caribbean and Central American karst. In *Karst Terrains: Environmental Changes and Human Impact*, edited by P.W. Williams. Cremlingen-Destedt: Catena Verlag

Troester, J.W., Back, W. & Mora, S.C. 1987. Karst of the Caribbean. In *Geomorphic Systems of North America*, edited by W.L. Graf, Boulder, Colorado: Geological Society of America

COLONIZATION

Colonization of a cave by an animal or plant species may be either primary or secondary. Primary colonization may be defined as the establishment of a population in an area not previously occupied by the species. A cave open to primary colonization may lack a species because of recent origin (e.g. during the karstification process), recent catastrophes (e.g. lava flows, glaciations), or geographic location (e.g. isolation). Secondary colonization is more characteristic of small locally confined areas less isolated from colonizing populations; in these areas the species disappeared following minor environmental alterations (e.g. marginal areas of a karstic massif after the retreat of a glacier, or polluted areas after recovery). Both primary and secondary colonization may be considered as continuous processes; nevertheless, colonization rates may increase or decrease owing to adverse climatic conditions, depletion of resources, high competition rates, or strong predation pressure.

In the biospeleological literature the term colonization refers mainly to the colonization of a cave by a surface population; however, subterranean organisms are active and apparently cognitive subjects, and their dispersal in new habitats, and hence the colonization of new caves, may be considered a relatively common event.

Considering the modalities of colonization, biospeleologists put great emphasis on active and passive colonization mechanisms. Usually active colonization is understood in the sense of colonization by means of the locomotory apparatuses and is the opposite of passive transport. However, some biospeleologists do not use the term "active" to refer to the mechanical method used to reach the subterranean environment, and view active colonization as synonymous with voluntary colonization. Indeed, one of the most common questions on colonization addressed by biospeleologists is not how epigean species colonized the subterranean domain, but why they did so. This is an old question due to the interpretation of the subterranean environment as a special one (the refugium model): caves have long been considered as refugia against climatic vicissitudes for the ancestors of troglobites, especially those of Pleistocene animals in temperate zones. Rouch & Danielopol (1987) challenged this idea, stating that there are no compelling forces to colonize the subterranean environment: the active colonization model replaced the refugium model. The refugium model is considered incompatible with the accumulated evidence that subterranean organisms occur on a worldwide scale, even in the absence of unfavourable climatic conditions; moreover, it does not explain the contemporary colonization processes of subterranean habitats and the fact that many hypogean animals are not relicts. The active colonization model proposes a unique scenario to explain colonization of subterranean habitats; for this reason, it was criticized by Botosaneanu & Holsinger (1991). They reported several examples for and against hypogean relicts and the refugial character of subterranean environments, stating that no unique explanation can be found to explain the colonization events. Finally, Stoch (1995) formulated an "adaptive zone model", suggesting that surface populations invade subterranean habitats to exploit a new set of resources; colonization may be followed by niche specialization and adaptive radiation. Most modern biospeleologists do not support the old hypothesis that the origin of subterranean animals must differ in certain instances from that of other animals: instead it is argued that caves are no more than one of several types of environments that ecologists deal with.

Several colonization models have been developed to explain the origin of groundwater fauna, some of them dealing primarily with interstitial fauna (see Interstitial Habitats: Aquatic). Among several others, the following models are mentioned herein: regression model (Stock, 1980); zonation model (Iliffe, 1986); two-step model (Notenboom, 1991); "modèle biphase" (Coineau & Boutin, 1992); three-step model (Holsinger, 1994); and adaptive zone model (Stoch, 1995). Some of these models can also be applied to terrestrial fauna; however, detailed colonization models dealing separately with terrestrial habitats were developed for caves (three-step model, Juberthie, 1984) and lava tubes (adaptive shift, Ashmole, 1993).

Coineau & Boutin (1992) synthesized most of the models outlined above for groundwater fauna. According to these authors, the stygobites living in inland subterranean waters belong to two groups: limnicoid stygobites, the marine ancestors of which lived in surface freshwaters before groundwater colonization, and thalassoid stygobites, which colonized continental groundwaters from the marine environment through the littoral interstitial zone during a marine regression. The two groups of stygobites settled into the inland groundwaters after two ecological and geographical transitions ("modèle biphase" or "two-step model") involving a "vertical transition" (active colonization of the littoral interstitial environments) and a "horizontal transition" (active or passive colonization of the continental interstitial groundwaters) (Figure 1).

Holsinger (1994) developed a "three-step model" of colonization of continental groundwaters; these steps (designated A, B, and C) are best regarded as components in a graded series or phases in a transition and not as rigid stops and starts. Species at step A are epigean benthic forms with little or no apparent troglomorphy, living in freshwater or in shallow littoral or sublittoral sea zones and demonstrate little specialization; most of these taxa remain potential epigean ancestors of stygobites. Species at step B are semi-hypogean organisms, with strong preadaptation to life in subterranean waters; eyes and pigment are typically rudimentary. Species at step C are stygobites; either limnostygobites or thalassostygobites, commonly troglomorphic and restricted to life in subterranean waters.

Ashmole (1993) summarized the theories on the origin of subterranean terrestrial species in volcanic islands, distinguishing between "ecological colonization" and "evolutionary colonization". Ecological colonization of lava flows commences as soon as the rock is cool; populations of "lavicolous" scavengers build up both on the surface and in cracks and dry caves, exploiting the "biological fallout", mainly windborne insects. On the other hand, evolutionary colonization of the deep cave zone (both mesocaverns and saturated caves) is the process by which some individuals from one type of habitat move into a different habitat that imposes new selection pressures, thus leading to adaptive divergence and speciation of the colonizing population. This "adaptive shift" may occur when preadapted epigean individuals accidentally penetrate cavities and form populations that are poorly or rarely influenced by gene flow. The shift involves rapid adaptive evolution in morphological and physiological characteristics.

Juberthie (1984) illustrated the theories on colonization of cave habitats by terrestrial species, dealing mainly with the role of the MSS (Mesovoid Shallow Substratum: see Interstitial Habitats: Terrestrial). Juberthie's scenario is a three-step model. During the first step, some populations inhabiting soil litter or mosses colonize the lower layer of forest soils, becoming preadapted to the subterranean environment. Next, they invade the MSS: during this step, the colonizing populations inhabit both compartments. Finally, MSS populations invade the "deep hypogean compartment" (e.g. caves); the isolation of cave-dwelling populations is related to geographical location, degree of karstification, and history (e.g. glaciations).

Finally Stoch (1995) clearly separated the colonization process from the speciation process (see Speciation), and developed

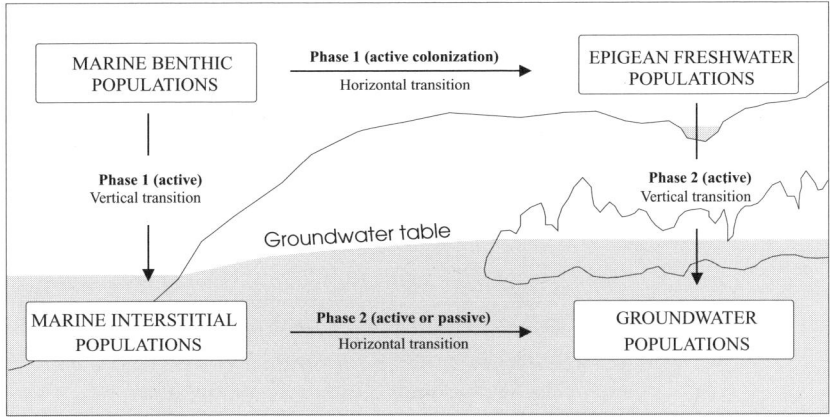

Colonization: Figure 1. The two ways of colonization of continental groundwaters according to the "two-step model" (modified after Coineau & Boutin, 1992): transition via surface freshwaters (limnicoid stygobites) and transition via littoral interstitial waters (thalassoid stygobites).

a multiple step model (called "adaptive zone model") to explain the origin of subterranean biodiversity following two different methods: colonization-speciation-adaptive radiation (e.g. rapid phylogenetic diversification via niche differentiation) or multiple colonization-speciation (when radiation does not occur) (Figure 2). Both active and passive colonization mechanisms are allowed by the adaptive zone model, but the refugium model is clearly ruled out. Following this model, climatic fluctuations and other events such as marine regressions clearly enhanced colonization rates, but speciation is not a consequence of colonization and an increase in the rate of colonization is not necessarily followed by increased speciation rates. Generalist colonizers accomplish the expansion of their range due to efficient dispersal adaptations and, after arrival, the establishment of a viable population at the new site. Repeated colonizations into the same site are likely to occur and colonization rates may be high. But generalist colonizers do not appear to speciate: the efficiency of colonization effectively prevents the accumulation of genetic differences from the source, and a speciation event in this case is unlikely. Following the adaptive zone model, only stochastic colonization by a founder with preadaptation to the new site is an event conducive to speciation.

Several hypotheses have been proposed to date the colonization processes, for example marine regressions, the Messinian salinity crisis for the Mediterranean area (Hsü, Ryan & Cita, 1973), and glaciations for temperate zones. Unfortunately, in most models based on historical events the age of colonization is confused with the cause of colonization. For example, some biospeleologists advocated that the Mediterranean salinity crisis (when most of the Mediterranean Basin dried up) forced isopods and amphipods to colonize groundwaters where they remained "stranded" ("regression model" developed by Stock, 1980). Following this model, the age and cause of colonization are determined by the same historical event, but an alternative explanation is possible. According to the adaptive zone model, Stoch (1995) supposed that the Messinian event simply interrupted the genic flux between brackish water and marine populations; epigean brackish water species may have colonized surface freshwaters and subsequently cave waters following the development of karstic areas. Therefore, these stygobitic species may be considered a recent limnicoid stygobite instead of older thalassoid stygobites, and an active colonization model may be advocated opposed to a refugium model.

Unfortunately, most of the theories dealing with the colonization process remain highly speculative and untested: for this reason, the debate between active colonization and refugium model supporters is still a central focus in the current biospeleological literature.

FABIO STOCH

See also **Adaptation; Evolution of Hypogean Fauna; Speciation**

Works Cited

Ashmole, N.P. 1993. Colonization of the underground environment in volcanic islands. *Mémoires de Biospéologie*, 20: 1–11

Botosaneanu, L. & Holsinger, J.R. 1991. Some aspects concerning colonization of the subterranean realm—especially of subterranean waters: A response to Rouch and Danielopol, 1987. *Stygologia*, 6(1): 11–39

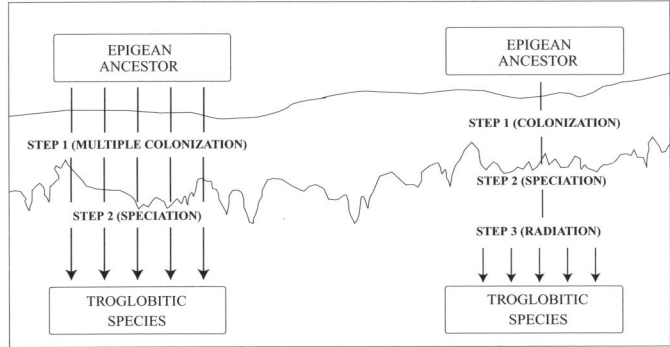

Colonization: Figure 2. Origin of troglobitic species according to the "adaptive zone model" (Stoch, 1995); two different ways are possible: multiple colonization followed by speciation, and colonization by a single ancestor followed by speciation and adaptive radiation.

Coineau, N. & Boutin, C. 1992. Biological processes in space and time. Colonization, evolution and speciation in interstitial stygobionts. In *The Natural History of Biospeleology*, edited by A.I. Camacho, Madrid: Museo Nacional de Ciencias Naturales

Holsinger J.R. 1994. Pattern and process in the biogeography of subterranean amphipods. *Hydrobiologia*, 287: 131–45

Hsü, K.J., Ryan, W.B.F. & Cita, M.B. 1973. Late Miocene desiccation of the Mediterranean. *Nature*, 242: 240–244

Iliffe, T.M. 1986. The zonation model for the evolution of aquatic faunas in anchialine caves. *Stygologia*, 2(1/2): 2–9

Juberthie, C. 1984. La colonisation du milieu souterrain: theories et modèles, relations avec la spéciation et l'évolution souterraine. *Mémoires de Biospéologie*, 11: 65–102

Notenboom, J. 1991. Marine regressions and the evolution of groundwater dwelling amphipods (Crustacea). *Journal of Biogeography*, 18: 437–54

Rouch, R. & Danielopol, D.L. 1987. L'origine de la faune aquatique souterraine, entre le paradigme du refuge et le modèle de la colonisation active. *Stygologia*, 3(4): 345–72

Stoch, F. 1995. The ecological and historical determinants of crustacean diversity in groundwaters, or: Why are there so many species? *Mémoires de Biospéologie*, 22: 139–60

Stock, J.H. 1980. Regression model evolution as exemplified by the genus *Pseudoniphargus* (Amphipoda). *Bijdragen tot der Dierkunde*, 50: 104–44

Further Reading

Camacho, A.I. (editor) 1992. *The Natural History of Biospeleology*, Madrid: Museo Nacional de Ciencias Naturales

Culver, D.C., Kane, T.C. & Fong, D.W. 1995. *Adaptation and Natural Selection in Caves: The Evolution of Gammarus minus*. Cambridge, Massachusetts: Harvard University Press

Gray, A.J., Crawley, M.J. & Edwards, P.J. 1987. *Colonization, Succession and Stability*. Oxford and Boston, Massachusetts: Blackwell Scientific

COMMUNICATIONS IN CAVES

On deep or wet pitches, communicating by means of whistle-blasts is an established practice and, in many situations, this or perhaps a loud voice will be sufficient. However, for cave rescue, expedition management or the execution of projects such as photography in a large chamber, communication by radio or telephone is often essential.

Possibly the first use of a telephone in a cave was during the exploration of Lamb Leer Cavern (Mendip, United Kingdom) in 1880 (Williams, 1995). Despite the disadvantage of having to lay a cable, telephones continue to be used because they are simple and robust. A variety of other cable-based systems exist—single-wire telephone, "guide wire" radio, and optical fibre. True "wireless" communication is difficult because rock, being conductive, absorbs radio waves. For line-of-sight work within cave passages, high-frequency (HF) walkie-talkies and CB radios are used, but magnetic induction equipment is usually necessary for communication through the rock itself. The technique of "earth current injection" was developed during World War I but is rarely used now although an enhancement, using a low-frequency (LF) carrier, is the basis for the latest high-performance induction radio/earth-current "hybrids" now used in the United Kingdom and Europe.

The single-wire telephone (SWT) (also known as an earth-return telephone) uses the conductivity of the ground to provide a return path for the current. The obvious advantage is that only half the weight of cable has to be carried. The devices used by cavers feature electronic amplification and will often operate without any specific earthing other than through the caver's body. SWTs are cheap, rugged, easy to build by amateurs, and are preferred to the traditional army field telephone. They are popular with expeditions and rescue groups throughout the world, and are also convenient for providing flood warnings to cavers.

If, instead of audio, an LF radio signal is passed along an SWT cable, it will work without any earthing, because the signal is able to jump gaps by a capacitive coupling effect, implying that the handset does not need to touch the cable either. Commercial equipment based on this principle is used for mine rescue. HF radio can also be guided along a cable placed in close proximity, although the principle of operation is quite different. For commercial use at VHF, "leaky-feeder" cable presents yet another mode of operation. Confusingly, all these systems are sometimes referred to as "guide wire" radio.

Rock is electrically conductive, which causes it to absorb radio waves, so HF radio is mostly limited to use within cave passages. Rescue teams use HF walkie-talkies (27 or 49 MHz) to communicate on pitches while manhandling an accident victim in a stretcher. Photographer Gavin Newman has used a modified CB radio to coordinate photography in Illu River Cave, Irian Jaya. He also found the radios useful to warn of impending flooding in this huge cave system (Newman, 1993). HF radio will penetrate a few metres into rock and can be used for locating dig points on the surface, although this normally requires specialized equipment (see separate entry, Radiolocation). In 2001, a CB radio was used to verify the surface location of the top of the Titan shaft in Peak Cavern (Derbyshire), with communication through 10 m of fill. Radio waves can travel for some distance along mineralized fault lines or dry well-jointed rock, so occasionally quite deep penetration of a cave is possible.

Low frequencies are attenuated less than high frequencies—long wave broadcasts, such as BBC Radio 4 (198 kHz) can be detected at the bottom of deep caves. Unfortunately, detecting a radio signal is easier than transmitting it—a 200 kHz signal would require an antenna some 750 m long, which is normally only feasible for mining installations. From the early 1960s, cavers began to experiment with loop antennas. Because of their small size (typically 1 m diameter) these radiate very little "true" radio energy, and mainly generate a magnetic field, which couples to the receiver loop by the principle of magnetic induction, and operates over a limited range of just a few hundred metres. Early systems used "baseband" audio but it was soon realised that there were benefits to using radio frequency modulation. Experimental induction systems have covered 27–185 kHz (i.e. most of the long-wave broadcast band), and most have used single-sideband (SSB) operation. The most notable designs have been Bob Mackin's commercially produced Molefone, dating

Communications in Caves: Using an induction radio with a ribbon-cable antenna in Ogof Ffynnon Ddu. (Photo by Mike Bedford)

from 1979, which has seen widespread use in Britain and Ireland, and Ian Drummond's CB Transverter, used mainly in the United States and Canada.

Two of the latest designs are John Hey's HeyPhone and the French "Nicola System" which was developed as an aid to flood warning in the Gouffre Berger. Both operate at 87 kHz SSB and are published designs. They are intended to work with induction loops or earth-current antennas, thus reviving the early–1900s trench-communications technique. With earth electrodes, the Nicola system has been demonstrated to work through 1000 m of rock. In 2001, working with the Cave Radio & Electronics Group (CREG)—a special interest group of the British Cave Research Association—the British Cave Rescue Council issued over 50 HeyPhones to British cave rescue teams (Bedford, 2001). This is the first time that a national organization has systematically equipped its rescue teams with such advanced equipment. The South Wales team soon had occasion to test their HeyPhones. A caver dislocated his shoulder some 5 km into Daren Cilau—a formidable cave protected by a daunting 520 m entrance crawl and requiring an underground camp to explore the far reaches. While the injured caver slowly made his way to the camp, two of his companions made a strenuous exit and raised the alarm. Some 28 hours later, the medical team was able to communicate with the surface to report that the casualty had been attended to, and would not require a stretcher carry (which, it had been estimated, would take up to three days to complete). Use of the HeyPhones (and of an SWT in the entrance crawl) considerably eased the logistics of this rescue, which involved 63 personnel (*Caves and Caving*, 2001).

Cave radios have been coupled, experimentally, to mobile telephones and HF radios on the surface, and messages relayed automatically over long distances. System development seems likely to continue, with the aim of providing extra functionality, for example medical data logging, text messaging, and digital image transmission.

DAVID GIBSON

See also **Accidents and Rescue; Exploring Caves; Radiolocation**

Works Cited

Bedford, M. 2001. Introducing the HeyPhone. *Caves and Caving*, 91: 15–17
Caves & Caving. 2001. Welsh News: Daren Cilau. *Caves and Caving*, 90: 10
Newman, G. 1993. The Caves of Thunder. *Descent*, 110: 26–27
Williams, B. 1995. Lamb Leer Cavern 1880–90: The lake and the talking machine. *Proceedings of the University of Bristol Speleological Society*, 20(2): 135–51

Further Reading

Note: The *Cave Radio and Electronics Group* is a special-interest group of the British Cave Research Association.
Bedford, M. 1994. A directory of cave radio designs. *Cave Radio and Electronics Group Journal*, 18: 3–4
  A list of experimental cave radio designs, most of which were short-lived or only ever saw local usage. Some references.
Bedford, M. 1999. Back to basics with cave communications. *Cave Radio and Electronics Group Journal*, 37: 6–10
  A fairly comprehensive introduction to the various methods of cave communications; no circuit diagrams, but references to other sources of information.
France, S. 1995. Induction radio or telephone? *Cave Radio and Electronics Group Journal*, 20: 7
  A discussion of the pros and cons of single-wire telephones versus induction radios.
Gibson, D. 1997. Single wire telephones for mines rescue. *Cave Radio and Electronics Group Journal*, 27: 7–11
  A comprehensive review of commercial SWT equipment. This issue of the CREG journal was a "special feature" on SWT and guide-wire radio (six articles).
Gibson, D. 2002. Design directory: Radiolocation and cave radio. *Cave Radio and Electronics Group Journal*, 48: 22–23
Mackin, R.O. 1984. Communications. In *Caving Practice and Equipment*, edited by D. Judson, Newton Abbot: David and Charles
  An introduction to the various methods of cave communications; includes simple telephone circuits and a brief description of radiolocation.

Useful Websites

David Gibson's Cave Radio page at http://www.caves.org.uk/radio/ includes a list of references to practical projects for telephones and radio, including the HeyPhone and Nicola systems, as well as links to sites that discuss communications and radiolocation.
The British Cave Research Association's Cave Radio & Electronics Group maintains a website at http://www.bcra.org.uk/creg/
The US National Speleological Society has a section for cave electronics, with a website at http://www.caves.org/section/commelect/

CONDENSATION CORROSION

Condensation corrosion occurs when a chemical reaction takes place between condensed water vapour and bedrock or speleothems in caves and can be classified as a form of chemical weathering (Fairbridge, 1968). The process has been proposed as a mechanism by which caves are enlarged and as a factor in the degradation of speleothems. Speleothems are an important natural resource and tourist attraction, hence the majority of studies of condensation corrosion have been aimed at establishing the rate of speleothem degradation in tourist caves. Evidence for the existence of condensation corrosion is greatest in caves in karst rocks.

Features attributable to condensation corrosion are particularly noticeable in cave entrances and in passages with unstable microclimates. Some of the features characteristic of the process include etching of wall rock (James, Jennings & Dyson, 1982), small pits only centimetres in diameter on flat roofs, and much larger bell holes (Tarhule-Lips & Ford, 1998). The degradation of speleothems, although not stand-alone evidence for the process, is to be expected where aggressive condensation waters are found. For instance, the deposition of calcite as cave coral, associated with areas of solution of both wall rock and speleothems, is regarded as evidence for the process (Hill & Forti, 1997). The same features are mimicked on a micron scale. Scanning electron micrographs of speleothems degraded by condensation corrosion reveal micro etching and pitting. Other karst processes, particularly erosive ones such as bio-karst processes and salt weathering, produce similar features.

In order for corrosion to take place, the water vapour in the cave atmosphere must first condense. Condensation can be observed as clouds in large chambers, sparkling droplets on roofs and walls, and as a thin film of moisture on surfaces. For water to form, the temperature of mixed air has to drop below the dew point. This takes place either when cold surface air flows into caves and encounters the warm wet air, a winter phenomenon, or when hot wet surface air comes into contact with the colder cave air, a summer phenomenon. Once a mist has formed, it can wet any surface that it comes into contact with. Mixing of two air masses is not essential, however, as water vapour in warm air will directly condense onto cold bedrock and speleothems. The drier the air, the greater the temperature drop required for it to reach its dew point.

Condensation is possible in all caves. The probability of condensates forming is greatest where the surface temperature range is large. Goede *et al.* (1990) stated that evidence of condensation corrosion is particularly noticeable in cave entrances in arid and semi-arid zones, where there is a large diurnal temperature change. In this case, moist air, at the average temperature of the area from within the cave, condenses on encountering cold night surface air.

Whether condensation corrosion takes place deep inside caves is controversial, for the essential temperature differential between air and rock is nonexistent or small at depth. Despite this, many of the features attributable to condensation corrosion are found deep in caves. Far from the cave entrance, the cave air is likely to be saturated with water vapour; consequently only small temperature changes are required for condensation to occur.

Many authors have referred to the phenomenon of condensation corrosion, but few have examined the mechanism beyond qualitative interpretations of calcite dissolution reactions. Two sets of experimental data are required to assess the corrosive potential of condensation waters on speleothems and thus model the rate of corrosion for sites within a cave. The condensing waters have to be collected and analysed, two difficult procedures because of the small quantities of water available and the exceptionally low concentrations of chemical species in them. Studies at Jenolan Caves (New South Wales, Australia) showed that condensation waters contained dissolved salts and gases. From these analytical results, saturation indices for calcite were calculated and led to the conclusion that condensing waters can be exceptionally aggressive to speleothems. Additional modelling showed that such waters are even more aggressive to calcite when they contain dissolved acid gases or salts of non-common ions. To complete the model, a budget of how much water condenses is needed. Experiments to assess the volumes and patterns of condensation have been made in the Glow-worm Cave, Waitomo, New Zealand (Schmekal & de Freitas, 2001).

An alternative approach is to measure the actual corrosion. Because of the roughness of the corroding surfaces, use of a micro-erosion meter is not practical. A method trialled by Avramidis *et al.* (2001), using slabs of Carrara marble as a substitute for speleothem calcite, has proved to be promising. In their initial experiments, Avramidis *et al.* were able to compare rates of condensation corrosion at a number of sites, and showed that in a three-month experiment, condensation corrosion was measurable with a maximum erosion rate of 1000 μm $(100 \text{ a})^{-1}$ and varied significantly between sites. Kinetics of calcite dissolution are slow, so there has been some debate as to whether the residence time of the aggressive condensate on speleothems is long enough to allow dissolution. The above experiments, and those at Jenolan Caves where condensation waters collected from speleothems contain more dissolved calcite than the condensing waters, confirm that the residence time is adequate for dissolution.

Most caves have enhanced levels of the acid gas, carbon dioxide (see Carbon Dioxide-Enriched Cave Air). The presence of carbon dioxide above atmospheric levels is believed to substantially increase corrosion (Pearce, 1999). Carbon dioxide-enhanced condensation corrosion can be viewed in many caves and is dramatically demonstrated in Odyssey Cave, Bungonia, New South Wales, Australia (James, 1977). Speleothems, in areas with atmospheres containing 2–3 v/v% carbon dioxide, exhibit the features of condensation corrosion listed above. Where levels are continuously above 4 v/v%, all speleothems have disappeared and the wall rock is deeply etched.

JULIA M. JAMES

Works Cited

Avramidis, P., Hong, J., Barnes, C. & James, J.M. 2001. A new method of measuring condensation corrosion. *Proceedings of the 13th International Congress of Speleology*, Brasilia: Speleological Society of Brazil (CD-ROM)

Fairbridge, R.W. 1968. Condensation corrosion. In *Encyclopedia of Geomorphology*, New York: Reinhold, pp.204–205

Goede, A., Harmon, R.S., Atkinson, T.C. & Rowe, P.J. 1990. Pleistocene climatic change in Southern Australia and its effect on speleothem deposition in some Nullarbor caves. *Journal of Quaternary Science*, 5: 29–38

Hill, C. & Forti, P. (editors) 1997. *Cave Minerals of the World*, 2nd edition, Huntsville, Alabama: National Speleological Society

James, J.M. 1977. Carbon dioxide in the cave atmosphere. *Transactions of the British Cave Research Association*, 4: 417–29

James, J.M., Jennings, J.N. & Dyson, H.J. 1982. Mineral decoration and weathering of the caves. In *Wombeyan Caves*, edited by H.J. Dyson, Sydney: Sydney Speleological Society

Pearce, F. 1999. Acid attack: Heavy breathing puts limestone caves at risk. *New Scientist*, 2186: 23

Schmekal, A.A. & de Freitas, C.R. 2001. *Condensation in Glow-worm Cave, Waitomo, New Zealand: Management Guidelines*, Wellington: Department of Conservation

Tarhule-Lips, R.F.A. & Ford, D.C. 1998. Morphometric studies of bell hole development on Cayman Brac. *Journal of Cave and Karst Studies*, 25: 119–30

Further Reading

Dublyansky, V.N. & Dublyansky, Y.V. 1998. The problem of condensation in karst studies. *Journal of Cave and Karst Studies*, 60: 3–17

Hoyos, M., Soler, V., Cañaveras, J.C., Sánchez-Moral, S. & Sanz-Rubio, E. 1998. Microclimatic characterization of a karstic cave: Human impact on microenvironmental parameters of a prehistoric rock art cave (Candamo Cave, Northern Spain). *Environmental Geology*, 33: 231–42

Jameson, R.A. & Alexander, E.C. 1991. Hydrology and chemistry of condensation waters in Snedgar's and Greenville salt peter caves, West Virginia. *Journal of Cave and Karst Studies*, 53: 60

Pudilo-Bosch, A., Martínez-Rosales, W., López-Chicano, M., Rodríguez-Navarro, C.M. & Vallejos, A. 1997. Human impact in a tourist karstic cave (Arcanea, Spain). *Environmental Geology*, 31: 142–49

Villar, E., Fernandez, P.L., Gutierrez, I., Quindos, L.S. & Soto, J. 1986. Influence of visitors on carbon dioxide concentrations in Altamira Cave. *Cave Science*, 13: 21–23

CONE KARST

Cone karst (Kegelkarst) landscapes represent some of the most spectacular and complex terrain developed on carbonate rocks, particularly in the humid tropics. The generally conical carbonate hills (cones) may be isolated from each other visually or share lower ground surfaces such as pedestals or ridge remnants. In some cone karst areas discrete cones protrude from a near-planar surrounding corrosion surface, which may be alluviated and traversed by allogenic rivers that flow from outside the karst area. In other areas the cones have overlapping flanks and/or are separated by more-or-less enclosed depressions (cockpits), such that the cones form deeply serrated ridges surrounding depressions in a polygonal arrangement that has been described as "egg box" topography (Ford & Williams, 1989).

Some of the best-known cone karst landscapes are in southern China, particularly around Guilin in Guangxi and in Guizhou (see Photo). Other notable examples occur throughout Southeast Asia, particularly in the Gunung Sewu (Thousand Hills) of Java (see Sewu entry), elsewhere throughout Indonesia, and in Malaysia, Papua New Guinea, Burma, Vietnam, Thailand, Cambodia, Laos, and the Philippines. There are also extensive cone karst areas in tropical areas of southern Mexico, in Belize and Guatemala, in Central America, and throughout the Caribbean Greater Antilles, in Cuba, Jamaica, Puerto Rico, and Hispaniola.

The cones themselves may dominate the landscape or may be subsidiary to pronounced enclosed depressions or to an extensive planar surface. Individual cones usually arise from a continuous carbonate surface, although this may be obscured by alluvium or by other sediments. In some instances cones are seated on non-carbonate rocks, for example in the Rio Frio area of western Belize, where they rest upon a granite base, but this is uncommon. The morphology of cones may be extremely variable, ranging from steep-sided peaks to subdued hemispherical hills or cupolas. Some cones attain a height of over 100 m and a diameter up to 500 m, yet most others are measured in the tens of metres. Cones, or conical hills, are of variable morphology and they have been classified on the basis of their height/width ratio by Balazs (1973) and others. This approach was extended by Day (1981) to incorporate cone shapes into an overall classification of tropical karst landscape styles. This approach produced a simple tripartite framework. In type I landscapes enclosed depressions are dominant and interspersed with subdued hills; in type II landscapes enclosed depressions and residual hills attain approximately equal prominence; and in type III landscapes isolated residual hills dominate intervening near-planar surfaces.

Quantitative approaches to cone karst landscapes are relatively straightforward where individual cones are isolated, but far more problematic where their bases are common or intersecting (Day, 1981). There is no definitive distinction between cone karst and tower karst, since both contain assemblages of residual hills of variable morphology. Karst towers are usually discrete, steep-sided individual hills set in a corrosional plain and representing one end of the spectrum; individual, closely spaced karst cones are intermediate; and coalescing conical hills separated by enclosed depressions and valleys represent the other extreme. It is this last assemblage that represents the classical Kegelkarst, as

Cone Karst: Conical hills in the Zhenning karst of Guizhou, China. (Photo by John Gunn)

described by Lehmann and later termed cockpit or polygonal karst. The Cockpit Country of Jamaica provides a typical example, in which conical hills dominate the skyline but the intervening depressions represent the active foci of dissolution.

Establishing a meaningful, objective classification of cone karst landscapes has proven problematic, although it is clear there are, in fact, underlying morphologic patterns. The seminal morphometric studies of Williams (1971, 1972) were the first to identify organization within cone karst landscapes, in particular serving to illustrate their essential polygonal morphology and spatial organization, and the essential similarity of their hydrological function to that of other types of fluvial drainage basin. Field measurement of cone karst is arduous and time-consuming, but maps, air photographs, and satellite imagery generally do not provide sufficiently detailed information at a suitable scale to be acceptable substitutes. Comparative data on the density and spacing of residual hills is relatively easily assembled, but meaningful measurements of height and three-dimensional shape, for example, are far more problematic. Where field data are available, usually for relatively restricted areas, a variety of approaches to measurement and classification have been employed. For example, in the context of general or overall landscape morphometry, surface roughness has been utilized as a discriminator of cone karst landscape styles, and wavelength variance within cone karst has been analysed by use of Fourier models. Development of more rigorous GIS models may be expected to contribute to further mathematical analysis of cone karst landscapes.

Various names have been ascribed to the cones themselves and to the overall landscape. In China, groups of residual hills sharing a common exposed bedrock base are known as fengcong (peak cluster) and isolated residual towers on a plain are termed fenglin (peak forest); intermediate categories include peak-cluster-depression, where the cones enclose discrete depressions, and fengcong-valley karst, where the cones are separated by valley networks (Ford & Williams, 1989; Smart et al., 1986). Elsewhere throughout the world residual carbonate hills, whether conical or not, are known as mogotes, Kuppen, pitons, pepinos, or hums.

Although some cones approach true conical symmetry, most are actually asymmetrical in at least one directional plane, reflecting structural or other geological influences and a degree of intrinsic heterogeneity. Asymmetry of the mogotes of northern Puerto Rico was considered by Monroe (1966, 1968) and others to be the result of directional preference in surface induration or case-hardening related to predominating trade winds, but was shown by Day (1978) to be directionally inconsistent and related rather to structural control and basal undercutting. In most instances, degree of symmetry is a function of scale and perspective; most approximately conical hills stripped of their vegetative cover prove to have highly asymmetrical slopes. As the cones themselves are erosional remnants, it is the intervening depressions and lowlands that are the focus of current karstic dissolution and fluvial geomorphic activity and, by comparison, the cones themselves are relatively inactive, although they are subject to a range of mechanical slope processes. Cone summits are rarely accordant, although in some areas there is a suggestion of dissected plateau remnants, and it is not unusual for conical hills in cockpit karst areas to be aligned in a pattern suggesting that they are the remnants of interfluvial ridges, although this is less clear in locations where isolated cones are set in a corrosional plain. Lehmann (1936) identified aligned or directed residual-bordered cockpits and depressions, which he termed Gerichteter karst.

Regional drainage in cone karst landscapes is largely autogenic, with individual hillslopes contributing runoff and throughflow to discrete depressions. Although this centripetal and vertical drainage may seem chaotic, and indeed the surface component is highly ephemeral, each depression and the surrounding hills constitute an organized drainage basin, with integrated epikarstic components. Locally there may be allogenic inputs from adjacent non-karst terrain, and many cone karst areas are fringed by springs and pocket valleys which debouche the regional underground drainage. Individual conical hills set in alluvial plains often have their own set of peripheral springs, whose behaviour can be altered radically by removal or modification of the natural vegetation cover.

Cone karst slopes and hilltops are highly irregular, often with a very patchy soil cover that is deepest in fractured pockets. Slopes may consist of combinations of small vertical cliffs, inclined bedrock surfaces, "staircases", or talus accumulations. By comparison, the intervening depressions and lowlands may have deep accumulations of colluvial sediment. Cone karst is developed in a wide range of carbonate rocks that are able to support slopes of variable steepness and continuity. Softer, more friable carbonates often have a surface induration, often referred to as case-hardening and in some cases exceeding a metre in thickness (Monroe, 1966). Bedding, jointing, faulting, and lithological variation also influence cone morphology.

Cave systems associated with cone karst include vertical shafts and small centripetal systems that evacuate drainage from individual cones and depressions, deeper regional conduits, and largely abandoned systems riddling the cones at higher elevations. Some summits contain large vertical shafts, which, given the restricted present surface catchment areas, are presumably relict features inherited from prior phases of the regional surface and underground drainage system. Paleomagnetic studies of sediments from caves in karst towers are potentially valuable in deciphering the chronology of cone karst landscape evolution (Williams, 1987), suggesting that they are time-transgressive landforms, with their summits being older than their bases.

Spectacular cone karst landscapes attracted the attention of early Chinese travellers such as Xu Xiake, and Grund (1914) adapted the landscape as the "mature" stage in his karst evolutionary cycle. It was not until the mid-20th century, however, that the first widely available and systematic geomorphological studies of cone karst were undertaken in Asia and the Caribbean by Herbert Lehmann (1936, 1954) and others. Some of the most detailed observations of cone karst were made in the 1960s by Conrad Aub (1969a, b), who drew particular attention to the intrinsic variability of the slopes and to the fact that rainwater throughfall was greater in the enclosed depressions than on the surrounding hill summits.

Several models of cone karst formation and evolution have been proposed. Although it is now generally accepted that preferential near-surface dissolution is the primary mechanism involved, with surface collapse playing only a minor role, the influence of regional groundwater levels and occasional surface inundation by floodwaters or upwelling groundwater is still not fully understood. Of particular concern is whether isolated karst cones represent a sequential evolution from previously linked conical hills, or whether the two develop independently under differing geological and environmental conditions; while both are possible, most evidence supports the latter hypothesis. In Jamaica, for example, it is clear that so-called "degraded" cone karst has developed independently where insoluble sediments are most voluminous, and not by modification of pre-existing "normal" cone karst. Another contentious issue is the extent to which surface drainage systems, developed prior to extensive karstification, have influenced cone karst morphology. In some cone karst areas, such as Puerto Rico, Belize, and Guatemala, there is a clear imprint of regional paleodrainage patterns, but in other areas this inheritance has been rendered undetectable by subsequent karstification.

Although cone karst is most often associated with humid tropical climates, relict examples occur in locations where equivalent climatic conditions formerly prevailed. Such relict cone karst has been recognized in central Europe. Similar near-conical residual hills have also been identified in temperate karst areas, and non-carbonate analogs exist in the inselbergs, bornhardts, and other residual hills of non-karst landscapes.

Cone karst landscapes have been subjected to a wide range of human land uses, many of which have proven to be unsustainable. Because of their ruggedness and remoteness, some cone karst areas have remained relatively unscathed by human influences, and represent significant biological reservoirs. Many cone karst landscapes contain significant archaeological remains, and they are also important in terms of agriculture, forestry, and tourism. Portions of many significant cone karst areas have been demarcated as protected areas, although this has infrequently involved explicit recognition of their karstic nature. Elsewhere, human encroachment has resulted in geomorphological and biological degradation, particularly as a result of uncontrolled forest clearance, unwise agricultural practices, urbanization, and industrial activities, including quarrying.

MICK DAY

See also **China (Figure 4); Cockpit Country Cone Karst; Morphometry; Sewu Cone Karst; Tower Karst**

Works Cited

Aub, C.F.T. 1969a. The nature of cockpits and other depressions in the karst of Jamaica. *Proceedings of the 5th International Congress of Speleology*, Paper M15, edited by Verband der Deutscher Höhlen und Karstforscher e.V.

Aub, C.F.T. 1969b. Some observations on the karst morphology of Jamaica. *Proceedings of the 5th International Congress of Speleology*, Paper M16, edited by Verband der Deutscher Höhlen und Karstforscher e.V.

Balazs, D. 1973. Relief types of tropical karst areas. In *IGU Symposium on Karst Morphogenesis*, edited by L. Jakucs, Szeged: Attila Jozsef University: 16–32

Day, M.J. 1978. Morphology and distribution of residual limestone hills (mogotes) in the karst of northern Puerto Rico. *Bulletin of the Geological Society of America*, 89: 426–32

Day, M.J. 1981. Towards numerical categorization of tropical karst terrains. *Proceedings of the 8th International Congress of Speleology*, edited by B.F. Beck, Huntsville, Alabama: National Speleological Society, vol. 1: 330–32

Ford, D.C. & Williams, P.W. 1989. *Karst Geomorphology and Hydrology*, London and Boston: Unwin Hyman

Grund, A. 1914. Der geographische Zyklus im Karst. *Ges. Erdkunde*, 52: 621–40 [The geographical cycle in the karst]. Translated and reprinted in *Karst Geomorphology*, edited by M.M. Sweeting, Stroudsburg, Pennsylvania: Hutchinson Ross, 1981

Lehmann, H. 1936. *Morphologische Studien auf Java*, Stuttgart: Engelhorns

Lehmann, H. 1954. Der tropische Kegelkarst auf den grossen Antillen. *Erdkunde*, 8: 130–39

Monroe, W.H. 1966. Formation of tropical karst topography by limestone solution and reprecipitation. *Caribbean Journal of Science*, 6: 1–7

Monroe, W.H. 1968. The karst features of northern Puerto Rico. *Bulletin of the National Speleological Society*, 30: 75–86

Smart, P.L., Waltham, A.C., Yang, M. & Zhang, Y. 1986. Karst geomorphology of western Guizhou, China. *Transactions of the British Cave Research Association*, 13(3): 89–103

Williams, P.W. 1971. Illustrating morphometric analysis of karst with examples from New Guinea. *Zeitschrift für Geomorphologie N.F.*, 15: 40–61

Williams, P.W. 1972. The analysis of spatial characteristics of karst terrains. In *Spatial Analysis in Geomorphology*, edited by R.J. Chorley, London: Methuen

Williams, P.W. 1987. Geomorphic inheritance and the development of tower karst. *Earth Surface Processes and Landforms*, 12(5): 453–65

Further Reading

Jennings, J.N. 1985. *Karst Geomorphology*, 2nd edition, Oxford and New York: Blackwell: 120–22 and 201–14

Miller, T.E. 1987. Fluvial and collapse influences on cockpit karst of Belize and eastern Guatemala. In *Karst Hydrology: Engineering and Environmental Applications*, edited by B.F. Beck & W.L. Wilson. Rotterdam: Balkema: 53–58

Sweeting, M.M. 1972. *Karst Landforms*, London: Macmillan and New York: Columbia University Press

CONSERVATION: CAVE BIOTA

Many cave biota have very restricted ranges, and limited opportunities for dispersal. Consequently they are strongly affected by even moderate environmental variability or pollution. The main threats come from a destruction of the habitats by quarrying; increased visitor numbers and vandalism; pollution by heavy metals, herbicides, or pesticides; excess nutrient input; or invasion by an epigean species. The majority of subterranean fauna is not considered in existing national or international protection laws. Only five world conventions are applicable to cave biota and habitats: the Ramsar Convention (1971) in which 12 karst and subterranean hydrological systems are included in 1150 designated wetland areas (see Ramsar Sites); the UNESCO World Heritage List (1972) which includes 13 caves (just 1.8% of the known world caves; see World Heritage Sites); the Washington

Convention on International Trade in Endangered Species of Wild Fauna and Flora (1973) where some subterranean species are listed (bats, guácharo or *Steatornis caripensis, Proteus anguinus*); the Rio Convention (1992) for the protection of biodiversity—although this does not include a detailed list of threatened species; and the International Union for Conservation of Nature and Natural Resources (IUCN) which published a "Red List" of 95 threatened cave biota (Baillie & Groombridge, 1996). There is no specific conservation list for cave flora. Despite these conventions, protection of species from subterranean habitats remains extremely difficult and requires measures to reduce or mitigate the negative impacts of human activities. The main conservation procedures for cave biota are: limitation of visitors in sensitive caves; closure of the most valuable caves; reduction of vandalism; ending pollution from external sources and removal of internal pollution left by cave explorers; rehabilitation of degraded karst ecosystems; and public education and information.

Limestone quarrying has a direct impact on landforms, as well as impacting on karst processes, groundwater systems, and the epikarstic biospace of cave fauna. In process terms the principal impact is the removal of stone with the potential for the complete destruction of caves in some areas, but it may also include depositional components such as the creation of spoil heaps. Gunn and Gagen (1989) have estimated that during the 20th century quarrying has been responsible for the removal of over 900 million tonnes of limestone from the Peak District (England), about the same as has been removed by natural processes over the whole of the Holocene. To reduce these major impacts on subterranean sites as well as on ecosystems, it is vitally important to set up a databank of the most interesting karst areas (from the geological, biological, and archaeological perspectives), and to limit industrial activity to the less important areas. Moreover, it is essential to oblige quarry companies to restore affected landscapes so as to redress damage to the environment and to the countryside.

All visitors to caves—occasional cavers, tourists, and scientists—have an impact on the cave itself and on cave biota. The main effects are compaction of the sediments, introduction of energy sources (mud, food residues, faeces, and lint), and the destruction of biotopes. These impacts are enhanced by increased visitor numbers and intentional vandalism or souvenir gathering of geological features (mainly speleothems) and cave biota (mainly bats, but also several terrestrial or aquatic invertebrates). A conservation policy must include controls on the number of visits and visitors by gating of cave entrances; the creation of protected areas that encompass the catchments of the most interesting caves with troglobitic taxa; the repression of vandalism; and the education of cavers and visitors. It is necessary to explain the importance of cave biota for scientific studies (evolution, speciation, population genetics, adaptation) using periodicals, books, the internet, and interpretation tours within selected caves. Degraded ecosystems could be restored by depollution campaigns including the removal of waste material (such as spent carbide) and speleothem repair (see Restoration of Caves and Speleothem Repair).

It is well known that industrial activities such as iron and steel plants produce dust containing heavy metals such as cadmium, chromium, copper, nickel, lead, and zinc. Karstic sites do not escape the problem of dust deposition and pollution of groundwater by heavy metals. For example, heavy metal pollution has been found in the Sloans Valley Cave System, the third longest cave system in Kentucky (United States), endangering the fragile ecosystem which contains a rare troglobitic species, the blind fish *Typhlichthys subterraneus* (see Pisces: Amblyopsidae).

In addition to industry, agricultural activities and residential and commercial developments have been shown to impact on groundwater quality in limestone and karst terrains and hence on cave biota. Of particular importance in intensively farmed karstlands is disruption of the ecological balance by the introduction of agrichemicals such as herbicides into groundwater. The most commonly used herbicides are Atrazine, Alachlor, Metolachlor, Glyphosate, 2-D, Imazaquin, Imazerahir, and Pendimethalin. Positive correlations between agricultural land and herbicide concentrations in springs and groundwater have been found in the United States and England. Pesticides such as Aldicarb, Carbaryl, Carbofuran, Dianizon, and Parathion, and toxic chemicals such as petroleum also impact subterranean ecosystems. It has been estimated that in the Unites States 3.5–21 million pounds weight of pesticide reach ground or surface water before degradation each year, and a significant amount of this is in karstlands. These products are carried by water through the soil into groundwater through the processes of infiltration, leaching, and percolation. All of these substances persist in the environment, being slowly, if at all, degraded by natural processes; in addition, all are toxic to life if they accumulate in any quantity. Residential and commercial development may increase the amount of the chemicals released into soil and water. In November 1981, 10 000 Salem cave crayfish (*Cambarus hubrichti*) and 1000 Southern cavefish (*Typhlichthys subterraneus*) were stressed or killed in Meramec Spring, Missouri, because of a fertilizer pipeline failure that released 80 000 litres of liquid ammonium nitrate.

The problem of an excess nutrient input into groundwater and cave streams is worldwide. For instance, the US Environmental Protection Agency (1998) reported that agricultural activity (primarily the application of fertilizers) is the leading source of pollution threatening the water quality of rivers and lakes in the United States. However, agriculture is not the only human activity affecting groundwater: an estimated 3 billion cubic metres of sewage (mainly human waste) and wastewater are discharged to surface waters every year in the United States, increasing the nutrient input in groundwater. Eutrophication is commonly defined as a process that increases nutrients, especially nitrogen and phosphorous forms in aquatic ecosystems. This frequently leads to an increase in algae populations and a decrease in biodiversity in surface waters. Nutrient pollutants alter the oligotrophic nature of groundwater ecosystems and severely alter groundwater food webs. The introduction of organic pollution can extirpate the indigenous fauna or completely replace the community with epigean fauna. Excess organic loadings create a biological oxygen demand that can quickly deprive fauna of dissolved oxygen. To fight excessive organic loading, legislation to control wastewater must be strengthened and waste treatment plants developed, especially in karst areas.

Infestation of epigean species in the subterranean environment, and their potential competition with the cave biota, fortunately seems rare. Indeed, the harsh ecological conditions of caves restrict the invasion of new species. Most of the cases are linked directly or indirectly with human activities. The fauna introduced by human activities is especially noticeable on islands, and sometimes this highlights important problems for native communities. For example, several species have invaded lava tubes in the Canary Islands. In show caves, visitors introduce lint, skin flakes, soil particles, food scraps, microflora, and fungi. These provide a rich source of food which benefits some cavernicolous species, but may also permit colonization by pre-adapted epigean species. The most important environmental alteration permitting new species is the use of artificial light, which allows the growth of several species of algae, mosses, and ferns that do not normally occur in the cave system (see Tourist Caves: Algae and Lampenflora). A very interesting case of a recent invasion of an exotic species occurred in the United States. The Red Imported Fire Ant (*Solenopsis invicta*) (Hymenoptera: Formicidae) was inadvertently introduced from Brazil in the 1930s and has gradually spread to cover 300 million acres in 12 southern states. These ants use their stings for not only defence but also for prey capture and pheromone dispersal. Elliott (1992) reported that *S. invicta* began invading the caves of central Texas in 1988. The colonies live in soil mounds, but the workers invade caves during summer and forage intensively for moisture and food in the caves. They can kill the Texas cave crickets (*Haedenoecus subterraneus*) and carry off other cave dwellers. They also attack the cave crickets outside the caves during their nocturnal foraging period. There is no efficient method to specifically protect the subterranean system from this invasion and the only technique is to progressively eradicate the external colonies.

There are many case studies from karst areas throughout the world that show how limited legal, contractual policies, or spontaneous initiatives are in protecting the subterranean environment, including cave biota, which is increasingly deteriorating as a consequence. An important tool to assist conservationists is the development of data banks showing the distribution of troglobites and particularly vulnerable cave biotypes. These should receive the highest level of protection both in terms of habitat preservation (prevention of mineral extraction and controls on access) and pollution control.

RAYMOND TERCAFS

See also **Environmental Impacts Assessment; Floral Resources; Tourist Caves**

Works Cited

Baillie, J. & Groombridge, B. (editors) 1996. *1996 IUCN Red List of Threatened Animals*, Gland, Switzerland: IUCN and Washington, DC: Conservation International; see also http://www.redlist.org/

Elliott, W.R. 1992. Fire ants invade Texas caves. *American Caves*, 5: 13

Gunn, J. & Gagen, P.J. 1989. Limestone quarrying as an agency of landform change. In *Resource Management in Limestone Landscapes: International Perspectives*, edited by D.S. Gillieson & D.I. Smith, Canberra: Department of Geography and Oceanography, University College, Australian Defence Force Academy (Special Publication 2)

US Environmental Protection Agency. 1998. *1996 National Water Quality Inventory*, Section 305(b) Report to Congress, Washington DC: US Environmental Protection Agency, Office of Water (also available at http://www.epa.gov/305b/)

Further Reading

Al-Assiuty, A.I.M. & Khalil, M.A. 1996. Effects of the herbicide atrazine on *Entomobrya musatica* (Collembola) in field and laboratory experiments. *Applied Soil Ecology*, 4: 139–46

Dickson, G., Briese, L., & Giesy J. Jr. 1979. Tissue metal concentrations in two crayfish species cohabiting a Tennessee cave stream. *Oecologia*, 44: 8–12

Eddy, F. & Williams, E. 1987. Nitrite and freshwater fish. *Chemical Ecology*, 3(1): 1–38

Gillieson, D. 1996. *Caves: Processes, Development and Management*, Oxford and Cambridge, Massachusetts: Blackwells

Hamilton-Smith, E. & Eberhard, S. 2000. Conservation of cave communities in Australia. In *Subterranean Ecosystems*, edited by H. Wilkens, D.C. Culver & W.F. Humphreys, Amsterdam and New York: Elsevier

Poulson, T.L. 1990. Developing a protocol for assessing groundwater quality using biotic indices in the Mammoth Cave region. In *Proceedings of the First Mammoth Cave National Park Karst Conference*, Mammoth Cave, Kentucky: National Park Service

Sasowsky, I.D., Fong, D.W. & White, E.L. (editors) 1997. *Conservation and Protection of the Biota of Karst*, Charlestown, West Virginia: Karst Water Institute (Special Publication 3)

Tercafs, R. 2001. *The Protection of the Subterranean Environment: Conservation Principles and Management Tools*, Luxembourg: PSP

Tronvig, K.A. & Belson, C.S. Top Ten List of Endangered Karst Ecosystems, http://www.karstwaters.org/TopTen3/topten3.htm

CONSERVATION: PROTECTED AREAS

There have always been protected areas in the natural landscape, some of which have probably been in existence since the Stone Age and are now often perceived as sacred sites. The popular idea that the concept of protected areas commenced with Yellowstone (United States) is a fallacy—probably the only genuine innovation at Yellowstone in 1872 was the use of the term "National Park". Nevertheless, the area has served as a flagship for modern nature protection.

The latter part of the 20th century saw an immense proliferation of protected areas, and almost as much proliferation of formal titles for such areas. Recognizing the confusion that has arisen from the complex patterns of nomenclature, the IUCN (International Union for Nature Conservation) established a world classification in an endeavour to provide an effective basis for comparability. There is only space here for a brief summary of this classification (see Table), and its proper application demands reference to the published manual (IUCN, 1999).

An obvious difference in national usage arises in relation to the use of the term "national park" in the countries of Europe. For instance, most of the UK National Parks are probably best placed in the IUCN Categories of Protected Landscapes (V) or Managed Resource Protected Areas (VI). This does not imply

Conservation: Protected Areas: IUCN Categories and Management Objectives of Protected Areas.

Category Ia: Strict Nature Reserve
Protected area managed mainly for science
Category Ib: Wilderness Area
Protected area managed mainly for wilderness protection
Category II: National Park
Protected area managed mainly for ecosystem protection and recreation
Category III: Natural Monument
Protected area managed mainly for conservation of specific natural features
Category IV: Habitat / Species Management Area
Protected area managed mainly for conservation through management intervention
Category V: Protected Landscape / Seascape
Protected area managed mainly for landscape / seascape conservation and recreation
Category VI: Managed Resource Protected Area
Protected area managed mainly for the sustainable use of natural ecosystems

---

they are of any less importance, but simply that they are based upon a totally different set of objectives and criteria to those, say, in the United States.

A further dimension which does not provide any further legal protection, but which strengthens the political dimension of protection, arises from international recognition of various national protected areas. The first of these was the Man and the Biosphere (MAB) programme launched in 1968 and which led to the declaration of Biosphere Reserves. These have a threefold role. Each reserve usually includes a protected area that meets conservation objectives, but the reserve as a whole also provides for development activities on a culturally and ecologically sustainable basis. Finally, it is expected to provide a basis for continuing research, education, and information exchange on the issues around the conservation-development nexus.

The Ramsar Convention (first agreed in 1971; see Ramsar Sites) provides for the international recognition of wetlands and fosters their conservation and sustainable use. Although some surface wetlands in karst were recognized, it was not until 1996 that the concept of subterranean wetlands was accepted, and now provides new opportunities for the recognition of karst sites. The World Heritage convention followed in 1972, and focussed upon furthering the protection and conservation of both the natural and cultural heritage of the world. World Heritage Areas are discussed in a separate entry and will not be given further attention here other than to emphasize that the convention does not, in itself, confer legal protection—that remains the responsibility of host governments.

There are some specific problems in the declaration of protected areas in karst regions:

1. The focus of popular attention upon caves led to many of the early protected areas in karst simply containing a single cave—or perhaps only the entrance area of a cave. These reserves fall within the National Monument category, but are usually totally inadequate in karst protection. Much of the subterranean karst may well fall outside of the karst system of which the cave is only a part, and the immense importance of the watershed is totally neglected.
2. Similarly, protected area boundaries may well be determined on the basis of surface divides, as the subterranean divide is often unknown or unrecognized, and so the system remains vulnerable to impacts generated outside of the protected area boundary.
3. Much of the community, and hence political, perception of the values to be maintained in protected areas is based upon the visible and aesthetically attractive higher animals and plants. This position is strongly reflected in the staffing of protected area management agencies—most employ biologists, but it is rare to find geologists or other earth scientists. So, protected area decision making focuses upon fur, feathers, and leaves! Only the most monumental scenery has been readily acceptable as a rationale for the protection of geological characteristics and values.
4. Even when karst is protected, the general public and most managers over-emphasize the importance of cave speleothems; the often virtually invisible ground water and cave biota are neglected. An even bigger problem is that the floors of caves, which may well be the most scientifically important element of a cave system, are frequently neglected, trampled, or dug away to improve access.

Virtually all protected areas face new threats of increasing visitor numbers without an equivalent increase in the resources to provide for adequate environmental protection. This is compounded by decreasing budgets and the demand that parks should earn their own budget through visitor charges and special events, which in turn distorts the central objectives of the park. Karst parks are forced to give priority to the development of tourist caves and the ways in which income derived from commercialization of these may help solve the fiscal problems.

There are also new demands being made of protected areas, notably by forestry, mining, grazing, and other exploitive industries that are increasingly seeking access to the resources "locked away" in parks. All too often this is proposed under the ambit of so-called "wise use" without any real recognition that this is contrary to the very basic objectives of protected areas. At the same time, a new interest in Integrated Conservation and Development Projects is emerging, often with real leadership or partnership being offered by the more responsible industrial leaders.

Another social dimension of protected areas is the rapidly growing call for collaborative management, particularly in joining with indigenous or oppressed people. Protected area managers have come to realize the extent to which, in the past, they may have neglected the rights of these people in trying to ape the "Yellowstone model" and in so doing lost an immense treasure of inherent knowledge and understanding of the natural environment and its meanings.

ELERY HAMILTON-SMITH

Further Reading

IUCN Commission on National Parks and Protected Areas 1999. *Guidelines for Protected Area Management Categories*, Gland,

Switzerland: IUCN (produced by a group of European organizations and published for IUCN)

Watson, J. *et al.* 1997. *Guidelines for Cave and Karst Protection*, Cambridge: IUCN (for the World Commission on Protected Areas Working Group on Cave and Karst Protection)

Worboys, G., Lockwood, M. & De Lacy, T. 2001. *Protected Area Management: Principles and Practice*, Melbourne and New York: Oxford University Press

Parks: A journal published by IUCN. See in particular issues
9(2): Protected areas program
10(2): Non-material values of protected areas
11(1): Biosphere reserves
11(2): ICDPS: Working with parks and people

See also the Best Practice Protected Area Guidelines Series of Handbooks. These and other publications are available through the IUCN Publications Service at http://www.iucn.org/bookstore/

CONSTRUCTION ON KARST

The dissolution features of karst provide problems for all forms of construction activity. Built structures demand special treatment on karst, in many respects similar to the approaches used for highways. Dams and reservoirs encounter a different set of problems related to potentially major leakage, while caves create special hazards in the excavation of tunnels (see Dams and Reservoirs on Karst; Highways on Karst; and Tunnelling and Underground Reservoir Construction on Karst). For surface structures, karst provides hazards and problems that fall into three groups, related to rockhead, to caves, and to dolines (sinkholes).

Rockhead creates problems due to its irregularities in most karst. In immature karst terrains, installation of piles may require longer elements for some parts of a site, reinforced ground beams can be designed to span any small new dolines that may develop, or raft foundations may be incorporated as a precaution. In karst that is more mature, rafts or groundbeams may still be appropriate for bridging cavities (see Figure 1). In Florida, either rafts or preparatory grouting are preferred where new dolines are recorded locally at rates above $0.05 \text{ km}^{-2} \text{a}^{-1}$. Heavy geogrid (a planar mesh manufactured from polymeric materials) stabilizes a soil profile, and can be designed to span a potential void in order to reduce the impact of any subsequent catastrophic collapse. Grouting of soils over a highly fissured rockhead, before installation of spread footings within the soil profile, may require works on a large scale. A site of $10\,000 \text{ m}^2$ on karst in Pennsylvania took 1200 m^3 of compaction grout through 560 boreholes that averaged 9 m deep; this was still more economical than piling to rockhead that was mostly 6–12 m deep.

Pinnacled rockheads in tropical karst terrains generally require that structures are founded on sound limestone, either by piling to rockhead at whatever depth it is encountered or spanning foundations between sound pinnacle tops. Driven piles may be bent, deflected, or poorly founded on unsound pinnacles (Sowers, 1986). Bored piles are preferred, though they can be difficult to place on steeply inclined rockhead. The founding of each pile tip is probed to ensure integrity; probing depths for cavity detection are outlined below. Undersized or unstable pinnacles may require assessment by probes splayed at 15° from the vertical from the founding point. A light structure can bear safely on the soil profile over a deeply pinnacled rockhead, after due care with respect to drainage modification and precautionary placement of geogrid where appropriate. Rockhead pinnacles up to 50 m high, which occur in some tropical karsts, offer some of the worst ground conditions possible for heavy structures that have to be founded on bedrock. Every pile location has to be treated as an individual ground investigation, and the contractor has to adapt to unique ground conditions as they are revealed by excavation.

Caves provide difficult foundation conditions largely because it is impossible to predict both the number and the size of caves beneath any given karstic site. Each site has to be assessed individually within the context of its geomorphology, and engineer-

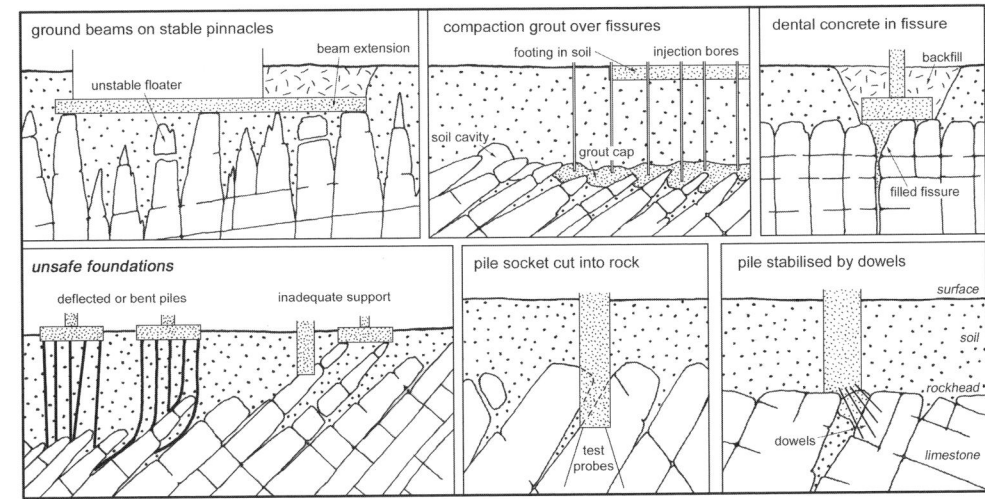

Construction on Karst: Figure 1. Various ways of achieving safe or unsafe foundations for buildings on soil-covered karst (largely after Sowers, 1986).

ing works must respond to the local conditions. Local records and observations may indicate the typical and maximum cave sizes previously encountered. The cave size determines the engineering philosophy whereby a defined minimum of sound rock should be proven by drilling beneath every foundation pad and pile tip. For typical karst in strong limestone, a cover thickness of intact rock that is 70% of the cave width ensures integrity. This guideline covers limestone with a normal density of fractures and bedding planes; local zones of heavy fissuring may reduce cave roof integrity, and thicker cover may be required in some softer limestones.

Large caves at shallow depths provide the major hazard. Statistically they are rare, but their absence may have to be proved by extensive probe drilling. Caves with open entrances are best assessed by direct exploration. Small shallow cavities may be collapsed by a monitored surcharge or by dynamic compaction, especially in chalks and weak limestones. Where caves are found at critical locations beneath planned foundations, the normal remedy is to fill them with mass concrete. Grout injection through boreholes may incur considerable losses by flowage into karstic cavities that extend far off site, and perimeter grout curtains may therefore reduce total costs. Alternatively, creating access to a cave may allow installation of shuttering and removal of any weak floor sediment before filling. Relocation of footings is usually an expensive option, but can prove essential over complex caves (Waltham, Vandenven & Ek, 1986). Bored piles can be placed through a cave to sound footing in the rock below; geotextile sleeves can be used to cast the concrete through the cave, but total cave filling is often preferred for its simplicity, at costs that may be little different.

Dolines are a significant hazard to structures founded on the soils that overlie karstic limestone. They are commonly known as sinkholes by engineers, especially when their floors are actively subsiding or collapsing, and the key to the minimization of sinkhole failures in any karst is the proper control of groundwater flows. Design codicils appropriate for any karst site include a ban on soakaway drains (dry wells) in or adjacent to the site, use of flexible lines and junctions on all drainage elements, diversion of inbound surface flows, and avoidance of reducing soil thickness over shallow rockhead. Dry wells can be employed where they are sealed onto open fissures and cased below rockhead, and retention basins in dolines require lining if needed by flood control regulations.

Control of water abstraction is also critical, especially where the original water table is close above rockhead. Florida's Disney World stands on 20–30 m of soils over limestone, and all its wells are monitored so that pumping can be switched when a local water table decline is detected; sinkholes have not occurred on the site. Dewatering by quarrying incurs legal responsibilities, and has been stopped at some sites by legal action to prevent further ground failures.

Grout sealing of rockhead fissures, prior to foundation works, is appropriate on less mature karst, but pinnacled rockheads require more elaborate cap grouting with cement slurries after plugging known open fissures with more viscous grouts (Kannan & Nettles, 1999). Compaction grouting (with a viscous cement mix) forms a solid block within the soil that can bridge over fissures, densifies the adjacent soil, and can lift subsided structures. It has been used to remediate sinkholes in soils over pinnacled rockheads, but its use requires caution because the

Construction on Karst: Figure 2. The result of indiscriminate house building on thick soil cover over mature karst is the occasional loss of structures to the random development of destructive doline collapses. The garages of this house in Ripon, England, were lost when the ground collapsed above cavernous gypsum. (Photo by Tony Waltham)

grout flow cannot be controlled and its placement may not remedy the initial cause of a failure. Where rockhead cannot be seen or treated, sinkhole activity can be reduced by laying geogrid into the soil; this is additional to proper drainage control.

A covered sinkhole that collapses or subsides beneath an existing structure requires individual remediation. Such features may develop in any karst, though their width and depth are limited by the soil thickness. They may involve only modest soil subsidence, with no development to rockhead, but structural underpinning with friction piles generally fails again within a few years unless the underlying fissure is sealed at the same time. A permanent sinkhole repair requires exposure of rockhead and choking of the causative fissure or cave throat with chunk rock, covered with graded fill, with or without a concrete slab or geogrid mat. Whether such action is preventative before site development or remedial after a failure, depends largely on how well the ground has been investigated and how well the problems of karst are understood.

TONY WALTHAM

See also **Geophysical Detection of Caves and Karstic Voids**

Works Cited

Kannan, R.C. & Nettles, N.S. 1999. Remedial measures for residential structures damaged by sinkhole activity. *Hydrology and Engineering Geology of Sinkholes and Karst: Proceedings of the Seventh Multidisciplinary Conference on Sinkholes and Karst*, edited by B.F. Beck, A.J. Petit & J.G. Herring, Rotterdam: Balkema: 135–39

Sowers, G.F. 1986. Correction and protection in limestone terrane. *Environmental Geology and Water Science*, 8: 77–82

Waltham, A.C., Vandenven, G. & Ek, C.M. 1986. Site investigations on cavernous limestone for the Remouchamps Viaduct, Belgium. *Ground Engineering*, 19(8): 16–18

Further Reading

Beck, B.F. & Herring, J.G. (editors) 2001. *Geotechnical and Environmental Applications of Karst Geology and Hydrology: Proceedings of the Eighth Multidisciplinary Conference on Sinkholes and Karst*, Lisse: Balkema

Sowers, G.F. 1996. *Building on Sinkholes: Design and Construction of Foundations on Karst Terrain*, New York: ASCE Press

CREVICE CAVES

Crevice caves are narrow rectilinear crevices of natural origin; unitary or in somewhat rectilinear networks; with little or no dissolutional speleogenesis; large enough for human entry and investigation; and extending beyond daylight. In granite, coarse-grained sandstone ("grit"), basalt, and a few other rocks, some are curved instead of rectilinear. Many are an intermediate stage of breakup of competent rock masses. They occur in many types of competent rocks, including limestone, and also in glaciers where they are known as crevasses. Commonly other pseudokarstic and occasionally some karstic processes interact with crevice caves, producing multiprocess caves. Some spectacular examples in quartzites in tropical rainforests have spawned vigorous debates in recent decades.

Because of the variety of their genetic processes and the variety of terrains in which these caves exist (and also because of limited communication), local terminologies have arisen, including breakaway caves, cambering caves, closed joint caves, crack caves, crevasse caves, crevasses, earth cracks, earthquake cracks, eruptive fissures, fissure caves, gravity-sliding caves, gulls, gull caves, joint caves, mass displacement caves, mass movement caves, open joint caves, rift caves, rifting caves, rock topple caves, sliding fracture caves, slope movement caves, tectonic caves, tilting caves, toppling caves, and windypits. "Windypit" is a local British term for crevice caves formed by "valleyward sliding" of large masses of bedrock (Cooper, Ryder & Solman, 1982). In other parts of England, those formed by lateral spreading are called gulls or gull caves. "Camber" is a related, largely British, engineering term. In this entry, the term "crevice cave" is used to conform to definitions of the 4th edition (1997) of the American Geological Institute's *Glossary of Geology*.

Significance of Crevice Caves

Where karstic caves are readily accessible, all but the largest crevice caves are commonly ignored. Thus they are not nearly as rare as the English-language speleological literature seems to suggest. In Poland, where most of the country is non-karstic, 530 crevice caves are documented in flysch sediments alone and 17 of these are more than 100 m long (Klassek, 2000). In addition to being objects of curiosity and popular sites for strenuous recreation, crevice caves commonly shelter small to large forms of plant and animal life. Several in the United Kingdom have been designated Sites of Special Scientific Interest (Cooper, 1983). Especially in Europe and Asia, some have been locations of notable archaeological discoveries. A few (e.g. Hematite Cave, Oregon, United States) are locations of unusual mineralogical occurrences. Others (e.g. Pit H Cave in the Great Crack of Kilauea Volcano, Hawaii) permit observations of patterns of magma flow deep in volcanic rift systems (Figure 1) and in shallower eruptive fractures. Several are show caves. They are a common cause of engineering problems. Vertical piping of soil into crevices is a valuable warning of incipient catastrophic bedrock movement.

Classifications

Classifications of crevice caves reflect local nomenclature, local processes, local topography, and local lithology. None has yet encompassed the full range from littoral caves to rift zone fractures. Perhaps the simplest classifications consider whether the cave extends into a cliff or slope, or parallels it, or is formed along a bedding plane or other structural feature, or opens vertically (Figure 2). Finlayson (1986) classified those in granite as open joint caves and closed joint caves. Vitek (e.g. 1989) has published limited classifications of pseudokarstic caves in several areas. Overall he has specified "fissure-type" and "crevice-type" caves, "selective weathering" crevice caves on bedding planes (up to 100 m long), mass movement "crevasse" and "block" caves, and "combined type" caves. Cooper (1983) classified mass movement caves in terms of their relation to slope: plateau-top, head-of-slope, slope, cliff, scree (talus) slope, and detached block forms.

Distribution, Forms, and Speleogenetic Mechanisms

A variety of processes, somewhat beyond the scope of this entry, initiate and enlarge crevices, singly, or in combination. Some of these are frost and ice wedging and heaving, tectonic movements, minor dissolution along bedding planes, dykes, fractures and mineral veins, gravitogenic mass movement, block creep or glide on granular, slick or incompetent structures, the complex of littoral processes, and upward or lateral movement of magma.

Locations of crevice caves depend largely on the processes which form them. Excluding glacier crevasses, most crevices occur on or near cliffs or steep slopes, but some are found at the top of batholithic domes, on gentle to moderately sloping lava fields, and in eroded plains and other areas of comparatively low relief. In volcanic regions, circumferential crevices form around many craters and calderas; a few persist as crevice caves.

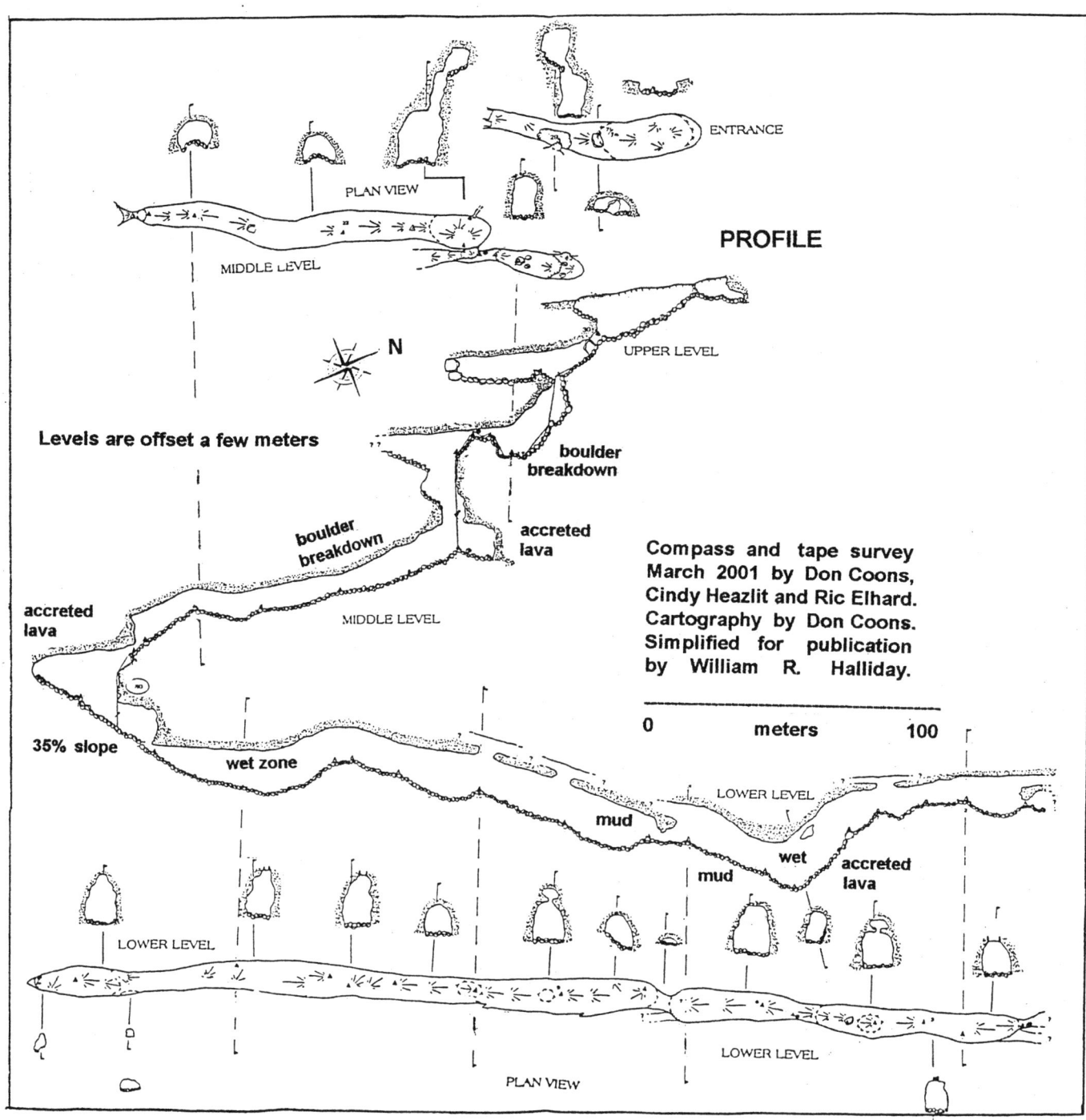

Crevice Caves: Figure 1. Plan and vertical section of Pit H Cave on the Great Crack of Kilauea Volcano, Hawaii. Courtesy Don Coons.

Crevice caves occurring parallel to cliffs tend to be especially impressive. In Montana (United States), Campbell (1968) described crevices and crevice caves over 60 m deep behind cliffs more than 300 m high, at distances up to 800 m from the cliff edge. Elsewhere in the United States, some in New York State and New England are especially well known.

Mass Movement or Gravity-Sliding Caves

Simple crevice caves are narrow roofed passages with flat, near-vertical walls. They occur singly or in small networks. Rooms tend to form at crevice junctions. If offsets are present along a crevice, they are angled sharply and "fit features" are evident unless erosion or dissolution has blunted them.

Crevice Caves: Figure 2. Vertical entrance of a crevice cave in sandstone and conglomerate, on Vancouver Island, Canada. (Photo by William R. Halliday)

These caves may be roofed by part of the original rock surface, by chockstones, by tilted blocks, and / or by soil and vegetation.

On hillsides and in terrains affected by lateral spreading, gravity eventually begins to widen such crevice caves and form new ones. This occurs more rapidly if an incompetent or slippery stratum or vulnerable bedding plane underlies a hard, dense formation. Other joints tend to form nearby, and enlarge into crevices. In some caves, they convert unitary crevices into networks. Block sliding, rotation, forward or backward tilting, and / or basal sapping play roles in determining the predominant form. In complex cases, major crevices commonly are parallel to slopes or cliff faces, with parallel and cross-crevices forming a network or an en-echelon pattern.

Under the influence of gravity, passage widths enlarge downslope, but the network pattern remains intact until blocks begin to rotate or slide independently. Especially on cross-crevices, "tilt features" then demonstrate misalignments, displacements, and / or rotation. At this stage, caves generally are roofed by offsets in fracture lines, slippage on bedding planes, chockstones, and / or soil and vegetation.

As displacement and rotation increasingly supplant tilting and sliding, erosional agents increasingly modify the caves, unroofing them, or converting them into talus caves or merely masses of talus. Sometimes, mapping of what appears to be an ordinary hillside talus cave reveals a persistent rectilinear pattern consistent with a sliding fissure. The rate of change commonly varies in different parts of such caves. The 1.6 km Bat Cave (North Carolina, United States) is a well-known combination of sliding fissure and talus cave.

Mass Movement and Other Crevice Caves in Basalt

Because of its brittleness, basalt tends to fracture into innumerable small cracks rather than large, cave-sized forms. However, some important crevice caves exist in volcanic areas. Curved crevice caves with accreted lava linings are present in the walls of craters on Japan's Miyake Island, and rectilinear examples (e.g. Roadcut Cave) exist hundreds of metres back from caldera cliffs on Hawaii's Kilauea Volcano. Some radial eruptive fissures on Italy's Mt Etna continue downslope as lava tube caves.

Along the Southwest Rift Zone of Kilauea Volcano, the Great Crack ("17 Mile Crack") is a narrow zone of en-echelon crevices, small grabens and crevice caves. Here, 600 m of passages have been mapped to a depth of 183 m in Pit H Cave, along three levels (Figure 2), and exploration of nearby crevice caves has barely begun. The levels of Pit H Cave are largely formed by chockstones; single rope techniques are needed between them. Individual chambers are many metres in width and length (Coons, 2001). At several levels, accreted lava demonstrates unconfined longitudinal magma flow. This is a different pattern from that in nearby Wood Valley Pit Crater, where lava tubes formed within a crevice cave. The Great Rift in Idaho (United States) also contains notable crevice caves. Preliminary study suggests that such crevices are self-propagating as lava is forced laterally into and through them.

Earth Cracks in Northern Arizona (United States)

Some of the partially open "Earth Cracks" in limestone in northern Arizona are associated with faults and grabens. Most show little or no evidence of dissolution ("Bottomless Pits" or Rio Frio Sink near Walnut Canyon National Monument is an exception). With a length of 452 m, Doney Fissure is considered the longest and 0.6 m of vertical displacement is present at the entrance. Displacement is only 5 cm at Sipapu Cavern, which is shorter but 152 m deep. Also in northern Arizona, Paiute Cave is a single, steeply sloping chamber 165 m deep without evidence of dissolution. Recent studies of karstic features in this general area suggest that at least some of these features are secondary to deep-seated evaporite karst. If so, they should be considered interface phenomena rather than purely pseudokarstic.

Karstic and Pseudokarstic Processes in Quartzite Caves

The role of dissolution in speleogenesis, karstification, and pseudokarstification of poorly soluble quartzite, conglomerate, and quartzose sandstones has been debated in recent years. Somewhat similar features in ferruginous metamorphics have also been discussed. All these areas are characterized by surface landforms, which appear more or less karstic, and by especially impressive crevice caves, some of which have been modified by karstic and / or other pseudokarstic processes. Their caves vary considerably in morphology and extent. Some are largely or entirely vertical. The world-famous pits ("shafts") of the Venezuelan rain forest plateaux are as much as 314 m deep and 350 m wide. From some of these, subhorizontal conduit caves extend as much as 2500 m, with cross sections from rectangular to oval (see Quartzite Caves of South America). Enormous volumes of clastic material have been removed by piping and by stream erosion so that the pits and conduits were not choked in their own debris. Laminar transport of colloidal silica presumably occurs as an inception mechanism.

Quartzite caves have been particularly studied in southern Africa, Venezuela, and Brazil. In some of these areas, underground drainage and conduit features correspond poorly to surface karstification. In the lapiazed quartzite area of the Mpumalanga Escarpment of southern Africa, underground drainage underlies less than 1% of the lapiazed surface (see Silicate Karst). Thus it is correct morphologically to refer to the Mpumalanga karst as a small quartzite karst within the large Mpumalanga piped crevice pseudokarst. Even there, however, the volume of rock removed by piping must have been much greater than that removed by intergrain dissolution. From the standpoint of process, therefore, it is correct to refer to the entire Mpumalanga lapiazed area as pseudokarstic.

This seeming conflict may be resolved by considering the caves as multiprocess features in an interface between karst and pseudokarst. A continuum exists between caves formed by piping and by dissolution. Even in very pure limestones, cave streams in karstic conduits carry small loads of insoluble residue freed from bedrock in the dissolution process. Other cavernous limestones contain horizons of sand, gravel, and even large boulders of poorly soluble material. After they have been freed during the process of dissolution, these materials are removed by active stream transport but the caves are dissolutional. Even some caves in limey sandstones and conglomerates show characteristics of phreatic speleogenesis although the dissolution may be primarily intergrain and thus the process is that of piping.

In clastic rocks with siliceous and ferric matrices, the continuum is not as clear. However, intergrain dissolution of even these binders has been documented in some quartzose localities (e.g. Striebel & Schäferjohann, 1997). This permits similar transport of siliceous particles by piping and streamflow in pseudokarstic environments, which morphologically may become karstic. For caves in the Venezuelan quartzite plateaus, the relative percentages of piping and of dissolution apparently have not been quantified. The amount of subsurface karstification may be considerably larger than in the Mpumalanga pseudokarst. Colloidal silica speleothems have been documented here.

Piping, however, is a misnomer for the process of erosive widening of vertical crevice caves. Instead, vertical crevice caves undergo an ordinary form of mass wasting common to bare exposures of many slightly soluble rocks. In basaltic cliffs in Hawaii and elsewhere, this process forms spectacular pseudokarstic karren hundreds of metres high and tens of metres wide.

Other Types of Crevice Caves

Many littoral caves are crevice caves subjected to hydraulic wedging and marine abrasion. Some are found far inland, on ancient shores of vanished lakes and seacoasts (see Littoral Caves). Among other unusual types of crevice caves are deep, extremely narrow stream slots in sandstone (southwestern United States), limestone (Mallorca, Spain) and other rocks. Here, overhanging walls of meandering channels shut out daylight for considerable distances.

WILLIAM R. HALLIDAY

See also **Piping Caves; Pseudokarst**

Works Cited

Campbell, N. 1968. The role of gravity sliding in the development of some Montana caves. *Bulletin of the National Speleological Society*, 30: 25–29

Coons, D. 2001. The Great Crack: An update on recent exploration in Hawaii. *Hawaii Speleological Survey Newsletter*. 9: 27–28. Also in *Cave Research Foundation Newsletter*, 29(3) (May 2001): 9–10

Cooper, R. 1983. Mass movement caves in Great Britain. *Studies in Speleology*, 4: 37–44

Cooper, R.G., Ryder, P.F. & Solman, K.R. 1982. The windypits in Duncombe Park, Helmsley, north Yorkshire. *Transactions of the British Cave Research Association*, 9(1): 1–14

Finlayson, B. 1986. The formation of caves in granite. In *New Directions in Karst: Proceedings of the Anglo-French Karst Symposium*, edited by K. Patterson & M.M. Sweeting, Norwich: Geo Books

Klassek, G. 2000. The caves in the Polish flysch Carpathians. *Newsletter of the Commission for Pseudokarst of the International Union of Speleology*, 5: 4–5

Striebel, T. & Schäferjohann, V. 1997. Karstification of sandstone in central Europe: Attempts to validate chemical solution by analysis of water and precipitates. In *Proceedings of the 12th International Congress of Speleology*, vol. 1, edited by J. Bauchle *et al.*, Basel, Switzerland: Speleo Projects

Vitek, J. 1989. Types of pseudokarst form (sic) in neovolcanic rocks in Carpathians (abstract). In *Proceedings of the 2nd Symposium on Pseudokarst*, Prague: Czech Speleological Society

Further Reading

Proceedings volumes of international symposia on pseudokarst, most recent volume being *Proceedings of the 6th International Symposium on Pseudokarst, Galyatető, Hungary, 1996*, edited by I. Eszterhás & S. Sárkozi

Newsletter of the Commission for Pseudokarst of the International Union of Speleology, edited by Istvan Eszterhás (www.clubs.privateweb.at/speleoaustria/pseudokarst.htm)

Earl, S. 1999. Selected caves of Kings Bowl, Wapi Lava Fields, and the Great Rift Volcanic Rift Zone. In *Idaho: The Underground Gem*, edited by T. Kline, Guidebook, 1999 Convention of the National Speleological Society, Huntsville, Alabama: National Speleological Society

Grimes, K. 1997. Redefining the boundary between karst and pseudokarst: A discussion. *Cave and Karst Science*, 24(2): 87–90

Sjöberg, R. 1986. A proposal for a classification system in granitic caves. In *Proceedings of the 9th Congress of Speleology*, vol. 2, Barcelona: IBYNSA: 25–29

Striebel, T. 2000. Probleme mit den Begriffen Karst und Pseudokarst. *Mitteilungen des Verbandes der Deutschen Höhlen- und Karstforscher*, 46: 99–105

Wray, R. 1997. A global review of solutional weathering forms in quartz sandstones. *Earth Science Reviews*, 47(3): 137

CRIMEA, UKRAINE

The Crimean peninsula in southern Ukraine joins the mainland by a narrow isthmus and juts out far into the Black Sea. The peninsula is mostly steppe plains gradually rising to the Crimean mountains in the south (the highest peak is Mt Roman-Kosh, 1549 m). In Crimea the Mountain, Foothill, and Flat karst areas are differentiated (Ivanov, 1961). Karst is developed to the greatest extent in the Mountain area, in a continuous sequence of Jurassic limestones 50 to 1300 m thick. These form the Main Ridge of the Crimean mountains, a massive scarp rising 600 m above the southern coast of Crimea, with more than 340 km^2 of karst. A distinctive feature of the Main Ridge is its alp-like plateaux. They are occupied by mountain meadow steppes, with thousands of karst sinks and dolines, and wide karren fields. Sink density approaches 90–100 per km^2 on Chatyrdag and Karabi. Along the edges of high scarps, karst fissures can reach up to 100 m in length at depths of 50 m.

The northern foothills of the Crimean mountains are two monoclinal cuestas with southern scarp faces up to 100 m high and gentle northern dip slopes. They are composed of Upper Cretaceous and Tertiary limestones and marls. The mountains contain small caves (the largest is Zmeinaya Cave, 320 m long, 5 m deep) and numerous Paleolithic and Neolithic sites. In Flat Crimea, limestone outcrops in the Tarkhankut peninsula area. At the base of limestone cliffs up to 30 m high, there are some sites of marine karst with submerged or marine caves, for example Tarkhankutskaya Cave (150 m long, 12 m deep).

The karst of the Crimean mountains was first described in 1915 by Alexander Kruber. The current views on Crimean karst were mainly developed by the 1970s, and form the theoretical basis of Ukrainian and Russian karstology. The research of karst in Crimea and Ukraine improved after B. Ivanov organized his karst expedition in 1958. The caving team of the expedition was headed by the young scientist Victor Dublyansky. Prior to the expedition 82 caves with total length of 1250 m were known in Crimea, and by the end of 1971 their number had increased to 800, and total length to 32 km.

The distribution of karst cavities on karst massifs is shown in the Table. In a morphogenetic classification of the karst cavities of Crimea (Dublyansky, 1977), the corrosive / gravitational, nival / corrosive, corrosive / erosive, and corrosive / abrasion classes are distinguished.

The corrosive / gravitational class includes karst cavities developed on fractures at the top of steep slopes and scarps. More than 30 caves of this type are known. They have a fissure morphology, and their depth reaches 60–100 m. At different depths they are choked by debris, and in many of them snow is retained in summer. The largest are Gremuchaya, Aryk-Bash, and Suyuru-Kaya, each 100 m deep.

The nival / corrosive class includes shafts formed as a result of corrosion by snow meltwater. There is a close correlation between these caves and stages of Quaternary glaciations. More than 360 such caves are known. The largest are the Kuryuch-Agai, Inzhenernaya, and Vodyanaya shafts, each nearly 100 m deep. In 40 of these caves there are long-term bodies of snow and ice with a total volume of up to 10 000 m^3.

The corrosive / erosive class is formed by underground streams. These form the longest karst hydrogeological systems, such as Soldatskaya-Karasubashi (21–26 km, Karabi), Beshtekne-Uzundzda-Chernaya (13 km, Ai-Petri), and Proval-Krasnaya (7–8 km, Dolgorukovsky), which are all unexplored but proved by fluorescin tracing. The systems start at caves and ponors, where permanent or temporary streams sink at the lower edge of limestone residual soil. Among the largest ponors are Soldatskaya Cave, the deepest cave in Crimea (1860 m long; 517 m deep), Molodezhnaya (261 m deep), and Kaskadnaya (400 m deep).

The karst hydrogeological systems terminate at resurgence caves, located at the bases of the karst massifs, at altitudes from 300 to 1300 m. Many of these are open horizontal caves, and today many are dry. The large cave springs are associated with

Crimea: Number and size of karst caves in Crimea. The massifs of the Main Ridge are listed from west to east from Ai-Petri to Karabi.

Massif	Number of caves	Total length, m	Total volume, m^3
Ai-Petri	327	24 741	293 400
Yalta	24	1134	17 020
Nikita	5	308	440
Babugan	32	1222	10 990
Chatyrdag	137	11 761	352 810
Dolgorukovsky	35	22 420	280 890
Demerddzi	16	1664	13 630
Karabi	256	18 968	743 111
Eastern Crimea	10	769	4990
Foothill Crimea	69	2842	34 890
Flat Crimean area	11	512	600
TOTAL	922	86 231	1 751 971

Crimea: Mramornaya Cave in the Chatyrdag karst massif. (Photo by Alexander Klimchouk)

karst massifs, where large catchment areas are available and there are conditions for long-term influx supply. The longest cave in the Crimea, Krasnaya Cave (20 km long, in the Dolgorukovsky massif) has developed mainly along tectonic faults. There are cascades and chambers up to 30–40 m high at the boundaries between tectonic units. In the entrance zone, cave galleries form at six levels and a large underground river flows in the cave. At the exit it forms a picturesque waterfall 18 m high. Maximum discharge of water is 22.7 $m^3 s^{-1}$.

At the western and eastern edges of the Crimean mountains and Tarkhankut peninsula, where the limestones emerge from under the Black Sea, and are abraded by wave action, there are both corrosional and littoral caves.

Both infiltration of surface waters and underground condensation are important in karst and cave development. The annual precipitation in Crimea ranges from 450 mm (Foothill and Flat Crimea) to 1200 mm (Main Ridge). Condensation in karst fissures can reach 3.8% of the mean precipitation, and during the summer and autumn, drainage of the underground rivers and the flows of karst springs is almost completely supported by condensation. On the basis of chemical analysis of 2300 water samples from 444 springs, the chemical denudation rate in the Mountain area has been estimated as $c.43 \times 10^{-6}$ m a^{-1}.

BORIS VAKHRUSHEV AND VICTOR DUBLYANSKY

Works Cited

Dublyansky, V. 1977. *Karst Caves and Shafts of the Crimea Mountains*, Leningrad: Nauka (in Russian)

Ivanov, B. 1961. *Principles of Karstological Zoning of Mountain Crimea*, Kiev: Kiev University (in Russian)

CRUSTACEA

The taxonomic status of crustaceans has long been debated by carcinologists (the scientists who study these arthropods). The group has been assigned to one of the phylum, subphylum, or superclass levels, with five, six, and even ten classes being recognized. The purpose of this essay is not to enter into a taxonomic debate, but to recognize that crustaceans are one of the oldest arthropod groups and certainly represent one of the largest, most diverse, and most successful groups of invertebrates on Earth, with some 40 000 species recognized (Hobbs, 2000). Although crustaceans occupy a truly impressive number of aquatic environments (freshwater, brackish, marine), some species have become successful on land as well (e.g. isopods). Crustaceans have a chitinous exoskeleton impregnated with calcium carbonate, with three major body divisions: the cephalon, thorax, and abdomen, and they possess paired antennules, antennae, mandibles, and maxillae.

Not only have crustaceans been successful in epigean (above-ground) habitats, but they have also moved into the hypogean (below-ground) realm, with six classes of crustaceans being recognized as inhabiting cave and other hypogean environments. Additionally, anchialine and interstitial habitats (although many of the latter are not actually associated with karst landforms) harbour troglomorphic crustaceans.

Like many other cavernicoles, many crustaceans are stygobites (aquatic obligate cavernicoles), while a few are troglobites (terrestrial obligate cavernicoles). Both groups exhibit various morphological, physiological, and behavioural adaptations to living in a dark spelean environment where food is in short supply. Their adaptations include reduction or loss of pigments and eyes, attenuated appendages, and lowered metabolic rates.

It is widely accepted that most obligate cavenicoles (including those other than crustaceans) evolved from epigean ancestors with some degree of pre-adaptation to living in hypogean environments and are thought to have evolved by either natural selection or by neutral mutation and genetic drift.

The six classes of crustaceans with subterranean representatives are Mystacocarida, Branchiopoda, Remipedia, Copepoda, Ostracoda, and Malacostraca. The primitive Class Mystacocarida is a meiobenthic group restricted to interstitial waters of intertidal sand beaches and subtidal sand substrates. It is most closely related to the Copepoda, and less than 20 species and subspecies are known from Africa, Europe, North and South America, and Australia. Because mystacocarids are not known from karstic settings, there will be no further discussion here of this group. Discussion of the remaining five classes of crustaceans follows. (The classes Copepoda and Ostracoda have their own entries in this Encyclopedia, as do three orders—Amphipoda, Isopoda, and Syncarida—of the Class Malacostraca.)

Class Branchiopoda

Only the subclass Diplostraca includes species inhabiting subterranean habitats (the hyporheos—or transition zone between streams and ground water, phreatic zone, caves). Although the Conchostraca (clam shrimps) and Cladocera (water fleas) are included in this subclass, only the latter order has hypogean representatives, with fewer than 100 of the approximately 450 recognized species occupying subterranean waters. These are small (0.2–2 mm), are laterally compressed, and there are only 20 + species that can be classified as stygobites. Subterranean cladocerans (particularly hyporheic species) are known from all continents and most demonstrate only minor troglomorphic adaptations.

Class Remipedia

Remipedia (Figure 1a) are among the most primitive of all extant crustaceans. Twelve species, placed in two families containing six genera (*Cryptocorynetes*, *Godzilliognomus*, *Godzillius*, *Lasionectes*, *Pleomothra*, and *Speleonectes*), are restricted to anchialine caves in the Bahamas, the Caicos Islands, Cuba, the Yucatán Peninsula (Mexico), the Canary Islands, and the Cape Range Peninsula (Western Australia). These small (up to 45 mm total length), colourless, blind, elongate, centipede-like predators are found in their highest densities in oxygen-deficient waters (<1 mg l^{-1}), and are often found associated with other stygobitic crustaceans such as caridean shrimps, cirolanid isopods, haziid amphipods, mysids, ostracods, and thermosbaenaceans.

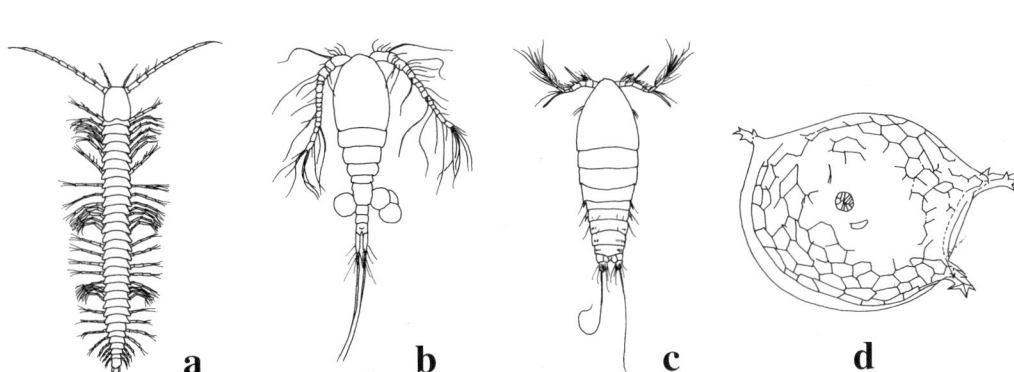

Crustacea: Figure 1. Line drawings of dorsal (d) and lateral (l) views of representative stygobitic crustaceans: **a.** Remipedia (d) *Speleonectes lucayensis* from the Bahamas (after Hobbs, 2000; total length 19.8 mm); **b.** cyclopoid copepod (d) *Kieferiella delamarei* from France (after Hobbs, 2000; length 740 μm); **c.** harpacticoid copepod (d) *Novocrinia trifida* from Belize (after Huys and Iliffe, 1998; length 505 μm); **d.** thaumatocyprid ostracod (l) *Danielopolina wilkensi* from the Canary Islands (after Kornicker and Iliffe, 1995; length 0.41 mm).

Class Copepoda

(See also Crustacea: Copepoda.) Copepods are a large and diverse group of crustaceans, and are the dominant forms of marine plankton. Presently, more than 14 000 species are placed in approximately 2300 genera in 210 families, yet the majority of subterranean species are placed in the following six orders: Platycopioida and Misophrioida (occur in anchialine habitats), Gelyelloida (found in caves in France and Switzerland), Calanoida, Cyclopoida, and Harpacticoida. Stygobitic species originated from surface, marine, and freshwater ancestors and reached different groundwater habitats mostly through interstitial and crevicular or karst routes.

Calanoid copepods are mainly planktonic filter-feeders, represented by some 2300 species, yet fewer than 20 are known from subterranean waters in a wide distribution ranging from the Americas to Madagascar and southeastern China. The cyclopoids (Figure 1b) are a poorly studied group, with approximately 150 species recognized from various caves, wells, anchialine habitats, and groundwater interstices. These characteristically blind crustaceans lack pigments and are known from North and South America, Asia, Africa, Europe, and Madagascar. The harpacticoids (Figure 1c) are represented by nearly 3000 species assigned to 40 families, at least 10 of which have stygobiont species. They are widely distributed in caves and various interstitial habitats, as well as in anchialine habitats. These vermiform crustaceans can be observed scraping food from various substrates rather than filtering water.

Class Ostracoda

(See also Crustacea: Ostracoda.) These bivalve crustaceans are fairly successful, being known from virtually all aquatic habitats, but mostly from the marine environment. Some 5700 benthic and planktonic species are recognized and more than 300 are endemic to subterranean waters, many assigned to the Order Podocopida. Although some species are known only from caves (e.g. *Candona jeanneli* in Marengo Cave, Indiana), most subterranean forms have been observed in springs, interstitial waters, and anchialine habitats (Figure 1d). Stygobites are reported from Afghanistan, Africa, Bahamas, Bermuda, Canary Islands, Central and Southern Europe, India, Jamaica, Japan, Java and Western Malaysia, Mexico, North, Central, and South America, and South Korea.

Class Malacostraca

Clearly this is the largest class of crustaceans, with approximately 23 000 species placed in six superorders, four of which contain hypogean species. The Superorder Syncarida (see Crustacea: Syncarida) is a poorly studied, freshwater group that is divided into three orders, all of which share the absence of a carapace or cephalic shield. The Order Anaspidacea (Figure 2a) is known only from Australia, including the island of Tasmania, and occupies a variety of aquatic habitats, including springs and caves, with one family (Psammaspididae) that is exclusively stygobitic. The Order Stygocaridacea (Figure 2b) occupies interstitial waters in epigean and hypogean environments in New Zealand and South America. Fewer than 10 species are assigned to four genera. The Order Bathynellacea (Figure 2c) has the greatest species richness of the three orders, with approximately 150 species recognized from interstitial epigean, well, and cave habitats. This ancient, relict group is known from Africa, Asia, Australia, Europe, Japan, Madagascar, New Zealand, and North and South America.

The Superorder Pancarida is represented by the Order Thermosbaenacea (Figure 2d). These are small (4 mm or less), eyeless (or with greatly reduced eyes), stygobites that evolved from shallow-marine ancestors and moved into anchialine habitats, caves, springs, thermal springs, wells, and occupy freshwater and saline environments. The group is widespread (found in Australia, the Bahamas, Cambodia, Costa Rica, Croatia, Cuba, France, Greece, Haiti, Italy, Lanzarote, Mallorca, Mexico, Morocco, Puerto Rico, Sicily, Spain, United States, and the Virgin Islands) and consists of at least 34 species assigned to six genera in two families.

The Superorder Peracarida consists of eight orders, six of which have known subterranean species. Common characters of this group include a carapace that is never fused to all the thoracomeres, although the carapace in isopods and amphipods secondarily has been lost, the first pair of thoracopods (thoracic somites with a pair of appendages) is modified as a pair of maxillipeds, the females have a brood pouch, and the pleopods lack

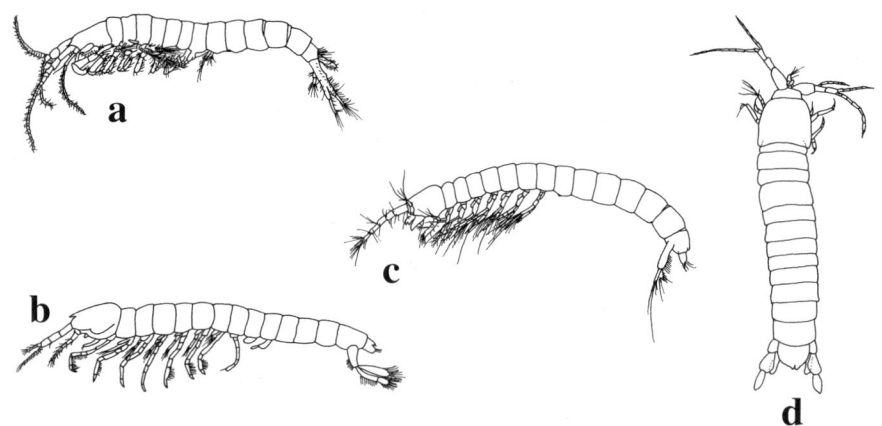

Crustacea: Figure 2. Line drawings of dorsal (d) and lateral (l) views of representative stygobitic crustaceans: **a.** anaspidacean syncarid (l) *Psammaspides williamsi* from Australia (after Hobbs, 2000; length 6.7 mm); **b.** stygocaridacean syncarid (l) *Stygocaris* sp. (after Hobbs, 2000; 500 μm); **c.** bathynellacean syncarid (l) *Notobathynella williamsi* from Australia (after Hobbs, 2000; length 1.46 mm); **d.** Thermosbaenacea (d) *Monodella relicta* from Israel (after Botosaneanu, 1986; length 2.6 mm).

an appendix interna (a slender appendage of the endopod). The Order Mysidacea is a diverse group of crustaceans with more than 1000 species widely distributed in marine, freshwater, and groundwater habitats. Stygobitic mysids (Figure 3a) occur in anchialine habitats, wells, and blue holes in Africa (Kenya, Zanzibar) the Bahamas, the Balearic Islands, Bermuda, Bosnia / Hercegovina, Canary Islands, Cuba, India, Jamaica, the Lesser Antilles, Mexico, Puerto Rico, southern Italy, and the Turks and Caicos islands. At least 28 stygobites, most of which are endemics, are recognized and the distribution of most species suggests that the majority colonized ground waters due to stranding and uplifting of their marine ancestors resulting from the regressions of the Tethys and Mediterranean seas. Other taxa have invaded ground waters more recently and are stygophiles.

The Order Tanaidacea is represented by some 850 marine benthic species. Only three species are characteristic inhabitants of marine caves in Bermuda, Koror Island (Palau), and Niue Island (South Pacific) and it is unclear whether or not any of these species are obligate cavernicoles.

The Order Cosinzeneacea was recently erected to house the stygobitic suborders Spelaeogriphacea and Mictacea. The Suborder Spelaeogriphacea is represented by three monotypic genera: *Spelaeogriphus lepidops* (South Africa—Figure 3b), *Potiicoara brasiliensis* (Brazil), and *Mangkurtu mityuula* (Western Australia). All have a small saddle-like carapace which is fused with the first thoracic somite and, anteriorly, is produced into a broadly triangular rostrum. The thoracopods and pleopods are reduced and the abdomen is elongate, often exceeding half the total body length. The Suborder Mictacea is represented by *Mictocaris halope* (Figure 3c), known from marine caves on Bermuda. This non-predatory crustacean swims, rests, or walks on the substrata of anchialine caves.

The Order Bochusacea was recently erected to accommodate three marine species belonging to the family Hirsutiidae, two of which are deep-water marine forms, while *Thetispelecaris remex* (Figure 3d) is a stygobite, inhabiting anchialine and submarine caves in the Bahamas.

The Order Isopoda (see Crustacea: Isopoda: Oniscidea) has more than 11 000 described terrestrial and aquatic species and is widely distributed across most habitats. These crustaceans lack a carapace, are dorsoventrally compressed, and have pleopods modified for respiration. Eight of eleven suborders have subterranean representatives: Anthuridea, Asellota, Calabozoidea, Cymothoidea, Microcerberidea, Oniscidea, Phreatoicidea, and Sphaeromatidea. Most anthurideans are marine, but approximately 20 species occur in interstitial sediments of rivers, beaches, wells, anchialine habitats, and caves in the Canary Islands, the Caribbean and Indian Ocean islands, Mexico, and South America. The Asellota (Figure 4a) is a diverse group that is divided into five superfamilies. These isopods inhabit ground

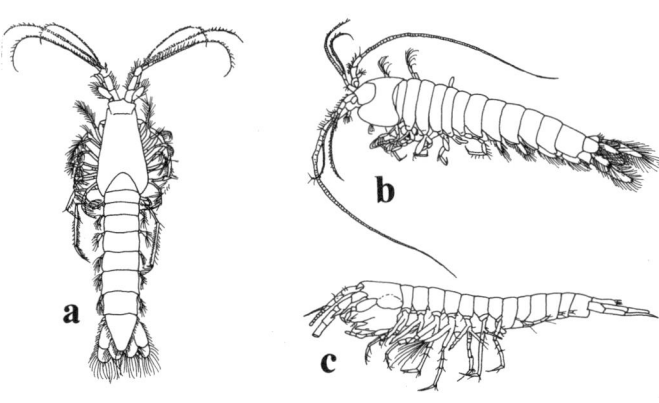
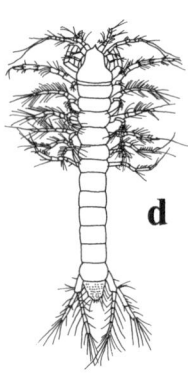

Crustacea: Figure 3. Line drawings of dorsal (d) and lateral (l) views of representative stygobiotic crustaceans: **a.** Mysidacea (d) *Speleomysis longipes* from India (after Hobbs, 2000; length 5.8 mm); **b.** speleogriphacean peracarid (l) *Spelaeogriphus lepidops* from South Africa (after Hobbs, 2000; length 6.0 mm); **c.** mictacean peracarid (l) *Mictocaris halope* from Bermuda (after Hobbs, 2000; length 3.0 mm); **d.** boshusacean peracarid (d) *Thetispelecaris remex* from the Bahamas (after Hobbs, 2000; length 1.2 mm).

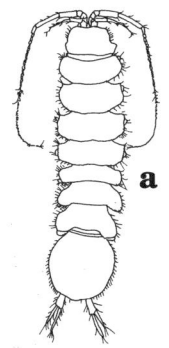
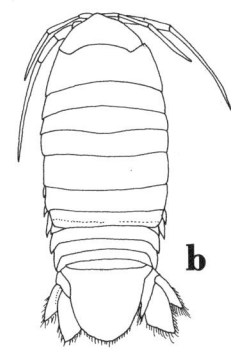
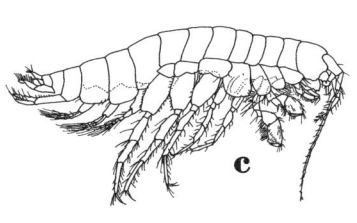

Crustacea: Figure 4. Line drawings of dorsal (d) and lateral (l) views of representative stygobitic crustaceans: **a.** asellid isopod (d) *Proasellus valdensis* from France (after Hobbs, 2000; length 8.0 mm); **b.** cirolanid isopod (d) *Anopsilana conditoria* from the Philippines (after Bruce & Iliffe, 1992; length 2.9 mm); **c.** amphipod peracarid (l) *Stygobromus exilis* from Kentucky, United States (after Hobbs, 2000; length 6.0 mm).

waters of caves, wells, springs, thermal springs, the hyporheos, and anchialine habitats in Africa, the Bahamas, Belize, Bonaire, Cuba, Curaçao, Europe, the Galapagos Islands, India, Indonesia, Japan, Madagascar, Malaysia, North and Central America, and Polynesia. The Calabozoidea is monotypic, restricted to a phreatic system in northern Venezuela, and this stygobite is closely related to the Asellota. Subterranean representatives of the cymothoideans are restricted to the mostly marine family Cirolanidae (Figure 4b), of which approximately 50 are stygobites. Microcerberideans are slender, blind, pigmentless stygobites of which about 60 species occur in interstices of marine beaches, wells, and caves near coastal areas of southern Africa, southeastern Asia, the Caribbean, Indian Ocean islands, Japan, and the Mediterranean. Oniscideans (see Crustacea: Isopoda: Oniscidea) are mostly terrestrial, although some stygobitic species (Trichoniscidae) inhabit cave waters. Although some 3500 species are known, only 14 families have subterranean representatives and are global in their distribution. The stygobite Phreatoicidea is the most ancient group of isopods, dating back to the Carboniferous. Prior to the breakup of Gondwana they invaded freshwater habitats, and currently most species are epigean, although some are known from wells and caves in Australia, India, New Zealand, and southern Africa. The sphaeromatideans are mostly marine crustaceans, but approximately 36 are stygobites that are restricted to subterranean waters in Southern Europe.

The Order Amphipoda (see Crustacea: Amphipoda) is characterized as being laterally compressed, lacking a carapace, and being generally an aquatic group consisting of more than 6700 species placed in three suborders: Caprellidea, Gammaridea, and Hyperiidea. Stygobitic amphipods are known only from the large Suborder Gammaridea (>5700 species) and the approximately 750 troglomorphic species are assigned to about 160 genera. Although there are more than 30 families, the Bogidiellidae, Crangonyctidae (Figure 4c), Hadziidae, and Niphargidae demonstrate the greatest biodiversity. They are found virtually in all parts of the globe with appropriate habitats, but Central and Southern Europe, the Mediterranean region, eastern and southern North America, and the Caribbean house the largest taxonomic diversity of stygobitic amphipods.

The Superorder Eucarida (see Crustacea: Decapoda) is characterized by a cephalothorax (carapace fused to all thoracomeres) and represents the most highly evolved group of malacostracans. Three orders are assigned to this superorder but only the Decapoda occupy subterranean environments. Decapods are the largest and most diverse order (nearly 10 000 species) and possess three pairs of maxillipeds and five pairs of functional pereiopods (hence the name Decapoda). Although widespread in distribution, the subterranean species are concentrated in Africa, Central and North America, and the Philippines and are divided among seven infraorders: Anomura, Astacidea, Brachyura, Caridea, Palinura, Stenopodidea, and Thalassinidea (the last three are not stygobites). The Anomura include crabs such as the marine hermit, porcelain, mole, and sand crabs, yet subterranean species are mostly known from two families: Aeglidae (*Aegla cavernicola*—Brazil) and Galatheidae (*Munidopsis polymorpha*—Lanzarote). Assigned to the Infraorder Astacidea are approximately 900 species of crayfishes and chelate lobsters. Only crayfishes have invaded subterranean waters, with 39 stygobitic species assigned to four genera (*Cambarus, Orconectes, Procambarus,* and *Troglocambarus*) in North America and Cuba. Brachyurans ("true crabs") are dorsoventrally compressed crustaceans, with about 4500 known species, of which only approximately 96 dwell in hypogean waters. They are widely distributed primarily in subtropical and tropical latitudes. The global Infraorder Caridea (shrimps) is quite diverse, with nearly 2000 species, yet only about 137 species dwell in springs, wells, caves, blue holes, cenotes, and anchialine habitats. They are concentrated in tropical latitudes but are known from Africa, Ascension Island, the Caribbean and western Atlantic islands, Georgia, India, the Indian Ocean islands, Indonesia, Japan, Madagascar, Melanesia, Mexico, Micronesia, the Middle East, Southern Europe, the United States, and Western Australia. A single stygophilic palinuran is known from a sea cave in New Zealand; the stenopodideans are represented by a stenoxene from a lava tube on Lanzarote, and the thalassinidean is known from a cave on Acklins Island, Bahamas.

Horton H. Hobbs III

Works Cited

Botosaneanu, L. (editor) 1986. *Stygofauna Mundi, A Faunistic, Distributional, and Ecological Synthesis of the World Fauna Inhabiting Subterranean Waters (Including the Marine Interstitial)*, Leiden: Brill

Bruce, N.L. & Iliffe, T.M. 1992. *Anopsilana conditoria*, a new species of anchialine troglobitic cirolanid isopod (Crustacea) from the Philippines. *Stygologia*, 7(4): 225–30

Hobbs III, H.H. 2000. Crustacea. In *Subterranean Ecosystems*, edited by H. Wilkens, D.C. Culver & W.F. Humphreys, Amsterdam and New York: Elsevier

Huys, R. & Iliffe, T.M. 1998. Novocriniidae, a new family of harpacticoid copepods from anchihaline caves in Belize. *Zoologica Scripta*, 27(1): 1–15

Kornicker, L.S. & Iliffe, T.M. 1995. Ostracoda (Halocypridina, Cladocopina) from an anchialine lava tube in Lanzarote, Canary Islands. *Smithsonian Contributions to Zoology*, 568: 1–32

Further Reading

Culver, D.C. & Wilkens, H. 2000. Critical review of the relevant theories of the evolution of subterranean animals. In *Subterranean Ecosystems*, edited by H. Wilkens, D.C. Culver & W.F. Humphreys, Amsterdam and New York: Elsevier

Gutu, M. & Iliffe, T.M. 1998. Description of a new hirsutiid (n.g., n.sp.) and reassignment of this family from Order Mictacea to the new Order Bochusacea (Crustacea, Peracarida). *Travaux du Muséum d'Histoire naturelle Grigore Antipa*, 40: 93–120

Juberthie, C. & Decu, V. (editors) 1994. *Encyclopaedia Biospeologica*, 2 vols, Moulis and Bucharest: Société de Biospéologie

Wagner, H.P. 1994. A monographic review of the Thermosbaenacea (Crustacea: Peracarida): a study on their morphology, taxonomy, phylogeny and biogeography. *Zoologische Verhandelingen* (Leiden), 1994: 3–338

Yager, J. 1981. Remipedia, a new class of Crustacea from a marine cave in the Bahamas. *Journal of Crustacean Biology*, 1: 328–33

CRUSTACEA: AMPHIPODA

Amphipods belong to the arthropod subphylum Crustacea and the order Amphipoda. They typically range in size from 2 to 50 mm, although a few are larger. Amphipods are common in aquatic ecosystems throughout many parts of the world, where they inhabit marine, brackish, and freshwater environments. A few species also live in terrestrial ecosystems. The order Amphipoda, which contains approximately 7000 described species, is divided into three and sometimes four suborders: Gammaridea, Caprellidea, Hyperiidea, and Ingolfiellidea. However, the latter may be highly specialized gammarideans that do not merit recognition as a suborder. Gammaridea, with more than 5500 described species, is the largest amphipod suborder and contains all of the freshwater and subterranean taxa. Approximately 21 superfamily groups, 95 families (or family groups) and more than 1000 genera are classified in this suborder (Holsinger, 1994).

Amphipods are important components of aquatic subterranean faunas in many parts of the world and are among the most abundant, widespread, and taxonomically diverse organisms found in subterranean groundwater ecosystems (Holsinger, 1993). The vast majority of subterranean amphipods are stygobites, which by definition are species restricted to subterranean groundwater habitats and characterized morphologically by loss or severe reduction of eyes and pigment, often accompanied by attenuation of the body and / or some appendages. Although body shape is variable among stygobitic amphipods (see Holsinger, 1994 for examples), many species generally resemble the species of *Stygobromus* shown in the Figure. Not all amphipods found in subterranean water are stygobites. Some species occur simultaneously in epigean (surface) and hypogean waters and do not exhibit the advanced level of morphological modification or troglomorphy that is typical of most stygobites. A few rare subterranean species in the family Talitridae are not strictly aquatic but are otherwise restricted to damp cave passages with high humidity.

Taxonomic diversity is significant at the family, genus, and species levels, and species richness is especially remarkable in several large subterranean genera, such as *Niphargus* and *Stygobromus*. Both of these genera contain over 100 described stygobitic species. On a global scale approximately 35 families (or family groups) and 166 genera of amphipods contain stygobites, and many of these genera and even some of the families are known only from stygobites. However, the vast majority of described stygobitic species (approximately 821 out of 866) are classified in 12 families and one unnamed family group. Four families with more than 100 described stygobitic species include Bogidiellidae (widespread), Crangonyctidae (northern hemisphere), Hadziidae (strongly circumtropical), and Niphargidae (Europe and Asia Minor).

The majority of subterranean amphipods live in freshwater, but some are found in brackish and even fully marine water. Subterranean amphipods, like other crustaceans that live in hypogean environments (e.g., thermosbaenaceans, isopods, and shrimps), occur in a relatively wide variety of groundwater habitats. Primary examples of these habitats include drip pools, streams, and occasionally phreatic lakes in caves; anchialine water in caves and crevices near coastal areas that is in contact with the nearby ocean through subterranean connections; phreatic aquifers accessed through wells and groundwater pumps; surface springs and seeps; and interstitial habitats in hyporheic zones beneath surface streams and elsewhere in coarse sediments saturated with groundwater. Probably the most important food source for subterranean amphipods are various forms of decomposing organic material and their associated microorganisms that

Crustacea: Amphipoda: *Stygobromus gracilipes*, a stygobitic amphipod (approximately 10 mm in length) from a stream in Endless Caverns, Virginia. (Photo by Lynda Richardson)

enter hypogean groundwaters from the surface. Much of this material is carried into subterranean channels primarily by filtration through cracks, crevices, sinkholes, and stream bottoms, and by running water that flows directly underground in sinking streams.

Subterranean amphipods are interesting biogeographically because of their taxonomic diversity, limited dispersal ability, and significant geographic isolation and restriction to groundwater aquifers. Isolation and restriction to groundwater aquifers has, in turn, apparently contributed significantly to the evolution of a large number of locally endemic species, which are common in large genera such as *Bogidiella, Niphargus, Pseudoniphargus*, and *Stygobromus*. Moreover, many taxa probably represent old phylogenetic lineages that have persisted in well-buffered groundwater habitats for long periods of time and are genuine relicts, both phylogenetic and distributional. However, not all stygobitic amphipods are old relicts. Some are apparently morphologically closely similar to epigean forms, suggesting a rather recent invasion and colonization of subterranean waters by the ancestors of these species.

It is commonly believed that stygobitic amphipods evolved from epigean ancestors, which had developed some degree of pre-adaptation for a hypogean lifestyle prior to invasion of subterranean waters. Putative ancestors of stygobites presumably gained access to and colonized hypogean groundwaters from both freshwater and marine environments. For example, the numerous stygobites in the exclusively freshwater family Crangonyctidae are believed to have evolved from putative freshwater ancestors that invaded groundwater from surface habitats. In contrast, stygobites in the Hadziidae complex currently live in brackish-marine waters or in freshwater caves developed in karst areas that were exposed to shallow marine transgressions in the Cretaceous and / or Tertiary. Thus the stygobites in this family complex are a mixture of those believed to be derived from marine ancestors that stranded in freshwater habitats following marine regressions and those derived more directly from marine ancestors that invaded and colonized anchialine habitats in concert with changes in sea level. Observations also suggest that stygobites can evolve from other stygobites through speciation in subterranean waters. Subterranean habitats provide many opportunities for the geographic isolation of local populations, which in turn should enhance speciation and ultimately result in the evolution of new subterranean taxa.

JOHN R. HOLSINGER

Works Cited

Holsinger, J.R. 1993. Biodiversity of subterranean amphipod crustaceans: global patterns and zoogeographic implications. *Journal of Natural History*, 27: 821–35

Holsinger, J.R. 1994. Amphipoda. In *Encyclopaedia Biospeologica* (vol. 1), edited by C. Juberthie & V. Decu, Moulis and Bucharest: Société de Biospéologie

Further Reading

Botosaneanu, L. (editor) 1986. *Stygofauna Mundi, A Faunistic, Distributional, and Ecological Synthesis of the World Fauna Inhabiting Subterranean Waters (Including the Marine Interstitial)*, Leiden: Brill
   Includes a series of 17 chapters on subterranean amphipods, with an introduction by J.H. Stock and contributions on various taxa by N. Coineau, J.R. Holsinger, G.S. Karaman, S. Ruffo, J.H. Stock, and W.D. Williams

Culver, D.C., Kane, T.C. & Fong, D.W. 1995. *Adaptation and Natural Selection in Caves: The Evolution of Gammarus*, Cambridge, Massachusetts: Harvard University Press
   The amphipod *Gammarus minus* Say is used as a model organism in this study on the evolution of cave animals.

Holsinger, J.R. 1994. Pattern and process in the biogeography of subterranean amphipods. *Hydrobiologia*, 287: 131–45

Notenboom, J. 1991. Marine regressions and the evolution of groundwater dwelling amphipods (Crustacea). *Journal of Biogeography*, 18: 437–54

CRUSTACEA: COPEPODA

The Copepoda—small aquatic crustaceans between 0.5 and 5 mm in length—are a large class of Crustacea, with over 12 000 species currently known worldwide. Most species are marine. The subclass is divided into ten Orders, four of which live in symbiosis with other organisms or are external parasites; only the larval stages are free-living. Six Orders (Platycopioida, Misophrioida, Calanoida, Harpacticoida, Gelyelloida and Cyclopoida) have free-living adults. The Calanoida and the Cyclopoida are by far the most dominant groups in freshwater and marine zooplankton.

Free-living copepods inhabit a wide variety of habitats, from the deep-sea floor to glacial pools at high altitudes or high latitudes. In subterranean environments, they occupy saturated and unsaturated zones in alluvial plains and karstic areas as well as anchialine caves. In many karstic caves, they are the most abundant inhabitants (Brancelj, 2002).

The name Copepoda is derived from the Greek words *kope* = "oar" and *podos* = "foot", referring to their broad oar- or paddle-like swimming legs. The English language uses the adopted Latin name—copepod(s). Others use translations of the Greek name (German: Ruderfusskrebse; Norwegian: Hoppekrebs; Russian: veslonogie rakoobraznye; Slovenian: veslonožci or ceponožci).

The body of free-living specimens is divided into two parts. The anterior part (cephalosoma) includes five somites, bearing antennulae, antennae, mouth appendages, and the first thoracic segment. This is followed by five thoracic segments, each bearing a pair of biramous swimming legs, and a genital segment with reduced legs. The rear part has four limbless abdominal segments, and an anal segment with a pair of appendages called caudal rami. Males differ from females in the structure of the antennulae, swimming legs, and genital segment. They are usually smaller, and some of their appendages are asymmetrical.

The swimming legs on the second to fifth thoracic segments are used for locomotion. In primitive forms, both rami are three-segmented. The adult or the copepodite stage (larva) always possess at least two pairs of swimming legs, and each pair is linked by a flat sclerite (plate), connecting the basal segments.

The Order Gelyelloida (with only two species) is known from karstic waters in France and Switzerland (Rouch & Lescher-

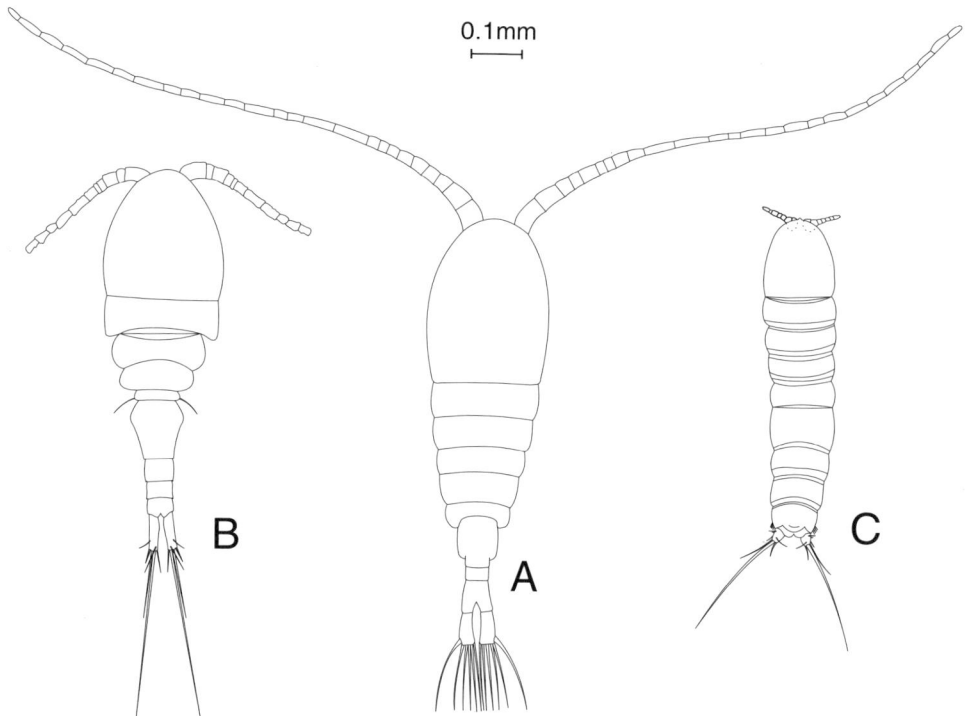

Crustacea: Copepoda: The most common representatives of freshwater Copepoda: (**A**) Calanoida (*Troglodiaptomus sketi* Petkovski, 1978); (**B**) Cyclopoida (*Diacyclops slovenicus* Petkovski, 1954); (**C**) Harpacticoida (*Paramorariopsis anae* Brancelj, 1992).

Moutoue, 1977; Moeschler & Rouch, 1988). The Order Platycopoidea (12 species) is known only from the Atlantic Ocean. Two stygobitic genera (*Nanocopia* and *Antriscopia*) occur in anchialine caves on Bermuda (Fosshagen & Iliffe, 1988). The Order Misophrioida (37 species) is entirely marine, and members occur also in anchialine caves on volcanic islands around the world. Most of the taxa of Misophrioida were discovered since 1980.

Members of the Calanoida, the Cyclopoida, and the Harpacticoida (see Figure) have worldwide distributions, inhabiting marine and freshwater habitats. In fresh water they are represented by about 2600 taxa. About one-quarter of these are subterranean and most of them were discovered in Europe.

The Order Calanoida is predominantly marine, but about 700 taxa are known from fresh water. About 10 taxa inhabit saturated zones in karst. They are known from the Crimea, the Balkan peninsula, France, China, and Mexico. *Speodiaptomus, Spelaeodiaptomus, Stygodiaptomus,* and *Troglodiaptomus* are exclusively stygobitic genera. Calanoida are an important element of specialized fauna in anchialine caves all over the world, being particularly common in the Caribbean area (Fosshagen & Iliffe, 1999).

The Order Cyclopoida has 12 families, all marine except one, Cyclopidae. This family is by far the most abundant and successful group of copepods in fresh water. About 180 taxa inhabit different types of groundwater, and the genus *Diacyclops* has the highest number of stygobitic taxa (Stoch, 2001), followed by *Acanthocyclops*. Strictly stygobitic genera are *Speocyclops, Graeteriella, Allocyclops, Austriocyclops,* and *Kieferiella*, and all but *Speocyclops* and *Allocyclops* are restricted to Europe.

The Order Harpacticoida has about 50 families, both freshwater and marine. They are primarily benthic organisms with worm-like bodies. A large number of taxa live associated with other organisms, usually inside their shells. In fresh water there are about 10 families with about 500 taxa living in groundwater. They have been found all over the world, including Australia (Lee & Huys, 1999). The genera with the largest number of stygobitic taxa are *Nitocrella* (about 50), *Elaphoidella* (about 120), and *Parastenocaris* (about 200).

Common adaptations of groundwater Copepoda are lack of pigmentation, reduction or absence of eye apparatus, reduction in the number of segments in the swimming legs and numbers of setae (genera *Speocyclops, Morariopsis, Parastenocaris*) which could also be a neotenic modification (*Speodiaptomus, Troglodiaptomus*). The number of eggs produced is usually smaller in comparison with their epigean relatives, but they are bigger. Body size is smaller and, particularly in Harpacticoida, worm-like, with the genus *Parastenocaris* as the most illustrative example. In predators the maxillipedes can be elongated (Brancelj, 2000).

There is no particular breeding season in groundwater Copepoda. Their lifespan is up to 15–20 times longer than epigean taxa and all developmental stages are prolonged (Rouch, 1968). During copulation, the male attaches an elongate spermatophore to the female, close to her genital opening. Fertilized eggs (one or a few) are carried by the female in two lateral or one central egg sac attached to the genital segment. In some genera, eggs are deposited directly on the substratum. A free-swimming larva (nauplium) is released from the egg and undergoes several moults before being transformed into a copepodite, and then into the adult after additional moults.

The Copepoda are the most common inhabitants of all types of groundwater, and specimens of some taxa, especially in caves, can be collected in large numbers. Nevertheless, the majority of taxa are restricted to small areas of distribution (endemics) (Brancelj, 2001). Some taxa in porous habitats (i.e. the phreatic zone) have very low population densities, and several hundreds or even thousands of litres of water must be filtered to collect one specimen.

Little is known about the feeding behaviour of some groundwater-dwelling groups such as Platycopioidea and Gelyelloida. Others are opportunistic gorgers (e.g. Misophrioida), feed on other animals, or are predominantly filter feeders (e.g. Calanoida). Representatives of the Cyclopoida and the Harpacticoida are omnivorous (feeding on different animal and plant material) or carnivorous. In groundwater there is no particular predator of the Copepoda, only Amphipoda (genus *Niphargus*) can occasionally prey on them.

ANTON BRANCELJ

Works Cited

Brancelj, A. 2000. *Parastenocaris andreji* n. sp. (Crustacea; Copepoda)—the first record of the genus in Slovenia (SE Europe). *Hydrobiologia*, 437: 235–39

Brancelj, A. 2001. Description of male of *Moraria radovnae* (Brancelj, 1988) (ex. *Moraria pectinata radovnae* Brancelj, 1988), a new endemic species from Slovenia and notes on other Slovenian endemic species of Copepoda. *Hydrobiologia*, 453/454: 513–24

Brancelj, A. 2002. Microdistribution and high diversity of Copepoda (Crustacea) in a small cave in central Slovenia. *Hydrobiologia*, 477(1): 59–72

Fosshagen, A. & Iliffe, T.M. 1988. A new genus of Platycopioida (Copepoda) from a marine cave on Bermuda. In *Biology of Copepods*, edited by G.A. Boxshall & H.K. Schminke, Dordrecht: Kluwer

Fosshagen, A. & Iliffe, T.M. 1999. Two cave-loving families of calanoids—the Ridgewayiidae and the Epacteriscidae. In *Program and Abstracts of the 7th International Conference on Copepoda*, Curitiba, Brazil

Lee, W. & Huys, R., 1999. A new genus of groundwater Ameiridae (Copepoda, Harpacticoida) from production bores in western Australia. In *Program and Abstracts of the 7th International Conference on Copepoda*, Curitiba, Brazil

Moeschler, P. & Rouch, R. 1988. Découverte d'un nouveau représentant de la famille des Gelyellidae (Copeda, Harpacticoida) dans les eaux souterraines de Suisse. *Crustaceana*, 55: 1–16

Rouch, R., 1968. Contribution à la connaissance des Harpacticides hypogés (Crustacés – Copépodes). *Annales de Spéologie*, 23: 5–167

Rouch, R. & Lescher-Moutoue, F. 1977. *Gelyella droguei* n. g. n. sp., curieux Harpacticide des eaux souterraines continentales de la nouvelle famille des Gelyellidae. *Annales de Limnologie*, 13: 1–14

Stoch, F. 2001. How many species of *Diacyclops*? New taxonomic characters and species richness in a freshwater cyclopid genus (Copeda, Cyclopoida). *Hydrobiologia*, 453/454: 525–31

Further Reading

Botosaneanu, L. (editor) 1986. *Stygofauna Mundi, A Faunistic, Distributional, and Ecological Synthesis of the World Fauna Inhabiting Subterranean Waters (Including the Marine Interstitial)*, Leiden: Brill

Dussart, B.H. & Defaye, D. 1995. *Introduction to the Copepoda*. Amsterdam: SPB Academic Publishing

Huys, R. & Boxshall, G. 1991. *Copepod Evolution*, London: Ray Society

Rouch, R. 1991. Copepoda. In *Encyclopaedia Biospeologica*, edited by C. Juberthie & V. Decu, Moulis and Bucharest: Société de Biospéologie

http://www.nmnh.si.edu/iz/copepod/ is a useful website with bibliographic, taxonomic, and research databases for all Copepoda

CRUSTACEA: DECAPODA (SHRIMPS, CRAYFISH, CRABS)

The malacostracans are the largest class of crustaceans, having approximately 23 000 species that are assigned to six superorders, four of which have species that live in subterranean environments. The Superorder Eucarida is one of these, and represents the most highly evolved group of malacostracans. Although three orders are assigned to the Eucarida, only the Decapoda are found in caves and other hypogean (below-surface) settings. Decapods are the largest and most diverse order of the Eucarida, with nearly 10 000 marine, freshwater, and semiterrestrial species, and are characterized by the possession of a well-developed carapace enclosing a branchial (gill) chamber, three pairs of maxillipeds (mouthparts), and five pairs of functional pereiopods (= 10 "legs", thus the derivation of "Decapoda"). These crustaceans are significant components of hypogean, primarily aquatic communities in temperate and tropical regions of the world dominated mostly by volcanic or karst terrains.

For most cavernicolous decapods, very little is known about their behaviour, ecology, evolution, and life history—clearly it is commonly difficult to assign a species accurately to one of the three categories of stygobite, stygophile, or stygoxene.

The order Decapoda is divided into two suborders: the Dendrobranchiata and the Pleocyemata. The Dendrobranchiata (about 450 species, most of which are penaeid and seregestid shrimps) is represented by a single stygoxenic penaeid shrimp found in an East African cave. All remaining decapods, including several types of shrimps, crabs, crayfishes, and lobsters, are assigned to the Pleocyemata. This suborder is represented in subterranean habitats by approximately 307 described species (numerous others remain undescribed) with 168 (54%) classified as stygobites that show varying degrees of troglomorphic adaptations (convergent behavioural, morphological, and physiological characteristics of obligate cavernicoles). Clearly the focus of this article is on the Pleocyemata, and the distinction between this suborder and the Dendrobranchiata is based primarily on gill structure.

Pleocyemate decapods are geographically widespread, and the obligate subterranean species are placed into one of four infraorders: Anomura, Astacidea, Brachyura, and Caridea. Additionally, the stygophilic *Jasus edwardsii* (Infraorder Palinura — the marine spiny and slipper lobsters) is known from an intertidal sea cave in southeastern New Zealand. A sixth infraorder, Stenopodidea, is represented by the stygoxenic *Stenopus spinosus* (restricted to a lava tube on Lanzarote, Canary Islands), as well as by *Odontozona addaia*, described from a marine cave on the island of Minorca, Spain. A seventh infraorder (Thalassinidea—

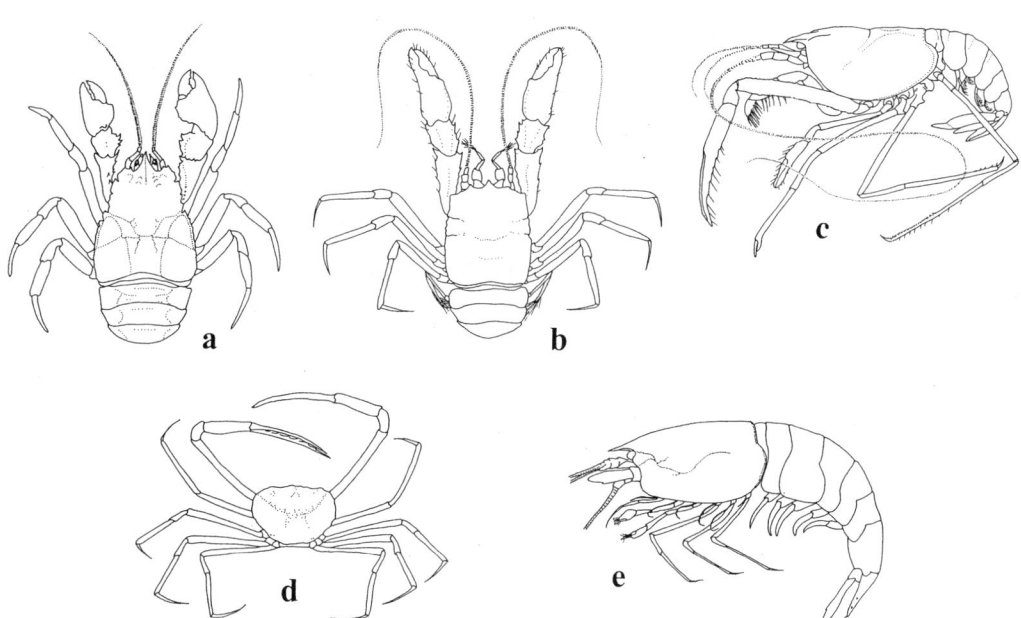

Crustacea: Decapoda: Figure 1. Line drawings of dorsal (d) and lateral (l) views of representative stygobiont decapod crustaceans: **a.** aeglid anomuran (d), *Aegla cavernicola*, from Brazil (modified from Hobbs et al., 1977; carapace length 15.0 mm); **b.** galatheid anomuran (d), *Munidopsis polymorpha*, from the Canary Islands (after Hobbs III, 2000); **c.** cambarid astacidean (l), *Troglocambarus maclanei*, from Florida, USA (after Hobbs & Franz, 1986; postorbital carapace length 13.0 mm); **d.** pseudothelphusid brachyuran (d), *Typholpseudothelphusa acanthochela*, from Belize (after Hobbs III, 2000; carapace width 37.5 mm); **e.** atyidid caridean shrimp (l), *Typhlatya pearsei*, from Mexico (after Hobbs et al., 1977; postorbital carapace length 5.2 mm).

Naushonia manningi, a probable stygophile) is known from an anchialine cave on Acklins Island, the Bahamas.

The Anomura are globally distributed, with approximately 1600 species recognized, yet only two species from two families have been described from subterranean waters. *Aegla cavernicola* (Aeglidae) (Figure 1a) is known from a stream in a single cave in São Paulo, Brazil, and little is known of the life history and ecology of this anomuran. *Munidopsis polymorpha* (Galatheidae) (Figure 1b) is found in a lava tube and wells (marine groundwater) on Lazarote, Canary Islands, and its life history, feeding, reproductive, and agonistic behaviours are well studied.

The Astacidea (including crayfish and chelate lobsters) is a diverse group of crustaceans (more than 500 species and subspecies of crayfishes are recognized) distributed over much of the globe (all continents except Africa). Only a single family (Cambaridae) is known to include obligate cave-dwelling species, yet numerous epigean species often invade subterranean waters worldwide (e.g. Astacidae, Cambaridae, Parastacidae). The following genera are represented by 39 stygobitic species (about 24% of stygobitic decapods) and subspecies restricted to North America (Florida Lime Sink, Appalachians, Interior Low Plateau, and the Ozarks), Mexico, and Cuba: *Cambarus* (11 species), *Orconectes* (eight species and subspecies) *Procambarus* (19 species and subspecies), and *Troglocambarus* (one species—Figure 1c). Most display reduced eyes and body pigments, attenuated appendages, low metabolic rates and reproductive potentials, and have long life histories. They are omnivorous opportunists, and basically consume anything organic—living or dead.

The Infraorder Brachyura (true crabs) is characterized by a cephalothorax that is flattened dorsoventrally and usually expanded laterally, a symmetrical but reduced abdomen that is flexed beneath the thorax, and uropods are rarely present. Most of the 7000 or so species are marine, but freshwater and tropical terrestrial species occur in epigean and hypogean environments. The following families have subterranean representatives located primarily in tropical settings: Gecarcinidae, Gecarcinucidae, Goneplacidae, Grapsidae, Hydrothelphusidae, Hymenosomatidae, Parathelphusidae, Potamidae, Potamonautidae, Pseudothelphusidae, Sundathelphusidae, Trichodactylidae, and Xanthidae. At least 31 species assigned to nine families are considered to be stygobites (the greatest number of species, nine, within Pseudothelphusidae—Figure 1d), whereas approximately 65 species in 11 families are either stygophiles or stygoxenes dwelling in caves, anchialine, and other subterranean habitats, and demonstrate a generally tropical, circumglobal distribution pattern.

The infraorder Caridea includes nearly 2000 species of caridean shrimps, which are characterized by variably enlarged chelate first or second pairs of pereiopods (except for the genus *Procaris*, which lacks chelation of any pereiopod) and a distinctly enlarged second pleuron overlapping the first and second pleura. The following families are diverse and have subterranean species: Agostocarididae, Alpheidae, Atyidae, Hippolytidae, Palaemonidae, Procarididae, and Rhynchocinetidae. Approximately 95 shrimps assigned to 40 genera and six families are classified as stygobites (approximately 56% of all stygobitic decapods), whereas 42 species in 16 genera and five families are either trogloxenes (terrestrial cave-dwellers unable to complete their life history in subterranean environments), stygoxenes, or stygophiles inhabiting a variety of hypogean habitats. Clearly the Atyidae (Figure 1e) are the most diverse, with at least 47 described stygobitic species, followed by the Palaemonidae with 25 stygobites.

The Red List of Threatened Animals (Baillie & Groombridge, 1996) listed 5205 animal species as threatened with extinction globally. This large number indicates that a significant proportion of the world's fauna is in trouble, and this includes stygobitic species of crayfishes (27) and shrimps (10).

HORTON H. HOBBS III

Works Cited

Baillie, J. & Groombridge, B. (editors) 1996. *1996 IUCN Red List of Threatened Animals*, Gland, Switzerland: IUCN and Washington, DC: Conservation International

Hobbs, H.H., Jr & Franz, R. 1986. New troglobitic crayfish with comments on its relationship to epigean and other hypogean crayfishes of Florida. *Journal of Crustacean Biology*, 6(3): 509–19

Hobbs, H.H., Jr, Hobbs III, H.H. & Daniel, M.A. 1977. A review of the troglobitic decapod crustaceans of the Americas. *Smithsonian Contributions to Zoology*, 244: 1–183

Hobbs III, H.H. 2000. Crustacea. In *Subterranean Ecosystems*, edited by H. Wilkens, D.C. Culver & W.F. Humphreys, Amsterdam and New York: Elsevier

Further Reading

Botosaneanu, L. (editor) 1986. *Stygofauna Mundi, A Faunistic, Distributional, and Ecological Synthesis of the World Fauna Inhabiting Subterranean Waters (Including the Marine Interstitial)*, Leiden: Brill

Hobbs, H.H., Jr, Hobbs III, H.H. & Daniel, M.A. 1977. A review of the troglobitic decapod crustaceans of the Americas. *Smithsonian Contributions to Zoology*, 244: 1–183

Juberthie, C. & Decu, V. (editors) 1994. *Encyclopaedia Biospeologica*, 2 vols, Moulis and Bucharest: Société de Biospéologie

Wilkens, H., Culver, D.C. & Humphreys, W.F. (editors) 2000. *Subterranean Ecosystems*, Amsterdam and New York: Elsevier

CRUSTACEA: ISOPODA (AQUATIC)

While land-based isopod crustaceans all belong to the Suborder Oniscidea, about 6000 species of isopods are found in aquatic habitats, mostly marine. Some 630 species inhabit fresh or brackish groundwaters: these are stygobites.

Stygobitic isopods occur in most of the subterranean biotopes: karsts, anchihaline caves, littoral marine sands, and alluvium of plains and rivers. Most large species of Sphaeromatidae, Asellidae, Stenasellidae, and Cirolanidae live in karst, while smaller species populate sediment interstitial water (see Interstitial Habitats: Aquatic): e.g. stygobite and phreatobite species of *Proasellus* (Figure 1). They have a wide ecological plasticity. Microisopods are strictly interstitial forms.

Isopods are crustacean Malacostraca without a carapace, with a dorsoventrally flattened body which comprises the head; an eight segment thorax, the first one being fused to the head (the remaining segments form the pereion); and an abdomen (or pleon) of six segments (or pleonites). One to five of the pleonites may be fused to the telson (the terminal part of the body) forming together the pleotelson. Antennae and mouthparts, pereiopods, pleopods, and uropods are appendages of the different body parts. Antennae are uniramous (single-branched). The first thoracopods are modified as maxillipeds. The uniramous pereiopods are ambulatory limbs; most pleopods serve a respiratory function, the second one is specialized for copulation in the male. Eggs are brooded in the marsupium formed by brood plates called oostegites.

Several groups exhibit five free pleonites: the Phreatoicidea, which have existed since the Carboniferous, possess a laterally compressed pleon and ambulatory uropods. *Phreatoicus* is one of the seven stygobitic genera. The Cirolanidae with 21 groundwater genera and 70 species (Botosaneanu, 2001) exhibit a convex and oval body with pleonites fused in varying degrees, and a tailfan (i.e. flat uropods acting with the telson for swimming). Volvation (ability to roll into a ball) may appear. The genus *Typhlocirolana* is highly diversified. The pleonites of the 40 stygobitic Sphaeromatidae are fused medially, the bulky body displays tubercles, or lateral spine-like protuberances in *Monolistra*. The Anthuridea (2.3–20 mm) have a cylindrical body ending as a flower bud, and the pleonites are free or fused (Figure 2).

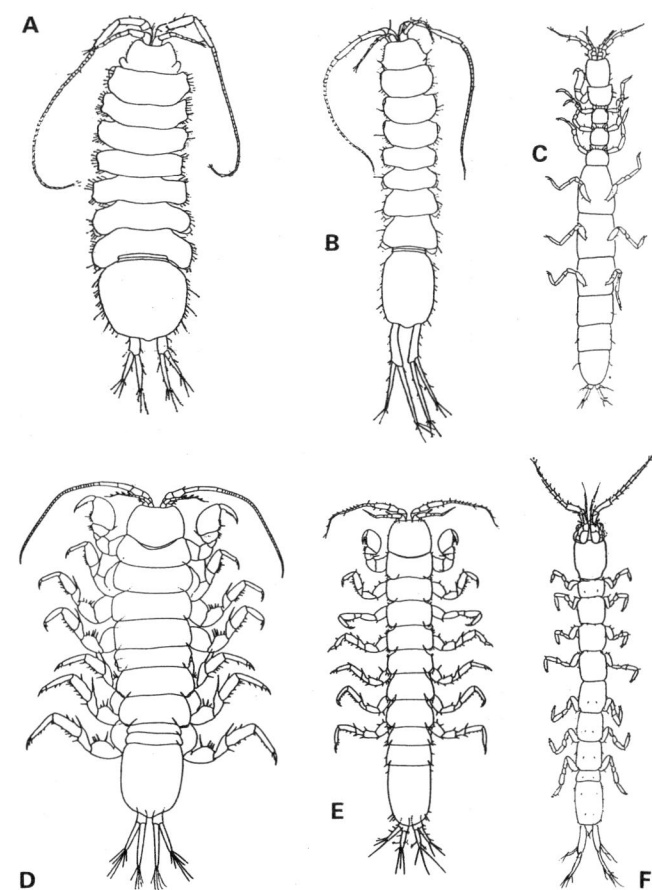

Crustacea: Isopoda (Aquatic): Figure 1. Large and wide karstic isopods: **A.** *Proasellus vandeli* Magniez et Henry, 1969 (Asellidae), 4 mm, from southwestern France; Tiny interstitial species: **B.** *Proasellus albigensis* (Magniez, 1965) (Asellidae), 5 mm, central France; **C.** *Microcerberus haouzensis* Coineau, Albuquerque, Boulanouar & Boutin, 1999 (Microcerberidae), 1.8 mm, from Morocco; **D.** *Stenasellus chapmani* Magniez, 1982 (Stenasellidae), 7.6 mm, from Sarawak. **E.** *Metastenasellus leysi* Magniez, 1986 (Stenasellidae), 3.3 mm, from southern Algeria; **F.** *Microcharon angelieri* Coineau, 1963 (Microparasellidae), 1.1 mm, from southern France and northern Spain.

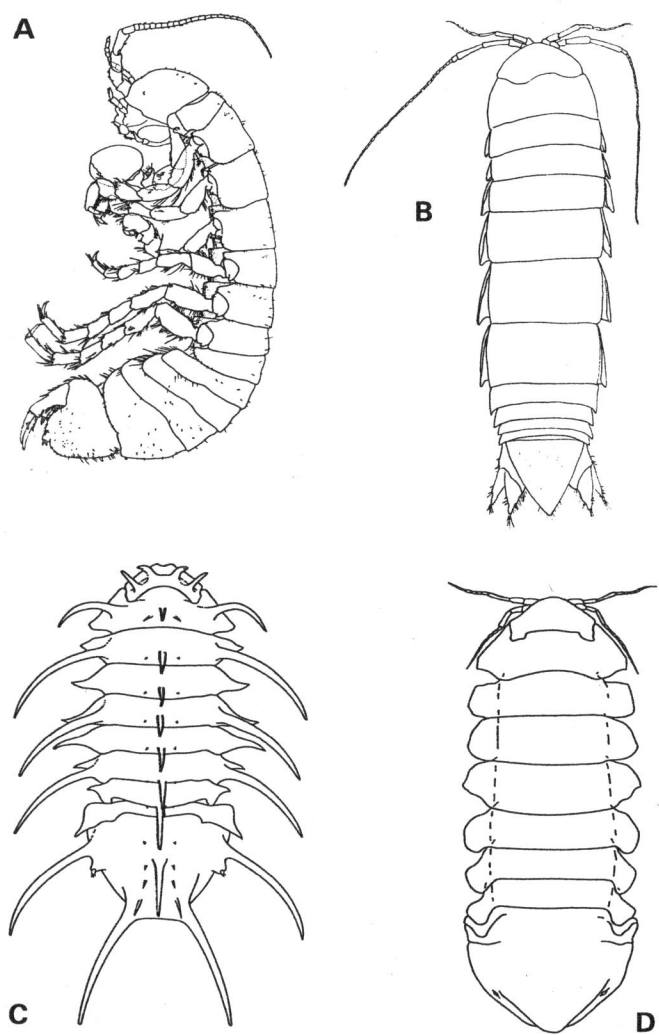

Crustacea: Isopoda (Aquatic) Figure 2. Large-sized groundwater isopods. **A.** *Crenoicus harrisoni* Nicholls, 1944 (Phreatoicidae), 12.7 mm, adult male from New South Wales, Australia. **B.** *Typhlocirolana haouzensis* Boutin, Boulanouar, Messouli et Coineau, 2001 (Cirolanidae), 11 mm, from Morocco. **C.** *Monolistra (Microlistra) spinosissima* Racovitza, 1929, 11 mm, female, from karsts of Slovenia. **D.** *Caecosphaeroma burgundum* Dollfus, 1898, 17 mm, male, from northeastern French karsts. After Wilson & Ho, 1996 (A), Boutin *et al.*, 2001 modified (B), Coineau *et al.*, 1994 (C, D).

The Asellota, characterized by one or two free pleonites and the styliform uropods, include the largest stygobite families. The Asellidae (260 species) have two free short and narrow pleonites. The eyes are more or less reduced. *Proasellus* with 118 taxa is the largest genus. The Stenasellidae (3.5–25 mm) display two wide pleonites, and an antenna exopod. The *Stenasellus* genus contains about 30 species. The tiny Microcerberidae (0.6–2.5 mm) show two long pleonites. *Coxicerberus* (50 species) and *Microcerberus* are the most diversified genera. The dwarf Microparasellidae (0.6–2 mm) have one free pleonite. *Microcharon* (85 species) displays long uropods (Figure 1) (Coineau *et al.*, 1994).

Most freshwater stygobitic isopods are derived from marine surface ancestors. The "two step model of colonization and evolution" provides one model of the establishment of marine surface ancestors in groundwaters (Coineau & Boutin, 1992). After active colonization of the littoral sands of the Tethys Ocean (first step), the ancestors passively settled into fresh interstitial groundwater due to marine regressions (second step). The distribution and origin of interstitial Cirolanidae, Anthuridae, Microparasellidae, and Microcerberidae result from such a model. The arrival of some species of *Microcharon, Microcerberus*, and *Typhlocirolana* ancestors in the groundwaters of Morocco is thought to be due to the regression of the Tethys at the end of the Turonian (Upper Cretaceous). Species from littoral karsts may also result from marine regressions: the Sphaeromatidae may have later actively colonized continental karsts. The oculated (furnished with eyes) marine ancestors of the Asellidae entered surface freshwaters (first step), and stygobites evolved from these epigean populations (second step). After the establishment of their marine ancestors in littoral karsts or interstitial habitats of the continental shields in Spain, the widespread *Stenasellus virei* species actively colonized alluvial river channels following Quaternary glaciations. Stenasellids also reached tropical aquifers in the same way (Magniez, 1999a, b).

Plate tectonics has also played a role in the evolution and distribution of ancient genera which occur in both the Old and the New Worlds. Their common ancestors occupied the littoral Tethys (*Microcerberus*, Coineau & Albuquerque, 2001) or freshwaters of continental shields (*Stenasellus*) before the break-up of Pangea and the opening of the Atlantic Ocean. In the same way, the ancestors of the Phreaoticidae were separated by the breakup of Gondwana (southern part of Pangea) and have evolved separately in India and Australia.

Groundwater isopods inhabit all continents. Many species occur in the Caribbean and the Mediterranean Basin. The Stenasellidae mainly occupy the northern tropical regions, in contrast to the Asellidae which live in a more northern area. Coastal Microcerberidae have a worldwide distribution while freshwater forms are limited to the southern United States, Cuba, Morocco, Central Europe, and South Africa. The Sphaeromatidae are restricted to southern Europe. Most Cirolanidae live in the Caribbean, the Mediterranean, and Africa.

Stygobites walk and sometimes burrow in river alluvium. They are mostly blind, but some asellids (*Proasellus*) are microphthalmic. Karstic species are large and wide, with long appendages (cf. *Stenasellus chapmani, Proasellus vandeli*). In contrast, elongation of the body and small size are adaptive traits to porous habitats (cf. *Metastenasellus leyi, Proasellus albigensis*); the dwarf microcerberids are thread-like with shorter antennae. The smallest interstitial species have a reduced number of pleopods and antenna articles; the microcerberid morphology results from progenesis (heterochrony of development due to precocious sexual maturity resulting in a small adult exhibiting the larval morphology of its ancestors). Microisopods lack an ovary, digestive glands, and optic lobes.

A K-reproductive strategy characterizes groundwater isopods as an adaptation to low energy budget, in contrast to surface relatives. Such evolution towards energy economy involves a reduced number of large eggs with abundant yolk: *Stenasellus* lays 15–60 eggs, *Proasellus walteri* 6–10, *Microcharon* two, and

Angeliera only one. Egg-laying occurs only twice a year in microisopods, once every two years in stenasellids. Lifespan is elongated with longer development, slower growth, and increased longevity ranging from two years in microparasellids to 15–20 years in stenasellids. Gametes, eggs, and juveniles are protected; *Typhlocirolana* and *Turcolana* are viviparous and there are relatively few breeding females. Reproduction occurs throughout the year without a discernible specific breeding season.

Stygobitic isopods are non-selective deposit feeders, ingesting bacteria, fungi, particulate organic deposits, and microdetritus. Furthermore, anthurids, cirolanids, and stenasellids may be carnivorous and prey upon copepods, oligochaetes, chironomids, or even congeners (smaller individuals of the same species). Small species may be prey for predators such as other large isopods, decapods, and fish that ingest sediments.

NICOLE COINEAU AND CLAUDE BOUTIN

Works Cited

Botosaneanu, L. 2001. Morphological rudimentation and novelties in stygobitic Cirolanidae (Isopoda, Cymothoidea). *Vie Milieu*, 51(1/2): 37–54

Coineau, N. & Albuquerque, F.E. 2001. Palaeobiogeography of the freshwater isopods Microcerberidae (Crustacea) from Caribbean and North America. *Proceedings of the 13th International Congress of Speleology, Brasilia (Brazil)*

Coineau, N. & Boutin, C. 1992. Biological processes in space and time: Colonization, evolution and speciation in interstitial stygobionts. In *The Natural History of Biospeleology*, edited by A.I. Camacho. Madrid: Museo Nacional de Ciencias Naturales

Coineau, N., Henry, J.P., Magniez, G. & Negoescu, H. 1994. Isopoda aquatica. In *Encyclopaedia Biospeologica*, vol 1, edited by C. Juberthie & V. Decu, Moulis and Bucharest: Société de Biospéologie

Magniez, G. 1999a. Isopodes Aselloïdes stygobies d'Espagne, IV-Stenasellidae: taxonomie, histoire évolutive et biogéographie. *Beaufortia*, 49: 115–39

Magniez, G. 1999b. A review of the family Stenasellidae (Isopoda, Asellota, Aselloidea) of underground waters. In Jan H. Stock Memorial issue, edited by D.L. Danielopol, K. Martens & J.C. von Vaupel Klein, *Crustaceana*, 72 (8): 837–48

Further Reading

Botosaneanu, L. (editor) 1986. *Stygofauna Mundi, A Faunistic, Distributional, and Ecological Synthesis of the World Fauna Inhabiting Subterranean Waters (Including the Marine Interstitial)*, Leiden: Brill

Boutin, C. 1993. Biogéographie historique des Crustacés Isopodes Cirolanidae stygobies du groupe *Typhlocirolana* dans le Bassin Méditerranéen. *Comptes Rendus de l'Académie des Sciences*, 316: 1505–10

Coineau, N. 2000. Adaptations to interstitial groundwater life. In *Subterranean Ecosystems*, edited by H. Wilkens, D.C. Culver & W.F. Humphreys, Amsterdam and New York: Elsevier

Juberthie, C. & Decu, V. (editors) 1994. *Encyclopaedia Biospeologica*, vol. 1, Moulis and Bucharest: Société de Biospéologie

Wägele, J.W. 1990. Aspects of the evolution and biogeography of stygobitic Isopoda (Crustacea: Peracarida). *Contributions to Zoology*, 60(3/4): 145–50

Wilson, G.D.F. & Keable, S.J. 2001. Systematics of the Phreatoicidea. In *Isopod Systematics and Evolution*, edited by B. Kensley & R.C. Brusca, Rotterdam: Balkema

CRUSTACEA: ISOPODA: ONISCIDEA (WOODLICE)

The Oniscidea (formerly called Oniscoidea) or terrestrial isopods represent a monophyletic suborder of the Order Isopoda (Crustacea, Malacostraca, Peracarida). They are the only group of crustaceans fully adapted to live on land. With about 4000 described species in 33 families, the Oniscidea represent the largest isopod suborder. Their common names are woodlice, slaters, sawbugs, and pillbugs. They occur in all terrestrial habitats, from littoral to high mountains, and from forests to deserts.

The Oniscidea have a dorso-ventrally flattened body, a cephalothorax with sessile compound eyes and uniramous antennulae and antennae, pereon (thorax) of seven segments, pleon (abdomen) of five segments, and a pleotelson. There are seven pairs of uniramous pereopods (legs), all more or less alike (hence "isopod"), and five pairs of biramous pleopods. Eggs and embryos are carried by the female in a ventral marsupium. The Oniscidea are characterized by (1) extreme reduction (1 to 3 articles) of the antennules; (2) endopods of pleopod 1 and / or 2 in males being elongated, and specialized as a copulatory apparatus; and (3) presence of a complex water-conducting system (Hoese, 1982a). In species best adapted to terrestrial life (e.g. Porcellionidae, Armadillidiidae, and Armadillidae families) the exopods of pleopods 1–2, 1–3, or 1–5 bear respiratory structures, called pseudotracheae or lungs (Hoese, 1982b). Terrestrial isopods possess general body morphologies correlated to their ecological strategies and behaviour, and can be grouped in four main categories (Schmalfuss, 1984): the "runners", with an elongated, slightly convex body and long pereopods; the "clingers", with a flat broad body and short strong pereopods; the "creepers", with small, convex, and elongated body, dorsum with longitudinal ribs, and slow movements; and the "rollers" (or conglobaters), with a highly convex body able to roll up into a ball.

The majority of terrestrial isopods need a high degree of air humidity for survival, due to their limited ability to preserve water. Hence, it is not surprising that a large number of Oniscidea have adapted to live in subterranean habitats in general, and in caves in particular. About 300 troglobitic species are known but many others are troglophilic or are occasionally found in caves. Many species also live in other endogean habitats such as the MSS (Mesovoid Shallow Substratum) and the nests of ants and termites. It is difficult to define a clear ecological boundary between the different subterranean environments, and many species are often found in caves as well as in the MSS. Troglobitic terrestrial isopods have been recorded from 16 families: Ligiidae, Mesoniscidae, Trichoniscidae, Styloniscidae, Olibrinidae, Scyphacidae, Philosciidae, Platyarthridae, Spelaeoniscidae, Porcellionidae, Trachelipodidae, Cylisticidae, Armadillidiidae, Scleropactidae, Eubelidae, and Armadillidae. The family with the largest number of troglobitic forms is the Trichoniscidae, ac-

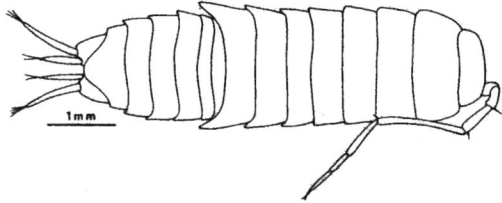

Crustacea: Isopoda: Oniscidea: Figure 1. *Haloniscus anophthalmus* Taiti & Humphreys, 2001 from groundwater calcretes in Western Australia.

counting for around 70% of all the cavernicolous terrestrial isopods with more than 50 genera in the northern hemisphere. Among the most interesting troglobitic species in this family are some forms which live in fresh water in caves. For example *Typhlotricholigioides aquaticus*, from the Cueva de Ojo de Aqua Grande, Vera Cruz, Mexico; *Mexiconiscus laevis* from some caves in San Luis Potosí Province, Mexico; and *Cantabroniscus primitivus* from several caves in Santander and Burgos Provinces, Spain. According to Vandel (1965) the systematic affinities between *Typhlotricholigioides* and *Cantabroniscus* prove their very ancient origin which can be dated back to the Carboniferous, before the fragmentation of Laurasia. It is not clear if aquatic life is a primitive condition for these species, as stated by Vandel, or if they have secondarily returned to water. The latter condition is most probably true for other Oniscidea occurring in subterranean waters, such as *Thailandoniscus annae* (Stylonicidae) from a cave in Thailand, and four species of *Haloniscus* (presently placed in Scyphacidae) from anchialine waters in a New Caledonian cave (*H. anophthalmus*) and from groundwater calcretes in Western Australia (*H. tomentosus, H. longiantennatus*, Figure 1, and *H. stilifer*, all described by Taiti & Humphreys, 2001). The Stylonicidae in the southern hemisphere and the Philosciidae in the intertropical belt also include many troglobitic species (about 5% of the global total each). The family Cylisticidae includes both troglobites (e.g. *Troglocylisticus cyrnensis* from Brando Cave, Corsica, Figure 2) and endogean forms (e.g. *Lepinisticus vignai* from Monti Lepini, central Italy, and most species of the *nasutus*-group of the genus *Cylisticus*). All the other families contribute only a very small percentage (in general less than 2% each) to the total number of troglobitic Oniscidea, with a few represented by single species, e.g. the Trachelipodidae with only *Trachelipus troglobius* from Movile Cave, Romania.

The adaptive traits to cavernicolous life present in oniscidean forms are typical of most troglobitic arthropods: reduction or absence of eyes and body pigment; development of long appendages, especially antennae and pereopods; and reduced metabolism and fertility with a small number of eggs in comparison with closely related epigean species. All ecological categories listed by Schmalfuss (1984) can be found in subterranean habitats, with a prevalence of "runners" in caves (e.g. Trichoniscidae of the subfamily Trichoniscinae, Stylonicidae, and Philosciidae) and "creepers" in the MSS (Trichoniscidae of the subfamily Haplophthalminae). Conglobating species are also common cave dwellers (e.g. Trichoniscidae of the subfamily Buddelundiellinae, Cylisticidae, Armadillidiidae, Scleropactidae, and Armadillidae). In some cases a few species belonging to conglobating families with a very convex body have secondarily developed a slightly convex body, most probably to penetrate narrow crevices, e.g. *Cylisticus (Platycylisticus) dobati* from a cave in Turkey, and *Armadillidium aelleni* from a cave on Malta.

In general, terrestrial isopods are considered to be very good biogeographic indicators due to their weak dispersal capacity. The troglobitic forms in particular, which at least in temperate regions include relict forms of ancient faunas, are very important from a biogeographical perspective. The areas of the world with the highest number of troglobitic Oniscidea are the southern United States and Mexico, southern Europe and the Mediterranean, and southwest Asia to the Caucasus. Most of these zones border the maximum extent of Quaternary glaciations, and the climatic changes occurring in the late Tertiary and in the Quaternary seem to have played a fundamental role in the adaptation of terrestrial isopods to cave life. Until recently only a small number of troglobitic terrestrial isopods were known from the tropics, which were considered to be particularly unfavourable for speciation in caves due to the limited climatic changes in the past. In recent years new investigations have been carried out and it has become clear that a rich oniscidean fauna is present in some tropical caves, for example in the lava tubes of the Hawaiian islands where four troglobitic species of *Hawaiioscia* and one of *Littorophiloscia* (Philosciidae) have been discovered (Taiti & Howarth, 1997; Rivera *et al.*, 2002).

Stefano Taiti

Works Cited

Hoese, B. 1982a. Der *Ligia*-Typ des Wasserleitungssystems bei terrestrischen Isopoden und seine Entwicklung in der Familie Ligiidae (Crustacea, Isopoda, Oniscoidea). *Zoologische Jahrbücher Abteilung für Anatomie und Ontogenie der Tiere*, 108: 225–61

Hoese, B. 1982b. Morphologie und Evolution der Lungen bei den terrestrischen Isopoden (Crustacea, Isopoda, Oniscoidea). *Zoologische Jahrbücher Abteilung für Anatomie und Ontogenie der Tiere*, 107: 396–422

Rivera, M.A.J., Howarth, F.G., Taiti, S. & Roderick, G.K., 2002. Evolution in Hawaiian cave-adapted isopods (Oniscidea: Philosciidae): Vicariant speciation or adaptive shifts? *Molecular Phylogenetics and Evolution*, 25: 1–9

Schmalfuss, H. 1984. Eco-morphological strategies in terrestrial isopods. *Symposia of the Zoological Society of London*, 53: 49–63

Taiti, S. & Howarth, F.G. 1997. Terrestrial isopods (Crustacea, Oniscidea) from Hawaiian caves. *Mémoires de Biospéologie*, 24: 97–118

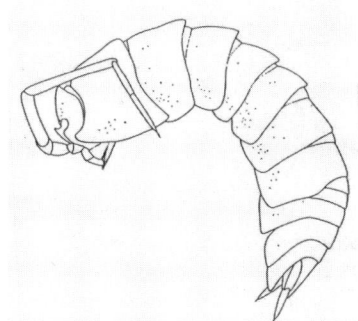

Crustacea: Isopoda: Oniscidea: Figure 2. *Troglocylisticus cyrnensis* Ferrara & Taiti, 1983 from Brando Cave, Corsica, France.

Taiti, S. & Humphreys, W.F. 2001. New aquatic Oniscidea (Crustacea, Isopoda) from groundwater calcretes of Western Australia. *Records of the Western Australian Museum*, Supplement 64: 133–51

Vandel, A. 1965. Sur l'existence d'Oniscoïdes très primitifs menant une vie aquatique et sur le polyphylétisme des isopodes terrestres. *Annales de Spéléologie*, 20: 489–518

Further Reading

Argano, R. 1994. Isopoda Terrestria: Oniscidea. In *Encyclopaedia Biospeologica*, edited by C. Juberthie & V. Decu, Moulis: Société Internationale de Biospéologie, vol. 1: 141–46

Manicastri, C. & Argano, R. 1989. An analytical synopsis of the troglobitic terrestrial isopods. *Monitore Zoologico Italiano* (Nuova Serie), Monografia 4: 63–73

CRUSTACEA: OSTRACODA

Musselshrimps, or Ostracoda, are small, bivalved Crustacea with over 300 subterranean representatives, mainly from interstitial habitats. Their calcified carapace has an average length of *c.*1 mm and completely envelops the body, which is largely non-segmented and has a reduced number of appendages compared to most other crustaceans. Ostracoda are divided into two subclasses, the Myodocopa and the Podocopa (Horne *et al.*, 2002). The myodocopans are exclusively marine and, since most of them have weakly calcified valves, are poorly represented in the fossil record. The podocopans are divided into three Orders: the exclusively marine Platycopida, the ubiquitous Podocopida (to date the most diverse group of Ostracoda, found in marine, brackish, and fresh waters), and the Paleocopida (diverse and widespread in the Paleozoic, but now represented only by the extremely rare marine superfamily Puncioidea). The Podocopida originated in the Ordovician (*c.*450 million years ago) and of the five extant suborders, three have representatives in freshwater environments, while the Bairdioidea and the Sigillioidea are entirely marine. Cypridocopina are by far the most diverse podocopidans; they comprise more than 80% of specific diversities on most continents (Martens, 1998). Most cyprid lineages are non-marine, but some are genuinely marine. Cytherocopina are mostly marine, but eight separate invasions of non-marine habitats have occurred. Darwinulidae are fully non-marine, but their extant diversity is low.

Ostracoda abound in the benthic and periphytic aquatic animal communities, but also in (semi-) terrestrial and subterranean environments. Ostracods constitute an important fraction of stygobitic fauna, especially in the interstitial domain. Danielopol & Hartmann (1986) listed 310 subterranean ostracod species from both marine and freshwater environments. However, only about 50 of these were reported from karst and caves, the majority being reported from truly interstitial habitats. This is surprising as ostracods have to calcify their valves after each moult (eight times), and hence would be expected to thrive in karst environments which are rich in carbonates. However, high concentrations of carbonates are not necessarily sufficient to create favourable conditions for ostracods, as the overall water chemistry and other factors such as temperature and food availability determine whether conditions are favourable for moulting and calcification of valves. Moreover, these conditions are thought to be genus- or species-specific. Karst and cave habitats can also be seen as intermediate systems between epigean and interstitial habitats. Various epigean species, not listed by Danielopol & Hartmann (1986), live in association with karstic water bodies. For example, in the dolomitic Molopo Oog area (South Africa), several tens of ostracod species from various genera generally considered to be epigean live in doline lakes, upwelling zones in lakes, or in their outflows. Finally, compared to epigean habitats, the interstitial zones of rivers, caves, and karst are relatively poorly studied.

Numerous podocopid and myodocopid ostracods are known from anchialine caves. For example, Maddocks & Iliffe (1986) reported 33 podocopid species from 24 Bermudan anchialine caves, one genus and ten species being new to science. Several other podocopid species belonging to the Bairdioidea, Cytheroidea, Candonidae, as well as Platycopida have been found from anchialine caves on the Galapagos Islands and Jamaica (Maddocks & Iliffe, 1993). Myodocopid anchialine ostracods belong to either Thaumatocyprididae or Halocyprididae. Based on phylogenetic studies, Danielopol (1990) and Danielopol, Baltanas & Humphries (2000) argued that the present-day troglobitic representatives of the genus *Danielopolina*, a genus of Thaumatocyprididae, originated from shallow benthic thaumatocypridinids that lived on the shelf of the Tethys Sea, rather than from deep-water species, as had previously been postulated.

All major freshwater podocopid lineages (cypridoids, cytheroids, and darwinuloids) have representative lineages in continental caves and karst. Of the Cypridoidea, the Candonidae are most successful in all stygobitic habitats, with representatives of at least ten genera being recorded. Some of these are highly endemic, for example *Danielocandona* spp. and *Caribecandona* spp. in wells from the West Indies, and *Namibcypris* spp. from karstic springs in Namibia (Martens, 1992). In Cyprididae, the smaller Cypridopsinae and, somewhat surprisingly, several relatively large Herpetocypridinae lineages are also able to colonize caves and riverbed sediments. The monospecific genus *Somalicypris* is thus far known only from Somalian wells, while the dozen or so species of *Humphcypris* are known from caves and springs throughout eastern and southern Africa, the middle East, and India (Martens, 1997). Several species in the Darwinuloidea appear to have developed a General Purpose Genotype and can survive in an exceptionally wide range of habitats (Van Doninck *et al.*, 2002), including caves and karst. For example, *Darwinula protracta* described from a cave on Mont Hoyo, Irumu, Zaire, is a synonym of the cosmopolitan and ubiquitous *D. stevensoni*. Most cavernicolous cytheroids belong to the entocytherid subfamily Sphaeromicolinae. Entocytherid ostracods are presently known to live as commensals (semi-parasites) on various kinds of other crustaceans (crayfish, amphipods, isopods), but species of *Sphaeromicola* are associated exclusively with subterranean isopods.

Subterranean ostracods show various morphological changes that have been interpreted as adaptations to stygobitic life, for example loss of sight, enlargement of chemosensorial aesthetascs, overall reduction in size, and simplification of limb chaetotaxy (mapping of setae on body segments). However, the signal from cave and karst ostracods is confusing. For example, Danielopol & Marmonier (1994) mentioned that loss of sight is not always compensated by enlargement of aesthetascs. Moreover, subterranean herpetocypridinids are generally around 2 mm long, which is relatively large for ostracods. Entocytherids show more adaptations to their commensal than to their subterranean way of life. This has led to the hypothesis that subterranean invasion, including caves and karstic aquifers, is an active process, by which ostracods can establish populations simply because the conditions are favourable. Subterranean invasion would not always be an adaptive process, and would generally be facilitated by preadaptions or by the existence of exaptations (i.e. structures that developed for other reasons than that for which they are presently seemingly adapted). An alternative hypothesis, the refugium model, holds that animals were forced to invade subterranean habitats under constraints, because of unfavourable epigean conditions, in which case subterranean faunas would have a relict character. Botosaneanu & Holsinger (1991) defend the latter view, but also propose a pluralistic approach, including both models (see also Colonization entry).

Uncertainty also exists about the age and antiquity of cave and karst faunas. Anchialine caves along the margins of the Tethys Sea are thought to have been colonized by Thaumatocypridinae since the end of the Jurassic (*c.*150 million years ago). A marine ancestry related to the Tethys Sea is also proposed for the commensal Sphaeromicolinae; here invasion of fresh groundwaters is postulated to have occurred during the Tertiary (or earlier) and the lineage became isolated there, at least in Europe, by the end of the Miocene (5–6 million years ago). Martens (1992) placed the African genus *Namibcypris* and the West Indian genus *Danielocandona*, both from subterranean habitats, in the same tribe (Namibcypridini). He also suggested that this lineage might have existed before the breakup of Africa and South America (65–70 million years ago), implying that this lineage would have lived underground for that period of time. However, Danielopol *et al.* (1994), following the above-mentioned preadaptation scenario, suggested that both groups invaded groundwaters independently from each other and much more recently.

KOEN MARTENS

Works Cited

Botosaneanu, L. & Holsinger, J.R. 1991. Some aspects concerning colonisation of the subterranean realm—especially of subterranean waters: A reply to Rouch & Danielopol, 1987. *Stygologia*, 6: 11–39

Danielopol, D.L. 1990. The origin of the anchialine cave fauna—the "deep sea" versus the "shallow water" hypothesis tested against the empirical evidence of the Thaumatocyprididae (Ostracoda). *Bijdragen tot de Dierkunde*, 60(3/4): 137–43

Danielopol, D.L., Baltanas, A. & Humphries, W.F. 2000. *Danielopolina korenickeri* n.sp. (Ostracoda, Thaumatocypridoidea) from a western Australian anchialine cave: Morphology and evolution. *Zoologica Scripta*, 29: 1–16

Danielopol, D.L. & Hartmann, G. 1986. Ostracoda. In *Stygofauna Mundi, A Faunistic, Distributional, and Ecological Synthesis of the World Fauna Inhabiting Subterranean Waters (Including the Marine Interstitial)*, edited by L. Botosaneanu, Leiden: Brill

Danielopol, D.L. & Marmonier, P. 1994. Ostracoda. In *Encyclopaedia Biospeologica*, edited by C. Juberthie & V. Decu, Moulis and Bucharest: Société de Biospéologie

Danielopol, D.L., Marmonier, P., Boulton, A.J. & Bonaduce, G. 1994. World subterranean ostracod biogeography: Dispersal of vicariance? *Hydrobiologia*, 287: 119–29

Horne, D.J., Cohen, A. & Martens, K. 2002. Biology, taxonomy and identification techniques. In *The Ostracoda: Applications in Quaternary Research*, edited by J.A. Holmes & A. Chivas, Washington DC: American Geophysical Union

Maddocks, R.F. & Iliffe, T.M. 1986. Podocopid Ostracoda of Bermudian caves. *Stygologia*, 2: 26–76

Maddocks, R.F. & Iliffe, T.M. 1993. Thalassocypridine Ostracoda from anchialine habitats of Jamaica. *Journal of Crustacean Biology*, 13: 142–64

Martens, K. 1992. On *Namibcypris costata* n.gen. n.sp. (Crustacea, Ostracoda, Candoninae) from a spring in northern Namibia, with the description of a new tribe and a discussion on the classification of the Podocopina. *Stygologia*, 7(1): 27–42

Martens, K. 1997. On two new crenobiont ostracod genera (Crustacea, Ostracoda, Herpetocypridinae) from Africa and Asia Minor, with the description of a new species from dolomitic springs in South Africa. *South African Journal of Science*, 93: 542–54

Martens, K. (editor) 1998. *Sex and Parthenogenesis: Evolutionary Ecology of Reproductive Modes in Non-marine Ostracods*, Leiden: Backhuys

Van Doninck, K., Schön, I., De Bruyn, L. & Martens, K. 2002. A general purpose genotype in an ancient asexual. *Oecologia*, 132: 205–12

CRUSTACEA: SYNCARIDA

Syncarida is a Superorder of Malacostraca in which the carapace is absent. Fossil Syncarida have been recorded in former brackish lagoons from the Carboniferous through to the Permian. Contemporary stygobitic taxa inhabit fresh interstitial groundwater, although some species still occur within marine beaches. Stygobitic syncarids have a wide ecological plasticity. They occur in astacid (freshwater decapod crustacean) galleries or coarse sediments (*Micraspides*) and in caves (*Koonunga*); psammaspidids, stygocarids, and bathynellids live in interstitial groundwater habitats—freshwater alluvial sediments associated with rivers and lakes in upland and lowland areas, mainly in the phreatic zone, and sometimes cave sediments.

The Syncarida body is long and almost cylindrical and is composed of the head; a thorax with seven or eight free segments—the first of which may be fused to the head; and the abdomen—composed of five free segments and the pleotelson, or of six segments before the telson. Both antennae are biramous, with a peduncle, a long exopod, and a short endopod with various segment numbers. Other appendages are the mouthparts; seven or eight biramous ambulatory thoracopods, the last pair being transformed into the male copulatory organ and reduced in the female of some groups; the pleopods (uniramous or biramous, reduced or absent, in variable number); and the biramous uropods which form a tail fan with the telson when they are foliated. The size ranges from 0.5 to 50 mm.

There are about 240 species belonging to three Orders, of which 21 species are fossils dating from Carboniferous up to Cretaceous times. The most primitive Order, the Palaeocaridacea, has 14 fossil genera and 19 species with stalked eyes, antennal statocysts (a balance organ), eight thoracic and six abdominal segments, eight ambulatory thoracopods, and six biramous pleopods, the two first being modified in a petasma for copulation. They have no furca (the last appendage of the body). The Anaspidacea Order exhibits seven thoracic segments (the first one being fused to the head), six abdominal segments, and a triangular telson. Eyes may be stalked, sessile, or absent. They possess an antennal statocyst, eight biramous or uniramous ambulatory thoracopods, and two to six pleopods (the male petasma is present). The oculated Anaspididae, with three genera, and Koonungidae, with seven genera (two species of which are Cretaceous and Triassic fossils) are the largest (7–56 mm) and exhibit a tail fan formed by the flat uropods grouped with the telson. The smaller interstitial Psammaspididae and Stygocarididae (1.4–14 mm), with two and four genera, have no eyes and no tail fan, uropods are bi-articulated. The Bathynellacea is the most diverse Order: two families, the Bathynellidae and the Parabathynellidae, contain 61 genera and about 200 small eyeless species (0.4–3.5 mm). The body is composed of eight thoracic segments, five abdominal segments, and the pleotelson. Thoracopods 1–7 are biramous, the eighth being transformed for copulation in the male and reduced in the female; furcae and uropods complete these stygobites. Numerous genera are monospecific (Serban, 1972; Schminke, 1973; Camacho, 1987; Coineau, 1996).

In contrast to swimming anaspidaceans from surface waters, all stygobitic syncarids walk within sediments using their thoracopods and their furca. They are blind and lack pigment. Their small size is an adaptation to interstitial life (cf. the tiny *Stygocarella, Nanobathynella, Leptobathynella*), which may be due to progenesis (precocious sexual maturity inducing small adults to retain some larval morphology). The minute *Acanthobathynella* exhibits the post-embryonal morphology of the less derived *Parvulobathynella*. All Bathynellacea display a zoea (larval) morphology (Schminke, 1981): they are paedomorphic, retaining some larval traits. The slender small syncarid species have no telson, and a reduced number of pleopods, appendage rami and segments. The thoracopod 7 of *Hexabathynella* and *Hexaiberobathynella* is absent. Digestive caeca and an ovary are also absent. Syncarids are K-strategists as an adaptation to the low-energy budget within subterranean environments. Eggs are spawned one after another; in anaspidaceans gametes are protected in spermatophores and spermateca; bathynellaceans lay only one egg with an abundant yolk in sediments. Syncarids also have a long embryonic development within their egg, hatchlings resemble the adult in structure, and have a slow growth for energy economy. They are relatively long lived, up to 2.5 years in *Paraiberobathynella* (Coineau, 2000).

Stygobitic syncarids are cold stenotherms. Bathynellids reside in groundwater at temperatures ranging from 8 to 11°C and do not inhabit water temperatures greater than 20°C. *Thermobathynella* occurs in thermal waters at 55°C. Syncarids are non-selective deposit feeders. Some parabathynellids may be carnivorous, preying upon copepods. The largest syncarids may be prey for amphipods or fishes in karstic waters.

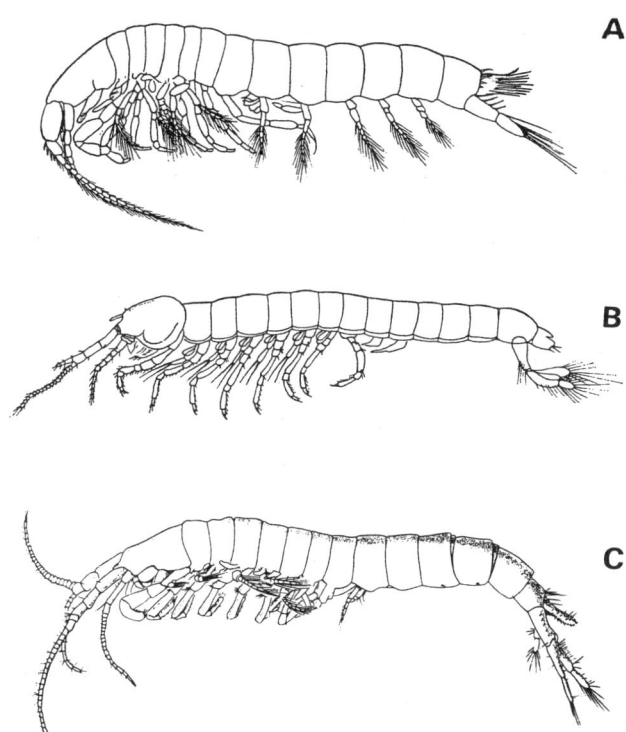

Crustacea: Syncarida: Figure 1. Anaspidacean syncarids: **A.** *Micraspides calmani* Nicholls (Koonungidae); **B.** *Parastygocaris andina* Noodt (2.5 mm, Stygocarididae); **C.** *Psammaspides williamsi* Schminke (6 mm, Psammaspididae).

Syncarids were already a diverse group in the Paleozoic, especially during the Carboniferous and the lower Permian on the Laurentia supercontinent, and later on Gondwana in the Permo-Triassic (Schram, 1981; 1984). The distribution of contemporary anaspidaceans covers only limited Gondwanian areas in Tasmania, Australia, New Zealand, Chile, and Argentina. The disjunct distribution of numerous bathynellid genera results from a Gondwanian origin. *Chilibathynella* and *Atopobathynella* occur in Chile, Australia, and New Zealand; the sister groups *Parvulobathynella* and *Acanthobathynella* are distributed in South America and the Ivory Coast; *Thermobathynella* is known from Amazonia and Central Africa, *Cteniobathynella* from central African area, Israel, and Brazil, and *Austrobathynella* from the southern parts of Africa and America. The history of the closely related genera *Texanobathynella* and *Iberobathynella* is related to the opening of the northern Atlantic. The *Iberobathynella* group is endemic to the Iberian Peninsula, while *Hexabathynella* exhibits a cosmopolitan distribution from Brazil, California, Iberian Peninsula, Bulgaria, Corsica, to Madagascar, Australia, and New Zealand. On a large scale, the Parabathynellidae display a worldwide distribution from temperate to tropical areas. The Bathynellidae occupy more limited regions of the globe, mainly in the temperate northern hemisphere as well as some local areas of the southern hemisphere. Therefore, plate tectonics has played a major role in the distribution and the evolution of the Syncarida through vicariance. According to Schminke (1981), the Syncarida have a freshwater origin. In contrast, other authors (Boutin &

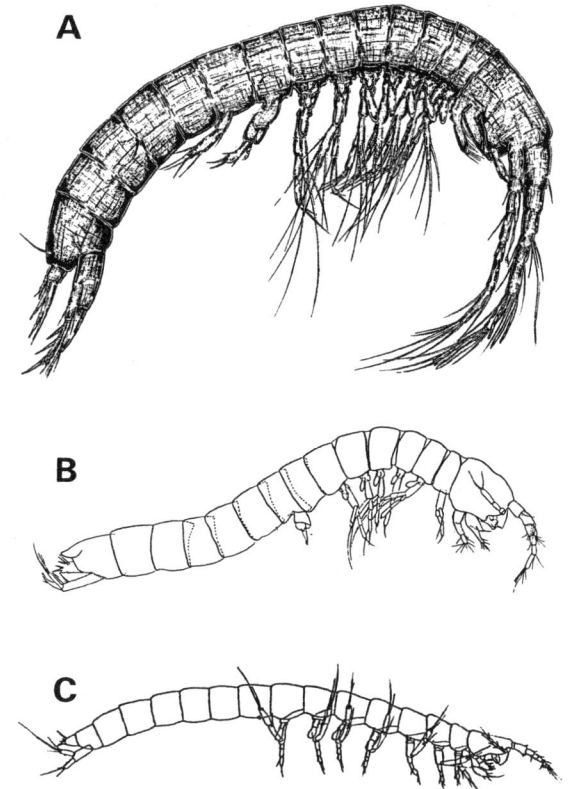

Crustacea: Syncarida: Figure 2. Bathynellacean syncarids:
A. *Bathynella (Bathynella) paranatans* Serban (0.9 mm, Bathynellidae);
B. *Hexabathynella knoepffleri* Coineau (1 mm, Parabathynellidae);
C. *Parvulobathynella riegelorum* Noodt (0.6 mm, Parabathynellidae).

Coineau, 1987; Camacho & Coineau, 1989; Coineau, 1996; Guil & Camacho, 2001) claim that freshwater Syncarida are derived from marine ancestors. These ancestors lived in the interstitial littoral of the Tethys Sea before the break-up of Pangea and of Gondwana, and colonized subterranean freshwater at the end of regressions mainly in the Jurassic and Cretaceous. Most of the extant species are located in areas formerly covered by Tethys gulfs and co-occur with other interstitial crustaceans of marine origin. Paleogeographic data also strongly indicate a marine origin. Tethys regressions, as well as orogenesis in some regions of the world, provide an understanding of the evolutionary history of the Syncarida.

NICOLE COINEAU AND ANA I. CAMACHO

Works Cited

Boutin, C. & Coineau N. 1987. *Iberobathynella* (Crustacea, Syncarida, Bathynellacea) sur le continent africain. Implications paléobiogéographiques. *Comptes Rendus de l'Académie des Sciences Paris*, 304: 355–58

Camacho, A.I. 1987. *La familia Parabathynellidae (Crustacea, Syncarida Bathynellacea) em la Peninsula Iberica. Taxonomia, filogenia y biogeografica*. Thesis University of Madrid

Camacho, A.I. & Coineau, N. 1989. Les Parabathynellacés (Crustacés Syncarides) de la Péninsule Ibérique. Répartition et Paléobiogéographie. *Mémoires de Biospéologie*, 16: 111–24

Coineau, N. 1996. Sous-classe des Eumalacostracés (Eumalacostraca Grobben, 1892), Super-Ordre des Syncarides (Syncarida Packard, 1885). In *Traité de Zoologie P.-P. Grassé, Crustacés*, edited by J. Forest, Paris: Masson, 7(2): 897–954

Coineau, N. 2000. Adaptations to interstitial groundwater life. In *Subterranean Ecosystems*, edited by H.K. Wilkens, D.C. Culver & W.F. Humphreys, Amsterdam and New York: Elsevier

Guil, N. & Camacho, A.I. 2001. Historical biogeography of *Iberobathynella* (Crustacea, Syncarida, Bathynellacea), endemic to the Iberian Peninsula. *Global Ecology and Biogeography*, 10: 487–501

Schram, F.R. 1981. Paleozoogeography of late Paleozoic and Triassic Malacostraca. *Systematic Zoology*, 26: 367–79

Schram, F.R. 1984. Fossil Syncarida. *Transactions of the San Diego Society for Natural History*, 20(13): 189–246

Schminke, H.K. 1973. Evolution, System und Verbreitungsgeschichte der Familie Parabathynellidae (Bathynellacea, Malacostraca). *Mikrofauna des Meeresbodens*, 24: 1–192

Schminke, H.K. 1981. Adaptation of Bathynellacea (Crustacea: Syncarida) to life in the interstitial ("Zoea Theory"). *Internationale Revue der Gesamten Hydrobiologie*, 66(4): 575–637

Serban, E. 1972. Bathynella (Podophallocarida, Bathynellacea). *Travaux de l'Institut de Spéologie Emile Racovitza*, 11: 11–224

Further Reading

Camacho, A.I. (editor) 1992. *The Natural History of Biospeleology*, Madrid: Museo Nacional de Ciencias Naturales

Juberthie, C. & Decu, V. (editors) 1999. *Encyclopaedia Biospeologica*, vol. 2, Moulis and Bucarest: Société de Biospéologie

Noodt, W. 1964. Naturliches System und Biogeographie der Syncarida (Crustacea Malacostraca). *Gewasser und Abwasser*, 37–38: 77–186

Noodt, W. 1970. Zur Eidonomie der Stygocaridacea, einer gruppe interstitieller Syncarida (Malacostraca). *Crustaceana*, 19: 227–44

Schram, F.R. (editor) 1986. *Crustacea*, Oxford and New York: Oxford University Press

Serban, E., Coineau, N. & Delamare Deboutteville, C. 1972. Recherches sur les Crustacés souterrains et mésopsammiques. I. Les Bathynellacés (Malacostraca) des régions méridionales de l'Europe occidentale. La sous-famille des Gallobathynellinae. *Mémoires du Muséum National d'Histoire Naturelle*, N.S., sér. A, Zool., 75: 1–107

Siewing, R. 1959. Syncarida. In *Bronn's Klassen und Ordnungen des Tierreichs*, edited by H. Weber, Leipzig: Akademische Verlagsgesellschaft, vol. 5(1) 4, 2: 1

CUBA

The Cuban archipelago, with an area of 110 992 km^2, is composed of the largest island of Cuba (104 km^2), the Isla de la Juventud, and more than 4000 smaller islands and cays in the Caribbean. In terms of geological structure, relief, and climate, Cuba can be divided into two regions: the Bahaman-Cuban plains (most of the island and the archipelago) and the mountains of the Oriente (in the southeast of Cuba).

About 60% of Cuba has carbonate rocks at outcrop or beneath thin soil cover: these are mainly of Mesozoic age and form very diversified karst terrains (Figure 1). The geologic structure of Cuba evolved from an archipelago of volcanic islands, with deposition of geosynclinal sediments. The main carbonate rocks were later folded during the Eocene orogeny. Subsequently the Cuban platform period was broken into tectonic blocks separated by faults and flexures. In the Neogene and Quaternary, shallow marine limestones were deposited on the marginal zones of the blocks, and now support a very distinctive littoral karst. Hydrothermal activity continues at mineral and thermal springs on the north coast (e.g. San Diego de los Baños), and some of the caves in this region are of hydrothermal origin.

The major caves of Cuba are extensive, sub-horizontal, multi-level systems. Sistema Cavernario Palmarito is the longest and most complex, with 48 km of passages on three main levels. It lies in the massif of Pan de Azucar, which has classical marginal poljes as well as cone and tower karst in the southern part of Sierra de los Organos. The highest karst massif is the Macizo de Guamuhaya, just to the north of Trinidad in central Cuba, where alternating calcareous schists and crystalline limestones rise in monoclines to a height of 300–700 m. In this region, near Topes de Collantes, Cueva Cuba Magiar (–390 m) is the deepest cave in Cuba.

During the Pleistocene, eustatic sea-level variations reached nearly 200 m in this part of the Caribbean and were of great importance for the development of karst over the whole Cuban archipelago. The most important consequences were the multiple levels of the cave systems, raised sea terraces with karstic and abrasional features, and intrusions of saline water reaching far into the interior, notably on the west coast of Pinar del Rio province, on the Zapata peninsula, and near Cabo Cruz in Oriente province. There are many cenotes in the coastal plains of the northwestern regions of Havana and Matanzas. In the inland phreatic zone below sea level there are extensive cave levels, draining to numerous submarine springs (mainly along the southern coast of western Cuba). Since karst waters are widely exploited in Cuba, saline water intrusions into the aquifers are among the most significant environmental problems. In the middle of the karst plains there are numerous flooded caves, shafts, and dolines.

Marine terraces along the coasts of Cuba are the site of some interesting karst processes. The terraces range from a few metres to 100 m above sea level and are characterized by zones of abrasional and karstic landforms. The lowest terraces are subject to mechanical erosion from waves and dissolution promoted by the mixing of saline water and rainfall, and by large-scale biochemical processes caused by algae and other organisms covering the coastal rocks. Among the landforms of littoral karren there are many bowl-shaped dolines, locally known as "casimbas". The finest littoral karst is in the coastal terraces of the Desembarco del Granma National Park, on the Sierra Maestra near Cabo Cruz, west of Santiago de Cuba. The National Park was established in 1986 (and later became a World Heritage Site) to ensure protection of the largest and the best-preserved system of littoral terraces in the world (from 180 m below sea level to 300 m above sea level) and also to protect the endemic sea fauna of the submarine terraces.

The climate of Cuba has annual dry and wet seasons with high amounts of rainfall. Up to 80% of the precipitation falls in the wet season from May till October, and annual totals are 1000–2000 mm. Mean air temperatures are 23–26°C. In these humid tropical climatic conditions, most of the bare surfaces of carbonate rocks within the Cuban archipelago are covered with corrosional microforms. The most common are fine networks

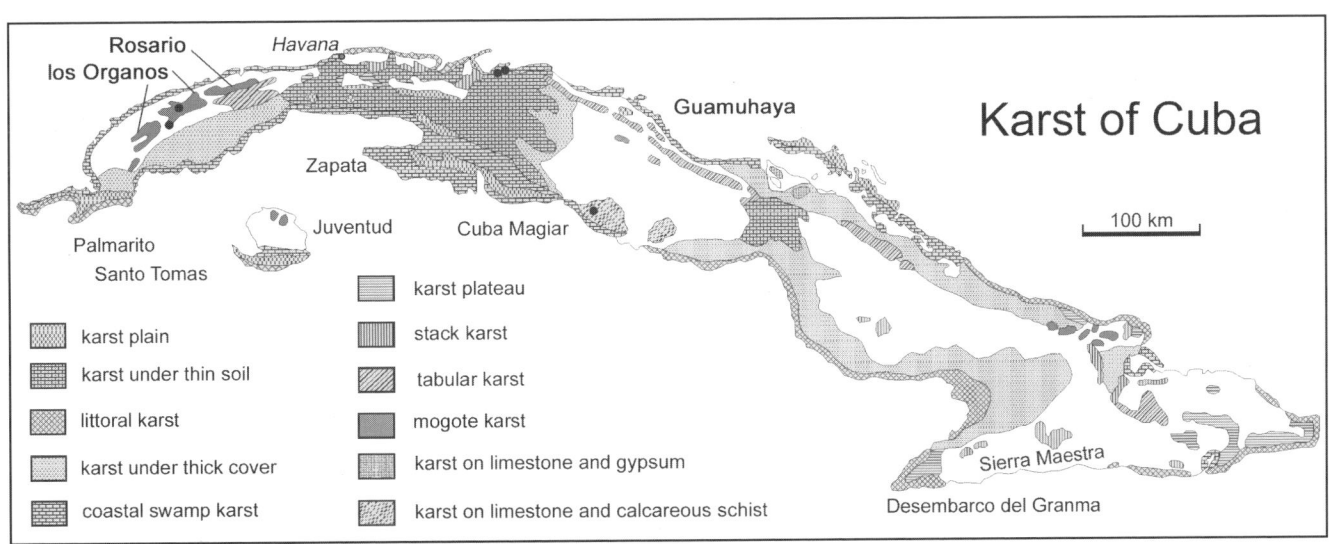

Cuba: Figure 1. Karst areas of Cuba (modified after Núñez Jiménez, 1984).

Cuba: Figure 2. The classic view across the mogotes of the Viñales valley in the Sierra de los Organos. (Photo by Tony Waltham)

of deep, cylindrical, flat-bottomed pits separated from each other by very sharp edges, locally known as "dientes de perro" (dog teeth).

The Guaniguanico mountain system in western Cuba has the most interesting humid tropical karst. Sierra de los Organos and the massif of Pan de Guajaibon (in the Sierra del Rosario) are some of the most extensive areas of cone and tower karst in the world. In the Sierra de los Organos, the most significant karst rock units are the Upper Jurassic to Lower Cretaceous Viñales massive and bedded limestones with flints and the Paleogene Ancon marly limestones and flysch. The dominant residual karst hills are called mogotes and have both cone and tower profiles (though their height/width ratios are lower than for towers of the Chinese karst). They form massifs within which there are small corrosion poljes (called hoyos). Between the massifs, both karstic and non-karstic, there are large karst-erosion poljes. The most famous polje of this type is the Valle de Viñales, which was listed as a World Heritage Site in 1999 due to its landscape values and traditional agriculture (mainly cultivation of tobacco) which has been unchanged for centuries (Figure 2).

The southeastern part of Sierra de los Organos is in the catchment of the Cuyaguateje River. This area of 840 km^2 is a fine hydrological system of mogote karst, with numerous central and border poljes as well as cave systems utilized by the river and its tributaries. In the massif of Sierra de Quemado there is one of the longest and best-known cave systems in Cuba—the Gran Caverna de Santo Tomás, with 46 km of mapped passages. The system has six recognizable relict and modern levels of the underground course of the Santo Tomás stream and its tributaries. It links the border poljes of Valle de Santo Tomás and Valle de Quemado (Figure 3). The cave developed in the dark Upper Jurassic limestones, and the border poljes developed on the contact zone with the Lower Jurassic San Cayetano shales and sandstones. The upper cave levels have abundant calcite speleothems, with many forms that are rare elsewhere. Notable are the many exceptionally large cones, some over a metre high, which are made of calcite rafts that fell to the floor of cave lakes when a frequent drip destroyed their surface tension support. The cave system is also known for its finds of cultural material left by the prehistoric inhabitants of Cuba, including the engravings in Cueva de Mesa and a human skeleton dated to 3000 years in Cueva de la Incógnita. The upper levels of the cave also have extensive accumulations of Pleistocene animal bones. For its multiple geological, paleontological, prehistorical, and historical values, the Gran Caverna de Santo Tomás was recognized as a national monument of Cuba in 1989.

Several caves in the Pinar del Rio, La Habana, and Matanzas provinces have been developed as tourist sites. These include Cuevas de Bellamar and Cueva Grande de Santa Catalina, both located in raised terraces in Matanzas. Both caves are also important for their speleothems, particularly for very large calcite crystals as well as large cones and mushrooms of fallen cave rafts.

ANDRZEJ TYC

See also **Caribbean Islands**

Further Reading

Nuevo Atlas Nacional de Cuba. 1988. La Habana: Instituto de Geografía de la Academia de Ciencias de Cuba

Núñez Jiménez, A. (editor) 1984. *Cuevas y carsos*, La Habana: Editoria Científico-Técnica; 2nd edition, 1990

Núñez Jiménez, A. 1990. *La Gran Caverna de Santo Tomás: Monumento Nacional*, La Habana: Ediciones Plaja Vieja

Panos, V. & Stelcl, O. 1968. Physiographic and geologic control in development of Cuban mogotes. *Zeitschrift für Geomorphologie*, 12(2): 117–73

Pulina, M. & Fagundo, J.R. 1992. Tropical karst and chemical denudation of Western Cuba. *Geographia Polonica*, 60: 195–216

Schenck, J.E., Porter, Ch. & Coons, D. 1999. NSS expedition to Cuba. *NSS News*, October 1999

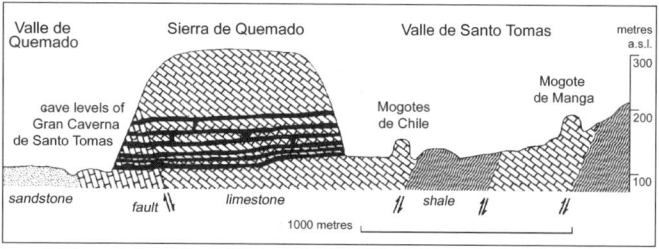

Cuba: Figure 3. Section of Sierra de Quemado and multiple cave levels in the Gran Caverna de Santo Tomás (modified after Núñez Jiménez, 1990, modified).

CUPP-COUTUNN CAVE, TURKMENISTAN

Cupp-Coutunn, the longest limestone cave in the former Soviet Union, is located in the Kugitang region of eastern Turkmenistan, close to the borders with Uzbekistan and Afghanistan. The cave is internationally famous for its mineralogy, both in the number of mineral species and in the great variety of its speleothem forms. In 1991, Cupp-Coutunn and its associated caves were placed on the Global Indicative List of Geological Sites (GILGES) by a working group of the World Heritage Convention.

The Cupp-Coutunn cave system (Figure 1; Maltsev & Self, 1992) is located on the southwestern flank of the Kugitangtau ridge, which runs north–south for some 50 km and attains an altitude of 3137 m. Low hills to the north connect the Kugitangtau ridge to the southwestern end of the Tien Shan mountain chain, from which it appears topographically as an outlier. The Kugitangtau is an anticlinal dome with a granite batholith intruded into its core. A major fault, the Eastern Kugitang Upthrust, along the north–south major axis of the anticline, separates the uplifted Kugitangtau from the Uzbek plain to the east. The Kugitangtau cuesta presents a scarp face over 1 km high, topped with a near-vertical wall of upper Jurassic limestone. The dip slope to the west is a gentle 7° to 15°, deeply dissected by canyons. This dip slope is cut by a second major fault, sub-parallel and of opposite throw to the Eastern Kugitang Upthrust, leaving the central spine of the ridge as a horst block.

The caves of southern Kugitangtau show a complex history. They developed as an extensive phreatic maze, following the then horizontal bedding in the limestone on several levels, probably in late Cretaceous times. These passages were subsequently abandoned as conduits for groundwater flow and infilled with argillaceous sediments. Neogene uplift and tilting of the strata rejuvenated the caves. Detrital deposits were locally eroded by groundwater introduced via a north–south fault set, though in places the consolidated remains of this ancient filling are retained as the walls or roof of the cave. New passage development was restricted to invasive tubes in the immediate vicinity of these faults. The air-filled passages were then invaded by thermal water entering from another (Chilgas) fault set, with chemical alteration of the cave wall rocks and sediments. Alteration effects include recrystallization of the limestone and replacement of carbonates by other minerals, mostly silicates and sulfides. Small amounts of hydrothermal calcite and fluorite were locally deposited within the cave, the calcite sometimes containing microscopic inclusions of metal sulfides. The thermal fluids do not appear to have been particularly aggressive, as no significant passage enlargement has been noted. Stable isotope analysis of rock and mineral samples suggests that the thermal water was an evolved basinal brine (Bottrell, Crowley & Self, 2001).

During the mid-Quaternary, there was some movement of the Chilgas faults. This caused the displacement of cave passages crossed by these faults, and collapse in some of the larger air-filled passages. The modern phase of development is marked by condensation corrosion of the cave walls and roof, the reworking of material from the thermal phase, and the deposition of speleothems. For the most part, the caves are independent of drainage from the surface, though some canyons have cut their beds deep enough to intersect the caves. However, the Kugitang region is semi-arid and significant quantities of water enter the caves about once per decade, locally introducing alluvium or reworking the older sediments.

The international importance of the Cupp-Coutunn cave system is in its speleothem mineralogy, which has been entirely deposited during the post-thermal phase of the caves' development. Some very rare minerals have been formed by the reworking of material from the thermal phase, whether this material was deposited in the cave itself or as vein minerals in fractures crossing the cave. Particularly interesting are zinc alumino-silicates, which are coloured green by trace quantities of nickel. Sauconite was the first to be identified, and then the very rare fraipontite (with only a few occurrences known worldwide, this is its first report as a cave mineral). Si and Al mobility is due to acidic gases released from a red, clay-like coating (ochre) on hydrothermally altered wall rock. Ochre is a variable but stratified material, with sulfate-reducing bacteria in the lower layers and oxidizing bacteria in the outer layers. It is more like a soil than a classic corrosion residue and various secondary oxide and silicate minerals have been identified within it (Maltsev, Korshunov & Semikolennykh, 1997). Hydrogen sulfide produced within the ocker is oxidized to sulfuric acid, which in turn attacks crystals of thermal phase fluorite. Gaseous hydrogen fluoride is released into the cave air, the evidence for this being tiny dark purple crystals of fluorite that have been found growing on calcite helictites and on euhedral gypsum crystals (Maltsev & Korshunov, 1998).

Cupp-Coutunn has spectacular displays of speleothems of the more common cave minerals. Where surface water infiltrates from canyon floors or from north–south faults, there are calcite stalactites, stalagmites, and even shields. A particularly beautiful banded flowstone (marble onyx) was commercially mined from 1970 until 1982, with three mine adits cut into the cave to provide easier access. The mining was finally halted following a campaign led by Moscow and Ashkhabad cavers. Aragonite is a main speleothem mineral in some parts of the cave, often in association with calcite. This association produces some of the most photogenic speleothems, such as quill anthodite bushes and multi-corallite splays (Figure 2). Helictites are abundant, sometimes with euhedral crystals of clear gypsum or blue celestite growing on their sides. Gypsum is found in most parts of the cave, as beards, flowers, crusts, chandeliers, and as large hollow stalagmites. Rare speleothem forms include solid crystalline stalagmites of gypsum, aragonite pseudo-stalactites, and macro-crystalline calcite stalactites.

Cupp-Coutunn / Promeszutochnaya (combined length 56 km) is part of a more extensive karst system that is still largely filled with ancient sediments. Nearby, and recognized as part of the Cupp-Coutunn system, is Geophyzicheskaya (length 4.5 km, with spectacular gypsum chandeliers) and Hashm-Oyeek (length 7 km, first reported by Diodorus Siculus c.40 BC). Dozens of other cave fragments have been found in neighbouring canyons throughout southern Kugitangtau. The area was once very popular with Russian cavers, but there have been few expeditions since Turkmenistan became an independent country.

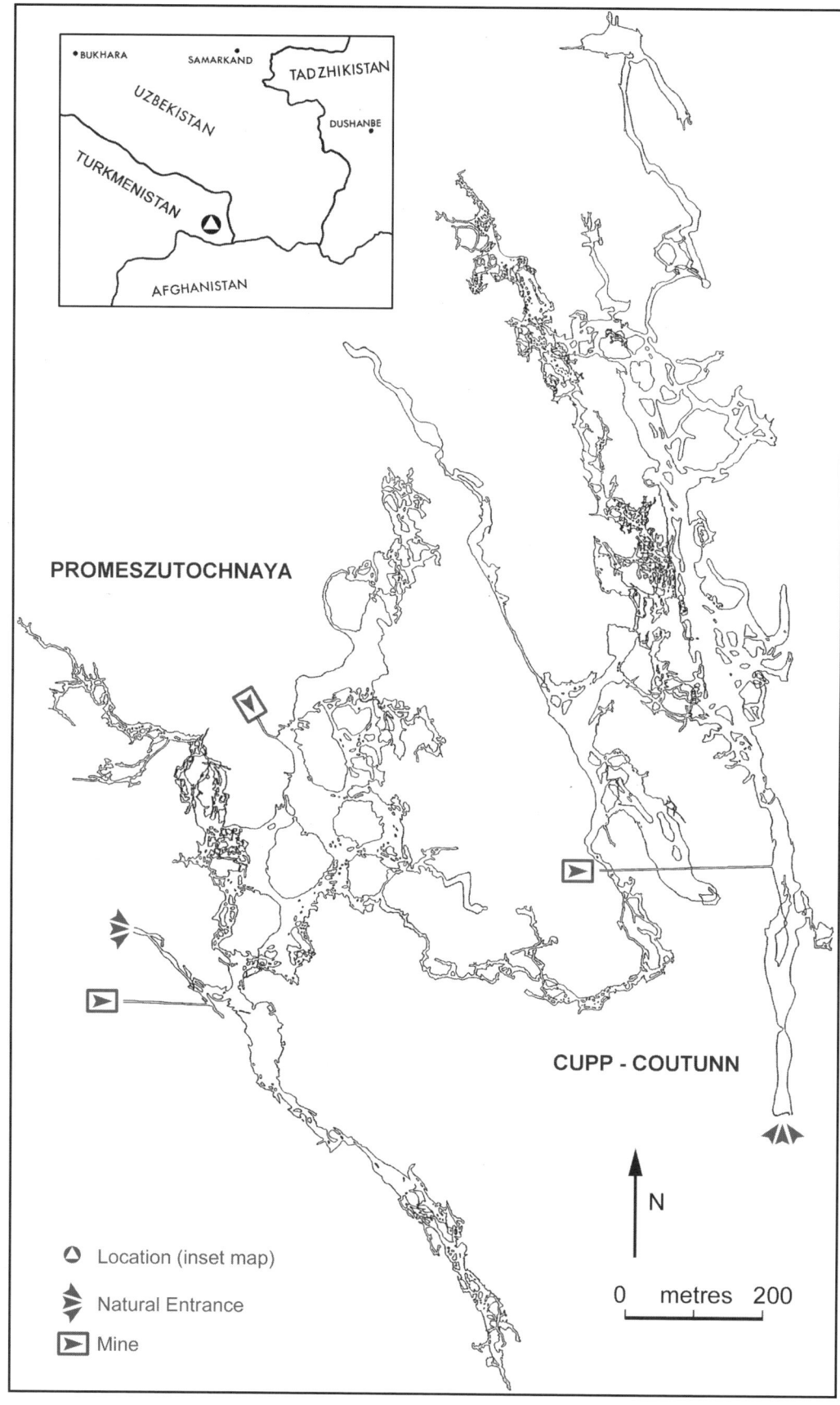

Cupp-Coutunn Cave, Turkmenistan: Figure 1. Cave survey (plan) and area map.

Cupp-Coutunn Cave, Turkmenistan: Figure 2. A calcite column has been overgrown by corallites (calcite), multi-corallites (calcite, aragonite, hydromagnesite), a branching helictite (aragonite), and a multi-corallite / spathite hybrid (the bent arm-like protrusion on the right). Aragonite spathites grow independently from the cave roof. (Photo by Charles Self)

Vladimir Maltsev's team (Moscow) was pre-eminent in both the exploration and the scientific study of the caves during the late Soviet period. Much of our modern understanding of ontogeny is based on speleothem observations made in these caves.

CHARLES ANTHONY SELF

Works Cited

Bottrell, S.H., Crowley, S. & Self, C.A. 2001. Invasion of a karst aquifer by hydrothermal fluids: evidence from stable isotopic compositions of cave mineralization. *Geofluids*, 1: 103–21

Maltsev, V.A. & Korshunov, V.A. 1998. Geochemistry of fluorite and related features of the Kugitangtou ridge caves, Turkmenistan. *Journal of Cave and Karst Studies*, 60(3): 151–55

Maltsev, V.A., Korshunov, V.A. & Semikolennykh, A.A. 1997. Cave chemolithotrophic soils. *Proceedings of the 12th International Congress of Speleology*, vol. 1, Basel, Switzerland: Speleo Projects: 29–32

Maltsev, V.A. & Self, C.A. 1992. Cupp-Coutunn cave system, Turkmenia, Central Asia. *Proceedings of the University of Bristol Spelaeological Society*, 19(2): 117–49

Further Reading

Maltsev, V.A. 1993. Minerals of the Cupp Coutann karst cave system, Southeast Turkmenistan. *World of Stones*, 1: 5–30
A detailed description in English with many colour photographs.

Maltsev, V.A. 1997. Cupp-Coutunn Cave, Turkmenistan. In *Cave Minerals of the World*, 2nd edition, C.A. Hill & P. Forti, Huntsville, Alabama: National Speleological Society
This book contains more than fifty other references to the cave in the main text.

http://fadr.msu.ru/~vvkor/maltsev/kugitang-caves.htm
A very well-constructed site with many photographs.

D

DAMS AND RESERVOIRS ON KARST

The nature of karst presents a great variety of risks associated with the construction of dams and reservoirs. The most frequent difficulties are caverns at dam sites, land subsidence at the bottom of reservoirs, water leakage at dam sites and from reservoirs, induced seismicity as a consequence of artificial storage, endangerment of endemic fauna, and changing of local groundwater balance. Because each karst region is different individual situations are very seldom, if ever, repeated.

Although the need for artificial water storage in karst has been recognized as important for a long time, dam construction projects were initially very rare and mostly resulted in failure. However, there is a long history of smaller-scale hydraulic works including cleaning the entrance and improving the flow capacity of ponors (in the Dinaric karst area, and in Greece, Turkey, and China), construction of small water storage ponds, and construction of water-driven mills at ponors. Dam construction dates from 1700 years ago (Iran, Sassanian era), but dam building in karst environments commenced only in the 20th century (Hales Bar, Tennessee, 1922; Camarasa, Spain, 1927; Wolf Creek, Kentucky, 1941). In the early 1950s the construction of reservoirs in the Dinaric karst region, former Yugoslavia, posed a special challenge. Notable examples were Bileća reservoir on the Trebišnjica River with a storage capacity of 1.28 billion m^3, reservoirs in Nikšičko Polje with 220 million m^3, and the Buško Blato Reservoir with 800 million m^3. After the 1950s a number of large reservoirs were built in other karst areas of the world.

Reservoirs in karst may fail to fill despite an intensive investigation programme and sealing treatment. In most cases the leakage occurs during the first filling. Among the largest-known losses were from Keban (Turkey), leakage 26 m^3 s^{-1}; Vrtac (Yugoslavia), 25 m^3 s^{-1}; Iliki (Greece), 13 m^3 s^{-1}; Camarasa (Spain), 11.2 m^3 s^{-1}; Lar (Iran), 10.8 m^3 s^{-1}; Marun (Iran), 10 m^3 s^{-1}; Mavrovo (FYR Macedonia), 9.5 m^3 s^{-1}; Great Falls (United States), 9.5 m^3 s^{-1}; and Canelles (Spain), 8 m^3 s^{-1}.

Persistent, time-consuming and step-by-step sealing treatment can in some cases resolve the problem and justify invested money. For example, the leakage was considerably reduced from Keban, Camarasa, Great Falls, Canelles, Mavrovo, and Marun. However in other cases the hydrogeological conditions were too complex for the available sealing technology and unacceptable leakage still exists at Vrtac, Lar, and Iliki.

According to the generalized geomorphological and hydrogeological features, dam sites and reservoirs in karst can be divided into three groups: dams and reservoirs in deep and narrow canyons; dams and reservoirs in river valleys; and dams and reservoirs in karst poljes. In the first case fluvial erosion is faster than the karstification process and karst features below the river bottom are rare as the water table is connected with the riverbed or is very close. Such riverbeds have limited leakage although the upper parts of the dam site and reservoir banks can be prone to leakage. A grout curtain in the riverbed and into both abutments is usually necessary but is technically and economically acceptable. In the second case the evolution of the karst aquifer proceeds at a similar speed to, or faster than fluvial erosion. Hence caverns and karst channels may be present at significant depth below the riverbed. The most unfavourable conditions occur in valleys with a hanging riverbed situated at high elevation. In such situations the water table is deep below the valley bottom. Deep and expensive grouting treatment and plugging of caverns are likely to be necessary. However, if the river valley is at base level, i.e. the discharge zone for the adjacent aquifer, the chances for safe water storage are better.

In the third case, karst poljes are hydrogeologically very complex places for dam and reservoir construction. Wide and very large zones of concentrated infiltration (ponors) are located in these depressions, particularly along the foothills, and the base of karstification is very deep. During the dry period the water table is deep below the polje bottom. During the wet period of the year, the water table rises abruptly and a considerable part of the floor is under the influence of a strong upwards recharge. The most common and unpredictable defects found during reservoir impounding and operation are collapses, funnels, and huge open cracks in the reservoir bottom. The treatment of karst poljes generally needs long and step-by-step remedial works, including deep and long grout curtains.

Large caverns at dam sites need a very detailed investigational approach using different methods (geology, geophysics, and speleology) and complicated treatment technology. During the construction of the Keban Dam (Turkey) about 30 caverns were treated. To fill the largest one about 605 000 m³ of limestone blocks, gravel, sand, and clay were thrown into the cave. The volume of the second largest cavern at the same dam site was 105 000 m³ (110 m below the river bed). At the Lar dam site (Iran) a few large caverns were discovered at depths of 250 m to 430 m below the riverbed. Some 117 000 m³ of fill was used, but reservoir losses are still very high (Djalaly, 1988). Cavities filled with soft clay at the Wujiangdu dam site (China; Figure 1) were treated down to a depth of about 250 m beneath the river bed (Zuomei & Pinshou, 1982). A grout curtain was used to isolate caverns on both sides of the Sklope dam site (Bozicević, 1971). At the Salman Farsi Dam (Iran) more than 20 caverns were discovered on both sides, along the grout curtain route, the largest being about 60 m long, 40 m high, and 20 m wide.

To reduce water losses from reservoirs in karst areas two alternative methods are applied. First, the surface can be made impermeable by encompassing large ponors with cylindrical concrete dams; isolating large ponor zones by dykes; closing estavelles by concrete plugs equipped with non-return valves; protecting sinking zones in alluvial overburden by surface compacting; protecting sinking zones in alluvial overburden by PVC foil; blanketing of karstified limestone by shotcrete; and dental treatment (filling of cavities) and various similar structures. Second, the underground medium can be sealed by grouting using different types of cement and cement mortar with additives and

Dams and Reservoirs on Karst: Figure 2. In the Puleirang doline, in Java's Gunung Sewu Karst, a hand-built concrete dam stands round the entrance to a shaft series that descends 200 m to the regional water table. Chunam concrete seals the limestone around the doline and its floor is a thick clay. The reservoir was a failure, as it drained out when a sinkhole developed in the clay soon after the first wet-season filling. (Photo by Tony Waltham)

cavern fillings (sand, gravel, chemicals, etc); polyurethane foam grouting; grouting with asphalt; construction of concrete plugs in karst channels; and construction of cut-off (diaphragm) walls. In special cases the "bath-tub" solution was applied, linking the dam site with an upstream aquaclude (e.g. Oymapinar, Turkey; El Cajon, Honduras). In some cases the grout curtains are very long (e.g. Buško Blato 6.77 km; Slano 6.2 km; Mosul (Iraq) 6 km) and deep (up to 300 m at Keban; Mratinje (Yugoslavia) 278 m; Wujiangdu 250 m).

The construction of dams and reservoirs in karst regions results in considerable environmental impacts. Mostly the impacts are positive and predictable (flood reduction, water supply improvement, power production, infrastructure improvement, and reduction of deforestation), but in some cases they are negative and unpredictable. The common negative environmental impacts are: severe changes to spring discharge, deterioration of groundwater quantity and quality, threats to endemic fauna, induced seismicity, and induced occurrence of collapse dolines and some other secondary uncertainties. For example, construction of the Trebišnjica Hydrosystem (Herzegovina) is expected to eliminate recharge through many ponors resulting in a loss of recharge to the East Herzegovina karst aquifers of around 6 billion m³ per annum. The low flow discharge of some springs has been increased but the annual average discharge at the large springs has decreased considerably. For example, the average yearly discharge of Ombla Spring was reduced from 33.8 m³ s^{-1} in natural conditions to 24.4 m³ s^{-1} after construction. Submergence of the Trebišnjica spring zone by Bileća reservoir prolonged the duration of floods in the Fatnicko Polje.

Induced seismic shocks, i.e. explosions of trapped "air pillows" in karst channels and siphons are recorded by nearby seismological observations. Subsidence induced by extensive water level fluctuation in man-made reservoirs may provoke considerable damage in the surroundings. For example at the Mavrovo

Dams and Reservoirs on Karst: Figure 1. A concrete buttress dam built in the limestone gorge of the Wu River in Chongqing District, China. (Photo by John Gunn)

DATING METHODS: ARCHAEOLOGICAL

In approximate order of durability, the buried archaeological materials that survive in caves are stone tools, teeth (including ivory), bone, pollen, metal, and vegetable material. The cave environment itself also determines how well objects survive and thus affects the viability of any direct dating attempt. After burial, stone tools survive for millions of years under most conditions. Bone and teeth survive very well in alkaline conditions but pollen does not. Speleothems and bone are both eroded and rotted by bat or avian guano due to acid leached from the decomposing guano. Pollen survives best in anoxic, acid environments and this is also true of wood and leather. Metals survive best if they are in a dry environment although impurities within most metal objects mean that they are liable to pitting. Most developments in dating methods have been aimed at applications to the time period 10 ka to greater than 3 Ma rather than to historical periods, due to the great interest in the evolution of humans and culture. To some extent, these applications have been paralleled by interest in dating the internal and external palaeoenvironment of the cave and in understanding the development of karst landscapes (see Paleoenvironments). Table 1 is a summary of common dating methods as applied to chronological studies of caves, archaeology, and the Pleistocene.

Stone tools cannot be dated directly except, occasionally, by the measurement of a small hydration rim that develops on obsidian, an amorphous (glassy) anhydrous silicate. As obsidian dating is mostly of concern to the relative dating of open-air Meso-American sites and is rarely met in cave archaeology the reader is referred to Friedman *et al.* (1997) for further information.

Stone tools that have been burnt in a hearth can be dated by thermoluminescence (TL). TL dating is a technique that measures the total radiation dose stored in trapped electrons in the mineral, as received from internal and external radioactive emission. The TL method relies on the assumptions that the mineral had no radiation dose upon burial or immediately after heating and that the average dose rate from burial to the present can be estimated from dose rates measured today. The dose acquired since the object was fired is measured by releasing the energy from the trapped electrons by progressive heating to produce a "glow curve". A dosimeter left in the burial environment

Dating Methods: Archaeological: K-Ar, potassium-argon; Ar-Ar = argon-argon. TL, thermoluminescence, OSL, optically stimulated luminescence, IRSL, infra-red stimulated luminescence. *Items in brackets show material that may not be as good for a reliable date. (After Aitken, 1990 and other sources)

Method	Age range	Material*
Dendrochronology	Up to about 10 ka	Wood
Radiocarbon	Up to 50 ka (uncalibrated)	Wood, charcoal, organics
K-Ar (Ar-Ar)	From 40ka to > 100 ka	Potassium rich minerals
Uranium Series (by mass-spectrometry).	Up to 500 ka	Speleothem, corals (tooth enamel, bone, peat)
Fission tracks	1 ka to > 100 ka	Volcanic glass and minerals
Luminescence (TL, OSL, IRSL)	Up to about 200 ka	Pottery, quartz grains, burnt flint (speleothem)
Electron spin resonance	Up to several Ma	Tooth enamel, quartz (speleothem).
Obsidian hydration	Up to several Ma	Obsidian
$^{10}Be/^{26}Al$	Up to several Ma	Silica sands and pebbles

for a year, typically, can be used to estimate the dose rate, and thus the age. The method was originally used to estimate ages of fired pottery, where the initial, archaeological dose is known to be zero, and it is frequently used to test for forgeries of antique pottery. One variation of the method relies on zeroing the past dose of quartz grains (and less reliably, feldspar grains) in sediments by sunlight, an effect known as bleaching, and is known as Optically Stimulated Luminescence (OSL). A second variant looks at radiation traps at the infrared end of the spectrum and is called IRSL. The technique can be used on single grains and it has the advantage that those grains that have been zeroed by sunlight can be distinguished and separated out from those within the same sample that have not been sufficiently bleached by sunlight. There seems to be no reason why these methods could not be used more frequently on cave sediments containing quartz grains. TL itself has been used a number of times on burnt flint from caves of the Levant (see Carmel Caves, Israel: Archaeology) where the zeroing occurred when flint was prepared in ancient cave hearths (Mercier et al., 1994).

Carbon-14 dating (Taylor, 1997) is regularly applied to the organic component of bone called collagen, including bones from caves, with varying degrees of success. Radiogenic ^{14}C is produced in the upper atmosphere by cosmic rays at a more-or-less constant rate. Living organisms take up ^{14}C as well as the stable isotopes, ^{13}C and ^{12}C. After death, there is no ^{14}C replenishment and the ^{14}C decays by beta particle emission at a rate determined by its half-life of 5568 ± 30 years. As the rate of production of radiogenic ^{14}C is not quite constant over time, a correction has to be made using a tree-ring master curve. The resulting date is quoted as radiocarbon calibrated ("cal" for short; e.g. 7680 ± 150 BP cal, where BP stands for "Before Present", and where "present" is 1950). The normal limit of the dating method without calibration is about 40–50 000 years, depending on the sample and upon the laboratory set-up. Many samples are now dated using an accelerator mass spectrometer (AMS) that measures the ratio of ^{14}C to ^{12}C isotopes rather than the number of beta decays. Beyond the range of tree-ring calibration, ^{14}C dates have been cross-calibrated using Uranium-Thorium (U–Th) dates and these indicate that a radiocarbon age of 18 000 years is too low by 2000 years and a radiocarbon age of 36 000 years may be too low by about 4000 years, though this latter remains to be confirmed.

Although bone is often well preserved in the slightly alkaline conditions of caves, it is porous and there may have been exchange with water in the sediment. Usually groundwater, particularly that coming into contact with limestone, has low ^{14}C compared to the other isotopes. This "dead carbon" has the effect of making the bone appear much older and it is difficult to make an allowance for it. This is also true of other organic materials that may become buried in a cave.

There are two other direct ways in which bone may be dated, the first being the U–Th method and the second the Electron Spin Resonance (ESR) method. It has been found that buried bone often takes up uranium, which is soluble in groundwater, soon after burial, but not the thorium daughter product, which is insoluble. Any thorium that is present in buried bone must have grown radiogenically from the uranium. For applications of U-series dating to archaeological materials see Schwarcz & Blackwell (1992) or Schwarcz (1997).

The U–Th method is more reliably used on speleothems (see entry on Paleoenvironments: Speleothems). Stalagmites and flowstones are sometimes associated with archaeological material lying above or below it so that U–Th analyses provide either maximum or minimum dates. A good example is the archaic *Homo sapiens* skull at Petralona cave, northern Greece, that was found attached to a wall by a thin layer of pink-brown flowstone (see Europe, Mediterranean: Archaeological and Paleontological Caves). By ESR and U–Th methods the calcite was found to be about 140–180 ka giving a minimum age for the skull (Latham & Schwarcz, 1992; Grün, 1996). As judged by comparative morphology, most paleoanthropologists think that the skull is about twice this age.

The ESR method (Grün, 2001) relies on measuring single trapped electrons that have accumulated at certain sites in bone phosphate due to the long-term damage by radiation. In this case, the radioactive emitters may be inside the bone, as in the case of U, Th, Ra and other decay daughters, or be in the nearby sediment or arise from energetic cosmic rays. Hence an estimate has to be made of internal and external dose rates to the material being measured and this is done either by direct analysis of the radioactive emitters or by using a dosimeter in the sediment. As for TL, the age of the material is then worked out from the total dose and the dose rate. One of the largest uncertainties is the effect of moisture over the burial time of the bone (or other object) because groundwater carries radioactive material and pore water also acts as a radiation absorber. Although the ESR method has been used on bone, it is thought that it is more reliable to use the enamel of teeth, as enamel, being made of hydroxyapatite, is non-porous and more resistant to the movement of U in or out of the sample.

Unfortunately, in the case of bone, we are often not quite certain that the uranium was always taken up immediately upon burial and, because of the porous nature of bone, that the system has remained closed to the migration of uranium. In finding out how and when the uranium was taken up in bone and teeth it is usual to consider two model scenarios (Grün, Schwartz & Chadam, 1988; Latham, 1997). In the first, we assume that the object has taken up U soon after burial and the U–Th age or ESR age is then known as the Early Uptake (EU) age estimate. Alternatively, it could be assumed that the U has been taken up at a constant rate from burial till now. This is a linear accumulation of U (LU) giving an LU age estimate. It is usual to compare the EU and LU age estimates to see how close they are and to see which, if any, is concordant with ages estimated by other methods. The models will not work if, however, there has been an unknown delay between burial and uptake of U and, in such a case, the U–Th and ESR ages will underestimate the true age.

Since the Earth's magnetic field reverses its polarity periodically, a magnetic reversal stratigraphy can be correlated to an independently derived timescale (see Paleoenvironments: Clastic Sediments). McFadden et al. (1979) used magnetic reversals at the Limeworks Cave, Makapansgat, in South Africa, to give age brackets of 2.0–3.2 Ma for a layer containing *Australopithecus africanus*. Magnetostratigraphic studies are underway at Sterkfontein and other Australopithecine caves in South Africa (see Africa, South: Archaeological Caves). The remains of earliest-known hominids in Europe are those found in the Gran Dolina infilling, Burgos, north Spain, and the reversed polarity of the sediments in which they lie dates them at more than 0.78 Ma.

The first attempts to date the famous Upper Palaeolithic rock art (parietal art) of caves of Spain and France was via a stylistic approach (see Art in Caves: History) in which progressions of style with time were thought to be recognisable. Now more direct methods have been used to date rock art, mostly by ^{14}C dating of pigments and binders containing charcoal, blood, and plant-derived material (Valladas et al., 1992). The dating of these pigments has been facilitated by the much smaller sample sizes needed in AMS over the older ^{14}C decay counting method. Studies in Cougnac Cave, Lot, France, showed that earlier paintings mostly of giant elk (*Megaloceros*) in red ochre were partly retouched in charcoal. The ^{14}C ages showed that the earliest paintings may have been Gravettian (28–23 000 BP) and the latest, in the same style, were as young as 14 000 BP, 10 millennia later, and of probable Magdalenian culture. The oldest dated paintings are those of Chauvet Cave, Ardèche, at 32–30 000 BP followed by those of the Mediterranean coastal cave of Cosquer at about 27 000 BP. Samples of bone from the "Red Lady" from Paviland Cave, Gower, South Wales, and associated with the use of ochre and art objects, date to between 25 250 and 27 450 BP (Aldhouse-Green, 2000).

In summary, it is no exaggeration to say that caves and rock shelters have frequently offered a unique physical and chemical environment for preserving a wide range of archaeological and environmental material and this has enabled the development and testing of a number of dating techniques. Beginning around 1980, these techniques have contributed in no small way to a new understanding of human evolution and culture.

ALF G. LATHAM

See also **Palynology**

Works Cited

Aitken, M.J. 1990. *Science-based Dating in Archaeology*, London and New York: Longman

Aldhouse-Green, S. 2000. *Paviland Cave and the "Red Lady": A Definitive Report*, Bristol: Western Academic and Specialist Press

Friedman, I., Trembour, F.W. & Hughes, R.E. 1997. Obsidian dating. In *Chronometric Dating in Archaeology*, edited by R.E. Taylor & M.J. Aitken, New York and London: Plenum Press

Grün, R. 1996. A re-analysis of ESR dating results associated with the Petralona hominid. *Journal of Human Evolution*, 30: 227–31

Grün, R., Schwarcz, H.P. & Chadam, J.M. 1988. ESR dating of tooth enamel: Coupled correction for U-uptake and U-series disequilibrium. *Nuclear Tracks and Radiation Measurements*, 14: 247–41

Grün, R. 2001. Electron spin resonance dating. In *Handbook of Archaeological Sciences*, edited by D.R. Brothwell & A.M. Pollard, Chichester and New York: Wiley

Latham, A.G. 1997. U-Series dating of bone by gamma-ray spectrometry: Comment. *Archaeometry*, 39: 217–19

Latham, A.G. & Schwarcz, H.P. 1992. The Petralona hominid site—U-series re-analysis of "Layer 10" calcite and associated palaeomagnetic analyses. *Archaeometry*, 34: 135–40

McFadden, P.L., Brock, A. & Partridge, T.C. 1979. Palaeomagnetism and the age of the Makapansgat hominid site. *Earth and Planetary Science Letters*, 44: 373–82

Mercier, N., Valladas, H., Jelinek, A., Meignen, L., Jonon, J.-L. & Reyss, J.-L. 1994. TL dates of burnt flints from Jelinek's excavations at Tabun and their implications. *Journal of Archaeological Science*, 22: 495–509

Schwarcz, H.P. & Blackwell, B. 1992. Archaeological applications. In *Uranium Series Disequilibrium: Applications to Earth, Marine and Environmental Sciences*, edited by M. Ivanovich & R.S. Harmon, 2nd edition, Oxford: Clarendon Press and New York: Oxford University Press

Schwarcz, H.P. 1997. Uranium Series Dating. In *Chronometric Dating in Archaeology*, edited by R.E. Taylor & M.J. Aitken, New York: Plenum Press

Taylor, R.E. 1997. Radiocarbon dating. In *Chronometric Dating in Archaeology*, edited by R.E. Taylor & M.J. Aitken, New York: Plenum Press

Valladas, H., Cachier, H., Maurice, P., Bernaldo de Quiros, F., Clotte, J., Cabrera Valdes, V., Uzquiano, P. & Arnold, M. 1992. Direct radiocarbon dates for prehistoric paintings at the Altamira, El Castillo and Niaux Cave. *Nature*, 357: 68–70

Further Reading

Aitken, M.J. 1998. *Introduction to Optical Dating: The Dating of Quaternary Sediments by the Use of Photon-stimulated Luminescence*, Oxford and New York: Oxford University Press

Aitken, M.J. 1997. Luminescence dating. In *Chronometric Dating in Archaeology*, edited by R.E. Taylor & M.J. Aitken, New York: Plenum Press

Aitken, M., Stringer, C.B. & Mellars, P. (editors) 1992. *The Origin of Modern Humans and the Impact of Chronometric Dating*, Princeton, New Jersey: Princeton University Press

Latham, A.G. 2001. Uranium-series dating. In *Handbook of Archaeological Sciences*, edited by D.R. Brothwell & A.M. Pollard, Chichester and New York: Wiley

Noller, J.S., Sowers, J.M. & Lettis, W.R. (editors) 2000. *Quaternary Geochronology: Methods and Applications*, Washington, DC: American Geophysical Union

Taylor, R.E. & Aitken, M.J. (editors) 1997. *Chronometric Dating in Archaeology*, New York: Plenum Press

DATING OF KARST LANDFORMS

Most of the absolute dating in karst has been made on speleothems (see Paleoenvironments: Speleothems entry) or clastic sediments (see Paleoenvironments: Clastic Sediments entry). However the absolute dating of karstic and pseudokarstic geomorphological surfaces is even more promising. Very little has been done on this important topic so far, with many karst researchers not aware that this is possible.

Precise dating of surface landforms is a very difficult task that requires measurements of very low concentrations in small samples, with high precision, which cannot be done with conventional techniques such as α- or β-counting or mass spectrometry. This was solved only when accelerator mass spectrometry (AMS) was introduced 20 years ago, for measurement of cosmogenic isotopes in rock samples. The most precise dating has been achieved with tandem accelerator mass spectrometry (TAMS) facilities.

Absolute dating of surface landforms can be done only by measuring cosmogenic isotopes (e.g. ^{36}Cl, ^{26}Al, ^{10}Be, ^{41}Ca, and ^{14}C) produced *in situ* in rocks at the Earth's surface. These isotopes are formed from secondary cosmic radiation, only in the surface few millimetres of the rock, by radiogenic nuclear reactions (Lal, 1988). There are three principal mechanisms by which cosmogenic isotopes are produced in terrestrial rocks: high-energy spallation by nucleons; neutron capture reactions; and muon-induced nuclear disintegrations. Cosmogenic isotope accumulation starts when the fresh rock surface is exposed to cosmic rays. ^{36}Cl, ^{26}Al, and ^{10}Be are formed with annual rates of 10–50 atoms $g^{-1} a^{-1}$ in appropriate rocks at sea level, four times faster at an altitude of 2000 m and 30 times faster at 5000 m. So the surface of a rock formed at sea level 1000 years ago can be dated by AMS in a sample of a few tens of grams. Concentration of cosmogenic isotopes depends on the irradiation site and geometry of the rock surface, so datable samples require knowledge of the geometric evolution of the rock.

The rock being dated should be a closed system in respect to its losing or gaining an isotope from the environment. To avoid such processes, it is best to use quartz grains for dating to avoid possible interference from atmospheric cosmogenic isotopes being absorbed into the rock matrix via seeping water, but other rocks without quartz can also be dated. Samples for dating can be divided into two types: samples whose past evolution is known a priori, and samples whose past evolution geometry is known qualitatively but the time constants are not known. The expected isotope concentrations due to *in situ* production can be determined if both the cosmic-ray intensity $I(t)$ and the sample geometry evolution $G(t)$ are known. Therefore, if one of these is known the other can be modelled; the degree of resolution on the parameter evaluated would depend on the number of isotopes studied.

Samples with well-known evolution and geometry—trees, monuments, rock surfaces polished by glaciers—give information about changes of the geomagnetic dipole field of the Earth; movement of the sample's latitude or altitude due to uplift, folding, or plate tectonic drift; and changes in the primary cosmic-ray flux due to solar modulation. Some non-characterized samples have a qualitative history of irradiation derived from geological or geophysical evidence, such as accreting sediments and eroding or denuding rocks. In such cases we can measure the rate of accumulation or erosion (or denudation) of the rock surfaces.

For example, we consider the expected concentration of an isotope C in a rock surface, eroding at a constant rate ϵ at a fixed location. Let us further assume that the rock formed from igneous activity at some time T in the past; in this case, the initial isotope concentration is zero. After T years of irradiation, we would expect the following depth profile for the concentration $C(x)$ of the isotope (Lal & Arnold, 1985):

$$C(x) = (1 - e^{-\lambda T})[P_o e^{-\mu X}/(\lambda + \epsilon\rho\mu)]$$

where P_o is the isotope production rate at the rock surface, λ is the disintegration constant of the isotope, ρ is the mean density of the rock, μ is the inverse of the mean absorption length for cosmic-ray particles in the rock, and the units of X, μ and ϵ are $g\,cm^{-2}$, $cm^2\,g^{-1}$, and $cm\,s^{-1}$, respectively. This equation assumes constant isotope production and erosion rates and no pre-irradiation history of the rock. If several metres of rock have been eroded away, the isotope concentration in the rock is in a quasi-steady state, and the surface concentration C_s becomes:

$$C_s(X = 0) = P_o/(\lambda + \epsilon\rho\mu),$$

with the concentration decreasing exponentially with depth X.

The resulting surface isotope concentration depends on the isotope half-life, the mean erosion rate, and the isotope production rate. We may define two irradiation time periods: the equivalent time T_{eq} (time taken to produce certain isotope concentration) and the effective time T_{eff} (time corresponding to the ratio of the concentration to production rate). We obtain:

$$T_{eq} = \ln(1 + \lambda/\epsilon\rho\mu)/\lambda,$$

$$T_{eff} = 1/(\lambda + \epsilon\rho\mu)$$

Erosion (corrosion) sets a new equilibrium value, lowering the mean lifetime of the isotope by a factor $1/(1 + \epsilon\rho\mu)$. This is equivalent to saying that erosion mean lifetime is $1/(\epsilon\rho\mu)$. The characteristic mean erosion period T_{ero} is defined by the ray attenuation length $(1/\mu)$, the density ρ, and the erosion rate ϵ:

$$T_{ero} = 1/(\epsilon\rho\mu).$$

The mean erosion period T_{ero} becomes comparable to one million years (the order of half-lives of ^{26}Al and ^{10}Be) when $\epsilon \sim 5 \times 10^{-5}\,cm\,a^{-1}$. For smaller erosion rates, the effective irradiation time will be determined by the isotope half-life, and for higher erosion rates, by the erosion rate. From a study of concentrations of isotopes of different half-lives, an exposed surface can be characterized and erosion rate determined. The erosion timescales accessible are simple functions of the isotope half-life (Lal, 1988). Such appropriate pairs of isotopes for limestone rocks are ^{14}C and ^{36}Cl or ^{41}Ca, but for dolomite rocks ^{10}Be can be used. ^{26}Al can also be used for dating surfaces of volcanic rocks, or limestones with high concentration of quartz (sand), because it is formed from silica-rich material. For silicate rocks, it is best to use ^{26}Al and ^{10}Be, but for rocks with high Ca or Fe concentration, ^{36}Cl and ^{41}Ca can also be used. Age determinations are possible only if two or more isotopes are used (Lal, 1987); unless the magnitude of erosion is known a priori, one has to determine both erosion rate and the time of exposure of the surface.

^{36}Cl has been used by Kubik *et al.* (1984) to date limestone landforms. Raisbeck and Yiou (1979) discussed the possible applications of ^{41}Ca for dating of $CaCO_3$ sediments.

Dating of Volcanic Caves

Absolute dating of volcanic caves may use two different types of methods depending on genesis of the caves. Volcanic lava tubes are formed during the cooling of the lava, so have the same age as the bedrock. This allows us to use the conventional isotopic disequilibrium dating methods that produce reliable dates (Faure, 1986). Some lava tubes in active volcanic regions (Hawaii, Iceland, Etna) are very young, but others in paleovolcanic regions (Rhodope Mountains, Bulgaria) are of Oligocene age (Shopov & Kolev, 1992). Appropriate methods are ^{230}Th/^{238}U dating for 1000 to 400 000 year old rocks, K/Ar for 100 000 to 2 billion year old rocks, or Rb/Sr dating for rocks older than 100 000 years. Many dating methods can be used for volcanic rocks of older age (Faure, 1986). Hekiniam, Chaigneau & Cheninee (1973) dated lava tubes with K/Ar dating.

Dating of denudo-erosional caves formed by weathering of volcanic rocks (Shopov & Kolev, 1992) and volcanic landforms requires use of *in situ* produced cosmogenic isotopes, as in the dating of karst landforms. Phillips *et al.* (1986) discussed the possibility of dating the young surface volcanic rocks by ^{36}Cl.

YAVOR Y. SHOPOV

Works Cited

Faure, G. 1986. *Principles of Isotope Geology*, 2nd edition, New York: Wiley

Hekiniam, R., Chaigneau, M. & Cheninee, J. 1973. Popping rocks and lava tubes from the Mid-Atlantic Rift Valley at 36 N. *Nature*, 245: 371–73

Kubik, P., Korschinek, G., Nolte, E., Ratzinger, U. & Ernst H. 1984. Accelerator mass spectrometry of 36-Cl in limestone and some paleontological samples using completely stripped ions. *Nuclear Instruments and Methods in Physics Research*, B5: 326–30

Lal, D. 1987. Cosmogenic nuclides produced in situ in terrestrial rocks. *Nuclear Instruments and Methods in Physics Research*, B29: 238–45

Lal, D. 1988. In-situ produced cosmogenic isotopes in terrestrial rocks. *Annual Review of Earth and Planetary Sciences*, 16: 355–88

Lal, D. & Arnold, J. 1985. Tracing quartz through the environment. *Proceedings of the Indian Academy of Science (Earth and Planetary Science)*, 94: 1–5

Phillips, F., Leavy, B., Jannik, N., Elmore, D. & Kubik, P. 1986. The accumulation of cosmogenic chlorine–36 in rocks: a method for surface exposure dating. *Science*, 231: 41–43

Raisbeck, G. & Yiou, F. 1979. Possible use of 41-Ca for radioactive dating. *Nature*, 227: 42–43

Shopov, Y. & Kolev, B. 1992. Volcanic Caves in Bulgaria. In *Proceedings of the 6th International Symposium on Vulcanospeleology*, edited by G.T. Rea, Huntsville, Alabama: National Speleological Society

Further Reading

Olsson, I.U. (editor) 1970. *Radiocarbon Variations and Absolute Chronology*, New York: Wiley

Wolfli, W. Polach, H.A. & Andersen, H.H. (editors) 1984. Accelerator mass spectrometry. Special issue of *Nuclear Instruments and Methods in Physics Research*, B5: 1–448

DENT DE CROLLES CAVE SYSTEM, FRANCE

The Dent de Crolles area is located in the Chartreuse Massif, in the French Northern Prealps (for location, see map in Europe, Alpine entry). At 2068 m in altitude, its summit rises more than 1500 m above the Isère valley (Figure 1). The mountain resembles a ship's prow, surrounded by high cliffs on three faces. The fourth side corresponds to a large cirque opening to the northwest, where the Guiers-Mort spring resurges, at 1332 m altitude. This narrow ridge extends to the north, to the Aup du Seuil, the Alpe and the Granier massifs, which also contain long and deep cave systems. The boat-shaped plateau surface shows bare karren, only lightly covered by discontinuous pine forest on the lower part, and with alpine meadows near the crest.

This original relief owes its shape to the geological structure, a perched syncline some 20 km long, running from the Dent de Crolles to Granier—high above the Isère valley between Grenoble and Chambéry. Each karst system is separated from its neighbours by transverse dextral faults. The heights of the cliffs correspond with the thickness of the Urgonian limestone, which reaches 450 m, overlying impervious Hauterivian marls. The syncline axis plunges south–north, so that water can concentrate in the trough, at the base of limestone in contact with the marls, and then reappear in the Guiers-Mort Cave.

Under a plateau only 1.8 km^2 in areal extent, runs a cave system more than 50 km long, which can be reached by seven different entrances: the Thérèse and P40 shafts on the plateau, the Glaz hole on the western face, the Annette, Chevalier, and Montagnards caves on the eastern face, and finally the Guiers-Mort spring (Figure 2). The caves are formed on four levels,

Dent de Crolles Cave System, France: Figure 1. Aerial view of Dent de Crolles in winter, from the south (photograph by P. Audra).

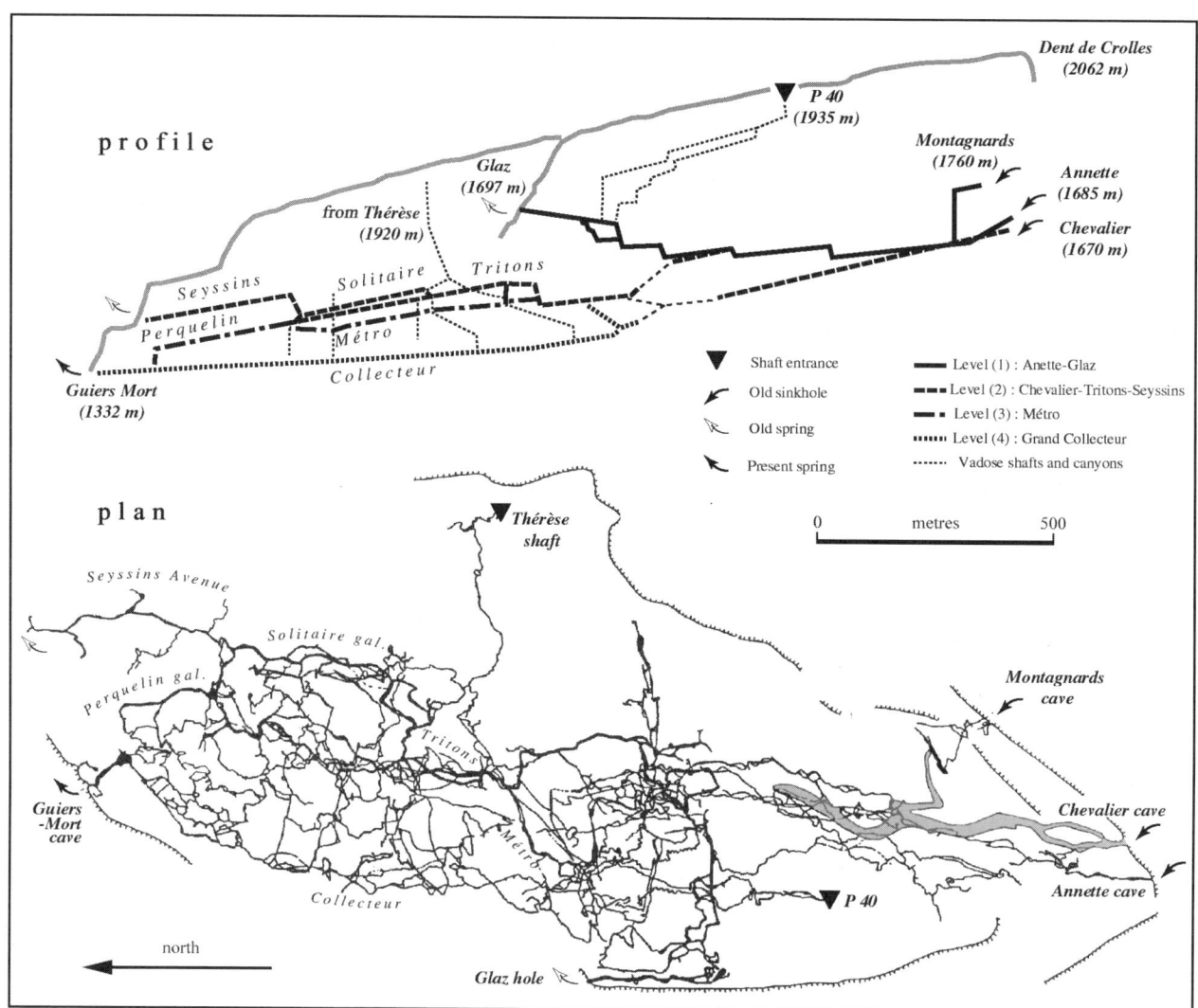

Dent de Crolles Cave System, France: Figure 2. Simplified plan and profile of the cave passages within the Dent de Crolles, France. The surface profile is drawn directly above the main mass of cave passages, and the Thérèse shaft lies higher on the plateau which also slopes down from east to west.

sloping to the north: Glaz-Annette (1700–1650 m), Chevalier-Tritons-Seyssins (1450–1550 m), Métro (1500–1400 m), and Collecteur (1450–1330 m). The three upper levels are perched and dry, but the lowest one, on the marl contact, harbours the main drain. Shafts and meander complexes link the plateau with different levels (Lismonde et al., 1997).

The Guiers-Mort and Glaz entrances have been known for a long time, for a distance of several hundred metres. E.A. Martel made a survey of them in 1899. During the 1930s the real exploration began (led by the Jarrets d'Acier—a local speleological club—and continued by Robert de Joly), which reached 119 m depth in 2.5 km of passage. De Joly made the first dye tracing between Glaz and Guiers-Mort, but it is Pierre Chevalier who will forever be connected with the history of the Dent de Crolles. From 1935 to 1947, under very difficult conditions and accompanied by Fernand Petzl, he explored a 17 km underground maze, connecting successively to the Guiers-Mort Glaz (1941), Annette (1946), and finally the P40 (1947), establishing a world depth record (-603 m). He also discovered the huge Chevalier gallery, only connected to the main system in 1984. These explorations were partly carried out during World War II, when travelling was hazardous and food supplies were heavily restricted. He related his discoveries with modesty in his famous work *Escalades souterraines* [Subterranean Climbers], one of the world's best-known caving books (Chevalier, 1948). From 1959 to 1971, the Tritons, a local speleological club led by M. Letrône, extended the known system to 36 km. They discovered the middle storeys (Tritons and Métro). Later explorations have extended the surveyed length to 55 km, with the main breakthroughs being the discovery of the Thérèse shaft, the system's upper entrance, extending the system depth to -668 m. It is now possible to go through Dent de Crolles from west to east, or from the plateau to the spring, following in Chevalier's footsteps.

Pierre Chevalier also recorded the first European observations distinguishing between vadose and phreatic forms (Chevalier, 1944). The formation of a huge labyrinth under such a small surface area could be explained by the combination of several factors: (1) a synclinal setting which concentrates water flow toward the fold axis; (2) allogenic flows from impervious neighbouring areas and also from glaciers; (3) gradual lowering of the water table, giving rise to the storeys or levels. A rise in the base level due to valley glaciation simply flooded former networks of caves; and (4) lowering of the water level in the looped galleries, giving rise to back-flooded mazes (Lismonde et al., 1997).

To date, three major phases of evolution have been identified (Lismonde et al., 1997; Marchand, 1985):

1. At the end of the Miocene, orogenic activity led to the present folded structure. An impervious depression developed at the site of the present Isère valley. Surface flows probably disappeared underground at the limestone contact, feeding the Annette, Chevalier, and Montagnards caves. The south part of the massif was drained toward the Glaz passages, to a base level located near the present-day 1700 m contour, creating the upper cave level. Scallop dimensions show that discharge was around several hundreds of litres per second. The northern part constituted an independent catchment towards an unknown spring.
2. During the Pliocene, uplift brought a progressive lowering of the water table. The upper level was drained and the Chevalier-Tritons-Seyssins level was formed, which drained both the northern and southern parts of the system. The position of the spring, located in the northern structural block, which is displaced eastwards, explains the location of this level in the Dent de Crolles eastern limb (and not along the axis of the syncline). Some mazes progressively drained this level into the Métro lower level, in a transitional period when both of the middle levels were active (Marchand, 1985).
3. A new water-table lowering divides the two structural blocks into two catchments: Chaos de Bellefont to the north and the present Dent de Crolles to the south. The Dent de Crolles main drainage system has developed in the axis of the syncline, on the contact with the marls, forming the lowest level.

The history of the cave system can be reconstructed using radiometric dating (Audra, 1994; Maire & Quinif, 1991). Level 2 was drained a long time before 250 ka. By 370 ka, Chevalier Cave (level 1) was already partly filled by clastic sediments covered by flowstones, but some flows, probably originating from the input of glacial meltwater along the eastern scarp, eroded the flowstones and contributed to the enlargement of this giant gallery during the Riss (Wolstonian) glaciation. This water input, combined with a glacial dam at the spring, led to major flooding and the deposition of glaciokarstic varves.

PHILIPPE AUDRA

See also **France: History**

Works Cited

Audra, P. 1994. *Karsts alpins. Genèse de grands réseaux souterrains. Exemples: le Tennengebirge (Autriche), l'Ile de Crémieu, la Chartreuse et le Vercors (France), Thesis*, Paris: Fédération française de spéléologie and Bordeaux: Association française de karstologie (Karstologia Mémoires, 5)

Chevalier, P. 1944. Distinctions morphologiques entre deux types d'érosion souterraine. *Revue de géographie alpine*, 3: 475–86

Chevalier, P. 1948. *Escalades souterraines*, Paris: Susse; as *Subterranean Climbers: Twelve Years in the World's Deepest Chasm*, London: Faber and Faber, 1951

Lismonde, B., Bohec, G., Caillault, S., Fouard, C., Gibon, M., Letrone, M. & Marchand, T. 1997. *La Dent de Crolles et son réseau souterrain*, Grenoble: Comité départemental de spéléologie de l'Isère

Maire, R. & Quinif, Y. 1991. Mise en évidence des deux derniers interglaciaires (stades 5 et 7) dans les Alpes françaises du Nord. *Spéléochronos*, 3: 3–10

Marchand, T. 1985. Le système souterrain de la Dent de Crolles, Isère. *Karstologia*, 5: 9–16

DI FENG DONG, CHINA

Di Feng Dong, Great Crack Cave, is part of a spectacular karst system situated near the town of Xin Long in Chongqing Province, China (see location in Figure 1 of China entry). The cave is still being explored by members of the China Caves Project, a collaboration between British cavers and the Institute of Karst Geology in Guilin.

Xin Long is located in an area of dissected cone karst approximately 35 km from the point where the Chang Jiang (Yangtze River) enters the Three Gorges. In this area, the geological structure comprises a series of plunging folds with axes oriented southwest to northeast. Impermeable sandstones and shales outcrop in high mountains to the southeast of Xin Long and provide a large catchment area that drains to the northwest into the limestones near Xin Long.

The limestones are remarkably monotonous. There are more than 1000 metres of finely laminated limestones ranging in age from Devonian to Jurassic, but there are almost no bedding planes, so the limestones give the impression of being one huge bed. Consequently, the only fissures providing easy routes for water are joints and faults. To the north, the Chang Jiang (Yangtze River) and its tributaries have been rapidly incising in response to uplift of the mountains and so all the rivers and caves in the region have experienced a rapid lowering of their base level. This has resulted in the incision of gorges and a series of underground river captures, mostly by caves eroding along the joints.

Within this geological setting the Di Feng karst system comprises four main parts: the Tien Jing Gorge, Di Feng Dong, Xio Zhai Tien Ken, and the Mie Gong He gorge (Figure 1).

The first part of the Di Feng karst system is the Tien Jing Gorge and is an extreme example of river incision in response to uplift. The valley is a broad box canyon with a more recent gorge entrenched in its floor. This gorge, Di Feng (The Great Crack), is almost hidden by trees because it is mostly less than

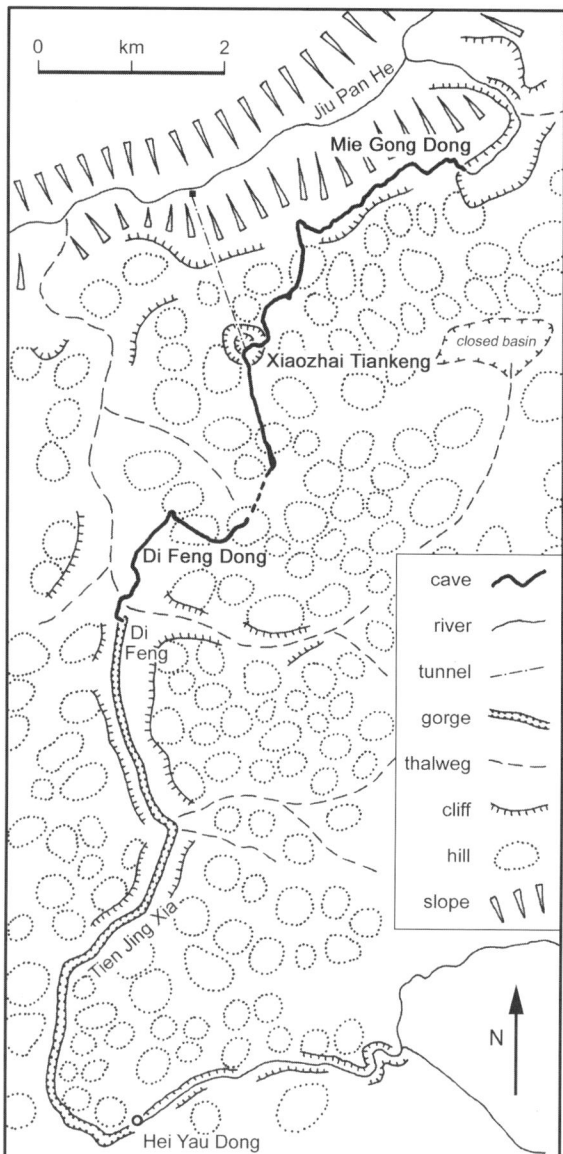

Di Feng Dong: Figure 1. Outline map of the Di Feng Dong cave system beneath the conical hills of the fengcong karst.

Di Feng Dong: Figure 2. The giant doline of Xiaozhai Tiankeng seen from the air, with the shadows of its inner doline hiding the entrances to the passages of Di Feng Dong. (Photo by Zhu Xuewen)

Di Feng Dong is strongly controlled by joints, and is typically 3–15 m wide with vertical walls at least 100 m high. Exploration is exciting, with lakes up to 250 m long, rapids, and waterfalls. Unfortunately, a small rise in water levels during exploration would prevent safe return and for this reason the exploration of Di Feng Dong and the connection with Xiaozhai Tiankeng is still in progress. The river's bedload of pebbles and cobbles clearly becomes suspended in flood, as stainless steel rigging bolts have been found sheared clean off and all equipment is stripped from the cave by annual floods.

The river in Di Feng Dong drains to the foot of the great doline of Xiaozhai Tiankeng (literally, the Big Sky Hole Behind

Di Feng Dong: Figure 3. The tall canyon passage carrying the main stream out into the base of the Xiaozhai doline midway through the cave system. (Photo by Tony Baker)

10 m wide. However, Di Feng reaches a depth of more than 200 m at the northern end where the cave of Di Feng Dong begins. In many places the gorge is only 2 m wide at floor level.

There are several active sinks at the head of the gorge, near Xin Long, and these take all the water except in times of very serious floods. Hei Yau Dong (Black Rock Cave) appears to be the last, major, active sink, so downstream of Hei Yau Dong the Tien Jing Gorge is normally dry apart from static lakes. Most of the water resurges through boulders and pebbles where the gorge continues into Di Feng Dong, so the cave contains an active river.

The second part of the system, the cave of Di Feng Dong, begins where the Tien Jing Gorge ends. The incision of the Great Crack in the floor of Tien Jing Gorge means that a 180 m abseil is required to reach the start of the cave. Like the gorge, the Small Village). This giant doline has mostly sheer rock walls, and is about 600 m across and 660 m deep. From the southern wall, the river from Di Feng Dong resurges in an awesome cave passage 15–20 m wide and more than 100 m high. At this point most of the water is captured by a tunnel which takes it to a hydro-electric power station on the Jiu Pan River about 1.5 km to the north. Consequently, the sink on the eastern side of the doline usually takes little or no water, and a traverse of this downstream section is easier. It begins with pitches of 12, 7, and 50 m, all into large and deep pools. The cave then turns north as a high rift passage with a series of pools, lakes, and pebble banks, and there appear to be larger high-level passages joining from the west in the roof. After 1500 m, the cave turns back on itself and then heads east towards the resurgence. More long lakes and short cascades lead into a narrower canyon, which ends at the lip of a beautiful cascade dropping 40 m into the Mie Gong He gorge.

When exploration of the upstream cave is completed, Di Feng Dong will offer a spectacular through trip 9 km long. The traverse down the great doline and out to the resurgence has already been made and involves a descent of 964 m. Combined with the Great Crack, this is one of the world's great karst sites.

KEVIN SENIOR

Works Cited

Senior, K.J. (editor) 1995. The Yangtze Gorges Expedition: China Caves Project 1994. *Cave and Karst Science*, 22(2): 53–89

DINARIC KARST

The Dinaric Mountains, which form the western part of the Balkan peninsula, take their name from Dinara mountain at the Croatian–Bosnian border. The Dinaric karst occupies the western part of the Dinaric Mountains, that is the whole Adriatic Sea littoral belt from the Trieste Gulf to the Drim River, and inner part from the Friuli Plain (Italy) to the valley of the Zapadna Morava River (Figure 1). The eastern and northeastern limits are less clear, because the carbonate rocks slowly and discontinuously give way to other rock formations. The inner limit of the Dinaric karst can be traced approximately along the line Tolmin–Novo mesto–Banja luka–Užice–Peć. In a straight line, it is about 600 km long and over 200 km wide, covering approximately 60 000 km², forming the largest continuous karstland in Europe.

The Dinaric karst developed in Mesozoic limestone. The youngest (Cretaceous, or even Tertiary) rocks form low relief along the coast and on the islands. In synclines, Eocene flysch was deposited and above there are "islands" of erosional relief with short streams, more soil, and vegetation. In some places there are rich bauxite deposits. Further inwards, the surface consists of Jurassic limestone, while in the east and northeast there are Triassic rocks, often dolomitic, but with less karst.

The climate alters from the Mediterranean coast, to moderate continental and mountainous in the interior. The whole of the Dinaric karst has over 800 mm of annual precipitation, with the highest amounts on the mountain barriers immediately above the coast, such as Crkvice (1070 m) above the Boka Kotorska bay, with over 4900 mm per year. On the coast, dry and hot summers and wet and fresh winters are typical. A strong downslope northeast wind bura (bora) is typical of the winter on the Adriatic side of the Dinaric barrier, especially at the bases of high plateaux and ridges. The vegetation cover corresponds

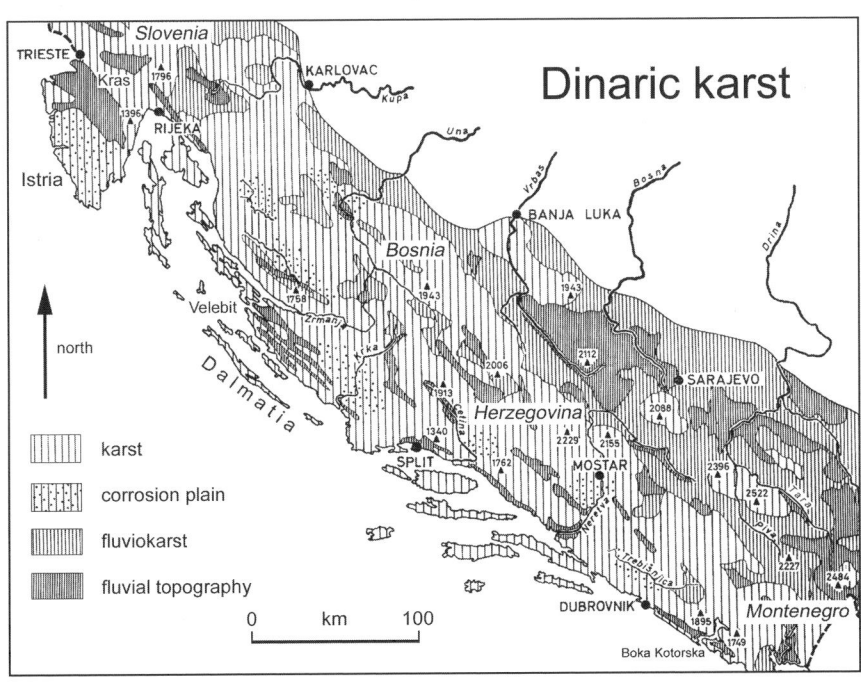

Dinaric Karst: Figure 1. The extent and morphological types of the Dinaric karst. From Roglić (1965).

to climate. Karst, referred to as a limestone desert, is found on the islands and along the coast, especially where the bura wind sprays salt water. Here there is practically no vegetation. The higher parts of the islands and coast-facing slopes have only grass and bush, while on the opposite slopes and inwards, there are mostly forests, some being so-called high (economically important) coniferous and mixed forests. The most common trees are Dinaric fir and birch (*Abieti-Fagetum dinaricum*). The deforestation is partly due to natural conditions (mediterranean climate), and partly due to thousands of years of grazing and burning to obtain pastures. After World War II, when goats were banned and the pressure of agrarian population decreased, forests began to gain ground and bushes now cover a lot of pastures and bare lands.

In spite of the high amount of precipitation, there are only a few surface streams on the Dinaric karst, mainly on impermeable terrain (flysch), or in karst poljes. However there are a few large rivers, with enough water to cut canyons through limestone belts: the Dalmatian Krka, Cetina, Neretva and Morača in the Adriatic basin, and the Kolpa, Una, Vrbas, Bosna, and Drina in the Black Sea basin. The canyon of the Tara River, a tributary of the Drina, is 78 km long and up to 1300 m deep. Sinking rivers, flowing underground from one polje to another, are very common. A typical example is the Ljubljanica River, draining 1100 km^2 of karst, which has a mean discharge of 56 m^3 s^{-1} and crosses five poljes with eight surface sections between caves.

Along the Adriatic coast, water often flows directly into the sea via undersea springs called vrulja, or to springs on or near the coast. Donjobrelska vrulja, at the foot of Mt Biokovo, has an estimated low flow discharge of 20–30 m^3 s^{-1}. The karst hydrology and the morphology of the shallow sea developed during the Pleistocene, when the sea level was more than 100 m lower than at present. Large conduits feeding karst springs are known from depths of 50 m below sea level, and Donjobrelska vrulja is at a depth of 38 m. Springs on the coast are often located at the bottoms of bays and were once used for water supply and energy (mills), this being the reason for location of some settlements (such as Bakar and Malinska). Along the coast there are some very short rivers with a high discharge, for example the Timavo, 2 km long, with a mean discharge of 30 m^3 s^{-1}, and the Ombla near Dubrovnik (at an altitude of 2 m) with a maximum discharge of 200 m^3 s^{-1}. The biggest springs in the Dinaric karst are the Buna (maximum discharge 440 m^3 s^{-1}), and the Ljuta (300 m^3 s^{-1}) in Boka Kotorska Bay. The most important and typical morphological features in the Dinaric karst are poljes (see Dinaride Poljes; Poljes). Other landforms include uvalas, dolines, and corrosional plains (zaravan).

According to the thickness of limestone above the impermeable base, the karst can be divided into deep and shallow types, similar to Jovan Cvijić's holokarst (deep, fully developed with karst poljes) and merokarst (shallow, with only a few karst features). There are over 20 000 explored caves, including deep shaft systems (e.g. Lukina Jama, 1392 m deep on Mt Velebit in Croatia) and long caves (e.g. Postojnska Jama—see separate entry; Krizna Jana (Figure 2); and others around the poljes of Cerkniško, Postojnsko, and Planinsko, in Slovenia).

The Dinaric karst is very important for the history of karstology and speleology, as well as in the development of karst science.

Dinaric Karst: Figure 2. The horizontal phreatic tunnel of Križna Jama (Slovenia) with its numerous deep lakes that make a visit to the cave an almost continuous boat journey. (Photo by Tony Waltham)

The term karst is derived from the Kras plateau in the northwesternmost part of the Dinaric karst (see separate entry, Kras). Similarly, the international terms polje, uvala, doline, kamenitza, and ponor originate from this region.

The history of exploration varies as the Dinaric karst has been divided into a number of political units (Slovenia was incorporated in Austria; Croatia in Hungary or in the Military Zone; Bosnia, Serbia, and Montenegro in the Ottoman Empire; Dalmatia and part of the coast in the Venetian Republic; and the Republic of Dubrovnik). For the history of the northwesternmost part (nowadays Slovenia), see the entry on Kras. The most important early descriptions of karst phenomena along the coast were written by Nikola Gučetić (in 1585), Alberto Fortis (in 1771), and Ivan Lovrić (in 1776). From the inner part of the Dinaric karst, one of the oldest mentions of the caves is in the work of the Moslem traveller, E. Čelebija (1660). Later, Austrian geologists carried out a lot of research in Vojna Krajina (Military Zone), with the aim of understanding the water supply. After 1878, Austrian geologists wrote the first modern studies about typical Dinaric karst in Bosnia, with an emphasis on karst poljes. Well-known karst hydrological theories were developed on this basis by A. Grund (1903) and F. Katzer (1909). One of the founders of the modern karstology, Jovan Cvijić, also based his works (1893, 1901) and theories on the study of Dinaric karst (see entries on Speleogenesis Theories).

ANDREJ KRANJC

See also **Dinaride Poljes; Kras, Slovenia; Poljes**

Further Reading

Alfirević, S. 1966. Hydrogeological investigations of submarine springs in the Adriatic. *Proceedings of the Congress of the International Association of Hydrogeologists*, Belgrade: 255–65

Gams, I. 1969. Some morphological characteristics of the Dinaric Karst. *Geographical Journal*, 135: 4
With an emphasis on karst poljes
Herak, M. 1972. Karst of Yugoslavia. In *Karst: Important Karst Regions of the Northern Hemisphere*, edited by M. Herak & V.T. Stringfield, Amsterdam and New York: Elsevier
With particular attention to the geology of the Dinaric karst
Petrović, B. & Prelević, B. 1965. Hydrologic characteristic of the Karst area of Bosnia and Herzegovina and a part of Dalmatia with special consideration of underground water connection. *Naše Jame*, 7(1–2): 79–87

Roglić, J. 1964. "Karst valleys" in the Dinaric karst. *Erdkunde*, B. 28: 2
Roglić, J. 1965. The delimitations and morphological types of the Dinaric karst. *Naše Jame*, 7(1–2): 79–87
Roglić, J. 1972. Historical review of morphologic concepts. In *Karst: Important Karst Regions of the Northern Hemisphere*, edited by M. Herak & V.T. Stringfield, Amsterdam and New York: Elsevier
Contains much data on the research history of Dinaric karst
Shaw, T.R. 2000. *Foreign Travellers in the Slovene Karst 1537–1900*, Ljubljana: Založba ZRC

DINARIC KARST: BIOSPELEOLOGY

The Dinaric karst covers more than 56 000 km^2. It is approximately 100 km wide and 600 km long in the region between the Adriatic Sea and the hills that separate it from the Pannonian Valley. It extends from 2000 m altitude to below sea level. Cave temperatures may be 16°C at the coast and 4°C in the Velebit Mountains or even close to zero around small perennial ice deposits in caves. Ecologically, the caves are very diverse, with nearly all the hypogean habitat types that are possible in temperate climates. Virtually all of the caves are ecologically heterogeneous, and have a rich and diverse fauna.

Richness and Structure of the Fauna

The first cavernicolous species in the world were described from the northwestern parts of the Dinaric karst in the 1830s to 50s. Later, some unusual discoveries were made in the central part (Dalmacija and Herzegovina) while the Slovenian section appears to be richest in aquatic species. Dinaric caves and interstitial waters are inhabited by more than 450 stygobitic species, which is by far the highest number in the world, when compared with other regions of corresponding size (Sket, 1999). It is nearly twice as high as any other region, except the east Balkan Province. A recent review of cave fauna revealed more than 790 terrestrial obligate cave species. The inclusion of aquatic and terrestrial subspecies would at least double the number of taxa (Sket, Paragamian & Trontelj, 2003).

The taxonomic composition at the level of order and class appears to resemble other cave faunas in southern Europe. Approximately 45% of the terrestrial species described are beetles from two groups: Cholevidae: Leptodirinae and Carabidae: Trechinae. Other insect groups are largely absent, but pseudoscorpions (Pseudoscorpiones) are comparatively numerous. Among the aquatic fauna, crustaceans are by far the most common, and stygobitic aquatic insects are absent. The Dinaric fauna is not particularly diverse at the higher taxonomic level. Its richness is mainly founded on the diversity within some genera such as *Niphargus*, *Proasellus*, *Monolistra*, *Anophthalmus*, and *Antroherpon*. Apart from their morphological diversity, they also show a high ecological diversity.

Distinctive Features of the Dinaric Hypogean Fauna

There are some Dinaric species that are particularly distinct morphologically. These include troglomorphies such as an accentuated (pseudo)physogastry (an apparently swollen abdomen) in terrestrial Coleoptera (*Leptodirus hochenwartii*, *Antroherpon* spp.), cuticular spines or outgrowths in aquatic Crustacea (*Niphargus balcanicus*, some *Monolistra* spp.) or outgrowths in Hirudinea (*Croatobranchus mestrovi*).

Unlike vertebrates, some invertebrate groups are unusually richly represented. A particularly rich stygobitic gastropod fauna has been found, with 130 species, a diversity that closely approaches that of all non-Dinaric hypogean gastropod species. The diversity of epizoic or parasitic Turbellaria: Temnocephalida, with 15 species, is also high. The mainly marine isopod family of Sphaeromatidae is represented by c.30 species and subspecies in the Dinarides, compared with less than 10 in other south European karst areas.

The most interesting species inhabiting Dinaric caves are unique stygobitic representatives of some higher taxa, for example *Eunapius subterraneus* (Spongillidae) is the only stygobitic freshwater sponge; *Velkovrhia enigmatica* (Hydrozoa: Bougainvilliidae) is the only known stygobitic cnidarian; *Congeria kusceri* Bole (Mollusca: Bivalvia: Dreissenidae) is the only definitively stygobitic clam; *Marifugia cavatica* (Polychaeta: Serpulidae) is the only stygobitic tube-worm; and *Proteus anguinus* (Amphibia: Proteidae) the only non-American stygobitic amphibian. The undescribed terrestrial planarian (cf. *Microplana*) is probably also the only troglobitic species of its group.

The three stygobitic species each for the Cladocera and Copepoda: Calanoida represent a comparatively high diversity, since these groups are extremely poorly represented in fresh waters elsewhere. In addition, the absence of otyglobitic fish species in the Dinaric karst is worth mentioning, although there are a number of endemic stygloxene fish species.

Ecological Diversity

A high ecological diversity (variety of habitats) is characteristic of most larger cave systems in the area, resulting in very rich local faunas. Of 16 caves worldwide containing 20 or more troglobitic and stygobitic species, six are in the Dinaric karst (Culver & Sket, 2000). By far the richest within the Dinaric karst is the Postojna–Planina Cave System (see separate entry), with approximately 50 stygobites and 30 troglobites, and close to 150 stygophilic or stygoxene species. Vjetrenica in Bosnia–Herzegovina (see separate entry) is nearly devoid of non-specialized species, while 40 stygobites and 32 troglobites are known within the system.

Some particularly interesting ecological entities in the Dinaric karst are worth mentioning.

1. An epizoic protozoan faunula of no less than seven Ciliata: Peritricha and two Suctoria were found on a single specimen of the isopod *Monolistra spinosissima*;
2. A hygropetric-like habitat (a thin film of water running down the rock) known from Vjetrenica in Her-

zegovina occurs in some other Dinaric (and south Alpine) caves, and is inhabited by some specialized beetles (Coleoptera: Catopidae: Leptodirinae) which are pholeuonoid in shape (with a spindle-shaped trunk and elongated appendages);
3. In very specialized conditions in the sink-cave Crnulja in Popovo polje, there are very dense and extensive colonies of the cave tube worm *Marifugia* and thick layers of its tubes have formed a type of hypogean tufa. This used to be inhabited by an interstitial fauna of oligochaetes, acarines, and some crustaceans. Flood prevention work in the polje prevented regular flooding of the cave, killing *Marifugia* and its associated fauna;
4. In the vicinity of hypogean ice, rich beetle faunas may be found, including some specialized oligostenothermal (limited to low temperatures) fauna, such as *Astagobius* spp.

Distribution Patterns (Biogeography)
The following distribution patterns (Sket, 1994) can be established for the hypogean Dinaric fauna:

1. widely distributed troglobitic or generalist species (for example, the west Palearctic *Asellus aquaticus*);
2. troglobitic taxa with relatives in the western and/or eastern karst areas (*Zospeum, Troglocaris, Sphaeromides, Niphargus* aggr. *rejici*);
3. holo-Dinaric or holo-Dinaric–South Alpine taxa (*Proteus, Marifugia, Monolistra, Titanethes*);
4. merodinaric taxa, i.e. limited to one of three subregions (northwest Dinaric: *Anophthalmus, Leptodirus, Hadziella*, subgenus *Microlistra*; southeast Dinaric: *Antroherpon, Plagigeyeria*, subgenus *Pseudomonolistra*; paralittoral (mainly along the coast): *Hadzia fragilis, Niphargus hebereri*);
5. narrowly endemic taxa.

In contrast to most other areas, a portion of the Dinaric anchialine fauna in the paralittoral area is of freshwater origin. The composition of the interstitial fauna within the Dinaric area differs from that of cave habitats.

Some Unresolved Problems
Although this is a relatively well-studied karst area, sometimes intensively so, it is far from being thoroughly investigated, even taxonomically. Some apparently well-known taxonomic groups such as *Niphargus, Monolistra,* and *Anophthalmus* continue to yield new taxa; but some groups have been almost completely neglected, e.g. Bacteria, Mycota, Protozoa, Nematoda, Acarina, and parasitic animals in general. Phylogenetic and phylogeographical relations within some large genera distributed across Europe (e.g. *Niphargus, Proasellus*) are completely unresolved.

Biochemical investigations on *Troglocaris* have confirmed the long-held hypothesis that at least some of what have been considered holo-Dinaric species may in fact be species aggregates. Because of the morphological similarity caused by convergent evolution, it will only be possible to investigate this using DNA analysis. An example of one of these taxa is the cave salamander, *Proteus*.

Recent biogeographical analyses have revealed that the ages of cave populations may differ vastly, even within one species. The surprisingly recent immigrations (in some *Monolistra* spp., and in *Proteus*; Sket, 1997) still await confirmation by other means.

The Dinaric Hypogean Fauna Is Endangered
The comparatively high diversity of taxa results in a greater vulnerability of hypogean fauna to disturbance. Fortunately, most phenomena which endanger cave fauna are also harmful to the human population (Sket, 1972). Although most obligate cave species seem to originate from euryoecious (generalist) ancestors, their own tolerances to changes of ecological parameters are largely unknown. In any case, the character of Dinaric cave fauna shows that the species may be highly specialized and diversified, therefore dependent on the habitat diversity. In addition, they are dependent on the environment's stability for their "slow life style" (K-strategy). This is particularly dangerous in combination with their limited distribution ranges. Due to the contiguity of the Dinaric karst it was nearly impossible to avoid endangering its hypogean biota either by draining waste and waste waters underground or by hydrotechnical interventions. Pollutant spills are not uncommon in the karst as well as in interstitial water within alluvial plains. The underground hydrography allows pollutants to spread in unknown and unseen directions. Cave fauna may be endangered either directly by the toxicity of the wastes or by competition with epigean species in organically polluted habitats. Both are common phenomena all over the Dinaric karst. Land drainage and water resource management, particularly in the southeast (Herzegovina), has changed the surface and subterranean hydrology to such an extent that some hypogean habitats have changed substantially. The drying out of the unique massive tubeworm colonies in the Crnulja in Popovo polje is one of the most severe results of such activities. Similar hydrological interventions have altered the connections between some systems, allowing mixing of previously isolated populations or races and therefore allowing harmful changes in the biodiversity.

BORIS SKET

See also **Postojna–Planina Cave System, Slovenia: Biospeleology; Vjetrenica, Bosnia–Herzegovina: Biospeleology**

Works Cited

Culver, D.C. & Sket, B. 2000. Hotspots of subterranean biodiversity in caves and wells. *Journal of Cave and Karst Studies,* 62(1): 11–17

Sket, B. 1972. Zaščita podzemeljske favne se ujema z življenjskimi interesi prebivalstva (Protection of subterranean life consonant with human interests). In *Zelena knjiga o ogroženosti okolja v Sloveniji*[The Green Book], edited by S. Peterlin, Ljubljana: Prirodoslovno Društvo Slovenije: 137–40; 164–65

Sket, B. 1994. Distribution patterns of some subterranean Crustacea in the territory of the former Yugoslavia. *Hydrobiologia,* 287: 65–75.

Sket, B. 1997. Distribution of Proteus (Amphibia: Urodela: Proteidae) and its possible explanation. *Journal of Biogeography,* 24: 263–80.

Sket, B. 1999. High biodiversity in hypogean waters and its endangerment—the situation in Slovenia, Dinaric karst, and Europe. *Crustaceana,* 72(8): 767–79

Sket, B., Paragamian, K. & Trontelj, P. 2003. A census of the obligate subterranean fauna in the Balkan Peninsula. In *Balkan Biodiversity: Pattern and Process in Europe's Biodiversity Hotspot,* edited by H.I. Griffiths & B. Krystufek, Amsterdam: Kluwer

Further Reading

Brancelj, A. & Sket, B. 1990. Occurrence of Cladocera (Crustacea) in subterranean waters in Yugoslavia. *Hydrobiologia*, 199: 17–20

Casale, A., Giachino, P.M. & Jalžić, B. 2000. Croatodirus (nov. gen.) bozicevici n. sp., an enigmatic new leptodirine beetle from Croatia (Coleoptera, Cholevidae). *Natura Croatica*, 9(2): 83–92

Guéorguiev, V.B. 1973. Sur la formation de la faune troglobie terrestre dans la péninsule balkanique durant le tertiaire. *Comptes rendus de l'Académie bulgare des Sciences*, 26(9): 1231–34

Sket, B. 1994. Yugoslavia (Bosnia–Herzegovina, Croatia, Macedonia, Montenegro, Serbia, Slovenia). In *Encyclopaedia Biospeologica*, vol. 1, edited by C. Juberthie & V. Decu, Moulis and Bucharest: Société de Biospéologie

Sket, B. 1997. Biotic diversity of the Dinaric karst, particularly in Slovenia: History of its richness, destruction, and protection. In *Conservation and Protection of the Biota of Karst*, edited by I.D. Sasowsky, D.W. Fong & E.L. White, Charles Town, West Virginia: Karst Water Institute

Sket, B. 1999. The nature of biodiversity in hypogean waters and how it is endangered. *Biodiversity & Conservation*, 8(10): 1319–38

Sket, B. & Karaman, G. 1990. *Niphargus rejici* (Amphipoda), its relatives in the Adriatic islands, and its possible relations to S.W. Asian taxa. *Stygologia*, 5(3): 153–72

Sket, B., Dovč, P., Jalžić, B., Kerovec, M., Kučinić, M. & Trontelj, P. 2001. A cave leech (Hirudinea, Erpobdellidae) from Croatia with unique morphological features. *Zoologica Scripta*, 30(3): 223–29

DINARIDE POLJES

Approximately 130 karst poljes are located in the Dinaric karst region (see separate entry), which straddles the borders of Bosnia, Herzegovina, and Montenegro (see map in Dinaric Karst). There are more than 50 large karst poljes in the area, generally elongated along the strike of the Dinaric Mountains, i.e. in a northwest–southeast direction.

The first comprehensive study of the Dinaric karst was made by Jovan Cvijić in 1900. In the early 1900s, local investigations focused mostly on draining of the temporarily flooded poljes, providing effective water intakes at the springs, and constructing small storage ponds. After 1945, karst research developed rapidly, stimulated by the need for detailed surveys for hydroelectric power projects in the region, as well as the initiation of reclamation projects.

West Bosnia and Herzegovina

Southwest Bosnia, Dalmatia, and west Herzegovina contain the world's largest concentration of well-developed karst poljes (Figure 1). Poljes are developed at four to five particular elevations, between 1200 and 20 m above sea level (Figure 2). A complex suite of processes governs their genesis in this part of the Dinaric karst, but faults of regional extent, mainly reverse faults, have played a key role, e.g. in Glamočko, Kupreško, and Livanjsko. Many small karst poljes are developed along the same overthrust, e.g. Mućko, Aržano, Vinica, and Rakitno poljes. The highest is Trebistovo Polje at 1278 m above sea level, and the lowest are Bokanjačko Polje and Vrgorac Lake, in Dalmatia, at only 20–25 m above sea level. A few of the very small poljes do not show any surface flow—Bokanjačko, Nadinsko, Dugo, and, without ponors, Posuško and Vir poljes.

Vukovsko and Ravno poljes are hydrologically connected via a short gorge. The longer axis of Vukovo follows an east–west fault and the underlying rock is dolomite. Ravno Polje occurs on limestone and is elongated along the Dinaric strike, northwest–southeast. Vukovsko and Ravno poljes never flood. Livanjsko Polje is the largest in the region, 65 km long and 6 km wide with a total area of 402 km². Its principal ponors are mostly located in the southwest foothills: Ćaprazlija, Kazanci, Veliki, and Opaki ponors in Livanjsko and Srdjevičko, and Sinjski, Proždrikoza, and Stara Mlinica ponors in the Buško Blato area. Mali Kablić Ponor is situated on the northeast border of the polje. In natural conditions, floods in the southern part of the polje (Buško Blato) lasted four to six months per year, ponding some 200 million cubic metres of water. For the purposes of hydroelectric power generation, all ponors have been shielded against surface leakage by grouting, plugging, and construction of local rock-filled dikes in front of ponor zones, and water from Buško Blato has been transferred to a man-made reservoir with a capacity of 800 million m³ s⁻¹.

Kupreško and Glamočko poljes both straddle the regional watershed between the Adriatic Sea and Black Sea catchments. Kupreško Polje occurs on a northnorthwest–southsoutheast fault and consists of three connected morphological units, with flooding being confined to the lower part of each. Glamočko has an upper polje and a lower polje. The flooding occurs for 6–10 months per year, and the volume of floodwater exceeds 80 million cubic metres. Kovači Ponor in Duvanjsko Polje is the largest ponor in this region, with a capacity of over 60 m³ s⁻¹.

Huge volumes of Neogene shale, conglomerate, clay, and sandstone have accumulated in these poljes. The lake deposits in Glamočko Polje are more than 700 m thick, they are 2300 m thick in Livanjsko Polje and more than 2500 m in Duvanjsko Polje. In the lower-lying poljes of Western Herzegovina (Posuško, Kočerinsko, Trnsko, and Imotsko poljes) the Neogene deposits are much thinner, e.g. only 145 m in thickness in Imotsko Polje. All of these deposits in the poljes function as hydrogeological barriers.

Eastern Herzegovina

With the exception of Nevesinjsko, all poljes in eastern Herzegovina are aligned along the Dinaric strike (Figure 1). Stepwise poljes are distributed from 950 m down to 50 m above sea level (Figure 2). Underground flows, up to 35 km long, dominate the drainage. Most of these poljes are tectonically controlled, and were developed along reverse faults or overthrusts. All poljes except Mokro are topographically enclosed. In natural conditions all of the poljes flooded periodically. In the rainy season inundation began with the lower poljes, and the highest poljes are the first to dry up.

Popovo Polje is a special karst landform (Figure 3). In its upper part, the alluvial cover is only 1–2 m thick, increasing to 15–20 m downstream. To date, more than 500 ponors, estavelles, and temporary springs have been identified in the polje. Temporary springs predominate in the upstream section, estavel-

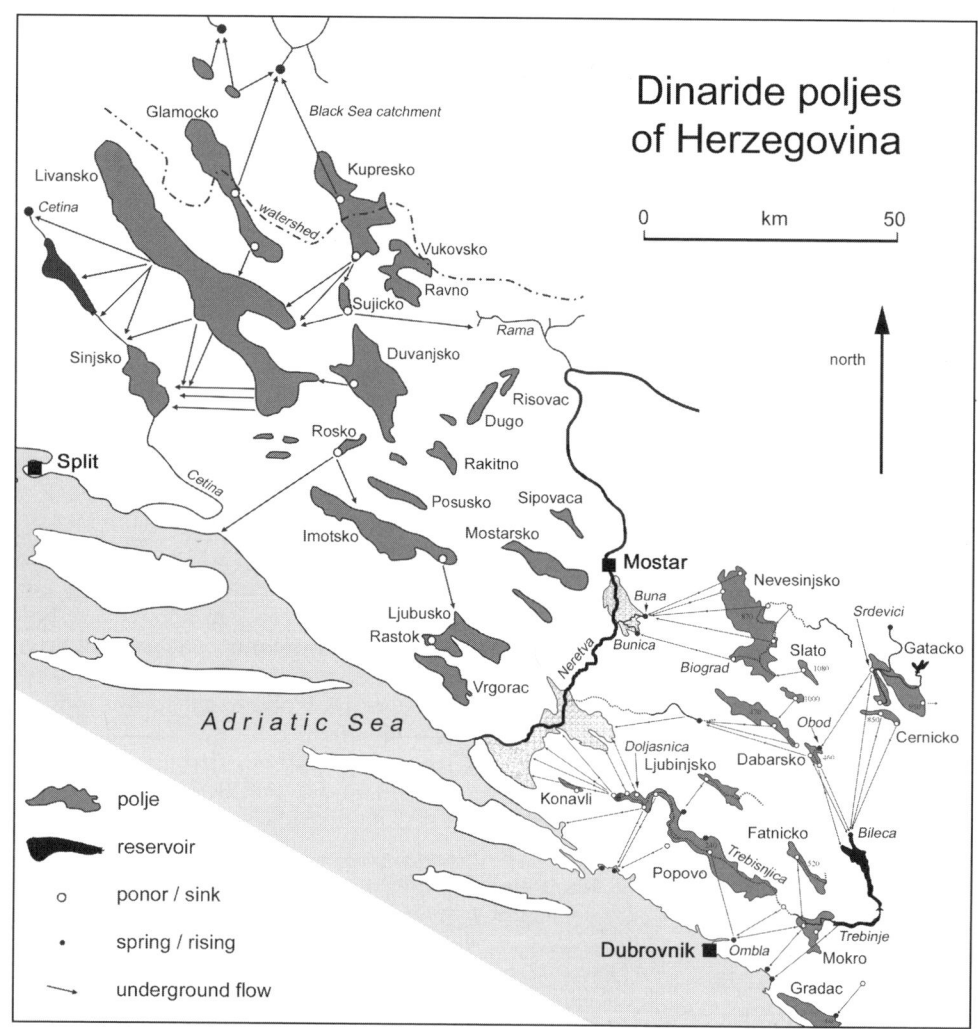

Dinaride Poljes: Figure 1. Simplified map of the main karst poljes in Bosnia and Herzegovina.

les in the middle, and ponors downstream. Water loss along the 65 km bed of the Trebišnjica River is more than 63 $m^3 s^{-1}$ in the dry season, and the total sinking capacity of the Popovo Polje ponors is more than 300 $m^3 s^{-1}$. The capacity of the largest single ponor (Doljašnica) is over 60 $m^3 s^{-1}$. However, the winter inflow to the polje is substantially greater and may exceed 1000 $m^3 s^{-1}$. The highest observed flow was 1362 $m^3 s^{-1}$ on 2 December 1903. Flooding under natural conditions reached a height of 40 m in the lowest section of the polje, and the average yearly flood duration was 253 days. As part of the construction of the Trebišnjica Hydrosystem, 62.2 km of riverbed was covered with shortcrete. Following the excavation of two large tunnels for the hydroelectric power-plant, the floods have practically been eliminated.

The Fatničko and Dabarsko poljes are developed along one overthrust of Cretaceous flysch on Eocene flysch. The depth of

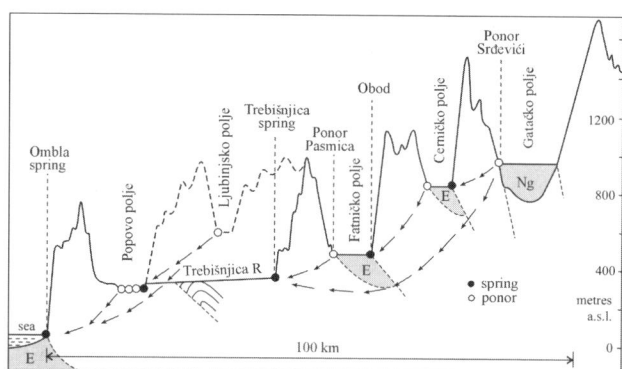

Dinaride Poljes: Figure 2. Simplified profile from the Gatačko Polje to the Ombla Spring.

Dinaride Poljes: Figure 3. Popovo Polje seen in profile, with the flat floor and sharply defined marginal slopes that characterize a true polje. (Photo by Tony Waltham)

limestone paleorelief below the surface is 150–200 m. Annual flooding in the Fatničko Polje lasts 79–213 days, and the highest water level can be as much as 38 m above the lowest point, Pasmica Ponor, which has an intake capacity of 25 m³ s⁻¹. During the maximum flood, the volume of accumulated surface water is 228 million cubic metres. The largest inflow into the polje is through the ephemeral Obod Estavelle (0–60 m³ s⁻¹), located at the overthrust contact. Dabarsko and Fatničko poljes are now interconnected by a tunnel 3.2 km in length.

In Nevesinjsko Polje two areas of water storage are formed, one in the northern part and a second, smaller one in the south, in front of Biograd Ponor. This ponor is the terminus of the Zalomka River and has an intake capacity of 110 m³ s⁻¹. The conglomerate deposits beneath the floor of the polje are 800 m thick. The water table in the dry season is more than 350 m below the surface of the polje. It fluctuates through a height range of 300–312 m.

The small Cerničko, Slato, and Lukavačko poljes are developed along a tectonic line where the limestones are overthrust on Eocene flysch. The Ključka River in Cerničko Polje is a good example of karst polje surface flow. The river originates in a large cave at the contact between flysch and limestone, and terminates about 300 m downstream in a ponor that has a recharge capacity of approximately 20 m³ s⁻¹.

Montenegro

Nikšićko Polje is composed of several different geomorphological and hydrological units: the Upper Nikšićko, Krupac, Slano, and Slivlje poljes. The thickness of the Tertiary sediments (clay and sand) over Cretaceous limestone and dolomite is 9–18 m. About 880 ponors and estavelles have been identified in the polje, with 851 of them being located along its southern perimeter. Slivlje Ponor is one of the biggest ponors known, with a sinking capacity of about 150 m³ s⁻¹ in the whole fissure zone, while the maximum capacity of the main opening itself exceeds 100 m³ s⁻¹. In the past 40 years the natural hydrological conditions have been altered enormously by a number of geotechnical projects, including cylindrical dams around ponors, the plugging of karst channels and more than 12 km of grout curtains.

PETAR MILANOVIĆ

See also **Cerknica Polje; Poljes**

Further Reading

Mijatović, B. 1984. *Hydrogeology of the Dinaric Karst*, Hannover: Heise

Milanović, P. 1981. *Karst Hydrogeology*, Littleton, Colorado: Water Resources Publications

Radulović, V. & Radulović, M. 1997. Karst of Montenegro. In *100 Years of Hydrogeology in Yugoslavia*, edited by Z. Stevanović, Belgrade: Faculty of Mining Geology, Institute of Hydrogeology (in Serbian, with English summary)

DISEASE

Most caves are healthy places, and most cavers are exceptionally healthy. Nevertheless, during the 20th century it became evident that some diseases are related to particular caves and their inhabitants. However, current concepts of cave diseases have nothing in common with the lurid tales of folklore.

Only in the 1920s was it suspected that local epidemics of rabies in tropical America might be transmitted by troglophilic vampire bats. This was proven a decade later, in Trinidad and in South America (Pawan, 1936; Torres & de Queiroz Lima, 1936, cited by Turner, 1975). Pawan isolated the rabies virus and developed an effective immunization vaccine.

Histoplasmosis, a lung disease caused by spores from a fungus, *Histoplasmosis capsulatum*, was next to be identified as a "cave disease". In 1947, an outbreak occurred among an Arkansas group that had explored an abandoned underground quarry. The first reports (e.g. Washburn, Tuohy & Davis, 1948) called it "cave sickness". It quickly entered speleological awareness (e.g. Halliday, 1949). Soon it emerged as a newly recognized disease of many parts of the world, both above and below ground. Small epidemics were identified in Mexico, Jamaica, parts of South Africa, Canada, and elsewhere. A few deaths have been reported. One conspicuous fatality was a British caver treated with high doses of steroids (John Frankland, personal communication, 2001).

Later, several cavers in Barbados and elsewhere developed a type of infectious jaundice called leptospirosis or Weil's disease. The infection is carried in rat urine, and cases have developed from contact with infected water in some caves in Barbados, Guam, the United Kingdom, the Irish Republic, and several other countries, especially in the tropics. The causative organism is endemic in rat populations worldwide.

In addition, a common but trivial disease of dusty caves was identified: cave dust pneumonitis or alveolitis. Its principal importance is as a warning that more serious lung disease may also be developing, with a longer incubation period.

With the vast expansion of 20th-century speleology, it became apparent that a host of other diseases may be acquired in caves, but the existence of the cave is relatively unimportant in their transmission. Some of these are the result of unusual cave environments, including but not limited to: mountain sickness in high-elevation caves, hypothermia (including frostbite), hyperthermia, hydrogen sulfide poisoning, and possibly nitrogen dioxide poisoning (silo filler's disease). Some authorities might include diseases resulting from the hazards of cave diving and the bites of poisonous snakes, some of which occasionally penetrate considerable distances into caves. Hydrogen sulfide poisoning in the rare caves with a high level of this chemical, and anoxic brain syndromes resulting from over-long exposure in hypoxic "bad air" caves are the only diseases in this group which are specific to caves. The controversy about radon in caves is discussed below and in a separate entry.

Additional diseases are related to pollution and/or contaminated water, in caves and elsewhere. Human pathogens in groundwater are discussed in the entry on Groundwater Pollution: Dispersed. Chronic carbon monoxide poisoning can result where vehicle exhaust fumes are permitted to enter caves. Poisonings by hydrocarbons, heavy metals, radioactive wastes, pesticides, and herbicides may occur as a result of unlawful disposal in dolines, pits, or cracks.

In dry caves, nests of rodents and birds carry a threat of several acute infectious diseases and should be avoided. These include plague, tularemia, hantavirus, Lassa and related haemorrhagic fevers, psittacosis, and possibly rabies (the incidence of rabies in rodents apparently is greatly underestimated). If the nest is moist with rat urine, it may contain the leptospirosis organism. Hantavirus, Lassa and related haemorrhagic fevers, however, occur primarily in parts of the world with few caves. Contrary to a recent best-selling book (Richard Preston's *Hot Zone*, 1995), the Ebola virus is not cave-related (it was briefly suspected in Kenya's Kitum Cave but this cave was reopened as a tourist attraction even before the book was published).

In addition to histoplasmosis, the fungi causing several other pulmonary diseases are at least potentially present in cave dust. These include coccidioidomycosis (San Joaquin Valley fever and its South American variant), cryptococcosis, actinomycosis, blastomycosis, aspergillosis, nocardiosis, and others. Cryptococcosis is especially likely in the presence of pigeon droppings. This group of diseases is a particular problem to persons with AIDS or other impairments of the immune system, and is rarely if ever seen in other cavers. In theory, another related hazard is the presence of some strains of atypical tuberculosis organisms in some soils.

Bites and stings of insects, spiders, and scorpions are of some concern, especially in the entrance zones of caves in areas where a known problem exists. The commonest effects of these are acute anaphylactic reactions to bee, wasp, or hornet stings, just as on the surface. Black widow spider bite is a regional American problem, sometimes presenting with findings suggestive of an acute abdominal catastrophe. Its diagnosis and treatment are more complicated than generally recognized, but fatalities are now rare. The bite of brown recluse spiders is also a regional American problem, but fortunately uncommon in caves. In addition to an acute systemic reaction, a subsequent necrosis of large areas of skin and muscle may result. Danger from scorpion sting is highly species-specific. A dangerous species occurs in the southwestern desert of the United States, and others exist in the Middle East and perhaps central Asia. In the United States, deaths now are rare. Danger from centipede sting is also species-specific. Some in Japan are said to be especially dangerous. The toxin exuded by millipedes is highly irritating to the eyes if rubbed after handling them.

A long list of serious diseases is transmitted by the bite of ticks, mosquitoes, and tsetse flies. All these vectors are occasionally found in the entrances of caves (and sometimes deeper, as in the case of ticks in some bat caves). But transmission of these diseases rarely if ever occurs in caves, so they are omitted here. Reduviid bugs (kissing bugs), however, not only transmit Chagas disease (a form of trypanosomiasis), but also cause some serious anaphylactic reactions. Cavers thus should use special caution when camping in dry caves in areas where they exist (e.g. the southwestern United States and some parts of Mexico). In a few parts of the world, botflies (and possibly a few related flies also occasionally found in cave entrances) not only bite their victims but lay eggs which hatch in the wound.

In addition to anaphylactic reactions to bee, wasp, hornet, or reduviid bug bites and stings, cavers are at risk from at least two hypersensitivity diseases: "Farmers' lung", caused by a thermophilic actinomycete, and avian protein anaphylaxis. These appear to be only theoretical risks.

Prolonged wetting can cause serious problems. The worst is immersion foot, which is a particular problem for caving expeditions in Sarawak and elsewhere in the tropics, despite well-planned precautions. Another is semi-humorously called "Kazumura crud". It has the form of small festering blisters of the chest and abdomen, seen after prolonged crawling in moist slimy caves in Hawaii.

Histoplasmosis is primarily an acute infection of the lungs. It becomes symptomatic a few days after exposure. In most children and some adults it is so mild that it is thought to be merely a "chest cold", but X-rays show an alarming pattern of small spots throughout both lungs, virtually identical to a rapidly spreading form of tuberculosis. Other patients with similar X-ray findings are acutely ill for several days and recover only gradually. Still others remain ill for months. A very few—especially those with compromised immune systems—develop serious complications: small abscesses in many parts of the body, dense infiltrations around the heart and base of the lungs which impair the circulatory system, or eye symptoms. The latter appear to be immunological reactions. A single attack provides at least limited immunity in most victims, but multiple infections have occurred in some cavers. The most publicized epidemics are those from caves, but the disease occurs widely in many parts of the world, both above and below ground. Its organism grows especially well in bat and bird droppings.

Outside the range of vampire bats, medical science's picture of rabies has changed greatly in the last century. In some aspects, opinions have changed repeatedly. The course of the disease is not as uniform as once believed, and it must be suspected in any person with an acute inflammatory disease of the brain or related abnormalities of muscle groups and nerves. A few people have now survived what formerly was an entirely fatal disease. Varieties of clinical courses are notably explicated in Brass (1994) which, however, fails to note that one of the two supposed victims of aerosol rabies in the Frio Cave of Texas probably scratched a rash on his neck with infected gloves (Constantine, 1967). Cases have now been found in much of the world without any evident exposure to any rabid animal, and research is seeking additional vectors. Most cases are still transmitted by the known bite of an animal, including insectivorous bats. In the United States, most current cases are caused by strains of rabies virus predominantly found in silver-haired bats, which do not live in caves. American and other cavers frequently come in contact with bats, but none of them have developed rabies. Aerosol transmission of rabies to humans is no longer considered a serious threat, and a recent "undetected bite" hypothesis of rabies transmission by insectivorous bats has become untenable. In 1999 the 29th North American Symposium on Bat Research concluded that there was no credible evidence supporting this concept, and it had "moved from hypothesis to fact without adequate testing". Cavers within the range of vampire bats, and those who must handle bats, should be immunized against the disease. Others should avoid them to the greatest degree possible.

Another concept which "moved from hypothesis to fact without adequate testing" is that of lung cancer from high levels of radon and its products, in caves and elsewhere. The anticipated epidemic of lung cancer in elderly American cavers has not materialized. Not even one excess case of lung cancer has been documented in non-smoking American cavers or employees of show caves. Agency silence about the uncertainty of inferred health risks from radon and its daughter elements (Aley, 2000) has blurred the issue, but clearly, lung cancer is not cave-related.

Perhaps the most bizarre of all cave diseases is acute salamander poisoning, reported only after excessive alcohol consumption. The mortality rate is about 50% (Bradley, 1981; Brodie, 1982). Swallowing enough live millipedes would probably have a similar effect!

WILLIAM R. HALLIDAY

See also **Groundwater Pollution: Dispersed; Radon in Caves**

Works Cited

Aley, T. 2000. Radon and radon daughters. In *Standard Handbook of Environmental Science, Health and Technology*, edited by J.H. Lehr, New York: McGraw-Hill
Bradley, S. 1981. A fatal poisoning from the Oregon rough-skinned newt (*Taricha granulosa*). *JAMA*, 246: 247–48
Brass, D.A. 1994. *Rabies in Bats: Natural History and Public Health Implications*, Ridgefield, Connecticut: Livia Press
Brodie, E.D. 1982. Toxic salamanders. *JAMA*, 247: 1408
Constantine, D. 1967. *Rabies Transmission by Air in Bat Caves*, Washington: US Government Printing Office
Halliday, W. 1949. Cave sicknesses. *National Speleological Society Bulletin*, 11: 28–30
Pawan, J. 1936. Rabies in the vampire bat of Trinidad, with special reference to the clinical course and the latency of infection. *American Journal of Tropical Medicine and Parasitology*, 30: 401–22
Turner, D. 1975. *The Vampire Bat: A Field Study in Behavior and Ecology*, Baltimore: Johns Hopkins University Press
Washburn, A., Tuohy, J. & Davis, E. 1948. Cave sickness: A new disease entity? *American Journal of Public Health*, 28: 1521–26

Further Reading

Anon. 1947–(annual editions). *Physicians' Desk Reference*, Montvale, New Jersey: Medical Economics Co.
Anon. 1992. *Technical Support Document for the 1992 Citizen's Guide to Radon*, Washington, DC: Radon Division, Office of Radiation Programs, US Environmental Protection Agency
Ashford, D. *et al.* 1999. Outbreak of histoplasmosis among cavers attending the National Speleological Society Annual Convention, Texas, 1992. *American Journal of Tropical Medicine*, 60: 899–903
Centers for Disease Control and Prevention. 1998. *Bats and Rabies: A Public Health Guide*, Washington: US Department of Health and Human Services; also available at www.cdc.gov./ncidod/dvrd/rabies
Cockrum, E.L. 1997. *Rabies, Lyme Disease, Hanta Virus and Other Animal-borne Human Diseases in the United States and Canada*, Tucson, Arizona: Fisher Books
Frankland, J.C. 1990. Expedition medicine and Histoplasmosis in Guangxi, China. *Cave Science*, 17(2): 85–86
Halliday, W. 1974. Cave medicine and first aid. In *American Caves and Caving: Techniques, Pleasures, and Safeguards of Modern Cave Exploration*, New York: Harper and Row
Lewis, W. 1989. Histoplasmosis: A hazard to new tropical cavers. *National Speleological Society Bulletin*, 51: 52–65
Tilton, B. 1999. First aid for trenchfoot. Website: Michael Hodgson's Adventure Network http://www.adventurenetwork.com/Healthsafe/Trench1staid.html

DISSOLUTION: CARBONATE ROCKS

To understand the evolution of karst in limestone terrains it is important not only to know how much limestone can be dissolved, but also how fast dissolution proceeds. The first topic is related to the equilibrium chemistry of the system H_2O–CO_2–$CaCO_3$, and the second to its reaction kinetics. The knowledge of dissolution rates in relation to a given geochemical situation is the key to understanding important processes in karst.

Depending on its saturation state, water containing carbon dioxide dissolves or precipitates limestone ($CaCO_3$) by three elementary chemical reactions, which proceed in parallel (Plummer, Wigley & Parkhurst, 1978). They are:

$$CaCO_3 + H^+ \leftrightarrow Ca^{2+} + HCO_3^- \quad (I)$$

$$2CaCO_3 + H_2CO_3 \leftrightarrow 2Ca^{2+} + 2HCO_3^- \quad (II)$$

$$CaCO_3 + H_2O \leftrightarrow Ca^{2+} + HCO_3^- + OH^- \quad (III)$$

For undersaturated solutions these cause dissolution rates that are given by the Plummer, Wigley & Parkhurst (PWP) equation, which was confirmed later by Chou, Garrels & Wollast (1989):

$$R = k_1(H^+)_s + k_2(H_2CO_3^*)_s + k_3 - k_4(Ca^{2+})_s (HCO_3^-)_s \quad (1)$$

The rate R is given in mmol cm^{-2} s^{-1}. The brackets denote the chemical activities at the surface of the mineral. k_1, k_2, and k_3, are rate constants of the forward reactions (I)–(III). k_4 comprises the corresponding backward reactions (I)–(III) (precipitation) and is therefore dependent on the CO_2 concentration in the solution. Reactions (I)–(III) can be summarized into:

$$CaCO_3 + H_2O + CO_2 \leftrightarrow Ca^{2+} + 2HCO_3^- \quad (IV)$$

Thus one molecule of CO_2 is consumed to dissolve one $CaCO_3$ unit. To react with $CaCO_3$, carbon dioxide must be converted to carbonic acid, H_2CO_3, and to H^+ and HCO_3^-

$$CO_2 + H_2O \leftrightarrow H_2CO_3 \leftrightarrow H^+ + HCO_3^- \quad (V)$$

The first reaction is slow (Usdowski, 1982). It takes several seconds to several minutes—depending on the pH of the solution and the temperature—until it comes to chemical equilibrium. Therefore conversion of CO_2 to H_2CO_3 plays an important role in the effective dissolution rates. When a solution of volume V interacts with limestone rock of reactive surface area A, the number of CO_2 molecules converted per unit time into H_2CO_3 must be equal to the number of $CaCO_3$ units released during this time:

$$V \times (d[CO_2]/dt) = A \times R \quad (2)$$

If the ratio V/A is sufficiently small, slow conversion of CO_2

becomes rate limiting. The Ca^{2+} and CO_3^{2-} ions released from the surface of the mineral must be transported away into the bulk of the solution by molecular diffusion. On the other hand, CO_2, H_2CO_3, and H^+ must migrate from the bulk to the mineral's surface. Schematics of these processes are shown by Figure 1. Dashed lines symbolize diffusional transport, and solid lines indicate chemical reactions. Due to diffusional transport, concentration gradients build up such that the concentrations in the bulk are different from those at the mineral surface. Therefore equation 1 must not be used to calculate the rates by inserting bulk concentrations. The correct rates have been obtained by involved models (Buhmann & Dreybrodt, 1985a, b; Dreybrodt & Buhmann, 1991), taking into account the PWP equation, conversion of CO_2, and diffusion. It turns out that the rates not only depend on the chemical composition of the solution but also on its hydrodynamics and the ratio V/A. Figure 2 shows typical results for a water layer with depth δ and stagnant or in laminar flow, which covers a plane surface of limestone ($\delta = V/A$ is given in cm on the curves). The solution is in equilibrium with an atmosphere containing CO_2 at a partial pressure $pCO_2 = 5 \times 10^{-3}$ atm. The dissolution rates are plotted against the calcium-concentration in the solution. Note: 10^{-8} mmol cm^{-2} s^{-1} corresponds to a retreat of bedrock by 1.17×10^{-2} cm a^{-1}. For very thin films $\delta < 5 \times 10^{-3}$ cm, the rates increase linearly with δ at a constant calcium concentration. This is the region where CO_2 conversion is rate limiting. For 5×10^{-3} cm $< \delta < 5 \times 10^{-2}$ cm the rates are only slightly dependent on δ. In this region both CO_2 conversion and diffusional mass transport control the rates. For $\delta > 5 \times 10^{-2}$ cm the rates drop with δ because mass transport takes over. The rates are then given by (Dreybrodt, 1988):

$$R = \alpha_{lim} (c_{eq} - c)/(1 + \alpha_{lim} \delta/3D) \quad (3)$$

where c is the actual concentration in mmol cm^{-3} and c_{eq} is

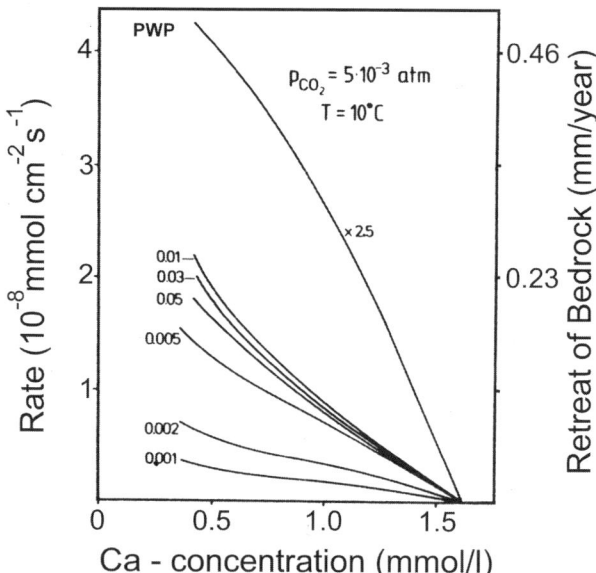

Dissolution: Carbonate Rocks: Figure 2. Dissolution rates in dependence of the Ca-concentration. A water film of depth δ either stagnant or in laminar flow covers a plane surface of calcite. δ is given in cm at the corresponding curve.

the equilibrium concentration of Ca with respect to calcite. The c_{eq} depends on the CO_2 concentration in the solution. The α_{lim} is the average slope for the curve with $\delta = 1 \times 10^{-2}$ cm; D is the coefficient of diffusion. The uppermost curve depicts maximal dissolution rates, which occur in fully turbulent flow with $\delta > 10$ cm, when CO_2 conversion is not rate limiting and the constant of diffusion due to eddies is higher by four orders of magnitude compared with the laminar flow. In this case rates are given by the PWP equation.

The significant influence of CO_2-conversion to dissolution and to precipitation rates has been shown by experiment. The enzyme carbon-anhydrase (CA) catalyses this reaction so effectively that even at μmolar concentrations the CO_2 conversion is effected instantly. Consequently, dissolution or precipitation rates are significantly enhanced upon addition of CA to the solution (Dreybrodt et al., 1996, 1997). When flow is turbulent the bulk becomes completely mixed and is separated from the mineral surface by a diffusion boundary layer (DBL). Mass transport into the bulk by molecular diffusion must then be effected through this layer. Its thickness ϵ depends on the hydrodynamic flow conditions. Figure 3 depicts the dissolution rates for a turbulent layer with depth δ of 1 cm for different values of the thickness ϵ of the DBL. For $\epsilon = 0$ the rates are maximal and entirely surface controlled. Since in this case the bulk concentrations of all species are equal to their surface concentrations, the rates can be obtained from the PWP equation. With increasing thickness of ϵ the rates drop and become almost independent of ϵ at values between 0.01 cm $< \epsilon <$ 0.03 cm. (Dreybrodt & Buhmann, 1991). All theoretical models have been verified experimentally (Dreybrodt et al., 1996; Liu & Dreybrodt, 1997).

When water is flowing in fractures the CO_2 cannot be replenished and is therefore consumed during dissolution. This can be modelled by a solution contained between two parallel planes

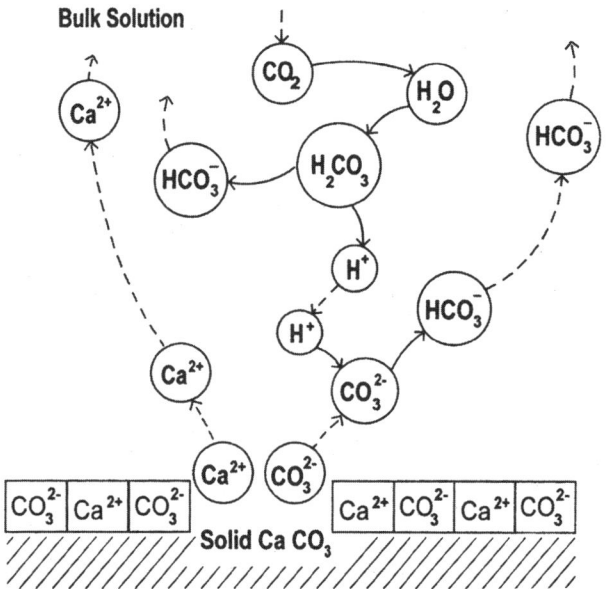

Dissolution: Carbonate Rocks: Figure 1. Schematic diagram of chemical reactions and transport during dissolution.

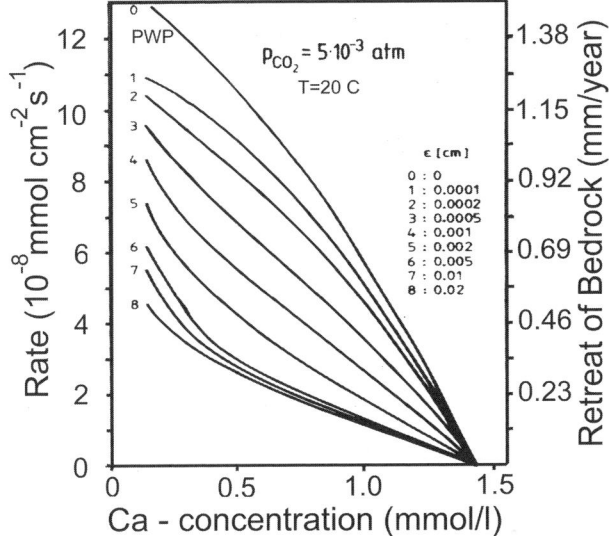

Dissolution: Carbonate Rocks: Figure 3. Dissolution rates against Ca concentration for turbulent flow. The depth δ of the water layer is 1 cm. The curves show rates for different values ε of the DBL. The rates decrease with increasing ε.

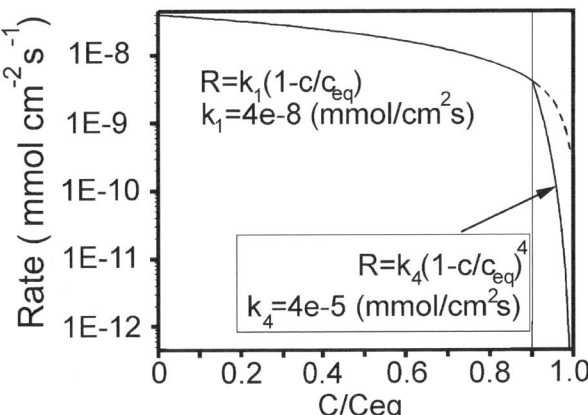

Dissolution: Carbonate Rocks: Figure 4. Switch to nonlinear kinetics. The dashed line shows the continuation of the linear rate law.

with a distance 2δ (Buhmann & Dreybrodt, 1985b). The results are similar to those shown above.

In all cases the dissolution rates can be approximated with an accuracy of about 10% by:

$$R = \alpha (c_{eq} - c) \qquad (4)$$

The rate constant α is in the order of 10^{-5} cm s^{-1}. It depends on temperature, pCO_2 in the solution, the value V/A, flow conditions and whether the system is open or closed with respect to CO_2. Values of α are reported by Buhmann & Dreybrodt (1985a, b), Dreybrodt (1988) for laminar flow, and for turbulent flow by Liu & Dreybrodt (1997). They can be used to estimate rates for many geological situations in karst (see Erosion Rates: Theoretical Models). Equation 4 is also valid for porous carbonate media with average pore diameter d (Baumann et al., 1985), provided that δ is replaced by $V/A = 0.15d$.

Close to equilibrium at concentrations c_t, inhibition of dissolution occurs for natural calcium carbonate minerals. The rates are then given by:

$$R = k(1 - c/c_{eq})^n, \; c \geq c_t \qquad (5)$$

where n is about 3 to 6 and c_t about 0.8 c_{eq}, depending on lithology and the chemical composition of the rock (Svensson & Dreybrodt, 1992; Eisenlohr et al., 1999). Due to this nonlinear behaviour the rates drop dramatically in comparison to those from equation 3. Therefore in this region the rates are controlled entirely by the surface reaction slowed down by inhibition, and consequently k does not depend on V/A and flow conditions, in contrast to the region of linear kinetics. This is of utmost importance in the early state of conduit evolution where water flows in narrow fractures with aperture widths of several tenths of a millimetre. The dissolution rates are then given by a linear rate law which switches to nonlinear kinetics at c_t. Figure 4 depicts these rates as a function of c. The drastic reduction of the rates and their nonlinear behaviour are the key to understanding the initial evolution of karst conduits (see Speleogenesis: Computer Models).

Dolomite, $MgCa(CO_3)_2$, dissolves by three parallel reactions involving H^+, H_2CO_3, and H_2O, but due to the structure of two-component carbonates the chemical reactions at the surface are far more complex. Busenberg & Plummer (1982) have suggested a rate law, valid for surface-controlled reactions far from equilibrium.

$$R^D = k_1^D (H^+)_s^{1/2} + k_2^D (H_2CO_3^*)_s^{1/2} + k_3^D - k_4^D (HCO_3^-)_s \qquad (6)$$

In contrast to calcite, the reactions with H^+ and H_2CO_2 exhibit rate terms with a reaction order $n = 1/2$, owing to complex processes at the surface. Since dolomite cannot crystal-

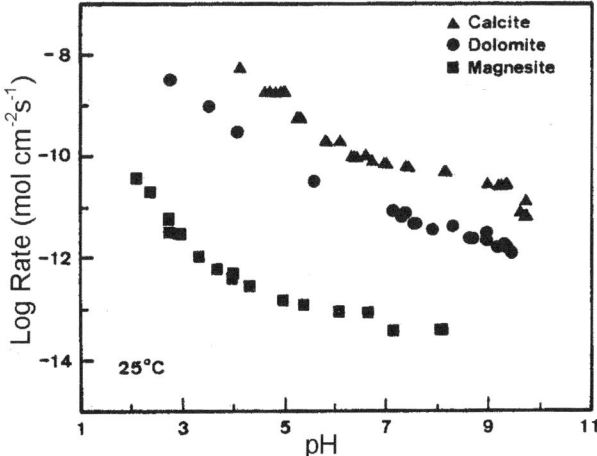

Dissolution: Carbonate Rocks: Figure 5. Comparison of the dissolution rates of calcite, dolomite, and magnesite (Chou, Garrels & Wollast, 1989).

lize from the solution, dissolution is irreversible and therefore equation (6) is not complete. The back-reaction term has been interpreted to result from HCO_3^- ions adsorbed at the surface and inhibiting dissolution. Chou, Garrels & Wollast (1989) found a similar rate law. The rates for dolomite at the same total hardness and pH of the solution are one order of magnitude smaller than for limestone. Figure 5 shows the experimental results of Chou, Garrels & Wollast, which compare the rates of dolomite dissolution with those of calcite at various pH-values, whereby total hardness of calcite and dolomite solutions are equal. Due to the much lower rates of dolomite dissolution, it also seems likely that in natural environments dissolution is surface controlled, or at least conversion of CO_2 and mass transport play a less important role as compared to calcite. However, investigations concerning this have not yet been reported.

Magnesite ($MgCO_3$) as a one-component carbonate shows the same reaction pattern as calcite, and therefore the rates are described by equation 1. The rate constants, however, are lower by about two orders of magnitude (Chou, Garrels & Wollast, 1989). Figure 5 shows the experimental rates in comparison to dolomite and calcite.

WOLFGANG DREYBRODT

See also **Carbonate Karst**

Works Cited

Baumann, J., Buhmann, D., Dreybrodt, W. & Schulz, H.D. 1985. Calcite dissolution kinetics in porous media. *Chemical Geology*, 53: 219–28

Buhmann, D. & Dreybrodt, W. 1985a. The kinetics of calcite dissolution and precipitation in geologically relevant situations of karst areas: 1. Open system. *Chemical Geology*, 48: 189–211

Buhmann, D. & Dreybrodt, W. 1985b. The kinetics of calcite dissolution and precipitation in geologically relevant situations of karst areas: 2. Closed system. *Chemical Geology*, 53: 109–24

Busenberg, E. & Plummer, L.N. 1982. The kinetics of dissolution of dolomite in CO_2–H_2O systems at 1.5 to 65°C and 0 to 1 atm p_{CO2}. *American Journal of Science*, 282: 45–78

Chou, L., Garrels, R.M. & Wollast, R. 1989. Comparative study of the kinetics and mechanisms of dissolution of carbonate minerals. *Chemical Geology*, 78: 269–82

Dreybrodt, W. 1988. *Processes in Karst Systems: Physics, Chemistry, and Geology*, Berlin and New York: Springer

Dreybrodt, W. & Buhmann, D. 1991. A mass transfer model for dissolution and precipitation of calcite from solutions in turbulent motion. *Chemical Geology*, 90: 107–22

Dreybrodt, W., Eisenlohr, L., Madry, B. & Ringer, S. 1997. Precipitation kinetics of calcite in the system $CaCO_3$–H_2O–CO_2: the conversion to CO_2 by the slow process $H^+ + HCO_3^- \rightarrow CO_2 + H_2O$ as a rate limiting step. *Geochimica et Cosmochimica Acta*, 61(18): 3897–904

Dreybrodt, W., Lauckner, J., Zaihua, L., Svensson, U. & Buhmann, D. 1996. The kinetics of the reaction $CO_2 + H_2O \rightarrow H^+ + HCO_3^-$ as one of the rate limiting steps for the dissolution of calcite in the system H_2O–CO_2–$CaCO_3$. *Geochimica et Cosmochimica Acta*, 60: 3375–81

Eisenlohr, L., Meteva, K., Gabrovšek, F. & Dreybrodt, W. 1999. The inhibiting action of intrinsic impurities in natural calcium carbonate minerals to their dissolution kinetics in aqueous H_2O–CO_2 solutions. *Geochimica et Cosmochimica Acta*, 63: 989–1002

Liu, Z. & Dreybrodt, W. 1997. Dissolution kinetics of calcium carbonate minerals in H_2O–CO_2 solutions in turbulent flow: The role of the diffusion boundary layer and the slow reaction $H_2O + CO_2 \Leftrightarrow H^2 + HCO_3^-$. *Geochimica et Cosmochimica Acta*, 61: 2879–89

Plummer, L.N., Wigley, T.M.L. & Parkhurst, D.L. 1978. The kinetics of calcite dissolution in CO_2–water systems at 5°C to 60°C and 0.0 to 1.0 atm CO_2. *American Journal of Science*, 278: 537–73

Svensson, U. & Dreybrodt, W. 1992. Dissolution kinetics of natural calcite minerals in CO_2–water systems approaching calcite equilibrium. *Chemical Geology*, 100: 129–45

Usdowski, E. 1982. Reactions and equilibria in the system CO_2–H_2O and $CaCO_3$–CO_2–H_2O. A review. *Neues Jahrbuch für Mineralogie Abhandlungen*, 144: 148–71

Further Reading

Dreybrodt, W. & Eisenlohr, L. 2000. Limestone dissolution rates in karst environments. In *Speleogenesis: Evolution of Karst Aquifers*, edited by A. Klimchouk, D.C. Ford, A.N. Palmer & W. Dreybrodt, Huntsville, Alabama: National Speleological Society

Morse, J.W. & Arvidson, R.S. 2002. The dissolution kinetics of major sedimentary carbonate minerals. *Earth Science Reviews*, 58(1–2): 51–84

DISSOLUTION: EVAPORITE ROCKS

Evaporites can form karst scenery, for example, gypsum or anhydrite karst covers large areas in the Ukraine, the United States, and Europe. Salt karst is observed on salt diapirs in arid climates (see Evaporite Karst).

To understand the processes in the evolution of karst in evaporite rocks, it is necessary to know the kinetics of dissolution as well as the equilibrium chemistry of the aqueous solutions involved. All evaporites dissolve by detachment of the ions constituting the crystal lattice and by diffusional mass transport from the surface into the solution. In contrast to carbonate rocks, no further chemical reactions complicate the kinetics of dissolution.

The most common evaporite is gypsum ($CaSO_4 \cdot 2H_2O$), which in an aqueous solution dissolves rapidly, with initial rates of dissolution being roughly several 10^{-5} mmol cm^{-2} s^{-1}. Its equilibrium concentration c_{eq} with respect to calcium at 20°C in pure water is 15.3×10^{-3} mmol cm^{-3}. Note that a dissolution rate of 10^{-5} mmol cm^{-2} s^{-1} corresponds with a retreat of the bedrock by 2.3 cm per year. Using rotating disc experiments, Jeschke, Vosbeck & Dreybrodt (2001) found that the surface reaction rates are given by:

$$R_s = k_s(1 - c_s/c_{eq}) \quad (1)$$

where c_s is the Ca concentration at the surface and $k_s = 1.1 \times 10^{-4}$ mmol cm^{-2} s^{-1}. The Ca^{2+} and SO_4^{2-} ions released from the mineral are transported away from the surface into the solution by molecular diffusion. Therefore concentration gradients exist and the surface concentration differs from that in the bulk. The transport rate R_D is given by:

$$R_D = k_D(1 - c_B/c_{eq}) \quad (2)$$

where k_D is the transport constant and c_B is the concentration of the bulk solution. Since, by mass conservation, R_S must be equal to R_D, one finds an effective rate law.

$$R = k_{eff}(1 - c_B/c_{eq}), \quad k_{eff} = (k_s \times k_D)/(k_s + k_D) \quad (3)$$

When $k_s \gg k_D$, k_{eff} becomes close to k_D and rates are controlled by diffusion. On the other hand if $k_s \ll k_D$, k_{eff} becomes k_s and the rates are surface controlled. The region where k_s and k_D are of similar magnitudes is called mixed kinetics.

When water flows are laminar in a film of thickness, δ, k_D is given by

$$k_D = 2D\, c_{eq}/\delta \quad (4)$$

(see Beek & Muttzall, 1975) where D is the coefficient of diffusion (10^{-5} cm^2 s^{-1}). For $\delta = 1$ mm one obtains $k_D = 3 \times 10^{-6}$ mmol cm^{-2} s^{-1} and the rates are controlled by diffusion.

When water flows turbulently the completely mixed bulk with concentration c_B is separated by a diffusion boundary layer of thickness ϵ from the surface of the mineral. Mass transport proceeds by molecular diffusion through this layer. In this case the transport constant is given by:

$$k_D = D\, c_{eq}/\epsilon \quad (5)$$

ϵ depends on the velocity of the flow and its geometry (e.g. flow in films with a free surface, flow in conduits) and decreases with increasing velocity in a complex manner (Beek & Muttzall, 1975). In natural situations ϵ varies between 10^{-2} cm and 10^{-1} cm (see also Dissolution: Carbonate Rocks). Therefore k_D becomes dependent on flow velocity and consequently also k_{eff}. This has been observed by James (1992) for gypsum and anhydrite.

In surface streams, gypsum dissolution rates between 1.5×10^{-5} and 5×10^{-5} mmol cm^{-2} s^{-1} have been observed, with the highest rates at the highest velocities (Navas, 1990). These rates are lower than k_s by about a factor of ten and are limited by diffusion across a boundary layer with $\epsilon \approx 10^{-2}$ cm.

Close to equilibrium at about 0.94 c_{eq}, and only in natural materials, the rate law switches, in a similar way to that in limestone, to higher order kinetics:

$$R = k_n (1 - c/c_{eq})_n, \quad c \geq 0.94\, c_{eq} \quad (6)$$

where $k_n = 3$ mmol cm^{-2} s^{-1} and $n = 4.5$. Figure 1 shows a plot of log R versus $\log(1 - c/c_{eq})$ for the experimental rates obtained in a batch experiment for selenite (Marienglas) (Jeschke, Vosbeck & Dreybrodt, 2001). Note that the concentrations increase from right to left. At low concentrations the linear rate law is represented by the straight line with slope $n_1 = 1.0$. At $c = 0.94\, c_{eq}$ the curve bends down into a steeper straight line with slope $n_4 = 4.5$. This region represents the nonlinear kinetics where, due to the inhibition of dissolution, the rates are controlled by surface reaction. Similar results have been obtained for alabaster and gypsum rock. This nonlinear behaviour is of utmost significance in the evolution of karst conduits (see Speleogenesis: Computer Models).

James (1992) has reported effective rate constants lower by a factor of ten in an experiment where transport control dominates (see equation 3).

Anhydrite (CaSO$_4$) is an evaporite rock with a higher thermodynamic equilibrium concentration $c_{eq} = 23.5$ mmol l^{-1} at 10°C compared with gypsum's 14.4 mmol l^{-1}. When anhydrite dissolves, the solution can become supersaturated with respect to gypsum, but remains still aggressive to anhydrite. Consequently, gypsum is precipitated until all the anhydrite is converted. Therefore dissolution rates can only be measured at concentrations below gypsum saturation. Experiments revealed the rate law:

$$R = k(1 - c/c_{eq})4.5, \quad k = 6 \times 10^{-6} \text{ mmol cm}^{-2} \text{ s}^{-1} \quad (7)$$

The initial rates are lower by a factor of 20 compared with gypsum, and are surface controlled (equation 1). These results are in contrast to the observations of James (1992), who reports a rate law with $n = 2$. In his experiments, however, the hydrodynamic conditions were poorly defined, and both surface rates and diffusion controlled the dissolution rates. It should be pointed out that this must be considered for each natural situation.

To compare the dissolution kinetics of anhydrite with those of gypsum, Figure 2 shows the evolution of Ca concentration

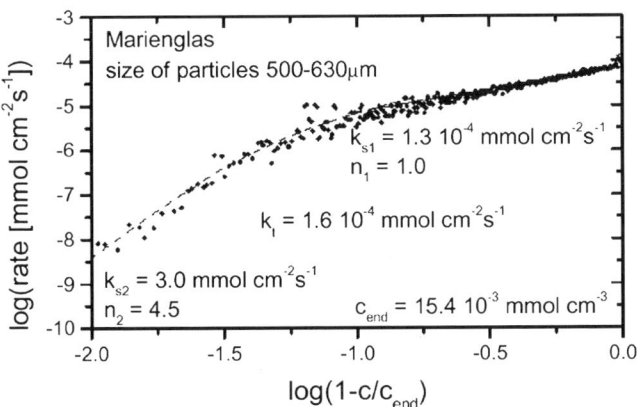

Dissolution: Evaporite Rocks: Figure 1. Surface-controlled rates as a function of Ca concentration.

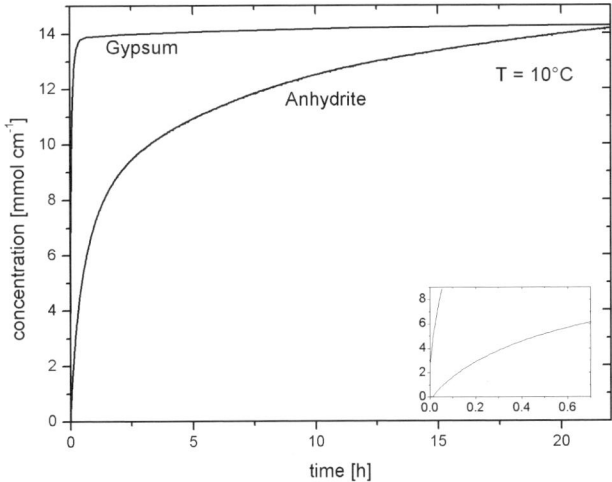

Dissolution: Evaporite Rocks: Figure 2. Comparison of the evolution of concentrations of gypsum and anhydrite. The inset shows the first 0.6 hours.

in a batch experiment with gypsum particles of 550 μm size in comparison to anhydrite particles of the same size in an identical experimental set-up. From such experiments the rate laws shown above are obtained. The experiment shows the much smaller increase of concentration for anhydrite.

The most soluble evaporite rock is rock salt with $c_{eq} = 5.4$ mol l^{-1}. Alkattan *et al.* (1997), using the rotating disc technique, have obtained a linear rate law with $k_s = 6.7 \times 10^{-1}$ mmol cm^{-2} s^{-1}. This extremely high rate constant in practically all cases of interest is large compared with transport constants k_D, and the dissolution of rock salt is controlled by diffusion. One extreme case, however, has been observed at extremely high floods in the salt caves of Mount Sedom, Israel, with a bedrock retreat of 0.2 mm s^{-1} (Frumkin & Ford, 1995). This is close to the maximal value of 0.17 mm s^{-1} derived from the value of Alkattan *et al.* (1997).

WOLFGANG DREYBRODT

Works Cited

Alkattan, M., Oelkers, E.H., Dandurand, J.-L. & Schott, J. 1997. Experimental studies of halite dissolution kinetics. 1. The effect of saturation state and the presence of trace metals. *Chemical Geology*, 137: 201–19

Beek, W.J. & Muttzall, K.M.K. 1975. *Transport Phenomena*, London and New York: Wiley

Frumkin, A. & Ford, D.C. 1995. Rapid entrenchment of stream profiles in the salt caves of Mount Sedom, Israel. *Earth Surface Processes and Landforms*, 20: 139–52

James, A.N. 1992. *Soluble Materials in Civil Engineering*, Chichester and New York: Ellis Horwood, p.434

Jeschke, A.A., Vosbeck, K. & Dreybrodt, W. 2001. Surface controlled dissolution rates in aqueous solutions exhibit nonlinear dissolution kinetics. *Geochimica et Cosmochimica Acta*, 65: 13–20

Navas, A. 1990. The effect of hydrochemical factors on the dissolution rate of gypsiferous rocks in flowing water. *Earth Surface Processes and Landforms*, 15: 709–15

Further Reading

Klimchouk, A. 2000. Dissolution and conversions of gypsum and anhydrite. In *Speleogenesis: Evolution of Karst Aquifers*, edited by A.B. Klimchouk, D.C. Ford, A.N. Palmer & W. Dreybrodt, Huntsville, Alabama: National Speleological Society

DISSOLUTION: SILICATE ROCKS

Quartz

The dissolution of quartz in water is essentially a simple process involving the formation of silicic acid by hydration (see Dove & Rimstidt, 1994, for more details):

$$SiO_2 \text{ (s)} + 2H_2O = H_4SiO_4 \text{ (aq)}$$

where the equilibrium constant at 25°C is:

$$K = \alpha H_4SiO_4 \text{ (aq)} = 1.1 \times 10^{-4}$$

Silicic acid (H$_4$SiO$_4$) is a very weak acid, barely ionically dissociated under normal karst conditions, and its solubility is low, being constant at about 10 mg l^{-1}. At pH values above nine, however, the quartz solubility increases markedly, due to the following ionic dissociation:

$$H_4SiO_4 \text{ (aq)} = H_3SiO_4^- + H^+$$

This highly alkaline condition is generally unusual in the karst environment, with the exception of rare cases.

Some authors (see references in Martini, 2000) have suggested that complexation of silicic acid with organic acids may increase the solubility of quartz, but mainly at basic pHs. Because such high pH values have not been observed in the dominantly acidic waters of quartzite karst, this effect is probably negligible, as supported by the low (<10 mg l^{-1}) silica concentration in these waters.

Another important aspect to consider is the extremely slow rate of quartz dissolution, which is 10^{-17} moles cm^2 s^{-1} at 25°C, and is entirely controlled by the rate of bond breaking and hydration of silica on the crystal surface. In the field this is exemplified by the high resistance of this mineral to weathering. The dissolution rate is considerably increased in the presence of organic acids, up to $10^{-15.5}$ moles cm^2 s^{-1}. The latter conditions are encountered in the organic-rich waters of quartzite terrains, and may play a role in karst development.

At 25°C the rate of quartz crystallization is also very slow, as exemplified by the evaporation of water containing silica, which is then not deposited in the form of quartz. During this process, the silica concentration can rise to several hundred mg l^{-1}, i.e. well above quartz solubility. In this range of high supersaturation, silica occurs dominantly in the form of polymers, such as H$_6$Si$_2$O$_7$, which was present in negligible quantities in the field of quartz undersaturation. Evaporation to dryness leads to the deposition of silica in an amorphous form, which is more or less hydrated (opal), and this later slowly decays into low-cristobalite or tridymite, and eventually to quartz. These silica polymorphs form preferentially due to their rates of deposition, which are higher than those for quartz. The deposition of amorphous silica is illustrated by the formation of opal popcorn in caves and at the surface, and the deposition of silcrete in semi-arid climates.

The very slow rate of quartz dissolution is the main factor controlling the speleogenesis of quartzite caves. As groundwater can remain undersaturated in silica, with respect to quartz, for a very long time, it is able to weather the rock penetratively by enlarging crystal junctions. Consequently, the quartzite loses its cohesion, and becomes friable and amenable to the formation of caves by piping, i.e. by the mechanical removal of quartz grains. Speleogenesis therefore starts by an initial phase (dissolution, forming karst), followed by a second phase (piping or pseudokarst) (see Pseudokarst; Piping Caves and Badlands Pseudokarst).

Minor cavities formed by a dissolution process comparable with the speleogenesis in limestone, gypsum, etc., i.e. dominantly by dissolution, have also been observed in quartzite. Cavities of this type are rare, and form at depth by a hydrothermal process at temperatures of several hundred degrees centigrade, or under conditions of very high pH. The rate of dissolution and the solubility of quartz are much higher in this environment, which explains the nature of these cavities.

Silicate Minerals

In contrast to quartz, the dissolution of silicate minerals is more complex. Silica produces the same non-ionic species below pH 9, but the cations—commonly K^+, Na^+, Ca^{2+}, Mg^{2+}, Fe^{2+}, Fe^{3+}, and Al^{3+}—form more ionically dissociated species, especially the monovalent and divalent ions. Since H^+ is fixed in nondissociated silicic acid, OH^- is produced to balance the cations, thus generating alkaline solutions. Therefore, under the chemical and physical conditions that usually prevail in the weathering zone, in contrast to quartz, the solubility of silicate minerals is strongly dependent on the pH: from neutrality the solubility increases in proportion to the acidity. Furthermore, the dissolution of most rock-forming silicates is complex and incongruent, i.e. it involves the subsequent formation of intermediate phases. The latter consist generally of minerals containing trivalent cations, like clay minerals, and aluminium and ferric hydroxides, which precipitate close to neutrality. These minerals do not form in extreme acidic or alkaline domains, where the dissolution is then congruent. An example is the dissolution of K-feldspar, which is a common rock-forming mineral:

$$KAlSi_3O_8(s) + 8H_2O = K^+ + Al^{3+} + 3H_4SiO_4 \text{ (aq)} + 4OH^-$$

Near neutrality, at room temperature, the solubility is a few mg l^{-1}, but increases to several hundred mg l^{-1} at pH values below four. Below this pH the dissolution is congruent, but above it kaolinite $[Al_2SiO_5(OH)_4]$ forms, which later decays into gibbsite $[Al(OH)_3]$. In the pH domain of low solubility, at 25°C, as for quartz, the dissolution rate is sluggish, i.e. $10^{-15.5}$ moles $cm^2 s^{-1}$.

Although small stream-caves have been reported in granite, the weathering of silicate rocks is less conducive to speleogenesis than that of quartzite, possibly due to the formation of clay minerals as micaceous particles that tend to clog the joints within the rock. Moreover, because silicate dissolution generally leads to oversaturation of silica with respect to quartz, in quartzite its presence as impurities may prevent the dissolution of the latter mineral. This delay in the dissolution of quartz may last for as long as the silicate minerals have not been leached out. Low-pH waters (<4) that have leached aluminosilicates then precipitate allophane (an amorphous hydrated Al-silicate) speleothems by evaporation or a fall in acidity due to the oxidation of organic acids and the release of CO_2.

JACQUES MARTINI

See also **Silicate Karst**

Works Cited

Dove, P.M. & Rimstidt, J.D. 1994. Silica–water interactions. In *Silica: Physical Behavior, Geochemistry and Materials Applications*, edited by P.J. Heaney, C.T. Prewitt & G.V. Gibbs, Washington, DC: Mineralogical Society of America

Martini, J.E.J. 2000. Dissolution of quartz and silicate minerals. In *Speleogenesis: Evolution of Karst Aquifers*, edited by A.B. Klimchouk, D.C. Ford, A.N. Palmer & W. Dreybrodt, Huntsville, Alabama: National Speleological Society

Further Reading

Bennett, P.C. 1991. Quartz dissolution in organic-rich aqueous systems. *Geochimica et Cosmochimica Acta*, 55: 1781–97

Helgeson, H.C., Murphy, W.M. & Aagaard, P. 1984. Thermodynamic and kinetic constraints on reaction rates among minerals and aqueous solutions. II. Rate constants, effective surface area, and the hydrolysis of feldspar. *Geochimica et Cosmochimica Acta*, 48: 2405–32

Holdren, G.R. & Speyer, P.M. 1986. Stoichiometry of alkali feldspar dissolution at room temperature and various pH values. In *Rates of Chemical Weathering of Rocks and Minerals*, edited by S.M. Colman & P.P. Dethier, London and Orlando, Florida: Academic Press

DIVING IN CAVES

Water is an essential agent in the formation of most limestone caves and many systems are still in a stage of development at which some parts remain permanently submerged. The desire to explore such underwater sections of cave passage is almost as old as recreational caving itself. A significant number of conventional cavers regard those who wish to dive into these dark flooded tunnels as the lunatic fringe. However, determined individuals who make the financial and time commitments to learn how to operate safely in this so-called "merciless environment", will experience one of the greatest opportunities for original exploration on Earth. The evolution of cave diving techniques has closely followed the development of diving equipment in general. Indeed, cave divers have contributed to the latter many times. This entry briefly reviews over 200 years of underwater achievements by many generations of cave explorers. It is inevitably only a summary of various milestone events from around the world; greater detail is available in Farr (2000), which provides essential reading.

Perhaps the first documented attempt to pass a section of underwater passage (or "sump") took place in 1773, at Peak Cavern in England. Mr Day tried to pass the Buxton Water Sump by holding his breath. He soon got into difficulty and was pulled from the water, on the verge of drowning, by one of the surface party who managed to catch hold of his arm. The explorer was (we are told) "speechless for some time". It has thus been observed that the event might also be considered as the first cave diving rescue! (Note that the earliest documented rescue attempt in a cave, by an equipped diver, probably took place in Austria's Lurloch in 1894.) A far more celebrated (and happily successful) free diving attempt occurred in 1922 when the French explorer Norbert Casteret passed a short sump in the Grotte De Montespan (French Pyrenees). He discovered a dry cave system beyond, which contained important archaeological remains; the scientific potential of cave diving was thus recognized some 80 years ago.

Very few explorers had access to the early brass-helmeted "Standard Diving Equipment", but some notable explorations did take place. The deep Fontaine De Vaucluse (France), the typical example of a "vauclusian resurgence", was explored to 23 m depth by Nello Ottonelli in 1878 (Cousteau, 1953). Two years later, Alexander Lambert (of Seibe Gorman Ltd) made some outstanding dives in the (artificial) flooded Severn Tunnel

(England). Although not strictly cave diving, his 300 m penetration certainly revealed the possibilities of the Standard Equipment and early "Fleuss" oxygen rebreather used (the latter efficiently recycling the exhaled gas). Another notable early use of Standard Equipment took place at Le Creugenat (Switzerland), where a 95 m sump was successfully passed in 1934.

Traditionally, in the United Kingdom, cave divers were ordinary cavers who learned to dive in order to pass flooded sections to discover further "dry" cave passages. (Note that this is in marked contrast with other parts of the world, such as Florida, where most practitioners are divers first, then extend their activities to underwater cave visits.) In 1934, Jack Sheppard used a home-made respirator in Swildon's Hole, possibly the first successful use of "self contained underwater breathing apparatus" (SCUBA) to pass a sump (Balcombe et al., 1990). The advantage of such equipment is that the diver is no longer reliant on a link with the surface for his breathing air supply. However, early SCUBA was not very reliable and exploratory work shortly afterwards (Balcombe, 1935) at Wookey Hole (England) still used Standard Equipment to "bottom walk" along the submerged course of the River Axe. On one notable dive, Graham Balcombe made the first ever live diving TV broadcast from inside his brass helmet, at Wookey in 1935. Those involved in these early UK explorations later formed the Cave Diving Group in 1946, one of the first non-military diving organizations in the world.

Of great significance to underwater exploration globally was the invention of the "Aqualung" in 1943, by Emile Gagnan and Jacques Cousteau. This was the first reliable "open circuit" SCUBA and for many cave dives, it still remains the most appropriate equipment. Although providing less total duration than "closed circuit" rebreathers, the Aqualung was lightweight, robust, and easy to transport through awkward cave passages. Cousteau's "Undersea Research Group" was also largely responsible for the widespread acceptance of free swimming with fins (as opposed to the bottom walking traditionally associated with the Standard Equipment). In 1946 they exploited their new equipment in a well-planned attempt to pass the Vaucluse sump. Unfortunately, the only way on was downwards; they reached a depth of over 60 m but almost died in the attempt due to carbon monoxide poisoning caused by a compressor fault (Cousteau, 1953). Significant explorations were also made soon afterwards in France, at Font Estramar and the Fontaine Des Chartreux; among those involved were Guy De Lavaur, plus Undersea Research Group members Farges and Morandiere. Elsewhere in France, Jean Alinat of the same team, successfully passed a 150 m long flooded section in the Gouffre Des Vitarelles.

The Aqualung helped many other countries develop their cave diving skills in the years following, one good example being the series of deep dives in the early 1950s at Florida's Wakulla Springs (United States). A few years later, the use of the Aqualung had spread to Australia, where underwater caves in the Mount Gambier and Jenolan regions were being explored. In the United Kingdom, postwar developments took a different direction as cave divers initially exploited the relatively easily available frogman's equipment (rebreathers and drysuits), rather than the Aqualung. Of particular note were Bob Davies' dives in Buxton Water Sump (Peak Cavern, England). Many observations of water levels in a "torricellian" airbell here prompted the publication of perhaps the first serious scientific paper, documenting work where scientific research (as opposed to cave exploration) was the main motive for cave diving (Davies, 1950).

From these early foundations, the list of increasingly audacious and successful cave diving explorations around the world grew exponentially in the 1960s, 1970s, and 1980s. Many major advances were made, particularly in the United States and in Europe, including the United Kingdom. The lightweight Aqualung proved ideal for exploration of remote underground sites, such as the downstream sumps of the Gouffre Berger in France, where Ken Pearce's UK team made notable progress on their expedition of 1963. The psychological barrier of being a long way from the nearest entrance was overcome, as witnessed by explorations at the Trou Madame, Emergence Du Ressel, Doux De Coly, and Fontaine Saint-Georges resurgences (France); The Blautopf (Germany); Keld Head (England); and Manatee Springs, Cathedral Canyon, and Wakulla 1987 project (United States). The great underwater systems of Australia's Nullarbor Plain and the submerged lava tube of Lanzarote's Atlantida Tunnel were pushed for impressive distances. Significant dives began to be performed in the Former Soviet Union, where the potential for large flooded cave systems was widely recognized. In the Bahamian Blue Holes, the early work of pioneers, such as George Benjamin, was followed up by various teams coordinated by Rob Palmer, who carried out much underwater scientific work (Palmer, 1989). The above are but a few examples, taken almost at random, from the extensive record of a superb era of exploration (Farr, 2000).

The other barrier, which was gradually overcome during the above period, was that of depth. Many French and Swiss systems explored were both deep and long, and certain explorations took deep diving to the absolute limits. The descents of Claude Touloumdjian and Jochen Hasenmayer (to depths of 143 m and 205 m, respectively) at Vaucluse, were outstanding, as was the Nacimiento Del Rio Mante (Mexico) dive to a depth of 267 m in 1989 (Exley, 1994). In the United Kingdom, Martyn Farr was the first person to take deep cave diving seriously; his 60 m deep exploration (1982) at Wookey Hole was significant (given the cold water and low visibility of British sumps) as was Rob Parker's 1985 push to a depth of 68 m here. Many of the above sites could not have been explored to such limits without the adoption of various synthetic breathing mixtures, in particular those based on the physiologically inert gas helium.

The 1990s have seen further advances at many of the sites mentioned above and elsewhere. Cave divers have made full use of advanced decompression software, multiple underwater scooters, the latest mixed gas rebreathers, underwater habitats (artificial gas-filled chambers for "dry" decompression stops) and submersible dive computers. Some recent ultra-long dives include the Woodville Karst Plain Project (WKPP) team's 4298 m penetration at Chip's Hole (United States) in 1996 and also Olivier Isler's 4352 m Doux De Coly (France) exploration of 1998 (since extended to 5000 m by a German team). Both of these are major achievements, particularly in view of the significant depth of much of the passages. However, the WKPP's 5506 m dive into Wakulla Springs (United States) in 1998, at an average depth of 87 m, represents a phenomenal technological advance.

Of particular note was a small British team's extension of the French Emergence Du Ressel underwater system. The first sump

Diving in Caves: Diving in Joint Hole, Yorkshire. Note the roof scallops. (Photo by John Cordingley)

(1995 m long and 80 m deep) was not passed until 1990, by Olivier Isler. However, in the second half of the 1990s, Rick Stanton and Jason Mallinson, in collaboration with Germany's Reinhard Buchaly, achieved a total penetration (in Sump 5) of over 3000 m. Unlike most of the expeditions referred to above, which had massive budgets, the most recent Ressel explorations were done quietly by a group of enthusiasts without major financial backing, in what has been described as "a triumph of adventure over technology". Conversely, most UK divers, whilst never achieving "record" lengths or depths on a world scale, have continued to develop specialized techniques for use at home. In the United Kingdom, the 1990s was the decade when underwater civil engineering skills were perfected, perhaps best exemplified by the extensive submerged excavations at Malham Cove Rising.

Throughout the last decade, the worldwide quest for ultimate depth has continued. In 1997, Pascale Bernabe reached a depth of 240 m in the Fontaine De Vaucluse (France) whilst Jim Bowden and Sheck Exley's deep explorations of 1993 and 1994, at Zacatón cenote in Mexico, attained a depth of 276 m. Exley, who was arguably the world's greatest cave diver, was sadly killed on this project. The deepest recorded cave dive, however, took place in South Africa, where Nuno Gomes (following on from an earlier deep dive by Exley) reached the bottom of Bushmansgat in 1996, at an astounding 282 m depth.

Increasingly throughout the 1990s, many very long underwater caves were explored in Mexico's Yucatán Peninsula (see separate entry). Here the water is clear and relatively warm, the passages are often spectacularly adorned with speleothems, and more significantly, the depths are usually minimal. All these factors have led to intense activity here in recent years. Many of these "cenotes" (a Spanish word derived from the Mayan word "dzonot", meaning a well) have been linked to give by far the longest underwater cave systems in the world including Ox Bel Ha system (107 km), Nohoch Nah Chich system (61 km), and the Dos Ojos system (56.7 km). The prospect of joining even these supersystems, to create a single network over 200 km in total length, reveals that cave diving has certainly come of age.

So what of the future? International cooperation on cave diving projects and interchange of ideas will continue to flourish. In recent years, an obvious trend has been the increasing reliance on electronic devices. This will continue in future, although sadly it will also eliminate from the cutting edge all but those with substantial financial resources. The three-dimensional high-speed sonar mapping of Wakulla Springs in 1999 is perhaps a pointer to the widespread use of technology that can be expected. There will be an increased utilization of remotely operated electronic devices, which will collect data about the underwater cave automatically. The French have already made excellent use of such vehicles (e.g. the "Telenaute", "Sorgonaute", and "Modexa") at Vaucluse, to explore beyond the limitations of the human diver. Perhaps one day, most underwater cave exploration will be done by such devices and there will no longer be any need for man to get in the water. In contrast, are the current experiments in the use of saturation diving techniques specifically for cave diving, and the development of the artificial gill, both of which will effectively allow man to stay underwater almost indefinitely.

JOHN CORDINGLEY

Works Cited

Balcombe, F.G. 1935. *The Log of the Wookey Hole Exploration Expedition, 1935* (Private publication by Balcombe)

Balcombe, F.G., Cordingley, J.N., Palmer, R.J. & Stevenson, R.A. 1990. *Cave Diving: The Cave Diving Group Manual*, Somerset: Castle Cary Press

Cousteau, J.Y. 1953. *The Silent World*, New York: Harper and London: Hamilton

Davies, R.E. 1950. Water at a depth of −5 ft discovered by diving in Peak Cavern. *Nature*, 166: 895

Exley, S. 1994. *Caverns Measureless to Man*, St Louis, Missouri: Cave Books

Farr, M.J. 2000. *The Darkness Beckons: The History and Development of Cave Diving*, 2nd edition, with a supplement: *Postscript—the 1990s*, London: Diadem Books and St Louis, Missouri: Cave Books

Palmer, R.J. 1989. *Deep Into Blue Holes: The Story of the Andros Project*, London: Unwin Hyman; Nassau, Bahamas: Media Publishing, 1997

Further Reading

Prosser, J. & Grey, H.V. (editors) 1992. *NSS Cave Diving Manual: An Overview*, Branford, Florida: Cave Diving Section of the National Speleological Society

Useful Websites

Cave Diving Group (UK), http://cavedivinggroup.org.uk
National Speleological Society Cave Diving Section (US), http://www.cavediver.org

DOLINES

A doline is a natural enclosed depression found in karst landscapes. Dolines are also sometimes known as sinkholes, particularly by engineers and especially in North America. They are usually subcircular in plan and tens to hundreds of metres in diameter, though their width can range from a few metres to about a kilometre. From the lowest point on their rim, their depths are typically in the range of a few metres to tens of metres, although some can be more than a hundred metres deep and occasionally even 500 m. Their sides range from gently sloping to vertical, and their overall form can range from saucer-shaped to conical or even cylindrical. Their lowest point is often near their centre, but can be close to their rim. Dolines are especially common in terrains underlain by carbonate rocks, and are widespread on evaporite rocks. Some are also found in siliceous rocks such as quartzite. Dolines have long been considered a diagnostic landform of karst, but this is only partly true. Where there are dolines there is certainly karst, but karst can also be developed subsurface in the hydrogeological network even when no dolines are found on the surface.

Doline derives from "dolina", a word of Slav origin meaning valley, possibly because these were the most common hollows in the landscape of the Dinaric karst, where there are few fluvial valleys. "Slepe dolina" means blind valley, where a stream disappears underground and so its valley ends in a steep face. The word doline entered international scientific literature largely through the writings of Cvijić (1893), despite the more accurate local term vrtača being current in the "classical" Karst of Slovenia. However, the usage of the word doline is so embedded in karst literature now that it would be fruitless to try to change it.

The term sinkhole is sometimes used to refer both to dolines (especially in North America and in the engineering literature) and to depressions where streams sink underground, which in Europe are described by separate terms (including ponor, swallow hole, and stream-sink). Thus the terms doline and sinkhole are not strictly synonymous. Hence, to avoid the ambiguity that sometimes arises in general usage, further qualification is required, such as solution sinkhole or collapse sinkhole. Indeed, the international terminology that is used to refer to dolines that are formed in different ways can also be very confusing. Table 1 lists the terms employed by different authors, the range of terms partly reflecting the extent to which genetic types are subdivided, and Figure 1 illustrates six main doline types.

It is widely recognized that enclosed depressions in karst can be formed by four main mechanisms: dissolution, collapse, suffosion, and regional subsidence (Table 2). However, in practice the complexity of natural processes often results in more than one mechanism being involved, in which case the doline is polygenetic in origin. A typical case is a depression formed initially by dissolution that later in its development is subject to collapse of its floor into an underlying cave, following a combination of downwards dissolution and upwards stoping of the cave roof. In such a case, the gentler upper slopes of the doline were formed by dissolution and the steeper lower slopes by collapse. Table 2 describes the different processes responsible for doline formation, the various styles of doline landforms produced, and the names used in English to refer to them. Beck (1984), Waltham (1989), White (1988), and Ford and Williams (1989) provide numerous examples of doline form and evolution.

Solution Dolines

The bowl-shaped form of a typical doline (Figure 2) indicates that more material has been removed from its centre than from around its margins. Where the principal process responsible for this is dissolution of the bedrock, it follows that there is a mechanism that focuses chemical attack. The amount of limestone that can be removed in solution depends upon two variables: first, the concentration of the solute and, second, the volume of the solvent (in this case the amount of water draining through the doline). Variations in either or both of these quantities could be responsible for the focusing of dissolution near the centre of the depression, but if local variation in solute concentration alone were sufficient to explain the occurrence of solution dolines, then they would be found on every type of limestone in a given climatic zone. This is not the case, as illustrated by comparison of landscapes formed on Devonian, Carboniferous, Jurassic, and Cretaceous limestones in England, where dolines are most frequently found on Carboniferous limestones and practically absent on Cretaceous and Jurassic limestones. It follows, therefore, that local spatial variations in water flow must be responsible for focusing corrosional attack.

The development of dolines of all kinds depends on the ability of water to sink into and flow through karst rocks to outlet

Dolines: Table 1. Doline / sinkhole English language nomenclature as used by various authors (modified from Waltham & Fookes, 2003).

Doline-forming Processes	Ford & Williams (1989)	White (1988)	Jennings (1985)	Bögli (1980)	Sweeting (1972)	Culshaw & Waltham (1987)	Beck & Sinclair (1986)	Other Terms in Use
Dissolution	solution	solution	solution	solution	solution	solution	solution	
Collapse	collapse	collapse	collapse	collapse (fast) or subsidence (slow)	collapse	collapse	collapse	
Caprock collapse		–	subjacent collapse		solution subsidence	–		interstratal collapse
Dropout	subsidence	cover collapse	subsidence	alluvial	alluvial	subsidence	cover collapse	
Suffosion		suffosion	cover subsidence				cover subsidence	ravelled, shakehole
Burial	–	–	–	–	–	buried	–	filled, paleo-

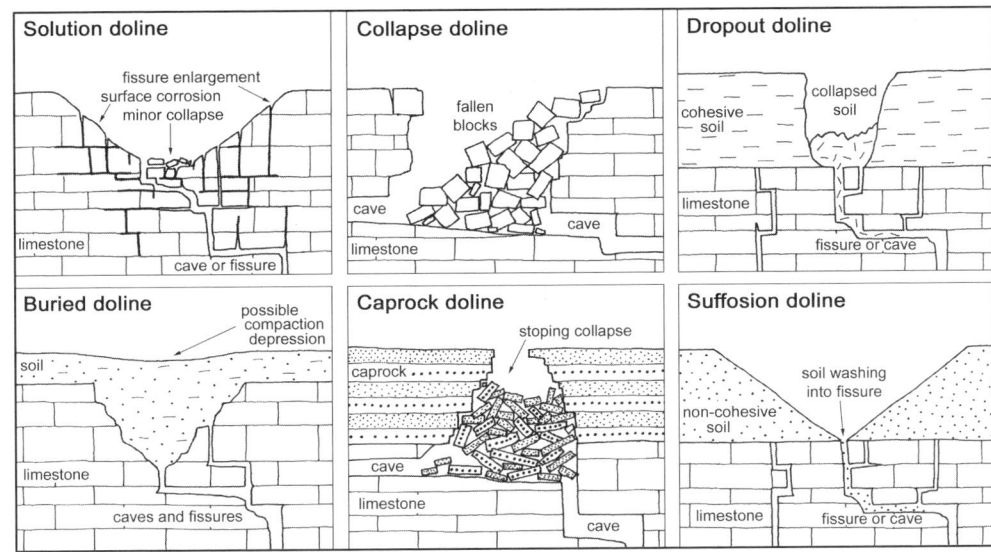

Dolines: Figure 1. Six main types of dolines (after Waltham & Fookes, 2003). Note that dropout dolines and suffosion dolines are two types of subsidence doline.

springs. The initiation of karst drainage is explained in Ford and Williams (1989) but, essentially, the exposure of limestones by erosion provides an input boundary for infiltration of water and a valley incised into the limestone provides an output boundary. The hydraulic gradient between the two sets up a groundwater circulation and dissolved limestone is discharged at springs. Streams and seepages flowing from non-limestone cover rocks (allogenic runoff) into exposed inliers of underlying limestone sink into the limestone. The recharge is centripetal, converging on the inlier, and so focuses solutional attack with the result that closed depressions are formed at such sites. Consequently, such depressions can be considered to be point recharge depressions. The floors of these dolines are usually in limestone, but their sides may be developed in clastic (non-carbonate rocks such as shales or sandstones) cover beds. These landforms are genetically transitional to stream-sinks.

When all cover beds have been removed, the exposed limestone is subject to diffuse recharge and solutional attack from direct precipitation (autogenic recharge). Rainwater is acidified in the atmosphere and infiltrating water is further acidified in the soil. On percolating downwards this water accomplishes most of its dissolutional work within 10 m of the surface. The highly corroded zone near the surface is termed the epikarst or subcutaneous zone. Within this zone fissures in the limestone are found to be especially enlarged by corrosion near the surface but taper with depth. Consequently, infiltration into the karst is rapid, but vertical water flow encounters increasing resistance with depth as fissures become narrower and less frequent. This produces a bottleneck effect after particularly heavy rain, resulting in temporary storage of percolation water in a perched epikarstic aquifer. Joints, faults, and bedding planes vary spatially within the rock because of tectonic history and variations in lithology, consequently the frequency and interconnectedness of fissures available to transmit flow also varies. Some fissures are more favourable for percolation than others, for example where several joints intersect, and as a result these develop as principal drainage paths. Water in the epikarstic aquifer flows towards them and as a result they are subjected to still more dissolution by a positive feedback mechanism and so vertical permeability is enhanced. The piezometric surface (water table) of the epikarstic aquifer draws down over the preferred leakage path similar to the cone of depression in the water table over a pumped well; streamlines adjust and resulting flow is centripetal and convergent on the drainage zone. By this means solvent flow is focused and, as the surface lowers, the more intensely corroded zones begin to obtain topographic expression as solution dolines. The diameters of neighbouring solution dolines are determined by the radii of intersecting draw-down cones (Figure 2). Small dolines develop in areas with high fissure frequency, typically in thinly bedded and closely jointed limestones, whereas particularly large dolines develop in rocks that are massive and have widely spaced joints.

Dolines formed by the focusing of dissolution either by recharge or by draw-down can occur in any climatic zone where water exists in a liquid state, provided there is unimpeded underground flow from recharge to discharge zones. Particularly large solution depressions often occur in the humid tropics where corrosion processes were uninterrupted by Pleistocene glaciations. In these places the term cockpit is sometimes applied to them after a particular style of landscape in Jamaica, where depressions are incised between intervening conical hills (see Cockpit Country Cone Karst, Jamaica).

Although small solution dolines have formed in 15 000 years or so in some mid to high latitude areas that were glaciated in the late Pleistocene, several tens to hundreds of thousands of years are required to develop large solution dolines in limestone. Once formed they may persist in the landscape for several million years provided there is sufficient thickness of limestone for their continued incision. Individual dolines may merge to form compound closed depressions (known as uvalas) and large dolines may subdivide internally into smaller second generation basins. Where all the available space is occupied by depressions, rather like an egg box, the landscape is termed polygonal karst, because the topographic divides of the adjoining solution depressions have a polygonal pattern when viewed in plan (Williams, 1971). However, not all dolines within polygonal karst are necessarily of dissolutional origin, because a small proportion of collapse

Dolines: Table 2. Types of doline, the processes that form them, and terms used in English to describe them.

Process	Karst Landforms Produced	Terms Used to Describe Them
1. Dissolution: Chemical dissolution (corrosion) of carbonate bedrock or physical solution of evaporite rocks.	Bowl-, saucer-, or funnel-shaped enclosed depressions in bedrock, usually with a soil cover. Depressions usually less than 1 km in diameter and often much smaller.	Solution doline / sinkhole Cockpit (in a humid tropical context)
2. Collapse: This can occur in unconsolidated coverbeds or in compact caprock, and may progress rapidly or slowly.		
(i) Sudden or progressive collapse of the roof of a cave into the underlying cavern. Collapse is entirely within karst rocks and may be propagated upwards from tens to hundreds of metres beneath the surface.	Cylindrical to steep-sided enclosed depressions in karst rocks with debris-filled floor, sometimes with sheer and even overhanging rock walls tens to hundreds of metres high. Some open near their base into the underlying cave; some contain lakes. Up to a few hundred metres in diameter, but often much less.	Collapse doline Cave-collapse sinkhole Cenote (only when they contain a lake)
(ii) Sudden or progressive collapse of non-karst caprock into an underlying cave. A ceiling collapse within a cave located in interstratified karst rocks stopes upwards into, and through, overlying consolidated caprock causing the surface to collapse. The upwards stoping may occur over tens to hundreds of metres.	Cylindrical to steep-sided enclosed depressions in caprock with debris-filled floor, sometimes but not always revealing underlying karst rock. Up to a few hundred metres in diameter, but often much less.	Caprock doline Subjacent karst collapse doline Interstratal collapse doline / sinkhole
(iii) Sudden failure in mechanically weak unconsolidated sediments overlying a subsurface cavity, giving rise to a depression at the surface. Sediments are evacuated downwards through solution pipes in the bedrock creating a void at the bedrock-sediment contact that enlarges by upwards propagation through the clastic coverbeds. The ceiling dome above the void eventually fails.	Steep-sided enclosed depressions (sometimes cylindrical when freshly formed) in unconsolidated cover sediments with debris-filled floor, usually a few metres to tens of metres in diameter. Such depressions are often found in sediment fill near the bottom of a solution doline.	Dropout doline Collapse doline Cover-collapse sinkhole
3. Suffosion and subsidence: Gradual subsidence of superficial coverbeds (as opposed to sudden collapse) giving rise to a depression at the surface. Formative processes involve the evacuation of superficial unconsolidated sediments through underlying solution pipes. Because the nature of cover deposits can vary considerably from uniform clays, silts, and sands through mixed alluvial facies to heterogeneous glacial deposits, the mechanisms involved in their evacuation also vary. Suffosion is one of the processes and involves the gradual physical and chemical downwashing of fines through underlying solution pipes in the karstified bedrock. Plastic flow of clays and silts can also occur with cover sediments sometimes being extruded as a wet slurry into underlying caves. Overlying sediments slowly settle and deform in response to gradual undermining. Coarse materials such as boulders and scree move downwards by gravity as underlying sediments are removed.	Dimpled surface of enclosed depressions in coverbeds, sometimes exposing windows of underlying bedrock beneath. Such depressions are often found in superficial sediments such as glacial drift, alluvium, loess, and sand. They are usually only a few metres in diameter, but larger forms can be produced if the removal of coverbeds exhumes buried dolines in bedrock.	Suffosion doline, especially in finer-grained coverbeds. Cover-subsidence sinkhole is used in the United States to describe depressions formed by the gradual settling of unconsolidated clastic coverbeds. Sometimes the sediments have completely buried a pre-existing doline, the form of which is being exhumed as the subsidence occurs. Shakehole is a term used in England to describe depressions in bouldery deposits such as glacial drift overlying limestone.
4. Regional subsidence: This process involves gentle, progressive settling of the ground surface over large areas as a consequence of gradual dissolution of deep, underlying interstratal evaporite beds by groundwaters.	Strike-aligned depressions are formed when beds are dipping, but otherwise gently sloping depressions are distributed across an undulating surface. Depressions are often of several square kilometres in area and so are very much larger than is generally understood by the term doline. Nevertheless, some smaller depressions are formed in this way.	Subsidence depression Solution trough

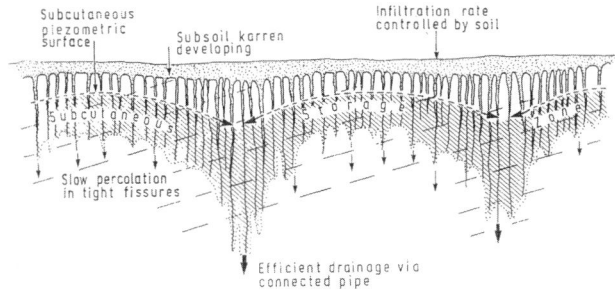

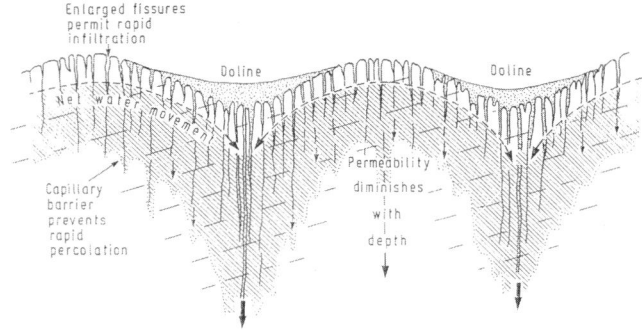

Dolines: Figure 2. Solution dolines develop in the uppermost weathered zone of the karst rock, termed the epikarst or subcutaneous zone. Drainage within that zone is achieved by leakage of water down fissures such as joints and faults. Major fissures capture most of the flow and therefore are the foci of solvent attack on the bedrock. This results in more rock being removed in solution from these locations than elsewhere, and this gradually gains topographic expression as enclosed depressions as the overall surface is lowered by chemical denudation (from Williams, 1983).

depressions is also likely to be included. The mesh sizes of different polygonal karsts can vary widely and is a direct reflection of the characteristic radii of individual solution dolines, in turn determined by the epikarstic processes described above. Dense fields of dolines constituting polygonal karsts are known from the mid latitudes (e.g. New Zealand and Tasmania) to the tropics, but none has been reported polewards of 50° latitude, probably because of the destructive influence of periglacial and glacial processes on surface landforms during the Pleistocene.

There are few examples of dolines that have survived severe glaciation, although large solution dolines at High Mark near Malham, England, clearly formed before the last glaciation and contain drift deposits in which many small suffosion dolines have developed. In contrast, there are numerous cases of dolines formed prior to the last glaciation and located beyond the ice limits that have been strongly subjected to processes other than dissolution. In many cases they have been partly filled by scree and others have been covered, or even completely buried, by loess. Post-glacial conditions have resulted in some of these materials being removed and in the re-expression and continued development of the underlying doline form. This is the case in the Sinkhole Plain of Kentucky, where many dolines were buried by fine clastic sediments. It follows, therefore, that solution dolines, like valleys in a fluvial landscape, can have long and complex histories over time periods when climates have changed and when the importance of dissolution compared to other geomorphic processes has waxed and waned (see also buried dolines below).

Collapse Dolines

Whereas most authors agree with the designation solution doline, there is considerable variation in nomenclature concerning depressions formed mainly by mechanical processes (Table 1). This is largely because of the variety of materials and processes involved, and the tendency of some authors to group types and of others to subdivide types. Collapse refers to rapid downward movement of the ground, whereas subsidence refers to gradual movement, sometimes without even ripping the surface. These processes can occur in karst bedrock (collapse dolines *sensu stricto*), in caprock that may stratigraphically overly it (caprock collapse dolines), and in veneers of unconsolidated sediments (dropout dolines). In all cases the collapse has to be preceded by dissolution of the karst rock to form a void into which material can fall. The movement may involve little water or may be promoted by lubrication by water and the kind of landforms produced depends upon which of the various materials and processes were involved.

Where collapse dolines form in karst bedrock the void is commonly part of a cave system. Collapse may occur following undermining from below as the roof of a cavity stopes upwards, ultimately causing the surface above to collapse or following dissolution from above that weakens the span of a cave roof, causing it to collapse. For example, solutional attack by drainage water near the bottom of a solution doline may combine with upwards stoping of an underlying cave roof to weaken a span from above and below, thereby causing the doline floor to collapse into a cave. The natural process of ground surface lowering inevitably means that caves will eventually be unroofed and, as the final stage involves failure of their thinning roof, the depressions produced must be regarded as a sub-type of collapse doline.

When limestones or evaporites are overlain by a caprock, a void may form by dissolution at the lithological junction between caprock and karst rock. Stoping may then work upwards through the cover rocks and lead to the formation of a caprock collapse doline, which may be entirely within rocks such as shales, sandstones, or even basalt (Figure 3). Such features propa-

Dolines: Figure 3. Caprock collapse doline in basalt overlying limestone near Timahdite, south of Fes, Morocco. (Photo by John Gunn)

gated from interstratal dissolution were termed subjacent collapse dolines by Jennings (1985), and good examples are found in Namurian sandstones around the North Crop of the South Wales coalfield (Thomas, 1974). While many of the South Wales dolines have formed following dissolution at the limestone–sandstone unconformity, some result from collapse propagated upwards from deeper caves that developed entirely within the limestone (Bull, 1980). Stoping has propagated through more than 1000 m of cover rocks at sites in Canada and Russia.

When recently developed, collapse and caprock collapse dolines are steep sided, even cylindrical in form, but over time their sides degrade and bottoms infill with debris, so that superficially they assume the bowl-shaped morphology of solution dolines for which they can be easily mistaken. Only excavation, drilling, or geophysical survey will reveal their true origin. Collapse dolines are on average smaller in diameter than solution dolines, although particularly large examples 700 m along their largest axis and up to 400 m deep are known in the Nakanai Mountains of New Britain (see Nakanai entry, with photo). In China, the name tiankeng (roughly meaning skyhole) is applied to some of these very large dolines, including Xiaozhai over Di Feng Dong (see Di Feng Dong, with photo) and those at Xingwen (see China entry). Progressive collapse has clearly been important in their genesis, but not all tiankengs are associated with appropriately massive river cave development, and there is debate over whether they are merely very large ordinary dolines or if they represent a different mechanism of doline formation.

Sometimes a collapse extends from a cave below the modern water-table level, in which case the collapse doline will contain a lake. Spectacular cylindrical collapses that descend below the water table are known as cenotes after the type-site in the Yucatán Peninsula of Mexico, although similar features are found elsewhere, such as in southeast Australia. The deepest known case of a collapse doline containing a lake is the Crveno Jezero (Red Lake) in Croatia, which is 528 m deep from its lowest rim, the bottom of the collapse extending 281 m below the modern level of the nearby Adriatic Sea. The collapse diameter at the surface is about 350 m and at lake level is about 200 m. Recent diving has found an active subterranean river that crosses the doline near its floor. The whole feature is thought to have formed by progressive upwards collapse of the cave roof, much of the collapse debris having been transported away by the underground river.

The capacity of a cave roof to resist collapse depends on the width of the roof span and on factors determining rock mass strength such as the thickness of beds, the composition, texture, and compressive strength of the rock, and the frequency and continuity of fissures. If stress exceeds the rock's strength then it will fail. In the simplest case where bedding is horizontal, each bed can be considered to act as a beam, but if it is fractured then it has less strength and acts as a cantilever. Collapse occurs where a critical span is exceeded for a given thickness and strength. White and White (1969) illustrate this with examples from West Virginia, where 0.5 m beds of carbonate rock can support roofs up to 25 m wide where they are unbroken beams, but only about 10 m where they act as cantilevers. Beams of more thickly bedded carbonates can support spans of 35 m or more, though cantilevered beds as massive as 5 m thick will fail with a span of 30 m or less. Massive reef limestones support the largest spans, which exceed 50 m in the Big Room of Carlsbad Cavern, New Mexico. In the Mulu karst of Malaysia, Sarawak Chamber is more than 250 m wide and about 500 m long. Its ceiling is actively collapsing and working upwards, but with few bedding weaknesses it is developing an arched profile that is stable in compression. Its ultimate evolution into an enormous collapse doline will be due more to surface lowering than to roof stoping.

Another process that increases the effective stress on rock arches and domes is removal of buoyant support by water-table lowering, which increases the effective weight on the span of the roof, resulting in its strength being exceeded and so in its failure and collapse. This occurs because in a fully saturated medium the buoyant force of water is 1 t m^{-3}, and if the water table is lowered by 30 m, the increase in the effective stress on the rocks is 30 t m^{-3} (Hunt, 1984). A gradual lowering of the water table occurs naturally as karstification proceeds, because of the increase in subterranean void space within the rock, but it occurs more rapidly when valley incision occurs, because springs are lowered too, and with them the level of the saturated zone that feeds them. More rapid still is the lowering of the level of saturation caused by sea-level lowering, a process that occurred frequently in the Pleistocene because of repeated glacio-eustatic (vertical movement of sea level caused by glaciation and deglaciation) fluctuations, although this only affected karsts well connected to the coast such as in Florida, southeastern Australia, and Yucatán, where it probably was a significant influence in the development of cenotes.

If unconsolidated coverbeds are drained by water-table lowering, then consolidation and compression occurs, leading to subsiding of the surface and collapse where clastic sediments span de-watered unsupported arches. This has been a common process in Florida where porous sandy formations overlie karstified limestones, but has been much exacerbated by groundwater pumping for water supplies, which has still further reduced buoyant support (Beck, 1984). This process and the resulting incidence of collapse attains dangerous hazardous proportions in karstified areas extensively de-watered by mining activities. These dolines in unconsolidated coverbeds have been referred to as cover collapse sinkholes (Table 1) by Beck and Sinclair (1986) and White (1988). They are genetically transitional to subsidence dolines.

Subsidence (Suffosion/Dropout) Dolines

When unconsolidated deposits such as alluvium, glacial moraine, loess, or sand mantle karstified rock, the sediments are sometimes evacuated downwards through corrosionally enlarged pipes in the underlying karst, resulting in gradual or rapid subsidence of the surface. Hence, the term subsidence doline is sometimes used for any doline in unconsolidated deposits although the term is also used for depressions formed by regional subsidence (see below). The main process by which the sediment moves is known as suffosion and involves the gradual winnowing and downwashing of fines by a combination of physical and chemical processes. The topographic consequence of this activity depends on whether the material is cohesive or non-cohesive. In cohesive sediments evacuation of material may proceed for some time without any surface expression. However, a void is formed that enlarges and stopes upwards resulting in a sudden, and sometimes catastrophic, failure of the ground surface. The depression thus formed is called a dropout doline or cover col-

Dolines: Figure 4. Dropout doline, Santang, Guizhou, China. (Photo by John Gunn)

lapse doline (Figure 4). As the final process is one of collapse, some authors group these dolines with those formed by collapse in bedrock (Tables 1 and 2).

Where the sediment is non-cohesive, the clayey fraction will tend to move as a slurry into underlying cavities, producing dejection fans and mudslides in cave passages, whereas the coarser fraction will tend to remain closer to the surface and line small bouldery depressions known as suffosion dolines that are usually only a few metres in diameter and depth. In Britain suffosion dolines formed in glacial boulder clay overlying limestone are widely referred to as shakeholes, although the term is sometimes applied to any doline. Similar features but in more uniform finer grained materials are referred to as cover subsidence sinkholes in the United States. Often a combination of processes is involved in the development of subsidence dolines, including corrosion and collapse of the underlying bedrock, as well as suffosion, mudflow, and void collapse in the mantling materials. Soil piping and surface collapse can also occur in unconsolidated non-karstic sediments in badland country, leading to a style of topography termed suffosional pseudokarst by White (1988).

Buried Dolines

Changes in the environment may result in bedrock dolines, however formed, becoming filled with sediment. There may be no evidence of such dolines at the ground surface, their presence only being revealed by geophysical prospecting or geotechnical survey. Good examples have been found in the central lowlands of Ireland where the lowest fills are of Tertiary age. Some buried dolines contain sediment of economic importance such as karst bauxites. Where they are no longer hydrologically active they may be regarded as paleokarst but if karstification continues, or is reactivated, in the underlying bedrock then there is likely to be gradual removal of material leading to surface subsidence.

Depressions Formed by Regional Subsidence

Particularly soluble rocks such as evaporites are often interbedded with clastic rocks. During the course of groundwater circulation the evaporite beds are dissolved away and gradual subsidence of the surface results. Waltham (1989) refers to this process as salt subsidence and provides a discussion of the mechanisms involved and the effects produced. When the beds are dipping, large depressions are produced along the strike. These can be many square kilometres in area and so are too large to be designated dolines and are best considered solution troughs. However, there is a continuum of forms and more localized subsidence can produce doline-sized features that could be termed subsidence dolines, although collapse is also sometimes implicated in their formation. Many have been mapped in the vicinity of Ripon in north Yorkshire, England. Ford and Williams (1989) provide further details and examples.

Mapping of Dolines

For various reasons we may wish to map dolines. This is particularly the case when karst collapses may be hazardous to human activity (Beck & Pearson, 1995). Hence, it is necessary to know how to recognize different genetic types of dolines and their boundaries. In the case of solution dolines, which depend for their development on hydrological and chemical processes in the epikarst, their perimeters are the topographic divides (watersheds), within which water drains centripetally towards the doline bottoms. In the case of collapse dolines, where the main forcing process is undermining of the surface by cavern collapse, the surface boundary is the outermost surface inflection or break of slope that marks the limit of zone of influence of the failure. This may not be static even in the short term, but can move outwards until stability is attained. The boundaries of suffosion dolines and subsidence depressions are also defined by topographic inflections and breaks of slope at the outer limits of influence of the subsurface processes that formed them. Bearing in mind the polygenetic nature of some dolines, it is possible to map nested features such as collapse depressions within the boundary of a larger solution depression.

Doline Hydrology

Solution dolines have a similar function in karst landscapes to the drainage basin in non-karstic lithologies: they channel water to an outlet at the lowest point in the doline. Lateral movement is by overland flow and throughflow if there is cover material over the bedrock and by subcutaneous flow. Vertical movement is by shaft flow, vadose flows, and vadose seeps, three points on a continuum of flow routes (Gunn, 1981). Suffosion dolines also form small drainage basins, but the hydrological function of collapse and caprock collapse dolines depends on their form. Where they are narrow and steep sided they may be virtually inert, particularly when they are formed in essentially flat landscapes. However, as they widen the sides usually become less steep and an internal drainage system may develop.

PAUL WILLIAMS

See also **Cone Karst; Karst Evolution**

Works Cited

Beck, B.F. (editor) 1984. *Sinkholes: Their Geology, Engineering and Environmental Impact*, Rotterdam: Balkema

Beck, B.F. & Pearson, F.M. (editors) 1995. *Karst Geohazards: Engineering and Environmental Problems in Karst Terrane*, Rotterdam: Balkema

Beck, B.F. & Sinclair, W.C. 1986. *Sinkholes in Florida: An Introduction*, Florida Sinkhole Research Institute Report 85–86–4

Bull, P. 1980. The antiquity of caves and dolines in the British Isles. *Zeitschrift für Geomorphologie*, Supplementband 36: 217–32

Cvijić, J. 1893. Das Karstphänomen [The karst phenomenon]. *Geographische Abhandlung*, 5(3): 218–329 (The section on dolines, pp.225–76, is translated into English in *Karst Geomorphology*, edited by M.M. Sweeting, Stroudsburg, Pennsylvania: Hutchinson-Ross, 1981)

Ford, D.C. & Williams, P.W. 1989. *Karst Geomorphology and Hydrology*, London and Boston: Unwin Hyman

Gunn, J. 1981. Hydrological processes in karst depressions. *Zeitschrift für Geomorphologie*, 25(3): 313–31

Hunt, R.E. 1984. *Geotechnical Investigation Engineering Manual*, New York: McGraw-Hill

Jennings, J.N. 1985. *Karst Geomorphology*, Oxford: Blackwell

Thomas, T.M. 1974. The South Wales interstratal karst. *Transactions of the British Cave Research Association*, 1: 131–52

Waltham, A.C. 1989. *Ground Subsidence*, Glasgow: Blackie and New York: Chapman and Hall

Waltham, A.C. & Fookes, P.G. 2003. Engineering classification of karst ground conditions. *Quarterly Journal of Engineering Geology and Hydrogeology*, 36: 101–18 in press

White, W.B. 1988. *Geomorphology and Hydrology of Karst Terrains*, Oxford and New York: Oxford University Press

White, E.L. & White, W.B. 1969. Processes of cavern breakdown. *Bulletin of the National Speleological Society of America*, 31(4): 83–96

Williams, P.W. 1971. Illustrating morphometric analysis of karst with examples from New Guinea. *Zeitschrift für Geomorphologie*, 15: 40–61

Williams, P.W. 1983. The role of the subcutaneous zone in karst hydrology. *Journal of Hydrology (Netherlands)*, 61: 45–67

DRAENEN, OGOF DRAENEN, WALES

Located in 1994, Ogof Draenen is Britain's most significant caving discovery in recent years. It is already one of the longest caves in Britain (70 km) and contains the longest continuous streamway (> 2.5 km) and some of the country's largest passages. It also has abundant gypsum flowers, rare gypsum needles and anthodites, plus significant bat-guano deposits. But it is its geomorphic interest that sets it apart and makes it worthy of recognition. Above and up dip from the streamway lie several tiers of abandoned cave passage, most of which are unrelated to modern hydrology. These preserve evidence that the cave system has behaved like a hydrological see-saw, with the flow switching from south to north and then south again in response to landscape evolution and valley incision.

Ogof Draenen lies under the interfluve between the Usk and Afon Lwyd valleys, on the northeastern margin of the South Wales coalfield (Figure 1). The entrance lies close to the hamlet of Pwll Ddu, six kilometres southwest of Abergavenny. First investigated by the Cwmbran Caving Club, the entrance to Ogof Draenen was a small, choked but draughting phreatic tube, almost buried under an old coal tip. The main digging breakthrough occurred on 6 October 1994; by 28 November, over 15 km of passage had been explored and surveyed (Kendall, 1994). Over the next few years, the rate of exploration averaged a remarkable 2 km per month, with many clubs and individuals contributing to numerous major discoveries. A comprehensive detailed survey, initiated by the Chelsea Spelaeological Society, had mapped over 68 km of passage by April 2002. The first geomorphological study of the cave was published by Simms *et al.* (1996) and expanded on in Waltham *et al.* (1997).

Ogof Draenen is developed entirely within the Carboniferous Dinantian Limestone, although some of the larger boulder chokes stope up into the overlying Namurian Millstone Grit. The limestone succession was deposited on a carbonate ramp close to the contemporary shoreline. This, coupled with pre-

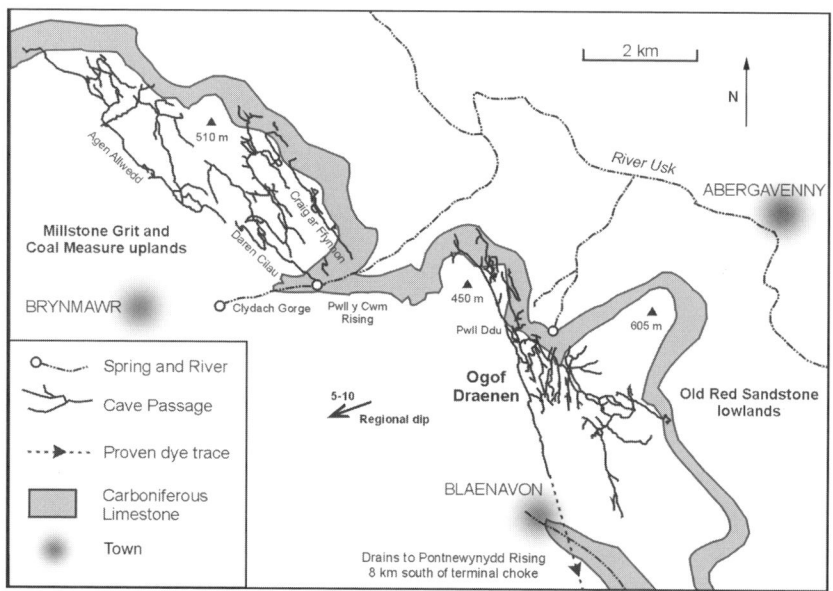

Draenen, Ogof Draenen, Wales: Figure 1. The Blaenavon area, South Wales, showing the location of Ogof Draenen and other major caves.

Draenen, Ogof Draenen, Wales: Figure 2. Schematic diagram showing the progressive evolution of Ogof Draenen. Stages 1a and 1b, initial drainage south to the Usk Valley; Stage 2, drainage reversal and flow orientated northwards to the Clydach Gorge; Stage 3, second drainage reversal and flow to the south again to the Afon Lwyd Valley.

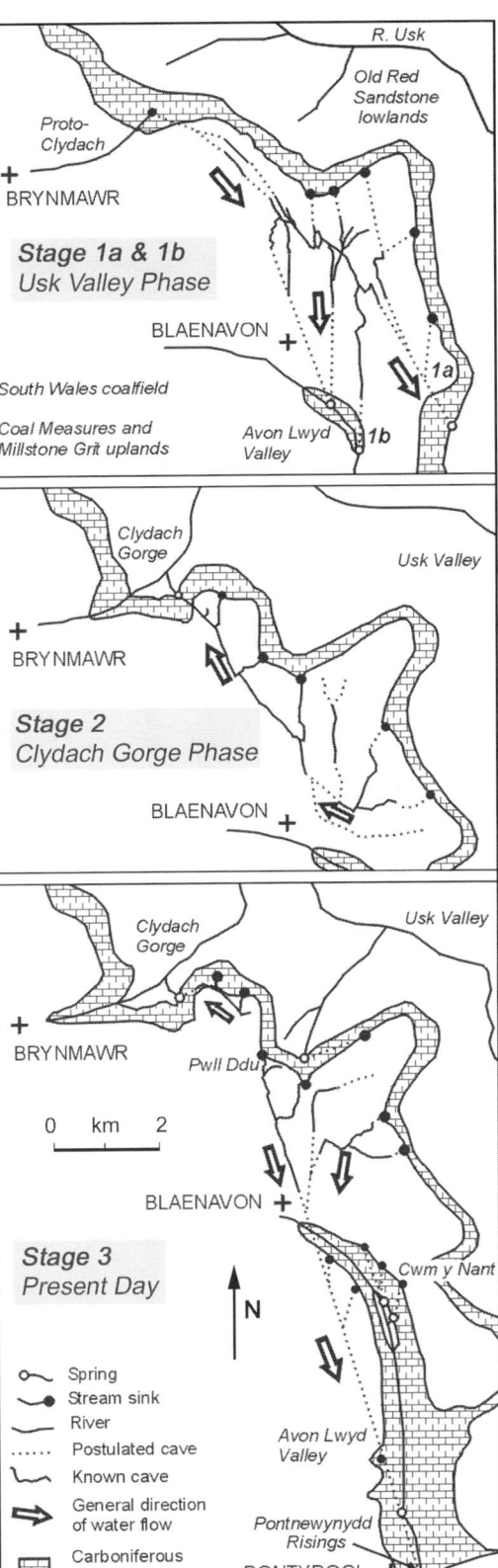

Draenen, Ogof Draenen, Wales: Figure 3. Solutional wall notches in the abandoned conduit of Megadrive survive on the left wall but have been lost by breakdown of the right wall. (Photo by Clive Westlake)

Namurian erosion and overstep by the overlying Millstone Grit, has created a wedge of limestone less than 25 m thick along the eastern margin of the escarpment, thickening westwards to c.80 m in the Pwll Ddu area. The limestone crops out in a narrow band around the escarpment overlooking the Usk Valley and as a re-entrant up the Afon Lwyd Valley. Much of the cave is overlain by less-permeable Millstone Grit, and, in some places, by the Coal Measures, which protect the system from erosion. The geological structure of the area is relatively simple. The rocks dip to the southwest at between 10° and 28°, and are cut by several minor faults, which can be identified underground by lines of boulder chokes. The dominant joint set which strongly influences cave development runs at 150–330°, slightly oblique to the strike and mirroring the main fault orientations. Some joints have been mineralized with barite and copper minerals.

Ogof Draenen has had a complicated multiphase origin. Evidence of at least four phases of cave development, associated with major reversals in flow direction and resurgence location, can be identified (Figure 2). Essentially, the system can be thought of as four separate vertically stacked cave systems, linked by more recent vadose inlets and passage intersections. Some passages are multiphase and are thus usually much larger than single-phase passages. Major sediment influxes have occurred throughout much of the cave's development, modifying and controlling passage genesis and evolution and creating some fine paragenetic passages. The major trunk conduits trend northwest–southeast along the dominant joint set. Because this coincides with the maximum hydraulic gradient, long unidirectional quasi-horizontal passage segments develop (Figure 3). This permits passage segments to rise or fall stratigraphically while remaining at a constant elevation. Draining into these are down-dip vadose tributaries. In places maze networks occur, usually in response to a change in passage orientation.

The earliest phase of cave development is represented by a major conduit system which extends southeastwards from a major northern sink at an elevation of at least 390 m, to a resurgence in the Usk Valley at c.370 m. Once this system was established, a series of down-dip captures began in response to incision in the Afon Lwyd Valley down dip to the west. This shift increased the down-dip component of underground drainage, causing a gradual migration in the conduit orientation to the southsoutheast. These can be identified as a series of maze networks which diverge off south from the initial conduit. The second phase of development occurred in response to rapid incision of the Clydach Gorge, effectively reversing the hydraulic gradient and causing a fundamental 180° re-orientation of the underground drainage. This created a joint-aligned conduit system draining northwest, fed by a dendritic series of down-dip inlets, resurging at around 300–310 m. Renewed valley incision, probably following glacial downcutting during the Devensian, in the Afon Lwyd Valley has resulted in a second fundamental 180° re-orientation of the underground drainage. Water tracing using fluorescent dyes has shown (Gascoine, 1995) that the water currently sinking around Pwll Ddu drains into the Main Streamway. From here it drains south to the Pontnewynydd Risings, 140 m lower and 8 km distant, in under 100 hours.

Although Ogof Draenen falls within a World Heritage Site, this designation is for the region's industrial heritage. However, the cave system is worthy of international recognition and is a proposed Site of Special Scientific Interest. Conservation has been a paramount consideration since the discovery of the system, and many kilometres of tape have been laid to conserve the undisturbed sediments, guano deposits, and gypsum crystals.

ANDREW FARRANT

Works Cited

Gascoine, W. 1995. Dye traces in Ogof Draenen. *Caves and Caving*, 69: 19–21

Kendall, A. 1994. Ogof Draenen. *Descent*, 119: 12

Simms, M.J., Farrant, A.R, & Hunt, J. 1996. Ogof Draenen. Where did it come from and where will it go? *Caves and Caving*, 71: 10–13

Waltham, A.C., Simms, M.J., Farrant, A.R. & Goldie, H.S. 1997. *Karst and Caves of Great Britain*, London and New York: Chapman and Hall

Further Reading

Ford, T.D. (editor) 1989. *Limestones and Caves of Wales*, Cambridge and New York: Cambridge University Press
Although published before Ogof Draenen was discovered, this book has much background information on the area.

Simms, M.J. 1998. *Caves and Karst of the Brecon Beacons National Park*, London: British Cave Research Association (BCRA Cave Studies, 7)
A guide to the classic karst areas of the region, including a section on Ogof Draenen.

Stevens, J., Millet, A., Stacey, P., McCombe, M., and members of the Chelsea Spelaeological Society. 1996. *Ogof Draenen Survey* (to BCRA Grade V)
Unpublished. Indispensable to anyone visiting the cave or undertaking any research.

EDWARDS AQUIFER, UNITED STATES

The Balcones fault zone section of the Edwards Aquifer is one of the most permeable and productive carbonate aquifers in the United States. The Edwards Aquifer extends from the Mexico–Texas border near Del Rio, eastward and then northeast for 350 km to near Waco (Maclay, 1995). The aquifer can be divided into four segments by groundwater or surface water divides, the most productive and utilized being the San Antonio segment. This is the primary water supply for about 1.7 million people, and supports a broad base of agricultural, municipal, industrial, and recreational uses. It extends between groundwater divides near Brackettville in the west and just north of San Marcos, a distance of about 280 km (Figure 1).

Southern Texas includes the southern portion of the Edwards Plateau, the Balcones fault zone, and the northern portion of the Gulf Coastal Plain. The Edwards Aquifer is named after and contained in the Lower Cretaceous Edwards Limestone, which was deposited in a complex of subtidal to supratidal, shallow-water marine to deep-basin environments. The limestone thickness is 130–270 m, with an average of 180 m. A series of faults in the Balcones fault zone has exposed the Edwards Limestone at the surface along the southern boundary of the Texas Hill Country. En echelon faulting has dropped the Edwards Limestone to great depth below the surface along the aquifer's southern zone. In some areas, fresh water can be found in the Edwards Limestone as far as 1000 m below sea level. The southern and eastern boundary of the aquifer is generally defined by the "bad water line", where total dissolved solids exceed 1000 mg l^{-1}. The Edwards Limestone is a noted oil producer in the Gulf Coastal Plain south of San Antonio.

Generally, the Edwards Aquifer is divided into three zones: the contributing or drainage area, the recharge zone, and the artesian zone (Figure 2). Surface streams forming on the contributing area (the dissected Edwards Plateau also referred to as the Texas Hill Country) flow south or east and cross the Edwards Limestone outcrop (the recharge zone), where in low flow conditions most surface water is captured by the aquifer. Rainfall on the recharge zone may also enter the aquifer. Groundwater in the artesian zone moves through the aquifer, generally from west to east, to discharge from a number of sites including Leona Springs (Uvalde County), San Pedro and San Antonio Springs (San Antonio), Hueco and Comal Springs (New Braunfels), and San Marcos Springs (San Marcos). Mean discharges for Comal Springs and San Marcos Springs, the two largest, are 8.3 and 5.6 m^3 s^{-1}. Residence time in the aquifer ranges from a few hours to many years depending upon depth of circulation, location, and other aquifer parameters.

The combined primary and secondary porosity and aquifer permeability of the Edwards Limestone is extremely high. Most wells do not fully penetrate the aquifer yet commonly have yields that exceed 3800 l s^{-1} with little or no drawdown. The Catfish Farm Well (near San Antonio), a 76 mm diameter flowing artesian well, discharged over 115 000 l s^{-1} to create a vertical column of water rising 10 m above the surface when first drilled. Well yields are generally limited by the size of the pump, and not by the yield of the aquifer.

The genesis of the Edwards Aquifer is complex and is the product of numerous geological processes including carbonate deposition, uplift, early karstification (soon after uplift), down faulting, volcanism, and current karstification (Oetting, Banner & Sharp, 1996). Meteoric waters from the recharge zone play an important role in dissolution in the aquifer, including the removal of limestone, dolomite, and gypsum. There is also evidence that mixing corrosion with hydrogen sulfide has resulted in significant permeability enhancement in the artesian portion of the aquifer. Many of the wells drilled south of the bad water line produce hydrogen sulfide.

Because of the extremely high transmissivity of the aquifer, recharge during rain events can be rapid. Monitoring wells in the recharge zone have risen as much as 50 m in response to large rainfall events. However, during drought conditions, aquifer levels in the region's index wells have dropped as much as 700 mm in one day. In July 2000, a pressure wave, created by reductions in pumping stress mandated during a drought, was recorded in wells over a 100 km length of the aquifer in under

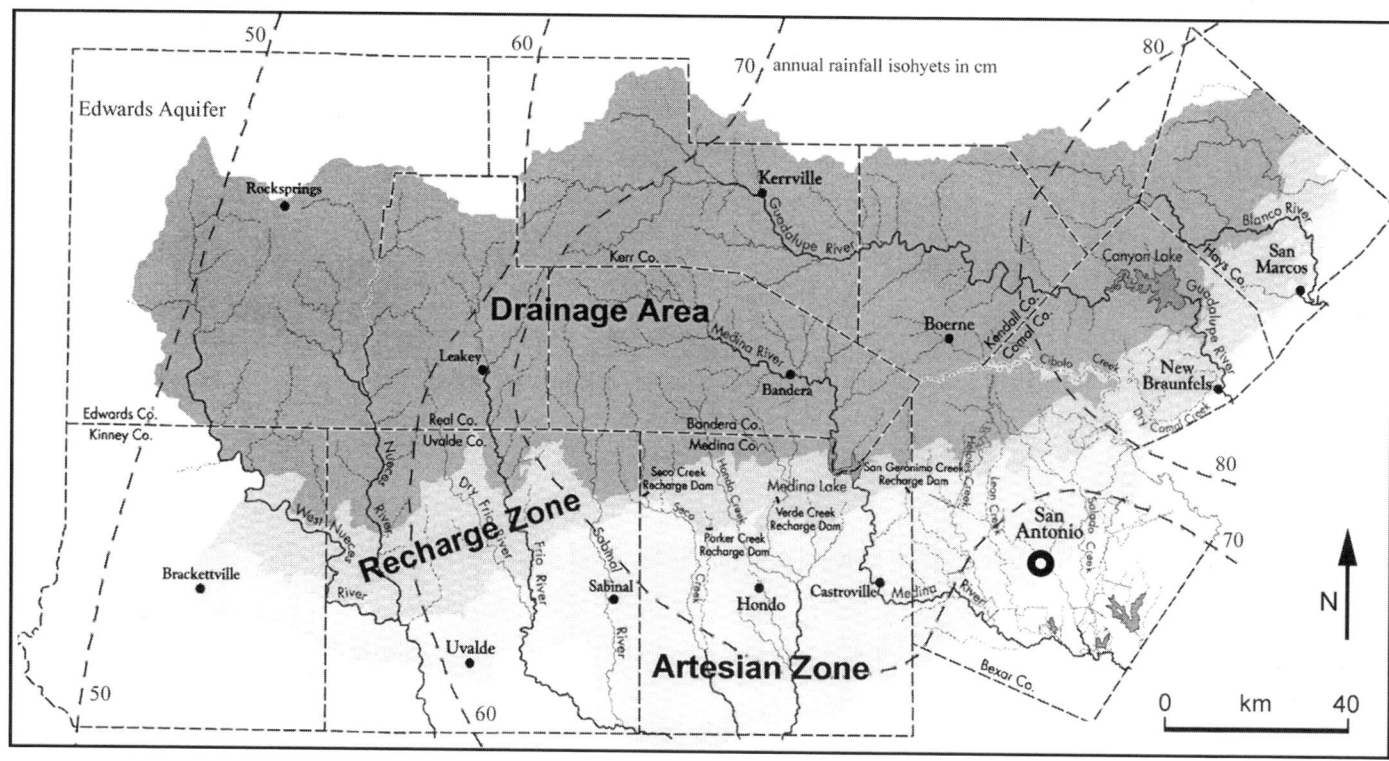

Edwards Aquifer: Figure 1. Outline map of the San Antonio segment of the Edwards Aquifer in Texas.

12 hours. Along with the extremely high well yields, this reflects the aquifer's very high transmissivity.

The San Antonio segment of the Edwards Aquifer is managed by the Edwards Aquifer Authority, with statutory ability to protect water quality and control groundwater withdrawals to protect springflow for the dependant aquatic communities. There is a withdrawal limit of $5.548 \times 10^8 \text{ m}^3 \text{ a}^{-1}$ to protect endangered species dependent upon Comal and San Marcos springs. The Authority uses a series of index wells to monitor the potentiometric surface of the aquifer on a daily basis across the region, and also records daily discharges from Comal and San Marcos Springs. On average, the Authority reserves about two thirds of the aquifer's water recharge for withdrawal by pumping, and leaves one third for discharge from the springs, thereby maintaining species habitat and downstream water rights.

Land use in the Edwards Aquifer recharge zone has historically been low-density cattle or goat ranching and, to a lesser extent, row cropping. Water quality in the Edwards Aquifer is generally very high, requiring very little treatment before use. However, urban and industrial development has encroached upon the recharge zone, and analyses do indicate some constituent increases, but all levels remain within drinking water quality standards. Nitrate concentrations are elevated in some areas, with the most likely sources being agricultural and residential use of fertilizers. Ammonium nitrate blasting agents, used in limestone mining, may also be a source of nitrate. Volatile or-

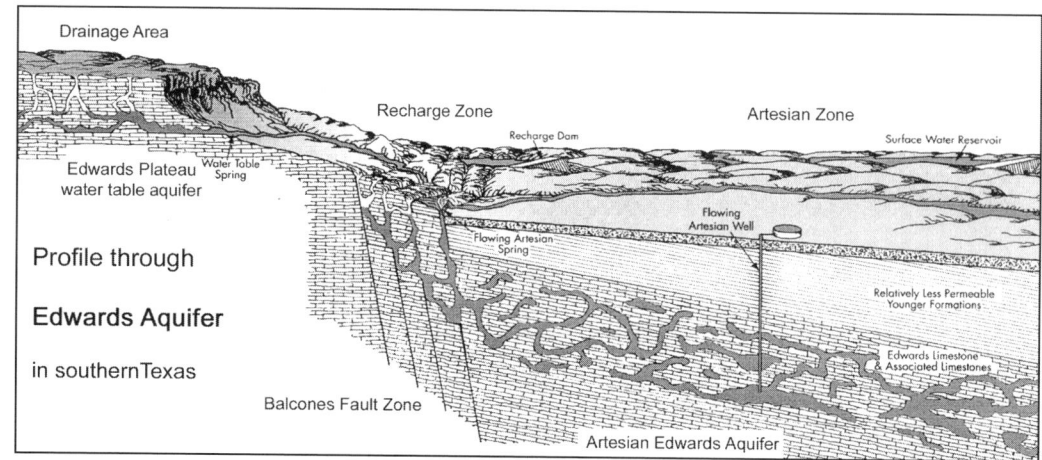

Edwards Aquifer: Figure 2. Cross section of the Edwards Aquifer near San Antonio.

ganic compounds have also been detected in the aquifer in some limited areas, and are most likely related to releases from petroleum underground storage tanks and dry cleaning facilities.

Known caves in the exposed Edwards Limestone associated with the Balcones Fault Zone are mainly small. Many are deeper than they are long, and jointing appears to play a strong role in controlling cave passage development. Genesis Cave, in Bexar County, one of the few caves that intersects the Edwards Aquifer, is 165 m long and 78 m deep. Indian Creek Cave (Uvalde) is the fourth longest cave in Texas with 6700 m of passage including an active streamway, which is very unusual for the area. Bracken Bat Cave (near New Braunfels) is only 130 m long but contains from 20 to 40 million Mexican Free-Tail bats (*Tadarida brasiliensis* mexicana), perhaps the largest collection of mammals in the world.

Drill bit drops, troughs in the water table, rapid response of groundwater elevations to storm events, and the presence of over 40 endemic species in the groundwater all indicate a very extensive network of underwater caves. However, their great depth and few access points leave very little potential for significant underwater exploration.

GEARY SCHINDEL, JOHN HOYT and STEVE JOHNSON

Works Cited

Maclay, R.W. 1995. *Geology and Hydrology of the Edwards Aquifer in the San Antonio Area, Texas*, Washington: United States Geological Survey (Water-Resources Investigations Report 95–4186)

Oetting, G.C., Banner, J.L. & Sharp, Jr., J.M. 1996. Regional controls on the geochemical evolution of saline groundwater in the Edwards Aquifer, Central Texas. *Journal of Hydrology*, 181: 251–83

Further Reading

Edwards Aquifer Authority, www.edwardsaquifer.org

Elliott, W.R. & Veni, G. 1994. *The Caves and Karst of Texas: NSS Convention Guidebook*, Huntsville, Alabama: National Speleological Society

Groschen, G.E. 1996. *Hydrogeologic Factors That Affect the Flowpath of Water in Selected Zones of the Edwards Aquifer, San Antonio Region, Texas*, Washington: United States Geological Survey (Water-Resources Investigations Report 96–4046)

Sharp, J.M., Jr. & Banner, J.L. 2000. The Edwards Aquifer: Water for Thirsty Texans. In *The Earth Around Us: Maintaining a Livable Planet*, edited by J.S. Schneiderman, New York: Freeman

EDWARDS AQUIFER, UNITED STATES: BIOSPELEOLOGY

The Edwards Aquifer is associated with the Balcones Escarpment in central Texas. The aquifer serves as the primary source of water for approximately 1.7 million people. It also supplies water for agricultural irrigation and for major springs. The springs provide the base flow of the Guadalupe River system and have been important in the development of the region historically. The portion of the Edwards Aquifer directly located on the escarpment is referred to as the Fault Zone Edwards Aquifer. There are two main areas, the confined (artesian) zone and the water-table zone. Water recharges the aquifer in the water-table (unconfined) zone. Numerous caves and sinks allow surface water to quickly flow to the aquifer. The aquifer is a porous, honeycombed, cavernous limestone made up of several formations. Water travels from the recharge zone downward to the artesian zone and then slowly to the north primarily through conduits that parallel the escarpment. The aquifer is about 280 km long, and from 8 to 64 km wide.

Extensive biological studies of this system began in the mid-1970s and have continued to the present. Sampling of the biota is done using fine mesh nets placed over the discharge of wells and springs throughout the region. The sampling began on a flowing artesian well at Southwest Texas State University in San Marcos. This well was drilled in 1895 to provide water for a large federal fish hatchery. The well penetrates a 1.5 m wide cavern in the Edwards at 59.5 m. The cavern is the source of water for the well. This well has been sampled almost continuously. This sampling, along with springs and other wells, has produced evidence of a very diverse aquatic community in the aquifer. Some of the deepest wells sampled are between 412 m and 610 m deep in San Antonio, Texas. These have produced two species of blind catfish, *Satan eurystomus* (Widemouth Blindcat) and *Trogloglanis pattersoni* (Toothless Blindcat), which are only found at these deep water levels. Numerous invertebrates are also found throughout the aquifer. There are 39 described invertebrates and several more unnamed. The aquifer has also produced 16 described vertebrates (all but two are Plethodontid salamanders in the genus *Eurycea*). Several more salamanders have been found and will probably be named as distinct species. This assemblage of 55 named species is highly diverse, possibly more so than any known aquatic subterranean ecosystem. The protozoa and other very small invertebrates have not been adequately sampled or described.

The organisms described from the aquifer range from flatworms to catfish. Turbellaria (free-living flatworms) are represented by *Sphalloplana mohri*. This colourless, eyeless planarian exhibits polypharyngy, the unusual condition of having many feeding tubes. It apparently feeds on decaying matter. Nematoda (roundworms) are represented by the species *Rhabdochona longleyi*, a parasite of the Texas Blind Salamander, *Eurycea rathbuni*. Hirudinea (leeches) are represented by the small leech, *Mooreobdella microstoma*, found on the Texas Blind Salamander. It is in the family Erpobdellidae.

Gastropoda (snails) are the most diverse invertebrate group in the aquifer, and represented by 13 different aquatic species in the family Hydrobiidae. The five genera are often less than 2 mm and may be interstitial as well as phreatic.

Crustaceans form about two thirds of the total aquatic fauna. Ostracoda are represented by *Sphaeromicola moria*, a member of the Entocytheridae. Isopoda are represented by four genera and five species. Amphipoda are the second most diverse group in the aquifer represented by 12 described species in eight different genera. These crustaceans are more diverse in this system than any other described surface or groundwater system worldwide. Several of these are marine relicts (having their closest relatives in the marine environment), and indicate colonization of freshwater habitats by a marine environment that last covered the area during the late Cretaceous or early Tertiary. The Thermosbaenacean *Monodella texana* represents another group of typically marine forms. The two described species of Decapoda are both shrimps in the family Palaemonidae.

Of the Coleoptera, three species of stygobitic beetles have been described from the Edwards Aquifer. *Haideoporus texanus* (Texas hell passer) is representative of the dytiscid beetles. It is a small, transparent, eyeless beetle that apparently spends its entire life cycle in the aquifer.

The vertebrates are represented by the two species of blind catfish mentioned earlier and 13 species of lungless salamanders. The most famous of these salamanders is the Texas Blind Salamander, *Eurycea rathbuni* (formerly *Typhlomolge rathbuni*), which is endemic to the aquifer. It is considered by many to be the most highly adapted vertebrate to subterranean aquatic life. It has long slender legs that would not support it on land. It has no functional eyes, only vestiges of retinas buried deep beneath the skin. It is colourless and is highly flattened dorsoventrally. It has adapted by having many special neurosensory structures over its head region for sensing food. Like most of the organisms found in the aquifer it exhibits a relatively long lifespan when compared to surface relatives. This salamander is found only near San Marcos, Texas.

The ecosystem exhibits amazing productivity in terms of the numbers of organisms issuing from wells and springs. There are likely two major energy pathways at work in the system. In the unconfined and nearby area of the aquifer the pathway is similar to that known in caves around the world. Organic matter deposited in the aquifer by recharge serves as food for fungi and other decomposers. Protozoa and larger invertebrates including the planarian, *Sphalloplana mohri*, feed on these organisms. The larger arthropods feed on the smaller organisms depending on their specializations. In the deep artesian area of the aquifer, isolated from organic debris, the source of energy is probably fossil organic matter (peat or oil) found in the Edwards formations. There is evidence of chemosynthesis. There is a saline–freshwater interface, and on the saline side, the water is without oxygen and large quantities of sulfur bacteria occur. The largest numbers of organisms occur at or near the interface. At the top of the food chain are the Widemouth Blindcat in the deep aquifer and the various blind salamanders in the unconfined or nearby areas of the aquifer.

The Texas Blind Salamander (see Plate 2i) was the first species listed as endangered in the Unites States in 1967; two other stygobitic species from the aquifer are listed as endangered (Comal Springs Dryopid Beetle, *Stygoparnus comalensis*, and Peck's Cave Amphipod, *Stygobromus pecki*), and the San Marcos Salamander as threatened. These species' very limited habitats are threatened by reduced spring flows caused by overpumping of groundwater. Overpumping could also cause encroachment of saline water into the aquifer. Another threat to the habitat of these species is the potential for groundwater contamination from sewage and urban expansion.

GLENN LONGLEY

See also **America, North: Biospeleology; Edwards Aquifer**

Further Reading

Hershler, R. & Longley, G. 1986. Phreatic hydrobiids (Gastropoda: Prosobranchia) from the Edwards (Balcones Fault Zone) Aquifer region, south-central Texas. *Malacologia*, 27(1): 127–72

Holsinger, J. & Longley, G. 1980. The subterranean amphipod crustacean fauna of an artesian well in Texas. *Smithsonian Contributions to Zoology*, 308: 1–62

Langecker, T.G. & Longley, G. 1993. Morphological adaptations of the Texas blind catfishes *Trogloglanis pattersoni* and *Satan eurystomus* (Siluriformes: Ictaluridae) to their underground environment. *Copeia*, 93(4): 976–86

Longley, G. 1981. The Edwards Aquifer: Earth's most diverse groundwater ecosystem? *International Journal of Speleology*, 11(1–2): 123–28

Longley, G. 1986. The biota of the Edwards Aquifer and the implications for paleozoogeography. In *The Balcones Escarpment: Geology, Hydrology, Ecology, and Social Development in Central Texas*, edited by P.L. Abbott & C.M. Woodruff Jr., San Antonio, Texas: Geological Society of America

Longley, G. 1992. The subterranean aquatic ecosystem of the Balcones Fault Zone Edwards Aquifer in Texas: Threats from over pumping. *Proceedings of the First International Conference on Ground Water Ecology*, edited by J.A. Stanford & J.J. Simons, Tampa, Florida: American Water Resources Association

Longley, G. 1994. The ecology and biota of the Balcones Fault Zone Edwards Aquifer in Texas. In *Toxic Substances and the Hydrologic Sciences*, edited by A.R. Dutton, Minneapolis, Minnesota: American Institute of Hydrology

Longley, G. 1996. Entries on Texas Blind Salamander, Toothless Blindcat, and Widemouth Blindcat. In *The New Handbook of Texas*, vol. 6, Austin: Texas State Historical Association

Stock, J.H. & Longley, G. 1981. The generic status and distribution of *Monodella texana* Maguire, the only known North American Thermosbaeneacean. *Proceedings of the Biological Society of Washington*, 94(2): 569–78

ENCANTADO, SISTEMA DEL RIO, PUERTO RICO

Sistema del Rio Encantado is situated in the North Coast karst belt on the island of Puerto Rico, 53 km southwest of San Juan and 3 km south of the town of Florida (Monroe, 1976). The cave is an underground tributary of the Rio Grande de Manati which flows north from the Cordillera Central Mountain Region to the Atlantic Ocean. The cave has been surveyed to a length of 17.1 km (with a vertical extent of 250 m), nearly all along a single continually traversable underground river (Figure 1).

Puerto Rico's North Coast karst belt occupies the northern third of the island, has an east–west extent of 135 km, and ranges in width from 10 to 23 km. The limestones form a rugged tropical karst topography. These lie within a thick sequence of platform carbonates and minor clastics ranging in age from middle Oligocene to Miocene. The basement rocks that underlie the limestone belt consist of Late Cretaceous and Early Tertiary volcaniclastic rocks and intrusives. The north-central area of the northern karst belt is known for its variety of karst landforms, which include cone karst and mogote (tower) karst, river caves, alluviated valleys, deep limestone gorges, and zanjones (vertical-walled trenches). The alignment of mogotes, dolines, zanjones, and river segments may be a function of local fracturing and faulting.

In the Rio Encantado area, the Lares Limestone gives rise to a distinctive cone karst with round cones that are capped by vertical cliffs forming small towers. Where jointing has influenced cone development, jagged saw-tooth cones and ridges have

Encantado, Puerto Rico: Outline map of the Sistema del Rio Encantado and associated caves beneath the cone karst of Puerto Rico; the darker areas are forested, and the lighter areas are cleared farmland.

formed. The younger Montebello Limestone forms ridges and cuesta scarps, and near Florida, forms steep cliffs on top of the Lares Limestone which results in a cliffed cone karst. The Mucarabones Sand (contemporary with the Lares Limestone) is present in the valleys between hills capped with Montebello limestone. The sands originated as debris from the volcanic interior, which were carried to the coastal area by rivers during periods between limestone deposition and gradual uplift of the island. These blanket sands have had a significant influence on karstification of the limestone by providing reservoirs of undersaturated water that dissolved the underlying limestone.

The Sistema del Rio Encantado is part of a groundwater basin within the Rio Grande de Manati watershed, one of six major watersheds along the south coast (Guisti, 1978). Tectonic tilt has affected the subterranean drainage systems such that springs are found predominantly on the west side of north-trending valleys. The Rio Grande de Manati forms a north-trending valley and the Manantial de las Aguas Frias, which is the resurgence for the Sistema del Rio Encantado, emerges from the west side of the river valley.

Allogenic streams flowing off the volcanic core formed the caves within the Rio Encantado groundwater basin. Concentrated recharge of this type forms linear cave systems with few side passages. This is apparent in the Sistema del Rio Encantado, which consists of a master river passage and very little base-level side passage development. The current allogenic catchment area for the groundwater basin encompasses only 4 km². It has been postulated that the recharge area was much larger in the past, but the original river may have been pirated either by the Rio Grande de Arecibo or the Rio Grande de Manati (Troester, 1992).

Dry valleys are common in the Encantado area, some containing intermittent streams after heavy rains. These valleys are lined with swallets that carry recharge underground. The autogenic recharge has formed the vertical entrances to the cave, and disjointed upper level passages and domes. The autogenic water also provides the source water that forms many unusual and spectacular speleothems.

The Sistema del Rio Encantado is formed in the Oligocene Lares Limestone. Upstream of the Encantado cave four segments of river passage are separated by collapse; Cueva Chica (50 m); Cueva Yuyu (500 m); an unnamed river segment (100 m); and Cueva Zumbo (1 km). A river draining the volcanics on the south side of the Manati River valley forms the headwaters of the conduit drainage. The river flows for a short distance on the Lares Limestone and then disappears into Cueva Chica.

Sistema del Rio Encantado has five entrances, is characterized by canyon passages (more than 20 m tall) and large phreatic boreholes, and has three rooms over 100 m across. Autogenic recharge has deposited flowstone cascades that cover the walls and floors and form dams in many places in the river. Large stalagmites and draperies abound and showerhead formations are common in the cave. There are four major sumps in the system. Multi-level floodwater mazes occur near the sumped areas of the cave and spectacular soda straw displays are common in these sections. Mazes also occur near collapsed areas of the system. In many places in the river, coral reef fossils stand out in relief against the water-sculpted limestone. Well-preserved dugong fossils have been observed in the walls of the cave.

Cueva Encantada reaches the main river via a walk-in entrance located at the bottom of a doline. Four kilometres downstream from the Encantada entrance is a short section of underwater passage that leads to the "Land Between the Sumps", an extraordinary section of the system consisting of large phreatic passages and tall canyons with many waterfalls. There are several sumps through the upstream end of Cueva Juan Nieves (with a 40 m vertical entrance series to the main river). The downstream section of the cave is a series of smaller pirated zones with older, less active, parallel passages. Ceiling heights reach over 30 m in places, until 600 m from the entrance the river sumps for the last time and reappears 400 m later in the Aguas Frias spring on the Rio Grande de Manati.

The karst surrounding the Sistema Rio del Encantado contains many smaller caves, which are developed at higher elevations than the active cave. These are paleo-segments of the system and are indicative of uplift of the island and sea-level fluctuations, which affect local base level. Surface erosion has lowered the land surface and collapses have dissected the paleo-cave passages leaving remnants of cave in the mogotes. A number of upper level caves have been found in the mogotes above the Rio Encantado. Cueva Viento, the largest and most complex of the paleo-segments, is 2000 m long. This cave has two vertical entrances near the top of a mogote and one horizontal entrance in the side of a dry valley. One of the vertical entrances is named

Volcan after the dark cloud of bats that leave the entrance every evening. A sinking stream enters the cave less than 100 m from the valley entrance. The cave consists of sediment-floored canyon passages, a series of complex mazes and loops, and very large breakdown rooms. The sinking stream appears in lower level passages, which extend for over 1 km before sumping. The water has been dye traced to the Agua Frias spring.

The caves of the Encantado karst are home to a variety of cave life. Amblypigids (tailless scorpions), cockroaches, freshwater crabs, and crayfish are common in the river passages, and ten different bat species occupy the caves. At dusk it is not uncommon to see boa constrictors waiting patiently in mogote cave entrances for the evening bat flights.

Prior to organized exploration, some of the caves within the Encantado groundwater basin were already known to locals who used the caves for a source of water. Taino petroglyphs in some of the dry caves indicate that human use of the caves goes back to pre-history.

Exploration and survey continues in the caves of Rio Encantado, but an expanding population and ever-growing development is posing a threat not only to Encantado but also to the entire North Coast karst belt. Commercial and industrial development projects involve extensive deforestation, landform obliteration, and water diversion and contamination. All of these are seriously impacting the karst regions of Puerto Rico. It is hoped that increasing public awareness of the island's karst areas will stimulate protection efforts and generate government response to the threats confronting the North Coast karst belt and its associated ecosystems.

PATRICIA KAMBESIS

Works Cited

Giusti, E., 1978. Hydrogeology of the karst of Puerto Rico, Washington: US Government Printing Office (US Geological Survey Professional Paper, 1012)

Monroe, W. H. 1976. The karst landforms of Puerto Rico, Washington: US Government Printing Office (US Geological Survey Professional Paper, 899)

Troester, J.W. 1992. The northern karst belt of Puerto Rico: A humid tropical karst. In *Hydrogeology of Selected Karst Regions*, edited by W. Back, J. Herman, H. Paloc, Hannover: Heise

ENTRANCE HABITATS

Two ecological factors of overriding importance in cave entrance habitats are the amount of solar radiation received and the availability of water. Most cave entrances or "thresholds" extend from a land surface fully exposed to sunlight to a depth where little light penetrates, merging, at a variable rate, into near darkness. For caves with high, open entrances, the light gradient is low, amounting to reductions of 1% or less per metre, while in the smallest entrances (<1m), the gradient may approach 100% per metre. Sometimes the cave entrance will be under water, or within a doline, and receives low illumination with indirect sunlight only. In many cases there are problems for ecologists in deciding where cave entrances begin, particularly where the rock outcrop and entrance is irregular. The limits of the cave threshold communities are likewise poorly defined. Some light penetrates deeply within caves, as it does in the ocean, although at levels undetectable by the human eye. Defining the threshold limit as the depth to which plants can grow is not satisfactory as different species have their own tolerance levels and some plants may contain chlorophyll and appear green, while growing heterotrophically (without photosynthesis). Oceanographers define a euphotic zone, where photosynthetically available light is >1% of its average surface value, followed by an aphotic zone where it is less than this. The scheme could be adapted for caves, but since light varies continuously and often irregularly in caves, the best approach may be to record a series of light levels as accurately as possible and consider the cave system as a continuum without resort to a rigid definition of the threshold. However, a reference level is sometimes useful when comparing caves and their biota, and a zone where >1% of the external surface light intensity occurs is suggested for the limit of the cave entrance habitat. In mid-latitudes, this level corresponds to a light intensity slightly exceeding 1000 lux ($c.18$ μE m^{-2} s^{-1}) when the external light intensity is optimal (midday, open sky) (where $1E = 6.02 \times 10^{23}$ photons).

The flora of cave entrances is often similar to that of local vegetation, differing only in minor respects where resurgences or air currents of different temperature occur. If the cave floor consists of moist sediment, a local forest flora is often represented as the illumination and humidity will approximate that of the woodland ground flora. Deeper within caves, the flora becomes restricted and a more or less marked zonation is often evident beginning with flowering plants, followed by ferns, mosses, lichens, liverworts, and algae. In European limestone caves, typical threshold genera of flowering plants are *Geranium*, *Urtica*, and *Viola*; ferns *Asplenium*, *Dryopteris*, and *Phyllitis*; mosses *Eucladium*, *Fissidens*, and *Thamnobryum*; lichens *Chrysothrix* and *Lepraria*; liverworts *Conocephalum*, *Plagiochila*, and *Porella*, and algae *Chlorococcum*, *Chroococcus*, and *Gloeocapsa*. In some cases, cave entrances provide refugia for plants where the terrestrial flora has been modified externally by man. Such sites have high conservation value and could provide a nucleus for future expansion when conditions outside the cave become more favourable. Most of the cave flora, excluding non-photosynthetic plants such as fungi, are facultative cavernicoles and a few species have been found only in caves. Plants with leaves (angiosperms, ferns, mosses, and some liverworts) show adaptations to low light levels. Leaves often become enlarged (to capture more light), leaf tissues thinner and stomata less numerous (due to high humidity). Stems become elongated, and branching less prolific. Sexual reproduction is often impaired at low light levels. A small group of cave mosses possess specially shaped cells, causing focusing of light onto the chloroplasts (*Mittenia* and *Schistostega*). Some algae show adaptations to low light by increasing the density of thylakoids in the chloroplasts. Cave soils provide a rooting medium for flowering plants but the flora is not confined to them. Rock crevices often provide moisture and support in addition to ferns and bryophytes. Lichens and algae are able to grow on bare rock walls, when the only source of water may be condensation from the atmosphere. Some ruderal species such as the nettle (*Urtica*) thrive in phosphorus-enriched soils where mammals use caves for shelter and deposit excreta. Aquatic communities, including flowering plants (e.g. *Potamogeton*) occur at wet cave entrances.

Invertebrates of cave thresholds often contain members of the woodland ground-fauna, with the flora and associated detritus supplying food and shelter. As with plants, a zonation is often apparent as the associated flora diminishes in abundance and diversity. Patterns are more complex for invertebrates since food and shelter can also be provided by allochthonous detritus passing through a cave and by structures within the cave. Consequently some animals are not dependent upon light and are often found more deeply within cave systems. Invertebrates of cave entrance habitats include several insects such as the mosquito *Culex* and caddis *Stenophylax* which fly short distances into cave entrances for diapause (a form of hibernation), entrances providing moderately equable climates throughout the year. While these species do not feed within caves, others do, and are fed upon by predators such as the Tetragnathid spider *Meta merianae* and gastropods such as *Oxychilus*. These species often form part of a well-defined wall-association, living in crevices or attaching themselves to the exposed rock surface. Cave entrances are also used by birds such as *Hirundo fulva* (cave swallow) and *Pyrrhocorax* spp. (chough), the latter especially in sea caves. Along with bats (e.g. *Barbastella barbastellus*), their excreta and nest detritus provide a rich food source for other animals such as rats, and nutrients for plants and bacteria.

Reptiles, amphibians, and mammals make considerable use of cave entrances. Brief occupations are made by numerous species during periods of inclement weather such as heavy rains, but regular use of caves is restricted to fewer species, many of which are carnivores. Hyenas and wolves make dens in caves where the young are raised. Some burrowing species such as the red fox and badger also make occasional use of caves while others, such as the brown bear, use them for hibernation, but usually venture beyond the cave threshold. Leopards and porcupines are also known to occupy them for considerable periods. During the Pleistocene, cave bears made regular use of caves as dens. Humans are perhaps the most significant users of cave entrances among the mammals. Many early humans made regular use of cave entrances for temporary shelter, permanent accommodation, and for religious or other cultural practices. Some caves remain in use today (see Human Occupation of Caves).

ALLAN PENTECOST

Further Reading

Chapman, P. 1993. *Caves and Cave Life*, London: HarperCollins; chapter 4: Flora and fauna

Claus, G. 1962. Data on the ecology of algae in Peace Cave, Hungary. *Nova Hedwigia*, 4: 55–79

Dobat, C. 1998. Flore (Lichens, Bryophytes, Pteridophytes, Spermatophytes). In *Encyclopaedia Biospeleologica*, vol. 2., edited by C. Juberthie & V. Decu, Moulis and Bucharest: Société de Biospéologie (in French)

Mason-Williams, A. & Benson-Evans, K. 1967. Summary of results obtained during a preliminary investigation of the bacterial and botanical flora of caves in South Wales. *International Journal of Speleology*, 2: 397–402

Vandel, A. 1965. *Biospeleology: The Biology of Cavernicolous Animals*, Oxford and New York: Pergamon Press (originally published in French, 1964)

ENVIRONMENTAL IMPACTS ASSESSMENTS

Karst areas are diversely rich in resources, yet are the world's most vulnerable landscapes to environmental impacts. The Earth's most productive springs and water wells occur in karst. The space provided by caves often provides habitats for rare and unusual species, serving as models of evolution and ecosystem development. Other species, such as bats, are internationally important as pollinators, seed dispersers, and controllers of crop and other insect pests. Hydrologically inactive caves offer stable environments that preserve many of the greatest archaeological and paleontological sites from the elements that would destroy them on the surface. Caves and surface karst terrains provide a wealth of recreational and scenic attractions. Paleokarst areas often contain tremendous deposits of oil, lead, zinc, and other fuels and minerals. The karstic bedrock itself, usually limestone, is often quarried for use in construction.

Historically, karst areas have not been heavily populated or utilized, and thus not heavily impacted. People have lived along the margins of karst, enjoying the benefits of springs, but most karst resources were difficult to access or of little direct or recognized economic benefit. However, technological advances in the 20th century changed this situation. Water and mineral resources are now easily extracted from karst, and many urban areas have grown onto karst areas. The underground spaces that offer protection and habitats also allow the consequences of surface activities, such as contaminants, to easily enter the subsurface and degrade those resources.

Environmental impacts assessments (EIA) are increasingly used to limit the adverse impacts of human activities on resources important to the health of human and non-human ecosystems. The unique attributes of karst areas require assessment methods specifically designed for those terrains. DRASTIC (Aller *et al.*, 1987), pump tests, monitoring wells, and other techniques that are effective in assessing and managing non-karst terrains, are often poorly effective to ineffective in karst. While several scientific disciplines need to be considered in assessing karst, a hydrogeologic EIA should first be conducted, since it defines the physical parameters of a study area and provides a conceptual model of karst development as a foundation for the other fields of research.

Three primary hydrogeologic EIA models for karst have been published and used. One or more can be used within a given investigation, based on the purpose, scale, and detail of study. Doerfliger, Jeannin & Zwahlen (1999) proposed EPIK to assess the vulnerability of karst terrains through the use of air photo interpretation, tracer tests, geophysical data, and geomorphological mapping in karst spring drainage basins. The EPIK system uses the resulting data to define an area's Epikarst (surface karst features), Protective cover (soil), Infiltration conditions, and

Karst drainage (degree of solutional conduit development). The characteristics of an assessed karst watershed area are matched to a category under each EPIK factor. The categories have numerical values which are multiplied by a constant for the EPIK factor, and the results are summed to identify the area's degree of sensitivity with respect primarily to groundwater contamination. This vulnerability mapping method has broad application for both local and regional level EIA.

A second karst hydrogeologic EIA model was proposed by the American Society for Testing and Materials (ASTM, 1995) and extended by the US Environmental Protection Agency (Schindel et al., 1997). A conceptual model of the karst aquifer under study is first developed based on existing information and modified as new information is collected. Initially, well and spring data are usually most available and used to examine water-table elevations, differences in permeability and chemistry, and responses to recharge. Karst-specific data are then considered, surveying the location, type, and geologic setting (fractures, strata, etc.) of caves and other karst features. The next step, in many investigations, is to design a series of tracer tests to determine groundwater drainage basins, flowpaths, travel times, storage, dilution, and dispersion. These factors change with groundwater discharge and should be conducted, if possible, during base flow and storm flow conditions. Geophysical techniques, such as electrical resistivity and microgravity, can be used where it is necessary to locate voids that are not accessible through known caves. This is usually done for relatively small areas since these techniques are labour, time, and cost intensive.

Veni (1999) proposed a third karst hydrogeologic EIA strategy for areas where tracer testing and geophysical techniques are not possible or available. This method examines geomorphological and hydrological characteristics of karst areas and features to assess their relative permeability and their potential vulnerability. The potential for doline (sinkhole) collapse is often best identified through such methods. This strategy can be applied regionally as well as to specific features, and is adjusted according to the dominant characteristics of karst within each study area. It can also be applied to develop superior conceptual models of the karst and / or better interpret the results of the tracer and other studies. Additionally, this strategy requires defining what level of impacts on karst resources is sustainable in order to truly determine the consequences.

All hydrogeologic karst EIA methods suffer from the failing that they identify areas and features of relatively greater or lesser vulnerability, and land managers not familiar with karst often interpret less vulnerable areas as not vulnerable. Scientists conducting EIA in karst should stress to land managers that by the nature of karst landscapes, all areas are vulnerable, only highly vulnerable areas are distinguished, and that karst cannot be adequately managed solely on a feature-by-feature basis. Studies on the relationship of water quality to non-natural impervious land cover (roads, parking lots, buildings etc.) show adverse environmental impacts significantly increase when impervious cover exceeds 15% of a surface watershed. Veni (1999) recommended a maximum of 15% impervious cover as an effective means of protecting karst aquifers when coupled with protection of critical areas identified by field surveys.

Rare, endangered, or otherwise important species occur in many karst areas. They include aquatic species such as invertebrates, fish and salamanders, and terrestrial species, such as bats and invertebrates. While these species may be protected by the above hydrogeologic EIA methods, specific biological EIA methods are needed where hydrologic protection measures are not present or are inadequate for karst fauna. For example, water quality protection measures may not protect bats as much as not disturbing them during their hibernation. The above geomorphological EIA strategy is applicable to biology, by delimiting areas where important faunas should occur and which species may be present.

Following a review of biological data, the first step of a biological EIA is a presence–absence survey for the fauna of concern in the study area. Some rare species will require multiple surveys to adequately determine if they are likely to be present. Others, such as migratory bats, can only be surveyed at certain times of year. A standard for the number and method of surveys may need to be established appropriate to the fauna. US Fish and Wildlife Service (2000) provides an example of karst biological survey standards. Monitoring of the fauna is also a key component of a karst biological EIA, to distinguish natural changes within the ecosystem from induced impacts and to determine what impacts are sustainable. While monitoring is usually not within the scope of a specific assessment, experience from monitoring studies with the species of concern should be incorporated into the karst biological EIA.

A significant portion of karst ecosystems may not be accessible or directly observable through caves and is thus difficult to assess in a karst biological EIA. This problem is especially important where quarries may largely or completely destroy the habitat and fauna, or where the habitat and fauna may be significantly altered and impacted by extensive urban and industrial development. A regional analysis of species localities relative to geologic factors may delineate the species' probable distribution and the degree of impacts they can withstand. Recovery plans for endangered karst invertebrate species have been based on such analyses (e.g. O'Donnell, Elliott & Stanford, 1994).

Karst EIA for cultural and paleontological materials also requires presence–absence surveys. Regional analysis can be helpful to identify areas where such materials are likely to occur but is less reliable than with similar biological studies. Materials present on the land surface may or may not be good indicators of materials in caves. Some past cultures and animals avoided caves while others were drawn to them. In either case, significant deposits may be present in caves due to deliberate or accidental accumulations. Both above and below ground, relatively recent sediments, free of cultural or paleontological materials, may bury or obscure older important deposits.

Cultural and paleontological karst EIA techniques commonly begin with detailed grid searches of the land surface. Such searches are also effective in hydrogeologic EIA in locating karst and other geologic features. Shovel tests are often necessary to evaluate potential deposits. Areas in caves that are likely to accumulate sediments, especially near entrances, should also be shovel tested, although such tests may be insufficient to characterize deep deposits. In some cases, geophysical methods can be used to evaluate a deposit's potential for significant materials and guide excavation or preservation strategies. In areas that may be lost to quarrying or other activities, evaluation methods may need to be modified into salvage operations if detailed studies are not feasible.

The complexity of karst terrains, their multiple and often difficult to define resources, and their high vulnerability to adverse impacts, requires prudence in EIA recommendations. There remains a lot to learn about how to best evaluate some attributes of karst and how to effectively protect and manage them. EIA recommendations should therefore be coupled with strategies recommended for the identification, delineation, and establishment of preserved or protected areas, appropriate management and maintenance of those areas, gating of caves and areas as necessary, and monitoring of resources within those areas, to determine the effectiveness of the recommended actions and the EIA process.

GEORGE VENI

See also **Conservation: Cave Biota; Conservation: Protected Areas; Groundwater Protection; Monitoring**

Works Cited

Aller, L., Bennett, T., Lehr, J.H., Petty, R.J. & Hackett, G. 1987. *DRASTIC: A Standardized System for Evaluating Ground Water Pollution Potential Using Hydrogeologic Settings*, Ada, Oklahoma: Office of Research and Development, Environmental Protection Agency (EPA-600/2-87-035)

American Society for Testing and Materials. 1995. *Standard Guide for Design of Ground-water Monitoring Systems in Karst and Fractured-rock Aquifers*, West Conshohocken, Pennsylvania: American Society for Testing and Materials (D 5717-95)

Doerfliger, N., Jeannin, P.Y. & Zwahlen, F. 1999. Water vulnerability assessment in karst environments: A new method of defining protection areas using a multi-attribute approach and GIS tools (EPIK method). *Environmental Geology*, 39(2): 165-76

O'Donnell, L., Elliott, W.R. & Stanford, R.A. 1994. *Recovery Plan for Endangered Karst Invertebrates in Travis and Williamson Counties, Texas*, Region 2, Albuquerque, New Mexico: US Fish and Wildlife Service

Schindel, G.M., Quinlan, J.F., Davies, G. & Ray, J.A. 1997. *Guidelines for Wellhead and Springhead Protections Area Delineation in Carbonate Rocks*, Atlanta, Georgia: Groundwater Protection Branch, Region 4, US Environmental Protection Agency (Report 904-B-97-003)

US Fish and Wildlife Service. 2000. *Terrestrial Karst Invertebrate Survey Protocols, July 7, 2000 version*, Austin Field Office: US Fish and Wildlife Service

Veni, G. 1999. A geomorphological strategy for conducting environmental impact assessments in karst areas. *Geomorphology*, 31: 151-80

Further Reading

Gillieson, D. 1996. *Caves: Processes, Development and Management*, Oxford and Cambridge, Massachusetts: Blackwells

Ogden, A.E. 1984. Methods for describing and predicting the occurrence of sinkholes. In *Sinkholes: Their Engineering and Environmental Impact*, edited by B.F. Beck, Rotterdam: Balkema

Sendlein, L.V.A. 1991. Analysis of DRASTIC and wellhead protection methods applied to a karst setting. In *Proceedings of the Third Conference on Hydrogeology, Ecology, Monitoring, and Management of Ground Water in Karst Terranes*, edited by J.F. Quinlan & A. Stanley, Dublin, Ohio: National Ground Water Association

Veni, G. & DuChene, H. (editors) 2001. *Living with Karst: A Fragile Foundation*, Alexandria, Virginia: American Geological Institute

Watson, J., Hamilton-Smith, E., Gillieson, D. & Kiernan, K. (editors) 1997. *Guidelines for Cave and Karst Protection*, Cambridge and Gland, Switzerland: International Union for Conservation of Nature and Natural Resources

EROSION RATES: FIELD MEASUREMENTS

In karst, as in most other terrains, both chemical and mechanical erosion processes operate and the total karst denudation rate is the sum of both processes. However, as high rock solubility is one of the main characteristics that gives rise to karst terrains, and as chemical erosion is easier to estimate than mechanical, most karst research, with the notable exception of Smith & Newson (1974) has focussed on dissolutional denudation rates. Recent work on the turbidity of karst springs (e.g. Bouchaou, Mangin & Chauve, 2002) provides the potential for more detailed analysis of mechanical erosion rates. The majority of studies of dissolutional denudation rates have been undertaken in carbonate karst and these are the primary focus of this essay. However, as Klimchouk *et al.* (1996) have noted, similar considerations apply to evaporite karsts, and the estimation of gypsum dissolution rates is particularly problematic because of the more rapid solution and the consequently greater spatial and temporal variability of dissolution.

Denudation rates are commonly expressed as mm per 1000 years (mm ka^{-1}), implying that all erosion contributes to surface lowering and that environmental conditions have remained broadly the same for millennia. The former assumption is incorrect, particularly in karst, while the latter is also highly questionable. The preferred unit is m^3 km^{-2} a^{-1} and 1 mm ka^{-1} is equivalent to 1 m^3 km^{-2} a^{-1}. Where surface lowering is measured directly then units of μm a^{-1} are appropriate.

Solutional erosion rates in carbonate karst may be estimated from knowledge of dissolution kinetics, runoff, carbon dioxide availability, and temperature (White, 1984), but there remains a need for field measurements to obtain actual values of regional denudation / transformation of relief; to compare denudation rates in contrasting environments and by different processes; to understand landform evolution; and to understand how the processes operate in a complex natural environment as opposed to the laboratory. When evaluating results from past studies it is important to understand what was actually measured and how the denudation rates were calculated. Although an early study by Spring & Prost (1883) was based on daily sampling, the vast majority of field measurements of solutional erosion rates in carbonate karst are based on spot samples, with denudation being estimated from the Corbel (1959) formula:

$$D = 4\ ET\ n/100$$

Where D is the limestone denudation rate (m^3 km^{-2} a^{-1}), E is runoff (precipitation-evapotranspiration, dm), T is the average CaCO$_3$ content of the water (mg l^{-1}) and $1/n$ is the proportion of the basin occupied by karst rocks. This formula suffers several problems, including the assumption that the bulk density of carbonates is 2.5 g cm^{-3} whereas it can range from 1.5 to 2.9; failure to consider hardness due to magnesium salts; and failure to consider sulfate salts. However, the three most important failings are that T is frequently the average of a few spot measurements, with the implicit assumption of a linear relationship between carbonate hardness and discharge; that carbonates present in solution only come from karst denudation whereas some can come from other sources; and that measurements are usually made at only one point, commonly the output of a drainage basin, with the implicit assumption that this is representative of conditions upstream.

Where water samples have been collected over a range of flow conditions it is apparent that the relationship between dissolved load and discharge is usually nonlinear and, particularly in small drainage basins, may be complicated by hysteresis effects (usually higher concentrations per unit discharge on the rising limb). In practice it is virtually impossible to correct for hysteresis, but by collecting samples over a range of discharges it is usually possible to construct a reliable discharge-concentration or discharge-load rating curve. This can then be applied to the discharge curve and the results summed to obtain the total annual solute load (Gunn, 1981). Greater accuracy may be obtained using a logging conductivity meter, developing a conductivity-concentration rating curve, and using this to predict the concentration at each measured discharge. Where discharge records cover only a relatively short time period they may be extended using rainfall-runoff relationships (Williams & Dowling, 1979).

Having computed the total solute load (TSL) at a point it is important to realise that this is made up of total corrosion of karst rocks by both autogenic (CKAu) and allogenic (CKAl) waters, less any deposition of previously dissolved material (D), together with corrosion of non-karst rocks by allogenic waters (CNK), solute accessions in rainfall and snowfall (AC), and any anthropogenic inputs such as fertilizers (AN). The gross karst solution is then (CKAu + CKAl) whereas the net karst solution is (CKAu + CKAl − D). Where precipitation of previously dissolved carbonates is minimal then gross and net solution will be similar, but elsewhere failure to account for deposition may result in a significant underestimate of gross denudation, which is the real measure of relief transformation. In contrast, failure to take into account the solution of non-karst rocks and solute accessions in precipitation will result in an overestimate of karst solution. For example, Williams & Dowling (1979) found that CNK and AC respectively made up 9.9% and 4.6% of the total solute load in the Riwaka Basin, New Zealand. Error in estimating erosion rates can arise from many sources and even in a careful study using hydrochemical budgeting and taking into account non-denudational components potential errors of around 25% are likely (Gunn, 1981).

Solutional erosion rates for whole drainage basins derived by sampling of water at the basin outlet are unlikely to be representative of any specific location within the basin. This information may best be obtained by an extension of the hydrochemical budgeting method discussed above. Water samples are collected from the full range of sites in the karst system—bare limestone surfaces, the soil zone, the subcutaneous zone, the main body of bedrock (sampled as vadose flows and seepages), and cave streams in both vadose and phreatic zones. These, together with estimates of the proportion of water following the various pathways through the system, permit the breaking down of the overall erosion budget (Williams & Downing, 1979; Gunn, 1981). Those few studies that have been made show that a high propor-

Erosion Rates: Field Measurements. Micro-erosion meter. (Photo by John Gunn)

tion of solution (50–85%) occurs within several metres of the surface in the soil (if present) and subcutaneous zone (uppermost bedrock). Caves account for very little of the erosion when averaged over the whole basin.

The principal drawback of the hydrochemical approach is that it requires frequent, ideally continuous, measurement of discharge and sufficient samples to establish the pattern and extent of variations in solute concentrations. This is not always possible and alternative methods that integrate erosion over a longer time period have therefore been derived. The two most commonly used in karst areas are the micro-erosion meter (Spate et al., 1985) which provides a direct measure of rock surface lowering (see Figure) and rock tablets (Trudgill, 1975). In contrast to the hydrochemical method these techniques are highly site-specific and may only be used to assess erosion rates on bare limestone surfaces, in the soil zone, at the soil–bedrock interface, and in cave streams. Trudgill (1975) suggested that tablets could be used to detect seasonal differences in erosion rates and raised the possibility of making reliable measurements over shorter time scales. However, when Crowther (1983) compared the rock tablet and hydrochemical methods in West Malaysia, he found that the tablets gave estimates two orders of magnitude less than those calculated using hydrochemical data. The most likely explanation is that natural rock surfaces come into contact with larger volumes of water than do isolated rock tablets, simply because of their greater lateral flow component (see also Erosion Rates: Theoretical Models). Thus, the two methods measure fundamentally different phenomena and the hydrochemical method provides the only reliable means of estimating solutional erosion rates on limestone surfaces. Different problems arise if tablets are placed in cave streams as they will project above the natural surface and as a consequence are likely to erode more rapidly. They are also likely to suffer from abrasion as well as corrosion, although this can be exploited by placing the tablets in nylon cages with differing mesh sizes and comparing the erosional losses suffered.

In some areas it may be possible to estimate long-term average erosion rates for two elements in the carbonate karst system—

bare limestone surfaces and cave passages. Surface lowering rates may be estimated from the height of limestone pedestals formed when part of a limestone pavement is protected from solution because it is capped with an erratic or a limestone block. The rate of erosion in cave streams may be estimated by an approach proposed by Gascoyne (1981) that makes use of uranium series dating of speleothems. If an old speleothem can be found in its growth position and near to an active streamway then its height above present stream level divided by its basal age provides an estimate of the mean maximum rate of passage lowering. The main shortcoming is that the routes of cave streams vary through time and if the streamway at the measurement point has not been occupied continuously since speleothem deposition the calculated erosion rate is no longer a maximum value. Care is also necessary in interpreting the results in terms of the overall distribution of solutional erosion since both solution and mechanical erosion are likely to be active in cave floor lowering. Direct comparison of long-term solutional erosion rates estimated from pedestal heights or speleothem ages with the results of modern process studies using hydrochemical or alternative methods is, strictly speaking, unsound. Nevertheless comparisons have been made and it is interesting to note that on the whole there is quite good agreement.

JOHN GUNN

Works Cited

Bouchaou, L., Mangin, A. & Chauve, P. 2002. Turbidity mechanism of water from a karstic spring: Example of the Ain Asserdoune spring (Beni Mellal Atlas, Morocco). *Journal of Hydrology*, 265: 34–42

Corbel, J. 1959. Vitesse de l'erosion. *Zeitschrift für Geomorphologie*, 3: 1–28

Crowther, J. 1983. A comparison of rock tablet and water hardness methods for determining chemical erosion rates on karst surfaces. *Zeitschrift für Geomorphologie*, 27: 55–64

Gascoyne, M. 1981. Rates of cave passage entrenchment and valley lowering determined from speleothem age measurements. In *Proceedings of the 8th International Speleological Congress, Bowling Green, Kentucky*, edited by B. Beck: 99–100

Gunn, J. 1981. Limestone solution rates and processes in the Waitomo District, New Zealand. *Earth Surface Processes and Landforms*, 6: 427–45

Klimchouk, A., Cucchi, F., Calaforra, J.M., Aksem, S., Finocchiaro, F. & Forti, P. 1996. Dissolution of gypsum from field observation. *International Journal of Speleology*, 25: 37–48

Smith, D.I. & Newson, M.D. 1974. The dynamics of solutional and mechanical erosion in limestone catchments on the Mendip Hills, Somerset. In *Fluvial Processes in Instrumented Watersheds*, edited by K.J. Gregory & D.E. Walling, London: Institute of British Geographers (Special Publication 6, 155–67)

Spate, A.P., Jennings, J.N., Smith, D.I. & Greenaway, M.A. 1985. The micro-erosion meter: Use and limitations. *Earth Surface Processes and Landforms*, 10: 427–40

Spring, W. & Prost, E. 1883. Etude sur l'eau de la Meuse. *Annales de la Société Géologique de Belgique*, 11: 123–220

Trudgill, S.T. 1975. Measurement of erosional weight-loss of rock tablets. *British Geomorphological Research Group, Technical Bulletin*, 17: 13–19

White, W.B. 1984. Rate processes: Chemical kinetics and karst landform development. In *Groundwater as a Geomorphic Agent*, edited by R.G. LaFleur, Boston: Allen and Unwin: 227–48

Williams, P.W. & Dowling, R.K. 1979. Solution of marble in the karst of the Pikikiruna Range, northwest Nelson, New Zealand. *Earth Surface Processes*, 4: 15–36

Further Reading

Smith, D.I. & Atkinson, T.C. 1976. Process, landforms and climate in limestone regions. In *Geomorphology and Climate*, edited by E. Derbyshire, London and New York: Wiley: 367–409

White, W.B. 2000. Dissolution of limestone from field observations. In *Speleogenesis: Evolution of Karst Aquifers*, edited by A. Klimchouk, D. Ford, A. Palmer & W. Dreybrodt, Huntsville, Alabama: National Speleological Society: 149–55

EROSION RATES: THEORETICAL MODELS

Lowering of the Earth's surface in karst areas of soluble rock is primarily caused by dissolution of the bedrock, in contrast to nonsoluble rocks, where mechanical erosion is the main process. Therefore knowledge of dissolution rates in various geological situations (see entries: Dissolution: Carbonate Rocks, Dissolution: Evaporite Rocks) allows estimation of denudation rates.

Direct observation of bedrock lowering (see Erosion Rates: Field Measurements) on bare limestone bedrock exposed to the atmosphere yields rates between 10 $\mu m\ a^{-1}$ and 50 $\mu m\ a^{-1}$. Dissolution on bare rock surfaces proceeds under the following conditions. A water film of about 1 mm thickness flows down the rock. Due to the roughness of the rock and raindrops impinging, eddies will occur and render the flow fully turbulent. The water is in equilibrium with CO_2 in the atmosphere. Figure 1 shows the rates for full turbulent flow obtained by modelling (Buhmann & Dreybrodt, 1985) for various thicknesses δ (denoted by the numbers on the curves) of the water film. The rates R (cm a^{-1}) can be approximated (see Dissolution: Carbonate Rocks) by

$$R = 1.17 \times 10^6\ \alpha(c_{eq} - c) \qquad (1)$$

where α is a constant in cm s^{-1}, c is the calcium concentration in mmol cm^{-3} in the solution, and c_{eq} the equilibrium concentration with respect to calcite. When a water film comes into contact with a limestone surface it attains 90% of c_{eq} within a time $T = 2\delta/\alpha$ (Dreybrodt, 1988). From Figure 1 one finds $T = 3000$ for a film of 0.1 cm. Therefore the water remains at low Ca-concentration and denudation is determined by the rates at about 20% of c_{eq}. From Figure 1 one reads a lowering of bedrock from about 400 $\mu m\ a^{-1}$ to 200 $\mu m\ a^{-1}$ depending on the thickness of the water film. If one assumes that water films are present only during 10% of the year, denudation rates between 20 $\mu m\ a^{-1}$ to about 40 $\mu m\ a^{-1}$ are obtained in agreement with observation. Rates for downcutting of rinnenkarren are expected to be of similar magnitude. Thus entrenchment of a few centimetres needs about 1000 years.

Water running down a vertical shaft in layers several millimetres thick is in supercritical laminar flow with velocities in the range of 1 to 10 m s^{-1}. Erosion rates range between 300 and 800 $\mu m\ a^{-1}$ (Figure 1) depending on the thickness of the water film. Therefore where water flows down in shafts continuously, for example Mammoth-Flint Cave, Kentucky, widening of the shaft diameter by 1 m takes about 1000 years.

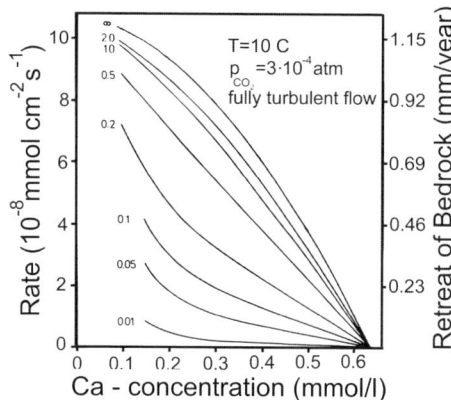

Erosion Rates: Theoretical Models: Figure 1. Dissolution rates for water films in fully turbulent flow. The solution is in equilibrium with atmospheric CO_2. The depth δ of the layer is denoted at the curves in units of cm.

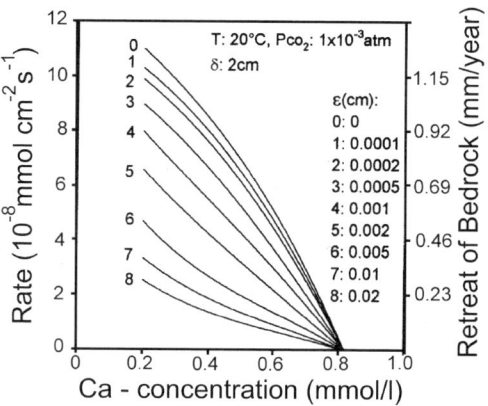

Erosion Rates: Theoretical Models: Figure 2. Dissolution rates of turbulently flowing streams with a depth deeper than 1 cm. The numbers denote the ϵ of the diffusion boundary layer.

A common procedure to measure denudation rates is to place bedrock tablets into the soil and measure their weight loss after a defined time. Observed rates are between 10 to 30 μm a^{-1}, similar to those on bare rock. This is somewhat unexpected because pCO_2 in soil is higher by at least a factor of 10. One has to consider, however, that dissolution proceeds by slowly moving water contained in the pores of the soil surrounding the limestone sample. The ratio of the volume V of water interacting with the limestone sample, to its surface area A is V/A-0.2d, where d is the pore diameter. This volume must now be taken for δ (Baumann et al., 1985). Soil water is in equilibrium with soil atmosphere with pCO_2 between 1×10^{-3} and 5×10^{-3} atm. From Figure 1 in entry: Dissolution: Carbonate Rocks, assuming c-0.5 c_{eq}, one finds rates of about 30 μm a^{-1} for pore sizes of 100 μm at a temperature of 10°C, and $pCO_2 = 5\times10^{-3}$ atm. For smaller pore sizes (likely in various soils) the rates drop linearly, e.g. to 16.5 μm a^{-1} at a pore size of 50 μm. The dissolution rates increase by roughly a factor of two when temperature rises by 10°C. They also increase with increasing pCO_2 (Dreybrodt, 1988). Since the rates depend critically on the saturation state of the water, knowledge of the Ca-concentration in the pore water is of utmost importance. Zambo and Ford (1997) have investigated the carbonate dissolution potential in soils of a doline in upland karst of Hungary. This potential is the amount of calcite which can be dissolved at prevailing conditions: pCO_2 in the soil and mean annual flux of infiltration. From this potential they estimated denudation rates in the range of 10 μm a^{-1} at the bottom of the doline. Dissolution of calcite in soil is thus a complex process compared to dissolution on bare rock surfaces, and direct comparison of these two rates is not appropriate.

Dissolution rates are much higher in karst streams and range between 100 μm a^{-1} and 800 μm a^{-1}. In streams the V/A ratio is large and rates are no longer controlled by CO_2 conversion, but by diffusion through a boundary layer. Its thickness ϵ decreases with increasing flow velocity (see Dissolution: Carbonate Rocks, Figure 3). For example, Figure 2 depicts rates for turbulent flow of water with $pCO_2 = 1\times10^{-3}$ atm as is common in stream water, for various values of ϵ. At low Ca-concentration of c-0.2 c_{eq} the rates vary from about 800 μm a^{-1} for high flow (low ϵ) velocities down to 230 μm a^{-1} when flow is slow (ϵ large). For higher Ca-concentrations the rates drop linearly. They depend only weakly on pCO_2 in the water (see Figure 3 in Dissolution: Carbonate Rocks) but they drop with decreasing temperature. The dependence of bedrock retreat on flow and consequently on ϵ has been observed by Lauritzen (1986). Dissolution rates in an isolated single cave conduit, with only one sink and one spring increased by almost a factor of ten from low discharge (1 m^3 s^{-1}) to high discharge at 10 m^3 s^{-1}.

Denudation rates in gypsum terrains have been observed by a variety of methods (Klimchouk et al., 1996). Direct retreat of ledges in gypsum and also observation of wall retreat of large gypsum boulders in rivers yield dissolution rates between 10–20 cm a^{-1} (James, 1992). Navas (1990) has observed rates of 20–40 cm a^{-1} on gypsum spheres suspended in rivers. Gypsum rates R (cm a^{-1}) can be estimated in such situations by

$$R = 2.3 \times 10^6 \frac{D}{\epsilon} \cdot (c_{eq} - c) \qquad (2)$$

(see Dissolution: Evaporite Rocks). $D = 10^{-5}$ cm^2 s^{-1} is the diffusion coefficient, ϵ [cm] is the thickness of the diffusion boundary layer in the order of 10^{-2} cm, and c, c_{eq} are the actual and the equilibrium concentrations in mmol cm^{-3}. For undersaturated water, $c = 0$, one obtains rates of about 18 cm a^{-1} for $\epsilon = 2\times10^{-2}$ cm. Depending on hydraulic conditions of flow, variations in ϵ are the reason for the observed variation of the rates. At extremely fast flow ϵ could become lower. For $\epsilon = 2\times10^{-3}$ cm rates are maximal at 180 cm a^{-1}. Such extreme values are reported by Klimchouk et al. (1996).

Measurements with micro-erosion meters on bare gypsum rock exposed to meteoric precipitation yield values from 0.2 up to 1.6 mm a^{-1} (Klimchouk et al., 1996). The rates are correlated linearly with annual precipitation. Conditions for dissolution on gypsum rock are similar to those for limestone discussed above. Flow is fully turbulent. Under such conditions gypsum dissolution is surface-controlled and in contrast to limestone dissolution is fast. (See equation 1 in Dissolution: Evaporite Rocks.) The time T to obtain 90% of saturation is short, in the order of 30 s. Therefore the solution flowing on a bare rock is close to saturation and the rates can be estimated by mass balance. At an annual precipitation W, in mm a^{-1}, $10^{-4}W$ litres of saturated solution with $c_{eq} = 2.5$ g l^{-1} flow from 1 cm^2 of bedrock. This yields a denudation rate of 1.1×10^{-3} W mm a^{-1}, which is close to values observed in Italy on exposed bedrock.

In summary, knowledge of the dissolution kinetics of limestone and gypsum is necessary to understand observations in the field.

WOLFGANG DREYBRODT

Works Cited

Baumann, J., Buhmann, D., Dreybrodt, W. & Schulz, H.D. 1985. Calcite dissolution kinetics in porous media. *Chemical Geology*, 53: 219–28

Buhmann, D. & Dreybrodt, W. 1985. The kinetics of calcite dissolution and precipitation in geologically relevant situations of karst areas: 1. Open system. *Chemical Geology*, 48: 189–211

Dreybrodt, W. 1988. *Processes in Karst Systems: Physics, Chemistry and Geology*, Berlin and New York: Springer

James, A.N. 1992. *Soluble Materials in Civil Engineering*, Chichester and New York: Ellis Horwood

Klimchouk, A., Cucchi, F., Calaforra, J.M., Aksem, S., Finocchiaro, F. & Forti, P. 1996. Dissolution of gypsum from field observation. *International Journal of Speleology*, 25: 37–48

Lauritzen, S.E. 1986. Hydraulics and dissolution of a phreatic conduit. *Proceedings of the 9th International Congress on Speleology*, vol. 1, Barcelona

Navas, A. 1990. The effect of hydrochemical factors on the dissolution rate of gypsiferous rocks in rocks in flowing water. *Earth Surface Processes and Landforms*, 15: 709–15

Zambo, L. & Ford, D. 1997. Limestone dissolution processes in Beke Doline, Aggtelek National Park, Hungary. *Earth Surface Processes and Landforms*, 22: 531–43

EUROPE, ALPINE

The Alpine regions of Europe are a complex system of high mountains that formed since Cretaceous times along convergent plate boundaries between the African and Central European plates. The main arc of the Alps stretches from the Mediterranean coast near the French–Italian frontier to the edge of the Vienna Basin and to the edge of the classical karst of Slovenia (Figure 1). There are also significant elements of the same tectonic structural unit, including the Cantabrians and the Pyrenees (and other chains within Spain), the Apennines, and the chain from the Dolomites to the Julian Alps (Figure 1). In eastern Europe, the Carpathians and Dinarides are also a product of the same Alpine orogenesis. The cores of all these mountain chains are schists and gneisses, but their rocks originated from Tethyan sediments deposited between the converging plates. Throughout the Mesozoic these included substantial areas and thicknesses of carbonate shelf oozes, which now form zones of Triassic, Jurassic, and Cretaceous limestones distributed along all the Alpine chains. Most are now strong limestones ideal for karst development, though some are dolomitized and others have been metamorphosed to marble. Evaporites are relatively rare.

There are many different types of alpine karst landscapes, and their variability is greater than in other regions of the world of similar extent. The prime causes of this variability, best seen in the main arc of the Alps, are the different ages of the carbonate rocks (mainly Jurassic and Cretaceous in the west, and Triassic in the east), the conditions of their sedimentation, and their different structures. Secondary causes are the different intensities and ages of tectonic deformation during the Alpine orogenesis

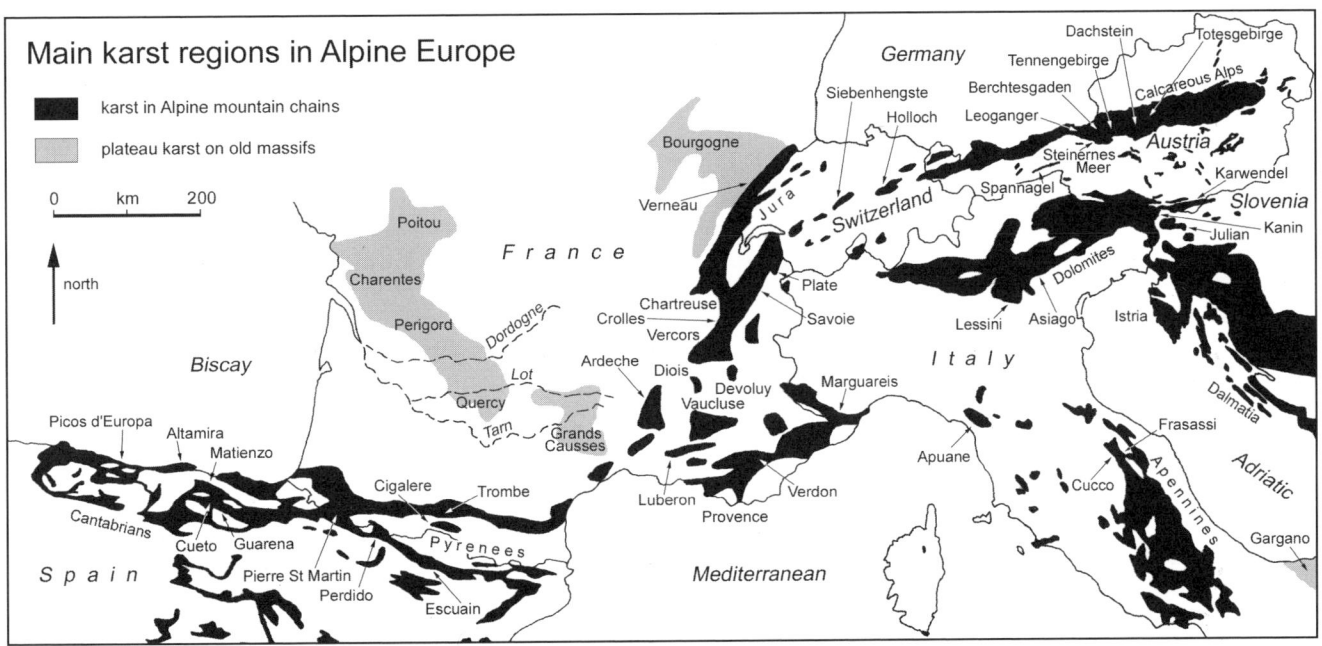

Europe, Alpine: Figure 1. Area location map.

(in general earlier in the Eastern Alps, and later in the Western Alps). A third important influence on the evolution of karst and caves in the Alps has been the very different climatic conditions in the past as well as today. The Atlantic climate in the western ranges, the Mediterranean climate on the southern border, the central European transition climate in the northeastern Alps, and the Illyrian climate in the Julian Alps are all different in the amount, intensity, and annual distribution of their precipitation. Temperatures decrease with altitude, so that the high alpine zones have nival and partially glacial karst. During the cold stages of the Ice Age, most karst terrains were covered by glaciers, so that many alpine regions now have polygenetic surfaces. Karst forms are not always dominant on the surface, where they were modified or destroyed by glacial erosion or are now covered by glacial deposits, but underground karstification is still intense.

Most of the slopes and plateaux up to altitudes of about 1700 m are green karst, covered by forests. Higher parts of the limestone massifs alternate between alpine meadows and zones of bare rock (naked karst). Only the very highest basins in the alpine limestone regions retain relatively small modern glaciers, notably on the Diablerets in Switzerland and on Hochkönig and Dachstein in Austria. Within the high alpine karst there are very few true poljes but the relief shows many traces of dry valleys inherited from Tertiary topographies that predate the main karstification.

River sinks are very rare in the Alps. The few great karst springs are fed only by rainfall and snowmelt, which enter the limestone outcrops directly through immense numbers of joints, fissures, and dolines. The catchments of the springs are not always easily defined, but their boundaries are typically independent of surface relief and topographic divides. Most of the karst springs are vauclusian risings, and the karstic cavities of the deep phreatic zone contain very large resources of water. In the French Alps, the best-known example is the Fontaine de Vaucluse. In the Eastern Alps, karst springs are very important for the supplies of drinking water—half the population of Austria (including the inhabitants of Vienna) utilize karst waters.

The importance of rock structure in the formation of specific karst forms is shown by comparison of the best examples of karren fields in the Alps. Within the Swiss Alps, karren are very common on the outcrops of the Cretaceous Schrattenkalk, notably on Schrattenfluh mountain and in the Glattalp-Märenberg region. In the Venetian Alps of northern Italy, the Jurassic Rosso Ammonitico forms the karst on Alpi Lessini and Monte Grappa. In the northern Calcareous Alps of Austria, the Upper Triassic Dachsteinkalk is carved into the alpine karst of the Steinernes Meer, Dachstein, and Totes Gebirge massifs. The characteristic forms of karren are different on these different limestones, and karren are very rare on other limestones within the alpine stratigraphic sequence.

The Dachsteinkalk, a strong limestone with a thickness of >1000 m, contains the longest and deepest known caves in Austria and also forms much of the Julian Alps. The main karst massifs form a chain along the northern Calcareous Alps (see separate entry), all with spectacular alpine karst, extensive karren field, and deep caves. Lamprechtsofen is currently the second deepest cave in the world (1632 m), and is unusual in that it was largely explored upwards from the resurgence. A concentration of karst caves has been explored in the northern part of the Dachstein massif, notably Hirlatzhöhle (87 km long, 1041 m deep) and Dachstein Mammuthöhle (58 km long, 1199 m deep). South of the main Alps, the karst of Styria has many caves including Lurhöhle (with 6 km of decorated galleries) and Drachenhöhle (with very large passages and well known for its rich finds of the fossil cave bear), all developed in Devonian limestones.

The Swiss Alps have a belt of Cretaceous limestones forming high escarpments with spectacular alpine karst from Leysin to Alpstein along the northern side of the Bernese Oberland. Within this belt, Hölloch and Siebenhengste (see separate entries) are the two major cave systems in the Schrattenkalk. Of the Jura mountain chains, only the zone straddling the border of Switzerland and France was strongly folded in the Alpine orogenesis. In these parts of the Swiss and French Jura, mostly forested limestone escarpments are distinguished by some very large karst resurgences. Some of the caves in the Jura are long but not deep, including the fine through-system of the Reseau Verneau with 32 km of passages in the French department of Doubs.

Karst massifs extend along the French Alps and show progressive changes in their geomorphology from the high alpine karsts near Mont Blanc to the softer fluviokarsts of Provence. The finest karst landscapes and the deepest caves are in the massive Urgonian facies of the Cretaceous limestones. North of Mont Blanc, the Désert de Platé is a bare karren field that is under snow for much of the year; its dipping limestone contains the world's first and fourth deepest caves (Mirolda, −1733 m; Jean Bernard, −1602 m), both distinguished by long narrow meandering canyons and numerous small shafts. Notable among the karsts around Grenoble, the three massifs of the Chartreuse each contain their own long and deep integrated network of multi-level caves—l'Alpe (60 km long), Granier (55 km long), and Dent de Crolles (50 km long, see separate entry). South of these lies the spectacular karst of the Vercors (see separate entry). Further south, the Vaucluse Plateau is a fluviokarst upland with a surface of broken limestone debris inherited from periglacial frost shattering and largely surviving in the modern dry climate. Caves including the deep shaft systems of Caladaire (−668 m) and Jean Nouveau (−579 m) drain to the famous Fontaine de Vaucluse where the water rises 308 m up the flooded shaft. The Grand Canon du Verdon is entrenched more than 800 m between plateaux of limestone distinguished by dolines, grassland, and few known caves.

On the French–Italian border, the Marguareis is a spectacular plateau of degraded glaciokarst with deep cave systems including the Piaggia Bella, 32 km long and 950 m deep with its low-level "collecteur" (main drain) partly entrenched into the green schists that underlie the limestone. Within the Southern Alps of Italy, the main carbonates occur east of Lake Garda. The Lessini and Asiago plateaus are pitted with numerous dissolution and collapse dolines in the finest of the karst landscapes in the Venetian Alps (see Asiago Plateau, Italy). Spluga della Preta (−985 m) on Lessini, and Abisso di Malga Fossetta (−974 m) on Asiago, are the deepest known shaft systems, and Buso de la Rana is a resurgence cave 24 km long at the foot of the Asiago. Further east, the Dolomites are distinguished by high white cliffs and spectacular glaciated landscapes, but known caves are few in the less karstified dolomite rock. In contrast, the Julian Alps cross the Italian–Slovenian border, where many vertical caves more than 1000 m deep have been explored in Monte Kanin (see separate entry).

The Apennines extend the length of Italy with isolated but significant areas of mountain karst. A long belt of gypsum karst lies along the Po Valley flank, and contains the Spipola-Aquafredda with 10.4 km of passages near Bologna. The Alpi Apuane have an alpine karst with extensive bare outcrops of variably metamorphosed limestones that include the famous Carrara marbles. One of the main groups of active quarries cuts into the side of the mountain drained by the complex cave system of Fighiera-Corchia, 52 km long and 1210 m deep. The mountains hold many other deep shaft caves, with five already explored to more than 700 m deep. Perugia has the high alpine karst of Monte Cucco, with the cave of the same name (31 km long and 945 m deep), and the caves of Fiume-Vento caves (23 km long) and Frasassi (see separate entry) at lower elevations towards the Adriatic coast.

The Pyrenees are a major westerly extension of Europe's alpine fold mountains. They have many areas of spectacular alpine karst with vast karren fields and numerous deep caves (Figure 2). Among the finest are in the karst around Pierre St Martin (see separate entry) with its many deep caves passing under the France–Spain frontier, and also the huge cave system of Felix Trombe / Henne Motre (with 101 km of mapped passages reaching a depth of 975 m). Much smaller but equally distinctive are the galleries of the Grotte de Cigalère, some of which are uniquely well decorated with calcite and gypsum, and the Grotte Moulis with its underground laboratories and its inner chambers decorated with aragonite. On the Spanish side of the Pyrenees, the finest alpine karst covers Monte Perdido, now in the Ordesa national park, due north of Huesca. The ice-filled galleries of Grotte Casteret lie just below the Brèche de Roland (with the Cirque de Gavarnie on the French side). The finest caves lie further east in the Escuain massif, where the Sistema Badalona can be followed through from sink to resurgence over a depth of 1149 m.

Alpine fold mountains continue into Spain as the Cantabrians. The Picos d'Europa (see separate entry) has the highest and most spectacular karst with an amazing number of deep cave systems, but there are other significant karst sites. The Matienzo depression extends over 26 km^2, largely drained into a ponor that lies 220 m below the lowest col through the perimeter mountains; the karst is underlain by numerous caves, of which >200 km have been mapped to date, with three systems each >25 km long. South of Matienzo, another karst plateau is broken by the 302 m deep Cueto shaft which continues down into large abandoned galleries and an active river passage through to the Coventosa resurgence, making a cave system 32 km long and 815 m deep. Further east, Ojo Guarena is the longest cave in Spain with a rambling maze of 99 km of mapped passages beneath a low escarpment.

Away from the main alpine fold mountains, the high Sierras of southern Spain are formed on huge tracts of limestone with karst landforms that are generally poorly developed in the hot and semi-arid climate. In the high Sierrania de Ronda, behind Malaga, the shafts and meanders of Sima GESM drop to a depth of 1098 m, while, along the coast to the east, the show cave of Nerja is noted for its very large stalactites.

Recent geomorphological and speleogenetic studies seem to confirm that the oldest generation of karst landforms was developed in the late Miocene. These include caves that survive only as tunnel fragments or as roofless caves. The Messinian regression rejuvenated the Western Alps with a lowering of the base level of karstification. In the Eastern Alps, it appears that a paleosurface developed by the early Oligocene has been preserved with minimal modification on some of the elevated karst plateaux. There are many detailed studies of the karst and caves in all parts of the Alps, but much remains to be explored and studied, and valuable results are coming from comparisons of the geomorphologies in the different areas of the region.

HUBERT TRIMMEL AND TONY WALTHAM

See also **Alpine Karst; Calcareous Alps, Austria**

Further Reading

Audra, P. 1994. *Karsts alpins. Genèse des grands réseaux souterrains. Exemples: le Tennengebirge (Autriche), l'Ile de Crémieu, la Chartreuse et le Vercors (France), Thesis,* Paris: Fédération française de spéléologie and Bordeaux: Association française de karstologie (Karstologia Mémoires, 5)

Avias, J. 1972. Karst of France. In *Karst: Important Karst Regions of the Northern Hemisphere,* edited by M. Herak & V.T. Stringfield, Amsterdam and New York: Elsevier

Courbon, P., Chabert, C., Bosted, P. & Lindsley, K. 1989. *Atlas of the Great Caves of the World,* St Louis, Missouri: Cave Books

Forti, P. & Sauro, U. 1996. The gypsum karst of Italy. *International Journal of Speleology,* 25(1–2): 239–50

Frisch, W., Kuhlemann, J., Dunkl, I. & Székely, B. 2001. The Dachstein paleosurface and the Augenstein Formation in the Northern Calcareous Alps: A mosaic stone in the geomorphological evolution of the Eastern Alps. *International Journal of Earth Sciences,* 90: 500–18

Kusch, H. & Kusch, I. 1998. *Höhlen der Steiermark: phantastische Welten,* Graz: Steirische Verlagsgesellschaft

Pavuza, R. (editor) 1993. *Akten des Symposiums über die Karstgebiete*

Europe, Alpine: Figure 2. Classic alpine glaciokarst in the high Pyrenees, with bare limestone lapiaz fretted with deep karren and broken by small dolines with soil floors. (Photo by Tony Waltham)

der Alpen: Gegenwart und Zukunft, Bad Aussee 1991, Vienna: Verband Österreichischer Höhlenforscher

Puch, C. 1998. *Grandes cuevas y simas de España*, Barcelona: Espeleo Club de Gràcia

Salzburger Höhlenbuch, 1975–96. 6 vols. Salzburg: Landesverein für Höhlenkunde in Salzburg

Wildberger A. & Preiswerk C. 1997. *Karst and Caves of Switzerland*, Basel: Speleo Projects

EUROPE, BALKANS AND CARPATHIANS

A long temperate karst belt extends through Romania, Moldova, Bulgaria, and former Yugoslavia (Figure 1). However, in the southernmost parts of Serbia and Bulgaria the climate is transitional between Mediterranean type and fully temperate continental type. Despite the complex geological background and the relatively low proportion of karst outcrops, more than 20 000 caves have been explored and mapped within this region. This figure may be due to a long tradition of speleological exploration and research. Notable scientists from the above-mentioned countries include Jovan Cvijić (see Geoscientists) and Emil Racoviţă (see Biospeleologists). In Bulgaria, where karst is widespread, the study of caves began early in 1878 when the Shkorpil brothers (Karel and Herman) wrote *Krazhki yavlenia*—the first book describing some of the Bulgarian caves. In Romania, by the end of the 1990s, a network of caving clubs had explored more than 12 000 caves. This shows that there may be great speleological potential in regions where there is only a relatively small area of exposed karst rocks.

The type of cave encountered in the Balkan region is related to the overall geological history of the area. Apart from the gypsum-karst area of Moldova, whose geological evolution is connected with a transition zone between the East European Platform and the Carpathian foredeep, the karst was strongly influenced by Tertiary uplift of the Carpathians and Balkans. This uplift has broken the unity of most stratigraphic units, thus breaking up the limestone outcrops between the high ridges and the slopes. In only a few cases was this continuity preserved, favouring deep drainage as within the Piatra Craiului Mountains of Romania, where Avenul din Grind (−540 m) is the deepest pothole in the region. For most mountain ranges in the area, an igneous or metamorphic core is surrounded by limestone. This setting favours the formation of relatively long caves with

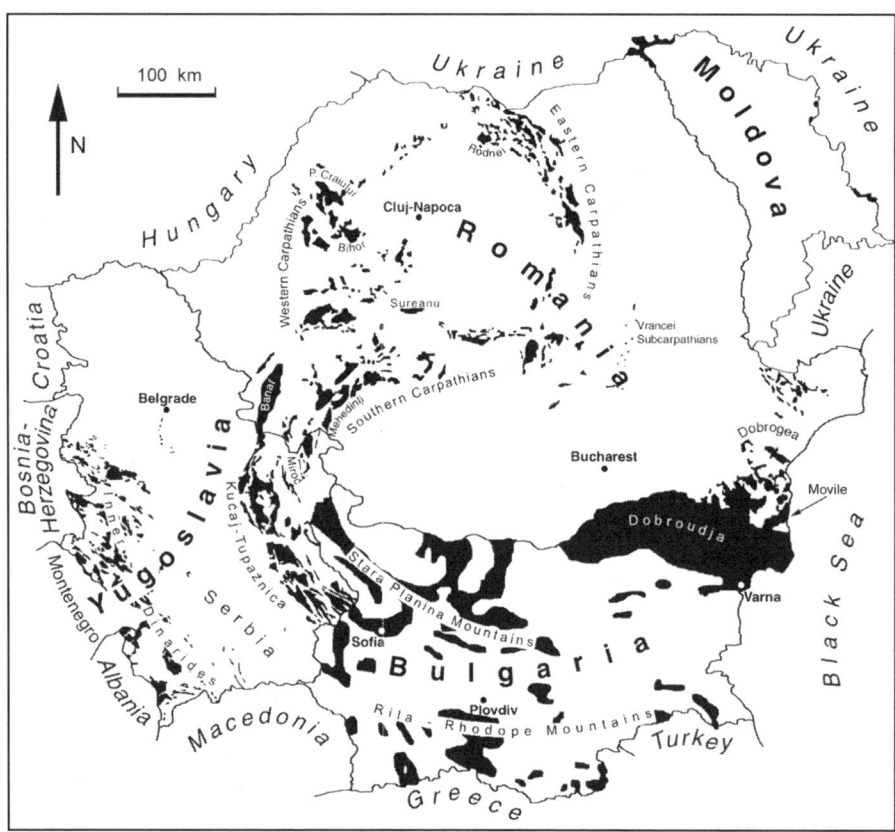

Europe, Balkans and Carpathians: Figure 1. Map of the major karst areas in southeastern Europe.

Europe, Balkans and Carpathians: Figure 2. Phosphate deposits in Ciclovina Cave, Romania. The upper level of the deposit is covered by flowstone. (Photo by Art Palmer)

Europe, Balkans and Carpathians The longest and deepest caves in southeastern Europe.

Length (m)			Depth (m)		
Romania					
1	Peștera Vântului	48 000	1	Avenul din Grind	540
2	Sistemul Humpleu–Poienița	39 000	2	Peștera Tăușoare	479 (−373; +106)
3	Peștera Hodobana	22 142	3	Avenul din Stanu Foncii	339
Serbia					
1	Velika Klisura	8500	1	Velika Klisura	310 (+296; −14)
2	Ušački Pećinski Sistem	6185	2	Rakin Ponor	285
3	Cerjanska Pećina	6025	3	Jama na Dubašnici	276
Bulgaria					
1	Douhlata	18 000	1	Raytchova Doupka	377
2	Orlova Chouka	13 437	2	Barkite 14	350
3	Jagodinskata	8501	3	Yamata	350
Moldova					
1	Zoloushka	92 000	1	Zoloushka	30

active passageways, often tiered and sometimes including mazes of relict passages richly decorated with speleothems.

Romania

About 2% of the exposed land surface of Romania consists of karst (Onac & Cocean, 1996), but this figure is misleading, as in many cases Neogene or Quaternary deposits overlie older formations, which include carbonate and evaporate karst-forming rocks. The most important karst rocks are Mesozoic limestones, but caves are also recorded in dolomites, halite, conglomerates, and sandstones. In Romania, karst is mainly developed in the Southern and Western Carpathians, with the largest limestone area being located in the Banat Mountains (adjacent to the border with the former Yugoslavia).

The karst plateaux contain karren, dolines, uvalas, and poljes, whose development was favoured by the great thickness of carbonate deposits (600 to 2000 m). These plateaus may be found at 800 to 1200 m altitude in the Pădurea Craiului, Mehedinți, and Bihor Mountains, and also at 1800 to 2200 m in the Retezat Mountains, which show typical alpine karst (Onac & Cocean, 1996). Bar Karst is mostly located on the slopes of the Carpathians, and consists of longitudal limestone bars (stripes), isolated during the uplift of the mountain range. Surface rivers, formed on the crystalline basement, cut through the limestone ridges in spectacular gorges or, in many cases, through long caves. The most typical karst systems of this type appear in the Piatra Craiului, Vîlcan, Mehedinți, and Șureanu Mountains of the Southern Carpathians.

The most important karst of Romania is located in the Western Carpathians, mainly in the Pădurea Craiului and Bihor Mountains. Bihor Mountains includes a high plateau with many 200–300 m deep potholes. Some of the potholes contain river passages that are several kilometres long; Avenul V5 is 225 m deep and 10 km long. The second longest cave of Romania is the Humpleu–Poienița system (39 km, −308 m), which includes a 5 km long underground stream with more than 30 sumps. The main features of this cave are the huge rooms (the Giants Room is 573 × 111 × 35 m), as well as the richness and the variety of the speleothems. Among these are the remarkable calcite scalenohedron crystals, 30–80 cm long, which may weigh over 50 kg. Oxygen-isotope ratios reveal that they are of hydrothermal origin. Valea Rea Cave (19 km long), in the western Bihor Mountains is considered to be one of the world's richest caves with respect to its mineralogy. The 35 minerals described from this cave include barite, celestine, quartz, dickite, metatyuyamunite, gypsum, and malachite, with most of these supporting a hydrothermal genesis. In addition, the Bihor Mountains also host five caves with perennial ice deposits. The most famous is Ghețarul de la Scărișoara, which contains an ice mass 75 000 cubic metres in volume. Pollen extracted from the lower part of the ice block gave an estimated age of 3500 years (Racovita & Onac, 2000).

The Pădurea Craiului Mountains have a karst terrain at altitudes of 400–800 m, where sinking streams follow long, branching, gently sloping caves, including Ciur Ponor (−203 m, 17 km) and Ponoraș (−211 m, 6 km). Peștera Vântului (Wind Cave), Romania's longest cave, is a resurgence with over 48 km of passages. It is basically a single-stream passage paralleled by three relict levels displaying outstanding horizontal and vertical meanders, and is also notable for its mineralogy (Onac, 1996), and geomicrobiology (Manolache & Onac, 2000).

In the Eastern Carpathians, the only significant karst area is the Rodnei Mountains—a network of alpine ridges. Here several Eocene limestone bodies contain influent caves, including Tăușoare Cave (470 m deep and 16 km long), famous for its strange limestone balls, gypsum speleothems, and rare mirabilite, leonite, konyaite, and syngenite minerals (Onac et al., 2001). Halite and gypsum occur in Miocene salt domes, diapirs, and massive salt beds protected by cap rocks. In the Vrancei Subcarpathians, the Peștera 6S de la Mânzălești (3234 m long) is the second-longest salt cave in the world. In the southeast of Dobrogea, Movile Cave (see separate entry) is a unique chemoautotrophically based cave ecosystem in which 34 invertebrate species are endemic.

In the Șureanu Mountains, a high karst plateau (at an altitude of 800 to 1200 m) is divided into two by the deep Strei Gorge. The main underground system is the 10 km long Șura Mare resurgence cave with a vertical range of 405 m. In the same area, Cioclovina Cave has been known since the late 19th century, when scientists visited it to search for cave bear bones and to investigate the 15–20-m thick phosphate deposit (Figure 2). Between 1912 and 1941 over 30 000 m^3 of guano phosphate were mined out and used as fertilizer, and thousands of cave bear skeletons from within the phosphate were ground up for the same purpose. In the upper layer of the phosphate deposit, Hallstadt ceramics, as well as skulls of *Homo sapiens fossilis* were found, and the cave is also famous for its many rare phosphate minerals, including crandallite, paratacamite, taranakite, collinsite, hydroxylellestadite, and berlinite.

On the Mehedinți Plateau in the southwestern Carpathians, the major cave system is Topolnița (20.5 km, −127 m), a three-level streamway that underlies a vast network of older passages. Spectacular karst landforms are concentrated at Ponoarele, including dissected limestone pavements, a natural bridge carrying a motor road, and the great lake of Zăton which is drained by the 5 km long Bulba Cave. In the same area, the richly decorated Cloşani Cave is used by the Institute of Speleology as an underground laboratory.

Yugoslavia (Serbia and Kosovo)

In Serbia, carbonate rocks crop out over a total area of 7803 km^2 in the Carpatho-Balkanides (eastern Serbia), and in the Inner Dinarides in the west (Figure 1). The most extensive karst, in Eastern Serbia, has swift-flowing springs (flow rates of 0.5–2 m^2 s^{-1}), which drain massifs of Mesozoic limestone. The greatest concentration of long and deep caves occurs within the Miroć carbonate platform—an impressive karst plateau cut by deep dolines, on the right bank of the Danube Gorge. Six small parallel streams, originating on the impervious basement, sink into the limestone, forming essentially vertical caves with some large active passages. Among them, Rakin Ponor is the second deepest cave in the country (−285 m) and Buronov Ponor (2.4 km, −187 m) is a beautiful cave, richly decorated, that intercepts the principal river with an average flow of $c.$ 1 m^3 s^{-1} (Zlokolica *et al.*, 1996). The resurgence lies 5 m below the present level of the Danube in the artificial lake of Iron Gates Dam.

The Kučaj–Tupažnica carbonate platform is the continuation of the Banat karst in Romania, and contains most of Serbia's caves. In the Kučaj Mountains, the Dubašnica pothole (−276 m) consists of large shafts, including a single drop of 120 m. In the same area, Bogovinska Pećina (5842 m long) is a complex resurgence cave developed on three levels. The vauclusian karst spring, Vrelo Krupac, at the foot of Mount Svrlijiške Planine, discharges 150 l s^{-1} and has been explored by diving down to −83 m. On Mount Kalafat, the third-longest Serbian cave, Cerjanska Pećina (6025 m), is an active swallet with extensive allogenic sediments.

In the Inner Dinarides, the limestones are Triassic in age, and important karst areas include Lelić and the Gilijeva mountain. The karst of Pešter plateau contains the Ušački Pećinski system (6185 m), which includes an active passage from sink to resurgence.

The western part of Kosovo seems to be the area with the greatest speleological potential—not yet explored due to political instability in the area. Velika Klisura (8500 m long, 310 m deep), is the most important major cave, situated close to the border with Montenegro.

Bulgaria

Bulgaria has outcrops of carbonate rocks across 24% of its territory, hosting over 5000 caves. Mesozoic carbonate rocks form the main karst in the Stara Planina Mountains. There is a marked distinction between the karst and caves of the Stara Planina, in the central part of Bulgaria, and those of the Rilo-Rhodopes Mountains in the south. The most important caves in Bulgaria, including the longest (Douhlata Cave which is 18 km long) and the deepest (Raitchova Doupka, −377 m) are located in the western and central part of Stara Planina. The Shopov cave system (in the western part of Stara Planina) is considered to be one of the world's most mineralogically rich caves, with 41 mineral species, 18 of these being described only from that locality (Shopov, 1988). The upper part of the cave has a hydrothermal origin, while the lower levels are formed by the dissolution of sulfuric acid produced by oxidation of iron sulfides in the upper part of the cave.

The large unbroken covered karst of the Dobroudja is characterized by a few short caves, dry valleys, and the presence of a sulfidic water aquifer at a depth of 400 m. Along the Danube Plain, Orlova Chuka Cave (near Ruse) in Lower Cretaceous limestone is 13 437 m long, and Magura Cave (2500 m long, near Vidin) is renowned for its wonderful Neolithic and bronze age guano paintings.

In the western part of the Rhodopes Mountains and the northern part of the Pirin Mountains, karst develops on the Proterozoic marbles. Of the 50 known caves in the area, more than 10 are of hydrothermal origin. Historically, the Romans mined these for lead ore. The marble sequence in the Pirin Mountains is over 2000 m thick, with the potential for hosting some of the world's deepest caves. The Pirin National Park was founded in 1963 and added to UNESCO's World Heritage List in 1983. It hosts diverse and unique landscapes, including some 70 glacial lakes, caves, waterfalls, and pine forests. It is also the home to hundreds of endemic and rare species, many of which are representative of the Balkan Pleistocene flora. Unfortunately cave ice seals all of the high-altitude caves. In the eastern part of Rhodopes Mountains, more than 250 caves are known. The largest karst area in the Rilo-Rhodopes is south of Plovdiv, where the Trigrad Plateau (at an altitude of 1500 m) is an old corrosion surface that is now breached by hundreds of dolines and cut by several canyons 300 to 400 m deep (Nicod, 1982). The largest cave in the region is Jagodinskata (8500 m long).

Paleokarst is widely documented in Bulgaria, mainly from boreholes and mining activities. Mineral deposits connected with paleokarst include metasomatic lead–zinc ores, bauxites, gypsum, kaolin, and iron ore.

Moldova

Karst research in Moldova has received little attention, because the rocks that are susceptible to karstification crop out only in small areas in the northwest of the country and along the Nistru (Dniestr) Valley in the east (Figure 1). The karst features develop on Miocene gypsum (10–40 m thick) (Andrejchuk & Klimchouk, 1996) and the development of caves began under deep-seated conditions during the Pliocene. Following the Prut Valley entrenchment into the gypsum during the Late Pleistocene, karstification processes were stimulated. Apart from the Zoloushka Cave (known as "Emil Racoviță" Cave by Moldova's speleologists), which is 92 km in length (and similar to the Ukraine gypsum caves), all of the other caves are small.

BOGDAN P. ONAC and SILVIU CONSTANTIN

See also **Dinaric Karst; Europe, Mediterranean**

Works Cited

Andrejchuk, V. & Klimchouk, A. 1996. Gypsum karst of the Eastern-European Plain. *International Journal of Speleology*, 25(3–4): 251–61

Manolache, E. & Onac, B.P. 2000. Geomicrobiology of black sediments in Vântului Cave (Romania): Preliminary results. *Cave and Karst Science*, 27(3): 109–12

Nicod, J. 1982. Canyons et grottes du plateau de Trigrad (Rhodope central, Bulgarie). *Revue de Géographie Alpine*, 70(3): 227–31

Onac, B.P. 1996. Mineralogy of speleothems from caves in Pădurea Craiului Mountains (Romania), and their palaeoclimatic significance. *Cave Science*, 24(3): 109–24

Onac, B.P. & Cocean, P. 1996. Une vue global sur le karst roumain. *Kras i Speleologia*, 8(17): 105–12

Onac, B.P., White, W.B. & Viehmann, I. 2001. Leonite [$K_2Mg(SO_4)_24H_2O$], konyaite [$Na_2Mg(SO_4)_25H_2O$] and syngenite [$K_2Ca(SO_4)_2H_2O$] from Tăuşoare Cave, Rodnei Mts, Romania. *Mineralogical Magazine*, 65(1): 103–09

Racovita, G. & Onac, B.P. 2000. *Scărişoara Glacier Cave Monographic Study*, Cluj-Napoca: Editura Carpatica

Shopov, Y.Y. 1988. Bulgarian cave minerals. *National Speleological Society Bulletin*, 50: 21–24

Zlokolica, M., Mandić, M. & Ljubojević, V. 1996. Some significant caves at the western rim of the Miroć karst (Yugoslavia). *Theoretical and Applied Karstology*, 9: 77–88

Further Reading

Bleahu, M.D. 1972. Karst of Rumania. In *Important Karst Regions of the Northern Hemisphere*, edited by M. Herak & V.T. Stringfield, Amsterdam and New York: Elsevier

Constantin, S. & Mitrofan, H. 1994. Romania. *The International Caver*, 12: 26–30

Djurović, P. (editor) 1999. *Speleological Atlas of Serbia*, Belgrade: Jovan Cvijić Geographical Institute of the Serbian Academy of Sciences (in Serbian and English)

Gavrilović, D. 1976. The karst of Serbia. *Memoires of the Serbian Geographical Society*, 13: 1–28

Goran, C. 1983. Les types de relief karstique de Roumanie. *Travaux de l'Institute de Spéologie "Emile Racovitza"*, 12: 91–102

Jalov, A. 1993. La spéléologie en Bulgarie. *Spelunca*, 52: 23–26

Stevanović, Z., Dragisić, V., Dokmanović, P., & Mandić, M. 1996. Hydrogeology of Miroć karst massif, eastern Serbia, Yugoslavia. *Theoretical and Applied Karstology*, 9: 89–96

EUROPE, CENTRAL

Central Europe covers Austria outside the Alps, Czechia, Hungary, Poland, and Slovakia. The region is a complex of karst areas covering the old Middle European Platform as well as the younger Western Carpathians and Carpathian Basin. Karst and cave areas are irregularly distributed (see Figure). The largest carbonate karst areas outcropping on the surface occur in southern Poland and in Slovakia (8000 km² and 2700 km², respectively). Although 15% of Austria is underlain by karstified rocks, only a tiny part of this karst lies outside the Alps. Carbonate rocks crop out over some 1.5% (1350 km²) of Hungary. In Czechia, limestone and dolomite outcrops support karsts of minor importance (less than 0.4% of territory, about 300 km²), but have been very well studied. The Krakow-Wielun Upland (Poland) with an area of 2500 km² and more than 1700 caves is the largest individual karst region in Central Europe.

The geological settings of karst in Central Europe range from Precambrian (marbles of Sudety Mountains) to late Quaternary (travertine of Bükk Mountains and Lower Tatra Mountains). The region is divided into two different tectonic units. Karst areas of Czechia and most of Poland are formed mostly by blocks of folded Paleozoic limestones, dolomites, and marbles belonging to the older Central European Platform. In South-Central Poland they are covered by sub-horizontal Mesozoic carbonate rocks. The southeastern part of the region has the small karst areas of the Western Carpathians and Carpathian Basin, on Alpine folded structures of Mesozoic carbonates.

In the Sudety Mountains (on the border of Poland and Czechia) karst is developed in crystalline limestones, Proterozoic marbles, and Carboniferous limestones. The longest cave in this area is the 2230 m Niedzwiedzia (Bear) Cave. In the Holy Cross Mountains in the north of the region, caves that developed in Devonian limestones are almost entirely filled with clastic sediments of a lower Triassic transgression (e.g. Chelosiowa Jama, a cave more than 3 km long). The surface karst landforms and some of the caves in this area are filled with sediments related to Pleistocene continental glaciations. Only the Holy Cross Mountains and the Silesia-Krakow highlands were within the

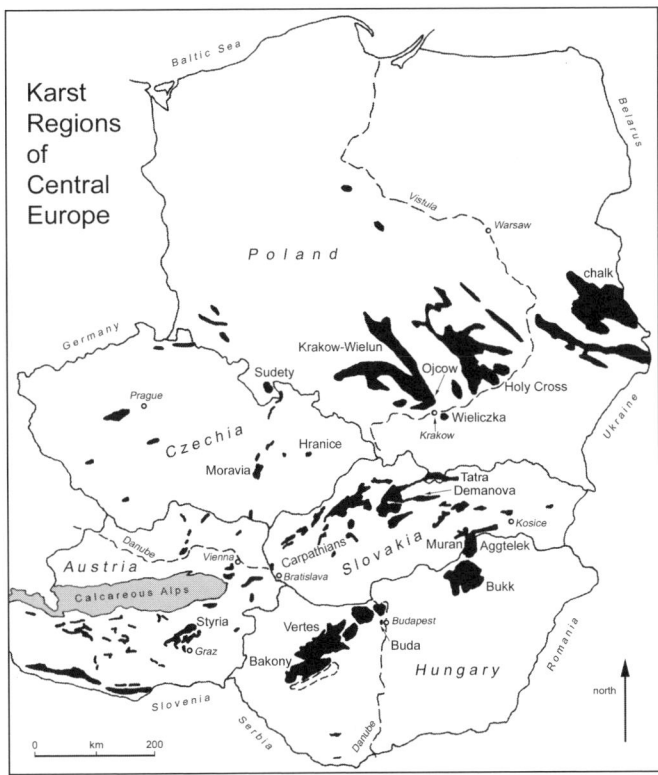

Europe, Central: The major karst regions of Central Europe.

Europe, Central: Table 1. The longest caves in Central Europe.

No.	Cave	Length (m)	Region	Country
1.	Amaterska-Punkva Cave System	32 500	Moravian Karst	Czechia
2.	Demanovsky Cave System	32 300	Lower Tatra	Slovakia
3.	Baradla-Domica Cave System	23 916	Aggtelek-Slovak Karst	Hungary/Slovakia
4.	Wielka Sniezna Cave System	22 000	Tatra	Poland
5.	Stratenska-Psie diery Cave System	21 800	Slovak Paradise	Slovakia

range of the Scandinavian ice sheet, and caves were developed by proglacial waters at the carbonate base; these include the Szachownica Cave in the Krakow area. Mesozoic rocks of the Silesia-Krakow Upland have strongly developed paleokarst phenomena, which are the result of hydrothermal processes in the middle Triassic carbonate rocks (including zinc and lead ore deposits). They also have relict well-developed karst in upper Jurassic limestones with large fossil poljes and fragments of large hydrothermal caves.

Czechia has karst areas developed in small lenses of Paleozoic carbonate rocks surrounded by non-karst rocks. The Moravian Karst, on Devonian limestones in contact with the Brno igneous massif and clastic culm rocks, has the largest Central European fluviokarst system—Amaterska-Punkva Cave System, which includes show caves. The Bohemian Karst, on Carboniferous limestones near Prague has strongly developed upper Paleozoic paleokarst, revealed during exploitation of limestones in quarries.

Different conditions of karst are found in the Western Carpathians and the Carpathian Basin, where folded Mesozoic carbonates form high-mountain and plateau karst. High mountain (alpine) karst occurs in a small area of the Tatra Mountains in the Czerwone Wierchy massif, at an elevation of more than 2000 m, where the deepest cave systems in the region can be found. An important feature of the high-mountain karst of the Czerwone Wierchy range was the proximity of the Scandinavian ice sheet in the Pleistocene and severe climatic conditions associated with the glaciation of the Tatras. The Wielka Sniezna System in the Tatra is the largest recognized autochthonous karst system in the region (explored cave systems and sections of underground flow proved by tracing). Other areas of high-mountain karst occupy the highest part of Belanske Tatra and the Lower Tatra Mountains, where Demanovsky Cave System is the best explored active fluvial cave system.

The Slovak Karst, the Muran Plateau, and the Slovak Paradise plateau karst lie in the central part of the Inner Carpathians. Parts of the North Hungarian Mountain Range, Aggtelek-Rudabánya, and Bükk Mountains have a similar geological and morphological setting, and contain some of the most interesting caves in Central Europe, including Ochtinska Aragonite Cave, Dobšina Ice Cave, and the Baradla-Domica cave system.

The Transdanubian Mountains contain the karst areas of Bakony, Vértes, Pilis, and Buda Mountains, low mountains and plateaus at altitudes of 250–700 m, built from Mesozoic and Tertiary carbonates. The Triassic Dachstein and Eocene limestones are the most important for karstification. Small patches of limestones were strongly karstified and denuded in the Cretaceous and Paleocene. The paleokarst features of the open-cast bauxite pits of the Bakony and Vértes Mountains are significant remains of this important regional karstification phase. Bauxite and manganese deposits are mined here in one of the most productive areas in Europe (see Bauxite Deposits in Karst).

In the Paleozoic basement of Central Europe small block structures occur, while the zones that separate them are sites of high tectonic activity. The results of this activity are both fossil and present-day hydrothermal phenomena, ore-bearing carbonate rock complexes, travertine deposition, gas emanation phenomena in caves, and specific hydrogeochemical environments. Although there are many significant fossil thermal water sites—several paleokarstic features within Paleozoic and Mesozoic carbonate rocks of the Middle European Platform in Czechia and Poland—some hydrothermal phenomena are still present in the Carpathians. The largest cave systems of thermal water origin are in the Buda Hills of Budapest, Hungary. These cave systems show strong tectonic control and have labyrinth plans. Most were created by the mixing corrosion effect of ascending warm waters and descending cold, meteoric waters. Pál-völgy Cave, with more than 7 km of network passage, is the longest cave in the area. A spectacular example of the influence of tectonic activity on the development of karst processes is represented by the Hranice karst in Czechia. A lens of Devonian limestone lies between the Carpathian Foredeep and the Bohemian Massif. Zbrašovské Aragonite Caves have features indicating development stages from cold phreatic, through hydrothermal, to the present-day vadose stage with emanation of juvenile CO_2. In the caves there are "ponds" of CO_2, with atmospheric concentration of up to 80%. Hranicka Chasm, in the same area, is 274

Europe, Central: Table 2. The deepest caves in Central Europe.

No.	Cave	Depth (m)	Region	Country
1.	Wielka Sniezna Cave System	814 (−807; +7)	Tatra	Poland
2.	Sniezna Studnia	763 (−726; +37)	Tatra	Poland
3.	Bandzioch Kominiarski Cave	562 (−546; +16)	Tatra	Poland
4.	Wysoka-Za Siedmiu Progami Cave	435 (−288; +147)	Tatra	Poland
5.	Stary hrad	432	Lower Tatra	Slovakia

m deep and filled with water that is hot at depth, with high contents of CO_2 and H_2S.

Because of their significant natural values most of the areas of carbonate karst in Central Europe are protected, with the national parks of Aggtelek, Slovak Paradise, Tatra and Lower Tatra Moutains, and Ojców, and regional parks in the Bohemian, Moravian, and Slovak karsts. In 2000, a unique underground conservation area was established within the Salt Mine in Wieliczka, the world's sole example of natural voids with halite crystals within salt-bearing formations. The caverns have volumes of 1000 m^3 (Upper Crystal Cave) and 700 m^3 (Lower Crystal Cave).

ANDRZEJ TYC

See also **Aggtelek and Slovak Karst; Calcareous Alps, Austria; Europe, Alpine**

Further Reading

Daranyi, F. 1972. Karst of Hungary. In *Important Karst Regions of the Northern Hemisphere*, edited by M. Herak & V.T. Stringfield, Amsterdam and New York: Elsevier

Glazek, J., Gradzinski, R. & Pulina, M. 1982. Karst and caves in Poland. In *Kras i Speleologia*, 4(XIII), 9–18, Katowice.

Gradzinski, M. 2001. Caves in Poland. *Polish Caving 1997–2001*, 3–5, Kraków.

Karszt es Barlang, Special Issue 1989. This special issue of the *Bulletin of the Hungarian Speleological Society* contains 21 papers on the caves and karst of Hungary.

EUROPE, CENTRAL: ARCHAEOLOGICAL CAVES

The caves of the Central European limestone formations have attracted archaeologists since the 1850s. One of the earliest and most important finds was made in August 1856 when workers employed by a local quarrying company cleared out a small cave known as the "Kleine Feldhofer Grotte" in the former Neanderthal (German for Neander valley). They found a number of bone remains and a local specialist, J.C. Fuhlrott, was called. When he arrived, the cave had already been emptied and no additional observations regarding find circumstances were possible. What remained was a cavity of 3 m width, 2.4 m height, and 4.5 m length together with the incomplete remains of a human skeleton which today is world-famous as the *Homo sapiens neanderthalensis* type fossil.

The Neanderthal used to be a small, deeply incised gorge west of Düsseldorf. The gorge, cut through a bed of Devonian limestone by the River Düssel, was about 50 m deep and some 800 m long. At least nine caves were known in the gorge, but they were all destroyed during intensive limestone quarrying during the late 19th and early 20th centuries. Recent excavations of the valley bottom have established the approximate location of the Feldhofer Grotte as well as the adjacent Feldhofer Kirche. Middle Paleolithic as well as Upper Paleolithic stone artefacts, faunal remains, and human remains, some of which could even be refitted with the 1856 skeletal fragments, were recovered from the redeposited cave sediments (Schmitz & Thissen, 2000).

The mid-19th century witnessed the onset of systematic archaeological investigation of many Central European cave sites. The primary aims of these initial investigations were partly paleontological: to examine the large quantity of faunal remains of now extinct Ice Age fauna found particularly in caves, and especially to investigate the possibility of the contemporary existence of an Ice Age human. In the Jurassic limestone formation of the French Jura, Swiss Jura, Swabian Alb, and Franconian Alb these aims were readily fulfilled and over the years numerous paleolithic remains have been unearthed from caves within this region.

The abundant caves and rock shelters of the Jurassic limestone formation are generally of karstic origin and they are still exposed to the eroding forces of running and percolating water. This means that walls and roof of the cavities are crumbling, leaving generally rough and irregular surfaces. Accordingly, parietal art is not preserved from the region (except perhaps for a few, albeit much debated, painted rock fragments reported from recent excavations). On the other hand, the caves of the Swabian Alb have yielded some of the oldest examples of mobile art objects as well as musical instruments belonging to the early Upper Paleolithic Aurignacian culture (approximately 30 to 34 000 BP) (Hahn, 1993).

The caves of Vogelherd, Geißenklösterle, and Hohlenstein Stadel have produced a series of delicate figurines carved of mammoth ivory. These include a number of fairly naturalistic animal plastics, a more schematic anthropomorphic engraving and an extraordinary lion-headed human figure (Figure 1). A flute made on a swan bone was also found in the Aurignacian layers of Geißenklösterle (Hahn & Münzel, 1995).

Many of the caves have been settled repeatedly throughout the Paleolithic, but during the extreme climatic conditions of the late Pleniglacial (approximately 20 to 23 000 BP) the area appears to have been almost deserted. Again the Swabian Alb stands out through a remarkable cluster of cave sites: out of a total of five Gravettian settlement sites within this region, four are located in the Ach valley (Figure 2, area A). Three of these sites (Brillenhöhle, Geißenklösterle, and Hohler Fels Schelklingen) are directly connected by refitting of lithic artefacts, while the fourth site (Sirgenstein) displays close typological, technological, and not least raw material affinities with the former three, and thus very probably belongs to the same settlement pattern, i.e. the same course of events (Scheer, 1993). These observations may be interpreted as the result of either true or limited contemporaneity involving the seasonal movements of a few, or perhaps only one, human group. The Gravettian of southwestern Germany is chronologically separated from the preceding Aurignacian as well as the succeeding Magdalenian by some 3000 to 5000 radiocarbon years. Thus, the intimate connection of the four Ach valley cave sites may perhaps indicate that in this area Gravettian settlement was a more or less isolated, event-like phenomenon.

During the Magdalenian (approximately 16 to 12 000 BP), however, the situation changed markedly. As climatic conditions improved a number of hunter-gatherer groups moved in and settled throughout the region. Archaeological evidence suggests

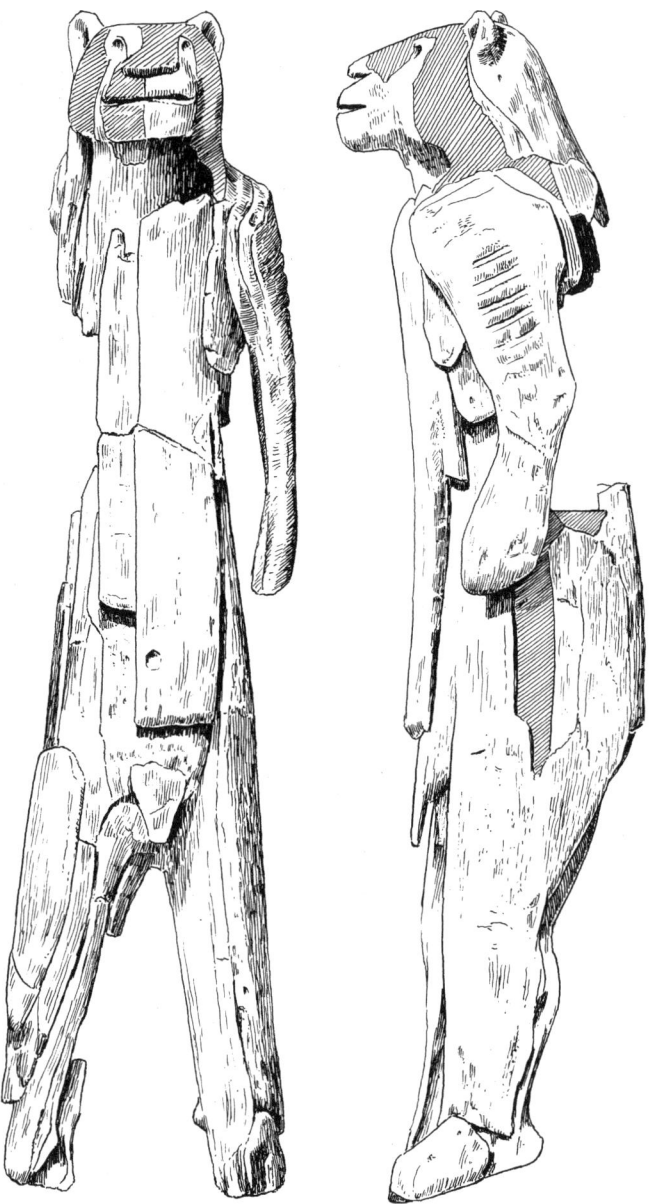

Europe, Central: Archaeological Caves: Figure 1. Human figure with lion-head, mammoth ivory. Found in early Aurignacian levels in Hohlenstein Stadel, southwest Germany, c.30 000–34 000 BP. Reproduced from Schmid (1989) by permission of Landesdenkmalamt Baden-Württemberg. Original in Ulmer Museum.

that they were flexible, mobile, and generally opportunistic hunters, who knew how to successfully exploit the rich and varied resources of their new environment. They were characterized by a complex settlement and mobility pattern which also involved numerous cave and rock shelter dwellings. In fact these sheltered sites by far outnumber the open-air settlement sites from the period in question, and most of the find-bearing caves within the region have indeed yielded Magdalenian artefacts.

However, there is an obvious methodological problem concealed in these observations, as cave sites and open-air sites are

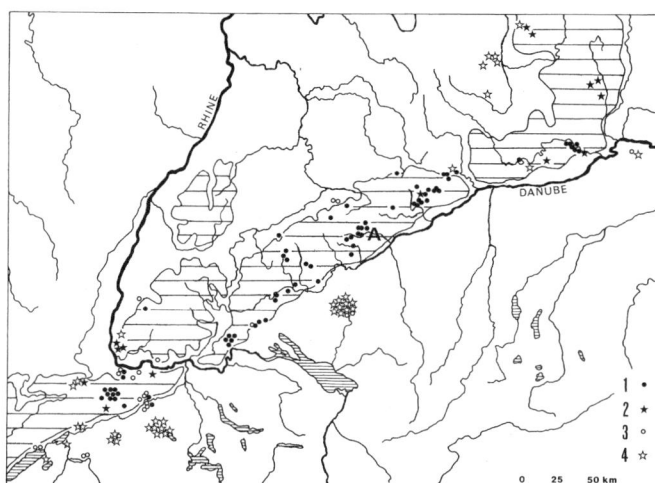

Europe, Central: Archaeological Caves: Figure 2. Late Pleistocene sites in southwest Germany and northwest Switzerland: 1: Magdalenian cave sites, 2: Late Paleolithic cave sites, 3: Magdalenian open-air sites, 4: Late Paleolithic open-air sites. A: the Ach valley. Hatched area: the mountainous region of the Jurassic limestone formation and the Black Forest.

subject to markedly different site formation and preservation processes. Owing to their visibility, caves are more likely to be re-occupied than most open-air sites, and they are thus also more likely to be discovered and investigated by archaeologists. In other words, the pattern of sites known within the region is most likely not in accordance with the prehistoric distribution of sites. The Magdalenian open-air sites are especially likely to be missing or at least under-represented.

Nevertheless, it is evident that caves and rock shelters were a popular habitation form during the early Late Pleistocene, and it is likely that the succeeding Late Paleolithic period witnessed a veritable migration from the sheltered sites in the deeply incised Jurassic valleys to the much more exposed open-air sites of the surrounding lake and river regions (Figure 2) (Eriksen, 1997). During the Late Paleolithic the climate had again improved and shelter was no longer the primary factor in locational decision making. This pattern of use pertained during the following prehistoric periods and although the caves of the region were often sought out for shelter or even defensive purposes, or in recent time for storage (as beer or wine cellars), they never regained the habitational importance of the late Upper Paleolithic.

BERIT VALENTIN ERIKSEN

Works Cited

Eriksen, B.V. 1997. Settlement patterns, cave sites and locational decisions in late Pleistocene Central Europe. In *The Human Use of Caves*, edited by C. Bonsall & C. Tolan-Smith, Oxford: Archaeopress

Hahn, J. 1993. Aurignacian art in Central Europe. In *Before Lascaux: The Complex Record of the Early Upper Paleolithic*, edited by H. Knecht, A. Pike-Tay & R. White, Boca Raton, Florida: CRC Press

Hahn, J. & Münzel, S. 1995. Knochenflöten aus dem Aurignacien des Geißenklösterle bei Blaubeuren, Alb-Donau-Kreis. *Fundberichte aus Baden-Württemberg*, 20: 1–12

Scheer, A. 1993. The organization of lithic resource use during the Gravettian in Germany. In *Before Lascaux: The Complex Record of the Early Upper Paleolithic*, edited by H. Knecht, A. Pike-Tay & R. White, Boca Raton, Florida: CRC Press

Schmid, E. 1989. Die altsteinzeitliche Elfenbeinstatuette aus der Höhle Stadel im Hohlenstein bei Asselfingen, Alb-Donau Kreis. *Fundberichte aus Baden-Württemberg*, 14: 33–118

Schmitz, R.W. & Thissen, J. 2000. *Neandertal: Die Geschichte geht weiter*, Heidelberg: Spektrum

Further Reading

Conard, N.J. & Floss, H. 1999. Ein bemalter Stein vom Hohle Fels bei Schelklingen und die Frage nach paläolithischer Höhlenkunst in Mitteleuropa. *Archäologisches Korrespondenzblatt*, 29: 307–16

Eriksen, B.V. 1991. *Change and Continuity in a Prehistoric Hunter-Gatherer Society: A Study of Cultural Adaptation in Late Glacial-Early Postglacial Southwestern Germany*, Tübingen: Archaeologica Venatoria

Hahn, J. 1986. *Kraft und Aggression: Die Botschaft der Eiszeitkunst im Aurignacien Süddeutschlands*, Tübingen: Archaeologica Venatoria

Hahn, J. 1988. *Das Geißenklösterle-Höhle im Achtal bei Blaubeuren I*, Stuttgart: Theiss (Forschungen und Berichte zur Vor- und Frühgeschichte in Baden-Württemberg, Band 26)

Wehrberger, K. (editor) 1994. *Der Löwenmensch: Tier und Mensch in der Kunst der Eiszeit*, Sigmaringen: Thorbecke & Ulmer Museum

EUROPE, CENTRAL: HISTORY

The caves of Central Europe have been explored since the Middle Ages; for example many well-organized visits to the Dragon's Cave (Drachenhöhle) in Styria (Austria) have been reported since 1387. Most of the explorations of this time were connected with mining activities or mineral prospecting. From the 16th century, there are well-documented cave visits with detailed descriptions. The local people from Amberg explored the Breitenwinnerhöhle (Franconia, Germany) in 1535. The first mention of the "Caverna Lunaris" (later Mondmilchloch) in the Pilatus mountain (Switzerland), in a publication of 1555, is followed in later centuries by many other descriptions of this cave. From Lower Austria, detailed reports (deposited at the Austrian National Library) from a well-planned official expedition, and a second exploration tour to the Geldloch in the Ötscher mountain, both of them in the autumn of 1592, precisely describe nearly one kilometre of passages. Smocza Jama (Dragon's Den) beneath the Wawel royal castle in Kraków, first written about in *c.*1190, became well known throughout Europe following its illustration in Münster's *Cosmographia* (1544 and many subsequent editions).

Since the 16th century, cave descriptions and also the first cave maps from Central European caves can be found in various books. In the Baroque period, the interest in curiosities and wonders of nature motivated explorations and special studies. The best documentation concerns the karst and caves of Slovenia and is published in the work *Die Ehre des Herzogthums Krain* by Johann Weichhard Freiherr von Valvasor (see entry on Speleologists). This is a historic, geographic, and cultural description of Krain, now part of Slovenia, published in 1689 in Laibach (now Ljubljana). From this important karst region, on the southeastern border of Central Europe, many further documents exist from the 17th and 18th centuries, the greatest interest being in Cerknica polje and its lake (see Cerknica Polje, Slovenia: History). During the same historic period, caves in the Harz mountains, especially the Baumannshöhle, as well as caves in Westphalia, such as the Kluterthöhle (both in Germany), were explored and described. Reports exist from caves in the Moravian karst (Czech Republic), in the central part of the Carpathian mountains (Slovakia and Hungary), and from many other regions.

By order of the Habsburg Emperor an important systematic research on the karst regions and caves in the Austro-Hungarian monarchy was made by Joseph Anton Nagel (see Speleologists), director of the imperial collections of natural history (the precursor of the Museum of Natural History in Vienna). He visited caves in Lower Austria (Geldloch) and Styria in 1747, and made two trips in 1748, the first in the Slovenian karst, and the second in the Moravian karst, where he visited the cave of Sloup and the abyss of Macocha. During a long trip to Hungary in 1751, he visited the cave of Demănova (Slovakia). He collected minerals and described his observations in long reports, with many illustrations made by artists participating on his travels.

Military interests during the wars between the Austro-Hungarian monarchy and the Turks resulted in documentation of the Veteranische Höhle near Orsova in a defile of the lower course of the Danube (Romania); 17 hand-drawn maps from 1788–1805 are preserved in the Austrian war archives. Other examples of activities at this time, are a further exploration of the Macocha abyss, with a map in 1784 and a description of the cave of Aggtelek (Hungary) published 1786; a plan of this cave exists from 1794. In the late 18th century, interest in paleontological finds grew and the caves of Franconia (Germany) have great importance for related discussions and studies—concerning in particular the cave bear.

In the Romantic period (first part of the 19th century), many caves in Central Europe were described in the guidebooks for travellers, while Ojców caves in Poland, popular since the 1780s, had possibly the world's first trained cave guides from *c.*1810. Many scientific observations and discoveries are documented from this time. The most important personality of this time is Adolf Schmidl (see Speleologists). He studied nearly all the karst and cave regions of the Austro-Hungarian monarchy (Classical karst, Lower Austria, Moravian karst, Hungaro-slovakian karst near Aggtelek), including the karst features in the Bihargebirge in Transylvania, now a part of Romania. In his publications, he introduced in 1850 the German term "Höhlenkunde" (cave science) still used today.

During the second part of the 19th century, systematic exploration of karst phenomena was developed in the Habsburg monarchy, and the term "karst" was first used as a scientific term (see Kras, Slovenia). The well-known work of Jovan Cvijić, *Das Karstphänomen* (1894), which summarized the results of his research, was published in Vienna (see Geoscientists). In the same year, also in Vienna, the first general overview on cave science in the German language, the book *Höhlenkunde* was published by Franz Kraus.

Europe, Central: History: National Speleological Federations in Central Europe (2002).

Austria	Verband österreichischer Höhlenforscher	http://www.hoehle.org
Croatia	Hrvatsko speleološko društvo	http://jagor.srce.hr/~mgarasic/hsd1.htm
Czech Republic	Česká speleologická společnost	http://www.natur.cuni.cz/~zeman/DOM6.htm
Germany	Verband der deutschen Höhlen-und Karstforscher	http://www.hfc-hersfeld.de/vdhk/vdhk.html
Hungary	Magyar Karszt- és Barlangkutató Társulat	http://www.fsz.bme.hu/mtsz/barlang/mkbt/mkbt.htm
Italy	Societá Speleologica Italiana	www.ssi.speleo.it
Poland	Speleologists included in the Polish Tourist Association	http://www.pza.org.pl
Slovakia	Slovenskej speleologičkej spoločnosti	
Slovenia	Jamarska zveza Slovenije	http://www.jamarska-zveza.si
Switzerland	Société Suisse de Spéléologie/Schweizerische Gesellschaft für Höhlenforschung	http://www.speleo.ch

In general, the speleology of the late 19th century is characterized in Central Europe by prehistoric and paleontological studies, as well as by discoveries and explorations in the classical karst regions. Excavations in the Moravian karst were made by Heinrich Wankel (1821–97); observations on ice formations in alpine caves were made by Eberhard Fugger between 1869 and 1886. In Vienna, the first speleological society (Verein für Höhlenkunde) in the world with a scientific background, was founded in 1879 (see Exploration Societies), followed by two sections of cave explorers, organized into the alpine societies of German and Italian languages in Trieste in 1883. In Germany, the first officially organized speleological club (Schwäbischer Höhlenverein) was founded in 1889. The Verein für Höhlenkunde in Österreich (Speleological Society of Austria), founded in 1907 in Graz (Styria), was later the first Federation of different clubs in that country. In Moravia, the Höhlensektion des Naturwissenschaftlichen Klubs (Cave Section of the Club of Natural Sciences) was founded in 1908 in Brno. In 1910, the Gesellschaft für Höhlenforschung in Krain (Speleological Society of Carniola, now Slovenia) was founded in Laibach (Ljubljana), a Komitee für Höhlenforschung in der Geologischen Kommission des Königreichs Kroatien und Slawonien (Speleological Committee in the Geological Commission of Croatia) in Agram (Zagreb), and a further Speleological Committee in the Geological Society of Hungary in Budapest.

In the years before World War I, advances in understanding the hydrology of the karst from Istria to Bosnia, were made by Alfred Grund, Franz Katzer, and many others, and the high alpine cave systems in the Northern Limestone Alps were explored.

In the 1920s, cave tourism began to be an important economic factor and many show caves were opened to the public. Cave sediments were a focus for researches with observations of deposits containing bones of Pleistocene cave bears in the Drachenhöhle near Mixnitz (Styria). Many studies in all parts of Central Europe show the importance of sediments for different scientific problems, including the first steps to a modern relative chronology. In Romania (1920) and in Austria (1923), official Institutes of Speleology were founded. At the University of Vienna, lectures on speleology were given by Georg Kyrle from 1924 up to his death in 1937. In Italy, an Institute of Speleology was established in connection with the administration of the Grotta di Postumia (up to 1918 called Adelsberger Grotte, since 1945 Postojnska Jama), managed by G.A. Perco (Perko). In the Moravian karst, important exploration and exploitation work in caves was made by Karel Absolon.

On 28 June 1928, a Federal Law concerning the protection of caves, supplemented by a series of decrees in the following year, was agreed by the Austrian Parliament. It was the first special law in the world to give rules for cave documentation, for administration and management of commercial caves, and for the education of cave guides.

After World War II, national (or regional) speleological federations were founded or re-founded in all countries of Central Europe, and international collaboration became more important. The third (Austria, 1961), fourth (Yugoslavia, 1965), fifth (Germany, 1969), and sixth (Czechoslovakia, 1973) International Congress of Speleology, all held with extended excursions, were held in Central Europe. Spectacular discoveries, great progress in the documentation of caves and karst phenomena, and new methods to determine the age of speleothems and development in karst water tracing have been the main fields of work in Central Europe from the 1960s to the present.

Hubert Trimmel

Further Reading

Saar, R. & Pirker, R. 1979. *Geschichte der Höhlenforschung in Österreich* [History of speleology in Austria], Vienna: Landesverein für Höhlenkunde in Wien und Niederösterreich

Shaw, T.R. 1992. *History of Cave Science: The Exploration and Study of Limestone Caves, to 1900*, 2nd edition, Broadway, New South Wales: Sydney Speleological Society

EUROPE, MEDITERRANEAN

Limestone crops out widely around the Mediterranean Basin, and many outstanding European karst areas are located relatively close to the Mediterranean coastline or are under the influence of Mediterranean climates in Spain, southern France, Italy, Malta, Slovenia, western Croatia, Bosnia-Herzegovina, Montenegro, Albania, Greece, and western Turkey. Karst in Mediterranean Europe is characterized by a heterogeneous patchwork of solutional landforms which shows few similarities, besides a clear predominance of bare rock features on the landscape.

The climatic patterns prevailing in Mediterranean Europe are characterized by high variability of annual precipitation, frequent severe summer drought periods, and the concentration of rainfall in irregular but intense downpours. The climate produces harsh conditions for plant communities and promotes enhanced denudation of soil and slope materials. In some areas annual rainfall exceeds 1000 mm, reaching 2000 mm in the mountains of Montenegro and Andalusia, Spain. However, over large areas different subhumid to semi-arid Mediterranean subclimates contribute to generating the great diversity of geomorphology and plant cover present in the Mediterranean Basin. Pleistocene glaciations did not directly affect most of the mediterranean areas during the repeated climatic changes of the Quaternary, although the higher mountains were subjected several times to cold, even periglacial, morphogenesis with some glaciers in Greece.

The Mediterranean climate is one of the most aggressive in respect of erosion, as a consequence of the great intensity of autumn storms that occur just after the severe summer drought period. Bioclimatic conditions are in general unfavourable for the maintenance of continuous vegetation cover over the karstified terrains. Furthermore, man has induced burning of the Mediterranean forests during the past 6000 years in order to obtain pasture and crop lands. Both factors encourage the dominance of different kinds of sclerophyllous shrubland and the retreat to more favourable places of the mature successional stages of plant cover, consisting basically of evergreen woods. Edaphic and climatic factors determine the distribution of the various types of shrubland formations, some of which are called *matorral* in Spanish and *garrigue* in southern France. These low, often sparse, scattered scrub plant communities are considered to be caused by deforestation processes, triggered by human impacts on limestone terrains under the influence of the harsh climatic regime. Because of the steep topography and the scant plant cover protection, Mediterranean soils are frequently affected by degradation, and even desertification, processes. However, it seems clear that the present barrenness of the Mediterranean karst has been artificially increased by human intervention throughout the ages.

Karst landforms dominate over large areas of some geographical regions including Istria, Dalmatia, Herzegovina, Apulia, the Greek coast, the island of Majorca and several mountain ranges and plateaux of Provence, Languedoc, Campania, and Andalusia. In all these representative Mediterranean karstlands, barren rock surfaces prevail over soil-covered ground, for the reasons described above. Alpine and Mediterranean limestone terrains have become the model for bare-karst landscapes, as opposed to the covered-karst types of central and northern Europe. In fact, the Dinaric karst, probably the best example of Mediterranean karst, has long been regarded in the geomorphological literature as a "classical" area for studying karst landforms.

Mediterranean caves are not very long, rarely exceeding 20 km in length, but generally have abundant speleothems, their growth being favoured by the prevailing bioclimatic conditions. Large chambers and relict caverns, partially filled with collapse boulders, clastic sediments, and speleothems are common throughout the region. Active drainage caves are locally important because of their lengths, as in the karsts of Languedoc and northern Catalonia (France), Sardinia (Italy), and southeastern Spain, where there are several caves longer than 10 km. Subsurface vertical shafts are abundant in the mountains, but only recently have explorations found karstic shafts deeper than 1000 m in Andalusia (Sima GESM; González-Ríos & Ramírez-Trillo, 1999) and Croatia (Kuhta & Bakšić, 2001). Some collapse landforms attain impressive dimensions, such as in the Crveno Jezero (Imotski, Croatia), a huge half-flooded chasm over 500 m deep and more than 200 m in diameter. Karren landscapes are especially remarkable in the Dinaric karst (Velebit range and Biokovo mountain) as well as in Majorca. Solution dolines stand out as the main surface features in the karsts surrounding the Adriatic coast (Slovenia, Dalmatia, Herzegovina, and Montenegro), and are also frequent in Spain and southern France. Poljes have a similar geographical distribution, and have been extensively described from Greece, Italy, Portugal, Spain, and especially the Dinaric karst, where some of the most famous poljes are located (see Dinaric Karst; Dinaride Poljes). The scarcity of water in these environments makes the karstic springs significant features for human settlements in the Mediterranean. Among several remarkable examples are the Ombla (Dubrovnik, Croatia) and the Fontaine de Vaucluse (Provence, France) which rank among the greatest springs of the world. Tufa deposits, karst corrosion surfaces, steep heads, and karst canyons are also frequent in the Mediterranean landscapes. As a whole, it appears that Mediterranean Europe contains many diverse karstic features, although rocky landscapes are the most widely represented.

Most of the Mediterranean countries are in relatively young orogenic belts, whose mountain chains run close to the coastline forming rugged terrains. The present configuration of the Mediterranean area consists of high and steep mountain ranges of the Alpine system, structured by complex folding and overthrusting. Regarding lithology, the carbonate rock substrates are more extensive than in most other regions of the world and give rise to some of the most typical soils and landscapes of the Mediterranean Basin. The majority of the mountain building corresponds to thick Mesozoic limestone outcrops, rising near to the sea shore and showing a rather sharp relief. Active tectonic uplift accounts for the high gradients and deeply dissected slopes observed in the coastal ranges.

Sea level acts as base level for many karsts surrounding the Mediterranean. Because of the proximity to the sea, coastal and karstic processes interact, producing specific littoral landforms and generating in those karst systems controlled by sea level, very unusual hydrological conditions. Caverns captured by marine abrasion, rock arches, the presence of steep-sided coves, and the drowning of large depressions and karst valleys as a result of postglacial sea-level rise, are examples of this kind of interacting geomorphological dynamics. The most striking features ob-

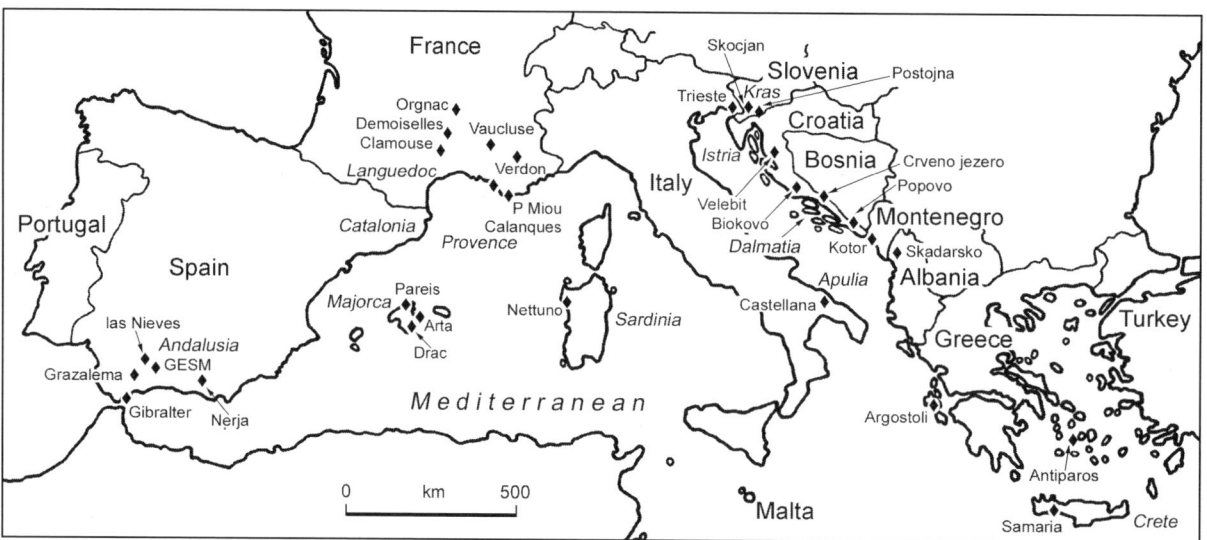

Europe, Mediterranean: Locations of karst sites in the Mediterranean referred to in the text.

served in these coastal karsts are the strong upwellings of fresh or brackish water that occur beyond the shore in the form of submarine springs. These springs, called *vrulje* in Dalmatia, are quite common all over the Mediterranean and in some cases the associated hydrostatic pressure changes can result in advance of salt-water intrusion and the existence of brackish springs inland. Besides the well-known littoral springs reported in the literature from France, Greece, and the Adriatic shore (including Istria, Dalmatia, and Montenegro), some other significant sites affected by sea-level change are the great flooded polje Skadarsko Jezero (on the border between Montenegro and Albania), the overall sea level-controlled landscape of Boka Kotorska (Montenegro) and Neretva Plain at Metković (Herzegovina), the *calanques* of Provence including the long submarine cave spring of Port Miou (explored by divers), and the recent discoveries in eastern Majorca of submerged cave systems exceeding 5 km in length.

Caves, karst springs, and diverse limestone landforms were related in different ways to old human settlements all around the Mediterranean Sea. Archaeological remains, from Paleolithic times to very recent historical events, are frequently found in the caves, containing fairly well-preserved evidence of the past. Ancient civilizations, especially Greek and Roman cultures, invested many karstic sites with religious and mythological significance (see Caves in History: The Eastern Mediterranean). The interest in caves and karst sites remained throughout the ages and was the main drive for the earlier explorations made in the 17th, 18th, and 19th centuries (Shaw, 1992). Many references to caves appeared in the books of travellers in the 19th century and Martel's expeditions incorporated knowledge on Mediterranean caves into modern speleology.

Since the mid-20th century many Mediterranean limestone terrains and their celebrated karst sites have become well known due to massive tourism development in this part of Europe. Wild landscapes, great springs, and well-decorated show caves (Aellen & Strinati, 1975) are natural tourist resources of great importance for several southern European countries. Some of the most visited show caves in the world are located in the Mediterranean region, including the Coves del Drac (Majorca, Spain), Grotta di Castellana (Apulia, Italy), Cuevas de Nerja (Andalusia, Spain), and the Aven d'Orgnac, Grotte de Clamouse, and Grotte des Demoiselles (Languedoc, France). Among many spectacular landscape and water sceneries are the Argostoli sea-mills (Cephalonia, Greece), the deep spring called Fontaine de Vaucluse (Provence, France), the tufa waterfalls of Krka river (Dalmatia, Croatia) and the karst canyons of Samaria (Crete, Greece), Torrent de Pareis (Majorca, Spain), and Gorges du Verdon (Provence, France). These karst locations can each attract more than 100 000 visitors per year.

Àngel Ginés

See also **Dinaric Karst; France, Southern Massif Central; Gibraltar Caves**

Works Cited

Aellen, V. & Strinati, P. 1975. *Guide des Grottes d'Europe occidentale*, Neuchâtel: Delachaux & Niestlé

González-Ríos, M.J. & Ramírez-Trillo, F. 1999. Las grandes cavidades de Andalucía: Historia de sus exploraciones. In *Karst en Andalucía*, edited by J.J. Durán & J. López-Martínez, Madrid: Instituto Tecnológico Geominero de España

Kuhta, M. & Bakšič, D. 2001. Karstification dynamics and development of the deep caves on the North Velebit Mt., Croatia. *Proceedings of the 13th International Congress of Speleology*, CD-Rom

Shaw, T.R. 1992. *History of Cave Science: The Exploration and Study of Limestone Caves, to 1900*, 2nd edition, Broadway, New South Wales: Sydney Speleological Society

Further Reading

Herak, M. & Stringfield, V.T. (editors) 1972. *Karst: Important Karst Regions of the Northern Hemisphere*, Amsterdam: Elsevier

EUROPE, MEDITERRANEAN: ARCHAEOLOGICAL AND PALEONTOLOGICAL CAVES

Greece has one of the world's most spectacular karst landscapes and was the subject of some of the earliest karstic studies from Thales and Aristotle to Pausanias. Limestone is the predominant rock in Greece, which in combination with a temperate climate that allows rapid infiltration of water and dissolution, has produced deep caves and large springs, attracting both animals and humans. Though Italy and Malta display smaller karstic landscapes, they contain caves with important fossiliferous deposits.

Ghar Dalam Cave was one of the earliest sites used by Neolithic man, who crossed to the Maltese islands around 7000 years ago. Even more remarkable than the presence of prehistoric man, are thousands of fossil animal bones. The cave, 145 m long, is formed in coralline limestone, 9 m above the Weid Dalam valley. It was first excavated by Arturo Issel in 1865 and excavations continued until 1922. Six distinct layers were revealed. The second lowest, about 125 000 years old, contained the remains of *Hippopotamus pentlandi* and *Hippopotamus melitensis*, as well as the dwarf elephant *Elephas mnaidriensis* and *Elephas falconeri*. The fourth lowest layer, around 18 000 years old, contains *Cervus elaphus* with brown bear, red fox, and wolf. The top layer is the Cultural layer, with pottery, "sling stones", and domestic animal remains. The fauna is Continental European and came to Malta via Sicily during a glacial phase. The dwarfism of some mammalian lineages started in Sicily and continued in Malta. Gigantism also appeared in some species, such as the dormouse *Leithia melitensis* and the giant lizard *Lacerta siculimelitensis*.

Guattari Cave is a small cave in Monte Circeo, Italy. Its entrance was discovered in 1939, having been closed for about 50 000 years. The cave is famous for a Neanderthal skull, with an apparently artificially enlarged hole at its base, found on the cave floor within a circle of stones. It caused a sensation as a case for ritual cannibalism but it was later shown that the cave was a hyaena den and that the circle was a natural arc of stones. The hole at the base of the skull had hyaena tooth marks, evidence that the hole was made by hyaenas, as in the case of *Homo erectus* skulls from the Zhoukoudian cave in China.

Lamalunge Cave in Altamura is another Italian cave, where one of the most complete European hominids was discovered in 1993. Dating estimates vary from 130 to 400 000 years BP. Because it is heavily encrusted in botryoidal (grape-shaped) speleothem, the skeletal remains have not been fully excavated. It is suggested that it is a pre-Neanderthal or an archaic *Homo sapiens*.

Corbeddu Cave near Oliena, Sardinia, is a karstic cave 110 m long, with four main chambers. Excavations started in 1982 by a Dutch–Italian team. It contains the oldest human bones from any Mediterranean island, dated at 20 000 years BP and 8750 years BP. The Paleolithic sediments also contain the endemic deer *Megaceros cazioti* and the ochotonid *Prolagus sardus*. The Neolithic layers are characterized by the presence of domesticated animal remains. The presence of small bone and enamel fragments between the teeth in the jaws of young deer, indicates the use of these jaws as tools.

Kokkines Petres (or Petralona Cave) in Chalkidiki (Greece), is one of the richest archaeological and paleontological caves in Europe. It was formed in late Jurassic limestone and was discovered by locals in 1959, who also discovered the skull of Petralona man in 1960. Small-scale excavations started in 1960 by J. Petrochilos. The skull was encrusted by a flowstone sheet, that had been hanging on the walls of the cave. This makes it difficult to find its position in the stratigraphic sequence. All dating studies by electron spin resonance (ESR) and U-series on material from the skull show an age of about 200 000 years BP. The skull presents a mosaic of *H. erectus*, archaic, and Neanderthal-like characteristics. The fauna found in the cave is divided into two groups. The older group is of Villafranchian age (< 700 000 years) and the younger is of the late middle and upper Pleistocene age. The Petralona skull belongs to the younger fauna. The older fauna includes taxa such as *Pliohyaena perrieri, Ursus deningeri, Canis lupus mosbachensis, Xenocyon cf. lycanoides*, and *Praemegaceros* sp. The younger fauna includes *Crocuta spelaea, Ursus spelaeus, Equus petraloniensis* n. sp., *Equus caballus piveteaui, Dicerorhinus hemitoechus*, and *Sus scrofa*. The two faunas were found mixed on the cave floor, owing to episodes of water flooding as evidenced from trampling of the bones. Coprolites on the cave floor show that the cave was a carnivore den. The cave entrance was closed during the upper Pleistocene, as the top of a talus cone reached the entrance.

Maara Cave is the biggest cave in Greece. It has a subterranean river, 13 km in length, that forms the springs of Aggitis river. The fossil cave fauna represents the whole Pleistocene. Taxa include *Archidiskodon meridionalis, Mammuthus trogontheri, Mammuthus primigenius, Cervus elaphus, Coelodonta antiquitatis*, and *Ursus spelaeus*. Mousterian artefacts have also been found.

Loutraki-Almopia Caves are located in a gorge, 120 km north of Salonica. The area has been designated as a "speleological park", with tens of caves formed in intensely karstified Maestrichtian limestone. Late Paleolithic fauna, with numerous juve-

Europe, Mediterranean: Archaeological and Paleontological Caves: *Panthera* sp. skeleton from Dyros Cave, Greece, the first panther fossil found in the Peloponnese. The cave has numerous bone deposits of extinct animals from the upper Pleistocene. (Photo by Costas Zoupis)

niles of *Ursus spelaeus, Crocuta spelaea, Panthera pardus, Vulpes* sp., *Capra ibex, Dama* sp., bovids, cervids, and foxes are found. Agios Georgios Cave in Kilkis, 50 km north of Salonica, is a hyaena den of *C. crocuta spelaea* with food remnants of *Cervus elaphus, Megaceros* sp., *Bos primigenius, Equus hydruntinus*, and *Equus caballus cf. germanicus*. Late Pleistocene fauna (12 200 ± 2500 years BP).

The Maronia Cave near Komotini, Thrace, has been identified by some as the cave in which, according to Homer, Odysseus blinded the Cyclops Polyphemus. It contains Neolithic and upper Pleistocene deposits with horses, cervids, bovids, and rhinos. Peristeri Cave, a recently discovered karstic cave 67 m deep, overlooks the Louros river in Epirus. It contains animal bones, stone artefacts, and hearths of middle Paleolithic. Theopetra Cave in Thessaly Plain contains a continuous sequence from middle Paleolithic deposits of about 70 000 years old to Chalcolithic, including Mesolithic deposits. Two partial skeletons were found, ^{14}C dated at about 9000 and 16 500 years BP respectively, and 46 000 year-old hearths and human footprints are also reported to have been found.

Kapsia Cave in Mantinia polje, Peloponnese, usually floods once a year due to winter rains. The cave was first explored in 1892, by E.A. Martel and N.A. Siderides. It is remarkable for its human remains which suggest that about 50 people were drowned in the cave, after a big flood took place in the 4th or 5th century AD. There has been human presence since Neolithic times. Apidima Cave in the Taenaron peninsula, a massive karstic landscape, has produced two Neanderthal-like skulls. Two ESR dates, on material associated with the two skulls, have given ages of 20 000 and 45 000 years BP respectively. The fauna appears to be much older and includes *Megaceros, Panthera pardus, Lynx lynx*, and *Hippopotamus amphibious antiquus*. Nearby are the Dyros Caves, a complex of caverns, lakes, and rivers, with Neolithic finds and remains of *Panthera* sp. (see Figure).

Charkadio Cave in Tilos Island, north of Rhodes, documents the latest survival of elephant fauna in Europe in the form of *E. falconeri* 3500 years ago. It arrived on the island about 50 000 years ago. Deeper sediments show that deer survived till 140 000 years ago. Vamos Cave at Chania, Crete is an underwater sea cave where the new species *Elephas chaniensis* has been found. Fossil deer and seal bones have also been discovered.

ANTONIS BARTSIOKAS

See also **Gibraltar Caves: Archaeology**

Further Reading

Bartsiokas A., Merdenisianos, C. & Zafiratos, C. 1982. The anthropological finds of Kapsia Cave, Tripolis and their history. *Bulletin de la Société Spéléologique de Grèce*, 18: 157–68

Bartsiokas, A. 2001. Peristeri I, a new Palaeolithic cave in Epirus: Palaeoanthropological and geophysical investigations. In *Archaeometry Issues in Greek Prehistory and Antiquity*, edited by Y. Bassiakos, E. Aloupi & Y. Facorellis, Athens: Hellenic Society of Archaeometry and Society of the Messenian Archaeological Studies

Grün, R. 1996. A re-analysis of electron spin resonance dating results associated with the Petralona hominid. *Journal of Human Evolution*, 30: 227–41

Kyparissi-Apostolika, N. (editor) 2000. *Theopetra Cave: Twelve Years of Excavation and Research 1987–1998*, Larissa, Greece: University of Thessaly

Petrocheilou, A. 1984. *The Greek Caves*, Athens: Ekdotike Athenon

Piperno, M. & Scichilone, G. 1991. *Il Cranio neandertaliano Circeo 1: studi e documenti / The Circeo 1 Neanderthal Skull: Studies and Documentation*, Rome: Istituto Poligrafico e Zecca dello Stato

Sondaar, P.Y., Klein Hofmeijer, G. & Sanges, M. 1989. The dramatic end of the Sardinian Paleolithic island economy. *B.A.R. International Series* 508, (I): 517–21

Spoor, F. 1999. The human fossils from Corbeddu Cave, Sardinia: A reappraisal. In *Elephants Have a Snorkel! Papers in honour of Paul Y. Sondaar*, edited by J.W.F. Reumer & J. De Vos, *Deinsea*, 7: 297–302

Stiner, M.C. 1994. *Honor among Thieves: A Zooarchaeological Study of Neandertal Ecology*, Princeton, New Jersey: Princeton University Press (for Italian caves)

Theodorou, G.E. 1988. Environmental factors affecting the evolution of island endemics: The Tilos example from Greece. *Modern Geology*, 13(2): 183–88

Tsoukala, E.S. 1991. Contribution to the study of the Pleistocene fauna of large mammals (Carnivora, Perissodactyla, Artiodactyla) from Petralona Cave (Chalkidiki, N. Greece). Preliminary report. *Comptes Rendus de l'Académie des Sciences Paris*, 312(serie II): 331–36

Tsoukala, E.S. 1992. The Pleistocene large mammals from the Agios Georgios Cave, Kilkis (Macedonia, N. Greece). *Geobios*, 25(3): 415–33

Tsoukala, E. 1999. Quaternary large mammals from the Apidima Caves (Lakonia, S Peloponnese, Greece). *Beiträge zur Paläontologie, Wien*, 24: 207–29

Zammit Maempel, G. 1989. *Ghar Dalam Cave and Deposits*, Malta: PEG Publications

EUROPE, NORTH

In Europe north of the Alps, geological or topographical factors restrict the development of extensive karst and deep caves (Figure 1). Most older rocks form complex basements where limestone formations are relatively thin—the one exception being the low plateaus of Carboniferous limestone. Most younger rocks form a relatively undeformed cover, with poorly lithified limestones, notably the Cretaceous chalk, that do not contain large cave systems. There are no young mountain ranges where folded limestones crop out over great ranges of altitude.

In Britain, the Lower Carboniferous limestones form significant areas of cavernous karst (Waltham *et al.*, 1997), notably in the Pennine Hills of northern England. The Peak District (see separate entry) in the southern Pennines has the largest single area of karst, with over 400 km^2 of rolling fluviokarst dissected by superimposed, dendritic, dry valleys, fresh since their periglacial modification in the Devensian. Further north, the Yorkshire Dales (see separate entry) sector of the Pennines was covered by Devensian ice and therefore has the best of Britain's glaciokarst, as well as being notable for its many fine and extensive cave systems. The outlying parts of the northern Pennines have even more extensive limestone pavements on smaller outcrops of various Carboniferous limestones. Some of these are Upper Dinantian limestones (often known by their old name of Yoredale limestones), each less than 30 m thick within clastic rock se-

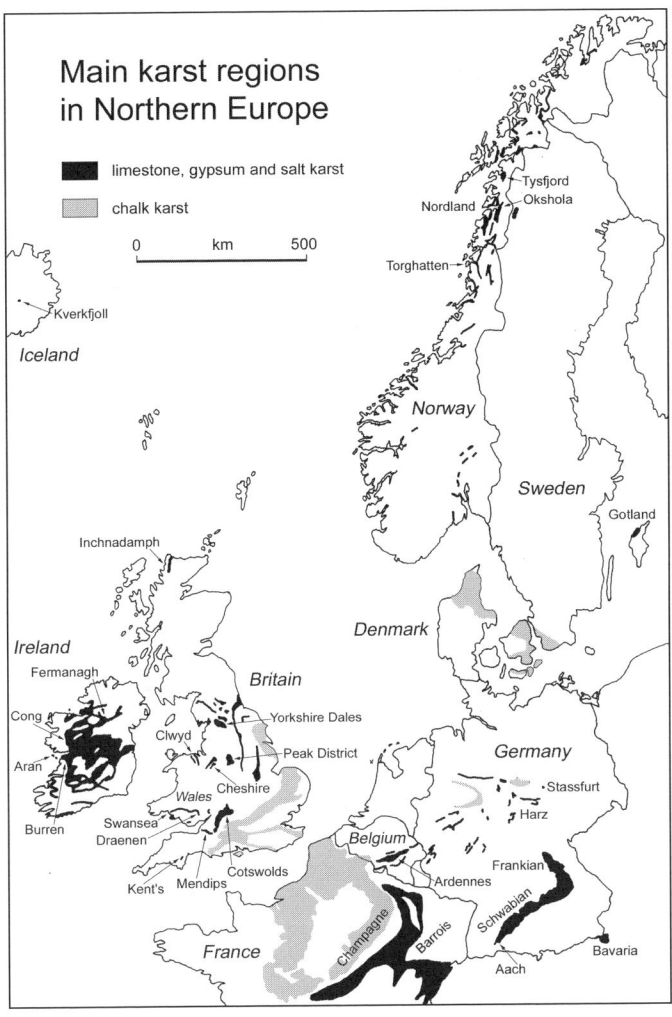

Europe, North: Figure 1. Location map of the main karst regions of northern Europe, including those described in their own entries.

quences. These also contain active and drained phreatic maze caves with kilometres of passages on joint networks with spacings of only a few metres. The Mendip Hills (see separate entry) of southern England were formed in thick, steeply dipping Lower Dinantian limestone with inclined cave networks, while Devon has a scatter of caves beneath isolated outcrops of folded Devonian limestones, including Kent's Cavern with its important Pleistocene bone sediments preserved under flowstone. The other British karst on pre-Carboniferous limestone is around Inchnadamph in Scotland, where stream caves underlie a glaciokarst on Durness dolomites.

Dinantian limestones also crop out in Wales (Ford, 1989). They form a broken low escarpment around the South Wales coalfield, where the gently dipping beds contain some very long cave systems. Under Mynydd Llangattwg, large phreatic conduits follow the strike of single bedding planes, but Pleistocene base-level lowering has allowed subsequent systems to establish further down-dip, and the older and younger caves have been linked by intersecting down-dip vadose inlets (Smart & Gardener, 1989). Ogofs Agen Allwedd and Daren Cilau each have over 30 km of passages, mainly beneath the cover of Namurian sandstone and therefore largely devoid of stalagmites and stalactites. Just to the south, Ogof Draenen (see separate entry) shows a comparable evolution but is a much longer cave. Just to the west, the sandstone cap of the Llangynidr Plateau is pocked with hundreds of large collapse dolines overlying extensive interstratal karst. Dan yr Ogof and Ogof Ffynnon Ddu lie on either side of the Swansea Valley; their long complexes of drained phreatic passages (17 and 51 km respectively), looping obliquely up and down bedding / joint intersections, have been rejuvenated by extensive vadose canyon development. Their evolution may span a million years, and some of the old passages are now well decorated with calcite. In North Wales a number of deep phreatic caves have been drained by mining activity, whose artificial drainage created a more recent version of the Pleistocene rejuvenations seen elsewhere in Britain. The limestone hills of Clwyd also host Britain's largest postglacial tufa deposits, at Caerwys (Pedley, 1987).

Ireland has large outcrops of Carboniferous limestone, a large proportion of which are now covered by glacial till and peat bogs. The best-exposed karst is close to the west coast. In Eire, the Burren glaciokarst (see separate entry) forms a large expanse of glaciated limestone pavements, replicated in the bare rock vistas of the windswept Aran Isles immediately offshore. Just south of the Burren, the shale-capped hills of County Clare provide allogenic drainage on to the limestone and into over 80 km of mapped caves, nearly all narrow twisting canyons cut below gently dipping bedding planes (Self, 1981). These include Doolin Cave, in which it is possible to walk dry beneath the Aille River while it is still flowing on the surface. The karst lowlands are noted for the turloughs—closed depressions that flood in most winters but have not yet evolved into poljes by lateral planation at a more mature and stable water table. Some permanent lakes drain underground. Among them is Lough Cong, with its boat canal that is normally dry because it was built across cavernous limestone by engineers not adequately familiar with karst. The highest limestone hills and therefore the deepest caves of Ireland are in County Fermanagh, in the Ulster sector. An excess of till restricts the extent of limestone pavements, but the glacial evolution of the karst is recorded in the many abandoned and rejuvenated caves.

Carboniferous limestone also occurs in Belgium, notably around the type localities of Dinant and Namur in the Ardennes. The Lesse River drains through a limestone hill at Han, where 10 km of cave passages form a cave network not yet fully connected (Figure 2). These are the only long caves in the Benelux countries, although dolines identify karst on many of the Ardennes plateaus.

Germany has a small area of spectacular glaciokarst in eastern Bavaria, where its border loops round a slice of the karst of Alpine Europe. Other than this, Germany's limestone karst is only significant in the low plateaus of the Schwäbischer and Frankischer Alber. These have over 6000 km^2 of fluviokarst terrains on soft Jurassic limestones. Most of the many caves are only small streamways, and few are longer than a kilometre. Where it crosses the western tip of the Schwäbischer Alb, the Danube loses about 6 m^3 s^{-1} of its flow into river-bed sinks, and this water rises from the Aach Spring, 12 km away and 170 m lower, at the head of a major tributary to the Rhine. There

Europe, North: Figure 2. Calcite stalagmites in Grotte Père Noel, the abandoned high-level tunnel that lies parallel to the river passage in the Grottes de Han, in the Belgian Ardennes. (Photo by Tony Waltham)

Europe, North: Figure 3. The dry valley of Millington Pastures in the Yorkshire Wolds (England), with the rounded slopes and sheep-grazed short grass that characterizes the chalk karst. (Photo by Tony Waltham)

are similarities between the Schwäbische karst and the Cotswold karst on equivalent Jurassic limestones in England, but the latter has almost no caves, although it constitutes an economically important aquifer with a major component of cavernous flow.

The most extensive karst in northern Europe is on the Cretaceous Chalk (Figure 3), which has large outcrops in England, France, Germany, and Denmark. The best of the chalk fluviokarst is in southern England and northern France, where the soft, rolling landscape contains shallow dolines and dendritic dry valleys largely floored by solifluction debris from Pleistocene stages of periglacial weathering. Underground drainage is partly by diffuse flow, but is largely by flow through networks of fissures rarely large enough to be entered. The chalk in France is more indurated than that further north, and has the only known long caves, with many small vadose streamways over a kilometre long.

Scandinavia contains only small areas of karst, and little outside the outcrops of Lower Paleozoic carbonates of Nordland in Norway and the island of Gotland in Sweden. Narrow bands of limestone and marble in the folded basement rocks of Nordland have many caves in excellent examples of stripe karst (see Stripe Karst). The metamorphic grade of the marbles increases towards the higher and more westerly of the Caledonian nappes. This commonly provides attractive, dipping foliation, distinctive minerals including garnets, and folded interbeds of non-carbonate rocks, which can be seen in the many clean-washed stream passages that underlie older, phreatic passages, some of which are well decorated with calcite. Large numbers of caves occur in the county of Nordland in Norway (St Pierre & St Pierre, 1985), and also across the border in Sweden. Tjoarvekrajgge (13 km long and 500 m deep) and Okshola / Kristihola (11 km) are the longest in Nordland, and Sweden's longest marble cave is Korallgrottan, with nearly 6 km of passages. Råggejavre-Raige, 580 m deep from its plateau sink to its fjord-side entrance, lies in a narrow band of steeply dipping marble, close to the wall of Tysfjord, Nordland. The island of Gotland in Sweden has a large outcrop of Silurian limestone that contains the show cave of Lummelundagrottorna (5 km long). Scandinavia has numerous elevated sea caves (including Torghatten, 160 m long and nearly 30 m high and wide right through a mountain ridge; Figure 4), talus and fissure caves, and tunnel caves on the Baltic coast that were excavated by the marine cobble abrasion but are now elevated up to 200 m altitude following isostatic rebound after the last glaciation. Iceland has no limestone, but has many small lava caves and ephemerally explorable glacier caves; the latter includes the Rivière de Kverkfjöll (2850 m long and 525 m deep) with subglacial stream trenches in volcanic rocks cut beneath an ice tunnel whose excavation was aided by fumarolic steam.

Metacarbonates (marbles) occur along the west coast of Spitsbergen and host relict caves and active hot springs. Almost all the caves are filled with ground ice, but their speleogenesis can be traced back to subglacial conditions as well as an important paleokarst legacy from the Devonian and to even earlier hydrothermal corrosion during the Caledonian. Active hot springs yield sulfide-containing waters of carbonate-sulfate-chloride types, with temperatures as high as 15°C. Abundant Permian evaporites (anhydrite) give rise to high sulfate content of groundwater, and gypsum precipitates in icings. Although the karst is presently in a permafrozen state there are active sub-permafrost aquifers, fed from the underside of polythermal glaciers. Hence, the area serves as a modern analog for subglacial karstification in northern Europe during the Pleistocene glaciations.

Europe, North: Figure 4. Daylight is silhouetted through the Torghatten cave in western Norway. The cave occupies the full width of the vertical band of marble that is visible in its roof. (Photo by Tony Waltham)

drained phreatic passages and large breakdown rooms. Gypsum karst creates significant engineering hazards in parts of Germany and England. The most important salt karst occurs in the Cheshire Plain, England, where natural dissolution has created many surface depressions that are now flooded as meres (lakes), typically a kilometre across. Brine abstraction has increased the rates of groundwater circulation, and has therefore accelerated surface subsidence at many sites, with a significant impact on the towns and infrastructure that stand on the drift-covered salt outcrops.

TONY WALTHAM

See also **Britain and Ireland**

Works Cited

Ford, T.D. (editor) 1989. *Limestones and Caves of Wales*, Cambridge and New York: Cambridge University Press

Kempe, S. 1996. Gypsum karst of Germany. *International Journal of Speleology*, 25(3–4): 209–24

Pedley, H.M. 1987. The Flandrian (Quaternary) Caerwys tufa, North Wales, an ancient tufa barrage deposit. *Proceedings of the Yorkshire Geological Society*, 46: 141–52

St Pierre, D. & St Pierre, S. 1985. Norway's longest and deepest caves. *Cave Science*, 12: 113–25

Self, C.A. 1981. *Caves of County Clare*, Bristol: University of Bristol Speleological Society

Smart, P.L. & Gardener, C.G. 1989. The Mynydd Llangattwg cave systems. In *Limestones and Caves of Wales*, edited by T.D. Ford, Cambridge and New York: Cambridge University Press

Waltham, A.C., Simms, M.J., Farrant, A.R. & Goldie, H.S. 1997. *Karst and Caves of Great Britain*, London and New York: Chapman and Hall

Further Reading

Downing, R.A., Price, M. & Jones, G.P. (editors) 1993. *The Hydrogeology of the Chalk of Northwest Europe*, Oxford: Clarendon Press and New York: Oxford University Press

Drew, D. 2001. *Classic Landforms of the Burren*, Sheffield: Geographical Association

Fogg, P. & Fogg, T. 2001. *Beneath Our Feet: The Caves and Limestone Scenery of the North of Ireland*, Belfast: Environment and Heritage Service, Department of the Environment

Gunn, J. (editor) 1994. *British Limestone Karst Environments*, London: British Cave Research Association (BCRA Cave Studies 5)

Lauritzen, S.-E. 1998. Karst morphogenesis in the Arctic: Examples from Spitsbergen. In *Global Karst Correlation*, edited by Yuan Daoxian & Liu Zaishua, New York: Science Press

Pfeiffer, D. & Hahn, J. 1972. Karst of Germany. In *Important Karst Regions in the Northern Hemisphere*, edited by M. Herak & V.T. Stringfield, Amsterdam and New York: Elsevier

A belt of Permian and Triassic evaporite rocks provides discontinuous outcrops extending from central England to northern Germany. The best of the gypsum karst is in Germany's Harz Mountains (Kempe, 1996). Outcrops are scored by valleys dry since the early Pleistocene, are pitted with dolines, and have deeply pinnacled rockheads (bedrock surface on which unconsolidated sediment is deposited) largely mantled by soils of insoluble residues. Five caves reach lengths of 1500–2500 m, with

EVAPORITE KARST

Evaporite deposits form in seas, lagoons, and internally drained lakes, where more water leaves the basin by evaporation than enters as rainfall, surface, and subsurface flow. Evaporite rocks are inorganic in origin and form by chemical precipitation in a concentrated solution. The most common evaporite rocks include gypsum, anhydrite, and halite, although other rarer salts can be significant locally in karst. Carbonates such as calcite and dolomite can also be of evaporitic origin; however, the term "evaporite karst" is normally employed to denote karst in the above-mentioned more soluble salts, most commonly in gypsum and halite. Under normal conditions, the solubility of gypsum is up to three orders of magnitude greater than that of calcite, but the solubility of salt is roughly 140 times greater than the solubility of gypsum. Where they are in contact with undersaturated water, evaporite rocks may dissolve at very high rates (see Dissolution: Evaporite Rocks).

Evaporite rocks underlie about 25% of the continental surface, although this fraction is greater in some major regions: 35–

40% for the entire United States and the same figure for the former Soviet Union. Major epochs of evaporite deposition occurred in the late Cambrian, the Permian, the Jurassic, and the Miocene. Lesser but still significant episodes took place in the Silurian, the Devonian, the Triassic, and the Eocene. Ancient platform and basin-wide evaporite or combined evaporite / carbonate / siliciclastic sequences often reach thicknesses exceeding 1000 m, and extend laterally across huge areas. Evaporite accumulation in the Quaternary was on a very much smaller scale. Penecontemporaneous evaporites are most often found in the arid and semi-arid zones, in the circumtropical desert areas that lie between 15 and 45° north and south of the equator. Modern evaporite accumulation locally occurs outside the major desert belts, in the strictly continental environment, in high mountain deserts, in the rain shadows of high mountain ranges, and in polar and circumpolar regions.

The variety of depositional settings in which evaporites can form ranges from deep-water, to subaerial, to lacustrine, so the mode of occurrence of evaporites and the associations of lithofacies are quite diverse. The picture is further complicated by the fact that evaporites are very susceptible to recrystallization, dissolution, and replacement. Gypsum deposited in the shallow subsurface is converted to anhydrite after burial, and back to gypsum as the sulfate re-enters the meteoric realm and is washed in cooler and fresher fluids. Sulfates and halides occur widely as rare to frequent interbeds in sequences containing carbonates and siliciclastic sediments, but also as massive beds. Individual beds of gypsum with thicknesses of 10–40 m are fairly common, although sequences of them may exceed 200–300 m. Massive salt units may reach more than 1000 m in thickness. When massive deposits of salt are buried by denser strata, they flow upward and approach the surface as diapirs or dykes. Diapirs may rise from original salt bodies occurring beneath several thousand metres of cover strata, disrupting and deforming them.

The karst that develops on evaporite deposits has many features in common with carbonate karst, but also has significant peculiarities imposed by the specific features of evaporite geology and dissolution processes. In characterizing evaporite karst, it is particularly important to refer to the overall settings of karst development, which are conveniently generalized by the evolutionary typology of karst types (Klimchouk & Ford, 2000). Types of karst are viewed as successive stages in geological / hydrogeological evolution, from deposition through burial to re-emergence, between which the major boundary conditions, the overall circulation pattern and extrinsic factors and intrinsic mechanisms of karst development appear to change considerably. The different types of karst are marked by characteristic styles of karst system development, which are particularly distinct in evaporites.

As evaporites are much more soluble than carbonates, they are much less common in outcrops. This is one of the reasons why evaporite karst has received relatively little appreciation in the past in the mainstream of karstology, and has commonly been considered of limited significance. However, since the 1970s it has been increasingly realized that karst processes in evaporites such as gypsum and salt operate extensively when they occur below various types of cover beds within the upper few hundred metres of sedimentary sequences, with this type of setting being widely distributed over the continents. Under dramatically increasing human impacts such as construction, mining, and the exploitation of groundwater and hydrocarbon resources, intrastratal evaporite karst suffers more severe environmental problems and hazards than karsts in carbonates.

Syngenetic Evaporite Karst

Karst processes readily commence in freshly deposited evaporites. Syngenetic karst in ancient evaporites, probably common in continental settings, is sometimes recognized among the paleokarst features. Modern syngenetic karst is limited to salt lakes, where it develops in lake-bottom evaporite deposits of various compositions. It is fairly prominent in many salt lakes in the pre-Caspian lowland and the Altai region in Russia, in Central Asia, on the Qinghai–Xizang Plateau of Tibet, and in the Middle East and the United States. In addition to various sculpturing forms created by rainfall and inflowing surface streams, the most typical karst features are "windows", "cauldrons", and holes in the lake-bottom salt masses created by currents of groundwater rising from the aquifers that occur below the lakes.

Intrastratal Evaporite Karst

After burial by younger sediments, karstification in evaporites may be initiated at any of the stages of intrastratal development, particularly at some stage *en route* back to the surface. Where poorly permeable clays or shales surround evaporites in a stratified sequence, intrastratal karst may not develop at all, and any considerable karstification will not commence until the soluble unit is exposed to the surface. The entire sequence of intrastratal karst types includes deep-seated karst, subjacent karst, and entrenched karst. They differ according to the degree to which the evaporite units are covered by overlying rocks, and hence according to the hydrogeological setting.

Deep-seated Evaporite Karst

Deep-seated karst is not evident at the surface, and the soluble rock is not exposed. Its development is favoured where the evaporite units are sandwiched between or intercalated with carbonate or siliciclastic aquifers (Klimchouk, 2000). Dissolution may be initiated by two different hydrogeological mechanisms. In the first, transverse hydraulic communication, usually ascending, across a gypsum unit, is established between the two surrounding aquifers, giving rise to maze cave patterns, provided that the proper structural prerequisites exist. Outstanding examples are the giant gypsum maze caves in the western Ukraine, which are presently relict where they were entrenched. Five of them occupy the top ranks in the list of the longest gypsum caves in the world (see table). In the deep-seated karst zone of this region, extensive cave systems still develop under confined settings through the transverse circulation mechanism, as evidenced by extensive data from exploratory drilling (see Ukraine Gypsum Caves and Karst). This mechanism is unlikely to occur in salt because in this rock the joints readily seal under lithostatic pressure. The second mechanism involves the attack of the base of the evaporite unit by solutions from the underlying aquifer. Once saturated, the water becomes heavier and returns to the aquifer, where it is replaced by less dense water. Such a system of natural convection may create vast isolated cavities in both gypsum and salt. Documented examples of this kind of speleogenesis in gypsum include numerous giant Schlotten-type caves in the Zechstein (the Upper Permian) gypsum in Germany (Kempe, 1996). Cavities of this type in both gypsum and salt are probably responsible for the initiation of vertical breakdown structures called "breccia

Evaporite Karst: The world's longest and deepest gypsum caves.

The longest gypsum caves in the world Name	Length (m)	Country	Age of Rock
Optimistychna	214 000+	Ukraine	Neogene
Ozerna	117 000	Ukraine	Neogene
Zoloushka	92 000	Ukraine	Neogene
Mlynki	27 000	Ukraine	Neogene
Kristal'na	22 000	Ukraine	Neogene
Kulogorskaja–Troja	14 300	Russia	Permian
Jester	11 800	USA	Permian
Spipola–Aquafredda	10 500	Italy	Neogene
Olimpijskaja–Lomonosovskaya	9 110	Russia	Permian
Slavka	9 100	Ukraine	Neogene
Agua, cueva de	8 350	Spain	Neogene
Verteba	7 820	Ukraine	Neogene
Cater Magara	7 300	Syria	Neogene
Park's Ranch	6 269	USA	Permian
Konstitutzionnaja	6 130	Russia	Permian
Kungurskaya Ledjanaja	5 600	Russia	Permian
Kumichevskaja	4 520	Russia	Permian
Zolotoj Kljuchik	4 380	Russia	Permian
Covadura	4 245	Spain	Neogene
The deepest gypsum caves in the world Name	Depth (m)	Country	Age
Tunel dels Sumidors	210	Spain	Triassic
Pozzo A	>200	Italy	Triassic?
Corall, sima del	130	Spain	Neogene
Triple Engle Pit	130	USA	Permian
Covadura	126	Spain	Neogene
Campamento, sima del	122	Spain	Neogene
Aguila, sima del	112	Spain	Triassic
Rio Stella–Rio Basino	100	Italy	Neogene
AB 6	100	Russia	Jurassic

Note: The list has been compiled using data provided by Belski (for United States), Calaforra (for Spain), Forti & Sauro (for Italy), Kempe (for Germany), Klimchouk (for Ukraine), Woigt & Schnadwinkel (for Syria), Malkov & Lavrov (for Russia), all references in Klimchouk et al. (1996).

pipes", which are abundant in many deep-seated evaporite karsts, as illustrated by studies from the United States, Canada, China, and Russia. They have diameters ranging from a few tens to over 100 m (sometimes even measured in kilometres) and propagate by upward stopping from depths as great as 1200 m. When several closely spaced breccia pipes reach the surface, they may give rise to large (measuring from several hundreds of metres to tens of kilometres) often elongated depressions, termed solution subsidence troughs or solution-induced depositional basins. The latter term applies if the rate of sedimentation is approximately equal to the rate of subsidence, and the depressions do not have a strong topographic expression. Such features are reported in Canada, the southwestern United States, Russia (in the Urals and Siberia), and in other regions. Dissolution of salt units may occur from one side if they are laterally in contact with thick aquifers with active groundwater circulation. Over a long geological time-scale the process causes the dissolution front to recede along the margins of the major intrastratal salt deposits, leaving some of the world's largest solution-induced basins, such as those in the Prairie Provinces of Canada and Montana (the Devonian salt) or in the south of the Siberian platform of Russia (the Cambrian to Devonian sulfates and salt). Prolonged dissolution of evaporite beds leads to the formation of areally extensive brecciated horizons and disconformities, a very common end result of intrastratal evaporite karst formation (Friedman, 1997). Disruption and brecciation of the carbonate beds due to intrastratal dissolution of sulfates or salt is an important factor of deep-seated carbonate karstification in complex interbedded sequences. Another important factor is the chemical interaction between carbonates and sulfates, which can greatly increase the solubility of dolomite, gypsum, and anhydrite (Palmer, 1995).

Deep-seated evaporite karst can be readily induced or enhanced by mining, drilling, or groundwater abstraction activities, which intensify groundwater circulation and enable undersaturated water to encounter and dissolve the evaporites (Klimchouk & Andrejchuk, 1996; Martinez, Johnson & Neal, 1998). This commonly results in the appearance of collapse and subsidence features, due to the re-activation of existing vertical breakdown structures or the formation of new ones. The solution mining of evaporite salts creates large cavities that occasionally collapse. Some of the largest sudden collapse dolines, up to a few hundreds of metres in diameter and in depth, occurred due to human-induced dissolution, either intentional or unintentional. Examples are abundant throughout the United States, Canada, Western Europe, western Ukraine, Russia, China, and other countries.

Subjacent Evaporite Karst

The re-emergence of evaporite units to the relatively shallow subsurface due to uplift causes local breaching of the hydrogeological confinement by erosion of the cover rocks and locally the evaporites. This represents the subjacent karst stage, marked by drastic changes in the hydrodynamics from confined through to semi-confined, to vadose and water table, the activation and localization of dissolution, and the distinct surface expression of various karst features. The latter include collapse and subsidence dolines (which form extensively at this stage due to a decrease in the hydraulic head and the removal of the buoyant support), dry valleys, and a range of hydrological karst phenomena, such as sinking streams, ponors, and springs. Subjacent karst in evaporites is well developed where substantial karstification is inherited from the preceding deep-seated stage. Examples include the gypsum karsts of the margins of the Zechstein basin in Europe (northeastern England and northern and eastern Germany); the southwestern part of the western Ukrainian gypsum karst in the Miocene rocks; the pre-Ural karst areas and the Pinega region in the Permian sulfates of Russia; the gypsum karst areas of the margins of the Permian Basin (eastern New Mexico, north-central Texas and western Oklahoma); and the predominantly salt karst of northeast Arizona in the United States. Because subjacent karst settings are transitional in terms of karst evolution, the respective karst areas are commonly adjacent to the areas of deep-seated or entrenched karst, or to both. They frequently lead to specific engineering problems, particularly the danger of subsidence.

Entrenched Evaporite Karst

Further erosional incision may create a situation where major valleys largely or entirely drain subhorizontal gypsum beds that remain capped with protective insoluble beds. This is entrenched karst, which is common on gypsum but rarely occurs on halite because most of the medium- to thin-bedded halite in the stratified sequences is dissolved during the previous stages. Entrenched gypsum karst may show well-preserved and accessible relict artesian caves (as in the Miocene gypsum of the western Ukraine, the Paleogene gypsum of the Paris Basin of France and the Permian gypsum of the pre-Urals region in Russia) and supports the formation of contemporary caves: mainly single-conduit or crudely branching caves that develop from swallow holes on the interfluves towards outlets on the valley slopes (as in the Permian gypsum of the western Oklahoma (United States) and the North Dvina region in Russia). A characteristic feature of many entrenched gypsum karsts are vertical solution pipes ("chimneys") that develop from the overlying bed, commonly limestone or dolomite, across the gypsum bed in areas of focused percolation. Their density may be as high as 300 per square km.

Open and Denuded Evaporite Karst

Areas of open and denuded karst in evaporites, with the rocks exposed to the surface, are limited in extent because the evaporites are highly soluble. Open karsts are found mainly in arid and semi-arid environments and where the evaporite units are fairly thick. Examples of open gypsum karst include Zorbas in

Evaporite Karst: A fine polygonal karst of dolines each about 100 m across in the gypsum karst of Sivas, Turkey. (Photo by Tony Waltham)

southern Spain; the Central Apennines and Sicily in Italy; the Diebel Nador and Oranais areas in eastern and western Algeria; the Ar Rabitat / Bir area in northwest Libya; the Sivas region of Turkey (see Figure and Turkey entry); central and northern Somalia; and some mountain areas in the northern Caucasus in Russia and in Central Asia. Many other smaller areas are associated with the gypsum cap rocks of salt diapirs. The extensive Gypsum Plain in western Texas and New Mexico, United States, probably falls into the denuded karst category, which is a former intrastratal karst. Exposed gypsum karsts exhibit many surface features similar to those of carbonates. Contemporary cave development in open settings favours the formation of linear or crudely dendritic caves, which commonly carry sink streams. Where the vadose zone is thick, vertical pitches form in gypsum caves, as, for example, in the Central Apennines and Sicily in Italy, Zorbas in Spain, and the Gypsum Plain in the United States.

Open karsts in salt are small, and are most commonly associated with rising diapiric structures. Where extrusion of salt is rapid, diapirs may support dramatic landscapes with a relief of up to 900 m, with large salt cliffs, deep cirques, crevasses, and canyons, and various karst features. Outstanding examples are the Khodja-Mumyn and Khodja-Sartis salt domes in the Pamir Mountains of Tadjikistan and the salt domes in the Zagros mountains of Iran. In the Tadjikistani sites, surface karst features are particularly abundant and include numerous dolines, deep collapses and shafts, various karren forms, pinnacles, and giant salt "mushrooms" capped with residual boulders of gypsum. Caves are also numerous, formed by captured ephemeral streams. They commonly begin with vertical shafts and continue as single subhorizontal passages, leading to outlets at base level. The largest known assemblage of salt caves is found in Mount Sedom near the Dead Sea in Israel (see Sedom Salt Karst entry), where over 100 caves with >20 km of passages have been surveyed, including the largest known salt cave, Malham Cave, explored for 5.7 km (Frumkin, 2000). Other noteworthy caves in halite are known from Spain, Romania, Hungary, Russia, Algeria, and Chile.

Mantled Evaporite Karst

Gypsum karst that has been mantled by glacial till, now partly exposed along the rims of river cliffs, has been described for Manitoba, Ontario, for the Maritime Provinces of Canada

(Ford, 1997), and for Pinega and the North Dvina regions of Russia. Characteristic landforms include densely packed small suffosion dolines, vertical solution pipes, pinnacles, shlotten, and karst trenches. Similar topographies are known above river cliffs in some entrenched gypsum karsts (pre-Urals, Russia). Where the gypsum is deeply mantled, as in other parts of the above-mentioned regions, karst features are represented mainly by collapse dolines and by small caves where the gypsum is locally incised by postglacial rivers. Areas within the major river valleys in the Erbo Basin of the Saragoza region of Spain show examples of the alluviated subtype of mantled gypsum karst.

ALEXANDER KLIMCHOUK

See also **Pinega Gypsum Caves; Speleogenesis: Deep Seated and Confined Settings; Syngenetic Karst**

Works Cited

Ford, D.C. 1997. Principal features of evaporite karst in Canada. *Carbonates and Evaporites*, 12(1): 15–23

Friedman, G.M. 1997. Dissolution-collapse breccias and paleokarst resulting from dissolution of evaporite rocks, especially sulfates. *Carbonates and Evaporites*, 12(1): 53–63

Frumkin, A. 2000. Speleogenesis in salt, the Mount Sedom area, Israel. In *Speleogenesis: Evolution of Karst Aquifers*, edited by A.B. Klimchouk, D.C. Ford, A.N. Palmer & W. Dreybrodt, Huntsville, Alabama: National Speleological Society

Kempe, S. 1996. Gypsum karst of Germany. In *Gypsum Karst of the World*, edited by A.B. Klimchouk, D. Lowe, A. Cooper & U. Sauro. *International Journal of Speleology*, Theme Issue 25(3–4)

Klimchouk, A.B. 2000. Speleogenesis under deep-seated and confined settings. In *Speleogenesis: Evolution of Karst Aquifers*, edited by A.B. Klimchouk, D.C. Ford, A.N. Palmer & W. Dreybrodt, Huntsville, Alabama: National Speleological Society

Klimchouk, A.B. & Andrejchuk, V.N. 1996. Environmental problems in gypsum karst terrains. In *Gypsum Karst of the World*, edited by A.B. Klimchouk, D. Lowe, A. Cooper & U. Sauro. *International Journal of Speleology*, Theme Issue 25(3–4)

Klimchouk, A.B. & Ford, D.C. 2000. Types of karst and evolution of hydrogeologic settings. In *Speleogenesis: Evolution of Karst Aquifers*, edited by A.B. Klimchouk, D.C. Ford, A.N. Palmer and W. Dreybrodt, Huntsville, Alabama: National Speleological Society

Martinez, J.D., Johnson, K.S. & Neal, J.T. 1998. Sinkholes in evaporite rocks. *American Scientist*, 86: 38–51

Palmer, A.N. 1995. Geochemical models for the origin of macroscopic solution porosity in carbonate rocks. In *Unconformities and Porosity in Carbonate Strata*, edited by A.D. Budd, A.H. Saller & P.M. Harris, Tulsa, Oklahoma: American Association of Petroleum Geologists

Further Reading

Carbonates and Evaporites 12(1), 1997 is a special issue containing papers presented on a Symposium on Evaporite Karst: Processes, Landforms, Examples, and Impacts

James, A.N. 1992. *Soluble Materials in Civil Engineering*, Chichester and New York: Ellis Horwood

Johnson, K.S. 1997. Evaporite karst in the United States. *Carbonates and Evaporites*, 12(1): 2–14

Klimchouk, A.B., Forti, P. & Cooper, A. 1996. Gypsum karst of the world: A brief overview. In *Gypsum Karst of the World*, edited by A.B. Klimchouk, D. Lowe, A. Cooper & U. Sauro. *International Journal of Speleology*, Theme Issue 25(3–4)

Klimchouk, A., Lowe, D., Cooper, A. & Sauro, U. (editors) 1996. Gypsum Karst of the World. *International Journal of Speleology*, Theme Issue 25(3–4)

Warren, J.K. 1989. *Evaporite Sedimentology: Importance in Hydrocarbon Accumulation*, Englewood Cliffs, New Jersey: Prentice Hall

EVOLUTION OF HYPOGEAN FAUNA

Hypogean fauna characterized by troglomorphisms have been used by virtually every school of thought in evolution in support of their theories, from 19th-century creationism to past and present neo-Lamarckism (including its derivatives of directional—"regressive"—evolution), and from the neutralist school to the selectionist / neo-Darwinian one. In fact, Darwin himself took a somewhat ambiguous position in the first edition of *On the Origin of the Species by Means of Natural Selection*, where he explained troglomorphism as a combination of both natural selection and use *vs.* disuse. In later editions he took a more neo-Lamarckian stance (for a review of these schools of thought, see Romero, 2001).

Evolution is opportunistic, and that is why life is ubiquitous on earth. The total number of troglomorphic species has been estimated to be between 50 000 and 100 000 (Culver & Holsinger, 1992). This number is considerable, given that most hypogean habitats: (1) are very reduced in space; (2) generally lack primary producers (plants); and, (3) have not been thoroughly explored, particularly in tropical regions where caves contain a very diverse fauna (Deharveng & Bedos, 2000).

Historically, most cave researchers have concentrated on hypogean animals that are troglomorphic. The hypogean fauna has thus been epitomized by animals displaying some kind of troglomorphism (see Table); however, not all hypogean faunas are troglomorphic. Surveys of cave fauna yield a large proportion of non-troglomorphic organisms; further, the proportion of troglomorphic organisms in caves is inversely proportional to latitude (Peck, 1988). The number of known hypogean fish species with troglomorphism is 86 (Romero & Paulson, 2001a) and 115 without troglomorphism (Poly, 2001). These data are consistent with the contention that a large proportion, perhaps even the majority, of hypogean animals worldwide are not troglomorphic. Therefore, colonization of the hypogean environment does not necessarily require morphological changes.

Some argue that hypogean environments are so "harsh" (i.e. poor in nutrients) that only "pre-adapted" organisms could survive in them (Holsinger, 2000). Available data do not support either generalization about the ecological conditions in caves or about the nature of the colonizing animals. First, animals that colonize caves can find food, reproductive niches, protection from predators, and a place for hibernation. Therefore, the hypogean environment can offer a number of ecological opportunities to many different species of many different taxa. That is why cave colonizers can undergo extensive adaptive radiations (leading to many differentiated populations and / or species) (e.g. Hoch & Howarth, 1999). Second, contrary to generalizations based on

Evolution of Hypogean Fauna: List of some features catalogued as "troglomorphic". (Not all troglomorphic organisms display all these characteristics at the same time.)

Morphological	Physiological	Behavioural
Reduced or lost		
Eyes, ocelli	Metabolism	Photoresponses
Visual brain centres	Circadian rhythms	Aggregation
Pigmentation	Fecundity	Response to alarm substances
Pineal organ	Egg volume	Aggression
Body size		
Cuticles (terrestrial arthropods)		
Scales (fish)		
Swim bladder (fish)		
Enlarged or increased		
Chemo- and mechano-sensors	Lifespan	
Appendages	Lipid storage	

studies of caves in temperate regions (e.g. Poulson & White, 1969), many caves are very rich in nutrients, particularly in tropical regions (e.g. Deharveng & Bedos, 2000), and some are even chemo-autotrophic, that is, rich in bacteria that produce organic matter by oxidizing sulfur (see Movile Cave, Romania). Both types of caves tend to be very rich in species, and some of those species have large populations.

"Pre-adaptations" have been described as features such as nocturnal habits and hyperdevelopment of sensory organs in epigean species that are considered useful and even necessary in hypogean environments (Holsinger, 2000; Langecker, 2000). However, a recent study of hypogean fishes failed to show that all or even most fish families with troglomorphic representatives were taxa characterized by pre-adaptive features (Romero & Paulson, 2001b). Furthermore, the Texas blindcat *Trogloglanis pattersoni*, found in deep aquifers, has rather minute barbels, which is unusual for other catfishes of its family (Ictaluridae) (Langecker & Longley, 1993). This is inconsistent with the notion that enlarged sensory organs are required to enhance survival potential in the hypogean environment. Therefore, "pre-adaptations" are neither necessary nor sufficient to that end.

Caves around the world have very variable conditions, with a wide range of temperature, water supply, and size. The only thing they have in common is that for most of their length no natural light is available. The two characteristics most closely correlated with light conditions are eyes and pigmentation. Animals that have been raised under conditions of total darkness display a lower degree of eye and pigment development (see Adaptation: Eyes); conversely, when troglomorphic animals are exposed during certain periods of time to light, they may redevelop, to a certain extent, both pigmentation and the visual apparatus.

This strongly suggests that many troglomorphic animals are derived from epigean species characterized by phenotypic plasticity. Phenotypic plasticity is the ability of a single genotype to produce more than one alternative form of phenotype in response to environmental conditions. The direction and degree of response to environmental factors is known as the reaction norm, which is genetically variable and subject to natural selection. Natural selection may favour those individuals with a higher capacity to express specific traits under appropriate conditions. Phenotypic plasticity often provides a reproductive advantage over a genetically fixed one because environmentally induced phenotypes have a higher probability of conforming to prevailing environmental conditions than genetically fixed ones.

Lack of light can trigger heterochrony, i.e. changes in the timing of development of features. Examples are paedomorphs (animals that do not reach morphological maturity [metamorphs] reproducing as juveniles), and neotenes (animals with an unusually slow rate of growth). Many cave organisms are either paedomorphic or neotenic. Most troglobitic salamanders are paedomorphic, and half of all known paedomorphic salamanders are troglobites. Salamander evolution into a paedomorphic condition can be quite fast. Individuals living in the hypogean environment gain an advantage by becoming paedomorphic because this condition gives them the flexibility to survive in an unpredictable environment. Paedomorphosis in *Eurycea neotenes* seems to be a response to selection for the ability to pass dry periods in hypogean aquatic refugia (Sweet, 1977; see also Amphibia). Also, neoteny in hypogean animals is well documented for reduced body size, loss of scales, fin modifications, and reduced ossification.

Natural selection that favours paedomorphs / neotenes fixes their paedomorphic / neotenic alleles in the cave population. Given that most cave populations are small and subject to very similar selective pressures within the same cave, this evolutionary process can take a relatively short period of time. In fact, paedomorphosis can be achieved via a major gene effect (a small genetic change generating a large phenotypic effect). Troglomorphic characteristics can arise via minor changes in developmental genes.

Only when there is a constant gene flow from the epigean environment can such changes be prevented. In this respect the recessive allele can be considered the "troglomorphic gene" because it manifests a morphologically and ecologically differentiated phenotype that is reproductively isolated from the epigean ancestor. This explanation is supported by the convergent nature of troglomorphic characteristics. Convergent evolutionary patterns are strong evidence of adaptation via natural selection. Isolation would later lead towards speciation (genetic differentiation from the epigean ancestor; see Speciation). Many troglomorphic organisms are believed to have recently invaded the hypogean environment, since their epigean ancestor is easily recognizable and can even interbreed with them to produce fertile hybrids.

The evolution of troglomorphic characteristics does not necessarily occur in parallel. This is because: (1) they are controlled by sets of genes independent of each other; (2) the degree of development of these characteristics is conditioned by their phylogenetic history; and (3) because the selective pressures behind each one of those characteristics may differ from cave to cave (Culver *et al.*, 1995; Romero & Paulson, 2001c). In addition to reduction / loss of phenotypic characteristics, many troglomorphic organisms exhibit enhancement of sensory systems (chemical and mechanical) that are favoured by natural selection, since these sensory systems increase the fitness (survival capabilities) by helping them to find food and mates. Complex, coordinated, and adaptive phenotypes can originate rapidly and with little genetic change, via correlated shifts in the expression of plastic traits. Composite characteristics, like those often observed among troglomorphic organisms, are produced by correlated phenotypic shifts that give the impression of a coevolved character set.

There are still grey areas in our understanding of the evolution of hypogean fauna. For example, we assume that behaviour plays a major role in the colonization of new niches, since behaviour is the most plastic part of the phenotype. Although changes in behaviour are well documented and present even in transitional forms, we do not know the role played by behaviour in the changes in other phenotypic characters. Behavioural changes usually precede external morphological evolution. Behavioural flexibility is thus the first condition for success in cave colonization. Only after colonization has taken place do morphological and physiological changes take place. Individual adaptability is the main target of selection. Behavioural invasion of a new environment by adults exposes the reaction norm of their progeny. Since hormonal production is closely linked to behaviour and hormones play a role in many developmental processes, there is a great potential for hormones to produce (or act as) developmental constraints. Therefore, the arrest in development of features could be due to diminishing hormonal production as a physiological consequence of the adaptation to the hypogean environment (i.e. many behaviours are no longer performed under conditions of darkness). This, in turn, could lead to the differential regulation of developmental genes.

The phenomenon of loss or reduction of phenotypic characteristics is not unique to troglomorphic organisms, and can be found among other animals such as parasites, deep-sea creatures, and the inhabitants of highly turbid waters. Also, limblessness and flightlessness are common among faunas living on small islands and high mountains. The loss of limbs among cetaceans is an example of a major evolutionary novelty by default. Even humans show loss or reduction of a number of characteristics inherited from their ancestors (Diamond & Stermer, 1999). Thus, troglomorphisms can be explained using well-known evolutionary mechanisms, without needing to resort to neo-Lamarckian explanations or terminology such as "regressive evolution". The study of this phenomenon has been largely neglected in mainstream evolutionary biology due to the sense that evolutionary novelties mean addition, not subtraction, of characteristics, and because of the way in which this biological phenomenon has been used by neo-Lamarckians to advance their own cause of either inheritance of acquired characteristics or the notion that evolution has some sort of directionality (Romero, 2001).

ALDEMARO ROMERO

See also **Adaptation; Biodiversity in Hypogean Waters; Colonization; Speciation in Caves and Karst**

Works Cited

Culver, D.C. & Holsinger, J.R. 1992. How many species of troglobites are there? *National Speleological Society Bulletin*, 54: 79–80

Culver, D.C., Kane, T.C. & Fong, D.W. 1995. *Adaptation and Natural Selection in Caves: The Evolution of* Gammarus minus, Cambridge, Massachusetts: Harvard University Press

Deharveng, L. & Bedos, A. 2000. The cave fauna of southeast Asia. Origin, evolution and ecology. In *Subterranean Ecosystems*, edited by H. Wilkens, D.C. Culver & W.F. Humphries, Amsterdam and New York: Elsevier

Diamond, J. & Stermer, D. 1999. Evolving backward. *Discover*, 19: 64–68

Hoch, H. & Howarth, F.G. 1999. Multiple cave invasions by species of the planthopper genus *Oliarus* in Hawaii (Homoptera: Fulgoroidea: Cixiidae). *Zoological Journal of the Linnean Society*, 127: 453–75

Holsinger, J.R. 2000. Ecological derivation, colonization, and speciation. In *Subterranean Ecosystems*, edited by H. Wilkens, D.C. Culver & W.F. Humphries, Amsterdam and New York: Elsevier

Langecker, T.G. 2000. The effects of continuous darkness on cave ecology and cavernicolous evolution. In *Subterranean Ecosystems*, edited by H. Wilkens, D.C. Culver & W.F. Humphries, Amsterdam and New York: Elsevier

Langecker, T.G. & Longley, G. 1993. Morphological adaptations of the Texas blind catfishes *Trogloglanis pattersoni* and *Satan eurystomus* (Siluriformes: Ictaluridae) to their underground environment. *Copeia*, 1993: 976–86

Peck, S.B. 1988. A review of the cave fauna of Canada, and the composition and ecology of the invertebrate fauna of caves and mines in Ontario. *Canadian Journal of Zoology*, 66: 1197–1213

Poly, W.J. 2001. Nontroglobitic fishes in Bruffey-Hills Creek Cave, West Virginia, and other caves worldwide. *Environmental Biology of Fishes*, 62: 73–83

Poulson, T.L. & White, W.B. 1969. The cave environment. *Science*, 165: 971–81

Romero, A. 2001. Scientists prefer them blind: The history of hypogean fish research. *Environmental Biology of Fishes*, 62: 43–71

Romero, A. & Paulson, K.M. 2001a. It's a wonderful hypogean life: A guide to the troglomorphic fishes of the world. *Environmental Biology of Fishes*, 62: 13–41

Romero, A. & Paulson, K.M. 2001b. Scales not necessary: The evolution of scalelessness among troglomorphic fishes. *Program Book and Abstract, Joint Meeting of Ichthyologists and Herpetologists, 81st Annual Meeting of the American Society of Ichthyologists and Herpetologists, 2001* (abstract) (unpublished; available at http://www.asih.org/meetings/2001/abstracts_2001.htm)

Romero, A. & Paulson, K.M. 2001c. Unparalleled evolution: Blindness, depigmentation, and scalelessness do not run hand in hand among troglomorphic fishes. *2001 NSS Convention. A Cave Odyssey. July 23–27, 2001, Mount Vernon, Kentucky, Program Guide*, p.64 (abstract) (unpublished; available at http://www.macalester.edu/~envirost/abstract_unparalleled_nss.htm)

Sweet, S.S. 1977. Natural metamorphosis in *Eurycea neotenes*, and the generic allocation of the Texas *Eurycea* (Amphibia: Plethodontidae). *Herpetologica*, 33: 364–75

EXPLORATION SOCIETIES

One hundred and fifty years ago, there were no cave exploration societies, some 100 years ago there were 15 or more, and now they are uncountable. This article is concerned with their development—why, when, and where they were formed, what effect they had on exploration and the publication of results, and how they spread out from the core of 19th-century European cave exploration activity. Recently the quantity of clubs has increased vastly, and an increasing proportion are devoted solely to the sporting aspects of caving. However, the motivated and progressive societies that achieve research results, both in exploration and cave study, are still similar in nature to those of 100 years ago.

The creation of cave exploration societies has had three main effects:

1. They bring together people with similar interests, stimulating deeper and longer-term involvement in the field;
2. By working together as a group, the members are able to undertake explorations that are technically more difficult and physically more demanding than those they could have done alone;
3. Their specialist cave publications not only increase the amount of speleological material published, but make it more readily available to people who have an immediate interest in it.

The locations of cave exploration societies are determined by the geography of the major karst areas, particularly those where karst scenery is widespread and spectacular. Another requirement is the presence of a sufficiently large educated population in or near the karst area.

The popularity of rock climbing and mountain exploration developed steadily in the 19th century, marked by the founding of the Alpine Club in 1857 and the Deutsche und Österreichische Alpenverein in 1862. Some of the earlier speleological groups were in fact offshoots of mountaineering clubs. The earliest known cave society was the Höhlenklub of Appenzell in eastern Switzerland, which existed in the early 1860s. Very little is known about it, for it evidently kept no records or minutes and did not publish its findings. Its members are known to have visited the caves of Wildkirchli and Dürr-Schrennen in 1863, for they wrote a short account of these in the visitors' book at the nearby inn. This Appenzell club was not heard of again and most certainly had no effect on the subsequent formation of the more scientific groups in Austria and Slovenia.

The first real speleological group was the Verein für Höhlenkunde, founded by Franz Kraus (1834–97). On 18 January 1879, he put a notice in a Viennese newspaper, inviting the attention of anyone who was interested. The response was good and after several months of preparation the Verein für Höhlenkunde was formally founded in Vienna on 19 December of the same year. Soon afterwards it published five issues of a bibliographical publication on caves, *Literatur-Anzeiger*, the first of which appeared on 17 February 1880. It was the world's first speleological periodical. Sixteen months after its formation, the group, which then had a membership of 44, became part of the Österreichischen Touristenclub as a separate Section für Höhlenkunde. Its new publication, the *Mittheilungen* of the Section, recorded the scientific exploration work of its members and was thus the earliest conventional speleological journal. A Karst Committee formed within the Section adopted a semi-professional approach to its work, which became increasingly associated with the practical problems of karst hydrology. Kraus succeeded in acquiring subsidies from two ministries in the Austrian government and also from the Südbahn Austrian railway company, and the Karst Committee's activities culminated in the sponsored cave work of Putick. The impetus spread over most of the Austro-Hungarian Empire, so that by the end of 1892 Kraus knew of 659 caves there.

Another very early cave group was the Section Küstenland des Deutschen und Österreichischen Alpenvereins. This "coastal section" had in fact been set up in 1873 but not with cave exploring in mind. It was only in 1883 that a cave section or Grottenabteilung was created, but from then on it was noted for the very difficult explorations that it carried out, including those of Škocjanske jame (Figure 1, and see separate entry) and Kačna jama. Its leaders were Anton Hanke (1840–91) and Jozef Marinitsch (1838–1915).

The first of the Italian cave societies also worked in Slovenia as well as that part of Italy close to the border. The Società degli Alpinisti Triestini was founded at Trieste on 23 March 1883. There were 73 founder members and the number soon grew to over 100. A subsection for caves, the Commissione Grotte, was formed and explored the Grotta Gigante, Trebiciano (329 m deep), the Abisso sopra Chiusa (227 m deep), and other deep caves (Figure 2). In 1885 this society merged with the still-existing Società Alpina delle Giulie. The total number of caves explored rose at an increasing rate, from four in 1883 to 230

Exploration Societies: Figure 1. Workers employed by the Deutsche und Österreichische Alpenverein to assist in the very difficult exploration of Škocjanske jame. A photograph taken between 1884 and 1892.

Exploration Societies: Figure 2. Žankajna jama in Croatia during exploration in 1925 when it was called Abisso Bertarelli, showing members of the Commissione Grotte of the Società Alpina delle Giulie (reproduced from their archives, with permission).

by 1900. The caves were recorded systematically and allocated catalogue numbers.

The influence of Austria and Italy was very evident in these first few groups working in Slovenia, which was then a part of Austria and close to the Italian border, but the Austrian influence was not quite as strong as it appears at first sight. Admittedly, the initial investigation came from Kraus, who continued to participate although he lived in Vienna, and the cave groups were administered as parts of Austrian societies. Many of the principal explorers and leaders, however, were local people: Hanke and Marinitsch were Slovenes and Hrasky was a Czech who lived and worked in Ljubljana; Putick was also a Czech.

It was in 1889 that the first purely Slav cave society was formed, a group called Anthron whose main discoveries were in the Postojna system. It cooperated closely with foreign explorers and E.A. Martel carried out his 1893 Postojna work with its members as well as with Kraus and Putick.

Of very great importance was the French Société de Spéléologie, founded in Paris by Martel in 1895 and concerned with specialist cave research as well as exploration. A great deal of the Society's influence was due to its publications, the *Bulletin* and the *Mémoires*. Together they comprised 98 numbers between 1895 and 1914, with a total of 4843 pages. Martel was the author of about 20 of the issues, including a special 810-page volume of annotated bibliography.

The wide scope of the Société de Spéléologie was signified in the first article of its original statutes, which read:

> The Speleological Society has been founded in order to engage in the exploration, facilitate the general study and aid in the control or utilisation of all kinds of subterranean cavities, whether known or unknown, natural or artificial; to encourage and give grants to investigations bearing in any way on caverns; and in short to popularise and develop researches of every kind into the interior of the earth from practical as well as theoretical, utilitarian as well as scientific points of view.

In England, the Yorkshire Ramblers' Club was much more oriented towards the sporting aspects of cave exploring, although it did do a lot of useful original work. It was formed in Leeds in 1892, primarily as a mountaineering club. In Gaping Gill, where Martel had made the first descent of the main shaft in 1895, the Yorkshire Ramblers' Club pursued the exploration of the entire passage system at the bottom, surveying a total of about 1100 m by 1906.

The rate of cave exploration increased greatly after 1900, even quite early in the century. It has been estimated that in many countries the number of caves known increased by a factor of ten between 1892 and 1911. In the Trieste and Istria region alone the number rose from 230 in 1900 to 510 in 1920, and by 1937 the total was 3500. No figures have been published for the number of caves known in the world today but it is likely to be more than a quarter of a million.

At the beginning of the century, the societies were mainly interested in the discovery and exploration of new caves. They took this seriously and for the most part published their results, often with plans of the caves. Of the 19th-century groups, only the Commissione Grotte of the Società Alpina delle Giulie and the Yorkshire Ramblers' Club still survive and both of them continue to publish. New societies founded in Italy shortly after the turn of the century included the Società Speleologica Italiana (1903), the Circolo Speleologico Romano (1904) and a Commissione Grotte in the Fiume (= Rijeka) section of the Club Alpino Italiano (1904). In England, the Yorkshire Speleological Association was founded in 1906; it explored and surveyed about 8 km of difficult caves but disbanded in about 1914.

Most of these societies were interested to some degree in the specialist scientific aspects of caves, as well as in accurate surveying and recording. All of them, however, were and are entirely amateur; a few of the members may be professional speleologists in institutes or universities and the societies sometimes receive grants or subsidies to aid in particular projects, but they remain amateur organizations. The link between the amateur's recreational cave exploring and the work of the professional scientist is often close and sometimes indistinguishable. Dr Walther von Knebel in 1906 said of it:

> Cave science depends upon the help of the layman for completion. It is not always possible for the expert to check and complete his studies. Cave sportsmen are as useful in cave science as, for example, Alpine sportsmen are for glaciology.

The amateur who enjoys working in caves for its own sake is often a willing collector of data, however time-consuming this may be. This type of society or group, with its combined interest in the scientific and recreational aspects of caves, has persisted throughout the 20th century. It includes or has included some of the most important national and local societies, for example the British Speleological Association, the Fédération Française de Spéléologie, the Société de Spéléologie, the National Speleological Society (United States), the South African Speleological Association, and the Sydney Speleological Society.

The very great growth in the numbers of cave societies or clubs formed in the 20th century has included many which are concerned solely with the sporting side of cave exploring. Some such groups, including the long-established Yorkshire Ramblers' Club, do seek new caves, often with notable success, and do publish their findings. In other cases, not even new discoveries are sought, but just the satisfaction of overcoming the physical difficulties of the exploration. The Leeds Ramblers' Club descended Gaping Gill in 1904 and continued to do so annually for many years, often without undertaking any original exploration there, and there are many later examples.

The vast increase in the number of societies is the result of many causes, including increased population, a larger proportion of people living in cities, the higher public profile of caving in the media, and the presence of even more clubs attracting recruits. An important influence in enlarging the total number of societies (although now masked by the causes given above) has been the growing internationalization of speleology. The entry for Speleologists in this Encyclopedia emphasizes Martel's role in spreading interest across Europe and even into the United States. Also, as the numbers of individual speleologists in the world has increased, their working abroad and their emigration has helped to start up similar activity in their new countries.

Two non-European countries are of particular interest in this respect, one whose cave work started surprisingly early, and the other surprisingly late. In Australia, where serious cave exploration and surveying was already going on in the 1880s, it was due not to amateurs in societies but to enthusiastic professionals working for the government, such as the surveyor, cave explorer, and writer Oliver Trickett. In the United States, on the other hand, despite Martel's already close links with American speleologists even before his visit there in 1912, it was not until 1939 that the first cave exploring society was founded: the Speleological Society of the District of Columbia, soon to be transformed into the National Speleological Society, which is now the world's largest cave society with over 12 000 members. There had been talk of forming a national society in the 1930s, but it did not come to anything. It seems that at that time individual academics and cave explorers were too thinly spread over that vast country to provide a sufficient nucleus in any one place to form a group. There was certainly no equivalent of the Australian government-led cave research.

To list the present-day societies and associations is difficult and has to be partly subjective, taking into account their size, new exploration, scientific work, publications, etc. The accompanying Table cannot be taken as listing the "top ten" or so, but certainly all those included are important and produce substantial results of a professional standard. In some countries (e.g. China), however, most exploration is institute-sponsored and, in others such as Poland, Spain, and Slovenia, the principal publications are from professional or academic institutions.

Exploration Societies: Some major present-day cave groups (arranged alphabetically). Specialist government and academic institutes are not included.

Association for Mexican Cave Studies	United States
British Cave Research Association	United Kingdom
Česka Speleogicka Společnost (Czech Speleological Society)	Czech Republic
Commissione Grotte "Eugenio Boegan", Società Alpina delle Giulie ("Eugenio Boegan" Cave Group of the Giulian Alpine Club)	Italy
Fédération Française de Spéléologie (Speleological Federation of France)	France
Jamarska zveza Slovenije (Speleological Association of Slovenia)	Slovenia
National Speleological Society	United States
New Zealand Speleological Society	New Zealand
Società Speleologica Italiana (Italian Speleological Society)	Italy
Société Suisse de Spéléologie (Swiss Speleological Society)	Switzerland
Spéléo-Club de Paris (Speleo Club of Paris)	France
Speleological Society of Japan	Japan
Speleološki odsjek HPD "Šeljezničar" (Cave section of the Croatian Mountaineering Club "Željezničar")	Croatia
Sydney Speleological Society	Australia
Verband der deutschen Höhlen- und karstforscher (Association of German Cave and Karst Researchers)	Germany
Verband Österreichischer Höhlenforscher (Association of Austrian Cave Researchers)	Austria

TREVOR SHAW

Further Reading

Guidi, P. 1994. *Cenni sull attività dei gruppi grotte a Trieste dal 1874 al 1900* [An Outline of Cave Club Activity at Trieste from 1874 to 1900]. *Atti e* Memorie della Commissione Grotte "Eugenio Boegan", 32 for 1994: 85–127

Shaw, T.R. 1992. *History of Cave Science: The Exploration and Study of Limestone Caves, to 1900,* 2nd edition, Broadway, New South Wales: Sydney Speleological Society
This gives more detail, with references to original sources.

EXPLORING CAVES

Human beings have been exploring caves for thousands of years, initially with nothing more than a hand-held light. Recreational caving, the exploration of caves for pleasure and as a sporting challenge, is generally held to have begun in the late 19th century with the explorations of Édouard-Alfred Martel, and the number of participants increased greatly from the 1950s with the general increase in leisure time and improved transport. As cave exploration became more popular, specialized equipment was developed. Equipment has changed considerably over this period with an escalation of changes in the last 50 years. This entry examines what a human being needs to explore a cave.

Lighting

A source of lighting has always been the most important piece of caving equipment. Lights were originally burning materials, such as flaming torches or candles. For most of the 20th century, carbide lamps burning acetylene gas were the standard caving light and they are still used for expedition caving. The 20th century also saw the introduction of electric lights which have gradually replaced carbide lights for most caving uses. The cap-mounted miner's lamp together with a belt-worn acid battery was adopted for caving, and more recently specialized caving lights have been developed. Modern electric caving lights are water resistant with two sizes of bulbs to give high or low illumination, and usually incorporating a spare bulb. Both rechargeable and disposable batteries are used. For expedition use a carbide light with an electric backup is the norm. Electric caving lights have progressed from simple tungsten bulbs to higher performance halogen or krypton bulbs and recently the trend is towards LED lights which are robust, rarely fail, and being more efficient, offer longer battery duration.

Protection from Cold, Water, and Abrasion

Cave explorers in 1900 wore many layers of everyday clothing, tweed suits being particularly popular in Europe. Abundant army surplus gear from World War I and (mainly) World War II kept cavers clothed and equipped for many years, but during the 1960s, cavers began to use clothing borrowed from other outdoor sports. Woolly underwear came from mountaineering but oversuits were simply mechanics' overalls. For very wet caves or those requiring complete immersion, wetsuits originally used for diving became common, often hand-made by the owner or highly modified with extra neoprene pads to provide protection for knees and elbows. A thin oversuit was often worn over these wetsuits in an attempt to reduce damage from abrasion. In the early 21st century the most usual caving clothing in temperate areas is thermal underclothing in one or two layers with a specialized oversuit over the top. These oversuits can be very tough and waterproof and are used in all but the wettest caves. There is, however, some variation in clothing around the world from hot to cold climates. Protection against abrasion is usually important so even in the hottest caves thin oversuits tend to be worn.

Footwear has also progressed. Originally miners' boots made of leather or rubber and with steel toecaps and nailed soles were used. During the 1980s specialized caving rubber wellingtons or walking boots became popular, often worn with neoprene bootees. More recently there has been a move towards more specialized footwear, such as water boots, normally used for canoeing or canyoning and in the future it is likely that highly specialized boots will be produced with rigid toecaps and wet-grip soles.

Originally helmets were worn as protection against hitting the roof or being struck by falling rocks but more recently they have also provided a means of mounting the light and freeing hands for climbing and carrying. Early helmets were made of leather, cardboard, or plaited bamboo, and bamboo helmets are still manufactured and used in mines in China. Since the late 1960s there have been several technological advances, including the development of plastic and fibreglass helmets designed for mining with lamp brackets and shock-absorbing internal cradles added. Modern caving helmets are lightweight, comfortable, and capable of absorbing a considerable shock.

Vertical Equipment

Equipment for climbing or descending pitches has progressed enormously since the 1950s. Before this time, exploration of shafts often took place by people being lowered down on a rope or winch, a method that had not changed for several hundred years. Free climbing was originally attempted with nothing but the occasional hand line, but this severely limited the caves that could be explored. Rope ladders developed quickly from the late 19th century and some large drops were descended by natural fibre rope ladders with wooden rungs. Wire rope ladders with aluminium rungs replaced natural fibre rope ladders and several climbs of over 300 m were made using this equipment, including Provatina and Epos chasm in Greece in the 1960s. Ladders though were still heavy and easy to fall from, thus requiring an additional safety line. The largest drops continued to be descended using hand-operated or powered winches.

The real advance came with the first readily available synthetic ropes. During the late 1960s North Americans started using a single rope to descend and ascend caves—Single Rope Techniques (SRT) were born. Low stretch, abrasion-resistant ropes were developed and SRT spread rapidly throughout the world during the 1970s. The rope passed through a friction device attached to a caver's seat harness and friction reduced the speed of the descent. These devices were originally borrowed from climbing and the "figure of eight" remains popular but devices specialized to descend long drops on single ropes soon began to appear. The rappel rack was invented in the United States and is still the American descender of choice. The rack consists of a bent metal frame in the shape of a giant paper clip

Exploring Caves: Figure 1. The first exploration of the deep shaft of Tripa tis Nifis in Greece; the main belay is a good thread, but the rock is thin bedded, so back up belays are onto the piton and the rope of the pitch above. (Photo by Tony Waltham)

with a number of aluminium friction bars like ladder rungs across it, which the rope passes back and forwards between. In Europe fixed pulley or "bobbin" devices were developed where the rope describes a tight "S" through the device to slow the descent by friction. Locking mechanisms have since been added.

To ascend a rope, short lengths of cord were originally tied with "prusik knots" around the rope which could be pushed up but would not slide down when under tension. Climbing a single rope has since been known as "prusiking" even though prusik knots which were cumbersome and slow were soon replaced by mechanical ascenders. The first mechanical ascenders—"Jumars"—revolutionized caving. Two distinct methods of climbing a rope developed: the sit / stand or frog, and the ropewalker. The frog uses two mechanical ascenders, one worn on the chest between a sit and chest harness, the other connected to the seat harness by a safety cord and with a long cord to a footloop. By standing on the long cord the chest ascender slides up the rope; then, while sitting back on the chest ascender, the second ascender can be lifted up the rope by hand. As both these devices are near hand / eye level, it is very simple to get the system on and off the rope allowing cavers to pass knots or easily change from one rope to another. In the United States, various forms of "ropewalking" developed. The ropewalker attaches one ascender to each foot with a third ascender or pulley at chest level to keep the body upright, and he ascends by a walking motion up the rope. This gives a considerable speed advantage over frog but is heavier and less manoeuvrable. Sit and stand systems are by far the most common techniques worldwide, although some people still use ropewalking systems, particularly for long pitches.

Vertical caving equipment has become more and more specialized with low stretch, highly abrasion-resistant ropes being further developed. Rope strengths have increased and thicknesses decreased: SRT began on 11 mm and 12 mm diameter ropes but by 2000 most exploration caving was being done on 8 mm to 9 mm rope. Most caving ropes are made from nylon, although for a time single metal wire ropes with special ascending and descending devices were used. These gained some popularity in the former Soviet Union due to the unavailability of adequate nylon rope.

For very wet caves or surface canyons, floating rope is often used, although the strength and abrasion resistance are not as high as other caving ropes. Life jackets are also frequently worn where a considerable amount of swimming is involved in cave exploration.

Ropes and ladders were originally anchored to natural rock belays but these are not necessarily available or in the right place for a good "hang" away from water or loose rock. Hence, in Europe, and increasingly elsewhere, rock bolts have largely superseded natural anchors. Rock bolts are also placed in suitable positions for cavers to abseil through caves pulling the rope after them where exit is possible through a second entrance. Several types of masonry anchor were initially used before cavers settled on the 8 mm self-drilling anchor or "spit". These have an expansion sleeve that has its own drilling teeth cut onto each anchor. They were initially placed using a removable handle and hammer but portable hammer drills have now largely replaced arm-power and it is common to use less bulky expansion sleeves that require a smaller hole. Larger bolts held in by resin with an integral ring hanger ("P-bolts") are often used in high traffic caves as they deteriorate only gradually with use.

The development of rock bolting has enabled long descents of difficult caves to be done safely. It has also enabled climbs upwards to be made much more easily. Since the 1950s a variety of ways of climbing upwards have been developed, including solid lightweight ladders, scaling poles, and more recently, small platforms. Using advanced bolting techniques, cavers have climbed hundreds of metres up vertical or overhanging rock walls to discover high-level roof passages. Titan Shaft in Peak Cavern, Derbyshire (England) is an example where cavers have climbed for over 140 m. The downside is that with bolts so physically easy to place, many unnecessary bolts have been inserted, leading to considerable damage to popular caves.

When SRT was first used the harnesses were borrowed from the climbing world and the body weight rested mainly on the groin (British cavers spent many uncomfortable years in Willans harnesses). Development in harnesses led to leg loops where the weight is taken on wide straps around the buttocks. Specialized caving harnesses designed to be sat in for long periods with increased abrasion resistance and minimal water absorption were developed. Chest harnesses have been produced to pull up an ascender and to keep the caver more upright. Chest harnesses can be sewn like brassieres or simply a length of tape passed around the chest and over the shoulders.

Miscellaneous Caving Equipment

To carry all this equipment a variety of bags and containers have been developed. Originally ordinary rucksacks were used but they were very expensive and were rapidly destroyed by abrasion. Hence, special bags were developed in very tough materials with minimal folds and loops to get abraded or caught on projections. Waterproof bags are often used to keep equipment dry and as flotation devices to aid swimming and transport heavy equipment through flooded passages.

With caves being explored to ever greater lengths, the need has increased for underground camping. In a large cave, camps can be much like those on the surface and in cold or wet caves

Exploring Caves: Figure 2. Exploring a wide passage with limited air space over a long lake in Gua Sodong in the Gunung Sewu karst of Java. (Photo by Andy Eavis)

tents may even be transported underground and erected. In small caves hammocks with or without covers are often used, fixed with rock bolts across or even along the passage. Underground food is generally kept as lightweight as possible and is heated on a normal camping stove. Sleeping bags must always be made from synthetic fibres as down loses its warmth when wet.

New caves are becoming harder to find in many parts of the world and most new discoveries are now made by digging, climbing, or diving. Cave digging ranges from simple excavation of loose sediments with bare hands to almost mining a way through. Battery-powered electric drills are commonly used for shot holes to place explosives for enlarging passages or removing boulders. Bolts are used extensively for climbing although scaling poles are still used for smaller climbs. Cave diving has become a sport in its own right (see Diving in Caves). Surveying known cave passages can often suggest sites for digging or climbing (see Surveying in Caves).

The scope for finding new caves and for making new discoveries in known caves is still enormous and there are still many lesser-developed nations that have seen very little cave exploration. As political conditions change, cavers become wealthier, and travel gets even easier, these countries will become centres of cave exploration.

ANDY EAVIS

See also **Recreational Caving**

Further Reading

Eyre, J. 1981. *The Cave Explorers*, Calgary, Alberta: Stalactite Press

Halliday, W.R. 1974. *American Caves and Caving*, New York: Harper & Row

Judson, D. (editor) 1984. *Caving Practice & Equipment*, Newton Abbot, Devon: David T. Charles

Lyon, B. 1983. *Venturing Underground: The New Speleo Guide*, Wakefield: EP

Marbach, G. & Tourte, B. 2000. *Techniques de la Spéléologie Alpine*, 3rd edition, Pont-en-Royan: Expé

Meredith, M. & Martinez, D. 1986. *Vertical Caving*, 2nd edition, Dent: Lyon Equipment

Montgomery, N.R. 1977. *Single Rope Techniques: A Guide for Vertical Cavers*, Sydney: Sydney Speleological Society (Occasional Paper no. 7)

Smith, B. & Padgett, A. 1996. *On Rope: North American Vertical Rope Techniques for Caving*, revised edition, Huntsville, Alabama: National Speleological Society

Thrun R. 1971. *Prusiking*, Huntsville, Alabama: National Speleological Society

Warild, A. 1994. *Vertical: A Technical Manual for Cavers*, 3rd edition, Sydney: Speleological Research Council; updated 2001 on CD-ROM only

EXTRATERRESTRIAL CAVES

As Earth is a water-dominated, volcanically active planet, there are numerous solutional and volcanic mechanisms always at work creating subsurface voids. Hence, Earth is cave-rich, but not the only body in the Solar System to have caves, as both the Moon (Oberbeck, Quaide & Greeley, 1969) and Mars (Figure 1) are known to have lava tube caves. There are probably other types of caves on many other bodies in our Solar System but direct evidence does not yet exist as exploration of the planets and smaller bodies is still at an early stage. Detection of subsurface voids has not yet been attempted on space missions.

Cave conditions within a planet's crust may be very different from those on the planetary surfaces that geologists have studied so far. On Earth, the cave environment often differs significantly from overlying surface conditions. For example, caves can maintain cool temperatures, saturated humidity, open pools, and sometimes permanent ice even when there is a hot desert on the surface. In most cases, the cave thermal regime will be more stable than at the surface because of insulation by surrounding material. For example, Mendell (in Horz, 1985) calculated that lunar lava tube interiors will remain at a constant $-20°C$ (at depths greater than 30 cm) despite mean surface temperature fluctuations of more than $300°C$ ($-233°C$ minimum at polar night to $+130°C$ at equatorial noon).

Classifications of extraterrestrial caves are based on knowledge of Earth caves and the fundamental physical processes responsible for their formation (see Table). This categorization helps geologists imagine possible cave types that could form under radically different extraterrestrial conditions, and to develop cave-detection protocols for future planetary missions.

Why are caves on other planets plausible? A primary mechanism for cave formation involves cracking of crustal material, often followed by other enlargement mechanisms. On any planet with a solid surface, application of an energy source (either internal or external) will result in cracking. Those cracks then provide the basis for cave formation by a variety of terrestrial and possibly other non-terrestrial mechanisms (see Table). Heat flow from the planetary interior drives the type of plate tectonics that we have on Earth, resulting in crustal movements and deformation. Uplift, faulting, and earthquakes can all result in surface and subsurface cracks ultimately leading to speleogenesis. In the case of planets like Mars with lower internal heat flow, little or no plate tectonic activity has ever existed. However, the majority of rocky planets, larger asteroids, and moons have an early history of significant heat flow and crustal melting. Other kinds of tectonism not involving large plate motions can occur, typically involving expansion and contraction and both producing cracking. Surface cracks can also result from impact cratering. Many other Solar System bodies are heavily cratered including Earth's Moon, and many moons of the gas giant planets, for example Ganymede and Callisto orbiting Jupiter, and Mimas and Dione orbiting Saturn. In the case of some of the moons around gas giant planets (e.g. the Jovian moon Io), tidal flexure, caused by a moon's close proximity to the enormous gravitational pull of its primary planet, has caused extremely active surface geological processes including extensive fissure formation (Moore *et al.*, 1999). Tectonic caves can be created by processes as simple as crumbling or non-fluid undercutting in faults, scarps, and other cracks (see Talus Caves). Subsequent enlargement can occur via freezing expansion of ice. Large blocks of igneous materials can crack off and slide, thus creating voids and passage. Mass wasting of boulders can also create labyrinths, for example perhaps disaggregating scarp crests along margins of sites like Vastitas Borealis on Mars.

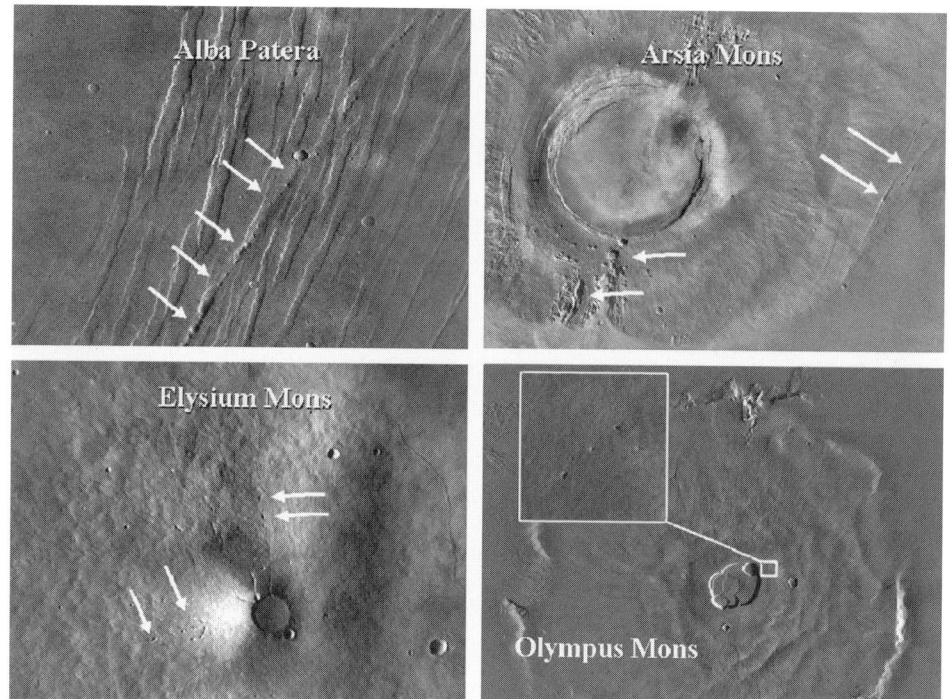

Extraterrestrial Caves: Mars images showing lava tubes (arrows) in a variety of volcanic landscapes. Viking Orbiter Camera Image; NASA.

A second major cave formation mechanism involves melting and resolidification of materials, as in the case of lava tube caves. Though the volcanically active eras of bodies such as Mars or the Moon are long past, very large lava tubes (3–10 times sizes typical for Earth) appear to be the norm due to the lesser gravity of those bodies (Bunnell, 1991) and possibly to their magmatic composition (Keszthelyi, 1995). The morphological evidence of lava tubes is easily seen in lunar images as strings of collapse features near the tube ends, for example at the western end of Mare Imbrium. Features on the flanks of Martian volcanoes like Olympus Mons also appear as lava tube strings radiating from central volcanic sources just as we see them on Earth (see Figure). Tidal flexure volcanism on Io may have produced lava tubes (Peale, Cassen & Reynolds, 1979). The history of Venus is also dominated by massive, global volcanism of unrivalled proportions (Sakimoto, Crisp & Baloga, 1997) and Venus may prove to have lava tubes (Zimbelman, 1998). On bodies with gravity less than Earth's, less lava flow trenching erosion will occur; thus formation of some lava tube types that we find on our planet (see Volcanic Caves) may be rare on bodies like the Moon and Mars.

Besides volcanoes that emit melted rock in the type of volcanism familiar on Earth, the cold, icy moons of Jupiter show evidence of volcanism involving supercooled water, other fluids, and ices known as "cryovolcanism" (Wilson, Head & Pappalardo, 1997). This process could produce caves that have no terrestrial counterparts, for example ice tubes, cryotumuli, or other as yet unimagined speleological objects. In addition, because of very low global temperatures, glacier caves could be permanent features on icy moons rather than the transient phenomena that they are on Earth.

Permafrost caves caused by ground ice suffosion (Ford, 1997; see also Piping Caves) and interaction with karst terrain (Pinneker & Shenkman, 1991) are known from the high latitude regions of Earth. Similar processes can be imagined in circumpolar regions of Mars, on icy satellites, or even icy asteroids or comets with periodic sublimation of ices occurring each time they pass perihelion. Ice sapping features are clearly visible on Mars, although the specific details of this process on a planet with a hyperarid surface (drier than any site on Earth), extremely low temperatures (lower than anywhere on Earth), and ground ice cannot be directly compared to related processes on Earth. Seasonal or longer-term cycles of thawing, gargantuan episodic floods, and subsequent sublimation of liquid into the tenuous atmosphere could have created very unusual subterranean passages.

Dissolution speleogenesis on Earth requires the presence of liquid water (often as weak carbonic acid or more aggressive sulfuric acid) and soluble rock. On cold planets and moons, non-water liquids, for example liquid nitrogen, liquid ammonia, or ethane, could conceivably dissolve rocky and icy materials depending upon the solubilities of those materials at the relevant temperatures. Whether carbonates and other evaporites exist on other planets is unclear. However, the early history of Mars as a wetter, warmer environment with a CO_2-dominated atmosphere has fueled speculation that carbonates may have formed abundantly during its early history and remain as deposits today. As Mars is a sulfur-rich planet, large standing water bodies would have had abundant sulfur to form evaporite minerals like gypsum. However, these minerals have not yet been directly detected on the Martian surface. Orbital spectroscopic detection of surface minerals has proven difficult beneath the well-mixed, globally distributed surface fine materials coating Mars. If caves did form in soluble materials during early wet Martian eras, they may have been preserved by subsequently dry and relatively inactive later periods. Thus, the average lifetime of any solution caves on Mars could greatly exceed their counterparts on Earth unless destroyed by impact or buried by wind-blown fines. In the latter case the caves may subsequently be revealed by wind-excavation.

Extraterrestrial Caves: Possible extraterrestrial (ET) cave types. Extraterrestrial caves can be formed by mechanisms that are common speleogenetic processes on Earth but there is an entire array of different processes and conditions that may produce caves of radically different properties than those we are familiar with on Earth.

Site Type	Possible Parent Material	Earth Formation Mechanisms	Possibile Unique ET Mechanisms
Solutional Caves (True "Karst")	Soluble rock	Dissolution of rock by solvent	
Epigenic caves	Limestone, dolomite, gypsum, other soluble rocks or minerals	Weak carbonic acid (groundwater or rain and dissolved CO_2), organic acids	Other non-water liquid solvents, e.g. liquid CO_2 on Mars, or liquid N_2 on Titan
Hypogenic caves	Limestone, dolomite, gypsum, other soluble rocks or minerals	Strong sulfuric acid (subsurface H_2S in solution at water table)	Sulfur-rich crust, e.g. on Mars, may make hypogenic caves more likely
Other karstic caves	Quartzite, sandstone, arkoses, opalinized silicates, etc.	Water dissolution of parent rock, decementation, tafoni formation	Other non-water liquid solvents, e.g. liquid CO_2 on Mars, or liquid N_2 on Titan
Pseudokarst Caves	Meltable or vaporizable solid	Melting or other phase transition	
Lava tubes or bubbles	Basalt, andesite, etc.	Molten rock with differential cooling of outer surface	Scale of tubes larger than Earth in lower gravitational fields, e.g. 0.38 g on Mars
Sub-ice volcanic caves	Ice masses overlaying volcanic terrain	Lava/ice or lava/permafrost interactions	Lava interactions with CO_2 ice or ice-clathrate interactions, ground ice sublimation
Glacier caves	Ice masses	Thermal and pressure-induced localized melting in water ice	Melting in super-cooled water ice, and other ices, e.g. CO_2, methane, or ammonia ice
Ice volcanism	Ice crusts with subsurface liquid and/or vapor	None known for Earth	Ice-covered bodies, e.g. Europa, Callisto, other outer planets' moons
Subsidence caves	Permafrost, ice/sediment complexes	Permafrost voids	Ground ice sapping and subsequent collapse and void formation on Mars or cold moons
Volatile labyrinths	Icy breccia or rubble	None known for Earth	Sublimation or boiling of ground ices, e.g. on perihelion passage by comets
Tectonic and Fracture Caves	Solid rock and ices	Faulting, rifting, other tectonic movements	Cratering and subsequent fracturing
Erosional Caves	Solid materials	Weathering effects	
Piping Caves	Unconsolidated materials	Subsurface erosion of particles by water, "suffosion" in arid environments	Subsurface erosion of particles by non-water solvents
Sea caves	Solid rock, welded tuff, ice	Water action (waves, floods) against rock	Massive flood events, e.g. on Mars Large bodies of water in the geological past
Wind-scoured caves	Solid rock, welded tuff, ice	Wind blasting of abrasive particles against rock	Global or regional dust storms on Mars Aeolian caves on Venus
Thermokinetic caves	Solid rock	Thermal expansion and contraction, spalling, exfoliation	Thermal regimes and unique physical effects much hotter and much colder than Earth.

On Earth, solutional caves are not restricted to limestones and evaporites but form in sandstones, arkoses, quartzites, granites, and other rock types, although the lower solubilities and greater resistance to weak acids of these rocks means that caves are much less common than in carbonate rocks. Typically, void is created by dissolution of cements between grains rather than in bulk rock. Cemented grain materials result from a sedimentary past. The status of such deposits on other planets is as yet unknown although recent images of Mars show apparently stratified cliff faces at sites like Chandor Chasma, possibly indicating sedimentary layers.

The scientific value of caves on other worlds could be immense. Apart from intrinsic interest in their very presence and mode of formation, they could also contain repositories of materials (e.g. frozen volatiles, unique minerals, or biological remains) extremely valuable for understanding formation and subsequent history of the bodies on which the caves may be found.

Penny Boston

See also **Pseudokarst**

Works Cited

Bunnell, D. 1991. Calculations of the relative size of lava tubes on the moon. Hawaii: The 6th International Symposium on Vulcanospeleology. *NSS News*, 49(12): 355

Ford, D.C. 1997. Principal features of evaporite karst in Canada. *Carbonates & Evaporites*, 12(1): 15–23

Horz, F. 1985. Lava tubes: Potential shelters for habitats. In *Lunar Bases and Space Activities of the 21st Century*, edited by W. Mendell, Houston, Texas: Lunar and Planetary Institute: 405–12

Keszthelyi, L. 1995. A preliminary thermal budget for lava tubes on the Earth and planets. *Journal of Geophysical Research—Solid Earth*, 100(10): 20411–20

Moore, J.M., Asphaug, E., Morrison, D., Spencer, J.R., Chapman, C.R., Bierhaus, B., Sullivan, R.J., Chuang, F.C., Klemaszewski, J.E., Greeley, R., Bender, K.C., Geissler, P.E., Helfenstein, P. & Pilcher, C.B. 1999. Mass movement and landform degradation on the icy Galilean satellites: Results of the Galileo nominal mission. *Icarus*, 140: 294–312

Oberbeck, V.R., Quaide, W.L. & Greeley, R. 1969. On the origin of lunar sinuous rilles. *Modern Geology*, 1: 75–80

Peale, S.J., Cassen, P. & Reynolds, R.T. 1979. Melting of Io by tidal dissipation. *Science* 203(4383): 892–94

Pinneker, E.V., & Shenkman, B.M. 1991. Karst hydrology in the zone of sporadic permafrost (taking the southern part of the Siberian Platform as the example). In *Proceedings of the 2nd International Symposium of Glacier Caves and Karst in Polar Regions,* University of Silesia

Sakimoto, S.E.H., Crisp, J. & Baloga, S.M. 1997. Eruption constraints on tube-fed planetary lava flows. *Journal of Geophysical Research—Planets*, 102(3): 6597

Wilson, L., Head, J.W. & Pappalardo, R.T. 1997. Eruption of lava flows on Europa: Theory and application to Thrace Macula. *Journal of Geophysical Research*, 102: 9263–72

Zimbelman, J.R. 1998. Emplacement of long lava flows on planetary surfaces. *Journal of Geophysical Research—Solid Earth*, 103(#B11): 27503–516

Further Reading

Cattermole, P. 1996. *Planetary Volcanism: A Study of Volcanic Activity in the Solar System,* 2nd edition, Chichester and New York: Wiley

Frankel, C. 1996. *Volcanoes of the Solar System,* Cambridge and New York: Cambridge University Press

Greeley, R. 1994. *Planetary Landscapes,* 2nd edition, New York: Chapman and Hall

Greeley, R. & Batson, R. 1997. *The NASA Atlas of the Solar System,* Cambridge and New York: Cambridge University Press; as *The Compact NASA Atlas of the Solar System,* 2001

Weissman, P., McFadden, L. & Johnson, T. (editors) 1991. *Encyclopedia of the Solar System,* San Diego and London: Academic Press

F

FILMS IN CAVES

The first cine film showing a cave was taken by Harry Short and released in Britain on 22 October 1896. *A Sea Cave Near Lisbon* was 80 ft long (one foot of film equated to about one second on screen) and was shot using daylight—capturing movement on film was an infant science, and suitable artificial light sources were not available. The crude cine cameras of the time, then as now, exposed a series of pictures on a length of film that was moved in front of the lens in a series of jerks. This effectively fixed the exposure and required very bright, continuous lights for filming underground. It was this factor, more than any other, that prevented early film makers from venturing underground.

However, an avid public expected and demanded entertainment from the new movie industry, and caves as a subject were experimented with many times. A series of suggestions to dramatize Rider Haggard's *King Solomon's Mines* in Cango Caves, a popular show cave in South Africa, failed due to a lack of finance, until in 1919 the Universal Film Manufacturing Co. included underground footage as part of a film financed by the Smithsonian Institute of Washington DC. The company used magnesium flares for lighting, but the fumes produced appalling conditions and "the smoke descended in clouds. The volunteers sent their flares flying and the party ran for it through the narrow going ... and helter skelter close on the heels of each other men, women and children ran for their lives ..." (Anon, quoted in Howes (1989), p.245.

Clearly, magnesium was far from suitable, yet for many years it remained the only option and, in 1924, several newsreels were made in Carlsbad Caverns, New Mexico, one of which concentrated on W.T. Lee's expedition for *National Geographic*. The cave—with its easily accessible, beautifully decorated passages—remained a prime location for many years, with productions usually linked to "travelogue interest" or "education". Other productions included Louis Mercanton's 1922 film *Phroso*, shot in the Grottes de St-Cézaire in France, and the 1928 Czech film *Demanovske Jeskyne*, these concentrating on show caves and elements such as stalactites and stalagmites, which were deemed to appeal to a paying audience.

Although education in a broad sense was instrumental in developing underground filming, cave exploration by amateur cameramen also played a part. Between 1925 and 1927, Russell Trall Neville made a 16 mm film, *In the Cellars of the World*, which was the first to depict a cave as being explored for fun. Neville's remarkable production culminated in a lengthy underground camp in Old Salts Cave in Kentucky.

Such films were expensive and difficult to produce. Notably, the only other early attempt was made in England by E.K. Tratman. His experiments began in 1933 and in 1937 around 5000 ft of film was shot, resulting in an edited 500 ft, 11-minute film of "real" caving on Mendip. For lighting, Tratman resorted to Tilley paraffin lamps, extending his exposures by filming at half speed, with cavers walking at half speed so that their movement appeared correct when projected.

Electric lights, coupled with greatly increased film sensitivity, aided further advances, but after World War II the influence of education and exploration waned in favour of Hollywood fiction, when caves were freely used as ready-made "sets". Carlsbad Caverns remained popular, and the excellent *Cave of Outlaws* (directed by William Castle) was shot there in 1951 (see photo). Other Carlsbad-based films include *King Solomon's Mines* (1950, Compton Bennett & Andrew Marton) and *The Spider* (1958, Bert I. Gordon). The latter is of interest as it was based partly on stills shot in the caverns, with the closing scenes (the spider impaled on a stalagmite) using a painting.

It is easy to recognize that commercial show caves, with their ease of access, will form the bulk of such locations although, when the influence of television is added to that of Hollywood, even these may be supplanted by mines, quarries, and artificial constructions purporting to be caves. Among "big name" films, the underground sequences in *Star Wars* (1977, George Lucas) were shot in the sandstone "caves" at Matmata in Tunisia, and in *Indiana Jones and the Last Crusade* (1989), Steven Spielberg's backdrop was the "caves" at Petra in Jordan. Bronson Caverns in California—where at least 50 films have been shot, including

Films in Caves: The making of *Cave of Outlaws* in Carlsbad Caverns in 1951. From the archives of the Carlsbad Caverns National Park.

representations of the Bat Cave from Batman movies—comprises a man-made tunnel; part of *Star Trek VI: The Undiscovered Country* (1991, Nicholas Meyer) was shot there.

Many classic books have been made into films, including Mark Twain's *The Adventures of Tom Sawyer* which, with its intrinsic cave scene, screened in 1930 and again in 1938 (Norman Taurog), this being perhaps the best of a series of remakes (*Tom Sawyer* was shot in 1973 in Howe Caverns in New York state, director Don Taylor). Jules Verne's *Journey to the Center of the Earth* (1959, Henry Levin) was shot in Carlsbad Caverns and received three Academy Award nominations. In 1953 *The Secret Cave*, based on Thomas Hardy's novel *Our Exploits at West Poley*, was filmed in Somerset for the Children's Film Foundation (John Durst). The inferior *Exploits at West Poley* was made at the Elstree studios (1985, Dirmuid Lawrence).

While the 1985 cave appeared convincing and real, the often false appearance of sets does nothing to improve any film; witness *Battle Beneath the Earth* (1967, Montgomery Tully) where the United States is to be invaded by Chinese armed forces through caves and tunnels which run under the Pacific. The 1960s television series of *Star Trek* is a classic example of cheaper production values, where flat floors and movable formations form an essential element of alien caverns. In *Grim* (1995, Paul Matthews) a team of "spelunkers" investigates monster-laden caves in Virginia—these being shot in Clearwell Caves (natural passages modified by iron miners) in the United Kingdom. The film lives up to its title.

If any generalizations can be made, fiction-based films inevitably fill their caves with formations or are subdued to the level of a mere backdrop. The plot invariably contains a sensuous female, such as in *The Cavern*, a war movie where six soldiers are trapped underground with Rosanna Schiaffino (1966, Edgar G. Ulmer), or an underground river that has to be free-dived to escape some awful fate, a theme borrowed from *King Solomon's Mines*. *Broken Arrow* (1996, John Woo) serves as an example, though with the decency to include a convincing, custom-made cave from Foam-Tec, while Sean Connery as James Bond made his underwater escape through Lucayan Caverns, Bahamas, in *Never Say Never Again* (1983, Irvin Kershner). A film with a different approach is *Ace in the Hole* (later retitled *The Big Carnival*: 1951, Billy Wilder), which found Kirk Douglas in a media circus culminating in the death of a trapped miner, a plot transparently based on the 1925 death of caver Floyd Collins in Kentucky's Sand Cave.

One arena ignored thus far is that of modern, specialist caver-made films which concentrate on exploration, usually for television consumption. Of the scores of active film makers, Sid Perou has gained international acclaim for his work, beginning with *Sunday at Sunset Pot* in 1967 and continuing with series such as *Beneath the Pennines*, *The World About Us*, *Realm of Darkness*, and *Hidden Depths*; he received an Emmy award for *Mysteries Underground* (1992), a film about Lechuguilla Cave in New Mexico. That an interest in caves exists outside the caving community is witnessed not only by the success of films such as those by Perou and Belgium-based Phillipe Axell (with his *Caves of Europe* series), but also with the 2001 release of *Journey into Amazing Caves* (Stephen Judson), an IMAX film concentrating on cave exploration and science, which became the highest-grossing large-format film of the year.

With the rise in small and light, high-definition digital video cameras and viable lighting units, technology has placed underground filming within the reach of all sport cavers. If a prediction is possible, rather than relying on recreating exploration for the sake of the camera, the future will see a rise in cavers recording their sport as it happens, presenting original discoveries directly to the audience.

Chris Howes

See also **Caves in Fiction; Photographing Caves**

Further Reading

Howes, C.J. 1989. *To Photograph Darkness: The History of Underground and Flash Photography*. Gloucester: Sutton. Chapter 10

FLORAL RESOURCES

The principal driving force for karst solution processes is the uptake of carbon dioxide and organic acids by percolation water. This occurs in the root zone of the vegetation that covers the limestone. Thus, changes to the vegetation on karst, as a result of clearance, fire, or climatic change, may provoke major changes in hydrology and limestone solution. For this reason alone, the conservation and management of the floral resources of limestone areas is of very high priority. In addition, the distinctive vegetation of limestone areas is a valuable resource for biodiversity conservation, for traditional medicines, and for wildlife habitat.

Soils on carbonate bedrock (see separate entry) are usually thin and deficient in most macronutrients, except for calcium and magnesium. In addition, trace elements may also be lacking. The rapid abstraction of underground water in karst areas means that plants and other organisms are subject to episodic and pro-

Floral Resources: Figure 1. Sparse limestone vegetation on the limestone slopes of the Cirque des Navacelles, France. (Photo by David Gillieson)

Floral Resources: Figure 2. A lone rowan tree survives on the bare limestone pavement of the Yorkshire Dales, England. (Photo by David Gillieson)

longed periods of drought. The sparse and often patchy vegetation (Figure 1) provides little shade, resulting in a greater range of soil temperatures. Those plants that are out-competed on other lithologies, and therefore survive in this harsh environment, are termed calcifuges. Some plants possess specific adaptations to the calcium-rich environment and show a high fidelity to limestone and other carbonate rocks; they are termed calciphiles. Many of the traditional Mediterranean herbs, such as rosemary and lavender, are calciphiles (Gams *et al.*, 1993). Exceptions to this are found in the shaded and damper habitats of limestone gorges, dolines, and crevices, in which more water-demanding species may find a refuge and flourish.

The specific conditions of the karst environment may produce physiological adaptations in plants which advantage them in this harsh environment. Reduced vegetation cover, and thus increased exposure to sunlight, enhance photosynthesis, often leading to an excess of carbon. This excess may be stored as protective structures such as thorns and woody stems, or as plant defensive compounds such as volatile oils (lavender and rosemary), phenols, and tannins. These render the leaves unpalatable to grazers, whether arthropods or vertebrates. These compounds may have value for pharmaceuticals or perfume manufacture.

Native plants growing on limestone are frequently displaced by exotic species, creating management problems. Most plant species can be classified according to their life strategies—their life-forms, resistance to disturbance, rates of seed production, and viability (Grime, 1977). Highly competitive species, such as blackberries, grow rapidly, resist disturbance, bear copious viable seed, and may inhibit the establishment and growth of other species near them, by poisoning the soil (allelopathy). Stress-tolerating plants, such as figs and eucalypts, have adaptations to ensure adequate water supply (deep roots, high root-shoot ratios), to resist fire (resistant bark, lignotubers, and vegetative reproduction), and to exploit low levels of nutrients (root fungi associations and high root hydrogen ion production) (Figure 2). Ruderal species, such as dandelions and St John's Wort, habitually colonize disturbed habitats. They have tuberous energy stores, very effective seed dispersal, and easily pollinated flowers. In general, stress-tolerating plants are more numerous on limestone soils with some ruderals present, but highly competitive species are rare. Those native species that are stress-tolerators are very susceptible to displacement by exotic species when the disturbance regime is altered radically.

The specific conditions of karst environments described above exert a strong ecological pressure on any colonizing species. These may respond in two ways; local extinction, or adaptation by accelerated evolution, often after an initial fall in the population. In the process, isolated populations in this highly fragmented environment are likely to develop into new species (Figure 3). Many of these limestone-adapted species simply can-

Floral Resources: Figure 3. A rare One Leaf Plant (*Monophyllea* sp.) in the entrance of Clearwater Cave, Mulu, Sarawak. This plant is endemic to the twilight zone of caves in Borneo. (Photo by David Gillieson)

not survive anywhere else. The extreme and varied conditions in limestone areas may result in unusual geographic affinities of a local flora. For example, in the Zhuang region of China, limestone areas support a much higher proportion of southern tropical species (88%) than adjoining non-limestone areas (Vermeulen & Whitten, 1999). In northern Thailand, limestone areas contain temperate plant species at the extreme southern limit of their range in Indo-China. In New Guinea, upper montane forests on limestone contain many plants of Gondwanan origin, while at lower elevations, on the same lithologies, the flora is dominated by Indo-Malayan species.

The most abundant lower plants on limestone are the bryophytes, or mosses and liverworts. These primitive plants can reproduce vegetatively, withstand dessication and have defensive chemical compounds to resist predation by most insects, snails, and bacteria. Many of the bryophytes on limestone are also found on buildings (cement and stone) and share affinities with those found on alkaline or calcareous soils elsewhere (Downing, Ramsay & Schofield, 1991). Some species curl up and enter suspended animation during droughts, expanding rapidly and becoming green after a rainstorm. Others survive in permanently cool and damp sites, especially around springs and cave entrances. In the tropics, a diverse microflora coats stalactites and rock surfaces in the twilight zone of caves. These may have a significant role in the limestone solution process. We know very little about the ecology of limestone microflora, especially in the area of biogeomorphology.

Most karsts have a low natural fire frequency, due to the shielding effects of limestone outcrops and reduced ground cover. In the impounded karsts of eastern Australia, natural fire frequencies are poorly documented but the fire interval may be 35 to 50 years or greater (Holland, 1994). Under these conditions, relict vegetation types may survive, for example, the vine thickets of North Queensland or the monospecific *Acacia* scrubs on the Bendethera karst, New South Wales. Limestone bedrock retains little water, so during drought the vegetation tends to dry out and become flammable. Fires sweeping up the slopes may burn the margins of karst vegetation or become accelerated and burn summit areas. This may not only destroy the vegetation but also the thin soil, leading to its erosion and abstraction down open joints. In Great Britain, there are rare fern species which are now largely restricted to limestone outcrops. In these karsts, soil erosion only occurs immediately after fires, with minimal erosion in the intervening periods.

Surface vegetation regulates the flow of water into the karst through interception, the control of litter and roots on soil infiltration, and the bacterial production of carbon dioxide in the root zone. The metabolic uptake of water by plants, especially trees, may regulate the quantity of water available to feed speleothems. Trees in particular are like large carbon dioxide pumps, releasing 20–25% of the atmospheric gas uptake through root respiration (Aley, 1994). The free vertical drainage of most limestone soils creates special conditions for evapotranspiration, gas exchange, and root penetration. Large eucalypts act as water pumps, taking up 250–270 litres per day for each tree of river red gum (*Eucalyptus camaldulensis*). Tree roots can penetrate to depths of 30–50 m in search of water, especially the figs in seasonally humid climates of northern Australia (Gillieson, 1996). The penetration of tree roots aids the enlargement of bedrock fissures in karst and ensures the high degree of secondary porosity characteristic of limestone terrains.

Floral Resources: Figure 4. A plant collector selling medicinal herbs and roots collected from the limestone forests of Mount Emei, China. (Photo by David Gillieson)

Large trees are usually sparse on limestone areas, thus they are of lesser interest to commercial forestry. Local people often harvest firewood and building timber from karst areas. Of more importance is an array of small forest products harvested for traditional medicines (Figure 4), for food (fungi and fruits), for handicrafts such as baskets, and for flowers. Some wild limestone plants, such a slipper orchids in Southeast Asia, are heavily collected and this may lead to local depletion or extinction. Many of these plants are easily cultivated so there is little justification for the practice.

DAVID GILLIESON

See also **Forests on Karst**

Works Cited

Aley, T. 1994. Some thoughts on environmental management as related to cave use. *Australasian Cave and Karst Management Association Journal*, 17: 4–10

Downing, A.J., Ramsay, H.P. & Schofield, W.B. 1991. Bryophytes in the vicinity of Jenolan Caves, New South Wales. *Cunninghamia*, 2(3): 371–84

Gams, I., Nicod, J., Sauro, U., Julian, E. & Anthony, U. 1993. Environmental change and human impacts on the Mediterranean karsts of France, Italy and the Dinaric region. In *Karst Terrains: Environmental Changes and Human Impact*, edited by P.W. Williams, Cremlingen: Catena

Gillieson, D. 1996. *Caves: Processes, Development and Management*, Oxford and Cambridge, Massachusetts: Blackwells

Grime, J.P. 1977. Evidence for existence of three primary strategies in plants and its relevance to ecological and evolutionary theory. *American Naturalist*, 111: 1169–94

Holland, E. 1994. The effects of fire on soluble rock landscapes. *Helictite*, 32: 3–9

Vermeulen, J. & Whitten, T. 2000. *Biodiversity and Cultural Property in the Management of Limestone Resources*, Washington DC: World Bank

Further Reading

Grime, J.P. 1979. *Plant Strategies and Vegetation Processes*, Chichester and New York: Wiley

Lichon, M. 1993. Human impacts on processes in karst terranes, with special reference to Tasmania. *Cave Science*, 20(2): 53–60

Trudgill, S.T. 1985: *Limestone Geomorphology*, London and New York: Longman

FLUVIOKARST

Fluviokarst may be defined as a landscape on carbonate rocks where the dominant landforms are valleys cut by surface rivers. Sweeting (1972, p.259) gives a broader definition, stating that "Fluviokarst is formed by the combined action of fluvial and karst processes". However, this is too broad since fluvial processes are involved in the development of virtually all karst landforms, including dolines and poljes which would not be considered as fluviokarst landforms. The term appears to have been first used in English by Roglic (1964) who used it in preference to the "halbkarst" (halfkarst) of Grund (1914) and the more widely used equivalent, "merokarst", introduced by Cvijić (1924). As originally used, a merokarst was envisaged as containing both fluvial and karstic elements as a result of having developed on thin sequences of limestones interbedded with other rocks (particularly marly beds) or on impure limestones. Mechanical erosion was considered to be more important than dissolution in the development of merokarst leading to a dominance of valley forms. Ford (1980) noted that limestones of low purity may display a low density of small dolines in an otherwise fluvial landscape. Such an assemblage can also develop in rocks with less than 50% carbonate minerals, such as calcareous sandstones, and Birot (1954) used the term "parakarst" for this type of development.

In addition to the term merokarst, Cvijić (1924) used "holokarst" to describe a fully karstic landform assemblage and "transitional karst" to describe a landform assemblage intermediate between holokarst and merokarst. His major example was the Causses of France (see France, Southern Massif Central) where there is a thick limestone sequence but "the karst basis formed by the impermeable bed is not so deep as in the holokarst and therefore the karstic evolution is more rapid; normal valleys are much more frequent than in the holokarst..." Sweeting (1972, p.253). Sweeting (1972) and Ford (1980) followed Roglic (1964) in considering parakarst, merokarst, and transitional karst as all being types of fluviokarst. On this basis, Sweeting suggests that many of the karstlands described in western and central Europe are fluviokarstic. However, in the majority of these karsts the valley forms are associated with allogenic rivers and take the form of through or blind valleys (see Valleys in Karst). Again this usage seems to be too broad and instead it is suggested that the term fluviokarst should only be used where the land surface over a wide area is largely dissected by valleys, the majority of which have their heads in the karst. There may be dolines in varying densities on the interfluves but these are not the dominant landform. As defined, fluviokarst may be seen as one end member in a continuous sequence of surface landform assemblages on carbonate rocks, the other being polygonal karst where the whole surface is pitted by dolines with little or no trace of any previous valley form. The latter represents an extreme form of a holokarst.

Active fluviokarst would be expected on karst rocks under two specific conditions. First, where the karst development is young and/or the hydraulic gradient is very low, the karst conduit system is not yet able to drain all of the surface and open channels remain above ground, at least in headwater areas. Such systems are not common but there are good examples in Canada. Second, in dry regions subject to seasonal torrential rains, there are regular dendritic valley patterns heading in the karst rocks and draining to their perimeter. These valleys are carved by channelled runoff at the height of storms but at all other times the epikarst can drain any surface water and there may be substantial caverns below. Dolines tend to be rare on interfluves or along the valleys and may be entirely absent. Examples include much of the highlands north and south of Jerusalem, the Edwards Plateau, Texas, and gypsum highlands on the south side of the Ebro valley in Spain.

Fluviokarsts that are dominated by dry valleys are more common than active fluviokarsts. The English Peak District and the chalklands of southern England provide good examples that are discussed below.

The geology and caves of the Peak District are described in a separate entry but briefly, Carboniferous (Dinantian) limestones crop out over some 520 km^2 and are surrounded by lithologies that support surface drainage, although only in the northern half of the area are these higher than the limestones. The basement rocks are several hundred metres beneath the surface and have had no influence on surface landform evolution and little, if any, influence on the inception and development of conduits and caves. Virtually no perennial surface drainage exists on the limestone and only two allogenic-fed rivers maintain flow across the karst. However, the limestone plateau is dissected by a dense dendritic system of valleys that are dry for all, or most, of the year (see photograph in Valleys in Karst entry). Most of these valleys have their heads entirely within the karst and either drain out of it or are tributary to the main allogenic rivers. The limestone sequence is thick (up to 1800 m) and the limestones are of high purity, in contrast to the common situation in a merokarst. Instead, it is generally accepted that these valleys have been inherited by the limestone after originating on an impermeable clastic cover. Their desiccation largely reflects gradual karstification of drainage, aided by regional water-table lowering in response to downcutting by major rivers. However, not all the desiccation was natural, as dewatering via mine soughs affects some rivers, most notably the Lathkill's middle reaches, which dry up every summer. Dry valleys were probably reactivated under Pleistocene periglacial conditions when the ground froze and extensive screes developed.

The Cretaceous Chalk crops out over large areas of southern England and northwestern France (see map in Europe, North entry). Although it is a relatively weak rock, the outcrops are surrounded by even weaker clays and sands, and consequently the Chalk generally forms relatively high ground, commonly as a series of escarpments. Surface flow is rare on the escarpments but dry valleys are common, forming steep coombes on the scarp faces and large dendritic systems on the dip slopes. Many valleys have springs at their lower ends, and winter rises of the water table create seasonal surface streams, known as bournes. The dry valleys are largely floored by solifluction debris from Pleistocene stages of periglacial weathering (Paterson, 1977), and formation under permafrost conditions is one of several hypotheses for valley origin (Smith, 1975). However, Smith argues that while the valleys would certainly have been reactivated under periglacial conditions, as in the Peak District, they were formed by surface streams and initial desiccation followed gradual karstification of drainage. Periglacial valley carving and/or deepening action was also important on the Mendip Hills (see separate entry) where Cheddar Gorge is a classic example of fluviokarst.

There is little doubt that in the early evolution of most, though by no means all, limestone areas there was surface drainage, and fluvial valley development. Where the limestone outcrop is relatively narrow and there are large allogenic rivers then karstification of drainage is likely to result in a landscape dominated by blind and through valleys with dolines on the interfluves. Where the outcrop is wider, and allogenic inputs are proportionately less, then a dendritic valley network is more likely to form on the limestone. Where recharge is dispersed then the valleys are likely to remain as the dominant landform following karstification of drainage and the landscape is best described as a fluviokarst. More concentrated recharge is likely to favour the growth of dolines that dissect the valleys (as is happening in the Mendip Hills) and may ultimately form a polygonal karst (see model for the Waitomo District of New Zealand in Gunn, 1986).

JOHN GUNN

Works Cited

Birot, P. 1954. Problèmes de morphologie karstique. *Annales de Geographie*, 63: 160–92

Cvijić, J. 1924. Types morphologiques de terrains calcaires. *Glasnik Geografskog drustva*, 10: 1–7

Ford, D.C. 1980. Threshold and limit effects in karst geomorphology. In *Thresholds in Geomorphology*, edited by D.R. Coates & J.D. Vitek, London and Boston: Allen & Unwin

Grund, A. 1914. Der geographische zyklus in karst. *Zeitschrift der Gesellschaft fur Erdkunde. Berlin*, 621–640

Gunn, J. 1986. Solute processes and karst landforms. In *Solute Processes*, edited by S.T. Trudgill, Chichester and New York: Wiley

Paterson, K. 1977. Scarp-face dry valleys near Wantage, Oxfordshire. *Transactions of the Institute of British Geographers*, 2: 192–204

Roglic, J. 1964. "Karst valleys" in the Dinaric karst. *Erdkunde*, 18: 113–116

Smith, D.I. 1975. The problems of limestone dry valleys implications of recent work in limestone hydrology. In *Processes in Physical and Human Geography*, edited by R. Peel, London: Heinemann

Sweeting, M.M. 1972. *Karst Landforms*, London: Macmillan and New York: Columbia University Press

FOLKLORE AND MYTHOLOGY

There are few features in the landscape that have so excited the imagination, and are as deeply embedded in the folklore of societies throughout the world, as caves. Associations range from the historical, or quasi-historical, to the fantastic and mythological and are recorded in orally transmitted traditions, legends, and place-names.

The historical association of a particular cave with a notable, or notorious, individual is a widespread phenomenon and well-authenticated cases are not uncommon. Bonnie Prince Charlie is reliably recorded to have sheltered in caves in Glenmoriston and on South Uist during his time as a fugitive after the Battle of Culloden, and Hardin County's cave in Cave-in-Rock State Park, Illinois, is associated with the notorious Harpe brothers' outlaw gang. Because of their interesting and romantic associations, such well-authenticated cases often become incorporated in community tradition and give rise to a number of quasi-historical associations with such figures. The fugitive Prince is traditionally associated with a large number of caves in the western Highlands, and caves on the islands of Arran, Jura, and Rathlin are each claimed as the hiding place of Robert the Bruce in 1306. The simple naming of caves as "The King's Cave", "The Spaniard's Cave", and "Smugglers' Lair" marks a further step away from historical reality, but one that has nevertheless some basis in fact.

A further step is marked by the association of caves with individuals whose historicity is uncertain but who have become traditional icons; Robin Hood's Cave in Derbyshire, England, and King Arthur's Cave in Gwent, Wales, are obvious examples. Such ahistorical beings have a partly mythological status, and it is in the realm of myth that caves make their most potent contribution to folklore.

Caves play many parts in the mythologies of the world, but three main roles are widespread. First, they are often seen as the route to the underworld; the Xibalba of the Central American Maya, the Tartarus of classical Greece, and Dante's Inferno. As such, they function as passages between worlds and across liminal or boundary zones, into which mortals pass and are transformed. The transforming nature of caves is a theme in its own right, although related to caves as passages to and from the underworld.

Secondly, caves are commonly seen as sources of power and creative energy. According to the ancient Greeks, Zeus was born in a cave on Crete, while ancient Japanese mythology attributed the fall of night to the retreat to a cave of the sun-goddess Amaterasu. According to some traditions, the Aztecs of Mexico City—the Tenochcas—originated in a group of caves before descending to establish their settlements on islands in the lakes. For many traditional societies in Australia, caves and rock shelters are associated with the heroic ancestors of the Creation Period, such as the Wandjina cult of the Western Kimberleys. These ancestral beings commonly feature in cave paintings that are traditionally believed to have been created by the Wandjina themselves, and which require repainting and restoration at intervals, in order to ensure the continuation of the social order and the natural resources on which the communities depend.

A related theme is the role of caves as a source of wisdom and knowledge. In ancient Greece and Rome, caves were the theatres for oracular pronouncements, as in the case of the Sybil of Cumae, Italy. In Christian Europe, from medieval to modern times, caves have been the scene of visionary experiences such as the manifestation of the Madonna in the grotto at Lourdes, France, or the Archangel Michael in the cave at Monte Saintangelo, Italy.

Thirdly, caves are widely seen as the abode of gods, demons, spirits, ancestors, and various malevolent forces often personified in the form of giants—such as the one-eyed cyclops Polyphemus of Homeric legend or the Polynesian "Giant with Teeth of Fire", or dragons. Before the advent of scientific paleontology the discovery of bones in caves was taken as proof of their use by dragons, and the Slovenian Olm, *Proteus*, a blind cave salamander, was regarded as dragon-fry.

The cave-dwelling monster is a powerful image in mythologies from many parts of the world, and the related myth of the dragon-slaying hero is widespread—examples spanning the globe from the Maori hero Pitaka, slayer of the taniwha (dragon) Kataoare, to the Nordic legend of Sigurd the Volsung (Richard Wagner's Siegfried), who killed the dragon Fafnir. The legend of Fafnir also provides an example of the theme of caves as places of transformation, for Fafnir was originally a giant but transformed himself into a dragon in order to guard the treasure that had been stolen from Andavi the dwarf. The subject of caves as repositories of gold, silver, and other valuables is a further worldwide aspect of cave mythology. Such myths probably owe their origins to the discovery of votive deposits within caves, or grave goods accompanying cave burials.

CHRIS TOLAN-SMITH

See also **Caves in Fiction; Caves in History: The Eastern Mediterranean; Religious Sites**

Further Reading

Bonsall, C. & Tolan-Smith, C. 1997. *The Human Use of Caves*, Oxford: Archaeopress

Chapman, P. 1993. *Caves and Cave Life*, London: HarperCollins

Layton, R. 1992. Rock art and indigenous religion. In *Australian Rock Art: A New Synthesis*, Cambridge and New York: Cambridge University Press

Leitch, R. 1987. Green Bottle Howffs: a pilot study of inhabited caves. *Vernacular Building*, 11:15–20

FOOD RESOURCES

Availability of food resources in hypogean environments is an important limiting factor for cave-dwelling animals, because it produces a strong selective pressure at the metabolic level on troglobitic and troglophilic species. However, availability of food resources does not on its own determine the distribution of species. Caves in the Mediterranean region are characterized by a deficiency of food resources while in tropical caves the trophic material is extremely abundant; yet, paradoxically, the former represents a true reservoir of relict or "living fossil" troglobites which are rare or absent in the latter. There are both ecological and biogeographical explanations for the situation, in which a variety of biotic and abiotic, and recent or geohistorical factors play essential roles.

Food chains in caves are substantially simple to describe. In fact, they are formed by few links or, in extreme cases, by only two links (Figure 1). From an energetic point of view, caves do not represent a closed system since there is an "exogenous" trophic energy flow from outside to inside the cave in addition to an "endogenous" flow, which takes place within the cave. The balance between energetic input and output makes cave populations stable. If caves were exclusively controlled by an endogenous flow, that is if they were a closed and isolated system, the energetic cycles would be likely to result in a natural and continuous energetic fall which would destroy the ecosystem in a short time. Therefore, even a low external input is required for a more or less constant trophic balance, although there are some exceptions. Movile Cave (southern Dobrogea, Romania) is a closed system where the primary energy is provided by hydrogen sulfide which is abundant in the cave water (see Movile Cave, Romania). A biofilm (see separate entry) composed of chemiolithic–autotrophic microbial material (sulfobacteria) is present at the water surface, where it represents the starting point of the cave trophic cycle.

The metabolic demands of certain stygobitic animals can be very low. Thus, the blind prawn *Typhlocaris salentina* is adapted to feed on guano (see below) that lies on the bottom of cave pools. Experimental results suggest that another stygobitic crustacean, *Spelaeomysis bottazzii*, is able to survive for years by scraping small amounts of imported or fossil biomass from the surface of soft calcareous rocks, in waters absolutely lacking primary production (Ariani, 1982) (Figure 2).

Inside caves the following food zones can be recognized: vegetation zone (with algae, bryophytes, and phanerogams); clay slime; water; air; and organic sediments (see Sediments: Biogenic). Each zone is distinct and, within them, qualitatively simple but well characterized biocoenoses (or communities of living creatures) can be described.

The vegetation zone is represented by the part of the cave entrance which is illuminated during the day. In this area there is an abundant trogloxenic fauna inhabiting shadowy cracks, mosses, rocks, and stagnant waters. Some of these species act as "shuttles" between internal and external cave environments. Troglobites rarely penetrate into this zone, generally only for temporary feeding necessities. In vertically developed caves, fallen vegetation provides the deepest biotopes with plenty of organic substances such as pieces of plants, dry leaves, pollen, and seeds, which on decomposing, feed a considerable number of bacteria and fungi.

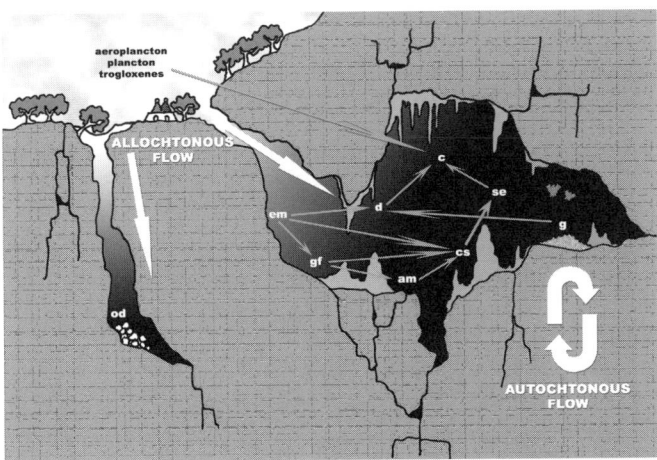

Food Resources: Figure 1. Scheme of trophic net in the cave ecosystem. *am*, endogenous autotrophic microflora; *c*, carnivorous; *cs*, clay-slime; *d*, detritivorous; *em*, eterotrophic microflora; *g*, guano; *gf*, growth factors (aminoacids, vitamins, oligo-elements, fatty acids, mineral salt); *od*, organic deposits; *se*, slime eater.

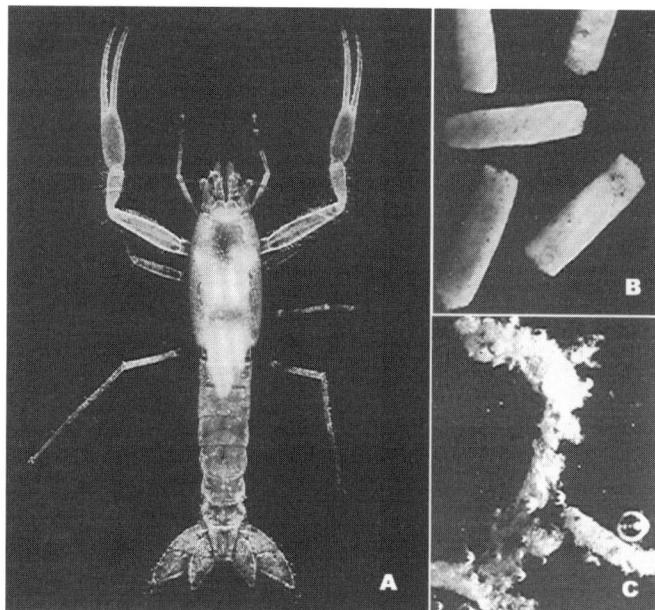

Food Resources: Figure 2. (A) Male individual of *Typhlocaris salentina* Caroli. (B) The faecal material of *Spelaeomysis bottazzi* Caroli, before HCl treatment and (C) after treatment. This technique demonstrated the presence of calcium carbonate. (From Ariani, 1982, reproduced with permission)

In horizontally developed caves, the (always passive) transport of organic material from the vegetation zone towards the interior, instead takes place through different vehicles such as draughts, rivulet waters, men, and animals (anemocorous, hydrocorous, or zoocorous transport, respectively). Clay slime, usually present in groundwaters, represents a biotope of great biological interest because it is very rich in trophic resources. Water contains different quantities of both microorganisms and oligoelements, partly of endogenous origin and partly coming from epigean environments. Underground water essentially consists of meteoric water infiltrated into the subsoil and occasionally contaminated by seawater intruding into the land mass.

The synthesizing activity of bacteria and other autotrophic organisms can be considered the first stage of an endogenous flow. Although the cave darkness implies the absence of green plants and photosynthesis, a microflora is nevertheless frequently present which is able to produce biological energy from inorganic substrata (CO_2, H_2S, FeO_2). Ferrobacteria are rather common in caves, where they extract energy from substrata rich in iron oxides. Nitrobacteria and Sulfobacteria draw nitrogen from the atmosphere and sulfur from sulphuric thermal waters, respectively, thus obtaining energy (see Microorganisms in Caves).

Bacteria and fungi are present everywhere inside caves, in soil, in clay mud, on rocks and speleothems, in the air, in the dripping water, in underground rivers and lakes, and obviously in decomposing vegetables and animals. A considerable quota of microorganisms has been reported from a kind of clayey-slimy concretion termed "foval" (Camassa, 1997). Fovals are brown-greyish formations that can be found in different positions in the caves, on the floor, or on the rocky walls (Figure 3). By microscopic observations (partly after laboratory culture) on foval materials, several microflora components have been identified, such as *Mucor mucedo*, *Mucor racemosus*, *Geotrichum* sp., and *Nitrospina gracilis*. The exceptionally high concentration of bacteria and micromycetes, the presence of nematodes and abundant organic scraps is indicative of an intense biological activity that probably produces organic substances and growth factors (aminoacids, proteins, vitamins, fatty acids) useful for cave biocoenoses (for example at Frasassi Caves, Italy).

Clayey sediment is absolutely essential for the colonization of cave pools by animals. Laboratory studies (Gounot, 1960) showed that reproduction in captivity of *Niphargus* and other aquatic amphipods was successful only after addition of cave clayey slime to the aquariums. In fact, the presence of this substratum assured the survival of the young *Niphargus*. Similarly, it is known that the proteus (*Proteus anguinus*) takes nourishment and growth factors from slime during the first year of life.

Marine and lake slime-eaters feed on both organic remains and microorganisms present in the sediment; similarly cave slime-eaters utilize not only exogenous organic substances present in the substratum but also living microorganisms in consequence of the discovery of autochthonous microflora in underground slimes (Jefferson, 1965). These autochthonous, autotrophic components include Thiobacteria that generate energy by oxidization of hydrogen sulfide, and Ferrobacteria (such as *Leptothrix*) oxidizing the hydrate ferrous carbonate. A third group of autotrophic bacteria (Nitrosobacteria) that utilize nitrates, is also supposed to be involved in energy production inside some caves. According to Caumartin (1964), Thiobacteria and Ferrobacteria are in competition with eterotrophic micro-organisms which come from the surface. Chodorowski (1959) reported bacteria performing such different functions as proteolysis, denitrification, and nitrogen fixation, and that play a role in sulfur and phosphor cycles in Polish caves. Clay slimes of a temporary lake in Baradla Cave (Hungary) were found to have a microflora comparable to that from farm soils of the same region, with nitrogen-fixing, nitrifiant, cellulose-lytic, and denitrifiant bacteria.

Among solid substrata in caves, particular biotopes are represented by speleothems and "montmilch" (or moonmilk). Their microfloras have been extensively studied as they were thought to be implicated in the alteration processes through which these formations are made (Mason-Williams, 1959; Pochon et al., 1964).

Aerial microflora are scarce because of the high degree of humidity that causes particles suspended in air to precipitate. This particle complex mainly or entirely consists of bacteria conveyed by air currents from outside or, in show caves, by visitors (Mason-Williams & Benson-Evans, 1958).

Besides bacteria, other microorganisms can be found in cave environments. Though the deepest parts of caves are marked by complete darkness, they are occasionally inhabited by autotrophic algae carried in by waters from epigean biotopes (Mason-Williams & Benson-Evans, 1958; Palik, 1960). It was also observed (Palik, 1960) that some algae are able to live in caves in eterotrophic conditions. Other important elements of cave biocoenoses are represented by Protista, often abundant in clayey slime or groundwaters, and by micromycetes which can be found as saprophytes on organic substrates of caves (decomposing animals and vegetables, and guano). All microbiological compo-

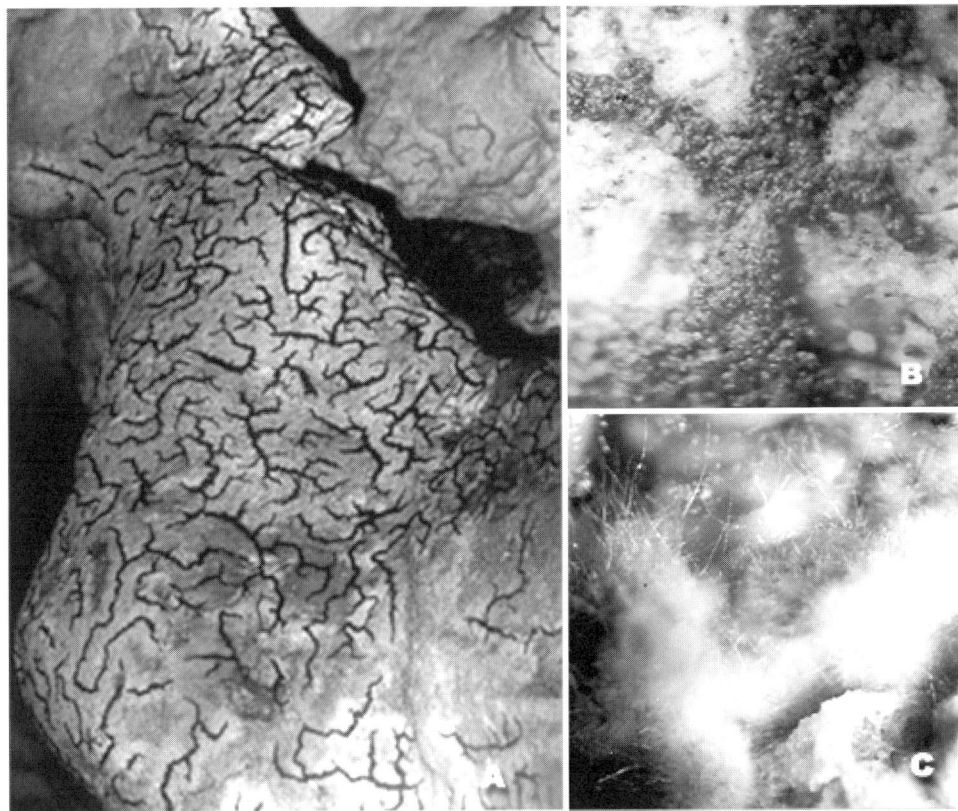

Food Resources: Figure 3. (A) Foval (a clay-slime vermiculation) in Zinzulusa Cave (southern Italy). (B) shows 20x and (C) 40x magnification. In (C) the *Geotrichum* hyphae are visible. (A, photo by N. Ciccarese; B and C, photo by M.M. Camassa.)

nents of caves, especially the autotrophic ones, can be considered fundamental to the preservation of a fragile ecological equilibrium in the underground ecosystem; they represent the base of the food pyramid, that is the primary producer of all trophic factors essential for most of the cave-dwelling species.

Caves open directly to the external environment are often inhabited by colonies of bats or by troglophilic birds such as *Collocalia* (a Chinese swallow) and *Steatornis* (the well-known "guácharo"). These animals generally produce a large quantity of faecal material (guano) which is deposited on the bottom of caves and partially along walls where the animals cling or nest. Sometimes bat populations are so numerous that guano layers reach a considerable thickness, up to several metres (see Guano). Guano represents a considerable trophic resource with a special biocoenosis. A large number of organisms such as protozoans, mycetes, nematodes, and insects (Collembola, Coleoptera, Diptera) are so connected to the presence of guano that they have attained specialized physiological and structural characteristics, and occupy a unique ecological niche. These organisms are termed "guanobites" and represent, where present, the second link of the food chain. The importance of this trophic resource inside a cave is immediately clear. Guano is produced by animals that normally go out of the cave to feed; this implies the introduction into the cave ecosystem of organic material rich in nourishing substances such as proteins, fats, carbohydrates, mineral salts, oligoelements, and vitamins, most of which could not be synthesized by autochthonous microflora. In addition, a great number of organisms, mainly parasites of bats or cave-dwelling birds, are deposited into the cave by the guano: these include protozoans, fungi, roundworms, flatworms, mites, and millions of bacteria.

MICHELE M. CAMASSA

See also **Microorganisms in Caves; Subterranean Ecology**

Works Cited

Ariani, A.P. 1982. Osservazioni e ricerche su *Typhlocaris salentina* (Crustacea, Decapoda) e *Spelaeomysis bottazzii* (Crustacea, Mysidacea). Approccio idrogeologico e biologico sperimentale allo studio del popolamento acquatico ipogeo della Puglia. [Remarks and investigations on *Typhlocaris salentina* (Crustacea, Decapoda) and *Spelaeomysis bottazzii* (Crustacea, Mysidacea). Hydrogeological and experimental biological approach to the study of the groundwater fauna of Apulia]. *Annuaio dell'Istituto e Museo di Zoologia dell'Universita di Napoli*, 25: 201–326

Camassa, M. 1997. Osservazioni biologiche in una grotta artificiale di Ginosa: ruolo delle foval nell'ecosistema cavernicolo. [Biological observations in an artificial cave of Ginosa (southern Italy): role of the foval in cave ecosystems]. *Thalassia salentina*, 23 (Suppl.): 189–91

Caumartin, V. 1964. Note sur la présence de dérivés de l'acide carbamique dans les grottes, son importance dans la corrosion des rochers et la réduction des oxydes metalliques. [Note about the presence of carbamic acid derivates, its importance in the rock corrosion and metallic oxides reduction]. *Spelunca Memoires*, 4: 17–23

Chodorowski, A. 1959. Les études biospéologiques en Pologne [Biospeleological studies in Poland]. *Biospeleologica Polonica*, 2: 122–44

Gounot, A.M. 1960. Recherches sur le limon argileux souterrain et sur son rôle nutritif pour les *Niphargus* (Amphipodes

Gammaridés). [Investigations on the underground clay slime and its role in the *Niphargus* feeding (gammarid amphipods)]. *Annales de Spéléolologie*, 15: 501–26

Jefferson, G.T. 1965. British cave faunas and the problem of their food supply. *Actes du IVe Congrès International de Spéléologie en Yougoslavie*, Ljubljana

Mason-Williams, A. 1959. The formation and deposition of moonmilk. *Transactions of the Cave Research Group of Great Britain*, 5(2): 133–38

Mason-Williams, A. & Benson-Evans, K. 1958. *A Preliminary Investigation into the Bacterial and Botanical Flora of Caves in South Wales*, Berkhamsted, Hertfordshire: Cave Research Group of Great Britain

Palik P. 1960. A new blue-green alga from the cave Baradla near Aggtelek (Biospeleologica Hungarika XII). *Annales of University of Science, Budapest, Biology*, 3: 275–86

Pochon, J, Chalvignac, M.A. & Krumbein, W. 1964. Recherches microbiologiques sur le mondmilch [Microbiological researches on montmilk]. *Comptes rendus de l'Academie de Sciences, Paris*, 258: 5113–15

Further Reading

Poulson, T.L. & Lavoie, K.H. 2000. The trophic basis of subsurface ecosystems. In *Subterranean Ecosystems*, edited by H. Wilkens, W.F. Humphreys & D.C. Culver, Amsterdam and New York: Elsevier

Vandel, A. 1965. *Biospeleology: The Biology of Cavernicolous Animals*, Oxford and New York: Pergamon Press (originally published in French, 1964)

FORESTS ON KARST

Many of the karst areas of the world are forested, or were forested prior to human activity. Since karst is widely dispersed around the world, the majority of the forest types present in the world exist, to at least some extent, on karst. With some notable exceptions, forests on karst are commonly less economically productive than nearby forests on non-karst areas. In many cases, this is because the autochthonous soils on carbonate karst (see Soils entry) are so shallow and/or rocky, that they have less moisture storage capacity than soils on nearby non-karst areas. In addition, soils in karst areas are often more prone to nutrient impoverishment than non-karst soils. The steep terrain also makes logging expensive and often hazardous. Gams (1991) concluded that the original meaning of the term karst was a forestless, stony, and waterless landscape. He further concluded that the absence of forests in the Dinaric karst region lasted for centuries and was a direct result of human activity, designed to clear the land for the pasturage of animals. Many other karst areas of the world have also experienced dramatic episodes of deforestation, with attendant damage to soil and water resources.

Karst areas, in regions with high precipitation distributed throughout the year, may support significant forests. Some spectacular temperate rainforest on karst exists in coastal portions of southeastern Alaska; this area receives about 2500 to 5000 mm of precipitation per year, and the precipitation is evenly distributed throughout the year. Much of this karst is developed on carbonate rocks with extremely high purity. The derived soils are predominantly organic and relatively shallow, but precipitation is high enough that moisture depletion of the karst soils seldom limits tree growth. The underlying karst groundwater system provides effective drainage for the overlying soils. In contrast, many adjacent non-karst areas are peat lands or have poor forest growth because the soils are waterlogged and lack effective natural drainage.

In some karst locations, much of the soil is derived from allogenic sources, such as alluvial and colluvial soils. Loess deposits (windblown glacial materials) of Pleistocene age blanket some karst areas and are often of sufficient thickness to support valuable forests, or (if cleared) good farmland. The majority of karstlands with good soils (and thus potentially good forests) are areas where much of the soil is derived from allogenic sources. For example, the King Country karst on North Island, New Zealand was originally thickly forested, although much of the forest has been logged and replaced by pasture. In this area, the limestone is covered by a thick mantle of volcanic ash and it is on this that the forest grew.

In most non-karst landscapes, runoff water flows primarily on the surface and by throughflow to streams; relatively little water moves directly downward into the groundwater system and laterally out of springs. In many karst landscapes, groundwater recharge predominates over surface runoff. Groundwater recharge can transport plant nutrients beyond the reach of tree roots in relatively short distances, whereas with surface runoff, more distance is available in which the landscape can detain the water-transported nutrients. As a result, forests in karst areas are commonly deficient in one or more plant nutrients, and are more likely to experience nutrient impoverishment than are forests in most non-karst areas. In addition, trace elements are often limiting or lacking.

Timber harvesting in some karst areas is likely to increase erosion and result in the introduction of sediment and organic matter into dolines and sinking streams (Aley & Aley, 1984). Where well-developed karst groundwater systems exist, sediment and both suspended and dissolved organic matter may adversely impact the water quality of springs. An illustration of these issues is provided by some karst areas in the Tongass National Forest in Alaska (Aley et al., 1993). Groundwater tracing has demonstrated that dolines in some areas, under consideration for timber harvesting, rapidly transport water to spring-fed streams, which provide essential spawning habitat for salmon. Pulses of sediment and organic matter, derived from timber harvest and timber access roads, can clog gravels in spawning areas and deplete dissolved oxygen concentrations critical to salmon eggs and fry. Under present forest policies, hydrological investigations are conducted in karst areas of the Tongass National Forest, prior to detailed planning for timber harvest. The results of the hydrological investigations have removed some areas from timber harvest and have added karst-specific constraints to some planned harvests.

Concerns have been raised about timber harvesting in karst areas on Vancouver Island, British Columbia, Canada. Many of the issues have been similar to those encountered on the Tongass National Forest; although the species composition is different, the harvest units in Canada have been larger than those in Alaska, and burning after harvest has been used at a few Canadian locations. Stokes and Griffiths (2000) prepared a valuable manual on karst inventory systems and principles for the Research Branch of the British Columbia Ministry of Forests. This manual includes an excellent summary of resource vulnerability classification and mapping approaches used in forested karst areas around the world. As is the case in Alaska, groundwater tracing investigations are recognized in British Columbia as a valuable part of resource vulnerability assessments.

Huntoon (1991, 1992) describes hydrologic and other impacts resulting from the massive deforestation that has occurred since 1958, in the subtropical south China karst. Approximately 100 million people live in the south China karst belt, and the greatest poverty in the nation is located in this region. Inadequate water supplies have long been a problem in south China, and the area is characterized by a severe annual flood–drought cycle. Deforestation, primarily for charcoal production for industrialization, has depleted the "green reservoir" of forests, which in turn has increased rates of surface and groundwater runoff and sediment production. Essential community water supply springs, at the base of karst towers, have ceased flowing and residents must now transport water from distant wells. The amplitude of the flood–drought cycles has been increased, which has in turn rendered useless many formerly productive fields in karst depressions.

Afforestation of karst areas may impact on underlying cave systems. For example, in some karst areas of Australia, eucalypt forests (which are relatively conservative of water) have been replaced with pine plantations. Clearance and replanting with pines was accompanied by soil erosion and the pines were planted at a greater density. In consequence, evapotranspiration rates increased and less water entered the karst groundwater system and associated caves. Similarly, at Lehman Caves, Nevada, decades of fire suppression has enabled trees, such as junipers and pinyon pines, to invade lands overlying the caves. The forest vegetation uses more water than the previous cover of shrubs (primarily sagebrush, *Artemesia* sp.), and an apparent result has been the drying of speleothems in the caves.

TOM ALEY

See also **Soils on Carbonate Karst; Soil Erosion and Sedimentation**

Works Cited

Aley, T. & Aley C. 1984. Effects of land management on cave and water resources, Dry Medicine Lodge Creek basin, Bighorn Mountains, Wyoming. Proceedings of the 1984 National Cave Management Symposium. *Missouri Speleology*, 25 (1–4): 79–92

Aley, T., Aley, C., Elliott, W.R. & Huntoon, P.W. 1993. *Karst and Cave Resources Significance Assessment, Ketchikan Area, Tongass National Forest, Alaska.* Report of the Karst Resources Panel to Tongass National Forest

Gams, I. 1991. The origin of the term karst in the time of transition of karst (kras) from deforestation to forestation. In *Proceedings of the International Conference on Environmental Changes in Karst Areas*, edited by U. Sauro, A. Bondesan & M. Meneghel, Padova: Department of Geography, University of Padova

Huntoon, P. 1991. Deforestation in the south China karst and its impact on stone forest aquifers. In *Proceedings of the International Conference on Environmental Changes in Karst Areas*, edited by U. Sauro, A. Bondesan & M. Meneghel, Padova: Department of Geography, University of Padova

Huntoon, P. 1992. Hydrogeologic characteristics and deforestation of the stone forest karst aquifers of south China. *Ground Water*, 30(2): 167–76

Stokes, T.R. & Griffiths, P. 2000. A preliminary discussion of karst inventory systems and principles (KISP) for British Columbia. British Columbia Ministry of Forests Research Program Working Paper 51

Further Reading

Huntoon, P.W. 1997. The case for upland recharge area protection in the Rocky Mountain karst of the western United States. In *Karst Waters and Environmental Impacts: Proceedings of the 5th International Symposium and Field Seminar on Karst Waters and Environmental Impacts, Antalya, Turkey, 10–20 September 1995*, edited by G. Günay & A. Johnson, Rotterdam: Balkema

FRANCE: BIOSPELEOLOGY

France is one of three countries where research on subterranean fauna began in the 19th century (the others being Slovenia and the United States). The first species of blind cave beetles and spiders were discovered in Pyrenean caves—which have one of the world's highest levels of subterranean biodiversity, including several of the most troglomorphic species such as the cave beetles *Aphaenops* (Figure 1) and *Hydraphaenops*.

In France about 650 exclusively subterranean (troglobitic and stygobitic) species are known. Most of them are recorded from the Pyrenees, the southern Massif Central, the Alps, and the Jura mountains. They inhabit a broad spectrum of habitats: karstic caves, interstitial MSS (Mesovoid Shallow Substratum), phreatic aquifers, and interstitial river habitats. Almost all are endemic; several *Speonomus* beetles show micro-endemism within an area of a few square kilometres; some Leptodirinae and Trechinae species are each only found in one cave. In the Pyrenees the range of terrestrial subterranean invertebrates is not limited to caves, but includes the whole karstic subterranean system and micro-voids of the MSS in rocks other than limestone. Some are "living fossils" from ancient groups now absent from surface habitats; others have actively colonized the cave habitats, especially the MSS and subterranean rivers.

Coleoptera with 186 troglobitic species represents the main terrestrial cave-dwellers; they consist of small populations of carnivorous Trechinae, *Aphaenops, Hydraphaenops, Trichaphaenops, Geotrechus, Duvalius,* and *Typhlotrechus* at the top of the food chain, and large populations of detritivorous Leptodirinae, with *Speonomus* and *Troglophyes* lines in the Pyrenees, *Diaprisius* in the southern Massif Central, and *Cytodromus* lines in Provence, Alps, and Jura with *Royarella, Troglodromus,* and *Isereus. Isereus xambeui* inhabits cold caves in Grande Chartreuse Mountain. Usually, Leptodirinae populations are larger than Trechinae, estimated at 30 000 to 50 000 individual Leptodirinae in caves and their interstitial systems in the Pyrenees.

Spiders are represented by 33 species, mainly Leptonetidae (*Leptoneta*) and Linyphiidae (*Troglohyphantes*), and a tropical

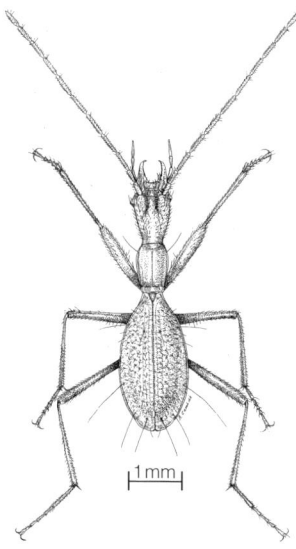

France: Biospeleology: Figure 1. *Aphaenops laurenti*, Grotte de Bordes de Crues à Seix en Couserans (Ariège).

relict, *Telema tenella*, in some caves of the Eastern Pyrenees. In the entrance and twilight zone the troglophilic *Meta* weave webs and cocoons. Three harvestmen families have troglobitic species in Pyrenees; Ischyropsalidae with *Ischyropsalis pyrenaea*, Travuniidae with blind and depigmented *Arbasus caecus* and reduced-eyed *Abasola sarea*, and Phalangodidae with *Scotolemon lucasi* in caves and the MSS in Ariège. Four species of extremely tiny Palpigradida *Eukoenenia* are found, each known by less than ten specimens. 33 species of troglobitic Pseudoscorpiones are known. The caves of the karst running along the south of the Massif Central are the habitat of *Chthonius* (seven species), *Neobisium* (three species), one *Spelyngochthonius*, and one *Roncochthonius*; nine species are known from Pyrenean caves, six from the Alps, four from Jura, and two in Corsica.

Considering the aquatic fauna, Crustacean Copepoda (60 species), Amphipoda (36 species), and Isopoda (29 species) are the most numerous. *Niphargus* inhabit all types of subterranean water and are the most representative of aquatic fauna in France (Figure 2); their ecology and biology have been extensively studied by the team at Lyon. Other Amphipoda, *Ingolfiella* and *Salentinella*, are mainly inhabitants of interstitial waters. The tiny Copepoda Cyclopidae and Harpacticidae are also well represented in karstic aquifers, groundwater, and interstitial water. *Gelyella droguei* found in Languedocian karst became the type of a new sub-order of Copepoda (Gelyelloida). Thousands of specimens are washed out of the karst through springs during flooding of cave streams, as for example in the karst system of Baget near Moulis; the subterranean aquatic populations are abundant but only a small proportion can be collected in caves.

Isopoda Stenasellidae *Stenasellus* with four relict species are found widely in karstic and interstitial habitats, whereas the asellid *Asellus aquaticus*—a surface species that occasionally invades groundwater and caves—is found in many caves in its more or less blind and depigmented form. Two Sphaeromatidae species are known: *Caecosphaeroma virei* in Jura and *C. burgundum* found in Jura and Charente. Two Cirolanidae species have been collected from caves, *Faucheria faucheri* in Languedoc and the eastern Pyrenees, and *Sphaeromides raymondi*, limited to some caves in Languedoc; Microparasellidae (*Microcharon*) inhabits interstitial waters. Syncarida, a primitive relict family, are mainly represented in interstitial water in southern France and Corsica by eight genera, including *Bathynella*, *Parabathynella*, and *Gallobathynella*. A single species of Decapoda, *Troglocaris inermis*, inhabits karstic water north of Montpellier; it has a phylogenetic relationship with Caucasian species.

Of the vertebrates, 12 species of bats regularly use caves to reproduce, hibernate, and rest during the day.

Several subterranean systems in France have been studied extensively over a long period of time: the Cent Fons system north of Montpellier; the system of Dorvan Massif in Jura; the interstitial aquatic habitats of the Rhône; and in the Pyrenees, the karst system of Baget, Le Goueil di Her, the interstitial habitat of Lachein, Nert streams near Moulis, Sainte-Catherine Cave near Moulis, Pigailh Cave, the MSS of Ravin de la Tir in Ariège, the cave fauna and MSS of Mt Canigou, and La Verna chamber of the Pierre Saint-Martin. The "Laboratoire souterrain du CNRS" with its two cave laboratories in the central Pyrenees at Moulis has been internationally important in biospeleological study. Since 1952 more than one hundred species from the Pyrenees and also from outside France have been reared in these cave laboratories. From the 1950s, experiments were carried out on the Slovenian cave salamander *Proteus anguinus*, and Moulis now has the only permanent and reproductive captive population in the world. Later species studied include the troglophilic *Euproctus asper*, a salamander endemic from the Pyrenees; from the Crustacea, several species of *Niphargus* (Amphipoda), *Stenasellus* and several terrestrial Onisciidae (Isopoda), aquatic Oligochaeta *Pelodrilus leruthi*, several species of Copepoda: Cyclopidae and Harpacticidae, one Mysidacea from Cuba, and the Decapoda: *Troglocaris*; from Arachnida, four spiders, four Opiliones *Ischyropsalis* and *Scotolemon*, one Amblypygi from Cuba; from myriapoda, two Diplopoda from the Pyrenees; and from Insecta, several Collembola, around 14 species of Coleoptera Leptodirinae (*Speonomus* and *Bathysciola* from the Pyrenees, *Closania* and *Drimeotus* from Romania), and one Orthoptera, *Dolichopoda linderi*. Adaptations to cave life have been demonstrated in these

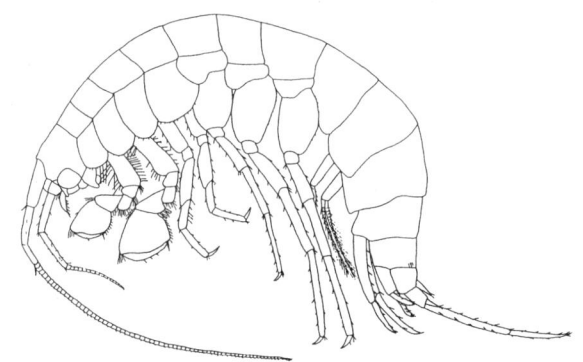

France: Biospeleology: Figure 2. *Niphargus virei* Chevreux (Niphargidae). Male (30.0 mm).

species in the laboratory: low level of fecundity, enlarged eggs, low speed of development, long adult lifespan (around 50 years for *Proteus*, 20 for *Troglocaris*, 10 for *Telema*), low resistance to dryness, disappearance of circadian rhythms, changes in behaviour, behavioural sexual barriers between species, development of chemo-hygroreceptors, and mechanoreceptors for low air flow in *Aphaenops*.

In 1979, the Société Internationale de Biospéologie was founded from Moulis and Lyon. The Society organizes a symposium on cave life and animals every two years, in different countries.

In France, the conservation of cave fauna is just beginning. All cave beetles—*Aphaenops, Hydraphaenops*, and *Trichaphaenops*—and Mollusca Hydrobiidae are protected; four caves and their fauna are protected National Nature Reserves, and another more extensive project is under consideration in Ariège. Bats and several of their cave habitats are protected by the European Bern and Bonn Conventions, and by French legislation.

CHRISTIAN JUBERTHIE

Further Reading

Jeannel, R. 1926. *La Faune cavernicole de la France*, Paris: Lechevalier

Juberthie, C. & Decu, V. (editors) 1994. *Encyclopaedia Biospeologica*, vol. 1; vol. 2 1998; vol. 3, 2001, Moulis and Bucharest: Société de Biospéologie

Vandel, A. 1965. *Biospeleology: The Biology of Cavernicolous Animals*, Oxford and New York: Pergamon Press (originally published in French, 1964)

Many data on French cave fauna are also published in the journals *Mémoires de Biospéologie, Annales de Spéléologie, Bulletin Museum National Histoire Naturelle Paris, Archives Zoologie expérimentale, International Journal of Speleology*, and *Spelunca*.

FRANCE: HISTORY

France has extensive karst, and caves have been described for a very long time by local historians, who took great pride in them. Caves close to large cities, at the foot of the massifs, or on the plateaus, attracted the first curious observers: the cave of Cuves de Sassenage (Isère) was first described in 1534. At the same time, caves and potholes were visited for practical reasons, such as for mineral exploitation, cheese maturation, and as a place of refuge. Historical remains have been found at the bottom of the Gouffre de Padirac (Lot) and the foot of Aven Noir (Aveyron).

The 1780s saw spectacular progress: several classic caves were discovered, such as the caves of Campan (Hautes-Pyrénées) and the caves of Osselles (Doubs), and explorations that were extremely difficult for the period were realized. Several known caves were extended in this way, such as the cave of La Balme (Isère) whose underground lake was explored, first by boat then by swimming, and several known but as yet unexplored caves were descended, such as the entrance shaft of the grotte des Demoiselles (Hérault) and Tindoul de la Vayssière (Aveyron).

In the 19th century it was first paleontologists, then archaeologists, who carried out most underground investigations. For example from 1880–82, the naturalist and paleontologist J.A. Lucante (1880, 1882) listed nearly two thousand caves and shafts in the whole of France. It was in this context that Édouard-Alfred Martel, who dominates the history of French speleology, appeared (see Speleologists entry). Born in 1859, his first love was the mountains, but in June 1888, he made two great exploration discoveries in caves: in the Grands Causses he discovered the lower level of the Dargilan cave (Lozère), and at Bramabiau (Gard) he carried out the first through-exploration of the underground river from end to end (Martel, 1889). The following year, he descended the large caves of Rabanel (Hérault) and Mas Raynal (Aveyron) and began to explore the underground Padirac River (Martel, 1890). His speleological career had begun: between 1888 and 1913 he undertook 26 annual campaigns throughout France, with a taste, however, for the medium-altitude massifs: Jura, Provence, Languedoc, Lot, and the Dordogne occupied him more than the Alps and the Pyrenees, where his visits were fewer and more specific.

Using more lightweight equipment than the Austrian explorers of the period he explored hundreds of caves, the descent of a large shaft taking only a few days of preparation. Many collaborators throughout the country (such as L. Armand, F. Mazauric, O. Decombaz, and E. Fournier) carried out their own extensive searches, and also accompanied Martel in the caves that they discovered. A small group of speleologists became organized in this way: Martel founded the Société de Spéléologie in 1895 and the journal *Spelunca* was launched the same year. However, World War I halted this impetus and ended speleology under Martel's leadership, though Martel himself continued to study and publish (e.g. Martel, 1921).

From 1920 to 1930, several great figures emerged in the wake of Martel. Norbert Casteret (see Speleologists entry) surveyed the central Pyrenees in several solo explorations: in 1935, he reached a depth of 243 m in the Gouffre Martel (Ariège), the deepest cave known in France at the time. Robert de Joly was interested especially in the Causses of Languedoc, discovering there the Aven Orgnac (Ardèche). In 1930, he created the Spéléo-Club de France, which in 1935 became the Société Spéléologique de France. French speleology progressed via a series of technological innovations (especially wire ladders), which meant obstacles that had stopped Martel could be overcome, such as the large shafts of Chourum Martin (Hautes-Alpes), descended by de Joly in 1929. It was also at this time that explorations of large cave systems began, such as at Henne-Morte (Haute-Garonne) and Dent de Crolles (Isère), which continued during World War II.

After 1945, speleological activity started again quickly and clubs were formed in Paris, Lyon, Nice, and Périgueux. The first congress of the International Speleological Union (UIS) took place in Paris in 1953. Many large expeditions were organized, and the world depth record was broken three times in France in less than ten years: in 1947, at Trou du Glaz (Isère), due to the connection of P.40 to Guiers-Mort Cave (−603 m); in 1953, at Pierre Saint-Martin (Pyrenees, −689 m); and in 1956 at the Gouffre Berger (Isère) where the depth of 1000 m was reached and exceeded for the first time (−1122 m). Casteret remained very active in this period, and his influence on the new generation of cavers was profound: in 1957, at the age of 60, he carried out another expedition to Henne-Morte.

This breadth of activity led in 1963 to the creation of the Fédération Française de Spéléologie, to which the great majority of French speleologists are affiliated. The first French speleological inventories were published, either by massif, or by local region. Single rope technique (SRT), a technological revolution established at the beginning of the 1970s, spread through the French school of caving and made several new discoveries possible: in 1979, Jean-Bernard Cave (Haute-Savoie) became the world's deepest known cave, replaced in 1998 by the neighbouring Mirolda Cave which was then 1616 m deep but was extended to 1733 m early in 2003 (see Morphometry of Caves). Around Henne-Morte Cave, the vast network of the Réseau Felix Trombe, whose length exceeds 100 km, was progressively explored and surveyed. Progress in cave diving also produced many discoveries and resolved some old speleological problems, such as the connection between Padirac Cave and the Finou resurgence, which was explored in September 1995.

Since 1980, French expeditions abroad have been equally divided between neighbouring countries (e.g. Spain and Austria) and more distant destinations: Algeria, China, and Papua New Guinea, where the first 1000 m deep cave in the southern hemisphere, Muruk cave, was explored by a French team. French speleology remains very active with more than 8000 cavers affiliated to over 600 clubs. The Fédération publishes exploration results and research through its two journals *Spelunca* and *Karstologia*, thus combining the sporting and scientific aspects of the discipline.

CHRISTOPHE GAUCHON

See also **Dent de Crolles; Pierre St Martin; Vercors**

Works Cited

Lucante, J.A. 1880, 1882. Essai géographique sur les cavernes de la France ... [A geographical study of the caves of France]. *Bulletin de la Société d'Études Scientifiques d'Angers*, 8 & 9: 81–156; 11& 12: 25–146

Martel, E.A. 1889. Sous terre (1eme campagne) [Under ground, 1st campaign]. *Annuaire du Club Alpin Français*, 15: 238–94

Martel, E.A. 1890. Sous terre (2eme campagne) [Under ground, 2nd campaign]. *Annuaire du Club Alpin Français*, 16: 100–44

Martel, E.A. 1921. *Nouveaux traité des eaux souterraines* [A new treatise on underground waters], Paris: Doin

Further Reading

Baring-Gould, S. 1894. *The Deserts of Southern France: An Introduction to the Limestone and Chalk Plateaux of Ancient Aquitaine*, 2 vols, London: Methuen and New York: Dodd
  Travels in the Causses [karst] of France, including its caves

Casteret, N. 1938. *Ten Years Under the Earth*, New York: Greystone and London: Dent (originally published in French as two books, *Dix ans sous terre* (1933) and *Au fond des gouffres* (1936)

Chevalier, P. 1951. *Subterranean Climbers: Twelve Years in the World's Deepest Cavern*, London: Faber (originally published in French, 1948)
  On the Trou de Glaz–Guiers-Mort cave system

Joly, R. de. 1975. *Memoirs of a Speleologist: The Adventurous Life of a Famous French Cave Explorer*, Teaneck, New Jersey: Zephyrus Press (originally published in French, 1974)

Martel. E.A. 1894. *Les abîmes* [The Depths], Paris: Delagrave
  On Martel's cave explorations in many countries

Minvielle, P. 1967. *La conquête souterraine* [The conquest of the underground world], Paris: Arthaud
  A history of cave exploration in Europe, with much on France

FRANCE, SOUTHERN MASSIF CENTRAL

Around the southern margin of the Massif Central, the Causses are a series of plateaus with very fine fluviokarst on gently folded limestones. The region is also notable for its caves with exceptionally fine aragonite speleothems, especially around the south of the Causses (particularly the Montagne Noire), in the Montpelier area, and in the foothills of the Pyrenees.

Causses

The Causses are extensive karstified limestone plateaus that are the remnants of ancient erosion surfaces. They are mainly undulating, with a dry stone-covered surface that is dissected by dry valley networks and by dolines ("sotches"), some of which contain siliceous deposits. Surface landforms include runiform relief ("pechs"), particularly on dolomites. The rocks form a monoclinal sedimentary series, bordered on the northern side by the Hercynian basement of the Massif Central. Marly Liassic rocks form an impermeable basement overlain by Jurassic and Cretaceous limestones and dolomites. Limestone is thought to have first become exposed in the late Jurassic, and evidence for a very ancient (late Cretaceous–early Tertiary) karst developed in a tropical climate is provided by traces of bauxite, sometimes in the shape of pisoliths, and by red clays, yellow sand, and quartz pebbles. Landform evolution continued through the Tertiary with phases of surface lowering and tectonic rejuvenation, particularly during the Oligocene and Miocene, with some local interruptions by volcanic activity. During the Pleistocene cold periods, the area was subject to severe nival conditions but was not glaciated. This accounts for the extensive frost-shattered debris on the Causse surfaces; some of this debris still forms sparsely vegetated felsenmeer, while elsewhere it is a significant component of the stony soils.

From north to south it is possible to distinguish the Périgord and Quercy Causses, the Grands Causses (Sauveterre, Méjean, Montagne Noir, and the Larzac which is one of the largest plateaus, reaching $>1000\,km^2$), and the Minervois and Garrigues Causses. The relief in parts of the Périgord-Quercy Causses resembles relict tropical cone karst, albeit altered by periglacial weathering during the Pleistocene, and Pelissier (1999) describes paleokarst in the Quercy Causses. The Grand Causses (850–1100 m above sea level) are bounded by major cuestas and separated by deep valleys such as the Lot, or by spectacular canyons, up to 600 m deep, such as those of the Tarn, Jonte, and Dourbie. All of the canyons and gorges have been formed by allogenic rivers although their discharge is added to by springs which drain the surrounding Causses plateaus. The area is noted for extremely heavy rainfall events particularly in the spring when 24-hour totals of up to 950 mm have been recorded.

The Causses contain many fine caves, including the Gouffre de Padirac (in the Causse de Gramat) with its large entrance shaft and a long extensive underground river lying beneath abandoned passages richly decorated with calcite. The Aven d'Orgnac (in the Ardèche) has a large chamber with many tall "plate" or "palm tree" type stalagmites and is also extremely well endowed with helictites, shields, and monocrystalline stalactites. Both caves featured prominently in early French exploration, E.A. Martel being associated with Padirac while Orgnac contains a vase with the ashes of its discoverer, Robert de Joly (see also France: History). More recently, the Doux de Coly has been explored by divers (see Diving in Caves) to reveal an active phreatic tunnel over 5000 m long beneath the Quercy Causse.

Aragonite Caves
Although aragonite is quite common as a cave mineral (see Speleothems: Carbonate) it is rarely found in such profusion as in a series of caves in the Montagne Noire. In several of these caves the aragonite is a beautiful blue colour due to copper mineralization. The Barrencs de Fournes (Blue Cave) is listed as one of the ten most important mineralogical caves in the world (Hill & Forti, 1997). It has been sealed since 1974 to ensure total protection but is the reference site for blue aragonite and is claimed to contain unique forms. More accessible is the Grotte de Limousis, a 2 km long show cave that contains "Le Lustre", a huge hanging aragonite cluster estimated to weigh about 9 tonnes and probably the largest single cluster known. A second tourist cave, the "Giant Hole", part of the 17 km long Grotte de Cabrespine, is also superbly decorated with both coralloid and acicular aragonite together with large calcite shields and a profusion of more common calcite speleothems. The smaller Grotte de l'Asperge contains spectacular, and possibly unique, acicular blue aragonite frostwork, blue coralloid aragonite, and a pale blue aragonite stalagmite, while the nearby PN77 cave contains brown and white-tipped tubular aragonite stalactites that have not been reported from anywhere else in the world. Also in the Montagne Noire region, the Grotte du Mont Marcou contains rare green aragonite speleothems (the colouration being due to nickel) and the Grotte Pousselière contains acicular aragonite and aragonite balls.

Outside the Causses, in the Montpelier region, the Grottes de la Clamouse is a popular tourist site that contains extensive deposits of coralloid and acicular aragonite, with hydromagnesite on some of the acicular tips, and a wide range of calcite speleothems, some of which have overgrowths of aragonite making the cave of particular interest. In the Pyrenees the most important sites for aragonite are the TM71 and André Lachambre caves. In TM71 the main aragonite is in a spectacular blue cluster and there are also unusual red speleothems and pure white monocrystalline triangular calcite stalagmites. Many of the 25 km of passages in André Lachambre are encrusted with pure white aragonite from floor to ceiling and this, together with cave pearls, hydromagnesite formations, and a unique vein of talc which crosses the cave passage, makes the cave one of the most spectacular in the world. The Gouffre d'Esparros also contains spectacular calcite and aragonite speleothems; part of the entrance series has been opened as a show cave with extensive instrumentation to measure tourist impacts on the cave climate. Two other notable sites in the Pyrenees are the Grotte de la Cigalère, a protected site containing rare gypsum speleothems, which include black gypsum, iron hydroxide and manganese hydroxide minerals, and large gypsum flowers and the Laboratoire souterrain du CNRS with its two cave laboratories at Moulis which has been internationally important in biospeleological study since 1952.

A common element to all the caves with aragonite is that they are formed in Jurassic dolomite or Paleozoic meta-dolomite, as aragonite is only deposited where the host rock contains over 7.5% Mg (Cabrol, 1979). Cabrol and Coudray (1982) found

France, Southern Massif Central: Figure 1. Aragonite and hydromagnesite, Réseau André Lachambre. (Photo by John Gunn)

France, Southern Massif Central: Figure 2. Water droplet at the end of fine aragonite needles, Réseau André Lachambre. (Photo by John Gunn)

that, as well as speleothems that contain one or more carbonate minerals as a result of diagenesis, there are some complex speleothems displaying successive deposition of calcite and aragonite minerals. They suggested that the spatial and temporal distribution of these speleothems with respect to their morphology and mineralogical features indicate that there were cycles of primary deposition and diagenetic alteration associated with changes in local climate driven by regional climatic changes.

JOHN GUNN

Works Cited

Cabrol, P. 1979. Trois types de concrétions d'aragonite très rares. *Spelunca*, 3: 119–21

Cabrol, P. & Coudray, J. 1982. Climatic fluctuations influence the genesis and diagenesis of carbonate speleothems in southwestern France. *NSS Bulletin*, 44: 112–17

Hill, C.A. & Forti, P. 1997. *Cave Minerals of the World*, 2nd edition, Huntsville, Alabama: National Speleological Society

Pelissier, T. 1999. Les phosphatières du Quercy. *Spelunca*, 73: 23–38

Further Reading

Cabrol, P. & Mangin, A. 2000. *Fleurs de pierres: les plus belles concrétions des grottes de France*, Lausanne: Editions Delachaux et Niestlé

Salomon, J.-N. 2000. Le Causse de Gramat et ses alentours: les atouts du paysage karstique. *Karstologia*, 35: 1–12

FRASASSI CAVES, ITALY

The Frasassi Caves are located in central Italy, on the eastern slopes of the Apennines some 40 km from the Adriatic. They developed in a small anticline ridge deeply dissected by the Sentino Gorge (Figure 1). The caves constitute one of the most famous Italian karst systems, partly because they host the most-visited show caves in the country. Speleological explorations of the area took place mainly from 1950 to 1980, but some large cavities have been known at least since Procaccini Ricci's 1809 *Memoria sulla Grottadi Frasassi nei dintorni di Fabriano*. The Frasassi system is of particular scientific interest, and is presently one of the best examples of a cave with sulfide water circulation: in its lower branches, hyperkarstic speleogenetic mechanisms are active, and an unusual ecosystem dominated by chemiosynthetic organisms has been documented.

Several caves, over 25 km in length, open along the steep walls of a gorge from 200 m above sea level (corresponding with the river bed) up to 500 m. They generally consist of networks of dendritic horizontal passages in which wide chambers alternate with smaller tubes: the Ancona Abyss, the largest chamber, has a volume of about one million cubic metres. The whole

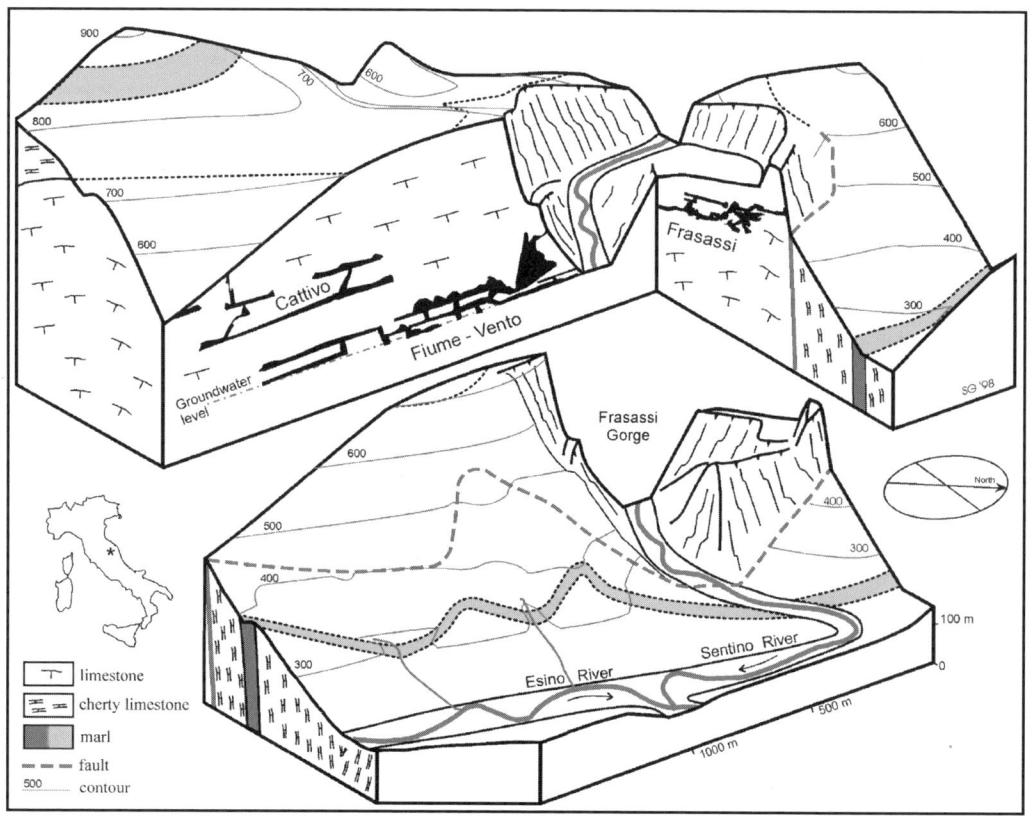

Frasassi Caves, Italy: Figure 1. Block diagram of the Frasassi karst system, with the caves of Buco Cattivo, Grotta del Fiume–Grotta Grande del Vento and Grotta del Mezzogiorno–Grotta di Frasassi in the limestone either side of the Frasassi Gorge of the Sentino River.

system consists of several superimposed and interconnected levels, the genesis of which is related to the different steps in the deepening of the base level.

The lower series, present mainly in the Fiume–Vento system, developed during the Middle and Upper Pleistocene, under geomorphological and hydrogeological conditions similar to the present day. It contains speleothems dated to 200 000 BP, and can be correlated with the alluvial deposits just outside the Frasassi Gorge. The upper levels, developed mainly in the Buco Cattivo and Mezzogiorno–Frasassi caves, show a different morphology, as a consequence of evolution under different hydrogeological settings.

The lower parts of the system are in contact with a sulfide aquifer, fed by rising waters from buried evaporitic formations and by seeping meteoric water. The phreatic waters are cold and rich in Cl^-, Na^+, and SO_4^{2-}, with a rather high H_2S content (up to 18 mg l^{-1}). The saline content varies throughout the year, being controlled by the mixing ratio between meteoric and sulfide waters: therefore higher values (about 2 g l^{-1}) occur at the end of the summer.

The presence of H_2S and its consequent oxidation allows secondary cave minerals to develop: gypsum is forming extensively in the lower part of the system, while thick gypsum deposits (up to 2 m or more, with a total volume of over 10 000 m³) are present in the upper dry series. Their prevailing occurrence is as white microcrystalline replacement crusts over the limestone walls, or as thick precipitated deposits over the floor (Figure 2). Large prismatic crystals of gypsum are common, and their genesis is related to the diagenesis of the microcrystalline deposits and to the seepage of supersaturated waters into clay deposits. Other cave minerals related to the sulfur cycle (halloysite, barite, and jarosite) have been detected in the Frasassi Caves.

Most of the redox processes develop in the interface between sulfide waters and the cave atmosphere, and they are responsible for the speleogenesis of the whole karst system. The hyperkarst corrosion to limestone is active in both the epiphreatic and the vadose zone, but gypsum deposition occurs only in the latter because supersaturation cannot be reached in the vadose groundwater. Where H_2S is released in the cave atmosphere, the limestone is highly corroded and partially covered by gypsum crusts associated with a few crystals of elemental sulfur. The intensity of this reaction has been measured using limestone tablets with a known surface area and weight: the average weight loss was about 15 mg cm^{-2} a^{-1} both in the groundwater and in the atmosphere.

The sulfur cycle is also responsible for the development of autotrophic microorganisms, which produce organic matter using the chemical energy released during sulfur oxidation (chemosynthesis). This organic matter is an important trophic support for a rich and complex community of invertebrates.

Bacterial colonies always cover the bottom of the flooded galleries, but they are also present in the aerated zone, where microbial films, some millimetres thick, may give rise to mucous stalactites (mucolites): the dripping water from these formations has a pH <1, being rich in H_2SO_4.

Measurements of C and N stable isotopes prove that the whole ecosystem in the sulfide caves is chemoautotrophic. In other areas, which are sparsely populated, the sulfide ones always show a noticeably high density of highly specialized fauna: seven new species have been detected so far, of which six have a distribution area restricted to the sulfide part of the system.

Frasassi Caves, Italy: Figure 2. One of the largest gypsum deposits in the Fiume–Vento Cave. (Photo by Sandro Galdenzi)

In 1974 a part of the cave was transformed into a show cave, the entrance of which is an artificial tunnel into the Ancona Abyss. The show cave has an average of 350 000 visitors per year and up to 6000 visitors per day. However, this heavy tourist traffic has had practically no impact on the cave microclimate, thanks to its very large volume and to three sliding doors in the artificial tunnel. The main environmental parameters have been monitored since 1980 at 10 different stations, and have been analysed periodically by a scientific committee, in order to determine the intervention needed to preserve the cave environment.

PAOLO FORTI and SANDRO GALDENZI

Further Reading

Galdenzi, S. & Menichetti, M. (editors) 1990. *Il carsismo della Gola di Frasassi*, Costacciaro: Memorie dell'Istituto Italiano di speleologia
  This has an exhaustive bibliography on every aspect of the Frasassi caves.
Sarbu, S.M., Galdenzi, S., Menichetti, M. & Gentile, G. (2000). Geology and biology of the Frasassi Caves in Central Italy: an ecological multi-disciplinary study of a hypogenic underground ecosystem. In *Subterranean Ecosystems*, edited by H. Wilkens, D.C. Culver & W.F. Humphreys, Amsterdam and New York: Elsevier
  This gives an update on the research inside the caves.

GEOPHYSICAL DETECTION OF CAVES AND KARSTIC VOIDS

Finding an underground opening is one of the most challenging exploration targets for a geophysicist. In addition to locating caves for exploration or scientific research, mapping hidden karst is necessary when engineering projects are planned in rock formations known to contain caves, because karstic voids and doline collapses can compromise the integrity of building foundations, dams, and bridges. Clay-filled voids in paleokarst can also jam rock-boring machines. Proper use of geophysical exploration for karstic voids can often help geotechnical engineers develop an effective programme of test borings and grouting, since if the bore spacing is greater than the cavity dimensions it is possible to miss it completely.

Geophysical methods employ indirect, non-intrusive observations to characterize and map variations in the physical properties of what lies concealed beneath the ground surface. All geophysical techniques require contrasts of some physical property (density, electrical resistivity, magnetic susceptibility, seismic velocity) between subsurface structures. Although void space in rock may represent an enormous contrast in physical properties that can be advantageous to an investigator seeking concealed caves, underground karst openings are frequently small, irregular targets whose effects are easily masked by those of surface irregularities. Techniques useful for deep exploration usually do so at the expense of resolution and accuracy; conversely, techniques capable of generating high-resolution images of shallow features are often based on high-frequency signals that are rapidly attenuated as they propagate through deeper soil and rock.

Bedrock voids, whether filled with air, water, or sediment, constitute "missing mass" in comparison to solid rock. The negative density contrast is identifiable as a local reduction in the Earth's gravitational acceleration (g) measured at the surface. Microgravity surveys conducted to locate bedrock voids involve densely spaced measurements, high precision elevation control, and sensitive gravity meters. Rybakov *et al.* (2001) used an automated gravity meter and measurements on a 5 m grid to survey about 2000 m^2 per day. Although the LaCoste and Romberg "G" meter (the industry standard) is read to a precision of 0.001 mGal (or about a millionth of the Earth's normal gravitational acceleration), true accuracy is seldom better than 0.01 mGal. To reduce the gravity observations to gravity anomalies, observed gravity must be separated from theoretical gravity, which varies according to latitude, elevation and surrounding topography, and geological material. Gravity on the Earth's surface decreases with elevation at a rate of about 0.197 mGal m^{-1}, so the elevation of each gravity station must be determined to an accuracy of 5 cm if reduced gravity variations of 0.01 mGal are to be significant. Tidal variations of as much as 0.3 mGals must be calculated or monitored, and the gravitational effects of neighbouring hills and valleys must be calculated and separated from the cavity's effect.

Interpretation of gravity anomalies is limited by ambiguities regarding shapes of structures which could be responsible for the anomaly (Figure 1). Moreover, voids may still be present where no anomaly is detected, due to a combination of additive and negative anomalies. Total volume of void space can be estimated, subject to assumptions or independent information regarding the average density of surrounding bedrock and the material filling caves and fractures (air-filled cavities create the largest anomaly; water-filled cavities create an anomaly effect of only 60% that of the same cavity containing air; and rubble or mud-filled cavities only about 40% that of air). Sophisticated computer analysis of data from closely spaced points allows anomalies to be distinguished by both their amplitude and their wavelength, which decrease and increase respectively with depth to a cavity. Data processing can apply filtering to separate the effects due to anomalies at different depths. Spectacular results have been obtained in surveys over old mines, where the voids form simple and recognizable shapes, but the random shapes and complexity of karstic cavities make them more difficult to identify in microgravity surveys: fractured rock around the cavity, and collapse features associated with caves can give an enhanced anomaly and a false picture of the size of the void

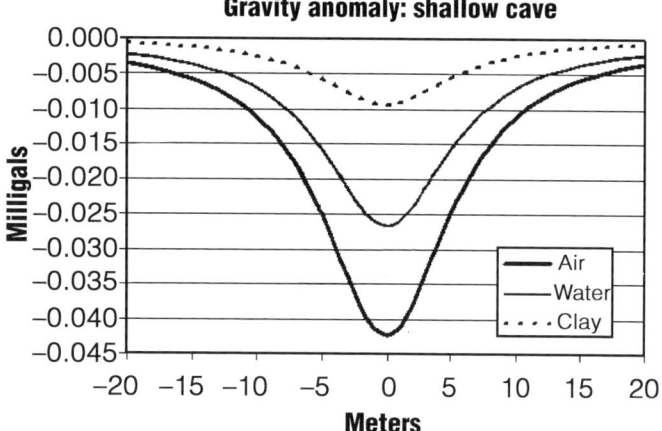

Geophysical Detection of Caves and Karstic Voids: Figure 1.
Gravity anomaly calculated for a horizontal cave, 3 metres in diameter and centered 6 metres under the surface, in dense (2700 kg m^{-3}) limestone. Forward models such as this example help the geophysicist determine grid spacing and minimum acceptable accuracy prior to designing an appropriate field exploration strategy.

(Chamon & Dobereiner, 1988). In karst, the old guideline that voids can only be recognized at depths less than twice their diameter is still appropriate. Gravity mapping of caves in mountainous areas is restricted by the difficulty in calculating with precision the large topographic effect of nearby mountains and valleys. Gravity mapping of soil-mantled karst requires use of an additional, independent method for determining the buried topography. However, gravity can be used in urban areas where cultural noise obscures electrical and magnetic signals. A negative gravity anomaly must be due to some form of missing rock, and this makes the data singularly accessible to the engineer who has to ensure safe ground for structural foundations.

Ground-penetrating radar (GPR) is one of the newer geophysical methods, and one of the most successful in detecting shallow caves (Chamberlain et al., 2000). Pulses of low-frequency (10 to 1500 MHz) electromagnetic energy penetrate the ground and are partially or totally reflected from rock or soil boundaries with contrasting electrical properties (notably their dielectric constants, or permittivities). Air-filled voids and layers of water-saturated sediment are strong radar reflectors. The reflected signals are detected on the ground surface and are collated by computer to produce the ground profiles; a series of parallel profiles can be combined to generate gridded 3-D images of the subsurface. GPR has been used to map joints enlarged by dissolution under thin soils, and may have the potential for mapping the cross-sectional shape of caves that are not too deeply buried by conductive overburden. The selection of radar frequency is probably the most important choice when undertaking a GPR survey, since lower frequencies (10–300 MHz) allow depth penetration to several tens of metres, but they will fail to detect small diameter anomalies. Frequencies of 500 MHz and greater provide excellent resolution, but they limit the subsurface depth penetration of the GPR to less than 5 m. Radar signals penetrate well into dry sandy soils and into carbonate rocks to map features at depths of several tens of metres. In contrast, clay-rich soils attenuate GPR signals and can restrict depth interrogation to depths of just 1 m. Lateral reflections limit the use of GPR in rugged terrain. Roadbeds are penetrated by GPR, and a transmitter / receiver towed by a car can be used to detect karstic voids developing under the highway. GPR is a relatively new technology that should become increasingly functional as field geophysicists discover ways to enhance its capabilities.

Electromagnetic induction measurements (EM) also use electromagnetic waves transmitted into the ground. Where the waves encounter electrical conductors in the ground, they induce electrical currents in these conductors, which in turn generate electromagnetic waves that can be collected at the surface by an antenna. The method can be useful for mapping karst where the voids in bedrock are filled with clay or water. Clay-filled fractures and caves partly filled with water conduct electrical signals far better than does carbonate bedrock, and often produce high-amplitude anomalies. EM measurements have proven more cost-effective than exhaustive test drilling for locating vertical clay-filled fractures prior to expansion of limestone quarries. On the other hand, air-filled fractures or voids are transparent to EM signals and are difficult to detect. However, caves that extend above the water table do not block the signal generated by conductive ground water, and so the water-filled part of the cave can be detected. The major limitation of EM measurements is their ambiguity. It may be difficult to isolate changes in depth to bedrock from lateral changes in electrical conductivity. It is sometimes possible to discriminate EM conductivity variations due to changes in depth to bedrock, from those due to water-filled karst, by using an instrument such as the Geonics EM34–3 and making conductivity measurements at three different frequencies in both the vertical and horizontal dipole modes. The EM34–3's lowest operating frequency is 400 Hz and its depth of exploration can vary from 60 to 100 m. EM instruments can provide more useful, albeit less detailed, information than can GPR in karst that is mantled by several metres of wet clay soil.

Electrical resistivity surveys also utilize contrasting electrical properties to characterize and map buried rock. Electrical current is transmitted directly into the ground through a pair of electrodes, which results in a voltage change measured between a second pair of electrodes. The apparent resistivity (ρ_a) of the ground can be calculated, and since low-porosity bedrock usually exhibits an electrical resistivity higher than overlying sediment, the buried topography can be interpreted. Electrical resistivity measurements should usually yield values consistent with EM conductivity, because apparent resistivity is simply the inverse of apparent conductivity, except when encountering subsurface voids. An air-filled cave is opaque to electrical currents but transparent to EM signals. It is unfortunate that air-filled caves and clay-filled caves have opposite resistivity characteristics, and some site investigation surveys have encountered problems where effects from a mixture of open and filled caves cancel each other out. When resistivity soundings indicate a highly resistive subsurface where an EM reading indicates a conductive subsurface, this discrepancy is probably due to a cave that extends both above and below the water table. Electrical resistivity profiles using arrays that simultaneously map lateral and vertical variations in apparent resistivity reveal caves even more clearly when data are interpreted using commercial inversion software on desktop computers.

Finding hidden caves or voids that might be large enough for direct human exploration has been less successful than profiling potential collapse areas in geotechnical investigations. Resistivity tomography has been successfully used to resolve known cavities, for example 20 × 30 m cavities at a depth of 10–20m at Harrison's Cave in Barbados (Shi, Morgan & Wharton, 1997), but proving the existence of as yet-undiscovered caves with resistivity remains elusive, perhaps because explorers may not be able to drill potential sites.

The induced polarization and self-potential methods have been tested as cave detection methods at Indian Echo Caverns, Pennsylvania (Sogade, Vichabian & Morgan, 1999; Vichabian & Morgan, 1999) and at Kartchner State Caverns, Arizona (Lange, 1999). The induced polarization method utilizes the difference between chargeability of air and water-saturated rock, while preferential infiltration of groundwater towards air-filled cavities in sedimentary rocks is thought to cause self-potential anomalies. Sogade, Vichabian and Morgan were able to resolve a 20 × 3 m cavity at a depth of 20 m. As with the resistivity method, detection of known cave passages is yet to be demonstrated.

Seismic reflection surveys can detect voids because the large negative reflection coefficient that exists between air and rock or water and rock generates an echo that is not only strong but which exhibits a phase reversal. A reflection from the overburden-bedrock contact is a compression, while that from a void begins as a dilation. Single sets of ground-based seismic data have often failed to produce interpretable results, but tomographic sections produced by computer analysis of larger banks of data (obtained from seismic velocities measured on hundreds of ray paths between multiple positions in boreholes) have proved more useful. The development of three-dimensional seismic tomography (3dT) is again dependent on powerful computer software, but has produced some impressive images of complex karstic voids in unseen ground (Figure 2). This is currently the most promising tool in geophysical searches for caves, but does represent a major investigation incorporating boreholes unless underground access is already available (as on a tunnel project).

Strategy

A prospecting strategy has to be developed for any specific set of potential karstic ground conditions and the choice of method depends on the degree of certainty required. Electrical dipole-dipole resistivity profiles are often the most reliable method for locating shallow air-filled caves. This method can approximate the size, shape, and depth of air-filled voids, but survey depth is usually limited to 20 m or less. Slopes where talus conceals cave entrances are usually too steep for other geophysical methods. Gravity surveys can distinguish heavily karstified zones from nearby areas with a lower overall void space, but gravity tells us little about the size and shape of individual voids. Seismic reflection and ground-penetrating radar usually image only the top of an air-filled void, and seismic tomography demands expensive exploration infrastructure and data processing. Locating hidden caves challenges even the experienced geophysicist, and non-specialists are unlikely to achieve success on their first attempts, despite having extensive technical backgrounds in related fields.

In locating buried topography of karstic terrains (such as dolines that are filled, covered, or concealed by sediment, usually glacial drift), microgravity profiles and EM measurements provide good reconnaissance tools, followed up by test drilling or by slower, more labour-intensive methods (electrical resistivity,

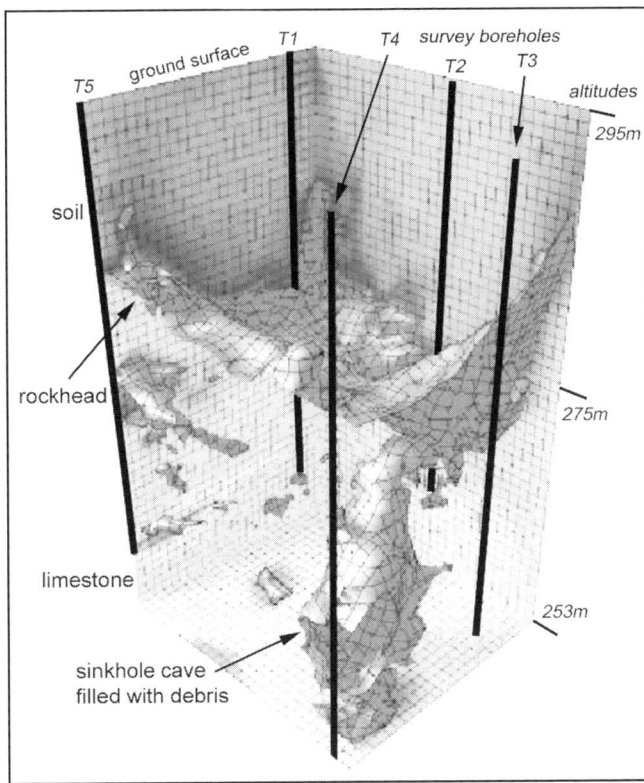

Geophysical Detection of Caves and Karstic Voids: Figure 2. An image of a doline beneath a road in Pennsylvania produced by 3-D seismic tomography between sources and receivers in an array of five boreholes. The debris-filled cave passage descending steeply through rock beneath a soil cover 20 m deep was subsequently verified by further drilling. (Courtesy of 3dT/NSA Engineering)

seismic reflection) capable of providing better subsurface images of identified conductivity highs and gravity lows. Geophysical exploration relies on sophisticated instruments and data analysis. The method selected is often determined by the needs of the investigator and the nature of the target, and even an experienced geophysicist might not know until field work begins which field strategies and instrument deployments will provide adequate information most efficiently in any given setting.

DONALD J. STIERMAN

Works Cited

Chamon, N. & Dobereiner, L. 1988. An example of the uses of geophysical methods for the investigation of a cavern in sandstones. *Bulletin of the International Association of Engineering Geology*, 38: 37–43

Chamberlain, A., Sellers, W.I., Proctor, C. & Coard, R. 2000. Cave detection in limestone using ground penetrating radar. *Journal of Archaeological Science*, 27: 957–64

Lange, A.L. 1999. Geophysical studies at Kartchner Caverns State Park, Arizona. *Journal of Cave and Karst Studies*, 61(2): 58–72; see also Discussion and Reply, 62(1): 27–29

Rybakov, M., Goldshmidt, V., Fleischer, L. & Rotstein, Y. 2001. Cave detection and 4-D monitoring: A microgravity case history near the Dead Sea. *Geophysics: The Leading Edge*, 20(8): 296–300

Shi, W., Morgan, F.D. & Wharton, A.E. 1997. Application of electrical resistivity tomography to image Harrison Caves in

Barbados, West Indies. *Annual Meeting Abstracts, Society of Exploration Geophysicists*: 350–353

Sogade, J., Vichabian, Y. & Morgan, F.D. 1999. Induced polarization in the detection of cave systems. In *Proceedings of the Symposium on the Application of Geophysics to Engineering and Environmental Problems*, edited by M.H. Powers, L. Cramer & R.S. Bell, Englewood, Colorado: Environmental and Engineering Geophysical Society

Vichabian, Y. & Morgan, F.D. 1999. Self potentials in cave detection. In *Proceedings of the Symposium on the Application of Geophysics to Engineering and Environmental Problems*, edited by M.H. Powers, L. Cramer & R.S. Bell, Englewood, Colorado: Environmental and Engineering Geophysical Society

Further Reading

Batayneh, A.T., Abdelruhman, A.A. & Moumani, K.A. 2002. Use of ground-penetrating radar for assessment of potential sinkhole conditions: An example from Ghor al Haditha area, Jordan. *Environmental Geology*, 41: 977–83

Bishop, I., Styles, P., Emsley, S.J. & Ferguson, N.S. 1997. The detection of cavities using the microgravity technique: Case histories from mining and karstic environments. In *Modern Geophysics in Engineering Geology*, edited by D.M. McCann, P.J. Fenning & G.M. Reeves, London: Geological Society

Branham, K.L. & Steeples, D.W. 1988. Cavity detection using high-resolution seismic reflection methods. *Mining Engineering*, 40(2): 115–19

Butler, D.K. 1984. Microgravimetric and gravity gradient techniques for detection of subsurface cavities. *Geophysics*, 49: 1084–96

Daniels, J. 1988. Locating caves, tunnels and mines. *Geophysics: The Leading Edge*, 7(3): 32–38, 52

Kearey, P. & Brooks, M. 1991. *An Introduction to Geophysical Exploration*, London: Blackwell

Kirk, K.G. & Werner, E. 1981. *Handbook of Geophysical Cavity-Locating Techniques*, Washington, D.C.: Federal Highway Administration (Implementation Package FHWA-IP-81-3)

Milsom, J. 1996. *Field Geophysics*, 2nd edition, New York: Wiley

Milanović, P.T. 2000. *Geological Engineering in Karst*, Belgrade: Zebra Publishing (Chapter 11: Applications of geophysical methods in karst)

Noel, M. & Xu, B. 1992. Cave detection using electrical resistivity tomography. *Cave Science*, 19: 91–94

Simpson, D. 2001. It's a vision thing. *Ground Engineering*, 34(12): 22–23

http://www.agiusa.com/2Dvoids.shtml—AGI collection of 2D resistivity imaging brochures: Cave and void detection

Proceedings volumes and the *Journal* of the Environmental & Engineering Geophysical Society include numerous case histories demonstrating the use of geophysical techniques to map and characterize karst.

GEOSCIENTISTS

Any listing of geoscientists who excelled in cave and karst science is subjective. At a given time, in all disciplines, a handful of workers appear to dominate their chosen field, yet identifying past giants can be controversial. Ideas initially seized as huge advances might later appear totally misdirected, even counter-productive—or vice versa. Moreover, not only brilliant theories and novel interpretations make individuals stand out, but their inspiration of peer scientists and stimulation of wider thinking. It is virtually inevitable that every practising geoscientist has his or her own list of worthy exemplars. Equally inevitably, each individual's list will be personal, even biased, for various reasons. Choices may be influenced by scientific background and peer grouping, by language and nationality, or by empathy (or lack of empathy) with a particular scientific specialisation or concept. Less subjectively, choices will also reflect the context of geoscientific achievement against the datum (and within the framework) of wider contemporary thinking. The writer and the Advisory Board consider that the scientists discussed below have made particularly important contributions to the development of the subject in the late 19th and the 20th centuries.

Jovan Cvijić (1865–1927)

Remembered for contributions to anthropology and ethnology as well as to geoscience, Cvijić was born at Loznica, in western Serbia. He was the foremost karstologist of his generation and, arguably, the greatest karst "generalist" of the 19th and early 20th centuries. After graduating from Belgrade University in 1888 he published his first paper, describing karst in eastern Serbia, in 1889, when he also began doctoral research in Vienna. His work, supervised by Alfred Penck, the only contemporary geomorphologist of comparable reputation to W.M. Davis (below), led to publication of "Das Karstphänomen" (1893).

Returning to Belgrade University as a professor, he worked there until his death. In 1947 the Jovan Cvijić Geographical Institute was founded in Belgrade, and in 1996 a memorial statue was erected in Loznica.

In karst terms Cvijić is best remembered for two pivotal publications. "Das Karstphänomen" provided the first major overview of karst terrains, clearly describing, with case-study examples, complex aspects of karstification. In "Hydrographie souterraine et evolution morphologique du Karst" (1918), Cvijić introduced the still valid concept of hydrographic zones. He also described karst surface lowering, and mentioned the resultant surface features now called unroofed caves. A third, less well-known, monograph contained vital new insights and revisions. Drafted in 1925, "La Géographie des terraines calcaires" was not published until 1960 (in French). Had publication been immediate, the emphasis of Davis's cave work [below] might have changed, as Cvijić demonstrated that erosional cyclicity could not dominate karst landscape development. He also argued that, with so many variables involved, no single, general rule covers the progression of all aspects of karstification, and stressed that, contrary to common expectation, knowledge of underground drainage is the key to understanding surface karst.

William Morris Davis (1850–1934)

Born in Philadelphia (United States), W.M. Davis became the dominating influence among peer researchers of his generation. Regrettably, he is best remembered as originator and champion of erosional cyclicity theories that subsequently lost favour, after dominating geoscientific thinking for decades. His miscalculation in promoting cyclicity too enthusiastically, too widely, and too dogmatically, overshadowed his many deep and valuable insights, including recognition that landscapes reflect geological structure, prevailing erosional or depositional processes, and development stage—a catechism that seems obvious today.

In karst terms, Davis's greatest achievement, published when he was 80 years old, was his *Origin of Limestone Caverns* (1930). Within this 150-page tome he discussed aspects of deep phreatic, shallow phreatic, and vadose cave development (see Speleogenesis Theories: Post–1890). Elements of his consideration of deep flow mechanisms and his recognition of extremely long speleogenetic timescales remain particularly valid today. However, some of his contemporaries were overwhelmed and misled by the attractiveness of his erosion cycle arguments and, like others since, they overlooked the immeasurable underlying value of his fundamental insights. Davis touched upon several ideas that survive within modern thinking about speleogenesis and related topics, such as the potential links between caves and mineralization. However, for many, his most perceptive deductions were obscured and devalued by the sheer volume of the enclosing text.

J Harlen Bretz (1882–1981)

Reputedly he was born Harley Bretz, in Michigan (United States), and changed his name to J. Harlen Bretz while at college, later discarding the full stop after the J! He was one of few genuinely intuitive and open-minded field geologists to contribute significantly to karst geoscience. Much of his career was based at the University of Chicago, whence he retired as an Emeritus Professor of Geology in 1947. His most controversial and celebrated work related to the "Scablands" of Washington State, formed (he believed) when catastrophic floods drained Glacial Lake Missoula. In a "uniformitarian" world his catastrophe ideas remained unacceptable to peer scientists for almost 50 years, until vindicated by photographs taken during the "Viking" Mars missions. After years of relentless, unwarranted criticism, his brilliance as an empirical scientist was belatedly acknowledged.

Based on observations of rocks of many ages across North America and the Caribbean, Bretz contributed a spectrum of ideas to karst science, and inspired much subsequent study. However, he is remembered above all for recognizing and documenting features diagnostic of phreatic or vadose conduit enlargement (Bretz, 1942). He also believed that his observations supported W.M. Davis's ideas, particular the fundamental and dominant role of deep-seated dissolution.

Otto Lehmann (1884–1941)

Lehmann grew up in Vienna, studying geography and history there and in Liepzig, tutored by eminent geographers including Penck and Brückner. After acquiring his doctorate in Vienna, Lehmann worked as an assistant lecturer in Leipzig, spending time (and hearing W.M. Davis lecture) in Paris before returning to Vienna. After wartime service as an artillery officer in Slovenia and Croatia he returned to geomorphology as a professor, first in Vienna (1920) and then in Zürich (1928), where he worked until his death.

His interests included karstology, climatology, human geography, and cartography, but he is remembered for his textbook *Die Hydrographie des Karstes* (1932). He described *estavelles* (features functioning either as springs or as swallow holes) and discussed the "karst hydrological contrast" whereby karst aquifers appear to have many inputs (sinks) but few outputs (springs). He also supported Martel and Katzer's "cave river" theories, and hence disagreed with Cvijić and Grund's "groundwater" models. In fact, both viewpoints have contextual validity, but for 40 years Lehmann's opinions dominated across most of Europe. Considering caves as pipes, Lehmann applied fluid mechanics concepts of conduit and open channel flow to speleogenesis studies, deducing that water movement can commence only in rocks with primitive perviousness (*hydrographische Wegsamkeit*), related to "initial cavities" (*Urhohlräume*) with a minimum 2 mm width.

In English-speaking countries Lehmann's ideas remain under-appreciated, perhaps because they were complex and inaccessible. More realistically, lack of subsequent interest might reflect his seeming failure to turn worthy, basic ideas into more comprehensive karst development models.

Georgij A. Maximovich (1904–79)

Maximovich, who became the foremost karst and cave scientist of the former USSR, was born in Warsaw. He graduated from the Dnepropetrovsk Mining Institute (Ukraine) in 1926 and completed his PhD (hydrogeology) in 1938. From 1934 to his death he was Head of Perm University's Dynamic Geology and Hydrogeology Department, one of the Soviet Union's leading karst study centres. Besides supervising 60 PhD projects, he established two scientific transaction series, *Peschery* (Caves) and *Gidrogeologija and Karstovedenie* (Hydrogeology and Karstology). He also founded and directed the All-Union Institute for Karstology and Speleology, a learned society with over 230 member scientists from all over the USSR.

Maximovich published widely on hydrogeology, hydrochemistry, and oil geology, but contributed most to karstology and speleology. Of his 535 publications, most involved cave and karst science, with 208 devoted to speleology. His two-volume *Principles of Karstology* (1963 & 1969) was a key reference source for generations of scientists. Maximovich's work covered cave study methodologies, speleogenesis, hydrology, hydrochemistry, cave sediments, cave microclimates, and the use of caves. Above and beyond this, however, his greatest contribution was to observation and description of regional speleology and karstology.

Alfred Bögli (1912–98)

Born in Bern, Switzerland, Alfred Bögli worked as a teacher in Bern and Luzern, before taking a geography degree at Friburg in 1939. His early studies included glacial morphology in the Alps, together with hydrology and karst geomorphology. Soon after World War II he began to study and survey the Hölloch cave system in Switzerland's Muota valley, a mammoth undertaking that was to dominate much of his life. Aside from his long involvement in exploring and mapping the Hölloch, Bögli's greatest of many contributions to karst geoscience was his recognition, development, and publication of ideas about *Mischungskorrosion*, or dissolution by mixing waters (Bögli, 1964). The mixture corrosion concept was accepted widely as the answer to a then current conundrum—how can meteoric water remain aggressive and be capable of dissolution deep underground, even after passing through great thicknesses of soluble rock? Another major contribution was *Karsthydrographie und physische Speläologie* (1978), translated into English in 1980. This widely based, well-balanced and, in many ways, understated textbook, remains a valued reference, reflecting Bögli's objective approach to the underlying source data and their interpretation.

Bernard Gèze (1913–96)

Bernard Gèze was, without doubt, a cave explorer and geoscientist *par excellence*. He was born in Toulouse (France) and became interested in caves and underground water during his late teens. His publication list eventually grew to more than 300 titles, and his contributions and related skills attracted wide recognition. He was a member of the French National Committee for Geology and the National Committee for Geodesy and Geophysics. He was President of the Speleological Commission of the National Scientific Research Centre and Honorary President of the International Speleological Union. Other attainments included presidencies of innumerable societies and committees, and the award of the Légion d'honneur (in 1961).

Gèze was distinguished as a caver (visiting more than 3000 caves) as well as a scientist, and joined many ground-breaking speleological explorations, including early work in the Henne-Morte and Padirac caves. His early cave and karst investigations produced interesting results, including ideas relating to structural influences on deep flow to vauclusian springs (e.g. 1939), appearing to contradict pre-existing ideas supported by Martel. In later years his work encompassed a wide spectrum of scientific disciplines and led to major publications, such as that on calcareous rock hydrology (1965), *La Spéléologie scientifique* (1965) and his lexicon of French speleological terms (1973). With Martel and de Joly he remains widely considered as one of the three founding fathers of French speleology and, specifically, he has been described as the father of scientific speleology.

Joseph Newell Jennings (1916–84)

"Joe" Jennings was born in Leeds (England). Reputedly his fascination with geomorphology was ignited by a school trip to nearby Malham. He took his first degree, essentially geomorphology and cartography, at Cambridge, followed by PhD research into the origin of the Norfolk Broads, but World War II intervened and it was years before he received his doctorate. After the war he lectured at Leicester University, until appointed Reader in geomorphology at the Australian National University in 1953. His first karst paper was published in 1956. People have commented upon the incongruity that he became an international karst authority while based in a country where karstic rocks are relatively scarce. Incongruous or not, he made the best of what is available in Australia—proving this to be far from insignificant—and spread his researches to Papua New Guinea, Malaysia, Canada, the United States, China, and New Zealand. His hundred or so karst related publications included the world-class texts *Karst* (1971) and *Karst Geomorphology* (1985). Publications aside though, his greatest contribution to karst research was a huge ability to inspire fellow workers, amateur and professional, in Australia and around the world. His entire philosophy, which could be a model for all scientists, is typified in his own words: "...avoid exclusiveness in method; collaborate if you can; predict and test; seek homologues; get previous theory out of your skin." (1978, pp.8–9).

Marjorie Mary Sweeting (1920–94)

Born in Fulham, London, Marjorie Sweeting gained her MA and doctorate at Cambridge, where she was a founder member of the University caving club. She moved to Oxford in 1951

Geoscientists: Marjorie Sweeting in the field, 1983. This is the top of the Watlowes dry valley near Malham, during the 1983 Anglo-French Karst Symposium. (Photo by Paul Williams)

and the rest of her professional life was spent there, culminating in her position as Emeritus Fellow at St Hugh's College from 1987 until 1994. She supervised many doctoral students, perhaps most notably (in the early 1960s) Derek Ford and Paul Williams, who have had a profound influence on the subsequent development of karst geomorphology and hydrology.

Of a plethora of publications, her textbook *Karst Landforms* (1972) and edited volume of benchmark papers *Karst Geomorphology* (1981) are best known. However, her field research was also groundbreaking in its time, including work relating Yorkshire Dales cave levels to erosion surfaces (Sweeting, 1950) and dissolution rates to limestone lithology (Sweeting & Sweeting, 1969). In 1977, as travel restrictions began to ease, Marjorie Sweeting was among the first western karstologists to revisit China's spectacular and varied karst landscapes, where she returned many times. Meeting and working with the country's leading karst researchers, she facilitated wide study by other western scientists and cave explorers, while also finding time to gather data for her final book—published posthumously in 1995.

DAVID J. LOWE

Works Cited

Bögli, A. 1964. Mischungskorrosion, ein Beitrag zum Verkarstungsproblem. [Mixture corrosion—a contribution to the karst problem]. *Erdkunde*, 18(2): 83–92

Bögli, A. 1978. *Karsthydrographie und physische Speläologie*, Berlin and New York: Springer; as *Karst Hydrology and Physical Speleology*, 1980

Bretz, J.H. 1942. Vadose and phreatic features of limestone caves. *Journal of Geology*, 50: 675–811

Cvijić, J. 1893. Das Karstphänomen. *Geographische Abhandlungen herausgegeben von A.Penck* [*Geographical Proceedings. Published by A. Penck*], 5(3): 218–329

Cvijić, J. 1918. Hydrographie souterraine et évolution morphologique du Karst [Underground hydrology and morphological evolution of karst]. *Recueil des Travaux de l'Institute de Géographie Alpine* [*Collected works of the Institute of Alpine Geology*], 4(4): 375–426

Cvijić, J. [1925] 1960. La géographie des terraines calcaires [Geography of calcareous terrains]. *Monographie*, vol. 341, Classe des sciences mathematiques et naturelles, 26: 1–212

Davis, W.M. 1930. The origin of limestone caverns. *Geological Society of America Bulletin*, 41: 475–628

Gèze, B. 1939. Les sources dites "vauclusiennes" [The springs referred to as "vauclusian"]. *Sciences Naturelles* [*Natural Sciences*], 1(6): 171–80

Gèze, B. 1965. Les conditions hydrogéologique des roches calcaires [Hydrological properties of calcareous rocks]. *Chroniques d'Hydrogeologie*, 7: 9–39

Gèze, B. 1965. *La Spéléologie scientifique* [Scientific Speleology]. Paris: Editions du Seuil

Gèze, B. 1973. Lexique des termes français de spéléologie physique et de karstologie [Lexicon of French physical speleological and karstological terms]. *Annales de Spéléologie*, 28(1): 1–20

Jennings, J.N. 1971. *Karst*, Cambridge, Massachusetts: MIT Press.

Jennings, J.N. 1978. Apostasy or apologetics? Confessions of a natural geographer. *Monash Publications in Geography*, 20. Melbourne: Monash University

Jennings, J.N. 1985. *Karst Geomorphology*, 2nd edition, Oxford and New York: Blackwell

Lehmann, O. 1932. Die Hydrographie des Karstes [Karst Hydrology]. In *Enzyklopädie der Erdkunde* [Encyclopedia of Geography]. Leipzig: Deuticke

Maximovich, G.A. 1963, 1969. *Principles of Karstology*, 2 vols, Perm: State Publisher (In Russian)

Sweeting, M.M. 1950. Erosion cycles and limestone caverns in the Ingleborough district. *Geographical Journal*, 115: 63–78

Sweeting, M.M. 1972. *Karst Landforms*, London: Macmillan and New York: Columbia University Press

Sweeting, M.M. (editor) 1981. *Karst Geomorphology*, Stroudsburg, Pennsylvania: Hutchinson Ross

Sweeting, M.M. 1995. *Karst in China: Its Geomorphology and Environment*, Berlin and New York: Springer

Sweeting M.M. & Sweeting, G.S. 1969. Some aspects of the Carboniferous limestone in relation to its landforms, with particular reference to northwest Yorkshire and County Clare. *Méditerranée*, 7: 201–09

Further Reading

Anon [many contributors], 1996. Bernard Gèze. *Spelunca*, 65: 21–44

Ford, T.D. 1995. Obituary: Marjorie Sweeting. *Caves and Caving*, 67: 36–37

Klimchouk, A.B., Ford, D.C., Palmer, A.N. & Dreybrodt, W. (editors) 2000. *Speleogenesis: Evolution of Karst Aquifers*, Huntsville, Alabama: National Speleological Society. [See especially: Lowe, D.J. "Development of speleogenetic ideas in the 20th century: The early modern approach"; Shaw, T.R. "Views on cave formation before 1900"; White, W.B. "Development of speleogenetic ideas in the 20th century: The modern period, 1957 to the present"]

Lowe, D.J. 1992. A historical review of concepts of speleogenesis. *Cave Science*, 19(3): 63–90

Spate, O.H.K. & Spate, A.P. 1985. Obituary. Joseph Newell Jennings 1916–1984. *Australian Geographical Studies*, 23(2): 325–37

GIBRALTAR CAVES, GIBRALTAR: ARCHAEOLOGY

The Rock of Gibraltar is an escarpment eroded in a klippe of Mesozoic carbonates and shales, situated on the south coast of the Iberian Peninsula. The limestones are heavily fissured by various faults and joints aligned northeast–southwest and northwest–southeast. The Rock forms a north–south asymmetric peninsula, 5.2 km long and 1.6 km wide, with a very steep eastern side and a more gently sloping western side. It rises to a maximum altitude of 426 m. The geology and morphology of Gibraltar fall into three main regions (Rose & Rosenbaum, 1991). An isthmus in the north is a sandy plain less than 3 m high formed as a Holocene tombolo that joins the Rock to the mainland. The Main Ridge of Jurassic limestones and dolomites forms a sharp crest with several peaks over 400 m. The Southern Plateaux are a succession of benches inclined gently southwards from 130 m to the present sea level.

The overall morphology of the Rock is controlled by tectonic structure, modified by subaerial and coastal marine erosion. Littoral processes have been especially active on the eastern face of the Rock, where the wave fetch is greater. Quaternary sea-level fluctuations have left raised features and controlled the evolution of slopes. Sediments that possibly range through the entire Quaternary lie over a large area of the surface and inside many caves and fissures (Rose & Hardman, 2000), providing the Rock with one of the most important and complete records in the Mediterranean. Several sets of terraced Quaternary marine deposits along the coast of Gibraltar range from 1 m to 210 m above sea level. Five steps have been identified as marine terraces at 1–25 m, 30–60 m, 80–130 m, 180–210 m, and above 210 m, where each terrace and its sediments is backed by a steep relict cliff. The most recent coastal landforms and sediments evolved in the last 250 ka, and are linked to Oxygen Isotope Stages (OIS) 1, 3, 5, and 7.

There are over 140 caves and fissures within the Rock of Gibraltar. Seven of these, all within the landforms of the last 250 ka, have revealed evidence of Neanderthal occupation, and

Gibraltar: Figure 1. The Rock of Gibraltar from the Strait of Gibraltar (southeast) with the Gorham's-Vanguard Cave complex at sea level and the staircased morphotectonic units from there to the highest peak (426 m).

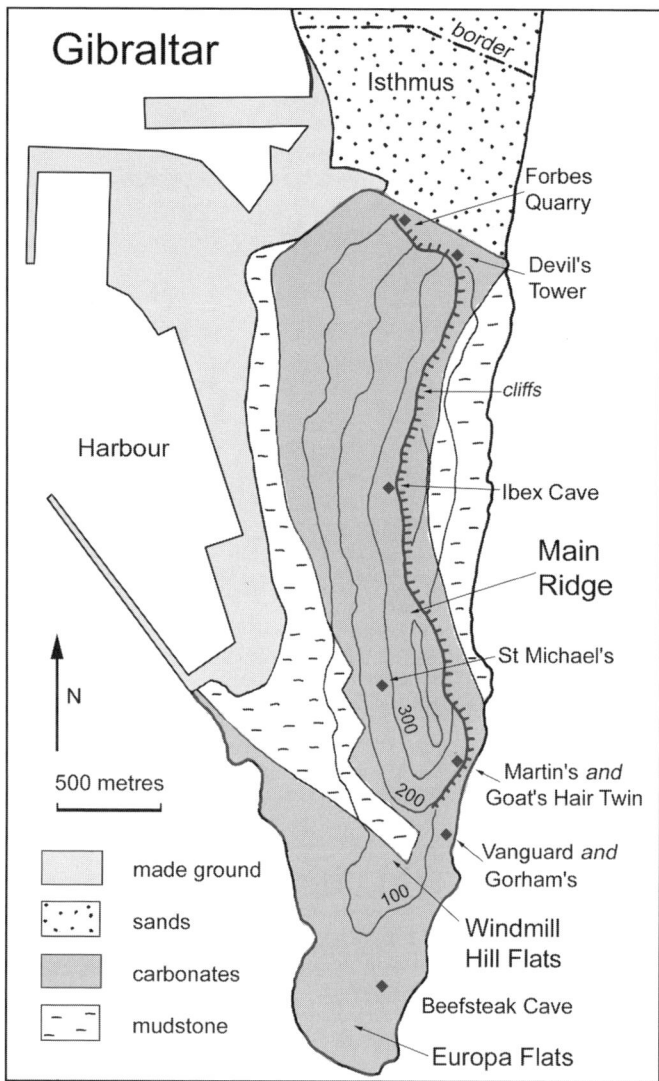

Gibraltar: Figure 2. Simplified morphotectonic map of the Gibraltar Peninsula. Contours at 100 m intervals.

the sites at Forbes's Quarry and Devil's Tower have produced Neanderthal fossil material. Gibraltar has been an important part of the Neanderthal story since 1848, when the first fairly complete fossil of these humans was found in Forbes's Quarry (Stringer, 1994). This cranium was exhibited by George Busk in September 1864 at the meeting of the British Association for the Advancement of Science, held in Bath. Hugh Falconer proposed that it should be made the type of a new human species *Homo calpicus* (after Calpe, the Roman name for Gibraltar). The proposal was, however, overtaken by the publication that same year of *H. neanderthalensis* by William King, who based his new species on the Neander Valley (Feldhofer) skeleton from Germany that had been discovered eight years after the Forbes's Quarry specimen (see Europe, Central: Archaeological Caves).

The Forbes's Quarry specimen remains one of the best preserved of all Neanderthals. A second important discovery was made in 1926 in the Devil's Tower Rock Shelter, close to Forbes's Quarry, which was systematically excavated by Dorothy Garrod, yielding fauna, artefacts, and charcoal. The Neanderthal remains consisted of parts of the cranium and jaws of a child, estimated to have died at the age of four. Also of great historical importance are the Rosia Bay bone breccias that were discovered when engineers were clearing the coastal cliffs in 1769. These fossils were studied by William Buckland and Georges Cuvier.

Active research continues in the Gibraltar caves, with work in Ibex, Gorham's, and Vanguard Caves on the eastern side of the Rock, ongoing since 1994. At Gorham's Cave recent work has revealed a late Neanderthal persistence as recent as 31 ka. The faunal and charcoal richness of the deposits at Gorham's and Vanguard, spanning the sequence OIS 5–1, is permitting a greater understanding of the relationship between Neanderthals and their ecology in one of their last refuges. Evidence from the higher part of the sequence at Gorham's Cave indicates a late arrival of modern humans to southern Iberia, probably after the Neanderthal extinction. The peak of Upper Paleolithic occupation occurs during the Solutrean, coinciding with the Last Glacial Maximum. In Gorham's Cave this is represented by parietal art depicting horse, ibex, aurochs, and red deer. The relationship between humans and the limestone of Gibraltar continued well after the Paleolithic. Of particular interest are the top levels at Gorham's Cave that have revealed a major Phoenician and Carthaginian shrine that was used from the 8th to the 3rd centuries BC. The Rock then became the northern Pillar of Herakles. From the 18th to the 20th centuries British engineers tunnelled over 60 km of roads inside the Rock, largely for military installations.

CLIVE FINLAYSON AND JOAQUIN RODRÍGUEZ-VIDAL

Works Cited

Rodriguez-Vidal, J. & Gracia, F.J. 2000. Landform analysis and Quaternary processes of the Rock of Gibraltar. In *Gibraltar during the Quaternary*, edited by C. Finlayson, G. Finlayson & D. Fa, Gibraltar: Gibraltar Government Heritage Publications

Rose, E.P.F. & Hardman, E.C. 2000. Quaternary geology of Gibraltar. In *Gibraltar during the Quaternary*, edited by C. Finlayson, G. Finlayson & D. Fa, Gibraltar: Gibraltar Government Heritage Publications

Rose, E.P.F. & Rosenbaum, M.S. 1991. The Rock of Gibraltar. *Geology Today*, 7: 95–101

Stringer, C. 1994. The Gibraltar Neanderthals. In *Gibraltar during the Quaternary*, edited by J. Rodríguez-Vidal, F. Díaz del Olmo, C. Finlayson & F. Giles, Seville: Asociación Española para el Estudio del Cuaternario (AEQUA Monografias, 2)

Further Reading

Finlayson, C., Finlayson, G. & Fa, D. (editors) 2000. *Gibraltar during the Quaternary*, Gibraltar: Gibraltar Government Heritage Publications

Stringer, C.B., Barton, R.N.E. & Finlayson, J.C. (editors) 2000. *Neanderthals on the Edge: 150th Anniversary Conference of the Forbes's Quarry Discovery, Gibraltar*, Oxford: Oxbow Books

GLACIER CAVES AND GLACIER PSEUDOKARST

Glacier pseudokarst is the suite of landforms analogous to limestone karst developed in glacier ice by the processes of melting, sublimation, and deformation. As ice melting by flowing water is analogous to dissolution of limestone, the features most commonly seen in glacier pseudokarst are caves and closed depressions, often with associated streams. The similarity is so compelling that many simply describe it as "glacier karst". However, the analogy is not exact. The major distinction is that glacier ice is more massive, generally mobile, and mechanically much weaker than most carbonate rocks. Hence there is limited matrix flow of water and openings at depth will tend to close quite rapidly if they are not being continually enlarged. Well-developed perennial glacier pseudokarst is restricted to ice that is not moving because it is too thin or flat lying.

Glacier ice consists of tightly-packed, large (cm scale), irregular crystals developed from permanent snow by sustained compression, melt and refreezing, and deformation. Although there may be weak flow of water through the "vein" system running around the crystals, ice may be considered impermeable to the bulk flow of water through it. Ice has peculiar physical properties; it resists deformation up to a certain stress, after which it deforms relatively rapidly. It also resists short-term stress, but creeps (flows) under sustained load. A glacier starts to move when the thickness and slope of a patch of ice are sufficient to develop the critical stress. Movement is by a combination of deformation, concentrated near the bed, and sliding at the bed. Typical surface velocities are of the order of tens of metres per year. Ice is weak under tension, and fractures to produce crevasses that may penetrate to the bed. Crevasses and the glacier bed constitute the preferred pathways taken by meltwater, although only the most actively enlarging openings can offset closure at depths beyond 50 m.

Most non-polar glaciers are "temperate", that is the ice is largely at its melting point, and host the most fully developed pseudokarst. "Polar" glaciers remain well below freezing, do not melt or slide easily, and show limited pseudokarst development. In cold regions such as Spitzbergen many of the glaciers are hybrid types, displaying a layer of cold ice above a temperate core and bed. In the upper reaches of glaciers and in perennial snow patches, pseudokarst can also develop in firn (or nevé), snow that has been partially converted to glacier ice. Firn is composed of millimetre-size crystals and is more porous than glacier ice. In some localities, it may reach tens of metres in thickness, at which point it may start to creep as a firn glacier.

Processes

Glacier ice can be expected to melt when it is at its melting point and heat energy is taken up. Solar and atmospheric energy result in the bulk of melting taking place in summer on the glacier surface, contributing the primary source of water for the development of glacier pseudokarst. Every litre of water can melt about 12 g of temperate ice per degree above zero Celsius. Flowing water also releases heat as it falls in elevation and every litre descending one metre can melt about 3 mg of ice. This is a small amount, but is important inside a glacier where other heat sources are absent. In geothermal settings, steam may act as a very effective, though localized cause of melting (e.g. Mt Rainier, Washington, United States, and the Vatnajökul, Iceland). There is also minor melting at the glacier bed caused by geothermal heat, sliding, and deformation of glacier ice.

Freezing may be considered the glacier analogue of calcite precipitation in limestone karst. It occurs wherever energy is removed from zero-degree water at the pressure freezing point (the freezing point of water is depressed in a thick glacier or ice sheet because of increased pressure, by $-0.7°C\,km^{-1}$). Ice varies from transparent to opaque as the crystals become smaller or if the boundaries between crystals contain air pockets. As water freezes, minerals and dissolved gases are ejected, resulting in dusty efflorescence (powder) and contained bubbles, respectively. Cooling is predominantly a surface process, occurring at night or in winter, resulting in cooling of surficial ice and freezing of exposed water. The distribution of freezing relative to the flow pattern of water determines whether stalactites, stalagmites, or sheet ice develop in glacier caves. Although similar in form to calcite speleothems, the weakness and plasticity of ice often result in the development of drooping and deformed shapes. In ponds, surface ice sheets are often abandoned or flooded as water levels change, producing complex ledges and chambers, often decorated by ice speleothems. Snow falling in cold moving water produces numerous floating "frazil" ice crystals that can build up to form rafts, ledges, and dams of white ice. Water rising up a glacier conduit may also consume heat energy, allowing frazil ice to form. Rising sections of internal glacier conduits are therefore likely to be very constricted.

Sublimation is the direct conversion of ice to water vapour and *vice versa*. Eight times more energy is required for sublimation than for melting, so it is quantitatively rather limited but, in contrast to melting, it is effective at sub-zero temperatures. As an erosional process, sublimation is effective where cold air becomes warmed in passing through an existing cavity. However, every degree of warming of a litre of air can only sublimate about 0.4 μg of temperate ice. Large scallops 10 to 200 cm in diameter seen on the walls of glacier conduits and crevasses are a product of erosion by sublimation. These features are analogous to limestone cave scallops but are formed by moving air rather than water. As a depositional process, sublimation produces hoar frost crystals, sometimes of spectacular size, in locations where moist air is chilled at sub-zero temperatures. Sublimation deposition allows more ordered composition of the molecules in ice, generating hexagonal plate crystals, sometimes 10 cm or more in diameter. These beautiful forms are often clustered in those locations favouring their growth, but they are fragile and prone to rapid melting and collapse under their own weight.

Tectonic processes are associated with ice collapse, deformation, and sliding. Melting temperate ice is particularly weak and tends to shatter as there is little cohesion between crystals, and water flow opens up numerous fractures. Tectonic cavities are often generated as a glacier slides over obstacles, or as thin ice buckles when forced against an obstacle. At depth and under compression, ice flow can rapidly close an opening, if there is no opposing water pressure. A glacier sliding over an irregular bed can respectively create and annihilate cavities on the lee and stoss (up-glacier) side of obstacles.

Few glaciers are free of sediment and surface melting is particularly sensitive to debris thickness. A thin dusting will enhance melting by absorbing more solar radiation than bare ice, whereas more than a few centimetres may prevent melting due to insulation. Glacier cave streams may have considerable difficulty transporting coarse sediments which can become trapped at narrowings and dips in conduits. Glacier ice may often become buried as blocks or *en masse* by rapid deposition of sediment, especially at the snout of glaciers where debris-rich streams emerge.

Hydrology and Drainage
Glacier drainage systems convey surface waters through the glacier to the snout, largely through discrete, dendritic (branching) conduits. They derive the bulk of their water from surface melt streams, or extra-glacial streams that usually enter the glacier at a distinct sink point. The "sink points" are often open shafts called "moulins" (see Figure), and the outlets are termed "portals". Moulins form where crevasses open downstream of a catchment on the glacier surface. Crevasses occupy zones of tension in the glacier. Individual crevasses migrate with the ice and as they pass out of the zone of tension, may close up. As new crevasses open, they capture the surface streams from their predecessors. Older crevasses thus often carry with them a series of abandoned moulins. These may persist as shallow shafts, but much of the deeper route may be restricted and transient because the weight of ice at depth forces the conduit to close if there is not very rapid erosion, or high water pressure. Distinctive stellar crystal clusters, up to a few metres in diameter, that are often seen in glacier ice form from freezing of water ponded in a sealed moulin.

Glacier hydrology is highly seasonal. During winter, the absence of meltwater permits wholesale closure of the deep drainage system. What remains is an ill-connected set of cavities reflecting the distribution of irregularities on the glacier bed. Spring melt produces abundant water on the glacier surface which is initially stored in the snow, but eventually collects to form supraglacial (surface) and marginal streams. If the stream encounters a crevasse, or reaches the margin of a glacier, it may feed down to the bed where water accumulates, applying pressure to the incompetent basal drainage system. The glacier can be locally lifted from its bed, permitting the water to spread and escape at the base. If there is enough water involved, melt-erosion can etch a channel in the ice at the bed which then becomes an active conduit, enlarging by melting and closing by deformation. The glacial conduit drainage system develops progressively up-glacier as the summer progresses, and gradually atrophies as winter closes in. Hence, the analogy with carbonate karst hydrology is more restrictive than generally perceived, applying most effectively to supraglacial surface streams that sink to follow both "vadose" and "phreatic" conduits through smaller and thinner glaciers, descending to the bed or resurging lower down the glacier. However, this is a rather seasonal, and spatially limited condition. Some moulins and areas of a glacier do not appear to be linked to conduits with rapid water flow and in these cases sinking water is considered to enter an undeveloped "linked cavity" system at the glacier bed. There does not seem to be much matrix recharge and flow through glacier ice so that storage of water is rather limited and largely associated with isolated cavities. In contrast, firn may sustain both conduit and diffuse flow systems as a result of its granular composition and lack of deformation.

Geomorphology
The similarity between karst and glacier discrete drainage systems results in corresponding morphological similarities: allogenic streams, blind valleys, stream sinks, vadose shafts, phreatic conduits, and pocket valleys (spring heads, reculées) are all present. In glacier caves, phreatic conduits are elliptical in section, vadose canyons meander, and vertical shafts are fluted and may widen at the base. At the glacier bed, channels may be carved up into the ice and down into till or bedrock. Channels cut up into ice are termed Röthlisberger channels; they are transient as they migrate with the ice, and can be annihilated by creep closure or being dragged over a bedrock obstacle. In contrast, channels cut into bedrock remain fixed in position, and may persist over winter. They are termed Nye channels, and where they exist may promote the rapid establishment of drainage systems. Subglacial streams flowing on unconsolidated materials like till also form channels. Their behaviour is poorly understood, but they may persist, providing they do not become choked on their own sediment.

The route that water follows through the glacier from the moulin to the portal reflects hydraulic laws, moderated by conditions within and at the base of the glacier. The pattern of flow depends on whether the conduits are water-filled or contain air. The former will tend to have circular cross sections, and will be directed by the hydraulic gradient towards the margins and the snout. Such conduits may lie well above the deepest part of a valley. Rising sections of conduit are expected to be wider and lower. If passages are air-filled, then the stream will follow the steepest bed slope to the lowest point, just like a surface river. However, the lack of water contact with the roof will make such channels persistently wide and low. Normally, there are more moulins than portals, and a dendritic (tree-like) network is implied, with fewer, larger channels down-glacier. However, many of these ideas are based on theory. Field investigations indicate that multiple channels exist with significant daily variation in the active flow route driven by the daily melt cycle.

Moulins are the most common visible glacier karst feature, and they have been explored to tens of metres in depth. Glacier caves develop very rapidly with abundant warm water, and so

Glacier Caves and Glacier Pseudokarst: An undescended moulin that swallows a large meltwater river into the Greenland ice sheet. (Photo by Tony Waltham)

can enlarge indefinitely to the limit of the mechanical strength of the ice. Where extra-glacial streams contact ice, conduits may be tens of metres in diameter. However, attempts at following deep glacier cave systems have seldom been successful as the passages become too narrow or low. Thermal energy is rapidly exhausted, and the rate of closure of an opening depends on size and depth below the surface.

The most common position for large open glacier caves is at shallow depth in stagnant ice, for example near the glacier margin (e.g. Paradise Ice Caves, Mt Rainier, Washington, United States), or the snout of a retreating glacier (e.g. Athabasca Glacier, Alberta, Canada). Here closure rates and deformation are negligible, and warm water from the valley sides may have opened up significant cave networks, many hundreds of metres in length. Significant cave passages may also develop where hydrothermal waters promote enlargement (e.g. Iceland), or where a glacier lake provides abundant water for a glacier outburst or "Jökulhlaup" (e.g. the drainage of Gornersee ice-dammed lake through the Gornergletscher in Switzerland, or overflow from a volcanic crater, e.g., Grimsvötn in Iceland). A significant cave may survive for many days after a flood, as closure rates are of the order of centimetres per day. Mechanical collapse is the greatest risk in large passages. The strength of glacier ice depends on its thickness and composition. Solid, cold ice is much stronger than warm ice with the crystal boundaries opened up by melting. The largest glacier conduit is estimated to have formed in an outburst of glacial Lake Missoula (Western Montana) some ten thousand years ago. The 1500 km^3 of water draining from the lake scoured over 5000 km^2 of eastern Washington State, and indicates that the final conduit diameter may have exceeded one kilometre.

Persistence of shallow subsurface conduits means that stagnant glacier ice commonly hosts spectacular surface karst with tessellated closed depressions, ponds, karst windows, arches, and springs. A debris cover seems to permit the most well-developed karst, possibly because it protects the underlying ice from decay, weakening, and mechanical collapse.

Tectonic caves are common in glacier ice. Many form in the lee of bedrock steps, where they often persist and at the margin may provide access to the glacier bed. Here the motion of the ice may be spectacularly revealed by the extrusion forms developed. Thin ice at the glacier snout may often form open-cored buckle folds, as it is forced against an obstacle. Crevasses are the most spectacular tectonic openings, and may reach tens of metres in depth, forming networks of fracture passages, often choked by ice blocks and arched by snow bridges to create a tectonic cave system suffused with cold blue light. Many perennial glacier caves show little evidence for active water flow, yet continue to enlarge slowly. This has been attributed to sublimation erosion occurring in winter when frigid external air is drawn in and warms up. The pattern of such erosion will reflect the meteorology of the cave air flow, with the strongest fluxes being driven by chimney effect winds.

Differential erosion of dirty ice results in the development of ice-cored mounds, debris-veneered ice slopes, and redistribution of the debris by mass movement, sometimes generating a distinctive pseudokarstic surface of closed depressions, blind valleys, and mud morasses (e.g. the many square kilometres of glacier pseudokarst at the terminus of the Malaspina Glacier, Alaska, or the lower Tasman Glacier, New Zealand). Streams cut caves beneath the surface, but they are frequently blocked by the relentless movement of debris, causing ponding and redirection of drainage. Glacier pseudokarst also develops in debris where buried ice melts *in situ*, developing closed depressions, in isolation known as "kettle holes", or assembled into extensive debris pseudokarst hillocks termed "hummocky moraine".

Glacier speleology or "Glaciospeleology" is concerned with the exploration and understanding of glacier caves. One of the earliest recorded descents was in 1898 when Vallot explored the Grand Moulin on the Mer de Glace on the French side of Mont Blanc, descending to a depth of 60.5 m. At that time, the explorers did not make any particular distinction between caves in glacier ice and those in rock. However, as caves in ice were almost impossible to tackle using existing techniques speleologists lost interest in this particular branch of caving. Interest was renewed in the 1980s when Polish speleologists began to explore the Hansbreen on Spitsbergen, Swiss speleologists the Gorner glacier and Icelandic glaciers, French speleologists revisited the Grand Moulin, and Italian spelogists first explored the Gorner and Miage glaciers in Switzerland with subsequent expeditions to the Karakorum, Tien Shan, Svalbard, and Patagonia. In recognition of this renewed interest the International Union of Speleology has set up a Glaciospeleology Commission, and the International Glaciospeleological Survey and La Venta group maintain relevant websites (http://glaciercaves.com/index.html and http://www.laventa.it/en/glacio.html). Vertical shafts ("moulins") have been explored to tens of metres, but have usually proven impenetrable at depth as the passages narrow to slots. In summer, moulins host violent cold streams; in winter closure of the basal drainage system causes them to pond up. Early winter would appear to be the optimum time for exploration, although abandoned moulins lying down-glacier may provide easier access to the subsurface system. Similar frustrations occur at outlets (portals). Typically, these rapidly become lower, broaden, and become flooded. The greatest success has been gained following moulins developed by warm extra-glacial streams, or by exploration of warm emergent waters, usually of geothermal origin. The most extensive, persistent caves are found at shallow depth in marginal, quasi-stagnant glacier ice. Glacier cave exploration requires some care due to the extremely cold water, rapid daily flooding, and the inherent weakness of glacier ice under tension. Snow-mantled glacier surfaces may conceal large crevasses that close relentlessly at depth. Rock debris on ice is also alarmingly unstable, and a fragment can turn crampons into skates.

Chris Smart

See also **Antarctica; Ice in Caves**

Further Reading

Badino, G. 1994. Phenomenology and first numerical simulations of the phreatic drainage network inside glaciers. *Proceedings of the 3rd International Symposium on Glacier Caves and Karst in Polar Regions, Chamonix*

Badino, G. 2001. Glacial karst phenomenology. *Proceedings of the 13th International Congress of Speleology*, Brasilia: Sociedade Brasiliera de Espeleologia (CD-ROM): 113

Benn, D.I. & Evans, D.J.A. 1998. *Glaciers and Glaciation*, London: Arnold and New York: Wiley

Fountain, A.G. & Walder, J.S. 1998. Water flow through temperate glaciers. *Reviews of Geophysics*, 36(3): 299–328

Hooke, R. Le B. 1989. Englacial and subglacial hydrology: A qualitative review. *Arctic and Alpine Research*, 21(3): 221–33

Krawczyk, W.E., Pulina, M. & Rehak, J. 1997. Similarity between the hydrologic system of the Werenskiold Glacier (SW Spitsbergen) and a karst. *Proceedings of the 12th International Congress of Speleology*, edited by J. Bauchle et al., Basel: Speleo Projects: vol. 1, 493–96

Paterson, W.S.B. 1994. *The Physics of Glaciers*, 3rd edition, Oxford: Pergamon

Piccini, L. & Badino, G. 2001. Moulins and contact caves in the Gornergletscher, Switzerland: Morphology and hydrology. *Proceedings of the 13th International Congress of Speleology*, Brasilia: Sociedade Brasiliera de Espeleologia (CD-ROM): 121

Smart, C.C. 1997. Hydrogeology of glacial and subglacial karst aquifers: Small River, British Columbia, Canada. In *Proceedings of the 6th Conference on Limestone Hydrology and Fissured Media, La Chaux de Fonds, Switzerland*, edited by P.-Y. Jeanin, Besançon, Switzerland: University of Franche-Comté: 315–19

Theakstone, W. 1988. Glacier caves and subglacial water in Nordland, Norway. *Cave Science*, 15: 121–27

GLACIERIZED AND GLACIATED KARST

Karst landscapes currently occupied by glaciers are considered "glacierized", although the glacier cover may obscure the karst inferred from considerations of geology, geomorphology, or the presence of hydrological features such as springs emerging from glacier mantled terrain. "Glaciated" karst has been previously occupied by glaciers. Such landscapes are more obviously karstic, with karst features and hydrology, while retaining erosional features and deposits from former glaciers. Modern glacierized karst is restricted to high latitude or high altitude sites, reflecting the limited distribution of current climatic conditions permitting perennial snow accumulation. However, during the Quaternary period, cooling of the global climate under 100 000 year cycles led to much more extensive glaciation as ice spread to lower altitudes and latitudes. Glacier retreat in the last 12 000 years has left a legacy of extensive glaciated karst, although this has been considerably modified over this interval. A minor phase of glaciation called the "Little Ice Age" reached its maximum in the late 19th century. Subsequent glacier retreat has exposed particularly fresh glaciated karst. Many glaciated karst landscapes have endured a number of glaciations over the last two million years, but evidence from more ancient glacial episodes is seldom preserved on eroded surfaces; deposits sequestered in caves may however be preserved through a number of glacial cycles. In this essay, the general character of glacier erosion and deposition is first described, and then applied to the understanding of glacierized and glaciated karsts at alpine and continental scales.

Glaciers develop from the accumulation of perennial snow to sufficient thickness and density as to permit movement by internal deformation and sliding at the bed. Movement carries the ice from relatively cold conditions, allowing snow accumulation, to warmer environments at lower altitudes or latitudes where melt dominates. As glacier ice moves, it erodes underlying material by abrasion (grinding) and plucking (removal of blocks). Abrasion leaves behind striated surfaces and produces finely ground rock particles. Plucking generates larger boulders, leaving fractured surfaces, mostly facing down-glacier. Glacial deposition is driven by melting, which releases an ill-sorted mixture of particles, broadly termed "till". However, till is frequently reworked by coincident meltwaters to generate gravelly river deposits and silty layers in lakes.

Glacier erosion and deposition are associated broadly with accumulation zones and melt zones, respectively. In an *alpine* setting, higher altitude erosion sculpts massive armchair-shaped valley heads termed "cirques", which juxtapose to generate sharp ridges and pinnacles, and feed ice into deep U-shaped valleys. Surfaces tend to be scraped bare of soil and striated (scratched). In contrast, thick till deposits occupy lower valleys and are often organized into ridges known as moraines. At the more extensive *continental* scale, ice builds up to form an ice sheet. Erosion occurs towards the centre and deposition around the margin, generating bare bedrock surfaces and a thick mantle of till respectively. This pattern of erosion and deposition expands and contracts as glaciers advance and retreat, resulting in superimposition of erosional and depositional forms.

Water is found at the glacier bed in cavities, in conduits, and as a thin film. Cavity water is isolated and may be at high pressure. Conduits may develop from cavities every year and are sustained by free surface or pressurized streams. The basal film is less than one millimetre in thickness and occupies much of the glacier bed. It is continually cycled through freezing and thawing. Glacier waters have little capacity for dissolution of limestone bedrock, as they lack the enhanced carbon dioxide generated in warm soils, and because solution is expended on crushed material rather than bedrock. However, much glacier meltwater contains so little dissolved material that it may dissolve very small amounts of limestone quite quickly. In contrast, high concentrations of dissolved minerals can be achieved as water freezes. Thus winter residual waters, and the freezing zone of the basal film can induce rapid local precipitation, normally of amorphous carbonate in the form of patches of white precipitates.

Glacierized alpine karst has received some study in the European Alps and the Canadian Rockies. The commonest features are sinking marginal streams near the glacier snout. However, hydrogeological analysis, water balance, and water chemistry of spring waters demonstrate that some subglacial water also appears in springs (e.g. Glacier de la Plaine Morte, Switzerland). Marginal and supraglacial meltwaters have been traced through the karst aquifer at Castleguard Mountain, Alberta, Canada. Castleguard Cave (see separate entry) penetrates the limestones beneath the Columbia Icefield and encounters glacier ice about 300 m below the glacier surface. Here seasonal conduit water is encountered in erosionally etched shafts in the cave, and stalactites are developing from seepage waters, indicating that karst aquifers may function beneath the upper reaches of alpine glaciers. There also appears to be an association between glacier and karst conduits, and between the basal film and karst seepage water. At Small River Glacier, British Columbia, combined drilling and dye tracing into a subglacial karst aquifer revealed that the aquifer rapidly captures much of the subglacial water, although the routing and timing of water delivery varies greatly as conduits and cavities open and close.

Crevasses determine the locations where abundant surface meltwaters can penetrate directly to erode the bed. They develop preferentially in zones of tension in thin ice, such as at the margins of glaciers, or where ice is deformed and fractured by underlying bedrock steps or hummocks. Subglacial ponors may consequently develop in post-glacially anomalous locations such as high up on bedrock valley sides and the crests of hillocks. Free surface subglacial streams tend to follow valley floors where preglacial ponors and depressions may exist. Pressurized channels are not constrained by gravity and may run along valley walls, or ascend opposing slopes, leaving tantalising traces once the ice has receded. Glacial erosion may expose existing cave passages which in turn may capture subglacial streams.

Stream sediments limit glacier–karst interaction by preferentially consuming solutional potential, and blocking conduits. As a result, cleaner up-glacier streams are more likely to enter and develop karst, compared to sediment-laden, down-glacier streams. These distinctions are not absolute, but depend on the activity of the glacier and the susceptibility of the rock to glacial erosion. Massive limestones appear to generate relatively little sediment, and therefore tend to remain open to subglacial waters. Thin, residual glaciers may also generate little sediment, allowing their meltwater to feed directly into underlying karst.

Glacier retreat reveals glaciated alpine karst. During an initial "paraglacial" period, unstable glacial deposits are mobilized by water and mass movement. Surface drainage basins are chaotic, reflecting the bedrock structure exploited by glacier erosion, and the pattern of glacial deposits. Karst openings and depressions with a sizable surface catchment may become choked, and even form ponds. Gradually more stable conditions develop as unconsolidated materials stabilize and vegetation cover is established. Karst pond linings may collapse and till surfaces develop suffosion depressions as sediment feeds into underlying karst openings (e.g. Castleguard Glaciers, Alberta, and Small River Glacier, British Columbia)

Karst surfaces exposed by glacier retreat retain evidence of former subglacial conditions, and of the destruction wrought on preglacial karst by glaciation (e.g. Glacier de la Tsanfleuron, Switzerland). For the first few years of exposure, delicate subglacial precipitate spicules, lobes, and veneers produced by basal freezing may linger, before succumbing to frost action. The sites of former crevasse recharge may be indicated by local etching of channels or narrow cave entrances on the crest of ridges and other positions anomalous in the postglacial landscape. Former subglacial channels remain, sometimes sculpted on walls, indicating former pressurized conditions. Cave systems may have been dissected, distorted, or unroofed by glacial erosion, and have apparently random shafts punched through where crevasse recharge has occurred. Erosive subglacial water following former percolation routes may dissolve speleothems developed when soil existed at the surface.

In erosive areas, extensive bare bedrock surfaces form: "Schichttreppen" (stepped) karst where bedding is near horizontal, and "Rundhöcker" (rounded mound, or roches moutonnées) karst where bedding is steeper. Pure limestone surfaces generate little soil and remain exposed to subaerial weathering. They often develop suites of 10 mm to 10 m scale solutional forms known as "karren" or "lapiès", and fractures become enlarged by solution, eventually producing the highly dissected classical limestone pavements typical of northern England (e.g. Malham, Yorkshire; see Figure) and the Burren, Ireland (see separate entry). Both carbonate and non-carbonate "erratic" boulders may remain, developing limestone pedestals beneath them as they protect the underlying limestone from dissolution.

Glacial sediments are often well preserved in caves. Openings may have glacial till injected, frustrating cave exploration, and often only isolated shafts in ridge tops remain open. Glaciers can impose a distinctive pattern of hydraulic head on the underlying karst, sometimes forcing flow up through former vadose shafts, so that deposits are inconsistent with postglacial flows. Similarly, glaciers or their deposits may block former outlets, inducing ponding. Gravel beds represent active flow and laminated clays and silts arise from still water. The exact association between these deposits and glacial conditions are not well understood. How does ponding indicated by clays develop in steep passages? At what stages are glacial sands and gravels introduced, reworked, and redeposited? And how are streams re-established in completely infilled passages? The association of sediments with the marginal glacial environment indicates that a majority of sediment introduction may occur only once near a glacier limit, but twice where the glacier overrides and subsequently retreats over the terrain, giving a puzzling dual deposit from a single glacial episode. Glacially derived sediments may also redirect internal drainage, and much paragenetic cave development has been in-

Glaciated Karst: The classic glaciated karst landscape above Malham in the Yorkshire Dales of northern England, with bare scars, dissected pavements, and dry meltwater valleys. (Photo by Tony Waltham)

duced by sustained hydraulic gradients through passages filled with glacial sediments (see Paragenesis entry).

Little is known of active glacierized continental karst; the closest analogues are in Spitsbergen (Svalbard) where karst springs emerge from below small ice sheets. Continental ice sheets are mobilized by their own topography rather than that of the buried landscape. The directions of ice flow and hydraulic gradient may bear little relationship to ground topography, being controlled largely by the ice surface slope. The flow in icesheets appears to have been focused in fast-moving ice streams that can cut deep glacial valleys, and construct enormous moraine complexes. At the frozen margins of icesheets, particularly steep hydraulic gradients arise from the steep glacier slope, and the accumulated meltwater derived from up-glacier. Groundwater may be driven through deep karst formations, destabilizing overlying beds and even erupting through spontaneous vertical shafts. The Earth's crust deforms under glacial loading and unloading leading to relatively short-term changes in base level and possibly inducing deep groundwater circulation impossible under static conditions. Similarly, water sequestered in glaciers causes significant falls in sea level (up to 130 m during the last glaciation), allowing drainage of karst aquifers, and formation of speleothems well below current sea level.

Continental glaciation can be utterly destructive to the karst landscape, erasing karst landforms in zones of erosion, and completely burying them in areas of deposition. Nevertheless, karst hydrological function can be maintained despite the loss of the surface karst landscape (e.g. carbonate areas of southern Ontario, and beneath the city of Winnipeg, Manitoba, Canada). Many prominent glaciated karst areas lack closed depressions and may have few surface outcrops of limestone. In some cases, caves have only been discovered during well drilling operations.

In summary, glaciation brings about a profound change in the controlling recharge, routing, and discharge of water in karst through moderating hydraulic head, water chemistry, and sediment load. Glaciers also remodel the surface landscape and cave systems through erosion and deposition. The relative transience and cyclicity of the glacial condition, and glacial sediments, makes interpretation of landscape and cave evolution more complex and subtle than in environments with a more consistent history.

CHRIS SMART

See also **Alpine Karst; Picos de Europa, Spain; Yorkshire Dales**

Further Reading

Anderson, N.L. & Hinds, R.C. 1997. Glacial loading and unloading: A probable cause of rock salt dissolution in the Western Canadian Basin. *Carbonates and Evaporites*, 12(1): 43–52

Bögli, A. 1964. Le Schichttreppenkarst. *Revue Belge de Géographie*, 88: 64–82

Ford, D.C. 1983. The effect of glaciations upon karst aquifers in Canada. *Journal of Hydrology*, 61: 149–58

Ford, D.C. (editor) 1983. Castleguard Cave and Karst, Columbia Icefields area, Rocky Mountains of Canada: A symposium. *Arctic and Alpine Research*, 15(4): 425–544

Ford, D.C. & Williams, P.W. 1989. *Karst Hydrology and Geomorphology*, London and Boston: Unwin Hyman: 472–89

Smart, C.C. 1984. Glacier hydrology and the potential for subglacial karstification. *Norsk Geografisk Tidsskrift*, 38: 157–61

Smart, C.C. 1984. Overflow sedimentation in an alpine cave system. *Norsk Geografisk Tidsskrift*, 38: 171–75

Smart, C.C. 1997. Hydrogeology of glacial and subglacial karst aquifers: Small River, British Columbia, Canada. In *Proceedings of the Sixth Conference on Limestone Hydrology and Fissured Media*, edited by P.-Y. Jeanin, La Chaux de Fonds, Switzerland: 315–19

Walder, J. & Hallet, B. 1979. Geometry of former subglacial water channels and cavities. *Journal of Glaciology*, 23(89): 335–46

GOLONDRINAS AND THE GIANT SHAFTS OF MEXICO

Few caves in the world provide the awe-inspiring impact of Sótano de las Golondrinas, in the state of San Luis Potosí, Mexico (see Figure). The entrance to the *sótano* (pit or shaft) is only 63 m × 49 m in diameter but the floor of the bell-shaped shaft measures 305 m × 134 m and lies more than 300 m below the surface. T.R. Evans made the first descent into the cave in 1967 (Raines, 1968). Experienced cave explorers agree that rappelling into this pit, sunlight streaming in from the surface and walls receding progressively further away from the caver, ranks as one of the most spectacular caving experiences. The shaft is large enough that clouds form within the entrance chamber, particularly in June when sunlight beams directly down the entrance shaft.

Tens of thousands of White-Throated Swifts (*Streptoprocne Zonaris*) and hundreds of Green Parakeets (*Aratinga holochlora*) spend each evening in the cave, making the pit a giant, natural bird cage. Each dawn, the swifts make a massive spiral flight up the shaft on their way to the surface to spend the day feeding. The flight typically lasts nearly an hour. The parrots follow in a more leisurely manner and often perch around the entrance throughout the morning. At dusk, the swifts return to the cave, each circling the entrance once or twice, then folding their wings and diving for their nests deep in the shaft.

The uneven shaft floor has 75 m of relief and is completely covered by bird guano. The deepest part of the room is 400 m below the uppermost point along the entrance. A tight fissure called "The Crevice" extends the depth of the cave to 512 m. However, the passage is unpleasant and explorers have not found any features to recommend it.

The Sierra Madre Oriental, the major mountain range along the eastern side of Mexico, contains many notable pits. A middle Cretaceous carbonate reef complex, commonly repeated by thrust faulting, makes up most of the range and provides the thick sequence of soluble rocks necessary for development of these features. While Golondrinas is arguably the most spectacular of the shafts, there are two other outstanding open-air pits, Hoya de Guaguas and Sótano del Rancho del Barro. Sótano del Rancho del Barro, more commonly called *El Sótano*, is a huge 455 m deep gash in the side of a cornfield-covered mountain west of Golondrinas. The entrance is about 450 × 200 m across and funnels to a floor approximately 100 × 200 m. Reaching the tree- and shrub-covered bottom requires rope and the 410 m

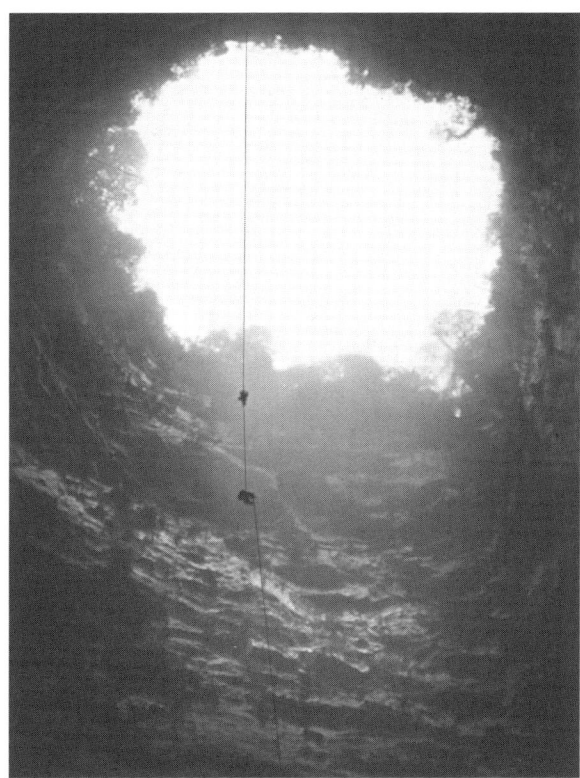

Golondrinas and the Giant Shafts of Mexico: Looking up the huge shaft of Sótano de las Golondrinas, with two climbers in tandem on the single rope. (Photo by Andy Eavis)

descent from the south side is nearly vertical. No known passages extend from the pit or its floor.

Closer to Golondrinas is another open-air "bird" pit, Hoya de Guaguas. The entrance is 50 m × 80 m and the mostly free-hanging drop to the bottom varies from 153–202 m, depending on the rigging point. The bottom is heavily vegetated and the pit provides a home to White-Throated Swifts (*Streptoprocne Zonaris*) and Green Parakeets. About 500 m of passage extends beyond the entrance pit to a water-filled passage. The total depth of the cave is 478 m and the total length is 725 m.

Another notable shaft is Sótano de la Cuesta, which sits on top of a remote section of the Sierra de El Abra. Although not particularly deep (174 m), the cave's visual impact makes it one of Mexico's "classic" pits. Cavers start their descent through a small hole less than a metre wide and about 1 m long and pass through a narrow slot for a couple of metres to where the ceiling extends out in all directions and the caver is suspended 170 m above the chamber floor with all walls more than 20 m away. A larger entrance nearby illuminates the room with an intense, midday beam of light. The floor of the pit is about 320 m long and 90 m wide and covered with flowstone, stalagmites, breakdown blocks, bird guano, and coral pipes. The cave has a total depth of 217 m and is formed in the middle Cretaceous El Abra reef complex.

Twenty six 200+ m air-filled shafts have been explored in the states of Chiapas, Jalisco, Oaxaca, Puebla, Queretaro, San Luis Potosí, and Veracruz (Sprouse, 2002). Most are entrance pits, like Golondrinas, El Sótano, Guaguas, and Cuesta. Others, like two pits inside Kijahe Xontjoa Cave in the state of Oaxaca, are parts of multiple-drop cave systems. At least two pits with depths in excess of 200 m are completely filled by deep phreatic flow; Nacimiento del Río Mante's Macho Pit (206 m) at the base of the El Abra, and Zacatón, on Rancho La Azufrosa west of Tampico. Zacatón has been plumbed to 329 m and is filled with warm, sulfur-rich, oxygen-depleted, slightly acidic water (Gary, 2001).

The origin of most of the giant shafts of Mexico is currently unclear. Some probably formed as underwater tubes lifting water along joints or bedding planes and others are swallet caves formed by aggressive, vadose water. The very large pits probably developed by stoping processes in which the collapsed material is dissolved and removed by aggressive waters passing through the breakdown.

LOUISE D. HOSE

See also **America, Central; America, South: History**

Works Cited

Gary, M. 2001. Los Cenotes de Rancho La Azufrosa. *Association of Mexican Cave Studies Activities Newsletter*, 24: 31–38

Raines, T.W. 1968. Sótano de las Golondrinas. *Association of Mexican Cave Studies Bulletin*, 2: 20

Sprouse, P. 2002. Deep pits of Mexico. *Association of Mexican Cave Studies Activities Newsletter*, 25: 18

Information about the Association of Mexican Cave Studies publications can be found at www.amcs-pub.org

Further Reading

Anon. 2001. Zacaton: The deep project: El Proyecto Buceo Espeleologico Mexico y America Central. Available at http://www.onr.com/user/zacaton/

Ford, D. 2000. Deep phreatic caves and groundwater systems of the Sierra de El Abra, Mexico. In *Speleogenesis: Evolution of Karst Aquifers*, edited by A.B. Klimchouk, D.C. Ford, A.N. Palmer & W. Dreybrodt, Huntsville, Alabama: National Speleological Society

Stone, B. & Raines, T. 1997. History of Mexican speleology to 1992. *Association of Mexican Cave Studies Activities Newsletter*, 22: 25–48

Walsh, M. & Woods, S. 2001. *Mexican Caving 2001: A Field Guide for the Exploration of the Caves of the Sierra Gorda*: Austin, Texas Mexican Caving 2001

GRAND CANYON, UNITED STATES

Grand Canyon National Park encompasses some 4900 km² in northern Arizona (Figure 1). The canyon dissects thick, regionally extensive Paleozoic carbonate units, which contain both paleokarst and active karsts. The active karst drains water from the plateaus and delivers it to the otherwise dry canyon environment below. Significant caves occur, some containing excellent records of paleoecology and human prehistory.

Two carbonate sections are exposed in the canyon walls, separated by a 350-m-thick Pennsylvanian–Permian clastic confining sequence. The upper carbonate section, the top of which forms

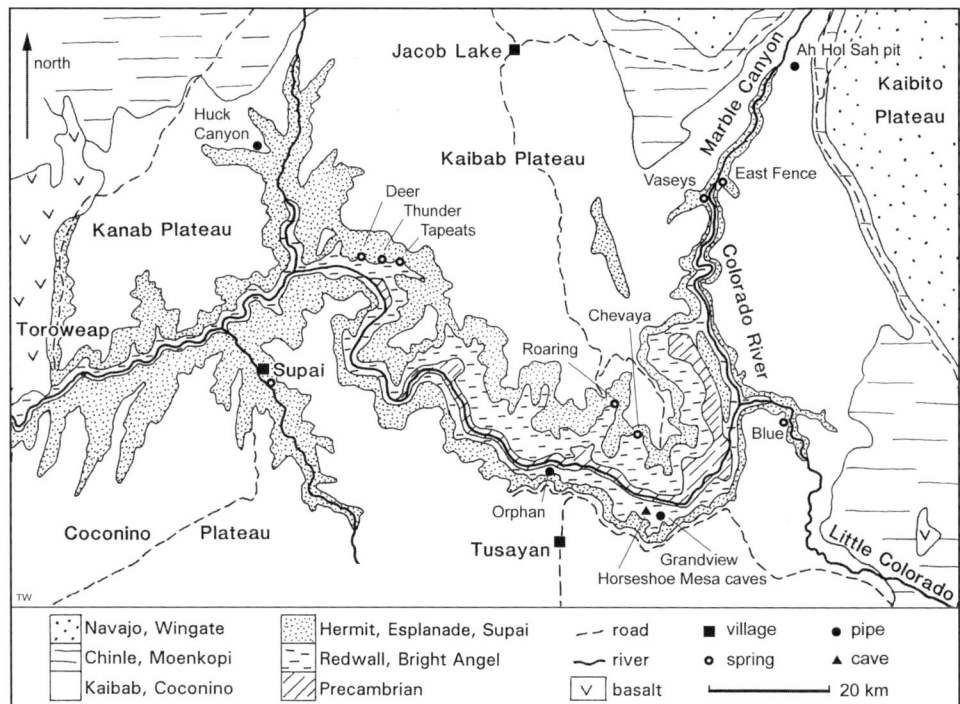

Grand Canyon: Figure 1. Geology and karst sites of the Grand Canyon of the Colorado River, Arizona. The rim of the canyon is the edge of the inlier of Hermit and older rocks, as the plateaus on each side of the canyon are capped by the Kaibab Limestone.

the surfaces of the surrounding plateaus, includes 225 m of Permian Kaibab and Toroweap limestones and dolomites, but most of the karst there occurs in laterally extensive gypsum facies. The lower Paleozoic carbonates include, from the top down, the massive Mississippian Redwall Limestone, the Devonian Temple Butte Formation, undivided Cambrian dolomites, and Cambrian thin-bedded, shaly Muav Limestone.

Paleokarsts developed on most of the several unconformities found in the carbonate sections. The best exposed is at the top of the Redwall Limestone. It developed in late Mississippian times during a regional emergence that subjected the upper part of the unit to a tropical environment. Dolines were abundant, and were laterally interconnected in the subsurface by integrated cave networks. The karst was drained by low-gradient westward-flowing streams that incised valleys into the otherwise flat surface. The depth of karstification increased westward in proportion to the depth of the valleys. A modern analogue is the Floridan karst. With subsidence, Mississippian and Pennsylvanian sediments covered the karst. Burial by several hundred metres of younger Paleozoic and Mesozoic sediments caused significant collapse and cementation. Now the paleokarst is expressed as infilled and buried karst topography at the top of the unit, as extensive karst breccia layers, and as filled or collapsed paleocaves. Most caves in the upper Redwall Limestone are remnants that have been re-excavated a short distance back from the cliff faces, but these are invariably filled in beyond this. Isolated, open cavities within the Redwall paleokarst were nucleation points for numerous breccia pipes that have subsequently developed throughout the region (Wenrich, 1985).

Triassic uplift south of the Grand Canyon appears to have activated early northward-directed artesian circulation in the Redwall aquifer (Huntoon, 1996). This caused dissolution of carbonate breccias in the incipient breccia pipes, creating more space, which allowed continued upward stoping. As the pipes pierced successive confining strata, upward circulation within their breccia fills, coupled with dissolution of the breccia, facilitated pipe growth, resulting in some pipes which propagated upward over 700 m through the remaining Paleozoic section and into Mesozoic strata. Some pipes became highly mineralized, and were first mined for copper (Grandview pipe) and later uranium (Orphan pipe, Hack Canyon pipes). Uranium production from the Grand Canyon pipes has been substantial.

Breccia pipes are still forming or being reactivated in the region as space is created by dissolution in the underlying lower Paleozoic artesian aquifers. Pits such as Dante's Descent (south of the canyon) and Ah Hol Sah are dramatic fresh examples on the surrounding plateaus. Openings in the reactivated pipes have been penetrated to depths reaching 160 m (Hose & Strong, 1981).

Late Cretaceous and early Tertiary uplift, coupled with canyon incision during Tertiary time, steepened hydraulic gradients, thereby facilitating karstic development, especially in the saturated lower Paleozoic carbonates. Exhumed phreatic caves in the Redwall Limestone on both sides of the canyon, some having 1000 m or more of passage, such as the Horseshoe Mesa caves, represent remnants of the earlier stages of this artesian karstification. These now dry caves are perched up to 1200 m above the Colorado River, whereas springs from the active parts of the Redwall artesian aquifer occur mostly where the Redwall Limestone lies at or below river level in structural basins adjacent to the Kaibab Upwarp. The three largest artesian springs are East Fence on the floor of Marble Gorge, Blue Springs in the Little Colorado River Canyon, and Havasu at the bottom of Cataract Canyon. Each drains a substantial artesian basin. Blue Springs

Grand Canyon: Figure 2. The tufa calcite of Travertine Falls, in a side canyon fed by springs rising from the Redwall Limestone far below the South Rim of Grand Canyon. (Photo by Trevor Ford)

have the largest discharge: a perennial flow of 6.5 $m^3 s^{-1}$. Massive travertine deposits downstream from the Blue and Havasu springs reveal active dissolution within the basins (Figure 2). The caves associated with the artesian karst systems are inaccessible owing to saturation. Elevated, exhumed remnant caves reveal that laterally extensive mazes predominate, often localized along extensional fractures. The most permeable parts of the aquifers are found within large but regionally localized extensional subsidence basins.

The adjacent elevated, unsaturated plateaus are drained by vadose conduits. The montane climate of the forested 2700-m-high Kaibab Plateau, north of the eastern Grand Canyon, resulted in the development of a karst surface characterized by numerous steep-walled dolines that appear to represent collapse into dissolving gypsum in the underlying Toroweap Formation. Large closed basins up to several km long, such as DeMotte Park, represent karstification localized along extensional faults. No perennial streams occur on the Kaibab Plateau, attesting to effective karst drainage. Horizontal caves in the Kaibab and Toroweap formations are virtually absent because the primary flow direction for recharge is vertical, through collector structures that feed into extensional fault zones.

Once in the fault zones, the water moves vertically through the Supai confining strata to the lower Paleozoic carbonates. It then circulates laterally to the canyon walls or floor via simple linear vadose drain conduits, primarily in the Muav Limestone, developed along the same fault zones concurrently with the incision of the lower Paleozoic carbonates by the Grand Canyon. The character of these simple conduits reflects steeper hydraulic gradients and lack of saturation in the elevated settings (Huntoon, 2000). The largest vadose caves, with up to 2000 m of surveyed passage in each cave, are Tapeats, Thunder, and Roaring Springs caves. The latter serves as the water supply for Grand Canyon National Park, feeding pipelines in which water is pumped up to both the north and south rim tourist facilities.

As an outstanding geological spectacle, Grand Canyon National Park was designated a World Heritage Site in 1979, but the karst constitutes only a small proportion of its scientific values. Access to its caves is restricted because of vulnerable features. These include Pleistocene paleontological deposits containing bones and dung from extinct species, excellently preserved prehistoric artefacts, including unique split-twig animal figurines dating from up to 4000 years ago in several Redwall caves, and Pueblo II pottery and other relics c.1000 years old. Bat Cave, in the far western Grand Canyon, hosts one of the most northerly colonies of Mexican free-tailed bats.

DONALD G. DAVIS AND PETER W. HUNTOON

Works Cited

Hose, L.D. & Strong, T.R. 1981. The genetic relationship between breccia pipes and caves in non-karstic terranes in northern Arizona. In *Proceedings of the 8th International Congress of Speleology*, edited by B.F. Beck, vol. 1, Huntsville, Alabama: National Speleological Society

Huntoon, P.W. 1996. Large-basin groundwater circulation and paleo-reconstruction of circulation leading to uranium mineralization in Grand Canyon breccia pipes, Arizona. *The Mountain Geologist*, 33: 71–84

Huntoon, P.W. 2000. Variability of karstic permeability between unconfined and confined aquifers, Grand Canyon region, Arizona. *Environmental and Engineering Geoscience*, 6: 155–70

Wenrich, K.J. 1985. Mineralization of breccia pipes in northern Arizona. *Economic Geology*, 80: 1722–35

Further Reading

Beus, S.S. & Morales, M. (editors) 1990. *Grand Canyon Geology*, New York: Oxford University Press

Euler, R.C. (editor) 1984. *The Archaeology, Geology and Paleobiology of Stanton's Cave, Grand Canyon National Park, Arizona*, Grand Canyon, Arizona: Grand Canyon Natural History Association

Huntoon, P.W. 1974. The karstic groundwater basins of the Kaibab Plateau, Arizona. *Water Resources Research*, 10: 579–90

Huntoon, P.W. 1989. Bat Cave guano mine, western Grand Canyon, Arizona. In *Geology of the Grand Canyon, Northern Arizona*, edited by D.P. Elston *et al.*, Washington, DC: American Geophysical Union (American Geophysical Union, 28th International Geological Congress Field Trip Guidebook T115/315)

Martin, P.S., Sabels, B.E. & Shutler, D. Jr 1961. Rampart Cave coprolite and ecology of the Shasta ground sloth. *American Journal of Science*, 259: 102–27

GROUNDWATER IN KARST

Definition of Groundwater: Aquifers and Aquicludes

Groundwater is the water present beneath the Earth's surface, stored in voids in bedrock and unconsolidated deposits. A body of rock containing and able to transmit significant quantities of water is termed an "aquifer"; an "aquiclude" is unable to transmit significant quantities of water, and generally acts as a barrier to the flow of groundwater. Aquifers are generally associated with particular geological formations or sequences, and generally adopt the respective formation name or stratigraphic age designation. The strong association between groundwater and geology has led the study of groundwater to be termed hydrogeology. Groundwater is replenished by "recharge", generally the deep percolation of soil water and (in karst aquifers) the inflow of sinking streams. The shallow subsurface and soil are commonly undersaturated; water coexists with air, and not all the openings in the rock are occupied by water. In soil-covered carbonate rocks dissolution is usually at a maximum immediately below the soil–bedrock interface forming a subcutaneous or "epikarstic" zone at the top of an unsaturated zone that eventually grades into a saturated zone where all voids are occupied by water. In karst hydrogeology, the terms "vadose" and "phreatic" are commonly used instead of unsaturated and saturated. The interface between these zones is the water table. Recharge water descends more or less vertically from the surface to the water table, travels sub-horizontally through the aquifer, and emerges as groundwater discharge in the form of seeps and springs where the water table reaches the surface. Many natural wetland areas are sustained by groundwater discharge.

Groundwater as a Reservoir: Resource and Water Balance

Groundwater is ubiquitous in favourable geological settings, and modest quantities may be readily obtained as the primary water supply for smaller communities and agricultural use. Only exceptionally productive aquifers can be exploited for large municipal supplies, because there is a limit to the quantity of water that can be withdrawn from an area.

An aquifer may be conceived as a reservoir or distinct body of water defined by its geological boundaries and quality. When the recharge and discharge are in balance, the aquifer is in equilibrium and its long-term volume does not change. If groundwater extraction causes the balance to go into deficit, aquifer water levels will decline. Computation of aquifer water balances is therefore a fundamental, but difficult aspect of groundwater management.

Natural water quality in an aquifer reflects the geological context, the solutes depending on the solubility of the minerals composing the aquifer. Many subsurface chemical processes are very slow, so the age of the water has a profound impact on quality, with older water becoming increasingly mineralized, depleted in oxygen, and soured by the presence of odorous hydrogen sulfide. Offsetting this deterioration with age is a decreasing risk of anthropogenic contamination in older waters. The age of groundwater is therefore of considerable interest, as it indicates the general position in an aquifer flow system, and the risk of pollution. It is common to determine the residence time of groundwater, perhaps based on the level of radioactive tritium or carbon, or by a simple ratio between aquifer volume and recharge or discharge rates. However, the average residence time is not a very useful concept in karst aquifers since they contain a mixture of waters of very different ages. Recharge through dolines or sinking streams directly enters the conduit network and may travel to springs at rates exceeding 1 km d^{-1}. Conversely, much of the recharge through thick soils or fine-grained sediments overlying the bedrock may take years to travel a few metres, and flow through the matrix of the rock can be exceedingly slow. Consequently, residence times in any karst aquifer will vary considerably with the flow path(s) that the water has followed.

Groundwater Dynamics

The flow of groundwater is energy-driven, with a progressive loss of energy from recharge to discharge. Rather than determining energies directly, it is more useful to describe energy indirectly as "hydraulic head", expressed in metres of elevation. The hydraulic head of any water surface is its elevation above datum. The head throughout a static water body is uniform; in a lake it is the surface elevation. This is because hydraulic head at a point is the sum of two components: elevation and pressure. The former is simply height above datum. The latter is a reflection of the force exerted by overlying water, and is represented by the height to which water would rise in a stand pipe. The balance of elevation and pressure head at any point in an aquifer can vary; as water goes deeper, pressure rises and elevation is lost. In this way, groundwater can descend to considerable depths, and yet still rise to discharge at the surface. The total hydraulic head, however, must always decline along a flow route.

The rate of loss of head along a flow route is termed the hydraulic gradient; it is a reflection of the rate of consumption of energy in driving the flow. Increased flow requires a greater hydraulic gradient, obtained by increasing water levels at the upstream end of the system. Counteracting the hydraulic gradient is the resistance to flow arising from fluid properties and rock form. In granular media such as sand or sandstone, the flow route follows the voids between the individual grains, and the ease of flow is termed the "hydraulic conductivity". All rocks have an inherent hydraulic conductivity ranging from 1 to 10^{-13} m s^{-1} for coarse gravel to marine clays respectively. The exceptional range of values of hydraulic conductivity has profound implications. First, strong contrast between hydraulic conductivities makes the distinction between aquifers and aquicludes realistic, but relative; silt in sand is an aquiclude, but in clay would be an aquifer. Second, it makes the actual flow route taken by water extremely sensitive to almost imperceptible geological variations. Third, it makes it practically impossible to accurately characterize the hydraulic properties of any aquifer in general and karst aquifers in particular.

The effects of hydraulic gradient and hydraulic conductivity on groundwater flow in porous media are summarized in Darcy's Law, which states that the flow per unit area is the product of the two factors, hydraulic gradient and hydraulic conductivity. Darcy's Law has seen widespread application, not least in the development of popular computer models designed to simulate groundwater flow based on interpolations of observations of hydraulic head and hydraulic conductivity (see Groundwater in Karst: Mathematical Models). Such simulations are very cost-effective ways of defining aquifer flow patterns, and exploring the impact of possible environmental change or resource use and abuse. However, groundwater may follow routes other than the pores in the rock matrix, most notably through fractures and solutionally enlarged karst openings. Hydraulic head in such aquifers remains the primary driving force for groundwater movement, but the medium can no longer be characterized by a hydraulic conductivity (a bulk property), but requires definition of the size, orientation, and linkage of openings. These data are seldom available, although explorable cave passages provide a sample of the upper end of the spectrum of openings. Flow in mapped cave streams can be described by a range of generalized open or closed conduit laws (e.g. Manning, Chèzy, and Darcy-Weisbach) that show flow to depend strongly on the aperture of the conduit. Flow routes intermediate between matrix flow and cave streams are tackled either by applying Darcy's law to an assumed "equivalent" hydraulic conductivity, or more recently by attempts at direct simulation of conduits networks to obtain a matching to observations of hydraulic head. While the Darcy approximation may provide an acceptable match to observations of hydraulic head, it is not known how realistic the flow pattern might be. In particular, Darcy simulations of karst aquifers significantly underestimate flow velocities, making the transmission of contaminants appear slower and more manageable than is really the case.

Hydraulic Head: Landscape Controls and Flow Systems

The correspondence between elevation and energy provides a simple rule of thumb for groundwater flow: recharge takes place at high ground, and discharge at lower elevations. In this sense groundwater flow can be viewed as obeying the same topographic rules as surface streams with corresponding divides and discharge along the axis of valleys. Strictly this rule applies to the topography of the water table, not the land surface, but the former can be provisionally considered a subdued analogue of the latter. The water table reaches the surface in discharge zones, but lies at greater depth elsewhere in recharge zones. Although there may be a number of natural features indicative of water-table level such as lakes, rivers, and springs, in many landscapes the thickness of the unsaturated zone is indeterminate, but may be many tens of metres in some karst areas. The water level in shallow wells provides the best indication of water-table elevation and the compilation of a water-table database using existing and custom wells is a primary task in hydrogeology. However, only shallow groundwater flow follows the local topography. At increasing depth, groundwater flow is influenced by wider scale topography up to regional scales of thousands of kilometres. The result may be multi-level flow systems at local, intermediate, and regional scales with frequent inconsistency of groundwater flow trajectories at different depths.

Hydraulic Conductivity: Geological Controls

Topographically directed flow systems develop in completely uniform porous media, and arise from consideration of only hydraulic gradient. The extreme natural variability of hydraulic conductivity often exerts greater control, especially in local redirection of groundwater. If the ground is considered to be composed of aquifers and aquicludes, then the geological extent of the aquifer dictates the storage and routing of groundwater.

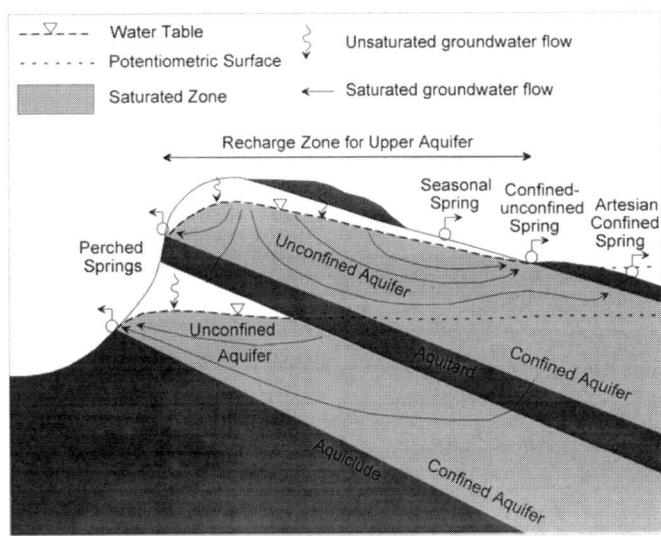

Groundwater in Karst: Figure 1. A simple escarpment composed of two aquifers sandwiched between three aquicludes. Recharge occurs at the surface exposures of the aquifers, and water is routed to respective springs or well based on the pattern of hydraulic head reflected in the water table in the unconfined aquifer, and the potentiometric surface in the confined aquifer.

Figure 1 is an idealized example of groundwater occupation of two porous medium aquifer units sandwiched between three aquicludes in an escarpment setting typical of moderately deformed sedimentary rocks. Geological exposure dictates the extent of the recharge area across the crest of the escarpment. Springs have developed at the lowest topographic points of each aquifer. The upper aquifer has filled beneath the overlying or confining aquiclude. An unconfined aquifer has a water table for its upper boundary; a confined aquifer has an aquiclude as its upper boundary. As a result, the water in a confined aquifer is pressurized and the hydraulic head is greater than the elevation of the upper boundary of the aquifer. It is convenient to consider a virtual water table termed a "potentiometric surface", the level to which water would rise if a well were drilled into the aquifer. It is quite possible for the potentiometric surface to be above the ground surface, in which case, artesian springs and flowing wells may occur. If recharge is highly seasonal, then the water table and potentiometric surface may also oscillate. The upper spring on the gentle dip slope may therefore be seasonal, only flowing when the water table reaches that elevation. The lower aquifer in Figure 1 is isolated from surface recharge and protected from surface-derived contamination. Recharge may occur by slow seepage from the upper aquifer, but the water would be expected to have a long residence time. Flow from the lower perched spring might be more steady, and the water more mineralized than for the upper perched spring.

Heterogeneity and Anisotropy

Much hydrogeological theory assumes a homogeneous aquifer in which hydraulic conductivity is uniform throughout. However, most aquifers are heterogeneous and geological contrasts result in variations in hydraulic conductivity. Groundwater is redirected or refracted when it encounters subtle changes in hydraulic conductivity. As a general rule, groundwater is focused through areas of high conductivity and away from areas of low conductivity. Some aquifers exhibit anisotropy (directional differences in hydraulic conductivity), for example in response to layering of the constituent rock materials, or the presence of fractures. In these cases, the groundwater trajectory tends to follow the higher hydraulic conductivity, rather than the direction of the steepest hydraulic gradient.

Fracture Aquifers

Intact igneous and metamorphic rocks are often heterogeneous and anisotropic. The intact rock has very low hydraulic conductivity, but fractures arising from tectonic, cooling, and decompression stresses may be present. Joints and bedding planes may similarly dominate the conductivity of some sedimentary rocks such as quartzites and massive limestones. The pathway taken by groundwater in such rocks depends on the relationship between the fracture pattern and hydraulic gradient. The latter is driven by topography, as expected. It is possible to simulate the flow of groundwater in fracture aquifers, but it is almost impossible to determine critical features of the fracture system such as the orientation, extent, connectivity, and aperture. In most circumstances, fracture aquifers are simulated with an equivalent porous-medium model.

Karst Aquifers

In rock types favouring dissolution by groundwater, the most active flow routes become enlarged over time, resulting in a highly heterogeneous and anisotropic karst aquifer. The process is particularly effective in fracture aquifers where groundwater flow is concentrated, but even porous soluble aquifers show development of solutional openings. Karst aquifers exhibit a more or less organized system of solutionally enlarged openings, typically directed and convergent towards a spring or springs. The hydraulics of enlarged openings favour water movement, so that the majority of flow may become concentrated in very few conduits, giving a dendritic (branching) network. Nevertheless, the pathways followed by karst groundwater are still directed by hydraulics, conveying higher elevation recharge to lower elevation springs. However, karst development leads to some hydrogeological peculiarities, and has evolved a distinctive terminology.

Karst Recharge

Where karst rocks are exposed at the surface, excess rain and snowmelt infiltrate through the soil and down into the bedrock to produce "autogenic" recharge. Surficial karst rocks become highly eroded by solution, producing many flow routes, and considerable water storage capacity, in the epikarst aquifer. The karst surface is commonly moulded into closed depressions that act to funnel recharge into more highly developed conduits. Passage through the epikarst tends to produce a steady flow of chemically enriched water. The general absence of surface rivers on karst demonstrates that autogenic karst recharge is much greater in magnitude than in other aquifers.

Less permeable materials such as shale, glacial till, or sandstone develop surface stream networks. Where such rocks are juxtaposed to, or overlie, karst, streams running onto the karst are prone to sink into the ground, constituting concentrated (focused) "allogenic" recharge. Typically, allogenic flow is variable, chemically dilute, and carries sediment, reflecting the properties of the corresponding surface catchment. Where the volume of water running onto the karst exceeds the capacity for recharge, the stream may flow for a considerable distance along an apparently conventional valley, before recharge is complete. The extent of such flow depends on the discharge of the stream, and is promoted by a common tendency for sink points to become choked by sediment and debris. In some cases, allogenic streams run completely across a karst landscape, and may have excavated a significant gorge. Where a permeable caprock overlies karst then dispersed allogenic recharge will occur, with the resulting dissolution leading to puzzling collapse dolines in the insoluble caprock.

The Karst Unsaturated or Vadose Zone

Flow through the unsaturated zone is dominated by gravity, so flow routes tend to be directed steeply downwards. Allogenic streams tend to develop discrete cave streamways with distinctive canyon passages and vertical shafts. Autogenic water is conveyed in seeps, trickles, and a hierarchy of widely distributed flows reflecting surface conditions, the degree of focusing by closed depressions, and the magnitude of storage in the epikarst. Apart from the epikarst, there appears to be relatively little storage of water in the unsaturated zone. The traditional term for unsaturated conditions is vadose, and this term is applied to the characteristic canyon and shaft style of cave system. In contrast to porous media, there is little capillary movement of water upwards through the unsaturated zone to sustain surface evaporation; during droughts, karst surface moisture is sustained by the epikarst aquifer.

The Karst Saturated or Phreatic Zone

Caves filled to the roof with water are termed phreatic, have distinctive rounded cross sections, and, with flow driven by pressure, may be directed both up and down. The transition from unsaturated to saturated or phreatic conditions technically occurs at the water table. In caves, the exact elevation of this transition depends on the flow of the stream and the geometry of the channel; local and temporary phreatic conditions often develop. Sections of cave passage which become temporarily filled with water are termed epiphreatic, and exhibit a mix of vadose and phreatic features. Phreatic conduits descending to great depths below the water table are termed bathyphreatic. They develop from a combination of geological control, and the presence of hydraulic gradients at depth.

Away from primary conduits many smaller flow routes exist, a more coherent regional water table can usually be defined, and a considerable volume of water can be stored. Flow in conduits requires very low hydraulic gradient compared to flow elsewhere in the aquifer, so the water table tends to be low and flat along the line of major conduits, and relatively steep in the surrounding aquifer. Under such "normal" conditions, water drains slowly from the aquifer to the nearest conduit which then rapidly conveys the water to the surface spring. Saturated zone storage is replenished not only by autogenic seeps, but also by water driven out of conduits pressurized during floods. Under the latter conditions, the karst water table is elevated along the line of the conduit, indicating flow out of the conduit and into the surrounding aquifer.

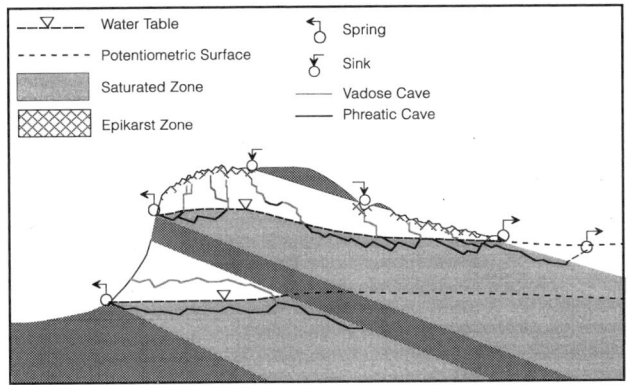

Groundwater in Karst: Figure 2. The same hydrogeological setting as Figure 1, rendered for a karst aquifer. The exposures of karst where autogenic recharge occurs show rugged topography with an underlying epikarst and vadose caves. The impermeable caprock has generated allogenic streams which enter the karst as vadose streamways, descending to the phreatic zone. The water table is sketched for the phreatic conduits and has flattened considerably compared to Figure 1. The deeper aquifers may contain conduits, but function much as conventional aquifers in terms of the potentiometric surface.

Figure 2 presents a karstic version of the dual aquifers represented in Figure 1. The overall governing topography is similar, but recharge is focused by closed depressions and at the margins of impermeable cover. In response, within the unsaturated zone, many vertically directed conduits (shafts) have developed, some under closed depressions, others at stream-sinks. As the land surface changes over time and the pattern of recharge evolves, some conduits become abandoned, but nonetheless remain as potential flow routes. In the saturated zone, the flow is directed towards the springs, but is constrained by geological structure resulting in the irregular path shown. Landscape changes also alter the position of the water table and springs may find lower elevation outlets, so that multiple level conduits exist. These have been omitted from the figure for clarity. In some karst aquifers, geological controls can sustain phreatic (water-filled) passages well above the regional water table, a phenomenon which may cause confusion in hydrogeological interpretation.

Summary

Overall, the physical principles guiding groundwater flow in karst aquifers are those applying to all groundwater flow. However, the peculiarities of solutional development of flow routes, and the possibility of exploring the larger of these flow routes has led to a distinctive form of permeability and a quite separate terminology. The resulting separation of karst from general hydrogeology has meant that the understanding of karst water resources has been compromised.

CHRIS SMART AND STEPHEN R.H. WORTHINGTON

See also **Karst Water Resources; Springs**

Further Reading

Bonacci, O. 1987. *Karst Hydrology with Special Reference to the Dinaric Karst*, Berlin and New York: Springer

Domenico, P.A. & Schwartz, F.W. 1998. *Physical and Chemical Hydrogeology*, 2nd edition, New York: Wiley

Ford, D.C. & Williams, P.W. 1989. *Karst Geomorphology and Hydrology*, London and Boston: Unwin Hyman

Price, M. 1996. *Introducing Groundwater*, 2nd edition, London and New York: Chapman and Hall

White, W.B. 1988. *Geomorphology and Hydrology of Karst Terrains*, Oxford and New York: Oxford University Press

Worthington, S.R. 2002. Test methods for characterizing contaminant transport in a glaciated carbonate aquifer. *Environmental Geology*, 42: 546–51

GROUNDWATER IN KARST: BOREHOLE HYDROLOGY

Boreholes are shafts of a few centimetres diameter drilled into the ground, typically for the extraction or observation of groundwater. In the former case, the borehole is pumped and draws water in from the surrounding aquifer (see Karst Water Resources). In the latter, the borehole remains passive in order to allow determination of undisturbed water levels in the aquifer, although observation boreholes are intermittently pumped to extract water samples or to evaluate aquifer properties. In karst, groundwater monitoring is conventionally located at springs and in caves, and complementary borehole data may prove difficult to reconcile with these observations. Yet many karst aquifers may have no enterable caves, and springs may not be accessible. For example, karst aquifers in weaker or younger limestones may not host caves; many karst aquifers are buried under alluvium and glacial till, or drowned by lakes or oceans; or landowners may deny access. In such cases, boreholes provide the only viable window on the karst aquifer. While boreholes may provide useful information on the aquifer, they need to be monitored using specialized techniques, and care should be exercised in evaluation of the data.

Wells drilled into karst aquifers differ significantly from those placed into porous-medium aquifers, such as sand or sandstone. They are often spectacularly productive, but show significant transient (short-term) degradation in quality, and are especially vulnerable to microbial contamination. It is important to identify and consider such differences, as treating a karst well with protocol devised for porous media may be inappropriate and compromise the user's safety.

A karst aquifer may be conceived as consisting of conduits arranged in a dendritic system placed within the carbonate rock mass. Large conduits are rare; smaller conduits are increasingly common, depending on the rock characteristics and aquifer history. The probability of a borehole intersecting a major, cave-sized conduit is only about 1–2%. Most boreholes in karst therefore only intersect small conduits, at most a few centimetres in diameter. These small conduits and solutionally-enlarged fractures, with apertures in the range 0.1 mm–1 cm, produce most of the permeability measured in karst boreholes and result in high productivity. Unfortunately, the enlarged fractures and conduits constitute an interconnected network through which not only water but also contaminants and bacteria can move rapidly and without filtration. If the aquifer or ground surface is contaminated, and there is uninterrupted connection to a supply conduit, then a well can become suddenly contaminated.

In conventional hydrogeology, borehole water level and water composition are generally taken to be characteristic of the aquifer at that location. Conduit flow and rapid transmission of water and impulses (e.g. from recharge or pumping) make water level data and water samples from karst boreholes potentially misleading. However, there are a number of useful tests in karst where information may be gathered from a single borehole, from a pair of boreholes, or from a large number of boreholes.

Useful single-borehole tests include drillers' records of bit drops and loss of circulation, flow meter tests, video logs, conductivity profiles, and water level and water chemistry monitoring. Bit drops indicate voids, and these may be isolated vugs or intersections with conduits. Loss of circulation is often associated with conduits. Flow meter tests involve measuring velocities up or down the well bore, and are most useful if done both in ambient flow and under pumping conditions. If the flow metering has sufficient resolution, these tests typically show active inflow and outflow at a small number of horizons. A less expensive method of electrical conductivity profiling supplements this information as well as revealing contrasts in the water chemistry of the different inflows. Downhole video is useful, especially if the camera has a capability of viewing sideways, since it can be seen if inflows are from discrete conduits or from fractures. Rapid changes in borehole water level and water chemistry are diagnostic of karst aquifers, and occur following rainfall or snowmelt events. Karst borehole water level time series often show consistent breaks in slope, which may be related to activation of overflows and underflows in the aquifer. However, since conventional bi-weekly sampling will not reveal such a pattern, continuous monitoring is recommended. Electrical conductivity is straightforward to measure and is proportional to total dissolved solids. Electrical conductivity, water temperature, and water level are often measured together. However, flow pattern in a borehole means the results of point sampling or monitoring may not be representative of the entire water column. Changes in these variables are dependent not only on the properties of the conduit network but also on the connectivity of this network to recharge. For instance, in glaciated areas there may be thick glacial sediments mantling the karst, resulting in the attenuation of the signal from recharging water.

Tests between boreholes form a second category. The probability of two boreholes encountering a common conduit is very low indeed, which means that natural gradient tracer tests seldom work in karst. Quinlan and Ewers (1989) exemplify the natural gradient problem with heavy-metal laden effluent from a sewage treatment plant in Kentucky, United States. Contamination was not picked up at any of 23 domestic wells in the down-gradient direction. Instead, the signature was picked up beyond the wells at a major spring. The effluent had flowed through a major conduit not intersected by any of the wells. The conduit formed an efficient, low-head pathway, resulting in convergent flow towards it. However, well contamination might have been induced by sustained pumping of the wells, or through flooding of the conduit.

Tracer tests between boreholes in karst are much more likely to be successful if a convergent flow test is carried out, with sampling at a pumping borehole. Such "induced gradient" tests are often successful over distances of tens or hundreds of metres. However, there is often considerable tracer loss, and the resulting velocities are not necessarily indicative of the peak velocities possible in a conduit.

A third category of borehole analysis is to use water-level data from a number of boreholes. An array of boreholes allows discrimination of general patterns of behaviour from idiosyncratic effects, unique to particular wells. The former may be taken as an "aquifer" signal, the latter as arising from the particular connectivity of the well. More controlled effects of this kind may be obtained using observation wells to define the pattern of drawdown around a pumping well. Karst aquifers frequently show marked departure from the radial symmetry predicted for the cone of depression in a porous medium. Elongation of the cone may be indicative of a conduit.

The classical multiple borehole study is of the Mammoth Cave area in Kentucky, where a water table map was compiled from water level in 1500 boreholes, supplemented by results from 500 tracer tests to springs and by maps of cave and surface streams. Care had to be taken to eliminate anomalous borehole water-level data arising from local pumping, poor connection, or storm runoff. The map showed that flow is convergent to distinctive troughs in the water table, and the troughs terminate at springs. The troughs are associated with major conduits and decrease in hydraulic gradient towards the spring. A decrease in gradient towards springs is a diagnostic feature of karst aquifers and is the opposite of the pattern in porous-medium aquifers.

The majority of work in karst hydrogeology has been undertaken from a speleological perspective. While this approach has been rewarding, it has not focused attention on the important unexplorable parts of karst aquifers, and has led to an implicit association of karst aquifers with a karst landscape. Where carbonates do not develop explorable caves, or they have been planed off and buried by glaciation, karst aquifers may still exist, and boreholes provide the primary window on these aquifers. Karst aquifers unfortunately compromise the interpretation of conventional pumping tests on boreholes, and karst hydrogeologists have been slow to develop alternative diagnostic techniques. Fortunately, these specialized methods are evolving and may provide important insight into water supply and aquifer management.

Chris Smart and Stephen R.H. Worthington

Works Cited

Quinlan, J.F. & Ewers, R.O. 1989. Subsurface drainage in the Mammoth Cave area. In *Karst Hydrology: Concepts from the Mammoth Cave Area*, edited by W.B. White & E.L. White, New York: Van Nostrand Reinhold

Further Reading

Milanović, P.T. 1981. *Karst Hydrogeology*, Littleton, Colorado: Water Resources Publications

Quinlan, J.F. & Ray, J.A. 1981. Groundwater basins in the Mammoth Cave Region, Kentucky. Mammoth Cave, Kentucky: Friends of the karst (Occasional Publication 1)

Smart, C.C. 1999. Subsidiary conduit systems: A critical hiatus in aquifer monitoring and modelling. In *Karst Modelling*, edited by A.N. Palmer, M.V. Palmer & I.D. Sasowsky, Charles Town, West Virginia: Karst Waters Institute

Worthington, S.R.H. 1999. A comprehensive strategy for understanding flow in carbonate aquifers. In *Karst Modelling*, edited by A.N. Palmer, M.V. Palmer & I.D. Sasowsky, Charles Town, West Virginia: Karst Waters Institute

GROUNDWATER IN KARST: CONCEPTUAL MODELS

Groundwater is present in virtually all rocks but the defining characteristic of karst aquifers is the presence of water in an interconnected network of solutionally enlarged fractures. These enlarged fractures are sometimes called channels, especially where apertures are less than 1 cm. The vast majority of enlarged fractures are of this size. Where apertures exceed one centimetre then turbulent flow is likely, and the enlarged fractures are called conduits. Where people can enter the conduits they are called caves. Small channels tend to have elliptical shapes, being elongated along fractures. As aperture increases there is a progression towards a circular shape, as long as the conduit is below the water table. Conduits above the water table tend to be eroded downwards and hence are usually much taller than they are wide.

The three fundamental requirements for a karst aquifer to develop are a soluble rock, flow through the rock of a fluid capable of dissolving the rock, and sufficient time. This discussion considers only carbonate rocks (limestones, dolostones, and metacarbonates), and water acidified by dissolved carbon dioxide from the atmosphere or soil, since these are the principal constituents in about 95% of carbonate aquifers. Dissolution occurs where meteoric water is able to infiltrate and flow through the rock. This situation includes all unconfined carbonate aquifers and some confined aquifers. High concentrations of dissolved solids ($>10\,000$ mg l^{-1}) in an aquifer are an indication that flow through the rock is sluggish and karstification is likely to be slow. Conversely, potable water in a carbonate aquifer is an indication of active flow and such an aquifer is almost certainly karstified. The third requirement for karstification is time. The evidence from field studies and from numerical modelling indicates that conduit networks can be created in as short a period as a few thousand years where there is a combination of steep hydraulic gradients, large initial apertures, and short flow pathways. However, periods of up to a million years are required in most aquifers. Most carbonate rocks were formed tens to hundreds of millions of years ago and many have been exposed at the surface for at least some millions of years, so there is almost always ample time for karstification to have taken place in any carbonate rock exposed at the surface.

Definition and evaluation of flow in carbonate karst aquifers has been subject to much speculation in the past. Only in the last 30 years has a coherent understanding emerged of how karst aquifers develop, as a result of progress in three main areas. First, experiments in the 1970s and 1980s showed that dissolution rates of calcite and dolomite drop precipitously as chemical saturation approaches. The implication is that fractures *throughout* a limestone aquifer can become enlarged, rather than just in the upper few metres, as earlier experiments had implied. A second development since the 1990s has been the use of computer modelling to integrate this new knowledge of dissolution kinetics with hydraulic equations to show the rate and distribution of conduit development in carbonate aquifers. Third, detailed measurements have shown that in most carbonates flow into boreholes occurs at just a few locations, and that these locations are fractures that have been greatly enlarged by dissolution. Thus it is the dissolution of the bedrock (i.e. karstification) that results in carbonate rocks becoming productive aquifers. It is now possible to combine these three developments to understand the processes and products of karstification.

Carbonate rocks of all ages evolve into karst aquifers. For instance, at one extreme there are caves in Archean marbles more than 2 billion years old near Lake Baikal in Russia, and at the other extreme there are caves in limestones less than 100 000 years old in the Bahamas. Young carbonates are unlikely to have been deeply buried below the surface, and so retain a high porosity. On the other hand, compaction has usually resulted in the loss of most of the intergranular porosity of pre-Cenozoic carbonates, although fracture openings develop and persist. Karstification in the older, more compacted carbonates has been better studied than in younger rocks. The conceptual model presented for flow in these aquifers is based on the relationship between porosity components (the fraction of void space) and permeability (the ease of flow through the rock). This standard model is followed by an exploration of the differences in young aquifers.

Most pre-Cenozoic carbonate aquifers have well-developed bedding planes and joints, and these intersecting fractures result in fracture porosity and permeability in these aquifers. The matrix blocks between the fractures yield a second type of porosity and permeability, and the linear, interconnecting conduits form a third order of porosity and permeability. Together, these three types of permeability result in a triple porosity (or triple permeability) aquifer. Tables 1 and 2 give examples of these three permeability types in four contrasting carbonate aquifers. The rock matrix accounts for almost all the porosity (usually $>95\%$) in carbonate aquifers, but the pores are usually small and often poorly connected, and so matrix permeability is low. The fractures typically have apertures of $0.01–0.1$ mm, and collectively have a much lower porosity than the matrix. However, because the fractures are connected together they result in moderate permeability. The third porosity element consists of the interconnected network of channels and conduits arising from solution. The network of solution channels varies from hierarchic, convergent tributary systems to braided maze patterns. In both cases there are large numbers of small channels (<1 mm in aperture) feeding larger channels and conduits, with the largest conduits being in the down gradient parts of the aquifers close to the springs. The conduit network dominates the transport of water through the aquifer although it makes up only a small fraction of the porosity of carbonate aquifers (Table 2).

Young carbonate aquifers are somewhat different from older aquifers. Porosity is much higher than in older aquifers because of the limited compaction, and matrix permeability can be high in these aquifers, in part due to the presence of unconnected voids, known as vugs. The effect of fracturing is more variable than in older aquifers. Conduits form both along fractures and also as vug-linking channels, which have apertures in the millimetre to centimetre range. In some young limestones there are zones where the vug-linking channels are so ubiquitous that the rock appears like a sponge. However, the lateral extent of these zones is uncertain, and the processes responsible for their development have yet to be fully understood.

Channels and conduits form the principal pathways for water movement through carbonate aquifers, and the average velocity from many tracer tests along large conduits is 1700 m day^{-1}. Where there are sinking streams recharging a karst aquifer and springs draining it, then tracer tests are commonly used to evaluate which sinks are connected to which springs. However, evaluation of conduits is much more difficult where recharge to a karst aquifer is only by percolation and where springs are not visible (e.g. where an aquifer discharges into large rivers, lakes, or along coasts). Boreholes are very useful in evaluating matrix and fracture flow, but are very problematic in evaluating conduit flow because of the low probability (0.01–0.02) of a borehole intersecting a conduit. Therefore, combining results from both boreholes and springs gives a much more comprehensive understanding of how a carbonate aquifer behaves than either alone can do.

The mode of recharge is important to aquifer evolution and a distinction is commonly made between allogenic and autogenic karst aquifers. The former receive some recharge from non-karst rocks, while the latter are fed solely by precipitation falling onto the area of outcrop or the overlying soil / sediments. In practice all "allogenic" aquifers also receive some autogenic recharge, although the proportion can be quite small, and some aquifers that are presently "autogenic" may have received allogenic recharge in the past. Allogenic recharge is usually concentrated and occurs where surface streams sink underground but recharge through a permeable non-carbonate caprock is also possible. In contrast, most autogenic recharge is dispersed and either enters the carbonate bedrock directly at outcrop or after infiltrating through the soil and any superficial deposits. However, concentrated autogenic recharge also occurs where dolines act to focus flow in the epikarst.

Sinking streams provide ideal tracer injection points, and many thousands of successful tracer tests have been conducted between sinking streams and springs that may be as far distant as several tens of kilometres. The springs in allogenic karst aquifers are known as resurgences, and are often characterized by substantial variation in water quality since they reflect the changing water quality of the sinking streams. Furthermore, streams often sink into accessible caves, and cave exploration and cave studies have provided substantial knowledge about allogenic karst aquifers.

Autogenic karst aquifers feed springs known as exsurgences. Where recharge is dispersed, the relatively slow percolation of water through the soil and epikarst zones into the conduits results in there being relatively little variation in water quality at exsurgences, and discharge variations are often less than at resurgences. However, where recharge is concentrated it is also more rapid and tracer tests from dolines in areas of polygonal karst have recorded velocities similar to those from sinking streams. The focusing of flow starts at the surface in dolines or in the epikarst zone with small conduits, which coalesce to form progressively larger conduits in a down gradient direction. Gaining knowledge about these conduits in the aquifer is difficult, however, as tracer testing is only possible where there is surface water (usually only during periods of high rainfall) or where water is introduced artificially from a tanker. Passable cave entrances in dolines or directly into the epikarst are less common than at sinking streams and most exsurgences are flooded and therefore accessible only by diving. Thus much less is known about the conduits in autogenic aquifers than in allogenic aquifers. Nevertheless, mathematical modelling has shown that autogenic karst aquifers should develop conduit networks, and this is substantiated by, for instance, exploration in areas of polygonal karst and in autogenic alpine karst aquifers. There are over 70 caves that have been explored to a depth of more than 1000 m and most are found high on mountains, lack sinking streams, have narrow entrance passages but have much larger cave passages at depth (often with substantial cave streams), and flow to flooded springs.

Cave streams in both allogenic and autogenic settings typically descend down dip along bedding planes and vertically via joints or, more rarely, faults to the water table. Once this is reached it is common for conduits to be below the water table for most of the distance to springs. This is indicated by studies of major relict conduits, most of which were below the water table for most of the time that water flowed through them, and by the low hydraulic gradients typically found in karst aquifers.

In thick carbonate aquifers conduit flow often extends to a considerable depth below the water table. This has been vividly demonstrated by the efforts of divers and several springs have been explored to depths of over 150 m. Evidence from thermal karst springs and from cavities intercepted in deep wells indicates karstification in some areas to depths exceeding 1000 m, such as in the Edwards aquifer in Texas. It has been found that the depth of conduit flow is proportional to the stratal dip and to the length of the catchment. Thus springs with large flows, which of necessity drain large catchments, often have very deeply descending conduits, as do conduits in steeply dipping Alpine aquifers. Conversely, platform carbonates often have low dips, and conduits in these aquifers often develop just a few metres below the water table.

STEPHEN R.H. WORTHINGTON and CHRIS SMART

Groundwater in Karst: Conceptual Models: Table 1. Matrix, fracture, and channel permeability in four carbonate aquifers (after Worthington, 1999)

Area	Hydraulic conductivity (ms^{-1})		
	Matrix	Fracture	Channel
Silurian dolostone, Smithville, Ontario, Canada	1×10^{-10}	1×10^{-5}	3×10^{-4}
Carboniferous limestone, Mammoth Cave, Kentucky, United States	2×10^{-11}	1×10^{-5}	3×10^{-3}
Cretaceous chalk, England	1×10^{-8}	4×10^{-6}	6×10^{-5}
Cenozoic limestone, Yucatán, Mexico	7×10^{-5}	1×10^{-3}	4×10^{-1}

Groundwater in Karst: Conceptual Models: Table 2. Matrix, fracture, and channel porosity in four carbonate aquifers (after Worthington, 1999)

Area	Porosity (%)		
	Matrix	Fracture	Channel
Silurian dolostone, Smithville, Ontario, Canada	6.6	0.02	0.003
Carboniferous limestone, Mammoth Cave, Kentucky, United States	2.4	0.03	0.06
Cretaceous chalk, England	30	0.01	0.02
Cenozoic limestone, Yucatán, Mexico	17	0.1	0.5

See also **Inception of Caves; Patterns of Caves; Speleogenesis (various entries)**

Further Reading

Dreybrodt, W. 1996. Principles of early development of karst conduits under natural and man-made conditions revealed by mathematical analysis of numerical models. *Water Resources Research*, 32: 2923–35

Farr, M. 2000. *The Darkness Beckons*, London: Diadem, and St Louis, Missouri: Cave Books

Gunn, J. 1986. A conceptual model for conduit flow dominated karst aquifers. In *Karst Water Resources*, edited by G. Günay & A.I. Johnson, Wallingford, Oxfordshire: International Association of Hydrological Sciences: 587–96

Palmer, A.N. 1991. Origin and morphology of limestone caves. *Geological Society of America Bulletin*, 103: 1–21

White, W.B. 1999. Conceptual models for karstic aquifers. In *Karst Modelling*, edited by A.N. Palmer, M.V. Palmer & I.D. Sasowsky, Charles Town, West Virginia: Karst Waters Institute (Karst Waters Institute Special Publication, 5): 11–16

Worthington, S.R.H. 2001. Depth of conduit flow in unconfined carbonate aquifers. *Geology*, 29: 335–38

Worthington, S.R.H. 1999. A comprehensive strategy for understanding flow in carbonate aquifers. In *Karst Modelling*, edited by A.N. Palmer, M.V. Palmer & I.D. Sasowsky, Charles Town, West Virginia: Karst Waters Institute (Karst Waters Institute Special Publication, 5): 30–37

GROUNDWATER IN KARST: MATHEMATICAL MODELS

Modern hydrogeology relies heavily on numerical modelling to provide a supplemental, less expensive alternative, to field investigations, and to explore various hypothetical conditions as an aid to planning or research. Mathematical models are based on various formulations of relationships between physical parameters (such as hydraulic conductivity or fracture geometry), usually expressed in computer code to exploit modern computational power and speed. Mathematical models of karst hydrogeology have only become viable as appropriate physical laws for groundwater flow and aquifer evolution have been formulated. As with all numerical models, the accuracy of the model relies on a well-constrained physical description from independent estimates of the aquifer and processes, and this remains a considerable barrier to practical implementation.

A wide variety of mathematical models are used in groundwater studies in carbonate aquifers. The most common types of model used are distributed parameter flow models, aquifer evolution models, lumped parameter flow models, and conduit models. The flow models are widely used, and are often known simply as groundwater models, numerical models, or computer models. The considerable challenge in modelling carbonate aquifers is to choose a mathematical representation of the flow system that is simple enough for implementation, yet embodies the essential elements of the aquifer so that meaningful results will be obtained. The ideal model would represent matrix flow in the differing lithologies within an aquifer; fracture flow in the joints, bedding planes, and faults which dissect the matrix blocks, and channel and conduit flow where the fractures have been enlarged by dissolution to apertures that may range from around 1 mm to more than 10 m on occasion. Provision of this level of detailed data for a real aquifer is impractical, and also places the modelling beyond current computer capability.

Distributed parameter models represent an aquifer by assigning values of parameters such as hydraulic conductivity and storage coefficients to many individual cells or nodes in an aquifer, of which there may be hundreds of thousands in complex models. One distributed parameter model, the MODFLOW program, has become the *de facto* standard for modelling three-dimensional flow in porous medium aquifers. Such a single continuum porous medium model can be adapted for karst aquifers by simulating the major conduits by lines of narrow cells within a thin layer in the model, although this would include neither smaller conduits, nor turbulent flow within the conduits. Several other types of model have been used for karst aquifers, including dual continuum (using two integrated, distinct porous media to represent the matrix and conduits), single or dual discrete fracture (using defined fracture networks in lieu of matrix and conduit components), and perhaps more realistic but difficult combinations such as porous continuum plus fracture or continuum plus conduit models.

In distributed parameter models the parameters required concern the physical properties assigned to each element such as the hydraulic conductivity and storage coefficient, the width and geometry of fractures, or the network, roughness, and size of conduits. Furthermore, hydraulic and hydrochemical processes require appropriate governing equations. Obtaining accurate values for the physical parameters over the entire aquifer would require impractical levels of field data, so simple "equivalent" values are substituted for actual variation. In addition, real-world processes are often more complex than those implemented in models and so many parameters are artificially "tuned" to generate appropriate behaviour from simpler approximations, such as by assuming a carbonate aquifer behaves as a porous medium and thus has no local variation in hydraulic conductivity.

Distributed parameter models can be implemented by either deterministic or stochastic realizations. In a deterministic model the best estimate for each parameter is assigned to each cell in the model, and these values are repeatedly adjusted until there is a satisfactory fit, usually between measured and simulated water levels, although spring flows are also used in some models. Although the tuning process is artificial, the single model generated is readily managed and visualized. A stochastic or probabilistic model selects randomly from a range of possible parameter values. Each resulting model has little significance in itself, but a large number of predictions can be pooled to provide an indication of the most likely range of results. While stochastic models avoid the problem of parameterization, designation of the appropriate range of values is just as difficult, and dealing with the pooled outputs is conceptually challenging.

There has been considerable discussion of the utility and accuracy of mathematical models in general, and of distributed parameter models in particular. Two fundamental problems arise with these models: they are inverse models and hence give a non-unique solution, and they are open systems. The inverse problem arises from taking aquifer flow data, such as water levels or spring flow, and working inversely to identify a combination of physical and geometric properties (e.g. hydraulic conductivity, fracture and conduit aperture, and connectivity) that result in the observed data. The resultant model is non-unique because there are many combinations of aquifer structure which could result in any given set of water levels or spring flows. The model is an open system because it exists in nature rather than in a laboratory; hence, there may always be some critical process or feature that has been missed, neglected, or dismissed as unimportant. This is particularly the case in carbonate aquifers, which are often modelled as equivalent porous media. Resultant predictions of flow velocity are usually too low by two to three orders of magnitude because of a failure to incorporate the conduit network and hydraulics resulting in rapid flow velocities.

A second class of models investigate aquifer evolution. Some fundamental theoretical studies were carried out in the 1960s, but comprehensive numerical modelling only started after laboratory experiments on simple scenarios in the 1970s and 1980s revealed that the dissolution rates of calcite and dolomite close to saturation are extremely slow. The starting point for these models is an arbitrary set of fractures with apertures much less than 1 mm, as is found in most bedrock aquifers. The fractures are usually arranged in a rectangular grid to represent the bedding planes and joints found in carbonate rocks. These models have two components, a hydraulics module and a geochemical module. The hydraulics module uses standard flow equations to calculate the flow through each fracture in the grid. The geochemical module then calculates the fracture enlargement which takes place at each iteration of the model. The two modules are run alternately, thus representing the positive feedback process whereby more flow occurs in the more open fractures, more dissolution takes place in these fractures which enlarges them further, etc. Modelling has now firmly established that this process inexorably leads to a conduit network in all carbonates where there is unconstrained recharge to, and discharge from, the bedrock, which includes almost all carbonate aquifers.

The early aquifer evolution models were one-dimensional or consisted of a simple two-dimensional grid, but more complex two- and three-dimensional grids are now being utilized. Other developments include the incorporation of turbulent flow equations and flow in the rock matrix as well as in fractures. Problems simulated include conduit development in the epikarst, conduit development in both allogenic and autogenic aquifers, development of both tributary and maze conduit networks, and the rapid development of conduits in the vicinity of dams.

A third class of models are known as lumped parameter flow models, or alternatively as grey box or black box models. This approach subsumes the flow processes within the aquifer into a small number of "response functions" which represent the overall effect of the aquifer on the properties of interest such as flow. The simplest models simulate output discharge at a spring in terms of measurable inputs such as precipitation and evapotranspiration. The matching of simulated to measured discharge is achieved by functions that provide for both lag and modulation of the input signal. Early models were an outgrowth of similar models used to simulate flow in rivers. More recent models sometimes simulate not only spring discharge but also properties such as water temperature or major ion concentrations. Lumped parameter transport models simulate the transmission of tracers (e.g. fluorescent dyes or tritium) or contaminants (e.g. gasoline). The tracers are interpreted in terms of advection, dispersion in conduits, temporary conduit storage, adsorption on mineral surfaces, and diffusion into the rock matrix.

The fourth class of models are conduit models, employing standard engineering formulae to focus on the hydraulic and geometric properties of conduits, and ignoring the matrix, fracture, and minor channel components of the aquifer. Such models are computationally straightforward, and may provide adequate representation of conduit component of flow in karst aquifers. They allow simulation of the distribution of hydraulic head in conduits, and the resulting flow routing patterns in response to floods. Alternatively, they can investigate flow processes or infer conduit morphology by inverting observations of discharge and head in a particular system. A particular application is in the interpretation of former flow through abandoned conduits, using measurements of the length of erosional scalloping on the conduit walls to obtain a paleovelocity. The product of velocity and conduit cross-sectional area in a cave system gives paleo-discharge and former flow routing.

Various hybrid models can be used to combine these model styles to provide the insight required within the constraints of available data and computing power. Simplified rendition of flow or flow frequency distributions can be linked with conduit models to predict landscape evolution, for example.

Distributed parameter single continuum models such as MODFLOW may be adequate for describing flow and transport in simple porous medium aquifers such as a homogeneous sand. However, flow, transport, and form in carbonate aquifers are far more complicated and cannot be adequately represented in MODFLOW. Unfortunately, it is possible to obtain acceptable simulation of observed hydraulic head with widely different parameter combinations, each model implying a "successful" simulation. However, in general such models will grossly under-represent groundwater and contaminant velocities and well capture zones. Even with more realistic models, karst aquifers are not amenable to being fully and accurately described, and so are unlikely to be accurately simulated by distributed parameter models. However, karst aquifer models remain valuable in allowing exploration of processes under conditions or in places precluding observation, such as in deep conduits, in the past, under exceptional flow, or in response to a development proposal such as damming, enhanced groundwater extraction, or recharge. Such unconstrained modelling, while revealing, is merely a form of disciplined speculation. Yet it is difficult to avoid accepting model predictions as fact, especially when accompanied by appealing visualizations.

STEPHEN R.H. WORTHINGTON AND CHRIS SMART

See also **Speleogenesis: Computer Models**

Further Reading

Gale, S.J. 1984. The hydraulics of conduit flow in carbonate aquifers. *Journal of Hydrology*, 70: 309–27

Hanna, R.B. & Rajaram, H. 1998. Influence of aperture variability on dissolutional growth of fissures in karst formations. *Water Resources Research*, 34: 2843–53

Jeannin, P.Y. 2001. Modeling flow in phreatic and epiphreatic karst conduits in the Hölloch Cave (Muotatal, Switzerland). *Water Resources Research*, 37: 191–200

Klimchouk, A.B., Ford, D.C., Palmer, A.N. & Dreybrodt, W. (editors) 2000. *Speleogenesis: Evolution of Karst Aquifers*,

Huntsville, Alabama: National Speleological Society [particularly the papers in section 4.2 "Modelling of initiation and propagation of single conduits and networks"]

Lauritzen, S.-E., Abbott, J., Arnesen, R., Crossley, G., Grepperud, D., Ive, I. & Johnson, S. 1985. Morphology and hydraulics of an active phreatic conduit. *Cave Science*, 12: 139–46

Oreskes, N., Shrader-Frechette, K. & Belitz, K. 1994. Verification, validation, and confirmation of numerical models in the Earth Sciences. *Science*, 263: 641–46

Palmer, A.N., Palmer, M.V. & Sasowsky, I. (editors) 1999. *Karst Modelling*, Charles Town, West Virginia: Karst Waters Institute (Karst Waters Institute Special Publication, 5)

Smart, C.C. 1988. A deductive model of karst landscape evolution based on hydrological probability. *Earth Surface Processes and Landforms*, 13: 271–88

Teutsch, G. & Sauter, M. 1998. Distributed parameter modelling approaches in karst hydrological investigations. *Bulletin d'Hydrogéologie*, 16: 99–109

GROUNDWATER POLLUTION: DISPERSED

Dispersed pollution, also known as nonpoint source pollution, comes from a wide variety of sources rather than from a single point source such as a pipe or injection well. As rainwater or snowmelt runs across the surface and percolates through the soil it picks up and carries away natural and man-made contaminants. The list of contaminants and sources is long, but can be categorized into some generalized groupings. Dispersed pollutants may include fertilizers, herbicides, and insecticides from agricultural and residential areas; oil, grease, and toxic chemicals from urban and highway runoff; sediment from crop and forest lands and construction sites; salt from irrigation and highway runoff; pathogens and nutrients from livestock, pet wastes, and faulty septic systems; and atmospheric deposition.

Contaminants from dispersed sources enter karst groundwater systems by allogenic or autogenic recharge. Allogenic recharge originates from nonkarst areas; surface streams flowing onto the karst land sink into the karst bedrock carrying any contaminants with the sinking water. Allogenic recharge undergoes very little or no natural filtration and contaminants in the sinking water directly contaminate the groundwater. Autogenic recharge originates from rain falling directly on the karst landscape. The recharge may be dispersed, infiltrating through the soil and percolating through the rock mass, or concentrated by overland flow and throughflow and entering at a point, often at the bottom of a doline (Gunn, 1983). Autogenic recharge waters pick up contaminants from the surface and during percolation through soil. Gibert (1990) describes the various forms of filtration and filtration interactions in karstic and porous aquifers. Autogenic recharge water percolating through soil undergoes natural filtration, but thin, karst soils with highly developed secondary porosity via macropores are limited in their filtration ability. Furthermore, solutionally enlarged joints at the soil–bedrock interface can rapidly transport draining water and contaminants to the groundwater system.

Although not the sole source of dispersed contaminants in groundwater, agriculture is often viewed as a major source of nutrients, organic chemicals, and pathogens. Nitrate is the primary nutrient of concern in groundwater because of its potential health risks such as methemoglobinemia (blue body syndrome), some cancers, and teratogenic effects. Agricultural sources of nitrate include inorganic fertilizers and animal wastes. A linear relationship has been shown between nitrate concentrations in karst springs and percent of land in agricultural land use in the Appalachian region of the United States (Kastrinos & White, 1986; Boyer & Pasquarell, 1995). Nitrate is an anion (negatively charged ion) and is not readily adsorbed to soil particles. Quickly percolating water on karst landscapes reduces the opportunity for denitrification processes to convert nitrate to the more stable ammonium cation and it quickly leaches to the groundwater system. Drinking water standard limits for nitrate nitrogen are set at $10\ mg\ l^{-1}$ in the United States and $11.3\ mg\ l^{-1}$ in Europe. Boyer and Pasquarell (1996) found nitrate concentrations in karst groundwater below animal grazing systems in the Appalachian region of the United States exceeded the drinking water standard on numerous occasions.

Pathogens have received an increasing amount of attention in recent years as a groundwater contaminant. Recent research emphasis has been placed on the human pathogens *Cryptosporidium parvum* protozoa and *Escherichia coli* O157:H7 bacteria, but there are potentially hundreds of pathogens that might be present in groundwater. Tests for faecal coliform bacteria are usually used to indicate the presence of faecal contamination and the possible presence of enteric pathogens. The lack of suitable natural filtration in karst systems has resulted in significant numbers of faecal coliform bacteria contaminating karst ground water as a result of agricultural activity (Pasquarell & Boyer, 1995; Gunn et al., 1998). Water in the United States and Europe does not meet drinking water standards if any total or faecal coliform bacteria are found.

Herbicides and pesticides have been found in some karst groundwaters (Hallberg, 1986; Pasquarell & Boyer, 1996). The appearance of herbicides in the karst groundwater of the intensively cropped Midwestern United States (Hallberg, 1986) is not surprising, but Pasquarell and Boyer (1996) routinely found atrazine in karst groundwater of the central Appalachian karst in the United States, even though the use of agricultural chemicals in that region is minimal because of the lack of widespread cropland. The results led the researchers to conclude that changes in agricultural management practices in that area that would even slightly increase reliance on agricultural chemicals could have dramatic effects on the karst groundwater quality. Drinking water quality standards in the United States place an upper limit of $0.003\ \mu g\ l^{-1}$ for atrazine; the limit on pesticides in Europe is generally $0.1\ \mu g\ l^{-1}$. Once karst groundwater resources are contaminated there may not be any quick remediation. Researchers in Slovenia found atrazine and alachlor still present in that country's groundwater seven years after their use was no longer permitted (Zgarjnar-Gotvajn, Zagorc-Koncan & Tisler, 2001).

Regulations for controlling dispersed pollution of groundwater in the United States are covered under Section 319 of the nation's Clean Water Act. Recent regulation requires the determination of the total maximum daily load (TMDL) of contaminants that individual surface streams can accommodate without exceeding water quality standards. TMDL regulations will only indirectly affect groundwater. In Europe the European Union Parliament and Council have agreed to a proposal by the European Commission for a Water Framework Directive. One objective of the legislation is the protection of all waters, surface and groundwater, and includes strict new standards for the use of fertilizers, manures, and pesticides. Economic incentives and voluntary programs are often used to attempt reduction of pollution from various sources. Best management practices to control diffuse pollution are commonly used. Practices such as nutrient budgeting to reduce the use of fertilizers and manures, composting of manures, storm water management, precision agriculture, and buffer strips are common. Such practices have not necessarily been developed for the special problems associated with karst terrain and may not be effective for protecting karst groundwater quality. Rapid percolation and lack of surface water on karst land means that source areas are expansive and little time is available for arresting contaminant movement to groundwater. Protection of karst groundwater from dispersed pollutants requires a holistic approach with a broad spectrum of treatments and practices (see Groundwater Protection).

DOUGLAS G. BOYER

Works Cited

Boyer, D.G. & Pasquarell, G.C. 1995. Nitrate concentrations in karst springs in an extensively grazed area. *Water Resources Bulletin*, 31: 729–36

Boyer, D.G. & Pasquarell, G.C. 1996. Agricultural land use effects on nitrate concentrations in a mature karst aquifer. *Water Resources Bulletin*, 32: 565–73

Gibert, J. 1990. Behavior of aquifers concerning contaminants: differential permeability and importance of the different purification processes. *Water Science and Technology*, 22:101–08

Gunn, J. 1983. Point recharge of limestone aquifers: A model from New Zealand karst. *Journal of Hydrology*, 61: 19–29

Gunn, J., Tranter, J., Perkins, J., & Hunter, C. 1998. Sanitary bacterial dynamics in a mixed karst aquifer. In *Karst Hydrology*, edited by C. Leibundgut, J. Gunn & A. Dassargues, Wallingford, Oxfordshire: International Association of Hydrological Sciences (IAHS Publication no. 247)

Hallberg, G.R. 1986. From hoes to herbicides: agriculture and groundwater quality. *Journal of Soil and Water Conservation*, 41: 357–64

Kastrinos, J.R. & White, W.B. 1986. Seasonal, hydrogeologic, and land-use controls on nitrate contamination of carbonate ground waters. *Proceedings of the Environmental Problems in Karst Terranes and Their Solutions Conference*, Dublin, Ohio: National Water Well Association

Pasquarell, G.C. & Boyer, D.G. 1995. Agricultural impacts on bacterial water quality in karst groundwater. *Journal of Environmental Quality*, 24: 959–69

Pasquarell, G.C. & Boyer, D.G. 1996. Herbicides in karst groundwater in southeast West Virginia. *Journal of Environmental Quality*, 25: 755–65

Zgajnar-Gotvajn, A., Zagorc-Koncan, J. & Tisler, T. 2001. Estimation of environmental impact of some of the most often occurring pesticides in Slovenian surface and underground water. *Water Science and Technology*, 44: 87–90

Further Reading

Drew, D. & Hötzl, H. (editors) 1999. *Karst Hydrology and Human Activities: Impacts, Consequences and Implications*, Rotterdam: Balkena

White, W.B. 1988. *Geomorphology and Hydrology of Karst Terrains*, Oxford and New York: Oxford University Press

Williams, P.W. (editor) 1993. *Karst Terrains: Environmental Changes and Human Impact*, Cremlingen-Destedt: Catena

Useful Websites

European Union Water Framework Directive, http://projects.dhi.dk/waterdir/Laws/index.htm

UK Department for Environment, Food & Rural Affairs, http://www.defra.gov.uk/environment/water/index.htm

US Environmental Protection Agency, http://www.epa.gov/OWOW/NPS/index.html

GROUNDWATER POLLUTION: POINT SOURCES

Point source pollution is contamination that originates from any discernable or discrete source such as a pipe, ditch, channel, well, container, concentrated animal feed operation, landfill, or industrial storage area. This is in contrast to dispersed or non-point sources such as contaminants in storm water runoff from agricultural or urban lands (see Groundwater Pollution: Dispersed). Many contaminants commonly occur as both point and non-point sources. For example, salt used for road de-icing and spread along roadways may be considered a non-point source, whereas contamination from a road salt storage facility would be considered a point source. Other contaminants that commonly occur as both point and non-point sources are fertilizers, herbicides, and pesticides: from broad application in agricultural and residential areas or as point sources from manufacturing, formulation, and disposal sites.

Point sources may enter the karst groundwater system by either an allogenic or an autogenic source. Allogenic point sources originate in a non-karst area from the discharge of contaminants into surface streams. These then flow onto karst and sink into the bedrock, where they impact groundwater. Autogenic point sources occur on the karst landscape and directly enter the groundwater system as a liquid, such as from a leaking underground storage tank, or may be carried by rainwater runoff (leachate) from an industrial storage or disposal site.

Point sources have been responsible for the most serious cases of groundwater contamination, affecting both human consumers of groundwater and the aquatic fauna. Contaminants detected in groundwater in karst systems have included a wide range of volatile organic compounds (benzene, TCE, PCE), heavy metals (lead, mercury, arsenic), pesticides (DDT, atrazine), and polychlorinated biphenyls (PCBs) (Crawford, 1998; Crawford & Groves, 1995). Particular problems are posed by non-aqueous phase liquids (NAPLs), which may be either dense (D) or light (L) depending upon their specific gravity. DNAPLs descend to, and accumulate at, the base of an aquifer whereas LNAPLs accumulate at the water table. Both DNAPLs and LNAPLs may act as a continuous source of karst groundwater contamination and may be transported relatively rapidly depending upon groundwater velocity and conduit morphology. LNAPLs may accumulate in conduits containing air as water flows from conduit-full to non-full conditions; in essence, the roof of the conduit acts as an oil / water separator. Accumulation of petroleum LNAPLs in conduits has created toxic and even explosive conditions in caves, homes, and schools, with some of the best documented cases occurring in the Bowling Green, Kentucky (United States) area. DNAPLs may collect in low areas in the bottom of active cave streams (or even dry conduits) and then be remobilized when groundwater velocities increase due to storm events.

Pathogens from human or animal waste frequently enter karst as a point source, both deliberately, such as where raw or inadequately treated sewage is discharged into a "soakaway" or drains from inadequate septic tanks, and accidentally such as from leaking sewer pipes. Numerous pathogen outbreaks have occurred in karst as little or no natural filtration occurs in the groundwater system. The Hidden River System in Horse Cave, Kentucky, was contaminated for many years with sewage discharged into a doline from a poorly designed and operated wastewater treatment plant. Most naturally occurring fauna were killed and sewage odours discharging from the cave made the downtown business district intolerable at times. Upon completion of a regional sewage treatment plant and piping of effluent to a surface stream, water quality in Hidden River Cave greatly improved and the cave has been reopened for tourism and recreational use (Gee & Heavers, 2002). More serious incidents in terms of human health include a *Cryptosporidium* (protozoa) outbreak in Braun Station (San Antonio, Texas) in 1984 which caused more than 200 people to become sick (D'Antonio *et al.*, 1995). The acute bacterial contamination (*E. coli* 0157:H7) of karst groundwater in Walkerton, Ontario (Canada) in May 2000 resulted in over 2000 people becoming sick and nine dying (Worthington, Smart & Ruland, 2002). While the impacts of pollution on humans are clear, impacts on cave fauna can easily go unnoticed. In Castleton, Derbyshire (England), leachate entering a doline dramatically reduced the abundance and diversity of aquatic invertebrate taxa although surface sites downstream of the cave resurgence were much less obviously affected (Wood, Gunn & Perkins, 2002).

Many factors control the occurrence and persistence of contaminants in groundwater once the source has been identified and controlled. If contaminants are introduced to groundwater through a thick column of soil, the soil may store the contaminant and release it slowly over time, or in large quantities during storm events. If the contaminant is directly injected into a conduit, little of the material may be held in storage in either the rock matrix or fill; therefore, the site may return relatively quickly to pre-contamination conditions in relation to non-karst aquifers (see also Groundwater Pollution: Remediation).

In the United States, point source contaminants have a long history of regulation through a number of federal and state programs, including:

- Clean Water Act (CWA)—including the national pollution discharge elimination system (NPDES) program which regulates end of pipe discharge for municipal and industrial waste into surface and groundwater systems through a state or federal permit system. A number of states regulate the disposal of wastes into sinkholes (dolines) through the NPDES program.
- Underground Injection Control (UIC) Program—regulates the disposal of hazardous materials into the ground through injection wells. Modified sinkholes may qualify as injection wells under the UIC program. Some cities, such as Bowling Green, Kentucky, are located on rolling sinkhole plains, and utilize injection wells to control storm water runoff in the urban environment. However, the injection wells commonly flush urban non-point runoff into the karst groundwater systems beneath the city.
- Safe Drinking Water Act (SDWA)—includes the source water protection program to protect public drinking water supply wells and springs through control of human activities around well and spring heads as well as in recharge areas. Development in the recharge zone of the Edwards Aquifer, Texas (see separate entry), is controlled through a state water quality protection program. This program requires the identification of all significant karst features and places design criteria such as minimum spacing requirements between karst features and potential pollution sources.
- Resource Conservation and Recovery Act (RCRA)—includes regulation of the transportation, storage, and disposal of hazardous wastes as well as municipal solid waste disposal. A number of hazardous waste sites and solid waste landfills exist in karst areas of the eastern United States and have resulted in widespread contamination in some aquifers and related surface water systems.
- Comprehensive Environmental Response, Compensation, and Liability Act (CERCLA or Superfund Act)—regulates abandoned hazardous waste sites. These are sites where the owner either can not be found or does not have the resources to either investigate or remediate soil and water contamination at the site. Generally, parties that have been identified as contributors to site contamination can be held liable for the site.

GEARY SCHINDEL AND JOHN HOYT

Works Cited

Crawford, N.C. 1988. Karst groundwater contamination from leaking underground storage tanks: Prevention, monitoring techniques, emergency response procedures, and aquifer restoration. In *Proceedings of the Second Conference on Environmental Problems in Karst Terranes and Their Solutions*, Nashville, Tennessee: Association of Ground Water Scientists and Engineers, National Water Well Association, p.213

Crawford, N.C. & Groves, C.G. 1995. Sinkhole collapse and ground water contamination problems resulting from storm water drainage wells on karst terrain. In *Proceedings of the Fifth Multidisciplinary Conference on Sinkholes and the Engineering and Environmental Impacts of Karst*, Gatlinburg, Tennessee, edited by B.F. Beck, Rotterdam: Balkema, p.257

D'Antonio, R.G., Winn, R.E., Taylor, J.P., Gustafson, T.L., Current, W.L. Rhodes, M.M., Gary Jr. G.W. & Zajac, R.A. 1985. A waterborne outbreak of cryptosporidiosis in normal hosts. *Annals of Internal Medicine*, 103: 886–88

Gee, J. & Heavers, D. 2002. *Exploring Caves and Karst: Cave and Karst Curriculum for Teachers*, Horse Cave, Kentucky: American Cave Conservation Association, can also be downloaded from www.cavern.org

Wood, P.J., Gunn, J. & Perkins, J. 2002. The impact of pollution on aquatic invertebrates within a subterranean ecosystem—out of sight out of mind. *Archiv für Hydrobiologie*, 115: 223–37

Worthington, S.R.H., Smart, C.C. & Ruland, W.W. 2002. Assessment of groundwater velocities to the municipal wells at Walkerton, *Proceedings of the 55th Conference of the Canadian Geotechnical Society, Niagara Falls, Ontario*, Canadian Geotechnical Society

GROUNDWATER POLLUTION: REMEDIATION

Groundwater contamination is the degradation of natural water quality as a result of human activities, and pollution occurs when contaminant concentration levels restrict the potential use of groundwater. Groundwater remediation aims to reduce contaminant concentrations to below the threshold standard for the intended use. Potential sources of karst groundwater contamination are discussed in the entries on Groundwater Pollution, and reasons why karst areas are often especially sensitive to contaminants at the land surface are discussed in the entries on Groundwater in Karst and on Groundwater Protection. Remediation of groundwater is always difficult and this entry examines the particular problems of remediating karst aquifers.

A key point for remediation is that karst aquifers commonly exhibit a triple porosity and triple permeability characteristic, with primary porosity occurring in the rock matrix, secondary porosity occurring in fractures and bedding plane partings, and tertiary porosity occurring in conduits. Each porosity type and its related permeability can be highly variable throughout an aquifer creating extremely heterogeneous and anisotropic conditions. Most groundwater storage occurs in the matrix and fractures but the majority of groundwater flow occurs in conduits. Hydraulic conductivity can range over 11 orders of magnitude in a very limited area creating an equally wide range of groundwater velocities within the same aquifer. This combination allows for the rapid movement of contaminants over long distances in karst. Hence there are problems in reactive response, with different parts of the aquifer requiring different remediation strategies. A contaminant release may largely have passed through the conduits before any reaction is possible, but may also slowly discharge from the matrix / fractures at or below detection limits for long periods of time. Computer models based on Darcy's Law are unable to accurately characterize karstic (turbulent) flow through solutionally enlarged fractures (see Groundwater in Karst: Mathematical Models). Consequently, the travel times predicted by most computer models are gross underestimates of true conditions, and the predicted contaminant occurrence shows much greater lateral and longitudinal dispersion (spread) than occurs in flow through a conduit.

Strategies for groundwater remediation are dependent upon a number of factors including the source (point or non-point), duration of release, site characteristics (such as soil thickness and type, depth to groundwater), physical properties of the contaminant, potential flow paths and receptors, health and environmental risks posed by the contaminant, and the resources available to address the problem. In developed countries, remediation of non-point sources, such as urban or agricultural storm water runoff, may be possible through programs to minimize generation of contaminants (e.g. education, limiting or banning the use of specific chemicals in recharge areas, street sweeping and cleaning, and limiting impervious areas in aquifer contributing or recharge areas). Aquifer protection techniques include implementation of best management practices during construction and operation, such as sediment control during construction, grass buffers around dolines, roughing filters on open ponors, storm water retention basins in urban areas, or hazardous materials spill traps on highways. Surface water may be treated at centralized collection points such as in storm water retention basins.

Generally, point sources such as industrial releases are best remediated as close to the source as possible to minimize the volume of water (and soil) to be treated and decrease remediation time. Groundwater remediation from point sources begins by controlling the contaminant source through either removal or containment. After source controls have been initiated, a site characterization is usually performed to determine exposure pathways, physical properties of the contaminants, and extent of migration of contaminants. Results from the site characterization are applied in a risk assessment to determine the potential for contaminants to impact public health and the environment. If a sufficient risk is present, a feasibility study is usually performed to determine the optimum remediation strategy. This may include a wide range of treatments including no action, removal of soils for treatment, soil vapour extraction, natural attenuation, pumping and treating of groundwater, *in situ* bioremediation, *in situ* groundwater treatment (neutralization or biodegradation), or a combination of methods. Groundwater treatment systems may be placed at or near the location of the spill, at the boundary of the property (regulatory point of compliance), at a spring or public water supply well (point of withdrawal), or at a domestic well (point of use). In some karst environments, remediation may be termed "technically impractical" because of the difficulties of defining and intercepting groundwater flow paths.

Dolines and sinking streams often provide targets for discharge of industrial and sewage waste. For example, the effluent (termed "dunder") from a rum distillery near the town of Wakefield (Jamaica) had choked a succession of stream-sinks, flooding an extensive area of karst, contaminating a series of irrigation and domestic wells and springs, and threatening the groundwater supply of much of northern Jamaica. Similarly, the discharge of heavy metals and sewage effluent into a doline in Horse Cave, Kentucky (United States), resulted in the contamination of the Hidden River Cave System (see Groundwater Pollution: Point Sources for detail). In both Jamaica and Kentucky no attempt was made to remediate the aquifers as such but when the contaminant inputs were finally controlled both groundwater systems recovered remarkably rapidly in quality and ecology. The implication seems to be that the conduit flow may limit the penetration of contaminants into the rock matrix and that rare karst organisms may survive in uncontaminated parts of the aquifer, ready to return when conditions permit. However, in other cases, contamination of karst has been severe and sustained.

Commonly there are intermittent problems with pathogens in karst groundwater supplies and immediately adjacent septic systems or manure accumulations are often responsible. In most cases the only effective remediation is treatment at the point of withdrawal although improved management may help to minimize the pathogen loading to the groundwater system. For example, nutrient (manure) management planning, development of buffer strips along streams and around stream-sinks, and installation of roughing filters in dolines can significantly reduce pathogens entering the aquifer. In addition, it is likely to be beneficial to use centralized sewage treatment rather than individual septic systems, and to construct storm water retention drains and retention ponds that limit contaminated storm water from entering the aquifer.

Pathogens can be derived from many places, and even the most careful genetic fingerprinting may fail to reveal a source. *Cryptosporidia* was first identified as a water-borne pathogen in the United States during an outbreak that occurred in the Edwards Aquifer, Texas, in 1984. Unfortunately, this pathogen often survives conventional water chlorination, and expensive filtration may be required. In Walkerton, Ontario (Canada), acute bacterial contamination of groundwater by *E. coli* 0157: H7 killed nine people and affected thousands of others in May 2000. The aquifer was not obviously karstic, and sources immediately adjacent to town wells were implicated. However, the pattern of contamination in domestic wells was quite irregular, and flow in wells focused in discrete horizons. Subsequent tracer testing revealed groundwater velocities almost two orders of magnitude greater than those suspected from computer modelling, indicating a karst aquifer (Worthington, 1999). As a result, there remains considerable uncertainty as to the source of contamination. No remediation has been attempted, but elaborate and expensive protection of the water supply has been undertaken.

The discharge of polychlorinated biphenyls (PCBs) into a pond and adjacent drainage ditch in Russellville, Kentucky (United States) resulted in contamination of a groundwater / surface water system and a ban on fish consumption was required along more than 100 km of the adjacent river (Kentucky Division of Water, 1986). Remediation of the site included removal of the pond, installation of a large and elaborate "pump and treat" system, and the excavation and removal of sediments along more than 1 km of surface stream channel. Groundwater remediation has been actively occurring for more than 10 years.

Remediation of non-aqueous phase liquids (NAPL) in conduits is extremely difficult. Attempts to develop remediation systems at discharge points (springs) are commonly limited by physical accessibility, surface water flooding, and sizing of treatment systems to address the high and low ranges in spring discharge. Some success has been obtained from traditional "pump and treat" systems and soil vapour extraction in the pollution source area but results can be highly variable depending on local site conditions. A major spill of industrial solvents near Nashville, Tennessee (United States) resulted in a LNAPL layer several metres in thickness. Sustained pumping and removal, followed by establishment of air stripping on the site moderated but did not solve the problem, and a film of solvents could be seen emerging from a local creek for several years afterwards (Quinlan, 1990). The implication is that remediation wells in karst are unlikely to adequately link into conduits and storage zones.

Gasoline, diesel fuel, and aviation fuel are common contaminants in karst. These are complex LNAPLs, with low solubility, but low pollution thresholds. Leaking underground storage tanks at gasoline stations or chronic spillage at airfields are the most common sources. However, it may not always be apparent which of many potential sites is the source, and the fractionation of the many compounds in common fuels makes them difficult to fingerprint in groundwater. Complex and difficult tracer testing may be required to identify the source as LNAPLs in karst are often routed idiosyncratically, especially in the highly weathered epikarst in the shallow subsurface. Some aquifers are chronically contaminated by fuels. Anticipatory regulation in the location, testing, and design of fuel storage and handling may be more effective than remediation.

GEARY SCHINDEL, STEVE JOHNSON, AND CHRIS SMART

Works Cited

Kentucky Division of Water, 1986. Report to Congress on Water Quality in Kentucky

Quinlan, J.F. 1990. Special problems of ground-water monitoring in karst terranes. In *Ground Water and Vadose Zone Monitoring*, edited by D.M. Nielsen & A.I. Johnson, Philadelphia: American Society for Testing and Materials (ASTM Special Technical Report 1053)

Worthington, S.R.H. 1999. A comprehensive strategy for understanding flow in carbonate aquifers. In *Karst Modeling*, edited by A.N. Palmer, M.V. Palmer & I.D. Sasowsky, Charles Town, West Virginia: Karst Waters Institute (KWI Special Publication 5)

Further Reading

Bedient, P.B., Rifai, H.S. & Newell, C.J. 1999. *Groundwater Contamination: Transport and Remediation*, Upper Saddle River, New Jersey: Prentice Hall

Driscoll, F.G. 1986. *Groundwater and Wells*, 2nd edition, St. Paul, Minnesota: Johnson Division

Fetter, C.W. 2001. *Contaminant Hydrogeology*, 4th edition, Upper Saddle River, New Jersey: Prentice Hall

Groundwater Monitoring and Remediation, National Ground Water Association, (www.ngwa.org).

1994. *Remedial Technologies Screening Matrix*, US Department of Commerce, National Technical Information Service (www.ntis.gov; PB95-10482, EPA/542/B-94/013)

GROUNDWATER PROTECTION

As almost 25% of world's population obtains its water from karst aquifers, it is vital that they are protected from degradation by contamination of the groundwater. More than 70 methods of mapping groundwater vulnerability have been either proposed or implemented. The methods range from statistical-mathematical, to parametric system models, to hydrogeologically based types. They also vary in terms of the purpose for which they were designed, for example land-use planning, protection zoning, and resource or source protection. Most methods for assessing the vulnerability of groundwater were designed with homogenous / isotropic aquifers in mind (for example, the American DRASTIC and the British GOD), and few make specific reference to karst terrains. [DRASTIC = Depth to water table, net Recharge, Aquifer media, Soil zone, Topography, Impact of vadose zone media, hydraulic Conductivity (Aller et al., 1987); GOD = Groundwater occurrence, Overall aquifer class, Depth to groundwater table (Foster, 1987)]. A modification to DRASTIC has been proposed which is designed to increase the sensitivity of the method to karst conditions: KARSTIC (Davis & Long, 2002). The terms A, R, S, T, I, and C are as per DRASTIC. K is a term describing the occurrence of swallow holes and fractures. The parameters are again rated and weighted. The overall effect is to markedly increase the vulnerability of areas adjacent to swallow holes and similar karst features.

Other methods, such as the method used in the Republic of Ireland, incorporate some recognition of the particularities of karst groundwater (Daly & Drew, 1999). The Irish method takes account of karst features, such as sinking streams, by incorporating a buffer zone of extreme vulnerability around them. In addition, risk assessment is applied: aquifers are subdivided according to degree of karstification, with greater protection being accorded to regionally important aquifers with a high degree of karstification than to (for example) poor aquifers with a low degree of karstification.

Resolving the difficulties involved in protecting the valuable water resources of vulnerable karst regions requires the use of a karst-specific methodology, which address the hydrogeological singularities of karst flow systems, such as thin soils, point recharge, minimal filtration in the unsaturated and saturated zones, and rapid flow rates. For example, in conventional hydrogeological settings, a water source (spring or borehole) has a protection zone designated over some or all of the catchment area for that source. In the protection zone, more stringent land-use regulation is applied, to lessen the risk of contaminants reaching the source. The outer limits of these protection zones are often defined in terms of time of travel and internationally they range from 10–400 days. Applying such a method to many karst aquifers would be impracticable, given the rapid rates of underground flow, as the resulting protection areas would be impossibly extensive. Attempts have therefore been made, particularly in Europe, to develop methods that can be used successfully in karsts. An important methodology is the EPIK system, developed in Switzerland and intended particularly for use in the karst of the Jura mountains (Doerfliger et al., 1999). The method maps intrinsic vulnerability and delineates source protection areas (the target is the source) rather than resource protection (in which the target would be the water table). Four parameters are taken into account: the degree of development of the Epikarst layer (e.g. dolines, karrenfields, swallow holes); the Protective cover; the Infiltration conditions (e.g. sinking streams); and the degree of Karst development of the aquifer. Protective cover is a relevant parameter, irrespective of rock type, but the other three parameters used in EPIK are of particular significance in karst areas. The assessed values for each parameter are mapped separately, rated, weighted and then combined to give a final numerical value (protection index) for each mapped unit. The index is then divided into four vulnerability classes and thence into four protection zones. In Slovakia, a variant on EPIK, termed REKS, was developed to suit local conditions. However, when EPIK has been applied outside of the region for which it was created, results have often been unsatisfactory.

A European-wide effort to address the problem of karst groundwater protection was the COST Action 65 1991–95 (COST means European Co-operation in the Field of Scientific and Technical Research) which was devoted to hydrogeological aspects of groundwater protection in karstic areas. Based on national studies, this action established an inventory of national regulations concerning the management of karst waters and made recommendations for management measures. A succeeding initiative, COST Action 620, (1997–2003) attempted to develop an objective methodology for intrinsic and specific vulnerability assessment and risk assessment in karstic environments. This approach was intended to be broad enough to encompass all the conditions found in European karst areas, but to be sufficiently flexible to be customized for individual karstic regions. The method consists of "core" and "peripheral" (bolt-on) components. Core components consist of a factor "O", expressing the degree of protection to contamination offered by those layers (soil, unconsolidated materials, non-karstified rocks, unsaturated zone of the karst), that may overlie the water table in the karstic aquifer and a factor "C" that describes the degree to which runoff is concentrated at or near the surface, into flow which then bypasses the protective layers, reaching the aquifer as point recharge via dolines or sinking streams. The two peripheral factors are a precipitation / recharge factor "P" which may be added to modify the "O" and "C" factors, in the light of the precipitation regime of a specific area and a factor "K" that describes the character of the sub-water table karst drainage system in terms of diffuse to conduit flow. The "K" factor is added where source protection rather than resource protection is required. Quality assurance / validation is applied via independent assessment of the whole karstic system (e.g. by spring hydrograph analysis) and comparison with the vulnerability maps. Variants on this pan-European approach are being developed to suit the needs of individual countries with karstified aquifers.

In the United States, protection of karst waters operates largely at state and local levels. States with extensive karst, such as Kentucky and Tennessee, have developed specific protection and legislative responses to karst conditions.

DAVID DREW and SUZANNE DUNNE

See also **Environmental Impacts Assessments**

Works Cited

Aller, L., Bennet, T., Lehr, J.H., Petty, R.J. & Hackett, G. 1987. *DRASTIC: A Standardized System for Evaluating Groundwater*

Pollution Potential using Hydrologic Settings, Ada, Oklahoma: Office of Research and Development, Environmental Protection Agency (EPA-600/2–87/035)

Daly, D, & Drew, D. 1999. Irish methodologies for karst aquifer protection. In *Hydrogeology and Engineering Geology of Sinkholes and Karst*, edited by B. Beck, A. Pettit & J. Herring, Rotterdam, Balkema

Davis, A.D. & Long, A.J. 2002. KARSTIC: A sensitivity method for carbonate aquifers in karst terrains. *Environmental Geology*, 42: 65–72

Doerfliger, N., Jeannin, P.Y. & Zwahlen, F. 1999. Water vulnerability assessment in karst environments: a new method of defining protection areas using a multi-attribute approach and GIS tools (EPIK method). *Environmental Geology*, 39(2): 165–76

Foster, S.S.D. 1987. Fundamental concepts in aquifer vulnerability, pollution risk and protection strategy. In *Vulnerability of Soil and Groundwater to Pollutants*, edited by W. van Duijvenbooden & H.G. van Waegeningh, The Hague: TNO Committee on Hydrological Research

Further Reading

Daly, D., Dassargues, A., Drew, D., Dunne, S., Goldscheider, N., Neale, S., Popescu, I.C. & Zwahlen, F. 2002. Main concepts of the "European Approach" to karst-groundwater-vulnerability assessment and mapping. *Hydrogeological Journal*, 10(2): 340–45

Drew, D. & Hötzl, H. (editors) 1999. *Karst Hydrogeology and Human Activities: Impacts, Consequences and Implications*, Rotterdam: Balkema

Gogu, R.C. & Dassargues, A. 2000. Current and future trends in groundwater vulnerability assessment. *Environmental Geology*, 39(6): 549–59

Vrba, J. & Zaporozec, A. 1994. *Guidebook on Mapping Groundwater Vulnerability*, Hannover: Heise

GUANO

Guano consists of the accumulated droppings of animals, usually bats, but occasionally of cave-dwelling birds, such as the swiftlets (*Collocallia*) of Southeast Asia, or the oilbirds (*Steatornis caripensis*) of South America. The term originates from the Quechua word *huana* for "dung", and originally applied to enormous accumulations of seabird excreta that sustained a 19th-century nitrate-mining industry on the islands of the Humboldt Current, off the coast of Peru. Bat guano accumulations can be equally extensive. At Carlsbad Caverns, New Mexico, some 92 million kg of guano were reportedly mined between 1903 and 1923 (Geluso, Altenbach & Kerbo, 1987). Niah Great Cave, Sarawak, contained at least 29 million kg of guano and related decomposition products (Wilford, 1951), while some 50 000 m³ of guano remain in the caves of Mona Island, West Indies, despite extensive mining of the thickest deposits (Peck & Kukalova-Peck, 1981). The original size of the deposits was estimated at 462 000 tonnes.

The size of guano accumulations is a function of the size of the bat colony, the length of time over which accumulation has taken place, and physical characteristics of the cave which determine rates of decomposition and removal. At Bracken Cave, Texas, a summer population of some 20 million Mexican Free-Tailed bats (*Tadaraida brasiliensis*) deposit an estimated 50 000 kg of guano annually (Barbour & Davis, 1969). At Eagle Creek Cave, Arizona, the same species of bat adds as much as 16 cm a year to the existing deposit (Altenbach & Petit, 1972), which shows no appreciable decomposition over timescales of at least 50 years.

Large guano deposits are not homogeneous, but undergo physical compaction, biological degradation, and chemical alteration with time and burial. At Niah Great Cave, Sarawak, Wilford (1951) identified an indistinct stratigraphy consisting of 0.7 m of "fresh" guano, supporting a thriving arthropod community, which merged into 3 m of finely textured "fossil guano", and finally 0.8 m of "rock phosphate". This sequence of deposition is typically accompanied by a progressive loss of organic content and a proportional increase in insoluble phosphatic residue. In very dry caves, such as Tramway Cave, Arizona, bat guano deposits may be preserved for more than 12 000 years without appreciable decomposition. This is possible because insectivorous bat guanos consist primarily of small fragments of insect chitin, which is highly resistant to chemical attack. Moist conditions are necessary to support decomposing micro-organisms capable of producing the enzyme chitinase.

Low rates of chitin decomposition make guano accumulations valuable repositories of paleoclimatic information. Guano chitin preserves a carbon stable isotope signature of the surface community from which it was derived, and this has been used to track climate change over thousands of years (Mizutani, McFarlane & Kabaya, 1992). The carbon content of guano chitin also makes it a convenient material for ^{14}C dating, and the dating of distinctive guano layers in cave sediments profiles has found use in studies of paleontological remains from caves.

Decomposition of bat guano and the bat urine associated with it produces free ammonia, which in Cueva del Tigre, Mexico, can reach 1200 ppm—far above the level of human tolerance (McFarlane, Keeler & Mizutani, 1995). In moist caves, some of this ammonia may be converted to nitric acid by bacterial action, which in turn reacts with the underlying bedrock to produce a suite of nitrate minerals (see Sediments: Biogenic).

Guano deposits associated with fruit-eating bats are common in tropical regions of the world. They are quite different to insectivorous guanos, consisting of dropped and excreted seeds, leaves, and twigs. Many of these seeds germinate in the deposit, producing spindly, pale seedlings, which die after a few weeks from lack of light for photosynthesis. Fruit bat guanos rarely accumulate to great depths or volumes, because decomposition of plant debris is much more rapid than the decomposition of insect chitin.

Unique to the neotropics, vampire bats (*Desmodus, Diaemus,* and *Diphylla*) are cave- and hollow tree-dwellers that feed exclusively on the blood of vertebrates, taking as much as 40% their own bodyweight at a feeding. Blood is a bulky food to carry in flight, so the bats excrete much of the water content within 20–30 minutes of feeding. The guano of these animals that subsequently accumulates in caves is a sticky, tar-like material of concentrated and partially digested blood solids, with a distinctive, ammonia-rich odour.

The South American Oil-Bird, *Steatornis*, is a fruit eater whose guano is very similar to that of fruit-eating bats. A large colony in Cueva del Guácharo, Venezuela, has been famous since it was first visited by Alexander von Humboldt in 1799 (De Bellard Pietri, 1957). In contrast, many caves in Southeast Asia support very large populations of cave swiftlets. Bird guano accumulations can be extensive in these caves, but differ little from insectivorous bat guano deposits that often occur in the same caves.

Actively decomposing bat guano accumulations support thriving communities of invertebrates. Detritus-feeding mites are typically present in enormous numbers, preyed upon by predatory mites and pseudoscorpions. These micro-arthropods support rich populations of macro-invertebrates, of which cockroaches, crickets, and predatory tailless whip scorpions predominate. Temperate zone bat guano ecosystems tend to support a lower diversity of macro-invertebrates. In tropical guano ecosystems, productivity may be high enough to support small vertebrates, usually frogs.

DONALD A. MCFARLANE

See also **Aves (Birds); Organic Resources in Caves; Sediments: Biogenic**

Works Cited

Altenbach, J.S. & Petit, M.G. 1972. Stratification of guano deposits of the Free-Tailed bat, *Tadarida brasiliensis*. *Journal of Mammalogy*, 53: 890–93

Barbour, R.W. & Davis, W.H. 1969. *Bats of America*, Lexington: University Press of Kentucky

De Bellard Pietri, E. 1957. El Guácharo. *Boletin de la Sociedad Venezolana de Ciencias Naturales*, 18: 1–4

Geluso, K.N., Altenbach, J.S. & Kerbo, R.C. 1987. *Bats of Carlsbad Cavern National Park*, Carlsbad, New Mexico: Carlsbad Caverns Natural History Association

McFarlane, D.A., Keeler, R.C. & Mizutani, H. 1995. Ammonia volatization in a Mexican bat cave ecosystem. *Biogeochemistry*, 30: 1–8

Mizutani, H., McFarlane, D.A. & Kabaya, Y. 1992. Carbon and nitrogen signatures of bat guanos as a record of past environments. *Mass Spectroscopy*, 40(1): 67–82

Peck, S.B. & Kukalova-Peck, J. 1981. The subterranean fauna and conservation of Mona Island, Puerto Rico. *National Speleological Society Bulletin*, 43: 59–68

Wilford, G.E. 1951. The phosphate deposits in the Niah Caves. In *Annual Report of the Geological Survey Department*, Kuchin, Sarawak: Government Printing Office (for the Geological Survey Department, British Territories in Borneo)

Further Reading

Hutchinson, G.E. 1950. Survey of existing knowledge of biogeochemistry. 3. The biogeochemistry of vertebrate excretion. *Bulletin of the American Museum of Natural History*, 96: 1–554

GUNPOWDER

For hundreds of years, gunpowder was a mixture of which the main ingredient, saltpetre (KNO_3), was sometimes extracted from caves. Although the Chinese were probably familiar with it, the black powder was unknown in Europe until Roger Bacon (1214–94) of England, discovered the formula about 1248. The other components, sulfur and charcoal, together comprised a fourth of the compound (Pierce, 1952; Lewis, 1956).

There have been other uses, such as a preservative, but the primary importance of saltpetre has been its utilization in gunpowder. The dirt in dry caves is one source of saltpetre, which can be obtained by leaching the sediments with water. The source of nitrates in the saltpetre earth is groundwater seeping into a cave from the surface: dry caves act as receptacles where the soluble nitrate can accumulate rather than be leached away. Other sources for saltpetre have been soils of certain warm countries, especially India's Bahar Province, long controlled by England; cellar walls, stables, the dirt underneath old buildings; and artificial composting beds of decaying animal and vegetable matter (Faust, 1949). In most cases, the soil or product had to be leached in hoppers and the resulting solution treated with an alkali to cause calcium to be precipitated and potassium to be substituted.

It was learned very early on that saltpetre could be found in caves. In 1490, a German cave was explored in search of it, and the Italian Niccolo Machiavelli (1469–1527) noted that one place saltpetre could be obtained was in "desolate caves where raine can not come in" (Hill & Forti, 1997; Faust, 1949). Although mining of saltpetre from caves is best documented in North America, caves in many other parts of the world have also been mined or described as containing saltpetre. These caves or localities include Pulo di Molfetta, Italy; Grotte des Eyzies and Grotte des Espelunges, Departments of Dordogne and Hautes-Pyrenees, France, mined in 1793; Germany, near Hamburg; Minas Gerais and other areas of Brazil; Jamaica; Sri Lanka (formerly Ceylon) "where 22 nitriferous caverns are mentioned" (from Antisell, 1852, p.392); the island of Mindanao, Philippines; St John's in the Cape Verde Islands; and Guizhou Province of China.

The most widespread use of caves for saltpetre occurred in the United States, for about 125 years, beginning in the mid–1700s. Within this time the activity intensified during the wars of 1775–83, 1812–15, and 1861–65. Even though cave saltpetre was noted by a 1700 French Mississippi River expedition (Shaw, 1992), its presence in the American colonies was generally unknown until the limestone regions of western Virginia were settled. A few caves were probably mined there by 1740 (Faust, 1964).

The munition needs of the Continental forces, during the American Revolution, led to increased mining of frontier caves, although their overall contribution remained small. Most of the powder and its elements were imported, and possibly only 3 to 5% of the saltpetre used was derived from caves. Even so, beginning in 1775, Charles Lynch and others made 11 000 pounds (about 5000 kg) from a single southwestern Virginia cave (Jefferson, 1982 [1787]). After the war, Thomas Jefferson estimated that 50 caves were worked in the Greenbrier region of what is now West Virginia (Jefferson, 1982 [1787]). Although that, and other Virginia locales, remained productive, by the early 1800s the centre of America's saltpetre industry moved west to Kentucky and Tennessee.

The early 19th century was the second epoch of concentrated saltpetre mining in US caves. This was especially true from 1808 to 1815, when foreign supplies were disrupted by British and French orders, a series of congressional acts, and war with England. Importation of cheap Indian saltpetre fell, and large powder mills, such as DuPont in Delaware, sought domestic supplies, causing the price to rise dramatically. Lexington, Kentucky, and Nashville, Tennessee, became central markets for trade in this commodity (O'Dell, 1995). The incomplete 1810 manufacturing census reported that the areas annually making the most saltpetre were Kentucky, middle Tennessee, and Virginia, with 202 000, 145 000, and 59 000 pounds, respectively. During the War of 1812, the output increased, with Kentucky alone contributing 400 000 pounds a year.

Samuel Brown (1769–1830), a Lexington saltpetre trader, wrote an article in 1805, in which he estimated the potential yield of several Kentucky caves, including Great Saltpetre Cave. He also explained how saltpetre was obtained from sandstone rock shelters, which often produced ten to twenty pounds per bushel (about 130–260 kg m^{-3}), compared with one to three pounds per bushel of cave earth. The most ambitious cave mining efforts, during the early 1800s, were at Great Saltpetre and Mammoth in Kentucky, and Big Bone in Tennessee. These were very elaborate works, employing 20 to 100 workers. Great Saltpetre is thought to have out-produced all caves of that day, while in 1813, Big Bone yielded 500 pounds per diem (George, 1988).

When peace returned in 1815, the US saltpetre market collapsed. It again became cheaper for the large eastern powder mills to import Indian saltpetre than to pay transportation costs from inland America. The scale of saltpetre mining was greatly reduced and though it lingered for decades, by 1860 the practice had practically ceased. However, the supply-starved Southern Confederacy during the American Civil War of 1861–65 inspired one last massive mining of caves for saltpetre. In April 1862, after initial efforts by Confederate and some state governments proved inadequate, a Nitre Bureau was established to more efficiently work the caves and to build nitre beds. The South was divided into fourteen districts, with a superintendent for each. Major Isaac M. St John (1827–80) was appointed chief, with headquarters in Richmond, Virginia, and a division office was established at Augusta, Georgia, for the better distribution of materials (Smith, 1990). The Bureau directly worked over 55 caves at one time or another. Sauta Cave, Alabama, was the largest operation, with up to 195 employees. Contracts for saltpetre were made with perhaps 700 individuals. Although the majority of these contractors leached dirt from underneath old houses and barns, undoubtedly they also mined several hundred more caves (Smith, 1990).

The Confederacy's total output of saltpetre is unknown. At the end of the third quarter, 1864, 1 735 000 pounds were reported, of which nearly 29%, or 512 000 pounds, came from caves (United States War Department, 1901). The Bureau continued to function until the end of the war, April 1865, when saltpetre mining from caves in America ended.

MARION O. SMITH

See also **Sediments: Biogenic**

Works Cited

Faust, B. 1949. The formation of saltpetre in caves, *NSS Bulletin*, 11: 17–23

Faust, B. 1964. *Saltpetre Caves and Virginia History*. In *Caves of Virginia*, edited by H.H. Douglas, Falls Church, Virginia: Virginia Cave Survey

George, A.I. 1988. Interim chronology of historic events at Great Saltpetre Cave, Rockcastle County, Kentucky, *The Journal of Spelean History*, 22(2): 7–11

Hill, C.A. & Forti, P. 1997. Nitrates. In *Cave Minerals of the World*, 2nd edition, Huntsville, Alabama: National Speleological Society

Jefferson, T. 1787. *Notes on the State of Virginia*, edited and with an introduction by W. Peden, Chapel Hill, North Carolina: University of North Carolina Press, 1982

Lewis, B.R. 1956. *Small Arms and Ammunition in the United States Service*, Washington, DC: Smithsonian Institution

O'Dell, G.A. 1995. Saltpeter manufacturing and marketing and its relation to the gunpowder industry in Kentucky during the nineteenth century. In *Historical Archaeology in Kentucky*, edited by K.A. McBride, W.S. McBride and D. Pollock, Frankfort, Kentucky: Heritage Council

Pierce, J.C. 1952. History of explosives, *Compressed Air Magazine*, 57(1): 2–8

Shaw, T.R. 1992. Cave exploration in America and cave exploration elsewhere in the world. In *History of Cave Science: The Exploration and Study of Limestone Caves, to 1900*, 2nd edition, Broadway, New South Wales: Sydney Speleological Society

Smith, M.O. 1990. *Saltpeter Mining in East Tennessee*, Maryville, Tennessee: Byron's Graphic Arts

United States War Department, 1901, *The War of the Rebellion: A Compilation of the Official Records of the Union and Confederate Armies*. 70 vols, Washington: Government Printing Office, 1880–1901 (ser 4, vol. 3: 698)

Further Reading

NSS Bulletin 43(4), 1981 is a special issue on American saltpetre caves

Faust, B. 1955. Saltpetre mining tools used in caves, *NSS Bulletin* 17: 8–18

George, A.I. 1988. Pre-1815 demise of the domestic saltpeter industry in Kentucky. *The Journal of Spelean History*, 22(2): 15–20

George, A.I. 2001. *The Saltpeter Empires of Great Saltpetre Cave and Mammoth Cave*, Louisville, Kentucky: HMI Press

O'Dell, G.A. 1989. Bluegrass powdermen: A sketch of the industry, *Kentucky Historical Society Register*, 87(2): 99–117

Smith, M.O. 1983. The Sauta Cave Confederate niter works, *Civil War History*, 29(4): 294–315

Smith, M.O. 1997. In quest of a supply of saltpeter and gunpowder in early Civil War Tennessee, *Tennessee Historical Quarterly*, 56: 96–111

H

HA LONG BAY, VIETNAM

From Haiphong to the Chinese border, at the northern end of Vietnam's long coastline, thick, folded Carboniferous and Permian limestones are carved into a host of spectacular and rocky islands. The finest of these are in and around Ha Long Bay, where nearly 2000 limestone islands essentially represent a drowned extension of the great karst terrains of China (see China entry). The hundreds of rocky islands that form the most beautiful and famous landscapes in the bay are individual towers in a classic fenglin karst where the intervening plains have been submerged by the sea (Figure 1). Many of the towers have vertical walls with openings into truncated caves. The smaller numbers of larger islands are clusters of steep limestone hills that are typical of fengcong karst—fragments of a spectacular karst that had not evolved through to the fenglin landforms. The largest block of fengcong karst forms Cat Ba Island on the western side of the bay.

Virtually all the limestone hills are covered by low but dense vegetation, mostly rooted in organic soils within dissolutionally opened fissures. Some of the fissures drop into caves, but most are too small for human access. All exposed rock surfaces are covered with blue-green algae, and are fretted into karren runnels between razor-sharp ridges. Deep dolines lie between the fengcong hills on the larger islands, and some contain saltwater lakes that are linked to the open sea by the caves.

This mature karst landscape has been modified by marine erosion since being invaded by the sea. The most obvious effect is the incision of deep and extensive sea-level notches. These are mainly dissolutional as they are equally well developed on cliffs that are sheltered from any major wave action, though the deepest notches do lie along the exposed south coasts. Notch expansion continues the original karstic process of vorfluter and lateral planation, and therefore creates the most spectacular of the limestone towers. The process is active, and rockfalls are numerous along the cliffs. Many of the karstic limestone hills have been thinned to uncharacteristic arête ridges by cliff retreat in the face of marine erosion. There are fewer foot caves than might be expected in such a mature karst. Smaller islands are only narrow towers with domed tops with minimal soil and no fissures, so that rainfall washes over the tower without significant penetration. Many of these have neither notching nor foot caves. Larger islands, with soil cover and vegetation, generally have well-developed notches and many foot caves (mostly fissures too narrow to enter). This suggests that mixing corrosion is critical to the development of both notches and foot caves, where only the larger islands gather enough fresh water from direct rainfall to promote significant mixing with the salt water. On the small tower islands, notches (and cliff retreat) develop faster than foot caves.

A few of the marine notches extend back into foot caves that reach through the limestone ridges. The finest of these can be followed through, by boat at low tide, into lakes that occupy drowned dolines within the fengcong karst of the larger islands. The Ho Ba Ham caves in the western side of Dau Be island (Figure 1) have at least three cave tunnels that provide access to three beautiful lakes (Waltham, 2000). These lakes are otherwise isolated within the inhospitable terrain of thorn vegetation and sharp karren on the surrounding karst hills.

True karstic foot caves are more numerous at higher levels, where they are abandoned from earlier phases of lateral dissolution on alluviated inland plains. Some of these also extend through the islands, but many close down to smaller or choked tubes. The islands also house many fragments of old phreatic cave systems. These are recognized by the vertical relief within their undulating passages, though many do have large flat ceilings planed by dissolution across the geological structure. They lie at up to 50 m above sea level, and some are truncated in the sides of dolines within the fengcong karst; the longest caves yet found each have only about 300 m of passage.

The remnants of phreatic caves are one part of the evidence of a long karstic evolution of Ha Long Bay before it was invaded and modified by the sea. Fluvial channels entrenched in the drowned karstic plain, and now recognizable as channels of deep water between some of the islands, appear to date from rejuvenation when sea level was lower during Pleistocene cold stages.

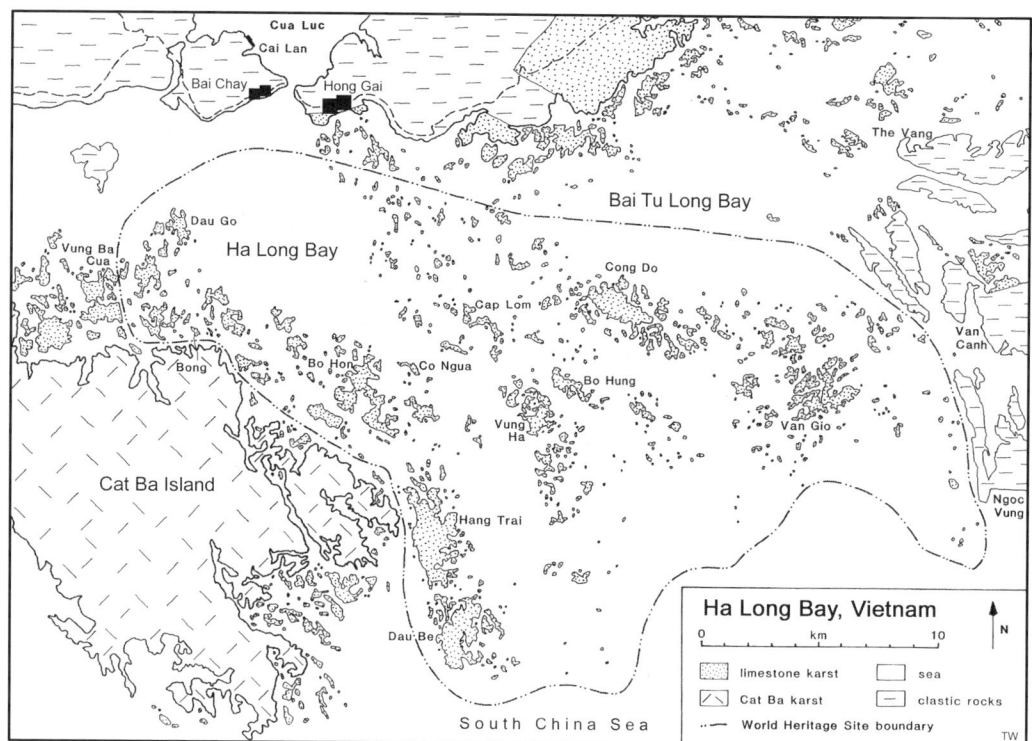

Ha Long Bay: Figure 1. Outline map of Ha Long Bay, Vietnam. Many of the smaller islands cannot be shown at this scale.

How much the karst was modified by the sea during Pleistocene stages of high sea level remains open to debate until the tectonic history of the area is better understood.

Though primarily important for its karst geomorphology, Ha Long Bay is also notable for its biology, and there is a particularly rich population of snails including many species adapted to the cave environment. A single sample from one cave included 14 different species—a truly remarkable level of biodiversity. Most of the islands have only relatively small areas of vegetation but the flora also demonstrates a considerable biodiversity. This is largely due to the great diversity of small habitats each with a distinctive topography and microclimate. There is also significant archaeological material in the bay, notably in foot caves. Limited archaeological research to date has already demonstrated that the islands were probably inhabited from the early Stone Age, and certainly since the Neolithic Age. During the rise of Buddhism, Ha Long Bay became recognized as an important centre and so a number of significant cultural sites have been identified from this period.

Ha Long Bay is a listed World Heritage Site, and appropriate conservation measures are now in place. It is a major tourist site, based on the resort town of Bai Chay (Figure 1). Most visitors take boat tours around the bay, calling at sites that are singularly beautiful (Figure 2). Visitation is therefore easily managed, and a selection of caves has been developed with boardwalks for tourist access. There is also a large indigenous population living on boats in sheltered parts of the bay. They catch seafood that is renowned for its quality, but their destructive habit of gathering and selling coral has now been almost completely stopped.

Hong Gai is the main city within the bay, and is the main port for shipping that takes coal from huge open-pit coal mines just inland. There is a continuing need to reduce pollution of the bay from out-dated coal handling procedures, and the expansion of Hong Gai into a major container port requires sensitive planning. The shipping traffic that is an essential and unavoidable component of the regional economy can only have a detrimental impact on the bay's marine biology, but fortunately has almost no impact on the karst geomorphology. The sight of an ocean-going freighter gliding between the limestone towers is one of the enduring images of Ha Long Bay.

TONY WALTHAM AND ELERY HAMILTON-SMITH

Ha Long Bay: Figure 2. Limestone tower islands in Ha Long Bay framed by the entrance of Bo Nau Cave. (Photo by Tony Waltham)

Further Reading

Waltham, T. 2000. Karst and caves of Ha Long Bay. *International Caver*, 2000: 24–31

HAWAII LAVA TUBE CAVES, UNITED STATES

The Hawaiian Islands are the world's most important area for the study of lava tube caves. A high proportion of the islands' voluminous lava flows passed through these tubes before they solidified. The volcanology of the islands is primarily related to the mid-Pacific "hot-spot" magma plume or plumes, over which the ocean floor moves slowly northwestward, carrying the string of islands with it. The islands' predominant rock types thus become progressively older northwestward.

Kauai (the most northwesterly of the major islands) is primarily composed of lavas 5 to 7 million years old. In the middle of the island chain, the oldest rocks of the island of Oahu are about 2 million years old. At the active southeast end, "The Big Island"—Hawaii Island proper—is still growing. About 95% of it is pseudokarst. Only one of its five volcanoes is considered extinct (Kohala), three are currently active or have recently been active (Mauna Loa, Kilauea, and Hualalai), and one older volcano may or may not have shut down (Mauna Kea). In addition, however, a poorly understood rejuvenation phase of volcanism produced significant lava tube caves on some of the older islands a few hundred thousand years ago.

East Maui has several extensive lava tube caves. One—Ka Eleku Cavern—has been minimally developed for ecotourism. The most recent cave-forming flow on this island was about 1790 AD, but its principal caves were formed in early Recent or Pleistocene times. There is no complete agreement on the ages of the various tube-forming flows on east Maui, and one—at the bottom of a deeply eroded gully—may be of Tertiary age. Contrary to many popular accounts, its famous Waianapanapa Caves are inland littoral caves, not lava tube caves.

With one possible exception, all the lava tube caves of Oahu are about two million years old. The longest is only about 100 m. The spatter features in at least one of the caves appear to be brand new.

Kauai's lava tube caves also are comparatively short and are in rejuvenation-phase lava. Also present are three sizeable, much-visited caves near sea level, which are either littoral caves in the older lava or short lengths of very ancient lava tube caves eroded to the point where they have lost their distinguishing characteristics of lava cooling and lava deposition.

Major lava tube caves exist on all the Big Island's volcanoes except Kohala (which apparently has only erosional caves). With about 200 km mapped in several cave systems, the Ailaau flow field of Kilauea Volcano is currently the world's most important. Here, Kazumura Cave alone has over 65.5 km mapped (at December 2001), making it the longest lava tube cave in the world (Figures 1 and 2). Both Mauna Loa and Hualalai volcanoes have caves with more than 10 km mapped.

Mauna Kea volcano has recently been found to have several shorter, unusually steep lava tube caves. Located in a gullied rain forest, all are invaded intermittently by torrential high-load streams, which have eroded them considerably. The largest is so impacted that one school of thought considers it entirely an erosional phenomenon. In stark contrast, some of the lava tube caves of the island's dry leeward side contain thick deposits of soluble volcanic minerals: gypsum, mirabilite, thenardite, and others. Even here, however, some caves show clear-cut depositional evidence of invasion by occasional flash floods: a serious potential public health problem in populated, beach, and fishing areas. Unlawful dumping of raw sewage and toxic and hazardous wastes in caves and dolines is very common in Hawaii.

The Big Island's Hawaii Volcanoes National Park, a World Heritage Site, is entirely pseudokarstic and contains many caves. Because of extensive early Hawaiian use of caves and also because of the rough, remote terrain, much of it is uninvestigated speleologically. At the other extreme, 120 m of its famous Thurston Lava Tube is electrically illuminated, and is the world's most visited lava tube cave. About 445 m long, it is a relatively featureless upper-level corridor entered through a jameo (a lava collapse

Hawaii Lava Tube Caves: Figure 1. Outline map of the Kazumura lava tube down the eastern slope of the Kilauea volcano on Hawaii's Big Island.

Hawaii Lava Tube Caves: Figure 2. The main tunnel of Kazumura Cave, which provides many kilometres of easy walking beneath the eastern slope of Kilauea. (Photo by Tony Waltham)

sink / doline extending through more than one level of cave (Halliday, 1993)). Permits are required for most of the other caves in the park. Some are notable for petroglyphs or for thickets of lava dripstone.

The caldera of Kilauea volcano has both "ordinary" lava tube caves and numerous small caves formed by subcrustal injection of fluid lava, followed by drainage. Some of these are largely featureless tubes. Others are tubular but have sunken ceilings that sagged while still plastic. Still others are hollow "whaleback" lava tumuli several metres high and several times as long, or long, sinuous hollow lava tumuli. A few are perimeter passages around large "lava rises" with sunken centres (Walker, 1991). Others resulted from drainage of sheet-flow units (Hon et al., 1994). Many of the caves in the caldera are hyperthermal and require special exploration techniques.

Also within Hawaii Volcanoes National Park, the Ailaau vent was the supply source for Kazumura and many other major caves in its flow field. It is about 350 years old. Most of the actual vent structure collapsed into Kilauea Iki crater, which developed more recently. Kazumura Cave now begins outside the park; part of its upper end was unknowingly destroyed by state highway construction. It originally extended from a point near Thurston Lava Tube to the 65-m contour near the ocean. Its remaining passages still have a unique unbroken descent of 1101 m in a linear distance of 34 km. Its overall floor plan is that of a long crooked pole with small bunches of spaghetti strung along it at intervals. Many entrances are present, with and without extensive secondary ceilings that form multiple levels. The upper, steeper part tends to be high and narrow, with some cutbanks and slip slopes that show downcutting by flowing lava. Lava falls up to 15 m high are present, some with large plunge-pool chambers at their bases. Farther downslope, the cave characteristically is wide and comparatively low. Innumerable individual rheogenic and spatter features are present, mostly undescribed. Many are formed of a reddish or brown late-stage lava which contrasts with the dark grey tube linings. About one-third of the cave is floored by breakdown, but two sections with long lengths of walking passage have undergone minor development for ecotourism.

Kaumana Cave is a county park in a suburb of the city of Hilo. Its more accessible sections are a very popular local attraction. This cave is in an 1881 Mauna Loa flow which nearly reached the ocean at Hilo. In 2000, a hydrostatic overload blew out a segment of its roof, on the edge of a suburban street. It is an intermittent but important part of Hilo's aquifer system, and initial hydrological studies are under way. Exploration and study of remote portions of this cave have been hampered by illegal dumping of used diapers, garbage, and other hazardous substances.

Located on the Southwest Rift Zone of Mauna Loa volcano, the Kula Kai cavern system contains the island's best-known show cave. One of its caves is 16 km long. Because of the presence of soluble minerals, cavern walls here tend to be much lighter-coloured than most caves in Hawaii, and the cave's speleogenesis appears to be unusually unconstrained by confining lava, with a tendency to form mazes.

The northwest and southwest slopes of Mauna Loa volcano contain numerous shorter caves, some of large volume. Their individual characteristics appear to be determined by the parameters of the flows that contain them. One is a notable demonstration of formation of a lava tube cave by wedging of crusts floating on top of a steep lava river—an oft-cited mechanism rarely seen in Hawaii. A few contain notable lava dripstone, and some contain Native Hawaiian fortifications and other evidence of use (Sinoto, 1992).

Hualalai is a much smaller volcano, but it also has large, varied cave systems up to at least 11 km long. Its Laniakea Cave was the first cave to be considered "celebrated" in Hawaii, but has been dreadfully neglected in recent decades. Others have impressive lava falls, and some have parallel passages running downslope. One Hualalai cave is especially notable for metre-high "sand bars" of topsoil washed in by flash floods.

The lava flow which formed Maui's 2724 m Ka Eleku Cavern is much blacker and was much more viscous than most of those on the Big Island. In Ka Eleku, lava speleothems characteristically are so dark and massive that commonly it is difficult to distinguish them from the walls of the cave.

WILLIAM R. HALLIDAY

See also **Hawaiian Islands: Biospeleology; Volcanic Caves; Vulcanospeleology: History**

Works Cited

Halliday, W. 1993. Kaluaiki and Thurston Lava Tube: an unrecognized jameo system? In *Proceedings of the 3rd International Symposium on Vulcanospeleology, Bend. OR, July 30–August 1, 1982*, edited by W. Halliday, Seattle: International Speleological Foundation

Hon, K. *et al.* 1994. Emplacement and inflation of pahoehoe sheet flows. *Geological Society of America Bulletin*, 106: 351–70

Sinoto, Y. 1992. Hawaiian use of lava tube caves and shelters. In *Proceedings 6th International Symposium on Vulcanospeleology, Hilo, HI, August 1991*, edited by G.T. Rea, Huntsville, Alabama: National Speleological Society

Walker, G. 1991. Structure, and origin by injection of lava under surface crust, of tumuli, "lava rises", lava rise pits", and "lava inflation clefts" in Hawaii. *Bulletin of Volcanology*, 52: 546–58

Further Reading

Calvari, S. & Pinkerton, H. 1998. Formation of lava tubes and extensive flow fields during the 1991–93 eruption of Mt. Etna. *Journal of Geophysical Research*, 103: 27 281–301

Ellis, W. 1823. *Journal of William Ellis*. Reprinted 1963, Honolulu: Advertiser Publishing

Forti, P. (editor) 1998. Proceedings, 8th International Symposium on Vulcanospeleology. Nairobi, February 1998. *International Journal of Speleology*, 27B (1–4): 1–163

Greeley, R. 1987. The role of lava tubes in Hawaiian volcanoes. In *Volcanism in Hawaii*, edited by R.W. Decker, T.L. Wright & P.H. Stauffer, Washington: Government Printing Office (US Geological Survey Professional Paper 1350)

Halliday, W. editor. 1993. *Proceedings, 3rd International Symposium on Vulcanospeleology. Bend, OR. July 30–August 1, 1982*. Seattle: International Speleological Foundation

Kauahikaua, J. *et al.* 1998. Observations on basaltic lava streams in tubes from Kilauea volcano, island of Hawaii. *Journal of Geophysical Research*, 103: 21 303–323

Kirch, P.V. 1985. *Feathered Gods and Fishhooks: An Introduction to Hawaiian Archaeology and Prehistory*, Honolulu: University of Hawaii Press

Macdonald, G., Abbott, A. & Peterson, F. 1983. *Volcanoes in the Sea: The Geology of Hawaii*, 2nd edition. Honolulu: University of Hawaii Press

Rea, G.Y. (editor) 1992. *Proceedings, 6th International Symposium on Vulcanospeleology, Hilo, HI, August 1991*, Huntsville, Alabama: National Speleological Society

HAWAIIAN ISLANDS: BIOSPELEOLOGY

The Hawaiian Islands are the emergent summits of massive volcanoes in the North Pacific Ocean. At more than 3500 km from the nearest comparable land masses containing caves, they are the most isolated group of high islands in the world. The eight high islands at the southeast end of the chain range in age from Hawaii, at less than one million years, to Kauai, at nearly six million years. Relatively few organisms colonized the islands, but these evolved into a diverse array of native species. Given the isolation and youth of the islands, the extreme youth of the lava tubes, the absence of taxa typically inhabiting continental caves, and the rarity of tropical cave species in general, obligate cave species or troglobites were not expected to occur in Hawaii. However, since 1971, more than 70 terrestrial troglobites have been discovered. Their study has led to a revision of theories on the evolution and ecology of cave animals (Howarth, 1980; 1993). Although no less interesting, the aquatic cave fauna of Hawaii (Kensley & Williams, 1986; Maciolek, 1986) is restricted to the coastal zone, where it is associated with anchialine ecosystems (see Anchialine Habitats).

Three types of caves support terrestrial troglobites in Hawaii: lava tubes, limestone caves, and piping caves. The most common caves in Hawaii are lava tubes (see Hawaii Lava Tubes). Limestone caves were formed by dissolution in elevated reefs and lithified sand dunes on the older islands of Oahu and Kauai. Piping (or suffosion) caves formed when water erosion plucked out softer material from under a layer of harder material. In Hawaii, they are best developed beneath montane rain forests on Molokai, where a lava flow covers an ash layer. Sea (littoral) caves and talus caves occur on all high islands but rarely support troglobites unless they intersect more extensive cavernous rock strata.

The main energy sources in Hawaiian caves are plant roots, especially those of *Metrosideros polymorpha*, the dominant pioneering tree on lava flows. Additional food energy comes from organic material washing or falling into crevices, surface animals wandering underground, and probably chemo-autotrophic bacteria. Native cutworm moths once roosted in caves in huge numbers, but the group has become rare. Cave bats and continental cave crickets do not occur in Hawaii. Cixiid planthoppers and their nymphs feed on living roots by sucking sap with piercing mouthparts. The blind flightless adults wander through subterranean voids in search of mates and roots (Hoch & Howarth, 1993; 1999). Caterpillars of noctuid moths prefer to feed on succulent new growth at root tips, but they also scavenge on a variety of foods. The pale adults are flightless or can fly only weakly, and stay underground. Tree crickets, terrestrial amphipods, and isopods are omnivores but feed extensively on roots. Cave rock crickets are also omnivorous, as well as being opportunistic predators. Millepedes, springtails, and flies are found feeding on rotting organic material and associated micro-organisms. Terrestrial water treaders suck juices from long-dead arthropods. The blind predators include spiders, pseudoscorpions, rock centipedes, thread-legged bugs, and beetles. Most of the cave predators will also scavenge on dead animal material (Howarth, 1987; 1991).

Unlike soil, which acts as a filter trapping nutrients and water, the voids in cavernous rock strata act as conduits that transport organic resources deep underground. Young basaltic flows, like cavernous areas generally, contain a vast anastomosing system of medium- to large-sized voids. The deeper voids (that is, the deep and stagnant air zones; see Subterranean Habitats) are stressful environments for most surface organisms, and the food resources there are unavailable to them. Rather than being relicts stranded in caves by changing climate, obligate cave species are highly specialized to exploit the abundant energy resources within these deeper voids and can colonize cave passages that are accessible to humans only where the environment is similar to that found in deeper voids; i.e. moist substrates, calm air, saturated with water vapour, and sometimes high levels of decomposition gases, especially carbon dioxide. Hawaiian troglobites appear to have evolved following an adaptive shift that allowed them to exploit underground environments. Three different surface habitats (moist forests, barren rock, and marine littoral) served as the source for ancestors of the cave species. In many cases (Figure 1A & B), their closest surface relative is still extant in one of these neighbouring habitats, corroborating the theory that cave species evolved following an adaptive shift, and indicating that the relictualism displayed by some troglobites may be secondary (Howarth, 1993). The adaptations displayed by Hawaiian troglobites are truly remarkable, including such anomalies as blind underground tree crickets, planthoppers, flightless flies, terrestrial water treaders and amphipods, and the epitome of adaptive shifts: the no-eyed, big-eyed hunting spider (*Adelocosa anops*, Figure 1C); other members of this family have large, well-developed eyes, hence their common name. Howarth & Mull (1992) show colour photographs of many related cave and surface species pairs.

Cave species evolved independently on five islands, indicating that cave adaptation is a general process and that they can evolve wherever there have been suitable habitats and available resources for a sufficient time. The degree of convergence in morphology, behaviour, and physiology among the independently derived fauna on different islands is striking, but can be understood as adaptations to cope with the stressful underground environment. Surprisingly, the degree of cave adaptation is not correlated with either the age of the island or the age of the cave, but with environmental factors and the size of the available habitat (Howarth, 1993; Taiti & Howarth, 1998; Hoch & Howarth, 1999). In fact, Hawaii, the largest and youngest island, supports the largest number of cave-adapted species. The diversity of cave species may be much higher than currently recognized. Recent morphological, behavioural, and molecular studies of several troglobitic groups on Hawaii Island have demonstrated that each cave may harbour unique populations or species. Hoch and Howarth (1993) showed that each cave population of *Oliarus polyphemus* studied has a unique mating song and small but consistent differences in morphology. Even the mating calls in neighbouring lava tubes were distinct, and some had diverged sufficiently to appear to be reproductively isolated (i.e. distinct biological species). Otte (1994) confirmed that the rock cricket *Caconemobius varius* also represented a complex of species. Different cave populations of moths and millipedes can also be distinguished morphologically, but the species have not been formally described.

Hawaiian caves are island-like habitats within islands, and like other Hawaiian environments, cave communities are threat-

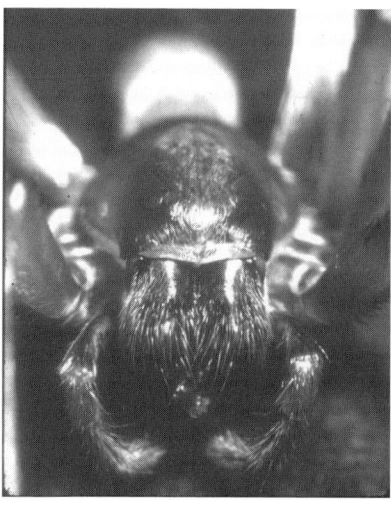

Hawaiian Islands: Biospeleology: Lycosid spiders. A & B. Comparison of a cave- and surface-adapted species pair. A. big-eyed hunting spider (*Lycosa* species) living on barren lava flows on Hawaii Island. B. Small-eyed, big-eyed hunting spider (*Lycosa howarthi*), living underground in young volcanic substrates on Hawaii Island. These two species differ only in their adaptations to their environments. C. No-eyed, big-eyed hunting spider (*Adelocosa anops*) from remnant lava tubes on Kauai Island. This species evolved its adaptations to caves independently of *L. howarthi* (B), via a different ancestor. (Photos by W.P. Mull)

ened, both by direct disturbance from visitors and indirectly from activities on the surface. Caves are used as refuse dumps; vegetation over caves is cleared; the rocks surrounding the caves are mined; groundwater is polluted; and alien plants and animals disrupt cave ecosystems. Human visitors may break plant roots, kill blind animals, trample bones, and damage the cave itself. Only two Hawaiian cave species currently enjoy formal protection as federally listed endangered species: the no-eyed, big-eyed hunting spider (Figure 1C) and the terrestrial sandhopper, both of which are known only from a small area of remnant caves on Kauai. Many other species deserve conservation initiatives. Most Hawaiian caves remain biologically unsurveyed, and many more endemic animals remain to be discovered, especially since many apparently widespread taxa may in fact represent species complexes. Effective conservation is predicated on accurate systematics studies. There is a dilemma: publicity featuring caves can increase visitation rates, leading to increased damage in caves; but if caves and their value are not made known, their resources may be destroyed through ignorance during changes in land use (Howarth, 1983). If these cave ecosystems had been destroyed before discovery, it would be difficult to speculate that they ever existed.

FRANCIS G. HOWARTH

See also **Colonization**

Works Cited

Hoch, H. & Howarth, F.G. 1993. Evolutionary dynamics of behavioral divergence among populations of the Hawaiian cave-dwelling planthopper *Oliarus polyphemus* (Homoptera: Fulgoroidea). *Pacific Science*, 47: 303–18

Hoch, H. & Howarth, F.G. 1999. Multiple cave invasions by species of the planthopper genus *Oliarus* in Hawaii (Homoptera: Fulgoroidea: Cixiidae). *Zoological Journal of the Linnean Society*, 127(4): 453–75

Howarth, F.G. 1980. The zoogeography of specialized cave animals: a bioclimatic model. *Evolution*, 34: 394–406

Howarth, F.G. 1983. The conservation of cave invertebrates. In *Proceedings of the First International Cave Management Symposium Held at Murray, Kentucky, July, 1981*, edited by J.E. Mylroie, Murray State University, Kentucky: privately published

Howarth, F.G. 1987. Evolutionary ecology of aeolian and subterranean habitats in Hawaii. *Trends in Ecology and Evolution*, 2: 220–23

Howarth, F.G. 1991. Hawaiian cave faunas: macroevolution on young islands. In *The Unity of Evolutionary Biology, Proceedings of the Fourth International Congress of Systematic and Evolutionary Biology*, vol. 1, edited by E.C. Dudley, Portland, Oregon: Dioscorides Press

Howarth, F.G. 1993. High-stress subterranean habitats and evolutionary change in cave-inhabiting arthropods. *American Naturalist*, 142: S65–S77

Howarth, F.G. & Mull, W.P. 1992. *Hawaiian Insects and Their Kin*, Honolulu: University of Hawaii Press

Kensley, B. & Williams, D. 1986. New shrimps (families Procarididae and Atyidae) from a submerged lava tube on Hawaii. *Journal of Crustacean Biology*, 6: 417–37

Maciolek, J.A. 1986. Environmental features and biota of anchialine pools on Cape Kinau, Maui, Hawaii. *Stygologia*, 2(1/2): 119–29

Otte, D. 1994. *The Crickets of Hawaii: Origin, Systematics and Evolution*, Philadelphia: The Orthopterists' Society

Taiti, S. & Howarth, F.G. 1998. Terrestrial isopods (Crustacea: Oniscidea) from Hawaiian caves. *Mémoires de Biospéologie*, 24: 97–118

Further Reading

Ahearn, G.A. & Howarth, F.G. 1982. Physiology of cave arthropods in Hawaii. *Journal of Experimental Zoology*, 222: 227–38

Camacho, A.I. (editor) 1992. *The Natural History of Biospeleology*, Madrid: Museo Nacional de Ciencias Naturales

Culver, D.C. 1982. *Cave Life: Evolution and Ecology*, Cambridge, Massachusetts: Harvard University Press

Howarth, F.G. 1981. Community structure and niche differentiation in Hawaiian lava tubes. In *Island Ecosystems: Biological Organization in Selected Hawaiian Communities*, edited by D. Mueller-Dombois, K.W. Bridges & H.L. Carson, Stroudsberg, Pennsylvania: Hutchinson Ross

Howarth, F.G. 1983a. Ecology of cave arthropods. *Annual Review of Entomology*, 28: 365–89

Howarth, F.G. 1983b. Bioclimatic and geologic factors governing the evolution and distribution of Hawaiian cave insects. *Entomologia Generalis*, 8:17–26

Howarth, F.G. 1988. The evolution of non-relictual tropical troglobites. *International Journal of Speleology*, 16: 1–16

Wilkens, H., Culver, D.C. & Humphries, W.F. (editors) 2000. *Subterranean Ecosystems*, Amsterdam and New York: Elsevier

HIGHWAYS ON KARST

To build good and safe highways on karst, and at the same time to identify and protect the natural heritage, it is necessary to take into account the special features of karst at the planning stage and in monitoring of construction (Knez & Slabe, 2001). Hence, karst geomorphologists, hydrologists, and speleologists, frequently contribute to decisions on the choice of right-of-way, methods of construction, the stabilizing of caves, and cost-benefit analyses of different methods of construction. Decisions on the best alignment for a highway through karst should be preceded by a complete study of the karst, considering data from various sources (e.g. a database of cave and water features) together with targeted fieldwork. As a rule, the more important karst features, whether surface landforms or caves and waters, are best avoided. The presence of voids in the karst rocks and the permeability of the surface, which can both influence the construction of highways, may be deduced, in part, from geological and geomorphological mapping, speleological investigation, and study of aquifer characteristics. The presence of voids and rockhead features may also be revealed by core drilling. Geophysical approaches such as geoelectric and ground-penetrating radar surveys, seismic reflection and refraction profiling, tomography and microgravimetry surveys may also be used (see Geophysical Detection of Caves and Karstic Voids). However, whether regarding the rock, the hydrogeology and climate, or the karst cover, each individual karst region has special characteristics and it is best to adapt the approach to these factors in each case. These methods are used before construction as well as during construction when the karst surface has been stripped and levelled and caves have been exposed.

The aim of the engineer is to prevent the possibility of collapse and of rapidly developing dolines. As a rule, the demands on construction are greater on younger carbonate rock, and where possible construction should be avoided on highly soluble gypsum, anhydrite, and salt formations. Special attention must be devoted to the effect of the construction and use of highways on karst waters. Impermeable road surfaces prevent the direct influence of runoff on underground waters (Stephenson & Beck, 1995), but ditches and culverts should be installed along the edges of highways to conduct the runoff into oil collectors, and only after being cleaned should the water be allowed to flow back onto the permeable karst surface. Highway drainage running onto superficial deposits overlying karst must also be directed so that it does not increase the frequency of subsidence dolines.

Many karst phenomena, particularly caves, are discovered only during construction. The removal of soil and vegetation from the karst surface, and major earthworks undertaken during the digging of road cuts, tunnels, and bridge foundations, also reveal surface and underground features, epikarst, and caves. These features form part of the natural heritage, and karst specialists can advise on methods of their preservation and on surmounting construction obstacles. Various features filled with soil and alluvium commonly dissect the karst surface, and where these are intersected during highway construction the sediments are normally removed. If they are not, erosion by water can cause the formation of subsidence dolines. The floors of larger depressions, where shafts and open fissures are common, must be reinforced with concrete and stone arches and the depressions may then be filled with layers of gravel (Figures 1 and 2). Larger shaft mouths are sealed with concrete lids.

The presence of voids is the most distinctive feature of construction work in karst regions. Cave ceilings collapse during blasting, and cross sections of passages are left in the slopes of cuts. More than 300 caves were opened over a 50 km stretch of an expressway across Slovenia's classic karst region. On covered karst, the possibilities for doline formation, and the location of inaccessible parts of caves should be determined by direct exploration, where possible, with additional geophysical surveys, and with downhole camera and acoustic scanning. Remedial work appropriate for the cave and the type of rock should follow, with the goal of preserving as many caves as possible. Easiest to preserve are shafts whose smaller entrances can be sealed with

Highways on Karst: Figure 1. One of many dolines encountered during construction of an expressway across Slovenia's classic karst region. (Photo by John Gunn)

Highways on Karst: Figure 2. Remedial work at the base of a doline on the route of the Slovenian expressway. (Photo by John Gunn)

concrete slabs. Old caves with solid ceilings can also be preserved and can remain hidden below the road where appropriate (beneath a concrete cover) but badly weakened caves must be filled. The most interesting and well-preserved caves can be completely protected, and even under highways they can be left accessible through concrete culverts. Well-preserved caves in the slopes of cuttings may be left open for viewing, but it is normally necessary to construct rock walls to close the entrances to caves where the ceilings are too weakened, or where they contain sediment fills that could be flushed onto the highway by water. Direct dynamic loading with vibration-rollers has often proved useful in testing the firmness of the ground, since even the process of compacting gravel on a roadbed can open previously undiscovered caves. The effects of blasting on nearby known caves must be monitored, since they may affect further construction work and the preservation of karst phenomena.

Monitoring the influence of highways on karst regions, for example on caves under a highway or the possibility of dolines developing, is essential. The water that runs off highways must be analysed regularly. Too often, wastewater cisterns are only tested during the first periods of extremely high water that can flush sediments onto the karst surface if the cisterns are too small. Runoff water from highways is usually polluted, primarily with solid particles, has higher COD (chemical oxygen demand) values, contains hydrocarbons, lead, and cadmium as well as chlorides in winter, and is particularly polluted following long periods of drought (Kogovšek, 1995).

Finally, karst geomorphologists may be able to provide input to highway landscape architects in designing the roadway landscape, so that the natural appearance of the karst region is as well preserved as possible.

MARTIN KNEZ AND TADEJ SLABE

Works Cited

Knez, M. & Slabe, T. 2001. Karstology and motorway construction. In *Proceedings of the 14th IRF Road World Congress*, Paris: International Road Federation and Union Routière de France

Kogovšek, J. 1995. Detailed monitoring of the quality of the water that runs off the motorway and its impact on karst water. *Annales, Annals for Istrian and Mediterranean Studies*, 7: 149–54

Stephenson, J.B. & Beck, B.F. 1995. Management of the discharge quality of highway runoff in karst areas to control impacts to groundwater: A review of relevant literature. In *Karst Geohazards: Engineering and Environmental Problems in Karst Terrane: Proceedings of the Fifth Multidisciplinary Conference on Sinkholes and the Engineering and Environmental Impact of Karst*, edited by B.F. Beck, Rotterdam: Balkema

Further Reading

Beck, B.F. & Herring, J.G. (editors) 2001. *Geotechnical and Environmental Applications of Karst Geology and Hydrogeology*, Rotterdam: Balkema

Benson, R.C. & Yuhr, L. 1993. Spatial sampling consideration and their applications to characterizing fractured rock and karst systems. In *Applied Karst Geology: Proceedings of the Fourth Multidisciplinary Conference on Sinkholes and the Engineering and Environmental Impacts of Karst*, edited by B.F. Beck, Rotterdam: Balkema

Bosák, P., Knez, M., Otrubová, D., Pruner, P., Slabe, T. & Venhodová, D. 2000. Palaeomagnetic research of a fossil cave in the highway construction at Kozina, SW Slovenia. *Acta Carsologica*, 29(2): 15–33

Fischer, J.A., Fischer, J.J. & Green, R.W. 1993. Road design in karst. *Environmental Geology*, 22: 321–25

HÖLLOCH, SWITZERLAND

Exploration of Hölloch cave was initiated by local people in the 1880s. By 1905, the cave was more than 6 km long making it the longest in Switzerland, and exploration accelerated after World War II. The length of Hölloch in 2002 is about 186 km, with a total vertical range of 941 m. Exploration in the cave was so fruitful that cavers did not start systematically to explore the surrounding karst until the 1970s. Several large caves have been found and could be connected. They include the Silberen System which is about 35 km long and 896 m deep. Several other caves are known in the area and the total length of mapped caves is about 240 km.

Because the cave system is very well developed, many scientific investigations were carried out by Alfred Bögli in and above Hölloch between 1946 and 1982. Hölloch was used as an example in support of some speleogenetic theories (Bögli, 1964, 1980; Ford & Williams, 1989; Palmer, 1991). The terminology of karren types developed in this region by Bögli (1960) is still used as a basis for the classification of karst surface geomorphology. Since the 1980s research has focused more on karst hydrology, using the opportunity to observe and measure groundwater flow and heads at various places within the karst system. Hölloch is an excellent example of the way a karst conduit network floods (Jeannin, 2001).

Hölloch underlies 7 km² of the 32 km² groundwater catchment of the Schlichenden Brünnen spring (Figure 1), which forms part of a larger karst in central Switzerland. The Schlichen-

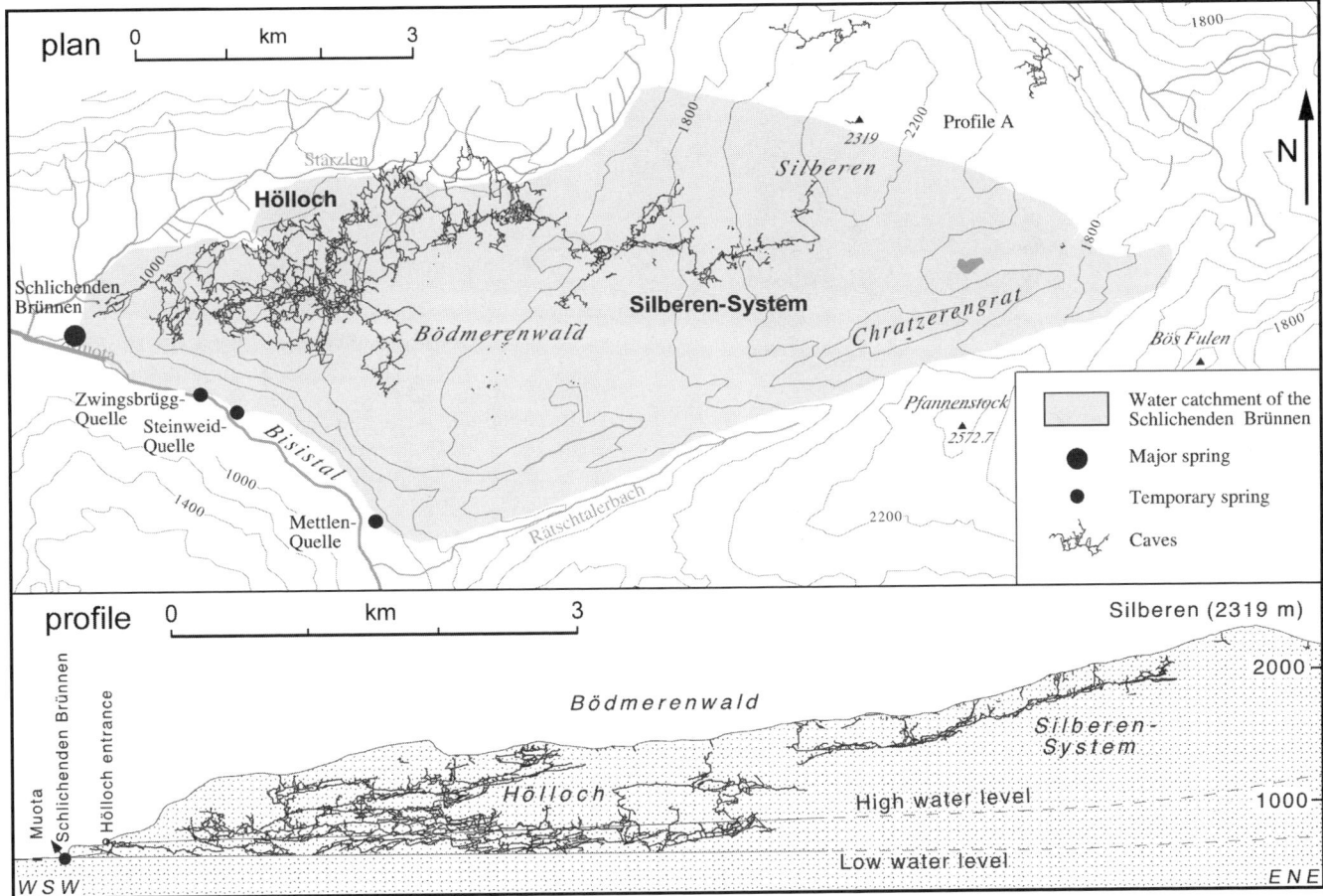

Hölloch, Switzerland: Figure 1. Plan and profile of the cave passages of Hölloch and the adjacent Silberen System.

den Brünnen spring emerges at an elevation of 638 m, where soils and vegetation cover the limestone. Between 1000 and 1700 m altitude, in the Bödmeren area, soils are thinner and karrenfields occur, but they are usually covered by forest. At higher elevations, karrenfields are bare and can look like crevasses and glaciers. The highest summit within the catchment reaches an altitude of 2319 m.

Mean annual precipitation in the area is about 2600 mm with a significant part falling as snow. The discharge at Schlichenden Brünnen spring ranges between $0.02-15$ m^{3}s^{-1}. Despite a 200–1000 m thick unsaturated zone where flow is mostly vertical (shaft flow), horizontal flow in Hölloch is mainly constricted and the main active conduits are located 10–100 m below the water table. Additional conduits, which occur above the water table, flood during high water events (Figure 1). Water-table fluctuations range from 40 m close to the spring to more than 220 m in the upstream part of the cave, located 5200 m away from the spring.

The karst system extends mainly within the Cretaceous limestones of the Helvetic nappe. In this region the limestones have been folded and overthrusted on top of other sedimentary rocks. In the Hölloch area the Cretaceous limestone series is repeated four times, to build a total thickness of about 1000 m of limestone. However, some shaly and sandy beds are present as discontinuous strata within this structure.

Hölloch can readily be divided into several zones. The lowest galleries, which are always submerged, are part of the phreatic zone. The passages of the intermediate zone are regularly flooded by high water, and lead to the upper level, where relict conduits are present as well as some active stream passages feeding into the lower phreatic conduits. Because the geologic structures dip to the north-northwest, it is in this direction that the lowermost galleries are found, while the uppermost conduits are located to the south–southeast.

The submerged zone has been explored only in a few places, where a maximum depth of 88 m below the water table has been reached. During heavy rains or snow melt, the water table can rise and submerge up to about 90 km of galleries. During large floods, the lowermost of the four cave entrances (95 m above Schlichenden Brünnen) becomes an overflow spring where flow can exceed 4 m^3 s^{-1}.

Most of the conduits in Hölloch display a perfect elliptical cross-section (Figure 2). From this morphology we can infer that most of the conduits developed under constricted phreatic flow conditions. Several stages of conduit development are identifiable. It seems that one stage developed within a vertical range of

Hölloch, Switzerland: Figure 2: Conduit with elliptic cross-section: a very common type of conduit in Hölloch. (Photo by Urs Widmer)

50–150 m, forming switchback looping passages. The various cave levels were mainly controlled by the position of the outlet, i.e. by the deepening of the Muota Valley and by the geological setting. The levels are all connected, forming a network of conduits developed in three dimensions.

The Silberen System lies in the highest portion of the Schlichenden Brünnen catchment. The seven entrances are characterized by series of shafts of about 200 m, leading into a complex network of phreatic passages. The largest underground chamber in Switzerland (the Mammutmünster, 77 m high, 86 m wide, and 100 m long in places) is in this cave. This system has mainly developed below the bare karst, while Hölloch is located below the forest. It is noticeable that only a few speleothems are present in both Hölloch and Silberen System, and many of those that do occur are yellow-brown or red coloured.

Karst in the Hölloch area is not adequately protected today, but it has been proposed as a site of national significance for natural purposes and should receive strengthened protection status in the future.

PIERRE-YVES JEANNIN AND ANDRES WILDBERGER

Works Cited

Bögli, A. 1960. Kalklösung und Karrenbildung. *Zeitschrift für Geomorphologie*, supplement 2: 4–21
Bögli, A. 1964. La Corrosion par mélange des eaux. *International Journal of Speleology*, 1(1–2): 61–70
Bögli, A. 1980. *Karst Hydrology and Physical Speleology*, Berlin and New York: Springer (original German edition, 1978)
Ford, D.C. & Williams, P. 1989. *Karst Geomorphology and Hydrology*, London and Boston: Unwin Hyman
Jeannin, P.-Y. 2001. Modeling flow in phreatic and epiphreatic karst conduits in the Hölloch cave (Muotatal, Switzerland). *Water Resources Research*, 37(2): 191–200
Palmer, A.N. 1991. Origin and morphology of limestone caves. *Geological Society of America Bulletin*, 103: 1–21

Further Reading

Bögli, A. 1970. *Le Hölloch et son karst / Das Hölloch und sein Karst*, Neuchâtel, Switzerland: Baconnière
   Although quite old this is probably the best overview about Hölloch, for those reading French or German.
Bögli, A. 1980. *Karst Hydrology and Physical Speleology*, Berlin and New York: Springer (original German edition, 1978)
   Textbook in which Hölloch cave is widely used as an example.
Bögli, A. & Harum, T. (editors) 1981. Hydrologische Untersuchungen im Karst des hinteren Muotatals (Schweiz). *Steirische Beiträge zur Hydrogeologie*, 33: 125–264 (special issue)
Wildberger, A., & Jeannin, P.-Y. 1995. Traçage des eaux souterraines dans la région entre le Bisistal et le Klöntal (cantons de Schwytz et Glaris), résultats des essais effectués entre 1992 et 1994. *Stalactite*, 45(2): 113–29
Wildberger, A. & Preiswerk, C. 1997. *Karst and Caves of Switzerland*, Basel, Switzerland: Speleo Projects
   Four-language edition giving an nice illustrated overview about karst and caves in Switzerland.

HONGSHUI RIVER FENGCONG KARST, CHINA

The middle reaches of the Hongshui River pass through a huge area of rugged and very spectacular fengcong that constitutes one of the world's finest karst landscapes (see location map in China entry). Upstream, the Hongshui drains from the less dissected karst of the Guizhou Plateau (at about 1200 m altitude), and downstream it flows into the fenglin karst of the Guangxi lowlands (at 300 m). The heart of the karst, around Qibailong in Dahua County, has been designated a new national park for its magnificent limestone landscapes.

Around Dahua, carbonates of Devonian to Triassic age are nearly 3000 m thick and have insoluble contents of less than 1.25%. They lie in box-folds and monoclinal structures with dips ranging from horizontal to 70°. Paleo-climates were tropical, hot and humid in the Tertiary Period, when coal measures were deposited and land surfaces were strongly levelled to form the Yunnan-Guizhou Plateau peneplain during a time of tectonic stability. Subsequent neotectonic movements include the mid-Tertiary Darong Phase and the end-Tertiary Tianyang Phase with several intermittent episodes of regional uplift. Dahua lies in the southern subtropical climatic zone, with a mean temperature of 19.1°C and annual precipitation of 1700 mm. Dissolutional erosion rates reach 80–300 mm ka^{-1}.

The Qibailong area has the finest karst landscapes characterized by high fengcong cones and deep depressions. Many karst hills are now bare as a consequence of deforestation and there is serious desertification within the region. Mean depth of the depressions is about 104 m, with the deepest depression at Gangfeng descending nearly 300 m. In one studied area of 203 km^2, there are 1038 deep depressions, occupying 18.3% of the total area of the Qibailong karst. Farming is only possible on the depression floors, and the local people say that only one tenth of the ground is soil. Many depressions are linear due to development along the main faults, and some of the fengcong massifs are defined as fault-bound blocks. Some of the marginal depressions are small poljes with well-defined corrosion levels dissected into stone teeth that are revealed by localized soil erosion.

PLATE 1i

Gypsum chandeliers, Lechuguilla Cave, New Mexico, USA
(photo: Art Palmer)

World's tallest stalagmite, Cueva Martin Infierno, central Cuba
(photo: Kevin Downey)

Tube in Source du Rautely, Herault, France
(photo: Kevin Downey, Urs Widmer, Phillipe Crochet)

Sarawak Chamber, Lubang Nasib Bagus, Mulu, Sarawak, Malaysia
(photo: Jerry Wooldridge)

PLATE 2i

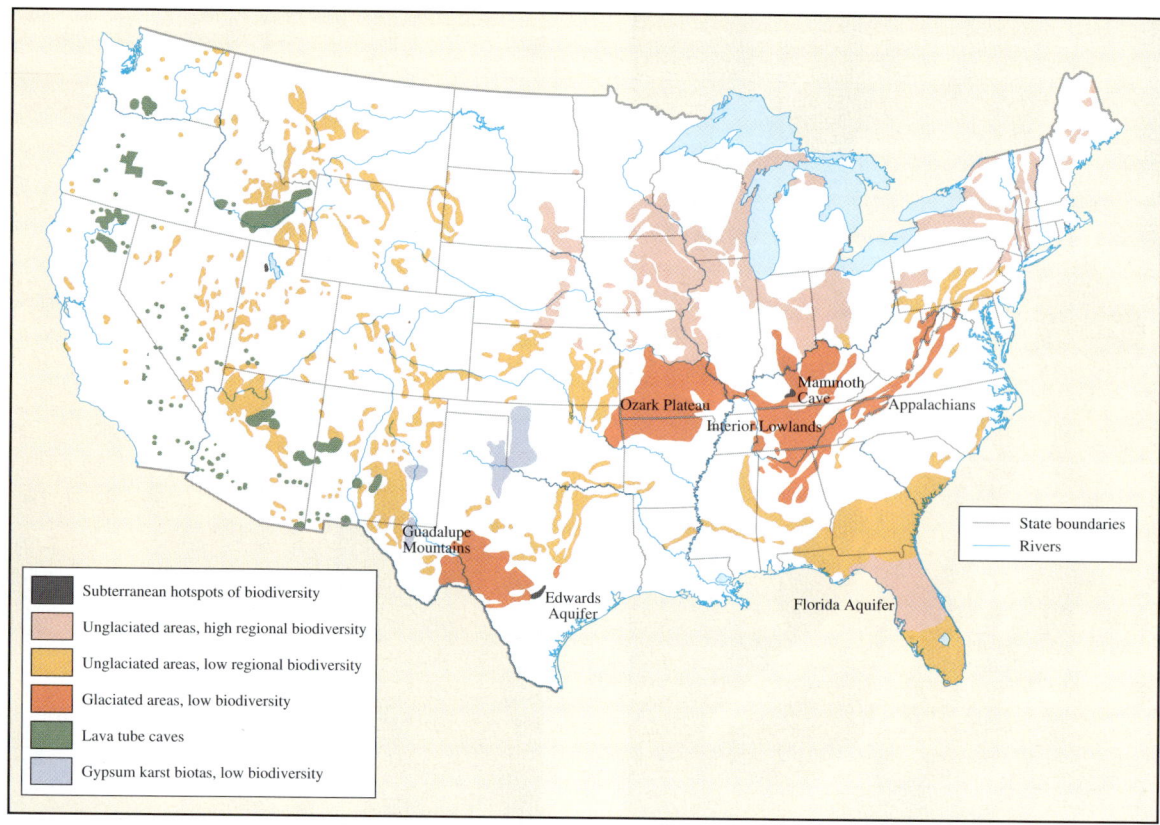

Levels of biodiversity within the karst terrains of USA (From Ricketts, *et al.,* 1999)

Typhlomolge rathbuni (Texas blind cave salamander), troglobite endemic to Edwards Aquifer, USA (photo: Chip Clark)

Amblyopsis spelaea (Northern cavefish), troglobite, USA (photo: Chip Clark)

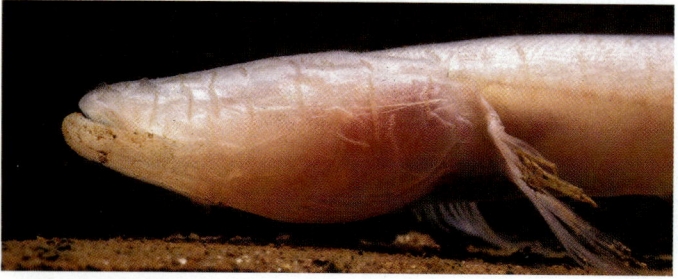

Head of *Typhlichthys subterraneus* (Southern cavefish), troglobite, USA (photo: Chip Clark)

Cave bat, *Miniopterus australis,* Deer Cave, Mulu, Sarawak, Malaysia (photo: Tony Waltham)

Opossum in cave nest, Anne's Cave, Waitomo, New Zealand (photo: John Gunn)

PLATE 3i

Loboptera troglobia, a cave-dwelling cockroach, Canary Islands, Spain (photo: Pedro Oromí)

Kentucky cave shrimp *Palaemonias ganteri,* troglobite endemic to Mammoth Cave, Kentucky, USA (photo: Chip Clark)

Collartida tanausu, a predatory troglomorphic thread-legged bug endemic to La Palma, Canary Islands, Spain (photo: Pedro Oromí)

Orconectes pellucidus, troglobitic eyeless crayfish, Mammoth Cave, Kentucky, USA (photo: Chip Clark)

Rhadine reyesi, a predatory troglobitic beetle, USA (photo: Jean Krejca, Steve Taylor)

Maiorerus randoi, a troglobitic harvestman endemic to a cave on Fuerteventura, Canary Islands, Spain (photo: Pedro Oromí)

Yellow mould covers a bat carcass in Carlsbad Cavern, New Mexico, USA (photo: Diana Northup)

PLATE 4i

Red Bull, Grotte de Lascaux, Vézère, France (photo: Taylor and Francis and its licensors. All rights reserved)

Great Hall of the Bulls, Grotte de Lascaux, Vézère, France (photo: Taylor and Francis and its licensors. All rights reserved)

Painting of Grotta Azzurra, Isle of Capri, Italy (painting by Yves Gianni)

Taino petroglyphs, Cueva Indios, northern Puerto Rico (photo: Kevin Downey)

Guihai tablets, Longuin Dong, Guilin, Guangxi, China (photo: Tony Waltham)

Banded luminescent calcite with uranyl content (photo: Yavor Shopov)

Luminescent hydrothermal calcite with Pb-Mn, Carlsbad Caverns, USA (photo: Yavor Shopov)

PLATE 5i

Selenite crystals, cave within the Naica Mine, Chiuhuahua, Mexico (photo: Kevin Downey)

Calcite encrustations, Cueva de Bellamar, Matanzas, Cuba (photo: Kevin Downey)

Spherical and oblate cave pearls, Baume de Postres, France (photo: Kevin Downey)

Green calcite containing nickel, cave in Montagne Noir, France (photo: Kevin Downey)

Blue aragonite, cave in Montagne Noir, France (photo: John Gunn)

Calcite coloured red by organic acids, Silvertip Cave System, Montana, USA (photo: Art Palmer)

Calcite and aragonite, Cupp-Coutunn Cave, Turkmenistan (photo: Charlie Self)

Blue-green calcite, cave in Montagne Noir, France (photo: John Gunn)

PLATE 6i

Rift passage, Ogof Ffynnon Ddu, South Wales, UK (photo: Jerry Wooldridge)

Big Room, Carlsbad Caverns, New Mexico, USA (photo: Art Palmer)

Tower of Pisa, Lechuguilla Cave, New Mexico, USA (photo: Art Palmer)

Main passage, Beimo Dong, Bama, Guangxi, China (photo: Jerry Wooldridge)

Flooded passage in Ingleborough Cave, Yorkshire Dales, UK (photo: John Cordingley)

PLATE 7i

Side passage tube, Hang Phong Nha, Quang Binh, Vietnam
(photo: Howard Limbert)

White River Series, Peak Cavern, Peak District, UK
(photo: Clive Westlake)

Chuan Yan, Li Jiang, Guangxi, China (photo: Andy Eavis)

Main chamber, Maomaotau Yan, Guilin, Guangxi, China
(photo: Andy Eavis)

River passage notch, Gua Air Jernih, Mulu, Sarawak, Malaysia
(photo: Jerry Wooldridge)

Plateau Mountain Ice Cave, Alberta, Canada (photo: Tony Waltham)

PLATE 8i

Resurgence of Dabong River, Guizhou, China (photo: Tony Waltham)

Solutional fissures in epikarst exposed by a quarry in Indiana, USA (photo: Art Palmer)

Fluviokarst, Lathkill Dale, Peak District, UK (photo: John Gunn)

Towers of degraded karst now in a desert, Farafra, Egypt; note the person on the right for scale (photo: Tony Waltham)

Doline karst on gypsum, Sivas, Turkey (photo: Tony Waltham)

Cone karst, Gunung Sewu, Java, Indonesia (photo: Tony Waltham)

Meltwater gorge in glaciokarst, Gordale Scar, Yorkshire Dales, UK (photo: Tony Waltham)

Fengcong karst, Caoping, Guangxi, China (photo: Tony Waltham)

Evolution of the karst has been characterized by multiple phases of uplift in competition with karstic erosion. Uplift created an original plateau, and a fengcong-depression karst evolved from the fenglin landscape on the paleo-peneplain. The basal plain of the fenglin was transformed into a fengcong landscape, with stream-sinks, dolines, and shafts developing in the new depressions as new underground river systems expanded beneath. The evolutionary sequence is matched by the succession of landscape styles down the length of the Hongshui River, from the Yunnan-Guizhou Plateau to the Guangxi Basin. These are, in order downstream: the Guizhou karst plateau with residual karstic hills, then fenglin valley or basin, then fengcong-valley (or with shallow depressions), then fengcong with deep depressions or gorges, and finally the Guangxi fenglin plain.

The fengcong-depression karst around Dahua has a double structure to its morphology. The upper parts of the limestone hills stand above a zone of rock terraces that survive as fragments of the old peneplain. Mean elevation difference between the summits of the conical hills and the rock terraces is about 90 m, which might be interpreted as the height of the old fenglin towers. The lower part of the fengcong-depression landscape lies below the terraces. Around Qibailong, two fossil peneplains remain. The higher is at 1000–1100 m altitude, developed from Late Oligocene to Middle Miocene. With elevation of the Yunnan-Guizhou Plateau from Late Miocene to Early Pleistocene, a second level of peneplain, at altitudes of 800–950 m, developed surrounding the older peneplain fragments. Later, new depressions developed in the second peneplain, so that the modern depression floors lie at altitudes of 500–600 m.

A low erosion level is represented by an exceptionally fine corrosion plain along both banks of the Hongshui River around the Disu resurgence. It is now entrenched by the main river and extends downstream as far as a small gorge carved through an anticline, the crest of which appears to have controlled the corrosion plain level in the past.

Abandoned high-level cave systems formed in multiple stages, as they generally have 3–7 levels. Commonly the highest cave level is 320 m above the depression floors, with major lower levels at heights of 170 m and 70 m. The modern underground rivers are developed beneath the depression floors. Their gra-

Hongshui River Fengcong Karst, China: A concrete boat, housing pumps for water abstraction, floats in a flooded karst window within the rugged karst terrain of the Disu catchment. (Photo by Tim Fogg)

dients are typically lower in the upper reaches and higher in the lower reaches, these ungraded profiles being a consequence of Pleistocene tectonic uplift. As an example, the upper reach of the Nansuo underground river has a slope of 1.2–2.0% beneath the plateau stage, but it is 5–6% towards its gorge resurgence. On the north side of the Hongshui, the Disu karst system drains 1000 km² of fengcong to the Chin Shui spring, with some water travelling at least 57 km underground. The rising has a base flow of $4 \text{ m}^3\text{s}^{-1}$, which rises to $540 \text{ m}^3\text{s}^{-1}$ in flood. Caves in the basin are mostly flooded, and five sites have been dived to large conduits at depths of >75 m.

Flooded shafts in the depression floors and in the corrosion plain are windows into the Disu phreas. They are invaluable as water resources in the almost waterless holokarst. Water levels can vary by more than 20 m from wet to dry seasons, so concrete boats are used to support pump houses that float up and down on the fluctuating lakes within some of the shafts (see photograph).

SONG LINHUA AND LIN JUNSHU

See also **Cone Karst; Tower Karst**

HUANGLONG AND JIUZHAIGOU, CHINA

Situated in the Minshan Mountains of Sichuan Province, China, these tourist attractions are renowned for their mountain scenery, forming a backdrop to spectacular waterfalls and travertine-dammed lakes. While the Jiuzhaigou site has some of the largest travertine dams in the world, the travertine pools of Huanglong are almost certainly the world's most beautiful. Both sites are inscribed in the World Heritage list.

Huanglong (yellow dragon), so called because the winding stream deposits travertine coloured yellow by microscopic algae, is located in a tectonically active region of alpine karst. The waters issue from a fault-guided thermal spring at an altitude of 3580 m and descend a broad valley for a distance of 3.5 km to the Fu River, at an altitude of 3115 m. Travertine is deposited extensively throughout the valley, consisting of travertine-dammed pools often beautifully decorated with rimstone, small cascades and extensive travertine terraces, and spring crusts up to 170 m in width (Figure 1). Exceptionally beautiful pools, with water coloured by algae, dominate the upper third and the extreme toe of the travertine run. Rounded cascades form a continuous cascade in the steeper middle section, and the swathe of yellow gives the site its name. The extent of exposed active travertine is nearly matched by fossil material covered by grass, scrub, and woodland. Isolated trees also grow on the lower parts of the active travertine. A boardwalk up the edge of the active travertine provides controlled visitor access, and Buddhist temples beside the highest pools attract crowds to an annual festival. Situated among snow-covered mountains, the site is perhaps the most spectacular of its kind in the world. Mean air temperature is 1.1°C and the region is extensively snow-covered during winter, so visits are normally only possible during the warmer months.

Huanglong and Jiuzhaigou: Figure 1. Terraced pools behind travertine dams in the middle section of the "yellow dragon" at Huanglong. (Photo by Tony Waltham)

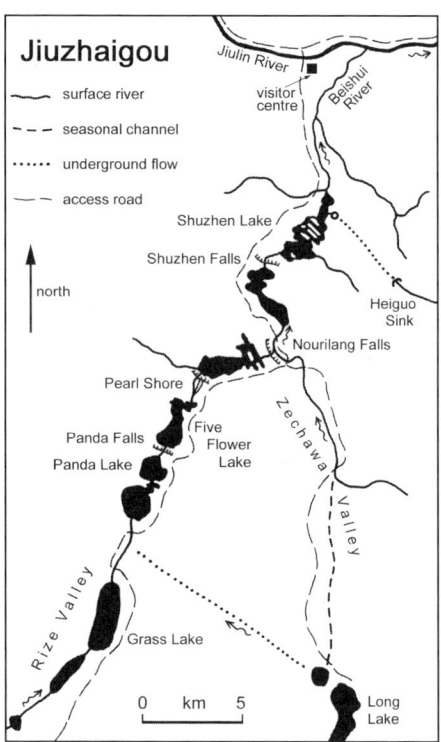

Huanglong and Jiuzhaigou: Figure 2. Outline map of the lakes behind the tufa dams of Jiuzhaigou, China.

The Huanglong travertines have been the subject of much research directed toward the relationships between the deposition of travertine, water chemistry, and hydrodynamics. Travertine deposition rates rarely exceed 5 mm a^{-1} and have been found to be strongly dependent upon the hydrodynamics of the system. Where water flows rapidly, such as over the lips of travertine dams, deposition rates are about four times higher than those in the stagnant water of pools. This results from the reduced boundary layer thickness when flow is rapid, affecting the rate of carbon dioxide escape to the atmosphere (see separate entry, Travertine). Although algae are abundant on the travertine, a biological component to travertine deposition has not been identified to date. The site is also notable for deposition of a rare mineral form of calcium carbonate (vaterite), requiring a high supersaturation with respect to Ca^{2+} and CO_3^{2-} ions. Ths mineral is unstable and soon reverts to calcite.

Jiuzhaigou is situated 50 km north of Huanglong, at a slightly lower altitude, much of it within magnificent forests of the Minshan karst, dominated by Huangshan (Yellow Mountain). This system arises in two separate valleys (Rize and Zechawa), fed by springs at an altitude of about 3000 m, meeting at a point about 15 km south of Juizhaigou village, where they form the Beishui River (Figure 2). Long Lake, in the Zechawa Valley, drains underground for a distance of 8 km, descending 200 m along a band of steeply dipping Palaeozoic limestone to springs in the western Rize Valley. Further downstream, the Heigou tributary also sinks into a cave, which drains along the strike of the limestone to a resurgence beside the Beishui. The Y-shaped stream system, about 30 km in total length, contains one of the largest series of travertine-dammed lakes in the world. The Juizhaigou World Heritage Site covers an area of 620 km^2 and contains a total of 114 lakes. Many of the largest lakes (up to 3 × 0.8 km) occur about 10 km south of Jiuzhaigou village, where most of the travertine is deposited. All of them are dammed by large accumulations of travertine, which form waterfalls of impressive size. Panda Lake is remarkable for its clear turquoise-blue water, containing stromatolitic travertine reefs smothered in diatoms and resembling coral in its depths. The lake is terminated at its lower end by the highest falls in the system (70 m). Below, Pearl Shore is an extensive area of turbulent water, flowing over a gently dipping mass of actively forming travertine. It is reached by a series of wooden walkways passing across the river that forms a complex series of runnels with water depth of just a few centimetres. At Pearl Falls, travertine deposition rapidly increases, forming a series of cascades 20 m high and 310 m in length. Below Pearl Falls is Nourilang Lake and Falls, one of the largest single travertine waterfalls of the system, over 20 m high and brightly coloured by algae and bryophytes. Next comes Shuzhen Lake, which is noteworthy for its complex morphometry, consisting of a series of breached, part-drowned and part-exposed travertine dams, up to 10 m high and only 1–2 m in width. At the upper end of the lake are Shuzhen Falls and Shore, a series of small but remarkably intricate and varied travertine cascades, and broad well-vegetated slopes covered with bryophytes and small trees.

Less is known about the chemistry of the Juizhaigou system, but the waters appear to be similar to those of Huanglong. It is probable that, in common with other travertine dammed lakes, calcite precipitation is augmented by the long residence times of water in the lakes and photosynthesis by the limnic flora. The rich, aquatic flora has been well studied. Over 200 species of algae have been found in the waters, with diatoms (*Achnanthes, Cymbella*) and cyanobacteria (*Calothrix, Schizothrix*) particularly well represented on the travertine. Bryophytes are conspicuous, especially on the travertine dams, the commonest species being *Bryum setchwanicum, Fissidens grandifrons,* and *Palustriella commutata*. The two former species have been found to deposit travertine between infoldings of the leaves, leading to the suggestion of biologically induced calcite precipitation. If supported by further evidence, this observation would be of significance in travertine dam formation, since bryophytes colonize the tops of dams and may be instrumental in their formation and maintenance.

ALLAN PENTECOST

See also **Travertine**

Further Reading

Pentecost, A. & Zhang, Z. 2000. The travertine flora of Jiuzhaigou and Munigou, China, and its relationship with calcium carbonate deposition. *Cave and Karst Science*, 27: 71–78

Pentecost, A. & Zhang, Z. 2001. A review of Chinese travertines. *Cave and Karst Science*, 28: 15–28

Zaihua, L., Svensson, U., Dreybrodt, W., Daoxian, Y. & Buhmann, D. 1995. Hydrodynamic control of inorganic calcite precipitation in Huanglong Ravine, China: Field measurements and theoretical prediction of deposition rates. *Geochimica et Cosmochimica Acta*, 59: 3087–97

HUAUTLA CAVE SYSTEM, MEXICO

The remarkable Sistema Huautla ranks as the 10th deepest (-1475 m) and 32nd longest (56 km) cave system in the world. It is the second deepest explored cave in the Americas and fourth longest in Mexico. Seventeen entrances in the Municipio de Huautla de Jimenez, a remote region of the Sierra Mazateca, Oaxaca, Mexico, provide access to the system. Dyed water from the cave exited at a spring in Santo Domingo canyon, about 1700 m lower and south of the explored Huautla system and the potential relief for the cave is 1850–1900 m (see Figure). Exploration continues, attempting to establish a connection between the upper Sistema Huautla and the lower, little explored, Huautla Resurgence. Current limits of exploration at both ends requires logistically challenging dives through water-filled passages (Stone, 1997).

The earliest record of exploration in the Huautla caves dates to the Classic Period (200–900 AD). Artefacts representing ceremonial use lie in chambers well beyond daylight. Modern cave exploration began in 1966 when American and Canadian cavers began exploring the region's caves by entering Sótano de San Agustín, which remains the most commonly used entrance to the system. The 1966 team established an American depth record of -606 m in short time under increasingly hostile surface conditions. Through an unfortunate stroke of fate, the Huautla area had become a popular destination of hippies seeking the indigenous psilocybin mushrooms, and foreigners were aggressively discouraged from visiting the area. One rope was cut during a caver's ascent to the surface (miraculously resulting in only minor injuries). Cavers abandoned the area until 1976.

From 1976 through the late 1980s, a mostly American team explored and mapped the system, extending the known cave to 53 km long and 1353 m deep. Extending the depth along several routes required entering water-filled passages and the team dived to the extent that standard scuba techniques allowed. They also explored caves and springs in the Santo Domingo canyon approximately 10 km to the south.

An expedition in spring 1994 culminated with a closed-circuit (rebreather) equipped dive through two sumps totalling 655 m of water-filled tunnels at the -1353 m level in Sótano de San Agustin. After the tragic death of one diver, two of the team broke into a giant, dry passage and extended the Huautla system to its present length and depth (Stone & am Ende, 1997; Stone, am Ende & Paulsen, 2002). Their progress was stopped at Sump 9, described as "one of the largest underwater tunnels yet discovered in Mexico" (Stone, 1997, p.172). As of spring 2003, no one else has explored the two kilometres of passage beyond the first sump at the bottom of Sótano de San Agustín.

Sistema Huautla developed in an extensive Cretaceous carbonate sequence that has not yet been studied in detail. Within the east-verging, Early Tertiary Laramide fold-and-thrust belt, intensely faulted and folded structures control the hydrologic flow paths and, hence, cave passage development. Mylonitic marbles crop out in some deep sub-surface exposures along fault zones. As determined in other major cave systems in Mexico's Cretaceous carbonates, including Sistema Purificación and Sistema Cheve (Hose, 1996; 2000), passages probably formed along north–south-trending structural discontinuities such as axial cleavage and primary and auxiliary fault planes.

The various entrances and upper passages in Sistema Huautla initially formed at the edge of clastic caprock contacts that have retreated towards the west. Hence, the eastern passages are older than those to the west. The extensive vertical development of the currently explored passage above the sumps formed in the

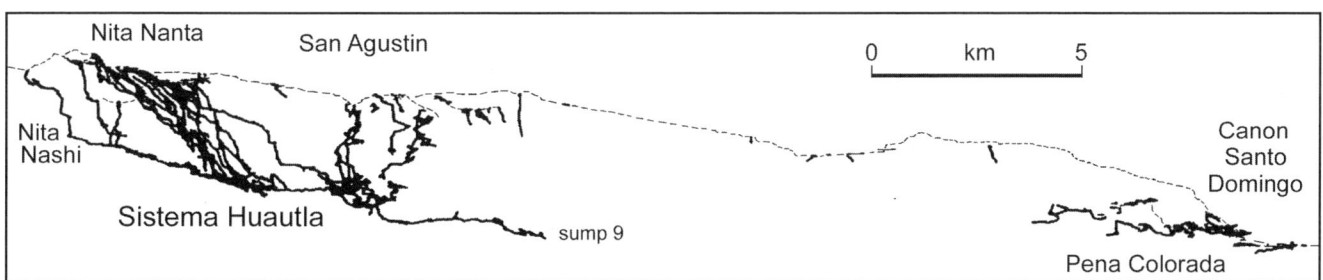

Huautla Cave System, Mexico: Line plot of caves in the Sistema Huautla area viewed from the southwest. Solid lines represent mapped cave passage. Broken lines show surface surveys. Sótano de San Agustín provides the most commonly used access to the main Huautla cave system (after compilation by Stone, 1997).

unsaturated zone by aggressive water flowing into the caves after flowing over the clay-rich, clastic terrains. Horizontal, commonly large passages in the lower portion of the system probably formed at the top of the regional phreas. As the river in the Santo Domingo canyon cut down and lowered the regional base level, those passages have mostly drained.

Exploration continues in the area, but the frontiers lie in extremely remote locations. Most of the diving leads require decompression dives beyond the usual limits of scuba technology. Hence, cavers have developed light and compact closed-circuit rebreathers suitable for transporting to and through extremely remote cave passages (see entry, Diving in Caves).

The Huautla project has spawned other major caving projects in Oaxaca. A higher, more remote region to the east called the Cerro Rabon attracted cavers. They found a virgin rain forest and a wealth of caves, including Mexico's fourth deepest cave, the 1223 m deep Kijahe Xontjoa, and caves rich in archaeological material. South of Huautla, across the Santo Domingo canyon, lies Sistema Cheve with a proven depth potential of 2540 m (Hose, 2000), where explorers have already mapped over 25 km to a depth of −1484 m. Making it North America's deepest explored cave

Louise D. Hose

See also **America, Central**

Works Cited

Hose, L.D. 1996. Geology of a large, high-relief, sub-tropical cave system: Sistema Purificación, Tamaulipas, México. *Journal of Cave and Karst Studies*, 58(1): 6–21

Hose, L.D. 2000. Speleogenesis of the Cheve cave system, Mexico. In *Speleogenesis: Evolution of Karst Aquifers*, edited by A.B. Klimchouk, D.C. Ford, A.N. Palmer & W. Dreybrodt, Huntsville, Alabama: National Speleological Society

Stone, B. 1997. The 1995 Río Tuerto Expedition. *Association of Mexican Cave Studies Activities Newsletter*, 22: 163–72

Stone, B. & am Ende, B. 1997. The 1994 San Agustin Expedition. *Association of Mexican Cave Studies Activities Newsletter*, 22: 44–64

Stone, W., am Ende, B. & Paulsen, M. 2002. *Beyond the Deep: The Deadly Descent into the World's Most Treacherous Cave*, New York: Warner

Further Reading

Stone, B. 1995. History of Huautla exploration. *Association of Mexican Cave Studies Activities Newsletter*, 21: 17–30

Stone, B. & Raines, T. 1997. History of Mexican speleology to 1992. *Association of Mexican Cave Studies Activities Newsletter*, 22: 25–48

HUMAN OCCUPATION OF CAVES

There are few aspects of the human experience that are such worldwide phenomena as the use of caves and few natural features in the landscape that have so excited the imagination (see also Folklore and Mythology). Cave use is an activity that spans the full temporal range of *Homo sapiens*, extending over half a million years from the present. Some finds hint at cave use by other members of the hominid family, though the context and biological status of the early humans represented in some caves by their fossils or artefacts is rarely unambiguous (see entries, Africa, South: Archaeological Caves and Atapuerca Caves, Spain). It is interesting to note that while cave use is a behavioural trait of many animals it does not appear to be one widely shared by humans' closest relatives, the other higher primates, though this may be because caves are not usually a feature of their preferred habitats. If we wish to compare human cave use with similar behaviour by other mammals we have to turn to the social carnivores, the wolves and hyaenas, both of which regularly use caves. Bears can provide another point of comparison though their use of caves is mainly as secure quarters for hibernation. Although the use of caves is not a uniquely human trait, some aspects of that use certainly are.

The wide range of different activities undertaken in caves by humans over several hundred thousand years fall broadly into two categories. Some activities may be described, rather loosely, as economic. The most frequently encountered activity is residence, either short or long term, and the word "troglodytes" or "cave-dwellers" from the Greek *troglodutes*, is a common English-language term, often used in a derogatory sense. Numerous examples exist of caves being used as workshops, often for "industrial" activities that needed to be segregated from other aspects of everyday life, such as the activities of the Roman bronze worker who established himself in Poole's Cavern, Buxton, England, or the small community of ropemakers who occupied the entrance to the Peak Cavern, Derbyshire, England until the middle of the 20th century. A more recent example is provided by the cave of Bedeilhac in the French Pyrenees which was used as an aircraft factory during World War II. Caves have also been used for storage and their use as repositories for rubbish is witnessed by the hundreds of prehistoric middens found in the mouths of caves along the Atlantic coasts of Europe from Cantabrian Spain to northern Scotland. The ancient Chinese used caves as sources of raw materials such as minerals, water, and chemicals, and swiftlet nests, the essential ingredient of bird's nest soup, are still harvested along with bat guano fertilizer at Niah Cave in Sarawak (see Organic Resources in Caves). It is some aspects of behaviour of this kind that humans have in common with many other animals. Wolves and hyaenas live in caves and dispose of their food debris there; elephants and antelopes have been reported entering caves in search of minerals and water, and many small mammals cache food stores in caves.

It is the second broad aspect of cave use that is uniquely human, the use of caves for ritual purposes. Two activities can be described under this heading, the use of caves as theatres for ritual, usually evidenced by cave art (see Art in Caves) and / or the presence of votive deposits, and the use of caves as burial vaults (see Burials in Caves). However, not all human remains found in caves are the result of formal burial and the victims of rockfalls can probably be classified along with the remains of bears that died in hibernation, while some human fragments should probably be treated as the remains of another cave-using carnivore's meal. Formal burial is, of course, a form of ritual but not all rituals carried out in caves involved human remains.

The distinction between economic and ritual behaviour may be partly false as it arises from the application of a 21st-century rationalist perspective which may not be entirely appropriate in other contexts. We know from ethnography, ethnohistory and everyday experience that many aspects of economic behaviour have a ritual dimension, while ritual behaviour often has an economic aspect.

The use to which a cave can be put depends on a number of factors of which the most important are its size and shape. A single burial can be squeezed into a small, dark crevice but residential activity requires space and light, while the more arcane of rituals seek obscurity and concealment. In terms of their suitability for human use it is possible to classify caves according to whether they are open (or day-lit) chambers which in some cases are indistinguishable from rock shelters, or deep fissure caves, although both types are often found as part of the same system. The former can be, and frequently are, used for both economic and ritual activities, whereas deep caves are rarely used at all and then only for ritual purposes. A third category consists of cavities formed within rockfalls (see Talus Caves), although these have more in common with rock shelters than true caves and are not discussed further.

Although open caves exhibit evidence for both economic and ritual use, this rarely appears to have been simultaneous and at many caves it is possible to document a change from one to the other. A frequently encountered change is from economic to ritual use, although in some cases caves went through several cycles of change. For example, at St Columba's Cave on the west coast of Scotland, a period of late prehistoric residential and industrial use preceded a phase of Early Christian ritual activity associated, tenuously, with the eponymous saint, while in the following Norse period the cave became the scene of industrial activity only to revert to being used as a burial ground in the 13th century. It appears to be a worldwide phenomenon that open caves, initially used for economic purposes, often become the scenes of ritual activity that may or may not include human burial. The timing of this change is, of course, not synchronous in an absolute sense but does seem to coincide with other, fundamental changes of an economic and social kind. The most common example is the change from hunting and gathering to farming. The view that formerly mobile hunters and gatherers, on adopting a more sedentary food-producing mode of subsistence, nevertheless retained proprietorial, emotional, and ideological ties with the more tangible aspects of their former existence is persuasive. In cases where people regarded as farmers are found engaged in the economic use of caves they are often nomadic pastoralists engaged in long distance transhumance, the caves serving as temporary shelters. Gypsies, travellers, and tinkers have habitually used caves throughout Europe in recent centuries and might be advanced as an exception to this rule, but these groups usually have more in common with hunter-gatherers or pastoralists than they have with sedentary farmers, and they often maintain a degree of mobility in their lives in which natural shelters could be seen as a convenient alternative to frail temporary structures (see Figure 1). The common factor that the economic use of caves by humans has is mobility. When people decide to adopt a sedentary mode of existence they usually come out of their caves and the use of these natural shelters may be given over to ritual. An exception to this rule is provided by the conversion of caves into houses

Human Occupation of Caves: Figure 1. The Tinker's Cave near Wick, from an early 20th-century postcard. Taken from Tolan-Smith (2001).

with formally built facades, a phenomenon still widely found throughout southern Europe but also persisting into the early 20th century in artificial caves in the English Midlands.

Deep caves—without natural light, difficult to access, damp, and often poorly ventilated—have rarely been the scene of economic activities, except in extreme circumstances of social unrest when they could offer a measure of security and the use as caves as places of refuge in times of conflict is widely reported. However, a cave system could as often as not turn out to be a trap as in the case of the massacre, by a force of MacLeods, of 195 members of Clan MacDonald in St Francis' Cave on the Hebridean island of Eigg in 1577 and the trapping of thousands of Japanese soldiers by US Marines in the Biak Caves, Irian Jaya in 1944.

Generally, the use of deep caves is confined to ritual activity. Reports of the penetration of truly deep caves before the development of modern speleology are rare but examples are known from the United States, Central America, southwest Europe (e.g. Ardèche Caves, Vezère Caves), and the Urals. Many caves are comprised of both "open" and "deep" components, and there are numerous archaeological examples of economic activity having

Human Occupation of Caves: Figure 2. The Cliff Palace in Mesa Verde National Park (Colorado, United States), a Pueblo Indian village occupied between 1050 and 1300 AD in the shelter of a sandstone cave that is a fluvial undercut in a canyon wall. (Photo by Tony Waltham)

taken place in cave entrances while the deeper zones bear witness to ritual activity, represented by art or votive deposits.

The importance of adopting a contextual approach to the study of human cave use applies at many scales of analysis, not just in the case of the juxtaposition of different activities. For example, our appreciation of the wonders of cave art is greatly enhanced by attempts to reconstruct the circumstances under which these images may have been viewed originally, taking account of light, shade, and viewpoint and acknowledging a potential role for other sensory stimuli such as sound. Taking a somewhat broader view, in the realm of ritual behaviour caves are not the only natural features imbued with special significance and many societies attribute ideological meaning to a range of natural features such as rocks, trees, streams, waterfalls, and chasms. Ritual cave use needs to be considered within this wider context if it is to be understood.

Similarly, the economic use of caves did not take place within a vacuum and it is unlikely that there has ever existed a society of exclusively cave dwellers. On the contrary, where we have any evidence, cave use often seems to have been an exception and many cave dwellers appear to have spent at least part of their time using other kinds of shelters, which they built themselves.

It would be inappropriate to ignore the social context of cave use even though this dimension of human behaviour is notoriously difficult to appreciate in periods before the emergence of documented history and recorded ethnographic observation, though periods for which such sources do exist illustrate the range of possibilities. In economic terms, caves are usually regarded as low status facilities, and the derogatory sense in which the term "troglodyte" is often used has already been referred to. Among most settled societies cave dwellers are usually considered to be social outcasts, although there are exceptions to this rule such as the cave-dwelling Guanche "noble women" of the Canary Islands. Caves also participate in the wider social arena. The study of votive deposits in Latvian caves dating from the post-medieval period has documented a clear correlation between increasing socio-economic stress and the intensification of ritual behaviour as evidenced by fluctuating rates of coin deposition. Many caves are destinations of pilgrimage as in the case of the grotto of St Bernadette at Lourdes, France, visited each year by thousands of Catholics and Batu Caves, Kuala Lumpur, Malaysia where hundreds of thousands of Hindus participate in the Thaipusam festival (see Religious Sites). These examples offer provocative insights into the social context of cave use regardless of the period under review.

In many parts of the world caves continue to be used in traditional ways but such is their emotive power that their attraction as tourist destinations is posing serious threats to their survival and the practices that take place within them. It has been necessary to restrict, or curtail altogether, visitor access to some of the most spectacular painted caves in France and Spain such as Lascaux, Pech Merle, Niaux, and Altamira while the exploitation for tourism of mummy caves such as those of the Ibaloi at Kabayan, Philippines (see Burials in Caves), is leading to an exponential increase in the rate of decay of the mummies themselves. Tourism may be seen as a 21st-century parallel to traditional pilgrimage and is a pattern of cave use in itself. As such, usage on this scale has to be managed for all and a number of famous caves have been designated World Heritage Sites.

CHRIS TOLAN-SMITH

See also **Art in Caves; Burials in Caves; Religious Sites**

Further Reading

This topic was discussed by specialists from five continents at a major international symposium held in Newcastle, England in 1993 and the publication of the conference proceedings provides the most comprehensive review of the subject. This entry has drawn extensively on material in that volume and readers seeking further information are referred to contributions by Bahn (context of cave art), Branigan (cave workshops), Eddy (the Guanche of the Canary Islands), Flood (aboriginal cave use in Australia), Gofer and Tsuk (cave burials in Israel), Sieveking (the context of cave art), Stone (Maya cave use), and Urtans (Latvian offering caves).

Bonsall, C. & Tolan-Smith, C. (editors) 1997. *The Human Use of Caves*, Oxford: Archaeopress

Chapman, P. 1993. *Caves and Cave Life*, London: HarperCollins

Kempe, D. 1988. *Living Underground: A History of Cave and Cliff Dwelling*, London: Herbert Press

Leitch, R. & Smith, C. 1993. Archaeology and ethnohistory of cave dwelling in Scotland. *Scottish Studies*, 31: 101–08

Tolan-Smith, C. 2001. *The Caves of Mid Argyll: An Archaeology of Human Use*, Edinburgh: Society of Antiquaries of Scotland

HYDRAULICS OF CAVES

Cave hydraulics concerns the behaviour of water in solution conduits, even those that are too small for human entry. Turbulent flow through these conduits is the most diagnostic feature of karst. Turbulence prevails at large flow velocities and conduit sizes, and involves rapid mixing between adjacent flow lines (for details see White, 1988, or any hydraulics text). Conduits are surrounded by myriad fractures, partings, and intergranular pores having little, if any, solutional enlargement. Most are too narrow to permit any but laminar flow (i.e. little or no mixing between adjacent flow lines). In fissures, turbulent flow occurs at Reynolds numbers (N_r) above 1000, where

$$N_r = \rho v w / \mu \quad (1)$$

and ρ = fluid density, v = flow velocity, w = fissure width, and μ = dynamic viscosity of fluid. As fissures widen and flow rate increases, their flow evolves from laminar to turbulent. As a result, dissolution rates increase (mainly in evaporite rocks), and transport of sediment becomes significant. Depending on temperature and gradient, this transition generally takes place when the fissures grow to ~2–20 mm wide.

Conduits receive groundwater recharge in several ways, arranged here in order of decreasing quantity and variability of discharge: (1) sinking streams, which drain large catchment areas and feed large flows directly into conduits; (2) karst depressions, mainly dolines, with small catchment areas that contribute numerous small trickles to the karst aquifer; (3) dispersed inputs through the epikarst; and, in places, (4) diffuse seepage through neighbouring insoluble rocks. The relative contribution of each depends on the geologic setting. Spring outlets are much fewer than recharge sources, and so convergent flow is typical of most karst aquifers.

Discharge is usually the independent variable in conduit flow, depending only on upstream conditions—i.e. catchment area and runoff rate—and so the following expressions are arranged to indicate the hydraulic gradient required to transmit that discharge. During their early phases of development, incipient conduits contain laminar flow, which in planar fissures is expressed by

$$i = (12 Q \mu)/(w^3 b \gamma) \quad (2)$$

where i = hydraulic gradient (head loss/flow distance); Q = discharge; w = fissure width (narrow dimension of fissure cross section); b = long dimension of fissure cross section; and γ = specific weight of water. Other geometries show similar relationships. Note the extreme sensitivity to fissure width. Wide openings require little gradient to transmit their flow, and therefore the head within them tends to be very low compared to that around them, and flow converges toward the large conduits from surrounding smaller ones.

In turbulent flow, hydraulic gradient is expressed as

$$i = (Q^2 f) / (4 \pi^2 r^5 g) \quad (3)$$

where f = friction factor (depends on N_r and wall roughness; usually ~0.05 in conduits), r = conduit radius, and g = gravitational field strength. Here too the gradient is very sensitive to passage size, and conduits have lower heads than do surrounding smaller openings, which favours convergent flow. Branchwork caves are the typical result (see Patterns in Caves, and Palmer, 1991). This equation can be modified to fit any configuration of closed-conduit or open-channel flow. In conduits that are temporarily or perennially water-filled, flow resembles that of pipes, with velocities that vary directly with the amount of discharge. Above the water table (or potentiometric surface), open-channel flow in cave passages has the mean velocity and discharge of surface streams of similar catchment area. Open-channel velocities vary with discharge, but changes in water depth absorb much of the effect of discharge variations. Both conduit types have leaky walls that allow exchange of water with the surrounding zones of laminar flow.

As discharge fluctuates, conduit flow responds differently from the surrounding laminar flow (Figure 1). During low flow, conduits at, or below, the water table provide lines of low head within the aquifer, as do surface rivers, and are fed mainly by dispersed recharge. During floods, most flow enters conduits rapidly through sinking streams and dolines (Figure 2). The dispersed, laminar input remains large but has much lower velocities and broader flood peaks. As conduit flow increases, the head in the conduits rises above that of neighbouring laminar-flow fissures, reversing the local gradients and flow directions (Figure 1). Passages that ordinarily carry open-channel flow may temporarily become water-filled.

This flow reversal is accentuated by several characteristics of turbulent flow. Equation (3) applies only to straight segments

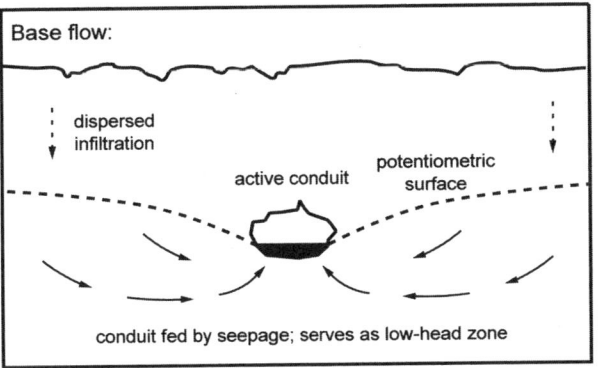
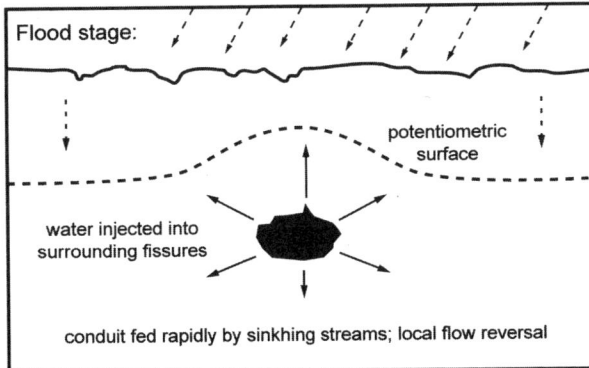

Hydraulics of Caves: Figure 1. Conduit flow at or below the potentiometric surface and its relation to water in the surrounding aquifer during low and high flow. Modified from Palmer (1984).

Hydraulics of Caves: Figure 2. Point-source recharge directly into a conduit at the peak of a severe flood (Cave Disappointment, New York State, United States). During low flow, only a trickle enters the cave at this point, and most of the discharge in the conduit below is supplied by laminar flow from the surrounding aquifer. (Photo by Art Palmer)

of uniform cross section. Bends and constrictions in a conduit cause local steepening of the gradient. In turbulent flow, the gradient increases with the square of the discharge, so the head increases greatly during floods. Finally, because of the enormous sensitivity of gradient to conduit radius (Equation 3), high flows tend to pond upstream from passage constrictions, even those that transmit low flow without difficulty. Rises in water level of more than 100 m have been observed in such areas.

In caves fed by rapid runoff, water is injected into openings in the surrounding aquifer in the same manner that bank storage is delivered by rising surface rivers. As the floodwaters subside, and throughout the period of base flow, water drains back into the conduits from the surrounding aquifer, helping to sustain the flow of cave streams. This frequent injection of solutionally aggressive floodwater produces peripheral mazes, pockets, and blind fissures superimposed on the original passages (see Patterns in Caves).

Above the water table, flow patterns are more complex. Passage floors are leaky, even during low flow, and solutional enlargement of floor fissures by this leakage tends to pirate increasing amounts of flow, eventually creating diversion routes that leave the original paths high and dry. During low flow, vadose seepage through the epikarst is mainly vertical through many small openings. As the recharge rate increases, some openings are unable to transmit all the flow delivered to them. Much of the flow is shunted along less steeply inclined routes to relatively few major drains, which drain water into the heart of the aquifer.

Because they conduct water so easily, one might expect conduits in the vicinity of a pumping well to behave as fixed-head boundaries. In other words, drawdown from a well should terminate at conduits and not extend beyond them. But a conduit's contact with the aquifer is not extensive or intimate enough for it to serve as more than a narrow band of enhanced recharge through relatively few open fissures. The influence of conduits on well drawdown (e.g. during pumping tests) is usually subtle or undetectable, which has often led traditional groundwater hydrologists to entirely ignore the presence or importance of conduits. Perhaps this view is justified from the standpoint of well yield, but it overlooks the essential role of conduits in determining natural patterns of head and groundwater flow within the aquifer, not to mention the rapid transmission of contaminants and their injection into surrounding openings during floods.

ARTHUR N. PALMER

Works Cited

Palmer, A.N. 1984. Geomorphic interpretation of karst features. In *Groundwater as a Geomorphic Agent*, edited by R.G. LaFleur, Boston and London: Allen and Unwin
Palmer, A.N. 1991. Origin and morphology of limestone caves. *Geological Society of America Bulletin*, 103: 1–21
White, W.B. 1988. *Geomorphology and Hydrology of Karst Terrains*, Oxford and New York: Oxford University Press

Further Reading

Atkinson, T.C. 1977. Diffuse flow and conduit flow in limestone terrain in the Mendip Hills, Somerset (Great Britain). *Journal of Hydrology*, 35: 93–110
Bögli, A. 1980. *Karst Hydrology and Physical Speleology*, Berlin and New York: Springer (original German edition, 1978)
Bonacci, O. 1987. *Karst Hydrology, with Special Reference to the Dinaric Karst*, Berlin and New York: Springer
Ford, D.C. & Williams, P.W. 1989. Karst Geomorphology and Hydrology, London and Boston: Unwin Hyman
Gale, S.J. 1984. The hydraulics of conduit flow in carbonate aquifers. *Journal of Hydrology*, 70: 309–24
Gunn, J. 1983. Point recharge of limestone aquifers—a model from New Zealand karst. *Journal of Hydrology*, 61: 19–29
Klimchouk, A., Ford, D.C., Palmer, A.N. & Dreybrodt, W. (editors) 2000. *Speleogenesis: Evolution of Karst Aquifers*. Huntsville, Alabama: National Speleological Society
Lauritzen, S.-E., Abbott, J., Arnesen, R., Crossley, G., Grepperud, D., Ive, A. & Johnson, S. 1985. Morphology and hydraulics of an active phreatic conduit. *British Cave Research Association Transactions*, 12(4): 139–46
Palmer, A.N. 1972. Dynamics of a sinking stream system, Onesquethaw Cave, New York. *National Speleological Society Bulletin*, 34(3): 89–110
Williams, P.W. 1983. The role of the subcutaneous zone in karst hydrology. *Journal of Hydrology*, 61: 45–67

HYDROCARBONS IN KARST

An important class of hydrocarbon reservoirs is related to karst features, especially paleocave systems. Some of the areas where paleocave-related hydrocarbon reservoirs have significant production are the Lower Ordovician, Siluro-Devonian, and Permian carbonates of west Texas, Lower Ordovician carbonates of Oklahoma, lower Cretaceous carbonates of Mexico, lower Ordovician carbonates of Bohigh Bay, China, and Devonian to lower Carboniferous carbonates in Kazakhstan. As examples of the size of this resource, the Lower Ordovician Ellenburger Group in west Texas contains 3703 million barrels of oil in place (Holtz & Kerans, 1992), and the Golden Lane field in Mexico has estimated ultimate reserves of 1900 million barrels of oil equivalent (Greenlee & Lehmann, 1993). One well from the Golden Lane field, the Juan Casiano No 7, had an initial production of 35 000 barrels per day and produced 70 million barrels of oil (Viniegra & Castillo-Tejero, 1970). Another well, the Potrero del Llano No 4, went out of control, blowing more than 100 000 barrels of oil per day, and is credited with having produced more than 95 million barrels of oil.

A paleocave system is defined as a cave system that is no longer related in time or space to the active karst system that formed it (Loucks, 1999). The paleocave system is commonly characterized by extensive collapse of former passages and cementation of clasts (angular pieces of rock) into lithified breccias. A paleocave hydrocarbon reservoir is a reservoir where the pore network consists of karst-related cavernous porosity and / or of cave-collapse-related interclast and fracture porosity. Generally the porosity does not consist of a pore network related to a single cavern, but consists of a pore network resulting from the coalescing of collapsed caverns within a cave system or multiple cave systems.

Paleocave hydrocarbon reservoirs have complex histories of formation and are a product of near-surface karst processes (epigenic caves) and later burial compaction and diagenesis. Some caves (hypogenic caves) form from fluids originating in the deeper subsurface, but these are relatively rare (Palmer, 1991) and have not been documented as having formed hydrocarbon reservoirs. Near-surface cave development processes include dissolutional excavation of passages, breakdown of passages, and cave-passage sedimentation (Figure 1). The most extensive reservoirs appear to form in areas that had composite unconformities. Composite unconformities form over periods of a few million to tens of millions of years, where the host strata stay near the Earth's surface during several rises and falls of sea level. The resulting near-surface cave systems are three-dimensional, cavernous, megapore networks containing varying amounts of breccias and sediments. Between the passages is seen relatively tight host rock with minor fracturing resulting from stress release near caverns.

Near-surface dissolutional excavation and cave sedimentation terminate as cave-bearing strata are buried into the deeper subsurface. Extensive mechanical compaction is initiated, resulting in the collapse of remaining passages, rebrecciation of existing large blocks and slabs, and redistribution of porosity types (Figure 1). Loucks (1999) analysed 19 buried paleocave systems, to identify the relationship of pore type with depth of burial. He found that caverns could remain intact down to approximately 3000 m. As cave passages collapse, new breakdown breccias form from their ceilings and walls. The volume of interclast porosity increases, whereas cavernous porosity decreases. The areal cross-sectional extent of brecciation and fracturing may be greater than that of the original passage because of the collapse of the cave ceiling and walls (Figure 1) into the former cavern. Sag features and faults commonly form over the collapse passages (Figure 1). Continued burial leads to more extensive mechanical compaction of previously formed breakdown clasts, causing

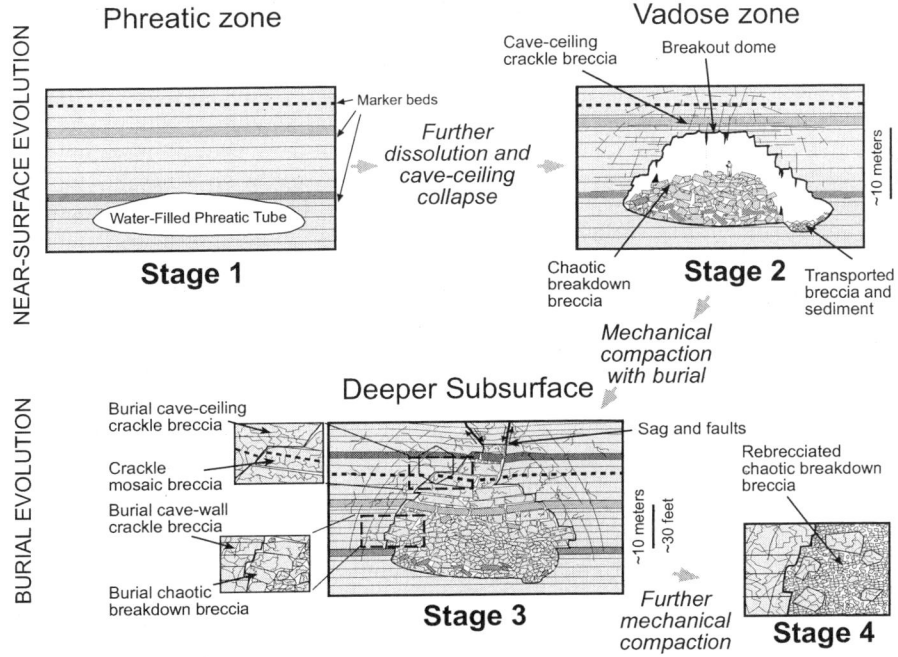

Hydrocarbons in Karst: Figure 1. Evolution of a cave passage during base-level drop, followed by deeper burial and compaction. Each stage of evolution has been verified through the study of modern cave systems and outcrops of paleocave systems. Modified from Loucks & Handford (1992) and Loucks (1999).

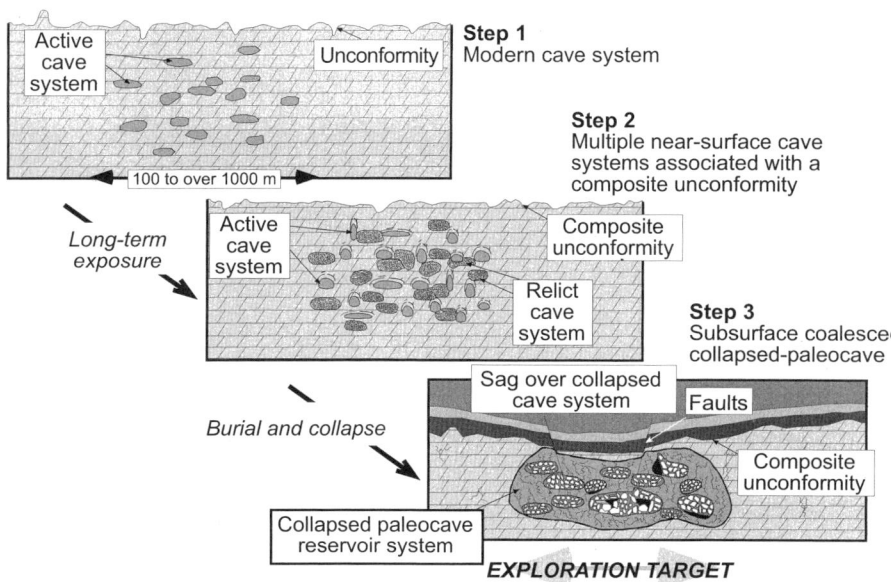

Hydrocarbons in Karst: Figure 2. Evolution of a cave system with burial and compaction. Faults and sags associated with the collapsed paleocave system are based on outcrop exposures and seismic analysis. Modified from Loucks (1999).

blocks with large void spaces between them to fracture, brecciate and pack closer together. As burial compaction continues, more fracturing occurs in the surrounding host rock, as well as in the chaotic breccias. This mechanical process expands the volume of crackle breccia and fractures farther out into host rock. The resulting brecciated zone is not as simple or well defined as the original passage. It now consists of a zone of compacted chaotic breccias, commonly overprinted with stress fractures. Radiating out from this larger pod or tube of compacted chaotic breccias, is a zone of crackle breccias and fractures. The end product is a coalesced collapse-paleocave system (Figure 2). This coalesced system has a pore network much finer and more diffuse than the original near-surface cavernous cave system.

The pore network, which provides the storage capacity and delivery system for the hydrocarbon reservoir, is correspondingly modified during burial. The pore network in a near-surface cave system is composed of caverns, large interclast voids, fractures in clasts, walls, and ceilings, and interparticle matrix in cave-sediment fill. Caverns can be preserved down to several thousand metres of burial but eventually collapse, forming smaller interclast pores and crackle and mosaic breccia fractures. Cave-sediment fill commonly is tightly cemented during burial diagenesis. Hydrocarbon reservoirs in shallowly buried paleocave systems are composed of cavernous, interclast, and fracture pores. In deeply buried reservoirs, the pore networks are composed mainly of crackle-breccia and fracture pores with some finer interclast pores.

A core description through a paleocave reservoir is presented in Figure 3. Three collapse-cave passages are filled with muddy cave sediment fill and clast-supported, matrix-rich chaotic breccias. The oil-producing pore network consists of open fractures in crackle breccia. The chaotic breccias are relatively tight because of the mud matrix.

Several concepts and tools can be used to explore for karst-related hydrocarbon paleocave reservoirs. Cave systems form in carbonate intervals, associated with major composite unconformities. These are the intervals in the stratigraphic column where

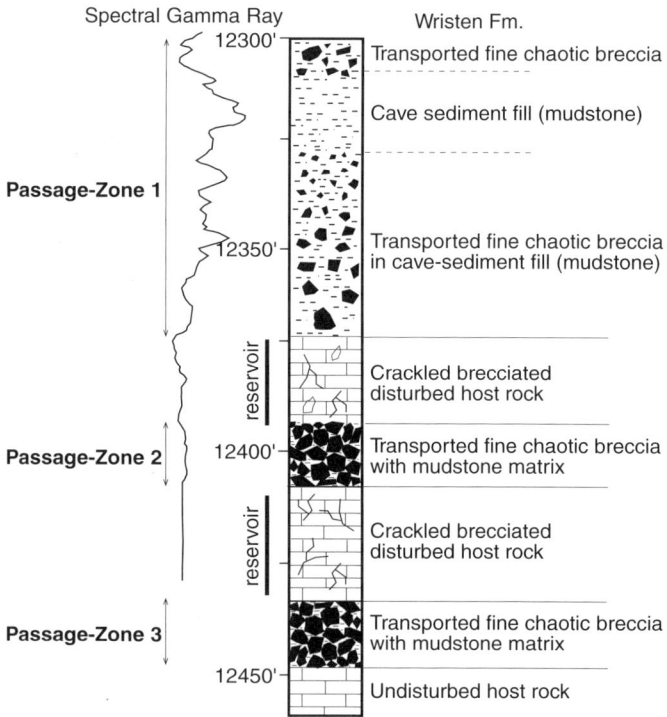

Hydrocarbons in Karst: Figure 3. Core description of a paleocave reservoir from the Siluro-Devonian Wristen Formation in the Exxon J P Ryan Trustee #1 well, Gaines County, Texas. Three levels of collapsed cave passages are filled with cave sediment and breccias. Response of the spectral gamma-ray log reflects argillaceous cave sediment fill. The reservoirs are in the crackled brecciated disturbed host rock. Entzminger & Loucks (1992) and Loucks (2001).

Hydrocarbons in Karst: Figure 4. Drilling for hydrocarbons near Carlsbad, New Mexico. (Photo by John Gunn)

exploration should be focused. At unconformities in the subsurface, evidence of karstification can be obtained from wireline logs, cores, bit drops, and two-dimensional and three-dimensional seismic surveys. Discontinuous argillaceous cave-sediment fills can be identified from conventional well logs, such as spontaneous-potential and gamma-ray logs. These cave fills can be used to map out potential paleocave trends. Interpreted seismic-scale features, such as missing reflections associated with structural sags, cylindrical faults, and anomalous thickening in shallower strata, can be used to identify possible coalesced paleocave systems.

ROBERT G. LOUCKS

Works Cited

Entzminger, D.J. & Loucks, R.G. 1992. Paleocave reservoirs in the Wristen Formation at Emeral field, Gaines-Yoakum Counties, Texas. In *Paleokarst, Karst Related Diagenesis and Reservoir Development: Examples from Ordovician-Devonian Age Strata of West Texas and the Mid-Continent*, edited by M.P. Candelaria & C.L. Reed, Texas: Permian Basin Section, Society of Economic Paleontologists and Mineralogists

Greenlee, S.M. & Lehmann, P.J. 1993. Stratigraphic framework of productive carbonate buildups. In *Carbonate Sequence Stratigraphy; Recent Advances and Applications*, edited by R.G. Loucks & Rick Sarg, Tulsa, Oklahoma: American Association of Petroleum Geologists

Holtz, M.H. & Kerans, C. 1992. Characterization and categorization of West Texas Ellenburger reservoirs. In *Paleokarst, Karst Related Diagenesis and Reservoir Development: Examples from Ordovician-Devonian Age Strata of West Texas and the Mid-Continent*, edited by M.P. Candelaria & C.L. Reed, Texas: Permian Basin Section, Society of Economic Paleontologists and Mineralogists

Loucks, R.G. 1999. Paleocave carbonate reservoirs: origins, burial-depth modifications, spatial complexity, and reservoir implications. *American Association of Petroleum Geologists Bulletin*, 83(11): 1795–1834

Loucks, R.G. 2001. Modern analogs for paleocave-sediment fills and their importance in identifying paleocave reservoirs. *Gulf Coast Association of Geological Societies Transactions*, 71: 195–206

Loucks, R.G. & Handford, C.R. 1992. Origin and recognition of fractures, breccias, and sediment fills in paleocave-reservoir networks. In *Paleokarst, Karst Related Diagenesis and Reservoir Development: Examples from Ordovician-Devonian Age Strata of West Texas and the Mid-Continent*, edited by M.P. Candelaria & C.L. Reed, Texas: Permian Basin Section, Society of Economic Paleontologists and Mineralogists

Palmer, A.N. 1991. Origin and morphology of limestone caves. *Geological Society of America Bulletin*, 103: 1–21

Viniegra, O.F. & Castillo-Tejero, C. 1970. Golden Lane fields, Veracruz, Mexico. In *Geology of Giant Petroleum Fields*, edited by M.T. Halbouty, Tulsa, Oklahoma: American Association of Petroleum Geologists

Further Reading

Candelaria, M.P. & Reed, C.L. (editors) 1992. *Paleokarst, Karst Related Diagenesis and Reservoir Development: Examples from Ordovician-Devonian Age Strata of West Texas and the Mid-Continent*, Texas: Permian Basin Section, Society of Economic Paleontologists and Mineralogists

Fritz, R.D., Wilson, J.L. & Yurewicz, D.A. (editors) 1993. *Paleokarst Related Hydrocarbon Reservoirs*, Tulsa, Oklahoma: Society of Economic Paleontologists and Mineralogists

Kerans, C. 1988. Karst-controlled reservoir heterogeneity in Ellenburger Group carbonates of West Texas. *American Association of Petroleum Geologists Bulletin*, 72(10): 1160–83

Mazzullo, S.J. & Chilingarian, G.V. 1996. Hydrocarbon reservoirs in karsted carbonate rocks. In *Carbonate Reservoir Characterization: A Geologic–Engineering Analysis, Part II*, edited by G.V. Chilingarian, S.J. Mazzullo & H.H. Rieke, Amsterdam and New York: Elsevier

Roehl, P.O. & Choquette, P.W. 1985. *Carbonate Petroleum Reservoirs*, New York: Springer

ICE IN CAVES

There has been confusion regarding the definition of "ice caves" as opposed to "glacier caves", described separately in this encyclopedia. Ford and Williams (1989, pp. 351–55) define ice caves as caves in rock containing both seasonal and perennial ice. Ice caves have also been referred to as *glacières* (Balch, 1900). The title here removes the confusion by its general reference to ice of any kind found in caves, but "ice cave" is used in the sense of Ford and Williams.

Seven types of ice have been recognized in caves, each resulting from distinctive mechanisms of formation (see Table). Specific environments are required for these types to form. For example, how does one explain the existence of perennial cave ice where the mean annual temperature is above zero? While in a number of cases ice caves may persist due to earlier climatic conditions extending over a few years to millennia (Silvestru, 1999), current meteorological conditions may allow ice to form in temperate climates.

The terms static and dynamic have been used to divide ice caves by their meteorological behaviour. For example, caves with only one entrance, leading down into a closed passageway with little air circulation, might be referred to as static, whereas multi-entrance caves experiencing significant airflow are dynamic. However, such a division may not be useful, as every cave is dynamic to varying degrees, and ice types cannot necessarily be assigned to one meteorological framework or another.

Five mechanisms of cave ice formation have been recognized and, whereas one may dominate at a given site, many ice caves are thought to be a combination.

Trapping of Cold Air

Balch (1900) recognized the cave trap type where dense, cold winter air displaces lighter warm air as it sinks to the lowest point in the cave, maintaining freezing temperatures there (static environment). The effect is commonly found in downward sloping caves with a bottleneck or funnel entrance, but existing accumulations of snow or ice are often required. Halliday (1954) provides several examples throughout the United States and Martin and Quinn (1991), in a lava ice cave near Trout Lake, Washington state, United States, give quantitative support to this meteorological principle of cave air movement. Lava caves appear effective in retaining ice, as they lack the fissured nature

Ice in Caves: Figure 1. Massive ice in Coultard Cave, Crowsnest Pass, Alberta. (Photo by Ian Drummond)

Ice in Caves: Ice types in caves.

Ice Type	Description	Formation	Climate
Drip or Flowstone	Stalactites, stalagmites, flowstone. Polycrystalline, clear to opaque	Freezing of infiltration water	Temperate seasonal
Recrystallized Snow	Opaque to bluish, layered	Accumulation of snow in cave traps which densifies and recrystallizes with infiltration component	Temperate perennial <6°C
Massive Ice	Clear or coarse polycrystalline ice. Occasionally with bubbles	Ponded water that freezes from the top downward. Can incorporate infiltration water and falling hoar ice	Temperate: seasonal or perennial
Hoarfrost	Needles, rosettes and hexagonal plate crystals up to 0.5 m. Small, tapering prismatic crystals	Humid air subliming onto cave walls below 0°C.	Seasonal or perennial, the latter in the permafrost zone
Extrusive	Curving fibrous crystals, similar to gypsum flowers up 20 cm long. Rare	Supercooled water forced through microfissures in rock below 0°C which freezes as it emerges from the cave walls	Margins of the permafrost zone
Intrusive	Massive ice subliming on the caveward face or with hoarfrost formation	Glacier ice intruded into cave passages at the glacier/rock contact	Temperate glaciers
Ice in clastic sediments	Intra-particle aggregations as irregular masses, lenses, needles or soil-like extrusions	Freezing of moist sediments at different rates (fast to slow respectively, see previous formation)	Temperate or permafrost depending on rate of freezing

of limestone and so tend to have limited air circulation (see Figure 3).

Cooling by Unidirectional Ventilation

During the winter, multi-entrance caves promote gravity drainage of dense outside air into the cave, which can cool the rock to below 0°C. In the summer, air within the cave, being cooler than outside, drains to lower entrances drawing in warmer air from outside. For ice to form in these circumstances, mean annual temperatures need to be below 0°C and while this condition is met in some regions, such as in mountainous western North America, it is likely that cave trap conditions with perennial ice are also required. Silvestru (1999) argues that no temperate ice caves are known in which perennial ice is sustained by this mechanism.

Trapping of Snow

The shape of some cave entrances allows snow to be captured during the winter months. Snow can accumulate on steep slopes below the entrance, away from direct sunlight. Steeply descending entrances also allow cold air drainage across the snow slope, aiding snow preservation. Percolation water assists the densification and recrystallization of the snow, which then tends to form perennial, layered deposits conforming to floor profiles. The Volcano Room in Q5, Vancouver Island, has accumulated a large cone of glacier-like ice, which has implications for Holocene climate reconstruction. Occasionally, caves in a warm temperate climate can preserve snow for a few years, as in Eldon Hole, England, which held snow for about two years following the bitter winter of 1947 (Myers, 1962).

Permafrost

Caves at high latitude or altitude may currently be within a permafrost zone, thus ensuring the cave walls to be at freezing temperatures. Ice found in such caves originates from the incursion of moist summer air freezing out on the cave surfaces as hoar frost. Furthermore, massive ice found in such caves appears to incorporate a significant proportion of the summer hoar, which has fallen to the floor (Yonge and MacDonald, 1999). The freezing out of the summer vapour can be so effective that it causes aridity in passages further into the cave, such as in Grotte Valerie in the Canadian Northwest Territory at 61.5°N, with mean annual temperatures of −8°C. In some cases, evaporite minerals, such as gypsum, may form.

Evaporative Cooling

Wigley and Brown (1976) theoretically predicted that winter air could cool further as it enters a cave before rising to the ambient temperature inside. This can be explained by the warming of winter air above the dewpoint temperature by the cave walls. Moisture from the walls can then evaporate into the unsaturated air and the latent heat absorbed cools the cave walls. A

Ice in Caves: Figure 2. Hoar ice in Plateau Mountain Ice Cave, Alberta (Photo by Ian Drummond).

Ice in Caves: Figure 3. Ice stalagmites in South Ice cave, a lava tube cave near Bend, Oregon, United States. (Photo by John Gunn)

"relaxation length", based on the characteristics of airflow (laminar or turbulent), predicts the dimensions of the cold zone. The cold zone is important in explaining the large ice deposits found at the entrance of caves. Castleguard Cave in the Canadian Rockies exhibits winter ice (the cave floods in summer) extending about 600 m into the cave. Similarly the Eisriesenwelt in Europe has extensive ice, but these represent maximum incursions. Scarisoara Glacier Cave in Romania reportedly has a "cold halo" resulting from evaporative cooling (Silvestru, 1999). The term "dynamic" may be appropriate for such caves, as substantial airflows are required for the ice to form.

Because ice caves contain perennial ice, there is interest in its climatic implications. Marshall and Brown (1974) analysed massive ice from Coultard Cave in the southern Canadian Rocky Mountains, thinking it to be glacier ice, which is climatically significant. However, crystallographic study of the ice showed it to be unstrained and therefore not to have been glacially intruded into the cave.

Yonge and MacDonald (1999) measured stable isotope concentrations (deuterium and oxygen-18) of ice types in caves throughout western North America, eventually coming up with an isotopic model for warming and cooling trends within permafrost caves. Surprisingly, these trends required a reverse interpretation to that of glacier ice, although snow accumulation caves are most likely similar to glaciers. Given an understanding of the mechanisms of cave ice formation, combined with the presence of datable organic material, stable isotope climate research has some potential, especially in the recent past. For example, Lauritzen has combined carbon-14 dating with stable isotopes to produce a climate profile in massive, banded ice in Svarthammarhola, Norway. Yonge and Macdonald (1999) have similarly analysed a 970-year-old, massive ice plug in Serendipity Cave in the Canadian Rocky Mountains. Silvestru (1999) has carbon-14 dated wood fragments from upper ice layers in Scarisoara Cave at 870 years BP and suggests that lower layers may have formed during the last glacial stage.

CHARLES J. YONGE

See also **Glacier Caves and Glacier Pseudokarst**

Works Cited

Balch, E.S. 1900. *Glacières or Freezing Caverns*, Philadelphia: Allen, Lane and Scott; reprinted with an introduction by W.R. Halliday, New York: Johnson Reprint Company, 1970

Ford, D.C. & Williams, P. 1989. *Karst Geomorphology and Hydrology*. London and Boston: Unwin Hyman

Halliday, W.R. 1954. Ice caves of the United States. *National Speleological Society Bulletin*, 16: 3–28

Marshall, P. & Brown, M.C. 1974. Ice in Coultard Cave, Alberta. *Canadian Journal of Earth Sciences*, 11(4): 510–18

Martin, K. & Quinn, R.R. 1991. Meteorological observations at Ice Cave, Trout Lake, Washington. *National Speleological Society Bulletin*, 52: 45–51

Myers, J.O. 1962. Cave physics. In *British Caving: An Introduction to Speleology*, 2nd edition, edited by C.H.D. Cullingford, London: Routledge and Keegan Paul

Silvestru, E. 1999. Perennial ice in caves in temperate climate and its significance. *Theoretical and Applied Karstology*, 11–12: 83–93

Wigley, T.M.L. & Brown, M.C. 1976. The physics of caves. In *The Science of Speleology*, edited by T.D. Ford & C.D.H. Cullingford. London and New York: Academic Press

Yonge, C.J. & MacDonald, W.D. 1999. The potential of perennial cave ice in isotope palaeoclimatology. *Boreas*, 28: 357–62

Further Reading

Geiger, R. 1965. *The Climate Near the Ground*, 4th edition, Cambridge, Massachusetts: Harvard University Press (original German edition, 1950)

Racovita, G. & Onac, B.P. 2000. *Scărişoara Glacier Cave Monographic Study*, Cluj–Napoca: Editoria Carpatica

INCEPTION OF CAVES

Inception: *The earliest stage of speleogenesis.*
Inception horizon: *Any lithostratigraphically controlled element of a rock sequence that, by virtue of physical, lithological, or chemical deviation from the predominant facies within the sequence, passively or actively favours the localized inception of dissolutional activity.*

As defined above (Lowe & Gunn, 1997), inception marks the transition from void-free rock to rock with dissolutional voids that allow transmission of groundfluids through the primitive rock mass, especially within bedded successions. This definition is deliberately non-specific as regards the chemical or physical changes causing void formation, the mechanisms driving fluid flow, or the timescales involved. Most models of early speleogenesis (inception) in carbonate rocks are based on water movement through physically formed voids. The alternative view presented here focuses on what might be termed chemical inception within, or at the boundaries of, particular beds (inception horizons).

Inception can commence during sedimentary diagenesis, when sediment changes into rock, but this is neither inevitable nor essential. Some rocks, including weakly cemented oolites and the English Chalk, exhibit primary intergranular permeability. In the Chalk, diffuse, porosity-related permeability does not support widespread cave development. However, conduits do form in these rocks, many guided by bedding-related factors rather than primary porosity (Lowe, 1992). Reef limestones do not undergo a soft sediment phase. Voids enclosed during reef growth are isolated or connected only across limited areas. In such rocks inception of *widespread* dissolutional permeability is inevitably post-diagenetic.

Gestation (Lowe & Gunn, 1997) follows inception, as increased dimensions of interconnected void systems permit extensive laminar flow. For any specific incipient conduit, gestation is complete when turbulent flow begins, but such conduits remain linked to extensive incipient void systems, where inception conditions persist. Whereas in an absolute sense inception is the earliest speleogenetic phase, at any point in time a rock mass may contain a continuum from voids at inception stage, through active and enlarging cave passages, to voids that are being abandoned or destroyed.

During diagenesis inception proceeds slowly throughout depositional basins. Early changes in permeability are inconspicuous, but these minuscule changes eventually produce noticeable effects. Inevitably, such changes imprint within sedimentary sequences even before lithogenetic fissures form and before any secondary tectonic adjustment. Thus, initial inception features relate to sedimentological variation, not to cross-stratal fractures. This has major significance to speleogenesis and potential speleogenetic frameworks.

The Inception Horizon Hypothesis or IHH (Lowe & Gunn, 1997; Lowe, 2000) grew from many field observations and from re-evaluation of earlier studies, emphasizing the fundamental role of stratigraphical constraints. Broadly, the IHH relates inception to specific, favourable, horizons—the inception horizons—within any particular rock succession. Though reworking and amalgamating concepts and underlying observations from many workers, including Waltham (1971) and Ford (1971), the foundation of the IHH is unlike that of earlier theories. Seemingly paradoxically, it stresses that most potentially cavernous successions are initially virtually impermeable. Inception horizons are suggested as being the key to permeability development and, once established, they exert guidance on speleogenesis during the lifetime of the rock mass.

Prior to tectonic uplift, potentially cavernous beds with no surface outcrop lie deeply buried by relatively less permeable or more permeable rocks. In either case, it is generally accepted that acidic meteoric water cannot be the initial agent of speleogenesis in deeply buried rocks. However, hydrocarbon explorations have provided abundant undoubted evidence of deep speleogenesis. Deep drilling in rocks of all ages all over the world confirms dissolutional void development well below the depth of any obvious landscape-related underground drainage systems. Hence, processes other than carbonic acid dissolution must initially be involved, together with drive mechanisms unrelated to gravitational flow. Hydrothermal activity, involving overall upward migration of hot and potentially aggressive fluids via

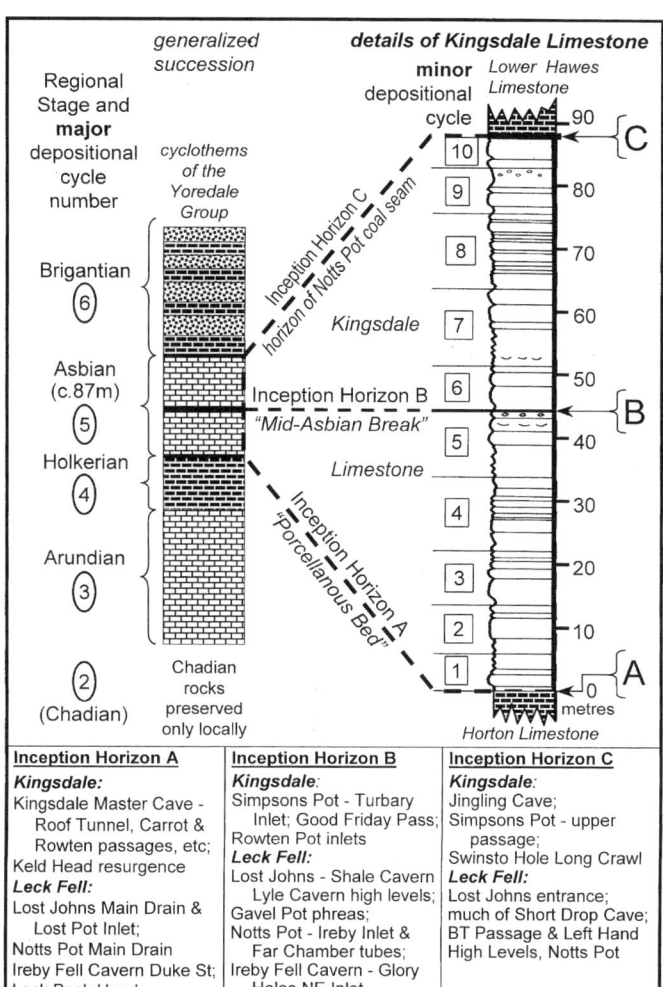

Inception of Caves: Figure 1. The Kingsdale Limestone of the western Yorkshire Dales (England) comprises a single major Carboniferous depositional cycle, containing 10 minor depositional cycles. Of 53 observed bedding planes within or bounding the formation, 11 represent cycle boundaries. Three of these are regionally important inception horizons, which have guided much of the bedding-related speleogenesis in this part of the succession. Selected examples from one area are noted; many other examples occur west and east of Leck Fell / Kingsdale.

fractures and / or bedding-related voids, may play a part in this and cannot be treated as a totally unrelated topic. Lateral fluid migration under diagenetic overburden pressure is also well known (e.g. Neuzil, 1994). Such migrations, where concentrated within or between specific beds, inevitably establish basin-wide weaknesses that are incipient drains. Additionally, earth tides (Davis, 1966), seismic pumping (Sibson et al., 1975), and local ionic diffusion can drive fluid motion in pre- and post-tectonic conditions.

There is no such thing as "homogeneous" sedimentary rock. Even rocks that seem homogeneous incorporate subtle variations of composition, texture, and chemistry. Inception horizons reflect locally significant extremes, many of which are found at depositional cycle boundaries (Figure 1; Table 1). Various types include:

Inception of Caves: Table 1. Relative elevations (metres above Ordnance Datum) and vertical difference between four major inception horizons in five cave systems in the western Yorkshire Dales, England (see also Figure 1). Close similarities of some elevations were formerly assumed to relate to surface "erosion levels". Inter-horizon differences vary across the outcrop, and are not corrected for geological dip or the effects of faulting

Position of inception horizon	Easegill Caverns	Lost Johns	Ireby Fell Cavern	West Kingsdale	Gaping Gill
[Base of Brigantian rocks] Horizon of Notts Pot Coal *[Top of Asbian rocks]*	Probably present but not yet identified	345 m OD	390 m OD	370 m OD	390 m OD
strata between	–	65 m	60 m	55 m	30 m
Mid-Asbian "break"	280 m OD	280 m OD	330 m OD	315 m OD	360 m OD
strata between	52 m	60 m	60 m	50 m	50 m
[Base of Asbian rocks] Porcellanous Bed *[Top of Holkerian rocks]*	228 m OD	220 m OD	270 m OD	265 m OD	310 m OD
strata between	–	–	–	–	50 m
? Mid-Holkerian "break" *[boundary between Cove and Kilnsey limestones]*	Open passages not yet explored at this horizon in these systems; beds may be absent locally, or active conduits may be in the flooded zone.				260 m OD

- obvious lithological contrasts between adjacent beds such as limestone and dolomite, or more subtle contrasts, such as micritic against sparry limestones, or finely against coarsely crystalline gypsum;
- evaporite-rich beds with high physical solubility, deposited under hypersaline conditions late in carbonate depositional cycles. These provide solutions that enhance carbonate dissolution. Buried sulfates may be reduced to sulfides and hence oxidized to acids;
- beds rich in sulfides (particularly pyrite), which can oxidize to produce acids;
- beds with greater primary permeability than adjacent rocks, including sandy beds, oolites, or coarsely crystalline horizons within more compact, finely crystalline, or tightly cemented beds.

Some inception horizons represent discontinuities between contrasting lithologies and hence have virtually zero thickness. For example, the boundary between the Porcellanous Bed (fine-grained, hypersaline limestone), and underlying sparite in the western Yorkshire Dales, England (Figure 1; Table 1) guides cave passage elevations regardless of topographical level. The speleogenetic efficiency of this boundary probably reflects what is missing rather than what remains and it is hypothesized that dissolution has removed late-cycle evaporites that are absent from the idealized cyclic successions. Among many lesser springs, the presence of two major resurgences (Leck Beck Head and Keld Head) at this horizon but at different elevations provides compelling evidence for the role of inception horizons. Farther east, major risings at Black Keld and Malham Cove, also at different elevations, are guided by the Cove / Kilnsey limestones boundary, an inception horizon that corresponds to the Mid-Holkerian Break (Table 1). Thin beds of atypical lithology, such as clay partings or coal seams, also occur in the western Dales (Figure 1; Table 1). A good example is the Notts Pot coal seam, rich in pyrite that provides a source of sulfuric acid. Similar conditions are found at contacts between thicker "clay" beds (terrigenous shales, seatearths, and volcanic tuffs) and carbonate beds. The Mid-Asbian Break (Figure 1; Table 1) is marked locally by a 2 m thick pyrite-rich shale that infills depressions in the highly uneven paleokarstic surface below. Argillaceous intervals are also implicated in guiding early speleogenesis in some evaporitic successions.

In the Forest of Dean (England), many cave passages lie at the upper and lower boundaries of the Crease Limestone. The inception horizons reflect depositional variation across the Arundian / Chadian and Chadian / Courceyan cycle boundaries (Lowe, 1993). Speleogenesis focused along both horizons at least as long ago as Permian times, and iron ore was deposited in significant cave systems during the Triassic. Inception almost

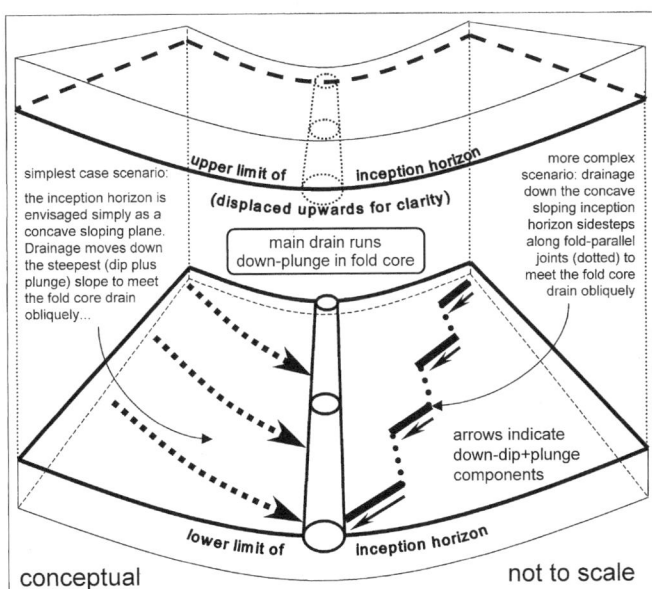

Inception of Caves: Figure 2. Exploded view of how, under "open-system" conditions, many underground main drains guided by unfaulted inception horizons tend to occupy the cores of plunging synclines. Whereas the figure shows one main drain, it can also be visualized as illustrating how two (or more) superimposed main drains can be guided by vertically separated inception horizons within the same folded succession.

certainly commenced during late Dinantian times, c.330 million years ago, long pre-dating the several tectonic episodes that uplifted the area before the onset of current landscape development.

Results of research projects designed to test the IHH are still awaited, but parallel investigations present potentially supportive evidence, and inception horizons have been described in many parts of the world. Strong support for the role of "non-traditional" dissolution derives from the Classical Karst of Slovenia. Here, Pezdič et al. (1998) demonstrated the potential inception role of previously unconsidered chemical–mineral interactions at contacts between limestone and dolomite units. Complex reactions include a process of de-dolomitization in the presence of clay minerals and dissolved silica, and one of the reaction products is free carbon dioxide, a potential source of excess acid.

Objection to the IHH has been raised by results presented by Knez (1996), also in the Classical Karst. Knez studied rock and rock discontinuities at Škocjanske jama in Slovenia, at a level of detail rarely, if ever, achieved elsewhere. He demonstrated conclusively that all "original channels" are concentrated along 3 (of 62 available) bedding planes, which he described as "slip surfaces". This could be viewed as proof that "early cave development" (inception) relates to void-forming fractures, not to stratigraphy. However, it should be noted that the rock mass slipped only on 3 out of 62 bedding planes and it is possible that these 3 planes had been subject to inception prior to the folding and compressional slippage. Hence, they could already have been weakened or may have provided lubricated movement surfaces. Alternatively, the chemical / physical / mineralogical make-up of these beds may have made them particularly susceptible to slip, i.e. the slip potential might have been enhanced by the same factors that support chemical inception.

The examples from England provide strong evidence that atypical beds, or contrasts between them, can provide potential foci for early establishment of laterally extensive discontinuities within untectonized sedimentary sequences. Later, usually following relative uplift, parts of these incipient discontinuities, linked by new tectonic fractures, guide dissolutional conduit development, eventually forming integrated cave systems. Before

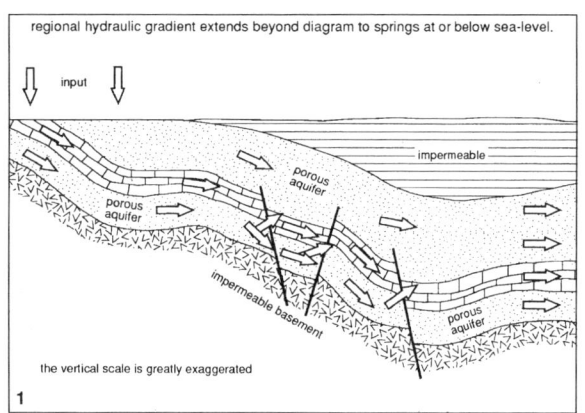
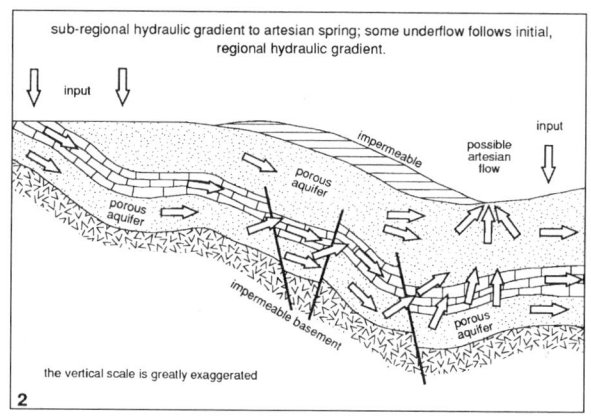
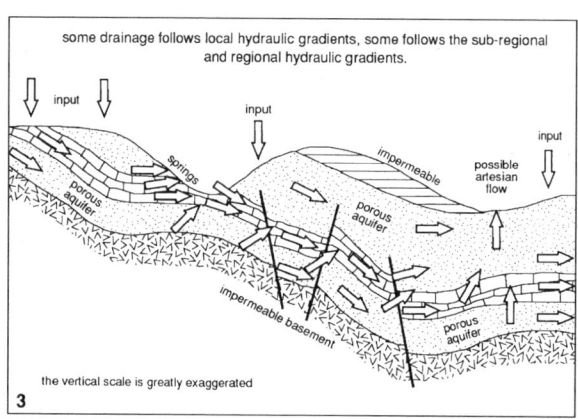
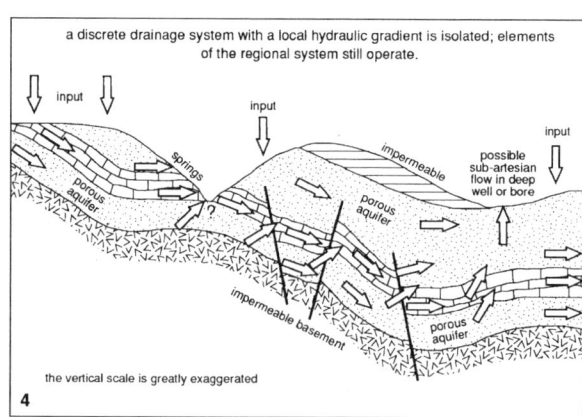

Inception of Caves: Figure 3. Schematic view of the stages in the establishment of local hydraulic gradients in a carbonate rock unit and adjacent porous aquifers confined by impermeable rocks: (1) all underground drainage follows extended routes along regional hydraulic gradients under artesian or deep phreatic conditions; (2) confinement may be broken by erosion allowing establishment of upward movement under artesian conditions; (3) deeper surface erosion intersects the carbonate aquifer to establish the earliest local hydraulic gradient, but some drainage still follows the regional hydraulic gradient; (4) elements of the carbonate bed are isolated from the regional situation and have only local hydraulic gradients, but surviving parts of the regional hydraulic gradient continue to function. (Redrawn from, and used with permission, *Acta Carsologica*, 26(2): 470)

uplift there can be no landscape-related hydraulic gradient as traditionally described. Underground fluid transfer must, however, follow a potential gradient between a start point and a point, or points, where fluids exit the sequence, passing into adjacent more porous rocks or along basin-marginal faults. This applies whether fluids enter the rock mass intra-stratally or originate as formation water within it. Later the situation changes. Pristine rock beds are folded, fractured, and uplifted, imparting the fundamental property of stratal dip. Horizontal beds survive only as local inflexions between dipping rocks, and most so-called horizontal sequences display significant dip when viewed regionally. Additionally, cross-stratal fractures (joints, faults, and thrust planes) inevitably develop. Some have a displacing effect, leading to truncation or isolation of potentially cavernous beds, forming barriers to fluid transfer. Other fractures form inception links between favourable beds at different absolute or stratigraphical levels.

Before uplift, underground fluids are driven along sub-horizontal, bedding-related, inception voids during diagenetic dewatering, as described above. After tectonism, remnant formational water, augmented by meteoric water if new geological and topographical relationships permit, follows combinations of pre-existing voids to move up- or down-dip under artesian pressure or under gravity. Bedding-guided segments, connected locally by lateral, upward, or downward fracture links, follow an overall gradient towards output points that might still be intra-stratal (described above) or at the surface (described below).

Immediately after uplift, before surface (and also submarine) output points are targeted, fluids migrate under closed-system or near closed-system conditions, and may move up-dip (but still down-gradient) under artesian pressure within inception horizon voids. Where structure permits, flow concentrates within inception routes in anticlinal fold axial zones, just as anticlines provide favoured migration routes and host accumulation of hydrocarbons and related water (itself a heretical concept a hundred years ago: Gries, 2002). For example, in Derbyshire (England), routes within buried anticlinal crests provide the favoured means for deep waters to reach thermal springs where inception horizons in fold crests are intersected by the land surface. More easily appreciated is the subsequent, open-system, situation. Gravitational drainage to surface outputs dominates, and many main drains follow plunging synclinal cores, even where fold limb dips are virtually imperceptible. This is displayed *par excellence* on Leck Fell in the western Dales, United Kingdom (Waltham, 1974). Two main drains are "superimposed", separated by c.130 m vertically, within the axial zone of a gentle syncline. Both are guided by inception horizons, the higher (Short Drop Cave) by the "Notts Pot Coal" and the lower (Lost Johns) by the "Porcellanous Bed" (Figure 1; Table 1). Dendritic tributaries join the main drains obliquely, down the dip of the inception horizons on the fold limbs (Figure 2). Analogues occur in adjacent systems such as Easegill Caverns, Pippikin Pot, Notts Pot, and Ireby Fell Cavern.

Newly rejuvenated and tectonized rock masses, incorporating dipping horizons and fractures cutting and / or linking them, are affected rapidly by erosion. Superimposed streams incise valleys into the rocks, cutting through any imprinted regional inception frameworks. Springs form where valleys intersect active inception horizons or fracture links at progressively lower levels (Figure 3). Thus, pre-existing drainage systems along relatively inefficient and extended regional hydraulic gradients are cut into shorter systems with local hydraulic gradients. New local drains rapidly become far more efficient than the former regional routes and capture a greater proportion of local underground drainage (consequently developing and decaying more rapidly). Drainage in segments down-dip of valley intersections might change direction to flow up-dip towards the newly exposed output points. Underflow drainage also continues along deeper regional routes below surface incision level. Even these deeper levels are eventually intercepted, and higher levels become largely abandoned sequentially.

DAVID J. LOWE

See also **Speleogenesis entries**

Works Cited

Davis, S.N. 1966. Initiation of ground-water flow in jointed limestone. *National Speleological Society Bulletin*, 28: 111–18

Ford, T.D. 1971. Structures in limestone affecting the initiation of caves. *Transactions of the Cave Research Group of Great Britain*, 13: 65–72

Gries, R. 2002. Who dares, wins. *Geoscientist*, 12(6): 4–5 & 10–11

Knez, M. 1996. Vpliv lezik na razvoj kraških jam: primer Velike doline, Škocjanske jame. [Bedding plane impact on development of karst caves: an example from Velika Doline, Škocjanske jame caves.], Ljubljana: Znanstvenoraziskovalni centre SAZU

Lowe, D.J. 1992. Chalk caves revisited. *Cave Science*, 19(2): 55–58

Lowe, D.J. 1993. The Forest of Dean caves and karst: Inception horizons and iron ore deposits. *Cave Science*, 20(2) 31–43

Lowe, D.J. 2000. The Speleo-inception concept. In *Speleogenesis: Evolution of Karst Aquifers*, edited by A.B. Klimchouk, D.C. Ford, A.N. Palmer & W. Dreybrodt, Huntsville, Alabama: National Speleological Society

Lowe, D.J. & Gunn, J. 1997. Carbonate speleogenesis: An inception horizon hypothesis. *Acta Carsologica*, 26(2): 457–88

Neuzil, C.E. 1994. How permeable are clays and shales? *Water Resources Research*, 30: 145–50

Pezdič J., Šušteršlć, F. and Misić, M. 1998. On the role of clay-carbonate reactions in speleo-inception: A contribution to the understanding of the earliest stage of karst channel formation. *Acta Carsologica*, 27(1): 187–200

Sibson, R.H., Moore, J.McM. & Rankin, A.H. 1975. Seismic pumping—a hydrothermal fluid transfer mechanism. *Journal of the Geological Society of London*, 131: 653–59

Waltham, A.C. 1971. Controlling factors in the development of caves. *Transactions of the Cave Research Group of Great Britain*, 13: 73–80

Waltham, A.C. 1974. Speleogenesis of the caves of Leck Fell. In *The Limestones and Caves of North-west England*, edited by A.C. Waltham, Newton Abbot: David and Charles

Further Reading

Klimchouk, A. 2000. Speleogenesis in gypsum. In *Speleogenesis: Evolution of Karst Aquifers*, edited by A.B. Klimchouk, D.C. Ford, A.N. Palmer & W. Dreybrodt, Huntsville, Alabama: National Speleological Society

Osborne, R.A.L. 1999. The Inception Horizon Hypothesis in vertical to steeply-dipping limestone: Applications in New South Wales, Australia. *Cave and Karst Science*, 27(1): 5–12

Osborne, R.A.L. 2001. Halls and narrows: Network caves in dipping limestone, examples from eastern Australia. *Cave and Karst Science*, 28(1): 3–14

INDIAN SUBCONTINENT

The Indian subcontinent includes the countries of India, Pakistan, Afghanistan, Nepal, and Sri Lanka. The north of the region was once fringed with Mesozoic and Tertiary Tethyan limestones, from Mozambique in the west to Thailand in the east. This originally continuous belt of limestone was interrupted by tectonism as the Indian plate moved north. The limestones that survive form a garland of isolated blocks of cavernous karst within a tectonically active belt. This stretches from the Baluchistan Plateau (partly in Iran, Afghanistan, and Pakistan), via the foothills of the Hindu Kush, through the entire Himalaya of India, Nepal, and Bhutan, to where it bends sharply south into Burma and the Chittagong Hills of Bangladesh (Gebauer, 1983).

The karst areas of the Indian continental block are on limestones of late Precambrian to Triassic ages. The oldest cavernous rocks are the Archaean limestones in the Arvalli Range on the border of Rajasthan and Madhya Pradesh (Figure 1). In sharp contrast, India also has caves in some very young rocks, including recent calcrete, on the Saurashtra coast of Gujarat. These rocks contain fragments of burned bricks, which can be expected to date to the late Renaissance, when the Portuguese started to colonize the region. Nepal has a cavernous limestone only 500 years old.

There are great climatic variations across the region, ranging from equatorial rain forest in southern India and Sri Lanka, to continental interior deserts in Afghanistan and Baluchistan, to frozen alpine regions in the Himalaya. This great climatic variation has significant impact on the development of karst across the region, and also on the scatter of caves that are known. The summit of Everest consists of limestone (though with no known karst features) and Rakhiot Peak Cave appears to be a solutionally enlarged fissure cave in marble, at an altitude of 6767 m on the avalanche-swept east face of the southern arête of Nanga Parbat.

Man-made caves, and caves used by man, are found all over the Indian sub-continent and are comparatively well researched and described. Hardly explored at all, are the many hundreds of underground towns and communities with up to a dozen levels of cave houses, that were excavated from 1300 BC onwards, in remote locations at high elevations in the Himalaya and Hindu Kush. Culturally, the Indian region is distinctive in

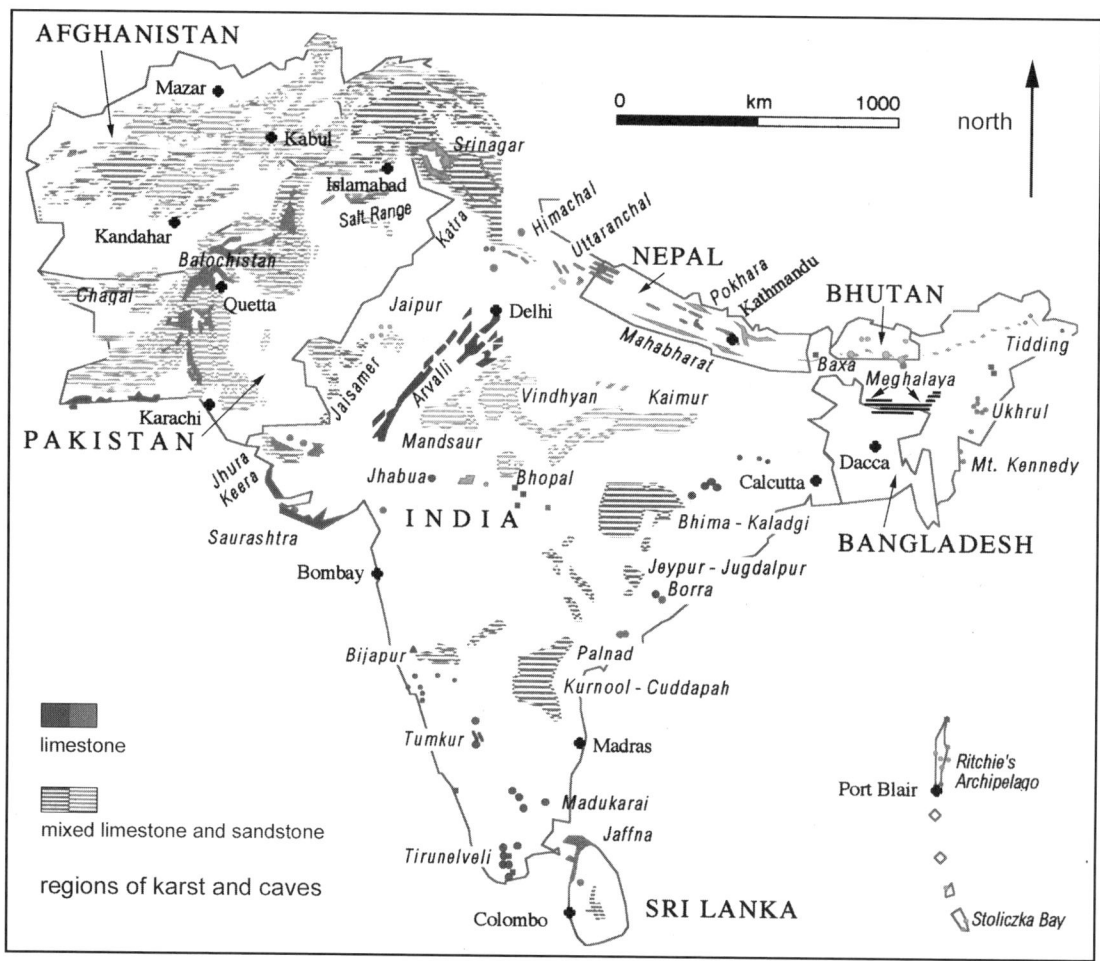

Indian Subcontinent: Figure 1. The main areas of limestone within the Indian sub-continent (compiled by Daniel Gebauer).

that almost every local inhabitant is familiar with a cave experience. In Hindu mythology, any cave is the centre of the universe and no humanity would be possibly without it. Just like an altar is the focus of attention in a Christian church, the "cave" (in Sanskrit: *gupha*) is the centre of every Hindu temple, though that cave may be natural or man-made.

Afghanistan

Within its dry mountainous terrain, Afghanistan has some significant areas of limestone and karst, with a few known caves. In the north, near Mazar-e-Sharif, areas of massive limestone lie in an adversely dry climate that may hinder cave and karst development. The Mar Koh, Safed Koh, and Sayed Koh ranges, above both banks of the Kabul River, contain a mixture of limestone, sandstone, and conglomerate. Both north and west of Kabul, there is limestone and karst that contains Afghanistan's largest caves yet found. Table Rock Cave was explored by American cavers for 492 m, and a French and Spanish team explored the Salang and Bolan Baba massifs of the Hindu Kush, finding Ab Bar Amada, 1220 m long. In the interior desert, the spectacular gour lakes of Band-i-Amir spread down 15 km of a fine canyon (Figure 2). They are fed by waters that rise from limestone springs and then sink into alluvium downstream of the ten travertine barriers, the larger of which are each over 20 m high. The travertine is largely of Pleistocene age, as are many other dry and abandoned deposits in the adjacent mountains. There is karst near Kandahar, with caves of which the longest is Shamshir Ghar (225 m). Caves may also exist in the Paktia region near to the border with Pakistan. In addition to these natural caves, there are numerous man-made "caves" cut into soft conglomerates and sandstone, that have served as religious sites, homes, storerooms, and military strongholds; they include the infamous Tora Bora "caves" in eastern Afghanistan. There is certainly some significant karst waiting to be searched for caves, but political instability renders most of the country inaccessible. A group of Pakistani cavers, based in Baluchistan, conducted some investigations in the Kandahar area in the later part of 2000.

India

India covers a vast area that has disproportionately little karst when compared to its neighbours. Meghalaya, in the northeast, is the notable exception, which has so far yielded over 200 km of surveyed caves, with much more cave and karst remaining to be explored. Further south, Tertiary limestones in the large basins of Kurnool, Cuddapah, Bhima, and Kalagdi, crop out across almost 2000 km^2 on the Deccan Plateau. This area, like almost all Indian karst, is only partly explored (Gebauer & Abele, 1985), but does contain some abandoned river caves, the longest of which is Belum Guhalu at 3225 m. The caves are largely infilled by clastic sediments, except those that have developed into a senile stage of cave destruction (Ruggieri, 1998). In the north, limestone is widespread throughout the Indian Himalaya, but karst development is limited. The Vale of Kashmir has some major karst resurgences (with temples built around them) and the famous Hindu shrine of Amarnath Cave is a rock shelter in thick limestone north of the Vale, but there is no known major cave development. The hills between Simla to Chakrata are known for their many potholes, including Lower Swift Hole that reaches a depth of 70 m.

The northeastern states of India contain the most significant karst. Assam has limestone and karst in the Kopili Valley and the Mikir Hills, in which caves are known, and a few have been explored. Manipur has limited outcrops of Cretaceous limestone, in which some caves have been reported and noted primarily for their prehistoric interest. Mizoram has limestone in which 61 caves have been explored to date, though few exceed 200 m in length. Nagaland has small limestone outcrops in the Tuensang and Phek districts, and there are reports of some cave and karst features. Meghalaya is the state with the most extensive tracts of limestone, which have a depth potential of up to 500 m a hot monsoon climate, and world record rainfall. These ideal factors have combined to create large areas of karst landscape with a wealth of impressive river caves. The limestone, of Cretaceous to Miocene age, forms a more or less continuous outcrop along the southern margin of Meghalaya, encompassing the Garo, Khasi, and Jaintia Hills, and also the Mikir and North Cachar Hills in Assam. The Garo Hills in the west have elevations of 450–600 m, and the land steadily rises to the east where the Khasi and Jaintia Hill rise to 1200–1500 m. All contain significant caves.

Underground exploration in Meghalaya began as early as 1827, largely by surveyors and engineers who noted cave and karst features while investigating exploitable mineral resources, and Kemp and Chopra, from the Indian Museum in Calcutta, made a comprehensive scientific study of Siju Cave in the Garo Hills in 1922. From 1992 onwards, European cavers have worked in cooperation with the Shillong-based Meghalaya Adventurers Association, to explore a substantial amount of caves. Over 840 caves are now recorded and 296 of them have been mapped to date, yielding a total passage length of 226 km. This includes South Asia's longest cave, Krem Kotsati-Umlawan, 21.5 km long (Figure 3), and two more caves are over 10 km long. Most of these caves have active trunk passages 3–12 m high and wide, commonly with sandstone ceilings or floors, where they were initiated along the boundaries of the gently dipping limestone beds (Figure 4); some of the caves carry large streams beneath sandstone outcrops, between inliers of limestone. The largest group of caves is in the Jaintia Hills, around Lumshnong and Sutnga, and this has been the focus of activities in recent years.

Indian Subcontinent: Figure 2. The giant rimstone dam made of travertine nearly 10 m high that retains one of the series of lakes at Band-i-Amir in Afghanistan. (Photo by Bill Renshaw)

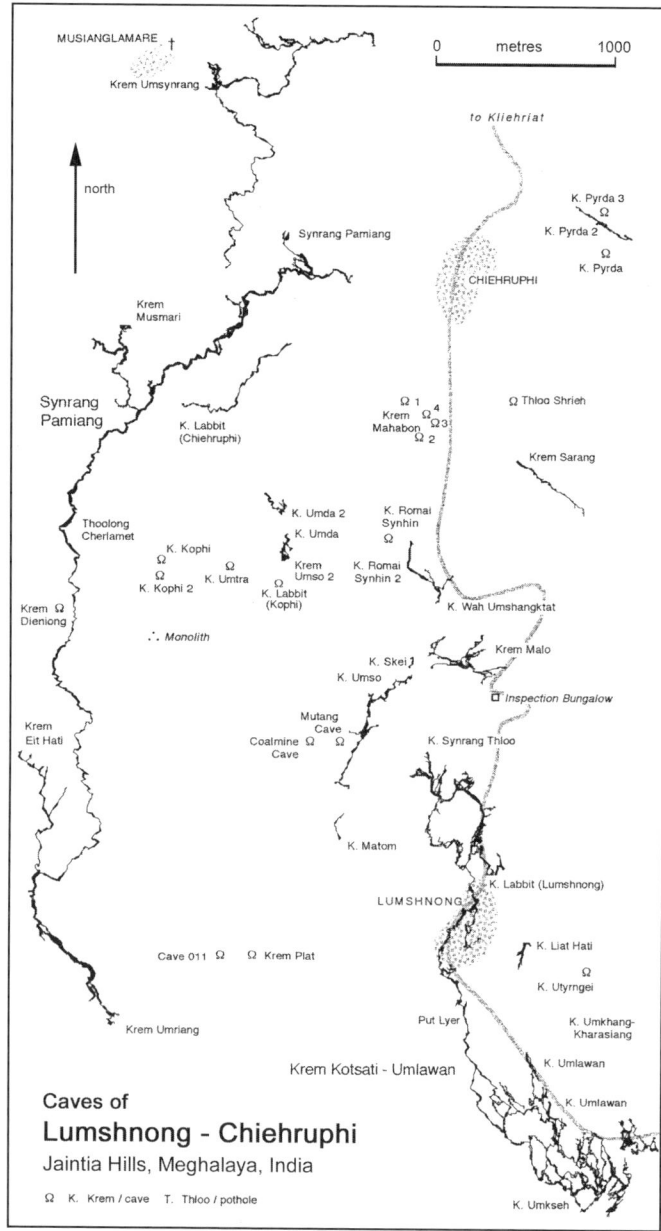

Indian Subcontinent: Figure 3. Outline map of the known caves between Lumshnong and Chiehruphi in the karst of Meghalaya, India (compiled by Daniel Gebauer).

Nepal

Nepal contains huge amounts of limestone but very little karst within the Himalaya. The bulk of Annapurna and Dhaulagiri are formed of the thick Nilgiri Limestone, but it is not pure, and slow dissolution in the cold, high altitudes has created minimal caves and karst landforms. Younger Tethyan limestones lie mainly on the dry north slopes (including the summit of Everest), and the few caves known in the Kali Gandaki zone, are no more than narrow shafts and short rift passages, usually covered or blocked by snow and ice. Far more important are the caves, dolines, gorges, and karrenfields near Pokhara, including Patale Chhango Gupha, with 2960 m of large passages that carry a small river (Waltham, 1996). These are all formed in an impure, conglomeratic limestone, only 500 years old, that originated as a debris flow from massive landslides of the Nilgiri Limestone faces of the Annapurna Himal. The cave probably originated as a piping failure in the poorly consolidated debris, before karstic dissolution features developed as the host material became better cemented. The gorge at its downstream end is a fine example of major cavern collapse. Small caves are scattered through isolated beds of limestone throughout the Himalayan southern foothills; the longest is a 1250 m long phreatic maze in the Chobar Gorge at Kathmandu.

Pakistan

Pakistan stretches from the Arabian Sea up to the high mountains of Central Asia. Much of the country is mountainous, with the Karakorum in the north and the Sulaman (or Suleiman) Range in the south and west. Within these long chains of mountains, some significant areas of karst are formed on limestones of Triassic to Eocene ages. Between the villages of Passu and Sust, in the Hunza Valley through the Karakorum, very hard marble contains some small caves, seldom more than a few tens of metres long, and also some significant karstic resurgences. The Chitral district, further northwest, has limestone but only a few very small caves. Immediately north of Islamabad, the limestone Margella Hills are known to contain over a dozen small single-chamber caves, while north of Islamabad, between Muree and Abbottabad, a large block of limestone also contains small caves. North and south of Peshawar, the tribal areas of Karran and Waziristan both contain extensive tracts of limestone and some recorded caves, including those in the Khyber Limestone that forms the walls of the infamous Khyber Pass.

Pakistan's largest areas of limestone and karst are found in the semi-arid state of Baluchistan. These include limestone surrounding the former hill station of Ziarat, mountainous limestones around the provincial capital of Quetta and the limestones of the Kalat Plateau further to the south, and these contain the

Indian Subcontinent: Figure 4. Main stream passage with flat sandstone roof in Siju Dobhakol Cave, India. (Photo by Simon Brooks)

Indian Subcontinent: Figure 5. Massive calcite speleothems in Moghul Well Gharra in the Baluchistan karst of Pakistan. (Photo by Simon Brooks)

country's major caves (Figure 5). Pir Ghaib Gharra, in the Bolan Pass, is the longest with 1270 m of passage, and Kach Gharra, near Ziarat, is the deepest, reaching a depth of 127 m (350 m long). Depth potential is considerable, and Kach Garra lies at an altitude of over 2200 m, near the top of the limestone, that is well over 1000 m thick. Much of the cave exploration in Pakistan has been since 1990, by British groups working with Pakistani cavers and mountaineers based in Quetta (Brooks, 2001). To date, there are 104 recorded caves in Pakistan with a combined passage length of 5700 m. In October 2000, the Pakistan Cave Research Association was formed to further cave exploration and research; it is based in Quetta, with close links to the University of Baluchistan and the Geological Survey of Pakistan.

Sri Lanka

Previously known as Ceylon, this island has an area of 65 500 km² and variations of climate across lush jungle, rolling hill country, tea plantations, semi-arid areas, and palm-fringed beaches. There are 45 recorded caves on Sri Lanka, most of which seem to be formed in non-karstic granite and gneiss and consist of just one or two chambers, that were commonly used as dwellings in prehistoric times. The karst areas of Sri Lanka are on Miocene limestones in two separate regions. On the northern tip of the island, the barren Jaffna peninsula has a low-level doline karst containing many small shafts and a few short, well-decorated caves, while tidal shafts have been explored by Czech cave divers to a depth of 56 m without reaching a floor. Further south, narrow bands of limestone and marble extend from Trincomalee on the east coast, through the central highlands, to Matara on the southwest coast. It is in this area that the largest and most complex caves on the island are located. The two Istripura caves, reputed to be around 600 m and 150 m long and containing large underground galleries and lakes, are now believed to be drowned under the Victoria Reservoir. Vava Pena (Bat Cave, also known as Wavulpane), situated near to Pallabeda, is a 300 m long stream cave that is home to a large bat colony.

SIMON BROOKS AND DANIEL GEBAUER

Works Cited

Brooks, S.J. 2001. Cave exploration in Pakistan. *Caves and Caving*, 90: 24–28

Gebauer, H.D. 1983. *Caves of India and Nepal*, Schwäbisch Gmünd & Ölmühle

Gebauer, H.D. & Abele, A. 1985. Kurnool 1984: Report of the speleological expedition to the district of Kurnool, Andhra Pradesh, India (*Abhandlungen zur Karst und Höhlenkunde*, 21)

Ruggieri, S. 1997. Cosa succede nel mondo: India. *Speleologia*, 18(36): 127

Waltham, A.C. 1996. Limestone karst morphology in the Himalayas of Nepal and Tibet. *Zeitscrigt für Geomorphologie*, 40: 1–22

Further Reading

Brooks, S. 2001. Caving in the abode of the clouds. *International Caver*, 2001, 40–44

Brooks, S.J. & Gebauer H.D. 1998. *Caving in the Abode of the Clouds, part 2: The Caves and Karst of Meghalaya, North East India*, Derbyshire: Caving in the Abode of the Clouds Project

Fort, M. 1976. Quaternary deposits of the Middle Kali Gandaki valley. *Himalayan Geology*, 6: 499–507

Gebauer, H.D., Mansfield, R., Kusch, H. & Chabert, C. 1995. *Speleological Bibliography of South Asia*, Kathmandu: Armchair Adventure Press

Lapparent, A.F. de 1966. Les Depots de travertins des montagnes afghanes a l'ouest de Kaboul. *Revue de Géographie Physique et de Géologie Dynamique*, 8: 351–57

Pearman, H. 2001. Caves of Afghanistan. *William Pengelly Cave Studies Trust Newsletter*, 84

Waltham, A.C. Limestone karsts of the Annapurna region, Nepal Himalayas. *Cave Science*, 18: 99–104

Wilson, J. 1988. Caves in southern Sri Lanka. *Caves and Caving*, 42: 22–23

INSECTA: APTERYGOTA

The small, primitively wingless Hexapoda, often considered as a subclass of Insecta under the name of Apterygota, include the orders Collembola, Protura, Diplura, Microcoryphia, and Zygentoma. Although their phylogenetic relationships are still controversial, they do not constitute a natural grouping (Bitsch & Bitsch, 2000), and Apterygota is only used today as a practical term. Apterygota are widely distributed worldwide, with many subterranean species, except Protura and Microcoryphia, which are strictly edaphic (i.e. soil-dwellers). Most Apterygota are decomposers living in soil and vegetation. Only the japygid Diplura and some Collembola are predators.

Collembola, or springtails, are small Hexapoda (0.2 to 8 mm) with a number of peculiar morphological features: mouthparts enclosed in the cephalic capsule (entotrophy), four-segment antennae (sometimes secondarily subdivided), a six-segment abdomen (sometimes partly fused), seven-joint legs, and the presence of a ventral tube on the first abdominal segment (see Figure). Many species are able to jump, thanks to the furcal apparatus: a specialized ventral structure of the third and fourth abdominal segments. A post-antennal organ of unknown function, plus eight ocelli, are often present on each side of the head. However, eyes tend to be lost in subterranean and deep soil species.

Springtails, the oldest known Hexapoda (*Rhyniella praecursor* is of Devonian age), are present on all continents. Because of their minute size, they are less well known than other Hexapoda, although knowledge of their taxonomy and ecology has progressed considerably in the last decade. Of the 7000–8000 extant species, more than 400 are known only from subterranean habitats, and two-thirds of them are limited to the Mediterranean basin. Tropical areas, still poorly known, seem to have a less diversified fauna of troglobites, in contrast to a higher diversity of troglophilic and guanobitic springtails.

The families most successful at colonizing subterranean habitats are Hypogastruridae, Onychiuridae, Entomobryidae, Paronellidae, Tomoceridae, and Arrhopalitidae, which include 20 to 200 troglobitic or guanobitic–troglobitic species each. The large families Neanuridae and Isotomidae and the small families Oncopoduridae, Sminthuridae, Spinothecidae, and Neelidae include only a few subterranean species. Cave Entomobryidae (specially the large genus *Pseudosinella*), and to a lesser degree Arrhopalitidae, are widespread on all continents; Hypogastruridae, Onychiuridae, and Tomoceridae are mainly diversified in northern temperate caves, while most Paronellidae are restricted to the tropics.

Collembola are abundant in all terrestrial habitats, from permanent snow to deserts, and from caves to forest canopy. They are usually the most numerous arthropods in subterranean habitats and second only to Acari in soils. They play a prominent role in subterranean communities, at the base of terrestrial invertebrate food webs, where they are preyed upon by a wide range of organisms, including gamasid mites, spiders, pseudoscorpions, and predatory beetles.

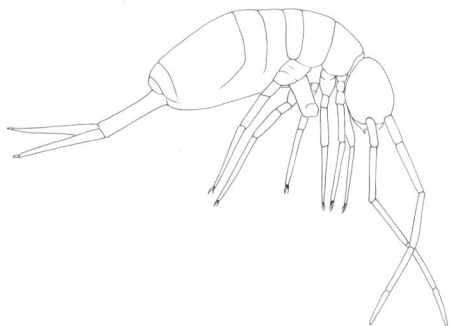

Insecta: Apterygota: *Troglopedetes longicornis*, a troglomorphic Collembola (Family Paronellidae) from a cave in northern Thailand (Mae Hong Son province). Body length: 2 mm.

Troglobitic springtails often exhibit morphological and biological modifications in relation to subterranean life. They tend to have reduced or absent eyes and pigment, elongated legs, claws and antennae, and a larger body size than their epigean relatives (Christiansen, 1965). Biological traits have been documented for a few troglobitic species: fecundity is lower than in related epigean species, development longer, ability to regulate water exchange weaker, respiratory metabolism slower, and resistance to starvation higher, associated with an increase in the lipidic fraction of the tissues (Thibaud & Deharveng, 1994). All these features are usually considered as evolutionary adaptations under the strong selective pressure of a thermally buffered habitat with water-saturated atmosphere and low food-supply.

Cave guano accumulations host a specialized springtail fauna, which has to cope with highly concentrated food resources and stable microclimatic conditions. A few species are adapted to this habitat, where they live in dense oligospecific assemblages. In contrast to troglobitic forms, guanobitic Collembola exhibit only slight morphological modifications in relation to subterranean life.

Diplura are represented underground by probably over a hundred troglobitic species (Bareth & Pagès, 1994). They are relatively large, elongate Hexapoda (body 2 to 9 mm long), of uniform general morphology, white or pink, always anophthalmous, flattened dorso-ventrally, with medium to long legs and antennae. The Campodeidae have two filiform and multi-articulate terminal cerci, while the Japygidae have short, uniarticulate, forcep-like, sclerotized cerci. Campodeids are decomposers linked to humid habitats, mainly edaphic and subterranean, where they may live in dense populations. Most subterranean species belong to the Campodeinae and live in temperate regions, especially in Europe, where they have been relatively well studied. Tropical caves have more recently provided a few troglobitic species of another subfamily, the Lepidocampinae (Condé, 1989). Several campodeids show spectacular troglomorphic traits, especially very long appendages and claws (e.g. *Oncinocampa* sp. from Picos de Europa in Spain). In contrast, japygids, which prey on small arthropods, occur as isolated individuals in caves and are only known from a few localities in the world. None of them exhibits a strong adaptation to underground life.

Microcoryphia (or Archaeognatha) and Zygentoma, formerly grouped in the order Thysanura, are now considered as two independent orders. These medium-sized arthropods (with a body length of up to 2 cm) are usually covered with scales and have the abdomen terminated by two cerci and a long median filament.

Microcoryphia, or bristletails, are characterized by a convex body, two big contiguous compound eyes, small abdominal styli, and the ability to jump by a rapid downward bending of the abdomen. No species is restricted to subterranean habitats, although they may be rather frequent at cave entrances in Europe.

Zygentoma, or silverfish, are dorso-ventrally flattened, with eyes much less developed than those of Microcoryphia, or absent. All are decomposers, some being strictly myrmecophilous (i.e. having a symbiotic relationship with ants and their nests). Only one of the five families of this order, the Nicoletiidae, has representatives in caves, either troglobites or guanobites. Nicoletiidae comprise about 80 species, 20% of which are cave-restricted (Mendès, 1994, modified). Subterranean species are known from a few localities in Mediterranean and tropical regions, and include some highly troglomorphic forms.

LOUIS DEHARVENG

Works Cited

Bareth, C. & Pagès, J. 1994. Diplura. In *Encyclopaedia Biospeologica*, vol. 1, edited by C. Juberthie and V. Decu, Moulis and Bucharest: Société de Biospéologie (in French)

Bitsch, C. & Bitsch, J. 2000. The phylogenetic interrelationships of the higher taxa of apterygote hexapods. *Zoologica Scripta*, 29(2): 131–56

Christiansen, K.A. 1965. Behavior and form in the evolution of cave Collembola. *Evolution*, 19(4): 529–37

Condé, B. 1989. Prodromes d'une évolution souterraine dans le genre *Lepidocampa* Oudemans (Diplura Campodeidae) [Prodromes of a subterranean evolution in the genus *Lepidocampa* Oudemans (Diplura Campodeidae)]. *Mémoires de Biospéologie*, 16: 153–56

Mendès, L.F. 1994. Thysanura. In *Encyclopaedia Biospeologica*, vol. 1, edited by C. Juberthie and V. Decu, Moulis and Bucharest: Société de Biospéologie (in French)

Thibaud, J.M. & Deharveng, L. 1994. Collembola. In *Encyclopaedia Biospeologica*, vol. 1, edited by C. Juberthie and V. Decu, Moulis and Bucharest: Société de Biospéologie (in French)

Further Reading

Cassagnau, P. 1990. Des Hexapodes vieux de 400 millions d'années: les Collemboles [400 million years old Hexapoda: the Collembola]. *Année Biologique*, 29: 1–69

Hopkin, S.P., 1997. *Biology of the Springtails (Insecta: Collembola)*, Oxford and New York: Oxford University Press

INSECTA: COLEOPTERA (BEETLES)

The Coleoptera (beetles) are the largest order within the class of insects, and part of the subclass Pterygota. Due to their importance they are considered separately from Insecta: Pterygota. Coleoptera are holometabolous insects (i.e. they possess a nymph stage), with hard and case-like anterior wings, and membranous posterior wings. The name of the order comes from Greek (*koleos* = elytrum or sheath + *pteron* = wing), meaning insects with anterior wings transformed into hard sheaths, or elytra, that cover the delicate posterior wings.

Of over 300 000 Coleoptera species known to date, the majority are terrestrial, with only about 9000 being aquatic. Approximately 2110 are troglobitic or stygobitic (aquatic), and most of these (83%) belong to the Carabidae: Trechinae and to the Cholevidae: Leptodirinae (Bathysciinae).

The strictly subterranean Coleoptera species belong to 14 families and are widespread on all continents, but the majority are concentrated in the Western Palearctic (especially in the Atlantic–Mediterranean area) and the Eastern Palearctic regions (81%), and in the south of the Nearctic region (13%). The troglophilic (mostly guanophilic) and the subtroglophilic species predominate in the tropical zone, and belong to the Tenebrionidae, to some groups of Carabidae and Cholevidae, to the Histeridae, Scarabaeidae, Ptiliidae, Colydiidae, Cryptophagidae, Hylophilidae, etc. (For details, see Decu & Juberthie, 1998, 2001; Peck, 1998; Newton Jr, 1998.)

Diversity and Distribution

The most important troglobitic and troglomorphic coleopteran taxa are:

Coleoptera aquatica: Only 32 stygobitic taxa are known at present. Most of these were discovered in wells (predominantly), caves, and springs, in areas between 45 degrees latitude north and south in both hemispheres, and they belong to the families Dytiscidae, Noteridae, Elmidae, Dryopidae, and Hydrophilidae. The genera *Morimotoa* (a Japanese genus, with four taxa, e.g. *M. phreatica*) and *Siettitia*, from France (e.g. *S. balsetensis* in Figure 1B, representing the first stygobitic Coleoptera discovered, in 1904); the *Phreatodytes* genus (a Japanese genus, with six taxa, e.g. *P. relictus*); and *Troglelmis laleupi*, from the former Zaïre are all typical of this group.

Carabidae: Of the 12 subfamilies of Carabidae (or ground beetles), the Trechinae show the most impressive adaptations, have a high diversity of species, and show "anophthalmic" and "aphenopsian" morphological types. Of the roughly 2000 taxa, more than 1100 are troglobites or troglomorphs, distributed in the West Palearctic province, the East Palearctic province (southeast China, Korea, and Japan), the Nearctic and Neotropic provinces, Australia, and New Zealand.

Some of the most widely distributed taxa, the most troglomorphic taxa, or the most scientifically investigated taxa are: *Tasmanotrechus* and *Idacarabus*, from Tasmania; *Trechiama* (with about 120 species), *Rakanatrechus* (with 20 species) and *Kurasawatrechus* (with 31 species), from Japan; *Sinaphaenops, Dongodytes, Cathaiaphaenops*, and *Shenaphaenops*, from China; *Duvaliomimus*, from New Zealand; *Paratrechus* (with 30 species) from Central America; *Mexaphaenops*, from Mexico; *Subilsia*, from Morocco; *Aphaenops* (e.g. *A. pluto*, Figure 1C), from France and Spain; *Pheggomisetes*, from Bulgaria, Orotrechus (with 40 species), from Italy and Slovenia; *Duvalius* (with >240 species), from the European Mediterranean coast including the Caucasus, the Maghreb region; and *Anophthalmus* (with 35 species), from Italy, Slovenia, and Croatia.

The Pterostichinae represent the second subfamily of Carabidae, important for the subterranean domain, among which the following are of particular note: *Jujiroa*, from Japan; *Rhadine*, from North America; *Ifridites mateui*, from Morocco; *Speluncarius* and *Wolltinerfia*, from the Eastern Pre-Alps; and *Miquihuana rhadiniformis*, from Mexico.

Cholevidae: After the Carabidae, the Cholevidae family is the one best represented in subterranean habitats. Approximately 746 troglobitic taxa, among which over 90% belong to the subfamily Leptodirinae, have been recorded. Within this family, three morphological types have been distinguished which show progressive stages in adaptation to subterranean environments: the "bathyscioid" type, present in the muscicolous or endogeous species, the less specialized forms, from the genus *Bathyscia, Speonomus* (Figure 1H) or *Diaprysius*; the type "pholeuonoid", from more specialized species, like *Pholeuon* or *Antrocharis*; and the type "leptodiroid", which is characteristic of highly specialized

Insecta: Coleoptera: Subterranean Coleoptera: diversity and distribution (approximate synthesis) (after Decu & Juberthie, 1998, partially modified).

Families	Number of troglobite and stygobite taxa/families	Proportion of subterranean Coleoptera/ families	Zoogeographical regions						
			I	II	III	IV	V	VI	VII
			West Palearctic (principal Atlantic–Mediterranean)	East Palearctic (Southeast China, Korea, Japan)	Nearctic, Neotropical	Australia, New Zealand	Ethiopian	Indian–Malaysian	Melanesian and Polynesian
Aquatic Coleoptera	32	1.52%							
Carabidae	1183	56.3%							
Cholevidae	746	35.2%							
Pselaphidae	82	3.8%							
Staphylinidae	27	1.3%							
Curculionidae	22	1.05%							
Histeridae	12	0.53%							
Other families (Ptiliidae, Scydmaenidae, Merophysidae, Tenebrionidae)	6	0.3%							
Proportion of subterranean Coleoptera by zoogeographical region (%)			68.75	13.06	13.55	2.34	1.4	0.66	0.24

troglobite species, from the genus *Leptodirus* or *Antroherpon* (Figure 1F). In latter case, the subelytral cavity, which has a role in regulating the hydric equilibrum of the body, is extremely large (i.e. it shows false physogastry).

The Leptodirinae troglobites are widespread in most of Europe, south of 45 degrees north, in Turkey, the Caucasian mountains, and some Mediterranean islands. They do not occur in Africa. A single species has been described from Korea (*Coreobathyscia solivaga*). Of the most important leptodirine troglobitic taxa we should mention the following: *Speocharis* (with 36 species) is recorded from Spain; *Speonomus* (with 80 species) from the Pyrenees and Sardinia, comprising the species most comprehensively studied from an ecological and biological point of view (*S. delarouzeei, infernus, longicornis, diecki, pyrenaeus*, etc.); *Isereus xambeui*, which is recorded from France, as well as *Pholeuon glaciale* from Romania or *Bathysciomorphus globosus* from Slovenia and Croatia, are species which populate ice caves at temperatures of 1 to 6°C; *Boldoria* (with seven species) and *Pseudoboldoria* (with 11 species), are recorded from Italy; *Antroherpon* (e.g. *A. dombrowskii* (Figure 1F), with 26 species), is a highly specialized genus, with extremely elongated appendages, and is recorded from Bosnia–Herzegovina, Croatia, Albania, and Montenegro; *Leptodirus hohenwarti*, a highly specialized genus and the first known troglobitic coleopteran (it was discovered in 1832) is recorded from Slovenia, Croatia, and Italy; *Hadesia vasiceki*, recorded from Bosnia–Herzegovina, possesses an oral apparatus specialized for filtering food from the sediment deposited by the flowing water; *Radziella styx* from Croatia, has an extremely elongated body; *Remyella scaphoides* is recorded from Serbia, and some specialists have proposed a fourth morphological type, the "scaphoid" form (Greek *skaphe* = boat), to accommodate it. *Drimeotus* and *Pholeuon* are recorded from Romania; *Bureschiana* is a genus found in Bulgaria and Greece; *Elladoherpon* from Greece and Macedonia is highly specialized, with a leptodiroid form; *Karadeniziella omodeoi* and *Hutheriella* are recorded from Turkey.

Staphylinidae: The approximately 27 known troglobitic species of this family are concentrated in the Canary Islands and Morocco (17), the rest occur in Algeria, Madeira, Spain, Italy, and Romania. The majority belong to the subfamilies Paederinae and Aleocharinae. The Paederinae include *Pignostygus galapagoensis*, a lavicolous relict from the Galapagos Islands, of neotropical origin, and the genus *Domene* with 11 relict species, all subterranean, found in the Canary Islands (e.g. *D. vulcanica*; see photograph in Canary Island: Biospeleology), Morocco (e.g. *D. aurouxi*, Figure 1E), and Spain. Among the Aleocharinae, *Apteranopsis* is present with six relict species in the Canary Islands (e.g. *A. tanausui*) and *Anopsapterus bordati*, with aphenopsian form from Algeria.

Pselaphidae: Most of the approximately 150 troglobitic and eutroglophilic taxa which mainly inhabit networks of cracks and in the Mesovoid Shallow Substratum (MSS) were discovered in the Occidental Palearctic (71) and Oriental faunal provinces, in the United States (41), and only a few were discovered in the intertropical areas of Africa, South America, and Asia. Most of them belong to the subfamilies Goniacerinae (the Palearctic genera *Bryaxis* and *Linderia*; *Tychobythinus*, a Holarctic genus with 14 cavernicolous species), and to the subfamily Batrisinae (with the Holarctic genus *Batrisodes*; *Seracamaurops* (e.g. *S. grabowskii*, Figure 1A); and *Troglamaurops* with nine troglobitic species from the former Yugoslavia, etc.); *Decumarellus sarbui* (Pselaphinae) was discovered in Romania, in Movile Cave (which contains mesothermal, sulfurous water). The sexual dimorphism of the Pselaphidae is extremely marked: there are microphthalmic and apteric males and winged and eyed males which, unlike the females (apteric, microphthalmic, or anophthalmic) may leave the caves (in the case of species of *Linderia* or *Glyphobythus*, for example).

Histeridae: The approximately 12 troglobitic species of Histeridae belong to three subfamilies and are to be found in the southwestern Palaearctic regions and in the south of the Neotropical (Mexico) region. According to Vomero (1998), the small number of troglobitic taxa is caused by extreme trophic specialization and diversification of life forms. Among the Acritinae, it is worth

Italy). They represent rhizophagous (root-eating) or saproxylophagous (rotten wood-eating) forms, anophthalmic, with elongated legs and rostrum, and a smooth, shiny integument.

Other families: The other 24 families of Coleoptera which include some troglobitic and troglomorphic species include *Malkinella cavatica* (Ptiliidae), from South Africa, which show anophthalmy, depigmentation, and apterism (absence of wings) (ADA) features; *Scydmaenus aelleni* (Scydmaenidae), from New Scotland, with the same features; *Gomya troglophila* (Merophysidae), microphthalmous species from Fiji.

Ecology and Adaptations

Habitats and abiotic factors: Subterranean Coleoptera populate all terrestrial and aquatic subterranean environments: caves, and networks of cracks (where most of the individuals live), the MSS, lava tubes, mines and some quarries, wells, springs, and subterranean rivers. They live in total obscurity at a range of temperatures, depending on altitude, between 1 and 25°C, and at close to 100% humidity.

Food: The subterranean environment represents an ecosystem dominated by detritivores. In general, the subterranean Coleoptera have maintained the eating habits of their epigean ancestors. Some of these, adults and larvae (except for the larvae of the Family Cholevidae) are saprophagous, guanophagous, humiphagous, microphagous, mycophagous (Cholevidae, Noteridae, Ptiliidae, Tenebrionidae), or rhizophagous, xylophagous, and cletrophagous (seed-eating) (e.g. Curculionidae, Dryopidae, and Elmidae). Others, adults and larvae, are predators (most of the Carabidae, Dytiscidae, Pselaphidae, Histeridae, Scydmaenidae, and Staphylinidae). There are also subterranean Coleoptera which show trophic specialization: e.g. *Neaphaenops tellkampfi*, which prefers the eggs and nymphs of *Hadenoecus*: Orthoptera (in Mammoth Cave, Kentucky), or *Rhadine subterranea* which eats the eggs of *Ceuthophilus*: Orthoptera (in Texan caves).

Population size: The populations of carnivorous carabid are small; in contrast, the populations of detritivorous leptodirines are large (approximately 1 million individuals, in the case of two species of *Speonomus*, at the MSS site of Ravin de la Tir, in France). A small proportion of these inhabit caves, but most inhabit the cracks connected with the caves. The seasonal migrations from caves to cracks and vice versa, related to climatic and trophic factors, have led to oscillations in the number of cavernicolous populations.

Endemicity: Almost all cave beetle species are endemic. Some are very scarce or are known only from one or two caves. Usually, the range of troglobitic species is one of the smallest known.

Predators and parasites: Two genera of Opilions (*Holoscotolemon* and *Scotolemon*) eat the leptodirines *Closania* and *Speonomus*. Also *Neobisium spelaeum* (a pseudoscorpion) consumes the leptodirine *Leptodirus hohenwarti*. The fungus Hyphomycetes: *Troglobiomyces guignardi* may attack leptodirines (dead or alive) of the genera *Troglocharinus*, *Speonomus*, *Closania*, etc.

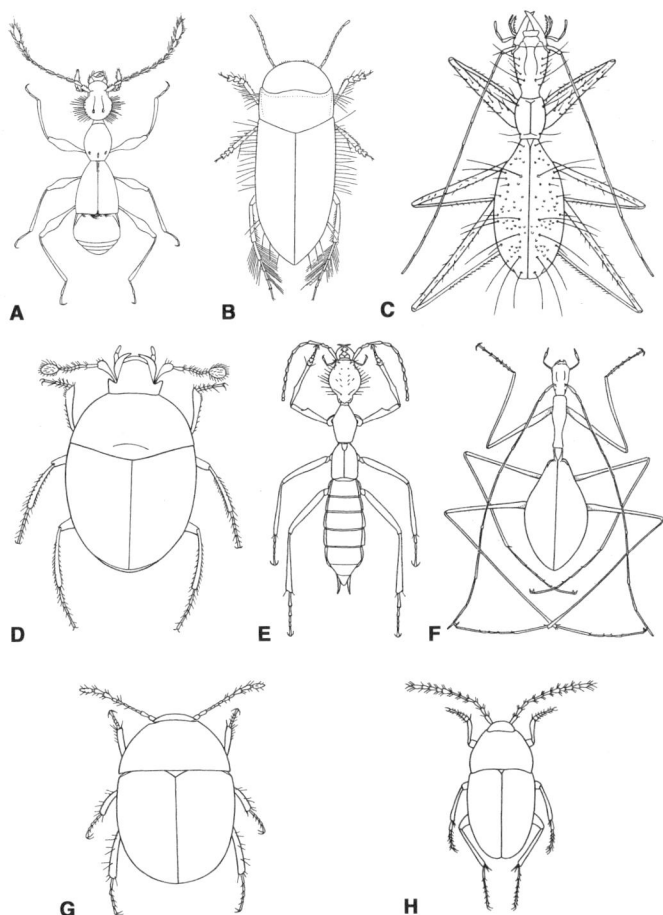

Insecta: Coleoptera: Figure 1. Troglobitic and stygobitic Coleoptera.
A. *Seracamaurops grabowskii* (G. Muller, 1926) (Pselaphidae);
B. *Siettitia balsetensis* (Abeille de Perrin, 1904) (Dytiscidae);
C. *Aphaenops pluto* (Dieck, 1869) (Carabidae: Trechinae);
D. *Spelaeacritus anophthalmus* (Jeannel, 1934) (Histeridae);
E. *Domene aurouxi* (Español, 1970) (Staphylinidae);
F. *Antroherpon dombrowskii* (Apfelbeck, 1907) (Cholevidae: Leptodirinae);
G. *Mehadiella paveli* (Frivaldszky, 1880) (Cholevidae: Leptodirinae);
H. *Speonomus colluvii* (Delay et al., 1983) (Cholevidae: Leptodirinae). (G and H: see entry on Interstitial Habitats [Terrestrial]) (From Jeannel, also 1928; Decu & Juberthie, 1998)

mentioning *Spelaeacritus anophthalmus* (Figure 1D) from Turkey, the first troglobitic hysterid described (in 1934), and *Iberacritus vivesi* from Spain; among the Abraeinae, *Spelaeabraeus agazzi*, from Italy; among the Bacaninae, the *Troglobacanius* genus (from Mexico), with four species, representing the most well-adapted troglobitic hysterids in the New World.

Curculionidae: Taxa discovered only in caves (approximately 30 species) belong to the subfamilies Polydrusinae (most of them) and Molytinae, and were collected in Central Europe (more than half in Italy) and Algeria. Over 22 of these belong to the genus *Troglorhynchus* (e.g. *T. vailati* or *T. amicalis*, from

Morpho-anatomical and ecophysiological adaptations are necessary for life in the subterranean environment, which is one of the most selective. These are common for troglobiotic Coleoptera:

Anophthalmy, depigmentation, apterism (ADA) (Figure 2): In the gradient from soil species to cave species, the eyes progressively disappear and are completely absent in most adapted troglomorphic species. The melanin pigment of the integument disappears, but a yellowish-fawn colour persists. Membranous wings are lacking or are reduced.

Sensory apparatus and length of appendages: A strongly modified sensory apparatus is present, particularly in the carnivorous Carabidae: Trechinae. Presumably, the soil species of the genus *Trechus* possesses the basic specialized sensory equipment of the Trechinae. The selective pressures of the cave environment tend to favour trechine species with a dominant olfactory system on the antennae and long trichobotria on the elytra. The abundance of chemoreceptory sensillae, the surface of olfactory sensillae, and the length of the ampullacea sensillae, increase from soil species, such as *Trechus*, to highly specialized species such as *Aphaenops*, *Hydraphaenops*, and *Sardaphaenops* (Figure 2). In Clivinini species, Nitzu & Juberthie (in Casale, Vigna Taglianti & Juberthie, 1998) demonstrated a gradual increase in the chemosensory functions of the antennae, from soil-dwellers towards neo- and paleotroglobite cave species. Within the spatial constraints of soil microspaces and limestone cracks, burrowing species are characterized by dominant mechanoreceptory equipment on the antennae and chemoreceptory equipment on the maxillary palps, which inform the animal by contact about its immediate environment. In contrast, in species inhabiting large spaces such as caves, olfactory detectors are concentrated on the antennae because these species have an overriding need to sense olfactory information from longer distances. The degree and type of modifications differ from one phyletic line to another.

The increase in appendage length in many carabid and leptodirid beetles is not due to a positive allometric (i.e. differential growth of different body parts) effect, but is simply a result of adaptation to cave life. The increase in antennae length shows a significant correlation with an increase in the number of olfactory receptors.

Reproduction and development: A reproductive K-strategy is well developed in Cholevidae: Leptodirinae with:

1. Reduction in the number of eggs, and production of larger eggs, particularly in the most adapted cave species, with only one egg every month.
2. Reduction of larval instars from three to one (a contracted cycle).
3. In species with a contracted cycle, the larval stage develops with proteic, lipidic, and glucidic products stored in the egg.
4. Protection against predators; the sole larval instar in a contracted cycle spends its entire development in a clay hollow which it has excavated.
5. Loss of reproductive seasonality. However, when the development of cave carabid is known, this shows few features of K-selection, except for in the species *Aphaenops cerberus*. Indeed, the larvae present a normal, three-instars development (for instance, *Trichaphaenops gounellei*).

Increased longevity: The adults of some Leptodirinae (*Speonomus*) and Sphodrina are notably long lived, with their lifetime being estimated at four to six years, which is observed both in the field and in laboratory conditions.

Survival: Like other troglobites, cave beetles store lipids, so they can survive prolonged periods of starvation.

Sensitivity to air dryness: An increase of transpiration through the integument and a decreased lifespan in dry air in one specialized species of *Aphaenops*, is recorded.

Circadian rhythms: Progressive disappearance of circadian locomotory activity is recorded in the transition from troglophilic species with more or less reduced eyes to blind troglobitic species. In microphthalmic species, the maintenance of a circadian clock plays a role in the entrainment of daily light / dark cycles of locomotory activity. In contrast, in blind species, the circadian cycles and the circadian clock system disappear completely.

VASILE DECU AND CHRISTIAN JUBERTHIE

Works Cited

Casale, A., Vigna Taglianti, A. & Juberthie, C. 1998. Coleoptera Carabidae. In *Encyclopaedia Biospeologica*, vol. 2, edited by C. Juberthie & V. Decu, Moulis and Bucharest: Société de Biospéologie (in English)

Decu, V. & Juberthie, C. 1998. Coleoptera (Généralités-synthèse). In *Encyclopaedia Biospeologica*, vol. 2, edited by C. Juberthie & V. Decu, Moulis and Bucharest: Société de Biospéologie

Jeannel, R. 1928. Monographie des Trechinae. Morphologie et distribution géographique d'un groupe de Coléoptères. *L'Abeille*, 35: 1–808

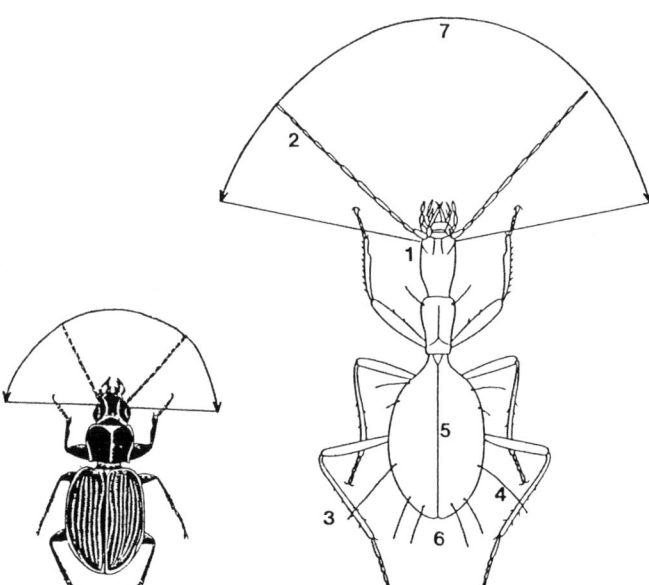

Insecta: Coleoptera: Figure 2. Morphological changes between a soil-adapted *Trechus* and a cave-adapted *Hydraphaenops* Carabidae: Trechinae. 1. no eyes; 2. long antennae; 3. long legs; 4. long trichobotria (conical protuberances with sensory hairs); 5. depigmented; 6. modified body shape; 7. increase in the area able to be examined by the antennae (after Casale *et al.*, 1998).

Newton Jr, A.F. 1998. Phylogenetic problems, current classification and generic catalog of the subfamily Leptodirinae. In *Phylogeny and Evolution of Subterranean and Endogean Cholevidae (= Leiodidae, Cholevinae)*, edited by P.M. Giachino & S.B. Peck, Turin: Museo Regionale di Scienze Naturali

Peck, S.B. 1998. Phylogeny and evolution of subterranean and endogean Cholevinae (= Leiodidae, Cholevinae): an introduction. In *Phylogeny and Evolution of Subterranean and Endogean Cholevidae (= Leiodidae Cholevinae)*, edited by P.M. Giachino & S.B. Peck, Turin: Museo Regionale di Scienze Naturali

Vomero, V. 1998. Coleoptera Histeridae. In *Encyclopaedia Biospeologica*, vol. 2, edited by C. Juberthie & V. Decu, Moulis and Bucharest: Société de Biospéologie (in English)

Further Reading

Bordoni, A. & Oromi, P. 1998. Coleoptera Staphylinidae. In *Encyclopaedia Biospeologica*, vol. 2, edited by C. Juberthie & V. Decu, Moulis and Bucharest: Société de Biospéologie (in French)

Decu, V., Juberthie, C. & Nitzu, E. Coleoptera (Varia). 1998. In *Encyclopaedia Biospeologica*, vol. 2, edited by C. Juberthie & V. Decu, Moulis and Bucharest: Société de Biospéologie (in French)

Giachino, P.M., Decu, V. & Juberthie, C. 1998. Coleoptera Cholevidae. In *Encyclopaedia Biospeologica*, vol. 2, edited by C. Juberthie & V. Decu, Moulis and Bucharest: Société de Biospéologie (in French)

Osella, G. & Zuppa, A.M. 1998. Coleoptera Curculionidae. In *Encyclopaedia Biospeologica*, vol. 2, edited by C. Juberthie & V. Decu, Moulis and Bucharest: Société de Biospéologie (in English)

Poggi, R., Decu, V. & Juberthie, C. 1998. Coleoptera Pselaphidae. In *Encyclopaedia Biospeologica*, vol. 2, edited by C. Juberthie & V. Decu, Moulis and Bucharest: Société de Biospéologie (in French)

Spangler, P. & Decu, V. 1998. Coleoptera aquatica. In *Encyclopaedia Biospeologica*, vol. 2, edited by C. Juberthie & V. Decu, Moulis and Bucharest: Société de Biospéologie (in English)

INSECTA: PTERYGOTA

The Pterygota are characterized by the presence of wings on the meso- and metathorax in adults. They are a diverse group with a number of subterranean species within the different orders. Those with troglobitic representatives are outlined in detail in this entry. Coleoptera are described in a separate entry.

Blattodea (cockroaches)

The Blattodea is a widely distributed order of easily recognized terrestrial insects, which are known from the Carboniferous and have maintained primitive characteristics, more so than the related Orthopteroid groups. It has been estimated that 4000 species exist or occur, but this figure is almost certainly an underestimate. The cockroaches range from 2 mm to 10 cm, with dorsoventrally compressed bodies, cursorial legs, and forewings—when present—modified into tegmina. However females are often wingless. Cockroaches are primarily nocturnal, and are usually found near the ground, hiding under bark, logs, or stones during the day. Some are adapted to arid conditions, but the majority of the species live in tropical forests or in damp habitats, and also in human habitations. Cockroaches are a common feature in tropical caves where they often show gregarious habits. However, only 32 species can be categorized as troglobites (obligate cave-dwellers) and display adaptive characters to an obligate hypogean life, such as depigmentation, blindness, absence of wings, and the structural reduction or absence of pulvilli and arolias (adhesive pads on the pretarsus). Cavernicolous cockroaches are known from Australasia, tropical Africa, Cuba, Mexico, and the Galapagos Islands, but are absent from most of the Holoarctic region with the exception of Canary Islands. In the latter, 13 cave species have been recorded, five of which (*Symploce microphthalma* and four species of the endemic *Loboptera*; see photograph in Canary Islands: Biospeleology) are true troglobites.

Orthoptera (grasshoppers, locusts, crickets)

The Orthoptera comprise more than 20 000 species of well-known heterometabolous insects. They occur over all but the coldest parts of the Earth's surface, being most developed in the tropics. They are often abundant, forming a characteristic and striking component of the fauna in many parts of the world. The Orthoptera range in size from 5 mm upwards and include the largest living insects, attaining 12 cm in length and up to 24 cm wingspan. They are winged (brachypterous) or apterous, with a large pronotum extending posteriorly over the mesonotum. Forewings are thickened and leathery (termed tegmina), hindwings fanlike, occasionally modified for stridulation (production of sounds) or camouflage, and hindlegs are usually large and adapted for jumping. The well-known "singing" is produced by stridulation in the males, usually by rubbing specially modified forewings together, or the hind femur against the tegmen. Most Orthoptera are herbivores, but many are omnivorous and some are even predatory.

Orthoptera are classically divided into two suborders, the Caelifera (grasshoppers) and Ensifera (crickets and tettigonids). Only the Ensifera, characterized mainly by their long antennae and their long female ovipositor, have successfully colonized caves. The most representative groups are some tettigonids (Gryllacridoidea: Rhaphidophoridae) and crickets (Grylloidea). These groups include many cavernicolous species, which represent a typical feature of both temperate and tropical cave ecosystems. Of the 300 Rhaphidophoridae species, more than half are cave-dwellers. It is thought that the origins of this family can be traced to a temperate zone of the Gondwana paleo-continent and roughly dated from the upper Carboniferous to the Jurassic. A typical Paleantarctic distribution is reflected in the subfamily Macrophatinae which is widely distributed, with several genera, in Australia, Tasmania, and New Zealand, and with the sole relictual genera *Heteromallus* and *Spelaeiacris* in Patagonia and the Cape Town region in South Africa, respectively.

The subfamily Rhaphidophorinae is widely distributed in Southeast Asia with the genera *Rhaphidophora*, *Tachycines*, and *Diestrammena* (one of the few cricket genera to include completely blind species) and in the Mediterranean region with the genus *Troglophilus*. The most widespread and numerous cave-dwellers in temperate areas are the genera *Dolichopoda* in southern Europe and *Hadenoecus* (see photo in Mammoth Cave: Biospeleology) and *Euhadenoecus* in the eastern United States, each including several troglomorphic species (showing morphological adaptations to cave life). Finally the Nearctic subfamily Ceutophilinae includes several genera in North and Central America characterized by different levels of troglomorphism.

A similarly complex picture occures in the true crickets (Grylloidea), where adaptation to cave life has taken place several times independently, even in a given tribe. Cavernicolous crickets belong to five families: the Phalangopsidae (e.g. *Paracophus, Longuripes, Arachnomimus, Malgasia, Endacusta*) found in the tropics worldwide, the Gryllidae (e.g. *Acroneuroptila, Discoptila, Petaloptila*) in the Mediterranean region, the Trigonidiidae in the Pacific islands, the Pentacentridae (e.g. *Nemobiopsis*) in Cuba, and the Oecanthidae (e.g. *Caconemobius, Thaumatogryllus*) in Hawaii. Some of them are true troglobites, others colonize caves and other natural cavities such as tree hollows or burrows.

Dermaptera (earwigs)

Nearly 1800 species of Dermaptera have been described from all parts of the world (except the polar regions), but only 15 species included in the suborders Arixenina and Forficulina are cave-dwellers. The earwigs are elongate heterometabolous insects, winged or apterous, with a mobile telescopic abdomen and cerci modified into terminal forceps. They range in length from approximately 7 to 50 mm, and vary in colour from buff to black. They favour damp, confined spaces and, although nocturnal, are attracted to lights. Their food consists of a wide range of living and dead plant and animal matter.

The Arixenina are cave-dwellers, although poorly adapted to cave life. They are dark brown in colour, wingless, with long antennae and legs, and their compound eyes are small but functional. They occur in large numbers in some caves in the larger Indonesian islands forming seething masses over guano: *Xeniaria esau* is a facultative ectoparasite of bats in the Niah Caves of Sarawak, while *Arixenia jacobsoni* behaves as a predator of arthropods in caves in Java.

Most of the Forficulina show nocturnal habits, hiding by day in crevices or other dark shelters. Although they prefer darkness, moisture, and crawl beneath all kind of debris on the ground, few species have been recorded from caves. Some of them show enhanced troglomorphic features such as *Anataelia troglobia* from La Palma (Canary Islands), *Diplatys milloti* from New Guinea, and *Anisolabis howarthi* from Hawaii. Other Forficulina species recorded in caves do not show troglomorphic adaptations. Some species are troglophilic since they have also been recorded in subtropical forests. Other species are colonizers of recent lava tubes.

Psocoptera (Psocids or booklice)

Psocoptera (also known as Copeognatha or Corrodentia) are a small order of heterometabolous insects consisting of about 3800 species from all regions of the world. They range from less than 1 to almost 10 mm in length and have a characteristic appearance due to their round, mobile head; long antennae; enlarged pterothorax; and the wings held like a roof over the abdomen. The adults of most species are winged, but alary polymorphism occurs, and apterism is common in one or both sexes. The Psocids are free-living land insects that primarily live on foliage or branches of trees and shrubs, on or under bark, on fences and walls, in leaf litter, under stones, on rocks, in caves, in human habitations, and in stored products. They feed on unicellular algae, lichens, fungal hyphae, spores, and fragments of plant or insect tissue.

Cave-dwelling species have been reported from Europe, North America, Cuba, Yucatán, Venezuela, Malaysia, Australia, and Congo. Probably none of them can be regarded as true troglobites, but at least ten species show troglomorphic features such as long antennae, pale colour, reduced eyes, and short or rudimentary wings. Some species are characterized by a typical structure of the hypopharynx that allows them to uptake water from humid air, favouring hypogean life. Occurrence of parthenogenesis is also reported in some cave species. *Prionoglaris stygia* is a troglomorphic species distributed in caves of France, Switzerland, Belgium, and Crete. *Troglotroctes ashmoleorum* is another troglomorphic species occurring in lava tubes and in underground barren volcanic rubble in the Ascension Islands. The genus *Speleketor* includes three species, of which only *S. flocki* from Arizona and Nevada is a true cave-dweller.

Hemiptera (bugs, leafhoppers, planthoppers)

The Hemiptera are the dominant group of heterometabolous insects. They range in length from about 1 to 100 mm, and comprise insects with a great range of different structural features and occupying a wide range of different environments. The most relevant feature characterizing the order is the structure of the mouthparts, consisting of hinged stylets resting in a dorsally rostrate labium. The hemiptera feed by suction, most are phytophagous, a few are invertebrate or vertebrate blood-sucking, and some aquatic species are carnivorous.

The suborder Heteroptera includes about 35 000 species, which are more abundant in the tropics than in the temperate zones. The group is characterized by having hemelytra: the first pair of the wings have a parchment-like basal and a membranous apical portion. The second pair form the membranous, flying wing. The suborder Homoptera includes about the same number of species; they are abundant all over the world, and are all phytophagous insects. They have membranous wings. Alary polymorphism is widespread; sometimes both sexes are dimorphic or polymorphic, and in some instances only the female is wingless.

About 70 species found in caves are reported among Heteroptera. While some are probably accidental visitors, others are guanobitic or obligate troglophiles such as in some lygaeids from Trinidad and Peru and some emesine reduviid species from Fiji and Hawaii. In addition the family Cimicidae includes troglophilic species that feed on bats living in the dark zone of caves. So far only four troglobitic/stygobitic Heteroptera species are known. These are the mesoveliid *Cavaticovelia aaa* (Hawaii), the emesines *Collartida anophthalma* (Hierro, Canary islands) and *Nesidiolestes ana* (Hawaii), and the water scorpion *Nepa anophthalma* from Movile Cave (Romania). All these species exhibit various troglomorphic features such as long legs and antennae, strongly reduced or absence of eyes and ocelli, depigmentation, and absence of wings.

The Homoptera appear unlikely candidates for a permanent life underground due to their obligate phytophagous diet. However, subterranean Homoptera species are now known in five families of the Fulgoridea, a large group of worldwide leafhoppers. Up to 60 species with different levels of adaptation to cave life have been described, geographically focusing on Hawaii, the Canary Islands, and Australia. Others have been recorded from the Galapagos Islands, Western Samoa, New Caledonia, New Guinea, New Zealand, the Balearic islands, southern France, the Azores, Jamaica, Mexico, and Reunion. Cave-dwelling planthoppers feed on roots penetrating the cave ceiling. They differ strongly from their epigean relatives with a reduction of compound eyes, ocelli, wings, and body pigmentation, and the most highly modified taxa are completely eyeless, flightless, and unpigmented. As shown in the troglobitic Hawaiian species *Oliarus polyphemus*, courtship and communication in these cave planthoppers is based on vibratory signals transmitted along root curtains.

Trichoptera (caddis-flies, caddises)

Trichoptera are holometabolous insects resembling small hairy moths: in fact they are phylogenetically related to butterflies and moths.

Cavernicolous Trichoptera are represented in the holoarctic region mainly by species from the family Limnephilidae, while in the southern hemisphere the family Hydropsychidae is most common in caves. Most species appear to use caves as a temporary refugial habitat, mainly as diapausing adults. Pre-imaginal instars often occur in flooding caves as result of passive drifting. In the cave ecosystem caddis flies are typical elements of the parietal community, resting on walls not only in the twilight zone, but also in deep, remote passages of caves. Many faunal inventories of caves all over the world list several Trichoptera species, however it is still unclear how many of them are regular cave-dwellers. *Wormaldia subterranea* is a noteworthy species discovered in Slovenian caves where it apparently completes its entire life cycle. However no morphological adaptation allows us to classify *W. subterranea* as a true troglobite.

Lepidoptera (Butterfies and moths)

The Lepidoptera are a relatively homogeneous group of holometabolous insects, including around 165 000 species. As typically phytophagous insects Lepidoptera are poorly cave-adapted organisms. Several moths and a few butterfly species are frequent visitors to caves where they rest on walls for winter or summer diapause (a resting phase of metabolic activity). However several Tineid species, especially in the genera *Monopis* and *Tinea*, complete their entire life cycle in caves where larvae feed on bat guano. In Hawaii several moths of the genus *Schrankia* (Noctuidae) complete their life cycle in the lava tubes, exploiting roots penetrating medium-size voids even within young lava, where they are among the first colonizing organisms.

Diptera (flies, mosquitoes, gnats, and midges)

Diptera are an order of holometabolous insects with one pair of membranous mesothoracic forewings and a metathoracic pair of club-like hindwings called halteres (which function as balancing organs during flight). They have mouthparts adapted for sponging, sucking, or lapping: mandibles, maxilles, and hypopharynx are variously modified as stylets in parasitic and predatory groups; the labium forms a proboscis that in the higher families forms a sponge-like pad with food canals. Their size ranges from about 1 mm to around 55 mm in length with a 10 cm wing span. The Diptera form one of the largest insect orders, with at least 150 000 species described. They occur from subarctic regions to subantarctic regions, but the greatest diversity of species occurs in the tropics.

Diptera are common in the subterranean environment where they occupy a wide spectrum of ecological niches with some species showing adaptations, for example the blood-sucking, spider-like, wingless bat flies (Nycteribiidae) and the bioluminescent, prey-attracting glow worms *Arachnocampa luminosa* (Mycetophilidae) of New Zealand caves. However, in comparison to other cave arthropods Diptera show a remarkably low number of troglobites. Noteworthy are *Allopnyxia patrizii* (Sciaridae) endemic to an Italian cave, *Crumomyia absoloni* and *Spelobia tenebrarum* (Sphaeroceridae) from Herzegovina and eastern North America respectively, and *Mormotomyia hirsuta*, a guanophagous, highly modified, hairy, spider-like fly (the only representative of Mormotomyiidae) from a cave in Kenya.

VALERIO SBORDONI AND MARINA COBOLLI

Further Reading

Baccetti, B., Cobolli, M., Minelli, A. & Ruffo, S. 1991. Phylum Artropodi. In *Zoologia (Trattato Italiano)*, edited by AA, Bologna: Editoriale Grasso

Brusca, R.C. & Brusca, G.J. 1990. *Invertebrates*, Sunderland, Massachusetts: Sinauer Associates

CSIRO 1991. *The Insects of Australia*, 2nd edition, Carlton, Victoria: Melbourne University Press

Vandel, A. 1965. *Biospeleology: The Biology of Cavernicolous Animals*, Oxford and New York: Pergamon Press (originally published in French, 1964)

Badonnel, A. & Lienhard, C. 1994. Psocoptera. In *Encyclopaedia Biospeologica*, vol.1, edited by C. Juberthie & V. Decu, Moulis and Bucharest: Société de Biospéologie

Bouvet, Y. 1994. Trichoptera. In *Encyclopaedia Biospeologica*, vol. 1, edited by C. Juberthie & V. Decu, Moulis and Bucharest: Société de Biospéologie

Brindle, A. & Oromi, P. 1994. Dermaptera. In *Encyclopaedia Biospeologica*, edited by C. Juberthie & V. Decu, vol. 1, Moulis and Bucharest: Société de Biospéologie

Desutter-Grandcolas, L., Di Russo, C. & Sbordoni, V. 1998. Orthoptera. In *Encyclopaedia Biospeologica*, vol. 2, edited by C. Juberthie & V. Decu, Moulis and Bucharest: Société de Biospéologie

Harvey, M.S., Shear, W.A. & Hoch, H. 2000. Onychophora, Arachnida, Myriapods and Insecta. In *Subterranean Ecosystems*, edited by H. Wilkens, D.C. Culver & W.F. Humphreys, Amsterdam and New York: Elsevier

Hoch, H. 1994. Homoptera (Auchenorrhyncha Fulgoroidea). In *Encyclopaedia Biospeologica*, vol. 1, edited by C. Juberthie & V. Decu, Moulis and Bucharest: Société de Biospéologie

Hoch, H. 2000. Acoustic communication in darkness. In *Subterranean Ecosystems*, edited by H. Wilkens, D.C. Culver & W.F. Humphreys, Amsterdam and New York: Elsevier.

Izquierdo, I. & Oromi, P. 1994. Dictyoptera–Blattaria. In *Encyclopaedia Biospeologica*, vol. 1, edited by C. Juberthie & V. Decu, Moulis and Bucharest: Société de Biospéologie

Juberthie, C. 2000. Conservation of subterranean habitats and species. In *Subterranean Ecosystems*, edited by H. Wilkens, D.C. Culver & W.F. Humpreys, Amsterdam and New York: Elsevier

Maldonado Capriles, J. 1994. Hemiptera—Heteroptera. In *Encyclopaedia Biospeologica*, vol. 1, edited by C. Juberthie & V. Decu, Moulis and Bucharest: Société de Biospéologie

Matile, L. 1994. Diptera. In *Encyclopaedia Biospeologica*, vol. 1, edited by C. Juberthie & V. Decu, Moulis and Bucharest: Société de Biospéologie

Sbordoni, V. & Forestiero, S. 1998. *Butterflies of the World*, 2nd edition, Buffalo, New York: Firefly Books

Turquin, M.-J. 1994. Lepidoptera. In *Encyclopaedia Biospeologica*, vol. 1, edited by C. Juberthie & V. Decu, Moulis and Bucharest: Société de Biospéologie

INTERSTITIAL HABITATS (AQUATIC)

The aquatic interstitial habitat constitutes groundwater-filled spaces between grains of unconsolidated and saturated sediments. Such porous biotopes occur in littoral sandy or gravelly sea bottoms and beaches, within freshwater lakes and river margins, alluvial plains, river bars, the river bed, and the hyporheic and phreatic zones. Ward *et al.* (2000), Malard, Ward & Robinson (2000), and Gibert, Danielopol & Stanford (1994) define the hyporheic zone as the porous aquifer extending up to several metres vertically beneath the river bed, and laterally beyond the channel for hundreds of metres. The phreatic zone is the deep portion of an aquifer that does not experience exchange with stream water, in contrast to the hyporheic zone. In the latter, the sediment layer is filled with river water, and just below, there is a mixture of both surface and groundwater. Light is attenuated rapidly with depth in unconsolidated sediments. The stability of other environmental variables (temperature, water flow, dissolved oxygen, organic matter) increases with depth from the river bed. Grain size is the most significant factor because permeability, porosity, dissolved oxygen, organic matter distribution, and water circulation tightly depend on granulometry. Grain size also determines interstitial space size and habitat suitability for organisms within clay, sand, gravel, or pebble substrates. Coarser sediment with low silt / clay content provide larger habitat spaces for stygobites and better groundwater flow. Generally, alluvial deposits exhibit larger grain size than more selected sea shore sediments. Most of the energy inputs are allochtonous. The hyporheic zone of rivers is viewed as a dynamic ecotone (or interface) with surface water–groundwater interactions (i.e. surface water downwellings, groundwater upwellings), and a high bioproduction and energetic transformation. At a microscale, interstices are interconnected, and the porous medium is heterogeneous and constitutes a mosaic of microenvironments.

The granular medium is inhabited by interstitial stygobites. These organisms move within spaces with a minimum of grain disturbance in contrast to burrowers which remove the substrate elements. In spite of the constrained habitat due to darkness, limited energetic sources, and reduced space a great number of phyla have interstitial representatives; for example Protozoa, Cnidaria, Rotifera, Gastrotricha, Loricifera, Tardigrada, Turbellaria, Nematoda, Oligochaeta, Polychaeta, Mollusca, Crustacea, Acarina, and Collembola. Many groups of Crustacea such as Copepoda, Mystacocarida, Ostracoda, Isopoda, Amphipoda, Syncarida, and Thermobaenacea have evolved a number of both marine and fresh groundwater species. Within the hyporheic zone a mixture of strict stygobites and epigean organisms, such as insect larvae which spend the earliest part of their life cycle within sediments, co-exist.

Interstitial stygobites present a set of morphological, behavioural, and physiological traits resulting from an evolutionary adaptation to life in the constrained granular milieu: they are phreatomorphic (Coineau, 2000). Phreatomorphology involves regressive characteristics such as pigment and eye loss. Such regressive evolution may result from the accumulation of neutral mutations, or from mutation and directional selection; it may also result from the non-synchronization of the expression of the *eyeless* master gene and stages of a slow development. Small size is characteristic and there is a strong correlation between grain size and stygobite size. Nematodes, Protozoa, crustacean ostracods, harpacticoid copepods, and bathynellids are generally less than 1 mm long and occur in fine sediments. Polychaeta, Mollusca, Acarina, crustacean amphipods, isopods, and Thermosbaenacea are generally slightly longer (1–3 mm) and live in coarser sediments. Miniaturization in some crustaceans such as Syncarida or Thermosbaenacea may result from progenesis (i.e. heterochrony of development due to precocious sexual maturity inducing small adults to display a larval morphology). Paedomorphy of minute interstitial forms includes the lack of body or appendage segments and rami, fewer extremities, and reduction of article number and armature. An adaptive trait to granular

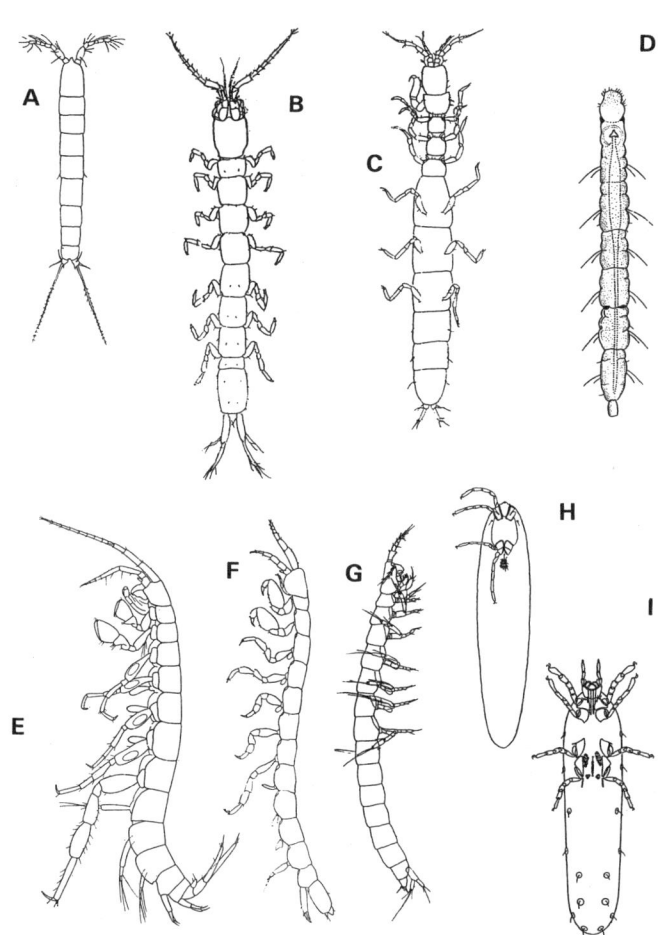

Interstitial Habitats (Aquatic): Eyeless, slender, and elongated minute interstitial organisms. A. Harpacticoid copepod (0.4–0.7 mm). B. *Microcharon angelieri* (isopod, 1.2 mm). C. *Microcerberus haouzensis* (isopod, 1.8 mm). D. *Potamodrilus fluviatilis* (worm Aphanoneura, 1 mm). E. *Bogidiella (Orchestigidiella) orchestipes* (amphipod, 1.8 mm). F. *Ingolfiella (Gevgeliella) petkovskii* (amphipod, 1.3 mm). G. *Parvulobathynella riegelorum* (syncarid parabathynellid, 0.7 mm). H. *Stygolimnochares elongata* (Acarina, 1.3 mm). I. *Wandesia vermiformis* (Acarina hydrachnellid, 2.5 mm). After Dussart, 1967 (A); Coineau, 1962 (B); Coineau *et al.*, 1999 (C); Dumicka & Juberthie, 1994 (D); Ruffo & Vigna Tagliantí, 1977 (E); Karaman, 1959 (F); Noodt, 1964 (G); Schwoerbel, 1986 (H, I).

sediments is elongation of the body, appendages, and sense organs compared to surface or cave relatives. Species of ciliates and Foraminifera living in sand are filiform (thread-like). Harpacticoid copepods, many isopods such as *Microcharon*, *Microcerberus*, stenasellids, and the amphipods Bogidiellidae and Ingolfiellidae exhibit a thread-like body with elongated posterior legs and uropods (see Figure). The water mites *Wandesia* and *Stygolimnochares* display a long posterior body. Long claws, legs, and uropods are an advantage for pushing the animal through the labyrinth of narrow channels. A high number of specialized sensory setae such as aesthetascs and plumose setae on antennae and other appendages allows the detection of chemical gradients, prey, or the opposite sex. Mucus glands along the body allow slipping while a thin and transparent integument, a strong muscular system, greater exoskeleton articulation providing body flexibility, and thigmotactism (keeping in contact with the substrate particles), represent other adaptations to interstitial life. Anatomical reductions include the absence of one ovary, a reduced number of testicular utricula and digestive caeca, and simple and linear cerebron and gut structures. Such reductions probably require developmental gene modifications. The relatively stable and oligotrophic interstitial environment has selected K-reproductive strategies and limited energy loss: few large eggs are produced, development is slow, spawning events are widely spaced in time, and reproduction is more or less continuous; longevity is increased, from 7–10 months (asellids) to about 15–20 years (amphipod *Niphargus* and isopod stenasellids). Late hatching; protection of gametes, eggs, and embryos; precocious fertilization; viviparity; and dormancy during dry or cold periods (tardigrades, nematodes) are other features enhancing energy economy. In addition, some organisms are able to regulate their relatively low oxygen consumption.

At a local–micro scale, hydrodynamism, hydrological and environmental factors, as well as biotic interactions explain the patchiness and the distribution patterns of interstitial organisms. Interstitial biotopes serve as pathways between different aquatic ecosytems, which increases biodiversity together with the tendency to speciate and to produce endemism in taxa of high ancestry (Danielopol, 2002). At the evolutionary scale, interstitial biotopes represent highways of fresh groundwater colonization by marine surface ancestors. The two-step and the three-step models of colonization and evolution provide an understanding of both the ecological change and the evolutionary processes involved before and during these invasions (Boutin & Coineau, 1990; Holsinger, 1994). Vicariant events, such as plate tectonics, Tethys regressions, as well as orogenesis have played a major role in the evolution and the distribution of old lineages. Many relictual forms including living fossils, especially among crustaceans, reside in interstitial habitats.

ELAINE F. ALBUQUERQUE AND NICOLE COINEAU

Works Cited

Boutin, C. & Coineau, N. 1990. "Regression Model", Modèle biphase d'évolution et origine des micro-organismes stygobies interstitiels continentaux. *Revue de Micropaléontologie*, 33(3/4): 303–22

Coineau, N. 2000. Adaptations to interstitial groundwater life. In *Subterranean Ecosystems*, edited by H. Wilkens, D.C. Culver & W.F. Humphreys, Amsterdam and New York: Elsevier

Danielopol, D.L. 2002. Taxonomic diversity of groundwater Harpacticoida (Copepoda, Crustacea) in southern France. A contribution to characterise hotspot diversity sites. *Vie Milieu*, 52(1): 1–15

Gibert, J., Danielopol, D.L. & Stanford, J.A. (editors) 1994. *Groundwater Ecology*, New York: Academic Press

Holsinger, J.R. 1994. Pattern and process in the biogeography of subterranean amphipods. In *Biogeography of Subterranean Crustaceans: The Effects of Different Scales*, edited by D.C. Culver & J.R. Holsinger, special issue of *Hydrobiologia*, 287(1): 131–45

Malard, F., Ward, J.V. & Robinson, C.T. 2000. An expanded perspective of the hyporheic zone. *Verhandlungen Internationale Vereinigung Limnologie*, 27: 431–37

Ward, J.V., Malard, F., Stanford, J.A. & Gonser, T. 2000. Interstitial aquatic fauna of shallow unconsolidated sediments, particularly hyporheic biotopes. In *Subterranean Ecosystems*, edited by H. Wilkens, D.C. Culver & W.F. Humphreys, Amsterdam and New York: Elsevier

Further Reading

Botosaneanu, L. (editor) 1986. *Stygofauna Mundi, A Faunistic, Distributional, and Ecological Synthesis of the World Fauna Inhabiting Subterranean Waters (Including the Marine Interstitial)*, Leiden: Brill

Delamare Deboutteville, C. 1960. Biologie des eaux souterraines littorales et continentales. *Vie Milieu*, suppl. 9: 740 pp.

Camacho, A.I. (editor) 1992. *The Natural History of Biospeleology*, Madrid: Museo Nacional de Ciencias Naturales

Juberthie, C. & Decu, V. (editors) 1994. *Encyclopaedia Biospeologica*, vols 1 & 2, Moulis and Bucharest, Société de Biospéologie

INTERSTITIAL HABITATS (TERRESTRIAL)

The terrestrial interstitial habitat or MSS (Mesovoid Shallow Substratum or Milieu Souterrain Superficiel) is the compartment of the terrestrial subterranean biome located between the base of the mineral soil horizon and bedrock. It is composed of a network of small voids, which may be found in one of two types of lithic aggregate: the screes on valley sides (colluvium), or, in volcanic areas, scoriaceous layers; and cracks and interconnected mesovoids in the superficial zone of the rock, including within lava flows (see Figures). The MSS is isolated from the surface by a mantle of soil; it may or may not connect with larger natural and artificial voids; and its depth varies from some tens of centimetres to several metres.

The MSS was first described by Juberthie et al. (1980; 1981) in the Central Pyrenees and South and West Carpathians, in some screes at the base of noncalcareous rocks (schist, gneiss, granite) and beneath limestone cliffs, and by Oromí, Medina and Tejedor (1986) and Howarth (1981) in cracks and mesovoids of the volcanic rocks of the Canary Islands and Hawaii. Ueno (1981) described what he termed "the upper hypogean habitat" along the banks of streams in Japan. Subsequently, the MSS has been described in the Pyrenees, Alps, Apennines, the Dinaric karst, the South and West Carpathians, the Stara Planina and Rhodope mountains in Bulgaria, Greece, Turkey, Japan, Taiwan, and China. It is thought to be present at low and mid

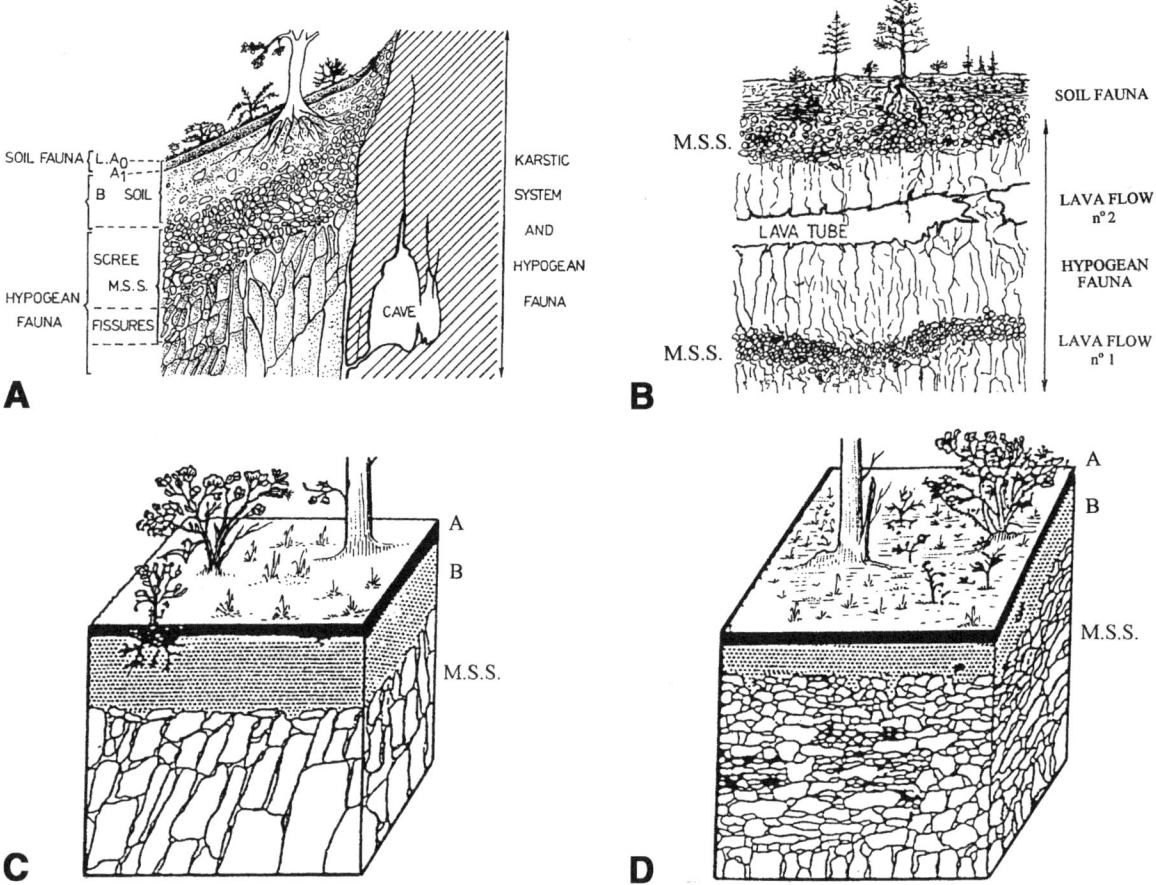

Interstitial Habitats (Terrestrial): (A) Diagram of the MSS in a calcareous zone in a scree at the base of a limestone cliff, connected with a cave system (after Juberthie, 2000); (B) Diagram of a volcanic MSS (after Juberthie, 2000); (C) Block-diagram of a MSS in the cracks and interconnected mesovoids of a superficial rock zone (after Juberthie, 1983); (D) Block-diagram of a MSS in a scree of noncalcareous rock (after Juberthie, 1983).

altitude in all mountains in the temperate zone, but seems to be absent in tropical areas, where instead the voids are filled with laterite and clay. The MSS has a discontinuous distribution, which is determined by geomorphological barriers, climatic, and structural factors.

The MSS can be formed in two ways: in the first, frost fragments nonvolcanic rocks into small stones of various sizes that accumulate on the slopes of valleys or at the base of fissures. When soil has covered the screes, a new MSS is available for colonization by cave or soil invertebrates. The cracks in the superficial zone of the rocks have the same origin, and are enlarged by mechanical expansion of the rock. The second process is volcanic: MSS is formed on some types of cracked lava flow or scoriaceous layers, or sometimes a combination of the two.

The physical environmental factors in the MSS are similar to those affecting caves in the same area. The major difference consists in the greater range of seasonal temperature variation—up to 10°C or more. In winter in the Pyrenees, at an altitude of around 1000 m, the average temperature in the MSS is 2–4°C; in the West Carpathians, at the same altitude, between April and October, the average temperature is 9.8°C. As in caves, the humidity is close to 100%.

The trophic sources in the MSS are allochthonous, and originate at the surface. As in the case of caves, trophic chains are short, with detritivores predominating. The carnivorous species consume worms, apterygotan insects, and the juvenile stages of other invertebrates (mainly Diptera, for example, Phoridae: *Tryphleba hyalinata*, a characteristic species of the Pyrenean MSS).

Two faunal communities inhabit the MSS; one specific to the MSS, and the other composed of soil dwellers. The first community comprises three types of species. The first inhabit only the MSS, for example, Cholevidae–Leptodirinae (e.g. *Speonomus colluvii* and *Mehadiella paveli*, see figures in Insecta: Coleoptera, and *Sophrochaeta globosa*), the second is made up of troglobitic and troglophilic species that inhabit both caves and the MSS (e.g. Isopoda: *Mesoniscus graniger*; Araneae–Linyphiidae; Diplopoda: *Typhloblaniulus*; Coleoptera–Carabidae: *Aphaenops, Duvalius*; Cholevidae–Leptodirinae: *Speonomus hydrophilus* and *S. zophosinus*, *Sophrochaeta*, *Drimeotus*;

Cholevidae– Cholevinae: *Catops, Choleva*), and the third is made up of typical troglobitic and troglophilic cave species, which rarely enter the MSS and then only in small numbers. The second community is composed of soil-dwellers that actively enter the MSS for food and humidity, or drift in with the flow of meteoric water. Some are the food supply for zoophagous species (for example Collembola and Diptera), while others, such as Chilopoda, Araneae, Pseudoscorpiones, or some Coleoptera, are predators, or parasites such as Hymenoptera. Depending on the temperature and humidity, MSS species move vertically upwards toward the soil horizon (mainly in spring and autumn) or downwards toward the deeper network of fissures within the bedrock (in winter and partially in summer). Characteristic MSS faunas present the same morphological (lack of eyes, loss of pigmentation, and absence of wings) and physiological adaptive features as eucavernicolous faunas.

The evolution of the MSS is cyclic, with three phases: (1) a juvenile stage, with colonization by soil- or cave-dwellers; (2) a mature stage, with an equilibrium community; and (3) an old stage, characterized by collapsed voids and disappearance of the fauna. The average time for one of these cycles ranges from ten thousand years to perhaps a hundred thousand years. In the Central Pyrenees, the most recent cycle involving the genesis of MSS in screes began 12 000–13 000 years ago, when forests replaced the post-Würm (post-Devensian) steppe cover (Juberthie, Dupré & Jalut, 1990).

It has been demonstrated that the route for cave colonization by the ancestors of cave-dwellers is not only via the karst surface and cave entrances, but also via the MSS. In volcanic areas, new lava tubes are colonized by existing troglobites from the volcanic MSS and older lava tubes.

CHRISTIAN JUBERTHIE AND VASILE DECU

Works Cited

Howarth, F.G. 1981. Non-relictual terrestrial troglobites in the tropical Hawaiian caves. In *Proceedings of the 8th International Congress of Speleology, Bowling Green, Kentucky*, vol. 1, edited by B.F. Beck, Huntsville, Alabama: National Speleological Society: 539–41

Juberthie, C. 1983. Le milieu souterrain: étendue et composition. *Mémoires de Biospéologie*, 10: 17–64

Juberthie, C. 2000. The diversity of the karstic and pseudokarstic hypogean habitats in the world. In *Subterranean Ecosystems*, edited by H. Wilkens, D.C. Culver & W.F. Humphreys, Amsterdam and New York: Elsevier

Juberthie, C., Delay, B. & Bouillon, M. 1980. Extension du milieu souterrain en zone non calcaire: description d'un nouveau milieu et de son peuplement par les Coléoptères troglobies. *Mémoires de Biospéologie*, 7: 19–52

Juberthie, C., Delay, B., Decu, V. & Racovitza, G. 1981. Premières données sur la faune des microespaces du milieu souterrain superficiel de Roumanie. *Travaux de l'Institut de Spéologie Emile Racovitza*, 20: 103–11

Juberthie, C., Dupré, E. & Jalut, C. 1990. *Aphaenops rebereti*, espèce "endogée" du sous genre *Geaphaenops* présente dans le MSS de la vallée d'Ossau, daté 12000 B.P. *Mémoires de Biospéologie*, 17: 181–90

Oromí, P., Medina, A.L. & Tejedor, M.L. 1986. On the existence of a superficial underground compartment in the Canary Islands. In *Proceedings of the 9th International Congress of Speleology, Barcelona*, vol. 2: 147–51

Ueno, S.I. 1981. A remarkable new trechine beetle found in a superficial subterranean habitat near Tokyo, Central Japan. *Journal of the Speleological Society of Japan*, 6: 1–10

Further Reading

Juberthie, C. & Decu, V. 1994. Structure et diversité du domain souterrain; particularités des habitats et adaptations des espèces. In *Encyclopaedia Biospeologica*, vol. 1, edited by C. Juberthie & V. Decu, Moulis and Bucharest: Société de Biospéologie

INVERTEBRATES: MINOR GROUPS

While molluscs, annelids, arthropods, and vertebrates include groups that may be abundant in hypogean habitats, particularly caves, some groups of invertebrates are less numerous (at least in caves), less important, or simply less noticeable. This entry will cover all these in limited detail, and therefore includes a very heterogeneous assemblage of fauna, mostly non-segmented (Porifera and Ameria), but with some oligomerous or even chordate groups (Oligomeria and Chordonia, or Deuterostomia).

Sponges (Porifera or Spongiaria) are sessile, marine, or freshwater organisms. Their bodies are bag-shaped and consist of a system of channels and chambers that filter water to obtain suspended particulate food. Marine or freshwater sponges often occur at high densities in dark or cave-like habitats, and even in true caves, where there is no competition for space with algae. In shallow-marine caves some deep-sea or relict elements have been recorded, e.g. the tiny glass sponge (Hexactinellida) *Oopsacas minuta*, the uniquely carnivorous sponges *Asbestopluma* sp. (Cladorhizidae), and a number of different species of calcareous sponges (Calcarea) from the group Pharetronida (Vacelet, 1996). Some probably stygobiotic species of Demospongiae, such as *Cinachyra subterranea*, have been described from euhaline anchialine caves in the Bahamas (Van Soest & Sass, 1981). Another anchialine species, *Higginsia ciccaresei*, has been described from anchialine cave waters in southern Italy (Pansini & Pesce, 1998). The only true stygobiotic freshwater species (Demospongiae: Spongillidae), *Eunapius subterraneus*, was found in the Dinaric caves of western Croatia, near Ogulin, where at least two races occur (Sket & Velikonja, 1986).

Flatworms (Platyhelminthes: Turbellaria) are mostly leaf-shaped, soft-bodied, hermaphroditic organisms with a ciliate skin, which creep along the bed of water bodies, feeding as predators. They range in size from a few millimetres to several centimetres long. Small species of different taxonomic groups inhabit the marine mesopsammal zone, and different families or even higher taxa may be characteristic of particular ecological zones. Planarians (Tricladida), of which over 150 stygobiotic species are known, are the most common in caves. The genus *Phagocata* has a number of cave species in Europe, Asia, and North America; *Dendrocoelum* has even more species, but is apparently limited to Europe, while species of the family Kenkiidae, mainly *Shalloplana*, are limited to North America and eastern Asia (Gourbault, 1994). A terrestrial species has been found in a cave in Slovenia, but has not yet been described. Cave tricladids are mostly 10–20 mm long.

Another interesting turbellarian group is the epizoic or parasitic Temnocephalida. While only one species is known from surface waters in Europe, a series of species were found to live on decapod shrimps and one on amphipods in the caves of the southern European Dinarides. They were also found on cave shrimps of the Caucasus. All European species belong to the family Scutariellidae (e.g. genera *Scutariella*, *Troglocaridicola*, *Bubalocerus*), they are tiny, on average 1 mm long, and are characterized by two diversely shaped tentacles on the anterior end, and a simple or complex sucker on the posterior end.

Cnidaria (jellyfish, hydra, sea anemones, and corals) are primarily sessile aquatic organisms with bag-shaped bodies called polyps. During their life cycle they undergo a free-swimming stage called medusa. However, each of the stages may be absent (e.g. the medusa stage is absent in sea anemones and corals.) Around the apical mouth in the polyp, as well as along the outer rim of the medusa, are tentacles armed with stinging intracellular structures called cnidae or nematocysts, which are mainly used to capture and kill the animal prey. Specimens of the freshwater *Hydra* spp. are often swept into caves or interstitial waters. Only *Velkovrhia enigmatica* (Hydrozoa: Bougainvilliidae), a tiny colonial species from the Dinaric caves, is known to be stygobiotic (Matjašič & Sket, 1971). On the polyps are budding vestigial, permanently fixed, medusae (gonophores) with male or female gonads. The dark walls of marine caves are often adorned with different non-specialized cnidarians, in addition to sponges and bryozoans. A number of tiny species inhabit the marine mesopsammal habitat. *Halammohydra* spp. (Hydrozoa: Halammohydridae) are reduced hydroid medusae consisting of a long mouth tube (manubrium) with tentacles around the umbrella, which is reduced to a small sucker on the proximal end of the manubrium. *Stylocoronella* spp. (Scyphozoa: Lucernariida) are so-called polypo-medusae, i.e. polyps with their apical part developed into a medusoid shape and structure.

Ribbon worms (Nemertini or Nemertea) are creeping animals, sometimes superficially resembling very narrow turbellarians; at least this is the case for all hypogean species. Their main characteristic is an eversible proboscis, which may be armed with stylets and poison glands for capturing living prey. A number of species inhabit the marine mesopsammal zone, the most numerous of which are the virtually thread-like *Ototyphlonemertes* spp., which are on average 10 mm long. Cave species of the freshwater genus *Prostoma*, occurring in Western Europe and Dinarides (Pust, 1990) are much broader although not much longer.

Roundworms or nematodes (Nematoda) constitute an extremely rich and ecologically diverse group, but paradoxically with extreme morphological conservativism. Nearly all are more or less elongate spindle-shaped, covered by a smooth and thick cuticle. They have a variety of feeding habits: detritus, bacteria, plant rootlets, and live animals, and approximately half the known species are parasitic. While some parasitic forms are large, all free-living species are smaller (from a few microns up to 2 mm). Most inhabit soil or littoral or freshwater benthic sediments. A number of species have been found only in caves, although this does not assure their stygobiotic nature, since the group is very poorly studied. *Desmoscolex aquaedulcis* had been found in a cave in Slovenia (Stammer, 1935) and was considered to be a marine relict. A number of other species of this morphologically very aberrant (with a characteristic ring structure) and presumably marine interstitial genus were later found in soils in other parts of Europe.

Arrow-worms or chaetognaths (Chaetognatha) are primarily predatory marine planktonic animals. Their bodies are elongate spindle-shaped, mostly transparent, and usually less than 2 cm long. They possess two groups of grasping bristles beside the terminal mouth which can be used to capture prey, and two pairs of horizontal lateral fins and a caudal fin. They are hermaphroditic. The few benthic species are shorter, with an otherwise unchanged morphology. *Paraspadella anops*, from an euhaline anchihaline cave in the Bahamas (Bowman & Bieri, 1989), is eyeless and can be considered to be stygobiotic, while some probably less specialized species have been found in other marine caves.

Interstitial Fauna

A number of other invertebrate groups inhabit either solely interstitial waters or surface and interstitial waters, occurring only rarely in cave waters. Gnathostomulids (Gnathostomulida) (Ax, 1956) resemble very slender turbellarians. Most of these are less than 1 mm long. However, their epithelial cells are monociliated, and in the pharynx they have cuticular jaws. There are about 80 species which inhabit the marine mesopsammal and a number of them are limited to layers with very little or no oxygen.

Gastrotrichs (Gastrotricha) are usually microscopic in size, the body is skittle- or tap-shaped, and covered with a cuticle that is often scaly or spiny, with ventral ciliated belts and long cilia on the head. The worm-like body of the group Macrodasyoida has adhesive tubes scattered all over the body surface; the group is almost exclusively marine interstitial. The mainly skittle-shaped Chaetonotoidea also live in freshwater benthic habitats and possess only one pair of caudal adhesion tubes. The diversity of the marine mesopsammal gastrotrichs is therefore particularly high. Few of the species appear to be freshwater psammobionts.

Rotifers (Rotatoria = Rotifera) are usually microscopic in size; the body can be very diverse in shape, but with a characteristic ciliary organ called the corona. This is usually a circular or bicircular group of cilia in the mouth region, which beat in turn, driving particulate food to the mouth or propelling the organism when it is not attached. The rotifers are a predominantly freshwater benthic group, but a number of species live interstitially in the marine as well as in the freshwater mesopsammal zone. The few presumed cave species are most probably not completely restricted to such habitats.

Roundworms (Nematoda) have already been mentioned. Although few, if any, cave-limited species exist, the group is extremely richly represented and diverse in the marine mesopsammal zone. A comparatively large number of these species show very aberrant shapes. The number of species is high and if the habitat is supplied with large amounts of detritus, the biomass of the nematodes may be twice as high as the rest of the fauna.

Kinorhynch (Kinorhyncha) are skittle-shaped organisms < 1 mm in length. All are benthic marine taxa, and some are restricted to the marine mesopsammal zone. Their chitinous cuticle is divided into rings; the anterior region (introvert), which can be protruded or withdrawn, is protected by a neck region with bristle-shaped appendages called scalids.

Loricifera are an entirely marine mesopsammal group first described in 1983 (Kristensen, 1983), the first of these species being *Nanaloricus mysticus*. The organisms are barrel-shaped and strongly armoured, with a large anterior introvert, armed with rings of very diversely shaped scalids. The adults will adhere to sand grains, while the so-called Higgins larvae are mobile. At less than 0.5 mm in length, their bodies consist of more than 10 000 cells.

Priapulida are sausage-shaped marine animals with a somewhat globular anterior introvert, densely set with short hooks; differently shaped vesicular appendages may be present on the posterior end. Some long species (up to 30 cm) burrow in benthic sediments, but a few species, about 1 mm long, live interstitially.

Kamptozoa (= Entoprocta) are goblet-shaped marine (rarely freshwater) animals up to 1 mm in length; tentacles on the upper end surround all of the openings, including the mouth and anus. Most are sessile and many are colonial, but some are solitary and mobile in spite of their polypoid appearance, for example *Loxosoma isolata*, which is a mesopsammobiont found in the northern Adriatic. Other mesopsammal species may also exist.

Water bears (Tardigrada) are tiny animals, mostly less than 1 mm long, with a stout cylindrical body with four pairs of unjointed legs carrying different numbers of claws. They are aquatic, although the water film in moss cushions can accommodate numerous individuals. Some species seem to be restricted to fresh interstitial waters, e.g. *Macrobiotus longipes* from the far north of Sweden. There is a rich assemblage of marine mesopsammal species which may be very aberrantly shaped; for example *Batillipes* spp., which have "fingers" with adhesive discs on their legs, or *Florarctus salvati* which have a broad membranaceous border around their trunks.

Sipunculans (Sipuncula) are sausage-shaped marine animals with a thick cuticle; the terminal mouth is surrounded by short tentacles and the terminal part of the body can invert itself up to the anus, which opens on the neck. Most species are several centimetres long. *Aspidosiphon exiguus* (which is about 3 mm long) seems to be restricted to marine or brackish interstitial waters in the Caribbean.

Bryozoans (Bryozoa = Ectoprocta) are sessile and colonial marine or freshwater animals. The body of a single specimen (zooid) consists of a box-like cystid, covered with cuticula which may even be calcified, and a polyp-like polypid which may retract into the cystid. The terminal mouth is surrounded by ciliated tentacles, and the anal opening is on its neck. Out of the few marine mesopsammal species, *Monobryozoon* spp. are the only solitary members of this group.

Lamp shells (Brachiopoda) are marine animals resembling bivalve molluscs, but with dorsal and ventral valves (rather than right and left valves). The organism itself has a paired and often spirally coiled holder of ciliated tentacles. Some species can be found attached to the walls of marine caves, while the tiny (1 mm or less) *Gwynia capsula* also occurs in the mesopsammal zone.

Sea cucumbers (Echinodermata: Holothurioidea) are mostly sausage-shaped marine animals related to starfish and sea urchins, a feature which is reflected in their internal structure. The mouth and anal openings are at opposite ends; the former is surrounded by tentacles. They can reach 45 cm in length, and the few interstitial species are only 2–5 mm long. *Leptosynapta minuta* (5 mm in size) has a distribution ranging from the western Mediterranean to the North Sea.

Sea squirts or ascidians (Tunicata = Urochordata: Ascidia) are sessile, solitary, or colonial marine animals, ranging from a few millimetres to several centimetres in length. Their bag-shaped bodies are covered by a thick layer of tunicine (similar to cellulose in plants), with two openings or siphons; at the oral siphon, the water enters a wide respiratory and filtering basket, which is the modified fore-gut. The tadpole-shaped larva has a tubular nerve cord, and a notochord in its tail, but the latter later degenerates. Some millimetre-long species of sea squirt inhabit the interstitial waters of very coarse sands.

Boris Sket

Works Cited

Ax, P. 1956. Die Gnathostomulida, eine raetselhafte Wurmgruppe aus dem Meeressand. *Abhandlungen der Mathematisch–naturwissenschaftlichen Klasse, Akademie der Wissenschaften und der Litteratur in Mainz*, 8: 1–32

Bowman, T.E. & Bieri, R. 1989. *Paraspadella anops*, new species, from Sagittarius Cave Grand Bahama Island, the second troglobitic chaetognath. *Proceedings of the Biological Society of Washington*, 102(3): 586–89

Gourbault, N. 1994. Turbellaria, Tricladida. In *Encyclopaedia Biospeologica*, vol. 1, edited by C. Juberthie & V. Decu, Moulis and Bucharest: Société de Biospéologie

Kristensen, R.M. 1983. Loricifera, a new phylum with Aschelminthes characters from the meiobenthos. *Zeitschrift fuer zoologische Sytematik und Evolutionsforschung*, 21: 163–80

Matjašič J. & Sket, B. 1971. Jamski hidroid s slovenskega krasa (A cave hydroid from Slovene karst). *Biološki Vestnik*, 19: 139–45

Pansini, M. & Pesce, G.L. 1998. *Higginsia ciccaresei* sp. nov. (Porifera: Demospongiae) from a marine cave on the Apulian coast (Mediterranean sea). *Journal of the Marine Biological Association of the UK*, 78: 1083–91

Pust, J. 1990. Untersuchungen zur Systematik Morphologie und Oekologie der in westphalischen Hoehlen vorkommenden aquatischen Hoehletiere. *Abhandlungen aus dem Westphalischen Museum fuer Naturkunde*, 52(4): 1–188

Sket, B. & Velikonja, M. 1986. Troglobitic freshwater sponges (Porifera, Spongillidae) found in Yugoslavia. *Stygologia*, 2(3): 254–66

Stammer, H.-J. 1935. *Desmoscolex aquaedulcis* n.sp., der erste suesswasserbewohnende Desmoscolecide aus einer slowenischen Hoehle (Nemat.). *Zoologischer Anzeiger*, 109: 311–18

Vacelet, J. 1996. Deep-sea sponges in a Mediterranean cave. In *Deep-sea and Extreme Shallow-water Habitats: Affinities and Adaptations*, edited by F. Uiblein, J. Ott & M. Stachowitsch, Vienna: Österreichische Akademie der Wissenschaften

Van Soest, R.W.M. & Sass, D.B. 1981. Marine sponges from an inland cave on San Salvador Island, Bahamas. *Bijdragen tot de Dierkunde*, 51(2): 332–44

Further Reading

Botosaneanu, L. (editor) 1986. *Stygofauna Mundi, A Faunistic, Distributional, and Ecological Synthesis of the World Fauna Inhabiting Subterranean Waters (Including the Marine Interstitial)*, Leiden: Brill

Higgins, R.P. & Kristensen, R.M. 1986. New Loricifera from southeastern United States coastal waters. *Smithsonian Contributions to Zoology*, 438: 1–70

Juberthie, C. & Decu, V. (editors) 1994. *Encyclopaedia Biospeologica*, 2 vols, Moulis and Bucharest: Société de Biospéologie

IRAN

Iran is geologically a part of the Alpine–Himalayan orogenic belt. Five major structural zones can be distinguished in Iran (Stocklin, 1968): Zagros Range; Sanandaj-Sirjan Range; Central Iran; East and Southeast Iran; and the Alborz and Kopet-Dagh Ranges. The Zagros is sub-divided into the three structural zones of Khozestan Plain, the Simply Folded Zone, and the Thrust Zone. The Zagros folded zone, with a width ranging from 150 to 250 km, is a sequence of late Precambrian to Pliocene shelf sediments about 12 km thick, mainly consisting of limestone, marl, gypsum, sandstone, and conglomerate. The Zagros Thrust Zone is composed of crushed limestone, radiolarite, and ultrabasic and metamorphic rocks. The Sanandaj-Sirjan Range consists mainly of granite, diorite, and metamorphic rocks. Central Iran is composed of Precambrian metamorphic basement overlain by shelf sediments and a chain of volcanoes. The Alborz Range consists of volcanics, sandstone, shale, siltstone, and limestone, and the Kopet Dagh Range consists of Mesozoic–Tertiary limestone, shale, and sandstone.

Karstic carbonate formations cover about 185 000 km^2 (which is 11% of Iran's land area) of which 55% is in the Zagros, 24% in Central Iran, 15% in the Alborz, and 5% in East and Southeast Iran (Figure 1). Most of the carbonate rocks are Cretaceous and Tertiary in age. The most important karst features in the Zagros Range are karren, grikes, springs, and to a lesser extent, caves and dolines. The main source of recharge is direct rainfall and snowmelt. Most of the springs are permanent (see photo of Margoon Spring in Springs entry) and high percentages of the spring waters are baseflow. Karst water in the Zagros Range is usually of good quality, and it is one of the most important sources of drinking water in the area. The Zagros fold zone is characterized by a repetition of long and regular anticlinal mountains. Most of the karst formations are sandwiched between two impermeable formations, so that they form independent highland aquifers. The general direction of groundwater flow is mostly parallel to the strike, towards cross-cutting gorges that determine local base levels.

Though there is an ever-increasing interest in cave research in Iran, only a few attempts have been made to establish a complete published inventory of the caves that have been explored so far. Most of the cave studies concentrate merely on visiting, photography, and / or mapping the explored caves. Sometimes the complete information about a cave has not been published or it cannot easily be found. The speleological committee of Iran has started to collect cave data, but this has not yet been published. Marefat (1994) reported 258 caves, but the report does not cover all the caves in Iran, and has only partial informa-

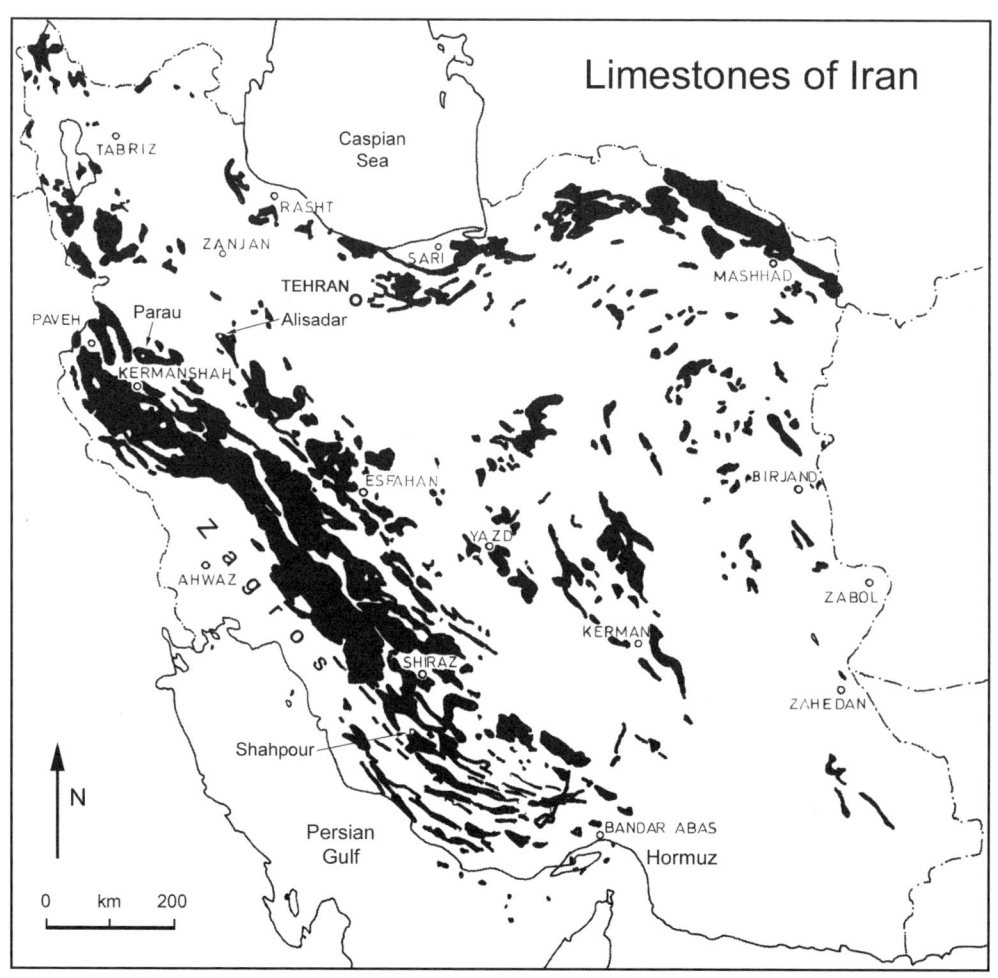

Iran: Figure 1. Distribution map of carbonate karst formation in Iran (Raeisi & Kowsar, 1997).

tion about location, altitude, survey, and length of some of the caves; neither geological settings nor hydrologic features are reported.

In spite of the large extent of karst in Iran, the number and lengths of caves are less than might be expected. There are many highland karst aquifers without any cave systems yet known, while big springs emerge at their bases of erosion. Several levels of caves are expected in the karst mountains of Iran, as a result of the rapid rates of uplifting and local valley incision. There are two main reasons for the small number of known caves. First, most of the karst areas are high mountains with steep slopes, so that many cave entrances have been filled in by talus or transported sediments or have been blocked by entrance breakdown in the high-risk earthquake regions of Iran. Second, many springs with high discharges are of the vauclusian type, with no explorable dry cave systems. Conduit systems of these springs, which have developed at higher levels in the past, cannot presently be seen on the surface. The short length of caves is also to be expected as the rapid rates of uplift and incision reduce the time for cave development at any one level. Water in the cave conduits leaks to lower levels and prevents further cave development. Also, recharge is mainly diffuse flow in the majority of the karst sites, so cave development is only initiated where branches of diffuse flow join each other beneath the surface.

The deepest cave in Iran is Ghar Parau, 751 m deep and 1365 m long, located just north of Kermanshah (Judson, 1973). It is developed in Cretaceous limestone below a sinkhole entrance at an elevation of 3050 m. It has 26 shafts along a small steeply descending canyon passage that has not been followed beyond a sumped section in a small local syncline. Most of the deep caves in Iran are located in Kermanshah province.

Ghar Alisadar is developed in Jurassic limestone which overlies schist and sandstone in the Sanandaj-Sirjan Zone (Laumanns et al., 2001). It is the longest (11 440 m) and most visited cave in Iran (400 000 visitors per year). More than 4 km of the cave passages lie at the water table, forming lakes of clear water up to 15 m deep, and the public underground boat tour is 2010 m long. Calcite and aragonite deposits are found in ledges a few metres above the present lake surface in most parts of the cave, implying that water levels in the cave were at fixed elevations for long periods in the past. Oscillation of the water table due to climate changes created up to nine of these ledges, at different elevations. The cave most likely developed under the water table by the influence of ascending volcanic CO_2. The average temperature, electrical conductivity, pH, and dissolved oxygen are 12°C, 270 μmhos cm^{-1}, 8.3, and 7.4 mg l^{-1} respectively. Ghar Sarab, 7 km south of Ghar Alisadr, also contains spectacular rimstone deposits (Figure 2) and has been subject to recent exploration (Brooks, 2002).

The Katelahkor cave, 2500 m long, lies 155 km south of Zanjan, and is also open as a tourist cave. It is an anastomatic multilevel cave, formed in Tertiary limestone. Ghori-Ghaleh, located near Paveh, is formed at the contact of the Jurassic limestone and conglomerate. It is a cave with a single passage about 1205 m long and mostly 2–3 m high. An underground stream flows through the whole length of the cave and emerges from the cave entrance with discharges that vary from 23 to 3300 l s^{-1}. This cave is under study for making it accessible as a tourist attraction. Shapour Cave, 80 km south of Shiraz, is an archeological and touristic attraction. It is a single anastomatic tiered cave 1229 m long. At the initial stage of subaerial exposure of the karstic Tertiary Asmari limestone, the Shapour River started entrenching a valley, and leakage from the river through joint and bedding plane fissures enlarged the cave (Raeisi & Kowsar, 1997). Remarkable salt caves in the Hormoz region have recently been surveyed by Czech speleologists, amongst them the 5010 m long Tri Nahaci Cave (Bosák et al., 1999).

Ezzat Raeisi

Works Cited

Bosák, P., Bruthans, J., Filippi, M., Svoboda, T. & Smid, J. 1999. Karst and caves in salt diapirs, SE Zagros Mts. (Iran). *Acta Carsologica*, 28(2): 41–75

Brooks, S. 2002. Iran 2001. *International Caver 2001* 82–83.

Judson, D.M. 1973. The discovery and exploration of Ghar Parau, Iran. *Transactions of the Cave Research Group of Great Britain*, 15(1): 19–26

Laumanns, M., Brooks, S., Dorsten, I., Kaufmann, G., Lopez-Correa, M. & Koppen, B. 2001. *Speleological Project Ghar Alisadr (Hamadan/Iran)*, Berlin: Speläoclub Berlin (Berliner Höhlenkundliche Berichte, 4)

Marefat, A. 1994. *Kuhha va Gahrhai Iran* [Mountains and Caves of Iran], Tehran: Goli (in Farsi)

Raeisi, E. & Kowsar, N. 1997. Development of Shahpour Cave, southern Iran. *Cave and Karst Science*, 24(1), 27–34

Stocklin, J. 1968. Structural history and tectonics of Iran: A review. *American Association of Petroleum Geologists Bulletin*, 52(7): 1229–58

Further Reading

Afrasibian, A. 1998. *Field Trip Guide: 2nd International Symposium on Karst Water Resources*, Tehran: Iran Karst Research Center, Ministry of Energy

Judson, D. 1973. *Ghar Parau*, London: Cassell and New York: Macmillan

Waltham, A.C. & Ede, D.P. 1973. The karst of Kuh-e-Parau, Iran. *Transactions of the Cave Research Group of Great Britain*, 15: 27–40

Iran: Figure 2. Calcite rimstone deposits from an older higher lake level form this mushroom island in the cave of Ghar Sarab. (Photo by Simon Brooks)

JEITA CAVE, LEBANON

Jeita Cave with its subterranean river is the most famous cave in Lebanon and has a history of exploration stretching back to 1836. The natural entrance of the cave is 18 km north of Beirut. The cave's main passage drains west along the northern side of the Nahr el Kelb (the Dog River). In 1971, a tunnel was excavated at Daraya to make a second access to the cave, at its upstream end (see map).

The first written document about the Jeita springs and their probable underground source is a letter sent by Father Fromage in 1736 that includes a remarkable hydrogeological description of the springs and mentions that there is a big cave with a huge lake inside the rocks. The first documented speleological activity was in 1836, when William Thomson progressed some 50 m until he reached an underground lake. Explorations between 1856 and 1902 reached 1000 m into the cave. Their main objective was assessment of the hydraulic potential of the underground river—an important water source for the city of Beirut. Between 1926 and 1927 two expeditions progressed a further 730 m and the cave was not extended further until the start of the Lebanese period of exploration in 1946 when a team reached 1950 m from the entrance. Further expeditions followed, and by 1958 they had reached the end of the cave and discovered several side galleries, giving about 9 km of total development (see map). Only minor discoveries have been made since.

Jeita Cave is completely formed in the Jurassic Kesrouane Formation, which is 1000 m thick. Fossiliferous micritic limestone makes up this formation but the lower part features later dolomitization, and most of Jeita Cave is formed within these dolostones.

Orogenesis in the Upper Jurassic led to a period of emergence and aerial exposure, when carbonate was fractured and karstified before burial in the Cretaceous. This early karstification was reactivated in the Neogene and subsequently when the present topography evolved. The western mountain chain is characterized by steep flanks and a maximum altitude of 3083 m, between the Mediterranean Sea and the Bekaa valley. This chain is cut by deep east–west valleys, one of which is Nahr el Kelb where the Jeita Cave is located. The steep topography together with an annual precipitation exceeding 1200 mm, mostly falling within six months of the year, have contributed in intensifying the recent karstification.

The major underground river in Jeita Cave has an average discharge of $2.3 \text{ m}^3 \text{ s}^{-1}$ and is used to supply about one million people with potable water. The cave plays the role of a lateral collector in the deeply karstified Jurassic aquifer. Many vertical caves lie in the limestone mass above Jeita Cave, but no connection has yet been made. The cave system drains a large area, extending beyond the hydrological basin of the Nahr el Kelb. Dye tracing has proven flow from the southern adjacent hydrological basin, from caves that are 18 km away, including Fouar Dara and Qattine Azar. The cave's outlet is on the stratigraphic contact with less permeable younger formations (Upper Jurassic volcanic rocks, Lower Cretaceous sand) forming a hydrogeological boundary that forced the underground river to rise to the surface. This barrier is also responsible for the large chambers (with a maximum height exceeding 60 m) in the cave.

The general plan of the cave is a sinuous line with the river flowing westward (see map). At the end of the cave, two branches from the inland extremity of the cave, each end with a siphon. The total slope gradient between the terminal siphon and the cave entrance is around 1/100. This illustrates the general flat passages characterizing the cave and its subterranean river, but small cascades and rapids break this monotonous topography. Jeita Cave begins with a big hall within a zone of large-scale meanders. The main passages of the cave then become more or less straight and two more large halls follow. There, the overall dimension becomes narrower through some rapids. Then the cave turns into a tunnel-shaped gallery for hundreds of metres. It becomes even larger with the Thompson's Cavern (250 m long and 60 m wide), before the Grand Chaos, a chamber over 500 m long where large collapsed blocks form the floor. This continues into the Castle of a Thousand and One Nights, a

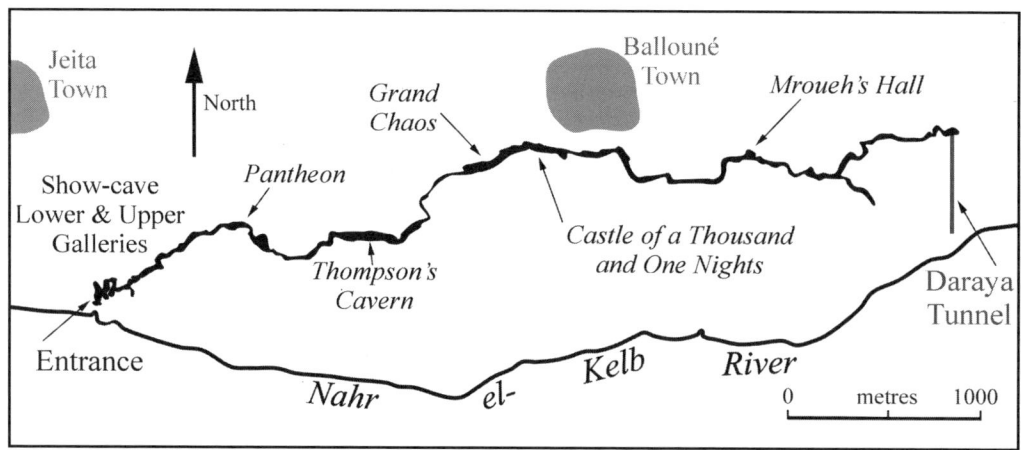

Jeita Cave, Lebanon: Outline map (surveyed by the Spéléo-Club du Liban)

well-decorated hall. A series of lakes connected by small cascades lead to Mroueh's Hall (200 m long and 50 m wide) also featuring large collapsed blocks. This is followed by a series of rapids to a broad junction floored with gravel. The northern branch is the longer and contains a beautiful sandy section before a series of lakes ends with a big terminal sump.

In January 1969, the Upper Gallery was opened to the public and became a major attraction of Lebanese tourism at that time. This new section is the most beautiful part of the cave with huge and delicate speleothems. Following the beginning of the 1970s, the Lebanese strife started, and in the early 1970s the show cave was closed and not reopened until 1995, five years after the end of the Lebanese war. The Lower Gallery can be visited in electric boats for a distance of 500 m. The 800 m long Upper Gallery lies more than 60 m above this lake and here tourists are left to wander on an elegantly designed path and to enjoy a wide variety of large speleothems. In some places, one can see down to the waters of the Lower Gallery. Conservation measures taken by the cave management include closing for one month per year, non-continuous and non-heating lighting, and electric boats. Since its re-opening, the show cave has received around 280 000 visitors per year.

FADI NADER

Further Reading

Bulletin of the Spéléo-Club du Liban Al Ouat'Ouate, 5 (1990)
    Special issue to commemorate the 50th anniversary of the first Lebanese expedition
www.jeitagrotto.com

JOURNALS ON CAVES

By the end of the 20th century, there were at least 2000 caving journals worldwide, but just 120 years earlier none existed. Their history is therefore far shorter than that of caving books (see related entry) and began only at the end of the 19th century, when the first caving clubs were formed in Austria, Italy, and France. Until that time, papers on caves were scattered in non-specific periodical publications.

The best-known early journal, completely devoted to speleology, was the French *Spelunca*, the bulletin of the Société de Spéléologie printed since 1895. This was followed a year later by *Mémoires de la Société de Spéléologie* (see Figure), which started in 1896. Both still exist, though they have had some gaps and / or changes in format; moreover, *Mémoires* changed its title twice, becoming *Annales de Spéléologie* (1946–75), then *Spelunca*, and finally *Karstologia* (1983). These two bulletins, the only ones started in the 19th century, dealt with all the fields of speleology and, in their first issues, hosted several papers by the French caver Édouard Alfred Martel, commonly regarded as the founder of modern speleology (see Speleologists).

In the first decades of the 20th century, only a few other journals appeared, all printed in Europe: in Italy *Rivista Italiana di Speleologia* (1903–04), *Mondo Sotterraneo* (1904), *Le Grotte d'Italia* (1927), and *Memorie dell'Istituto Italiano di Speleologia* (1931); in Hungary *Barlangkutatás* (1913–44); in Austria *Speläologisches Jahrbuch* (1920–36). The *Proceedings of the University of Bristol Spelaeological Society* started in 1920 and is still being produced. Six years later, *Mitteilungen über Höhlen und karstforschung* (1926–40) started in Germany.

In 1940, the first two caving journals were printed outside Europe: *NSS (National Speleological Society) News* and the *Bulletin of the National Speleological Society*, which took this name from the second issue (1941), the first being named *Bulletin of the Speleological Society of the District of Columbia*. The *NSS Bulletin* later became the *Journal of Cave and Karst Studies*.

In the next two decades, tens of new caving journals appeared each year so that, in the early 1960s, over 500 caving bulletins were printed in five continents. Some of them had a tremendous impact on the development of speleology in their own countries, but many disappeared after a relatively short time. Among the most famous of such journals were: *Ceskoslovenski Kras* (Czechoslovakia, 1948–90), *Rassegna Speleologica Italiana* (Italy, 1948–73), *Speleon* (Spain, 1950–83), *Travaux de L'Institute de Spéléologie Émile Racovitza* (Romania,1963–97), and *Serie Espeleologica* (Cuba, 1967–74).

The exact number of current caving journals is not known, due to their fast dynamics and normally short lifetime, but it

Journals on Caves: Addresses of the UIS Documentation Centres

Argentina
Library "Dr Emilio Maury", Grupo Espeleológico Argentino, Heredia 426 (C1427CNF) Buenos Aires. Email gea@mail.retina.ar

Austria
Speläologisches Dokumentationszentrum des Institutes für Höhlenforschung, c/o Naturhistorisches Museum, Burgring 7, A-1014 Wien

Belgium
Centre Documentation, Union Belge de Spéléologie UBS/SSW, Maison de la spéléologie, rue Belvaux 93. B-4030. Email caving.service@speleo.be

France
Documentation Fédération Française de Spéléologie, 28 rue Delandine, F-69002 Lyon. Email FFS.biblio@wanadoo.fr

Germany
Bibliothek des Verbandes der Deutschen Höhlen- und Karstforscher Dechenhöhle 5, D-58644 Iserlohn. Email dechenhoehle@t-online.de

Italy
Centro documentazione speleologica, Isituto Italiano di speleologia, c/o Università, Via Zamboni 67, I-40127 Bologna. Email ssibib@geomin.unibo.it

Japan
Natural Science Museum, c/o Dr Uéno, Hyakunin-cho 3–23–1, Shinjuku, Tokyo 160

Portugal
Biblioteca Sociedade portuguesa de espeleologia, rua Saraiva de Carvalho 233, P-1350 Lisboa. Email spe.nacional@clix.pt

Library of "Kras I speleologia", Laboratory of Research and Documentation of Karst Environment, University of Silesia, ul. Bedzinska 60, 41–200 Sosnowiec. Email atyc@us.edu.pl

Romania
Institut de spéléologie, c/o V Decu, 11 rue Frumoasa, R-78114 Bucharest 12. Email alex.petrulescu@dataline.ro

Slovenia
Institut za Raziskovanje Krasa ZRC SAZU, Titov trg 2, 6230 Postojna. Email kranjc@zrc-sazu.si

Spain
Centre de documentació espeleologica, Ap.C.32110, E-08080 Barcelona 6

Switzerland
Centre de Documentation UIS, c/o Bibliothèque de la Société Suisse de spéléologie, CH – 1614 Granges. Email ssslib@vtx.ch

United Kingdom
British Cave Research Association Library, c/o Roy Paulson, Holt House, Holt Lane, Lea, Matlock, Derbyshire, DE4 5GQ. Email librarian@bcra.org.uk

United States
National Speleological Society Library, 2813 Cave Avenue, Huntsville, Alabama 35810. Email nss@caves.org

Venezuela
Biblioteca Sociedad Venezolana de espeleologia, Apartado 47.334, Caracas 1041-1. Email kghneim@yahoo.com

certainly exceeds 2000. They may be grouped into three different categories. By far the largest number is that of the local or regional bulletins issued by caving clubs and regional Federations, which represent over 95% of the whole. Their audience is normally limited to the members of the group editing the journal and only few tens of copies circulate outside that group. The papers printed in these journals deal with every aspect of speleology but are rarely of more than local interest. However, some of these are worth noting, due to their outstanding scientific level: *Atti e Memorie* of the Commissione Grotte E. Boegan in Trieste (Italy, 1962), *Proceedings of the University of Bristol Spelaeological Society, Endins* of the Speleological Group of Mallorca (Spain, 1974), and *Journal of Sydney University Speleological Society* (Australia, 1950).

The second category consists of the journals (40–60) edited by the national speleological organizations: most countries in Europe, North America, and Oceania have their own bulletin, while in Africa, Asia, and Central and South America, only a few countries with a well-developed speleology have an official national bulletin. For example, in South Africa (*The Bulletin of SASA*, 1956); in Asia, Lebanon (*Al Ouat'Ouate*, 1955), Israel (*Nikrot Zurim*, 1980), China (*Carsologica Sinica*, 1982) and Japan (*Journal of the Speleological Society of Japan*, 1975); and in South America, Venezuela (*Boletin de la Sociedad Venezolana de Espelelologia*, 1967) and Brazil (*Espeleo Tema*, 1970). The target of the national journals are the cavers of that country and, therefore, their circulation may be minimal (a hundred or few hundred copies) but it may exceed 10 000 (such as *NSS News*, United States, the speleological journal with the greatest print run in the world), ranging normally between 1000 and 2000 copies. Some tens or hundreds of copies of these journals circulate outside their country of origin, mainly reaching the other national speleological societies. Sometimes the national organization prints more than one journal: in this case the second is always a scientific bulletin in which studies and research of general interest are printed. Some of the most renowned are: *Caves and Caving* (1937) in England, which was one of the parents of the present *Cave and Karst Science* (the other being *Transactions of the British Cave Research Association*), *Acta Carsologica* (1955) in Slovenia, *Theoretical and Applied Karstology* (1984) in Romania, and the already cited *Karstologia* in France, *Le Grotte d'Italia* in Italy, and *NSS Bulletin* in the United States.

The third and last category corresponds to the international caving journals: only four are currently printed and all are official journals of international societies. Three of them are journals of the International Union of Speleology (UIS) and cover different caving fields: *UIS Bulletin* (1970) reports the relevant information for the activity of UIS itself, the *International Journal of Speleology* (1964) accepts only scientific papers in all the speleological fields, and *Speleological Abstracts* (1964) is a bibliographic bulletin. The last international journal is *Mémoires de Biospéologie*, edited since 1978 by the Société Internationale de Biospéologie. Finally, it must be noted that an attempt to print a private international magazine on cave exploration was first made in Canada (*Caving International*, 1978–92) and then in England (*International Caver*, 1991–97): these attempts were well accepted in the speleological world but survived only 14 and 25 issues respectively due to lack of financial support. However,

Journals on Caves: Title page of one of the first issues of the *Mémoires de la Société de Spéléologie*.

International Caver has published an irregular yearbook since 1997.

Although, or perhaps because it falls into none of the above categories, special mention should be made of *Descent*, which is the only commercial caving journal that is independent of any club or national body. First published in 1969 by Bruce Bedford, the magazine is now produced by Wild Places Publishing of Cardiff, United Kingdom, under the editorship of Chris Howes. There are six issues per year.

How to get information

The extremely high number of caving journals, together with their average very low circulation, irregularity in publication or frequent sudden death, causes extreme difficulty in searching for articles on a given caving topic. For this reason, as soon as the number of journals started to increase dramatically, just after World War II, several national speleological associations begun to print bibliographical bulletins, the first of which was *Internationale Bibliographie fur Speläologie* (1950–60), edited annually by the Austrian Speleological Society, as a supplement to their official journal *Die Höhle* (1950). In 1958, the National Speleological Society started printing *Speleo Digest*, in which the most important papers appearing in the local bulletins and journals of the United States were reprinted. From 1961 to 1991, the British Cave Research Association printed *Current Titles in Speleology*, while other national associations such as in Belgium (*Bibliographie Spéléologique Belge*) and Spain (*Bibliografia Espeleològica Hispànica*) published similar bibliographical lists, even if for a shorter timespan. The defect of all these publications was that they were not truly "general", reflecting the home territory of the editor, and moreover suffered from lack of worldwide distribution.

The UIS decided to solve this problem by supporting the Swiss Speleological Society in printing *Speleological Abstracts*, which, since its first issue, tried to be truly international. Presently this bibliographic magazine has over 70 collaborators from around 20 different countries and reports on about 5000 papers scattered in over 1000 journals. From 2002 it is possible to consult its last six issues directly through the net. However, knowledge of the existence of a paper is not useful unless one can obtain copies of it. The scarcity of available issues for many of the caving journals of the world precludes the possibility of having most of them at least in the national speleological libraries. For this reason, the UIS set up a network of *Documentation Centres* (see Table), covering all the main speleological areas of the world: they are, upon request, ready to supply copies from the journals present in their library.

PAOLO FORTI

See also **Exploration Societies**

K

KAIJENDE ARÊTE AND PINNACLE KARST, PAPUA NEW GUINEA

The arête and pinnacle karst of Mt Kaijende is located in the Western Highlands district of Papua New Guinea, between the settlements of Laiagam and Porgera (5° 30′S). It was first brought to the attention of scientists by Jennings and Bik (1962) and investigated further by Williams (1972). Karst in the region is developed in Lower Miocene limestone of more than 1000 m stratigraphic thickness. Exposures in cliff faces show it to be extremely massively bedded, with some beds of the order of 150 m. Mt Kaijende is a triangular block, faulted on its northwest, southwest, and eastern sides, and tilted gently to the southsoutheast. Fault scarps vary in height, some reaching almost 1500 m high. The region was probably first uplifted and exposed to denudation in the early Pliocene.

The arête and pinnacle karst occurs near the summit of Mt Kaijende, which rises to about 3500 m. It covers an area of 8–10 km^2, although the extent of bare karst is less. The arêtes are naked, reticulated, saw-topped ridges with spires, with practically vertical side slopes $c.120$ m high. They are crudely aligned, dominantly northnortheast–southsouthwest. Photolineaments, presumably representing master joints, strike at 5°, 25°, 85°, 115°, and 155°. These have been enlarged and deepened by solution, and their intersection has produced a network of deep chasms that isolate massive, sheer-faced, polyhedral blocks as much as 20 ha in area at their base. The retreat of their walls has reduced the immense intervening joint blocks to a series of spired ridges with pinnacles. The walls are corrugated by near-vertical rock drains typically 2–3 m in semicircular diameter, tens of metres in length and often terminating in gaping holes. The converging heads of these solution gutters from different flanks of the blocks impart a sinuosity to the arête ridges. The bare pinnacle tops are probably sharp and fluted with rillenkarren, but investigation has not been close enough to determine details. Vertical perforations within the blocks have expanded into irregularly fluted cylindrical depressions, about 70–100 m wide at their top, and perhaps exceeding 100 m deep.

Although Jennings and Bik named the morphology "arête-and-doline karst", Williams considered it misleading to imply that the enclosed depressions resemble dolines as they are normally understood, because of their very steep rock sides and intimate connection to the arête ridges. Therefore he preferred to omit the term doline from the description of the terrain, which he described as "arête and pinnacle karst" (see photo).

The rock surfaces are essentially bare on their crests, being steep and in an exposed, hostile environment near the upper limit of montane forest (the regional tree-line is at about 3700 m). However, the limestone faces gradually gain an increasingly dense plant cover as the surfaces descend into the more sheltered

Kaijende Arête and Pinnacle Karst: The spectacular arête and pinnacle karst on the slopes of Mt Kaijende, seen from the air. (Photo by Paul Williams)

environment of joint canyons, along which the bottoms of closed depressions are aligned. Thus most of the enclosed depressions within the arête and pinnacle terrain are vegetated at their base. However, authoritative detail is imprecise, because the topography has been studied from the air and approached from below, but, because of its extreme inaccessibility, has never been explored very far or been the subject of a field survey. Consequently, there are no measurements of the height of the arête ridges, but they appear to be of the order of 100 m or more. At an elevation of about 3200 m, thick, virtually impenetrable moss-forest, swirling in mist, clings like a sodden cloak to the rugged slopes. The daytime temperature at that height was measured as 11°C, and water dripping from the moss-draped branches had a pH of 3.9. The upper slopes of Mt Kaijende summit are almost always cloud-covered, so fog-drip must make a large contribution to the annual precipitation. Rainfall at Porgera, which is 8 km to the west and at 2200 m in the nearby valley, has been measured at about 3700 mm, so it is likely to be considerably more at the summit. Solutional denudation rates have not been measured, but must be at the upper end of international estimates.

Consideration of the lapse rate indicates that the mean annual temperature of the summit of Mt Kaijende is about 10°C, and the tropical location implies little seasonal variation. Nevertheless, there is a significant diurnal variation, with a range of about 12°C being measured at 2800 m. Night-time frosts are common above 2430 m, so freezing conditions must often affect the arête and pinnacle terrain. Frost shattering was probably common during the last glacial maximum, because at that time climatic conditions suitable for glacier ice formation existed from about 3600 m upwards.

On the lower slopes of the mountain, at around 2900 m, arête and pinnacle karst progressively gives way to a plateau incised with polygonal karst of relatively subdued relief. This is completely clothed in rain forest, except for the bottoms of some of the larger depressions which are covered with coarse grassland (locally called "kunai") and tree-ferns, the forest having being excluded by frost drainage.

Further examples of arête and pinnacle karst have been reported but not described on mountains to the west. Similar landforms are found in the tsingy of Madagascar, the pinnacles of Mount Api in Mulu National Park, Sarawak, and the stone forest of Lunan in China, although the Kaijende pinnacles are much higher and more dramatic than those of the stone forest (Salomon, Ford & Williams, 1996).

PAUL W. WILLIAMS

See also **Karren; Madagascar; Mulu; Shilin Stone Forests**

Works Cited

Jennings, J.N. & Bik, M.J. 1962. Karst morphology in Australian New Guinea. *Nature*, 194: 1036–38
Salomon, J.-N., Ford, D.C. & Williams, P.W. 1996. Les forêts de pierres. *Karstologia*, 28(2): 25–40
Williams, P.W. 1972. Morphometric analysis of polygonal karst in New Guinea. *Bulletin of the Geological Society of America*, 83: 761–96

KANIN MASSIF, SLOVENIA–ITALY

The Kanin (Canin) Massif (see map), with its high concentration of deep cave systems, some of the world's deepest shafts, and its spectacular glaciokarstic landscape, is one of the finest examples of alpine karst in the world (see also photograph in Alpine Karst entry). The massif straddles the border between Slovenia and Italy, and is bounded by several fluvioglacial valleys. It forms part of the western Julian Alps, the highest point being Mt Kanin at an altitude of 2587 m.

The massif basically comprises an Upper Triassic series of Norian dolomite (the Main Dolomite) overlain by up to several hundred metres of Dachstein Limestone, which is calcareous dolomitic or totally calcareous in composition. The main geological structure is an east–west-trending anticline, cut by an east–west fault dipping southwards at an inclination between 60° and 80°. In addition, the anticline has been dissected into blocks by four sets of subvertical faults.

The neighbouring valleys contain several karst springs at altitudes between 350 m and 870 m. Their mean discharges are up to $1000\ l\,s^{-1}$, with maxima up to $100\ m^3\,s^{-1}$. The altitude difference between the highest cave entrances of the massif and the lowest part of the caves at the base of the mountain is about 2000 m. This makes Kanin potentially a massif with some of the world's deepest cave systems. Several water-tracing experiments, in which dyes were injected into the cave streams, have proven some of the drainage directions; sinks on Kaninski podi drain to the Boka, Mala Boka and Zyltior risings, while sinks on Goričica drain to Zyltior and Glijun (Audra, 2000).

Serious cave exploration on both sides of the border started in the 1960s, when numerous caves were investigated. In the 1970s a depth of 920 m was reached in Abisso Michelle Gortani, on the Italian side of the border. The longest system within the massif is Complesso del Foran del Muss—also in Italy. It was first explored in the early 1970s. Many deep and independent caves have been investigated, and some of the latter have been found to be connected to the main system, the explored portion of which now reaches a depth of 1140 m, extends over 20 km and has 23 entrances (Benedetti & Mossetti, 2000). Probably the most explored system is the Complesso del Col delle Erbe, including the Abisso Gortani. Its present length is 15 km, with a depth of 935 m. At least 15 independent caves or cave systems on the Italian side are deeper than 600 m, and four of these are longer than 7 km.

On the Slovenian side of the massif, a depth of 911 m was reached in Skalarjevo Brezno in 1988. At the beginning of the 1990s three caves deeper than 1000 m were explored on the Goričica Plateau. Two of these, Črnelsko brezno and Čehi 2 were investigated by Italian teams, and Vandima was explored by Slovenians. Recent exploration in Renejevo brezno, on the Kaninski podi plateau, have revealed passages down to a depth of 1068 m. In the 1990s, two extremely deep single shafts were explored; Brezno pod velbom has a 501 m deep entrance shaft in a cave 852 m deep, and Vrtiglavica is a single shaft 643 m deep.

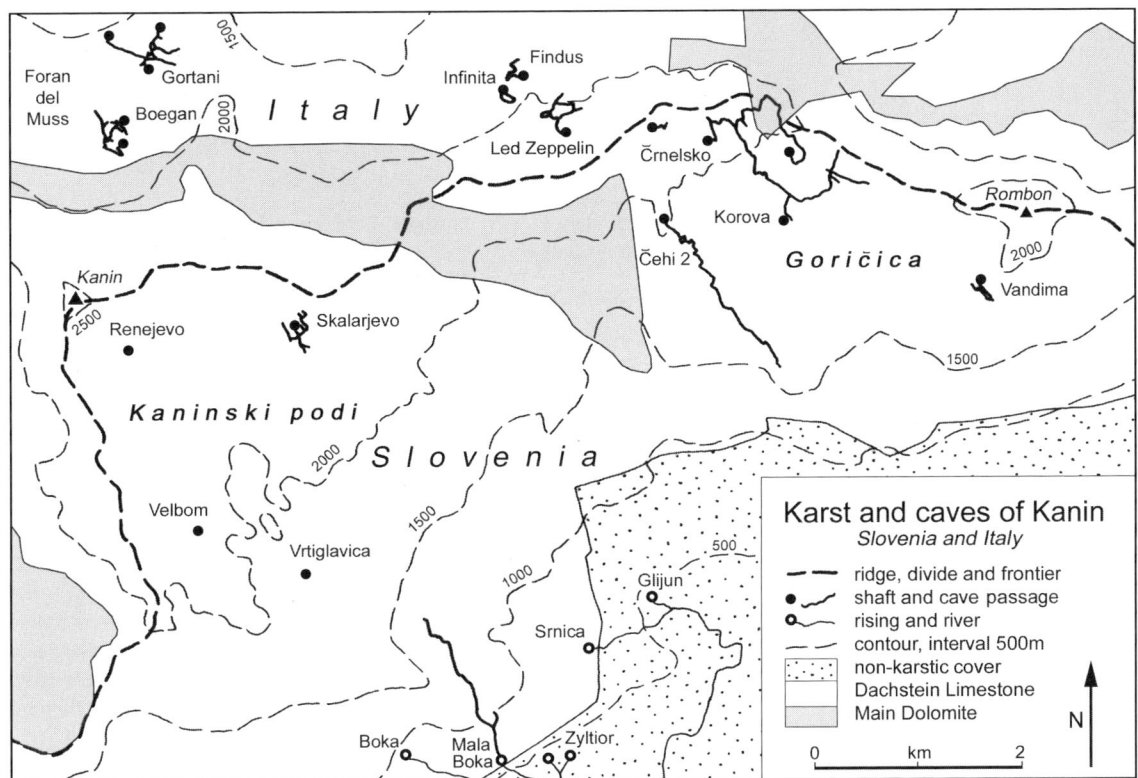

Kanin Massif, Slovenia–Italy: Map of the main features of the Kanin massif (after compilation by Philippe Audra). The cave passages of Complesso del Foran del Muss and Renejevo brezno are not marked; Gortani is the main entrance to the Complesso del Col delle Erbe.

Different parts of the massif exhibit different cave morphologies. Systems of phreatic channels at 1400–1800 m altitude are typical for large caves on the northwest, Italian side (Complesso del Foran del Muss, Complesso del Col delle Erbe). On the other hand, phreatic features are rarely observed on Pala Celar (on the Italian northeastern side) and the Kaninski podi plateau (on the Slovenian southwestern side). There the caves are guided by the main subvertical tectonic structures. Shafts connected by straight and meandering canyons (known as pitch–ramp series) are a characteristic pattern. Many deep single shafts were explored in the area.

The deepest caves of the massif are located on Goričica plateau (on the southeast, Slovenian side). Pitch–ramp series down through the Dachstein Limestone terminate at the contact with the Main Dolomite. From there, in Čehi 2 and Črnelsko brezno, vadose canyons dip down the contact to depths of 1533 m and 1380 m respectively. A similar but less pronounced pattern can be observed in Abisso Led Zeppelin on the Italian side.

Several caves have abandoned passages of phreatic origin at different levels. The position and orientation of these passages provide a record of the ancient water tables. The passages are pre-Quaternary in origin and were formed in hydrological settings completely different to those prevailing at the present day. Some contain speleothems and clastic sediments. An attempt was made to U–Th date the speleothems, but they proved to be too old (> $c.$350 ka) for this method to be used. Some varves and car-

Kanin Massif, Slovenia–Italy: The deepest cave systems of the Kanin massif (as at the end of 2002)

Cave	Country	Depth (m)	Length (km)
Čehi 2	Slovenia	1533	5
Črnelsko brezno	Slovenia	1380	9
Vandima	Slovenia	1182	3
Sistema Foran del Muss	Italy	1140	20
Renejevo brezno	Slovenia	1068	2
Abisso Led Zeppelin	Italy	960	4
Complesso del Col delle Erbe	Italy	932	15
Skalarjevo brezno	Slovenia	911	3
Brezno pod velbom	Slovenia	852	1

bonates in Črnelsko brezno exhibited reverse polarity in their paleomagnetism, and are therefore older than 780 000 years (Audra, 2000).

FRANCI GABROVŠEK

See also photo in **Alpine Karst**

Works Cited

Audra, P. 2000. Le karst haut du Kanin (Alpes Juliennes, Slovénie–Italie). *Karstologia*, 35: 27–38

Benedetti, G. & Mosetti, A. 2000. Il Complesso del Foran del Muss (Monte Canin–Friuli–Venezia Giulia). *Speleologia*, 43: 3–21

Further Reading

Cassagrande, G., Cucchi, F., Manca, P. & Zini, L. 1999. Deep hypogean karst phenomena of Mt. Canin (Western Julian Alps): a synthesis of the state of present research. *Acta Carsologica*, 28(1): 57–69

Gabrovšek, F. 1997. Two of the world's deepest shafts on Kaninski podi plateau in western Julian Alps, Slovenia. In *Proceedings of the 12th International Congress of Speleology*, vol. 4, La-Chaux-de-Fonds, Switzerland: International Union of Speleology

KARREN

Limestone that outcrops over large areas as bare and rocky surfaces is furrowed and pitted by characteristic sculpturing landforms that generate a distinctive karstic landscape. These solutional forms, ranging in size from less than 1 mm to more than 30 m, are collectively called karren, an anglicized version of the old German word Karren (the equivalent of the French terms lapiés and lapiaz). The terms karren and lapiés were formerly used to describe both the great array of individual karren forms and the great exposures of solutional features superimposed on bare limestone terrains. Currently, these groups of complex karren forms tend to be called karrenfields or Karrenfelder, in order to differentiate such large-scale exokarst landforms from their smaller karren components (see Table).

Several attempts at karren classification were made during the second half of the 20th century, especially after the pioneer works on the alpine and Dinaric karsts by Eckert (1902) and Cvijić (1924). More recent classifications of karren features (e.g. Bögli, 1980; Ford & Williams, 1989) include the process that generated each karren form, the size of the resulting features, the topographical pattern of their distribution over the rock surface, the presence or absence of a soil cover, and the kind of water action on the limestone (direct rainfall, standing water accumulation, sheet wash, channellized flow, melting of glacier or snow patches, and percolating water). Recent advances in karren studies are related to experimental simulations using plaster of Paris (Dzulynski, Gil & Rudnicki, 1988) and to investigations of the morphometrics of individual features (Ginés, 1996; Goldie & Cox, 2000).

Several different weathering processes may produce microkarren over limestone surfaces. Some of the microkarren features, such as biokarstic borings, are the result of specific solutional processes induced by cyanobacteria, fungi, algal coatings, and lichens. At this scale, many different patterns of minute hollows and pits are common, especially in arid environments, because the occasional wetting of the rock produces irregular etching, frequently coupled with biokarstic action (Fiol, Fornós & Ginés, 1996). Microrills are the smallest karren form showing a distinctive rilling appearance. Microrills consist of very tiny and sinuous runnels, 0.5–1 mm wide, rarely more than 5 cm long; they are caused by dew and thin water films, enhanced in coastal locations by supralittoral spray. Some other specific karren features develop near the coastline.

The majority of etched surfaces in semiarid environments display a rather complex microtopography that rarely presents linear patterns, the only exception being microrills. The general trend is a chaotic and holey limestone surface in which focused corrosion dominates, without any kind of integration in drainage patterns. These solutional features related to focused corrosion, give rise to depressions of different sizes, more or less circular in plan, such as the rainpit and the kamenitza karren types. Rainpits are small cup-like hollows, sub-circular in plan and nearly parabolic in cross section, whose diameter ranges from 0.5–5 cm and rarely exceed 2 cm in depth; they appear clustered in groups, or even packed by coalescence. The kamenitza karren type (Figure 1) consists of solution pans, generally flat-bottomed, from a few square centimetres to several square metres in size, that are produced by the solutional action of still water that accumulates after rainfall; their borders, frequently elliptical or circular in plan, are overhanging and may have small outlet channels.

Many types of karren are linear in form, controlled by the direction of channelled waters flowing along the slope under the effect of gravity. Most of the conspicuous furrowed appearance shown by karren terrains is due to this group of longitudinal landforms. The smaller ones are called rillenkarren (Figure 2) and are easy to distinguish from solution runnels or rinnenkarren (see below) by their trough width, which rarely exceeds 4 cm. Rillenkarren can be defined as narrow solution flutes, closely packed, less than 2.5 cm in mean width, consisting of straight grooves separated by sharp parallel ribs, that are initiated at the rock edges and disappear downwards. Rillenkarren are remarkable for the regularity of the herringbone pattern that they form on the top of the rocks and by their individual linear rills, parabolic in cross section and up to 60 cm long, and whose shape is constant along their whole length. Rillenkarren are produced by direct rainfall and their limited extent seems to be explained by the increase of water depth attaining a critical value that inhibits further rill growth downslope. Neither dendritic patterns nor tributary channels can be recognized in rillenkarren flutes, as opposed to the normal (or Hortonian) erosional rills.

Solution runnels are not as straight and regular in form as rillenkarren, being greater and more diversified in shape and origin. Solution runnels or rinnenkarren are normal (Hortonian) rills and develop where threads of runoff water are collected into channels. Classification of solution runnels is difficult because of the great diversity of topographic conditions, the complex processes involved, and the specific kind of water supply feeding the channel. Rinnenkarren is the common term to describe the

Karren: Classification of karren forms. Shaded areas indicate karren forms developed under soil cover. Faint-line upper frame encloses free karren single forms. Bold-line lower frame encloses complex large-scale landforms.

Solutional Environment	1 mm	1 cm	10 cm	1 m	10 m	100 m	1 km	Synonyms
Biokarstic	Borings							
Surface Wetting		Irregular Etching						
Thin Water Films		Microrills						Rillenstein
Storm Showers			Rainpits					Solution Pits
Direct Rainfall			Rillenkarren					Solution Flutes
				Solution Runnels				Rinnenkarren
Channelled Water Flow					Wall Karren			Wandkarren
				Decantation Runnels				
				Meandering Runnels				Mäanderkarren
Standing Water				Kamenitza				Solution Pan
				Solution Bevels				Ausgleichsflächen
Sheet Wash Water Flow				Trittkarren				Heelsteps
		Cockling Patterns						
			Solution Ripples					
				Trichterkarren				Funnel Karren
Snow Melt				Sharpened Edges				Lame Dentate
				Decantation Runnels				
				Meandering Runnels				
Ice Melting					Meandering Runnels			Mäanderkarren
Infiltration				Grikes				Kluftkarren
				Rundkarren				*Rounded Runnels*
				Smooth Surfaces				
Soil Percolation Water				*Subsoil Tubes*				*Bodenkarren, Subcutaneous Karren*
				Subsoil Hollows				
				Cutters				
Complex Processes				*Undercut Runnels*				*Hohlkarren*
				Clints				Flachkarren
					Pinnacles			Spitzkarren
						Pinnacle Karrenfield		
						Limestone Pavement		Karrenfeld
							Stone Forest	
							Arête Karst	
								Lapiés

equivalent of Horton's first-order rills on soluble rocks; they result from the breakdown of surface sheetflows that concentrate into a channelled way and they are also wider than rillenkarren. These solution runnels are sculpted by the water runoff pouring down the flanks of the rocks and have distinctive sharp rims separating the channels; their width and depth range from 5–50 cm, being very variable in length (commonly from 1–10 m, but in some cases exceeding 20 m long). Rundkarren are rounded solution runnels developed under soil cover; they differ from rinnenkarren in the roundness of the rims between troughs and can be considered good indicators of formerly soil-covered karren. Many transitional types from rundkarren to rinnenkarren can be found, due to deforestation and re-shaping of the rocks after subsequent soil removal by erosion. Undercut runnels or hohlkarren are associated with semi-covered conditions, as suggested by the bag-like cross sections of the channel, resulting from enhanced corrosion at the soil contact. Decantation runnels are rills, which reduce in width and depth downslope because the solvent supply is not directly related to rainfall, but corresponds to overspilling stores of water, such as moss clumps, small snow banks, or soil remnants. On steeper rock outcrops, the runnels are nearly parallel along the slope, but on moderate slopes some kind of dendritic or wandering patterns can be found. Wall karren are the typical straight runnel forms developing on sub-vertical slopes, but meandering runnels are more frequent on moderately inclined surfaces or where some kind of decantation feeding occurs over flat areas or gentle slopes. Wall karren, fed from local decantation points, as well as some proglacial ice-melting meandering runnels, may attain remarkable dimensions exceeding 30 m in length. Obviously, transitional forms of runnels are abundant in the majority of karren outcrops, with the exception of areas with arid climates.

Other types of karren features are linear forms controlled by fractures. Grikes or kluftkarren are solutionally widened joints or fissures, whose widths range from 10 cm to 1 m, being deeper than 0.5 m and several metres long. Grikes are one of the commonest and widespread karren features and separate limestone blocks into tabular intervening pieces, called clints in the British literature and Flachkarren in German. For this reason, clint and grike topography is the most typical trend in the limestone pavements, such as the Burren (Ireland; see separate entry) and Ingleborough (northwest England; see Yorkshire Dales entry). The

Karren: Figure 1. Kamenitza, Malta. (Photo by John Gunn)

Karren: Figure 3. Typical assemblage of karren features at the Serra de Tramuntana mountains (Mallorca, Balearic Islands). The vertical solution runnels are intensely fluted by conspicuous rillenkarren 1 to 2.5 cm wide. Note the stepped pattern horizontally carved into the rock slope. (Photo by Àngel Ginés)

term "cutters" is commonly used in North America as a synonym for grike, although it is best applied to a variety of grike that develops beneath soil cover. Giant grikes, larger than 2 m wide to over 30 m deep, are called bogaz or corridors. Corridor karst or labyrinth karst constitutes the greatest expression of this type of fracture-controlled karrenfield. Splitkarren are similar smaller-scale features, resulting from solution of very small weakness planes, being less than 1 cm deep and 10 cm long.

Finally, there is a group of karren features closely related to the solutional action of unchannelled washing by water sheets. Many of them, particularly trittkarren and solution ripples, show a characteristic trend that is transverse to the rock slope. At the foot of rillenkarren exposures, subhorizontal belts of unchannelled surfaces can be observed; they are called solution bevels and appear as smoothed areas flattened by sheet water corrosion. More distinctive forms are trittkarren or heelsteps, which are the result of complex solutional processes involving both horizontal and headward corrosion resulting from the thinning of water sheets flowing upon a slope fall. The single trittkarren consists of a flat tread-like surface, 10–40 cm in diameter, and a sharp back-slope or riser, 3–30 cm in height. On steep surfaces, high velocity shallow flows produce cockling patterns, some of them more or less transversal to the slope direction. Pulsating flow can also produce several kinds of solution ripples, apparently related to eddies of the water flow.

A wide variety of peculiar karren forms are produced by special conditions, such as where solution takes place in contact with snow patches or damp soil. Trichterkarren are funnel-shaped forms that resemble trittkarren, but are formed at the foot of steep outcrops where snow accumulates. Sharpened edges or "lame dentate", as funnel karren features, are developed beneath snow cover. Rounded smooth surfaces, associated with subsoil tubes and hollows are very common subcutaneous forms, due to the slow solution produced in contact with aggressive water percolating through the soil.

In Bögli's classifications, two kinds of complex karren forms are recognized: clints or flachkarren, and pinnacles or spitzkarren. These latter, three-dimensional forms, range from 0.5–30 m in height and several metres wide, and are formed by assemblages of single karren rock features, being the constituents of larger-scale groups of complex forms, the karrenfields or karrenfelder. Clints or flachkarren are flattened blocks of rock, outcropping more or less parallel to the bedding, that become isolated by the solutional widening of intervening joints. Pinnacles or spitzkarren are pyramidal blocks characterized by sharp edges, resulting from the solutional removal of rock from their sides, as well as from cutting through furrow karren features. Pinnacles

Karren: Figure 2. Rillenkarren, northwest Nelson, New Zealand. (Photo by John Gunn)

are exceptionally well developed in the tropics, where spectacular landscapes constituted by very steep ridges and spikes have been reported. In some cases, such as the Shilin or Stone Forest of Lunan (see Shilin Stone Forest), the presence of transitional forms, evolving from subsoil dissected stone pinnacles sometimes called "dragons' teeth" to huge and rilled pinnacles more than 30 m in height, can be observed. The particular karren assemblage that develops on limestone coasts is considered in the entry on Coastal Karst.

Karrenfields are bare, or partly bare, extensions of karren features, from a few hectares to a few hundred square kilometres. At the end of the 19th century, karrenfields were described in alpine and mediterranean karst environments, but recent explorations of karstified terrains in the tropics have documented many impressive karrenfield landscapes, such as the Tsingy Bemaraha (see Madagascar entry) and the Gunung Mulu National Park (Sarawak, Malaysia; see Mulu entry). Additional work is needed to clarify the relation between karren assemblages and climate, on the basis of the current knowledge accumulated in the last decades from arctic, alpine, humid-temperate, mediterranean, semiarid, and humid-intertropical karsts. Some well-known karrenfields correspond to formerly glaciated areas, such as the Gottesackerplateau (Allgäu Alps, Germany), the Désert de Platé (Haute Savoie, France), the Lapis de Tsanfleuron (Valais, Switzerland), and Glattalp (Schwyz, Switzerland) in the Alps; as well as the limestone pavements of the Burren (County Clare) in Ireland and the Hutton Roof Crag (Westmorland) in England. Karrenfields are also a major constituent of the typical bare-rock landscapes that are associated with deforested mediterranean karst, as occurs in the Velebit mountains (Dalmatia, Croatia) and the pinnacle karrenfields of the Tramuntana ranges (Mallorca Islands, Spain). Some extreme examples of karrenfields are the so-called crevice karst and arête karst, described from locations such as Mt Kaijende (New Guinea; see separate entry), where pinnacles can attain more than 40 m. Many celebrated karst sceneries are in fact karrenfields, being considered a significant tourist resource. This is the case of the Ubajara National Park (Ceará, Brazil) and especially the famous Shilin of Lunan (Yunnan, China), known in the literature as the Shilin Stone Forest, which attracts more than 1 million visitors each year.

ÀNGEL GINÉS

Works Cited

Bögli, A. 1980. *Karst Hydrology and Physical Speleology*, Berlin and New York: Springer (original German edition, 1978)

Cvijić, J. 1924. The evolution of lapiés: A study in karst physiography. *Geographical Review*, 14: 26–49

Dzulynski, S., Gil, E. & Rudnicki, J. 1988. Experiments on Kluftkarren and related Lapies forms. *Zeitschrift für Geomorphologie*, 32(1): 1–16

Eckert, M. 1902. Das Gottesackerplateau. Ein Karrenfeld im Allgäu. Studien zur Lösung des Karrenproblems. *Zeitschrift des D.U.O. Alpenvereins, Wissenschaftliche Ergänzungshefte*, 1(3): 1–108

Fiol, L.A., Fornós, J.J. & Ginés, A. 1996. Effects of biokarstic processes on the development of solutional rillenkarren in limestone rocks. *Earth Surface Processes and Landforms*, 21: 447–52

Ford, D.C. & Williams, P.W. 1989. *Karst Geomorphology and Hydrology*, London and Boston: Unwin Hyman

Ginés, A. 1996. Quantitative data as a base for the morphometrical definition of rillenkarren features found on limestones. In *Karren Landforms*, edited by J.J. Fornós & A. Ginés, Palma de Mallorca: Universitat de les Illes Balears

Goldie, H.S. & Cox, N.J. 2000. Comparative morphometry of limestone pavements in Switzerland, Britain and Ireland. *Zeitschrift für Geomorphologie*, Suppl.-Bd. 122: 85–122

Further Reading

Ford, D.C. & Lundberg, J. 1987. A review of dissolutional rills in limestone and other soluble rocks. *Catena*, Suppl.8: 119–40

Fornós, J.J. & Ginés, A. (editors) 1996. *Karren Landforms*, Palma de, Mallorca: Universitat de les Illes Balears

Ginés, A., 1995. Deforestation and karren development in Majorca, Spain. In Environmental Effects on Karst Terrains, edited by I. Bárány-Kevei, Szeged: Attila József University (special issue of *Acta Geographica Szegediensis*)

Trudgill, S.T. 1985. *Limestone Geomorphology*, London and New York: Longman

KARST

Karst is terrain with distinctive hydrology and landforms arising from the combination of high rock solubility and well-developed solution channel (secondary) porosity underground. Aqueous dissolution is the key process. It creates the secondary porosity and may be largely or wholly responsible for a given landform on the surface. Where it is not quantitatively predominant, it is the essential (trigger) process that permits others to operate, e.g. where a doline forms by mechanical collapse of insoluble strata into a solution cavity below. In hydrogeology, a karst aquifer is one modified from earlier granular and / or fracture aquifer conditions by development of interconnected solutional channels ("caves" where big enough for human entry): their larger aperture and efficient interconnection usually imply that water will circulate much more rapidly than in an unmodified aquifer. In the study of landforms, the karst geomorphic system is distinguished from the fluvial, glacial, eolian, coastal, and other systems because of the leading role of dissolution, which results in water circulating underground rather than running off at the surface in river channels.

Derivation of the Term

"Karst" is a germanicization of a regional place name, "Kras" (Slovenian) or "Il Carso" (Italian) given to the hinterland of Trieste Bay in the northwest Dinaric area (Gams, 1993). It is believed to derive from a pre-Indo-European word "karra" meaning stone, although this has been disputed by Hromnìc (2001). Already in Roman times (Latin = carsus) it defined "stony ground". This particular terrain was bare and stony because deforestation, followed by overgrazing with sheep and goats, had caused loss of all soil on its limestones into underlying solutional cavities. The name was promulgated by travellers in the 17th and 18th centuries and in the 19th century became widely adopted to describe the similar limestone landscapes extending from north Italy to Greece. The "father" of karst studies,

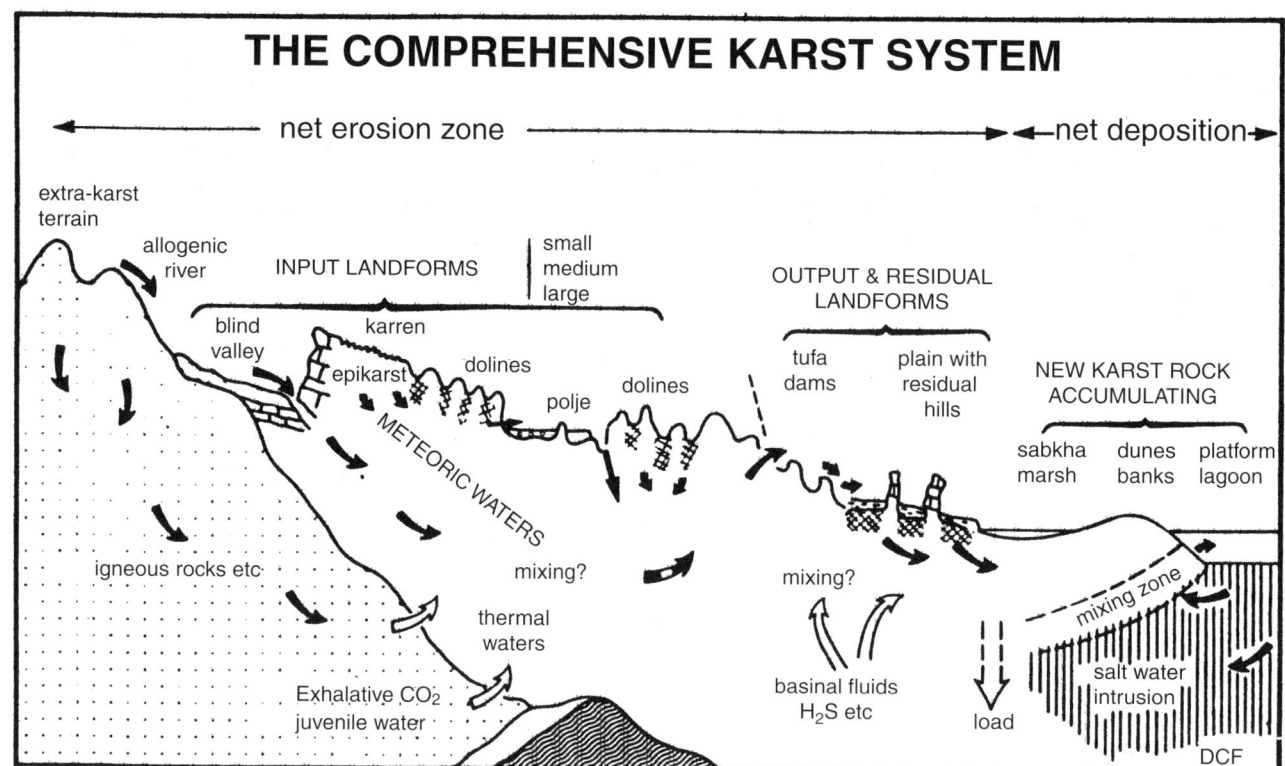

Karst: A schematic illustration of the components of the karst system in carbonate rocks and their interrelationships. From Ford & Williams (1989).

Jovan Cvijić, confirmed it when he entitled the first major Western monograph on solutional landscapes *Das Karstphänomen* in 1893. Sawicki (1909) expanded it as the global term for such topographies when describing tropical sites he had visited. It is the accepted term in China, the other great historic centre of dissolutional landscape studies (Yuan, 1991).

The Karst System

On Earth the principal karst rocks are, in descending order of solubility, salt, sulfates (gypsum and anhydrite), and carbonates (limestone and dolostone). There is limited development of karst features and hydrologic behaviour on some sandstones, quartzites, and even on granites.

The greatest range of form and development occurs on the carbonate rocks. Components of the karst system found in them are illustrated schematically in the diagram. Karst dissolution occurs during deposition of the youngest rocks, such as modern coral reefs, and extends to dolostones 3.4 billion years or more in age containing the earliest fossil traces of life. Most karst is created by meteoric water that fell as rain or snow, but intruding sea waters, trapped interstitial waters being expelled by compression (basinal fluids), and juvenile waters released from consolidating magmas (lavas) are also effective locally. On the Earth's surface it is useful to distinguish between forms created where water passes underground ("input landforms" such as dolines) and those where it returns ("output" forms such as sapped gorges or deposits of travertine). Underground, dissolution is known to occur to depths of five kilometres or more, where "pressure solution" may remove as much as 40% of the original thickness of a consolidated limestone, expelling it in the basin fluids.

Classifying Karst

There are many classifications in the international literature, serving different aims. The most widely used include the following. "Holokarst" describes terrain where all water passes underground within short distances, prohibiting development of surface stream channels except inside closed depressions. Where such karst discharges onto adjoining insoluble terrain it may be termed "karst barré. Cvijić (1893) limited "holokarst" to terrains extending to the sea coast; i.e. there are no river channels downstream of them. "Merokarst" or "halbkarst" were terms for mixed karst and stream channel topography found on less pure limestones but are not now in common use. "Fluviokarst" (see separate entry) describes the case where large rivers flowing off other rocks ("allogenic rivers") are able to maintain surface courses across a karst because of their magnitude; it is also used occasionally in the sense of merokarst.

Many authors classify karst assemblages by climate. At the extreme, nine distinctive types have been recognized—"temperate" (rain all year), "Mediterranean" (summer dry), "tropical humid", "arid", "semi-arid", "glacial" or "alpine", "periglacial" or "nival", "coastal tropical", and "coastal temperate" (Jennings, 1985).

At the surface a karst may be "bare", "subsoil" where covered by its insoluble residuum, or "mantled" where covered by transported detritus such as glacial till (Quinlan, 1972). Where it is overlain by later consolidated but insoluble rocks, there is "intrastratal" dissolution along bedding planes and fractures within the soluble rock or "interstratal" dissolution if that unit is entirely removed. This is "subjacent" karst where the water is of direct meteoric origin, and "covered" if it is expressed by collapse dolines, sinking streams, or other features at the surface. The distinctive caves and karst created by ascending thermal waters are "hypogene" features.

"Paleokarst" describes ancient surface and underground karst that has been buried by later rocks and is now inert; "buried karst" is equivalent (Bosák et al., 1989). Where there is renewed groundwater flow with dissolution through such karst, it is "rejuvenated". If erosion exposes it at the surface it is "exhumed". "Relict" karst describes surface forms (usually towers) created under past climatic conditions, surviving in the present landscape. "Relict" caves have been abandoned by their formative streams but remain open cavities. "Fossil karst" can mean any of these conditions.

"Pseudokarst" describes karst-like forms created by processes other than dissolution. The chief forms are caves and collapse dolines (sinkholes). Chief processes are lava flow ("volcanokarst"), melting of glaciers or ground-ice masses ("thermokarst"), and mechanical washout of pipes in silt, sand, or gravel ("piping").

DEREK FORD

See also **Asia, Northeast: History; Carbonate Karst; Evaporite Karst; Kras, Slovenia; Paleokarst; Pseudokarst**

Works Cited

Bosák, P., Ford, D.C., Glazek, J. & Horacek, I. (editors) 1989. *Paleokarst: A Systematic and Regional Review*, Amsterdam and New York: Elsevier

Cvijić, J. 1893. Das Karstphänomen. *Geographisches Abhandlung*, 5(3): 218–329
Ford, D.C. & Williams, P.W. 1989. *Karst Geomorphology and Hydrology*, London and Boston: Unwin Hyman
Gams, I. 1993. Origin of the term "karst", and transformation of the classical karst (kras). *Environmental Geology*, 21(3): 110–14
Hromnĭc, C.A. 2001. Karst, Kras or Karaśattu: Whence the name? *Cave and Karst Science*, 28(2): 79–88
Jennings, J.N. 1985. *Karst Geomorphology*, 2nd edition, Oxford and New York: Blackwell
Quinlan, J.F. 1972. Karst-related mineral deposits and possible criteria for the recognition of paleokarsts. *Proceedings of the 24th International Geological Congress, Montreal*, Montreal: McGill University
Sawicki, L. 1909. Ein Beitrag zum geographischen Zyklus im Karst. *Geographisches Zeitschrift*, 15: 185–204
Yuan D. 1991. *Karst of China*, Beijing: Geological Publishing House

Further Reading

Gillieson, D. 1996. *Caves: Processes, Development and Management*, Oxford and Cambridge, Massachusetts: Blackwells
Milanović, P.T. 1981. *Karst Hydrogeology*, Littleton, Colorado; Water Resources Publications
Salomon, J.-N. 2000. *Précis de Karstologie*, Pressac: Presses Universitaires de Bordeaux
Sweeting, M.M. 1972. *Karst Landforms*, London: Macmillan, and New York: Columbia University Press, 1973
Shaw, T.R. 1992. *History of Cave Science: The Exploration and Study of Limestone Caves, to 1900*, 2nd edition, Broadway, New South Wales: Sydney Speleological Society
White, W.B. 1988. *Geomorphology and Hydrology of Karst Terrains*, Oxford and New York: Oxford University Press

KARST EVOLUTION

Karst evolution concerns the way in which surface and subterranean features of karst rocks develop over time. These rocks have the common property of being especially soluble in water, and include evaporites (rock salt, anhydrite, and gypsum) and carbonates (limestone, marble, and dolomite). Karst terrains are also subject to erosion by other natural processes, but solution (sometimes called corrosion) has the greatest effect on them, and karst landforms are the direct or indirect consequence of these solution processes. Rocks that do not develop karst features, such as granite and schist, are also subject to solution, but it has only a minor effect on them as compared with other natural processes. The evaporate rocks are so soluble that in humid climates they are reduced to a very low relief; whereas the moderate solubility and greater physical strength of the carbonate rocks gives them more resistance and hence they are the most globally significant host rocks for karst. Nevertheless, not all carbonates develop significant karst, especially if they are argillaceous (muddy) or extremely porous (like coral). Karst develops best on pure, dense, thick limestones and marble. Carbonate rocks in general outcrop or are close to the surface over about 12% of the ice-free continental area, and well-karstified carbonates cover about 7–10%. Surface and subsurface evaporite rocks occur beneath approximately 25% of the continental surface, although they show fewer exposures than carbonate rocks (Ford & Williams, 1989).

Karst takes its name from the Dinaric Mountains beside the Adriatic Sea, inland from Trieste and centred on Slovenia, where the terms "carso" (Italian) and "kras" (Slovenian, Czech, Slovakian, and Polish) are used to describe the local landscape (see Kras, Slovenia). The term "karstification" is applied to the combination of processes, chemical and physical, that give rise to the features of karst both below and above ground.

To understand the evolution of karst it is best to imagine a geologically simple situation of dense, well-bedded limestones covered in places with remnants of still uneroded clastic caprock, and incised by a few deep valleys. The thick limestones extend well below sea level and dip gently towards the coast. The climate is humid and mild enough for water to be in a liquid state for all or most of the year. The natural processes that lead to the development of karst in such a terrain can be conceptualized as an open system comprising two interacting hydrological and geochemical subsystems. The hydrological cycle provides the main source of natural energy that powers the evolution of karst, because water is the solvent that dissolves karst rocks and then carries them away in solution. Geochemical processes control the rate of dissolution (the speed with which solid rock is converted into ions in solution), which in a carbonate karst context depends very strongly on the extent to which the water has become acidified by dissolved carbon dioxide (which produces carbonic acid) during its passage through the atmosphere and soil layer before making contact with the limestone. The concentration of carbon dioxide (CO_2) in the open atmosphere is about 0.03% by volume, whereas it is commonly 2% in the soil, and can even reach 10%. A factor of 100 increase in the concentration of CO_2 results in a roughly fivefold increase in the solutional denudation rate (White, 1984). Although this is important, the amount of rainfall is even more significant, the wettest places in the world having the fastest rate of limestone solution. For example, limestone denudation by solution processes has been estimated to be as high as 760 $m^3 a^{-1} km^{-2}$ in very wet places such as parts of Papua New Guinea, where rainfall can reach 12 000 $mm\ a^{-1}$, but as low as 5 $m^3 a^{-1} km^{-2}$ in some arid zones like the Nullarbor Plain in southern Australia, with a rainfall of less than 350 $mm\ a^{-1}$. The amount of solutional attack on the limestone rock therefore depends on the concentration of the solute (determined by geochemical processes) and the volume of solvent (determined by the rainfall).

Evolution of the Hydrological System

Karst will only evolve if water can get underground and dissolve caves, which provide conduits for the evacuation of materials in solution (as well as insoluble residues) that were dissolved at and near the surface. Critical in the evolution of karst, therefore, is the development of a plumbing system that permits water to sink underground and to circulate through the karst rocks. This first involves the enlargement of fissures in the rock into tiny interconnected passages sufficient to permit the passage of water from the highlands where it sinks underground to the bottom of neighbouring valleys. The small passages that develop are of the order of millimetres in diameter. They can be envisaged as proto-caves, and their development is explained more fully in the Speleogenesis entries. It is sufficient to note here that this is the first and essential step in karst evolution.

The development of karst hydrology also depends on the manner in which water enters the karst. Rainwater that falls directly onto the limestone outcrop is known as autogenic recharge. It infiltrates diffusely into the rock via countless fissures. By contrast, rain that falls on to impervious non-karstic caprock but later flows on to the karst is known as allogenic recharge. It runs off as organized streams, which sink underground soon after encountering the limestone. It contrasts with the diffuse nature of autogenic recharge by being high-volume point recharge. Not only that, but it has had a different geochemical history, because in its runoff path it encountered different vegetation, soils, and rock mineralogy. Thus the chemical aggressivity (power to dissolve rock) of autogenic and allogenic waters towards limestone often differs. Autogenic waters are mainly acidified by carbonic acid, but allogenic waters may have flowed from peat bogs (and hence contain organic acids) or have encountered sulfide minerals when draining from shales, and hence contain sulfurous acid. They also tend to have a greater mechanical load, which can abrade limestone and help to incise cave floors.

Given the existence of proto-cave conduits from the highland surface to springs in the valleys, point recharge then enlarges some of these pathways into caves in which sinking rivers establish their subterranean flow paths. Sinking streams along the allogenic input boundary converge underground and emerge at springs at the output boundary of the system, thereby establishing dendritic subterranean drainage networks. This process was modelled physically by Ford and Ewers (1978) and is explained in detail by Ford and Williams (1989).

Even if there were no cover beds shedding allogenic water, diffuse autogenic recharge alone is capable of developing protoconduits and establishing a karst groundwater circulation system, although the conduits formed are not as large as those developed by point recharge from allogenic streams. Diffuse recharge subjects the entire surface to solution, but since most of the acidification of the infiltrating water is achieved in the soil zone, most of the solutional attack is just beneath the soil, where the percolating water has its greatest aggressivity and first encounters the bedrock. Up to 90% of the total solution can be achieved in the top 10 m or so of the limestone outcrop, and since water mainly penetrates the rock by means of joints and faults, these fissures become more widened by corrosion near the surface than they are at greater depth. The surface of the karst is therefore very permeable, but permeability (the capacity to transmit water) in the rock decreases with depth. This highly corroded superficial zone is termed the epikarst or subcutaneous zone (discussed more fully in the entry on Dolines).

Evolution of Surface Landforms

As soon as subsurface drainage is established, karst landforms can develop on the surface. Whereas the most characteristic subterranean features of karst are caves, the most typical surface landforms are closed depressions, especially dolines, which are enclosed bowl- or saucer-shaped depressions of usually a few hundred metres in diameter and some tens of metres deep. When dolines occupy all the available space, the surface has a relief like an egg-box and is known as polygonal karst, but this does not always develop and often dolines are dispersed or in clusters across an undulating surface. For a fuller discussion of the development of dolines and polygonal karst see the entry on Dolines.

The solution doline has a role in the karst landscape that resembles that of the valley in the fluvial landscape: it collects rainwater and discharges it from the surface. Solution dolines develop in the subcutaneous zone and drain water centripetally to enlarged fissures that discharge it vertically to the deep groundwater system. Small allogenic streams also develop enclosed depressions, termed stream-sinks or swallow holes, where they disappear underground. Large allogenic streams penetrate farther into the karst in well-defined valleys before they sink, and they produce landforms known as blind valleys, because their valleys usually terminate abruptly in a cliff or steep slope. The sinking streams give rise to caves, and if their roof is close to the surface it sometimes collapses, producing a cylindrical or crater-like depression termed a collapse doline. Subterranean rivers at a shallow depth beneath the surface can have their courses revealed by lines of collapse dolines, and in cases where collapse has proceeded for a long time the cave can be almost entirely unroofed, producing a gorge of cavern collapse (although not all gorges in karst are produced in this way).

As solution depressions evolve, some enlarge laterally and coalesce, producing compound closed depressions known as uvalas. Between the depressions are residual hills. Where the rate of vertical incision of dolines is significantly greater than the rate of solutional denudation of the surrounding land, the interdoline areas develop into hills. This is particularly common in humid tropical karsts, where the residual hills can be so well developed that they visually dominate the landscape, which is sometimes described as cone karst, a well-known example being the Gunung Sewu in Java (see Sewu Cone Karst). "Fengcong" is the term used in China to describe such karsts, with "fengcong-depression" recognizing both the positive and negative elements of the landscape. In Jamaica, the depressions in such karsts are known as cockpits (see Cockpit Country Cone Karst).

When vertical denudation eventually reduces the bottom of dolines / cockpits to the level of the regional water-table, they can incise no farther, so instead they widen their floors, with the result that the residual hills between them become isolated. Sometimes the lower slopes of these hills become over-steepened by undercutting and collapse at their base, a process brought about by the corrosional attack of swamp waters—made particularly vigorous if allogenic rivers periodically flood the intervening plains. The landscape is then transformed into tower karst (see separate entry), superb examples being known in southern China (where it is called "fenglin") and Vietnam. In this process the caves are drained and dismembered and their remnant passages are left at various elevations within the towers. Eventually, even the residual hills are removed by solution, and only a corrosion plain is left. In the Kinta Valley in Malaysia, the alluvial veneer over a corrosion plain has been removed in the process of alluvial tin mining to reveal the pinnacled corroded surface beneath. Another superb example of a corrosion plain is the Gort lowland of counties Clare and Galway in western Ireland, where Pleistocene glaciers have stripped away the mantle of residual soil, alluvium, and loose rock to reveal the karstified bedrock beneath.

If tectonic uplift were to elevate an old karst surface, then its erosion systems would be rejuvenated and new caves and landforms would develop. But whereas in the first cycle of karstification the rock was unweathered and had only primary porosity, in the second cycle there is an inheritance of landforms on the surface and secondary porosity underground. Thus a new phase of karst evolution would exploit the inherited features and develop them further.

In practice, many karst areas have developed on rocks that have been folded and faulted. These tectonic influences considerably complicate karst evolution and are of major significance in guiding groundwater flow and denudation of the surface. Faulted terrains often provide the conditions in which the largest enclosed karst depressions—known as poljes—are developed, some exceeding 100 km² in area (see entry on Poljes).

Modelling Karst Evolution

Various attempts have been made to model the processes described above. Early conceptual models of karst landscape development were presented by Grund (1914) and Cvijić (1918) (translations into English are available in Sweeting, 1981), but it was not until the late 20th century that models became quantitative. Ford and Ewers (1978) used a physical laboratory model to elucidate the development of proto-caves and successive flow paths. White (1984) developed a theoretical expression showing the relationship between chemical and environmental factors in the solutional denudation of limestones that convincingly demonstrated the relative importance of the factors involved. Subsequent progress in this area is discussed in Erosion Rates: Theoretical Models, and modelling of conduit development is discussed in Speleogenesis: Computer Models. Ahnert and Williams (1997) developed a three-dimensional model of surface karst landform development (see Figure) that started with a ter-

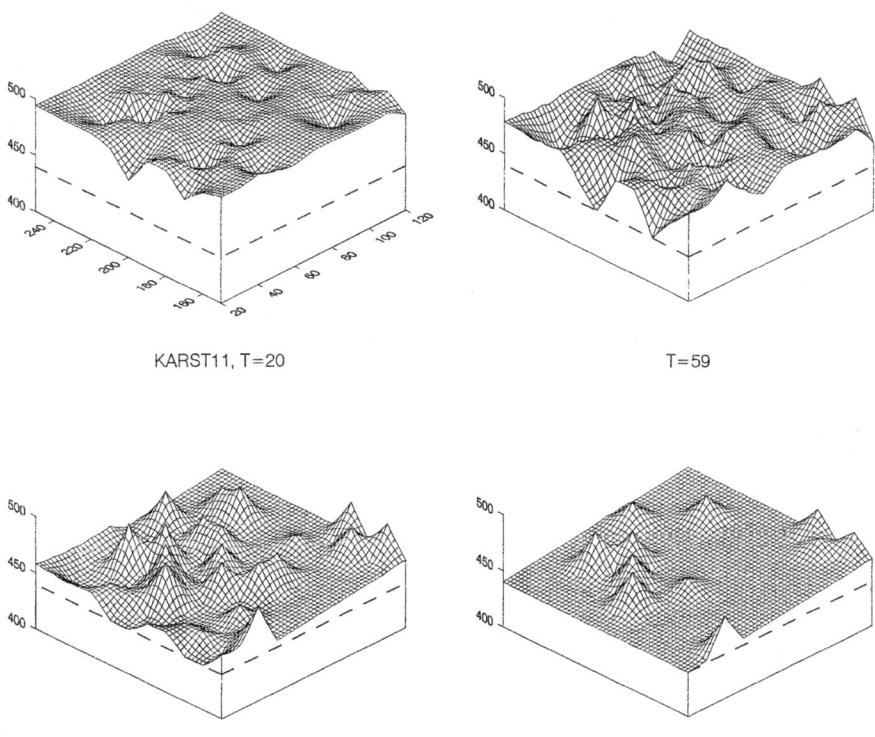

Karst Evolution: Model KARST 11 from Ahnert and Williams (1997). The water table in the karst slopes down a hydraulic gradient towards baselevel ($z = 440$) at the left edge (the outflow margin) of the block diagram. Doline karst is developed by the 20th iteration of the computer model ($T = 20$) and doline bottoms reach the water-table level by $T = 59$. Doline floors widen and converge at water-table level, producing a corrosion plain with the same slope as the hydraulic gradient, the beginning of which is evident by $T = 98$.

rain in which proto-conduit connections were already established and then showed how the relief might develop given different assumptions about starting conditions, such as randomly disposed sites of greater permeability or random variations in initial relief. This model illustrated sequential steps in the development of doline and polygonal karst and revealed the importance of divergent and convergent flow paths for giving spatial variations in solutional denudation that were sufficient to explain the development of residual cones between incising depressions.

PAUL W. WILLIAMS

See also **Speleogenesis**

Works Cited

Ahnert, F. & Williams, P.W. 1997. Karst landform development in a three-dimensional theoretical model. *Zeitschrift für Geomorphologie*, Supplementband 108: 63–80

Cvijić, J. 1918. Hydrographie souterraine et evolution morphologique du karst. *Receuil des Traveaux de l'Institut Géographie Alpine*, 6(4): 375–426

Ford, D.C. & Ewers, R.O. 1978. The development of limestone cave systems in the dimensions of length and breadth. *Canadian Journal of Earth Science*, 15: 1783–98

Ford, D.C. & Williams, P.W. 1989. *Karst Geomorphology and Hydrology*, London and Boston: Unwin Hyman

Grund, A. 1914. Der geographische Zyklus im Karst. *Ges. Erdkunde*, 52: 621–40

Sweeting, M.M. (editor) 1981. *Karst Geomorphology*, Stroudsburg, Pennsylvania: Hutchinson Ross
Translations of the Cvijić, Grund, and other classic papers.

White, W.B. 1984. Rate processes: Chemical kinetics and karst landform development. In *Groundwater as a Geomorphic Agent*, edited by R.G. LaFleur, Boston and London: Allen and Unwin

KARST HYDROLOGY: HISTORY

Caves are known to have been of importance to humans in prehistoric times and karst springs would also have been important for water supply. In historic times, probably the earliest record of karst hydrology is the visit, in 1100 BC, of Tiglath Pileser, King of Assyria, to a stream cave close to one of the springs that form the sources of the Tigris in Kurdistan (Shaw, 1992). The earliest example of what may be termed "karst water resource management" took place around 700 BC when natural cave passages and shafts beneath Jerusalem were adapted and enlarged so that water from the Gihon spring could be brought into the city. There is a major area of karst terrain in southern Greece, and karst phenomena made a great impression on the Greek philosophers as well as being important in the development of their civilization (Crouch, 1993). The earliest known statement on the origin of springs is that of Anaxagoras (500–428 BC), who supposed that they originated from large underground reservoirs supplied by rainfall. The term "katavothron" was applied by the ancient Greeks to the point at which a surface stream disappears underground (see Ponors) and Sophocles (496–406 BC) was the first to report on a katavothron on the river Inachos. Plato (427–347 BC) put forward two quite different hypotheses on the source of springs, one involving water within the Earth, the other, less well known, involving rainfall. Aristotle (384–322 BC), one of the earliest writers to specifically mention underground rivers in the Peloponnese, believed that springs originated from rainfall. Eratosthenes (275–194 BC), as reported by Strabo in his Book 8, described the connection between katavothra in the Pheneos polje to Ladon Spring in the Peloponnese. He also correlated spring rains and the discharge of karst springs. There was probably parallel work in China, and a book on caves in North China containing a description of cave hydrography is reputed to have been written about 221 BC, although there are some doubts over this.

About 30 BC, an extensive shallow artificial tunnel system in indurated calcareous dune sands was used to siphon off fresh water from salt water at a Palace at Mersah Matruh in the coastal zone just west of Alexandria, Egypt. This system is still in use and was the first development of water from a freshwater lens in a coastal karst aquifer.

Strabo (63 BC–20 AD) devoted the eighth of his 17 books on *Geography* to the "Karst Phenomena of the Poljes" and other karst phenomena, and in 37 AD the Jewish historian Josephus Flavius recorded, in *The History of the Jewish War*, the study of a water source from the Jordan and the use of chaff by Tetrach of Trachonitis to trace an underground stream. Seneca (4 BC–65 AD) was perhaps the most important Roman philosopher and in his book III, *Naturales Questinones*, he describes solution processes and the development of large caves, and he explains the disappearance and reappearance of streams in karst. The first known map locating a karst feature was by a Roman cartographer who marked the location of the Fonte Timavi, or Timavo karst springs near Trieste.

"An Encyclopedia of Knowledge" was compiled in 52 epistles around 970 AD by members of the Arabian Order of the Brothers of Purity (Ikhwanus Safa). In these documents Arabian monks wrote about caves inside mountains and springs discharging the water stored in caves. They contributed many new concepts relating to hydrology and geology. An example is the writings of the famous Arabian Abdul Hasan Ali Masudi on hydrogeology, geological cross sections, artesian conditions, and karst concepts.

The Hongshan Karst Spring in Jiexiu County, Shanxi Province, China has been in use since the Song Dynasty (1000 AD). In 1040 AD the spring's discharge was measured as 3 m^3 s^{-1} and separated into three channels to irrigate nearly 400 km^2.

The first recorded attempt to prove a connection between the Škocjanske jama stream-sink and the Timavo karst springs in Slovenia was made about the end of the 16th century by Imperato. During the 17th and 18th centuries a great variety of qualitative investigations were undertaken in European karst areas. For example, Athanasius Kircher, in his *Mundus Subterraneus* (1665) interpreted fluctuations of water in a polje as due to the seasons, and theorized on the connection of underground streams. Melchior Goldast reported on the Blautopf, one of Germany's largest springs, and between 1747 and 1748, Johannes Antonius Nagel, a mathematician, was assigned by the Hapsburgs to study poljes and caves. Between 1778 and 1779, Balthasar Hacquet, in a four-volume work, described many karst hydrologic phenomena, including limestone dissolution and the relationship between streams in a karst area, and poljes.

Toward the end of the 19th century, Jovan Cvijić (see Geoscientists) provided a systematic treatment of karren, dolines, karst rivers, karst valleys, and poljes in the Adriatic coastal area. This work and that from early researchers in Germany, France, and Italy formed the basis for the earliest quantitative testing in karst. Large scale karst-water tracing was accomplished by injecting sodium fluorescein and potassium chloride in swallow holes of the Danube in 1877, and in 1907, Truseus and Bartman, using lithium chloride and pitchblende, carried out tracing studies in the Istrian Karst. In 1926 the first plant spores were used for tracing; owing to their small size, and the fact that they formed an emulsion of solid bodies, they were thought to provide an ideal tracing medium although subsequently they have largely been abandoned in favour of fluorescent dyes (see Water Tracing: History). In the early 20th century, A. Grund recognized the zone of saturation in karst during studies in the Adriatic region where he noted that sea level was the base level for the karst hydrologic system.

By the later part of the 19th century, quantitative methods for determining groundwater velocity and permeability had been developed, although these were only applicable to water moving through homogeneous granular material with infinite character and isotropic conditions. These conditions do not usually apply in karst areas but in 1935, Victor Stringfield used a potentiometric map to interpret regional groundwater flow in the Floridan limestone aquifer. Similarly, in 1959, Cooper reported on the dynamic balance between fresh water and salt water in the Biscayne aquifer in Florida, and applied pumping test methods using the Theis equation during a large-scale pumping test on the Floridan aquifer at Fernandina, Florida. These methods were then successfully applied in the Huntsville, Alabama area with major pumping tests in hard dense, karstified limestones of Mississippian age. Subsequently they were applied to storm flow and recession analysis, and to a great variety of hydrological problems associated with quarrying, landfill siting, and water supply development.

With the many new instruments and techniques that have evolved during the later part of the 20th century, it is possible to describe more precisely the physical character of karst geological systems enabling quantitative methods to be applied to fractured rocks with solution conduits. More meaningful results can also be obtained from pumping tests on karst aquifers (see Groundwater in Karst: Borehole Hydrology).

PHILIP E. LAMOREAUX

See also **Speleogenesis Theories: Early**

Further Reading

Back, W. & LaMoreaux, P.E. (editors) 1983. V.T. Stringfield Symposium, Processes in karst hydrology. *Journal of Hydrology*, 61(1–3): 1–355

Crouch, D.P. 1993. *Water Management in Ancient Greek Cities*, Oxford and New York: Oxford University Press

Drew, D. & Hötzl, H. (editors) 1999. *Karst Hydrogeology and Human Activities: Impacts, Consequences and Implications*, Rotterdam: Balkema

LaMoreaux, P.E., Prohic, E., Zötl, J., Tanner, J.M. & Roche, B.N. (editors) 1989. *Hydrology of Limestone Terranes, Annotated Bibliography of Carbonate Rocks, Volume Four*, Hannover: Heinz Heise

Milanović, P.T. 2000. *Geological Engineering in Karst. Dams, Reservoirs, Grouting. Groundwater Protection, Water Tapping, Tunneling*, Yugoslavia: Zebra

Pfeiffer, D. 1963. Die geschichtliche Entwicklung der Anschauungen uber das Karstgrundwasser [The history of karst hydrology]. *Beihefte zum Geologischen Jahrbuch*, 57: 1–111

Shaw, T.R. 1992. *History of Cave Science: The Exploration and Study of Limestone Caves to 1900*, 2nd edition, Broadway, New South Wales: Sydney Speleological Society

Soliman, M.M., LaMoreaux, P.E., Memon, B.A., Assaad, F.A., & LaMoreaux, J.W. 1998. *Environmental Hydrogeology*, Boca Raton, Florida: Lewis Publishers

White, W.B. & White, E.L. 1989. *Karst Hydrology*, New York: Van Nostrand Reinhold

Zötl, J. 1974. *Karsthydrogeologie*, Vienna and New York: Springer

KARST RESOURCES AND VALUES

Karst areas offer a remarkable number of economic, scientific, and cultural-human values. The overview which appears here is largely based upon that which first appeared in the IUCN / WCPA Guidelines for Cave and Karst Protection (Watson *et al.*, 1997), and relates also to many other entries in this encyclopedia dealing with archaeology, conservation, flora, forests, human occupation of caves, religious sites, soils, speleotherapy, and tourist caves.

Most of the world's population is dependent upon agriculture, and agriculture is ultimately dependent upon the upper few centimetres of the Earth's surface. Some karsts offer rich and highly productive soils that are utilized for both general and specialized agriculture. Millions of people live in karst areas, but karst soils are often particularly vulnerable due to degradation by a variety of karst-specific processes that add to the usual pressures on soil. Caves are sometimes used for some specialized forms of agriculture and industry, including fish breeding, mushroom growing, and cheese production. In Southeast Asia, the natural occurrence of cave swiftlets provides a multimillion dollar industry in harvesting of nests for bird's-nest soup.

It is estimated that at least one quarter of the world's population gain their water supplies from karst, either from discrete springs or from karst groundwater. Thus in some karsts settlement patterns have been strongly influenced by sources of water. Ancient Mayan people made extensive use of caves and cenotes. On a world scale, major engineering works have been, and continue to be, undertaken to provide for the effective utilization of spring or ground waters. An interesting example is the ongoing redevelopment of the water supply system in Iran where karst ground waters are being utilized to replace the traditional qanat (ghanat) system that had been very seriously damaged by military action. Irrigation and the generation of hydroelectric energy are other major uses to which karst waters are put. Water supply may be particularly difficult to obtain in karst areas upstream of major springs, either for agriculture or human consumption.

In some karsts, major forest resources exist, or have previously existed. Where they have been removed by clear felling, they are rarely replaced as the stresses of soil erosion and regular periods of aridity prevent natural regeneration. However, where the soils are particularly suitable for forestry, plantation forests—often of exotic species—have replaced the natural forests. Many of these offer high levels of productivity.

Limestone is an important resource with application in many areas of agriculture and industry, for example, as a flux in steel making, and it is also used to reduce some forms of industrial pollution, for example, removal of sulfur dioxide gases. Limestone is also extracted for building construction, agriculture, road construction, cement manufacture, or other industrial purposes. Important mineralization has occurred in some karsts, and mining of these other minerals is often extremely profitable. Some limestones are also hosts to hydrocarbons (oil and gas).

Tourism is a major economic activity in some karsts, including the use of both developed and undeveloped caves, and surface scenery, thereby generating local employment. Each year many millions of people visit tourist caves globally, with some single cave areas receiving as many as two million people. It is sometimes difficult to neatly define tourist caves, but using the wider contemporary definition of such caves as including, for instance, the many temple caves of Asia, there are probably several thousand worldwide. Remote appreciation is also possible by means of films, videos, and photographic volumes, the production of which can be a significant component of some local economies. Such media also reinforce the value of caves and karst for tourism and as environments that need caring for.

In some parts of the world caves are used for various forms of speleotherapy and as sanitoria generally for respiratory ailments, especially where hot springs are also present, as in Budapest (Hungary). Some caves are still used for permanent residence. Others are used for occasional shelter, and as sites of refuge from air raids or for military activities. However, virtually all of the so-called caves that received considerable publicity in the recent conflict in Afghanistan consisted of artificially constructed underground tunnels and rooms; a similar example is the extensive underground tunnels in Vietnam.

It is commonly said with a high degree of truth, that "Caves are the books in the library of the history of the earth." Karst offers bedrock geologists clear exposures of lithological units, geological structures, and minerals. The fossils contained in the limestone are often readily visible and so offer paleontologists ready access to important sites. Geomorphologists derive insight into landform evolution and climate change over broad areas from the morphology of particular caves and the study of cave sediments. The cave floors with their layers of sediment contain invaluable and commonly readily dated evidence of geoclimatic and other earth history. Clastic fragments often provide evidence of tectonic and other geomorphological change. Although speleothems provide much of the beauty in caves, they also provide evidence of the past.

Some of the world's most important, and best-preserved archaeological and paleontological material has been found in floors of caves. Such deposits have provided much of our knowledge about evolutionary patterns that gave rise to our contemporary mammal fauna and even the human race. Thus the floors are amongst the most important components of any cave system.

Karst provides an extremely important habitat for a wide range of specialized plant and animal species, both on the surface and underground. To commence with the surface, karst provides an environment that favours plants that can tolerate seasonal aridity, prefer alkaline soils, and have low fire resistance. The result is that a number of species are found only on karst. Further, climatic change may lead to some vegetation communities being isolated and continuing to survive on karst even though they no longer occur on neighbouring lands. In turn, a number of animal species are often associated with those plants and so are most commonly found on karst. A small number may even occur only on karst.

A major factor that underlies the character of the surface environment is that the surface of karst is often highly irregular with a multitude of small-scale erosional landforms and this often provides a great diversity of micro-climates and a variety of small soil pockets. These demonstrate a very high degree of endemicity and biodiversity. This is most evident amongst the snail populations, where any one hill in a tower karst area may house a number of species found only on that hill.

The underground environment has a very high proportion of specialized species found only in the subterranean environment. These range from microscopic species found only in tiny fissures and in the groundwater to various invertebrates that inhabit the meso-cavernous voids that cannot be entered by scientists. However, invertebrates from meso-cavernous voids do enter caves and can be studied there along with other species found only in the caves. Again, there is a very high level of endemicity and localization of species. Thus karst provides an ideal environment for the study of adaptive radiation and other aspects of evolutionary change.

Karst appears to have virtually always been a place of importance and interest to people and to have exerted a considerable influence on cultural patterns. Thus cave and karst sites have long provided for particular modes of dwelling, for spiritual religious functions, for aesthetic appreciation and artistic creativity, and for recreational purposes.

The use of caves as dwellings commenced in the pre-human hominoids, remained dominant through the Stone Age and continues in a few cultures to the present. The early use of caves as dwellings provides, of course, for their present archaeological importance. Caves may also be used in emergencies as a place of shelter or even as hospitals, or for other special purposes.

In many parts of the world caves and other limestone landforms are recognized as important sacred sites or temples by, for example, the Mayan people of Central America, and the Hindu, Buddhist, and Christian societies. Some Buddhist communities also build temples that mimic caves, as with the great temple of Sokkurum in South Korea and temples built in Yangoon for the World Buddhist Conference. In some cases spiritual values relate to underground waters. Mayan priests prayed for assistance in water management to the water god Chac. Certain karst springs, such as those at Muktinath in Nepal, are sacred to both Buddhists and Hindus. Few western tourist caves lack a "cathedral chamber", further emphasizing the spiritual connections some feel with cave environments. Around the world caves continue to be used as burial sites, and places of worship continue to be erected amid karst, for example in the karst towers of southern China and Malaysia.

Many of the world's most scenic environments owe much of their appeal to karstic phenomena, including many mountain areas that not only attract cavers, but also walkers, climbers, photographers, artists, poets, novelists, and nature lovers. Classical Western art includes a multitude of cave images, as sacred sites or as places of beauty and fantasy. The tower karst of China is commemorated in many thousands of art works and pervades the written literature, appearing even in the poetry of Mao Tse-tung.

Karst is also a major resource for both mass tourism and for popular recreation, ranging from opportunities for family picnics to the most extreme physical challenges of cave exploration and cave diving. Speleology provides many extremely valuable and successful examples of partnership between amateurs (the recreational speleologists) and professional scientists. Organized speleology in itself supports a large number of major local and national organizations as well as several international bodies, all of which sponsor regular conferences dealing with issues in cave management, conservation, and research.

The various economic, spiritual, and scientific values of karst are often readily demonstrated in a compact area, and commonly make caves and karst areas splendid examples for education. The ecological chains of cause and effect, and the complex determinants of both environmental integrity and the relationship between the environment and human settlements, are rarely so clearly evident. Surprisingly, the use of karst parks for this purpose is still in its infancy but some have now developed professionally staffed education centres with on-site accommodation and learning facilities.

Cultural resource management is often an important consideration in karst areas. Some springs and caves have long served as foci for human settlement or activity and now contain valuable records of the evolution of societies layer by layer, in sediment or in art upon the cave walls. The prehistoric legacy found in some caves is well known and has contributed in a major way to knowledge of our ancestors. The historical archaeology of some karsts is also important, including such features as water reticulation systems established in some Chinese karsts. Again, these provide a wonderful resource for learning about human cultural history.

Considerable heritage value can be attached to the built environment in some karst areas, ranging from some prehistoric constructions in caves to some cave resorts in Europe and the distinctive cave-associated tourist hotels of Australia and the United States. Probably few other kinds of landscapes or land systems attract such a remarkable diversity of human and cultural interest, while also serving important economic and scientific purposes. Many of the authors contributing to this volume are people who have spent many years of their life in the pursuit of speleological inquiry. Often this commenced as recreation, but in due course led to a fully developed professional career. But whether as amateurs or professionals, most share an immense joy in and commitment to cave experience and cave inquiry. This encyclopedia in itself is, in fact, a major symbol of the many values which people place upon caves and karst. Caves can be virtually all things, to somebody, somewhere. Further, many of those with a lifetime experience of caves follow a series of more-or-less parallel interests and live surrounded by not only working materials, but an accumulation of literature, images, and sounds from the whole spectrum of caving interests.

ELERY HAMILTON-SMITH

Works Cited

Watson, J., Hamilton-Smith, E., Gillieson, D. & Kiernan, K. (editors) 1997. *Guidelines for Cave and Karst Protection*, Cambridge and Gland, Switzerland: International Union for Conservation of Nature and Natural Resources

Further Reading

Vermeulen, J. & Whitten, T. 1999. *Biodiversity and Cultural Property in the Management of Limestone Resources*, Washington, DC: World Bank

KARST WATER RESOURCES

Large numbers of people live in areas underlain by carbonate rocks and it has been estimated that up to a quarter of the global population is supplied largely or entirely by karst waters, including over 100 million people in China and a significant proportion of the population in many European countries. However, water supply in karst regions is often a limiting factor in human occupation and development, because it is so variable in both quality and accessibility in time and space. Paradoxically, karst water is often abundant deep underground, but successful exploitation requires equipment and capital. Utilizing karst water resources therefore requires an understanding of the location of the water and consideration of the technology available; location determines accessibility and hydrogeological factors determine the quality and quantity of water; technology determines whether the water is practically available and potable.

The primary control on water resources is climate, and as karst occurs in all climatic zones, this dictates the overall water resource conditions. The high permeability of karst rocks often results in little surface water, but enhanced groundwater recharge. Surface runoff is occasionally retained in "dew ponds" in locally impermeable depressions, but these have limited capacity and duration, and are prone to contamination by surface activities. In the shallow subsurface the epikarst aquifer may retain water for many months, sometimes sustaining small, temporary springs in the side of depressions. High permeability also results in a low-gradient water table, which translates into a thick unsaturated zone. Recharge water descends through the unsaturated zone to the water table, beneath which it is stored in the matrix and fractures, travelling along a limited number of conduits to emerge briefly at karst windows, or permanently at springs. Karst aquifers also receive recharge from allogenic sinking streams. In contrast to autogenic recharge, allogenic water tends to be variable in quality and quantity, in part because disappearing streams seem to lend themselves so naturally to waste disposal! A disconcerting paradox of karst water resources is that spring water is not necessarily pure, but may become contaminated with sediment, pathogens, and pollutants, especially during high flows. Some karst regions are traversed by surface rivers, often occupying deep gorges, making the water inaccessible.

Readily accessible water resources are therefore restricted in their distribution and reliability (see Table), but may be somewhat enhanced by development. Dew ponds can be artificially constructed, deepened or sealed, or runoff can be retained in cisterns. Excavations in the epikarst can tap the shallow aquifer.

Karst Water Resources: Karst water resource classification.

Location	Resource	Origin	Quality	Quantity
Surface	dew ponds, cisterns	surface runoff	soft, vulnerable	limited transient
	epikarst springs	epikarstic aquifer	moderate, vulnerable	small, short-lived
	springs/karst windows	autogenic aquifer	hard, reliable	abundant, steady
		allogenic aquifer	variable, vulnerable	variable
Underground	rivers	surface runoff aquifers	variable, vulnerable	variable
	adits	aquifer	good-poor	good
	dug wells	aquifer		variable
	drilled wells	aquifer		variable

In all cases the objective is provision of sufficient storage to sustain supply through dry periods. In contrast, many settlements have developed around karst springs that can provide much more abundant and reliable water supply, sustained by the karst aquifer. However, springs vary a great deal in quantity and quality of water supply, reflecting the mode of recharge and the storage capacity of the associated aquifer. Allogenic recharge routed through a discrete conduit to an overflow spring can result in intermittent flooding by contaminated water, whereas a large, confined, autogenic aquifer may provide sustained excellent quality water supply. Sometimes, karst water reserves can be enhanced by damming springs and rivers. In successful cases, the resulting elevation of the water table significantly amplifies the storage capacity beyond that of the apparent surface reservoir. However, the permeability of karst rock means that any form of surface barrage tends to be bypassed quite quickly by leakage and only at great expense can effective grout barriers be installed in the subsurface (see Dams and Reservoirs on Karst).

Access to subsurface karst water has sometimes been possible through caves, and in some cases horizontal tunnels have been driven to tap the aquifer. The great thickness of the karst unsaturated zone makes for demanding construction of vertical dug wells, and the rubbly, eroded character of weathered carbonates can make modern rotary and air track drilling somewhat difficult due to loss of circulation pressure and collapse of the bore wall. Drilled wells revolutionize exploitation of karst water because they permit more arbitrary access to water, freeing settlements from limited surface occurrences. However, karst aquifers are heterogeneous, and this results in very diverse production from wells; those encountering exclusively matrix storage have very low yields, whereas those encountering conduits can have massive yields, sometimes sufficient for cities or industries with demands beyond the normal viability of groundwater exploitation. Most wells have intermediate performance in terms of yield, or can have their yield enhanced by "development" using intensive pumping or pressurized injection of hydrochloric acid or dry ice. A simple strategy, occasionally used in regions with explored caves, is to drill directly into a cave stream, although this demands quite precise mapping. Some success has been gained by drilling along surface fracture traces, or fracture intersections, because these zones are taken to indicate the likely presence of enhanced solutional development in the subsurface. There has been relatively little success with the use of subsurface geophysics to locate reliable karst water supply.

The theory of groundwater exploitation through wells is based on easily defined porous or equivalent porous media in which the hydraulic head in a well is reduced by pumping, inducing a radially symmetrical inward flow, and "drawdown" (lowering) of the water table in a characteristic cone of depression centred on the well. The depth and extent of the cone of depression depends on the size of the well, the pumping rate, and the hydraulic properties of the rock. The theory has been extended in recent years to define "capture zones" for wells: the area from which water is drawn to provide the flow from the well. Such definition is important, not just in defining the practical limits of well spacing, but in determining the possible risk or source of groundwater contamination.

Drawdown and capture zones in karst aquifers are much more difficult to determine because of heterogeneity. Most wells intersect only small conduits, a few centimetres or less in aperture, and have capture zones that may be elongated in the direction of the conduit. Occasionally a large conduit is intersected by a well; such a conduit acts essentially as an extension of the well. The drawdown is focused not just on the well, but also on the conduit, drawing water from the adjacent fractures and matrix. The resulting cone of depression will be highly eccentric, being dependent as much on the conduit(s) as on the forcing well. Some karst aquifers show metres or even tens of metres of water table rise in response to a rain storm, and the resulting flow field in the aquifer and well is rather complex. Capture zones for karst wells are therefore not only highly eccentric, but will vary with time, frustrating application of standard management practices.

Many aquifers are currently compromised or threatened by over-exploitation or contamination. Excessive withdrawal of water from karst aquifers was not possible with self-regulating springs, but pumped wells can more completely drain an aquifer

Karst Water Resources: Sulphur Spring, a major rising exploited for its water resources in the lowland karst of central Florida. (Photo by Tony Waltham)

and its springs. It is therefore common to monitor groundwater levels using passive observation or monitoring wells. The high-permeability connection of major conduits to springs results in lower heads in the conduits than in the surrounding fractured bedrock. The water table is thus not a simple planar surface, and requires a large number of well-water levels for adequate definition. Contamination of karst aquifers is often rapid in entry and transmission, and poorly attenuated compared to porous media. Contaminants may be difficult to detect and track in monitoring wells, so that *a priori* planning may be required in anticipation of possible problems.

A number of specialized techniques for monitoring karst groundwater have been developed. Foremost has been the use of tracers for defining groundwater trajectories, and defining catchment areas, prior to occurrence of contamination (see Water Tracing). Second has been use of springs to provide a monitoring point which integrates flow from a large area. These methods work well in karst where sinking streams and springs are accessible. However, many karst aquifers lack such convenient access points, or information may be required for areas removed from primary conduits. Monitoring wells provide the key here, and appropriate techniques have been slowly developed. Collecting data from both springs and wells gives a more complete understanding of a carbonate aquifer than either can alone; springs provide information on the conduit network and wells (in a rather idiosyncratic manner) provide information on matrix and fracture hydrogeology. The behaviour of karst aquifers often varies dramatically with flow. It is therefore essential to implement continuous or high-frequency monitoring and sampling if such conditions are to be identified and characterized. Idiosyncratic short-term responses of wells to recharge events arise from the connectivity of the particular well to the various levels in the aquifer, and may misrepresent broader aquifer conditions.

Karst groundwaters are a major resource around the world, but in many cases the water is poorly managed due to the absence of regulation, or the application of inappropriate methods of waste management and water use. In general, it is best to take a precautionary approach and presume that carbonate aquifers are karstic, and assume that there may be rapid recharge and rapid flow through the aquifer via a conduit network. In the worst case such groundwater has similar properties to surface river water with its attendant problems of turbidity, pathogens, and chemical contaminants, and requires appropriate filtration and sterilization treatment.

Chris Smart and Stephen R.H. Worthington

See also **Groundwater in Karst**

Further Reading

Domenico, P.A. & Schwartz, F.W. 1998. *Physical and Chemical Hydrogeology*, New York: Wiley

Drew, D.P & Hötzl, H. 1999. *Karst Hydrology and Human Activities: Impacts, Consequences and Implications*, Rotterdam: Balkema

Ford, D.C. & Williams, P.W. 1989. *Karst Geomorphology and Hydrology*, London and Boston: Unwin Hyman

Price, M. 1996. *Introducing Groundwater*, 2nd edition, London and New York: Chapman and Hall

Quinlan, J.F. & Ewers, R.O. 1989. Subsurface drainage in the Mammoth Cave area. In *Karst Hydrology: Concepts from the Mammoth Cave Area*, edited by W.B. White & E.L. White, New York: Van Nostrand Reinhold

White, W.B. 1988. *Geomorphology and Hydrology of Karst Terrains*, Oxford and New York: Oxford University Press

Worthington, S.R.H., Davies, G.J. & Ford, D.C. 2000. Matrix, fracture and channel components of storage and flow in a Paleozoic limestone aquifer. In *Groundwater Flow and Contaminant Transport in Carbonate Aquifers*, edited by C.M. Wicks & I.D. Sasowsky, Rotterdam: Balkema

KHAMMOUAN, LAOS–VIETNAM

The Khammouan Karst is 290 km long and 40 km wide in Central Laos, continuing for 40 km into Vietnam, with a 60 km wide northern branch and a 15 km wide southern branch. In Laos, it is mainly an arcuate, northwest–southeast-trending faulted anticlinorium following the Thakhek Fault, one of the largest in Southeast Asia. In Vietnam, synclines are more developed.

The 1100 m thick main carbonate sequence is mainly middle Carboniferous to lower Permian, consisting of limestones, dolomitic to a variable extent, and of dolomites. Advanced diagenesis makes the rock extremely hard. Devonian, and some Visean carbonates are also present. Middle Triassic Indosinian tectonic events generated huge erosion. Karst formed and was subsequently buried under thick Mesozoic fluvial sandstones and shales. A new uplift started at the beginning of the Tertiary and still continues. Erosion has stripped off 2700 m or more of cover rocks in parts of Khammouan. Present karst features result from this long history.

The karst is drained west into the Mekong River and east to the Tonkin Gulf toward the China Sea, with a divide close to the political border. The Mekong flows at an altitude of 140 m at low waters and the karst uplands are mainly between 600 and 800 m, with some summits up to 1146 m. The monsoon-related, tropical climate has a dry season from November to mid-May and a rainy season with a peak in August. Annual rainfall is at least 1900 to 2750 mm. Temperatures are 25–26°C near the Mekong River and on the Vietnam coast. The main allochthonous streams have a varying perennial flow. Specific river discharge is 52 l s^{-1} km^{-2} on the reference Xé Bang Fai River, Laos. Calculated flows in Xé Bang Fai Cave are 68 m^3 s^{-1} on average (600 maximum, 2.2 minimum) and 11 m^3 s^{-1} on average in Nam Hin Boun Cave (100 maximum, 0.4 minimum), on a daily computation basis. At the Phong Nha Cave outlet, Vietnam, 10 m^3 s^{-1} are reported at the dry season. Evidence for a 40–50 m water rise in at least 40 × 40 m cross-section passages has been reported from both countries.

Flash floods do occur in binary karsts, especially where sandstone and shale cuestas dominate the karst boundary, as to the northeast in Laos. In Nam Non Cave, enormous quantities of sandstone boulders (average size 20 cm and up to 60 cm) are found throughout the 5 km wide karst crossing. Karst drains are commonly sub-horizontal in most of the area, but vauclusian

Khammouan, Laos and Vietnam: Figure 1. Rugged limestone hills form complex massifs between alluviated basins in the Khammouan karst on the Laotian side of the border with Vietnam. (Photo by Tony Waltham)

springs (dived to a depth of 54 m) are present along main valleys and near the Mekong River, where karst aquifers have been successfully drilled.

Surface morphology combines fengcong karst, tower karst and plains, plus cone karst in Vietnam. Massifs are bounded by steep slopes and long cliffs, 200 and even 400 m high (Figure 1). The holokarst surface is extremely irregular with dry valleys, corridors, and many dolines, with extremely developed karren features, including tsingy pinnacles up to 10 m high with razor-sharp edges. A few open shafts are up to 50 m deep. Shaliness and dolomite content variations determine several types of landscape, with somewhat different landforms and vegetation.

Impressive poljes show different stages of maturity. Mature hyperpoljes are comprised of high carbonate cliffs surrounding

Khammouan, Laos and Vietnam: Figure 2. The Hin Boun River Cave which local people navigate on fishtail-powered longboats to travel between villages in the karst basins of Laos. (Photo by Tony Waltham)

a core of substratum rocks, as in the Nam Pha Thène basin. Cliff foot dissolution and subsequent collapse generate lateral enlargement. Polje drainage may include peripheral springs, underground rivers below polje floors, and caves with sinking streams. Ponor lakes, emissive in the rainy season, are found in both poljes and karst plains, some being fed by moderately enlarged fissures.

Active caves usually open at the base of cliffs, though a number of relict caves are known up to 300 m above plains and polje floors. Caves are commonly huge, especially when they collect allochthonous water, such as the 50×50 m passage sections in Hang Vom, Phong Nha, and Xé Bang Fai Caves, or the 50 to 100 m wide sections in Nam Hin Boun Cave and many fossil galleries. Several large chambers are known: 260×240 m in Tham En, and 210×155 m in Tham Koun Dôn. The largest cave entrance is 215 m wide and 30 m high (Tham En). Relict entrances are commonly closed by boulders and the remaining section may be small. Phreatic and vadose features co-exist, the phreatic ones maybe representing either older evolutionary stages or a morphology acquired during high water periods. Sloping relict conduits (up to 30°) in several caves are probably drains of formerly drowned areas. Long sumps are known, some being open in the dry season. Shafts are rare and are usually not deep, with so far two exceptions in Laos; several are located in thick sand and boulder passage fills and formed under pressure variations associated with fluctuating water level in underlying conduits.

Cave morphology reflects various flowing conditions: ceiling cupolas (phreatic conditions), scallops, potholes, and rocky floors (fast flows), underground karren on walls and floors (decreasing water flow), etc. Water pressure during large floods destabilizes walls (as in Kagnung Cave) and removes large boulders. Fast allochthonous streams generate relatively linear caves, as the Xé Bang Fai and the Nam Hin Boun (Figure 2). Sandstone boulders and pebbles from flash floods are common and may be imbricated, even in the 30 m wide Nam Non galleries. Gravel and sands are common and clay is encountered near sumps and remaining pools. Antidunes, sand waves, current ripples, and climbing ripples characterize flows of decreasing energy in the passages. Meander bars exist in Nam Hin Boun Cave. Speleothems are frequent in relict passages and more developed in the higher, older ones. They are present in the upper sections of active passages with only occasional flooding. All classical speleothem types are found, and shields and common cave pearls are also found. Rimstone pools are present in several cave streams (such as Hang Lanh).

The longest caves are the Nam Non (22.1 km), Hang Khe Ry (18.9 km), the longest part of Hang Vom (15.5 km) and Nam Hin Boun (12.4 km). Surveyed development is 130 km in Laos, 44.5 km for the Phong Nha System, and 32 km for the Hang Vom System, Vietnam. The deepest cave is Hang Khe Ry (141 m deep) and the second deepest Tham En (122 m deep). Tham Phi Seua in Laos gently rises up to $+315$ m.

CLAUDE MOURET

See also **Asia, Southeast for location map**

Further Reading

Limbert, H. 2002. Vietnam 2001, Ha Giang, Cao Bang and Quang Binh provinces. *International Caver 2001*, 60–65
Milner, S. 1999. Vietnam '99. *Australian Caver*, 149: 16–19
Mouret, C. 2001. Les grands poljés du karst du Khammouane, Laos Central. [The large poljes of Khammouan karst, Central Laos]. *Proceedings of the 4th European Exploration Speleology Congress, May 2000*, Profondeville, Belgium: Union Belge de Spéléologie: 83–89
Mouret, C. 2001. Le karst du Khammouane au Laos Central. Dix ans d'exploration. *Spelunca*, 84: 7–32
Waltham, A.C. & Middleton, J. 2000. The Khammouan karst of Laos. *Cave and Karst Science*, 27(3): 113–20

KRAS, SLOVENIA

The Kras is a karst plateau to the northeast of Trieste Bay (Adriatic Sea, Mediterranean), on both sides of the Slovene–Italian border (see map in Dinaric Karst entry). From the name of the plateau—Kras in Slovene, Karst in German, Carso in Italian, comes from the root "kar/gar", meaning rock—the international term karst has developed. The region is elongate in the northwest–southeast Dinaric trend, and is 40 km long and up to 13 km wide, covering about 550 km². The rocks, which form the Kras, are Cretaceous and Tertiary limestone and dolomite, being carbonate deposits of shallow, warm-water shelf environments. They range in age from the lower Cretaceous to Eocene, and were followed by flysch deposition.

The terrain is dominated by two ridges in the Dinaric direction, with the great depression of Dol between them. Dol is probably a dry valley of the Reka River or may even have a tectonic origin. The plateau has very little large-scale relief but has numerous collapse and dissolutional dolines (Figure 1). There are huge expanses of rocky terrain, with exposed white limestone that is *in situ* or broken, where a sparse vegetation of grasses and shrubs is rooted in the fissures that contain the only soil. Dolines are so numerous that much of the ground is a polygonal karst. This is the classical karst of "rocky ground".

The annual precipitation is 1400–1650 mm, but on the Kras there are no surface streams. A considerable part of the rainwater sinks very fast and feeds the deep karst aquifer. Allogenic drainage is important for aquifer recharge, especially by the River Reka sinking in Škocjanske Jame (Škocjan Caves, see separate entry) and also by losses from the rivers Vipava, Soča, and Raša, and some other superficial streams from the non-karstic border. The Reka River reappears in the Timavo springs on or near the Adriatic coast, after an underground flow length of about 41 km. One part flows to the north of a dolomitic barrier and the other through the Abisso Trebiciano (Labodnica) to the south of the barrier. The flows merge again before reappearing in the springs. Comparison of the lowest discharges of the Reka (at drought below 1 m³ s⁻¹) and the Timavo springs (9.1 m³ s⁻¹), shows that the aquifer is substantially fed by other sources. Due to a high degree of karstification and consequently almost immediate infiltration, and due also to the relatively high amount of precipitation, the autochthonous infiltration of rainwater is important, contributing about 65% of annual spring flow.

Kras, Slovenia: A large dissolutional doline is a component of the landscape of broken rocks, patchy grass cover, and stands of woodland that characterize the Kras. (Photo by Tony Waltham)

The Kras inner structure is heterogeneous: a complex system in which the primary drainage channels are interconnected with smaller interconnected channels and fissures. In the main passages, water drains by fast flow (300 m h^{-1}) but the karst aquifer shows typical heterogeneity. Its exploitation is complicated by the great depth down to the phreatic zone. In the eastern part of the aquifer, the water is found during low water level at altitudes of 292 m and 210 m (Škocjanske Jame). The water level lowers towards Kačna Jama (156 m) and Abisso Trebiciano (12 m), while the Timavo springs are at sea level. Heavy rainfall is followed by a rapid increase in this level, with underground water level increasing by 30 m in the western part and by more than 100 m in the east. In such conditions, the Reka has a discharge of about 300 m^3 s^{-1} and the Timavo springs yield 127 m^3 s^{-1}. The mean flows of the Reka River and Timavo springs are 8.3 m^3 s^{-1} and 30.2 m^3 s^{-1}, respectively. The first tracing of the Reka was realized in 1864, when researchers used not only hydrological observations (discharge, observations of water level, measurements of rainfall) but also a series of tracing methods, including eels, salts, and dyes. The Timavo springs were renowned for being the most suitable for ships' water supply in the 4th century BC, and these springs and those of Brojnice, at Aurisina, provide the water supply to Trieste. At Brestovica there is a borehole where water is pumped from the phreatic zone at sea level and pumped to the Kras plateau, up to 500 m higher.

Some caves of Kras are important paleolithic and prehistoric sites. Cave tourism started as early as the 17th century at Vilenica Cave. People descended to the bottom of Velika Dolina (collapse doline) of Škocjanske Jame in the 18th century; in 1808 Pečina na Hudem letu was made a show cave, and in 1819 the visitor's book of Škocjanske Jame was introduced. There are now ten show caves on Kras.

Speleological research began in 1839 with the first attempts to follow the Reka underground. In 1841, the bottom of Abisso Trebiciano was reached at a depth of 329 m, making it the world's deepest explored cave for about 50 years. From 1884 to 1890, Škocjanske Jame was surveyed to the sump (dived in 1991). In 1889 Kačna Jama was discovered with its 200 m entrance shaft, and followed to reach the Reka in 1972. Among thousands of caves on Kras, Škocjanske Jame (5.8 km long, 250 m deep), Brezno pri Risniku (230 m deep), Kačna Jama (12.5 km long, 280 m deep), Abisso Trebiciano (329 m deep), and Lazaro Jerko (0.5 km long, 380 m deep) are notable. Grotta Gigante and Škocjanske Jame both have very big underground chambers: Martel's Hall, in the latter, has a volume of 2.2 million m^3.

Andrej Kranjc

See also **Dinaric Karst (location map); Škocjanske Jame, Slovenia**

Further Reading

Civita, M., Cucchi, F., Garavoglia, S., Maranzana, F. & Vigna, B. 1995. The Timavo hydrogeologic system: an important reservoir of supplementary water resources to be reclaimed and protected. *Acta Carsologica*, 24: 169–86

An explanation and up-to-date data on the Timavo hydrological system, relating to the hydrology of Kras.

Habič, P., Knez, M., Kogovšek, J., Kranjc, A., Mihevc, A., Slabe, T., Šebela, S. & Zupan, N. 1989. Škocjanske Jame speleological revue, *International Journal of Speleoleology*, 18(1–2): 1–42

Detailed physical speleological description of Škocjanske Jame, including the history of exploration and arrangements for tourists.

Kranjc, A. 1994. About the name and the history of the region Kras. *Acta Carsologica*, 23: 82–90

Explanation of the name Kras and how the name has been transformed into the international term "karst".

Kranjc, A. (editor) 1997. *Slovene Classical Karst*, Ljubljana, Slovenia: Kras

Detailed description of Kras from different points of view, from natural history to human, with emphases on karst morphology and hydrology.

KRUBERA CAVE, GEORGIA

At the dawn of the new millennium, Krubera (Voronja) Cave in the Arabika Massif, western Caucasus, became the deepest known cave in the world, with a depth of 1710 m, although it lost the record to France's Gouffre Mirolda early in 2003.

The Arabika Massif is one of the largest limestone massifs of the western Caucasus. It is located in Abkhasia, a republic currently within Georgia (for location, see maps in Russia and Ukraine and in Caucasus entries). The massif, with its strongly pronounced glaciokarstic landscape at elevations ranging between 1900 and 2500 m (the highest peak, the Peak of Speleologists, rises to an altitude of 2705 m), is composed of Lower Cretaceous and Upper Jurassic limestones. In the central part of Arabika the Cretaceous beds are retained only in a few ridges and peaks, as well as in small outcrops along the synclinal cores. The central part of the massif is composed of Upper Jurassic strata that dip continuously to and beneath the Black Sea shore. The massif is highly tectonized, with a fault-block structure strongly controlling both cave development and groundwater flow systems. To the northwest, northeast, and east, Arabika is bordered by the deeply incised canyons of the Sandripsh, Gega, and Bzyb rivers. The latter separates Arabika from the adjacent Bzybsky Massif, another area with major speleological prospects in the western Caucasus, including Snezhnaja–Mezhonogo (−1370 m) and Pantjukhina (−1508 m), among many other significant caves (see Caucasus, Georgia).

Among the several hundred caves known in the Arabika Massif, some deep caves explored during the 1980s stand out, including the Iljukhina system (−1240 m), the Arabikskaja system (Kujbyshevskaja–Genrikhova Bezdna; −1110 m), Dzou Cave (−1080 m), Moskovskaja Cave (−970 m), and Cherepash'ja Cave (−650 m). In 2001 Sarma Cave, previously explored to −700 m, was pushed to a depth of 1530 m. The deepest cave, Krubera (Voronja), is located in the Ortobalagan glaciated trough valley, some 300 m to the southeast of and 60 m above Kujbyshevskaja Cave, the main entrance to the Arabikskaja system. Although Krubera Cave is not connected directly with the Arabikskaja system, it is most probably part of a single linked hydrological system.

An open-mouthed 60 m shaft was first documented by Georgian researchers in the early 1960s, who named it after Alexander Kruber, a founder of karst science in Russia. The early exploration was stalled by an impassable squeeze in a meandering passage which led off from the foot of the entrance shaft. During the 1980s the speleological club of Kiev pushed the cave to −340 m by breaking through several critically tight meanders between the shafts. During this time the cave received its alternative name Voronja (Crow's Cave), owing to the number of crows nesting in the entrance shaft. Meanwhile, other deeper caves of the Ortobalagan Valley diverted the main effort, and the exploration of Krubera Cave was suspended. In August 1999 the Ukrainian Speleological Association expedition, led by Yury Kasjan, recommenced work in Krubera Cave and made a major breakthrough by finding a deep continuation, explored to −750 m, from a window in the wall of a pitch at −220 m. Further expeditions explored the new branch to −1410 m in 2000 and to −1710 m in 2001 (see profile).

The cave is developed in the thickly bedded and massive Upper Jurassic limestones, in the fold zone of the Berchil'sky

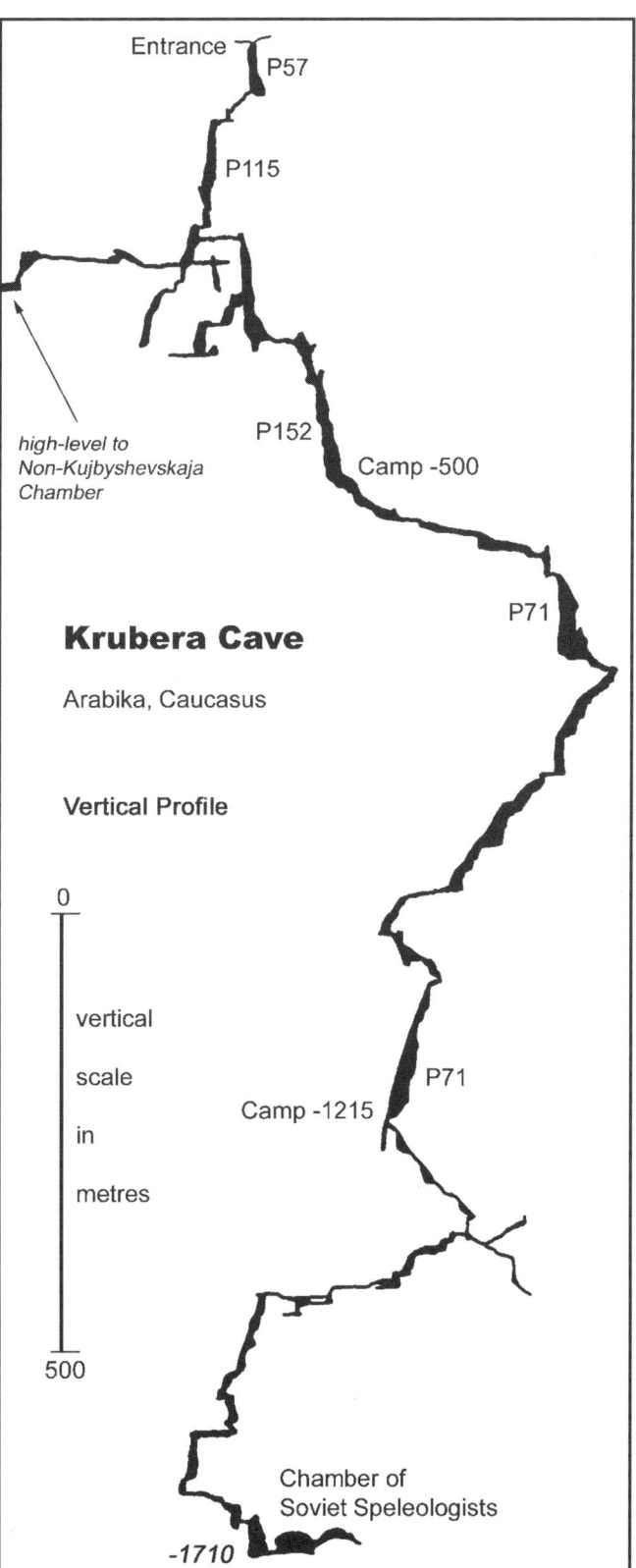

Krubera Cave, Georgia: Vertical profile through the Krubera Cave.

anticline entrenched by the Ortobalagan trough valley. The main branch descends steeply in vertical pitches, separated by short meanders, while shifting slightly towards the southern slope of the anticline. Apart from the "non-Kujbyshevskaja" branch, which stretches for almost 500 m to the northwest, the cave develops within quite a small area (400 m by 400 m), remaining within a small tectonic block and not extending beyond the southern margin of the trough valley. There is a strong tectonic control of cave development in plan view. Some segments of the major caves stretch along faults, other sections twist within major tectonic blocks and reflect back inside the blocks when intercepted by a fault. The main branch of the Krubera Cave turns many times and drops steeply via vertical pits, separated by short meanders. Through both the degree of morphological development and its hydrological system, the cave seems to be a tributary to the adjacent Kujbyshevskaja Cave. A small water flow (up to 1 s^{-1}) appears in the cave at a depth of about 340 m. This flow disappears and reappears at various levels, but never increases significantly.

Major karst springs with individual average discharges of 1 to 4 $\text{m}^3 \text{ s}^{-1}$ are located along the fringes of the massif at altitudes above sea level ranging from 1 m (Reproa Spring) to 540 m (Gegsky Vodopad). Submarine springs are also known, emerging from the floor of the Black Sea at depths of 20–40 m (and probably greater). Some boreholes located along the shore of the Black Sea yield karstic groundwater from depths of 40–280 m below sea level. An outline of the hydrogeological structure of the massif and its true speleological potential were revealed in the 1980s, when spectacular progress was made in deep cave exploration. Two dye-tracing tests during 1984 and 1985 proved connections between the major caves and springs. In particular, tracers injected in the Kujbyshevskaja Cave were detected in Kholodnaja Rechka (1.5 $\text{m}^3 \text{ s}^{-1}$, 50 m above sea level) and Reproa (2.5 $\text{m}^3 \text{ s}^{-1}$; 1 m above sea level) springs along the seashore. Tracer was also detected in a borehole which yields groundwater at a depth of 40 m below sea level, located between these two springs. This gave a reason to identify the major karst hydrological system as being potentially the deepest in the world at that time, with a vertical range greater than 2300 m. The system comprises the majority of the southeastern flank of the major Arabika anticline.

At its present lowest explored point of -1710 m (530 m above sea level) Krubera Cave neither enters a master river passage nor shows any signs of major flooding (which could indicate close proximity to the base-level of a main collector). These features, together with the proven connection of the Arabikskaja caves to large springs on the shore of the Black Sea, suggest a good potential for further deepening the cave system of the Ortobalagan valley. Equally realistic is the possibility of connecting caves with entrances at higher elevations into the Krubera or Kujbyshevskaja caves. The best prospects can be found in the nearby Berchil'skaja Cave (-500 m), the entrance of which is some 150 m above, and Martel's Cave, located some 80 m higher up. Hence the future possibility of locating a 2000 m+ system in the area is exceptionally good.

ALEXANDER KLIMCHOUK

See also **Caucasus, Georgia**

Further Reading

Klimchouk, A. & Kasjan, Ju. 2001. Krubera (Voronja): The deepest cave in the world. In a search of -2000 m in the Arabika Massif, Western Caucasus. *Caves & Caving*, 91: 34–41. The same article is also published in *Spelunca*, 82: 15–24 and *NSS News*, 59(9): 252–57

LIMESTONE AS A MINERAL RESOURCE

Limestones have been worked for many thousands of years—initially for building stone and, more recently, for a wide range of constructional and industrial uses. In Europe and North America, the industrial revolution of the 19th century led to a rapid increase in the demand for limestone, which was one of the main commodities (together with coal) carried by the early canals and railways. More recently there has been a global increase in the demand for limestone for use in cement making. In some countries, such as Great Britain and the United States, it is also of major importance as an aggregate. Indeed, 70% of the stone produced in the United States is limestone. Limestone has a very large range of uses that may be grouped under seven main headings.

Aggregate: The principal uses of aggregate are as roadstone (which is often coated with asphalt), as a component of concrete, and as ballast (large material, e.g. railways). The suitability of a particular limestone for use as an aggregate is governed by its physical properties, most notably density and porosity, and its mechanical properties. Density should be >2.65 g cm^{-3}; hence chalk is not suitable. Porosity controls water absorption which should be low ($<1\%$) so as to minimize damage from frost and from salt crystallization. In addition, a porous limestone will absorb more bitumen in the coating process, thereby increasing costs. Dolomites tend to have higher porosity and hence are less suitable as roadstones. The mechanical properties control the response of limestone to external stimuli such as shear stress or compressive impact.

Cement: Cement is strictly a manufactured product, made by blending different natural and synthetic materials, in order to achieve rather precise chemical proportions of lime, silica, alumina, iron, and sulfate in the finished product. Over 80% of the raw material is limestone and the manufacture of cement is the main user of limestone in many countries, and is second only to aggregates in others. Low magnesium limestones ($<3\%$ $MgCO_3$) are necessary and specific cements may have more stringent requirements. The limestone and other raw materials are finely ground and blended to an exact chemical specification prior to being fed into large rotary kilns for conversion into cement "clinker". The clinker is then cooled and ground, and at this stage a small amount of gypsum ($CaSO_4$) is added to control the setting time.

Lime (quicklime): Lime is made by burning (calcining) limestone or dolostone in a furnace, thereby driving off CO_2. Lime has been manufactured for several hundred years, initially in small wood-fuelled kilns, through a series of increasingly sophisticated designs to the large modern rotary kilns. The initial stimulus was the need for agricultural lime, which is applied directly to the soil. Non-agricultural uses of lime include the manufacture of mortar, plaster, calcium silicate brick, aerated insulation blocks, lime washes, and a variety of external renderings. It is also reacted with water to produce hydrated (slaked) lime which has a variety of chemical and industrial uses.

Chemical processing: Limestones and lime are used in many chemical-processing industries. For example, in the making of glass, limestone or lime may be used to flux silica sand, to form chemically fused calcium silicates. They also render the glass less soluble, reduce its brittleness, and improve its appearance by enhancing its lustre. High calcium stone is preferred for window glass and high magnesium stone for glass containers and tumblers, as the magnesium oxide provides greater resistance to etching by chemical solvents and acids. Limestone is also important in papermaking, as hydrated lime is reacted with SO_2 to form calcium bisulfite, which is heated as a liquor in which wood pulp is digested, to dissolve all constituents, except those which are cellulose based. The end product is then refined and pressed as the initial basis for papermaking. A high calcium limestone is required but there is an increasing trend to replace limestone by magnesia, ammonia, or soda ash as the alkaline agent to react with SO_2. Finely ground limestone fillers are used in the manufacture of paint, ceramics, rubber, plastics, and a variety of floor-covering materials. Whiting is a specific category of fine-grained filler derived by fine grinding such that $>95\%$ is <44 μm. Chalk is a particularly good source material.

Limestone as a Mineral Resource: Production of limestone (including dolomite and chalk) in Great Britain by end use (data from British Geological Survey Annual Mineral Statistics)

	1981		1988		1992		1996	
	Mt	%	Mt	%	Mt	%	Mt	%
Roadstone	30.0	42.5	52.1	43.5	52.9	49.0	37.1	38.6
Other aggregate	27.0	38.3	49.4	41.3	36.8	34.1	40.1	41.7
Cement	8.8	12.4	10.3	8.6	8.9	8.3	9.7	10.1
Agriculture	1.4	2.0	1.3	1.1	1.4	1.3	1.4	1.5
Iron and steel	0.9	1.3	2.2	1.9	2.2	2.0	3.0	3.1
Fillers	0.8	1.1	1.1	0.9	1.3	1.2	1.6	1.7
Glass making	0.3	0.5	0.3	0.2	0.2	0.2	0.0	0.0
Building stone	0.1	0.1	0.2	0.2	1.4	1.3	0.2	0.2
Other uses	1.1	1.7	2.8	2.3	2.8	2.6	3.0	3.1
TOTAL	70.4		119.7		107.8		96.1	

Ferrous and non-ferrous metal processing: Limestone is used as a flux in the processing of a variety of metals including copper, lead, zinc, and most notably iron. Limestone is added to the blast furnace where it reacts with the fuel and impurities in the ore and forms a molten slag that is removed gravimetrically.

Building / dimensional stone: Not all limestones are suitable for building, only those that have a high mechanical strength. Where the rock is to be used outside, or in other wet applications, it must also have a high durability (soundness), that is an ability to resist wetting / drying and freezing / thawing cycles. Marble has long been prized for statuary and ornamental uses (see Ornamental Use of Limestone), but in the dimensional stone trade the term is commonly applied to those limestones that will take a polish, whether or not they have actually been metamorphosed.

Environmental protection: An increasingly important use of limestone is in the removal of pollutants from air and water. Limestone aggregate is used in sewage works, where it provides a surface upon which colonies of bacteria can become established. These bacteria feed upon the bacteria already present in the sewage. Hydrated lime is used to balance the pH of domestic water from surface sources, which are often quite acidic. However, groundwater from limestone aquifers in general, and chalk in particular, may need to have its pH reduced to avoid build up in pipes. Problems associated with "acid rain" became a major issue during the 1980s and one proposed solution was the fitting of flue gas desulfurization (FGD) equipment to power stations and other major consumers of fossil fuels. There are over 200 methods of FGD but one of the most popular uses limestone, which is reacted with sulfur dioxide to produce gypsum.

The proportion of limestone used for each of these purposes varies from country to country, but in general the greatest use of limestone is as an aggregate in those countries where there is a shortage of easily exploitable hard rocks or gravels. However, in those countries where there are alternative sources of aggregate, the greatest use of limestone is in the production of cement.

An exception would be the United States, where limestone is commonly preferred as an aggregate, even where other rocks are available. The table shows the production of limestone by end use in Great Britain over the period 1981–1996.

JOHN GUNN

Further Reading

Carr, D.D., Rooney, L.F. & Freas, R.C. 1994. Limestone and dolomite. In *Industrial Minerals and Rocks*, edited by D.D. Carr, 6th edition, Littleton, Colorado: Society for Mining, Metallurgy, and Exploration
  A very good general reference that gets about as non-technical as a technical reference can get!
North, F.J. 1930. *Limestones, Their Origins, Distribution, and Uses*, New York: Van Nostrand, and London: Thomas Murphy
Prentice, J.E., 1990. *Geology of Construction Materials*, London and New York: Chapman and Hall
  This book does not refer only to limestone but it does have a good overview of the dimension stone and aggregate industries—including limestone.
Raymond, F. (editor) 1994. *Mendip: Limestone Quarrying, A Conflict of Interests*, Tiverton: Somerset Books
  Proceedings of a seminar to discuss aspects of quarrying on Mendip. Includes discussion of uses of limestone, minerals planning policy, and environmental impacts.
Smith, M.R. (editor) 1999. *Stone: Building Stone, Rock Fill and Armourstone in Construction*, London: Geological Society (Geological Society Engineering Geology Special Publication No. 16)
Smith, M.R. & Collis, L. (editors); revised by Fookes, P.J., Lay, J., Sims, I. & West, G. 2001. *Aggregates: Sand, Gravel and Crushed Rock Aggregates for Construction Purposes*, 3rd edition, Bath: Geological Society (Geological Society Engineering Geology Special Publication No. 17; originally published 1985)
Stanier, P. 1985. *Quarries and Quarrying*, Shire Publications
  Discussion of the history of quarrying in the UK. Includes 7 pages related to limestone and a list of 20 books for further reading.

LITTORAL CAVES

Littoral caves, more commonly known as sea caves, are found throughout the world, actively forming along present coastlines and as relict sea caves on former coastlines. Some of the best-known sea caves are European. Fingal's Cave, on the Scottish island of Staffa, is a spacious cave some 70 m long, formed in columnar basalt. The Blue Grotto of Capri, although smaller, is famous for the apparent luminescent quality of its water, imparted by light passing through openings underwater (see picture in colour insert and entry Art Showing Caves). The Romans built a stairway in its rear and a now-collapsed tunnel to the surface. The Greek islands are also noted for the variety and beauty of their sea caves. Numerous sea caves have been surveyed in England, Scotland, and in France, particularly on the Normandy coast. The largest sea caves are found along the west coast of the United States and in the Hawaiian islands.

True littoral caves are those formed by processes other than dissolution. They should not be confused with dissolutional caves that pre-dated the wave action but were then intersected and revealed as the cliff line was eroded back, or with the dissolutional voids formed in the littoral zone on tropical islands (see Speleogenesis: Coastal and Oceanic Settings). In some regions, such as Ha-Long Bay, Vietnam (see separate entry), caves in carbonate rocks are found in littoral zones but were formed by dissolution. Littoral caves may be found in a wide variety of host rocks, ranging from sedimentary to metamorphic to igneous, but caves in the latter tend to be larger due to the greater strength of the host rock.

In order to form a sea cave, the host rock must first contain a zone of relative weakness. In metamorphic or igneous rock, this is typically either a fault (Figure 1), as in the caves of California's Channel Islands, or a dyke, as in the large sea caves of Kauai's Na Pali Coast (Hawaii). In sedimentary rocks, this may be a bedding-plane parting or a contact between layers of different hardness (Figure 2). The latter may also occur in igneous rocks, such as in the caves on Santa Cruz Island (California), where waves have attacked the contact between the andesitic basalt and the agglomerate.

The driving force in littoral cave development is wave action. Erosion is ongoing anywhere that waves batter rocky coasts, but where sea cliffs contain zones of weakness, rock is removed at greater rate along these zones. As the sea reaches into the fissures thus formed, they begin to widen and deepen due to the tremendous force exerted within a confined space, not only by direct action of the surf and any rock particles that it bears, but also by compression of air within. Blowholes (partially submerged caves that eject large sprays of sea water as waves retreat and allow rapid re-expansion of air compressed within), attest to this process. Adding to the hydraulic power of the waves is the abrasive force of suspended sand and rock. Most sea-cave walls are irregular and chunky, reflecting an erosional process where the rock is fractured piece by piece. However, some caves have portions where the walls are rounded and smoothed, typically floored with cobbles, and result from the swirling motion of these cobbles in the surf zone.

Rainwater may also influence sea-cave formation. Carbonic and organic acids leached from the soil may assist in weakening rock within fissures. As in solutional caves, small speleothems may develop in sea caves. The largest that the author has seen are stalactites and sheets of flowstone in sea caves formed in calcareous sandstone.

Life within sea caves may assist in their enlargement as well. For example, sea urchins drill their way into the rock, and over successive generations may remove considerable bedrock from the floors and lower walls.

Sea-cave chambers sometimes collapse leaving a "littoral sinkhole". These may be quite large, such as Oregon's Devil's Punchbowl or the "Queen's Bath" on the Na Pali coast. Small peninsulas or headlands often have caves that cut completely through them, since they are subject to attack from both sides, and collapses of sea-cave tunnels often leave free-standing "sea stacks"

Littoral Caves: Figure 1. Lady's Harbor Cave, Santa Cruz Island, California, is formed on a prominent fault visible along the right wall. (Photo by Dave Bunnell)

Littoral Caves: Figure 2. Littoral caves below the vertical cliffs of South Wales; the rock is limestone but there is almost no karstification and the caves have been excavated by wave action on the bedding planes and joints. (Photo by Tony Waltham)

Littoral Caves: Long sea caves of the world (compiled by D. Bunnell and C. Self).

Rank	Cave Name	Location	Region	Country	Length (m)
1	Painted Cave	Santa Cruz Island	California	USA	374
2	Waiahuakua Cave	Kauai	Hawaii	USA	352
3	Cueva Tres Pisos	Baja California	California	USA	316
4	Waiwaipuhi Cave	Kauai, Hawaii	Hawaii	USA	290
5	Sunbeam-by-the-Sea Cave	Punta Banda, Baja	Baja	Mexico	270
6	Catacombs Cave	Anacapa Island	California	USA	246
7	Cathedral Cave	Anacapa Island	California	USA	241
8	Forbidden Fissures	Santa Cruz County	California	USA	241
9	Sandside Head Cave #2	Thurso, Highlands	Scotland	UK	230
10	Sea Maze	San Luis Obispo County	California	USA	229
11	Waialoha Cave	Kauai, Hawaii	Hawaii	USA	225
12	Virgin's Spring	Lundy Island, Devon	England	UK	225
13	Sand Hill Bluff Cave	Santa Cruz Co.	California	USA	206
14	Jumbo Gumbo	Santa Rosa Island	California	USA	205
15	Parallel Tunnels	Punta Banda, Baja	Baja	Mexico	204

along the coast. The Californian island of Anacapa is thought to have been split into three islets by such a process.

Most sea caves are small in relation to other cave types. A compilation of sea-cave surveys by the author showed three over 300 metres, 15 over 200 metres, and 86 over 100 metres in length. The Table presents details of the 15 caves with surveyed lengths exceeding 200 metres. In Norway, several apparently relict sea caves exceed 300 metres in length, but are not included in the above compilation of actively forming sea caves. There is no doubt that many other large sea caves exist but are unsurveyed due to their remote locations and hostile sea conditions.

Several factors contribute to the development of relatively large sea caves. The nature of the zone of weakness itself is surely a factor, although difficult to quantify. A more readily observed factor is the situation of the cave's entrance relative to prevailing sea conditions. At Santa Cruz Island, the largest caves face into the prevailing northwest / west swell conditions—a factor which also makes them more difficult to survey. Caves in well-protected bays sheltered from prevailing seas and winds tend to be smaller, as are caves in areas where the seas tend to be calmer.

The type of host rock is important as well. All of the largest sea caves are in basalt, a relatively strong host rock compared to say, sedimentary rock. Basaltic caves can penetrate far into cliffs where most of the surface erodes relatively slowly. In weaker rock, erosion along a relative zone of weakness may not greatly outstrip that of the cliff face.

Time itself is another factor—i.e. the length of time that a cave has been actively enlarging. The active littoral zone changes throughout geological time by an interplay between sea-level change and regional uplift. Recurrent ice ages during the Pleistocene have changed sea levels within a vertical range of some 200 metres. Significant sea caves have formed in the Californian Channel Islands that are now totally submerged by the rise in sea levels over the last 12 000 years. In regions of steady uplift, continual littoral erosion may produce sea caves of great height—Painted Cave is almost 40 m high at its entrance.

Finally, caves that are larger tend to be more complex, as discussed below. By far the majority of sea caves consist of a single passage or chamber. Those formed on faults tend to have canyon-like or angled passages that are very straight. In Seal Canyon Cave on Santa Cruz Island, entrance light is still visible from the back of the cave 189 m from the entrance. By contrast, caves formed along horizontal bedding planes tend to be wider with lower ceiling heights. In some areas, sea caves may have dry upper levels, lifted above the active littoral zone by regional uplift.

Sea caves can prove surprisingly complex where numerous zones of weakness—often faults—converge. In Catacombs Cave on Anacapa Island (California), at least six faults intersect. In several caves of the Californian Channel Islands, long fissure passages open up into large chambers beyond. This is invariably associated with intersection of a second fault oriented almost perpendicularly to that along the entrance passage.

DAVID BUNNELL

Further Reading

Bunnell, D. 1983. The amazing caves of Santa Cruz Island. *NSS News*, 41(2): 86–91
Bunnell, D. 1988a. *Sea Caves of Santa Cruz Island*, Santa Barbara, California: McNally and Loftin
Bunnell, D. 1988b. The 1987 Na Pali Coast Sea Caves Expedition. *NSS News*, 46(2): 446–53
Bunnell, D. 1993a. *Sea Caves of Anacapa Island*, Santa Barbara, California: McNally and Loftin
Bunnell, D. 1993b. Sea caving in the Channel Islands. *NSS News*, 51(6): 150–59
Bunnell, D. 1995. Preliminary list of the long (>100m) sea caves of the world. *Geo^2*, 21: 38–39
Bunnell, D. 1998. California's coastal sea caves. *NSS News*, 56(10): 292–300
MacCulloch, D.B. 1975. *Staffa*, 4th edition, North Pomfret, Vermont: David and Charles
Moore, D.G. 1954. Origin and development of sea caves. *National Speleological Society Bulletin*, 16: 71–76
Petrocheilou, A. 1984. *The Greek Caves*, Athens: Ekdotike Athenon
Rodet, J. 1983. Karst et littoral du Bec de Caux. *Karstologia*, 2: 23–32
Sjöberg, R. 1981. Tunnel caves in Swedish Archean rocks. *Transactions of the BCRA*, 8(3): 159–67
http://www.pipeline.com/-caverbob/seacave.html Long Sea Caves of the World

M

MADAGASCAR

The island of Madagascar lies in the Indian Ocean, about 500 km from Mozambique. The fourth largest island in the world at 594 000 km², it is of continental origin, having broken away from the African mainland c.160 million years ago. It stretches 1600 km north to south and rises to an altitude of 2876 m. While much of the surface is lateritic sandstone over crystalline bedrock, there are extensive belts of karst, especially in the west, besides smaller areas of cave-bearing lava, quartzite, and granite (Figure 1). There was sporadic cave exploration during the French colonial period, particularly from the 1930s to 1960s, but major international interest in Madagascar's karst dates from the easing of restrictions on access to the country in the early 1980s, since when a number of British, French, and German expeditions and cavers from the United States, Australia, and Italy have documented many caves. The principal karst areas from north to south are as follows.

Ankarana

This area of some 200 km² of highly karstified Jurassic limestone lies about 70 km south of the northern city of Antsiranana (Diego Suarez). Initial exploration and documentation was undertaken, virtually single-handedly, by Jean Radofilao, but the area was really placed on the map by British expeditions in 1981 and 1986 (Wilson, 1990). The limestone, which rises up to 200 m in sheer walls on the west, has been deeply eroded under a strongly seasonal tropical climate, giving rise to a distinctive landscape of small pinnacle karst fretted by rillenkarren (see separate entry, Karren) and known locally as "tsingy" (derived from the ringing sound emitted when struck). The karst is split by a series of deep, narrow fault-controlled canyons, some of which give access to caves (Figure 2). These sheltered canyons are inhabited by many endemic species of lemur, while the tsingy plateaus are sparsely colonized by succulent xerophytic plant species. The caves are perhaps best known for their watercourses and as one of the few places where crocodiles (*Crocodylus niloticus*) can be found underground; large eels are also present and probably provide the crocodiles' main food source. Bats include

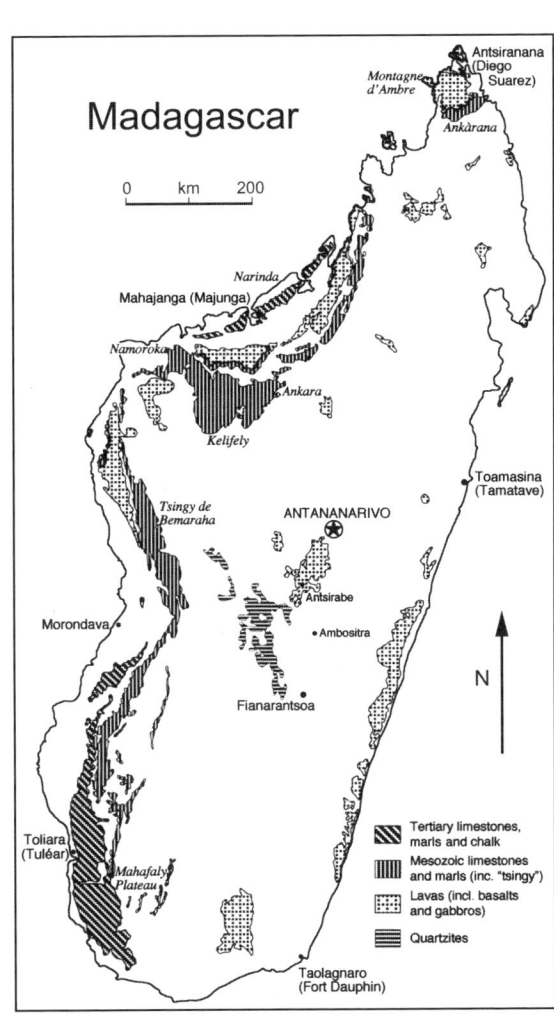

Madagascar: Figure 1. The main karst areas in Madagascar.

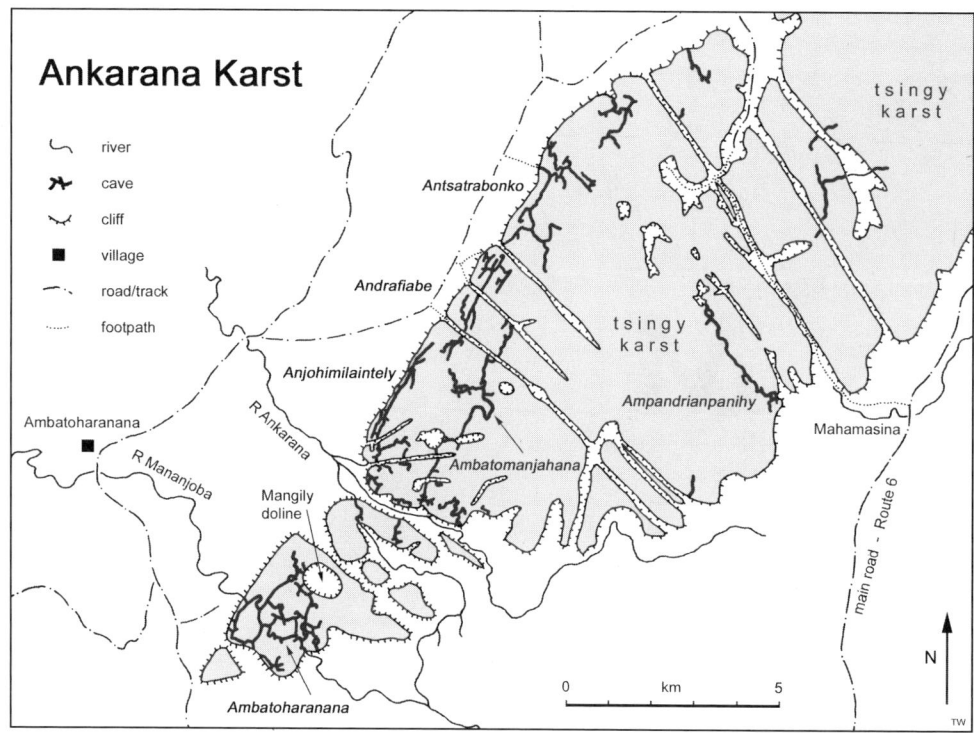

Madagascar: Figure 2. Outline map of the main caves within the Ankarana karst, Madagascar.

both the insectivorous microchiroptera and the large fruit-eating species, *Eidolon dupraenum* and *Rousettus madagascariensis*. The more significant caves are Ambatoharanana (Grotte des Crocodiles) >18 000 m (longest in Madagascar), Ambatomanjahana (Grottes des Rois) 2200 m, Andrafiabe 11 460 m, Antsatrabonko 10 400 m, Antsiroandoa 1100 m, Lavaka Fanihy (Grotte des Chauve-Souris) 4460 m, and Milaintety 8300 m. Most of the limestone is within the Ankarana Special Reserve, under the management of Assoc. Nationale pour la Gestion des Aires Protegées (ANGAP); an entry fee is payable and permits are required for scientific collecting. Some caves have been used as tombs and access to these by foreigners is normally restricted.

Narinda

Tertiary limestone and dolomite is found northeast of Majunga on the west coast. The topography is largely cone karst with rounded hills rising 20 to 40 m above a palm savannah. Caves are not known to be numerous, but Anjohibe (place of the big cave), previously known as Grotte de Andranoboka, with a length of over 5.3 km (Laumanns & Gebauer, 1993), was developed as a show cave during the colonial period (probably in the 1940s). Only a few rusty ladders and electrical insulators remain. Nearby is Anjohikely (place of the small cave), that is over 2.1 km long, well decorated and less trafficked than Anjohibe. To the west of Majunga, the karst continues as a plain with caves beneath.

Namoroka

This area comprises ~180 km² of Jurassic limestone, south of the Betsiboka estuary and the Andranomavo River. This area is very remote and has seldom been visited by cavers (Laumanns & Gebauer, 1993; Middleton, 1998b). The largest known cave is the complex maze of Anjohiambovonomby (Zebu Well Cave, 4630 m). The karst is similar to Ankarana, but has less relief and is more broken by canyons. Notable are some large bogaz that are flooded for several months each year and remain dry and bleached white for the other months.

Kelifely

Even though this area is contiguous with Namoroka and Ankara, the Kelifely Plateau—about 50 × 70 km—is treated separately. It is approached from the north or east and in the south rises in a 400–500 m escarpment and is bisected by the Mahavavy River. Access problems and threat of bandits have limited speleological investigation in recent years. Notable are Abadie's work in the 1940s (Abadie, 1950), Rossi's detailed geomorphological studies in the 1970s (Rossi, 1986), and Peyre's attempt in 1982 (Peyre, 1983). The plateau exhibits a variety of karst landscapes, with canyons, dolines, poljes, fluviokarst, cockpit karst, tsingy, and areas of volcano-karstic interaction. Some streams have part of their course underground; the Tondraka sinks and rises a number of times, flowing about 12 km underground before rising in the bed of the Mahavavy (Abadie, 1950). Peyre (1983) describes the main sink as "needing a rope" (he did not enter). Abadie reported three caves in the Kassijie Forest (Peyre's guides were too frightened of a local tribe to take him to them), one at the foot of Doany Hill and one (Anjohibe) near Ambararata village.

Tsingy de Bemaraha

This large, forested tropical karst extends for ~100 km north–south and up to 15 km east–west in the central west of Madagascar, 150 km north of Morondava. The natural values of the region were recognized as early as 1927, when the Reserve Naturelle Integrale was established over a large part of the karst. This was expanded in 1966, to cover all of the karst north of

Madagascar: Figure 3. Well-developed tsingy of karren-fluted pinnacles in the Tsingy de Bemaraha. (Photo by John Middleton)

the Manambolo River; at 152 000 ha, it is the largest protected area in Madagascar. In 1990, the area was the first in Madagascar to be inscribed on the World Heritage List (Bosquet & Rabetaliana, 1992) and it is now managed with assistance from UNESCO. It was listed on account of both its superlative natural phenomena—particularly the spectacular pinnacle karst known as tsingy (Figure 3) and areas of exceptional beauty, and because it contains important natural habitats for conservation of biological diversity. Caves have been documented along the Manambolo Gorge, the river providing the only easy access into the area, and in less accessible parts of the karst (e.g. Dobrilla & Wolozan, 1994; Middleton, 1996, 1998a). North of the river, the caves consist of networks of joint-controlled passages that may be very extensive. Very old pottery may be encountered in the caves, which hopefully will not be disturbed.

Tulear Region

This is the most southerly extension of the western limestone belts, extending from ~300 km north of Tuléar (Toliara) to about 200 km south. The major feature in the north is the Mickoboka Plateau (where there are reports of pits—locally termed "avens"—up to 165 m deep) and, in the south, the Mahafaly Plateau, where there are numerous small caves and large pits up to 100 m in depth (Middleton, 1999). Underground waters in the region are inhabited by an endemic, eyeless white fish, *Typhleotris madagascariensis*.

The main documented region of lava tubes is around Montagne d'Ambre, just south of Antsiranana, and particularly to its west, south of Andranofanjava (Middleton, 2000). Caves have also been reported in quartzite at Mt Ibity (or Ibinty) in the central highlands (152 m long Grotte Albert) at an altitude of 1900 m, the highest cave in Madagascar.

GREGORY MIDDLETON

Works Cited

Abadie, Ch. 1950. Note sur la région du Kelifely. *Bulletin de l'Académie Malgache*, new series 28 (for 1947–48): 12–16

Bosquet, B. & Rabetaliana, H. 1992. *Site du patrimoine mondial des Tsingy de Bemaraha et autres sites d'intérêt biologique et écologique du fivondronana d'Antsalova*, Paris: Unesco

Dobrilla, J.-C. & Wolozan, D. 1994. Spéléologie sous les Tsingy de Bemaraha, Madagascar 1993. *Etudes et Documents de l'ADEKS*, 4 (1994): 1–62

Laumanns, M. & Gebauer, H.D. 1993. Namoroka 1992: Expedition to the karst of Namoroka and Narinda, Madagascar. *The International Caver*, 6: 30–36

Middleton, G. 1996. The 1995 Australo-Anglo-Malagasy Speleo-Ornitho-Malacological Expedition: Tsingy de Bemaraha, Western Madagascar. *Journal of the Sydney Speleological Society*, 40(9): 141–58

Middleton, G. 1998a. The 1996 International Speleo-Ornitho-Geo-Malaco-logical Expedition: Northern Tsingy de Bermaraha, Western Madagascar. *Journal of the Sydney Speleological Society*, 42(1): 3–18

Middleton, G. 1998b. Narinda and Namoroka Karst Areas: Madagascar 1997. *Journal of the Sydney Speleological Society*, 42(10): 231–43

Middleton, G. 1999. Mahafaly Plateau and Onilahy Valley Expedition: Tulear Karst Region: Madagascar 1998. *Journal of the Sydney Speleological Society*, 43(7): 157–67

Middleton, G. 2000. Madagascar 1999: A preliminary reconnaissance of the lava caves of Andranofanjava region. *Journal of the Sydney Speleological Society*, 44(10): 335–42

Peyre, J-C. 1983. Au bout de la paperasse et des demarches administratives: encore l'aventure. *Spéléologie*, 122: 14–18

Rossi, G. 1986. Aspects morphologiques du karst du Kelifely (Madagascar). In *New Directions in Karst: Proceedings of the Anglo-French Karst Symposium September 1983*, edited by K. Paterson & M.M. Sweeting, Norwich: Geo Books

Wilson, J. 1990. *Lemurs of the Lost World: Exploring the Forests and Crocodile Caves of Madagascar*, London: Impact

Further Reading

Decary, R. & Kiener, A. 1971. Inventaire schématique des cavités de Madagascar. *Annales de Spéléologie*, 26: 31–46

De Saint Ors, J. 1959. Les phénomens karstiques à Madagascar. *Annales de Spéléologie*, 14: 275–91

Middleton, J. & Middleton, V. 2002. Karst and caves of Madagascar. *Cave and Karst Science*, 29(1): 13–20.

Rushin-Bell, C.J. 1998. Caving in Madagascar. *NSS News*, 56(9): 260–61

Protected Areas Programme
http://www.wcmc.org.uk/protected_areas/data/wh/bemaraha.html

MAMMOTH CAVE REGION, UNITED STATES

Mammoth Cave (Kentucky, United States) is the longest cave in the world, with about 557 km of mapped passages. The cave consists of several large sections, explored separately through various entrances and connected by later discoveries (Figure 1). It is located mainly within Mammoth Cave National Park, established in 1941, and is administered by the National Park Service. The Park, which encompasses 214 km^2, was designated a World Heritage Site in 1981, on the basis of its geological, archaeological, and biological significance. It was designated an International Biosphere Reserve in 1990. The Park receives an average

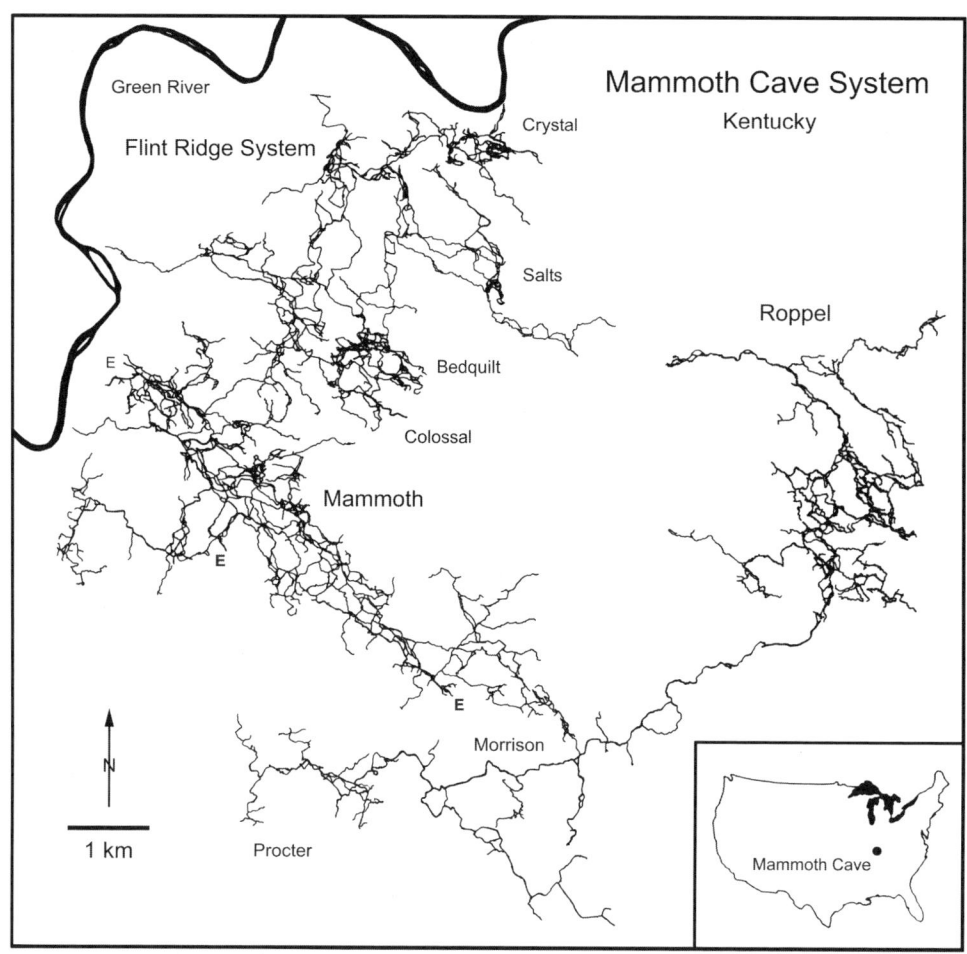

Mammoth Cave Region, United States: Figure 1: Map of the Mammoth Cave System, Kentucky (based on surveys by the Cave Research Foundation and Central Kentucky Karst Coalition). Names are shown for caves explored separately but connected by later discoveries. E = entrances to tour routes in the main part of Mammoth Cave.

of 1.7 million visitors per year. About 15 km of trails in the cave are open to the public, through 13 different guided tours. Other parts of the cave, which lie within the Park, are accessible only under permit for research or mapping. Mammoth Cave extends beyond the Park boundaries, but access to these sections is restricted. Exploration and surveying of the system are carried out by the Cave Research Foundation and the Central Kentucky Karst Coalition (White & White, 1989).

The cave is located in a low-relief plateau of Mississippian (lower Carboniferous) limestones and dolomites, locally capped by low-permeability sandstones and shales (Figure 2). The climate is temperate and humid, with a mean annual rainfall of ~1300 mm, and both surface and subsurface karst processes are very active. The average stratal dip is about 0.3° toward the northwest. The capped region (Chester Upland) rises to an altitude of c.250 m. Stream erosion has breached the caprock, forming irregular flat-topped ridges separated by karst valleys. Most recharge to the system is through dolines in the valley floors, and from the adjacent Pennyroyal Plateau, a vast karst plain to the southeast, where the clastic rocks have been completely removed by erosion. Its surface lies about 45–60 m below the ridge crests of the Chester Upland. The karst area of the Pennyroyal Plateau and Chester Upland covers thousands of square kilometres, extending through western Kentucky and into adjacent Indiana and Illinois. All explored passages of the Mammoth Cave System lie in the Chester Upland, but dye traces show considerable subsurface inflow from ponors and dolines up to 10 km away in the Pennyroyal (Quinlan & Ewers, 1989).

Throughout the erosional history of the region (since the late Carboniferous, with few interruptions), the karst landscape has gradually migrated northwestward toward the centre of the Illinois Basin, as the overlying clastic rocks have been stripped away. Superimposed on this broad trend are irregular fluctuations in stream entrenchment rate, interspersed with occasional aggradation. A simplified review of the landscape evolution with time can be obtained simply by traversing the region from northwest to southeast, i.e. from downdip regions, where the caprock is still intact, to updip regions where the limestone is completely exposed and ultimately removed entirely by erosion.

In most places, the Pennyroyal is dimpled by an almost continuous array of dolines, but the overall surface has very little relief. Along the foot of the Chester Upland the surface is partly buried by thick colluvial, alluvial, and residual sediment. Remnants of alluvial gravel and sand persist at high elevations. Karst is also subdued in the far updip regions, because the carbonates are shaly and less deeply dissected. Surface drainage from this area sinks where it encounters purer carbonates or greater erosional relief. The Pennyroyal surface is a stratally discordant erosion surface formed during a period of rather stable base level (Miotke & Palmer, 1972).

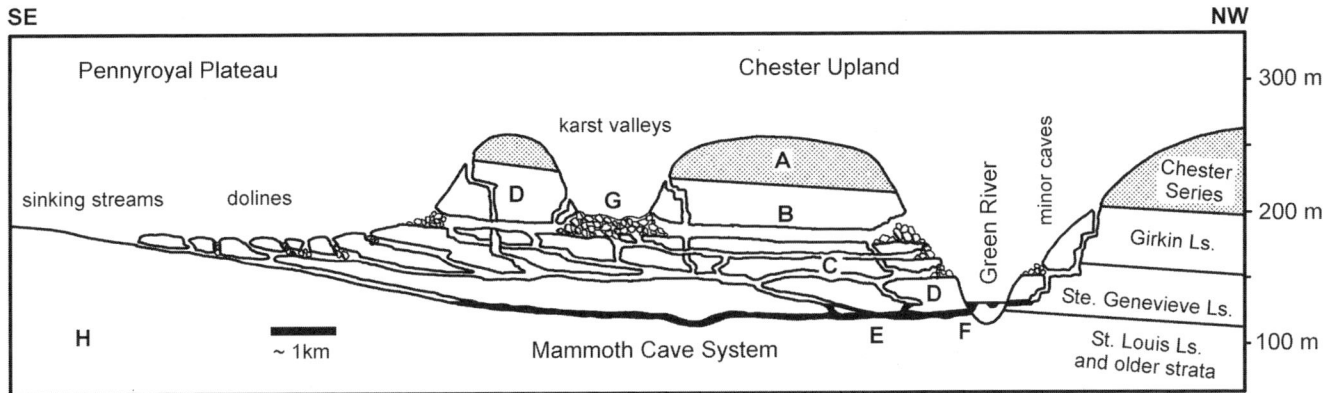

Mammoth Cave Region, United States: Figure 2: Geologic cross section through the Mammoth Cave region. A = caprock composed mainly of sandstone and shale, with minor limestone units (late Mississippian–mid-Carboniferous age); B = large upper-level canyons and tubes of late Tertiary age; C = tubular passage levels of Pleistocene age; D = vadose infeeders of various ages; E = passages flooded by late Pleistocene rise in base level; F = springs partly blocked by alluvium; G = truncation of upper-level passages by karst valleys; H = impure limestones that support surface drainage.

The Green River, one of the few permanent surface rivers in the region, is the local base level for cave development at Mammoth Cave. The river now lies only 100 m below the ridge crests, but late Pleistocene alluvial fill has raised its level about 15 m. The Green River is a tributary to the extensive Ohio River, and so the passage layout and sediments of Mammoth Cave hold clues to the erosional history of the entire east-central United States.

Typical passages in Mammoth Cave consist of narrow vadose canyons that change downstream to tubular phreatic conduits. Canyons are fed by, or alternate with, vertical shafts up to 45 m deep. The local carbonate rocks are prominently bedded, with few conspicuous faults and joints, and so nearly all passages follow the bedding. Canyons have nearly flat ceilings and are almost invariably oriented directly down the local dip, following local irregularities in the partings along which cave inception took place (Palmer, 1981). Most canyons are only 1–3 m wide and highly sinuous, as the result of local variations in dip direction, which in turn are caused by minor irregularities (mainly depositional) in bedding-plane partings. Canyons in the uppermost levels of the cave are large, up to 20 m wide and 30 m high, and are generally far less sinuous than their narrower counterparts (Figure 3). Most vadose entrenchment involves headward retreat of waterfalls and rapids, imposed by massive or relatively insoluble beds.

Tubular passages have lenticular cross sections elongated along the bedding (Figure 4). They are not as sinuous as the typical narrow canyon, because their greater width engulfs all but the largest bedding irregularities. Most tubes form at or below the water table, although some minor ones (usually having rather flat ceilings) are perched vadose conduits. Fissure passages and linear passage segments are rare.

Shafts are generally round or oval in horizontal cross section and nearly all have developed in stages by successive downward dissolution from bed to bed, as shown by sequential drains at many elevations in their walls. Very few shafts descend in single drops along major fractures. Each drain leads to narrow canyons

Mammoth Cave Region, United States: Figure 3: Dyer Avenue, a remnant of the 183–189 m level of late Pliocene age in the Crystal Cave section of the Mammoth Cave System. Remnants of a former tourist trail are visible. (Photo by Art Palmer)

Mammoth Cave Region, United States: Figure 4: Turner Avenue, a tubular passage at the 168 m level of early Pleistocene age in the Flint Ridge section of the Mammoth Cave System. (Photo by Art Palmer)

or small stratally perched tubes that interconnect in tangled complexes. The abundance of sequential shaft drains is partly responsible for the great interconnectivity between various parts of the cave. Most shafts are clustered around ridge boundaries, where they receive diffuse seepage through the thin eroded edge of the clastic caprock.

Passages in Mammoth Cave are scattered over a vertical range of ~120 m. Vadose canyons can form anywhere above base level, and roughly half of them are still active today. But tubular passages cluster at only a few elevations, which indicate significant pauses in base-level lowering. Vertical sinuosity in the tubes indicates that while forming, they extended as much as 23 m below the water table. Former base levels, that controlled the vertical positions of the tubes, can be determined by the elevations of points where there is a change in the original downflow direction from downdip vadose canyons to tubes with irregular, low gradients and no systematic relation to the dip. Many tubes are oriented almost along the strike of the beds, following shallow routes along the same beds that guide the incoming vadose canyons. These vadose-phreatic transition points cluster at elevations of 210, 183–189, 168, and 152 m. These four levels were roughly dated from geomorphic evidence (Miotke & Palmer, 1972; Palmer, 1989) and by paleomagnetic analysis of sediment (Schmidt, 1982). Since then, much more precise dates have been obtained from $^{26}Al / ^{10}Be$ ratios in quartz sediment (Granger, Fabel & Palmer, 2001). While the sediment is at the surface, these two isotopes are produced in tiny quantities by cosmic radiation. When the sediment is carried underground, the isotopes decay at different rates, and their ratio indicates the duration of burial.

The two upper levels (above 168 m) are wide canyons and tubes, filled to various depths with stream-borne quartz silt, sand, and gravel. Some are completely filled, but in others, most of the sediment has been removed by stream erosion. The greatest sediment thickness is about 20 m. Passages at these levels have sediment dates of 2.3–3.5 million years ago and, of course, the passage origins predate the sediment. These passages represent slow fluvial entrenchment during the late Tertiary, which was interrupted by periodic aggradation, probably in response to climate changes. The rather flat Pennyroyal Plateau surface formed at this time. At 2.3 million years ago, there was widespread aggradation to as much as 30 m, which filled all upper-level passages, some partially and others completely. Aggradation also took place on the Pennyroyal surface. This event appears to coincide with the first major North American glaciation.

The two lower levels contain smaller passages, partly because groundwater recharge had been fragmented into many sub-basins by that time, and also because pauses in base level were shorter. They represent fairly rapid Pleistocene entrenchment of the Green River. Prior to that time, karst was probably sparse on the Pennyroyal, because of limited entrenchment. The rapid drop in base level during the Pleistocene caused nearly all surface drainage on the Pennyroyal to be diverted underground, and the extensive doline-pocked surface of today began to form. Cave levels at 168 and 152 m are well-developed tubes with little sediment fill. Sediments at these levels are, respectively, about 1.5 and 1.2 million years old. Passages at different elevations are mainly shafts, small canyons, and crawlways, mostly of vadose origin. Some large tubes have formed below the 152 m level, but their elevations are not consistent. They include local tube segments fed by many vadose infeeders, as well as passages perched on relatively insoluble beds.

The sudden abandonment of the large upper-level passages and development of many small passages at lower levels, apparently resulted from major changes in the patterns of surface drainage (see Palmer, 1981). Until the early Pleistocene, the Ohio River was rather small, not significantly larger than the Green River. Most of the drainage from the east followed a more northerly route (the so-called Teays River), which drained directly into the Mississippi River. The first major glacial advance into the region blocked the Teays and diverted its flow into the Ohio. Suddenly the Ohio became the largest river in the eastern United States. It began to entrench its valley rapidly and, as a tributary of the Ohio, the Green River did the same. Minor adjustments in drainage pattern may account for the periodic pauses in valley incision that resulted in the tubular passage levels. Diversion of the Teays drainage into the Ohio took place about 1.5 million years ago, accounting for the rapid entrenchment below the upper two cave levels.

At any time during the evolution of the cave, the pattern of active passages has been crudely dendritic. On cave maps, this pattern is obscured in several ways. For example, there are many superimposed levels of old, abandoned passages overlying more recent active ones. Even many active vadose passages cross over active or abandoned passages at lower levels. Post-glacial alluvial fill in the valley has flooded the lowest 15 m or so of the cave, reactivating many formerly abandoned passages.

Secondary mineral deposits in Mammoth Cave are highly varied. Calcite is rare because of the rather low caprock permeability, but evaporite minerals, such as gypsum and epsomite, are abundant in the driest passages.

The Mammoth Cave region contains hundreds of smaller caves, some of which are huge in their own right. Fisher Ridge Cave consists of 161 km of mapped passages northeast of Mammoth Cave, and there is almost certainly, as yet undiscovered, a connection between them. The Martin Ridge Cave System, to the southwest, contains more than 53 km of mapped passages. The Pennyroyal Plateau also contains many caves with one or two levels of passages. Some are very large, and most are highly flood-prone. An example is the Hidden River Cave System, east of Fisher Ridge Cave, which contains several tens of kilometres of explored passages.

The astonishing length of Mammoth Cave results from several coinciding factors. Its drainage basin is large, nearly 300 km². Prominent bedding, with many lithologic contrasts, allows diversion of vadose water along many successively lower partings. The resistant caprock has protected many of the uppermost passages from erosional destruction. Finally, the many small but discrete adjustments in base level have fostered the development of numerous, independent cave levels.

Mammoth Cave is also of great archeological, biological, and historical interest. As early as 4000 years ago, native humans ranged through many kilometres of the cave on a quest for minerals or adventure, using only reed torches for light. Artefacts, paleofaeces, and several mummified bodies have been discovered in the cave (see America, North: Archaeological Caves). Cave biota include the rare Kentucky cave shrimp and eyeless fish (see Mammoth Cave: Biospeleology). Sediment in the main passage of Mammoth Cave was mined for nitrates to make gunpowder during the "War of 1812" (see Gunpowder). Since then the cave has been more or less continually open to the public, first by a series of private owners, and later by the National Park Service.

ARTHUR N. PALMER

See also **Speleogenesis: Unconfined Settings**

Works Cited

Granger, D.E., Fabel, D. & Palmer, A.N. 2001. Pliocene-Pleistocene incision of the Green River, Kentucky, determined from radioactive decay of ^{26}Al and ^{10}Be in Mammoth Cave sediments. *Geological Society of America Bulletin*, 113(7): 825–36

Miotke, F.-D. & Palmer, A.N. 1972. *Genetic Relationship among Caves and Landforms in the Mammoth Cave National Park Area*, Würtzburg: Böhler

Palmer, A.N. 1981. *A Geological Guide to Mammoth Cave National Park*, Teaneck, New Jersey: Zephyrus Press

Palmer, A.N. 1989. Geomorphic history of the Mammoth Cave System. In *Karst Hydrology: Concepts from the Mammoth Cave Area*, edited by W.B. White & E.L. White, New York: Van Nostrand Reinhold

Quinlan, J.F. & Ewers, R.O. 1989. Subsurface drainage in the Mammoth Cave area. In *Karst Hydrology: Concepts from the Mammoth Cave Area*, edited by W.B. White & E.L. White, New York: Van Nostrand Reinhold

Schmidt, V.A. 1982. Magnetostratigraphy of Sediments in Mammoth Cave, Kentucky. *Science*, 217: 827–29

Thornbury, W.D. 1965. *Regional Geomorphology of the United States*, New York: Wiley

White, W.B. & White, E.L. (editors) 1989. *Karst Hydrology: Concepts from the Mammoth Cave Area*, New York: Van Nostrand Reinhold

Further Reading

Harmon, R.S., Schwarcz, H.P. & Ford, D.C. 1978. Stable-isotope geochemistry of speleothems and cave waters from the Flint Ridge—Mammoth Cave system, Kentucky: implications for terrestrial climate change during the period 230,000 to 100,000 years B.P. *Journal of Geology*, 86: 373–84

Hess, J.W. & White, W.B. 1993. Groundwater geochemistry of the carbonate karst aquifer, south-central Kentucky, USA. *Applied Geochemistry*, 8: 189–204

Palmer, A.N. 1987. Cave levels and Their interpretation. *National Speleological Society Bulletin*, 49: 50–66

White, W.B., Watson, R.A., Pohl, E.R. & Brucker, R.W. 1970. The central Kentucky karst. *Geographical Review*, 60: 88–115

MAMMOTH CAVE, UNITED STATES: BIOSPELEOLOGY

Introduction

In order to understand the biospeleology of Mammoth Cave, it must be viewed in the context of the South-Central Kentucky Karst, where there are two historical and four functioning ecosystems. In pre-settlement times, prairie and savannah maintained by fire were prevalent. These ecosystems were converted to agriculture over the past two centuries, but forest, river, and cave ecosystems are relatively intact. Mammoth Cave biota is among the most diverse in the world (Culver *et al.*, 1999), with approximately 130 regularly occurring species, roughly divided among troglobites, troglophiles, and trogloxenes (Barr, 1967; Poulson, 1992).

Interconnections and Habitats

Functionally, since sinking streams and cave streams are tributaries of base-level rivers by way of springs, they are all part of the river continuum, with the important distinction that the middle section is underground. These distinct but connected aquatic ecosystems are energetically supported by in-washed organic debris from the forest and former prairie / savannah ecosystems. Food transport is usually down gradient, but natural back-flooding from the river into cave streams is also important.

Cave aquatic habitats can be roughly divided on the basis of water quantity. Ephemeral pools occur in rimstone dams, near terminal breakdowns, in passages rarely flooded, and in shafts. These may feed nearby shallow streams, tributaries to master shaft drains grading into base-level streams, which feed springs on Green River. As the river lowers its channel, cave streams follow and leave dry upper levels.

These upper levels are habitat for the terrestrial cave ecosystem, also dependent upon forest and agricultural land for food. The import of food is mostly accomplished by cave crickets, and to a lesser degree, by wood rats (*Neotoma magister*). Also bats, which feed in the forest, use caves for refuge, depositing guano. Crickets' eggs within caves also support separate communities. Raccoons, entering caves to feed on bats and cave crickets, may leave significant quantities of scat. Habitats here are determined by proximity to entrances with variable temperature and humidity, which in turn determines the species depositing guano. Leaf litter falling into entrances, and flood debris deposited on passage surfaces, are also locally important to the terrestrial cave ecosystem.

The Biota and Their Interactions

One prevalent aspect of aquatic life in the Mammoth Cave System is habitat partitioning by similar species within a taxonomic group. Two stygobitic species each of planaria (*Sphalloplana percocea* and *S. buchanani*), amphipods (*Stygobromus vitreus* and *S. exilis*), isopods (*Caecidotea stygius* and *C. bicrenata*), and fish (*Typhlichthys subterraneus* and *Amblyopsis spelaea*) occupy upstream and downstream habitats, respectively. Stygobites found only in base-level streams include a snail (*Antroselates spiralis*), and the endangered Kentucky Cave Shrimp (*Palaemonias ganteri*, see Figure 1). The more adaptable cave crayfish (*Orconectes pellucidus*) occupies habitats ranging from base level to tiny streams, and can travel out of water if necessary. The stygophilic amphipod *Crangonyx packardi*, crayfish *Cambarus tenebrosus*, the sculpin *Cottus carolinae*, and the spring fish *Chologaster agassizi* often occur in organically rich situations. With the exception of sculpin, fish and crayfish are predators and the remaining species are primarily grazers.

Common organisms living within the sediments of Mammoth Cave streams are nematodes (undescribed), copepods (*Maraenobiotus*, *Moraria*, *Nitocra*, and *Parastenocaris*), tardigrades (*Macrobiotus*), and oligochaete worms (*Aeolosoma*). Worm casts and tracks are also visible on mud banks of cave streams, and these organisms are preyed upon by troglobitic beetles, e.g. *Pseudanophthalmus striatus*, *P. menetriesii*, and *Neaphaenops tellkampfi*. This zone forms an ecotone between aquatic and terrestrial cave ecosystems. As part of the community dependent upon flood-deposited organic films, the springtails *Folsomia candida* and *Pseudosinella* are preyed upon by the troglobitic harvestman, *Phalangodes armata*. Another major ecotone exists at cave entrances, where litter from surface vegetation accumulates via gravity and wood rats. The collembolans *Tomocerus*, *Hypogastrura*, *Sinella*, and *Arrhopalites* are found in the entrance zone. Predators include the beetle, *Pseudanophthalmus*, and a rhagidid mite (Poulson, 1992).

The cave cricket (*Hadenoecus subterraneus*, Figure 2) buries its eggs in sandy passages with moderate moisture in the constant temperature zone, and *Neaphaenops tellkampfi* is especially skilled at finding those eggs. After cave crickets, this beetle has the highest density of any species in Mammoth Cave, and a small community subsists on beetle faeces (Poulson, 1992). The springtail *Arrhopalites* and the dipluran *Litocampa* are consumers, which are preyed upon by the mite *Arctoseius*, the spider *Anthrobia*, and the pseudoscorpion *Kleptochthonius*. These latter two are in turn preyed upon by *N. tellkampfi*.

Mammoth Cave, United States: Biospeleology: Figure 1. The Kentucky Cave Shrimp (*Palaemonias ganteri*) is an endangered species found only in the Mammoth Cave area. (Photo by Chip Clark).

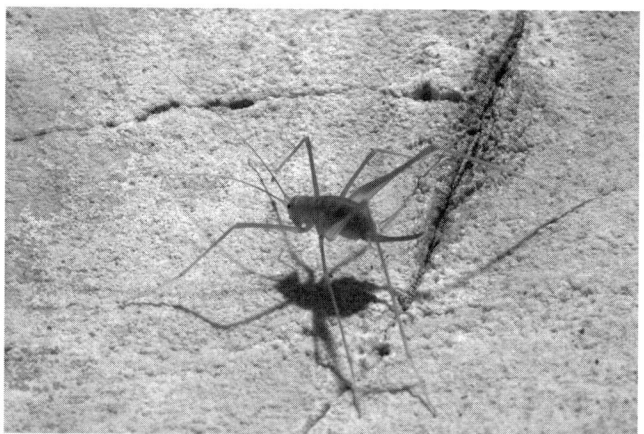

Mammoth Cave, United States: Biospeleology: Figure 2. The Cave Cricket *Hadenoecus Subterraneus* is a keystone species in the terrestrial cave ecosystem due to many species dependent on its eggs and guano. (Photo by Rick Olson)

In addition to their eggs, *Hadenoecus* guano is also important as a food source. Crickets feed in surface habitats at night and return to the cave to roost. Here their guano supports the millipedes *Scoterpes* and *Antriadesmus*, the springtails *Hypogastrura*, *Arrhopalites*, *Pseudosinella*, *Tomocerus*, and *Lepidocyrtus* plus the bristletail *Litocampa*, the beetles *Ptomophagus hirtus* and *Batrisodes henroti*, the snail *Carychium stygius*, and the mites *Ceratozetes* and *Belba*. These in turn are preyed upon by the pseudoscorpion *Kleptochthonius*, the beetle *Pseudanophthalmus menetriesii*, the larval dipteran *Macrocera nobilis*, and the spider *Phanetta*. The spider *Meta ovalis* and the cave salamander *Eurycea lucifuga* are also present and prey upon crickets.

Woodrats and raccoons were formerly abundant in Mammoth Cave, and though today reduced, their faeces support specialized communities. Latrines of the Eastern Wood Rat *Neotoma magister* sustain larva of the fly *Psychoda* and fungus gnat *Bradysia*, and the beetle *Ptomaphagus hirtus*, which are preyed upon by the rove beetle *Quedius* (Richards, 1990). Raccoon faeces support a similar community, with the exception that cave crickets may preempt fly larvae, most notably *Spelobia* and *Amoebelaria*.

Due to low populations, bat guano in Mammoth Cave is today negligible as an energy source, but would have been highly significant during pre-settlement times, since Mammoth Cave was formerly one of the largest bat hibernation sites in the world. Indiana Bats (*Myotis sodalis*) and to a lesser extent Gray Bats (*M. grisescens*) were prominent species in Mammoth Cave only 150 years ago, but are today listed as endangered. Little Brown Bats (*M. lucifugus*) were also abundant, with the Big Brown Bat (*Eptesicus fuscus*) and Eastern Pipistrelle (*Pipistrellus subflavus*) being less common (Toomey *et al.*, 1998). These, and more rare bat species, such as *M. leibii* and *M. septentrionalis*, have been estimated to have had a total population of 9–12 million just in the Historic Section (Tuttle, 1997). Ecological restoration of this portion of Mammoth Cave, and the facilitation of the return of bats is an ongoing effort (Olson, 1996).

Major Conservation Issues

In addition to correcting major ecological distortions to the Historic Section of Mammoth Cave, lamp flora are a problem as in all show caves, and elimination via wavelength selection has been achieved on a small scale (Olson, 2002). On a karst landscape scale, contamination of groundwater recharge is a major issue, especially from Interstate Highway 65 (Olson & Schaefer, 2002). A problem also exists downstream on Green River in the form of Lock and Dam Number 6, which ponds water up into Mammoth Cave and degrades the habitat for the endangered Kentucky Cave Shrimp (Olson & Leibfreid, 1999).

RICK OLSON

Works Cited

Barr, T. 1967. Ecological studies in the Mammoth Cave System of Kentucky, I. The Biota. *International Journal of Speleology*, 3: 147–203

Culver, D., Master, L., Christman, M. & Hobbs, H. 1999. Obligate cave fauna of the 48 contiguous United States. *Conservation Biology*, 14: 386–401

Olson, R. 1996. This old cave: The ecological restoration of the historic entrance ecotone of Mammoth Cave, and mitigation of visitor impact. *Proceedings of the Fifth Annual Mammoth Cave National Park Science Conference*, Mammoth Cave National Park, Kentucky: National Park Service

Olson, R. 2002. Control of lamp flora in Mammoth Cave National Park. In *Proceedings of the International Conference on Cave Lighting, November 2000, Budapest, Hungary*, edited by T. Hazslinszky, Budapest: Hungarian Speleological Society: 131–33

Olson, R. & Leibfreid, T. 1999. The importance of inventory and monitoring data sets in resolving ecosystem management problems at Mammoth Cave National Park. In *On the Frontiers of Conservation: Proceedings of the 10th Conference on Research and Resources Management in Parks and on Public Lands*, edited by D. Harmon, Hancock, Michigan: George Wright Society

Olson, R. & Schafer, J. 2002. Planned spill retention and runoff filtration structures on Interstate 65 in the South-Central Kentucky Karst. *Proceedings of the 2001 National Cave and Karst Management Symposium, Tucson, Arizona*, edited by T. Rea, Tucson, Arizona: US Department of Agriculture, Forest Service

Poulson, T. 1992. The Mammoth Cave ecosystem. In *The Natural History of Biospeleology*, edited by A.I. Camancho, Madrid: Museo Nacional de Ciencias Naturales

Richards, P. 1990. The effects of predation on invertebrate community structure in the cave rat fecal latrine. Cedar Falls, Iowa: Cave Research Foundation (Cave Research Foundation 1990 Annual Report)

Toomey, R., Colburn, M., Schubert, B. & Olson, R. 1998. Vertebrate paleontological projects at Mammoth Cave National Park. In *Proceedings of Mammoth Cave National Park's Seventh Science Conference*, Mammoth Cave National Park, Kentucky: National Park Service

Tuttle, M.D. 1997. A Mammoth discovery. *Bat Conservation International*, 15(4): 3–5

Further Reading

Barr, T. 1985. Cave life of Kentucky. In *Caves and Karst of Kentucky*, edited by P.H. Dougherty, Lexington: Kentucky Geological Survey

MARINE CAVE HABITATS

Marine (or littoral) caves form special environments within rocky shores. Although their walls and ceiling are of hard substrate, the bottom substrate can be either soft or hard. Marine caves can be as large as many cubic metres; however small voids and crevices of several cubic centimetres are also included in this habitat, which can ecologically be termed as the mesolithial. The biocenosis (or community of living creatures) in the mesolithion is comparable to that in the marine mesopsammon (sandy-bottomed habitats) including the interstitial fauna between grains of sand. Therefore, when considering the mesolithion the fauna between or under boulders and pebbles should be included.

The larger sea caves which lie at the water surface and therefore include air domes are called grottos and passages, and those which lie below the water surface are called bag caves (if they have only one entrance; from the German *Sackhöhle*) or tunnels (Riedl, 1966). All these kinds of sea caves have a constant euhaline salinity level, in contrast to anchialine caves where salinities decrease from their sea entrance, to a brackish or freshwater environment (see Anchialine Habitats). Holes or crevices within sea caves form caves within caves. They may shelter mobile animals from being visible if they are very narrow; piddock holes can completely contain the inhabitants (Abel, 1959).

Three main abiotic parameters influence the composition of the biocenosis of sea caves: light, currents, and constancy of substrate. Light decreases to 10% of surface illumination in cave entrances (reflected light zone), and to 1% and lower (scattered light zone) in the inner cave sections (Riedl, 1966). This reduction is caused by distance from the entrance, by substrate inclination, and, most importantly, by absorption of light with depth. The important consequence is that photosynthesis, and therefore the existence of plants, ceases in this habitat. The greater the depth of sea caves, where light absorption can attain lower values than 1% of surface light on the open substrate, the less significant they are ecologically.

Currents of at least 2–10% of the strength of surface currents are needed in order to guarantee a constant influx of plankton into sea caves; these are then prey for the many sessile filter feeders. With increasing distance from the cave entrance, the number of passive filter feeders increasingly exceeds the number of active ones. Where currents are absent or weak, plankton cannot penetrate the boundary layer of the substrate and "empty sections", where sessile fauna cannot exist, are found (Gili, Riera & Zabala, 1984).

The mobility of the substrate is of importance when considering small voids such as the holes and crevices between boulders and pebbles. The latter must be heavy enough to remain in position for at least two months in order to give colonizing animals the chance to grow and mature. Where voids change in size or location with a periodicity of less than two months they provide shelter only for mobile fauna.

Sea caves characteristically consist of five sections (Riedl, 1966; see Figure):

1. Entrance section, where crusted red algae may still have a chance of existence.
2. Foreground section, where hydroids and bryozoans dominate.
3. Central section, where the densest populations of sponges, oysters, barnacles, zoantharians, and corals occur, partially overlain by higher bryozoans and hydroids, gorgonians, and finger-shaped sponges.

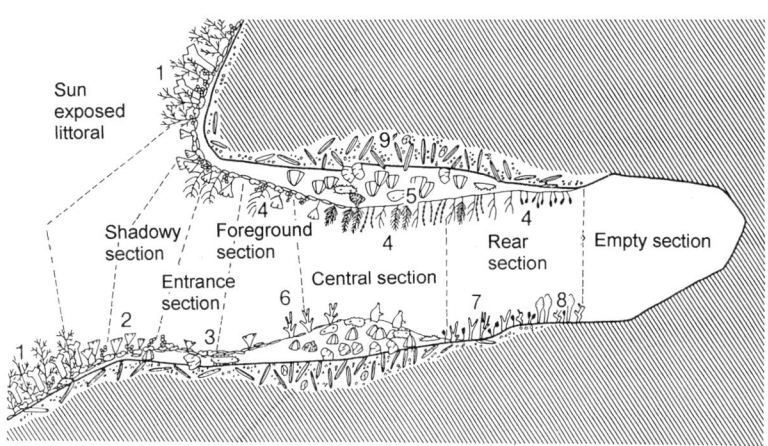

Marine Cave Habitats: A characteristic bag cave and its five sections, and the zones outside the cave. 1: light-exposed algae; 2: shadow-flora; 3: crusted flora; 4: hydroids; 5: sponges, oysters, barnacles; 6: bryozoans; 7: gorgonarians; 8: finger-like sponges; 9: boring mussels and sponges. (According to Ott, 1996)

4. Rear section, where the density of organisms, which are predominantly sponges, decreases.
5. Empty section, where no sessile organisms can exist.

Generally, the biocenosis of sea caves is characterized by species which prefer currents (rheophilic) but avoid light (photophobic). These are predominantly sessile, filter-feeding groups such as sponges, barnacles, mussels, ascidians, and polychaetes, but also comprise anthozoans and hydrozoans. These groups make up to 80–90% of the total population. The remaining 10–20% are mobile animals which are animal grazers, deposit feeders, or predators such as gastropods, crustaceans, also apparently decapods, and fish. Blenniid fish (blennies) prey on sponges and other "aufwuchs" (the plants and animals that are attached to or move about on the surfaces of submerged plants or debris) (Zander, 1990). There are no troglobitic species found exclusively in sea caves, but many troglophilic species exist which search for prey during the day and can even leave caves. Several shrimp and fish species hide in caves during the day, but leave them and become active at night.

Some special adaptations are characteristic of cave-dwelling organisms. Many species are conspicuous by their red or bright yellow colouring. This is true of many sponges, anthozoans which can dominate in the central section, crustaceans, ascidians, and fish. The red colour is advantageous because of its invisibility in the scattered light zone (below 1% of surface light). Red-coloured fish of Mediterranean caves include, for example, the suprabenthic *Anthias anthias* (Serranidae) and *Apogon imberbis* (Apogonidae), as well as the epibenthic *Lipophrys nigriceps* (Blenniidae) and *Tripterygion melanurus* (Tripterygiidae) (Zander & Heymer, 1976). *Parablennius zvonimiri* (Blenniidae) changes from red in the cave to chocolate brown outside (Zander, 1990). It is apparent that these species are diurnally active, in contrast to dusty-coloured species like *Oligopus ater* (Bythitidae) or *Phycis phycis* (Gadidae) which are active nocturnally.

Diurnally active troglophilic fish have enlarged eyes which enable them to use the reduced light. The epibenthic *Lipophrys nigriceps* additionally possess bigger lenses and cones, as well as a more advantageous relation of rods and cones than related species from the free littoral (Zander, 1982). In contrast, nocturnally active fish have rather reduced eyes.

The adaptation of sessile organisms to life in caves are short planktonic periods: the parenchymula larvae of sponges, planula of cnidarians, and cyphonautes of bryozoans settle out and become sessile within hours in order to use the substrate as quickly as possible. Mobile cave-dwellers such as crustaceans or echinoderms have relatively long larval periods with durations of several months. Life-spans of one year or longer are an adaptation of sessile animals to prolong the period of distribution (e.g. Cnidaria, Echinodermata; Riedl, 1966).

C. DIETER ZANDER

See also **Anchialine Habitats**

Works Cited

Abel, E.F. 1959. Zur Kenntnis der Beziehungen der Fische zu Höhlen im Mittelmeer. *Pubblicatione della Stazione Zoologia Napoli*, 30 (Suppl.): 519–28

Gili, J.M., Riera, T. & Zabala, M. 1984. Physical and biological gradients in a submarine cave in the western Mediterranean coast (north-east Spain). *Marine Biology*, 90: 291–98

Ott, J. 1996. *Meereskunde: Einführung in die Geographie und Biologie der Ozeane*, Stuttgart: Ulmer

Riedl, R. 1966. *Biologie der Meereshöhlen: Topographie, Faunistik und Ökologie eines unterseeischen Lebensraumes*, Hamburg: Parey

Zander, C.D. 1982. Morphological and ecological investigations on sympatric *Lipophrys* species (Blenniidae, Pisces). *Helgoländer Meeresunter-suchungen*, 34: 91–110

Zander, C.D. 1990. Benthic fishes of sea caves as component of the mesolithion in the Mediterranean Sea. *Mémoires de Biospéologie*, 17: 57–64

Zander, C.D. & Heymer, A. 1976. Morphologische und ökologische Untersuchungen an den speleophilen Schleimfischartigen *Tripterygion melanurus* Guichenot, 1850 und *T. minor* Kolombatovic, 1892 (Perciformes, Blennioidei, Tripterygiidae). *Zeitschrift für Zoologische Systematik und Evolutionsforschung*, 14: 41–59

MENDIP HILLS, ENGLAND

The smallest of the four main karst areas in Britain, the Mendip Hills of southwest England, are of special interest because of the unusual complexity of their geological structure, history of erosion, burial and exhumation, and the occurrence of periglacial features. Although there are only about 200 known caves, with an aggregate length of no more than 60 km, these systems display many classic geomorphological features, and are often cited as the type examples for cave development in geologically unconfined settings (Ford & Ewers, 1978) (see also Speleogenesis: Unconfined Settings).

The Mendips form an elongate plateau 40 km long and 8 km wide, rising abruptly to a maximum of 250 m above the surrounding lowlands at or close to modern sea-level. Structurally they consist of four *en echelon* asymmetrical periclines, each with a central core of Devonian sandstone flanked by Carboniferous limestone. Marginal dips are high, mostly between 20° and 70°; the steepest dips occur on the northern limbs, which in places are overturned. The Carboniferous limestone sequence comprises about one kilometre thickness of regular, medium- to thick-bedded platform carbonates, with a thin but hydrologically important sequence of calcareous shales (the Lower Limestone Shale) at the base (Waltham et al., 1997). In deep structural basins either side of the Mendips, these strata are overlain by sandstones and Coal Measures, themselves partially buried by Mesozoic strata. Folding and thrusting have created complex joint and fault patterns in the rocks.

Permian uplift and erosion created a rugged desert topography incised by many ravines. The sandstone core of the western pericline was breached and gutted, leaving the limestones as cuestas. Regional extension and karst solution created numerous fissures. The Hills then began to subside. Ravines were filled with scree and fan deposits, creating a cemented calcareous breccia known locally as the "Dolomitic Conglomerate". The rugged crest was subject to planation. Fissures were infilled with terrestrial Triassic and marine Jurassic deposits, and the Mendips became buried under thick Jurassic and Cretaceous deposits. There was extensive lead and zinc mineralization during the burial. Paleokarst features are now exposed in quarries, including marine surfaces with borings, filled caves, and collapses, but they do not appear to have influenced modern karst development to a significant extent.

Tertiary and Quaternary erosion by eastward scarp retreat has partially exhumed the former Triassic landscape beneath, exposing the limestone and allowing the development of cave and karst systems, initially in the west and then progressively eastwards during the past two million years or more. The Hills were never glaciated, but ice reached their northern margins and there were severe periglacial conditions periodically. During this time, base-level continued to fall, permitting multiphase caves to form. The largest of these are found in central Mendip, because cave development has been active there for a long period but surface lowering and dissection have not yet destroyed them. Caves on western Mendip are generally older truncated fragments of relict phreatic systems, many of which are important archaeological sites. Conversely, eastern Mendip is characterized by numerous immature sinks and springs but generally few caves. Exceptions are the extremely well-decorated caves intersected by Fairy Cave Quarry and the Stoke Lane Slocker system with its active streamway and large relict chambers.

The principal surface karst landforms are dolines, dry valleys, and gorges (Ford & Stanton, 1968). The central plateau is indented with regular dendritic patterns of dry valleys that deepen into the gorges of Cheddar, Ebbor, and Burrington Coombe as the steep flanks of the Hills are approached. Cheddar is the largest limestone gorge in Britain. The origin of the valleys is attributed to inheritance from the non-karstic cover rocks, which was succeeded by episodic entrenchment during periglacial stages or when surface runoff could be generated during very severe storms. The longer dry valleys are all indented by dolines, but local runoff can flow along them during the most severe storms.

The Hills contain numerous solution dolines. These are generally small, but can be up to 20 m in depth and 100 m in diameter. Stream-sink dolines have allogenic stream-flow from the Limestone Shales or cover rocks. Dolines are also scattered along dry-valley floors, and on slopes and interfluves, but there is never the complete colonization required to form polygonal karst. Larger, shallow closed basins drain to one or more central small dolines. They are infilled with clays, loessic or thermoclastic deposits, with modern suffosion dolines draining through them. Most have an overflow outlet in the perimeter, indicating that they filled episodically as lakes; one has a lake-edge corrosion platform 25 m wide, with a low limestone cliff at its back (Ford & Stanton, 1968). These basins are taken as evidence of effective permafrost blockage of the smaller catchments; larger catchments supplied by allogenic streams maintained groundwater intake through the most severe cold. There are collapse dolines through thin covers of Liassic limestones or marls resting on the Carboniferous limestones.

The steeply dipping limestone has produced a characteristic style of cave development, with most influent caves developed at the stratigraphic base of the limestone. Typically, an "invasion" vadose streamway descends rapidly down dip to the local base-level, whereupon the passage continues as an undulating phreatic tube, ultimately reaching the resurgence after rising stratigraphically through the limestone sequence. Depending on its orientation, the phreatic passage takes on either a looping profile following bedding planes down dip and ascending stratigraphically up joint or fault risers, or develops along strike as a series of shallower loops or quasi-horizontal elliptical tubes along bedding partings. Most systems display a mixture of both. The deepest loop so far explored is 60 m deep in Wookey Hole. Due to the depth that the active phreatic conduits reach, and the sediment that accumulates in them to obstruct the loops, no Mendip cave has been explored completely from sink to spring. Most known caves are either vadose influent systems, resurgence systems, or relict fragments truncated by surface entrenchment or lowering. The study of these caves has stimulated development of the "four-state model" for cave long section genesis and also modern understanding of the network-linking rules that govern the plan patterns of a majority of caves forming where there is unconfined groundwater circulation (Ford, 2000).

There are substantial deposits of clastic sediments in all but the youngest caves, ranging from pebble and cobble sizes down to sand, silt, and clay. Most relict passages become blocked by

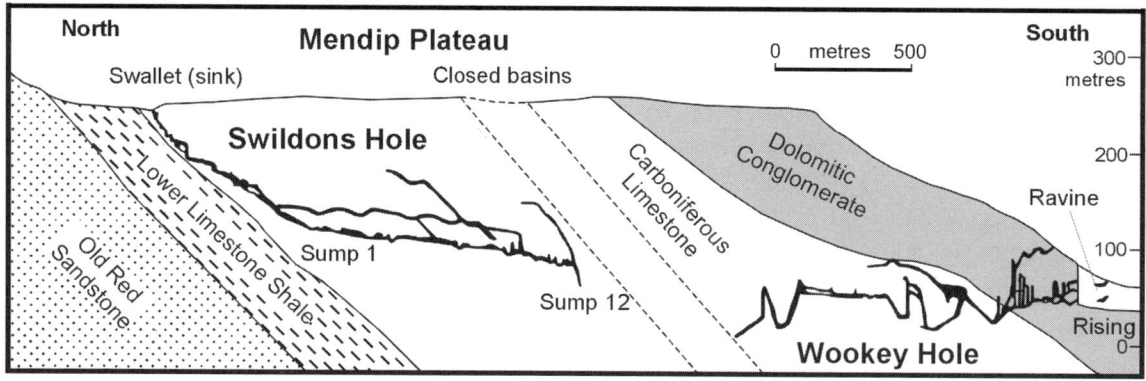

Mendip Hills: Figure 1. Profile of the known passages in Swildons and Wookey caves, beneath the Mendip Hills. The vertical scale is exaggerated by 5, and there is a gap of 2300 m between the nearest points in the two caves.

them, and elsewhere they have aided the development of paragenetic features. Many of these clastic fillings are interbedded with speleothem layers that have been dated to the warm phases of the past 350 000 years. This suggests that the clastics were deposited chiefly during colder phases when forest cover was reduced or eliminated.

At 9.1 km in length, the longest cave on Mendip is Swildons Hole, a classic influent cave system (Figure 1). The streamway initially cascades steeply down a small vadose canyon to a siphon, "Sump 1", the location in 1935 for a pioneering diving attempt using a home-made respirator. Beyond it the passage gradient slackens and the streamway changes to a series of phreatic loops. The troughs are marked by ten further siphons before becoming impassable at a twelfth. Between the sumps, the stream flows through vadose canyons entrenched into the loop crests. Above lie series of abandoned passages at several levels, representing former courses of the stream that demonstrate a complex sequence of passage captures.

St Cuthbert's Swallet nearby is a similar influent cave, with numerous vadose canyons developed either side of a plunging anticline uniting at depth into a single phreatic looping streamway. Extensive collapse, coupled with major sediment influxes and paragenetic development, have modified much of the original system, which is further disturbed by the influx of tailings from lead-mining operations.

Both caves have been traced to Wookey Hole, a resurgence which has been known since Roman times, a show-cave, and the site of many pioneering cave dives. Here, the River Axe flows through a series of deep phreatic loops within the Carboniferous Limestone, before reaching daylight via a channel with shallower sumps developed in the overlying Triassic Dolomitic Conglomerate.

To the west, the Charterhouse Caves (GB Cavern, Charterhouse, Manor Farm, and Longwood Swallets) are further classic examples of complex influent caves all draining to Cheddar Gorge. Relative to their size, this group of caves, especially GB Cavern (Figure 2), is one of the most intensively studied in the world, with much pioneering work on geomorphology (Ford, 1964; Atkinson, 1967; Smart *et al.*, 1984), landscape evolution, and paleoclimate (Atkinson *et al.*, 1978). All four major caves have similar morphology: vadose streamways with up to four vertically stacked abandoned phreatic conduits which occur at the same elevations in each cave, suggesting that they experienced a uniform response to base-level changes. Complex sequences of thick clastic sediments interbedded with stalagmites have been dated using U-series, ESR, and paleomagnetic methods.

The Charterhouse waters resurge at springs at the mouth of Cheddar Gorge, the lowest place at which limestone is exposed along the southern front of the main plateau. There is a series of abandoned conduits above them. The largest is Gough's Cave, which intercepts the active conduit. Discovered in 1898, this show cave is also one of Europe's most important Upper Paleolithic sites and the home of "Cheddar Man", one of several skeletons found in the cave. Mitochondrial DNA from this skele-

Mendip Hills: Figure 2. Very white calcite dripstone in the old high level of Bat Chamber in GB Cavern. (Photo by Tony Waltham)

ton was found to match that of a schoolteacher living nearby today.

Quarrying has had a major impact, especially on eastern Mendip, where much of the outcrop has been quarried away or has planning permission for quarrying. Some quarries are now sub-water-table operations and have intercepted groundwater conduits between the sinks and resurgences. The sub-water-table quarrying may affect the Bath Hot Springs 20 km to the northeast.

ANDY FARRANT AND DEREK FORD

See also **Speleothems: Carbonate for photo from Shatter Cave**

Works Cited

Atkinson, T.C. 1967. The geomorphology of Longwood Swallet, Charterhouse-on-Mendip. *Proceedings of the University of Bristol Spelaeological Society*, 11: 161–85

Atkinson, T.C., Harmon, R.S., Smart, P.L. & Waltham, A.C. 1978. Paleoclimatic and geomorphic implications of ^{230}Th/^{234}U dates on speleothems from Britain. *Nature*, 272: 24–28

Ford, D.C. 1964. On the geomorphic history of GB Cave, Charterhouse-on-Mendip, Somerset. *Proceedings of the University of Bristol Spelaeological Society*, 10: 149–88

Ford, D.C. 2000. Speleogenesis under unconfined settings. In *Speleogenesis, Evolution of Karst Aquifers*, edited by A.B. Klimchouk, D.C. Ford, A.N. Palmer & W. Dreybrodt, Huntsville, Alabama: National Speleological Society

Ford, D.C. & Ewers, R.O. 1978. The development of cave systems in the dimensions of length and breadth. *Canadian Journal of Earth Sciences*, 15: 1783–98

Ford, D.C., & Stanton, W.I. 1968. Geomorphology of the south-central Mendip Hills, *Proceedings of the Geologists' Association*, 79(4): 401–27

Smart, P.L., Moody, P.D., Moody, A.A.D. & Chapman, P.R.J. 1984. Charterhouse Cave: Exploration, geomorphology and fauna. *Proceedings of the University of Bristol Spelaeological Society*, 17: 5–27

Waltham, A.C., Simms, M.J., Farrant, A.R. & Goldie, H.S. 1997. *Karst and Caves of Great Britain*, London and New York: Chapman and Hall

Further Reading

Barrington, N. & Stanton, W.I. 1977. *Mendip, the Complete Caves and a View of the Hills*, Cheddar: Cheddar Valley Press
A very thorough guidebook, indispensable for anyone interested in Mendip caves and the karst landscape

Farrant, A.R. 1999. *Walks around the Caves and Karst of the Mendip Hills*, London: British Cave Research Association (BCRA Cave Studies, 8)

Friederich, H. & Smart, P.L. 1982. The classification of autogenic percolation waters in karst aquifers: A Study in GB Cave, Mendip Hills, Somerset. *Proceedings of the University of Bristol Spelaeological Society*, 16: 143–60

Gunn, J. (editor) 1994. *An Introduction to British Limestone Karst Environments*, London: British Cave Research Association

Halstead, L.B. & Nicoll, P.G. 1971. Fossilized caves of Mendip. *Studia Speleologica*, 2(3): 93–102

Hanwell, J.D. & Newson, M.D. 1970. The great storms and floods of July 1968 on Mendip. *Wessex Cave Club Occasional Paper*, 2(2)
Describes the events and aftermath of the 1968 floods.

MICROBIAL PROCESSES IN CAVES

Microorganisms are the most widespread form of life on Earth and represent the largest biomass grouping of organisms. They are adapted to many modes of life, unfamiliar to us as dwellers on the planet's surface. In common with ourselves, some microorganisms use atmospheric oxygen to respire an organic substrate (food) to generate energy for vital processes. However, many are adapted to life in the subsurface, where oxygen is usually absent (i.e. conditions are anoxic). These utilize other types of respiration, most commonly reduction of nitrate, or oxidized forms of iron and manganese or sulfate; processes which are less energy-efficient than oxygen respiration, but allow bacteria to colonize and thrive in a wide range of environments. All of these bacteria are heterotrophs—that is, they require an organic substrate for respiration, and cell growth and division. Other bacteria are chemotrophs and can gain energy from the oxidation of inorganic substrates, such as sulfide and methane. Some bacteria are also autotrophs, having the ability to convert inorganic forms of carbon (such as carbon dioxide or bicarbonate dissolved in karst groundwaters) into organic carbon for cell growth.

The range of subsurface environments colonized by bacteria is huge. In soils, microbes play a central role in the degradation of plant and other organic material, under both oxic and anoxic conditions. Microbes and their activity have been identified at depths of hundreds of metres, in deeply buried aquifer sediments (Krumholtz, *et al.* 1997). In deep marine sediments, active bacteria have been identified at depths >700 m below the sea floor and the frequent recovery of biodegraded oil from reservoirs of even greater depth indicates that microorganisms are active deeper still.

Microbial activity is crucial to the origin and development of most "normal" limestone caves. Carbon dioxide, which forms the carbonic acid that corrodes limestone, is derived principally from respiration of soil microorganisms (and plant roots). The atmosphere contains only 0.035% CO_2, which provides minimal corrosive power via carbonic acid formation, whereas soil gases contain 0.1 to over 1% CO_2. The influence of soil bacteria in this respect is clear from the distribution of the world's largest limestone cave passages (in the tropics where the warmer soils are more productive) to seasonal differences in karst spring-water chemistry in temperate zones, due to differing seasonal rates of soil microbial activity.

In cave sediments, microbial activity has been shown to degrade organic matter in detrital sediments washed into cave systems (Humphreys, 1991; Bottrell, 1996). The nature and rates of these processes are similar to surface soils (Bottrell, 1996) and the microbial populations are presumably introduced with the surface-derived sediment. These microbes are, however, important in their ability to break down the complex surface-derived organic material in the sediment and thus form the basis of subterranean food chains.

Microbial reduction and oxidation of nitrogen species is implicated in the origin of nitrate deposits and nitrate-rich waters in caves. Sources of nitrogen may be indigenous guano or ammonium ions leached in by percolation from soil horizons overlying the caves (Hill, 1981). Bacteria such as *Nitrosomonas* spp. oxidize ammonium to nitrite, and then *Nitrobacter* spp. carry out the final stage of nitrification, the oxidation of nitrite to nitrate. These nitrates can crystallize as a variety of mineral forms in dry caves. An alternative view (Lewis, 1992) suggests that nitrogen-fixing microorganisms may exist in cave sediments, but to date only nitrifying bacteria have been identified from caves (Fliermans & Schmidt, 1977).

Iron and manganese are the two metallic elements that most commonly undergo redox transitions in natural environments, which can be microbiologically mediated. Whilst reactions involving iron are likely prevalent, the distinctive black / purple colourations associated with manganese oxide deposits often make them a focus for study. Most iron and manganese formations and deposits in caves result from the microbiological oxidation of soluble Mn(II) and Fe(II) species in inflowing water to precipitate insoluble Mn(IV) and Fe(III) hydroxides (e.g. Peck, 1986; Manolache & Onac, 2000). Microorganisms, capable of such reactions, have also been found in cave clay deposits (Morehouse, 1968).

The microbial system that has received the most attention in caves and karst is that involving reduction and oxidation of sulfur species. The stable sulfur species under earth surface conditions is sulfate, but in anoxic environments a wide range of microbes are able to gain energy by using sulfate to oxidize organic matter (anaerobic respiration), usually reducing the sulfate to hydrogen sulfide. This process has been documented from several cave environments (e.g. Bottrell, *et al.*, 1991; Lauritzen & Bottrell, 1994). A larger variety of studies describe sulfidic waters entering the shallow, oxidized zone of karst aquifers, where hydrogen sulfide can be oxidized. This reaction represents a very significant energy source and again, a wide range of sulfur-oxidizing microbes are adapted to make use of it. The oxidative part of this cycle also has the ability to produce acidity that could be involved in development of significant secondary porosity and cave development (e.g. Morehouse, 1968; Hill, 1990). One example is Cueva de Villa Luz, Mexico (see Villa Luz entry). A particularly interesting case is the discovery of a previously closed cave system in Romania (Movile Cave, see separate entry), where a whole ecosystem has developed, based on chemoautotrophic microbial mats. These mats gained energy by oxidation of sulfide from thermal waters and could fix inorganic carbon for growth. This ecosystem was thus truly subterranean, completely independent of surface-derived photosynthetic organic matter.

SIMON BOTTRELL

See also **Biofilms; Guano; Microorganisms in Caves; Sediments: Biogenic**

Works Cited

Bottrell, S.H., Smart, P.L., Whitaker, F. & Raiswell, R. 1991. Geochemistry and isotope systematics of sulphur in the mixing zone of Bahamian blue holes. *Applied Geochemistry*, 6: 97–103

Bottrell, S.H. 1996. Organic carbon concentration profiles in recent cave sediments: Records of agricultural pollution or diagenesis? *Environmental Pollution*, 91:325–32

Fliermans, C.B. & Schmidt, E.L. 1977. *Nitrobacter* in Mammoth Cave. *International Journal of Speleology*, 9: 1–19

Hill, C.A. 1981. Origin of cave saltpeter. *National Speleological Society Bulletin*, 43: 110–26

Hill, C.A. 1990. Sulphuric acid speleogenesis of Carlsbad Cavern and its relation to hydrocarbons, Delaware Basin, New Mexico and Texas. *American Association of Petroleum Geologists Bulletin*, 74: 1685–94

Humphreys, W.F. 1991. Experimental re-establishment of pulse-driven populations in a terrestrial troglobite community. *Journal of Animal Ecology*, 60: 609–23

Krumholz, L.R., McKinley, J.P., Ulrich, G.A. & Suflita, J.M. 1997. Confined subsurface microbial communities in Cretaceous rock. *Nature*, 386: 64–66

Lauritzen, S.-E. & Bottrell, S.H. 1994. Microbiological activity in thermoglacial karst springs, south Spitzbergen. *Geomicrobiology Journal*, 12: 161–73

Lewis, W.C. 1992. On the origin of cave saltpeter: A second opinion. *National Speleological Society Bulletin*, 54: 28–30

Manolache, E. & Onac, B.P. 2000. Geomicrobiology of black sediments in Vantului Cave (Romania): Preliminary results. *Cave and Karst Science*, 27: 109–12

Morehouse, D.F. 1968. Cave development via the sulphuric acid reaction. *National Speleological Society Bulletin*, 30: 1–10

Peck, S.B. 1986. Bacterial deposition of iron and manganese oxides in North American caves. *National Speleological Society Bulletin*, 48: 26–30

MICROORGANISMS IN CAVES

Microbiologists study five main groups of organisms: bacteria, fungi, algae, protozoa, and viruses. The property that unites all microorganisms is their minute size. Prokaryotic microorganisms can be as small as 0.2 μm in diameter and as big as 750 μm in diameter in the newly discovered colourless sulfur bacterium, but the average bacterium is 1 by 3 μm. Eukaryotic cells are usually much larger, ranging from 2 μm to more than 200 μm in diameter. Typical bacterial shapes are cocci (spherical), rods, filaments, and spirals, but one can see bacteria that resemble braided or twisted rope-like shapes, stars, beads-on-a-string, etc. and fungal spores take on many unusual shapes, including some that look like ancient urns. The calcifying and silicifying algae take on intricate shapes with highly ornamented structures. Protista (algae and protozoa) and fungi are eukaryotic cells, similar to human cells in having a "true nucleus" ("eu" = true, "karyon" = nucleus). Bacteria and Archaea are prokaryotic ("pro" = before) and lack a true membrane-bound nucleus. Viruses are not even cells, just genetic material surrounded by a protein coat, and are incapable of independent existence. We know almost nothing about viruses in caves, with the exception of the rabies virus associated with bats.

Within caves, microbes cycle nutrients such as carbon and nitrogen, provide nutrition to invertebrates, and participate in the formation of secondary minerals (speleothems). Microbes may also produce some growth factors, such as vitamins, needed by higher organisms. Microorganisms in caves worldwide have been studied since the end of the 19th century, with many pioneering studies done in Europe (Caumartin, 1963; Juberthie & Decu, 1994).

Like all organisms, microorganisms require water, an energy source, and nutrients such as carbon and nitrogen, in order to grow. There is a broad range of physical conditions that microbes can tolerate, allowing them to occur in many habitats, including those hostile to humans. Microorganisms, particularly bacteria, are metabolically very diverse in what they can use for energy sources. Microbes can be divided into metabolic classes relating to the sources of energy they use. The three groups are heterotrophs, which utilize organic substrates as an energy source (these are also called chemoheterotrophs); photoautotrophs, which obtain energy from light; and chemolithotrophs, which obtain energy from inorganic compounds. Carbon for cell synthesis is obtained from organic substrates; however some microbes, including the photoautotrophs, fix CO_2. Some microorganisms in cave entrances and twilight zones use sunlight for energy, but most use chemical energy sources. In general, chemoheterotrophs are found where there is sufficient organic matter, and chemolithotrophs are found in the absence of organic matter where inorganic sources of chemical energy exist, such as manganese and iron.

Microorganisms can enter caves by flowing water, gravity, or on air currents, travelling through fissures, porous rocks, and tiny voids in the overlying rock, and can be carried in on humans or animals entering caves. Studies of microbes in caves have identified a wide variety of different microorganisms present; how they interact with and adapt to the cave environment; and their role in creating and destroying secondary mineral deposits. A new area of study is the broad field of geomicrobiology. Microbes are also seen as important in food-limited cave environments because they recycle organic material into microbial biomass as the first step in the food chain, or as primary producers (see Food Resources). Other microbes can be agents of disease, such as the fungus causing histoplasmosis.

Microbiological Methods

Basic microbiological techniques can be used to study karst microbes. Typically, we want to find out which microbes are present, their abundance, and what they are doing. Light microscopy with appropriate staining techniques (e.g. Gram's Stain) can be used to observe microorganisms and for enumeration. Staining with metabolic dyes such as INT has been used to indicate respiratory activity. Electron microscopy (e.g. scanning or transmission, see Figure 1) allows greater magnification and examination of the interaction of microbes with their environment.

Traditional methods of culturing microorganisms from the environment (e.g. spread plating) can be useful, although it has been estimated that only 0.1% can be cultured using standard techniques. Pure cultures of one kind of organism can be tested for biochemical and physiological characteristics for identification. The types and proportion of fatty acids (lipids) in cell membranes of microorganisms can be used to provide a "fingerprint" for identification. Newer applications do not require the microbes to be cultured. For example, fluorescent monoclonal antibodies are being developed for identification, but have had limited applications to cave microbes. Methods based on the microbial genetic sequences of DNA and RNA are being employed to identify cave microorganisms and to evaluate microbial diversity. Stable isotopic ratios are very useful in determining the contribution of microbes to the cave food-chain and their role in the production of secondary mineral deposits.

Major Groups of Microorganisms

Bacteria and Archaea

We are only now becoming aware of the complex range of prokaryotic microbes found in caves, and a tremendous amount of

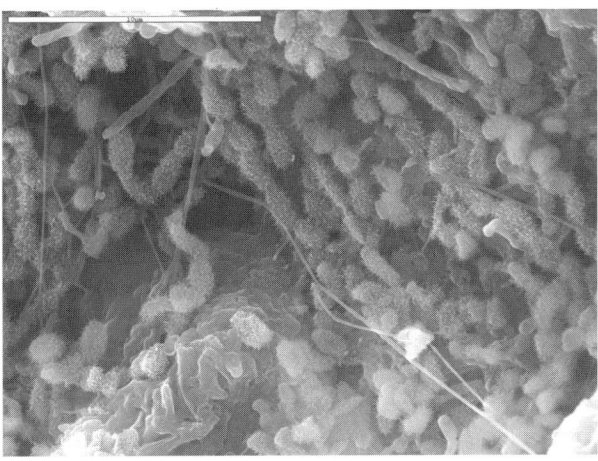

Microorganisms in Caves: Figure 1. Scanning electron microscopy reveals the diversity of microorganisms present on the wall of a lava tube in the Cape Verde Islands. (Photomicrograph by M. Spilde, D. Northup, and P. Boston)

work remains to be done. Archaea, including those that produce methane, are just beginning to be studied in caves.

One of the most visible groups of bacteria seen in caves are colonies of actinomycete bacteria. They are evident as reflective white (sometimes pink or gold) dots on moist limestone or lava. These actinomycetes are responsible for the distinctive odour of caves and soils. Filamentous in nature, actinomycetes may be widespread in caves due to lower temperatures and high humidity. They have been implicated in the biodeterioration of cave paintings studied extensively in Altamira and other Spanish caves (Groth & Saiz-Jimenez, 1999). Two of the more commonly reported genera are *Streptomyces* and *Nocardia*. Actinomycetes are often very dense on the walls and ceilings of lava tube caves.

Bacteria are involved in all phases of the nitrogen cycle in caves. Studies have concentrated on the role of nitrifying bacteria, such as *Nitrosomonas* and *Nitrobacter*, in the generation of saltpetre, a component of gunpowder. The former oxidizes ammonia to nitrite and the latter oxidizes nitrite to nitrate. Bacteria that fix atmospheric nitrogen are also found in caves. Thus, the processes of ammonification, nitrification, denitrification, and nitrogen fixation have all been documented in caves.

Studies of sulfur bacteria have rapidly expanded in the last decade as several new sulfur-containing caves have been discovered, including Movile Cave (see separate entry), whose food web is based on chemolithotrophic processes involving sulfur and other compounds (Sarbu, Kane & Kinkle, 1996). Isotopic studies have established a microbial role in the production of sulfur compounds and in the enlargement of passages in caves (reviewed in Northup & Lavoie, 2001). Both sulfide / sulfur oxidizers (*Thiobacillus*, *Beggiatoa*, and *Thiothrix*) and sulfate-reducers are found in caves.

Several studies have proposed microbial participation in the formation of cave manganese and iron deposits. Manganese- and iron-oxidizing bacteria have been reported from caves, including the genera *Leptothrix*, *Gallionella*, and *Clonothrix*. Most studies have reported only on the presence of these bacteria, but some studies have established the production of manganese and iron oxides in cultures of organisms from caves.

Cyanobacteria, formerly classified as algae (i.e. blue-green algae), are photosynthetic bacteria that produce oxygen. They are frequently found in the entrance and twilight zones of caves, alone, or as one partner in the lichen symbiosis with fungi. Some are adapted to very low-light conditions, and some have the ability to fix atmospheric nitrogen, allowing them to colonize low-nitrogen areas. Two cyanobacteria described from caves, *Geitleria calcarea* and *Scytonema julianum*, can be encrusted by calcium carbonate. Genera most often noted include *Oscillatoria*, *Phormidium*, *Gleocapsa*, and *Lyngbya*.

Algae
Algae are a diverse group of phototrophic, eukaryotic organisms that contain chlorophyll. Several algae are adapted to the low light levels of cave entrances, and may produce chlorophyll in darkness. Some algae may grow chemoheterotrophically in total darkness. Green algae (Chlorophyta) and diatoms (Bacillariophyceae) have been extensively reported from caves. Along with cyanobacteria, algae can be significant nuisance organisms in areas of artificial lighting in caves. Both groups can also figure prominently in studies of speleothem formation in entrance and twilight regions, and in studies of the biodeterioration of rock. Both groups also serve as food sources for other organisms, acting as locally important primary producers in cave entrances.

Fungi
A wide variety of fungi have been documented from caves (Dickson & Kirk, 1976; Rutherford & Huang, 1994), but studies of their activities have been limited. Caumartin (1963) believed there were no true indigenous cave fungi in the typical low-food cave; however, the richness of the media generally used to isolate fungi may prevent the discovery of cave-adapted (oligotrophic) forms. In caves, fungi are found on any source of organic material, such as wood, carcasses, and guano, often as large mycelial mats, or growths with bright colours (Figure 2). Fungi are important decomposers in caves and produce a variety of extracellular enzymes (lipases, proteinases, and chitinases) that degrade organic detritus. Most fungi found in caves are moulds (filamentous fungi) in the Zygomycetes (bread moulds such as *Mucor*) or Deuteromycetes (fungi imperfecti, including, most commonly, *Penicillium*, *Aspergillus*, *Fusarium*, and *Trichoderma*). Mushrooms (Basidiomycetes) are also found in caves on dead matter, and mycorrhizal fungi have been documented from rootlets in a cave stream. Besides their decomposer role, fungi may be pathogens. The causative agent of histoplasmosis, *Histoplasma capsulatum*, grows on bat guano. Filamentous fungi have been implicated in the formation of micritic laminae in speleothems, and can serve as nucleation sites (Went, 1969).

Slime Moulds
Slime moulds are eukaryotes that share characteristics with both fungi and protozoa. They produce spores, feed on bacteria, and can move across surfaces fairly rapidly. Perhaps the least-studied microorganisms in caves, cellular slime moulds in the genera *Dictyostelium* and *Polysphondylium* have been documented in caves, where they may be most active in bacteria-rich soils.

Protozoa
Protozoa are mobile, one-celled eukaryotic microorganisms that do not have cell walls. They are all chemoheterotrophs. Some protozoa are parasitic on cave vertebrates and invertebrates, while others are free-living on organic matter, bacteria, and other protozoa. They, in turn, serve as food for other cave biota. When protozoa find conditions unfavourable for growth they may form a dormant cyst. The first paper on cave protozoa appeared in 1845 and since that time more than 350 species have been identified from aquatic habitats and from cave biota (Gittleson & Hoover, 1969). Cave protozoans correspond well with those found in forest litter, and most are considered to be troglophilic. Protozoans that are parasitic on troglobitic cave fauna are themselves considered troglobitic.

Microbes as Agents of Destruction

Microorganisms have been shown to be significant agents of dissolution of rock (Northup & Lavoie, 2001) and destruction of paintings in caves. For example, some sulfur-oxidizing bacteria produce sulfuric acid as a waste product, which reacts with limestone to form gypsum, a mineral that is highly soluble in water. An example of the destruction wrought on prehistoric cave paintings by microorganisms is the "Maladie Verte" of Lascaux, where human-transported microorganisms contaminated cave paintings and human activities raised the temperature in the cave, increasing microbial growth (see also Tourist Caves: Algae and Lampenflora).

Human Impacts on Microbes

Human activities can have a major impact on microbes. The input of human faecal pollution brings in organic matter that stops the growth of native microbes adapted to normal low-food conditions in caves. Polluting bacteria may pose a threat of infection to cavers. Faecal contamination in the Red Lake area of Lechuguilla Cave in New Mexico has persisted for at least five years. Northup and Boston have advocated low-impact caving such as wearing clean caving clothes, packing out all wastes, and leaving areas undisturbed as microbial nature preserves, as ways for us to minimize our impact on cave microorganisms.

Microbes and Caver Health

Water in caves may be heavily polluted from farmland grazed by sheep or cattle, or from leaking septic tanks, and may contain

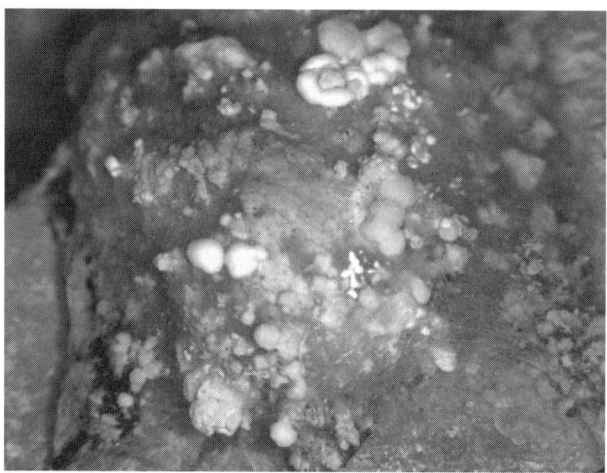

Microorganisms in Caves: Figure 2. Yellow mould covers a bat carcass in Carlsbad Cavern, Carlsbad Caverns National Park, New Mexico. (see also in colour insert; photo by D. Northup)

high concentrations of faecal indicator bacteria, posing a risk to caver health. For example, Gunn *et al.* (1998) report faecal coliform concentrations of up to 14 400 CFU / 100 ml and faecal streptococci concentrations up to 440 CFU / 100 ml in springs draining recreational cave systems at Castleton, England. Polluted water in caves may exceed standards for water quality for recreational uses, and it is not advisable to drink any cave waters. Other microbial dangers to cavers are the Histoplasmosis fungi, usually encountered growing on deposits of bat guano. Often a whole group of cavers is infected from the same exposure. Rabies from bats is a potential source of risk to cavers, although no cases are known. (See also Disease.)

Future Studies

Many microbes identified from deep caves are similar to surface forms, being non-residents transported into caves by water, air, sediment, and animals. However, most microbiological studies have focused on typical heterotrophic microbes known from surface studies and have missed microorganisms that are difficult to culture. Culture-independent, molecular phylogenetic techniques that compare genetic sequences of organisms have since shown that many novel organisms can be found in caves (Angert *et al.*, 1998). Multidisciplinary studies and the application of techniques from surface studies are now greatly expanding our knowledge of the roles of cave microorganisms.

DIANA E. NORTHUP AND KATHLEEN H. LAVOIE

See also **Biofilms; Microbial Processes in Caves**

Works Cited

Angert, E.R., Northup, D.E., Reysenbach, A.-L., Peek, A.S., Goebel, B.M. & Pace, N.R. 1998. Molecular phylogenetic analysis of a bacterial community in Sulphur River, Parker Cave, Kentucky. *American Mineralogist*, 83: 1583–92

Caumartin, V. 1963. Review of the microbiology of underground environments. *National Speleological Society Bulletin*, 25: 1–14

Dickson, G.W. & Kirk, P.W. 1976. Distribution of heterotrophic microorganisms in relation to detritivores in Virginia caves (with supplemental bibliography on cave microbiology and mycology). In *The Distributional History of the Biota of the Southern Appalachians*, edited by B.C. Parker & M.K. Roane, vol. 4: Algae and Fungi, Charlottesville, Virginia: University Press of Virginia

Gittleson, S.M. & Hoover, R.L. 1969. Cavernicolous protozoa: Review of the literature and new studies in Mammoth Cave Kentucky. *Annales de Spéléologie*, 24(4): 737–76

Groth, I. & Saiz-Jimenez, C. 1999. Actinomycetes in hypogean environments. *Geomicrobiology Journal*, 16(1): 1–8

Gunn, J., Tranter, J., Perkins, J. & Hunter, C. 1998. Sanitary bacterial dynamics in a mixed karst aquifer. In *Karst Hydrology*, edited by C. Leibundgut, J. Gunn & A. Dassargues, Wallingford, Oxfordshire: International Association of Hydrological Sciences (IAHS Publication no. 247)

Juberthie C. & Decu, V. (editors) 1994. *Encyclopedia Biospeologica*, vol. 1, Moulis and Bucharest: Société de Biospéleologie

Northup, D.E. & Lavoie, K.H. 2001. Geomicrobiology of caves: a review. *Geomicrobiology Journal*, 18(3): 199–222

Rutherford, J.M. & Huang, L.H. 1994. A study of fungi of remote sediments in West Virginia caves and a comparison with reported species in the literature. *National Speleological Society Bulletin*, 56(1): 38–45

Sarbu, S.M., Kane, T.C. & Kinkle, B.K. 1996. A chemoautotrophically based cave ecosystem. *Science*, 272: 1953–54

Went, F.W. 1969. Fungi associated with stalactite growth. *Science*, 166: 385–86

MILITARY USES OF CAVES

Throughout human history, caves have been used in military situations, and many karst landscapes have been arenas for significant military conflict. The nature of karstlands, particularly the irregular topography, restricted surface water supply, and the suitability of caves for refuge and ambush, affords strategic offensive and defensive advantages to native combatants familiar with the surface and underground terrain and poses tactical problems for unfamiliar foreign forces. Caves and karstlands are especially well suited to guerrilla warfare by small, mobile local units, and partisans have hidden and lived in caves since time immemorial (Kempe, 1988).

Caves have been employed historically for a multitude of specific military purposes, both offensive and defensive. Aggressive activities include planned ambushes and strategic entrapment, troop concealment, training and deployment, munitions storage, weapons testing, command and communications facilitation, imprisonment, and execution. Caves have been important sources of military materials including lead and other metals used to manufacture ammunition, saltpetre, which is a principal component of gunpowder, and even flint, from which were crafted early knives, arrowheads, flintlocks, and other weapons.

Defensive strategies include troops and civilians seeking refuge from conflict, both from land invasion and from air raids, establishing armed defensive positions, and caching essential survival supplies and potential spoils of war such as precious metals, jewelry, rare books, priceless artwork, and secret documents. Caves have also been used as medical facilities for treatment of military and civilian casualties, and as military burial grounds.

The impacts of military activities in caves and karstlands may be profound and long lasting. Many caves have been extensively modified or expanded for military purposes, their resources exploited and their ecosystems disrupted or destroyed. Karst groundwaters have been contaminated, both deliberately and by accident, and surface landforms disfigured or obliterated by tunnelling, heavy equipment use, or bombardment. Indigenous flora and fauna likewise have been devastated. Notwithstanding, military impact on karstlands has received relatively little attention.

Prehistoric conflict involving caves is speculative, although it seems reasonable to envision early hominids hiding from aggressors in caves and seeking underground flint for weapon making. Reliable documentation of military use of caves increases through human history, with the best-known examples from the 20th century. An early example is from China, where the emperor Sun Quan (222–252 AD) ordered a cave in Jiangsu Province to be explored as a potential route for the invasion of neighbouring provinces (LaMoreaux, 1999). Even earlier, during the first century BC, the Indian King Kharavela recorded his military exploits by means of inscription in the monastic caves of Orissa.

Caves may serve as useful refuges, but equally they can represent deathtraps. In 1577 nearly 400 members of the Clan Ranald, a scion of the MacDonalds, were suffocated in the Cave of Francis (or St Francis' Cave) on the Scottish Hebridean island of Eigg when they were found and trapped there by a force of their enemies, the MacLeods, who lit a large fire in the cave entrance (Kempe, 1988). Such tactics are not always successful, as in a 1923 incident in which Free State troops tried to dislodge seven IRA men from the Clashmealcon Caves in County Kerry, Ireland. Examples of fortified cave entrances include the Erasmus Castle at the entrance to Predjama Cave in Slovenia and the Covolvo del Butistone in the Venetian Prealps (Gams et al., 1993). During the Middle Ages the caves beneath Buda, now incorporated into Budapest, were enlarged and connected for military purposes; in the 1930s the labyrinth was converted into a shelter large enough to accommodate 10 000 people, and during the Cold War it served as a secret military installation. The Veterani Cave in Romania was extensively fortified by troops of the Austro-Hungarian Empire in the 17th and 18th centuries during conflicts with Turkish forces (Patay, 1997). Caves, both natural and man-made, also featured as military installations during the Crimean War (1853–56).

Cave use for military storage, construction, training, and testing is also well documented. During World War I, the Bethlehem Steel Company used one of the Reddington Caves, Pennsylvania, as an artillery firing range (Folsom, 1956). An aircraft assembly plant was set up in the Bedeilhac Cave in France in 1940 (Nicod et al., 1996), and during 1944 a Heinkel aircraft factory was established in a cave near Vienna. Also during World War II a German ammunition dump in Postojna Cave (Slovenia) was blown up by Yugoslav partisans (Kempe, 1988). Nuclear missiles are said to have been stored in Cuban caves during the Cuban missile crisis, and in Russia and China during the Cold War.

Perhaps the most bizarre military operation involving caves was the World War II "Bat bomb" project, in which a US military scientist proposed attaching small incendiary devices to Mexican free-tailed bats that were to be dropped over Japan. Thousands of bats were collected from caves in Texas and stored under conditions resembling hibernation, guarded by US Marines. The project was ultimately abandoned in favour of the atomic bomb (Folsom, 1956; Couffer, 1992).

Caves have played a significant role in the military history of several islands, including Gibraltar. In 1704 Spanish troops concealed themselves overnight in St Michael's Cave while preparing unsuccessfully to attack the British and Dutch defenders. The island's cave systems, extensively modified, were used as civilian air raid shelters and hospitals during World War II. Similarly, caves played important roles during the 1690–1796 Maroon Wars in Jamaica.

World War I saw extensive military activity in caves and karst throughout Europe. Northeastern Italy was affected particularly (Sauro, 1987). Over 100 Slovenian caves were modified for military use (Kepa, 2001); in Postojna Cave Russian prisoners were used to build the famous "chasm footbridge", which is still used by tourists. Examples of caves being used as military hospitals and burial sites are legion. In 1769 Russian soldiers and sailors apparently were both hospitalized and buried in caves in Minorca following battles with the Turks. Similar underground hospitals have been reported in caves ranging from Gibralter to Eritrea.

Caves have been used as military execution sites and for the disposal of executed victims. Large numbers of Armenians reportedly were asphyxiated in caves by Turkish troops during 1915, repeating an earlier Turkish action in the Melidoni Cave in Crete in 1822 (Nicod et al., 1996). During World War II Yugoslav partisans are reputed to have thrown thousands of Italians and others to their deaths in karst chasms known as *foibes* (Pizzi, 1998). In March 1944, in reprisal for partisan actions, German troops massacred 335 Italian civilians in the Ardeatine Caves near Rome. In 1997 some 300 victims were recovered from a cave near Hrgar, in Bosnia.

Some of the most bloody military actions involving caves took place in the Pacific during World War II when advancing Allied troops encountered tenacious Japanese defenders entrenched in heavily fortified cave entrances and passages. Intense conflict in 1943–45 consumed caves and karst on Tarawa, Okinawa, Iwo Jima, Guam, and Peleliu, among others, with the flame-thrower playing a pivotal role in clearing the caves. The battle for Iwo Jima resulted in the deaths of 22 000 Japanese defenders and nearly 6000 US Marines. Thousands of Japanese soldiers were trapped by US Marines in the Biak Caves of Irian Jaya in 1944; in Japan itself the Akiyoshi caves were extensively fortified and heavily defended.

There are numerous other instances of caves being used temporarily during battle, for example in South Africa's Transvaal and Zululand, on the Northwest Frontier of India, and during the American Civil War in the 1800s. In February 1944, during the Allied landings on the Anzio beachhead, there occurred a decisive but costly action known as the Battle of the Caves. On a more permanent basis, various guerrilla forces have used caves as military bases and sanctuaries. The Sohoton caves and the surrounding karst were a stronghold of native partisans during the Philippine–American war until they were subdued in 1901. Mao Tse-tung and his revolutionary forces made extensive use of caves prior to and during the Long March of 1934, and Fidel Castro and his followers, including Che Guevera, based their 1959 Cuban Revolution around caves (Núñez Jiménez, 1986). The caves of Cao Bang were the initial military headquarters and training grounds for Ho Chi Minh's Viet Minh during the 1940s. During World War II Tito's partisans were headquartered in caves at Drvar, Bosnia, and during the Vietnam War (1961–75) Viet Cong forces made offensive and defensive use of both natural and man-made caves and tunnels, in which fierce hand-to-hand fighting subsequently occurred.

Most recently, caves have figured prominently in wars in Afghanistan. Mujaheddin fighting Russian occupying troops between 1979 and 1989 used caves, particularly in the Zhawar region, as bases for guerrilla activities and proved impossible to defeat. In 2001–02 these same artifically enlarged "caves", including the Tora Bora cave and tunnel complex in the White Mountains, were used by the Al-Qaeda network to evade and resist Allied military forces.

MICK DAY and JEFF KUENY

See also **Gunpowder**

Works Cited

Couffer, J. 1992. *Bat Bomb: World War II's Other Secret Weapon*, Austin: University of Texas Press

Folsom, F. 1956. *Exploring American Caves: Their History, Geology, Lore, and Location: A Spelunker's Guide*, New York: Crown Publishers

Gams, I., Nicod, J., Julian, M., Anthony, E. & Sauro, U. 1993. Environmental change and human impacts on the Mediterranean karsts of France, Italy and the Dinaric Region. In *Karst Terrains: Environmental Changes and Human Impacts*, edited by P.W. Williams, Cremlingen: Catena

Kempe, D.R.C. 1988. *Living Underground: A History of Cave and Cliff Dwelling*, London: Herbert Press

Kepa, T. 2001. Karst conservation in Slovenia. *Acta Carsologica*, 30(1): 143–64

LaMoreaux, P.E. 1999. The historical perspective. In *Karst Hydrology and Human Activities: Impacts, Consequences and Implications*, edited by D. Drew & H. Hötzl, Rotterdam: Balkema

Nicod, J., Julian, M. & Anthony, E. 1996. A historical review of man-karst relationships: Miscellaneous uses of karst and their impacts. *Rivista Geografica Italiana*, 103: 289–338

Núñez Jiménez, A. 1987. *Geografia y Espeleologia en Revolucion*, La Habana: Imprenta Central de las Fuerzas Armadas Revolucionarias

Patay, K. 1997. Abbildungen der Veterani-Hohle aus dem 17–18 Jahrhundert. *Acta Carsologica*, 26(2): 149–57

Pizzi, K. 1998. Silentes Loquimur: Foibe and border anxiety in post-war literature from Trieste. *Journal of European Studies*, 28(3): 217–29

Sauro, U. 1987. The impact of Man in the karstic environments of the Venetian Prealps. In *Karst and Man: Proceedings of the International Symposium on Human Influence in Karst*, edited by J. Kunaver, Ljubljana: Department of Geography, University E. Kardelj

Further Reading

Collins, J.M. 1998. *Military Geography for Professionals and the Public*, Washington, D.C.: National Defense University Press

Peltier, L.C. & Pearcy, G.E. 1966. *Military Geography*, Princeton, New Jersey: Van Nostrand

O'Sullivan, P. & Miller, J.W. 1983. *The Geography of Warfare*, London: Croom Helm and New York: St Martins Press

Winters, H.A., Galloway, G.E., Reynolds, W.J. & Rhyne, D.W. 1998. *Battling the Elements: Weather and Terrain in the Conduct of War*, Baltimore: Johns Hopkins University Press

MINERALS IN CAVES

A cave mineral is any secondary mineral formed within a natural subterranean cavity, fissure, or tube that is human-size or larger and that extends past the twilight zone. A "secondary" mineral is derived by a physicochemical reaction from a primary mineral in bedrock or detritus, and is deposited because of a unique set of conditions within a cave (i.e. the cave environment has influenced the mineral's deposition). Secondary mineral deposits formed in this way in caves are called speleothems. However, a "cave mineral" is definitely not the same as a "speleothem", although speleothems are composed of cave minerals. The term "speleothem" refers to the mode of occurrence of a mineral in a cave (i.e. its morphology, or how it looks), not to its composition. For example, calcite is not a speleothem, but a calcite stalactite in a cave is a speleothem. The speleothem type "stalactite" can be composed of many minerals besides calcite (e.g. aragonite, gypsum, halite, ice, etc.). While only about 40 speleothem types have been recognized worldwide, a total of 255 "official" secondary minerals are known to occur in caves (Hill & Forti, 1997). Cave minerals can have a number of different origins and depositional settings. Figure 1 shows a generalized depositional scheme for all the different classes of cave minerals discussed in this section.

Cave minerals, like minerals in the outside world, can be organized according to the classification scheme of *Dana's System of Mineralogy* (Gaines et al., 1997), where grouping is by chemical class.

Native elements: Sulfur (S) is the only native element known to have a secondary origin in caves. Sulfur forms in "abnormal" cave situations where: (1) fumarole activity exists in volcanic caves; (2) primary pyrite and / or marcasite are oxidized in mine caves; or (3) reduced sulfur (H_2S) is oxidized in sulfuric acid caves. This last mechanism is responsible for most large cave sulfur deposits, e.g. those in Lechuguilla Cave, New Mexico, United States.

Sulfides: Sulfide minerals in caves almost always have a primary origin (i.e. they pre-date the cave and are only exposed by the cave), but in rare circumstances they can form as secondary mineral deposits—usually either as sulfide inclusions within hydrothermal calcite or as sulfide minerals produced by the anaerobic bacterial reduction of iron sulfates within cave clay. For more information on this topic refer to Sulfide Minerals in Karst.

Oxides and hydroxides: Common oxide–hydroxide cave minerals fall into three categories: ice (H_2O), manganese oxides, and iron oxides. Caves with ice speleothems in them can be limestone or gypsum caves with an average or seasonal temperature that falls below freezing, lava tube caves where the insulating effect of lava rock enhances the survival of ice, and caves within glacial ice itself (see Ice in Caves). The most famous caves containing ice speleothems are those in the limestone mountains of Europe (e.g. Eisriesenwelt and Riesenheishöhle of the Austrian Alps).

Manganese oxides are introduced into caves by stream water and, less frequently, by dripping water or associated with ore mineralization. The soluble Mn^{2+} in water is oxidized to the insoluble Mn^{4+}, thus causing the precipitation of manganese oxide minerals, a process probably aided by bacteria. Manganese minerals, such as birnessite [$(NaCa)Mn_7O_{14} \cdot 3H_2O$], usually occur as black layers coating stream cobbles.

The most common iron oxide–hydroxide mineral is goethite [$FeO(OH)$], often referred to by the catch-all term "limonite". Goethite usually occurs within cave sediment, but it can also take the form of dripstone and flowstone. Goethite in caves is also probably related to the activity of bacteria.

Halides: Halide minerals are relatively rare in caves, with halite (NaCl) being the most common, but only where halite evaporite rock exists in the overburden and where the climate is (or has been) arid. Halite is known to form seeping water

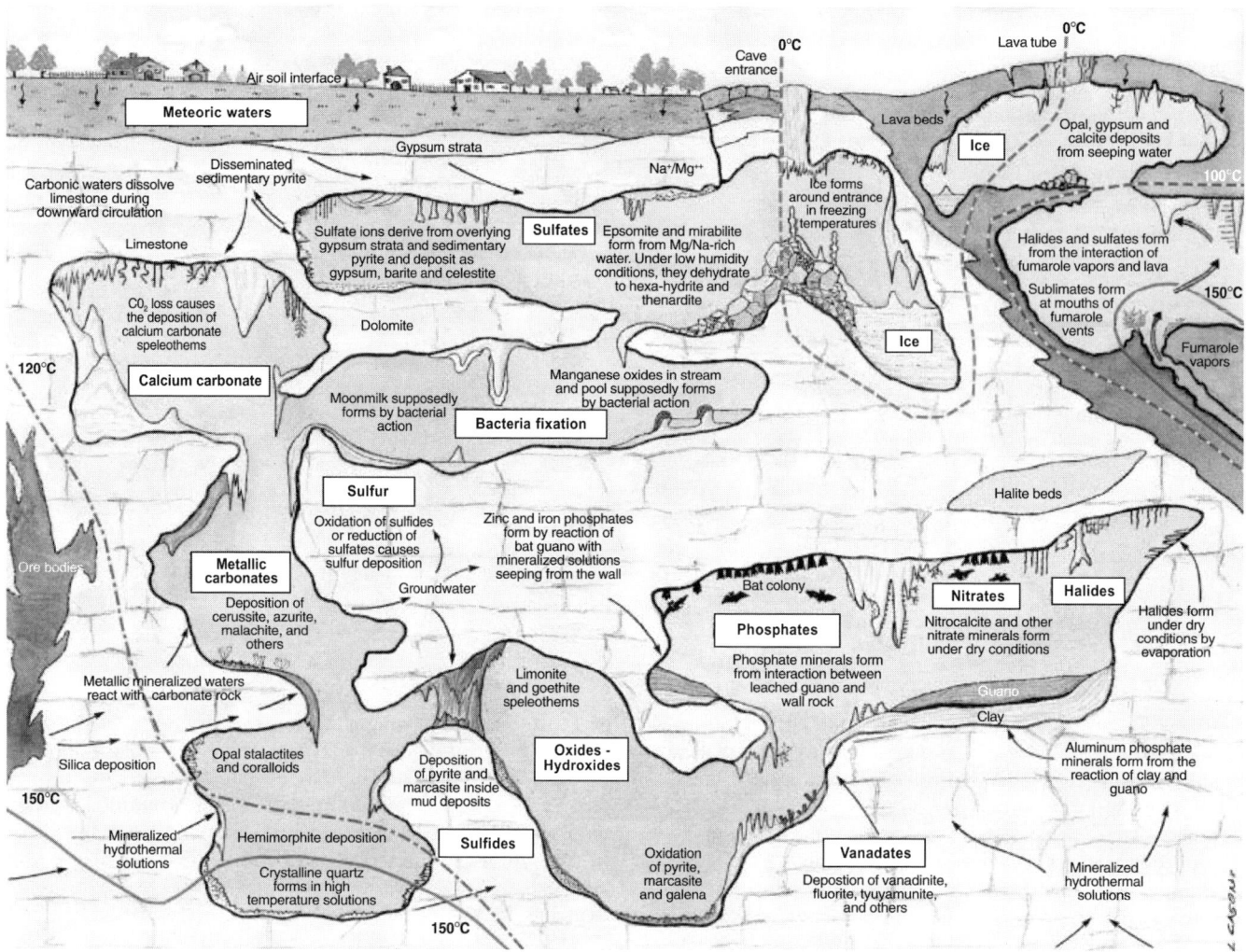

Minerals in Caves: Figure 1. Different depositional settings for different classes of cave minerals. From *Cave Minerals of the World*, Hill & Forti, 2nd edition. © 1997, National Speleological Society, Inc. Used with permission. Drawing by Luciano Casoni.

speleothems such as crusts, hair, cotton, and flowers, and dripping water speleothems such as stalactites and stalagmites. It also forms as euhedral spar crystals. Halide minerals usually form by the simple mechanism of groundwater dissolution of halite bedrock and reprecipitation in an underlying cave, but they can also derive from bat guano. Some, such as fluorite, can have a hydrothermal origin. A classic example of caves formed in halite rock, and containing halite minerals and speleothems, are the Mount Sedom caves of Israel (see Sedom Salt Karst).

Arsenates: Arsenate minerals are very rare in caves, and where they do occur are usually associated with arsenic-rich ore.

Borates: Only one borate mineral is known from a cave occurrence, associated with the borax deposits of the Mojave Desert, California, United States.

Carbonates: By far the most important class of cave minerals are the carbonates, with the two most common cave minerals, calcite ($CaCO_3$) and aragonite ($CaCO_3$), making up >95% of all speleothem deposits. Carbonate speleothems are abundant because most caves are formed in limestone rock, which readily supplies the necessary calcium and carbonate. Since this is such an important class of cave minerals, a separate Encyclopedia entry has been written on this topic (see Speleothems: Carbonate for a more in-depth discussion). In summary, the deposition of carbonate minerals begins when rainwater picks up carbon dioxide in the air and soil zone above a cave to form a weak carbonic acid. This acid dissolves limestone bedrock as groundwater percolates downward toward a cave; in this way, the water can become saturated with calcium carbonate. The precipitation of carbonate minerals happens when this carbonate-bearing groundwater reaches a cave: the exchange of carbon dioxide in the water with the cave atmosphere causes the precipitation of calcite (and other carbonate minerals). Carbon dioxide exchange is the primary cause of carbonate mineral precipitation in caves, but less often it can be caused by evaporation, the common-ion effect, and / or pressure–temperature changes.

Besides calcite and aragonite, a number of other carbonate minerals also exist in caves. Usually these minerals are either ore-related (e.g. azurite, cerussite, hydrozincite, malachite), or related to the evolution of cave waters with respect to magnesium content, evaporation, and / or carbon dioxide loss (e.g. hydromagnesite, huntite, nesquehonite, monohydrocalcite). In dolostone caves, where there is a general trend of increasing magnesium in solution with precipitation of carbonate species, the deposition of calcite, and then aragonite, huntite, and hydromagnesite is a common sequence. Dolomite also occurs as a secondary mineral in some caves, but primarily as a replacement of other carbonate minerals.

Nitrates: Nitrate minerals are not common in caves, but they do occur where cave conditions are dry enough, and the relative humidity low enough, for these very soluble minerals to crystallize. Nitrate minerals are both hygroscopic (they can absorb moisture from the air) and deliquescent (they can dissolve in that moisture). Thus, as the relative humidity of a cave seasonally oscillates, these minerals can alternately disappear (become deliquescent and sink into cave sediment or walls) and reappear (effloresce).

Nitrocalcite [$Ca(NO_3)_2 \cdot 4H_2O$] is historically interesting in that it is the nitrate constituent of "saltpetre earth", which was mined as an ingredient for gunpowder in the Revolutionary War, the War of 1812, and the Civil War in the eastern United States (e.g. Mammoth Cave, Kentucky; see Gunpowder). The source of nitrate to the saltpetre earth is groundwater seeping into a cave from the surface: dry caves act as receptacles where the soluble nitrate can accumulate rather than being leached away. In addition, bat guano can supply nitrate for cave mineralization (see entries on Guano and Sediments: Biogenic). Nitrocalcite, nitrammite, nitre, and nitromagnesite are all derived directly from bat guano in the caves of southern Africa (e.g. Chaos Cave, South Africa), where these minerals can form as crusts, flowers, and small stalactites.

Phosphates: Over 50 different phosphate minerals are known to form in caves, and almost all of these derive from bat guano. Bat guano is rich in both nitrogen and phosphorus; decomposition by leaching involves a process by which the very soluble nitrogen is first removed, leaving the relatively insoluble phosphate. This process often creates a complex stratified suite of phosphate minerals within the guano. In limestone caves, the most common phosphate minerals are brushite, carbonate-hydroxylapatite, and hydroxylapatite. All of these minerals contain Ca^{2+}, which is derived from limestone bedrock in contact with the bat guano.

Phosphate minerals are not usually recognizable because they form mainly as indistinct fine-grained powders within or near bat guano. However, rarely phosphates can also form as dripstone or flowstone deposits, or they can be associated with ore deposits. An unusual occurrence of brushite ($CaHPO_4 \cdot 2H_2O$) exists in Kartchner Caverns, Arizona (United States): a 2 m long, ivory-yellow deposit of moonmilk-flowstone can be seen on the side of a breakdown block covered on top with fresh bat guano.

Silicates: Silicate minerals in caves fall into three categories: opal / quartz; clay minerals; and ore minerals. Opal ($SiO_2 \cdot nH_2O$) is the most common non-clay silicate mineral in caves, and can be locally abundant in lava tubes where basaltic rock supplies the silica. Clay minerals (e.g. montmorillonite) usually reconstitute within the cave from detrital (residual) sediment washed into the cave by surface streams.

Some silicate minerals are useful indicators of precipitation conditions within a cave. Opal forms under low-temperature conditions, whereas quartz (SiO_2) forms under high-temperature conditions (hydrothermal waters). Endellite—a colourful, waxy, hydrated halloysite; $Al_2Si_2O_5(OH)_4 \cdot 2H_2O$—has been used as an indicator mineral for sulfuric-acid cave development in the Guadalupe Mountains, New Mexico, United States. Endellite is a mineral that is known to form in low pH, sulfuric-acid rich waters.

Sulfates: The second most important class of cave minerals, after the carbonates, is the sulfates. The third most common cave mineral, gypsum ($CaSO_4 \cdot 2H_2O$), is in this class, as are two other fairly common minerals, epsomite ($MgSO_4 \cdot 7H_2O$) and mirabilite ($NaSO_4 \cdot 10H_2O$). Like the carbonates, a separate entry has been written about this important class: Speleothems: Evaporite.

Sulfate cave minerals fall into five general categories depending on origin: (1) those that form in a "normal" limestone or gypsum rock setting (e.g. gypsum); (2) those that form in lava tubes (e.g. thenardite); (3) those that derive from bat guano (e.g. arcanite); (4) those that derive from ore bodies (e.g. chalcantite); and (5) those that are associated with fumarole activity (e.g. voltaite). Because of this variety of possible origins, more different sulfate minerals are known to exist in caves than any other crystal class (Hill & Forti, 1997, list 64 different sulfate cave minerals). Gypsum can occur in all five settings.

Sulfate speleothems can assume many of the same forms as carbonate speleothems (e.g. stalactites, stalagmites), but more frequently they display a fibrous habit (e.g. crusts, flowers, cotton). A spectacular form of gypsum is "chandelier" stalactites, such as those that occur in Lechuguilla Cave, New Mexico (Figure 2). Epsomite and mirabilite (like the nitrates) are deliquescent and effloresce only under relatively low-humidity cave conditions.

Vanadates: Like the arsenates, vanadates are rare in caves and occur mostly in ore settings. However, the vanadate mineral

Minerals in Caves: Figure 2. Crystal chandelier, Lechuguilla Cave. (Photo by David Harris)

tyuyamunite $[Ca(UO_2)_2(VO_4)_2 \cdot nH_2O]$ can be an indicator mineral for sulfuric-acid cave development. Hydrogen sulfide (which oxidizes to sulfuric acid) precipitates uranium (and vanadium) in groundwater. When these constituents are remobilized in the vadose zone, bright-yellow tyuyamunite crystals form as crusts, such as in Carlsbad Cavern, New Mexico, United States.

Many cave minerals are rare or fragile. Others may be unique to a single cave or location within a cave. Therefore, the collection of cave minerals and speleothems is almost never justified, and even scientific sampling should be kept to a minimum or avoided if sampling will cause the complete destruction of a unique mineral occurrence. Cave minerals belong in, and should stay in, caves!

CAROL A. HILL AND PAOLO FORTI

See also **Speleothems**

Works Cited

Gaines, R.V. *et al.* 1997. *Dana's New Mineralogy: The System of Mineralogy of James Dwight*, 8th edition, Chichester and New York: Wiley, 3 vols

Hill, C.A. & Forti, P. 1997. *Cave Minerals of the World*, 2nd edition, Huntsville, Alabama: National Speleological Society

Further Reading

Hill, C.A. & Forti, P. 1997. *Cave Minerals of the World*, 2nd edition, Huntsville, Alabama: National Speleological Society
The authors recommend their book for further reading because it is the only extensive book available on this subject. Colour photographs of many cave minerals and of each major speleothem type and subtype are included, as are over 5000 references to cave minerals worldwide.

MINERAL DEPOSITS IN KARST

A variety of different mineral deposits are found in the infilling of paleokarst and neokarst. Karst-related mineral placer deposits are divided into four genetic types: sedimentary deposits, weathering deposits, hydrothermal deposits, and fluids. The metallic ore deposits include gold, tin, wolframite, columbite, and tantalite placers, bauxites (see separate entry), iron, manganese, lead-zinc, copper, uranium, mercury, and vanadium. Non-metallic ore deposits include phosphorites, diamond, ruby, sapphire, and spinel placers, barite, antimony, fluorite, onyx marble, clays, clay pigments, sands, coal, and peat. The fluid deposits are oil, gas, water, and brines.

Karst mineral deposits can also be divided by the source of the useful components as follows: (1) autochthonous deposits, where the components are formed due to destruction of the host rocks; (2) allochthonous deposits, where the components were brought from outside; and (3) parautochthonous deposits, where the components are foreign to the host rocks, but the same as their close surroundings (Zuffardi, 1976). The karst-related placers are divided into alluvial, deluvial, proluvial, and eluvial types by their genesis. Karstified surfaces of carbonate rocks act as traps for heavy minerals and also protect the placers from erosional destruction. Karst depressions and pockets create extremely favourable conditions for weathering of ore-bearing rocks, and for extraction of valuable components.

Diamond placers in a karst environment are known from the Siberian Platform, Russia. Quaternary alluvial placer deposits, occurring on Cambrian karstified dolomites with pockets and depressions, have been mined since 1997 in the Ebelyakh River basin. Neogene and Cretaceous alluvial placers in paleokarst depressions are also known there. A collapse karst placer, filling a deep karst hollow in Silurian and Ordovician carbonate rocks near the Aikhal kimberlite pipe, was exhausted in the 1960s. Buried middle Carboniferous alluvial placers in paleokarst depressions in Ordovician carbonate rocks have been explored in the Morkoka River basin. Buried alluvial and residual placers of upper Triassic–lower Jurassic Age were prospected in paleokarst depressions in lower Ordovician carbonate rocks near the Nakynskaya and Botuobynskaya kimberlite pipes. Quaternary alluvial diamond placers in karst erosional depressions have been exploited during the last 40 years in the Koyva and Vishera river basins in the Ural Mountains, Russia. They formed by rewashing of ancient placers.

Deluvial diamond karst-associated placers were mined in Bakwanga, Congo Republic. These placers formed due to destruction of the nearby kimberlite pipes. Diamondiferous material was redeposited downslope and accumulated in karst dolines up to 80 m deep. Breccia 20–80 m thick, containing 60% sandstones, dolomites, and weathered kimberlites, filled the lowest parts of the dolines. Deluvial karst diamond placer was also mined in the Lichtenburg area, South Africa. The placer developed due to rewashing of Precambrian and Cretaceous diamondiferous deposits. Diamond-bearing deluvium had accumulated up to 45 m deep within the karst depressions.

Mineral Deposits in Karst: Working a ruby mine in Mogok, Burma. The miners are digging the sediment from between the limestone pinnacles of rockhead that they have exposed, as the best stones are usually found at the bottom of the fissures. (Photo by Tony Waltham)

There are three types of gold placers in karst: (1) polygenetic placers in karst erosional depressions; (2) alluvial placers, which occur on karstified carbonate rock with rugged relief; and (3) eluvial chemical weathering crusts, which develop in contact zones of sulfidized ore occurrences and carbonate rocks (Kropachev, 1972). The first and the last are numerous in the Urals, Russia. Gold-bearing layers occur in karst erosional depressions, which are a few kilometres in length and 50–60 m deep. Alluvial placers have been mined recently in the Tommot, Yakokit, Seligdar, and Maly Yllymakh river basins (Yakutia), and in the Birusa River basin, East Sayan (Russia).

Tin and wolframite alluvial karst placers are mined in Malaysia and Laos in the Nam-Phatene River basin, and in Vietnam at Tin-Tac polje. Kinta Valley within the Malayan peninsula is the world's most productive area. It lies between the Main and Kledang ranges, which consist of granite masses, while the bedrock of the valley floor is generally limestone. The tin-bearing alluvium lies on karstified limestone with a highly irregular surface of pinnacles and depressions. These act as ripples to concentrate the cassiterite brought down by rivers. The alluvium varies from 1–2 m to over 60 m thick.

Sapphire and ruby are extracted from karst placers in Sri Lanka (the Pelmadulla deposit). Ruby and spinel are obtained from alluvial and deluvial karst-associated placers in Burma (Myanmar, the Mogok deposit; see Figure).

It is thought that the Mississippi Valley-type deposits, the world's largest productive lead-zinc resources, are of hydrothermal karst genesis (Dźulyński & Sass-Gutkiewicz, 1989). They are known in Upper Silesia (Poland), in the southern Appalachians (US), and in the eastern Alps (see Sulfide Minerals in Karst).

Iron ore deposits in karst and paleokarst depressions are numerous. They accumulated as weathering residua. Currently mined deposits include the Qui-Xa (Vietnam), Alapaevskoye and Akkermanovskoye (Urals, Russia), and deposits in northeast Bavaria (Germany). Nickel deposits were formed in the karst pockets, developed along the contact between carbonate and ultrabasic rocks in the Urals, Russia. Nickel, iron, manganese, and chromium, the weathering products of ultrabasic rocks, migrate into karst pockets where they form compound precipitates. Low-temperature hydrothermal barite deposits in karst cavities have been exhausted in the Tyuya Muyun (Kirgizstan). Residual barite accumulations in karst depressions are known from Strawczynek and the Holy Cross Mountains (Poland), and from Missouri (US). The Missouri deposit yielded barite from weathered barite veins and leached limestones. Paleokarst-related uranium deposits have been exploited in Tyuya Muyun (Kirgizstan), Orphan Mine (Arizona, US), Bakouma (Central African Republic), Le Vigan (France), and in the Vise region (Belgium).

Phosphorites of infiltrational metasomatic genesis often fill large depressions and pockets in karstified rocks. Numerous deposits were prospected in East Sayan (Russia), in Tennessee and Florida (US), and in Liège (Belgium). Phosphatic cover deposits exist in the high carbonate islands (such as Nairu) in the Pacific Ocean. The ores are trapped on and within karst.

Approximately one-half of the world production and reserves of reservoir oil and gas are contained in carbonate rocks (see Hydrocarbons in Karst). Remarkable resources of hydromineral raw materials, connected with deep salt karst, are accumulated within the Siberian Platform. In addition to common macroelements and microelements such as strontium, rubidium, caesium, and lithium, the unique concentrated brines of deep horizons contain rare elements (yttrium, tantalum, niobium, europium, cerium, thorium, zirconium, molybdenum, and tungsten).

Andrei G. Filippov

Works Cited

Dźulyński, S. & Sass-Gutkiewicz, M. 1989. Pb–Zn ores. In *Paleokarst: A Systematic and Regional Review*, edited by P. Bosák, D.C. Ford, J. Glazek & J. Horácek, Amsterdam and New York: Elsevier

Kropachev, A.M. 1972. K geneticheskoy klassifikatsii otlozheniy v karstovykh polostyakh i depressiyakh [Genetic classification of deposits in karst cavities and depressions]. In *Karst Sredney Azii i gornykh stran* [Karst of Central Asia and mountainous regions], Tashkent: Uzbekgidrogeologiya

Zuffardi, P. 1976. Karst and economic mineral deposits. In *Handbook of Strata-Bound and Stratiform Ore Deposits*, vol. 5, edited by K.H. Wolf, Amsterdam and New York: Elsevier

Further Reading

Amstutz, G.C., El Goresy, A., Frenzel, G., Kluth, C., Moh, G., Wauschkuhn, A. & Zimmermann, R.A. (editors) 1982. *Ore Genesis. The State of the Art*, Berlin and New York: Springer

Bosák, P., Ford, D.C., Glazek, J. & Horácek, I. (editors) 1989. *Paleokarst: A Systematic and Regional Review*, Amsterdam and New York: Elsevier

Kutyrev, E.I. & Lyakhnitsky, U.S. 1982. Rol' karsta v formirovanii mestorozheniy medi, svintsa, zinka, sur'my, rtuti i flyuorita [Role of karst in the origin of lead, zinc, antimony, mercury, and fluorite deposits]. *Litologiya i Poleznie Iskopaemie*, 2: 54–69

Laznicka, P. 1985. *Empirical Metallogeny: Depositional Environments, Lithologic Associations and Metallic Ores*, vol. 1: *Phanerozoic Environments, Associations and Deposits*, Amsterdam and New York: Elsevier

Prokopchuk, B.I., Metelkina, M.P. & Shofman, I.L. 1985. *Drevniy karst i ego rossypnaya minerageniya* [Ancient karst and its placer metallogeny], Moscow: Nauka

Quinlan, J.F. 1972. Karst-related mineral deposits and possible criteria for the recognition of paleokarsts: A review of preservable characteristics of Holocene and older karst terraines. In *Proceedings of the 24th International Geological Congress, Montreal*, vol. 6

Tsykin, R.A. 1985. *Otlozheniya i poleznye iskopaemye karsta* [Deposits and raw materials in karst], Novosibirsk: Nauka

MOLLUSCA

Subterranean habitats worldwide are permanently inhabited by several hundred molluscan species. Most terrestrial species belong to the Gastropoda: Pulmonata (snails and slugs), whereas aquatic species primarily belong to the operculate Gastropoda: Prosobranchia (freshwater snails), and a few to the Bivalvia (mussels).

Approximatively 100 species of terrestrial troglophilic and troglobitic snails are known. The entrance zone of caves, being mostly dark, cool, wet, and rich in organic debris, is a common habitat for terrestrial molluscs inhabiting similar surface habitats in (troglophiles). In the deeper cave zones we find some mollusc species unique to hypogean habitats (troglobites). These may be relicts from early climatic changes, for example some molluscs known from European caves are survivors of a subtropical fauna, still found today in the subtropics and tropics but extinct in surface habitats in Europe; other molluscs survived the Quaternary glaciations beneath ice-free "massifs de refuge". Often these relicts possess an adaptation to subterranean life, for example no or little pigmentation, loss of eyes, a specialized radula for carnivorous feeding, or a respiratory system allowing an amphibious life. The diet of troglobitic species may contain a higher proportion of organic matter from invertebrates and their secretions and a lower proportion of vegetal organic matter than that of surface-dwelling species.

More than 100 species of snails live in subterranean waters, mostly in phreatic aquifers. Living specimens are occasionally found in cave streams, but usually they are only accessible at wells, in groundwater outlets (artesian and karstic springs), by drilling and pumping, or when trapped in water pools deep in caves after violent flooding. Aquatic subterranean snails (stygobites) are depigmented, eyeless, and particularly small. They normally live in fresh water, although certain species are also known from brackish underground waters (such as Movile Cave, Romania). Certain aquatic snails that live only in springs (crenobionts) can penetrate and live underground; other aquatic snails live in interstitial habitats.

The first cavernicolous mollusc, *Carychium spelaeum* (now known as *Zospeum*), was discovered in Adelsberger Grotte (now Postojnska Jama) in 1839. *Zospeum schaufussi* from caves in the Spanish Pyrenees and *Carychium stygium* from Mammoth Cave, Kentucky, were described in 1862 and 1897 respectively. Many stygobitic genera and species were discovered during the 19th century in alluvial sediments, in wells (for example, *Avenionia* in 1882 in France), and in springs. A few discoveries were made inside caves: e.g. *Hydrobia quenstedti* in 1873 (southern Germany), *Vitrella tschapecki* in 1878 (southern Austria), and *Lartetia virei* in 1903 (northern Italy) (these are now all incorporated in the genus *Bythiospeum*). In the early 1900s, zoologists such as Wagner, Kuscer, and Sturany described many new species and genera from caves and karstic springs in the Dinaric karst (see Dinaric Karst: Biospeleology). In the second half of the 20th century, numerous subterranean snails were discovered worldwide in subterranean waters as well as in caves (in Europe, North and South America, New Zealand, Australia, Japan, China, Caucasus, and North Africa).

The correct taxonomic classification of molluscs needs knowledge of their anatomy. The shells, particularly of the aquatic Hydrobiidae, often show great variability in shape, so efforts were made to find living subterranean snails in order to examine their anatomical structures. On this basis, our knowledge about taxonomic classification and biogeographical relationships has greatly improved.

The majority of subterranean snails are found in the western Palearctic zone, particularly in the Balkan Peninsula. The tropical and subtropical zones have fewer troglobitic and stygobitic snails. Some genera living in hypogean habitats in the Balkan Peninsula represent relicts of a molluscan fauna which has otherwise been extinct in Europe since the Tertiary. Examples are *Pholeoteras euthrix* in Herzegovina and Croatia, and *Pholeoteras zilchi* in Greece (Prosobranchia: Cyclophoridae); *Hydrocena cattaroensis* (Prosobranchia: Hydrocenidae) in Montenegro, Croatia, and Albania; *Sciocochlea collasi* and *Sciocochlea nordsiecki* (Pulmonata: Clausiliidae) in Greece; and *Congeria kusceri* (Bivalvia: Dreissenidae) from cave streams in Croatia.

Most terrestrial snails found in tropical and subtropical caves belong to the following Prosobranch families: Hydrocenidae (e.g. *Georissa papuana* from Papua New Guinea, *Georissa pangianensis* from Sumatra); Assimineidae (e.g. *Cavernacmella kuzuuensis* from Japan, *Anaglyphula minutissima* from Sumatra); and Cyclophoridae (e.g. *Opisthostoma mirabilis* from Borneo). Most terrestrial subterranean snails in the western Palearctic are Pulmonata, belonging primarily to the Zonitidae family (genera: *Oxychilus*, *Spelaeopatula*, *Gyralina*, *Lindbergia*, and others). Monotypic troglobitic species are *Meledella werneri* (Mljet Island, Croatia), *Troglaegopis mosorensis* (Croatia), and *Troglovitrea argintarui* (Romania). A few single genera or monotypic species of Pupillidae, Orculidae (e.g. *Speleodentorcula beroni* from Greece), Clausiliidae, and Cochlicopidae are restricted to the caves of the Balkan Peninsula, particularly the Dinaric karst. *Cryptazeca spelaea* and *C. elongata* (Subulinidae) are known only from Cantabrian caves. In the Carychiidae family, cavernicolous snails belong to the genus *Zospeum*—all species being blind (*Z. spelaeum* and related species in the Dinaric karst, northeastern Italy, and southern Austria; *Z. schaufussi* and related species in the French–Spanish Pyrenees). *Carychium stygium* and *C. exile* are widespread in subterranean habitats throughout the southeastern United States. An interesting representative of a troglobitic slug is *Troglolestes sokolovi* (Trigonochlamydia) from Caucasian caves.

Aquatic snails belong mostly to the Prosobranch rissoacean family Hydrobiidae. Aquatic snails restricted to phreatic habitats are widespread. Species and genera found and described from cave waters (streams or pools) include, for example, *Catapyrgus spelaeus* and *Opacuincola coeca* (New Zealand); *Pseudotricula eberhardi* (Tasmania); *Selmistomia beroni* (Papua New Guinea); *Akiyoshia uenoi* (Japan); *Antroselates spiralis*, *Antrobia culveri*, and *Holsingeria unthanksensis* (all from the United States); *Andesipyrgus sketi* (Colombia and Ecuador); *Bythiospeum [Paladilhiopsis] grobbeni* (Slovenia); *Heleobia [Semisalsa] dobrogica* (Movile Cave, Romania); *Alzoniella feneriensis* (Italy); *Paladilhia umbilicata*, *Palacanthilhiopsis vervierii*, *Moitessieria lescherae*, *Alzoniella pyrenaica*, and *Palaospeum bessoni* (all from France).

The majority of stygobitic snails have been discovered in karstic springs or by drilling wells, for example the remarkable molluscan fauna of Texan aquifers (*Phreatodrobia*, *Balconorbis*, *Phreatoceras*, *Stygopyrgus*, and *Texapyrgus* genera). Recent new discoveries of stygobites include, *Alzoniella cornucopia* and *Plagigeyeria stochi* (Italy); *Sardopaladilhia plagigeyerica* and *Sardohoratia sulcata* (Sardinia); *Kerkia brezicensis* (Slovenia); *Atebbania bernasconii* and *Heideella makhfamanensis* (southern Morocco).

RENO BERNASCONI

Further Reading

Bernasconi, R. 1995. Two new cave prosobranch snails from Papua New Guinea: *Selmistomia beroni* n.gen. n.sp. (Caenogastropoda: Hydrobiidae) and *Georissa papuana* n.sp. (Archaeogastropoda: Hydrocenidae) (Zoological results of the British Speleological Expedition to Papua New Guinea 1975). *Révue Suisse de Zoologie*, 102(2): 373–86

Bodon, M., Cianfanelli, S., Manganelli, G., Girardi, H. & Giusti, F. 2000. The genus *Avenionia* Nicolas, 1882, redefined (Gastropoda, Prosobranchia, Hydrobiidae). *Basteria*, 64: 187–98

Gittenberger, E. & Maassen, W.J.M. 1980. *Hydrocena (H.) cattaroensis* (L. Pfeiffer, 1841) (Prosobranchia Neritacea) und ihre Lebensbedingungen [*H. cattaroensis* and its conditions of life]. *Basteria*, 44: 9–10

Haase, M. 1995. The stygobiont genus *Bythiospeum* in Austria: A basic revision and anatomical description of *B. cf. geyeri* from Vienna (Caenogastropoda: Hydrobiidae). *American Malacological Bulletin*, 11(2): 123–37

Hershler, R. & Holsinger, J.R. 1990. Zoogeography of North American hydrobiid cavesnails. *Stygology*, 5(1): 5–16

Hershler, R. & Thompson, F.G. 1992. A review of the aquatic gastropod subfamily Cochliopinae (Prosobranchia: Hydrobiidae). *Malacological Review, Supplement* 5: 1–140

Manganelli, G., Bodon, M. & Giusti, F. 1995. The taxonomic status of *Lartetia cornucopia* De Stefani, 1880 (Gastropoda, Prosobranchia, Hydrobiidae). *Journal of Molluscan Studies*, 61: 173–84

Morton, B., Velkovrh, F. & Sket, B. 1998. Biology and anatomy of the "living fossil" *Congeria kusceri* (Bivalvia: Dreissenidae) from subterranean rivers and caves in the Dinaric karst of the former Yugoslavia. *Journal of Zoology*, 245: 147–74

Saul, M. 1966. Shell collecting in the limestone caves of Borneo. *Conchologist's Newsletter, London*, 19: 128–30

MONA, PUERTO RICO

Isla de Mona is a carbonate island some 55 km² in area, located midway between Puerto Rico and Hispaniola. Part of the Commonwealth of Puerto Rico, it is a National Park and the largest uninhabited island in the Caribbean. The island has been tectonically uplifted to a maximum elevation of 90 m, and consists of the Mio-Pliocene Lirio Limestone and underlying Mona Dolomite, with a small Pleistocene limestone coastal plain along the southwest coast. Sheer cliffs plunge 40–80 m from a flat "mesata" or plateau surface directly to the sea on all sides, except where the Pleistocene coastal plain is located. The cliffs contain numerous cave openings, which are concentrated at the contact of the Lirio Limestone and the Mona Dolomite (Figure 1). The caves are the world's largest known examples of "flank-margin caves", a type of cave that develops in carbonate coastlines as a result of sea-water and freshwater mixing inside the coastline (see also Speleogenesis: Coastal and Oceanic Settings).

The interior of the island contains very few horizontal caves. There are, however, numerous pits with a depth range of 5–20 m, in concentrations of several hundred per square kilometre (Frank *et al.*, 1998). These pits commonly lead to a single chamber, but no extensive horizontal cave. The mesata contains a few large closed depressions. The central depression, Bajura de los Cerezos, shows evidence that sinking streams occur during high rainfall. Other depressions, such as Cuevas del Centro and Los Corrales de los Indios, contain short caves in the walls of the depression. A few large collapse chambers are known on the mesata, but none leads to significant lateral cave passages.

Isla de Mona's flank-margin caves are spectacular. They are extensive, with spacious, well-decorated chambers and complex passage configurations. The largest system, Sistema del Faro, is located at the southeastern end of the island, and links the caves of Lirio, Faro, and las Losetas into a continuous series of passages with 19.1 km of survey (Figure 2). The cave has a very maze-like quality, and numerous entrances exist where it has been intersected by the retreat of the cliffs. Because the cave developed by mixing of sea water, and fresh water from within the island, the cave is extensive parallel to the coast, over a range of 2.5 km, but penetrates into the island a maximum distance of only 220 m. The cave has many large chambers adjacent to each other, such that there are numerous connections between them. This complexity gives the cave its maze-like character, and 19.1 km surveyed does not mean 19.1 km of continuous passage, because surveying the numerous chambers has required a large number of survey lines.

North from Sistema del Faro along the east coast, and west along the north coast, the caves become smaller and less frequent, as the north coast is the more elevated side of the island, and the Lirio Limestone has been thinned by erosion. Where small patches of Lirio Limestone remain, the entire patch is commonly underlain by a single cave, such as Cueva de Frio. West from Sistema del Faro along the south coast, many caves exist where

Mona, Puerto Rico: Figure 1. Southeast corner of Isla de Mona. Lirio limestone is dark upper layer, Mona dolomite is the lighter lower layer. Note cave entrances at the contact, cave collapse entrances on the mesata surface, and the lighthouse in the distance for scale. All visible cave entrances are part of the 19.1 km Sistema del Faro.

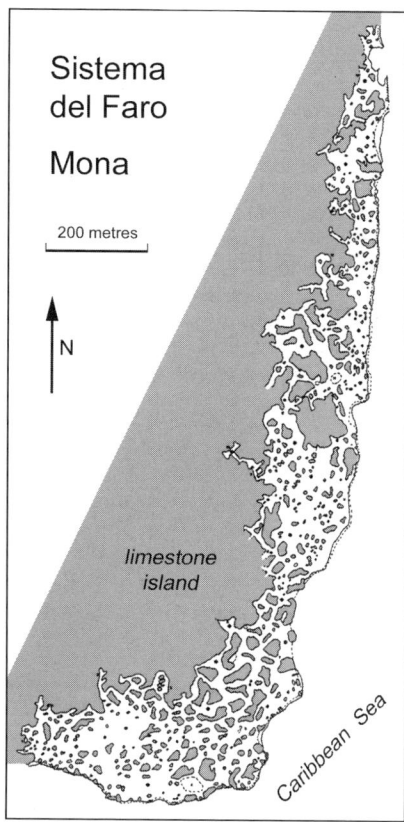

Mona, Puerto Rico: Figure 2. Outline map of Sistema del Faro, the largest coastal cave on Isla de Mona. Based on a survey made in 1998–2000, from a map drawn up by Marc Ohms. The cave is laterally extensive but penetrates inland no more than 220 m, which is consistent with the flank-margin model.

the Lirio Limestone is thicker. Cueva de los Parajos, 2 km southwest of Sistema del Faro, is another extensive cave with over 10 km of passages and several large chambers (50 m across and 15 m high). Numerous caves with survey lengths of roughly 1 km or more are known along the southwest coast, where the coastal plain makes access to the cliffs possible. Where the island cliffs swing north along the west coast, there are a number of large caves, including Cueva del Diamante, Cueva del Esqueleto, and Ceuva del Capitan, each with several kilometres of passages.

The flank-margin caves commonly have several levels, which are believed to reflect different sea-level positions while the cave was forming. The caves also have many subaerial calcite speleothems, which show evidence of being subjected to phreatic dissolution. Again, changes in sea level, and therefore the position of the water table, are thought to have caused this dissolutional attack on the speleothems. Paleomagnetic analysis has shown one cave to be at least 1.6 million years old. The large size of the Isla de Mona flank-margin caves is due to fact that they were formed initially before the onset of glacial changes in sea level at the start of the Pleistocene (Frank *et al.*, 1998). Therefore, sea level stayed constant for a long period of time, and the caves grew very large. Then the caves were uplifted and drained by tectonic activity, so that subaerial speleothems could grow. Glacial sea-level changes began, which allowed water to re-enter the caves, enlarging them and also dissolving the speleothems. Tectonic uplift of the island, and its caves, has continued, so that today most of the caves are far above any interaction with the sea, regardless of how sea-level changes due to glaciation–deglaciation cycles. The caves are dry and dusty much of the year, but during major rainfall events, such as tropical storms, the caves gather large amounts of water, which drips from the ceilings.

The caves of Isla de Mona were mined for guano fertilizer from 1887 to 1927. Most of the cave earth was removed, and elaborate trails and causeways were built by the miners to move the earth to the cave entrances. Ore carts, iron track, and other relicts from mining are common in some of the caves. It is interesting to note that the miners took care to avoid damaging the cave speleothem formations, even though it meant more work to lay the trails. The mining operation resulted in clear-cutting of the native forest for fuel and building material, upsetting the hydrological balance of the island. Visited by Columbus in 1493 for water, and listed on old British naval charts as a watering locality, visitors must now bring all their fresh water with them.

JOHN MYLROIE

See also **Blue Holes of Bahamas; Caribbean Islands; Speleogenesis: Coastal and Oceanic Settings**

Works Cited

Frank, E.F., Mylroie, J., Troester, J., Alexander, E.C., & Carew, J.L. 1998. Karst development and speleogenesis, Isla de Mona, Puerto Rico. *Journal of Cave and Karst Studies*, 60(2): 73–83

Further Reading

Briggs, R.P. & Seiders, V.M. 1972. *Geologic Map of the Isla de Mona Quadrangle, Puerto Rico*, Reston, Virginia: US Geological Survey (US Geological Survey Miscellaneous Geological Investigations, Map 1–718)

Frank, E.F. 1998. History of the guano mining industry, Isla de Mona, Puerto Rico, *Journal of Cave and Karst Studies*, 60(2): 121–25

Gonzalez, L.A., Ruiz, H.M., Taggart, B.E., Budd, A.F. & Monell, V. 1997. Geology of Isla de Mona, Puerto Rico. In *Geology and Hydrogeology of Carbonate Islands*, edited by H.L. Vacher & T.M. Quinn, Amsterdam and New York: Elsevier

Journal of Cave and Karst Studies, 60(2): 1998 is a special issue dedicated to the caves of Isla de Mona

Mylroie, J.E., Carew, J.L., Frank, E.F., Panuska, B.C., Taggart, B.E., Troester, J.W. & Carrasquillo, R. 1995. Comparison of flank margin cave development: San Salvador Island, Bahamas and Isla de Mona, Puerto Rico. In *Proceedings of the Seventh Symposium on the Geology of the Bahamas*, edited by M.R. Boardman, San Salvador Island: Bahamian Field Station

MONITORING

Human use and interaction with cave and karst systems can readily lead to adverse impacts on the quality and integrity of both the environment and visitor experience. Monitoring involves the systematic sampling, measurement, and recording of characteristics reflecting environmental and social conditions. It is a management tool that identifies actual and emerging problems related to impacts. Monitoring also assists planning and priority setting, and provides a means to evaluate the effectiveness of management actions in preventing or reducing impacts. This entry briefly reviews the monitoring process and its applications in cave and karst management. The main steps involved in developing a monitoring programme are given in the Table.

In cave and karst systems, impacts can be generated internally, or on adjacent non-karst lands, and it is difficult to predict the pathways that impacts will follow. This is due to the complex and variable conduit system linking surface and subterranean areas. The integrity of the karst system depends on the interrelationship among bedrock, soils, water, gases, landforms and biota, and damage to any one of these can affect the entire system. Hence, karst managers need to develop a holistic understanding of this web of linkages and incorporate this knowledge into the design of monitoring programmes.

Traditionally, monitoring programmes in caves and karst have focused on accumulated visitor impacts in show caves, and in the surface areas immediately adjacent. Accordingly, most karst monitoring programmes have been based on visitor impact management approaches. These include the Recreational Opportunity Spectrum (Clark & Stankey, 1979), Visitor Activities Management Process (Parks Canada, 1985), Limits of Acceptable Change (Stankey *et al.*, 1985), Visitor Impact Management (VIM: Kuss, Graefe & Vaske, 1990), and Visitor Experience and Resource Protection (Belnap *et al.*, 1997). Monitoring programmes, using these approaches, describe environmental and social conditions, identify visitor impacts, measure and assess condition, and develop management actions to reduce impacts.

At Jenolan Caves, New South Wales (Australia), the VIM approach has been modified into the Social and Environmental Monitoring (SEM) programme (Hamilton-Smith, 2000). This programme recently moved away from a purely visitor impact approach, to consider any process that may cause impact. This is more akin to sustainability models that consider the resource as a whole rather than starting from the issue of specific impacts. The SEM programme now includes consideration of impacts and issues not directly related to visitors, such as exotic species invasion or wildfire. Given that karst systems are highly interconnected, a holistic approach is a more suitable basis for future monitoring programmes than impact management approaches.

In keeping with a holistic approach, monitoring of both show caves and wild caves must be considered. Unfortunately, impacts in wild caves are poorly documented as they are difficult to quantify, consequently monitoring efforts have been limited. Recent research in New Zealand caves successfully developed and tested techniques for monitoring indicators of impacts, such as sediment erosion, compaction, and tracking (Bunting, 1998). A higher priority for similar research will ensure that wild caves are included in future monitoring programmes.

The identification of indicators is an important consideration for karst monitoring programmes (see Table). For any given issue, more than one indicator may be required, and indicators developed for the same issue in one karst area will not necessarily be appropriate for other karst areas. Some indicators, such as variables relating to air and water quality, are well established and relatively easy to measure.

Identifying indicators for other aspects of the karst system are more problematic. For example, selection of indicator species in subterranean invertebrate faunas is very difficult, as they are

Monitoring: Summary of main steps in the monitoring process.

Step	Notes
1. Identify issues and/or objectives	Issues may be identified from a range of sources including plans of management, legislation, management experience, or stakeholder concerns. This step enables the monitoring programme to stay focused.
2. Identify indicators	Key indicators are specific, measurable environmental, or social variables that reflect overall condition. Indicators should be relevant and specific to prominent environmental or social issues, and sensitive to changes in condition.
3. Establish standards	A standard is the minimum acceptable condition for each indicator. Setting of standards is intrinsically subjective, and their selection needs to consider the degree of rigour and reliability required for assessing condition.
4. Develop monitoring plan	Planning should include identifying the location, methods, frequency, and timing of measurements, analysis and reporting procedures, and management strategies for dealing with impacts.
5. Compare existing conditions with standards	Standards provide a baseline against which existing conditions can be compared. When indicators approach or exceed a standard, this signals a problem requiring management attention.
6. Identify causes of impact	Causes of impact may be complicated and should not rely on guesswork, but should be based on monitoring results to avoid making decisions based on false assumptions.
7. Develop management strategies	The collection of monitoring data is meaningless if data cannot be translated into actions to prevent violation of standards or to address indicators that are out of compliance with standards.
8. Continue ongoing monitoring and regularly review	Monitoring programmes need to be dynamic. Analysis of results may require readjustments in the methods, frequency, and location of data collection, or the revision or rejection of indicators.

generally cryptic, patchily distributed, and often occur in low numbers. To address this problem, researchers at Mammoth Cave, United States, developed an Index of Biological Integrity (IBI) for monitoring purposes (Poulson, 1992). This index combines habitat data with population and community data, using information on all species and habitats present. Instead of assessing the response of individual indicator species, the IBI develops cave community "signatures" for different classes of impacts. This example demonstrates that detailed research and innovative thinking are required before indicators can be selected for use in karst monitoring.

Conditions in karst can change rapidly, so the timing and frequency of data collection is an important consideration, and will vary for each indicator. Diurnal, seasonal, or annual changes in indicators occur regardless of impacts, and baseline data should be collected to determine these patterns. However, event-based monitoring should also be undertaken, as impacts related to events may not be detected when regular sampling intervals are used. Where events, such as periods of high visitation, are predictable, continuous sampling before, during, and after an event will give the most reliable information on the movement of impacts through a karst system. Unpredictable events, such as chemical spills, should be continuously monitored immediately after the event has occurred.

Current environmental and social conditions in a karst area are determined by comparing indicators against standards. Some standards have been formalized as part of best-practice management, or may be set within legislation. For example, national and international health authorities have established standards for acceptable levels of many pollutants, while occupational health and safety legislation in many countries establishes standards for acceptable exposure levels of workers.

Published research and expert opinion may also be the source of standards for a wide range of issues, including visitor perceptions of crowding or the quality of habitat required to sustain endemic or endangered species. Such standards may only apply to the country or region in which the research occurred, and may need to be modified to suit local conditions. For many site-specific issues, established standards may not be appropriate or may be lacking, and on-site research, coupled with management and stakeholder input, may be required to develop appropriate standards.

Monitoring programmes are now well established in many cave and karst areas across the world. The information provided by these programmes is an important tool for identifying impacts and developing management actions to maintain social and environmental conditions. Given the important social, economic, scientific, and environmental values of karst areas and their susceptibility to impact, continued improvement of the monitoring process should remain a high priority well into the future.

MIA THURGATE

See also **Tourist Caves entries**

Works Cited

Belnap, J., Freimund, W.A., Hammett, J., Harris, J., Hof, M., Johnson, G., Lime, D.W., Manning, R.E., McCool, S.F. & Rees, M. 1997. *VERP: The Visitor Experience and Resource Protection Framework: A Handbook for Planners and Managers*, Denver: National Park Service

Bunting, B. 1998. The physical impacts of recreational users in caves: Methods currently in use for assessing recreational impacts in two New Zealand caves. In *Cave and Karst Management in Australia XII*, edited by D.W. Smith, Carlton South, Victoria: Australasian Cave and Karst Management Association

Clark, R.N. & Stankey, G.H. 1979. *The Recreation Opportunity Spectrum: A Framework for Planning, Management and Research*, Portland, Oregon: Pacific Northwest Forest and Range Experiment Station (General Technical Report PNW–98)

Kuss, F.R., Graefe, A.R. & Vaske, J.J. 1990. *Visitor Impact Management*, vol. 2: *The Planning Framework*, Washington, DC: National Parks and Conservation Association

Hamilton-Smith, E. 2000. Managing for Environmental and Social Sustainability at Jenolan Caves, New South Wales, Australia. *Acta Geographica Szegedensis*, 36: 144–52

Parks Canada. 1985. *The Management Process for Visitor Activities*, Ottawa, Ontario: National Parks Directorate, Visitor Activities Branch

Poulson, T.L. 1992. Assessing groundwater quality in caves using indices of biological integrity. *Proceedings of the Third Conference on Hydrogeology, Ecology, Monitoring and Management of Ground Water in Karst Terranes*. Dublin, Ohio: National Ground Water Association

Stankey, G.H., Cole, D.N., Lucas, R.C., Petersen, M.E. & Frissell, S.S. 1985. *The Limits of Acceptable Change (LAC) System for Wilderness Planning*, Ogden, Utah: Intermountain Forest and Range Experiment Station (General Technical Report INT–176)

Further Reading

Acta Carsologica 31(1), 2002: 5–177 is a special issue in which papers from an international workshop on monitoring karst caves are published

Drew, D. & Hötzl, H. (editors). 1999. *Karst Hydrogeology and Human Activities: Impacts, Consequences and Implications*, Rotterdam: Balkema

Gillieson, D. 1996. *Caves: Processes, Development and Management*, Oxford and Cambridge, Massachusetts: Blackwell

Nilson, P. & Tayler, G. 1997. A comparative analysis of protected area planning and management frameworks. In *Proceedings—Limits of Acceptable Change and Related Planning Processes: Progress and Future Directions*, edited by S. McCool & D.N. Cole, Ogden, Utah: Rocky Mountain Research Station (General Technical Report INT-GTR–371)

Watson, J., Hamilton-Smith, E., Gillieson, D. & Kiernan, K. (editors) 1997. *Guidelines for Cave and Karst Protection*, Cambridge: International Union for Conservation of Nature and Natural Resources

Williams, P.W. (editor) 1993. *Karst Terrains: Environmental Changes, Human Impact*, Cremlingen, Germany: Catena

MORPHOLOGY OF CAVES

Caves are formed as three-dimensional features. Their morphology can be described by their plan patterns or vertical sections. The principal plan patterns that develop within a given phase (level) are angular or sinuous branchworks and reticulate or anastomosing mazes. The patterns become more complex where there are sequences of such passages at different levels, complexly interconnected, for example in Mammoth Cave, Kentucky. Ford (2000) outlined a conceptual model that describes four stages of the vertical development and morphology of caves according to the fractures available where the caves formed (see Speleogenesis: Unconfined). There can be many combinations of these cave types, and with development of the karst they can be further modified. The principal types of modification are vadose entrenchment, paragenesis (see separate entry), and bypass galleries. Entrenchment occurs in the upper parts of phreatic loops if the piezometric level lowers. Paragenetic passages develop when sediments protect the lower part of the passage by directing the flow and dissolution between the fill and the ceiling under phreatic conditions. If the entire ceiling is lifted, a paragenetic canyon develops. Bypasses form new pathways for water near obstructions or deep phreatic loops.

In the vadose zone, drawdown caves develop when the effective porosity of the karst increases, thus lowering the piezometric level. Early conduits drain towards the lowered water table forming drawdown caves. The original phreatic forms may still be preserved, but the majority of the passage has been shaped in the vadose conditions. Invasion vadose caves form when a new stream, or water collected within the rock, invades a cave that was formed in a previous phase of speleogenesis. They develop where the effective porosity is high or where the resistance in fissures is low. This is usually connected with young mountain systems. Most alpine caves are of this type, consisting of fluted shafts connected by short meandering canyons.

Passage Cross-Section Morphology

The morphology of cave passages is best described by their cross sections which result from the activity of dissolutional and mechanical erosion processes during the development of the cave system. Some cross sections can be explained by geological structures and the solubility of the rock, but most are polygenetic. The geological setting influences morphology by providing secondary porosity, rock strength, and susceptibility to dissolution and erosion (Lauritzen & Lundberg, 2000). Fractures and bedding plane surfaces or their intersections are the most important structures for the initiation and shaping of passages. Symmetric profiles develop along horizontal bedding planes or vertical joints; dipping strata or joints produce asymmetric cross sections. Rock strength influences the size of the gallery. Most carbonate rocks can support large cave galleries, but this is not the case with some evaporite rocks. In homogeneous rock, the hydrodynamic factors of the water flow determine the form of the passage. If the solubility varies through a stratigraphic column, less soluble parts or insoluble particles, nodules, or layers can protrude from the walls and become significant features in the cave profile.

Passage shape is influenced by the hydrologic zone in which it is situated and so by the flow properties. The profile may be phreatic (Figure 1) or vadose (Figure 2) or a combination of the two. Phreatic passages are, or were originally, completely filled with water such that dissolutional processes could act evenly on all cave walls. In these conditions passages with rounded cross sections develop. Elliptical passage profiles are common due to development along guiding bedding planes or joints. Passages in homogeneous rock or passages with changing flow conditions can have irregular profiles but they retain characteristic smooth circular forms. The profiles of phreatic passages can reach dimensions of several tens of metres. In the unsaturated (vadose) zone dissolution only occurs in that part of the cave that contains water. Passage enlargement is downwards and forms underground canyons. These canyons are often sinuous and like surface rivers form cave meanders. Vadose passages vary from small features to large underground canyons such as the 95 m high and 10–20 m wide canyon in the Škocjan Cave, Slovenia (see photo in Škocjan entry). Narrow vadose canyons have often developed at the bottom of phreatic passages. The characteristic combined profile with a phreatic tube in the upper part and a vadose canyon is called a "keyhole" profile. A combined profile can be the result of the draining of the phreatic zone due to a change in the external base level, the developing conductivity of the karst and the subsequent lowering of the piezometric level in it, or a drop in the quantity of the water in the passage.

Vadose shafts are vertical or nearly vertical passages formed by falling water originating from sinks on the surface, from the

Morphology of Caves: Figure 1. Peak Cavern, Derbyshire, England. The phreatic tube has been drained by a large vadose canyon into which the present misfit stream drains. (Photo by John Gunn)

Morphology of Caves: Figure 2. One of the larger vadose canyons in the Yorkshire Dales caves, Hensler's Main Drain in the Gaping Gill System. (Photo by Tony Waltham)

epikarst, or at drops from one cave level to another. Most of the falling water does not touch the walls, corroding mostly at the bottom and deepening the shaft. Shafts often show a lenticular cross section in their upper parts and a more rounded cross section at the bottom due to the spray of the water (see Figure 3 in United States of America entry). The largest shafts are several hundred metres deep and most of the deepest caves in the world consist of a series of shafts connected by short sections of vadose canyons.

Cave sediments are important for cave morphology. They can fill and modify the profile of a passage, and they can also influence cave formation. Fluvial sediments transported by underground rivers can contribute to the erosion of passages in vadose conditions or can fill phreatic channels causing the paragenetic development of the passage (see Paragenesis).

Erosional Sculpture Within Passages (Rock Relief)

Bedrock features that dissect the walls of caves are combined under the term "cave rock relief" (Slabe, 1995). They are smaller than the wall (lateral surface) of a cave. According to their shapes, which frequently reflect the mode of their formation, such features may be classified as: channels and groups of flutings; depressions (scallops, pockets (Figure 3), bell holes, potholes, and cups); notches and bevels; and protuberances (pendants, rock knives, spikes, and crags). They may occur as isolated individuals, linked together in networks, or packed together and overlapping, reflecting many factors operating simultaneously or contributing to their formation in sequence. The texture and composition of the rock often have a decisive influence on the origin and shape of these features (Slabe, 1995). Standard "scallops" (Figure 4) form due to accelerated dissolution where the chemically saturated boundary layer of the water is locally detached by fixed eddies in the flow along the rough surface of the rock. The form of scallops (spoon-shaped depressions of various sizes with steepened upstream surfaces, overlapping each other in dense patterns) reflects the shape and hydraulic conditions in passages (Knez & Slabe, 1999). Length and breadth reflect primarily the velocity of the water; the faster the flow the shorter the scallop (Curl, 1974). Bedrock composition also influences their shape and size, and scallops form best on homogeneous, fine-grained rocks. Flutes (*sensu* Curl, 1966) are a rare form that exhibit scallop asymmetry but are elongated across the direction of flow and represent very stable conditions of detachment. Paleohydraulic conditions can be reconstructed from scallop measurements in passages long abandoned by flowing water. Scalloping by sublimation in flowing air is common

Morphology of Caves: Figure 3. A fine example of a solution pocket developed along a joint in the ceiling of Grutta de Torrinha, Brazil. (Photo by John Gunn)

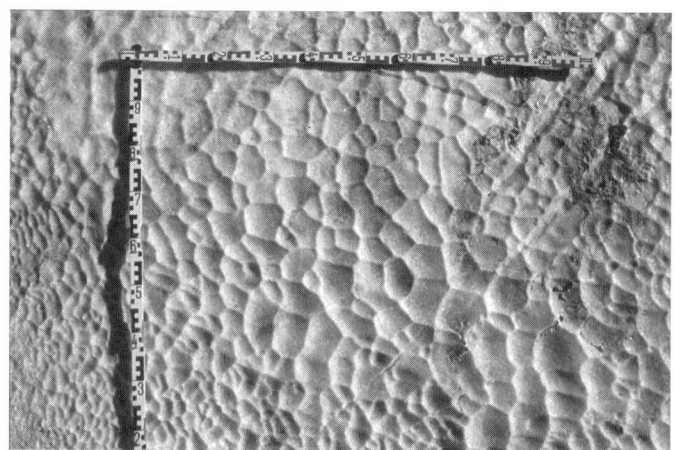

Morphology of Caves: Figure 4. Scallops in marble. Surface stream channel, Svartisen, North Norway. (Photo by Stein-Erik Lauritzen)

in ice and snow caves, the scallops being much longer than in water at the same velocity due to the lower viscosity of air.

Ceiling and wall pockets reveal a great diversity of dissolutional forms. Their origin and varied shapes are consequences of eddies in flow caused by the characteristics of the rock, the shape of the passages, and / or by hydrodynamic conditions. Eddies may form around the entire circumference of a passage but their effects are preserved primarily on their ceilings and upper walls. Isolated individual ceiling pockets may form due to mixing corrosion where tributary water enters from a fissure, the mixing of fresh and salt water, differing temperatures of water and rock driving cellular convection, particularly in hydrothermal caves (Forti, 1996), and air trapped under ceilings in flooding caves (Mücke, Völker & Wadevitz, 1983). One particular type of pocket is the result of convectional cellular circulation established in very slowly flowing waters due to solute density differences and is particularly evident in gypsum caves due to their greater solubility (Kempe et al., 1975).

Where vadose channel gradients are steep and the rock is hard, stream potholes (French—"marmites") may develop. These are circular in cross section and may have a depth up to several times their diameter. They occur in all strong rocks, being drilled when grinder boulders are trapped and spun (rock mills) in any small hole in the channel bed. However, they are most frequent and display the most regular form in soluble rocks because dissolution in the swirling water reinforces the grinding and may replace it entirely (Ford, 1964). Diameters range from a few centimetres to several metres.

Wall and ceiling flutings are characteristic on the sloping surfaces of doline shafts through which large quantities of water may flow. In caves formed in gypsum, ceiling channels may be carved by fresh water with a lower specific density. Underfit floor channels form due to the flow of small quantities of water across the floors of passages that are largely abandoned by their formative flow. Rock pillars form due to the eddying of water in vertical fissures dissecting walls. Knife-like blades or spikes occur where the rock is fisssured more densely. Residual pinnacles are frequently found on the floors of passages. Horizontal notches carved into walls or irregular ceilings and planar, horizontal dissolutional bevels in ceilings form where the water surface is stable for a long period and flow is slow. Cellular density currents are established that remove denser, solute-rich water from the surface, replacing it with lighter, chemically aggressive water (Ford, 1989).

Distinctive cave rock features occur along the contacts with fine-grained sediments. Paragenetic above-sediment channels and anastomoses are formed by water flowing over sediments and dissolving upwards into the ceilings of nearly-filled passages; if later dissolution reshapes them, only pendants may remain. Small, cup-like solutional pits form at the contact with wet sediments, and scallops may develop where water flows along a permeable contact with overhanging walls. Pockets occur due to water flowing through fissures at the contact point with sediment. Sub-sediment dissolutional sculpture is characteristic of periodically flooded passages in which water deposits smaller quantities of fine-grained sediments. Flutes are formed on sloping walls where water trickles from the sediment. Small notches are characteristic of vertical and overhanging walls, as are small pendants on ceilings (Slabe, 1995). Pits also occur on sloping walls and deep niches may be carved into the rock walls where water flows in meandering channels over fine-grained sediments (Bretz, 1942).

Where there is frequent percolation of water from the surface, flutes may form along slopes, isolated scallops on vertical and overhanging walls, and small pendants on ceilings (Slabe, 1995). Small channels, single or parallel in groups, that are frequently found descending from slightly opened bedding planes and fissures are the result of small, acidic discharges of water from them, often during floodwater recession. Ceiling pockets form due to the dispersion of water that reaches the ceiling through fissures (Dreybrodt & Franke, 1994). Dripping water carves cup-like pits. Ceiling pockets, large and shallow scallops, and ceiling channels form due to the condensation of moisture from the air onto the rock (see Condensation Corrosion). Some bell holes may have formed in a similar fashion (Tarhule-Lips & Ford, 1999). Unusual cave condensation forms are found in thermal caves (Cigna & Forti, 1986) and condensed moisture often etches the surfaces of the rock in cave entrance zones in all climates. Dissolutional notches and pits form under ice masses accumulating in cold caves.

Biokarstic processes most frequently pit or dissect the rock at much smaller scales. Tiny pits and pendants form under lichen; exposure to light particularly influences their shapes. Larger animals can also leave their traces on the rock. Rock can be weathered several centimetres deep by sediments, condensing moisture, trickling water, freezing, and microorganisms (Zupan Hajna, 2001). Such small-scale sculpture is a frequent and significant clue to the formation and development of karst caves. Distinctive features appear in every type of cave, and the traces of a variety of processes can be interwoven in them. The processes may operate simultaneously to create the diversity, or they can be distributed over various periods of development, where younger traces more or less distinctly cover the older.

Breakdown

Breakdown is the term for the collapse of caves and the debris it produces (see Sediments: Autochthonous). Breakdown plays a role in the evolution and enlargement of many cave passages. It can occur at any time during the evolutionary history of a cave but is most frequent during the enlargement and decay phases. Geological control is apparent in the process of breakdown where the original hydrodynamically created form is modified by mechanical degradation. In most karst rocks, it is the bedding planes and joints that fail under mechanical stress, so the spacing of joints and bedding planes dictates the occurrence and form of breakdown. The stable form for a breakdown passage is a dome. The most stable symmetric domes occur in well-bedded and horizontally bedded rocks. Asymmetric and less stable domes develop where beds dip. Within the bedrock, breakdown is most likely to lie at passages where prominent joint scts intersect. The main reason for all collapses is mechanical breakdown inside a bedding plane, between bedding planes, or between fissured blocks. Some breakdowns can be triggered by earthquakes but this is probably a rare occurrence. The largest known collapse chamber in the world is Sarawak chamber in Borneo, which developed at the axis of an anticline and a thrust fault and along the contact between the impermeable Mulu Formation and the Melinau Limestone (see Mulu Karst and Caves).

ANDREJ MIHEVC, TADEJ SLABE, and STANKA ŠEBELA

Works Cited

Bretz, J.H. 1942. Vadose and phreatic features of limestone caverns. *Journal of Geology*, 6(2): 675–811

Cigna, A.A. & Forti, P. 1986. The speleogenetic role of air flow caused by convection. *International Journal of Speleology*, 15: 41–52

Curl, R.L. 1966. Scallops and flutes. *Transactions of the Cave Research Group of Great Britain*, 7(2):121–160

Curl, R.L. 1974. Deducing flow velocity in cave conduits from scallops. *NSS Bulletin*, 36(2): 1–5

Dreybrodt, W. & Franke, H.W. 1994. Joint controlled pockets (Laugungskolke) in ceilings of limestone caves: A model of their genesis, growth rates and diameters. *Zeitschrift für Geomorphologie N. F.*, 38: 239–45

Ford, D.C. 1964. Stream potholes as indicators of erosion phases in caves. *NSS Bulletin*, 26(4): 67–74

Ford, D.C. 1989. Characteristic of dissolutional cave system in carbonate rocks. In *Paleokarst: A Systematic and Regional Review*, edited by P. Bosák, D.C. Ford, J. Glazek & I. Horácek, Amsterdam: Elsevier

Ford, D.C. 2000. Speleogenesis under unconfined settings. In *Speleogenesis: Evolution of Karst Aquifers*, edited by A.B. Klimchouk, D.C. Ford, A.N. Palmer & W. Dreybrodt, Huntsville, Alabama National Speleological Society, 319–324

Forti, P. 1996. Thermal karst systems. *Acta Carsologica*, 25: 99–117

Kempe, S., Brandt, A., Seeger, M. & Vladi, F. 1975. "Facetten" and "Laugdecken", the typical morphological elements of caves developing in standing waters. *Annales de Spéléologie*, 30(4): 705–08

Knez, T. & Slabe, T. 1999. Škocjanske jame, Slovenia: Development of caves related to rock characteristic and rock relief. In *Contribución del estudio científico de las cavidades kársticas al conocimiento geológico* [The scientific study of karstic cavities as a contribution to geological knowledge], edited by B. Andreo, F. Carrasco, & J.J. Durán, Málaga: Patronato de la Cueva de Nerja, Instituto de Investigación

Lauritzen, S-E. & Lundberg, J. 2000. Solutional and erosional morphology. In *Speleogenesis: Evolution of Karst Aquifers*, edited by A.B. Klimchouk, D.C. Ford, A.N. Palmer & W. Dreybrodt, Huntsville, Alabama: National Speleological Society

Mücke, B., Völker, R. & Wadevitz, S. 1983. Cupola formation in occasionally inundated cave roofs. *Proceedings of the European Regional Conference on Speleology*, vol. 2, Sofia: Bulgarian Federation of Speleology

Slabe, T. 1995. *Cave Rocky Relief and its Speleogenetical Significance*, Ljubljana: Zbirka ZRC 107

Tarhule-Lips, R. & Ford, D.C. 1999. Morphometric studies of bellhole development on Cayman Brac. *Cave and Karst Science*, 25(3): 119–307

Zupan Hajna, N. 2001. Weathering of cave walls in Martinska jama, SW Slovenia. *Proceedings of the 13th International Congress of Speleology*, Brasilia: Sociedade Brasiliera de Espeleologia (CD-ROM)

MORPHOMETRY OF CAVES

A number of parameters may be used to characterize and compare dimensions and shapes of caves and their patterns. They are derived from data reduction of cave surveys or from resultant cave maps. The pattern of the cave can be represented and measured by building a traverse framework, comprising a combination of traverse (centre) lines for each individual passage, based on tape, compass, and clinometer readings. The cave shape may be represented by a complex surface in three dimensions, to characterize which additional measurements are made, tied to the framework.

Caves are generally compared by their length and depth. The length of a cave is the combined length of all the centreline segments that constitute a framework of the entire integrated passage system. It is necessary to distinguish between the plan length, that is the length of passage projections on the horizontal plane, and the true length. The difference can be substantial for inclined caves, and can be very large for a system in which pitches and vertical shafts predominate. In the extreme case of a single vertical shaft, it will have no plan length but the true length will be equal to the depth. The distribution of caves by length varies between regions, due to both the variable degrees of exploration and differing local geological and hydrogeological conditions. Smooth distributions are characteristic for well-studied regions with a large number of explored caves. All populations are truncated at the short end, commonly at 10 or 30 m, because of the lower limits of documentation and inventory. Extremely long caves are not common. By March 2003, there were 13 caves in the world with lengths exceeding 100 km (Table 1) and 44 caves with lengths exceeding 50 km. The listing of the longest caves is rather conservative as their growth relies on systematic exploration and mapping over many decades. The number of caves of lesser lengths increases dramatically.

The majority of those caves that have a substantial vertical component were explored "top-down" from an upper entrance or entrances. Their depth may be defined as the vertical range between the altitudes of the entrance (the highest entrance if there is more than one) and the lowermost point reached in the cave. Less commonly, a cave may be explored upwards from a low entrance or entrances without any connection being made to a higher point on the surface. In this case it is more appropriate (from an exploratory perspective) to define the "height" of the cave as the vertical range between the altitude of the lowest entrance and the highest point reached in the cave. If a cave has sections that extend higher than the highest entrance (or lower than the lowest entrance in a "bottom-up" system), then the total vertical range of the cave is measured by the amplitude between the altitudes of the extreme upper and lower points. Depth may be represented by "−" and height by "+", but is often taken as the vertical range, since the positions of the accessible entrances are not significant to the total morphology of the cave. For example, in Table 2, the depth (vertical range) of Boj Boluk is 1415 m, of which 1158 m is below the highest entrance and 257 m is above. As the depth potential for direct cave exploration is limited by the vertical extent of continuous carbonate rocks, deep caves are concentrated in the mountain regions. Most of the world's deepest caves (Table 2) are in the young mountain belt that stretches across Europe from the Pyrenees (Spain) in the west to the Caucasus (Georgia) in the east. This reflects a bias in exploration efforts and the last two decades of the 20th century witnessed a rapid increase in deep cave explora-

Morphometry of Caves: Table 1: The longest caves of the world (as of March 2003)

	Cave	Country	Length (km)
1	Mammoth Cave System	USA, Kentucky	556.9
2	Optymistychna (in gypsum)	Ukraine	214.0
3	Jewel Cave	USA, South Dakota	205.6
4	Hölloch	Switzerland	186.0
5	Lechuguilla Cave	USA, New Mexico	180.0
6	Wind Cave	USA, South Dakota	173.3
7	Fisher Ridge Cave System	USA, Kentucky	169.1
8	Siebenhengste-Hohgant	Switzerland	145.0
9	Ozernaja	Ukraine	117.0
10	Gua Air Jernih	Malaysia Sarawak	111.0
11	Ox Bel Ha (under water)	Mexico	107.1
12	Toca de Boa Vista	Brazil	102.0
13	Réseau Felix Trombe	France	101.0
14	Ojo Guarena	Spain	99.3
15	Sistema Purificacion	Mexico	94.3
16	Zoloushka (in gypsum)	Moldavia–Ukraine	92.0
17	Hirlatzhöhle	Austria	86.6
18	Bullita Cave System	Australia	81.9
19	Raucherkarhöhle	Austria	78.6
20	Ease Gill Cave System	Great Britain	75.0
21	Friar's Hole Cave	USA, West Virginia	73.3
22	Ogof Draenen	Great Britain	70.0
23	Kazumura Cave (in lava)	USA, Hawaii	65.5
24	Organ Cave	USA, West Virginia	63.6
25	Nohoch Nah Chich (under water)	Mexico	61.0
26	Réseau de l'Alpe	France	60.2
27	Red del Silencio	Spain	60.0

Morphometry of Caves: Table 2: The deepest caves of the world (as of March 2003)

	Name	Country	Depth (m)
1	Gouffre Mirolda	France	1733
2	Krubera (Voronja)	Georgia	1710
3	Lamprechtstofen	Austria	1632
4	Réseau Jean-Bernard	France	1602
5	Torca del Cerro	Spain	1589
6	Cehi 2 "la Vendetta"	Slovenia	1533
7	Sarma	Georgia	1530
8	Vjacheslava Pantjukhina	Georgia	1508
9	Sistema Cheve	Mexico	1484
10	Sistema Huautla	Mexico	1475
11	Sistema del Trave	Spain	1441
12	Boj Bulok	Uzbekistan	1415
13	Puertas de Illamina	Spain	1408
14	Lukina Jama	Croatia	1392
15	Evren Gunay düdeni (Peynirlikönü)	Turkey	1377
16	Sneznaja–Mezhennogo	Georgia	1370
17	Réseau de la Pierre Saint-Martin	France–Spain	1342
18	Siebenhengste	Switzerland	1340
19	Slovacka jama	Croatia	1320
20	Cosa Nostra Loch	Austria	1291
21	Gouffre Berger	France	1271
22	Gouffre Muruk	Papua New Guinea	1258
23	Pozo del Madejuno	Spain	1255
24	Torca de los Rebecos	Spain	1255
25	Abisso Paolo Roversi	Italy	1249
26	Iljukhina	Georgia	1240
27	Sotano Akemati	Mexico	1226
28	Kijahe Xontjoa	Mexico	1223
29	Schwersystem	Austria	1219
30	Abisso Olivifer	Italy	1215
31	Gouffre Gorgothakas	Greece, Crete	1208

tion in other regions of the world. Table 2 lists the deepest caves known at March 2003. However, the fast pace of exploration changes the list substantially within periods of a few years. Frequent changes in ranking of caves occur due to small differences, and new deepest caves may appear quite often. All the deepest caves are combinations of vertical shafts and inclined passages, some constituting very complicated three-dimensional systems. The deepest single pits explored so far are in the Slovenian side of the Monte Kanin massif (see separate entry): Vrtiglavica (643 m deep) and Brezno pod velbom (501 m deep).

Caves can also be compared by the area occupied by passages and chambers, and by their combined volume. The former parameter can be measured from cave maps, whereas to determine volume, it is necessary to sum up volumes of individual elements of the cave, determined from original survey measurements of lengths, widths, and heights. If this was not done routinely throughout all the mapping history of a complex cave, then its volume can be only roughly evaluated. Data on cave areas and volumes in the literature are scarce. Areas of the largest caves are of an order of hundreds of thousands of square metres and rarely exceed one million square metres. Volumes of some longest caves can be as large as several millions of cubic metres. It is noteworthy that the volume of some of the largest individual underground chambers, such as Sarawak Chamber in the Mulu karst in Malaysia (12 million m^3), is greater than the volumes of some of the longest caves that integrate tens or hundreds of kilometres of passages. Even greater are volumes of some huge and deep dolines or pits, such as Luse (50 million m^3) and Minye (26 million m^3), in the Nakanai karst of Papua New Guinea.

Caves are also characterized by the area of cave fields, which is the area of a polygon that reasonably closely delineates a plan array of mapped passages, and by the volume of cave block, which is the volume of rock that contains a cave. The latter parameter can be obtained by multiplying the area of the cave field by the vertical amplitude of a cave.

For the purposes of speleogenetic analysis and hydrogeological and engineering characterization of karstified rocks, some specific parameters that can be derived from basic cave measures and from cave field and block measures are used. Specific volume (the cave volume / length ratio) characterizes an average size of cave passage in the cave system. Passage network density is characterized conveniently by using the ratio of the cave length to the area of the cave field (km km^{-2}). Areal coverage is the area of the cave itself divided by the area of the cave field, expressed as a percentage. It refers to plan-view cave porosity density, which is equivalent to the probability of a drill hole encountering the cave. Cave porosity is a fraction of the volume of a cave block occupied by mapped cavities. It is the volume of the cave divided by the volume of cave block, expressed as a percentage.

Morphometric analysis of caves or their particular components (form) includes various exercises aimed at determining their characteristics and the relationship between cave morphology and structural, lithological, and hydrogeological factors. This can shed more light on the origin of caves or their particular forms. Recent studies suggest that at least some of the above specific parameters can be indicative of whether the caves evolved under an unconfined setting or a confined setting. Analysis of samples of typical caves formed in the respective settings shows that average passage network density for unconfined settings is 16.6 km km^{-2} while this parameter for confined settings is 177.6 km km^{-2}. Average areal coverage for unconfined speleogenesis is 6.16%, while it is 32.8% for confined speleogenesis. Cave porosity for unconfined speleogenesis is 0.54% but it is an order of magnitude greater for confined speleogenesis (5.4%).

ALEXANDER KLIMCHOUK

See also **Morphology of Caves; Patterns of Caves; Speleogenesis**

Further Reading

Courbon, P., Chabert, C., Bosted, P. & Lindsley, K. 1989. *Atlas of the Great Caves of the World*, St. Louis, Missouri: Cave Books

MORPHOMETRY OF KARST

Morphometry is the measurement and analysis of form (shape). In the context of geomorphology, defined as the study of earth surface processes and landforms, morphometry may be utilized in the quantitative analysis of individual landforms and their landscape assemblages. In this context, general geomorphometry is "the measurement and analysis of those characteristics of landform which are applicable to any continuous rough surface. This is distinguished from "specific geomorphometry", the measurement and analysis of specific landforms" (Evans, 1972, p.18).

Morphometric studies in karst terrains have a long history, beginning with the doline shape analyses conducted by Cvijić (1893). Other, sporadic, morphometry was applied to karst in the first half of the 20th century, notably by Cramer (1941), who made a detailed worldwide inventory of thousands of dolines, their numbers, sizes, and densities, and by Meyerhoff (1933), who examined doline size and distribution using measures of relief and spacing. However, it was not until the "quantitative revolution" of the 1960s that morphometry became widely utilized in karst studies.

Although karst morphometry has focused on surface landforms and landscapes, it has also been applied to caves (see Morphometry of Caves), where quantitative data on shapes and sizes of cave passages, specific erosional and depositional features, and even entire systems are valuable evidence of speleogenic origins. Geometric analyses have been conducted on cave passage cross-sections, passage meanders, and scallops, but quantitative treatments of cave morphology in general are sparse.

Pioneering work on karst morphometry, initially evolving from fluvial drainage basin studies, was carried out by Williams (1969), who treated swallets (stream-sinks), and later enclosed karst depressions (dolines) as discrete catchment areas, and conducted quantitative analyses of their shapes, sizes, and drainage components, concluding that fluvial morphometric relationships also applied in the karst landscape. Williams (1971; 1972a,b) and others subsequently extended these analyses to tropical karst terrains, which had previously been regarded as chaotic assemblages of depressions, hills, and valleys. Williams showed that these were, in fact, organized landscapes with distinct landscape patterns reflecting predictable modes of closed depression development. In particular, he identified so-called polygonal karst, in which enclosed depressions are surrounded by residual hills whose summits could be connected by lines to produce a polygonal pattern. Focusing on the enclosed depressions, Williams revealed that they could be considered viably as individual drainage basins, and that their dispersion pattern tended towards uniformity, reflecting general slope and structural controls.

Morphometric analyses of dolines have been legion. A large number of studies have focused on spatial distribution, pattern, and density, correlating doline density, for example, with carbonate lithology, mantle thickness, and limestone purity. Lavalle (1968) employed doline parameters as a measure of landscape dissection, and several studies have used doline numbers and sizes to assess degree of karstification, for example employing an index of pitting. Other studies have attempted to illuminate doline evolution, for example using spatial relations, historical sequences, or using space-time substitution. Clustering of "daughter" dolines around individual "mothers" apparently occurs in some areas, although not others. Some dolines elongate through time, and Day (1983) showed that, on the raised reef terraces of Barbados, doline numbers increased through time until they attained a density that was structurally controlled. Depth / diameter ratios have also been employed, for example to illuminate differences between doline populations.

Other workers have expanded morphometric analysis to other landscape components, including regional valley systems. Again following prior studies in fluvial landscapes, morphometry of karstic valley systems has focused on such issues as valley ordering, density, pattern, and orientation. Several studies have examined the relationships between valley systems and karst depressions, which are, in a sense, "competitive". Other studies have employed morphometry at a smaller scale, focusing, for example on the components of limestone pavements (e.g. Goldie & Cox, 2000).

Morphometry has also been applied to residual carbonate hills (cones or towers), which are of variable morphology and which have been classified on the basis of such dimensions as their heights and diameters. This approach was extended by Day (1981) to incorporate both enclosed depressions and residual hills into an overall classification of tropical karst landscape styles. This approach produced a simple tripartite framework as follows: in type I landscape enclosed depressions are dominant, and interspersed with subdued hills; in type II landscape enclosed depressions and residual hills attain approximately equal prominence; and in type III landscape isolated residual hills dominate intervening near-planar surfaces. Morphometric approaches to karst landscapes are relatively straightforward where individual landforms are clearly defined, but are far more problematic where landforms coalesce or intersect.

Field measurement of karst landscape is arduous and time-consuming, but maps, air photographs, and satellite imagery generally do not provide sufficiently detailed information at a suitable scale to be acceptable substitutes. Comparative data on the density and spacing of individual landforms is relatively easily assembled, but meaningful measurements of height and three-dimensional shape, for example, are far more problematic. Where field data are available, usually for relatively restricted areas, a variety of approaches to measurement and classification have been employed.

Other studies have gone beyond individual landforms to consider general karst landscape morphometry, which should provide overall indices of landscape morphology, although this has until recently been less fruitful than might be hoped. McConnell and Horn (1972) employed quadrat analysis to analyse doline density, revealing patterns suggestive of other than single random processes or contagion, but suggestive of multiple random processes, such as dissolution and collapse. Surface roughness has also been utilized as a discriminator of tropical karst landscape styles by Day (1979) and, with greater sophistication, by Brook and Hanson (1991) who used double Fourier series analysis to analyse landscape wavelength variance within doline and cockpit karst landscapes in Jamaica. They were able to demonstrate statistically the greater complexity and roughness in the latter, and illuminated the differing roles of fracture sets on landscape development. More recently, GIS modelling has been applied to karst landscapes, an approach that holds great promise.

One incentive for morphometric studies in karst geomorphology has been the hope that they would reveal explicit linkages between karst landforms and landscape assemblages on the one hand and suites of karst process variations on the other; that they would serve to forge a link between disparate information about forms and processes, especially in differing climatic zones. In this context morphometry has not proven to be as useful a tool as had been hoped. Nevertheless, quantification of form has made inter-regional comparison more rigorous. It has also helped in the development of meaningful indices of landscape morphology and it has clarified the role of lithological, structural, and other factors in influencing karst landform development. Development of more rigorous mathematical modelling techniques and the employment of digital elevation models and related landscape data in geographic information systems (GIS) may be expected to contribute to more productive mathematical analysis of karst landscapes.

MICK DAY

See also **Cone Karst; Dolines; Morphometry of Caves; Tower Karst**

Works Cited

Brook, G.A. & Hanson, M. 1991. Double Fourier series analysis of cockpit and doline karst near Browns Town, Jamaica. *Physical Geography*, 12(1): 37–54

Cramer, H. 1941. Die Systematik der Karstdolinen. *Neues Jahrbuch für Mineralogie, Geologie und Paleontologie*, Beilage-Band, Abt B 85: 293–382

Cvijić, J. 1893. Das Karstphänomen. *Geographische Abhandlungen Wien*, 5(3): 218–329

Day, M.J. 1979. Surface roughness as a discriminator of tropical karst styles. *Zeitschrift für Geomorphologie N.F., Supplementbande*, 32: 1–8

Day, M.J. 1981. Towards numerical categorization of tropical karst terrains. *Proceedings of the 8th International Congress of Speleology*, vol. 1, edited by B.F. Beck, Huntsville, Alabama: National Speleological Society

Day, M.J. 1983. Doline morphology and development in Barbados. *Annals of the Association of American Geographers*, 73: 206–19

Evans, I.S. 1972. General geomorphometry, derivatives of altitude, and descriptive statistics. In *Spatial Analysis in Geomorphology*, edited by R.J. Chorley, London: Methuen

Goldie, H.S. & Cox, N.J. 2000. Comparative morphometry of limestone pavements in Switzerland, Britain and Ireland.

Zeitschrift für Geomorphologie N.F., Supplementbande, 122: 85–112

Lavalle, P. 1968. Karst morphology in south central Kentucky. Geografiska Annaler, 50: 94–108

McConnell, H. & Horn, J.M. 1972. Probabilities of surface karst. In Spatial Analysis in Geomorphology, edited by R.J. Chorley, London: Methuen

Meyerhoff, H.A. 1933. Texture of karst topography. Journal of Geomorphology, 1(4): 279–95

Williams, P.W. 1969. The geomorphic effects of groundwater. In Water, Earth and Man, edited by R.J. Chorley, London: Methuen

Williams, P.W. 1971. Illustrating morphometric analysis of karst with examples from New Guinea. Zeitschrift für Geomorphologie N.F., 15: 40–61

Williams, P.W. 1972a. Morphometric analysis of polygonal karst in New Guinea. Bulletin of the Geological Society of America, 83: 761–96

Williams, P.W. 1972b. The analysis of spatial characteristics of karst terrains. In Spatial Analysis in Geomorphology, edited by R.J. Chorley, London: Methuen

Further Reading

Goudie, A.S. (editor) 1990. Geomorphological Techniques, 2nd edition, London and Boston: Unwin Hyman

MacGillivray, C.M.I. 1997. A review of quantitative techniques for measuring karst terrain, with examples from the Caribbean. Caribbean Geography, 8(2): 81–95

MOVILE CAVE, ROMANIA

An artificial shaft dug in 1986 near the Black Sea, in the limestone platform of the Romanian Dobrogea, intercepted a cave with thermal sulfurous water. Movile Cave, which has no natural entrance, connects to a short groundwater system and shelters a rich biological community with 48 species of aquatic and terrestrial invertebrates, many of which are new taxa and 34 are endemic to this ecosystem. Stable isotope data shows that the Movile Cave ecosystem derives its organic carbon solely from *in situ* chemoautotrophic production, the cave being the first terrestrial ecosystem known to be independent of solar energy (Sârbu, Kane & Kinkle, 1996); mineral-rich deep-sea hydrothermal vents are the only other known non-photosynthetic ecosystems on Earth. The Movile discovery supports the Deep Hot Biosphere theory and the expectation of exobiologists that chemoautotrophic (also known as chemolithoautotrophic) life forms may be found beneath the surface of Mars, in volcanic caves.

Movile Cave is located near the city of Mangalia in southeastern Romania, 3 km from the Black Sea coast, at the eastern border of the Dobrogea Plateau. The upper part of the plateau consists of fossiliferous, oolitic limestones c.12.5 Ma in age. The plateau is locally disturbed by solutional karstic features, such as dolines, dry valleys, and cliffs (in Romanian, "Movile" means hillocks). Except for Limanu, a 4 km long maze cave, the area previously offered little interest for cave explorers. However, in 1986, a power-plant construction project began near Mangalia and Cristian Lascu, a speleologist at the "Emil Racoviţă" Institute in Bucharest, suggested that the proposed location should be investigated with a few drill holes, in order to avoid possible karstic collapses. During the survey of an exploratory shaft he found the cave at a depth of 18 m beneath the surface.

Movile Cave is a horizontal maze cave 240 m long, with two levels. The upper level is dry, with a phreatic morphology including rounded and elliptical sections, 1–2 m in diameter. Red and yellowish fine clay is abundant. There are no speleothems and the only secondary mineral deposits are a few calcite crusts associated with millimetric aragonite needles and gypsum (Diaconu & Morar, 1993). The lower level has a clastic morphology that indicates a more recent origin. It consists of a 25 m succession of four underwater passages, 0.5–2 m wide and 1–3 m deep, separated by three small airbells. A very soft, grey, organic sediment lies on the bottom. Secondary gypsum needles are present on the walls above the water table. A 5 m shaft from a pool connects the two cave levels.

Cave Environment

Movile is a low-grade thermal cave. The air temperature is 20°C and humidity is 100%. There is no detectable air movement (Boghean & Racoviţă, 1989). The upper level's atmosphere contains c.20% oxygen and around 1% carbon dioxide, and the atmosphere in the airbells becomes progressively poorer in oxygen (19–16–7%) and enriched in carbon dioxide (1.5–2–3.5%) due to biological activity (Sârbu, 2000). The airbells also contain small amounts of hydrogen sulfide and methane. Moisture on the walls has a pH of 3.5–4. The groundwater contains 30 mg l^{-1} H$_2$S, with a pH of 7.2 at a constant temperature of 21°C. Although the water contains dissolved oxygen in a 1 mm layer at the air–water interface it is anoxic below this layer. The water contains no organic compounds from the surface and is probably derived from a deep (180–600 m) thermomineral aquifer in the Mesozoic limestones. A slight current was detected in the underwater passages.

Cave Life

The lower level of Movile Cave hosts a rich and diverse biological community. On the walls of airbells it is possible to see simul-

Movile Cave: Figure 1. *Haemopsis caeca* devouring a worm.

taneously tens of spiders, Isopoda, centipedes, and Coleoptera. Near the water level, every square decimetre contains nearly 100 specimens of the gastropods *Semisalsa* and *Paladilchia*, 1–3 mm long snails. The water surface in the airbells is covered with a layer of white, cream-like organic matter. This mat is a mixture of fungi and sulfide-oxidizing micro-organisms including *Thiomicrospira* sp., *Beggiatoa* sp., and *Thiobacillus thioparus* (Vlăsceanu, Popa & Kinkle, 1997; see also Biofilms entry). The terrestrial fauna is composed of 30 species of cave-dwelling invertebrates, including 23 endemic species (Sârbu, 2000; Sârbu & Popa, 1992). Of particular interest are species of pseudoscorpions, spiders, and centipedes that are new to science. *Agroecina cristiani*, a 20 mm long arachnid that is a close relative of a cave spider found inhabiting lava tubes in the Canary Islands, is very abundant. There are also numerous specimens of *Criptops anomalans*, a voracious centipede that reaches up to 100 mm in length. Its favourite prey are the isopods *Trachelipus* and *Armadillidium* and the coleopterans *Clivina, Medon*, and *Tychobitinus*.

The aquatic fauna is represented by 18 species that are new to science, with 11 endemics. Here there are even more peculiar creatures. The water scorpion *Nepa anophthalma*, which grows up to 25 mm long, is the first cave-adapted aquatic hemipteran known in the world. Hidden below the surface it extends two pincer-like front legs, probing the water for signs of shrimps or worms. To breathe, the water scorpion arches its hollow tail upwards into the air, drawing air through this built-in snorkel. A new species of the predatory leech, *Haemopis caeca*, the only species of cave leech known, preys on worms and *Niphargus*, a common amphipod. *H. caeca* also grazes the organic mats of bacteria and fungi.

The cave fauna generally displays morphological features typical for species with a long evolutionary history of isolation in caves, and there is paleogeographical evidence to suggest that the ancestors of some species may have invaded the subterranean system as early as the late Miocene (5.5 million years ago), when the climate was much warmer. Most of the cave's invertebrate species show a reduction or loss of eyes. In compensation they present an enlargement of appendages and extraoptic sensitive structures. Another troglomorphic adaptation is the loss or reduction of pigmentation and some of the cave fauna are white or translucent.

Energetics of the Movile Ecosystem

In most instances life in caves is based on food that originated from the surface. Vegetal detritus and animal excreta, such as bat guano, represent rich food resources and percolation water brings small amounts of organic matter along limestone cracks. However, at Movile Cave a deposit of clay overlying the limestone seals the cracks and pores and several different experiments have proved that the subterranean community can receive little, if any, organic input from the surface. In particular, whereas the soil, plants, surface sediments, and other caves in the area contain the radionuclides ^{137}Cs and ^{90}Sr as a result of the Chernobyl nuclear accident, no trace of them has been detected in Movile Cave sediments. The dehydrogenase (a bacterial enzyme) activity in samples of sediment from the limestone strata overlying the Movile passages is very low, proving the scarcity of external microbes and organic compounds (Sârbu, 2000).

The presence in Movile Cave of a large population of invertebrate species and abundant predators, with a high metabolic activity, suggests a large energy base for the ecosystem. Studies of stable isotope ratios indicate that the community is totally dependent on *in situ* chemoautotrophic production within the bacterial mat. Carbon and nitrogen stable isotopes are fractionated by organisms during metabolic processes, the lighter isotopes being preferentially selected. All forms of inorganic carbon in the cave are isotopically light. The Movile Cave atmosphere contains CO_2 as a mixture of heavier CO_2 from limestone sulfuric corrosion and lighter CO_2 presumably from endogenetic methane oxidation. The nitrogen in the bacterial mat in the cave is enriched in the lighter nitrogen isotope. The organisms in Movile Cave consuming these mats are isotopically lighter both in carbon and nitrogen when compared with similar organisms sampled from the nearby Limanu cave and some other surface locations. If the Movile Cave ecosystem were dependent on allochthonous input of organic material from the surface, the organism would be isotopically heavier, for both carbon and nitrogen. The stable isotope analyses demonstrate the absence of photosynthetically derived carbon in the cave but also suggest that the carbon of the microbial mat is solely of chemoautotrophic origin (Sârbu, Kane & Kinkle, 1996). Since the reduced carbon in organic molecules metabolized by Movile microbiota does not take part in the external carbon cycle, the energy used by microorganisms to fix inorganic carbon both from atmospheric CO_2 and from methane must result from an internal source, the oxidation of hydrogen sulfide from the thermomineral groundwater:

$$H_2S + 2O_2 = SO_4 + 2H + 798.2 \text{ kJ mol}^{-1}$$

Chemical energy resulting from the oxidation of reduced compounds such as sulfide, methane, iron, or ammonium is the alternative source to photosynthesis for manufacture of carbohydrates and cells in aphotic habitats such as sulfurous caves, deep-sea vents, and underground thermomineral waters.

Although most of the oxygen in the cave probably comes from the surface atmosphere, part of it may be produced by iron reduction bacteria in a sulfidic environment. An alternative

Movile Cave: Figure 2. The spider *Agroecina cristiani* (new endemic species), a 20 mm long arachnid.

Movile Cave: Figure 3. Phreatic tunnel in the upper level of Movile Cave.

source for oxygen in the cave could be hyperoxidized iron compounds from the red clay. The results are fine grains of pyrothine and greigite: iron sulfide compounds that form spheroidal aggregates resembling a blackberry. Such features and associated magnetic bacteria have been detected in Movile Cave.

Origin of Movile Cave and the Mangalia Karst

Movile Cave is a recent, shallow, and limited segment of the extensive and deep karstic system of southern Dobrogea. The Dobrogea Platform consists of Mesozoic and Cenozoic limestones up to 800 m thick, and is bordered by the Danube to the west and by the Black Sea to the east. This large limestone body experienced several karstification periods related to oscillations in the level of the Black Sea which have favoured the formation of large karstic networks. A study based on analysis of 52 boreholes has proved that at least three different karstic levels were formed. The deepest level is located at an average depth of 200 m and was formed during the Upper Miocene, 5.5 million years ago. At that time, the Black Sea, as throughout the Mediterranean Basin, experienced a dramatic hydrological and climatic event known as the Messinian Crisis. Due to a northward movement of the African plate, the Gibraltar Strait closed, separating the Mediterranean Basin from the Atlantic Ocean. In a few millennia the Mediterranean Sea lost nearly all of its water by evaporation, leaving a few hypersaline lakes. Consequently, the satellite seas of the former Tethys, climatically and hydrologically dependent on the mother-sea, experienced a drop in their levels of several hundred metres (Hsu, 1978). The study of shallow-water deposits in boreholes in the Danube Delta has shown that the Black Sea level was probably 200 m lower (Lascu, Popa & Sârbu, 1994). At this stage, the Dacian Lake, another remnant basin located in the west of Dobrogea, drained towards the Black Sea through large karstic passages carved in the Dobrogea carbonate platform. Some of these karst conduits seem to be still active, circulating huge amounts of water from the Danube River towards the Black Sea (up to 10×10^6 l s^{-1}). Several 200 m deep submarine canyons in the Black Sea are thought to be remnants of the Messinian karstic development. Six hydrogeological boreholes which penetrate this level provide large amounts of water ranging between 110 and 225 l s^{-1}; the water is sulfidic, low-grade thermal (24–25°C), at a pressure of some 1.7 atm.

Later oscillations of the Black Sea level led to several other karstification periods. A middle-Quaternary speleogenetic episode may be responsible for the formation of the upper level of Movile Cave. Another karstification episode was contemporary with the Würm II glacial, c.15 000 years ago, when the Black Sea level was about 50–60 m lower. This initiated collapse processes that produced large dolines in this area. The Movile doline collapse cut the pre-existing phreatic conduits and water drainage was lowered to its present level.

Speleogenesis of the modern cave may have been influenced by microbial activity as the presence of sulfide-oxidizing bacteria could have generated small amounts of H_2SO_4, increasing the rate of carbonate dissolution. According to the mechanism proposed by Egemeier (1981), the acid environment of Movile Cave, which includes gypsum deposits, is actively undergoing enlargement.

CRISTIAN LASCU

Works Cited

Boghean, V. & Racoviţă, G. 1989. Données préliminaires sur le topoclimat de la Grotte "Peştera de la Movile" (Mangalia-Dobrogea de Sud-Roumanie). *Miscellanea Speologica Romanica*, 1: 19–31

Diaconu, G. & Morar, M. 1993. Analyses mineralogiques dans la grotte "Pestera de la Movile" (Mangalia-Dobrogea de Sud). *Karstologia*, 22: 15–20

Egemeier, S.J. 1981. Cavern development by thermal waters. *National Speleological Society Bulletin*, 43: 31–51

Hsu, K.J. 1978. When the Black Sea was drained. *Scientific American*, 238: 52–63

Lascu, C., Popa, R. & Sârbu, Ş. 1994. Le Karst de Movile (Dobrogea de Sud) (I). *Revue roumaine de géographie*, 38: 85–94

Sârbu, Ş.M. 2000. Movile Cave: A chemoautotrophically based groundwater ecosystem. In *Subterranean Ecosystems*, edited by H. Wilkens, D.C. Culver & W.F. Humphreys, Amsterdam and New York: Elsevier

Sârbu, Ş.M., Kane, T.C. & Kinkle, B.K. 1996. A chemoautotrophically based groundwater ecosystem. *Science*, 272: 1953–55

Sârbu, Ş.M. & Popa, R. 1992. A unique chemoautotrophically based cave ecosystem. In *The Natural History of Biospeleology*, edited by A.I. Camacho, Madrid: Museo Nacional de Ciencias Naturales

Vlăsceanu, L., Popa, R. & Kinkle, B.K. 1997. Characterization of *Thiobacillus thioparus* LV43 and its distribution in a chemoautotrophically based groundwater ecosystem. *Applied and Environmental Microbiology*, 63: 3112–27

Further Reading

Sârbu, Ş.M. & Kane, T.C. 1995. A subterranean chemoautotrophically based ecosystem. *National Speleological Society Bulletin*, 57: 91–98

MULU, SARAWAK

The Gunung Mulu National Park covers a large slice of preserved rain forest in the northeast corner of Sarawak, Malaysia. Within its boundaries, the mountains of Api and Benarat are formed of steeply dipping, very massive, Miocene Melinau Limestone. Annual rainfall of 5000–10 000 mm, and temperatures rarely outside 20–30°C, create the classic hothouse environment, where karst matures most rapidly. Drainage from the sandstone mountain of Gunung Mulu feeds onto the Api-Benarat limestone outcrop—which therefore contains some of the world's largest caves—and through to a terraced alluviated plain, eroded into the limestone and overlying shales. Serious cave exploration started in 1978, since when a series of British expeditions have mapped over 320 km of passages within the Park.

Gunung Api is the largest limestone block (Figure 1). It is largely drained by Gua Air Jernih (Clearwater Cave), over 110 km long. Its large passages give the southern end of Api a cavernous permeability close to 3%, an unusually high figure and one that takes no account of further caves not yet explored. The lowest level carries the river (Figure 2), whose base flow of 0.1 $m^3 s^{-1}$ rises to about 5 $m^3 s^{-1}$ on many afternoons, in response to daily rainstorms. Its upstream section has lengths of canyon separated by flooded loops, but the downstream 2000 m forms a singularly impressive, clean canyon out to the resurgence boulder choke. It once carried all the Melinau River from sinks in its gorge, but these are now buried and almost sealed.

The abandoned high-level passages of Air Jernih are huge tunnels along the strike, forming a sequence at successive levels on just a few major dipping bedding planes. Dip-oriented inlets include phreatic risers more than 100 m deep, now abandoned as dry shafts. Many of these mature passages are graded to their contemporary outlet levels. They originated as phreatic routes with varying extents of switchback looping, guided by joints across their host bedding planes. Downloops were removed by paragenetic roof dissolution associated with massive sedimentation, while canyons were cut through the uploops. Huge thicknesses of clastic sediments survive in many passages. The pattern of stream sinks around Api has changed over time, as alluvial fans from the sandstone of Gunung Mulu have built up against the limestone, with streams shifting easily across their convex surfaces. The caves have evolved in response, with complete reversals of flow in some of the strike-oriented passages that lie across the regional drainage direction.

Wall notches, typically 2 m high and cut 4–5 m into vertical walls, are a feature of the Mulu caves; a few are active, most are abandoned. They are formed by dissolutional undercutting at river levels, which are controlled by the resurgences onto the alluvial plain. The cave rivers rise and fall by the notch height, with the almost daily flood events. The main notches form downstream of large sinks at the alluvial plain base level, and the finest active examples are in the Cave of the Winds along the large streamway from the Melinau Paku to the Melinau.

A sequence of 20 notch levels in Gua Air Jernih, spread over 300 m of altitude and dated by their associated stalagmite and clastic sediments, records a base level that has fallen at a mean rate of 0.19 m per 1000 years, for 700 000 years (Farrant *et al.*, 1995). Phases of interglacial gravel aggradation alternated with base-level lowering during Pleistocene cold stages. Over about 10 Ma, the limestone hills of Mulu have evolved by remaining

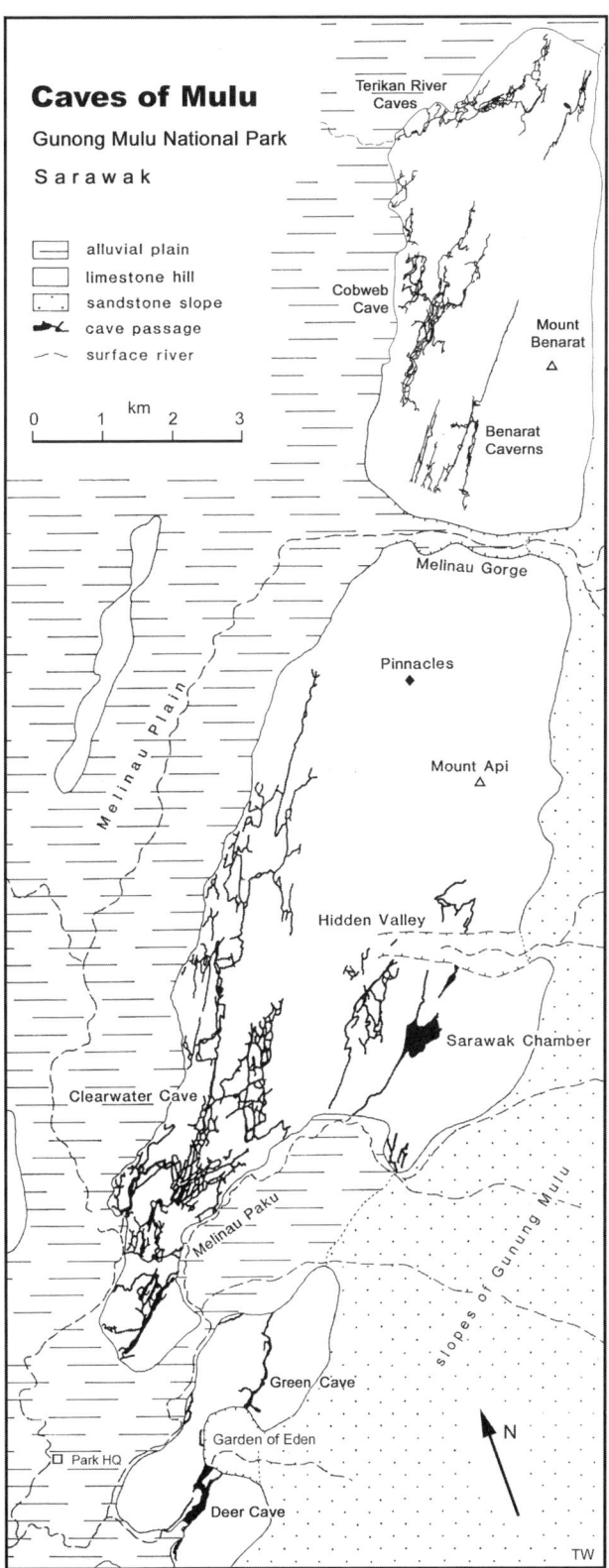

Mulu, Sarawak: Figure 1. Mapped cave passages in the limestone hills of the Gunung Mulu National Park, Sarawak (compiled from surveys by the British expeditions 1978–2001).

Mulu, Sarawak: Figure 2. The main river passage in Gua Air Jernih (Clearwater Cave) in Gunung Api. (Photo by Tony Waltham)

Mulu, Sarawak: Figure 3. The largely abandoned tunnel that is the main passage of Deer Cave, probably the largest passage in the world. The roof is more than 100 m above the show-cave path that winds across the floor. (Photo by Tony Waltham)

almost uneroded, while the adjacent plains and non-carbonate slopes have been intermittently lowered. Combined with steady tectonic uplift, this has had the effect of the limestone hills rising slowly from an almost stationary plain.

In the east side of Api, Lubang Nasib Bagus carries a second river underground, from Hidden Valley south to a resurgence beside the Melinau Paku. It drains through the base of Sarawak Chamber, which is 700 m long and 300–400 m wide, with an arched roof 100 m high. By far the world's largest cave chamber, this only survives due to the very massive nature of the Melinau Limestone. It appears to have originated as a wide inclined slot cut along a major dipping bedding plane by a powerful river; this migrated laterally, probably in part by waterfall retreat, and was matched by roof blockfall to create a single stable arch above.

South of Api, a smaller hill is breached by Deer Cave, with a passage 100 m high and wide for more than a kilometre of length—making it the world's largest cave passage (Figure 3). It is one of a series of larger, older caves that lie close to the eastern edge of the limestone. Between it and Green Cave, the Garden of Eden is an asymmetrical doline, 800 m across, with an inlet stream off the sandstone. The doline has grown as the stream sink has shifted its course off its own alluvial fan; wall scallops indicate that it flowed out through Green Cave at one stage. There is debate over how much the doline originated by collapse of a cave passage or chamber, but there is no evidence for the scale of individual collapse events within a progressive process. The concept of a single very large collapse does become real when the doline's size is compared to that of Sarawak Chamber, not far to the north.

North of Api, Gunung Benarat contains more very large caves. Cobweb Cave (30 km long) has a complex of old high levels, but with few long graded passages, unlike the layered, more mature tunnels of Benarat Caverns (8 km long) and the various caves of the Terikan River. North of Benarat, Gunung Buda is another smaller limestone hill, containing another 30 km of caves, explored mainly by American groups; they include Gua Kulit Siput, whose phreatic ramps reach 470 m above the entrances at plain level.

The surface karst of Mulu is dominated by forested pinnacles. Very steep slopes and high cliffs form the perimeter of the limestone mountains, whose high ridges are scored by deep dolines. All surfaces, steep and gentle, are fretted into pinnacles and blades each 1–10 m high, now shrouded in forest vegetation. The Pinnacles on Gunung Api are up to 50 m tall and rise clear of the forest canopy (Figure 4), but they are just the largest and best known of a widespread landform. Most pinnacles are razor-sharp subaerial forms, but rounded versions are evolving beneath

Mulu, Sarawak: Figure 4. The Pinnacles on Gunung Api with blades of limestone rising clear of the forest canopy. (Photo by Tony Waltham)

the thicker organic soils. The Melinau River carries the largest single flow off Gunung Mulu, and has maintained a surface course across the limestone outcrop between Api and Benarat; its gorge has vertical cliffs up to 500 m high, which cleanly truncate the ancient phreatic tubes of Benarat Caverns.

In addition to its extraordinary karst geomorphology, the Gunung Mulu National Park hosts an amazing variety of fauna and flora, both above and below ground. Deer Cave has over two million bats that roost on its high roof every day, and emerge most evenings in one of the world's most spectacular bat flights. In most of the Mulu caves, bats and swifts provide the guano that supports a pyramid of cave fauna, topped by large spiders and centipedes.

The Gunung Mulu National Park was established in 1975, to preserve a sample of Borneo's rain forest. The caves were intended to provide some visitor interest and financial support, but their sheer scale has turned them into a major tourist attraction. In 2001, the Park was designated as a World Heritage Site, on the strength of both its surface and underground features, which should support appropriate management and conservation of this magnificent karst.

TONY WALTHAM

Works Cited

Farrant, A.R., Smart, P.L., Whitaker, F.F. & Tarling D.H. 1995. Long-term Quaternary uplift rates inferred from limestone caves in Sarawak, Malaysia. *Geology*, 23: 357–60

Further Reading

Eavis, A.J. 1985. *Caves of Mulu '84*, Bridgewater: British Cave Research Association
Smart, P.L. & Willis, R.G. (editors) 1982. Mulu '80 expedition. *Cave Science*, 9: 55–164
Waltham, A.C. & Brook, D.B. (editors) 1978. *Caves of Mulu*, London: Royal Geographical Society
Waltham, T. 1997. Mulu—the ultimate in cavernous karst. *Geology Today*, 13: 216–22
Wilford, G.E. 1964. The geology of Sarawak and Sabah caves. *Bulletin of Geological Survey of Borneo Region, Malaysia*, 6: 1–181

MUSIC IN AND ABOUT CAVES

The amazing sonority and acoustic qualities of certain cave chambers have appealed to musicians since prehistoric times when footprints of dancers circling clay sculptures were made on the floor of caves in the French Pyrenees. Later religious ceremonies accompanied with instruments and chant were performed in Mayan caves of Central America and in Buddhist temple caves of Southeast Asia.

In historic times the earliest known performances of secular music in caves date to the end of the 18th century. In Peak Cavern (Derbyshire, England) around 1778, choral groups would climb around to the niche overlooking Roger Rain's House and sing for famous and wealthy travellers (Bray, 1778). Somewhat earlier in the Franche-Comté region of France, the provincial royal intendant organized a very successful festive banquet in the Grotte d'Osselle, involving music and dancing illuminated with hundreds of torches (Minvielle, 1970).

In 1822, dances were being held in the world-famous Postojnska Jama (then known as Adelsberger Grotte). Here on Whit Monday the Tournier Platz (now Congress Hall) was lit with rustic chandeliers and the locals would dance to traditional music (Russell, 1828). These dances continued annually throughout the 19th century. Farther back in a magnificent chamber the management later installed a stage and seating area where symphony concerts are still held occasionally.

In the latter half of the 20th century the idea of holding concerts in show caves during the summer caught on in several countries. In France there are jazz and folk concerts in the Aven Armand, Lozère and pan flute concerts in the Grotte de Lombrives, Ariège. In St Michael's Cave, Gibraltar, band concerts predominate. Once a year in Romania musicians hike for an hour carrying their instruments to Peştera Româneşti in the Poiana Rusca mountains to play classical concerts. On the isle of Kephalonia in Greece, classical and choral music are performed in Drogarati Cave. Contemporary and *concrète* music concerts were held in a relict gallery of Jeïta Cave in Lebanon before the outbreak of the war there. There are also winter venues such as the annual Christmas carol concert in the entrance chamber of Peak Cavern, England, which is attended by several hundred people.

Some show caves hold regular yearly festivals where world-class ballet troupes, symphony orchestras, and famous soloists come to perform. Theatre, choral groups, and jazz musicians have performed in the Balver Höhle in the Sauerland, Germany since 1949 (Allhoff-Cramer, 1996). Starting in 1960, the Cueva de Nerja near Malaga, Spain set up a large stage beneath a splendid speleothem display where many prestigious ballet troupes and soloists have performed (Ortega, 1970). The Baradla Barlang in northeastern Hungary has also hosted many fine concerts every summer (see photograph) and in South Africa the Cango Caves have been staging choral concerts for 40 years.

The European practice of holding concerts in caves was not carried over to North America. Instead, during the hot summer months there, the early colonists would take to their local caves

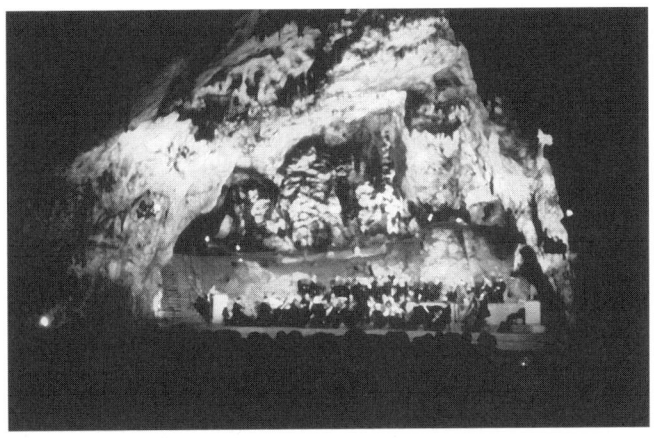

Music in and about Caves: Concert of cave-inspired classical and contemporary music in Baradla Barlang, Hungary, in July 1996.

and dance to live music. In 1839 hundreds of candles lit Weyer's Cave (now Grand Caverns) in Virginia and a band played for square dancing on the clay floor of the Ball Room. Then in 1878, with the opening of Luray Caverns, Virginia, a wooden floor was laid down, a band played, and anyone who wished to dance was charged 25 cents (Gurnee, 1978). As in many other caves worldwide, the cave guides would play simple tunes on a row of stalactites. This inspired Leland Sprinkle to create a "stalacpipe organ" in Luray Caverns by harnessing an organ console to various cave formations which when struck by rubber-tipped plungers would sound the desired notes. However, this entailed some damage to the cave where a few stalactites had to be ground down to achieve the proper pitch. The first concert where the cave itself became a musical instrument was in 1957.

During the 1920s and 1930s some caves in the southern United States proved ideally suitable for establishing speakeasy nightclubs with ballroom dancing. At Lost River Cave in Bowling Green, Kentucky; at Bangor Cave in Alabama; at Longhorn Caverns in Burnet, Texas; and at Wonderland Cave in Bella Vista, Arkansas, the customers danced to the music of name orchestras, but all these cave clubs had closed by the 1950s. Today nightclub-discotheque caves with recorded or live dance music can be found in Bermuda at Prospero's Cave, in Cuba at the Cueva del Pirate in Varadero, and in the Dominican Republic at Meson de la Cava in Santo Domingo.

Damage to the cave environment due to sustained use by nightclubs and discotheques is deplorable. However, the occasional use of cave spaces for concerts or dancing has been found acceptable in most cases. Hearing a great concert with the superb acoustics and surroundings of a natural cave chamber is a truly memorable experience surpassing imagination.

This special sonority combined with the emotions of wonder and mystery experienced in caves have inspired many musicians to compose music about caves. This music inspired by specific, actual caves around the world ranges through many categories including classical, contemporary, folk, country, jazz, and New Age. One of the first cave musical works and certainly the best known worldwide was the "Fingal's Cave Overture", a masterpiece of early Romantic music composed by Felix Mendelssohn Bartholdy following his visit in 1829 to this celebrated cave in the Hebrides of Scotland. Later in the 19th century two cave pieces for classical guitar were written: one by a Czech composer, Johann Mertz, entitled "Fingals Höhle", and the other by a French composer, Napoléon Coste, "La Source du Lyson (sic)", dedicated to the famous resurgence cave in the Doubs department.

Works of contemporary art music in the 20th century paid homage to various caves. In 1968, Zsolt Durkó composed a piece for concert orchestra and choir celebrating the renowned Altamira Cave in Spain. The *concrète* music composer, François Bayle, wrote an electro-acoustical work for the spectacular Jeïta Cave in Lebanon where sounds recorded in the cave were modified electronically and mixed with synthesizer music. In the realm of experimental music, Mariolina Zitta recently improvised a series of unusual pieces by tapping (without damaging) on the formations in caves of Sardinia and Liguria, Italy.

In the folk music field many short traditional works were inspired by Scottish caves: a bagpipe chant for the Cave of Gold at Harlosh, Isle of Skye, a march for the ubiquitous Piper's Cave, and another march for Huntley's Cave near Grantown. In the United States, following the tragic death of Floyd Collins in Sand Cave, Kentucky in 1925, six different event ballads were written and recorded in his honour. Jazz cave music seldom appears but in 1956 Raymond Scott composed, "Blue Grotto in Capri", and in the 1980s Kristian Blak wrote two long works, "Concerto Grotto" and "Antifonale", for two caves in the Faroe Islands. Recently New Age music composers have been finding caves very inspiring. Mathias Grassow set down three pieces dedicated to the famous lava tube caves on Lanzarote, Canary Islands. The British caver, Steve Thomas, composed a collection of synthesizer pieces all revolving around the cave diving experience ("More people have been to the Moon", 1997).

The timeless, magical surroundings of caves continue to provide ideal venues for musical performances and, hopefully, the awesome beauty of caves will inspire many more compositions.

DAVID N. BRISON

Works Cited

Allhoff-Cramer, A. 1996. *Die Festspiele Balver Höhle*, Balve: Festspiele Balve Höhle

Bray, W. 1778. Cited in Woodall, B. 1976. *Peak Cavern*, Buxton: Brian Woodall

Gurnee, R.H. 1978. *Discovery of Luray Caverns, Virginia*, Closter, New Jersey: Gurnee

Minvielle, P. 1970. *Guide de la France Souterraine*, Paris: Tchou

Ortega, E. Rodriquez. 1970. *La Cave de Nerja*, Nerja: Patronato de la Cueva de Nerja

Russell, J. 1828. *A Tour in Germany and Some of the Southern Provinces of the Austrian Empire in 1820, 1821, 1822*, Edinburgh: Constable; cited in Shaw, T.R. 1980. British and American travellers in the Cave of Postojna before 1865, *Naše Jame*, 22: 101

Further Reading

Brison, D. 1966. Caves on phonograph records, *York Grotto Newsletter*, 7(5): 65–71, 7(6): 79–84

Brison, D. 1997. Les grottes dans la musique et la chanson enregistrées, *Grottes et Gouffres*, 144: 19–28; see also same text in English: Caves celebrated in recorded music and song, *Al-Ouat'Ouate*, 11, 1997–1998: 96–107

Brison, D. 2001. The early Floyd Collins ballads. *Journal of Spelean History*, 34 (3): 3–24

Mills, M. 1986–1989. Caves in music, *Shepton Mallet Caving Club Journal*, 8(1): 22–37, 8(4): 145–68, 8(6): 244–84

MYRIAPODA (CENTIPEDES AND MILLIPEDES)

Myriapoda is an artificial superclass of about 15 000 species, containing all terrestrial Arthropods having antenna, mandibles, and above all more than ten pairs of legs (terrestrial Isopoda, Arachnida, and Insecta have respectively seven, four, and three pairs). The two subclasses centipedes (Chilopoda) and millipedes (Diplopoda) are the best represented in the subterranean world. Others (Pauropoda, Symphyla, and Penicillata) are microscopic edaphic (soil-dwelling) myriapods that are occasionally cave-dwelling but rarely true troglobites.

Centipedes (about 2500 species distributed in 4 orders) have a flattened body, with a supple leathery integument, only one pair of legs per segment, and three pairs of mouth parts. Most (except the worm-like Geophilomorpha) are very agile and all are predators in which the first pair of legs is modified into poison-claws. Only about 50 species (mostly Lithobiomorpha) can be considered as troglobitic. Geophilomorpha (more than 25 pairs of legs) inhabit leaf litter or soil fissures, and are rarely cavernicolous (though there are examples from Cuba, France, and Movile Cave, Romania).

Within the Scolopendromorpha order (21 pairs of legs), only one family (Cryptopidae) has a few species showing adaptations such as depigmentation or elongated appendages (Cuba, United States) or species known only from caves (Cuba, Italy, and Yugoslavia). Scutigeromorpha (antenna and 15 pairs of considerably elongated legs) has some species inhabiting caves in Mediterranean countries (e.g. the trogloxene *Scutigera coleoptrata*). Lithobiomorpha (15 pairs of legs, usually elongated) are mostly represented in European and circum-Mediterranean caves by about 40 species belonging to the genus *Lithobius*. The majority of these species are endemic having a very limited geographical distribution, and are dispersed from the Pyrenees to the Carpathians and Balkans. Only two cave-dwelling species are known from the Far East, and none in the New World. Of the characteristics usually associated with adaptation to subterranean habitats, loss of eyes and pigmentation are the most frequent expression. Less frequent is the elongation of appendages, which gives an unusual "scutigeroid" (or Scutigera-like) aspect in some Lithobius, such *L. drescoi* (Spanish Basque province). An increase in the number of antennal articles is even more rare, though *L. sbordonii* (Sardinia) has 111.

Millipedes (about 10 000 species distributed in 12 orders) have two pairs of legs per segment and only two pairs of mouth parts. Segments are usually cylindrical, often bearing paranota (lateral projections), each one concealing a pair of repugnatorial glands. Generally slow-moving and eating dead plant material, their relatively permeable integument confines them to hygrophilous and cryptophilous habitats. A large number are accidental inhabitants of caves (trogloxene), and although the number of troglobites is greater than in centipedes (about 300–400 species), only five orders have true cavernicolous forms.

If the classic adaptive characteristics such as depigmentation, loss of eyes, and decalcification of integuments are frequent, other characteristics such as elongation of appendages or increased body size, are not the rule. In orders having a variable number of segments (such Iulida and Lysiopetalida), this number increases, but it stays unchanged (or rarely decreases) in orders having a fixed number of segments (such as Glomerida, Polydesmida, and Chordeumida). A special adaptation concerns the mouth parts (enlarged lamellae which act as a filter) in various millipedes inhabiting waterlogged caves.

Glomerida are characterized by their small number of segments (11–12) and their ability to roll up in a sphere. Cavernicolous species are generally smaller than epigean ones, and are blind and unpigmented. The true troglobites are distributed in a dozen endemic genera (such as *Speleoglomeris*) which are all European, distributed from the Pyrenees to the Caucasus and Balkans. The genus *Trachysphaera* (previously *Gervaisia*), while not exclusively troglobitic, is remarkable for its decorative integument and its distribution from the Cantabric Mountains to the Caucasus.

Polydesmida are remarkable for their segments (usually numbering about 20) having usually strong lateral projections, and containing glands that often secrete cyanide. All are blind and many species are troglophilic or troglobitic. These last are divided into two categories: those whose epigean close relatives are sympatric, such as the poorly modified European *Polydesmus* (= *Brachydesmus*) and the Japanese *Epanerchodus*; and those (all small or microscopic forms in size) whose epigean close relatives are allopatric and are relics of a former tropical fauna—they are distributed in eight small genera in Europe (e.g. *Trichopolydesmus*), and five in North America (e.g. *Speodesmus*).

Chordeumida (= Craspedosomida) are relatively agile millipedes living in the temperate countries of the world. Small or moderate in size, their segments (usually numbering about 30), without repugnatorial glands, always bear three pairs of dorsal hairs. They possess preanal spinnerets secreting silk. The endemism rate is very important in this order, and thus troglophilic and troglobitic forms are particularly numerous. Some genera contain both epigean and troglobitic species; the latter show only poor morphological adaptations (e.g. the European *Nanogona* and US *Cleidogona*), modifications in the sense of reduction of size and number of ocelli (e.g. *Opisthocheiron* in southern France), or more important modifications in the sense of elongation of appendages (e.g. *Psychrosoma* from Spain). But usually species having the full range of adaptations to the subterranean habitat are distributed in some small endemic genera scattered in all karstic countries of Europe, North America, and the Far East.

The body of Iulida is usually very elongated in shape, composed of numerous cylindrical segments, and their repugnatorial glands secrete quinones. Palearctic species are represented by numerous troglophilic and troglobitic taxa that are mainly members of two families: the rather eastern Iulidae and the rather western Blaniulidae. All have a similar shape caused by the loss of pigmentation and elongation of body and appendages. Four genera have a relatively large area of distribution, such as *Mesoiulus* (discontinuous area from Turkey to Spain), *Blaniulus* (= *Typhloblaniulus*, common in caves of southern France), and *Metaiulus* (terricole or soil-inhabiting in Great Britain, troglophilic in south France). The latter, and also a dozen endemic small genera having restricted distribution (in western Europe and North Africa) have affinities with their epigean (and sympatric) homologues. But some of them (such as the Spanish *Paratyphloiulus*) belong to a line that possesses allopatric epigean homologues living in Nepal and Southeast Asia. Cave-dwelling species are rarer in palearctic Asia and in North America: they belong to other families (Paraiulidae and Mongoliulidae). In tropical and subtropical countries, the family Glyphiulidae, remarkable for its multicarenate segments, inhabits numerous caves from south of China, India, and Southeast Asia. In Central America, the genus *Jarmilka* (from Belize) has a number of segments particularly low for Iulida (18). In South America, only the genus *Pseudonannolene* has some species presumed troglobitic (in Brazil).

Callipodida (= Lysiopetalida) are iulida-like in shape (but slightly compressed laterally) and have preanal spinnerets producing silk and repugnatorial glands secreting phenols. Many are carnivorous and live in the hottest of the temperate zones of the northern Hemisphere, often in hypogean habitats. There are no notable modifications to the external morphology of the European cave-dwelling genera, such as *Apfelbeckia* (former

Yugoslavia) or *Callipus foetidissimus* (cryptophilous in southern France, troglobitic in the catacombs of Paris). However, *Tetracion* (United States) and *Paracortinus* (China and Vietnam) are unpigmented, and show elongated appendages and reduced eyes.

JEAN-PAUL MAURIÈS

Further Reading

Ginet, R. & Decou, V. 1977. *Initiation à la Biologie et à l'écologie souterraines*. Paris: Delage

Hopkin, S.P. & Read, H.J. 1992. *The Biology of Millipedes*, Oxford and New York: Oxford University Press

Juberthie, C. & Decu, V. (editors) 1994. *Encyclopaedia Biospeologica*, vol. 1, Moulis and Bucharest: Société de Biospéologie
For Myriapodes, see pages 249–66, which includes a bibliography of 128 titles, 12 figures, and 3 maps (in French)

Lewis, J.G.E. 1981. *The Biology of Centipedes*, Cambridge and New York: Cambridge University Press

Shear, W.A. 1969. A synopsis of the cave Millipeds of the United States, with an illustrated key to genera. *Psyche*, 76(2): 126–43

NAHANNI KARST, CANADA

Nahanni National Park (the first UNESCO World Heritage Site) extends 230 km along the South Nahanni River, where it crosses the central and eastern ranges of the remote Mackenzie Mountains in northern Canada (see Canada Karst and Caves, Figure 1, for location). Valley to summit elevations range from 200 to 2600 m. The history of glaciation is complex—the western half experienced repeated alpine glaciation and the eastern quarter continental glaciation, but a central corridor was always ice-free, due to its aridity. The modern climate is subarctic, mean annual temperatures being $-4°$ to $-12°C$ and precipitation 400–600 mm, roughly half of which falls as snow.

In the western park, a granitic batholith of Cretaceous age is intruded into mixed carbonate and clastic strata, deforming them. At the contact are Rabbitkettle Hotsprings, calcite travertine deposits, which are the largest in Canada. "North Mound" is a classic tiered wedding-cake structure, Holocene in age, 250 m in basal diameter and very colourful. The springs are perennial, with temperatures $\sim 21.5°C$ and total hardness of 475–550 mg l^{-1} as $CaCO_3$. Nearby (and at Yohin Lake downstream) are dozens of large piping sinkholes (see Pseudokarst entry) in river terrace silts resting on coarse gravels.

The principal karst is in the eastern park, on 140–180 m of thick, massively bedded Devonian limestones and >1000 m of weaker, banded dolostones beneath them. The limestone is overlain by black shales. The strata are folded into broad, anticline domes rising to an altitude of 1500–1600 m. The South Nahanni and its tributaries have carved meandering, antecedent canyons through them, with walls as high as 1200 m (Ford, 1991). The karst forms a belt up to 20 km wide between the South Nahanni River First Canyon dome and Ram Canyon dome 50 km further north. Waters flow into it from the crests of the domes and from flanking shale lowlands. The underground drainage appears to be relatively simple. Groundwater flows south to perennial springs in First Canyon that are $c.520$ m below the stratigraphic top of the dolostones, or north to springs emerging where the limestone dips under the shale cover. The active cave channels are inaccessible but dye tracing suggests flow rates of 20–30 m h^{-1} along major routes. The systems are very dynamic, floodwaters quickly rising 50 m in some big dolines. The waters have a bicarbonate composition, with a total hardness of 100–200 mg l^{-1} at the springs and the estimated regional dissolution rates are equivalent to 18–27 mm of lowering per 1000 years (Brook & Ford, 1982).

The outstanding landforms are dissolutional corridors ("streets") in the limestone that follow major vertical fractures created by the doming (Brook & Ford, 1978). Individual corridors are 30–100 m deep, 15–100 m wide, and up to 6 km in length. For a distance of 13 km they intersect to form a natural labyrinth. The walls recede from frost shattering, causing some parallel corridors to amalgamate into broader closed depressions, like squares in a pattern of city streets; the greatest measures 800

Nahanni Karst: Raven Lake, a doline within a karst corridor, Nahanni Karst. The doline is 300 m in length. Cliffs rise 150 m above the waterline, which itself may rise a further 50 m. The lake drains dry in winter. (Photo by Derek Ford)

× 400 m. Isolated towers are preserved within them. Floor profiles of corridors and squares are highly irregular, with local streams sinking into depressions between talus accumulations or into bedrock shafts.

At the north end of the labyrinth the shale cover and glacifluvial sands encroach to reduce the limestone outcrop to a narrow spillway with three small (<2.5 km^2) but elegant poljes developed in it. There are many collapse and suffosion dolines in the shales and sands. In the labyrinth and elsewhere on the limestone are large, vertical-walled dolines (see photo) and smaller, elliptical dissolutional shafts. Many trap the water of successive melt seasons, depth increasing slowly until pressure bursts an ice plug below and the feature drains with catastrophic rapidity.

Flanks of the domes are incised by consequent streams, creating canyons up to 800 m deep. Ancient glacial moraines blocked the downstream ends of seven of these streams, which now drain underground into the limestone or dolostone. Lafferty Canyon, 24 km in length, is an intermediate case; all waters sink into the dolostone except during occasional summer flash floods, when overflow into South Nahanni River rolls 50-tonne boulders along the bed.

More than 200 relict caves are known but most are sealed off by ground ice or permafrozen silts within a few metres. Grotte Louise–Grotte Mickey (*c.*3 km long) is a multilevel series of phreatic passages exposed in limestone cliffs at the junction of First Canyon and Lafferty Canyon, 200 m above the South Nahanni River: it has complex clastic fills, seasonal and perennial ice (Schroeder, 1977). Grotte Valerie (*c.*2 km long) opens in south-facing cliffs 300 m above the river. It was created by influents from the surface 60 m overhead draining to a major bedding plane where a sequence of shapely arched phreatic passages developed. In summer, cold air drains from a low exit, drawing in warm air to replace it through an entry 40 m higher. This creates (1) a "warm entrance cave" (+6 to +1°C) with active speleothem growth, supplying moist air to (2) a "cool exit cave" (0 to −1.5°C) covered with ice and hoar frost: behind and below both is (3) a "permafrost cave" (−2.5°C) receiving only the cold air of winter, dry and dusty, without speleothems or ice and preserving the remains of 80 or more wild sheep. It is a spectacular example of cave climatic zonation.

The Nahanni is the most accentuated periglacial karst yet reported from the Arctic or Subarctic. Its development in a cold, rather dry climate is attributed to the young domal fracturing and the fact that the main belt has not been over-ridden by glaciers recently, perhaps not for 350 000 years or more.

DEREK FORD

See also **Alpine Karst; Glacierized and Glaciated Karst**

Works Cited

Brook, G.A. & Ford, D.C. 1978. The nature of labyrinth karst and its implications for climaspecific models of tower karst. *Nature*, 275: 493–96

Brook, G.A. & Ford, D.C. 1982. Hydrologic and geologic controls of carbonate water chemistry in the sub-Arctic Nahanni karst, Canada. *Earth Surface Processes and Landforms*, 7(1): 1–16

Ford, D.C. 1991. Antecedent canyons of the South Nahanni River. *The Canadian Geographer*, 35(4): 426–31

Schroeder, J. 1977. Les formes de glaces des grottes de la Nahanni, T.N.O. *Canadian Journal of Earth Sciences*, 14(5): 1179–85

Further Reading

Ford, D.C. 1984. Geology; geomorphology; hydrology. In *Nahanni National Park Reserve: Resource Description and Analysis*, Winnipeg, Manitoba: Parks Canada

Hartling, R.N. 1993. *Nahanni: River of Gold, River of Dreams*, Hyde Park, Ontario: Canadian Recreational Canoeing Association

NAKANAI CAVES, PAPUA NEW GUINEA

The Nakanai mountains are located on the island of New Britain, east of Papua New Guinea (see the map in the Asia, Southeast Islands entry). They are world famous for their megadolines, which carry huge underground streams (Table), in the middle of one of the world's last tropical rain forests. The island's Miocene limestone platform lies on an old paleogene volcanic arc located along the south coast. An active volcanic arc lies along the northeast coast, and the limestones were covered by a thick Pliocene volcano-sedimentary series.

The Nakanai mountains include a large plateau (5000 km^2 in area), with a high point at an altitude of 2185 m in the north, which descends to the south coast and is cut by canyons more

Nakanai Caves, Papua New Guinea: Major caves and dolines of Nakanai

Cave	Depth	Exploration team	Observations	Low-water discharge
Muruk	−1178 m	France (1985, 1995, 1998)	Through-cave to Berenice, 17 km long,	2 m^3 s^{-1}
Gamvo	−478 m	United Kingdom (1985)	Stream-sink cave	5 m^3 s^{-1}
Arcturus	−475 m	France (1988)	Muruk system (not connected)	
Minye	−468 m	Australia (1968), France (1978, 1985)	Megadoline 26 M m^3, 410 m shaft	20 m^3 s^{-1}
Ka 2	−414 m	France (1978, 1980)	Kavakuna system (not connected)	20 m^3 s^{-1}
Bikbik Vuvu	−414 m	France (1978, 1980)	Megadoline 141 m shaft	
Nare	−415 m	France (1978, 1980), United Kingdom (1985)	Megadoline 4.7 M m^3, 238 m shaft	20 m^3 s^{-1}
Kavakuna	−392 m	France (1978, 1980)	Megadoline	7 m^3 s^{-1}
Ora	−270 m	Australia (1972–1973)	Megadoline	
Luse	−224 m	France (1980)	Megadoline 61 M m^3	

Nakanai Caves, Papua New Guinea: Figure 1. Nare megadoline (300 m), with the Vaisseau Fantôme river flowing at the bottom (15–20 m³ s⁻¹). (Photo by J.-P. Sounier)

than 1000 m deep (e.g. Galowe, Wunung, and Bairaman). Since the beginning of the late Pliocene uplift, the volcano-sedimentary deposits have been weathered and eroded, bringing the Yalam limestones to the surface and allowing karst processes to take place. The clay sediments presently exist as a covering of detritus, only a few metres thick, overlying the karst relief. This combination results in the formation of rounded hills and deep depressions, connected by small valleys.

The mountains have an oceanic monsoon climate, with considerable rainfall (6 m per year on the coast, 10 to 12 m per year in the mountains). This significant and continuous humidity, combined with high temperatures all year round, allows the growth of rain forest.

The underground karst benefits from three types of hydrological input. Diffuse drainage across thick soils and tiny fissures allows saturation and gives rise to calcite deposition in shallow caves (upstream, Muruk Cave). Rapid drainage into larger fissures causes flow-path enlargement at moderate depths. Allogenic aggressive drainage first concentrates in thalwegs or canyons (such as Kavakuna) during high water and is then absorbed into stream sinks. High water is permanent during the rainy season and frequent during the "dry" season, so solution can easily occur in the deep and remote parts of the caves.

In the "dry" season the karst is fed essentially by diffuse drainage, with occasional episodes of turbid runoff entering via the surface gully system, but these are limited in both time and volume (two to three times the low-water discharge). The forest / soil system plays a remarkable role. Just one or two days without rain are enough to drain the soils, which are then able to absorb up to 50 mm of rain without causing any runoff. But if the soil is saturated from previous rainfall, 20 mm of rain are sufficient to produce runoff. During heavy rainfall, in some large catchments, discharge can reach several $m^3 s^{-1}$ (10 times the low-water discharge), probably up to 100 $m^3 s^{-1}$ during the highest water, and 1000 $m^3 s^{-1}$ during exceptional flooding (50 times the low water discharge in the Kavakuna Matali system).

Karst processes are influenced by exceptional conditions, with pure limestones being subjected to solution. The high rainfall supplies large amounts of water, which is concentrated on the clay floors of valleys before being injected into the karst, and the rain forest, with a highly productive biomass, provides abundant carbon dioxide and humic acids. In addition there is a steep topographic gradient due to the high tectonic activity. For these reasons, karstic denudation is estimated at a world record of 400 $m^3 km^{-2} a^{-1}$ (Audra, 2001; Maire, 1990). Nakanai can be considered as a "hyperkarst", since so many factors favouring karst processes are pushed to their extreme. The Nakanai caves are therefore on a dramatic and very large scale, and 80 km of conduits have been surveyed by 11 expeditions.

The formation of the megadolines is due to the upward stoping of the ceilings of large passages (up to 150 m high) below the floors of dolines (up to 150 m deep). If the drains are at a depth of around 350 m (e.g. Nare, Minye), the thickness of rock in the roof can be quite thin. If it collapses, it can form

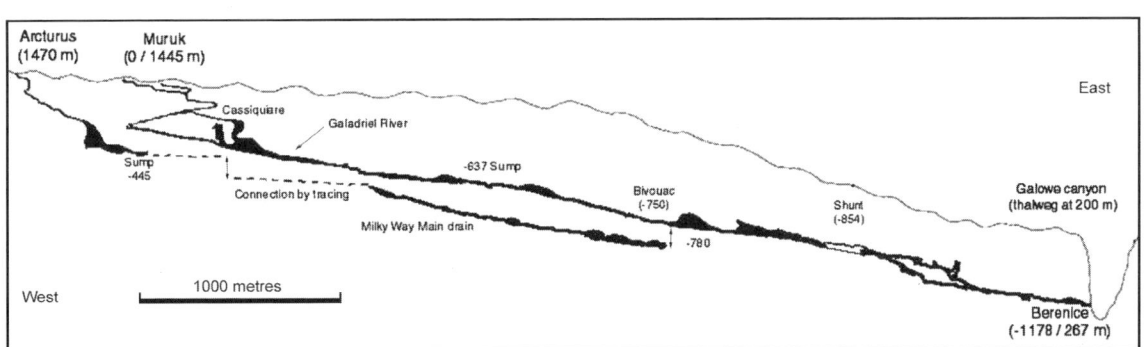

Nakanai Caves, Papua New Guinea: Figure 2. Profile of Muruk system.

Nakanai Caves Papua New Guinea: Figure 3. The main passage in Nare. (Photo by Dave Gill)

open windows to the subterranean rivers below (Maire, 1981; Figure 1). Where the depth of the drain is deeper (>450 m in Muruk), collapses do not exist (Audra, 1999). Such large passages develop in soft chalky limestone, by downcutting linked to a high flow rate.

The Muruk system began as a tube sloping gently from a stream-sink to a spring (Figure 2). It evolved by entrenchment into a canyon hollowed out by giant potholes, and downstream into large passages (>50 m in diameter). It is a monophase system, which is exceptional in a mountain environment, with no perched passages except in the vicinity of the spring. The Berenice resurgence is perched 50 m above the Galowe river, which is mainly fed by the Mayang spring (flowing at 20 m^3 s^{-1}). This shows the lag time between karst processes as compared with fluvial entrenchment, which is the result of the rapid uplift. The morphology is typical of a juvenile karst system.

Paleomagnetic dating of fluvial sediments originating from reworked soils shows normal polarities (i.e. an age of <780 ka). Uranium-series dating on speleothems has only provided ages younger than 50 ka. The Muruk system was probably initiated around 100 to 200 ka, as shown by the Berenice perched spring at 50 m above base-level. The efficiency of the karst processes, which led to such extraordinary dimensions, is peculiar to the Nakanai mountains.

Cave genesis starts with water flowing over the clay cap and entering the limestone through the stream-sinks. It then follows a nearly straight line to the springs through limestone that was not karstified in its deeper parts. Subsequently, the cave system evolved by entrenchment and enlargement caused by torrential flows, mainly in the vadose zone. The initial cave outline was not greatly changed by subsequent evolution. This differs greatly from the common karst model, where shafts drop vertically through the vadose zone and horizontal galleries are located near the water table. Muruk can be regarded as a juvenile system model. As a case study it can aid in the understanding of other more ancient and generally more complex systems with similar profiles.

PHILIPPE AUDRA AND RICHARD MAIRE

Works Cited

Audra, P. 1999. Genèse des grands vides souterrains du réseau de Muruk, influence des séismes sur le karst (Montagnes Nakanaï, Nouvelle-Bretagne, Papouasie-Nouvelle-Guinée). *Etudes de géographie physique*, 28 (suppl.): 37–42

Audra, P. 2001. Karst solution and denudation assessment in Nakanai mountains. In *Nakanai, 20 Years of Exploration*, Antibes: Hemisphere Sud

Maire, R. (editor) 1981. Papouasie-Nouvelle Guinée. *Spelunca*, 3 (suppl.)

Maire, R. 1990. *La haute montagne calcaire*. Thesis, Paris: Fédération française de spéléologie and Bordeaux: Association française de karstologie (Karstologia Mémoires, 3)

Further Reading

Audra, P., Lauritzen, S.E. & Rochette, P. 1999. Datation de sédiments (U / Th et paléomagnétisme) d'un hyperkarst de Papouasie-Nouvelle-Guinée (Montagnes Nakanaï, Nouvelle-Bretagne). *Etudes de géographie physique*, 28 (suppl.): 43–48

Audra, P., de Coninck, P. & Sounier, J.-P. 2001. *Nakanai, 20 Years of Exploration*, Antibes: Hemisphere Sud

Gill, D.W. (editor) 1988. *The Untamed River Expedition 1985*, Stockport: Gill

Hache, P., Hobla, F., Philips, M., Sessegolo, D. & Sounier, J.-P. 1995. Muruk, hémisphère sud, premier −1000. *Spelunca*, 60, 35–54

Pernette, J.-F. 1981. *L'abîme sous la jungle*, Grenoble: Glénat

Sounier, J.-P. 1995. *Nakanaï: Dans les gouffres géants de Papouasie*, Marseille: Spelunca Librairie

Sounier, J.-P. 1996. Muruk: the Southern Hemisphere's first −1000. *International Caver*, 17: 17–25

Sounier, J.-P. 2000. Muruk: l'épilogue? *Spelunca*, 77: 15–22

NEW ZEALAND

The main rocks supporting karst in New Zealand are Oligocene limestones and Ordovician marble (Figure 1). During the mid-Oligocene limestones were deposited over most of the Northland peninsula and along the east and west coasts of what are now both North and South Islands. However, plate tectonic deformation along the Pacific–Australian plate boundary, associated uplift since the Miocene, and considerable erosion has reduced the outcrop to discontinuous patches widely distributed throughout the length of the country. Miocene to Plio-Pleistocene limestones also occur, especially along the east coast of North Island, although they tend to be poorly cemented coquina limestones (limestones composed mainly of fossil debris) with high porosity and little karstification beyond widened fissures. Ordovician marble is found mainly in the northwest of the South Island

New Zealand: Figure 1. Map of outcrops of limestone and marble in New Zealand. Most carbonate rocks have karst development, especially in their groundwater systems, although surface karst landforms and caves are best developed on marble and dense crystalline limestones. Occurrences of lava caves near Auckland are also indicated. Marble outcrops at Takaka, Mt Arthur, and Mt Owen.

(northwest Nelson). Caves are also found in Quaternary basalts, especially in the vicinity of the city of Auckland in North Island, where there are about 60 short lava tube caves (see photo in Pseudokarst entry).

North Island

The Northland peninsula has many small limestone outcrops (Figure 1), some with caves, although much is too argillaceous to support well-developed karst. The most important karst is near Waipu, about 100 km north of Auckland, where doline karst with karren fluted outcrops covers a few square kilometres. The limestones exceed 100 m in stratigraphic thickness and are pure but thinly bedded. Two Tone Cave is developed at the unconformable contact of the Oligocene limestone and underlying Mesozoic greywacke.

The main karst area in North Island is in the west between Kawhia and Aria, in a district referred to locally as the King Country (Figure 1). The karst occurs at altitudes below 400 m and currently experiences a moist, warm temperate climate. The mean annual temperature of caves and mean annual rainfall range from 13°C and 1600 mm at 100 m above sea level, to about 11.5°C and over 2300 mm in the hills at around 350 m. About 20% of the original podocarp-hardwood evergreen rainforest remains, usually in reserves. Carbon dioxide production in volcanic soils overlying the karst has been investigated by Gunn and Trudgill (1982).

Karst is developed in 100 m thick, step-faulted, Oligocene limestones blanketed by thick volcanic ash. Prominent north–south faulting results in the limestones ascending from Waitomo towards the west. As a result of Upper Tertiary and Quaternary tectonic and erosional processes, the limestone outcrops and subcrops now present a fragmented mosaic and in total cover about 800 km^2 of the region. Dips are gentle, seldom exceeding 10–15°. The limestones are part of the Te Kuiti Group (White & Waterhouse, 1993), which comprises calcareous siltstones and sandstones near its base and thinly bedded limestones above. Occasionally the lower clastic formations are underlain by coal measures, but sometimes both are absent and the limestones rest directly and unconformably on Mesozoic greywackes. The limestones are divided into a lower Orahiri Limestone and an upper Otorohanga Limestone, in places separated by the calcareous Waitomo Sandstone. The limestones are particularly notable for their thinly bedded flagginess, which is partly a consequence of pseudo-bedding along stylolites, and most beds are of the order of only decimetres in thickness, just occasionally exceeding a metre. Associated with this thin bedding is closely spaced jointing. Thus the limestones have a very high fissure frequency, which is exploited by hydrogeological processes.

Karstification commenced in the upper Tertiary prior to the burial of many dolines by mid-Pleistocene ignimbrites, many of which have since been exhumed. The southern shore of Kawhia Harbour along the edge of the Rakaunui Peninsula has karst that descends below sea level, with partly drowned dolines containing remnants of ignimbrite fill. There are also well-developed intertidal karren and, in places, shallow multiple solution notches reflecting the position of former sea levels. These reach up to about 2 m above sea level. In the larger patches of limestone around Waitomo, the landscape is notable for very good examples of polygonal karst, with doline densities of 55 per km^2. Most are solution dolines, although impressive collapse dolines in bedrock and suffosion dolines (usually in tephra infills within solution dolines) also occur. Hydrological processes in solution dolines have been described by Gunn (1981b).

There are numerous caves in the King Country, the longest being Gardners Gut Cave (>12 km; Figure 2), a totally vadose dendritic river cave that can be traversed from stream-sinks to resurgence beside the Waitomo Stream. It probably commenced to form about 250 000 years ago as the overlying cap of Miocene mudstone was eroded to reveal inliers of underlying limestone. The rate of incision of the Gardners Gut cave stream has been estimated by Williams (1991) as 0.13 m ka^{-1} by dating the basal calcite of a flowstone (118 ka BP) and speleothems (12 ka BP) at various heights above the modern cave stream. This is of the same order as the rate of uplift of the region. Rates of karst denudation in the uplands of the district have been measured by Gunn (1981a) as 69^{+18}_{-7} m km^{-2} a^{-1} (or mm ka^{-1}), which is about half the rate of uplift.

Many caves in the region have spectacular glowworm displays (Pugsley, 1984). Some are of international significance and of importance to the tourist industry, especially in the Waitomo

New Zealand: Figure 2. Calcite speleothems in Gardner's Gut Cave in the King Country Karst. (Photo by Andy Eavis)

Glowworm Cave, the climate of which has been studied by de Freitas et al. (1982) and de Freitas and Littlejohn (1987). Waitomo is also the original home of Blackwater Rafting, which commenced as a tourist venture in Ruakuri Cave.

Extensive limestone outcrops occur in a discontinuous belt over about 450 km along the East Coast of North Island. The limestones rise to 1300 m in places and form distinctive scarps. The carbonates are mainly Upper Tertiary to early Pleistocene in age, and vary in composition from dense crystalline limestone to friable coquina. Particularly important from the point of view of karst development is the Whakapunake Limestone, of mid-Pliocene age, a cream-coloured, well-bedded, hard barnacle plate coquina up to 300 m thick. Despite the young age of many limestones and the recent uplift, there is significant karstification including cave development and active groundwater networks, especially in the more indurated limestones. The most significant caves are in the Wairoa district of northern Hawke Bay. Especially important is the Te Reinga Cave system (4 km long, 134 m deep) developed in Whakapunake Limestone. However, thick sandstone coverbeds over the limestones inhibit the development of solution dolines, most closed depressions originating by progressive collapse of clastic sediments into developing caves.

The Te Aute Limestone south of Hawke Bay is often less well cemented and more porous than at Te Reinga, so karstification is less advanced. For example the Te Mata Peak, Maraetotara Plateau, and Kahuranaki uplands are composed of a soft friable coquina limestone that varies from 30–330 m thick. Early stages of karstification are evident along joints and there are occasional closed depressions, springs, and calcareous tufa deposits. The groundwater circulation is unstudied, but probably mainly diffuse.

South Island

Extensive areas of cavernous karst occur in the northwest corner of the South Island, in northwest Nelson, and northern Westland (Figure 1). Tertiary limestones occur as a discontinuous strip for over 250 km along the northwest coast, and in places extend inland where they have sometimes been uplifted to 1000–1500 m, for example in The Haystack and Thousand Acre Plateau of the Matiri Range north of the Buller Gorge. The limestones are very variable in thickness and in lithology, from thin and sandy to thick and crystalline. In the Paturau district the Takaka Limestone is 25–50 m thick, whereas at Tarakohe it is 80 m thick but sandy in its lower half. Near Charleston and Punakaiki in north Westland, the Tertiary carbonate sequence is highly variable in thickness but in places exceeds 700 m. The Potikohuna Limestone is particularly important for karst development, being a dense polyzoan biosparite. It is stylobedded and has a platy appearance in weathered outcrop, giving rise to the Pancake Rocks where an uplifted last interglacial marine platform (34–36 m) has been partly stripped of gravelly beach sediments by surf and spray. The flaggy layering appears related to the differential exploitation of insoluble clays and micas formed along stylolites by pressure solution. Inland from the coast at Punakaiki, the limestones underlie a plateau at 300–400 m that is heavily dissected by solution dolines, forming the southernmost polygonal karst in New Zealand. Dolines, karren, stream-sinks, springs, and gorges are well developed. Because of the copious rainfall (up to 3 m annually), there is a high density of caves, some of which have rich fossil avifauna deposits of international significance, Honeycomb Hill Cave being a particularly important site (Worthy & Mildenhall, 1989). The area also contains excellent examples of underground river capture, Bullock Creek-Cave Creek being the most well-known case (Figure 3). The capacity of the stream-sink zone is 15–20 $m^3 s^{-1}$ and the combined peak discharge of outflow springs is of the order of 30–40 $m^3 s^{-1}$, the largest flood flow from any karst system in New Zealand. Most of the area is rugged and covered in dense evergreen rainforest with essentially intact natural vegetation making it virtually impenetrable. This is important for the conservation of many endemic species of plants and invertebrates.

A greater diversity of karst features is found in the Arthur Marble, which crops out extensively in northwest Nelson (Figure 1). It is found mainly in the mountains to 1875 m, but extends below sea level in Golden Bay. The marble is Ordovician in age, about 1000 m thick, and occurs in a discontinuous belt for about 90 km. It is largely covered by undisturbed natural vegetation, some in the alpine zone, and is mostly within two National Parks. Mt Arthur and Mt Owen were both glaciated in the Pleistocene and provide the best examples of glacio-karst in the Southern Hemisphere with dry cirque basins, karrenfeld, and limestone pavements (Williams, 1992a). They contain the five deepest caves in the country and the three longest (see Table). Nettlebed Cave, the deepest (889 m), and Bulmer Cave, the longest (50.1 km), both have a development history that is likely to exceed 1 million years. The Takaka valley contains New Zealand's largest spring, the Waikoropupu Spring, with an average discharge of $c.15$ $m^3 s^{-1}$ and an age spectrum of outflowing waters of 2–20 years (Williams, 1992b). Water clarity measurements have shown it to have the third clearest water yet recorded in the world and the aquatic ecology of the spring has international significance. The spring also has local cultural significance to the Maori people. However, much of the catchment area is under farmland with intensifying landuse and is not protected.

The Riwaka River catchment in northwest Nelson has an average rainfall of 2518 mm and the solution denudation was calculated as 100 ± 24 $m^3 km^{-2} a^{-1}$ by Williams and Dowling

New Zealand: The five deepest and six longest caves in New Zealand. All but Gardner's Gut Cave are in northwest Nelson in the South Island.

Cave	Area	Depth (m)	Length (m)	Rock
Nettlebed Cave	Mt Arthur	889	24 252	Ordovician marble
Ellis Basin System	Mt Arthur	775	28 730	Ordovician marble
Bulmer Cavern	Mt Owen	749	50 125	Ordovician marble
Bohemia Cave	Mt Owen	713	9300	Ordovician marble
HH Cave	Mt Arthur	710	1906	Ordovician marble
Megamania Cave	NW Nelson	<200	14 800	Oligocene limestone
Honeycomb Hill Cave	NW Nelson	<200	13 712	Oligocene limestone
Gardner's Gut Cave	Waitomo	<70	12 197	Oligocene limestone

(1979) who estimated that ~80% of the solution was accomplished by rain falling directly onto the marble outcrop with most corrosion occurring in the uppermost 10–30 m of the rock beneath the soil. The remaining 20% comes mainly from conduit (cave) enlargement by allogenic streams that originate on neighbouring non-carbonate rocks. Thus the overall rate of surface lowering in the marble catchment is about 80 mm ka^{-1} under present conditions. Within the estimated error terms, the solution denudation rate for Ordovician marble in the Riwaka basin is about the same as that calculated by Gunn (1981a) for Oligocene limestones at Waitomo, despite the entirely different carbonate lithologies. These solutional denudation rates are moderately rapid on a world scale.

Rates of downcutting of cave streams may also be compared. Basal ages of stalagmites in Metro Cave in northern Westland indicate an incision of Tertiary limestones by the cave stream at a rate of about 0.27 m ka^{-1} (Williams, 1982). This is twice as rapid as the rate determined for Waitomo, but less than that calculated for Nettlebed Cave beneath Mt Arthur, where the rate of water-table lowering in the marble was estimated by the same means as about 0.44 m ka^{-1}. Whereas the rate of solutional denudation is determined principally by the annual rainfall, the rate of cave stream incision is determined by at least two factors: (1) the erosive power of the stream (sum of mechanical and chemical erosion) and (2) the rate of uplift of the land (which increases potential energy). If the erosive power of the stream is high, the rate of cave passage downcutting can equal the uplift rate. This is probably the case for the caves investigated. At each site the rate of tectonic uplift is probably <1 m ka^{-1}.

Fiordland is part of the Te Wahipounamu–South West New Zealand World Heritage site. It is an exceptionally rugged and inaccessible area with mountains to over 2200 m and deeply glaciated valleys and fiords. The vegetation of the region is largely thick Southern Beech forest, which gives way to alpine herbfield above the treeline at about 1100 m. Mean temperatures at sea level range between 18° (summer) and 2°C (winter) and annual precipitation on windward west-facing slopes sometimes exceeds 10 m. Although caves, dolines, and karrenfeld are known, these areas are little explored. Karst occurs in narrow bands of marble in Paleozoic metamorphic rocks. Very steeply dipping Permian limestones to 750 m thick have been mapped near the Hollyford valley, but have not been investigated for karst features. Patches of Oligocene limestones also occur in the region and are known to contain caves, the best investigated being Aurora Cave on the shore of Lake Te Anau, which is 6.4 km long and 267 m deep and contains an excellent dated record of glacial and interstadial deposits (Williams, 1996).

PAUL WILLIAMS

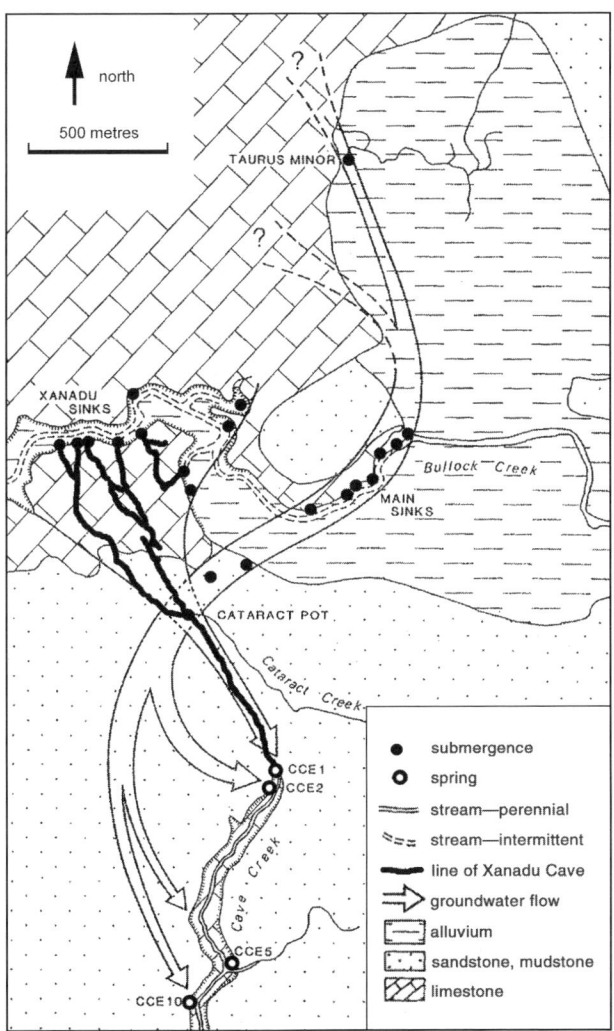

New Zealand: Figure 3. Subterranean river capture of Bullock Creek, near Punakaiki, northern Westland, South Island (from Williams, 1992a after Crawford, 1994).

See also **Speleothems: Evaporite** (photo from Puketiti Flower Cave, King County Karst)

Works Cited

de Freitas, C.R., Littlejohn, R.N., Clarkson, T.S. & Kristament, I.S. 1982. Cave climate: Assessment of airflow and ventilation. *Journal of Climatology*, 2: 383–97

de Freitas, C.R. & Littlejohn, R.N. 1987. Cave climate: Assessment of heat and moisture exchange. *International Journal of Climatology*, 7: 553–69

Gunn, J. 1981a. Limestone solution rates and processes in the Waitomo district, New Zealand. *Earth Surface Processes and Landforms*, 6: 427–45

Gunn, J. 1981b. Hydrological processes in karst depressions. *Zeitschrift für Geomorphologie*, 25(3): 313–31

Gunn, J. & Trudgill, S.T. 1982. Carbon dioxide production and concentrations in the soil atmosphere: A case study from New Zealand volcanic ash soils. *Catena*, 9: 81–94

Pugsley, C. 1984. Ecology of the New Zealand Glowworm, *Arachnocampa luminosa* (Diptera: Keroplatidae), in the Glowworm Cave, Waitomo. *Journal of the Royal Society of New Zealand*, 14(4): 387–407

White, P.J. & Waterhouse, B.C. 1983. Lithostratigraphy of the Ke Tuiti group: A revision. *New Zealand Journal of Geology and Geophysics*, 36: 255–66

Williams, P.W. 1982. Speleothem dates, Quaternary terraces and uplift rates in New Zealand. *Nature*, 298: 257–60

Williams, P.W. 1991. Tectonic geomorphology, uplift rates and geomorphic response in New Zealand. *Catena*, 18: 439–52

Williams, P.W. 1992a. Karst in New Zealand. In *Landforms of New Zealand*, 2nd edition, edited by J.M. Soons & M.J. Selby, Auckland: Longman Paul

Williams, P.W. 1992b. Karst hydrology. In *Waters of New Zealand*, edited by M.P. Mosley, Wellington: New Zealand Hydrological Society

Williams, P.W. 1996. A 230 ka record of glacial and interglacial events from Aurora Cave, Fiordland, New Zealand. *New Zealand Journal of Geology and Geophysics*, 39: 225–41

Williams, P.W. & Dowling, R.K. 1979. Solution of marble in the karst of the Pikikiruna Range, northwest Nelson, New Zealand. *Earth Surface Processes*, 4: 15–36

Worthy, T.H. & Mildenhall, D.C. 1989. A late Otiran-Holocene palaeoenvironment reconstruction based on cave excavations in Northwest Nelson, New Zealand. *New Zealand Journal of Geology and Geophysics*, 32: 243–53

Further Reading

Crossley, P. (compiler) 1988. *The New Zealand Cave Atlas: North Island*, New Zealand Speleological Society

Main, L. (editor) 1993. *The New Zealand Cave Atlas, vol. 2: South Island*, 2nd edition, New Zealand Speleological Society www.massey.ac.nz/~sglasgow/nzss/ the NZ Speleological Society web site

NULLARBOR PLAIN, AUSTRALIA

The Nullarbor Plain is a remarkably flat, nearly treeless arid karst plain developed on the Tertiary limestones of the Eucla Basin. It lies on the southern border of the continent between the states of Western and South Australia, and forms a biogeographical barrier between west and east (see map in Australia entry). This is Australia's largest karst (200 000 km²) and perhaps the most intensively studied. The region has an arid to semi-arid climate with annual rainfall between 150 and 250 mm, with a maximum in the winter months. Annual potential evaporation is 1250–2500 mm. Relief is low, less than 5 m, with slope angles of 1°–3°. The Nullarbor Plain has no surface drainage, but relict stream channels are found on the northern and western flanks of the plain. These have meandering traces and are infilled with alluvium and eolian deposits. Elsewhere there are many aligned depressions with clay pans; these are of structural origin (Lowry & Jennings, 1974). These clay pans, locally termed "dongas", act to channel the slight runoff and direct it into dolines and blowholes. Dolines are sparse, steep sided, and show evidence of collapse into underlying cavities (Figure 1). Blowholes are very numerous, fed by deep solution pipes of complex origin. The caves are extensive, much modified by salt wedging and collapse processes, and commonly descend gently to near-static pools and lakes of brackish to saline water. A low-gradient water table underlies the plain and its depth ranges from 30 m in the north to 120 m in the south. The

Nullarbor Plain, Australia: Figure 1. Collapse doline on the northeastern part of the Nullarbor Plain, South Australia. The depression leads into Koonalda Cave. (Photo by John Gunn)

Nullarbor Plain: Figure 2. Looking out of Koonalda Cave into the collapse doline. Passage continues at this size for some 500 m. (Photo by John Gunn)

flooded tunnels of Cocklebiddy Cave are more than 6 km long but most of them can only be explored by divers. Withdrawal of hydrostatic support at times of lower sea level has led to collapse into these water table caves, allowing entry into large caverns such as Koonalda (Figure 2), Abrakurrie, Mullamullang, and Weebubbie Caves (Figure 3). All of these caves are characterized by having spectacularly grand old trunk passages that retain a 20–40 m diameter for several kilometres.

These deeper caves of the Nullarbor may intersect the water table, producing underground lakes, as at Weebubbie, Koonalda, and Tommy Graham's caves. These lakes are remarkable for their exceptional clarity, depth, and high salinity. The large

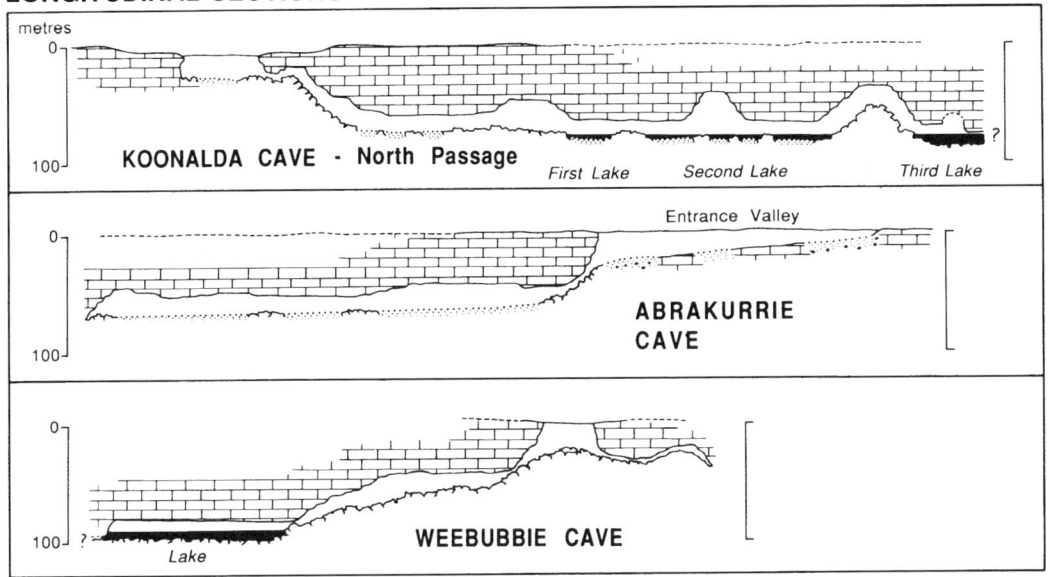

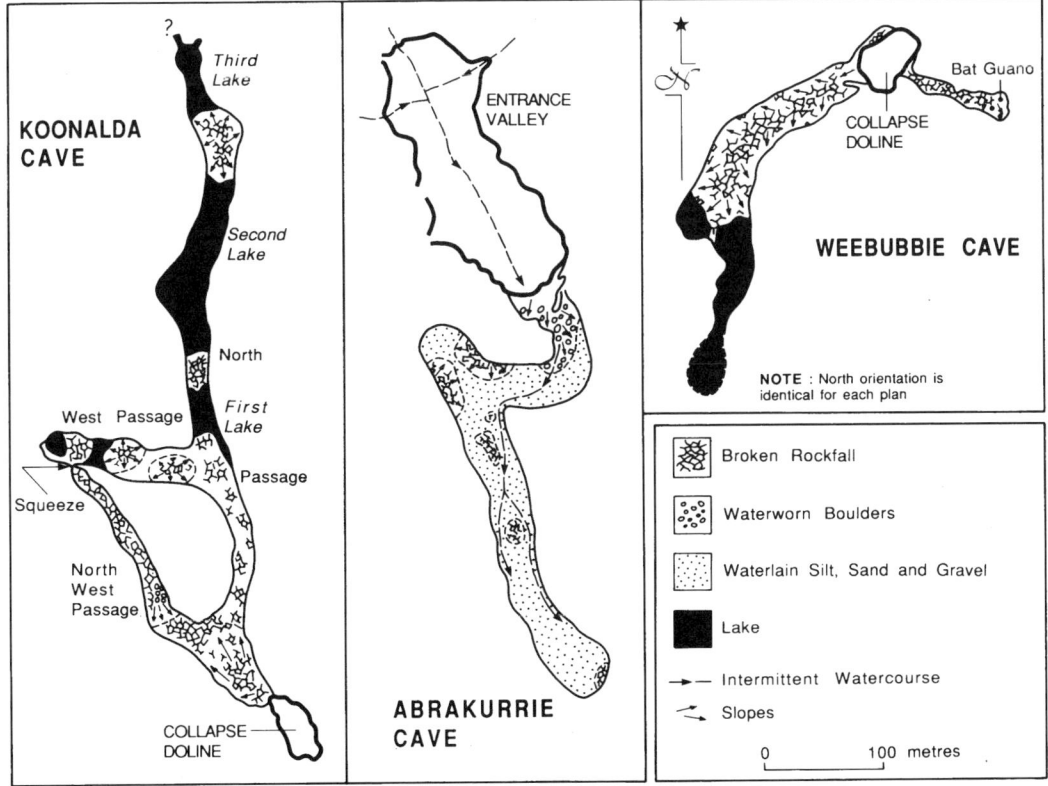

Nullarbor Plain, Australia: Figure 3. Plans and sections of three deeper caves on the Nullarbor Plain, Australia.

cave passages are typically flattened collapse arches with extensive breakdown blocks on the floor. Individual collapse events have been recorded on a time-scale of decades, especially near entrances. Cones and ridges of coarse sandy weathering debris are found in many caves, one of the best known being "The Dune" in Mullamullang Cave. However, the majority of the caves on the Nullarbor Plain are shallow and contain a variety of passage forms, dominated by breakdown and also including phreatic passages, e.g. Old Homestead Cave. Many of these shallow caves are entered via cylindrical blowholes up to 30 m deep. Networks of anastomosing half-tubes are common in the upper zone of the limestone and create a large air volume, producing strong air movements in response to variations in atmospheric pressure. Stream-cut passages are uncommon in Nullarbor caves, and running water is absent except during flash flooding into cave entrances. In several caves the exceptionally dry microclimate has produced mummified wallabies, and, in one, the preserved remains of a Tasmanian tiger (*Thylacinus cynocephalus*), dating from 3500 years BP. A particular feature of the Nullarbor caves is the abundance of halite, which produces both speleothems (see photo in Speleothems: Evaporite) and many weathering forms through wedging during recrystallization. Gypsum speleothems are also abundant, with calcite speleothems less obvious, largely because of destruction through salt wedging. In Kelly Cave basal calcite speleothems are overlain by gypsum speleothems, and these in turn by halite. Speleothem history has much to contribute to the study of past climates and the history of the caves, which is at present unknown (Lowry & Jennings, 1974).

When the long-term operation of karst processes in the arid zone is considered, a fundamental dilemma emerges: are the present features the product of slow, low-intensity processes in a variable but essentially arid climate, or do they reflect inheritance from a past, wetter climate under which processes operated at a faster rate? There is some evidence for wetter conditions on the Nullarbor Plain in the distant past: first, the speleothem sequences mentioned previously; second, the dark-brown to black colours of calcite speleothems which have been demonstrated (Caldwell *et al.*, 1983) to be due to organic compounds such as humic acids, fulvic acids, and phenolics, suggesting more effective leaching of soil organic matter; and third, the presence of relict solution tubes and spongework in Thampanna and Old Homestead caves and the presence of solution scallops (spoon-shaped hollows) in the flooded tunnels of Weebubbie and Cocklebiddy caves (Grodzicki, 1985) suggest a more vigorous fluvial regime in the geological past. Thus the cave systems of the Nullarbor Plain may have originated in a more humid climate phase, and probably span the entire Quaternary and part of the Tertiary. Since the onset of the present arid climate, these features have been heavily modified by salt wedging and collapse processes (Lowry & Jennings, 1974).

Under present climatic conditions, exposed outcrops are merely pitted by raindrops, and no solution rills are present. Some solution of limestone does occur in the coastal fringe of the Nullarbor Plain, and mixing corrosion may be important in this context (James, Rogers & Spate, 1990). The large caves of the Nullarbor may have been formed initially by enhanced solution at the mixing zone of fresh and saline waters. The groundwater of the Eucla Basin has a high salinity and some solution may also be occurring at great depth in the limestone. Inland periodic low-intensity rainfall may cause some solution on the margins of clay pans, and calcretes are forming, but overall rates of solution must be very low. Of more significance are the intrusions of rain depressions of tropical origin, which occur twice or three times per decade. These have the capacity to dump large amounts of water on the western Nullarbor, flooding clay pans and dolines, and perhaps maintaining epiphreatic passage forms in several caves such as Thampanna and Old Homestead.

During the last interglacial–glacial cycle the sea level dropped at least twice, steepening hydraulic gradients. This prompted a shift in the isohyets, allowing the saltbush shrubland of the interior plain to extend across terrain which today lies closer to the coast and supports eucalypt woodland. Conversely, during the last interglacial, eucalypt woodland would have extended farther inland than it does today and this may have enhanced solutional processes. These changes in the vegetation cover may have acted to destabilize the continental dunefields to the north of the Nullarbor and so allowed reddened quartz sands to accumulate downwind in caves and dolines on the eastern portion of the plain. Cave and doline infills on the eastern margin of the Nullarbor are dominantly red eolian quartz sand with well-developed silica skins from pedogenesis. At least three different pedogenic cycles are involved, with some loessic additions. These sands have their origin in the Great Victoria Desert and suggest a wind regime different to the present predominantly southerly to southwesterly airflow. The distribution of these sands, in a narrow belt around the Diprose Caves, suggests that they may have moved as sand streaks across the gravelly plain. The most recent disturbance of these sands, dated by thermoluminescence at <360 years BP, may reflect the increased fire frequency consequent on pastoralism and the spread of rabbits across the Nullarbor Plain. Recent environmental impacts on the Nullarbor and other arid karsts are reviewed by Gillieson (1993).

DAVID GILLIESON

Works Cited

Caldwell, J.R., Davey, A.G., Jennings, J.N. & Spate, A.P. 1983. Colour in some Nullarbor Plain speleothems. *Helictite*, 20: 3–10

Gillieson, D. 1993. Environmental change and human impact on arid and semi-arid karsts of Australia. In *Karst Terrains: Environmental Changes and Human Impact*, edited by P.W. Williams, Cremlingen: Catena

Grodzicki, J. 1985. Genesis of the Nullarbor Plain caves in Southern Australia, *Zeitschrift für Geomorphologie*, 29: 37–49

James, J.M., Rogers, P. & Spate, A.P. 1990. The role of mixing corrosion in the genesis of the caves of the Nullarbor Plain, Australia. *Proceedings of the 10th International Congress of Speleology*, edited by A. Kosà Budapest Hungarian Speleological Society: 263–65

Lowry, D.C. & Jennings, J.N. 1974. The Nullarbor karst, Australia. *Zeitschrift für Geomorphologie*, 18: 35–81

Further Reading

Gillieson, D.S. & Spate, A.P. 1992. The Nullarbor karst. In *Geology, Climate, Hydrology and Karst Formation: IGCP 299 Field Symposium in Australia*, edited by D.S. Gillieson, Canberra: Department of Geography and Oceanography, Australian Defence Force Academy

Gillieson, D.S., Wallbrink, P., Murray, A. & Cochrane, A. 1996. Estimation of wind erosion using GIS modelling and caesium-137 on the Nullarbor Plain karst, Australia. *Zeitschrift für Geomorphologie Supplement*, 105: 73–90

Jennings, J.N. 1983. The disregarded karst of the arid and semiarid domain. *Karstologia*, 1: 61–73

ORGANIC RESOURCES IN CAVES

A wide range of organic materials is commonly found in caves. Some of these enter caves by being washed or blown into the cave, seeping through cracks in the bedrock, or simply by falling into an entrance. The living fauna and flora within or visiting caves is a further significant contributor. Tree roots and other vascular plant materials, algae, fungi, mosses and bryophytes (particularly in the twilight zone), birds, bats, and other vertebrates all make up a significant biomass. In turn the biota provide their excreta and their dead remains. All of these sources may interact with each other and with the rocks and minerals of the cave to produce a range of both further organic compounds and minerals. Many organic deposits have been mined or excavated as economic resources, and these are the subject of this entry. Hydrocarbons in karst are the subject of a separate entry.

Probably the greatest total of organic deposits are provided by the accumulated guano deposited by bats or birds (see Guano entry). These deposits provide an immensely rich food resource for other species, particularly invertebrates, and a substrate for the growth of fungi. They are often the basis for the genesis and development of special suites of minerals and in many regions their spontaneous combustion at intervals provides a still further source of chemical reactions. Guano, most commonly that from insectivorous bats, was mined extensively from caves in the 19th and early 20th centuries in Puerto Rico, the Bahamas, Mexico, and the southern United States for use as fertilizer (being rich in nitrogen and phosphate). The invention of chemical fertilizers has dramatically reduced the need for guano, and bat conservation measures have closed some operations. However, small-scale mining continues in a number of tropical caves, e.g. in Malaysia, the Caribbean, and Latin America, and there is a "cottage industry" level of extraction for the home garden market even in modern industrialized countries. Bone deposits in caves are also rich in phosphates and mining was carried out at the extensive vertebrate fossil deposits at Wellington Caves (New South Wales) from 1913–19.

Until the end of the 19th century nitrate deposits (saltpetre), formed largely as a result of seepage from the surface above the cave but often enriched by bat or rat guano, were commonly mined in order to manufacture gunpowder or other explosives (see Gunpowder entry).

Although the organic mineral guanine, which provides luminescence to liquid cosmetics, was originally found in bird guano in 1846, it has never been commercially produced from guano of any kind. Other organic minerals, occurring as a component of stalactites and crusts on the walls of caves containing bat guano deposits such as at Murra-el-elevyn Cave, Nullarbor Plain, Western Australia, appear to have no commercial value.

The nests of *Collocalia* and *Aerodramus* swiftlets are without doubt the most economically important cave products. These nests are comprised largely of dried saliva, and cemented to the cave walls. They are collected, cleaned and processed, then used to make bird's-nest soup. The processed nests command a very high price in the (Asian) gourmet market, now ranging up to $US 4000 per kilogram (Vermuelen & Whitten, 1999: 32–35, 59–60). The management of the industry varies from one country to another. Traditional family harvesting had, for many centuries, ensured the sustainability of the industry, but with disruption of indigenous cultures and rising profitability, violent takeovers of some areas have occurred and poaching is becoming widespread (Valli & Summers, 1990). There is now an increasing demand for governmental or other regulation to re-establish sustainability (Casellini, Foster & Hien, 1999).

ELERY HAMILTON-SMITH

See also **Sediments: Biogenic**

Works Cited

Casellini, N., Foster, K. & Bui Thi Thu Hien. 1999. *The "White Gold" of the Sea: A Case Study of Sustainable Harvesting of Swiftlet Nests in Coastal Vietnam*, Hanoi: IUCN Vietnam Programme

Valli, E. & Summers, D. 1990. *The Nest Gatherers of Tiger Cave*, London: Thames and Hudson, and Charlottesville, Virginia: Eastman Kodak

Vermeulen, J. & Whitten, T. 1999. *Biodiversity and Cultural Property in the Management of Limestone Resources*, Washington, DC: World Bank

Further Reading

Chapman, P. 1993. *Caves and Cave Life*, London: Harper Collins

Francis, C. 1987. *The Management of Edible Bird's Nest Caves in Sabah*, Kota Kinabalu, Malaysia: Wildlife section, Sabah Forest Department

ORGANISMS: CLASSIFICATION

All classifications are somewhat arbitrary. Furthermore, the groups that are proposed necessarily reflect the views of the classifier, since the latter chooses both the discriminative characteristics and the methodology used for the formation and the ordering of groups. The classification of living creatures is a well-known example that illustrates this. Classifications of organisms were successively utilitarian and anthropomorphic, typological, evolutionary, and more recently phenetic (based on global similarity) and finally phylogenetic, aiming for greater objectivity. Subterranean organisms include representatives of most large groups of the animal kingdom plus some groups of bacteria, fungi, and plants. These organisms live in a great variety of subterranean habitats and any reasonable classification of them must necessarily be based on ecological criteria that reflect the wide variety of habitats and life histories utilized. Several different classifications have been proposed, although they cannot always be easily compared because the authors frequently use similar names but employ different definitions.

Terrestrial *versus* aquatic species: The diversity of subterranean habitats results in a great diversity of potential niches suitable for terrestrial animal species using mainly atmospheric oxygen, as well as for aquatic species using dissolved oxygen for respiration. There is presently no common term referring to the whole terrestrial subterranean fauna, whereas the aquatic subterranean fauna is known as the stygofauna (Botosaneanu, 1988), the Styx being the main subterranean river of Hades in Greek mythology. The stygofauna includes all subterranean aquatic fauna, whether continental or marine, living in open water, in caves and crevices, or in interstitial water within sediments. As the prefix "troglo" (from the Greek word *troglê*) means hole, cavity, or cave, all kinds of animal species observed in caves or other subterranean cavities form the troglofauna, and this includes both terrestrial and aquatic species. However, since the introduction of the prefix stygo- for aquatic subterranean organisms, the prefix troglo- has increasingly been used specifically for terrestrial cave fauna.

Organisms within caves may occur on the ground or in scree, on the walls and stalagmites, or on the roof and on stalactites. Others, the stygobites, occur in subterranean lakes or rivers utilizing open water biotopes in the dark environment of caves. When considering the terrestrial and aquatic fauna within caves, most speleobiologists recognize three different types of inhabitants: troglobites, troglophiles, and trogloxenes.

Troglobites are species permanently and exclusively living in caves or other subterranean cavities such as that of the "superficial underground compartment" (Camacho, 1992; Humphreys, 2000), also called the MSS (mesovoid shallow substratum; Juberthie, 2000), or in fissures or interstices providing large spaces compared to the animal body size. They are sometimes considered as the true cave-dwelling organisms and are often called troglobites (Holsinger & Culver, 1988). The adjectives "troglobiontic" or "troglobitic" are used, but compound names including troglobite are generally preferred by many authors. Troglobites generally present troglomorphic (or "troglobiomorphic") characteristics *sensu* Christiansen (1962), including morphological traits such as reduction or loss of eyes and tegument pigmentation, thinning of cuticle, and elongation of appendages, especially antennae, in arthropods. In addition, they frequently have biological, ecological, or behavioural traits such as a low metabolism and fecundity associated with a K-reproductive strategy, and a tendency to the loss of circadian and seasonal rhythms. This set of characteristics is called "troglomorphy" (or "troglobiomorphy") and is considered an adaptation to subterranean environments (see the entries on Adaptation). Troglobites exhibiting marked troglomorphic traits are generally considered as "paleotroglobites" as their ancestors colonized subterranean habitats in ancient times, from millions to hundreds of million years ago. When the ancestral group has disappeared from all surface habitats, the paleotroglobites, being the sole representative of the group, are often called "living fossils" or relict species. Other troglobites are genuinely limited to caves and differ from all closely related surface (epigean) species in respect to some specific traits but may exhibit few or no troglomorphic traits. These species are called "neotroglobites", having relatively recently colonized the subterranean habitat. Conversely, some epigean organisms may appear to possess troglomorphic characteristics even though they do not inhabit subterranean environments.

Troglophiles are species that can establish and maintain subterranean populations but also occur in surface habitats that may have little similarity with the cave environment. They are able to support viable reproductive hypogean populations and may be common in caves, but are not obligate subterranean inhabitants. They may exhibit only some troglomorphic traits but these are generally less marked than in troglobites. They often look very similar to closely related surface species. Sometimes their status is uncertain, such as when they are theoretically able to live in surface habitats but are unknown outside of caves.

Trogloxenes are species that sometimes occur in caves but belong to surface communities, usually living and feeding in epigean habitats. They have sometimes been called "xenocaval" and as some are found in caves only by chance they have also been called "tychocaval" organisms or "accidentals" (Holsinger & Culver, 1988). However, the term trogloxene is by far the most commonly used and following Racovitza (1907) it is usual to distinguish the "regular trogloxenes" (such as the bats that usually reside in caves but feed outside of the cave) and the accidental or "occasional trogloxenes", corresponding to the tychocaval or accidentals, including many surface forms exceptionally and passively drawn into some caves (surface fishes or aquatic invertebrates) or temporarily resting in caves (snakes, birds, rodents, or many terrestrial invertebrates).

Some authors have proposed other categories. For example, Ginet and Decou (1977) added a fourth category to the three categories discussed above, the "subtroglophiles", for animal species such as bats that frequent caves only during one part of their life cycle such as the diapause period (a form of hibernation). These taxa do not feed in the cave whereas the troglophiles are able to feed and spend their entire life cycle within cave habitats. This distinction has not been widely adopted and most authors would consider bats to be trogloxenes. Christiansen (1962) focused more attention on the level of adaptations to subterranean life than to the different habitats; he considered four categories of cave animals: the "trogloxenes" (*sensu* accidental trogloxenes), the "troglomorphs" (ancient and well-adapted troglobites exhibiting most troglomorphic traits), the "ambimorphs" which exhibit only some traits related to caves or other subterranean habitats (probably some recent troglobites and some troglophiles), and the "epigiomorphs" which live in caves but have a phenotype still similar to that of surface animals (many guanobitic species, and some troglophiles or recent troglobites).

Aquatic fauna living in groundwater within caves, fractured rocks, and pores can be placed in all the categories proposed above. Thus the stygofauna *sensu lato* includes "stygobites", "stygophiles", and "stygoxenes" with definitions similar to those for the three main troglofauna categories but limited to aquatic organisms.

Some classifications result from the utility of defining some particular communities. For example, bat, or sometimes swiftlet, guano is often a major source of food for terrestrial cave species that are termed guanophages. Some authors also recognize "guanobites", "guanophiles", and "guanoxenes" (see Humphreys, 2000) but such categories cannot be the basis for a rational classification because it is not always possible to identify definite feeding groups of species. Moreover, many subterranean species are polyphagous, with complex variations of diet and habitat use in the same cave. Other tentative classifications are based on the spatial and abiotic characteristics of different parts of the subterranean ecosystems (see Hypogean Habitats). Some cave biologists have distinguished between the entrance (cave threshold) communities, the wall (or parietal) communities, and the deep and clayey soil communities, but it is impossible to recognize a finite number of clearly defined communities.

More interesting is the separation of stygobites into "troglostygobites" (mainly karstostygobites) and "phreatostygobites". The first live in open water of cave rivers and lakes whereas the latter live in the groundwaters of alluvial sediments, i.e. in interstitial waters, where the space available for animals is limited (see Interstitial Habitats entries). In each habitat, the selective pressure of the environment acting on the evolution of the body size and morphology is quite different as are the resulting stygobites: larger animals (cave fishes, amphibians, or crayfishes) with hypertelic appendages (e.g. some cave shrimps with very long antennae) only occur as troglostygobites, but very small species, with more elongated shapes, short antennae and appendages, and phreatomorphic traits (Coineau, 2000) mainly occur as phreatostygobites (see comparative figures of different kinds of stygobitic isopods in Crustacea: Isopoda Aquatica). Similar differences in morphological traits have been observed in terrestrial invertebrates between cave terrestrial troglobites and deep soil "edaphobites" (Humphreys, 2000). The latter generally have a smaller and thinner body, with shorter appendages than the closely related troglobitic species.

Other classifications of stygobites have been proposed that are from an evolutionary and biogeographical perspective: "limnostygobites" inhabit fresh continental subterranean waters, whilst "thalassostygobites" inhabit either marine sediments or marine caves, or coastal biotopes with a marine influence. Some groups of Crustacea (for example the micro-isopod crustaceans *Microcharon* spp. and *Microcerberus* spp.) have interstitial species belonging to the same genus in the two groups of stygobites and both exhibit a clear phreatomorphy. The thalassostygobites include many troglobitic species living either in marine caves (marine troglobitic species) or in anchialine caves—littoral caves always in connection with marine water but displaying some continental influence, and characterized by spatial variation of ecological factors, especially salinity (Sket, 1986; see also Anchialine Habitats). There is a great variety of anchialine habitats and a rich anchialine fauna including three categories of stygobites. Some taxa belong to groups usually continental and limnostygobitic (e.g. the numerous species of the cirolanid isopod genus *Typhlocirolana* discovered first in brackish waters of Balearic caves), whereas others belong to marine groups (e.g. the galatheid crab *Munidopsis polymorpha* from an anchialine lava tube in the Canary Islands). A third kind of taxa occur solely in anchialine caves and thus are characteristic of this habitat in all locations, for example, *Speleonectes* spp. and other genera of Remipedia, a group discovered only 20 years ago. They have a very plesiomorphic trunk, but very apomorphic head appendages and are generally considered to be one of the most primitive classes of Crustacea.

Some limnostygobites (such as the above micro-isopods) have no closely related species in surface freshwaters. They are clearly related to marine groups and are derived from marine ancestors that have directly colonized the coastal and continental groundwaters and are called "thalassoid" limnostygobites. Other limnostygobites, originating from surface freshwater ancestors which colonized groundwaters and interstices in alluvial sediments or the subterranean waters of caves, are called "limnicoid" limnostygobites (Coineau & Boutin, 1992; Notenboom, 1991). Groundwater cirolanid isopods are thalassoid limnostygobites whereas the subterranean isopods Asellidae or the cave amphipods *Gammarus* (Culver, Kane & Fong, 1995) are limnicoid limnostygobites.

CLAUDE BOUTIN

Works Cited

Botosaneanu, L. (editor) 1986. *Stygofauna Mundi, A Faunistic, Distributional, and Ecological Synthesis of the World Fauna Inhabiting Subterranean Waters (Including the Marine Interstitial)*, Leiden: Brill

Camacho, A.I. 1992. A Classification of the aquatic and terrestrial environments and their associated fauna. In *The Natural History of Biospeleology*, edited by A.I. Camacho, Madrid: Museo Nacional de Ciencias Naturales

Christiansen, K. 1962. Proposition for the classification of cave animals. *Spelunca*, 2: 76–78

Coineau, N. 2000. Adaptations to interstitial groundwater life. In *Subterranean Ecosystems*, edited by H.K. Wilkens, D.C. Culver & W.F. Humphreys, Amsterdam and New York: Elsevier

Coineau, N. & Boutin, C. 1992. Biological processes in space and time: Colonization, evolution and speciation in interstitial stygobionts. In *The Natural History of Biospeleology*, edited by A.I. Camacho, Madrid: Museo Nacional de Ciencias Naturales

Culver, D.C., Kane, T.C. & Fong, D.W. 1995. *Adaptation and Natural Selection in Caves: The Evolution of Gammarus minus*. Cambridge, Massachusetts: Harvard University Press

Holsinger, J.R. & Culver, D.C. 1988. The invertebrate cave fauna of Virginia and a part of Eastern Tennessee: Zoogeography and ecology. *Brimleyana, Journal of the North Carolina State Museum*, 14: 1–163

Humphreys, W.F. 2000. Background and glossary. In *Subterranean Ecosystems*, edited by H.K. Wilkens, D.C. Culver & W.F. Humphreys, Amsterdam and New York: Elsevier

Juberthie, C. 2000. The diversity of the karstic and pseudokarstic hypogean habitats in the world. In *Subterranean Ecosystems*, edited by H.K.Wilkens, D.C. Culver & W.F. Humphreys, Amsterdam and New York: Elsevier

Notenboom, J. 1991. Marine regressions and the evolution of groundwater dwelling amphipods (Crustacea). *Journal of Biogeography*, 18: 437–54

Racovitza, E.G. 1907. Essai sur les problèmes biospéologiques. *Archives de Zoologie Expérimentale et Générale*, 6: 371–478

Sket, B. 1986. Ecology of the mixohaline hypogean fauna along the Yugoslav coasts. *Stygologia*, 2(4): 317–38

Further Reading

Gibert, J., Danielopol, D.L. & Sanford, J.A. (editors) 2000. *Groundwater Ecology*, San Diego: Academic Press

Ginet, R. & Decou, V. 1977. *Initiation à la Biologie et à l'écologie souterraines*, Paris: Delarge

Juberthie, C. & Decu, V. (editors) 1994, 1998, 2001. *Encyclopaedia Biospeologica*, 3 vols, Moulis and Bucharest: Société de Biospéologie

Vandel, A. 1964. Biospéologie: La Biolgre des animaux cavernicoles, Paris: Gauthier-Villars

Wilkens, H., Culver, D.C. & Humphreys, W.F. (editors) 2000. *Subterranean Ecosystems*, Amsterdam and New York: Elsevier

ORNAMENTAL USE OF LIMESTONE

Limestone has great value as a resource for many types of construction, and depending on the specific properties and characteristics of the particular rock, its use can be interpreted as ornamental. There are many popular uses of cut (dimensional) quarried limestones, including exterior and interior building surfaces and monuments. Limestones may be used ornamentally in both their weathered and unweathered states. Many unweathered limestones, freshly cut, are extremely attractive and interesting by virtue of their rich and varied textures and constituents, often involving visible fossil content. Polishing enhances this attraction but can also be a source of confusion in that the building trade uses the term "marble" to describe any limestone, and sometimes non-limestone, rocks that can be polished whereas strictly it should be applied only to metacarbonates.

In addition to cut or dimensional stone, blocks of natural weathered rock from surface limestone outcrops are very attractive to many people for ornamental purposes, providing a sculptured appearance which can appear very dramatic and "modern". Such weathered limestone blocks, with their wide variety of attractive karren features, have been used ornamentally in various ways for several centuries. In Great Britain, solutionally sculpted limestone pavement blocks have been used in garden rockeries since the mid-19th century, and in gateways and wall toppings for even longer. Unfortunately, due to the popularity of these blocks a very high proportion of the pavement in the country has been quarried, and the extraction has caused considerable environmental damage to surface limestone outcrops. The clash between landscape conservation and resource exploitation has been the focus of considerable discussion and action, including legislation, to prevent the damage and to conserve the remaining natural limestone landscape features. In particular, the Wildlife and Countryside Act passed by the UK Parliament in 1981 provided a mechanism on which to base protection of the limestone pavements. This is the only case in the UK where a specific landform has been protected, as opposed to protection for areas or sites of special scientific interest.

There are many cases around the world of similar ornamental use of weathered limestone blocks, for example features taking advantage of attractive patterns of runnels are used in gardens in China, Slovenia, and Switzerland. Large intact Flachkarren, or clints, which have been scrubbed clean are sold as "sculpture" in the Jura area of Switzerland, and near to Mammoth Cave in Kentucky, United States, clints have been modified and in some cases painted to form a "sculpture park". The striking solutional sculpture of humid tropical areas (e.g. southern China) is also naturally ornamental and has been copied in artificial materials in Disneyland, Florida, as well as being used in its natural state in China, for example, in the Imperial Palace in Beijing.

Quarrying of limestones for ornamental use has a very important impact on the visual environment. This may be adverse at

Ornamental Use of Limestone: Dimensional stone quarry, Borba, Portugal. Blocks of limestone are cut with a diamond saw providing a cross section through the karst. (Photo by John Gunn)

the point of extraction, although some dimensional stone quarries provide fascinating cross sections through the karst (see photo in Quarrying of Limestone entry), but is usually positive in the case of the construct using the stone. Most dimensional stone quarries are smaller in both scale and extent than quarries producing aggregates.

Many of the world's most famous buildings use some type of limestone or marble in their construction. Examples include the Taj Mahal, in which Makrana marble from Rajasthan is used, and many 17th-century buildings in Rome constructed from, or decorated with, a marl limestone named the Cottanello "marble". Another famous limestone is Portland Stone, extracted from quarries on the Isle of Portland on the south coast of England. This is a Jurassic oolitic limestone in which the texture is very even, which means that it can be sculpted or cut in any direction. It became famous in the 17th century after Wren chose it for the reconstruction of St Paul's Cathedral in London. Many famous buildings have been constructed from it since, including the United Nations building in New York. The Indiana Limestone of Lawrence County, Indiana, has also been used in the construction of well-known buildings including the Empire State Building and the Pentagon in the United States. This limestone is very easily worked, has an attractive pale buff colour, and makes up about 80% of the dimensional limestone used in the United States.

Limestones from France have been very widely used for ornamental purposes, the most well-known examples being in the great Gothic cathedrals of the north, more modern examples including the extensive and widely marketed use for tiling, especially of floors. Greece and Portugal are two other sources of such materials. In Greece limestone has been used ornamentally since Classical times and is seen in many of the famous buildings such as the Parthenon in Athens. Many other famous structures of the ancient civilizations of the Middle East and Mediterranean are composed of limestones, for example the Sphinx in Egypt. Another very striking example of a limestone used ornamentally is the Globigerina Limestone of Malta. This is a relatively soft yellow limestone which hardens on exposure to the atmosphere, and its softness renders it easy to carve into shapes before hardening occurs. It is worked in this way for both public buildings and private residences on both Malta and Gozo, and the ornamentation is very varied.

In Central America, a further example of the ornamental use of quarried limestone in buildings from an old civilization comes from the Mayan city of Nakbe in Guatemala. The limestone used in the most impressive buildings was quarried at Nakbe, and excavations in these quarries were important for understanding Mayan culture.

A final example of a particularly striking ornamental use is of redeposited limestone, or tufa. Pieces of this material are used in grottoes and gardens, in the former case to create a rough, pocked, or bony surface, deemed attractive in the 18th and 19th centuries in England and copying the contemporary ideas of the appearance of cave interiors (see Art Showing Caves). In horticulture, pieces of tufa are used to provide a microhabitat for lime-loving plants, creating a mini-rock garden such as in an extensive tufa rockery at the Lakeland Horticultural Society gardens at Holehird in Cumbria, United Kingdom. Harder tufas and travertines are used ornamentally in buildings, for example in the Tubingen area of southwest Germany and at the Bagni di Tivoli and other sites in and near Rome, including the colonnades around St Peter's Square. More recent applications include the Basilica of the National Shrine of the Immaculate Conception in Washington, DC, where the South Vestibule and Narthex (passageway) are being clad with travertine and marble following the installation of a new marble relief sculpture believed to be the largest of its type in the world.

Thus, ornamental use of limestones is widespread, varied, and sometimes curious, exploiting the attractiveness and interest of the weathered and unweathered, cut and uncut, polished and unpolished sufaces of the endlessly varied rocktype.

HELEN GOLDIE

Further Reading

Smith, M.R. (editor) 1999. *Stone: Building Stone, Rock Fill and Armourstone in Construction*, London: Geological Society (Engineering Geology Special Publication No. 16)

Useful websites

www.swgfl.org.uk/jurassic/portland.htm provides details on the uses and quarrying of Portland stone.

www.limestonecountry.com/Limestone.html provides details of Indiana stone.

http://www.imiweb.org/stonemagazine/dimston1.htm is the website of *Stone*, the trade magazine for specifiers, fabricators, and suppliers of dimensional stone.

P

PALEOENVIRONMENTS: CLASTIC CAVE SEDIMENTS

Caves make excellent sediment traps, with nearly all caves having some form of clastic sediment within them. These deposits, and any fossils or artefacts they contain, are often preferentially preserved compared to surface deposits and can provide valuable clues to the genesis and evolution of the cave system and to the changing environment outside. Furthermore, many cave sediments can be dated, allowing this information to be put into a chronological context. However, clastic cave sediments are often overlooked, even though they are an essential part of the evolutionary history of the host cave.

Clastic sediments can occur throughout a cave system, but an important distinction can be made between cave entrance and interior facies. Cave entrance deposits are generally much more complex and often contain valuable archaeological deposits. Archaeological artefacts or fossils may be dated using radiocarbon, thermoluminescence, or amino acid racemization dating techniques, providing information on the confining clastic sediments (see Dating Methods: Archaeological). Interior deposits may be devoid of significant archaeological material but often provide valuable information about the genesis and evolution of the cave system.

Clastic cave sediments fall into two main categories: autochthonous and allochthonous. Autochthonous clastic sediments are generated within the cave by endogenetic processes operating at the site of deposition, although such sediments may be subsequently transported further into the cave. The commonest sediments of this type are the products of roof collapse and breakdown. Most clastic cave sediments, however, are allochthonous; generated outside caves and transported into them. Two main transport pathways into caves occur: either vertically by collapse, slumping, and translocation (gravitational fills); or laterally by fluvial action, mudflows, solifluction, or more rarely by ice and wind, although water is by far the dominant transport agent. Aeolian deposits are rare in caves, usually only occurring in entrance passage, but reworked loess is a common component of cave sediments in northern Europe and America. Similarly, glacial deposits rarely penetrate far underground, even where cave systems extend under ice caps such as Castleguard Cave, Canada. However, glacially derived sediments are often transported long distances underground by meltwater.

Many of the processes, bedforms, and deposits found in caves are identical to those seen in surface fluvial environments. As with surface clastic sediments, the range of depositional environments is immense, although in general cave sediments are often very chaotic with rapid lateral facies changes and complex stratigraphic relationships. This is partially because underground the normal laws of superposition do not apply. Although sediments in a particular passage normally young upwards, sediments in overlying passages are normally older. Furthermore, within single sediment sequences, repeated cut and fill episodes, collapse, subsidence, and re-sedimentation can generate very complex stratigraphies. Additional complexities arise when sediments cemented by flowstone are subsequently partially eroded, leaving false floors. Moreover, several unusual transport mechanisms occur in caves. Under pipe-full conditions during flood or mass movement events, sediment may be emplaced as a single, fluidized, self-perpetuating sliding mass of sediment similar to that proposed for esker formation. This "sliding bed" facies enables sediment bodies to penetrate far into a cave system (Gillieson, 1986). At the other extreme, Bull (1981) suggested that translocation of mud down joints and fissures can accumulate to create "cap mud" deposits, often infilling caves to roof level.

Influxes of large amounts of clastic sediment can also affect cave development by protecting passages, floors, and walls from further dissolution, concentrating any erosion upwards (see Paragenesis entry) and modifying passage morphology. This is a fundamental process that is vastly underestimated in cave development. Many caves show at least some evidence of modification in this way, especially in low-gradient systems with an allochthonous sediment source.

Clastic cave sediments and the sedimentary structures they contain can provide clues on the paleoenvironment and paleohy-

Paleoenvironments: Clastic Cave Sediments: Figure 1. Clastic sediments infilling a cave in Eldon Hill Quarry, Castleton, Derbyshire, England. These sediments consist of coarse sandstone and shale derived from the adjacent Namurian Millstone Grit and Edale Shales (see Peak District entry). Some of the clasts have chatter marks suggestive of a glacial origin. Paleomagnetic dating of the fine-grained cap muds and U-series dating of intercalated speleothem deposits suggest these sediments may be up to 900 000 years old. (Photo by John Gunn)

drology at the time of deposition. Furthermore, clast lithology, surface texture, chemistry, and mineralogy may give clues as to the sediment source area, catchment area, transport mechanism, relative age, and transport distance. Many caves in temperate latitudes contain enormous quantities of coarse, poorly sorted, gravelly allochthonous sediment. Analysis of such sediments in South Wales using clast analysis and examination of their microscopic surface textures, demonstrated that these were not deposited under current interglacial conditions, but are consistent with a fluvioglacial origin (Bull, 1981). Analysis of sediments in the Peak District, England, demonstrated that sediments exhibited distinctive mineralogical and chemical characteristics dependent on age and source (Bottrell, Hardwick & Gunn, 1999). In Clearwater Cave, Sarawak, systematic changes in clast lithology downstream give an indication of transport distance underground.

Crucially, clastic cave sediments can be dated in various ways, permitting any paleoenvironmental or sedimentological inferences to be put into a chronological context. Indirectly, paleontological or archaeological dating methods can be applied if fossils or artefacts are present within the sediment, although most bones and artefacts often occur near cave entrances. Moreover, they can be reworked and do not necessarily date the cave or the host sediment. Pollen grains and phytoliths can occur widely in sediments (see Palynology). The abundance, type, and preservation of pollen grains can indicate surface environments at the time of deposition and potentially constrain the age of the sediment.

Paleomagnetism is the most widely used method for directly dating clastic sediments, although only clays and silts can be reliably dated, often only as far back as the first few polarity reversals. Records of secular magnetic variation can be obtained from caves and compared to other dated sections such as lake sediments, although curve matching can prove problematic in older sediments. Alternatively, samples can be obtained throughout a vertically stacked series of cave passages to identify periods of reversed polarity sediments. Such an approach has been undertaken in many caves, including Mammoth Cave, Kentucky (Schmidt, 1982) and the Mulu Caves, Sarawak (Farrant *et al.*, 1995). From this technique, rates of base-level lowering can be estimated. However, without independent dating control, the correlation between observed sediment polarity and the paleomagnetic timescale becomes increasingly tenuous beyond the first few reversals. Other paleomagnetic techniques such as susceptibility and anisotropy have not been extensively used on cave sediments.

Absolute dates can be obtained using radiometric methods. Although clastic cave sediments cannot be dated using U-series or Electron Spin Resonance methods, speleothem deposits interbedded within clastic sediment sequences can. Studies in many caves have utilized this method to constrain the age of sediment sequences. Examples are the studies of Quinif and Maire (1998) in the Gouffre de Pierre Saint-Martin and of Williams (1996) in Aurora Cave on the slopes above Lake Te Anau in Fiordland, New Zealand. The latter cave descends steeply from a formerly glaciated valley and drains to the Te Anau tourist cave on the lakeshore. Sequences of glaciofluvial sediments interbedded with speleothems in the cave are evidence of the number and timing of glacial advances and the status of intervals between them. Twenty-six U-series dates on speleothems underpin a chronology of seven glacial advances in the last 230 000 years, with the peak of the late Otira glaciation (Last Ice Age) at *c.*19 000 years BP. With five advances in the Otiran, the last glaciation is more

Paleoenvironments: Clastic Cave Sediments: Figure 2. Extensive gravels of glaciofluvial origin were deposited in Crag Cave, County Kerry, Ireland between *c.*114 000 and 65 000 years ago. (Photo by John Gunn)

complex than previously recognized. Comparison of the record from the Aurora Cave sequence with that interpreted from other onshore deposits is very convincing. Glacial deposits on slopes above the cave may be evidence of "missing" glacial events in the mid to early Pleistocene that are not preserved at other sites due to their obliteration by successive glacial advances.

Recent advances in the use of terrestrial cosmogenic isotopes as dating tools provide another promising method of dating clastic cave sediments. Cosmic ray reactions in the uppermost few metres of the Earth's crust produce minute quantities of the rare isotopes ^{26}Al, ^{10}Be, and ^{36}Cl, whose accumulation can be used to determine surface exposure ages, measure erosion rates, and trace sediment production and dispersal. As these elements are only produced near the ground surface, the isotope ratios can be used to date the burial age of surface-derived allogenic sediments. Such methods have been used to date fluvial sediments in the upper levels of Mammoth Cave, Kentucky and hence determine the rates of incision of the Green River during the last 3.5 million years (Granger, Fabel & Palmer, 2001). However, the technique is in its infancy and problems determining the rate of terrestrial cosmogenic nuclide production arise as their production varies spatially and temporally due to the shielding influence of the Earth's atmosphere and with changes in geomagnetic field intensity.

ANDREW FARRANT

See also **Sediments: Allochthonous Clastic; Sediments: Autochthonous Clastic**

Works Cited

Bottrell, S., Hardwick, P. & Gunn, J. 1999. Sediment dynamics in the Castleton Karst, Derbyshire, UK. *Earth Surface Processes and Landforms*, 24: 745–59

Bull, P.A. 1981. Some fine grained sedimentation phenomena in caves. *Earth Surface Processes and Landforms*, 6: 11–22

Farrant, A.R., Smart, P.L., Whitaker, F.F. & Tarling, D.H. 1995. Long term Quaternary uplift rates inferred from limestone caves in Sarawak, Malaysia. *Geology*, 23: 357–60

Gillieson, D. 1986. Cave sedimentation in the New Guinea Highlands. *Earth Surface Processes and Landforms*, 11: 533–43

Granger, D.E., Fabel, D. & Palmer, A.N. 2001. Pliocene–Pleistocene incision of the Green River, Kentucky, determined from radioactive decay of cosmogenic ^{26}Al and ^{10}Be in Mammoth Cave sediments. *Geological Society of America Bulletin*, 113: 825–36

Quinif, Y. & Maire, R. 1998. Pleistocene deposits in Pierre Saint-Martin Cave, French Pyrenees. *Quaternary Research*, 49: 37–50

Schmidt, V.A. 1982. Magnetostratigraphy of sediments in Mammoth Cave, Kentucky. *Science*, 217: 827–29

Williams, P.W. 1996. A 230 ka record of glacial and interglacial events from Aurora Cave, Fiordland, New Zealand. *New Zealand Journal of Geology and Geophysics*, 39: 225–41

Further Reading

Bosák, P. (editor) 1989. *Paleokarst: A Systematic and Regional Review*, Amsterdam: Elsevier

Bull, P.A. 1980. Towards a reconstruction of timescales and paleoenvironments from cave sediment studies. In *Timescales in Geomorphology*, edited by R.A. Cullingford, D.A. Davidson & J. Lewin, Chichester and New York: Wiley

Ford, D.C. & Williams, P.W. 1989. *Karst Geomorphology and Hydrology*, London and Boston: Unwin Hyman

Ford, T.D. 2001. *Sediments in Caves*, London: British Cave Research Association (BCRA Cave Studies Series, 9)
A short introduction to cave sediments for the educated caver.

Karstologia Memoires No. 2: Quinif, Y. 1990 *Remplissages karstiques et paleoclimats*, Paris: Fédération française de spéléologie and Bordeaux: Association française de karstologie

Karstologia Memoires No. 5: Audra, P. 1994. *Karsts alpins*, Paris: Fédération française de spéléologie and Bordeaux: Association française de karstologie
These provide a good introduction to much previous work published in French.

Kranjc, A. 1989. *Recent Fluvial Cave Sediments, Their Origin and Role in Speleogenesis*, Ljubljana: SAZU, ZRC

Smart, P.L. & Francis, P.D. (editors) 1991. *Quaternary Dating Methods: A User's Guide*, Cambridge: Quaternary Research Association
A specialist technical guide to dating methods.

White, W.B. 1988. *Geomorphology and Hydrology of Karst Terrains*, Oxford and New York: Oxford University Press

PALEOENVIRONMENTS: SPELEOTHEMS

Caves are amongst the longest-lived components of landscapes, surviving for millions of years, or longer where preserved as paleokarst. They function as giant sediment traps, sampling all rock, chemical, and organic waste products that are mobile in the outside environment. They are important in many types of paleoenvironmental research (Ford, 1997). A stalagmite may be likened to a stone tree that can grow for many thousands of years. Calcite stalagmites, plus flowstones and underwater crystal linings, are now being studied intensively. Aragonite and gypsum speleothems attract less attention because of their rarity and comparative instability.

Dating Speleothems

Speleothems are dated by several different methods. They are among the most important deposits for establishing precise ages of events during the last 500 000 years.

^{14}C *method:* Carbon-14 is created by cosmic radiation in the atmosphere and has a half-life of 5730 years. It is precipitated together with the stable isotopes, ^{12}C and ^{13}C, in most types of speleothem calcite and can be detected by mass spectrometry for 8–10 half-lives, i.e. 45 000–57 000 years Before Present (BP). Unfortunately, the proportion of "dead" carbon (dissolved from ancient rock in which all ^{14}C has decayed in the speleothem can range 5–38% or more. This uncertainty, plus the short time range, limits the utility of the method.

U-series methods: These are the most important at the present time. Uranium isotopes are common trace components in rocks, especially shales. They weather readily, forming $UO_2(xCO_3)$ ions in solution that contribute U atoms to precipitating speleothems. The U decays by emitting ^{4}He nuclei, electrons, and X-rays, until stable as ^{206}Pb or ^{207}Pb. The early decay products,

^{230}Th and ^{231}Pa, are insoluble in waters of normal pH and thus will not be precipitated (Ivanovich & Harmon, 1992).

The decay of ^{234}U (half-life = 245 000 years) to ^{230}Th (half-life = 75 700 years) is the principal method. Figure 1 gives the graph of the dating equation, which is complex because decay of ^{238}U must also be considered. With modern mass spectrometer techniques it is possible to date a sample 500 000 years in age with an analytical error of only ±15 000 years (2 standard deviations). Decay of ^{235}U (half-life = 713 million years) to ^{231}Pa (half-life = 34 000 years) is a similar method but with a range of only ~200 000 years: it is little used because of the scarcity of these isotopes.

The chief problems encountered with these methods are:

1. insufficient uranium. Acceptable results can be obtained with as little as 0.02 ppm U but >0.1 ppm is preferred. Most calcite speleothems have 0.02–2.0 ppm U; aragonite is often much richer;
2. presence of ^{230}Th carried on detritus (chiefly clay particles) in the calcite. This produces too great an age and affects many speleothems, particularly at archaeological sites. ^{232}Th can derive only from detritus, thus where ^{230}Th / ^{232}Th >20, the sample is considered "clean". Dirty calcites can often be tackled by repeated extractions or datings;
3. high porosity, which permits leaching of uranium. It is a serious difficulty in evaporitic calcite but can be avoided in other types.

Despite these problems, several thousand speleothem ^{230}Th/^{234}U dates of high analytical quality have been published.

Decay of ^{238}U to ^{234}U permits dating back to ~1 500 000 years. However, these isotopes precipitate together in calcite and their initial ratio cannot be calculated where ^{230}Th is in equilibrium (>600 000 years). In Figure 1, one speleothem has an initial ratio always close to 3.35 and might be used for a reasonable ^{234}U/^{238}U estimate beyond 600 000 years. The other sample in the figure has a widely varying ratio and could not be used.

Uranium–Lead dating: The half-life of ^{238}U is 4.7 billion years, similar to the age of the Earth itself. Its decay to stable ^{206}Pb is the principal means of dating the oldest rocks on the planet. Most speleothems are much younger and thus contain little ^{206}Pb derived from decay, making it difficult to differentiate from background concentrations of non-radiogenic ^{206}Pb present in all speleothems. Samples with large variations of ^{238}U concentration within them are required in order to construct a ^{238}U/^{204}Pb v ^{206}Pb/^{204}Pb linear trend ("isochron") that estimates the age: such speleothems appear to be rare (Richards *et al.*, 1998).

Paleomagnetism: Most speleothems contain small quantities of precipitated or detrital grains of magnetite, which is common in all environments. When deposited, the grains orient (declination) and tilt (inclination) towards the Earth's magnetic poles and become locked in position by subsequent calcite accumulation. The declination, inclination, and intensity of the Earth's magnetic field change continuously by small amounts and sometimes shift abruptly or reverse. This behaviour has been dated using quick-setting lavas. In principle, the age of a speleothem may be determined by comparing its paleomagnetic record to the lava record. The method requires large volumes of calcite; in speleothems it is used chiefly to detect the Brunhes / Matuyama reversal at ~780 000 years and some older excursions and reversals (Latham & Ford, 1993).

Amino acid racemization: Many speleothems contain trace quantities of soil organic matter (see below). The protein amino acids in them decay slowly from an L to a D configuration, at a rate that is constant where the temperature is constant. Lauritzen *et al.* (1994) measured the decay in extracts from an ancient Norwegian speleothem, obtaining a fair correlation with its U-series ages and suggesting that racemization might be used to extend dating a little beyond the ^{230}Th / ^{234}U dating range.

Environmental Isotope Records

Many stable elements, including hydrogen, oxygen, carbon, and sulphur involved in the dissolution and deposition of calcite, aragonite, and gypsum, occur in two or more isotopic configurations, where the "heavier" isotope has one or more extra neutrons in its nucleus. When vapour condenses or a solid is precipitated, the heavy isotope is slightly more abundant ("enriched") in the denser phase. The amount of enrichment may be controlled entirely by the ambient temperature ("equilibrium fractionation") or determined by dynamic mechanisms such as evaporation ("kinetic fractionation"). These effects are potent tracers of environmental processes (Clark & Fritz, 1997) that become locked into a speleothem when its $CaCO_3$ or $CaSO_4 \cdot 2H_2O$ are precipitated.

The fractionation of $^{18}O:^{16}O$ is the most used. There are three competing controls:

1. because of ice sheet growth during the Quaternary Ice Ages the oceans become slightly enriched in ^{18}O (one part per thousand, or "permil");
2. cave interior temperatures are close to mean annual values outside; ^{18}O of calcite, etc. increases 0.24 permil per degree C (°C) of cooling; but
3. less ^{18}O is evaporated from the oceans during cold periods and it precipitates out of the clouds after shorter distances, reducing the proportion of ^{18}O falling as inland rains.

Many published studies have shown that effect 3 most often predominates, i.e. speleothems are a little depleted in ^{18}O during cold periods (Figure 2), but there may be enrichment in coastal areas or the effects may cancel each other elsewhere. All such signals may be negated where there is strong evaporation during deposition.

Measurement of the $^{18}O:^{16}O$ ratio in calcite, aragonite, or gypsum can only show whether any temperature change was "warmer" or "cooler". The actual change (°C) requires that the water of precipitation (trapped in microscopic fluid inclusions in speleothems) also be measured. $^{1}H:^{2}H$ (the D / H ratio) is preferred because oxygen ratios may be disturbed by interacting with the calcite, etc. Work on the problem continues (Dennis, Rowe & Atkinson, 2001).

Part of the carbon in a speleothem derives from dissolution of limestone and other rocks, where $^{13}C:^{12}C = 0 \pm 5$ permil compared to VPDB (a global standard), and part from

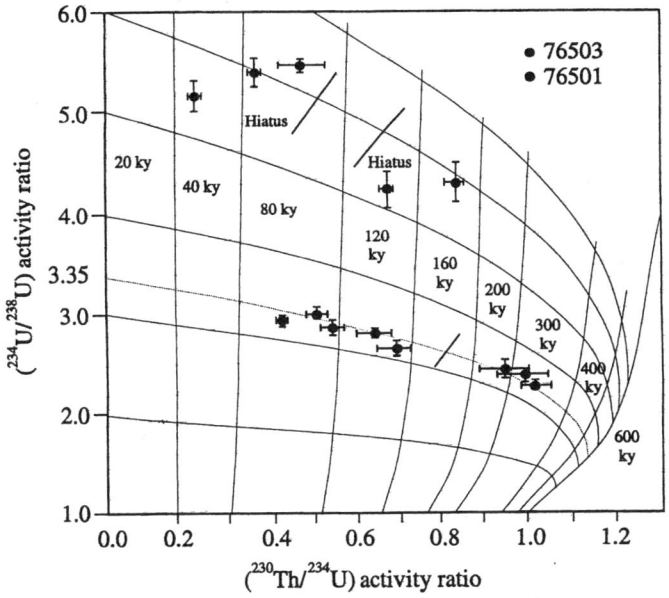

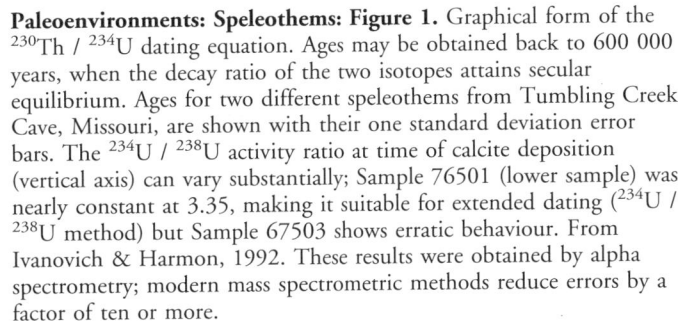

Paleoenvironments: Speleothems: Figure 1. Graphical form of the ^{230}Th / ^{234}U dating equation. Ages may be obtained back to 600 000 years, when the decay ratio of the two isotopes attains secular equilibrium. Ages for two different speleothems from Tumbling Creek Cave, Missouri, are shown with their one standard deviation error bars. The ^{234}U / ^{238}U activity ratio at time of calcite deposition (vertical axis) can vary substantially; Sample 76501 (lower sample) was nearly constant at 3.35, making it suitable for extended dating (^{234}U / ^{238}U method) but Sample 67503 shows erratic behaviour. From Ivanovich & Harmon, 1992. These results were obtained by alpha spectrometry; modern mass spectrometric methods reduce errors by a factor of ten or more.

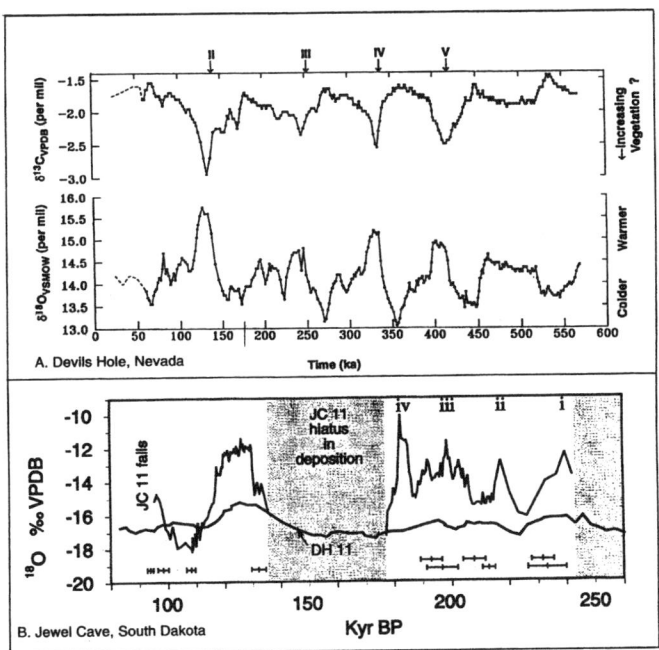

Paleoenvironments: Speleothems: Figure 2. A. The δ^{13}C and δ^{18}O records of Sample DH11, Devil's Hole, Nevada. The sample is a 36 cm core from a thermal water calcite spar coating in the cave, which is part of the spring outlet of a major regional aquifer. Dating control is by 21 mass spectrometer ^{230}Th / ^{234}U and ^{234}U / ^{238}U ages obtained at regular intervals along the core. From Coplen et al., 1994. **B.** For comparison, the δ^{18}O record from the upper part of a vadose speleothem, JC11, that grew in Jewel Cave, South Dakota, until it fell about 90 000 years ago. Dating control by ten mass spectrometer ^{230}Th / ^{234}U ages is shown, with two standard deviation error bars. This location is colder than the Devil's Hole region. The local vadose zone feedwater is vulnerable to cold, as indicated here, where speleothem deposition ceases during the coldest periods. Note that the amplitude of ^{18}O cyclic variation is much greater than in the Devil's Hole thermal water, where the climatic signal is muted by effects of mixing and slow groundwater circulation.

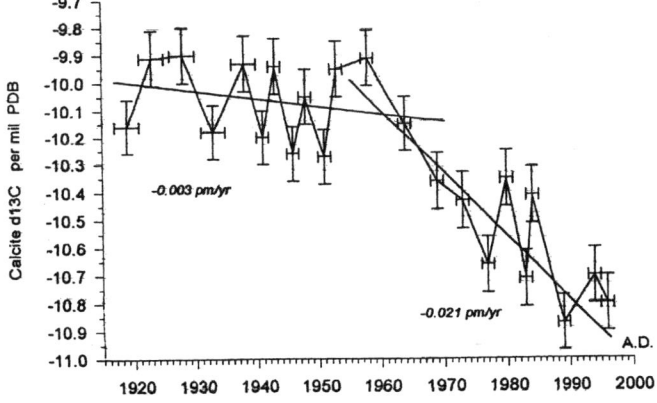

Paleoenvironments: Speleothems: Figure 3. An example of current studies exploiting annual depositional bands detected in some speleothems. The speleothem is from southern France. Two trends in the ^{13}C / ^{12}C ratio can be seen: (1) between 1920 and the 1950s, the "Suess Effect", the slow dilution of ^{13}C in the atmosphere caused by the burning of fossil fuels; (2) the "Bomb Effect", rapid dilution caused by detonation of nuclear bombs in the atmosphere (Genty et al., 1998).

atmospheric CO_2 $^{13}C:^{12}C$ is approximately −7 permil in CO_2 in the open air, reducing to around −14 permil in soil air under rich grasslands, and to −26 permil under dense forest. As a consequence, $^{13}C:^{12}C$ ratios may record processes in vegetation and soil above a cave or aquifer. In Figure 2A, it is seen that the carbon signal is negatively correlated with the oxygen signal in the DH11 phreatic calcite spar, suggesting slight increase in vegetal activity during interglacials in this desert setting: the amplitude of the speleothem $^{13}C:^{12}C$ oscillation here is about 1.4 permil. The forest-grassland–desert transition shifts back and forth across Jerusalem during a glacial climatic cycle, where the amplitude in vadose zone speleothems has been found to be as great as 12 permil (Frumkin, Ford & Schwarcz, 2000).

Organic and Other Cyclic Traces in Speleothems

Under magnification, many speleothems display micrometric layering of apparently cyclic kinds. This includes growth microterminations (with or without some dissolution) that suggest seasonal flooding or drying up of the feedwater, and even seasonal alternation of calcite and aragonite deposition. The most widespread, however, is probably colour banding in calcite without growth terminations. The banding usually has a couplet form–lighter to darker–with a colour range between yellow and dark brown. This is caused by variation in trace concentrations of fulvic acids (chiefly), plus humic acids, and fine particulate organic matter, incorporated in the crystal lattice (van Beynen et al., 2001). Shopov (1987) showed that the banding is best seen when fluoresced with ultraviolet light and suggested that, in many instances, each couplet represents one climatic year of deposition (see Speleothems: Luminescence). This has now been confirmed by many studies where the annual accumulation can be checked by field measurement, historic records, etc. (Baker et al., 1993). In temperate climates, the darker component is usually deposited in winter, or during the spring thaw if the ground freezes (Genty & Quinif, 1996). There are sometimes correlative variations in the abundance of common trace elements such as Ba, Mg, and Sr deposited in speleothems (Fairchild et al., 2000).

Speleothems may also incorporate pollen grains, enhancing paleovegetal information (see Palynology). However, their abundance is generally too low for conventional palynological ecologic reconstructions. Bacteria, fungi, algae, and mites (chiefly arthropods) have also been found in trace amounts.

This work is opening up a fruitful field for very high-resolution paleoenvironmental studies, similar to that of tree rings but extending much further back in Quaternary time. Investigations are expanding geographically to speleothems in hot desert and alpine regions. The stable isotopic composition of annual banding of known age is permitting precise tracking of human impacts on the environment, as shown in the example in Figure 3.

DEREK FORD

See also **Dating of Karst Landforms**

Works Cited

Baker, A., Smart, P.L., Edwards, R.L. & Richards, D.A. 1993. Annual growth banding in a cave stalagmite. *Nature*, 304: 518–20

Clark, I.D. & Fritz, P. 1997. *Environmental Isotopes in Hydrogeology*, Boca Raton, Florida: CRC Press

Coplen, T.B., Winograd, I.J., Landwehr, J.M. & Riggs, A.C. 1994. 500 000-year stable carbon isotope record from Devil's Hole, Nevada. *Science*, 26; 361–65

Dennis, P.F., Rowe, P.J. & Atkinson, T.C. 2001 The recovery and isotopic measurement of water from fluid inclusions in speleothems. *Geochimica et Cosmochimica Acta*, 65(6): 871–84

Fairchild, I., Borsato, A., Tooth, A.F., Frisia, S., Hawkesworth, C.J., Huang, Y., McDermott, F. & Spiro, B. 2000. Controls on trace element (Sr-Mg) compositions of carbonate cave waters: Implications for speleothem climatic records. *Chemical Geology*, 166; 255–69

Ford, D.C. 1997. Dating and paleo-environmental studies of speleothems. In *Cave Minerals of the World*, 2nd edition, C.A. Hill & P. Forti, Huntsville, Alabama: National Speleological Society

Frumkin, A., Ford, D.C. & Schwarcz, H.P. 2000. Paleoclimate and vegetation of the last glacial cycles in Jerusalem from a speleothem record. *Global Biogeochemical Cycles*, 14(3): 863–70

Genty, D. & Quinif, Y. 1996. Annually laminated sequences in the internal structure of some Belgian stalagmites—importance for paleoclimatology. *Journal of Sedimentary Research*, 66(1): 275–88

Genty, D., Vokal, B., Obelic, B. & Massault, M. 1998. Bomb ^{14}C time history recorded in two modern stalagmites. *Earth and Planetary Science Letters*, 160: 795–809

Ivanovich, M. & Harmon, R.S. 1992. *Uranium-series Disequilibrium: Applications to Earth, Marine and Environmental Sciences*, 2nd edition, Oxford: Oxford University Press

Latham, A.G. & Ford, D.C. 1993. The paleomagnetism and rock magnetism of cave and karst deposits. In *Applications of Paleomagnetism to Sedimentary Geology*, edited by D.M. Aissaoni, D.F. McNeill & N.F. Hurley, Tulsa, Oklahoma: Society of Economic Paleontologists and Mineralogists

Lauritzen, S.-E., Haugen, J.E., Lovlie, R. & Gilje-Nielsen, H. 1994. Geochronological potential of isoleucine epimerization in calcite speleothems. *Quaternary Research*, 41: 52–58

Richards, D.A., Bottrell, S.H., Cliff, R.A., Strohle, K.-D. & Rowe, P.J. 1998. U-Pb dating of a Quaternary-aged speleothem. *Geochimica et Cosmochimica Acta*, 62(23–24): 3683–88

Shopov, Y.Y. 1987. Laser luminescent microzonal analysis: A new method for investigation of the alterations of climate and solar activity during the Quaternary. In *Problems of Karst Study in Mountainous Countries*, edited by T. Kiknadze, Tbilisi, Georgia: Metsniereba

Van Beynen, P.E., Bourbonniere, R., Ford, D.C. & Schwarcz, H.P. 2001. Causes of color and fluorescence in speleothems. *Chemical Geology*, 175(3–4): 319–41

Further Reading

Bradley, R.S. 1999. *Paleoclimatology: Reconstructing Climates of the Quaternary*, 2nd edition, London and New York: Academic Press

Ford, D.C. & Williams, P.W. 1989. *Karst Geomorphology and Hydrology*, London and Boston: Unwin Hyman

PALEOKARST

Although a great range of geological structures and landform features of variable ages have been described as paleokarst, the term is difficult to define. The most inclusive definition is to say that paleokarst is evidence for karst processes acting in the past. This definition intentionally includes both karst landforms formed in the past and the deposits that fill them. The essential characteristic of paleokarst is that it is inert, i.e. there is no longer any significant gain or loss of matter in it as a result of karst processes. Every feature of karst that we recognize today, from towers and giant maze caves to microkarren and helictites, existed in the past and may be preserved in the geological record by burial or may have survived in the surface or underground landscape through isolation from weathering and erosion.

As paleokarst *sensu stricto* is inert it should be distinguished from fossil karst and relict karst. Fossil karst is correctly used to describe karst features that are not in equilibrium with modern landscape process, but are not inert from a karst perspective. Relict karst is isolated from the karst excavation processes that formed it, but still is subject to modification for example by weathering, breakdown, and speleothem deposition. However, the terms paleokarst and fossil karst are often used interchangeably, while cavers frequently, and erroneously, refer to "fossil passage" when they are actually describing relict passage that is above the present zone of active erosion by running water.

Paleokarst is a worldwide phenomenon that is found on all continents, including Antarctica, and in the geological record from the early Proterozoic to the Holocene. Very young paleokarst, including filled caves, occurs in raised coral reefs on the north shore of Oahu, Hawaii. Paleokarst has been reported in a range of soluble rocks including limestone, dolostone, chalk (England), gypsum (Ukraine and Canada), salt (Canada), and carbonatites (Canada).

Diagenetic Settings

For many limestones the process of diagenesis, which converts sediments into rock, begins with a period of subaerial exposure and karst development (see Syngenetic Karst). Consequently many karsts have undergone at least one previous period of karstification. Large filled caves and dolines preserved at grand unconformities are the most outstanding forms of paleokarst, but the most frequent type of paleokarst occurs when epikarst (ectogenetic karst) is preserved. This often occurs at minor disconformities in carbonate sequences when short periods of subaerial exposure are followed by a marine transgression and further carbonate sedimentation. For this reason some carbonate petrologists consider paleokarst to be a facies. Karst can develop in lithified soluble rocks (telogenetic karst) wherever and whenever they are exposed to karst processes. This can occur at or near the Earth's surface and deep underground where hydrothermal, thermal, or artesian processes can dissolve the rock, producing endogenous karst. This means that most karsts, particularly those in Mesozoic or older rocks, have been subjected to more than one period, and often many periods, of karstification. Multiple or polycyclic karstification is more often the rule than the exception. Some karst features, such as caves and dolines, may be buried and exhumed a number of times. For this reason paleokarst should not be thought of as "dead" karst. The real death of karst comes when the rock mass is completely removed by solution. It is caves without roofs, not filled caves, that are really on their deathbed (see photo).

Preservation of Paleokarst

There are three main ways in which paleokarst can be preserved: burial, isolation, and cessation of process. When karst landscapes are covered by sediment, buried karsts are produced. Depressions in the karst surface, and caves, are filled, or partially filled, with sediment. The degree of preservation of surface solution features depends on the energy with which the sediment is deposited. Buried karsts are protected from surface karst processes while they remain isolated from the surface groundwater system. Subjacent karst may develop in buried karst rocks if they are close to the surface. However, buried karsts are not protected from *per ascensum* karst processes. Many karsts remain buried, their presence being discovered only by drilling or geophysical investigation. In other cases, natural processes of uplift and erosion have exposed buried karst rocks at the Earth's surface revealing, and / or exhuming, the paleokarst. When karst landforms, particularly caves, become isolated from karst processes, ancient karst features may survive as relict karst. This can occur, for example, when caves become disconnected from the local hydrological system in regions with low erosion rates or where climatic changes convert wet tropics to deserts. Relict caves can also be preserved after *per ascensum* processes cease. Thermal caves, formed at depth in localities where there is no longer rising warm water, and the giant gypsum caves of the Ukraine, from which artesian waters have now drained, belong to this category.

Rises in sea level can flood both surface karst and caves, causing karst processes to cease. As a result paleokarst is an important feature of limestone coastlines. Caves below sea level may still carry fresh water to the sea, producing submarine springs. The Sea Mills of Argostoli, Kephallonia, Greece are a curious flooded-karst feature. Sea water sinks into a doline on one side of the limestone island, only to rise, slightly diluted, just above sea level on the other side of the island, ten kilometres away.

Plate tectonics is significant when thinking about all types of paleokarst: buried, relict, and exhumed. In parts of the world where the present landscape has ancient origins, such as Australia and South Africa, relict karst forms, including caves, may have formed in the ancient continent of Gondwana, within the Antarctic Circle. Similarly, relict karst landforms in Europe may have formed under tropical conditions.

Caves without roofs are an important paleokarst landform which has only recently attracted scientific attention. Caves without roofs are produced when lowering of the karst surface intersects, and removes the roof of a cave. A cave without a roof may be a cave floor or a depression consisting of cave walls and floor, exposed at the surface. They can be considered as paleokarst because they are now under the influence of surface lowering, rather than karst processes. Often caves without roofs are filled with, or contain a considerable quantity of, cave sediments. Interest in caves without roofs arose when motorway construction in the Kras (Slovenia) intersected a considerable number of these features (see photo).

In geologically active parts of the world, a mass of soluble rock can subside, become buried, and then be uplifted again on a number of occasions. Each time the rock is at the surface it

Paleokarst: Roofless cave discovered during construction of an expressway through the Classical Karst of Slovenia. (Photo by John Gunn)

is possible for karst to develop. The same rock mass may also be intruded by igneous rocks resulting in hydrothermal or thermal karst, or may form part of an artesian basin resulting in artesian karstification. At some localities, four, six, and ten periods of karstification have been recognized. With increased scientific interest in this field, it is likely that recognition of many periods of karstification will become more common. Multiple karstification can effect the whole range of paleokarst features, from large caves and dolines to cases where modern epikarst is invading and re-activating ancient epikarst.

Economic Deposits

Paleokarst is of great economic significance: large lead-zinc ore bodies, oil resources, bauxite, uranium, iron, and clay deposits occur in paleokarst. Paleokarst aquifers are important sources of water and can present a significant flooding hazard to underground mining operations. Mississippi Valley-type ore bodies are one of the most significant forms of paleokarst ore deposits. These consist of large bodies of galena and sphalerite emplaced in limestone (see Sulfide Minerals in Karst). Many of these bodies have plan shapes similar to those of caves and some ore occurs as speleothems. The three-kilometre long zinc-lead ore body at Nanisivik, Baffin Island, Canada has been interpreted as a large keyhole-shaped cave passage in which sulfide ores were emplaced. Ancient karst surfaces, paleokarst unconformities, and cavernous zones in limestone are significant traps and reservoirs for petroleum (see Hydrocarbons in Karst). Paleokarst is important in the great areas of carbonate-hosted petroleum such as the Middle East and Texas. Paleokarst bauxites account for 10% of the world's resources of aluminium. They develop in deeply weathered sediments filling ancient karst depressions and in some cases caves. Karst bauxites include large deposits in Jamaica and the majority of bauxite deposits in Europe, such as those of Greece and Hungary (see Bauxite Deposits in Karst). The phosphate deposits of Pacific islands such as Nauru have developed in and on karst surfaces. The complex shape of karst surfaces makes them ideal traps for dense minerals forming placer deposits (see Mineral Deposits in Karst). Some filled dolines and caves contain economic concentrations of alluvial gold, diamonds, cassiterite (tin ore), rare earths, and precious stones. Paleokarst aquifers and submarine springs can both be significant resources for water supply.

Scientific Significance

Paleokarst can be an important source of scientific information. It can preserve information about events and environmental conditions in the past which is not preserved elsewhere in the geological record. This is particularly important in areas where there are significant gaps in the stratigraphic record. In some cases the only record of a particular event, or the only means of dating an event, is provided by paleokarst. Evidence for a marine transgression in the Czech Republic during the Miocene is provided by a tiny deposit of fossil-bearing limestone, adhering to the wall of Zbrašovské Aragonite Cave. Paleokarst exposed in a small quarry in Tasmania provides the principal evidence for the age of a major Devonian folding event in eastern Australia. With continuing improvements in dating techniques, paleokarst has great potential for providing a window into times in the past about which we presently have little or no information. Palaeokarsts, including caves without roofs, are a rich source of vertebrate fossils. Examples include the hominid sites of the Transvaal, South Africa and the Tertiary marsupial deposits of Riversleigh, Queensland, Australia. Dinosaur fossils have been found in paleokarst in Belgium, Romania, and Slovenia. Modern caves are complex sedimentary environments in which many different types of sediments (e.g. sand, mud, bone, guano) and minerals are deposited. These are preserved in paleokarst, and frequently become transformed into rock. Paleokarst deposits are complex and present petrologists and sedimentologists with a fascinating range of unusual rock types and sedimentary structures to study. The variety of their economic and scientific significance means that scientists from a range of backgrounds are interested in paleokarst. As a consequence, information about paleokarst is published in a great range of literature.

Finding and Recognizing Paleokarst

Evidence for karst processes in the past occurs in two main forms, morphological and material. Morphological evidence includes surface and underground karst structures (buried, exhumed, or preserved) which are indicative of past processes, while material evidence includes earth materials (minerals, sediments, and rocks) deposited in, on, or over karst. Paleokarst is most easily recognized where large (but not too large) buried or filled karst features are intersected and exposed. Sections through filled cave passages in cliffs or quarry faces, such as those in Czatkowice Quarry, Poland, are quite compelling. For this reason, much paleokarst research has focussed on filled caves and dolines exposed in quarries (see photo in Paleoenvironments: Clastic Sediments). However, it is doubtful whether researchers would recognize some of the world's largest caverns as paleokarst if they were filled with sediments and then exposed in section (e.g. Sarawak Chamber, Mulu, Sarawak: 700 m long, 400 m wide, and 280 m high). On a different scale, oil and mining geologists tend to see paleokarst exposed in drill cores. Their work often centres on the microfabrics produced in carbonate rocks by karst processes.

In some areas, such as the Kras of Slovenia, where paleokarst deposits are exposed at the surface, paleokarst has not yet been found in caves. Paleokarst deposits are most likely to be intersected by younger caves in steeply bedded impounded karsts where there are a limited number of flow paths through the limestone, or where recent caves have developed from below (*per ascensum*) by thermal or artesian processes. Recognizing paleokarst in caves can be difficult. Some caves, such as those in the Peak District, England, intersect ancient open cavities (see photo in Peak District entry), while others, such as Jenolan Caves, New South Wales, Australia and Beremend Crystal Cave, Hungary, intersect ancient sediment-filled caves. In addition, some caves that are now open and accessible were once filled with sediments or ore bodies. These exhumed caves can sometimes be recognized because their morphology is unrelated to modern hydrological conditions, or because they contain tiny remnants of the material that once filled them but many probably go unrecognized. Modern caves may contain passages, chambers, and speleogens that are remnants of older cave systems. A modern stream cave may intersect cupolas that were part of an ancient thermal cave. If polycyclic karst is common, it must be common for caves to be made up of sections of differing ages, formed by different processes. Recognizing the different paleokarst elements and unravelling the history of complex caves is a challenging, four-dimensional, puzzle.

ARMSTRONG OSBORNE

Further Reading

Bosák, P., Ford, D.C., Glazek, J. & Horácek, I. (editors) 1989. *Paleokarst: A Systematic and Regional Review*, Amsterdam: Elsevier

Ford, D.C. 1986. Genesis of paleokarst and strata-bound zinc-lead sulfide deposits in a Proterozoic dolostone, northern Baffin Island, Canada—a discussion. *Economic Geology*, 81: 1562–66

Ford, D.C. 1995. Paleokarst as a target for modern karstification. *Carbonates and Evaporites*, 10(2): 138–47

James, N.P. & Choquette, P.W. (editors) 1988. *Paleokarst*, New York and London: Springer

Jones, B. & Smith, D.S. 1988. Open and filled features on the Cayman Islands: Implications for the recognition of paleokarst. *Canadian Journal of Earth Sciences*, 25: 1277–91

Korpas, L. 1998. Paleokarst studies in Hungary. *Occasional Papers of the Geological Institute of Hungary*, 195: 1–139

Kranjc, A. (editor) 1999. Papers presented at the 7th International Karstological School "Roofless Caves." Postojna, 1999. *Acta Carsologica*, 28(2): 15–210

Olson, R.A. 1984. Genesis of paleokarst and strata-bound zinc-lead sulfide deposits in a Proterozoic dolostone, northern Baffin Island, Canada. *Economic Geology*, 70: 1056–1103

Osborne, R.A.L. 2000. Paleokarst and its significance for speleogenesis. In *Speleogenesis: Evolution of Karst Aquifers*, edited by A.B. Klimchouk, D.C. Ford, A.N. Palmer & W. Dreybrodt, Huntsville, Alabama: National Speleological Society

Osborne, R.A.L., 2002. Paleokarst: cessation and rebirth. In *Evolution of Karst: From Prekarst to Cessation*, edited by F. Gabrovsek, Ljubljana: Zalozba ZRC

Quinlan, J.F. 1972. Karst-related mineral deposits and possible criteria for the recognition of paleokarsts: A review of preservable characteristics of Holocene and older karst terranes. In *Proceedings of the 24th International Geological Congress, 1972*: Section 6: 156–68

Wrenrich, K.J. & Sutphin, H.B. 1994. Grand Canyon caves, breccia pipes and mineral deposits. *Geology Today*, May-June 1994: 97–104

Wright, V.P. (editor) 1991. *Palaeokarsts and Palaeokarstic Reservoirs*, Reading: Postgraduate Research Institute for Sedimentology, University of Reading (Occasional Publication Series, 2)

PALEONTOLOGY: ANIMAL REMAINS IN CAVES

Deprived of the life-giving energy of the sun, caves are an environment that, in principle, is hostile to life. Few mammals regularly use the deeper areas of the cave, exceptions being bats that can find their way in total darkness, and cave bears, now extinct (Pinto & Andrews, 2003). On the other hand, many animals dwell near the entrances of caves, making use of the shelter provided. However, the majority of animal remains that are found in caves have been introduced by other means, human agency being one of them. This is an account of the ways that animal bones come to be in caves and how they may be interpreted in the archaeological record to take account of the biases introduced by these different processes.

Bone Accumulation in Caves

Animal remains accumulate in caves in four basic ways: the animals live and die in the cave; they fall in by accident and cannot get out again; they are taken in by predators; or their bones are transported in after death by other means (Figure 1). Cave size and shape, altitude, and relation to water determines its usefulness for animal occupation and whether animals may be trapped, carried in, or washed in.

Common dwellers of caves include bats, porcupines, and many species of mammalian and avian predator. Some tropical caves have large accumulations of bat remains mixed with the deep layers of bat guano. Large mammals may seek shelter in caves and may be trapped, for example, by heavy snowfalls. Caves in England have been observed to be home to foxes and badgers, and in Africa, leopards have been found living in caves, e.g. the Mount Suswa caves in Kenya (Brain, 1981). The lair of the well-known man-eating lions of Tsavo was eventually located in a nearby cave. In the past, human beings and their ancestors frequently lived in caves, not only in temperate regions but in Africa as well (Balch, 1914; Klein, 1975; Deacon, 1979). It has been inferred that many fossil animal assemblages found in caves were derived from animals actually living in the caves. For example, it is probable that the bats found in some abundance in one unit of the caves at Westbury-sub-Mendip, Somerset (England), were living in the cave, for where the bats occurred there were few other mammals (Andrews, 1990). Similarly, extremely rich amphibian remains were found in wet clay deposits in a cave at Draycott, Somerset, and again these were the only animals present. In the past also, there is good evidence that cave bears occupied and died in large numbers in caves, and this has been documented for the Westbury caves. Occupation over thousands of years has resulted in huge quantities of bones being accumulated in some European caves.

The second way animals get into caves is by accident, trapped in a pit fall chamber or in the soft mud of the bottom of a pit,

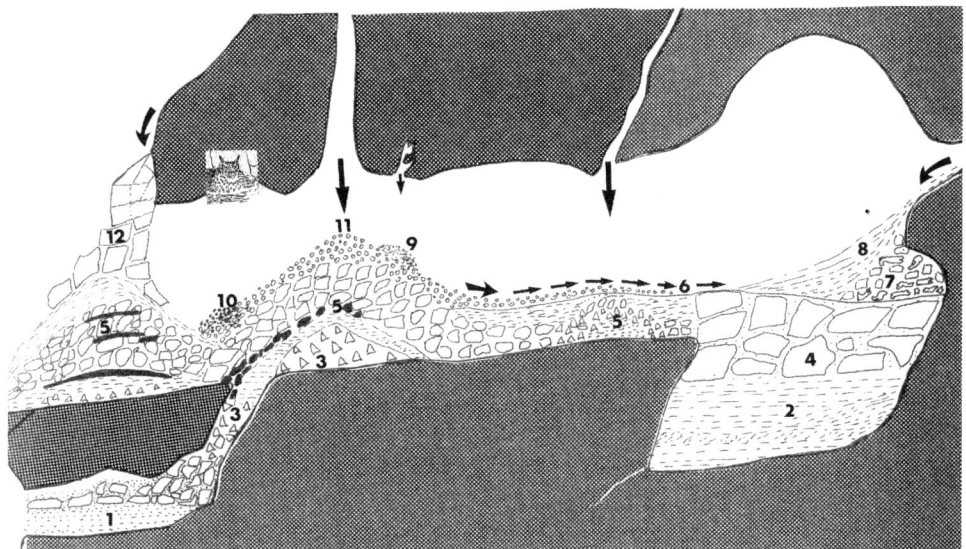

Paleontology: Animal Remains in Caves: Figure 1. Generalized section of a cave. The numbers indicate different forms of accumulation of bones and sediment: **1.** Accumulation of cave earths in a closed chamber; **2.** Water-lain silts; **3. & 11.** Breccia formation beneath an aven, and secondary transport into the lower chamber; **4.** Roof fall; **5.** Cave entrance breccias with bone; **6.** Breccias flattened by trampling of large animals; **7.** Den accumulation of bones; **8.** Water-transported muds from further inside the cave; **9.** Accumulation of bat remains beneath roosting area in the cave roof; **10.** Accumulation of small mammals beneath owl roosting/nesting area in the cave roof; **12.** Closing of cave entrance by collapse of the cliff space.

from where they cannot escape. Usually the structure of the cave provides an indication, if this is the case. For example, the Sima de los Huesos accumulation of human and cave bear bones in Spain may have been a natural trap, at least for the cave bears, and the bears found in the deep pit of Somiedo cave likewise (Pinto & Andrews, 2003). Both these caves are in Spain, but there is also the famous Natural Trap cave in Wyoming, United States (White et al., 1984). Elephants and rhinoceroses may enter caves seeking water in karstic crevices, and mineral salts. Remains of these animals have been found in caves on Mount Elgon, Uganda, after meeting with accidents inside the caves (Andrews, 1990). Rodents have been found living (and dying) at the bottom of cave (and mine) shafts, where they are able to survive for a time on food debris blown in. For example, rodents have been found at the bottom of a 30 m vertical shaft in the Eastwater Cave, Somerset. Finally, animals may be trapped in caves, not by the cave formation itself, but by inclement weather such as heavy snowfalls (Sutcliffe, 1955). A feature of such bone accumulations is that animals are represented by whole skeletons, as there is no way for the animals to be removed, dead or alive. Carnivores are sometimes attracted into such caves, if not too deep, but their meals are dearly bought when they cannot get out again.

The third way of entry of bones into caves is when transported by predators. The predators either live in the caves and accumulate prey remains in their pellets or scats, carry prey individuals into the cave and discard bones uneaten, or they may live just outside the cave and their prey remains either fall in or are carried in by another process. For example, partly eaten remains of bovids have been found in African caves, left over from leopard kills and rabbit remains have been found in Devon caves, brought in by foxes. In one unusual case, the abundant remains of porpoises were found in a cave formerly inhabited by striped hyenas in Qatar (Andrews & Whybrow, 2003). Some species of owl and diurnal raptors also shelter in caves, and their droppings contain extremely rich remains of their small mammal prey. The remains of the predators themselves may be present in the caves in which they lived, but their absence cannot be taken as evidence for absence of these predators. For example, several different predators have been identified as living at different times in the caves at Westbury-sub-Mendip, Somerset, by analysis of their prey remains (Andrews & Ghaleb, 2000).

The fourth way of entry is transport by water, gravity, etc. A model for this was provided for Swildons Hole, a modern cave in Somerset (Andrews, 1990). Bones from a bovid were found in a blind chamber on top of a talus cone in the process of accumulation beneath an aven, which communicated to the surface above through a narrow crack. On the surface was a deep doline, caused by the gradual movement of soil down into the chamber below, and it is inferred that the bones were carried with the soil over a period of 850 years. Movement can be much more rapid, as in the cavern collapse at GB cave, Somerset, when heavy rain caused the collapse of about 4000 m^3 of surface sediment and soil into an underground chamber. Water transport of sediment and bone into caves has also been documented for several levels in the Westbury sequence.

Bone Modification in Caves

The method of entry into caves clearly affects the degree to which bones may be modified. In the first case listed above, of animals living and dying in caves, it is apparent that initially they will be preserved as whole skeletons. Where there is free access to the cave, however, scavengers may enter the cave (some may even live there) and they may destroy much of the material. This is generally the case with large mammals dying in caves; for example, hyenas scavenged the leopard kills in the East African caves at Mount Suswa, but the mammals brought into the South African caves by leopards were still intact, since no scavenger had found them. Smaller animals may be scavenged as well, and even in the case of the large bat accumulations mentioned above, some of the bones may be modified, even destroyed, by insect action and fungal / bacterial decay. For small mammals, however, it remains the case that natural accumulations such as this may be distinguished from predator assemblages by their greater completeness and relative lack of modification. For example, in Wookey Cave, Somerset, a barn owl assemblage of-dispersed rodent bones, some showing evidence of digestion (see below), was found with some incomplete but still articulated bat skeletons which showed no evidence of digestion. The latter were clearly derived from bats living in the cave and dropping to the floor of the cave when they died and their bones becoming mixed with the owl prey bones (Andrews, 1990).

There may be other factors operating as well to disperse or otherwise modify the bones from natural deaths in caves. By their very nature, caves have restricted areas compared with land surfaces above, and animals moving about the cave, whether scavengers or later entrants into the cave, trample, break, and scatter the bones of earlier deaths. Rock falls and sediment

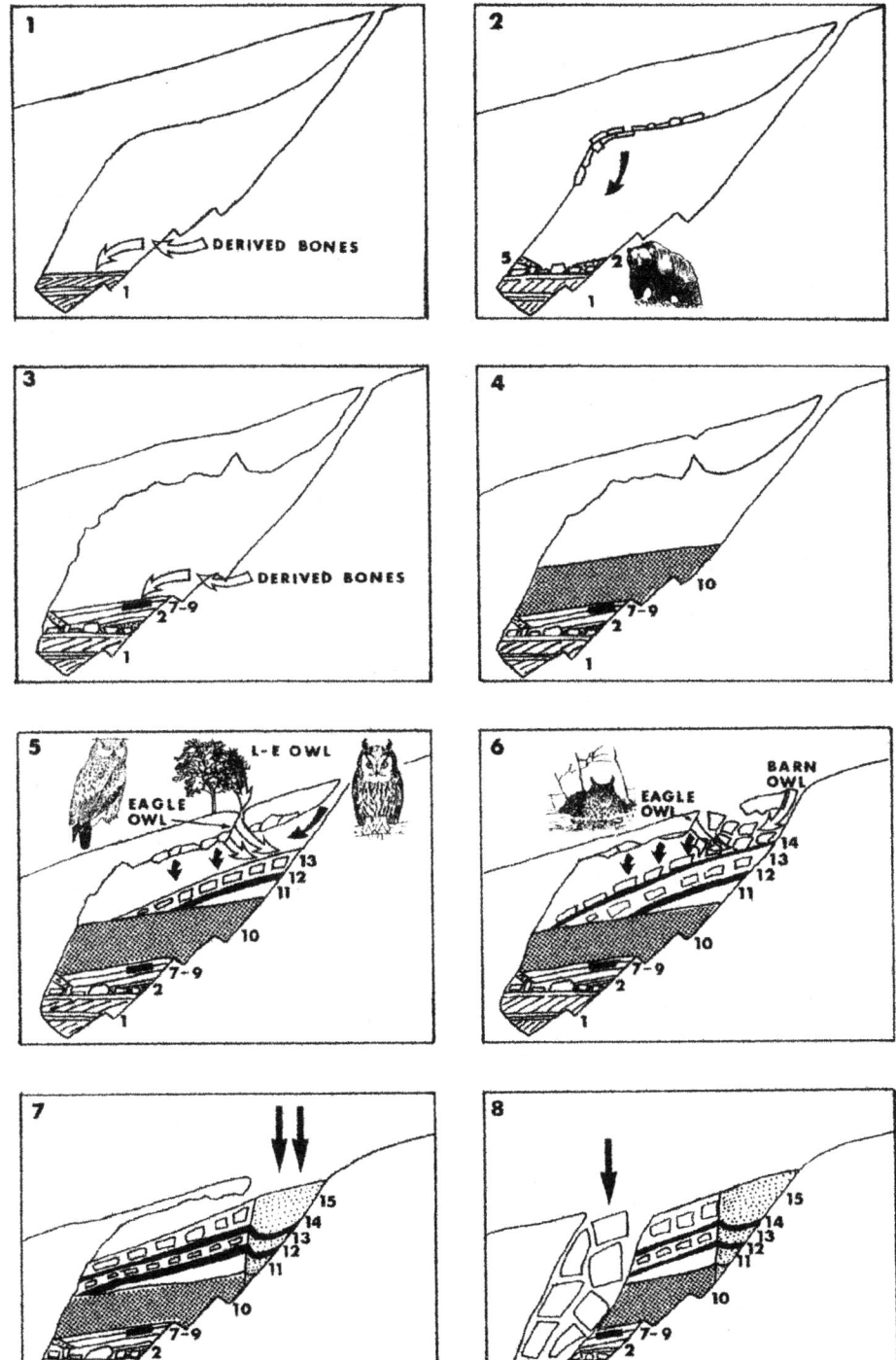

Paleontology: Animal Remains in Caves: Figure 2. Summary of the taphonomy of one of the caves at Westbury-sub-Mendip, shown in 8 stages from bottom to top, numbered 1 to 8 top left of each figure. **1.** Deposition of water-transported sands and gravels, generally lacking fossils. **2.** Major collapse of roof-fall blocks onto the surface of the eroding sands; occupation of the chamber by all-male groups of cave bears in units 2 and 5. **3.** Further accumulation of water-transported sediments with local pockets of water-transported fossils in unit 8. **4.** Silts and sands of unit 10, with dispersed fauna throughout and concentration of bat remains at the top, derived from bats living in the cave. **5.** Formation of cave breccias of units 11 to 13, with multiple sources of sediments (black arrows) and fossils (open arrows); large mammals may have been transported into the cave along with the sediment, but the small mammals were the product of eagle owl and long-eared owl predation. **6.** Formation of cave breccias of unit 14 and lower subunits of unit 15 infilling space between roof fall blocks; origin of sediments and fossils as before, with the predators being barn owls and eagle owls; at this time the roof of the chamber was largely destroyed. **7.** Formation of the upper subunits of unit 15, now probably exposed to the open air, so that the fossil accumulations cover a greater area. **8.** The final stage came with the collapse of the sediments along one side of the cave, down into a chamber below, and the infill of the void thus left by loosely consolidated infill breccias, indicated by the vertical arrow. The numbers refer to unit numbers, the black arrows indicate source and direction of the sediments and the open arrows indicate source and direction of the fossil bones.

pressure may also produce breakage, and the result is that most bones in caves become broken and scattered, even if originally they were present in the cave as entire animals. The very high humidity within caves and its changes, the acidity or alkalinity of water, and dripping water may also further corrode the bones.

Animals trapped by pitfalls also enter the cave entire, and in this case they more often remain in one piece since the structure of the cave prevents entry by scavengers. This is not always the case, however, for scavengers may be tempted in and then cannot get out again, and in this case the damage to bones may be extreme as the scavengers are driven by hunger to consume as much of the skeleton as they can. The other factors mentioned above come into play as well: trampling, sediment movement, and corrosion.

Predation is both the most important cause of death in mammal faunas and the cause of their greatest modifications. Many predators live in caves, and numerous examples are known of owls roosting in caves (Andrews, 1990, Avery, 2001), and larger predators such as hyenas and leopards living and hiding their prey in caves (Maguire et al., 1980). The amount of bone present and the species representation in prey assemblages vary from predator to predator but are always biased with respect to the community from which they came. The important issue is how to identify the predator and hence the probable bias, but the relationship of prey to predators is highly complex and is dependent on several unrelated factors. Predator size, for example, is related to prey size, but it varies with seasonal and annual cycles of climate, habitat, and prey populations, but predator adaptability and ability to learn may offset the impact of these cycles. Independent of these factors is the potential difference in activity patterns between predator and prey, so that nocturnal and diurnal predators hunting the same area may produce completely different prey assemblages. Because of these variations in predator / prey relationships, neither the size of the prey nor the species preyed upon are diagnostic of any single predator species.

The biases produced by predators as a result of their hunting and selective behaviour can be interpreted and allowed for if the identity of the predator is known (Andrews, 1990). It has been found that different predators produce specific patterns of bone modification of their prey, so that it is sometimes possible to predict predator species from the consideration of modifications to the bones making up faunal assemblages. Bone breakage in prey assemblages may be diagnostic of certain predators but not of others; but bone digestion is the most diagnostic feature, although here too it is not possible to distinguish all predator species. Degree of digestion also varies between predators according to their state of hunger and their maturity.

Several Pleistocene cave faunas have been analysed based on these criteria, in order to identify their modes of accumulation (Figure 2). The long cave sequence at Westbury-sub-Mendip has been shown to have faunas accumulated by barn owls, eagle owls, and long-eared owls (Andrews, 1990). The equally good sequence at Dolina cave, one of the Atapuerca caves in Spain, has similarly been shown to have faunas accumulated by an eagle owl, tawny owl, and an unidentified species of small mammalian carnivore (Fernandez-Jalvo & Andrews, 1992). In both cases there were additional modifications that went some way towards obscuring these patterns, but despite additional breakage, extensive cave corrosion, and in one case weathering (at the top of the Westbury sequence where the roof had fallen in), the original evidence of the predator damage was still apparent.

In the final method of entry of bones into caves, the association of bones with water-lain sediments and the presence of abrasion of bone surfaces, provide evidence of transport. Large mammal bones have been found deep inside caves in Wales and in Swildon's Hole, Somerset, where they were washed in by streams flowing through the caves. An example from the fossil record comes from the Westbury cave again (Andrews, 1990). Two of the major units were formed as part of a series of mudflows brought into the cave from outside, and the mammal fossils are dispersed in it. Some of the bones entered the cave with the mudflow, and some were the result of cave bear occupation in or near this part of the chamber. Carnivore damage is moderately common, with some of the bones being highly modified. The bone is rounded and extremely broken. Human presence is indicated by cut marks on a few of the bones.

PETER ANDREWS AND ANA C. PINTO LLONA

Works Cited

Andrews, P. 1990. *Owls, Caves and Fossils: Predation, Preservation, and Accumulation of Small Mammal Bones in Caves, with an Analysis of the Pleistocene Cave Faunas from Westbury-sub-Mendip, Somerset, UK*, Chicago: Chicago University Press

Andrews, P. & Ghaleb, B. 2000. Taphonomy of the Westbury Cave bone assemblages. In *Westbury Cave*, edited by P. Andrews, J. Cook, A. Currant & C. Stringer, Bristol: Western Academic and Specialist Press

Andrews, P. & Whybrow, P. 2003. A striped hyaena den site in Qatar, United Arab Emirates. *Préhistorie, Anthropologic Méditerranéenes*

Avery, D.M. 2001. The PlioPleistocene vegetation and climate of Sterkfontein and Swartkrans, South Africa, based on micromammals. *Journal of Human Evolution*, 41: 113–32

Balch, H.E. 1914. *Wookey Hole, its Caves and Cave Dwellers*, London and New York: Oxford University Press

Brain, C.K. 1981. *The Hunters of the Hunted? An Introduction to African Cave Taphonomy*, Chicago: University of Chicago Press

Deacon, H.H. 1979. Excavations at Boomplaas Cave: A sequence through the Upper Pleistocene and Holocene in South Africa. *World Archaeology*, 10: 241–57

Fernandez-Jalvo, Y. & Andrews, P. 1992. Small mammal taphonomy of Gran Dolina, Atapuerca (Burgos) Spain. *Journal of Archaeological Science*, 19: 407–28

Klein, R.G. 1975. Middle Stone Age man-animal relationships in southern Africa: Evidence from Die Kelders and Klasies River Mouth. *Science*, 190: 265–67

Maguire, J., Pemberton, D. & Collett, N.H. 1980. The Makapansgat limeworks grey breccia: Hominids, hyaenas, hystricids or hillwash. *Palaeontologia Africana*, 23: 75–98

Pinto, A.C. & Andrews, P. 2003. *Taphonomy and Palaeoecology of Bears from N. Spain*, Oviedo, Spain: Fundacion Oso de Asturias

Sutcliffe, A.J. 1955. A preliminary report on the reindeer remains from Banwell Bone Cave. *Journal of the Axbridge Cave Group Archaeological Society*, 2: 36–40

White, J.A., McDonald, H.G., Anderson, E. & Soiset, J.M. 1984. Lava blisters as carnivore traps. In *Contributions in Quaternary Vertebrate Paleontology: A Volume in Memorial to John E. Guilday*, edited by H.H. Genoways and M.R. Dawson, Pittsburgh: Carnegie Museum of Natural History

Further Reading

Stiner, M.C. 1994. *Honor Among Thieves: A Zooarchaeological Study of Neandertal Ecology*, Princeton, New Jersey: Princeton University Press

Kurtén, B. 1976 *The Cave Bear Story: Life and Death of a Vanished Animal*, New York: Columbia University Press

PALEOTECTONICS FROM SPELEOTHEMS

Over the last few decades, seismotectonic studies of speleothems have proved that broken speleothems and actively growing stalagmites are the most powerful tools for the quantitative and chronological reconstruction of seismotectonic events over the last 500 000 years (Forti, 1999). The first person to suggest that speleothems may record earthquake activity was Becker (1929), who proposed the idea after studies of Bing Cave in Germany. Nevertheless, until the 1980s, geophysicists did not give serious consideration to the method because it was impossible to discriminate between the fracturing of speleothems by earthquake-induced effects and that caused by other mechanisms. However, improvements in sampling and statistical analysis now allow the paleotectonics of karst areas to be reconstructed, using not only broken speleothems but also actively growing stalagmites. They may be used as a tool for the detection of ancient earthquakes, and the relative and absolute dating of seismotectonic activity, the determination of its magnitude, and for improving general seismic hazard evaluation. The dating of speleothems is described in Paleoenvironments: Speleothems.

Detection of Ancient Earthquakes using Broken Speleothems

Caves in seismically active areas may contain broken and collapsed speleothems, but before it can be determined that broken speleothems are really related to earthquakes, all other factors must be discounted. The more common natural causes of speleothem breakage are: (1) An increase in the weight loading of stalactites growing from porous or highly fractured ceilings; (2) Sliding of stalagmites, columns, and flowstones along unconsolidated walls and floors; and (3) The presence of ice tongues during glaciation.

The morphology of the breakages can help to determine their cause: several types of breakage are caused only by tectonic stresses (Figure 1), the most characteristic being the perfect breakage of stalagmites along subhorizontal planes. This unusual type of breakage can be explained by the resonance induced by high-frequency seismic waves. Another sign of seismic activity is the presence of consistent speleothem breakage in certain directions. The preferential azimuths of the collapsed stalagmites normally coincide with the main structural directions in the cave area. However, it is often difficult to use this type of analysis, because it is virtually impossible to be sure that broken speleothems still retain their pre-earthquake orientation. Breakages resulting from seismic activity may also be distinguished from non-seismic breakages by a statistical analysis of their ages as the dates of earthquake-induced breakages tend to be grouped together at around the same time.

Estimating Earthquake Magnitude

Some seismic-induced breakages (e.g. A and B of Figure 1) may also provide data for evaluating earthquake accelerations, thus making it possible to define both the epicentre and the magnitude of the event. The method is similar to that used by geophysicists studying the effect of earthquakes on tombstones (Postpischl et al., 1991). The stalagmite itself is considered to be a homogeneous cylinder, perfectly connected to the floor, with its breakage being caused by resonance induced by the earthquake, and with the ratio of the diameter / height of the broken stalagmite being related to the horizontal acceleration of the seismic waves (which, in turn, is affected by the distance from the epicentre and the magnitude). Although this appears to be simple from a theoretical point of view, performing such a study is very complex because real stalagmites are very different from the model of a homogeneous cylinder, perfectly attached to the floor of the cave.

Actively Growing Stalagmites

Seismotectonic information may also be extracted from stalagmites that are still being deposited. Schillat (1977) considered the stalactite–drip-stalagmite system to be a "recording pendulum", in which the stalagmite, with its successive growth layers, acts as a recorder of the "actual verticality". A polished section along the growing axis of a stalagmite normally shows well-marked, symmetrical growth layers. Ideally, this axis records the vertical direction, which, if stable over time, should be a rectilinear segment. In reality, progressive and / or sudden variations in the growth axis can be found in many stalagmites, associated with environmental conditions in the cave itself, or associated with tectonic events. However, there are also many speleological factors which may affect the orientation of the stalagmite growth axis, for example: (1) permanent air currents; (2) migration of the dripping point along a fracture; and (3) gravity sliding of the stalagmite over unstable material.

In order to rule out these local effects, it is necessary to perform a statistical analysis of a large number of stalagmites. A three-dimensional analysis of the stalagmite growth axis enables the reconstruction of the tectonic movements experienced by

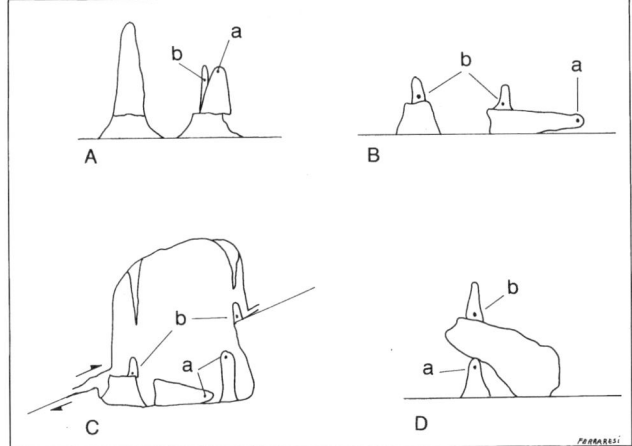

Paleotectonics from Speleothems: Figure 1. Characteristic breakage of speleothems, induced by seismic stress. Resonance-induced stalagmite fractured along a subhorizontal plane: (A) the upper part is still standing on its base, being only slightly translated and / or rotated from its original position; (B) the broken upper part lies on the floor close to its base; (C) stalagmite collapse caused by the displacement of the adjacent wall; (D) new stalagmite growing over a fallen rock, which covered an older stalagmite. Positions (a) and (b) indicate characteristic sampling points for absolute (U / Th and / or ^{14}C) dating of deposits which occurred just before (a) or after (b) the seismic event.

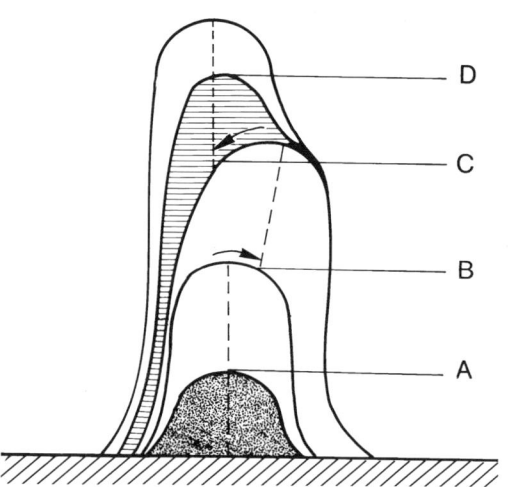

Paleotectonics from Speleothems: Figure 2. Evidence of earthquakes as seen in the inner structure of a stalagmite: i.e. sudden and sharp vertical changes in the stalagmite axis (B & C) and abrupt variations in the texture, colour, and chemical composition of the growing layers (A, C, & D) may be induced by seismic shocks.

the speleothem, and the order in which they occurred. This may be converted into a chronology using absolute dating.

The above-mentioned data on tectonic movements are not the only information that can be obtained from stalagmites—they may also provide records of the major earthquakes that hit the cave area. In many cases the vertical polished section of a stalagmite shows some clear discontinuities in the growth axis (Figure 2). Such discontinuities may be abrupt vertical variations and / or sharp changes in colour or texture, both of which can reveal the occurrence of an earthquake if local factors are excluded. By applying this method to two different gypsum caves, Forti & Postpischl (1986) reconstructed the seismic history of Bologna (Italy) over the last 1200 years with an average error of ±11 years.

Seismic Hazard

The evaluation of the seismic hazard for a given area is normally based on its seismic history, which can rarely be known for more than two thousand years. Clearly this span of time is too short to determine the possible magnitude of the strongest earthquake that might hit a particular region in future. By using seismic analyses of speleothems it is possible to recognize and date strong paleo-earthquakes up to the limits of radiometric dating methods, which presently exceed 500 000 years.

PAOLO FORTI

Works Cited

Becker, H.K. 1929. Höhle und Erdbeben. *Mitteilungen Über Höhlen- und Karstforschung*, 14: 130–33

Forti, P. 1999. Evidencias tectónicas y sísmicas a partir del estudio de espeleotemas: conocimiento actual y desarrollo futuro. In *The Scientific Study of Karstic Cavities as a Contribution to Geological Knowledge, Proceedings of the Geology Symposium Nerja Cave (Malaga, Spain), June 17–20*, edited by B. Andreo, F.Carrasco & J.J. Durán, Nerja, Malaga: Patronato de la Cueva de Nerja, Instituto de Investigación

Forti, P. & Postpischl, D. 1986. May the growth axes of stalagmites be considered as recorders of historic and prehistoric earthquakes? Preliminary results from the Bologna karst area. *Proceedings of the International Symposium on Engineering Problems in Seismic Areas*, vol. 1, Bari, Italy: Istituto di geologia applicata e geotecnica, Università di Bari

Postpischl, D., Agostini, S., Forti, P. & Quinif, Y. 1991. Palaeoseismicity from karst sediments: the "Grotta del Cervi" cave case study. *Tectonophysics*, 193: 33–44

Schillat, B. 1977. Conservation of tectonic waves in the axes of stalagmites over long periods. *Proceedings of the 7th International Speleological Congress*, edited by T. Ford, Sheffield: BCRA

Further Reading

Quinif, Y. (editor) 1998. Karst & Tectonics: Proceedings of the International Symposium—Han '98, *Spéléochronos* hors-série, Fac. Polytecnique de Mons, Belgium

PALYNOLOGY

Palynology, the study of microscopic organic matter known as palynomorphs (Traverse, 1988), offers enormous potential in karst research since it can provide important evidence about ancient environments. The most commonly studied palynomorphs are pollen and spores, but also important are algal microfossils, dinoflagellate cysts, fungal spores, cyanophyte phycomas, amoeboid cysts, and a variety of other remains which can be lumped under the general name of kerogen and analysed using palynofacies analysis. Apart from the limestones and evaporites themselves, which sometimes contain palynomorphs coeval with their deposition, two important types of sites are available for palynological study in karst terrains: closed depression fills and cave fills.

Understanding of taphonomy (the production, transport, deposition, and preservation of fossils) in karst depositional systems is still developing. Sediments in karst systems may be highly oxidized and/or subject to repeated wetting and drying. These are inimical to the preservation of organic matter. Organic matter in karst depositional systems is therefore often poorly preserved. However, good to adequate preservation of organic matter will occur where the oxygen supply is limited through rapid burial or waterlogging, or where desiccation prevents bacterial activity.

Closed depression (doline, uvala, or polje) fills offer what is often the only available evidence for past vegetation patterns in many karst areas. Often the closed depression contains, or has contained, permanent water, thus enabling palynomorphs to be preserved by waterlogging and anoxia. Palynological studies of doline fills have been carried out for many karst areas, for instance in Ireland (Coxon & Flegg, 1987). These doline fills provide virtually the only evidence for late Tertiary and early Pleistocene vegetation in Ireland and hence are of enormous importance. In arid countries, for instance in the Libyan desert, pollen was preserved by desiccation in a closed depression fill at Grerat-es-Salam (Gilbertson *et al.*, 1994).

Palynomorphs enter caves in a diversity of ways. The flora in cave mouths is usually distinctive and of low diversity, often dominated by ferns, and this flora has an impact on cave pollen assemblages. In the entrance facies of most caves and throughout those caves where there is a vigorous circulation of air, much pollen arrives through the air (Genty, Diot & Oyl, 2002). In caves where there is a diurnal circulation, fern spores, which are shed nocturnally, are carried into the cave at night by air currents entering the cave at low level. This, plus the dominance of ferns in many cave entrance floras, may account for the high incidence of fern spores in many cave deposits (Coles *et al.*, 1990). Palynomorphs may also be recycled by wind, for instance as a component in loess; they may be recognized as having been reworked by being ecologically and chronologically mixed.

Other mechanisms are also important. Some percolation water is reputed to carry significant amounts of pollen into caves, although rigorous experiments by a number of authors (e.g. Genty, Diot & Oyl, 2002) has shown that this is comparatively unusual. Pollen has been documented to have been carried over 5 km in streamways and it is also likely that pollen sometimes enters caves via mudflows. Pollen may be carried by mammals and birds on their fur/feathers or in their gut contents. For example, a single vole may carry several thousand pollen grains in its fur during the flowering season. Pollen has been found in animal dung in a number of caves, for instance in the American Midwest and Libya. Insects may also carry significant amounts of pollen into caves. Ground-nesting bees carry significant amounts of Compositae pollen into some southern European cave fills and in Sarawak burrowing wasps introduce enormous quantities of pollen into some cave deposits.

Once it is within the cave system, pollen may be preserved in clastic sediments and flowstones. Very often rapid deposition of clastics or flowstone dilutes the pollen so that it is extremely sparse, often no more than five pollen grains per gram. It is clear from a number of studies that, once the effects of entrance floras have been allowed for, cave pollen assemblages are similar to those of the immediately adjacent local pollen rain or to Quaternary deposits of equivalent age (e.g. Coles *et al.*, 1990).

The first palynological study of cave sediments was undertaken by Schutrumpf (1939). Subsequently, cave pollen studies have been carried out in many karst regions. Many of the early studies, including Schutrumpf's, were undertaken to correlate cave sediments with external sequences and thus provide a measure of dating. With the rise of radiometric dating, the objective of many studies became explicitly paleoenvironmental, although correlation using palynology is still undertaken where datable materials are absent, chronological resolution is poor, or the sediments are beyond the range of radiometric tools. A good example is the use of palynology to date paleokarst fills in southwest England as being Rhaetian in age. Notable research on pollen in speleothems, with complementary dating by radiometric methods, has been carried out in Belgium by Bastin (1990). Palynology is an integral part of many archaeological investigations in caves, since it provides a picture of the flora close to the cave and thus enables environmental and climatic reconstruction. Occasionally, palynology provides important behavioural evidence. The most famous case of this is still Arlette Leroi-Gourhan's (1975) palynological evidence for the burial of bunches of flowers with the body of a young Neanderthal girl in Shanidar Cave, Iraq, around 70 000 years ago (see Shanidar Cave: Archaeology).

CHRIS HUNT

Works Cited

Bastin, B. 1990. L'analyse pollinique des concrétions stalagmitiques: méthodologie et résultats en provenance des grottes belges. *Karstologia Mémoires*, 2: 3–10

Coles, G.M., Gilbertson, D.D., Hunt, C.O. & Jenkinson, R.D. 1990. Pollen taphonomy in caves. *Cave Science*, 16: 83–89

Coxon, P. & Flegg, A. 1987. A Late Pliocene / Early Pleistocene deposit at Pollnahallia, near Headford, Co. Galway. *Proceedings, Royal Irish Academy*, 87B: 15–42

Genty, D., Diot, M.-F. & Oyl, W. 2002. Sources of pollen in stalactite dip water in two caves in southwest France. *Cave and Karst Science*, 28(2): 59–66

Gilbertson, D.D., Hunt, C.O., Feiller, N.R.J. & Barker, G.W.W. 1994. The environmental consequences and context of ancient floodwater farming in the Tripolitanian pre-desert. In *Environmental Change in Drylands: Biogeographical and Geomorphological Perspectives*, edited by A.C. Millington & K. Pye, Chichester and New York: Wiley: 229–51

Leroi-Gourhan, A. 1975. The flowers found with Shanidar IV, a Neanderthal burial in Iraq. *Science*, 190: 562–64

Schutrumpf, R. 1939. Die pollenanalytische Datierung der Altsteinzeitlichen Funde von Mauren. *Das Ahnenerbe, bericht über die Kieler Tagung 1939 der Forschungs und Lehgemeinschaft*

Traverse, A. 1988. *Paleopalynology*, Boston: Unwin Hyman

Further Reading

Baker, A., Caseldine, C.J., Hatton, J., Hawkesworth, C.J. & Latham, A. 1997. A Cromerian complex stalagmite from the Mendip Hills, England. *Journal of Quaternary Science*, 12(6): 533–37

Bastin, B. 1978. L'analyse pollinique des stalagmites: une nouvelle possibilité d'approche des fluctuations climatiques du Quaternaire. *Annales de la Société Géologique de Belgique*, 101: 13–19

Carrión, J.S., Munuera, M., Navarro, C., Burjachs, F., Dupré, M. & Walker, M.J. 1999. The palaeoecological potential of pollen records in caves: The case of Mediterranean Spain. *Quaternary Science Reviews*, 18(8–9): 1061–73

Hunt, C.O. & Gale, S.J. 1986. Palynology: A neglected tool in British cave studies. In *New Directions in Karst*, edited by K. Paterson & M.M. Sweeting, Norwich: GeoBooks: 323–32

Navarro, C., Carrión, J.S., Navarro, J., Munuera, M. & Prieto, A.R. 2000. An experimental approach to the palynology of cave deposits. *Journal of Quaternary Science*, 15(6): 603–19

Marshall, J.E.A. & Whiteside, D.I. 1980. Marine influence in the Triassic "uplands". *Nature*, 287: 627–28

PAMUKKALE, TURKEY

The Pamukkale hydrothermal field is located in western Turkey, a region of intense vertical tectonism that commenced in the Upper Miocene with the formation of horst and graben structures (see map in Turkey entry for location). Since later tectonic movements have modified the fracture patterns and enhanced the anisotropic permeability, there are four reservoirs in the Pamukkale hydrothermal system: Paleozoic marble, Mesozoic crystallized limestone, Pliocene limestone, and Quaternary travertine. Pliocene units of claystone, siltstone, marl, and sandstone act as an impermeable caprock in the system (see Figure).

The drainage area of the Pamukkale thermal springs is about 102 km^2 and the annual mean rainfall has been estimated to be 881 mm using rainfall-altitude relationships. The isotopic characteristics of the thermal waters show that they possess almost the same deuterium composition as that of local meteoric water. Hence, the water constituting the thermal fluid is of a predominantly meteoric origin. The cold fluid, circulating through faults and fractures, is heated by magmatic intrusions at great depth, and ascends from deep reservoirs to the surface. Discharge takes place along fault systems that appear to follow the northern margin of the Çürüksu graben, but some thermal springs are also present in the Çürüksu valley. Hydrochemical surveys have shown that the chemical and physical characteristics of the thermal waters are affected by cooling, dilution, and mineral deposition as they ascend, because the interaction with relatively cold groundwater becomes prominent in the near-surface zone (Figure). The combined discharge of the Pamukkale thermal springs is about 385 l s^{-1}. However, recession curve analysis indicates an estimated active annual reservoir capacity of 16 x 10^6 m^3 and this corresponds to a discharge rate of 510 l s^{-1}. Günay *et al*. (1997) suggested that the excess (125 l s^{-1}) could be abstracted from the aquifer for recreational purposes at the site or to enhance travertine formation.

The thermal springs are issued at 35°C and a pH of 5.6. They exhibit a calcium–bicarbonate–sulfate composition with total dissolved ions of 2300 mg l^{-1}. The temperature and chemistry of individual thermal springs are more or less constant through time, indicating that thermal systems develop chemical compositions in equilibrium with the reservoir rocks. The thermal springs have very high carbon dioxide concentrations and the source of this gas is thought to be the decomposition of marine carbonate rocks. Experiments on calcium carbonate precipitation kinetics at the travertine site show that as the thermal water runs over the surface rapid carbon dioxide outgassing occurs, and as a result the water becomes highly supersaturated with respect to calcium carbonate, which then precipitates as the travertine terraces for which Pamukkale is famous. Age estimations show that travertine deposition started during the late Pleistocene.

The five principal morphological types of travertine formation at Pamukkale are (1) terraced mounded travertine, (2) fissure-ridge travertine, (3) range-front travertine, (4) eroded-sheet travertine, and (5) self-built channel travertine (Altunel & Hancock, 1994). The total area of these travertine types is about 10 km^2. The beautifully ornate architecture of the travertine terraces consists of a curving and steeply raked, or overhanging, ramp of overlapping cup-shaped pools on all scales from a few centimetres to a few metres. Small-scale terraces tend to be initially constructed where thermal water becomes turbulent as it flows over or around the obstructions, such as at breaks of slope, leaves,

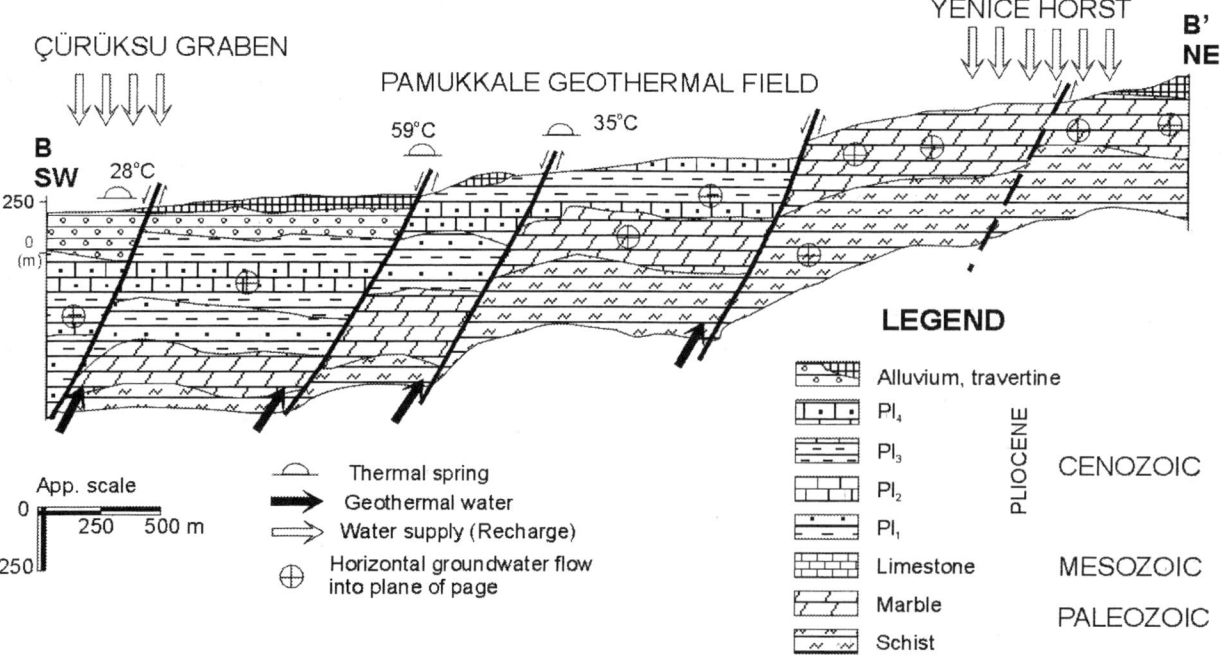

Pamukkale, Turkey: Hydrogeological cross-section of the Pamukkale geothermal field (vertical scale bar represents the topographic relief).

twigs, pebbles, or wall-like self-built channels. As long as water continues to flow over the terraces they grow higher and their maximum development occurs where there is a series of pools separated by rims of travertine (Altunel & Hancock, 1994). The morphology of the rimstone pools was studied by Ekmekçi et al. (1995) who found that flow regime strongly affects the grading from micro-terraces to flowstone.

The oxygen-18 and deuterium stable isotope values of the thermal waters range from -9.42 to -8.65 ‰ and -59.7 to -57.1 ‰, respectively. The tritium concentration varies between 0.8 and 4.5 tritium units (TU) because of mixing with shallow cold groundwater. Since all the thermal waters are immature, an indication of the reservoir temperature is provided by the solubility of silica minerals. Chalcedony solubility gives an equilibrium temperature in the range 63–93°C.

The existence of the thermal springs and white travertine terraces, together with the cultural ruins of the ancient city of Hierapolis, attract many visitors to Pamukkale and in 1988 UNESCO declared the area a Natural and Cultural World Heritage Site and inscribed it on the World Heritage list. Unfortunately, with the increasing numbers of tourists, and the development of tourist facilities, serious environmental problems became apparent. The main causes of damage were uncontrolled walking over the delicate travertine surface; swimming in thermal springs; intensive and uncontrolled excavations including wells; vehicle entry to the site; sewage leaking from non-isolated cesspools; and waste storage in karstic features. In an attempt to reduce the damage the Pamukkale Preservation and Development Plan was published in 1992 by the Governorship of Denizli, Ministry of Culture, and UNESCO. The scope of this plan was to create more suitable protection activities for the natural and archaeological assets (Dilsiz, 2002).

The main objectives of this ongoing project are to determine the nature of the hydrologic and thermal flow systems; to describe the hydrochemical characteristics of the springs; to identify carbonate mineral dissolution and precipitation reactions; to perform environmental studies to prevent pollution in and around the thermal springs and travertine area; and to increase both the yield of the thermal waters and the area of white travertine. Covered concrete channels were constructed to prevent the loss of thermal water and further pollution along the water path, and water outlets in the thermal spring zone were protected by concrete structures. A series of concrete terraces, imitating the natural morphology, were constructed through the old asphalt road with the aim of encouraging new deposition of travertine (Dilsiz, 2002). Despite the many studies that have been conducted or are still in progress, the site is still at risk. Hence, it is essential that special regulations and laws should be enacted soon to ensure permanent protection of the natural assets in Pamukkale.

CÜNEYT DILSIZ AND GÜLTEKIN GÜNAY

See also **Travertine**

Works Cited

Altunel, E. & Hancock, P.L. 1994. Morphology and structural setting of Quaternary travertines at Pamukkale, Turkey. *Geological Journal*, 28: 335–46

Dilsiz, C. 2002. Environmental issues concerning natural resources at Pamukkale protected site, southwest Turkey. *Environmental Geology*, 41: 776–84

Ekmekçi, M., Günay, G. & Şimşek, Ş. 1995. Morphology of rimstone pools at Pammukale, western Turkey. *Cave and Karst Science*, 22: 103–06

Günay, G., Şimşek, Ş., Ekmekçi, M., Elhatip, H., Yeşertener, C., Dilsiz, C. & Çetiner, Z. 1997. Karst hydrogeology and environmental impacts of Pamukkale thermal springs. In *Karst Waters & Environmental Impacts*, edited by G. Günay & A.I. Johnson, Rotterdam: Balkema

PARAGENESIS

Paragenesis is the term introduced by Renault (1968) to describe the processes by which cave genesis within the phreatic zone is modified by influxes of sediment. True paragenesis occurs in the phreatic zone when influxes of sediment armour the floor and lower walls of a passage, concentrating dissolution upwards and thereby creating a distinctive passage morphology.

In the absence of any clastic sediment, the process of dissolution will produce a continuous increase in the cross-sectional area of a cave passage, unlike a surface stream excavated in sediment in which the channel will expand in area to some limit (normally the "bankfull flood" stage) and then stabilize. Paragenesis is, in a sense, the equivalent happening in a phreatic conduit after it has undergone the basic expansion needed to handle flow up to "bankfull equivalent". As a phreatic passage enlarges by dissolution, the cross-sectional area increases and the average flow velocity decreases, favouring deposition of any transported sediment. Thus the average flow velocity is held constant at the threshold of sediment transport. Since sediment deposition generally occurs on the passage floor, the passage enlarges upwards (and often migrates laterally) over time, creating a high phreatic canyon, terminating only at the water table (Ford & Ewers, 1978). Although most dissolution occurs upwards, Groves, Vaughan & Meiman (1998) demonstrated that acidic water and dissolution still occur within the saturated sediment, although the total solute flux is much reduced. Dissolution takes place within the sediment and on the bedrock floor and walls, but it is very small in comparison with that in unshielded flow.

A clear distinction should be made between paragenesis *sensu stricto* and vadose entrenchment followed by alluviation. Both processes may result in sediment-filled canyon passages but the geomorphic and hydrological implications of these two possible origins are radically different. Palmer (2000) argues the case for vadose incision followed by aggradation in Mammoth Cave (Kentucky), as many of the passages observed do not fit the criteria for true paragenesis. However, even alluviated passages often display some minor paragenetic modification, including small anastomosing half-tubes cut into passage walls and ceilings by water flow within the saturated sediment. Indeed, some passages may originate as vadose canyons, become alluviated, and continue to evolve paragenetically, creating a two-phase passage. Where such passages can be observed it is because the sediment

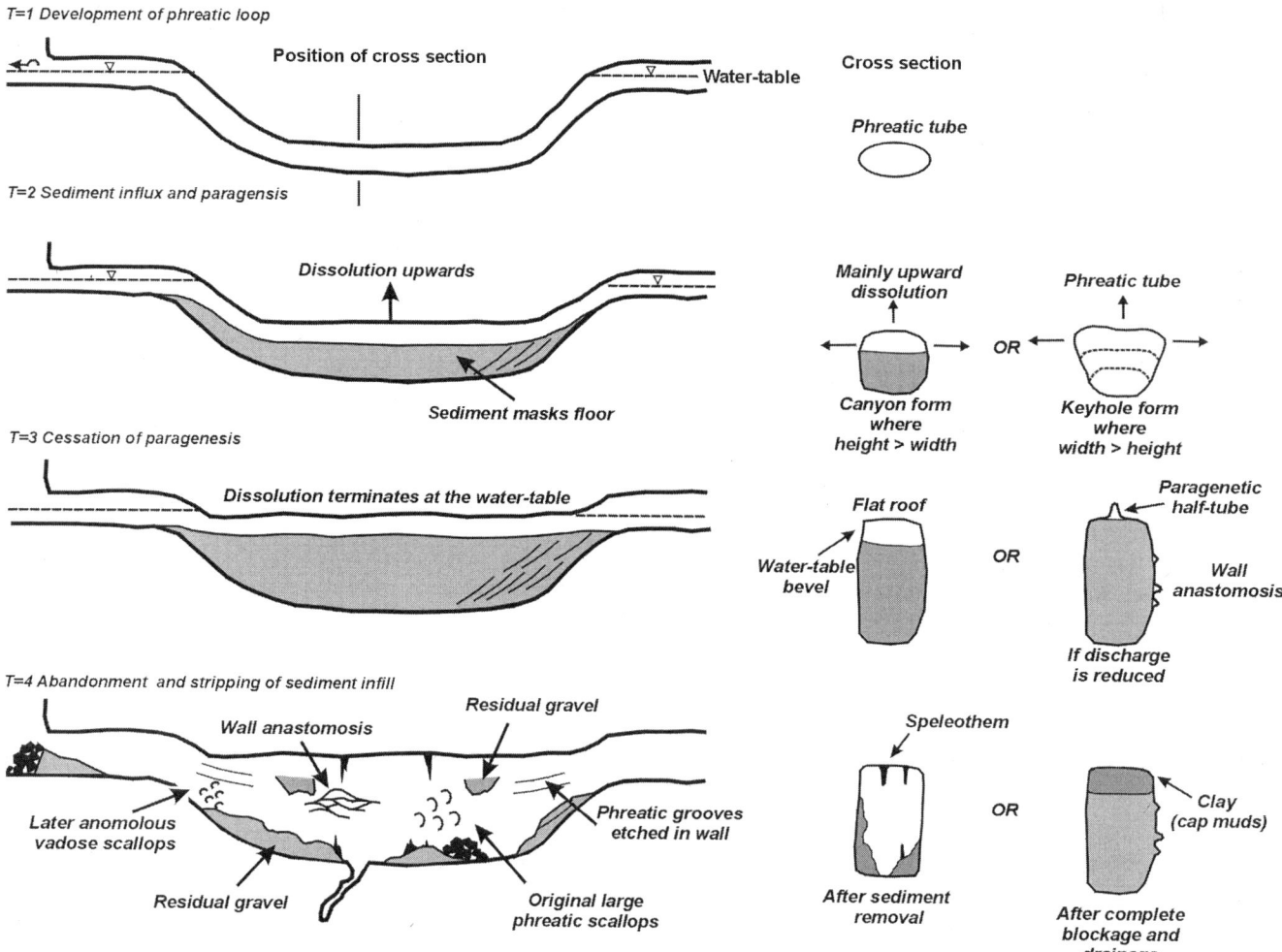

Paragenesis: Schematic paragenetic evolution of a phreatic conduit (elevation and cross section) under conditions of high sediment flux and on abandonment. T = 1: Development of a phreatic loop prior to sediment aggradation. T = 2: Sediment influx and the initiation of alluviation and paragenetic erosion. T = 3: Continued sediment influx and termination of upwards paragenetic erosion at the water table. T = 4: Abandonment and subsequent stripping of sediment by vadose dripwaters, revealing passage morphology.

was subsequently flushed out to render the passage accessible. The downstream portion of Thomas Avenue in the Mammoth Cave system is an excellent example where canyon infilling terminated, with the stream carving very regular meanders into bedrock at a new water-table level. Several cycles of sediment aggradation, paragenesis, and sediment flushing may occur in a given cave, in response to base-level or climatic fluctuations, or to avulsions in stream channels on clastic rocks upstream of the caves.

Paragenetic passages can be recognized by a characteristic suite of features which distinguish them from either vadose canyons or more classic phreatic (i.e. "syngenetic") passages. In particular, paragenetic canyons may superficially resemble vadose canyons. However, they are morphologically and genetically distinct and can be distinguished from vadose canyons by several criteria. Vadose canyons often form part of a keyhole profile where the initial phreatic opening along a bedding plane is at roof level. In paragenetic canyons the opposite is true; the guiding fracture is at floor level. Paragenetic canyons meander in a similar fashion to vadose canyons except that the axis of the meander shifts progressively *upstream* against the direction of flow (Lauritzen & Lauritsen, 1995). Furthermore, there is often evidence for a complete sediment fill, with pockets of sediment preserved in alcoves or cemented by speleothem deposits. In addition, there are many instances where bedrock passage roofs rise and fall in synch with sediment bars on the floors underneath, the passage enlarging upwards to match the surface of the sediment profile beneath to maintain cross-sectional area.

As well as the gross passage form, several erosional morphological features can be identified, including paragenetic anastomoses, pendants, and half-tubes. Morphologically and genetically different from bedding plane anastomoses, these dissolutional features are formed by water flowing along the sediment–rock interface. Unlike bedding plane anastomoses they are not confined to a single bedding plane and can occur on vertical passage walls and ceilings. Paragenetic pendants result from an extreme form of anastamosis where the sediment carrying stream becomes braided and flow becomes confined into a series of channels etched into the roof, leaving isolated pendants between the channels. Paragenetic wall notches form at the level of the sediment floor and have the same gradient as the passage, and can either form in the vadose zone as a result of alluviation, or in the phreatic zone during true paragenesis. The latter often display a looping profile, mimicking the looping profile of the sediment floor, governed by the local channel bedload dynamics (Allen, 1970). In some passages, several vertically stacked grooves may be etched in the side of a single passage, marking successive positions of the sediment floor. Examples of vadose notches can be seen in many caves. Excellent examples occur in the Caves Branch caves of Belize (Ford, 2000b) and the Mulu caves, Sarawak. Here, notches relate to lateral incision following sediment deposition controlled by external alluvial fan aggradation at the resurgence (Farrant *et al.*, 1995). Another superb example occurs in the Nanisivik zinc-lead mine, Baffin Island (Ford, 1986; Olson, 1984) where a horizontal paragenetic notch 1 m high, 400 m wide, and 800 m long is host to sulfide ore deposits.

As described above, paragenesis can modify the internal passage morphology (Lauritzen & Lundberg, 2000), but can also modify cave patterns (Palmer, 2000), and long section geometry (Ford, 2000a). In looping cave systems, sediment accumulates in the base of phreatic loops where paragenetic dissolution of the ceiling erodes the passage upwards. This process combined with vadose incision over loop crests work together above and below the base level to reduce passage amplitude and create a water-table cave as the ultimate end product. Ford (2000a) described a classic example of this in the Swildons Hole streamway (see Mendip Caves). Bypass passages or anastomotic mazes can develop where downward loops become choked, increasing the hydraulic head across the obstruction, enabling otherwise insignificant fissures to be opened up.

The causes for sediment influx are varied and depend on the geomorphic, tectonic, and climatic location of the cave. Moreover, caves are dynamic features and many display evidence of one or more phases of sediment infilling. Sediment accumulation may be caused by a rapid decrease in stream transport capacity as a stream enters a low gradient cave; ponding in phreatic loops; external aggradation at the resurgence; or increased sediment flux into the cave following climatic change (or simply upstream channel avulsion, as noted above). In temperate caves, especially in northern Europe, massive sediment influxes are typically associated with a climatic deterioration at the beginning of a glacial stade and the onset of periglacial conditions. Interglacial weathering products are quickly stripped off and washed underground. Glacial deposition may block the outlets of preglacial caves, causing sediment aggradation and abandonment. In tropical regions, aggradation may be a result of climatic fluctuations, for example creating increased rainfall, increased erosion, and alluvial fan build up. Even slight rises in base level may cause alluviation and ultimately paragenesis of cave passage.

Clearly, the ability to observe these features requires the sediment fill to be partially or totally removed, either by natural geomorphic processes or by human actions such as quarrying or excavation in the search for new passage. Many caves display some evidence of either paragenetic development or subsequent modification. Its recognition is crucial as vadose and paragenetic passages have very different geomorphic and paleoclimatic implications, yet the true extent and significance of paragenesis and alluviation in speleogenesis and development is probably underestimated. This is partially because by their very nature, most paragenetic caves are flooded or choked with sediment and thus remain unexplored.

ANDREW FARRANT

See also **Morphology of Caves**

Works Cited

Allen, J.R.L. 1970. *Physical Processes of Sedimentation*, London: Allen and Unwin and New York: Elsevier

Farrant, A.R., Smart, P.L., Whitaker, F.F. & Tarling, D.H. 1995. Long term Quaternary uplift rates inferred from limestone caves in Sarawak, Malaysia. *Geology*, 23: 357–60

Ford, D.C. 1986. Genesis of paleokarst and strata-bound zinc/lead sulfide deposits in a Proterozoic dolostone, Northern Baffin Island—a discussion. *Economic Geology*, 81(6): 1562–63

Ford, D.C. 2000a. Speleogenesis under unconfined settings. In *Speleogenesis: Evolution of Karst Aquifers*, edited by A.B. Klimchouk, D.C. Ford, A.N. Palmer & W. Dreybrodt, Huntsville, Alabama: National Speleological Society

Ford, D.C. 2000b. Caves branch, Belize and the Baradla–Domica system, Hungary and Slovakia. In *Speleogenesis: Evolution of Karst Aquifers*, edited by A.B. Klimchouk, D.C. Ford, A.N. Palmer & W. Dreybrodt, Huntsville, Alabama: National Speleological Society

Ford, D.C. & Ewers, R.O. 1978. The development of limestone caves in the dimensions of length and depth. *Canadian Journal of Earth Sciences*, 15: 1783–98

Groves, C.G., Vaughan, K. & Meiman, J. 1998. Carbonate chemistry of interstitial fluids within cave stream sediments. *Proceedings of the International Meeting of the International Geological Correlation Program, Project 379: "Karst Processes and the Global Carbon Cycle"*, Bowling Green, Kentucky

Lauritzen, S.-E. & Lauritsen, Å, 1995. Differential diagnosis of paragenetic and vadose canyons. *Cave and Karst Science*, 21: 55–59

Lauritzen, S.-E. & Lundberg, J. 2000. Solutional and erosional morphology. In *Speleogenesis: Evolution of Karst Aquifers*, edited by A.B. Klimchouk, D.C. Ford, A.N. Palmer & W Dreybrodt, Huntsville, Alabama: National Speleological Society

Olson, R.A. 1984. Genesis of paleokarst and strata-bound zinc-lead sulphide deposits in a Proterozoic dolostone, northern Baffin Island Canada. *Economic Geology*, 70: 1056–1103

Palmer, A. 2000. Speleogenesis of the Mammoth Cave system, Kentucky, USA. In *Speleogenesis: Evolution of Karst Aquifers*, edited by A.B. Klimchouk, D.C. Ford, A.N. Palmer & W. Dreybrodt, Huntsville, Alabama: National Speleological Society

Renault, P. 1967–68. Contribution à l'étude des cations méchaniques et sédimentologiques dans la spéléogenese. *Annales de Spéléologie*, 22: 5–21 & 209–67; 23: 259–307 & 529–96; 24: 317–37

Further Reading

Pasini, G. 1975. Sull'importanza speleogentica dell' "erosione antigravitativa". *Le Grotte d'Italia*, 4: 297–326

Simms, M.J. 2001. *Exploring the Limestone Landscapes of the Burren and the Gort Lowlands*, Belfast: Burrenkarst.com

PATAGONIA MARBLE KARST, CHILE

The karsts of the Chilean Patagonia archipelago are the most southerly (50–52°S) and the most inhospitable on the planet, due to the extreme rainfall and strong winds. Alpine karst is well developed on the islands of Madre de Dios and Diego de Almagro, in the province of Ultima Esperanza (see America, South map). The archipelago is uninhabited except near Guarello Quarry (which has exploited lime since 1948 for the steel works at Talcahuano). Pure limestones are rare in Chile, and these partly calcareous islands were recognized in 1930–50 within an inventory of mineral resources (Biese, 1956). Cave and karst exploration, however, only began in 1995, with further expeditions in 1997 and 2000 (Maire, 1999).

The sedimentary rocks of the archipelago form part of the pre-Jurassic basement of the Andean cordillera. The limestones are bounded to the west by the Pacific Ocean and to the east by the Patagonian granite batholiths dating from the early Cretaceous, and are interbedded with volcanic sediments. The limestones and marbles of the Tarlton Formation, which is more than 500 m thick, date from the Carboniferous and Permian. They were formed as coral reefs, which settled on underwater volcanic intraoceanic mounts, forming atolls surrounded by bioclastic limestone formations.

The archipelago has an isothermic subpolar climate. The confrontation between the tropical anticyclones and the southern low pressures accounts for the climate of the "roaring forties". Precipitation reaches 7330 mm per year (80% rain, 20% snow) at the Guarello station, with an average of 611 mm per month. In the shelter of rock dolines and at the bottom of the cliffs, the magellanic forest of *Nothofagus*, inhabited by hummingbirds, constitutes one of the last primary forests in the world. It is similar to the equatorial cloud forests of New Guinea at altitudes of 3000–4000 m. The combination of wind and rain is the origin of a widespread hydro-aeolian karren not previously described. Tapered, parallel ridges 1–4 m long and 10–30 cm high, formed in the shelter of volcanic blocks deposited by glaciers (Figure 1). These forms of lateral differential dissolution are located on cols that are exposed to the northwesterly winds, where horizontal karren grooves are cut between them by dissolution in surface water pushed by the wind.

On the rock slopes, rillenkarren form small catchment areas with small deep canyons from 5 to 10 m deep, a true natural laboratory for studying regressive erosion by dissolution related to the slope and flow of water. Each stream leads to a stream-sink cave, whose exploration is hazardous due to the frequent flood events. Littoral karren are very well developed, and there are remarkable staged notches of erosion, corresponding to glacio-isostatic uplift; 10–12 m on Diego de Almagro, and 4–8 m on Madre de Dios.

The rate of surface dissolution is 3 mm in 50 years (60 m^3 km^{-2} a^{-1}). It has been calculated with precision on Guarello Island, which was exploited for lime. Traces of paint dating back to 1948 show a raised relief of 3 mm. Elsewhere, small dykes of basalt and lamprophyre are highlighted by dissolution, and stand proud by 600–1000 mm. On the col of Madre de Dios, an inclined marble pillar, protected by a piece of basalt, measures more than 1.5 m high. A 10 m broad dyke projects by 4–6 m, due to differential dissolution of the surrounding marble which was weakened by contact metamorphism.

Patagonia Marble Karst, Chile: Figure 1. Hydro-aeolian karren on Madre de Dios island. These aeolian ridges of limestone occur on the leeward sides of glacial volcanic rock fragments due to erosion by high-velocity wind and heavy rain. (Photo by R. Maire)

Patagonia Marble Karst, Chile: Figure 2. The Avenir Sinkhole is a 50 m waterfall at the contact of volcanoclastic rocks and marbles, located on Diego de Almagro Island. (Photo by R. Maire)

The underground karst is very developed, but explorations are still in a preliminary phase. In the western part of Madre de Dios, an important karst system was reached through two stream-sinks. La Perte du Futur (376 m deep) is a cave with an alpine morphology, fed by a sink, and constituting the upstream part of the system. La Perte du Temps, 2.65 km long, forms the intermediate part of the system; it is fed by another large sink (estimated $2\ m^3\ s^{-1}$) located at the contact between the limestone and the volcanic tuffs. The resurgence is to the west on the Pacific coast. This active karstic system shows some older parts with a few eroded concretion formations.

On the island of Diego de Almagro, la Perte de l'Avenir (130 m deep, 1.2 km long) constitutes a young cave system, probably formed during and after the melting of the glaciers within the last 20 000 years (Figure 2). It is located at the contact between the marble and volcanic sediments, at the lip of a glacial cirque occupied by two lakes. The lakes are drained by a 50-m waterfall (0.2–$3\ m^3\ s^{-1}$) into the cave, which continues as an underground canyon through a marble dome, and then continues on the surface as a roofless canyon. Above the canyon, an abandoned meltwater canyon can be observed. Between the underground canyon and the glacial canyon, a network of phreatic tubes has developed by subglacial and proglacial meltwater erosion.

Bordering the sea, dry caves with glacial infill and massive stalagmites have been observed. Older caves, truncated by glacial erosion on the side of certain valleys, indicate the existence of several generations of karst. Resurgences are all located at the coast or below sea level. The most significant resurgence known from diving is the siphon of Lobos (49 m deep, discharge $2\ m^3\ s^{-1}$) on Madre de Dios. On the archaeological side, the 2000 expedition for the first time highlighted the cave burials of the Alakalufs, a maritime people, whose bones have been carbon-dated as nearly 4000 years old.

RICHARD MAIRE

Works Cited

Biese, W.B. 1956. Uber Karstvorkommen in Chile. *Die Höhle*, Dezember 1956: 91–96
Maire, R. 1999. Les "glaciers de marbre" de Patagonie, Chili: un karst subpolaire de la zone australe. *Karstologia*, 33: 25–40

Further Reading

Fage, L.-H., Maire, R. & Pernette, J.-F. 1997. L'expédition Ultima Esperanza en Patagonie chilienne: les karsts de l'extrême. *Proceedings of the 12th International Congress of Speleology*, vol. 6, La-Chaux-de-Fonds, Switzerland: International Union of Speleology / Swiss Speleological Society
Hobléa, F., Jaillet, S. & Maire, R. 2001. Erosion et ruissellement sur karst nu: les îleś subpolaires de la Patagonie chilienne. *Karstologia*, 38: 13–18

PATTERNS OF CAVES

Solutional caves exhibit several distinct patterns (Figure 1). "Branchwork caves" consist of converging tributaries that merge downstream into fewer but generally larger passages. They are the most common type, representing >60% of all solutional caves. "Maze caves", in contrast, contain many closed loops in which the passages form simultaneously. "Network mazes" are angular grids of intersecting fissures guided by fractures. "Anastomotic mazes" consist of curvilinear tubes that intersect in a braided pattern and usually follow bedding-plane partings. "Spongework mazes" are irregular cavities enlarged from primary pores, which interconnect in a three-dimensional array like the holes in a sponge. "Ramiform mazes" resemble inkblots in map view, with branches extending outward from irregular central rooms. Some caves include a combination of patterns. For example, most ramiform caves contain zones of spongework or network mazes. Many small caves have only single passages, or are erosionally dissected fragments of larger caves, but they share the origins of the types described above.

Cave patterns reflect the local hydrogeologic setting (Palmer, 1975; 1991). Branchwork caves are formed by groundwater recharge through many discrete inputs, particularly dolines. Each source contributes a small amount of water, capable of forming a passage, or at least an incipient passage. Where bedding is prominent, passages are sinuous and curvilinear. Where fractures are prominent, passages are rectilinear and meet at sharp angles. Branchworks usually possess more than one tier or storey, and the combination of active and relict passages can mask or confuse the branching pattern on maps. Among many examples described in separate entries are Mammoth Cave (United States), Friar's Hole Cave (see Appalachian Mountains entry), caves in the Mendip Hills (England), and Ogof Draenen (see Draenen, Ogof Draenen, Wales).

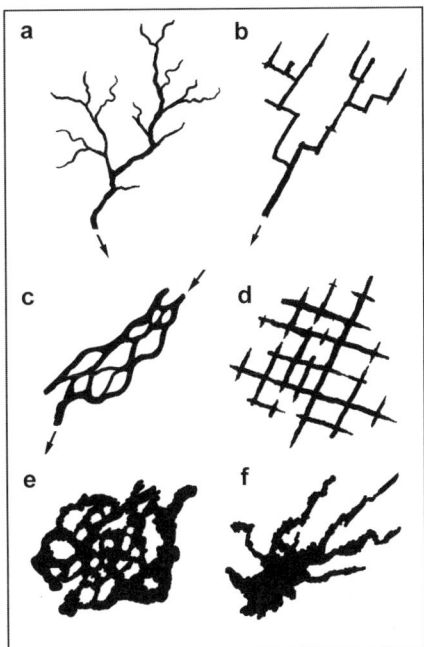

Patterns of Caves: Figure 1. Cave patterns in plan view (adapted from Palmer, 1991). **a** = branchwork in prominently bedded strata; **b** = branchwork in prominently fractured strata; **c** = anastomotic maze; **d** = network maze; **e** = spongework maze; **f** = ramiform pattern. Arrows show dominant flow direction. In many caves, the pattern is obscured by rudimentary development, multiple levels, or superposition of more than one pattern.

Maze caves require simultaneous enlargement of many competing paths. This is achieved by water with a steep hydraulic gradient and/or short flow paths from where the solutionally aggressive water first encounters the soluble rock (Palmer, 1991). Their specific patterns are controlled by the mode of groundwater recharge and by local structural conditions.

Periodic floodwaters delivered by allogenic recharge (e.g. sinking streams) pond in the caves under great pressure, injecting aggressive water into surrounding openings and enlarging them all simultaneously (see Hydraulics of Caves). Where fractures are prominent, local networks and blind fissures are superposed on the initial cave pattern (e.g. see Sof Omar Cave, Ethiopia entry). Where low-angle bedding planes or thrust faults are prominent, anastomotic mazes form (e.g. parts of Hölloch, in Switzerland, see separate entry). In massive rocks with intergranular pores, local spongework mazes can form (Palmer, 1975). Where fractured soluble rock receives uniform aggressive seepage through an overlying or underlying insoluble formation, two-dimensional network caves are formed (e.g. see Ukrainian Gypsum Caves entry). Three-dimensional network mazes are usually formed by the mixing of two waters of contrasting chemistry, for example high-CO_2 water rising from depth mixing with shallow low-CO_2 water. Wind and Jewel Caves (United States, see separate entry) may have formed in a combination of these two ways. Speleogenesis by sulfuric acid, generated by oxidation of hydrogen sulfide, can produce either network or ramiform patterns. The latter tends to develop along major flow routes where generation of sulfuric acid is intense (e.g. Carlsbad Cavern and Lechuguilla Cave (United States), and Cupp-Coutunn Cave, Turkmenistan, see separate entries). It is also theoretically feasible for three-dimensional networks to be formed by cooling of thermal water ascending along multiple fractures (Palmer, 1991), but most examples in the literature seem instead to have formed by mixing. Many spongework mazes are products of mixing of fresh water and saline water along seacoasts, in young limestones of high primary porosity (see Speleogenesis: Coastal and Oceanic Settings).

Artesian conditions are commonly thought to produce maze caves. However, there is no inherent tendency for this to happen. Although many mazes formerly experienced artesian conditions, their patterns result instead from one of the above processes.

The vertical profile of a cave depends on a combination of geologic conditions and geomorphic history. Ford (1971) explained the evolution of cave profiles with time (see Figure in

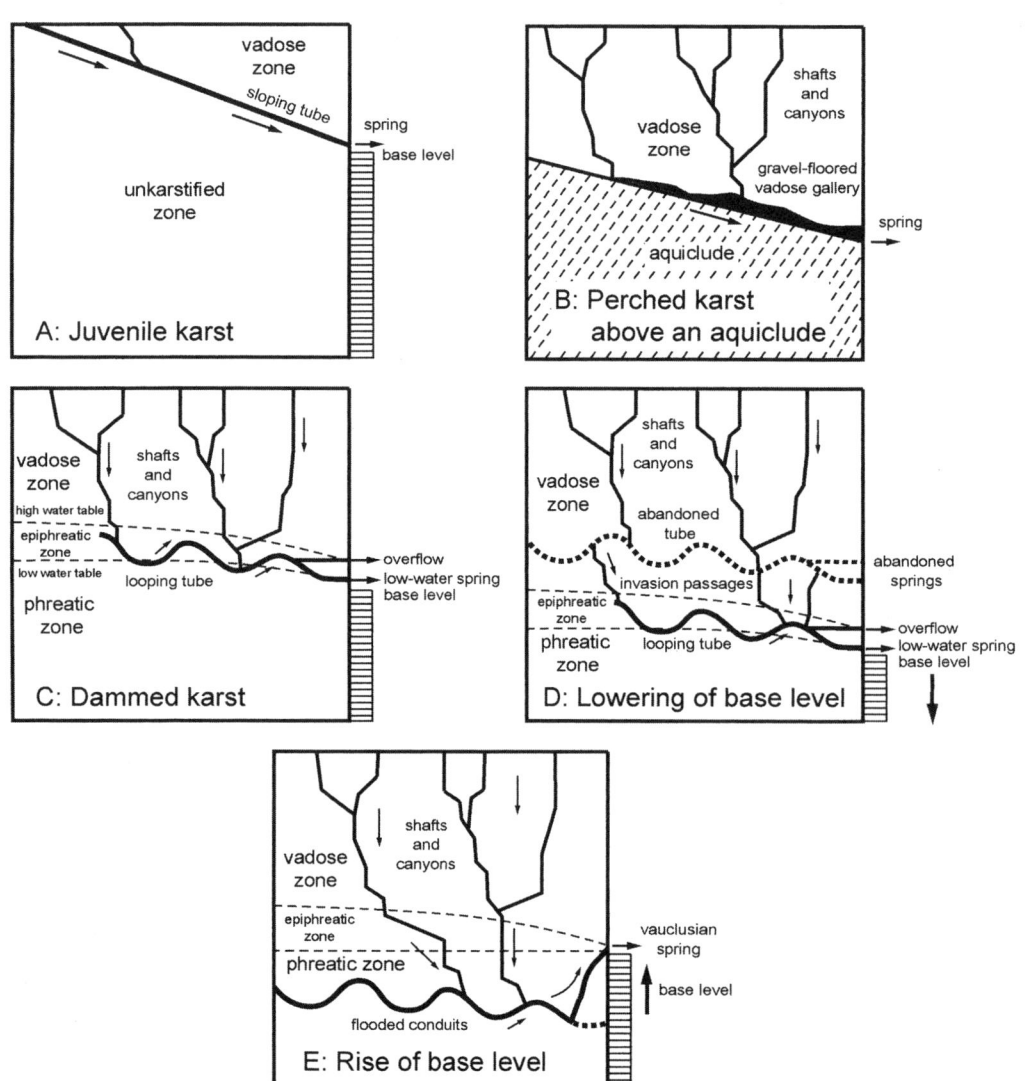

Patterns of Caves: Figure 2. Evolution of cave profiles, based on concepts in Audra (1994). See text for details.

Speleogenesis: Unconfined Settings entry). Early in an aquifer's history, fissure frequency is low. Caves contain deep phreatic loops that descend to considerable depth below the water table (piezometric surface). With time, fissure frequency increases due to stress release, and later passages (often successively lower levels) have a greater number of phreatic loops but with lower amplitude. Where the fissure frequency is great enough, cave passages follow the water table with no significant phreatic loops. From empirical evidence, Worthington (2001) suggested that the depth of initial phreatic cave development in cave systems >3 km long is determined by decreasing viscosity of water as temperature rises with depth. Flow depth is proportional to flow-path length and stratal dip.

Audra (1994) presented an alternate view. "Juvenile karst" (Figure 2a) prevails when soluble rocks are first exposed by uplift and removal of any impermeable cover. Because of sparse fracturing, the water table is steep and located high above fluvial base level. Phreatic paths are later entrenched by vadose water as the water table drops. These are common in young, rapidly developing karst (e.g. see entry on Nakanai, Papua New Guinea). Geologic structure determines the aquifer development in several ways. Where the aquifer is perched above base level on an underlying aquiclude, there is no significant phreatic cave development (Figure 2b). Shafts and canyons converge on conduits at the aquiclude top and feed springs along hill slopes (e.g. see entry on Dent de Crolles System, France). Mechanical erosion, aided by detrital sediment, quickly enlarges the main routes. Major springs open into the heads of blind valleys. In "dammed karst" (Figure 2c) the karst aquifer extends below the spring outlet, which is determined by a fluvial or structural base level. Major passages either follow the water table or loop below it. During high flow, water rises in phreatic lift tubes and emerges at overflow springs (as in Castleguard Cave, Canada, see scparate entry). In this model, high-amplitude looping passages form in the epiphreatic zone and are enlarged by aggressive high flows. Their amplitudes can exceed 200 m. During low flow the water follows lesser openings at lower elevations.

As fluvial base level drops with time, successively lower phreatic passages develop (Figure 2d). Pauses in base-level lowering produce cave levels that correlate with river terraces (see Mammoth Cave Region). Vadose development extends down to the new water table. Former conduits and springs are abandoned and partly filled with floodwater sediments and secondary minerals. Perched in the vadose zone, old phreatic conduits are cut by new shafts and canyons that feed active conduits. Base-level rises (Figure 2e) cause flooding of conduits, although they often preserve their original morphology. Some become sediment-filled, but the main flow lines remain active. New ascending routes, or reactivation of relict conduits, may form vauclusian springs (e.g. Fontaine de Vaucluse, France).

ARTHUR N. PALMER AND PHILIPPE AUDRA

See also **Morphology of Caves; Speleogenesis**

Works Cited

Audra, P. 1994. *Karsts alpins. Genèse de grands réseaux souterrains. Examples: le Tennengebirge (Autriche), l'Ile de Crémieu, la Chartreuse et le Vercors (France), Thesis*, Paris: Fédération française de spéléologie and Bordeaux: Association française de karstologie (Karstologia Mémoires, 5)

Ford, D.C. 1971. Geologic structure and a new explanation of limestone cavern genesis. *Transactions of the Cave Research Group of Great Britain*, 13(2): 81–94

Ford, D.C. & Ewers, R.O. 1978. The development of limestone cave systems in the dimensions of length and depth. *Canadian Journal of Earth Sciences*, 15: 1783–98

Palmer, A.N. 1975. The origin of maze caves. *National Speleological Society Bulletin*, 37: 56–76

Palmer, A.N. 1991. Origin and morphology of limestone caves. *Geological Society of America Bulletin*, 103: 1–21

Worthington, S.R.H. 2001. Depth of conduit flow in unconfined carbonate aquifers. *Geology*, 29(4): 335–38

Further Reading

Bögli, A. 1980. *Karst Hydrology and Physical Speleology*, Berlin and New York: Springer (original German edition, 1978)

Ford, D.C. & Williams, P.W. 1989. *Karst Geomorphology and Hydrology*, London and Boston: Unwin Hyman

Gillieson, D. 1996. *Caves: Processes, Development, and Management*, Oxford and Cambridge, Massachusetts: Blackwell

Klimchouk, A., Ford, D.C., Palmer, A.N. & Dreybrodt, W. (editors) 2000. *Speleogenesis: Evolution of Karst Aquifers*, Huntsville, Alabama: National Speleological Society

PEAK DISTRICT, ENGLAND

The Peak District is situated in the centre of England, and is one of four main caving regions in Britain (see map in Europe, Northern). Lower Carboniferous (Dinantian) limestones crop out over an area of some 520 km^2, which is known as the "White Peak" and forms c.30% of the Peak District National Park. The largest unbroken karst area in Britain, it measures 40 km north–south and up to 20 km west–east (Figure 1). The surrounding Triassic sandstones (in the south) and the Namurian rocks of Millstone Grit facies support surface drainage, but only in the northern half of the area are these topographically higher than the limestones. The limestones, ranging from a few hundred metres to over 1800 m in thickness, were deposited in three main environments: lagoonal, reefal, and basinal. Subsequent folding has produced dips of up to 20° in lagoonal facies rocks, with much steeper dips in reefal rocks. Bedding planes are rare in the reef limestones but common elsewhere, with individual beds from 0.5 m to over 5 m thick. Three types of bedding discontinuity are recognized: contrasts in limestone lithology that reflect relatively rapid changes in carbonate deposition; pyrite-rich shale bands that mark short episodes of dominantly terrestrial sedimentation; and 1 mm–1 m thick green clay bands, known locally as "wayboards", that originated as falls of volcanic dust. Dinantian vulcanism also produced basaltic lavas and tuffs, locally interbedded with the limestones, and intrusive dolerites ("toadstones") also occur. Wayboards, toadstones, and volcanic rocks have influenced subsequent conduit / cave development.

Locally preserved paleokarstic features indicate at least one period of subaerial conditions during the Carboniferous (Ford, 1984). Following burial beneath later sediments, some paleokarstic voids were exploited as migration routes for hydrothermal fluids, becoming infilled by mineral deposits, mostly during the late Carboniferous. Mineralization is a key feature of the Peak

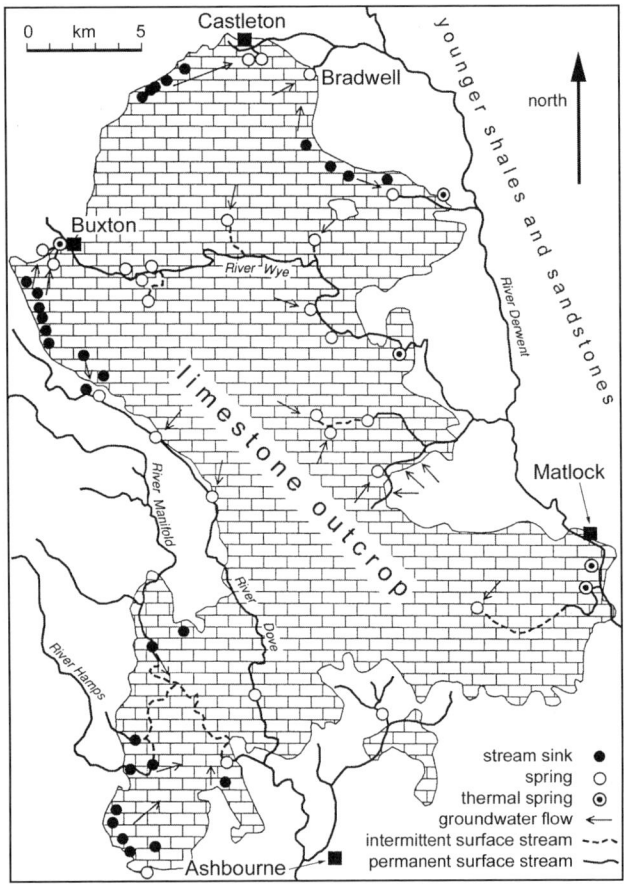

Peak District: Figure 1. Location of the White Peak, showing regional geological setting.

District karst, which is traversed by hundreds of veins. Most known veins were mined for lead between Roman times and the mid-20th century and some are still exploited for fluorspar. One form of fluorspar, Blue John, which is of high value as an ornamental stone, occurs only within paleokarstic voids on Treak Cliff, near Castleton, in the north of the White Peak.

Drainage always presented a problem to lead miners, and increasingly so as time progressed. Pumps were employed in many mines, but a later widespread solution was the construction of drainage adits ("soughs"), to dewater the mines. Soughs were driven from river level to lower the natural water levels and drain the phreatic zone. This was so successful that around Matlock, in the south of the limestone outcrop, no significant natural springs remain, and the whole area drains via a few major soughs. Lead miners encountered large, isolated dissolution cavities within mineral veins, faults, and fractures. Some cavities carried torrents of water and were described as "Great Swallows", whereas others were termed "Large Opens". In the Castleton area a notable concentration of these "vein-related cavities" is thought to represent a primitive phase of phreatic karst development. They provided inception routes for early water movement through the limestone, thereby stimulating later phreatic and vadose, bedding-related cave development (Ford, 1984, 2000).

There are over 210 caves in the Peak District, with a total length of $c.50$ km, but only nine individual systems are over 1 km long (Gunn, Lowe & Waltham 1998). Most of the caves, and all those over 1 km long, are located where escarpments of Namurian Millstone Grit strata are close enough to the karst, and high enough, for surface streams to cross the shale and sink into the limestone. Castleton's Peak–Speedwell cave system is the longest ($c.15$ km) and has both the greatest vertical range ($c.290$ m) and the deepest explored sump (Speedwell Main Rising, 70 m and continuing) in England. The system has a 8.4 km^2 autogenic catchment with dispersed recharge and a 5 km^2 allogenic catchment, within which 13 streams flow across solifluction debris-covered slopes and sink in marginal reef limestones. Eight of the stream sinks are associated with accessible influent caves, including the Giants–Oxlow system (5 km long; 240 m deep). The influent caves consist primarily of narrow meandering vadose streamways that drain to sumps, the deepest of which are at a similar altitude to the input sumps in Speedwell Cavern. Within Peak–Speedwell, the majority of the passage length consists of long phreatic tubes developed on three horizons within the gently dipping limestones and modified to varying degrees by vadose trenching and collapse. However, the most unusual feature of the cave system is the number, and vertical extent, of its mineral vein related cavities. At least 57 have been identified, with a cumulative vertical extent of over 2600 m and a volume of about 185 000 m^3. Most spectacular is the recently discovered Titan, Britain's largest natural shaft, with a vertical range of some 219 m, including a free-hanging drop of 142 m (Figure 2). At the top of the cavity the east–west extent is $c.100$ m and the north–south width is 15–20 m. A surface dig is in progress to connect with the shaft via a passage near the top that is presently blocked by a boulder choke.

Eldon Hole, south of, and connected hydrologically to, the Peak–Speedwell cave system, is the largest open pothole in the Peak District, being 34 m long and 6 m wide at the surface and over 60 m deep. It was first explored around 1600 (Shaw, 1992), but the first properly documented descent was by Lloyd (1772), who also provided a survey. The nearby Eldon Hill Quarry exposes several narrow dissolutional passages up to 10 m high, apparently aligned along north–south joints. Their relationship to the vein-related cavities remains uncertain, but paleomagnetic dating of glaciofluvial infill (see photo in Paleoenvironments: Clastic Cave Sediments entry) indicates ages of around 900 000 years (Ford, 2000). Glaciofluvial sediment sequences older than 730 000 years have also been found near Matlock, where some cave segments follow mineral veins, and mining has re-exposed paleokarstic features. These sediments are the earliest evidence of Quaternary glaciation in Britain outside East Anglia. Evidence for even earlier cave development is provided by Early Pleistocene bones recovered from Victory Quarry Fissure near Buxton (Bramwell, 1977) and speleothem dating from the Olduvai event (1.67–1.87 Ma) at Elderbush Cave in the Manifold Valley (Rowe, Austin & Atkinson, 1989). It is generally assumed that the first significant breaching of the clastic cover took place in the Early Pleistocene and marked the beginning of significant cave development. However, limestone was exposed in the south of the region during Permo-Triassic times, when cavities were enlarged in the Brassington area (Walsh *et al.*, 1972) and conduit inception must have been ongoing for millions of years beneath the "impermeable" cover. Evidence for deep circulation of meteoric water in hydrothermal convection systems that are allogenic-fed rivers, the Dove and the Wye, maintain perennial flow across the karst. However, the most notable feature of the limestone plateau is that it is dissected by a dendritic system of valleys ("dales") that are dry for all, or most, of the year. Thus, the Peak District is an internationally important example of a relict fluviokarst (see Photo in Valleys in Karst).

The valleys are thought to have been inherited by the limestone after originating on an impermeable clastic cover. Their desiccation largely reflects gradual karstification of drainage, aided by regional water table lowering in response to downcutting by major rivers. However, not all the desiccation was natural, as dewatering via mine soughs affects some rivers, most notably the Lathkill's middle reaches, which dry up every summer. Dry valleys might have been active again under Pleistocene periglacial conditions when the ground froze and extensive screes developed. Solution dolines, subsidence dolines, and a few small collapse dolines pit the limestone surface, but they are mere details within the dominantly fluviokarstic landscape. Additionally, the courses of most lead veins are marked by depressions that were either modified or created by lead miners. Quarrying, a major local industry, has also impacted on the natural landforms and more limestone was removed by quarrying during the 20th century than by natural solutional processes acting over the whole of the Holocene (the last 10 000 years).

JOHN GUNN

See also **Europe: North; Fluviokarst; Quarrying in Limestone; Radon in Caves; Sulfide Minerals in Karst**

Works Cited

Bramwell, D. 1977. Archaeology and paleontology. In *Limestones and Caves of the Peak District*, edited by T.D. Ford, Norwich: Geo Abstracts

Ford, T.D. 1984. Palaeokarst in Britain. *Cave Science*, 11(4): 246–64

Ford, T.D. 2000. Vein cavities: an early stage in the evolution of the Castleton Caves, Derbyshire, UK. *Cave and Karst Science*, 27(1): 5–14

Gunn, J., Lowe, D.J. & Waltham, A.C. 1998. The karst geomorphology and hydrogeology of Great Britain. In *Global Karst Correlation*, edited by Yuan Daoxian & Liu Zaihua, New York: Science Press

Lloyd, J. 1772. An account of Elden Hole in Derbyshire. *Philosophical Transactions of the Royal Society*, 61(1): 250–56

Rowe, P.J., Austin, T.J. & Atkinson, T.C. 1989. The Quaternary evolution of the south Pennines. *Cave Science*, 16(3): 117–21

Shaw, T.R. 1992. *History of Cave Science: The Exploration and Study of Limestone Caves, to 1900*, 2nd edition, Broadway, New South Wales: Sydney Speleological Society

Walsh, P.T., Boulter, M.C., Ijtaba, M. & Urbani, D.M. 1972. The preservation of the Neogene Brassington Formation of the southern Pennines and its bearing on the evolution of upland Britain. *Quarterly Journal of the Geological Society of London*, 128: 519–59

Further Reading

Cave Science, 18(1), 1991 is a special issue on the Peak–Speedwell Cave System

Ford, T.D. (editor) 1977. *Limestones and Caves of the Peak District*, Norwich: Geo Abstracts

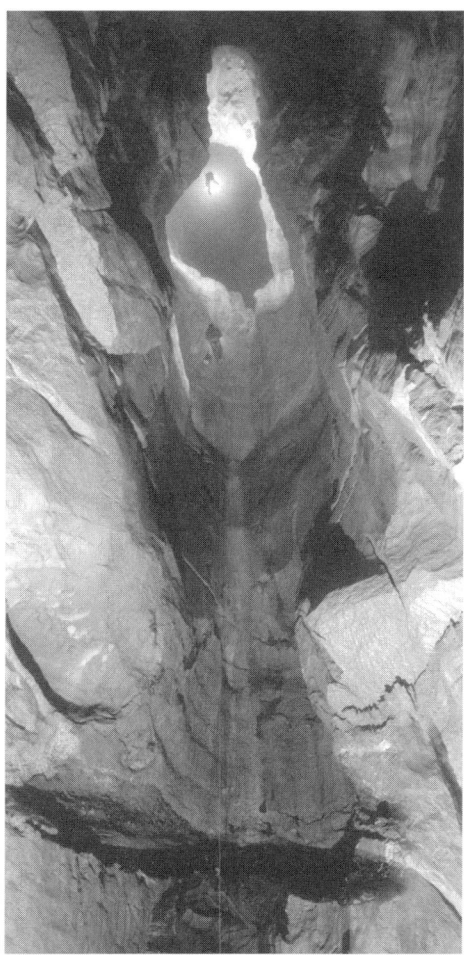

Peak District: Figure 2. Looking up the Titan Shaft, in the inner reaches of the Peak–Speedwell cave system, Castleton, Derbyshire. (Photo by Paul Deakin)

thought to be primarily within the carbonate sequence, is provided by the warm springs that issue at Buxton, Stoney Middleton, Bakewell, and Matlock. Water-tracing experiments in the central White Peak also suggest that there is a large body of slow-moving groundwater, in contrast to the margins where the water from sinking streams moves very rapidly through the conduit systems.

A notable feature of the Peak District caves in general, and the Castleton caves in particular, is the high concentration of radon gas. Ventilation systems have been installed in each of the four show caves in the Castleton area to ensure the safety of visitors and employees, but Giants Hole is thought to have the highest measured radon gas concentration of any natural limestone cave in the world.

Some cave deposits confirm that the Peak was ice-covered during at least one early glaciation, but Quaternary glaciers are not thought to have influenced major river valley incision significantly, and the area lies outside the most recent (Devensian) ice limits. Consequently there is virtually no bare limestone, although soils are generally thin ($c.1$ m) and developed largely from loessic material rather than from weathered limestone. Virtually no surface drainage exists on the limestone, and only two

Ford, T.D. & Gunn, J. 1992. *Caves and Karst of the Peak District: An Excursion Guidebook to the Karst Landforms and Some Accessible Caves within the Peak District*, London: British Cave Research Association

Ford, T.D. & Rieuwerts, J.H. (editors) 2000. *Lead Mining in the Peak District*, 4th edition, Ashbourne: Landmark Publishing

Gill, D.W. & Beck, J.S. 1991. *Caves of the Peak District*, Clapham: Dalesman

Gunn, J. 1985. Pennine karst areas and their Quaternary history. In *The Geomorphology of North-west England*, edited by R.H. Johnson, Manchester: Manchester University Press

Harrison, D.J. & Adlam, K.A.McL. 1985. *Limestones of the Peak: A Guide to the Limestone and Dolomite Resources of the Peak District*, London: HMSO (British Geological Survey Mineral Assessment Report 144

PHOTOGRAPHING CAVES

The history and development of cave photography is closely linked to two factors: technology and motivation. When the invention of photography was announced in 1839 it brought widespread excitement, but the early process required a cumbersome camera and tripod, and also that the sensitized materials were prepared, exposed, and developed within a few minutes. Restrictions such as these and the long exposure times required meant that few photographers—who were, in the main, motivated by commercial portraiture—ventured far from their studios. Of those that did, John Dillwyn Llewelyn photographed cave entrances in South Wales between 1852 and 1856, but photography in darkness awaited better light sources than the prevalent candles, gas lamps, and oil lamps.

The turning point came with the commercial preparation of magnesium in 1863. When burned, the metal produces light which is rich in the actinic (blue) part of the spectrum that photographic plates were sensitive to. To illustrate its photographic power, on 27 January 1865 Alfred Brothers, a professional photographer of Manchester, England, photographed a chamber in the Blue John Caverns in Derbyshire—the first successful underground cave photograph.

Following the American Civil War there was a perceived need to attract tourists to Kentucky's Mammoth Cave. A Belgian immigrant, Charles L. Waldack, was employed to take stereo photographs for sale to an avid public. Stereo pictures were taken using a double-lensed camera, simulating the slightly different views seen by each eye and giving an impression of depth—such pictures were a popular commodity. By 24 June 1866 Waldack had produced seven good images—returning for a second, three-month-long expedition in July. Spending up to 35 hours underground, he echoed Brothers by burning strips of magnesium bound into a taper—up to 120 tapers were used for each picture. Subsequently, a set of 42 images was placed on sale.

Waldack's meticulous techniques and pictures were widely reported and brought universal acclaim. The photographic press exclaimed that: "It is a greater triumph of photography than anything yet accomplished . . . Mr. Waldack deserves the thanks of the world of science and art." (Edward L. Wilson, "Photography in the Mammoth Cave", 1866; reproduced in Howes, 1989, pp. 68–69)

During his experiments, Waldack found that light sources close to the camera produced uninteresting pictures which lacked relief or shadows and, in addition, the cave air was so humid that light reflected back into the lens, degrading the image. He surmounted these problems by placing the magnesium burners to one side of the camera, producing greater detail in his pictures. Waldack frequently had no room to manoeuvre in narrow passages, so for the second expedition he chose a wider-angle lens, conferring more freedom to compose his picture and placing him closer to the subject. Basic tenets such as these—using a wide-angle lens with separate lighting positions—remain true today.

This commercial motivation—benefiting both the photographer and the cave owners—remained the controlling factor behind cave photography for much of the 19th century. Under-

Photographing Caves: Figure 1. Waldack's 1866 stereo photograph of the Deserted Chamber in Mammoth Cave clearly shows clouds of magnesium fumes drifting overhead. (Photo by Charles L. Waldrack)

ground photography increased alongside tourism, fuelled in the United States by the spread of the railroad which, in some cases, was deliberately laid near commercial caves. The rail companies frequently funded cave photography for advertising (typified by the work of W.R. Cross and W.F. Sesser during the 1880s), to encourage tourists to use their services.

However, technology did not stand still. Glass plates were improved and exposure times were reduced, and the "wet" process gave way to dry plates, which removed the need to prepare these in the field. Recording red colours in black and white tones remained difficult, producing virtually black areas, but no solution to this was found until panchromatic film was introduced in 1906. For now, cave photographers avoided taking pictures containing ochre-rich mud or iron-stained formations.

Experiments were made with alternative light sources to magnesium, such as limelight (used by A. Veeder around 1877 at Howe's Cave near New York) and electricity (used in 1881 by C.H. James in Luray Caverns, Virginia). These pictures were characterized by their bland lighting and static nature: exposure times were too lengthy to include people in the scene as they could not stand still for the duration of the exposure. Magnesium ribbon remained the only realistic choice, although it was less than perfect because of the copious clouds of magnesium oxide ash that obscured the scene within seconds and prevented a second exposure.

In 1887 the invention of flashpowder was announced and the new compound (consisting of finely ground magnesium mixed with an oxidizing agent—potassium chlorate) was soon used underground when, in 1888, Max Müller built and deployed a "magnesium blitzlicht" lamp in the Hermannshöhle in Germany. His techniques were explained in a monograph, published with 20 prints. Flashpowder offered cave photographers a valuable new tool, which "instantaneously" illuminated the scene—the powder was both powerful and portable. The problem of fumes remained, but pictures could be taken with far greater ease, and subtly the nature of artificial light photography changed. The inclusion of people within the scene became more common as the likelihood of blurring was reduced by the shorter exposure times, while shadows became harsher.

With flashpowder in widespread use, by the 1890s cave photography was well served by established techniques, but a change in emphasis was under way. Cave exploration for its own sake began, and photographers such as Ben Hains in the United States and Charles Henry Kerry in Australia recorded the caves that they explored. In France, Édouard Alfred Martel required photographs to support his explorations, and encouraged his companions to develop new photographic techniques and equipment, writing the first book on the subject in 1903: *La Photographie souterraine* (Underground Photography). As a motivating force, depicting the cave as a place of exploration had at last outstripped commercial interests, although, in France, Martel's dictum that lighting should simulate daylight from behind, above, and to one side of the lens, restrained the discipline from reaching maturity.

As part of the rise in cave exploration during the closing years of the 19th century and the first decades of the 20th, many amateur and professional photographers worked underground, but they were outshone by one man: James Henry Savory (universally known as Harry Savory). A professional photographer, between 1910 and the mid-1920s Savory produced an outstanding series of pictures from the caves of the Mendip hills of southwest England. By the time that Savory concluded his work the change in emphasis was complete: cave photography now belonged to dedicated photographers who held a genuine interest in the underground world.

Cave photography also had a profound effect during the 1920s. At Carlsbad, New Mexico, Ray V. Davis had solved many of the difficulties in photographing the immense and beautiful chambers in the nearby Bat Cave (today known as Carlsbad Caverns). The pictures were influential in persuading President Coolidge to inaugurate the caves as a national monument (later to become a national park). Davis was keenly involved with preserving the cave for posterity and his photographs are perhaps some of the most important tools ever used to support conservation.

Photographers such as Savory and Davis produced stunning pictures because, in part, they were so skilful in lighting their subjects. Light behind a subject produces a silhouette; with light to one side there is strong relief which is suited to showing texture. By combining more than one flash, the effects are often startling.

In 1929 the flashbulb (aluminium burned in oxygen sealed within a glass globe) removed the problem of magnesium fumes, but, initially, flashbulbs were large and expensive. Just as Brothers and Waldack had helped popularize magnesium, in 1952 it fell to Ennis Creed "Tex" Helm to use 2400 flashbulbs in a single exposure in Carlsbad Caverns, thus helping his sponsor—Sylvania Electric Products—to produce a new and appealing advertisement. Flashbulbs, being waterproof, also enabled Luke Devenish to make the first stumbling attempts at photography in sumps at Wookey Hole, England in 1955, by which time cave photographic techniques were well known. Even so, it was not until smaller and cheaper bulbs became available that cave photographers finally turned away from flashpowder in the late 1950s and early 1960s.

Other advances in cave photography during the latter half of the 20th century relied on better and cheaper electronic components, resulting in the proliferation of electronic flashguns. These differ from bulbs in that they have a shorter duration of flash and typically emit light over a narrower angle. The effect is striking. Moving water lit with a bulb is blurred and appears to flow, while an electronic flash "freezes" movement and produces harder shadows. Cavers, then as now, mixed the light sources to obtain the results that they required. All flash sources can be assigned a guide number, which is relative to the sensitivity of the film; higher numbers indicate a more powerful flash. Using the guide number and the formula: Guide Number = Distance from flash to subject multiplied by the lens Aperture ($GN = D \times A$) it is relatively easy, even with multiple flash situations, to determine the correct settings to use on the camera.

Despite advances in cameras, film, and lighting, techniques had changed little for many years. Typically, the camera was mounted on a tripod and, with all lights extinguished and the shutter open to admit light to the film, single, or multiple flashes were manually fired by the photographer or assistants. As long as the camera remained steady and everyone stayed still, the ensuing picture was sharp. Therein lay the difficulty: if the camera or subject moved, the picture was blurred. The unwieldy tripod did not fit in well with modern exploration into remote areas of new caves, and the restrictions produced very static, staid poses.

The answer lay with caver-designed slave units, which have become widely used since the late 1970s. Slave units are attached

Photographing Caves: Figure 2. Pictures by R.V. Davis, such as this one taken in The Big Room in Carlsbad Caverns, influenced the decision to inaugurate the national park. (Photo by R.V. Davis)

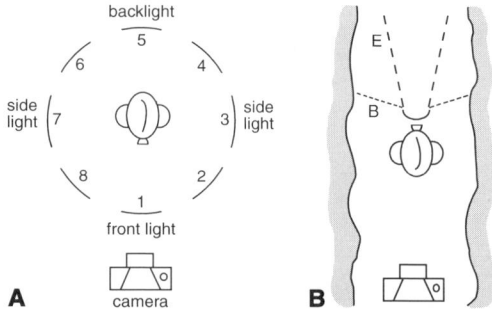

Photographing Caves: Figure 3. A: Lighting details can be recorded using a simple code system, where the letters E and B represent an electronic or bulb flash and a number indicates the direction of flash relative to the camera. A side-lit photograph taken with a single flashbulb would be designated B3 or B7, for example. Other photographs in this section are coded in this manner.
B: A caver firing a flashgun away from the camera but from within the picture area produces a silhouette. A flashbulb (B) can have an advantage in some situations, as the more directional light from an electronic flash (E) may not illuminate the nearby passage walls.

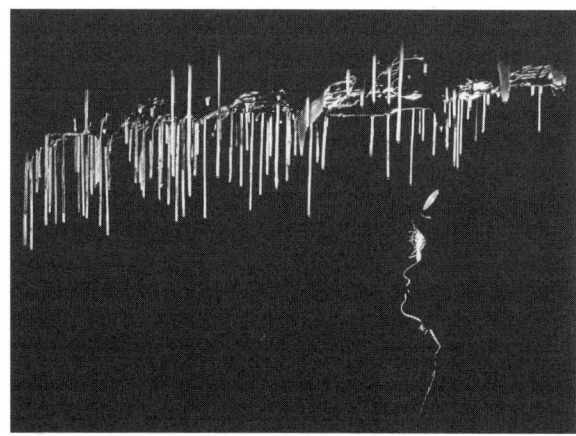 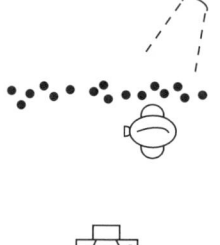

Photographing Caves: Figure 4. Back-lighting—as with these straw stalactites in Dan yr Ogof, Wales (E5)—produces dramatic results characterized by strong highlights and dense shadows. In the diagram the caver is placed between the flash and the camera, ensuring that no direct light reaches the lens. (Photo by Chris Howes)

Photographing Caves: Figure 5. Cave photographers use the characteristic long-duration light from a flashbulb—about 1/30th second rather than the 1/10 000th second duration of electronic flash—to simulate flowing water by allowing it to blur. (Arête Chamber, Ogof Ffynnon Ddu, Wales, B6; Photo by Chris Howes)

to a flashgun and detect a pulse of light to act as an electronic switch. By firing a master flash attached to the camera the photographer can therefore trigger any additional number of flashes without using intervening cables. Because all the flashes fire together, the camera can be hand-held and subjects are able to move—the short-duration flash of modern flashguns prevents blurring.

In addition, most caver-built slave units use infrared wavelengths. Film is insensitive to infrared light, so by using an appropriate filter over the master flash no light from the camera position is recorded in the picture. This further increases the photographer's creative control as pictures with strong back- or side-lighting can readily be taken. These slave units are often sensitive at distances in excess of one kilometre which, although this is greater than required, is a major advantage because weaker pulses of reflected light can be detected in shadow areas and around passage corners. Slave units confer the freedom to compose a picture exactly as the photographer envisages—a tremendous advantage in that cavers can be more easily and quickly photographed in action.

In the latest technological advances, digital cameras are further revolutionizing cave photography. Less light is required to record an image, which is immediately viewable on an LCD screen so that errors in lighting and exposure can be corrected and the picture retaken. An age-old photographic technique is to build up a cave picture using separate flashes to fill in different portions of the scene, but digital techniques can extend this theme. By taking a series of pictures from the same viewpoint with different camera settings or lighting positions, portions of each can be assembled on a computer using graphic-editing software. The requirement that everything is perfectly exposed in a single shot is no longer valid.

Purists might grumble as pictures produced using this technique become more prevalent. Perhaps the early pioneers would wince with displeasure—they struggled with heavy and unmanageable equipment, preparing and developing their glass-plate negatives underground. In comparison, using highly sensitive films, digital cameras and light-weight, small yet powerful flashguns which can be placed and fired at will, today's cave photographers have an easy time indeed.

Yet cave photography is not totally encapsulated and controlled by technology—it is a creative discipline conducted within a harsh, unyielding environment. In 1866 Charles Waldack wrote that: "You will agree with me that photographing in a cave is photographing under the worst conditions" (Charles L. Waldrack, "Photographing in the Mammoth Cave", 1866; reproduced in Howes, 1989, p. 66). His statement was true then and, arguably, is more so now that explorers penetrate ever further into the Earth. Technology may ease aspects of the challenge, but the impetus to produce an outstanding picture remains one that is very personal to the photographer. Ultimately, motivation and creativity remain more important than technology.

Chris Howes

Further Reading

Howes, C.J. 1989. *To Photograph Darkness: The History of Underground and Flash Photography*, Gloucester: Sutton and Carbondale: Southern Illinois University Press

Howes, C.J. 1997. *Images Below: A Manual of Underground and Flash Photography*, Cardiff: Wild Places

Martel, É.A. 1903. *La Photographie souterraine*, Paris: Gauthier-Villars; reprinted Marseille: Laffitte, 1989

PHYTOKARST

Phytokarst encompasses a range of small-scale landforms, produced by the action of plants on limestone surfaces. Such landforms have been identified since the early 1970s, when Folk, Roberts & Moore (1973) published observations from a site at Hell, Grand Cayman Island. Most phytokarst features reported in the literature are thought to be erosional features, produced through dissolution of limestone by organic acids, although depositional features produced by the reprecipitation of calcium carbonate, encouraged by plant activity, could also be defined as phytokarst. The term phytokarst is rather restrictive as it implies plant action only, and biokarst is a better, broader term for organically-influenced karst features (see Biokarstification). Phytokarren and photokarren are other terms which have been used to described plant-influenced landforms in karst areas.

Most phytokarst features, described in the literature, are centimetres to metres in scale and consist of a series of pits and upstanding areas. In some cases they form fields of pinnacles oriented towards a source of light, and in other cases they form randomly orientated spongework nested on larger-scale topography. Notable examples of light-oriented pinnacles have been reported from tropical cave entrances (e.g. as recorded by Waltham & Brook (1980) and Bull & Laverty (1982) from Mulu, Borneo), and temperate coastal caves (as discovered by Simms (1990) along the Burren coast, Co. Clare, Ireland). Random spongework forms were first identified by Folk, Roberts & Moore (1973) from near-coastal limestone terrain in Grand Cayman Island in the Caribbean, received more detailed study by Jones (1989) and are common on young limestone deposits on many tropical islands. Other examples of phytokarst forms include algal and lichen-induced pitting, observed on many coastal and terrestrial limestone surfaces, in tropical and temperate areas. The term phytokarst has also been applied to the constructional role of plants and micro-organisms growing on speleothems on vertical slopes and cave entrances (such as tower karst foot caves) in tropical karst. Such speleothems are sometimes orientated away from the vertical wall or cave entrance, and have been given the term "aussen stalactiten" by some authors (e.g. Viles, 1984). Their spongy, plant-covered and often tufa-like appearance marks them out from other, less organically influenced stalactites.

Despite the dramatic appearance of many phytokarst forms, it has proved difficult to determine the exact role of biological agents in their formation. Using scanning electron microscopy, it is possible to show that micro-organisms and plant roots are boring into limestone surfaces at the micron to millimetre scale. However, the challenge remains to try and link such microscale processes with the production of landforms at the centimetre to metre scale.

HEATHER VILES

See also **Biokarstification; Coastal Karst; Tourist Caves: Algae and Lampenflora**

Works Cited

Bull, P.A. & Laverty, M. 1982. Observations on phytokarst. *Zeitschrift für Geomorphologie*, 26: 437–57

Folk, R.L., Roberts, H.H. & Moore, C.H. 1973. Black phytokarst from Hell, Cayman Islands, British West Indies. *Bulletin of the Geological Society of America*, 84: 2351–60

Jones, B. 1989. The role of microorganisms in phytokarst development on dolostones and limestones, Grand Cayman, British West Indies. *Canadian Journal of Earth Sciences*, 26: 2204–13

Simms, M.J. 1990. Phytokarst and photokarren in Ireland. *Cave Science*, 17: 131–33

Viles, H.A. 1984. Biokarst: Review and prospect. *Progress in Physical Geography*, 8: 523–43

Waltham, A.C. & Brook, D.B. 1980. Cave development in the Melinau limestone of the Gunung Mulu National Park. *Geographical Journal*, 146: 258–66

Further Reading

Ford, D.C. & Williams, P.W. 1989. *Karst Geomorphology and Hydrology*, London and Boston: Unwin Hyman

Jennings, J.N. 1985. *Karst Geomorphology*, 2nd edition, Oxford and New York: Blackwell

PICOS DE EUROPA, SPAIN

The Picos de Europa, in northwest Spain, hosts the largest concentration of deep caves in the world. Covering an area of *c.*500 km², the Picos are located on the northern flank of the Cantabrian Mountains and contain the highest peaks in this range, rising to an altitude of over 2500 m. The Picos are dissected from north to south by the spectacular gorges of the Cares and Duje rivers, which, flowing at altitudes of from 300 to 400 m, divide the Picos into three sub-massifs (Figure 1). Declared in 1918, the Picos de Europa was the first national park in Spain and one of the first in Europe.

Geological Background

The Picos are composed almost exclusively of a 1500 m thick sequence of Carboniferous (Visean to Moscovian) marine limestones. This sequence starts with the nodular limestones and radiolarites of the Alba Formation (50 m) and the dark, thin-bedded limestones of the Barcaliente Formation (250–300 m). Both formations were deposited in a deep platform setting over most of the Cantabrian zone. In the Picos de Europa, overlying the Barcaliente, are the deposits of a large steep-margined carbonate platform, dominated by the massive limestones of the

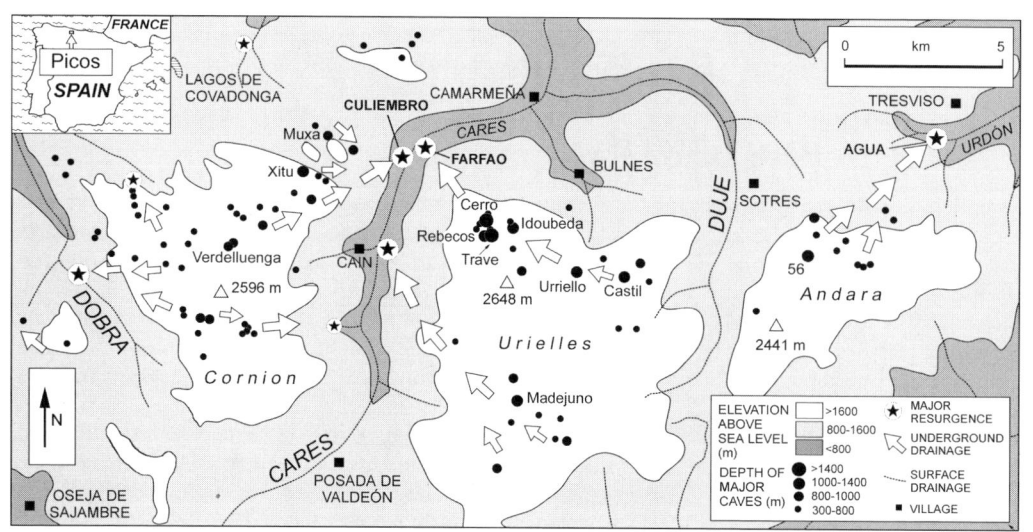

Picos de Europa, Spain: Figure 1. Outline map of the three massifs of the Picos de Europa, with sites of the main caves and known underground drainage patterns.

Valdeteja (300 m) and Picos de Europa Formations (up to 1000 m). In areas surrounding the Picos, deposits coeval to this carbonate platform mainly consist of siliciclastics.

During the late Carboniferous, the Cantabrian zone was incorporated into the Variscan orogeny and the Picos limestones were deformed into an imbricate system of thrust sheets. The strike of the thrusts is roughly east–west, and their dip increases from south to north. Although pre-Carboniferous siliciclastics are found locally at the base of some thrust sheets, most are formed almost exclusively of limestones. As a consequence, the stacking of thrusts has produced a huge vertical accumulation of limestone.

In the Picos de Europa and adjacent areas, a Permo-Mesozoic, siliciclastic-dominated sedimentary cover was deposited over the Hercynian basement. The current relief of the Cantabrian Mountains, including the Picos de Europa, is related to Alpine uplift during the Eocene to Miocene. As a consequence, the Permo-Mesozoic cover was eroded from the western Cantabrian Mountains, including the Picos de Europa. The fluvial gorges dissecting the Picos run perpendicular to the structures, suggesting that the present drainage was inherited from the cover and superimposed onto the Carboniferous basement during the Tertiary.

Quaternary glaciation extensively affected the Picos de Europa (Smart, 1986). Glaciers covered most of the highlands and occasionally, the Cares and Duje valleys. The largest ice cap developed on the Western Massif, covering an area of $c.50$ km^2 and locally reaching 300 m in thickness. Gale & Hoare (1997) have recognized five glacial phases, the most extensive probably being latest Tertiary.

The rugged Picos highlands are composed of plateaus, usually at altitudes ranging from 1500 to 2000 m, and ridges and horns that typically rise $c.500$ m above the plateaus. The plateaus have no surface drainage and their bare bedrock surfaces are covered by karren and dolines (Figure 2). However, the most notable feature is the presence of closed depressions up to 1000 m across, which are glacial modifications of pre-Quaternary karst depressions (Smart, 1986).

Thousands of caves are known in the Picos, and hundreds are explored every summer. Caves are predominantly vertical vadose systems located in the highlands. The majority are small shafts that become impenetrable or blocked at depths of a few tens of metres. However, a significant number of caves attain considerable depths: 96 caves exceed 300 m and 9 caves exceed 1000 m in depth (see Table). Such a concentration of deep caves is related to the paucity of non-carbonate lithologies and the large differences in elevation between the highlands and the valley floors, resulting in an exceptionally thick vadose zone.

Most of the deep vadose Picos caves consist of shafts systems interconnected by narrow canyons. Shafts are guided by Alpine faults or Variscan thrusts. An impressive example of the latter is the Trave system which, between −500 and −1000 m of depth, has several series of shafts following the dip of a major thrust (Vidal, 1990). Where shafts are developed in massive limestones (Valdeteja and Picos de Europa Formations) they can reach spectacular proportions, and four single pitches exceeding 300 m of depth are known.

Many of the large vadose Picos caves are partially subglacial in origin. Glaciers provided large catchments and concentrated meltwater into certain sinks, explaining large shafts at points

Picos de Europa, Spain: Figure 2: Typical landscape of the Picos highlands, showing the bare bedrock surfaces of the Vega Huerta Plateau (to the left) and Canal de la Duernona (to the right) on the Western Massif. Glacial rounding and karren are common. The area shown in the lower part of the image contains abundant shafts, including the 949 m deep Pozo del Llastral and the 420 m deep Pozo de la Garita Cimera. (Photo by Carlos Rossi)

which currently receive little modern drainage (Smart, 1986). Additionally, some cave morphologies, particularly the "pitch-ramp" systems, are best explained by the action of glacial meltwater (Senior, 1987). The pitch-ramp systems, characteristic of many Picos canyons, are inclined bedrock terraces ascending from the base of each pitch to the head of the next, and are vadose features that are formed by pitch retreat (Senior, 1987).

Except in the case of the resurgence caves, only the deepest caves reach the water table, which is defined by segments of low-gradient streams interrupted by inundated phreatic passages. Overall, horizontal water-table passages are uncommon, particularly in the resurgence caves such as Culiembro and Cueva del Agua. Instead, these caves consist of active and relict looping phreatic tubes, suggesting that a significant part of the present drainage is through phreatic caves.

Relict phreatic conduits are uncommon in the influent Picos caves. However, some deep caves intersect large, looping, strike-oriented phreatic tubes in their lower sections (typically up to 100–150 m above the present water table). Excellent examples are found in the M2 cave, Pozu Jultayu, Sil de Oliseda, and Asopladeru de la Texa (Western Massif), and in the Torcas Urriellu, Castil, del Cerro, and Idoúbeda (Central Massif). At higher elevations, relict phreatic conduits are scarcer and smaller, and most have been invaded by vadose streams, resulting in deep canyons. Some phreatic tubes are found at altitudes as high as 1000 m above the water table. Based on the rate of base-level lowering deduced from uranium-series dates of speleothems, some of these high-level phreatic conduits may be as old as 3 million years (Smart, 1986). In caves of the Western and Eastern Massifs, some relict phreatic conduits preserved siliciclastic

Picos de Europa, Spain: Caves in the Picos de Europa more than 900 m deep.

Cave	Depth (metres)	entrance altitude (m)	massif	resurgence name	gorge	altitude (m)
Torca del Cerro del Cuevón (T-33)	1589	2019	Central	Farfao	Cares	320
Sistema del Trave	1441	2042	Central	Farfao	Cares	320
Torca de los Rebecos (T27)	1255	2083	Central	Farfao	Cares	320
Pozu Madejuno (Omega 45)	1255	2425	Central	Molinos	Cares	450
Sima 56	1169	1975	Eastern	Agua*	Urdón	465
Torca Idoúbeda	1168	1856	Central	Farfao	Cares	320
Sistema del Xitu	1148	1652	Western	Culiembro*	Cares	390
Torca Urriello	1022	1860	Central	Farfao	Cares	320
Torca Castil (PC-15)	1028	2000	Central	Farfao*	Cares	320
M2 (Pozo de Cuetalbo)	972	1990	Western	Capozo*	Cares	900
Pozo del Llastral (Beta 3)	949	1950	Western	Reo Molin*	Dobra	930
Torca Tortorios (CT-1)	943	?	Central	Farfao	Cares	320
Pozu Cabeza Muxa	939	1504	Western	Culiembro*	Cares	390
Torca del Jou de Cerredo	910	2325	Central	Farfao	Cares	320
Sima del Jou de la Canal Parda	903	2215	Western	?	?	?

* indicates that the resurgence has been dye traced from the sink cave.

sediments related to the erosion of the Permo-Triassic cover (Smart, 1986; Fernández et al., 2000).

The water drained by the caves resurges in a few springs at the bottom of the gorges. Underground drainage is in a broadly east–west direction, following the strike of beds, thrusts, and post-Hercynian faults. Because the dip is towards the north, no resurgences exist in the southern front. In the Western Massif, the most important resurgences are Culiembro in the Cares gorge (see below) and Reo Molín in the Dobra gorge (Figure 1). In the Central Massif, underground drainage is predominantly towards the west–northwest, and the Farfao and Molinos springs at the Cares gorge drain the southern and northern parts of the massif, respectively. In the Eastern Massif, the most important resurgence is Cueva del Agua, located in the Urdón gorge.

The Culiembro and Farfao systems are examples of important and relatively well-known underground drainage systems. The Cueva de Culiembro, located at an altitude of 390 m, in the Cares gorge, and with a summer discharge of 1 m^3 s^{-1}, drains the northeastern part of the Western Massif (Figure 3). The Culiembro drainage system has two branches (Figure 1). The north branch drains to the southeast, following the strike of a series of thrust sheets. In this area, three caves (Pozu les Cuerries, 545 m deep, Cabeza Muxa, 906 m deep, and Asopladeru de la Texa, 837 m deep) descend to segments of a low-gradient streamway, interrupted by sumps up to 28 m deep, that has been dye traced to Culiembro. The altitude of these sumps decreases towards Culiembro, defining the water table. To the south, a second drainage branch is developed within a series of northeast-oriented thrust sheets, draining to the northeast, and containing several major caves, including the 1148 m deep Sistema del Xitu. Its terminal sump has been dye traced to Culiembro, 1 km distant from and 115 m below the Xitu sump. The nearby Pozu Jultayu (820 m deep, c.12 km long) contains a large horizontal streamway, which has been also dye traced to Culiembro, 2 km

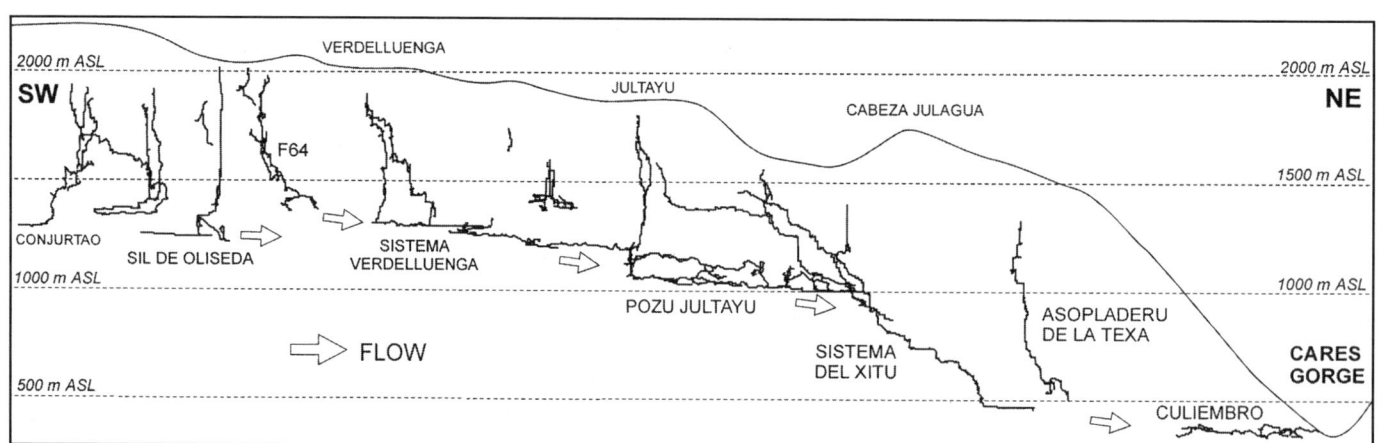

Picos de Europa, Spain: Figure 3: Profile through the major caves in the northern sector of the Cornion Massif, the western part of the Picos de Europa. Survey data from Oxford University Cave Club home page.

distant from and 600 m below the downstream end of Jultayu. Upstream, the Jultayu river is fed by the low-gradient streamways found at the bottoms of the 644 m deep, 8 km long Sistema Verdelluenga and the 806 m deep Sil de Oilseda. Other major caves, located 1–2 km further to the west, may be also related to the Culiembro system. They are the Canal Parda (903 m deep), Porru la Capilla (863 m deep) Conjurtao (658 m deep), Jorcada Blanca (590 m deep), and F20 (582 m deep) systems, all developed along a common fault plane and ending in sumps located at similar altitude (1300–1350 m). This altitude is similar to those of the low gradient streams and sumps of Verdelluenga, Sil de Oliseda, and Jultayu, suggesting a hydrologic connection and that the water table in this sector is 600–800 m higher than elsewhere in the Culiembro drainage area (Figure 3; Roberts, 1986).

The Farfao resurgence, located at an altitude of 320 m, in the Cares gorge, drains the northern Central Massif (Figure 1), producing 3 m^3 s s^{-1} in summer. The area drained by Farfao consists of two major thrust sheets, and contains six caves exceeding 1000 m in depth, including the 1589 m deep Torca del Cerro, one of the deepest in the world. Four of these caves (Trave, Cerro, Urriellu, and Castil) have low-gradient streams at their bottoms, interrupted and/or ending in sumps. The coincidence in elevation of streams and sumps indicates that they are at the water table. The elevation of the terminal sumps decreases progressively to the northwest, indicating general drainage towards Farfao. This has been further confirmed by dye tracing in Torca Castil, which is 8 km distant from Farfao.

CARLOS ROSSI

Works Cited

Fernández, E., Calaforra, J.M. & Rossi, C. 2000. Karst in the Picos de Europa massif (North Spain). In *Speleogenesis: Evolution of Karst Aquifers*, edited by A.B. Klimchouk, D.C. Ford, A.N. Palmer & W. Dreybrodt, Huntsville, Alabama: National Speleological Society

Gale, S.J., & Hoare, P.G. 1997. The glacial history of the northwest Picos de Europa of northern Spain. *Zeitschrift fur Geomorphologie*, 41(1): 81–91

Roberts, S. 1986. Combined survey of the top camp caves. *Proceedings of the Oxford University Cave Club*, 12: 55

Senior, K.J. 1987. Geology and speleogenesis of the M–2 Cave System, Western Massif, Picos de Europa, Northern Spain. *Cave Science*, 14(3): 93–103

Smart, P.L. 1986. Origin and development of glacio-karst closed depressions in the Picos de Europa, Spain. *Zeitschrift fur Geomorphologie*, 30(4): 423–43

Vidal, B. 1990. Picos de Europa. Sistema del Trave: −1441; T–31: −570. *L'Aven*, 50: 63–87

Further Reading

Alonso, V. 1998. Covadonga National Park (Western Massif of Picos de Europa, NW Spain): A calcareous deglaciated area. *Trabajos de Geología*, 20: 167–81

Beniot, P. (editor) 1985. *Les Picos de Europa. Spelunca* supplement, 19, 64 pp

Maragliano, D., Muñoz, J. and Estévez, J.A. 1998. Récord de España en la Torca del Cerro del Cuevón (Macizo Central de los Picos de Europa, Asturias). *Subterránea*, 10: 20–29

Marquínez, J. & Adrados, L. 2000. La geología y el relieve de los Picos de Europa. *Naturalia Cantabricae*, 1: 3–10

Sánchez, J. & Cerdeño, R. 1999. Resumen de las exploraciones de la Torca de la Nieve—estudio del nivel freático. *Subterránea*, 11: 36–41

Sefton, M. 1984. Cave explorations around Tresviso, Picos de Europa, Northern Spain. 1975–1983. *Cave Science*, 11(4): 199–237

Smart, P.L. 1984. The geology, geomorphology and speleogenesis of the Eastern Massifs, Picos de Europa, Spain. *Cave Science*, 11(4): 238–45

PIERRE SAINT-MARTIN, FRANCE–SPAIN

The French–Spanish Pyrenean massif of Pierre Saint-Martin is of interest both historically, and for its superb geology and speleology. Situated 50 km southwest of Pau, this alpine karst area of 140 km^2, at altitudes between 430 m (Illamina and Bentia springs) and 2504 m (Pic d'Anie), is affected by a montane climate—temperate and marine and very humid (with a rainfall of 2500 mm a^{-1}). The area is characterized by typical alpine topography: rillenkarren and lapiés, rock benches, snow shafts, and underground glaciers, depressions, small valleys, and dolines of glaciokarstic origin. The first speleological investigations known in the area (of the Napia Cave) go back to 1818. Towards 1880, some local naturalists began to explore the shafts beneath Sainte-Engrâce and towards the col of Pierre Saint-Martin. They were joined from 1892 to 1909 by Édouard Alfred Martel, Eugene Fournier, and their companions, who carried out several reconnaissances and explored the canyons of Kakouetta, Holzarté, and Olhadubie.

Between 1930 and 1949, Max Cosyns and Norbert Casteret undertook many descents and explorations of canyons and shafts in the area. In 1950, Georges Lépineux discovered the entrance of the Pierre Saint-Martin cave. More than 320 m deep, the Lépineux shaft was descended for the first time in 1951, but in 1952 exploration was halted by the death of Marcel Loubens in an accident caused by a faulty winch. In 1953, the bottom of La Verna chamber was reached at −734 m. This was the world depth record at the time, and it was the largest underground chamber then known (4.5 million cubic metres).

From 1951 to 2002, more than 350 km of shafts and passages were explored and surveyed and in 1996 the Association pour la Recherche Spéléologique Internationale à la Pierre Saint-Martin (ARSIP) was formed to coordinate research. At the end of 2001, the number of caves explored is impressive: 49 deeper than 300 m, and four deeper than 1000 m (Table 1). The regional map of the caves shows a remarkable density of underground systems (Figure 1). Thanks to underground tracings with fluorescein indicating the hydrological relations between the networks, three large drainage basins are recognized. Two karstic systems drain northwest. The Bentia Basin (2.37 m^3 s^{-1} mean flow, 30 km^2 in area) includes the Pierre Saint-Martin, Lonné Peyret, and Soudet. The Illamina Basin (5.64 m^3 s^{-1}, 80 km^2) lies to the south, largely in Spain, with the Partages and BU56 caves draining to the d'Arrestialeko (58 km long). The third basin is Issaux

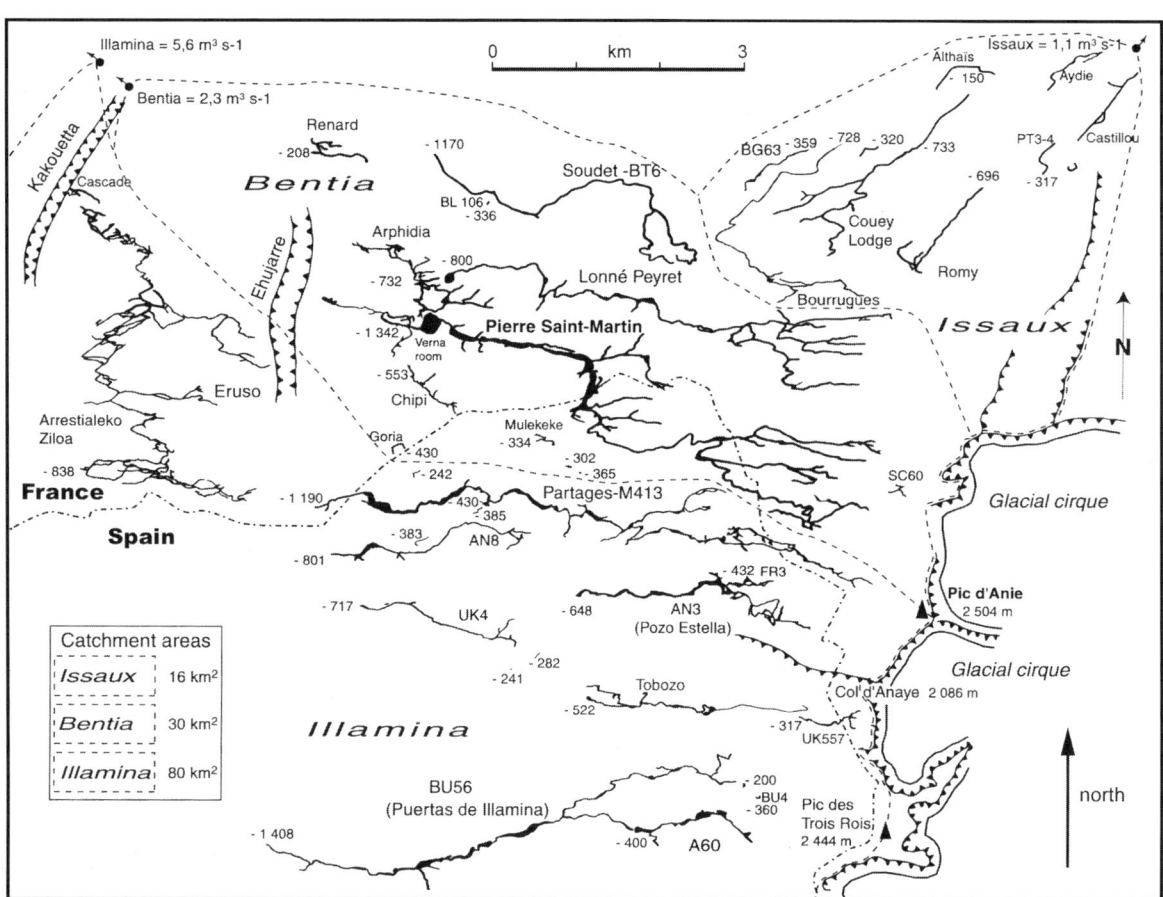

Pierre Saint-Martin France–Spain: Figure 1. Outline map of the cave passages in the Pierre St Martin and adjacent cave systems, beneath the France–Spain border. The numbers indicate the depths in metres of each cave below its own top entrance.

($1.15 \text{ m}^3 \text{ s}^{-1}$, 16 km²), to the northeast, which drains the Bourrugues, BG 63, Couey Lodge, and Romy caves. A fourth basin to the east, not indicated on the map, is Lees Athas ($1.2 \text{ m}^3 \text{ s}^{-1}$, 17 km²), but it has few known caves of importance.

The underground and surface karstification is due to favourable geological and geomorphological conditions. Upper Cretaceous limestones—350 m thick and very fractured—directly overlie Paleozoic basement in the axis zone of the Pyrenees. Since the major Miocene uplift, erosion of the overlying impermeable flysch and shales made it possible for karstification to begin in the underlying limestones. Between 20 and 6 million years ago, under subtropical climate conditions, the sedimentary cover was

Pierre Saint-Martin, France–Spain: Table. The deepest caves of Pierre Saint-Martin Massif (in 2002).

Cave	Depth (m)	Length (km)	Location
Las Puertas de Illamina (BU56)	1408	14.5	Isaba (Spain)
Gouffre de la Pierre Saint-Martin	1342	52.5	Arette et Sainte-Engrâce (France), Isaba (Spain)
Réseau de Soudet (BT6)	1170	10.3	Arette et Sainte-Engrâce (France)
Gouffre des Partages (M413)	1097	23.9	Arette (France), Isaba (Spain)
Arresteliako Ziloa	838	58.1	Sainte-Engrâce (France)
Gouffre AN8	810	7.2	Isaba (Spain)
Gouffre Lonné Peyret	807	24.3	Arette et Sainte-Engrâce (France)
Gouffre du Couey Lotdge	733	8.7	Arette et Osse-en-Aspe (France)
Gouffre des Bourrugues (B3)	728	7.5	Arette et Osse-en-Aspe (France)
Sima de Ukerdi Abajo (UK4)	717	4.6	Isaba (Spain)
Grotte d'Arphidia	712	22.3	Sainte-Engrâce (France)

subjected to intense weathering, indicated by the pockets of red earth (with micaceous clays, smectite, chlorite, and goethite) eroded by the modern karst drainage.

The evidence of lowering of limestone surfaces and of the erosion of karstic closed basins is numerous. The upstream edge of the massif exposes cavities in walls, relict unroofed caves, and large surface stalagmitic masses. However, the calcite deposits are not currently being formed at this altitude, in the absence of soil and vegetation. These old speleothems exposed by erosion are probably pre-Quaternary, and were formed at a time when the massif was less high and the vegetation and soil were more abundant. U–Th dating puts the age at more than 350 ka. These paleo-cavities lost their catchment areas when Pleistocene glacial erosion destroyed the higher cirques.

There are also inactive caves within the limestones, 80–120 m above the Paleozoic basement, but at different altitudes, such as in the caves SC60 (length 1930 m), UK4 (1600 m), AN8 (1500 m), and Chipi Joseteko (1300 m). These dry horizontal conduits, with phreatic morphology, were intersected by more recent caves, and they formed in relation to old base levels which progressively changed with deepening of the valleys. The current base level ranges between 435 and 445 m altitude at the height of the Illamina and Bentia springs. At the downstream end of the Bentia karstic system, 22 km of passages in the Arphidia Cave have five levels between 1100 and 600 m altitude. This situation, low down in the system, is due to a combination of two factors: Pleistocene uplift and the presence of Devonian limestones in the Paleozoic basement. These basement limestones enabled the formation of drains beneath the Cretaceous limestones, due to lowering of the base level. The sediments in the Aranzadi passages—passages which are perched 100 m above La Verna chamber—indicate that the underground river formerly flowed in this gallery, but a collapse captured the river in the Paleozoic basement. Significant voids were formed by infiltration in the Devonian floor, and the chamber there evolved by collapse. La Verna chamber (255 × 245 × 180 m) is thus related to a hydrological capture by an underground version of a collapse doline.

The Aranzadi passage is accessible by the old EDF (Electricité de France) tunnel, constructed between 1956 and 1960 with the aim of capturing the underground river, but, because of the variability of the flow (from $0.05 \text{ m}^3 \text{ s}^{-1}$ up to an estimated $5-15 \text{ m}^3 \text{ s}^{-1}$), the project was abandoned. This passage contains a sequence of sediments over 20 m in thickness, the most important known in the alpine karst environment (Maire, 1990; Quinif & Maire, 1998). Three unmatched detrital assemblages are observed: a lower ensemble of blocks and scree; a main sequence of silts and calcareous varved clays (Figure 3); a series of fluvial terraces encased in the lower and main sequence. About 50 U–Th dates have been determined on the speleothems. The lower unit probably dates from a former cold period before 330 ka (isotope stage 10). The principal sequence, which ranges from 225 000 to 300 000 years BP, is allotted to glacial stage 8, from which the external moraine deposits have disappeared. The fluvial terraces are covered with uneroded stalagmites, of which the oldest is 190 000 years BP. The oldest stalagmite from La Verna is older than 210 000 years; it is located at the same level as the Aranzadi passage and was eroded by the old river. Consequently, the abandonment of the Aranzadi passage corresponds with the evolution of La Verna between 194 000 and 211 000 years BP.

Pierre Saint-Martin, France–Spain: Figure 2. The underfit streams in the main passage of Gouffre Pierre Saint-Martin between Puits Lépineux and Salle de la Verna. (Photo by Tony Waltham)

Seismotectonic indicators are widespread in the caves. In BU56, the walls of the meandering passage between 380 m and 450 m depths are broken by a series of tear faults, related to the reactivation of a major east–west fault after the cave passage was formed. Within the Pierre Saint-Martin caves there are numerous collapses on faults, which are evidence of seismotectonic activity due to the compression of the Iberian plate against the European plate.

The massif of Pierre Saint-Martin appears to be a natural laboratory for further detailed study, in space and time, of the different karstification parameters and their complex interactions in an alpine karst situation. The construction of a ski

Pierre Saint-Martin, France–Spain: Figure 3. The fluvioglacial deposits in Aranzadi gallery (Pierre Saint-Martin Cave). The main unit of laminated clay, with carbonate-rich varves devoid of pollen, correlates with stage 8. The river terraces inset into this unit formed during stage 7 before collapse of Verna Room between 194 000 and 211 000 years BP. (Photo by Yves Quinif)

station during the 1960s caused a degradation of the surface karst and pollution of the caves in the Issaux basin, but a project to open La Verna chamber for tourism could become a reality in the future.

RICHARD MAIRE

Works Cited

Maire, R. 1990. *La haute montagne calcaire*, Paris: Fédération française de spéléologie and Association française de karstologie (Karstologia Mémoires, 3)

Quinif, Y. & Maire, R. 1998. Pleistocene deposits in Pierre Saint-Martin Cave, French Pyrenees. *Quaternary Research*, 49: 37–50

Further Reading

Douat, M. (editor) 2002. *Bulletin n.17 de l'ARSIP*: 1–250
Pernette, J.-F. 1983. *Rivières sous la Pierre*, Paris: Fernand Nathan
Tazieff, H. 1952. *Le gouffre de la Pierre Saint-Martin*, Paris: Arthaud

PINEGA GYPSUM CAVES, RUSSIA

Permian gypsum forms a long north–south outcrop west of the Urals, in northern Russia. The main bed of soluble evaporites is a complex of gypsum and anhydrite, that thickens to ~50 m around the small town of Pinega, due east of Arkhangelsk (see area map in Russia and Ukraine). Low plateaux in the almost horizontal gypsum are dissected by the Pinega River and its tributaries. The rugged karst landscape is overgrown with taiga forest of birch and spruce. It is broken by rocky dolines, and large areas are surfaced in spectacular schlottenkarst, with funnel-shaped dolines spaced on 3–4 m centres, tapering into vertical shafts, that are usually plugged with vegetation, snow, or soil just a few metres down. These features are larger than those that are common on gypsum elsewhere, and they constitute an excellent form of polygonal karst. In addition, scattered vertical shafts drop into bedding plane caves.

The plateau edges drop 50 m to the intervening valley floors, and many are broken by long scars and crags of bare gypsum, with some separated pinnacles, sharpened up from the tops of schlotten divides, so that they resemble a small-scale pinnacle karst. Perennial streams flow along many of the valley floors, but some thalwegs are dry—mostly where their streams have invaded older cave passages. There are many sinks, karst lakes, and resurgences, mainly where the thalwegs with more irregular profiles are almost broken into chains of dolines.

Numerous caves are truncated in the valley walls, but most cave entrances are choked by rock debris. New caves, both in the crags and in the doline floors, are often only found when they are exposed by collapse. Some of the new collapses are formed by inwashing of soil and debris, but others are due to rapid dissolution of the gypsum by snowmelt water. Face retreat of the gypsum has been measured at 20 mm a^{-1}, where it is subject to a constant flow of unsaturated surface water. Over 50 km of cave passages have been mapped in the Pinega gypsum, and 22 caves are each longer than 1 km.

A notable feature of the Pinega caves is that a large proportion have dendritic passages that carry, or have carried, significant underground streams. The finest group of caves is at Zheleznye Vorota (Iron Gates), just northeast of Pinega (see Figure 1). The two main caves of Olympyskaya and Lomonosovskaya are connected by a sump, which has been dived to establish a single system with over 9 km of passages. A stream can be followed through the length of the system. Part of it is in shallow vadose canyons, part is in drained phreatic tubes with almost no entrenchment, and it follows a shallow phreatic loop between the

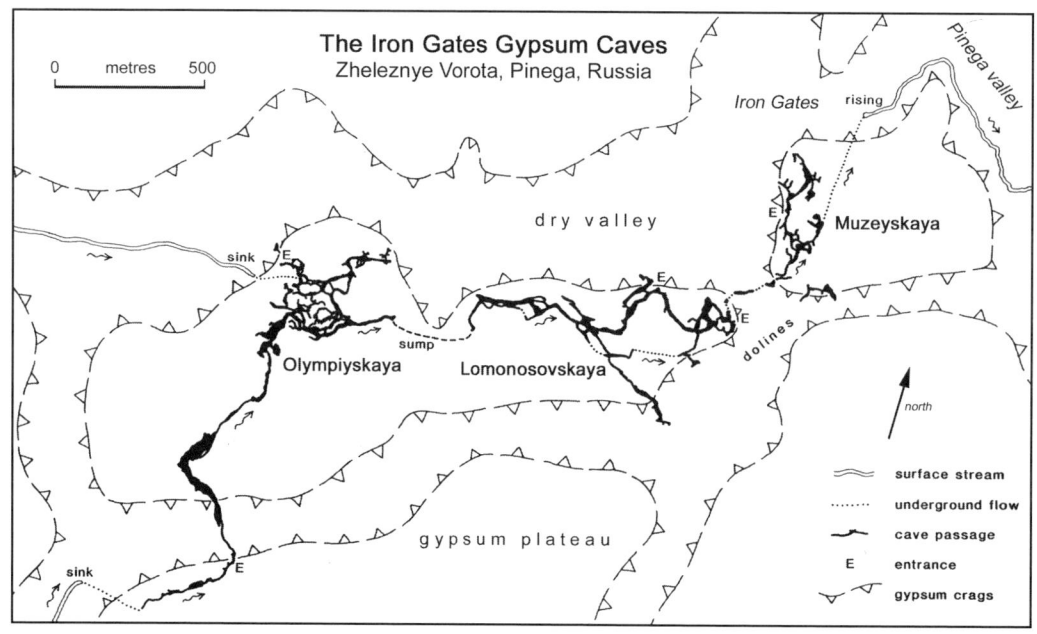

Pinega Gypsum Caves, Russia: Figure 1. Outline map of the gypsum caves in the Iron Gates Nature Reserve, just above Pinega, Russia (after surveys by Arkhangelsk Geologia).

Pinega Gypsum Caves, Russia: Figure 2. The main tunnel in Lomonosóvskaya, an elliptical phreatic tube in the gypsum, liberally decorated with ice that forms every winter and largely melts away each summer. (Photo by Tony Waltham)

two drier cave systems. The main passages have very low gradients. They were initiated on bedding planes, of which some are clay-bearing paleokarst horizons and others are on thin interbeds of solutionally-resistant dolomite. Many of the largest phreatic tubes are within white gypsum but have ceilings of beautiful, pale blue anhydrite.

At least two higher levels of passages can be recognized. These also follow the bedding, but are now heavily collapsed. Some have thick curved slabs that are slowly peeling away from the ceiling by plastic deformation, which is aided by the hydration of anhydrite to gypsum. Kulogorskaya is another dendritic system with multiple levels close to the main river just northeast of Pinega; it has over 16 km of mapped passages.

Other caves are drained phreatic mazes of joint-controlled passages. Symphoniskaya (3.2 km long) and Golubinskaya (1.6 km long) are both maze caves that carry no modern streams and lie abandoned at higher levels in the gypsum plateaux. Their long, narrow, fissure passages locally widen, and some link into bedding-guided chambers. The mazes are comparable in appearance to those in the gypsum caves of the Ukraine; it is unclear if they were formed in similar conditions of slow, rising phreatic flow, or at a river-level water table beneath multiple drainage inputs from the schlotten fissures.

In contrast, the larger passages in the main dendritic cave systems have acted as major karst conduits. Phreatic tubes 10–20 m wide and 3 m high, have carried streams much larger than those that now occupy some of them as underfits. The segments of large old passage, now abandoned above the modern streams, may have been formed by subglacial meltwater during the Pleistocene. The braided system of surface valleys, cut with irregular profiles into the gypsum plateau at the Iron Gates (see Figure 1), is also characteristic of subglacial phreatic drainage.

Pinega lies just outside the Arctic zone of permafrost. The ground never freezes to depths of more than a few metres, and the larger cave streams continue to flow through the winter, beneath a landscape with a thick blanket of snow for nearly half the year. However, freezing air blows through the caves that have multiple entrances, or just sinks into those with single entrances. All percolation seepage, therefore, freezes as it enters the caves, and creates huge ice cascades and spectacular icicles (Figure 2). Lakes freeze over to form underground skating rinks, even where water continues to flow beneath. In addition, water vapour freezes into giant hoarfrost of ice crystals that are plastered over the walls and ceilings of all passages near the entrances. The total effect is very beautiful, and nearly all the ice displays disappear each summer and form anew every winter. Only below some of the larger open entrances, does blown-in snow accumulate beyond what can melt, so that some small cave glaciers are created; the oldest ice yet found has been dated to 3000 years ago.

The finest of the gypsum karst landforms, and many of the longer caves, now lie protected within the Pinega National Park and the Iron Gates Nature Reserve. Both these sites are primarily known for their populations of deer, moose, and black bears that roam unmolested through the taiga-covered karst.

TONY WALTHAM

Further Reading

Malkov, V.N., Nikolaev, J.I. & Kuznetsova, V.A. 1986. Regionalisation of the Pinega and Kuloj basins according to settings and intensity of exogenous geological processes. In *Geologija i poleznye iskopaemye Arkhangelskoj oblasti*, Moscow: 154–74

Malkov, V.N., Nikolaev, J.I. & Luskan, V.F. 1988. Types of karst caves of the Piezhje. In *Peschery: peschery v gipsakh i angidritakh*, Perm University Publishing: 46–50

Waltham, T. 1994. Gypsum caves of Pinega. *International Caver*, 10: 15–22

Waltham, T. & Cooper, A. 1998. Features of gypsum caves and karst at Pinega (Russia) and Ripon (England). *Cave and Karst Science*, 25: 131–40

PIPING CAVES AND BADLANDS PSEUDOKARST

Introduction

Grain-by-grain removal of particles by groundwater flow produces caves and smaller conduits in some poorly soluble rocks and unconsolidated material. Nonkarstic hydrologists and engineering geologists consider these phenomena on the basis of mindsets with radically different terminologies and methodologies. If read carefully, however, their literatures are perceived to achieve largely similar conclusions.

The traditional engineering geology mindset is expressed especially by Parker *et al.* (1990), and the general viewpoint of academic nonkarstic hydrologists is explicated elegantly by Dunne (1990). Dunne seems more concerned with movement of water through micropores and intergranular and interpedal voids than with depressions, caves, and smaller conduits, but includes them in his overall construct. The editors of the authoritative volume which brought together these divergent papers deliberately avoided attempting to arbitrate these semantic and conceptual conflicts (Higgins & Coates, 1990, p.vii). An additional European terminology of suffosion pseudokarst has not been widely accepted elsewhere.

Parker's concepts and terminology are congruent with those of the current mainstream of English-language speleology. Consequently, Parker's terminology and concepts are employed in this entry, and extended to more consolidated rocks than those he studied. For details of the inception stage of piping, interested readers are referred to Dunne (1990) and numerous references cited therein.

Definitions: Piping Caves

Piping caves are natural underground spaces; large enough to enter and investigate; extending beyond daylight; formed by grain-by-grain subterranean removal of particles by channelled flow of groundwater; primarily from poorly soluble rock or granular material; and with little or no dissolution. Other than the cited human module, no clearcut line of differentiation can be specified between large pipes and small piping caves, and an interface or continuum exists between piping caves and dissolutional caves in rocks of various solubilities. Terms used in the literature for piping caves include clay caves, claystone caves, gully caves, loess caves, mud caves, mudstone caves, peat caves, shale caves, slutch caves, suffosion caves, tunnel caves, and tunnels.

Definitions: Badlands

This term is not the equivalent of the Spanish-language "malpais" which generally is applied to rough flowfields of lava. Badlands topography consists of a complex of bare, moderately steep slopes intricately dissected by intermittent stream runoff, with narrow interfluves and generally with a resistant caprock. Its type locality is Badlands National Park, South Dakota (United States). Especially notable examples also are exhibited in Petrified Forest National Park, Arizona (United States). Here, expansion and contraction of clays with rainfall and drought repeatedly roof and unroof many small closed conduits (Mears, 1968). Badlands topography develops especially in shales, mudstones, claystones, and siltstones, but also occurs in some thin-bedded calcareous and arenaceous rocks, and in volcanic ash. Mears and Parker et al. (1990) (Figure 2) and others have neatly depicted mechanisms which produce piping and other pseudokarstic landforms in such topography. The terms avalanche pseudokarst, loess pseudokarst, mudflow pseudokarst, piping pseudokarst, and suffosion pseudokarst have been used to describe badlands pseudokarst.

Significance

Some planetary geologists (e.g. Malin & Edgett, 2000) have suggested that certain landforms on Mars are piping features. Thus the possibility of extraterrestrial piping caves must be given serious consideration as potential sources of subterranean ice (see also Extraterrestrial Caves). On Earth, however, piping caves are primarily of interest as geological curiosities and recreational sites. Piping itself is a serious, often overlooked cause of engineering problems.

Types of Piping

Piping appears in different physical forms and originates in different ways, depending on climatic and geological conditions. Basically, it is accomplished by one of two specific mechanisms. One is a form of mass fluidization of soil with failure of lithostatic resistance. This produces "heave", "sand boils", and the like, without any well-defined conduit and hence is not considered further in this entry. The other mechanism is an enclosed process forming more or less stable "pipes" in poorly soluble rocks and unconsolidated materials. These intergrade into dissolution conduits in karstifiable rocks. This process occurs through eluviation, sapping, seepage face erosion, "tunnel scour", and other erosive mechanisms. Primarily it occurs in drylands settings, especially in granular media and in rocks predisposed to slake, e.g. mudstones and claystones. Large examples tend to occur in these rocks where shattered in avalanches and mudslides. In humid regions, this form of piping is said to occur only in soils and peat (e.g. Jones, 1990), but this fails to consider recent speleological literature including that on caves in quartzose rocks.

Classical Piping in Drylands

Mostly relying on English-language sources, Parker et al. (1990) have provided a global overview of more than a century of studies of drylands piping and the pseudokarstic features which it produces, including subsurface drainage with caves and smaller conduits, funnel-shaped sinks and swallets, dry valleys, natural bridges, and shallow circular lakes. Some of these features resemble karstic and other pseudokarstic landforms (e.g. thaw lakes on permafrost pseudokarsts). Classically, the process begins with laminar transport of fine particles through sand or gravel, or through tiny cracks in coherent rocks. A limited amount of solution of matrix or of particles may occur (Striebel & Schäferjohann, 1997). Once a continuous channel is established, transport becomes turbulent and scour and other erosive mechanisms enlarge it. As the pipe grows larger, the volume of flow increases. Slumping and local roof collapse may permit entry of additional volumes of runoff, enlarging it further, or may dam it with surface debris, leading to development of a tortuous conduit pattern. The resulting pipes may propagate vertically or at grade, upslope or downslope. Some develop braided or dendritic networks. Perhaps in many cases, the incipient pipes develop at the bottom of shrinkage or other crevices rather than de novo between sand- or gravel-sized particles. This occurs especially in clayey shales and other rocks whose physical properties change with their state of hydration. Alternating swelling and shrinkage repeatedly form large and small cracks.

Some clays are especially slippery when wet, with formation of dislocation blocks, large and small block slippages and rotation, extensive spalling and even underground slides, all of which contribute materially to erosion and pipe enlargement. Numerous authors cited by Dunne report that a high exchangeable sodium content deflocculates such clays, concentrating flow of water in crevices and thus providing additional lubrication. Piping is not restricted to such rocks, however. It also occurs in quartzose rocks. Here, corrosion and limited dissolution may modify crevice features into multiprocess caverns with features typical of karst (see entries on Pseudokarst and Crevice Caves). Further, the superficial parts of some pyroclastic flows and ash fall deposits lose cohesiveness when wet, and fracture in large blocks (some of which form the roof of fractures). Further, they also tend to become slurries of wet sand which flow through newly-formed streamway pipes one to a few metres below the surface at the bottom of roofed fractures (Halliday, 1986).

Badlands Pseudokarst

Some badlands topography is riddled with pipes, piping caves, funnel-shaped dolines, dry valleys, and other features of centripetal subsurface drainage, commonly observed in karstic terrain. Locally, these form specific landscapes. Because their bedrock tends not to be notably coherent, most individual forms are short-lived, but the general landscape evidently persists throughout long periods of scarp retreat. Some features in Triassic shales or claystones in Arizona's Petrified Forest National Park (United States) seem to be especially stable, but elsewhere, each storm may replace old features with new examples.

Piping Caves

Because they are short-lived, only a few classical pipes in poorly consolidated material persist long enough to reach the size of caves as defined in this volume. Perhaps simplest are the much-visited caves in claystones and mudstones in Southern California's arid Arroyo Tapiado area. Rainfall here is rare and scant, but is sufficient to cause numerous piping caves in narrow gorges in Pliocene lake deposits through the mechanisms of block slumping, disaggregation, and outwash. Except where they are partially or completely filled with disaggregated slump, the gorges are deep, narrow meandering stream slots with vertical walls. Most of the caves are unitary. Blind valleys leading to pipe swallets are common. Crumbly slumped blocks form the roofs of caves as much as 110 m long. Multiple cave levels and collapse entrances to caves are not uncommon. Cave rooms are up to 25 m high and 10 m wide. In normal weather, the caves are dry and dusty and "dry waterfalls" up to 15 m high have been reported. Apparently similar caves are present in Saskatchewan's Big Muddy Valley (Canada).

Disruption of structural stability during avalanching evidently facilitates piping speleogenesis in claystones and mudstones. Officers Cave in Eastern Oregon (United States) is especially well known (Parker, 1964). The 345 m cave is in a rotated landslide mass of Oligocene / Miocene John Day siltstone which contains montmorillonite and ash, and is slippery when wet. Desiccation cracks are numerous, with innumerable small fracture sets, slickensides on the sides of small blocks, extensive spalling, and block fracturing. A dendritic drainage conducts disaggregated sand-sized and smaller particles to an intermittent streamway into which the cave sags with headward erosion of its scarp. Small vertical pipes perforate nearby hillsides.

In semi-arid western Colorado (United States), more than 100 claystone and mudstone caves have been found, especially near the town of Grand Junction. Most of these also are in landslide and slump masses. The largest is the Anvil Points Claystone Cave, where 620 m of passages have been mapped (Davis, 2001). Here, a chaotic mixture of clay, silt, sand, and angular blocks of sandstone sags intermittently into another dendritic stream network. The upper part of the rectilinear Catarina-Confusion Cave System in Texas' Palo Duro Canyon (United States) also is in a landslide mass. Large pseudokarstic sinks are reported elsewhere in this canyon, and approximately 50 smaller piping caves have been noted. In Cameron County, Texas (United States), Model T Cave was formed by runoff of irrigation water which cut a short subterranean passage through an old trashpile and the surrounding soil. Its small central room was roofed by the hood of a Model T Ford; the walls and floor are soil. Archaeological discoveries included a steering wheel and a sparkplug.

Loess Caves

Loess pseudokarst is a special case. The study of loess is an active subscience, with international symposia and an extensive literature entirely separate from speleology. Numerous pseudokarstic features form in loess and loess-like silt, especially in vast areas of China. Although Jakucs (1977) attributed much of such piping to decalcification accompanied by pedological alteration, caves as defined in this volume are perhaps the least of these features. Cavernous features have been described, but the vast majority in China are artificial habitations and shelters. Some of these are artificially enlarged pipes and a few such pipes evidently are large enough to qualify as caves.

Piping in Humid Lands

Jones (1990) has provided a limited global overview of piping in humid lands. Such piping tends to occur in or at the top of thick, saturated soils rather than in bedrock or alluvium as in drylands. Such pipes are generated by high through-flow discharges (Jones, 1990) but rarely remain open long enough to form caves.

In the United States, wetlands piping occurs in badlands topography in humid eastern Oklahoma and a variety of other settings. The 804 m long Christmas Canyon Cave (Washington state), is a low (<0.5 m), braided channel in volcanic ash atop a mudflow deposit beneath a recent lava flow in Mount St Helens National Volcanic Monument (Figure 1). In the 1980 blast zone on the other side of Mount St Helens, extensive but very short-lived pseudokarsts developed in thick ash-cloud pyroclastic accumulations. Much of their volume was soon piped through crevices, some of which were in parallel groups (Figure 2).

In quartzites in tropical rainforests, groundwater flow through pipes has removed enormous volumes of clastic particles, forming conduits indistinguishable from karstic dissolution conduits. This is one extreme of the limestone–quartzite dissolutional spectrum. In New Zealand, the 8 km Bohemia Cave was largely formed by groundwater erosion in phyllites underlying marble.

Peatlands, and particularly upland blanket bogs, are very susceptible to piping erosion and pseudokarst landscapes with stream-sinks, closed depressions and risings are common. Small pipes (<10 cm diameter) occur at different depths through the profile but larger pipes may form by upwards erosion at the interface with the underlying bedrock. These pipes often reach accessible dimensions but are only rarely long enough to qualify as true caves. Pipes are common in all areas of upland blanket bog in Britain and Ireland, and caves have been explored in the Derbyshire Peak District (where they are known as slutch caves and are up to 50 m long), on the slopes of Kippure outside Dublin, and on Cuilcagh Mountain in County Fermanagh, Northern Ireland, where Poll na Mona has a length of 150 m. "Soil caves" in tropical Ecuador (Funkhowser, 1951) may have had a similar origin in leached lateritic soil.

Interfaces and Multiprocess Caves

Concomitant rockfall and piping form caves up to 300 m long in clean, friable, and poorly cemented sandstones in Arkansas (United States). In Minnesota (United States), low, wide maze caves and tubular caves up to 350 m long exist in the similar St

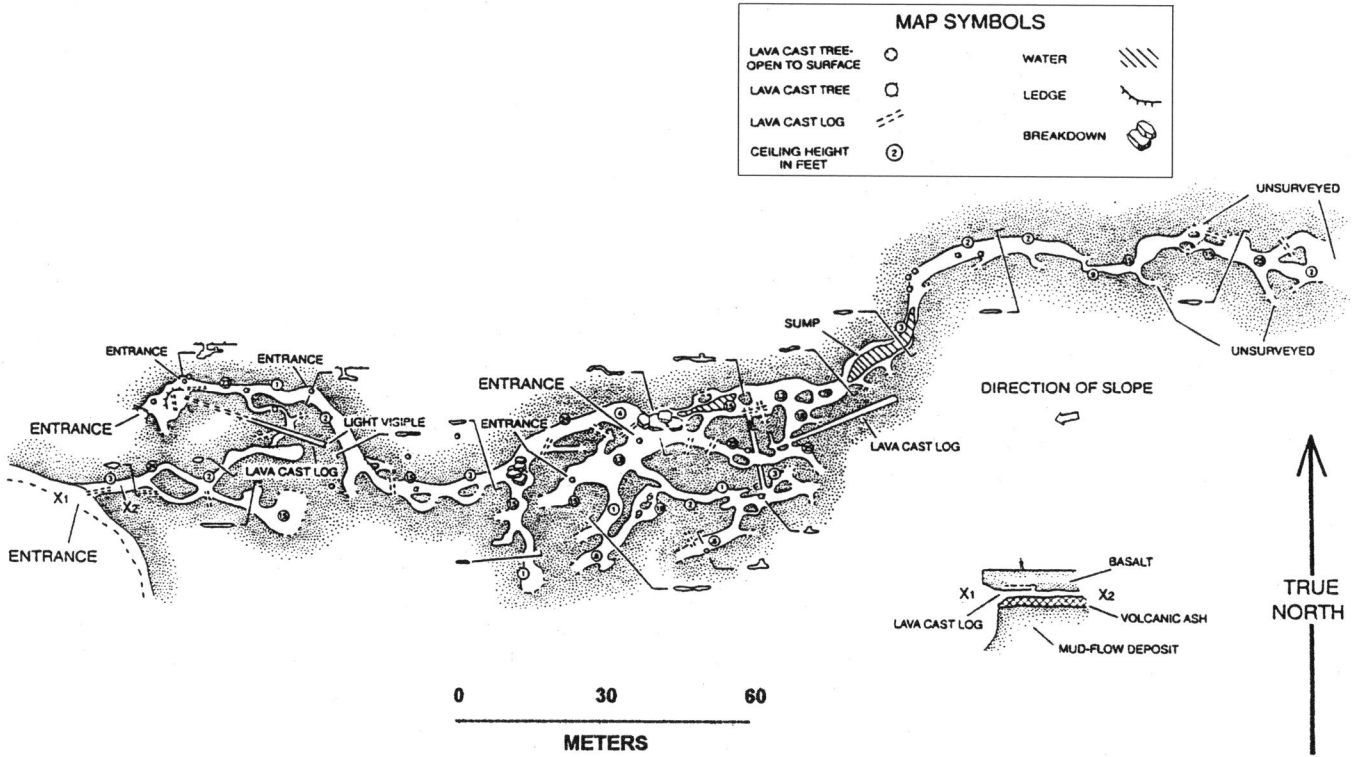

Piping Caves and Badlands Pseudokarst: Figure 1. Map of Christmas Canyon Cave, a long piping cave in volcanic ash beneath a basalt flow, Washington state (United States). Courtesy Larry King.

Peter sandstone. One maze cave under downtown Minneapolis underlies an entire city block.

In central Kenya, Gigglers Caves appear to fall within an interface between crevice caves, dissolutional caves, and piping, much like multiprocess caves in quartzite, discussed in the entries on Pseudokarst and Crevice Caves. They are developed along a series of small crevices in a hard granular tuff. Their passages appear primarily phreatic, with rounded conduits and small rounded chambers which fill in flood season. Bedding plane development is prominent, and only a few linear crevice passages are present. In northwestern Kenya, voluminous Kitum Cave, Makningen Cave, and some others on Mt Elgon appear to be large multiprocess caves formed primarily by removal of enormous quantities of particulates by disaggregation and periodic stream transport. Other processes include considerable dissolution of complex bedrock. Makningen Cave is primarily a dead-end borehole passage up to 70 m wide and 10 to 20 m high, but it also has a high, rounded chimney more than 5 m in diameter. These caves were formed in little-studied lakebed deposits quite unlike those in the Arroyo Tapiado area. The Mt Elgon lakebed complex contains agglomerate, tuff, organic components, and evaporites. Soluble salts provide underground salt licks for both wild and domestic animals.

Extraterrestrial Piping and Piping Caves

Malin and Edgett (2000) have observed several types of Martian terrain which may be analogous to various badlands and piping pseudokarst features (see also Extraterrestrial Caves). Especially intriguing evidence of conduits exists in two or three specific terrains: alcoves associated with resistant layers in cliffs, landslide or slumped material, and "perhaps colluvium". On Earth, piping occurs in all three such terrains, and piping caves in at least two of them. Characteristically, Martian channel heads and interiors appear not to be clogged with clastic debris, implying efficient transport through conduits. Present-day conditions on Mars are hyper-arid beyond any conditions on earth, but some hyper-

Piping Caves and Badlands Pseudokarst: Figure 2. 18 May 1980 pyroclastic deposits in the Spirit Lake pseudokarst, Mt St Helens, Washington state (United States). Pits are along crevices that are intermittently roofed by slumped blocks, with piping at the bottom of the crevices. (Photo by William Halliday)

Piping Caves and Badlands Pseudokarst: Figure 3. Makningen Cave, Kenya, a voluminous piping cave in partially soluble lakebed deposits. Note the large chimney above the caver. (Photo by William Halliday)

arid terrestrial regions may have piping caves that may be useful as models for Mars. Limited photodocumentation of the Oumou Caves in the Republic of Chad (Dossal, n.d., c.1995) suggests such a possibility. The Cave of Swimmers and others in sandstone on Gilf Kebir in Egypt also may be analogs but are even less known geologically.

WILLIAM R. HALLIDAY

Works Cited

Davis, D. 2001. Anvil Points Claystone Cave Complex: An exceptionally large "mud cave" system. *NSS News*, 59(11): 331–34

Dossal, D. (photographer) n.d. (c.1995) Untitled photograph. In *Expedition 1994 "Tibesti—Terre Interdite"*, edited by P. Courbon et al., Carcassonne: Imp. Gabrielle, p.36

Dunne, T. 1990. Hydrology, mechanics, and geomorphological implications of erosion by subsurface flow. In *Groundwater Geomorphology: The Role of Subsurface Water in Earth-Surface Processes and Landforms*, edited by C.G. Higgins & D.R. Coates

Funkhowser, J. 1951. Soil caves in tropical Ecuador. *NSS News*, 9: 4

Halliday, W. 1986. Caves and other pseudokarstic features of Mount St. Helens: 1980–1985 observations. In *Mount St Helens: Five Years Later*, edited by S.A.C. Keller, Cheney, Washington: Eastern Washington University Press

Higgins, C.G. & Coates, D.R. (editors) 1990. *Groundwater Geomorphology: The Role of Subsurface Water in Earth-Surface Processes and Landforms*, Boulder, Colorado: Geological Society of America (GSA Special Paper 252)

Jakucs, L. 1977. *Morphogenetics of Karst Regions*, New York: Wiley

Jones, J. 1990. Piping effects in humid lands. In *Groundwater Geomorphology: The Role of Subsurface Water in Earth-Surface Processes and Landforms*, edited by C.G. Higgins & D.R. Coates

Malin, M. & Edgett, K. 2000. Evidence for recent groundwater seepage and surface runoff on Mars. *Science*, 288: 2330–35

Mears, B. 1968. Piping. In *Encyclopedia of Geomorphology*, edited by R.W. Fairbridge, New York: Reinhold

Parker, G. 1964. Officers Cave, a pseudokarstic feature in altered tuff and volcanic ash of the John Day Formation in Eastern Oregon. *Bulletin of the Geological Society of America*, 75: 393–407

Parker, G. et al. 1990. Piping and pseudokarst in drylands. In *Groundwater Geomorphology: The Role of Subsurface Water in Earth-Surface Processes and Landforms*, edited by C.G. Higgins & D.R. Coates, Boulder, Colorado: Geological Society of America

Striebel, T. & Schäferjohann, V. 1997. Karstification of sandstone in central Europe: Attempts to validate chemical solution by analysis of water and precipitates. In *Proceedings of the 12th International Congress of Speleology*, vol. 1, edited by J. Bauchle et al., Basel, Switzerland: Speleo Project

Further Reading

Péwé, T., Liu Tungsheng, Slatt, R.M. & Li Bingyuan. 1995. *Origin and Character of Loesslike Silt in the Southern Qinghai-Xizang (Tibet) Plateau, China*, Washington DC: US Government Printing Office (US Geological Survey Professional Paper 1549)

Rideau, D. & Hansen, A. (editors) 1993. *Disorder and Granular Media*, New York: North Holland

PISCES (FISHES)

A Chinese account from 1541 was probably the first record of a "blind, white fish", the description given of the depigmented, often eyeless, fishes which are found in caves and other subterranean habitats. Although this stygobitic species, and others, were already known, the first formally described species was *Amblyopsis spelaea* from Mammoth Cave, Kentucky (United States) in 1842. By 1969, when the seminal work by Georges Thinès (*L'Évolution régressive des poissons cavernicoles et abyssaux*) was published, the total number of stygobitic fish described was about 50. Since this date, due to a large increase in cave exploration, particularly in China, Southeast Asia, and Mexico, the total has nearly tripled to 92 described species. Five countries alone (China, Mexico, Brazil, United States, and Thailand) contain almost 50% of the total, although subterranean fishes have been recorded in 30 countries. The trend of description fits an exponential curve and it is possible to predict that there could be 250 known species by 2050. Stygobitic fishes seem to be restricted to tropical and subtropical regions with limits of 40°N and 25°S. Many species of fishes are known from caves outside this range but none are obligate. Two hypotheses may explain this: (1) glaciation in the northern hemisphere in Pleistocene times rendered extinct any pre-existing stygobitic fishes and there has been insufficient time for more to evolve; (2) the lower temperature in temperate regions leads to a reduced food supply so that stygobitic fish cannot evolve.

Nearly three-quarters of all hypogean fishes belong to the Superorder Ostariophysi, but as this group comprises over 60% of all freshwater fishes this is not surprising. The principal families containing hypogean fish are the Cyprinidae and Balitoridae (with 37 species) and various siluriform (catfish) families (with 24 species). Eighteen families have stygobitic representatives and only a few species are of marine origin. The genera *Lucifuga*, with five species (all restricted to hypogean waters in the Caribbean), *Ogilbia* (two species, both Bythitidae), and *Milyeringa* (Gobiidae) are the main marine invaders. In 17 cases two or three species exist in the same hydrological system. Stygobitic fishes account for only 0.3% of all Teleosts (the large group of fishes with bony skeletons).

Most species live in vadose streams (e.g. *Ancistrus cryptophthalmus* in Brazil) but some are known only from phreatic regions. Perhaps the most extreme hypogean environment is that within the Edwards Aquifer, San Antonio, Texas, United States (see Edwards Aquifer: Biospeleology). This aquifer exists to a depth of at least 600 m and has a very diverse fauna including two fishes, *Satan eurystomus* and *Trogloglanis pattersoni* (both Ictaluridae). These fishes show adaptations to the extreme hydrostatic pressure (Langecker & Longley, 1993). Other fishes live in shallow phreatic aquifers and the aberrant, and probably ancient, *Phreatobius cisternarum* (Pimelodidae) appears to inhabit water sheets underlying the mouth of the Amazon River.

Hypogean fishes are very restricted in distribution and 50 are known from only one site each. Even where species are known from more than one site, they may exist as sub-populations between which there is little, if any, demographic or genetic exchange. Only two species are widely distributed, *Typhlichthys subterraneus* in the United States and *Lucifuga spelaeotes* in the Bahamas, and both may consist of several, genetically distinct, sibling species.

Very few of the known species have been properly studied. The best known is the cave form of *Astyanax fasciatus* from Mexico. Since its discovery in 1936, and the important observation that the troglomorphic cave form and the normal epigean form can interbreed, many observations and experiments have been undertaken (Wilkens, 1988). It is likely that the troglomorphic populations are actually a separate species, and they were once considered as a separate genus (*Anoptichthys*). The four stygobitic species from the family Amblyopsidae are also relatively well known (see Pisces: Amblyopsidae). Other species which have been relatively well studied are *Pimelodella kronei* (Pimelodidae) (e.g. Trajano, 1991), *Caecobarbus geertsii* (Cyprinidae) (e.g. Heuts & Leleup, 1954), the cave form of *Poecilia mexicana* (Poeciliidae) (e.g. Parzefall, 2001) and *Milyeringa veritas* (Gobiidae) (e.g. Humphreys, 2001).

Typical troglomorphic features include absence of eyes and melanin pigment; elaboration of extra-ocular sensory structures such as lateral line organs and canal neuromasts; behavioural changes (slowing of swimming speed, loss of aggressive behaviour); and extreme K-selection with lower growth rate, increased longevity, delayed reproduction, and fewer and bigger eggs with more yolk. The lost characters are termed "regressive" characters, the others are termed "constructive" since they allow the fishes to exist in the usually exacting cave environment (see Evolution of Hypogean Fauna). The most advanced regressive and constructive features are found in phylogenetically old fishes, those evolving longest in the subterranean environment (e.g. Wilkens, 1988).

Many other species of fishes have been recorded from hypogean habitats (see e.g. Poly & Boucher, 1996). Most of these are accidentally present and will not be able to support viable populations within the cave / subterranean environment. While in the cave they may function as predators or prey, or when dead, as a substrate for decomposition thereby providing potential food for other cave animals. Some species can feed and reproduce successfully underground (stygophilic species). Very few of these have been studied but they could provide information on the mechanisms of cave colonization and the acquisition of troglomorphy. One species from China (*Varicorhinus macrolepis*) appears to be a habitual trogloxene. It migrates into the warmer cave water in winter, existing on stored fat, but leaves the caves in summer to feed and breed (Zhang, 1986). There are also a number of species in the Dinaric karst which behave as habitual trogloxenes.

Most subterranean fishes are opportunistic feeders, taking whatever food they can obtain. Some are generalist predators but only a few are thought to be the top predators in a system (possibly *Satan eurystomus* and *Ogilbia pearsei*). Given the sporadic nature of food supplies within caves these animals are adapted to withstand long periods without food and build up large, to massive, deposits of fat and adipose tissues within the body.

Subterranean fishes are among the world's most endangered animals. Since around 50 taxa are known from one location only, they are susceptible to extinction following anthropogenic disturbances, e.g. a toxic spill. The international "Red List" contains three subterranean fishes on the critically endangered category and two Chinese species, *Sinocyclocheilus hyalinus* (the fish recorded in 1541 above) and *Triplophysa gejiuensis* have been described as "nearly extinct" by Chinese workers. Many other species are probably vulnerable and require special consideration.

GRAHAM S. PROUDLOVE

Works Cited

Heuts, M.J. & Leleup, N. 1954. La géographie et l'ecologie des grottes du Bas- Congo: les habitats de *Caecobarbus geertsii* Blgr. *Annales Musée Royale du Congo Belge*, 35: 1–71

Humphreys, W.F. 2001. *Milyeringa veritas* (Eleotridae), a remarkably versatile cave fish from the arid tropics of northwestern Australia. *Environmental Biology of Fishes*, 62: 297–313

Langecker, T & Longley, G. 1993. Morphological adaptations of the Texas blind catfishes *Trogloglanis pattersoni* and *Satan eurystomus* (Siluriformes, Ictaluridae) to their underground environment. *Copeia*, 1993: 976–86

Parzefall, J. 2001. A review of morphological and behavioural changes in the cave molly, *Poecilia mexicana*, from Tabasco, Mexico. *Environmental Biology of Fishes*, 62: 263–75

Poly,W.J. & Boucher, C.E. 1996. Nontroglobitic fishes in caves: Their abnormalities, ecological classification and importance. *American Midland Naturalist*, 136: 187–98

Trajano, E. 1991. Population ecology of *Pimelodella kronei*, troglobitic catfish from southeastern Brazil (Siluriformes, Pimelodidae). *Environmental Biology of Fishes*, 30: 407–21

Wilkens, H. 1988. Evolution and genetics of epigean and cave *Astyanax fasciatus* (Characidae, Pisces). Support for the neutral mutation theory. *Evolutionary Biology*, 23: 271–367

Zhang, C. 1986. On the ecological adaptation and geographical distribution of the barbine fish *Varicorhinus macrolepis* (Bleeker). *Acta Zoologica Sinica*, 32: 266–72

Further Reading

Environmental Biology of Fishes 62(1–3), 2001 is a special issue on hypogean fishes.
Many papers providing a modern perspective on cave fish biology.

Parzefall, J. 1998. Behavioural adaptations of cave fishes. In *Encyclopaedia Biospeologica*, vol. 2, edited by C. Juberthie & V. Decu, Moulis and Bucharest: Société Internationale de Biospéologie
  Describes the various behaviour changes that subterranean fishes have undergone during their adaptation.
Proudlove, G.S. 2001. The conservation status of hypogean fishes. *Environmental Biology of Fishes*, 62: 201–13
  Describes why cave fishes are so vulnerable and provides an IUCN threat category for 85 species.
Thinés, G. 1969. *L'évolution rgressive des poissons cavernicoles et abyssaux* [The regressive evolution of cave and deep sea fishes], Paris: Masson
  The best review of the biology of cave fishes.
Thinés, G. & Proudlove, G.S. 1986. Pisces. In *Stygofauna Mundi*, edited by L. Botosaneanu, Leiden: Brill
  Lists all troglomorphic species, including those from non-cave habitats (e.g. river rapids).
Weber, A., Proudlove, G.S. & Nalbant, T.T. 1998. Morphology, systematic diversity, distribution, and ecology of stygobitic fishes. In *Encyclopaedia Biospeologica*, vol. 2, edited by C. Juberthie & V. Decu, Moulis and Bucharest: Société Internationale de Biospéologie
  Contains a list of subterranean fishes and some details of their biology.

PISCES (FISHES): AMBLYOPSIDAE

This freshwater family, in the Order Percopsiformes, is distributed in the southern and eastern (unglaciated) United States. It is characterized by the presence of a flattened head, a strongly protruding lower jaw, a jugular vent (anus), and small embedded cycloid scales, except on the head, which is naked. Individuals have rows of sensory papillae on the head, body, and tail. Their eyes range from small (microphthalmic) in the epigean and stygophilic species, to vestigial (remnant eye tissue under the skin) in the stygobitic ones. Stygobitic species are also characterized by: (1) depigmentation (they have a pinkish colour due to the blood vessels showing through the translucent skin, with only a few, mostly nonfunctional, melanophores); (2) low metabolism; (3) low fecundity; and (4) increased swimming efficiency, tactile receptivity, and longevity. The systematics of this family need revision since genetic studies have shown that they are much more complex that previously believed (Bergström et al., 1995). The six species of this family demonstrate the transition from epigean (surface) to hypogean waters: *Chologaster cornuta* is epigean, *Forbesichthys agassizi* is a stygophile or facultative cavernicole, and *Typhlichthys subterraneus*, *Amblyopsis spelaea*, *Amblyopsis rosae*, and *Speoplatyrhinus poulsoni* are all stygobites, with increasing troglomorphy from *T. subterraneus* through to *S. poulsoni* in the sequence above (see Figure). Comparative characteristics are summarized in the Table. The family is most closely related to another Percopsiform, *Aphredoderus*, but it may merit its own order, the Amblyopsiformes.

Swamp fish, *Chologaster cornuta*. This species is characterized by being dorsally brown and ventrally creamy white, with three dark stripes on each side. It is found in swamps, ponds, ditches, and slow streams in the Atlantic Coastal Plain from southeast Virginia to central Georgia. It feeds mostly at night, on small crustaceans and aquatic insects. It spawns in March and April and may live up to two years. Although locally common, individuals are hard to spot because they are largely nocturnal and found in heavily vegetated waters.

Spring cavefish, *Forbesichthys agassizi*. This species is characterized by being dorsally dark brown to nearly black, grading to lighter brown laterally. It is ventrally cream-yellow, often with a thin yellow stripe along each side. It is found in central and western Kentucky (west to the Tennessee River) to southern central Tennessee and west across southern Illinois to southeastern Missouri. The Missouri population may have been isolated from the others for 2000 years when the Mississippi River changed its course. Individuals are active in springs at night (feeding on crustacea, insect larvae, and oligochaetes) and they usually retreat underground during the day. The few individuals which venture into the spring portions of their habitat hide under rocks or debris. They prefer highly oxygenated water, and show scotophilia (i.e. they respond to light by moving away from it), thigmotaxis (orienting themselves using clues in the substrate), and a wide range of temperature tolerance. Little is known of their breeding habits, but spawning probably takes place underground in the winter.

Southern cavefish, *Typhlichthys subterraneus*. This species is probably a mosaic of unrelated (paraphyletic) populations (Bergström et al., 1995). It is found in the subterranean waters of two major disjunct ranges separated by the Mississippi River: the Ozark Plateau of central and southeastern Missouri and northeastern Arkansas, and the Cumberland and Interior Low plateaux of northwest Alabama, northwest Georgia, central Tennessee and Kentucky, and southern Indiana. It inhabits deep pools and streams, where individuals feed mostly on copepods. Breeding probably occurs in late spring in association with rising water levels and individuals are long-lived, slow-growing, and

Pisces (Fishes): Amblyopsidae: Summary information for amblyopsid species.

Species	Maximum size (standard length, SL, mm)	Eyes	Pigmentation	Number of rays in fins			Pelvic fins	Number of rows of papillae in the caudal fin
				Dorsal	Anal	Caudal		
C. cornuta	68	Microphthalmic	Yes	9–12	9–10	9–11	Absent	0–2 (branched)
F. agassizii	75	Microphthalmic	Yes	9–11	9–11	11–16	Absent	0–2 (branched)
T. subterraneus	75	Vestigial	No	7–10	7–10	10–15	Absent	0–2 (branched)
A. spelaea	110	Vestigial	No	9–11	8–11	11–13	Absent/reduced	4–6 (branched)
A. rosae	65	Vestigial	No	7–9	8	9–11	Absent	4–6 (branched)
S. poulsoni	72	No vestiges?	No	9–10	8–9	21–22	Absent	4 (unbranched)

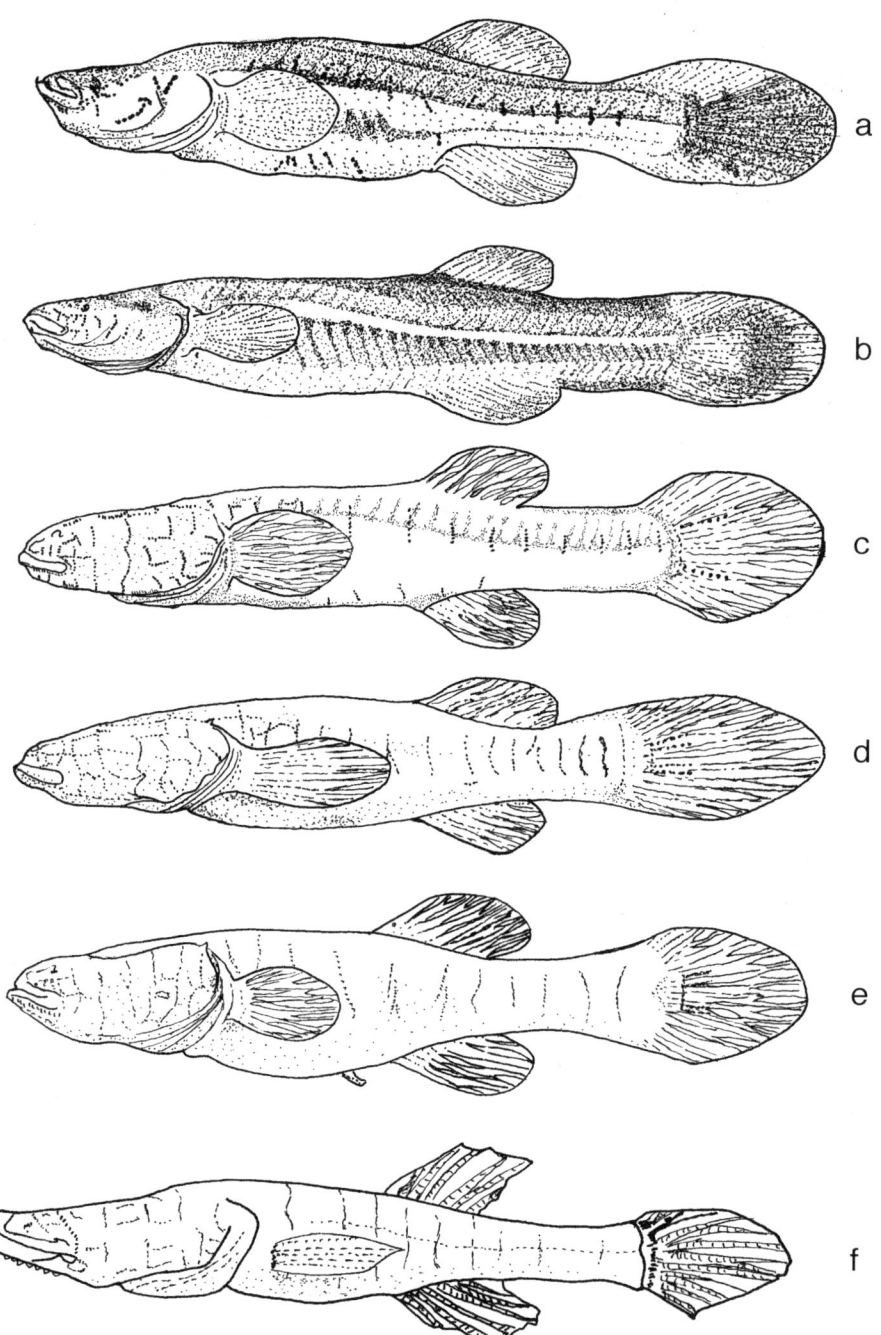

Pisces (Fishes): Amblyopsidae: Drawings of (a) *Chologaster cornuta*, (b) *Forbesichthys agassizi*, (c) *Typhlichthys subterraneus*, (d) *Amblyopsis rosae*, (e) *Amblyopsis spelaea*, and (f) *Speoplatyrhinus poulsoni*. Drawing by John Ellis.

do not respond to light. It is classified as "Vulnerable" in the Red List of the International Union for the Conservation of Nature and Natural Resources (IUCN) (Romero, 1998b).

Ozark cavefish, *Amblyopsis rosae*. This species is made up of at least four genetically distinct populations (Bergström *et al.*, 1995), and is found in 41 sites occurring on the Springfield Plateau, in seven counties of three states: southwest Missouri (20 sites), northwest Arkansas (10 sites), and northeast Oklahoma (11 sites). The verified historic range was larger (Willis & Brown, 1985). Individuals are found mostly in small cave streams with chert or rubble sediments, in pools over silt and sand sediments, and in karst windows or wells. Most of their diet is copepods but they also eat small salamanders, crayfish, isopods, amphipods, and young of their own species. Their breeding habits are not well understood, but they have an extended spawning season with a peak in late summer. They are classified as "Vulnerable" by IUCN and "Threatened" by the US Fish & Wildlife Service (USFWS) (Romero, 1998a).

Northern cavefish, *Amblyopsis spelaea*. This species is found in about 45 caves in Kentucky and about 17 caves in southern Indiana (Keith, 1988). The distribution may be limited by competition from the Southern Cavefish. Typical habitats are caves and subterranean passages in well-developed karst terrain where it is a top predator in both habitats (Poulson, 1963). Breeding occurs during high water from February to April. Females carry eggs in their gill cavities until hatching and then carry young until they lose their yolk sacs—a total period of four to five months. Young appear in late summer and early fall. They are scotophilic. This was the first stygobitic species of fish described in the scientific literature. It is classified as "Vulnerable" by the IUCN (Romero & Bennis, 1998).

Alabama cavefish, *Speoplatyrhinus poulsoni*. This species is characterized by its extremely elongated and anteriorly depressed head, which comprises one-third of the standard length in adults, with a laterally compressed snout and terminal mouth. It is found only in Key Cave, Lauderdale County, Alabama, on the north bank of the Tennessee River. Its habitat is not well understood but probably consists of phreatic groundwater. Its total population size is estimated to be less than 100 individuals, which would make it one of the most endangered fish species in the world. It is classified as "Critically Endangered" by the IUCN and "Endangered" by the USFWS (Romero, 1998c).

The classic studies of Poulson and co-workers have elucidated many aspects of the biology of this unique family, but much remains to be learned. However, all of the stygobitic species are threatened to some degree and it is vitally important that the need for conservation is taken into account in future studies.

ALDEMARO ROMERO

See also **Adaptation: Morphological**

Works Cited

Bergström, D.E., Noltie, D.B. & Holtsford, T.P. 1995. Ozark Cavefish genetics: the phylogeny of Missouri's Ozark Cavefish (*Amblyopsis rosae*) and Southern Cavefish (*Typhlichthys subterraneus*), Missouri: Missouri Department of Conservation

Keith, J.H. 1988. Distribution of Northern Cavefish, *Amblyopsis spelaea* Dekay, in Indiana and Kentucky and recommendations for its protection. *Natural Areas Journal*, 8: 69–79

Poulson, T.L. 1963. Cave adaptation in Amblyopsid fishes. *American Midland Naturalist*, 70: 257–90

Romero, A. 1998a. Threatened fishes of the world: *Amblyopsis rosae* (Eigenmann, 1842) (Amblyopsidae). *Environmental Biology of Fishes*, 52: 434

Romero, A. 1998b. Threatened fishes of the world: *Typhlichthys subterraneus* (Girard, 1860) (Amblyopsidae). *Environmental Biology of Fishes*, 53: 74

Romero, A. 1998c. Threatened fishes of the world: *Speoplatyrhinus poulsoni* Cooper and Kuehne, 1974 (Amblyopsidae). *Environmental Biology of Fishes*, 53: 293–94

Romero, A. & Bennis, L. 1998. Threatened fishes of the world: *Amblyopsis spelaea* De Kay, 1842 (Amblyopsidae). *Environmental Biology of Fishes*, 51: 420

Willis, L.D. & Brown, A.V. 1985. Distribution and habitat requirements of the Ozark cavefish, *Amblyopsis rosae*. *American Midland Naturalist*, 114: 311–17

Further Reading

Environmental Biology of Fishes, 62 (2001) is a special issue on hypogean fishes and contains a number of papers on the Amblyopsidae. Many of the recent studies described in the essay text are included in this volume.

Bechler, D.L. 1983. The evolution of agonistic behavior in Amblyopsid fishes. *Behavioral Ecology and Sociobiology*, 12: 35–42

Eigenmann, C.H. 1909. *Cave Vertebrates of America: A Study in Degenerative Evolution*, Washington DC: Carnegie Institute of Washington.

This is a major work, which still gives the best description of the eyes of this family; it is beautifully illustrated.

Poulson, T.L. 1961. *Cave Adaptation in Amblyopsid fishes*, Ph.D. Dissertation. Ann Arbor: University of Michigan

Rosen, D.E. 1962. Comments on the relationships of the North American cave fishes of the family Amblyopsidae. *American Museum Novitates*, 2109: 1–35

Smith, P.W. & Welch, N.M. 1978. A summary of the life history and distribution of the Spring Cavefish, *Chologaster agassizi*, Putnam, with population estimates for the species in southern Illinois. *Illinois Natural History Survey, Biological Notes*, 104

US Fish and Wildlife Service 1989. *Ozark Cavefish Recovery Plan*, Atlanta: US Fish and Wildlife Service

US Fish and Wildlife Service 1990. *Alabama Cavefish, Speoplatyrhinus poulsoni Cooper and Kuehne 1974, Recovery Plan*, Jackson, Mississippi: US Fish and Wildlife Service

Woods, L.P. & Inger, R.F. 1957. The cave, spring, and swamp fishes of the family Amblyopsidae of central and eastern United States. *American Midland Naturalist*, 58: 232–56

PLITVICE LAKES, CROATIA

The Plitvice Lakes are located in the Dinaric karst region of Croatia, between the mountain massifs of Mala Kapela in the southwest and Plješivica in the east. In the late 19th century they were described as the "only cataract lakes in Europe". The natural dams that separate the lakes form waterfalls, for which the lakes are justly famous. The dams themselves are constructed of travertine or calcareous tufa (porous hard limestone), precipitated from the lake water. This process is still going on, with new travertine barriers, curtains, stalactites, stalagmites, channels, and cascades being formed and existing ones altered. The water continues to breach the travertine barriers at different locations, creating new lakes and barriers (Srdoč et al., 1985; Plenković, Marčenko & Srdoč, 1989).

In the hard waters of the Plitvice Lakes, in areas of strong aeration and under favourable temperature conditions, calcium carbonate ($CaCO_3$) precipitates in considerable quantities. Some hydrobionts (bacteria, algae, and mosses) retain the particles of calcite, and in this way gradually form tufa barriers (Matoničkin & Pavletić, 1967). The precipitation of calcium carbonate from lake water requires the interaction of abiotic and biotic factors, along with the interglacial climatic conditions that prevail at the present day. The basic abiotic factors are: pH between 8.2 and 8.4, $CaCO_3$ saturation index over 3.0, and dissolved organic matter concentration less than 10 mg l^{-1}. Biotic factors influencing precipitation include the presence of bacteria and algae on the aquatic plants growing in the waterfalls. Phyto-

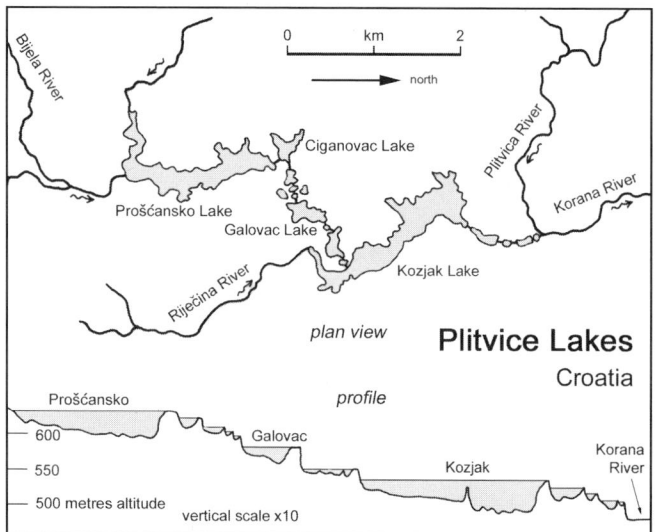

Plitvice Lakes, Croatia: Plan and profile of the tufa-dammed lakes of Plitvice in Croatia. Note that the vertical scale is greatly exaggerated.

plankton and periphyton are also important in the formation of lacustrine sediment. Furthermore, as tufa is deposited, it coats the bottoms and sides of the watercourses, petrifying everything in the water—fallen trees, stones, and even entire beaches (Stilinović & Božičević, 1998).

The natural beauty and ecological importance of the Plitvice Lakes led to their designation as the first National Park in Croatia, in 1949. In 1979 they were added to the UNESCO list of World Heritage Sites (WHS). The National Park is 29 482 ha in area, of which 22 300 ha (75%) is covered by forests. The surface area of the 16 lakes is only 192 ha. The remainder of the park is grassland and other vegetation. There are a total of 16 lakes (Riđanović & Božičević, 1996), which are geographically an integral part of the Korana River (Figure 1). The distance between the first lake (Prošćansko, at an altitude of 637 m) and the last lake (Novakovića Brod, at an altitude of 503 m) is 8300 m. Eleven of the lakes have surface areas of more than 1 ha. The two deepest and largest lakes are Prošćansko (area 682 720 m^2, depth 37 m) and Kozjak (815 060 m^2 in area, 46 m deep). In total, the lakes store about forty million cubic metres of high-quality fresh water. The lakes can be divided into two subsystems: the Upper Lakes (12 lakes) which lie on dolomite bedrock, and the Lower Lakes (4 lakes) which lie on limestone bedrock. At the end of the last lake, the Novakovića Brod, the Plitvica River flows out through a magnificent waterfall with a height of 28 m.

The Upper Triassic strata in this area comprise 400 to 600 m of predominantly dolomite rocks underlain by shales whose presence was essential in the formation of the lakes. The Jurassic strata comprise alternating well-bedded limestones of varying thickness, containing irregular dolomite intercalations.

The Cainozoic biogenic calcareous tufa deposits have been badly damaged by human activity. In the Lower Lakes, unconsolidated muddy sediments have covered the lake bottoms, making the bedrock impermeable. The estimated thickness of calcareous tufa varies from 30 to more than 50 m, covering a zone several hundred metres in width in the Upper Lakes section (Božičević, 1990).

The exact hydrological catchment area and boundaries of the Plitvice Lakes are not known, although numerous investigations have been carried out. Determination of the exact catchment area is problematical, due to strong, direct, complex, and unpredictable interactions between groundwater and surface water flowing from the different karst subregions. The average annual temperature in the lake area is 8–9°C and the annual rainfall varies from 1100 to 1700 mm, with an average of about 1400 mm. The climate is continental.

The annual mean inflow into the first Prošćansko Lake varies from 1.8 to 2.9 $m^3 s^{-1}$. Outflow from the last Novakovića Brod Lake varies from 2.4 to 2.8 $m^3 s^{-1}$. Minimum inflow is 0.579 $m^3 s^{-1}$ and minimum outflow is 0.605 $m^3 s^{-1}$, while maximum inflow is 12.5 $m^3 s^{-1}$. Since 1980 there has been a general decrease in discharges.

Scores of algae and non-vertebrate species have been found in the lakes, indicating that the water quality is oligotrophic or mesotrophic. However, some other species indicate a continuous increase in the trophic state values (towards eutrophication) of the water in the lakes. Of the fish species, the autochtonous lake trout (*Salmo trutta lacustri*) and the river trout (*Salmo trutta fario*) are worthy of note. The river crayfish disappeared entirely from the lakes in 1958, but reappeared in the Lower Lakes in October 1995.

Ognjen Bonacci and Tanja Roje-Bonacci

Works Cited

Božičević, S. 1990. Hidrogeološke zanimljivosti Plitvičkih jezera [Hydrogeological characteristics of the Plitvice Lakes], *Ekološki glasnik*, 7–8: 28–33

Matoničkin, I. & Pavletić, Z. 1967. Hidrologija protočnog sistema Plitvičkih jezera i njegove ekološko biocenološke značajke (Hydrology of the Plitvice Lakes flow system and its ecological biocenological characteristics), *Krš Jugoslavije—Carsus Iugoslavie*, 5: 83–126

Plenković, A., Marčenko, E. & Srdoč, D. 1989. Periphyton on glass slides in aquatic ecosystems of the National Park Plitvice Lakes. *Periodica Biologica*, 91(1): 88

Riđanović, J. & Božičević, S. 1996. Geographic–physical conditions of runoff and hydrogeological characteristics of the Plitvice Lakes. *Acta Geographica Croatica*, 31: 7–26

Srdoč, D., Horvatinčić, N., Obelić, B., Krajcar, I. & Sliepčević, A. 1985. Procesi taloženja kalcita u krškim vodama s posebnim osvrtom na Plitvička jezera (Calcite deposition processes in karst waters with special emphasis on the Plitvice Lakes). *Krš Jugoslavije—Carsus Iugoslavie*, 11(4–6): 101–204

Stilinović, B. & Božičević, S. 1998. The Plitvica Lakes—a natural phenomenon in the middle of the Dinaric Karst in Croatia. *European Water Management*, 1(1): 15–25

Further Reading

Chafetz, H.S. & Folk, R.L. 1984. Travertines: depositional morphology and the bacterially constructed constituents. *Journal of Sedimentary Petrology*, 54(1): 289–316

Lorah, M.M. & Herman, J.S. 1988. The chemical evolution of a travertine depositing stream: geochemical processes and mass transfer reactions. *Water Resources Research*, 24(9): 1541–52

Ford, T.D. 1989. Tufa—the whole dam story. *Cave Science*, 16(2): 39–49

Plitvice National Park, http://www.archaeology.net/plitvice/

POLJES

In a strict sense, poljes can be defined as depressions in limestone karst. They are generally elliptical, with bottoms that slope relatively gently from the inflow (spring) zone to the outflow (often swallow hole) zone (Bonacci, 1987b). They are the largest and most conspicuous type of surface depression found in carbonate rock megarelief. Poljes are frequently aligned along tectonic and fold axes, often forming flat alluvial valleys bordered by relatively steep, bare limestone ridges, ranging from almost a kilometre to several kilometres in width (see photo in Dinaride Poljes entry). They are somewhat elongated, often with surface streams flowing along the longer axis. Poljes vary in size from less than 0.5 km^2 to more than 500 km^2 in area.

The formation of depressions in karst terrain results from the successive and simultaneous action of a number of environmental factors, including underlying geology and paleoclimate. According to Trudgill (1985), there are two important factors that affect the formation of poljes: surface lateral planation at the average height of water-table levels in the sediment, and the accumulation of the sediments themselves. In the case of poljes, the erosion is lateral (Sweeting, 1972)—where sediments and water come in contact with the limestone massif. Neogene and Quaternary superficial deposits such as terra rossa often tend to accumulate on the floors of poljes. Poljes show complex hydrological and hydrogeological features and characteristics, such as permanent and temporary springs, permanent and lost rivers, swallow holes and estavelles (openings that may function as either a sink or a spring).

Karst poljes can be found in various parts of the world, but are particularly common in the Mediterranean countries (Greece, Italy, France, Spain, Morocco, Tunisia, Slovenia, Croatia, Bosnia and Herzegovina, and Montenegro). There are a few in Asia (China), a large number in Cuba, Jamaica, Canada (in the Nahanni area), and at least one (Bögli, 1980) in the United States in Tennessee (Grassy Cove). The Dinaric karst of Slovenia probably contains the highest concentration of classic poljes (see Cerknica Polje, Slovenia: History), but they also occur in Croatia, Bosnia and Herzegovina (see Dinaride Poljes) and Montenegro. Thus, the term has passed from Croatian, Serbian, and Slovenian into other languages, and is now used internationally. As Sweeting (1972) pointed out, over time "polje" has acquired a more specific scientific meaning than simply a flat overgrown or cultivated area—firstly to indicate a surface karst landform and secondly to mean all such extensive karst plains surrounded by limestone hills.

In the Dinaric karst (and similarly elsewhere), poljes represent the only areas with conditions favourable for human habitation. Surrounded by bare rocky terrain, they tend to be covered with arable soil and have either permanent or temporary springs and rivers. Although relatively small in size, they are thus significant from an economic and social standpoint.

From the hydrological point of view, a polje is only part of a wider system. It cannot and should not be treated as a complete system, but only as a subsystem in the process of surface and groundwater flow through and over the karst massif. Consequently, poljes cannot be studied adequately without establishing measurement points within the surrounding karst mass, and in the poljes of higher and lower groundwater horizons that connect to the subsystem being analysed.

Karst poljes flood regularly in the cold and wet periods of the year (in Dinaric karst from October to April) and in summer they have a tendency to dry up (Bonacci, 1985; Bonacci & Plantić, 1997). According to their hydrological regime, inflows, and outflows, they can be classified into four basic types: (1) closed poljes; (2) upstream open poljes; (3) downstream open poljes; and (4) upstream and downstream open poljes. Flooding is caused by the limited capacity of outlet structures and / or by a high groundwater level in the surrounding area. Poljes thus play an important role in the hydrological balance of karst areas.

Some anthropogenic changes have already been made to improve the hydrological regimes of polje areas, and more will be done in the future. However, some environmental damage has also been done and in some cases this has exceeded the benefits. From a hydrotechnical standpoint, it should be borne in mind that karst poljes are often linked to other adjacent upstream and downstream poljes—this being especially true for the Dinaric karst. Anthropogenic influences on the hydrological regimes of poljes can be divided into four categories: water storage, an increase in the capacity of outlet structures, surface hydrotechnical work, and the pumping of groundwater (Bonacci, 1987a).

Water storage in karst is not just affected by water retention in the poljes themselves—most water storage schemes involve damming river sections, in some cases either permanently flooding adjacent parts of the poljes or raising water table levels. Often human activity makes it necessary to preserve the fertile parts of the poljes as oases, only permanently flooding the less valuable areas of the karst. Building reservoirs in karst has a major influence on hydrotechnical relations in the poljes, and this effect is often difficult to assess (see Dams and Reservoirs on Karst). Spring discharges tend to increase in downstream areas, but there may be some additional negative consequences that cannot be amended easily. In upstream areas, drainage systems formerly inactivated by downward karstification may be reactivated (Milanović, 1986).

To prevent the flooding of karst poljes, attempts have been made to enlarge the capacities of ponors (see Ponors). These have usually failed because the capacities of ponors depend to some degree upon the size of inflows. In the last hundred years, it has been necessary to drill relief tunnels, and the connections between adjacent poljes have had to be enlarged.

In general, surface hydrotechnical construction work in poljes involves hydrological regulatory work in open streams, construction of channels for surface drainage, large land reclamation projects, and other measures. It is generally assumed that surface drainage does not have a great influence on the underground water level, which essentially governs outflow processes in karst. However, some forms of construction work may adversely affect the hydrological regime.

The pumping of groundwater from a karst aquifer for irrigation is a common practice, and since the 1980s it has been gaining prominence as a method of supplying water in karst areas. When groundwater is abruptly extracted, this can lead to fracturing of the upper layers of unstable lithologies, because of the presence of interlinked fissures within the karst mass. This can create new ponors and activate existing ones. Pumping groundwater from the karst also significantly lowers the local water table.

OGNJEN BONACCI

See also **Cerknica Polje, Slovenia: History; Dinaride Poljes**

Works Cited

Bögli, A. 1980. *Karst Hydrology and Physical Speleology*, Berlin and New York: Springer (original German edition, 1978)

Bonacci, O. 1985. Flooding of the poljes in karst. In *Papers Presented at the 2nd International Conference on the Hydraulics of Floods and Flood Control, Cambridge, 24–26 September 1985)*, Cranfield: BHRA

Bonacci, O. 1987a. Karst hydrology and water resources—past, present and future. In *Water for the Future: Hydrology in Perspective*, edited by J.C. Rodda & N.C. Matalas, Wallingford: International Association of Hydrological Sciences

Bonacci, O. 1987b. *Karst Hydrology with Special Reference to the Dinaric Karst*, Berlin and New York: Springer

Bonacci, O. & Plantić, K. 1997. Hydrology of the Drežničko Polje in the karst (Croatia). In *Karst Waters and Environmental Impacts: Proceedings of the 5th International Symposium and Field Seminar on Karst Waters and Environmental Impacts, Antalya, Turkey, 10–20 September 1995*, edited by G. Günay & A. Johnson, Rotterdam: Balkema

Milanović, P. 1986. Influence of the karst spring submergence on the karst aquifer regime. *Journal of Hydrology*, 84: 141–56

Sweeting, M.M. 1972. *Karst Landforms*, London: Macmillan and New York: Columbia University Press

Trudgill, S. 1985. *Limestone Geomorphology*, London and New York: Longman

Further Reading

Breznik, M. 1998. *Storage Reservoirs and Deep Wells in Karst Regions*, Rotterdam: Balkema

Drew, D. & Hötzl, H. (editors) 1999. *Karst Hydrogeology and Human Activities: Impacts, Consequences and Implications*, Rotterdam: Balkema

Ford, D.C. & Williams, P.W. 1989. *Karst Geomorphology and Hydrology*, London and Boston: Unwin Hyman

Milanović, P.T. 1981. *Karst Hydrogeology*, Littleton, Colorado: Water Resources Publications

Yuan D. 1991. *Karst of China*, Beijing: Geological Publishing House

PONORS

Ponors or swallow holes are fissures in the karst mass through which water sinks underground. Field (1999) cites two definitions:

1. A hole or opening in the bottom or side of a depression where a surface stream or lake flows either partially or completely underground into the karst groundwater system. A sea ponor is where sea water flows or is drawn into an opening by a vacuum in karstified rock;
2. A hole in the bottom or side of a closed depression, through which water passes to or from an underground channel.

Ponors play an important role from the hydrological–hydrogeological standpoint of water circulation in karst. According to their hydrological function, ponors can swallow water permanently or can function partly as ponors and partly as springs, i.e. as estavelles. Very often ponors are situated close to the margins of the floor of a polje while estavelles are more central (see Poljes; Cerknica Polje; and Dinaride Poljes). Water circulation in karst terrain is characterized by the existence of a strong inflow–outflow relationship. Ponors generally serve as the main inflow networks.

From the morphological standpoint (Milanović, 1981) ponors can include: large pits and caves; large fissures and caverns; systems of narrow fissures; or alluvial ponors. Jamas (vertical or steeply inclined shafts) most frequently function as ponors and present paths for the direct contact of surface water with the underground water in the karst mass. Ponors often provide entry points to the underground karst for cavers, and their investigation can reveal the positions, dimensions, and interactions of surface and underground karst features and water flow in the karst and on its surface.

Marine (sea) ponors are a rare phenomenon in karst. Most sea ponors are short-lived and actually represent the functioning of a vrulja (submarine spring) immediately after it dries up. The only permanent sea ponor in the world is the sea mill of Argostoli, located on Kefalonia Island in the Ionian Sea (Greece) (Glanz, 1965).

When an open stream flows through a karst terrain, the surface water frequently sinks underground and water losses occur through small ponors, which are found in karst fissures. In some karst regions, rivers sink underground through huge ponors or a well-developed ponor zone, and reappear at a large resurgence.

The capacity of a ponor to swallow water depends only upon the water level in the pre-ponor retention if the flow in the main karst conduit is not under pressure. A pre-ponor retention is a temporary water accumulation in the surface depression just above the ponor. The dimensions of these pre-ponor retentions vary greatly, and are a function of surface morphology. When the flow comes under pressure the ponor swallow capacity depends exclusively upon the difference between the level of water in the pre-ponor retention and the average level of the spring exit. When the groundwater level in the surrounding karst mass is higher than the water level in the pre-ponor retention, the ponor acts as an estavelle (Bonacci, 1987). In order to determine the swallow capacity of independent, large ponors, it is possible to use specially designated measurement devices based upon the principle of measuring velocities and / or pressure changes at a few points. More frequently there are numerous ponor zones with several large and a great number of small ponors in one area, and particularly in poljes. In that case, in order to define the swallow capacity of each ponor zone, it is necessary to carry out measurements of the groundwater levels in an appropriate part of the karst mass, since they influence the swallow capacity of that ponor zone.

With the objective of flood prevention in karst areas, particularly in poljes, attempts have been made to increase the capacities of ponors. Such attempts have usually failed because the capacity of ponors depends on the conduit system to which they drain as well as upon their size. Where the ponors do not have sufficient capacity as outlet structures, other hydrotechnical structures are often built to deliver water from the flooded areas.

Sometimes occasionally flooded karst areas are transformed into permanent storage basins (see Dams and Reservoirs in Karst). To prevent water losses the bottom and the sides of the new storage basin must be rendered impermeable but this is problematic where the reservoir is located on a permeable karst terrain with many ponors (Breznik, 1998). There have been more unsuccessful attempts to isolate or surface seal ponors than successful ones and water losses from a reservoir can result in large increases in the discharge of karst springs (Bonacci & Jelin, 1988).

OGNJEN BONACCI

See also **Dams and Reservoirs on Karst; Hydraulics of Caves for Photo; Poljes**

Works Cited

Bonacci, O. 1987. *Karst Hydrology with Special Reference to the Dinaric Karst*, Berlin and New York: Springer

Bonacci, O. & Jelin, J. 1988. Identification of a karst hydrological system in the Dinaric karst (Yugoslavia). *Hydrological Sciences Journal*, 33(5): 483–97

Breznik, M. 1998. *Storage Reservoir and Deep Wells in Karst Regions*, Rotterdam: Balkema

Field, M.S. 1999. *A Lexicon of Cave and Karst Terminology with Special Reference to Environmental Karst Hydrology*, Washington DC: US Environmental Protection Agency

Glanz, T. 1965. Das Phänomen der Meermühlen von Argostolion [Phenomenon of the sea mills of Argostoli], *Steierische Beiträge zur Hydrologie*, 17: 113–27

Milanović, P. 1981. *Karst Hydrogeology*, Littleton, Colorado: Water Resources Publications

Further Reading

Drew, D. & Hötzl, H. (editors) 1999. *Karst Hydrogeology and Human Activities: Impacts, Consequences and Implications*, Rotterdam: Balkema

Ford, D.C. & Williams, P.W. 1989. *Karst Geomorphology and Hydrology*, London and Boston: Unwin Hyman

Kogovšek, J. 1996. Sinking rivers quality—the Pivka case study. *Acta Carsologica*, 25: 145–56

POSTOJNA–PLANINA CAVE SYSTEM, SLOVENIA

Postojnska Jama (Postojna Cave) is the longest known cave system in Slovenia (20 570 m in length, 115 m deep). Between Postojna Cave and the Planina Cave system there is still about 500 m of unexplored passages. If both caves are connected, the cave system will be about 30 km long. Postojnska cave system has several entrances and consists of: Postojnska Jama, Črna Jama, Pivka Jama, Magdalenska Jama, and Otoška Jama (Figure 1).

People have lived in the entrance parts of the cave system for at least 10 000 years. The oldest inscriptions on cave walls date from the 13th and 14th centuries. The Slovenian historian and naturalist Baron Valvasor described Postojnska Jama in 1689 as one of the most spectacular caves in the world that was known at the time, even though just a small part of today's cave systems was then known. In 1748 Joseph Nagel made the first map of the cave, and in 1781 Tobias Gruber inferred the underground water connection between Postojnska Jama and Planina Cave system (see entry, Cerknica Polje, Slovenia: History). The Postojnska cave system was also important for the development of biospeleology, being the location of the first cave fauna specimens described scientifically (see Postojna–Planina Cave system: Biospeleology).

On 14th April 1818, the discovery by cave guide Luka Čeč of extensive new parts of the cave marked the beginning of official cave tourism in Postojnska Jama and 5 km has now been opened as a show cave. In 1997, the caves were visited by about 400 000 tourists. Between the years 1818 and 1992 Postojnska Jama were visited by 26 million people, the highest number in a single year being over 900 000 in 1985. Postojnska Jama has one of the oldest tradition in guidebooks; from 1821 to the present, 110 guidebooks have been published. The first printed guidebook was written by Girolamo Agapito in 1823.

The River Pivka sinks into Postojnska Jama at 511 m altitude from the impermeable Eocene flysch of its basin, and resurges from the Planina Cave system (Figure 1) as a spring of the Unica River. The Pivka is an interesting river in that the main drainage is to the Black Sea, but part of the river before entering the Postojnska system flows to the Adriatic (Habič, 1989).

The Postojnska caves are developed on two principal levels. The higher level starts at an altitude of 529 m and comprises dry passages which are partly filled with cave sediments, flowstone, and collapse blocks. A great number of these passages have been reshaped through collapses, with many of the largest chambers being formed by collapse. The lower level comprises the active underground channels of the Pivka River which are situated west of the higher levels. The two levels are connected by side passages.

The caves are developed in the c.800 m thick Upper Cretaceous bedded limestones. The Cenomanian and Turonian limestones are more thinly bedded and include chert lenses or beds while the Senonian limestones are thick bedded to massive. The caves are situated between two regionally important Dinaric-trending faults, the Idrija fault to the north and Predjama fault to the south. The principal folding deformation in Postojnska Jama is the Postojna anticline. Cave passages are developed in both flanks of the anticline, and follow strike and dip of the bedding planes, especially those with interbedded slips. Sections of the underground River Pivka between Otoška Jama and Pivka Jama follow the dip of bedding planes and the northeast–southwest fault zones.

Sediments from Postojnska Jama were studied by Gospodarič (1976) who determined 10 principal development stages of the cave system by absolute dating of flowstone and by the relative age of cave sediments; according to Gospodarič (1976) the oldest sediments originate from an interglacial period (Mindel-Riss). In 1999 sediments from Postojnska Jama were studied by paleomagnetic analyses (Šebela & Sasowsky, 1999) and it was found that most were deposited during the Brunhes Normal Epoch (younger than 0.73 Ma). However, samples from a small natural

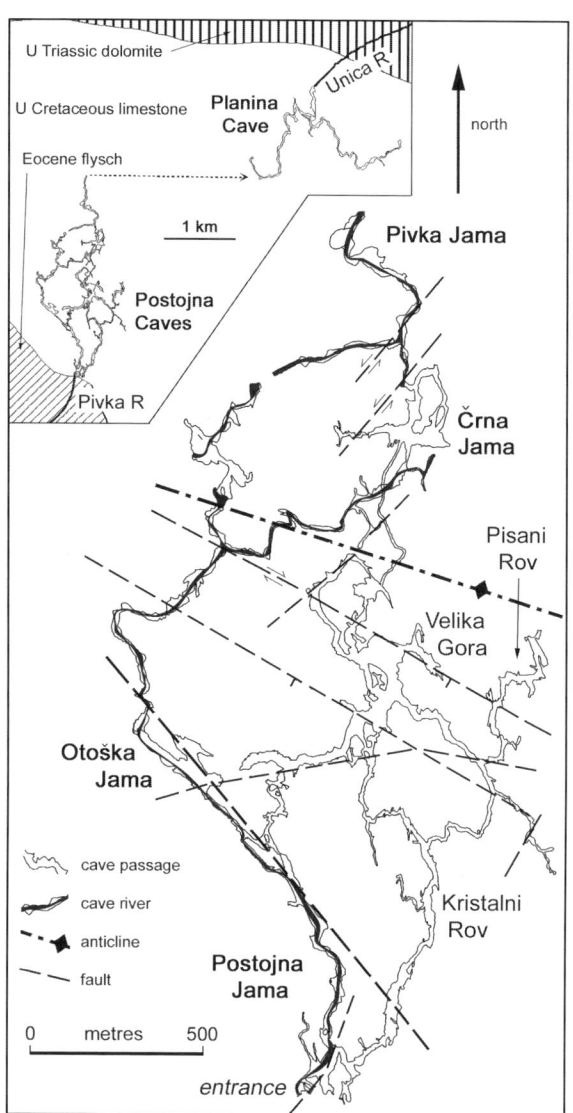

Postojna-Planina Cave System, Slovenia: Figure 1. Main elements of the structural geology of the limestones containing the Postojna caves, Slovenia; the inset shows the relationship of the Postojna and Planina caves.

passage which is accessible from the tunnel between Postojna and Črna caves showed reverse polarity and are at least 0.73–0.90 million years old. These cave sediments are cut by a younger cross-Dinaric fault zone with horizontal displacement. This is the best example of neotectonic activity in Postojnska caves.

The average annual temperature in Postojna is 8.4°C and most of the average annual rainfall of 1579 mm infiltrates directly into the karstified limestone. This water appears in the caves as permanent or seasonal trickles that contain up to 250 mg l^{-1} $CaCO_3$. Measurements over several years have shown that the precipitation amount is among the most important factors for dissolving limestone and consecutive karstification. The percolation water is oversaturated and deposits flowstone. Permanent trickles in Pisani rov (Gams & Kogovšek, 1998) with discharge of about 100 ml min^{-1} deposit carbonate up to 18–39 g m^{-3} in one year.

Postojna-Planina Cave System, Slovenia: Figure 2. Calcite deposits in a high-level gallery of Planina Jama. (Photo by Tony Waltham)

Monitoring of discharge and water tracing tests with fluorescent dyes (Kogovšek, 2000) has shown that water percolating from the surface through the 100 m thick roof into Kristalni rov reaches a velocity of 4.5 m h^{-1} through the main, most permeable conduit (maximum discharge through the conduit, $Q_{max} = 1.5$ l min^{-1}) and 0.7 m h^{-1} through less permeable ones ($Q_{max} < 0.18$ l min^{-1}). To determine what would happen if a pollutant was accidentally discharged on the surface above the cave, 6 m^3 of water containing Uranin dye was poured on the surface and gave a flow velocity of 80 m h^{-1} through the most permeable conduit (almost 20 times faster than under natural conditions) and 0.05 m h^{-1} through the less permeable fissure.

STANKA ŠEBELA and JANJA KOGOVŠEK

Works Cited

Gams, I. & Kogovšek, J. 1998. The dynamics of flowstone deposition in the caves Postojnska, Planinska, Taborska and Škocjanske, Slovenia. *Acta Carsologica*, 27(1): 299–324

Gospodarič, R. 1976. The Quaternary cave development between the Pivka basin and polje of Planina. *Acta Carsologica*, 7: 8–135

Habič, P. 1989. Pivka karst bifurcation on Adriatic-Black Sea watershed. *Acta Carsologica*, 18: 233–64

Kogovšek, J. 2000. How to determine the way of percolation and transport of substances by water tracing test in natural conditions. *Annales: Annals for Istrian and Mediterranean Studies*, 10: 133–42

Šebela, S. & Sasowsky, I.D. 1999. Age and magnetism of cave sediments from Postojnska jama cave system and Planinska jama cave, Slovenia. *Acta Carsologica*, 28(2): 293–305

Further Reading

Čar, J. & Gospodarič, R. 1984. About geology of karst among Postojna, Planina and Cerknica. *Acta Carsologica*, 12: 91–106

Gams, I. 1965. H kvartarni geomorfogenezi ozemlja med Postojnskim, Planinskim in Cerkniškim poljem. *Geografski Vestnik*, 37: 61–101

Gams, I. 1966. Factors and dynamics of corrosion of the carbonatic rocks in Dinaric and Alpine karst of Slovenia. *Geografski Vestnik*, 38: 11–68 (In Slovenian, with an English summary)

Habe, F. 1986. *The Postojna Caves*, 4th edition, Ljubljana: Tone Tomšič

Šebela, S. 1998. *Tectonic Structure of Postojnska Jama Cave System*, Ljubljana: Založba ZRC (In Slovenian and English)

Kogovšek, J. 1999. New knowledge about the underground water drainage in the northern part of Javorniki Mountains (High Karst). *Acta Carsologica*, 28(1): 161–200

Shaw, T.R. & Čuk, A. 2002. Royal and other notable visitors to Postojnska jama 1819–1945. *Acta Carsologica*, 31(1) supplement 1: 1–107

Vokal, B., Obelić, B., Genty, D. & Kobal, I. 1999. Chemistry measurements of dripping water in Postojna Cave. *Acta Carsologica*, 28(1): 305–20

POSTOJNA–PLANINA CAVE SYSTEM: BIOSPELEOLOGY

The Postojna–Planina Cave System (Postojnsko–planinski jamski sistem, or PPCS) comprises the hypogean bed of the river Pivka, between its sink at Postojna and its resurgence at Planina, along with the adjacent relict cave corridors. Upstream passages more than 20 km in length (Postojnska jama, Otoška jama, Magdalenska jama, Črna jama, Pivka jama, and Lekinka) include approximately 4 km of river sections. Most of the 6 km of the downstream section (Planinska jama) is hydrologically active, and includes more than 2 km of the Pivka River in addition to its confluences. A phreatic section, so far unexplored, with a length of 500 m (measured in a straight line), connects both sections.

Since 1818, more than 5 km of relict passages in Postojnska jama have been utilized as tourist caves. The River Pivka is now being polluted by effluents from the town of Postojna. A biospeleological station was installed in Postojnska jama and has been intermittently active since the 1930s.

Scientific Importance of the Cave System

These caves are sometimes referred to as "the cradle of speleo biology", since they are the type localities of some of the first cave animals described scientifically, in the 19th century. In 1797 the cave salamander (*Proteus anguinus*) was first seen here in its primary cave environment. The caves are also the type locality of the first troglobitic beetles discovered (*Leptodirus hochenwartii*), pseudoscorpions (*Neobisium spelaeum*), spiders (*Stalita taenaria*), cnidarians (*Velkovrhia enigmatica*), and even of the trogloxene crickets (e.g. *Troglophilus cavicola*). In fact the cave system is the type locality for more than 60 invertebrate species. It is faunistically by far the richest cave system in the world, with 49 stygobitic and 35 troglobitic species (Culver & Sket, 2000). A high number of troglophilic and accidental terrestrial and aquatic species have also been recorded. The scientifically most interesting are populations of those stygophilic species, which exhibit clines (i.e. a gradual increase) in their degree of troglomorphism between the sink and the resurgence. Along the hypogean part of the Pivka River, for example, the sponge *Ephydatia muelleri*, the snail *Ancylus fluviatilis*, the leech *Dina krasensis*, the amphipod *Synurella ambulans*, and the isopod *Asellus aquaticus* all show this type of variation.

The relatively intensive ecological studies of this cave system have resulted in the discovery of a characteristic relationship between specialized (starvation-resistant but less active) stygobites and generalist stygophilic (more active but more food-demanding) species. When organic pollution increases, the surface species gain a competitive edge, pushing the stygobites further underground (Sket, 1973). Postojnska jama was also the site of a detailed study of "Lampenflora" (see Tourist Caves: Algae and Lampenflora).

Ecological Conditions

The normal cave temperature of the system varies between 9 and 10°C; the monthly mean temperature of the hypogean Pivka varies between 2 and 17°C in Postojnska jama and between 6 and 12°C in Planinska jama; daily variations may exceptionally reach 3°C in Planinska jama, but only during high water discharges. The discharge of the Pivka River varies between negligible values and in excess of 40 m^3 s^{-1}. At high discharges, water takes only seven hours to pass through the 9 km of cave corridors between the sink and the resurgence, but five days or more in dry periods. Being influenced by the river to a large degree, the air temperatures in the system are dependent on the precipitation/discharge fluctuations and the regional air temperature regime in a particular year. Therefore, the mean yearly temperature within the cave may deviate slightly from the mean air temperature outside the cave.

Aquatic Fauna

The benthic fauna of Pivka River in Postojnska jama consists largely of epigean animals. Some of them are restricted to the sink region, while others inhabit the entire hypogean river bed. These include, in addition to those clinally variable populations outlined above, the water-flea *Chydorus sphaericus*, stonefly larvae *Nemoura* sp. and *Brachyptera tristis*, and some mayfly larvae (family Leptophlebiidae). The isopod *Asellus aquaticus* is represented in Postojnska jama by pigmented or slightly depigmented specimens of the type subspecies *A. aquaticus*. With increasing distance from the sink, more troglomorphic subpopulations of stygophiles and some stygobitic species prevail. Examples include tiny hydrobioid snails, which may reach densities of more than 10 000 specimens per square metre. The amphipod *Niphargus spoeckeri* is among the most common species on the riverbed surface, while the smaller *N. aquilex dobati* is present within deeper interstitial layers in Planinska jama. The cave shrimp *Troglocaris anophthalmus*, and the highly troglomorphic isopod *Asellus aquaticus cavernicolus* (which may be a separate species) are present in pools. The amphibian *Proteus* is rare in the Tartarus in Postojnska jama, while it may be seen regularly in the bed of the Pivka River in Planinska jama. However, no very young or very old specimens have been found there, forcing

Briegleb (1962) to consider the hypogean river as a marginal ecotope for this animal.

During periods of increased organic pollution (in the 1950s to 1980s) the less troglomorphic species or subpopulations increased in density, even in the remote parts of the hypogean stream. Prior to 1950, and following the construction of a purification plant in Postojna in the 1980s, the stygobites were and have become more abundant.

The fauna of the Rak branch in Planinska jama includes a greater proportion of stygobitic fauna. In addition to *Proteus*, cave shrimps and troglomorphic asellids, and a larger number of tiny stygobitic snail species (*Neohoratia subpiscinalis* and *Acroloxus tetensi*) are common. The tiny sessile cnidarian *Velkovrhia enigmatica* may also reach high densities.

The fauna of the percolation waters are distinctive. *Niphargus stygius* is the characteristic amphipod species in pools, accompanied by a number of copepod species. Some harpacticoid copepods (e.g. *Elaphoidella* spp.) have been found within drips of percolating waters from the crevicular system in the ceiling. Amphipods are represented by the small *Niphargus wolfi* and *Niphargobates orophobata*; the latter was filtered out from only one such drip in Planinska jama, and its only known relative was later discovered on Kriti Island (Greece).

Terrestrial Fauna

The fauna in the non-modified entrance areas of the caves is characteristic of most of the Slovenian Dinaric karst. On the walls, resting specimens of the moths *Triphosa dubitata* and *Scoliopteryx libathrix* may be seen. Trogloxene crickets, *Troglophilus* spp., may be present deeper within caves, particularly during the winter. In the organic debris on the cave floor there may be numerous edaphic–troglophilic animals such as the millipede *Brachydesmus subterraneus*, the troglophile beetle *Laemostenes cavicola*, and Collembola. Among the troglobites, the large amphibious isopod *Titanethes albus* may be particularly numerous. Occasionally, tiny snails (two *Zospeum* spp.) may be found on the damp walls. Spiders and pseudoscorpions are even rarer, while troglobitic beetles, including *Leptodirus*, three *Anophthalmus* species and others, are the species most regularly caught using bait.

BORIS SKET

See also **Dinaric Karst: Biospeleology**

Works Cited

Briegleb, W. 1962. Zür Biologie und Oekologie des Grottenolms (*Proteus anguinus* Laur. 1768). *Zeitschrift fuer Morphologie und Oekologie der Tiere*, 51: 271–334

Culver D.C. & Sket, B. 2000. Hotspots of subterranean biodiversity in caves and wells. *Journal of Cave and Karst Studies*, 62(1): 11–17

Sket, B. 1973. Gegenseitige Beeinflussung der Wasserpollution und des Hoehlenmilieus. In *Proceedings of the 6th International Congress of Speleology*, vol. 5, Olomouc: International Union of Speleology

Further Reading

Brancelj, A. 1986. Rare and lesser known harpacticoids (Copepoda Harpacticoida) from the Postojna–Planina Cave System (Slovenia). *Biološki Vestnik*, 34(2): 13–36

Brancelj, A. 1987. Cyclopoida and Calanoida (Crustacea, Copepoda) from the Postojna–Planina Cave System (Slovenia). *Biološki Vestnik*, 35(1): 1–16

Kranjc, A. & Malečkar, F. 1988. *Postojnska jama: 170 let odkrivanja, raziskovanja in turističnega razvoja* [Postojna Cave: 170 years of discovery, research, and tourist development], Postojna: Kraška muzejska zbirka pri IZRK ZRC SAZU

Turk, S., Sket, B. & Sarbu, S. 1996. Comparison of some epigean and hypogean populations of *Asellus aquaticus* (Crustacea: Isopoda: Asellidae). *Hydrobiologia*, 337: 161–70

PSEUDOKARST

Introduction

Large areas of the Earth's land surface are characterized by concentrated subsurface drainage, or by interconnected voids caused by processes other than dissolution. Where these processes occur in loess, lava flows, glaciers, permafrost regions and elsewhere in poorly soluble rocks, terrains characterized by dolines, caves, sinking streams and / or other features commonly observed in karst are termed pseudokarst.

Definitions

A working session of the 1997 12th International Congress of Speleology concluded that "pseudokarsts are landscapes with morphologies resembling karst, and / or may have a predominance of subsurface drainage through conduit-type voids, but lack the element of long-term evolution by solution and physical erosion" (Kempe & Halliday, 1997). Watson Monroe and many others have said it more simply: karst-like morphology produced by some process other than solution. Pseudokarstic caves are natural underground spaces, large enough to enter and investigate, extending beyond daylight, and produced by some process other than dissolution.

Among especially common types of pseudokarstic caves are lava tube caves, talus caves ("boulder caves") on mountainsides and in stream gorges (called purgatory caves in some parts of the United States), crevice caves, glacier caves, and littoral caves ("sea caves"), each of which has a separate entry in this Encyclopedia. Few stream-cut voids are large enough to meet the basic definition of "cave", and by this definition, rock shelters and tafoni are not considered caves in this volume.

Significance of Pseudokarst

Resources and values of pseudokarsts are dependent on the processes which formed them, their features, and their extents. Their geological values differ considerably from those of karsts. They provide access for study of a wide range of geological features, some of which are of special interest to planetary geologists (see Extraterrestrial Caves). They contain a wider range of minerals than do karstic caves. Their archaeological and paleontological values largely depend on their age and location. Their biological resources vary widely. Because pseudokarsts tend to be smaller than karsts, development of troglobitic species tends to be limited. Some in Hawaii and the Canary Islands, however, are so large and so fortuitously located that these have developed nonetheless (see Hawaiian Islands: Biospeleology and Canary Islands: Biospeleology).

Just as in karsts, educational, recreational, wilderness, and other cultural resources and values are highly site-specific. Recreational caving in some stream-smoothed granite talus caves is exceptionally enjoyable. While comparatively few show caves have been developed in pseudokarsts, their educational values are at least as great as those of their karstic counterparts. Some (e.g. parts of Hawaii's Kazumura Cave) are as aesthetically pleasing as all but the greatest karstic caves. During much of the 20th century, the Paradise Ice Caves of Mount Rainier were among the most noted visitor attractions of the United States' Pacific Northwest states.

Hazards

Outburst floods from normothermic or hyperthermic glacier caves are perhaps the most spectacular hazard of pseudokarst. A large literature exists on engineering problems caused by crevice pseudokarsts, isolated crevice caves, piping, and piping caves (see entries on Crevice Caves and Piping Caves). Sudden collapse of overloaded cave roofs is especially a hazard of pseudokarsts containing lava tube caves. Public health hazards exist where such caves function jointly as floodwater or perennial conduits and unlawful disposal sites. In a few pseudokarstic caves, the water flow is formidable; an alpine torrent in a talus cave caused the recent death of an experienced, well-equipped Colorado caver (see entries on Talus Caves and Volcanic Caves).

History

Pseudokarstic forms were observed in loess, perhaps 2300 years ago (Liu *et al.*, cited by Péwé *et al.*, 1995). These became known outside China late in the 19th century; a 1879 description by von Richthofen is often cited. Roman writings mentioned lava tube caves on Mt Etna. The presence of large caves in or beneath some glaciers also must have been known locally from early times. During the early 20th century, the specific terms "pseudokarst" and "pseudokarstification" originated several times, in several languages, and for several types of features. Reports accumulated from a wide variety of terrains. Most of them described features remote from centres of learning and their writers were not academics. Commonly they were in obscure publications and many were in languages that were not widely read. Locally invented terminologies baffled "outsiders", even of the same nationality. Especially confusing were diligent efforts to apply karstic concepts and terminology to phenomena which only looked karstic. Yet an impressive body of knowledge gradually accumulated.

The German geologist von Knebel (1908, p.171) apparently was the first to use any of these terms. Seeing Icelandic streams disappear into fractured basalt, he recorded that "in many lava areas, rivers are features of the subsurface, but it is proper to consider this only as pseudokarstification ("pseudoverkarstung" in his German-language account).

Beginning around 1927, Russian scientists pioneered the study of karst-like features in permafrost and poorly soluble rocks. In 1931 and 1935, F.P. Savarensky wrote about karst-like phenomena in clayey sediments and loess, calling them "clay karst" and "loess karst". In 1947, N.A. Gvozdetskij recommended qualified use of the term "pseudokarst", correctly pointing out that the processes are real, not "pseudo".

Especially after World War II, these papers became known in the portion of Europe under Soviet domination, and elsewhere to a lesser degree. Meanwhile, a significant paper was published in Italy, specifically referring to "A pseudokarstic phenomenon ('fenomeno pseudocarsico') in clay" (Floridia, 1941), and Malaurie used the term in the title of a short geological note in French in 1948. These reports, however, were not in widely read journals.

In the 1950s, English began to be a common scientific language. In 1950, Kukla appended an English summary to a report on sizeable sandstone caves in Bohemia and Kunský (1957) discussed types of pseudokarstic caves in a notable English-language paper. Innumerable reports have appeared since, but many continue to be in little-read languages and / or journals. Proceedings volumes of international symposia on pseudokarst, however, tend to unify central European concepts.

Controversies

In part because of its divergent origins, the concept of pseudokarst is not universally accepted, and agreement on what should be included is less than complete. Several leading dissenters did not attend the 1997 IUS working session which agreed upon the cited definition (Kempe & Halliday, 1997). William B. White specifically considered glacier caves and related features to be "karst-like" rather than either pseudokarst or karst; Eraso and Pulina (1992) and some others consider them to be karstic despite the lack of dissolution. Jennings and some others have accepted the term "volcanokarst" for volcanic pseudokarst, but Grimes pointed out that this term previously was established for a dissolutional form in volcanic ash. A similar controversy exists about the term "thermokarst". Even in 1972, Marjorie Sweeting noted that "to include such forms is to make the definition of karst too wide and thus to lose much precision". Other controversies have arisen because of misunderstanding. In the 1950s and 1960s, William E. Davies repeatedly identified certain American closed depressions as pseudokarstic; these were shown as such in the 1970 US National Atlas. Subsequently it was found that these were caused by karstification at depth. Ervin Otvos, a coastal geologist, correctly pointed out (1976) that these are karstic phenomena, not pseudokarstic, and Otvos also proposed restricting the term "pseudokarst" to "only processes and forms involving predominantly piping and thermokarst". The conclusions of the IUS 1997 working session, however, reflect broad agreement that this proposed limitation was too narrow. Controversies about quartzite karst and pseudokarst have diminished as a result of recent increased emphasis on specific features and processes evident in various tropical areas of quartzite and other poorly soluble rocks (see entry on Silicate Karst).

Classification of Pseudokarst

Pseudokarsts may be classified by morphology, by process, and by lithology. Most classifications have been morphological, or a combination of morphology with other factors. Nearly all have been regionally oriented (e.g. central Europe, western United States, north-central Sweden, etc.) or limited to specific terrains (e.g. volcanic, glacier, etc.). On a global basis, the 1997 IUS working session specifically identified "pseudokarst on lava", "pseudokarst on ice", "pseudokarst on permafrost", "pseudokarst on talus", and "pseudokarst on unconsolidated sediments and volcanic ash". At least two additional types were mentioned but not discussed. Based largely on that working session, the following classification is used in this entry:

1. rheogenic pseudokarsts (pseudokarsts on lava flows);
2. glacier pseudokarsts;
3. badlands and piping pseudokarsts;
4. crevice pseudokarsts including littoral pseudokarsts;
5. talus pseudokarsts;
6. permafrost pseudokarsts;
7. consequent pseudokarsts.

Rheogenic Pseudokarsts and Their Caves

Rheogenic pseudokarsts are features of certain types of lava flows, primarily pahoehoe basalts. In these flows, individual caves up to 65.6 km long have been explored. In addition to specific rheogenic features, many have cross sections (Figure 1), speleothems (see photo in Volcanic Caves entry), speleogens and petromorphs comparable to those of karstic caves. Many have multiple entrances. Some of these occur in patterns reflecting braided cavern passages within the flows. Patterns consistent with extraterrestrial lava tube caves have been identified on Mars, Venus, and Io, and a NASA-funded world database on lava tube caves is maintained at Arizona State University (United States). Pseudokarst in other volcanic terrains is discussed as crevice pseudokarst and badlands pseudokarst (q.v.).

Glacier Pseudokarsts and Their Caves

Glacier caves provide unique observation points for study of features and mechanics of glacial flows. Until recently, however, their values and their hazards commonly were overlooked. The devastating 1892 outburst flood from Switzerland's Tête Rousse Glacier may have been the first which was recognized immediately as having come from a previously unknown glacier cave.

Glacier caves vary greatly in size. The world's largest waterfilled cavern chamber is probably the enormous space beneath Antarctica's Ross Ice Shelf, although some have proposed the Antarctic cavern which contains Lake Vostok. At least until recent ablation of its piedmont lobe, one of the world's longest subterranean rivers ran (or runs) beneath part of Alaska's Malaspina Glacier, a distance of some 50 km. More than 19 km of passages were mapped in the Paradise Ice Caves of Mount Rainier (Washington state, United States) before its glacier melted completely. This large system and most other large glacier snout caves were largely formed by modification of small subglacial and intraglacial conduits by sublimation and evaporation. Fountain and Wilder (1998) have provided an excellent overview of this process. Other glacier caves are formed by plastic arching of ice in the lee of rocky obstructions and by snow bridging of crevasses. Still others (moulin caves) form by meltwater enlargement of crevasses. To date, moulin caves have been explored to depths of nearly 200 metres. This permits sampling of ice nearly 100 000 years old. The enormous, much-depicted moulin of New Zealand's Fox Glacier is believed to remain unexplored because of the formidable cliffside waterfall which created it.

Crevice Pseudokarst and Its Caves

Crevice pseudokarst and its caves reach their maximum expression in four terrains: volcanic and other terrains which have undergone extensive fracturing; littoral pseudokarst; cliff and mass movement terrains; and glaciers. Mass movement terrains often are not recognized as such, but locally cause serious engineering problems due to gravity-driven sliding or tilting. Most of the island of Hawaii (the "Big Island" of the Hawaiian Archipelago) consists of crevice pseudokarst, but only a few of its crevices are caves (the remainder of the pseudokarst of the island consists of rheogenic pseudokarst). Perhaps most visible of all crevice forms are crevasses in glaciers. A few crevice caves have commercial value as show caves. Only rarely are cavernous littoral areas extensive enough to comprise a landscape, but at least one exists: at Ballybunion, on the west coast of Ireland.

Talus Caves and Their Pseudokarsts

In Sweden and some other areas where granitic and metamorphic rocks predominate, talus caves are as important as karstic caves. Most of these caves are in talus accumulations too small to be considered landscapes. However, Colorado's Lost Creek Pseudokarst (United States), formed in granite, is comparable in features and size to sections of Puerto Rico's famous karst (Hose, 1996). In the northeastern United States, large blockfields (boulder fields) contain lengthy, tortuous maze caves. Boulder accumulations large enough to contain caves have been identified on Mars (Malin & Edgett, 2000).

Badlands (Piping) Pseudokarst and Its Caves

Badlands and other forms of piping pseudokarst are best known for causing serious engineering problems but several sizeable piping caves are important individual features (e.g. Officers Cave, Oregon, Anvil Points Claystone Cave, Colorado, and Christmas Canyon Cave, Washington state, United States). While piping originally was described in loess, it occurs in many forms of poorly consolidated material and some rocks, including quartzite, where it transports some of the clastic debris generated in the development of multiprocess caves. Some small-scale features on Mars suggest recent piping on that planet (Malin & Edgett, 2000).

Permafrost Pseudokarst and Its Caves

Permafrost pseudokarst is formed by a combination of thawing and piping in areas of tundra and taiga where permafrost under-

Pseudokarst: Figure 1. A drained lava tube showing cross section similar to phreatic tube in limestone. Wiri Cave, Auckland, New Zealand. (Photo by John Gunn)

lies the ground surface. Such terrains occupy more than 10% of the Earth's surface. In summer, parts of these terrains are conspicuous for circular to oval thaw ponds and drained, steep-walled depressions up to 10 km in diameter and 1 to 40 m deep, with flat bottoms. Locally they may occupy as much as half of the surface, pockmarking the landscape much as in the case of some karstic plateaux. These depressions are formed by local collapse of thawing soil layers which contain ice wedges and permafrost polygons, with subsequent piping. Related karst-like features include funnel-shaped pits, ponors, and dry valleys. Small caves form by melting of ice veins in subsurface polygons and by piping in earthy walls of depressions. Especially in Europe, the term "thermokarst" has been applied to these terrains, but there is nothing dissolutional in the process which formed them. Sweeting has decried the term, pointing out its confusing similarity to the established term "thermal karst".

Consequent Pseudokarsts and Their Caves

Istvan Eszterhas has developed the concept of consequent pseudokarst: karst-like terrains formed by the action of natural processes interacting with mines, underground quarries, and other subsurface works of mankind. Their surface features tend to be rectilinear, and commonly reflect serious engineering problems (Figure 2). Some consequent collapse areas contain extensive caverns formed by natural stoping. These are bounded on all sides by talus or fracture surfaces, much as in the case of karstic or lava tube caves in which breakdown has filled the original space and the present-day cave is entirely above the original cave.

Interfaces and Multiprocess Caves

Because of the variety of processes and lithologies that form pseudokarsts, numerous interfaces and multiprocess caves exist. In Brazil, "canga caves" have been identified beneath limonite-cemented, haematite-rich surface debris. Some are believed to be solutional, others corrasional, and perhaps some are a combination of these processes. More study is needed. Several geologists have discussed small closed depressions caused by dissolution on horizontal or gently sloping surfaces of poorly soluble rocks. While large quantities of insoluble particles are removed by piping or streamflow, this phenomenon generally is considered to be karstic. Similarly, karren (lapiès) also are found on a variety of soluble and poorly soluble rocks. Examples of the latter are the much-photographed granite boulders of the Seychelles Islands. Most lapiès are karstic, but some in lava tube caves clearly are the result of thermal erosion and are pseudokarstic phenomena. Dolines penetrating thin mantles of loess and other poorly soluble material overlying karstified bedrock are also karstic, not pseudokarstic. Because dissolution of calcareous materials in loess is part of the process forming its karst-like features, some geologists have considered them also to be karstic. This, however, is so minor a part of the process that, in this entry, loess features are considered part of badlands pseudokarst. Grimes has differentiated between laterite karst (in which silicate minerals are removed in solution from within a deep weathering profile) and laterite caves in western Australia which appear to have been caused by piping.

Karstic and Pseudokarstic Processes in Quartzite Caves

The role of dissolution in speleogenesis and karstification of poorly soluble quartzites and related rocks has been much debated in recent years. The spectacularly pitted Sarisarinana and other rainforest quartzite plateaus of Venezuela, the gentler Chimanimani area of Zimbabwe, and some regions of South Africa have received especially intensive consideration (see Silicate Karst). Somewhat similar features in ferruginous metamorphic rocks also have been discussed.

In calcareous rocks, karstification requires both dissolution and physical removal of varying amounts of clastic debris originally present in the bedrock. Even in very pure limestones, laminar or turbulent flow must carry small loads of insoluble residue freed from bedrock by dissolution, with karstic conduits acting as pipes. Other cavernous limestones contain horizons of sand, gravel, and even large boulders of poorly soluble material (e.g. Espluga de Francoli, Tarragona, Spain). These similarly are freed by dissolution, and choke incipient conduits if not actively removed by stream transport. Further, some caves in limy sandstones and conglomerates (e.g. Cova de Salnitre, Catalunya, Spain) also have features characteristic of phreatic speleogenesis despite an even greater proportion of clastic material. Thus, it appears that there is a continuum between piping and dissolution caves in carbonate rocks, including rocks with a mere carbonate matrix.

In clastic rocks with siliceous and ferric matrices, such a continuum is not as clear. However, dissolution of these non-carbonate binders has been documented in some quartzose localities (e.g. Striebel & Schäferjohann, 1997). Such dissolution permits similar transport of siliceous particles by piping and stream flow.

Some but not all caves in quartzites in the humid tropics contain dissolutional speleogens, and appear to be part of this continuum. Photodocumentation of passages in these caves shows patterns typical of phreatic conduits. But comparatively enormous volumes of clastic material must have been removed so that the pits and conduits were not choked by their own debris. In such a multiprocess continuum, no specific percentage of processes can be said to separate karstic, interface, and pseudokarstic caves. It is of some interest that colloidal silica speleothems have been documented in Brazilian and Venezuelan caves,

Pseudokarst: Figure 2. Consequent pseudokarst atop an abandoned coal mine in Wyoming (United States). US Geological Survey aerial photo by C.R. Dunrud published in US Geological Survey Circular 876. Scale is shown by roads.

but actual laminar transport of colloidal silica as an inception mechanism apparently has not been studied. On the other hand, published maps, descriptions, and photographs of the quartzite Chimanimani Caves of Zimbabwe and some in South Africa do not document similar evidence of dissolution. Like caves near Ellenville, New York (United States), the Chimanimani Caves appear to be rectilinear crevice caves with chambers developed vertically, without evidence of significant dissolution or piping. It therefore appears that some major caves in quartzose rocks are largely or entirely pseudokarstic crevice caves while others are multiprocess caves that have become predominantly karstic.

Martian Analogues of Terrestrial Pseudokarsts

Malin and Edgett (2000) have reported several Martian features which may be analogous to terrestrial pseudokarsts (see also Extraterrestrial Caves). One consists of talus accumulations where alcoves are "littered" with boulders "several metres to several decimetres" in diameter. These are directly downslope from egresses of recent outbursts of water, some of which evidently entered the talus. On Earth, talus accumulations with these parameters commonly contain talus caves, and terrestrial talus acts as a baffle to subsurface atmospheric movement. As a result, some talus caves function as natural deep-freezes, and contain unseasonal snow and ice. On Mars, they may protect recent spelean ice accumulations from ablation. In addition, the open conduits above these Martian talus accumulations may be analogous to terrestrial piping caves. Malin and Edgett merely referred to badlands topography in polar pits, without explanation. But they presented detailed evidence of piping or other open conduits in perhaps three specific Martian terrains: alcoves associated with resistant layers in cliffs, in landslide or slumped material, and "perhaps in colluvium". On Earth, piping caves occur in all three types of terrain.

Present-day Martian terrains are hyper-arid beyond any conditions now present on Earth. Some extremely hyper-arid terrestrial terrains, however, may sufficiently approach Martian conditions that their piping caves may serve as useful models of Martian conduits. Speleological observations are largely lacking in such forbidding parts of the Earth. But a published photograph of the Oumou Caves, Republic of Chad, suggests that they should be investigated as possible terrestrial equivalents in clastic sedimentary rock. The Cave of Swimmers and other sandstone caves reported on Egypt's Gilf Kebir also may be equivalents, but except for archaeological features, little is known of these. In a humid region, Christmas Canyon Cave (Washington state, United States) is a low but extensive piping cave in volcanic ash beneath a resistant layer of basalt, seemingly replicating the Martian stratigraphy. Major piping caves (e.g. Officers Cave, Oregon and Anvil Points Claystone Cave, Colorado, United States) also may have features usefully analogous to Martian conduits in landslide or slumped material (see Piping Caves and Badlands Pseudokarst).

WILLIAM R. HALLIDAY

See also **Crevice Caves; Glacier Caves and Glacier Pseudokarst; Littoral Caves; Piping Caves; Talus Caves**

Works Cited

Eraso, A. & Pulina, M. 1992. *Cuevas en Hielo y Rios Bajo los Glaciares*, Madrid: McGraw-Hill

Floridia, G. 1941. Un particolare fenomeno pseudocarsico manifestato da algune argile. *Bolletino della Societa di Sciencio Naturale ed Economiche di Palermo*, 23: 10–19

Fountain, A. & Wilder, J. 1998. Water flow through temperate glaciers. *Reviews of Geophysics*, 36(3): 299

Hose, L. 1996. The Lost Park Reservoir Project, Park County, Colorado. In *The Caves and Karst of Colorado, 1996 NSS Convention Guidebook*, edited by R. Kolstad, Huntsville, Alabama: National Speleological Society

Kempe, S. & Halliday, W. 1997. Report of the discussion on pseudokarst. In *Proceedings of the 12th International Congress of Speleology*, vol. 6, Basel, Switzerland: Speleo Projects: 107

Knebel, W. von 1908. *Höhlenkunde mit Berücksichtigung der Karstphänomene*, Braunschweig: Friederich Vieweg und Sohn

Malin, M. & Edgett, K. 2000. Evidence for recent groundwater seepage and surface runoff on Mars. *Science*, 288: 2330–35

Otvos, E. 1976. "Pseudokarst" and "pseudokarstic terrains": Problems of terminology. *Geological Society of America Bulletin*, 87: 1021–22

Péwé, T., Liu Tungsheng, Slatt, R.M. & Li Bingyuan. 1995. *Origin and Character of Loess-like Silt in the Southern Qinghai-Xizang (Tibet) Plateau, China*, Washington: Government Printing Office (US Geological Survey Professional Paper 1549)

Striebel, T. & V. Schäferjohann. 1997. Karstification of sandstone in Central Europe: attempts to validate chemical solution by analyses of water and precipitates. In *Proceedings of the 12th International Congress of Speleology*, vol. 1, Basel, Switzerland: Speleo Projects: 473–76

Further Reading

Proceedings volumes of international symposia on pseudokarst, most recent volume being *Proceedings of the 6th International Symposium on Pseudokarst*, Galyatető Hungary, 1996, edited by I. Eszterhás & S. Sárkozi

Newsletter of the Commission for Pseudokarst, International Union of Speleology. (bilingual), edited by Istvan Eszterhás (www.clubs.privateweb.at/speleoaustria/pseudokarst.htm)

Halliday, W. 1974. *American Caves and Caving: Techniques, Pleasures, and Safeguards of Modern Cave Exploration*, New York: Harper and Row, pp.46–70

Kosack, H.-P. 1952. Die Verbreitung der Karst- und Pseudokarsterscheinungen über die Erde. *Petermanns Geographische Mitteilungen*, 96 Jahr, 1st Q.: 1ff.

Kunský, J. 1957. Thermomineral karst and caves of Zbrašov, northern Moravia. Sbor. Čs. Spol. zeměpis., 62, 4 : 306–351.

Parker, G. & Higgins, C. 1990. Piping and pseudokarst in drylands. In *Groundwater Geomorphology: The Role of Subsurface Water in Earth-surface Processes and Landforms*, edited by C. Higgins and D. Coates, Boulder, Colorado: Geological Society of America

White, W., Jefferson, G. & Haman, J. 1966. Quartzite karst in southeastern Venezuela. *International Journal of Speleology*, 2: 309–14

QUARRYING OF LIMESTONE

The extraction of limestone has a long history, going back quite literally to the Stone Age. Three primary groups of extraction techniques may be identified: mechanical techniques; techniques that use low explosives (blackpowder); and modern techniques that use high explosives, often in combination with ANFO (Ammonium Nitrate and Fuel Oil). In general terms, there is a technological progression, but non-explosive and low-explosive techniques are still used for the extraction of dimensional (building) stone.

The earliest forms of quarrying utilized human muscle to remove limestone from free faces. For example, drawings of early English quarries show labourers suspended from ropes prising loose stones from valley sides. Large blocks were reduced by labourers with sledgehammers. Similar techniques are still practised in some developing countries where there is an abundance of labour. There are various modern non-explosive techniques, including the use of drilling, feathers and wedges, mechanical shovels on weaker limestones, and diamond saws to cut dimensional stone, particularly marble (Figure 1).

Explosives were introduced into quarrying in the 19th century. The essential property of any explosive is that, on detonation, it is converted as rapidly as possible into gases which occupy many times the original volume of the explosive. In high explosives the gases are produced almost instantaneously at very high temperatures and pressures and are accompanied by an intense shock wave. Blackpowder (gunpowder) is slower in action and the gases are released at much lower pressures. This difference in explosive property determines the amount of rock liberated on detonation, together with the resulting end-form of the blasted face.

Blackpowder was the primary explosive used in quarrying limestone until the mid-20th century when it was largely replaced by high explosives. It is only rarely used in modern quarrying, usually where there is a need to minimize fracture damage to the rock, as in the extraction of dimensional stone. In the older quarries the ground above the face was cleared of overburden prior to blasting, so as to reduce the amount of washing needed to clean the stone. Initially this was done by teams of quarrymen, a process known as piking (hand picking). Piking located suitable fractures and joints into which black powder could be poured and fired almost at random. These blasts were sometimes augmented by header tunnels, which were excavated beneath the face and packed with explosives. Teams of men with sledgehammers were employed to break down oversized rocks and to remove loose or overhanging rocks from the face, a process known as face-dressing. Blackpowder was replaced

Quarrying of Limestone: Figure 1. Tunstead Quarry, Derbyshire, England showing a typical rock face blasted using ANFO/emulsion explosives (top) and the same face after restoration blasting and habitat reconstruction (bottom).

firstly by high explosives such as TNT and nitro-glycerine and then by a combination of high explosives and ANFO. To accommodate the new explosives and take advantage of new technologies there were also changes to drilling and blasting designs. The aim of the drilling and blasting design is to excavate safely the maximum amount of stone of the desired fragmentation into an easily removed blast pile, while leaving the rock face in a condition suitable for further blasts to take place.

Limestone quarrying has three main environmental impacts: aesthetic, geomorphological, and hydrogeological. The aesthetic impacts are obvious, and to most people quarrying represents a visual intrusion into the landscape, although it has been argued that quarry faces provide features of interest in areas otherwise devoid of natural cliffs. In the past, dust was a major associated problem but in most countries this has been much reduced by the imposition of stringent controls.

Geomorphologically, quarrying represents the most dramatic impact of human activity on karst landscapes. It is initially erosional, through the removal of materials from the ground, but may also include depositional components such as spoil heaps and tailings lagoons. In process terms quarrying represents an extremely potent erosive agent. For example, during the 20th century, quarrying removed more limestone from the English Peak District than had been removed by natural processes during the entire Holocene period, some 10 000 years. This process activity results in the destruction and modification of surface and underground landforms including hills, valley sides, dolines, and cave systems. The latter has been a cause of particular concern, where the caves contain deposits of archaeological, paleoenvironmental or paleontological interest, or where there is a rich subterranean fauna.

When extraction ceases, a hole remains and, unless it is infilled, the perimeter slopes will begin to evolve under the influence of natural processes. Ultimately it may become virtually indistinguishable from the surrounding landscape, the time involved being a function of the excavation processes, geology (structure and strength), and the intensity of natural processes. In the English Peak District, a characteristic sequence of landforms develops in abandoned blackpowder quarries (Figure 2). In contrast to blackpowder quarries, there are just two primary landforms on faces in high explosive / ANFO quarries: blast fracture cones and buttresses. Blast fracture cones are doline-like features with a lateral extent of 3–5 m, while rock buttresses accord with the position of drilled shot holes and are easily identified on recently blasted faces by the prominent explosives scorch marks. They project out from the quarry face and increase in size and lateral extent towards the quarry or bench floor. Their vertical extent is half to two-thirds of the face height, in contrast to the buttresses produced by blackpowder blasting, which are frequently full face height.

Quarrying is often associated with groundwater pollution, principally by fine material and also, in some cases, by fuel oil. In some quarries, pumping of groundwater allows working to extend sub-water table but this may result in the drying up of springs and surface streams. Hence it is important that any new quarries are located in areas where potential impacts may be minimized.

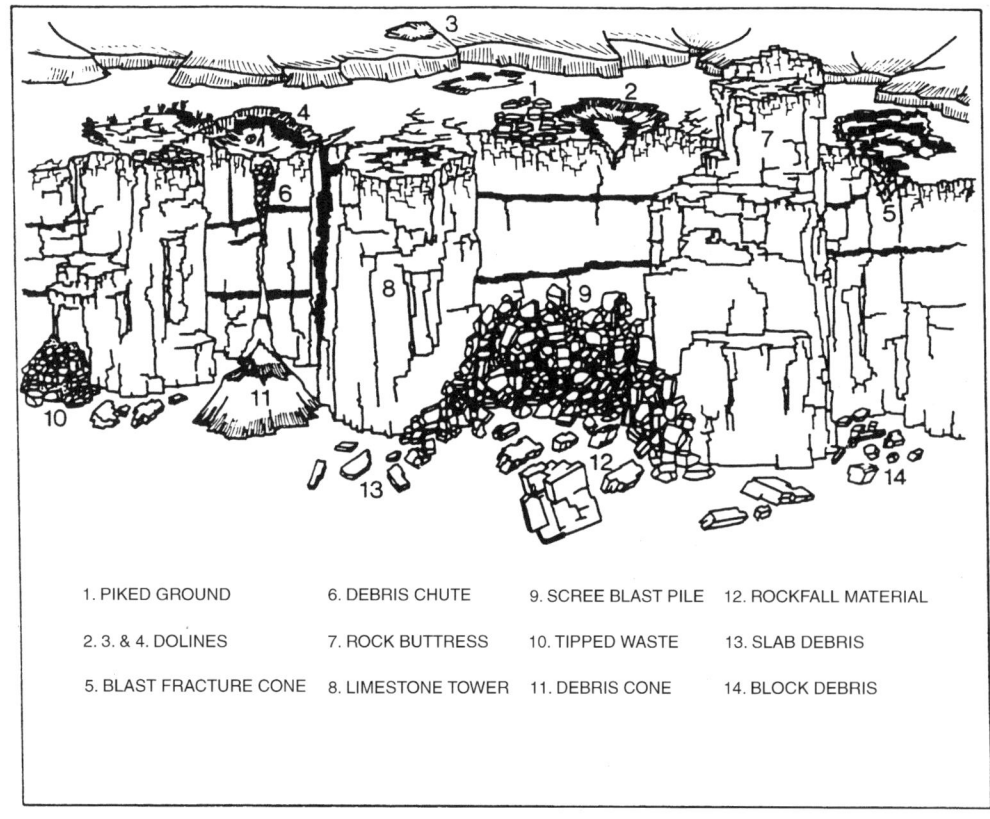

1. PIKED GROUND
2, 3. & 4. DOLINES
5. BLAST FRACTURE CONE
6. DEBRIS CHUTE
7. ROCK BUTTRESS
8. LIMESTONE TOWER
9. SCREE BLAST PILE
10. TIPPED WASTE
11. DEBRIS CONE
12. ROCKFALL MATERIAL
13. SLAB DEBRIS
14. BLOCK DEBRIS

Quarrying of Limestone: Figure 2. Typical landform assemblage in an abandoned quarry in the English Peak District excavated by blackpowder blasting.

A principal concern, in areas where limestone is extracted, is the restoration of sites once quarrying has ceased. The small scale and extent of the blackpowder blasted quarries, together with the manner in which their margins evolve under the influence of natural processes, mean that they will eventually be assimilated into the natural landscape, over periods of tens to hundreds of years. By contrast, the scale of modern quarries and the methods used in their excavation are such that they will continue to intrude upon the natural landscape for many centuries. In order to speed up the process of assimilation a new technique, Landform Replication, was developed in Britain. The first step is to identify the key elements in a natural limestone landform sequence. These are then replicated using restoration blasting and habitat reconstruction. In the Peak District, restoration blasting aims to replace the engineered appearance of a production blasted quarry face by a sequence of constructed landforms, whose scale and extent mimics that of a natural limestone valley side. Habitat reconstruction aims to establish vegetation, similar to that on natural valley sides, on the constructed landforms. Landform replication techniques have been applied on the Niagara escarpment in Canada and other techniques for quarry reclamation, particularly of the quarry floor, have been pioneered at the Banburi Quarry near Mombasa, Kenya and at the Lune River Quarry in Tasmania.

John Gunn

See also **Limestone as a Mineral Resource**

Further Reading

Brashaw, P. (editor) 2002. *Reclamation of Limestone Quarries by Landform Simulation: Summary of Lessons Learnt from Trial Sites*, London: Department for Transport, Local Government and the Regions

Drew, D. & Hötzl, H. (editors) *Karst Hydrogeology and Human Activities: Impacts, Consequences and Implications*, Rotterdam: Balkema. (Section 6 contains a number of papers on the Impacts of extractive industries).

Ekmekci, M. 1993. Impact of quarries on karst groundwater systems. In *Hydrogeological Processes in Karst Terrains*, edited by G. Günay, International Association of Hydrological Sciences Publication no. 207

Gillieson, D. & Houshold, I. 1999. Rehabilitation of the Lone River Quarry, Tasmanian Wilderness World Heritage Area, Austrialia. In *Karst Hydrogeology and Human Activities: Impacts, Consequences and Implications*, edited by D. Drew & H. Hötzl, Rotterdam: Balkema

Gunn, J. 1993. The geomorphological impacts of limestone quarrying. In *Karst Terrains: Environmental Changes & Human Impact*, edited by P.W. Williams, Cremlingen-Destedt: Catena

Gunn, J. & Bailey, D. 1993. Limestone quarrying and quarry reclamation in Britain. *Environmental Geology*, 21(3): 167–72

Gunn, J., Bailey, D. & Handley, J. 1997. *The Reclamation of Limestone Quarries using Landform Replication*, London: Department of the Environment, Transport and the Regions

Harrison, D.J., Buckley, D.K., & Marks, R.J. 1992. *Limestone Resources and Hydrogeology of the Mendip Hills*, Keyworth: British Geological Survey (BGS Technical Report WA/92/19)

Hobbs, S.L. & Gunn, J. 1998. The hydrogeological impacts of quarrying karstified limestone, options for prediction and mitigation. *Quarterly Journal of Engineering Geology*, 31: 147–57

Mineral Planning, 2001–2002. Three papers entitled Limestone landform simulation in quarry restoration were published in issues 89 (pp.28–32), 90 (pp.13–17) and 91 (pp.16–19). The journal publishes many short papers on aspects of limestone quarrying and quarry restoration.

Walton, G. & Allington, R. 1994. Landform replication in quarrying. *Transactions of the Institute of Mining and Metallurgy* (Section A: Mining industry), 103: A55–66

QUARTZITE CAVES OF SOUTH AMERICA

While not well endowed with carbonate karst South America hosts the world's best-developed quartzite karst. Extensive quartzite outcrops (see map in America, South), in varied climatic and geomorphic settings, allow the existence of a large number of caves of impressive extent and depth. Although exploration and research in quartzite areas is still in its early stages, there are now at least 11 caves over 1 km in length (Table 1), and 16 over 200 m in depth (Table 2), including the deepest and longest quartzite caves in the world.

The study of quartzite caves and karst is a relatively recent field of karst science (see separate entry, Silicate Karst). The model for formation of caves in quartzite was initially based on the quartzite caves of South Africa (Martini, 1979), but has found wider applicability on the quartzite caves of South

Quartzite Caves of South America: Table 1. The longest quartzite caves of South America.

Cave	Location	Country	Length (m)
1. Gruta do Centenário	Inficionado Peak	Brazil	3790
2. Gruta da Bocaina	Inficionado Peak	Brazil	3220
3. Sima Auyán-tepui Noroeste	Auyán-tepui	Venezuela	2950
4. Gruta das Bromélias	Ibitipoca	Brazil	2750
5. Sima Aonda Superior	Auyán-tepui	Venezuela	2128
6. Sima Aonda	Auyán-tepui	Venezuela	1690
7. Lapão	Chapada Diamantina	Brazil	1600
8. Sima Acopán 1	Acopán	Venezuela	1376
9. Sima de la Lluvia	Sarisariñama	Venezuela	1352
10. Sima Menor de Sarisariñama	Sarisariñama	Venezuela	1179

Quartzite Caves of South America: Table 2. The deepest quartzite caves of South America.

Cave	Location	Country	Depth (m)
1. Gruta do Centenário	Inficionado Peak	Brazil	481
2. Gruta da Bocaina	Inficionado Peak	Brazil	404
3. Sima Aonda	Auyán-tepui	Venezuela	383
4. Sima Auyán-tepui Noroeste	Auyán-tepui	Venezuela	370
5. Sima Aonda 3	Auyán-tepui	Venezuela	335
6. Sima Aonda 2	Auyán-tepui	Venezuela	325
7. Sima Auyán-tepui Norte	Auyán-tepui	Venezuela	320
8. Sima Mayor de Sarisariñama	Sarisariñama	Venezuela	314
9. Sima Auyán-tepui Norte 2	Auyán-tepui	Venezuela	297
10. Sima Aonda Este 2	Auyán-tepui	Venezuela	295

America. Current knowledge of quartzite karst in South America allows the recognition of at least two major types of caves: (1) vertical, fissure-like caves related to high plateaux; and (2) horizontal or gently sloping caves in flat lying areas of "cuesta" landscape.

Fissure vertical caves were recognized initially in the Precambrian quartzites of the Roraima Group, in the Gran Sabana area of southern Venezuela. In this ever wet region, where precipitation can reach as much as 4000 mm a^{-1}, a number of isolated flat-topped towers, known locally as tepuis, some over 1 km high, rise abruptly from the dense rainforest. The tops of the tepuis are criss-crossed with joints, some of them (mainly the ones close to the scarps) opening into vertical shafts over 100 m deep, that give access to joint-controlled passages at the bottom. Many of these passages contain underground streams, although it has not been possible to connect the passages to resurgences observed at the scarps. The difficulty of access has acted as a major limitation to exploration in this area. To date, some of the most visited plateaux are Sarisariñama, Auyán-tepui, Roraima, Autana and Kukenan, but there is still much work to be done in these areas, while many other tepuis await exploration. The Sarisariñama plateau contains massive open shafts, the largest, Sima Mayor, reaching as much as 350 m in diameter and 314 m in depth, holding an internal volume of c.12 million m^3 (Figure 1). The deepest caves to date have been explored at the Auyán-tepui. Sima Aonda and Sima Auyán-tepui Noroeste are, respectively, 383 m and 370 m deep. The latter is also the longest cave in the area, reaching 2950 m in length. The proximity of these caves to the tepui scarp has led to the suggestion that pressure relief has aided in the enlargement of the joints (Galán, 1991). Surface karst features include swallets, resurgences, pinnacles, towers, and solution basins (kamenitza). A small portion of the Roraima quartzite karst has been included in the National Park of Cainama, established in 1962 and now a World Heritage Site. The National Park protects mainly the Auyán-tepui, including its main attraction, the 970 m high Angel Falls.

In south–central Brazil, recent research has yielded another area of outstanding potential. The Inficionado Peak is one of a series of peaks in the Caraça Range, composed of Precambrian quartzites of the Caraça Group. Although not a plateau like the Venezuelan tepuis, it rises abruptly over the surrounding lower area, possessing a 1 km high scarp in its southeast-facing slope. The top of the Inficionado Peak covers only 0.9 km^2, but it presents a number of joints, some of them opening into deep and narrow crevasses that lead into rectilinear passages containing streams. The general character of the Inficionado karst resembles those of the tepuis, with the exception that there is no surface drainage, and it is sometimes possible to traverse the caves to entrances situated at the face of the scarp. The deepest and longest quartzite cave in the world, and currently the deepest cave in South America, is the 481 m deep and 3790 m long Gruta do Centenário (Figure 1). Gruta da Bocaina (404 m deep and 3220 m long) is another significant cave. Surface karst features are poorly developed, except for large collapse dolines and resurgences. Due to its small area, the potential at the Inficionado Peak appears limited, although there are still many open joints awaiting exploration. Other high quartzite peaks elsewhere in Brazil may hold similar potential for deep and long quartzite caves.

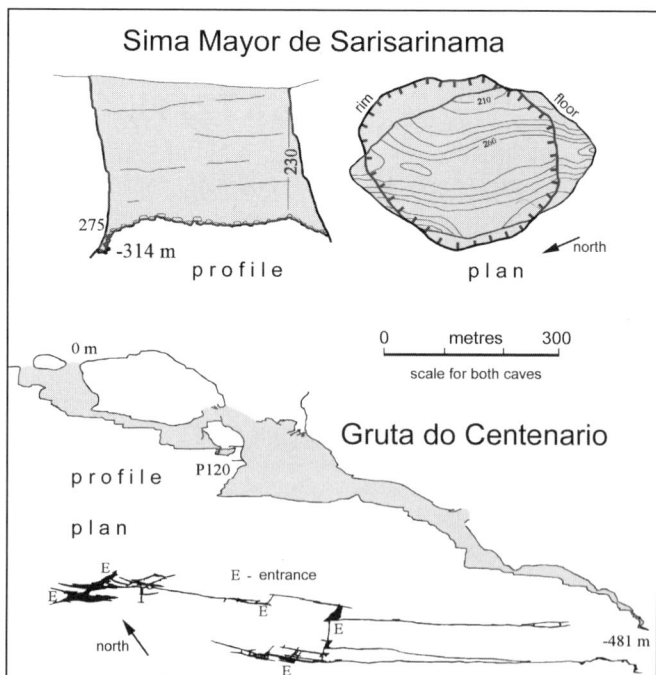

Quartzite Caves of South America: Figure 1. Plans and profiles of two major caves in South American quartzite: Sima Mayor de Sarisariñama and Gruta do Centenário (courtesy of Grupo Bambuí de Pesquisas Espeleológicas).

Quartzite Caves of South America: Figure 2. Typical passage morphology in Lapão Cave (Lencóis) showing breakdown modification of vadose canyon. The 1.6 km cave is developed in metasandstones and conglomerates of the Tombador Formation and is the fourth longest cave in sandstones and quartzites in Brazil. (Photo by John Gunn)

The second type of quartzite cave, commonly found in South America, comprises caves occurring in the gently rolling cuesta-type of relief typical of the Andrelândia Group rocks of southern Minas Gerais state, Brazil (Corrêa Neto, 2000). These caves tend to follow favourable quartzite horizons (parallel to the cuesta gentler slope) and develop along active passages with frequent tributary junctions. Many of these caves end in restrictions, due to the accumulation of sand, although some can be followed to another entrance. The largest concentration of this type of cave occurs at the Ibitipoca State Park, where the Gruta das Bromélias, with 2750 m of mapped passages, is the longest cave known. Other areas, with a large number of caves, are the Luminárias and Carrancas areas, also in southern Minas Gerais State, and the Espinhaço Range, extending across the states of Minas Gerais and Bahia. The Chapada Diamantina in central Bahia State is the northern extension of the Espinhaço orographic system. Here, Lapão Cave is the fourth largest cave in sandstones and quartzites in Brazil. Along the 1600 m of passage the dominant morphology is of breakdown modification of vadose conyons but in some parts the remnant elliptical passage cross sections suggest modification of phreatic conduits (Figure 2). The cave is also notable for opan speleothems.

AUGUSTO AULER

See also **America, South; Silicate Karst**

Works Cited

Corrêa Neto, A.V. 2000. Speleogenesis in quartzite in southeastern Minas Gerais, Brazil. In *Speleogenesis: Evolution of Karst Aquifers*, edited by A.B. Klimchouk, D.C. Ford, A.N. Palmer & W. Dreybrodt, Huntsville, Alabama: National Speleological Society

Galán, C. 1991. Disolución y génesis del karst en rocas carbonáticas y rocas silíceas: Un estudio comparado [Dissolution and genesis of karst in carbonate and siliceous rocks. A comparative study]. *Munibe*, 43: 43–72

Martini, J.E.J. 1979. Karst in Black Reef quartzite near Kaapsehoop, eastern Transvaal. *Annals of the Geological Survey of South Africa*, 13: 115–28

Further Reading

Chalcraft, D. & Pye, K. 1984. Humid tropical weathering of quartzite in southeastern Venezuela. *Zeitschrift für Geomorphologie*, 28(3): 321–32

Galán, C. & Lagarde, J. 1988. Morphologie et evolution des cavernes et formes superficielles dans les quartzites du Roraima (Venezuela) [Morphology and evolution of caves and surface forms in the Roraima quartzites (Venezuela)]. *Karstologia*, 11–12: 49–60

Martini, J.E.J. 2000. Dissolution of quartz and silicate minerals. In *Speleogenesis: Evolution of Karst Aquifers*, edited by A.B. Klimchouk, D.C. Ford, A.N. Palmer & W. Dreybrodt, Huntsville, Alabama: National Speleological Society

Szczerban, E. & Urbani, F. 1974. Carsos de Venezuela. Parte 4: Formas carsicas en areniscas precambricas del Territorio Federal Amazonas y Estado Bolivar [Venezuelan Karst. Part 4. Karst forms in Precambrian sandstones in Amazonas Federal Territory and Bolivar State]. *Boletin de la Sociedad Venezolana de Espeleologia*, 5(1): 27–54

Urbani, F. 1978. Les karsts gréseux du Venezuela. Une spéléologie entre ciel et terre [The Venezuelan sandstone karst. A speleology between the sky and Earth]. *Spelunca*, 18(1): 24–8

Urbani, F. 1986. Notas sobre el origen de las cavidades en rocas cuarciferas precambricas del Grupo Roraima, Venezuela [A note on the origin of caves in precambrian quartzite of the Roraima Group, Venezuela]. *Interciências*, 11(6): 298–300

Wray, R.A.L. 1997. A global review of solutional weathering forms on quartz sandstones. *Earth Science Reviews*, 42: 137–60

R

RADIOLOCATION

Radiolocation, in the speleological context, is the procedure of determining the position and depth of an underground radio transmitter beacon, by making measurements on the surface using a radio receiver. The name is perhaps a misnomer, because the underlying physical principle owes more to the phenomenon of magnetic induction than to "true" radio. Using radiolocation, an underground aven or choked inlet passage can be correlated with a surface feature quicker than by conventional cave surveying methods (see Surveying Caves). Such fixes are useful, not only for exploration, but as an aid to cave communication using induction radios, which generally require the shortest possible communications path (see entry, Communications in Caves). Rescue teams maintain maps showing the location of these surface and underground stations, thus allowing communications to be set up swiftly in the event of an incident. Radiolocation is also used to verify underground surveys—the point on the surface immediately above a survey station is radiolocated and its position measured by an accurate surface survey or GPS reading.

Radiolocation in its present form dates from the mid-1950s, when transistorized equipment started to become readily available to experimenters and the mid-1960s also saw a spate of research by the US Bureau of Mines. The method of radiolocation used by cavers has changed little since its inception. A transmitter antenna, consisting of many turns of wire on a loop of typically 500 mm diameter, is placed horizontally underground with the aid of a spirit level (Figure 1). An amplifier drives the antenna with a very low frequency (VLF) signal, causing it to generate an alternating magnetic field. Typically, a power of 10 W is used at 1–3 kHz. At such low frequencies there is virtually no electromagnetic radiation—the power is dissipated as heat in the antenna. The magnetic field induces a signal in a receiver antenna, which is usually of similar construction to the transmitter. If the receiver loop is orientated such that no magnetic field lines pass through it, then it will not detect any signal. Establishing this "null" orientation at several locations allows the operator to triangulate a "ground zero" point directly above the underground transmitter, at which location the field lines are vertical (Figure 2). The method is described by Glover (1976) and France (2001).

With ground zero (GZ) fixed, there are several methods of determining the depth. The simplest, and one that has been used by the mining industry and for commercial pipe and sewer location, is to measure the field strength using calibrated equipment and to calculate the depth from the knowledge that the field strength decays as the inverse cube of distance. Cavers normally use a different method that does not depend on accurate measurement of signal strength. At a measured distance x from GZ, the angle α of the field line to the horizontal is determined using the "nulling" procedure outlined above (Figure 3). The depth d below GZ is then given (Glover, 1973) by

$$\frac{x}{d} = \frac{\sqrt{(8 + 9 \tan^2 \alpha)} - 3 \tan \alpha}{2}$$

The careful surveyor will take several readings at different distances, plot a graph, find a best-fit curve and thereby arrive at a good approximation to the depth. In practice, a couple of rough-and-ready approximations are adequate—for the field line at 45°, we have, $x/d \approx 0.562$, so the depth is approximately twice the distance x and, when $x/d = 1$, the field angle is $\alpha \approx 18.4°$.

The formula assumes that the field lines are in a "bar magnet" pattern. Obvious exceptions are if a magnetic ore body distorts the field or if the transmitter is tilted (Gibson, 1998). With a levelled antenna, an experienced operator can achieve an accuracy for GZ of 1 m for a 50 m depth (2%) and a depth accuracy of perhaps 5%. With a 5° tilt, geometry dictates additional errors of 3%. If the surface terrain is uneven, this must be taken into account or further errors will result. If the transmitter is grossly tilted, then the technique described here will not work. However, the methods of radiolocation used by the mining industry do allow for this situation, where body-worn transmitters can allow prone or unconscious miners to be located. An adaptation

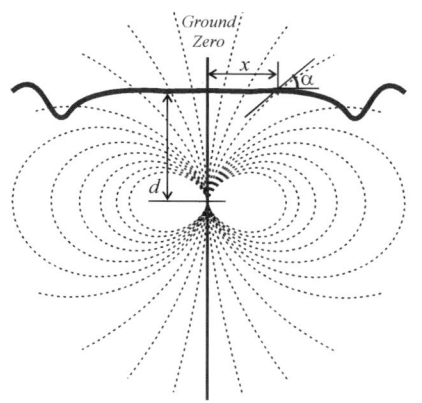

Radiolocation: Figure 1. Levelling a radiolocation transmitter loop prior to use, in Valley Entrance, Kingsdale Master Cave, North Yorkshire. (Photo by Mike Bedford)
Radiolocation: Figure 2. The magnetic field lines from an underground transmitter form the familiar "bar magnet" pattern. Measurements of field angle (α) and distance from Ground Zero (x) on the surface allow the depth (d) to be calculated.
Radiolocation: Figure 3. Using a nulling technique to measure field angle—note the clinometer attached to the side of the receiver loop antenna. (Photo by Mike Bedford)

of the technique using vertical antennas allows cave-to-cave location.

Derivations of depth from measurements on different bearings from GZ will sometimes differ markedly, especially if the transmitter is deeply buried, or if $x \gg d$. If the transmitter is known to be level, and if the presence of ore bodies can also be discounted, then the likely cause is that the ground is anisotropic with respect to electrical conductivity—which is only to be expected for a bedded and jointed rock such as limestone. Indeed, an adaptation of this radiolocation technique can be used to measure the electrical conductivity of the ground by a non-contact means.

The fact that the electrical conductivity of the rock can affect what is generally assumed to be a purely magnetic measurement is an important observation, demonstrating that magnetic induction is not the only phenomenon at work. It is the predominant effect only when the separation of the transmitter and receiver is much less than a "skin depth"; this being a figure of merit used to describe the extent to which electromagnetic waves penetrate into a conductor. If this condition cannot be met, the "secondary field" caused by induced currents becomes significant and prevents accurate measurement of α. Additionally, neither the depth equation nor the inverse-cube law hold true in these conditions. A useful GZ measurement is still possible, but depth determination is prone to errors of around 20% at two skin depths. It is for this reason that the accuracy of deep radiolocations must be called into question.

Close to a transmitter, skin depth does not have the usual physical interpretation, but is nevertheless defined, formulaically, in the way familiar to geophysicists. In the presence of a water table or conducting overburden, skin depth can be very low. It may vary from under 10 m (conductivity 0.01 Ω^{-1} m^{-1}, frequency 100 kHz) to over 500 m (0.001 Ω^{-1} m^{-1}, 1 kHz). Techniques for correcting the errors due to deep radiolocation are being developed (Pease, 1997).

When radiolocation leads to new cave entrances being opened, this may not always be for the best. In Dale Barn Cave (East Kingsdale, Yorkshire), divers radiolocated an aven that was then excavated in an impressive engineering operation. This opening of the far reaches of a cave to non-divers—with the attendant question of cave conservation—is not unique. Several points in the far reaches of the Gaping Gill system (Yorkshire) are known to be very close to the surface but so far remain unexcavated. In Wales, radiolocation aided the opening of a "back door" entrance to Ogof Draenen and this led to fierce arguments about the need to balance the needs of cave explorers and rescuers with the need for cave conservation (Lovett, 1999). With a new understanding of how magnetic fields propagate in rock, it is likely that location systems will develop further, becoming more widespread and easier to use. Although a useful aid to surveying, the need for conservation must be borne in mind.

David Gibson

Works Cited

Note: The *Cave Radio & Electronics Group* and the *Cave Surveying Group* are special-interest groups of the British Cave Research Association.

France, S. 2001. Cave surveying by radio-location—2. *Cave Radio and Electronics Group Journal*, 44: 21–23
Gibson, D. 1998. Radiolocation errors arising from a tilted loop. *Compass Points (BCRA Cave Surveying Group Newsletter)*, 21: 18–20
Glover, R.R. 1973. Cave depth measurement by magnetic induction. *Lancaster University Speleological Society Journal*, 1(3): 66–72
Glover, R.R. 1976. Cave surveying by electromagnetic induction. In *Surveying Caves*, edited by B. Ellis, Bridgwater: British Cave Research Association
Lovett, B. 1999. Making an entrance, stage left... *Descent*, 147: 24
Pease, B. 1997. Determining depth by radiolocation: An extreme case. *Cave Radio and Electronics Group Journal*, 27: 22–25

Further Reading

Gibson, D. 1996. How accurate is radio-location? *Cave and Karst Science*, 23(2): 77–80

A discussion of some of the geophysical causes of errors in radiolocation.

Gibson, D. 2000. A channel sounder for sub-surface communications: Part 2—computer simulation of a small buried loop. *Cave Radio and Electronics Group Journal*, 41: 29–32
Some theoretical background to recent work on radiolocation at extreme depth, using formulations given by Wait (1982).

Gibson, D. 2001. Cave radio notebook. 46: The role of skin depth in cave radio. *Cave Radio and Electronics Group Journal*, 45: 28
A note explaining that for radiolocation, skin depth does not have the physical interpretation usually given.

Gibson, D. 2001. Cave surveying by radio-location-1. *Cave Radio and Electronics Group Journal*, 43: 24–26
An introduction to a series on radiolocation that collects together existing knowledge and outlines new techniques. It includes references to practical articles on beacon construction.

Wait, J.R. 1982. *Geo-Electromagnetism*, New York: Academic Press
This is a highly mathematical book, but it describes in detail the behaviour of electromagnetic fields in a conducting medium and is the basis of current work in the field (e.g. see Gibson, 2000).

Historical Further Reading

Brooks, N, & Ellis, B. 1956. An independent check on the survey of Ogof Ffynnon Ddu. *South Wales Caving Club Newsletter*, 16: 1–2
This is one of the earliest reports of cavers using radiolocation. No detail is given.

Mixon, W. 1966. Locating an underground transmitter by surface measurements. *National Speleological Society News*, 24(4): 61, 74–75
This article covers the principles in a similar manner to Glover (1973, 1976).

Useful Websites

David Gibson's Cave Radio page at http://www.caves.org.uk/radio/ includes links to sites that discuss cave communications and radiolocation.

The British Cave Research Association's Cave Radio & Electronics Group maintains a website at http://www.bcra.org.uk/creg/

The US-based National Speleological Society has a section for cave electronics, with a website at http://www.caves.org/section/commelect/

RADON IN CAVES

Radon, the heaviest naturally occurring gaseous element, is inert, colourless, and odourless. Although it has 20 known radioactive isotopes, the most significant is ^{222}Rn and the remainder of this article is concerned solely with this radioactive gas. ^{222}Rn is a decay product of ^{226}Ra (radium) which is itself part of the decay series of uranium (^{238}U), an element which is widely distributed in the Earth's crust, though generally in low concentrations. When radon in turn decays, it forms very small solid particles of other radioactive substances, the short-lived isotopes of lead (^{214}Pb), polonium (^{218}Po and ^{214}Po), and bismuth (^{214}Bi), which are collectively known as radon progeny or "daughters", before decaying into the stable element ^{206}Pb. The ratio of radon progeny to radon gas is called the equilibrium factor (F).

When first formed the radon progeny exist as free ions known as the unattached fraction. Some of the progeny attach themselves to ambient aerosols and are then called the attached fraction; others attach themselves to surfaces where decay takes place, a process known as plateout. As radon is a noble (inert) gas it is almost completely exhaled after inhalation with no effects on health. However, the radon progeny are very reactive, and if they are inhaled they are likely to be deposited in the lungs where the α-radiation from their decay can damage tissue. Long-term studies of uranium miners (e.g. Lubin *et al.*, 1994) have shown that exposure to high concentrations of radon progeny increases the risk of lung cancer and may also increase the risk of other cancers such as acute myeloid leukaemia. The majority of researchers are of the opinion that long-term exposure to lower concentrations of radon progeny in domestic housing also significantly increases the risk of lung cancer (Lubin & Boyce, 1997; Darby, 1998). However, there are still some workers who regard this link as unproven (e.g. Aley, 2000) and there is considerable debate over the risk to caver health from exposure to radon. This is considered later, after discussion on measurement methods and measured concentrations in caves.

The SI unit of radiation activity is the Becquerel (Bq) and the concentration of radon gas is expressed in Becquerels per cubic metre of air (Bq m^{-3}) although in North America the older unit of activity, the Curie, is still used and concentrations are often expressed in picoCuries (pCi, where 100 pCi $\simeq$ 3.7 Bq). The Potential Alpha Energy Concentration (PAEC), effectively the α-energy produced by radioactive decay of radon progeny, is usually measured in Working Levels (WL) and exposure levels are often expressed in working level hours (WLh, where 1 WLh is exposure to 1 WL for 1 hour) or Working Level Months (WLM, where 1 WLM = 170 WLh). In most studies of radon in caves and mines the radiation dose accrued by an individual is calculated as a function of exposure time in a radiation environment (potential α-energy exposure) and is expressed in milliSieverts (mSv). There are many complexities in the conversion of measured radon gas concentrations into an estimate of radon progeny concentration and further problems in equating the amount of time exposed to progeny into an estimate of dose. However, the commonly used approximations are that 5 mSv $\simeq$ 68 WLh and that 3700 Bq m^{-3} of radon gas at equilibrium (F = 1) is equivalent to 1 WL.

Track etch detectors are the most widely used method of measuring radon gas concentrations in cave air. These are passive devices containing an α-sensitive film in a sealed chamber that are left in a location for a period of time (7 days to 3 months depending on how high the concentrations are). They are then removed, etched, and the impacts counted. The activated carbon monitors that are commonly used in buildings have been less successful in caves due to absorption of water in the high humidity conditions usually present underground, and Friend (2000) has also reported problems with water ingress into some track etch detectors. Spot measurements of radon progeny may be made using the Kusnetz method whereby an air sample is drawn through a filter and the activity on the filter is counted. This is a simple and effective method routinely employed by the author in show caves in Britain and Ireland but it has the disadvantage that the activity must be counted between 40 and 90 minutes after sampling, precluding use at any distance from an entrance.

There are also a variety of proprietary instruments for obtaining spot measurements of progeny but most are bulky and expensive, a notable exception being the Thompson & Nielson "Radon Sniffer" which has been used in caves by the author and others.

Concern over the risk to the health of cavers from exposure to radon first arose in the mid-1970s and since then there have been many thousand measurements in caves across the world, particularly those open to the public. Mean concentrations for caving regions range from 200 to over 3000 Bq m^{-3} but averages for an individual cave system may be more than ten times higher (Hyland & Gunn, 1994). The highest measured concentration in a natural limestone cave (as opposed to a mine) is thought to be the 155 000 Bq m^{-3} recorded by Gunn *et al.* (1991) at Giants Hole in Derbyshire, England. The concentration of radon at any point will be a complex function of emanation, exhalation, and the cave climate. Emanation is the net radon production rate in the rock walls of a cave, or in unconsolidated sediment within the cave, and ultimately depends on the concentration of uranium (which then decays to radium) in the material. Exhalation is the proportion of radon produced that is released into the atmosphere. As radon is an inert gas it is able to diffuse away from the place where it is produced and some atoms travel from within the rock / sediment grain to a pore and thence to a crack and the cave atmosphere. Exhalation is influenced by geological factors (Ball *et al.*, 1991) and by climate, especially barometric pressure. Radon is a soluble gas and hence may be added to, or less commonly removed from, the cave atmosphere by water. Once in the cave the radon gas concentration will be modified by natural ventilation (if any) and this will also influence the equilibrium factor and hence the concentration of radon progeny. The combination of all these factors means that radon / radon progeny concentrations exhibit marked temporal variations (hourly, daily, and seasonally) and spatial variations (from region to region, from cave to cave within a region, and from site to site within a cave) which makes prediction very difficult. For example, a study in Britain during 1991–92 found that mean concentrations in the Peak District were almost eight times higher than some 90 km away in the Northern Pennines (Hyland & Gunn, 1994). Concentrations are generally lower in winter than in summer because there is a strong draught into most caves in winter whereas in summer the air tends to be more static or to draught out (see Climate of Caves).

Returning to the question of risk to cavers' health, it is important to distinguish between recreational users of caves and those who are employed to work underground as guides in show caves or adventure caves, as instructors at outdoor pursuits centres, or in some other capacity. In many countries legislation controls the amount of radiation that an employee may receive while at work. For example, in the United Kingdom the Ionising Radiations Regulations (1999) sets a maximum exposure of 15 mSv a^{-1} and states that anyone receiving more than 6 mSv a^{-1} must be designated as a "Classified Employee". Initial measurements showed that in some show caves the guides were receiving high radiation doses and it was necessary to install forced air ventilation systems to reduce the radon concentrations. This is very much a last resort when doses cannot be reduced by controlling exposure through changes to working patterns, as although the ventilation has proved to be very effective in reducing the radon concentrations it also changes the cave microclimate, possibly to the detriment of speleothems and cavernicoles. In Britain, and probably also in other countries, the radiation dose that individuals may accrue during recreational caving is not proscribed by law, although in Britain the National Caving Association (1996) has endorsed a recommendation from the National Radiological Protection Board (Kendall, 1995) that recreational cavers should not exceed an exposure of 10^6 Bq h m^{-3} (*c*.3 mSv) in any one year. On the basis of a survey reported by Hyland & Gunn (1994), this represents just over 110 hours underground in the Peak District but almost 900 hours in the North Pennines, emphasizing the need for more data on the spatial and temporal variability in radon and radon progeny concentrations in individual caves.

While most recent publications have adopted the position that cavers should take precautions to reduce, or keep to a minimum, their radiation dose it should also be noted that several medical practitioners and experts in the field of radiation protection who are also active cavers are sceptical as to the risk that exposure to radon progeny poses to recreational cavers. These sceptics cite the facts that cavers are, in general, healthy individuals and that there are few, if any, known cases of cavers who were not also smokers dying of lung cancer. It has also been suggested that the cave environment may be particularly subject to plate out and to a low attached fraction of radon progeny although this has not been proven and some initial results suggest that the opposite may be the case. In the light of all the adverse publicity that has been given to radon, and particularly radon in caves, it may come as some surprise to learn that "Radon is used for therapeutic purposes in many medical facilities around the world. Bathing in radon water and *radon exposure in caves* [my italics] are the most widely employed forms of application" (Falkenbach *et al.*, 2002). This paper further states that there is a large body of evidence showing the beneficial long-term effects of radon therapy in rheumatic diseases and discusses "speleotherapeutic" radon exposure in Austria.

John Gunn

Works Cited

Aley, T.J. 2000. Radon and radon daughters. Section 15.3 in *Standard Handbook of Environmental Science, Health and Technology*, edited by J.H. Lehr & J.K. Lehr, New York: McGraw-Hill

Ball, T.K., Cameron, D.G., Coleman, T.B. & Roberts, P.D. 1991. Behaviour of radon in the geological environment: A review. *Quarterly Journal of Engineering Geology*, 24: 169–82

Darby, S. 1998. Risk of lung cancer associated with residential radon exposure in south-west England: A case-control study. *British Journal of Cancer*, 78: 394–408

Falkenbach, A., Kleinschmidt, J., Soto, J. & Just, G. 2002. Radon progeny activity on skin and hair after speleotherapeutic radon exposure. *Journal of Environmental Radioactivity*, 62: 217–23

Friend, C.R.L. 2000. Contribution of airflow to the control of seasonal variation in radon concentrations in Ogof Fynnon Ddu, Penwyllt, South Wales. *Cave and Karst Science*, 27: 101–08

Gunn, J., Fletcher, S. & Prime, D. 1991. Research on radon in British limestone caves and mines, 1970–1990. *Cave Science*, 18: 63–67

Hyland, R. & Gunn, J. 1994. International comparison of cave radon concentrations identifying the potential alpha radiation risks to British cave users. *Health Physics*, 67: 176–79

Kendall, G.M. 1995. Limiting radon exposure in caves and abandoned mines. *Radiation Protection Bulletin*, 165: 16–17

Lubin, J.H., Boice, J.D., Edling, C. *et al.* 1994. *Radon and Lung Cancer Risk: A Joint Analysis of 11 Underground Miners Studies*,

Rockville, Maryland: Institutes of Health (National Institute for Health Publications 94/3644)

Lubin, J.H. & Boise, J.D. 1997. Lung cancer risk from residential radon: Meta analysis of eight epidemiological studies. *Journal of the National Cancer Institute*, 89: 49–57

National Caving Association 1996. *Radon Underground* (report of the NCA Radon Working Group; summary available at http://web.ukonline.co.uk/nca/radonsum.htm)

Further Reading

Friend, C.R.L. 1996. Radon exposure during underground trips: A set of guidelines for cave and mine exploration in Britain. *Cave and Karst Science*, 23: 49–56

International Commission on Radiological Protection (ICRP). 1993. *Protection Against Radon 222 at Home and at Work*, Oxford: Pergamon Press (Annals of the ICRP, 23(2); ICRP Publication 65)

RAMSAR SITES—WETLANDS OF INTERNATIONAL IMPORTANCE

The Convention on Wetlands is an intergovernmental treaty adopted in 1971 in the Iranian city of Ramsar, and hence is commonly known as the Ramsar Convention. It was the first global treaty on conservation and wise use (sustainable use) of wetlands as natural resources. "Wetlands of International Importance" (Ramsar Sites) are sites designated under the Ramsar Convention on Wetlands and represent those wetland areas within a country, or areas shared by two or more countries, that are important internationally, but also nationally and locally. The criteria for inclusion in the List fall into two types: sites containing representative, rare, or unique wetland types; and sites of international importance for conserving biological diversity. Ramsar Sites are designated "To develop and maintain an international network of wetlands which are important for the conservation of global biological diversity and for sustaining human life through the ecological and hydrological functions they perform." (Vision for the Ramsar list, from the 7th Meeting of the Conference of the Parties, May 1999.) Ramsar Sites cover a number of different habitat types based on the Ramsar Classification System which includes 42 categories grouped into marine and coastal wetlands, inland wetlands, and human-made wetlands. All three groups include "karst and other subterranean hydrological systems".

Although the importance of wetlands for their biodiversity values and for the well-being of their human communities has long been recognized, they are still among the most threatened ecosystems globally. Appropriate management of the whole catchment area (including both surface water and groundwater), and the wise use, that is sustainable use, of wetlands at national level and through international cooperation are essential for the conservation of dynamic and unstable wetland ecosystems. Technical and policy guidelines have been developed to assist countries in preparing their national wetland policies based on these principles.

Wetlands and karst systems have some key factors in common. They both depend on water and the hydrological system of the catchment area. The quality and quantity of water, in combination with other factors, are crucial for maintaining the form and structure of karst and also for maintaining the ecological character of wetlands. Both are extremely fragile systems: changes in one component can affect the ecosystem quality and considerably change its character. To conserve and maintain wetlands, including karst and other subterranean hydrological systems, requires good knowledge of their main characteristics and an understanding of the processes that form and sustain them.

In accordance with Article 2.2 of the Ramsar Convention, which deals with the Ramsar List, "wetlands should be selected for the List on account of their international significance in terms of biology, botany, zoology, limnology or hydrology". From this point of view, the principal wetland conservation values of karst and other subterranean hydrological systems are:

1. rarity of karst phenomena or functions;
2. inter-dependency and fragility of karst systems and their hydrological and hydrogeological characteristics;
3. rarity of these ecosystems and endemism of their species;
4. importance for conserving particular taxa of fauna and flora.

Additionally, the functions and values range from maintaining—on a sustainable basis—high-quality water for drinking and other human uses (water for grazing animals or agriculture, tourism, and recreation) to supporting life in cave systems.

The main goal of including subterranean wetlands in the Ramsar List is to assist in the conservation and wise use of subterranean wetland functions and values and in implementation of Ramsar principles and strategic guidelines. In general terms, many "living" karst areas are wetlands, both surface and subterranean. Both direct (e.g. visitors to caves, researchers) or indirect (e.g. pollution—particularly water pollution; dumping of solid waste or sewage; development of infrastructure; water abstraction, retention in reservoirs, and other uses) development pressures are increasing. Appropriate management, including conservation and sustainable use, is crucial to maintain the functions and values of the interacting karst surface and subterranean hydrological systems in the whole catchment area and to prevent or mitigate threats to karst wetlands. The Ramsar Convention can help first by fostering conservation and wise use of subterranean wetland systems in general, and second by ensuring that examples of the most characteristic karst wetlands are considered and added to the List of Wetlands of International Importance.

Karst Ramsar Sites

There are presently (August 2002) 133 Contracting Parties to the Convention, with 1180 wetland sites, covering 103.2 million ha, designated for inclusion in the Ramsar List of Wetlands of International Importance. Those karst and other subterranean hydrological systems (type Zk) included in the List are shown in the Table. Since the introduction of the subterranean wetland type (Zk) in 1996, 12 Ramsar Sites that include subterranean wetlands have been added to the List. Ten are in carbonate rocks, but for only six areas have karst subterranean hydrological systems been ranked remarkable or dominant. The three European sites (see Table) are also listed as World Heritage Sites. Skocjanske Jame (see separate entry) is an underground river cave system developed in the area of the "classical" karst area of Kras, Slovenia. The main hydrological characteristics are the

Ramsar Sites: Type Zk Sites (karst and other subterranean hydrological systems) included in the List of Wetlands of International Importance. The table shows all of the wetland types present in the site and the subdivision of type Zk, where Zk(a) designates marine and coastal karst wetland; Zk(b) inland karst wetland; and Zk(c) anthropogenic karst or other subterranean wetlands. All except Sites 2 and 3 are underlain by carbonate rocks; Sites 2 and 3 are "other subterranean hydrological systems". The list was correct in May 2002. Source: Ramsar Convention Bureau: Information Sheets on Ramsar Wetlands, provided by Dineke Beintema, Wetlands International, Wageningen, The Netherlands (www.wetlands.org).

No.	Country	Sitename	Designated	Wetland Types	Type Zk
1	ALGERIA	La Vallée d'Iherir	02.02.2001	N,P,Ss,Y,**Zk**	Zk(b)
2	ALGERIA	Chott Ech Cherg, Sad'da	02.02.2001	Sp,Ss,Tp,Xf,Y,Zg,**Zk**	Zk(b)
3	ALGERIA	Oasis de Tamantit et Sid Ahmed Timmi	02.02.2001	R,Y,**Zk**	Zk(c)
4	CUBA	Ciénaga de Zapata	12.08.2001	A,B,C,D,E,F,G,L,M,N,O,P,Q,R, Sp,Ss,Tp,Ts,U,W,Xf,Xp,Y,**Zk**	Zk(b)
5	HUNGARY	Baradla Cave System and Related Wetlands	14.08.2001	**Zk**	Zk(b)
6	GUATEMALA	Parque Nacional Laguna del Tigre	26.06.1990	M,N,O,P,R,Sp,Ss,Tp,Ts,Xf,Y,**Zk**	Zk(b)
7	MADAGASCAR	Lac Tsimanampetsotsa	25.09.1998	Q,**Zk**	Zk(b)
8	MEXICO	Dzilam (reserva estatal)	07.12.2000	A,H,J,**Zk**	Zk(b)
9	NICARAGUA	Cayos Miskitos y Franja Costera Immediata	08.11.2001	A,B,C,E,F,G,H,I,J,K,M,N,P,Tp, Ts,W,Xf,Zg,**Zk**	Zk(b)
10	PAPUA NEW GUINEA	Lake Kutubu	25.09.1998	M,N,O,Tp,Xf,**Zk**	Zk(b)
11	SLOVAK REPUBLIC	Domica	02.02.2001	**Zk**,Ts,	Zk(b)
12	SLOVENIA	Škocjanske jame	21.05.1999	**Zk**	Zk(b)

extremely high fluctuations of groundwater level, flowing river currents fed by rainwater, and pools of stagnant water. The underground river has a discharge ranging from 0.05 to >400 m^{3}s^{-1} and water levels in the cave have a 130 m range. The area contains typical karst phenomena and karst features developed at the contact between permeable and impermeable rocks and in the limestones. The cave system also provides a habitat for numerous endemic and endangered animal species. The Baradla–Domica Cave System (see entries on Aggtelek and Slovak Karst and on Aggtelek Caves, Archaeology), shared between Hungary and Slovakia, is the largest subterranean hydrological system of the karst plateau in the territory of the two countries. The site is characterized by a permanent subterranean river, ponds, many speleothems, and diverse representatives of subsurface fauna as well as rich archaeological remains. The Aggtelek and Slovak Karst provides a habitat for more than 500 species of troglobite, troglophile, and trogloxene animals including endemic species (such as *Niphargus aggtelekas*), as well as species first described from this region. The most important archaeological sites are the settlements of Bükk culture both inside and in front of the cave entrance, with charcoal drawings unique in Central Europe.

The Zk wetland type is still to be further elaborated and it is intended that more surface and subterranean karst wetlands will be included. Additions to the list in the 1990s include Lake Tsimanampetsotsa in Madagascar where caves and underground rivers adjoin the lake on its eastern side, and Lake Kutubu in Papua New Guinea which has major subterranean inflow and outflow. The remainder of this article briefly describes five areas that have the potential to be Zk sites.

The karst catchment area of the Ljubljanica River in Slovenia is a complex site that includes a series of intermittent lakes in karst poljes and caves with underground rivers and is a good example of the interaction and interdependency between surface and subterranean wetlands. In addition to the landforms (see Cerkniško Polje, Slovenia: History), there are more than 300 bird species known in the area and 11 fish species, some of which are uniquely adapted to the intermittent character of karst lakes. The area also has an extremely rich and endemic aquatic and terrestrial subterranean fauna (see Postojna–Planina Cave System, Slovenia: Biospeleology).

The Otway Basin of Southern Australia provides an excellent example of a subterranean wetland with two aquifers totally separated by an impermeable layer of clay sediments. Although the upper aquifer has become severely polluted as a result of human settlement, the lower continues to provide high-quality water for domestic and other purposes. In the southern part of the basin numerous cenotes penetrate the impermeable layer, but the separation of the aquifers is maintained by thermoclines. The most important biota consists of a diversity of freshwater stromatolites, the continuing survival of which is threatened by eutrophication.

One of the most famous and spectacular subterranean wetlands occurs at Waitomo in the North Island of New Zealand. The walls and roof of a large cave with a flowing river are covered with the larvae of a fly, commonly termed glow-worms because of the luminescence of their caudal segments. These live by using a drapery of sticky threads to catch small gnats or flies, which originate from the upstream waters exposed on the surface.

The small and very isolated mid-Atlantic island of Bermuda contains more than 150 known caves, many of which contain anchialine pools in their interior. The caves have been found to contain 75 stygobitic species (see Walsingham Caves, Bermuda: Biospeleology). Due to their limited distribution, the fragile nature of the anchialine cave habitat, and severe water pollution and/or development threats, 25 of these species are listed as critically endangered.

Caves and karst features are common in nearly all parts of Mexico's Yucatán Peninsula, where the world's largest known underwater caves carry water to the Caribbean Sea along the east coast (see Yucatán Phreas). The Yucatán State Secretary of Ecology has inventoried more than 3000 cenotes although less than 100 of these have been explored by divers or otherwise scientifically investigated. Biological investigations of these caves have revealed a rich stygobitic fauna that is primarily marine relict species. Deep-well injection of sewage near Merida in the interior of the Peninsula and at new resort developments along the Caribbean coast is adversely impacting water quality within the caves and groundwater.

GORDANA BELTRAM

See also **Conservation: Protected Areas; World Heritage Sites**

References and Further Reading

Direction générale des forêts. 2001. *Atlas des zones humides algériennes d'importance internationale*, Alger: Direction générale des forêts, Ministère de L'agriculture.
Ramsar Convention Bureau. 1999. Resolution VII.13: Guidelines for identifying and designating karst and other subterranean hydrological systems as Wetlands of International Importance, *People and Wetlands: The Vital Link*, San José, Costa Rica (Ramsar Convention on Wetlands, Conference Proceedings)
Ramsar Convention Bureau. 1996. Resolution VI.5: Inclusion of subterranean karst wetlands as a wetland type under the Ramsar Classification System, Brisbane, (Ramsar Convention on Wetlands, Conference Proceedings, vol. 4/12: 16)
Ramsar Convention Bureau. 1998. Report of the Sub-regional Workshop (21–23 September 1998, Postojna, Slovenia)
Ramsar Convention Bureau. 2000. *Ramsar Handbooks for the Wise Use of Wetlands*, vols 1–9, Gland, Switzerland: Ramsar Convention Bureau
Ramsar Convention Bureau. 2000. *Strategic Framework and Guidelines for the Future Development of the List of Wetlands of International Importance*, vol. 7, Gland, Switzerland: Ramsar Convention Bureau (Ramsar Handbooks for the Wise Use of Wetlands, vol. 7)
UNESCO. 1994. *Ramsar Convention on Wetlands*, convention text, Certified copy, Paris, 13 July 1994
For detailed information on the Ramsar Convention see also http://ramsar.org/.

RECREATIONAL CAVING

The earliest cave exploration was curiosity-led, most visits being made for the advancement of science, particularly archaeology. Inevitably some of these scientists developed an interest in exploring caves for their own sake, and individuals emerged for whom the primary motivation was exploration. However, most of the early explorers also developed a strong interest in one or more areas of cave science (speleology), the classic example being Édouard Alfred Martel (1859–1938).

Cave exploration and documentation remained the realm of scientists and speleologists until the end of the 19th century. However, since the latter part of the 20th century there has been a steady increase in what is now called recreational caving. This form of caving activity is one of "going caving purely for the joy of caving" similar to the way that others enjoy walking or any other outdoor leisure pursuit. Thus, by the start of the 21st century, speleology, cave exploration, and recreational caving were all well established throughout the world with a huge range of clubs (see Exploration Societies) to cater for their different needs. The popularity of caves and caving is indicated by the large number of links to sites on the web. Advances in techniques for exploring caves have been such that clubs are less important than they once were as a source of communal equipment, but they remain important for training in the practical aspects of caving and in instilling a respect for the cave environment.

Recreational caving, as a separate activity from cave exploration, probably began with the development of outdoor pursuit groups such as the scouting (formed in 1908) and girl guide (formed in 1910) movements. These groups began a trend that saw visits to adventurous outdoor areas such as wild rivers, mountains, and caves as places to develop the "character" of individuals and groups. This form of activity was taken to new levels when companies were formed in the late 1980s to provide small through to large corporations with activities that developed "management" and "team-building" skills for staff members. Some of these activities were conducted in inappropriate caves resulting in detrimental impacts on the cave environment.

Another aspect of recreational caving occurs when individuals accidentally locate a cave entrance and with no prior experience or knowledge "visit the cave". Generally these visitors are ill-informed about the cave environment and therefore put themselves as well as the cave at risk from their activities. Some may go on to join exploration societies and become experienced cavers, but regrettably others indulge in acts of cave vandalism.

By the 1980s a significant quantity of cave exploration had been conducted in areas that were easily accessible from major centres of population. This meant that continued exploration would require greater technical effort or travel to more remote areas. Such high levels of exploration did occur but there was also a significant growth in the number of cavers who enjoyed visiting known and fully explored caves for their beauty, serenity, or purely the thrill of the sporting experience. This also added to the high levels of recreational caving throughout the world. An inevitable consequence of the increased numbers of recreational cavers has been significant impacts on popular caves. The lack of new caves or extensions to older systems in these areas of high visitation has led to significant changes in caving habits, including the development of speed caving and "speleo olympics". However, the majority of caver visits around the world are still for the pure joy of going caving and showing visitors or first-time cavers the fun, beauty, and tranquility of the cave environment. Unfortunately not everyone understands the fragility of the cave environment and terrible vandalism, both deliberate and accidental, has been inflicted on many caves. Examples of this vandalism can be seen on the Cave Vandalism website at http://wasg.iinet.net.au/vandals.html.

Concern over the impacts of recreational cavers on sites in Australia led to the development of a Minimal Impact Caving

Code (MICC). Codes of Practice or Ethics had been developed previously but the recreational impacts of cavers was still known to be significant (Spate & Hamilton-Smith, 1991). After consulting with cavers from around the world (using the internet) during the early 1990s, the cavers drafting the Australian MICC drew on a wide range of knowledge and experience in formulating the code that was finally adopted by the Australian Speleological Federation in 1995.

Other countries have followed the Australian lead with MICCs being developed to suit the requirements of caves in the United States in 1995 and Great Britain in 1999. These techniques are designed to reduce caver impacts by encouraging cavers to think carefully about every caving trip. Impacts can be minimized by track marking delicate areas of cave where trail widening can have significant impacts, or marking a single path, where a number of passages lead to the same location, to reduce impacts on the other trails. The development of MICCs does not ensure that cavers abide by them or even put them into practice. The strong support of both cave managers and cavers is required if using minimal impact caving techniques is to reduce the everyday impact of cavers.

Karst management throughout the world has been forced to change as outdoor recreational organizations and companies placed demands for access to cave and karst resources upon them. Karst management plans have to balance the needs for access with the management desire to conserve karst resources.

The major challenges that modern recreational cavers face are the increasing restrictions being placed upon them by cave managers who are attempting to protect the cave resource they are managing. These include the requirements for recognized leader qualifications and for protection of landowners, fellow cavers, and committees from possible legal action by providing public liability insurance. Further restrictions are likely to be applied by lawmakers as they deal with health and safety concerns.

The most recent development in recreational caving is the construction of commercial artificial caves, similar to artificial climbing walls, where groups can enjoy the "fun" of underground activity without the impact on the "natural" environment. These artificial caves can combine several of the adventurous activities associated with caving, such as oozing through mud, squeezes, water, climbing, and descending pitches all into one "cave". Although a "natural" cave may contain these features, they are likely to be more widely distributed. There has even been a genuine rescue from one such cave in Britain when a participant became trapped in a squeeze!

RAULEIGH WEBB

Work Cited

Spate, A. & Hamilton-Smith, E. 1991. Cavers impacts—some theoretical and applied considerations. In *Proceedings of the 9th Biennial Conference of the Australasian Cave and Karst Management Association*, edited by P. Bell, Margaret River, Western Australia: Australasian Cave and Karst Management Association

Useful Websites

Speleo Link http://hum.amu.edu.pl/~sgp/spec/links.html
Australian Speleological Federation Standards and Guidelines, http://www.caves.org.au/standards/set_standards.htm
US Minimum Impact Caving Code, http://www.caves.org/committee/conservation/www/b_caving/caving_rules.htm
National Caving Association of Great Britain Minimum Impact Caving Code, http://web.ukonline.co.uk/nca/canda/mimpcode.htm
TeamBuilding website example http://www.teambuildinginc.com/retreats_a_caving.htm
Outdoor Adventures–Caving http://www.mountainmayhem.com/caving.html

RELIGIOUS SITES

Mountains are visible and powerful elements in the landscape, and many have become important focal points for mythology (Bernbaum, 1997). In contrast, caves are hidden and hence perhaps unlikely candidates for veneration. Yet many sacred mountains contain caves that form part of their story, and caves share many of the attributes that have underpinned the attribution of sacredness to mountains, including their capacity to evoke a sense of mystery and the eternal. Hence, caves and other karst features have acquired profound spiritual significance in many parts of the world. Some karstic religious sites bear physical testimony to adoption by successive traditions, while some remain shared between faiths.

The religious significance of caves and karst varies across and within traditions. Christianity, professed by $c.33\%$ of the world's population, is arguably the most anthropocentric religion the world has seen, and rejects animism and pantheism in favour of a monotheistic outlook. Nevertheless, numerous karst sites are considered holy by Christians, as at the Massabielle Grotto and spring at Lourdes, where the Catholic girl Bernadette is said to have witnessed apparitions of the Virgin Mary in 1858. Worship of a landform would represent unacceptable idolatry within the rapidly growing Islamic tradition (covering 22% of the world's population). However, caves are very significant within the traditions of Hinduism ($c.15\%$), Buddhism ($c.6\%$), traditional Chinese religions including Taoism and Confucianism (4%), primal and indigenous faiths (3%), and a wide variety of smaller traditions. Neither are caves necessarily irrelevant for some among the $c.14\%$ of the world's population who profess no preference for any particular religion, not all of whom are atheists or agnostics.

Various types of karst landforms are considered sacred. Limestone summits such as the world's highest peak, Chomolungma or Sagarmatha, Goddess Mother of the Earth (Everest), and karst towers elsewhere, are or have been important to a number of traditions, including local Buddhism and Chinese Taoism and Confucianism. Followers of animist traditions in southeast Asia still make offerings to the spirits at curiously shaped limestone outcrops. The identification of the nurturing Earth as female by many cultures is linked to a perception that caves are womb-like, and the phallic connotations of speleothems have also been recognized (Dunkley, 1995). Some sacred caves are entirely artificial. The pyramids of Egypt have sometimes been interpreted as artificial sacred mountains inside which caves were deliberately constructed. Pilgrimage is an important part of the experience of some sites. Thus, visits to the Elephanta caves in India entail three journeys: across water, up a mountain, and into the cave, which serve to promote abandonment of the everyday world. Some karstic springs are also objects of veneration, and healing or divinatory properties may be attributed to their waters.

Sites Around the World
Flower pollen in a Neanderthal burial in Shanidar Cave, northern Iraq (see separate entry), suggests the use of caves to give expression to the spiritual 60 000 years ago. Although their meaning is unclear, ice-age paintings in caves such as Lascaux in France (see Vézère Caves, France: Archaeology) are among other early physical legacies that suggest caves were imbued with religious significance. The mixing of human blood with ochre used for artwork in Wargata Mina, Tasmania, emphasizes the probability of spiritual meaning.

Votive offerings have been found in a cave on Mt Ida in Crete, said to be the birthplace of Zeus, the father of the gods in Greek mythology. Among the many sacred caves of Crete are Eileithyia Cave, used for cult rituals from the Neolithic until the 5th century BC and sacred to Eileithyia, the goddess of childbirth; Kamares Cave, sacred since the Minoan period and possibly also dedicated to Eileithyia; and the cave of Agia Parajkevi Skotinou, a site of religious importance in antiquity and Christian times. Caves were also used for rites connected with the Phrygian mother goddess Cybele.

A cave at the World Heritage-listed Delphi site beneath sacred Mt Parnassus was sacred to Pan and the nymph Corycia. Remains of two fountains that date from the Archaic period and Roman era occur at the Castalian Spring, where niches cut in the rock face may have held offerings to the nymph Castalia. The female deity of the Earth was worshipped at Delphi during the Mycenaean period. In the early 3rd century BC the site became dominated by the Aetolians, and in 191 BC it was conquered by the Romans. It lost its religious significance with the spread of Christianity.

In the Christian faith, the parish church of St Paul's in Rabat, Malta, includes a grotto where St Paul is said to have lived during a three-month stay on the island in 60 AD following a shipwreck. Due to its supposed miraculous properties, soft limestone from this cave continued to enjoy popularity as a cure for fevers long after cave carbonate ceased to be widely used medicinally in the mid–18th century (Shaw, 1992). A cave near Cavadova in the Cantabrian Mountains of Spain is said to have provided refuge for King Pelayo who led the Christians in a battle that saw the first victory of the Christians over the Moors, a victory of great symbolic significance in the Christian conquest of Spain. A chapel in the entrance to Cavadonga Cave dates from the 8th century. Probably the most celebrated of Christian sites today is at Lourdes, where three chapels have been built, with the main Basilica of St Pius X accommodating 30 000 worshippers. Christian ceremonies are still occasionally conducted within caves, some of which, such as Lucas Cave at Jenolan, Australia, have been formally consecrated as churches.

Hindu societies have established many shrines in caves. The Batu Caves in Malaysia (see photograph in Asia: Southeast entry) attract over 800 000 devotees during the festival of Thaipusam. The Ajanta and Ellora caves near Bombay are significant for three major Indian religions, and include both carved temples and wall paintings. Reliefs, sculptures, and a temple at Elephanta are dedicated to the Hindu god Lord Shiva and probably date back to the Silhara kings of the 9th–12th centuries AD.

The Hindu site of Amarnath Cave at 3888 m altitude in the Kashmir Himalayas attracts pilgrims who worship at an ice stalagmite. Some caves formed in ice are also of religious significance. Examples occur at c.4200 m in the central Himalayas, where the Bhagirath River, which discharges from two meltwater outflow caves in the snout of the Gangotria Glacier, joins with the Alakanada River that discharges from a glacier in the next valley to form the sacred Ganges River.

At Kusma, west of Pokhara in Nepal, lies the famous Hindu shrine of Gupteswary Cave. Other shrines occur south of Pokhara in the Mahabharat Hills and at Halesi Cave, on the east bank of the Dudh Kosi–Sun Kosi confluence, which is used for a festival on the birthday of Ram. Other Nepalese Hindu shrines include Goraknath Cave in the Bagmati Valley and Shivaji Cave, Bhojpur. The karstic spring complex at Muktinath at c.3800 m in the Annapurna Himal, close to the Tibetan border, is sacred to both Hindus and Buddhists. The sight of pilgrims gathering waters from the springs and fountains in the more developed Hindu sector recalls images of Christians collecting Holy Water from Lourdes. The darkness within a curtained, cave-like recess beneath the altar in a simple Buddhist gompa is lit by a natural fire burning gas that issues from rock crevices beside the emerging springwater.

In China the divine power or spirit of the Tao permeates all things and beings, animate and inanimate. Karst sites of religious significance are widespread. For example, in Guangxi Province, the walls of caves in the hill that contains the heavily visited Longuin tourist cave in the city of Guilin bear inscriptions that

Religious Sites: A black statue of Buddha on a shrine in Sriwilai Cave, Tha Pra District, Thailand. (Photo by John Gunn)

span 77 different dynasties, with 93 inscriptions from the Sung dynasty alone (see colour page 4). More than 200 statues, mostly 50–200 cm high, have been carved into the limestone outcrops of Western Hill in Guilin. They date from the spread of Buddhism into southern China during the Tang dynasty.

Kham altars and inscriptions from the 9th and 10th centuries occur in Phong Nha Cave, north–central Vietnam. Cave shrines are also found amid the Huong Tich Mountains. In the Marble Mountains near Danang, one karst tower is a pilgrimage site and Buddhist temples have been established in its caves, together with some Confucian shrines. Some of these caves were Hindu shrines during the reign of the Champa.

In Thailand, where c.95% of the population profess Buddhism, at least 200 caves are currently used for worship, meditation, or retreat (see photograph). Some have been utilized for over 1000 years, and on the frontiers of settlement the religious use of new caves continues to be initiated (Munier, 1998; Dunkley, 1995). Statuary and other relics also occur in numerous caves in Burma (Myanmar), including Pindaya Cave, Shwe Ohm-min Cave, and Mimehtu Cave in southern Shan State, and the Bingyi Caves and Kogun Cave near Moulemein (Mawlamyine) (Dunkley et al., 1989). In Japan, many legends are associated with lava tube caves on sacred Mt Fuji. Shotoku Taishi, a 6th-century prince, is said to have descended the crater into a vast cavern where he spoke to a fire-breathing dragon who transformed into the Buddha of All-Illuminating Wisdom, dwelling in the cave palace to save all sentient beings. In the 17th century the founder of the Fuji-ko sect is said to have taken up residence in one sacred cave where he stood immobile in a meditative bid to restore stability to a nation riven with unrest.

In Central America, caves and cenotes have played a major role in the religious traditions of Maya peoples for over 2000 years and were the location of elaborate rites (Bower, 1998). Classic Maya settlements constructed from AD 250–900 appear to have been placed strategically over caves that had great religious and political significance. There is evidence to suggest the alignment of major structures with cave passages underneath them, including a large cave under the El Duende Pyramid, Guatemala. Offerings and relics likely to have been used by shaman have been reported from at least three caves in southern Belize (see also America, Central: Archaeological Caves).

In Oceania, lava tube caves are significant for indigenous people in Hawaii who follow traditions associated with the fiery goddess of volcanoes, Pele. For Australian Aborigines, natural landmarks that form part of an extended kinship system are the centres for religion and ceremony. One gaping cave entrance at Uluru is considered to be the wailing mouth of a grief-stricken mother of the carpet-snake people, whose son was struck down by a venomous snake warrior, while another cave important to the hare wallaby men is still used for male initiation. Only in a few cases are deeper karst caves known to have been entered.

Management

Notwithstanding the sacred status with which karstic and other cave sites have been imbued, many remain vulnerable to damage and land-use decisions. Some karstic religious sites have been deliberately targeted during political and military conflicts in order to demoralize or assert dominance over devotees, as in the defacing of statuary and inscriptions on karst hills in China during the ascendancy of Mao's Red Guard. During the undeclared war in Laos in 1969, an unmarked but probably American aircraft fired a single rocket into Tham Phiu, killing 400 villagers who had fled into the sanctuary of this religious site. In some cases the sense of security and protection felt at sacred sites may encourage defenders to base themselves in cave temples, which may place the site at risk. The entrance to Phong Nha Cave bears a legacy of heavy air raids during the American war against Vietnamese forces who stored munitions in the cave. Similarly, bullet-pocked masonry amid the cave shrines of the Marble Mountains attests to the heavy fighting that occurred there during the American war.

Extensive physical modification of the entrances to sacred caves is common, as at Ajanta. Cave walls may be modified through the execution of wall inscriptions and, in a few cases, stalagmites have been carved to enhance their similarity to religious icons. Some sites are closely safeguarded from harm and even the eyes of the uninitiated, but others attract a multitude of devotees, creating severe environmental pressures. The small grottoes that formed the initial focus of worship are sometimes overwhelmed by construction of major temple complexes, as at Koanoi, Thailand, and at Lourdes. Concerns about the degradation of the natural environment that might be felt by a karst geo-ecologist may hold little sway in decisions regarding a feature that for devotees may—in becoming sacred—have lost its original status as a landform and have been transformed into an inherently incorruptible embodiment of the Divine.

However, the solitude, tranquility, and peace that facilitated the initial attribution of sacredness may be lost entirely if crowding or physical transformation go too far, and many practical difficulties may also arise. The increasingly large numbers of pilgrims reaching Muktinath since access conditions eased in the late 1970s have resulted in severe deforestation to obtain fuel for heating and cooking, and the development of accommodation blocks has seriously detracted from its aesthetic qualities.

Some cave and karst shrines have been formally designated for natural or cultural heritage conservation, such as the World Heritage Ajanta Caves. But in other cases a major extractive industry can occur very close to a religious site, as at Batu Caves, where nearby limestone quarrying has been a significant management issue. In Hawaii, lava tube caves of cultural significance have been polluted by the dumping of refuse and the diversion of urban stormwaters. In 1988 the spring at Lourdes was found to have been contaminated by infiltration from a rubbish tip 7 km distant. Human-induced climate change also has management implications, most obviously for sacred caves in rapidly retreating glaciers such as those at the source of the Ganges.

KEVIN KIERNAN

See also **Burials in Caves; Folklore and Mythology**

Works Cited

Bernbaum, E. 1997. *Sacred Mountains of the World*, Berkeley: University of California Press

Bower, B. 1998. Sacred secrets of the caves (Mayan archaeology). *Science News*, January 24, 1998

Dunkley, J.R. 1995. *The Caves of Thailand*, Sydney: Speleological Research Council

Dunkley, J.R., Sefton, M., Nichterlein, D. & Taylor, J. 1989. Karst and caves of Burma (Myanmar). *Cave Science*, 16(3): 123–31

Munier, C. 1998. *Sacred Rocks and Buddhist Caves in Thailand*, Bangkok: White Lotus

Shaw, T.R. 1992. *History of Cave Science: The Exploration and Study of Limestone Caves, to 1900*, 2nd edition, Broadway, New South Wales: Sydney, Speleological Society

Further Reading

Fergusson, J. & Burgess, J. 1880. *Cave Temples of India*, London: Allen; reprinted New Delhi: South Asia Books, 1969

Flood, J. 1983. *Archaeology of the Dreamtime*, Sydney: Collins; revised edition, New Haven, Connecticut: Yale University Press

Hayden, D. 1987. Caves. In *Encyclopedia of Religion*, edited by M. Eliade, vol. 3, New York: Macmillan

Kiernan, K. 1995. Genesis 1:31 versus managing the devout: On the spiritual and religious use of karst. *Cave Management in Australasia*, 11: 208–21

Scully, V. 1962. *The Earth, The Temple, and the Gods: Greek Sacred Architecture*, New Haven, Connecticut: Yale University Press

RESTORATION OF CAVES AND SPELEOTHEM REPAIR

Human visitation can have negative impacts on caves and their contents. Cave restoration efforts serve to remediate or repair damage caused by carelessness, inadvertent actions, or intentional vandalism. Project planning for cave restoration often begins with thorough inventory and documentation of the features within. Prior to restoration and repair, the cave fauna, flora, microbiota, habitat, clastic sediments, speleothems, hydrological and geological features, as well as archaeological and paleontological resources should be evaluated. The extent of damage, proposed restoration strategies, and potential future impacts should be recorded. Careful planning and implementation is required to minimize alteration of cave resources and the first objective of cave restoration should always be: *Do no harm*.

Cave restoration planning should center on one of three objectives. Decisions will focus on 1) restoration to a former natural condition; 2) restoration to a previous historic period; or 3) simple improvement of the aesthetic state without harming resources. In a show cave with decades of accumulated impact, partial restoration may be the best achievable option whereas restoration to a former natural condition is more likely to be successful in a wild cave that is not as ecologically disturbed. Doline and stream-sink cleanups are common in karst terrains where these features are often used as trash dumps. Removal of garbage and debris mitigates pollution and improves groundwater quality.

Caving groups volunteer thousands of hours annually to cave and karst restorations, donating time, labour, materials, and expertise for conservation efforts. On public and private lands in the United States and in many other countries, cavers provide significant volunteer value for cave and karst protection projects. When funding is available, cost-shares, partial reimbursements, cooperative agreements, and project contracts with caving groups provide support for ongoing restoration projects.

Cave Restoration Tasks

Cave restoration projects employ a variety of tasks, tools, and skills. For example, projects might include rubble and debris removal, mud removal, spray paint cleaning, lint picking, speleothem cleaning and repair, doline and stream-sink clean-out, or habitat restoration. Large projects removing rubble, artificial fill, and debris often require multiple project days scheduled over several years. Other projects require minimal time and human resources. For example, sponging away a single muddy footprint left on pristine flowstone can prevent permanent calcification of the imprint. Smaller, more delicate restoration projects on gypsum or cave pearls may require only a few hours of attention from one or two cavers (Figure 1).

Cavers should follow stringent minimum-impact guidelines when restoring recently discovered chambers. Pristine environments deserve careful restoration strategies and require changing to fresh, clean garments. Sensitive areas with suspected microbial significance might require more specialized clean-room techniques or sterile procedures if restoration tasks are necessary (Hildreth-Werker & Werker, 1999). In caves that have had high levels of visitation or historic commercial use, negative impacts are less likely to occur from restoration efforts. A good example is Moondyne, a severely damaged former show cave in Australia that had been closed since 1959. Labour-intensive restoration

Cave Restoration and Speleothem Repair: Figure 1. Clad in a tyvek suit, this caver uses clean room technique to restore cave pearls found in the Pearlsian Gulf of Lechuguilla Cave in New Mexico. (Photo by Val Hildreth-Werker)

in the 1990s involved removal of all rubbish, pathways, and other infrastructure, and cleaning of the walls and speleothems. The cave was restored to a virtually pristine condition and now serves as a guided education site (Bell, 1993).

After evaluation and documentation of the site, restoration typically begins with collection of litter and removal of contemporary graffiti. Consultation with scientists as well as historical and cultural preservation experts is always appropriate before erasing graffiti and removing trash. The history of an area may be literally written on the cave walls and should be protected. Pictographs, petroglyphs, and historic signatures may be layered under contemporary graffiti. Also, historically important mud glyphs may be easily overlooked. Significant historical, prehistorical, geological, mineralogical, climatological, or biological resources may be present, but visible only to the trained eye. Removing contemporary trash and graffiti or emphasizing historic and scientific significance can result in increased visitor respect for cave resources (Goodbar, 2003).

Restoration projects can facilitate species recovery. Restoring entrance features, airflow, or hydrological conditions may support rehabilitation of species and habitats (Aley, 1989). Old trash and wood piles may be providing habitats for particular cave species and should be removed in stages to allow fauna to migrate to new areas and biofilms to recover. Some restoration involves carefully removing lint and dust (garment fibres, epidermal matter, hair, and small particles of debris) from along tour trails where it accumulates, discolouring formations and providing a habitat for opportunistic organisms (Jablonsky, Kraemer & Yett, 1995).

Groups of cavers working multi-day sessions are needed to restore speleothems covered with excavation sediments or to remove construction debris and artificially deposited clastic sediments added in show cave chambers to make flat floors. For example, in the late 1980s, tonnes of clay and rock were removed to restore natural floors at Wisconsin's Mystery Cave (Netherton, 1993). Sediment restoration also takes place at Caverns of Sonora in Texas where 50 or more cavers gather for annual bucket brigade projects. Teams remove tonnes of blast rock from trail construction, yet recognize and leave the natural breakdown in place (Veni, 1998).

Restoration leaders should evaluate the degree of previous impact in a cave passage and plan methods to avoid creating new damage. Trails through cave passages can be clearly marked to help confine visitor impact. Even in wild caves, travelling on durable surfaces and previously compacted routes will aid in preservation of sediments, small floor speleothems, and invertebrate populations. Unfortunately, if footprints are visible beyond the designated paths, others will tend to follow and trails will expand. If footprints remain outside of delineated trails, they can be erased to restore the natural appearance of clastic sediments. Gentle combing motions with lightweight nylon brushes will erase footprints and avoid stirring up dust (Hildreth-Werker & Werker, 1997). Deep footprints in rock flour or sand can be camouflaged with natural sediments taken from alongside the trail. Removing visible traces of human travel tends to mitigate future damage.

Tools for Cave Restoration
Appropriate tools for restoration are chosen according to the task. Some projects require toothbrushes or tweezers, while others require shovels or power drills. In show caves, industrial skills and equipment are required to accomplish infrastructure-related tasks such as removal or remediation of wiring, guardrails, walkways, or electrical transformers that may be contaminated with PCBs. It is wise to contact hazardous waste removal experts if toxic waste is encountered during cave or doline restoration. For the cleaning and repair of delicate speleothems, different types of tools and expertise are necessary. For restoration in wild caves, lightweight, compact tools are typically preferred. Scrub brushes with nylon bristles, soft absorbent sponges, vinyl gloves, flexible plastic scrapers, buckets, hand-held sprayers, water, and human perseverance are effective tools for cleaning speleothems and many cave surfaces. Catchments, towels, or sponge dams are placed downslope to capture restoration runoff, which is then removed from the cave and properly disposed.

Studies at the Jenolan Caves, Australia, first indicated that pressurized water is safe for cleaning durable speleothems (Bonwick, Ellis & Bonwick, 1986). However, more recent analysis with a scanning electron microscope found that the cave surfaces were suffering some damage (Spate & Moses, 1993). Current practice is to minimize washing, to use low-pressure water, and to sometimes use a wet-dry vacuum cleaner. Caution must be exercised to avoid harming fragile speleothems, splattering nearby cave features, or destroying cave life with streams of high-pressure water. Water from within the cave, ideally percolation water, should be used for restoration because water from the surface may be chemically aggressive and lead to speleothem damage.

Opportunistic photosynthetic organisms often grow near electric lights in show caves (see Tourist Caves: Algae and Lampenflora). Sodium hypochlorite solutions and other chemical agents have been used to control lamp flora, but must be used only at a minimum concentration, applied only where required, and rinsed carefully to avoid indiscriminate killing of cave biota. Manufactured chemicals are not recommended for cave cleaning tasks since fumes, residues, and by-products can be harmful to biota and cave systems. Anthropogenic agents typically produce toxic by-products through out-gassing and degradation. Recommendations for improved control of undesirable algae and microflora include advanced lighting technologies and wavelengths selected to inhibit photosynthesis. Recent experience demonstrates that with properly designed, located, and controlled lighting, lamp flora can be eliminated.

Speleothem Repair Tools and Materials
Specialized products that are relatively safe for long-term applications in cave systems are used for repair tasks. Recommended products include archival epoxies and museum-grade cyanoacrylate adhesives. Degradation and out-gassing characteristics are reduced and longevity characteristics are strengthened in these products (Werker, 2003). High tensile strength cement products that contain minimal calcium hydroxide are recommended if cement is required for cave projects. High-austenitic stainless steels are resistant to corrosion and are more suitable to cave environments than many other construction materials. For stalagmite and stalactite repair, austenitic stainless round stock or threaded rod is installed as a supporting rod to pin the speleothem pieces back in place (Figure 2). Fracture sites and repair seams can be filled with a mixture of archival epoxy and rock dust (Veni, 1997). Draperies, soda straws, helictites, gypsum

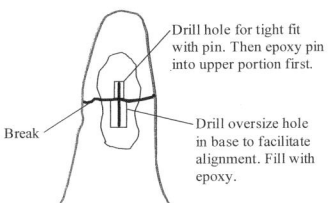 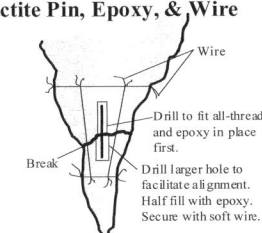

Cave Restoration and Speleothem Repair: Figure 2. Repair of stalactites and stalagmites using pin and epoxy method. (Drawings by Jim C. Werker)

crust, flowstone, rimstone dams, and even large speleothems can be repaired (Werker, 2003). Poulter (1987) and others have utilized similar technology in Australia.

Research is needed to define the safest materials for use in subterranean systems. Cave biologists and chemists should confirm the safety of product characteristics before manufactured chemicals are placed in cave environments, whether for construction of infrastructure or for repair and restoration (Spate *et al.*, 1998). Cave conservation and management will benefit from materials data gained through laboratory analyses and field monitoring to document the degradation, out-gassing, and longevity characteristics of materials placed in cave systems.

As speleological research increases and as we understand more about the ecological processes of cave and karst systems, restoration and repair techniques will continue to evolve and become less intrusive. In the future, identification of cave-safe materials will help define improved practices for cleaning and speleothem repair. Cavers who perform restoration and repair are actively defining low-impact caving techniques. If visitor and caver ethics improve to avoid unnecessary impacts to cave passages, the need for restoration and repair may decline (see minimal impact codes in Recreational Caving entry).

JIM C. WERKER and VAL HILDRETH-WERKER

Works Cited

Aley, T. 1989. Restoration and maintenance of natural cave microclimates. *National Speleological Society News*, 47(2): 39–40

Bell, P. 1993. Moondyne Cave: Development to guided adventure. *Cave Management in Australasia*, 10: 9–11

Bonwick, J., Ellis, R. & Bonwick, M. 1986. Cleaning, restoration, and redevelopment of show caves in Australia. In *Proceedings of the Ninth International Congress of Speleology*, vol 2: 221–23

Goodbar, J. 2003. Cave graffiti: The writing is on the wall. In *Cave Conservation and Restoration*, edited by V. Hildreth-Werker & J.C. Werker, Huntsville, Alabama: National Speleological Society

Hildreth-Werker, V. & Werker, J.C. 1997. Lechuguilla restoration: Techniques learned in the Southwest. *National Speleological Society News*, 55(4): 107–13

Hildreth-Werker, V, & Werker, J.C. 1999. Restoration, trail designation, and microbial preservation in Lechuguilla Cave. In *Proceedings of the National Cave Management Symposium 1997*, edited by R.R. Stitt, Bellingham, Washington: National Cave Management Symposium Steering Committee

Jablonsky, P., Kraemer, S. & Yett, W. 1995. Lint in caves. In *Proceedings of the National Cave Management Symposium 1993*, edited by D.L. Pate, Carlsbad, New Mexico: National Cave Management Symposium Steering Committee

Netherton, W. 1993. Mystery Cave trails: Past, present and future. In *1989 National Cave Management Proceedings*, edited by J.R. Jorden & R.K. Obele, New Braunfels, Texas: Texas Parks and Wildlife Department

Poulter, N. 1987. The restoration of the Jewel Casket, Yallingup Cave, W.A. *Australian Caver*, 115: 6–8

Spate, A. & Moses, C. 1993. Impacts of high pressure cleaning: A case study at Jenolan. *Cave Management in Australasia*, 10: 45–48

Spate, A., Hamilton-Smith, E., Little, L. & Holland, E. 1998. Best practice and tourist cave engineering. *Cave Management in Australasia*, 12: 97–109

Veni, G. 1997. Spelothems: Preservation, display, and restoration. In *Cave Minerals of the World*, 2nd edition, edited by C.A. Hill & P. Forti, Huntsville, Alabama: National Speleological Society

Veni, G. 1998. 1998 Sonora restoration project. *The Texas Caver*, 43(6): 87–89. Reprinted in *SpeleoDigest*, 1999: 301–02

Werker, J.C. 2003. Speleothem repair techniques. In *Cave Conservation and Restoration*, edited by V. Hildreth-Werker & J.C. Werker, Huntsville, Alabama: National Speleological Society

Further Reading

Hildreth-Werker, V. & Werker, J.C. (editors) 2003. *Cave Conservation and Restoration*, Huntsville, Alabama: National Speleological Society

RUSSIA AND UKRAINE

Russia and Ukraine have large areas of karst and the Western Ukraine contains one of the largest gypsum karst areas in the world. The karst may be considered in three areas.

European Sector

This covers the large Precambrian craton that occupies most of the Ukraine and the European part of Russia. It extends from the Arctic Ocean to the Black Sea. Karst is developed mainly in various intrastratal settings (deep-seated, subjacent, entrenched, and denuded) in carbonates (limestones, dolomites, and chalk) and evaporites (gypsum and salt) of different ages. In many regions, karst develops in intercalating sulfate-carbonate successions. Most of the significant caves in the region are in gypsum.

The Northern Russian province includes the Pinega (see Pinega Gypsum Caves) and North Dvina karsts in the Lower Permian sulfates, some of the largest integrated gypsum karst regions in Europe. The gypsum-anhydrite sequence, in places intercalated with dolomites and limestones, is commonly 40–60 m thick and lies at shallow depth. It supports a spectacular variety of karst landforms and hydrological features. More than 360 gypsum caves are now known in these regions, with a total length of over 100 km (Malkov & Gurkalo, 1999).

The Central Russian province occupies the centre of the Eastern European Plain, which corresponds to the southern part of the structural Moscow depression. Karst is developed in the Carboniferous and Upper Devonian limestones and dolomites. Drilling has proved that deep-seated karst in artesian conditions is present in many localities. Caves are rare (the longest being Poneretka Cave in the Valday Upland, with 1430 m of passage)

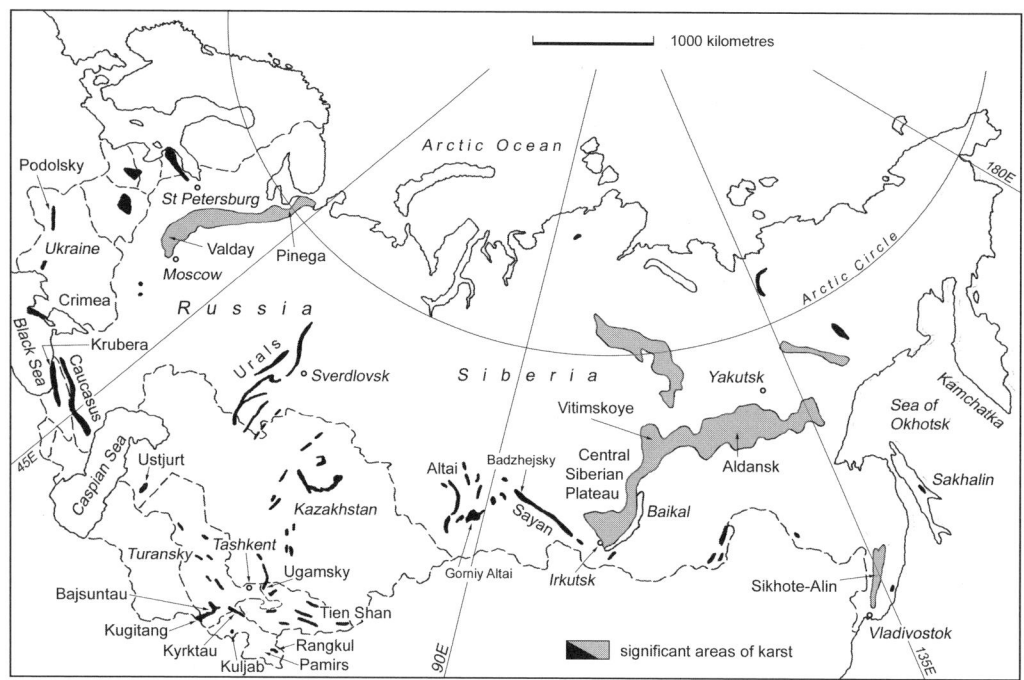

Russia and Ukraine: Generalized map of the major areas of karst in Russia, Ukraine, and the Asian and Baltic republics, including karst on both carbonate and evaporite rocks. The larger areas, in the grey tone, include complex smaller outcrops of non-karstic rocks.

but surface karst landforms, such as dolines and dry valleys, are common in the areas of subjacent and entrenched karst.

The Kama-Middle Volga province lies between the Volga River and the Ural Mountains, mainly within the Kama River catchment. Karst is developed in Permian sulfates, dolomites, and limestones and in Carboniferous and Devonian carbonate rocks. Collapse and subsidence dolines are common in many areas, originating from active karstification at various depths. The most intensely karstified areas are those within paleo-valleys and alluvial plains, where karst breakdowns may propagate through unconsolidated cover up to 100 m thick. Caves are abundant in the areas of entrenched karst, particularly in the gypsum karst areas in the easternmost, Kamsko-Ufimsky region. There are more that 200 known gypsum caves in the latter region, the largest being the 5700 m long Kungur Ice Cave.

The Fore-Caspian province lies in the southeastern part of the Eastern European Plain, north of the Caspian Sea. Over 1500 salt diapirs are known in the region, some close to the surface and some being denuded. They support open salt and gypsum karst, and gypsum caves are known in the caprocks (e.g. Baskunchakskaja, 1438 m long).

The Moldavian-Ukrainian province includes, within the borders of Ukraine, the Podol'sko-Bukovinsky region of gypsum karst, the Prichernomorsky region of limestone karst, the Krivorozhsky region of quartzite and carbonate karst, and the Donetzk region of gypsum and carbonate karst. The Podol'sko-Bukovinsky region in the Western Ukraine is internationally renowned for its giant maze caves in Neogene gypsum. It is a model example of artesian speleogenesis (Klimchouk, 2000), and contains the five longest gypsum caves in the world, accounting for well over half of the total known length of gypsum caves worldwide (see Ukraine Gypsum Caves and Karst). Few caves are known in the calcareous sandstones (Stradchanska, 360 m long and Studencheskaja, 242 m long). The Prichernomorsky lowland is located north of the Black Sea, with karst developed in the Neogene limestones. Isolated fissure-like caves of artesian origin are occasionally seen in old limestone mines around Odessa, the largest being the caves of Novorossijskaja (1404 m) and Natalina (1292 m). Caves up to 150 m long, partially submerged beneath the sea, are known in the coastal cliffs of the Tarkhankut peninsula, Crimea. In the Krivorozhsky region caves have been found in Precambrian dolomites and ferruginous quartzites.

The Carpathians extend as an arc from Slovakia through Ukraine to Romania, and the Ukrainian segment is composed mainly of flysch. Small caves are known in the exotic massifs of Upper Jurassic and Lower Cretaceous limestones (Druzhba, 256 m long), but interesting caves are also found in conglomerates (Krasny Kamen', 900 m long and 56 m deep) and in sandstones (Prokhodnoj Dvor, 520 m long, 40 m deep).

The Crimean Mountains lie in the south of the Crimean peninsula in Ukraine and structurally continue as the Caucasus Mountains along the southern border of Russia. Both mountain regions belong to the Alpine folded system. The Crimean karst is in a chain of low plateau-like massifs of Upper Jurassic limestones, with remarkable development of open karst, and almost 1100 caves are documented (see Crimea, Ukraine).

Urals Region

The Urals region includes the Hercynian folded system of the Ural Mountains and the foredeep that separates the Urals from the adjacent Eastern European craton. The mountains, which extend north to south for almost 2000 km, are eroded and rounded, and largely covered with forests. Gypsum karst with a variety of karstic landforms and some caves (Ishcheevskaja Cave, 1000 m long) occurs widely in the Fore-Ural. In the Ural fold mountains carbonate karst predominates, developed in Paleozoic limestones and dolomites. Because of stripe tectonics, contact karst is common. The largest vauclusian karst springs are located in the Western Urals (the Goluboje Ozero spring, dived to −56 m, and the Krasny Kljuch Spring, dived to −38 m). Caves are numerous, particularly in the southern part of the Western Ural province. The largest caves are the Sumgan-Koutuk Cave (9860 m long and 130 m deep), Kinderlinskaja (7900 m long, 185 m deep, with a 230 m long and 48 m deep siphon) and Kizelovskaja-Viasherskaja System (7600 m long, 43 m deep). Of special archaeological interest are the Kapova Cave (2640 m long, 103 m deep) with Paleolithic rock paintings, as well as the Ignatjevskaja and Staromuradymovskaja caves in the Central Urals.

Siberia and the Far East

In Siberia, a huge taiga-covered territory that lies east of the Ural Mountains, karst is not very common, but some important karst regions are studied in the mountains of southern Siberia, in the Sikhote-Alyn' range of the Far East, and on the island of Sakhalin (see Siberia entry). The mountains of Altaj and Sajan of southern Siberia are both large ranges with flat tops and locally jagged ridges, rising above 2200 and sometimes above 3000 m, but they differ in their geological evolution and tectonics. The Altaj are Hercynian, the Sajan are Caledonian, and the Salair-Kuznetzk forms a transitional tectonic zone. In the Altaj more than 210 caves have been explored in the Rifean and Devonian limestones, including Altajskaja (4175 m long, 240 m deep), Tutkushskaja (1400 m long, 200 m deep), and Kektash (1780 m long, 340 m deep). In the Sajan, particularly in the Eastern Sajan range, caves are abundant and varied. Some large caves occur in the Cambrian limestones (e.g. Zhenevskaja, 6000 m long, 65 m deep, Kubinskaja, 3000 m long, 274 m deep and Torgashynskaja, 1560 m long, 75 m deep). In Lysanskaja Cave (2400 m long) eight siphons have been explored up to 95 m long and 20 m deep. The most remarkable speleological features of the region are caves in the Badzhejsky area, developed in Ordovician conglomerates. These include the Bol'shaja Oreshnaja Cave (47 000 m long, 155 m deep), the world's largest conglomerate cave and the second longest cave in Russia, and some other significant caves (e.g. Badzhejskaja, 5900 m long, 240 m deep). In the Salajr-Kuznetzk province caves occur mainly in the Lower Cambrian limestones, where more than 120 caves are documented, including the Jaschik Pandorry Cave, a complex 3-D system, 10 000 m long and 180 m deep.

ALEXANDER KLIMCHOUK

See also **Asia, Central; Caucasus, Georgia; Crimea, Ukraine; Krubera Cave; Pinega Gypsum Caves, Siberia; Ukraine Gypsum Caves and Karst**

Works Cited

Klimchouk, A.B. 2000. Speleogenesis of great gypsum mazes in the Western Ukraine. In *Speleogenesis: Evolution of Karst Aquifers*, edited by A.B. Klimchouk, D.C. Ford, A.N. Palmer & W. Dreyboldt, Huntsville, Alabama: National Speleological Society

Malkov, V.N. & Gurkalo, E.N. 1999. Speleological regionalization and distribution of caves in the Arkhangelsk region. *Peschery (Caves): Inter-university Scientific Transactions*, Perm: Perm University: 10–15 (in Russian)

Further Reading

Chikishev, A.G. 1979. *Problems of the Karst Studies of the Russian Plain*, Moscow: Moscow University (in Russian)

Dublyansky, V.N. & Dublyanskaya, G.N. 1992. *Mapping, Regionalization and Engineering-Geological Assessment of Karstified Terrains*, Novosibirsk (in Russian)

Gvozdetsky, N.A. 1962. Issues of the geographic regionalization of karst on the territory of the USSR. In *Obshchie voprosy karstovedenija*, Moscow: Academy of Science (in Russian)

Gvozdetsky, N.A. 1981. *Karst*, Moscow: Mysl. (in Russian)

Ivanov, B.N. 1972. Carbonate karst of the Ukraine and Moldavia. In *Karst in Carbonate Rocks*, Moscow: Moscow University (in Russian); Transactions of the Moscow Society of Naturalists, vol. 47 (in Russian)

Klimchouk, A.B. & Dublyansky, V.N. 1993. The state of the speleological investigations of the territory of the former USSR. *Svet (The Light)*, 1–2 (7–8): 24–27 (in Russian)

Maximovich, G.A. 1962a. Tectonic regularities of the karst distribution on the territory of the USSR. In *Obshchie voprosy karstovedenija*, Moscow: Academy of Science: 40–54 (in Russian)

Maximovich, G.A. 1962b. Distribution and regionalization of karst of the USSR. *Gidrogeologija and karstovedenije*, vol. 1, Perm: Perm University (in Russian)

Maximovich, G.A. & Kostarev, V.P. 1973. Karst regions of the Ural and Fore-Ural. In *Voprosy fisicheskoj geografii Urala*, Transactions of Perm University, vol. 308 (in Russian)

S

SALUKKAN KALLANG, INDONESIA: BIOSPELEOLOGY

The impressive cone karst of Maros lies in South Sulawesi (Indonesia) between 4°7′ S and 5°1′ S. It is a rugged limestone plateau of Eocene to middle Miocene age, with basaltic and dioritic intrusions of middle Miocene to late Pliocene age. These dykes and sills have strongly influenced the karstification; in particular they are at the origin of a network of narrow and deep corridors which constitute an unusual feature of the karst landscape in Maros (Brouquisse, 1986). Among the more than 150 caves known in this region (Bedos et al., 1994), the caves of the system Gua Salukkan Kallang-Towakkalak (SKT) are outstanding for their beauty and their length. SKT contains more than 24 km of underground passages, including over 10 km of a large underground river, in five different caves: Gua Wattanang (440 m), Gua Tanette (9.7 km), Gua Batu Neraka (750 m), Lubang Kabut (1145 m), and Gua Salukkan Kallang (12.5 km) (Figure 1). Several other caves probably correspond with dismantled relict levels of the system, including the giant shafts of Lubang Tomanangna (−190 m) and Lubang Kapa Kapasa (−210 m). The resurgence is an undived sump at an altitude of 40 m, more than 10 km from the easternmost river sinks that are thought to feed it.

The fauna of the SKT system is patchily distributed. Stygobitic fauna is only found in small endogenous inlets, whereas the main river is populated by epigean species of fish and invertebrates. For most of their length, the SKT galleries appear at first sight to be totally devoid of terrestrial life. Terrestrial animals are concentrated around flood debris along the river banks, and on bat, swiftlet, and cricket droppings scattered in the system. Guano accumulations, typical of many caves in Southeast Asia, are rare and small in SKT. In spite of this severe shortage of favourable habitats, the SKT subterranean fauna is more diverse than that of any other Southeast Asian and even tropical cave system studied until now (Deharveng & Bedos, 2000) (see Asia, Southeast: Biospeleology).

Ninety-three species have been recorded in the SKT system, excluding bats and swiftlets which have never been studied in detail in the Maros karst. Non-troglobitic species are the most numerous. Most are trogloxene species, brought into the system by the annual floods of December to February, especially strong in Maros. This dominance of outside species is in line with observations elsewhere in the tropics. Guano communities are also diverse, with 19 recorded species. Some are guanophiles also found outside, but several are known only from caves, like the Cambalopsidae millipedes, which are among the most numerous arthropods in SKT. The cave-restricted component of the SKT fauna (guanobitic–troglobitic, troglobitic, and stygobitic species) represents 30% of the total richness.

The most conspicuous invertebrates of SKT are the giant arthropods, typical of Southeast Asian caves. They are associated with energy-rich habitats and are frequently seen on cave walls near guano piles or cave entrances. The giant arthropod community consists of a dense population of a large rhaphidophorid cricket, which is preyed upon by heteropodid spiders, amblypygids, and centipedes. Rhaphidophorid crickets may disperse their droppings far from their feeding place, providing food resources for troglobites in the energy-poor areas of the caves. Giant arthropods do not exhibit any morphological adaptation to cave life, although they are mainly or exclusively found in caves. They have fairly wide distributions, i.e. neither restricted to the SKT system nor to the Maros karst.

The terrestrial meso- and micro-arthropod groups present in virtually any cave of Southeast Asia also occur in SKT: schizomids, several families of small spiders (Ctenidae, Pholcidae, and Ochyroceratidae in SKT), cambalopsid millipedes, entomobryid Collembola, and nocticolid cockroaches; these troglobitic or guanobitic–troglobitic species are often modified in relation to subterranean life (eyes and pigment reduced or lost, elongated appendages).

The stygobitic fauna, far less rich than the troglobitic fauna in SKT, includes representatives of two groups regularly found in Southeast Asian caves: *Dugesia leclerci*, a flatworm of a widespread tropical genus, and an undetermined Bogidiellidae—an

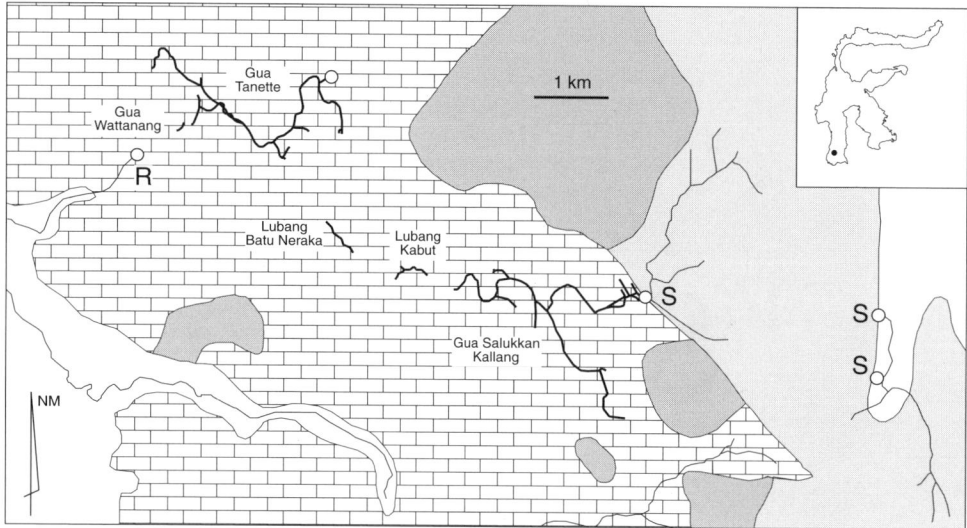

Salukkan Kallang, Indonesia: Biospeleology: Figure 1. Map of the Gua Salukkan Kallang–Towakkalak underground system in the Maros karst. White: alluvial deposits; bricks: limestone; pale grey: sedimentary and volcanic rocks of the Camba formation; dark grey: basalt and diorite; R: resurgence of Towakkalak; S: sink; thick lines: cave passages; thin lines: epigean rivers.

amphipod family frequent in many subterranean habitats worldwide. Like most troglobites, these stygobites are local endemics, with related species in other Southeast Asian caves.

The SKT system also hosts more unexpected animals. The pterostichid beetle, *Mateuellus troglobioticus*, is the only subterranean representative in the world of the tribe Abacetini. Several taxa exhibit a high level of troglomorphy. Among terrestrial animals, the palpigrad *Eukoenenia maros* is the second troglomorphic species of this genus for tropical Asia, whereas many cave species are known from Mediterranean regions. The campodeid *Lepidocampa hypogaea* is the second troglobitic species of the widespread and speciose tropical genus *Lepidocampa*. The aquatic fauna has comparatively more remarkable species. An undescribed microphthalmic fish of the genus *Bostrychus* occurs at very low density in the SKT system. At least two atyid shrimps are present in SKT: one is microphthalmic with a size similar to outside species of the family; the other is much larger and totally blind. But the most interesting species in SKT is a tiny aquatic crab: *Cancrocaeca xenomorpha* (Ng, 1991, Figure 2), which represents the first and the only known case of adaptation to cave life for the primarily marine family Hymenosomatidae. It also shows the highest degree of eye regression in the world for a cave crab.

Biogeographically, SKT has a typical Southeast Asian cave fauna, enriched by elements of disputable origin (such as *Mateuellus troglobioticus*), possibly derived from local species. The hymenosomatid crab *Cancrocaeca xenomorpha* is the only indication of a possible relationship with the Australasian region, as several species of this family live in the fresh waters of Australia and Papua New Guinea (Ng, 1991). At a finer scale, biogeographical affinities remain obscure, given our very poor taxonomic knowledge of subterranean tropical fauna, and the absence of reference phylogenetic frameworks for any of the taxa concerned.

New hotspots of subterranean biodiversity will be discovered in the future, given the huge extension of undocumented karst areas in the tropics. But investigations in the last decade in tropical America and Southeast Asia have not modified the current figure: the SKT system remains the richest hotspot of tropical cave biodiversity (Deharveng & Bedos, 2000), followed by the Air Jernih cave system of Mulu (Malaysia: Sarawak), the species richness of which is described in detail by Chapman (1984). The conservation of the SKT biological richness is therefore a concern at a world level. The SKT system and its surface drainage area, partly included in the Nature Reserve of Karaenta, remain relatively undisturbed, but the growing human pressure on the surroundings is a real problem. The lowlands are heavily populated, and deforestation is progressing in the wildest parts of the karst. The opening of two new big limestone quarries recently brought unprecedented threats to the local subterranean ecosystems. The uniqueness of the Maros karst regarding landscape, geodiversity, and archaeology is, however, increasingly recognized and may help in the coming years in proposing relevant management measures to protect this exceptional biodiversity hot spot.

LOUIS DEHARVENG and ANNE BEDOS

See also **Asia, Southeast: Biospeleology**

Works Cited

Bedos, A., Brouquisse, F., Deharveng, L., Leclerc, P. & Rigal, D. 1994. Grandes cavités du karst de Maros [Big caves of the Maros

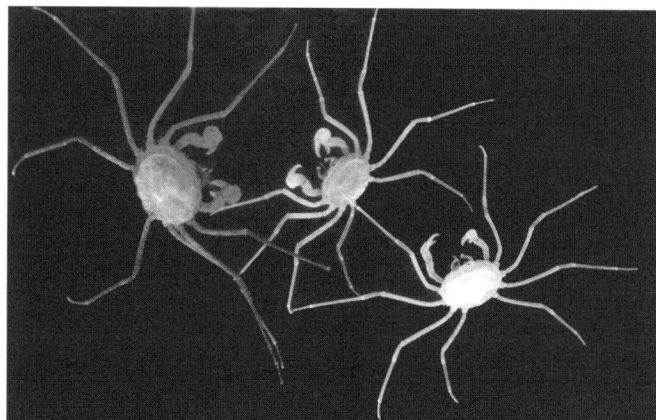

Salukkan Kallang (Indonesia): Biospeleology: Figure 2. The troglomorphic crab *Cancrocaeca xenomorpha*.

karst]. In *Indonésie 92, rapport spéléologique*, Toulouse: Association Pyrenéenne de Spéléologie

Brouquisse, F. 1986. Cadre géologique [Geological context]. In *Expédition Thaï-Maros 85, rapport spéléologique et scientifique*, Toulouse: Association Pyrenéenne de Spéléologie

Chapman, P. 1984. 1. The invertebrate fauna of the caves of Gunung Mulu National Park. *The Sarawak Museum Journal*, 30(51): 1–18

Deharveng, L. & Bedos, A. 2000. The cave fauna of Southeast Asia: Origin, evolution and ecology. In *Subterranean Ecosystems*, edited by H. Wilkens, D.C. Culver and W.F. Humphreys, Amsterdam and New York: Elsevier

Ng, P.K. 1991. *Cancrocaeca xenomorpha*, new genus and species, a blind troglobitic freshwater hymenosomatid (Crustacea: Decapoda: Brachyura) from Sulawesi, Indonesia. *Raffles Bulletin of Zoology*, 39(1): 59–73

SEDIMENTS: ALLOCHTHONOUS CLASTIC

A good deal of the material to be seen in any cave is derived from its catchment, transported into the cave by water, wind, mass movement, or even glacial processes. These sediments have an allochthonous origin, in comparison to the sediments formed within the cave, which have an autochthonous origin. This transport from surface to underground may be episodic, and material may have been temporarily stored in small terraces or floodplains prior to entering the cave. Thus the material may have already been altered by weathering or sorting since leaving its source area in a catchment.

Over human timescales earth surface processes tend to be quiescent, punctuated by a very few high-energy events due to floods, landslides, dust storms, or storm surges. Over the longer timescale of a cave's life, often measured in millions or tens of millions of years, such events are more commonplace. During these extreme geomorphic events a great deal of material can be brought into a cave, and often the accompanying flows of water will rework the existing sediments quite profoundly. A flood may bring in coarser sands and gravels that will be deposited as thick beds, often partly eroding then sealing in older deposits. If the cave floods, then slow settling from a deep pool will produce thick deposits of fine clays, often banded, and containing organic layers which indicate either discrete phases in flooding or the successive contributions of tributary valleys feeding the cave system. In a lake deposit we generally have a more complete record of all events in the catchment. In a cave these may have been partially or totally eroded, leaving a fragmented record that is very hard to interpret. Abrupt lateral facies change, unconformities, and stratigraphic reversals may all be present, and require detailed study of sedimentary architecture to unravel. Significant conceptual advances have been made in basin sedimentation models using three-dimensional analysis of sediment facies derived from vertical sections, drill holes, and remote sensing techniques. The application of these techniques to cave sediments is in its infancy, but offers very good prospects. The confined nature of most cave sediments, and the haphazard exposure of good sections, will necessitate the more intensive excavation of sections and the use of non-invasive techniques such as seismic reflection, resistivity, and ground-penetrating radar. There is also an ethical question regarding the information gain relative to the destruction of cave sediment stratification; this scarce resource can only be excavated once.

Caves can be seen as underground gorges and floodplains, in which sedimentation proceeds in modes similar to those of surface streams. This provides a conceptual scheme for sedimentation processes where water is the transporting agent. The major difference between the surface and underground streams is that in the latter, the water and sediment are confined within a conduit. This results in two main effects. First, dramatic fluctuations in water level due to either flood stage or to passage morphology result in steep gradients in energy along a cave passage. There is thus a greater range of sediment textures per unit length of channel than on the surface. This affects both estimation of past flow velocities from sediments, and also stratigraphic correlation. Second, subsequent flows of water down a particular cave passage may wholly or partially remove the sediment deposited by a prior event. The resistance of an individual parcel of sediment to this process of reworking will depend on its texture and on the passage geometry at the site.

The degree of this reworking will largely depend on the texture of the sediment. Thus, the large particles of boulder size, and the very fine cohesive clays will both be resistant to reworking once emplaced in a cave. In Westmoreland Cave (Mole Creek, Tasmania) dolerite boulders of Last Glacial age are wedged in the passages and appear little altered by weathering. In contrast, sand-sized sediments will be readily moved and reworked. This is due to the velocity of erosion being higher than the velocity of transport for very coarse and extremely fine particles. In Mammoth Cave (Kentucky) successive floods may move backflooding deposits of silt from the underground stream to higher elevations. After every flood, a thin layer of clay is deposited over all submerged passages. Some parcels of sediment may be shunted into side passages during very high floods and will remain there unaltered beyond the reach of successive flow events. Other parcels may be sealed in by rapid flowstone growth. Finally, sediments may persist in caves for a long time when they are resistant to water erosion. Murphy (1999) has described clays derived from relict glacial rock flour overlain by more modern goethite cemented cobbles in a flooded Yorkshire cave.

In many karst areas sediments enter the cave by the agency of wind or gravity fall, the rate being dependent on sediment supply and the trapping effects of surface vegetation as well. Where the entrance is a shaft or tube, such as at Naracoorte Caves in South Australia, then the sediments and animal bones will accumulate by gravity fall as a cone or pile at the base of the shaft (see photo). Some sediment will be transported further down the slope of the cone to form a colluvial fan. Both the cone and the fan will have rills or channels where water has eroded the accumulating sediments. Tabular layers of clean washed sand interspersed with red sand or clay layers provide further evidence of sorting during water flows. Small slumps or mass movements are likely to occur when high rates of sediment supply block the entrance shaft, allowing water to pond. The extent to which this slumping occurs depends on the nature of the sediment, its water content, and the hydraulic head of water above the deposit. This periodic slumping also causes some alignment of bones and spreads them over the cave floor. Cave sediments may thus be deposited by either gravity-fall or water

Allochthonous Clastic Sediments: Huge sand cone in Sand Cave, Joanna (near Naracoorte Caves). (Photo by Steven Bourne)

transport processes. The distinction between these becomes blurred when we consider such processes as turbidity currents sliding down steep sediment banks into a cave pool, or the injection of fluidized mudflows into tropical cave passages by landslides (Gillieson, 1986).

Once emplaced into a cave, any sediment is existing under conditions of total darkness, near constant high humidity, and near constant temperature. This reduces the amount of chemical alteration that can occur. However, with time some migration of solutes into and out of the sediment will occur. This may be as the result of wetting and drying cycles due to floods. In this context the porosity and mean grain size of the sediments has a great influence on the degree of diagenesis. Spectacular blue and red banded clay sediments in Selminum Tem, New Guinea, owe their banding to alternate layers of very fine clay and slightly coarser silt (Gillieson, 1986). Reducing conditions are maintained in the fine clays, with a dominance of ferrous iron salts, while in the silts the increased porosity allows oxidation and a dominance of ferric iron. Sediment banks are commonly cemented by iron oxide bands or by calcite from drips. Truly ancient sediments have reaction rims, which may penetrate into the adjoining rock surfaces. At Bungonia and Jenolan Caves, New South Wales, secondary dolomite and pyrite are common constituents of the paleokarst facies. This can make their identification and analysis very difficult, but scientists are beginning to realize the great extent and value of ancient sediments as paleokarst indicators in caves (Osborne, 1995).

DAVID GILLIESON

See also **Paleoenvironments: Clastic Sediments**

Works Cited

Gillieson, D.S. 1986. Cave sedimentation in the New Guinea Highlands. *Earth Surface Processes and Landforms*, 11: 533–43

Murphy, P.J. 1999. Sediment studies in Joint Hole, Chapel-le-dale, Yorskhire. *Cave and Karst Science*, 26(2): 87–90

Osborne, R.A.L. 1995. Evidence for two phases of late Palaeozoic karstification, cave development and sediment filling in south-eastern Australia. *Cave and Karst Science*, 22(1): 39–44

Further Reading

Bull, P.A. 1982. Some fine-grained sedimentation phenomena in caves. *Earth Surface Processes and Landforms*, 6: 11–22

Ford, D.C. & Williams, P.W. 1989. *Karst Geomorphology and Hydrology*, London and Boston: Unwin Hyman, Chapter 8: Cave Interior Deposits, pp.316–30

Ford. T.D. 2001. *Sediments in Caves*, London British Cave Research Association (BCRA Studies 9)

Gillieson, D. 1996. *Caves: Processes, Development and Management*, Oxford and Cambridge, Massachusetts: Blackwells

Moriarty, K.C., McCulloch, M.T., Wells, R.T. & McDowell, M.C. 2000. Mid-Pleistocene cave fills, megafaunal remains and climate change at Naracoorte, South Australia: Towards a predictive model using U/Th dating of speleothems. *Palaeogeography, Palaeoclimatology, Palaeoecology*, 159: 113–43

Osborne, R.A.L. 1984. Lateral facies change, unconformities and stratigraphic reversals: their significance for cave sediment stratigraphy. *Cave Science*, 11(3): 175–84

SEDIMENTS: AUTOCHTHONOUS CLASTIC

Caves become filled with a variety of materials both during development and after their formation. These include chemical deposits (speleothems such as stalactites and flowstone) and clastic sediments (broken rock material). The clastic sediments may be "allochthonous" (originating outside of the cave, for example from sinking streams bearing alluvium), or may be "autochthonous" (originating within the cave). The latter are perhaps the least studied of all cave deposits, but are a diverse group of materials that form by a variety of processes.

The most familiar of these materials is breakdown (from the process by the same name or as "incasion") which forms by the collapse of the walls or ceiling of the cave. The breakdown is classified by size, ranging from blocks to chips. Blocks are composed of more than one original bed, and can be quite large; volumes of greater than 25 000 m³ have been documented. Slabs are fragments of a single bed, generally a few cubic metres in volume. Chips are derived from the destruction of a bed, and have volumes in the cubic centimetre range.

Collapse occurs when the mechanical strength of the rock mass is exceeded. Breaks usually happen along existing zones of weakness; joints, bedding planes, or faults. Although the heterogeneous nature of natural rock masses makes prediction complicated, it is possible to model the process of breakdown using reasonable assumptions. The simplest case (White, 1988) is for a cave passage developed in flat-lying rocks. Under this instance, overlying beds of limestone may be considered as mechanical beams. The critical thickness of the beam (the thickness at which it will fail under its own weight) is given by:

$$t = \frac{\rho l^2}{2S}$$

For this case, in consistent units, t = thickness of beam, ρ = rock density, l = beam length (roof span), and S = flexural stress.

The process of natural breakdown is only rarely observed directly; usually only the result is seen. Davies (1951) suggested that most breakdown found in caves fell soon after the cave was initially drained of water (transition from phreatic to vadose conditions). The loss of buoyant support to the wall and roof promotes this collapse. Although it is found throughout many caves, breakdown is more likely to occur at spots where support of the ceiling is lessened. Passage intersections are one such point. Others are places where one passage overlies another. Sometimes the process of breakdown is the only mechanism to connect two passages. Some of the most voluminous cave chambers were formed (at least partially) by breakdown processes. Conversely, many formerly continuous passages are now blocked, at least in terms of human exploration, by massive breakdown piles (see photo). This autochthonous sediment may be removed from the cave by ongoing stream action, by both chemical and mechanical erosion.

The size of the breakdown is greatly affected by the characteristics of the existing rock, specifically the bedding thickness, joint spacing, and type of limestone. Thick bedding and wide joint spacing promote large block breakdown. The specific mechanical or chemical process that is operating may also affect the size.

Processes other than removal of underlying support can also act to promote breakdown of the cave walls. In those regions where winter temperatures reach freezing, water within the wall rock near the entrance of the cave may freeze. This generates a tremendous pressure that may shatter the rock in a process known as frost wedging (or gelifraction). A similar process is thought to occur via mineralogical processes. Secondary gypsum or halite grows in the bedding planes of some cave walls and the pressure of crystallization may then break the material apart (White & White, 1969; White & White, 2000: p.428). Halite may also crystallize in the pores of rock, resulting in the creation of detrital piles of sand-sized salt and bedrock particles.

Direct observation of collapse events that form breakdown is documented, but rare. Some have been massive, such as a 35 ton event in the Rotunda Room in Mammoth Cave, Kentucky (United States), in 1994 (USNPS, undated). No one was injured. The collapse was caused by frost wedging brought on by an unusual cold period. In June 1984 a massive collapse (c.150 tonnes) occurred in Valhalla Pit, Alabama (United States), resulting in the death of two cavers (Knutson, Sims & Sims, 1985). No specific triggering event was identified. Earthquakes have been postulated as initiators of breakdown but Davies (1951) noted that the New Madrid earthquake (1811, Richter magnitude 8) did not cause any collapses at Mammoth Cave (Kentucky), even though it was just 250 km away.

Weathering earth (detritus) is the term assigned to the insoluble material that remains after dissolution of the limestone while the cave is being formed. It is usually fine-grained (clay to sand size), and difficult to distinguish from allochthonous sediments that have been washed in to the cave. Sometimes it clings to moist walls of the cave, forming a pattern known as mud vermiculations. Coarse material may also result from autogenic processes. If the host rock for the cave is rich in chert, insoluble nodules of this material may collect on the floor of the cave in substantial quantities.

In caves formed by ascending sulfur-rich waters, a peculiar type of clastic / chemical sediment may be found. Alunite is an aluminosulfate mineral that forms by reaction of existing detrital clays with acidic water. It has proved valuable as a dating tool in some areas (Polyak *et al.*, 1998). Also in these settings, reaction of limestone with sulfuric waters may create thick rinds of gypsum that fall to the floor of the cave after it has been drained.

Autochthonous clastic sediments have the potential to provide other sorts of useful information about the cave and its surrounding environment. As our understanding of these deposits and processes increases, they are certain to become a focus of additional scientific studies (see Paleoenvironments: Clastic Sediments). Biologically derived materials may also, arguably, be considered as autochthonous sediments but these are discussed in three separate entries on Sediments: Biogenic, Guano, and Organic Deposits in Caves.

IRA D. SASOWSKY

Works Cited

Davies, W.E. 1951. Mechanics of cave breakdown. *NSS Bulletin*, 13: 36–43

Knutson, S., Sims, M. & Sims, L. 1985. American caving accidents. *NSS News*, 43(11) pt. 2: 355–67

Polyak, V.J., McIntosh, W.C., Güven, N. & Provencio, P. 1998. Age and origin of Carlsbad Cavern and related caves from 40Ar / 39Ar of Alunite. *Science*, 279: 1919–22

United States National Park Service, Undated. *Rockfall*. Electronic publication available at http://www.nps.gov/maca/rockfall.pdf

White, W.B. 1988. *Geomorphology and Hydrology of Karst Terrains*, Oxford and New York: Oxford University Press

White, E.L. & White, W.B. 1969. Processes of cavern breakdown. *NSS Bulletin*, 31(4): 83–96

White, E.L. & White, W.B. 2000. Breakdown morphology. In *Speleogenesis: Evolution of Karst Aquifers*, edited by A.B. Klimchouk, D.C. Ford, A.N. Palmer & W. Dreybrodt, Huntsville, Alabama: National Speleological Society

Further Reading

Bögli, A. 1980. *Karst Hydrology and Physical Speleology*, Berlin and New York: Springer

Davies, W.E. 1949. Features of cavern breakdown. *NSS Bulletin*, 11: 34–35

Autochthonous Clastic Sediments: Passage blocked by chip breakdown, Greenbrier River Cave, West Virginia (United States). (Photo by Ira Sasowsky).

Ford, D.C. & Williams, P.W. 1989. *Karst Geomorphology and Hydrology*, London and Boston: Unwin Hyman.
Section 7.12 discusses Breakdown in caves.

Ford, T.D. 2001. *Sediments in Caves*, London: British Cave Research Association (BCRA Cave Studies Series, 9)
A short introduction to cave sediments for the educated caver.

SEDIMENTS: BIOGENIC

Biogenic sediments in caves can be divided into three major types: detrital (external material directly transported from the surface), internal (excrement or guano produced by animals inside the cave), and secondary (material formed inside the cave).

Detrital sediments in caves are formed from biogenic material directly transported from the surface, including soil, dung, peat, woody matter, suspended organic material, animal bodies, and garbage flushed into the cave by rain water or air flow. Their composition does not differ significantly from that of the same material on the surface. Sometimes peat forms small stalactites (Hill & Forti, 1997). The rest of the material forms sediments on the floor of the caves.

The main internal sediment is excrement (guano; see separate entry), which sometimes forms huge deposits inside caves. It is very good manure, rich in nitrates and phosphates, and so is sometimes mined in caves. It is even the main part of the export of Guanaco Island (northern Peru). Guano frequently reacts with the bedrock or clays and forms many cave minerals, as described below.

Secondary biogenic sediments, sometimes referred as biogenic speleothems (Forti, 2001), are of much greater interest because they differ significantly from surface sediments. They are produced by biomineralization, which can be divided into three types: (i) induced minerals formed by biogenic mineralization, (ii) matrix minerals formed by transformation of organic bodies, and (iii) nucleation minerals formed by microorganisms as centres of nucleation.

Induced mineralization

Biogenic mineralization produces 88 biogenic cave minerals classified by origin to guanogenic, microbiogenic, osteogenic, and anthropogenic minerals, and also organic deposits (e.g. mumijo) which do not have a definite chemical composition. Seventy-three guanogenic cave minerals are formed by the reaction of bat guano and animal excrement with cave bedrock and minerals. Phosphates, nitrates, chlorides, sulfates, and organics are derived from the guano or excrement to form cave minerals. Some of the guanogenic minerals are found only in caves. Some carbonate minerals also have a guanogenic origin. Two minerals are known to be formed only by combustion of cave guano (Martini, 1994).

Microbiogenic minerals can be formed directly by microorganisms by biomineralization through enzymes (Northup *et al.*, 1997), or by producing substances that lead to the precipitation of minerals (e.g. by changing the pH of their surroundings). Microbial processes in caves (see separate entry) often involve redox reactions produced by aerobic (chemolithotrophic) microorganisms, which obtain energy directly from the oxidation of inorganic compound or by anaerobic (heterotrophs) organisms which obtain energy from the oxidation of organic matter and reduce inorganic compounds. The "sulfur cycle" (see Microbial Processes in Caves) is such a process, whereby both sulfur-oxidizing and sulfate-reducing bacteria are involved. They produce a wide variety of cave minerals: most frequently native sulphur, gypsum, and iron oxides-hydroxides, but other less common forms such as celestite, fibroferrite, opal, endellite, pyrolusite, and marcasite have been reported in the literature (Hill & Forti, 1997). In total, 19 cave minerals are known to be formed by microorganisms.

Deposition of saltpetre (KNO_3) and all other cave nitrates is driven by nitrifying bacteria. Moreover, at least a part of the chemical reactions which lead to the mineralization of guano are surely controlled by microorganisms.

Fungi, algae, and bacteria have all been implicated in the precipitation of carbonate dripstones but it has not been proved if these microorganisms were actively involved in precipitation or whether they were just accidentally buried there. Algae can trigger the precipitation of calcium carbonate from solution, and may subsequently trap and bind the particles to carbonate speleothems. This may occur as the algae take up carbon dioxide during photosynthesis causing $CaCO_3$ to precipitate. Algae will only contribute to carbonate deposition at the entrance and twilight region of caves. Finally, bacteria that utilize carbon dioxide (such as *Thothrix* in the sulfur cycle) have been proved to cause accelerated carbonate speleothem growth (Forti, 2001).

Moonmilk is a precipitate that seems to be originated by microbiological reactions. In fact, two of the four possible mechanisms for the evolution of moonmilk (Hill & Forti, 1997) are biochemical corrosion of the bedrock by organic acid produced by microorganisms (*Arthrobacter*, Flavobacterium, and *Pseudomonas*), and active precipitation of moonmilk by bacteria (*Macromonas bipunctata*).

Osteogenic cave minerals are formed by the reaction of bones with mineralizing solutions in caves and only two such minerals are formed in this way.

Two cave minerals are known to result from human activity. Vaterite forms on remains of carbide introduced by cavers (Hill & Forti, 1997) and one large calcite flowstone in Italy has been described as derived by the living activity of larvae of a troglobitic insect (*Tricoptera wormaldia*) (Poluzzi & Minguzzi, 1998). Its complex genesis has been referred to the large community of larvae living inside the cave on the surface of wide anthropogenic organic deposits. It is thought that the larval respiration caused the production of large amounts of carbon dioxide, which in turn reacted with water saturated by gypsum, thus causing the deposition of calcite just around the larvae.

Organic formations that do not have a fixed chemical composition include pigotite, amberat, mumijo, and asphalt. Pigotite is formed by leaching of granitic or quartzitic rocks in the presence of organic matter and an acidic environment. Amberat is composed of the dung and urine of cave rats. Mumijo is a black, very soluble substance formed from the excrement of small animals (rabbits or mice) in dry caves of Central Asia by organic reactions (with the help of chemolithotrophic bacteria), which allows slow migration of the soluble substances of the excrement to form mumijo. It has strong and spicy smell and is used in traditional medicine. Asphalt is derived from petroleum-bearing rocks (Sasowsky & Palmer, 1994). Sometimes the large amount of organic matter produced in the "sulfur cycle" allows the evolution of speleothems (pseudo-stalactites) consisting of a single organic mat (mucus), which are normally called "mucolites".

Matrix mineralization

Matrix biogenic mineralization produces minerals formed by transformation of organic bodies, most commonly represented by rootsicles. When roots enter cave voids their surface may become a preferential area for the flow of seeping water and for calcium carbonate deposition. Speleothems growing over roots are normally called "rootsicles" (Hill & Forti, 1997). In wet tropical environments, the root complexes of large trees may become the main driving factor for the evolution of those peculiar speleothems, called "showerhead".

In show caves, where light is artificially supplied, plants may halt calcium carbonate deposition or even cause the corrosion of speleothems, due to the acid secretion of their roots (see Tourist Caves: Algae and Lampenflora)

Nucleation biomineralization

Microorganisms also can be centres of nucleation. Fungal hyphae may act as nuclei for crystallization and as a site for attachment for crystals. Algae can trigger the precipitation of calcium carbonate, subsequently trapping and binding the particles to carbonate speleothems. Algae respire CO_2 consequently causing precipitation of $CaCO_3$.

YAVOR Y. SHOPOV

See also **Guano; Microbial Processes in Caves; Organic Resources in Caves**

Works Cited

Forti P. 2001. Biogenic speleothems: An overview. *Abstracts of the XVth International Symposium of Biospeleology*, 3–6

Hill, C.A. & Forti, P. 1997. *Cave Minerals of the World*, 2nd edition, Huntsville, Alabama: National Speleological Society

Martini, J. 1994. Two new minerals originated from bat guano combustion in Arnheim cave, Namibia. *South African Speleological Association Bulletin*, 33: 66–69

Northup, D.E., Reysenbach, A.L. & Pace, N.R. 1997. Microorganisms and speleothems in *Cave Minerals of the World*, 2nd edition, edited by C.A. Hill & P. Forti, Huntsville, Alabama: National Speleological Society

Poluzzi, A. & Minguzzi, V. 1998. Un caso di biocostruzione in un ambiente di grotta. *Memorie dell Instituto Italiano di Speleologica*, 10: 93–100

Sasowsky, I.D. & Palmer, M.V. (editors) 1994. *Breakthroughs in Karst Geomicrobiology and Redox Geochemistry*, Charlestown, West Virginia: Karst Waters Institute

Further Reading

Ehrlich, H.L. (editor) 1996. *Geomicrobiology*, 3rd edition, New York: Dekker

Frye, G. (editor) 1984. *Encyclopedia of Mineralogy*, Stroudsburg, Pennsylvania: Hutchinson Ross

Shopov, Y.Y. 1989. Genetic classification of cave minerals. *Proceedings of the 10th International Congress of Speleology*, vol. 1, edited by A. Kosà, Budapest: Hungarian Speleological Society

SEDOM SALT KARST, ISRAEL

Mount Sedom is the world's best-studied salt karst region, containing the largest known salt cave. It is also the lowest and one of the most arid karst regions worldwide (annual precipitation c.50 mm). The caves have been explored systematically since 1981.

The exposed head of a diapir, composed of Neogene salt beds, forms the elongated (11 by 1.5 km) ridge of Mount Sedom within the Dead Sea graben. The diapir ranges in height from 400 m below sea level to 160 m below sea level. The mean estimated diapir rising rate is about 6–7 mm per year (Frumkin, 1996) and it is still rising. Salt karst has developed in the upper part of the diapir, where salt beds are vertical or steeply inclined. They often yield under the shear stress induced by salt tectonics and develop minor faults. Horizontal fissures are rarely found in Mount Sedom. Before its subaerial extrusion, the top of the rising diapir suffered dissolution by groundwater. Residual, relatively insoluble anhydrite, shale, and dolomite accumulated above the salt, forming a cap-rock up to 50 m thick. The top of the cap-rock is a roughly tabular desert surface.

Some 107 caves have been explored in Mount Sedom with a total length of c.20 km (Figure 1). The caves formed during the Holocene and the oldest cave passage has been dated to c.8000 years BP (Frumkin et al., 1991) by radiocarbon dating of wood twigs carried and deposited by cave streams. The lower levels of the caves still carry floodwater during infrequent rainfall events. The floodwater flows from surface runoff from many small (up to 0.7 km²) ephemeral catchments, developed on the cap-rock. The flow is captured into fissures which enlarge to form shafts. Sodium chloride concentration in flood flow increases dramatically within the caves, from $c.10$ g l^{-1} at stream sinks up to 200–350 g l^{-1} (Frumkin, 1994). Most floodwaters remain chemically aggressive while flowing rapidly within cave passages. Apart from dissolution, cave development is promoted by physical erosion attributed to fast-flowing water, cutting into the soft salt. After a short inception period, salt caves develop mainly by open-channel flow. The cave retains a downstream slope, supporting gravitational sediment-rich flow. The larger caves consist of a vertical shaft at the sink point or slightly downstream, and a salt passage draining the shaft. Shafts with a surface catchment area <200–300 m² usually do not have a draining passage; the small discharge infiltrates in the bottom of the shaft towards the water table. The typical passage in Mount Sedom caves is a vadose canyon, developed along a joint, bedding plane, or fault within the salt. Young passages are often steep, incised in bedrock, and follow the initial fissure closely. Older passages are often more sinuous, moderately inclined (a few %), and their floor is shielded by alluvium. Some wide passages with highly developed meander notches are indicative of flow without much recent downcutting; lateral migration of meanders sometimes causes destruction of earlier canyon walls. Downward development of passages is achieved either by incision of a passage floor or by a stream capture to a lower-level passage. The older caves in Mount Sedom have several tiered levels above the modern channel.

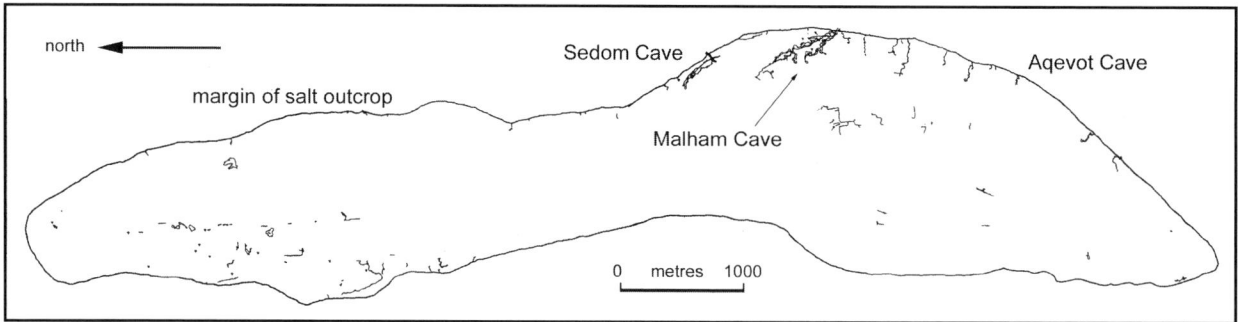

Sedom Salt Karst: Figure 1. Outline map of the known cave passages in the Mount Sedom salt dome, Israel.

Cave passages are either ingrowing vadose canyons or wide low passages with flat ceilings and corrosion bevels, developed by paragenetic dissolution (Frumkin, 1998). Some cross sections have been deformed by Holocene tectonics. The development of a cave profile and morphological features such as meanders and notches is sometimes constrained by less soluble layers (dolomite, shales, anhydrite) interbedded within the salt. Nickpoints in a passage profile are often associated with insoluble beds. Dolomite beds are the most prominent, as they are much more resistant to both dissolution and erosion, relative to the salt.

Many caves are integrated systems—they can be physically traversed along the full distance from sink to outlet (Figure 2). The largest caves, Malham Cave (the longest known salt cave; 5591 m long, 135 m deep) and Sedom Cave (1799 m long) drain to the south basin of the Dead Sea, the hydrologic base level of the region. They are branchwork caves, each with several stream passages joining underground. Residence time of floodwater in these caves is short, measured in minutes up to a few hours in a single flow event. Integrated cave systems are developed along the margins of the rising diapir, where conditions are favourable: hydraulic gradients are high, and fissures are long, open, and abundant because of lithostatic stress release.

Caves in the central portions of the mountain, terminating above the water table, appear to have no distinct outlet. These are referred to as "inlet" caves. The downstream parts of inlet caves often contain silt and clay banks with no continuation of explorable passages. Water ponded in the bottom of these caves has been found to become saturated with salt within a few hours. This evidence suggests that the limit of exploration is also a hydraulic limit between two sequential modes of water flow: rapid turbulent floods prevailing from the sink to the cave bottom, and diffuse infiltration below the bottom of the explorable passage, down to the output boundary of Mount Sedom. Three of the studied caves in northern Mount Sedom supported perennial brine ponds at least throughout the period 1984–95. The ponds are perched some tens of metres above the water table, and their levels seems to be controlled by fissure widths which decrease with depth. Water level in each lake also differs from the others by tens of metres. Both dissolution and precipitation features are observed on walls bordering ponds, as well as on cave walls where ponds have dried out. Horizontal notches indicate levels of aggressive water temporarily diluting the pond during floods. Large secondary halite crystals on some cave walls indicate supersaturation of the ponded water between successive floods. The ground plan of inlet caves range between three end members: elongated conduit, chamber, and maze, depending on fissuring properties of the rock and the hydraulic head applied by flash floods.

Most integrated cave systems originated in the past as inlet caves, created by a capture of subaerial channel into a cap-rock

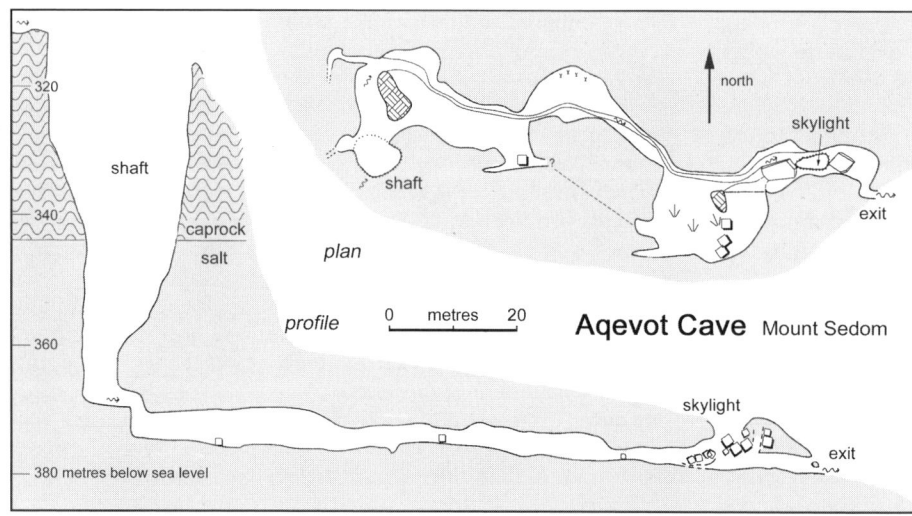

Sedom Salt Karst: Figure 2. Plan and profile of Aqevot Cave.

fissure. Diversionary routes are common where earlier conduits are blocked by alluvium, until a connection is established with an output point and a stable condition is achieved in ground plan. If a connection to the output boundary is not established, the cave continues to evolve as an inlet cave.

AMOS FRUMKIN

Works Cited

Frumkin, A. 1994. Hydrology and denudation rates of halite karst. *Journal of Hydrology*, 162: 171–89
Frumkin, A. 1996. Uplift rate relative to base level of a salt diapir (Dead Sea, Israel), as indicated by cave levels. In *Salt Tectonics*, edited by G.I. Alsop, D.J. Blundell & I. Davison, London: Geological Society
Frumkin, A. 1998. Salt cave cross sections and their paleoenvironmental implications. *Geomorphology*, 23: 183–91
Frumkin, A., Magaritz, M., Carmi, I. & Zak, I. 1991. The Holocene climatic record of the salt caves of Mount Sedom, Israel. *The Holocene*, 1(3): 191–200

Further Reading

Frumkin, A. 1994. Morphology and development of salt caves. *Bulletin of the National Speleological Society*, 56: 82–95
Frumkin, A. 1996. Determining the exposure age of a karst landscape. *Quaternary Research*, 46: 99–106
Frumkin, A. 1996. Structure of northern Mount Sedom salt diapir (Israel) from cave evidence and surface morphology. *Israel Journal of Earth Sciences*, 45: 73–80
Frumkin, A. 1997. Liquid Crystal Cave, Israel. In Cave Minerals of the World, edited by C. Hill & P. Forti, Huntsville, Alabama: National Speleological Society
Frumkin, A. 2000. Speleogenesis in salt — the Mount Sedom area, Israel. In *Speleogenesis: Evolution of Karst Aquifers*, edited by A.B. Klimchouk, D.C. Ford, A.N. Palmer & W. Dreybrodt, Huntsville, Alabama: National Speleological Society
Frumkin, A. & Ford, D.C. 1995. Rapid entrenchment of stream profiles in the salt caves of Mount Sedom, Israel. *Earth Surface Processes and Landforms*, 20: 139–52
Frumkin, A. & Raz, E. 2001. Collapse and subsidence associated with salt karstification along the Dead Sea. *Carbonates and Evaporites*, 16(2): 117–30

SELMA PLATEAU CAVES, OMAN

Huge cave passages lie hidden beneath the nearly barren, remote highlands of the Selma Plateau in northern Oman. Local legend states that God honoured a brave shepherd girl, Selma, by sending down seven stars that created the seven vertical shafts on the plateau today. Three of the shafts drop into one of the world's largest cave rooms, Majlis Al Jinn (Meeting Hall of the Spirits). The other four link into a cave system 11.5 km long and 385 m deep. The dimensions and vertical development of these caves seem remarkably incongruous with the current arid climate of the region.

The three shafts that break through the roof of Majlis Al Jinn, provide access and sunlight into the Meeting Room of the Spirits. An X-shaped hole, Khonshilat beyn Al Hiyool, is the largest entrance and drops 136 m near the south side of the room below. The slightly smaller Khonshilat Maqandeli (the name local people also apply to the cave below) drops the shortest distance, 118 m, from a nearly circular hole 20 m in diameter. The smallest entrance, Khoshilat Minqod, provides the most dramatic entrance with a 158 m descent into the centre of the massive chamber. The cave consists of a single chamber approximately 300 m x 200 m x 100 m (Davison, 1985); its measured volume is 3.9 million m³. The chamber is evenly rounded in plan and has a gently domed roof. All three shafts are on joint intersections, which have almost no influence on the chamber slope. The ceiling is nearly devoid of speleothems, although some dripstone hangs from alcove walls. Sand, silt, and clay sediments mostly cover the floor. A cracked mudflat fills most of the northern quarter of the floor. Sparse breakdown lies on the floor, mostly on the eastern side of the room, but conspicuously less abundant than might be expected in a chamber of this size. Apparently relict flowstone, cave pearls, and large coral pipes cover parts of the west-central floor and provide some hints to an earlier, wetter time in the cave's history. Dripping water from the ceiling initially formed coral pipes as pillars of sediment under protective caps of more resistant material (in the same manner as "hoodoos" on the surface). The pillars and cap rocks became calcified and coated with splatter cave coral. Much of the sediment has been eroded from under the coral pipes in Majlis Al Jinn, commonly leaving behind mostly hollow coral pipe shells. No known passages extend from the chamber.

The other large shaft entrances of the Selma Plateau are Bayn Halayn (Arch Cave), Kahf Khasha (Funnel Cave), Kahf Aqabat Khushil (Seventh Hole), and Three Window Cave (Figure 1). All four shafts lie in the floors of wadis cut 5–20 m deep across the plateau surface. They carry floodwater in storm events. The four caves all descend joint-guided shafts for about 250 m, then drain down the dip of 5–10° and converge into a cave system before resurging (Thomas & Robinson, 1997; Ganter, 1998). Seventh Hole (Figure 2) has broken shafts dropping 250 m into the head of the Canyon Room, 200 m long and 80 m wide. Its outlet streamway descends to meet larger and older passages with eroded stalagmites and cemented fills. These continue to the now intermittent resurgence of Kahf Tahry, truncated in its canyon wall. Three Window Cave offers a superb through trip, 6 km long dropping 385 m with 39 pitches of 3–34 m, many over large gour barriers, sometimes into pools of crystal-clear water. Its long streamway enters the base of the Canyon Room below Seventh Hole. The other tributary passages are smaller, but Funnel Cave has a clean shaft of 170 m just inside its entrance.

The caves of the Selma Plateau developed in the Middle Eocene Seeb Formation, a massive, argillaceous, bioclastic, basinal limestone with subordinate clastic units. An upper, unnamed marly member probably forms the ceiling of Majlis Al Jinn. Most of that chamber and all of the other caves lie in the stratigraphically lower, cliff-forming Leyh Member of the Seeb Formation. The caves immediately drop through the typically resistant, bioclastic calcarenite of the upper and middle layers, forming low-gradient trunk passages in the lower parts of the formation, which are interbedded marl, clayey limestone, and

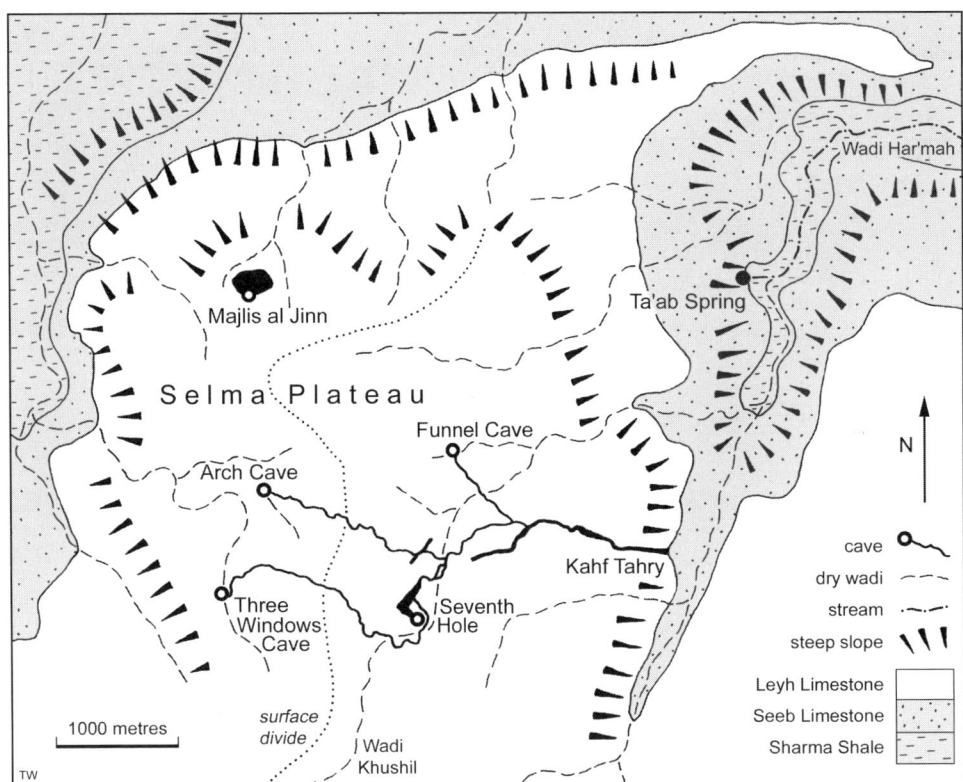

Selma Plateau Caves, Oman: Figure 1. Outline map of the Selma Plateau, with the Selma Cave System (with the four sink entrances and the Kahf Tahry outlet) and also the cave chamber of Majlis al Jinn.

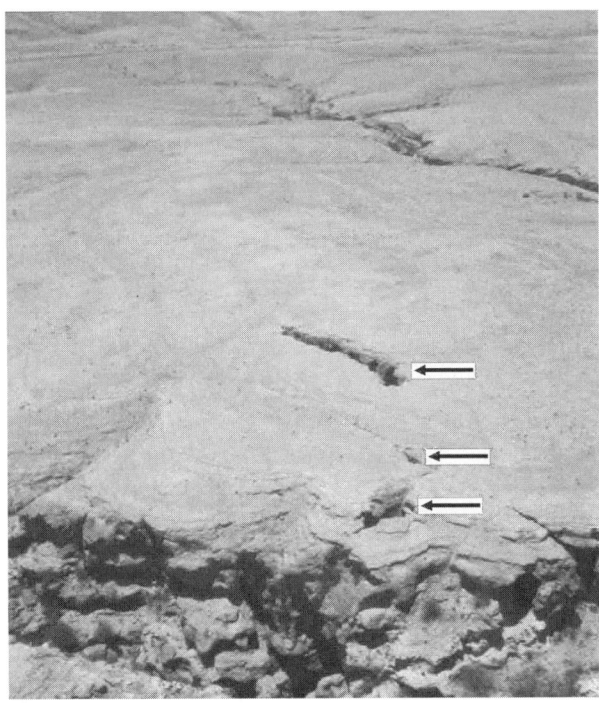

Selma Plateau Caves, Oman: Figure 2. Arrows point to the entrances of Seventh Hole, as seen from the air southsoutheast of the cave. Note the lack of surface drainage entering these entrances, yet the largest chamber of the Selma cave system lies at the base of the entrance drops. (Photo by Louise Hose)

nodular bioclastic limestone. Kahf Tahry also formed in this lower unit, suggesting that a stratigraphic barrier may have perched cave development.

In the Miocene, the Seeb Formation experienced a northeast–southwest compression phase of Alpine deformation that uplifted the Oman Mountains. The Selma Plateau gently arched up without significant faulting. Uplift of a few tens of metres continued into the Quaternary.

While recognized as a spectacular karst region, the Selma Plateau has received very little speleological investigation beyond initial cave exploration. Far more mysteries remain than clear understanding of the regional speleogenic history. Majlis Al Jinn appears to be the product of hypogenic waters rising from depth, dissolving the huge chamber beneath a structurally strong dome ceiling, leaving only insoluble sediments behind. If so, the cave formed when the water table was much higher than it is today, probably millions of years ago. No attempt has yet been made to determine the age of the cave or the validity of the hypogenic origin hypothesis.

The large size of the rooms and passages in the main cave system demands some explanation. The Canyon Room and the four entrance shafts could represent former conduits of rising, hypogenic, acidic groundwater, within an environment comparable to that which formed Majlis Al Jinn. When erosion exposed these previously buried conduits, they were invaded by surface waters—the wadis beyond the sinks are much reduced in size. Alternatively, Seventh Hole and the Canyon Room may simply be a much older sink with an outlet to the large lower passages now obscured.

LOUISE D. HOSE

Works Cited

Davison, W.D. Jr. 1985. Majlis Al Jinn Cave, Muscat: Sultanate of Oman, Public Authority for Resources (Report PAWR 85-20)

Thomas, V. & Robinson, L. 1997. Oman '97. *Caves and Caving*, 76: 25-28

Further Reading

Bechennec, F., Le Metour, J. & Roger, J. 1992. Geological map of Tiwi Sheet (scale 1:100 000) with explanatory notes. Muscat: Sultanate of Oman, Ministry of Petroleum and Minerals, Directorate General of Minerals

Burns, S.J., Fleitmann, D., Matter, A., Neff, U. & Mangini, A. 2001. Speleothem evidence from Oman for continental pluvial events during interglacial periods. *Geology*, 29: 623-26

Burns, S.J., Matter, A., Frank, N. & Mangini, A. 1998. Speleothem-based paleoclimate record from northern Oman. *Geology*, 26: 499-502

Ganter, J. 1998. Caving on the Selmeh Plateau, Sultanate of Oman. *NSS News*, April 1998: 100-108

Hanna, S. & Belushi, M.A. 1996. *Introduction to the Caves of Oman*, Muscat: Sultan Qaboos University

SEWU CONE KARST, JAVA

Gunung Sewu means Thousand Hills, but there are probably ten times that number of small rounded limestone hills in an unbroken karst of over 1000 km^2. Lying along the south coast of Java, just southeast of the city of Yogyakarta, the Sewu karst is formed in a belt of Miocene reef limestones. These lie on the flank of the chain of active volcanoes that makes up the core of Java. The climate is equatorial, with little seasonal or diurnal temperature variation from a mean of 26°C, and an annual rainfall of 2000 mm, all of which occurs between November and April.

The major feature of Gunung Sewu is the surface topography. The limestone hills are remarkably uniform in size and shape, each being about 200 m in diameter and 50 m high. They were first described by Herbert Lehman as the type example of kegelkarst, or cone karst. However, the hills are actually hemispherical (Figure 1). They do not have the sharp summits of the truly conical hills now known in the fengcong cone karsts of South China. Gunung Sewu certainly constitutes a classic karst landscape, but is perhaps better regarded as the type area for hemispherical karst, which is matched elsewhere in the world, notably in Sulawesi's Bohol karst and in Jamaica's Cockpit Country.

The rounded profile of the hills is probably created by a greater degree of rock head (the rock surface beneath unconsolidated deposits) dissolution beneath thicker soils retained on the hilltops. The origin of this soil is therefore critical and diagnostic, and in Gunung Sewu it may owe its longevity at least in part to regular renewal by airfall ash from the nearby volcanoes. When soil is stripped from the hills by erosion, slope profiles adjust to the gradient and the karst matures into a landscape of conical hills, as in China.

The conical hills of Gunung Sewu have broadly concordant summit levels that rise to about 400 m above sea-level in the centre of the karst. Between them, depressions are floored by thick layers of smectite-rich soils, derived largely from volcanic ash. The overall pattern of these depressions reveals the dendritic systems of the valleys, but these are now broken into strings of dolines. It appears that valleys were superimposed on to the limestone on a surface close to the level of the summit trend, before being segmented by karstic capture. Combined with the small dimensions of the passages in most of the caves, this suggests that the karst landscape is fairly young, even though the hills show a uniformity indicative of maturity.

There are no long surface stream courses, as all drainage finds a sink within a very short distance. The few surface rivers that drain into the karst from inland catchments also sink into open caves. There is little recognizable structure within the limestones, and therefore little geological influence on the topography. Some areas of softer, more chalky limestones have slightly more gentle hill slopes, and the chalk of the Wonosari Plateau, north of Gunung Sewu, has a gently rolling topography. Its diffuse groundwater flow beneath a shallow water table feeds into the limestone karst at an interdigitated facies boundary, and rapidly descends to depth in the more permeable cavernous limestone.

Gunung Sewu has a high rural population, with many small villages of farmers, who tend every bit of soil on the doline floors and also struggle to terrace and farm some of the hill slopes. In the long dry season there is a serious shortage of water, and many farmers have to carry water from the few accessible cave sources, often involving time-consuming candle-lit underground scrambling. A series of deep boreholes in the karst proved to be almost dry, and yielded no usable water resources.

Consequently, two small teams of British cavers systematically explored the caves of central and western Gunung Sewu in 1982–83. Their purpose was to discover better supplies of underground water that could be reached by shallow wells or deeper (targeted) boreholes. They visited 257 sinks and mapped 216 caves, most of which were shafts; 28 reached more than

Sewu Cone Karst, Java: Figure 1. The landscape of seemingly endless conical hills that makes the Gunung Sewu karst. It is clear that most hills are actually hemispherical and not as uniformly conical as those in the Chinese Karst. (Photo by Tony Waltham)

Sewu Cone Karst, Java: Figure 2. The shaft of Gua Lebak Bareng, 140 m deep. (Photo by Andy Eavis)

100 m deep, and nine extended to more than a kilometre of passages (Willis et al., 1984). Most caves drop steeply or vertically (Figure 2) to sump pools on the regional water table, which rises inland with a mean gradient of 1:150. A minority of longer caves have either youthful meandering canyons or sections of larger partially drained phreatic tunnels. There is no consistent pattern of sequentially developed old caves, and few dry high-level passages are known. Nearly all the caves in central Gunung Sewu drain to a single choked resurgence on the beach at Baron Bay. This lies in a recess between cones that are spectacularly undercut and truncated along the ocean coastline.

The largest underground flows are the two small rivers that are fed from the Wonosari Plateau. The Kali Suci drains through a series of short but large caves, before passing across the floor of the dramatic Grubug shaft, 65 m deep below its small daylight eyehole, and descending a flight of cascades to a sump. This marks the water table in a trough defined by its efficient conduit flow through to the Baron resurgence on a gradient of only 1:750. The Bribin Cave carries a small river through 3 km of lakes in a tunnel mostly 20 m high and wide.

The eastern end of the Gunung Sewu karst was explored by another team of cavers in 1984 (Stoddard, 1985). This area has fewer deep shafts and more long stream caves, including the splendid Luweng Jaran with 12 km of passage including a very long streamway and some beautifully decorated high-levels.

TONY WALTHAM

See also **Cone Karst**

Works Cited

Stoddard, S. 1985. Anglo-Australian Speleological Expedition to Java, 1984. *Cave Science*, 12(2): 49–60
Willis, R.G., Boothroyd, C. & Briggs, N. 1984. The caves of Gunung Sewu, Java. *Cave Science*, 11(3): 119–53

Further Reading

Lehman, H. 1936. Morphologische Studien auf Java. *Geographische Abhandlungen*, Serie 3, 9: 1–114
Waltham, A.C., Smart, P.L., Friederich, H., Eavis, A.J. & Atkinson, T.C. 1983. The caves of Gunung Sewu, Java. *Cave Science*, 10(2): 55–96

SHANIDAR CAVE, IRAQ: ARCHAEOLOGY

Shanidar Cave overlooks a tributary of the Tigris River in northeastern Iraq. It is famous for its nine Middle Paleolithic burials known as the Shanidar Neanderthals (1–9), found by Ralph Solecki in excavations between 1953 and 1960. The diseased state of one of the skeletons and the "flower burial" of another of these Neanderthals has changed our view of Neanderthals from ape-like and brutal creatures to beings showing altruism and compassion.

The cave, eroded in the Middle Cretaceous limestone of the Zagros Mountains, has yielded the partial skeletal remains of two infants and seven adult Neanderthals from the Middle Paleolithic, 28 representatives of anatomically modern humans from the Holocene, and a wealth of floral, faunal, and cultural data. The Neanderthal sample is the largest from the Middle East, providing an almost continuous sequence of human evolution, from the appearance of Neanderthals through to the appearance of modern humans and the domestication of plants and animals. The Neanderthal sample is divided into two groups: the typical Neanderthals (Shanidar 1, 3, 5) ^{14}C-dated at more than 45 000 years BP and the Early Neanderthals (Shanidar 2, 4, 6, 7, 8, 9) tentatively dated between 65 000 and 100 000 years BP.

Solecki excavated only the central part of the cave down to bedrock, and found five major layers: A, B1, B2, C, and D. The sediments consist of a loamy soil and numerous rocks. The rocks came from the roof of the cave after at least five major rock falls. Some of the Neanderthal skeletons were found badly crushed under the debris. The most convincing evidence for a death caused by a rock fall comes from Shanidar 5. His pelvis was beneath a rock, his legs' anterior surfaces were facing down and his trunk was bent backwards so that the skull was at the other side of the rock next to the pelvis.

The skeletal remains from Shanidar Cave are described in a monograph by Eric Trinkhaus. Shanidar 1 sustained a massive blow to the right side that badly damaged his right arm, foot, and leg, and a crushing fracture to the left eye that would have rendered his left eye blind. These injuries, which had occurred long before his death, are merely evidence that he could not have been an effective hunter, not necessarily that he was treated with compassion. Shanidar 4 became famous as the "flower burial". It is the best example of an intentional burial from the Middle Paleolithic. Shanidar 4, 6, 8, and 9 were found superimposed on each other in a natural rocky crypt. Usually, crypts and pits are considered as evidence of intentional burial. The covering of the Shanidar 4 skeleton by rocks may have prevented carnivores from exhuming the body. Indeed, skeletal remains of carnivores have been found in the cave, such as wolf, jackal, fox, and brown bear. Therefore, the completeness of Shanidar 4 skeleton is attributable to its covering by rocks and not to the absence of carnivores from the site. Besides, no comparatively well-preserved animal skeletons have been found in the cave.

The Shanidar 4 skeleton was found in a "fetal position", i.e. with his legs flexed. This may imply either an intentional burial or simply a desire to accommodate the corpse in the small confines of the rocky crypt. Flexed skeletons are generally considered as evidence of intentional burial, even though one cannot distinguish a ritual or religious burial from that of a corpse disposal. The presence of exceptional "grave goods", such as flowers, is another line of evidence for an intentional ritual burial. Solecki had already collected some 50 soil samples from the occupational deposits. When he uncovered the remains of Shanidar 4, he collected some representative samples from the crypt area below Shanidar 4 for pollen analysis. He noticed that there were two distinct layers—one just above and one just below the skeleton of Shanidar 4. The soil of the upper layer was a loose light brown sandy loam, and it was apparently disturbed by rodents. The lower soil was a very tough dark-brown organic loam interlayered with flowstone sheets.

Palynological analysis of the upper soil by Leroi-Gourhan showed that it had only isolated pollen grains, as did the samples from the rest of the cave. In the lower soil, however, there were extraordinary amounts of herbaceous and arboreal plants. There were not just isolated pollen grains but compact masses of fossilized pollen, many in the form of anthers—145 of them. In addition, the lower soil contained a large amount of noncarbonized (unburned) wood (oak, pine, juniper, ash, and fir). In fact, there was a whole layer of organic material full of wood fragments and flower anthers that could have been used as a sort of bedding. This gives a special character to the whole cave and shows that Shanidar 4 is an intentional burial. This does not mean that the flowers had the same meaning for the dead Neanderthals as they have today for us. For instance, the flower anthers were found beneath the skeleton and one of the flowers is covered with lots of spikes.

A criticism of the hypothesis that Shanidar 4 had been buried intentionally with flowers was based on the following assertions, that:

1. There were no grave pits in the cave;
2. There was no good stratigraphic evidence;
3. The superimposition of the skeletons shows temporal separations, not contemporaneity;
4. The rocky crypt was a shelter;
5. Only one anther was found;
6. The wind and rodents may have carried the tree branches and the flowers into the burial;
7. There was no sampling for pollen analysis from the other areas of the cave;
8. The rock fall preserved the pollen, unlike the other areas of the cave where no large concentrations of pollen were found;
9. No stratum of pollen was identified;
10. The pollen was not fossil, but contamination at the time of excavation; and
11. The intact nature of the skeleton of Shanidar 4 was due to the lack of carnivores in the cave.

In view of the fact that the above arguments are unsound, it is concluded that Shanidar 4 was at least an intentional burial. A thick layer of blossomed herbaceous and arboreal plants had been placed beneath the body of Shanidar 4.

ANTONIS BARTSIOKAS

Further Reading

Constable, G. 1980. *The Neanderthals*, New York: Time-Life Books
Dettwyler, K.A. 1991. Can paleopathology provide evidence for "compassion"? *American Journal of Physical Anthropology*, 84: 375–84
Gargett, R.H. 1989. Grave shortcomings. The evidence for Neanderthal burial. *Current Anthropology*, 30(2): 157–90
Leroi-Gourhan, A. 1975. The flowers found with Shanidar IV, a Neanderthal burial in Iraq. *Science*, 190: 562–64
Perkins, D. 1964. Prehistoric fauna from Shanidar, Iraq. *Science*, 144: 1565–66
Solecki, R.S. 1963. Prehistory in Shanidar Valley, Northern Iraq. *Science*, 139: 179–93
Solecki, R.S. 1971. *Shanidar, The First Flower People*, New York: Knopf.
Solecki, R.S. 1975. Shanidar IV, a Neanderthal flower burial in Northern Iraq. *Science*, 190: 880–81
Solecki, R.S. 1977. The implications of the Shanidar cave Neanderthal flower burial. *Annals of the New York Academy of Sciences*, 293: 114–24
Stringer, C. & Gamble, C. 1993. *In Search of the Neanderthals: Solving the Puzzle of Human Origins*, London and New York, Thames and Hudson
Trinkaus, E. 1983. *The Shanidar Neandertals*, New York: Academic Press

SHILIN STONE FOREST, CHINA

Areas of spectacular pinnacle karst in southern China have so many tall, closely-packed limestone pinnacles, that for 2300 years they have been known as shilin—which translates as stone forest. They differ from pinnacle karst elsewhere in their complex geological evolution. The largest and best known is the type example at Shilin, in Lunan County, Yunnan Province (Figure 1), but there are many more similar stone forests, both in Yunnan and elsewhere in China.

The stone forests occur in exposed beds of massive limestone, which dip at less than 10°. The finest occur in the Permian Maokou Limestone, but some are on other Permian and Carboniferous beds. The limestone is broken by sub-vertical joints, which have been eaten out by dissolution to form deep fissures, slots, and canyons between the tall and narrow blocks of limestone that remain. The limestone is fretted by rainwater into pinnacles and arêtes, which are scored by long deep rinnenkarren,

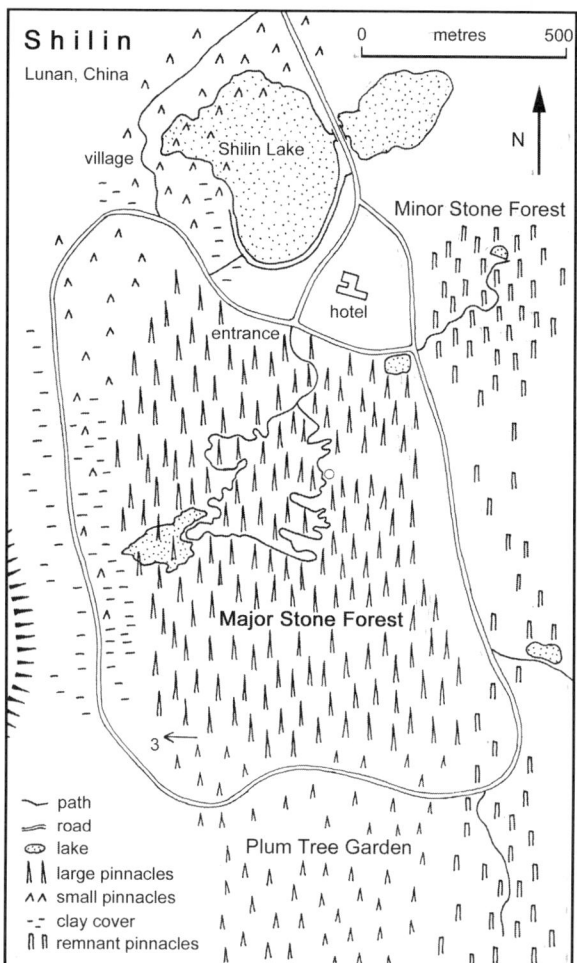

Shilin Stone Forest, China: Figure 1. Outline distribution of the main types of pinnacle karst at Shilin, the type stone forest at Lunan, China.

Shilin Stone Forest, China: Figure 2. The central part of the stone forest at Shilin, with the pagoda-style temple perched on the pinnacles to provide a viewpoint for visitors who are otherwise trapped on footpaths between the pinnacle bases. (Photo by Tony Waltham)

with razor-sharp crests on all the ridges and summits (Figure 2). Pinnacles are mostly 10–30 m tall, but some reach 50 m; bedding planes in the limestone generally prevent the formation of any taller pinnacles. Collapses and toppling failures do occur, but most pinnacles are stable; the low bedding dip and nearly vertical jointing are critical factors in this respect. The pinnacles are bare rock and there is little or no vegetation in the intervening fissures. It is possible to walk around in much of the network of fissures and canyons between the pinnacles; the Shilin site at Lunan has had tourist trails through it since 1614, during the Ming dynasty.

As in all well-developed pinnacle karst, the modern morphology is largely the result of a long period of dissolutional erosion by plentiful rainwater, charged with carbon dioxide in a warm climate. The sharp crests and intervening rillenkarren are evidence of erosion by direct rainfall. Pinnacle walls that have evolved under vegetation are more fretted, with random channels etched into them. Nearly all rock surfaces are dark grey, due to a crust rich in blue-green algae. The lower parts of many pinnacles have smoother faces with rounded ribs, and it is clear that these surfaces have evolved under water or beneath saturated masses of organic soil. Horizontal swamp notches indicate the role of gently flowing water at the local base level, while other pinnacles can be seen still half buried in soil, where sub-soil rockhead dissolution has been significant.

A transect across the Lunan Shilin reveals the recent geomorphological evolution of the landforms. The Major Stone Forest occupies an area of about 100 ha; it slopes and dips gently to the west towards a clay-floored basin. Within this, proto-pinnacles that project from the clay as stone teeth a few metres high have well-rounded surfaces that have clearly not long been exposed to rainwater corrosion. On the eastern side of the main stone forest, the Minor Stone Forest and a field of isolated, shattered pinnacles, up to 20 m tall, together represent an old age, degraded version of the shilin landform. Further east, the entire limestone bed has been removed by subaerial erosion.

This evolutionary sequence accounts for the main landforms of Shilin, and can be repeated in many other pinnacle karsts. The special factor of the Chinese shilin is its much longer evolutionary history. The Maokou limestone was indurated and then uplifted soon after its deposition. Its surface was first exposed to karstic erosion in the late Permian. It was then covered by the basalts and tuffs of the Upper Permian Ermei Volcanics. Subsoil dissolution continued beneath the tuffs, creating a pinnacled rockhead. This was then partly exposed and further carved into a small-scale version of pinnacle karst, in a Mesozoic tropical environment. Regional subsidence caused a second burial of the limestone, as Eocene red clays accumulated in lakes that gradually extended across the karst. Only in the Quaternary did uplift initiate re-exposure of the limestone. The newly uncovered surfaces were already eroded into pinnacles, but these have been considerably deepened and sharpened by rainwater within the modern environment.

Within Lunan County, a number of separate smaller stone forests provide the evidence for this long evolution. Different sites have pinnacles half covered by basalt lavas, Permian tuffs and Eocene red clays; some also have pinnacles that are half submerged in modern lakes. In each case, the pinnacles can be seen at some stage of evolution towards their modern form. It is the inherited fossil karst features that distinguish the stone forests of shilin from the areas of pinnacle karst recorded elsewhere, notably at Mount Kaijende and Mulu (see separate entries).

TONY WALTHAM

See also **Karren**

Further Reading

Ford, D.C., Salomon, J.-N. & Williams, P.W. 1996. "Les forêts de Pierre" de Lunan (Yunnan, Chine). *Karstologia*, 28(2): 25EJ 40

Song L. 1986. Origination of stone forests in China. *International Journal of Speleology*, 15: 3–13

Song L., Waltham, T., Cao N. & Wang F. (editors) 1997. *Stone forest, a treasure of natural heritage*, Beijing: China Environmental Science Press

SIBERIA, RUSSIA

There are more than 1400 caves known in Siberia, being widely spread along the south of the area. Gorniy Altai, the East Sayan Upland, the southern part of Central Siberian Upland, the Primorsky Range and the Priolkhonskoye Plateau near Lake Baikal, Sikhote-Alin' in the Far East, and Sakhalin Island are the most investigated speleologically (see map in Russia entry). Karst caves are the most numerous, and although crevice caves are very widespread on the Siberian Platform, only a few have been explored. Caves formed by abrasion are very common on the coasts of the Pacific Ocean, Lake Baikal, and reservoirs. Weathering caves, grottos, and niches are rather numerous, but they attract speleological investigation relatively rarely, owing to their small sizes. More than 30 volcanic caves have been mapped on the Kamchatka Peninsula and Kuril Islands, and a glacier cave has been described in the Baikalskiy Range.

The karstic caves are found in rocks from a wide range of geological ages—Archean, Proterozoic, all divisions of the Paleozoic, Triassic, and Jurassic—and are usually developed in calcitic and dolomitic marbles, limestones, dolomites, and carbonate rocks of mixed composition. But caves in other karst rock types also occur, for example, caves in carbonatites, calcyphyres, skarns, gypsum, gypsum anhydritic rocks, carbonate conglomerates, dolomitized gravelites, and sandstones.

The Gorniy Altay Mountains contain more then 400 caves. The largest are the Altayskaya (4175 m long, 248 m deep), Këktash (2300 m long, 350 m deep), Tutkushskaya (1400 m long, 218 m deep), and Soantekhnicheskaya (900 m long, 215m deep) caves.

More than 100 karst caves are known in Upper Proterozoic and Cambrian carbonate rocks in Kuznetskiy Alatau and Gornaya Shoriya. The largest are the Korolëva (5070 m long, 171 m deep), Fantaziya (6200 m long, 272 m deep), and Yashchik Pandory (10 100 m long, 182 m deep) caves. Yashchik Pandory is in massive Lower Cambrian limestone. The upper level is rich in ancient calcite speleothems. The lower level lies at the water table, where cave lakes occupy one-third of the passages in dry seasons.

174 karst caves have been explored in the East Sayan Upland. A few significant karst caves, Bol'shaya Oreshnaya (47 000 m long, 197 m deep), Badzheyskaya (6000 m long, 170 m deep), Tëmnaya (2500 m long, 55 m deep), Rucheynaya (1300 m long, 85 m deep), are developed in Ordovician carbonate conglomerates. Bol'shaya Oreshnaya is the longest conglomerate cave in the world. It is a deep and spacious ancient phreatic labyrinth, which consists of inclined and horizontal passages, with chambers, pits, crevices, cellular parts, lakes, streams, and a large flooded zone. Karbonatitovaya Cave is the only one in carbonatites known in the world.

There are 26 karst caves on the south of the Central Siberian Upland. The largest are Botovskaya Cave (57 260 m long, 6 m deep), the longest maze in Siberia and Russia, developed in Lower Ordovician limestones and sandstones, and Argarakanskaya Cave (8300 m long, 56 m deep) in Lower Cambrian limestones.

In the middle part of the Central Siberian Upland, 14 small horizontal solution caves in Ordovician gypsum are known at the Vilyey River. The largest, Oyusutskaya-9 Cave, has a length of 95 m.

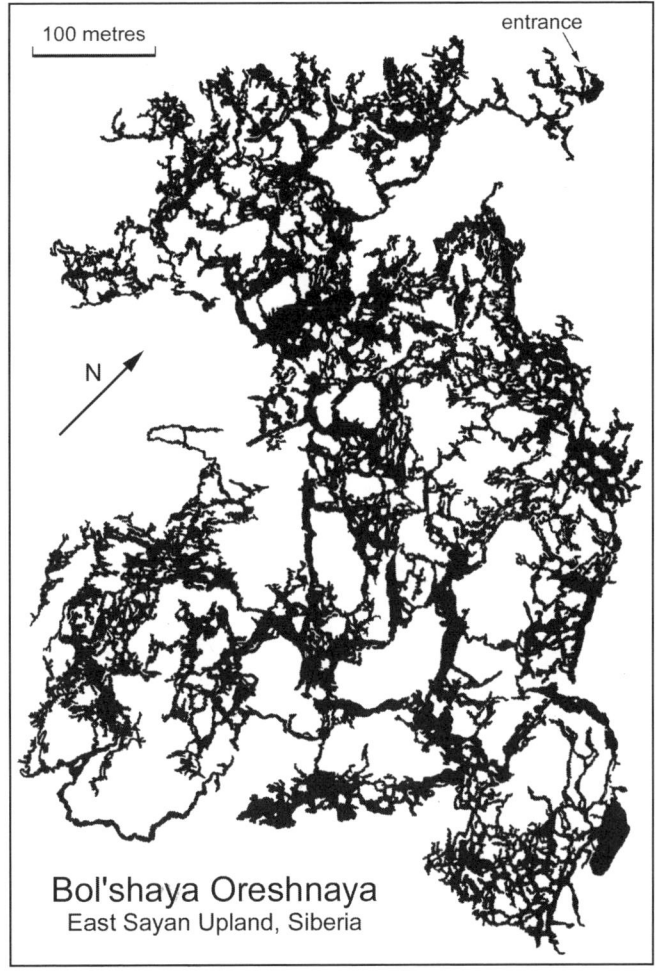

Siberia, Russia: Bol'shaya Oreshnaya Cave, the longest conglomerate cave in the world (survey by the Divnogorsky Speleo-Club), showing 36 km of surveyed passages as of 1987 (presently the cave length is 47 km).

At the east of the Siberian Platform, 12 karst carbonate caves with a total length of 2245 m were found in the Yudoma–Mayskiy Foredeep. The largest, Abagy–Dzhie (1400 m long, 8 m deep), is a horizontal water table labyrinth in Riphean dolomitized limestones. Lakes cover 70% of the cave floor.

The Khamar–Daban Range has only seven caves developed in Lower Proterozoic marbles and limestones. The Primorsky Range and Priolkhonskoye Plateau are the most speleologically investigated among the mountains surrounding Lake Baikal. They contain 55 and 24 caves respectively. Ancient phreatic caves, developed in Upper Archean–Lower Proterozoic graphite marbles, scarns, and calcyphyres on the Priolkhonskoye Plateau, are the most interesting. Some of them were formed in pre-Middle Miocene time (Aya, Ryadovaya, Oktyabr'skaya), some in pre-Eocene time (Mechta, Bol'shaya Baydinskaya, Shamanskaya Caves).

The other ranges (Baikalskiy, Barguzinskiy, Ikatskiy, Morskoy), situated around Lake Baikal, contain single karst, weathering, glacier, and abrasion caves. Kal'tsitovaya Cave, developed in Upper Proterozoic limestone in the Morskoy Range, has a hydrothermal genesis. It has tube-shaped inclined passages, encrusted with hydrothermal calcite 20 cm thick.

The Stanovoye Upland, which includes the Severo–Muyskiy, Delyun–Uranskiy, and Aglan–Yan Ranges, contains 35 caves in Cambrian limestones. Most caves are relict corrosional / erosion cavities. A few caves in Aglan–Yan Range are active, formed by recent water streams. The karst shaft Klyuch in Severo–Muyskiy Range is the deepest, at 57 m.

Six karst caves, connected with Upper Riphean–Lower Cambrian tremolite limestones, have been studied on the Vitim Upland. The largest, Dolganskaya Yama (5120 m long, 125 m deep), is a compound volume labyrinth of steeply inclined, vertical, and horizontal passages. Multiple branched systems of large organic tubes (spongework) are in the ceilings. Calcite flowstones and corallite cones are widespread. There are also two siphon lakes.

Among 30 caves known in Selenginskaya Dauriya, there are only two formed in limestones, the rest being weathering cavities. They were developed in various effusive, intrusive, and sedimentary rocks of the Lower Cambrian, Upper Paleozoic, Mesozoic, and Miocene ages. All of them are small.

The Mongolo–Okhotskiy folded belt has been explored very slightly. In the western part, in the Onon and Gazimur River basins, only small karst caves are described (Lurgikanskaya, Monasytuyskaya, Soktuyskaya, etc.) developed in the Riphean, Lower Cambrian, Silurian, and Devonian marbles and limestones. In the eastern part of the belt there are only four small karst caves. The Bureinskiy Median Mass includes 65 karst caves, with a total length of 2473 m, in Proterozoic carbonate rocks. Most of them are situated in the Lesser Kningan Range. The largest is Ledovaya Cave (385 m long, 64 m deep).

Cambrian, Carboniferous, Permian, and Triassic limestones are very widespread in Sikhote–Alin'. Two hundred karst and four volcanic caves, with a total length of 11 700 m have been investigated. The largest are the Proshchal'naya (3200 m long, 73 m deep), Spasskaya (2220 m long, 16 m deep), and Solyanik (425 m long, 125 m deep) caves.

The karst caves of the Sakhalin Island are in Upper Jurassic reef limestones in the East Sakhalin Mountains. During the last 20 years, 56 karst caves with a total length of 1415 m have been discovered. The most significant are the Kaskadnaya (208 m long, 123 m deep) and Vaydinskaya (300 m long, 64 m deep) caves.

Paleolithic sites have been studied in Strashnaya, Bukhtarminskaya, Ust'-Kanskaya, and Denisova caves in Gorniy Altay, and in Dvuglazka and Grot Proskuryakova Caves in Kuznetskiy Alatau, Dyuktaiskaya Cave in Yakutia, and the Geographicheskogo Obshchestva in the Far East. Mesolithic sites and younger cultural horizons have been studied in Eleneva Cave near Krasnoyarsk city and in Bol'shaya Ludarskaya Cave near Lake Baikal. Neolithic sites were excavated in the Aydashinskaya Cave in Kuznetskiy Alatau, in the Boro-Khukhan, Shamanskaya, Uzurskaya, Skriper, Obukheikha, and Tonta caves near Lake Baikal, and in the Chërtovy Vorota, Letuchaya Mysh, and Vereshchagina caves in the Far East. Bronze Age sites have been explored in Vereshchagina, Denisova, and Aydashinskaya Caves. Iron Age, Middle Age, and Contemporary Ethnography sites have been discovered in the numerous caves in Gorniy Altay, in the Yenisey River Basin, near Lake Baikal, and in the Far East.

Paleontological investigation of Siberian caves has continued for over 150 years. Vertebrate fossils from more than 100 caves have been studied in different regions of southern Siberia and Sakhalin Island. The Upper Pleistocene and Holocene fauna remains are the most studied (Ovodov, 1980). Only a few caves contain more ancient bones. The most ancient fossils were found in Aya Cave near Lake Baikal. Numerous Middle Miocene remains of mammals (Lagomorphs of the Amphilaginae family, and rodents of the *Gobicricetodon* genus / Zapodidae family), a tortoise of the *Trionyx* genus, and an extinct snake-headed fish of the *Channa* genus were extracted (Filippov et al., 2000). The bones of two extinct species of bears have been discovered in Siberian caves: the small cave bear, *Ursus (Spelaearctos) rossicus Borisjak*, from the Strashnaya Cave in Gorniy Altay, and *Ursus (Selenarctos) sp.*, close to the Himalayan species, from Botovskaya Cave (Ovodov & Filippov, 2000). Middle Pleistocene small mammal fauna have been described in the Stariy Zamok cave in Eastern Siberia. Nine species of bats are found in the caves of the Far East, six species in Transbaikalian and Sakhalin caves, and eight species in Pribaikalia and Krasnoyarsk territory. They belong to the genera *Myotis, Plecotus, Murina*, and *Ambliotus*.

ANDREI FILIPPOV

Works Cited

Filippov, A.G., Erbajeva, M.A. & Sychevskaya, E.K. 2000. Miotsenovye otlozheniya v peshchere Aya na Baikale [Miocene deposits in the Aya Cave near Baikal]. *Geologiya i geofizika* [Geology and Geophysics], 41(5): 754–63

Ovodov, N.D. & Filippov, A.G. 2000. Vymershie medvedi Sibiri [Extinct bears of Siberia]. In *Paleogeografiya kamennogo veka. Korrelyatsiya prirodnykh sobytiy i arkheologicheskikh kul'tur paleolita Severnoy Azii i sopredel'nykh territoriy* [Paleogeography of the Stone Age. Correlation of Natural Events and Archaeological Paleolithic Cultures of Northern Asia and Surrounding Territories], Krasnoyarsk

Ovodov, N.D. 1980. Peshchernye mestonakhozhdeniya ostatkov mlekopitayushchikh Sibiri i Dal'nego Vostoka [Cave Sites of Fossil Mammals in Siberia and the Far East]. In *Karst Dal'nego Vostoka i Sibiri* [Karst of the Far East and Siberia], Vladivostok: Academy of Sciences of the USSR, Far East Scientific Center

Further Reading

Bersenev, Y.I. 1989. *Karst Dal'nego Vostoka* [Karst of the Far East], Moscow: Nauka

Filippov, A.G. 1993. Peshcheri Irkutskoi oblasti [Caves of the Irkutsk Region]. In *Peshchery. Itogi issledovaniy* [Caves. Results of Investigations], Perm: Perm University Publishing House

Filippov, A.G. 1993. Peshery Buryatii [Caves of Buryatia]. In *Peshchery. Itogi issledovaniy* [Caves. Results of Investigations], Perm: Perm University Publishing House

Filippov, A.G. 1997. Gravity caves of the Siberian Platform. In *Proceedings of the 12th Interational Congress of Speleology*, 1, Switzerland, La Chaux-de-Fonds

Filippov, A.G. 2000. Speleogenesis of the Botovskaya Cave, Eastern Siberia. In *Speleogenesis: Evolution of Karst Aquifers*, edited by A.B. Klimchouk, D.C. Ford, A.N. Palmer and W. Dreybrodt, Huntsville, Alabama: National Speleological Society

Molodin, V.I., Bobrov, V.V. & Ravnoushkin, V.N. 1980. *Aydashinskaya peshchera* [Aydashinskaya Cave], Novosibirsk: Nauka

Tsykin, R.A. 1990. *Karst Sibiri* [Karst of Siberia], Krasnoyarsk: Krasnoyarsk University Publishing House

http://www.cavingclub.fegi.ru has a list of the Far East and Sakhalin Island caves

SIEBENHENGSTE, SWITZERLAND

The cave region of Siebenhengste is one of the world's major speleological sites. It includes more than 280 km of passages over a vertical range of 1500 m. The multiphase cave system is one of the best examples of the relationship between surface morphology and cave genesis, and may help to elucidate the morphological evolution of the northern rim of the Alps from the Mio-Pliocene to the Pleistocene.

The Siebenhengste is situated in the northwestern part of the Alps, adjacent to the Molasse basin. From Lake Thun at 558 m altitude, the system extends up to the Schrattenfluh, a massif that lies beyond the deeply incised valley of the Emme. The entire chain forms a southeast-dipping slope, cut to the northwest by steep cliffs. The upper parts, between 1700 m and 2000 m altitude, are either largely denuded and composed of limestone pavement, or still covered with sandstone and grassy soil. At lower altitudes, sandstone is predominant, and firs grow on the swampy ground. The annual precipitation is between 1500 and 2000 mm.

The cave system known as Réseau Siebenhengste-Hohgant (149 km long and 1340 m deep) is composed of the labyrinth of Siebenhengste, the F1 at Hohgant, and the Faustloch (see Figure). Other well-known caves are St Beatus Cave (12 km long), a tourist cave, and the scientifically interesting Bärenschacht (60 km long), which represents the downstream part of the Siebenhengste (Funcken et al., 2001). The other most important caves are A2 (14 km) and K2 (16 km). Since the cave genesis is multiphase, it is impossible to indicate a general cave morphology. The Siebenhengste and A2 labyrinth consists of phreatic looping galleries draining to the northeast, which are interlaced with younger shafts and meandering canyons crossing them and continuing towards the Zone profonde. There, the main phreatic tube coming from F1 continues to Faustloch and to the almost fully phreatic Bärenschacht. F1, K2, and St Beatus Cave show huge meanders in their upper parts.

On top of a siliceous Lower Cretaceous limestone the 40- to 50-m thick Drusberg marls normally form the impermeable base of the karstic system. Above this lies the 150–200-m thick Schrattenkalk, the main karstic aquifer, where most of the caves are found. The Kieselkalk to Schrattenkalk sequence is lower Cretaceous in age. The Upper Cretaceous, consisting of about 20 m of Gault sandstone and Seewen limestone, is only found in thin layers in the extreme southeast. The overlying Hohgant series of sandstones is of Eocene age, up to 200 m thick. Karstification may occur in its calcareous layers, creating several cave conduits superimposed on the Schrattenkalk system, but largely unconnected with it.

The general geological structure is a monoclinal slope, dipping to the southeast at about 15–30°, which is interrupted by a large normal fault, extending from Lake Thun up to the Schrattenfluh (Hohgant–Sundlauenen fault). The throw of the fault is around 150 m in the Hohgant region and increases to more than 1000 m in Sundlauenen, thus interrupting the continuous dip of the Cretaceous and Eocene sediments. Observations mainly in Bärenschacht indicate that the main fault was active during sediment deposition in the Lower Cretaceous and continued its activity into the Eocene. Related to this, a karst void was filled with Upper Cretaceous sandstone (Häuselmann et al., 1999). Another karstification took place during the Upper Cretaceous to the Eocene transgression, but only traces survive.

The cave region is divided into two catchment areas, the smaller one being the St Beatus cave system, west of the main fault (see map). The other one extends from the springs of Bätterich and Gelberbrunnen beside Lake Thun through Bärenschacht, Faustloch, Siebenhengste, and Hohgant, up to the Schrattenfluh, thereby draining three massifs.

Soon after cave exploration allowed the connection of several caves, the first observations suggested a genesis in distinct phases. Today, we can recognize 13 distinct phases (Jeannin et al., 2000), the 13th one being connected with the present-day water table. The genesis of these phases may be linked to the height of the respective springs by following the phreatic tubes and observing vadose–phreatic transitions. Since there are no indications that the springs were dammed by surface processes, we can conclude that they were located on the valley floors. Therefore, the succession of the phases indicates a series of valley deepenings.

The uppermost phases of cave genesis have been found mainly in Siebenhengste, and to a lesser extent also in the Hohgant area. The altitudes of the springs were 1900, 1800, 1720, 1585, and 1505 m successively. The overall morphology indicates that the springs were located in the Eriz Valley. We postulate that at this time the Aare Valley (Lake Thun) did not yet exist and that the Eriz may be the river-bed of the paleo-Aare. The next series of phases had their springs in the Aare Valley.

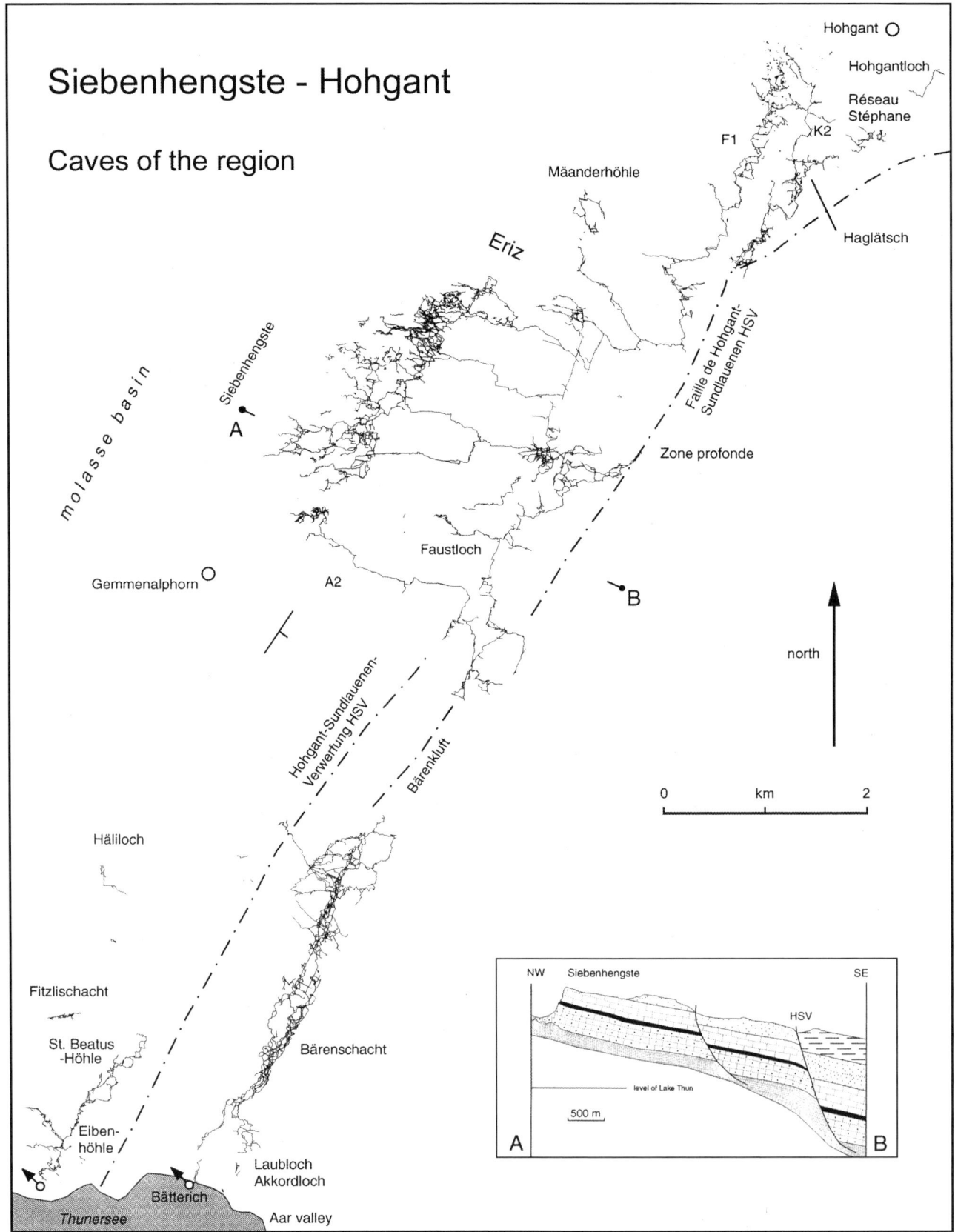

Siebenhengste, Switzerland: Outline map of the known cave passages in the region of Siebenhengste, Switzerland.

Therefore, an important geomorphic event has to be placed between the phases of the 1505 m and 1440 m springs, turning the flow direction 180° and drying up most of the Eriz springs.

The springs of the lower phases are located at 1440, 1145, 890, 805, 760, 700, 660, and 558 m. It appears that the phases reflect floodwater conditions, with inclinations of around 2°. Observations of gallery morphology indicate a time of long stability for each phase, followed by a rapid base-level lowering and readjustment of the flow-paths to the new conditions. The hydrological connection to the neighbouring Schrattenfluh Massif is thought to have occurred when the spring was at the 660 m level.

Since the Aar Valley was subject to Quaternary glaciations, the base-level lowering of the last eight phases is thought to been caused by the glaciers. A relative chronology of erosional and depositional events has been established, and some cave deposits have been dated, indicating both a base-level deepening by glaciers, as well as ages of more than 350 000 years ago for the phase with the 760-m spring. So we have to assume that the oldest galleries, in upper Siebenhengste, are at least early Pleistocene in age, but a Pliocene or even Miocene genesis is also possible, since there is morphological and sedimentological evidence that the first galleries were formed before glaciation began.

Both the speleological exploration and the scientific investigation are still continuing, although technical difficulties in the shafts, the danger of floods, and narrow passages all hinder rapid progress.

PHILIPP HÄUSELMANN AND PIERRE-YVES JEANNIN

Works Cited

Funcken, L., Moens, M. & Gillet, R. 2001. Bärenschacht— l'interminable exploration. *Stalactite*, 52(1): 9–22

Häuselmann, Ph., Jeannin, P.-Y. & Bitterli, T. 1999. Relations between karst and tectonics: the case-study of the cave system north of Lake Thun (Bern, Switzerland). *Geodinamica Acta*, 12: 377–88

Jeannin, P.-Y., Bitterli, T. & Häuselmann, Ph. 2000. Genesis of a large cave system: the case study of the North of Lake Thun system (Canton Bern, Switzerland). In *Speleogenesis: Evolution of Karst Aquifers*, edited by A.B. Klimchouk, D.C. Ford, A.N. Palmer & W. Dreybrodt, Huntsville, Alabama: National Speleological Society

Further Reading

Bitterli, T. 1988. Das Karstsystem Sieben Hengste-Hohgant-Schrattenfluh, Versuch einer Synthese. *Stalactite*, 38(1–2): 10–22

Bitterli, T. 1990. Das Réseau des Lausannois. *Höhlenforschung im Gebiet Sieben Hengste–Hohgant*, 2

Bitterli, T., Funcken, L. & Jeannin, P.-Y. 1991. Der Bärenschacht— der alte Traum von −950 m Tiefe. *Stalactite*, 41(2): 71–92

Funcken, L. 1994. Bärenschacht—plus de 36 kilomètres post-siphon. Une exploration hors du commun. *Stalactite*, 44(2): 55–81

Funcken, L. 1997. Explorations récentes au Faustloch, réseau des Sieben Hengste, Suisse. *Proceedings of the 12th International Congress of Speleology*, vol. 4, La Chaux-de-Fonds: International Union of Speleology

Gerber, M., Bitterli, T., Jeannin, P.-Y. & Morel, Ch. 1994. A2 Loubenegg. *Höhlenforschung im Gebiet Sieben Hengste–Hohgant*, 3

Häuselmann, Ph. 1999. Le Bärenschacht (Berne, Suisse): Etude de l'aval du Réseau des Sieben Hengste. *Etudes de géographie physique, suppl.* XXVIII: 113–16

Häuselmann, Ph., Schafheutle, M., Huber, L. & Brandt, C. 2000. Prospektion Laublochgegend–Waldegg. *Höhlenforschung im Gebiet Sieben Hengste–Hohgant*, 4

Hof, A. 1997. Labyrinthes et Sieben Hengste. *Proceedings of the 12th International Congress of Speleology*, vol. 1, La Chaux-de-Fonds: International Union of Speleology

Hof, A., Rovilles, Ph., & Jeannin, P.-Y. 1984. Le Réseau. Höhlenforschungim Gebiet Sieben Hengste–Hohgant, 0

Jeannin, P.-Y. 1991. Mise en évidence d'importantes glaciations anciennes par l'étude des remplissages karstiques du Réseau des Siebenhengste (chaîne bordière helvetique). *Eclogae Geologicae Helveticae*, 84(1): 207–21

Back issues 2–4 of *Höhlenforschung im Gebiet Siebenhengste-Hohgant* are best obtained from SpeleoProjects (http://www.speleoprojects.com/): 0–1 are out of print.

SILICATE KARST

This entry covers karst development in siliceous rocks, dominantly quartzites, but also lithologies such as granite and basic rocks. In the past, these rocks were regarded as unsuitable for karst development, although surface morphologies similar to those on carbonate rocks (e.g. karren, pavements of clints delineated by grikes and fields of pinnacles) were observed on quartzite, sandstone, and on nearly all other silica and silicate lithologies. In the 1970s, the exploration in Venezuela (especially the Roraima quartzite in southern Venezuela) of spectacular caves, potholes, swallow holes, resurgences, dolines, and poljes, showed scientists that important karstic features can develop in quartzite. The best silicate karst developments have been reported from Venezuela and Brazil (see Quartzite Karst of South America), and from South Africa and Australia. Sandstones with a carbonate matrix, which host important cave systems, are not treated here, since their formation is controlled by the same weathering process as carbonate rocks.

Exokarst

Weathering of exposed siliceous rocks commonly leads to the formation of a range of karren forms, particularly kamenitzas and rinnenkarren. Pavements showing grikes are less common, and these are usually less well developed than comparable features produced on carbonates. The most spectacular features are produced by weathering under soil cover—a process which is conducive to progressive formation of sand and clay in quartzite and felsic rocks along joints and bedding planes. By erosion of the soft, deeply altered material, unaltered or only moderately weathered rock is left standing as pinnacles and towers in the case of quartzite, and as tors, domes, and heaps of large rounded boulders in the case of granite. These features are ubiquitous worldwide.

In the figure, this process is illustrated for the case of quartzite, which is the most favourable lithology for siliceous karst development. In this case, the weathering involves the dissolution of quartz along crystal boundaries, a process leading to progressive

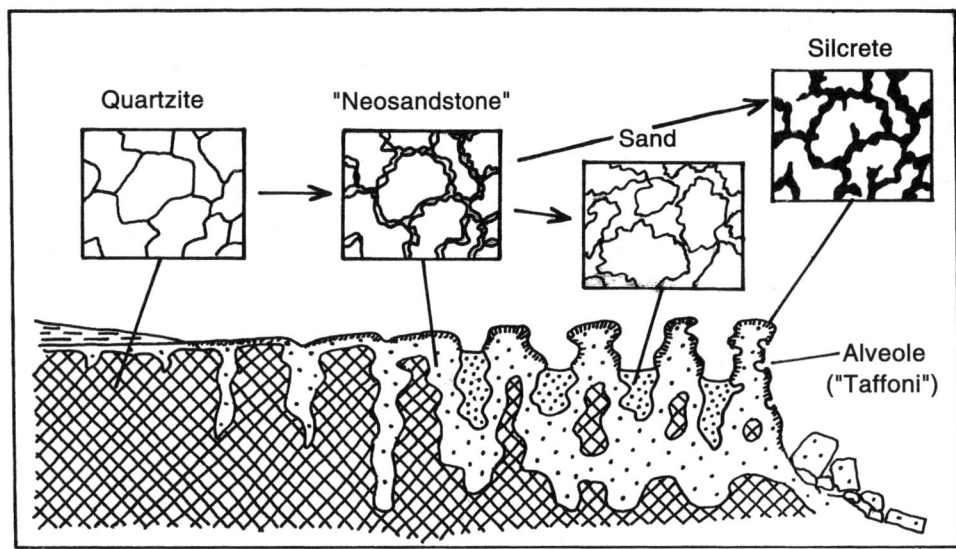

Silicate Karst: Cartoon illustrating the weathering and development of pinnacle fields in quartzite, showing the progressive disintegration of the rock into sand, which is then removed by erosion, thus exhuming the pinnacles of moderately weathered or unweathered quartzite. In humid climates with dry seasons, during periodic evapotranspiration, amorphous hydrated silica (opal) precipitates and forms a hard surficial crust (silcrete) by cementing the interstices in the weathered quartzite. Alveoles (or tafoni) form by disintegration of the softer weathered quartzite underneath, in places where the silcrete did not develop well, which is mainly in the hollows. The term "neosandstone" is suggested as the moderately weathered quartzite has been misinterpreted as sandstone in many instances by field geologists.

weakening of the rock into a "neosandstone", and eventually into loose sand (see Dissolution: Silicate Rocks). Very often this weathered quartzite has been misinterpreted as a sandstone by field geologists. The grains constituting the neosandstone are generally very jagged and interlocked, a texture inherited from the parent quartzite. In some cases, slabs of this weathered rock may display an odd flexibility, due to articulation of the grains, a quartzite variety that has been called "itacolumite". Especially for climates comprising dry seasons, the pinnacles are often indurated at the surface by an opal matrix deposited by evaporation (see Figure).

The time necessary to produce quartzite pinnacle fields, including the endokarsts that may be associated with them, has been estimated from calculations in some instances where the parameters are known: volume of rock removed, surface drained, rainfall, recharge, and silica in groundwater. In the case of the Berlin system, developed in a relatively thin quartzite bed (see the section below on the silicate karst of southern Africa), it appears that it took no more than two million years to form, perhaps less. This order of magnitude also seems to be applicable to other systems in southern Africa, which show a shallow development. Longer periods were required to develop larger systems, comprising spectacular tall tower fields and massive caves, as for example in Venezuela and north Australia, where the weathering was much deeper. In these cases, the weathering period may have been over 10 million years. Such long periods of weathering can develop preferentially if the topography is smooth and more or less peneplaned, since in rugged terrains mechanical erosion can supersede the chemical process. Therefore it is likely that such tower fields are exhumed after periods of continental uplift and subsequent renewal of erosion.

Endokarst

Although the surface morphology outlined above is suggestive of karst scenery, the fluvial drainage remains surficial in most cases—large exokarst landforms (dolines and poljes) and features diagnostic of endokarst (potholes, swallow holes, caves, and resurgences) are rare. For instance, in South Africa the latter features are developed on less than 1% of the quartzite outcrops and are generally grouped in very localized systems. The cave networks are often located close to escarpments, particularly where the former open at the base of cliffs on the down-dip side of inclined plateaux. Usually these systems do not extend more than a few hundred metres from the resurgences, but exceptionally are more than one kilometre long. They do not develop into the central parts of large plateaus, in contrast with the classic carbonate karst where entire massifs are affected by endokarstic dissolution where the climate is sufficiently humid. Typically, the quartzite caves occur in mountains with high rainfall regimes.

The caves are often developed at the contact with impervious rocks such as shale, siltstone, and schist. The base level can also be controlled by the depth to which weathering can occur, as is the case where the quartzite formation is very thick. Typical keyhole passages and canyons have been observed, although they are not very common. The passages generally display square cross sections or show ceilings arched over a flat floor of impervious rock that forms the riverbed, sometimes making up the bulk of the cavity volume. The fissure-like passages that developed along vertical joints are often observed where the base level is not lithologically controlled. A characteristic feature of some caves is the extreme variation in the size of a single passage, which in a downstream direction can narrow from a 10 m diameter tunnel into a narrow tube impenetrable to a caver. This irregularity is controlled by variations in the degree of quartzite weathering. The majority of the quartzite caves are vadose and still active, although relict levels have been observed in some complex systems, generally only slightly above the active sections.

One unusual type of quartzite cave has developed along linear rifts recently formed by gravity tectonics. These caves run parallel to the edges of plateaus and include some of the deepest pothole-type caves in quartzite. They have similarities to crevice caves (see separate entry) but are considered here because they capture small surface streams that contribute to speleogenesis. Such caves have been reported in Zimbabwe, South Africa, Brazil, and Venezuela.

In silicate rocks other than quartzite, the karst phenomena are less typical. Caves in granite are relatively frequent and have been reported, for example in Australia, California, Zimbabwe, and Swaziland. They often consist of shallow active caves developed along relatively important streams, disappearing into valley-floor accumulations of rounded granite boulders, up to several metres across. Generally, these boulders are nuclei of unweathered rock freed from their saprolitic matrix by surface erosion. The distance between the swallow hole and the resurgence may be over one kilometre in some cases, as in California and in Swaziland. Due to the chaotic nature of the boulder accumulation and the volume of the flow, in many cases the underground streams cannot be explored. Rare cases of short passages entirely developed in granite have been reported (see section below, Silicate karst of Australia).

A special mention must be made here of quartzite–dolerite caves, which are developed in the upper parts of deeply weathered dolerite sills intruded into quartzite. They are well developed in South Africa. The ceilings of the caves are flat and consist of hard quartzite, but the passages themselves are entirely developed in the red clay derived from the dolerite.

Subsequent to their formation, caves of volcanic origin sometimes capture surface streams, which contribute somewhat to their enlargement (e.g. Mouret, 1993). As the caves proper are not of karst origin, they are not considered here (see Volcanic Caves entry).

Silicate Karst of Southern Africa

Three main quartzite cave areas have been identified in this subcontinent and are described here as examples of silicate karst (see map in Africa, Subsaharan entry). The characteristics differ considerably from one place to another.

Cape Peninsula

This small district is situated immediately south of Cape Town, where about 100 caves are known in a fairly pure, tabular quartzite of Ordovician age, which is several hundred metres thick. They occur mainly in the rounded summit areas of mountains, but only extend to relatively shallow depths. Most of them are active, vadose, and controlled by a network of vertical joints, along which the quartzite had initially been weathered into a very friable material which disintegrates easily into white sand, as observed in the caves. Sometimes this joint network gives the cave systems a "pseudo-phreatic" pattern. The base levels seem to be controlled by harder, less-weathered quartzite at depth. The longest cave is Ronan's Well (800 m long). In the same area a few caves are developed along linear rifts parallel to the edge of tabular plateaux. One of them, the Bat–Giant System, reaches a total length of 1650 m.

Mpumalanga Escarpment

In the northeast corner of South Africa, this escarpment marks the geographical transition from the High-Feld to the Low-Feld region and geologically coincides with the eastern termination of the Transvaal Basin, which is filled with Late Archean to Early Proterozoic sediments. About 60 caves have been reported in the fairly pure quartzite of the Wolkberg Group and of the Black Reef Formation (both uppermost Archean), and of the Pretoria Group (Lower Proterozoic). The strata are affected by static regional-contact metamorphism, but are undeformed and monoclinal, dipping westwards.

The Berlin karst systems, near Nelspruit, are the most spectacular both in their surface expression (pinnacles and dolines) and in their caves. They developed within an inclined plateau of Black Reef quartzite, 25 m thick, with a 30 cm thick layer of shale-like tuff interstratified close to its base. The caves are grouped into two systems totalling about 2.5 km of vadose passages. Surface water is collected by large dolines, and flows through the caves towards resurgences located down dip. The great majority of the passages are developed on the thin tuff layer, which everywhere constitutes the impervious base level. The cross sections are flat to equidimensional and show extreme variations in size along the same passage: large chambers, which formed in zones of deeply weathered, friable, quartzite can narrow down to impenetrable channels where the alteration of the rock has been relatively mild.

Another important system is developed in the Daspoort Formation of the Pretoria Group. About 50 small dolines, and swallow holes occur on a 25 m thick quartzite forming a plateau inclined to the north at 5 to 10°. A few of them give access to caves developed at shallow depths below the surface. The water captured by dolines flows through the caves and reappears at resurgences on the down-dip edge of the plateau. The most important system is Magnet Cave, developed at depths of 5 m or less under the plateau, but in which 2.4 km of passages have been mapped. Access to the cave is through a large number of dolines generated by roof collapse. The shallow depth of the system is demarcated by a 10–40 cm thick shaly tuff forming an impervious barrier, which is visible on the floor of all the passages. Most of the passages are flat. In this cave, an unusual phenomenon is the frequent diffluences of active streams into side passages, some of which remain active even during periods of low flow. This is facilitated by the very flat shaly floor, which widely spreads the flow. The phenomenon produced a complex passage pattern reminiscent of a phreatic cave.

Quartzite–dolerite caves have been reported in the same area. They are developed in the Wolkberg Group, the Black Reef Formation, and the Pretoria Group, and are almost entirely vadose. The most important of them is Mogoto Cave (1.6 km), which is developed in a 3 m thick dolerite sill that intrudes the Black Reef quartzite, forming a plateau inclined at 25° to the west.

Chimanimani Mountains

This area is located on the border between Zimbabwe and Mozambique. The karst phenomena are restricted to a small (<1 km^2) plateau perched at an altitude of 2200 m, receiving a high rainfall and underlain by several hundred metres of quartzite belonging to the Umkondo Group (Early Proterozoic). The plateau is criss-crossed by a network of rift-type potholes collecting small surface streams, the deepest of them reaching 305 m below the surface. No horizontal collector passages were reached at the bottoms, but a resurgence was located at about 500 m below the plateau, giving access to a small cave, which is also rift related. In the caves the quartzite is variably friable, indicating weathering.

Silicate Karst of Australia

Remarkable exokarst scenery has been observed in Upper Proterozoic siliceous cemented sandstone forming tabular plateaux in Arnhem Land, Northern Territory (Jennings, 1983; Wray, 1997). The best portion is the Ruined City, characterized by a "two stories" morphology. The lower storey is developed in a massive sandstone dissected into large rectangular blocks by a network of narrow linear gorges, up to 30 m deep, controlled by a joint system. On the flat top of the blocks, the upper storey is developed in a thinly bedded sandstone. It forms a field of towers reminiscent of the limestone karst of southern China, but smaller in size, as the pinnacles reach only a maximum height of about 20 m. The sandstone has been surficially hardened by a cement, probably composed of opal and iron oxides, under which the rock is soft and friable.

Caves appear to be rare, consisting of arches and short tunnels with several entrances. As the area is part of an Aboriginal tribal land, speleological investigations have been limited. The climate is relatively dry, with seasonal rainfall. Caves have been reported several kilometres to the northwest, in the East Kimberley, and also in Proterozoic quartzite. The most important is Whalemouth Cave, close to Turkey Creek. It formed by the disappearance of a stream flowing in a shallow valley on a quartzite plateau. It consists of a 220 m long tunnel with a large diameter, up to 60 m high and 45 m wide at the exit in the cliff face bordering the plateau.

In the Girraween National Park, southeast Queensland, caves have been reported in Triassic granite (Finlayson, 1982). All are active, generated by the disappearance of streams. They dominantly consist of narrow gorges developed along vertical joints by stream erosion from the surface, and subsequently more or less covered by large granite boulders, which form the ceiling of the caves. Nevertheless, in three places the passages formed entirely in granite from initial horizontal joints, but over short distances, for instance 25 m in Goebel's Cave. In River Cave—the 50 m long active passage developed by a surface stream entrenching itself obliquely along a joint inclined at 20°—the floor and the ceiling are therefore in solid granite, but the upper side-wall is a boulder choke. In all the caves the rock is unweathered, and any saprolite, if present, has been washed away by stream erosion.

Speleogenesis

Particularly in Venezuela and South Africa, cave formation was by piping through zones of deeply weathered quartzite initially transformed into very friable neosandstone, by the dissolution process outlined above (see also Dissolution: Silicate Rocks). This dual speleogenetic model: weathering first, followed by mechanical removal of sand, has been more or less accepted since the 1960s. The piping process starts from a spring, where the local water table intersects the surface when the quartzite has lost enough cohesion. The first channels formed in this way are small-diameter pipes branching upstream. Where this regressive process reaches the surface, it may trap storm water or capture a stream, which results in enhanced erosion and enlargement of the pipes into accessible passages. At a more advanced stage the caves are segmented into several tunnels and bridge, by ceiling collapse and complete removal of the deeply weathered quartzite, and eventually are entirely transformed into subaerial canyons.

There is no accurate estimate of the time necessary for a quartzite cave to develop, except that the process must be geologically very fast, as observations of the rapid formation of cavities by suffosion in more or less sandy material have shown. It may be that a large majority of the quartzite cave systems are not older than a few thousands or tens of thousands of years, which contrasts with the much longer time required to weather the rock. Therefore, it appears that quartzite caves must be geologically ephemeral, with possible exceptions such as the Cueva del Cerro Autana in Venezuela. The fact that the majority of the caves are still active supports the theory that the caves are ephemeral in contrast to carbonate rocks, which may host very ancient relict caves.

In the case of gravity-tectonic rifts at the edges of plateaux, the formation of cavities is due to a variable extent to a process not controlled by water. However, water may play a role in secondary enlargement. Moreover, at the inception stage, rifting may allow deep penetration of water thereby promoting the weathering of quartzite at depths greater than usual. Like the other quartzite caves, the accessible voids are relatively young.

An important condition for the formation of quartzite caves is a flow fast enough to transport sand grains. This means that high gradients are essential and explains their preferential location in rugged terrains, their very dominant vadose nature, and their restricted location, i.e. caves cannot form everywhere in quartzite. This is perhaps the most important difference to the carbonate karsts, where solute concentrations are usually inversely related to flow rate and caves can be formed by water moving very slowly.

The preferential location of quartzite caves in humid areas is due to greatly increased infiltration into the ground, and to a thicker vegetation cover which slows surficial erosion and protects the deeply weathered rock. The disappearance of the cave systems may be more rapidly completed after a climatic change from humid to arid or semi-arid, due to sparser vegetation. This might be the case for the karst areas of northern Australia, where the climate is drier than in the other type area treated above, and where caves seem to exist only as relicts.

Speleogenesis in other silicate rocks may follow a similar two-stage process to that in quartzites. However, caves are less well developed, probably due to the incongruent dissolution of most silicates, like feldspar, a process that leaves impervious clayey residues, which are less amenable to piping (see entry, Dissolution: Silicate Rocks). According to Finlayson (1982), the Australian granite caves were initiated by tensional opening of horizontal joints following pressure release when the rock was exhumed as a result of erosion. An opening 1 or 2 cm wide might be sufficent to allow a flow fast enough to move sediment capable of mechanically eroding a cave passage. However, this process is only possible over short distances due to loss of head. Alternatively, the joint may have been enlarged by an initial solution process but this hypothesis seems less likely as no significant pervasive weathering has been observed along the joints, and because the clay minerals have a clogging effect which impedes piping. Quartzite–dolerite caves are an exception, and in this case the quartzite is practically unweathered, but the dolerite is altered to red clay. Due to the irregular compaction of the weathered dolerite, thin planar voids developed between the sill and the quartzite and were subsequently considerably enlarged by surface water. The caves formed exclusively by erosion of the clay, whereas the quartzite acted only as a rigid roof protecting the passages from collapse.

There are exceptions to the two-phase speleogenetic model, as, in some rare cases, cavities are formed in quartzite directly by dissolution along joints, as in carbonates. They have been observed in mines, however, and formed under unusual conditions at several 100°C, in a hydrothermal environment. Under these conditions the solubility and the dissolution rate of quartz are both much higher, thus explaining the formation process.

Pseudokarst Versus Karst

In this review the term "karst" has been applied to silicate. However, there is some controversy about this appellation, as some scientists are of the opinion that karst morphology must essentially be due to dissolution. For them, the term "pseudokarst" should be used in the cases where the morphology resembles a karst, but the genesis is not to be accounted for by dissolution (see Pseudokarst entry). As some of these authors think that dissolution does not play a significant role in silicate rocks, they use the term "pseudokarst" instead of karst. Other authors support the use of the term "karst" for silicate rocks, arguing that solution plays an important role, especially in the case of quartzite, where dissolution of quartz is congruent. Some of the "karst" proponents add that the difference between matter removed as ions and molecules, or as larger particles, is minimal, and does not justify two appellations for a very similar process and for a comparable morphological result. It was pointed out that in the case of limestone caves, speleogenesis implies not only transport of solutes, but also of detritus in variable proportions. This indicates that mechanical erosion can play an important role in impure limestone, and suggests a continuum between carbonate and silicate karst (see "karstic and pseudokarstic processes in quartzite caves" in the Pseudokarst entry). It was proposed that the definition of karst should be revised into: "a morphology due to ground water circulation and characterized by the more or less complete disappearance of the surficial drainage systems" (Martini, 1987). It is mainly on this basis that the Slovenian karst has been selected as the type area: the diagnostic features would then be dolines, poljes, swallow holes, resurgences, and caves, provided that their origin is bound to water circulation. If this definition were to be adopted then the pavements and pinnacle fields would not be considered to be typically part of a karst morphology. The term "pseudokarst" would then apply to a morphology where underground cave drainage is missing, as for instance in the cases of volcanic cavities, deflation pans, dune morphology, etc., which are not generated by underground drainage.

Jacques Martini

Works Cited

Finlayson, B. 1982. Granite caves in Girraween National Park, Southeast Queensland. *Helictite*, 20: 53–59

Jennings, J.N. 1983. Sandstone pseudokarst or karst? In *Aspects of Australian Sandstone Landscapes*, edited by R.W. Young & G.C. Nanson, Wollongong, New South Wales: University of Wollongong

Martini, J. 1987. Les phénomèmes karstiques des quartzites d'Afrique du Sud. *Karstologia*, 9: 45–52

Mouret, C. 1993. Formation of underground streams in volcanic rocks: Tahiti, French Polynesia, and Bali, Indonesia. *Proceedings of the 11th International Congress of Speleology*, Supplement, edited by Zhang Shouyue, Beijing: Secretariat, XI International Congress of Speleology

Further Reading

Chabert, C. & Courbon, P. (editors) 1997. *Atlas des Cavités non Calcaires du Monde*, Paris : Union Internationale de Spéléologie

Marker, M. & Swart, P. 1995. Pseudokarst in the Western Cape, South Africa. *Cave and Karst Science*, 22(1): 31–38

Martini, J.E.J. 2000. Quartzite caves in Southern Africa. In *Speleogenesis: Evolution of Karst Aquifers*, edited by A.B. Klimchouk, D.C. Ford, A.N. Palmer & W. Dreybrodt, Huntsville, Alabama: National Speleological Society

Wray, R. 1997. Quartzite dissolution: karst or pseudokarst. *Cave and Karst Science*, 24(2): 81–86

Includes references to most of the significant earlier papers on sandstone karst / pseudokarst in Australia

Younger, P.L. & Stunnell, J.M. 1995. Karst and pseudokarst: an artificial distinction. In *Geomorphology and Groundwater*, edited by A.G. Brown, Chichester and New York: Wiley

ŠKOCJANSKE JAMA, SLOVENIA

Škocjanske Jama is the second largest show cave in Slovenia, with 55 000 visitors in 2001. The influent cave is 5800 m long and 250 m deep, and is situated on the southeast side of the Kras plateau in western Slovenia (see entry, Kras, Slovenia). Because of its large chambers, the grand dimensions of the river gallery, the magnificent sink of the River Reka, the deep collapse dolines, and the heroic history of its exploration Škocjan is a World Heritage Site on the natural heritage list of UNESCO, while the caves and the karst above are protected as a regional park.

The Škocjanske caves are developed in thickly-bedded Cretaceous rudist limestone and thinly-bedded Paleocene dark limestone, dipping southwest (Habič et al., 1993). In the Cretaceous limestone, where most of the cave is developed, 15–125 m thick beds of rock are characteristic between tectonized bedding planes, with most proto channels developed along these structures. Fractures and fracture zones (north–south and northwest–southeast) are the other significant morphological controls. The fracture zones can be up to a few dozen metres wide. Phreatic loops, large chambers, and some collapse dolines are developed along them.

The cave was formed by the sinking of the River Reka flowing from Eocene flysch rocks, which are mostly composed of sandstone. Large variations of the discharge (mean annual discharge is 8.18 $m^3 s^{-1}$, maximum discharge measured was 387 $m^3 s^{-1}$) cause floods within the cave. Heights of the usual annual floods are about 20 m, exceptional floods are about 100 m, and the highest recorded flood reached a height of 132 m in the cave.

The Reka River enters the cave 80 m below Škocjan village, at an altitude of 317 m (Figure 1). After 200 m of gallery, the river re-emerges and flows through two collapse dolines divided by a short gallery. The Velika doline is c.160 m deep, with its floor at an altitude of 270 m (Figure 2). The Reka then flows into the main part of the cave. This is a massive canyon 20–30 m wide and about 30–110 m high, heading mainly northwest (Figure 3). This continues as Hanke's Passage, c.1 km long, 10–15 m wide, and 95 m high. The underground river passage widens in places, forming large chambers. At 308 m long, 123 m wide, and 146 m high, Martel's Chamber, with a volume of 2 100 000 m^3, is the largest chamber in the cave, and in the Kras plateau. After Martel's Chamber, the dimensions of the gallery are smaller, and lead to a sump, 20 m deep and 60 m

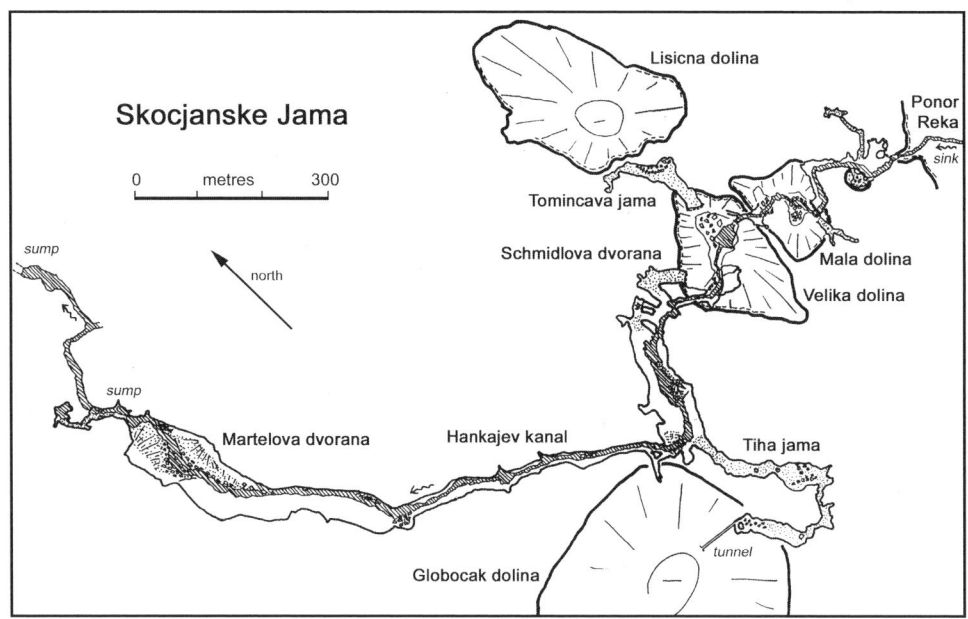

Škocjanske Jama: Figure 1. Plan survey of the Škocjanske Jama and its adjacent dolines, Slovenia.

long, that was dived in 1991. Beyond, there are some rapids with the terminal sump at an altitude of 190 m. Reka River then flows to Kačna Jama, about 900 m distant and 12 km long, after which its underground course splits into several separate flows that finally reappear at the coastal springs of the Timavo River, 40 km away.

High above the underground river there are some abandoned galleries. Most important is Tiha Jama (Silent Cave) discovered

Škocjanske Jama: Figure 2. The entrance dolines of Škocjanske Jama, with the Reka River flowing through the deep shadows and into the main cave entrance almost directly beneath the camera. (Photo by Tony Waltham)

Škocjanske Jama: Figure 3. The massive vadose canyon of Hankajev kanal with the River Reka far below the bridge that carries the tourist trail out of the abandoned Tiha jama gallery on the left. (Photo by Tony Waltham)

in 1904 and now connected by an artificial tunnel to the collapse doline of Globočak. The other gallery is Tominčeva Jama, with its entrance in the north wall of the Velika doline collapse; it is blocked by collapse material from one of the neighbouring dolines.

There are several phases of evolution evident in the cave. Large galleries have paragenetic levelling of their roof at an altitude 320 m. A canyon, 90 m deep was later entrenched into them, after a regional decline of the water-table level in the plateau. However, the oldest part of the cave system can be seen on the surface as a section of a large unroofed cave, 1.8 km in length. Karst denudation exposed remains of the cave with fluvial sediments deposited in cave environment, flowstone, and stalagmites. These cave remains are probably of Pliocene age.

There are several collapse dolines near the cave. They are several hundred metres wide and up to 160 m deep. The largest is Sekelak, with a volume of 8.5 million m^3.

There are some archaeological remains in the entrance parts of the cave, the oldest dating to the Neolithic period. Above the entrance there was a Bronze Age settlement. The caves were mentioned in antiquity and the first attempt of water tracing was done by F. Imperato in 1599. The entrance part to the collapse dolines was explored when Egenhafner swam through in 1816. Exploration down the lakes and rapids of the Reka were performed by J. Svetina in 1839, and then by A. Schmidl in 1851, but his boats were destroyed by a high flood. In 1884, a caving branch of the alpinist club from Trieste, guided by A. Hanke, and later J. Marinitsch, continued with the exploration. In the first year they passed the sixth waterfall, the main obstacle to previous explorers, and reached the terminal sump in 1890.

The high level part of the cave, Tiha Jama, was discovered in 1904, but only in 1991 was the terminal sump dived and 600 m of new passages explored.

Since 1823 the entrance parts of the caves have been used for tourism. Between 1884 and 1905 some bridges (Figure 3) and several kilometres of pathways were made, most of them cut by hand in vertical or overhanging walls high above the Reka. In 1933, the tunnel from the Globočak collapse doline was excavated into Tiha Jama, providing the loop for the tourist cave trail which ranks as one of the finest in the world. In 1959 the cave was electrified and in 1986 the elevator and tourist centre at the entrance to the cave was built.

ANDREJ MIHEVC

Works Cited

Habič, P. Knez, M., Kogovšek, J., Kranjc, A., Mihevc, A., Slabe, T., Šebela, S. & Zupan, N. 1989. Škocjanske Jame speleological revue, *International Journal of Speleoleology*, 18(1[-]2): 1–42

Further Reading

Kranjc, M. 2001. Škocjanske jame, an addition to bibliography. *Acta Carsologica*, 30/1: 213–28

Kranjc, M., Knez, M., Mihevc, A., Kogovšek, J., Slabe, T., Šebela, S. & Hajna, N.Z. 1998. Typical karst area: Škocjanske jame, Slovenia. In *Global Karst Correlation*, edited by Yuan Daoxian & Lui Zaihva, New York: Science Press

Mihevc, A. 2001. *Speleogeneza Divaškega krasa*, [Speleogenesis of the Divača karst] Ljubljana: Založba ZRC SAZU

http://www.wcmc.org.uk/protected_areas/data/wh/skocjan.html
From the World Conservation Monitoring Centre, a useful compilation of geographic and conservation information

SOF OMAR CAVE, ETHIOPIA

Sof Omar Cave is one of the longest known caves in Africa (>15 km), and is an outstanding example of a cave formed by a large river, of a meander cutoff cave, and of a flood water maze cave. The cave is located at the village of Sof Omar in Bale Province, Ethiopia, some 300 km southeast of Addis Ababa, and is formed in the thick-bedded Jurassic Antalo Limestone. The limestone is overlain by Tertiary basalts which form a plateau in the area. The Webi Gestro river has incised a 150 m deep gorge in the plateau, cutting through the basalts to expose the underlying limestone. The source of the river is in the 4300 m high Bale Mountains 120 km upstream of the cave, and the catchment area for the river is 3800 km^2.

Sof Omar Cave has formed at a prominent meander in the river (Figure 1). The groundwater flow path across the neck of the meander offered a shorter flow path and a steeper hydraulic gradient than flow on the surface around the meander loop. The capture of surface water apparently occurred in two stages. In the first stage a cave passage was formed which bypassed the downstream 600 m of the dry valley. This now-dry passage has a substantial fill of sediments. In the cave it terminates upstream at a large circular breakdown chamber, the Astrodome. In the second stage the river was captured underground at its current sink point, which is 700 m further upstream from the original sink point. The current underground river passage is 2000 m long and averages 20 m in width and height (Figure 2). There are several sections of relict river passage of similar size, which were abandoned when the river migrated to its present course. One of these relict passages is blocked by limestone and basalt boulders where a roof collapse stopes up to the plateau above, where there is a large doline. The modern underground river passage is substantially larger than the relict route via the Astrodome, suggesting that the river has followed the modern route for a longer period than the Astrodome route.

In the dry season, between November and March, the river in the cave averages 1 m in depth and is confined to just one passage. However, in the wet season the river rises at least 7 m and floods several kilometres of smaller passages. Steep hydraulic gradients and chemically aggressive waters have combined to produce similar dissolution rates along many different pathways, resulting in a flood water maze. There are two major sets of joints in the cave, with strikes of 20° and 110°, and both sets have been used by flood waters to create a gridiron pattern of passages about 1 m in width. Passages have formed either along these joints or at the intersection between these joints and near-horizontal bedding planes. At the downstream end of the cave there are two superimposed mazes, each having formed at the intersection of a prominent bedding plane with the major joints.

The cave is an Islamic Holy Place with a shrine in a dry passage close to the upstream entrance, and the areas close to the entrances have probably been visited for a considerable

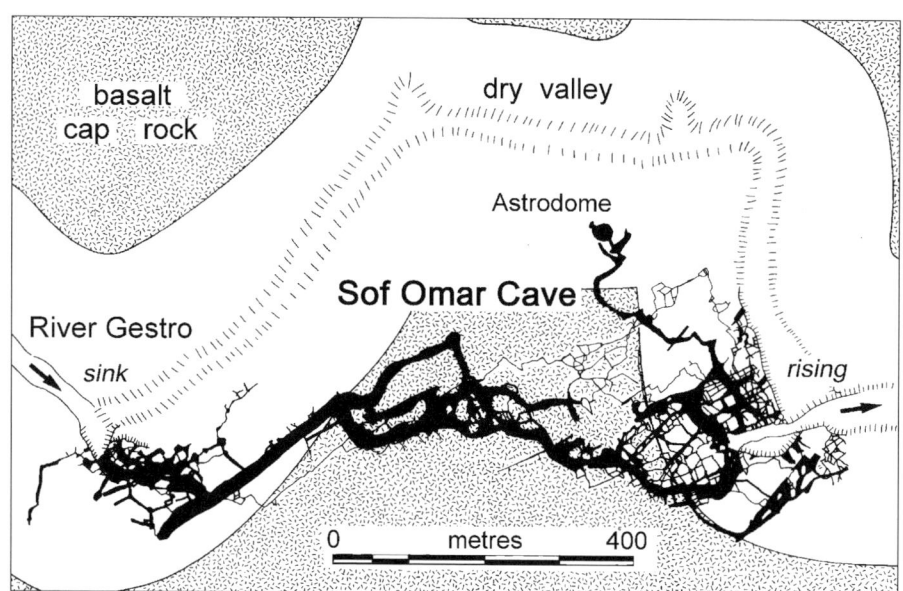

Sof Omar, Ethiopia: Figure 1. Map of Sof Omar Cave. The unshaded areas along the valley are underlain by limestone.

Sof Omar, Ethiopia: Figure 2. The Great Dome and the end of Safari Straight, Sof Omar cave. (Photo by John Gunn)

period. The main river passage and several side passages were mapped in 1966, and the cave was comprehensively explored and mapped in 1972, revealing a total length of 15 km of passage. The spectacular nature of a surface river going underground for a distance of 1 km and the easy nature of the cave have made it a tourist destination, though the cave is a long way from major population centres.

Steve Worthington

Further Reading

Amatt, S.N., Catlin, D., Ramsden, P.R., Ramsden, R.W., Raynor, T. Renvoize, T. & Worthington, S.R.H. 1973. The caves of Ethiopia. *Transactions of the Cave Research Group of Great Britain*, 15: 107–68

Brown, L., Gunn, J., Walker, C. & Williams, O. 1998. *Cave Ethiopia '95 and '96 Expedition Report*, Huddersfield: Limestone Research Group, University of Huddersfield

Palmer, A.N. 1975. The origin of maze caves. *NSS Bulletin*, 37: 56–76

Robson, G.E. 1967. Caves of Sof Omar. *Geographical Journal*, 133: 344–48

SOILS ON CARBONATE KARST

Soils on carbonate karst may be autogenic (authigenic) or allogenic. Autogenic soils are formed in place from the insoluble impurities left over following dissolution and leaching of the bulk of the carbonate parent material. Allogenic soils are formed from superficial deposits transported and deposited over the carbonate rocks by gravity, water, glaciers, or wind.

Soils of allogenic origin are common on carbonate areas throughout the world. Materials derived from higher elevation, noncarbonate areas and deposited over lower elevation carbonate rocks may form colluvial or alluvial soils. Colluvial soils are frequent on the Central Appalachian carbonate region of the United States and one common example is the Clarksburg silt loam, a gray-brown podzolic soil of the order Alfisols (complete descriptions of the US soils mentioned in this report can be viewed at http://www.statlab.iastate.edu/soils/nssc/). This soil forms in colluvium accumulated by water action and soil creep. Alluvial soils form along stream terraces and stream channels incised into the carbonate bedrock. Alluvial soils also form in dolines where materials derived from carbonate bedrock are transported by water and deposited as alluvium in the doline bottom. The alluvium then weathers to form an alluvial soil like the Lindside silt loam (order Inceptisol) and the Huntington silt loam (order Mollisol), both common to the Central Appalachian carbonate region. Chavies fine sandy loam (order Alfisol) is an example of an alluvial soil along stream terraces and channels. It is formed in alluvium washed chiefly from upland areas underlain by acid sandstone and shale. Other allogenic sources of soil-forming materials are glacial till and morainic deposits, aeolian deposits (loess), and volcanic ash. These drift materials can result in thick soil covers over karst bedrock. Loess soil depths exceeding 10 m occur over carbonate bedrock in the Midwestern United States.

Soils formed in Quaternary loessic parent material over limestone are found in the Peak District (England) and glacial drift material deposited over carbonate bedrock is common in Britain and Ireland as well as in the United States. The Central Plain of Ireland is a large, low-lying region dominated by limestone rocks and covered in soils formed in glacial drift up to 60 m thick. Thick drift deposits obscure karst expressions in the underlying carbonate bedrock. Diffuse infiltration and greater water storativity is characteristic of soils formed on thick drift deposits thus limiting karst development in the bedrock. Soils developed in tephra (volcanic ash) deposited over limestone are common in Japan and in the King Country New Zealand (see New Zealand), where the Pleistocene tephra is up to 10 m thick and forms a yellow-brown loam. Development of an ironpan in the B horizon concentrates subsurface drainage and encourages dissolution of the underlying limestone.

Soils of authigenic origin exhibit many common characteristics around the world, but differences do occur due to characteristics of the carbonate bedrock from which they are derived, climate, leaching, topography, and history of land use. The primary source of authigenic soil-forming materials over carbonate bedrock is the insoluble residual material resulting from weathering and dissolution of the carbonate bedrock. The world's most well-developed karst regions occur on nearly pure limestone with less than 10% insoluble material. In regions with a warm, moist, climate soil formation is rapid on most lithologies (e.g., less than 50 000 years to form one metre of soil), but in areas with carbonate bedrock the low volume of soil-forming insoluble residual material results in shallow soil depths. For example, Yuan et al. (1991) estimated that 250 000 to 850 000 years are required to form one metre of soil from limestone in the tropical Guangxi region of south China. The bedrock frequently crops out in karst landscapes, and soil depths can vary, in short distances, from zero to several metres where soil has accumulated in low points on the highly undulating bedrock surface.

The shallowness of most autigenic soils over carbonate bedrock, coupled with their well-drained and free drainage into the bedrock, make them highly susceptible to drought and this makes establishment and maintenance of continuous plant cover difficult, resulting in accelerating erosion and a lack of organic matter accumulation in the surface soil layer. Seasonally dry periods coupled with the droughty nature of karst soils in the Mediterranean region and in Australia severely dry plant canopies and fires are frequent. Fires leave the soil surface unprotected and severe erosion has led to complete denudation of the soil cover on some sloped areas in the Mediterranean region.

The chemical and physical characteristics of soil developed over carbonate rocks are primarily determined by amount of leaching and the local climate. White (1988: chapter 8) states that the primary modification of the insoluble soil-forming materials is the preferential leaching of silica. In the tropics, high leaching rates and warm temperatures cause the removal of virtually all of the silica leaving behind aluminium, which is immobile. High-aluminium tropical soils are known as bauxite and are mined for aluminium (see Bauxite Deposits in Karst). Cooler temperatures reduce the rate of removal of silica and clay and iron minerals hydrate. Where temperatures are warmer there is more leaching of silica and clays, which tend to accumulate in deeper soil layers, and dehydration of iron minerals. The clays in these soils are often red and are regionally referred to as "red clays" in the United States and "terra rossa" in Europe. Two US examples of these soils are the southern and central Appalachian Frederick and Frankstown silt loams (order Ultisol) and the Cumberland silt loam (order Alfisol) common to the Mammoth Cave, Kentucky area.

Thick karst soils such as those developed on allogenic drift materials deposited over limestone have an ability to store large volumes of water in contrast to the thin autogenic karst soils. Higher water storage capacity tends to even out flows and create a diffuse infiltration to the bedrock. Thin autogenic karst soils, in combination with high infiltration rates, create flashy flows and rapidly dry out between storms. The droughtiness associated with thin autogenic karst soils minimizes the opportunity to accumulate organic materials in the soil, which can substantially increase soil water-holding capacity and reduce soil erodability.

Soils are usually the first defence against movement of contaminants into groundwater. The ability of soils to act as buffers depends on their chemical nature, permeability, and thickness. Autogenic karst soils are generally thin and well drained, thus limiting their ability to physically filter and detain contaminants and pathogens long enough for natural purification and neutralization to take place. Droughty conditions found in thin, well-drained karst soils limit actual evapotranspiration, which can be an important mechanism for removal of some contaminants by vaporization. For further discussion of herbicide, pesticide, and fertilizer leaching through karst soils see Groundwater Pollution: Dispersed.

Soils are fundamental to successful agricultural pursuits. The thin and uneven nature of autogenic karst soils makes them difficult or impossible to till in many karst regions. Agricultural systems that do not depend on tillage for success are often the systems of choice on thin karst soils. Examples of those systems are forages in grazed pasture systems, viticulture, and trees as part of forestry or orchard operations. Grazed pasture systems on autogenic and allogenic karst soils are well known throughout the world with some of the most productive grassland systems in the world occurring in the United Kingdom and Europe, United States, New Zealand, and Australia. The Appalachian region of the United States has about 18% agricultural land (Boyer & Pasquarell, 1995), but more than one-third of the region's agricultural output occurs on karst land and is made up mostly of livestock systems on grazed grasslands and orchard production of apples, peaches, and nuts. Much of the rest of the region's karst land is forested and produces some of the world's finest quality hardwoods for furniture production. Successful viticulture depends on well-drained soils that experience a dry season toward the end of the growing season. Wines produced from grapes grown on karst soils of the Mediterranean region are famous and among the best quality wines worldwide. Other important regions for viticulture are New York State and southern Australia where limestone soils are common.

Soils over carbonate bedrock present technical and environmental challenges for construction projects. Thin soils, uneven bedrock surfaces, and removal of soil through subsurface drainage into the karst bedrock often cause failures in highways and building foundations (see Construction on Karst). Many of the clays found in karst subsoils are subject to extreme shrinking and swelling further complicating competent construction. Landfills and sewage lagoons depend on thick soil mantles in order to operate properly. The thin, well-drained soils common to karst land are not generally suitable for either construction. Even when synthetic liners are used the uneven nature of the carbonate bedrock and drainage from below can cause catastrophic failure. Thick soils over carbonate bedrock are also subject to failure because of the under-drained nature of karst. Several sewage lagoons holding domestic sewage and liquid swine manure have failed on the thick soils covering carbonate bedrock in southeastern Minnesota (United States). Although those structures were constructed with heavy synthetic liners, soil was removed from under the liners by subsurface drainage into the carbonate bedrock and catastrophic failures occurred introducing contaminants and pathogens into the local groundwater and springs.

Conservation and protection of fragile soil resources over karst carbonate bedrock is a daunting task, which is discussed in Soil Erosion and Sedimentation. Briefly, practices that encourage plant growth and discourage unnatural accumulation of running water are essential for the protection of karst soils.

DOUGLAS G. BOYER

Works Cited

Boyer, D.G. & Pasquarell, G.C. 1995. Nitrate concentrations in karst springs in an extensively grazed area. *Water Resources Bulletin*, 31: 729–36

White, W.B. 1988. *Geomorphology and Hydrology of Karst Terrains*, Oxford and New York: Oxford University Press

Yuan D. *et al.* 1991. *Karst of China*. Beijing: Geological Publishing House

Further Reading

Curtis, L.F., Courtney, F.M. & Trudgill, S.T. 1976. *Soils in the British Isles*, London and New York: Longman

Drew, D.P. 1983. Accelerated soil erosion in a karst area: The Burren, Western Ireland. *Journal of Hydrology*, 61: 113–24

Trudgill, S.T. 1985. *Limestone Geomorphology*, London and New York: Longman (pp.66–70 and 116–22 discuss soil formation and erosion)

Useful Websites

UK Soil Survey and Land Research Centre, http://www.silsoe.cranfield.ac.uk/sslrc/

US National Soil Survey Center, http://www.statlab.iastate.edu/soils/nssc/

US Department of Agriculture, Natural Resources Conservation Service, http://www.nrcs.usda.gov/

SOIL EROSION AND SEDIMENTATION

Soil erosion is the detachment and transport of soil by water, wind, ice, or gravity. Soil erosion caused by natural and geological processes tends to be slow, but episodic, and in equilibrium with soil-forming processes. The influences of man's activities on soil erosion are evident all around us. Urbanization, agriculture, deforestation, construction, and recreation are among the most important activities of man that often accelerate rates of soil erosion.

In order for soil erosion to take place, soil particles must first be detached from the soil surface. The energy required to accomplish this is most commonly provided by raindrop splash or energy supplied by movement of accumulated water either as sheet flow across the soil surface or as concentrated flow in rills. Plant cover is the most effective natural means of controlling soil erosion. Plant canopies effectively reduce the kinetic energy of falling raindrops, thus reducing detachment of soil particles by raindrop splash. Plants also inhibit the accleration of flowing water and often improve the infiltration capacity of soil, effectively reducing the opportunity for surface water to accumulate in detention hollows. Plants also contribute the organic matter found in surface soil layers. Organic matter in surface soil is important from an erosion standpoint because it increases the water-holding capacity of the soil and improves infiltration capacity by maintaining soil aggregates. The shallow depth of many autogenic karst soils, coupled with their well-drained nature and free drainage into the bedrock, make these soils highly subject to drought. This in turn makes establishment and maintenance of continuous plant cover difficult resulting in accelerated erosion and a lack of organic matter accumulation in the surface soil layer. Seasonally dry periods coupled with the droughty nature of karst soils in the Mediterranean region (Gams *et al.*, 1993) and in Australia (Gillieson, 1993) severely impact plant canopies and fires are frequent. Fires leave the soil surface unprotected and severe erosion has led to complete denudation of the soil cover on some steep karst areas in the Mediterranean region (Gams *et al.*, 1993). Karst soils are further discussed in the entry Soils on Carbonate Karst.

There are many published examples of soil erosion occurring on karst lands as a result of deforestation and agricultural exploitation. Nearly every essay in the book entitled *Karst Terrains: Environmental Changes and Human Impact* (Williams, 1993) contains at least one example of severe soil erosion as a result of deforestation and agriculture. Highly variable soil depths and exposed bedrock make many karst soils difficult to cultivate so grazing of improved or natural pastures is often the landuse of choice. Poor management and overgrazing lead to reduced plant cover, exposed soil surfaces, and compacted soil—all factors that accelerate soil erosion.

Many studies have found that runoff increases following deforestation in proportion to the area of land cleared. Hence, soil erosion is the primary consequence of deforestation and is directly related to hillslope length and gradient and the amount of precipitation and runoff. Harding and Ford (1993) studied the effects of clearcutting on soil erosion and forest regeneration on limestone slopes in Vancouver Island, British Columbia. They found soil erosion was most severe on the slopes that were too steep to have developed a dense epikarst adequate for trapping eroding soil particles. Very little forest regeneration had taken place on bare limestone slopes with little trapped sediment. Indeed, on the best of sites, only 20% of the original volume of timber had grown 75 years after logging. The study led the authors to conclude that glaciated and karstified slopes in British

Columbia are vulnerable to soil erosion and desertification that is permanent in terms of human history. Ford (1987) found that clearcutting on karstified limestone and dolomite plains in Ontario, Canada, also led to complete loss of soil and litter over broad areas. However, the epikarst was well developed, resulting in efficient trapping of eroding soil particles, nutrients, and water in the karren troughs and microcaves. Hence the forest was able to re-establish itself, but with a higher proportion of undesirable tree species growing to a much lower height than the original forest.

In the Burren plateau karst of Ireland (see separate entry) thin soils, large areas of bare rock, and sparse vegetation have been attributed to glacial erosion. However, accumulations of older brown-earth soils in grikes and layers of reddish-brown mineral soils under ancient walls (in contrast to modern walls resting directly on bare rock surfaces or thin layers of rendzina soil) led Drew (1983) to argue that the Burren was once covered by an extensive layer of soils formed in glacial drift. During the Bronze Age, deforestation resulted in soil erosion that removed most of the extensive mineral soil leaving the area bare and unproductive. Since the Bronze Age a sparse residual rendzina soil has been developing and exists in small patches of 1–5 cm thickness between vast areas of bare limestone pavement. Deforestation and early agriculture with resultant soil erosion have also been blamed for the bare appearance of karst in the Mediterranean, Yugoslavia, and the northern Pennines of England.

Stone forest aquifers in the south China karst belt (see Shilin Stone Forest) have suffered reduced recharge as a result of deforestation and soil erosion on upgradient karst slopes. The stone forest aquifer consists of a deep (up to 100 m), well-developed epikarst infilled with externally derived sediments, soils, karst breccias, and residual clays (Huntoon, 1992). Massive soil erosion from the low-lying karst plains left a virtual "forest" of rock spires and towers that at one time made up a highly developed epikarstic zone hidden by a continuous soil mantle. The remaining network of roofless dissolution-widened fissures, cavities, and tubes now create a shallow aquifer with low water storage capacity. A high degree of lateral permeability rapidly drains the epikarst aquifer. At one time the hills surrounding the karst plains were forested and supplied a steady recharge of water. Nearly complete deforestation since 1958 has created a loss of what the Chinese call their green reservoir and the amplitude of the flood / drought cycle has been intensified. Desertification has commenced on drier sites, and the hills are characterized by scanty vegetation, parched and highly eroded red soils, and bare limestone: the water-holding capacity of the limestone hills has been severely compromised. The lack of plentiful and good-quality water in the region has made this one of the poorest economic regions of China.

Many pollutants and human pathogens are known to attach themselves to sediments. Increases in sediment loads may lead to associated increases in contaminants. Phosphorus is one important example of a contaminant that binds tightly to soil particles. Increased soil erosion and rapid transport of sediment and phosphorus through karst systems can lead to eutrophication in downstream surface water bodies. Accelerated sediment loads also lead to higher costs of maintaining water treatment facilities and filling in of water-holding reservoirs. Increased sediment loads in karst systems may also lead to plugging of conduit systems thereby altering local hydrology, which then leads to urban flooding, surface collapses, and clogged water systems. Severe soil erosion from road-building sites in Papua New Guinea (James, 1993) covered karst surfaces creating an unstable surface. Infilling of caves with sediments caused subsurface drainage to become surface drainage with increased flooding and springs dried up. Effects of externally derived sediments on karst systems are discussed in the entry on Sediments: Allochthonous Clastic. Suitable soil conservation strategies need to be developed specifically for the special conditions that exist on karst.

DOUGLAS G. BOYER

See also **Forests on Karst**

Works Cited

Drew, D.P. 1983. Accelerated soil erosion in a karst area: The Burren, Western Ireland. *Journal of Hydrology*, 61: 113–24
Ford, D.C. 1987. Effects of glaciation and permafrost on the development of karst in Canada. *Earth Surface Processes and Landforms*, 12: 507–21
Gams, I., Nicod, J., Sauro, M., Julian, E. & Anthony, U. 1993. Environmental change and human impact on the Mediterranean karsts of France, Italy and the Dinaric Region. In *Karst Terrains: Environmental Changes and Human Impact*, edited by P.W. Williams
Gillieson, D. 1993. Environmental change and human impact on karst in arid and semi-arid Australia. In *Karst Terrains: Environmental Changes and Human Impact*, edited by P.W. Williams
Harding, K.A. & Ford, D.C. 1993. Impacts of primary deforestation upon limestone slopes in Northern Vancouver Island, British Columbia. *Environmental Geology*, 21: 137–43
Huntoon, P.W. 1992. Hydrogeologic characteristics and deforestation of the stone forest aquifers of south China. *Ground Water*, 30: 167–76
James, J.M. 1993. Burial and infilling of a karst in Papua New Guinea by road erosion sediments. *Environmental Geology*, 21: 144–51
Williams, P.W. (editor) 1993. *Karst Terrains: Environmental Changes and Human Impact*, Cremlingen-Destedt: Catena

Further Reading

Hardwick, P. & Gunn, J. 1990. Soil erosion in cavernous limestone catchments. In *Soil Erosion on Agricultural Land*, edited by J. Boardman, I.D.L. Foster & J.A. Dearing, New York: Wiley
Morgan, R.P.C. 1995. *Soil Erosion and Conservation*, 2nd edition, Harlow, Essex: Longman, and New York: Wiley

SOLUTION BRECCIAS

Solution breccias are common and widespread geographically and throughout the geological record. The brecciation may occur entirely within one soluble rock (usually limestone or dolostone) or where more soluble evaporite rocks are removed in mixed evaporite–carbonate or evaporite–clastic rock sequences. It may be extended upwards by mechanical failure (stoping) into overlying insoluble strata. For example, dissolution of $c.180$ m of salt has propagated through 400 m of carbonates overlain by $c.650$ m of clays, mudstones, and sandstones at a site in Saskatchewan, Canada.

Brecciation can occur during the earliest stages of diagenesis (eogenesis), when evaporites are dissolved in supratidal carbonate sequences, or where caves in case-hardened carbonate sand dunes collapse. It is common at interstratal sites during deep burial (mesogenesis), caused by the expulsion of trapped sea water and other basinal fluids or by invading hydrothermal waters, and it may extend to pressure solution (stylolite) depths of 4 km or more. It is also common in superficial karst terrains, where circulating meteoric waters are known to brecciate carbonate rocks at depths as great as 2 km.

Breccia fabrics and matrix

Stanton (1966) recognized three principal fabrics (Figure 1):

1. Crackle breccia, where beds sag apart and crack upon dissolutional removal of support, but there is little displacement;
2. Pack breccia, where large fragments (clasts) support each other in a pile. The clasts have usually dropped and vary from partly rotated to completely disorganized (chaotic brecciation) in orientation. Globally, sizes of such clasts in a pile range from small pebbles to "cyclopean breccias" of blocks of hundreds of cubic metres, but are normally more limited within any given breccia;
3. Float breccia, where the larger fragments are separated from each other and "float in" (are supported by) a matrix of fine material.

A genetic association of these fabrics is often found, consisting of crackle breccias at the top and around the perimeter of a body, pack breccia within it where beds are thicker, and float breccias at the base or where beds are thin. In some Mississippi Valley Type (MVT) lead–zinc deposits, these may be underlain by a basal "trash" zone of insoluble residua (see Figure 1; Sangster, 1988; Dźulyński & Sass-Gutkiewicz, 1989).

Clast-supported breccias may consist simply of clasts and voids, with little or no matrix or cement. This is most common in breccias in dolines and caves that are actively developing today. Matrix is usually present in eogenetic and mesogenetic breccias, consisting of carbonate fines and insoluble residua that have filtered down or (more rarely) washed-in sands and clays. Older breccias of all types will normally display partial or complete cementation. Calcite is the principal cement, but aragonite, dolomite, and gypsum are also common. In MVT the chief cements are dolomite, pyrite, galena, and sphalerite.

Form and location of the principal types of breccia bodies

Solution breccias are common on the surface and underground in modern karst terrains, chiefly where dolines, river cliffs, or shafts and chambers in caves have collapsed. Individual collapses may be millions of cubic metres in volume. Where surficial

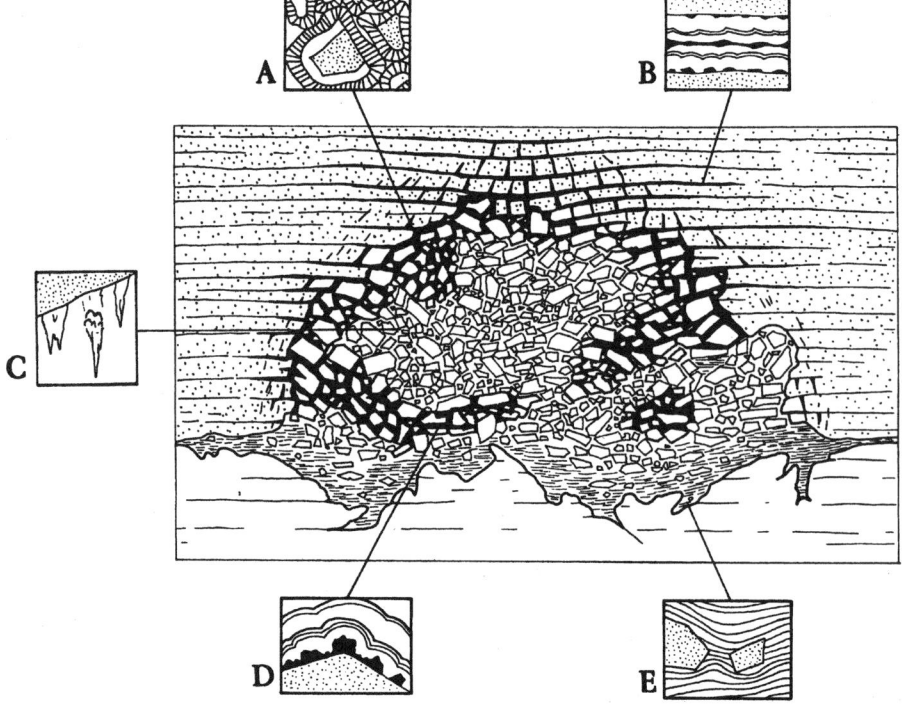

Solution Breccias: Figure 1. Model for a breccia-hosted lead-zinc deposit in limestone or dolostone, illustrating "crackle brecciation" around the perimeter; "chaotic pack breccia" with void-filling cements (black) in the centre and "float breccia" in a matrix of "trash" (insoluble residua) at the base. Enlargements **A–D** show the typical microstratigraphy of calcite or dolomite cements at different places in the breccia; **E** details float stratigraphy in the trash. From Dźulyński & Sass-Gutkiewicz (1989).

Solution Breccias: Figure 2. A breccia pipe of solutionally undermined and collapsed limestone blocks with an open cavity at the top where void migration is still continuing: exposed in an island cliff in Ha Long Bay, Vietnam. (Photo by Tony Waltham)

collapses are preserved in buried paleokarsts, Choquette and James (1988) term them "mantle breccias".

Extensive brecciation is usually present where the proximal margins of buried salt bodies on the continents are subject to dissolution by groundwaters penetrating the cover rocks, creating a receding "salt slope" (Ford & Williams, 1989, p.460). In Manitoba and Saskatchewan, the salt slope of a Devonian deposit is c.800 km in length and now at a mean depth of 200 m below the surface. It has receded (been "subroded") westwards for an average of 130 km since burial during the Devonian Period. Breccia pipes (Figure 2), also called "geological organs", "breakthroughs", or "prismatic bodies" by various authors, are the most widely reported breccia bodies, with thousands of examples being described in the world literature (Bosák et al., 1989). Diameters range from <10 m to >10 km, with a majority being 20–250 m. Most are plumb-vertical, with reported heights ranging from 20 m to more than 1000 m. Higher examples have usually stoped upwards through one or more cover formations, which may include siliciclastics, coal measures, and extrusive volcanic rocks. The upward termination may be in undisturbed rock, in downfaulted but not brecciated strata, in a closed depression at the surface, or even upstanding as a firmly cemented and now-resistant "castille" above an erosional plain (as described by Hill, 1996, in New Mexico). Most breccia pipes originate in point dissolution of salt (occasionally of gypsum or anhydrite) above a fracture junction, anticlinal crest, or buried reef that channels groundwater. They may be targets for the later precipitation of economic ore, such as the well-known uranium pipes of the Grand Canyon, Arizona (Huntoon, 1996).

Other forms are described in most detail from MVT deposits (e.g. Sangster, 1988). Breccia domes typically 30–100 m in diameter are similar to breakdown rooms in modern caves or transitional to breccia pipes. Tabular or sinuous forms are wide but shallow breccias extending for 1 km or more, often along paleoreef margins. Straight linear features ("runs") and reticulate mazes include individual breccia-filled corridors up to 30 m high, 150 m or more wide, and one or more km in length. Most of these features occur at relatively shallow depths beneath long-exposure paleokarst surfaces, and are believed to be genetically associated with them. Some are attributed to brecciation and subsequent sulfide ore (cement) deposition from invading hydrothermal solutions.

DEREK FORD

See also **Bear Rock Karst; Sulfide Minerals in Karst**

Works Cited

Bosák, P., Ford, D.C., Glazek, J. & Horácek, I. (editors) 1989. *Paleokarst: A Systematic and Regional Review*, Amsterdam: Elsevier

Choquette, P.W. & James, N.P. 1988. Introduction. In *Paleokarst*, edited by N.P. James & P.W. Choquette, New York and London: Springer

Dźulyński, S. & Sass-Gutkiewicz, M. 1989. Pb–Zn ores. In *Paleokarst: A Systematic and Regional Review*, edited by P. Bosák, D.C. Ford, J. Glazek & J. Horácek, Amsterdam: Elsevier

Ford, D.C. & Williams, P.W. 1989. *Karst Geomorphology and Hydrology*, London and Boston: Unwin Hyman

Hill, C.A. 1996. *Geology of the Delaware Basin, Guadalupe, Apache and Glass Mountains, New Mexico and Texas*, Midland, Texas: Society for Sedimentary Geology

Huntoon, P.W. 1996. Large-basin ground water circulation and paleo-reconstruction of circulation leading to uranium mineralization in Grand Canyon breccia pipes, Arizona. *The Mountain Geologist*, 33(3): 71–84

Sangster, D.F. 1988. Breccia-hosted lead–zinc deposits in carbonate rocks. In *Paleokarst*, edited by N.P. James & P.W. Choquette, New York and London: Springer

Stanton, R.J. 1966. The solution brecciation process. *Geological Society of America, Bulletin*, 77: 843–48

SOVIET UNION: SPELEOLOGICAL HISTORY

There are numerous early mentions, going back to the 4th century BC, of caves in the territories that later composed the Russian Empire and Soviet Union. However, it was during the 18th century that a scientific approach to the study of karst phenomena, and caves in particular, emerged in Russia. Five main periods can be broadly distinguished in the subsequent history of cave exploration and study (Gorbunova, 1988, 1990; Gorbunova & Dublyansky, 1999; Dublyansky, 1999). The distinction between the periods is based on changes in political and economical systems, predominant concepts about caves and their significance, motives for exploration, and ways in which cave exploration and study developed.

18th century

The conquest and development of huge territories of Ural and Siberia in the far north of the Russian empire was accompanied by a series of expeditions commissioned by the state to study and explore natural resources. Their reports contained many significant data about karst and caves. The first detailed field manual on the study of caves was compiled by Professor Johan Gmelin for the Kamchatka Expedition (also known as the Great North Expedition) of the St Peterbourg Academy of Science, carried out between 1733 and 1743. In the second half of the 18th century, the great Russian naturalist, Michail Lomonosov, expressed many important theories about the nature of karst phenomena and caves, cave formations, and microclimate. It is noteworthy that many of the earlier scholars realized that caves in both limestone and gypsum had similar features.

19th century to early 20th century

This period is noted for remarkable growth in data about the karst and caves of the Russian Empire. Such data were supplied by numerous regional geographical and geological expeditions and by missions that investigated the Empire's outlying districts, such as Crimea, Caucasus, Ural, Central Asia, Western and Eastern Siberia, and the Far East. At the beginning of the 19th century, V. Severgin published the first countrywide review of caves in the Russian state. In the second half of the century systematic efforts were made to study the natural resources of the state's huge territories, stimulated by the burst in the industrial development of the central and far east regions and by construction of railroads, and generating many cave descriptions and maps.

Specialized work in some caves included studies of cave paleontology (A. Ivanov), ice formations (E. Fedorov), microclimate (Ju. Listov), and archaeology (A. Kirkor and G. Ossovsky). In 1887 Listov published a detailed field manual of his cave studies, which covered aspects of geology, morphology, hydrogeology, formations and sediments, microclimate, speleogenesis, and local folklore. In the 1890s, the Crimean-Caucasian Mountaineering Club was established in Crimea and its members made many explorations and studies of caves. It was one of the first examples of the involvement in cave studies of enthusiastic amateurs organized in a public society, a phenomenon that emerged at the same time in Western Europe as a result of initiatives by E.A. Martel, and later developed into what is now termed speleology.

From the beginning of the 20th century, the number and significance of regional karst cave studies continued to grow exponentially in all the main regions of the Russian Empire, particularly in relation with systematic geological mapping. In 1900, a fundamental book *About Karst Phenomena in Russia* was published by Alexander Kruber. In this and later publications, Kruber not only gave a painstaking characterization of many karst regions and caves, but he also laid some important theoretical principles of karst and cave science. His role in karstology and speleology in Russia and Ukraine is somewhat comparable with that of E.A. Martel in Western Europe. The important distinction, however, is that Kruber did not found amateur speleological societies, as did Martel. Martel visited Russia in 1903, at the invitation of the Minister of Agriculture and Russian Domains (Cigna, 1977).

1917–1957

The revolutions of 1917 and the subsequent civil war and terror of Stalin's regime interrupted many common activities, including cave and karst studies. Scientific and applied geological research were resumed in many regions during the 1920s, with incidental caves studies. There were many karst research projects during the subsequent decades (with the natural exception of the World War II years) because the demands of industrial development in many karst terrains stimulated regional investigations as well as progress in the theory and methods of karst studies. The All-Union karstological conferences, held in 1933, 1947, and 1956, reviewed the advances and outlined the tasks for future researchers. The latter conference, held in Moscow, was attended by over 2000 participants representing 284 institutions.

By 1957, several karstological schools, led by eminent scientists (N. Gvozdetsky, B. Ivanov, G. Lykoshin, G. Maksimovich, I. Popov, D. Sokolov, N. Sokolov, A. Stupishin and others) had formed in various scientific institutions. However, exploration and study of caves had not advanced to the same extent during this period. Unlike many countries of Western Europe and North America, where the main progress in cave exploration during the 20th century was linked with amateur speleological activity organized in clubs and societies, no such speleological movement evolved in the Soviet Union during the first half of the century. This was due to continued economic hardships and the nature of the totalitarian regime. Instead, the state, concerned with military use of caves for missile sites and underground constructions, made efforts to organize the exploration and study of caves through governmental organizations. Such activities were shrouded in secrecy, especially between 1942 and 1956. Later, inventory and documentation of natural caves was included in the duties of the Ministry of Geology. However, the state-driven efforts were not too successful in terms of the number and size of caves explored. By the end of 1957, no more than 1000 caves, mainly easily accessible and small, were known in the entire Soviet Union. The longest cave remained Kungur Ice Cave in Ural (4.7 km), known since the 18th century, and the deepest cave was the Bottomless Pit in Crimea, where a depth of 100 m was reached in 1927. In view of the remarkable achievements made at that time in cave exploration in Western Europe and North America, it seemed that the Soviet Union was irreversibly behind. The weak state of speleological exploration and research was recognized by leading karst scientists, and was reflected in the resolution of the All-Union Karst Conference held in Moscow in 1956. This conference also pointed out the need for involving amateurs in cave exploration through the development of caving movements.

1958–1991

It is believed that modern speleology in the Soviet Union was born in 1958. There were several coinciding reasons that provided a boost to cave exploration and studies. In that year, the Interdepartmental Karst Commission was created under the Academy of Sciences to coordinate and advance karst and cave studies. The state undertook measures to promote and support outdoor sports, such as mountaineering, hiking, and wild rafting, directing them to exploration; caving found a natural place in such a system. The recognition by leading karst scientists of the significance of cave studies motivated them to encourage amateur exploration activity. Last but not least, publication in Russian of mass circulation books, by Norbert Casteret (in 1956 and 1959) and William Halliday (in 1962), had encouraged many young people to devote themselves to cave exploration.

In 1958 a scientific institution, called the Complex Karst Expedition, based in Crimea, was formed under the Ukrainian Academy of Sciences. In subsequent years, it established a sound practice of cooperation between scientists of various fields and sporting cavers, while carrying out the systematic study of caves in the Crimean Mountains and the rest of Ukraine. Encouraged by eminent karstologist Dr Boris Ivanov, Victor Dublyansky and Vladimir Iljukhin were instrumental in developing the organizational structure of the caving movement throughout the Soviet Union. In 1962 the first national field caving seminar was held in Crimea, where principles of exploration techniques and methods of cave study were laid down, as well as the structure of the training system. Amateur caving groups and clubs appeared in many regions and their enthusiastic exploration activity generated massive new data on caves. The pace of cave exploration in the USSR, since 1958, can be illustrated by the following facts:

1. The length of mapped passages in two caves, Ozerna and Optimistychna in Ukraine, passed 100 km in 1974 and 1975 respectively;
2. The threshold of 1000 m depth was reached in 1976 in Kievskaja (KILSI) cave in Tien Shan and by 1991 there were 5 caves over 1000 m deep in the country, the deepest being Pantjukhina Cave in the Caucasus (1508 m);
3. By 1991, the number of explored caves in the USSR exceeded 7500 (cf. less than 1000 explored caves in 1957)

In 1965, Soviet speleologists attended the International Union of Speleology (UIS) Congress for the first time and in 1977, the National Association of Soviet Speleologists (NASS) joined the UIS. The growth in cave exploration in almost every region of the country stimulated considerably the development of karst and cave science. Between 1958 and 1991, about 85 major scientific conferences and symposia were held, over 50 monographs or thematic collections of papers were published and several tens of PhD dissertations were defended on karst and caves. Soviet speleology became an important part of the international speleological scene.

1992 to the Present

After the breakup of the USSR at the end of 1991, the Ukrainian Speleological Association (established in 1992) and the Russian Speleological Union (established in 1996) maintained lively exploration activity in many regions of the former Soviet Union. Among the most remarkable achievements of the recent period are: (1) surveying of Optimistychna Cave in the Western Ukraine to 214 km; (2) exploration of Botovskaya Cave in Siberia for over 32 km; (3) exploration of the deepest cave in the North (Russian) Caucasus, Gorlo Barloga, to a depth of 870 m; (4) exploration in January 2000 of what was then the deepest cave in the world, Krubera in the Arabika Massif, Georgia; and (5) exploration of a 1530 m deep cave, Sarma, in Arabika.

ALEXANDER KLIMCHOUK

See also Asia, Central; Caucasus, Georgia; Crimea, Ukraine; Krubera Cave, Georgia; Pinega Gypsum Caves, Russia; Russia and Ukraine; Siberia; Ukraine Gypsum Caves and Karst

Works Cited

Cigna, A. 1977. Martel's voyage to Russia in 1903. *International Journal of Speleology*, 26(3–4): 79–87

Dublyansky, V.N. 1999. From the history of domestic speleology (XVIII century). *Peschery* [Caves], Perm: Perm University: 127–55 (in Russian)

Gorbunova, K.A. 1988. From the history of domestic speleology (the first half of the XX century). *Peschery* [Caves], Perm: Perm University: 96–104 (in Russian)

Gorbunova, K.A. 1990. From the history of domestic speleology (XIX century—the beginning of the XX century). *Peschery* [Caves], Perm: Perm University: 96–105 (in Russian)

Gorbunova, K.A. & Dublyansky, V.N. 1999. From the history of domestic speleology (the first half of the XX century). *Peschery* [Caves], Perm: Perm University: 116–26 (in Russian)

Further Reading

Gusev, A. (editor) 1997. *35 Years of the Speleology Club MGU (Moscow State University) 1961–1996*, Moscow: Speleology Club (in Russian)

Dublyansky, V.N. & Ilyukhin, V.V. 1987. Speleology in the USSR. *Studies in Speleology*, 7: 5–15

SPANNAGEL CAVE, AUSTRIA

The Central Alps of Austria (see Europe, Alpine for location map) are occupied by crystalline rocks and karst features are restricted to the presence of carbonates (mainly marbles) and carbonate-bearing schists. Cave systems are typically located at altitudes in excess of 2000 m and are commonly spatially associated with glaciated areas. At the head of the Tux Valley in the western Zillertal Alps of Tyrol in western Austria, a number of high-altitude caves are present in the vicinity of the Spannagel Hut (2528 m). The largest of these caves is Spannagel Cave, the longest cave in the province of Tyrol, with a total surveyed length of 9 km. The cave developed in upper Jurassic calcite marbles of the Hochstegen Formation, which form a 20–30 m thick unit sandwiched between gneisses and dipping north–northwest. The structural setting in conjunction with the vicinity to the Hintertux Glacier are key factors controlling cave formation in this area. Marginal moraines of the glacial advance during the Little Ice Age show that even fairly recently some parts of the cave systems were in a subglacial position.

Initial conduits formed along the well-developed bedding planes and the prominent east–west and north–south-trending fractures (Jacoby & Krejci, 1992). The most common passage types are vadose canyons, typically no more than 1–2 m wide, which extend down to the base of the marble. Meandering streams are episodically active in these canyons and their peak discharges are only a few litres per second even during snow melt. The stream beds show pothole-like erosion features cut into the underlying gneiss and filled by sandy gravel. The modern streams are incapable of creating such features, thus

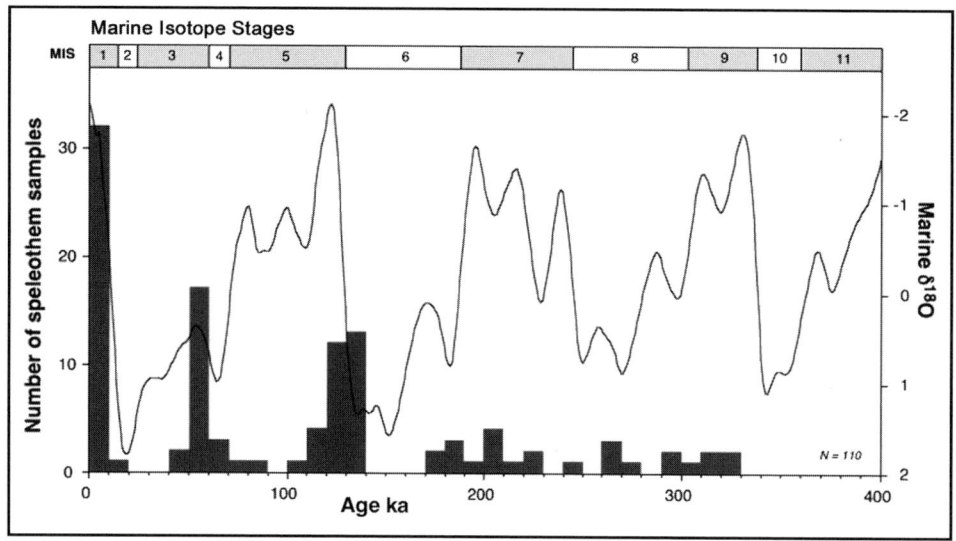

Spannagel Cave, Austria: Histogram of dated speleothems from Spannagel Cave correlated with late Pleistocene climatic variations.

indicating their ancient origin. Phreatic passages are preserved in the interior portions of both the western and the northern branch of the cave system. These passages show circular to oval cross sections and abundant large scallops on both walls, floors, and ceilings. Many were developed in the highest possible position within the marble, i.e. just beneath the overlying granitic gneiss. These phreatic tubes are commonly cut by vadose canyons giving rise to keyhole profiles. Many passages of Spannagel Cave show active destruction due to collapse, particular those located close to the surface. As a result, these passages develop elongate near-rectangular cross sections.

Spannagel Cave is currently unaffected by meltwaters from the nearby glaciers, but the cave most probably acted as a subglacial drainage system for meltwaters during the Pleistocene glaciations. Well-rounded allochthonous gneiss cobbles reaching a maximum diameter of 500 mm, derived from the nearby glaciated areas, are present even in the most distant parts of the cave system, attesting to their entrainment by high-energy streams.

The climate of Spannagel Cave is well known due to a number of continuous temperature records obtained by data loggers in various parts of the cave system. With the exception of the northernmost section close to the main entrance—which is operated as a show cave—the remaining cave system is characterized by an air temperature between +1.2 and +2.4°C. The higher temperature within this range is restricted to the westernmost branch of the cave, where overburden reaches 190 m. None of the interior sites show seasonal temperature variations. Air flow in Spannagel Cave is strongly controlled by the outside temperature (and pressure): during the cold season air ascends through the relatively warm cave and exits at the main entrance, i.e. at the highest position within the system. During the warm season, typically commencing in May, air flow reverses and warm air enters the exterior parts in the south. During October the difference between outside pressure and in-cave pressure diminishes, giving rise to stagnant conditions, which are slowly superseded by the establishment of a new upward-oriented winter flow regime, typically by early November. Relative humidity inside the cave is in excess of 96%, regardless of season. Likewise, the partial pressure of carbon dioxide in the cave air, which ranges from 280 to 325 ppmv, does not reveal a seasonal pattern.

A noteworthy feature of Spannagel cave system is the presence of calcite speleothems, some of which are active today, despite cave air temperatures only slightly above freezing. Speleothems include flowstones, stalagmites, stalactites, soda straws, spar, and helictites composed of low-Mg calcite, as well as gypsum encrustations of walls, ceilings, and calcite stalactites. Ancient speleothems, however, are more abundant than modern ones and commonly show evidence of subsequent erosion and / or dissolution. Due to the fact that seepage waters have to migrate through the overlying granitic gneiss, both dripwaters and speleothems are exceptionally rich in U (up to 218 ppm). Mass spectrometric U–Th disequilibrium dating on a large number of samples from this cave reveals a wide range of ages, from more than 400 000 years to essentially modern (Spötl et al., 2003), thus demonstrating (a) the high age of this cave system, and (b) that the environmental conditions at this high-alpine site repeatedly permitted speleothem growth during the Quaternary period. The distribution of U–Th ages mimics the well-known glacial-interglacial climate pattern in that the majority of the dates fall within interglacials (marine isotope stages 1, 5e, 7, and 9), but there is also evidence that stalagmites and flowstones formed during pronounced warm intervals within glacials, e.g. during isotope stage 3 (see Figure; Spötl & Mangini, 2002).

Christoph Spötl

See also Calcareous Alps, Austria

Works Cited

Jacoby, E. Krejci, G. 1992. Die Höhle beim Spannagelhaus [The Cave near Spannagelhaus, Tyrol, Austria]. *Wissenschaftliche Beihefte zur Zeitschrift "Die Höhle"*, 26: 1–148

Spötl, C., Burns, S.J., Frank, N., Mangini, A. & Pavuza, R. 2003. Speleothems from the high-alpine Spannagel Cave, Zillertal Alps (Austria). In *Studies of Cave Sediments*, edited by I.D. Sasowsky & J. Mylroie, Dortrecht: Kluwer, in press

Spötl, C. & Mangini, A. 2002. Stalagmite from the Austrian Alps reveals Dansgaard-Oeschger events during isotope stage 3: Implications for the absolute chronology of Greenland ice cores. *Earth Planetary Science Letters*, 203: 507–18

SPECIATION

Speciation, the process by which one species evolves into a different species or splits over time into two or more new descendant species, has been effectively reviewed in cave animals by Barr & Holsinger (1985), Sbordoni *et al.* (2000), and in aquatic cavernicoles by Coineau & Boutin (1992) and Holsinger (2000). Readers are referred to these works for extensive analysis of the general models and data used by researchers in the field.

A difficulty with all discussions of speciation is that there is usually an underlying assumption that the species being discussed conform to the biological species model. This model assumes that genetic isolation is the key factor associated with the origin of new species and that gene exchange is the cement that holds a species together. In some cases, where a given bisexual species is generally confined to a region where underground movement between different cave habitats occurs regularly, it is possible this model may apply. This appears to be true for cave beetles in many cases. Often the model is either questionable or clearly inapplicable. First, gene flow is not essential to maintaining species identity—as is shown by the existence of species in asexual organisms, which are quite comparable to those seen in sexual species. A second problem lies in the assumption that spatially separated cave systems which harbour the same species of cave organisms but have physical barriers to underground gene flow, are maintained as one species by occasional interchange via surface populations. There is very strong evidence that even when large surface populations of the same species exist there is rarely significant gene interchange between the surface and hypogean populations (Kane & Culver, 1992; Wilkens, 1988). Furthermore, since the biological species concept depends on gene flow and cave species, with very few exceptions, are determined on phenotype, using the biological species concept requires similarity or pattern between the phenotypic and biological species. The evidence so far available rarely supports this similarity of pattern (Kane & Culver, 1992; Sbordoni *et al.*, 1987; 2000).

Which model or models to use is thus an area fraught with controversy. Kevin de Queiroz (1998) proposes that the conflict among concepts of the species stems largely from a confusion between species concepts and species criteria. He says that most of this argument concerns the criteria by which species are determined, originated, or maintained and that there is a general agreement concerning species concepts ". . . that species are segments of population-level evolutionary lineages" (de Queiroz 1998, p.63).

If we accept this concept of species then speciation becomes the process of developing such segments and the problem becomes one of defining the "segments". A good argument can be made for defining the segment phenotypically since phenotypic "segments" are those subject to selection, and selective forces have no way of determining genotype directly. Furthermore, there is some evidence that while the forces of selection have much to do with morphological or phenotypic speciation, they have little to do with genetic speciation (Sbordoni *et al.*, 2000, p.469). Genetic speciation studies almost never involve genes associated with morphological features of adaptive importance and thus cannot be subject to selection, which is the final determinant of lineages.

Since the Barr & Holsinger review (1985), some workers have emphasized the suitability of the Wrightian adaptive shift model for cave speciation, where adaptation is an active adaptive shift to novel habitats rather than a negative reaction of an ancestral population to unfavourable surface conditions (see Howarth, 1987; Culver, 1987), but most earlier works and many recent works support a primarily or even a strictly allopatric model for cave animal speciation, where species formation arises from common ancestors separated by geographical boundaries. Some investigators have offered evidence supporting parapatric (involving contiguous populations) (Howarth, 1987; Peck, 1990) or peripatric (Sbordoni *et al.*, 1985) speciation. Holsinger (2000) makes an important and clear distinction between two types of speciation in cave organisms. The first type (phase 1) involves the evolutionary changes a species makes when it first successfully invades caves. The second (phase 2) involves changes made in established cave lineages. These two evolutionary processes are very different, both in mechanism and products. The first produces often unclear changes in physiology, little or no troglomorphy, and has a heavy association with preadaptations. Thus, in any group of organisms, many genera or families never make the transition from phase 1 to phase 2 while many others do. The second phase usually involves clear morphological, physiological, and behavioural changes and often results in increased troglomorphy. Phase 1 speciation has been the subject of most speciation controversy and for terrestrial faunas, this has largely focused on the question of what environmental conditions are responsible for, or at least associated with, the initial cave invasions and cave adaptation. The first and most widely applied theory was the Pleistocene climatic shift or climatic relict model. This model views the development of cave faunas as a result of invasion of caves and subsequent isolation in these by external climatic changes eradicating the ancestral surface populations. This extinction of surface faunas could be a result of rising or lowering of temperatures so that there are warm and cold relict species or lineages. There appears to be substantial evidence for both types of relicts in temperate karst regions in Europe and North America, which were the first to be well studied. As a result of increased knowledge of tropical cave faunas the adaptive shift theory or local habitat shift model was developed (Howarth, 1973; 1987). This theory was elaborated upon and extended by other workers and proposes that troglobites evolve from preadapted forms which actively invade caves to exploit new niches and are evolved from native fauna (Holsinger, 2000). Their evolution is regarded as a continuing process and not an episodic one as in the relict theory. In addition, by implication the speciation is parapatric rather than allopatric. Sbordoni *et al.* (2000) point out that many tropical environments went through major environmental shifts during the Pleistocene. In addition they point out that the degree of troglomorphy in cave faunas is high where the surface conditions make movement between caves improbable and troglomorphy is low or absent where such movement should be easy. They say ". . . it is hard to deny the prevalent relictual character of both terrestrial and aquatic cave communities . . .".

Whether or not aquatic cave communities are predominantly relictual, they pose some problems not seen in terrestrial organisms. The first of these is the problem of how they got into the caves. There are two excellent reviews by Holsinger (2000) and Coineau & Boutin (1992) detailing the processes. In summary they involve invasion into caves via interstitial and crevicular habitats from marine habitats, evolution from organisms stranded by regression of marine habitats, and evolution from freshwater ancestors largely via stream capture, spring failure, or lowered water tables. The invasion of caves by marine animals involves a double adaptation since the species have to develop freshwater tolerance and either preadapted or adaptive for subterranean environments.

Whether terrestrial or aquatic, the species which become successful cave lineages in the deeper reaches of caves are a small percentage of those which have the opportunity of doing so. The most fundamental unanswered question remains: what physiological and behavioural characteristics allow certain groups to readily make this transition and not others? Many forms become successful troglophiles, often dominant where there is little or no competition from highly adapted or troglomorphic species, but fail to produce any derivative highly adapted species. While there appear to be obvious (usually unsubstantiated) reasons for this in some cases, usually the reasons are not apparent.

Speciation within troglobitic lineages is almost certainly primarily due to allopatric speciation (Sbordoni *et al.*, 2000) although the fact that frequent cases of species pairs exist, frequently involving species with different levels of troglomorphy, may offer some suggestion that other speciation modes may sometimes play a role. The spatially limited distribution of troglobite species, the degree of subdivision of populations, and isolation of these from each other all lend themselves well to the process of allopatric speciation. Furthermore the distribution pattern of troglobitic and primarily cavernicole troglophiles further supports the idea that allopatric speciation is dominant (Christiansen & Culver, 1987).

KENNETH CHRISTIANSEN

See also **Adaptation; Colonization; Evolution of Hypogean Fauna**

Works Cited

Barr, T.C., Jr. & Holsinger, J.R. 1985. Speciation in cave faunas. *Annual Review of Ecology and Systematics*, 16: 313–37
Christiansen, K. & Culver, D. 1987. Biogeography and distribution of cave Collembola. *Journal of Biogeography*, 14: 459–77
Coineau, N. & Boutin, C. 1992. Biological processes in space and time colonization, evolution, and speciation in interstitial stygiobionts. In *The Natural History of Biospeleology*, edited by A. Camacho, Madrid: Museo Nacional de Ciencias Naturales
Culver, D.C. 1987. The role of gradualism and punctuation in cave adaptation. *International Journal of Speleolology*, 16: 17–31
De Queiroz, K. 1998. The general concept of species, species criteria and the process of speciation. In *Endless Forms, Species and Speciation*, edited by D.J. Howard & S.H. Berlocher, Oxford and New York: Oxford University Press
Holsinger, J. 2000. Ecological derivation, colonization, and speciation. Population genetic structure, speciation and evolutionary rates in cave-dwelling organisms. In *Subterranean Ecosystems*, edited by H. Wilkens, D.C. Culver & W.F. Humphreys, Amsterdam and New York: Elsevier
Howarth, F.G. 1973. The cavernicolous fauna of Hawaiian lava tubes. I Introduction. *Pacific Insects*, 15: 139–51
Howarth, F.G. 1987. The evolution of non-relictual tropical troglobites. *International Journal of Speleology*, 16: 1–16
Kane, T. & Culver, D. 1992. Biological processes in space and time, analysis of adaptation. In *The Natural History of Biospeleology*, edited by A. Camacho, Madrid: Museo Nacional de Ciencias Naturales
Peck, S.B. 1990. Eyeless arthropods of the Galapagos Islands, Ecuador: Composition and origin of the cryptozoic fauna of a young, Tropical Oceanic Archipelago. *Biotropica*, 22(4): 366–81
Sbordoni, V., Allegrucci, G. Cesaroni, D., Sbordoni, M. & de Matthaeis, E. 1985. Genetic structure of populations and species of Dolichopod cave crickets: Evidence of peripatric divergence. *Bolletim di Zoologia*, 52: 139–56
Sbordoni, V., Allegrucci, G., Caccone, A., Carchini, G. & Cesaroni, D. 1987. Microevolutionary studies in Dolichopodinae cave crickets. In *Evolutionary Biology of Orthopteroid Insects*, edited by B.M. Baccetti, Chichester: Horwood and New York: Halstead Press
Sbordoni, V., Allegrucci, G. & Cesaroni, D. 2000. Population genetic structure, speciation and evolutionary rates in cave-dwelling organisms. In *Subterranean Ecosystems*, edited by H. Wilkens, D.C. Culver & W.F. Humphreys, Amsterdam and New York: Elsevier
Wilkens, H. 1988. Evolution and genetics of epigean and cave *Astynax fasciatus* (Characidae, Pisces): Support for the neutral mutation theory. *Evolutionary Biology*, 23: 271–367

Further Reading

Claridge, M.F., Dawal, H.A. & Wilson M.R. (editors) 1997. *Species: The Units of Biodiversity*, London and New York: Chapman Hall
Harrison, R. 1998. Linking evolutionary pattern and process. In *Endless Forms, Species and Speciation*, edited by D.J. Howard & S.H. Berlocher, Oxford and New York: Oxford University Press
Smith, A.B. 1994. *Systematics and the Fossil Record: Documenting Evolutionary Patterns*, Oxford and Cambridge, Massachusetts: Blackwell Science
Sbordoni, V. 1982. Advances in speciation of cave animals. In *Mechanisms of Speciation*, edited by C. Barrigozzi, New York: Liss
Slobodchikoff, C.N. (editor) 1976. *Concepts of Species*, Stroudsburg, Pennsylvania: Dowden, Hutchinson and Ross
Wiley, E.O. 1978. The evolutionary species concept reconsidered. *Systematic Zoology*, 27: 17–26
Wilson, R.A. (editor) 1999. *Species: New Interdisciplinary Essays*, Cambridge Massachusetts: MIT Press

SPELEOGENESIS

Etymologically, speleogenesis is the origin of caves. In a wider sense the term means not only the actual origin but also the entire life history of caves from gestation to obliteration (complete infilling or decay). Ideally speleogenetic study should explain comprehensively why and how caves originate and develop, and which factors guide these processes. Clearly an object must be defined before its genesis can be studied, but the notion of cave genesis makes little sense when applied strictly to the set of objects that conform to the common anthropocentric definition of caves (see Caves entry). Much of the data important for speleogenetic analysis derives from direct observations in explorable caves. However, restrictions imposed by the human entry requirement make the limits bounds of the objects under study (and hence their structure and shapes) accidental, artificial, and discontinuous. Thus they differ greatly from those of the natural systems that originated to perform a specific hydrogeological function, and from which caves developed. Moreover, the formative processes begin to operate and the forms themselves begin to be created with cavity sizes much smaller than those that may be entered by humans. The problem of speleogenesis in its broadest sense is further complicated methodologically because speleogenetic processes are so diverse in nature that no attempt to contrive an integral theoretical approach to the problem seems feasible. The solution lies in viewing each apparent genetic class of cave separately, and setting up criteria to define them based on the essential characteristics and functions of caves as natural systems. This approach is realized most comprehensively with regard to solution (karst) caves. These caves have been created principally by dissolution of bedrock by water circulating through pre-existing networks of openings such as fissures and pores. These caves are most abundant and important scientifically and practically, and they are the principal concern of karstology and geospeleology. Speleogenesis in the narrow sense followed here concerns the origin of solution caves and knowledge of related issues. A general overview of cave origin by other processes is provided in the entry on Caves. For greater detail see also Crevice Caves, Glacial Caves, Littoral Caves, Piping Caves, Volcanic Caves, and Pseudokarst.

Solution caves form where subsurface water flow is strong enough to remove dissolved bedrock and to keep undersaturated water in contact with the rock. As mobile groundwater is the principal agent of speleogenesis, speleogenetic study is closely linked to hydrogeology. Dissolutional enlargement of earlier porosity gives rise to a conduit structure organized to facilitate fluid circulation in a downgradient direction. Therefore, from the hydrogeological perspective speleogenesis is the creation and evolution of organized conduit permeability in soluble rocks, the principal concern of karst hydrogeology. An understanding of the principles of speleogenesis and its most important controls is indispensable to the proper comprehension of the evolution and behaviour of karst systems in general and of karst aquifers in particular.

No single speleogenetic model can be applied to a wide variety of geological and hydrological settings (see Speleogenesis Theories: Post–1890). Different dissolutional mechanisms operate on different lithologies and in different settings to produce caves. Evaporite rocks such as gypsum and common salt dissolve by simple two-phase dissociation, and dissolution rates are controlled solely by diffusion. Significant dissolution of carbonate rocks relies on more complex mechanisms involving additional sources of acidity such as carbonic acid produced by carbon dioxide, hydrosulfuric acid produced by hydrogen sulfide, and sulfuric acid produced by the oxidation of H_2S or of metallic sulfides. Some processes may enhance or rejuvenate dissolutional aggressiveness, e.g. the mixing of waters of disparate chemistry, cooling of water, sulfate reduction, and de-dolomitization. Carbonate dissolution by carbonic acid groundwater dominates overwhelmingly in near-surface environments, but other mechanisms are more important in deep-seated settings.

Modern speleogenetic views appreciate the crucial importance of the preparatory (inception), initiation (gestation), and early development (rapid growth) stages in building up the pattern of conduits that evolve into explorable cave systems. An understanding of the initiation mechanisms comes from modelling based on the combined consideration of equilibrium chemistry, dissolution kinetics, and flow dynamics (see Speleogenesis: Computer Models). Such studies have revealed the basic functional relationships between conduit growth and other variables. The stage of initiation of the original flow path, during which it conducts nearly saturated water, is slow, covering geologically lengthy timespans. A positive feedback loop operates between discharge and the growth rate, making the mechanism self-accelerating. Widening of an initial opening causes increased flow through it. Thus, dissolution rates increase along the entire flow path, and so on. The rapid enlargement stage begins when water can pass through the entire conduit while preserving considerable undersaturation. This represents the breakthrough event, resulting in a boost of the growth rate up to a certain limit (about $0.01-0.1$ cm a^{-1} in limestones), if hydrogeological settings permit increasing discharge. After breakthrough, conduit enlargement is almost independent of discharge. At typical hydraulic gradients the switch to rapid dissolution kinetics coincides approximately with the laminar flow–turbulent flow transition, and with the onset of sediment transport. This combined threshold, the most significant in speleogenesis, occurs within the growing conduit's aperture range from about 5–15 mm, and indicates the birth of a proper solution cave.

Understanding of cave pattern formation requires that the mechanisms of initiation and early development be viewed in a broad geological and hydrogeological context. The evolutionary typology of karst considers the entire life cycle of a soluble formation, from deposition (syngenetic karst) through deep burial and re-emergence (the group of intrastratal karst types: deep-seated, subjacent, entrenched, and denuded) to complete exposure. The "pure line" of exposed development is represented by the open karst type, which is karst evolved solely when soluble rock has been exposed at the surface, or developed in formations that have not experienced burial at all. Different types of karst, which concurrently represent the stages of karst development, are marked by characteristic associations of the structural prerequisites for groundwater flow and speleogenesis, flow regime, recharge mode and recharge/discharge configurations, groundwater chemistry, and degree of inheritance from earlier conditions. To generalize further consideration and help to emphasize the primary importance of the principal hydrogeological conditions, three major speleogenetic settings are distinguished: (1) coastal and oceanic; (2) deep-seated and confined; and (3) unconfined (see entries on each of these for more details)

Although coastal and oceanic settings are commonly characterized by unconfined circulation, they are treated separately because of the specific conditions for speleogenesis determined by the dissolution of porous, poorly indurated carbonates by mixing of waters of contrasting chemistry at the halocline. Spongework cave patterns are most common in these settings.

The distinctions between confined and unconfined settings influence speleogenesis in many ways. Most aggressive recharge to unconfined karst aquifers comes from the surface, whereas speleogenesis in confined settings relies on aggressiveness supplied to soluble rocks by recharge from adjacent, commonly underlying, formations. Carbonic acid dissolution predominates

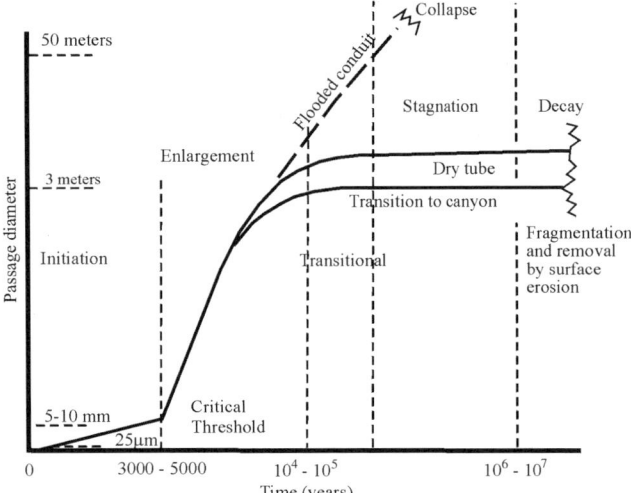

Speleogenesis: The evolutionary history of a single conduit (from White, 1988).

overwhelmingly in unconfined carbonate aquifers, whereas under confined settings speleogenesis occurs through a great variety of dissolutional mechanisms. In terms of the conduit initiation and early development model, differences in the hydraulic controls of growing conduits are of primary importance. Under unconfined phreatic conditions the resistance of conduits themselves, particularly of their narrowest parts, governs discharge through them. Discharge increases with the growth of the conduit at the expense of piracy from less favourable routes. Simultaneously the positive feedback loop between discharge and enlargement rate gives rise to strong competition and selective speleogenetic development. The result is that sparse branchwork cave patterns form most commonly in unconfined settings. By contrast, in confined aquifer systems several factors act to suppress the selective, competitive development of conduits at the early stage, but to favour formation of pervasive maze cave patterns where structural prerequisites are appropriate. Discharge through conduits in confined aquifer systems is constrained at certain levels by the hydraulic properties of adjacent feeding beds (inflow control) or confining beds (outflow control). Recharge that enters soluble rock from adjacent permeable but insoluble rock is commonly dispersed and available equally to many original openings. Last but not least, transverse circulation through a soluble unit sandwiched between other permeable rocks, a common relationship in sedimentary basins, imposes rather short flow distances that favour the growth of many openings at compatible rates.

Somewhat specialized speleogenetic conditions occur where hypogene fluids, commonly enriched in H_2S, flow upwards at sedimentary basin margins, to enter adjacent carbonate massifs and mix with oxygenated meteoric waters. The resultant sulfuric acid causes a burst of aggressiveness that generates complex 3D ramiform or network patterns. Details of cave morphologies depend on the structure of the host limestones, the extent (depth) of the mixing zone, the presence of the confining cover beds, and on whether hypogene water inflow is localized or dispersed. Replacement of calcite with gypsum and subsequent dissolution of the latter may also play a part. Somewhat similar patterns form where hypogene fluids, either juvenile or connate, are heated and enriched with CO_2. Aggressiveness is generated by the cooling of water rising from depth. Such mechanisms are commonly distinguished, respectively, as sulfuric acid speleogenesis and hydrothermal speleogenesis, or collectively as hypogenic speleogenesis.

Confined aquifers contain water at pressures greater than atmospheric, and the head lies above the base of the confining beds. In contrast, unconfined settings offer a variety of hydrodynamic zones (saturated zone below the water table, unsaturated zone above it, and the zone of intermittent saturation near the water table). Each zone displays distinct groundwater chemistry and flow regimes. Up to the 1960s there was much debate on whether caves form preferentially in the phreatic or vadose zones, or near the water table (see: Speleogenesis Theories: Post–1890). Modern approaches integrate a drainage basin wide analysis of geomorphic history and an analysis of optimum hydraulic path development in varying structural conditions. Much speleogenesis commences in the phreatic zone, although with lowering of the water table it continues in all zones, each of which leaves characteristic imprints on cave morphology. Varying recharge conditions, local structural peculiarities, and, especially, multi-phase geomorphic development may give rise to complex cave systems where branchwork patterns superimpose and interconnect, or combine with clusters of maze patterns.

Thus far only the initiation and enlargement of solution caves has been considered. When caves are decoupled from their formational settings, they become largely or entirely fossilized (see Paleokarst). This is a stage of stagnation, dominated by breakdown fragmentation and infilling by chemical and clastic sediments. The complete life history of a cave (see Figure) includes the decay or obliteration stage. This may be related to various geomorphic processes, such as erosional incision, lateral undercutting, and unroofing by overall denudational lowering of the land surface. The life of caves commonly spans a few million years, but may be much shorter (in intensively glaciated areas, in shallow settings, or in evaporite rocks) or far longer. In extreme cases caves have survived since Paleozoic or even late Proterozoic times.

ALEXANDER KLIMCHOUK

See also **Inception of Caves**

Further Reading

Bögli, A. 1980. *Karst Hydrology and Physical Speleology*, Berlin and New York: Springer (original German edition, 1978)
Gabrovšek, F. 2000. *The Evolution of Early Karst Aquifers: From Simple Principles to Complex Models*, Ljubljana: Založba, ZRC
Gillieson, D. 1996. *Caves: Processes, Development and Management*, Oxford and Cambridge, Massachusetts: Blackwells
Klimchouk, A.B., Ford, D.C., Palmer, A.N. & Dreybrodt, W. (editors) 2000. *Speleogenesis: Evolution of Karst Aquifers*, Huntsville, Alabama: National Speleological Society. Particularly the following papers:
Dublyansky, Y. Hydrothermal speleogenesis: Its settings and peculiar features.
Hill, C. Sulfuric acid hypogene karst in the Guadalupe Mountains.
Klimchouk, A.B. Speleogenesis under deep-seated and confined settings.
Klimchouk, A.B. & Ford, D.C. Types of karst and evolution of hydrogeological settings.
White, W.B. 1988. *Geomorphology and Hydrology of Karst Terrains*, Oxford and New York: Oxford University Press

SPELEOGENESIS THEORIES: EARLY

When the book of Genesis was regarded as a geological textbook, the context in which people thought about speleogenesis was very different to the accumulated knowledge that forms its background now. The age of the Earth was believed to be very young, therefore fast processes of cave formation had to be sought. Thus, it is not surprising that some of the earlier hypotheses took advantage of tectonics (to leave natural cavities), rocks that were still soft and so could be eroded rapidly, and the vast erosive power of the waters of the biblical Flood draining down to the Abyss beneath. Alternatives and variants of these included the suggestion that soluble rock salt had once occupied the places that became caves, and a logical argument that the bodies of animals drowned in the Flood and buried in the deposited mud must have decomposed, with the resulting gases forming cavities before the mud hardened.

Although these ideas were not then as foolish as they now appear, none of them was even at the start of the trail that has led to the still-developing knowledge of today. They form the protohistory of the subject. But there were elements of some of these rapid processes that play an important part in modern theories. The power of water in the Deluge drains was recognized, and the faults resulting from tectonic movements do provide planes of weakness in which caves can begin to develop.

Early modern ideas about speleogenesis developed in several stages. First came the recognition that the Earth was not so new after all, and so time was available for caves to be formed by the slow removal of solid material. At first, no distinction was made between erosion and solution, and erosion, being easier to understand, was introduced several centuries earlier. Then the process of solution became recognized as a separate agency, with its action eventually being attributed to an acid from the air that was later identified as carbon dioxide. For a long time all this activity was assumed to take place when it could be seen, in streams that were the underground equivalent of surface streams. These successive stages did not occur neatly as a steady progress from the simple to the advanced. They overlapped: some writers were unaware of others' work, and some continued to hold on to the more primitive ideas because they did not sufficiently examine the evidence or because they were under the influence of religious conviction.

The process of erosion was of course no different from the erosion involved in some of the Flood hypotheses; it was the recognition that it could be carried out by normal streams that constituted the advance. The realization that hard rock could be eroded in this way, at a rate sufficient to account for existing caves, did not occur until the 18th century. However, a long time before that there were intermediate ideas such as removal of soft inclusions in the rock by erosion and the action of water as a modifying agent after a tectonic origin. John Hutton, in 1780, suggested another such intermediate idea, that streams of present-day size had formed cave passages by eroding the rock while it was still new and hence relatively soft.

The idea of limestone being gradually dissolved away by solution rather than erosion in water was mainly a 19th-century concept arising from the more detailed and more precise examination of caves at that time. In a few cases the water was supposed to be hot or fiercely acidic; in others, before the presence of carbon dioxide in rain-water was realized, it was thought to be ordinary pure water.

The concept of solution in especially aggressive water persisted in various forms throughout the 19th century, but it had no direct effect on the subsequent understanding of normal speleogenesis. What it did do was to emphasize to later workers the fact that solution in an acid was often the best or the only way of explaining certain characteristics of caves.

A real understanding of this was not possible until the existence of phlogiston was disproved and oxygen identified in the 1780s as an essential element in combustion, by which carbon dioxide was formed. Observation might, and did, suggest solution before that time, and indeed the earlier concepts of unspecified aerial acid did allow some approach towards understanding the chemistry involved.

At first there were some who accepted the fact that limestone was dissolved in carbonic acid to the extent necessary to account for stalactites, but could not bring themselves to believe that this effect could wholly explain the formation of large cavities and caves.

In 1830 the importance of dissolved carbon dioxide in rendering the cave water acidic was properly appreciated by Charles Lyell and by [C.]E. Thirria and soon afterwards the function of joints as the initiators of the solution process was recognized. At first solution was thought of as occurring in the vadose zone, i.e. where trickling and flowing water could be actually seen; it was not until 1870 that phreatic solution was suggested and only in the 1890s and afterwards was it advocated as a major factor in speleogenesis. The 1870 recognition that much cave

Speleogenesis Theories, Early: Figure 1. Dr Franklen Evans, who in 1870 was the first to state that some caves are formed in the phreatic zone.

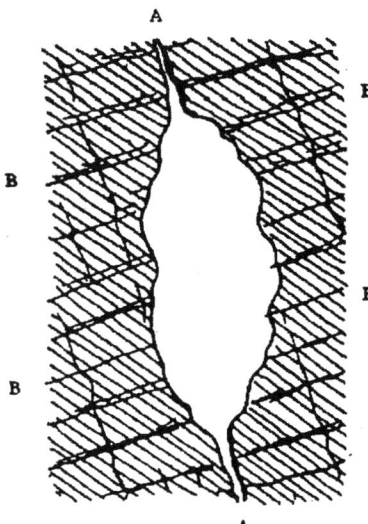

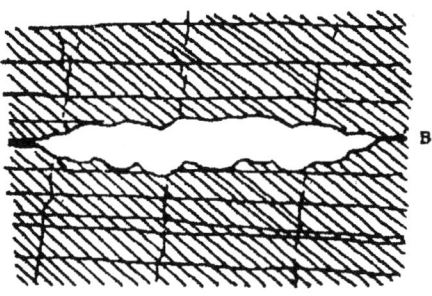

Speleogenesis Theories, Early: Figure 2.
Diagrams by Dupont in 1894 showing typical cross sections of joint- and bedding-determined cave passages formed by solution when completely full of water. AA is a joint; BB are bedding planes.

formation occurred below the water table was made in a little-known paper by Franklen Evans, a medical doctor in Wales (Figure 1).

By the turn of the 20th century therefore, three main theories of speleogenesis existed, each of them correct in certain circumstances. Armand Flamache maintained that erosion alone was sufficient to form caves; E.A. Martel and others believed that a combination of both vadose solution and erosion provided the complete explanation; while Edouard Dupont (Figure 2) and Alfred Grund considered that phreatic solution was the principal factor. Each of these schools maintained its own point of view as if it were applicable in all cases and hence two major arguments arose, one between the advocates of erosion and solution and the other between the vadose and the phreatic solutionists. Gradually, very gradually in some cases, both disagreements were resolved by the realization that the three processes were all valid and that they operated in combination, either together or successively, in varying proportions according the local circumstances.

The disparity of viewpoints had been due largely to the varied experience of their exponents and the different regions with which they were familiar.

TREVOR SHAW

Further Reading

Shaw, T.R. 1992. *History of Cave Science: The Exploration and Study of Limestone Caves, to 1900*, 2nd edition, Broadway, New South Wales: Sydney Speleological Society
This gives much more information, with references to the original sources.

Shaw, T.R., 2000 Views on cave formation before 1900. In *Speleogenesis: Evolution of Karst Aquifers*, edited by A.B. Klimchouk, D.C. Ford, A.N. Palmer & W. Dreybrodt, Huntsville, Alabama: National Speleological Society
A fuller treatment than in this entry, with references.

SPELEOGENESIS THEORIES: POST–1890

Recognition of noteworthy speleogenesis (cave development) theories is inevitably time-related and subjective. Seemingly insignificant factors may become crucial as keys to broader theories, and ideas that appeared misguided against the groundswell of contemporary science may later gain credibility. Among a plethora of ideas, some ludicrous or bizarre, many strands of later speleogenetic thinking existed before 1900. Major theories that matured in the early 20th century grew from the more reasonable of the earlier ideas.

Towards the end of the 19th century geomorphological investigation of surface karst led to increased speculation about sub-surface karst and underground water. Cvijić (1893) discussed surface development in the Dinaric karst and later (1918) linked karst landforms to sub-surface hydrology. He also described dry, transitional and saturated hydrographic zones (now better known as vadose, shallow phreatic, and deep phreatic zones), fuelling much subsequent controversy. Cvijić considered the transitional zone to be the major speleogenetic focus, with groundwater movement concentrated at the water table.

Early in the 20th century two pre-eminent geomorphologists, A. Penck and W.M. Davis, studied the Classical and Dinaric karsts. Prejudiced by their near-patriarchal assertions, two conflicting models began to evolve, typified in the work of Grund (1903) and Katzer (1909). Grund believed that karst aquifers divide into two broad zones, a lower zone saturated by stagnant water, and a higher zone where water circulates in open caves above a regional water table. In contrast, Katzer denied the existence of a saturated zone and a water table, but assumed all karst water follows open cavities from surface sinks to topographically lower springs. He considered caves to be mutually independent, because the intervening rock is virtually impermeable. From these early ideas, three outwardly conflicting general theories eventually emerged, identifying the speleogenetic focus as the deep phreatic zone, the shallow phreatic zone, or the vadose zone.

Davis himself was the main advocate of deep phreatic development, publishing "The origin of limestone caverns" in 1930, when he was 80 years old. Following Grund's basic theme, he considered that speleogenesis begins when freshwater replaces primitive brines trapped during sedimentation and retained during limestone diagenesis. He visualized slow dissolutional initiation of groundwater routes, possibly at great depth, and distinguished between this and subsequent void enlargement. Emphasizing the potentially vast timescales of cavern formation, he predicted deep-seated processes lasting tens or hundreds of millions of years, with shallower but still sub-water-table growth progressing more quickly. Flooded passages originating at depth are eventually drained following relative uplift. This two-cycle model explains caves that must have originated below a water table but which clearly occupy levels well above contemporary water tables.

By implication, ideas in Lehmann's (1932) publication "Die Hydrographie des Karstes" may also relate to deep phreatic speleogenesis. This was the only European contribution to the contemporary debate to gain serious recognition, but was overlooked or ignored by most North American and British researchers. Basing his reasoning on the then current understanding of fluid mechanics, Lehmann believed speleogenesis would not commence unless "initial cavities" (*Urhohlräume*) at least 2 mm wide existed within the rock. A minimal primary permeability (*hydrographische Wegsamkeit*) was also necessary, as water ingress and egress to and from the rock are required to initiate flow. Both aspects now appear to relate most closely to the phreatic, particularly deep phreatic, processes predicted by Grund and Davis. However, Lehmann's view that karst cavities are mutually independent and unrelated, has more in common with the views of Katzer.

Extensive field observations allowed Bretz (1942) to classify cave features as diagnostic of passage growth either in the phreatic or vadose zones. He supported and enhanced the broad Davisian model, postulating a third phase between Davis's phreatic and vadose cycles. He believed that, during this intermediate phase, flooded voids below the water table are progressively filled by red clay.

Early shallow phreatic theories were typified by Swinnerton's (1932) ideas, which re-emphasized Cvijić's threefold hydrographic zone division. Swinnerton acknowledged limited deep dissolution, but believed that most dissolutional growth takes place within a fluctuating shallow phreatic or water-table zone. Subsequently, following uplift and drainage of formerly flooded passages, cave modification and growth continue in the vadose zone.

Ideas presented by Gardner (1935) are commonly quoted as exemplifying vadose speleogenesis theories, but actually include aspects relating to early sub-water-table development. Gardner described dominantly down-dip passage formation commencing in relatively thin "carrier beds" or "aquifers", which somehow favour cave development. He suggested that, although static groundwater exists at depth, surface valley incision (Figure 1) is required to initiate gravitational down-dip flow and concomitant dissolution along intersected carrier beds. Continued downcutting eventually intersects lower carrier beds, leaving upper levels abandoned as caves forming along lower beds capture their drainage. Whereas Gardner's theory undeniably described vadose modification within stratigraphically determined zones, pre-drainage dissolution along the carrier beds must, by definition, have been a phreatic process.

Another aspect of vadose cave development was introduced by Malott (1937), within an abstract describing his "invasion theory of cavern development". He suggested that after uplift and water-table lowering, percolating meteoric water gathers in pre-existing primitive voids within the rock and channels downwards towards the water table, selectively enlarging preferred routes. As with Gardner's ideas, Malott's abstract related mainly to vadose modification, and the origin of the primitive phreatic cavities that guided the process was barely discussed.

From these brief descriptions it is clear that overlap existed between theories that seemed to champion either deep phreatic, shallow phreatic, or vadose-dominated speleogenesis. Considering the heat supposedly generated by this controversy, it is interesting to revisit a statement made by Martel in 1896, long pre-dating the theories:

> No theory about the origin of caves is universal: those that have been put forward have generally claimed to be too inclusive; almost all of them are partially correct; the whole truth lies sometimes in their combination, sometimes in the application of a particular one in a particular case. (Martel, 1896, pp.53–54).

With hindsight, each major school of thought accommodated aspects of cave origin or modification in all three hydrographic situations. Rather than the viability of each aspect being in dispute, it was simply a question of which was dominant. As pointed out by Martel, this would actually vary temporally and locally.

Partial corroboration of Martel's vision was provided by Rhoades and Sinacori (1941), who published a mainly mathematical consideration of underground flow. By assuming that water behaviour in limestone terrains fits idealized flow patterns deduced for isotropic aquifers, they demonstrated that both deep phreatic and shallow phreatic processes do operate, and can co-exist at the same locality. Significant deep dissolution occurs during the early drainage of an uplifted rock mass towards a specific discharge point. Over time voids near the discharge point enlarge preferentially and a master conduit cuts back headwards, following the local water table. Eventually, shallow processes dominate over deep-seated alternatives (Figure 2).

During the next 20 years or so interest turned away from hydrographic zones, towards more detailed considerations of underlying processes and mechanisms. No further generic works of major stature appeared, yet many specific contributions were published, mostly reflecting an overall improved knowledge of actual caves, or of chemical and hydrological factors involved in their development. Without doubt, as in the past, valuable ideas were published and overlooked in languages unfamiliar to the leading scientists, whatever their nationality, and effort was duplicated in different countries. This is nowhere better exemplified than by Laptev's (1939) Russian publication on mixing water dissolution, which was overlooked in the West and later revisited independently by Bögli.

Some other significant contributions pre-dating the next major advance must at least be mentioned. Howard (1964) questioned how meteoric water could remain aggressive and capable of dissolution when, theoretically, it should be neutralized within 10 mm of entering the rock. His conclusion that acid must be created within the rock if early voids are to form has only relatively recently found wider favour. Also in 1964, Bögli

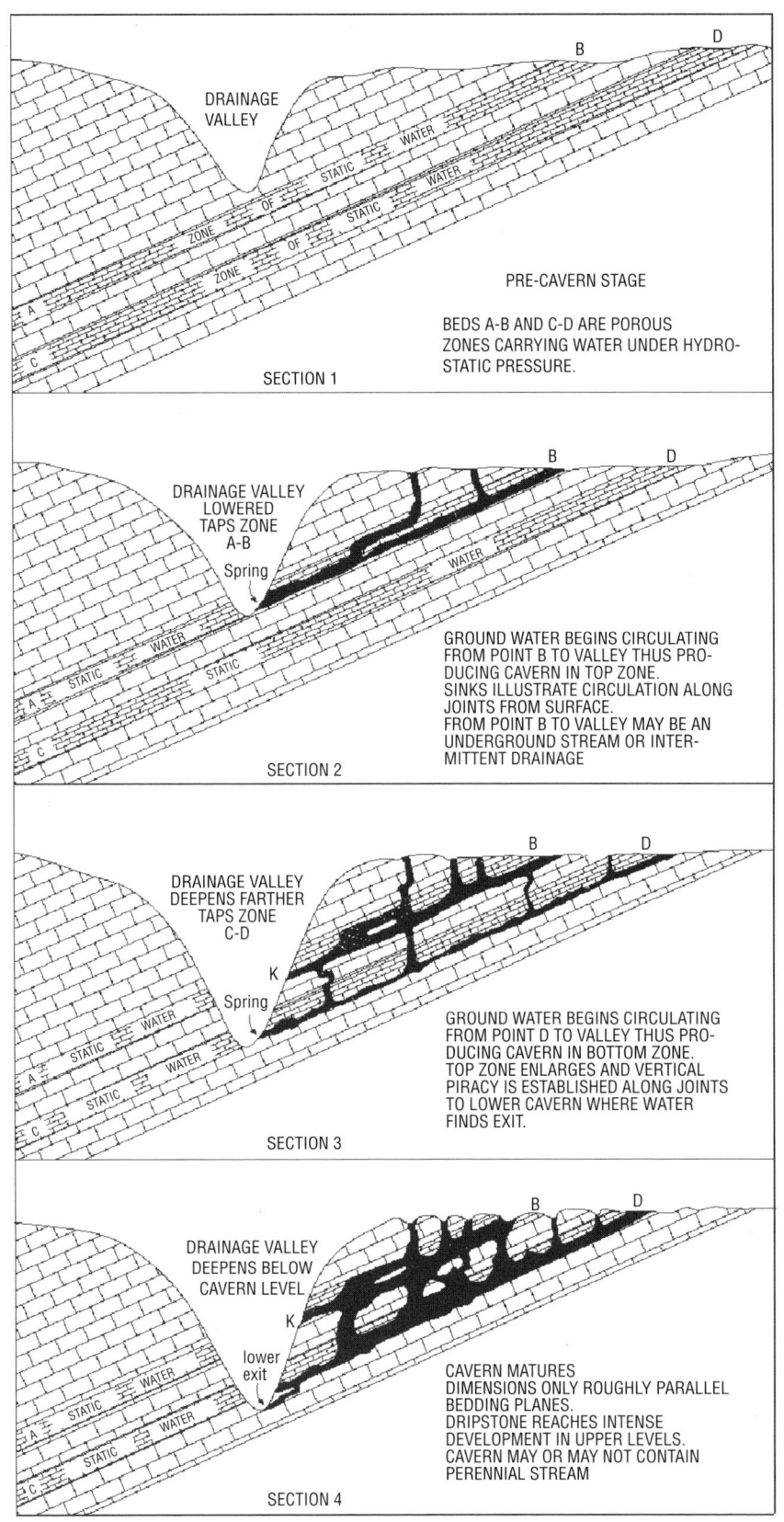

Speleogenesis Theories: Post–1890: Figure 1. Sketch sections illustrating the origin and development of a large limestone cavern (redrawn after Gardner, 1935, from Lowe, 2000).

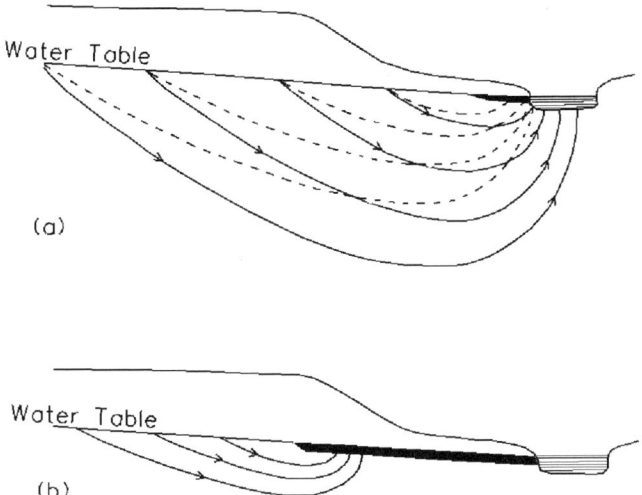

Speleogenesis Theories: Post–1890: Figure 2. Groundwater flow development stages in limestone according to Rhoades and Sinacori. Broken lines in (a) indicate the changes produced by backward growth of a water table cave from the discharge point. Diagram (b) shows the theoretical effect after significant conduit growth (redrawn after Rhoades & Sinacori, 1941, from Lowe, 2000).

described the possibility of "mixture corrosion", whereby two saturated waters can combine and become capable of additional dissolution. This has considerable impact upon cave modification in a variety of contexts. Ford (1965) introduced a new cave development model, which eventually evolved into the Ford-Ewers model, discussed below. In 1966 S.N. Davis questioned how and why water could move along rock joints that are barely more permeable than the surrounding rock. Concluding that mechanisms other than gravity must be involved, he suggested that "groundwater pumping" related dominantly to earth tides was implicated. In 1975 Palmer published "The origin of maze caves", a seminal work that received and has retained widespread acceptance. All of these and other, arguably equally pivotal, ideas from authors including S.A. Durov, C.A. Kaye, P.K. Weyl, J.V. Thrailkill, O. Langmuir, J.N. Jennings, and A.C. Waltham, all contributing to knowledge of aspects of cave development, are traceable via the sources of further reading listed below.

In the context of the earlier controversy, White and Longyear (1962) suggested that "The multitudinous theories . . . are neither correct nor incorrect in the general case, they are irrelevant . . .". This was quoted by Ford (1965) when presenting his new cave origin model, based largely upon study in the Mendip Hills. Ford's ideas, with a strengthened geological component, grew into the more widely applicable Ford-Ewers model (1978), which negated earlier controversy. The possibilities and explanations presented within the model were accepted widely, and White (1988) asked ". . . 'Do caves form above, at or below the water table?' To all of these possibilities, the Ford-Ewers model answers, 'yes!'."

The Ford-Ewers model, sometimes referred to as the fissure frequency hypothesis or the four-state model, forms the common ground of most current international speleogenetic thinking. Its applicability is described in various modern texts (see Further Reading) and details need not be examined here. Broadly, however, explanations for cave frameworks, including deep phreatic loops, sub-horizontal vadose streamways alternating with water-filled tubes, and deep shaft systems, are related to the availability of fissures (including bedding plane discontinuities) within the rock mass.

Since the appearance of the Ford-Ewers model, views of some aspects of cave development have changed, and advances have been made in various branches of speleogenetic study. Some of these are discussed further in the entries on Inception of Caves; Speleogenesis; Speleogenesis: Coastal and Oceanic Settings; Speleogenesis: Computer Models; Speleogenesis: Deep-seated and Confined Settings; and Speleogenesis: Unconfined Settings.

DAVID J. LOWE

See also **Karst Hydrology: History**

Works Cited

Bögli, A. 1964. Mischungskorrosion—ein Beiträg zur Verkarstungsproblem [Mixture corrosion—a contribution to the karst problem]. *Erdkunde*, 18: 83–92

Bretz, J.H. 1942. Vadose and phreatic features of limestone caves. *Journal of Geology*, 50: 675–811

Cvijić, J. 1893. Das Karstphänomen [The karst phenomenon]. *Geographische Abhandlung*, 5(3): 218–329

Cvijić, J. 1918. Hydrographie souterraine et évolution morphologique du Karst [Underground hydrology and morphological evolution of karst]. *Recueil des Travaux de l'Institute de Géographie Alpine*, 4(4): 375–426

Davis, S.N. 1966. Initiation of ground-water flow in jointed limestone. *National Speleological Society Bulletin*, 28: 111–18

Davis, W.M. 1930. Origin of limestone caverns. *Geological Society of America Bulletin*, 41: 475–628

Ford, D.C. 1965. The origin of limestone caverns: A model from the central Mendip Hills, England. *Bulletin of the National Speleological Society of America*, 27: 109–32

Ford, D.C. & Ewers, R.O. 1978. The development of limestone cave systems in the dimensions of length and depth. *Canadian Journal of Earth Sciences*, 15: 1783–98

Gardner, J.H. 1935. Origin and development of limestone caverns. *Bulletin of the Geological Society of America*, 46: 1255–74

Grund, A. 1903. Die Karsthydrographie: Studien aus Westbosnien [Karst hydrology: Studies from West Bosnia]. *Geographische Abhandlungen*, 7: 103–200

Howard, A.D. 1964. Processes of limestone cave development. *International Journal of Speleology*, 1: 47–60

Katzer, F. 1909. *Karst und Karsthydrographie. Zur Kunde der Balkanhalbinsel* [Karst and Karst Hydrology. For Understanding the Balkan Peninsula], Sarajevo: Kajon

Laptev, F.F. 1939. Aggressive action of water on carbonate rocks, gypsum and concrete. *Trudy Spetgeo*, vol. 1, Moscow and Leningrad: GONTI

Lehmann, O. 1932. Die Hydrographie des Karstes [Karst Hydrology]. In *Enzyklopädie der Erdkunde* [Encyclopedia of Geography], edited by O. Kende, Leipzig and Vienna: Franz Deuticke

Malott, C.A. 1937. Invasion theory of cavern development [abstract]. *Proceedings of the Geological Society of America* for 1937, p.323

Martel, E.A. 1896. Applications géologiques de la spéléologie. *Annales des Mines*, Paris, sér- 9, 10: 53–54

Palmer, A.N. 1975. The origin of maze caves. *National Speleological Society Bulletin*, 37: 57–76

Rhoades, R. & Sinacori, N.M. 1941. Patterns of groundwater flow and solution. *Journal of Geology*, 49: 785–94

Swinnerton, A.C. 1932. Origin of limestone caverns. *Geological Society of America Bulletin*, 43: 662–93
White, W.B. 1988. *Geomorphology and Hydrology of Karst Terrains*, Oxford and New York: Oxford University Press
White, W.B. & Longyear, J. 1962. Some limitations on speleogenetic speculation imposed by the hydraulics of water flow in limestone. *National Speleological Society, Nittany Grotto Newsletter*, 10: 155–67

Further Reading

Bögli, A. 1980. *Karst Hydrology and Physical Speleology*, Berlin and New York: Springer (original German edition, 1978)
Ford, D.C. & Williams, P.W. 1989. *Karst Geomorphology and Hydrology*, London and Boston: Unwin Hyman
Lowe, D.J. 1992. A historical review of concepts of speleogenesis. *Cave Science*, 19(3): 63–90
Lowe, D.J. 2000. Development of speleogenetic ideas in the 20th century: The early modern approach. In *Speleogenesis: Evolution of Karst Aquifers*, edited by A.B. Klimchouk, D.C. Ford, A.N. Palmer & W. Dreybrodt, Huntsville, Alabama: National Speleological Society
Palmer, A.N. 1991. Origin and morphology of limestone caves. *Geological Society of America Bulletin*, 103: 1–21
Shaw, T.R. 1992. *History of Cave Science: The Exploration and Study of Limestone Caves, to 1900*, 2nd edition, Broadway, New South Wales: Sydney Speleological Society
Shaw, T.R. 2000. Views on cave formation before 1900. In *Speleogenesis: Evolution of Karst Aquifers*, edited by A.B. Klimchouk, D.C. Ford, A.N. Palmer & W. Dreybrodt, Huntsville, Alabama: National Speleological Society
Watson, R.A. & White, W.B. 1985. The history of American theories of cave origin. In *Geologists and Ideas: A History of North American Geology*, edited by E.T. Drake & W.M. Jordan, Boulder, Colorado: Geological Society of America (Geological Society of America, Centennial Special vol. 1)
White, W.B. 2000. Development of speleogenetic ideas in the 20th century: The modern period, 1957 to the present. In *Speleogenesis: Evolution of Karst Aquifers*, edited by A.B. Klimchouk, D.C. Ford, A.N. Palmer & W. Dreybrodt, Huntsville, Alabama: National Speleological Society

SPELEOGENESIS: COASTAL AND OCEANIC SETTINGS

Cave development, or speleogenesis, in coastal or oceanic settings is very different from that found in the interior of continents. These differences depend on three main factors that are a direct result of the proximity to the coast: the soluble rocks are young; sea-water and freshwater mixing occurs; and sea-level change has occurred. First, in most parts of the world, the coastal limestones and dolomites (hereafter collectively called "carbonates") are geologically young, commonly less than 25 million years old. They have not been transported far from their place of origin, and have not undergone the alteration or diagenesis caused by deep burial over long periods of time. As a result, they are not as crystallized and dense as the ancient carbonates found in continental interiors. This type of youthful, porous carbonate is called "eogenetic". Second, these young carbonates contain a freshwater lens, the result of denser sea water intruding from the coast beneath the less-dense fresh water (Figure 1A). The term "lens" refers to the shape of this freshwater body, which is thin at the coastline but thickens inland. Within the island, where the freshwater body continues to the coast on the far side, it thins again, such that, when seen in cross section, the water body takes the shape of a lens (Vacher, 1988). Where sea water and fresh water come into contact within the carbonate rock, their chemistry becomes mixed and they are able to dissolve the rock, even though each water body separately might have been fully saturated with carbonate (Back et al., 1986). Third, the coastal location means that the freshwater lens will migrate if sea level changes, as the lens is floating on the sea water. If sea level changes, and the freshwater lens changes position as a result, then the site of carbonate dissolution and speleogenesis also changes. All carbonate islands and coastlines around the world have experienced sea-level change during the Pleistocene, caused by changing ice volumes on the continents, called "glacio-eustatic" sea-level changes. Some islands, such as Guam (Mylroie et al., 2001) or Isla de Mona (Frank et al., 1998), have also experienced tectonic movement, which causes "local" sea-level change, in that it does not affect coasts elsewhere. Such local changes overprint the record of glacio-eustatic sea-level change for these islands. The synthesis of the characteristic features of island karst into a coherent conceptual framework has been called the Carbonate Island Karst Model (Mylroie et al., 2001).

Karst development can also be influenced by the nature of the carbonate rocks. There are many depositional environments for carbonates, some of which can be spatially very close together, such as reef, lagoon, beach, and dune settings. Because the rock is young (or eogenetic) and has not been recrystallized and altered, the initial differences in the type of carbonate deposited are much more obvious than in older rocks. These differences can influence water flow and therefore dissolution rates and pathways.

Carbonate islands create a speleogenetic environment which differs from that which occurs along the carbonate coastline of a continent. For example, small carbonate islands like Bermuda are very different from large coastlines such as the Yucatán of Mexico. On an island, all the water present must be derived from meteoric precipitation falling on the carbonate rocks of the island; this recharge is called autogenic water. There is no water inflow from adjacent areas, as can occur where the continental interior collects water and discharges it through carbonate rocks along the coast (known as allogenic recharge). For small islands especially, the amount of water available can be very limited.

The simplest way to show the unique cave development found in carbonate islands and coasts is to look at a small island made entirely of carbonate rocks. Carbonate rock dissolves in the lens, preferentially in two locations. Water from rainfall enters the ground from and moves downward through the vadose zone to the top of the freshwater lens at the water table. At that point, the vadose water mixes with the phreatic water at the top of the lens. Even though both waters might be saturated with respect to carbonate, when they mix an undersaturated solution may result, and the water becomes capable of dissolving more rock (Bögli, 1980). In small islands, single- or multiple-chamber

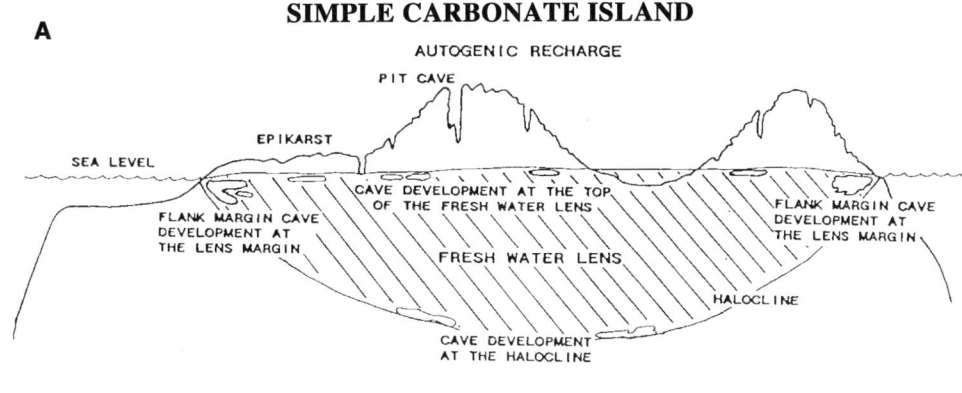
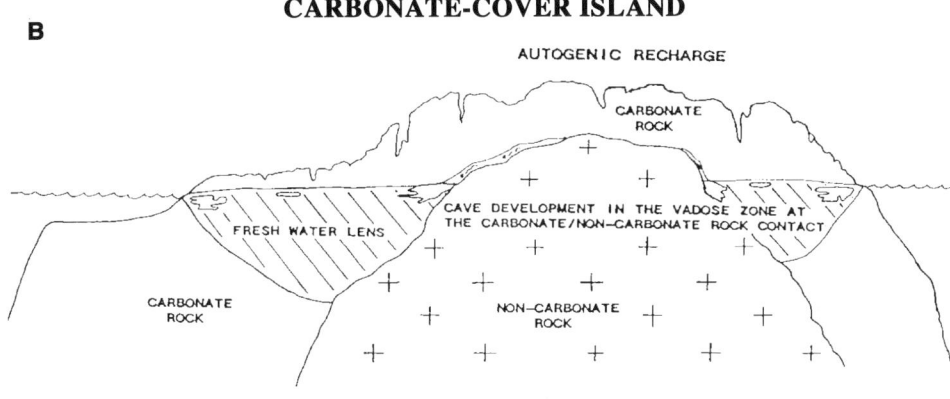
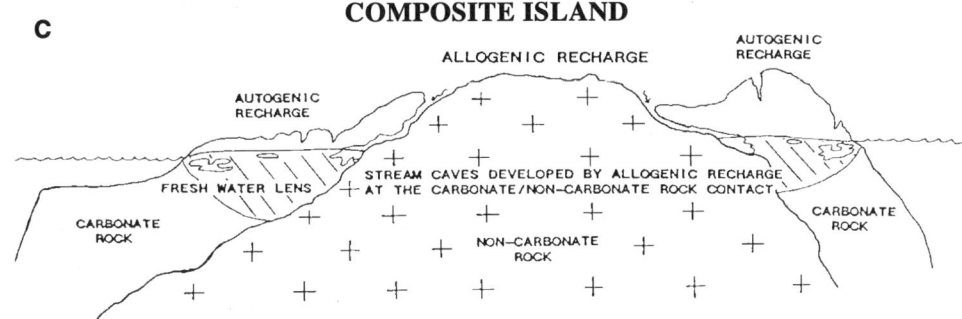

Speleogenesis: Coastal and Oceanic Settings: Figure 1. Sea-level and basement relationships for carbonate islands (with some vertical exaggeration). (A) The simple carbonate island, with no noncarbonate rock within the region of the freshwater lens. (B) The carbonate-cover island, where noncarbonate rock at depth can deflect vadose flow and distort the freshwater lens. (C) The composite island, where noncarbonate rock influences both surface and subsurface flow.

caves may develop at the top of the lens (Figure 1). These caves tend to be low and wide, and of an oval to suboval plan. In islands with low relief, such as the Bahamas, these voids will have thin roofs and will be prone to collapse. Such collapse voids are called "banana holes", and can occur in densities of up to 3000 km^{-2} (Mylroie, Carew & Vacher, 1995). At the base of the lens, fresh water is in contact with sea water, and as noted earlier, when these waters mix, the resulting combination is capable of dissolving more rock. Small caves may develop here as well. But in both cases, at the top and bottom of the lens, the caves formed are not true conduits, but only mixing chambers. They contain no flow markings indicating turbulent flow or other features found in stream caves. In addition, both the top of the lens and the base of the lens constitute a density interface. Organic material may collect at such interfaces. When the organic material decomposes and oxidizes, it creates extra carbon dioxide to help drive more dissolution of the rock. If the organic material is abundant, then the system may use up all the oxygen to become anoxic. Under such conditions, materials like hydrogen sulfide and other acids may form, greatly increasing the dissolution of the rock (Bottrell et al., 1991).

At the margin or edge of the lens, where it is close to its discharge point to the sea, the two favourable mixing zones for rock dissolution are superimposed on each other (Figure 1). The lens is also decreasing in cross-sectional area at this point, so the water discharge is faster as it is forced through a smaller area. Reactants come together quickly, and the products are removed quickly. All these factors are cumulative, and lead to the production of a special type of cave called a flank-margin cave (Mylroie, Carew & Vacher, 1995). The term "margin" refers to the position of the cave at the margin of the lens, and the term "flank" refers to the cave's position under the flank of the high ground that forms the island. These caves are the largest of those found on small carbonate islands. They are also not true conduits, but rather are mixing chambers sitting in the diffuse flow path at the edge of the freshwater lens. Flank-margin caves can be very large: some with over 1 km of surveyed passages are known from the Bahamas, and on Isla de Mona over 19 km of passages have been mapped for a single cave. These numbers are a little misleading, in that all these caves tend to be a collection of oval and suboval chambers and short, straight passage segments, in a very maze-like pattern (Figure 2; see also survey in Mona entry).

In addition, carbonate islands can be subdivided into three categories (Figure 1) based on the relationship between the sea level and any carbonate / noncarbonate contact. Many carbonate islands are built on volcanic pedestals, over which the carbonate rock has been deposited in layers of variable thickness. Simple carbonate islands (Figure 1A) have no noncarbonate rocks exposed at the surface or inside the island within the range of glacio-eustatic sea-level change. Islands with a carbonate cover (Figure 1B) have noncarbonate rocks beneath a carbonate veneer, and the contact between them is within the position of the freshwater lens for all or part of a glacio-eustatic cycle. Composite islands (Figure 1C) contain carbonate and noncarbonate rocks exposed on the surface.

On islands with a carbonate cover, water descending through the vadose zone is shunted along the carbonate / noncarbonate contact, producing stream caves carrying water to the freshwater lens. In composite islands, this process is augmented by the development of sinks and insurgences at the surface expression of the carbonate / noncarbonate contact. The noncarbonate outcrop acts as a large catchment funnelling dissolutionally aggressive allogenic water on to the adjacent carbonate rocks. In the phreatic zone of carbonate cover and composite islands, the lens can be subdivided into the "basal zone", where the base of the fresh water forms the transition zone to the underlying sea water, and the "parabasal zone", where the base of the fresh water rests on noncarbonate rock (Mink & Vacher, 1997).

The surface of carbonate islands contain a characteristic epikarst, with unusual morphology as a result of the youthful age of the carbonates and the pervasive presence of salt spray. In the absence of allogenic catchments on adjacent noncarbonate terrain, sinking streams, blind valleys, and springs are rarely found. Closed depressions are common, but many represent tectonic features, or constructional features produced by deposition, such as fossil lagoons or swales between dune deposits. These depressions, while internally drained by dissolution pathways, were not created by dissolutional excavation. Vadose flow along the contact between the carbonates and noncarbonates, however, can undercut the overlying carbonates, producing large collapse voids that may prograde to the surface, as observed on Bermuda (Mylroie, Carew & Vacher, 1995) and Guam (Mylroie et al., 2001).

In many islands, caves have developed during the Pleistocene in Pleistocene-aged carbonate rocks. In some cases, it appears that the cave development occurred almost simultaneously with the deposition of the rocks; these caves are called "syngenetic caves" (Jennings, 1968; see also Syngenetic Karst). As an example, in the Bahamas, carbonates are deposited when glaciers melt and sea level rises on to the flat island platforms. Carbonate sediment production begins, and the sediments are washed up into beaches and then blown up into dunes and form rocks

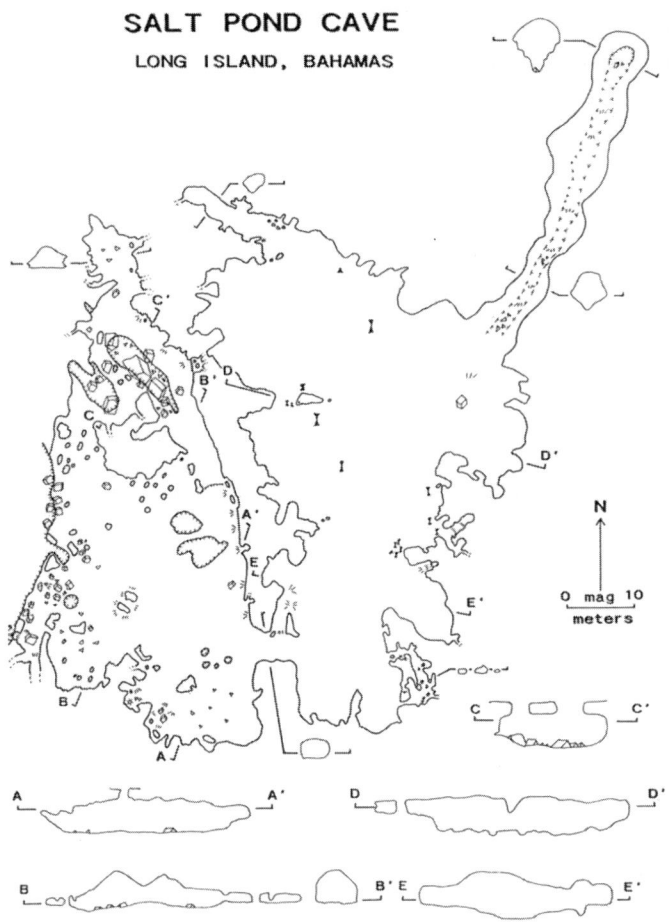

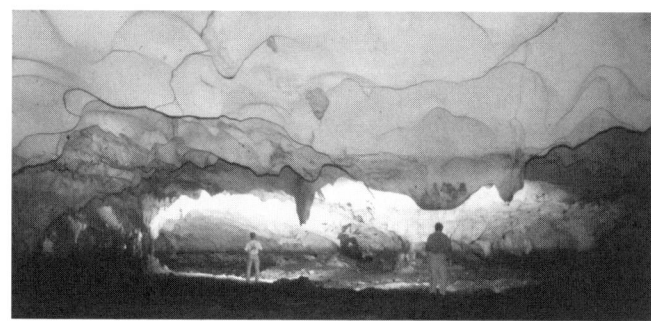

Speleogenesis: Coastal and Oceanic Settings: Figure 2. Salt Pond Cave, Long Island, Bahamas. (A) Map. (B) Photograph of the main chamber, looking north. Note the wide, low aspect of the chamber, and the dissolution surfaces on the floor and ceiling. The cave was mined for guano in the 1800s and the dark stain line represents the original guano level. (Photo by John Mylroie)

called "eolian calcarenites" or "eolianites". Sea level commonly continues to rise in this situation, a freshwater lens is created within the eolianites, and flank-margin cave development begins. If sea level falls, then the caves are drained. Throughout the Bahamas, there are numerous flank-margin caves that formed during the last interglacial 131 000 to 119 000 years ago, when the glaciers melted back a little more than in the present interglacial and sea level was 6 m higher. Some of these caves are formed in rocks that were deposited earlier in the Pleistocene (Mylroie, Carew & Vacher, 1995), and some are in rocks deposited early in the same sea-level event that then later continued to rise and produce the caves.

These caves can have chambers that are 14 000 m³ in volume, and 1 km of passages. They all formed in a small time window of 10 000 to 12 000 years, when sea-level was higher than at present, during the last interglacial (131 000 to 119 000 years ago). The extremely large caves on Isla de Mona are as much as 2 million years old (Panuska et al., 1998). Their large size is a result of their development prior to the glaciations of the Pleistocene, such that sea level stayed constant for a long time and mixing dissolution was able to excavate very large caves. Exploration by submarines in the Bahamas has shown that flank-margin caves also developed when sea level was lower during glaciations (Mylroie, Carew & Vacher, 1995).

When islands become large, or when there are extensive limestone coastlines, then the freshwater lens will change from diffuse flow to conduit flow. As an island becomes larger (as when sea level falls during glaciation) an ever-increasing amount of water must go through a proportionally smaller perimeter, and under these conditions, conduit flow begins. The lens is distorted by conduit flow, as it can now discharge laterally to conduits. Flank-margin cave development is inhibited as a result. In the Bahamas, large conduit systems are found at depths that equate to a sea level that would expose the shallow banks, dramatically increasing the size of the island. At today's sea levels, the islands are much smaller and only flank-margin caves are forming.

JOHN MYLROIE

See also **Blue Holes of Bahamas; Caribbean Islands; Littoral Caves; Mona, Puerto Rico**

Works Cited

Back, W., Hanshaw, B.B., Herman, J.S. & Van Driel, J.N. 1986. Differential dissolution of a Pleistocene reef in the ground-water mixing zone of coastal Yucatan, Mexico. *Geology*, 14: 137–40

Bögli, A. 1980. *Karst Hydrology and Physical Speleology*, Berlin and New York: Springer (original German edition, 1978)

Bottrell, S.H., Smart, P.L., Whitaker, F. & Raiswell, R. 1991. Geochemistry and isotope systematics of sulphur in the mixing zone of Bahamian blue holes. *Applied Geochemistry*, 6: 97–103

Frank, E.F., Mylroie, J., Troester, J., Alexander, E.C. & Carew, J.L. 1998. Karst development and speleologenesis, Isla de Mona, Puerto Rico. *Journal of Cave and Karst Studies*, 60(2): 73–83

Jennings, J.N. 1968. Syngenetic karst in Australia. In *Contributions to the Study of Karst*, edited by P.W. Williams & J.N. Jennings, Canberra, Australia, Australian National University Research School of Pacific Studies

Mink, J.F. & Vacher, H.L. 1997. Hydrogeology of northern Guam. In *Geology and Hydrogeology of Carbonate Islands*, edited by H.L. Vacher & T.M. Quinn, Amsterdam and New York: Elsevier

Mylroie, J.E., Carew, J.L. & Vacher, H.L. 1995. Karst development in the Bahamas and Bermuda. In *Terrestrial and Shallow Marine Geology of the Bahamas and Bermuda*, edited by H.A. Curran & B. White, Boulder, Colorado: Geological Society of America

Mylroie, J.E., Jenson, J.W., Taborosi, D., Jocson, J.M.U., Vann, D.T. & Wexel, C. 2001. Karst features of Guam in terms of a general model of carbonate island karst. *Journal of Cave and Karst Studies*, 63(1): 9–22

Panuska, B.C., Mylroie, J.M., Armentrout, D. & McFarlane, D. 1998. Magnetostratigraphy of Cueva del Aleman, Isla de Mona, Puerto Rico and the species duration of Audobon's Shearwater. *Journal of Cave and Karst Studies*, 60(2): 96–100

Vacher, H.L. 1988. Dupuit–Ghyben–Herzberg analysis of strip-island lenses. *Geological Society of America Bulletin*, 100: 580–91

Further Reading

Carbonates and Evaporites 10(2), 1995 is a special issue devoted to the 1995 Paleokarst Field Conference held on San Salvador Island, Bahamas in February 1995. It contains a number of papers dealing with island and coastal karst

James, N.P. & Choquette, P.W. (editors) 1988. *Paleokarst*, New York: Springer

Journal of Cave and Karst Studies 60(2), 1998 is a special issue devoted to caves on Isla de Mona, Puerto Rico

Palmer, R.J. 1989. *Deep Into Blue Holes: The Story of the Andros Project*, London: Unwin Hyman

Purdy, E.G. & Bertram, G.T. 1993. *Carbonate Concepts from the Maldives, Indian Ocean*, Tulsa, Oklahoma: American Association of Petroleum Geologists

Vacher, H.L. & Quinn, T.M. (editors) 1997. *Geology and Hydrogeology of Carbonate Islands*, Amsterdam and New York: Elsevier

SPELEOGENESIS: COMPUTER MODELS

Many models of speleogenesis have been proposed on the basis of observations in caves. However, since the early 1980s efforts have been made to understand speleogenesis beginning from basic principles of fluid flow in fractures and conduits and the physical chemistry of limestone or gypsum dissolution (see Dissolution: Carbonate Rocks, and Dissolution: Evaporite Rocks). In the first models, simple scenarios of the evolution of a single karst conduit were investigated by digital computation. These provided basic elements for further research and are therefore discussed here in some detail.

We consider a one-dimensional fracture with aperture width a and length L. Water aggressive to limestone is driven through it by a hydraulic head h. As water flows from the entrance down the hydraulic gradient it dissolves the rock and the fracture widens. Widening is fastest at the entrance but slows down when the solution approaches equilibrium further downstream. To calculate the amount of widening by dissolution one needs information on bedrock retreat as a function of the chemical composition of the solution. Early knowledge of calcite dissolution suggested a linear rate law where the rate R_1 is given by

$$R_1 = k_1 (1 - c/c_{eq}) \tag{1a}$$

and k_1 is a constant in mmol cm^{-2}s^{-1}, c is the calcium concentration and c_{eq} is the equilibrium concentration with respect to calcite or gypsum, respectively. However, this rate law predicts that conduit water would become saturated close to the entrance and therefore caves could not originate at all. The apparent contradiction was resolved by Palmer (1984) who suggested that close to equilibrium dissolution must be inhibited to such an extent that the water can penetrate deeply into the rock without becoming saturated. It was found later by experiment that above about 90% of the equilibrium concentration c_{eq} a nonlinear rate law is valid both for limestone and gypsum:

$$R_n = k_n (1 - c/c_{eq})^n \quad (1b)$$

where k_n is the higher order rate constant, and n is between 3 and 6 depending on lithology. With such a rate law applied digital modelling proved successful and results were obtained which could explain early conduit evolution in scales of both length and time. For example, take a fracture 1 m wide and 1000 m long, with an initial aperture width of $a = 0.03$ cm. The hydraulic head at its entrance is 10 m and zero at its exit. Figure 1a illustrates the profiles sculptured by dissolution for various times. At the onset of karstification a funnel-like shape is created at the fracture entrance while further along widening is even but very slow. The small increase of aperture width at the exit triggers a positive feedback loop. The flow through the fracture increases with the third power of the exit aperture width. Increase of flow rates enhances dissolutional widening and vice versa. This can be seen from Figure 1b which depicts the flow through the fracture. The open circles denote the times of the profiles shown in Figure 1a. Flow and correspondingly widening at the exit are initially slow but accelerate until within a very short time span a dramatic increase in flow and in fracture width is observed. After the breakthrough flow becomes turbulent. The conduit widens rapidly and evenly until the hydraulic head can no longer be supported. Flow and further evolution of the cave will therefore be controlled by recharge and the conduit will then widen evenly, and eventually become vadose. It is interesting to note that even though this evolution model of a single conduit seems complex an analytical estimation for the breakthrough time T can be given (Dreybrodt, 1996; Dreybrodt & Gabrovšek, 2000; Gabrovšek, 2000) by the relation

$$T = \tau \cdot \left(\frac{L^2}{h}\right)^{\frac{n}{n-1}} \cdot a^{-\frac{2n-1}{n-1}} \cdot (k_n)^{\frac{1}{n-1}} \cdot (c_{eq})^{-\frac{n-1}{n-1}} \quad (2)$$

where T is in years, a is the initial aperture width, units are in cm, mol cm^{-3}, mol cm^{-2}s^{-1} respectively, and $\tau = 10^{-13}$. This equation is a consequence of the positive feedback loop by which breakthrough occurs when the exit width has been opened to about three times its initial value (cf. Figure 1a). Equation 2 reveals how the parameters determining cave evolution relate to T in a simple way.

Modelling of cave evolution has been extended to two-dimensional nets of fractures with similar boundary conditions: inputs of constant heads at the upper boundary and output points at zero head at base level (Dreybrodt & Siemers, 2000). Figure 2 represents an example. The aquifer consists of a limestone bed 1 m thick, 1 km long, and 500 m wide. It is dissected into two orthogonal sets of fractures spaced at 10 m apart. The fracture aperture widths are statistically distributed (lognormal) with average of 0.02 cm, and 20% of fractures have been omitted. The hydraulic head acting at the input points at the left side is 10 m and zero at the base level to the right. The upper and lower margins are impermeable. Figure 2 illustrates the evolution of conduits and the head distribution in the aquifer. A competition of several conduits is observed until one breaks through. After breakthrough the hydraulic head acting on that channel breaks down and flow is restricted to constant recharge of 1000 cm^3 s^{-1}. Consequently the hydraulic head drops and flow of the nearest competing channel is directed towards the leading conduit such that a cave system becomes integrated. This simulation confirms the Ewers speleogenetic high-dip-model (see Ford & Williams, 1989). It is important to note that breakthrough times for two-dimensional nets with boundary conditions of constant head show an analogous dependence on the parameters as in equation 2, provided L is taken as the length of the aquifer (Siemers & Dreybrodt, 1998). Many different scenarios for cave evolution have been simulated for constant head conditions, such as Ewers' low-dip-model, where several rows of input points with different distances are subjected to equal heads. Those inputs closest to the output first experience breakthrough. Thereafter conduits start to grow from the more distant inputs and connect to inputs from which channels have already succeeded in breaking through. By this way a cave system integrates headwards. Differing lithologies of bedrock in aquifers can also be simulated. They can cause isolated cave systems without entrances and exits or caves with entrances but no exits (Dreybrodt & Siemers, 2000). Two-dimensional modelling enables one also to investigate the role of mixing corrosion as a cave-forming mechanism. Although in combination with linear dissolution mixing corrosion is insignificant as a cave-forming

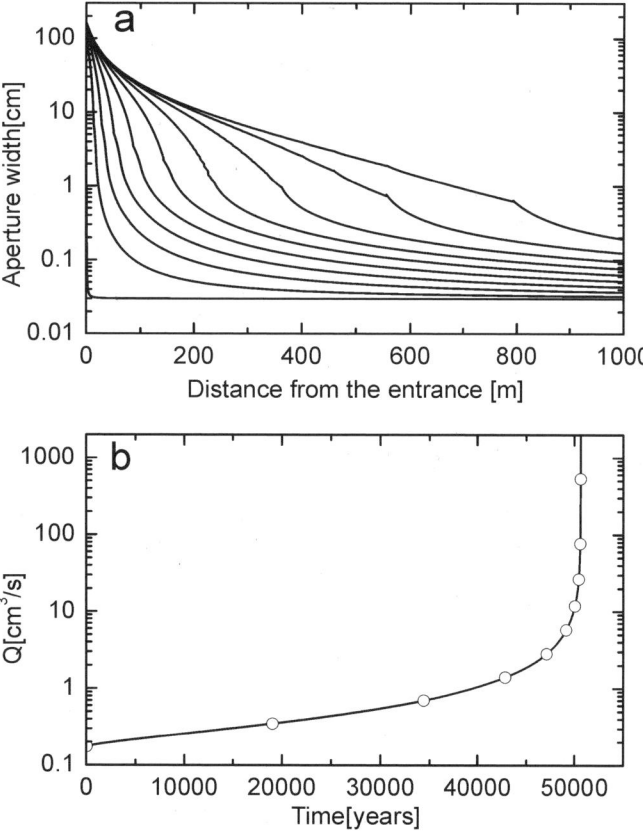

Speleogenesis: Computer Models: Figure 1. Evolution of a single conduit. a) Profiles of aperture widths. b) Flow through conduit in dependence on time. The open circles mark the times for which the profiles in a) are shown.

process, in combination with nonlinear kinetics (see equation 1b) it can play an important role in the evolution of cave systems under constant head conditions (Gabrovšek & Dreybrodt, 2000). Subterranean sources of CO_2, either from volcanic sources or the production of CO_2 by microbial action can also enhance karstification (Gabrovšek, Menne & Dreybrodt, 2000).

Where dam sites are constructed in limestone or gypsum terrains unnaturally steep hydraulic gradients arise. The question of whether karstification under such conditions can be intensified to such an extent that the structure might be endangered during its lifetime has been investigated by modelling (Dreybrodt, Romanov & Gabrovšek, 2002). Although such models must be treated with care it cannot be excluded that karstification below dam sites can lead to serious problems.

So far all models presented have boundary conditions of constant head in common. Those are most likely to be valid in the early evolution of conduits. The later state of natural karstification after breakthrough is governed by recharge from the catchment area and by turbulent flow in the major conduits. In this case one has to model constant recharge at the input points.

The further evolution of karst aquifers after breakthrough has been modelled by employing a combined continuum and discrete pipe network concept. This approach is based upon hydraulic coupling of a pipe network of larger prominent conduits (e.g. those after breakthrough) to a continuum with given hydraulic conductivity. This continuum represents the system of unwidened narrow fissures constituting most of the aquifer's storage. A constant recharge is supplied to the continuum, and also directly into the pipes. Exchange of water between the pipe system and the continuum is proportional to the head difference between the pipe-flow system and the continuum at the particular nodes, where they intersect. A further assumption is that the solution entering from the continuum into the pipe system is close to equilibrium with respect to calcite (e.g. 90% of c_{eq}). Models restricting dissolutional widening to the pipe systems only have been presented (Clemens et al. 1997; Sauter & Liedl, 2000). This modelling approach has been extended by Kaufmann and Braun (2000) to karst aquifer evolution in fractured, porous rocks by finite element techniques, with much higher spatial resolution. This approach, although basically similar to that of Clemens et al. (1997) avoids some parameters which are difficult to specify. Figure 3 illustrates an example of their results. Figure 3a shows the modelling domain, where the thin lines denote fractures and the dots are input points. The fractures are embedded into a porous matrix with conductivity of 10^{-5} m s^{-1}. A river at the bottom fixes the head at 1 m. All other boundaries are impermeable. Initial width of all fractures is 0.2 mm. Recharge of 400 mm a^{-1} is evenly distributed. Figure 3b shows the head distribution at time zero. After 2800 years some fractures have enlarged and grown headwards (Figure 3c). The distribution of hydraulic heads has changed. After 8000 years a net of conduits has developed headwards.

Models presented so far deal with the evolution of karst aquifers in the dimensions of length and breadth. Recent modelling of the evolution in the dimensions of length and depth has been reported by Gabrovšek and Dreybrodt (2001). The modelling domain consists of the vertical cross section of a small karstic plateau, 200 m long and 30 m high as illustrated by Figure 4. Recharge Q is evenly distributed across the surface and seeps through the fractures of the aquifer down to base level,

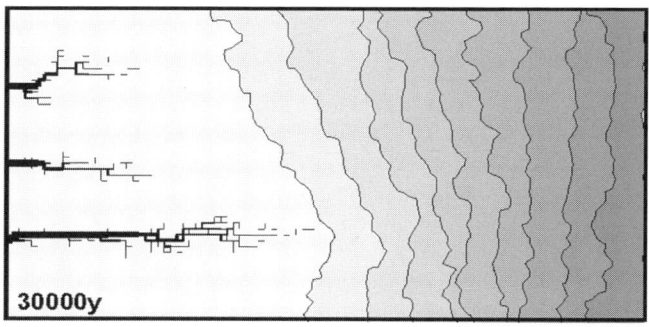
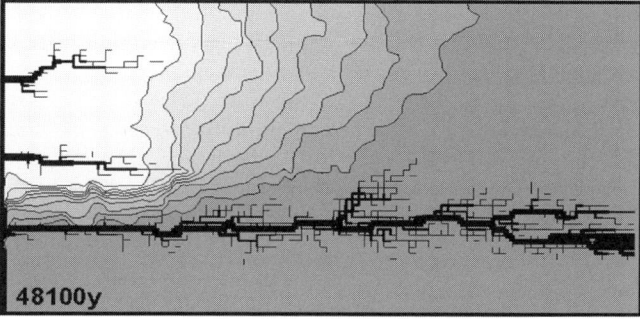
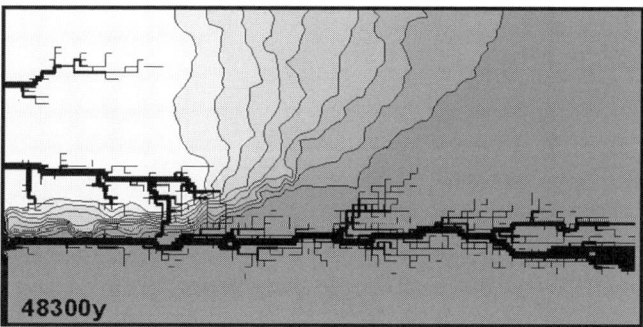
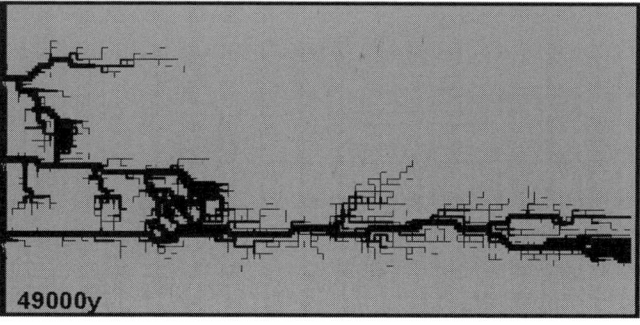

Speleogenesis: Computer Models: Figure 2. Evolution of conduits and distribution of heads for a two-dimensional aquifer under constant head conditions. The isolines of the head are given in steps of 1 m.

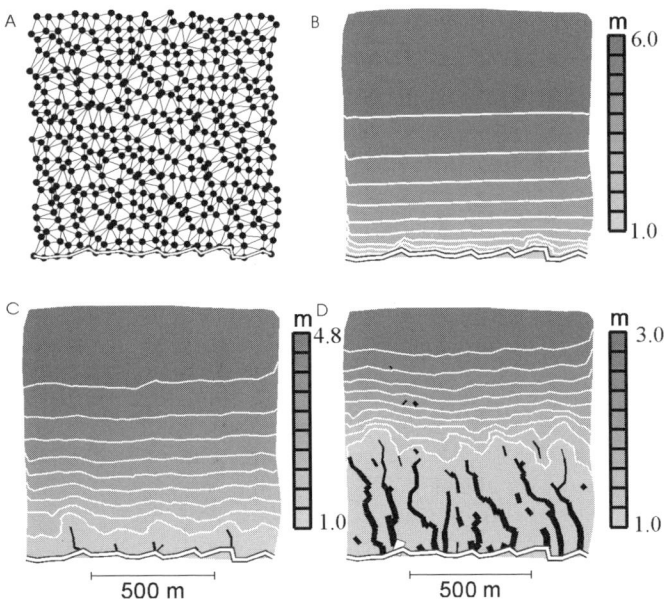

Speleogenesis: Computer Models: Figure 3. Evolution of an aquifer in fractured porous rocks with a constant recharge of 400 mm a^{-1}. The white lines show the head distribution. The aquifer drains into the river. (From Kaufmann & Braun, 2000)

represented by a river, or emerges from the seepage face of the cliff on the right hand side. All other boundaries are impermeable. The rock is fissured by some wide prominent fractures with aperture widths $a_p = 0.02$ cm. These are embedded into a net of narrow fractures with aperture widths a_v or a_h of 0.005 cm, and a vertical spacing $L_v = 0.5$ m. Horizontal spacing L_h is 2 m. First model scenarios omitted all prominent fractures to obtain a homogeneous aquifer, where water enters at the plateau and seeps downwards to the water table. Due to dissolution of the rock in the vadose zone its calcium concentration at the water table is about 90% of equilibrium concentration. Dissolution in the phreatic zone is primarily restricted to a narrow fringe at the water table, where permeability increases such that water flow is directed along the water table. The water table drops continuously until it reaches the river. Flow is concentrated towards the spring and a conduit grows headwards along the water table declining to become horizontal at base level. The final result is a water table cave as suggested by Swinnerton (1932). Figure 5a illustrates this evolution of the water table. If in addition to recharge an input region of constant head h_{in} is added, the water table is kept stable, because increasing permeability

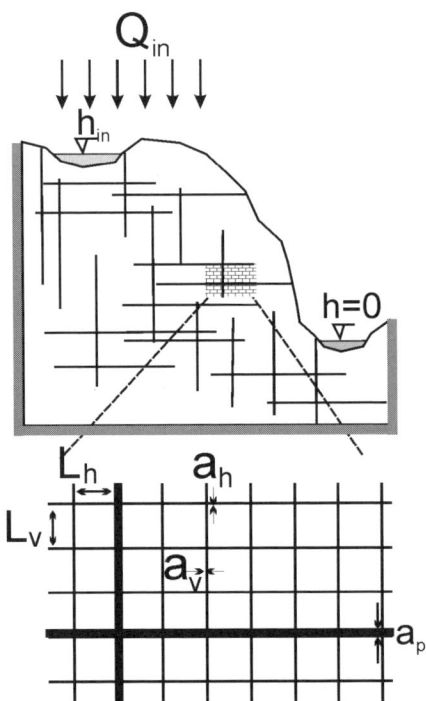

Speleogenesis: Computer Models: Figure 4. Modelling domain for evolution of an aquifer in dimension of length and depth. (From Gabrovšek & Dreybrodt, 2001)

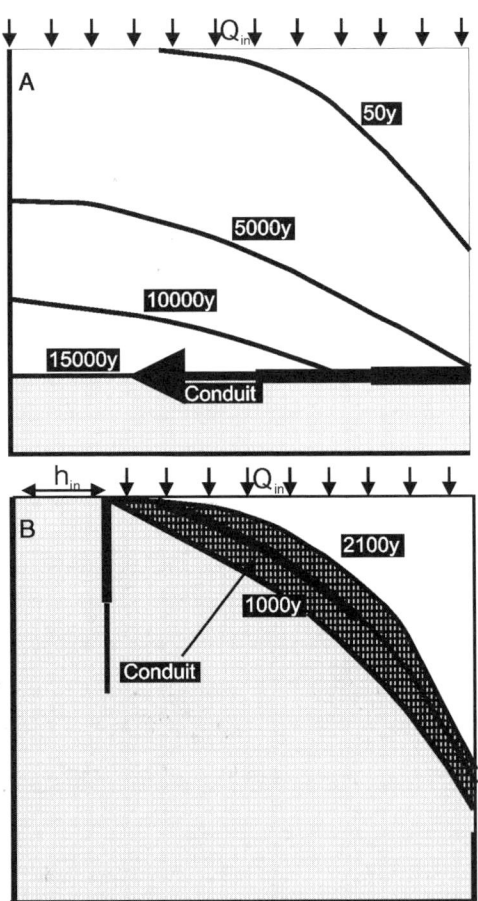

Speleogenesis: Computer Models: Figure 5. A) Evolution of the water table for an aquifer under conditions of constant recharge. B) A region of constant head h_{in} is added. First the water table is low (1000 years) and determined by recharge. As a conduit grows from the region of constant head h_i it rises. The dark shaded area is the region where dissolution is most active.

of the rock is compensated by increasing flow from the input. Conduits grow along the water table towards base level until breakthrough occurs. This is illustrated by Figure 5b. Vadose flow in the cave will determine its further evolution. Such a behaviour was predicted by Rhoades and Sinacori (1941) from field observations. If one incorporates a net of prominent fractures with significantly larger aperture widths, deep phreatic loops grow below the water table as predicted in Ford's four-state-model (see Ford & Williams, 1989).

In summary, modelling has been successful in explaining empirical theories of cave genesis. In the future it will be of utmost importance to a deeper understanding of the way by which physical and chemical processes govern karst and its evolution.

WOLFGANG DREYBRODT and FRANCI GABROVŠEK

See also **Groundwater in Karst: Mathematical Models; Speleogenesis; Speleogenesis: Deep-Seated and Confined Settings; Speleogenesis: Unconfined Settings**

Works Cited

Clemens, T., Hückinghaus, D., Sauter, M., Liedl, R. & Teutsch, G. 1997. Modelling the genesis of karst aquifer systems using a coupled reactive network model. In *Hard Rock Hydrosystems*, edited by T. Pointet, Wallingford, Oxfordshire: IAHS

Dreybrodt, W. 1996. Principles of early development of karst conduits under natural and man-made conditions revealed by mathematical analysis of numerical models. *Water Resources Research*, 32: 2923–35

Dreybrodt W. & Gabrovšek, F. 2000. Dynamics of the evolution of a single karst conduit. In *Speleogenesis: Evolution of Karst Aquifers*, edited by A.B. Klimchouk, D.C. Ford, A.N. Palmer & W. Dreybrodt, Huntsville, Alabama: National Speleological Society

Dreybrodt, W. & Siemers, J. 2000. Cave evolution on two-dimensional networks of primary fractures in limestone. In *Speleogenesis: Evolution of Karst Aquifers*, edited by A.B. Klimchouk, D.C. Ford, A.N. Palmer & W. Dreybrodt, Huntsville, Alabama: National Speleological Society

Dreybrodt, W., Romanov, D. & Gabrovšek, F. 2002. Karstification below dam sites: A model of increasing leakage in reservoirs. *Environmental Geology*, 42: 518–24.

Ford, D.C. & Williams, P.W. 1989. *Karst Geomorphology and Hydrology*, London and Boston: Unwin Hyman

Gabrovšek, F. 2000. *Evolution of Early Karst Aquifers: From Simple Principles to Complex Models*, Ljubljana, Slovenia: Zalozba ZRC

Gabrovšek, F., Menne, B. & Dreybrodt, W. 2000. A model of early evolution of karst conduits affected by subterranean CO_2 sources. *Environmental Geology*, 39: 531–43

Gabrovšek, F. & Dreybrodt, W. 2000. Role of mixing corrosion in calcite-aggressive H_2O-CO_2-$CaCO_3$ solutions in the early evolution of karst aquifers in limestone. *Water Resources Research*, 36: 1179–88

Gabrovšek, F. & Dreybrodt, W. 2001. A model of the early evolution of karst aquifers in limestone in the dimensions of length and depth. *Journal of Hydrology*, 240: 27–34

Kaufmann, G. & Braun, J. 2000. Karst aquifer evolution in fractured, porous rocks. *Water Resources Research*, 36: 1381–92

Palmer, A.N. 1984. Recent trends in karst geomorphology. *Journal of Geological Education*, 32: 247–53

Rhoades, R. & Sinacori, M.N. 1941. The pattern of ground-water flow and solution. *Journal of Geology*, 49: 785–94

Sauter, M. & Liedl, R. 2000. Modelling karst aquifer genesis using a coupled continuum-pipe flow model. In *Speleogenesis: Evolution of Karst Aquifers*, edited by A.B. Klimchouk, D.C. Ford, A.N. Palmer & W. Dreybrodt, Huntsville, Alabama: National Speleological Society

Siemers, J. & Dreybrodt, W. 1998. Early development of karst aquifers on percolation networks of fractures in limestone. *Water Resources Research*, 34: 409–19

Swinnerton, A.C. 1932. Origin of limestone caverns. *Geological Society of America Bulletin*, 34: 662–93

Further Reading

Dreybrodt, W. 1990. The role of dissolution kinetics in the development of karstification in limestone: A model simulation of karst evolution. *Journal of Geology*, 98: 639–55

Palmer, A.N. 1991. The origin and morphology of limestone caves. *Geological Society of America Bulletin*, 103: 1–21

SPELEOGENESIS: DEEP-SEATED AND CONFINED SETTINGS

For a long time, traditional theories on the origin of caves and karst were largely concerned with shallow and unconfined settings, ultimately implying a close hydrological and morphogenetic relationship between the surface and subsurface. The common view was that karst processes and speleogenesis commence only when soluble rocks are exposed, whether immediately after deposition or after prolonged burial, and receive recharge from the overlying surface. Features at considerable depth were commonly treated as paleokarst. However, exposure of a soluble formation at the surface represents only a minor, although important, part of its karstification history that may commence during the deep-seated intrastratal stage that precedes exposure. The last decades of the 20th century witnessed a rapid increase in the recognition of speleogenesis with no apparent relation to the surface.

When related to speleogenesis, the term "deep-seated" does not carry a genetic meaning, but implies considerable spatial remoteness and separation from the surface. The term "hypogenic speleogenesis" is more genetically specific, indicating that cave-forming mechanisms rely on aggressiveness that has been produced at depth beneath the surface. "Confined" and "artesian" refer to hydrodynamics, and imply that groundwater is under pressure in a bed or stratum confined by less-permeable rocks or sediment above it. The hydraulic heads in this type of aquifer lie above the base of the confining beds. As far as groundwater dynamics is concerned, the vast majority of deep-seated and hypogenic speleogenesis occurs in confined settings, or in settings that are unconfined but paragenetic or subsequent to confinement. Hence, the terms deep-seated, hypogenic, and artesian speleogenesis refer to closely related and overlapping, although not entirely equivalent, concepts.

The "classic" concept of artesian flow assumes that recharge of confined aquifers occurs only in limited areas, where they crop out at the surface (usually at basin margins) and that groundwaters move laterally through separate aquifers within the area of confinement. These simplistic assumptions have created a major problem in interpreting artesian speleogenesis. After moving for a considerable distance, and for a long time, through a soluble rock unit, water should be incapable of significant dissolution in the confined flow area. However, modern hydrogeologists are well aware that there are virtually no impervious rocks or sediments, just large contrasts in permeabilities. Hydraulic continuity in basins, and close cross-formational communication between aquifers, are common characteristics of artesian settings (Shestopalov, 1981; Tóth, 1995). Artesian aquifers commonly receive significant proportions of their recharge from vertically adjacent formations, not only from exposed marginal recharge areas. Adoption of these views into karstology has allowed the reconciliation of many controversial points of view regarding the interpretation of speleogenesis in confined settings (Klimchouk, 2000).

Artesian basins containing carbonate and sulfate formations are widespread throughout cratonic and foreland regions. Within confined areas, cross-formational flow is predominantly ascending, being more intense in areas underlying prominent topographic lows, such as large river valleys. Most of the aggressive recharge to soluble units in confined settings comes from the underlying aquifer formations. Recharge can be dispersed across wide areas, which favours the formation of maze patterns, or can be focused locally, along high-permeability pathways such as fault zones. Cross communication between formations of different lithology and zones with contrasting geochemical environments or different physical conditions supports the operation of a wide variety of dissolutional mechanisms which, under deep-seated and confined settings, may lead to the formation of caves.

Aggressiveness in some cases represents an original undersaturation of groundwater with respect to the solid phase that is being entered, such as in the case of low-sulfate waters from underlying carbonates or sandstones entering a gypsum bed. It can also reflect the acquisition of new sources of acid (e.g. by oxidation of hydrogen sulfide), or be due to a number of mechanisms that rejuvenate the dissolutional capacity of fluids, such as the mixing of groundwaters of contrasting chemistry, the cooling of water, sulfate reduction and dedolomitization.

Carbonic acid dissolution, which dominates overwhelmingly in unconfined carbonate aquifers, also operates as a hypogenic agent, although the origin of the acidity is different. It can be related to CO_2 generated from igneous processes, to thermometamorphism of carbonates, or to thermal degradation and oxidation of deep-seated organic compounds by mineral oxidants. Creation of significant caves by hypogenic carbonic acid depends mainly upon the rejuvenation of aggressiveness by mixing, or by a drop in temperature. The latter mechanism is distinguished as hydrothermal speleogenesis, occurring in high-gradient zones where ascending flow is localized along faults.

The dissolution of carbonates by hydrosulfuric acid is another important speleogenetic process in deep-seated anoxic environments where there are sufficient sulfate sources for reduction, and where the H_2S generated can escape from the reducing zones—settings typical of the margins of sedimentary basins containing evaporate formations. In shallower conditions, where H_2S-bearing waters rise to interact with oxygenated meteoric groundwaters, sulfuric acid dissolution can be a very strong speleogenetic agent. Substantial sulfuric acid dissolution can also be caused by oxidation of metallic sulfides such as pyrite, where it is localized in ore bodies or along certain horizons or bedding planes.

Dissolution in deep-seated and confined settings is commonly slow, due to the prevailing sluggish circulation and, hence, to mass-balance restrictions. However, being operative over prolonged periods of geological time, it is generally important for cave inception, that is the opening up of pathways for further, more effective, circulation (Lowe & Gunn, 1997). Creation of significant caves occurs when continuing uplift brings stratified artesian aquifers closer to the eroding surface, or where local, highly effective flow-paths discharge deep-seated fluids into shallower aquifers. Both situations serve to activate groundwater circulation. Cave patterns depend mainly upon structural conditions, mode of recharge, and degree of confinement, rather than upon particular dissolution mechanisms.

In cases of transverse circulation across a uniformly fissured soluble bed enclosed within a stratified artesian system, dispersed aggressive recharge and short flow distances favour the formation of maze patterns. Discharges through fissures, and hence the rates of fissure enlargement, are kept uniform by restrictions imposed by the hydraulic properties of the feeder bed or the confining bed. Examples include some of the longest caves in both limestones (Wind and Jewel caves, see separate entry and Figure 1, in the Lower Carboniferous Madison Limestone of South Dakota, United States, and Botovskaja Cave in the Lower Ordovician limestones of Siberia) and gypsum (Optimistychna, Ozerna, and other giant mazes in the Miocene gypsum of the western Ukraine; see Ukraine Gypsum Caves). Smaller but no less significant caves are Fuchslabyrinth and Moestrof caves in the Muschelkalk limestones of Germany and Luxemburg, Knock

Speleogenesis: Deep-Seated and Confined Settings: Figure 1. Wind Cave, South Dakota, is a network maze consisting mainly of irregular fissure-like passages. The cave was formed along zones of paleokarst and former sulfate zones by mixing of at least two groundwater sources, probably including deep-seated waters. (Photo by Art Palmer)

Speleogenesis: Deep-Seated and Confined Settings: Figure 2. Geophyzicheskaya Cave, one of the large hypogene caves in the Karljuksky region, Kugitang Range. (Photo by Alexander Klimchouk)

Fell Caverns in the Carboniferous limestones of the United Kingdom, Estremera Cave in the Neogene gypsum of Spain, and Denis Parisis cave in the Paleogene gypsum of France. High passage density, strong fissure guidance and the presence of numerous feeder channels at the bottom, and outlet features at the top, of the layered passage systems are characteristic of such caves.

A specific mechanism of cave development driven by natural convection in confined settings is exemplified by the vast rooms intercepted by many mines in the Harz region of Germany. Aggressive water from the underlying aquifer attacks the bottom of the massive Zechstein (Upper Permian) gypsum bed. Once saturated, the water becomes denser and returns into the aquifer, being replaced by the less-dense aggressive water (Kempe, 1996).

Focused rising of deep-seated fluids into massive carbonate rocks commonly creates irregular large rooms or more complex ramifying patterns of merged rooms and smaller maze-like or separately rising passages diverging from them. Cave development occurs due to different mechanisms where aggressive fluids enter carbonates or where aggressiveness peaks (e.g. in the high thermal gradient zone in cases of hydrothermal speleogenesis, or in the zone of mixing of H_2S-bearing waters with the oxygenated meteoric waters in cases of sulfuric acid speleogenesis). Such caves are most likely to occur in carbonate build-ups at the margins of sedimentary basins (e.g. Carlsbad Caverns and Lechuguilla Cave, see separate entry, other caves in the Guadalupe Mountains, New Mexico, United States, the Cupp–Coutunn system, see separate entry, and other caves in the Karljuksky region of Turkmenistan, Figure 2) or in deep fault zones in tectonically active regions (e.g. Vento Cave, Frasassi [see Frasassi entry] and caves in the Buda Hills in Budapest, Hungary). An outstanding example of a giant cave created by, and still filled with, thermal water under high pressure is the cavern intercepted by deep drilling in Archaean marbles in the Rhodope mountains of Bulgaria (Dublyansky, 2000). This cavern, with an estimated volume of 237.6 million m^3 and a maximum intersected vertical dimension of 1341 m, is by far the largest known (although not directly explored) cave, by volume, in the world.

ALEXANDER KLIMCHOUK

See also **Inception of Caves; Patterns of Caves; Speleogenesis (various entries)**

Works Cited

Dublyansky, V.N. 2000. A giant hydrothermal cavity in the Rhodope Mountains. In *Speleogenesis: Evolution of Karst Aquifers*, edited by A.B. Klimchouk, D.C. Ford, A.N. Palmer & W. Dreybrodt, Huntsville, Alabama: National Speleological Society

Kempe, S. 1996. Gypsum karst of Germany. In *Gypsum Karst of the World*, edited by A.B. Klimchouk, D. Lowe, A. Cooper and U. Sauro, *International Journal of Speleology*, Theme issue 25(3–4)

Klimchouk, A.B. 2000. Speleogenesis under deep-seated and confined settings. In *Speleogenesis: Evolution of Karst Aquifers*, edited by A.B. Klimchouk, D.C. Ford, A.N. Palmer & W. Dreybrodt, Huntsville, Alabama: National Speleological Society

Lowe, D.J. & Gunn, J. 1997. Carbonate speleogenesis: An inception horizon hypothesis. *Acta Carsologica*, 26(2): 457–88

Shestopalov, V.M. 1981. *Natural Resources of Underground Water of Platform Artesian Basins of the Ukraine*, Kiev: Naukova Dumka (in Russian)

Tóth, J. 1995. Hydraulic continuity in large sedimentary basins. *Hydrogeology Journal*, 3(4): 4–15

Further Reading

Dublyansky, Y. 2000. Hydrothermal speleogenesis: Its settings and peculiar features. In *Speleogenesis: Evolution of Karst Aquifers*, edited by A.B. Klimchouk, D.C. Ford, A.N. Palmer & W. Dreybrodt, Huntsville, Alabama: National Speleological Society

Ford, D.C. & Williams, P.W. 1989. *Karst Geomorphology and Hydrology*, London: Unwin Hyman

Ford, T.D. 1995. Some thoughts on hydrothermal caves. *Cave and Karst Science*, 22 (3): 107–18

Hill, C. 2000. Sulfuric acid hypogene karst in the Guadalupe Mountains. In *Speleogenesis: Evolution of Karst Aquifers*, edited by A.B. Klimchouk, D.C. Ford, A.N. Palmer & W. Dreybrodt, Huntsville, Alabama: National Speleological Society

Klimchouk, A.B., Ford, D.C., Palmer, A.N. & Dreybrodt, W. (editors) 2000. *Speleogenesis: Evolution of Karst Aquifers*, Huntsville, Alabama: National Speleological Society

Lowe, D.J. & Gunn, J. 1995. The role of strong acid in speleo-inception and subsequent cavern development. In *Environmental Effects on Karst Terrains*, edited by I. Barany-Kevei special issue of *Acta Geographica*, 34: 33–60

Palmer, A.N. 1991. Origin and morphology of limestone caves. *Geological Society of America Bulletin*, 103: 1–21

Palmer, A.N. 1995. Geochemical models for the origin of macroscopic solution porosity in carbonate rocks. In *Unconformities and Porosity in Carbonate Strata*, edited by A.D. Budd, A.H. Saller & P.M. Harris, Tulsa, Oklahoma: American Association of Petroleum Geologists

SPELEOGENESIS: UNCONFINED SETTINGS

Dissolution caves in unconfined settings are created by meteoric water circulating at relatively shallow depths. The flow between sinkpoints and springs is contained within the soluble strata. It does not rise into them from below and does not tend to pass so deep beneath insoluble and impermeable confining strata that it is geothermally heated to a significant degree. The great majority of explored caves longer than 100 m or so belong to this class. Transitions to the types developed in young coastal rocks or in deep-seated and confined settings do occur.

In most cases the passages (conduits) of unconfined caves have developed along fissures such as bedding planes, joints, and faults, rather than through interlinked pores in the matrix of the rock. Fissure control of cave patterns is overwhelmingly predominant in the carbonate and sulfate rocks. Due to its greater solubility, salt can display important matrix (intergranular) permeability and caves developed in it may have patterns similar to those of coastal caves in young, porous limestones.

The great, fissure-guided cave systems of the world are three-dimensional complexes, usually with multiple "levels" created to drain to springs at successively lower elevations. For genetic analysis, plan patterns are considered first, then development in depth, considering, to begin with, only initial caves (the earliest level).

Development of the Plan Patterns of Caves

The simplest situation is that of a single conduit formed by one sinking stream—the case shown in Figure 1a. There are innumerable examples in nature, such as underground captures across the necks of entrenched river meanders. The controls which determine whether speleogenesis can occur, and the rate of development where it can, are the aperture of the fissure or linked sequence of fissures to be penetrated, their length, the hydraulic head, and the solvent capacity of the water. The kinetics of the carbonate and sulfate minerals in normal waters are non-linear—groundwater rapidly dissolves up to about 90% of its solvent capacity and thereafter approaches thermodynamic equilibrium (chemical saturation) asymptotically. Physical simulations and computer models show that there is initial competition between dissolutional microconduits, each slowly extending its 90% saturation position towards the output boundary, which will be a spring line or a pre-existing passage (Dreybrodt, 2000). One, or a few get ahead, gaining a hydraulic head ("equipotential") advantage (Figure 1a, frames 2 and 3). "Breakthrough" occurs when the rapid (~90%) front reaches the output. The winning conduit will be only ~1 cm in diameter initially, but will enlarge rapidly and capture most of the flow from its competitors, greatly reducing their rates of extension.

Sinuous, branching patterns, as shown, are most common in bedding planes. The form tends to be more direct in joints and faults. Minimum continuous apertures for effective penetration to occur are believed to be between 10 and 100 microns.

More often, a cave will have multiple inputs. Figure 1b shows the inputs in one rank, such as occurs where allogenic streams sink along a contact with limestone. In this particular model, the distance to the output boundary is the same for all inputs, as are the hydraulic head, amount of water available, and its solvent capacity. Nevertheless, because initial apertures chance to be slightly more advantageous there, the right-hand microconduit gets ahead, distorting the equipotential pressure field in its favour and eventually breaking through first (1). The others then break through to it in the sequence 2–3–4, linking up between the nearest branches. This creates a dendritic (tree-like) network of channels, feeding one trunk conduit to one spring. The sinuosity in the figure is characteristic of genesis in bedding planes or low-angle thrust faults. Where joint control is dominant, linking patterns are more angular.

Where initial fissure apertures are similar at all inputs also, equidimensional mazes of conduits may develop, with rectilinear patterns in joints and anastomoses in bedding planes. They can aggregate many kilometres when mapped but rarely have long overall flowpaths (more than 1 km or so). Many are floodwater mazes (Palmer, 2000) formed in the entrance zones of longer

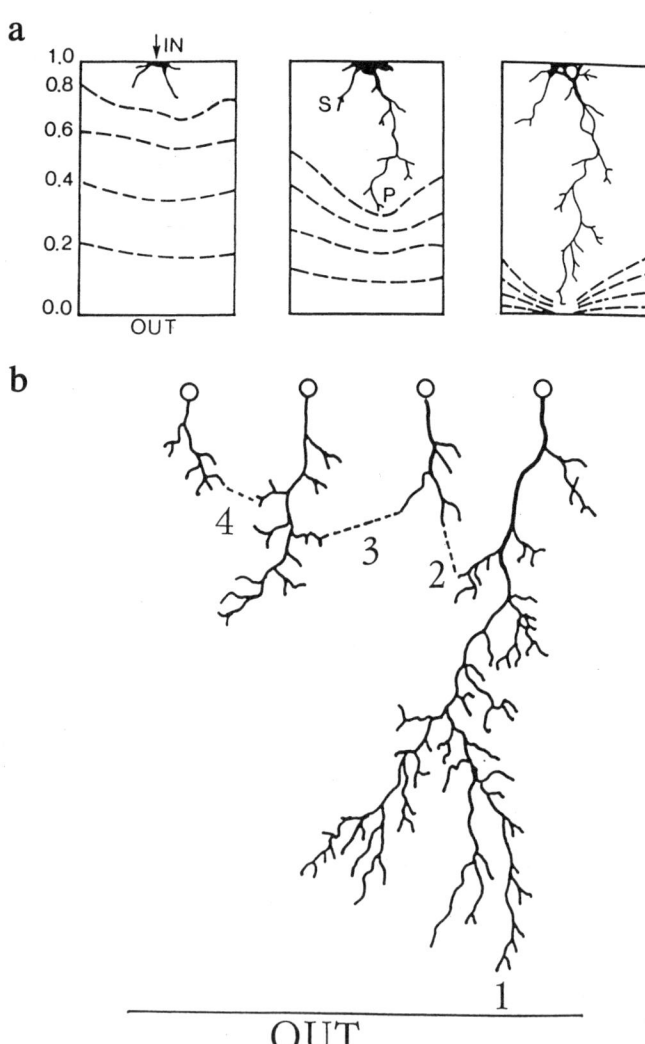

Speleogenesis: Unconfined Settings: Figure 1. (a) The competitive extension of dissolutional micro-conduits across a fissure with anisotropic porosity. "P" = principal or victor conduit; "S" = secondary conduits. Dashed lines are hydraulic equipotentials. (b) Competitive extension where there are multiple inputs in one rank. Numbers and dashed lines indicate the predicted sequence of breakthrough connections that will occur. Both figures are drawings of actual dissolution experiments with plaster models—see Ford and Williams (1989), pp.249–261, for details.

caves, because allogenic floods impact strongly there and surface erosion has unloaded the jointing, permitting it to gape; the mazes discharge into dendritic collectors downstream.

In nature, aperture, distance, hydraulic head, amount, and solvent capacity of water all tend to differ between inputs, however. Many caves are destined to have more distorted patterns than that of Figure 1b from the onset. In particular, in dipping strata, potential spring lines are often along the geological strike (i.e. on the left or right side of the figure), giving flowpath length advantages to one or a few inputs, as at Hölloch and Siebenhengstehöhle in Switzerland (see separate entries).

Broad areal karsts, such as limestone pavements, doline fields, or cockpit country, drain into many ranks of inputs, often with highly irregular patterns. Nevertheless, the principles of conduit initiation, competition and link-up remain the same as above. Broadly dendritic patterns of dissolution passages are created, beginning with the inputs closest to potential springpoints or otherwise most effectively connected to them. The surface forms themselves (dolines, cockpits, etc.) cannot develop to significant size until their conduit drains have broken through to join the cave.

Development in Length and Depth

The patterns that unconfined cave systems should display, when viewed in long section between the sinkpoints and springs, have been the subject of much debate, with different speleologists advocating development chiefly above the water table, below it, or along it (see Speleogenesis Theories: Post 1890; Ford, 1998). A resolution proposed by the author is presented here.

There can be three groundwater zones in unconfined karst settings:

1. an uppermost zone that is wholly vadose; i.e. water flows by gravity alone, through fissures in aerated rock. When it is very wet, the fissures may fill briefly but this does not affect cave development significantly;
2. a zone which is waterfilled initially but drains progressively as its local microconduits break through to springs or pre-existing caves and become enlarged, drawing the water table down;
3. a phreatic zone of water-filled fissures beneath a water table (piezometric surface). It may include an epiphreatic (seasonally or episodically flooded) zone of lateral transmission.

Caves in Zone 1 tend to be sequences of shafts down the steepest fractures, linked together by short, often sinuous, stream canyons. Gravitational control of flow is predominant. These caves are best developed in young mountain terrains where rapid uplift and stressing opens vertical fractures wide and to substantial depths. The majority of the world's caves that are deeper than 1 km or so, gain most or all of their depth in this zone. Krubera (Voronja) Cave in Georgia (see separate entry), until 2003 the depth record holder, is a fine example. Where the karst rock formations are relatively thin and / or were little stressed during uplift, however, Zone 1 conditions may never have existed.

Cave passages in Zone 2 (the "drawdown vadose zone"—Ford, 2000) display initial phreatic forms such as elliptical cross sections in bedding planes, but with subsequent gravitational entrenchments beneath them. This combination can be found locally in Zone 1 also, where linking passages are perched on shale bands or other obstructions. The proportion of phreatic to vadose form may increase with depth. In vertical shafts, early phreatic features tend to be destroyed by the splashing water. Once drained, Zone 2 may be invaded by new streams if surface input points happen to be relocated, e.g. by glacial action; this introduces younger, wholly vadose forms—"invasion vadose" morphology.

Entire cave systems, from sinkpoints to springs, can develop wholly within Zone 1 and / or Zone 2 settings, especially where karst strata are perched on insoluble rocks above regional base levels. There can be deep gravitational entrenchment into the insolubles; e.g. some "contact caves" in Greenbrier County, West Virginia, have 90% + of their volume in erodible shales beneath the limestones that hosted the initial passages. However, most extensive cave systems have substantial Zone 3 (phreatic or water-table) sections. Their length is usually greater than that of the vadose parts.

System geometry becomes more complex and varied in Zone 3. The four basic possibilities are best understood where strata dip steeply, as shown schematically in Figure 2—the "four state

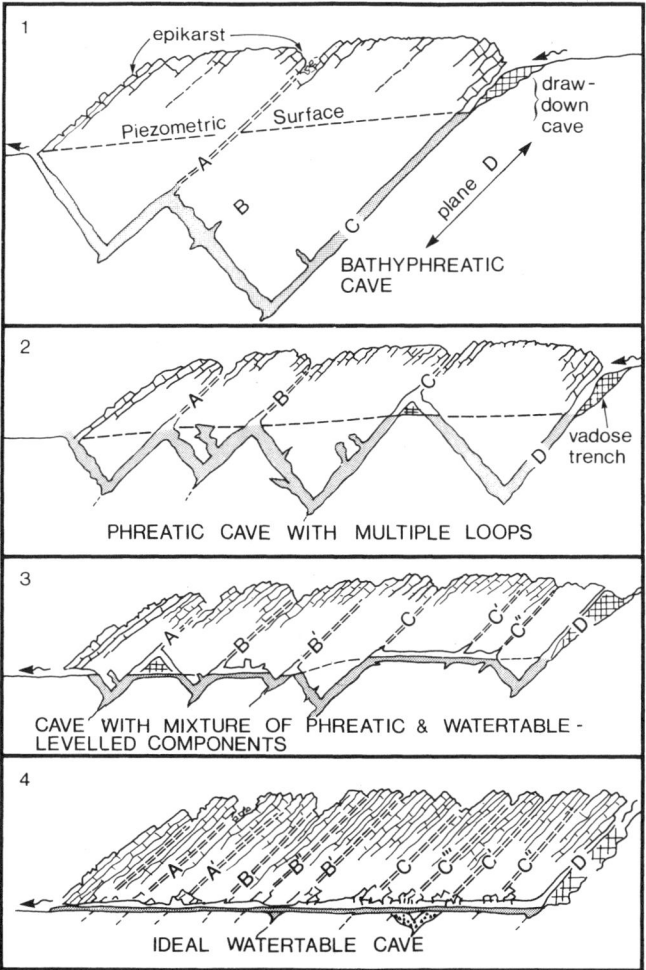

Speleogenesis: Unconfined Settings: Figure 2. The Four State Model that differentiates the system geometry of unconfined phreatic and water-table caves. See Ford and Williams, 1989, pp.261–271, for details.

model". If the density of penetrable and interconnected fissures is very low, geologic structure may force the microconduits to follow courses below the elevation of the spring ("phreatic loops"). A State 1 system passes from Zone 2 to the spring via one sub-water-table loop. In State 2 there is a sequence of loops whose crests fix the local elevations of the water table. Where frequency and interconnection of penetrable fissures is greater, the cave may display a mixture of loops and gently graded passages at the water table (State 3). Very high frequency permits continuous, gently graded, development along the water table (State 4), similar to caves in young, porous limestones. There is probably phreatic looping to depths of 1000 m or more in some State 1 caves. Individual loops greater than 250 m deep are common in State 2.

States may vary in different sectors of extensive caves, reflecting changes in geologic conditions. In some multilevel caves there is more State 3 or 4 morphology in the lower, younger levels, as a consequence of dissolution progressing widely in the aquifer. In addition to frequency, fissure orientation can also be important; e.g. in Figure 2, one exceptionally open vertical fracture along the plane of the page between sink and spring could yield a State 4 cave instead of the deep State 1 loop. These variables underline the fact that phreatic cave patterns can rarely be predicted in detail.

By definition, a cave is a void large enough for human entry. Many unconfined systems will include considerable lengths of passages that are too small for humans but, nevertheless, originate and function in the same manner as the enterable systems. At the other end of the scale are active river passages (vadose, water-table, and phreatic) that are tens of metres in diameter; White (1988, p.287) discusses evolutionary histories for such large conduits.

Multi-level (Multi-phase) Caves

As noted, most extensive cave systems have two or more "levels" that developed to drain to successively lower springs—"level" denoting the historical succession but not implying that the galleries must be near-horizontal; often, there is a sequence of State 2 or 3 geometries. In vadose caves, the lower levels may be simple entrenchments beneath the older passages. Where there is water table or phreatic development, the new springs are usually offset laterally tens to hundreds of metres or more. Distributary patterns are sometimes seen. The new springs steepen the hydraulic gradient in the downstream section of the old cave, which then adjusts to its new "level" in a sequence of breakthrough under-captures (*French*—"soutirages") that move the hydraulic steepening progressively upstream along the trunk and tributaries. Portions of individual capture links are incorporated into the final dendritic pattern of the new level but others are left redundant, becoming drained relicts or silted backwaters; see Ford and &Williams 1989, p.275, for an example.

Superimposition of successive levels, redundant links, and intruding invasion vadose caves from new sinkpoints, make maps of great systems such as Mammoth Cave, Kentucky (see separate entry), more complicated in appearance than almost any other geomorphic or hydrogeologic phenomena. Explorers can never be sure what will be seen around the next corner or up the next climb.

DEREK FORD

See also **Groundwater in Karst: Conceptual Models; Inception of Caves; Morphology of Caves; Morphometry of Caves; Patterns of Caves; Speleogenesis: Deep-Seated and Confined Settings**

Works Cited

Dreybrodt, W. (editor) 2000. Part 4: Theoretical Fundamentals of Speleogenetic processes. In *Speleogenesis: Evolution of Karst Aquifers*, edited by A.B. Klimchouk, D.C. Ford, A.N. Palmer & W. Dreybrodt, Huntsville, Alabama: National Speleological Society

Ford, D.C. 1998. Perspectives in karst hydrogeology and cavern genesis. *Bulletin d'Hydrogéologie*, 16: 9–29

Ford, D.C. (editor) 2000. Part 5.3 Speleogenesis Under Unconfined Settings. In *Speleogenesis: Evolution of Karst Aquifers*, edited by A.B. Klimchouk, D.C. Ford, A.N. Palmer & W. Dreybrodt, Huntsville, Alabama: National Speleological Society

Ford, D.C. & Williams, P.W. 1989. *Karst Geomorphology and Hydrology*, London and Boston: Unwin Hyman

Palmer, A.N. 2000. Hydrogeologic control of cave patterns. In *Speleogenesis: Evolution of Karst Aquifers*, edited by A.B. Klimchouk, D.C. Ford, A.N. Palmer & W. Dreybrodt, Huntsville, Alabama: National Speleological Society

White, W.B. 1988. *Geomorphology and Hydrology of Karst Terrains*, Oxford and New York: Oxford University Press

Further Reading

Bögli, A. 1980. *Karst Hydrology and Physical Speleology*, Berlin and New York: Springer (original German edition, 1978)

Gillieson, D. 1996. *Caves: Processes, Development and Management*, Oxford and Cambridge, Massachusetts: Blackwells

SPELEOLOGISTS

It has been exceedingly difficult to select people to be included in this biographical entry as the most significant speleologists of all time. The ones selected are people who have had the greatest influence on the subject overall, whether by making extensive regional studies of the karst at an early period, thus setting a standard of professionalism (for example, Xu Xiake, see Asia, Northeast: History entry, or Valvasor and Nagel); by extending the range of cave studies into a multidisciplinary subject, including fauna, flora, environment, and high-quality mapping (for example, Schmidl); by establishing international collaboration and specialist publication on caves, for example, Martel; or by extensive popular writing about their own and other explorations, thus making the subject known all over the world and inspiring many to take up caving for sport or for science (for example, Casteret). In every case these people have done much more than what is mentioned above, most notably vast amounts of original exploration. It has been necessary to leave out many others who have also made notable explorations or otherwise

Speleologists: Figure 1. J.W. Valvasor aged about 48, in 1689.

Speleologists: Figure 3. Martel in 1895.

extended the knowledge of karst. However, some of them are mentioned in the regional history entries.

Johann Weichard Valvasor (1641–93)

Baron Valvasor (Figure 1) was born in Ljubljana, Slovenia, and actively explored caves in that region from 1678 to 1689. He published descriptions of 13 of them, many for the first time, and his plan of Podpeška jama is only the second known printed map of a natural cave. His observations on the associated underground rivers are also of importance. The complex behaviour of the Cerknica intermittent karst lake caused him to attempt to explain it by a system of underground lakes and reservoirs (see entry or Cerknica Polje). This was published first as a paper in the *Philosophical Transactions* (of the Royal Society of London) in 1687, and as a result he was elected a member of that Society, among the most eminent men in Europe. Valvasor's wider reputation rests on his monumental topographical book on Slovenia, *Die Ehre des Herzogthums Crain* ... (1689), which comprised 2872 large pages in four volumes. Many chapters in this are concerned with karst, and all his own cave explorations are contained in it. The cost of publication exhausted his considerable fortune and he had to sell his castle and his library. He died soon afterwards—a martyr to his enthusiasm and dedication.

The significance of his work on karst and caves lay very largely in the fact that he described his own original observations. The

Speleologists: Figure 2. Adolf Schmidl.

Speleologists: Figure 4. Casteret aged 52, photographed at Montréjeau in 1949 by Winifred Hooper and reproduced by permission.

span of years over which he studied the karst and the detail of his descriptions all contributed to his importance, but most significant was his realization, for the first time, that caves and subterranean streams were an integral part of the whole system of karst hydrology. He was thus the first true speleologist.

Joseph Anton Nagel (1717–1800)

Nagel was born at Rittberg in Westphalia (Germany), but spent most of his life in Austria, where he was attached to the Emperor's court as a mathematician. After this he held posts in the University of Vienna and was keeper of the royal scientific collections (the basis of the present natural history museum). It was by command of the Emperor Franz I that Nagel undertook his travels in the cave areas of the old Austro-Hungarian Empire, and his lengthy manuscript reports, illustrated by a series of drawings, were lodged in the Royal Library in Vienna. His Austrian visit took place in 1747 and in the following year he made two extensive journeys of several weeks each in Slovenia, followed by a tour of some of the karst regions of Moravia (now in the Czech Republic). His cave visits in Austria were of relatively minor importance, but in Moravia he penetrated 416 m into Sloup Cave (measuring the distance remarkably accurately as 406 m) and explored part of Císařská jeskyně. In Slovenia he produced the first plans of Postojnska jama, Vilenica, and Socerbska jama; ventured 660 m into Planinska jama; and descended Črna jama to the underground Pivka. His many other visits there include what seems to be the first one to Željnske jame near Kočevje, of which he also made a plan.

Nagel is significant for his determined examination of caves in several totally different areas. He was the first to do this, thereby anticipating the better known 19th-century figures such as Schmidl, Kraus, and Martel. He did not, however, discuss the regional characteristics of each set of caves.

Adolf Schmidl (1802–63)

Schmidl (Figure 2) was described by Martel as "the real originator of speleology or the scientific study of caves". He was born on 18 May 1802 at Königswart (now Kynžvart) in the Czech Republic, and came to Vienna in his youth, where he studied philosophy and law from 1819 to 1825. His cave explorations (between 1850 and 1856) were made when he was working in Vienna. From 1857 until his death on 20 November 1863 he was professor of geography at the Polytechnic at Budapest.

Nearly all Schmidl's cave exploration took place in Slovenia, Austria, and Hungary. He set out with the avowed intention of "establishing the exact topography of the caves" of Slovenia, exploring and recording meticulously and having accurate surveys made by his companion Ivan Rudolf. His accounts of the several caves of the Postojna system, and at Predjama, Škocjanska jama, Križna jama, and other caves provided the first exact descriptions of them. In Postojnska jama itself, his major achievement was the discovery of nearly half a kilometre more of the underground river Pivka. The most important of his findings were published in his book *Die Grotten und Höhlen von Adelsberg, Lueg, Planina und Laas* (1854), together with a discussion on cave fauna, cave meteorology, and other scientific aspects of cave research. In the opinion of Martel, this book was "the real starting point of speleology". In 1855 Schmidl turned his attention to Austria. In particular he explored the Geldloch and published a plan. His account is very detailed and pays particular attention to temperatures and to barometric pressures as a means of determining altitudes. It was in the same paper that he drew attention to Strein's 16th-century exploration of the cave, and printed Strein's manuscript in its entirety. In August of the following year he made a thorough investigation of the Aggtelek cave in Hungary, which, with its length of 8667 m, remained the longest cave in Europe until 1893. Once again he included tables of temperatures and also a note on cave fauna.

Schmidl made a conscious effort to bring the various branches of cave study together, both in research and in publication, and it is significant that the earliest word for this purpose in any language, Höhlenkunde, meaning cave study, was introduced by Schmidl in 1850. It is this breadth of his interests in cave studies, coupled with the extent of his discoveries and the technical difficulties that he surmounted, that justify Schmidl's reputation.

Édouard Alfred Martel (1859–1938)

Martel (Figure 3) was born at Pontoise in northern France and practised law in Paris from 1887 until 1899, after which he devoted himself entirely to caves and to writing. Martel's explorations were the basis of all his cave work. Without them he would have had nothing with which to influence people, and he would not have had the data to support his ideas on underground water flow and other scientific matters. Indeed, his main objectives were to explore, survey, record, and publish, which he did to an impressive extent over many years. The exact number of explorations during his entire career has never been worked out, but has been estimated to be about 1500. Even by the end of 1893 (after only six years of exploration) he had visited 230 caves. Of the 110 vertical caves included in this number, 90 were previously undescended; of 120 mainly horizontal caves, 30 were previously unentered, and 45 had not been explored fully. Also by this time, he had surveyed 28 km of cave passages himself, while his collaborators had surveyed another 22 km.

Most of his discoveries were made in the course of 26 annual "campaigns", from 1888 to 1913, although he continued his cave studies and writing after World War I and up to his death in 1938. The campaigns took place in seventeen countries: France, Belgium, Germany, Ireland, England, Norway, Switzerland, the Czech Republic, Slovakia, Hungary, Slovenia, Greece, Italy, Spain, Turkey, Russia, and the United States. All the discoveries were fully published at the time. Experienced as he was in exploring many different kinds of ever more difficult caves, it is only to be expected that Martel developed some of the basic techniques involved. In 1889 he and Gaupillat were the first to use telephones in caves, where they found them useful on deep shafts, especially any that were more than 100 m deep. Their apparatus was fairly light for the time, being about 8 cm in diameter and weighing 480 g, and they used it successfully with as much as 400 m of wire. Martel's other claim for originality in equipment was in the cave use of the folding canvas canoe. The volume and quality of his publications are as impressive as his explorations. He produced some 20 books and 780 papers, many of them in scientific journals of the highest quality. Many of his writings were also translated and published abroad.

Martel was more than just a cave explorer, cave researcher, and writer. He also consciously caused the study of caves to spread into countries outside the European core where it was already flourishing. He was a leader who inspired and encouraged people to investigate the caves and karst problems of their own lands. The Société de Spéléologie, founded by Martel in Paris in 1895, enjoyed high scientific standing from the outset, and it was one means by which he contrived the extension of cave study into an international subject. Foreign membership of the Société was remarkably high. Some 21% of the founder members lived outside France, indicating the close links already existing before 1895. Between 1895 and 1904 the proportion rose to 29%; in addition three foreign cave societies were members. Many papers by foreign contributors, most of them members, were published in the Société's journal, *Spelunca*, comprising between 14% and 50% of the papers printed in individual years.

There can be no doubt about the extent of Martel's personal links with speleology in other countries. Of his 26 annual campaigns of exploration, 19 went outside France in what are now 20 nations and, in addition, he made lecture tours and other visits abroad. At least 61 of his own publications on caves appeared elsewhere in his lifetime. Many of these were papers presented to learned societies, and there were also popular articles and the texts of public lectures; other were simply translations or reprints of work already published in France, showing the interest with which this was regarded abroad.

Norbert Casteret (1897–1987)

Born at Saint-Martory in Haute-Garonne (France), a soldier in World War I, and an enthusiastic skier and swimmer, Casteret (Figure 4) approached cave exploration more as a dedicated sportsman than a scientist. His achievement of over 2000 caves explored, including many very major ones, was done mostly from 1922 till when he was over 60, but his 35 m rope climb in and out of the Gouffre de Planque was made at the age of 15. Six of his new discoveries, including Montespan, contained prehistoric art. Other specially notable explorations were in the Grotte de la Cigalère, the Gouffre Martel (then the deepest in France), and the Henne-Morte. Although the discoveries that he made during his long period of active exploration were extensive and important, speleology was already an established subject in his lifetime, and similar explorations of equal difficulty were being made elsewhere.

What sets Casteret apart is his writing, conveying his enthusiasm as well as his results to the general public and hence inspiring many young people to take up cave exploration. His 47 books have been translated into 16 languages and some have been published in Braille. The diversity of Casteret's achievements is reflected in the medals that he received, which include the Croix de Guerre, 4 gold medals from geographical and sporting organizations, and 3 life-saving medals (one of them earned when he was 78).

TREVOR SHAW

See also **Archaeologists; Biospeleologists; Geoscientists**

Further Reading

André, D. (editor) 1999. *L'homme qui voyageait pour les gouffres. Actes du colloque tenu à Mende les 17 et 18 octobre 1997* [The man who travelled for caves], Mende: Archives départementales de la Lozère Proceedings of a symposium in honour of Martel

Casteret, N. 1943. *E.A. Martel, explorateur du monde souterrain* [E.A. Martel, Explorer of the Underground World], Paris: Gallimard

Chabert, J. 1986. *Norbert Casteret: bibliographie*, Paris: Gandini and Spelunca Librairie (with supplement, *Compléments et rectificatifs*, by M. Casteret and J. Chabert, 1997)

Chabert, C. & de Courval, M. 1971. *E.A. Martel 1859–1938 bibliographie*, Paris: Spéléo-Club de Paris

Jolfre, J. 1992. *Norbert Casteret explorateur d'abîmes* [Norbert Casteret Cave Explorer], Toulouse: Milan

Reisp, B. 1983. *Kranjski polihistor Janez Vajkard Valvasor* [Carniolan polymath J.W. Valvasor], Ljubljana: Mladinska Knjiga

Salzer, H., 1931. *Die Höhlen- und Karstforschungen des Hofmathematikers Joseph Anton Nagel* [The Cave and Karst Researches of the Court Mathematician J.A. Nagel], *Speläologisches Jahrbuch*, 10–12: 111–21 & Taf. VII–XI

Shaw, T.R. 1978. Adolf Schmidl (1802–1863) the father of modern speleology? *International Journal of Speleology*, 10: 253–67

Shaw, T.R. 1992. *History of Cave Science, the Exploration and Study of Limestone Caves, to 1900*, 2nd edition, Broadway, New South Wales: Sydney Speleological Society

Gives extra information on all except Casteret

SPELEOTHEM STUDIES: HISTORY

As in the case of speleogenesis history, explanations of how speleothems formed were influenced by the wider beliefs of the time. As late as the 18th century it was commonly thought that rocks, especially crystals and mineral ores, grew: that they possessed a form of life lower than that of plants but rather similar. So it is no surprise to find that this idea was once applied to stalactite growths. Explanations of speleothem growth can be divided into three classes: those in which the material grows as a living substance; those in which it is deposited in some way from underground vapours; and those in which dripping water is either congealed into stone or deposits stony material that it has carried with it.

The belief that rocks and minerals grow like plants was applied to speleothems in the 17th and early 18th centuries. It was subsequently developed and persisted even into the 19th century. It was first advocated by John Beaumont, who studied caves in the Mendip Hills of England and presented his conclusions to the Royal Society in 1676. Like most of his contemporaries, Beaumont believed that fossil shells had grown in the rock where they were found, so the application of the idea to stalagmites was a logical extension of that belief. Better known is J.P. Tournefort's visit to the cave in the Greek island of Antiparos in August 1700 (see Figure in Caves in History: The Eastern Mediterranean). The exploration was moderately difficult for the time, involving rope descents and the use of a rigid ladder brought in for the purpose, but Tournefort's main interest was in the origin of speleothems. He noted what looked like the growth rings of trees, and stated that "These stems of marble must certainly vegetate".

The vapours occurring in caves, either from normal underground humidity or derived from deep in the earth, were held by some to condense or deposit to form stalactites or else to cause the ordinary percolating water to solidify. In one theory, vapours were believed to provide the nourishment that enabled vegetative growth to proceed.

Dripping water was most commonly regarded as the source of speleothem material. In some cases it was believed to congeal of itself from a variety of causes; in others it was supposed to consist of or contain a "lapidifying juice", whose special nature enabled it to turn into stone. A natural development of this idea considered that the petrifying material was a stony substance already present in the water, and gradually the distinction was made between material in suspension and in solution. Slowly, from the beginning of the 18th century, it came to be realized that the presence of this material was made possible by the water containing an acid of some sort, which enabled it to dissolve part of the surrounding rock. The recognition of the atmospheric origin of this acid, and finally, in 1812, of the chemical reactions involved, brings us to the present day.

Typical of early statements at the start of the pathway to modern thought is this one of 1655 by the Danish professor, Olaus Worm. He wrote that speleothems are:

> Formed by deposition from water which has rock-forming properties because it carries within itself finely divided mineral matter.... Dripping down from a high crack, it adheres where it can in the shape of a cone ... (quoted in Shaw, 1992)

Although he and several others spoke of the water depositing its particles or salts, they did not say what they thought caused the deposition. The first person explicitly to attribute the deposition of stalactitic matter to the effect of evaporation of water was Karl Nicolaus Lang, in 1708. But as yet there was no suspicion that it was an acid that caused the solution of these "salts" and that it was evaporation of this acid that caused the deposition.

The explanations incorporating such an acid can be divided into three groups. First come those (1700 to 1742) in which the solution of rocky material is aided by some unspecified acid in the atmosphere, which may be what is now recognized as carbon dioxide. Then comes another group (1782 to 1794) in which the atmospheric acid seems to be equivalent in every way to carbon dioxide, although recognition of its chemical composition was still delayed by the waning concept of phlogiston. And finally, from 1812 onwards, the role of carbon dioxide was fully appreciated. The unspecified acid was described as the "acid spirit of the atmosphere" or, more often, simply as "the aerial acid". J.G Lesser in 1735 suggested that rock material was dissolved by the acid, but that this material was deposited in caves by the evaporation of water. A few years later, however, in 1742, Louis Bourget associated the deposition with the release of an "air" from the water.

In 1812 the distinguished French naturalist, Georges Cuvier, suggested that deposition occurred as the acid gas escaped from the water, and he was the first person positively to identify this gas as carbonic acid:

> Certain waters, after dissolving calcareous substances by means of the superabundant carbonic acid with which they are impregnated, allow these substances to crystallize after the acid has evaporated; and, in this manner, form stalactites, and other concretions. (Quoted in Shaw, 1992)

This acid solution theory became generally, although not universally, accepted in the first half of the 19th century, in contrast to its sporadic and partial appearance before 1812. Two factors probably brought this about. The clear and authoritative statements in English by the very influential scientists Cuvier and Benjamin Silliman made it available in at least two languages, and the demise of the phlogiston theory enabled the chemical action of carbon dioxide in the solution of limestone to be more clearly understood. The carbon dioxide involved in the processes of solution and deposition was at first either implied to come from atmospheric air, or else its origin was ignored altogether. It was left to Justus von Liebig in about 1840 to associate the production of the gas by rotting vegetation with its concentration in the soil and the rock beneath. Liebig also appreciated for the first time that it was only the excess of carbon dioxide over and above the equilibrium quantity that was released in the cave.

Since 1900 there has been a large growth in interest in mineralogy and types of speleothems (see entries on Speleothems).

TREVOR SHAW

Further Reading

Shaw, T.R. 1992. *History of Cave Science: The Exploration and Study of Limestone Caves, to 1900*, 2nd edition, Broadway, New South Wales: Sydney Speleological Society
This gives much more information, with references to the original sources.

Shaw, T.R. 1997. Historical introduction. In *Cave Minerals of the World*, by C. Hill & P. Forti, 2nd edition, Huntsville, Alabama: National Speleological Society
A much fuller treatment than in this entry, with references and including also non-calcite speleothems.

SPELEOTHEMS: CARBONATE

A speleothem is a secondary mineral deposit formed in caves (Moore, 1952). The term refers to the mode of occurrence of a cave mineral; i.e. its morphology, or how it looks, not to its composition. This section covers carbonate speleothems. For a discussion of the various other mineral classes that form speleothems in caves, refer to the entry on Minerals in Caves.

The carbonates, a group that has the $(CO_3)^{2-}$ anion as its essential component, are the most important class of cave minerals. The two most common cave minerals, calcite $(CaCO_3)$ and aragonite $(CaCO_3)$, belong to this class and together probably comprise >95% of all mineral deposits in caves. Hydromagnesite $[Mg_5(CO_3)_4(OH)_2 \cdot 4H_2O]$ is the third most common carbonate cave mineral, and is probably the most common constituent of moonmilk in dolostone caves. Carbonate speleothems have been used by humans for medicinal and other purposes for over 2000 years.

Practically all of the various forms that speleothems display in caves are represented by the carbonate minerals, with the exception of fibrous speleothems, which is the main form taken by the sulfate minerals (see Speleothems: Evaporite). Hill & Forti (1997) classified speleothems into 38 "official" types, based primarily on morphology and secondarily on origin and

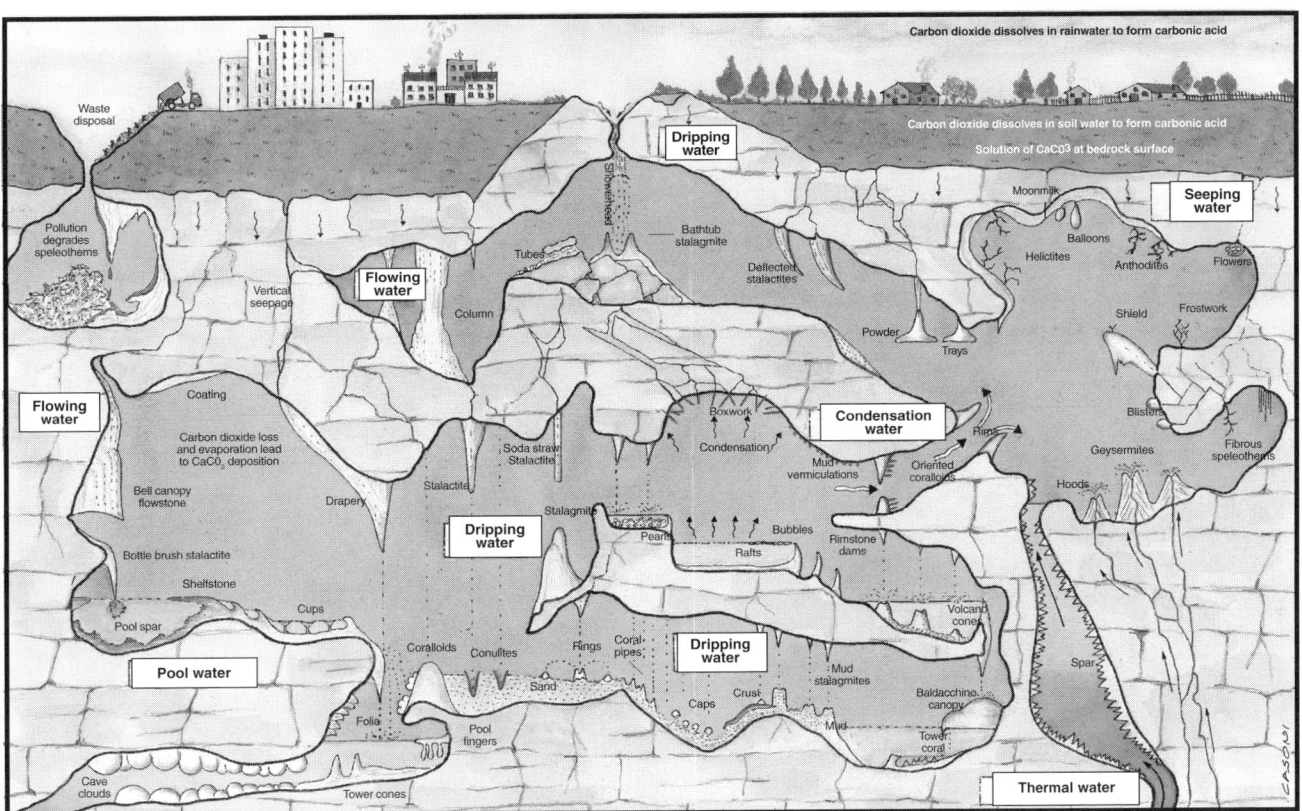

Speleothems: Carbonate: Figure 1. Diagram of various carbonate speleothem types and subtypes formed by flowing water, dripping water, pool water, seeping water, condensation water, and thermal water. From *Cave Minerals of the World*, Hill & Forti, 2nd edition. Drawing by Luciano Casoni.

crystallography. These various types form from flowing water, dripping water, pool water, seeping water, condensation water, and thermal water (or a combination of two or more of these processes) (see Figure 1 and photos in colour plate section). Carbonate speleothem types (and subtypes) formed primarily from dripping water are stalactites, stalagmites, columns, coralloids, conulites, coral pipes, showerheads, cave rings, and cave caps. Those formed from flowing water are flowstone, draperies, and canopies. Types and subtypes formed from pool water (at or near the surface of the pool) are cave rafts, shelfstone, rimstone dams, cave pearls, folia, pool fingers, cave bubbles, cave cups, "bottle-brush" stalactites, tower coral, cave cones, and pool spar; those formed in deep pools or in the "phreatic zone" (zone of saturation) are cave clouds and cave mammillaries. Seeping water can create a variety of speleothems: helictites, moonmilk, boxwork, frostwork, anthodites, shields, coatings, crusts, cave blisters, cave balloons, cave tubes, and powder. Condensation water forms rims, cave trays, and oriented coralloids, and thermal water forms large spar crystals and geysermites. Variation of colour and luminescence in these speleothem types is due to impurities in solution (see Speleothems: Luminescence).

The chemical evolution of cave waters with regard to the precipitation of carbonate speleothems can be divided into three stages (Holland *et al.*, 1964): (1) carbonation in the soil zone, where rainwater picks up carbon dioxide to form a weak carbonic acid (Figure 1); (2) dissolution of carbonate bedrock, where the carbonic acid dissolves limestone (and dolostone) bedrock; and (3) carbonate precipitation, where carbon dioxide exchange with the cave air, and to a lesser degree evaporation, causes carbonate speleothems to precipitate.

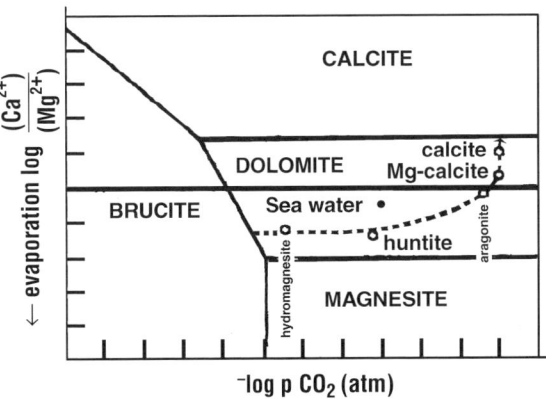

Speleothems: Carbonate: Figure 2. A graph showing the evolution of cave water with respect to increasing evaporation and carbon dioxide loss. After Lippman (1973). From *Cave Minerals of the World*, Hill & Forti, 2nd edition.

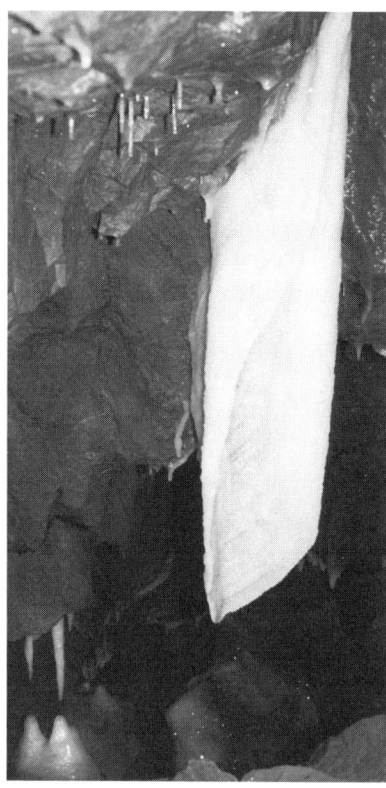

Speleothems: Carbonate: Figure 3. The Angel's Wing, a pure white calcite curtain in Shatter Cave, Mendip Hills, England. (Photo by John Gunn)

In the last few decades, carbonate speleothems (in particular stalagmites and flowstone) have proved important to many fields of research such as absolute dating (Ford, 1997), paleoclimatology and paleoenvironmental reconstruction (see Paleoenvironments: Speleothems); and paleoseismology (see Paleotectonics from Speleothems). Hill & Forti (1997) list hundreds of published papers dealing with these categories.

Because carbonate (and other) speleothems are often rare or fragile, every effort should be made to preserve them within the cave environment. Conservation efforts in both wild and show caves have included: (1) limiting access to mineralogically sensitive caves by gating or other techniques (Veni, 1997); (2) monitoring the cave environment with respect to temperature, humidity, carbon dioxide, and lighting levels (Cabrol, 1997; Veni, 1997); (3) effecting repair of broken speleothems (Veni, 1997); and (4) restricting collection of speleothems and cave minerals, even for scientific sampling. Cave minerals and speleothems belong in, and should stay in, caves!

CAROL A. HILL and PAOLO FORTI

See also **Minerals in Caves; Speleothems: Evaporite; Speleothems: Luminescence; Speleothem Studies: History**

Works Cited

Cabrol, P. 1997. Protection of speleothems. In *Cave Minerals of the World*, 2nd edition, edited by C.A. Hill & P. Forti, Huntsville, Alabama: National Speleological Society

Ford, D.C. 1997. Dating and paleo-environmental studies of speleothems. In *Cave Minerals of the World*, 2nd edition, edited by C.A. Hill & P. Forti, Huntsville, Alabama: National Speleological Society

Hill, C.A. & Forti, P. (editors) 1997. *Cave Minerals of the World*, 2nd edition, Huntsville, Alabama: National Speleological Society

Holland, H.D., Kirsipu, T.W., Huebner, J.S. & Oxburgh, U.M. 1964. On some aspects of the chemical evolution of cave waters. *Journal of Geology*, 72: 36–67

Lippman, F. 1973. *Sedimentary Carbonate Minerals*, Berlin and New York: Springer

Moore, G.W. 1952. Speleothem—a new cave term. *National Speleological Society News*, 10(6): 2

Veni, G. 1997. Speleothems: preservation, display, and restoration. In *Cave Minerals of the World*, 2nd edition, edited by C.A. Hill & P. Forti, Huntsville, Alabama: National Speleological Society

Further Reading

Hill, C.A. & Forti, P. (editors) 1997. *Cave Minerals of the World*, 2nd edition, Huntsville, Alabama: National Speleological Society

The relative amount of carbon dioxide loss, evaporation, and the amount of magnesium in solution determines the carbonate mineral species that will precipitate (Figure 2, follow dashed line and arrow). As groundwater entering a cave degasses carbon dioxide, calcite is precipitated so that the magnesium ion increases relative to the calcium ion (i.e. the Mg / Ca ratio increases). After calcite, magnesium-enriched calcite (Mg-calcite) precipitates, then aragonite (Mg / Ca ratio >2.9), then huntite, then hydromagnesite (Mg / Ca >16). Aragonite, the polymorph of calcite (both have the same chemical composition but different crystal structures), can usually (but not always) be distinguished from calcite because it displays a needle-like habit.

SPELEOTHEMS: EVAPORITE

A speleothem is a secondary mineral deposit formed in caves (Moore, 1952). The term refers to the mode of occurrence of a cave mineral, i.e. its morphology, or how it looks, not to its composition (Hill & Forti, 1997). Evaporite speleothems are those where deposition is controlled primarily by evaporation, even though a number of different mechanisms may be involved in the evaporative process (see Figure 1).

The two main classes of evaporite minerals in caves are the sulfates and halides (for other cave mineral classes, see Minerals in Caves). Sulfates represent the largest class, with gypsum ($CaSO_4 \cdot 2H_2O$) being by far the most common cave sulfate and the third most common cave mineral after calcite and aragonite (it is estimated that gypsum represents $c.2–3\%$ of all cave deposits). Halide speleothems are much less abundant than sulfates, usually forming where halite (NaCl) bedrock exists in the overburden, but also, rarely, forming from bat guano. The best and widest display of halide speleothems has been described from the Mount Sedom caves of Israel (Frumkin & Forti, 1997).

The depositional environments under which evaporite speleothems form are varied: (1) a limestone, gypsum, or halite rock

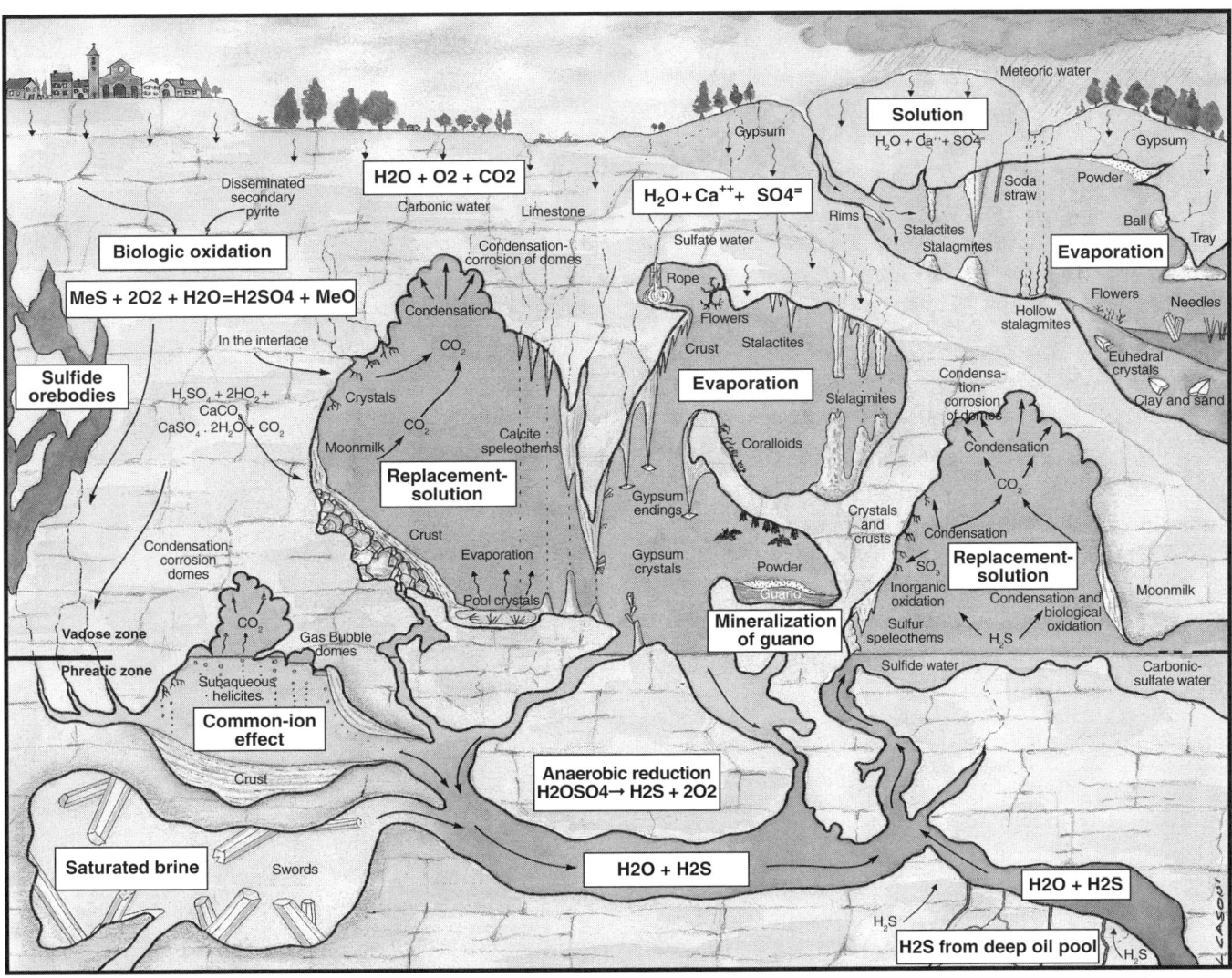

Speleothems, Evaporite: Figure 1. All mechanisms causing sulfate deposition in caves. From *Cave Minerals of the World*, Hill & Forti, Second Edition. Drawing by Luciano Casoni.

setting; (2) lava tubes; (3) bat guano; (4) associated with ore bodies; and (5) associated with fumarole activity. Evaporite speleothems can assume almost all the same forms as carbonate speleothems (see Speleothems: Carbonate), but they usually display a fibrous or filamentary habit. Fibrous speleothems can be divided into four subtypes, depending on the length of the crystal fibres and the way in which the fibres intertwine with each other: (1) single crystals (hair); (2) bunches of single crystals (cotton) (Figure 2); (3) fibres matted together (rope) (Figure 3); and (4) fibres that split (flowers). Often one or more of these subtypes will occur together; e.g. all four types coexist in Valea Rea Cave, Romania (Onac et al., 1995).

Even where evaporite speleothems do form as dripstone (Figure 4), the distinct genetic mechanism under which these form (i.e. supersaturation due to evaporation) causes them to differ in morphology from carbonate speleothems. For example, sulfate stalactites are typically more contorted or multibranched than carbonate stalactites, and their growth depends, almost exclusively, upon surficial percolation water rather than upon water that feeds through a central tube. This commonly results in the central tube being absent, or partially (if not completely) obstructed. Furthermore, the effect of permanent air currents with respect to carbonate and evaporite stalactites is exactly the opposite. In the carbonate case, growth is controlled by CO_2 diffusion, and because such diffusion is not influenced by air currents, the stalactite is deflected in the direction of air movement (i.e. in the same direction that the water droplets are deviated before they fall). However, in evaporite speleothems, the inverse effect dominates, and so stalactites deviate towards the source of the air current where maximum evaporation is occurring.

Compared with the carbonates, evaporite minerals form helictites only rarely and with extreme difficulty. This is because the process of evaporation leads to a greater obstruction of the helictite's feeding capillary. Also, it is much easier for wind-related speleothems, such as rims and trays, to develop in an evaporative environment. Euhedral crystal occurrences are also more common: gypsum crystals may range from a few microns

Speleothems, Evaporite: Figure 2. A pile of selenite crystals in the Zhucaojing cave, in the Xingwen karst of China, where pyrite beds overlying the limestone provide a ready source of sulfur and sulfate. (Photo by Andy Eavis)

(gypsum powder) up to several metres (gypsum swords) in length.

Since the minerals making up evaporite speleothems are often highly soluble, these speleothems can therefore have shorter "lifespans" in caves than carbonate speleothems. For example, thenardite (Na_2SO_4) speleothems are only deposited in the volcanic caves of Mount Etna, Italy during the initial hot and dry time

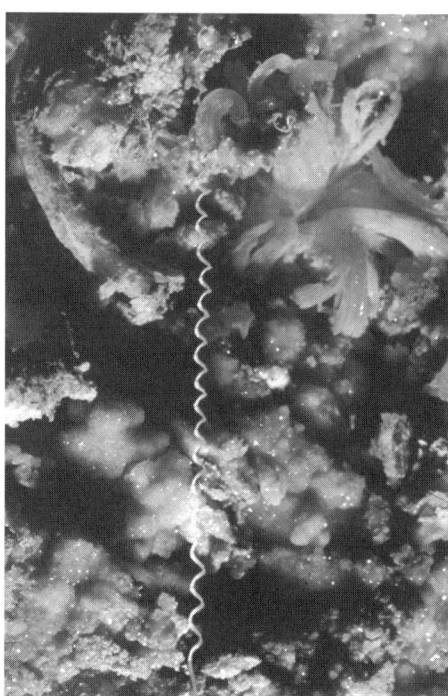

Speleothems, Evaporite: Figure 3. "The Spring", a gypsum rope in Puketiti Flower Cave, New Zealand. (Photo by John Gunn)

Speleothems, Evaporite: Figure 4. Salt stalagmite, Thamapana Cave, Nullarbor Plain, Australia. (Photo by John Gunn)

of lava tube formation. This thenardite rapidly transforms into mirabilite as water seeps into the cave and the temperature falls; then it completely dissolves when the cave reaches equilibrium conditions and the seeping water stops evaporating. Often evaporite speleothems are seasonally ephemeral, like the mirabilite and epsomite flowers present in the gypsum caves of Bologna in Italy and the gypsum powders found in the Pinega (Russia) and New Mexico (United States) gypsum caves (Forti, 1996).

Evaporite speleothems—in particular gypsum—have proved to be useful tools for paleoclimatic studies. The abundance of calcite and gypsum speleothems in caves formed in gypsum rock depends strictly on the climate of the area in which the cave is located (Calaforra & Forti, 1999). Therefore, where there is a sudden change in the depositional record (from dominant calcite to dominant gypsum, or vice versa, or even to no deposition), it is reasonable to suppose that the climate of the area had changed during this time.

Extra effort is required to preserve evaporite speleothems, due to their fragile crystal form (e.g. hair, cotton, flowers) and to their unstable nature outside the cave environment (i.e. many evaporite minerals disintegrate into powder when removed from a cave). Like carbonate speleothems and all cave minerals, these belong in, and should stay in, caves!

PAOLO FORTI and CAROL A. HILL

See also **Minerals in Caves**

Works Cited

Calaforra, J.M. & Forti, P. 1999. May the speleothems developing in gypsum karst be considered paleoclimatic indicators? *Abstracts, International Union for Quaternary Research, 15th International Congress, August, 1999, Durban, South Africa*, Durban: International Union for Quaternary Research

Forti, P. 1996. Speleothems and cave minerals in gypsum caves. *International Journal of Speleology*, 25(3–4): 91–104

Frumkin, A. & Forti, P. 1997. Liquid Crystal Cave. In *Cave Minerals of the World*, 2nd edition, C.A. Hill & P. Forti, Huntsville, Alabama: National Speleological Society

Hill, C.A. & Forti, P. 1997. *Cave Minerals of the World*, 2nd edition, Huntsville, Alabama: National Speleological Society

Moore, G.W. 1952. Speleothem—a new cave term. *National Speleological Society News*, 10(6): 2

Onac, B.P., Bengeanu, M., Botez, M. & Zeh, J. 1995. Preliminary report on the mineralogy of Valea Rea Cave (Bihor Mountains, Romania). *Theoretical and Applied Karstology*, 8: 75–9

Further Reading

Hill, C.A. & Forti, P. 1997. *Cave Minerals of the World*, 2nd edition, Huntsville, Alabama: National Speleological Society

SPELEOTHEMS: LUMINESCENCE

Many speleothems exhibit luminescence when exposed to ultraviolet (UV) light sources or other high-energy beams. Nearly 50 minerals known in caves have the capacity to exhibit luminescence, but so far, only 17 are actually observed to be luminescent in speleothems. For detailed explanation of luminescence types and properties of minerals, refer to Shopov (1997). Most known luminescent centres in calcite are inorganic ions of Mn, Tb, Er, Dy, U, Eu, Sm, and Ce. Claims that Sr causes luminescence in carbonates and Cu in calcite and aragonite are in error, as are suggestions that visible luminescence in calcite may be Pb-activated, because Pb emits only ultraviolet light. Minerals contain many admixtures. Usually several centres activate luminescence of the sample and the measured spectrum is the sum of the spectra of two or more of them. Conventional luminescent research methods have a number of disadvantages when applied to speleothems, so several special methods have been developed (see Table).

Luminescence of minerals formed at normal cave temperatures is due mainly to molecular ions and sorbed organic molecules. Luminescence of uranyl ions (Figure 1) is also very common in such speleothems. Before using a speleothem for any paleoenvironmental work, it is necessary to determine that all of its luminescence is due to organics.

Calcite speleothems frequently display luminescence, which is produced by calcium salts of fulvic and humic acids (Shopov, Dermendjiev & Buyukliev, 1989) derived from soils above the cave (White & Brennan, 1989; van Beynan et al., 2001). These acids are released by the roots of living plants, and by the decomposition of dead vegetative matter. Root release is modulated by visible solar radiation via photosynthesis, while rates of decomposition depend exponentially upon soil temperature. Soil temperatures are controlled chiefly by solar infrared and visible radiation (Shopov et al., 1994) where soils are bare or grass-covered, and by air temperatures where soils are covered by forest or bush. In the first case, microzonality of luminescence of speleothems can be used as an indirect Solar Activity (SA) index, and in the second case, as a paleotemperature proxy.

A time series of a Solar Activity (SA) index "Microzonality of Luminescence of Speleothems" can be obtained by Laser Luminescence Microzonal Analysis (LLMZA) of cave flowstones (see Table).

The luminescing organics in speleothems can be divided into four types: (1) calcium salts of fulvic acids; (2) calcium salts of humic acids; (3) calcium salts of huminomelanic acids; and (4) organic esters. All these four types are usually present in a single speleothem, as hundreds of different compounds with similar

Speleothems: Luminescence: Special speleothem luminescence research methods.

Method	Authors	Obtainable information
I. Impulse Photography of Luminescence (IPL):	Shopov & Tsankov (1985)	Diagnostics of minerals; registration of colour and zonality of fluorescence, phosphorescence and its spectra; UV photography; extraction of single mineral samples; chemical changes of the mineral-forming solution; climate and solar activity variations during the Quaternary.
1. Photography of phosphorescence (IPP)		
2. Photography of fluorescence & phosphorescence (IPFP).	Shopov & Grynberg (1985)	
II. Laser Luminescent–MicroZonal Analysis (LLMZA)	Shopov (1987)	Microzonality of luminescence; changes of the mineral-forming conditions. High resolution records of climate and solar activity variations. Reconstruction of annual rainfall and annual temperature in the past. Estimation of past cosmic rays and galactic cosmic rays. Speleothem growth hiatuses.
III. Colour Slide Spectrophotometry (CSS)	Shopov & Georgiev (1989)	Wide-band spectra of phosphorescence, fluorescence, and diffuse reflectance of minerals; spectra of quick processes.
IV. Autocalibration Dating (ACD)	Shopov et al. (1991)	High precision speleothem dating of speleothems of any age, climatic and solar activity cycles, variations of the speleothem growth rate.
V. Time Resolved Photography of Phosphorescence (TRPP)	Shopov et al. (1996)	Determination of the lifetime of the luminescent centre. Uplift of the region. Past mixing of surface and epithermal or hydrothermal waters during mineral growth. Estimation of the temperature of deposition, plus all information obtainable by IPP.

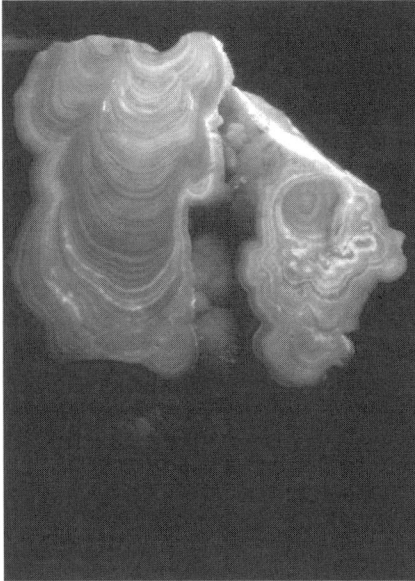 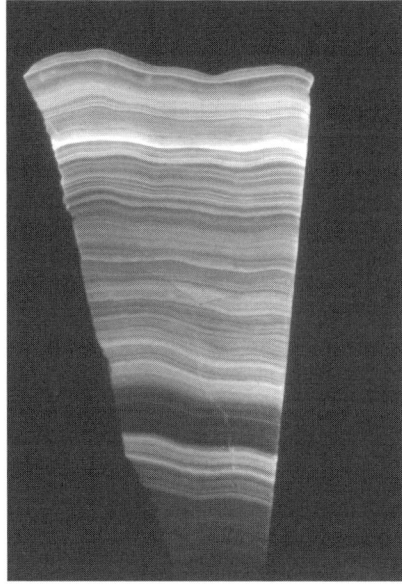 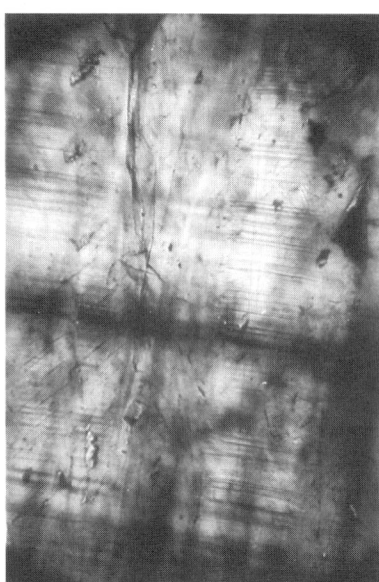

Speleothems: Luminescence: Figure 1. (left) Annual banding of luminescence of uranyl ions in a calcite flowstone, due to variations in pH of the water (also in colour plate section).

Speleothems: Luminescence: Figure 2. (middle) Luminescence of calcite growing over glacial and interglacial periods with variations of the luminescence colour due to changes in vegetation growing over the cave.

Speleothems: Luminescence: Figure 3. (right) Annual banding of luminescence of organics in a flowstone cross section. Microscopic magnification × 50. Dark parts are clay inclusions. Fluid inclusions are visible as bubbles and some crystal surfaces are also visible.

chemical behaviour but different molecular weights. The concentration distribution of these compounds (and their luminescence spectra) depends on the type of soils and plants over the cave, so study of the spectra can yield information about paleosoils and past vegetation. Colour changes in the visible portion of the luminescence spectrum imply major changes in plant ecology and thus organic matter decomposition, but are rarely observed (Figure 2).

Speleothem growth rate may vary significantly within a single speleothem (Shopov et al., 1994). Where there are no growth interruptions (hiatuses), such variations represent rainfall variation. Luminescence techniques can visualize annual microbanding on a scale of 20–60 μm that is not visible in normal light (Figure 3). If this banding is visible in normal light, or the luminescence curves have sharp profiles or jumps (e.g. as in Baker et al., 1993), it suggests that speleothem growth ceased for a certain period during the year and that such time series are not useful for deriving rainfall proxy records for paleoclimate studies.

Luminescence in high-temperature hydrothermal minerals is mainly due to cations, because any molecular ions and molecules are destroyed. It signifies the hydrothermal conditions under which the minerals form. Minerals deposited from low-temperature hydrothermal solutions display short-lived fluorescence, due to the cations, and longer duration phosphorescence due to molecular ions. Where calcite has only orange-red, short-lived phosphorescence, it is an indicator of very high-temperature hydrothermal solutions (>300°C). But if it has also long-time phosphorescence, then it is a lower-temperature deposit. The minimum temperature for the appearance of orange-red luminescence is estimated to be between 46°C and 60°C. Luminescence of hydrothermal calcite formed at lower temperatures than these is similar to the standard meteoric speleothem luminescence shown in Figure 1.

The tectonic uplift history of an area can be deduced from luminescence in combination with absolute dating (Shopov et al., 1996), where the luminescence is due to epithermal mineralizing solutions in the older parts of a speleothem deposit, but the mixing of these waters with meteoric waters containing organics appears in the younger parts.

Finally, luminescence may be used to estimate the absolute age of the speleothem itself by the autocalibration dating method (Shopov et al., 1991), which has been shown to be the most precise method for samples younger than 2000 years (Shopov et al., 1994).

YAVOR Y. SHOPOV

See also **Paleoenvironments: Speleothems**

Works Cited

Baker, A., Smart, P.L., Edwards, R.L. & Richards, D.A. 1993. Annual growth banding in a cave stalagmite. *Nature*, 304: 518–20

Shopov, Y.Y. & Grynberg, M.A. 1985. A new method for direct photography of luminescence. *Exped. Ann. Sofia Univ.* 1: 139–45

Shopov, Y.Y. & Tsankov, L.T. 1985. A photographic apparatus for luminescent analysis (B. Patent 40439 from 30 Aug. 1985). *Bulletin of Institute for Inventions and Rationalizations*, 12: 1–7

Shopov, Y.Y. 1987. Laser luminescent microzonal analysis: A new method for investigation of the alterations of climate and solar activity during the Quaternary. In *Problems of Karst Study of Mountainous Countries*, edited by T. Kiknadze, Tbilisi, Georgia: Meisniereba

Shopov, Y.Y. & Georgiev, L.N. 1989. CS-Spectrophotometry: A new method for direct registration of spectra of variable processes. *Abstracts of XXVI Colloquium Spectroscopicum Internationale*, vol. 3, edited by D.L. Tsalev, Sofia, Bulgaria

Shopov, Y.Y., Dermendjiev, V. & Buyukliev, G. 1989. Investigation on the variations of climate and solar activity by a new method—LLMZA of cave flowstore from Bulgaria. *Proceedings of the 10th International Congress of Speleology*, vol.1, edited by A. Kosà, Budapest: Hungarian Speleological Society

Shopov, Y.Y. Dermendjiev, V. & Buyukliev, G. 1991. A new method for dating of natural materials with periodical macrostructure by autocalibration and its application for study of the solar activity in the past. In *Proceedings of the International Conference on Environmental Changes in Karst Areas*, edited by U. Sauro, A. Bondesan & M. Meneghel, Padova: Dipartimento di Geografia, Universita di Padova

Shopov, Y.Y., Ford, D.C. & Schwarcz, H.P. 1994. Luminescent micro-banding in speleothems: High-resolution chronology and paleoclimate. *Geology*, 22: 407–10

Shopov, Y.Y., Tsankov, L., Buck M. & Ford, D.C. 1996. Time resolved photography of phosphorescence: A new technique for study of thermal history and uplift of thermal caves. In *Climate Change: The Karst Record*, edited by S.-E. Lauritzen, Charlestown, West Virginia: Karst Waters Institute

Shopov, Y.Y. 1997. Luminescence of cave minerals. In *Cave Minerals of the World*, 2nd edition, edited by C.A. Hill & P. Forti, Huntsville, Alabama: National Speleological Society

van Beynan, P.E., Bourbonniere, R., Ford, D.C. & Schwarcz, H.P. 2001 Causes of colour and fluorescence in speleothems. *Chemical Geology*, 175(3–4): 319–41

White, W.B. & Brennan, E.S. 1989. Luminescence of speleothems due to fulvic acid and other activators. *Proceedings of the 10th International Congress of Speleology*, vol. 1, edited by T. Hazslinszky & K. Bolner-Takacs, Budapest: Hungarian Speleological Society

SPELEOTHERAPY

Speleotherapy is a special type of medical treatment, involving subterranean environments and their specific microclimate. It is a therapeutic method, which is used for rehabilitation in curing respiratory diseases, asthma, and allergies (Horvath, 1989). It has been documented from prehistoric times, when humans were using caves for both shelter and ritual. The modern application of speleotherapy began in the 1950s, when speleotherapeutic centres were established in some Eastern and Central European countries, bringing together experts in speleology, medicine, especially pulmology, and natural sciences. In the United States and United Kingdom, speleotherapy is not known to be practised.

During World War II, the inhabitants of Ennepetal, Germany, used to take shelter in Kluterthöhle. Dr Karl Hermann Spannagel (1909–86) was surprised by the improvement in the health of asthma sufferers while they visited the cave. Soon after the end of the war, he began to study the therapeutic effects of caves in treating bronchial asthma, chronic bronchitis, and pertussis (whopping cough) (Schmidt, 1989). He described the properties of the environment of Kluterthöhle that gave rise to the therapeutic effects, and in 1949 published medical reports, praising the therapeutic effects of the cave climate on certain illness of the respiratory tract.

Following his pioneering work, modern speleotherapy was initiated in the karst caves of Hungary and Slovakia. In order to find out the beneficial factors of the cave environment, intensive research was carried out by national medical institutions to evaluate the pulmonary parameters of their patients. The remarkable findings resulted in the creation of an international society, which was established to provide international links, therapeutic protocols, and a scientific approach in treating an increasing number of adults and children suffering from asthma and other allergic diseases. The Speleotherapy Commission of the International Speleological Union (UIS) was set up in 1969 by representatives from Austria, Czechoslovakia, France, Germany, Hungary, and Italy.

The effects of speleotherapy are clearly shown in improved pulmonary function, reduction of non-specific bronchial hyperactivity, and the improvement of the humoral and the cellular response of the immune system. After the cure is completed, the aggravation of disease is rare and drug consumption is reduced. Furthermore, this leads to a reduction in the cost of treatment in hospital or at home, and also to a reduction in absences from school or work. The quality of life of these patients is highly improved.

Patients stay in the cave for four hours a day or more, depending on the microclimatic conditions and the medical facilities available. The patients perform physical activities, and take part in sports and breathing exercises. Educational programmes are normally organized by medical staff, giving information about lung diseases, self-help, and relaxation. Medical surveys are regularly undertaken throughout the whole course of therapy. In most speleocentres the therapy consists of three treatments which last for three weeks and are repeated every six or eight months. Speleotherapy as an additional method of healing enhances the rehabilitation of patients. The effects of gymnastics, breath rehabilitation, training, and musical therapy are intensified in the cave (Horvath, 1989).

The agreeable environment of caves chosen for speleotherapy is felt as one enters the underground space, due to so-called passive or indirect healing effects. In these caves the air temperature is fairly constant, and natural ventilation provides an optimal exchange of the air in the cave. The cave provides a comfortable stress-free climate, which has a strong relaxation effect (Horvath, 1989). The absence of inorganic, organic, and biological pollutants favours the treatment of allergies. Due to the absence of pathogenic micro-organisms, the atmosphere is sterile. The absence of air pollution is one of the most important passive curative factors of the cave microclimate. In spite of the fact that some small amounts of external pollutants can enter the cave on patients' clothes, the microclimate will purify the air of the cave, ensuring its biological and mechanical cleanliness. This extraordinary characteristic is attributed to high relative humidity (>90%), aerosols, radioactivity, and ionization.

Aerosol particles in the cave atmosphere contain dissolved substances, including very high concentrations of calcium and magnesium together with sodium, potassium, iodine, fluorine, and trace elements. Inhaled into the respiratory tract, these ions can exert a local anti-inflammatory, mucolytic effect and can stimulate removal of mucus from the lungs. Aerosols, together with the radioactivity and ionization, create a very complex system, which is perhaps the most important factor from the point of view of the biological activity of the cave climate.

The CO_2 level can be many times higher in the caves than in the outdoor air (see Carbon Dioxide-Enriched Cave Air). This stimulates deeper breathing and has a direct anti-spasmodic effect. The CO_2 also increases the ionization of calcium. The O_2 concentration is slightly lower than on the surface, but this difference has no biological significance. Pollutant gas components (SO_2, NO_2, NH_3, etc.) and O_3 are absent from the air of caves chosen for speleotherapy.

The radioactive gas radon is present in all caves (see Radon in Caves), but concentrations are usually low in sites used for speleotherapy, and the radiation dose never exceeds the levels permitted by national regulations and laws. Radon is thought to result in ionization, a bio-positive effect called hormesis, and the stimulation of the immune system. The bio-positive effect can be observed in the induction of DNA repair enzymes, the effect of detoxification, and the induction of regulative polypeptide synthesis. A combination of all these speleoclimatic parameters is thought to be responsible for the healing effect.

Karst caves, artificial caves, or operating and closed mines are only suitable for speleotherapy if they fulfil specific physical, chemical, and biological criteria. There are three categories of speleotherapy sites: cold caves, rock salt and potash mines, and natural and artificial thermal caves. In cold caves the average temperature ranges from 6°C to 10°C, and the relative humidity from 80 to 100%. These caves are used particularly for the treatment of respiratory diseases. In Central European countries, as well as in the Caucasus, speleotherapy centres have been established predominantly in cold caves.

In Germany there are over 12 speleotherapy centres, including the Kluterthöhle (Ennepetal), which was one of the first caves to be listed by the German Balneologic society (for the medicinal use of water baths). In Austria, speleotherapy centres have been established in four natural karst caves. In Sežana, Slovenia, there is a speleotherapy centre located in an artificial cave used as a shelter during World War II. In this locality only adult patients are treated. In Slovakia, speleotherapy for children has a long tradition and extraordinary results in Bystra cave. In the Czech republic there are centres in four caves and in Zlaté Hory, an old gold mine. Hungary has a 40-year tradition of speleotherapy, with centres in Tapolca, Miskole, Budapest and in the Baradla–Domica cave system.

Rock salt and potash mines have a distinctive microclimate with temperatures of 13–20°C and a relative humidity of 45–70%. This speleoclimate is characterized by its strong aerosol effect. Apart from respiratory diseases, it is also used for the treatment of cardiovascular diseases, bacterial and atypical dermatitis, neurodermitis, psoriasis and burns. In Eastern European countries, there are extensive rock salt and potash deposits and some large speleotherapy centres have been set up 100–400 m underground. The favourable temperature permits long-term visits either during the day or through the night. Different treatment methods are applied for different illnesses. This type of speleotherapeutic treatment has been practised for more than 30 years in the Ukraine (Starobin salt mines), the Minsk region (Belarus), the Perm region of the Western Urals (Russia), in Wieliczka (Poland), and in Romania. Special environments have been created in artificial chambers, which provide nearly the same conditions as in the cave, with an emphasis on aerosol concentration. These are now widely used in Uzghorod and Kiev, Ukraine and Perm, Russia.

The third category of speleotherapy sites includes natural and artificial thermal caves with warm or hot air temperatures, with or without elevated radioactivity. Rheumatic and other illnesses can be treated by speleotherapy in these caves. In Central Europe the Gastein Heillstollen is one of the leading centres in this field, and has been operating since 1952. The cave's climate is characterized by a radon concentration (^{222}Rn) of 4.5 nanocuries per litre of air, with temperatures of 38–41.5°C and relative humidities of 70–95%.

Taking care of patients undergoing speleological treatment in caves also involves a concern for the cave microclimate, speleothems, and the entire underground ecosystem. Only intact, well-preserved, and protected caves can offer the beneficial effects needed for speleotherapy, which is why this activity is closely connected with geological and speleological research.

VANJA DEBEVEC

Works Cited

Horvath, T. 1989. Possibilities of speleotherapy in the treatment of children's asthma. In *Allergie and Immunologie, Interasma '89. Proceedings of Symposium on Speleotherapy*, edited by J. Přibyl & D. Řičny, Brno: Institute of Geography of CSAS 126: 7–11

Schmidt, H. 1989. Speleotherapy in the Klutert Cave. In *Allergie and Immunologie, Interasma '89. Proceedings of Symposium on Speleotherapy*, edited by J. Přibyl & D. Řičny, Brno: Institute of Geography of CSAS 126: 13–14

Further Reading

Beamon, S. 1998. Taking the healing airs. *Descent*, 144. 22–23

Říčný, D., Sandri, B. & Trimmel, H. 1994. Beiträge zu Speläotherapie und Höhlenklima *Proceedings of the 10th International Symposium of Speleotherapy, Bad Bleiberg, 1992*, Vienna: Speleotherapy Commission of the International Union of Speleology: vol. 2: 306

International Symposium of Speleotherapy Solotvino, Abstracts. Ukraine, 1998 Speleotherapy Commission of the International Union of Speleology.

Proceedings of the 11th International Symposium of Speleotherapy, Zlaté Hory: EDEL, 1999

Slovak Environmental Agency (editors) 1997. *Conference on Protection and Medical Utilisation of Karst Environment*, Banská Bystrica, Slovakia: DALI

SPRINGS

Karst springs are those places where karst groundwater emerges at the surface. Karst spring discharge ranges over seven orders of magnitude, from seeps of a few millilitres per second to large springs with average flows exceeding 20 m^3 s^{-1} (Table 1). Flow may be steady, seasonal, periodic, or intermittent, and may even reverse. Karst springs are predominantly found at low topographic positions, such as valley floors, although they may be concealed beneath alluvium, rivers, lakes, or the sea (vrulja). Some karst springs emerge at more elevated positions, usually as a result of geological or geomorphological controls on their position.

Springs in non-karst rocks may result from the convergence of flow in a topographic depression or from the concentration of flow along open fractures such as faults, joints, or bedding planes. Flow in porous media is limited by hydraulic conductivity, so that associated springs almost always have very small flow, often discharging over an extensive "seepage face." Larger springs are possible in fractured rocks such as basalt, where flow may be concentrated along open or weathered fractures. What distinguishes karst springs is that they are the output points from a dendritic network of conduits, and therefore tend to be both larger and more variable in discharge and quality than springs arising from coarse granular or fractured media.

Springs: Table 1. Large springs of the world (after Ford & Williams, 1989 and Milanović, 2000, modified).

Spring	Flow (m^3 s^{-1})	Basin (km^2)
Tobio, Papua New Guinea	>100	–
Matali, Papua New Guinea	90	350
Trebišnjica, Herzegovina	80	1140
Bussento, Italy	>76	–
Dumanli, Turkey	50	2800
Galowe, Papua New Guinea	40	–
Ras el ain, Syria	39	–
Tisu, China	38	1000
Ljubljanica, Slovenia	39	1100
Stella, Italy	34	–
Ombla, Croatia	34	600
Chingshui, China	33	1000
Spring Creek, Florida	33	>1500
Oluk Köprü, Turkey	>30	>1000
Timavo, Italy	30	>1000
Frio, Mexico	28	>1000
Yedi Miyarlar, Turkey	>25	>1000
Mchishta, Georgia	25	–
Buna, Herzegovina	24	110
Coy, Mexico	24	>1000
Liu longdong, China	24	–
Kirkgozler, Turkey	24	–
Silver, Florida	23	1900
Rainbow, Florida	22	>1500
Vaucluse, France	21	1100
Sinjac (Piva), Yugoslavia	21	500
Bunica, Herzegovina	20	510
Grab-Ruda, Croatia	20	390
Trollosen, Spitzbergen	20	–

In general, karst springs can be considered in terms of their hydrological function, their geological position, and their karstic drainage or "plumbing". These controls exert an influence on spring hydrology and hydrochemistry, allowing a diagnosis of aquifer form and function from measurements of spring water. This is particularly important in evaluation of spring water as a resource and in assessment of the overall condition of the aquifer. Karst springs have been classified in many different ways, and Table 2 shows eight different attributes. In theory, these attributes could be combined to describe a spring. For instance the spring at Sof Omar Cave, Ethiopia could be described as a "perennial, full-flow, gravity resurgence". In practice, most karst springs are described in terms of their most important attribute, depending upon the interest of the observer and the context of the application.

General Hydrology

Springs represent the outlet of water recharged at higher elevations, and stored and routed (conveyed) through the karst aquifer. In general, the larger the body of stored water relative to aquifer flow, the greater is the residence time of the water, and the more steady (consistent) will be spring flow. It is convenient to consider the energy of water at any point in an aquifer in terms of "hydraulic head": the elevation to which water would rise up a standpipe opening at that point. Groundwater always flows from high to low head, at a rate driven by the rate of fall in head, and governed by the resistance of the flow route. Thus a steady perennial spring will exhibit consistent distribution of head in the aquifer whereas seasonal and intermittent springs will show marked variation in the hydraulic head in the aquifer. Reversing springs, or "estavelles", occur where head in the aquifer is sometimes lower than head in a connecting water body such as a river or lake, forcing surface water into the aquifer.

Hydraulic head is the sum of local elevation and water pressure. Springs in which water is derived from open passages with free surface streams are driven by elevation head and termed gravity springs. In contrast, springs where water rises from depth are pressure springs. The most profound pressure springs are termed "Vauclusian" springs after the Fontaine de Vaucluse, France, which has been dived to a depth of 315 m. In some cases, the internal hydraulic head in an aquifer greatly exceeds that required to drive the flow of water. Springs with excess hydraulic head exhibit a marked upwelling or jet, and are sometimes termed "artesian" springs.

Geological Controls on Karst Springs

The location and form of karst springs is determined primarily by the distribution of karst rocks, and the pattern of potential flow paths (fractures) in the rock. Where karst rocks are intermixed with impermeable rocks, the latter act as barriers to groundwater flow, and karst springs tend to develop as "contact springs" where the boundary between the karst and impermeable rock is exposed at the surface. Where the impermeable unit underlies the karst, it enhances the elevation of the karst water, and the spring (and aquifer) is considered "perched", as it lies above the topographically optimum discharge point. Where the impermeable unit overlies the karst aquifer, it enhances the pressure of karst water, and springs are then described as "confined"

Springs: Table 2. Classification of karst springs according to different attributes.

Parameter	Type		Name for spring
Flow duration	continuous	steady	perennial: base flow
		variable	perennial
		periodic	ebbing and flowing, rhythmic or episodic
	discontinuous		intermittent
	non-existent		paleo- or abandoned spring
Reversing flow	reversing		estavelle
	non-reversing		(no specific term)
Conduit type at spring	open conduit (with airspace)		gravity spring
	water-filled conduit		pressure, artesian or Vauclusian spring
Geology	aquiclude underlies spring		contact spring
	aquiclude overlies spring		confined or artesian spring
	spring rises through alluvium		alluviated spring
	fault		fault-controlled spring
	joint		joint-controlled spring
	bedding plane		bedding plane-controlled spring
Topographic position	at valley floor		local base level or graded spring
	above valley floor		hanging (rejuvenated) spring
	below valley floor		buried spring
Relationship to bodies of surface water	at sea or lake level		base level spring
	below sea level		submarine spring (vrulja)
	below lake level		sublacustrine spring
	between high tide and low tide		intertidal spring
	above bodies of surface water		hanging (rejuvenated) spring
Distributaries	distributary with most variable or intermittent discharge (usually the highest one)		overflow spring
	distributary with steadiest discharge (usually the lowest one)		underflow spring
	intermediate distributary		underflow-overflow spring
	no distributaries		full-flow spring
Recharge	autogenic recharge		exsurgence
	allogenic recharge		resurgence
Chemistry	dilute water		fresh water spring
	enriched water		mineralized spring
	saline water		saline/brackish spring
	saturated with carbonates (tufa)		petrifying, tufaceous spring
	saturated with hydrogen sulfide		sulfur spring
	saturated with iron hydroxide		chalybeate (ochre) spring
	hot water		thermal spring
Culture/exploitation	religious		shrine
	fish culture		hatchery spring
	livestock and irrigation		watering spring
	bottled water		bottling spring
	brewing, distilling		brewery, distilling spring

and are often artesian. Confined springs tend to benefit from greater water storage than perched springs, and so exhibit more sustained flow. In many cases, karst springs may be concealed beneath permeable but insoluble materials such as river alluvium or talus. The term "quicksand" describes sandy material that is continually boiling in an alluviated spring orifice.

The primary flow routes in most karst aquifers are determined by the presence of discontinuities such as joints, bedding planes, and faults. In many cases, fractures direct karst conduits away from the lowest elevations favoured by consideration of hydraulic gradients, resulting in vertical and horizontal displacement of the spring from the expected topographic outlet point. Such springs are described as being "controlled" by the respective geological feature.

Geomorphological Controls on Karst Springs

Water tends to follow the steepest hydraulic gradient to a discharge point, so that springs tend to occur in valleys, and on the outside of river meander bends. Changes in landscape influence springs through erosional deepening and depositional infilling of valleys. If an underground karst stream is exposed by erosion, then a new spring may develop at the breach and the original downstream passage will be abandoned. The effect of valley deepening may be to leave an existing spring above the valley floor, and to permit a lower outlet to develop. The original spring tends to be progressively abandoned as lower outlets develop, often resulting in a suite of immature underflow springs and intermittent overflows. Valley infill may block former springs, rejuvenating former high level outlets and creating alluvial springs in the valley floor. Changes in sea level have a similar redirectional effect. Many submarine karst springs (vrulja) that presumably developed during Pleistocene low sea levels still function following postglacial rises in sea level.

Karst Controls on Springs

The quality and magnitude of flow from a karst spring reflects the form and function of the karst aquifer, and in particular the recharge processes and the conduit network. Springs deriving much of their water from allogenic surface catchments are known as resurgences. They exhibit properties closely related to those of the surface catchment: typically variable flow and unreliable water quality. Springs in autogenic aquifers, which receive the bulk of their recharge from a karst surface, are known as exsurgences, and exhibit less variability in discharge and composition. Where recharge to the limestone aquifer is moderated by thick unconsolidated deposits exsurgences are even less variable. For example, in Florida the overburden provides significant storage and buffers the underlying aquifer from fluctuations in recharge and water quality. Variable springs are most highly developed where the karst aquifer has massive, bare limestone exposed at the surface, limited storage, and small extent. Karst springs draining aquifers with substantial storage, or relatively isolated from surface recharge by overlying strata, will tend to exhibit more steady behaviour. In the past, such flow behaviour has been attributed to distinctive "diffuse' and "conduit" Karst aquifers, but it is now recognized that recharge or underflow-overflow effects are responsible, and that a diffuse karst aquifer is an oxymoron.

Most karst drainage systems are dendritic in form like surface drainage channels; thus smaller tributaries come together to generate a progressively larger stream, emerging at a single, integrative spring. However, geomorphic history and geology endow many karst aquifers with multiple levels of conduits, often punctuated by vertical shafts, so that karst flow systems may exhibit three-dimensional features uncommon or impossible in surface drainage systems. Often a single conduit discharges through a number of distributary springs. Both valley deepening and infilling can result in distributaries. Aquifers experiencing variations in hydraulic head (i.e. internal water level) as flow varies may show flow-dependent distributaries. Where distributary springs occur across a range of elevations, they fall into a vertical hierarchy. The lowermost members are "underflows" and tend to exhibit steady flow. In contrast, the overflow springs are intermittent, or show much greater variability in flow. Such effects may not necessarily be associated with multiple springs; in some cases internal overflows develop, for example in a high-level bypass to an obstruction in a streamway. However, the karst spring may exhibit sudden changes in composition and flow as such internal overflows are activated.

A few karst springs show remarkable periodicity in their flow, with a typical period of minutes to hours. In general, this is attributed to the existence of an internal siphon which progressively fills and drains. However, periodicity may be restricted to specific flow conditions, or the form of the pulse may not match the hydraulics of a simple siphon, and more complex "plumbing" has to be invoked. Periodicity in hydrothermal springs is seen in geysers. The key feature of geysers is the warming of a pressurized body of water to boiling point and the explosive spontaneous boiling occurring as pressure is released.

Many karst springs occur adjacent to or beneath the surface of rivers, lakes, or the sea; the majority are likely unacknowledged. The interaction between the aquifer and the external water body rests on the hydraulic head distribution and the pattern of connections (springs, sinks, and estavelles) that exists. The relationship is most complex where the timing of peak flow (and head) differs between the aquifer and water body. Often the aquifer response lags that of the surface water body. The result is a phase of inflow to the aquifer, followed by an expulsion of the recently influent water, followed by the original aquifer water, chased by actual storm water. A reach of river channel

Springs: Figure 1. Giant Springs lie on the banks of the Missouri River at Great Falls. Montana, United States. They produce a steady flow of 3–6 m^3 s^{-1} from a deep aquifer karst with an extensive catchment to the south. (Photo by Tony Waltham)

Springs: Figure 2. Margoon Waterfall Spring in the Zagros Mountain Range, Fars Province, Iran. This perennial spring has an average discharge of 800 l s^{-1} and emerges 58 m above the base of the cliff. Water emerges at 2200 m above sea level from Asmari Formation limestone (Oligo-Miocene) on a nearly vertical normal fault. (Photo by John Gunn)

encountering multiple karst conduits may show delightfully complex patterns of sinking, transmission, and rising in successive pools.

Hydraulic head is primarily dependent on elevation and pressure, but density of water may have an influence too. The greater density of seawater means that it has slightly greater hydraulic head than an equivalent fresh water body. This is most conspicuous in the visible discharge mound of many submarine springs (vrulja) as the fresh water rises to the surface. However, it may be more significant in driving subtle groundwater circulation in marine settings. Deeper openings in a submarine distributary may act as inflow points for saline water. Within the aquifer, the seawater rises, mixes with freshwater and emerges as brackish springs, often slightly above sea level. Similar density effects may develop in hydrothermal waters, but have not been extensively investigated.

Where karst spring water is supersaturated, calcareous tufa deposits develop at the orifice and downstream. Such petrifying springs mantle all objects in calcite, and often build up distinctive mounds and barrages in areas of peak precipitation (see Travertine). A peculiar property of such springs is "self-damming", arising from enhanced precipitation in the thinner, more turbulent free surface flows over the lip and outer surface of the barrier. Spectacular rimstone dams may occasionally reach many metres in height, but are prone to leakage and sudden abandonment fuelled by the enhancement of hydraulic head that they engender.

Hydraulic Influences on Karst Spring Discharge

The relationship between karst spring behaviour and the hydraulics of the aquifer permits analysis and simulation to aid research and management. Attempts to derive the internal plumbing of the aquifer are often ambiguous and constrained by the necessity of substituting a limited number of simplified equivalent components for the likely complexity of the system. Hydraulic modelling indicates that open channels accommodate changes in flow with limited change in water level, compared to closed (water-filled) conduits. In other words once a cave streamway fills at a point, relatively large head (water level) changes will occur. A second important feature is that during floods, restrictions result in disproportionate rises in head. Large heads (i.e. high water levels) and rapid changes in head are common during floods. The former may induce flow-dependent routing of water to distributary springs. Spring discharge is often moderated by the water stored and released by large head changes. Both phenomena have startled and trapped cave explorers.

Hydraulic head in a conduit changes with the balance between inflow and outflow, with the resulting volume causing more or less head change depending on the form of the conduit. In complex karst aquifers, all of the various inlets, stores, overflows, and outlets develop their own local head, depending on their form and flow conditions. This can result in significant internal rerouting of water under variable flow conditions. Moreover, the rerouting effects may not appear to be reproducible under superficially similar conditions. Development of simple spring-rainfall-response models has been an objective of karst hydrogeologists for many years. Such idiosyncratic nonlinearity as arises from rerouting effects has been a major cause of frustration for such modelling attempts.

As the aquifer drains, and spring flow declines, water is withdrawn from aquifer storage. The resulting flow recession may be modelled by a simple exponential dependence of flow on time. The inverse of the exponential parameter has units of time, and can be taken as an indicator of the residence time of water in the aquifer. Integration of the exponential model provides estimates of aquifer volume.

Quality of Karst Spring Waters

The water emerging from a karst spring consists of a mixture of water from various recharge routes and storage zones. As the environment and duration of recharge, and storage vary, so too will the resulting composition of the water. For example, allogenic recharge water will tend to be more turbid and chemically dilute than autogenic recharge. Long-term storage may result in depletion of the dissolved oxygen in the water, and deep flow may lead to warming or mineralization. In principle, these natural tracers should allow the source and routing of karst spring water to be derived. However, many of these characteristics (e.g.

temperature, turbidity, dissolved oxygen, hardness) do not have fixed values associated with particular environments, they are not conserved in transit, and mixing with other waters may induce chemical reactions. Nevertheless, it is often possible to develop empirical characterization for particular karst aquifers to allow discrimination of the proportion of basic water sources contributing to spring flow. The chemical composition of spring waters often results in distinctive deposits, biota, and exploitation, allowing a chemical classification.

Karst Springs as a Water Resource

The sustained, accessible supply of good quality water at karst springs has made them an important traditional water resource, especially in those regions where surface water is scarce. However, many karst springs have proven unreliable in both quality and quantity of supply, and have been superseded by wells and pumps. Karst springs are now regarded as valuable integrators of the aquifer, and, compared to observation wells, provide comprehensive monitoring sites for assessment of contamination and supply. The hydraulic and chemical principles described above, coupled with groundwater tracing allow catchment area, origin, storage, and routing to be determined. However, such interpretation requires considerable simplification, and seldom acknowledges the nonlinear and apparently idiosyncratic behaviour that is often revealed by sustained, comprehensive monitoring of springs. Nevertheless, the water quality of springs has a profound influence on their use, both as water sources and cultural resources.

Chris Smart and Stephen R.H. Worthington

Further Reading

Bonacci, O. 1987. *Karst Hydrology, with Special Reference to the Dinaric Karst*, Berlin and New York: Springer

Bonacci, O. 2001. Analysis of the maximum discharge of Karst springs. *Hydrogeology Journal*, 9: 328–38.

Ford, D.C. & Williams, P.W. 1989. *Karst Geomorphology and Hydrology*, London and Boston: Unwin Hyman

Mangin, A. 1969. Etude hydraulique du mécanisme d'intermittence de Fontestorbes (Ariège). *Annales de Spéléologie*, 24(2): 255–98

Milanović, P.T. 2000. *Geological Engineering in Karst*, Belgrade: Zebra

Newson, M.D. 1971. A model of subterranean limestone erosion in the British Isles based on hydrology. *Transactions of the Institute of British Geographers*, 54: 55–70

Ray, J.A. 1997. Overflow conduit systems in Kentucky: A consequence of limited underflow capacity. In *The Engineering Geology and Hydrogeology of Karst Terranes: Proceedings of the Sixth Multidisciplinary Conference on Sinkholes and the Engineering and Environmental Impacts of Karst*, edited by B.F. Beck & J. B. Stephenson, Rotterdam: Balkema: 69–76

Smart, C.C. 1988. Artificial tracer techniques for the determination of the structure of conduit aquifers. *Groundwater*, 26: 445–53

Smart, C.C. & Ford, D.C. 1986. Structure and function of a conduit aquifer. *Canadian Journal of Earth Sciences*, 23: 919–29

Worthington, S.R.H., Davies, G.J. & Quinlan, J.F. 1992. Geochemistry of springs in temperate carbonate aquifers: Recharge type explains most of the variation. *Proceedings of the Fifth Conference on the Hydrology of Limestone and Fractured Aquifers*, edited by P. Chauve & F. Zwahlen, Besançon: University of Besançon, vol. 2: 341–48

STAMPS AND POSTCARDS

Philately is one of the most popular hobbies in the world. Postage stamps and cancellations (marks that void a stamp through the use of a bar, wavy line, picture, or special wording) are unique, and at times valuable records of history, geography, politics, art, and numerous other aspects of human civilization. From the earliest years of the hobby, most philatelists have preferred to collect by country. However, since the mid–1950s, many philatelists have become increasingly interested in topical stamp collecting, on a wide variety of specific themes or subjects, for example caves and bats. A third and recent subcategory of stamp collecting is deltiology—the study and collecting of postcards.

The First Stamps and Postcards

In 1837, English schoolmaster and civil servant Rowland Hill proposed the idea of the adhesive postage stamp as one of many postal reforms in Britain. Through Hill's efforts, on 1 May 1840, Britain issued the world's first official adhesive postage stamp, a one-penny denomination universally referred to as the Penny Black. By 1860 most nations had adopted postage stamps. Designs at first imitated those of Britain, with portraits usually depicting heads of state and symbols, or artistic designs, generally being national in character. Pictorial designs were increasingly used toward the end of the 19th century, including those of caves; cave fauna such as bats, salamanders, and blind fish; cave scientists and speleologists; rock and cave art; and minerals.

The first stamp with an illustrated bat was designed in 1894 by the China Custom Post to commemorate the birthday of the Empress Dowager. The 1-candareen (Chinese monetary value) and the 9-candareen stamps both have five stylized bats (called "wu fu") in a ring surrounding the tree of life. These symbolize the five great happinesses—health, wealth, long life, good luck, and tranquility (Figure 1). The first true picture of a bat appeared on three different denominations of a 1948 Chile stamp. It depicted the Yellowed-shouldered Bat, *Sturnia lilium*, and was printed in three colours (Figure 2).

Stamps depicting caves first appeared in 1925, when Lebanon issued a postage-due stamp depicting Pigeon Grotto, and next in 1930, when they issued a second Pigeon Grotto stamp used for regular mail. From then on, nations around the world issued stamps referencing caves, including depictions of entrances; rooms with many formations; stalagmites and stalactites; cave streams; and tourists (Figure 3). Minerals, rock art, and natural bridges are other areas that can be collected on stamps, of which the 1899 natural bridge stamp issued by Tasmania depicting Tasman's Arch is an example (Figure 4).

The earliest known illustrations of caves on postcards are Kuhstall in Saxony (used in 1887), Einhornhöhle in the Harz mountains (1890), and the Blue Grotto in Capri (1893). Established "show" caves throughout the world began publishing large numbers of cards and, by 1910, millions were sold at cave souvenir shops, "promoting" the cave by mailing to friends and relatives, or collected as personal remembrances (Figure 5).

Postcards and Stamps: Figure 1. China (Scott No. 16) Five bats surrounding the Tree of Life **Figure 2.** Chile (Scott No. 255m) Yellow-shouldered Bat **Figure 3.** Austria (Scott No. 1496) Entrance to Peggau Cave, Styria **Figure 4.** Tasmania (Scott No. 89) Tasman's Arch **Figure 5.** Grotte de Remouchamps 1925 postcard **Figure 6.** France (Scott No. B703) Norbert Casteret **Figure 7.** A 1875 cancellation from Weyers Cave Station on the Baltimore and Ohio Railroad, United States.

Cave scientists and speleologists have also been pictured on stamps. In 2000, France honoured three speleologists: Norbert Casteret (Figure 6), Haroun Tazieff, and Jacques-Yves Cousteau, for their exploration of caves around the world, and for furthering knowledge in the cave sciences both underground and underwater. Emil Racoviţă (the father of biospeleology), Lazarro Spallanzani (who discovered echo-location in bats), and Dr Joseph Pawan (a Trinidad bacteriologist who identified the rabies virus in the vampire bat) are just a few of the cave scientists depicted on stamps.

The final collector's frontier is speleological cancellations. The cancellation can be from a city or town that has the word "cave, grotto, cavern" in its postmark (Figure 7), or a postmark with speleological reference, such as a bat or a caver's helmet and lamp. Cancellations can also commemorate a significant event like a Regional Speleological Meeting, the Annual NSS Convention in the United States, or an Annual Music Festival like the Concert Simponic held in Peştera Româneşti-Timis in Romania. Cancellations and postmarks date back to the early 1800s. This area of collecting is the most difficult as there are usually no checklists and the postmarks are not catalogued. Often philatelists who collect in certain areas of interest write the handbook and checklist so that collectors with similar interests can have a starting point. The American Philatelic Research Library in Pennsylvania is one of the best in the world for this type of research.

Organizations and Publications

The American Topical Association (ATA), in Arlington, Texas, is one of the specialized organizations of stamp collectors in the United States. It publishes a bi-monthly magazine, *Topical Times*, as well as special handbooks. The ATA has checklists for many cave and cave-related topics, one of which is *Handbook 128*, entitled *Bats in Philately: A Comprehensive Illustrative Study and List of Bats on Stamps*. The largest general organization for stamp collectors in the Western Hemisphere is the American Philatelic Society (APS), in State College, Pennsylvania, which publishes a monthly journal, *The American Philatelist*. Since 1863, the Scott Publishing Company has produced the Scott Standard Postage Stamp Catalogue, which is the most widely used by North American collectors. It is published annually in a six-volume set, which lists every adhesive postage stamp ever issued and their current value. In Europe similar references for stamps are the *Stanley Gibbons Catalogues, Yvert et Tellier's Catalogues*, and *Michel's Catalogue*. An affiliate of the National Speleological Society is the Speleophilatelic Section, which publishes a quarterly journal: *The Underground Post*. Its European counterpart, published at least three times a year, is *Speleophilately International*.

Collecting Procedures

One of the attractions of stamp and postcard collecting is the ease of getting started. With access to enough incoming mail, especially from abroad, a person can build a collection at little or no expense. Tens of thousands of items, including many older issues, can be bought for pennies. However, the harder to find and rare items can cost upwards of tens of thousands of dollars. Little special equipment is required. A collector needs only an album to house the collection, hinges, or other types of mounts to attach the stamps to the pages, and a pair of stamp tongs with which to handle them. Stamps and accessories can be readily purchased from a professional stamp dealer in nearly every city, or from those who operate exclusively by mail. A search of the Internet for websites that specialize in stamps, postcards, and cancellations will surprise you. One only needs to join a stamp club and enjoy caving from the top of the kitchen table.

THOMAS LERA

See also **Art Showing Caves**

Internet References

For the American Philatelic Society: http://www.stamps.org/
For the American Topical Association: http://home.prcn.org/~pauld/ata/index.html

For over 4200 Philatelic Resources on the Internet, including references for all catalogues: http://www.myexecpc.com/~joeluft/resource.html

Further Reading

Irwin, D.J. 1986. Picture postcards of caves and caving. *American Spelean History Association Journal*, 20(1): 1–29
Lera, T.M. 1995. *Bats in Philately*, Albuquerque, New Mexico: American Topical Association
Lera, T.M. 2000. Archaeology and cave art. *NSS News*, 58(11): 314–15 and 326
Lera, T.M. 2001a. A journey into Earth. *Scott Stamp Monthly*, 19(1): 70–72
Lera, T.M. 2001b. Cave exploring. *Topical Times*, January / February: 22–24
Stepanek, V.J. & Friedrich, W.-P. (editors) 1998. Special issue. *Nyctalus* (N.F.) 6: 3–108

STRIPE KARST

Stripe karst forms along narrow outcrops of steeply dipping karstic rock beds. This is an extreme case of contact karst, where the allogenic contact perimeter is very large relative to the area of the karst outcrop. Stripe karst is the dominant karst found in metamorphic marble outcrops of the Scandinavian Caledonides, and is named the "Norwegian karst type", as it was first described there (Horn, 1937). Analysis of the geometric properties of a stripe suggests that stripe karst can be defined as a narrow karst outcrop with length to width ratio greater than 3, and is fully developed when the ratio reaches 30. The absolute width is equal to or less than twice the penetration length of allogenic contact karstification. In most cases this width limit is some hundred metres. Impermeable and insoluble rocks, forming aquicludes, surround and isolate individual karst stripes (Lauritzen, 2001). Two or more stripe outcrops may exist side by side as independent aquifers with no hydrological communication except for surface runoff.

Stripe Karst: Figure 1. Marble stripe karst outcrop at Svartisen, north Norway. The steeply dipping stripe is only 20–30 m thick, but extends laterally for almost 10 km. This outcrop hosts the Pikhåg Caves. (Photo by Stein-Erik Lauritzen)

Stripe karst develops characteristically in metamorphic carbonates where multiple folding results in very complicated outcrop patterns. Subsequent erosion exposes these beds at various angles of dip and intersection with the land surface.

Karstification is intensive at lithological contacts between marble and mica schists, but also at those between different marble types. Mica schist contacts often contain iron oxides and sulfides, giving rise to sulfuric acid speleogenesis; many caves were initiated along such contacts.

Stripe karst contacts are classified into three main types. At gentle dips, karstification can take place at the confined upper contact, at the perched lower contact, or at both. Even in very steeply dipping beds caves tend to concentrate along one of the contacts, most frequently at the hanging wall contact. In many cases, speleogenesis starts under phreatic conditions at the upper, confining contacts (which may also carry sulfide mineralization), but caves may cut down towards the lower contact under vadose conditions and become perched.

Due to metamorphic recrystallization, primary voids are absent, and speleogenesis is entirely dependent on fractures formed subsequently in regimes of brittle deformation. Commonly, two orthogonal sets of fracture occur, which together with the allogenic contact plane form a "box" unit. This may display various attitudes depending on stratal dip and fracture orientation, which fall into six cases when high and low stratal dips are included.

Stripe karst cave systems form four main morphotypes: subvertical, tiered phreatic networks or mazes; low dip phreatic networks or mazes; looping systems with vadose trenches; and linear drainage systems. The various morphotypes do show systematic dependence on stratal dip of the allogenic stripe contacts, of the type of contacts, and to a lesser extent, the fracture patterns (Lauritzen, 2001).

Karstification depth in stripes can penetrate beyond 100 m below the land surface. Some areas display a high density of grikes at the surface, which may be taken as a kind of epikarst. Due to the high erosion rates prevailing during the Quaternary,

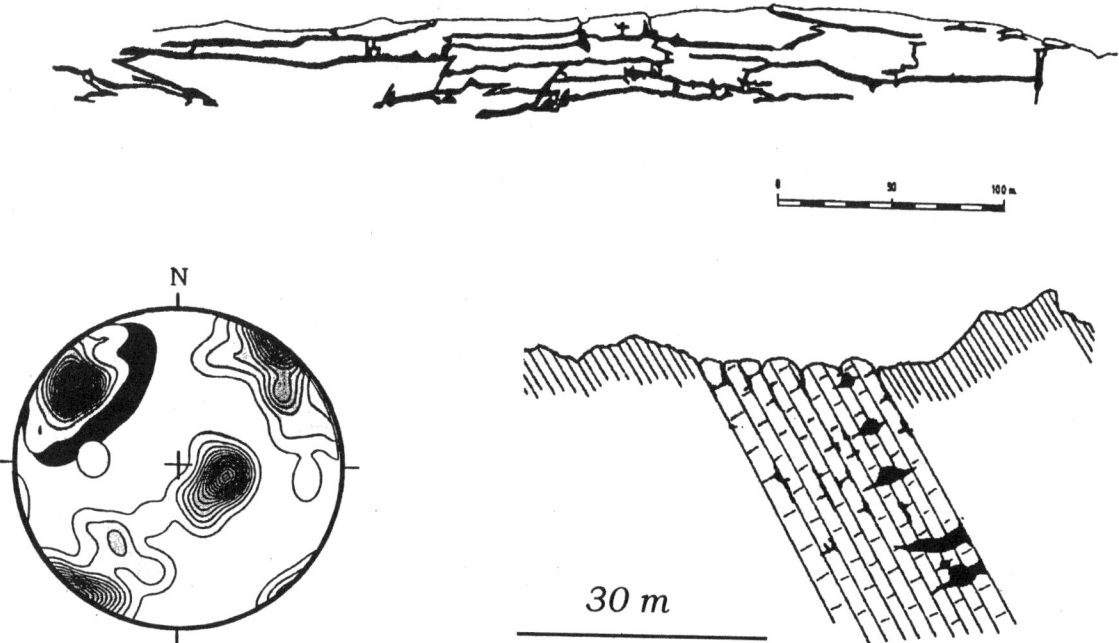

Stripe Karst: Figure 2. The Pikhåg Caves, Svartisen. Top: vertical section of the tiered network, forming 2000+ m of phreatic tubes. Lower left: Stereonet pole-plot of guiding fractures (contours) and marble / mica schist interface (black). Lower right: cross section through the stripe, showing cave concentration at and near the overhanging wall contact. From Lauritzen (2001).

stripe karsts in Scandinavia are relatively young: although many caves are several glacial cycles old, no unequivocal evidence of pre-glacial, or Tertiary caves has yet been recognized. A typical example of stripe karst is the Pikhågan outcrop and caves in Norway (Figures 1 and 2).

STEIN-ERIK LAURITZEN

Works Cited

Horn, G. 1937. Über einige Karsthöhlen in Norwegen. *Mitteilungen für Höhlen und Karstforschung*, 1–15

Lauritzen, S.E. 2001. Marble stripe karst of the Scandinavian Caledonides: An end-member in the contact karst spectrum. *Acta Carsologica*, 30: 47–79

SUBTERRANEAN ECOLOGY

Subterranean ecology is concerned with the study of the interactions that determine the distribution, abundance, and characteristics of the biological communities within habitats beneath the land surface. There are a wide range of subterranean habitats covering most latitudes and encompassing terrestrial and aquatic environments (freshwater, brackish, and marine) including caves, interstitial groundwater habitats, and the open voids connected by cracks and fissures within karstic environments (the Mesovoid Shallow Substratum or MSS; Juberthie, 2000). Many factors influence the communities within individual subterranean habitats including resource availability (particularly food), interactions between organisms of the same or different species (intraspecific and interspecific competition), the degree of isolation from the surface (epigean) and other subterranean (hypogean) habitats, and the impact of human activity. Significant advances regarding our understanding of the evolution and natural selection of organisms within subterranean environments have been made, although much has still to be learnt.

Abiotic Environment

The physical and chemical characteristics of individual subterranean ecosystems may be highly variable and locally significant. Two factors are of primary importance in determining the structure and functioning of subterranean biotic communities: the absence of light, and thermal regime. Permanent darkness is the single factor that separates all subterranean habitats from epigean systems. However, it is important to recognize that significant interactions occur between surface and subterranean ecosystems in threshold zones (ecotones) where light and the influence of epigean primary productivity may be attenuated due to the presence of plant roots or the transport of dissolved and particulate allochthonous organic material short distances underground.

The thermal characteristics of an environment are a primary determinant of the rates of chemical and biological processes. Temperature within subterranean ecosystems is strongly influenced by both latitude and altitude and is usually close to the mean annual surface air temperature in the epigean environment (see Climate of Caves). Subterranean air temperature is usually relatively constant throughout the year, although cave morphology may exert a strong influence, causing a cold trap in some caves, while in others with large bat populations temperature may be elevated to between 30–45°C. In most caves humidity is usually high (relative humidity >95%) and the distinction between aquatic and terrestrial realms is not always clear. Water temperature typically displays a low amplitude variation but may experience relatively slow to medium-term changes depending on the volume of water and speed of transfer from epigean environments. In salinity-stratified anchialine caves, there is a steep thermal gradient associated with depth. This may reflect both evaporation of water at the surface and geothermal warming from below.

Trophic Characteristics

Hypogean ecosystems are food limited when compared to photosynthetic epigean systems and as a result most subterranean food-chains and webs are relatively simple with few trophic links (see Food Resources). In the absence of direct sunlight energy, subterranean food webs are almost exclusively heterotrophic and largely dependent upon the transport of allochthonous primary productivity from surface habitats. However, compared to epigean food webs energy transfer is highly efficient between each trophic level and there may be considerable variability in both the volume and quality of the organic material input into individual systems.

Energy limitation is a primary determinant of community abundance in most subterranean ecosystems. The primary consumers of dissolved and particulate organic material (DOM and POM) are bacteria, protozoa, and fungi. Micro- and macroinvertebrate taxa and higher organisms (e.g., amphibians and fish) consume these organisms if present. However, most taxa do not usually have specialized diets and will utilize a wide range of food resources. In many terrestrial subterranean systems the distribution of organic material is patchy and its input episodic. As a result the distribution, abundance, and diversity of terrestrial taxa may strongly reflect the predictability of the input of coarse and fine particulate organic matter (CPOM and FPOM), including faecal material (guano) from birds, bats, or large arthropods, or animal carcasses (Poulson & Lavoie, 2000).

Hypogean systems in direct contact with sinking surface streams have been studied in greater detail than most other subterranean ecosystems. Typically these systems have a greater input of CPOM, such as leaf litter and woody debris, than systems fed by autogenic percolation water. The pulsed nature of the input and the highly variable quality of the material strongly influences both the structure and composition of the aquatic community. In systems with sinking streams a large potential energy input may occur.

Research on the trophic basis of subterranean ecosystems has indicated a wide variety of energy resources utilized by organisms at a range of spatial and temporal scales, including communities supported by guano transported into caves by birds and bats in tropical caves and near-surface cave systems dependent on plant and tree roots (Wilkens, Humphreys & Culver, 2000). The discovery of a chemoautotrophically driven cave system, independent from sunlight energy was arguably one of the most important ecological discoveries since the identification of faunal communities associated with deep-sea thermal vents (see Movile Cave). The use of stable isotope tracing has begun to clearly identify the feeding linkages and hierarchy of subterranean food webs in a number of locations, although many feeding relationships are still poorly understood (Pohlman, Cifuentes & Iliffe, 2000).

Biological and Ecological Characteristics

Given the comparative scarcity of resources in subterranean habitats it is not surprising that the faunal communities are typified by lower diversities of taxa and abundances of individuals than epigean habitats. In addition, due to the non-uniform distribution of resources the spatial distribution of organisms is patchy. The threshold / transition zone between the epigean and hypogean environment is usually the most biologically diverse and supports greater abundances of organisms than deeper subterranean habitats. The degree to which individual organisms are adapted to hypogean environments is variable. Historically all cave organisms were given the prefix "troglo-" but the term stygofauna is now widely used for all subterranean aquatic fauna, whether or not they are found in caves *sensu stricto*. There is no equivalent common term referring to the whole terrestrial subterranean fauna. However, since the introduction of the prefix stygo- for aquatic organisms, the prefix troglo- is increasingly being used specifically to describe terrestrial cave organisms (see Organisms: Classification). Most contemporary authors distinguish between three groups of organisms in both aquatic and terrestrial subterranean habitats. Trogloxenes (terrestrial) and stygoxenes (aquatic) are organisms that do not normally occur in subterranean enviornments but may accidentally, or in some instances actively, enter hypogean habitats in search of food. They are unable to complete their life cycle underground but may play an important role in the functioning of subterranean ecosystems, contributing significant inputs of organic material in the form of faeces and carcasses, and taking the roles of predators and / or prey of subterranean organisms. Troglophiles and stygophiles encompass taxa that are able to live, exploit resources, and successfully reproduce and maintain populations within the subterranean environment, but are not exclusively confined to them. Stygophilic organisms have been subdivided into three groups: (1) permanent hyporheic taxa that may be present in the subterranean environment during any stage of their life history; (2) amphibitic taxa whose life cycles require the use of both surface and groundwater environments; and (3) occasional hyporheic taxa that actively utilize the resources in the subterranean environment and may seek refuge there during unfavourable conditions (e.g. floods or droughts) (Gibert *et al.*, 1994). Troglobites and stygobites are obligate occupants of subterranean habitats that do not normally occur in epigean environments, although they may be recorded in shallow hypogean environments.

Many troglobitic and stygobitic organisms, and some troglophilic and stygophilic populations, may be geographically restricted to a limited number of sites or even endemic to individual cave systems. Many of the organisms display morphological, physiological, and / or behavioural characteristics related to the physical limitations of the subterranean environment. The evolution and natural selection of subterranean organisms has stimulated considerable interest since the work of Charles Darwin (see Evolution of Hypogean Fauna). However, the mechanism(s) and time required for the development of adaptive (e.g., enhanced development of sensory organs) and regressive (e.g, reduction or absence of eyes) troglomorphic characteristics has stimulated much debate (Culver, Kane & Fong, 1995) (see the six Adaptation entries).

The morphological characteristics of subterranean fauna (troglomorphism, see Adaption: Morphological entries) include the widely reported absence of pigmentation and regression or complete absence of eyes (e.g., the European cave salamander *Proteus anguinus* and many troglophilic Crustacea). Some stygophilic and troglophilic arthropods have longer appendages (e.g., antennae and legs) compared to other epigean organisms with close affinities (e.g., cave crickets from the family Rhaphidophoridae). Many hypogean terrestrial arthropods have thin and less waxy cuticles to facilitate the removal of water through the integument due to the high relative humidity within many subterranean habitats (Juberthie, 2000). Some groundwater organisms are typically longer and thinner (vermiform) than their surface water counterparts, probably to facilitate movement between interstitial spaces. Other morphological characters indicative of subterranean organisms include the reduction or absence of scales in fish (e.g. *Sinocyclocheilus hyalinus*: Cyprinidae), the modification of feet and claws (e.g. subterranean planthoppers), and the development of highly sensitive chemical and mechanical receptors to provide subterranean organisms with detailed information about their surrounding environment.

A range of troglomorphic physiological and behavioural characteristics have been recorded in subterranean organisms. In response to the comparatively stable environment and absence of light many taxa do not respond, or have reduced responses, to daily cycles (circadian rhythms). Many taxa have slow metabolic processes and a greater resistance to starvation in response to the scarce energy resources (spatially and temporally) (Hervant, Mathieu & Barre, 1999). Growth and reproductive rates (including number of eggs and offspring per brood) of stygobitic and troglobitic taxa are usually lower than epigean taxa, and both development time to maturity and longevity are usually greater. However, in some instances when environmental conditions fluctuate, some populations, particularly troglophilic and stygophilic organisms, may take on the characteristics of epigean populations. A range of behavioural adaptations have been recorded in subterranean environments including a decrease in the aggregation of some terrestrial taxa (e.g., Collembola); a reduced response to natural alarm substances in cave fish (e.g., *Astyanax fasciatus* (Characidae) and *Caecobarbus geertsii* (Cyprinidae)) and an increased sensitivity to vibration and reduced aggression between individuals of the same species (intraspecific competition) (Parzefall, 1992). The role and development of these behavioural characteristics is providing a greater understanding of the evolution of hypogean organisms and the compensatory benefits that they give the organism within subterranean environments.

Disturbance and Conservation

Within ecology, disturbance is acknowledged as a primary force of structural change. It is widely assumed that subterranean ecosystems, and the communities that occupy them, are relatively stable when compared to epigean systems. Anthropogenic activities and pollution constitute the most serious threat of disturbance to the natural functioning of both aquatic and terrestrial subterranean ecosystems (see also Conservation: Cave Biota). However, since the impacts of human activities on subterranean ecosystems are largely unseen they have historically been poorly documented. In addition, the diffuse nature of many pollution sources has meant that it has not always been possible to quantify the impacts on subterranean communities. Agricultural activities in epigean catchments may lead to a significant increase in the volume of nutrients entering hypogean environments and as a result change the structure of the subterranean food web. In extreme instances microbial communities may proliferate and the indigenous subterranean community may be displaced by epigean taxa (Notenboom, Plénet & Turquin, 1994). Management activities, such as the closure of cave entrances, have led to the reduction and even extinction of some bat populations. In open cave systems, the number of human visitors, particularly to show caves, may lead to significant changes to the microclimate and the cave atmospheric chemistry. Visitors and recreational cavers may inadvertently carry and deposit organic matter within caves and potentially introduce epigean taxa. In addition, algae and higher plants may be able to colonize show caves where artificial illumination occurs (see Tourist Caves: Algae and Lampenflora). This primary productivity potentially changes the trophic basis of the cave and may be utilized by hypogean taxa, but may also allow the colonization and the development of populations of epigean taxa.

The organisms within subterranean environments may provide valuable information regarding the evolution and functioning of natural ecosystems. However, our greatest challenges may involve the protection of hypogean habitats and organisms from the deleterious impacts associated with anthropogenic activity so that future generations can obtain a greater understanding of ecosystem processes and subterranean ecology.

PAUL J. WOOD

See also **Biology of Caves; Food Resources**

Works Cited

Culver, D.C., Kane, T.C. & Fong, D.W. 1995. *Adaptation and Natural Selection in Caves: The Evolution of* Gammarus minus, Cambridge, Massachusetts: Harvard University Press

Gibert, J., Stanford, J.A., Dole-Oliver, M.-J. & Ward, J.V. 1994. Basic attributes of groundwater ecosystems and prospects for research. In *Groundwater Ecology*, edited by J. Gibert, D.L. Danielopol & J.A. Stanford, San Diego and London: Academic Press

Hervant, F., Mathieu, J. & Barre, H. 1999. Comparative study on the metabolic responses of subterranean and surface-dwelling amphipods to long-term starvation and subsequent refeeding. *Journal of Experimental Biology*, 202: 3587–95

Juberthie, C. 2000. The diversity of the karstic and pseudokarstic habitats in the world. In *Subterranean Ecosystems*, edited by H. Wilkens, W.F. Humphreys & D.C. Culver, Amsterdam and New York: Elsevier

Notenboom, J., Plénet, S. & Turquin, M.-J. 1994. Groundwater contamination and its impact on groundwater animals and ecosystems. In *Groundwater Ecology*, edited by J. Gibert, D.L. Danielopol & J.A. Stanford, San Diego and London: Academic Press

Parzefall, J. 1992. Behavioural aspects in animals living in caves. In *The Natural History of Biospeleology*, edited by A.I. Camacho, Madrid: Museo Nacional de Ciencias Naturales

Pohlman, J.W, Cifuentes, L.A. & Iliffe, T.M. 2000. Food web dynamics and biochemistry of anchialine caves: A stable isotope approach. In *Subterranean Ecosystems*, edited by H.Wilkens, W.F. Humphreys & D.C. Culver, Amsterdam and New York: Elsevier

Poulson, T.L. & Lavoie, K.H. 2000. The trophic basis of subsurface ecosystems. In *Subterranean Ecosystems*, edited by H. Wilkens, W.F. Humphreys & D.C. Culver, Amsterdam and New York: Elsevier

Wilkens, H., Humpreys, W.F. & Culver, D.C. (editors) 2000. *Subterranean Ecosystems*, Amsterdam and New York: Elsevier

Further Reading

Camacho, A.I. 1992. *The Natural History of Biospeleology*, Madrid: Museo Nacional de Ciencias Naturales

Chapman, P. 1993. *Caves and Cave Life*, London: HarperCollins

Gibert, J., Danielopol, D.L. & Stanford, J.A. (editors) 1994. *Groundwater Ecology*. London and San Diego: Academic Press

Gibert, J., Mathieu, J. & Fournier, F. 1997. *Groundwater / Surface Water Ecotones: Biological and Hydrological Interactions and Management Options*, Cambridge and New York: Cambridge University Press

Wilkens, H., Humpreys, W.F. & Culver, D.C. (editors) 2000. *Subterranean Ecosystems*, Amsterdam and New York: Elsevier

SUBTERRANEAN HABITATS

Hypogean or subterranean habitats can be defined as habitable places within the hypogean realm, this being a comparatively closed space below the soil or below the barren rock surface (see Figure). Hypogean habitats are dark, their environmental parameters are comparatively stable, food input is limited, and so is the potential for colonization. Apart from darkness and environmental stability, lack of food—caused by the absence of photo-autotrophs and barriers to external inputs—is one of the most influential ecological factors. The endogean (edaphic) environment (i.e. the soil zone immediately below the surface), although dark, is not classified as a hypogean habitat because it contains rich and varied food resources, in contrast with the deeper zones. There are also some habitats related to the edaphic ones, known collectively as the aquatic Mesovoid Shallow Substratum (MSS), including the hyporheic habitat with *Niphargus* spp., and water-saturated soils in Brazilian llanos (grassy soils), which contain a rich copepod fauna. These habitats can be considered as ecotones (transition or threshold habitats) to various epigean habitats, rather than as hypogean habitats. This same approach has been used to identify other ecotones such as anchialine pools. However, since these habitats have traditionally been studied by biospeleologists, they have been included in this review.

Several attempts have been made to classify hypogean habitats, biotopes, or communities, for example the comparatively loose classifications of Vandel (1965) or Camacho (1992); a classification limited to the aquatic environment by Vandel (1968) and Husmann (1970); and a very detailed classification by Juberthie (1983). The classification presented here is based on large-scale habitat differences; the trophic base and small-scale habitat differences, which together (in aphotic habitats) are the equivalent of the composition of the plant community in epigean habitats; and biogeographical factors (with the proviso that the presence of other species also defines each species' environment) or ecological factors underpinning the composition of the animal community. The classification follows the Council of Europe directives (Devilliers & Devilliers-Terschuren, 1996) and takes into account the main ideas of the previous authors. For the sake of clarity it is presented hierarchically. The Encyclopedia has separate entries on Anchialine, Entrance, Interstitial, Marine, and Thermal Water Habitats.

Like any epigean patch of land, a cave may include a number of habitats. The following classification could assist in deciding whether to protect formally (legally) a cave system based on the diversity of the habitats it contains. It can also give an explanation of the high (or low) biotic diversity in a system, along with the ecological basis for its conservation.

Terrestrial Hypogean Habitats

The Terrestrial Interstitial Environment (Terrestrial MSS) is a systems of interconnected channels in deposits of scree (or dry gravel) not infilled by soil. Such deposits are usually covered by layers of more or less compact soil. This habitat is present worldwide in rocky areas. The terrestrial interstitial habitat may be considered as an ecotone between the soil (edaphic) and the crevicular or cave environment. Climatic conditions resemble those in caves, while the food input may be comparatively high. Terrestrial interstitial habitats may be subdivided into:

1. Terrestrial interstitial habitats in carbonate (karst) territories (equivalent to "terrestrial epikarst") which are sometimes inhabited by highly troglomorphic troglobites (particularly beetles) and by edaphic animals. They probably form an important route for migration of future troglobites into caves.
2. Terrestrial interstitial habitats in nonkarst lithologies, primarily utilized by edaphic animal species.

Cave Entrance and Crevicular Habitats are parts of caves or crevices where daylight penetrates. The depth of light penetration depends on the position, general direction, and shape of the void, and of the width of its opening to the exterior. The light gradually diminishes in its intensity and loses its spectral constituents with distance from the opening; other climatic factors gradually attain their hypogean values. A selection of phototrophs are present, disappearing gradually in this order: flowering plants, ferns, mosses, green algae, and finally cyanobacteria. These habitats are present worldwide and may be considered as ecotones between the surface and dark cave habitats. The subordinate categories below (shallow and wider voids) may be divided further by a combination of the following characteristics:

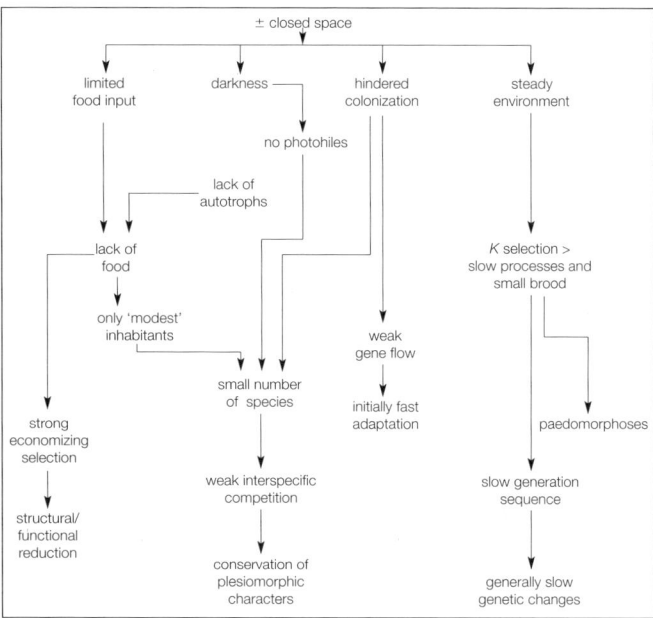

Hypogean Habitats: Interrelations between some characteristics of typical hypogean habitats. (Reproduced from: *Sket, B. 1996*)

regions poor in troglobites vs. regions rich in troglobites;
biogeographical/faunistic regions;
wooded areas, vs. open humid areas, vs. arid areas;
the presence of trogloxene/troglophile invertebrate assemblages (Rhaphidophoridae, Opiliones, Gastropoda, etc.);
habitats of trogloxene vertebrates;
containing hibernating and/or estivating (i.e. resting during hot periods) animal colonies.

1. Shallow crevices are narrow spaces in the fractured rock which may be connected with deeper (dark) crevicular or cave systems. They may be inhabited by selected epigean sighted species, particularly spiders and pseudoscorpions.
2. Habitats in the entrance parts of caves (or artificial cavities) are wider voids in the rock, which may continue into dark cave habitats. They mainly contain trogloxene or accidental animal species, sometimes constituting characteristic parietal faunas of resting (hibernating or estivating) species (particularly some Lepidoptera, Trichoptera, and Diptera), of photophobic migrants feeding outside daily (particularly Rhaphidophoridae crickets); various accidentals and troglobites are also regularly present. These habitats have developed worldwide in karst and volcanic regions.

Dark Cave and Crevicular Habitats are dark voids in the rock, which can be inhabited by small animals. They are potentially present worldwide in karst and volcanic regions, but in general are inhabited only in temperate to tropical areas. The subordinate types below may be further subdivided on the basis of rock type (karst vs. lava), climate (tropical vs. warm temperate vs. cold temperate regions), and biogeographical region.

Dark cave terrestrial habitats with a high input of allochthonous organic matter can be considered as ecotones with surface terrestrial habitats, particularly with endogean ones. If present, organically enriched deposits may be inhabited by a random assemblage of non-troglomorphic surface (mainly soil) animals. The wider surroundings of these deposits are often comparatively densely populated by troglobites, if they exist in the region. This habitat may be further subdivided according to the origin and composition of food deposits:

1. Input of food (wood, foliage with soil) by abiotic means (streams, wind, or falling into the habitat). Soil-like deposits may be inhabited by a selection of accidental migrants from the soil;
2. Input of food (guano, bird droppings) by animal vectors of particular groups or species (e.g. bat colonies, the oilbird *Steatornis*, swiftlets (*Aerodramus*), and cave crickets). In temperate regions the deposits may be similarly inhabited as above, while in tropical caves they may be particularly rich in Microlepidoptera larvae and in cockroaches (Blattaria). In the case of the latter, predators (like Amblypygi) may be very abundant around them;
3. Input of food by plant growth (e.g. penetrating tree rootlets) is particularly common in lava caves because they usually have a relatively thin rock cover. Species feeding on roots are mainly weevils (Coleoptera: Curculionidae) or homopterans (Homoptera: Cixiidae);
4. Input of food from autochthonous (chemoautotrophic) production in the cave water may be limited to deep parts of the cave. The taxonomic composition of the animal community on such food deposits (which mainly comprise bacterial and fungal mats) may depend strongly on the randomness of the local immigration.

Energetically poor, dark cave habitats are inhabited by scarce trogloxene through to troglobitic faunas. Since all cave organisms are bound to the substratum, rather than the void itself, moderately wide crevices are essentially the same habitat as wider cave corridors. They may be classified further by a combination of biogeographical position and any set of climatic and substratum characteristics, as described below:

1. In cold areas they are inhabited by a small population of trogloxenes, and only exceptionally by troglobites (e.g. the beetle *Glacicavicola*);
2. In warm temperate areas they may be inhabited by locally rich troglobitic faunas; around perennial ice deposits in caves in temperate areas, with temperatures close to 0°C and a wet substratum, there are often diverse and dense beetle faunas, with some species (e.g. *Astagobius* spp.) being specialists of this environment;
3. In the tropics they may be inhabited by a few troglobites, as well as scarce edaphic and other lucifugous (light-avoiding) forest animals;
4. On rocky walls, which are usually covered by a thin layer of clay that is regularly wetted, dark cave habitats may be inhabited or visited by snails (*Zospeum* spp.), pseudoscorpions, and beetles (however, their complete habitat often includes clay deposits for breeding);

5. Clay and silt deposits are often inhabited by millipedes and some beetles;
6. The cave hygropetric habitat, which is a rocky wall covered by a thin film of trickling water, is inhabited by some amphibious, originally terrestrial or originally aquatic animals. Some highly troglomorphic leptodirine beetles (e.g. *Hadesia vasiceki*) are specialized for this habitat, while other animals (e.g. the amphipod *Typhlogammarus mrazeki*) are facultative. This habitat is known only from some caves in the Dinaric karst and the Southern Calcareous Alps.

Aquatic Hypogean Habitats

Aquatic Habitats in Unconsolidated Sediments (Interstitial Waters) are water-filled systems of interconnected voids between grains of unconsolidated sediments—mainly gravel and sand. The interstitial space within one sediment deposit is normally contiguous: the length of the channels is immense, while their width depends on the size of the smallest sediment particles. The habitat in any region may be continuous over long geological periods, although it may (together with streams) gradually change its position and its character. If the substrate cover above the interstitial water body is thin, the temperature may vary daily in the upper layers, and variations may be extreme if the substrate is dark in colour. These habitats are generally inhabited by a selection of benthic animals and by a rich array of particularly small and slender specialized troglomorphic animals. Interstitial habitats have developed worldwide, in non-karstic and karstic regions, and even within caves. The subordinate types below may be further classified in combination with sediment grain size (stygopsammal in sand, stygopsephal in gravel); average temperature; high-order biogeographical position; and type of pollution.

Marine and coastal interstitial habitats include the marine mesopsammal zone, which is developed worldwide in sandy parts of the shallow sea bottom. It is an ecotone with different types of the marine benthic habitats. The characteristic inhabitants include Nematoda, Turbellaria, Copepoda: Harpacticoida and Ostracoda; in coarser sediments, some larger crustaceans: Amphipoda and Isopoda. The habitat may be further subdivided according to detritus content and depth. Secondly, coastal or mixohaline interstitial waters are present worldwide along sandy or gravelly sea coasts where an influx of fresh water from the continental side occurs (similar habitats have developed in dry landlocked areas; they are often regarded as marine relicts, but may in fact be secondarily saline). The remarkable spatial and temporal variability of most climatic parameters (temperature, salinity, and oxygen) is particularly characteristic of this type of habitat. The fauna consists largely of non-troglomorphic and troglomorphic Nerillidae (Annelida) and crustaceans (Copepoda, Isopoda, and Amphipoda). It may be further subdivided by salinity level and salinity fluctuation.

Freshwater phreatic interstitial habitats are potentially present worldwide on land along stream channels, and in sediment-filled depressions. They are often excellent sources of potable water. Benthic animals may be scarce in these habitats and they may be inhabited by a rich assemblage of small troglomorphic and/or stygobitic Oligochaeta, Copepoda, Amphipoda, and Isopoda. In cold climates, the interstitial fauna may be limited to non-stygobites in the hyporheal. The hyporheal and phreatic subordinate classes below may be further subdivided by average temperature, and location (non-karst, karst surface, or within a cave). These to a large extent define the characteristics of the fauna.

The hyporheal zone is by its position and ecological parameters an ecotone below the stream benthic zone. Climatic parameters are variable and the food input is high; the oxygen concentration may gradually diminish towards the phreatic zone. It is inhabited by small benthic animals and some interstitial stygobites; Hydracarina and Oligochaeta from both categories may be particularly numerous. This layer may be important as a refugium of the benthal fauna during dry periods.

The phreatic zone comprises layers of interstitial water below the hyporheic ecotone or away from the stream bed. Here, the ecological and faunistic influences from the epigean environment diminish. The fluctuations in ecological parameters are extremely low, particularly daily; the parameters may also be spatially homogeneous. The surface layers are comparatively well supplied with organic matter, and therefore have rich bacterial floras and locally rich stygobite faunas. Deeper layers are almost devoid of food, which has been exhausted in the surface layers. The deep phreatic zone is thus inhabited only by rare and particularly specialized stygobites. Both layer types may be further subdivided by the combination of depth and character of the cover above the water body (which also influences the food input); water-body exchange time; current speed; and degree of oxygenation.

Aquatic Habitats in Porous Rocks are waters in rock crevices, karst conduits (whether the voids are accessible to humans or not), or in volcanic caves. Each of the sub-categories below may be further subdivided according to climatic zone, biogeographical position, and geological character.

Hypogean beds of perennial sinking streams (a perennial sinking river is a stream on the surface with an established flora/vegetation and fauna, which sinks underground) are ecotones with the epigean river habitats. Epigean plants and animals may in part drift underground to form a potential food source, while some animals may penetrate underground as competitors to the obligate hypogean biota. In the hypogean parts of these streams, the communities are composed of stygoxene, stygophile, and stygobite organisms, the latter gradually prevailing with the distance from the ponor (swallow hole). In a population of a stygophile species, troglomorphism may increase clinally (gradually) in the same direction. Climatic parameters may to some degree (depending on the distance from the sink, water mass, and the current) vary daily and yearly. As a result, food input may also vary, depending on the season and the intensity of the precipitation. Such habitats have developed worldwide in karst regions. The environment may be further subdivided, according to the distance from the sink, which defines the volume and quality of food; the composition of local surface fauna and local cave fauna; and whether it is absolutely dark, or illuminated from the sink or from the resurgence.

Hypogean beds of intermittent sinking streams only superficially resemble the habitat of perennial sinking streams; the intermittent rivers lack an established vegetation and fauna; however, terrestrial organisms (alive or dead) which drift, will enrich the food supply underground. Therefore, the hypogean parts of the stream may be inhabited by dense populations of local stygobitic species. The variability of climatic parameters and food input are as for perennial sinking streams. These habitats have developed worldwide in karst regions.

Cave waters with autochthonous energy resources are mineral waters with nutrients (such as sulfur and hydrocarbons) on which chemoautotrophic bacteria may flourish, building the nourishing substratum for heterotrophic bacteria and fungi on which higher animals may feed. Very few caves containing this type of habitat have been reported, but Movile Cave in Romania (see separate entry) is a notable example. Cave waters with autochthonous energy resources, which are also open to the surface, are regularly invaded by epigean animal species, while stygobites are scarce or absent. Cave waters with autochthonous energy resources that are not open to the surface may be inhabited by very diverse, sometimes rich, stygobitic faunas. Owing to isolation these fauna may be of a very different taxonomic composition compared with other local hypogean faunas. This type of habitat is particularly rare (or little known) and could be further subdivided by the type of energy source.

Percolation waters are bodies of water originating from the more or less diffuse trickling of water through mostly narrow crevices in the rock. The lack of a food supply is particularly characteristic—it is limited mainly to particles brought from the surface or imported secondarily (underground) from other habitats. In general, climatic parameters are particularly stable in these habitats. They are inhabited by highly specialized troglomorphic species, and Crustacea predominate. They have developed worldwide in karst, but are less common in volcanic rocks. Where wider crevices are present, the percolation water may be ecologically identical with the intermittent sinking streams.

Waters in fissure systems in the rock (mainly in the aeration zone) have primarily been studied in the ceiling layers above caves in epikarst, although they can develop away from any cave. The narrow space is characteristic—in some the water may be held by capillary pressure. Their fauna primarily consists of Copepoda (particularly Harpacticoida) and smaller Amphipoda; facultatively amphibious terrestrials (e.g. Collembola) may also be present.

Cave waters mainly originating from percolation may be classified further according to habitat configuration: trickles, streamlets, puddles, rimstone pools, pools, lakelets; by the position within the cave system: at the water table or in the aeration cave layer ("zone"); and by connections to, and food input from, other habitats. These habitats may be inhabited by a rich assortment of stygobitic animals, particularly crustaceans, and sometimes fish and amphibians; some extraordinary taxa (e.g. Polychaeta, Porifera) may also be present.

Anchialine (anchihaline) habitats are waters in voids near the coast, where sea water is responsible for their mixohalinity (differnt mixtures of fresh and marine waters) and/or the presence of marine-derived animal species. Hydrographical connections to the sea—as well as to the continent—are hypogean, while the contact of the water body with the open air is restricted to at least some degree. Habitats can potentially develop along all the sea coasts, but anchialine fauna have been found only in tropical to warm temperate regions. The subordinate categories can further be divided according to a combination of characters: lava vs. karst; average temperatures; and biogeographical regions.

Euhaline (marine) anchialine cave habitats are very deep sea-caves or the euhaline parts of mixohaline anchialine caves. This constitutes an ecotone between hypogean habitats and sea caves. Salinity, temperature, and the low food input may be very stable, and the habitat is sheltered from elevated summer temperatures (particularly if the channel is directed downwards). They are inhabited by some slow-moving, or sessile, marine stygophiles, including some deep-sea animals (e.g. sponges: Pharetronida and Hexactinellida; fish: Ophidiidae) as well as some stygobites (e.g. Crustacea: Remipedia, some Isopoda).

Open anchialine pools are coastal pools connected to the sea by cave corridors. The water body is stratified according to salinity. There is a general similarity to shallow-marine habitats, particularly in the presence of phototrophic algae and fluctuation of climatic parameters. The cave-like character of these habitats is exhibited by their limited accessibility to invading biota, as well as by the presence of some stygobites. Apart from a number of marine biota, some stygophiles and stygobites inhabit anchialine pools; most characteristic are taxonomically diverse red shrimps, amphipods: Hadziidae, some copepods, and ophidiid fishes. They may be further classified by distance from and degree of connection to the sea, the salinity regime, and the distance from the equator.

Mixohaline and freshwater anchialine caves are real caves with an anchialine regime, but which are not able to be invaded by epigean marine biota. They may also be dark corridors connected to anchialine pools. A salinity stratification prevents mixing of water layers, and more or less strongly de-oxygenated layers are characteristic. The caves are primarily inhabited by euryhaline stygobites of marine origin (e.g. a number of amphipods: Hadziidae, shrimps: *Typhlatya*, and fish: Ophidiidae), some of limnic origin (e.g. some amphipods: *Niphargus* and some copepods); a few euryhaline marine species may also be present. Habitats may further be classified by the salinity regime.

Thermal hypogean waters are hypogean waters with a temperature higher than the average yearly temperature of the area. The actual habitat comprises the ecotonal area close to the spring where allochthonous food is available. It may be inhabited by a very scarce fauna of relict stygobites, particularly the crustacean family Stenasellidae. Some are eurythermic and others are polystenothermic. The habitat is potentially present everywhere, but the stygobites that live there are known only from tropical to temperate regions. To become thermal, the hypogean waters have usually descended to a great depth in the Earth's crust and are devoid of organic matter (food); however, they may be inhabited by chemoautotrophic bacteria. Thermal habitats may be divided further according to temperature; the absence or presence of chemoautotrophic bacteria; the presence of higher mineral concentrations; and their biogeographical position.

BORIS SKET

See also **Anchialine Habitats; Entrance Habitats; Interstitial Habitats (Aquatic); Interstitial Habitats (Terrestrial); Marine Cave Habitats; Thermal Water Habitats**

Works Cited

Camacho, A.I. 1992. A classification of the aquatic and terrestrial subterranean environment and their associated fauna. In *The Natural History of Biospeleology*, edited by A.I. Camacho, Madrid: Museo Nacional de Ciencias Naturales

Devilliers, P. & Devilliers-Terschuren, J. 1996. *A Classification of Palearctic Habitats*, Strasburg: Council of Europe (Nature and Environment, 78)

Husmann, S. 1970. Weitere Vorschläge für eine Klassifizierung subterraner Biotope und Biocoenosen des Süsswasserfauna. *Internationale Revue der Gesammten Hydrobiologie*, 55(1): 115–29

Juberthie, C. 1983. Le milieu souterrain: étendue et composition. *Mémoires de Biospéologie*, 10: 17–65

Sket, B. 1996. The ecology of the anchihaline caves. *Trends in Ecology and Evolution*, 11(5): 221–25.

Vandel, A. 1965. *Biospeleology: The Biology of Cavernicolous Animals*, Oxford and New York: Pergamon Press (originally published in French, 1964)

Vandel, A. 1968. Le milieu aquatique souterrain. *Bulletin de la Société Zoologique de France*, 93(2): 209–25

Further Reading

Bou, C. 1974. Les méthodes de récolte dans les eaux souterraines interstitielles. *Annales de Spéléologie*, 29(4): 611–19

Chappuis, P.A. 1950. La récolte de la faune souterraine. *Notes Biospéologiques*, 5: 1–29

Delamare-Deboutteville, C. 1960. *Biologie des eaux souterraines littorales et continentales* [Biology of littoral and continental hypogean waters]. Paris: Hermann

Ginet, R. & Decou, V. 1977. *Initiation à la biologie et à l'écologie souterraines* [Introduction to the biology and ecology of underground]. Paris: Delarge

Howarth, F.G. 1981. Lava tube ecosystem as a study site. In *Island Ecosystems: Biological Organization in Selected Hawaiian Communities*, edited by D. Mueller-Dombois, K.W. Bridges & H.L. Carson, Stroudsburg, Pennsylvania: Hutchinson

Juberthie, C. & Delay, B. 1981. Ecological and biological implications of the existence of a superficial underground compartment. *Proceedings of the 8th International Congress of Speleology*, Bowling Green, Kentucky: 203–06

Orghidan, T. 1959. Ein Lebensraum des unterirdischen Wassers: der hyporheische Biotop. *Archiv für Hydrobiologie*, 55: 392–414

Sarbu, S. & Popa, R. 1992. A unique chemoautotrophically based cave ecosystem. In *The Natural History of Biospeleology*, edited by A.I. Camacho, Madrid: Museo Nacional de Ciencias Naturales

Sket, B. 1977. Gegenseitige Beeinflussung der Wasserpollution und des Hoehlenmilieus. *Proceedings of the 6th International Congress of Speleology 1973*, vol. 5, Olomouc: International Union of Speleology

Sket, B. 1986. Ecology of the mixohaline hypogean fauna along the Yugoslav coast. *Stygologia*, 2(4): 317–38

SULFIDE MINERALS IN KARST

Limestone is the host rock for two of the most important classes of economic sulfide mineralization, and thus karst terrains host many of the world's economic sulfide ore deposits. Many karst areas have long mining histories, extending back in Europe to at least Roman times and, in some cases, the Bronze Age. Exploitation of sulfide mineral resources has had a profound effect on the cultural and economic development of many karst regions, but mining and associated mineral processing has also caused environmental degradation in some areas. One notable impact has been drainage of the host limestone to exploit the sulfide minerals. For example, over a large part of the English Peak District (see separate entry) the karst hydrogeology has been profoundly influenced by the driving of lead mine drainage levels (soughs).

From a genetic perspective, sulfide deposits in limestone fall into two broad categories: "syngenetic" deposits, which formed at the same time as the limestones themselves were deposited or lithified; and "epigenetic" or "hydrothermal" deposits, which formed within the limestone after lithification. In the latter group, pre-existing karstic porosity and permeability may play an important role in determining or influencing the location and style of mineralization. Much more rarely, "secondary" sulfide minerals form within the cave environment, either as a result of hydrothermal processes or via microbiological processes in cave sediments.

The best known of the syngenetic limestone-hosted sulfide deposits are the "Irish-type" lead-zinc ores. In the Irish localities, after which the group is named, these formed at the same time as, or soon after, deposition of the host Lower Carboniferous limestone sequence. Lead isotopic and other studies show that lead (and, by analogy, other metals) was derived from the underlying lower Paleozoic basement by brines driven by hydrothermal convection (e.g. Mills *et al.*, 1987). Fault-hosted mineralization of the appropriate style, age, and lead isotopic composition has indeed been found in the Paleozoic basement rocks (Haggerty *et al.*, 1996). Sulfide was derived predominantly from bacterial reduction of seawater sulfate within the host limestone. Where the ascending metal-rich brines met the sulfide-rich limestone sediment pore waters a perfect environment for sulfide ore deposition was created, resulting in some of the world's major lead-zinc ore deposits.

The classic limestone-hosted epigenetic sulfide deposits are the Mississippi Valley-type lead-zinc deposits of North America. However, these are one example of a wide range of types of limestone-hosted lead-zinc deposits that are common worldwide. Other well-known and broadly similar deposits occur in the Carboniferous limestone of the Derbyshire Peak District, Yorkshire Dales, North Wales, and Mendip areas of Britain. Other classic localities worldwide are found in Silesia, the Eastern Alps, Anatolia, Russia, and Canada. In these deposits the lithified limestone provides a host for sulfide deposits formed in faults and other fractures as replacement deposits—where limestone has been dissolved by the ore-forming fluid and the porosity refilled—and as deposits in pre-existing karstic porosity.

The role of the limestone host rock in the formation of these deposits has long been debated and may be complex. Limestones are often massive and relatively brittle rocks and may thus have extensive fracture networks, which promote deep groundwater flow and assist the emplacement of mineralizing fluids. Pre-existing karstic porosity in buried limestones will clearly enhance this effect. Limestones may also play a chemical role in the localization of mineralization in these deposits. It has been suggested that limestone can neutralize brines carrying H_2S and dissolved metals, resulting in sulfide mineral deposition. Clearly, in replacement-type deposits, there has been corrosion of limestone associated with ore emplacement, but in many other deposits these relationships do not exist.

The role of pre-existing karstic porosity in the formation of these deposits has often been significant or even crucial. Sass-Gustkiewicz (1983) showed that the style of mineralization in Silesian deposits was controlled by its relationship to pre-existing karstic porosity. At two major deposits in Canada (Pine Point and Nanisivik), mineralization is focused into zones of pre-existing karstic porosity. In the classic Derbyshire (England) localities, deep groundwater flow along fault systems allowed dissolutional development of passages that were later infilled with calcite, along with galena and sphalerite. However, even where ores are deposited in solution cavities away from the main faults, good evidence of replacement textures is rare and often equivocal (e.g. Ford & King, 1965). Solution cavity generation may be significantly separated in time from the later emplacement of ore minerals, even in apparently "replacive" deposits.

Perhaps the best-described example of "secondary" hydrothermal sulfide minerals in caves, are sulfide mineral inclusions in sparry calcite in the Cupp-Coutunn Cave system, Turkmenistan (see separate entry). These minerals formed when pre-existing cave passages were invaded by hot basinal brines; the same brines formed epigenetic fault-hosted galena mineralization in limestone at greater depths. Cinnabar and metacinnabar (sulfides of mercury) are known from the Gaudakskoy caves of the former USSR, where again they originate from thermal waters (Lazarev & Philenko, 1976).

Secondary cave sulfide minerals can also form at low temperatures, as a result of reactions of hydrogen sulfide produced by sulfate-reducing bacteria (see Microbial Processes in Caves). Bacterial sulfate reduction is ubiquitous in anoxic environments in marine systems, where sulfate is abundant, but is generally less significant in terrestrial aquatic environments, where sulfate concentrations are usually much lower. Iron sulfide mineral coatings formed in this way on cave walls, have been reported from the freshwater-marine mixing zones of the Bahamas (Bottrell *et al.*, 1991). Here, hydrogen sulfide is produced in a biologically active zone, localized within the density gradient of the mixing zone. Most of the sulfide is reoxidized in shallower groundwater, but some reacts with detrital iron oxides to form solid iron sulfide. Similarly, pyrite has been reported in sediments from phreatic caves in Florida (Martin & Harris, 1993). Such processes are likely to be less significant in caves with fresh waters. Seeman (1970) reported pyrite and marcasite (both forms of FeS_2) in cave sediments, the sulfur source being sulfate evaporites, and Bottrell *et al.*, (1999) reported sulfide mineralization of wood fragments in freshwater sediments in the Speedwell Mine / Cave system of Derbyshire, England. Lauritzen & Bottrell (1994) describe iron sulfide spherules from thermoglacial karst springs in southern Svalbard, though these most likely originate from the deeper, thermal part of the cave system, where waters have higher salinity and have undergone bacterial sulfate reduction.

SIMON BOTTRELL

See also **Mineral Deposits in Karst**

Works Cited

Bottrell, S.H., Smart, P.L., Whitaker, F. & Raiswell, R. 1991. Geochemistry and isotope systematics of sulphur in the mixing zone of Bahamian blue holes. *Applied Geochemistry*, 6: 97–103

Bottrell, S.H., Hardwick, P. & Gunn, J. 1999. Sediment dynamics in the Castleton karst, Derbyshire, UK. *Earth Surface Processes and Landforms*, 24: 745–59

Ford, T.D. & King, R.J. 1965. Layered epigenetic galena-barite deposits in the Golconda Mine, Brassington, Derbyshire, England. *Economic Geology*, 60: 1686–1701

Haggerty, R., Cliff, R.A. & Bottrell, S.H. 1996. Pb-isotope evidence for the timing of episodic mineralization on the Llanrwst and Llanfair-Talhaiarn orefields, north Wales. *Mineralium Deposita*, 31: 93–97

Lauritzen, S.-E. & Bottrell, S.H. 1994. Microbiological activity in thermoglacial karst springs, south Spitzbergen. *Geomicrobiology Journal*, 12:161–73

Lazarev, J.S. & Philenko, G.D. 1976. Geologo-mineralogiceskie osobenosti Guardakskoy karstovoy peschery. *Peschery*, 16: 45–63

Martin, H.W. & Harris, W.G. 1993. Mineralogy of clay sediments from three phreatic caves of the Suwanee River Basin. *NSS Bulletin*, 54: 69–76

Mills, H., Halliday, A.N., Ashton, J.R., Anderson, I.K. & Russell, M.J. 1987. Origin of a giant orebody at Navan, Ireland. *Nature*, 327: 233–35

Sass Gustkiewicz, M. 1983. Zinc-lead ore structures from the Upper Silesian region in the light of solution transfer. In *International Conference on Mississippi Valley Type Lead-Zinc Deposits*, edited by G. Kisvarsanyi, S.K. Grant, W.P. Pratt & J.W. Koenig, Rolla: University of Missouri

Seeman, R. 1970. Pyritfunde in der Dachstein-Mammuthohle (Oberosterreich). *Die Hohle*, 21: 83–89

SURVEYING CAVES

Cave surveying is an attempt to accurately record the form of a cavity and, like much else in speleology, can involve a great deal of dedication and discomfort if it is to be done well. Cave surveys (or cave maps in US terminology) serve a range of purposes, from the most basic of describing the shape of the cave, through detailed mapping to enable further exploration or scientific work, to full 3D visualization techniques for the placing of archaeological remains.

History

The first known cave plan was published in *De ortu et causis subterraneorum* (Georg Agricola, 1546, p.146) and is of the Stufe di Nerone (Nero's Oven), Pozzuoli, near Naples (Italy). However this is a man-made cavern in tufa deposited from ancient hot springs. The first drawing of a natural cave was a very approximate sketch of the Baumannshöhle, made by Von Alvensleben in 1656, but it was never published and now resides in the archives of Magdeburg (Germany). Next was a rather more realistic sketch of Long Hole, Cheddar (United Kingdom) made before 1680 by John Aubrey, however this was not published until 1992 (Boycott, 1992).

The first published cave survey was of Pen Park Hole, Bristol (United Kingdom) in 1683 (Mullan, 1993 and see figure in Britain and Ireland: History). The survey (plan and elevation)

was made during the second descent of the cave in 1682 by Captain Greenville Collins (Southwell, 1683). This survey, whilst not accurate, gives a reasonable idea of the form of the cave, and as the captain was from a survey ship this was probably the first survey to use any instruments (for depth measurement at least).

There were several more surveys in the 18th century, of somewhat variable accuracy; an elevation of Demänova cave in Czechoslovakia (1719, by Juraj Buchholtz), better surveys of both Pen Park Hole (1775, by William White) and Baumannshöhle (1702, by Hardt), Postojna Cave, Slovenia (J.A. Nagel, 1748), Aggtelek, Hungary (1794, by József Sartory), a 1781 sketch of Madison's Cave by Thomas Jefferson (Jefferson, 1782, pp.34–36) (the first American cave map), and the caves of Sloup, Czechoslovakia (Süsz, 1800). Whilst some of these are quite good representations of the cave, the first genuinely accurate survey is a plan of the Grotte de Miremont, Rouffignac (France) in 1765, by Nicolas Thomas Bremontier, a civil engineer and inspector-general of highways. This plan and 27 cross sections must have been made with compass and tape or chain, and compares very well with the survey made by E.A. Martel 128 years later.

From 1800 on, surveying became intimately associated with cave exploration. Activity in the two fields expanded together—you can't tell what you've found unless you survey it, and the survey is often the key to further discoveries. Thus explorers started to make good quality surveys as they discovered new parts of a cave system. Perhaps the best example is the work to piece together the parts of the Postojna system in Slovenia between 1748 and 1852 (Shaw, 1992).

Surveying continued to develop in the 19th century, notably in Belgium, Slovenia, Australia, and the United States. In the United States, the main drive for surveying was as an aid to the exploitation of saltpeter deposits. The first measured survey there was an 1805 compass and chain map of Great Saltpeter by John James DuFour. This was followed by numerous surveys of Mammoth Cave, Kentucky, by far the most accurate of which was the one in 1834–35 by Edmund F. Lee, a professional surveyor, which took several months. Later surveyors were restricted by the showcave management, who obtained an injunction against publication of one survey as it would show that the cave extended beyond their land, potentially allowing another entrance to be opened and reduce their takings. Such surveys would also have spoiled the extravagant claims the management made for the length of the system (between 40 and 50 km by 1952, but claims for a length of 600 miles were made!).

The first underground theodolite (or transit) surveys were made by mine surveyors as the theodolite had become standard for mine surveys by 1832. The first definite theodolite cave survey was by a Mr Hodgeson in Ingleborough Cave, Yorkshire (United Kingdom) in 1838, under the direction of the landowner James W. Farrer. It seems possible that Edmund Lee's Mammoth Cave survey of 1835 was also done using such instruments, which would make it the first.

Techniques

Details of the techniques used in early surveys are not often given so it is difficult to be specific about exact techniques. However a lot can be inferred by comparing with modern surveys and reading contemporary reports. The overall impression is that the techniques used have changed remarkably little since the end of the 18th century. A notebook with compass or transit / theodolite and tape or chain have been the tools of the trade up till the present day. Accuracy has improved with better equipment and technique as surveyors progressed from pacing through to marked ropes, tapes, chains, and towards the end of the 19th century the measurement of slope as well as direction. Over this time the size and weight of the instruments has improved significantly from heavy miner's dials in big wooden boxes to modern aluminium-bodied compasses and clinometers, introduced in the late 1960s. The earliest compass used a succession of concentric wax rings on which the needle directions were inscribed. This allowed the survey to be re-created on the surface. This was certainly used for mine surveying and was probably used in natural caves too. There have been very few instruments designed specifically for caving so available items have been pressed into service—such as mining instruments (compass, transit), theodolite, the Abney level, and the forester's clinometer.

The first detailed description of cave surveying techniques is by E.A. Martel (Martel, 1894, pp.24–28). He describes the construction of a canvas-covered notebook 12 × 17 cm composed of alternate pages of 1 mm / 1 cm grid and blank, and with a notch cut out in the top right hand corner for a small compass to be attached. He worked with one assistant who carried a light and moved ahead until he was about to become obscured. Martel aligned the notebook using the compass, recorded the reading and then drew the corresponding line on the page, drawing the cave around this line, carefully keeping the notebook properly aligned with the compass. This is similar to the modern plane table technique. Distances were measured by pacing except where the terrain was too difficult, when a marked rope was used. Martel generally didn't record slight

Surveying Caves: The original compass-and-tape survey of Sarawak Chamber in the Mulu karst; only when the survey was drawn up in the cave was it realized that the seemingly endless survey over the breakdown pile had made a loop through the world's largest underground chamber. (Photo by Andy Eavis)

slopes, but noted that boulder slopes always rest at between 33 and 36 degrees.

Since the early days, cave surveying has consisted of marking points in the cave called stations, and measuring legs (US: shots) between them, although advanced electronic instruments may finally change this fundamental method in the 21st century. A leg is defined by the length, angle, and inclination between two stations. The development of cave surveying has really just been advances in the instruments used to make these measurements.

Arthur Butcher's seminal book, including CRG (Cave Research Group of Great Britain) grades based on instruments used, defined modern technique (Butcher, 1950). This prompted a great deal of surveying in the United Kingdom in the 1950s. William E. Davies and Lang Brod covered similar ground in the United States in the 1950s and 1960s (Brod, 1962). Bryan Ellis later modified the CRG grades to produce the BCRA (British Cave Research Association) grades defined in terms of accuracy, which have become widely used throughout the world (Ellis, 1976).

The standard method since the 1950s has been to use compass, clinometer, and tape although many less accurate surveys, especially in largely horizontal caves, miss out the inclination reading and some still use pacing or a chain for length. For higher accuracy a theodolite or transit is used: a transit is just a very precise compass, but a theodolite measures relative angles from a baseline at the entrance of the cave, so errors will accumulate as the survey length increases. Such high-accuracy surveys are rarely justified—usually only on main passages of important caves or if starting an expensive project such as blasting a new entrance or connection.

The aluminium-bodied Suunto compass and clinometer became popular in the 1960s, and although some American surveyors still persist with the classic Brunton pocket Transit compasses (invented c.1898), the Suunto and Sisteco / Silva equivalents have become standard the world over. Fibreglass tapes are preferred to metal tapes as they are much easier to handle underground.

In the 1970s and 1980s (in France and Switzerland particularly) the topofil became quite popular—this is a device that reels out thread to measure the distance, and incorporates the inclinometer and compass. It was quick, if slightly less accurate than conventional compass and tape, but its popularity has declined in recent years. Some groups of Russian and Ukranian surveyors use a water level to measure height difference, rather than angle. This has the advantage that it can be used round corners between any two stations the pipe will reach, and the height measurements can be done on a separate trip from the compass and tape survey. Other popular instruments used up until the 1970s include the Abney level to measure inclination up to about 60 degrees.

Drawing was done by simply transferring the direction and length measures to the page at a suitable scale (and allowing for slope), until around the 1960s when using calculators or computers to do the simple trigonometry became feasible. The resulting coordinates are plotted on graph paper. Recent advances in software and technology have enormously speeded up this drawing process so that the surveyor merely enters the data and immediately gets a finished plot of the legs, stations, and even passage size.

The advent of computers also made it possible to deal with survey errors systematically. Loops in caves are common and there will always be a discrepancy in the position of a point measured by two different routes. Mathematical techniques, originally developed by land surveyors and commonly referred to as "least squares loop closure", find the best fit positions, assuming there are no gross errors ("blunders"). Techniques to identify blunders (which can be remarkably common in survey data, see Fish, 1999) are also now found in cave survey software. These can only be applied to blunders within loops.

The first cave survey software was written on university mainframes and the first survey to be produced with the aid of such software was of the Fergus River Cave, Ireland, in 1964 (Hanna, 1964). Since then innumerable programs have been written. In 1980 David McKenzie had a program called ellipse that would not only process the centreline but also plot the walls, and adjust them to fit as loop closures moved the centreline. This was not repeated on personal computers until Toporobot gained this functionality in about 1995. Toporobot, started in 1972 by Martin Heller, is one of the most long-lived programs, and is still popular and being developed today. It is unusual in that it defined and requires a particular surveying methodology to get the best out of it. Most other software instead reflects existing techniques and does its best to cater for them. SMAPS by Doug Dotson was very popular for many years in the 1980s, but ceased to be maintained in 1995 and has been superseded. Compass by Larry Fish is probably currently the most popular cave survey software in the world, having had long and consistent development with good documentation since the mid 1980s. The multi-platform, multilingual Survex by Olly Betts has also become popular since its creation in 1990. Many other programs are used, often being popular in a particular area. Modern software can generate impressive 3D models with rendered walls, overlay elevation details and maps, and do flythroughs as well as importing data from a number of sources.

Representing a 3D void, often of very complex shape, in 2D on paper has always been problematic and many techniques have arisen over the years to try to represent the cave as well as possible. Plans may be coloured to show multiple levels, elevations may be drawn as "extended elevations" where the passages are "unfolded" into one plane to better describe the slopes and feel of the cave at the expense of the representation of above / below relationships. Cross-sections and isometric views show the passage shape. Numerous symbols are used to indicate floor sediment type, speleothems, drops, climbs, slopes etc. Numerous standard symbol sets have been defined over the years, the latest being by the International Union of Speleology in 1999 (UIS, 1999).

Cave surveyors have always made use of new technology as soon as it becomes sufficiently robust, and at the turn of the 21st century electronic devices are becoming available which will probably mean significant changes to the cave surveying method over the next couple of decades. Martin Sluka of the Czech Republic made what is probably the first practical electronic cave surveying device in 1996—a laser distance-measuring device attached to a laser point, and electronic inclinometer with a conventional compass (Sluka, 1999). This took conventional readings but without the user having to move to the end point of the leg. Other researchers are working on completely electronic devices that are simply pointed at the next station and the readings internally recorded for later downloading (Auriga project, Martin Melzer http://home.nikocity-de/andymon/hfg/avriga4-htm). Inertial based devices that are simply carried through the cave recording a track are also possible; these produce data fundamentally different from the station + legs concept. Indeed, the Wakulla II project (in Florida) produced a device that worked in this way in 1998. It had a rotating scanning sonar head which traversed the (underwater) cave on a dive-scooter, automatically generating a helical 3-D scan of points describing the wall position, but it did weigh nearly 200 kg out of the water.

Underwater surveying is a particularly difficult task due to the poor visibility and lack of time. A depth gauge is used instead of measuring inclination and a dive line pre-marked with tags is used for distances. The compass is a wrist-compass, typically marked in 5 degree graduations. If the survey is done carefully (preferably in both directions along the line to make a loop) then quite good accuracies of 5% are obtainable. This has been standard technique since the 1950s (Lloyd, 1970). During the 1990s more accurate techniques have been developed by John Cordingley, made practical by better diving equipment allowing longer, warmer dives. A compass is screwed to an A4 slate and a fibreglass tape fitted with clothes pegs is used for distance. A dive computer is used for depth. All these improve accuracy so that 1% errors on 1 km loops are achievable (Cordingley, 1997). Cyrille Brandt has developed equally accurate techniques by two-man teams in European sumps.

Radiolocation is a technique to fix an underground point to the corresponding point on the surface, usually to find out where the extremities of a cave lie and reduce cumulative errors in the survey (see Radiolocation). Induction radio is used with a carefully levelled transmitter underground. A surface receiver can be used to find the point directly above the transmitter by searching for "nulls"—the plane in which a vertically held receiver loop receives no signal from the underground loop. Lines along nulls will intersect over the transmitter. The depth of the transmitter can be determined either using signal strength measurements, or by finding the point at which a null occurs when the receiver is held at 45 degrees to the perpendicular. It is also possible (but more difficult) to use the technique between two caves to help determine relative position.

WOOKEY

See also **Radiolocation**

Works Cited

Boycott, T. 1992. Cave references in John Aubrey's *Monumenta Britannica*. University of Bristol Spelaeological Society Newsletter, 8(3): 5–9
Brod, L. 1962. Cave mapping: A systematic approach. *Missouri Speleology*, 4(1–2)
Butcher, A.L. 1950. *Cave Survey*, Warwickshire: Cave Research Group of Great Britain
Cordingley, J. 1997. Surveying underwater caves. *Compass Points*, 16: 9–12
Ellis, B. (editor) 1976. *Surveying Caves*, Bridgwater: British Cave Research Association
Fish, L. 1999. How common are blunders in cave survey data? *Compass Points*, 25
Glover, R.R. 1976. Cave surveying by electromagnetic induction. In *Surveying Caves*, edited by B. Ellis, Bridgwater: British Cave Research Association
Hanna, K. 1964. Survey. In D.J. Patmore & F.H. Nicholson, The Fergus River Cave, County Clare, Ireland. *Proceedings of the University of Bristol Spelaeological Society*, 10: 285
Jefferson, T. 1782. *Notes on the State of Virginia*, Paris, privately printed (many later editions)
Lloyd, O.C. 1970. An underwater cave survey. *Transactions of the Cave Research Group*, 12(3): 197–99
Martel, E.A. 1894. *Les abîmes* [The Depths], Paris: Delagrave
Mullan, G.J. 1993. Pen Park Hole, Bristol: A reassessment. *Proceedings of the University of Bristol Spelaeological Society*, 19(3): 291–311
Shaw, T. 1992. *History of Cave Science: The Exploration and Study of Limestone Caves, to 1900*, 2nd edition, Broadway, New South Wales: Sydney Speleological Society
Sluka, M. 1999. Surveying with digital instruments. *Compass Points*, 23
Southwell, R. 1683. A description of Pen Park Hole in Glocestershire. *Philosophical Transactions of the Royal Society*, 13: 2–6
UIS 1999. Cave symbols: The official UIS list. www. karto.ethz.ch/neumann-cgi/cave-symbol.pl

Further Reading

Day, A.J. 2002. *Cave Surveying*, British Cave Research Association
Dasher, G.R. 1994. *On Station*, Huntsville, Alabama: National Speleological Society
Grossenbacher, Y. 1991. *Topoghraphie Souterrain*, Editions du Fond, Société Suisse de Spéléologie

SYNGENETIC KARST

Syngenetic karst is a term coined by Jennings (1968) for karst features, including caves, that form within a soft, porous, soluble sediment at the same time as it is being cemented into a rock. Thus, speleogenesis and lithogenesis are concurrent.

Jennings was describing the active karst geomorphology of the Quaternary dune calcarenites of Australia. Concurrent studies by sedimentologists of paleokarst horizons at unconformities in the stratigraphic record used the related concept of eogenetic diagenesis: processes that affect a newly formed carbonate or evaporite sediment when it is exposed to subaerial weathering and meteoric waters (Choquette & Pray, 1970). The resulting eogenetic karst (or "soft-rock karst") is distinguished from telogenetic ("hard-rock") karst, that has developed on hard indurated limestones that have been re-exposed after a deep burial stage.

The terms syngenetic and eogenetic overlap but involve different viewpoints. The former is best used for geomorphological studies of modern soft-rock karsts; whereas the latter is best retained for diagenetic studies of paleokarst porosities, where the sequence of dissolution and cementation events is much more complex. Some, but not all, paleokarst is eogenetic: the separation of eogenetic, mesogenetic (burial), and telogenetic features

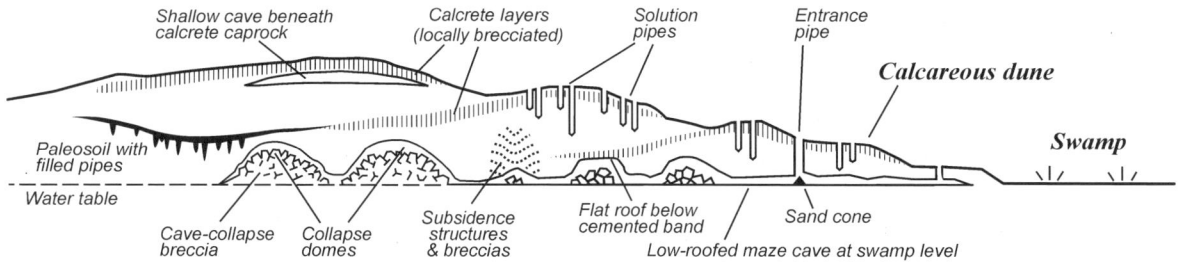

Syngenetic Karst: Figure 1. Features of syngenetic karst developed on a calcareous dunefield.

requires a detailed study of cement morphology, mineralogy, chemistry, and related dissolutional and brecciation features, at both the microscopic and macroscopic scale.

Some workers have applied the term "syngenetic cave" to lava tubes. Although that is an etymologically valid use of "syngenetic", it is unrelated to the present topic.

Syngenetic karst has several distinctive features, as well as many that are shared with classical (telogenetic) karst. In the following discussion, dune calcarenites in a Mediterranean climate are used as an initial example (Figure 1). In calcareous dunes, percolating rainwater gradually converts the unconsolidated sand to limestone by dissolution and redeposition of calcium carbonate. This initially produces a cemented and locally brecciated calcrete layer near the surface (Figure 4). Terra rossa soils may also develop. Below this, the downward percolating water dissolves characteristic vertical solution pipes (Figure 2), and simultaneously cements the surrounding sand. Early cementation tends to be localized around roots, to form distinctive rhizomorphs or rhizocretions. Cementation can occlude the primary inter-granular porosity, but dissolution can generate localized secondary porosity of a mouldic, vuggy, or cavernous character.

Mixing corrosion occurs where percolation water meets the water table, which is typically controlled by the level of a nearby swampy plain, which also provides acidic water. Near the coast, water levels fluctuate with changing sea levels, and further complexity results from a thin freshwater lens floating above sea water, which results in two mixing zones, above and below the lens (see Speleogenesis: Coastal and Oceanic Settings).

In the early stages of solution, the loose sand subsides at once into any incipient cavities, possibly forming soft-sediment deformation structures. Once the rock is sufficiently hardened to support a roof, caves can develop. Horizontal cave systems of low, wide, irregular, interconnected chambers and passages (see Salt Pond Cave figure in Speleogenesis: Coastal and Oceanic Settings) form, either in the zone of maximum solution at the water table, or by subsidence of loose material from beneath stable calcrete layers. Flat cave roofs are common; either marking the limit of solution at the top of the water table, or where collapse has reached the base of an indurated (cap rock) zone. Where a shallow impermeable basement occurs, its topography may concentrate water flow along buried valleys to form linear caves.

Sizable caves can form in less than 100 000 years. Surface dissolutional sculpturing is rare, as there is little solid rock for it to act upon. However, some sculpturing can occur on exposed calcrete layers.

The subsidence of partly-consolidated material can form a variety of breccias and sag structures; these can be further cemented as diagenesis continues (Figure 3). Mantling breccias can occur as part of the surface soil (Figure 4). Within the caves, breakdown of the soft rock is extensive. In many cases, rubble-filled collapse domes largely supplant the original dissolutional cave system at the water table. Subsidence may reach to the surface to form dolines. In paleokarst exposures, these collapse areas would appear as both discordant and concordant (intrastratal) breccias. In extreme cases, mass subsidence of broad areas can generate a chaotic surface of tumbled blocks and fissures.

Variations can occur in different climates or sediments. For example, calcrete is supposedly best developed in semi-arid climates, whereas dissolution and brecciation are more abundant in wet climates. Sequences of marine sediments undergoing cyclic

Syngenetic Karst: Figure 2. Solution pipes (or, more strictly, dissolution pipes) are distinctive features of syngenetic karst (Lyndberg & Taggart, 1995). They are vertical cylindrical tubes with cemented walls, typically 0.3 to 1 m in diameter, which can penetrate down from the surface as far as 20 m into the soft limestone. The pipes may contain soil and calcified roots (and root growth may have occurred hand-in-hand with dissolution of the pipe). They occur as isolated features, or in clusters with spacings as close as less than a metre. (Photo by Ken Grimes)

Syngenetic Karst: Figure 3. Subsidence structure in syngenetic karst. Thin horizontal beds of a beach calcarenite were partly cemented into individual plates that then subsided as dissolution undermined them. Continuing cementation stabilized the tilted beds before the present cave formed. (Photo by Ken Grimes)

emergence can develop syngenetic breccia layers and karst surfaces at the top of each cycle. In coarse-grained sediments, preferential dissolution of aragonite fossils (e.g. coral) can form a coarse mouldic porosity. Where soluble evaporites are interbedded with carbonates, they may be removed completely to undermine the overlying carbonate beds and form extensive intrastratal brecciated layers (see separate entry, Evaporite Karst). However, such breccias can also form in later mesogenetic and telogenetic settings, so are not necessarily eogenetic.

Dissolutional porosity generated during the eogenetic stage of paleokarsts can direct water flow and further dissolution during the later mesogenetic and telogenetic stages (see Inception in Caves), and also host ore minerals or hydrocarbons. (See Sulfide Minerals in Karst; Hydrocarbons in Karst.)

KEN G. GRIMES

See also **Evaporite Karst; Paleokarst; Speleogenesis: Coastal and Oceanic Settings**

Works Cited

Choquette, P.W. & Pray, L.C. 1970. Geologic nomenclature and classification of porosity in sedimentary carbonates. *American Association of Petroleum Geologists Bulletin*, 54: 207–50

James, N.P. & Choquette, P.W. (editors) 1987. *Paleokarst*, New York: Springer
The introduction has useful summaries of meteoric diagenesis and breccias. Many of the paleokarsts described in this book are at least partly eogenetic in origin.

Jennings, J.N. 1968. Syngenetic karst in Australia. In *Contributions to the Study of Karst*, edited by P.W. Williams & J.N. Jennings, Canberra: Research School of Pacific Studies, Australian National University

Lyndberg, J. & Taggart, B.E. 1995. Dissolution pipes in northern Puerto Rico: An exhumed paleokarst. *Carbonates and Evaporites*, 10(2): 171–83
A review and case study of "solution pipes"—a distinctive feature of syngenetic karst.

Further Reading

Esteban, M. & Klappa, C.F. 1983. Subaerial exposure environment. In *Carbonate Depositional Environments*, edited by P.A. Scholle, D.G. Bebout & C.H. Moore, Tulsa, Oklahoma: American Association of Petroleum Geologists
A well-illustrated review of karst and related meteoric diagenetic features. Many, but not all, of the examples are of syngenetic karsts.

Klimchouk, A.B. & Ford, D.C. 2000. Lithological and structural controls of dissolutional cave development. In *Speleogenesis: Evolution of Karst Aquifers*, edited by A.B. Klimchouk, D.C. Ford, A.N. Palmer & W. Dreybrodt, Huntsville, Alabama: National Speleological Society
Page 56 describes breccias in interbedded sulphates & carbonates, but not all these are syngenetic. Page 60 discusses diagenetic stages, including the eogenetic stage.

Moore, C.H. 1989. *Carbonate Diagenesis and Porosity*, Amsterdam and New York: Elsevier
A detailed review of carbonate and evaporate sediments and the evolution of their secondary porosity. Chapter 3 describes the tools used to determine the nature and timing of solutional and cementation events. Chapter 5 discusses evaporites. Chapter 7 discusses meteoric diagenesis (both eogenetic and telogenetic).

Mylroie, J.E., Jenson, J.W., Taborosi, D., Jocson, J.M.U., Vann, D.T., & Wexel, C. 2001. Karst features of Guam in terms of a general model of carbonate island karst. *Journal of Cave and Karst Studies*, 63(1): 9–22
Their model involves eogenesis (i.e. syngenesis) in young, porous limestones exposed by uplift or sea-level change.

White, S. 2000. Syngenetic karst in coastal dune limestone: A review. In *Speleogenesis: Evolution of Karst Aquifers*, edited by A.B. Klimchouk, D.C. Ford, A.N. Palmer & W. Dreybrodt, Huntsville, Alabama: National Speleological Society
A recent case study from Australia.

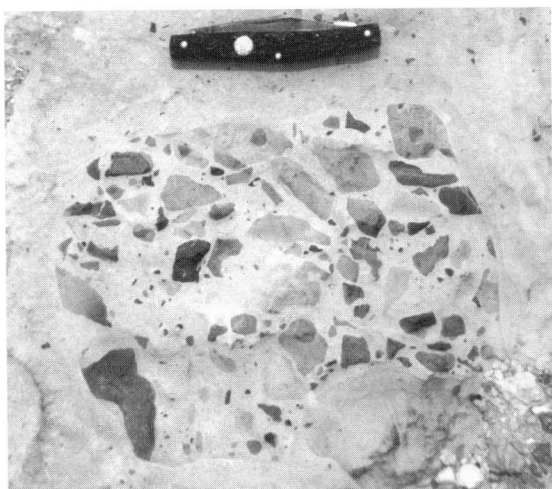

Syngenetic Karst: Figure 4. Calcreted, multi-generation, mantling breccia in dune calcarenite. The large, 20 cm clast contains at least two earlier generations of smaller clasts. Note the blackened pebbles: some authors have suggested that these may indicate carbon derived from fires. (Shinn & Lidz, pp.117–31 *in* James & Choquette, 1987)

TALUS CAVES

Talus caves are interconnected spaces between rocks, large enough for humans to enter and investigate, and which extend beyond daylight; or, a single such space beneath one or more boulders. They are produced by a variety of processes and commonly interface with crevice caves, and through erosion, with multiprocess caves. Alternative names in the literature include avalanche caves, block-field caves, boulder caves, purgatory caves, rockpile caves, rockslide caves, scree caves, and slope failure caves.

Significance

Many speleologists consider talus caves to be inconsequential, but Sweden's Bodagrottorna has more than 2800 m of passages. These are the commonest type of cave in Sweden, and a growing literature in several languages is centred around proceedings volumes of international symposia on pseudokarst convened in Europe. Sjöberg (1989a) found that about 15% of Swedish examples have high scientific and / or recreational values. Some in temperate climates are natural deep-freezes, containing unseasonal snow and forming ice *in situ* through mechanisms which may be of special importance on Mars (see Extraterrestrial Caves). Many others provide microclimates favourable to specialized flora and fauna, including relict populations. Others, in arid regions, protect running or ponded water from evaporation. Some talus caves in California (United States) provide delightful recreational caving, and others have been developed successfully as show caves (e.g. the widely advertized Lost River Caves and Polar Caves, New Hampshire, United States). In Pinnacles National Monument (California), Bear Gulch Cave is an especially popular tourist attraction and an important bat roost. In southeastern California and elsewhere, talus caves served as human habitations as late as the 20th century. In Europe, others are sites of cultural and archaeological significance. In Sweden, some talus caves are considered to be important indicators of recent tectonic activity, while others in present and past littoral zones are indicators of abrasive capacity of small seas like the Baltic. Because cavers commonly underestimate them, however, some talus caves are especially dangerous to recreational cavers. One of Colorado's Lost Creek caves has claimed the life of an experienced, well-equipped caver, and another nearly died in a similar cave in California's Thunder Canyon.

Distribution of Talus Caves

Talus caves are found worldwide; almost everywhere that competent rocks in steep terrains fracture into large angular blocks or undergo spheroidal weathering. In the United States they are found from Maine and South Carolina to California and Washington state. In the northeastern part of the United States they occur especially in anorthosite, but others are found in granite, gneiss, schist, marble, and other metamorphic rocks. They also occur in some volcanic and sedimentary rocks including limestone, where a few small examples are near-end features of the karst cycle.

Types and Classifications

In Europe, talus caves have been studied intensively, and have been classified by process, by lithology, and by a combination of these. Many writers have mentioned frost wedging and erosion in their formation. Most classifications have been applied to relatively small regions. In Sweden, Sjöberg (1989b) found four types of talus ("boulder") caves: glacial boulder caves; abrasive boulder caves; one boulder cave formed by frost wedging; and neotectonic boulder caves. Sjöberg's glacial boulder caves include caves in coarse moraines and caves in erratics. The latter include both boulder piles and caves in single split blocks. His abrasive boulder caves are littoral features (see Littoral Caves). The single frost-wedged cave was formed during a sudden 19th-century event when a large rock slab separated and slid down a cliff (similar caves are present in canyons and along cliffs in other parts of the world, including Hawaii).

Unfortunately, Sjöberg gave no information on the size of these caves. Clearly the largest and most important type are in the group he termed "neotectonic boulder caves". These are in Archean rocks largely located in the Swedish section of the

Fennoscandinavian uplift. Sjöberg subdivided these as talus caves in split roches moutonnées, and talus caves in collapsed mountain slopes. Sjöberg and others found that roches moutonnées containing talus caves were "perfectly formed" or "perfectly rounded". They were striated by glacial flow prior to tectonic activity which formed the caves during late deglaciation. These chaotic boulder piles include Bodagrottorna, the largest non-karstic cave in Sweden, with more than 2800 m of passages formed *in situ*. Most of the Swedish caves "in collapsed mountain slopes" are within a belt of high neotectonic activity. These are within voluminous talus accumulations and consist of "more or less vertical caves with bigger grottoes connected by narrow passages". In Sweden, length of passages in this type of cave reaches several hundred metres. Sjöberg also found one unusual cave in a displaced mountain top. It is a maze cave beneath a summit block, 110 by 50 by 10 m, which slid two or three metres along a slope of 15 to 20 degrees. Rounded forms in part of the cave indicated that a smaller, weathered cave existed before sliding occurred.

Striebel (1999) identified other types of talus caves and specific processes which formed them. Among them are caves formed by several types of erosion, frost splitting, and rock movement. Specific types included: (1) "woolsack" and "mattress" caves in tors and other features resulting from spheroidal weathering of granite (the names derive from the general shape of the boulders forming the caves); (2) "gorge bottom caves" (called "purgatory caves" in some parts of the United States); and (3) "erosion boulder caves" or "boulder fragment caves" (gorge bottom caves in which a labyrinthine conduit has been eroded between boulders).

In the United States, the commonest types of talus caves are rockslide / rockpile caves and purgatory caves ("gorge bottom caves").

Rockslide / Rockpile Caves

Rockslide or rockpile caves are slope failure features found at the base of cliffs and on slopes where they form boulder fields. A variety of mechanisms is involved: block glide, grusification, and others. On steep slopes, some represent a stage of disintegration of crevice caves during mass movement. Caves formed partially as crevices and partly in talus are common in rock masses undergoing different rates of downslope movement. Many maps of what initially are believed to be talus caves reveal patterns of gravity-sliding crevice caves. Others develop more dramatically. Especially well known are the Polar Caves (New Hampshire, United States) where chemical weathering of feldspars and mica along semi-vertical joints has loosened large blocks on a receding cliff 70 m high. Together with frost action, instability of the oversteepened cliff has dislodged blocks up to 12 m in diameter, forming jumbled talus caves at its base.

Other well-known US rockslide / rockfall caves include Yosemite Falls Indian Cave, California; Chuckanut Mountain Caves, Washington state; TSOD Cave, New York state; MDBATHS Cave, New Hampshire; and Rockhouse Cave, South Carolina. Entirely within an area 90 by 250 m and with a relief of 50 m, TSOD Cave is listed as 4 km long. Few parts of it are more than 6 m from one of 355 entrances (Narducci, 1991). However, figures cited for some multi-kilometre rockpile caves in the northeastern United States have been challenged. This is because some measurements attributed to a single cave may have been made in several adjoining caves together with uncovered spaces between them.

Purgatory Caves

The term "purgatory cave" is in common use only in California and parts of the northeastern United States but conforms to usage in the American Geological Institute's *Glossary of Geology*. No other term for these caves is in common use in English-language speleology. In part, this reflects the distribution of these gorge bottom caves that are formed by talus partially filling narrow, steep-walled gorges and a few narrow grabens. Even in comparatively arid regions, large examples commonly contain streams and small ponds (Figure 1). Characteristically their features change almost from metre to metre, with large and small rooms, abrupt overhanging drops, and tight squeezes. Many contain evidence of extensive mechanical abrasion; the Greenhorn Cave System is a notable California example, 1.8 km long, 152 m deep and segmented at only two locations. Its abrasion-smoothed walls are scenic delights and provide wonderful recreational caving. Others in California contain sheets of travertine and other speleothems.

Interfaces and Multiprocess Caves

Well-known examples of caves intermediate between talus and crevice caves include North Carolina's 1.6 km Bat Cave (United States) and Hungary's 370 m Csorgo-Lyuk, the longest non-karstic cave in that country. Combined purgatory-rockpile caves are not uncommon in the northeastern United States, and Thunder Canyon Cave is a purgatory cave formed along a series of crevices in California. In Llano County, Texas (United States), 350 m Enchanted Rock Cave was roofed by granite talus after fracturing, weathering, grusification, and "mechanical suffosion" (Elliott & Veni, 1994). Colorado's Grand Meeander Cave (United States) is an extraordinary example of multiprocess speleogenesis. It became a cave when an accumulation of fallen rock blocked much of the overhang of a huge sandstone alcove formed by groundwater sapping (Wright, 1994), with particulate transport by piping and stream erosion (Figure 2). Washington state's Boulder (Boulder Creek) Cave (United States) is a smaller example in volcanic rock. The impressive Lost Creek

Talus Caves: Figure 1. This granite talus cave ("purgatory cave") in the southern California desert has a small perennial stream. (Photo by William R. Halliday)

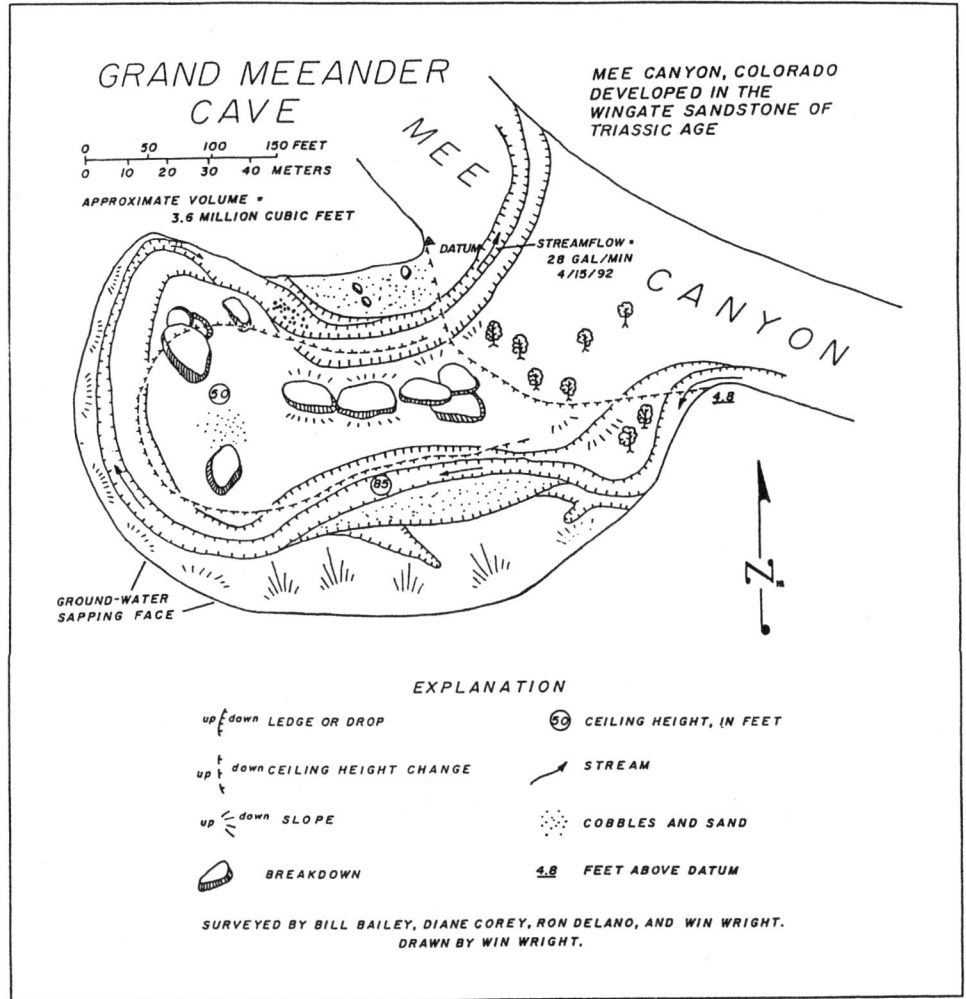

Talus Caves: Figure 2. Grand Meeander Cave, Colorado (United States), a multiprocess alcove which rockfall converted into a cave. It is one of the very few meander niches which extends beyond daylight. Courtesy Richard Rhinehart.

Pseudokarst (Colorado, United States) is an erosional feature so unusual that it is not universally accepted as a talus pseudokarst. It apparently is the result of multiprocess speleogenesis largely *in situ*. Located at an altitude of 2700 m in the Rocky Mountains, it combines alpine talus accumulation and piping and erosion by a mountain torrent and its tributaries. Enormous quantities of granitic sand and debris have been cleared from caves and giant sinkholes in this system, which is located within a narrow 5 km dendritic complex of large pseudokarstic windows, flat-stacked boulders, and sizeable ridges of partially grusified granite. Lost Creek disappears into swallet caves many times, reappearing from resurgence caves to flow across flat-bottomed dolines. It passes through ridges up to 600 m wide and 60 m high (Hose, 1996). The relative amounts of overroofing by tumbling and sliding of boulders, by slumping into the pseudokarstic conduits, and of speleogenesis *in situ* is not clear. About a century ago, engineers built a dam in this pseudokarst without realizing that the planned reservoir would immediately empty through the talus. Striebel (1999) and Sjöberg (1989a; 1989b) have documented smaller talus pseudokarsts formed by in-situ fragmentation and erosion.

Talus Caves as Glacières

Some talus and crevice caves in temperate zones are efficient natural deep-freezes containing unseasonal snow and ice. Ice usually is found in one of two locations: where cold air settles to an impenetrable layer, with residual snow and ice at and near the bottom, or where settling occurs along an inclined layer or back and forth through rocks, with ice near an egress at the bottom.

Mechanisms that have been proposed as explanations for this disproportionate trapping of snow and cold air are adiabatic cooling, evaporation from a very large surface area, and multiple small obstacles serving as baffles. In the United States, recorded examples are especially common in the northeastern states and in West Virginia.

Martian Analogues of Terrestrial Talus Caves

Malin and Edgett (2000) have reported several Martian features which may be analogous to terrestrial pseudokarsts (see also Extraterrestrial Caves). One consists of talus accumulations where cliff alcoves are littered with boulders "several metres to several decimetres in diameter". These are immediately downslope from locations of presumed recent outbursts of water which presumably flowed into or through the boulder piles. On Earth, talus accumulations with these parameters commonly contain talus caves. On Mars, they may protect residual ice from ablation.

WILLIAM R. HALLIDAY

Works Cited

Elliott, W. & Veni, G. (editors) 1994. Enchanted Rock Cave. In *The Caves and Karst of Texas: NSS Convention Guidebook*, Huntsville, Alabama: National Speleological Society

Hose, L. 1996. The Lost Park Reservoir Project, Park County, Colorado. In *The Caves and Karst of Colorado: 1996 NSS Convention Guidebook*, edited by R. Kolstad, Huntsville, Alabama: National Speleological Society

Malin, M. & Edgett, K. 2000. Evidence for recent groundwater seepage and surface runoff on Mars. *Science*, 288: 2330–35

Narducci, M. (editor) 1991. *Guide to the Caves of the Northeast: NSS Convention Guidebook*, Huntsville, Alabama: National Speleological Society

Sjöberg, R. 1989a. Caves as indicators of neotectonics in Sweden. In *2nd Symposium on Pseudokarst: Proceedings of the Symposium in Janocicky near Broumov 1985*, Prague: Czech Speleological Society

Sjöberg, R. 1989b. An inventory of caves in the county of Vasternorrland, north Sweden. In *2nd Symposium on Pseudokarst: Proceedings of the Symposium in Janocicky near Broumov 1985*, Prague: Czech Speleological Society

Striebel, T. 1999. Working meeting "Caves in Sandstone and in Granite". *Mittteilungsheft der Höhlenforschungsgruppe Blaustein* (Germany), 15: 45–53

Wright, W. 1994. The hydrogeology of Grand Meeander Cave. *Rocky Mountain Caving*, 11: 18–19

Further Reading

Proceedings volumes of the international symposia on pseudokarst, most recent volume being *Proceedings of the 6th International Symposium on Pseudokarst, Galyatető, Hungary, 1996*, edited by I. Eszterhás & S. Sárkozi

Newsletter of the Commission for Pseudokarst of the International Union of Speleology, edited by Istvan Eszterhás (www.clubs.privateweb.at/speleoaustria/pseudokarst.htm)

Balch, E.S. 1900. *Glacières, or Freezing Caverns*, Philadelphia: Allen, Lane and Scott; reprinted 1970 with a new introduction by W.R. Halliday, New York: Johnson Reprint Company

Bunnell, D. & Richards, B. 1986. Granite corrasional caves in California. In *Proceedings of the 9th International Congress of Speleology*, vol. 2, Barcelona: IBYNSA: 11

Engel, T. 1993. *Caves of the Eastern Adirondacks*, 2nd edition, privately published

Evans, J. *et al.* (editors) 1979. *An Introduction to the Caves of the Northeast: Guidebook for the 1979 NSS Convention*, Huntsville, Alabama: National Speleological Society

Perry, C. 1946. *New England's Buried Treasure*, New York: Stephen Daye Press

THERMAL WATER HABITATS

Those thermal waters categorized as hypogean habitats comprise all hypogean waters with temperatures higher than the annual average for the locality in which they occur (thermal springs, designated as *Umraumfremde Quellen* by Vouk, 1953, also constitute part of this habitat). According to Schwabe, thermal waters are among the oldest freshwater habitats on Earth, and, as they have shown great stability over geological time, they may harbour relict faunas. Since these waters have mostly been warmed in the deeper layers of the Earth's crust, they are largely devoid of organic material (except where chemo-autotrophic bacteria proliferate). Higher temperatures lower the solubility of gases and therefore the dissolved oxygen content; higher temperatures also enhance the metabolism of poikilothermal (or cold-blooded) animals. Therefore, thermal waters are generally harsher environments than non-thermal hypogean ones. Faunal populations in the hypogean areas of thermal waters may be extremely sparse. However, in the threshold habitats of springs, which may be enriched with food from the surface, these waters may be inhabited by rich populations of stygobitic animals (Sket & Velkovrh, 1981). Stygobites, with their typically very low metabolism, are generally well adapted for life in these habitats. Taxonomically, the inhabitants of thermal waters may be simply populations of eurythermal (tolerant of a wide range of temperatures) species, or stenothermal (specialized for a defined temperature range) subspecies or species.

By far the most common inhabitants of thermal hypogean waters are crustaceans, and the only ubiquitous inhabitants are species of the Family Stenasellidae. Near the formerly ice-covered areas of southern Europe, thermal springs (15–28°C) are almost the only known localities supporting stenasellids, such as *Balkanostenasellus skopljensis* in Slovenia and *Protelsonia hungarica* in Croatia. These populations are probably relicts of thermophilic animals from the pre-Pleistocene period. But even in the subtropics and tropics, some stenasellids occur in thermal springs: i.e. *Stenasellus pardii* and *S. costai* in Somalia at 29–31°C, *Johanella purpurea* in Algeria at 29°C, and *Mexistenasellus coahuila* in Mexico at 24–34°C. Their ecology indicates that the group originated in warmer geological periods. *M. coahuila* and *S. costai* both inhabit high-temperature waters, even those with less than 2 mg l^{-1} oxygen and in the presence of hydrogen sulfide. In an exceptional case, the adaptation of stenasellids may be in the opposite direction, for example, some populations of *Stenasellus virei hussoni* live at very low temperatures (less than 6°C) in the Pyrenees.

The most widely known extreme thermostygobite (specialized for warm hypogean waters) is the thermosbaenacean *Thermosbaena mirabilis* (described by Monod, 1924). Its dense population inhabits cyanobacterial mats in the baths of El Hamma, Gabes, Tunisia, at 45°C. Since it also occurs in dark hypogean thermal waters, displays troglomorphism, and is a member of an exclusively stygobitic group, this is almost certainly a stygobitic species, which also flourishes in illuminated areas that contain more food but are without direct competitors. The most extreme thermostygobite appears to be the bathynellacean *Thermobathynella adami*, which has been recorded from a spring in Zaïre with a temperature of 55°C.

Only in Slovenia has the regional fauna of thermal hypogean waters been investigated systematically. This fauna consists of 16 stygobitic species (up to five in one locality) of which only two appear to be polystenothermal (specialized to higher temperatures): the sphaeromatid isopod *Monolistra (Microlistra) calopyge* and the gastropod *Hadziella* sp., while the others (e.g. some *Niphargus* spp., the interstitial *Parabathynella stygia*) also occur in cold hypogean waters. Animals were found at temperatures of 15–28°C (normal temperatures in the continental parts of Slovenia are around 10°C) and they were usually found in waters with low oxygen concentrations of 5.5–7.5 mg l^{-1}. The faunistically richest thermal system known is at the springs at Cuatro Cienagas in Coahuila, Mexico; the water temperature is 29.5–34°C, and the oxygen concentration is 1.7–4.1 mg l^{-1} (often below 50% saturation). Their stygobitic fauna consists of at least five species of gastropods (*Paludiscala caramba, Coahuilix hubbsi*, and others), two amphipods (*Mexiweckelia* spp.), three cirolanid isopods (*Speocirolana thermydronis, Sphaerolana* spp.), and the stenasellid *Mexistenasellus coahuila*.

Only a few preliminary studies regarding the biology of stygobites from thermal waters have been undertaken. *Balkanostenasellus skopljensis thermalis*, which occurs in a Bosnian spring at 24°C has been kept successfully for longer periods at 16°C. *Protelsonia hungarica thermalis* can be found in springs and interstitial water in western Croatia at 18.2–24.4°C, while it can survive a temperature range of 13–28°C and shows an optimum at 20°C in the laboratory. It definitely could not survive at the normal local temperature of approximately 10°C.

BORIS SKET

Works Cited

Monod, T. 1924. Sur un type nouveau de Malacostrace: *Thermosbaena mirabilis* nov. gen. nov. sp. [About a new type of Malacostraca: *Thermosbaena mirabilis* nov. gen. nov. sp.]. *Bulletin de la Société Zoologique de France*, 49: 58–68

Sket, B. & Velkovrh, F. 1981. Podzemeljske živali v termalnih vodah [Subterranean animals in thermal waters]. *Biološki vestnik*, 29(2): 91–120

Vouk, V. 1953. Umraumfremde Quellen [Surrounding alien springs]. *Archiv für Hydrobiologie*, 48(1): 129–33

Further Reading

Barker, D. 1959. The distribution and systematic position of the Thermosbaenacea. *Hydrobiologia*, 13(3): 209–35

Lattinger-Penko, R. 1970. Vérification expérimentale de l'interdependance de la distribution de *Stenasellus hungaricus thermalis* (Crustace, Isopode) et les conditions de température du biotope) [Experimental verification of the interdependence between the distribution of *Stenasellus hungaricus thermalis* (Crustacea, Isopoda) and the temperature conditions of the biotope]. *Annales de Spéléologie*, 25(2): 319–34

Magniez, G. 1974. Données faunistiques et écologiques sur les Stenasellidae (Crustacea, Isopoda, Asellota) des eaux souterraines [Ecological and faunistic data on the Stenasellidae (Crustacea, Isopoda, Asellota) of subterranean waters]. *International Journal of Speleology*, 6(1): 1–80

Taylor, D.W. 1966. A remarkable snail fauna from Coahuila, Mexico. *Veliger*, 9(2): 152–228

Zilch, R. 1972. Beitrag zur Verbreitung und Entwicklungsbiologie der Thermosbaenacea [Contribution to the distribution and the developmental biology of Thermosbaenacea]. *Internationale Revue der Gesammten Hydrobiologie*, 57(1): 75–107

TOURISM AND CAVES: HISTORY

The overall human relationship to caves probably comprises three major phases. The first is marked by extensive use of caves for shelter, habitation, and ritual purposes. Caves were not feared or avoided, but often appear to have provided desirable residential locations. The transition to the next phase seems to mark a significant step in cultural evolution. Although some cave sites continued to be utilized in various ways, many were abandoned, and there seems to have been a flowering of legends which portrayed caves as places to be feared, often because they were reputed to be the abode of various imagined monsters or evil spirits. Although climatic change and other physical explanations have often been evoked, it could be argued that this transition marks the beginning of curiosity, and the asking of questions about the natural world. Caves, being difficult to explain within the framework of everyday experience, became explicable only in terms of such myths.

Progressively, at different times in different cultures, this curiosity was translated into the final phase of travel and investigation, and the beginnings of the natural sciences. Shaw (1992: pp.7–8) commences his seminal history with the travels of Assyrian kings some 3000 years ago—and there were doubtless earlier travellers who failed to leave any record for us. The great rise in investigatory and other more-or-less individualized travel came with the 15th and 16th centuries, particularly in Europe. Many travellers have since visited and described caves (see, for example, Shaw, 2000). Perhaps one of the more remarkable of these early travellers was Xu Xiake (1586–1641) of China, who is said to be the only Chinese to have travelled throughout his own country for his own satisfaction rather than for official or military purposes, and who spent 34 years doing so (see Asia, Northeast: History).

Tourism is perhaps defined as the progressive democratization and commercialization of travel and could be considered to have arisen out of the early years of the industrial revolution. The distinction between travellers and tourism is a fuzzy one, of course. Herodotus refers to paying guides in some sites of particular interest *c*.450–400 BC, and certainly travellers to Vilenica Cave in Slovenia were being charged a fee for access at least by the early part of the 17th century. For this reason, Vilenica is now generally recognized as one of the first "Show Caves".

The event which best marks the beginning of modern tourism took place on 5th July 1841, when Thomas Cook's first excursion left Leicester for an outing to Loughborough. By this time, some of the great tourist caves of the world were open to the public. Postojnska (Postojna, Adelsberg) Cave had been gradually commercialized from the beginning of the 19th century; Wookey Hole in Britain, known for some 2000 years, similarly underwent progressive commercialization at about the same time; and commercial tours commenced at Mammoth Cave (United States) in 1816.

The greatest development of cave tourism occurred in the second half of the 19th century. New caves were continually being discovered, and many of these were selected for commercial development. An especially interesting development occurred in Australia, where both the Naracoorte and Jenolan Caves attracted a rapidly growing number of visitors from the 1860s onwards. By the 1880s, Jenolan was undeniably the best-known and one of the most popular tourist destinations of the country. Its success was such that following the economic depression of the 1890s, state governments commenced to employ surveyors or prospectors to search for other caves that might also provide a similar economic resource for the tourism industry. Many of the other show caves of the nation were discovered and developed during this period. Jenolan was also of considerable interest in that it was the site of pioneering development of technology in cave tourism, with extensive use of clockwork-driven magnesium ribbon lamps for visitor display, the first experimental electric lighting in caves (1880), and the first use of hydroelectric generation in Australia. Electric lighting proved to be a dominant technology in cave tourism, with early permanent installations at Kraushohle (1883), Postojnska (1884), and Jenolan (1887).

One interesting aspect of these early caves was the extent to which travellers and discoverers, together with early managers or guides, became famous personalities. Examples include Stephen Bishop at Mammoth Cave, William Gough at Cheddar (Somerset, Great Britain), and Jeremiah Wilson at Jenolan. It would be easy to name many others. Many appear to have been extroverts who cultivated this reputation as a means of making their site more attractive to visitors. Even to this day, they remain as heroes of exploration, fabled raconteurs or comics, and guides of exaggerated wisdom. Some, who were initially men of little education, took up an interest in reading and literature in order to be able to converse in more positive ways with tourists, who were often people of relatively higher status and greater learning.

Eventually, the cave tourism experience became remarkably stereotyped, with visitors being grouped into parties, each led through brightly lit caves along fixed pathways by a talking guide. As a result, the immense enthusiasm for caves which characterized the 19th century industry gradually faded. Early attempts to recover the quality of cave tourism and to more effectively compete with the growing range of alternative tourist sites tended to concentrate on the improvement of infrastructural arrangements such as pathways and lighting. Certainly as technology has improved, this has led to an improvement in presentation and display. However, uniformity of the cave tourism experience and boredom are not changed by the caves being more brightly lit. More adequate attempts to diversify and enhance the quality of the personal experience itself, by changing the very character of the tour have generally only developed since 1990. Some exciting innovations are also to be found in countries entering into mass tourism for the first time and offering experiences growing out of their own cultural patterns rather than merely copying previous eurocentric models.

The other important development, particularly since the 1980s, has been the increasing role of both research and monitoring in the management of cave tourism—both the quality of the visitor experience and the impact of tourism on the cave and surrounding environment. Monitoring of environmental factors such as air quality and air-borne debris are covered in separate entries, as is research on the growth of algae and lampenflora, induced by the artificial lighting. The entry on Tourist Caves describes contemporary environmental and visitor monitoring and assessment.

ELERY HAMILTON-SMITH

Works Cited

Shaw, Trevor R. 1992. *History of Cave Science: The Exploration and Study of Limestone Caves, to 1900*, 2nd edition, Broadway, New South Wales: Sydney Speleological Society

Shaw, Trevor R. 2000. *Foreign Travellers in the Slovene Karst 1537–1900*, Ljubljana: ZRC Publishing and Karst Research Institute at ZRC SAZU

TOURIST CAVES

Tourist caves can be simply defined as those displayed to the general public in return for a fee or other financial consideration. The history of this tradition dates back to at least the early 17th century (see Tourism and Caves: History). Tourist caves may be managed by governmental agencies, non-governmental organizations (e.g., speleological clubs, religious communities), and individuals or corporations operating as a commercial business. People have always been fascinated by caves for a wide variety of reasons, such as mystery and curiosity, fantasy, and appreciation of beauty, so it is not at all surprising that visiting caves became commercialized. In the words of a 15th-century commentator on village markets, "No sooner does one man find a way of enjoying himself, than another finds a way to make a penny from it." Owners and managers have a wide range of objectives. Some doubtless have a purely financial motivation. Others wish to share their enthusiasm for the beauty and wonder of caves with other people; to use caves as a wonderful opportunity for environmental appreciation or education, or as a site for spiritual experience, prayer, or meditation. This article will deal in turn with the following aspects of tourist caves:

1. Visitor experiences.
2. Approaches adopted by management to shape these experiences to meet their own assumptions, values, or objectives.
3. The physical infrastructure used to facilitate access to the caves, including possible guidelines for its development.
4. Real and potential impacts of tourism in karst areas.
5. Assessment and monitoring of both environmental change and visitor experience in order to guide corrective action in conservation or tour management.

Visitor experiences are largely shaped by management practices and the kinds of opportunities made available to the public. Because most visitors have little understanding of possible alternatives in tour management, the result is a closed and self-reinforcing circle, where visitor expectations reflect what is already offered, rather than seeking new and different kinds of experience. All too often, the experience is a stereotyped one, where visitors are allocated to groups and led through a brightly lit cave along a (usually) concrete pathway by a guide who talks at the group, giving his or her version of what they are seeing. The visitor experience thus ranges from boredom to a general level of satisfaction in which the key dimension is often the personality and approach of the guide rather than any quality of the cave itself. Guides may recognize this and give considerable attention to customer relations, or alternatively, offer a dramatic or comic performance to their visitors. However, a gradually widening diversity of opportunities is offered today in the attempt to enhance visitor experience and retain market share in an increasingly competitive tourism marketplace.

It is perhaps useful at this point to consider the experiences which are least often provided. Being alone in the cave is only very rarely provided, and then usually unintentionally. Yet those who have been alone in a cave often speak of this as a very special and highly valued experience. One young woman who was able to walk alone into a cave and then turned her light off to sit in the dark said afterwards, "I could hear the Earth thinking!" Even if not alone, there is rarely a genuine opportunity for peaceful meditation, although some may find this during an unguided tour. Personal inquiry and investigation is all too rare, although some guides with a sense of creativity are able to turn even a group tour into a personalized quest and experience of inquiry. Even some so-called "adventure" tours are simply another version of the stereotyped tour, distinguished only by the absence of constructed paths and fixed lighting.

An increasing problem for guiding practice is the multilingualism of visitors. Some countries give attention to linguistic skills in selection of guides and may provide printed materials in a wide range of languages. Alternatively, they may seek the support of international tourism agencies in provision of interpretation. Others rely entirely upon self-guided tours, sometimes with signs highlighting special features, audio commentary, or other displays. Still others simply leave all communication to the unfortunate guides, who may simply shout louder in the hope that somehow, this may enhance understanding.

Talking to visitors is usually termed interpretation, though this extremely simplistic view of interpretation is based on the straightforward communication of the guide's cognitive perspective on the cave. The emotive and affective experience of the cave is either ignored or forgotten. A more radical view of the problem that is gaining increasing support is that the interpretation, even if well done, endeavours to compensate for poor presentation of the cave experience. From this perspective, the whole manner of presentation must be reviewed to make the cave experience much more self-evident to the visitor.

Contemporary approaches to the enhancement or improvement of visitor experience include:

1. Training of those guides who provide the traditional guided tour to offer a more interesting presentation and to increase the actual involvement of visitors in shaping the cave experience.
2. Asking guides to accompany rather than lead the party, joining in conversation with visitors rather than delivering a formal presentation.
3. Altering the lighting arrangements within the cave to provide an ambience of drama or mystery, or highlighting special features while leaving other parts of the cave in virtual darkness.
4. Offering unguided (often called self-guided or self-timing) tours where visitors move through the cave at their own pace. Effective tours of this kind normally still provide a staff presence within the cave for both safety and visitor comfort reasons and as an information source in responding to visitor questions. Particularly in Asia, many sacred sites and temple caves offer what amounts to an unguided tour in which visitors move through at their own pace. Some of these may provide the opportunity for prayer or meditation. Many have elaborate shrines or other constructed development to enhance the religious experience (see Religious Sites).
5. Offering unguided tours of caves with pathways and possibly some other infrastructure but without fixed lighting. Each visitor is provided with a hand-held light to carry out their own search for interesting features.
6. Offering what are commonly termed adventure tours. Visitors are provided with helmets, overalls, and appropriate footwear (if necessary) and are conducted through either totally undeveloped caves or caves where there has been only minimal development to improve safety.
7. Providing walking routes through surface karst features. These are increasingly popular, but rarely is the opportunity taken to genuinely demonstrate the relationship between surface and subterranean features.
8. Providing the opportunity for guides to work as field assistants in research programs at the cave or park for which they have responsibility so that they are better able to communicate the nature of scientific research to visitors.
9. Providing for current researchers or other cave scientists to undertake leadership of tours as guest guides on special occasions.
10. Undertaking restoration work. Restoration of caves is becoming increasingly important, and a restored cave may even become a special feature in itself (e.g., Bell 1994; see also Restoration and Speleothem Repair). Further, restoration often provides an opportunity to involve members of the general public in work teams, and hence to immensely deepen their appreciation of the cave environment.
11. Providing various forms of artificial entertainment such as the screening of video tapes, holographic presentations, use of sound and light or dramatised presentations (see Music in Caves), staged performances, and even the use of robots. Unless these are of outstanding merit, they probably do more to detract from the natural values of the cave than to accentuate them.
12. Assisting movement through the cave with some form of vehicular transport. The most long-standing and well-known example is probably the miniature train at

Postojnska in Slovenia. Other examples include boats, buses with a transparent top, or other motorized vehicles. If these are done tastefully and appropriately they may even enhance the experience rather than simply be a practical convenience. Certainly the spectacular boat journey in the lower levels of Jeita Cave (Lebanon) or the voyage through Ghar Alisadr (Iran) are regarded as two of the highlights of contemporary cave tourism experience.

The in-cave experience may be supported with various presentations on the surface. Sometimes these are totally irrelevant to the cave, but they often include visitor centres with a diversity of interpretive information or other scientific displays. A few of these are now providing clear information on conservation and tour management to enhance visitor understanding of the opportunities and constraints in cave tourism. The more elaborate may include museum-level presentation (e.g. Waitomo Museum of Caves, New Zealand). At the Naracoorte World Heritage Area (Australia), visitors to the Wonambi Fossil Centre see computerized robots of extinct species within a reproduction of a Pleistocene forest and swamp. Alternatively, they may visit the Bat Interpretation Centre where they view real-time video imagery of the immense population of bats in one of the caves.

For many years, guide books of varying quality were made available to visitors. Currently, they are more likely to find only a choice of brightly coloured postcards, folders or brochures, or photographic books. However, there are still outstanding examples where visitors are able to obtain quality guides and environmental books that relate either to the cave or to the region as a whole. A small number of sites also offer videotapes or CD presentations. A diversity of souvenirs and gifts may also be available but rarely contribute to the understanding or appreciation of the cave itself. Regrettably, in some countries speleothems taken from the caves are themselves sold as souvenirs, thus giving respectability to continuing vandalism.

A major opportunity for improved practice is provided by the various regular or occasional conferences of cave and karst managers. Regional conferences have been held regularly in Australasia (since 1972) and the United States (since 1975), while international conferences are now convened (since 1990) by the International Show Caves Association (ISCA). Australia and New Zealand also have a continuing professional association: the Australasian Cave and Karst Management Association (ACKMA). The published proceedings of these conferences provide a growing body of significant literature on cave management.

The physical infrastructure in caves may have been in place since the 19th century and thus acquired an historical value. Many of the more recent installations have repeated the old pattern of extensive modification to the natural structure of the cave, neglect—and often destruction—of the floor with its extensive deposits and other evidence of regional geoclimatic history, and the use of insensitive and / or excessive lighting intensity. There are very few explicitly published guidelines to preferable infrastructure designs. Various engineering or accessibility standards may set structural and dimensional requirements, for example for stairs or pathways, but these are generally based upon building codes and not sensitive to natural environment or conditions. The outstanding examples of environment-sensitive codes come from Australia, where the *Burra Charter* (ICOMOS, 1999) provides guidelines for conservation management of cultural sites and the *Australian Natural Heritage Charter* (IUCN, 1996) provides similar guidelines for natural sites. Where a present installation has attained historic significance, any redevelopment should consider the principles provided in the *Burra Charter* while any cave development of infrastructure should recognize the Natural Heritage Charter. A number of guidelines are provided and some of those which have particular relevance to cave presentation and conservation management worldwide include:

1. In so far as possible, conservation management should only exercise minor interference or modification to the natural environment and only in so far as this is necessary to ensure conservation. Additions should only occur if they do not alter the natural characteristics of the site.
2. These additions must be clearly identified in one way or another, so that no confusion will arise between natural and introduced matter.
3. Any additions to the site, for instance pathways for access purposes, should be designed and constructed in such a way that they can be readily removed, allowing for complete reinstatement of natural conditions. (Obviously this precludes the traditional poured concrete pathways of many tourist caves.)
4. Comprehensive and detailed records should be kept of any modifications so that if necessary those modifications can be reversed.

Charters such as the Australian Natural Heritage Charter are also accompanied by publications such as the *Natural Heritage Places Handbook* (IUCN, 1998) as a more detailed practical guide to action. Cave managers in Australia have also done considerable work towards the development of practical guidelines (e.g., Spate *et al.*, 1998; Hamilton-Smith *et al.*, 1998). An international guidelines document designed to further improve practice in conservation management of tourist caves and other sites on karst lands is in preparation.

Tourism generates a wide range of impacts on karst sites, and particularly upon tourist caves themselves. On the surface, the most obvious impacts relate to the changes in vegetation and increase in soil erosion that result solely from visitors walking about the environment. Other issues include the increased demand on water reserves, problems of both solid and liquid waste disposal, increased vandalism, and ecological changes due to the disappearance of some species and the entry of invasive species. Every effort should be made to minimize these impacts but usually it will not be possible to eliminate them. The most damaging impacts within caves are currently a result of management actions in the development of inappropriate infrastructure and other destructive actions. These may well be reduced significantly at the planning, design, and construction stages but otherwise may demand very significant redevelopment if they are to be dealt with properly.

One of the implementation issues in conservation management is the common practice of many work crews leaving behind minor debris resulting from their work. Metal fragments from fabrication of guardrails, or cuttings of electric wiring, often introduce materials toxic to cave fauna. Small clippings of copper will generate compounds toxic to invertebrates, while the cadmium impurities in galvanizing are toxic to microbiota and so will inevitably damage the integrity of cave soils. Probably the most intractable of impacts, resulting directly from the presence of visitors is the accumulation of lint, consisting of fibres from clothing, dust carried in by visitors and flakes of human skin (see Tourist Caves: Airborne Debris). Visitors may also leave behind less visible evidence of their presence in the cave, including invasive species, some of which may be microbiota. Other impacts such as changes in temperature, relative humidity, and the level of carbon dioxide in the cave (see, Tourist Caves: Air Quality) may have a significant impact in low-energy caves. In conjunction with artificial lighting, they may have the further and more drastic effects of encouraging the phenomenon of lampenflora (algal and other plant growth; see Tourist Caves: Algae and Lampenflora). Again, lampenflora is best prevented at the design, planning, and construction stage. Its occurrence is simply evidence of poorly designed, excessive and / or wrongly located lighting. When it has occurred it should be removed as its continuing presence actually damages the rock substrate, and this can be most easily accomplished by using a very dilute sodium hypochlorite solution.

A vital part of good conservation management in any tourist area is the assessment and monitoring of change (see Monitoring). It is useful to emphasize the distinction that assessment is the process of making judgements about the significance and necessary action to deal with changes; monitoring is simply the act of consistent measurement of those changes in order to provide a database for assessment. A number of basic parameters should be subject to continuous monitoring simply to provide comprehensive knowledge of the environment including both seasonal changes and any extraordinary changes. The parameters which might be included in both surface and underground monitoring include temperature, relative humidity, carbon dioxide levels, the levels of radon and radon daughters, water quality and flow levels, and wind and other air movement patterns. Careful consideration must be given to measurement sites and such questions as the extent to which specific microclimates might demand attention. Further monitoring may be necessary to deal with identified problems or potential problems. This may well be on a continuous basis but where peak events (e.g. flooding) are of significance then special measurement and monitoring may need to be undertaken during these events. Monitoring also needs to relate to, and often be supported by, genuine research. Common assumptions about the nature and cause of problems may well be incorrect and if these problems are to be dealt with, it will be necessary to carry out research in order to delineate the causative processes.

While there is a genuine awareness of the need for environmental monitoring and it is commonly carried out in many karst areas, the quality of the visitor experience has not been subject to the same level of research and enquiry. Given the dependence of many tourist caves upon a continuing flow of visitors, it is vital that the assessment of visitor experience be systematically carried out. Perception and assessment of environmental quality by visitors is as important as that by managers. Visitors are often sensitive to issues which managers take for granted, while at thesame time, managers may well be sensitive to issues of which visitors are totally unaware. Thus, the program of assessment and monitoring needs to integrate both environmental and experiential monitoring (Hamilton-Smith, 2002). Systematic participant observation properly coupled with informal interviewing is a very powerful tool in monitoring of experience. In one simple example, observation showed that acts of vandalism that are of concern to managers are generally carried out by visitors at the rear end of tour groups, possibly because they cannot hear the guide and have become bored. Another successful example specifically examined motor coach tours to a major Australian cave destination. The researcher rode with bus tours to the destination, shared with passengers over lunch, joined with their cave tour, and then rode the bus back to the starting point. This study not only identified a number of organizational problems that were detracting from the experience, but also identified a major but previously unrecognized problem in the telling of inappropriate (often anti-environmental) jokes and anecdotes by tour drivers. However, observation alone is not enough and there should also be regular surveys of visitor values, attitudes, and experience, perhaps accompanied by occasional special studies that focus in more depth on revealed problems.

An interesting example of a survey study was carried out at Waitomo, New Zealand by Doorne (1998) in an examination of perceived crowding. Actual crowding was self-evident, and was the most common source of complaints from visitors. Doorne found that visitors of European origin defined crowding as being tightly forced into a small space with many other people as happened during most cave tours. Asian visitors, by contrast, defined crowding as having to join a queue in order to purchase tickets, and then again to actually join a tour. This greater understanding enabled the managers to develop more effective strategies to deal with the problem than was previously the case.

The most potent approach to minimizing undesirable impacts is undoubtedly ensuring the best attainable practice in planning, designing, and developing tourist caves and related tourism opportunities. Many problems are predictable and can be greatly reduced by effective planning in advance, and by not repeating the mistakes of the past.

This article can only be a brief review of issues and contemporary trends in tourist cave management. There are, in addition to the specific references provided below, an immense number of sources of information available. At the descriptive level, many countries publish directories of tourist caves, or even major guidebooks covering all caves in the country. These now seem to be less published and utilised than in the past, probably because the Internet provides a cheaper and more flexible access to the same information (http://www.showcaves.com/ provides an international listing of tourist caves).

ELERY HAMILTON-SMITH

Works Cited

Australia ICOMOS. 1999. *The Burra Charter*, Melbourne: Australia ICOMOS (also available at http://www.icomos.org/burra_charter.html)

Australia IUCN. 1996. *Australian Natural Heritage Charter*, Canberra: Australian Heritage Commission and Australian Committee for IUCN (also available at http://www.ahc.gov.au/infores/publications/anhc/)

Australia IUCN. 1998. *Natural Heritage Places Handbook*, Sydney: Australian Committee for IUCN (also available at http://www.ahc.gov.au/infores/publications/nhnames/)

Bell, P. 1994. Moondyne Cave: Development to guided adventure. *Cave Management in Australasia*, 10: 9–11

Doorne, S. 1999. *Visitor Experience at the Waitomo Glowworm Cave*, Wellington: Department of Conservation

Hamilton-Smith, E. 2002. Management assessment in karst areas. In Proceedings of the meeting on monitoring of karst areas, edited by A. Kranjc, *Acta Carsologia*, 31(1): 13–20

Hamilton-Smith, E., McBeath, R. & Vavryn, D. 1998. Best practice in visitor management. *Cave and Karst Management in Australasia*, 12: 85–96

Spate, A., Hamilton-Smith, E., Little, L. & Holland, E. 1998. Best practice and tourist cave engineering. *Cave and Karst Management in Australasia*, 12: 97–109

Further Reading

Gillieson, D. 1996. *Caves: Processes, Development and Management*, Oxford and Cambridge, Massachusetts: Blackwell

Watson, J., Hamilton-Smith, E., Gillieson, D. & Kiernan, K. (editors) 1997. *Guidelines for Cave and Karst Protection*, Cambridge: International Union for Conservation of Nature and Natural Resources

For further information on tourist cave conservation and visitor management, see the various series of conference proceedings, such as *Proceedings of the National Cave Management Symposium* (NSS), *Proceedings of the Australasian Cave and Karst Management Association* (ACKMA), and *Proceedings of the International Show Caves Association* (ISCA).

TOURIST CAVES: AIR QUALITY

The air quality in tourist caves has to be maintained, to safeguard the resource against corrosive gases, and to protect the health and comfort of tourists and staff. Tourists can create great changes in the cave microclimate and the composition of its atmosphere. Cigna (1993) regarded the three major impacts as being increases in carbon dioxide (CO_2) and temperature and changes in humidity. To these is added a reduction in oxygen (O_2) and increases in anthropogenic pollutant gases. Studies have had diverse aims, such as the protection of speleothems (Baker & Genty, 1998), the preservation of prehistoric paintings (Cabrol, 1997), and the health of sensitive cave fauna.

Recommended techniques for the analysis of cave air are nondispersive infrared spectroscopy (NDIR) and gas chromatography, both of which allow *in situ* continuous monitoring. NDIR instruments are recommended for analysis of CO_2. Gas chromatography instruments allow the concentration of other gases such as nitrogen, O_2, carbon monoxide, and methane to be obtained simultaneously with CO_2, but are expensive and require helium as the carrier gas. Grab samples taken in Tedlar bags can be measured in the laboratory, where trace gases can be identified and quantified by gas chromatography mass spectroscopy. The latter technique has been used to identify 71 trace gases present in the atmosphere of Jenolan Caves, Australia.

Carbon dioxide is a critical component in the solution and precipitation of calcium carbonate (calcite or limestone); these processes are illustrated in the following equation:

$$\text{Solution} \Rightarrow$$
$$CaCO_3(s) + CO_2(g) + H_2O(l) \leftrightarrows Ca^{2+}(aq) + 2HCO_3^-(aq)$$
$$\Leftarrow \text{Precipitation}$$

CO_2 added to the system causes calcite solution, and its removal causes calcite precipitation. Thus, to corrode active calcite speleothems, the concentration of CO_2 must be higher in the cave atmosphere than in calcite-depositing waters. Increases of CO_2 in the cave air will initially reduce the rate at which active speleothems form and finally will cause their solution. The equation above supports a conclusion by Baker and Genty (1998), that speleothems depositing from solutions with high calcium concentrations can support greater increases in CO_2. These processes are temperature dependent and increases in temperature will alter the degree of CO_2 damage and rate at which it takes place.

Experimental data shows that there are no simple rules governing the behaviour of tourist-generated CO_2 within a cave or between caves. CO_2 accumulation from tourist respiration will depend on size and frequency of the cave tours, the energy the tourists expend in traversing the cave and the proportions of the caverns visited. The rate at which CO_2 levels return to background after tours finish, the relaxation time, depends on what mechanisms are available to remove CO_2 (see Carbon Dioxide-Enriched Cave Air).

Tourist-exhaled CO_2 is superimposed on a background of CO_2 from other sources. The background level of CO_2 is hard to establish and the assumption that it is the same as a nearby undeveloped cave is not valid. Background levels can be obtained experimentally during visitor-free periods. They are not constant and can vary by orders of magnitude. Despite difficulties in establishing reliable figures, over five years, a steady increase in background CO_2 has been observed at Jenolan Caves, Australia. O_2 to CO_2 ratios and $^{13}\delta C$ measurements show the increase is accompanied by an increase in the proportion of background CO_2 attributable to the respiration of micro-organisms. This is not surprising, as tourists add micro-organisms, and food for them, to the cave environment.

Fast relaxation times to background levels are desirable because the slow transfer of carbon dioxide from air to water is a necessary step in the solution of calcite. At most sites, ventilation controls relaxation times. Studies at Jenolan Caves, in the same cave system for the same time period, show that well-ventilated sites reach background in hours, whereas sites with poor ventilation take days.

If action is to be taken to reduce the levels of CO_2 in a cave, thresholds should be established above which management action is necessary. In caves with active speleothems, two thresholds can be calculated by using aqueous geochemical models, a conservative corrosion threshold and a higher dissolution threshold, which takes into account the kinetics of calcite solution. Comprehensive air and water experimental data are required for threshold calculations. An unpublished study at Waitomo Caves, New Zealand, established a corrosion threshold of 2500 ppm CO_2. This figure has been used to manage tourist CO_2 in the Glow-Worm Cave. In the Jenolan Caves investigations, corrosion thresholds were found to range from 2700–28 000 ppm CO_2. In all cases, corrosion thresholds were not exceeded by the maximum CO_2 measured. In caves with inactive speleothems, there are no calcite saturated solutions to buffer increases in CO_2. Thus corrosive condensates (see Condensation Corrosion), produced by tourists, are particularly harmful.

Automobile gases may be drawn into caves, if approach roads and car parks for tourist caves are poorly located (James et al., 1998). Up to 200 litres of CO_2 are produced for every litre of gasoline. Automobile CO_2 can be identified from O_2 to CO_2 ratios and $^{13}\delta C$ studies. Automobiles also produce SO_x (sulfur oxides), NO_x (nitrogen oxides), VOCs (volatile organic compounds), and particulates. Speleothems have no protection against SO_x and NO_x gases that produce strong acids. VOCs are not likely to cause damage, although Young (1996) suggests that hydrocarbons may provide a food source for micro-organisms. At Jenolan Caves, the main access road passes through a cave and thousands of tourist vehicles use it each year. An investigation of air quality in the associated cave system found that gases attributable to automobile emissions had penetrated over a kilometre into the cave passages. Fortunately, even close to the road, SO_x, NO_x and VOCs concentrations were exceptionally low.

Health aspects of cave air quality are also important. The occupational health and safety limits for CO_2, O_2, and radon (see, Radon in Caves) in the work place are often controlled by legislation. Many of the gases in automobile emissions also have recommended levels, for example, that for benzene is 1 ppm over 8 hours. Osborne (1981) believes that a limit of 1000 ppm of CO_2 is necessary to avoid symptoms of hypercarnapia in tourists with poor health undergoing exercise. Despite this, he recommends a threshold limit of 0.5 v/v% CO_2 well below 1.25 v/v% CO_2 (Australian mine standard). Methods most frequently used to control air quality in tourist caves are reductions in visitor numbers and artificial ventilation. However, James (1994) has argued that control by forced ventilation could actually damage a cave. Thus, when developing caves for tourism, plans for maintaining air quality, without resorting to artificial ventilation, are a priority.

JULIA M. JAMES

Works Cited

Baker, A. & Genty, D. 1998. Environmental pressures on conserving cave speleothems: Effects of changing surface land use and increased cave tourism. *Journal of Environmental Management*, 53: 165–75

Cabrol, P. 1997. Protection of speleothems. In *Cave Minerals of the World*, 2nd edition, edited by C.A. Hill & P. Forti, Huntsville, Alabama: National Speleological Society

Cigna, A.A. 1993. Environmental management of tourist caves. *Environmental Geology*, 21: 173–80

James, J.M. 1994. Carbon dioxide in tourist cave air. *Comptes Rendus du Colloque International de Karstologie à Luxembourg*, 27: 187–95

James, J.M., Antill, S.J., Cooper, A. & Stone, D.J.M. 1998. Effect of automobile emissions on the Jenolan Caves. *Acta Carsologica*, 27(1): 119–32

Osborne, R.A.L. 1981. Towards an air quality standard for tourist caves: Studies of carbon dioxide-enriched atmospheres in Gaden-Coral Cave, Wellington Caves, NSW. *Helictite*, 19: 48–56

Young, P. 1996. Mouldering monuments. *New Scientist*, 2054: 36–38

TOURIST CAVES: AIRBORNE DEBRIS

Caves that are used for tourist activities show obvious ill effects from the deposition of airborne particles. However, all caves that are entered by humans suffer from particle contamination, and caves also have particles transported from the surface by natural processes. Airborne particles in cave atmospheres, and deposits that result from them, have been studied with varied perspectives. Went (1970) investigated condensation nuclei (0.03 μm diameter); these are generated mainly by combustion processes and persist for several days. The effects of cave tourism have been investigated by Jablonsky, Kraemer & Yett (1993) and there have been extensive studies of dust deposition in temple caves in China (Christoforou, Salmon & Cass, 1996).

The material deposited in caves generally consists of mineral particles, mainly clay minerals, as well as organic material; textile fibres, human dander (skin flakes) and hair, industrial dust and soot, and natural organic material; spores, pollen, plant dust, and insect bodies. The nature of the deposits in any cave depends on the sources of the dust and the processes that generate it within, or transport it into, the cave. Dust particles are often measured by their aerodynamic properties; here the aerodynamic diameter is used to describe particles. A particle of aerodynamic diameter 1 μm settles under the influence of gravity at the same rate as a spherical particle of 1 μm diameter and a density of 1 g cm^{-3}. The Table shows the settling rates of particles. Settling is the main process for deposition of larger particles. Particles less than 1 μm also are significantly moved by Brownian motion and will even attach to the underside of horizontal surfaces.

Measurements of airborne dust in caves, on the surface in a country area and in a city, show similar diameter distributions. The Figure is a measurement of air-borne dust in Jenolan Caves (New South Wales, Australia), during a cave tour. By plotting particle surface area against diameter, a distribution results that is more normally distributed than plotting particle numbers or particle mass against diameter, which produce skewed distributions.

Natural dust sources are aided by wind, which is the major agent for generating airborne dust from soil particles, pollen, spores, plant and insect material, as well as smoke and ash from wildfires and volcanoes. Natural dust is transported into caves by airflow from the surface, and caves with "chimney" circulations can transport air laden with dust to great depths, a possible cause of the mobile dust described by Holsinger (1962), in Boundless Cave, Virginia. Barometric "breathing" has deposited a massive sand dune in Mullamullang Cave, Nullarbor Plains, Australia. Slow, bi-directional air flow near the entrances of

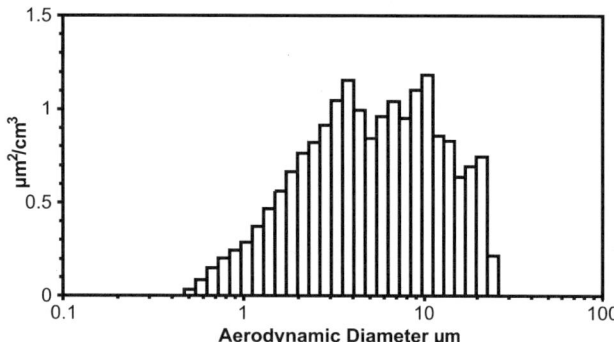

Tourist Caves: Airborne Debris: Diameter distribution and concentration of dust particles in the air, as a cave tour passed a point in Jenolan Caves, New South Wales, Australia. Before the tour group arrived the concentration was very low.

Particle Diameter μm.	Brownian motion displacement in 10sec. (mm. RMS)	Settling Velocity m s^{-1}	Time to settle one metre
0.01	1.0	6.95×10^{-8}	166 days
0.1	0.12	8.65×10^{-7}	13.4 days
1	2.3×10^{-2}	3.48×10^{-5}	7.98 hours
10	7.0×10^{-3}	3.06×10^{-3}	327 sec.
100	2.2×10^{-3}	0.261	3.86 sec.

Tourist Caves: Airborne Debris: The Brownian motion displacement and settling rates of particles of different aerodynamic diameter (see Baron & Willeke 1993).

"static" caves will transport airborne dust significant distances from the entrance before the time needed for the dust to settle elapses. Aeolian deposits are to be expected in many caves in all climates, but are rarely reported.

Anthropogenic dust sources are mainly mining, agriculture, and transport, and in third-world countries, cooking fires are significant. Humans themselves are a source of hair and dander, while the clothes that they wear shed textile fibre fragments. Anthropogenic dust is transported into caves by the same processes as natural dust, but in addition, a major mechanism for transport of dust involves the textiles used for clothing. The total surface area of typical textile clothing is in the order of 150 m^2 kg^{-1}. This area is quite efficient at capturing airborne particles outside a cave. The flexible nature of textiles and the movement of the wearer then enable the particles to be dislodged from the clothing and re-entrained into the air inside the cave. A typical tourist in a show cave releases 1 μg s^{-1}.

Methods of measurement are many, but commonly dust can be captured by drawing air through a filter, it can be measured in a stream of air by optical means, or it can be collected on sample plates as it deposits. The data in the Figure were acquired by a TSI Aerodynamic Particle Sizer, which draws an air sample through two laser light beams, while the air stream is decelerating. The transit time determines the aerodynamic diameter of each particle. A simple method in caves is to use glass Petri dishes placed on horizontal surfaces, to sample the dustfall for a period of weeks (Michie, 1997). The dust film can then be examined microscopically, the light transmission of the dish can be measured, or the material can be removed for chemical or physical analysis.

Management of dust in tour caves is by design and control of the infrastructure (Michie, 1999). In a well-managed tour cave, the dust deposited by each visitor may be tens of thousands of times less than that deposited by "wild" cavers with dirty clothes. Re-entrainment of particles, which have fallen on walking surfaces, will occur if floors are not kept quite clean. Ideal paths are elevated above the cave floor, have solid side barriers up to knee height, are frequently cleaned of deposited material and are made of materials that will not fragment. The environment outside a tour cave should be kept dust free so that material will not be trapped in visitors' clothing and carried into the cave, or be carried into the cave by air exchange. Any form of combustion in a cave releases very fine particles (mainly carbon) which have a disproportionately large surface area and are extremely efficient at penetrating deep into caves and discolouring all surfaces.

The effects of anthropogenic dust on caves may be subtle, but threaten many cave values. In all show caves, the colour of decorations is degraded as they become covered with dust. In many caves this effect has not been noticed but there is awareness of the accumulation of textile fibres (only a minor component of dust) which are picked out by hand. At Jenolan Caves, one hundred years as a show cave resulted in severely degraded decorations that were only restored by steam cleaning or, as is now preferred, water spray cleaning (Bonwick & Ellis, 1985). This treatment is not without residual damage (Spate & Moses, 1993) and is impossible for many types of speleothem, so degradation may be irreversible.

Dust has a profound effect on the life forms in caves. The minerals and organic material may overwhelm the original food chain, and new opportunistic species may invade the cave, although natural dust may be an essential part of the food chain. Organic material is decomposed by fungi and bacteria, whereas natural textile fibres (cotton, wool, etc.) decompose to add to the food chain, but synthetic fibres endure to be cemented into surfaces. Fibre dyestuffs, which may be 10% of the mass of a textile, often contain heavy metals or undesirable chemicals that could cause damage to cave fauna or flora when the fibre degrades. The organic decomposition causes corrosion of mineral surfaces.

NEVILLE A. MICHIE

Works Cited

Baron, P.A. & Willeke, K. 1993. Gas and particle motion. In *Aerosol Measurement: Principles, Techniques, and Applications*, edited by K. Willeke & P.A. Baron, New York: Van Nostrand Reinhold

Bonwick, J. & Ellis, R. 1985. New caves for old: Cleaning, restoration and redevelopment of show caves. In *Proceedings of the 6th Australasian Conference on Cave and Karst Management, Waitomo Caves, New Zealand*, edited by D.R. Williams & K.A. Wilde, Carlton South, Victoria: Australasian Cave and Karst Management Association

Christoforou, C.S., Salmon, L.G. & Cass, G.R. 1996. Fate of atmospheric particles within Buddhist cave temples at Yungang, China. *Environmental Science and Technology*, 30(12): 3425–34

Holsinger, J.R. 1962. Shifting sands noted in Boundless Cave, Virginia. *NSS News*, 20(5): 57–58

Jablonsky, P., Kraemer, S. & Yett, W. 1993. Lint in caves. In *Proceedings of the 9th National Cave Management Symposium, Carlsbad, New Mexico*, edited by D.L. Pate, Carlsbad, New Mexico: National Cave Management Symposium Steering Committee

Michie, N.A. 1997. The threat to caves of the human dust source. In *Proceedings of the 12th International Congress of Speleology*, vol. 5, La Chaux-de-Fonds, Switzerland: International Union of Speleology/Swiss Speleological Society

Michie, N.A. 1999. Management of dust in show caves. In *Cave Management in Australasia XII: Proceedings of the 13th Australasian Conference on Cave and Karst Management, Mount Gambier, South Australia*, edited by K. Henderson, Carlton South, Victoria: Australasian Cave and Karst Management Association

Spate, A. & Moses, C. 1993. Impacts of high pressure cleaning, a case study at Jenolan. In *Proceedings of the 10th Australasian Conference on Cave and Karst Management, Auckland, New Zealand*, Carlton South, Victoria: Australasian Cave and Karst Management Association

Went, F.W. 1970. Measuring cave air movements with condensation nuclei. *NSS Bulletin*, 35(1): 1–9

Further Reading

Willeke, K. & Baron, P.A. (editors) 1993. *Aerosol Measurement: Principles, Techniques, and Applications*, New York: Van Nostrand Reinhold
A comprehensive textbook for modern aerosol science, measurement, and problems.

TOURIST CAVES: ALGAE AND LAMPENFLORA

Visible growths of algae, cyanobacteria (formerly known as blue-green algae), mosses, and moss protonema (green filamentous structures arising from an asexual spore of moss) are common within electrically lit caves. In some caves lichens and ferns are also locally encountered. All such growths are termed lampenflora because of their association with electric lights.

Lampenflora change the natural appearance of cave features and, if not promptly treated, damage speleothem surfaces due to the production of organic acids which corrode the surface. In caves with prehistoric art (such as Lascaux) lampenflora are extremely destructive (Ruspoli, 1986). Management programmes to minimize and control lampenflora typically involve two strategies (Aley, Aley & Rhodes, 1984). First, light intensity / light duration thresholds are identified below which little or no visible plant growth will develop; light intensities on sensitive cave surfaces are then mostly kept below these thresholds. Second, lampenflora growth which does occur is treated with 5.25% sodium hypochlorite solution (bleach). However, the use of bleach solutions for control of lampenflora in caves with prehistoric art is likely to damage features of significance and is thus not likely to be a viable management strategy for such caves.

Lampenflora management strategies should not be predicated on an assumption that algae did not exist in the cave prior to the introduction of electric lighting for visitors. Studies by Claus (1962; 1964), Hajdu (1966), Kol (1967), and others demonstrate that many genera and species of algae and cyanobacteria grow in the perpetually dark portions of many caves. Such growth is generally not very obvious, is typically black in colour, and is limited to frequently or perennially wet cave surfaces. Some of the algal and cyanobacteria species found growing in the perpetually dark portions of caves become components of the lampenflora if the cave is electrically lit; many of these species are distributed widely around the world.

The composition of lampenflora varies substantially among caves in the United States which the author has studied. In 1984 in Carlsbad Caverns, New Mexico, lampenflora growths consisted of about 70% cyanobacteria, 20% green algae, and 10% moss protonema. In addition, there were diatoms present in about 25% of all of the clusters of lampenflora. Yellow-green algae were also present, but they were found in very few locations. Lampenflora were associated with 43% of the lights inspected in the Caverns, and the total number of lampenflora species present in the cave was estimated at 200. Carlsbad Caverns is in a semi-arid region and moisture availability is limiting even where lampenflora exist. In 1985 lampenflora growths in Oregon Caves, Oregon, consisted of about 15% mature moss, 10% moss protonema, 40% cyanobacteria, and 35% green algae. Diatoms were present in about half of the clusters of lampenflora, and the total number of lampenflora species present was estimated at 100. Oregon Caves has a Mediterranean climate with a dry summer. Most cave surfaces are routinely wet, and lampenflora were associated with every fixed light in the cave.

Visible lampenflora are usually limited to moist or wet surfaces. Soft surfaces (such as cave sediments and moonmilk) provide more moisture storage than hard surfaces (such as found on actively depositing speleothems). As a result, soft surfaces are more prone to the development of lampenflora and especially to the existence of luxuriant growths. However, hard surfaces often have adequate moisture to support lampenflora, especially in humid regions.

The total light energy received, a function of the intensity of the light and the duration of the lighting, is a critical factor for the establishment and growth of lampenflora. Theoretically, a period of continuous lighting for a given number of hours per day should yield more lampenflora growth than short periods of lighting which total the same number of hours per day. This is because plants make a number of chemical and physiological changes between light phase and dark phase conditions, and these changes require some time and plant energy. The management significance of this is that switching lights on and off multiple times during the course of a day in addition to reducing the total lighting period helps prevent or reduce lampenflora.

In general, the light necessary to produce lampenflora consisting of algae and cyanobacteria is less than the amount required for the establishment and growth of moss protonema. The establishment and growth of moss protonema requires less light than does the growth of mature moss, and ferns and lichens require still more light. In Carlsbad Caverns, 85% of the lampenflora received incident light intensities of 3.6 footcandles (39 lux) or more; a similar estimate for Oregon Caves was 4.2 footcandles (45 lux). Both of these caves were usually illuminated for most or all of the time that the cave was open for tours.

Light intensity, rather than the type or colour of the light, is the important factor. A notable exception is green light, since the photosynthetic pigment found in lampenflora primarily absorbs red and blue light and reflects green light, thus reducing photosynthetic efficiency. However, for aesthetic reasons, green cave lighting is an impractical strategy. Yellow lighting is a possible alternative, since algae do not strongly absorb yellow light. Due to filters that absorb parts of the white light spectrum, coloured lights produce lower incident light intensities on cave surfaces than do uncoloured lights of the same type and wattage; this accounts for the common observation that there is less lampenflora growth around lights using coloured filters than around white lights. For the same incident light intensities on moist cave surfaces, there is no detectable difference in lampenflora growths between fluorescent and incandescent lights.

Growth plots testing the extent of lampenflora growth on various cave substrates were established in both Carlsbad Caverns and Oregon Caves. Lint and other detritus from visitors and their clothing produced the most extensive and rapid lampenflora growths. Laundry products contain phosphates and residual phosphates are present in lint. Phosphates enhance algal growth in aquatic systems.

Many chemical agents have been tested for their ability to kill lampenflora. For most situations, 5.25% sodium hypochlorite solution applied as a light mist proves the most effective control while concurrently minimizing adverse impacts on cave features. More dilute solutions can be used, but they typically require the use of much more solution. The solution oxidizes the plant material and ruptures plant cells. One to two weeks after treatment fungal growth on the dead plant material will usually make it possible to wash the dead material from speleothems without scrubbing. Calcium hypochlorite solutions should not be used since they leave a calcium residue which is difficult to remove. Hydrogen peroxide is ineffective in controlling lampenflora.

Sodium hypochlorite should be carefully applied as a light mist on surfaces with lampenflora; multiple treatments should be used in areas with dense growth. Adverse impacts are minimized by treating areas as soon as the lampenflora is visible. Drippage or runoff of the treating solution should be captured; burlap or other fabric which will be oxidized by the treating solution can be placed adjacent to the target areas. While cave fauna could be killed if they come in direct contact with the sodium hypochlorite solution, the oxidizing ability of the solution (and thus its toxicity) diminishes rapidly in the presence of materials (such as lampenflora, cave lint, and burlap) which are subject to oxidation. Careful application and control of the sodium hypochlorite is the key to protecting cave fauna while conducting lampenflora control work (see Restoration of Caves and Speleothem Repair).

Recent advances in lighting, such as the use of fibre optics and low energy lamps, can permit more precise use of light in caves. If appropriately used, this enhanced control has the potential to permit cave lighting with reduced lampenflora growth.

TOM ALEY

See also **Biofilms**

Works Cited

Aley, T., Aley, C. & Rhodes, R. 1984. Control of exotic plant growth in Carlsbad Caverns, New Mexico. Proceedings of the 1984 National Cave Management Symposium. *Missouri Speleology*, 25(1–4): 159–71

Claus, G. 1962. Data on the ecology of the algae of Peace Cave in Hungary. *Nova Hedwigia*, 4(1): 55–79

Claus, G. 1964. Algae and their mode of life in the Baradla Cave at Aggtelek II. *International Journal of Speleology*, 1: 13–20

Hajdu, L. 1966. Algological studies in the cave at Maytas Mount, Budapest, Hungary. *International Journal of Speleology*, 2: 137–49

Kol, E. 1967. Algal growth experiments in the Baradla Cave at Aggtelek. *International Journal of Speleology*, 2: 457–74

Ruspoli, M. 1986. *The Cave of Lascaux: The Final Photographs*, New York: Abrams and London: Thames and Hudson

Further Reading

Aley, T.J. 1996. Caves in crisis. *1997 Yearbook of Science and the Future*, edited by D. Calhoun, Chicago: Encyclopaedia Britannica

Gillieson, D. 1996. *Caves: Processes, Development and Management*, Oxford and Cambridge, Massachusetts: Blackwells

Giordano, M., Mobili, F., Pezzoni, V, Hein, M.K. & Davis, J.S. 2000. Photosynthesis in the caves of Frasassi (Italy). *Phycologia*, 39(5): 384–89

Hazslinszky, T. (editor) 1985. *International Colloquium on Lamp Flora*, Budapest: Hungarian Speleological Society

Olson, R. 2003. Control of lamp flora in developed caves. In *Restoration and Conservation of Caves*, edited by V. Hildreth-Werker and J. Werker, Huntsville, Alabama: National Speleological Society

TOWER KARST

Tower karst landscapes represent perhaps the most spectacular terrain developed on carbonate rocks. Broadly, they consist of relatively isolated, often steep-sided, residual carbonate hills (towers) protruding from a near-planar surrounding corrosion surface, which is often alluviated and traversed by allogenic rivers that flow from outside the karst area. Some of the best-known tower karst landscapes are in southern China, particularly around Guilin in Guangxi (see Figure) and in Guizhou. Other notable examples occur throughout Southeast Asia, particularly in Malaysia, Thailand, Indonesia, and Vietnam; in Central America; and in the Caribbean Greater Antilles.

The towers, which dominate the landscape, are steep-sided to hemispherical residual hills, some over 200 m tall and 500 m in diameter. Various names have been ascribed to the towers and the overall landscape: in Chinese, tower karst is known as fenglin (peak forest) where towers are isolated, and fengcong (peak cluster) where groups of residual hills share a common exposed bedrock base; in Spanish the towers are termed mogotes; in French they are tourelles or pitons; in German Turmkarst; elsewhere they are known as "haystacks" and pepinos.

The towers are of variable morphology. Balazs (1973) suggested that four classes could be recognized on the basis of their height / width ratio: the Yangshuo, Organos, Sewu, and Tual types, named after their "type examples" in China, Cuba, Java, and Kai-Ketjil Island (Indonesia) respectively. Individual towers usually rise from a continuous carbonate surface, although this is often obscured by alluvium, and in some cases individuals are visibly connected to one or more of their neighbours with which they share a common basal "footprint". In the vicinity of Alligator Pond, in southern Jamaica, towers rise from a continuous unmantled carbonate rock base. Other towers rise from a surface incorporating non-carbonate rocks, such as in the Sierra de los Organos, Cuba and in Chillagoe, Queensland, Australia. In Malaysia and in the Rio Frio area of the Cayo District, Belize, limestone hills are founded on granite, which forms the floor of the Rio Frio Cave. In Puerto Rico, the mogotes protrude through a cover of "blanket sands"; in Ha Long Bay, Vietnam, and elsewhere, drowned marine towers rise from the sea. Ford and Williams (1989) suggest that there are four major genetic types of tower karst: (1) residual hills protruding from a planed carbonate surface veneered by alluvium; (2) residual hills emerging from carbonate inliers in a planed surface cut mainly across non-carbonate rock; (3) carbonate hills protruding through an aggraded surface of clastic sediments that buries the underlying karst topography; and (4) isolated carbonate towers rising from steeply sloping pedestal bases of various lithologies.

Although some towers approach conical symmetry, most are actually asymmetrical, reflecting structural or other influences. Asymmetry of the mogotes of northern Puerto Rico was considered to be the result of directional preference in surface induration or case-hardening, but was shown by Day (1978) to be directionally inconsistent and related rather to structural control and basal undercutting. As the towers themselves are erosional remnants, it is the surrounding lowlands that are the focus of current karstic and fluvial geomorphic activity, and there exists a landscape duality in which the carbonate tower environment and the surrounding alluvial environment are quite distinct.

Tower slopes and hilltops are highly irregular, often with only a minimal soil cover. Slopes consist of combinations of vertical cliffs, inclined bedrock surfaces, "staircases", or talus accumulations. Occasional rockfall episodes constitute a hazard in populated areas. By contrast, the intervening lowlands often have deep sediment covers produced by alluviation. Regional drainage is largely allogenic, although there may be significant autogenic inputs via springs at tower bases. Most tower karst is developed in mechanically competent carbonates that are able to support steep slopes (Day, 1982). Elsewhere, more friable carbonates have a surface induration, often referred to as case-hardening and in some cases exceeding a metre in thickness. Bedding, jointing, faulting, and lithological variation also influence tower morphology (Panos & Stelcl, 1968).

Cave systems associated with tower karst include active river caves dissecting the towers at the level of the surrounding lowlands, undercut marginal meander or foot caves or "swamp notches" (Jennings, 1976; McDonald, 1979), and largely abandoned formerly phreatic systems riddling the towers at higher elevations. Paleomagnetic studies of sediments from these caves have proven extremely valuable in deciphering the chronology of tower karst landscape evolution (Williams, 1987), indicating that they are time-transgressive landforms, with their summits being considerably older than their bases.

Tower Karst: A perfect example of a limestone tower beside the Jingbao River in the fenglin karst of Guangxi. (Photo by Tony Waltham)

Spectacular tower karst landscapes attracted the attention of early Chinese travellers such as Xu Xiake, and Grund (1914) described the landscape as the "senile" stage in his karst evolutionary cycle. It was not until the mid-20th century, however, that the first widely available and systematic geomorphological studies of tower karst were undertaken in Asia and the Caribbean by Herbert Lehmann (1954) and others. Some of the most detailed observations of towers were made in the 1970s by McDonald (1976a,b; 1979), who drew particular attention to their perimeters and to the role of basal slope development, including undercutting by rivers.

Several models of tower karst formation and evolution have been proposed. Of particular concern is whether isolated karst towers (fenglin) represent a sequential evolution from previously linked fengcong hills, or whether the two develop independently under differing geological and environmental conditions. While both are possible, most evidence supports the latter hypothesis. Although tower karst is most often associated with humid tropical climates, relict examples occur in locations where equivalent climatic conditions formerly prevailed. Such relict tower karst has been recognized in central Europe (Bosák et al., 1989). Analogous carbonate karst landscapes have also been identified in the Kimberley ranges of Western Australia and in the labyrinth karst of the Northwest Territory of Canada (Brook & Ford, 1978). A sandstone analog exists in Arnhem Land in the Northern Territory of Australia (see Silicate Karst).

MICK DAY and TAO TANG

See also **Cone Karst; Yangshuo Karst**

Works Cited

Balazs, D. 1973. Relief types of tropical karst areas. In *IGU Symposium on Karst Morphogenesis*, edited by L. Jakucs, Szeged: Attila Jozsef University: 16–32

Bosák, P., Ford, D.C., Glazek, J. & Horácek, I. 1989. *Paleokarst: A Systematic and Regional Review*, Amsterdam and New York: Elsevier

Brook, G.A. & Ford, D.C. 1978. The origin of labyrinth and tower karst and the climatic conditions necessary for their development. *Nature*, 275: 493–96

Day, M.J. 1978. Morphology and distribution of residual limestone hills (mogotes) in the karst of northern Puerto Rico. *Bulletin of the Geological Society of America*, 89: 426–32

Day, M.J. 1982. The influence of some material properties on the development of tropical karst terrain. *Transactions of the British Cave Research Association*, 9(1): 27–37

Ford, D.C. & Williams, P.W. 1989. *Karst Geomorphology and Hydrology*, London and Boston: Unwin Hyman

Grund, A. 1914. Der geographische Zyklus im Karst. *Ges. Erdkunde*, 52: 621–40 [The geographical cycle in the karst]. Translated and reprinted in *Karst Geomorphology*, edited by M.M. Sweeting, Stroudsburg, Pennsylvania: Hutchinson Ross, 1981

Jennings, J.N. 1976. A test of the importance of cliff-foot caves in tower karst development. *Zeitschrift für Geomorphologie Supplementbande*, 26: 92–97

Lehmann, H. 1954. Der tropische Kegelkarst auf den grossen Antillen. *Erdkunde*, 8: 130–39

McDonald, R.C. 1976a. Limestone morphology in south Sulawesi, Indonesia. *Zeitschrift für Geomorphologie Supplementbande*, 26: 71–91

McDonald, R.C. 1976b. Hillslope base depressions in tower karst topography of Belize. *Zeitschrift für Geomorphologie Supplementbande*, 26: 98–103

McDonald, R.C. 1979. Tower karst geomorphology in Belize. *Zeitschrift für Geomorphologie Supplementbande*, 32: 35–45

Panos, V. & Stelcl, O. 1968. Physiographic and geologic control in development of Cuban mogotes. *Zeitschrift für Geomorphologie*, 12(2): 117–73

Williams, P.W. 1987. Geomorphic inheritance and the development of tower karst. *Earth Surface Processes and Landforms*, 12(5): 453–65

Further Reading

Monroe, W.H. 1968. The karst features of northern Puerto Rico. *Bulletin of the National Speleological Society*, 30: 75–86

Sweeting, M.M. 1995. *Karst in China: Its Geomorphology and Environment*, Berlin and New York: Springer

Yuan D. 1991 (editor). *Karst of China*, Beijing: Geological Publishing House

TRAVERTINE

This term is applied to spring-deposited calcium carbonate, derived from the *Lapis tiburtinus* (rock of the Tiber), where it was quarried extensively as a building stone in Roman times. Travertine at the type locality, Bagni di Tivoli, is deposited by thermal springs permeating marine limestones, but travertines are deposited from a wide range of springs containing abundant Ca, and usually, abundant carbon dioxide. A related term is *calcareous tufa*, often used to describe the softer varieties of travertine, which were too weak for construction purposes. This term is still used widely, especially in Europe, though apt to be confused with the Italian *tufo*, a term (now largely defunct) for rocks of volcanic origin. Travertine has a broad definition encompassing calcareous tufa and cave calcite (speleothem), whose mechanism of formation is similar.

Most travertines are formed from waters containing calcium and bicarbonate ions through the following reaction:

$$Ca^{2+} + 2(HCO_3)^{2-} \rightleftharpoons CaCO_3 + CO_2 + H_2O$$

The reaction proceeds to the right when carbon dioxide is removed from the water by evasion to the atmosphere or by photosynthesis. Atmospheric evasion only occurs if the partial pressure of CO_2 in the water exceeds that of the atmosphere. This occurs in most springwaters, because CO_2 is added to water from a soil atmosphere rich in CO_2. However, travertine will only be precipitated if the Ca concentration is high enough to exceed the solubility product of the minerals calcite or aragonite. Typically, this requires water derived from a catchment containing limestone or gypsum, where the dissolved Ca exceeds about 1 mM l^{-1} (40 ppm). The removal of excess CO_2 is enhanced by water turbulence and warming. Sunny cascades are efficient at CO_2 removal and many of the best-known travertines are found on these. Bright sunshine also increases photosynthesis by aquatic plants, removing CO_2 below the atmospheric equilibrium level. The relative importance of the two processes, evasion and photosynthesis, depends upon springwater discharge, area of stream bed, stream morphometry, and climate. In general, travertines are deposited in the upper reaches of streams, and where discharge is high most CO_2 loss is by evasion. In small seeps in regions of low rainfall and high illumination, photosynthesis is probably more important. A small group of travertines, described as invasive meteogenes, result from the direct reaction of atmospheric CO_2 with a highly alkaline groundwater. They are formed on a small scale around lime-burning sites and in regions containing the rare mineral portlandite, both of which produce localized groundwaters containing calcium hydroxide.

Travertines show a wide range of morphologies due to development on different terrains. Most common are cascades and barrages. Cascades often become coated in travertine in karst regions and reach impressive dimensions. Well-known examples are the falls of Urach, Germany; Huangguoshu, China; Turner's Falls, Oklahoma; and the Reotier travertine cascade in the French Alps (Figure 1). At these sites, the rate of travertine erosion keeps pace with deposition, so the travertines form an inverted parabola shaped by the falling curtain of water. Where deposition exceeds erosion, large prograding aprons occur, such as those at Pamukkale, Turkey (see separate entry) and Mammoth Hot Springs, Wyoming. Occasionally, spectacular tall narrow channels are produced, e.g. the Gutersteiner Falls, Germany and the famous Pamukkale "self-built channels". Related to cascades are travertine dams. These differ only in impounding water behind them to form ponds or lakes and thus grow above the water line. Erosion exceeds deposition, and often a large series of apparently self-perpetuating dams are built across a river, leading to spectacular lake and waterfall scenery. The best known is Plitvice, Croatia (see Plitvice Lakes entry) but there are equally impressive sites in Afghanistan at Band e Amir (see photo in Indian Subcontinent entry), and in China (see separate entry, Huanglong and Jiuzhaigou). Dams may form in series, because the intervening lakes alter water chemistry by increasing residence time and photosynthesis, extending the precipitation process. The examples above originate from springs feeding rapidly descending watercourses, but where the terrain is flat, or deposition rapid, precipitation is localized around the spring orifice, forming a travertine mound. Mounds can reach impressive dimensions, such as Solomon's Prison in Iran (69 m) and the mound springs of the Lake Eyre Basin of Australia, but most are smaller, ranging from 5 m to less than half a metre in height. Many are fed by artesian springs. In Britain, the spring mounds of Great Close Mire in Yorkshire are noted for their rich flora. Other morphologies include fissure ridges, which are linear mounds forming along fault-springs, stream crusts, and cemented gravels. Stream crusts consist of travertine linings to the stream bed, which are of variable thickness (0.5–100 cm or more), often laminated and merging with cascades on steeper sections. In low to moderate flow rates ($c.20-100$ cm s^{-1}) travertine often develops around mobile fragments of gravel to form

Travertine. Figure 2. Water cascading into the Mediterranean Sea from the Antalya travertine terraces, which are among the largest freshwater carbonate deposits in the world. The cliffs are about 30 m high. (Photo by John Gunn)

rounded, laminated pebbles, termed oncoids or pisoids. Regular movement by the current permits all-round growth and eventual accumulation of oncoids into shoals or bars along some stream sections. Similar pebbles sometimes occur in the littoral zone of karstic lakes. Cemented gravels, similar to calcrete, are common where calcareous waters penetrate coarse alluvium. This can occur over broad floodplains and the resulting deposit, as hard as concrete, may weather out as travertine terraces. Such terraces are common around the Mediterranean Sea, for example at Antalya, Turkey (Figure 2). Other travertines erode soon after their formation, and can be redeposited as valley-fills or behind travertine-dammed lakes. These clastic travertines are usually soft, containing travertine "sand", oncoids, and fragments of "phytoherm" (incrustations on plants such as algae, bryophytes, and flowering plant stems). They are sometimes quarried for marl. Apart from their economic value, the deposits have been used widely in paleoecological and paleoenvironmental studies. Fossils of invertebrates, particularly Mollusca, Ostracoda, Vertebrata, and the impressions and pollen of plants, are often well preserved and have been used to reconstruct Quaternary environments.

Actively forming travertines are often rich in flora and fauna. The latter often harbours specialists, such as *Pericoma* larvae producing setae upon which the travertine is deposited, making travertines important for conservation. Dating of travertine using ^{14}C is often problematic due to the "hard water effect" but successful results have been achieved using Uranium series. Useful environmental data have also been obtained from their stable isotope geochemistry.

ALLAN PENTECOST

Travertine: Figure 1. The Reotier travertine cascade in the French alps showing rapid prograding carbonate deposition over a blue-green algal / diatom biofilm. (Photo by Allan Pentecost)

Further Reading

Bouyx, E. & Pias, J. 1971. Signification geologique des travertins d'Awpar (Afghanistan). *Compte Rendu Hebdomadaire des Seances et Mémoires de l'Academie des Sciences, Paris*, 273: 2468–71

Chafetz, H.S. & Folk, R.L. 1984. Travertines: depositional morphology and the bacterially constructed constituents. *Journal of Sedimentary Petrology*, 54: 289–316

Goudie, A.S., Viles, H.A. & Pentecost, A. 1993. The late-Holocene tufa decline in Europe. *The Holocene*, 3: 181–6

Pedley, M. 1993. Sedimentology of the late Quaternary barrage tufas in the Wye and Lathkill valleys, north Derbyshire. *Proceedings of the Yorkshire Geological Society*, 49: 197–206

Pentecost, A. 1995. The Quaternary travertine deposits of Europe and Asia Minor. *Quaternary Science Reviews*, 14: 1005–28

TUNNELLING AND UNDERGROUND DAMS IN KARST

Man-made underground structures in karst include tunnels, large chambers, dams, and reservoirs. Excavation of tunnels, in the form of qanats, in arid regions such as Persia, Iraq, Egypt, and Libya is more than 3000 years old. In Iran, some were excavated in karstified limestone. Tunnels or galleries as a method of exploiting groundwater were well known for centuries throughout the Mediterranean and Middle East. A 1600 m long water transmission tunnel in limestone on the island of Samos (Greece) was excavated using hammer and chisel about 530 BC. Another purpose for ancient tunnel excavation in karst regions was drainage of temporarily flooded karst poljes (see Karst Hydrology: History).

Successful tunnel excavation requires reliable identification and mapping of cavernous zones in order to adjust the tunnel route to the actual geological conditions. The common investigation methods are detailed geological mapping at the surface, drilling, water level monitoring, geophysical surveys (from the surface and from within the tunnel), and speleological exploration. During excavation of tunnels and large chambers in karst, the major problems arise from the presence of faults and large caverns, and from groundwater intrusion. Of these, only caverns may be of such dimensions to represent a problem for which there may be no technical solution other than relocation. Tunnels appear to be the structures most vulnerable to damage or failure in karst, and this is especially true of high-pressure tunnels for hydroelectric power plants. If a karst conduit is encountered on the tunnel route, excavation progress can be slowed because of many different difficulties. These include excessive free space (which must be filled, bridged, or by-passed), caves which are filled and masked by sediments of unknown dimensions, geotechnical features that may require a change of excavation method to avoid complicated repair work, groundwater discharges into the tunnel, and undiscovered caverns close to the tunnel which can cause collapse during excavation or subsequent operation.

If, during excavation, a cavern is discovered on the tunnel route, its extreme dimensions may be a serious barrier to further progress. In certain cases, the only good solution is to construct a bypass or to relocate the tunnel route. Examples are the headrace tunnel at Capljina (Herzegovina), the traffic tunnel through the Velebit Mountain (Croatia), a tunnel on the Rome–Naples (Italy) railway line, and over 20 railway tunnels in China. Along the route of the Ucka road tunnel in Croatia, more than 1300 m of karst channels were investigated, including a cavern 175 m long and 70 m wide (Hudec, Bozicević & Bleiweiss 1980).

Especially unfavourable conditions occur when the groundwater level is (permanently or temporarily) above the tunnel axis, since excessive water inflow through karst conduits can provoke sudden tunnel flooding. Human lives and tunnelling equipment can be seriously endangered, particularly if there is no drainage by gravity. After six years of operation of the 6170 m long Sozina railway tunnel (Yugoslavia), it was flooded by an abrupt groundwater intrusion through karst channels, at a rate of 6.5 m^3 s^{-1}. A 1750 m long drainage tunnel, 2 m lower, and 15 m from the main tunnel was excavated to resolve the problem. Without drainage by gravity, any groundwater intrusion of more than 100 l s^{-1} is an extremely serious problem. In those cases horizontal pilot boreholes and consolidation grouting ahead are

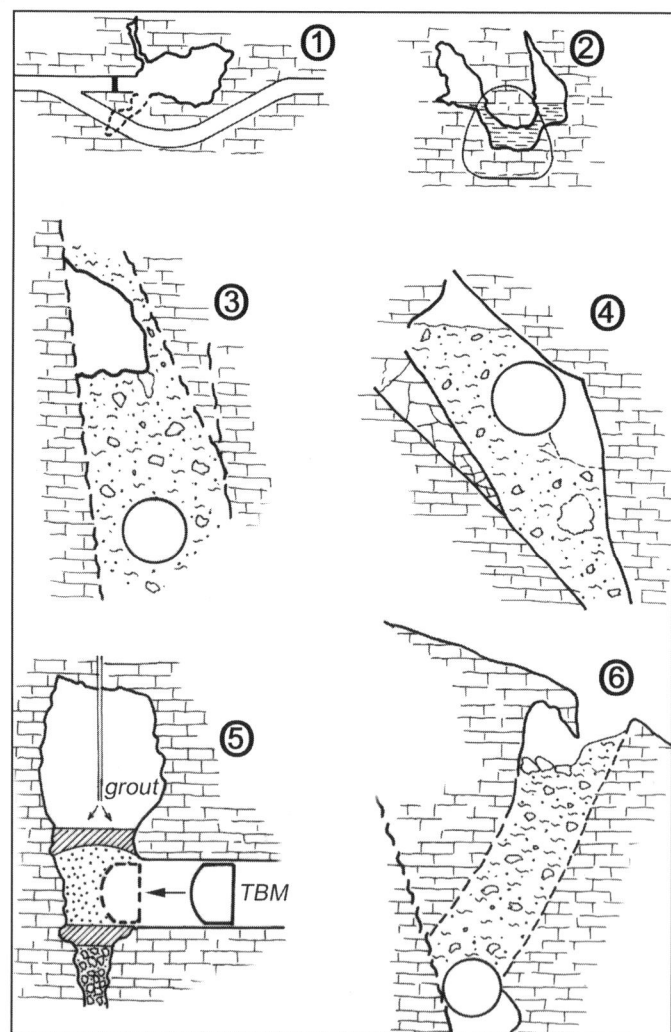

Tunnelling in Karst: Examples of caverns found on a tunnel route. 1. Deviation of tunnel to bypass large empty cavern. 2. Cavern at water table level. 3 & 4. Caves filled with sediment. 5. Empty cavern ahead of TBM, requiring prior grouting. 6. Collapse at surface caused by tunnel tube excavated through cave fill.

employed prior to driving the tunnel. Floods in the tailrace tunnel of the Capljina reversible power plant (Herzegovina) and in the Kouhrang transmission tunnel (Iran) necessitated complex remedial works as there was no scope for gravity drainage.

Undiscovered or unsatisfactorily treated caverns around the tunnel lining can cause cracking of the lining and collapse during the tunnel operation. This problem is especially common in tunnels under pressure for hydroelectric power. The opposite problem may occur in water-tapping galleries which aim to intersect joints, joint systems, and zones containing active karst conduits. Gallery-tapping structures are successfully used for water supply in different parts of the Mediterranean region, including the Dinaric karst. For example, a 300 m long gallery collects the water supply of the town of Trogir, Croatia (Mijatović, 1984). At many sites the gallery is a water-collecting channel itself, but the water inflow is increased by drainage holes or wells driven from the gallery, as at the Lez Spring (France) and the Zvir Spring (Croatia).

Special care is needed when using a tunnel-boring machine (TBM) in karstified rock as TBM technology is not as flexible as conventional tunnelling. Major groundwater intrusions, large caverns, and caverns filled with clay can be complex technical problems for TBMs. Because of the vulnerability of TBM technology, very detailed geological, hydrogeological, geotechnical, and geophysical investigations are needed ahead of excavation so that any karstified zones (with possible large caverns) may be avoided by changing the tunnel alignment. If a TBM penetrates a clay-filled cavern, the usual consequence is sinking of the TBM head and in the worst case the TBM head can be completely lost. Because of this, all clay has to be removed and the cavern filled with concrete up to the level of the tunnel, or bridging technology has to be applied. Good examples of problems with TBM usage in a highly karstified zone are the Zakucac II headrace tunnel (Croatia), and the Fatnica-Bileca transmission tunnel (Herzegovina)

The problems of excavating large chambers in karstified rock are mostly related to groundwater intrusion and rock mass stability. Experience has shown that the treatment of large natural caves is easier than treatment of similar problems in tunnels, because of the larger working space. If a chamber is endangered by underground water the voids and cavities can be permanently stabilized by cleaning, filling by concrete, and grouting. In the case of flowing water or water under pressure, special treatment techniques have to be used. The ultimate condition for successful treatment is to stop the groundwater flow by prior grouting. In the vadose zone, treatment of empty caves is not necessary and only faults and large joints have to be protected by rock bolts and grouting. Large chambers in karst are mostly excavated for military purposes (shelters for submarines, ships, and planes) and for hydroelectric plants. One of the largest is the chamber for the Capljina power plant which is 98 m long, 29 m wide, and 77m high and is excavated in highly karstified rock 45 m below the groundwater level. The basic problems during excavation were the intersection of active karst channels resulting in intrusion of groundwater under pressure.

An artificial underground reservoir is a portion of an aquifer that has been improved to control the discharge regime of artificially stored water. Extensive cave systems and large natural chambers can provide significant storage. More than 20 underground reservoirs, with a storage capacity of between 1×10^5 and 1×10^7 m^3, have been created in the karst regions of China (Lu, 1986). A large underground reservoir was formed in Linlang Cave in Qiubei, Yunnan, in 1955–60. The annual average discharge of the karst conduit is 23.8 m^3 s^{-1}, and the maximum discharge exceeds 100 m^3 s^{-1}. Artificial storage in the karst channel system was achieved by constructing a 15 m high underground dam. A headrace tunnel provides 109 m of head for a power station with an installed capacity of 25 MW. In Guizhou, 16 underground dams have been constructed for irrigation. Five underground reservoirs have been designed on Miyako Island (Japan). The first one, Minafuku Dam (grout curtain 500 m long and 16.5 m high with a storage capacity of 720 000 m^3), was completed in 1979 (Yoshikawa & Shokohifard, 1993). An underground reservoir was constructed in the spring zone of Nevesinjsko polje (Herzegovina) by plugging a conduit in karstified Eocene conglomerate. However, because of inadequate manipulation and an abrupt increase of water pressure in the cave system the overburden above the underground reservoir collapsed.

A massive concrete plug, 10 m high and on average 3.5 m wide, was constructed to plug the channel outlet of the Obod estavelle at Fatnicko polje (Herzegovina). The peak discharge of the Obod is 60 m^3 s^{-1} and the water pressure in the underground karst channel system increased rapidly to 1000 kN m^{-2}. Several tens of new springs appeared on the hill slopes 80–100 m above the plug level, and because of strong seismic shocks a road on the hillside started sliding and many houses at a distance of 250 to 300 m above the estavelle were damaged. To alleviate the pressure, blasting was used to create a 1.53×1.40 m opening in the concrete plug and within 6 hours all the artificially stored water was discharged.

The Ombla power plant (Croatia) has one of the largest underground damming and storage projects. The recorded discharge of the Ombla Spring is 2.3 112.5 m^3 s^{-1} and the deepest cave conduit is about 150 m below sea level, and about 200 m behind the main spring outlet which is at sea level. The position of the flysch and the elevation of overflow springs limit the underground dam at about 100 m above sea level. Due to this height and the required overburden of the alignment, the arch-like underground dam will need to be located at least 200 m behind the spring outlet and it will be necessary to construct grout curtains and to plug karst channels.

Artificial underground reservoirs may be classified into three basic groups: storage within one karst channel; storage in all types of porosity and the aeration zone, including inactive channels; and underground storage which is physically coupled to surface reservoirs. Depending on geomorphological, hydrogeological, and geotechnical features, five general concepts can be identified for underground reservoir construction: (1) surface dam in front of the main discharge zone; (2) diaphragm wall connected to impervious or basement rocks; (3) plugging of individual caves to enable water storage in the system of karst conduits; (4) grout curtain favourable for construction of a wide watertight screen connected to impervious rocks; and (5) combination of these underground geotechnical structures.

In comparison with surface storage, underground water storage has many advantages: it does not flood fertile land; the problem of relocating infrastructure, residential facilities, and historical monuments does not exist; negative effects of evaporation, thermal stratification, and hydrodynamic, biological, and ecological effects are eliminated; sedimentation is negligible; failures of underground dams cannot be catastrophic; and destruction of landscape is minimal. However, the plugging of karst channels can cause floods in upstream depressions, the development of collapse and subsidence dolines at the surface, and induced local seismicity.

PETAR MILANOVIĆ

See also **Dams and Reservoirs on Karst**

Works Cited

Hudec, M., Bozicević, S. & Bleiweiss, R. 1980. Support of cavern roof near Tunnel Ucka. *Proceedings of the 5th Yugoslav Symposium for Rock Mechanic and Underground Works, Split, Yugoslavia*, Yugoslav Society for Rock Mechanics

Lu Y. 1986. *Some Problems of Subsurface Reservoirs Constructed in Karst Regions of China*, Beijing: Institute of Hydrogeology and Engineering Geology, Chinese Academy of Geological Sciences

Mijatović, B. 1984. *Hydrogeology of the Dinaric Karst*, Hannover: Heise

Yoshikawa, M. & Shokohifard, G. 1993. *Underground Dams: A New Technology for Groundwater Resources Development*, Proceedings of the International Karst Symposium, Shiraz, Iran, edited by A. Afrasiabian, Iran: Ministry of Energy

Further Reading

Guoliang, C. 1994. Prediction of covered karst halls. *Proceedings of the 7th International IAEG Congress*, edited by R. Oliveira, L.F. Rodrigues, A.G. Coelho & A.P. Cunha, Rotterdam: Balkema

Milanović, P. 1997. Tunneling in karst: Common engineering-geology problems. In *Proceedings of the International Symposium on Engineering Geology and the Environment*, edited by P.G. Marinos, G.C. Koukis, G.C. Tsiambaos & G.C. Stournaras, Rotterdam: Balkema

Milanović, P. 2000. *Geological Engineering in Karst*, Belgrade: Zebra

Milanović, P. 2000. Engineering problems in karst. In *Present State and Future Trends of Karst Studies: Proceedings of the 6th International Symposium and Field Seminar*, Paris: UNESCO

Yuan D. 1990. *The Construction of Underground Dams on Subterranean Streams in South China Karst*, Guilin: Institute of Karst Geology (also available at http://www.karst.edu.cn/igcp/igcp299/1991/part3–4.htm#341)

TURKEY

Limestone, dolomite, and other carbonate rocks cover about one third of Turkey's 250 000 km², and most are strongly karstified. The largest and most important limestone karst forms most of the Taurus Mountains, and there are also significant karst terrains in Southeast Anatolia (along the Syrian border), Central Anatolia (south and west of Ankara), and along the Anatolian Black Sea coast around Zonguldak and in Thrace, west of Istanbul (Figure 1). There are important regions of gypsum karst in eastern Anatolia around Sivas.

Temucin Aygen founded Turkish speleology when he invited foreign cavers to explore the caves of Anatolia in the mid-1960s and established the Turkish Speleological Society in 1964. Most early explorations were by foreign expeditions, but the country now has several active caving groups and a Karst Water Resources Centre (UKAM) at Hacettepe University in Ankara.

The Taurus Mountains, which formed by lateral and vertical tectonic movements upon rock units of various ages from Cambrian to Recent, form a continuous karst belt 200 km wide and 2000 km long across southern Turkey. They rise from near sea level to elevations of 2500–3750 m with high mountains, sharp peaks, deep valleys, narrow gorges, and rugged plateaus. The Taurus Mountain range is an eastern extension of the Alpine orogenic belt, and many karst features follow structural lineaments. The famous travertine terraces of Pamukkale (see separate entry) are of hydrothermal origin whereas travertines of freshwater karstic origin around Antalya are amongst the largest travertine deposits in the world (see photograph in Travertine entry). The Antalya terraces lack the beautiful pools of Pamukkale but end in a series of cliffs 30 m high with waterfalls dropping straight into the Mediterranean. Very fine caves and karst landforms are spread across almost the entire range, and include the following sites of particular note.

Manavgat Gorge, east of Antalya, is about 1000 m deep and at its foot is perhaps the most beautiful cave system in Turkey,

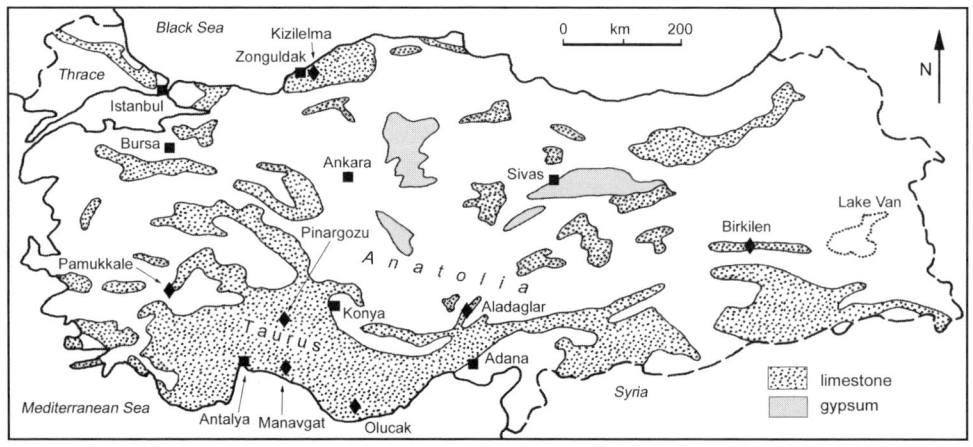

Turkey: Figure 1. The main outcrops of limestones (including dolomites) and gypsum in Turkey. Many of the soluble rocks are interbedded with shales, and some of these outcrops have minimal karst development.

Altinbesik Cave with 4500 m of passages. During winter and spring water level in the cave rises by more than 80 m, completely flooding most of the known system, and all exploration has been during summer when the outflow ceases. The source of this, and several other springs in the gorge, is thought to include water that sinks on the southern shores of Lake Beysehir some 50 km to the north. The cave has been explored via a series of climbs, lakes, and through one sump of 120 m to a final sump, as yet undived. The nearby Martal Cave, which was discovered during a drought in 1998, contains over 1500 m of very large passage, including a cavern over 100 m in diameter. Further down the gorge, the Dumanli spring, now submerged by 120 m of water following construction of the Oymapinar Dam, had an estimated mean discharge of 50 $m^3 s^{-1}$ making it possibly the largest karst spring in the world issuing from a single opening (Karanjac & Günay, 1980). Some 30 km to the northwest, the Oluk Köprü Springs have a minimum discharge of 30 $m^3 s^{-1}$ most of which discharges from the Köprüçay Conglomerate in a narrow canyon a kilometre long.

Tilkiler Düdeni is the second longest cave in Turkey (6650 m), and was found during gallery excavations for the Oymapinar Dam on the Manavgat River. It is the largest of several caves formed in a Miocene conglomerate made up of limestone clasts in a carbonate matrix (Figure 2; see also Değirmenci *et al.*, 1994).

Kocain Cave, some 20 km from Antalya, is notable for its large entrance (35 x 70 m), a large chamber (600 m long, 50–60 m high), and for archaeological remains from Roman times.

Pinargözü Cave is 11 km west of Yenisarbademli in Isparta Province. It is the longest cave in Turkey with 15 km of passages which have been explored upwards from the resurgence via active streamways and cascades. The cave rises 720 m above the resurgence, and its water temperature is only 4–5°C.

Olucak is a plateau of 400 km^2 in İçel Province, the nearest town being Anamur. Miocene limestones, rising to 1800 2000 m above sea level, contain the two deepest caves in Turkey, Çukurpınar Cave (−1192 m, 3500 m long) and the 1377 m deep Peynirlikönü Cave (also known as Evren Gunay Cave). The entrances to these two caves are only 500 m apart but they trend in different directions. Both are typical Taurus streamsink caves with series of narrow, wet shafts.

Kirkgözler Springs. This group of closely spaced small magnitude springs 30 km north of Antalya formed along a fault that juxtaposes Taurus Mountain with impermeable Cretaceous ophiolites that underly the Antalya travertine plateau. They have a combined mean discharge of 24 $m^3 s^{-1}$ and are an important

Turkey: Figure 2. Rift passage in Kurukopru Cave, one of the longest caves in the Köprüçay Conglomerate. (Photo by John Gunn)

water resource. Cave divers have explored two large water-filled conduits beyond the springs, Kirkgöz-Suluin which contains a water-filled chamber, the "Stadium", that is about 100 m wide, 150 m long, and 35 m deep, and Kirkgöz-1. Underwater speleothems were discovered in Kirkgöz-Suluin indicating that the

Turkey: The deepest and longest caves in Turkey.

Cave	Location	Depth (m)	Length (m)
Peynirlikönü Dudeni	Anamur	1377	
Çukurpınar Dudeni	Anamur	1192	3500
Kuyukule	Dedegöl	832	
Pinargözü	Yenierbademli	720	15 000
Sütlük Dudeni	Pozanti	640	
Subatagi Düdeni	Camlica	670	
Tilkiler Düdeni	Manavgat	159	6650
Kızılelma Mağarasi	Zonguldak	145	6630

Turkey: Figure 3. Kizoren Obruk, one of many large collapse dolines in the Konya basin, part of the Central Anatolian karst. It has been important as a water resource for centuries, as shown by the old caravanseri on the far side of the obruk; water is still pumped for irrigation and potable supply (left of picture). (Photo by John Gunn)

cave was not always flooded. Deep flooded conduits have also been explored at Dudenbasi (-60 m) and at Finike-Suluin (-120 m) which also has underwater speleothems.

Aladaglar is an extensive limestone massif north of Adana, with glaciokarstic surfaces and cave entrances at altitudes from 1200 m to over 3300 m. Karst springs with individual mean discharges of $1-8 \text{ m}^3 \text{ s}^{-1}$ and a total discharge of about $32 \text{ m}^3 \text{ s}^{-1}$ are grouped in four sites on the eastern flank of the massif, at elevations of 450–1100 m. There are also some large springs on the western flank and the high degree of flow concentration suggests that well-developed conduit systems are present. Structural and hydrogeochemical considerations indicate that most of the groundwater is recharged from the high-altitude karst surfaces and drained to the eastern flank springs (Bayari & Gurer, 1993). The depth potential for the underground drainage is thus well over 2000 m, and perhaps up to 2500–2900 m, so Aladaglar may contain the deepest karst hydrological system in the world, though water tracing is still in progress to prove this. Sütlük Düdeni (-640 m) and Subatagi (-670 m) are the deepest caves yet explored in the high karst.

Southeast Anatolia has clayey limestones around Gaziantep and Urfa that show poor karstification, but karstic features are widespread in Cretaceous and older limestones that crop out along the marginal fold belt. A karst ridge to the north contains the Birkilen Caves, including the splendid river passage carrying a major headwater of the River Tigris for 860 m from sink to resurgence and often therefore known as the Tigris Tunnel (see Karst Hydrology: History for an account of early exploration).

Central Anatolia is a closed basin bounded by high mountains. The average elevation is around 1200 m and the lowest point is Tuz Gölü (Salt Lake). There are karst zones in the Mesozoic recrystallized limestones of the Taurus belt along the southern margin of the basin, and also in Neogene lacustrine limestone (Obruk Limestone) in the centre of the basin. The limestone is named for the many large, steep-sided dolines (obruks, with maximum depth of 145 m) some of which are permanently flooded and provide an important water resource in an otherwise semi-arid region (Figure 3). The floor of the Konya basin consists of older karstified rocks and most drainage is underground, with the regional hydraulic gradient in the Neogene limestone towards Tuz Gölü.

The Black Sea coast of northwest Anatolia and Thrace has karst in lenticular limestone blocks that extend over limited areas. The Thrace karst has formed in Eocene limestone 100–150 m thick that extends as a thin belt nearly parallel to the Istranca massif. The reef limestones are indurated, hard, porous, thick bedded or massive, with many solution cavities. The rolling upland karst inland from Zonguldak is distinguished by a series of long cave systems with large flood-prone streamways and some very fine calcite decorations.

Gypsum karst extends over about 9000 km^2 in the Upper Kizilirmak basin of central and eastern Anatolia. Miocene gypsum, which contains rock salt (halite) interlayers, is the most extensive unit, underlying about half the basin. Karst features include dolines, ponors, poljes, and some small caves, but the most spectacular feature is the extensive polygonal karst east of Sivas. This has hundreds of adjacent rounded solution dolines and also a scatter of very large collapse features (Waltham, 2002). Recession coefficients and consistency of the spring discharges indicate that the gypsum aquifer has a large storage capacity and that groundwater flow occurs slowly, primarily through enlarged joints and fractures.

JOHN GUNN AND GÜLTEKIN GÜNAY

Works Cited

Bayari, C.S. & Gürer, I. 1993. Use of hydrologic, hydrochemical and isotopic data in identification of groundwater flow patterns in Lower Zamanti Basin (Eastern Taurids-Turkey). DOGA: *Turkish Journal of Earth Science*, 2: 49–59

Değirmenci, M., Bayari, C.S., Denizman, C. & Kurttas, T. 1994. Caves in conglomerates, Köprüçay basin: western Taurides-Turkey. *NSS Bulletin*, 56: 14–22

Karanjac, J. & Günay, G. 1980. Dumanli spring, Turkey: The largest karstic spring in the world? *Journal of Hydrology*, 45: 219–31

Waltham, T. 2002. Gypsum karst near Sivas, Turkey. *Cave and Karst Science*, 29(1): 39–44

Further Reading

Bayari, C.S. & Ozbek, O. 1995. An inventory of karstic caves in the Taurus Mountain Range (Southern Turkey): Preliminary evaluation of geographic and hydrologic features. *Cave and Karst Science*, 21: 81–92

Eroskay, S.O. & Günay, G. 1980. Tectogenetic classification and hydrogeological properties of the karst regions of Turkey. *Proceedings of 1st International Symposium on Karst Hydrogeology, 9–19 October 1979, Oymapinar*, Antalya, Turkey: DSIR Publication

Kaçaroğlu, F., Degirmenci, M. & Cerit, O. 2001. Water quality problems of a gypsiferous watershed: Upper Kızılırmak Basin, Sivas, Turkey. *Water, Air and Soil Pollution*, 128: 161–80

Kincaid, T.R., 2000. Storage-Dominated vs Flow-Dominated Caves: A Hydraulic Model for Cave Development. Proceedings: Underwater Science and Technology Meeting—SBT2000, December 2–3, 2000. Middle East Technical University, Ankara, Turkey.

Kincaid, T.R., 2000. Mapping and Modeling the Morphology of Underwater Caves in the Taurus Mountains and Antalya Travertine Plateau, Southern Turkey. Proceedings: Underwater Science and Technology Meeting—SBT2000, December 2–3, 2000. Middle East Technical University, Ankara, Turkey.

Kincaid, T.R., 2000. Speleogenesis in the Kirkgozler Region of the Taurus Mountains, Southern Turkey. Proceedings: Underwater Science and Technology Meeting—SBT2000, December 2–3, 2000. Middle East Technical University, Ankara, Turkey.

The International Research and Application Center for Karst Water Resources (UKAM), Hacettepe University, Ankara, have published four field guides that contain a great deal of useful information, including road logs.

UKAM, 1985. Guide Book for Field Trip to Central Anatolian and Taurus Karst Regions, 13–19 July, 1985, International Symposium on Karst Water Resources, Ankara-Antalya, Turkey

UKAM, 1990. Guidebook for International Symposium and Field Seminar on Hydrogeologic Processes in Karst Terranes, 7–17 October 1990, Kemer-Antalya- Ankara.

UKAM, 1995. Guidebook for International Symposium and Field Seminar on Karst Waters & Environmental Impacts, 10–20 September 1995, Beldibi-Antalya, Turkey

UKAM, 2000. Guidebook for International Symposium and Field Seminar on Present State and Future Trends of Karst Studies, September 2000, Marmaris, Turkey.

UKRAINE GYPSUM CAVES AND KARST

The extensive gypsum karst in the western Ukraine is renowned for its giant maze caves. It is internationally important as a model example of artesian speleogenesis (Klimchouk, 2000b). The region contains the five longest gypsum caves in the world. The host gypsum bed, ranging from a few metres to more than 40 m in thickness, is the main component of the Miocene evaporite formation that girdles the Carpathians to the northeast. The gypsum occurs on the southwestern edge of the Eastern European platform, where it extends along the Carpathian Foredeep for over 300 km in a belt ranging from several kilometres to 40–80 km wide (Figure 1). It occupies over 20 000 km², together with some separated areas that occur to the northeast of the unbroken belt.

Most Miocene rocks along the platform margin rest on the eroded terrigenous and carbonate Cretaceous sediments. The Miocene gypsum bed is variable in structure and texture. Most commonly it grades from microcrystalline massive gypsum in the lower part through to variably grained bedded gypsum in the middle, to megacrystalline rock in the upper horizon. A layer of evaporitic and epigenetic limestone, locally called "Ratynsky", commonly overlies the gypsum. This layer ranges from half a metre to more than 25 m in thickness. The gypsum and the Ratynsky limestone comprise the Tyrassky Formation, which is overlain by the Upper Badenian unit, represented either by argillaceous and marly limestones and sandstone or, adjacent to the foredeep, by marls and clays of the Kosovsky Formation. The total thickness of the capping marls and clays ranges from 40–60 m in the platform interior to 80–100 m or more in the areas adjacent to the regional faults that separate the platform edge from the foredeep.

The present distribution of Miocene formations and the levels of their denudation exposure vary in a regular manner from the platform interior towards the foredeep. The Tyrassky Formation dips 1–3° towards the foredeep and is disrupted by block faults in the transition zone. To the south and southwest of the major Dniester Valley, large tectonic blocks drop down as a series of steps, the thickness of the clay overburden increases, and the depth of erosional entrenchment decreases. Along the tectonic boundary with the foredeep the Tyrassky Formation drops to a depth of 1000 m or more. This variation—he result of differential neotectonic movement—played an important role in the hydrogeological evolution of the Miocene aquifer system, and resulted in the differentiation of the platform edge into four zones. The gypsum was entirely removed by denudation within the first zone, but the other three zones represent distinct types of karst: entrenched, subjacent, and deep seated (Klimchouk, 2000b). The gypsum bed is largely drained in the entrenched karst zone, is partly inundated in the subjacent karst zone, and remains under artesian confinement in the deep-seated karst zone.

In hydrogeological terms the region represents the southwestern portion of the Volyno–Podolsky artesian basin (Shestopalov,

Ukraine Gypsum Caves and Karst: Figure 1. Location of the gypsum karst of the Western Ukraine.

1989). The Sarmatian and Kosovsky clays and marls serve as an upper confining sequence. The lower part of the Kosovsky Formation and the limestone bed of the Tyrassky Formation form the original upper aquifer (above the gypsum), and the Lower Badenian sandy carbonate beds, in places together with Cretaceous sediments, form the lower aquifer (below the gypsum), the latter being the major regional one. The hydrogeologic role of the gypsum unit has changed with time, from initially being an aquiclude, intervening between two aquifers, to a karstified aquifer with well-developed conduit permeability (Klimchouk, 1997a, 2000a, b). The regional flow is from the platform interior, where confining clays and the gypsum are largely denuded, toward the large and deep Dniester Valley and the Carpathian foredeep. In the northwest section of the gypsum belt the confined conditions prevail across its entire width. In its wide southeast section the deeply incised valleys of Dniester and its left-hand tributaries divide the Miocene sequence into a number of isolated, deeply drained interfluves capped with the clays (Podol'sky area). This is the entrenched karst zone where most of the explored, presently relict, maze caves are located. To the south–southeast of the Dniester (Bukovinsky area) the gypsum remains largely intact and is partly inundated (the subjacent karst zone). Further in this direction, as the depth of the gypsum below the clays increases and entrenchment decreases, the Miocene aquifer system becomes confined (the deep-seated karst zone). In this zone the groundwater flow pattern includes a lateral component in the lower aquifer (and in the upper aquifer, but to a lesser extent) and an upward component through the gypsum in areas of potentiometric lows, where extensive cave systems develop, as evidenced by numerous data from exploratory drilling.

Eleven large caves over 2 km in length are known in the region (see Table). Most of these caves are located north of the Dniester River. Two other large caves, Zoloushka and Bukovinka, occur in the Bukovinsky region, near the Prut River and the border with Moldova and Romania, generally in the area of artesian flow within the Miocene aquifer system but within local, particularly uplifted blocks, where entrenchment into the upper part of the gypsum caused unconfined (water-table) conditions to be established in the Holocene. Most of the caves have only one entrance, either through swallow holes at the interfluves or from gypsum outcrops in the slopes of the major valleys. Some caves and their entrance series were known to local people from time immemorial (e.g. Ozernaya, Kristal'naya, Mlynki, Verteba), but others were discovered by cavers via digs (e.g. Optimistychna, Slavka, Atlantida). Two caves (Zoloushka and Bukovinka) became accessible when opened by gypsum quarries. Systematic cave exploration and mapping in the region began in the 1960s.

All the large gypsum caves in the region are mazes developed along vertical and steeply inclined fissures arranged into multi-storey laterally extensive networks. Aggregating passages form lateral two- to four-storey systems that extend over areas of up to 1.5 km^2 (Figure 2). A notable feature of the mazes is the exceptionally high passage network density, which is characterized conveniently by using the ratio of a cave length to an area occupied by a cave system. This parameter varies from 118 (Verteba Cave) to 270 (Gostry Eovdy Cave) km km^{-2}, with the average value for the region being 164 km km^{-2}. Values of areal coverage and cave porosity (fractions of the total area and volume of the rock within a cave field, occupied by passages) vary for individual caves from 17.5 to 48.4% (average 29.5%) and from 2 to 12% (average 4.5%) respectively, being roughly an order of magnitude greater than these characteristics for typical unconfined caves. Optimistychna Cave (Optimisticheskaya in Russian spelling) is the longest gypsum cave, and the second-longest cave of any type known in the world, with more than 214 km of passages surveyed. By area and volume the largest caves are Ozernaya (330 000 m^2 and 665 000 m^3) and Zoloushka (305 000 m^2 and 712 000 m^3), followed by Optimistychna Cave (260 000 m^2 and 520 000 m^3).

Maze caves in the region were developed under confined conditions, due to upward transverse groundwater circulation between aquifers below and above the gypsum (Klimchouk, 1992, 2000b) (see Speleogenesis: Deep Seated and Confined Settings). According to the morphology, arrangement and hydrologic function of the cave mesoforms during the main (artesian) speleogenetic stage, three major components can be distinguished in the cave systems (Figure 3):

1. **Feeding channels**, the lowermost components in a system: vertical or subvertical conduits through which water rose from the sub-gypsum aquifer to the master passage networks. Such conduits are commonly separate but sometimes they form small networks at the lowermost part of the gypsum. The feeding channels join

Ukraine Gypsum Caves and Karst: Table. Morphometric parameters of large gypsum caves in the Western Ukraine.

No.	Cave	Length (km)	Average cross-sectional area (m^2)	Density of passages (km km^{-2})	Areal coverage (%)	Cave porosity (%)
1	Optimistychna	214.0	2.8	147	17.6	2.0
2	Ozernaya	117.0	6.0	150	44.6	5.0
3	Zoloushka	92.0	8.0	142	48.4	3.8
4	Mlynki	27.0	3.3	141	37.6	3.4
5	Kristal'naya	22.0	5.0	169	29.2	6.0
6	Slavka	9.1	3.7	139	27.6	3.4
7	Verteba	7.8	6.0	118	34.7	12.0
8	Atlantida	2.52	4.5	168	30.0	4.0
9	Bukovinka	2.4	2.5	120	21.5	4.4
10	Ugryn	2.12	3.8	177	33.3	5.7
11	Gostry Govdy	2.0	1.7	270	17.5	4.0

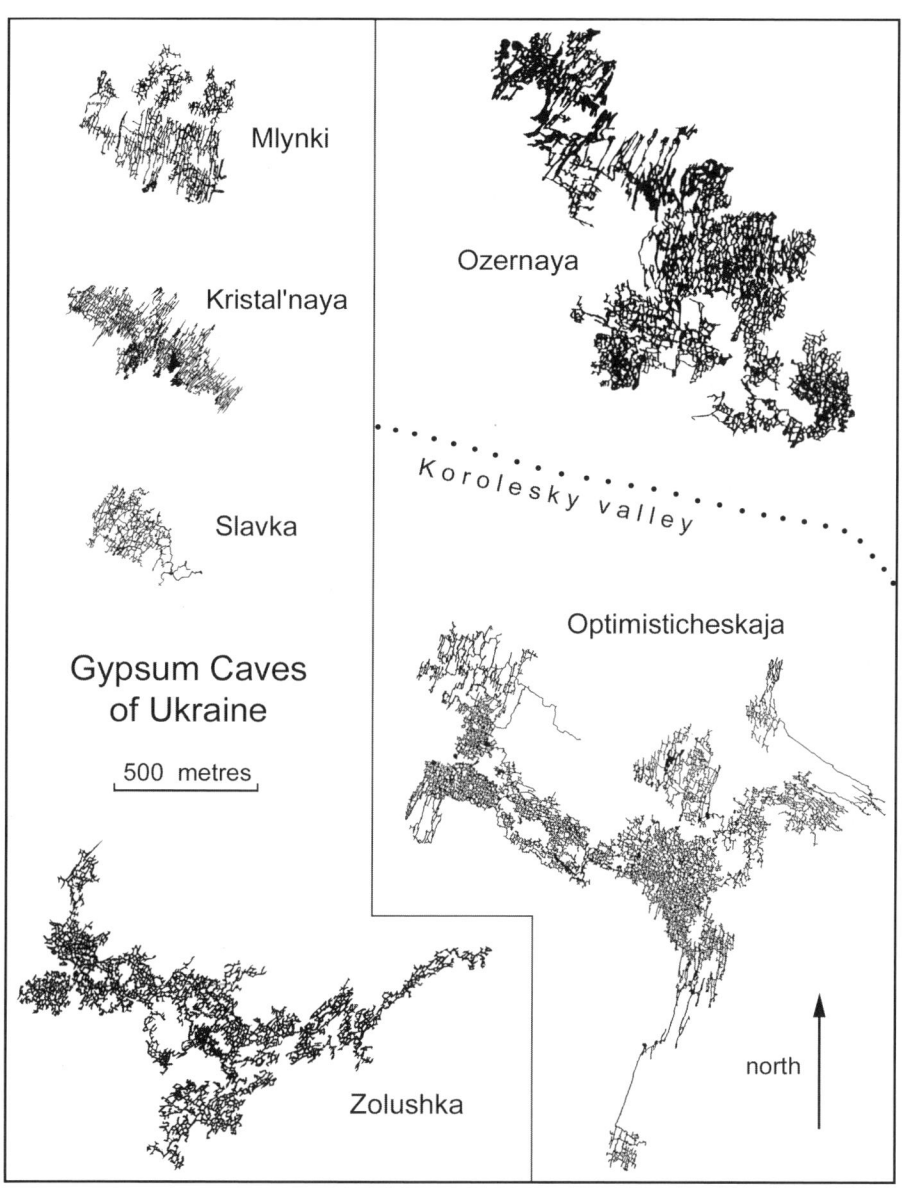

Ukraine Gypsum Caves and Karst: Figure 2. The very long gypsum caves of Ukraine drawn to the same scale. Ozernaya and Optimisticheskaja (Optimistychna) are shown in relation to each other, but the others are each at separate sites (after surveys by the cave clubs of L'vov, Ternopol, Chernivtsky, Kiev, and others).

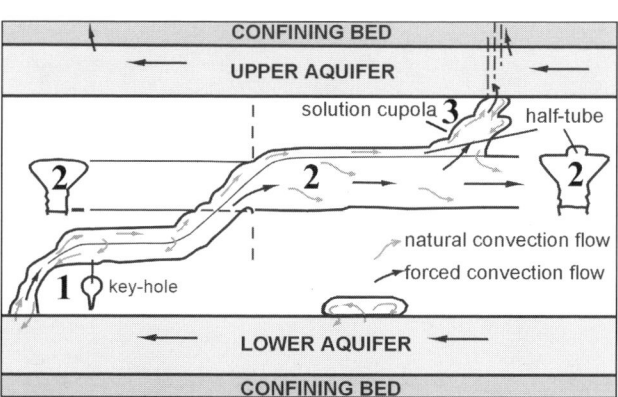

Ukraine Gypsum Caves and Karst: Figure 3. Main morphogenetic features of maze cave systems in the western Ukraine shown according to their hydrologic functionality. 1 = feeding channels, 2 = master passages, 3 = outlet features.

master passages located at the next upper level and scatter uniformly through their networks

2. **Master passages**: horizontal passages that form laterally extensive networks within certain horizons in the middle part of the gypsum bed (Figure 4). They received dispersed recharge from numerous feeding channels and conducted flow laterally to the nearest outlet feature

3. **Outlet features**: domes, cupolas, and vertical channels (dome pits) that rise from the ceiling of the master passages to the bottom of the overlying bed. They discharged water from cave systems to the overlying aquifer.

The predominant sediments in the maze caves of the region are successions of fine clays, with minor beds of silty clays. These fill passages to a variable extent and can reach 5–7 m in thickness. Breakdown deposits are also common. They include chip, slab, and block breakdown material from the gypsum, as well as more massive breakdown from the overlying formations. Calcite speleothems (stalactites, stalagmites, flowstones, and helictites)

Ukraine Gypsum Caves and Karst: Figure 4. "Master passage" in Ozernaya Cave. (Photo by John Gunn)

occur locally in zones of vertical water percolation from overlying formations. Gypsum crystals of different habits and sizes are the most common cave decorations. They are of largely subaerial origin. Hydroxides of Fe and Mn occur as powdery layers within the clay fill of many caves, indicating repeated transitional cycles from a reducing to an oxidizing geochemical environment. Massive deposition of Fe / Mn compounds in the form of powdery masses, coatings, stalactites, and stalagmites has occurred in Zoloushka Cave, where a rapid dewatering caused by groundwater abstraction during the last 50 years gave rise to a number of transitional geochemical processes, some of which appear to show considerable microbial involvement (Andrejchuk & Klimchouk, 2001).

The Western Ukrainian maze caves provide the most outstanding and unambiguous evidence for the transverse artesian speleogenetic model. The artesian speleogenesis in the Podol'sky region took place mainly during the late Pliocene through to the middle Pleistocene. It was induced by incision of the Dniester valley and its left-hand tributaries into the confining clays, and respective activation of the upward transverse groundwater flow within the underlying artesian system. Breaching of artesian confinement and further incision of the valleys during the middle Pleistocene caused substantial acceleration of groundwater circulation within the Miocene artesian system. The majority of passage growth probably occurred during this transitional period. Where the water table was established in the gypsum for a prolonged time, further widening of passages occurred due to horizontal notching at the water table. Eventually, with the water table dropping below the lower gypsum contact, cave systems in the entrenched karst zone became largely relict. Cave development under confined or semi-confined conditions continues today within the zones of deep-seated and subjacent karst.

There are large bioepigenetic deposits of native sulfur in the pre-Carpathian region, within the deep-seated karst zone, associated with the Miocene gypsum bed. Sulfur is embedded in epigenetic calcite that partially (at the top) or wholly replaces the gypsum. The artesian "ascending" speleogenesis in the gypsum layer played a fundamental role in the origin of the sulfur deposits (Klimchouk, 1997b). This is not only because it provided the large amounts of dissolved sulfates needed to fuel the large-scale sulfate reduction, but also because speleogenesis opened pathways for the flow of groundwater between the lower and upper aquifers through maze cave systems in the gypsum. Such a flow pattern and speleogenetic evolution within the gypsum provided the spatial and temporal framework within which the sulfur cycle processes took place, as well as controlling the geochemical environments, and the migration of reactants and reaction products between them.

ALEXANDER KLIMCHOUK

See also **Evaporite Karst; Speleogenesis: Deep Seated and Confined Settings**

Works Cited

Andrejchuk, V.N. & Klimchouk, A.B. 2001. Geomicrobiology and redox geochemistry of the karstified Miocene gypsum aquifer, Western Ukraine: a study from Zoloushka Cave. *Geomicrobiology*, 18(3): 27–95

Klimchouk, A.B. 1992. Large gypsum caves in the Western Ukraine and their genesis. *Cave Science*, 19(1): 3–11

Klimchouk, A.B. 1997a. The role of speleogenesis in the Miocene gypsum in the Western Ukraine in groundwater circulation in the multi-storey artesian system. In *Karst Waters and Environmental Impacts: Proceedings of the 5th International Symposium and Field Seminar on Karst Waters and Environmental Impacts, Antalya, Turkey, 10–20 September 1995*, edited by G. Günay & A. Johnson, Rotterdam: Balkema

Klimchouk, A.B. 1997b. The role of karst in the genesis of sulfur deposits, Pre-Carpathian region, Ukraine. *Environmental Geology*, 31: 1–20

Klimchouk, A.B. 2000a. Speleogenesis under deep-seated and confined settings. In *Speleogenesis: Evolution of Karst Aquifers*, edited by A.B. Klimchouk, D.C. Ford, A.N. Palmer & W. Dreybrodt, Huntsville, Alabama: National Speleological Society

Klimchouk, A.B. 2000b. Speleogenesis of great gypsum mazes in the Western Ukraine. In *Speleogenesis: Evolution of Karst Aquifers*, edited by A.B. Klimchouk, D.C. Ford, A.N. Palmer & W. Dreybrodt, Huntsville, Alabama: National Speleological Society

Shestopalov, V.M. (editor) 1989. *Water Exchange in Hydrogeological Structures of the Ukraine: Water Exchange Under Natural Conditions*, Kiev: Naukova dumka (in Russian)

Further Reading

Klimchouk, A., Lowe, D., Cooper, A. & Sauro, U. (editors) 1996. *Gypsum Karst of the World*, International Journal of Speleology, Theme issue 25 (3–4)

UNITED STATES OF AMERICA

The United States contains a wide variety of soluble rocks and karst types (Figure 1). It contains five of the world's seven longest caves (see Table), but because the thickest soluble rocks in the country are located in low-relief regions, its solution caves have limited depth. However, some of its volcanic caves on Hawaii have the greatest vertical extent of any in the world. (US sites described in more detail elsewhere in this volume are also shown in Figure 1.)

The most extensive karst regions of the United States are located in the Interior Low Plateaus (Figure 1) of the east-central states (Kentucky, Tennessee, Missouri, and neighbouring states). Carbonates of mainly Cambrian–Ordovician, Silurian–Devonian, and Mississippian (lower Carboniferous) age are exposed in broad plains and plateaux dissected by river valleys to depths rarely exceeding 200 m (Figure 2). Although continuous sections of soluble rocks are only 100–200 m thick, their low average dip of less than half a degree causes them to be exposed over large areas. Where carbonates are exposed, the karst consists of doline plains bordered by ponors that receive allogenic water from less soluble rocks. Farther downdip, the karst consists of dissected carbonate uplands capped by resistant clastic rocks. Caves are predominantly branchworks on several levels tied closely to the history of river entrenchment. The most notable, Mammoth Cave (Kentucky), is the longest cave in the world, with 557 km of mapped passages (see Mammoth Cave entry). Farther west in this region, in the Ozark Plateaux of Missouri and Arkansas, chert is so abundant in some of the carbonate rocks that residual chert fragments have accumulated to depths as much as 40 m, subduing much of the karst relief. Nevertheless, caves and large karst springs are abundant. Missouri alone contains more than 5400 documented caves and is second only to Tennessee (with more than 6300 caves) as the most cave-rich state.

The carbonate strata thicken eastward in the Appalachian Mountains (Figure 1; see Appalachian Mountains entry; Kastning & Kastning, 1991). The western half of the region is a broad plateau in which extensive karst occurs in bands in erosional escarpments and along entrenched valleys that have breached the overlying Pennsylvanian (late Carboniferous) caprock. The region contains many caves that typically consist of multilevel stream passages fed by vertical shafts up to 180 m deep (Figure 3). Farther east the Appalachians are strongly folded and faulted, producing long, linear karst belts in valleys and along ridge flanks. Most caves in these rocks are long strike-oriented stream passages. The deepest cave in the eastern part of the continent is the Omega Cave System, Virginia, with a vertical range of 384 m. Complex network mazes form where water enters through a thin permeable sandstone cap. Along the eastern border of this region is a broad valley floored by highly deformed Cambrian–Ordovician carbonates up to 4 km thick in nearly continuous sequences. Subsurface drainage is prevalent, but karst topography is subdued by the low relief. Accessible

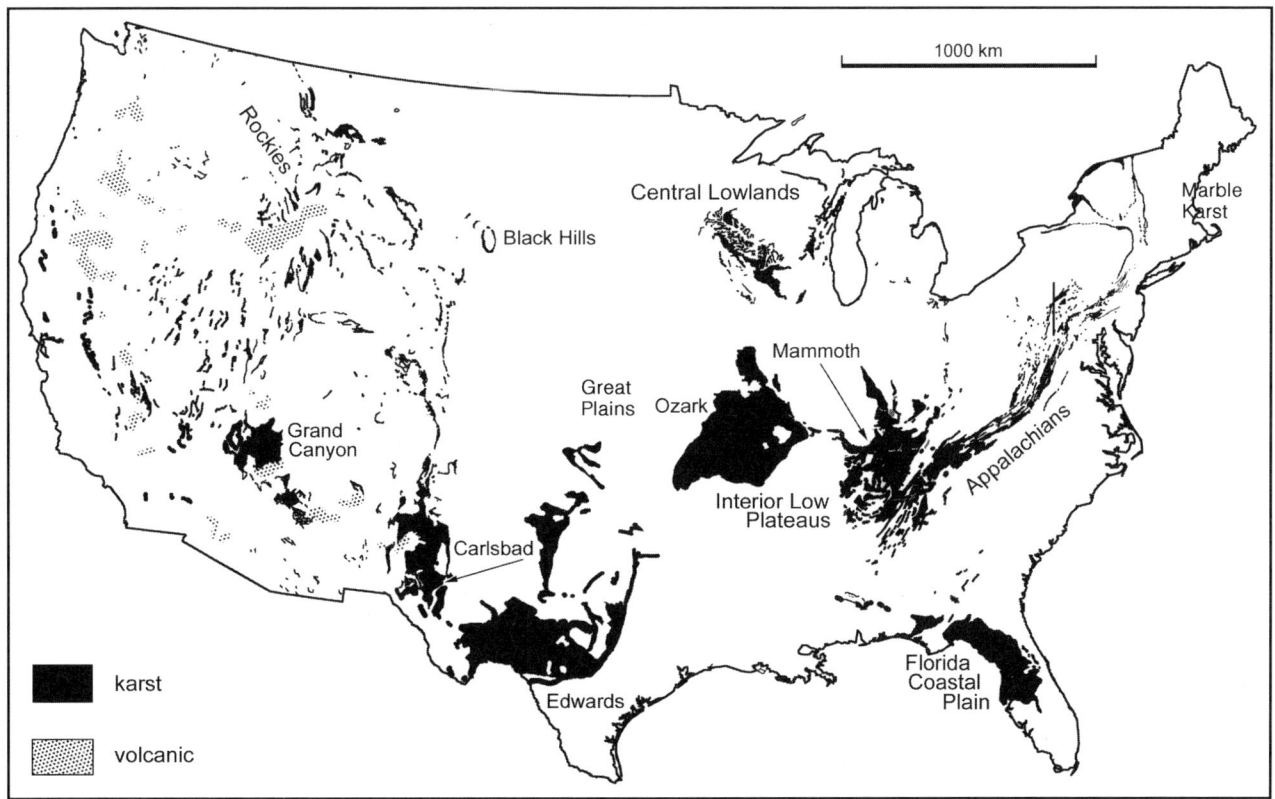

United States of America: Figure 1. The main karst regions of the "lower 48" states of the United States. Not shown are Alaska (with its island karst areas) and the Hawaiian Islands (which are entirely volcanic).

United States of America: Longest caves in the United States.

Cave	State	Surveyed length (km)	Surveyed depth (m)
1. Mammoth Cave System	Kentucky	557	116
2. Jewel Cave	South Dakota	206	193
3. Lechuguilla Cave	New Mexico	180	478
4. Wind Cave	South Dakota	173	202
5. Fisher Ridge Cave System	Kentucky	169	109
6. Friars Hole Cave System	West Virginia	73	188
7. Kazumura Cave (Lava tube)	Hawaii	66	1101
8. Organ Cave System	West Virginia	64	148
9. Blue Spring Cave	Tennessee	53	71
10. Martin Ridge Cave System	Kentucky	52	96

caves are restricted mainly to residual hills. Luray Caverns is the best-known example. Some caves in the folded Appalachians have little relationship to the surrounding topography or drainage patterns, which, together with maze-like patterns and mineralization by metallic ores, suggests a now-inactive hypogenic origin.

Along the eastern margin of the Appalachian Mountains, especially in the northeastern states, the strata have been highly metamorphosed, and karst and small caves are located in isolated patches of Precambrian and Paleozoic marble. The foothills of the Adirondack Mountains, New York, contain some of the world's oldest marbles, more than 1.2 billion years old. Its caves are apparently all of Pleistocene age, but there are scattered remnants of Precambrian paleokarst filled with basal Cambrian sandstone.

Lapping onto the eroded eastern and southern edges of the Appalachians in the southeastern United States are undeformed Tertiary strata of the Coastal Plain (Figure 1). These include porous carbonate rocks that form broad lowlands extending throughout much of Florida and parts of adjacent states. These strata are among the world's most productive aquifers. Solution and collapse dolines are common and are the cause of considerable property damage. Caves are abundant, mostly water-filled, and still actively enlarging. A great quantity of water flows through the caves to springs along major rivers or the seacoast. Much of the flow to conduits is contributed by diffuse seepage through the permeable surrounding rock. Some relict caves are complex labyrinths, but most are rambling tubular passages with

United States of America: Figure 2. Low-relief karst plains studded with dolines typify much of the Interior Low Plateaux. The Mitchell Plain of Indiana is shown here, with an outlier of sandstone-capped limestone in the background. (Photo by Art Palmer)

United States of America: Figure 3. Caves in the Appalachian Mountains are strongly guided by folds and faults. Vertical shafts up to 180 m deep are common in the southern Appalachians. This fault-controlled shaft in Ellison's Cave, Georgia, is the deepest known in the region (view looking upward). (Photo by Art Palmer)

crudely dendritic patterns. In general the water chemistry resembles that of most other temperate-climate karst areas, although dissolution is locally enhanced by seawater mixing along the coast and by oxidation of rising hydrogen sulfide. Late Pleistocene sea-level rise has flooded many caves and dolines, although in this low-relief region most cave development was already below the water table. Exploration of water-filled caves by divers indicates that the potential for discovery has barely been touched. The Leon Sinks Cave System in northwestern Florida is the longest known underwater cave in the United States, with a surveyed length of 30.5 km.

The broad glaciated Central Lowlands of the north-central United States (Figure 1) contain scattered karst areas on early Paleozoic carbonates where glacial sediment is thin or absent. Coldwater Cave in Iowa and Mystery Cave in Minnesota are the largest caves in the region, each more than 20 km long. Some are conduit systems, like those of the extensive karst plateaux to the south, but others, such as Mystery Cave, are angular fissure networks enlarged by water from river meander cutoffs. Many small fissure caves appear to be hypogenic, as they contain deposits of sulfide and other ore minerals. Deep-seated dissolution of evaporites has produced collapse breccias in some areas. Analysis of speleothems and sediments in caves of this region, as well as in those of the northeastern states, has helped to decipher local aspects of Pleistocene continental glaciation.

The Great Plains of the west-central United States are almost devoid of soluble rocks at the surface, except at the southern end, where nearly horizontal Cretaceous limestone and Permian gypsum are exposed, and in small uplifted fault blocks of Ordovician limestone. In the limestones, surface karst is subdued by the dry climate, although there are many high-discharge springs and large, well-decorated caves. Highly permeable Cretaceous carbonates are exposed along the Balcones Fault zone of central Texas and provide most of the region's water supply (see Edwards Aquifer; Edwards Aquifer: Biospeleology). Broad karst plateaux extend northwestward from the fault zone and include many notable caves. Honey Creek Cave, the longest in Texas (31 km), consists mainly of active stream passages requiring lengthy swims. The delicate helictite displays in the Caverns of Sonora are considered some of the world's most attractive. Bracken, Frio, and Ney Caves contain the nation's largest bat colonies. Bracken Cave alone has a population of 20–40 million Mexican free-tail bats. Some of the country's most productive petroleum reservoirs are located in Permian and Ordovician carbonates of west Texas, where much of the porosity is provided by deeply buried paleokarst, collapse of former cave systems, and deep-seated dissolution.

Farther west and north, the Permian gypsum extends across large parts of Texas, New Mexico, Oklahoma, and Kansas (Johnson, 1996). The country's largest cave in gypsum is Jester Cave in Oklahoma, which is more than 10 km long. Much groundwater recharge to the gypsum takes place through ephemeral sinking streams, but most caves can be followed only a short distance before they become impenetrably small. Apparently these inputs feed large systems of diffuse flow that emerge at regional springs.

The Black Hills of South Dakota and Wyoming consist of an elongate dome that rises from the Great Plains as an easterly outlier of the Rocky Mountains (Figure 1). Limestone and scattered gypsum crop out in the foothills that ring the central core of igneous and metamorphic rocks. Surface karst is limited to a few dolines and sinking streams, but the region includes some of the world's largest, oldest, and most complex caves (see Wind and Jewel Caves).

The southwestern United States consists of arid fault-bounded plateaux and mountains, many of them containing Paleozoic limestones and dolomites. Surface karst is virtually absent in this dry region, but many caves have formed by deep-seated processes, including H_2S oxidation. Caves of this origin are especially abundant in the Permian reef complex of the Guadalupe Mountains, New Mexico (see Carlsbad Cavern and Lechuguilla Cave) and in Mississippian limestone exposed in the Grand Canyon and its tributaries (see Grand Canyon entry). The Guadalupes include some of the world's most exotic caves. Lechuguilla Cave, more than 170 km long, is widely considered the world's most beautiful cave, owing to its profuse and unusual mineral deposits.

Western mountain ranges contain hundreds of small exposures of Paleozoic limestone and dolomite. At high altitudes they have been exposed to alpine glaciation, which has destroyed some karst features, but in some areas karst and cave development has been enhanced where glacial features such as moraines and cirques concentrate groundwater recharge into the carbonate rocks. Many caves and sinking streams are located high above nearby valley bottoms, but their host rocks are rarely more than a few hundred metres thick. Therefore, deep caves are rare except where they follow tilted strata for long distances. Karst is common along the sedimentary flanks of most of the Rocky Mountain ranges. Columbine Crawl (473 m deep) and Great Expectations ("Great X") Cave (429 m deep) in Wyoming each once held the US depth record. Some caves in the Rockies show evidence for hypogenic origin, and certain maze caves were probably formed by mixing of deep and shallow water (e.g. Cave of the Winds in Colorado). The Sierra Nevada contains notable

United States of America: Figure 4. Exhumed mid-Carboniferous paleokarst is pervasive throughout much of the western United States. This example shows formerly sediment-filled caves intersected by the Bighorn River in Wyoming. (Photo by Art Palmer)

marble karst, and Lilburn Cave and Bigfoot Cave in California are two of the largest marble caves in the world. The rainy islands along the coast of southeastern Alaska have extensive karst. El Capitan Pit in Prince of Wales Island, Alaska, is the deepest known solution shaft in the United States, at 182 m.

Paleokarst is abundant in the United States, and among several significant horizons are two that extend throughout much of the country (Palmer & Palmer, 1989). The lower horizon is developed in early Ordovician limestones and is best exposed in the southeastern United States, although it extends (mainly in the subsurface) across nearly the entire country. It includes a great deal of carbonate breccia and is host for economic-grade sulfide minerals, especially in Tennessee and Missouri, and petroleum in parts of the southwest. The upper paleokarst is developed on Mississippian (early Carboniferous) strata and is part of the widespread unconformity that forms the basis for dividing the Carboniferous of North America into two periods. This paleokarst is well exposed throughout the Rocky Mountain region and extends mainly in the subsurface as far as the east-central states (Figure 4). This is the paleokarst so prominent in the Black Hills.

Volcanic pseudokarst extends over broad areas of the western states, particularly Idaho, California, Oregon, Washington, and New Mexico (Halliday, 1960), but by far the most important volcanic region is the "big island" of Hawaii, the largest and southernmost in the Hawaiian island chain. Currently active volcanism supplies large flows that develop extensive caves, including the deepest and longest known lava caves in the world (see Hawaii Lava Tube Caves). Some Hawaiian lava flows have potential for caves exceeding 2000 m in vertical extent.

ARTHUR N. PALMER

Works Cited

Halliday, W.R. 1960. Pseudokarst in the United States. *National Speleological Society Bulletin*, 22(2): 109–13

Johnson, K.S. 1996. Gypsum karst in the United States. In *Gypsum Karst of the World*, edited by A. Klimchouk, D. Lowe, A. Cooper & U. Sauro, *International Journal of Speleology*, 25 (3–4)

Kastning, E.H. & Kastning, K.M. (editors) 1991. *Appalachian Karst: Proceedings of Appalachian Karst Symposium (Radford, Va.)*, Huntsville, Alabama: National Speleological Society

Palmer, M.V. & Palmer, A.N. 1989. Paleokarst of the USA. In *Paleokarst: A Systematic and Regional Review*, edited by P. Bosák, D. Ford, J. Glazek & I. Horácek, Amsterdam: Elsevier

Further Reading

Halliday, W.R. 1976. *Depths of the Earth: Caves and Cavers of the United States*, New York: Harper & Row

Veni, G. & DuChene, H. (editors) 2001. *Living with Karst: A Fragile Foundation*, Alexandria, Virginia: American Geological Institute

V

VALLEYS IN KARST

The term "karst valley" was introduced by Cvijić (1893) who provided a fourfold classification into pocket, blind, semi-blind, and dry valleys. Blind and dry valleys were subdivided into primary valleys, formed by allogenic rivers flowing from impermeable rock, and secondary valleys, formed in the bed of a normal karst river. Roglic (1964) argued that the term "karst valley" should be discontinued because strictly valleys are the result of water flowing overly the surface and are formed by fluvial, as opposed to karstic processes. Sweeting (1972, p.103) suggested that many would find this an overly pedantic view but that karst valleys should be recognized as "the most important of the landforms occurring in the karst which are not produced by true karst processes". Her fourfold classification into through (allogenic), blind and half-blind, pocket, and dry valleys is followed in this review. Valleys are a feature of most karst regions, the exceptions being areas of polygonal karst where the whole surface is pitted by dolines, although even here there may be vestiges of former valley networks. However, there are also some karst regions where the surface is dissected by a dense network of valley systems forming a fluviokarst (see separate entry).

Through valleys are formed by rivers that have their origins on non-karst lithologies and which maintain perennial flow through the karst to the output boundary. The valley is incised by both dissolution and mechanical erosion. Most through valleys are steep-sided, and gorges are more common in karstic rocks than in other lithologies, partly because most carbonate rocks are mechanically strong and partly because of a general absence of surface runoff and consequent reduction in mass wasting. Antecedent gorges form where uplift occurs at a rate less than the river's capacity to incise. There are four main reasons for the development of through valleys. First, karstification may not yet be sufficiently advanced; that is, the input from outside the karst exceeds the present capacity of the limestones to absorb it. In this case the river will usually be influent, with discharge decreasing both downstream and progressively over time. Second, the allogenic river may transport and deposit sufficient clastic material onto the karst to render the river bed virtually impermeable. In the third situation, the riverbed is rendered impermeable by permafrost but downcutting continues during summer melt periods. A fourth situation is where the hydraulic gradient is low and the river is at the local base level for drainage. In this case the river will usually be effluent, with discharge increasing downstream due to inputs from springs and direct recharge through the bed. The rivers Dove and Wye in Derbyshire, England, and the Green River in Kentucky, United States, are good examples of the fourth sub-type.

Some influent rivers lose water to the karst gradually over a long reach; the upper Danube in Germany and the Takaka in New Zealand are good examples. However, it is much more common for flow to be lost at a point, or series of points. The processes of dissolution and transport of clastic sediment underground result in a gradual lowering of the bed at these sink points, and downstream of them the river has less erosive power. Hence, over time an upward step develops at the sink point. Underground, the capacity of the conduits increases as they are enlarged by erosion and ultimately the lowest sink may be able to accommodate the entire base flow. This is the first stage in the formation of a blind valley, but as the sink is overtopped at discharges greater than base flow it is commonly called a half-blind or semi-blind valley. The conduit system may later enlarge sufficiently for the sink to take even the highest of flood flows forming a true blind valley (Figure 1). During the intervening time the valley below the sink may become progressively vegetated and increasingly difficult to recognize as ever having carried a river. If the sink-point migrates upstream then the height of the closure at the end of the blind valley may be just a few metres, however, if a large river continues to sink at the same point for many years, and the hydraulic gradient is high, the closure may grow to several hundred metres.

Pocket valleys (or steepheads) are the reverse of blind valleys, since they occur in association with large springs close to the margins of karst areas. Most of them are short and may form

Valleys in Karst: Figure 1. Blind valley where Fell Beck sinks into Gaping Gill on the slopes of Ingleborough, Yorkshire, England. (Photo by John Gunn)

Valleys in Karst: Figure 2. Dry valley, Lathkill Dale in the Peak District fluviokarst, England. (Photo by John Gunn)

by headward recession, as water from the spring undermines the rock above it, or by cavern collapse. Evidence for cavern collapse may be provided by a natural bridge, as at Marble Arch in County Fermanagh, Northern Ireland.

Dry valleys, lacking stream channels in their floor, are found on many lithologies but usually only close to the headwaters. An exception is where a dry valley forms as a consequence of river capture, a process that can occur on all lithologies. Long, well-developed dry valleys, are found in many karst areas, particularly where there are, or were, allogenic inputs, and are commonly similar in form to through valleys (Figure 2). Three major groups of hypotheses have been suggested for their formation: (1) differing climates in the past, with either greater rainfall or permafrost; (2) superimposition from non-karst strata followed by karstification of drainage; and (3) a fall in the level of the water table due to uplift of the land mass, incision of major valleys, or scarp recession. To these should be added the progressive desiccation of a through valley as the sink-point migrates upstream. Over time the floor of a dry valley may become dissected by dolines and the original fluvial form may be lost completely, as has happened in the Waitomo area of New Zealand. Alternatively, a substantial increase in surface discharge, following climate change or blockage of underground conduits by sediment deposition, may result in previously relict dry valleys becoming re-activated.

JOHN GUNN

Works Cited

Cvijić, J. 1893. Das Karstphänomen. *Geographische Abhandlungen Wien*, 5(3): 218–329
Roglic, J. 1964. "Karst valleys" in the Dinaric karst. *Erdkunde*, 18: 113–16
Sweeting, M.M. 1972. *Karst landforms*, London: Macmillan and New York: Columbia University Press, 1973 (Chapter 6 on Karst Valleys provides many examples)

Further Reading

Smith, D.I. 1975. The problems of limestone dry valleys–implications of recent work in limestone hydrology. In *Processes in Physical and Human Geography*, edited by R. Peel, M. Chisholm & P. Haggett, London: Heinemann: 130–47
Warwick, G.T. 1964. Dry valleys in the southern Pennines, England. *Erdkunde*, 18: 116–23

VERCORS, FRANCE

The Vercors extends over 1350 km² and is the largest alpine karst area in the Northern French Prealps (for location, see map in Europe, Alpine entry). The mean altitude is 1200 m and the highest peak is Grand Veymont (2341 m). The Vercors is bounded by high scarps and cut by deep gorges: Bourne cross-valley, Cholet reculée, Écouges canyon, and Combe Laval (Figure 1). The fold structures create stratigraphic ridges and valleys. Meadows and villages occur on the synclines in the Alpine molasse, whereas the anticlines of Urgonian limestone have produced a landscape of forest-covered hills. Altitudes rise eastward towards the High Plateau (1000–1600 m), reaching 1800–2000 m on the eastern scarps, with slopes carved into bare clints. This wilderness is protected in the largest French conservation area (17 000 ha). The folds follow a north–south trend, with a transverse saddle occupied by the Bourne gorge, toward which most of the underground drainage converges. Arbois is the third largest French spring, with a mean discharge of 8 m³ s⁻¹, and drains a 230-km² catchment. During exceptional floods, several overflows act with the siphon d'Arbois, the Bournillon resurgence (80 m³ s⁻¹) and the Luire cave (60 m³ s⁻¹), giving a total discharge of about 190 m³ s⁻¹! The Grotte de la Luire, located 20 km from the springs, flows from the entrance following a 450 m deep backflooding—the highest recorded in the world.

Vercors contains a large number of major cave systems, including the Gouffre Berger, the first cave in the world to be explored to a depth of −1000 m (Figure 2). Geological structures determine the underground drainage pattern, which may follow the troughs of perched synclines (Clot d'Aspres systems), anticline limbs (the Berger cave), edges of dammed synclines (Trou qui Souffle, Luire), or recumbent fold saddles (Tonnerre). Shafts and canyons take most cave systems directly down through the 300–400 m thickness of cavernous Urgonian limestones, sometimes as a single shaft, such as Pot 2 (302 m deep). Water then concentrates along the underlying Hauterivian marls, where it is entrenched in huge galleries (such as in the Gournier, Figure 3, and the Berger). Perched karst branchworks (including the Grotte Brudour) are drained by springs located at the marl contact, at the head of reculées (such as the Cholet). Dammed karst systems form complex three-dimensional mazes by backflooding, with level storeys left by progressive base-level lowering (Trou qui Souffle, Luire). Vercors is the best-known karst in the Alps, due as much to the number of deep caves as to the many publications discussing their genesis (Audra, 1994, 1995a, 2001). Eocene Pyrenean movements created the folded structures, with which some paleokarst features are associated, including huge pockets filled with weathered soils.

Miocene sea level was located near the present-day 1000–1200 m altitude. The relief was moderate, and often corresponded with the axes of the anticlines. Some tunnel caves linked the synclinal poljes. These caves are now completely disconnected, truncated, or even unroofed by erosion (Grotte de Pré-Létang). The oldest ones predate the orogeny (>12 Ma) and were formed according to the Miocene sea base level; others were formed in Upper Miocene as the area was uplifted (Antre

Vercors, France: Figure 2. The Hall of the Thirteen in the main passage of the Gouffre Berger, here clearly aligned on massive inclined fractures as it follows roughly down the dip. (Photo by Tony Waltham)

de Vénus, Coufin upper level). The Messinian fall in the base level (110 m below the present sea level in the Rhône valley) occurred during the main Rhodanian uplift (c. 5–6 million years ago). It caused valley entrenchment and the subsequent development of vertical cave systems that adapted to the new geological structure. The Zanclean transgression caused a prolonged sea-level highstand (lasting 1.5 million years). It ended the general deepening of both karst and valleys, and probably flooded the Messinian vadose systems that now appear as vauclusian springs (Taï, Diable). During this long period of stability, the largest and highest levels of major cave systems—located 200 m above the present base level—could have developed (Gournier, Bournillon, upper levels of Trou qui Souffle, Luire, Vallier); they were mostly related to synclinal poljes.

In these old cave systems, sediments are composed of residual quartz, heavy minerals, and clays (particularly kaolinite). They derive from the weathering of Upper Cretaceous sandstones and molasse in warm and humid conditions, producing an unconsolidated soil cover during the Tertiary (Audra, 1995b). This material was later removed and trapped in the karst. These sediments always constitute the oldest beds in the large alpine cave systems. They are covered by massive speleothems, frequently corroded by later inflows.

Vercors, France: Figure 1. A road winds along the massive cliff of Urgonian Limestone that forms the rim of the Combe Laval. (Photo by Tony Waltham)

Vercors, France: Figure 3. The lake in the entrance chamber of the Grotte de Gournier. (Photo by Tony Waltham)

The Pliocene transgression created a morphogenetic break. However, the subsequent uplift finally stopped this morphogenic quiet period. Moreover, after the late Pliocene (c.2.3 million years ago), a climatic cooling during the Pleistocene dominated the morphogenetic processes. During each glaciation, a morphogenic hiatus occurred. Consequently, the base level was lowered step by step, and poljes opened towards the Bourne Valley. However, karst evolution also depended on the position of the massif with respect to glaciers and base level. In western Vercors, where springs were located downstream from the glacier front, successive underground levels developed (Gournier, Coufin, Trou qui Souffle, Luire). In contrast, the evolution of areas located close to the large glaciers has nearly ended, leaving only some local entrenchment or invasion shafts in the vadose zone. Gouffre Vallier contains sediments linked to the first Plio-Quaternary glaciations (Audra & Rochette, 1993).

Two types of sediments correspond with Quaternary environments: gelifracts are characteristic of periglacial environments. They form talus fans at the bottom of shafts and outwash deposits in galleries (Delannoy, 1998). Glaciokarstic varves are silty, laminated sediments composed largely of calcite flakes. This material results mainly from glacial abrasion on rocky surfaces, and from mechanical erosion by torrential flow in the vadose zone. During the summer melt, suddenly large amounts of water were supplied to the karst system and could not be discharged when the outlets were blocked by moraines or glaciers (Maire, 1990). Flooding then occurred, sometimes over depths of several hundreds of metres. The important suspended load of glacial flour settled out in the calm, quiet lacustrine environment, thereby plugging the karstic voids. Water flow ceased in winter, and the systems drained slowly until the next seasonal cycle that produced a new lamina of sediment.

Up to now, theories of karst evolution have strongly favoured glacial processes. However, field evidence shows that glacial processes play a very limited role in the evolution of alpine caves, mostly causing plugging in epiphreatic conditions (Bini *et al.*, 1998). Nevertheless, on the surface, glacial tongues have carved out the surface landscape, producing typical glaciokarstic forms.

PHILIPPE AUDRA

Works Cited

Audra, Ph. 1994. *Karsts alpins. Genèse de grands réseaux souterrains. Exemples: le Tennengebirge (Autriche), l'Ile de Crémieu, la Chartreuse et le Vercors (France)*, Thesis, Paris: Fédération française de spéléologie and Bordeaux: Association française de karstologie (Karstologia Mémoires, 5)

Audra, Ph. 1995a. Looking at alpine karst speleogenesis, with examples in France (Vercors, Chartreuse, Ile de Crémieu) and in Austria (Tennengebirge). *Cave and Karst Science*, 3: 75–80

Audra, Ph. 1995b. Signification des remplissages des karsts de montagne. *Karstologia*, 25(2): 13–20

Audra, Ph. 2001. French Alps karsts: study methods and recent advances. In *Cave Genesis in the Alpine Belt*, edited by Ph. Hauselmann & M. Monbaron, Freiburg: Institut de Géographie Université de Fribourg (UKPIK Rapports de recherches, vol. 10)

Audra, Ph. & Rochette, P. 1993. Premières traces de glaciations du Pléistocène inférieur dans le massif des Alpes. Datation par paléomagnétisme de remplissages à la grotte Vallier (Vercors, Isère, France). *Compte-rendu à l'Académie des Sciences*, 2(11): 1403–09

Bini, A., Tognini, P. & Zuccoli, L. 1998. Rapport entre karst et glaciers durant les glaciations dans les vallées préalpines du Sud des Alpes. *Karstologia*, 32: 7–26

Debelmas, J. 1983. *Alpes du Dauphiné*, Paris: Masson

Delannoy, J.-J. 1998. Apport de l'endokarst dans la reconstitution morphogénique d'un karst. Exemple de l'antre de Vénus (Vercors, France). *Karstologia*, 31: 27–41

Maire, R. 1990. La haute montagne calcaire. Thesis, *Karstologia Mémoires*, 3

Further Reading

Audra, Ph., Delannoy, J.-J. & Hobléa, F. 1993. Signification paléogéographique des réseaux perchés des Préalpes françaises du nord: exemples en Chartreuse et Vercors. *Études de géographie physique*, 22: 3–17

Delannoy, J.-J. 1991. Vercors, histoire du relief. Carte géomorphologique commentée. Lans-en-Vercors: Parc Naturel Régional du Vercors

Delannoy, J.-J. 2000. Les paysages karstiques du Vercors. *Spéléo*, 21: 26–30

VÉZÈRE ARCHAEOLOGICAL CAVES, FRANCE

The Vézère river in the Périgord (Dordogne), southwest France, flows through limestone country which is riddled with caves and rock shelters, hundreds of which were occupied intensively during the last Ice Age by Neanderthals and modern humans, and about 25 of which were decorated. The Vézère caves became a World Heritage Site in 1979.

Scientific investigation of the Vézère sites began in the 1860s, when numerous shelters were excavated, and evidence was found for Ice Age occupation, the existence of Ice Age portable art, and the co-existence of people with extinct animals such as mammoths. For example, in 1864 an engraving of a mammoth on a piece of mammoth ivory was unearthed by Edouard Lartet in the rock shelter of La Madeleine. Some of the shelters in the valley became world-famous: La Madeleine gave its name to the Magdalenian period, the final cultural phase of the French Ice Age; Le Moustier gave its name to the Mousterian, the period of the Neanderthals; while the shelter of Cro-Magnon, below which some Ice Age burials were discovered in 1868, gave its name to anatomically modern humans.

Of the decorated caves, by far the most famous is that of Lascaux, which was discovered by four boys in 1940, and which contains the most spectacular collection of Ice Age wall art yet found (see colour plate section). It is best known for its 600 magnificent paintings of aurochs (wild cattle), horses, deer, and signs, but it also contains almost 1500 engravings dominated by horses. The decoration is highly complex, with numerous superimpositions, and clearly comprises a number of different episodes. The best-known feature is the Hall of the Bulls, containing several great aurochs figures, some of them 5 m in length, the biggest figures known in Ice Age art; the hall also contains an enigmatic figure, baptised the unicorn. One remarkable painted figure at the end of the Axial Gallery, dubbed the falling horse, is painted around a rock in such a way that the artist could never see the whole figure at once, yet when the figure is flattened out with photographs it proves to be in perfect proportion.

A shaft features a painted scene of what seems to be a bird-headed man with a wounded bison and a rhinoceros, which has often been interpreted in shamanistic terms, though with little justification. The narrow Cabinet of the Felines forms the farthest extremity of the cave, and is filled with engravings, including a remarkable horse seen from the front, as well as the eponymous felines. It was in a shaft in this narrow corridor that a piece of Ice Age twisted rope was found.

Stone tools for engraving were found in the cave's engraved zones. Many lamps were also recovered (one of them a particularly well-carved specimen in red sandstone), as well as 158 fragments of pigment, and colour-grinding equipment, including crude mortars and pestles, stained with pigment, and naturally hollowed stones still containing small amounts of powdered pigment. There are scratches and traces of use-wear on many of the mineral lumps. It was found that there were sources of ochre (red) and of manganese dioxide (black) within 500 m and 5 km of the cave respectively. The shades vary considerably—the colour of ochre is modifed by heat, and Ice Age people clearly knew this. At Lascaux they also mixed different powdered minerals and were apparently experimenting with different combinations and heating procedures.

Scaffolding was clearly used in some galleries to reach the upper walls and ceiling—one gallery preserves the actual sockets for beams that must have supported such a scaffold. Much of the cave floor was lost when the site was adapted for tourism in the 1940s, but the site was probably never a habitation, being visited briefly for artistic activity or ritual. Charcoal fragments from the cave floor have provided radiocarbon dates around 15 000 BC and in the 7th millennium BC. The cave was closed to tourists in 1963 owing to pollution—a "green sickness" consisting of a proliferation of algae, and a "white sickness" of crystal growth; it was possible to reverse the effects of the green sickness and arrest the development of the white; but to ensure the survival of the cave's art, it has been necessary to restrict the number of visitors drastically. As compensation, a facsimile, Lascaux II, is now open nearby.

Among the other caves, three played an important role in the discovery and acceptance of the whole phenomenon of Ice Age cave art. It was at La Mouthe, in 1895, that the owner decided to remove some fill and exposed an unknown gallery. Engravings and paintings were discovered on its walls, and since there were Ice Age deposits in the blocking fill, it was clear that the images must also be from that period. In 1899 an Ice Age lamp was unearthed in the cave, carved in red sandstone and with an ibex engraved on it. This was the first evidence of a lighting system for cave art.

In 1901, engravings were found in the caves of Les Combarelles, near Les Eyzies; the principal cave, a long narrow corridor, contains hundreds of images of horses, bison, deer, mammoths, and humanoids, as well as rarer species such as bear, rhino, and big cat. They are attributed to the mid-Magdalenian period, *c.*14 000 BC. A few days later, the rich art was found by local schoolteacher Denis Peyrony in the nearby cave of Font de Gaume—this cave contains 230 figures, including no less than 82 bison among which are many polychrome specimens. There are also famous images of horse, deer (including a rare scene, a male reindeer licking the forehead of a female reindeer), mammoths, and rhino, all attributable to the mid-Magdalenian period.

The sheer density and richness of Ice Age occupation in the Vézère valley led to Les Eyzies being dubbed the "Capital of Prehistory". Analysis of faunal material from many sites has indicated that in many phases this region was primarily occupied during the winter months.

PAUL G. BAHN

Further Reading

1984. *L'Art des cavernes: Atlas des grottes ornées paléolithiques françaises* [Cave art: Atlas of the French palaeolithic decorated caves], Paris: Ministère de la Culture

Bahn, P.G. 1994. Lascaux: Composition or accumulation? *Zephyrus*, 47: 3–13

Bahn, P.G. & Vertut, J. 1997. *Journey Through the Ice Age*, London: Weidenfeld and Nicolson and Berkeley: University of California Press

Barrière, C. 1997. L'art pariétal des grottes des Combarelles, *Paléo*, hors série (May)

Capitan, L., Breuil, H. & Peyrony, D. 1910. *La caverne de Font-de-Gaume aux Eyzies (Dordogne)*, Monaco: A. Chêne

Capitan, L., Breuil, H. & Peyrony, D. 1924. *Les Combarelles aux Eyzies (Dordogne)*, Paris: Masson

Laming-Emperaire, A. 1959. *Lascaux: Paintings and Engravings*, Harmondsworth, Middlesex and Baltimore, Maryland: Penguin

Leroi-Gourhan, A. & Allain, J. (editors) 1979. *Lascaux Inconnu*, Paris: Centre National de la Recherche Scientifique

Ruspoli, M. 1987. *The Cave of Lascaux: The Final Photographic Record*, London: Thames and Hudson and New York: Abrams

VILLA LUZ, CUEVA DE, MEXICO

An extraordinary cave with a remarkable ecosystem lies nestled in the coastal foothills of southern Mexico. Cueva de Villa Luz is about 3 km south of the village of Tapijulapa in the Municipio de Tacotalpa, Tabasco, Mexico, and local guides to the cave are readily available in the town's central plaza. The indigenous people have recognized Cueva de Villa Luz (also known as Cueva de las Sardinas) as unique since before historical memory. They noted a startling abundance of small, cave-adapted fish unaffected by surface droughts. When food became scarce, the cave continued to provide fish. The Zoque respected the sacredness of the cave and only harvested fish once or twice a year after solemn prayers and offerings to their gods. Today, *La Pesca de la Sardina* is re-enacted every Palm Sunday weekend and the fish continue to thrive in a robust ecosystem, unparalleled by any other cave in the world.

Nearly every remarkable feature in Villa Luz results from the dozens of hypogenic (deep) water inlets within the cave. These small, subterranean springs bring warm (28°C) solutions from hypogenic sources into the 2 km long air-filled cave. Abundant hydrogen sulfide and carbon dioxide gases are released immediately into the cave atmosphere. Episodic, intense gas releases cause extraordinary changes in local atmospheric compositions. Within only minutes, the hydrogen sulfide level can dramatically increase or the oxygen level rapidly decline to lethal amounts (Hose *et al.*, 2000). Some passages consistently contain dangerously high concentrations of carbon dioxide. The gases mix with water vapour in the cave air and moisture on the walls, forming sulfuric and carbonic acids. Some visitors have received chemical burns from drips or prolonged contact with the walls. The scientists studying this cave have developed unique techniques and equipment for ensuring their safety, including wearing acid gas masks, continuous portable electronic gas monitors, goggles to protect their eyes from dripping acid, and miniature scuba tanks for safe evacuation. While these conditions lead to a hazardous environment for speleologists, they also provide the necessities for other life beyond the reaches of light.

Ubiquitous microbes in the cave called chemoautotrophic bacteria are capable of using the energy of chemical reactions; they facilitate the oxidation of the hydrogen sulfide and use the resultant heat energy in the same manner as plants use light energy for photosynthesis. Although several reactions may take place, a typical one is the simple oxidation of hydrogen sulfide to produce sulfuric acid. This reaction releases 798 kJ mole^{-1} of energy, which the bacteria use with the readily available carbon dioxide and water to manufacture cells. Sulfur-oxidizing bacteria, abundant in the water and on the walls and ceiling of the cave, are the producers in the Villa Luz ecosystem.

The most notable bacteria are ephemeral, rubbery, white deposits that resemble stalactites, have the texture of mucous, and drip sulfuric acid. These "snottites" (see Figure) comprise colonies of diverse microbes and their drips have pH of 0–3 (Hose & Pisarowicz, 1999). They grow quickly, up to a centimetre in length a day, and cluster in areas of the cave with the highest atmospheric hydrogen sulfide concentrations. Researchers have identified *Thiobacillus* (a sulfur-oxidizing bacteria) as the genus of the dominant clones (Hose *et al.*, 2000).

Mites, worms, and other small organisms live within the snottites. Small, flying insects (most common is *Tendipes fulvipilus*), whose larvae graze on the bacteria, are so abundant in some areas that they fill the passages in buzzing clouds. Large populations of spiders, representing several species, capture the midges in their webs for their food. Many other insects inhabit the cave. At least three species of cave-adapted fish fill the top few centimetres of the streams. *Poecilia mexicana*, a mollie, is the dominant species. Most of the mollies lack any pigmentation. They are pink due to abnormally high levels of haemoglobin and mostly feed on bacteria along with lesser amounts of insects (Langecker *et al.*, 1996). A yet unidentified *Hemipterin* (bug) preys on the fish. At least four species of bats, including one vampire, also inhabit the cave. The varied and abundant organisms of the cave form a complex, but so far little investigated, ecosystem.

Cueva de Villa Luz has formed in a Cretaceous limestone sequence along the strike of the northwest limb of an anticline.

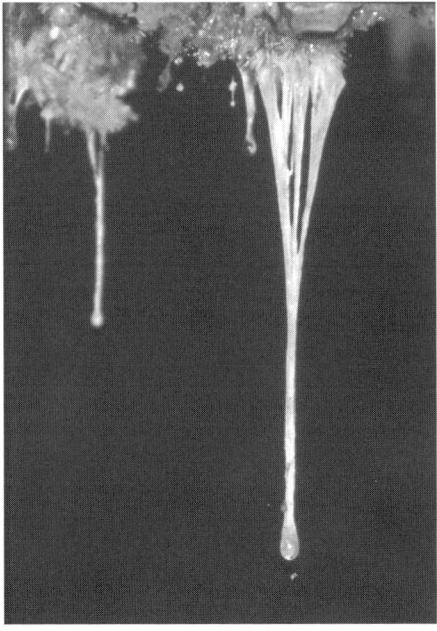

Cueva de Villa Luz, Mexico: Snottites in Snot Heaven. Sulfuric acid drops at their tips have pHs of 0–3. (Photograph by Louise Hose)

The floor of the cave seems to represent the local water table level. It appears to be a very young cave with extraordinarily fast cave-forming processes at work. Highly aggressive sulfuric acid along with the milder carbonic acid are rapidly destroying the limestone bedrock and enlarging the cave.

Calcite exposed to sulfuric acid undergoes a replacement reaction that results in gypsum and carbonic acid. Since the air in Villa Luz is permeated with sulfuric acid, gypsum coats nearly all of the walls and ceiling in a variety of forms, including delicate selenite crystals and microcrystalline masses that resemble toothpaste. Gypsum does not react with the atmospheric sulfuric acid but does create a buffer between the vulnerable limestone bedrock and the acidic atmosphere. The gypsum coatings typically harbour sulfuric acid and record pH values of 1–3. Many microbes live within the nearly ubiquitous gypsum paste, probably facilitating the production of the sulfuric acid.

The gypsum frequently drops from the ceiling and overhanging walls into the streams, which rapidly dissolve the highly soluble gypsum and remove it from the cave. Limestone exposed after the gypsum peels off the wall is again exposed to the sulfuric acid vapours, converts to gypsum, and the cycle continues as cave passages enlarge (Palmer & Palmer, 1998).

Sulfur folia and other elemental sulfur coatings drape walls near some of the most sulfur-rich springs. Deposits of sulfur folia have not been reported from any other cave in the world. The folia cover selenite (gypsum) crystals and appear to represent sub-aqueous deposits of sulfur now lining walls well above a flowing stream. Their origin is not clearly understood.

LOUISE D. HOSE

See also **Microbial Processes in Caves**

Works Cited

Hose, L.D. *et al.* 2000. Effects of geomicrobiological processes in a hydrogen sulfide-rich, karst environment. *Chemical Geology: Special Geomicrobiology Issue,* 169: 399–b423
Hose, L.D. & Pisarowicz, J.A. 1999. Cueva de Villa Luz, Tabasco, México: Reconnaissance study of an active sulfur spring cave. *Journal of Cave and Karst Studies,* 61(1): 13–21
Langecker, T.G., Wilkens, H. & Parzefall, J. 1996. Studies on the trophic structure of an energy-rich Mexican cave (Cueva de las Sardinas) containing sulphurous water. *Mémoires de Biospéléologie,* 23: 121–25
Palmer, A.N. & Palmer, M.V. 1998. Geochemistry of Cueva de Villa Luz: An active H_2S cave [abstract]. *Journal of Cave and Karst Studies,* 60(3): 88

Further Reading

Hose, L.D. 2001. Cave of the snotties. In *Caves: Exploring Hidden Realms,* M.R. Taylor, Washington, DC: National Geographic
Hose, L.D. 2001. La Pesca de la Sardina. *Association of Mexican Cave Studies Activities Newsletter,* 24: 50
Hose, L.D. 1999. Cave of the sulfur eaters. *Natural History Magazine,* (April 1999): 54–61
Pisarowicz, J.A. 2001. The acid test: Cueva de Villa Luz. *Association of Mexican Cave Studies Activities Newsletter,* 24: 48–54

VJETRENICA, BOSNIA-HERZEGOVINA: BIOSPELEOLOGY

Vjetrenica is a cave with approximately 8 km of active and relict passages. It opens some metres above Popovo polje, near Zavala in southeast Herzegovina (Bosnia–Herzegovina), and drains into the polje without swallowing its waters. It is located some distance from any major sources of industrial or urban pollution. In the summertime, a strong wind up to 3 m s^{-1} blows continually from the entrance, giving the cave its name (*vjetar* = wind). However, the wind ceases after any significant rain, when a lake approximately 1 km inside the cave rises and closes a siphon-like structure. The direction of the wind during the winter is not known with certainty. If conditions are windless this does not alter the other meteorological parameters (temperature and humidity) in the main cave passage, at least not for a few days.

In summer, the temperature in the main passage is between 11 and 11.5°C, and the relative humidity is close to 100% right up to the entrance when the wind is blowing out of the cave. The summer temperature of all streams is slightly above 11°C. Climatic conditions in Vjetrenica appear to be fairly but not entirely constant, since the temperatures of the bottom sediment and of the lakes in the main passage have been measured down to 10.2°C. However, there are no measurements available for the winter months.

The lower parts of Vjetrenica are hydrologically active, and the upper parts have a modest network of small streams and well-aerated pools. The calcium content in all waters is between 43 and 50 mg l^{-1}, with sodium between 1.7 and 2.3 mg l^{-1}, while the potassium content is highly variable. Even in the remotest parts of Vjetrenica, dead plant material may be found, particularly in the streams. This material is most probably brought in by very high-energy streams of percolating water, falling from numerous cracks in the ceiling, after rain. The substrate of one small stream in the deep interior appears to be somewhat polluted with organic material. Trogloxene invertebrates are comparatively scarce even close to the entrance, and they cannot contribute significantly to the food resources of the cave. No bats have been observed in Vjetrenica.

Vjetrenica has attracted the interest of biologists for some time. It was highly praised by the Czech, Karel Absolon, and was later visited by the first expedition of the revived Slovenian caving club in 1932. It was frequently visited by foreigners (e.g. Remy, 1940), and after World War II there were numerous expeditions by biospeleologists from Ljubljana.

Vjetrenica is one of the most faunistically rich caves in the world (Culver & Sket, 2000), due to its biogeographical position in the Dinaric karst, its size, and its ecological heterogeneity. Vjetrenica is inhabited by more than 30 troglobites and 40 stygobites. There are only a few troglophiles or trogloxenes, and there are virtually no unspecialized aquatic inhabitants.

The threshold zone of the main passage and its side branches contain some entrance fauna characteristic of the region. The parietal fauna (on the cave walls) consists of resting moths *Triphosa sabaudiata* and large numbers (up to 10 specimens per square metre) of the dipteran *Limonia nubeculosa*. Cave crickets *Troglophilus* spp. and *Dolichopoda araneiformis* and the large odoriferous centipede *Apfelbeckia* sp. are scarce. *Trogulus torosus*, a large-bodied harvestman species and a regional endemic, is rare. Young crickets and centipedes can be found up to 400 m from the entrance.

The most common troglobite of the main corridor and remote parts is the large and highly troglomorphic beetle *Antroherpon apfelbecki*; some other beetles (*Speonesiotes* spp., *Neotrechus* spp., *Aphaenopsis* spp.) are less common to extremely rare. The glomeridellid centipede *Typhloglomeris caeca* is limited to larger clay deposits. Single specimens of the harvestman *Travunia vjetrenicae* (Laniatores: Travuniidae) can be found in the active corridor known as the Absolonov canal.

Particularly interesting is the hygropetric-like habitat (a thin film of water flowing down the rock) on walls with extensive flowstone, inhabited by the specialized leptodirine beetle *Hadesia vasiceki* and also by the large amphipod *Typhlogammarus mrazeki*.

The animal communities in the cave waters are very diverse. In the past, shallow pools in the main corridor contained abundant populations of shrimps (*Troglocaris* spp.) and amphipods *Hadzia fragilis*. However, it appears that this fauna was destroyed indirectly by activities related to tourism.

In the lakes of Donja Vjetrenica (Lower Vjetrenica), the large, spiny, and extremely troglomorphic amphipod *Niphargus balcanicus* is particularly characteristic, and this is also the only known locality for the mysid *Troglomysis vjetrenicensis*. The rapidly flowing small stream in the Absolonov canal is particularly rich, with—among others—the predatory amphipod *Typhlogammarus*, rich colonies of tiny gastropods *Iglica absoloni*, and occasionally *Proteus anguinus*. The Veliko jezero (Great Lake) is particularly characterized by the specialized digger amphipod *Niphargus trullipes* and the less specialized but similarly large *N. vjetrenicensis*. Shrimps are also common. It should be mentioned that as many as three, and maybe even four, species of Atyidae shrimps are present in Vjetrenica (*Spelaeocaris pretneri*, *Troglocaris* cf. *anophthalmus*, *T. hercegovinensis*, and an undescribed *Troglocaris* sp.).

Tiny hydrobioid gastropods, *Lanzaia vjetrenicae*, and the serpulid worm, *Marifugia cavatica*, are characteristic of small streams in remote parts of Vjetrenica.

Biogeographical Relationships of the Fauna

A number of genera and species have a holo-Dinaric distribution: *Proteus*, *Marifugia*, the terrestrial gastropod *Zospeum amoenum*, the cockle *Congeria kusceri*, two species of *Troglocaris* shrimps, and others. *Titanethes hercegowinensis*, *Monolistra hercegoviniensis*, the shrimp *Spelaeocaris*, the leech *Dina absoloni*, and all of the beetle genera, have southeast mero-Dinaric distribution.

A number of *Niphargus* species, some isopods, the beetle *Hadesia*, and the centipede *Typhloglomeris* are narrow endemics even within the southeast Dinarides. For the time being, some species may be regarded as endemics of the Vjetrenica Cave. Two species are particularly enigmatic; the amphipod *Hadzia* and the mysid *Troglomysis* are indisputably species of coastal marine origin, but are incorporated here into a freshwater fauna not related to any recent or ancient seas.

Need to Protect the Faunas

Although far from obvious centres of pollution, the rich fauna of Vjetrenica is not immune from harm. Used batteries and similar waste left behind by tourists who have visited the cave in the past few years have already caused some damage. The use of any chemicals in the limited farming above the cave could cause more general problems, with the most potentially harmful being caused by the use of insecticides. The soil is very thin on the surface, and streams of percolating water are active in the cave just a few hours after rainfall. It is reasonable to assume that a large portion of the chemicals applied to the land could be washed into the caves via this route.

BORIS SKET

See also **Dinaric Karst: Biospeleology**

Works Cited

Culver, D.C. & Sket, B. 2000. Hotspots of subterranean biodiversity in caves and wells. *Journal of Cave and Karst Studies*, 62(1): 11–17

Remy, P. 1940. Sur le mode de vie des *Hadesia* dans la grotte Vjetrenica. *Revue Française d'Entomologie*, 7: 1–8

Further Reading

Sket, B. 2001. The hygropetric habitat in caves and its inhabitants. *Abstracts of the XVth International Symposium of Biospeleology*, Société Internationale de Biospéologie

VOLCANIC CAVES

The significance of volcanic caves is dependent on their size, type, contents, and location. Shield volcanoes generally owe their existence to the tendency of flowing lava to form tubes, in which it is insulated against convective and radiative heat loss. These tubes may conduct molten lava for tens of kilometres with a temperature loss of only 0.5 to $1.0°C$ km^{-1}. If drained, the largest of these tubes may persist as lava tube caves. An understanding of their origin, development, and features is important for comprehension of many volcanic locales. In addition, some lava tube caves are hydrologically important.

Hawaii's Kazumura Cave (US) is the longest on record (65.6 km). It is 32 km from end to end and has a vertical extent of 1100 m. Its floor plan is basically sinuous, with local braiding on one or more levels. It is notable for plunge pools up to about 20 m wide (Figure 1), more than 60 entrances and is never more than about 20 m below ground. Other types of volcanic caves contain cave art, habitations, and other cultural features, or are notable sites for recreation. Some are successful show caves and important nature reserves.

Volcanic caves (often called lava caves) also include hollow hornitos, hollow tumuli, lava rise and flow lobe caves, many moulds of trees, and eruptive fissures that intergrade into lava tube caves. Open vertical volcanic conduits are also included. Not included, however, are littoral and other crevice caves in volcanic rock (e.g. Fingal's Cave, Scotland), nor are piping caves in pyroclastics (see entries on Crevice Caves, Littoral Caves, and Piping Caves).

Lava Tube Caves and Lava Tubes

In the geological literature, the term "lava tube" has numerous contradictory meanings. For clarity it should be restricted to

Volcanic Caves: Figure 1. Eureka Falls in the upper part of Kazumura Cave, Hawaii, with the rims of a drained lava plunge pool at its foot. (Photo by Kevin Allred)

roofed conduits of flowing lava, either active, drained, or plugged. Such conduits are from a few centimetres to tens of metres in diameter, and from perhaps 0.1 m to many kilometres in length. Most lava tubes are too small or too full of lava to be considered caves. Advancing flow lobes, rifts carrying unconfined flow of lava, and lava trenches are not lava tubes. Shelly pahoehoe tubes are a special thin-skinned subtype found near rapidly degassing vents. Rarely, basalt dykes drain in the form of lava tubes (e.g. Cueva de la Fajanita, La Palma, Canary Islands).

Origins and Development of Lava Tube Caves

Late in the 20th century, it became apparent that most lava tube caves form by two basic processes: crusting over of surface rivers of lava, and / or drainage of still-molten lava from beneath solidified crusts. Features of large lava tube caves (e.g. Kazumura Cave, Hawaii) show that they characteristically formed and developed within lava flow fields, as a result of a complex sequence of events.

In vent regions, currents of preferential flow develop within crusted flows. In gullied terrain (e.g. Ape Cave, Washington State), preferential flow follows the course of pre-existing gullies. In more uniform terrain (especially on less than 4 degrees of slope), sinuous currents develop in the flow field where lava can advance with minimal energy loss. Lava temperature and plasticity are initially uniformly high. With cooling of the flow field, these currents become increasingly demarcated from the surrounding lava, and become walled-off tubes. Where lava forms pools, braiding and distributary tubes form. Small surface tubes develop independently, or at points where lava escapes to the surface. Meltdown of lava between subparallel tubes may form large tube segments. If the supply of lava increases while the crust is still plastic, the original tube may become distended in one of several patterns discussed below. Drainage by thermal erosion and other mechanisms, such as backcutting at lavafalls, produces "master tubes". Downcutting produces passages with volumes greater than that of the lava supply, creating free-surface lava rivers. It also deprives some braids of their lava supply and may drain them. This partial drainage also removes buoyant support and initiates breakdown. Except at low gradients, much of this material is carried down-tube. Blockages can cause breakouts, feeding surface flows containing additional small tubes (see below), much as in the inception phase.

On comparatively steep slopes (2–4 degrees), thermal erosion by comparatively shallow lava rivers commonly produces canyon passages with cutbanks and slipoff slopes. Some of these lava rivers eventually flow metres below their original bed, and create cavities much larger than their maximum flow. Where ceiling collapse introduces cold air, crusting of the surface of the lava river forms secondary ceilings. Below one degree of slope, tubes tend to be comparatively low and wide, with most or all of the tube filled with flowing lava rather than a free-surface river.

On gentle slopes, the molten interiors of advancing flow fronts are restrained by resistance of rapidly cooling peripheral lava. Similar to the situation at the vent, lava currents develop at points of low resistance. Repeated small breakouts occur and flow fronts advance by breakout and expansion of comparatively small flow lobes with thin "skins". Commonly these small lobes pile up rapidly in isolated segments of the flow front, then occur in another area, and then another. Under these conditions, the interior of each small lobe may remain hot for many days, and their "skins" characteristically break down, forming homogeneous flow fronts. Sometimes, however, breakouts from a single cooling lobe, or a complex of several partially homogenized lobes, may cause drainage beneath a stable crust, forming a flow lobe cave. Characteristically, these are elongate spaces 10–20 m wide and 1–2 m high, occurring singly, in sequences, or as three-dimensional clumps (nests). The longest recorded to date (Christmas Cave, Hawaii) has 632 m of passage and chambers on two levels.

Drainage, after pressurized injection of lava beneath a slightly plastic crust (subcrustal injection), may produce somewhat similar caves at the flow front and elsewhere in the flow field. These include two forms of hollow tumuli (domes or ridges projecting above the general surface of the flow), lava rise caves, and caves composed of a mixture of these forms. Sinuous hollow tumuli may be a few hundred metres long, with a chamber as much as 15 m wide. Cavernous lava rises are of two types. One is a low, rounded dome with a second or collapsed centre tens of metres in diameter, with cave remnants in much or all of its perimeter ridge; some also have short patent drainage and or feeder tubes in addition. A few open into other drained features nearby. The other type of lava rise cave consists of somewhat tubular spaces following the course of long, narrow lava rises with a flat top. Characteristically, these caves are originally wide and low, but become narrowed by the irregular sagging of their plastic roof. Locally this may produce multiple parallel passages 1–2 m high, or only boundary ridge passages may remain. Some of these caves are hundreds of metres long and have features consistent with minor conduit flow. These may be regarded as primitive lava tube caves.

Volcanologists studying rivers of flowing lava have observed speleogenesis by crusting of their molten surfaces and by wedging of floating plates of lava. At least one cave on a steep slope on Mauna Loa volcano is roofed by wedged lava plates, but roof patterns of most recorded lava tube caves are not consistent with this type of speleogenesis. Further study is needed.

Lava Tube Caves in Higher Viscosity Lavas

Lava tube and related caves form primarily in low-viscosity pahoehoe basalt (and possibly carbonatite). Some also have been reported in higher-viscosity lavas, including clinkery basalt (aa lava), and a few in andesite (e.g. Nishiyuura Ana, Japan) and rhyolite. Most of Italy's Mt Etna is covered by aa lava, and many of its numerous lava tube caves are beneath or within aa lava. Although these have been extensively studied, details of the origin and development of lava tube caves in high viscosity lavas are not well understood.

Lava Tube Caves as Groundwater Conduits

Some lava tube caves can function as perennial, seasonal, or floodwater conduits for groundwater, with consequent public health implications. They function as leaky pipes, and underlying volcanic ash or dense unfractured basalt layers may contribute to such transport. Except at water tables, very large reservoir spaces with large adsorptive surfaces commonly underlie lava tubes elsewhere, and may minimize public health problems.

Underwater Lava Tube Caves

Extensive submarine lava flows have been deposited through submarine lava tubes. In addition, subaerial lava tube caves can be submerged by rising sea level. The best-known penetrable example is probably Tunel de la Atlantida (the submarine portion of the Cueva de Los Verdes System, Lanzarote, Canary Islands).

Ages of Lava Tube Caves

Most lava tube caves are much younger than their calcareous analogs, but a few have ages measured in millions of years. One near the bottom of a gorge on the island of Maui (Hawaii) is in a flow mapped as Tertiary, and another in mid-Pacific is considered to be 8 million years old. In central Europe, short fragments of lava tube caves in Paleogene volcanics still have lava stalactites *in situ*.

Open Vertical Volcanic Conduits and Related Features

Open vertical volcanic conduits occur in several forms (Skinner, 1993). Especially important are bottle-shaped pits and "inverted goblet" pits. At 204 m, Iceland's Thrinukagigur is the deepest subaerial example (Figure 2). Hawaii's Na One crater contains a similar pit which begins on a ledge near the bottom of the crater; the combined depth is 268 m. Divers have reached a depth of 123 m in the inner conduit of the underwater portion of Hawaii's Kauhako Crater; the bottom was not visible at that depth. In Japan, deep, extensive crevice caves in thick scoria have accreted linings. On Mt Etna, several eruptive fissures with accreted linings are roofed and continue downslope as ordinary lava tube caves.

Lava Mould Caves (Tree Mould Caves)

Occasionally, logjams entrapped in flowing lava have formed complexes of tree moulds, with some individual trees more than 2 m in diameter and 20 m long. An especially complex area is in Japan's Yoshida-tanai area. A comparatively small group exists in the Mt St Helen's National Volcanic Monument (US). Here a few upright tree moulds have charred wood in their stumps, suitable for ^{14}C dating. More rarely, petrified wood is present in older caves of this type. Similar moulds of animals are known.

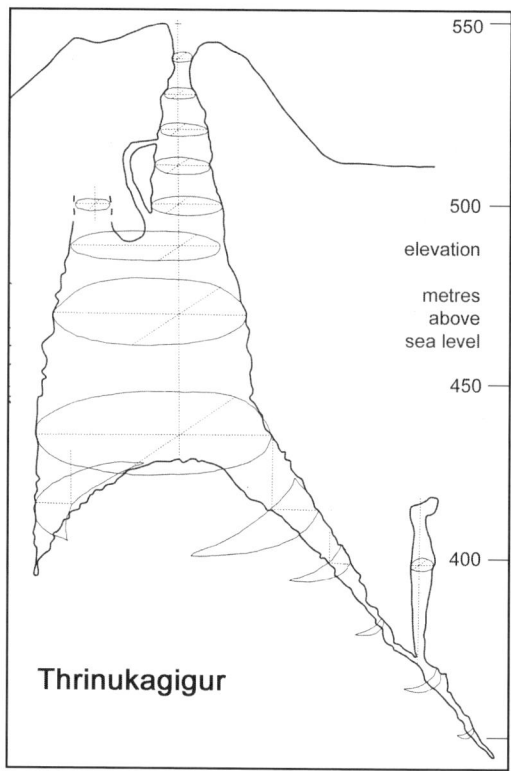

Volcanic Caves: Figure 2. Vertical section of Thrinukagigur, Iceland, the world's deepest open volcanic conduit (survey by Arni Stefansson).

A cave consisting of a mould of a Tertiary rhinoceros can be entered near Blue Lake in Washington state (United States). Moulds of elephants are present on the flank of Nyiragongo volcano in central Africa. Interiors of a few buildings, overrun by high-viscosity lava on Mt Etna, have been found preserved in much the same way.

Specific Features of Volcanic Caves

In addition to nonspecific features, such as breakdown, inwash, and tectonic crevices, four principal types of features exist in volcanic caves: speleothems, speleogens, petromorphs, and rheogenic features. Larson (1993) has brought considerable order into their terminology, condensing more than 1000 published English-language names into 174 terms, most of which are now widely used. Of these, approximately 20 describe common features.

The range of minerals deposited in volcanic caves is considerably greater than in karstic caves (see separate entry, Cave Minerals). Small calcite stalactites are not uncommon in lava tube caves, and large calcite speleothems develop where limestone or calcareous sand dunes overlie or are adjacent to well-watered lava tube caves. Elsewhere, silica dripstone and flowstone speleothems are more common. In a few locations, they form large stalactitic masses. Gypsum and other sulfate speleothems are also very common. They may appear while lava tubes and crevices are still very hot and their composition evolves with cooling and changes in humidity. Tiny siliceous microgours are seen on some vertical walls.

The original definition of the term "speleothem" did not consider the possibility of lava dripstone. In practice, the term commonly has been extended to include speleothem-like forms composed of lava. Stalactitic types are especially notable for variety of form and content. Some of the most attractive consist of a shiny form of pitchstone. In contrast, some of dacite are broad, thin, granular, and dull black.

With a maximum length of 2–3 m, tubular lava stalactites are the commonest form in Hawaii, Iceland and some other areas, and the most studied. Their basic form and size is similar to "soda straw" calcite stalactites. They are extruded by differential "filter-pressed" segregation of crystals within cooling lava (Allred & Allred, 1998). Larger tubular stalactites with a patent canal are much less common. While still plastic, some lava tubular stalactites are deflected by wind currents. Others ("vermiforms") twist for other reasons, and some have helictitic or coralloidal extensions. Still others flatten into "pipestem stalactites". Tapered and "shark's tooth" lava stalactites are also common. Tapered lava stalactites form from slumping of glaze, or as a result of simple dripping or streaming of lava. "Dip-layered" stalactites have multiple concentric layers caused by rise and fall of molten lava flowing beneath them, and are much less common. Where new lava pours into overhanging entrances of older caves, stalactitic columns may exceed 5 m in width and height and 1 m in thickness. Some have an attractive feathery surface.

Except at the base of lavafalls or other overhangs, where they may have the form of tall, symmetrical towers, lava stalagmites are characteristically tall thin accumulations of lava droplets a few centimetres to 2 m high (Figure 3). They tend to occur in rows, beneath drip lines. The size and shape of individual droplets vary from small globules to elongated teardrops more than 1 cm long, but tend to be consistent at each location. Where dripping lava is especially fluid, "worm nests" form instead of stalagmites. Rarely the droplets are poorly cohesive and form piles of partially agglutinated nodules instead of stalagmites. Lava spindles (stretched lava projections) up to about 2 cm develop where viscous lava surfaces pull apart. In areas of especially turbulent air currents (e.g. Kenya's Mt Suswa caves), thin lava strands have formed complexes resembling matted yarn. Complexes of spatter are common on walls, ceilings, and floors.

Many speleogens of lava tube caves resemble those of karstic caves. Solutional speleogens may occur in caves in lavas with unusually high sulfate and carbonate content (Larson, 1993). Others are the result of thermal erosion or plastic deformation. Probably the most impressive are plunge pools. These are as much as 20 m wide and several metres deep. Characteristically, large examples have collapsed crusts where underlying lava has drained. Usually they are found at the head of large chambers with vertical headwalls and tapered down-slope extensions. Lava karren, stalactitic pendants, cutbanks, and slipoff slopes are evidence of thermal erosion. Grooves and scratches cut by solid fragments in moving lava are common.

Rheogenic Features

Many types of rheogenic features result from lava flow in lava tube caves. Photographs in Larson's (1993) glossary depict these and other significant features. Primary ceilings commonly contain important clues to speleogenesis. These include wedged plates of lava that originally floated downstream on a lava river (Figure 4), and longitudinal seams where levees arched completely across such a river. Ceiling channels and longitudinal grooves and ridges caused by movement of lava against the ceiling are much more common. Cupolas are the result of upward pressure while lava still was plastic. Some extend upward to the underside of a hornito or short overflow level. Accreted linings are from 0.1 to over 20 cm in thickness, occurring singly or in multiples, which represent pulses of lava. Some uniformly coat the entire diameter of the passage. Others are thick at floor level (where they may merge with the floor) and taper upward, or curl outward into the passage, forming scrolls several centimetres in diameter. In a few caves which originally were wide at floor level and narrow near their ceiling, a succession of thick accreted

Volcanic Caves: Figure 3. An unusually fine group of stalagmites of solidified dripped lava glass in Apua Cave on Hawaii. (Photo by Tony Waltham)

Volcanic Caves: Figure 4. A ball of congealed lava caught between levee ledges in Ape Cave, United States. (Photo by John Gunn)

linings has completely filled the upper part of the tube with what are seen as vertical lamina. Small fingers of accreted linings are often squeezed into cavities of extratubal material. Flow lines are markers of flowing lava at grade whereas strand lines are horizontal and are remnants of lava pools. Diagonal "drag lines" were cut and / or deposited while the level of a lava river was falling or rising.

Perhaps the simplest rheogenic features are patterns or textures along the centre or the entire width of the passage floor, with or without a longitudinal bulge. Of special beauty is a feathery dendritic pattern termed arborescent lava. "Tubes-in-tube" form where the lower part of a shallow intratubal flow has drained without its chilled upper surface sagging or collapsing. Where segmental collapse interrupts tubes-in-tube, residual bridges occur. Where tubes-in-tube are especially large, their ceilings may be confused with secondary ceilings. Where most of a tube-in-tube is lacking, levees and gutters may be prominent. Ledges and benches may result from incomplete development of large tubes-in-tube or secondary ceilings, from localized downcutting, or from simple accretion at the edges of a stable stream of flowing lava. Aprons are present where lava congealed while flowing downward into an underlying cavity. During turbulent flow, spatter, splashing, wind currents, and thermal erosion may combine to create fimbriated globular complexes on floors and on the edges of ledges, levees, and other overhangs.

WILLIAM R. HALLIDAY

See also **Hawaii Lava Tube Caves**

Works Cited

Allred, K. & Allred, C. 1998. Tubular lava stalactites and other related segregations. *Journal of Cave and Karst Studies*, 60(2): 131–40

Grimes, K.G. 1995. Lava caves and channels of Mount Eccles, Victoria. *Vulcon Proceedings 1995*, Melbourne: Victoria Speleological Association

Halliday, W.R. 1998. Sheet flow caves of Kilauea Caldera, Hawaii County, Hawaii. *International Journal of Speleology*, 24B (1/4): 108

Larson, C. 1993. An illustrated glossary of lava tube features. *Western Speleological Survey Bulletin*, 87

Skinner, C. 1993. Open vertical volcanic conduits: a preliminary investigation of an unusual volcanic cave form with examples from Newberry Volcano and the central High Cascades of Oregon. In *Proceedings of the 3rd International Symposium of Vulcanospeleology 1982*, edited by W. Halliday, Seattle, Washington: International Speleological Foundation

Further Reading

Barone, N., Mirabella, O. & Licitra, G. (editors) 1987. *Atti, IV Symposium Internazionale di Vulcanospeleologia, Catania (Italy)*, Catania: Centro Speleologico Etneo

Forti, P. (editor) 1998. Vulcanospeleology: Proceedings of 8th International Symposium on Vulcanospeleology, Nairobi, Kenya, *International Journal of Speleology*, 27B(1/4): 1–163

Halliday, W.R. (editor) 1976. *Proceedings of the (First) International Symposium on Vulcanospeleology and its Extraterrestrial Applications, White Salmon, Washington (US)*, Seattle: Western Speleological Survey

Halliday, W.R. (editor) 1993. *Proceedings of the Third International Symposium on Vulcanospelelogy, Bend, Oregon (US)*, Seattle: International Speleological Foundation

Journal of Geophysical Research, 103(B–11), 1998: special section on Long Lava Flows

Oromi Masoliver, P. (editor) 1996. *Proceedings, 7th International Symposium on Vulcanospeleology, Santa Cruz de La Palma, Canary Islands*, Sant Ciment de Llobregat (Barcelona): Forimpres, S.A.

Rea, G.T. (editor) 1992. *Proceedings, 6th International Symposium on Vulcanospeleology, Hilo, Hawaii (US)*, Huntsville, Alabama: National Speleological Society

Walker, G.P.L. 1991. Structure, and origin by inflation under surface crusts, of tumuli, "lava rises", "lava rise pits", and "lava-inflation clefts" in Hawaii. *Bulletin of Volcanology*, 53: 546–58

VULCANOSPELEOLOGY: HISTORY

Vulcanospeleology is the study of caves in volcanic rocks—primarily lava tube caves and open vertical volcanic conduits. By some criteria, it is the fastest growing subdivision of speleology (see Volcanic Caves).

The history of human exploration of volcanic caves probably began on the lower slopes of Italy's Mount Etna and Japan's Mount Fuji. Natural human curiosity has drawn numerous visitors to these spectacular volcanoes from nearby centres of population and culture, throughout recorded time. A long continuum of written observations exists at Mount Etna. In the 1st century BC, Latin poet Titus Lucretius Carus wrote fancifully about "siliceous caves full of air and wind" on its slopes. In 1591 and 1698, some of its caves were mentioned in influential books by Fileoteo and Kircher. At Mount Fuji, a specific investigation of a lava tube cave was documented in 1203 and another in 1678. No true continuum developed here, however, and Japanese vulcanospeleologists had to begin anew after World War II.

In general, the volcanic caves of oceanic islands were the next to be investigated. Vikings and other early European voyagers encountered Iceland, and then other volcanic cave areas in the Atlantic Ocean. The world-famous Surtshellir ("Cave of the demon Surt") is mentioned in Icelandic sagas from about 1000 years ago. By 1757 it was the site of the first published map of any lava tube cave (Figure 1). In 1991 a historic first successful descent of the world's deepest known open vertical volcanic vent—Thrinukagigur, 204 m deep—was accomplished by Arni Stefansson.

In the Canary Islands, major explorations of Cueva del Viento and Cueva de San Marcos were documented in 1774 and 1776. Cueva de Los Verdes became a centre of local culture, and in 1857, Georg Hartung described it in considerable detail. Unfortunately, one picturesque illustration (Figure 2) misled science about the basic nature of lava tube caves for more than a century. The illustrator incorrectly depicted its overall form as that of a long, straight railroad tunnel. This misconception persists today.

Here and elsewhere, explorations and investigations eventually became more systematic and scientific. In 1965, Joaquin Montoriol-Pous began a notable wave of Spanish-language vulcanospeleology. From the Canary Islands, this wave reached Iceland, the Galapagos Islands, and Rwanda. Included was the

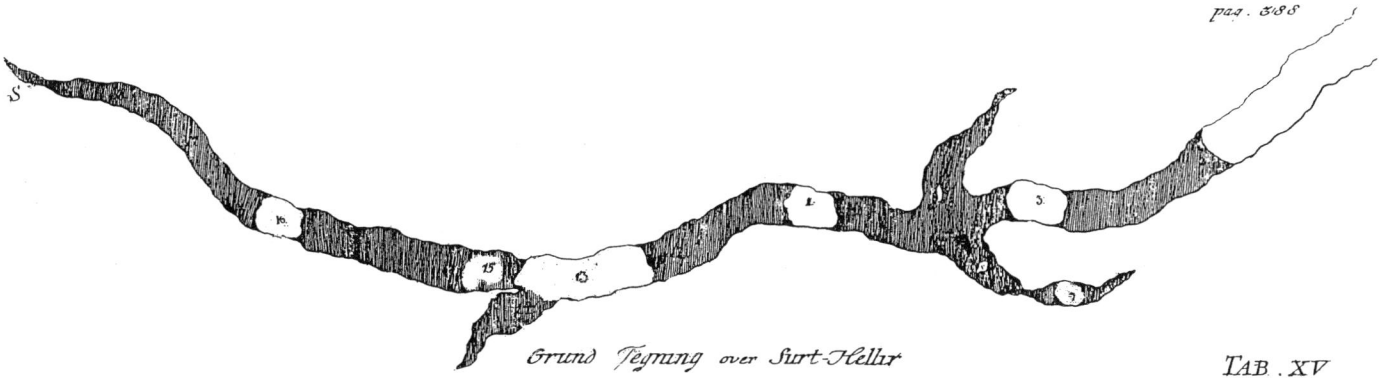

Vulcanospeleology, History: Figure 1. The first map of a lava tube cave. Plan of Iceland's Surtshellir system, published in 1757. Courtesy Jan Paul van der Pas.

prolonged exploration and documentation of Cueva del Viento, now recognized as the world's second-longest lava tube cave. Because of language barriers, however, this Spanish wave of vulcanospeleology was essentially isolated from rapid developments in vulcanospeleology elsewhere.

Shortly after the 15th-century settlement of the Azores, caves were discovered in many areas, including what became downtown Ponta Delgada and pasture land just outside Angra do Heroismo. Notable 19th-century accounts appeared in 1821 and 1873. In the last decades of the 20th century, Os Montanheiros led in systematic mapping and other documentation of caves throughout the archipelago.

In the Indian Ocean, English and French military expeditions and settlement resulted in considerable 18th- and 19th-century documentation of lava tube caves on the islands of Mauritius and Réunion. Theories about their origin were first published in 1804. From these pioneer works, a continuum of data entered the 20th-century French mainstream of speleology but not that of the English language. In the 1930s and 1940s, similar caves were reported on Grand Comoro Island and in northern Madagascar, but, in the chaos of decolonization, follow-up foundered until recent studies by Greg Middleton, there and in Mauritius.

In the Pacific Ocean, Montoriol-Pous and other Spanish vulcanospeleologists followed Charles Darwin to lava tube caves in the Galapagos Islands after a very long gap in observations. Perhaps surprisingly, repeated reports of lava tube caves on the isolated Easter Island began almost a century earlier. These were followed up by Thor Heyerdahl's archaeological investigations in the 1950s.

Hawaii, however, was the principal focus of world interest in the mid-Pacific during the 18th and 19th centuries. This strongly influenced the development of its vulcanospeleology. Certain Hawaiian lava tube caves had long been an integral part of native Hawaiian life. "Pre-contact" Hawaiian knowledge of these is better recorded in published recollections of newly Christianized 19th-century Hawaiians than in traditional chants. Around 1820, one of the first missionary houses in Hawaii was built at the entrance of Laniakea Cave, one of Hawaii's most celebrated lava caves. This and several other caves were described in some detail in 1823. Several broadly educated Christian missionaries soon observed lava tube caves in all stages of development. Some of them published accounts of what they saw, and began to record the processes observed in their formation. In 1849 James Dana was the first of many pioneer geologists to build on their work.

In 1912 Thomas Jaggar came to Hawaii and founded the Hawaiian Volcano Observatory on the rim of Kilauea Caldera. Jaggar knew several pioneer academic American speleologists, and his writings are full of references to caves. With publisher Lorrin Thurston and other Hawaiians he began the first real wave of Hawaiian speleology. In mid-century, Kenneth Emory's archaeological discoveries in Hawaiian caves expanded it significantly. In 1991, his pupil and successor Yosihiko Sinoto linked it to modern vulcanospeleology. The latter began in 1955, however, with an article reprinted three years later in the *Bulletin of the National Speleological Society*. Planetary geologists, and

Vulcanospeleology, History: Figure 2. 1857 artist's perception of Cueva de Los Verdes (Lanzarote Island, Spain). This engraving in a Swiss scientific periodical was the first widely disseminated illustration of a lava tube cave. Unfortunately it long misled the scholarly community into believing that such caves were little more than railroad tunnels. William R. Halliday Collection.

British, German, French, Japanese, and American speleologists, began increasingly intensive studies of Hawaiian caves. Following similar biological discoveries in Japan, in Hawaii important biospeleological discoveries began in the 1970s (see Hawaiian Islands: Biospeleology). Ultimately the Hawaii Chapter of the National Speleological Society and the Hawaii Speleological Survey emerged in leadership roles. Kevin and Carlene Allred directed the mapping of the 65.5 km Kazumura Cave, the world's longest lava-tube cave.

Continental lava tube caves generally have been the last to be explored and / or studied. Even fragmented tubular caves in Oligocene lava in southern Bulgaria have been recognized as lava tube caves only recently. In Africa, apparently only the tuff caves of Mount Elgon were known to 19th-century Europeans, and the history of vulcanospeleology in Africa largely is that of the Cave Exploration Group of East Africa in the latter half of the 20th century. Among projects led by Jim Simons, systematic exploration and mapping of Leviathan Cave in Kenya yielded a length of 10.5 km.

A somewhat similar pattern of discovery and exploration occurred in Australia. Some lava tube caves in the state of Victoria were known to Europeans by the mid-1880s, with some important reports as early as 1866. In eastern Australia, however, most volcanic caves were remote, and their exploration and study tends to be entirely 20th century.

Until late in the 19th century, the lava tube caves of the western United States were also remote and little known. After the American Civil War (1861–65), reports began to accumulate. Within a few years of the 1941 founding of the National Speleological Society (NSS), Erwin Bischoff brought some of them into the speleological mainstream. Vulcanospeleology then began to flower in the United States in the 1960s and 1970s, largely contemporaneously with its flowering in other parts of the world and initially independent of it. Publication of *Caves of California* and *Caves of Washington* (Halliday, 1962, 1963) systematized information on several especially important areas and introduced a body of terminology still in wide use. The Cascade and Oregon Chapters of the NSS became leaders in the systematic exploration and mapping of caves at Mount St Helens and elsewhere in the Pacific Northwest. The caves near Mount St Helens, and others near Bend, Oregon, soon were "discovered" by Ronald Greeley and other planetary geologists who applied their findings to extraterrestrial phenomena. Staff geologists of the US Geological Survey began to contribute influential papers. With NASA funding, Ronald Greeley created a world database on lava tube caves and their formation, located at Arizona State University.

In 1972 the NSS convened the first international symposium on lava tube caves, near Mount St Helens. Subsequently, others have been held in Italy (three in Catania), Japan, Kenya, the Canary islands, and two more in the United States (Bend, Oregon and Hilo, Hawaii). In 1993 the International Union of Speleology created a Commission on Volcanic Caves which facilitates communications between leaders of the field and organization of the international symposia.

WILLIAM R. HALLIDAY

See also **Hawaii Lava Tube Caves; Volcanic Caves**

Works Cited

Halliday, W. 1962. *Caves of California*, Seattle: Western Speleological Survey

Halliday, W. 1963. *Caves of Washington*, Washington: Washington State Department of Conservation (Washington Division of Mines and Geology Information Circular No. 40)

Further Reading

The published *Proceedings of the International Symposia on Vulcanospeleology* contain much historical information. To date, *Proceedings* volumes have been published for the First, Third, Fourth, Sixth, Seventh, and Eighth symposia.

Banti, M. (editor) 1993. *Proceedings of the International Symposium on the Proto-history of Speleology*, Citta di Castello (Italy): Nuova Prhomos

Borges, P.A.V. & Silva, A. (editors) 1994. *Actas do III Congreso Nacional de Espeleologia et do I Encontro Internacional de Vulcanoespeleologia das Ilhas Atlanticas, 30 September–4 October 1992*, Acores, Portugal: Angra do Heroismo

Hernandez Pacheco, J.J., Izquierdo Zamora, I., Martin Esqivel, J.L., Medina Hernandez, A.L. & Oromi Masoliver, P. 1995. *La Cueva del Viento*, Tenerife: Consejería de Política Territorial del Gobierno de Canarias

Montoriol-Pous, J. & Montserrat, A. 1984. Vulcanoespeleologia Espanola: 1910–1977. *Sota Terra*, 5: 2–14

The *Journal of Spelean History*, published by the American Spelean History Association (now the history section of the National Speleological Society, Huntsville, Alabama) also contains many articles on the history of volcanic caves.

WALSINGHAM CAVES, BERMUDA: BIOSPELEOLOGY

Bermuda and its extensive anchialine (coastal marine) caves are of exceptional biological and biogeographical significance due to their isolated mid-ocean location, unique geological history, and remarkably rich and diverse stygobitic fauna (Sket & Iliffe, 1980; Iliffe, Hart & Manning, 1983; Iliffe, 1994). Indeed, Bermuda's caves qualify as a biodiversity hot-spot of global importance. At least 78 endemic, cave-dwelling species, mostly crustaceans, have been identified from Bermuda caves, including two new orders, one new family, and 15 new genera. In order of abundance, Bermuda's anchialine fauna includes: 19 species of ostracods, 18 species of copepods, eight species of amphipods, six species of shrimp, four species of isopods, and four species of mites. Many of these species are found only in a single cave or cave system. Due to their limited distribution, the fragile nature of the marine cave habitat, and severe water pollution and / or development threats, 25 of these species have been listed as critically endangered (Baillie & Groombridge, 1996).

Bermuda is a volcanic seamount that formed in the Mid-Atlantic about 50–60 million years ago. Plate tectonics and sea-floor spreading have maintained Bermuda's location relative to North America (about 1000 km off the Carolinas), while increasing its distance from Europe and Africa as the Atlantic Ocean enlarged. Thus, Bermuda has never been part of, or closer to, a continental landmass.

Coral-reef derived limestone, first deposited as coastal sand dunes, caps most of present-day Bermuda. Approximately one million years ago, limestone caves began forming during glacial periods when sea level was as much as 100 m lower (Palmer, Palmer & Queen, 1977). Later, as postglacial sea levels rose, encroaching sea water drowned large portions of the caves. Continuing collapse of overlying rock into the large solutionally formed voids created the irregular chambers and fissure entrances that are commonly seen in Bermuda's caves. Extensive networks of submerged passageways, developed primarily at depths between 17 and 20 m below present sea level, interconnect otherwise isolated cave pools. These passages, only accessible to divers, are well covered at all depths with impressive stalactites and stalagmites, confirming that the caves must have been dry and air-filled for much of their history.

The sea-level, brackish pools located in the interior and / or entrances of Bermuda's caves are classified as anchialine habitats. The term anchialine was coined by Holthuis (1973) to describe "pools with no surface connection to the sea, containing salt or brackish water, which fluctuates with the tides." Bermuda's cave pools have a thin, brackish layer at the surface, overlying fully marine waters at depth (Sket & Iliffe, 1980; Iliffe, Hart & Manning, 1983). In the caves bordering Harrington Sound, subterranean waters tidally exchange with the sea through coastal springs. Caves farther inland typically contain slowly moving or near-stagnant waters. The input of food in most caves is primarily derived from the sea itself, although chemosynthethic bacteria may provide an additional source of organic matter to the anchialine cave ecosystem (Pohlman, Cifuentes & Iliffe, 2000).

Even on a small island like Bermuda, caves are not evenly distributed. Most of Bermuda's 150 known caves are located in the Walsingham district, a kilometre-wide isthmus separating Castle Harbour and Harrington Sound (see map). This part of Bermuda consists of hilly, wooded terrain underlain by highly karstified limestone containing numerous caves and dolines. While most of these caves initially appear to be relatively small and end in tidal saltwater pools, diving explorations have shown them to be highly integrated. The caves consists of large, sub-sealevel chambers, floored with breakdown and profusely decorated with speleothems. The Walsingham Caves consists of two larger and mostly underwater cave systems, Walsingham (1300 m long with seven known entrances) and Palm (500 m long with five entrances), plus many smaller caves.

Bermuda's anchialine caves, especially the Walsingham Caves, are inhabited by an unexpectedly high diversity of unique and previously unknown marine invertebrates. Included among this fauna are extremely ancient relict organisms that can be

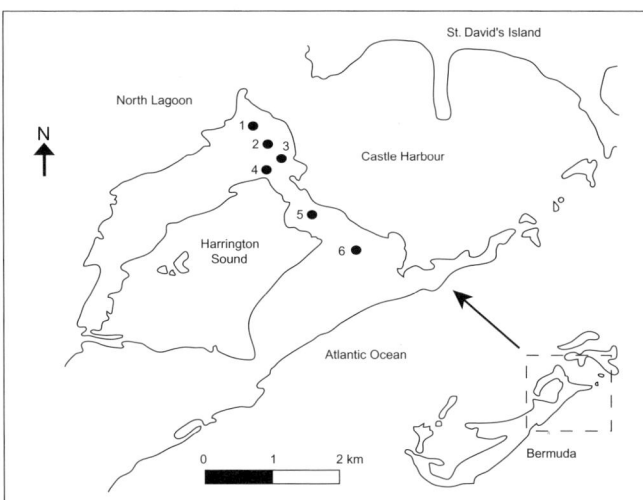

Walsingham Caves, Bermuda: Location of the Walsingham caves district, Bermuda.

legitimately referred to as "living fossils". As examples, the copepod *Erebonectes* is one of the most primitive of known calanoids, while the *Antrisocopia* agrees in many ways with the description of a theoretical ancestral copepod. Some of Bermuda's cave-dwelling species have close affinities with Old World cave and groundwater fauna, and probably colonized the subterranean habitats on Bermuda early in the island's history when the Atlantic was much narrower. The amphipod genus *Pseudoniphargus*, which has two species in Bermuda caves, was previously known only from caves and groundwater around the Mediterranean, the Azores, and the Canary Islands. Other animals inhabiting Bermuda caves have close relatives occurring in caves on other isolated oceanic islands from both the Atlantic and Pacific. The misophrioid copepod *Speleophriopsis* includes cave species from Palau, Bermuda, and the Balearic Islands. Finally, some animals are closely related to deep-sea species. The order Mictacea, for example, includes cave species from Bermuda and the Bahamas, plus deep-sea species from the Atlantic and Indo-Pacific. Thus, Bermuda's cave species are providing important clues about the evolution and dispersal of present oceanic species.

Belying the considerable age of both caves and the obligate cave-dwelling organisms that inhabit them, this environment is one of the rarest and most delicate on Earth. The potential impact of human activity on caves is profound—over time, even just visiting of caves can result in irreparable damage. The four primary threats to Bermuda caves are: (1) construction and quarrying activities, (2) water pollution, (3) dumping and littering, and (4) vandalism (Iliffe, 1979). Quarrying has destroyed numerous significant caves, particularly at Government Quarry in the Walsingham district. Construction of luxury town homes directly on top of Church and Bitumen Caves may destroy their endangered anchialine fauna. The Karst Waters Institute twice named these two caves to their list of the Top Ten Most Endangered Karst Ecosystems on Earth. The anchialine pool of Bassett's Cave, once said to be the longest and geologically most instructive cave in Bermuda (Nelson, 1840), was used by the United States Navy as a cesspit for disposal of raw sewage and waste fuel oil. Many of Bermuda's caves have been used as dumping sites. The bulldozing of large piles of partially burned rubbish into the anchialine pool of Government Quarry Cave resulted in depletion of dissolved oxygen and anaerobic production of hydrogen sulfide. Groundwater circulation transmitted this pollution to at least five other caves as much as half a kilometre or more away (Iliffe, Jickells & Brewer, 1984). In these polluted caves, all stygobitic species have disappeared. Since many of Bermuda's cave species are endemic and are often restricted to only one cave or cave system, pollution or destruction of these habitats can result in the extinction of entire species. Finally, few of Bermuda's larger caves have escaped the effects of vandals maliciously breaking and removing fragile stalactites and stalagmites or defacing cave walls with their names.

THOMAS M. ILIFFE

See also **Anchialine Habitats; Marine Cave Habitats**

Works Cited

Baillie, J. & Groombridge, B. (editors) 1996. *1996 IUCN Red List of Threatened Animals*, Gland, Switzerland: IUCN and Washington, DC: Conservation International

Holthuis, L.B. 1973. Caridean shrimp found in land locked saltwater pools at four Indo-West Pacific localities (Sinai Peninsula, Funafuti Atoll, Maui and Hawaii Islands), with the description of one new genus and four new species. *Zoologische Verhandelingen*, 128: 1–48

Iliffe, T.M. 1979. Bermuda's caves: a non-renewable resource. *Environmental Conservation*, 6: 181–86

Iliffe, T.M. 1994. Bermuda. In *Encyclopaedia Biospeologica*, vol. 1, edited by V. Decu & C. Juberthie, Moulis and Bucharest: Société de Biospéléologie

Iliffe, T.M., Hart, C.W. Jr & Manning, R.B. 1983. Biogeography and the caves of Bermuda. *Nature*, 302:141–42

Iliffe, T.M., Jickells, T.D. & Brewer, M.S. 1984. Organic pollution of an inland marine cave from Bermuda. *Marine Environmental Research*, 12:173–89

Nelson, R.J. 1840. On the geology of the Bermudas. *Geological Society of London Transactions, Series 2*, 5:103–23

Palmer, A.N., Palmer, M.V. & Queen, J.M. 1977. Geology and origin of the caves of Bermuda. In *Proceedings of the 7th International Congress of Speleology*, edited by T. Ford, Sheffield

Pohlman, J.W., Cifuentes, L. & Iliffe, T.M. 2000. Food web dynamics and biochemistry of anchialine caves: a stable isotope approach. In *Subterranean Ecosystems*, edited by H. Wilkens, D.C. Culver & W.F. Humphreys, Amsterdam and New York: Elsevier

Sket, B. & Iliffe, T.M. 1980. Cave fauna of Bermuda. *Internationale Revue der Gesamten Hydrobiologie*, 65: 871–82

Further Reading

Iliffe, T.M. 1992. Anchialine cave biology. In *The Natural History of Biospeleology*, edited by A.I. Camacho, Madrid: Museo Nacional de Ciencias Naturales
Review article on anchialine biodiversity, biogeography and evolution

Iliffe, T.M. 2000. Anchialine cave ecology. In *Subterranean Ecosystems*, edited by H. Wilkens, D.C. Culver & W.F. Humphreys, Amsterdam and New York: Elsevier
This review article contains water quality and tide data from Bermuda caves, as well as information on anchialine habitat, ecology, biodiversity, and evolution.

Sterrer, W. 1992. *Bermuda's Marine Life*, Bermuda: Bermuda Natural History Museum and Bermuda Zoological Society
A well-illustrated introductory guide to 225 marine species, including bacteria, plants, invertebrates, fishes, reptiles, birds, and mammals.

Useful Websites
Anchialine Caves and Cave Biology, http://www.cavebiology.com. This website contains detailed information on the anchialine caves and cave fauna of Bermuda, the Bahamas, and the Yucatán Peninsula.

WATER TRACING

Water tracing can be defined as the use of natural or induced properties to label a body of water, allowing detection of that water at some point downstream, thereby gaining understanding of the character of the flow path followed by the water. In karst environments, water tracing finds particular value in defining the path followed by inaccessible underground streams. The primary application for this is in catchment definition and related water and land use management, but it finds applications amongst hydrogeologists for clarifying conduit network structure, and is increasingly used in litigation concerning karst groundwater rights or contamination. The following review is intended to present the primary elements of karst water tracing, but particular attention should be paid to the legal and ethical issues associated with tracing public and private water supplies.

Tracers
Natural tracing exploits existing properties of water, whereas artificial tracing implies deliberate alteration of the water. Tracers can be classified as physical, chemical, isotopic, and biological, each with its corresponding analytical method. Physical traces may be natural or artificial physical impulses (e.g. flood or dam release pulses; Ashton, 1966; Williams, 1977) or changes in temperature or turbidity. In general, flood impulses travel faster than the water originally comprising the flood, especially in closed conduits. Microscopic inert particles (e.g. *Lycopodium* spores) are a specialized form of physical tracer, sometimes used to simulate transmission of bacteria. Chemical tracers can be simple ionic soluble materials like salt, or specialized fluorescent dyes like uranine (fluorescein sodium) or Rhodamine WT (e.g. Smart & Laidlaw, 1977). Stable and radioactive isotopes are highly detectable and may vary naturally, or in response to a deliberate injection. The isotopes of water (^{2}H, ^{3}H, ^{18}O) are particularly valuable, as they can be considered to mimic water most effectively. Biological tracers are often present inadvertently, as in the detection of coliform bacteria in wells. However, artificial bacteriophages have been used (Kennedy, 2000), and DNA-based tracing is now possible (Sabir *et al.*, 2000).

Natural tracing includes any natural chemical species, such as dissolved calcium, but more particularly refers to the stable isotopes of the oxygen and hydrogen in water, sulfur, carbon and chlorine, and the radioactive isotopes of hydrogen and carbon. These environmental isotopes are widely used in groundwater studies, but their success depends on the existence and consistency of marked contrasts in source composition. In contrast, artificial tracing allows greater control on the magnitude of the tracer signal and the specificity of the site to be traced, and is widely used in karst hydrogeology.

Apart from the issue of mimicking the movement of water as closely as possible, tracer selection depends on practical issues of cost, danger, sensitivity of analysis, characteristics of background (naturally occurring) concentrations, ease of handling and risk of contamination, and ability to discriminate a range of tracers at one time. Fluorescent dyes are the predominant tracer used at present. There are a number of "safe" dyes (e.g. uranine, eosine) which can be detected at the parts per trillion level, although they suffer to some extent from retardation, photo-decay and variable background (e.g. Käss, 1998). Analytical sensitivity should not be the primary criterion in tracer selection; it is important to consider all aspects of tracer performance.

Applications and Relevance
Tracing in karst is most widely used in "point-to-point" mode, to define the trajectory taken by underground water. Typically, this implies the identification of the destination spring of a sinking stream. Successful establishment of a rapid link between a sinking stream and a spring means that there is a conduit link between the point of injection and the point of recovery. A series of point-to-point tracer tests can be used to establish a regional network of underground flow routes analagous to the network of a surface river system. Most karst systems are dendritic systems with a number of tributaries feeding one trunk conduit. However, parallel conduits and distributary systems, where a single conduit feeds a number of springs, are also found, e.g. the Mendips in the United Kingdom (Atkinson, 1977) and Central Kentucky karst (Quinlan & Ewers, 1989). Replication of tracer tests under different flow conditions may show different flow routing, often arising from adoption of underground overflow routes by the floodwater. Monitoring the time required for a tracer to reappear generates travel time and, with an estimate of (linear) distance, groundwater velocity. Conduit velocities in karst conduits are usually between 100 m d^{-1} and 10 km d^{-1} with averages of 1700 m d^{-1} (Worthington, Davies & Ford, 2000). There have been many successful traces over distances as great as several tens of kilometres.

Tracing provides a major tool in the hydrogeological characterization of karst aquifers for water resource studies. Defining the regional network allows the catchment areas associated with springs or wells to be determined, permitting appropriate land use management to be developed. A major feature of karst land use schemes is pre-emptive definition of contaminant trajectories; the velocity of karst groundwater in conduits is so rapid that it does not permit reaction in the event of contamination. The way in which a tracer is modified, in passing through the aquifer from the initial spike injected, provides an analogy to the dilution and dispersion of potential contaminants. In addition, tracing also allows systematic tracking of the source of contaminants appearing in karst springs or wells. Effective tracing provides unequivocal evidence of groundwater trajectories. Flow rates revealed by tracing may be orders of magnitude faster, and contaminant concentrations much higher, than predicted by conventional groundwater models. In mantled karst, tracing may be quite challenging, and recourse to borehole injection methods may be necessary.

Approaches
A variety of styles of tracing are available. The simplest methods are inexpensive, but may be aesthetically unacceptable, and yield less compelling information than more sophisticated methods. Qualitative tracing involves nothing more than the identification of the tracer in the water. Visual detection generally requires high concentrations of tracer and is considered unacceptable for ecological, aesthetic, and legal reasons. Normally, a water sample should be analysed to confirm the presence of tracer. Qualitative tracing provides point-to-point and associated routing and catchment information. Semi-quantitative tracing involves defining the concentration of the tracer in the water over time. The resulting time-concentration breakthrough curve provides unequivocal evidence of the tracer, and can be used in determining characteristics of the traced route. Quantitative tracing combines concentration measurements with flow determinations, permitting the compilation of mass-breakthrough curves. The area under the mass-breakthrough curve indicates the total mass of tracer collected at a site. Comparison of this value to the mass of tracer injected allows definition of tracer recovery. This may be used to identify the presence of other flow routes, but can also arise from more insidious loss of the tracer by decay or adsorption on surfaces.

Injection
Successful injection of tracer requires total and immediate dissolution or dispersion in the receiving water. Clean injection is easier with liquids than powders, and may require use of an injection hose and flushing with a large volume of water to deliver the tracer effectively. In addition to streams, tracer may be injected into boreholes, closed depressions, fissures, or spread on the ground surface, with increasing likelihood of failure or massive loss of tracer as the tracer is increasingly removed from major conduits in the karst. In general, the mass injected should allow a coherent signal above background fluctuations, but should not compromise safety or aesthetics. The quantity of tracer for a conduit trace can be based on calculations (Field, 2003; Worthington & Smart, 2003). Traces using wells or on the ground surface will typically require much larger quantities.

Sampling and Sample Analysis
Personal detection of a tracer by eye, smell or taste is simplest, but provides the least coherent information. Convenient integrative sampling of fluorescent dyes is possible with granular activated charcoal packets (fluocapteurs) deployed in a range of sites and replaced at intervals of hours to weeks. The dyes are eluted from the charcoal with an alkaline-alcohol mixture, but the resulting mix of compounds may prevent coherent interpretation. Discrete water samples can be collected manually or by automatic water samplers. While laborious, this allows construction of time-concentration curves at the resolution of the sampling interval. Continuous sampling and analysis provides the most detailed and immediate tracer data, but increases the field logistic costs. Examples include *in situ* detection of common salt tracer with a conductivity meter and detection of fluorescent dyes using a field fluorometer. The sampling method is selected based upon the tracer used, the application and the resources available.

The mode of analysis depends on the same considerations. Thus fluorescent dyes can be detected visually, with a colorimeter, a filter fluorometer or a spectrofluorometer, with increasing discrimination and sensitivity. Good protocol demands the development of analytical standards, replicates, and a range of controls in the laboratory and in the field.

Interpretation
Point-to-point tracing provides simple positive and negative tracer results that can be interpreted to generate the "hydrogeography" of an aquifer showing linkages, networks and variable routing. In practice, this is often complicated by equivocal or inconsistent results and dependence on flow conditions in the aquifer. Features such as crossovers, convergence, divergence, and conditional routing may be identified. Catchments may be defined by aggregating and interpolating the traced routes to a particular destination. Again, results may be disjointed, fuzzy, or variable, compared to the discrete catchments associated with surface rivers.

Tracer travel times can be converted to "linear" velocities using the distance between the injection and sampling points. Velocity is an indicator of relative conduit openness. The relationship of velocity to discharge can demonstrate whether the conduit is predominantly closed (water filled), giving a velocity-discharge power function with an exponent of 1; an open channel will yield a power function with an exponent <1. This is because changes in discharge may only be accomplished by varying the velocity in a closed conduit, whereas an open channel may also vary in cross-sectional area. If the power function has an exponent >1, either the data are corrupt, or there has been a routing switch involving multiple conduits.

Time concentration and mass curves allow more sophisticated analysis of the hydraulics and behaviour of the tracer. Tracer concentration typically declines more gradually than it rises, because of storage in backwater areas (dead zones) or adsorption of the tracer. Compound peaks can arise from multiple routes or unsteady flow effects (temporary dilution or redirection of the tracer). A well-defined breakthrough curve can be used to define time to peak, mean travel time, and a dispersion coefficient for the conduit. If 100% recovery of tracer is assumed, then the ratio of the mass injected to the area under the time concentration curve provides the discharge from the system. A rapidly developing field is the computer analysis of tracer breakthrough curves to allow determination of parameters (average properties) of the traced system; for example, the conduit roughness, dispersion coefficient, and dead zone volume (e.g. Field & Pinsky, 2000).

Practice and Ethics of (Dye) Tracing
Effective groundwater tracing requires maximizing the likelihood of a correct positive while respecting the impact that the tracing may have on people and the environment. Loss of a tracer leaves uncertainty as to the explanation. In some cases tracer is held in storage until a subsequent flood, in other cases failure to monitor the correct site or delayed or infrequent sampling may be the cause. The resulting ambiguity means that the particular tracer may not be used again with confidence in that area. False positive results are similarly confounding, although the error may not be apparent. The main cause of false positives is misinterpreted background or contamination. Background arises from substances that can be mistaken for the tracer; an example is the fluorescence arising from organic matter that can be mistaken as a tracer dye. Contamination arises from the presence of the tracer from sources other than the trace. This

can arise from environmental presence or release of the tracer, for example the green fluorescent dye uranine is widely used in antifreeze fluids, and is ubiquitous in industrialized regions. Autocontamination occurs when the operator transfers tracer to the samples, a particular problem with activated charcoal and powdered dyes. A false positive can also arise from uncoordinated tracing in the same region.

Maximizing the likelihood of a correct positive trace requires knowledge of the site, good protocol, and experience. Once regional geology, hydrology, and local knowledge have been compiled, it is best to undertake a simple, even self-evident trace as a proof of practice. More sophisticated tracing can thus be built on local experience.

There is much sensitivity to the deliberate contamination of water supplies, not only for aesthetics and health, but because labelled water may not be usable. For example, colorimetric determination of residual chlorine in drinking water is difficult in heavily dyed water. Future tolerance and support of tracing rests on local experience. In much of the world, tracing is regulated, not only to prevent unnecessary or ill-conceived tracing with inappropriate materials, but also to ensure that the results of bona fide tracer tests are duly reported and archived.

Where tracing is undertaken as an aspect of litigation, then particularly stringent protocols must be adopted. The appropriate procedures depend on the regional and legal context and are best left for professional practitioners. One particularly difficult issue is demonstrating that the tracer detected is indeed the material injected. Most tracers are not unique, and background and contamination can occur in any trace. A time-concentration curve provides less equivocal evidence than a positive charcoal detector, but a fully quantitative mass recovery is more compelling.

Conclusion

Tracing remains the primary tool of the karst hydrogeologist, and it provides essential information on groundwater flow. However, it is not the only tool available, and it is much more effective when employed along with appropriate hydrochemical, geological and hydrological techniques.

CHRIS SMART AND STEPHEN R.H. WORTHINGTON

Works Cited

Ashton, K. 1966. The analyses of flow data from karst drainage systems. *Transactions of the Cave Research Group of Great Britain*, 7(2): 161–203

Atkinson, T.C. 1977. Diffuse flow and conduit flow in limestone terrain in the Mendip Hills, Somerset (Great Britain). *Journal of Hydrology*, 35: 93–110

Field, M.S. & Pinsky, P.F. 2000. A two-region nonequilibrium model for solute transport in solution conduits in karst aquifers. *Journal of Contaminant Hydrology*, 44: 329–51

Field, M.S. 2003. A review of some tracer design equations for tracer-mass estimation and sample-collection frequency. *Environmental Geology*, 43: 867–881

Worthington, S.R.H., and Smart, C.C., 2003, Empirical determination of tracer mass for sink to springs tests in karst. In: *Proceedings of the ninth multidisciplinary conference on sinkholes and the engineering and environmental impacts of karst* (Ed. B. Beck). American Society of Civil Engineers, Reston, Virginia., pages 287–295

Käss, W. 1998. *Tracer Techniques in Geohydrology*, Rotterdam: Balkema (original German edition published 1992)

Kennedy, K.G. 2000. Bacteriophages as particle migration indicators in subsurface environments. In *Tracers and Modelling in Hydrogeology*, edited by A. Dassargues, Wallingford, Oxfordshire: IAHS

Quinlan, J.F. & Ewers, R.O. 1989. Subsurface drainage in the Mammoth Cave area. In *Karst Hydrology: Concepts from the Mammoth Cave Area*, edited by W.B. White & E.L. White, New York: Van Nostrand Reinhold

Sabir, I., Haldorsen, S., Torgersen, J., Aleström, P., Gaut, S., Pedersen, T.S., Colleuille, H. & Kitterød, N.-O. 2000. Synthetic DNA tracers: Examples of their application in water related studies. In *Tracers and Modelling in Hydrogeology*, edited by A. Dassargues, Wallingford, Oxfordshire: IAHS

Smart, P.L. & Laidlaw, I.M.S. 1977. An evaluation of some fluorescent dyes for water tracing. *Water Resources Research*, 13: 15–33

Williams, P.W. 1977. Hydrology of the Waikoropupu Springs: A major tidal karst resurgence in northwest Nelson, (New Zealand). *Journal of Hydrology*, 35: 73–92

Worthington, S.R.H., Davies, G.J. & Ford, D.C. 2000. Matrix, fracture and channel components of storage and flow in a Paleozoic limestone aquifer. In *Groundwater Flow and Contaminant Transport in Carbonate Aquifers*, edited by C. Wicks & I. Sasowsky, Rotterdam: Balkema

WATER TRACING: HISTORY

The use of tracers to determine karst groundwater flow reportedly began in 10 AD when Philippus used chaff to trace the spring that is the source of the Jordon River (Whiston, 1957). Mayer used chaff and sawdust as tracers in southwest Germany during the 17th century, and Hagler used salt (sodium chloride) to trace a sinking stream to the water supply spring for Lausen, Switzerland in an experiment to find the source of a typhoid epidemic in 1872 (Käss, 1998). Three tons of sodium chloride were used in 1899 to trace the sinking at Malham Tarn in Yorkshire, England (Ford & Williams, 1989).

Fluorescein dye, discovered in 1871 and combined with salt to increase its fluorescence and solubility, was first used by Knop to trace the stream sinks of the upper Danube River to Aach Spring in Germany in 1877 (Knop, 1878). Visual observation of the green dye, sodium fluorescein (uranine, CI Acid Yellow 73), at springs was used as the primary karst tracing technique until the late 1950s. There were, however, problems associated with the visual detection method, including: (1) large quantities are required for detection and low dye concentrations may be missed; (2) springs and downstream water become visually impaired; (3) a large amount of labour is necessary for monitoring; and (4) dye may resurge at a spring or springs not monitored.

A major breakthrough solving most of the above problems occurred when Dunn (1957) demonstrated that sodium fluorescein could be adsorbed by granular activated coconut charcoal and then eluted from the charcoal by a solution of potassium hydroxide in ethanol. With this technique, granular charcoal packets in a wire or fibreglass screen mesh are placed in all springs where injected dye could possibly resurge. These charcoal packets (called dye receptors or fluocapteurs) are exchanged at regular intervals and then, after being eluted, dye extracted from the charcoal can be observed with a bright light. Visual detection of the dye in the charcoal elutant is possible even though it is not visually detectable in the resurgent water since dye adsorbs onto the charcoal and accumulates, resulting in higher concentrations as the dye cloud flows past.

In the early 1970s, Smart and Brown (1973) demonstrated that rhodamine WT (CI Acid Red 388), a red dye, could also be adsorbed onto charcoal granules and eluted with a solution of ammonium hydroxide, 1-propanol, and distilled water. The technique of detecting sodium fluorescein and rhodamine WT on charcoal dye receptors resulted in numerous dye traces performed by researchers in Europe and North America. However, trace results remained unreliable when insufficient dye was injected. Background fluctuation was also a problem, and interpretation under a bright light was very subjective.

The next dye tracer technique breakthrough was the use of a fluorometer for analysis, an instrument that quantitatively measures dye concentrations in water samples and charcoal elutant (Käss, 1967; Wilson, 1968). When fluorescent materials are irradiated, they emit light, and the emitted or fluoresced light always has a longer wavelength than that which is absorbed during irradiation. The fact that each fluorescent dye exhibits its own combination of excitation and emission spectra when analysed by a fluorometer, greatly increased the detectability and objectivity in the detection of fluorescent dyes.

Glover (1972) introduced the use of commercial optical brighteners to trace groundwater. Small bundles of surgical quality cotton, not treated with any brightening agent, are used as receptors and changed at regular intervals. Adsorbed dye is visually detected by its characteristic light blue fluorescence when observed under an ultraviolet light. Eight fluorescent dyes that could be detected using charcoal or cotton dye receptors were evaluated by Smart and Laidlaw (1977), and by the mid-1970s numerous dye traces were being performed, primarily using three dyes, often injected simultaneously at three different locations: sodium fluorescein, rhodamine WT, and optical brightener.

At the same time that these powerful dye tracer techniques were being developed in the mid 1970s, Lambert (1976) and Quinlan and Ray (1981) demonstrated that the water table could be contoured in karst aquifers by measuring the water level elevation in uncased water wells in the Bowling Green and Mammoth Cave, Kentucky, United States areas. A combination of dye tracing and potentiometric surface investigations permitted the delineation of karst groundwater basins and the approximation of the actual flow routes of subsurface streams through karst aquifers.

Technological advances in the late 1980s brought karst researchers the scanning spectrofluorophotometer, an instrument that provides the lowest detection limits and the most reliable dye analysis. A synchronous scan is performed with the excitation and emission monochrometers kept at a fixed wavelength separation during the scan. The emission spectra from the synchronous scan is then displayed on a computer monitor. Scanning spectrofluorophotometers allow four to six dyes to be injected simultaneously and then detected in water and charcoal elutant samples. The dyes usually used, because of less overlap between spectra, are: (1) tinopal CBS-X (CI Fluorescent Brightener 351); (2) sodium fluorescein (CI Acid Yellow 73); (3) eosine (CI Acid Red 87); and (4) sulphorhodamine B (CI Acid Red 52). Use of the spectrofluorophotometer permits dye detection for several dyes at concentrations of about 1 part per trillion, and if charcoal dye receptors are used to accumulate and concentrate the dye, it can be detected in the elutant even though it never reached a concentration of 1 part per trillion in the water that flowed past the receptor.

Dye tracing has evolved from a simple visual research tool for determining the resurgence for a sinking stream to a highly technical one used to provide definitive results concerning groundwater flow in karst aquifers for regulatory purposes. Dye traces and potentiometric surface mapping combined with geologic information about stratigraphy and structure usually permit the construction of a site conceptual hydrogeological model. This type of investigation is often required by groundwater regulatory agencies in the United States as part of a groundwater monitoring and remediation plan for karst aquifers. It is therefore imperative that dye traces be performed following strict adherence to procedures and protocols (Crawford & Roach, 2001). The author estimates that 6000 to 8000 dye traces have been performed to trace karst groundwater flow in the United States since 1970. Most of them have been performed by private consulting firms to determine the flow of contaminated groundwater or to establish groundwater flow for monitoring purposes.

Other Tracers

Spores of the Canadian club moss, *Lycopodium clavatum* L., were used for groundwater tracing by Mayr (1953) and further developed by Maurin and Zötl (1959) and Dechant (1959). Lycopodium spores, with an average diameter of only 33 μm, are usually dyed, often with fluorescent dyes, previous to injection. With spores dyed different colours, four to six simultaneous traces can be performed. Nylon plankton nets with a sieve opening of 25 μm are used to collect the spores at emergent springs, and analysis is performed by counting the spores under a microscope. Numerous spore drift traces were performed in the alpine karst and German Uplands of Europe. However, problems associated with the time and expense of dyeing the spores, potential spore contamination influencing the results, siltation of the plankton nets, and time and potential operator error in identifying and counting the spores under a microscope, have resulted in spore drift tracing being largely replaced by the superior fluorometric techniques (Smart *et al.*, 1986).

Fluorescent microspheres are occasionally being used as tracers, but they tend to have the same problems as the spore drift technique. However, since they are similar in size and have similar transport behaviour as bacteria, they are sometimes used for hygienic evaluations. Bacteria and other micro-organisms have also been used as well as bacteriophages, but the considerable amount of work involved in their preparation, handling, and examination have limited their use.

Salts, such as sodium chloride and lithium and potassium chlorides have been used as tracers as well as non-fluorescent dyes, but the large quantities required have limited their use (Brown *et al.*, 1972). Radioactive isotopes have been used, and for some species the public health disadvantages may be overcome by post-sampling activation analysis, but very specialized facilities are required (Kruger, 1971). Radioactive isotope tracers tend to be costly and hazardous, and they need skilled handling by personnel supported by atomic laboratories.

An additional method of tracing karst waters that does not require the use of tracer substances is known as flood pulse analysis. The pulse may result from a natural event such as a storm or the melting of snow / glacier ice, or it may be created artificially by the breaching of a small temporary dam or by opening the sluice gates on a permanent structure (Ashton, 1966). Flood pulses were used successfully in 1879 as part of a series of notable early experiments to trace the source of the River Aire in Yorkshire, England (see summary in Smith & Atkinson, 1977). These traces are rarely used, but they do serve to provide point-to-point connections for karst aquifers where dye tracing is difficult, such as confined karst aquifers that discharge at artesian springs. For example, Williams (1977) showed that pressure pulses produced by release from a hydroelectric dam in the Takaka Valley, New Zealand, took only 9–11 hours to travel over 20 km through a marble aquifer to an artesian spring. Tritium measurements at the spring suggest a minimum flow through time of 2–4 years (Stewart & Down, 1982).

NICHOLAS C. CRAWFORD

Works Cited

Ashton, K. 1966. The analyses of flow data from karst drainage systems. *Transactions of the Cave Research Group of Great Britain*, 7(2): 161–203

Brown, R.H., Konoplyantsev, A.A., Ineson, J. & Kovalevsky, V.S. (editors) 1972. *Groundwater Studies: An International Guide for Research and Practice*, Paris: UNESCO

Dechant, M. 1959. Das Anfärben von Lycopodiumsporen. *Steirische Beiträge zür Hydrologeologie*, 11: 145–49

Dunn, J.R. 1957. Stream tracing: Mid-Appalachian region, *National Speleological Society Bulletin*, 2: 7

Ford, D.C. & Williams, P.W. 1989. *Karst Geomorphology and Hydrology*, London and Boston: Unwin Hyman

Glover, R.R. 1972. Optical brighteners—a new water tracing reagent. *Transactions of the Cave Research Group of Great Britain*, 14(2): 84–88

Käss, W. 1967. Erfahrungen mit Uranin bei Farbversuchen. *Steirische Beiträge zür Hydrologeologie*, 18/19: 123–340

Käss, W. 1998. *Tracing Technique in Geohydrology*, Rotterdam: Balkema (originally published in German, 1992)

Knop, A. 1878. Über die hydrographischen Beziehungen zwischen der Donau und der Aachquelle im badischen Oberlande. *Neues Jahrbuch für Mineralogie*, 350–63

Kruger, P. 1971. *Principles of Activation Analysis*, New York: Wiley

Lambert, T. Wm. 1976. *Water in a Limestone Terrane in the Bowling Green Area, Warren County, Kentucky*, Kentucky Geological Survey, Report of Investigations 17, Series 10, 43

Maurin, V. & Zötl, J. 1959. Die Untersuchung der Zusammenhänge unterirdischer Wässer mit besonderer Berücksichtigung der Karstvarhältnisse. *Steirische Beiträge zür Hydrologeologie*, 11: 5–184

Mayr, A. 1953. Blütenpollen und pflanzliche Sporen als Mittel zur Untersuchung von Quellen und Karstwässern. *Anzeiger der Österreichische Akademie der Wissenschaften, Mathom – naturwiss. Klass*, 90: 94–98

Quinlan, J.F. & Ray, J.A. 1981. *Groundwater Basins in the Mammoth Cave Region, Kentucky*, map, Mammoth Cave, Kentucky: Friends of the Karst

Smart, P.L. & Brown, M.C. 1973. The use of activated carbon for the detection of the tracer dye Rhodamine WT. *Proceedings of the 6th International Speleological Congress*, Praha: Academica, vol. 4: 285–92

Smart, P.L. & Laidlaw, I.M.S. 1977. An evaluation of some fluorescent dyes for water tracing. *Water Resources Research*, 13: 15–23

Smart, P.L., Atkinson, T.C., Laidlaw, I.M.S., Newson, M.D. & Trudgill, S.T. 1986. Comparison of the results of quantitative and non-quantitative tracer tests for determination of karst conduit networks: An example from the Traligill basin, Scotland. *Earth Surface Processes and Landforms*, 11: 249–61

Smith, D.I. & Atkinson, T.C. 1977. Underground flow in cavernous limestone with special reference to the Malham area. *Field Studies*, 4: 597–616

Stewart, M.K. & Downes, C.J. Isotopic hydrology of Waikoropupu Springs, New Zealand. In *Isotopic Studies of Hydrologic Processes*, edited by E.C. Perry & C.W. Montgomery, DeKalb, Illinois: Northern Illinois University Press

Whiston, W. 1957. *The Life and Works of Flavius Josephus*, Philadelphia: Winston

Williams, P.W. 1977. Hydrology of the Waikoropupu Springs: A major tidal karst resurgence in northwest Nelson (New Zealand). *Journal of Hydrology*, 35: 73–92

Wilson, J.F. 1968. Fluorometric procedures for dye tracing. In *Techniques of Water-Resources Investigations*, Book 3, Chapter A12, Washington: United States Geological Survey

Further Reading

Crawford, N.C. & Roach, S.D. 2001. *Karst Groundwater Investigation Research Procedures*, Bowling Green, Kentucky: Crawford Hydrology Laboratory, Center for Cave and Karst Studies, Department of Geography and Geology, Western Kentucky University, www.dyetracing.com

Käss, W. 1998. *Tracing Technique in Geohydrology*, Rotterdam: Balkema

WILDERNESS

There are two widely different definitions of wilderness—the legal and the social. At the legal extreme, wilderness can be narrowly defined as "an area of undeveloped land which retains its primeval character and influence, which appears to have been affected primarily by the forces of nature, and with the nearly undetectable imprint of man's work substantially unnoticeable" (Wilderness Society, 1984, p.27). The United States Wilderness Act of 1964, Section 2(c), recognizes a wilderness "as an area where the earth and its community of life are untrammeled by man, where man himself is a visitor who does not remain" (Lera, 2001). At the social extreme, it is whatever people perceive natural wilderness to be—potentially the entire universe.

Wilderness Management

In theory there are two alternatives to actually managing a wilderness. Firstly, all use could be prohibited. However, most wilderness philosophy believes that wild places should be used and enjoyed. Secondly, a *laissez-faire* attitude could be adapted, which is also in violation of the wilderness philosophy requiring protection and perpetuation (preservation), and would likely ensure that most wilderness would vanish. The focus clearly should be on preserving the natural integrity of the wilderness environment, while still providing for human use and enjoyment, without needless restrictions to protect the area.

The many benefits derived from wilderness depend on the preservation of its undisturbed natural integrity. "Wilderness", "National Park", and "World Heritage Site" may seem to be mutually contradicting terms. If wilderness areas were to be simply placed off-limits to human use, they would probably remain natural and unimpaired as they have for centuries. However, wilderness is a multiple-use human resource, addressing needs for: watershed protection, scientific research, habitat for threatened or endangered plant and animal species (as well as game and non-game fish and wildlife), and many types of non-motorized recreation. Without some sort of monitoring and control, many outstanding wilderness areas would begin to lose those very values for which they were designated. Wilderness management, therefore, is not the management of the physical and biological resources themselves, but of the human use of those resources. Some wilderness managers find themselves forced to manage resources, as invasive species have no respect for human-defined boundaries.

Major Issues Facing Cave Wilderness

Several issues must be resolved when caves are considered for wilderness designation. Each significant cave needs to be inventoried for its unique list of wilderness values, uses of the land above the cave and impacts of those uses, definition of appropriate wilderness boundaries, and an established methodology to manage the cave wilderness. Cave wilderness boundaries should be defined in a more flexible manner than surface wilderness boundaries, as caves are best measured linearly rather than areally. The surface area under which a cave lies may be known at the time of wilderness designation, but allowances must be made to include additional cave passages within the wilderness area as they are explored, even if they extend under land managed by different agencies.

Cave features are very different from surface features and require a different management approach. However, the public's use of a cave wilderness for recreation, research, education, or inspiration are analogous to similar uses of surface land wilderness areas. Determining the cave's carrying capacity or limits of acceptable change will be a key—and difficult—issue in developing cave wilderness management guidelines. The problem is exacerbated by the fact that whereas on the surface the environment is largely self-renewing—given adequate control over visitor numbers and behaviour—the same is not true of caves. In areas with increasing pressures on recreational resources, the successful management of cave wilderness areas is of particular importance. The welfare of such resources should not be neglected—irrespective of location or political motivation—because of the nature and number of impacts that may occur, many of them irreversible. Any wilderness disturbance must be carefully planned and controlled in order to optimize the resource.

Defining the Boundaries of Underground Wilderness

The US Federal Cave Resource Protection Act provides the opportunity to begin wilderness assessment for all significant caves on federal lands (Lera, 2000). In 2001, the National Park Service defined their authority over cave wilderness, as:

> ... all cave passages located totally within a surface wilderness boundary will be managed as wilderness. Caves that have entrances within wilderness, but contain passages that may extend outside the surface wilderness boundary, will be managed as wilderness. Caves that have multiple entrances located both within, and exterior to, the surface wilderness boundary, will be managed consistent with the surface boundary; those portions of a cave within a wilderness boundary will be managed as wilderness. (National Park Service, 2001)

Lechuguilla Cave, New Mexico, for example, is located within the boundaries of the Carlsbad Caverns Wilderness Area, and is managed as wilderness by the National Park Service.

Karst ecosystems may be disturbed either directly through physical intrusion into, or direct interference with, the cavern environment (e.g. effluent disposal), or indirectly through the alteration of the natural surface overlying the cave and of the entire water catchment area of the cave. The disturbance of underground wilderness areas in karst ecosystems is a complex process, which varies in significance according to the nature of the impact. Where disturbance is totally uncontrolled, it is the most detrimental and therefore least desirable. Landscape management of karst ecosystems will protect and preserve cave wilderness. There are several cave World Heritage Sites but not all would qualify as true wilderness. Perhaps the best example is the Nahanni Karst in Canada (see separate entry). The Tasmanian Wilderness World Heritage Area (see Australia entry) has a number of caves within its boundaries, but they are managed to the full spectrum of levels—from major tourist caves to totally unvisited and unmanaged caves. The Caves of the Aggtelek and Slovak Karst World Heritage site (see separate entry) is similar in that it contains major tourist caves and is also affected by tourism and quarrying.

As Frank Elger observed, "Ecosystems are not only complex, but more complex than we think." (Elger, 1977, p.3). With this in mind, we should be cautious about consciously manipulating the last of our true wilderness—the underground wilderness.

THOMAS LERA

See also **Conservation: Protected Areas; Ramsar Sites; World Heritage Sites**

Works Cited

Elger, F.E. 1977. *The Nature of Vegetation, its Management and Mismanagement: An Introduction to Vegetation Science*, Norfolk, Connecticut: Aton Forest Publishers

Lera, T.M. 2000. *A Summary of Legislation and Organizations Involved in the Preservation of Caves and Bats*, available at http://www.caves.org/section/ccms/bat2k/index.htm

Lera, T.M. 2001. Can underground wilderness remain a reality? *NSS News*, 59(3): 64–66

National Park Service. 2001. Wilderness preservation and management: 6.3.11.2 Caves. In *2001 NPS Management Policies*, Washington DC: National Park Service, Department of the Interior; available at www.nps.gov/refdesk/mp/index.html

Wilderness Society. 1984. *The Wilderness Act Handbook*, Washington, DC: Wilderness Society

Further Reading

Agee, J.K. & Johnson, D.R. (editors) 1989. *Ecosystem Management for Parks and Wilderness*, Seattle: University of Washington Press

Green, M.J.B. & Paine, J. 1997. State of the world's protected areas at the end of the twentieth century. Paper presented at the *IUCN World Commission Symposium on Protected Areas in the 21st*

Century: From Islands to Networks, 24–29 November 1997, Albany, Australia (unpublished; available at http://wcpa.iucn.org/pubs/pdfs/AlbanyConfReport.pdf)

Hendee, J.C., Stankey, G.H. & Lucas, R.C. 1991. *Wilderness Management*, 2nd edition, Golden, Colorado: North American Press

Willers, W.B. (editor) 1999. *Unmanaged Landscapes: Voices for Untamed Nature*, Washington, DC: Island Press

World Conservation Monitoring Centre and the IUCN Commission on National Parks and Protected Areas. 1994. *1993 United Nations List of National Parks and Protected Areas*, Gland, Switzerland and Cambridge: IUCN

World Conservation Monitoring Centre and the IUCN Commission on National Parks and Protected Areas. 1998. *1997 United Nations List of Protected Areas*, Gland, Switzerland and Cambridge: IUCN

WIND AND JEWEL CAVES, UNITED STATES

Caves of the Black Hills, South Dakota, are complex, multi-tiered network mazes, with an unusually complex history, spanning 300 million years (Palmer & Palmer, 1989). The Black Hills uplift is an elongate structural dome with a core of Precambrian igneous and metamorphic rocks, with dipping sedimentary rocks exposed along its flanks. Dips are typically 3–5 degrees, but in local areas are much steeper. Caves and karst are located almost exclusively in the Mississippian (lower Carboniferous) Madison Limestone, which has an average thickness of 100–150 m. The two largest caves, Jewel and Wind Caves, are among the world's longest, with 205 and 173 km of surveyed passages, respectively (Figure 1). The cave areas are located in a dry climate, with rainfall averaging about 450–500 mm yr^{-1}, and surface karst is limited to a few scattered dolines and sinking streams.

The Madison originally contained substantial sulfate deposits (Sando, 1985). While the formation was still exposed at and near sea level, early diagenetic changes included dissolution, mobilization, reduction, and calcite replacement of the original sulfates (Palmer & Palmer, 1995). Early voids, up to two metres in diameter, formed as the result of sulfate reduction to hydrogen sulfide, which oxidized as it rose into higher strata to form sulfuric acid. Near the end of the Mississippian Period (c.300 million years ago) the Madison was exposed to surface karst development, which produced dolines and fissures, and some additional cave passages in the upper strata. Sands and clays of the Pennsylvanian (late Carboniferous) Minnelusa Formation filled all the dolines and most early cave passages. Deep burial, by as much as two kilometres of mainly detrital sediments, continued through the Cretaceous Period. Uplift of the Black Hills during the Laramide Orogeny (late Cretaceous—early Tertiary) exposed the Madison to erosion, and the onset of rapid groundwater flow during the early Tertiary formed the present caves. The present caves mainly follow earlier voids and diagenetically altered zones of Mississippian age. The conspicuous fracture pattern, extending radially outward from the centre of the Black Hills uplift, seems to indicate a post-Laramide age—but the diagenetic and paleokarst features follow those same trends, indicating that the Black Hills represented an active positive area, even during the Carboniferous. This resulted from reactivation of Precambrian structures during the early Carboniferous Antler Orogeny of western North America (Palmer & Palmer, 1989).

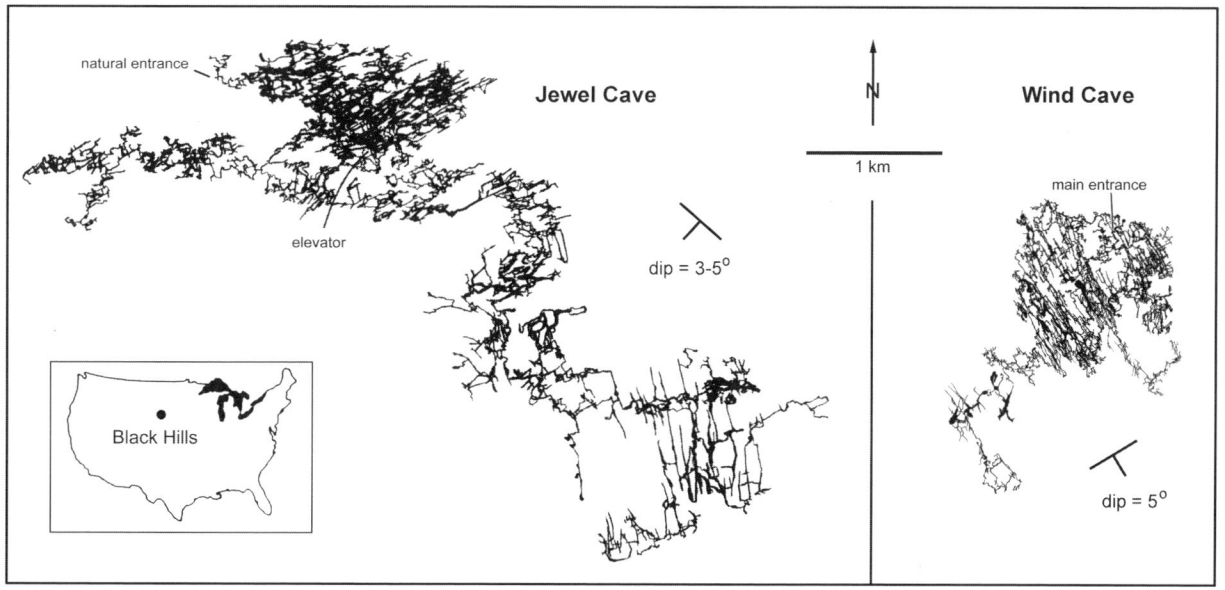

Wind and Jewel Caves, United States: Figure 1: Plan-view maps of Jewel Cave and Wind Cave, South Dakota. Jewel Cave is located in the southwestern flank of the Black Hills, and Wind Cave is located 30 km away in the southeastern flank. Based on maps by National Park Service staff and others.

Sandwiched between sandstones and shales, the Madison is one of the foremost artesian aquifers of the United States. Most groundwater flow feeds springs by rising along fractures in the overlying Minnelusa Formation. However, the caves reach the water table in only a few places and give no indication of extending indefinitely downdip into artesian regions. The caves do not extend to the bottom of the Madison, and in only a few places do they reach the base of the overlying Minnelusa Formation (mainly sandstone and shale).

The uppermost cave passages are irregular rooms that form a crudely planar zone, slightly discordant to the strata. Lower passages are mainly tall fissures following joints and minor faults (Figure 2; see also photograph in Speleogenesis: Deep-Seated and Unconfined). Intermediate levels are located within prominently bedded strata and consist of low, wide, rambling passages. The abundant cave tiers, at least six in places, are sub-parallel to the strata and show no horizontal levels (note contrast with Carlsbad Cavern and Lechuguilla Cave in New Mexico). Although a few passages contain small trickles of running water, there is no indication that discrete cave streams have contributed significantly to cave origin. Except for exhumed cave relics and sinking streams, surface karst is very scant, and the caves do not appear to be related to present hydrologic patterns. Entrances are accidental intersections of the caves by surface erosion. Sediments consist mainly of sand and clay from redistribution of paleofills. Manganese oxides are common in Jewel Cave, either derived from paleofill or precipitated by oxidation from rising anoxic water.

Displacement of continuous rock spans indicates that some of the faults have been reactivated after cave development. Many cave passages intersect filled paleokarst voids as much as 100 m below the top of the Madison, exposing bright red and yellow sand and clay that is only partly indurated. All of Jewel Cave, and the lower half of Wind Cave, possess a lining of calcite spar crystals up to 15 cm thick, except where removed by later dissolution or collapse.

The exact cave origin is unclear. Various authors have used the caves as examples of either artesian or thermal cave development. However, the layout of the caves makes it clear that mixing between waters of differing chemistry was responsible (Palmer & Palmer, 1989). Three sources are feasible: groundwater entering as sinking streams in the outcrop belt; rising thermal water from adjacent basins or from thermal convection; and diffuse infiltration through the overlying Minnelusa Formation. The fact that the caves are mostly concentrated beneath thin caps of sandstone, suggests that diffuse infiltration may have been more important than is generally recognized. Mixing of any of these three sources, which would differ in carbon dioxide content, would produce solutionally aggressive fluids. Widespread calcite wall coatings postdate cave development (Figure 2). They may represent stagnant thermal conditions, owing to blockage of springs around the perimeter of the Black Hills by the Oligocene White River Group. This group once extended over most of the cave region, and its remnants still persist in lowlands around the caves. The position of these remnants indicates that the present topography around the caves has not changed significantly from the late Eocene. Maximum cave enlargement is most likely of Paleocene–Eocene age.

The present caves contain abundant evidence of their lengthy geologic history (Palmer & Palmer, 1995). Early diagenetic features include bedded breccias and breccia dykes, H_2S-related porosity, calcite replacement of primary sulfates, rinds of quartz, calcite, and secondary dolomite, calcite veins, and intense alteration of bedrock to a porous quartz-rich mass between calcite veins. Deep-burial deposits include clear scalenohedral calcite void coatings and euhedral quartz crystals. Boxwork in the caves, especially Wind Cave, is the result of exposure and weathering of early diagenetic zones, in which sulfuric acid altered clay-rich bedrock to a removable mass, while not affecting the early veins of gypsum, which were later replaced by calcite. The boxwork seen today is a combination of the relatively resistant calcite veins protruding from the highly weathered intervening bedrock. Early diagenetic alteration zones in the bedrock have been turned into friable, powdery fluff or sandy material, that is easily removed by gravity and by water during periodic water-table rises.

ARTHUR N. PALMER

Wind and Jewel Caves, United States: Figure 2: Tour route in Jewel Cave, showing irregular passage pattern, paleokarst remnants in ceiling, calcite wall coating, and sediments rich in manganese oxides (lower right wall). (Photo by Art Palmer)

See also **Paleokarst; Speleogenesis: Deep-seated and Confined Settings**

Works Cited

Palmer, A.N. & Palmer, M.V. 1989. Geologic history of the Black Hills caves, South Dakota. *National Speleological Society Bulletin*, 51: 72–99

Palmer, A.N. & Palmer, M.V. 1995. The Kaskaskia paleokarst of the Northern Rocky Mountains and Black Hills, northwestern U.S.A. *Carbonates and Evaporites*, 10(2): 148–160

Sando, W.J. 1988. Madison Limestone (Mississippian) paleokarst: a geologic synthesis. In *Paleokarst*, edited by N.P. James & P.W. Choquette, New York and London: Springer

Further Reading

Bakalowicz, M.J., Ford, D.C., Miller, T.E., Palmer, A.N. & Palmer, M.V. 1987. Thermal genesis of dissolution caves in the Black Hills, South Dakota. *Geological Society of America Bulletin*, 99: 729–38

Conn, H. & Conn, J. 1977. *The Jewel Cave Adventure: Fifty Miles of Discovery under South Dakota*, Teaneck, New Jersey: Zephyrus Press

Deal, D.E. 1968. Origin and secondary mineralization of caves in the Black Hills of South Dakota, U.S.A. *Proceedings of the 4th International Congress of Speleology, Ljubljana, 1965*, vol. 3, Speleological Society of Yugoslavia

Ford, D.C., Lundberg, J., Palmer, A.N., Palmer, M.V., Schwarcz, H.P. & Dreybrodt, W. 1993. Uranium-series dating of the draining of an aquifer: The example of Wind Cave, Black Hills, South Dakota. *Geological Society of America Bulletin*, 105: 241–50

Howard, A.D. 1964. Model for cavern development under artesian ground water flow, with special reference to the Black Hills. *National Speleological Society Bulletin*, 26: 7–16

WORLD HERITAGE SITES

In the 1950s, international concern over the potential destruction of the Abu Simbel temples in Egypt by a proposed reservoir led to a massive campaign under the auspices of UNESCO resulting in reassembly of the temples on another site. Further international campaigns led to international cooperation in the protection of other cultural monuments, and in due course UNESCO, with the support of the International Council on Monuments and Sites (ICOMOS), prepared a draft convention on the protection of cultural heritage. Meanwhile, international concerns over the protection of outstanding examples of natural heritage were also mounting. The 1965 White House Conference, a meeting of the International Union for Conservation of Nature (IUCN) in 1968, and the 1972 UN Conference on the Human Environment all called for the establishment of a World Heritage Trust. It was seen that this would lead to international cooperation to protect the world's superb natural and scenic areas and historic sites for the present and the future of the entire world citizenry. In 1972 the UN General Conference brought together these two strands of concern, and agreed upon a Convention Concerning the Protection of the World Cultural and Natural Heritage. Basically, this convention provided for: the identification of such sites and their inscription upon a register; the responsibility of the state party for continuing protection of site integrity and provision of access by all peoples; and international support as necessary in restoration and protection of sites.

Criteria for assessment of nominated sites have been separately developed for natural and cultural heritage. Those for natural heritage are that the sites should (from Operational Guidelines for the Implementation of the World Heritage Convention):

1. be outstanding examples representing major stages of earth's history, including the record of life, significant ongoing geological processes in the development of land forms, or significant geomorphic or physiographic features; or
2. be outstanding examples representing significant ongoing ecological and biological processes in the evolution and development of terrestrial, fresh water, coastal and marine ecosystems and communities of plants and animals; or
3. contain superlative natural phenomena or areas of exceptional natural beauty and aesthetic importance; or
4. contain the most important and significant natural habitats for in-situ conservation of biological diversity, including those containing threatened species of outstanding universal value from the point of view of science or conservation.

Cultural heritage criteria demand that the site should:

1. represent a masterpiece of human creative genius; or
2. exhibit an important interchange of human values, over a span of time or within a cultural area of the world, on developments in architecture or technology, monumental arts, town-planning, or landscape design; or
3. bear a unique or at least exceptional testimony to a cultural tradition or to a civilization which is living or which has disappeared; or
4. be an outstanding example of a type of building or architectural or technological ensemble or landscape which illustrates (a) significant stage(s) in human history; or
5. be an outstanding example of a traditional human settlement or land-use which is representative of a culture (or cultures), especially when it has become vulnerable under the impact of irreversible change; or
6. be directly or tangibly associated with events or living traditions, with ideas, or with beliefs, with artistic and literary works of outstanding universal significance (the Committee considers that this criterion should justify inclusion in the List of sites only in exceptional circumstances and in conjunction with other criteria cultural or natural)

Furthermore, and in relation to both natural and cultural heritage, each site is expected to meet a series of conditions in respect to integrity. It is sufficient here to summarize those that relate to natural heritage as an example:

1. boundary issues, to ensure that the whole of any one phenomenon is included (e.g., the watershed of a karst area), and that the size is sufficient to demonstrate the totality of any one process or to genuinely protect an ecosystem, and generally, to provide for an adequate buffer zone around the area of universal value, either within or external to the boundary of the site

2. protection issues, to ensure adequate long-term legislative, regulatory, or institutional protection
3. management issues, including the development and implementation of an adequate plan of management.

Underlying all of the detail is the fundamental assumption that a World Heritage site will truly be of universal value, not simply of national importance. This is increasingly tending to be seen, at least in relation to natural heritage, as going hand-in-hand with the notion of representation, and demanding a degree of unique character in any one site. The process by which a site comes to be recognized usually commences with local or international concern being expressed, leading to a nomination being prepared and forwarded to UNESCO by the host government (usually referred to as the State Party). This leads in turn to assessment by IUCN (for natural heritage), or ICOMOS (cultural site), or both (mixed sites). The recommendations arising from the assessment are then placed before the annual meeting of the World Heritage Committee for decision.

The Table lists the karst sites that were inscribed as World Heritage as of July 2002. Many karst sites are of both natural and cultural value, but as the Table shows, only a small number have been assessed and designated as mixed sites, even though at least some probably should have been. Furthermore, on at least one occasion, both the nomination and assessment processes were based entirely upon surface features and biodiversity and totally failed to recognize that a large part of the site was karst. In others, although the site was recognized as including karst, no account was taken of the subterranean sector with its own biodiversity. Thus a number of the important karst sites included in the register of properties were not inscribed as a result of karst phenomena, even though these may be of outstanding importance.

World Heritage Sites: Karst sites currently inscribed as World Heritage. (N = Natural heritage, C = Cultural heritage, M = Mixed sites, * = site has a separate entry in this encyclopedia.)

Australia			**New Zealand**	
The Tasmanian Wilderness	M		Te Wahipounamu	N
Riversleigh & Naracoorte fossil mammal sites	N		**Philippines**	
Shark Bay	N		Puerto-Princesa Subterranean River National Park	N
Greater Blue Mountains	N		**Russia**	
Lord Howe Island	N		Western Caucasus	N
Bulgaria			**Solomon Islands**	
Pirin National Park	N		East Rennell	N
Canada			**Slovenia**	
Nahanni National Park	N*		Škocjan Caves	N*
Rocky Mountain Parks	N		**South Africa**	
China			The Fossil Hominid Sites of Sterkfontein, Swartkrans, Kromdraai, and Environs	C
Huanglong	N*		**Spain**	
Jiuzhaigou Valley	N*		Altamira Cave	C*
Wulingyuan	N		Atapuerca Caves	C*
Zhoukoudian	C		**Sweden**	
Croatia			Södra Ölands Odlingslandskap	C
Plitvice Lakes National Park	N*		**Thailand**	
Cuba			Thungyai-Huai Kha Khaeng wildlife sanctuary	N
Desembarco del Granma National Park	N		**Turkey**	
Viñales Valley	C		Pamukkale	M*
Alejandro de Humboldt National Park	N		**United Kingdom (Pitcairn Islands)**	
France			Henderson Island	N
Caves of the Vézère	C*		**United States**	
France & Spain			Grand Canyon National Park	N*
Pyrenees-Mount Perdu	M		Mammoth Cave National Park	N*
Hungary & Slovakia			Carlsbad Caverns National Park	N*
Aggtelek and Slovak Karst	N*		**Venezuela**	
Indonesia			Canaima National Park	N
Lorentz National Park (Irian Jaya)	N		**Vietnam**	
Lao PDR			Ha Long Bay	N*
Luang Prabang	C		**Yugoslavia (Montenegro)**	
Madagascar			Durmitor National Park	N
Tsingy Bemaraha	N			
Malaysia				
Gunung Mulu National Park	N*			
Mexico				
Chichen Itza	C			
Sian Ka'an	N			

N.B. Mogao Caves, Longmen Grottoes, Ajanta Caves, Ellora Caves and the Cave of Sokkuram are also inscribed on the WHS list but they are artificially excavated sites.

Thus there are special problems in the assessment and full recognition of karst. These arise in part from the organizational relationship between natural and cultural heritage, but much more from the emphasis upon biodiversity and lack of awareness of the special nature of karst. Indeed, karst has received little recognition, partly because it is commonly assumed to be a permanent characteristic of any landscape and not threatened, and partly because of the disciplinary focus on biological sciences. Similarly, much of the important, remarkably diverse, and often endemic invertebrate fauna of karst areas remains unrecognized in assessment because it lacks fur or feathers.

Many karst areas have a potential capacity to meet the natural heritage criteria simply because the very nature of karst embodies these characteristics. One of the special features of karst is the dynamic and complex interaction of the various elements that comprise it and more attention might well be paid during assessment to the extent to which any one site visibly demonstrates this. Perhaps one of the most difficult special problems related to karst is the extent to which many karst areas which have been, or might be, recognized as world heritage are only a very small component of the total karst system, sharing both the watershed area and the drainage system of the whole. This means that the karst area is often impacted by activities external to the management boundary and sometimes occurring a considerable distance from the site. Sediment and chemical pollution may be transmitted through the groundwater system over very long distances. In particular, agriculture, viticulture, and similar land uses commonly lead to eutrophication (or enrichment) of the groundwater. In turn this often leads to algal blooms in exposed waters and to a variety of impacts on aquatic fauna. Although it is unlikely that an adequately managed World Heritage area will in itself generate undesirable downstream impacts, it is possible that this may happen. For instance, if a natural water flow had been altered to better serve downstream agriculture and the natural flow was restored to better meet World Heritage standards of integrity then this may well be seen as a highly undesirable impact by those affected.

The continuing management of a karst World Heritage site provides significant challenges. Currently, there is a considerable rise of interest in effective monitoring and assessment of both environmental changes and visitor experience (Hamilton-Smith, 2000; Hamilton-Smith & Ramsay, 2001; Kranjc, 2002). One aspect of this problem is that in making some of the great caves of World Heritage areas accessible to the public as tourist caves, often outmoded systems of infrastructure and display are still in use, or even still being developed. One of the current challenges which is currently receiving examination with a view to developing guidelines is the strategies and methods for effective presentation of World Heritage values in karst to the public.

ELERY HAMILTON-SMITH

See also **Ramsar Sites**

Works Cited

Hamilton-Smith, E. 2000. Managing for environmental and social sustainability at Jenolan Caves, New South Wales, Australia. In *Essays in the Ecology and Conservation of Karst*, edited by I. Barany-Kevei & J. Gunn, special issue of *Acta Geographica Szegedensis*, 36: 144–52

Hamilton-Smith, E. & Ramsay, A. 2001. Social and environmental evaluation at Jenolan Caves, New South Wales. *Evaluation Journal of Australia*, n.s., 1: 60–65

Kranjc, A. (editor) 2002. Proceedings of the meeting on monitoring of karst areas. *Acta Carsologica*, 31(1)

Further Reading

Wong, T., Hamilton-Smith, E., Chape, S. & Friederich, H. (editors) 2002. *Proceedings of the Asia-Pacific Forum on Karst Ecosystems and World Heritage*, available at www.heritage.gov.au/apfp/pubs/Karst-full.pdf

World Heritage Centre at http://www.unesco.org/whc/

World Conservation Monitoring Centre at http://www.unepwcmc.org/

International Union for Conservation of Nature at http://www.iucn.org/

International Council on Monuments and Sites at http://www.icomos.org/

YANGSHUO KARST, CHINA

Located some 60 km south of Guilin, in the province of Guangxi, southern China, Yangshuo is a small town at the centre of the most spectacular tower karst for which Guilin and Guangxi are famous. The terrain is a classic example of one of the four relief types of tropical karst defined by Balazs (1971). The Yangshuo type is characterized by a peak base diameter that is less than 1.5 times its height. The other three types have lower hill profiles, with base / height ratios of 1.5–3.0 (Organos type, from Cuba), 3–8 (Sewu type, from Indonesia), and more than 8 (Tual type, from Indonesia). These differences appear to be dependent on lithological features, as well as on their geomorphological evolution. The Yangshuo type is developed on hard and compact Devonian carbonates, while the others are on younger and more porous carbonate rocks.

Geologically, Yangshuo is at the southern end of a 90 km long, north–south, arcuate synclinal structure, protruding to the west near Guilin (Figure 1). Basement rocks crop out on all sides around the Guilin-Yangshuo basin, at 1936 m Haiyan Mountain to the east, at 1215 m Jiaqiao Mountain to the west, and at 2142 m Yuecheng Mountain to the north. Overlaying the basement, above a strong angular unconformity of Caledonian orogeny, are upper Paleozoic and Mesozoic rocks, with some scattered Tertiary outliers. The main carbonate rocks are 4600 m thick, ranging from middle Devonian to lower Carboniferous age. These occur in the central part of the synclinorium and form the basis of the tower karst formation from Yangshuo to Guilin. The thickly bedded, pure limestone of the Devonian Rongxian Formation, 400 m thick, is especially important for the development of the finest karst landforms.

The main Pleistocene deposits are poorly sorted boulder beds, that appear as many low mounds on the dissolutional karst plain. Most of the cobbles are sandstone or quartzite, derived from the lower Paleozoic outcrops, with a general diameter of 100–300 mm, though boulders up to 1 m are not unusual. Whether these are moraines, fluvioglacial deposits, or debris-flow deposits, remains controversial. Holocene alluvial deposits form terraces alongside the modern Li Jiang (Li River), through the central part of the basin. These are well-sorted gravels, covered by silty sand, with a total thickness of about 10 m, but may be thicker over buried karstic depressions. Recent drilling for the many bridge sites on the Li Jiang have revealed a trough, 30–50 m

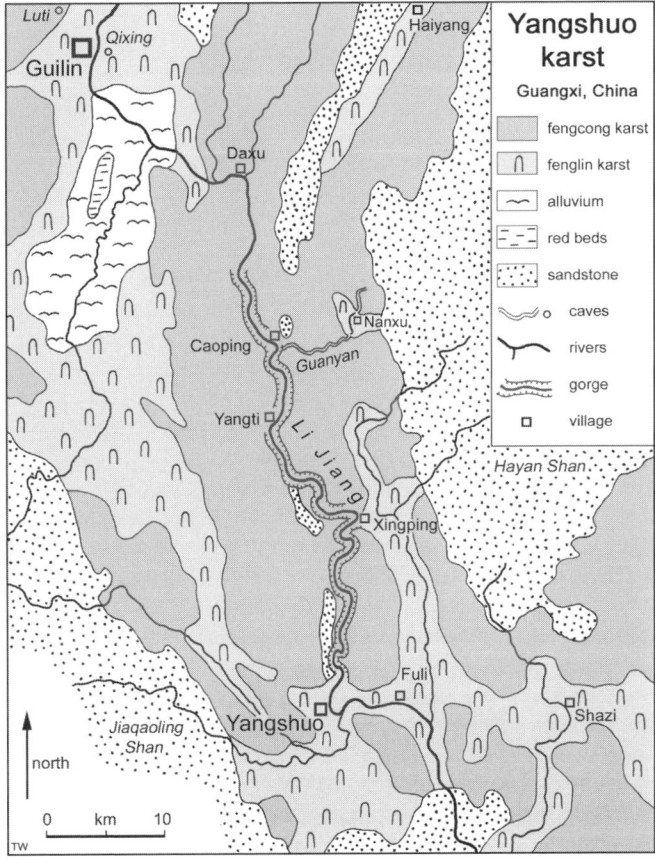

Yangshuo Karst, China: Figure 1. Outline map of the geomorphology of the karst between Yangshuo and Guilin.

Yangshuo Karst, China: Figure 2. The panorama of steep limestone cones and towers breached by the Li Jiang, that are seen on the justly famous boat trip between Guilin and Yangshuo. (Photo by John Gunn)

Yangshuo Karst, China: Figure 4. Clustered towers and steep cones form the fengcong karst to the west of Caoping. (Photo by Tony Waltham)

deep, infilled by Pleistocene boulder beds under the Holocene alluvium. This trough is considered to be a feature incised during Quaternary stages of lower sea levels, and may be associated with some of the cave development in the karst aquifer beneath the dissolutional plain.

Influenced by the Asian monsoon circulation, Yangshuo's climate is characterized by wet summers with frequent storms, but dry autumns and winters. Annual mean precipitation ranges from 1570 mm at Yangshuo to 1990 mm at Guilin, but 62% of the total rainfall is concentrated in the rainy season from April to August. Mean temperatures are 18–19°C. Based on data from many years of monitoring, the mean limestone denudation rate in the karst is around 90 mm ka^{-1}. The Li Jiang is the main drainage system with its base level at 103 m at Yangshuo and 141 m at Guilin. The catchment above Yangshuo is 5520 km^2, giving an annual mean discharge of 215 $m^3 s^{-1}$, with flood peaks at 6330 $m^3 s^{-1}$. Carbonate rocks in the central part of the catchment cover 3464 km^2, or 62.8% of the total. Allogenic water from the non-carbonate outcrops washes down into the karst basin and plays an important role in the formation of the fenglin karst of the region (Figure 2).

There are two types of tower karst in Yangshuo—the fenglin (peak forest; Figure 3) and the fengcong (peak cluster; Figure 4). The fengcong is defined as groups of rocky hills or peaks rising from shared limestone foot-slopes. Closed depressions lie between the peaks, so the landscape is sometimes described as "peak cluster depression". The fenglin is defined as limestone towers or peaks that are isolated from each other by flat limestone surfaces, which are generally covered by a thin layer of loose sediment. The peaks are usually completely surrounded by a karst plain, so the landscape may be called a "peak forest plain". The relative heights of the peak surface are generally 30–80 m above the plain in the middle of the basin, but rise to 300–500 m in the mountainous regions of the fengcong.

The profiles of the peaks are columnar towers where the limestone is gently dipped in the trough of the syncline, but form steep, broken escarpments on both flanks of the fold, where the dip is steeper. No strict zonal distribution of the fenglin and fengcong can be established, either in plan or profile, though this was claimed by Zhongjian (1935) and Gellert (1962) for the fenglin karst at Baisha, 10 km northwest of Yangshuo. The two types are mixed in their areal distribution (Figure 1). However, it is evident that most of the fenglin towers occur in those areas where they are subjected to strong lateral erosion by allogenic rivers, either the Li Jiang or its tributaries. In contrast, the fengcong hills lie beyond the zones of allogenic fluvial erosion, as in areas west of Guilin and in blocks on both sides of the Li Jiang gorge between Caoping and Yangshuo (Figure 1).

About 600 caves have been recorded within the area. Caves in the fenglin are generally short because of the limited size of the limestone towers. Within a population of 80 of the isolated peaks, only 24 have caves totalling more than 200 m. Most of these are foot caves with inflow from the alluvial plains indicated by the scallops on their walls. Cave development plays a significant role in the recession of cliffs around the peaks, and therefore in the formation of the fenglin landscape.

Caves in the fengcong karst are generally longer. There are more than 23 stream caves, each with a length of more than 1 km, on both sides of the Li Jiang gorge. Those on the eastern side receive allogenic recharge from Haiyang Mountain and discharge into the Li Jiang. The longest known caves are in the Guanyan system, east of Caoping, which has a catchment of 80 km^2 and flows of 0.3–8 $m^3 s^{-1}$. A total length of 10 530 m of passages have been mapped (Waltham, 1986), but the central section of active phreatic passages has not yet been explored.

Yangshuo Karst, China: Figure 3. The classic landscape of fenglin tower karst just to the southeast of Yangshuo. (Photo by Tony Waltham)

The resurgence end of the caves has already been opened for tourists, who cruise on an underground stream from a boat dock on the Li Jiang. Other tourist caves in the area include the well-decorated Luti Dong (Reed Flute Cave), the 1300 m long Qixing Dong (Seven Star Cave) with its large abandoned phreatic passages, and some small caves around Yangshuo. Many caves have fossils of the Pleistocene Ailuropoda-Stegodon fauna, as well as archaeological relics.

Dating and stable isotopic studies of a 1200 mm tall stalagmite from Panlong Cave, 30 km northwest of Yangshuo, revealed paleoclimatic changes over the past 36 ka that included the last glaciation, the Younger Dryas and the Holocene optimum, while an analysis of the clay beneath stalagmites in Shuinan Cave showed paleoclimatic change during the Jaramillo subchron at 1.0–0.9 Ma (Li et al., 1998). A paleomagnetic study of flowstone from a cave on Tunnel Hill, indicated a maximum lowering rate for the karst plain and the Li Jiang of 23 mm ka^{-1} (Williams, 1987).

The initiation and early evolution of the Yangshuo tower karst is unclear. Late Cretaceous Atopochara plant fossils are found both in the Red Beds on the floor of the basin and in many small patches of red breccia scattered on peaks 300–600 m higher. This correlation shows that the Cretaceous Red Beds used to cover a remarkable area, and the fenglin karst could only develop after their removal. It appears that much of the modern landscape may have evolved through the Quaternary, after the Asia monsoon became seasonal (Yuan, 1987).

Karst dominates the environment in the Guilin-Yangshuo basin, and groundwater abstraction from the karst aquifer is increasing, especially as tourism grows. Daily extraction is about 150 000 m^3. Karst collapses happen frequently, especially in those areas where groundwater is overpumped from limestone beneath alluvial soils. Local records include 148 incidents of collapse, with the earliest dating back to 1498 AD in the Ming Dynasty. Drilling of new wells is now controlled by municipal, environmental, and water protection regulations.

YUAN DAOXIAN

Works Cited

Balazs, D. 1971. Relief types of tropical karst areas. Presented at European Regional Conference of *IGU Symposium on Karst Morphogenesis*

Li B., Hauge, F. & Lovlie, R. 1998. Paleomagnetic record from Shuinan cave sediments, Guilin, and its bearing on paleoclimatic changes. *Journal of Geomechanics*, 4(4): 50–57

Gellert, J.F. 1962. Der Tropenkarst in Sud-China im Rahmen der Gebirgsformung des Landes. *Verhandlungen des Deutschen Geographentages*, 33: 376–84

Waltham, A.C. (editor). 1986. *China Caves '85: The First Anglo-Chinese Project in the Caves of South China*, London: Royal Geographical Society

Williams, P.W. 1987. Geomorphic inheritance and the development of tower karst. *Earth Surface Processes and Landforms*, 12: 453–65

Yuan D. 1987. New observations on tower karst. In *International Geomorphology*, vol. 2, edited by V. Gardiner, New York: Wiley

Zhongjian Y & Chardin, T. 1935. Cenozoic geology of Guangxi and Guangdong. *Bulletin of the Geological Society of China*, 14(2): 179

Further Reading

Silar, J. 1965. Development of tower karst of China and North Vietnam. *National Speleological Society Bulletin*, 27(2): 35–46

Sweeting, M.M. 1995. *Karst in China: Its Geomorphology and Environment*, Berlin: Springer

Williams, P.W., Lyons, R.G. & Wang X. 1986. Interpretation of the paleomagnetism of cave sediments from a karst tower in Guilin. *Carsologica Sinica*, 5(2): 119–25

Yuan D. (editor) 1991. *Karst of China*, Beijing: Geological Publishing House

Zhu X. 1988. *Guilin Karst*, Shanghai: Scientific and Technical Publishers

YORKSHIRE DALES, ENGLAND

The central part of England's Pennine Hills is better known as the Yorkshire Dales, after the deep glaciated valleys (dales) running through it, and this is also the name of the National Park that contains the finest glaciokarst landscapes in Britain.

The Lower Carboniferous Great Scar Limestone is nearly 200 m thick, and lies with a very gentle northerly dip on an impermeable basement. The dales descend gently from the north, in the direction of Pleistocene ice flow, so that some cut through the entire thickness of limestone. Between the dales, hills that rise 150 m above the limestone are formed of shales (with thin limestones and sandstones), and provide allogenic streams draining onto the limestone. The broken carbonate platform is known as the Craven Uplands, and is truncated to the south by the Craven Faults, where the limestone is downthrown to depth beneath the Craven Lowlands further south. Between the shale caps, the dale floors and the faults, the total area of limestone outcrop is about 320 km^2 (Figure 1). Erosive Devensian glaciation followed earlier stages of Pleistocene ice cover, and stripped much of the limestone to clean bare outcrops that constitute excellent glaciokarst. During interglacial stages, the karst matured with extensive cave development in a mainly temperate climate. The conspicuous elements of the glaciokarst are extensive limestone pavements, thousands of dolines and some deep meltwater gorges.

Perhaps the best-known karst occupies the limestone benches that surround Ingleborough Hill (Figure 2 and see also photo in Floral Resources entry). Devensian ice flowed along both sides of the hill, leaving limestone pavements that are terraced on strong beds of limestone and separated by ice-plucked scars. The pavements are dominated by postglacial rundkarren, rounded beneath a cover of lichen, feeding into deep grikes (kluftkarren). Away from the pavements, the limestone is covered by glacial till, which is thickest and most extensive in the southwestern lee of the hill. The till is pocked by hundreds of suffosion dolines, locally known as shakeholes, most of which drain into narrow fissures in the underlying limestone. Larger potholes (vertical shafts up to 10 m across and 100 m deep) are scattered across the karst. Some are active stream sinks fed by streams off the shale. Others, located along interglacial positions of the retreating shale margin, are almost dry, and many are widened by wall collapse. They include Gaping Gill (see photo in Valleys in Karst), which still takes Fell Beck down its 95 m deep shaft

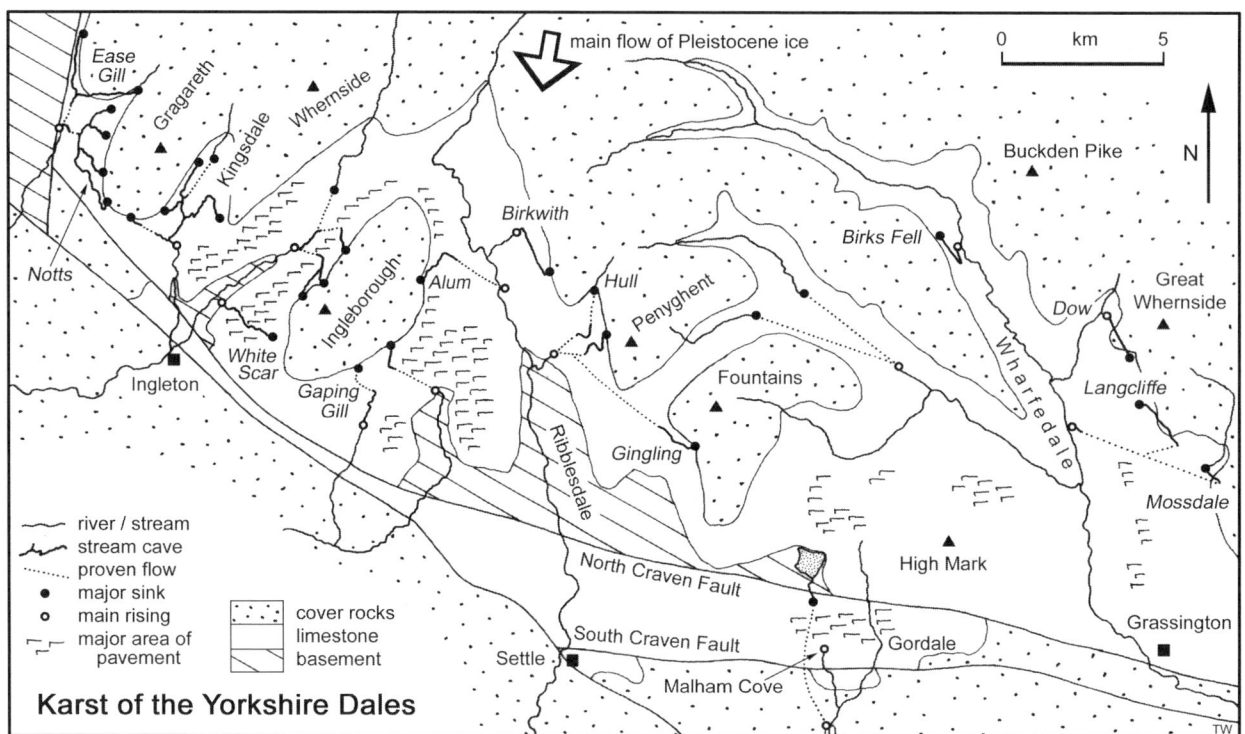

Yorkshire Dales, England: Figure 1. Main features of the geology, karst, and glacial geomorphology of the Yorkshire Dales, with the major underground drainage routes through known cave passages or proven by dye tracing.

through the roof of the main chamber. A complex of 18 km of active and abandoned caves, mostly phreatic, link beneath the meltwater gorge of Trow Gill and drain to the Ingleborough Cave resurgence.

East of Ingleborough, the glaciokarst above Malham also has extensive pavements, but is noted more for its entrenched proglacial meltwater features. The Watlowes dry valley ends at the lip of the 80 m high cliff of Malham Cove. The width of the cove

Yorkshire Dales, England: Figure 2. Limestone pavements on the glacially stripped top surface of the limestone, with the summit mass of Ingleborough formed of shales and sandstones in the background. (Photo by Tony Waltham)

Yorkshire Dales, England: Figure 3. Stalagmites stand on banks of clastic sediment in the abandoned phreatic tunnel that leads to Gour Hall, Pippikin Pot, in the Ease Gill Cave System. (Photo by Tony Waltham)

suggests that it may have originated by glacial plucking, before its modification as a meltwater cascade. The adjacent Gordale carries a tufa-depositing, underfit stream, and deepens between cliffs over 100 m high at its mouth in the fault scarp, where it leaves the Craven Uplands. East of Malham, the High Mark plateau is a polygonal karst with gently rounded dolines 200–800 m across, which appear to have been overridden by the Devensian ice.

Over 1400 caves are known in the Yorkshire Dales, with 50 of them having more than 1000 m of passage. A total of 330 km of cave passages have been mapped, but the deepest cave reaches only 211 m. The typical Dales cave has a vadose streamway only a few metres high and wide, cut beneath a bedding plane or thin shale bed; it drains roughly downdip, to the north (see photo in Morphology of Caves). Its meandering canyon is interrupted by vertical waterfall shafts on major joints, where the cave drops to lower bed horizons. Some caves are tall rifts along the joints. At the level of the resurgence, the cave continues as a phreatic tube, looping down joints and up the bedding planes, towards the south, against the dip. Swinsto Hole is the type example (Waltham et al., 1981), with a kilometre of streamway leading into a kilometre of flooded tube, explored the whole way from sink to resurgence.

Swinsto is also typical in that its modern drainage route intersects and invades segments of older passages. These include large phreatic tunnels, now well above the water table, partly choked with mud and partly decorated with stalactites and stalagmites (Figure 3). They were interglacial and preglacial trunk conduits, abandoned when new resurgences were established at the lowest outlets where the deepening dales crossed the Craven Faults. In

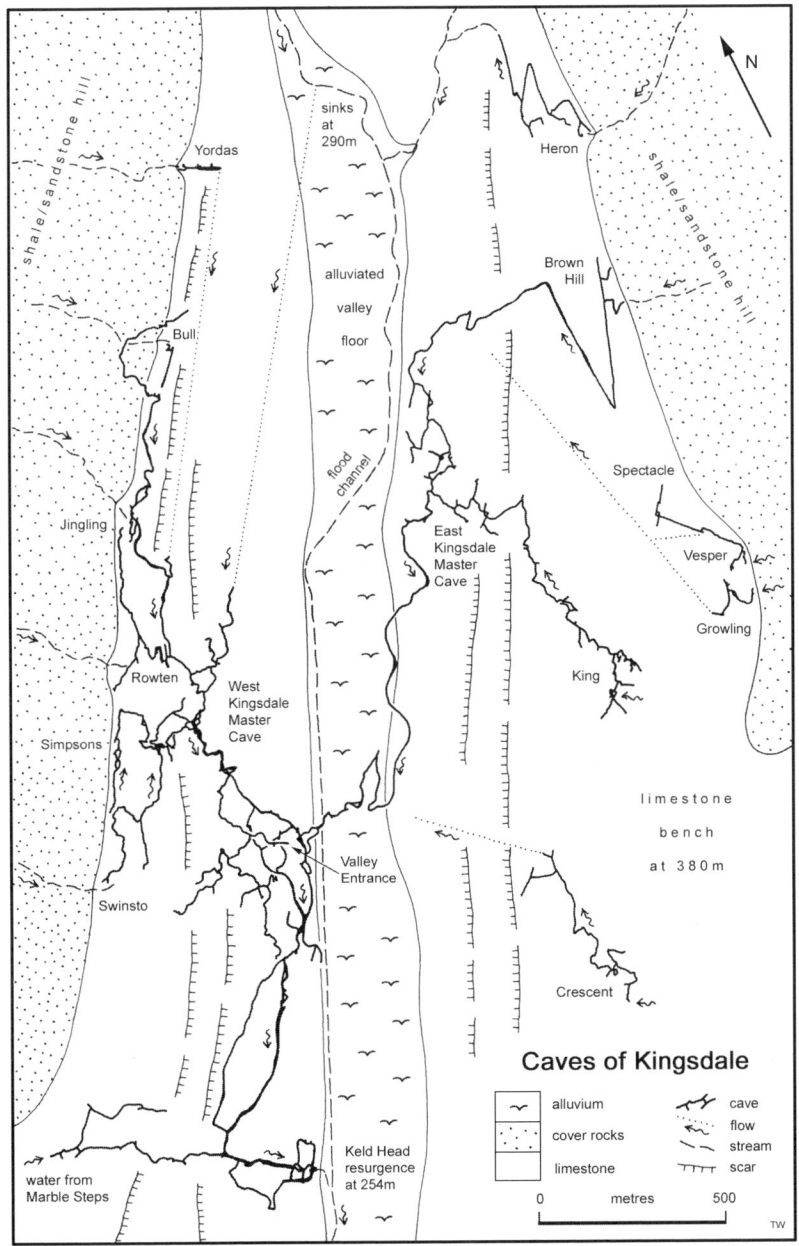

Yorkshire Dales, England: Figure 4. Known passages in the Kingsdale Cave System, Yorkshire (compiled from surveys by various caving clubs).

many caves, fragments of these old passages survive. Some can be traced back to abandoned sinks, which are now the larger, partly-collapsed dolines and potholes away from the shale margins, and others are truncated in the walls of the glaciated dales. Stalagmites dated from some of the old caves, show periodic deposition during interglacial stages of the Pleistocene (Gascoyne & Ford, 1984). Sequential levels of passage abandonment, dated by their oldest stalagmites, indicate the drainage of the karst as the dales were excavated. The lower half of their depth has been achieved within the last 400 000 years. Valley deepening can be broadly correlated with shale margin retreat during stripping of the limestone benches, as the karst evolved through alternating stages of fluvial and glacial erosion that must reach back into the early Pleistocene.

West of Ingleborough, Gragareth Hill has more than 100 km of mapped cave passages beneath its largely till-covered limestone flanks. The longest single cave system is currently Ease Gill, with 75 km of mapped passages extending under both sides of the normally dry valley of the same name. On the north side, numerous, down-dip, canyon streamways capture flows from various parts of the Ease Gill stream, and feed them all into a parallel master cave, then up-dip to a flooded resurgence back in the same valley. The main drain lies beneath a large abandoned passage, parts of which are very well decorated. On the south side, caves drain from sinks on the high-level shale margin of Gragareth. The modern drains are descending vadose canyon tributaries into the long, low-level, vadose Lost Johns Master Cave just above resurgence level. A parallel, old, phreatic master cave, with equally numerous inlets in Notts Pot, now carries an underfit stream in a floor canyon and is a relict feature of the interglacial karst drainage. On the eastern flank of Gragareth, the Kingsdale caves collect inlet streams from both sides of the valley (Figure 4). Both Swinsto Hole and King Pot demonstrate the northerly downdip vadose drainage before turning south in the phreas. The flooded tube carrying the King Pot drainage passes beneath the valley floor and continues, without air space, to the resurgence where it is truncated in the side of the glaciated dale.

Mossdale and Langcliffe are two distinctive caves that lie east of Malham. Each has 10 km of passages formed within the Brigantian Middle Limestone, a band just 30 m thick within the clastic sequence above the Great Scar Limestone. The small stream passages have angular courses that switch between two steps of joint rifts with sandstone floors. They trend downdip until they can break through into the underlying Great Scar Limestone by descending faults that cut through the sandstones and shales. Only in Langcliffe Pot has the breakthrough point yet been reached by exploration.

Nearly all the karst and caves lie within the Yorkshire Dales National Park. Tourism sustains many of the villages in the dales, and the white limestone scars attract hill walkers. The upland is grazed by sheep that maintain the open aspect of the grassland fells, by keeping down new trees and shrubs. Aggregate quarries are a localized threat to the limestone, but most of the spectacular glaciokarst remains in good condition.

TONY WALTHAM

Works Cited

Gascoyne, M. & Ford, D.C. 1984. Uranium series dating of speleothems: Results from the Yorkshire Dales and implications for cave development and Quaternary climates. *Cave Science*, 11: 65–85

Waltham, A.C., Brook, D.B., Statham. O.W. & Yeadon, T.G. 1981. Swinsto Hole, Kingsdale: A type example of cave development in the limestone of northern England. *Geographical Journal*, 147: 350–53

Further Reading

Gascoyne, M., Schwarz, H.P. & Ford, D.C. 1983. Uranium series ages of speleothems from northwest England: Correlation with Quaternary climate, *Philosophical Transactions of the Royal Society of London*, B301: 143–64

Waltham, A.C. 1970. Cave development in the limestone of the Ingleborough district. *Geographical Journal*, 136: 574–85

Waltham, A.C. 1990. Geomorphic evolution of the Ingleborough karst. *Cave Science*, 17: 9–18

Waltham, A.C., Simms, M.J., Farrant, A.R. & Goldie, H.S. 1997. *Karst and Caves of Great Britain*, London and New York: Chapman and Hall

YUCATÁN PHREAS, MEXICO

The Yucatán Peninsula is a low-lying limestone platform, extending over 75 000 km^2, between the Caribbean Sea and the Gulf of Mexico. The limestone ranges in age from Paleocene in the interior, to Quaternary on the coastal margins. There are numerous roughly circular cenotes (dolines; from the Maya word d'zonot meaning a well) that lead to the water table. Cenotes in the interior are steep walled with large chambers, extending below the water table to depths in excess of 60 m, on the perimeter of central piles of rock breakdown. These rarely lead to more than 500 m of accessible passage, exploration being impeded by silt, breakdown rock, and the depth limits of scuba diving. In contrast, cave diving explorations, which accelerated in the mid–1980s along the Caribbean coast, have identified at least 91 horizontally extensive submerged cave systems, with a combined length greater than 500 km and an average maximum depth of −16 m (see Table). Among these are the three longest flooded cave systems in the world and 13 other systems more than 5 km long, firmly establishing this region as the leading site for explored underwater cave systems.

The peninsula is underlain by a density-stratified aquifer, with a thin freshwater lens recharged by rapid infiltration of rainwater. The shallow underlying saline water flows coastward, in response to mixing and entrainment by freshwater outflow above the halocline interface. There is a compensatory influx of deeper saline water from the coastal margins. Unlike the blind flank margin cave development in island karst settings (see separate entry, Speleogenesis: Coastal and Oceanic Settings), channel networks have been explored here to distances 12 km inland from the Caribbean coast. The channel networks drain to coastal springs that are often located at the headward end of coastal inlets known as caletas. Quantitative dye tracing, conducted over 5 km in Sistema Nohoch Nah Chich, showed that 99.7% of

Yucatán Phreas, Mexico: Summary of completely submerged cave systems longer than 10 km of the Yucatán Peninsula. Yaxchen and Aerolito are discontinuous and the passages are broken by several open cenotes.

Sistema/Cueva	Length (km)	Max Depth (m)	Number of cenotes
Ox Bel Ha	107.1	33.5	60
Nohoch Nah Chich	61.0	73.6	n/a
Dos Ojos	56.7	119.1	24
Sac Actun	24.0	25.0	22
Naranjal	21.1	34.7	8
Aerolito	19.5	23.5	3
Yaxchen	18.3	22.3	40
Ponderosa	15.0	16.5	18
Nohoch Kiin	13.9	19.8	12

the freshwater flow occurs in the cave channels (Worthington, Ford & Beddows, 2000). As a consequence of the extensive cavernous porosity, the aquifer is highly transmissive, such that 40% of the Caribbean microtide (30 cm amplitude) is transmitted to cenotes 5 km from the coast. The hydraulic gradient is also extremely flat (7–10 mm km^{-1}) as measured in the northwest of the peninsula by Marín and Perry (1994). No coastal cave passages have yet been reported along the north or west coastlines. However, very large springs at the intersection of the "Ring of Cenotes", a concentration of cenotes that overlies the deeply buried Chicxulub Cretaceous–Tertiary impact crater (Perry et al., 1995), attest that the peninsula is highly karstified throughout.

The caves formed at some time(s) before our present sea-level high stand, as indicated by the great abundance of drowned vadose speleothems. Many of the shallow passages are now wholly within the freshwater lens or intersected by the halocline which is found at 18 m at Cenote Car Wash (in Sistema Aktun Ha), 9 km from the coast, and at shallower depths as the coast is approached. Mixing corrosion is thought to be an important speleogenic process in these horizontally extensive systems but bacterially mediated reactions, associated with hydrogen sulfide layers, may also enhance dissolution (Stoessell, Moore & Coke, 1993). Good examples of vertically extensive fracture-guided passages are found predominantly within 2 km of the coastal zone, such as in Sistemas Abejas and Aak Kimin. The deep pit cenotes of the peninsula interior, combined with deeper tiers of passage completely in the saline zone along the Caribbean coast, such as at The Pit (-119 m in Sistema Dos Ojos), the Blue Abyss (-74 m in Sistema Nohoch Nah Chich), and Sistema Aak Kimin (-69 m), indicate that extensive cavernous development occurred at lower sea levels. It may extend to the 120 m Quaternary sea minimum, or deeper in the older buried limestones.

Sistema Ox Bel Ha (Maya for "Three Paths of Water") is characterized by inter-linked northwest–southeast parallel main passages, with an average depth of 15 m, and four near-shore springs that connect to the Caribbean Sea (Figure 1). Exploration of Sistema Esmerelda (the precursor to Ox Bel Ha) only began in May 1998. The poetic name "Three Paths of Water" was adopted because of the connection with Sistema Chikin Ha, and an anticipated southern link to Sistema Yaxchen. Promising

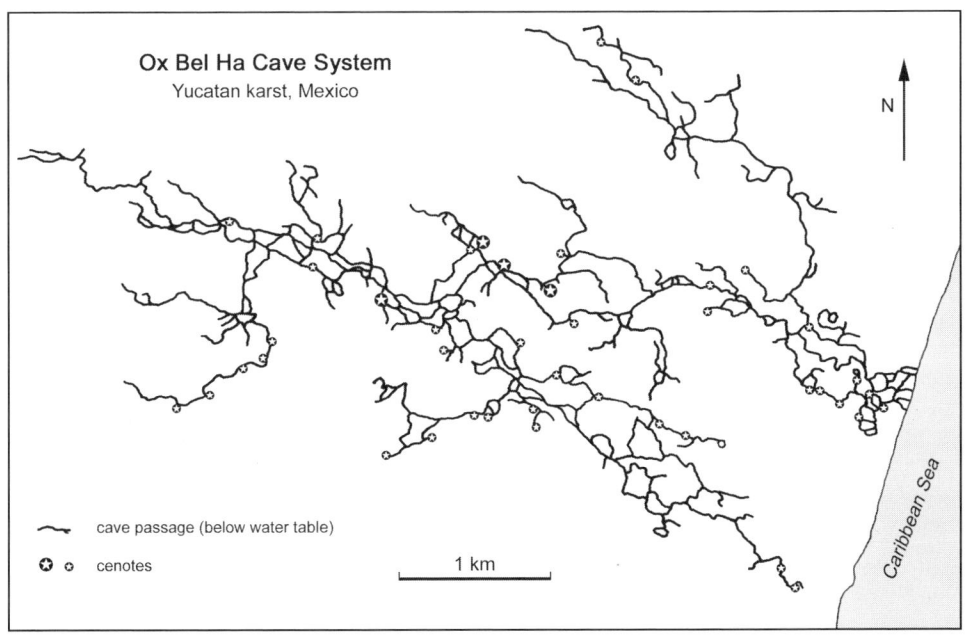

Yucatán Phreas, Mexico: Figure 1. Outline map of the known underwater cave passages of Ox Bel Ha, the longest of the caves beneath the Yucatán karst of Mexico (cave map by Grupo de Exploracion Ox Bel Ha).

Yucatán Phreas, Mexico: Figure 2. A diver passes subaerial calcite deposits in one of the flooded phreatic passages near Cenote of the Sun (in Sistema Naranjal). (Photo by Jill Heinerth)

leads northwest towards Sistema Naranjal (20 km long) are being explored by the Grupo de Exploración Ox Bel Ha.

Sistema Nohoch Nah Chich (Maya for "Giant Bird House"), with its main entrance 3 km from the coast, was first dived in 1987. The Cedam cave diving team effectively used jungle camps to survey the amazingly decorated upstream passages, generally more than 5 m wide, filled with crystal-clear fresh water at depths of less than 10 m. Exploration downstream, under the paleo-beach ridge, encountered restricted passages that were highly unstable with low visibility. Nonetheless, Sistema Nohoch Nah Chich was explored from 6 km inland, through to the Caribbean Sea via the near-shore spring at Cenote Manati / Tankah / Casa, and was the longest underwater cave system in the world, until displaced by Sistema Ox Bel Ha in 2000. Many leads exist, including the potential to locate a more northerly discharge point identified by the divergence of 50% of the flow at 3 km inland. Most tantalizing is the potential connection with Sistema Dos Ojos, which lies as little as 200 m away at one point.

Sistema Dos Ojos (Spanish for "Two Eyes") was first dived through the "two eyed" double collapse cenote in 1986. Exploration increased in the early 1990s, with expeditions that have resulted in the survey of 56.7 km of passage and the notable discovery of The Pit in 1994. Ongoing exploration has found The Pit Cenote to be the deepest known site in the Caribbean coast caves, at a depth of 119m. At present it is not certain which coastal spring is fed by Sistema Dos Ojos, but a portion of flow is likely to exit via Caleta Xel-Ha.

Many cave passages are known to contain Maya pottery, fire pits, carvings and human remains below the present water table, highlighting the long-standing cultural and religious importance of these sites to the indigenous Maya. At present, the cenotes and caletas are used extensively for recreational water activities and are favoured locations for small and large-scale tourism developments and nature parks, such as Caleta Xel-Ha, Caleta X-Caret, and Gran Cenote. The population of the Yucatán Peninsula is also wholly dependent on the freshwater lens for drinking water. However, there is justified concern that groundwater resources and the anchialine cave habitat (known to host at least 37 stygobitic species) may be degraded by disposal wells that pump sewage effluent into the saline zone, while cess pits and garbage dumps leach into the freshwater lens from above. It is certain that almost all water (including contaminants) will eventually pass through the cave systems to be discharged onto the sensitive barrier reef system. The phenomenal growth of Cancun and other coastal resorts demands that regional and multidisciplinary understanding of the aquifer is developed in order to maintain quality and achieve sustainable use of groundwater resources. It is hoped that ongoing cave diving exploration will continue to raise awareness and concern for the sub-aquatic environment, and therefore assist in conservation.

PATRICIA A. BEDDOWS

See also **Blue Holes of Bahamas; Caribbean Islands; Diving in Caves; Speleogenesis: Coastal and Oceanic Settings**

Works Cited

Marín, L.E. & Perry, E.C. 1994. The hydrogeology and contamination potential of northwestern Yucatán, Mexico. *Geofísica Internacional*, 33(4): 619–23

Perry, E., Marín, L., McClain, J. & Velázquez, G. 1995. Ring of cenotes (sinkholes), northwest Yucatan, Mexico: Its hydrogeologic characteristics and possible association with the Chicxulub impact crater. *Geology*, 23(1): 17–20

Stoessell, R.K., Moore, Y.H. & Coke, J.G. 1993. The occurrence and effect of sulfate reduction and sulfide oxidation on coastal limestone dissolution in Yucatan cenotes. *Ground Water*, 31(4): 566–75

Worthington, S.R.H., Ford, D.C. & Beddows, P.A. 2000. Porosity and permeability enhancement in unconfined carbonate aquifers as a result of solution. In *Speleogenesis: Evolution of Karst Aquifers*, edited by A.B. Klimchouk, D.C. Ford, A.N. Palmer and W. Dreybrodt, Huntsville, Alabama: National Speleological Society

Further Reading

Back, W., Hanshaw, B.B., Herman, J.S. & Van Driel, J.N. 1986. Differential dissolution of a Pleistocene reef in the ground-water mixing zone of coastal Yucatan, Mexico. *Geology*, 14(2): 137–40

Barton, H. 2000. The Pit. *Underwater Speleology*, 27(3): 10–17

Gerrard, S. 2000. *The Cenotes of the Riviera Maya*, Tallahassee, Florida: Rose Printing

Martin, J.B., Wicks, C.M. & Sasowsky, I.D. (editors) 2002. *Hydrogeology and Biology of Post-Paleozoic Carbonate Aquifers*, Charlestown, West Virginia: Karst Waters Institute

Matthes, A.W. 2001. Yucatan Deep Speleological Dive Team: Proyecto de investigacion de Cenotes y Grutas Yucatan 2000. *Underwater Speleology*, 28(1): 6–8

Phillips, B.A. 2001. The exploration of Sistema Ox Bel Ha. *Underwater Speleology*, 28(4): 4–8

www.caves.org/project/qrss/ Official site of the Quintana Roo Speleological Survey (QRSS) with statistics on the long and deep cave systems on the Caribbean coast, Yucatan Peninsula

www.mexicocavediving.com/ is the official webpage of the Grupo de Exploracion Ox Bel Ha

www.cavebiology.org/ Anchialine Caves and Cave Fauna of the World webpage contains a very useful summary of Yucatán Peninsula biospeleology

Notes on Contributors

Alberquerque, Elaine F.
Universidade Santa Ursula, Instituto de Ciencias Biologicas e Ambientais, Rio de Janeiro, Brazil.
  Interstitial Habitats (Aquatic)

Aley, Tom
Founder and President, Ozark Underground Laboratory, Protem, Missouri, USA. Certified forester, professional hydrogeologist and a licensed professional geologist. Past president of the American Cave Conservation Association. Has written over 80 professional publications, most of which have focused on caves and karst.
  Forests on Karst; Tourist Caves: Algae and Lampenflora

Andrews, Peter
Professor, Department of Paleontology, Natural History Museum, London, UK. Author of *Owls, Caves, and Fossils: Predation, Preservation, and Accumulation of Small Mammal Bones in Caves* (1990). Contributor to *Miocene Hominoid Fossils: Functional and Phylogenetic Implications*, edited by D. Begun (1997) and to the journals *Journal of Zoology, Nature, Philosophical Transactions of the Royal Society of London, Science*.
  Paleontology: Animal Remains in Caves

Audra, Philippe
Departement de géographie, Université Nice-Sophia-Antipolis, Nice, France. Contributor to *Cave and Karst Science, Journal of Cave and Karst Studies, Karstologia*.
  Dent de Crolles Cave System, France; Nakanai Caves, Papua New Guinea; Vercors, France

Auler, Augusto
Researcher at CPMTC, Departamento de Geografia, Universidade Federal de Minas Gerais, Belo Horizonte, Brazil. Co-author of *As Grandes Cavernas do Brasil* (2001). Contributor to *Cave Minerals of the World*, edited by C.A. Hill & P. Forti (1997); *Global Karst Correlation*, edited by Yuan Daoxian (1998), and to the journals *Cave and Karst Science, Journal of Cave and Karst Studies, Karstologia, Proceedings of the University of Bristol Spelaeological Society, Quaternary Research*, and *Zeitschrift für Geomorphologie*.
  America, South; America, South: History; Bambuí Karst, Brazil; Boa Vista, Toca da Boa Vista, Brazil; Quartzite Caves of South America

Badino, Giovanni
Researcher, Dipartimento di Fisica Generale, Università di Torino, Italy. Author of *Gli Abissi Italiani* (1984); *Tecnica di Grotta* (1988); *Tecniche di Grotta* (1992); *Fisica del Clima Sotterraneo* (1995); *Grotte e Forre: Tecniche Speciali* (1997); *Il Fondo di Piaggia Bella* (1999). Former President, Società Speleologica Italiana. Member of the steering board and of the scientific board of Associazione La Venta, devoted to underground exploration all over the world.
  Antarctica

Bahn, Paul G.
Freelancer researcher, and non-executive director, Wetland Archaeology and Environments Research Centre, University of Hull, UK. Author or co-author of *Easter Island, Earth Island* (1992); *Journey through the Ice Age* (1997); *The Cambridge Illustrated History of Prehistoric Art* (1998); *Mammoths* (2nd edition, 2000); *Archaeology: Theories, Methods and Practice* (3rd edition, 2000). Editor of *The Story of Archaeology* (1996); *The Cambridge Illustrated History of Archaeology* (1996); *The Atlas of World Archaeology* (2000); *The Archaeology Detectives* (2001); *The Penguin Archaeology Guide* (2001). Contributor to *Past Worlds: The Times Atlas of Archaeology*, edited by C. Scarre (1988); *The Oxford Companion to Archaeology*, edited by B.M. Fagan (1996), and to journals on the topics of Ice Age art, prehistoric art, Easter Island and the history of archaeology. Contributing editor of *Archaeology* magazine, advisory editor to *Antiquity* and *The Rapa Nui Journal*.
  Altamira Cave, Spain: Archaeology; Ardèche Caves, France: Archaeology; Art in Caves; Art: Cave Art in Europe; Atapuerca Caves, Spain: Archaeology; Vézère Caves, France: Archaeology

Bárány-Kevei, Ilona
Head of Department of Climatology and Landscape Ecology, University of Szeged, Hungary. Author of *Biogeography* (in

Hungarian, 1991); *Theory and Experience of Geoecological Mapping* (in Hungarian, 1997); *Soil Geography* (in Hungarian, 1998); *Methods of Field and Laboratory Research in Geography* (in Hungarian, 1998); *Structure and Function of Karst Ecological System* (in Hungarian, 2002). Contributor to *Proces et Mesure de l'érosion*, edited by A. Godard & A. Rapp (1984); *Karst et Évolutions climatiques*, edited by J.N. Salomon & R. Maire (1992); *New Perspectives in Hungarian Geography* (1992); *Physico-Geographical Research in Hungary*, edited by A. Kertész & F. Schweitzer (2000). Editor of *Environmental Effects on Karst Terrains* (1995); *Urban Climate* (1997); *Essays in the Ecology and Conservation of Karst* (1999, with John Gunn); *Acta Climatologica* (1997, 1999). Contributor to *Acta Carsologica, Cave and Karst Science, Mittelungen des Verbandes der deutschen Höhlen- und Karstforscher, Mittelungen der Österreichischen Geographischen Gesellschaft, Studia Carsologica,* and *Zeitschrift für Geomorphologie*.
  Bauxite Deposits in Karst

Bartsiokas, Antonis
Assistant Professor, Department of History and Ethnology, Democritus University of Thrace, Greece, and Director, Department of Speleology, Anaximandrian Institute of Human Evolution, Voula, Greece. Author of *The Palaeontology of Cythera Island* (1988). Contributor to *Bulletin de la Société Spéléologique de Grèce, Journal of Archaeological Science, Journal of Human Evolution, Proceedings of the Royal Society, Science,* and *The Anatomical Record*.
  Europe, Mediterranean: Archaeological and Paleontological Caves; Shanidar Cave, Iraq

Beddows, Patricia A.
PhD student, School of Geographical Sciences, University of Bristol, UK. Contributor to *Speleogenesis: Evolution of Karst Aquifers*, edited by A.B. Klimchouk, D. Ford, A. Palmer & W. Dreybrodt (2000).
  Yucatán Phreas, Mexico

Bednarik, Robert G.
President, International Federation of Rock Art Organisations (IFRAO), Melbourne, Australia. Author of *Rock Art Science: The Scientific Study of Paleoart* (2001). Co-author of *Indian Rock Art and its Global Context* (1997); *Nale Tasih: Eine Floßfahrt in die Steinzeit* (1999). Contributor to *Rock Art and Posterity: Conserving, Managing and Recording Rock Art*, edited by C. Pearson & B.K. Swartz (1991); *Ancient Images, Ancient Thought: The Archaeology of Ideology*, edited by S. Goldsmith, S. Garvie, D. Selin & J. Smith (1992); *Preservation of Rock Art*, edited by A. Thorn & J. Brunet (1996); *Indian Archaeology in Retrospect, Vol. 1, Prehistory*, edited by S. Settar & Ravi Korisettar (2000). Contributor to many journals, including *Anthropologie, Antiquity, Cambridge Archaeological Journal, Current Anthropology, Helictite, Journal of World Prehistory, Studies in Speleology*. Editor of *Rock Art Research*, the *AURA Newsletter* and *CARA Newsletter*; editor of the *AURA Occasional Monographs*; editor-in-chief of the *IFRAO-Brepols Rock Art Series*.
  Art in Caves: History; Art: Cave Art in Australasia

Bedos, Anne
Laboratoire d'Ecologie Terrestre, Université Paul Sabatier, Toulouse, France. Contributor to *Subterranean Ecosystems*, edited by H. Wilkens, D.C. Culver & W.F. Humphreys (2000).
  Salukkan Kallang, Indonesia: Biospeleology

Beltram, Gordana
Counsellor to the Director, Nature Conservation Unit, Environment Agency of the Republic of Slovenia, Ljubljana, Slovenia. Co-editor of *Convention on Biodiversity: National Report of the Republic of Slovenia* (1997). Contributor to *Educating for Sustainable Tourism: Proceedings of the International Conference held in Slovenia* (1992); *Nature for East and West: Proceedings of the SAEFL*, edited by M. Kütel & G. Thélin (1998); *Ramsarska konvencija in slovenska mokrišča* [The Ramsar Convention and Wetlands in Slovenia], edited by A. Sovinc (1999); *Mediterranean Wetlands at the Dawn of the 21st Century*, edited by Th. Papayanis & T. Salathé (1999); and to the journals *Coastline: EUCC Magazine* and *Proteus*.
  Ramsar Sites

Bernasconi, Reno
Retired from governmental management of the quality control of pharmaceuticals, Switzerland. Author of *Revision of the Genus Bythiospeum (Mollusca Prosobranchia Hydrobiidae) of France* (1990). Contributor to *Proceedings of the 12th International Congress of Speleology* (1997); *Fauna Helvetica 2: Mollusca. Atlas der Mollusken der Schweiz und Liechensteins* (1998); and to the journals *Documents Malacologiques* and *Revue Suisse de Zoologie*. Former Editor of *Speleological Abstracts* (1975–97).
  Mollusca

Bonacci, Ognjen
Professor and Head of Hydrology Department, Civil Engineering Faculty, University of Split, Croatia. Author of *Karst Hydrology* (1987); *Rainfall Main Input into Hydrological Cycle* (in Croat, 1994). Co-editor of *Regionalization in Hydrology* (1990). Contributor to *European Water Management, Hydrogeology Journal, Hydrological Processes, Hydrological Sciences Journal, Journal of Hydrology, Regulated Rivers: Research and Management*.
  Plitvice Lakes, Croatia; Poljes; Ponors

Boston, Penny
Director, Cave and Karst Research Program, New Mexico Institute of Mining and Technology. Editor of *Astrobiology Journal* and four proceedings volumes. Contributor to *Encyclopedia of Environmental Microbiology* (2001) and to *Chemical Geology, Journal of Cave and Karst Studies, NSS News*.
  Biofilms; Extraterrestrial Caves

Bottrell, Simon
Contributor to *Speleogenesis: Evolution of Karst Aquifers*, edited by A. Klimchouk, D. Ford, A. Palmer & W. Dreybrodt (2000), and to the journals *Cave and Karst Science, Cave Science, Earth Surface Processes and Landforms* as well as many geochemical journals.
  Microbial Processes in Caves; Sulfide Minerals in Karst

Boutin, Claude
Professor, Laboratoire d'Ecologie Terrestre, Université Paul Sabatier, Toulouse, France. Contributor to *The Natural History of Biospeleology*, edited by A.I. Camacho (1992) and to the journals *Crustaceana, Hydrobiologia, Mémoires de Biospéologie, Stygologia*.
  Crustacea: Isopoda, aquatic Organisms: Classification

Boyer, Douglas G.
Research Hydrologist, Appalachian Farming Systems Research Center, US Department of Agriculture, Agricultural Research Service, Beaver, West Virginia, USA. Contributor to *Agricultural and Forest Meteorology, Earth Surface Processes and Landforms, Journal of the American Water Resources Association, Soil Science.*
 Groundwater Pollution: Dispersed; Soils on Carbonate Karst; Soil Erosion and Sedimentation

Brady, James E.
Professor, Department of Anthropology, California State University at Los Angeles, USA. Co-editor of *In the Maw of the Earth Monster: Studies of Mesoamerican Ritual Cave Use* (in press). Contributor to *Archaeology of Ancient Mexico and Central America: An Encyclopedia*, edited by S.T. Evans & D.L. Webster (2001) and to the journals *Archaeology Magazine, American Antiquity, Journal of Cave and Karst Studies.*
 America, Central: Archaeological Caves

Brancelj, Anton
Head of Department for Freshwater and Terrestrial Ecosystems Research, National Institute of Biology, Ljubljana, Slovenia. Contributor to *Cladocera: The Biology of Model Organisms*, edited by A. Brancelj, L. De Meester & P. Spaak (1997), and to the journals *Hydrobiologia, Biološsi vestnik.*
 Crustacea: Copepoda

Brison, David N.
Retired television cameraman, France 2 Television. Contributor to *Grottes et Gouffres, Subterránea, Flash* (NSS Cave Photography Section).
 Music about and in Caves

Brook, George A.
Professor, Department of Geography, University of Georgia, Athens, Georgia, USA. Contributor to *Evolution of Environments and Hominidae in the African Western Rift Valley*, edited by N.T. Boaz (1990); *Quaternary Deserts and Climatic Change*, edited by A.S. Alsharhan, K.W. Glennie, G.L. Whittle & C.G. St. C. Kendall (1998); *Paleoenvironmental Reconstruction in the Arid Lands*, edited by A.K. Singhvi & E. Derbyshire (1999); and to the journals *Holocene, Journal of Archaeological Science, Palaeogeography, Paleoclimatology and Palaeoecology, Zeitschrift für Geomorphologie.*
 Africa, Sub-Saharan

Brooks, Simon
UK. Co-editor of *Caving in the Abode of the Clouds; The Caves and Karst of Meghalaya, N.E. India* (1994); *Caving in the Abode of the Clouds, Part 2: The Caves and Karst of Meghalaya, North East India* (1998). Contributor to *Caves and Caving, Descent, International Caver, Stalactite, Speleologia.*
 Indian Subcontinent

Bulog, Boris
Associate Professor, Department of Biology, Biotechnical Faculty, University of Ljubljana, Slovenia. Co-author of *Proteus: The Mysterious Ruler of Karst Darkness* (1993). Contributor to *Acta Biologica, Acta Carsologica, European Journal of Physiology, Journal of Morphology, Mémoires de Biospéologie, Water, Air and Soil Bulletin.*
 Amphibia: Proteus

Bunnell, David
USA. Author of *Sea Caves of Santa Cruz Island* (1988); *Sea Caves of Anacapa Island* (1993). Editor and contributor, *NSS News.*
 Littoral Caves

Camacho, Ana I.
Científico Titular CSIC, Departamento de Biodiversidad y Biología Evolutiva, Museo Nacional de Ciencias Naturales, Madrid, Spain. Editor of *The Natural History of Biospeleology* (1992). Contributor to *Grandes Cuevas y Simas de España*, edited by C. Puch (1998) and to the journals *Archiv für Hydrobiologie, Hydrobiologia, Mémoires de Biospéologie, Stygologia, Trav. Inst. Spéol. "Émile Racovitza".*
 Crustacea: Syncarida

Camassa, Michele M.
Istituto Sperimentale di Biologia del Sottosuolo "P. Parenzan", Latiano (Brindisi), Italy. Contributor to the journals *Mémoires de Biospéologie, Environmental Biology of Fishes, Thalassia Salentina.*
 Food Resources

Chamberlain, Andrew T. (Adviser)
Senior lecturer, Department of Archaeology and Prehistory, University of Sheffield, UK. Author of *Human Remains* (1994); *A Catalogue of Quaternary Fossil-Bearing Cave Sites in the Plymouth Area* (1994); *Earthly Remains: The History and Science of Preserved Human Bodies* (2001). Contributor to *Companion Encyclopedia of Archaeology*, edited by G. Barker (1999); *Human Osteology in Archaeology and Forensic Science*, edited by M. Cox & S. Mays (2000); *Evolution and the Human Mind*, edited by P. Carruthers & A. Chamberlain (2000); *Handbook of Archaeological Sciences*, edited by D.R. Brothwell & A.M. Pollard (2001). Contributor to *Holocene, Journal of Archaeological Science, Nature.*
 Britain and Ireland: Archaeological Caves

Chenoweth, Sean
Department of Geography, University of Wisconsin-Milwaukee, USA.
 Cockpit Country Cone Karst, Jamaica

Christiansen, Kenneth
Professor Emeritus of Biology, Grinnell College, Grinnell, Iowa, USA. Author or co-author of *The Collembola of North America North of the Rio Grande* (1980; revised 1998); "Collembola" in *Soil Biology*, edited by D. Dindal (1990); *The Collembola of Hawaii* (1992); "Cave life in the light of modern evolutionary theory" in the *Natural History of Biospeleology*, edited by A. Camacho (1992). Contributor to many journals, including *Annales de Spéléologie, Evolution, Journal of Biogeography, NSS Bulletin.*
 Adaptation: Morphological (External); Speciation

Cigna, Arrigo A.
Professor, Società Speleologica Italiana, Cocconato (Asti), Italy. Editor of *Proceedings of the 1st Congress of the International Show Caves Association; Proceedings of the 3rd Congress of the International Show Caves Association* (2000); *International Journal of Speleology.*

Contributor to *Karst Terrains: Environmental Changes and Human Impact*, edited by P. Williams (1993); *Cave Minerals of the World*, edited by C.A. Hill & P. Forti (2nd edition, 1997), and to the proceedings of national and international speleological congresses, and the journals *Environmental Geology, International Journal of Speleology, Le Grotte d'Italia, NSS News, VIA*. Honorary Member, National Speleological Society, USA.
 Climate of Caves

Ciochon, Russell L.
Professor, Department of Anthropology, University of Iowa, USA. Contributor to *Acta Anthropologica Sinica, American Scientist, Archaeology, Natural History, Nature*.
 China: Archaeological Caves

Cobolli, Marina
Department of Human and Animal Biology, La Sapienza University, Rome, Italy.
 Insecta: Pterygota

Coineau, Nicole
Université Pierre et Marie Curie, Observatoire Océanologique de Banyuls, Laboratoire Arago, Banyuls-sur-mer, France. Contributor to *Stygofauna Mundi*, edited by L. Botosaneanu (1986); *The Natural History of Biospeleology*, edited by A.I. Camacho (1992); and to the journals *Crustaceana, Mémoires de Biospéologie, Stygologia*.
 Crustacea: Isopoda (Aquatic); Crustacea: Syncarida; Interstitial Habitats (Aquatic)

Constantin, Silviu
Senior researcher, Department of Geospeleology, "Emil Racovita" Institute of Speleology. Bucharest, Romania. Co-author of *Karst of Southwestern Romania* (1996); *Sisteme carstice majore din zona Resita-Moldova Noua* [Major karstic systems from Resita-Moldova Noua area] (1996). Editor of *The Great Romanian Caves Atlas* (in preparation). Contributor to *Bulletin Speleogic, International Caver, Journal of Quaternary Science, Theoretical and Applied Karstology*. Managing Editor, *Theoretical and Applied Karstology*, and former Editor, *Speleotex*.
 Europe, Balkans and Carpathians

Cordingley, John
British Cave Rescue Council Diving Officer and UK Cave Diver (CDG Qualified), UK. Author of *The Peak Cavern System, A Cavers' Guide* (1986); "Peak Cavern, a climb beyond The Far Sump" in *The Last Adventure*, edited by A. Thomas (1989); *The Peak District Sump Index 1994; The Peak District Sump Index Update 1997*. Co-editor of *The Cave Diving Group Manual 1990*. Contributor to *Cave and Karst Science, Compass Points (Journal of the BCRA Cave Survey Group), UIS International Cave Diving Magazine*, and a large number of contributions on cave exploration to the major British caving journals over 25 years.
 Diving in Caves

Crawford, Nicholas C.
Professor, Center for Cave and Karst Studies, Applied Research and Technology Program of Distinction, Department of Geography and Geology, Western Kentucky University, Bowling Green, Kentucky, USA. Author of over 200 articles and technical reports dealing primarily with groundwater contamination of karst aquifers. Co-author of *Living with Karst* (2001). Honorary Member, National Speleological Society, USA.
 Water Tracing: History

Crothers, George M.
Director, William S. Webb Museum of Anthropology and Assistant Professor, Department of Anthropology, University of Kentucky, Lexington, Kentucky, USA. Contributor to *Caving Basics: A Comprehensive Guide for Beginning Cavers*, edited by G.T. Rea (3rd edition, 1992); *Formation Processes in Archaeological Context*, edited by P. Goldberg, D.T. Nash & M.D. Petraglia (1993); *Fleeting Identities: Perishable Material Culture in Archaeological Research*, edited by P.B. Drooker (2001); *The Woodland Southeast*, edited by D.G. Anderson & R.C. Mainfort (2002); and to the journals *Journal of Archaeological Science, Missouri Cave and Karst Conservancy Digest, NSS Bulletin, Tennessee Anthropologist*.
 America, North: Archaeological Caves

Culver, David C.
Professor, Department of Biology, American University, Washington, DC, USA. Author of *Cave Life* (1982); co-author of *Adaptation and Natural Selection in Caves* (1995). Contributor of articles to *Conservation Biology*, edited by M. Soule (1986); *Groundwater Ecology*, edited by J. Gibert, D. Danielopol & J. Stanford (1994); *Terrestrial Ecosystems of North America*, edited by T.H. Ricketts *et al.* (1999); *Subterranean Ecosystems* (2000). Co-editor of *Subterranean Ecosystems* (2000). Contributor to *Conservation Biology, Journal of Biogeography, Journal of Cave and Karst Studies, The Sciences, Stanford Journal of Environmental Law*. Honorary Member, National Speleological Society, USA.
 Adaptation: Genetics

Davis, G. Donald
Denver, Colorado, USA. Contributor to *Lechuguilla: Jewel of the Underground*, edited by M.R. Taylor (1991); *Caving Basics*, edited by T. Rea (3rd edition, 1992), and to the journals *Alpine Karst, Geo2, Journal of Caves and Karst, Journal of Spelean History, NSS Bulletin, NSS News, Rocky Mountain Caving*. Honorary member, National Speleological Society, USA.
 Grand Canyon, United States

Davis, Emily (Adviser)
Owner, Speleobooks, Schoharie, New York, USA. Co-editor of *Guide to Speleological Literature of the English Language 1794–1996* (1998).
 Accidents and Rescue

Day, Mick
Professor, Department of Geography, University of Wisconsin-Milwaukee, USA. Contributor to *Earth Surface Processes and Landforms, Bulletin of the Geological Society of America, Zeitschrift für Geomorphologie, Cave and Karst Science, Journal of Cave and Karst Studies*.
 America, Central; Aves (Birds); Cockpit Country Cone Karst, Jamaica; Cone Karst; Military Uses of Caves; Morphometry; Tower Karst

Debevec, Vanja
General Secretary of the International Union of Speleology (UIS) Permanent Commission of Speleotherapy, and Head of Department for Research and Development, Park Škocjanske jame, Slovenia.
    Speleotherapy

Decu, Vasile
Senior Researcher and Chief of the Biospeleological Department of the Speleological Institute "Emil Racovitza", Romanian Academy, Bucharest, Romania. Author or co-author of *Recherches sur les grottes du Banat et d'Olténie* (1967); *Subterranean World* (in Romanian, 1971); *Caves of Romania* (in Romanian, 1976); *Initiation à la biologie et a l'écologie souterraines* (1977); *Chiroptera of Romania* (in Romanian, in press). Co-editor of *Résultats des expéditions biospéologiques cubano-roumaines à Cuba* (1973); *Fauna hipogea y hemiédafica de Venezuela y de otros paises de América del Sur* (1987); *Soil Fauna of Israel* (1995); *Encyclopedia Biospeologica* (vol. 1, 1987, vol. 2, 1998, vol. 3, 2001). Contributor to many journals, including *Acta Zoologica Cracoviensia, Annales de Spéléologie, Archiv für Hydrobiologie, Mémoires de Biospéologie, NSS Bulletin, Travaux de l'Institut de Spéologie "Emile Racovitza"*.
    Insecta: Coleoptera; Interstitial Habitats (Terrestrial)

Deharveng, Louis
Directeur de Recherches, CNRS, Laboratoire d'Ecologie Terrestre, Université Paul Sabatier, Toulouse, France. Contributor to *Subterranean Ecosystems*, edited by H. Wilkens, D.C. Culver & W.F. Humphreys (2000), and to many journals including *Mémoires de Biospéologie*. Co-editor *Mapping Subterranean Biodiversity: Cartographie de la biodiversité souterraine* (2001).
    Asia, Southeast: Biospeleology; Insecta: Apterygota; Salukkan Kallang, Indonesia: Biospeleology

Dilsiz, Cuneyt
Faculty of Earth and Life Sciences, Vrije University, Netherlands. Contributor to *Environmental Geology*.
    Pamukkale, Turkey

Drew, David
Senior Lecturer, Geography Department, Trinity College, Dublin, Ireland. Author of *Man–Environment Processes* (1983); *Karst* (1989); *The Burren Karst* (2001); *The Karst of Ireland* (2001) and contributor to *Karren Landforms*, edited by J. Fornos & A. Gines (1996); *Gambling with Groundwater: Physical, Chemical and Biological Aspects of Aquifer-Stream Relations*, edited by J. van Brahana (1998); *Hydrogeology and Engineering Geology of Sinkholes and Karst, 1999*, edited by B.F. Beck, A.J. Pettit & J.G. Herring (1999); *Groundwater in the Celtic Regions*, edited by N.S. Robins & B.D.R. Misstear (2000); *Oxford Companion to the Earth*, edited by P.L. Hancock & B.J. Skinner (2000). Co-editor and author of articles in *Karst Hydrogeology and Human Activity: Impacts, Consequences and Implications* (1999). Contributor to *Environmental Geology, Proceedings of the Geological Association*.
    Burren Glaciokarst, Eire; Groundwater Protection

Dreybrodt, Wolfgang
Professor, Institute of Experimental Physics, University of Bremen, Germany. Author of *Processes in Karst Systems: Physics, Chemistry and Geology* (1988). Co-editor of *Speleogenesis: Evolution of Karst Aquifers* (2000). Contributor to *Global Karst Correlation*, edited by Yuan Daoxian (1998); *Karst Modeling* (1999) and to the journals *Cave and Karst Science, Chemical Geology, Environmental Geology, Geochimica et Cosmochimica Acta*, and *Zeitschrift für Geomorphologie*.
    Carbonate Minerals: Precipitation; Dissolution: Carbonate Rocks; Dissolution: Evaporite Rocks; Erosion Rates: Theoretical Models; Speleogenesis: Computer Models

Dublyansky, Victor
Professor, Department of Geology, Perm State University, Russia. Author of *Karst Caves and Shafts of the Crimea Mountains* (1977, in Russian); *Karst Caves of the Ukraine* (1980, in Russian); *Hydrogeology of Karst in Alpean Folded Zone of Southern USSR* (1984, in Russian); *Karst and Groundwaters of the Mountain Massifs of the Western Caucasus* (1985, in Russian); *Microclimate of Karst Caves in the Crimean Mountains* (1989, in Russian); *An Intriguing Speleology* (2000, in Russian). Contributor to *Speleogenesis: Evolution of Karst Aquifers*, edited by A. Klimchouk, D. Ford, A. Palmer & W. Dreybrodt (2000), and to the journals *Journal of Cave and Karst Studies, Studies in Speleology*.
    Crimea, Ukraine

Dunkley, John Robert
Consultant, Dunkley Consulting Pty Ltd, Canberra, Australia. Author of *The Exploration and Speleogeography of Mammoth Cave, Jenolan* (1971); *Jenolan Caves: As they Were in the Nineteenth Century* (1986); *A Bibliography of the Jenolan Caves* (1988); *The Caves of Thailand* (1995). Co-editor of *Caves of the Nullarbor* (1967); *Karst of the Central West Catchment, NSW: Resources, Impacts and Management* (2000). Contributor to *Cave Science, Helictite*.
    Asia, Southeast: History

Dunne, Suzanne
Postgraduate student, Department of Geography, Trinity College, Dublin, Ireland.
    Groundwater Protection

Eaves-Johnson, K. Lindsay
Masters candidate, Department of Anthropology, University of Iowa, USA.
    China: Archaeological Caves

Eavis, Andy
UK. Leader, China Caves project. President, Speleological Federation of the European Community. Vice President, International Union of Speleology (UIS).
    Exploring Caves

Eriksen, Berit Valentin
Department of Prehistoric Archaeology, University of Aarhus, Denmark. Author of *Change and Continuity in a Prehistoric Hunter-Gatherer Society: A Study of Cultural Adaptation in Late Glacial-Early Postglacial Southwestern Germany* (1991). Co-editor of *Humans at the End of the Ice Age: The Archaeology of the Pleistocene–Holocene Boundary* (1996); *As the World Warmed: Human Adaptations across the Pleistocene–Holocene Boundary* (1998). Contributor to *The Human Use of Caves*, edited by C. Bonsall & C. Tolan-Smith (1997).
    Europe, Central: Archaeological Caves

Fairchild, Ian
Professor of Earth Surface Processes and Dean of Natural Sciences, School of Earth Sciences and Geography, Keele University, UK. Over 100 published contributions, including to the journals *Chemical Geology, Geochimica Cosmochimica Acta, Journal of the Geological Society, London, Science.*
  Chemistry of Natural Karst Waters

Farrant, Andrew
British Geological Survey, Keyworth, Nottingham, UK. Author of *Walks around the Caves and Karst of the Mendip Hills* (1999) and co-author of *Karst and Caves of Great Britain* (1997). Contributor to *Cave and Karst Science, Geology, Proceedings of the University of Bristol Spelaeological Society.*
  Draenen, Ogof Draenen, Wales; Mendip Hills, England; Paleoenvironments: Clastic Sediments; Paragenesis

Filippov, Andrej G.
Formerly, East Siberian Research Institute of Geology, Geophysics and Mineral Resources, Irkutsk, Russia.
  Mineral Deposits in Karst; Siberia Caves, Russia

Finlayson, Clive
Director of the Gibraltar Museum, Gibraltar and co-director of the Gibraltar Caves Research Project. Author of *Birds of the Strait of Gibraltar* (1992). Co-editor of *Neanderthals on the Edge: 150th Anniversary Conference of the Forbes's Quarry Discovery, Gibraltar* (2000) and *Gibraltar during the Quaternary* (2000). Contributor to *Neanderthals on the Edge* (2000) and *Studies in Honour of D.A.E. Garrod*, edited by R. Charles & W. Davis (1998) and to the journal *Antiquity*.
  Gibraltar Caves: Archaeology

Ford, Derek (Adviser)
Emeritus Professor, School of Geography and Geology, McMaster University, Hamilton, Ontario, and Adjunct Professor of Earth Sciences, University of Waterloo, Canada. Co-author of *Karst Geomorphology and Hydrology* (1989): Co-editor of *Paleokarst: A World Regional and Systematic Review* (1989); *Geomorphology sans frontières* (1995); *Speleogenesis; Evolution of Karst Aquifers* (2000); *Present State and Future Trends of Karst Studies* (2001). Contributor to many journals including most recently *Bulletin d'Hydrogeologie, Chemical Geology, Quaternary Research*. Honorary Member, National Speleological Society, USA.
  Aggtelek and Slovak Karst, Hungary–Slovakia; Bear Rock Karst, Canada; Canada; Carbonate Karst; Castleguard Cave, Canada; Karst; Nahanni Karst, Canada; Paleoenvironments: Speleothems; Solution Breccias; Speleogenesis: Unconfined Settings

Forti, Paolo
Professor of Geomorphology and Speleology, University of Bologna, Italy. Co-author of *Le cavita' naturali dell'Iglesiente* (1982); *I cristalli di gesso del Bolognese* (1983); *La cavita' naturali della Repubblica di San Marino* (1983); *Cave Minerals of the World* (1986; 2nd edition 1997). Editor or co-editor of *L'idrogeolia del bacino minerario dell'Iglesiente* (1983); *Morfologie Carsiche e Speleogenesi—diapositive didattiche* (1993); *Grotte marine d'Italia* (1994); *Volcanospeleology* (1998); *Karst Geomorphology* (1999). Contributor of articles to *Karst Terrains: Environmental Changes and Human Impact*, edited by P. Williams (1991); *La Vena del Gesso*, edited by U. Bagnaresi et al. (1994); *Il lago del Fucino e il suo emissario*, edited by E. Burri & G. Tavano (1994); *Gypsum Karst of the World*, edited by A. Klimchouk et al. (1997). Contributor to *Atmospheric Environment, International Journal of Speleology, Marine Geology, NSS Bulletin, Tectonophysics*.
  Books on Caves; Frasassi Caves, Italy; Journals on Caves; Paleotectonics from Speleothems; Speleothems: Carbonate

Frankland, John
MD, Lancaster, UK. Cave rescue doctor for the busiest cave rescue team in the world (the Cave Rescue Organization in Yorkshire, England) for 30 years. Co-author of *Race Against Time* (1988), a detailed history of the Cave Rescue Organization's first 50 years.
  Accidents and Rescue

Frumkin, Amos
Senior Lecturer in Geography and Director of Cave Research Center, Department of Geography, The Hebrew University of Jerusalem, Israel. Contributor to *Salt Tectonics*, edited by I. Alsop, I., D. Blundell & I. Davison (1996); *Cave Minerals of the World*, edited by C.A. Hill & P. Forti (1997) and to the journals *Earth Surface Processes and Landforms, Israel Journal of Earth Sciences, Journal of Caves and Karst Studies, Quaternary Research, The Holocene*.
  Sedom Salt Karst, Israel

Gabrovšek, Franci
Researcher, Karst Research Institute, The Scientific Research Center of the Slovenian Academy of Sciences and Arts, Slovenia. Author of *Evolution of Karst Aquifers: From Simple Principles to Complex Models* (2000), and contributor to *South China Karst*, edited by Chen Xiaoping et al. (1998); *Speleogenesis: Evolution of Karst Aquifers*, edited by A. Klimchouk, D. Ford, A. Palmer & W. Dreybrodt (2000). Editor of *Evolution of Karst: From Prekarst to Cessation* (2002). Contributor to *International Caver, Proceedings of the 12th International Congress of Speleology*.
  Kanin Massif, Slovenia-Italy

Galdenzi, Sandro
Instituto Italiano di Speleologia, Ancona, Italy. Co-editor *Il carsismodella Gola di Frasassi* (1990). Contributor to *Subterranean Ecosystems*, edited by H. Wilkens, D.C. Culver & W.F. Humphreys (2000) and *Journal of Cave and Karst Studies, Environmental Geology*.
  Frasassi Caves, Italy

Gauchon, Christophe
Maitre de Conférences, Agrégé de Géographie, Laboratoire de Géographie de l'Université de Savoie, France. Author of *Des Cavernes et des Hommes* (1997). Contributor to *Spelunca*. Editorial board member of *Karstologia*. Secretary of karst phenomena Commission (French National Committee of Geography).
  France: History

Gebauer, Daniel
Author of *Caves of India and Nepal* (1983); *Kurnool 1984* (1985); *Speleological Bibliography of South Asia* (1996); *Unexplored Caves and Limestone Areas of Arunachal Pradesh, Assam, Manipur, Mizoram and Tripur (North-east India)* (1997); *Caves of Mizoram (North-east India)* (1999). Contributor to the journals *British Caving, Caves and*

Caving, Grottes et Gouffres, International Caver, International Journal of Speleology, Spelunca, and *Stalactite.*
  Indian Subcontinent

Gibson, David
Consultant Electronic Design Engineer (self-employed). Technical Editor of the BCRA's *Cave Radio & Electronics Group Journal.* Contributor to *Advanced Signal Processing for Communication Systems,* edited by T. Wysocki, M. Darnell & B. Honary (2002), and to the journals *Cave and Karst Science, Cave Radio & Electronics Group Journal, Compass Points, Electronics World.*
  Communications in Caves; Radiolocation

Gillieson, David (Adviser)
Professor of Geography and Head of School, School of Tropical Environment Studies and Geography, James Cook University, Cairns, Australia. Author of *Caves: Processes, Development and Management* (1996) and co-author of *Guidelines for Cave and Karst Protected Areas* (1997). Contributor to *Global Karst Correlation,* edited by Yuan Daoxian (1998); *Karst Hydrogeology and Human Activities: Impacts, Consequences and Implications,* edited by D. Drew & H. Hötzl (1999); *Oxford Companion to the Earth Sciences,* edited by B. Skinner & I. Stewart (2000); *Australia Underground,* edited by E. Hamilton-Smith & B. Finlayson (2003). Contributor to *Earth Surface Processes and Landforms, Environmental Geology, International Journal of Speleology, Proceedings of the Linnean Society of New South Wales,* and *Zeitschrift für Geomorphologie.*
  Asia, Southeast Islands; Chillagoe and Mitchell-Palmer Karsts, Australia; Floral Resources; Nullarbor Karst, Australia; Sediments: Allochthonous Clastic

Ginés, Àngel
Associate Professor of Ecology, Department of Biology, Universitat de les Illes Balears, Palma de Mallorca. Co-editor of *El carst I les coves de Mallorca* (1995); *Karren Landforms* (1996). Contributor to *Late Quaternary Sea-level Changes in Spain,* edited by C. Zazo (1987); *El karst en España,* edited by J.J. Durán & J. López-Martínez (1989); *The Natural History of Biospeleology,* edited by A.I. Camacho (1992); *Environmental Effects on Karst Terrains,* edited by I. Bárány-Kevei (1995); *Karst and Agriculture of the World,* special issue of *International Journal of Speleology* (1999), and to the journals *Acta Carsologica, Endins, Earth Surface Processes and Landforms, Geodinamica Acta, Quaternary Science Reviews.*
  Europe, Mediterranean; Karren

Glover, Ian
Institute of Archaeology, University College London, UK. Author of *Archaeology in Eastern Timor, 1966–67* (1986); *Early Trade Between India and Southeast Asia—A Link in the Development of a World Trading System* (1990). Contributor to *South Asian Archaeology,* edited by N. Hammond (1973); *Early South-East Asia: Essays in Archaeology, Historical Geography,* edited by R. Smith & W. Watson (1979); and to the journal *World Archaeology.* Dr Glover has excavated many caves in Indonesia.
  Asia, Southeast: Archaeological Caves

Goldie, Helen
Department of Geography, University of Durham, UK. Contributor to *Karst and Caves of Great Britain,* edited by A.C. Waltham, M.J. Simms, A.R. Farrant & H.S. Goldie (1997), and to the journals *Acta Geographica Szegediensis, Cave and Karst Science, Environmental Geology,* and *Zeitschrift für Geomorphologie.*
  Ornamental Uses of Limestone

Grimes, Ken G.
Consultant geologist, Regolith Mapping, Australia. Co-author of *Field Guidebook to Karst and Volcanic Features in Southeast South Australia and Western Victoria* (1999). Contributor to *Australian Cave and Karst Management Association Journal, Cave and Karst Science, Environmental Geology, Helictite, Memoirs of the Queensland Museum.* Editor of *Helictite*: The Journal of Australasian Speleological Research.
  Syngenetic Karst

Günay, Gültekin
International Research and Application Center for Karst Water, Hacettepe University, Ankara, Turkey. Co-editor of *Karst Water Resources* (1996); *Karst Waters and Environmental Impacts* (1997). Contributor to many journals including *Cave and Karst Science, Environmental Geology, Journal of Hydrology.*
  Pamukkale, Turkey; Turkey

Gunn, John (Editor)
Professor, Limestone Research Group, Department of Geography, University of Huddersfield, UK. Co-author of *Caves and Karst of the Peak District* (1990); *The Reclamation of Limestone Quarries using Landform Replication* (1997). Co-editor of *Karst Hydrology* (1998); *Essays in the Ecology and Conservation of Karst* (2000) and Editor, *An Introduction to British Limestone Karst Environments* (1994). Authored over 80 papers in journals and edited books. Chairman of the International Geographical Union's Karst Commission for the period 2000–2004. Joint Editor of *Cave and Karst Science,* Editorial Advisory Board member of *Environmental Geology.*
  Erosion Rates: Field Measurement; Fluviokarst; France, Southern Massif Central; Limestone as a Mineral Resource; Peak District, England; Quarrying of Limestone; Radon in Caves; Turkey; Valleys in Karst

Halliday, William R. (Adviser)
Nashville, Tennessee, USA. Author of *Adventure is Underground* (1959); *Caves of California* (1962); *Caves of Washington* (1963); *Depths of the Earth* (1966); *American Caves and Caving* (1974), and contributor to *Mount St. Helens Five Years Later,* edited by S.A.C. Keller (1986). Editor of *Proceedings, 1972 International Symposium on Vulcanospeleology and its Extraterrestrial Applications* (1972); *Selected Caves of the Pacific Northwest: Guidebook of the 1972 National Speleological Society Convention* (1972); *Proceedings, 1982 Third International Symposium on Vulcanospeleology* (1982); *Introduction to Hawaii Caves: Field Guide for the 6th International Symposium on Vulcanospeleology* (1991). Contributor to *Geological Newsletter, GP Magazine, NSS Bulletin/Journal of Cave and Karst Science, Science, Studies in Speleology.* Honorary President, Commission on Volcanic Caves, International Union of Speleology. Honorary Member, National Speleological Society, USA.
  America, North: History; Caves in History: The Eastern Mediterranean; Crevice Caves; Disease; Hawaii Lava Tube Caves, United States; Piping Caves and Badlands Pseudokarst; Pseudokarst; Talus Caves; Volcanic Caves; Vulcanospeleology: History

Hamilton-Smith, Elery (Adviser)
Chair, IUCN/WCPA Task Force on Cave and Karst Protection. Adjunct Professor, Cave and Karst Management, Charles Sturt University, New South Wales, Australia. Fellow of both the Australian Speleological Federation and the Australasian Cave and Karst Management Association. Co-author of *South-east Karst Province of South Australia* (1993); *Guidelines for Cave and Karst Protection* (1997); *Karst Management Considerations for the Cape Range Karst Province Western Australia* (1998). Co-editor of *Proceedings of the Asia-Pacific Forum on Karst Ecosystems and World Heritage* (2002). Contributor to *Essays in the Ecology and Conservation of Karst*, edited by I. Barany-Kevei & J. Gunn (2000); *Subterranean Ecosystems*, edited by H. Wilkens & D. Culver (2000) and to the journals *Evaluation Journal of Australia, Helictite, Natura Croatica*.
    Conservation: Protected Areas; Karst Resources and Values; Organic Resources in Caves; Tourism and Caves: History; Tourist Caves; World Heritage Sites

Häuselmann, Philipp
PhD student, Institute of Geography, University of Fribourg, Switzerland, and Postdoctoral fellow, Earth and Atmospheric Sciences, Purdue University, USA. Co-editor of *Speleogenesis in the Alpine Belt* (2001). Contributor to *Speleogenesis: Evolution of Karst Aquifers*, edited by A. Klimchouk et al. (2000); *Speleogenesis in the Alpine Belt*, edited by P. Häuselmann & M. Monbaron (2001), and to the journals *Al Ouat'Ouate, Etudes de Géographie Physique, Geodinamica Acta, Jahrbuch vom Thuner- und Brienzersee, Stalactite*.
    Siebenhengste, Switzerland

Hervant, Frédéric
Hydrobiologie et Écologie souterraines, Université Claude-Bernard Lyon I, France. Contributor to *Aquatic Sciences, Archiv für Hydrobiologie, Hydrobiologia, Journal of Experimental Biology, Mémoires de Biospéologie*.
    Adaptation: Physiological

Hildreth-Werker, Val
Hillsboro, New Mexico, USA. Co-editor of *Cave Conservation and Restoration* (2003), Co-Chair of the Conservation Division for the National Speleological Society.
    Restoration of Caves and Speleothem Repair

Hill, Carol A.
Adjunct Professor, Geology Department, University of New Mexico, Albuquerque, USA. Author of *Cave Minerals* (1976), *Geology of Carlsbad Cavern* (1987), and *Geology of the Delaware Basin* (1996). Co-editor of *Cave Minerals of the World* (1986; 2nd edition 1997). Contributor to *American Association of Petroleum Geologists Bulletin, Environmental Geology, Journal of Cave and Karst Studies, Journal of Geology*, and *Science*.
    Minerals in Caves; Speleothems: Evaporite

Hobbs, Horton H., III
Professor, Department of Biology, Wittenberg University, Springfield, Ohio, USA. Co-author of *The Crayfishes and Shrimp (Palaemonidae, Cambaridae) of Wisconsin* (1988). Editor of *A Study of Environmental Factors in Harrison's Cave, Barbados, West Indies* (1995). Contributor to *Biodiversity of the Southeastern United States: Aquatic Communities*, edited by C.T. Hackney, S.M. Adams & W.H. Martin (1992); *Conservation and Protection of the Biota of Karst*, edited by I.D. Sasowsky, D.W. Fong & E.L. White (1997); *Encyclopedia Biospeologica*, edited by C. Juberthie & V. Decu (1998); *Subterranean Ecosystems*, edited by H. Wilkens & D. Culver (2000); *Ecology and Classification of North American Freshwater Invertebrates*, edited by J.A. Thorp & A.P. Covich (2nd edition, 2001); and to the journals *American Zoologist, Conservation Biology, Crustaceana, Hydrobiologia, International Journal of Speleology, Journal of Crustacean Biology, Smithsonian Contributions to Zoology*. Honorary Member, National Speleological Society, USA.
    Crustacea; Crustacea Decapoda

Holsinger, John R.
Professor of Biological Sciences, Old Dominion University, Norfolk, Virginia, USA. Author of *The Freshwater Amphipod Crustaceans (Gammaridae) of North America* (1972); *Descriptions of Virginia Caves* (1975), and contributor to *Stygofauna Mundi*, edited by L. Botosaneanu (1986); *Encyclopedia Biospeologica* (vol. 1), edited by C. Juberthie & V. Decu (1994); *Subterranean Ecosystems*, edited by H. Wilkens & D. Culver (2000). Co-editor of *Biogeography of Subterranean Crustaceans* (1994). Contributor to *Hydrobiologia, Journal of Natural History, Proceedings of the Biological Society of Washington, Proceedings of the 12th International Congress of Speleology*. Honorary Member, National Speleological Society, USA.
    Crustacea: Amphipoda

Hope, Jeanette
River Junction Research, New South Wales, Australia.
    Australia: Archaeological and Paleontological Caves

Hose, Louise D.
Director, National Cave and Karst Research Institute, Carlsbad, New Mexico, USA. Author of articles in *Encyclopedia of Environmental Issues*, edited by C.W. Allin (2000); *Speleogenesis: Evolution of Karst Aquifers*, edited by A.B. Klimchouk, D. Ford, A. Palmer & W. Dreybrodt (2000); *Caves: Exploring Hidden Realms*, edited by M.J. Taylor (2001). Contributor to *Astrobiology, Chemical Geology, Geology, Journal of Cave and Karst Studies, NSS Bulletin, Revista Mexicana de Ciencias Geológias*. Editor of *Journal of Cave and Karst Studies* (1996–).
    Golondrinas and the Giant Shafts of Mexico; Huautla Cave System, Mexico; Selma Plateau Caves, Oman; Villa Luz, Cueva de, Mexico

Howarth, Francis G.
L.A. Bishop Chair of Zoology, Department of Natural Sciences, Bernice P. Bishop Museum, Honolulu, Hawaii, USA. Author of *Hawaiian Insects and their Kin* (1992). Co-editor of *Balancing Nature: Addressing the Impact of Importing Non-native Biological Control Agents, An International Perspective* (2001). Contributor to *The Unity of Evolutionary Biology*, edited by E.C. Dudley (1991); *The Conservation of Insects and their Habitats* (1991); *Problem Snake Management: The Habu and Brown Treesnake*, edited by G. Rodda, Y. Sawai. Chzar & H. Tanaka (1999); *Encyclopedia of Biodiversity*, edited by S. Levin (2000); *Measures of Success in Biological Control*, edited by G.M. Gurr & S.D. Wratten (2000), and contributor to the journals *American Naturalist, Bioscience, Evolution, International Journal of Speleology, Mémoires de Biospéologie, Science*, and *Trends in Ecology and Evolution*.
    Hawaiian Islands: Biospeleology

Howes, Chris
Freelance author and photographer, Cardiff, UK. Author of *Cave Photography: A Practical Guide* (1987); *To Photograph Darkness: The History of Underground and Flash Photography* (1989); *Images Below: A Manual of Underground and Flash Photography* (1997); *The Spice of Life: Biodiversity and the Extinction Crisis* (1997); *Radical Sports: Caving* (2002). Editor of *Descent*, the magazine of underground exploration. Contributor to *A Man Deep in Mendip*, edited by J. Savory (1989); *On Caves and Cameras*, edited by N. Thompson (2002), and to the journals *Cave Science, New Scientist, Proceedings of the University of Bristol Spelaeological Society*. Peter M. Hauer Spelean History Award (1993). Tratman Award for Speleological Literature (1997). Giles Barker Award for excellence in cave photography (1997). Spelean Arts and Letters Award for contributions to underground photography (1998). Outdoor Writers' Guild Award for Photographic Excellence (1998).
    Films in Caves; Photographing Caves

Hoyt, John
Aquifer Sciences Program Manager, Edwards Aquifer Authority, San Antonio, Texas, USA.
    Edwards Aquifer and the Texas Karst, United States; Groundwater Pollution: Point Sources

Humphreys, William Frank
Senior Curator, Terrestrial Invertebrate Zoology, Western Australian Museum, Australia. Editor or co-editor of *The Biogeography of Cape Range, Western Australia* (1993); *Proceedings of the XII International Congress of Arachnology* (1993); *Subterranean Ecosystems* (2000); *Subterranean Biology in Australia 2000* (2001). Author of articles in the above volumes and in *Encyclopedia Biospeleologica* vol. 3 (2001); *The Biology of Hypogean Fishes*, edited by A. Romero (2001): *Australia Underground: A Tribute to Joe Jennings* (2002). Contributor to *Comparative Biochemistry and Physiology, Crustaceana, Journal of Animal Ecology, Journal of Crustacean Biology, Journal of Zoology, Mémoires de Biospéologie, Nature, Zoological Journal of the Linnean Society, London*.
    Australia: Biospeleology

Hunt, Chris
Senior Research Associate, Division of Geographical Sciences, University of Huddersfield, UK. Contributor to *Archaeometry, Cave and Karst Science, Journal of Archaeological Science, Journal of Arid Environments, Proceedings of the Prehistoric Society*.
    Palynology

Iliffe, Thomas M.
Associate Professor, Department of Marine Biology, Texas A&M University at Galveston, USA. Contributor to *Galápagos Marine Invertebrates*, edited by M.J. James (1991); *Diversidad Biológica en la Reserva de la Biosfera de Sian Ka'an, Quintana Roo, México*, edited by D. Navarro & E. Suárez-Morales (vol. 2, 1992); *Natural History of Biospeleology*, edited by A. Camacho (1992); *Encyclopaedia Biospeologica*, edited by C. Juberthie & V. Decu (vol. 1, 1994); *Subterranean Ecosystems*, edited by H. Wilkens, D.C. Culver & W.F. Humphreys (2000); and to many journals including *Acta Carsologica, Crustacean Research, Crustaceana, Hydrobiologia, Internationale Revue der Gesamten Hydrobiologie, International Journal of Speleology, Mémoires de Biospéologie, Nature, Science, Stygologia*.
    Walsingham Caves, Bermuda: Biospeleology

James, Julia
Emeritus Professor, School of Chemistry, University of Sydney, New South Wales, Australia. Contributor to many caves and karst science books, journals, and conference proceedings.
    Accidents and Rescue; Carbon Dioxide-enriched Cave Air; Condensation Corrosion; Tourist Caves: Air Quality

Jeannin, Pierre-Yves
Hydrogeologist, Centre d'Hydrogéologie in Neuchâtel and Director of the Swiss Institute for Speleology and Karst Studies, La Chaux-de-Fonds, Switzerland. Co-editor *Modelling in Karst Systems* (1998). Contributor to *Speleogenesis: Evolution of Karst Aquifers*, edited by A.B. Klimchouk, D.C. Ford, A.N. Palmer & W. Dreybrodt (2000) and to many journals including *Cave and Karst Science, Environmental Geology, Ground Water, Stalactite, Water Resources Research*.
    Hölloch, Switzerland; Siebenhengste, Switzerland

Johnson, Steve
Senior Hydrogeologist, Edwards Aquifer Authority, San Antonio, Texas, USA.
    Edwards Aquifer and the Texas Karst, United States

Juberthie, Christian
Past Director of CNRS Laboratory, Laboratoire souterrain du CNRS, Moulis, France. Coeditor of *Encyclopaedia Biospeologica* (3 vols, 1994, 1997, 2001). Editor-in-chief *Mémoires de Biospéologie*.
    France: Biospeleology

Judson, David
Architect and surveyor who surveyed one of Britain's longest caves, Dan-yr-Ogof (1967–1977), was a founder member of the British Cave Research Association (BCRA) in 1973 and the co-founder of the Ghar Parau Foundation, the speleological equivalent of the Mount Everest Foundation (1974). He is currently Insurance Manager of BCRA and Legal and Insurance Officer of the National Caving Association.
    Britain and Ireland: History

Kambesis, Patricia
Graduate student (karst geology), Department of Geography and Geology, Hoffman Environmental Research Institute, Western Kentucky University, Bowling Green, Kentucky, USA. Co-author of *Deep Secrets: The Discovery and Exploration of Lechuguilla Cave* (1999). Editor of *Association for Mexican Cave Studies Report* (1990). Contributor to *Lechuguilla: Jewel of the Underground*, edited by M.R. Taylor (1991), and to the journal *Clay and Clay Minerals*.
    Encantado, Sistema del Rio, Puerto Rico

Kashima, Naruhiko
Professor Emeritus, Ehime University and part-time instructor Junior College, Matsuyama Shinonome Gakuen, Ehime, Japan. Co-author of *A Primer of Speleology* (1981, in Japanese); *Karst* (1996, in Japanese). Editor of *A Guidebook for the Nature of Ehime* (1997). Contributor to *Annals of the Speleological Research Institute of Japan*,

International Journal of Speleology, Journal of the Speleological Society of Japan.
    Akiyoshi-dai, Japan; Cheju-do Lava Caves, South Korea

Kiernan, Kevin
Research Associate, School of Geography and Environmental Studies, University of Tasmania, Australia. Author of *Lake Pedder* (1986); *The Management of Soluble Rock Landscapes: An Australian Perspective* (1988); *Caves, Karst and Management at Mole Creek, Tasmania* (1989); *An Atlas of Tasmanian Karst* (1995); *The Conservation of Coastal Landforms* (1996); *The Conservation of Landforms of Glacial Origin* (1997). Contributor to *The South-West Book*, edited by H. Gee & J. Fenton (1978); *Caves of Northwest Thailand*, edited by J.R. Dunkley & J.B. Brush (1986); *Hydrogeology of Selected Karst Regions*, edited by W. Back, J. Herman & H. Paloc (1992); *Archaeology of Aboriginal Australia*, edited by T. Murray (1998) and to the journals *Australian Archaeology, Australian Journal of Earth Sciences, Cave and Karst Management in Australia, Helictite, Journal of the Sydney Speleological Society, Nature,* and *Zeitschrift für Geomorphologie.*
    Australia; Religious Sites

Klimchouk, Alexander (Adviser)
Department of Hydrogeological Problems, Institute of Geological Sciences, National Academy of Science, Kiev, Ukraine. Co-editor of *Gypsum Karst of the World* (1996) and *Speleogenesis: Evolution of Karst Aquifers* (2000). Author of more than 180 papers on geospeleology and karstology, including more than 70 papers published in major international geoscience and speleological journals and proceedings. Vice-President of the International Union of Speleology (IUS) and President of the UIS Commission on Karst Hydrogeology and Speleogenesis. Honorary Member, National Speleological Society, USA.
    Asia, Central; Caucasus, Georgia; Caves; Evaporite Karst; Krubera Cave, Georgia; Morphometry of Caves; Russia and Ukraine; Soviet Union: History; Speleogenesis; Speleogenesis: Deep-Seated and Confined Settings; Ukraine Gypsum Caves and Karst

Knez, Martin
Scientific researcher, Karst Research Institute, Scientific Research Center of the Slovenian Academy of Sciences and Arts, Postojna, Slovenia. Author of *The Bedding-Plane Impact on Development of Karst Cave: An Example of Velika Dolina* (1996) and co-author of *Minerals in the Slovene Karst Caves* (1992). Co-editor of *South China Karst* (1998). Contributor to *Slovene Classical Karst*, edited by A. Kranjc (1997); *Field Guide of Karst in Slovenia*, edited by A. Kranjc (1997); *Global Karst Correlation*, edited by Yuan Daoxian (1998); *South China Karst*, edited by Chen Xiaoping et al. (1998), and to the journals *Acta Carsologia, Annales, Carbonates and Evaporites, Environmental Geology, International Journal of Speleology, Materials and Geoenvironment*, and *Zeitschrift für Geomorphologie.*
    Highways on Karst

Kogovšek, Janja
Professional Adviser, Karst Research Institute ZRC SAZU, Postojna, Slovenia. Co-editor of *South China Karst* (1998). Contributor to *Kras: Slovene Classical* Karst, edited by A. Kranjc (1997); *Karst Hydrogeological Investigations in South-western Slovenia*, special issue of *Acta Carsologica* (1997); *Global Karst Correlation*, edited by Yuan Daoxian (1998); *South China Karst*, edited by Chen Xiaoping et al. (1998), and to the journals *Acta Carsologia, Acta Geologica Sinica, Environmental Geology,* and *International Journal of Speleology.*
    Postojna Planina Cave System, Slovenia

Kranjc, Andrej
Scientific Adviser, Karst Research Institute, ZRC SAZU, Ljubljana, Slovenia. Author of *Recent Fluvial Cave Sediments, Their Origin and Role in Speleogenesis* (1989); co-author of *Proteus, The Mysterious Ruler of Karst Darkness* (1993). Editor of *Kras: Slovene Classical Karst* (1997). Contributor to *Encyclopaedia Biospeologica*, edited by C. Juberthie & V. Decu (1994); *Enciklopedija Slovenije*, edited by M. Javornik et al. (1996); *Global Karst Correlation*, edited by Yuan Daoxian & Liu Zaihua (1998); *Karst Hydrogeology and Human Activitioes: Impacts, Consequences and Implications*, edited by D. Drew & H. Hötzl (1999), and to the journals *Acta Carsologica, Cave and Karst Science, International Journal of Speleology, Slovensky Kras.* Editorial board member of *Acta Carsologica, Acta Geographica, Annales, Karstologia, Kras I Speleologia, Slovensky Kras.*
    Dinaric Karst; Kras, Slovenia

Kryštufek, Boris
Associate Professor and Senior Curator, Department of Vertebrates, Slovene Museum of Natural History, Ljubljana, and Institute for Biodiversity Studies, Scientific and Research Centre of the Republican Slovenia, Koper, Slovenia. Author of *Mammals of Slovenia* (1991); *Fundamentals of Conservation Biology* (1999); *Mammals of Turkey and Cyprus* (2001). Co-editor of the *Atlas of European Mammals* (1999); *Key to Vertebrates of Slovenia* (1999). Contributor to *European Bat Research 1987*, edited by V. Hanának, I. Horáček & J. Gaisler (1988); *Wild Sheep and Goats and their Relatives. Status Survey and Conservation Action Plan for Caprinae*, edited by D.M. Shackleton (1997); *Mousterian Bone Flutes and Other Finds from Divje Babe I Cave Site in Slovenia,* edited by I. Turk (1997); *Prague Studies in Mammalogy*, edited by I. Horáček & V. Vohralik (1992); *Mustelids in a Modern World: Management and Conservation Aspects of Small Carnivore:human Interactions*, edited by H.I. Griffiths (2000); and to the journals *Folia Zoologica, Journal of Biogeography, Journal of Zoology, Mammalian Biology, Mammalia, Myotis.*
    Chiroptera (Bats)

Kueny, Jeff
Department of Geography, University of Wisconsin-Milwaukee, USA. Contributor to *Journal of Cave and Karst Studies, Caribbean Geography.*
    America, Central; Military Uses of Caves

LaMoreaux, E. Philip
Senior Hydrogeologist, P.E. LaMoreaux & Associates, Inc, Tuscaloosa, Alabama, USA. Author or co-author of *Hydrology of Limestone Terranes; Annotated Bibliography of Carbonate Rocks* (1970); *Hydrogeology and Management of Hazardous Waste by Deep Well Disposal* (1989); *Environmental Hydrogeology* (1997). Co-editor of *Guide to the Hydrology of Carbonate Rocks* (1984); *Hydrology of Limestone Terranes; Annotated Bibliography of Carbonate Rocks* (vol. 3, 1988; vol. 4, 1989, vol. 5, 1993); *Springs and Bottled Waters of the World: Ancient History, Source, Occurrence, Quality and Use*

(2001). Contributor to *Sinkholes: Their Geology, Engineering, and Environmental Impact: Proceedings of the First Multidisciplinary Conference on Sinkholes* (1984); *Karst Hydrogeology and Karst Environment Protection: Proceedings of the 21st Congress of the IAH* (1988); *Proceedings of the Third Multidisciplinary Conference on Sinkholes and the Engineering and Environmental Impacts of Sinkholes and Karst* (1989); *Selected Papers on Aquifer Overexploitation: from the 23rd International Congress of IAH* (1991); *Proceedings of the International Conference On Karst-Fractured Aquifers: Vulnerability and Sustainability* (1996) and to the journals *Bulletin of the Association of Engineering Geologists, Geotimes, Journal of Hydrology, KWI Conduit, Professional Geologist*. Editor-in-Chief, *Journal of Environmental Geology*.

Karst Hydrology: History

Lascu, Cristian
Geologist, "Emil Racoviță" Speleological Institute, Cluj, Romania.
Movile Cave, Romania

Latham, Alf G.
Lecturer in Archaeological Science, School of Archaeology, Classics and Oriental Studies, University of Liverpool, UK. Contributor to *Uranium Series Disequilibrium: Applications to Earth, Marine and Environmental Sciences*, edited by Ivanovich & Harmon (1992); *Geoarchaeology: Exploration, Environments, Resources* (1999); *The Cenozoic of Southern Africa*, edited by T.C. Partridge & R.R. Maud (2000); *Handbook of Archaeological Sciences*, edited by D.R. Brothwell & A.M. Pollard (2001), and to the journals *Archaeometry* and *Cave Archaeology and Palaeontology Research Archive*.

Africa, South: Archaeological Caves; Carmel Caves, Israel: Archaeology; Dating Methods: Archaeological

Lauritzen, Stein-Erik
Geologisk Institutt, University of Bergen, Norway. Editor of *Climate Change: The Karst Record* (1997). Contributor to *Acta Carsologica, International Journal of Speleology, Journal of Cave and Karst Studies, Theoretical and Applied Karstology*. Honorary Member, National Speleological Society, USA.

Stripe Karst

Lavoie, Kathleen H.
Professor of Biology and Dean, Faculty of Arts and Science, Plattsburgh State University of New York. Co-author of *Introduction to Speleology* (2002). Contributor to *A Guide to Speleological Literature of the English Language 1794–1996*, edited by D.E. Northup, E.D. Mobley, K.L. Ingham III & W.W. Mixon (1998); *Subterranean Ecosystems*, edited by H. Wilkens, D.C. Culver & W.F. Humphreys (2000), and to the journals *American Midland Naturalist, Astrobiology Journal, Comparative Biochemistry and Physiology, Geomicrobiology Journal, Journal of the Helminthological Society of Washington, Journal of Mammology*, and *Microbial Ecology*.

Microorganisms in Caves

Lera, Thomas
Falls Church, Virginia, USA. Author of *Bats in Philately* (1995). Contributor to the journals *American Philatelist, Journal of Cave and Karst Studies, Journal of Spelean History, NSS News, Scott Stamp Monthly*. Editor of *The Underground Post* and Contributing Editor to *Speleophilately International*. Former conservation editor of *NSS Bulletin*.

Stamps and Postcards; Wilderness

Llona, Ana C. Pinto
Laboratorio de Prehistoria, Asturias, Spain. Co-author of *Taphonomy and Palaeoecology of Bears from N. Spain* (2002).

Paleontology: Animal Remains in Caves

Longley, Glenn
Director, Edwards Aquifer Research and Data Center and Professor of Aquatic Biology, Southwest Texas State University, San Marcos, Texas, USA. Author of articles on the salamander and blindcat in *The New Handbook of Texas* (1996). Contributor to *Copeia, Malacologia, Proceedings of the Biological Society of Washington, Smithsonian Contributions to Zoology*.

Edwards Aquifer, United States: Biospeleology

Loucks, Robert G.
Bureau of Economic Geology, John A. and Katherine G. Jackson School of Geosciences, University of Texas, Austin, USA. Co-editor of *Silicilastic Diagenesis and Fluid Flow: Concepts and Applications* (1996); *Carbonate Sequence Stratigraphy: Recent Advances and Applications* (1993). Contributor to *Clastic Diagenesis*, edited by D.A. McDonald & R.C. Surdam (1984); *Carbonate Sequence Stratigraphy*, edited by R.G. Loucks & R. Sarg (1993); *Petroleum Geology of North Africa*, edited by D.S. MacGregor, R.T.J. Moody & D.D. Clark-Lowes (1998); *Modern and Ancient Carbonate Eolianites: Sedimentology, Sequence Stratigraphy, and Diagenesis*, edited by F.E. Abegg, P.M. Harris & D.D. Loope (2001); and to the journals *American Association of Petroleum Geologists Bulletin, Gulf Coast Association of Geological Societies Transactions*, and *Journal of Applied Geophysics*. Associate editor for the *American Association of Petroleum Geologists Bulletin*.

Hydrocarbons in Karst

Lowe, David J. (Adviser)
Field geologist and, lately, geoscientific data project manager, British Geological Survey, Keyworth, Nottingham, UK. Joint author of *Dictionary of Karst and Caves* (1995, revised 2002). Co-editor of *Gypsum Karst of the World* (1996), and *Implications of Speleological Studies for Karst Subsidence Hazard Assessment* (2003). Joint Editor of *Cave and Karst Science*. Editorial Board member for *International Journal of Speleology, Naše jame, Theoretical and Applied Karstology*, and the *Virtual Journal of Speleogenesis and Evolution of Karst Aquifers*. Contributor to *Limestones and Caves of Wales* (1989), *Speleogenesis: Evolution of Karst Aquifers* (2000), *Paviland Cave and the 'Red Lady'* (2000) and to the journals *Acta Carsologica, Cave Science*. Visiting Research Fellow, Limestone Research Group, University of Huddersfield, UK; Research Associate, Karst Research Institute, Postojna, Slovenia.

Geoscientists; Inception of Caves; Speleogenesis Theories: Post–1890

Lundberg, Joyce
Department of Geography and Environmental Sciences, Carleton University, Canada. Contributor to *Speleogenesis: Evolution of Karst Aquifers*, edited by A.B. Klimchouk, D. Ford, A. Palmer & W. Dreybrodt (2000) and to the journals *Cave and Karst Science*,

Journal of Cave and Karst Science, Palaeogeography, Palaeoclimatology, Palaeoecology, Quaternary Research, The Holocene.
   Coastal Karst

Maire, Richard
Directeur de recherches au CNRS, Université Bordeaux, France. Author of *La haute montagne calcaire* (1990). Co-editor *Karsts de Chine Centrale* (1995). Editor-in-chief *Karstologia*.
   Patagonia Marble Karst, Chile; Pierre Saint-Martin, France-Spain

Martens, Koen
Researcher, Department of Freshwater Biology, Royal Belgian Institute of Natural Sciences and Guest Professor, University of Ghent, Belgium. Co-editor of *Speciation in Ancient Lakes* (1994); *The Evolutionary Ecology of Reproductive Modes in Non-marine Ostracoda* (1994); *Evolutionary Biology and Ecology of Ostracoda* (2000). Editor of *Sex and Parthenogenesis: Evolutionary Ecology of Reproductive Modes in Non-marine Ostracods* (1998). Contributor to *Crustaceana, Heredity, Hydrobiologia, Journal of Evolutionary Biology, Oecologia, Proceedings of the Royal Society, London, Zoological Journal of the Linnean Society*. Member of the editorial board of *Hydrobiologia, Phegea*.
   Crustacea: Ostracoda

Martini, Jacques
Formerly, Council for Geoscience (former Geological Survey), Pretoria, South Africa; currently retired in France. Contributor to *Speleogenesis: Evolution of Karst Aquifers*, edited by A. Klimchouk et al. (2000), and to the journals *Annals of the Geological Survey of South Africa, Karstologia, South African Speleological Society Bulletin*. Honorary life member of the Société Suisse de Spéléologie and of the South African Spelaeological Association.
   Dissolution: Silicate Rocks; Silicate Karst

Mathieu, Jacques
Hydrobiologie et Écologie souterraines, Université Claude-Bernard Lyon I, France. Co-editor *Groundwater / Surface Water Ecotones: Biological and Hydrological Interactions and Management Options* (1997). Contributor to *Groundwater Ecology*, edited by J. Gibert, D. Danielopol & J. Stanford (1994) and to the journals *Archiv für Hydrobiologie, Hydrobiologia, Mémoires de Biospéologie, Stygologia*.
   Adaptation: Physiological

Mauriès, Jean-Paul
Laboratoire de Zoologie-Arthropodes, Museum National d'Histoire Naturelle de Paris, Paris, France. Co-editor *Acta Myriapodologica* (*Mémoires du Museum National d'Histoire Naturelle*) (1996).
   Myriapoda

McFarlane, Donald A.
Associate Professor, Keck Science Center, The Claremont Colleges, Claremont, California, USA. Contributor to *Jamaica Underground: The Caves, Sinkholes and Rivers of the Island*, edited by A.G. Fincham (1997); *Extinctions in Near Time: Causes, Contexts and Consequences*, edited by R.D.E. MacPhee (1999), and to the journals *Biogeochemistry, Caribbean Journal of Science, Cave and Karst Science, Journal of Cave and Karst Studies*, and *Quaternary Research*.
   Guano

Messana, Giuseppe
Senior Researcher, CNR – Istituto per lo Studio degli Ecosistemi Sezione di Firenze, (formerly Centro di Studio per la Faunistica ed Ecologia Tropicali), Firenze, Italy. Contributor to *Encyclopaedia Biospeleologica, Archiv fur Hydrobiologie, Crustaceana, Italian Journal of Zoology, Mémoires de Biospéologie, Tropical Zoology*. President of Société de Biospéologie. Editor of the journal *Tropical Zoology*.
   Africa: Biospeleology

Meyer-Rochow, Victor Benno
Professor of Biology, Faculty of Engineering and Science, International University of Bremen, Germany. Author of *The New Zealand Glow-Worm* (1990). Contributor to *The Compound Eye and Vision of Insects*, edited by G.A. Horridge (1975); *Mechanisms and Phylogeny of Mineralization in Biological Systems*, edited by S. Suga & H. Nakahara (1991); *Atlas of Arthropod and Sensory Receptors*, edited by E. Eguchi & Y. Tominaga (1999); *Sensory Biology of Jawed Fishes*, edited by B.J Kapoor & T.J. Hara (2001); and to the journals *Biologist, Invertebrate Biology Journal of Insect Physiology, Physiology and Behaviour, Proceedings of the Royal Society of London*. Editorial board member of *Invertebrate Biology, Acta Neurobiologiae Experimentalis*, and *Entomologica Fennica*.
   Adaptation: Eyes

Michie, Neville
Freelance Cave Scientist, New South Wales, Australia. Contributor to several Conference Proceedings and to the journal *Helictite*.
   Tourist Caves: Airborne Debris

Middleton, Gregory
Manager Integrated Policies and Strategies, Resource Management and Conservation Division, Department of Primary Industries, Water and Environment, State of Tasmania, Hobart, Tasmania, Australia. Author or co-author of *Timor Caves, NSW* (1973); *Wilderness Caves of the Gordon-Franklin River System, Tasmania* (1979); *Oliver Trickett: Doyen of Australia's Cave Surveyors* (1991). Editor or co-editor of *Bungonia Caves* (1972); *Cave Management in Australasia II* (1997). Contributor to *Bungonia Caves* (1972), *Encyclopedia Biospeologica*, vol. 3, edited by C. Juberthie (2001), and to *Australasian Cave and Karst Management Association Journal, Journal of Spelean History, Journal of the Sydney Speleological Society, International Journal of Speleology, Proceedings of the 7th International Congress of Speleology, Proceedings of the 12th International Congress of Speleology Studies in Speleology*. Editor of *Australian Speleo Abstracts* (1970–79).
   Australia: History; Madagascar

Mihevc, Andrej
Lecturer and researcher at Karst Research Institute, Postojna, Slovenia. Author of *Notranjska A-Z: priročnik za popotnika in poslovega človeka* (1999); *Speleogeneza Divaškega krasa* (2001). Co-editor of *South China Karst* (1998). Contributor to *Karst Hydrogeological Investigations in South-western Slovenia*, special issue of *Acta Carsologica* (1997); *Global Karst Correlation*, edited by Yuan Daoxian (1998); *South China Karst* (1998); *Geografski atlas Slovenije: drzava v prostoru in casu*, edited by J. Fridl et al. (1998), and to the journals *Acta Carsologica, Environmental Geology, Geografia Fisica e Dinamica Quaternaria, Geografski Vestnik, International Journal of Speleology, Kras i Speleologia*.
   Škocjanske Jama, Slovenia

Milanović, Petar
Professor emeritus, Yugoslavia. Author of *Karst Hydrogeology* (1981); *Geological Engineering in Karst* (2000) and contributor to *Karst Hydrogeology and Water Resources* (1979); *Hydrogeology of the Dinaric Karst* (1984); *Karst Waters and Environmental Impacts* (1997) and to the journals *Environmental Geology, Episodes, Journal of Hydrology, Journal of International Geoscience.*
    Dams and Reservoirs on Karst; Dinaride Poljes; Tunnelling and Underground Dams in Karst

Miller, Rebecca
Archaeologist in the Université de Liège, Service de Préhistoire, Liège, Belgium.
    Belgium: Archaeological Caves

Moldovan, Oana
Institutul de Speologie "Emil Racovitza", Cluj, Romania. Contributor to *Mémoires de Biospéologie*.
    Adaptation: Morphological (Internal); Biodiversity in Terrestrial Cave Habitats

Mouret, Claude
France. President of Speleological Federation of European Community, member of the Bureau of International Union of Speleology and Vice-President of French Federation of Speleology (former President). Author of more than 200 articles on karst and caves and of a book on French karst. Guest author in congresses on Southeast Asia. Invited author for an Encyclopedia on Asian Caves. Editor of 5 books on karst and caves, including a review of Southeast Asian caves, a dictionary of Speleology and a study on caves and karst conservation. Contributor to *Spelunca and Karstologia Bulletins and Memoirs*.
    Asia, Southeast; Burials in Caves; Khammouan, Laos-Vietnam

Mueller, Bill
Department of Geography, University of Wisconsin-Milwaukee, USA. Contributor to *Journal of the Wisconsin Society for Ornithology*.
    Aves (Birds)

Mylroie, John
Professor of Geology, Department of Geosciences, Mississippi State University, USA. Co-author of *A Field Trip Guidebook of Lighthouse Cave, San Salvador Island, Bahamas* (1994); *Geology and Karst of San Salvador Island, Bahamas: A Field Trip Guidebook* (1994); *Field Guide to Sites of Geological Interest, Western New Providence Island, Bahamas* (1996); *Geomorphology and Quaternary Geology of the Inner Part of the Sognefjord Area and an Introduction to the Caves and Karst of Dummdalen* (1996); *The Geology of South Andros Island, Bahamas* (1998). Editor of *Western Kentucky Speleological Survey Annual Report* (1984); *Field Guide to the Karst Geology of San Salvador Island, Bahamas* (1988); *Proceedings of the Fourth Symposium on the Geology of the Bahamas* (1989); *Karst Landsforms and Caves of Nordland, North Norway* (1996); *Proceedings of the Ninth Symposium on the Geology of the Bahamas and other Carbonate Regions* (1999). Contributor of articles to *Unconformities and Porosity in Carbonate Strata*, edited by D.A. Budd *et al.* (1995); *Climatic Change: The Karst Record*, edited by S.-E. Lauritzen (1996); *Geology and Hydrogeology of Carbonate Islands*, edited by H.L. Vacher & T.M. Quinn (1997); *Speleogenesis: Evolution of Karst Aquifers*, edited by A. Klimchouk *et al.* (2000). Contributor to *American Scientist, Carbonates and Evaporites, Cave and Karst Science, Climate Research, Earth Surface Processes and Landforms, Geology, NSS Bulletin/Journal of Cave and Karst Studies*.
    Blue Holes of Bahamas; Mona, Puerto Rico; Speleogenesis: Coastal and Oceanic Settings

Nader, Fadi
PhD student at the physico-chemical geology lab, Katholieke Universiteit Leuven, Belgium. Former General Secretary, Spéléo-Club du Liban. Adjunct Secretary, International Union of Speleology (UIS). Contributor to *International Caver Magazine, Caves and Caving, Al Ouat'Ouate (SCL bulletin)*.
    Jeita Cave, Lebanon

Northup, Diana E.
Associate Professor, Centennial Science and Engineering Library, and Associate, Museum of Southwestern Biology, University of New Mexico, Albuquerque, New Mexico, USA. Co-editor and compiler of *A Guide to Speleological Literature of the English Language 1794–1996* (1998). Contributor to *Astrobiology Journal, Geomicrobiology Journal, Journal of Cave and Karst Studies, American Midland Naturalist*, and *Comparative Biochemistry and Physiology*.
    Microorganisms in Caves

Olson, Rick
Ecologist, Division of Science and Resources Management, Mammoth Cave National Park, Kentucky, USA. Co-author of *Living With Karst, A Fragile Foundation* (2001). Contributor to *Restoration and Conservation of Caves*, edited by V. Hildreth-Werker & J. Werker (2003).
    Mammoth Cave, United States: Biospeleology

Onac, Bogdan P.
Professor, Department of Mineralogy, University of Cluj and Speleological Institute "Emil Racovita", Cluj, Romania. Author of *Speleothems from Caves in Padurea Craiului Mountains: A Mineralogic, Crystallographic and Paleoclimatic Study* (1998); *Geology of Karst Terrains* (2000); *Scarisoara Glacier Cave* (2000). Editor of *Quaternary Studies in Romania: Achievements and Perspectives* (2000); *Karst Studies and Problems: 2000 and Beyond* (2000). Contributor to *Cave Minerals of the World*, edited by C.A. Hill & P. Forti (2nd edition, 1997); *Karst Processes and the Global Carbon Cycle*, edited by Yuan Daoxian (2001); and to the journals *Cave and Karst Science, European Journal of Mineralogy, Journal of Quaternary Science, Quaternary Research, Theoretical and Applied Karstology*.
    Europe, Balkans and Carpathians

Oromí, Pedro
Titular Professor of Animal Biology, Department of Animal Biology, University of La Laguna, Tenerife, Canary Islands, Spain. Author of *Los Apiónidos de las Islas Canarias* (1986); *Islas Galápagos: volcán, mar y vida en evolución* (1992); *Catálogo Espeleológico de Tenerife* (1995); and *Catalogue of the Coleoptera of the Canary Islands* (2000). Editor of *La Cueva del Viento* (1995); and *Proceedings of 7th International Symposium on Vulcanospeleology* (1996). Contributor to *The Unity of Evolutionary Biology*, edited by E.C. Dudley (1991);

The Natural History of Biospeleology, edited by A.I. Camacho (1992); *Encyclopaedia Biospeologica*, edited by C. Juberthie & V. Decu (1994 and 1998); and contributor to the journals *Evolution, Journal of Evolutionary Biology, Mémoires de Biospéologie, Proceedings of the Royal Society of London, Trends in Ecology and Evolution*.
    Canary Islands: Biospeleology

Osborne, Armstrong
School of Development and Learning, University of Sydney, New South Wales, Australia. Contributor to *Speleogenesis: Evolution of Karst Aquifers*, edited by A.B. Klimchouk, D. Ford, A. Palmer & W. Dreybrodt (2000); *Evolution of Karst: From Prekarst to Cessation*, edited by F. Gabrovšek (2002), and to *Acta Carsologica, Australian Journal of Earth Sciences, Cave and Karst Science, Helictite*.
    Paleokarst

Otte, Marcel
Université de Liège, Service de Préhistoire, Liège, Belgium.
    Belgium: Archaeological Caves

Palmer, Arthur (Adviser)
Professor of Hydrology and Director of Water Resources Program, Earth Sciences Department, State University of New York, Oneonta, New York, USA. SUNY Distinguished Teaching Professor of Hydrology, Geochemistry, and Geophysics. Author of *Geology of Wind Cave, Wind Cave National Park, South Dakota* (1981); *A Geologic Guide to Mammoth Cave National Park* (1981); *Jewel Cave—a Gift from the Past* (1984, revised 1995); *Wind Cave: An Ancient World beneath the Hills* (1988, revised 1995). Co-editor of *Karst Modeling* (1999) and *Speleogenesis: Evolution of Karst Aquifers* (2000). Contributor to *Acta Carsologica, American Association of Petroleum Geologists Memoirs, Carbonates and Evaporites, Geological Society of America, Journal of Cave and Karst Studies*. Honorary Member, National Speleological Society, USA.
    Carlsbad Cavern and Lechuguilla Cave, United States; Hydraulics of Caves; Mammoth Cave Region, United States; Patterns of Caves; United States of America; Wind and Jewel Caves, United States

Pavuza, Rudolf
Karst- und Höhlenkundl. Abteilung Naturhistorisches Museum, Vienna, Austria.
    Calcareous Alps, Austria

Pentecost, Allan
School of Health and Life Sciences, King's College London, UK. Contributor to *The Ecology of Cyanobacteria*, edited by B.A. Whitton & M.A. Potts (2000), and to the journals *Cave and Karst Science, Geology Today, Geomicrobiology Journal, Proceedings of the Geological Association, Quaternary Science Review*.
    Entrance Habitats; Huanglong and Jiuzhaigou, China; Travertine

Perritaz, Luc
Researcher, Institute of Geography, University of Fribourg, Switzerland. Contributor to *Karstologia, Zeitschrift für Geomorphologie*.
    Africa, North

Price, Liz
Kuala Lumpur, Malaysia. Author of Malaysian cave bibliography (up to 1997) (1998); *Caves and Karst of Peninsular Malaysia* (2001).
    Asia, Southeast: Archaeological Caves

Proudlove, Graham
Department of Zoology, The Manchester Museum, Manchester University, UK. Contributor to *Caving Practice and Equipment*, edited by D. Judson (1991); *Mapping Subterranean Biodiversity*, edited by D.C. Culver, L. Deharveng, J. Gibert & I.D. Sasowsky (2001); *A Cave and Mine Conservation Audit for the Masson Hill Area* (2001); and to the journals *Caves and Caving, Journal of the Craven Pothole Club*.
    Pisces (Fish); Britain and Ireland: Biospeleology

Raeisi, Ezzat
Department of Geology, Shiraz University, Iran. Contributor to *Carbonates and Evaporites, Cave and Karst Science, Journal of Cave and Karst Studies, Iranian Journal of Science and Technology, Journal of Engineering*.
    Iran

Reddell, James R.
Curator of arthropods, Texas Memorial Museum, University of Texas at Austin, USA. Author of many papers on caves, cave fauna, and conservation. Editor of *Studies on the Cave and Endogean Fauna of North America* (3 vols, 1986–2001). 2001 NSS Science Award for lifetime contributions to speleology.
    America, Central and Caribbean Islands: Biospeleology

Ribera, Carles
Facultat de Biologia, University of Barcelona, Spain. Author of more than 50 contributions on cavernicolous spiders.
    Arachnida; Arachnida: Acari; Arachnida: Aranae; Arachnida: Minor Groups

Rodríguez-Vidal, Joaquin
Department of Geology, University of Huelva, Spain.
    Gibraltar Caves: Archaeology

Roje-Bonacci, Tanja
Civil Engineering Faculty, University of Split, Croatia.
    Plitvice Lakes, Croatia

Romero, Aldemaro
Director and Associate Professor, Environmental Studies Program, Biology Department, Macalester College, St Paul, Minnesota, USA. Author of *Manual de Ciencias Ambientales* (1992); *Canaima* (1992); *Venezuela: Mágico pais de la biodiversidad* (1993); *Vida Verde* (1994); *How to Build an Environmental Academic Program* (2002). Editor of *The Biology of Hypogean Fishes* (2001), and *Environmental Issues in Latin America* (2002). Contributor of articles to *El Manejo de los Ambientes y Recursos Costeros en América Latina y el Caribe* (1990); *La gerencia de los 90* (1991); *Voices from the Environmental Movement: Perspectives for a New Era*, edited by D. Snow (1992); *Ambiente y Desarrollo Urbano* (1992). Contributor to *Biodiversity, Copeia, Environmental Biology of Fishes, Journal of Spelean History, NSS Bulletin, NSS News*.
    Adaptation: Behavioral Biospeleologists; Evolution of Hypogean Fauna; Pisces: Amblyopsidae

Rossi, Carlos
Professor, Departamento de Petrologia y Geoquimica, Universidad Complutense de Madrid, Spain. Contributor to *Speleogenesis: Evolution of Karst Aquifers*, edited by A. Klimchouk *et al.* (2000); *Quartz Cement in Sandstone Reservoirs*, edited by R. Worden & S. Morad (2000). Contributor to *AAPG Bulletin, Journal of Geochemical Exploration, Journal of Marine and Petroleum Geology, Journal of Sedimentary Research, Revista de la Sociedad Geológica de España*.
  Picos de Europa, Spain

Sabol, Martin
Assistant Professor, Department of Geology and Paleontology, Comenius University in Bratislava, Slovakia. Contributor to *Mineralia Slovaca, Slovak Geological Magazine, Slovenský Kras—Acta Carsologica Slovaca*.
  Aggtelek Caves, Hungary-Slovakia: Archaeology

Sambugar, Beatrice
Associate researcher, Museum of Natural History, Verona, Italy. Contributor to *Checklist delle specie della fauna italiana*, edited by A. Minelli, S. Ruffo & S. LaPosta (1995), and to the journals *Annales de Limnologie, Hydrobiologia, Journal of Zoology, Mémoires de Biospéologie*.
  Annelida

Sasowsky, Ira D.
Associate Professor, Department of Geology, University of Akron, Ohio, USA. Co-editor of *Breakthroughs in Karst Geomicrobiology and Redox Geochemistry* (1994); *Conservation and Protection of the Biota of Karst* (1997); *Karst Modeling* (1999); *Groundwater Flow and Contaminant Transport in Carbonate Aquifers* (2000). Contributor to *Clays and Clay Minerals, Geology, Geomorphology, Journal of Hydrology, Quaternary Research, Water Research, Water Resources Research*.
  Sediments: Autochthonous Clastic

Sauro, Ugo
Associate Professor of Physical Geography, Department of Geography, University of Padova, Italy. Co-editor of *Le Grotte del Veneto: paesaggi carsici e grotte del Veneto* (1989); *Proceedings of the International Conference on Environmental Changes in Karst Areas* (1991); *Altopiani Ampezzani: geologia, geomorfologia, speleologia* (1995); *Gypsum Karst of the World* (1996). Contributor to *Environmental Geology, Geomorphology, Zeitschrift für Geomorphologie*.
  Asiago Plateau, Italy

Sbordoni, Valerio
Department of Biology, Tor Vergata University, Roma, Italy. Editor-in-chief *International Journal of Speleology*. Author of many journal papers on evolutionary genetics and population biology, subterranean fauna, and Antarctic fauna. Contributor to *Encyclopedia Biospeologica*, edited by C. Juberthie & V. Decu (1998); *Subterranean Ecosystems*, edited by H. Wilkens *et al.* (2000).
  Insecta: Pterygota

Schindel, Geary
Chief Technical Officer, Edwards Aquifer Authority, San Antonio, Texas, USA. Author of several papers on karst, groundwater, and tracer testing.
  Edwards Aquifer and the Texas Karst, United States; Groundwater Pollution: Point-Source; Groundwater Pollution: Remediation

Šebela, Stanka
Higher scientific researcher at Karst Research Institute ZRC SAZU, Postojna, Slovenia. Author of *Tectonic Structure of Postojnska Jama Cave System* (in Slovenian and English, 1998). Co-editor of *South China Karst* (1998). Contributor to *Kras: Slovene Classical Karst*, edited by A. Kranjc (1997); *Karst Hydrogeological Investigations in South-western Slovenia*, special issue of *Acta Carsologica* (1997); *Global Karst Correlation*, edited by Yuan Daoxian (1998); *South China Karst*, edited by Chen Xiaoping *et al.* (1998), and to the journals *Acta Carsologia, Acta Geologica Sinica, Environmental Geology, Geological Journal, International Journal of Speleology, Studia Carsologica*.
  Postojna—Planina Cave System, Slovenia

Self, Charles Anthony
UK. Chairman, Genetic Mineralogy working group, Mineralogy Commission, Union International de Spéléologie. Editor of *Caves of County Clare* (1991). Contributor to *Cave Science, Cave Geology, Cave and Karst Science, Geofluids, Proceedings of the University of Bristol Spelaeological Society*.
  Cupp-Coutunn Cave, Turkmenistan

Senior, Kevin
Graduated in Geology and Physical Geography but has worked for IBM for most of his life. Member of numerous British caving expeditions to Spain, Uzbeckistan, Irian Jaya, Laos, and China.
  Di Feng Dong, China

Shaw, Trevor (Adviser)
Honorary Research Fellow, Karst Research Institute, Postojna, Slovenia. Author of *History of Cave Science: The Exploration and Study of Limestone Caves, to 1900* (2nd edition, 1992); *Foreign Travellers in the Slovene Karst, 1537–1900* (2000). Contributor of articles to *Festschrift Lurgrotte 1894–1994*, edited by R. Benischke (1994); *Jamaica Underground*, edited by A.G. Fincham (1997); *Cave Minerals of the World*, edited by C.A Hill & P. Forti (2nd edition, 1997); *L'Homme qui voyageait pour les gouffres*, edited by D. André & H. Duthu (1999); *Speleogenesis: Evolution of Karst Aquifers*, edited by A. Klimchouk *et al.* (2000). Contributor to many journals, including *Acta Carsologica, Cave and Karst Science, Helictite, International Caver, Journal of Caves and Karst Studies, Journal of Spelean History, Naše jame, Slovensky Kras, Studies in Speleology*. Honorary Member, National Speleological Society, USA; Peter M. Hauer Spelean History Award, 1985; Petrbok Medal of the Česká Speleologická Spolecnost, 1994. Editorial Board member of *Acta Carsologica*, and Advisory Board member of *International Journal of Speleology*.
  Archaeologists; Asia, Northeast: History; Caribbean Islands: History; Cerknica Polje, Slovenia: History; Exploration Societies; Speleogenesis Theories: Early; Speleologists; Speleothem Studies: History

Shopov, Y. Yavor
Senior Research Assistant, University Center for Space Research, University of Sofia, Bulgaria. Author of articles in *ESR Dating and*

Dosimetry, edited by M. Ikeya & T. Myki (1985); *Climatic Change: The Karst Record*, edited by S.-E. Lauritzen (1996); *Cave Minerals of the World*, edited by C.A. Hill & P. Forti (2nd edition, 1997); *Global Karst Correlation*, edited by Yuan Daoxian & L. Zaihua (1998); *Encyclopedia "World of the Earth Sciences"*, edited by L. Lerner & B. Lerner (2002). Contributor to *Acta Crystallographica, Annales Geophysicae, Geology*. Editor-in-chief of *Solar Eclipse Journal* (1999–), and editorial board member of *Theoretical and Applied Karstology* (2000–). President of the Royal Society of Bulgaria.

Dating of Karst Landforms; Sediments: Biogenic; Speleothems: Luminescence

Shrewsbury, Carolina
International SpeleoArt coordinator (originally based in the UK, now the USA). Contributor to *Illuminations* (publication of the NSS Arts and Letters Section), *Journal of the Sydney Speleological Society, Stalactite, Underground Photographer*.

Art Showing Caves

Simek, Jan
Professor, Department of Anthropology, University of Tennessee, USA. Author of *A K-Means Approach to the Analysis of Spatial Structure in Upper Paleolithic Habitation Sites* (1984). Co-editor of *Cave Archaeology in the Eastern Woodlands*, special issue of *Midcontinental Journal of Archaeology* (2001). Contributor to *American Antiquity, Antiquity, Journal of Archaeological Science, Journal of Human Evolution, Southeastern Archaeology*.

Archaeology of Caves: History

Sket, Boris (Adviser)
Professor, Department of Biology, Biotechnical Faculty, University of Ljubljana, Slovenia. Author of *Subterranean Life in Karst* (in Slovenian, 1979), and articles in *Fauna of Slovenia*, edited by F. Bernot *et al.* (1998), *Handbuch der Reptilien und Amphibien Europas*, edited by K. Grossenbacher & B. Thiesmeier (1999). Editor or co-editor of *Manual for Cavers* (in Slovenian, 1964); *Identification Keys for Animals of Yugoslavia* (in Slovenian, series: 1967–68); *Fauna of Slovenia* (in Slovenian, in preparation). Editor for biology and biotechnical topics for *Lexicon* (in Slovenian, 1973, 1976, 1985, 1988). Contributor to *Archiv für Hydrobiologie, Biodiversity and Conservation, Journal of Biogeography, Journal of Zoology, Proceedings of the Biological Society of Washington, Trends in Ecology and Evolution*. Editorial Board member of *Mémoires de Biospéologie*, Advisory Board member of *International Journal of Speleology*, Associate Board member of *Stygologia*, Associate Editor of *Zootaxa*.

Anchialine Habitats; Biodiversity in Hypogean Waters; Biology of Caves; Dinaric Karst: Biospeleology; Invertebrates: Minor Groups; Postojna-Planina Cave System, Slovenia: Biospeleology; Subterranean Habitats; Thermal Water Habitats; Vjetrenica, Bosnia-Herzegovina: Biospeleology

Slabe, Tadej
Higher scientific researcher, Karst Research Institute, Postojna, Slovenia. Author of *Cave Rocky Relief and its Speleogenetical Significance* (1995). Co-editor of *South China Karst* (1998). Contributor to *Kras: Slovene Classical Karst*, edited by A. Kranjc (1997); *Global Karst Correlation*, edited by Yuan Daoxian (1998); *South China Karst*, edited by Chen Xiaoping *et al.* (1998), and to the journals *Acta Carsologia, Annales, Atti mem. Comm. Grotte Eugenio Bpegan, Environmental Geology, International Journal of Speleology*.

Morphology of Caves

Smart, Chris
Associate Professor and Graduate Chair, Department of Geography, University of Western Ontario, Canada. Author of a number of papers on karst modelling, karst groundwater, and tracing, specializing in alpine and glaciated terrain, most recently in *Earth Surface Processes and Landforms, Environmental Geology, Hydrological Processes, Theoretical and Applied Karstology*.

Alpine Karst; Glaciated and Glacierized Karst; Glacier Caves and Glacier Pseudokarst; Groundwater in Karst; Groundwater in Karst: Borehole Hydrology; Groundwater in Karst: Conceptual Models; Groundwater in Karst: Mathematical Models; Karst Water Resources; Springs; Water Tracing

Smith, Marion O.
Retired; former Assistant Editor, *The Papers of Andrew Johnson*, University of Tennessee, Knoxville, Tennessee, USA. Author of *The Exploration and Survey of Ellison's Cave* (1977); *Letters from TAG, 1966–1969* (1992); *Saltpeter Mining in East Tennessee* (1990). Contributor to *Civil War History, Florida Historical Quarterly, Georgia Historical Quarterly, Journal of Spelean History, Tennessee Historical Quarterly*.

Gunpowder

Song Linhua
Professor, Institute of Geography, Chinese Academy of Sciences, Beijing, China. Co-editor of *The Pinnacle Karst of Stone Forest, Lunan, Yunnan, China, An Example of a Sub-jacent Karst* (1986); *Karst Landscape and Cave Tourism* (1993) and *Stone Forest: A Treasure of Natural Heritage* (1997). Author of many papers including in the journals *Acta Carsologica, Acta Geologica Sinica, Cave Science, Studies in Speleology*.

Hongshui River Fengcong Karst, China

Spötl, Christoph
Institut für Geologie und Paläontologie, Universität Innsbruck, Austria.

Spannagel Cave, Austria

Steward, Paul Jay
Production control and material specialist for Lockheed Martin, USA. Author of *Tales of Dirt, Danger and Darkness* (1998). Contributor to *American Caves, Central Jersey Caver, Illuminations* (publication of the NSS Arts and Letters Section), *NSS News*.

Caves in Fiction

Stierman, Donald J.
Associate Professor of Geophysics, Department of Earth, Ecological and Environment Sciences, University of Toledo, Toledo, Ohio, USA. Contributor to *Bulletin of the Seismological Society of America, Environmental Geology, Geoarchaeology, Journal of Geophysical Research, Pure and Applied Geophysics, Science, Tectonophysics*.

Geophysical Detection of Caves and Karstic Voids

Stoch, Fabio
Consultant and associate researcher, Museums of Natural History of Verona, Udine, and Trieste, Italy. Author of several book chapters

on karstic environments, including in *Studies in Crenobiology: The Biology of Sprigs and Springbrooks*, edited by L. Botosaneanu (1997); *Ponds and Pond Landscapes in Europe*, edited by J. Boothby (1999). Editor of *Grotte e fenomeno carsico* [Caves and Karst Phenomena] (2001). Contributor to several journals, including *Annales de Limnologie, Belgian Journal of Entomology, Crustaceana, Hydrobiologia, Mémoires de Biospéologie, Memorie dell'Istituto Italiano di Speleologia*.

    Colonization

Stone, Andrea
Professor, Department of Art History, University of Wisconsin-Milwaukee, USA. Author of *Images from the Underworld: Naj Tunich and the Tradition of Maya Cave Painting* (1995). Editor of *Heart of Creation: The Mesoamerican World and the Legacy of Linda Schele* (2002). Contributor to *Time and Space: Dating and Spatial Considerations in Rock Art Research*, edited by J. Steinbring et al. (1993); *The Human Use of Caves*, edited by C. Bonsall & C. Tolan-Smith (1997); and to *Journal of Cave and Karst Studies*.

    Art: Cave Art in the Americas

Taiti, Stefano
CNR Researcher, Istituto per lo Studio degli Ecosistemi, Firenze, Italy. Contributor to *Proceedings of the Second Symposium on the Biology of Terrestrial Isopods* (1989) and to the journals *Fauna of Arabia, Invertebrate Taxonomy, Journal of Natural History, Mémoires de Biospéologie, Zoological Journal of the Linnean Society*. Co-Editor of *Tropical Zoology*.

    Crustacea: Isopoda: Oniscidea

Tarhule-Lips, Roosmarijn
Lecturer, Department of Geography, University of Oklahoma, USA. Contributor to *Cave and Karst Science, Journal of Cave and Karst Studies, Journal of Physical Geography*.

    Caribbean Islands

Tao Tang
Department of Geography and Planning, State University of New York, College at Buffalo, USA. Contributor to *Earth Surface Processes and Landforms, Middle States Geographer, Chinese Journal of Geographic Studies*.

    Tower Karst

Taylor, Steven J.
Associate research scientist, Center for Biodiversity, Illinois Natural History Survey, Champaign, Illinois, USA. Contributor to *Annals of the Entomological Society of America, Entomological News, Florida Entomologist, Great Lakes Entomologist, Journal of Cave and Karst Studies*. Associate editor for life sciences, *Journal of Cave and Karst Studies*.

    America, North: Biospeleology

Tercafs, Raymond
Senior Research Associate of the Belgian Fund for Scientific Research, Institute of Zoology, Department of Animal Physiology, University of Liege, Belgium. Author or co-author of *Atlas de la vie souterraine: Les Animaux cavernicoles* (1972); entry on Biospéleologie in *Encyclopedia Universalis* (1974). Contributor to *The Natural History of Biospeleology*, edited by A. Camacho (1992); *Encyclopedia Biospeologica*, edited by C. Juberthie & V. Decu (1994, 2001), and to the journals *Annales de Spéléologie, Ecological Modelling, Environmental Conservation, International Journal of Speleology, Mémoires de Biospéologie*.

    Conservation: Cave Biota

Thurgate, Mia
Karst Resources Manager, Karst Resources Department, Jenolan Caves Reserve Trust, New South Wales, Australia. Contributor to *Acta Geographica Szegedensis, Records of the Western Australian Museum Supplement, Stalactite*.

    Monitoring

Tolan-Smith, Chris
Senior lecturer, Department of Archaeology, University of Newcastle upon Tyne, UK. Author of *Late Stone Age Hunters of the British Isles* (1992); *Landscape Archaeology in Tyneside* (1997); *The Caves of Mid-Argyll: An Archaeology of Human Use* (2001). Co-editor of *The Human Use of Caves* (1997). Contributor to *Contributions to the Mesolithic in Europe*, edited by P.M. Vermeersch & P. Van Meer (1990); *The Late Glacial in North-West Europe*, edited by N. Barton et al. (1991); *The Neolithic in No-Mans Land*, edited by P. Frodsham (1996); *Proceedings of the Berlin INQUA Symposium on Human Adaptations across the Pleistocene-Holocene Transition*, edited by L.G. Strauss & B.V. Eriksen (1998); *14C and Archaeology Acts of the 3rd International Symposium*, edited by J. Evin et al. (1999); and to the journals *Archaeologica Cambrensis, Proceedings of the Prehistoric Society, Scottish Studies*.

    Folklore and Mythology; Human Occupation of Caves

Tooth, Anna F.
Hydrologist with The Environment Agency, Worthing, West Sussex, UK. Contributor to *Journal of the Geological Society, London, Chemical Geology*.

    Chemistry of Natural Karst Waters

Trajano, Eleonora
Associate Professor, Departamento de Zoologia, Instituto de Biociências, Universidade de São Paulo, Brazil. Contributor to *Encyclopedia Biospeologica*, edited by C. Juberthie & V. Decu (1994); *Subterranean Ecosystems*, edited by H. Wilkens et al. (2000); *Fundação e a Produção Florestal do Estado de São Paulo* (2001), and to many journals including *Biological Rhythm Research, Biotropica, Entomologia, Environmental Biology of Fishes, Mémoires de Biospéologie, Revista Brasileira de Entomologia, Revista Brasileira de Zoologia*.

    America, South: Biospeleology

Trimmel, Hubert
Professor and retired Director of the Department of Karst and Cave Science, Museum of Natural History, Vienna, Austria. Lecturer, Institutes of Geography of the Universities of Salzburg and Vienna. Author of *Höhlenkunde* (1968). Editor of *Speläologisches Fachwörterbuch* (1965) and of the quarterly bulletin *Die Höhle* (1953–). Honorary President of the International Union of Speleology and of the Federation of Austrian Speleologists. Honorary member of the speleological federations of Germany, Hungary, and Italy.

    Europe, Alpine; Europe, Central: History

Tyc, Andrzej
Researcher, University of Silesia, Department of Geomorphology, Sosnowiec, Poland. Author of *Guide des terrains choisis des Sudety et Haut-Plateau de Silesie-Cracovie* (1987); *Anthropogenic Impact on Karst Processes in the Silesian-Cracow Upland* (1997, in Polish); *Development of Natural Processes on the Bratsk Reservoir's Banks* (2000, in Russian). Co-editor *Limestone Exploitation in Landscape Parks* (1998). Contributor to *Karst et evolution climatiques*, edited by J.N. Salomon & R. Maire (1992); *Karst Hydrogeology and Human Activities: Impacts, Consequences and Implications*, edited by D. Drew & H. Hötzl (1999); *Essays in the Ecology and Conservation of Karst*, edited by I. Barany-Kevei & J. Gunn (2000); and to the journals *Acta Carsologica, Acta Geographica, Annales Societatis Geologorum Poloniae, Kras i Speleologia, Studia Carsologica*.
  Cuba; Europe, Central

Vakhrushev, Boris
Dean of Geography Faculty and Manager of Speleology and Karst laboratory, Tavrichesky National University, Crimea, Ukraine. Co-author of *Karst and Groundwaters of the Mountain Massifs of the Western Caucasus* (1985).
  Crimea, Ukraine

Veni, George
Hydrogeologist, owner of George Veni and Associates, San Antonio, Texas, and Adjunct Professor for Center for Cave and Karst Studies, Western Kentucky University, Bowling Green, Kentucky, USA. Author of *Caves of Bexar County* (1988) and *Geomorphology, Hydrology, Geochemistry, and Evolution of the Karstic Lower Glen Rose Aquifer, South-central, Texas* (1997). Editor or co-editor of *Caves and Karst of Texas* (1994); *Speleology in Brazil* (special issue of *Journal of Cave and Karst Studies*, 1996); *Living with Karst* (2001). Contributor to *Images from the Underworld: Naj Tunich and the Tradition of Maya Cave Paintings*, edited by A. Stone (1995); *Cave Minerals of the World*, edited by C.A. Hill & P. Forti (2nd edition, 1997); *Conservation and Restoration of Caves*, edited by V. Hildreth-Werker & J. Werker (2003), and to the journals *Environmental Geology and Water Science, European Journal of Mineralogy, Geoarchaeology, Journal of Cave and Karst Studies*.
  Belize River Caves; Environmental Impact Assessment

Viles, Heather
School of Geography and the Environment, University of Oxford, UK. Editor of *Biogeomorphology* (1988). Editorial board member of *Zeitschrift für Geomorphologie*. Contributor to *Cave and Karst Science, Earth Surface Processes and Landforms, Progress in Physical Geography*.
  Biokarstification; Phytokarst

Waltham, Tony (Adviser)
Senior Lecturer in Engineering Geology, Department of Civil Engineering, Nottingham Trent University, UK. Author of *Caves* (1974); *The World of Caves* (1976); *Catastrophe: the Violent Earth* (1978); *Caves, Crags and Gorges* (1984); *Yorkshire Dales National Park* (1987); *Ground Subsidence* (1989); *Foundations of Engineering Geology* (1994, revised 2002). Co-author of *Caves of Mulu* (1978); *China Caves 1985* (1986); *The Underground Atlas* (1986); *Xingwen* (1993); *Karst and Caves of Great Britain* (1997). Editor of *Limestones and Caves of Northwest England* (1974). Contributor to *Quarterly Journal of Engineering Geology, Cave and Karst Science, Zeitschrift für Geomorphologie, Proceedings of Geologists' Association*. Editorial Board Member for *Geology Today, Mercian Geologist*. President of British Cave Research Association.
  Asia, Northeast; Asia, Southwest; China; Construction on Karst; Europe, North; Ha Long Bay, Vietnam; Mulu, Sarawak; Pinega Gypsum Caves, Russia; Sewu Cone Karst, Java; Shilin Stone Forests, China; Yorkshire Dales, England

Webb, Rauleigh
Australia. Speleologist, Computer Systems Analyst (self-employed). State Cave Recorder and Map Curator for Western Australia. Webmaster for Western Australian Speleological Group (WASG) and Australasian Cave and Karst Management (ACKMA). Numerous publications at both Australian Speleological Federation (ASF) and ACKMA Conference.
  Recreational Caving

Weber, Axel
Zoologisches Institut und Zoologisches Museum, Universität Hamburg, Germany. Contributor to *Encyclopedia Biospeologica*, edited by C. Juberthie & V. Decu (1998); *Subterranean Ecosystems*, edited by H. Wilkens & D. Culver (2000); and to the journals *Copeia, Mémoires de Biospéologie*.
  Amphibia

Werker, Jim C.
Hillsboro, New Mexico, USA. Co-editor of *Cave Conservation and Restoration* (2003), Co-Chair of the Conservation Division for the National Speleological Society.
  Restoration of Caves and Speleothem Repair

Wildberger, Andres
Swiss Institute for Speleology and Karst Studies, La Chaux-de-Fonds, Switzerland. Co-author *Karst and Caves of Switzerland* (1997).
  Hölloch, Switzerland

Williams, Paul (Adviser)
Professor of Geomorphology, School of Geography & Environmental Science, University of Auckland, New Zealand. Co-author of *Karst Geomorphology and Hydrology* (1989). Editor of *Karst Terrains: Environmental Change and Human Impact* (1993). Editorial board member of *Earth Surface Processes and Landforms* and *Zeitschrift für Geomorphologie*. Member of the World Commission on Protected Areas (WPCA) of the International Union for the Conservation of Nature (IUCN). Honorary Member, National Speleological Society, US.
  Dolines; Kaijende Arête and Pinnacle Karst, Papua New Guinea; Karst Evolution; New Zealand

Wood, Paul (Adviser)
Department of Geography, University of Loughborough, Loughborough, UK. Contributor to *Aquatic Conservation: Marine and Freshwater Ecosystems, Archiv für Hydrobiologie, Biological Conservation, Cave and Karst Science, Hydrobiologia*.
  Britain and Ireland: Biospeleology; Subterranean Ecology

Wookey
Software designer, Cambridge, UK. Contributor to *Caves and Caving*. Editor, *Cave Radio and Electronics Journal*.
  Surveying Caves

Worthington, Stephen R.H.
Principal hydrogeologist, Worthington Groundwater, Dundas, Ontario, Canada. Contributor to *Speleogenesis: Evolution of Karst Aquifers*, edited by A.B. Klimchouk, D. Ford, A. Palmer & W. Dreybrodt (2000), and to the journals *Earth Surface Processes and Landforms, Environmental Geology, Environmental Monitoring and Assessment, Geology*.

    Appalachian Karst, United States; Groundwater in Karst; Groundwater in Karst: Borehole Hydrology; Groundwater in Karst: Conceptual Models; Groundwater in Karst: Mathematical Models; Karst Water Resources; Sof Omar, Ethiopia; Springs; Water Tracing

Yuan Daoxian
Research professor and former director, Institute of Karst Geology, Chinese Academy of Geological Science, Guilin, China. Author of *Glossary of Karstology* (in Chinese, with English indices, 1988); *The Science of Karst Environment* (in Chinese, with English summary, 1988). Editor of *Karst of China* (1991); *Environmental Geology* (1997); *Global Karst Correlation* (1998). Contributor to *Hydrogeology of Selected Karst Regions*, edited by W. Back & H. Paloc (1992); *Karst Terrains: Environmental Change and Human Impact*, edited by P.W. Williams (1993), and to the journals *Carsologica Sinica, Environmental Geology, Journal of Hydrology, Journal of the Speleological Society of Japan, Zeitschrift für Geomorphologie*. Member of Editorial Advisory Board for *Environmental Geology*. President of Commission on Karst Geology, Geological Society of China.

    Yangshuo Karst, China

Yonge, Charles J.
President of Alberta Karst Consulting, and Adjunct Professor, Department of Physics & Astronomy, University of Calgary, Alberta. Author of *Under Grotto Mountain: Rats Nest Cave* (2001). Contributor to *Boreas, Cave and Karst Science, Cave Science, Chemical Geology (Isotope Geoscience Section), Journal of Hydrology, Nature, Proceedings of the National Academy of Sciences*.

    Ice in Caves

Zander, C. Dieter
Professor, Zoologisches Institut and Museum, University of Hamburg, Germany. Author of *Parasit-Wirt-Beziehungen. Einführung in die ökologische Parasitologie* (1997). Contributor to *Biology of Benthic Organisms*, edited by B.F. Keegan & O. Ceidigh (1977); *Biology and Ecology of Shallow Coastal Waters*, edited by A. Eleftheriou *et al.* (1996); *Das Mittelmeer, Fauna, Flora, Ökologie*, edited by R. Hofrichter (2001), and to the journals *Helgoländer Meeresunters, Mémoires de Biospéologie, Mitteilungen aus dem Hamburger Zoologischen Museum und Instituts, Zeitschrift für Zoologische Systematik und Evolutionsforschung*.

    Marine Cave Habitats

Index

Note: Main encyclopedia entries are indicated by **bold** type.

A2, Siebenhengste, Switzerland 647
AB 6, Russia 345
Abadie, Ch. 494
Abagy-Dzhie, Siberia 646
Abisso del Corno, Italy 117
Abisso del Nido, Italy 117
Abisso di Malga Fossetta, Italy 117, 118, 326
Abisso Gortani, Italy 468
Abisso Led Zeppelin, Italy 469
Abisso Michelle Gortani, Italy 468
Abisso Olivifer, Italy 525
Abisso Paolo Roversi, Italy 525
Abisso Trebeciano, Kras, Slovenia 486
Aborigines, Australia 124, 129, 624
Abosolon, Karel 336, 759
About Karst Phenomena in Russia (Kruber) 662
Abrakurrie Cave, Nullarbor Plain, Australia 120, 545
abrasion as coastal process 231
Abri Sassi system, Italy 117
Abu Dhabi
   geomorphology 115
   Kahf Hamam 115
Acari **74**, 143, 144
   Acari Parasiti 74
      Metastigmata 74
      Prostigmata 74
   Acari Terrestria 73
      Astigmata (Acaridida) 74
      Mesostigmata (Gamasida) 73
      Notostigmata (Opilioacarida) 73
      Prostigmata (Actinedida) 73
   Hydracarina 74
accelerator mass spectrometry dating 280, 281
accidentals 584
accidents in caves **1**
   See also rescue in caves.
Ace In The Hole (Wilder) 360
Achibakh Massif, Georgia 203
acid solution theory, speleothem formation 690
acoustic behaviour, behavioural adaptations **3–4**
Acta Carsologica 465
actinomycetes 507
actinomycosis 294
active colonization 235
Actun Chek, Belize 140
Actun Kan (Jobitzina), Guatemala Mexico 42
Actun Tunichil Muknal, Belize 140
Actun Tun Kul, Belize 37
adaptations
   Amblyopsidae 7
   Amphibia, morphological 9, 12
   Amphipoda 258

anophthalmy, depigmentation, apterism (ADA) 450
Arachnida 71
behavioural **3–4**
   acoustic behaviour 3–4
   aggregation/schooling 3
   aggression/antagonistic behaviour 3
   alarm substances, responses to 3
   circadian rhythms 3
   coprophagy 3
   echo-location 3
   feeding 3
   photoresponse 3
   reproductive behaviour 3
   scotophilia 3
Blattodea 451
Callipodia **536**
Callocalia spp. 131
Chilopoda **535**
Cholevidae 7, 447–448
Chordeumida **535**
Coleoptera **449–450**
Collembola 7, 8, 9, 446
Copepoda 260
Crustacea 11, 254
Diplopoda **535**
eyes **4–5**
flora 318, 362
Fulgoridea 452–453
genetic **6**
Homoptera 126
Hydracarina **74**
hypogean fauna, evolution of 347–348
internal **10–11**
interstitial habitats, aquatic 454, 455
Isopoda 264
littoral cave fauna **502**
morphological **7–9**
to MSS 180, 457
Orthoptera 452
physiological **11–12**
   dysoxic environments 11–12
   factors influencing 11
   metabolic rates 11, 12
   oxygen consumption 11, 12
   respiratory rates 12
   sensory functions 12
sensory apparatus, Coleoptera 450
speciation **665–666**
Stenasellidae 724
Syncarida 269
troglomorphism **7–9**, 548, 603

adaptive radiation 180
   hypogean fauna 347–348
   and karst resources 480
adaptive shift model of colonization **236**
adaptive shift theory of speciation 665
adaptive zone model of colonization **235, 236, 237**
Adelocosa anops 417, 418
Adelsberger Grotte, Italy 336, 533
Adventures of Tom Sawyer, The (Twain) 206, 360
Aeneid, The (Virgil) 205
aeolian deflation and cave formation 204
Afghanistan
   Band-i-Amir 443, 737
   Shamshir Ghar 443
   Table Rock Cave 443
   Tora Bora caves 443, 510
   travertine 443, 737
Afka Cave, Lebanon 208
Africa
   biospeleology **24–25**
   burials in caves 168
   North
     Algeria **14–15**, 620
       Anou Achra Lemoun 15
       Anou Boussouil 13, 14–15
       Anou Ifflis 13, 15
       Babors karst 15
       Boll Maza 13
       Chott Ech Cherg, Sad'da 620
       Dahredj Ghar Kef 15
       Djurdjura karst 15
       El Outaya 15
       Kef El Kaous 15
       La Vallée d'Iherir 620
       Oasis de Tamantit et Sid Ahmed Timmi 620
       Rahr Amalou 15
       Rhar Bou Maza 15
       Rahr Es Skhoun 15
       Rahr Medjraba 15
       Tafna underground river 15
       Tassili N'Ajjer 14
     Egypt **16**
     Libya **16**
       Umm al Masabih cave 16
     Morocco **13–14**
       Kef Aziza 13
       Kef Tikhoubaï 13
       Kef Toghobeit 13
       Rhar Chara 13
       Rhar Chiker 13
       Wit Tamdoun 13, 14
     Tunisia **15–16**
       Mine cave (Rhar Djebel Serdj) 15–16
       Rhar Ain et Tsab 15
   South
     Apocalypse Pothole 20
     archaeological caves **16–20**
       Apollo 11 Cave 20
       Border Cave 19, 82
       Coopers 19
       Drimolen 19
       Gladysvale 19
       Gondolin 19
       Klasies River Mouth Cave 19, 21–22, 82
       Kromdraai 17, 19
       Makapangsgat Limeworks Cave 17, 19
       Montagu Cave 19
       Nelson's Bay Cave 19
       Northern Cape 19
       Rose Cottage Cave 19
       Silberberg Grotto 17
       Sterkfonetein 17, 18
       Swartkrans 17, 18
       Wonderwerk Cave 19
     Archeulian culture 19
       Bushmangat 20, 21
       Cango Cave 20
       Cape Peninsula
         Bat-Giant System 651
         Ronan's Well 651
       Cave of Hearths 19
       DNA studies, hominid evolution **19**
       Howieson's Poort MSA technology **19, 20**
       Mpumalanga
         Magnet Cave 651
         Mogoto Cave 651
       quartzite caves
         Berlin karst systems 651
         Cape Peninsula 651
         Chimanimani Mountains 651
         Daspoort Formation 651
         Mpumalanga Escarpment 651
   Sub-Saharan
     caves of **20–22**
     cenotes **20–21**
     fauna **21**
     geomorphology **20**
     Kenya
       Banburi Quarry 611
       Elgon, Mount 22, 25, 562, 592
       Giggler's Caves 592
       Kitum Cave 22, 25, 294, 592
       Leviathan Cave 22, 766
       Makingen Cave 22, 592
       Suswa, Mount 22, 561, 562, 763
     Namibia
       Daleib Wanei tufa 23
       Dragon's Breath Cave 21
       Harasib Cave 21
       Isha Baidoa tufa 23
     Otjikoto Lake National Monument **21**
     Rand gold mining 22–23
     Rwanda
       Ubuvomo bwa Musanze 22
       Ubuvomo bwa Nyrabadogo 22
     Sof Omar, Ethiopia 20, **655–656**, 699
     Somalia
       Ail Afwein 23
       Las Anod 23
     surface karst **22–23**

Tanzania
    Hades Cave 22
Zambia
    Lusaka 22
Zimbabwe
    Big End Chasm 21
    Bounding Pot 21
    Jungle Pot 21
    Mwenga Mwena 20, 21
    Wonderhole, Cinhoyi Caves 21
Agapito, Girolamo 601
Agassiz, Jean Louis Rodolphe **152**
Agen Allwedd, Wales 166, 341
aggregate limestone, uses of 489
aggregation/schooling, behavioural adaptation **3**
aggression/antagonistic behaviour, behavioural adaptations **3**
Aggtelek Caves, Hungary/Slovakia **25–26**
    Aggtelek-Domica region 25–26
    Bükk culture 28–29
    Domica Cave 28–29
    geomorphology 25, 26, 28
    speleogenesis 26
Aggtelek-Domica Region, Hungary/Slovakia **25–27**
    Baradla-Domica Cave 26–27
    Beké Cave 26
    and cave surveying 715
    Dobšina Ice Cave 25, 27
    geomorphology 25, 26
    Kossuth Cave 26
    Ochtina Aragonite Cave 26
    Rákóczi Cave 26
    speleogenesis 26
    Stratenska Cave System 27
    Szabadsag Cave 26
    Vass Imre 26
Aggtelek National Park and Biosphere Reserve Hungary/Slovakia 25, 333, 335
Agia Parajkevi Skotinu, Greece 623
Agios-Georgios Cave, Greece 340
Agnew, John 88
agriculture soil erosion and 658
Agroecina cristiani 529
Agua, cueva de, Spain 345
Aguila, sima del, Spain 345
Ailaau flow field, Hawaii, United States 415, 416
Ail Afwein, Somalia 23
Aillwee Cave, Burren glaciokarst, Ireland 171
Ain Hit, Saudi Arabia 115
airborne debris and tourist caves
    defined **731**
    and life forms **732**
    management **732**
    measurement **732**
    particle settling **731**
    sources **731–732**
airflow in caves 229–230
air quality in tourist caves
    analysis of **730**
    CO2
        management of **730–731**

relaxation levels **730**
sources of **730, 731**
health aspects of **731**
human impact on **730**
speleothems and **730**
Ajanta Cave, India 623, 624
A Journey To The Center of the Earth (Verne) 205–206, 360
Akemati, Mexico 36
Akhshtyr Massif, Georgia 201
Akhtsy Massif, Georgia 201
Akhun Massif, Georgia 201
Akiyoshi Caves, Japan
    Akiyoshi-do Cave 30
    caves, classification of **29**
    Current-mark-no-ana Cave 30
    geomorphology **29**
    history of exploration 99
    Inugamori-no-ana Cave 30
    Kuzuga-ana Cave 30
    military use of caves 510
    Sano-ana Cave 30
    Taisho-da Cave 30
    Takaga-ana Cave 30
Akiyoshi Plateau, Japan 96
Akka-Do, Japan 96
Alabama cavefish *See* Speoplatyrhinus poulsoni.
alabaster, dissolution rates 299
alarm substances, responses to, behavioural adaptations **3**
Alejandro de Humboldt National Monument, Venezuela 53
Alek Massif, Georgia 201
Alekseeva Massif, Georgia 201
Alexander the Great and caves 207
algae **508**
    in biokarstification 147, 470
    control by chemical agents 734
    distribution 733
    entrance habitats 318
    light intensity and 733–734
    management of 733
    problems of 733
    on travertine 424
Algeria
    Anou Achra Lemoun 15
    Anou Boussouil 13, 14–15
    Anou Ifflis 13, 15
    Babors karst 15
    Boll Maza 13
    cave wall paintings of 14
    Chott Ech Cherg, Sad'da 620
    Constantine region 15
    Dahredj Ghar Kef 15
    Djurdjura karst 15
    El Outaya 15
    geomorphology 15
    gypsum caves 15
    history of exploration 14–15
    human use of caves 14–15
    Kef El Kaous 15
    La Vallée d'Iherir 620
    Oasis de Tamantit et Sid Ahmed Timmi 620

Algeria *(Continued)*
  Oran Meseta **15**
  Rahr Amalou 15
  Rahr Es Skhoun 15
  Rahr Medjraba 15
  Rhar Bou Maza 15
  salt caves **15**
  speleothems **15**
  Tafna underground river 15
  Tassili N'Ajjer 14
  Tellian Atlas **15**
Alice's Adventures in Wonderland (Carroll) 206
Aligheri, Danté (*The Divine Comedy*) 205
Alinat, Jean 302
alkalinity testing of cave and karst water 214
Allegory of the Cave, the (Plato) 206
Alligator Pond, Jamaica 735
allogenic runoff and doline formation 305
allopatric model of speciation 665, 666
Allpnyxia partizii 453
Allred, Carlene 766
Allred, Kevin 766
All-Union Karst Conference 662
Al Ouat'Ouate 465
Alpi Apuane, Italy 327
Alpine Club, England 350
alpine karst **31–33**, 198, 200, 201–203, 220
  caves 32
  dissolution 31
  exploration of 32–33
  frost shattering 31
  glaciated 388
  glaciers and 31
  hydrology 31
  landforms 31–32
  processes of 31
  resources of 33
Altajskaja, Siberia 629, 645
Altamira Cave, Spain **33–34**
  art in 34, 84, 85, 90
  fauna 34
  great hall of paintings 33–34
  history of discovery 33
  Horse's Tail art 33
  musical performances in 534
  occupation layers 34
  as tourist cave 428
Altinbesik Cave, Taurus Mountains, Turkey 740–741
Alu Caves, China 98, 151
aluminium ores in karst 136
alunite 194, 635
alveolitis 293
Alyn River Caves, Wales 166
Amarnath Cave, India 443, 623
Amaterasu 364
Amaterska-Punkva Cave System, Czechia 331
Ambatoharanana, Madagascar 494
Ambatomanjahana, Madagascar 494
amberat, formation of 636
Amblyopsidae **595–597**
  adaptations 12
  Amblyopsis rosae 595, 596
  Amblyopsis spelea 593, 595, 596, 597
  Chologaster cornuta 595, 596
  Forbesichthys agassizi 595, 596
  morphology 595
  North America, distribution of 46
  scotophilia in 595
  Speoplatyrhinus poulsoni 46, 595, 596, 597
  stygobitic species morphology 595
  thigmotaxis in 595
  Typhlichthys subterraneus 244, 595–596
Amblyopsis rosae 595, 596
Amblyopsis spelea 593, 595, 596, 597
Amblypygi **75–76**, 143
America, Central
  archaeological caves **40–42**
  Belize
    Actun Chek 140
    Actun Chichem Ha 42
    Actun Tunichil Muknal 140
    Actun Tun Kul 37, 140
    Barton Creek Cave 140
    Blue Creek Cave 140
    Caracol 37
    Caves Branch caves 571
    caves of **37, 140**
    Cebada Cave 37, 140
    Chiquibul Cave System 37, 40, 140
    Kabal Group 37
    Petroglyph-St Herman's Cave 140
  biospeleology **38–40**
  fauna, threats to 39–40
  geomorphology **34–35, 37**
  Guatemala
    Chiquibul Cave System 37, 140
    El Duende Pyramid 624
    Naj Tunich 37, 40, 92
    Parque Nacionale Laguna del Tigre, 620
    Sistema del Rio Candalaria 37
    Tikal 37
    Xibalba 37
  history of exploration **36**
  human impacts on karst 37
  karst, characterization of 35–36
  karst vegetation 37
  Mexico
    Akemati 36
    Balankanche 40, 42
    Cueva de Villa de Luz 38
    Cueva Encantada 40
    Dos Pilas 41
    Gruta del Rio San Geronimo 37
    Gruta de Palmito 36
    Grutus de Juxtlahuaca 37, 92
    Oaxaca, troglobitic fauna 38
    Ocotempa 36
    Pyramid of the Sun, Teotihuacan 40, 42
    Quintana Roo 37
    Rio Chontalcoatlan 37

San Miguel, hoya de 37
Sistema Cuetzalan 36
Sistema Purificacion 36, 525
Sumidero Yochib 37
Tabasco, troglobitic fauna 38
Veshtucoc system 36–37
Yucatán Phreas **787–788**
    caves of 787–788
    cenotes 786
    Chichen-Itza 37, 41
    geomorphology 786–787
    human impacts on 788
    Maya and 788
    Peninsula 39, 620–621
    Ramsar sites 620–621
    Sac Actun 787
    Sistema Aak Kimin 787
    Sistema Abejas 787
    Sistema Actun Ha 787
    Sistema Aerolito 787
    Sistema Chikin Ha 787
    Sistema Esmerelda 787
    Sistema Naranjal 787
    Sistema Nohoch Kiin 787
    Sistema Nohoch Nah Chich 786–787, 788
    Sistema Ponderosa 787
    Sistema Yaxchen 787
    speleogenesis 787
    stygobitic fauna 39
    troglobitic fauna 39
Nicaragua
    Cayos Miskitos y Franja Costera Immediata 620
pollution 37
protection 37
research 36
Sierra Madre Oriental, troglobitic fauna 38–39
Southern Mountains, caves of 36–37
speleogenesis 36
threats to karst 37
Veracruz, troglobitic fauna 38
America, North
    archaeological caves **43–45**
    bats 46, 500
    biospeleology **45**
    Canada **175–178**
    fauna, aquatic 46
    history of speleology **49–53**
    invertebrates 46
    stygobites 46
    troglobites 46
    United States **749–750**
    See also individual country by name.
America, South
    Argentina **56**
        Cueva de Doña Otilia 56
        Cueva de las Brujas 56
        Cueva de las Manos 91
        Cueva del León 56
        Sistema Cuchillo Cura 56

Bahia
    Chapada Diamantina 92
    Lapa da Sol 92
biospeleology **58–59**
    aquatic fauna, interstitial 58–59
    Chiroptera 58
    diversity of 57–58, 59
    history of 57
    Insecta 58
    invertebrates 58
    soil animals 58
    stygobites 58
    troglobites, terrestrial 58
Bolivia 55
Brazil 56–57
    Bambuí karst 133
    Barriguda, Toca da 156, 157
    Boa Vista, Toca da 55, 56, **156–157**, 525
    Boquierão 55
    Caverna Aroe Jari 57
    Caverna da Pedra Pintata 91
    Caverna dos Ecos 57
    Caverna Huayllas 55
    Chiflonkkakka 55
    Gruna da Água Clara 55
    Gruta Casa de Pedra 56, 57
    Gruta da Bocaina 55, 611, 612
    Gruta das Bromélias 611, 613
    Gruta de San Pedro 55
    Gruta do Janelão National Park 93, 134
    Gruta do Padre 55
    Gruta Planaltina 57
    Gruta San Miserato 55
    Gruta Umajalanta 55
    Gruto do Centenário 55, 611, 612
    Ibitipoca State Park 613
    Lapa do Angélica 55
    Lapa São Mateus III 55
    Lapa São Vicente I 55
    Lapão Cave 611, 613
    Minas Gerais 92, 93
    Peracu National Park 93
    Toca da Barriguda 55
    Toca das Confuspes 57
    Torotoro National Park 55
Chile 56
    Cueva del Valle de la Luna 56
    Cueva Mylodon 56
    Patagonia marble karst **572–573**
    Perte du Futur 55, 56
Colombia 54
    Cueva del Cunday 54
    Cueva del Indio 54
    Cueva de los Guácharos 54
    Cueva de los Guácharos National Park 54
    Hoyo del Aire 54, 60
    Sistema Hermosura 54
Cuba
    Cueva Pichardo 93

America, South *(Continued)*
　Ecuador 54–55
　　Cueva de Gallardo 55
　　Cueva del Cascajo 55
　　Cueva de los Cayos de Coangos 55
　　Cueva de San Bernadino 55
　　Cueva de Shimpiz 55
　geomorphology **53**
　history of speleology **59–60**
　Paraguay 55–56
　Peru 55
　　El Tragadero 55
　　Gruta de los Guácharos 55
　　Gruta Huagapo 55
　　Millpu de Kaukiran 55
　　San Andrés de Cutervo National Park 55
　　Sima Pumacocha 55
　　Uchkopisjo 55
　Uruguay
　　Cueva de Arequita 56
　Venezuela
　　Alejandro de Humboldt National Monument 53
　　Angel Falls, National Park of Cainama 612
　　Cueva Alfredo Jahn 53
　　Cueva del Cerro Autana 652
　　Cueva del Guácharo 53, 55, 59, 409
　　Cueva del Samán 53, 55
　　Cueva los Encantos 53
　　Cueva los Laureles 53
　　Cueva Sumidero La Retirada 53
　　geomorphology 53
　　Haitón del Guarataro 53
　　National Park of Cainama 612
　　Sima Acopán 1 611
　　Sima Aiyán-tepui Noroesti 55, 611, 612
　　Sima Aonda 55, 611, 612
　　Sima Aonda 2 55, 612
　　Sima Aonda 3 55, 612
　　Sima Aonda Este 2 612
　　Sima Aonda Superior 611
　　Sima Auyán-tepui Norte 2 612
　　Sima de la Lluvia 611
　　Sima Mayor 612
　　Sima Menor de Sarisariñama 611, 612
　See also individual country by name.
American Philatelic Research Library 704
American Philatelic Society 704
American Topical Association 704
amino acid racemization dating method 556
ammonia and guano 409
Amphibia **61**
　adaptations, morphological 9, 12
　characterization 61
　distribution 25, 144
　morphology 61
　neoteny 61–62
　Plethodontidae 61
　prey 61
　Proteidae
　Proteus anguinus **61–62**, 279, 289, 366, 370
　　adaptations in 6, 11, 62
　　discovery 148
　　distribution 61, 62
　　embryonic development 62–63
　　hearing 63
　　lateral-line sensory system 63
　　magnetic field sensitivity 63
　　morphology 62
　　neoteny 62
　　threats to 63
　　visual system 63
　troglobitic species of 61
Amphipoda **257–259**
　adaptations of 7, 12, 258
　Africa 24
　aquatic habitats 258
　Australia 126, 182
　Bogidiellidae 258, 631–632
　Britain and Ireland 163
　Crangoncytidae 259
　evolution of 259
　France 370
　Gammaridea 257
　Gammarus minus 6, 9, 12
　Hadziidea 259
　Ingofiella canariensis 181
　Niphargidae 258
　Niphargus aggtelekus 620
　Niphargus balcanicus 760
　Niphargus stygius 604
　North America 47
　Pseudoniphargus associatus 180
　Pseudoniphargus salinus 181
　Rhipidogamarus mivariae 180
　sandhoppers, terrestrial 180, 417
　South America 58
　Southeast Asia 110
　Spelaonicippe buchi, Lanazarote 179
　Stygobromus canadensis 200
　Talitridae 258
　taxonomic diversity of 257, 258
　Amud Cave, Israel 196
Anaspidacea 255, 268–70
anastomotic mazes 573, 574
Anataelia troglobia 452
Anaxagoras and karst hydrology 478
anchialine habitats **64–65**
　Amphipoda 65, 258, 260
　Belize 38
　Canary Islands 181, 182
　Cape Range, Australia 182
　continental 182
　Crustacea 65, 254, 256, 257
　Dinaric Karst 65
　defined 64, 712
　fauna of 65
　Hawaii 417
　stratification in 64–65
　thalassostygobites 549

thermal gradient in 707
    Walsingham Caves, Bermuda 767
Ancona Abyss, Frasassi cave, Italy 374, 375
andesite 762
Andrafiabe, Madagascar 494
Andrè Lachambre, France 373
anemometers, use in caves 230
Angel Falls, Cainama National Park, Venezuela 612
Anguila 190
anhydrite
    breccia pipes, formation of 661
    evaporite karst 343
    Pinega gypsum caves, Russia 588–589
    principle karst rock 474
    Spitsbergen 342
Anisolabis howarthi 452
Anjohiambovonomgy, Madagascar 494
Anjohibe, Madagascar 494
Anjohikely, Madagascar 494
Ankarana, Madagascar 493–494
Annah, Ireland 162
Annapurna Himal, Nepal 623
Anne-A-Kananda, Australia 122
Annelida **65–66**
    Clitellata 66
    Hirudinea 66
    Marifugia cavatica 66, 289, 290, 760
    morphology 66
    Oligochaetes 66
    Polychaeta 66
Annette cave, France 283
annual deposition band dating 557, 558
Anomura 261, 262
anophthalmy, depigmentation, apterism (ADA) 450
Anou Achra Lemoun, Algeria 15
Anou Boussouil, Algeria 13, 14–15
Anou Ifflis, Algeria 13, 15
Antalya terraces, Turkey 737, 740
Antarctica **68–69**
    Collins Glacier, King George Island 68
    crevasses 69
    glacier caves 68
    history of exploration 68
    ice, sublimation of 69
    moulins 68
    Ross Ice Shelf 606
Anthroherpon apfelbecki 760
Anticosti Island, Canada 175
Antifonale (Blak) 534
Antigua 190
Antiparos, Greece 208
Antrisocopia spp. 768
Antroherpon dombrowskii 448, 449
Antsatrabonko, Madagascar 494
Antsiroandoa, Madagascar 494
Anvil Points Claystone Cave, Colorado, United States 591, 606
Anza-Borrego, California, United States 53
A Passage to India (Forster) 206
Ape Cave, Washington, United States 761
Apidima Cave, Peloponnese 340

Apocalypse Pothole, South Africa 20
Apollo 11 Cave, South Africa 20
Appalachian Mountains, United States **69–70**
    Burnsville Cove, Virginia 69
    Butler-Sinking Creek Cave System 69
    Chestnut Ridge Cave System 69
    geomorphology 69–70, 749–750
    hypogean fauna in 142
Apterygota
    Collembola 445, 446 *See also individual entry for this order.*
    Diplura 47, 143, 182, 445, 446
    Microcoryphia 445, 446
    Zygentoma 445, 446–447
Aqevot Cave, Israel 638
Aqualung 302
aquatic hypogean habitats *See* habitats, aquatic hypogean
aquifers.
    conduits, evolution of 678, 679
    evolution models, mathematical 402
    length/depth modelling 679–680
    protection and remediation 406
    See also groundwater.
Arabika Massif, Georgia 201, 202
Arabikskaja cave system, Georgia 201
Arachnida **71–76** 369–370
    Acari **73–74**, 143
        Acari Parasiti 74
        Acari Terrestria 73
    adaptations 71
    Amblypigi **75–76**
    Aranae 24, **71–73**
        Araneomorpha 71, 72
        Liphistomorpha 71, 72
        Mygalomorpha 71, 72
    classification 71
    distribution 71
    diversity 143–144
    Opilionida **75**
        Cyphophthalmi 75
        Laniatores 75
        Palpatores 75
    Palpigradida **76**
        Eukoeneniidae 76
        Protokoeneniidae 76
    Pseudoscorpionida **75**
    Ricinulei **76**
    Schizomida **76**
    Scorpionida **74**
Arachnocampa luminosa 3, 4, 453
aragonite 690–691
    Aragonite Caves, France 373–374
    and breccia cementation 660
    Ghar Sarab, Iran 461
    Kugitangtau, Turkmenistan 273, 275
    Moulis, Grotte, France 327
    oxygen fractionation dating 556, 557
    precipitation of 194, 512, 692
    structure of 184
    Wyandotte Cave, Indiana, United States 44, 81
Aragonite Caves, France 373–374

Aranae **71–73**
   adaptations 144
   Agroecina cristiani 529
   Araneomorpha 71, 72
   biodiversity in caves 143, 144
   Canary Islands 180
   distribution 71
   Liphistomorpha 71, 72
   Mygalomorpha 71, 72
   North America 46
   South America 58
Araneomorpha **71, 72**
   Agelenidae 73
   Anapidae 73
   Austrochilidae 73
      Hickmania troglodytes 73
   Clubionidae 73
   Ctenidae 73, 631
   Dictynidae 46
   Dysderidae 72
   Gnaphosidae 73
   Leptonetidae 46
   Linyphiidae 72
      Phanetta subterranea 46
   Liocrandiae 73
   Lycosidae 73
   Meitdae 73
   Nesticidae 46, 73
   Ochyroceratidae 72, 631
   Oonopidae 73
   Pholcidae 72, 631
   Prodidomidae 73
   Telemidae 46
      Telema tenella 370
   Tetrablemidae 73
   Theraphosidae 38
   Theridiidae 73
   Theridiosomatidae 73
Aranzadi passage, Pierre Saint-Martin, France 587
Arbois spring, France 754
arborescent lava 764
archaeological caves
   Altamira Caves, Spain **33–34**
   America, North **43–45**
   Ardèche Caves, France **82–83**
   Asia, Southeast **107–109**
   Atapuerca Caves, Spain **119–120**
   Belgium **138–139**
   Europe, Central
      history of exploration 333
      human settlement of 333–334
   history of research 40–41
   Mount Carmel, Israel **195–197**
   South Africa **17–20**
   Vézère Caves, France 757
   See also burials in caves.
archaeological methods, dating **279–281**
archaeologists **76–79**
archaeology of caves, history of **80–82**
   Africa 82
   Europe **80**
   New World **81–82**
Archangel Michael 364
Archeulian culture, South Africa 19
Arch-Treasure Cave, Canada 178
Arcos/Pains, Bambuí karst, Brazil 133
Arctomys cave, Canada 51, 52, 177
Arctotherium brasiliense 157
Arcturus, Nakanai Caves, Papua New Guinea 538
Ardeantine Caves, Italy 510
Ardèche caves, France **82–83**
   archaeology 82–83
   art in 82–83
   history of exploration 427
Arecibo Observatory 190
Argarakanskaya Cave, Siberia 645
Argentina 56
   art in caves 91
   Cueva de Doña Otilia 56
   Cueva de las Brujas 56
   Cueva de las Manos 91
   Cueva del León 56
   history of speleology 60
   Sistema Cuchillo Cura 56
argon dating 194–195
Argostoli sea-mills, Greece 338
Aripo No. 1 Cave, Trinidad **190**
Aristotle and karst hydrology 478
Arixenia jacobsoni 452
Arixenina
   Arixenia jacobsoni 452
   Xeniaria esau 452
Armand, L. 371
Arresteliako Ziloa, Pierre Saint-Martin, France 586
Arrow Grotto, Feather Cave, New Mexico, United States 92
arrow-worms *See* Chaetognatha
Arroyo Tapiado, California, United States 591
arsenates
   minerals in caves **512**
ARSIP *See* Association pour la Recherche Spélélogique Internationale à la Pierre Saint-Martin (ARSIP).
artesian speleogenesis 681–682
Arthropoda *See individual orders by name.*
Arthur, George 129
Arthur, Mount National Park, South Island New Zealand 542
artificial caves 41
art in caves **83–84**
   Altamira Cave, Spain **34**
   Americas 44, **91–93**
      dark zone art 91, 92
      distribution of 91–92
      Maya art 92
      mud-glyph cave art 92
      Olmec art 92
      study of 91, 92–93
   Ardèche caves, France 82–83
   Australasia **88–89**
      Judds Cavern 129
      Koonalda Cave 88–89

Mount Gambier 89
Papua New Guinea 89
Britain 162
Caribbean Islands 92, 189
dating 281
Encantado, Sistema del Rio, Puerto Rico 318
Europe **90–91**
    acoustic properties, study of 91
    history of discovery 90
    interpretations 90–91
    subject matter 90
    techniques 90
history of discovery 85, 90
Indonesia 108
Kapova Cave, Urals 629
lampenflora and 733
Lascaux caves, France 85, 757
Magura Cave, Bulgaria 330
Maya 92
Naj Tunich, Guatemala 92
Niah caves, Sarawak 108
showing caves **86–88**
Southeast Asia 107
Thailand 107
Vézère caves, France 90
Vietnam 108
Wandjina cult, Australia 364
art showing caves **87–88**
    history of 86–87
    inspiration 87–88
    photography 88
Aruba 190
Aryk-Bash Cave, Ukraine 253
Asbestopluma sp. 457
A Sea Cave Near Lisbon (Short) 359
Asellota 256–257
Ash Cave, Kentucky, United States 168
Asia, Central **94–95**
    geomorphology 94
    Kazakhstan 95–96
    Kopetdag range
        Bakharden Cave 94
    Pamir Mountains 96
        Syjkyrdu Cave 94, 96
    Tien Shan 94–95
        Bojbulak 95
        Cupp-Coutunn Cave system 95
        Fersman's Cave 95
        Festival'naja-Ledopadnaja system 95
        Geofizicheskaya Cave 95
        Karljuksky Cave 94
        Khashim-Ojuk Cave 95
        Khodzharustam Cave 95
        Kievskaja Cave 95, 663
        Kjaptarkhana Cave 95
        Kun-Ee-Gout 95
        Kyrktau Plateau 95
        Kysyl Dzhar Mountains 95
        Pobednaja Cave 95
        Uluchurskaja Cave 95
        Zajdmana Cave 95
    Turansky Plain 94–95
        Sarykamyshskaya 94
    See also individual entry for each country.
Northeast **96**
    China **217–220**
    geomorphology 96
    history of exploration **97–100**
    Japan 96
        Akiyoshi caves **29–30**, 510
        Akiyoshi Plateau 96
        Akka-Do 96
        Byakuren-Do 96
        Gyokusan-Do 96
    Korea 96
        Namgamduk system, Korea 96
    Mongolia 96
    Taiwan 96
    See also individual country by name.
Asia, Southeast
    archaeological caves **107–109**
    Batu Lubang, Halmahera 110
    biospeleology **109–112**
        biogeographical distribution of fauna 110
        endemic species swarms 111
        evolution of fauna 110
        guano accumulations 112
        guanobites 111–112
        habitat, fragmentation of 111, 112
        history of research 109
        quarrying disturbance 112
        relict fauna 111
        taxonomic compositia 110
        troglomorphic species 111
    burials in caves 168
    Cambodia 100
    cone karst 104
    flora 100
    geomorphology 100, 102, 104
    Gunung Sewu cone karst, Java 104, 241, 278, **641–642**
    hong lakes 103
    hong of 103
    human use of caves and karst 103–104
    Indonesia 109, 631
    karst 102–103
    Khammouan karst, Laos **483–485**
    Laos 100
    Malaysia 100, 531–533
    minerals in karst 103
    Myanmar 100, 101
    paleokarsts 102
    Papua New Guinea 105
    Philippines 105
    pinnacle karst 104–105
    Puerto Princesa Subterranean River, Philippines 105
    speleogenesis 100–102
    speleological investigations, history of 112
    Thailand 100
    tower karst of 104

Asia *(Continued)*
   Vietnam 100
   See also individual country by name.
Asia, Southwest **114–115**
   Abu Dhabi 115
   Iran 115–116 **460–461**
   Iraq 115
   Israel 114
   Jordan 114–115
   Lebanon 114
   Oman 115
   Saudi Arabia 115
   Turkey **740–742**
      Altinbesik Cave, Taurus Mountains 740–741
      Birkilen Caves 742
      Cave of the Seven Sleepers 208
      Çukurptnar Cave, Taurus Mountains 741
      Dog Cave 208
      Dumanli spring 699, 741
      Evren Gunay düdeni (Peynirlikönü), depth of 525
      Keban Dam 278
      Kirkgöz–1 741
      Kirkgözler 699, 741
      Kirkgöz-Suluin 741
      Kizilelma Mağarasi, Taurus Mountains 741
      Kocain Cave, Taurus Mountains 741
      Kurukopru Cave 741
      Kuyukale, Taurus Mountains 741
      Martal Cave, Taurus Mountains 741
      Oluk Köprü 699, 741
      Pamukkale 187, **568–569**, 737
      Peynirlikönü Cave (Evren Gunay Cave), Taurus Mountains 741
      Pinargözü Cave, Taurus Mountains 741
      Pit of the Jinns Cave 208
      Plutonium Cave 208
      Sabatagi Düdeni, Taurus Mountains 741, 742
      Sütlük Düdeni, Taurus Mountains 741, 742
      Tuz Gölü 742
      Yedi Miyarlar 699
   Yemen 115
   See also individual country by name.
Asiago plateau, Italy **116–118**, 326
   cave types 117–118
   geomorphology 116–117
   hydrology 118
   speleogenesis 117–118
   urbanization, impact of 118
Asopladeru de la Texa, Picos de Europa National Park, Spain 583, 584
aspergillosis 294
asphalt 637
Assman hygrometers, measurement using 228
Association for Mexican Cave Studies 36, 38
Association pour la Recherche Spélélogique Internationale à la Pierre Saint-Martin (ARSIP) 585
Astacidea 261, 262
Astagobius spp. 290
Astyanax fasciatus 5, 38, 594, 708
Atapuerca Caves, Spain **119–120**
   burials in 119
   dating 119
   Homo heidelbergensis 119
   human occupation of 119
Atea Kananda, Papua New Guinea 105
Athabasca Glacier, Alberta, Canada 387
Athabasca Tar Sands, Canada 176
Atlantida, Ukraine 746
atmospheric pressure in caves 229–230
At The Earth's Core (Burroughs) 206
Atti e Memoire 465
Aub, Conrad 235, 242
Aubrey, John 714
Augensteine defined 173
Aurora Cave, Fiordland New Zealand 543, 554, 555
Austen, R.A.C. 80
Australasian Cave and Karst Management Association (ACKMA) 728
Australia
   Aborigines 124, 129, 624
   Abrakurrie Cave 120, 545
   Anne-A-Kananda 122
   archaeological and paleontological caves **124–125**
   art in caves
      Gambier, Mt 89
      Judds Cavern 129
      Koonalda Cave 88–89
      Wandjina cult 364
      Wargata Mina 124, 129
   Australian Fossil Mammal Sites World Heritage Area 124
   Ballawinnie Cave 122, 124, 129
   Bathurst district 127
   biospeleology **125–126, 181–183**
   Blanche Cave 128
   Breccia Cave 124, 127, 129
   Buchan Caves Reserve 129
   Bullita Cave System 123, 525
   Bungle Bungle range 123
   Bungonia caves 127, 634
   Cape Range
      biospeleology of **181–183**
      geomorphology 181
      karst areas of **121**
   Caveside Caves 129
   Chillagoe 89, 121, 192, **215–216**
   Cloggs Cave 124
   Cocklebiddy Cave 545, 546
   Darwin glass artefacts 124
   Diprose Caves 546
   Drum Cave 127
   Easter Cave 120
   endemicity of subterranean fauna 126
   Eugenana karst 120
   Exit Cave, Tasmania 122
   Eyrie, Lake Basin 737
   Fish River Caves 128
   Gambier, Mt, art in 89
   geomorphology 120
   Girraween National Park 652
   glaciation and troglobite evolution 125

Goebel's Cave 652
Greater Blue Mountains World Heritage Area, 121
Gregory National Park 123
history of exploration **127–129**
Hunter Island 123
Ice Tube Growling Swallet system 122
Jenolan Caves, New South Wales
   and cave tourism 726, 731, 732
   history of exploration 127
   Lucas Cave 623
   paleokarst of 561, 634
   restoration 626
   Social and Environmental Monitoring (SEM) programme 519
Judds Cavern 129
karst areas of 120–121
karst evolution 120
Kelly Cave 546
Kempsey, New South Wales 121
Kenniff Cave 124
Kimberley, Western Australia 121
Koonalda Cave 83, 88–89, 120, 544, 545
Koongine Cave 85, 89
Kutikina Cave 122, 124, 129
lava tube caves 123
Lawn Hill National Park 124
Limekilns Cave 127
Lucas Cave, Jenolan 623
Malangine Cave 85, 89
Mapala Rockshelter 125
McKeon's Caves 128
Milyeringa veritas 182, 183, 594
Mitchell-Palmer karst **215–216**
Monarch Cave 123
Moondyne Cave 625–626
Mosman's Cave 127
Mullamullang Cave 545, 546, 731
Murra-el-Elevyn Cave 547
Naracoorte Caves 124, 128, 633, 726
Naracoorte Caves Conservation Area 124, 128
Newdegate Cave, Tasmania 123
New Guinea 2, art in 89
Niggly Cave 122
Nombe rockshelter 125
Northern Territory/Queensland, karst areas of **123**
Nullarbor Plain **544–545**
Oakden caves 129
Old Homestead Cave 121, 546
Orchestra Caves, art in 89
Otway Basin 620
River Cave 652
Riversleigh 124
Rockhampton 129
Rocky Cape 123
Rocky Cape North Cave 127, 128
root mats 126
Ruined City 123, 651
Seton Rockshelter 124
silicate karst of **651–652**
Sthenerus spp., 124
Strongs Cave 120
stygobitic fauna **125, 126**
Tasmania, karst areas of 122–123
Tasmanian Wilderness World Heritage Site 122, 778
Tasman's Arch, Tasmania 704
Thampanna Cave 546
Tommy Graham's Cave 545
troglobitic fauna **125–126**
Tunnel Cave 127, 129
Victoria Cave, Naracoorte 121, 144
and vulcanospeleology 766
Waitomo Museum of Caves 728
Wargata Mina 122, 124, 129, 623
Weebubbie Cave 120, 545
Wellington Caves, New South Wales 124, 127, 128, 547
Westmoreland Cave, Tasmania 633
Wet Caves 129
Whalemouth Cave 123, 652
Wombeyan karst 120
Australian Fossil Mammal Sites World Heritage Area, Australia 124
Australian Natural Heritage Charter and conservation management 728
Australian Speleological Federation 622
Australopithecine caves, South Africa 17
Australopithecus africanis 16, 23, 82
Australopithecus robustus 17
Austria
   Bergerhohle-Bierloch 173
   Calcareous Alps **173–175**
   Cosa Nostra Cave 173, 525
   Dachsteinkalk 326
   Dachstein-Mammuthöhle 174, 326
   Dachstein massif 174
   Dachstein-Rieseneishöhle 174
   DÖF-Sonnenleiterschacht cave 174
   Drachenhöhle 326, 335
   Eastern Alps 173
   Eisriesenwelt cave 173, 174
   history of exploration 336
   Lamprechtstofen 173, 326, 525
   Lorloch 301
   Lurhöhle 326
   Raucherkarhöhle 174, 525
   Schwartzmooskogeleishöhle Gebirge 174
   Schwersystem 525
   Spannagel Cave **663–664**
   and speleotherapy 698
   St Lukas Guild 88
   Totes Gebirge 173, 174
   Verband österreichischer Höhlensforscher 336
   Verein für Höhlenkunde 336, 350
automata and cave art 87
Aven Armand, France 533
Aven Armand (Rudaux) 88
Aven d'Orgnac, France 338
Aven Noir, France 371
Aven Orgnac, France 371
Avenul din Grind, Romania 329
Avenul din Stanu Foncii, Romania 329

Avenul V5, Romania 329
Aves 1301–131
   Africa 25
   casual birds 130
   caves, use of 130
   cave species 130
   Collocalia spp. 131, 409
   echolocation 3, 131
   habitual birds 130
   Steatornis caripensis 57, 59, 130–131, 190, 366, 410
   terrestrial cave habitats, biodiversity in 144
   usual birds 130
Aya Cave, Siberia 646
Aydashinskaya Cave, Siberia 646
azurite 512

Babors karst, Algeria 15
Bacon, Roger 410
Bacson culture 108
bacteria
   in biofilms 145
   chemoautotrophic 150, 366, 375, 507, 758
   cyanobacteria 508, 733
   ferrobacteria 366, 506
   functions in cave environment 366, 505
   in groundwater 404, 405, 407
   nitrobacteria 366, 505–506
   sulfobacteria 366, 506
   thiobacteria 366
   in tufa deposition 147
   See also microorganisms.
Badalona, Sistema, Spain 327
badlands
   defined 590
   pseudokarst 591, 606
Badlands National Park, South Dakota, United States 590
bad water line defined 313
Badzhevskaja, Siberia 629, 645
Bahamas 189
   blue hole caves on 155–156, 189, 302
   Conch Blue Hole, North Andros Island 156
   Dean's Blue Hole, Long Island 155
   flank margin caves 189
   Lucayan Caverns 156, 360
   pit caves on 189
   Remipedia 156
   Salt Pond Cave 676
   Watling's Blue Hole, San Salvador Island 155
Bahia
   Chapada Diamantina, art in 92
   Lapa da Sol, art in 92
Baishuitai, China 218
Bakharden Cave, Central Asia 94
Balankanche, Mexico 40, 42
Balazs, Denes 113
Balch, E.S. (*Glacières, or Freezing Caverns*) 51
Balch, Herbert 166
Balcombe Graham 166, 302
Balkanostenasellus skopljensis 724
Ballawinnie Cave, Australia 122, 124, 129

Ballybunion, Ireland 606
Ballynamintra caves, Ireland 162
Balutopf, The, Germany 302
Balver Höhle, Germany 533
Bamazomus vespertinus 182
Bamboo Bottom, Jamaica 234, 235
Bambuí karst, Brazil **133–135**
   conservation of 135
   geomorphology 133
   Gruta do Janelão 93, 134
   karst areas 133–134
   limestones 134
banana hole caves, Jamaica 189
Banburi Quarry, Kenya 611
Band-i-Amir, Afghanistan 443
Ban Dong, China 218
Bandzioch Kominiarski Cave, Poland 331
Banff Hotsprings, Canada 177
Bangor Cave, Alabama, United States 534
Baradla Barlang, Hungary 533
Baradla-Domica Cave System, Hungary/Slovakia 28, 29, 331, 366, 620, 698
   geomorphology **26–27**
   speleothems **26**
   and speleotherapy 698
Barbados
   Bowmanston Cave 190
   geomorphology 190
   Harrison's Cave 190
Barbuda **190, 192**
Bärenschacht, Siebenhengste, Switzerland 647
barite
   mining of 514
   Valea Rea Cave, Romania 329
Barkite, Bulgaria 14, 329
Barlangkutatás 464
barnacle zone defined 231, 232
Barrencs de Fournes, France 373
Barriguda, Toca da Brazil 156, 157
Bárta, Juraj 28
Barton Creek Cave, Belize 140
Bar-Yosef, Ofer **196**
basalt
   clinkery 762
   and crevice caves 251
   pahoehoe 761
   volcanic caves **760–764**
baseband audio and cave communication 238
Basidiomycetes 508
Basilica of the National Shrine of the Immaculate Conception and tufa 551
Bassett's Cave, Bermuda 768
Bat Cave, New Mexico, United States 251
   archaeology of 43
   and cave photography 579
   See also Carlsbad Caverns National Park.
Bat Cave, North Carolina, United States 722
Bat-Giant System, Cape Peninsula, South Africa 651
bat guano *See* guano.
Bathurst district, Australia 127

Bathynellacea 255
Bathysciomorphus globosus 448
Batman 360
bats **225–227**
    See also Chiroptera.
Bat's Cave, Antigua 191–192
Bats in Philately: A Comprehensive Study and List of Bats on Stamps (*Handbook 128*) 704
Battle Beneath the Earth (Tully) 360
Batu Caves, Kuala Lumpur, Malaysia
    history of research 109
    ritual use of 103, 428, 623, 624
Batu Lubang, Halmahera Southeast Asia 110
Baturong, Malaysia 108
Baum, L. Frank (*Dorothy and the Wizard in Oz*) 206
Baumannshöhle, Germany 335, 714
Bauxite **135–136**
    alumina production 136
    bauxite belts, major 135, 136
    environmental damage by mining 136
    in karst deposits 135
    paleokarst 330, 332, 560
    production by country 135
    reclamation of mines 136
Bayle, François 534
Bayn Halayn (Arch Cave), Selma Plateau, Oman 639
Bear Gulch Cave, Pinnacles National Monument, United States 721
Bear Rock karst, Canada **137–138**, 178
bears See cave bears.
Beast in the Cave, The (Lovecraft) 206
Beatushöhle, Switzerland 87
Beaumont, John 689
Bedeilhac Cave, France
    industrial use of 426
    military use of 510
Bedford, Bruce 466
Bednarik, Robert 85
Beduzzi, Carlo 87
behavioural adaptations See adaptations, behavioural.
Beké Cave, Hungary/Slovakia 26
Belcher Islands, Hudson Bay Canada 175
Belgium
    archaeological caves **138–139**
    Bois Laterie Cave 138, 139
    Chaleux Cave 138
    Couvin Cave 139
    Engihoul Cave 139
    Fond-de-Forêt Cave 138, 139
    Furfooz Cave 138
    Grotte du Docteur Cave 138, 139
    Grotte Père Noel, Grottes de Han 342
    Grottes de Goyet Cave 138, 139
    Grotte Walou Cave 138, 139
    La Naulette Cave 138, 139
    Marche-les-Dames Cave 139
    Montaigle Cave 139
    Presle Cave 139
    Remouchamps Cave 139
    Trou Al'Wesse Cave 138, 139
    Trou Da Somme Cave 138, 139
    Trou Magrite Cave 138, 139
    Trou Sandron Cave 138, 139
Belisarius xambuei 74
Belize
    Actun Chek 140
    Actun Chichem Ha 42
    Actun Tunichil Muknal 140
    Actun Tun Kul 37, 140
    Aves 130
    Barton Creek Cave 140
    Blue Creek Cave 140
    Caracol 37
    Caves Branch caves 571
    Cebada Cave 37, 140
    Chiquibul Cave System 37, 40, 140
    Kabal Group 37
    karst areas of 140
    Petroglyph-St Herman's Cave 140
    Rio Frio Cave 735
    stream caves 140
    stygobitic fauna 38
    troglobitic fauna 38
Bellamar Cave, Cuba 191, 272
Belski, David 52
Belum Guhalu, India 443
Benarat Caverns, Gunung Mulu National Park, Malaysia 532
Beneath the Pennines (Perou) 360
Benjamin, George 155, 302
Bennett, Compton (*King Solomon's Mines*) 359
Berchil'skaja Cave, Georgia 487
Beremend Crystal Cave, Hungary 561
Bergerhohle-Bierloch, Austria 173
berlinite 329
Berlin karst systems, South Africa 651
Bermuda
    Bassett's Cave 768
    Bitumen Cave 768
    Church Cave 768
    Government Quarry Cave 768
    Prospero's Cave 534
    Walsingham Cave System 141, **767–768**
Bernabe, Pascal 303
Beshtekne-Uzundzda-Chernaya system, Ukraine 253
Besson, Jacques **151–152**
Betts, Olly 716
BG63, Pierre Saint-Martin, France 586
Biak Caves, Irian Jaya
    military use of caves 510
    strategic use of 427
biblical caves 208
Bibliografia Espeleològica Hispànica 466
Bibliographie Spéléologique Belge 466
Big Bone Cave, Tennessee, United States 411
Big End Chasm, Zimbabwe 21
Bigfoot Cave, California, United States 52, 752
Big Muddy Valley, Canada 591
Bigonda cave, Italy 117, 118
Bikbik Vuvu, Nakanai Caves, Papua New Guinea 538
Bileca reservoir, Yugoslavia 277
Bil-le-mot Cave, South Korea **212**

Bim Sejeune cave, Haiti **189**
Bing Cave, Germany 565
Bingyi Caves, Thailand 624
   biodiversity, protection of 244
   biodiversity hotspots in caves
   hypogean waters **141–142**
   terrestrial **143–144**
   bioerosion 231, 232
   biofilms **145–146**
      defined 145
      and dissolution/mineralization processes 145
      and evolutionary selection 146
      factors controlling 145
      formation of 146
      microbial mats 146
      morphology 145
      stromatolites 146
      tubercle formation and 145–146
   Biograd Ponor, Herzegovina 293
   bioherms 175
   biokarstification **147–148**
      and cave morphology 523
      and deposition 147–148
      and dissolutional processes 147
      and karren 470
      organisms involved in 147
      vs. phytokarst 581
      plant/soil involvement 148
   biological species model of speciation 665
   bioluminescence 3, 4
   biomineralization 636
   Biospeleologica (Chappuis and Jeannel) 149
   Biospéleologica (Jeannel and Racovitza) 154
   biospeleologists **151–154**
   biospeleology
      advances in 149
      Africa **24–25**
      America, Central **37–40**
      America, North **45–49**
      America, South **57–59**
      Asia, Southeast **109–111**
      Australia **125–126**
      biodiversity **141–144**
      biogeographic investigations 149
      Britain and Ireland **163–164**
      Canary Islands **179–181**
      Cape Range, Australia **181–183**
      Caribbean Islands **37–40**
      cave populations, genetic structure of 150
      chemoautrophic organisms 150, 366
      cladistics and 149
      classification of cave species 150
      conservation of species and 150, 243–245
      Dinaric karst **289–290**
      diversity of **149**
      ecology of subterranean habitats **706–707**
      Edwards Aquifer **315–316**
      evolution, cave-related 149–150
      France **369–371**
      Hawaii, United States **417–418**

      history of **148–149**
      karst hydrogeology and 150
      Mammoth Cave, Kentucky, United States **499–501**
      Postojna-Planina Cave System, Slovenia **603–604**
      Salukkan Kallang, Indonesia **631–632**
      Southeast Asia **109–112**
      Vjetrenica, Bosnia-Herzegovina **759–760**
      Walsingham caves, Bermuda **767**
      See also adaptation, colonization.
Biospéologie: La Biologie des Animaux Cavernicoles (Vandel) 154
Biosphere Reserves, declaration of 246
birds **130–131** *See* Aves.
bird's nest soup 131, 409, 479 547
Birkilin Great Cave, Turkey 207, 742
birnessite 511
Bischoff, Erwin 766
Bishop, Steven 87, 726
Bitumen Cave, Bermuda 768
Bivalvia
   Dreissenidae
      Congeria kusceri **516**
black widow spider bite 294
Blak, Kristian
   Antifonale 534
   Concerto Grotto 534
Blanche Cave, Australia 128
blastomycosis 294
Blatchley, W.S. 49, 51, 81
Blattaria
   Blattodea
      adaptations 451
      distribution 143, 144, 451
      Loboptera troglobia 180
      morphology 451
      Nocticollidae 25, 110, 631
   Loboptera troglobia 180, 451
bleaching 280
blind cave fish 98, 148, 151, 152, 182, 183, 594
See also Pisces.
Blind cave salamander 365
Blue Creek Cave, Belize 140
Bluefish Caves, Yukon, Canada 43
Blue Grotto (Grotta Azzura), Capri 87, 703
Blue Grotto in Capri (Scott) 534
blue hole caves 176, 189
   Bahamas 155–156, 302
   cenotes of 156
   definition of 155
   distribution of 155–156
   morphology 155
   reversing flow of and biological environment 156
   speleogenesis 155
Blue John Caverns, England 578
Blue Lake, Washington, United States 762
Blue Mountains, Jamaica 189
Blue Spring Cave, Tennessee, United States 70, 750
Boa Vista, Toca da, Brazil 55, 56, **156–157**
   depth of 525
   geomorphology 156–157

speleogenesis 157
speleothems 157
Boaz, Noel 222
Boca del Infierno cave, Dominican Republic 189
Bochusacea 256
Bock, H. 228
Boddagrottorna, Sweden 722
boehmite 135
Bögli, Alfred
    biography of **381**
    history of speleogenesis 671, 673
    Hölloch, Switzerland 420
Bogovinska Pećina, Serbia 330
Boguminskaya Cave, Georgia 203
Bohemia Cave, South Island, New Zealand 543, 591
Böhm, Jaroslav 28
Bohol karst, Philippines 104, 113
Boischatel stream cave, Canada 175
Bois Laterie Cave, Belgium 138, 139
Bojbulak, Turansky Plain, Central Asia 95
Boj Bulok, Uzbekistan 525
Boka Kotorska Bay, Dinaric karst 288, 338
Bokanjačko Polje, Bosnia/Herzegovina 291
Boletin de la Sociedad Venezolana de Espelelologia 465
Boll Maza, Algeria 13
Bol'shaja Oreshnaja Cave, Siberia 629, 645
Bol'shaya Baydinskaya Cave, Siberia 646
Bol'shaya Ludarskaya Cave, Siberia 646
Bonaire **190**
bone and dating methods 279, 280
bone accumulation in caves 562
Bonnechere stream cave 175
Bonnie Prince Charlie 364
Boquierão, Brazil 55
borates in caves **512**
Borbon, Cueva de, Hispaniola 192
Border Cave, South Africa 19, 82
bored piles and construction 248
borehole hydrology **397–398**
    analysis, tests, and methods 398
    conduits, aquifer and 397–398
    wells, use in 397
    Borneo 104
Boro-Khukhan, Siberia 646
Bosnia
    Dinaride poljes 291
    Dugo Polje 291
    Duvanjsko Polje 291
    Glamočko polje 291
    Imotsko Polje 291
    Kazanci ponor 291
    Kočerinsko Polje 291
    Kovači Ponor 291
    Kupreško polje 291
    Nadinsko Polje 291
    Opaki ponor 291
    Posuško Polje 291
    Proždrxikoza ponor 291
    Sinjski ponor 291
    Stara Mlinica ponor 291

    Trebistoho Polje 291
    Trnsko Polje 291
    Veliki ponor 291
    Vir Polje 291
    Vjetrenica 141, 289
        biospeleology **759–760**
        climate **759**
        conservation **760**
        fauna, distribution **760**
        geomorphology **759**
        hydrology **759**
Bosnia-Herzegovina
    Dinaric karst **287–288**
    hypogean fauna in 142, **289–290**
    poljes **291**
    Vjetrenica
        biospeleology **759–760**
Botosaneanu, L. (*Stygofauna Mundi*) 149
Botovskaya Cave, Siberia 645, 646, 663
Bottomless Pit, Crimea 662
Bottomless Pits, Arizona, United States 251
bottom walking and diving 302
Boulder Creek Cave, Washington, United States 722
Bounding Pot, Zimbabwe 21
Boundless Cave, Virginia, United States 731
bournes 363
Bowden, Jim 303
Bowmanston Cave, Barbados 190
Brachyura 261, 262
Bracken Bat Cave, Edwards Aquifer, United States 314
Bracken Cave, Texas, United States 409, 751
Brain, C.K. 17
Brambiau (Rudaux) 88
Branchiopoda 254, **459**
branchwork caves 573
Brandt, Peter Andreas 59–60
Brazil
    Arcos/Pains, Bambuí karst 133
    Bambuí karst **133–135**
    Barriguda, Toca da 156, 157
    Boa Vista, Toca da 55, 56, **156–157**, 525
    Boquierão 55
    Caverna Aroe Jari 57
    Caverna da Pedra Pintata 91
    Caverna dos Ecos 57
    Caverna Huayllas 55
    Chiflonkkakka 55
    Cordisburgo, Bambuí karst 133
    Gruna da Água Clara 55
    Gruta Casa de Pedra 56, 57
    Gruta da Bocaina 55, 611, 612
    Gruta das Bromélias 611, 613
    Gruta de San Pedro 55
    Gruta do Janelão 93, 134
    Gruta do Padre 55
    Gruta Planaltina 57
    Gruta San Miserato 55
    Gruta Umajalanta 55
    Gruto do Centenário 55, 611, 612
    Ibitipoca State Park 613

Brazil *(Continued)*
　Inficionado karst 612
　Lagoa Santa, Bambuí karst 133
　Lapa do Angélica 55
　Lapa São Mateus III 55
　Lapa São Vicente I 55
　Lapão Cave 611, 613
　Mambaí, Bambuí karst 133
　Minas Gerais, art in 92, 93
　Montes Claros, Bambuí karst 133
　Peracu National Park, art in 93
　Peruaçu, Bambuí karst 133, 134
　quartzite karst 612
　São Desidério, Bambuí karst 133
　São Domingos, Bambuí karst 133, 134
　Serra do Ramalho, Bambuí karst 133
　Terra Ronca State Park 134
　Toca da Barriguda 55
　Toca das Confuspes 57
　Torotoro National Park 55
breakdown 523
breakaway caves *See* caves, crevice.
breccia
　Mississippi Valley Type (MVT) 660
　pipes 344–345, **661**
　solution **660–661**
Breccia Cave, Australia 124, 127, 129
Breder, Charles Marcus **154**
Breitenwinnerhöhle, Germany 335
Bremontier, Nicolas Thomas 715
Bretz, J Harlan
　biography of **381**
　Caves of Missouri 52
　history of speleology 51, 671
Breuil, Henri Édouard Prosper **77, 79**, 85, 222
Brezno pod Velbom, Slovenia 469, 526
Brezno pri Risniku, Kras, Slovenia 486
Bribin Cave, Gunung Sewu, Java 642
bristletails *See* Microcoryphia.
Britain
　archaeological/paleontological caves **160–162**
　caves and karst 340–341
　chalk aquifers, fauna of 163
　distribution of fauna 163
　history of exploration 164–166
　protection of fauna 164
　stygobitic fauna of 163
　terrestrial fauna of 163
　threats to fauna 164
　troglobitic fauna of 163
　See also United Kingdom.
British Cave Rescue Council 239
British Cave Research Association (BCRA) 166, 466
British Columbia
　clearcutting and soil erosion in 658–659
　karst development 178
British Speleological Association (BSA) 166
Brixham Cave, England 162
Brodrick, Harold 166
Broken Arrow (Woo) 360

Bronson Caverns, United States 359–360
Brooke, G.A. 527
Broom, Robert 17
Brothers, Alfred 578
Brown, Samuel 411
brown spider bite 294
Brunnecker Cave, Austria 174
　brushite 513
　bryophytes 362
　entrance habitats 318
　on travertine 424
Bryozoa 459
BU56 cave, Pierre Saint-Martin, France 585, 586
Buchaly, Reinhard 303
Buchan Caves Reserve, Australia 129
Buchner, Bertold 158
Bucholz, Juraj 715
Buckfastleigh cave system, England 166
Buckland, William
　biography **77, 78**, 80
　and diluvial theory 164
　Gibraltar cave fossils 384
　Red Lady of Paviland 161
　Reliquae Diluvianae 164, 208
Buco Cattivo, Frasassi caves, Italy 374
Buda Hills, Hungary 332
Buddha of All-Illuminating Wisdom 624
Buddhism and ritual use of caves 622
Bukhtarminskaya, Siberia 646
Bükk culture, Aggtelek Caves, Hungary/Slovakia
　charcoal drawings 28–29
　Domica Cave 29
　pottery of 28
Bukovinka, Ukraine 746
Bulgaria
　Barkite 14, 329
　Douhlata 329, 330
　Jagodinskata 329, 330
　Magura Cave 330
　Orlova Chuka 329, 330
　Pirin National Park 330
　Raytchova Doupka 329, 330
　Shopov cave system 330
　Yamata 329
Bulletin of the National Speleological Society 464
Bulletin of the SASA 465
Bulletin of the Speleological Society of the District of Columbia 464
Bullita Cave System, Australia 123, 525
Bullock Creek-Cave Creek, South Island, New Zealand 542
Bulmer Cave, South Island, New Zealand 542, 543
Buna, Herzegovina 699
Buna spring, Dinaric karst 288
Bundera Sinkhole, Australia 182, 183
Bungle Bungle range, Australia 123
Bungonia caves, Australia 127, 634
Bunica, Herzegovina 699
Buontalenti, Bernardo (Grotta Grande) 86
Burggaillenreuth, Bavaria 77
burials in caves 81, **167–168**, 196
　Africa 168

Ash Cave, Kentucky, United States 168
Asia, Southeast 107, 168
Atapuerca Caves, Spain 119
Creag nan Uamh caves, Scotland 160
Cheddar Gorge, England 162
Déroc caves, France 167
Dordogne caves, France 167
Duyong Cave, Philippines 167
Foissac Caves, France 167
France 167
Hoabinhian culture 167
Indonesia 108, 109
Iraq 167, 643
Ireland 162
Israel 167
Khammouan karst, Laos 167
Malaysia 108
Mammoth Cave, Kentucky, United States 168
Manunggul Cave, Philippines 168
Matangkib Cave, Philippines 168
Maya 81, 84, 168
North America 44
Paleolithic, England 162
Papua New Guinea 105
Patagonia, Chile 573
Paviland Cave, England 161
Scotland 160
Shanidar Cave, Iraq 167, 643
Thailand 107
Uzbekistan 167
buried dolines 309
Burnsville Cove, Virginia, United States 69
Buronov Ponor, Serbia 330
Burra Charter, Australia and conservation management 728
Burren glaciokarst, Ireland **169–171**, 341
    Aillwee Cave 171
    Carren depression 170
    cave systems of 171, 581
    dating of 171
    Doolin cave system 171
    enclosed depressions 170
    erosion in 659
    geomorphology 169–170
    Glencurran Cave 171
    hydrology 170
    kamenitza 171
    limestone pavements 170–171, 473
    speleogenesis 170
Burroughs, Edgar Rice (*At The Earth's Core*) 206
Bushmangat, South Africa 20, 21, 303
Busk, George 162, 384
Buško Blato polje, Bosnia/Herzegovina 291
Buško Blato reservoir, Yugoslavia 277
Bussento, Italy 699
Busso della Neve di Zingarella, Italy 117
Butcher, Arthur 166, 716
Butler-Sinking Creek Cave System, Appalachian Mountains, United States **69**
Buxton Water Sump, England 301, 302
Byakuren-Do, Japan 96

Bystra Cave, Slovakia 698
Bzybsky Massif, Georgia 201–203

Cabinet of the Felines, Lascaux France 757
Caelifera 451
Caerwys tufa deposits, England 341
caesium 514
Caherguillarmore, Ireland 162
Caicos Islands 192
Caipora bambuiorum 157
cajunite 136
Calabash Cave, China 223
Calabozoidea 256, 257
Caladaire cave system, France 326
Calanoida 259, 260
Calcareous Alps, Austria **173–175**
calcifuges 361
calciphiles 361
calcite **187–188, 296**, 690
    annual deposition band dating 557, 558
    and biomineralization 636
    and breccias 660
    Cigalère, Grotte de, France 327
    Cueva de Villa Luz, Mexico 759
    dissolution 296, 677
    erosion rates **323, 324**
    formation of 512, 714
    GB Cavern, England 504
    Ghar Sarab 461
    Grotte Père Noel, Grottes de Han, Belgium 342
    Gouffre de Padirac 371
    Humpleu-Poienita system, Romania 329
    Kugitangtau, Turkmenistan 273, 275
    luminescence in 695, 696
    Mammoth Cave, Kentucky, United States 495
    Nordland marble stripe, Norway 342
    oxygen fractionation dating 556, 557
    precipitation of **187–188**, 512, 692, 730
    radiocarbon dating 557–558
    speleothems 690
    structure of 185
Calgeron cave (GB Trener Cave), Italy 117
Callipodia
    adaptations 536
    distribution 535–536
    morphology 535–536
Cambarus hubrichti 244
Cambodia
    geomorphology 100
    Roung Dei-Ho/Roung Thom Ken system 102
Camerons Cave, Australia 183
Campamento, sima del, Spain 345
Campan, France 371
Canada **175–178**
    Anticosti Island 175
    Arch-Treasure Cave 178
    Arctomys cave 51, 52, 177
    Athabasca Glacier, Alberta 387
    Banff Hotsprings 177
    Bear Rock Breccia karst **137–138**, 178

Canada *(Continued)*
  Belcher Islands, Hudson Bay 175
  Big Muddy Valley 591
  Cambrian shield 175
  Castleguard Cave 177, **198–200**
  Castleguard II 199
  Close to the Edge 52, 177
  Columbia Icefield 198
  Coultard Cave, Alberta 436, 437
  Crow's Nest Pass 177
  history of exploration
  Franklin Mountains 178
  Gargantua 177
  Goose Arm karst 177
  Grotte Louise-Grotte Mickey, South Nahanni karst 538
  Grotte Valerie, Northwest Territory 436
  Grotte Valerie, South Nahanni karst 538
  Mackenzie Mountains 178
  Maligne River 177
  Medicine Lake 177
  Mt. Tupper system 178
  Nahanni National Park 52
  Nakimu Caves 51, 52, 178
  Nanisivik, sulfide ores in 560, 571
  Niagaran dolostones 175
  Pine Point, Northwest Territories 176
  Precambrian Shield 175
  Rat's Nest Cave 177
  Serendipity Cave, Rocky Mountains 437
  South Nahanni karst 178, 537–538
  speleothems 178, 200
  Steam Caves, Banff 51
  stygobitic fauna of 200
  Thanksgiving Cave 178
  Volcano Room, Q5, Vancouver Island 436
  Western Cordillera 175, 177–178
  Wood Buffalo National Park 52, 176
  Yorkshire Pot 52, 177
Canadian Shield, composition of 175
Canalobre Cave, Spain 215
Canary Islands
  adaptations of fauna 179
  biospeleology **179–181**
  Cueva de Don Justo 179
  Cueva del Llano 179
  Cueva de los Roques 179
  Cueva de los Verdes 764
  Cueva del Viento 764
  Cueva de San Marcos 764
  Cueva de Todoque 179
  El Hierro 179, 180–181
  Felipe Reventón, Cueva de 179
  Fuerteventura 179, 181
  Gran Canaria 179
  Jameo del Agua 179
  La Gomera 179
  Lanzarote 179
  La Palma 179, 180
  lava tube caves 179
  Loboptera troglobia 180
  Maiorerus randoi 179, 181
  speleogenesis 179
  Tenerife 179–180
  troglobites, evolution of 179
  Tunel de la Atlantida, Lanzarote 179, 302, 762
  Verdes, Cueva de los 179
  volcanic origin of 179
Cancrocaeca xenomorpha 632
Candona jeanneli 255
Cango Caves, South Africa 20, 359, 533
Cao Bang caves, Vietnam 510
capacitive sensors, measurement using 228
Cape Peninsula, South Africa 651
Cape Range, Australia
  biospeleology of **181–183**
  geomorphology 181
  relict fauna 181
  World Heritage site potential 183
Caprazliha ponor, Bosnia/Herzegovina 291
Capri, Blue Grotto (Grotta Azzura) 87, 703
caprock collapse dolines 304, 305, 306, 307, 308
Carabidae 447
  Trechinae 289, 369
Caracol, Belize 37
carbon–14 (^{14}C) dating 279, 555
carbonate-covered island, development of 675
carbonate-hydroxylapatite 513
carbonate karst **184–186**
carbonate minerals **187–188, 512–513**
carbonate rocks
  classification 185
  dissolution
    calcite 296
    dolomite 297–298
    magnesite 297, 298
    PWP equation 295, 296
    rates
      control mechanisms of 296
      inhibition of 297
      turbulent flow 296–297
      undersaturated solution 295–296
carbonatites 186, 559
carbon dioxide
  carbonate rocks and 296–297, 658
  and cave air **183–184, 730–731**
  and evaporite speleothems 693
  hazardous levels of **184**
  and karstification 679
  and speleogenesis theories 669
  in tourist caves **730–731**
carbonic acid dissolution 682
carbon monoxide poisoning 293
carbon oscillation dating 558
Caribbean Islands **189–190**
  biospeleology 37–40
  geomorphology 189
  guano mining 189, 191, 192
  history of exploration 189, 191–192
  Huliba Cave 190
  Oropouche Cumaca Cave, Trinidad 190

paleontology 191
See also specific islands by name.
Caridea 257, 261, 262
Carlsbad Caverns, New Mexico, United States 51, 52, **192–194**
   Big Room 193, 194, 195, 308, 580
   and cave photography 579, 580
   films, use in 359
   geomorphology 193–194
   guano accumulations in 409
   guano mining 194
   history of exploration 51
   lampenflora and 734
   Main Corridor 193, 194
   minerals in 194
   speleogenesis 192
Carmel Caves, Israel **195–197**
   Es-Skhul Cave 195, 196
   Hayonin Cave 195
   Kebara cave 196
   Mousterian culture 195
   Natufian culture 195, 196
   Neanderthals 196
   Qafzeh Cave 196
   Tabun Cave 167, 195
   Wad, El 195
   Wadi el-Mughara 195, 196
Carpentaria Cave, Australia 216
Carrara marble, Italy 327
Carren depression, Burren glaciokarst, Ireland 170
Carroll, Lewis (*Alice's Adventures in Wonderland*) 206
Carsologica sinica 465
Cartailhac, Emile 82
Carus, Titus Lucretius 764
Carver's Cave, Minnesota, United States
   history of speleology 50
case-hardening of cone karst. 242
casimbas 271
cassiterite
   paleokarst, economic resources of 560
   Southeast Asia karst 103
Castalia and religious sites 623
Castellana, Grotta di, Italy 338
Casteret, Norbert 371, 662
   biography **687, 689**
   diving 301
   Pierre Saint-Martin, France 585
   on stamps 704
Castil, Picos de Europa National Park, Spain 583, 585
Castle, William (*Cave of Outlaws*) 359
Castleguard Cave, Canada 177, **198–200**
   Artesian Spring 199
   Central Cave 198, 199
   Columbia Icefield 198, 199
   Downstream Complex 198, 199, 200
   glaciated karst 388
   Headward Complex 198
   history of exploration 52
   hydrology 199–200
   ice deposits in 437
   speleogenesis 198

speleothems 200
stygobitic fauna 200
Castleguard II, Canada 199
Castlepook Caves, Ireland 162
Catacombs Cave, California, United States 491
Catarina-Confusion Cave System, Texas, United States 591
Cater Magara, Syria 345
Catfish Farm Well, Edwards Aquifer, United States 314
Cathedral Canyon, United States 302
Cathedral Cave, California, United States 491
Cathedral Formation, Canada **198**
Catherwood, Frederick 40
Caucasus, Georgia **200–203**
Cave, the (Warren) 206
Cave and Karst Science 465
   cave bears
      bones in caves 77, 119, 326, 329, 335, 339–340, 562, 646
      cave bear cult 82
      hibernation in caves 319
cave digging 355
Cave Diving Group, United Kingdom 302
cave dust pneumonitis 293
cave entrance/crevicular habitats 709–710
Cave-in-Rock, Illinois, United States 50, 364
Cave of Antiparos, Greece 208–209, 689
Cave of Bolanchen, Campeche, Mexico 42
Cave of Dionysos, Greece 208
Cave of Gold, Scotland 534
Cave of Hearths, South Africa **19**
Cave of Loltun (Thompson) 40
Cave of Marmarospilia, Greece 208
Cave of Outlaws (Castle) 359
Cave of Swimmers, Egypt 608
Cave of the Dragon, Greece 208
Cave of the Falling Star, Saudi Arabia 115
Cave of the Naiads, Greece 208
Cave of the Seven Sleepers, Turkey 208
Cave of the Winds, Gunung Mulu National Park, Sarawak, Malaysia 531
cave prawn, blind *See Typhlocaris salentina.*
Cave Radio & Electronics Group (CREG) **239**
cave radios and cave communication **239**
cave rescue *See rescue in caves.*
Cave Rescue Organization, France 1
Cave Research Foundation, United States 52
Cave Research Group (CRG), Britian 166
Cavern, The (Ulmer) 360
Caverna Aroe Jari, Brazil 57
Caverna dos Ecos, Brazil 57
Caverna Huayllas, Brazil 55
Caverna Lunaris, Switzerland 335
cavernous lava rises 761
Caverns of Copan (Gordon) 40
Caverns of Sonora, Texas, United States 626, 751
caves **203–204**
   accidents and rescue **1–2**
   air and carbon dioxide 183–184
   airflow in 229–230
   alpine **32**, 177
   archaeological **40–42** *See also* archaeological caves.

caves *(Continued)*
  archaeology, history of **80–82**
  artificial 41
  art in *See* art in caves.
  art showing caves **86–88**
  atmospheric pressure in 229–230
  banana holes 189, 675
  biblical 208
  biology of **148–150**
  blue holes 176, 189
  books on **158–159**, 205–206
  burials in **167–168**
  chambers 204
  classification of 203–205
  climate in **228–230**
  colonization of **235–237**
  communication in **238–239**
  conduction in 228
  construction on 247–248
  convection in 228
  corrosional 254
  crevice caves 204, **249–252**
  cultural views on 40
  decay/obliteration stage of 668
  defined **203**
  diseases associated with **293–295**
  diving **301–303**
  entrance habitats **318–319**
  erosion caves 204
  exploration of **350–352**
  exploring **352–355**
  extraterrestrial **355–357**
  films in **359–360**
  flank-margin 189, 675–677
  folklore and mythology **364–365**
  food resources in **365–367**
  fumarole **68**
  geophysical detection of **377–379**
  glacier 204, **385–387**, 606
  gravity sliding 250–251
  gravity tectonics 650, 652
  and gunpowder **410–411**
  history of exploration 350
    See also individual cave or region by name.
  human occupation of **426–428**
  hydraulics of **429–430**
  hypogenic 94, 157, 192
  ice caves **385–387**
  ice in caves **435–437**
  inception of **437–441**
  influent 504
  initiation stage 667
  journals on **464–466**
  lava mould 762
  littoral caves 204, 252, 254, **491–492, 501–502**
  loess 591
  mass movement 250–251
  maze 573–574
  microbial processes in **505–506**
  microorganisms in **506–509**
  military use of **509–510**
  mineral deposits in **511–514**
  monitoring of **519–520**
  morphology 204, 235, **521–523**
  morphometry of **524–526**
  multi-level (multiphase) **686**
  music in/about **533–534**
  nitrate mining in 51, 409
  organic resources in **547**
  paragenesis **569–571**
  passages 204
  patterns of **573–575**, 667–668, 684–685
  permafrost 356, 436
  photographing **578–581**
  piping caves 204, **589–593**
  pit 189
  postcards and 703–705
  profiles, evolution of 574–575
  pseudokarstic **604–608**
  quartzite 252, 607–608, 611–613, 652
  radiolocation **615–616**
  radon in **617–618**
  rapid enlargement stage 667
  recreational caving **621–622**
  relative humidity in **228–229**
  relict 559
  religious sites and **622–624**
  restoration of **625–627**
  ritual use of 40, **41–42**, 103, 112
  salt 346
  and saltpetre **410–411**
  solution caves **203–204**, 205, 667
  speleogenesis **666–668**
  speleotherapy 698
  stagnation stage of 668
  stamps and 703–705
  stream-sink 190
  sulfuric acid speleogenesis 668
  surveying **714–717**
  talus 606, **721–724**
  taphonomy 561–564
  temperature in 228
  thermal 559
  thermal waves in 228, 229
  tourism, history of **725–726**
  tourist *See* tourist caves.
  vertical caves 204
  volcanic caves 204, 282–283, **760–764**
    See also lava tube caves.
  without roofs **559, 560**
  See also speleogenesis.
Caves and Caving 465
Caves Branch caves, Belize 571
cave sickness 293
Caveside Caves, Australia 129
Caves of California (Halliday) 766
Caves of Missouri (Bretz) 52
Caves of Pennsylvania (Stone) 52
Caves of Virginia (McGill) 52
Caves of Washington (Halliday) 766

Caves of West Virginia (Davies) 52
caves without roofs 559, 560
Cave Vertebrates of North America (Eigenmann) 154
Caving International 465
caving lights 353
Cayman Islands 189, 192
Cayos Miskitos y Franja Costera Immediata
   Nicaragua, Central America 620
CB radios and cave communication 238
Cebada Cave, Belize 37
Čeč, Luka 148, 601
Čehi 2, Slovenia 468, 469, 525
Čelebija, E. 288
Celebrated American Caverns (Hovey) 51, 158
celestine 329
celestite 273, 636
cement, uses of 489, 490
cementation *See* diagenesis.
cenotes
   blue holes 156
   defined 303
   development of 308
   Yucatán 786
centipedes *See* Myriapoda.
centipede sting 294
Cerjanska Pećina, Serbia 329, 330
Cerknica Polje, Slovenia 210–211, 288
Černičko polje, Herzegovina 293
Cerro Rabon, Huautla Cave System, Mexico 426
Čertová pec Cave, Hungary/Slovakia 29
cerussite 512
Česká speleologická společnost, Czech Republic 336
Ceskoslovenski Kras 464
Ceuthophilus cunicularis 47
Chabot (Gard) cave, France 82, 85
Chad, Republic of
   Oumou Caves 608
Chaetognatha **458**
   anchialine habitats 65
   Paraspadella anops 458
Chagas disease 294
chalcantite 513
Chaleux Cave, Belgium 138
chalk
   cave development 438
   Cretaceous chalk in northern Europe 344, 363–364
   karst 185, 342
   paleokarst 559
Chandor Chasma, Mars 357
Chapada Diamantina, Bahia, Brazil 92
Chappuis, P.A. 149
Charkadio Cave, Tilos Island 340
Charterhouse Caves, Mendip Hills, England 504
Chartreuse massif, France 326
Chartreux, France 302
Chauvet, Jean-Marie 82–83
Chauvet Cave, Ardèche France 82, 85, 90, 281
Cheddar Gorge 162
Cheddar Man 504
Cheju-Do Lava Caves, South Korea **212–213**

   caves of 212–213
   geomorphology 212
   minerals in 212
Cheju Island, South Korea 212
Chelosiowa jama, Poland 331
Chelsea Spelaeological Society 310
chemoautotropic bacteria 348, 507
   Cueva de Villa Luz, Mexico 758
   Frassasi caves, Italy 375
   Movil Cave, Romania 366, 529–530
chert 185
Chërtovy Vorota, Siberia 646
Chestnut Ridge Cave System, Appalachian Mountains, United States 69
Chevalier, Pierre (*Escalades souterraines*) 284–285
Chevalier cave, France 283
Chevolidae
   Leptodirinae 289, 369
Chèzy's Law, hydraulics 395
Chichen-Itza, Yucatán, Mexico 37, 41
Chiflonkkakka, Brazil 55
Chile
   Cueva del Valle de la Luna 56
   Cueva Mylodon 56
   La Perte de l'Avenir, Patagonia 573
   La Perte du Temps, Patagonia 572
   Patagonia marble karst **572–573**
   Perte du Futur 55, 56, 573
   siphon of Lobos, Patagonia 573
Chillagoe, Australia 89, 192, **215–216**
Chilopoda **535**
   adaptations 535
   distribution 24, 143, 535
   morphology 534–535
   Scolopendromorpha 535
Chimanimani Mountains, South Africa 651
China **217–220**
   Alu Caves 98, 151
   archaeological caves of **221–224**
   Baishuitai 218
   Ban Dong 218
   bauxite production in 135
   Calabash Cave 223
   Chingshui 699
   Chongking province 219–220
   cone karst 218, 242
   Daji Dong 218
   deforestation and 659
   Di Feng Dong 220, **285–287**
   fengcong karst 217, 220, 242
   fenglin karst 217, 242
   Furong Dong 220
   Ganhe bridge 218
   Gebihe Dong 218
   Guangxi karst 217, 224, 241
   Hei Yau Dong 285
   history of research **97–99**
   Homo erectus 221, 222, 223
   Hongshan karst spring and karst hydrology 478
   Hongshui River fengcong karst **422–423**

China (Continued)
  Huangguoshu 736
  Huanglong and Jiuzhaigou **423–424**
  Liu longdong 699
  Liujiang Cave, karst 224
  Longtan Cave 222
  Longuin Cave 623–624
  Luti Dong (Reed Flute Cave), Yangshuo 783
  Mawang Dong 218
  Nyangziguan spring 220
  Panda Lake, Jiuzahaigou 424
  Panlong Cave, Yangshuo 783
  Panyang caves 218
  Qikeng Dong 220
  Qixing Dong (Seven Star Cave), Yangshuo 783
  Saguo Dong 218
  San Qiao 220
  Santang Dong 218
  Shilin stone forests **643–644**
  Shuinan Cave, Yangshuo 783
  Shuzen Lake, Jiuzahaigou 424
  Sichuan 218–219
  Solue Dong 218
  stamps depicting 704
  Tangshan 223
  Tien Jing Gorge 285, 286
  Tisu 699
  tower karst 217
  travertine 424, 736
  vaterite 424
  Wujia Dong 218
  Wujiangdu dam 278
  Xiaochai Tiankeng doline 286–287
  Xiaoyanwan doline 219
  Yangshuo karst 782–783
  Yanzi Dong 218
  Yunnan 218
  Zhoukoudian caves 220
  Zhucaojing Cave 219
China Caves Project 99
Chingshui, China 699
Chi Ni Tsu 98
Chipi Joseteko, Pierre Saint-Martin, France 587
Chip's Hole, United States 302
Chiquibul Cave System, Belize/Guatemala 37
Chiron, Léopold 82, 85
Chiroptera **225–226**
  America, North 46
  America, South 58
  biodiversity 144
  Carlsbad Caverns, United States 194
  description of 225–226
  flight mechanism 226
  Madagascar 493–494
  Megachiroptera 225
    Pteropodidae 225
    Rousettus spp. 225
  Megadermatidae 124
  Mexico 39
  Microchiroptera 225
  parasites of 74
  and rabies 294
  roost selection of 226
  Siberia 646
  and stamps 703–705
  Tandarida brasiliensis mexicana 314, 393, 751
  threats to **227**
  vampire bats 58
  vespertilionid 225
chlorite 587
Chologaster cornuta 595, 596
Chomolungma (Mt Everest, Sagarmatha), India/Tibet 622
Chongking province, China 219–220
Chordeumida 535
Chott Ech Cherg, Sad'da, Algeria 620
Chou Chhü-Fei (*Lin Wai Tai Ta*) 98
Chourum Martin, France 371
Christianity
  and ritual use of caves 622
  St Paul's Church, Rabat, Malta 623
Christmas Canyon Cave, Washington, United States 591, 592, 606, 608
Christmas Pot, Australia 216
Christol, Jules de 80
chromium 514
Chuckanut Mountain Caves, Washington, United States 722
Church Cave, Bermuda 768
Church Hole, England 162
Ciclovina Cave, Romania 329
Ciénaga de Zapata, Cuba 620
CI Flourescent Brightener 351 dye and water tracing 772
Cigalère, Grotte de, France 327
Ciliata
  Peritricha 289
cinnabar 714
Ciochon, Russell 222
circadian rhythms, behavioural adaptation 3
Circadian rhythyms, Coleoptera 450
Ciur Ponor, Romania 329
Cixius spp. 180
Cladocera 254, 289
Clamouse Cave, France 215, 338
Clarias cavernicola 21
Clashmealcon Cave, Ireland 510
clastic sediments and karst dating 281
Clean Water Act (CWA)
  pollution of groundwater, point sources **405**
Clearwater cave, Sarawk, Malaysia *See* Gua Air Jernih.
Clearwell Caves, England 360
climate
  in caves **228–230**
  and karst classification 474
  monitoring networks in caves 230
  paleoclimate studies 215
climatic relict model of speciation 665
clinometer 716
Clitellata
  Oligochaeta
    Hirudinea 66, 67
Cloggs Cave, Australia 124

Close to the Edge window, Canada 52, 177
Clottes, Jean 85
Cnidaria
　morphology 458
　Velkovrhia enigmatica 458, 603
CO_2 See carbon dioxide.
coastal karst 215, **231–233**, 674
Cobweb Cave, Gunung Mulu National Park, Sarawak, Malaysia 532
coccidioidomycosis 294
Cocklebiddy Cave, Nullarbor Plain, Australia 545, 546
Cockpit Country, Jamaica **233–235**, 242
cockpit doline 306
cockroaches See Blattaria.
Cogol dei Siori, Italy 117
Cogol dei Veci, Italy 117
Colani, Madeline 108
cold traps in caves 229
Coldwater Cave, Iowa, United States 751
Coleman, Jack 166
Coleoptera 290, 369, **447–450**
　adaptations 449–450
　Anthroherpon apfelbecki 760
　Carabidae 48, 447
　Cholevidae 7, 447–448
　circadian rhythyms 450
　Coleoptera aquatic 447
　Curculionidae 449
　Dinaric karst 289
　diversity 143–144
　distribution 25, 143, 144, 447–448
　Dysticidae 126
　Dytiscidae 25
　ecology 449
　endemicity 449
　food sources 449
　in France 364
　Histeridae 448–449
　Hydrophiidae 25
　Leodidae 48
　Leptodirinae 289, 448
　Merophysidae 449
　in North America 47–48, 500
　predators/parasites 449
　Pselaphidae 448
　Pterostichinae 447, 632
　Ptilidae 449
　reproduction 450
　Rhadine reyesi 47
　Scydmaenidae 449
　in South America 58
　Staphylinidae 180, 448
　in Tenerife 180
　Trechinae 447
Coleridge, Samuel Taylor (*Kubla Khan*) 205
collagen, dating of 280
collapse chamber 523
collapse dolines
　cave roof resistance, factors affecting 308
　described 305, 308

　nomenclature 304, 306
　speleogenesis 306, 307–308
Collartida anophthalma 181
Collartida tamausu 180
Collembola
　adaptations 7, 8, 9, 446
　biodiversity 143, 144
　distribution of 24, 39, 110, 446, 631
　morphology 446
　taxonomy of 445
　troglobitic species of 24, 446
Collins, Floyd 206, 360, 534
Collins, Greenville 715
Collins Glacier, King George Island, Antarctica 68
collinsite 329
Collocalia spp. 131, 409, 547
Colombia
　Cueva del Cunday 54
　Cueva del Indio 54
　Cueva de los Guácharos 54
　Hoyo del Aire 54, 60
　Sistema Hermosura 54
colonization **235–237**
　active colonization model 235
　adaptive shift model of colonization 236
　adaptive zone model of colonization 235, 236, 237
　of caves by microorganisms 507
　and clay sediment 366
　ecological 236
　evolutionary 236
　groundwater, colonization of 236
　horizontal transition colonization 236
　of lava tube caves 179
　marine regressions and 237
　modalities of 235
　from the MSS (mesovoid shallow substratum, milieu souterrain superficiel) 457
　primary defined 235
　rates of 237
　refugium model of colonization 235, 237, 268
　regression model of colonization 237
　secondary defined 235
　stygobites 236
　three-step model of colonization 235
　two-step model of colonization 235, 237, 263
　vertical transition colonization 236
　zonation model of colonization 235
Columbia Icefield, Canada 198
Columbine Crawl, Wyoming, United States 751
Coly, Doux De, France 302
Commissione Grotte, Società Alpina della Giulie, Italy 351
communication, wireless and cave communication 238
compaction See diagenesis.
compaction grouting and construction 248
compass
　Suunto 716
　Transit 716
Compass software 716
Complesso del Colle delle Erbe, Italy 468, 469
Complesso del Foran del Muss, Italy 468, 469

Complex Karst Expedition 663
composite island, development of 675, 676
Comprehensive Environmental Response, Compensation, and
       Liability Act (CERCLA)
   pollution of groundwater, point sources **405**
computers and cave surveying 716
Concerto Grotto (Blak) 534
Conch Blue Hole, Bahamas 156
condensation corrosion 194, **240–241**
conduction in caves 228
conduit models, mathematical 402
conduits
   aquifer and borehole hydrology 397–398
   aquifers, evolution of 678, 679
   closed conduit laws 395
   conduit initiation/early development model 668
   conduit models, mathematical 402
   development in caves 204
   evolution of 678, 679
   hydraulics of caves 429
   and karst evolution 297, 299
   paragenetic evolution, phreatic conduit 570
cone karst **242–243**
   China 218, 242
   Cockpit Country, Jamaica **233–235**
   Cuba 190
   features of 242
   formation models 242–243
   Goose Arm karst, Canada 177
   Gunung Sewu, Java **641–642**
   morphology of 242, 243
   morphometry 242, 527
   quantitative analysis of 241–242
   relict 243
   speleogenesis 242–243
   symmetry of 242
Confucianism and ritual use of caves 622
Congeria kusceri 289, 516
Cong Nuoc, Vietnam 102
Conjurtao, Picos de Europa National Park, Spain 585
consequent pseudokarst 607
conservation
   Australian Natural Heritage Charter 728
   Bambuí karst, Brazil 135
   Burra Charter, Australia 728
   cave biota, Canary Islands 150, **243–245**
   Cockpit Country, Jamaica 235
   ecology, subterranean 708
   entrance habitats 318
   flora 360
   France 371
   Hawaii lava tube caves, United States 417–418
   Integrated Conservation and Development Projects 246
   International Union for Conservation of Nature and Natural
       Resources (IUCN) 244, 245
   Korean Association for Conservation of Caverns 99
   legal issues 245, 246
   Mammoth Cave, Kentucky, United States 500–501
   and photographing caves 579
   procedures 244
   protected areas **245–246**
   Ramsar sites 619
   restoration and 627
   soils, carbonate karst 658
   speleothems 692
   and tourist caves 728–729
   travertine 737
   Vjetrenica, Bosnia-Herzegovina 760
   World Heritage Sites **779**
construction on karst **247–248**
   dams and reservoirs **277–279**
   highways **419–420**
convection in caves 228
Convention Concerning the Protection of the World Cultural and
       National Heritage 777
Convention on Wetlands *See* Ramsar Convention.
Convolvo del Butistone, Italy 510
Coopers Cave, South Africa 19
Cope, Edward Drinker 152
Copena culture and caves 44
Copepoda 255, **259–261**
   adaptations 260
   Calanoida 259
   Cyclopoida 259
   distribution 259
   Gelyelloida 259
   Harpacticoida 259
   Misophroida 179, 259, 260
   Walsingham Caves, Bermuda 767
coprolites 339
coprophagy 3
Coral, sima del, Spain 345
coral reefs 185
corals *See* Cnidaria.
Corbeddu Cave, Sardinia 339
Cordingley, John 717
Cordisburgo, Bambuí karst, Brazil 133
corrosion as coastal process 231
corrosion plains 477
corrosive/erosive class, karst cavities 253
corrosive/gravitational class, karst cavities 253
Corycia and religious sites 623
Corycian Cave, Greece 208
Cosa-Nostra-Bergerhöhle system, Austria 173–174
Cosa Nostra Cave, Austria 173, 525
Cosinzeneacea 256
cosmogenic isotopes, measuring 281–282
Cosmographia (Münster) 335
Cosquer cave, Mediterranean 281
COST action 65, groundwater protection 408
COST action 620, groundwater protection 408
Coste, Napoléon (*La Source du Lyson*) 534
Cosyns, Max 585
Cotal Cave System, Kentucky, United States 70
Cottanello marble and limestone 551
Cougnac Cave, France 281
Coultard Cave, Alberta, Canada 436, 437
Country Fermanagh, Ireland 341
County Clare, Ireland 232, 341
Cousteau, Jacques 302

Couvin Cave, Belgium 139
Cova de Salnitre, Spain 607
Covadura, Spain 345
cover collapse doline 308–309
cover-collapse sinkhole 306
cover-subsidence sinkhole 306
Coy, Mexico 699
Crabs 262, *See* Decapoda.
crack caves *See* crevice caves.
Cragonyctidae, evolution of 259
Craig ar Ffynnon, Ogof, Wales 166
Cramer, H. 526
crandallite 329
Craseonycteridae 225
    Craseonycteris thonglongyai 226
crayfish 262
    eyeless 5, 12
    North America 47, 55
Creag nan Uamh caves, Scotland 160
CREG *See* Cave Radio & Electronics Group (CREG).
Creswell Caves, England 162
Crete
    Vamos Cave 340
crevasse caves *See* crevice caves.
crevice caves 204, **249–252**
    classification of 249
    modification of 251
    processes in 249–252
    significance of 249
    speleogenesis of 249
crevice pseudokarst 606
crickets 451 *See* Orthoptera.
Crimea, Ukraine **253–254**
    Bottomless Pit 662
Crimean-Caucasian Mountaineering Club 662
Criptops anomalans 529
Črna Jama, Postojna-Planina Cave System, Slovenia 601
Črnelsko brezno, Slovenia 468, 469, 470
Crnulja cave, Herzegovina 290
Croatia
    Grab-Ruda 699
    hypogean fauna in 142
    Plitvice Lakes **597–598**
    Krka river, tufa waterfalls of 338
    Ombla 699
    Ombla power plant 739
    Prošćansko Lake, Plitvice Lakes 598
    Slovacka jama, depth of 525
Croatobranchus mestrovi 67
Crocodylus nitolicus, Madagascar 493
Crogan, John 87
Cross, W.R. 579
Crow's Nest Pass, Canada 177
Crumomyia absoloni 453
Crustacea **254–257**
    Amphipoda **257–259**
    Bochausacea 256
    Branchiopoda 254
    Britain and Ireland 163
    Cape Range, Australia 181, 182
    Cladocera 254, 289
    Copepoda 255, **259–261**
    Decapoda **261–262**
    Dendrobranchiata 261
    description of **254–257**
    Dinaric karst 289
    Diplostraca 254
    Eucarida 257, 261
    evolution of 254
    interstitial habitats 254
    Isopoda 256, **263–265**
    Malocostraca 24, 141, 255
    Mystacocarida 354
    Ostracoda 65, 255, **267–268**, 315
    Remipedia 65, 126, 156, 254
    Syncarida 255, **268–270**
    Tanaidacea 256
Crveno Jezero, Croatia 308
cryovulcanism 356
Cryptazeca elongata 516
Cryptazeca spelaea 516
cryptococcosis 294
crystallization/contraction and cave formation 204
Csorgo-Lyuk, Hungary 722
Cuatro Cienagas, Mexico 725
Cuba **271–272**
    Bellamar Cave 191, 272
    Ciénaga de Zapata 620
    Cueva del Pirate, musical performances in 534
    Cueva Pichardo, art in 93
    history of exploration 191
    karst areas of 189, 271
    mogotes of 190
    speleogenesis 271
Cucco, Monte, Italy 327
Cuertana Casas Archaeological Zone, Mexico 42
Cuervo, Father Romauldo 60
Cueva Alfredo Jahn, Venezuela 53
Cueva de Arequita, Uruguay 56
Cueva de Culiembro, Picos de Europa National Park, Spain 583, 584
Cueva de Doña Otilia, Argentina 56
Cueva de Don Justo, Canary Islands 179
Cueva de Frio, Mona, Puerto Rico 517
Cueva de Gallardo, Ecuador 55
Cueva de Nerja, Spain 327, 338, 533
Cueva de San Bernadino, Ecuador 55
Cueva de San Marcos, Canary Islands 764
Cueva de Shimpiz, Ecuador 55
Cueva de Todoque, Canary Islands 179
Cueva de Villa Luz, Mexico 38, **758–759**
    and biofilms 145, 146
    ecosystem 758
    minerals in 759
    speleogenesis 758–759
Cueva de las Brujas, Argentina 56
Cueva de las Pinturas, Guatemala 42
Cueva de los Cayos de Coangos, Ecuador 55
Cueva de los Guácharos, Colombia 54
Cueva de los Parajos, Mona, Puerto Rico 518

Cueva de los Roques, Canary Islands 179
Cueva de los Verdes, Canary Islands 179, 764
Cueva del Agua, Picos de Europa National Park, Spain 583, 584
Cueva del Capitan, Mona, Puerto Rico 518
Cueva del Cascajo, Ecuador 55
Cueva del Cerro Autana, Venezuela 652
Cueva del Cunday, Colombia 54
Cueva del Diamante, Mona, Puerto Rico 518
Cueva del Esqueleto, Mona, Puerto Rico 518
Cueva del Guácharo, Venezuela 53, 55, 59, 131, 409
Cueva del Indio, Colombia 54
Cueva del León, Argentina 56
Cueva del Llano, Canary Islands 179
Cueva del Pirate, Cuba 534
Cueva del Samán, Venezuela 53, 55
Cueva del Tigre, Mexico 409
Cueva del Valle de la Luna, Chile 56
Cueva del Viento, Canary Islands 764
Cueva Encantada, Sistema del Rio Encantado, Puerto Rico 40, 317
Cueva los Encantos, Venezuela 53
Cueva los Laureles, Venezuela 53
Cueva Magiar, Cuba 272
Cueva Mayor, Spain 119
Cueva Mylodon, Chile 56
Cueva Pichardo, Cuba 93
Cueva Sumidero La Retirada, Venezuela 53
Cueva Tres Pisos Cave, California, United States 491
Cueva Viento, Sistema del Rio Encantado, Puerto Rico 317
Cuevas de Bellamar, Cuba 191, 272
Cuevas del Centro, Mona, Puerto Rico 518
Çukurptnar Cave, Taurus Mountains, Turkey 741
cultural resource management 481
Culverson Creek Cave, West Virginia, United States 70
Cumberland Caverns, Tennessee, United States 70
Cunningham, Alan 127
Cupp-Coutunn Cave, Turkmenistan 95, **273–275**, 714
Curaçao 190, 192
Current-mark-no-ana Cave, Japan 30
Current Titles in Speleology 466
Cuves de Sassenage, France 371
Cuvier, Georges 76, 78, 80, 384, 690
Cvijić, Jovan
　　biography of **380**
　　and cave exploration 191, 288, 291, 304, 335
　　Das Karstphänomen 335
　　and karst classification 473–474
　　and quantitative karst testing 479, 526
　　speleogenesis theories and 670, 671
Cwmbran Caving Club, Wales 310
cyanobacteria 508
Cybele and religious sites 623
Cyclopoida 259, 260
Cymothoides 256
Cyrenaica karst, Libya 208
Czatowice Quarry, Poland 560
Czechia
　　Amaterska-Punkva Cave System 331
　　Czerwone Wierchy massif 331
　　Demänova, and cave surveying 715
　　Dobšina Ice Cave 331
　　Octinska Aragonite Cave 331
　　Sloup Cave 335, 688, 715
　　and speleotherapy 698
　　Zlaté Hory 698
Czerwone Wierchy massif, Czechia 331

Dabarsko polje, Herzegovina 293
Dachsteinkalk, Austria 326
Dachstein-Mammuthöhle, Austria 174, 326
Dachstein massif, Austria 174
Dachstein-Rieseneishöhle, Austria 174
Dahredj Ghar Kef, Algeria, 15
Daji Dong, China 218
Daleau, Francois 85
Daleib Wanei tufa, Namibia 23
dam construction on karst **277–279**
　　caverns and 278
　　classification of 277
　　common defects 277–278
　　environmental impacts of 278–279
　　history 277
　　problems with 277
　　underground fauna, effects on 279
　　unfavorable conditions for 277
Dana, James 765
Danger Cave, Utah, United States 43
Danielopolina spp. 183
Dan yr Ogof, Wales 166, 341, 580
Darcy's Law 394
Darcy-Weisbach Law, groundwater dynamics **395**
Daren Cilau, Ogof, Wales 166, 239, 341
Dargilan cave, France 371
dark cave/crevicular habitats 710–711
Dart, Raymond 16, 82
Darwin, Charles 765
　　biography of **152**
　　On the Origin of Species by Means of Natural Selection 347
Darwin glass artefacts, Australia 124
Das Karstphänomenon (Cvijić) 335, 380, 474
Daspoort Formation, South Africa 651
dating
　　accelerator mass spectrometry 280, 281
　　amino acid racemization dating method 556
　　annual deposition band dating 557, 558
　　archaeological methods **279–281**
　　argon 194–195, 279, 282
　　art in caves 281
　　calcified plate 224
　　carbon–14 279, 280, 555
　　carbon oscillation dating 558
　　collagen 280
　　cosmogenic isotopes, measuring 281–282
　　dendrochronology 279
　　developments in 279
　　electron spin resonance (ESR) 196, 279, 280
　　fission track 279
　　guano 409
　　karst landforms **281–283**
　　lava tubes 282, 762
　　magnetic reversal stratigraphy 280

optically stimulated luminescence 279
oxygen fractionation dating 556, 557
radiometric 285
speleothems 224, 279, 281, 554–558, 692
stone tools 279
surface isotope concentration and 282
tandem accelerator mass spectrometry (TAMS) 281
thermoluminescence (TL) 196, 279
U-series gamma ray spectrometry 196, 200, 224, 279, 280, 282
volcanic caves 282–283
Davies, Bob 302
Davies, William E. 605
Caves of West Virginia, Caves of Maryland 52
and cave surveying 716
Davis, Donald 52
Davis, Ray V. 579
Davis, S.N. 673
Davis, William Morris
biography of **380–381**
and history of speleology 49, 51, 670
Davis Blowout Cave, Texas, United States 226
Dawkins, William Boyd **77, 79**, 162
Day, Mr. 301
Day of the Cross 41
Dead Sea Scrolls 114, 208
See also Carmel Caves, Israel.
Dean's Blue Hole, Long Island, Bahamas 155
Decombaz, O. 371
Decapoda **261–262**
adaptations 261
Africa 24
Anomura 261
Astacidea 261, 261
Brachyura 261
Cave shrimp in Postojna Cave 603
Caridea 262v
crayfish in North America 47, 55, 500
Kentucky Cave Shrimp 500
Decou, V. (*Encyclopaedia Biospeologica*) 149
Deep Hot Biosphere Theory 528
deep phreatic development, speleogenesis theory 670–671
deep-seated evaporite karst 344–345
Deer Cave, Sarawak, Malaysia 532, 533
deforestation and soil erosion 658, 659
De Kerasoret 191
De Lavaur, Guy 302
De Levis, G.A. 228
Delphi, Greece 623
Demänova, Slovakia 335, 715
Demanovske Jeskyne 359
Demanovsky Cave System, Slovakia 331
Demoiselles, Grotte des, France 338, 371
Dendrobranchiata 261
dendrochronology dating 279
Denisova, Siberia 646
Denis Parisis, France 683
Dent de Crolles cave system, France **283–284**
history of exploration 284–285, 371
deposition, rates of 187
De rerun natura (Titus Lucretius Carus) 228

Dermaptera 452
Arixenina 452
Dyplatis milloti 25
earwigs 25, 180
Forfifculina 452
morphology 452
terrestrial cave habitats, biodiversity in 143, 144
Dermoptera **225**
Déroc caves, France 167
De Sautuola, Don Marcelino Santiago Tomás Sanz 85
Descent 466
Deschamps, Elliete Brunel 82
Desembarco del Granma National Park, Cuba 271
desertification and soil erosion 659
Desideri, Father 99
Desmocolex aquaedulcis **458**
Desmodus rotundas 227
Deuteromycetes 508
Devenish, Luke 579
Devil's Hole, Nevada, United States 52, 557
Devil's Tower Rock Shelter, Gibraltar 383, 384
dewpoint probes, measurement using 228
diagenesis
and carbonate karst 185–186
and cave inception 438
settings 559
diamond placers 514
diapause 319
diapirs 344
diaspore 135
dickite
and sulfuric acid speleogenesis 194
Valea Rea Cave, Romania 329
Die Fhre des Herzogthums Krain (Valvasor) 335, 687
Die Grotten und Höhlen von Adelsberg, Leug, Planina und Laas (Schmidl) 688
Die Höhle 466
Die Hydrographie des Karstes (Lehmann) 381, 671
dientes de perro defined 272
Di Feng Dong, China 220, **285–287**
Di Feng karst, China 285
Dina absoloni 67
Dinaric Karst **287–288**
deforestation of 288
history of exploration 288
Lukina Jama 288, 525
speleogenesis 287, 288
vegetation cover 287–288
Dinaric Karst, biospeleology **289–290**
adaptations of 289
biodiversity 143
biochemical investigation of 290
distribution of 290
ecological diversity of 289–290
as endangered species 290
taxonomic composition 289
Dinaride poljes **291–293**
Dionysus and caves 208
Diplatys milloti 452
Diplopoda 534

Diplopoda *(Continued)*
  adaptations 535
  biodiversity 143, 144
  Chordeumida 535
  Glomerida 535
  Iulida 535
  Lysiopetalida 535
  morphology 535
  Polydesmida 535
Diplostraca 254
Diplura
  Cape Range 182
  distribution of 143, 144, 446
  morphology 446
  taxonomy of 445
  troglobitic species of 446
Dipolopoda
  Cambalidae 38
  Cambalopsidae 631
  classification 534, 535–536
  distribution 24, 38, 47
  Trichopetalidae 38
Diprose Caves, Nullarbor Plain, Australia 546
Diprotodontia 225
Diptera 453
  Mormotomyiidae
    Mormotomyia hirsuta 453
  Mycetophilidae
    Arachnocampa luminosa 453
  Sciaridae
    Allpnyxia partizii 453
  Sphaeroceridae
    Crumomyia absoloni 453
    Spelobia tenebrarum 453
  terrestrial cave habitats, biodiversity in 143, 144
diseases and caves **293–295**
dissolution
  aqueous, and karst 473
  carbonate rocks **295–298**
  caves, dissolutional 184
  as coastal process 231
  evaporite rocks **298–300**
  karstic 242
  rates of 187
  silicate rocks **300–301**
  speleogenesis, role of 251
  water flow 194
dissolution speleogenesis, mechanism of 356
distributed parameter models 401, 402
Divine Comedy, The (Dante) 205
diving in caves **301–303**
  accidents and rescue 2
  Aqualung 302
  depth barrier 302, 303
  future of 303
  history of 52, 301
  International Underwater Cave Rescue and Recovery team 2
  SCUBA 302
  surveying 717

  techniques, evolution of 301
  technology, advances in 302–303
Djurdjura karst, Algeria 15
Dobšina Ice Cave, Hungary/Slovakia 25, 27, 331
DÖF-Sonnenleiterschacht cave, Austria 174
Dog Cave, Turkey 208
Dolganskaya Yama, Siberia 646
Dolichoiulus spp. 180
dolines **304–308**
  burial 304
  buried 309
  caprock collapse 304, 305, 306, 307–308, 309
  China 218
  collapse 304, 305, 306, 307–308
  and construction 248
  defined 304
  dropout 304
  hydrology of 309
  interstratal collapse defined 306
  mapping of 309
  New Brunswick 177
  Newfoundland 177
  Nova Scotia 177
  origin of term 288
  Pine Point 176
  remediation of 248
  solution 304–307
  speleogenesis 219, 304–305
  suffosion 304, 305, 306, 308–309
  term, origin of 304
  Xiaochai Tiankeng, China 286–287
  Xingwen, China 219
Doljašnica ponor, Herzegovina 291–292
dolomite
  and breccias 660
  dissolution of 297–298
  evaporite karst 343
  as mineral resource 489
  properties of 185
dolostone
  Dolomites, Italy 326, 327
  and karst aquifer development 399
  karst formation, role in 186
  paleokarst 559
  principle karst rock 474
  properties of 185
Domene vulcana 180
Domene vulcanica 180
Domica Cave, Hungary/Slovakia 28–29
Dominican Republic
  cave and karst areas of 189
  Meson de la Cava 534
Domus Aurea, excavation of 86
Doney Fissure, Arizona, United States 251
Dongryong-gul cave, Korea 99
Dong Xingren, China 222
Donjobrelska vrulja, Dinaric karst 288
Doolin cave system, Ireland 171, 341
Dordogne caves, France 167
Dorothy and the Wizard in Oz (Baum) 206

dosimeter 279
Dos Ojos system, Mexico 303, 787, 788
Dos Pilas, Mexico 41
Dotson, Doug 716
Douhlata, Bulgaria 329, 330
Drac, Coves del, Spain 338
Drachenhöhle, Austria 326, 335
Draenen, Ogof, Wales **310–312**, 341
   cave development 312
   depth of 525
   exploration of 166, 310
   geography of 310
   hydrology of 311–312
   multiphase origins 312
   and radiolocation 616
   speleogenesis of 310, 311, 312
Dräger gas analyzer 183
Dragon Bone Hill (Longgushan), China 221
dragon bones in medicine 97, 221
Dragon Bone Slope (Longgupo), China 223–224
Dragon Pool Cave (Longtangdong), China 222–223
Dragon's Breath Cave, Namibia, 21
Dragon's Den, Poland 335
dragons and folklore 61, 148, 158, 365
DRASTIC method, groundwater protection 408
drawdown/capture zones, karst aquifers, determining 482
Draycott, England 561
Drimolen, South Africa 19
drinking water standards
   nitrates 403
      pesticides/herbicides 403
      regulation of 404
Drogarati Cave, Greece 533
dropout doline defined 306, 308
Drum Cave, Australia 127
dry valleys 363
Dubašnica pothole, Serbia 330
Dublyansky, Victor 253, 663
DuFour, John James 715
Dugo Polje, Bosnia/Herzegovina 291
Dumanli, Turkey 699, 741
Dunstan Cave, Trinidad **190**
Dupont, Edouard 670
Durand, Jacques 5
Durkó, Zsolt 534
Durov, S.A. 673
Durst, John (*Secret Cave, The*) 360
Dust Cave, Alabama, United States 43
Duvanjsko Polje, Bosnia/Herzegovina 291
Duyong Cave, Philippines 167
Duzkhra Massif, Georgia **201**
Dvuglazka Cave, Siberia 646
Dydd Byraf cave system, Wales 166
dye tracing *See* water tracing.
Dyplatis milloti 25
Dyros Cave, Greece 339
Dyuktaiskaya Cave, Siberia 646
Dzhentu Massif, Georgia **201**
Dzilam State Reserve, Mexico 620
Dzou Cave, Georgia **201**

Eagle Creek Cave, Arizona, United States 409
Earle, Augustus 157
earth cracks *See* crevice caves.
earth current injection and cave communication 238
earthquake cracks *See* crevice caves.
earthquakes
   paleotectonics with speleothems **565–566**
earwigs *See* Dermaptera.
Easegill Caverns, England 441, 525
Easter Cave, Australia 120
Easter Island 83
Eastern Kugitang Upthrust, Turkmenistan 273
Eastwater Cave, England 561
Ebbou cave, France 82
Ebola virus 294
Echinodermata
   Holothurioidea 459
echolocation 131, 226
ecology, subterranean **706–708**
   characteristics of 706–707
   conservation 708
   fauna of 707–708
   food sources 365–367, 707
ecosystems, restoration of 244
Ecuador
   Cueva de Gallardo 55
   Cueva de San Bernadino 55
   Cueva de Shimpiz 55
   Cueva del Cascajo 55
   Cueva de los Cayos de Coangos 55
Edenvale caves, Ireland 162
Edwards Aquifer, Texas, United States **313–315**
   biospeleology **315–316**
   caves of 314–315
   chemosynthesis 315
   energy pathways of 315
   geography of 313
   geology 313
   hydrology 314, 315
   land use 314
   pollution of 405, 407
   saline-freshwater interface in 315
   speleogenesis 314
   zone classification 313–314
Edwards Aquifer Authority 314
egg box topography of cone karst 241, 305
Eglwys Faen, Wales 166
Egypt
   Cave of Swimmers 608
   geomorphology 16
   pyramids, religious views of 622–623
   White Desert 16
Eigenmann, Carl H.
   biography of 154
   Cave Vertebrates of North America 154
Eileithya cave shrine, Crete 208, 623
Einhornhöhle 703
Eisriesenwelt cave, Austria 173, 174
El Capitan Pit, Alaska, United States 752
El Duende Pyramid, Guatemala, Central America 624

El Outaya, Algeria 15
El Tragadero, Peru 55
Elderbrush Cave, Peak District, England 576
Eldon Hole, Peak District, England 164, 576
electrical resistivity surveys 378, 379
electroconductivity (EC) testing, karst water 214
electromagnetic induction measurements (EM) 378, 379
electron spin resonance (ESR) dating 196, 279, 280
Eleneva Cave, Siberia 646
Elephanta caves, India 623
Elgon, Mount, Kenya 22, 25, 562, 592
Ellipse software and cave surveying 716
Ellis, Bryan 716
Ellis Basin System, South Island, New Zealand 543
Ellora Cave, India 623
Elphas falconeri 340
Emergence Du Ressell, France 302–303
Emil Racoviță Cave Moldovia 330
Emory, Kenneth 765
Empire State Building and limestone 551
Encantado, Sistema del Rio, Puerto Rico 190, **317–318**
    art in caves 318
    biospeleology 318
    Cueva Encantada 317
    Cueva Viento 317
    geology 316, 317
    hydrology 317
    karst area 317–318
    speleogenesis 316–317
    taino petroglyphs in 318
    threats to 318
Enchanted Rock Cave, Texas, United States 722
Encyclopaedia Biospeologica (Juberthie and Decou) 149
endangered species
    Cape Range 183
    Edwards Aquifer, United States 315
    Hawaii lava tube caves, United States 417–418
    Myotis grisescens, Mammoth Cave, Kentucky, United States 500
    Myotis sodalis, Mammoth Cave, Kentucky, United States 500
    North America 48
    Palaemonias ganteri. 500
    Proteus anguinus 63
    Speoplatyrhinus poulsoni 595, 596, 597
    Synocyclocheilus hyalinus 594
    Triplophysa gejieunsis 594
endellite and biomineralization 636
endemic fauna
    Bermuda 767
    Canary Islands 179
    Cape Range, Australia 182
    Dinaric karst 289–290
    Edward Aquifer, Texas 315–316
    Vjetrenica 759–760
Endins 465
Endless Cavern, Virginia, United States 51
endokarst
    development 650
    diagnostic features of 650
    gravity tectonics caves 650, 652
    quartzite-dolerite caves 651, 652

en echelon faulting 313
energy fixation, chemoautotropic 182
Engihoul Cave, Belgium 139
Engis Caves, Belgium 139
English Chalk, United Kingdom 438
Ensifera 451
entrance habitats **318–319**
    animal use of 319
    biospeleology 319
    conservation of 318
    ecological factors of 318
    flora of 318
    human use of 319
    species adaptations 318
    threshold communities, limits of 318
entrenched evaporite karst 346
environmental impact assessments **319–320**
    cultural/paleontological survey 320
    history of human use 319
    hydrogeologic methods
        EPIK system 319–320
        geomorphological/hydrological technique 320
        geophysical techniques 320
        tracer testing 320
    karst areas, resources of 319
    karst biological EIA 320
    limitations of 320
    presence-absence survey 320
    recommendations 321
    regional analysis of species localities 320
    shovel tests 320
    uses of 319
eosine (CI Acid Red 87) dye and water tracing 772
EPIK system, groundwater protection 408
epsomite
    deposition of 694
    formation of 513
    Mammoth Cave, Kentucky, United States 495
Eptesicus fuscus 227
equilibrium constants and quartz dissolution 300
equilibrium fractionation 556
Eratothenes and karst hydrology 478
Erebonectes spp. 768
erosion
    caves 204
    rates
        bioerosion as coastal process 231
        bioerosion in warmer regions 232
        cave passages 322–323
        coastal karren 231
        conditions for 323
        Corbel formula 321
        field measurements **322–323**
        gross denudation, underestimating 322
        hydrochemical budgeting method 322
        hysteresis effects 322
        karst denudation rates, equations 321
        laminar flow, equation 323
        limestone 322–323
        limitations of 322

micro-erosion meter 322, 324
modelling 323–324
rock tablets 322
in soil 324
solutional erosion rates 322
and temperature 324
total solute load (TSL) 322
turbulent flow rates, equation 323, 324–325
and speleogenesis theories 669
Escalades souterraines (Chevalier) 284
Espeleo Tema 465
Esper, Johann Frederich 77
Espluga de Francoli, Spain 607
Essai sur les problèmes biospéleogiques (Racovitza) 149, 154
Es-Skhul cave, Israel 195, 196
Estremera Cave, Spain 683
Eszterhas, Istvan 607
Ethiopia
    Sof Omar 20, **655–656**, 699
Etna, Mount, Italy 762
Et-Tabun Cave, Israel 197
Eucalyptus spp. 181
Eucarida 257, 261
Eukoenenia maros 632
Eunapius subterraneus 289, 457
Euproctes aster
    adaptations of 12
Euproctus asper 370
Europaeische Hoelenfauna (Hamann) 149
Europe
    Alpine region **325–327**
    Austria
        Calcareous Alps **173–175**
        Dachsteinkalk, Austria 326
        Dachstein Mammuthöhle, Austria 326
        Drachenhöhle, Austria 326, 335
        Lamprechstofen, Austria 326
        Lurhöhle 326
        Spannagel Cave **663**
        Totes Gebirge 173, 174
    France
        l'Alpe cave, 326
        Caladaire cave system 326
        Cigalère, Grotte de 327
        Dent de Crolles **283–285**, 326, 371
        Felix Trombe/Henne Morte 327, 525
        French Alps, geomorphology of 326
        Gavarnie, Cirque de 327
        Granier cave 326
        Grotte Casteret 327
        Jean Nouveau 326
        Moulis, Grotte 327
        Pierre St Martin 327, **585–588**
        Platé, Désert de 326
        Reseau Verneau 326
        Vercors **754–755**
        Verdon, Gran Canon de 326
        Vaucluse Plateau 326
    Italy
        Abisso di Malga Fossetta 326
        Alpi Apuane 327
        Asiago plateau **116–118**, 326
        Badalona, Sistema 327
        Carrara marble 327
        Cucco, Monte 327
        Fighiera-Corchia 327
        Fiume-Vento caves 327
        Frasassi cave 327, **374–375**
        Lessini plateau 326
        Piaggia Bella 326
        Rana, Buso de la 326
        Spipola-Aqua-Fredda 327
        Spluga della Preta 326
    Spain
        Matienzo 327
        Nerja 327, 338, 533
        Ojo Guarena 327, 525
        Ordesa national park 327
        Perdido, Monte 327
        Picos d'Europa 327, **582–583**
        Sima GESM 327
    Switzerland
        Hirlatzhöhle 174, 326, 525
        Hölloch 326, **420–421**
        Siebenhengste 326, 525, **647–649**
    Balkans/Carpathians **328–330**
    Bulgaria
        Barkite 14, 329
        Douhlata 329, 330
        Jagodinskata 329, 330
        Magura Cave 330
        Orlova Chuka 329, 330
        Pirin National Park 330
        Raytchova Doupka 329, 330
        Shopov cave system 330
        Yamata 329
            exploration of 328
            hypogean fauna in 141, 142
    Moldovia, geology of 330–331
        Emil Racoviță 330
        Zoloushka 329, 330, 345, 525
    Romania 329–330
        Avenul din Grind 329
        Avenul din Stanu Foncii 329
        Cioclovina Cave 329
        Ciur Ponor 329
        Ghețarul de la Scărișoara 329
        Movile Cave 329, 365, 507, 516, **528–530**
        Peștera Hodbana 329
        Peștera 6S de la Mănzălești 329
        Peștera Vântului 329
        Sistemul Humpleu-Poienița 329
        Șura Mare cave 329
        Tăușoare Cave 329
        Topolnița 329
        Ušački Pićinski Sistem 329, 330
        Valea Rea Cave 329
    Serbia
        Bogovinska Pećina 330
        Buronov Ponor 330

Europe *(Continued)*
    Cerjanska Pećina 329, 330
    Dubašnica pothole 330
    Jama du Dubašnici 329
    Rankin Ponor 329
    Velika Klisura 329, 330
    Vrelo Krupac 330
    Yugoslavia, geomorphology 330
    See also individual country by name.
  Central **331–332**
    Aggtelek and Slovak karst **25**, 333, 335
    archeological caves **333–334**
    Baradla-Domica Cave System, Hungary/Slovakia **26–27**, 29, 331, 620, 698
    Czechia
      Amaterska-Punkva Cave System 331
      Czerwone Wierchy massif 331
      Dobšina Ice Cave 331
      Octinska Aragonite Cave 331
    Gastein Heillstolen 698
    history of exploration **335–336**
    Hranicka Chasm 332–333
    Hungary
      Pál-völgy cave 332
    Lower Tatra Mountains National Park 333
    Poland 331, 333
      Bandzioch Kominiarski Cave 331
      Chelosiowa jama 331
        Krakow-Wielun Upland 331
        Niedzwiedzia Cave 331
      Ojców National Park 333
        Salt Mine, Wieliczka 333
        Sniezna Studnia 331
        Szachownica Cave 331
        Tatra Mountains National Park 333
        Wielka-Snizena Cave System 331
        Wysoka-Za-Siedmiu Progami Cave 331
        Zabrašovké Aragonite Caves 332, 560
    protected areas of 333
    Slovakia
      Demanovsky Cave System 331
      Slovak Paradise National Park 333
      Stary-hrad 331
      Stratenska-Psie diery Cave System 331
    See also individual country by name.
  Mediterranean **337–338**
    archaeological/paleontological caves **339–340**
    Argostoli sea-mills, Greece 338
    Aven d'Orgnac, France 338
    Boka Kotorska, Montenegro 338
    Castellana, Grotta di, Italy 338
    Clamouse, Grotte de, France 338
    Cueva de Nerja, Spain 327, 338, 533
    Demoiselles, Grotte des, France 338
    Dinaric karst **287–288**, 337
    Drac, Coves del, Spain 338
    Fontaine de Vaucluse, France 338
    human settlement of 337–338
    Krka river, tufa waterfalls of, Croatia 338
    Mediterranean caves in history 208–209, 338
    Neretva Plain, Herzegovina 338
    Pareis, Torrent de, Spain 338
    Port Miou, France 338
    Samaria, karst canyons of, Greece 338
    Skadarsko Jezero, Montenegro/Albania 338
    and tourism 338
    Verdon, Gorges du, France 338
    See also individual country by name.
  North **340–341**
    Britain 340–341 *See also* United Kingdom.
      Cretaceous Chalk 340, 342, 363–364
        Mendip Hills **503–505**
        Ogof Draenen **310–312**
        Peak District **576–577**
        Yorkshire Dales **783–785**
    France
      Causses karst 372–374
    Germany 341–342, 343
    Ireland 341
      Burren **169–171**
    Norway 342, 705, 706
    Sweden 342, 722
    See also individual country by name.
europium 514
Eurycea rathbuni 315
euryoecious defined 290
eutrophication defined 244
Evans, Franklin 670
Evans, G.W. 127
Evans, T.R. 390
evaporite karst
  caves in 346
  classification of 344
  deep-seated 344–346
  distribution of 344
  entrenched 346
  features of 344
  intrastratal 344
  mantled 346–347
  open and denuded 346
  speleogenesis 343–344
  subjacent 346
  syngenetic 344, 676, **717–719**
evaporite rocks
  dissolution rates
    alabaster 299
    anhydrite 299–300
    control mechanisms 299
    equilibrium and rate laws 299
    gypsum 298, 299
    inhibition of 299
    rock salt 300
    surface reaction rates, equation 298
    transport rate, equation 298–299
    turbulent water flow, equations 299
evaporite speleothems **692–694**
evolution
  Amphipoda 259
  Crustacea 254
  human 196

hypogean fauna **347–349**
of karst *See* karst evolution.
MSS (mesovoid shallow substratum, milieu souterrain superficiel) 456, 457
Ostracoda 267
regressive 454, 455
speciation **665–666**
troglomorpic species 347–349
Evren Gunay düdeni (Peynirlikönü), Turkey 525
Ewers, R.O. 673
Ewers speleogenetic high-dip model 678
exaptations defined 268
Exit Cave, Tasmania 122
Exley, Sheck 52, 303
exokarst
 itacolumnite 650
 neosandstone 650
Exploits at West Poley (Lawrence) 360
exploration of caves *See* caves, exploring.
exploration societies **350–352**
 Australia, influence of **352**
 Austria/Italy, influence of 351
 data collection 351–352
 France, influence of 351
 history 350–352
 societies, partial list of 352
 See also specific society by name.
exsurgence springs 701
extraterrestrial caves **355–357**
 lava tube caves 355
 Mars 355, 592–593
 Moon 355
 plate tectonic activity 355
 pseudokarst 608
 speleogenesis 355
 types of, theoretical 357
eyes, adaptations **4–5**
 appearance, variability of 5
 degeneration of 5
 development of eyes 5
 lack of light 4
 regression 5
Eyrie, Lake Basin, Australia 737

F1, Siebenhengste, Switzerland 647
F20, Picos de Europa National Park, Spain 585
Fafnir, legend of 365
Fairy Cave Quarry, Mendip Hills, England 503
Falconer, Hugh 384
Fang Chengda 98
Fantaziya, Siberia 645
Faouar Dara, Lebanon 114
Faro, Sistema del, Mona, Isla de National Park, Puerto Rico 517–518
Farr, Martyn 302
Farrer, James W. 715
Fatničko polje, Herzegovina 293
Faulkner, Charles 82, 85
Faune cavernicole de la France (Jeannel) 154
Fauzan Cave, France 81

Fea, Leonardo 109
Feather Cave, New Mexico, United States
 Arrow Grotto 92
 art in 92
 ritual use of 44
Federación Argentina de Espeleología 60
Fédération Française de Spéléologie, France 372
Federov, E. 662
feeding, behavioural adaptation **3**
Felderhof Cave, Germany 81
Felipe Reventón, Cueva de, Canary Islands 179
Felix TrombeHenne, France 327, 525
felsenmeere 175
fengkong karst 217, 220, 242
fenglin karst 217, 242
Fergus River Cave, Ireland 716
ferrobacteria 366
Fersman's Cave, Turansky Plain, Central Asia 95
Festival'naja-Ledopadnaja system, Turansky Plain, Central Asia 95
Ffynnon Ddu, Ogof, Wales 166, 239, 341
fibroferrite 636
Fighiera-Corchia, Italy 327
Figuier cave, France 85
films in caves **359–360**
Fingal's Cave, Scotland 204
Fingal's Cave Overture (Mendelssohhn) 534
Fingal's Höhle (Mertz) 534
Fiordland
 Aurora Cave, New Zealand 543, 554, 555
fish **593–594**
 See Pisces.
Fish, Larry 716
Fisher Ridge Cave, Mammoth Cave, Kentucky, United States 52, 495, 525, 750
Fish River Caves, Australia 128
Fisht Massif, Georgia 201
fission track dating 279
fissure caves *See* crevice caves.
fissure frequency and karst development 186
Fiume-Vento caves, Italy 327
Flamache, Armand 670
flank-margin caves 189, 675–677
flatworms *See* Platyhelminthes.
Flavius, Josephus and karst hydrology 478
flintstone 185
flood pulse analysis and dye tracing 773
flora
 bryophytes 362
 classification of 361–362
 commercial forestry 362
 crevice caves 249
 distribution of 362
 entrance habitats 318
 environmental conditions 360–361
 and fire 362
 nutrients 360
 resources, conservation of 360
 water flow, regulation of 362
flowstone *See* speleothems.
Floyd Collins Crystal Cave, Kentucky, United States 159

flue gas desulfurization (FGD) and limestone 490
Fluess oxygen rebreather, diving 302
fluorescent microspheres and dye tracing 772
fluorescin tracing, in cave locating 253, 769, 771, 772
fluorite 273
fluorometer and water tracing 772
fluorospar 576
fluvial geomorphic activity and cone karst 242
fluviokarst **363–364**
   classification of 363
   Cretaceous Chalk, United Kingdom 363–364
   formation of 363
   Peak District, United Kingdom 342, 363, 561, **575–577**
flysch defined 288
Foissac Caves, France 167
Foley Caves, Ireland 162
folklore and caves **364–365**
Fond-de-Forêt Cave, Belgium 138, 139
Fontaine de Vaucluse, France 301, 303, 338
Fontaine Saint-Georges, France 302
Font de Gaume, France 757
Font Estramar, France 302
food resources **365–367**, 707
   clay sediment and 366
   chemosynthesis 145, 366, 507
   flow of energy 366
   food chains 365
   food zones 365
   guano 367
   microorganisms 366–367, 508
   montmilch 366
   organic material, transport of 366
   population, stability of 365
   slime-eaters 366
   speleothems 366
   vegetation zone 365
footings, regeography of 248
Forbesichthys agassizi 595, 596
Forbe's Quarry, Gibraltar 383
Forbidden Fissures, California, United States 491
Ford, Derek 52, 673
Ford-Ewers model, speleology theory 673
Ford, Trevor 166
Forel'naja Cave, Georgia 202
Forest of Dean, England 439
Forest Reserves, Cockpit Country, Jamaica 235
forests on karst **368–369**
   afforestation 369
   economic productivity of 368
   environmental impacts, logging 369
   groundwater recharge 368
   nutrient transport 368
   rainforests on 368
   soil, sources of 368
   timber harvesting 368–369
Forfifculina
   Anataelia troglobia 452
   Anisolabis howarthi 452
   Diplatys milloti 452
Forster, E.M. (*A Passage to India*) 206

For Whom the Bells Tolls (Hemingway) 206
Fouar Dara, Lebanon 463
Fourier models, cone karst analysis 242
Fournier, Eugene 371, 585
Four State Model 685–686
foval 366
Fox, Robert 109
fracture modelling 677, 678
fraipontite 273
France
   Andrè Lachambre 373
   Aragonite Caves 373–374
   Arbois spring 754
   Ardèche caves, history of exploration 427
   Aven Armand 533
   Aven d'Orgnac 338
   Aven Noir 371
   Aven Orgnac 371
   Barrencs de Fournes 373
   bauxite production in 135
   Bedeilhac, cave of 426
   BG63, Pierre Saint-Martin 586
   biospeleology **369–371**
   Cabinet of the Felines, Lascaux 757
   Caladaire cave system 326
   Campan 371
   Causses Karst 372–373
   Cave Rescue Organization 1
   Chabot (Gard) cave 82, 85
   Chauvet Cave, Ardèche 85, 90, 281
   Chartreuse massif 326
   Chourum Martin 371
   Cigalère, Grotte de 327
   Clamouse Cave 215, 338, 373
   Cuves de Sassenage 371
   Dargilan cave 371
   Demoiselles, Grotte des 338
   Denis Parisis 683
   Dent de Crolles cave system **283–285, 326**, 371
   Déroc caves, burials in 167
   Dordogne caves 167
   Ebbou cave 82
   Fauzan Cave 81
   Fédération Française de Spéléologie 372
   Felix Trombe/Henne Motre 327, 525
   Figuier cave 85
   Foissac Caves 167
   Font de Gaume 757
   Gavarnie, Cirque de 327
   Giant Hole, the 373
   Gouffre Berger 239, 302, 371, 525, 754–755
   Gouffre de Padirac 371, 373
   Gouffre de Pierre Saint-Martin 554
   Gouffre d'Esparros 373
   Gouffre Martel 371
   Gouffre Mirolda Cave 204, 525
   Gouffre Vallier 756
   Gournier cave system 755, 756
   Grand Veymont Peak 754
   Grotte Brudour 755

Grotte de Cabrespine 373
Grotte de Gournier 756
Grotte de la Cigalère 373
Grotte de la Luire 754
Grotte de l'Asperge 373
Grotte de Limousis 373
Grotte de Lombrives 533
Grotte de Miremont 715
Grotte de Pré-Létang 755
Grotte des Esplunges 410
Grotte des Eyzies 410
Grotte du Mont Marcou 373
Grotte Pousselière 373
Grotto des Remouchamps 704
Guiers-Mort Cave 283, 371
Henne-Morte 371
history of exploration **371–372**
Jean-Bernard Cave 372
Jean Nouveau 326
La Balme 371
La Chappelle-aux-Saints Cave 167
La Fontainguillère 167
La Mouthe cave 85
Lascaux
    art in 85, 757
    Hall of the Bulls 757
    as tourist cave 428
La Tête-du-Lion 82
Les Combarelles 757
Le Tuc d'Audoubert 84, 90
Luire cave 754
Mas Raynal 371
Mirolda Cave 372
Moulis, Grotte 327
Niaux 428
Osselles 371
Padirac Cave 372
Padirac River 371
Pair-non-Pair cave 85
Palabres Cave 167
Pech Merle 428
Pierre Saint-Martin
    alpine karst 327
    Aranzadi passage 587
    Arresteliako Ziloa 586
    BU56 cave 585, 586
    Chipi Joseteko 587
    Gouffre de la Pierre Saint-Martin 371, 525, **585–588**
    Gouffre des Bourrugues (B3) 586
    Gouffre des Partages (M413) cave 585, 586
    Gouffre du Couey Lotdge 586
    Gouffre Lonné Peyret 586
    Grotte d'Arphidia 586, 587
    La Verna chamber 585
    Réseau de Sondet (BT6) 586
    Romy cave 586
    SC60 cave 587
Piglaith Cave 370
PN77 cave 373
Port Miou 338

Rabanel 371
Reotier travertine cascade 736
Réseau de l'Alpe 525
Réseau Jean-Bernard 525
Rouffignac caves 167
Société de Spéléogie 351, 371, 688
St. Bernadette, Lourdes 428, 622, 623, 624
Tindoul de la Vayssière 371
TM71 373
troglobites 372
Trou du Glaz 371
Trou qui Souffle 755, 756
Vallier cave system 755, 756
Vaucluse 699
Vaucluse, Fontaine De 301, 302, 303, 338
Vercors **754–756**
Verdon, Gorges du 326, 338
Vézère caves
    archaeology **757**
    art in 90
    history of exploration 427
    religious views of 623
Frankenstein (Shelley) 206
Frankischer Alb, Germany 341
Franklin Mountains, Canada **178**
Frasassi caves, Italy 327, **374–375**
    Ancona Abyss 374, 375
    ecosystem 375
    fauna in 374, 375
    Grotta del Fiume-Grotta Grande del Vento 374
    Grotta del Mezogiorno-Grotta di Frasassi 374, 375
    history of exploration 374
    minerals of 375
    speleogenesis 375
    speleothems 375
    sulfide water cycle 374, 375
free-climbing and cave exploration 353
Freyre, Father 99
Friar's Hole Cave System, West Virginia, United States
    geomorphology 70
    hydrology 70
    length/depth of 525, 750
    speleogenesis 70
Friedrich, Caspar David 87
Frio, Mexico 699
Frio Cave, Texas United States 294
Fromage, Father and Jeita Cave 463
frost action as coastal process 231
frostbite 293
Frost Pot, Canada 200
Fuchslabyrinth caves, Germany 682
Fuerteventura Island, Canary Islands 179, 181
Fugger, Eberhard 336
Fuhlrott, J.C. 333
Fuji, Mt Japan 624
Fun Fun, Cueva Dominican Republic 189
fungi 508
Furfooz Cave, Belgium 138
Furong Dong, China 220

Gagnan, Emile 302
galena
  and breccias 660
  formation of 714
  paleokarst, economic resources of 560
Gallus, Alexander 85
Galowe, Papua New Guinea 699
Gambier, Mt Australia 89
Gammarus minus
  adaptations 9, 12
  genetic basis of adaptation 6, 9
Gamvo, Nakanai Caves, Papua New Guinea 538
Ganges River, India 623
Ganhe bridge, China 218
Gaping Gill, England 166, 522, 616, 783
García-Zambrano, Angel 41
Gardner, J.H. 671
Gardner's Gut Cave, North Island, New Zealand 541, 543
Gargantua, Canada **177**
garnets 342
Garrod, Dorothy 195, 196, 384
gas, mining of 514
Gaspe Peninsula, Quebec 177
Gastec gas analyzer 183
Gastein Heillstolen, Central Europe 698
Gastropoda **516**
  anchialine habitats 65
  Bivalvia **516**
  Carychiidae
    Zospeum spp. **516**
  in Edwards Aquifer, United States 315
  hypogean habitats 65
  Prosobranchia **516**
  Pulmonata **516**
Gastrotricha **458**
Gauntlett, George Edward 99
Gavarnie, Cirque de, France 327
GB Cavern, Mendip Hills, England 504
Gebihe Dong, China 218
Geißenklösterle caves, Germany 333
Geitleria calceria and calcium deposition 508
Gelyelloida 259
Genesis Cave, Edwards Aquifer, United States 314
genetic adaptations *See* adaptations, genetic.
genetic analysis, Cape Range fauna 182
geochemical processes and karst evolution 475
Geofizicheskaya Cave, Turkmenistan 95, 273
Geological Society of Hungary 336
geomorphometry 526
geophysical detection, caves and karstic voids **378–379**
  electrical resistivity surveys 378, 379
  electromagnetic induction measurements (EM) 378, 379
  gravity anomalies and 377–378, 379
  ground-penetrating radar (GPR) 378
  induced polarization method 379
  LaCoste and Romberg "G" meter 377
  negative density contrast 377
  prospecting strategies 379
  resistivity tomography 379
  seismic reflection surveys 379
  seismic tomography, three-dimensional 379
  self-potential methods 379
Georgia
  Akhshtyr Massif 201
  Akhtsy Massif 201
  Akhun Massif 201
  Alek Massif 201
  Alekseeva Massif 201
  Arabika Massif 201, 202
  Arabikskaja cave system 201
  Berchil'skaja Cave 488
  Boguminskaya Cave 203
  Bzybsky Massif 201–203
  Caucasus **200–203**
  Duzkhra Massif 201
  Dzhentu Massif 201
  Dzou Cave 201
  Fisht Massif 201
  Forel'naja Cave 202
  Gorlo Barloga Cave 201, 663
  Grafsky Proval Cave 202
  Great Caucasus 200–201
  Iljukhina Cave system 201, 487, 525
  Kaldakharsky Fault 201, 202
  Krestik-Turist Cave 201
  Krubera Cave 201, **487–488**, 525, 663
  Kujbeshevskaya-Genrikova Bezdna Cave 201, 488
  Majskaya Cave 201
  Martel's Cave 488
  Mchista 699
  Mchista Spring 202
  Moskovskaja Cave 201, 487
  Napra Cave 202
  Pantjukhina Cave 202, 203, 487, 663
  Parjashchaya Ptitsa Cave 201
  Pionerskaja Cave 202
  Rostovskaja Massif 201
  Sarma Cave 201, 487, 525, 663
  Sneznaja-Mezhennogo System 202, 487, 525
  Sochinsky artesian basin 201
  Vjacheslava Pantjukhina, depth of 525
  Vorontsovsky Massif 201
  Zagedanskaya Massif 201
  Zagedon Massif 201
geoscientists **380–382**
  See also individuals by name.
geotextile sleeves and construction 248
Germany
  Balver Höhle, musical performances in 533
  Baumannshöhle, and cave surveying 714
  Bing Cave 565
  Felderhof Cave 81
  Frankischer Alb 341
  Fuchslabyrinth caves 682
  Gutersteiner Falls 737
  Hermannshöhle and cave photography 579
  history of exploration 335
  Kluterthöhle 335, 697, 698

Moestrof caves 682
Schwäbischer Alb 341
Gesellschaft fur Höhlenforschung, Slovenia 336
geysers 701
Gèze, Bernard **382**
Ghar Alisadr, Iran 116, 461
Ghar Dalam, Malta 208, 339
Ghar Parau, Iran 115–116, 461
Ghar Sarab, Iran 461
Ghețarul de la Scârișoara, Romania 329
Ghori-Galeh, Iran 461
Giacominerloch, Italy 117
Giant Hole, the, France 373
Giants Hole, Peak District, England 577
Giants-Oxlow system, Peak District, England 576
Giant's Room, Sistemul Humpleu-Poienița, Romania 329
gibbsite 135, 301
Gibbs triangle, use of 184
Gibraltar Caves **383–384**
   archaeology 383–384
   Devil's Tower Rock Shelter 383, 384
   Forbe's Quarry 383
   geology 383
   Gorham's Cave 383, 384
   human uses of 384
   Ibex Cave 384
   morphology 383
   Rosia Bay 384
   St Michael's Cave 533
   Vanguard Caves 383, 384
gigantism, appearance of 339
Gigantopithecus blacki 223
Giggler's Caves, Kenya 592
GILGES *See* Global Indicative List of Geological Sites (GILGES).
Giniba Cave, Yemen 115
Girraween National Park, Australia 652
GIS models, cone karst analysis 242
glaciated karst **388–390**
   continental glaciation 390
   crevasses and 389
   distribution of 388
   glacial sediments and 389
   glaciers 388–389
   Rundhöcker 389
   Schichttreppen 389
   stream sediments and 389
   water 388–389
glaciation
   Calcarous Alps, Austria 174
   Canada 176–177
   and collapse doline formation 308
   and colonization 237
   effects of 175, 176
   and erosion 659
   karren and 473
   Spitsbergen, subglacial karstification 342
glacier caves 204, 606, **385–387**
   Antarctica **68–69**, 606
   closure/deformation of 387

development of 385
erosion, differential 387
exploration of 387
freezing process 385
geomorphology 386–387
glaciospeleology 387
hydraulic laws of water 386
hydrology and drainage 386
ice, physical properties of 385
Jökulhlaup 387
kettle holes 387
mechanical collapse 387
melting process 385
moulins 68, 386, 606
Nye channels 386
Röthlisberger channels 386
sediment and 385–386
sublimation 385, 387
tectonic caves 387
tectonic processes 385
glacier crevasses, geography of 249
glacière
   Dobšina Ice Cave, Hungary/Slovakia 27
Glacières, or Freezing Caverns (Balch) 51
glacier pseudokarst **387**, 606
glacio-eustatic fluctuations 308
glaciokarst **169–171**, 201, 341, 385, 387
glaciospeleology 387
Gladysvale, South Africa 19
Glamočko polje, Bosnia/Herzegovina 291
glass making and limestone 489
Glaz, France 284
Glencurran Cave, Burren glaciokarst, Ireland 171
Glennie, Aubrey 166
Global Indicative List of Geological Sites (GILGES) 273
Glomerida
   distribution 535
   morphology 535
   Trachysphaera spp. 535
   troglobitic species of 535
Glory of the Duchy of Carniola (Valvasor) 152
glowworms 3, 4, 453, 541–542, 620, 731
Gmelin, Johann 662
GOD method, groundwater protection 408
Goebel's Cave, Australia 652
goethite 587
Gogteik, Myanmar 103
gold, alluvial 560
gold placers 514
Goldast, Melchior and karst hydrology 478
Golding River Cave, Jamaica 190
gold placers 514
Golondrinas, Sótano de las Mexico **390–391**
   description of 36, 390, 391
   El Sótano 390
   exploration of 391
   fauna of 390, 391
   geology of 390
   Hoya des Guaguas 390, 391
   Sótano de la Cuesta 391

Golondrinas *(Continued)*
  Sótano del Rancho del Barro 390
  speleogenesis 391
Golubinskaya, Pinega gypsum caves, Russia 589
Goluboje Ozero spring, Urals 629
Gomera, La, Canary Islands 179
Gomes, Nuno 303
Gondolin, South Africa 19
Goose Arm karst, Canada 177
Goraknath Cave, Nepal 623
Gordale, England 785
Gordon, Bert (*The Spider*) 359
Gordon, George (*Caverns of Copan*) 40
Gorham's Cave, Gibraltar 383, 384
Gorlo Barloga Cave, Georgia 201, 663
Gort lowland, Ireland 477
Gostry Govdy, Ukraine 746
Gouffre AN8, Pierre Saint-Martin, Spain 586, 587
Gouffre Berger, France 239, 302, 371, 525, 754–755
Gouffre de la Pierre Saint-Martin, France/Spain 586
Gouffre de Padirac, France 371, 373
Gouffre de Pierre Saint-Martin, France 554
Gouffre des Bourrugues (B3), Pierre Saint-Martin, France 586
Gouffre d'Esparros, France 373
Gouffre des Partages (M413) cave, Pierre Saint-Martin, France 585, 586
Gouffre Des Vitarelles, France 302
Gouffre du Couey Lotdge, Pierre Saint-Martin, France 586
Gouffre Gorgothakas, Crete 525
Gouffre Lonné Peyret, Pierre Saint-Martin, France 586
Gouffre Martel, France 371
Gouffre Mirolda Cave, France 204, 525
Gouffre Muruk, Papua New Guinea 105, 525
Gouffre Vallier, France 756
Gough, William 726
Gough's Cave, Mendip Hills, England 162, 504
Gournier cave system, France 755, 756
Government Quarry Cave, Bermuda 768
Grab-Ruda, Croatia 699
Grafsky Proval Cave, Georgia 202
Graham Cave, Missouri, United States 43, 44
Grampian Speleological Group 166
Gran Canaria Island, Canary Islands 179
Grand Canyon National Park, Arizona, United States **392–393**
  Ah Hol Sah 392
  Blue Springs artesian spring 392–393
  breccia pipes, growth of 392
  Dante's Descent 392
  DeMotte Park 393
  East Fence artesian spring 392
  geology of 391–392
  Havasu artesian spring 393
  Horseshoe Mesa Caves 392
  hydrology 392–393
  mining of 392
  Redwall caves 392, 393
  Roaring Springs Cave 393
  speleogenesis 392
  Tapeats Cave 393
  Thunder Cave 393
Grand Caverns (Weyer's Cave), Virginia, United States 50, 534, 704
Grand Meeander Cave, Colorado, United States 722, 723
Gran Dolina, Spain 119, 564
Grand Veymont Peak, France 754
Granier cave, France 326
granite, caves in 301, 650
Grassow, Mathias 534
Grassy Cove, Tennessee, United States 599
gravity
  anomalies and geophysical detection, caves and karstic voids 377–378, 379
  effect on crevice caves 251
gravity-sliding caves *See* caves, crevice.
gravity springs 699
gravity tectonics caves 650, 652
Great Caucasus, Georgia 200–201
Great Cave, Niah, Malaysia 108
Great Close Mine, England 737
Great Crack, Hawaii, United States 251
Greater Antilles, The 189
Greater Blue Mountains World Heritage Area, Australia 121
Great Expectations (Great X) Cave, Wyoming, United States 751
Great Northern Peninsula, Newfoundland 177
Great North Expedition 662
Great Rift, Idaho, United States 251
Great Saltpetre Cave, Kentucky, United States
  and cave surveying 715
  and gunpowder 411
Greece
  Agia Parajkevi Skotinu, religious views of 623
  Agios-Georgios Cave 340
  Argostoli sea-mills 338
  bauxite production in 135, 560
  Cave of Antiparos 208–209, 689
  Cave of Dionysos 208
  Cave of Marmarospilia 208
  Cave of the Dragon 208
  Cave of the Naiads 208
  Corycian Cave 208
  Delphi, religious views of 623
  Drogarati Cave 533
  Dyros Cave 339
  Eileithya cave shrine 208, 623
  Gouffre Gorgothakas, Crete 525
  Hermes Cave 207
  history of exploration 339
  Ida, Mt 623
  Kamares Cave, religious views of 623
  Kokkines Petres (Petralona Cave) 339
  Loutraki-Almopia Caves 339–340
  Maara Cave 339
  Mallorca's Cave 208
  Panagia Eleousa Cave 208
  Samaria, karst canyons of 338
Greeks, ancient and caves 86, 207, 208
Greeley, Ronald 766
Green Cave, Gunung Mulu National Park, Malaysia 532
Greenhorn Cave System, California, United States 53, 722
Green River, Mammoth Cave, Kentucky, United States 497
Gregory National Park, Australia 123

greigite 530
Gremuchaya Cave, Ukraine 253
Grerat-es-Salam, Libya 567
grikes 216, 471
Grim (Matthews) 360
Grot Proskuryakova Cave, Siberia 646
Grotta del Fiume-Grotta Grande del Vento, Italy 374
Grotta dell'Elefante Bianco, Italy 117
Grotta del Mezogiorno-Grotta di Frasassi, Italy 374, 375
Grotta di Fiumelatte, Italy 228
Grotta Gigante, Kras, Slovenia 486
Grotta Grande, Italy 86
Grotta Pavese 86
Grotte Albert, Madagascar 495
Grotte Brudour, France 755
Grotte Casteret, Spain 327
Grotte d'Arphidia, Pierre Saint-Martin, France 586, 587
Grotte de Cabrespine, France 373
Grotte de Chauve-Souris, Madagascar 494
Grotte de Gournier, France 756
Grotte de la Cigalère, France 373
Grotte de la Clamouse, France 373, 338
Grotte de la Luire, France 754
Grotte de l'Asperge, France 373
Grotte de Limousis, France 373
Grotte de Lombrives, France 533
Grotte de Miremont, France 715
Grotte de Pré-Létang, France 755
Grotte des Crocodiles, Madagascar 494
Grotte des Esplunges, France 410
Grotte des Eyzies, France 410
Grotte du Docteur Cave, Belgium 138, 139
Grotte du Mont Marcou, France 373
Grotte Louise-Grotte Mickey, Canada 538
Grottenhof Residenz 86
Grotte Père Noel, Grottes de Han, Belgium 342
Grotte Pousselière, France 373
Grottes de Goyet Cave, Belgium 138, 139
Grottes des Rois, Madagascar 494
Grotte Valerie, Northwest Territory, Canada 436, 538
Grotte Walou Cave, Belgium 138, 139
Grotto des Remouchamps, France 704
Grotto of the Nativity, Israel 208
groundbeams, use in construction 247
ground beetles *See* Coleoptera.
ground ice suffosin and cave formation 356
ground-penetrating radar (GPR) and geophysical detection, caves & karstic voids 378
groundwater
    aquifer
        anisotropy 395–396
        fracture aquifers 396
        karst aquifers 396, 481
        phreatic zone 396–397
        residence times 394
        vadose zone 396
    borehole hydrology **397–398**
    chemistry **213–215**, 476
    continental, colonization of 236
    dynamics 394–396
    ecosystems 244, 246, 315–316
    exploitation 482
    fauna 236, 315–316
    flow
        aquifers, principle pathways of 399, 400
        laminar 187
        modelling 671, 673
        regulation of 362
        turbulent 187
    karst 201
        testing of 214, 400, 479
    karst, chemistry of 201, 213–215
    and lava tube caves 762
    management of 394
    models
        conceptual **399–400**
        mathematical **400–401**
    pollution
        dispersed **403–404**
        point sources **404–405**
        remediation **405–406**
    protection **408**
    recharge 368, 394, 481
    tracing *See* water tracing.
grouting, use in karst 247–248
Gruber, Tobias 601
    Briefe hydrographischen und physikalischen Imhalts aus Krain 211
Gruna da Água Clara, Brazil 55
Grund, Alfred 288, 478, 670
Grupo Espeleologico Mexicano 36
Gruta Casa de Pedra, Brazil 56, 57
Gruta da Bocaina, Brazil 55, 611, 612
Gruta das Bromélias, Brazil 611, 613
Gruta de los Guácharos, Peru 55
Gruta del Rio San Geronimo, Mexico 37
Gruta de Palmito, Mexico 36
Gruta de San Pedro, Brazil 55
Gruta de Torrinha, Brazil 522
Gruta do Janelão National Park, Brazil 93, 134
Gruta do Padre, Brazil 55
Gruta Huagapo, Peru 55
Gruta Planaltina, Brazil 57
Gruta San Miserato, Brazil 55
Gruta Umajalanta, Brazil 55
Gruto do Centenário, Brazil 55, 611, 612
grutus de Juxtlahuaca, Mexico 37, 92
Gua Air Jernih (Clearwater Cave), Sarawak, Malaysia 531, 532, 554
Gua Batu Neraka, Indonesia 631
Gua Cha, Malaysia 108
guácharo 130, 190, 191, 366
Guadakskoy Caves, USSR 714
Guadalupe caves, New Mexico, United States
    ramifying patterns of 193
    speleogenesis 194
    speleothems 193
Guadalupe Mountains, New Mexico, United States 192
Gua Gunung Runtuh, Malaysia 108
Gua Kulit Supit, Sarawak, Malaysia 532
Gua Lebak Bareng, Gunung Sewu, Java 642
Guangxi karst, China 96, 98, 217, 224, 241, 423

Guaniguanico mountain system, Cuba 272
guano **409–410**
   accumulations, size of 409
   ammonia and 409
   Carlsbad Caverns, New Mexico, United States
     accumulations in 409
   dating of 409
   fauna associated with 410
   food resources 367
   formation of 513
   fruit bat guano, characteristics of 409
   Mammoth Cave, Kentucky, United States 500
   mining 409
     Carlsbad Caverns National Park, United States 194
     Caribbean Islands 189, 191, 192
     Jamaica 235
     United States 194
     See also gunpowder.
   organic resources in caves 547
   and paleoclimatic information 409
   Salukkan Kallang, Indonesia 631
   stratigraphy 409
   term, origin of 409
   vampire bat guano, characteristics of 409–410
guanobites 366
Gua Salukkan Kallang, Indonesia 631
Gua Salukkan Kallang Towakkalak (SKT), Indonesia 631
Gua Tambun, Malaysia 108
Guatari Cave, Italy 339
Guatemala
   Actun Kan (Jobitzina) 42
   Chiquibul Cave System 37
   Cueva de las Pinturas 42
   El Duende Pyramid 624
   Lanquin Cave, 42
   Naj Tunich 37, 40, 92
   Parque Nacionale Laguna del Tigre 620
   Sistema del Rio Candaleria 37
   stygobitic fauna **38**
   Tikal 37
   troglobitic fauna **38**
   Utatlan 42
   Xibalba 37
Gua Wattanang, Indonesia 631
Gučetić, Nikola 288
gudgeon *See Milyeringa veritas.*
guide-wire radio and cave communication **238**
Guiers-Mort Cave, France 284, 371
Guizhou karst, China 96, 218, 423, 734
gull caves *See* crevice caves.
gunpowder **410–411**
   mining of 410–411
   in South America 59
Gunung Mulu National Park, Sarawak, Malaysia 468, **531–533**
   Benarat Caverns 532
   Cave of the Winds 531
   Cobweb Cave 532
   dating of 554
   Deer Cave 532, 533
   flora/fauna of 533
   geomorphology 105, 531, 532–533
   Green Cave 532
   Gua Air Jernih (Clearwater Cave) 531, 532, 554
   Gua Kulit Supit 532
   karren fields 473
   Melinau River 533
   morphology 531, 532–533
   paragenesis 571
   pinnacles, light-oriented 581
   Pinnacles, The 532
   Sarawak Chamber 204, 308, 526, 532
   speleogenesis 531–532
   wall notches of 531
Gunung Sewu cone karst, Java 104, 241, 278, **641–642**
   history of exploration 641–642
   human impact on 641
   hydrology 642
   speleogenesis 641
   topography 641
Gupteswary Cave, Nepal 623
Guri Cave, Philippines 109
Gutersteiner Falls, Germany 737
Guwa Lawa, Indonesia 108
Gvonzdetskij, N.A. 605
Gvozdetsky, N. 662
Gyokusan-Do, Japan 96
Gypsies and cave use 427
gypsum
   and bauxite filters, sewage treatment 136
   and biomineralization 636
   black 373
   breccia pipes, formation of 661
   and breccias 660
   Canada 176, 178
   Carlsbad Caverns, New Mexico, United States 194
   caves 345
   Cigalère, Grotte de, France 327
   Cueva de Villa Luz, Mexico 759
   deposits in caves 194
   dissolution, nonlinear rate law 678
   erosion rates **324**
   evaporite karst 220, 343, 344, 346
   formation of 513
   Kugitangtau, Turkmenistan 273
   Majskaya Cave, Georgia 201
   Mammoth Cave, Kentucky, United States 495
   Nullarbor Plain, Australia 546
   oxygen fractionation dating 556, 557
   paleokarst 330, 559
   Pinega gypsum caves, Russia 588–589
   principle karst rock 474
   speleothems 692
   Spitsbergen 342
   Tăuşoare Cave 329
   Valea Rea Cave, Romania 329
gypsum karst 95, 220, 343, 344, 346
   Germany 343
   Turansky Plain 95

Turkey 742
  See also evaporite karst.

habitats
  anchialine **64–65**, 712
    Amphipoda 65, 258, 260
    Belize 38
    Cape Range, Australia 182
    continental 182
    Crustacea 65, 254, 256, 257
    defined 64, 712
    distribution 64, 65
    fauna of 65
    Hawaii, United States 417
    Remipedia 65, 126, 156, 183
    stratification in 64–65
    thalassostygobites 549
    thermal gradient in 707
  aquatic hypogean **711–712**
    euhaline habitats 712
    hyporheal zone 711
    interstitial waters 711
    mixohaline/freshwater habitats 712
    percolation waters 712
    phreatic zone 711
    thermal hypogean waters 712
  aquatic interstitial **454**, 711
    adaptations 454, 455
    evolution, regressive 454, 455
    factors influencing environment 454
    fauna morphology 454–455
  classification **709–712**
  dark cave/crevicular 710–711
  entrance habitats **318–319**
  hypogean **709–712**
    biodiversity in 141–142
    classification of 709–712
    ecological factors influencing 706
    interrelations between 710
  phreatic 258, 711
  subterranean **709–712**
    classification 709
  terrestrial interstitial 179, 709
  thermal water 712, **724–725**
Hacquet, Balthasar and karst hydrology 478
Hacquet, F. 211, 478
Hadenoecus subterraneus 46, **500**
Hades Cave, Tanzania 22
Hadesia vasiceki 448
Haemopis caeca 66, 529
haemorrhagic fever 294
Hagengebirge karst system, Austria 173
Haggard, Rider, (*King Solomon's Mines*) 359
Hains, Ben 579
Haiti, cave and karst areas of 189
Haitón del Guarataro, Venezuela 53
Halesi Cave, Nepal 623
halides **511–512**
halite
  evaporite karst 343, 344, 345, 346

  formation of 511–512
  Nullarbor Plain, Australia 546
  speleothems 692
  Tăușoare Cave 329
Halla, Mount, South Korea 212
Halliday, William R.
  Caves of California 766
  Caves of Washington 766
  history of speleology 52, 663
Hall of the Bulls, Lascaux, France 757
halloysite
  and sulfuric acid speleogenesis 194
Ha Long Bay, Vietnam **413–414**
  archaeological research 414
  fauna 414
  flora 414
  geology 413
  Hong Gai 414
  human use of 414
  speleogenesis 413–414
Hamann, O. (*Europaeische Hoelenfauna*) 149
Hammamat Ma'in, Jordan 115
Hams-Oyeek, Turkmenistan 273
Han-dul Cave, South Korea **212**
Hang Khe Ry, Vietnam 102, 485
Hang Vom Cave, Khammouan, Laos/Vietnam 485
Hanimec Cave, Slovakia **183**
Hankajev Canal, Škocjanske jama, Slovenia 654
Hanke, Anton 350, 655
Hanson, M. 527
hantavirus 294
Haptolana spp. **183**
Harasib Cave, Namibia 21
hardness of water, testing 214
harnesses and cave exploration 354
Harpacticoida 259, 260
Harrison's Cave, Barbados 190
Harrisson, Barbara 108
Harrisson, Tom 108
Hartung, Georg 764
harvestman spiders *See* Opiliones.
Hasenmeyer, Jochen 302
Hatat Lohum, Oman 115
Hawaiian Volcano Observatory 765
Hawaii lava tube caves, United States **415–418**, 765, 766
  adaptations of fauna 417
  dumping, unlawful 415, 416
  evolution of fauna 417
  geology 415
  human influence on 417–418
  nutrients, sources of 417
  speleogenesis 415
  troglobitic fauna in 417
  and vulcanospeleology 765–766
Hawaii Volcanoes National Park, United States 415
Hayonim Cave, Israel 195
Haywood, John (*Natural and Aboriginal History of Tennessee*) 81
Hazelton, Mary 163, 166
heavy metals poisoning 294
Hei Yau Dong, China 285

Helens, Mount St, Washington, United States 591, 592, 762
helictites 194, 693–694
   Kugitangtau, Turkmenistan 273, 275
   See also speleothems
Hellbrun water garden 87
Heller, Martin 716
Hellyer, Henry 127
Helm, Ennis Creed "Tex" 579
helmets and cave exploration 353
Helmholtz resonator, caves as 230
Hematite Cave, Oregon, United States 249
Hemingway, Ernest (*For Whom the Bells Tolls*) 206
Hemiptera
   Heteroptera 452
   Homoptera 143, 452
   morphology 452
   Reduviidae 180
Hendea myersi cavernicola 4
Henderson, John 127
Henne-Morte Cave, France 371
herbicide poisoning 294
herbicides
   effects on cave biota 244
   pollution of groundwater 403
Hermannshöhle, Germany 579
Hermes Cave, Greece 207
Herodotus and cave tourism 725
Herzegovina
   Biograd Ponor 293
   Bokanjačko Polje 291
   Buna 699
   Bunica 699
   Buško Blato polje 291
   Caprazliha ponor 291
   Cerničko polje 293
   Dabarsko polje 293
   Doljašnica ponor 291–292
   Dugo Polje 291
   Duvanjsko Polje 291
   Fatničko polje 293
   Glamočko polje 291
   Imotsko Polje 291
   Kazanci ponor 291
   Ključka River 293
   Kočerinsko Polje 291
   Kovači Ponor 291
   Kupreško polje 291
   Lukavačko polje 293
   Nadinsko Polje 291
   Neretva Plain 338
   Nevesinjsko polje 293
   Obod Estavelle 293, 739
   Opaki ponor 291
   Popovo Polje 291–292
   Posuško Polje 291
   Proždrikoza ponor 291
   Sinjski ponor 291
   Slato polje 293
   Stara Mlinica ponor 291
   Trebišnjica 699
   Trebišnjica Hydrosystem 278, 292
   Trebišnjica River 291–292
   Trebistoho Polje 291
   Trnsko Polje 291
   Veliki ponor 291
   Vir Polje 291
   Vjetrenica 141, 289
      biospeleology **759–760**
      conservation 760
      geomorphology 759
      hydrology 759
   Zalomka River 293
Heslop, Linda 88
heterochrony 348
Heteroptera
   Cimicidae 452
Hexapoda *See* Insecta, Apterygota.
Heyden, Doris 40
Heyerdahl, Thor 765
HeyPhone and cave communication 239
HH Cave, South Island, New Zealand 543
Hickmania troglodytes 73
Hidden River Cave System, Kentucky, United States 405, 406, 498–499
Hierro, El Island, Canary Islands 179, 180–181
Higher Kiln Quarry cave system, England 166
high-frequency (HF) walkie-talkies and cave communication 238
highways on karst **419–420**
   construction methods 419
   discovery of caves 419
   effects of 419, 420
   obstacles to 419
   preservation of caves 419–420
   voids, prescence of 419
Hill, Rowland 703
Hillaire, Christian 82
Hill-Caves of Yucatan (Mercer) 40
Hinduism and ritual use of caves 622, 623
Hirlatzhöhle, Austria 174, 326, 525
Hirudinea **66**
   Croatobranchus mestrovi 67
   Dina absoloni 67
   in Dinaric karst 289
   Erpobdellidae 67
   Haemadipsidae 67
   Haemopsis caeca 529
   Mooreobdella microstoma 315
Hispaniola
   history of exploration 192
   karst areas of 189
Histoplasma capsulatum and histoplasmosis 293, 508
histoplasmosis 293, 294, 508, 509
Hoabinhian culture 107, 108, 167
Ho Ba Ham caves, Ha Long Bay, Vietnam 413
Hobbit, The (Tolkien) 206
Höhlensektion des Naturwissenschaftlichen Klubs, Moravia 336
Hohler Fels Schelklingen caves, Germany 333
Hölloch, Switzerland 326, **420–422**
   classification of 420, 421
   depth of 525

geography of 420–421, 422
geology of 421
history of exploration 420
protection of 422
speleogenesis 421–422
Homer (*Odyssey, The; Iliad, The*) 205
Homo antecessor 119
Homo calpilcus 384
Homo erectus 18, 19, 196
    China 221, 222, 223
    Greece 339
    Thailand 107
Homo ergaster 18, 224
Homo habilis 18, 224
Homo heidelbergensis 119
Homoptera
    Cixidae 25
    Fulgoridea 452–453
    Fulgoroidea 126
    Hypochthonellidae 25
    terrestrial cave habitats, biodiversity in 143, 144
Homo sapiens neanderthalensis 195, 333, 384, 642
Homo sapiens sapiens 19, 195, 196, 221, 224
Honduras
    Talgua Cave (Cave of the Glowing Skulls) 42
    Tauleve 42
Honeycomb Hill Cave, South Island, New Zealand 542, 543
Honey Creek Cave, Texas, United States 751
Hongshan karst spring, China and karst hydrology 478
Hongshui River fengcong karst, China **422–423**
    cave systems of 423
    Dahua 422, 423
    erosion of 423
    geology 422
    morphology 423
    Qibailong 422, 423
    speleogenesis 422, 423
    water resources of 423
Hoq Cave, Yemen 115
horizontal transition colonization 236
hormesis and speleotherapy 698
Horn, J.M. 527
Horse Cave, Kentucky, United States 405, 406
Hose, Louise 52
hotspot theory of geologic origin 179
hotspots of diversity
    hypogean waters **141–142**
    terrestrial **143–144**
Hovey, E.O. 49
Hovey, Horace C. 49, 51, 158
Howarth, Frank 110
Howe Caverns, New York, United States 360, 579
Howes, Chris 466
Howieson's Poort MSA technology, South Africa 19, 20
Hoyo del Aire, Colombia 54, 60
Hrasky, 351
Hronek, Clarence 52
Hrvatsko speleološko društvo, Croatia 336
Hualalai volcano, Hawaii, United States 415, 416
Huangguoshu, China 736

Huangguoshu Falls, China 218
Huanglong, China 187, **423–424**
    morphology 423
    travertines, research of 424
    vaterite 424
Huanglong Ravine, China 187
Huang Wanbo 222
Huautla Cave System, Mexico
    exploration, current 426
    history of exploration 425
    Sóntano de San Augustín 425
    speleogenesis 425–426
    Systema Cheva 425, 426
    Systema Puficicatión 425
Hudson Bay Lowlands, composition of 175–176
Huliba Cave, Caribbean Islands 190
Hulu Cave, China 223
Huludong, China 223
human impact on soil erosion 658
humans, anatomically modern *See Homo sapiens sapiens.*
human use of caves
    Activities, classification of 426
    Atapuerca Caves, Spain 119
    Baradla Cave, Hungary 28, 29
    Belgium 138
    Creag nan Uamh caves, Scotland 160
    deep caves 427
    economic use 426, 428
    factors determining use 427
    Ghar Dalam Cave, Malta 339
    Gibraltar 384
    Ha Long Bay, Vietnam 414
    Indian subcontinent 442–443
    Indonesia 109
    Kutikina Cave, Australia 122, 124
    Lascaux, France 757
    legitimizing settlement 41
    Mammoth Cave, Kentucky, United States 499
    military use **509–510**
    Niah Cave, Sarawak, Malaysia 108
    North America 43
    Paleolithic 195, 334
    residential use 426, 427
    ritual use 426, 427–428, 531–533
    social use 427, 531–533
    sources of raw materials, use of 426
    South America 59
    Southeast Asia 103–104
    as storage 426
    tourism 428, **725–726**
    as waste repositories 426
Humboldt, Alexander von 57, 131, 191, 410
    biography of **152**
    and Caribbean Islands 191
    Personal Narratives of Travels to the Equinoctial Regions of America during the years 1799–1804 59
Hume, Hamilton 127
humidity sensors, measurement using 228
Humpleu-Poienita system, Romania 329
hums 242

Hungary
  Aggtelek Caves **28–29**, 715
  Baradla Barlang 533
  Baradla-Domica Cave System **26–27**, 29, 331, 620, 698
  bauxite production in 135–136, 560
  Beké Cave 26
  Beremend Crystal Cave 561
  Čertová pec Cave 29
  Csorgo-Lyuk 722
  Dobšina Ice Cave 25, 27
  Domica Cave 28–29
  Kossuth Cave 26
  Pál-völgy cave 332
  Rákóczi Cave, 26
  and speleotherapy 698
  Stratenska Cave System 27
  Szabadsag Cave 26
  Vass Imre 26
Hunter Island, Australia 123
huntite 194, 513, 692
Huntley's Cave, Scotland 534
Hutton, John 669
Hyaena Den, England 79, 162
Hyatt, Alpheus **152**
hydra *See* Cnidaria.
Hydracarina **74**
  Eylaioidea 74
  Halacaroidea 74
  Hydrachnellae 74
  Hydryphantoidea 74
  Stygotrombioidea 74
  Trombioidea 74
hydration and cave formation 204
hydraulic action as coastal process 231
hydraulic gradient
  and doline formation 305
  laminar flow equation 429
  turbulent flow equation 429
hydraulics of caves **429–430**
  conduits, discharge of 429
  conduits, recharge of 429
  defined 429
  fissures, turbulent flow 429
  flow patterns above water table 430
  flow patterns of wells 430
  flow reversal in 429–430
  hydraulic gradient, laminar flow equation 429
  hydraulic gradient, turbulent flow equation 429
  rapid runoff fed caves 430
  Reynolds numbers, equation 429
hydrocarbon poisoning 294
hydrocarbons in karst **431–432**
Hydrocena cattaroensis 516
hydrochemical budgeting method, erosion rates 322
hydrogen sulfide 529
  in anchialine habitats 64, 181
  in Cueva de Villa Luz, Mexico 146
  in Movile Cave, Romania 529–530
  sulfur-oxidizing bacteria 506, 507, 713
  See also sulfuric acid speleogenesis

hydrogen sulfide poisoning 293
hydrogeology
  defined 394
  groundwater models, conceptual **399–400**
  groundwater models, mathematical **401–402**
    distributed parameter 401, 402
    MODFLOW 401, 402
    types of models 401
hydromagnesite
  described 690
  Kugitangtau, Turkmenistan 275
  precipitation of 692
hydrosulfuric acid dissolution 682
hydrothermal speleogenesis 668, 682
hydroxylapatite 513
hydroxylellestadite 329
hydrozincite 512
Hydrozoa
  Bougainvilliidae
    Velkovrhia enigmatica 289
hygogean habitats **709–712**
hygrometers, measurement using 228
Hymenoptera 25, 47, 143
  Solenopsis invicta 47, 245
Hymenosomatidae 632
Hyop-jae cave system, South Korea 212
hyperthermia 293
hypogenic speleogenesis 192, 668
  mechanisms of 355–356
  sulfuric acid 157, 192, 574, 668
hyporheic zone 454
hypothermia 1, 293

Ibex Cave, Gibraltar Caves, Gibraltar 384
Ibitipoca State Park, Brazil 613
Icaronyctaris index 225
ice caves 204, **385–387**
  ice in caves **436–437**
    climatic implications of 437
    Coultard Cave, Canada 436, 437
    Dachstein-Rieseneishöhle, Austria 174
    defined 435
    Eisriesenwelt, Austria 173, 174
    Grotte Valerie, Northwest Territory, Canada 436
    ice types in caves, list of 436
    isotopic model, warming and cooling trends 437
    Serendipity Cave, Rocky Mountains Canada 437
    speleogenesis 435–437
    static *vs.* dynamic 435
    Volcano Room, Q5, Vancouver Island, Canada 436
Iceland
  Rivière de Kverkfjöll 342
  Surtshellir 764, 765
  Thrinukagigur 762, 764
Ice Tube Growling Swallet system, Australia 122
Ida, Mt, Greece 623
Ignatjevskaja Cave, Urals 629
Iliad, The (Homer) 205
Iljukhin, Vladimir 663
Iljukhina Cave system, Georgia 201, 487, 525

Illu River Cave, Irian Jaya **238**
immersion foot 294
Imotsko Polje, Bosnia/Herzegovina 291
Imperato, F. 478, 655
In the Cellars of the World (Neville) 359
Inception Horizon Hypothesis (IHH) **438–440**
inception of caves **437–441**
    defined 437–438
    diagenesis 438
    drains 439, 441
    fluid migration 441
    gestation 438
    hydraulic gradients 440–441
    Inception Horizon Hypothesis (IHH) 438–440
    inception horizons 438, 439
    speleogenesis, evaporitic successions 439–440
    spring formation 441
Incógnita, Cueva de la Cuba 272
Index of Biological Integrity (IBI) 520
India **443–444**
    Ajanta Cave 623, 624
    Amarnath Cave 443, 623
    Belum Guhalu 443
    Chomolungma (Mt. Everest, Sagarmatha) 622
    Elephanta caves 623
    Ellora Cave 623
    Ganges River 623
    Ganges River, religious views of 623
    geology 443
    history of exploration 443
    human use of caves 442–443
    Krem Kotsati-Umlawan 443
    Lower Swift Hole 443
    Rakhiot Peak Cave 442
    Siju Cave 443
Indiana Jones and the Last Crusade (Spielberg) 359
Indian Cave, Tennessee River, United States 50
Indian Creek Cave, Edwards Aquifer, United States 314
Indian Echo Caverns, Pennsylvania, United States 379
Indohyinae, Cape Range 182
Indonesia
    archaeological caves 108–109
    burials in caves 108, 109
    Gua Batu Neraka 631
    Gua Salukkan Kallang 631
    Gua Salukkan Kallang Towakkalak (SKT) 631
    Gua Wattanang 631
    Guwa Lawa 108
    history of exploration 113
    human use of caves 109
    Leang Burung 2 109
    Lene Hara Cave 109
    Lubang Kapa Kapasa 631
    Lubang Tomanangna 631
    Mateullus troglobioticus 632
    Nature Reserve of Karaenta 632
    palynology 567
    Salukkan Kallang, biospeleology **631–632**
    Sulawesi, tower karst 105
    Uai Bobo 1 109
    Uai Bobo 2 109
    Ulu Leang 1 109
    Ulu Leang 2 109
induced polarization method 379
induction radio and cave communication 239
industrial/sewage releases 406
Inficionado karst, Brazil 612
Ingleborough Cave, England 715
Insecta
    America, South 58
    Apterygota 24, 47, 445–447
        See also individual entry for this order.
    Cape Range 181
    Coleoptera **447–450**
        See also individual entry for this order.
    Hadenoecus subterraneus 500
    Hymenoptera 25, 47, 143, 245
    hypogean fauna in 141
    North America 47
    Orthoptera **451**
        See also individual entry for this order.
    Pterygota 25, 143, **451–453**
        See also individual entry for this order.
    Siphonaptera 25, 143
    Southeast Asia 110
    Thysanura 24, 143
insect bites and stings 294
Integrated Conservation and Development Projects 246
International Biosphere Reserves 495
International Caver 465, 466
International Cave Rescue Commission 1
International Congress of Speleology 336
Internationale Bibliographie für Speläologie 466
International Glaciospeleological Survey 387
International Journal of Speleology 465
International Show Caves Association (ISCA) 728
International Society of Speleological Art 88
International Speleological Union (UIS) 371, 465
International Underwater Cave Rescue and Recovery team 2
International Union for Conservation of Nature and Natural Resources (IUCN) 244, 245
interstitial aquatic habitats **454**, 711
interstitial habitats, terrestrial 179, 709
intrastratal evaporite karst 344
Inugamori-no-ana Cave, Japan 30
invertebrates **457–459**
invertebrates, interstitial fauna **458–459**
Inzhenernaya cave, Ukraine 253
ionic disassociation 300
Ionising Radiation Regulations, United Kingdom 618
Iran 115–116, **460–461**
    cave research in **460–461**
    Ghar Alisadr 116, 461
    Ghar Parau 115–116, 461
    Ghar Sarab 461
    Ghori-Galeh 461
    Katelahkor 461
    Margoon Waterfall Spring 702
    mineral deposits **461**
    Salman Farsi Dam 278

Iran *(Continued)*
  Shapour Cave 461
  Solomon's Prison 737
  tourist caves **461**
  Tri Nahaci Cave 461
Iraq
  burials in caves 167, 643
  geomorphology 115
  Shanidar Cave 115
    archaeology **642–643**
    burials in 167, 643
    Neanderthals 642–643
    palynology 567, 643
    religious views of 623
Ireby Fell Cavern, England 441
Ireland
  Annah 162
  archaeological/paleontological caves **160–162**
  Ballybunion 606
  Ballynamintra caves 162
  biospeleology **163–164**
  burials in caves 162
  Burren glaciokarst **169–171**, 341, 473, 581, 659
    Aillwee Cave 171
    Doolin cave system 171
    Glencurran Cave 171
    Kilcorney-Meggagh depression 170
    Pol-an-Ionain 171
    Polldubh cave 171
    Poll na gCeim 171
    Pollnagollum cave system 171
    Vigo Cave 171
  Caherguillarmore 162
  Carren depression, Burren glaciokarst 170
  Castlepook Caves 162
  chalk aquifers, fauna of 163
  Clashmealcon Cave 510
  Country Fermanagh 341
  County Clare 232, 341
  distribution of fauna 163
  Doolin Cave 341
  Edenvale caves 162
  Fergus River Cave 716
  Foley Caves 162
  Gort lowland corrosion plains 477
  history of exploration **164–166**
  Kilgreany caves 162
  Killuragh Caves 162
  Lough Cong 341
  Newhall/Barntick caves 162
  palynology 566
  Poll na Mona 591
  protection of fauna 164
  Shandon caves 162
  threats to fauna 164
Irian Jaya
  Biak Caves
    strategic use of 427
  Kutierleruk Cave 107
Iron Gates Nature Reserve, Russia 588, 589

iron hydroxides 636
  Grotte de la Cigalère, France 373
  Ukraine 748
iron ore
  paleokarst 330, 560
  placer deposits 514
iron oxides
  and biomineralization 636
  formation of 511
iron sulfide minerals 530, 714
Isha Baidoa tufa, Namibia 23
Ishceevskaja Cave, Urals 628
Iskender-i-Birkilin cave, Turkey 207
Isla de Mona *See* Mona, Isla de.
Islam
  and ritual use of caves 622
  Sof Omar, Ethiopia 655–656
Isler, Olivier 302, 303
Isopoda **263–264**
  anchialine habitats 65
  Anthuridea 263
  Asellidae 263, 264
  Asellota 263
  Australia 182
  biodiversity 143
  Canada 200
  Cirolanidae 24, 46, 263, 264
  description of 263
  distribution of 263–264, 370
  Edwards Aquifer, United States 315
  Microcerberidae 263, 264
  Microparasellidae 263
  Phreatoicidea 263, 264
  reproductive strategy of 264–265
  Southeast Asia 110
  Sphaeromatidae 263, 264, 289
  Stenasellidae 24, 263, 724
    Balkanostenasellus skopljensis 724
    Protelsonia hungarica 724
  Trichonisidae 47
isotopic disequilibrium method, karst dating 282
Israel
  Aqevot Cave 638
  Amud Cave 196
  burials in caves 167
  Carmel Caves **195–197**
  Es-Skhul cave 195, 196
  Et-Tabun Cave 197
  geomorphology 114
  Hayonim Cave 195
  Jamal cave 195
  Kebara Cave 196
  Mousterian culture 195
  Natufian culture 195, 196
  Qafzeh Cave 196
  Qumran Caves 208
  salt caves 208
  Sedom, Mount 300, 512, **637–639**
  Tabun Cave 167, 195
  Wad, El cave 195

Wad, El terrace 196
Wadi el-Mughara 195, 196
Issel, Arturo 339
Istra, hypogean fauna in 142
Istripura caves, Sri Lanka 445
Italy 325–327
  Abisso del Corno 117
  Abisso del Nido 117
  Abisso di Malga Fossetta 117, 118, 326
  Abisso Gortani 468
  Abisso Led Zeppelin 469
  Abisso Michelle Gortani 468
  Abisso Olivifer 525
  Abisso Paolo Roversi 525
  Abri Sassi system 117
  Alpi Apuane 327
  Asiago plateau **117–118**, 326
  Bigonda cave 117, 118
  Buso de la Rana 326
  Bussento 699
  Busso della Neve di Zingarella 117
  Calgeron cave (GB Trener Cave) 117
  Carrara marble 327
  Castellana, Grotta di 338
  Cogol dei Siori 117
  Cogol dei Veci 117
  Complesso del Colle delle Erbe 468, 469
  Complesso del Foran del Muss 468, 469
  Cucco, Monte 327
  Fighiera-Corchia 327
  Fiume-Vento caves 327
  Frasassi caves 327, **374–375**
    Ancona Abyss 374, 375
    Buco Cattivo 374, 375
    ecosystem 375
    fauna in 374, 375
    Grotta del Fiume-Grotta Grande del Vento 374
    Grotta del Mezogiorno-Grotta di Frasassi 374, 375
    history of exploration 374
    minerals of 375
    redox processes 375
    speleogenesis 375
    speleothems 375
    sulfide water cycle 374, 375
  Giacominerloch 117
  Grotta dell'Elefante Bianco 117
  Grotta di Postumia 336
  Guatari Cave 339
  history of exploration 336
  Kanin, Mt. Italy/Slovenia 468
  Kanin massif **468–470**
  Lamalunge Cave 339
  Lessini plateau 326
  Obelix 117
  Pala Celar 469
  Piaggia Bella 326
  Pozzo A 345
  Pulo di Molfetta 410
  Rio Stella-Rio Basino 345
  Spaluga di Lusiana 117
  Spipola-Aquafredda 327, 345
  Spluga della Preta 326
  Stella 699
  Stufe di Neroni (Nero's Oven), and cave surveying 714
  Timavo 699
IUCN *See* International Union for Conservation of Nature and Natural Resources.
IUCN Red List 63
Iulida
  Blaniulidae 535
  Glyphiulidae 535
  Iulidae 535
  Jarmilka spp. 535
  Mongoliulidae 535
  morphology 535
  Paraiulidae 535
Ivanov, A. 662
Ivanov, Boris 253, 662, 663
IVPP *See* Institute of Veterbrate Paleontology.

Jabal Akhdar, Oman 115
Jackson's Bay Caves, Jamaica 190
Jägerbrunntroghöhle, Austria 173
Jagger, Thomas 765
Jagodinskata, Bulgaria 329, 330
Jama du Dubašnici, Serbia 329
Jamaica
  Alligator Pond 735
  Bamboo Bottom 234, 235
  banana hole caves 189
  bauxite production in 135
  bauxite mining 135, 236
  Blue Mountains 189
  Cockpit Country **233–235**, 242
  Golding River Cave 190
  guano mining 236
  history of exploration 191
  Jackson's Bay Caves 190
  karst areas of 189–190
  Maroon wars, military use of caves 510
  Quashies River Cave 189–190
  Riverhead Cave 191
  Runaway Bay Caves 191
  Sleeping Pools 190
  Swanswick Formation 233
  Troy-Windsor Trail 235
  Windsor Great Cave 234, 235
Jamal cave, Israel 195
Jamarska zveza Slovenije, Slovenia 336
Jameo del Agua, Lanzarote 179
James, C.H. 579
Japan
  Akiyoshi Caves **29–30**
    Akiyoshi-do Cave 30
    Current-mark-no-ana Cave 30
    Inugamori-no-ana Cave 30
    Kuzuga-ana Cave 30
    Sano-ana Cave 30
    Taisho-da Cave 30
    Takaga-ana Cave 30

Japan *(Continued)*
  Akiyoshi Plateau 96
  Akka-Do 96
  Byakuren-Do 96
  Fuji, Mt 624
  geomorphology 96
  Gyokusan-Do 96
  history of exploration and research 99
  hypogean fauna in 142
  military use of caves 510
  Minafuku Dam 739
Japan Caving Association 99
Japan Caving (Japan Caving Association) 99
Japanese Vulcanospeleological Society **212**
Jarmilka spp. **535**
Jarrets d'Acier club, France 284
Jaschik Pandorry Cave, Siberia 629
Java
  Bribin Cave, Gunung Sewu 642
  Gua Lebak Bareng, Gunung Sewu 642
  Gunung Sewu cone karst 104, 241, 278
  Luweng Jaran, Gunung Sewu 642
Jean-Bernard Cave, France 372
Jeannel, René Gabriel 149
  biography of **153, 154**
  Biospéleologica 154
  Faune cavernicole de la France (Jeannel) 154
Jean Nouveau, France 326
Jefferson, Thomas
  archaeology 81
  and cave surveying 715
  and gunpowder 410–411
  Notes on the State of Virginia 50
Jeita Cave, Lebanon 208, **463–464**
  history of exploration 463
  musical performances in 533
  speleogenesis 463
  speleothems 464
  Upper Gallery 464
jellyfish *See* Cnidaria.
Jenkins, Wally 52
Jennings, Joseph Newell 382, 673
Jenolan Caves, New South Wales Australia
  and cave tourism 726, 731, 732
  history of exploration 127
  Lucas Cave 623
  paleokarst of 561, 634
  restoration 626
  Social and Environmental Monitoring (SEM) programme 519
Jester Cave, Oklahoma, United States 345, 751
Jewel Cave National Monument, South Dakota, United States
  dating of 57
  geomorphology 775, 776
  history of exploration 52
  length/depth of 525, 750
  minerals in 776
  speleogenesis 775
Jia Lanpo 224
Jiuzhaigou, China 187, **423–424**
John Day country, Oregon, United States 53

joint caves *See* crevice caves.
Joint Hole, England 303
joints, density of 186
Joly, Robert de 284, 371, 373
Jorcada Blanca, Picos de Europa National Park, Spain 585
Jordan
  geomorphology 114–115
  Hammamat Ma'in 115
Jordas (Yordas) Cave, England 87
Journal of Cave and Karst Studies 464
Journal of the Speleological Society of Japan 465
Journal of the Sydney University Speleological Society 465
Journal (Speleological Society of Japan) 99
Journey Into Amazing Caves (Perou) 360
Judd, Henry 129
Judds Cavern, Australia 129
jumars and cave exploration 354
Jumbo Gumbo, California, United States 491
Jungle Pot, Zimbabwe 21

K2, Siebenhengste, Switzerland 647
Ka 2, Nakanai Caves, Papua New Guinea 538
Kabal Group, Belize 37
Kach Gharra, Pakistan 445
Kačna Jama, Slovenia 184, 350, 486, 654
Kadziella styx 448
Ka Eleku cave, Hawaii, United States 415, 416
Kahf Aqabat Khushil (Seventh Hole), Selma Plateau, Oman 639, 640
Kahf Hamam, Abu Dhabi 115
Kahf Hoti, Oman 115
Kahf Khasha (Funnel Cave), Selma Plateau, Oman 639
Kaijende, Papua New Guinea **467–468**
  arête and pinnacle karst 467
  exploration of 468
  morphology 467–468
  solutional denudation rates 468
  speleogenesis 467
Kalate Egeanda, Papua New Guinea 89
Kal'tsitovaya Cave, Siberia 646
Kamares Cave, Greece 623
Kamchatka Expedition 662
kamenitza 216
  origin of term 288
Kamptozoa 459
Kanchanaburi, Thailand 167
Kanin massif, Italy/Slovenia **468–480**
  cave morphology 468–469
  exploration of 468
  geology 468
  hydrology 468
  speleogenesis 469–470
Kaninski podi, Slovenia 469
Kanung Cave, Khammouan, Laos/Vietnam 485
kaolin 330
kaolinite 301
Kapova Cave, Urals 629
Kapsia Cave, Peloponnese 340
Karake style, cave art 89
Karaman, S. 149

Karbonatitovaya Cave, Siberia 645
Karljuksky Cave, Tien Shan Central Asia 94
karren **470–472**
   classification 470–471, 472
   corridor (labyrinth) karst 472
   decantation runnels 471
   defined 470
   erosion rates, coastal 231
   flachkarren 472
   hyro-aolian
      Patagonia marble karst, Chile 572
   kamenitza 470, 472
   karrenfields 473
   kluftkarren (grikes) 471
   limestone pavements 473
   microrills 470
   morphology 470
   pinnacles 472–473
   rainpits 470
   rillenkarren 216, 470, 472
   rinnenkarren 216, 470
   rundkarren 471
   solution bevels 472
   spitzkarren 472
   trichterkarren 472
   trittkarren (heelsteps) 472
   tropical 232
   wall karren 471
karst **473**, **653**
   alpine **31–32**, 198, 200, 201–203, 220, 388
   bauxite **135–137**, 330, 332
   biokarstification **147**
   carbonate karst **184–186**
   cavities, classification of 253
   chalk 342
   chemistry of groundwater 201, 213–215
      See also groundwater.
   classification of 288, **473–474**, 475
   closed depression fills, study of 566–567
   coastal 215, **231–233**, 674
   cone karst **242–243**
   construction on **247–248**
   covered 475
   development 178, 185, 186, 254, 559
   dissolution 473
      carbonate rocks **296–298**
      evaporite **rocks 298–300**
   dolines **304–309**
   dolomite 103, 297–298
   economic significance of 173, **479–481**
   environmental impact assessments **319–321**
   epikarst 94, 305, 306, 559, 676
   erosion rates
      field measurements **321–323**
      theoretical models **323–325**
   evaporite **343–347**
   evolution
      change, study of and 480
      conduits 297, 299
      hydrologic cycle 475, 476–477
      modelling 476, 477–478
      processes of 475–476
      surface landforms 477
   fengcong karst 217, 220, 242
   fenglin karst 217, 242
   floral resources of **360–362**
   fluviokarst **363–364**, 474
   forests on **368–369**
   fossil 475, 559
   freshwater 215
   geophysical detection of **377–379**
   Gerichteter karst 242
   glaciated **388–390**
   glaciokarst 201, **385**, **387**
   gypsum 95, 220, 343, 344, 346
   highways on **419–420**
      See also karst, construction on.
   holokarst 474
   hydrocarbons in **431–432**
   hydrogeology 150, 394, 399–400
   hydrology, history of 191, **478–479**
   landforms 178, **281–283**
   military impact on 509
   mineral deposits in **514–515**
   mogotes 190
   monitoring 483
   morphometry of **526–528**
   paleokarst 216, 221, 330, 332, 475, **559–561**
   parakarst 363
   phytokarst 189, **581–582**
   pinnacle 104–105, 218
   pollution and 244
   polygonal 233, 363
   pseudokarst 475, **605–608**, 653
   quifeng karst 218
   regions, protected areas in 246
   rejuventated 475
   relict karst 475, 559
   resources and values **479–481**
   salt karst 343, 344, 345, 346
   sandstone karst 103
   silicate karst **449–652**
   speleogenesis 474, 652
      See also Speleogenesis.
   springs **699–703**
   stripe karst 178, **705–706**
   subjacent 475
   sulfide minerals in **713–714**
   syngenetic **717–719**
   tower 217, 241, **734–736**
   tunnelling in **738–740**
   valleys in **753–754**
   water resource management 478, **481–483**
Karst Belt, Puerto Rico 190
Karst Dynamics Laboratory, China 221
karstification
   phases, examples of 173
   and water table 308
Karstologia 372, 464, 465
Karst Waters Institute list of endangered species, 768

Kartchner State Caverns, Arizona, United States 229
   cave detection methods 379
   minerals in 513
Kasjan, Yury 487
Kaskadnaya, Siberia 646
Kastning, Ernst 49, 52
Katelahkor, Iran 461
Katzer, F. 288, 670
Kauai, Hawaii, United States 415
Kauhako Crater, Hawaii, United States 762
Kaumana Cave, Hawaii, United States 416
Kavakuna, Nakanai Caves, Papua New Guinea 538
Kaye, C.A. 673
Kazakhstan
   bauxite production in 135
   geomorphology 95–96
Kazanci ponor, Bosnia/Herzegovina 291
Kazumura Cave, Hawaii, United States 204, 415, 416, 525, 750, 760–761
Keban Dam, Turkey 278
Kebara Cave, Israel 196
Kef Aziza, Morocco 13
Kef El Kaous, Algeria 15
Kef Tikhoubaï, Morocco 13
Kef Toghobeit, Morocco 13
kegelkarst *See* cone karst.
Keith, Sir Arthur 17
Këktash, Siberia 629, 645
Keld Head, England 302, 438
Kelifely, Madagascar 494
Kelly Cave, Nullarbor Plain, Australia 546
Kenniff Cave, Australia 124
Kent's Cavern, England 77, 78, 82, 341
Kentucky cave shrimp *See Palaemonias ganteri*.
Kenya
   Banbuti Quarry 611
   Elgon, Mt 22, 25, 592
   Giggler's Caves 592
   Kitum Cave 22, 25, 294, 592
   Leviathan Cave 22
   Makingen Cave 22, 592
   Suswa, Mount 22, 561, 562, 763
Kerry, Charles Henry 579
Kerschner, Irvin (*Never Say Never Again*) 360
Kessler, Hubert 28
Key Cave, Alabama, United States 597
K-feldspar 301
Khammouan, Laos/Vietnam **483–485**
   Hang Khe Ry 485
   Hang Vom 485
   Kanung Cave 485
   karst, burials in 167
   Nam Hin Boun river 104
   Nin Ham Boun 113
   Phong Na System 485
   speleogenesis 483
   Tham En, 102, 485
   Tham Koun Dôn 102, 485
   Tham Nam Thieng 102
   Tham Phi Seua 485

Kharavela, King and military cave use 509–510
Khashim-Ojuk Cave, Turansky Plain, Central Asia 95
Khéo-Phay, Vietnam 167
Khodja-Mumyn salt karst, Tadjikistan 96, 208
Khodzharustam Cave, Turansky Plain, Central Asia 95
Khonshilat beyn al Hiyool, Selma Plateau, Oman 639
Khonshilat Maqandeli, Selma Plateau, Oman 639
Khonshilat Minqod, Selma Plateau, Oman 639
Kievskaja Cave, Turansky Plain, Central Asia 95, 663
Kijahe Xontjoa Cave, Mexico 391, 426, 525
Kilauea volcano, Hawaii, United States 415, 416
Kilcorney-Meggagh depression, Burren glaciokarst, Ireland 170
Kilgreany caves, Ireland 162
Killuragh Caves, Ireland 162
Kim-nyong Cave, South Korea **212**
Kinderlinskaja Cave, Urals 629
kinetic fractionation 556
King, William 384
King Arthur's Cave, Wales 364
King Pelayo and Christianity 623
King Pot, Yorkshire Dales, England 786
King Salmanassar III and cave art 86
King Solomon's Mines (Bennett and Marton) 359
King Solomon's Mines (Haggard) 359
Kinorhynca 458
Kinta Valley, Malay 515
Kircher, Anathasius
   biography of **152, 153**
   and karst hydrology 210, 478
   Mundus subterraneus 152, 158
Kirkgöz–1, Turkey 741
Kirkgözler Springs, Turkey 699, 741
Kirkgöz-Suluin, Turkey 741
Kirkor, A. 662
Kitum Cave, Kenya 22, 25, 294, 592
Kiver, Eugene 53
Kizelovskaja-Viasherskaja System, Urals 629
Kizilelma Mağarasi, Turkey 741
Kjaptarkhana Cave, Turansky Plain, Central Asia 95
Klasies River Mouth Cave, South Africa 19, 21–22, 82
Klein-Feldhofer Grotte, Germany 333
Ključka River, Herzegovina 293
Kluterthöhle, Germany 335, 697, 698
Klyuch karst shaft, Siberia 646
Knebel, W. von 605
Knock Fell Caverns, England 682–683
Knop, A. 771
Ko, Robbie 113
Ko Hung (*Pao Phu Tzu*) 97
Koanoi, Thailand 624
Kocain Cave, Taurus Mountains, Turkey 741
Kočerinsko Polje, Bosnia/Herzegovina 291
Kogun Cave, Thailand 624
Kohala volcano, Hawaii United States 415
Kokkines Petres (Petralona Cave), Greece 339
Kolkbläser-Monsterhöhlen, Austria 173
Komitee für Höhleforschung in der Geologischen Kommission des Königreichs Kroatien und Slawonien, Croatia 336
Konstitutzionnaja, Russia 345
konyaite 329

Koonalda Cave, Australia 83, 88–89, 120, 544, 545
Koongine Cave, Australia 85, 89
Kopetdag range
    Bakharden Cave 94
Kopisch, August 87
Korea 96
    Dongryong-gul cave 99
    geomorphology 96
    history of research 99
    Namgamduk system 96
Korean Association for Conservation of Caverns 99
Korean Caving Association 99
Korolëva, Siberia 645
Kossuth Cave, Hungary/Slovakia 26
Kosswig, Curt 154
Kota Tampan, Malaysia 108
Kovači Ponor, Bosnia/Herzegovina 291
Kozjak Lake, Plitvice Lakes, Croatia 598
Kra Isthmus, Thailand 110
Krakow-Wielun Upland, Poland 331
Kras, Slovenia 288, **485–486**
    geology 485
    hydrology 485–486
    Škocjanske jama **653–654**
    speleological research 486
    tourism 486
Krasnaya Cave, Ukraine 254
Krasny Kljuch Spring, Urals 629
Kraus, Franz 335, 350
Kraushöhle 726
Krem Kotsati-Umlawan, India 443
Krestik-Turist Cave, Georgia 201
Kristal'na, Ukraine 345, 746
Krizna Jana, Dinaric karst 288
Krka river, tufa waterfalls of, Croatia 338
Kromdraai, South Africa 17, 19
Kruber, Alexander 253, 487, 662
    About Karst Phenomena in Russia 662
Krubera Cave, Georgia 201, **487–488**, 525, 663
Krupac Polje, Montenegro 293
Kubinskaja, Siberia 629
Kubla Khan (Coleridge) 205
Kugitangtau ridge, Turkmenistan 273
Kujbeshevskaya-Genrikova Bezdna Cave, Georgia 201, 488
Kula Kai cavern system, Hawaii, United States 416
Kulogorskaja-Troja, Russia 345
Kulogorskaya, Pinega gypsum caves, Russia 589
Kumichevskaja, Russia 345
Kun-Ee-Gout, Turansky Plain, Central Asia 95
Kungur Ice Cave, Russia 628, 662
Kungurskaya Ledjanaja, Russia 345
Kuppen 242
Kupreško polje, Bosnia/Herzegovina 291
Kurukopru Cave, Turkey 741
Kuryuch-Agai cave, Ukraine 253
Kusch, Heinrich 113
Kutierleruk Cave, Irian Jaya 107
Kutikina Cave, Australia 122, 124, 129
Kutubu, Lake Papua New Guinea 620
Kuyukale, Taurus Mountains, Turkey 741

Kuzuga-ana Cave, Japan 30
Kvira Massif, Georgia 203
Kyrktau Plateau, Turansky Plain, Central Asia 95
Kyrtle, Georg 336
Kysyl Dzhar Mountains, Turansky Plain, Central Asia 95

La Balme, France 371
Laboratoire souterrain du CNRS, Moulis, France 149, 154, 373
La Chappelle-aux-Saints Cave, France 167
LaCoste and Romberg "G" meter 377
Lac Tsimanampetsotsa, Madagascar 620
Lady's Harbor Cave, California, United States 491
La Fontainguillère, France 167
Lagoa Santa, Bambuí karst, Brazil 133
Lago Colony Cave, Aruba 190
Lake St. Martin 176
l'Alpe cave, France 326
Lamalunge Cave, Italy 339
Lamb Leer Cavern, England 238
Lambert, Alexander 301
La Mouthe cave, France 85
lampenflora **733–734**
    Control by chemical agents 734
    composition 733
    defined 733
    distribution 733
    light intensity and 733–734
    management of 733
    problems of 733
Lamprechtsofen, Austria 173, 326, 525
La Naulette Cave, Belgium 138, 139
Lancaster-Ease Gill cave system, England 166
Land Between the Sumps, Sistema del Rio Encantado, Puerto Rico 317
Landry Member, Bear Rock Breccia karst, Canada **137, 138**
Lang, Karl Nicolaus 690
Lang Cuom Cave, Vietnam 108, 167
Langcliffe Cave, England 786
Langmuir, O. 673
Laniakea Cave, Hawaii, United States 416, 765
Lankester, Edward Ray **153–154**
Lanquin Cave, Guatemala 42
Lanzarote, Canary Islands
    fauna of 179
    geomorphology 179
    Jameo del Agua 179
    Túnel de la Atlántida 179, 302, 762
Laos
    geomorphology 100
    Hang Khe Ry, Khammouan 485
    Hang Vom Cave, Khammouan 485
    Kanung Cave, Khammouan 485
    Khammouan 167, **483–485**
    Nam Hin Boun Cave 483, 485
    Nam Non Cave 102, 483, 485
    Pak Ou Cave 103
    Tham En, Khammouan 102, 485
    Tham Koun Dôn, Khammouan 102, 485
    Tham Phi Seau 102
    Tham Phi Seua 485
    Tham Phiu 624

Laos *(Continued)*
　Tham Pong 167
　Tham Sao Hin 102
　Tham Thing 113
　Tham Toutche 102
　Xé Bang Fai Cave 102, 483, 485
Lapa da Sol, Bahia, Brazil 92
Lapa do Angélica, Brazil 55
La Palma Island, Canary Islands 179, 180
Lapão Cave, Brazil 611, 613
Lapa São Mateus III, Brazil 55
Lapa São Vicente I, Brazil 55
La Perte de l'Avenir, Patagonia, Chile 573
La Perte du Futur, Patagonia, Chile 573
La Perte du Temps, Patagonia, Chile 573
La Photographie souterraine (Martel) 579
lapiés *See* karren.
La Recherche (Lorblanchet) 85
Las Anod, Somalia 23
Las Casas, Bartholomé de 191
Lascaux, France
　art in 85, 757
　Cabinet of the Felines 757
　Hall of the Bulls 757
　as tourist cave 428
La Source du Lyson (Coste) 534
Las Puertas de Illamina, Pierre Saint-Martin, Spain 586
Lassa fever 294
Last Man, The (Shelley) 206
La Tête-du-Lion, France 82
Laucer, Fernando Morban 85
Laurente Cave, Philippines 109
Laurenti, Josephi Nicolai 148, **152**
Lavaka Fanihy, Madagascar 494
Lavalle, P. 527
La Vallée d'Iherir, Algeria 620
lava mould caves 762
lava tube caves **415–416, 762–763**
　andesite 762
　Australia 123
　basalt, clinkery 762
　basalt, pahoehoe 761
　Canary Islands 179
　cavernous lava rises 761
　Cheju-do, South Korea 212
　dating of 282, 762
　features of 762–763
　and groundwater 762
　Hawaii, United States **415–416**
　in higher viscosity lavas 760–761, 761
　lava mould caves 762
　Madagascar 495
　minerals in 762
　MSS, colonization by 457
　rheogenic features 763–764
　rhyolite 762
　underwater caves 762
　See also volcanic caves.
La Verna chamber, Pierre Saint-Martin, France 585
Lawn Hill National Park, Australia 124

Lawrence, Dirmuid (*Exploits at West Poley*) 360
Lawson, William 127
Lazaro Jerko, Kras, Slovenia 486
Lazionectes exleyi 183
lead/zinc deposits 713
　Canada 176
　metasomatic 330
　mining of 514, 576
　Mississippi Valley Type (MVT) breccias 660
　placers 514
leaky-feeder cable and cave communication 238
Leang Burung 2, Indonesia 109
Lebanon
　Afka Cave 208
　Faouar Dara 114
　Fouar Dara 463
　geomorphology 114
　Jeita Cave 208, **463–464**, 533
　　description of **463–464**
　　history of exploration **463**
　　hydrology **463**
　　musical performances in 533
　　speleogenesis **463**
　　speleothems **464**
　　Upper Gallery **464**
　Pigeon Grotto 703
　Qattine Azar 114, 463
Lechuguilla Caves, New Mexico, United States **508**
　geomorphology 192, 193
　history of speleology 52
　length/depth of 525, 750
　management of 774
　minerals in 511, 513, 751
　speleothems 194
Le Creugenat, Switzerland 302
Ledovaya Cave, Siberia 646
Lee, Edmund F. 49, 715
Lee, W.T. (*National Geographic*) 359
Leech *See* Hirudinea.
　Haemopis caeca 529
Leeward Islands 190
Legal Institute for the Environment 181
Le Grotte d'Italia 464, 465
Lehman Caves, Nevada, United States 369
Lehmann, Herbert 235, 242, 641, 735
Lehmann, Otto **381**
　Die Hydrographie des Karstes 381, 671
Lene Hara Cave, Indonesia 109
Leoganger karst, Austria 173
leonite 329
Leon Sinks Cave, Florida, United States 751
Lepidocampa hypogaea 632
Lepidoptera 25, 143
　Alucitidae 25
　Noctuidae 25
　Oecophoridae 25
　Plutellinae 25
　terrestrial cave habitats, biodiversity in 143, 144
Lépineux, Georges 585
Leptodirus hochenwartii 603

Leptodirus hohenwarti 448
Leptonetidae 369
leptospirosis 294
Leptostraca 65
Leroi-Gourhan, Andrè 85, 91
Leroi-Gourhan, Arlette 567
Les Combarelles, France 757
Lesser, J.G. 690
Lesser Antilles, The 190
Lessini plateau, Italy 326
Letrône, M. 284
Le Tuc d'Audoubert, France 84, 90
Letuchaya Mysh, Siberia 646
Levantine Neanderthal-AMH successions 196
level, Abney 716
level, water and cave surveying 716
Leviathan Cave, Kenya 22
L'Évolution régressive des poissons cavernicoles et abyssaux (Thinès) 593
Libya 16
    Cyrenaica karst 208
    Grerat-es-Salam, palynology 567
    Umm al Masabih cave 16
lichen zone 231
Licinopsis spp. 180–181
Liebig, Justus von 690
LIFE *See* Legal Institute for the Environment.
lighting used in cave exploration 352–353
Lilburn Cave, California, United States 52, 752
Limbert, Howard 113
lime, uses of 489
Limekilns Cave, Australia 127
Limestone Caribees 190
limestones 474
    cave streams in 252
    in chemical processing 489
    and CO_2 enriched cave air 183
    composition of 184
    construction and 248, 550
    dissolution **295–296**, 678
    distribution of 185
    and doline formation 307–308
    erosion rates, measuring **322–323**
    and flora **360–362**
    and karst aquifer development 399
    karstification and 332
    kluftkarren (grikes) 471
    medicinal uses of 623
    as mineral resource **489–490**
    ornamental uses of **550–551**
    paleokarst 559
    pavements 170–171, 473, 783
    precipitation of 730
    quarrying 244, 550–551, **609–611**
    sulfide minerals in 713–714
    use as building/dimensional stone 490
Limeworks Cave, Makapansgat, Africa 280
Limits of Acceptable Change 519
Lin Shih-Chen (*Pên Tshao Kang Mu*) 98
Lin Wai Tai Ta (Chou) 98
Liphistomorpha 71, 72

Lister, Martin 152
Listov, Ju. 662
Li Thai (King of Sukothai) 112
lithium 514
lithology and cave morphology **235**
Little Foot skeleton, South Africa 17
littoral caves 204, 252, 254, **491–492**
    cave development 491, 492
    distribution of 491
    habitat 501
        fauna, factors influencing 501
        food sources 501
        sections of 501–502
        substrate mobility 501
        troglophilic fauna 502
    morphology 492
    speleogenesis 491
littoral zone processes 231–232
Liujiang Cave, China 224
Liujiang Man 224
Liu longdong, China 699
liverworts *See* bryophytes.
Ljubljanica, Slovenia 699
Ljubljanica River, Slovenia 288
Ljuta spring, Dinaric karst 288
Llangynidr Plateau, Wales 341
Llewelyn, John Dilwyn 578
Lloyd, John 164
Llyn Du cave system, Wales 166
Llyn Parc cave system, Wales 166
Loboptera troglobia, Canary Islands 180
local habitat model of speciation 665
loess caves **591**
Loewengreif, Jeršinovič von 148
Loltun Cave, Mexico 42, 84
Lomonosov, Michail 662
Lomonovskaya, Pinega gypsum caves, Russia 588, 589
London Chalk Basin, England 215
Longgong karst, China 218
Longgupo, China 223–224
Longgushan, China 221
Long Hole, England 714
Longhorn Caverns, Texas, United States 534
Long Rong Rien cave, Thailand 107–108
Longtan Cave, China 222
Longtangdong, China 222–223
Longuin Cave, China 623–624
Lorblanchet, Michel (*La Recherche*) 85
Lord of the Rings, The (Tolkien) 206
Lorenz National Park, Papua New Guinea 105
Loricifera 459
Lorloch, Austria 301
Los Corrales de los Indios, Mona, Puerto Rico 517
Lost Creek Caves, Colorado, United States 721
Lost Creek Pseudokarst, Colorado, United States 606, 722–723
Lost River Cave, Kentucky, United States 534
Lost River Caves, New Hampshire, United States 721
Loubens, Marcel 585
Lough Cong, Ireland 341
Loutraki-Almopia Caves, Greece 339–340

Lovecraft, H.P. (*The Beast in the Cave*) 206
Lovrić, Ivan 288
Lower Kane Cave, Wyoming, United States 184
Lower Swift Hole, India 443
Lower Tatra Mountains National Park, Europe, Central 333
Lubang Kapa Kapasa, Indonesia 631
Lubang Tomanangna, Indonesia 631
Lucante, J.A. 371
Lucas, George (*Star Wars*) 359
Lucas, John 128
Lucas Cave, Jenolan, Australia 623
Lucayan Caverns, Bahamas 156, 360
Ludwig II (king of Bavaria) 87
Luire cave, France 754
Lukavačko polje, Herzegovina 293
Lukina Jama, Dinaric karst 288, 525
luminescence of speleothems **695–696**
    causes of 696
        minerals, types of 695
Lummelundagrottorna, Sweden 342
lumped parameter flow models, mathematical 402
    Lunan, China, stone forest 473, 643
Lund, Peter Wilhelm 59–60
Lune River Quarry, Tasmania 611
lung cancer 294–295
Lungun Cave, Phillipines 168
Luray Caverns, Virginia, United States
    accessibility of 750
    and cave photography 579
    history of exploration 51
    musical performances in 534
Lurgikanskaya, Siberia 646
Lurhöhle, Austria 326
Lusaka, Zambia, Africa 22
Luse, Nakanai Caves, Papua New Guinea 538
Luse doline, Papua New Guinea 526
Luti Dong (Reed Flute Cave), Yangshuo, China 783
Luweng Jaran, Gunung Sewu, Java 642
Lyell, Charles 81, 669
Lykoshin, G. 662
Lynch, Charles 410
Lysanskaja Cave, Siberia 629

Maara Cave, Greece 339
Machiavelli, Niccolo 410
Macho Pit, Nacimiento del Rio Mante, Mexico **391**
Macizo de Guamuhaya, Cuba 271
Mackenzie Mountains, Canada 178
Macocha, abyss of, Moravia 335
Madagascar
    Ambatoharanana 494
    Ambatomanjahana 494
    Andrafiabe 494
    Anjohiambovonomgy 494
    Anjohibe 494
    Anjohikely 494
    Ankarana 493–494
    Ankarana Special Reserve 494
    Antsatrabonko 494
    Antsiroandoa 494
    Chiroptera 493–494
    Crocodylus nitolicus 493
    Grotte Albert 495
    Grotte de Chauve-Souris 494
    Grotte des Crocodiles 494
    Grottes des Rois 494
    Kelifely 494
    Lac Tsimanampetsotsa 620
    Lavaka Fanihy 494
    lava tube caves 495
    Milaintety 494
    Namoroka 494
    Narinda 494
    Peyre, J.-C. 494
    Radofilao, Jean 493
    Reserve Naturelle Integrale 494
    Rossi, G. 494
    speleogenesis 493
    Tsingy de Bemaraha 473, **494–495**
    Tulear Region 495
    Typholeotris madagascariensis 495
    Zebu Well Cave 494
Madison Cave, Virginia, United States 50, 715
Madison Limestone Formation, South Dakota, United States
    geomorphology 775, 776
    speleogenesis 775
Madonna, the 364
Mae Hong Son, Thailand 167
Magdalenian era 333–334
Magdalenska Jama, Postojna-Planina Cave System, Slovenia 601
magnesite
    dissolution of 297, 298
    formation of 513
Magnet Cave, Mpumalanga South Africa 651
magnetic induction equipment and cave communication **238**
magnetic reversal stratigraphy and dating methods 280
Magura Cave, Bulgaria 330
Magyar Karst- és Barlangkutató Társular, Hungary 336
Main Corridor, Carlsbad Caverns 193, 194
Main Ridge, Crimea, Ukraine 253
Maiorerus randoi, Canary Islands 179, 181
Majlis Al Jinn (Meeting Hall of the Spirits), Selma Plateau, Oman 639
Majskaya Cave, Georgia 201
Makapangsgat Limeworks Cave, South Africa 17, 19
Makingen Cave, Kenya 22
Maksimovich, G. 662
malachite
    formation of 512
    Valea Rea Cave, Romania 329
Malacostraca 255, 256
    Bathynellacea 24, 255
        Thermobathynella adami 724
    Thermosbaenacea 24
        anchialine habitats 65, 141
        Australia 126, 183
        description of 255, 256
        Halosbaena fortunata, Lanazarote 179
        Halosbaena tulki 182
        hypogean habitats 141

Monodella texana 315
Thermosbaena mirabilis 24, 724
Mala Karlovica Cave, Slovenia 210
Malangine Cave, Australia 85, 89
Malaspina Glacier, Alaska, United States 606
Malaysia
   archaeological caves 108
   Batu Caves, Kuala Lampour
      history of research 109
      ritual use of 103, 428, 623, 624
   Baturong 108
   burials in caves 108, 167
   geomorphology 100
   Great Cave, Niah 108
   Gua Cha 108
   Gua Gunung Runtuh 108
   Gua Tambun 108
   Gunung Mulu National Park, Sarawak 468, **531–533**
      dating of 554
      geomorphology 105, 531, 532–533
      Green Cave 532
      Gua Air Jernih (Clearwater Cave) 531, 532, 554
      Gua Kulit Supit 532
      karren fields 473
      Melinau River 533
      paragenesis 571
      pinnacles, light-oriented 581
      Pinnacles, The 532
      Sarawak Chamber 204, 308, 526, 532
      speleogenesis 531–532
      wall notches of 531
   history of exploration 113
   Kota Tampan 108
   Niah Cave, Sarawak
      guano accumulations in 409
      human occupation of 108
      industrial use of 426
   Painted Cave, Niah 108
   Tempurung Cave 102
Malham Cave, Mount Sedom, Israel 638
Malham Cove, England 438, 784
Malham Cove Rising, England 303
Maligne River, Canada 177
Mallinson, Jason 303
Mallorca's Cave, Greece 208
Malocostraca 255
Malott, C.A. 671
Malta
   Ghar Dalam 208
   St Paul's Grotto 208
   St Paul's Church, Rabat 623
Maltsev, Vladimir 273, 275
Mambaí, Bambuí karst, Brazil 133
Mammoth Cave, Kentucky, United States 191, 204, 206, 214, **495–498**
   active passages, patterns of 498
   art in 92
   bats 500
   biodiversity in 143
   and borehole analysis 398
   burials in 168
   and cave photography 578
   and cave surveying 715
   conservation 500–501
   dating of 554, 555
   ecosystems 499
   factors determing length of 499
   fauna 10, 499–500, 516
   Fisher Ridge Cave 52, 495, 525
   geomorphology 495–498, 521
   Green River 497
   guano 500
   and gunpowder 411, 499
   Hadenoecus subterraneus 499, 500
   Hidden River Cave System 498–499
   history of speleology 50, 52
   human use of 499
   Kentucky cave shrimp 500
   length/depth of 525, 750
   mineral deposits in 44, 498
   Myotis grisescens 500
   Myotis sodalis 500
   paragenesis 569, 570
   Rotunda Room 635
   sediments, allochthonous 633
   soils of 657
   speleogenesis 497
   stygobitic fauna 499–500
   surface drainage patterns 498
   Thomas Avenue, paragenesis 570
Mammoth Cave National Park 52, 495
Mammoth-Flint Cave, Kentucky, United States 323
Mammoth Hot Springs, Wyoming, United States 737
Mammutmünster, Hölloch, Switzerland 422
Mamo Kananda, Papua New Guinea 105
Man and the Biosphere programme (MAB) 246
manganese
   mining of 514
   paleokarst 332
manganese hydroxide
   Grotte de la Cigalère, France 373
   Ukraine 748
manganese oxides
   formation of 511
   Jewel Cave, South Dakota, United States 776
Manitoba, platform composition 176
Man-jang Cave, South Korea 212
Mannerism and cave art 86
Manning's Law, hydraulics 395
Mansfield, R. 113
Mansuy, Henri 108
Mantee Springs, United States 302
mantled evaporite karst 346–347
Manunggul Cave, Philippines 168
Mapala Rockshelter, Australia 125
marble *See* metacarbonate.
Marble Caves, Vietnam 103
Marcano, Vicente 60
marcasite
   and biomineralization 636
   formation of 714

Marche-les-Dames Cave, Belgium 139
Marc-René Marquis de Montalembert 152
Margoon Waterfall Spring, Iran 702
Marichard, Jules Ollier de 82
Marifugia cavatica 67, 289, 290, 760
marine caves *See* littoral caves.
marine regressions and colonization 237
marine terraces, raised 190
Marinitsch, Jozef 350, 655
Mark Twain Cave, Missouri, United States 206
marmites 522
Maronia Cave, Thrace 340
Mars
    Chandor Chasma 357
    pseudokarst analogies 608
    rheogenic pseudokarst 606
    and talus caves 721, 723–724
    Vastitas Borealis 355
    volcanic caves 355, 583–593
Martal Cave, Taurus Mountains, Turkey 741
Martel, Édouard Alfred 191
    and Anthron 351
    biography of 371, **687, 688–689**
    and cave exploration 352, 371, 621, 662
    and cave photography 579
    and cave surveying 715
    Dargilan cave 371
    in England 166
    in Ireland 166
    Kapsia Cave 340
    La Photographie souterraine 579
    Padirac Cave, France 373
    Pierre Saint-Martin, France 585
    publications of 158, 464
    and speleogenesis theories 670, 671
Martel's Cave, Georgia 488
Martel's Hall, Škocjanske jama, Slovenia 486, 653
Martin Ridge Cave System, Mammoth Cave, Kentucky, United
    States 495, 750
Marton, Andrew (*King Solomon's Mines*) 359
Mason, Revil 19
Mas Raynal, France 371
mass displacement caves *See* crevice caves.
mass movement caves *See* crevice caves.
Masuzo Uémo 99
Matali, Papua New Guinea 699
Matangkib Cave, Philippines 168
Mateullus troglobioticus, Indonesia 632
mathematical model, underground water flow 671, 673
Matienzo depression, Spain 327
matrix mineralization 637
Matthews, Paul (*Grim*) 360
Mauna Kea volcano, Hawaii, United States 415, 416
Mauna Loa volcano, Hawaii, United States 415
Mawang Dong, China 218
Maximovich, Georgij A. **381**
Maya
    art in caves **92**
    burials in caves 81, 84, 168

ritual use of caves 44–45, 624
Mazauric, F. 371
maze caves 573, 574
M2 cave (Pozo de Cueltabo), Picos de Europa National Park, Spain
    583, 584
McConnell, H. 527
McDonald, R.C. 735
McDougal's Cave, Missouri, United States **206**
McEnery, John **77–78**, 80, 162
Mchista, Georgia 699
Mchista Spring, Georgia 202
McKenzie, David 716
McKeon's Caves, Australia 128
McKinnon, Daniel 191
MDBATHS Cave, New Hampshire, United States 722
Mechta Cave, Siberia 646
medical use of speleothems 97, 98, 690
Medicine Lake, Canada **177**
Mediterranean, caves in history **207–209**
Megachiroptera
    Pteropodidae 225
    Rousettus spp. 225
Megaloceros cazioti 339
Megalonyx jeffersoni 51
Megamania Cave, New Zealand 543
Melidoni Cave, Crete 510
Melinau River, Gunung Mulu National Park, Sarawak, Malaysia 533
Memoire dell'Instituto Italiano di Speleologia 464
Mémoires de la Société de Spéléologie 464, 465
Mendell, W. 355
Mendelssohn, Felix (*Fingal's Cave Overture*) 534
Mendip Hills, England 341, **503–504**
    and cave photography 579
    Charterhouse Caves 504
    dolines of 503
    Fairy Cave Quarry 503
    Ford-Ewers model of speleogenesis 673
    fossils 561, 562, 563, 564
    GB Cavern 504, 562
    Gough's Cave 504
    Lamb Leer Cavern 238
    morphology 503, 504
    Pen Park Hole 164, 714–715
    quarrying and 504
    speleogenesis 503
    St Cuthbert's Swallet 504
    Stoke Lane Slocker system 503
    Swildon's Hole 302, 504, 562, 564, 571
    Westbury caves 162, 562
    Wookey Hole 79, 302, 503, 504, 562, 579
Mendoça Mar, Francisco de 59
Mercanton, Louis (*Phroso*) 359
Mercer, Henry (*Hill-Caves of Yucatan*) 40
Meregill Hole, England 2
merokarst 288, 363
Mersah Matruth and karst hydrology 478
Mertz, Johann (*Fingal's Höhle*) 534
Mesa, Cueva De, Cuba 272
Meson de la Cava, Dominican Republic 534

mesovoid shallow substratum *See* MSS (mesovoid shallow substratum, milieu souterrain superficiel).
Messinian salinity crisis **237**, 530
metacarbonate
    formation of 186
    and karst aquifer development 399
    New Zealand 541
    Spitsbergen 342
metacinnabar 714
metal processing and limestone 490
metatyuyamunite 329
Metro Cave, South Island New Zealand 543
Mexico
    Akemati 36
    aquatic fauna 38
    Balankanche 40, 42
    Cave of Bolanchen, Campeche 42
    Cerro Rabon 426
    Chichen Itza 37, 41
    Chiroptera 39
    Coy 699
    Cuatro Cienagas 725
    Cuertana Casas Archaeological Zone 42
    Cueva del Tigre 409
    Cueva de Villa Luz
        and biofilms 145, 146
        ecosystem **758**
        minerals in **759**
        speleogenesis **758–759**
    Cueva Encantada 40
    Dos Ojos system 303, 787, 788
    Dos Pilas 41
    Dzilam State Reserve 620
    fauna, distribution of 38
    Frio 699
    Golondrinas, Sótano de las 36, **390–391**
    Gruta del Rio San Geronimo 37
    Gruta de Palmito 36
    Grutus de Juxtlahuaca 37, 92
    Huautla Cave System 36, **425–426**, 525
    hypogean fauna in 142
    Kijahe Xontjoa Cave 391, 426, 525
    Loltun Cave 42, 84
    Macho Pit, Nacimiento del Rio Mante 391
    Maya and 788
    Nohoch Nah Chich 303, 525
    Oaxaca, troglobitic fauna 38
    Ocotempa 36
    Ox Bel Ha system 303, 525, 787
    Paralell Tunnels, Baja 491
    Pyramid of the Sun, Teotihuacan 40
    Quintana Roo 37
    Rio Chontalcoatlan 37
    Sac Actun, Yucatán 787
    San Luis Potosí 92
    San Miguel, hoya de 37
    Sistema Cheve 36, 525
    Sistema Cuetzalan 36
    Sistema Purificacion 36, 525
    Sótano de San Augustín 425
    Sotano Akemati 525
    Sumidero Yochib 37
    Sunbeam-by-the-Sea Cave, Baja 491
    Systema Cheva 425, 426
    Systema Puficicatión 425
    Talgua Cave (Cave of the Glowing Skulls), Honduras 42
    Tauleve, Honduras 42
    troglomorphic fish of 38
    Veshtucoc system 36–37
    Xochicalco 42
    Yucatán **786–788**
    Zacatón 391
Meyer, Nicholas (*Star Trek VI: The Undiscovered Country*) 360
Meyerhoff, H.A. 526
Michel's Catalogue 704
micrite 185
microbial mats 146, 529
Microcerberidea 256
Microchiroptera **225**
    Craseonicteridae 225
    Mystacinidae 225
    Natalidae 225
    Phyllostomidae 225
    Rhinolophidae 225
    Vespertilionidae 225
Microcoryphia 446
    morphology 446
    taxonomy 445
micro-erosion meters 322, 324
microkarren 470
micromycetes 366
microorganisms **505–508**
    actinomycetes 507
    algae 508
    bacteria 507
    Basidiomycetes 508
    biofilms **145–146**
    biokarstification and 147–148
    and carbon dioxide enrichment 184
    cave art and 508
    and cave development 505
    caver health and 508–509
    in cave sediments 505
    chemolithotrophs 507
    colonization of caves by 507
    cyanobacteria 508
    Deuteromycetes 508
    energy fixation 505
    energy sources 507
    environments, subsurface 505
    eukaryotic 506
    food resources 366–367, 508
    fungi 508
    Guadalupe caves 194
    heterotrophs 507
    human impacts on 508
    microbiological techniques 507
    and the nitrogen cycle 507
    and nitrogen cycling 505–506
    oxidation of iron and manganese deposits 506–507

microorganisms *(Continued)*
  oxidation of sulfur 506, 507
  photoautotrophs 507
  prokaryotic 506
  protozoa 508
  redox reactions, microbially mediated 194
  requirements of 506–507
  and rock dissolution 508
  role of in cave environments 506
  saltpetre and 507
  slime molds 366, 508
  and sulfur cycling 506
  Zygomycetes 508
micro-whipscorpion *See* Schizomida.
Mictacea 65, 256, 768
Middleton, Greg 765
Milaintety, Madagascar 494
Milieu Souterrain Superficiel *See* MSS (mesovoid shallow substratum, milieu souterrain superficiel).
military use of caves **509–510**
  Akiyoshi caves, Japan 510
  Ardeantine Caves, Italy 510
  Bat bomb project, World War II 510
  Battle of the Caves 510
  Bedeilhac Cave, France 510
  Biak caves, Irian Jaya 510
  burial sites 510
  Cao Bang caves, Vietnam 510
  Clashmealcon Cave, Ireland 510
  Convolvo del Butistone, Italy 510
  Crimean War 510
  Cuban Missle Crisis 510
  documentation of 509
  guerilla warfare 509
  hospitals 509, 510
  Iwo Jima 510
  Long March of 1934 510
  Maroon wars, Jamaica 510
  Melidoni Cave, Crete 510
  military execution sites 510
  1959 Cuban Revolution 510
  Postojna Cave, Slovenia 510
  Predjama Cave, Slovenia 510
  Reddington Caves, Pennsylvania, United States 510
  Sohoton caves, Phillipines 510
  source of materials 509
  St Francis cave, Scotland 510
  St Michael's Cave, Gibraltar 510
  storage facilities 509
  strategy 509
  Tora Bora caves, Afghanistan 510
  Veterani Cave, Romania 510
  Vietnam War **510**
Miller's Cave, Missouri, United States 168
millipedes *See* Myriapoda.
millipede toxin 294
Millpu de Kaukiran, Peru 55
Milyeringa veritas, Australia 182, 183, 594
Mimehtu Cave, Thailand 624
Minafuku Dam, Japan 739

Minas Gerais, Brazil 92, 93
Mine cave (Rhar Djebel Serdj), Tunisia 15–16
minerals, carbonate *See* carbonate minerals.
minerals in caves **512–513**
  arsenates 512
  borates 512
  carbonate 512–513, **690–692**
  classification of 511
  defined 511
  halides 511–512
  native elements 511
  nitrates 513
  oxides/hydroxides 511
  phophates 513
  silicates 513
  speleothems, carbonate **690–692**
  speleothems, evaporite **693–696**
  sulfates 513
  sulfides 511
  vanadates 513–514
  See also specific cave or mineral by name.
minerals in karst **514–515, 713–714**
mines and speleotherapy 698
Mingaria Massif, Georgia 203
  minimum-impact guidelines 621–622, 625–626
Miniopterus schreibersii 227
Minye, Nakanai Caves, Papua New Guinea 538
Minye doline, Papua New Guinea 526
mirabilite
  deposition of 694
  formation of 513
  Majskaya Cave, Georgia 201
  Mammoth Cave, Kentucky, United States 44
  Medicinal use 44
  Tăuşoare Cave 329
mircotidal environments, processes in **232**
Mirolda Cave, France 372
Misophrioida 259, 260
  Australia 126, 183
  Canary Islands 179
Mississippi Valley Type mineral deposits 514, 660
Mitchell, Thomas 124
Mitchell-Palmer karst, Australia **215–216**
Mitchell River, Australia 216
Mitteilungen über Höhlen und karstforchung 464
mixture corrosion, speleogenesis theory 671, 673
Mlynki, Ukraine 345, 746
Močulski (Motschoulsky), 148
modèle biphase of colonization 235
Model T Cave, Texas, United States 591
MODFLOW 401, 402
Moestrof caves, Germany 682
Moghul Well Gharra, Pakistan 445
mogotes 190, 242, 272
Mogoto Cave, Mpumalanga, South Africa 651
Moldovia
  Emil Racoviță 330
  Zoloushka 329, 330, 345, 525
Molefone and cave communication 238–239
Mollusca **516–517**

    Africa 24
    Bivalvia
        Dreissenidae
            Congeria kusceri Bole 289
        distribution of 516
        Gastropoda 516
        Hydrobiidae 516
        North America 46
        Subulinidae
            Cryptazeca elongata 516
            Cryptazeca spelaea 516
        taxonomic classification of 516
        trogobitic species 516–517
Molodezhnaya Cave, Ukraine 253
molybdenum 514
Mona, Isla de National Park, Puerto Rico 189, **517–518**
    Cueva de Frio 517
    Cueva del Capitan 518
    Cueva del Diamante 518
    Cueva del Esqueleto 518
    Cueva de los Parajos 518
    Cuevas del Centro 517
    flank-margin caves of 517–518, 674, 676, 677
    hydrology 518
    Los Corrales de los Indios 517
    mogotes of 190
    Sistema del Faro 517–518
    speleogenesis 518
Mona Island, West Indies 409
Monarch Cave, Australia 123
Monasytuyskaya, Siberia 646
Mondmilchloch, Switzerland 335
Mondo Sotterraneo 464
Mongolia
    geomorphology 96
    history of research 99
monitoring of caves **519–520**
    applications 519
    data collection 520
    event-based monitoring 520
    holistic approach to 519
    Index of Biological Integrity (IBI) 520
    indicators 519–520
    Limits of Acceptable Change 519
    process 519
    Recreational Opportunity Spectrum 519
    Social and Environmental Monitoring (SEM) programme 519
    standards 520
    Visitor Activities Management Process 519
    Visitor Impact Management (VIM) 519
    Vistor Experience and Resource Protection 519
monohydrocalcite 513
Monroe, Watson 604
monsters in caves 365
Montagnard cave, France 283
Montagu Cave, South Africa 19
Montaigle Cave, Belgium 139
Montanheiros, Os 765
Montenegro
    Boka Kotorska 338

    hypogean fauna in 142
    Skadarsko Jezero 338
Montes Claros, Bambuí karst, Brazil 133
Montespan, Grotte De, France 301
montmilch 366
montmorillonite 513
Montoriol-Pous, Joaquin 764, 765
Moondyne Cave, Australia 625–626
moonmilk
    and biomineralization 636
    constituents of 690
Morales, Andreas 191
Moravian Karst 227
More People Have Been To The Moon (Thomas) 534
Mormotomyia hirsuta 453
Morocco 13–14
    Kef Aziza 13
    Kef Tikhoubaï 13
    Kef Toghobeit 13
    Rhar Chara 13
    Rhar Chiker 13
    tufas of 14
    Wit Tamdoun 13, 14
morphological adaptations *See* adaptations, morphological.
morphology of caves **523–524**
    anastomoses 523
    biokarstic processes and 523
    breakdown 523
    bypasses, development of 521
    cave sediments 522
    drawdown caves, development of 521
    entrenchment 521
    flutes 522, 523
    modification, principle types of 521
    paragenetic passages, development of 521
    passage morphology 521–522
    phreatic 521, 522
    pockets 523
    rock relief, classification of 522
    scallops 522, 523
    solutional pits 523
    stream potholes 523
    vadose canyon 522
morphometry
    caves **524–526**
    cone karst 242
    karst **526–527**
        doline analysis 527
        field measurements 527
        history of 526–527
        regional valley systems analysis 527
        residual carbonate hills analysis 527
Mortillet, Gabriel de 82
Moskovskaja Cave, Georgia 201, 487
Mosman's Cave, Australia 127
Mossdale Caverns, England 2, 786
mosses *See* bryophytes.
Mother Grundy's Parlour, England 162
Mottle, Maria 28
Mouhot, Henri 113

Mouis, Grotte, France 327
Moulins 68, 386, 606
mountain sickness 293
Mouret, Claude 113
Mousterian culture, Israel 195
Movile Cave, Romania 329, 365, 507, 516, **529–530**
   adaptations of fauna 529
   cave environment 528
   chemoautotrophic energy production 529–530
   ecosystem of 528
   fauna of 73, 528–529
   geology 528
   minerals in 530
   morphology 528
   speleogenesis 530
Mpumalanga pseudokarst, South Africa 252, 651
Mramornaya Cave, Ukraine 253
MSS (mesovoid shallow substratum, milieu souterrain superficiel) 179, 236, 369, 709
   adaptations of fauna 457
   colonization of caves from 457
   defined 455, 456
   distribution of 455–456
   environment of 456
   fauna of 456–457
   nutrient sources in 456
   speleogenesis 456, 457
mucolites 637
mud-glyph cave art 92
Mud Glyph Cave, Tennessee, United States 82, 83–84, 85
Muhammed, Prophet and caves 208
Mullamullang Cave, Nullarbor Plain, Australia 545, 546, 731
Müller, Max 579
multi-level (multiphase) caves 686
multiprocess caves 591–592
multiway differential thermocouple, measurement using 228
Mulu **531–533** *See* Gunung Mulu National Park, Sarawak, Malaysia.
mumijo
   formation of 636–637
   medicinal use of 637
Mundewa Cave, Myanmar 102, 103
Mundus subterraneus (Kircher) 152, 158
Murchison, Roderick 164
Murra-el-Elevyn Cave, Nullarbor, Australia 547
Muruk cave, Papua New Guinea 105, 372, 525
music in/about caves **533–534**
   acoustics of 534
   cave environment, damage to 534
   concerts in 533–534
   history of performances 533
   inspired by 534
muskeg 175
mussel/echinoid zone **231, 232**
Mwenga Mwena, Zimbabwe 20, 21
Myanmar
   geomorphology 100, 101
   Gogteik 103
   Mundewa Cave 102, 103
   Peik Kyin Myaung Cave 103
   Pyin-Oo-Lwin 102

Mygalomorpha **71, 72**
Mynddyd Llangattock, Wales 141, 166
Myotis austroriparius 226
Myotis blythi 227
Myotis grisescens 227, 500
Myotis lucifugus 226
Myotis myotis 227
Myotis sodalis 227, 500
Myriapoda **534–536**
   biodiversity 143, 144
   Callipodia 535–536
   Cape Range 181
   Chilopoda 534–535
      Criptops anomalans 529
   Dipolopoda 534
   Pauropoda 534
   Penicillata 534
   Polydesmida 58
   Symphyla 534
Mystacinidae **225**
Mystery Cave, Minnesota, United States 751
Mystery Cave, Wisconsin, United States 626
mythology and caves 364–365

Nacimiento Del Rio Mante, Mexico 302
Nadinsko Polje, Bosnia/Herzegovina 291
Nagel, Joseph Anton 87, 211, 335, 601
   biography **687–688**
   and cave surveying 715
   and karst hydrology 478
Nahanni National Park, Canada 52
Naj Tunich, Guatemala 37, 40, 92
Nakanai Caves, Papua New Guinea **538–540**
   denudation 539
   geomorphology 538–539
   hydrology 539
   karst processes 539
   megadolines, formation of 539–540
   Muruk System 538, 539, 540
Nakanai Mountains, New Britain, Papua New Guinea 308
Nakimu Caves, Canada 51, 52, 178
Namgamduk system, Korea 96
Nam Hin Boun Cave, Laos 483, 485
Nam Hin Boun river, Khammouan, Laos/Vietnam 104
Namibia
   Daleib Wanei tufa 23
   Dragon's Breath Cave 21
   Harasib Cave 21
   Isha Baidoa tufa 23
Nam Non Cave, Laos 102, 483, 485
Namoroka, Madagascar 494
Namurian sandstones, Australia 308
Nang Nuan oil field, Thailand 102, 104
Nanisivik, Canada 560, 571, 714
Nanjing Institute of Paleontology and Geology (NIPG) **223**
Na One crater, Hawaii, United States 762
Napia Cave, Pierre Saint-Martin, France 585
Napra Cave, Georgia 202
Naracoorte Caves, Australia 124, 633, 726
Nare, Nakanai Caves, Papua New Guinea 538

Narinda, Madagascar 494
National Association of Soviet Speleologists 663
National Caving Association, United Kingdom 618
National Geographic (Lee) 359
National Monument of Japan, Taisho-da Cave 30
National Park of Cainama, Venezuela 612
national park term, use of 245–246
National Pollution Discharge Elimination System (NPDES) 405
National Radiological Protection Board 618
National Speleological Society 52, 352
National Speleological Society (NSS) Bulletin 465
National Speleological Society (NSS) News 464, 465
Native Americans 81, 82
native elements in caves 511
Natufian culture, Israel 195, 196
Natural and Aboriginal History of Tennessee (Haywood) 81
Natural Trap Cave, Wyoming, United States 562
Nature Reserve of Karaenta, Indonesia 632
Neanderthal man *See Homo sapiens neanderthalensis.*
Nelson, Nels 81
Nelson's Bay Cave, South Africa 19
Nematoda
 Desmocolex aquaedulcis 458
 distribution 458
 morphology 458
 Rhabdochona longleyi 315
 roundworms 458
Nemertea 458
Neobisium spelaeum 603
neoteny
 defined 348
 morphological (internal) adaptations 11, 61
Nepa anopthalma 529
Nepal
 Annapurna Himal 623
 Goraknath Cave 623
 Gupteswary Cave 623
 Halesi Cave 623
 Patale Chhango Gupha 444
 Shivaji Cave 623
 speleology 444
Neretva Plain, Herzegovina 338
Nerja, Cuevas de, Spain 327, 338
nesquehonite 513
Netherlands Antilles 190
Nettlebed Cave, South Island, New Zealand 542, 543
Never Say Never Again (Kerschner) 360
Nevesinjsko polje, Herzegovina 293
Neville, Russell Trall 51
 In The Cellars of the World 359
New Brunswick karst 177
Newdegate Cave, Tasmania 123
Newfoundland karst 177
New Guinea 2, Australia 89
Newhall/Barntick caves, Ireland 162
New Zealand **540–543**
 Fiordland
  Aurora Cave 543, 554, 555
  geomorphology 543
 geomorphology 540–541
 hypogean fauna in 142
 North Island 541–542
  geomorphology 541
  glowworm displays 541–542
  karstification 541
  King Country 541–542
  polygonal karst of 541
 South Island
  Arthur Marble 542
  Bohemia Cave 543, 591
  Bullock Creek-Cave Creek 542
  Bulmer Cave 542, 543
  Ellis Basin System 543
  HH Cave 543
  Honeycomb Hill Cave 542, 543
  Megamania Cave 543
  Metro Cave 543
  Mt Owen National Park 542
  Nettlebed Cave 542, 543
 Wiri Cave 606
Ney Cave, Texas, United States 751
Niagaran dolostones, Canada 175
Niah Cave, Sarawak, Malaysia
 art in 108
 guano accumulations in 409
 human occupation of 108
 industrial use of 426
Niaux Cave, France 428
Nicaragua, Cayos Miskitos y Franja Costera Immediata 620
Nicholson Expedition of 1930 51
nickel deposits 514
Nicola System and cave communication 239
Niedzwiedzia Cave, Poland 331
Nietoperek, Poland 227
Niggly Cave, Australia 122
Nikšićko Polje, Montenegro 293
Ningaloo fringing reef, Australia 183
Nin Ham Boun, Khammouan 113
niobium 514
NIPG *See* Nanjing Institute of Paleontology and Geology (NIPG).
Niphargus aggtelekus 620
Niphargus balcanicus 760
nitrammite 513
nitrates
 minerals in caves 513
 pollution of groundwater, dispersed 403
nitre 513
 See also gunpowder.
nitrobacteria 366, 505–506
nitrocalcite 513
nitrogen dioxide poisoning 293
nitromagnesite 513
nitrosobacteria 366
nival/corrosive class, karst cavities 253
nocardiosis 294
Nohoch Nah Chich, Mexico 303, 525
Nombe Rockshelter, Australia 125
non-aqueous phase liquids (NAPL's)
 pollution of groundwater, point sources **404–405**
 pollution of groundwater, remediation **407**

North Wales Caving Club 166
Northern Cape, South Africa 19
Northern cavefish See Amblyopsis spelea.
Norway
   Nordland
      marble stripe karst 186, 342
         Okshola/Kristihola 342
         Tjoarvkrajgge 342
      Pikhåg Caves 705, 706
      Torghatten sea cave 342
Notes on the State of Virginia (Jefferson) 50
Nothofagus forest, Patagonia, Chile 572
Notoptera 144
Notts Pot, England 441, 786
Nova Scotia karst, Canada 177
Novorossijskaja, Ukraine 628
nucleation biomineralization 637
Nullarbor Plain, Australia 302, **544–546**
   Abrakurrie Cave 120, 545
   Cocklebiddy Cave 545, 546
   Diprose Caves 546
   dolines of 544
   Gambier, Mt, art in 89
   halite 546
   Kelly Cave 546
   Koonalda Cave 83, 88–89, 120, 544, 545
   Mullamullang Cave 545, 546, 731
   Murra-el-Elevyn Cave 547
   Old Homestead Cave, 121, 546
   speleogenesis 546
   speleological significance of 128–129
   speleothems 546
   Thampanna Cave 546
   Tommy Graham's Cave 545
   underground lakes 545
   Weebubbie Cave 120, 545
nutrient input, excess 244
Nyangziguan spring, China 220
Nyári, Jenő 28
Nyiragongo volcano, Central Africa 762
nymphaea 86, 87

Oakden caves, Australia 129
Oasis de Tamantit et Sid Ahmed Timmi, Algeria 620
Oban Bay caves, Scotland 160
Obelix, Italy 117
Obod Estavelle, Herzegovina 293, 739
obruks 742
obsidian 279
Obukheikha, Siberia 646
ocher 273
Ochtina Aragonite Cave, Slovakia 26
Ocotempa, Mexico 36
Octinska Aragonite Cave, Czechia 331
Odysseus and caves 208
Odyssey, The (Homer) 205
Odyssey Cave, Australia 184
Officer's Cave, Oregon, United States 591, 606

Ohachue Massif, Georgia 203
oil
   in karst **431–432**
   paleokarst, economic resources of 560
oilbirds See Steatornis caripensis.
Ojców caves, Poland 335
Ojców National Park, Poland 333
Ojo Guarena, Spain 327, 525
Ok-san Cave, South Korea 212
Okshola/Kristihola, Norway 342
Oktyabr'skaya Cave, Siberia 646
Old Homestead Cave, Nullarbor Plain, Australia 121, 546
Old Salts Cave, Kentucky, United States 359
Oligochaeta
   Enchytraidae 67
   Haplotaxidae 67
   Hirudinea 66, 67
   Lumbricidae 67
   Lumbriculidae 67
   Naididae 67
   Ocnerodrilidae 67
   Parvidrilidae 67
   Phallodrillinae 67
   Tubificidae 67
Olimpijskaja-Lomonosovskaya, Russia 345
Olmec culture and art in caves 92
Olsen, John 129
Oluk Köprü, Turkey 699, 741
Olympyskaya, Pinega gypsum caves, Russia 588
Oman
   geomorphology 115
   Hatat Lohum 115
   Jabal Akhdar 115
   Kahf Hoti 115
   Quanaf Cave 115
   Selma Plateau **639–640**
      Bayn Halayn (Arch Cave) 639
      Kahf Aqabat Khushil (Seventh Hole) 639, 640
      Kahf Khasha (Funnel Cave) 639
      Khonshilat beyn al Hiyool 639
      Khonshilat Maqandeli 639
      Khonshilat Minqod 639
      Majlis Al Jinn (Meeting Hall of the Spirits) 639
      Three Window Cave 639
   Tawi Atayr 115
Ombla, Croatia 699
Ombla power plant, Croatia 739
On Canadian Caverns (Gibbs) 51
Oniscidea 256
On the Origin of Species by Means of Natural Selection (Darwin) 347
Onychophora 24
Opaki ponor, Bosnia/Herzegovina 291
opal
   and biomineralization 636
   Cheju-Do Lava Caves, South Korea 212
   formation of 513
   precipitation of 650
   and quartz dissolution 300
Opilionida 75
   adaptations 4

Cyphophthalmi
   Pettalidae 75
   Sironidae 75
France 372
Hendea myersi cavernicola 4
Laniatores
   Gonyleptidae 75
   Phalangodidea 75
   Phalangodinae 75
   Trabunidae 75
   Maiorerus randoi, Lanazarote 179
North America 46
Palpatores 75
South America 58
terrestrial cave habitats, biodiversity in 143, 144
Troglus torosus 759
optical brighteners and water tracing 772
optical fiber and cave communication **238**
optically stimulated luminescence 279
Optimistychna, Ukraine 345, 525, 663, 746, 747
Ora, Nakanai Caves, Papua New Guinea 538
Orchestra Caves, Australia
   art in 89
Order of the Brothers of Purity (Ikhwanaus Safa) and karst
      hydrology 478
Ordesa National Park, Spain 327
Oregon Cave National Monument, Oregon, United States 52, 733, 734
Organ Cave System, West Virginia, United States 70, 750
   depth of 525
   history of speleology 51
organic acids and dissolution rates 300
organic resources in caves **547**
organisms, classification of **548–549**
   limnostygobites 549
   stygofauna 548, 549
   thalassostygobites 549
   troglobites 548
   troglofauna 548
   troglophiles 548
   trogloxenes 548–549
Organos, Sierra de los Cuba 272
Orlova Chuka, Bulgaria 329, 330
Oropouche Cumaca Cave, Trinidad 190
Orthoptera **451–452**
   adaptations 452
   biodiversity 144
   Caelifera 451
   classification 451
   crickets in North America 46, 47
   crickets in South America 58
   distribution 451–452
   Ensifera 451
   Grylloidea 451
   morphology 451
   Phalangopsidae 25
   Rhaphidophoridae 25
   Rhaphidophorinae 451
Osselles, France 371
Ossovsky, G. 662

Ostariophysi 593
   Balitoridae 593
   Cyprinidae 593
osteodontokeratic culture 17
osteogenic cave minerals and biomineralization 636
Ostracoda 255
   adaptations of 267–268
   Bairdioidea 267
   Ciprididae
      Cypridopsinae 267
   Cypridocopina 267
   Cypridoidea
      Canonidae 267
   Cytheroidea 267
   Darwinulidae 267
   distribution of 24, 65, 267
   enthocytherids
      Sphaeromicrolinae 267, 268
   evolution of 267
   Halocyprididae 267
   morphology 267
   Myodocopa 267
   origins of 267
   Paleocopida 267
   Platycopida 267
   Podocopa 267
   Sigillioidea 267
   Sphaeromicola moria 315
   Thaumatocyprididae 267, 268
Otjikoto Lake National Monument, Namibia 21
Otoška Jama, Postojna-Planina Cave System, Slovenia 601
Ottonelli, Nello 301
Otvos, Ervin 605
Otway Basin, Australia 620
Oumou Caves, Chad 608
Out of Africa I migration 19, 196
Out of Africa II migration 19, 196
Owen, David 49
Owen, Louisa 49
Owen, Mount National Park, New Zealand 542
Owl Cave, Washington, United States 92
Ox Bel Ha system, Mexico 303, 525
oxygen fractionation dating 556, 557
Oyusutskaya–9 Cave, Siberia 645
Ozark blind salamander *See Thyphlotriton spelaeus.*
Ozark cavefish *See Amblyopsis rosae.*
Ozark plateau, United States 52
Ozerna, Ukraine 345
Ozernaja, Ukraine 525, 746, 747

Pacheco, Alonso 59
Packard, Alpheus Spring 51, 152–153
Packard, L. 149
Padirac Cave, France 372
Padirac River, France 371
Padrirac (Rudaux) 88
Padurea Craiului mountains, Romania 329
paedomorphosis 348
Painted Cave, California, United States 491
Painted Cave, Niah, Malaysia 108

Pair-non-Pair cave, France 85
Paiute Cave, Arizona, United States 251
Pakistan
   geology 444
   history of exploration 445
      Kach Gharra 445
      Moghul Well Gharra 445
      speleothems 445
Pakistan Cave Research Association 445
Pak Ou Caves, Laos 103
Palabres Cave, France 167
Pala Celar, Italy 469
Palacios-Vargas, José 38
Palaemonias ganteri, Mammoth Cave, Kentucky, United States 500
paleodrainage patterns of cone karst 243
paleoenvironments
   clastic sediments **553–555**
   speleothems **556–557**
      absolute dating 692
      amino acid racemization dating method 556
      annual deposition band dating 557, 558
      carbon–14 (^{14}C) dating method 555
      carbon oscillation dating 558
      equilibrium fractionation 556
      kinetic fractionation 556
      oxygen fractionation dating 556, 557
      paleoclimate studies and 215
      paleomagnetism dating method 556
      uranium-lead dating 556
      U-series dating methods 555–556, 557
Paleoindian occupation of caves 43
paleokarst 216, 221, 475, **559–561**
   and caves without roofs 559, 560
   characteristics of 559
   diagenetic settings 559
   economic resources of 560
   finding/recognizing 560–561
   fossil karst 559
   and hydrocarbons 176, **431–432**
   minerals in 330, 332
   and plate tectonics 559
   preservation of 559–560
paleomagnetism and sediment dating 554–555
paleomagnetism and speleothem dating 556
paleontology in caves **562**
Palmarito, Sistema Cavernario, Cuba 271
Palmer, Arthur 52
Palmer, Leo 166
Palmer, Rob 302
Palmer River, Australia **216**
Palmerville Fault, Australia **215**
Palo Duro canyon, Texas, United States 53
Palpigradida **76**
   Diversity 143–144
   Eukoeneniidae **76**
   France 371
   Indonesia 632
   Protokoeneniidae **76**
Pál-völgy cave, Hungary 332
palynology **566–567**
   applications 567
   closed depression fills 566–567
   defined 566
   palynomorphs, preservation of 566, 567
   palynomorphs, transport mechanisms 567
Pamir Mountains 96
   salt karst 96, 208
   Syjkyrdu Cave 94, 96
Pamukkale, Turkey 187, 208, **568–569**, 737
   environmental problems **569**
   hydrology **568**
   sinter terraces of 187
   thermal springs, composition of 568, 569
   travertine formations 568–569, 737, 740
Panagia Eleousa cave, Greece 208
Panama
   stygobitic fauna 38
   troglobitic fauna 38
Pan and religious sites 623
Panda Lake, Jiuzhaigou, China 424
Pan de Azucar massif, Cuba 271
Pan de Guajaibon massif, Cuba 271
Panlong Cave, Yangshuo, China 783
Pantjukhina Cave, Georgia 202, 203, 487, 663
Panyang caves, China 218
Pao Phu Tzu (Ko) 97
Papua New Guinea
   Arcturus, Nakanai Caves 538
   Atea Kananda 105
   bauxite production in 136
   Bikbik Vuvu, Nakanai Caves 538
   burials in caves 105
   cave art in 89
   Galowe 699
   Gamvo, Nakanai Caves 538
   geomorphology 105–106
   Gouffre Muruk, depth of 525
   Ka 2, Nakanai Caves 538
   Kaijende arête and pinnacle karst **467–468**
   Kalate Egeanda, cave art in 89
   Kavakuna, Nakanai Caves 538
   Kutubu, Lake 620
   Lorenz National Park 105
   Luse, Nakanai Caves 538
   Luse doline, depth of 526
   Mamo Kananda 105
   Matali 699
   Minye, Nakanai Caves 538
   Minye doline, depth of 526
   Muruk cave 105, 372
   Nakanai Caves **538–540**
   Nakanai Mountains, New Britain 308, 538
   Nare, Nakanai Caves 538
   Ora, Nakanai Caves 538
   Selminum Tem Cave 105, 634
   soil erosion 659
   Tobio 699
Paradise Ice Caves, Washington, United States 53, 204, 387, 605, 606
paragenesis **569–571**

defined 569
and dissolution 569
erosional features 570–571
looping caves systems and 571
paragenetic evolution, phreatic conduit 570
paragenetic passages, morphology 570
paragenetic pendants 571
sediment influx, causes 571
vadose canyons, morphology 570
and vadose entrenchment/alluviation 569
vadose notches 571
Parallel Tunnels, Baja, Mexico 491
Paranthropus robustus 17
parapatric speciation 665
paratacamite 329
Pareis, Torrent de, Spain 338
Parietal Markings Project 85
Parjashchaya Ptitsa Cave, Georgia 201
Parker, Rob 302
Park's Ranch, United States 345
Parque Nacionale Laguna del Tigre, Guatemala Central America 620
Parthenon, and limestone 551
Patagonia marble karst, Chile **572**
dissolution, rate of 572
geomorphology 572
hyro-aeolian karren 572
karst systems 573
Nothofagus forest 572
Patale Chhango Gupha, Nepal 444
pathogens
pollution of groundwater, **404, 405**
pollution of groundwater, remediation **407**
patterns of cave development **573–575**, 667–668, 684–685
Patterson, William 127
Pauropoda 534
biodiversity 144
Paviland Cave, Wales 80, 281
Pawan, J. 293
Peak District, England 342, 363, 561, **576–577**
bedding discontinuities 575
Blue John Caverns 578
Buxton sump 302
caves of 576–577
drainage 576
Elderbrush Cave 576
Eldon Hole 164, 576
geomorphology 575, 657
Giants Hole 577
Giants-Oxlow system 576
and limestone quarrying 610
mineral mining in 576, 713
Peak Cavern 238, 301, 302
morphology 521
musical performances in 533
ropemaking in 426
Peak-Speedwell system 166, 576
radon and 577, 618
Speedwell Cavern 576, 714
speleogenesis 575
Titan Shaft 576, 577

Treak Cliff 576
White Peak 575, 576
Pearce, Ken 302
Pech Merle, France 428
Pečina na Hudem Ietu, Kras, Slovenia 486
Peggau Cave, Syria 704
Peik Kyin Myaung Cave, Myanmar 103
Pei Wenzhong **79**, 224
Peking Man 79, 221
Pele and ritual cave use 624
Peloponnese
Apidima Cave 340
Kapsia Cave 340
Penck, Alfred 380, 670
Pengelly, William **77, 78–79**, 162
Pen Park Hole, England 164, 714–715
Pentagon, and limestone 551
Pên Tshao Kang Mu (Lin) 98
pepinos 242
Peracarida
Mysidacea 256
anchialine habitats 65
Heteromysoides cotti, Lanazarote 179
Troglomysis vjetrenicensis 760
Peracu National Park, Brazil 93
Perak Man 108
Perco, G.A. (Perko) 336
Perdido, Monte, Spain 327
Pericarida 255–256
peripatric speciation 665
Peristeri Cave, Greece 340
Peritricha, Dinaric karst 289
periwinkle zone 231, 232
permafrost pseudokarst 606–607
Perou, Sid
Journey Into Amazing Caves 360
Sunday at Sunset Pot 360
The World About Us 360
Personal Narratives of Travels to the Equinoctial Regions of America during the years 1799–1804 (Humboldt) 59
Perte du Futur, Chile 55, 56
Pertrified Forest National Park, Arizona, United States 590, 591
Peru
El Tragadero 55
geomorphology 55
Gruta de los Guácharos 55
Gruta Huagapo 55
Millpu de Kaukiran 55
San Andrés de Cutervo National Park 55
Sima Pumacocha 55
Sociedad Peruana de Espeleologia 60
Uchkopisjo 55
Peruaçu, Bambuí karst, Brazil 133, 134
Peştera Hodbana, Romania 329
Peştera Româneşti, Romania 533, 704
Peştera 6S de la Mănzăleşti, Romania 329
Peştera Tausoare, Romania 329
Peştera Vântului, Romania 329
pesticides
poisoning 294

pesticides *(Continued)*
    pollution of groundwater, dispersed **403**
    pollution of groundwater, point sources **404**
Petralona Cve, Greece 339
petrifying springs 702–703
Petrochilos, J. 339
Petroglyph-St Herman's Cave, Belize 140
Peynirlikönü Cave (Evren Gunay Cave), Turkey 741
Peyre, J.-C. 494
Peyrony, Dennis 757
phenotypic plasticity 348
philately 703–705
Philippines
    archaeological caves 109
    bauxite production in 136
    Bohol karst 104, 113
    burials in 167
    Duyong Cave 167
    Guri Cave 109
    history of exploration 113
    Laurente Cave 109
    Lungun Cave 168
    Manunggul Cave 168
    Matangkib Cave 168
    Puerto Princesa Subterranean River National Park 105
    Rebel Cave 109
    St Paul's Cave 113
    St Paul's Underground River 105, 106
    Tabon Cave 109, 168
    Tadyaw Cave 168
Philippines Mountaineering Federation 113
Philipus and dye tracing 771
Pho Binh Gia, Vietnam 167
Pholeoteras euthris 516
Pholeoteras zilchi 516
Pholeuon glaciale 448
Phong Na System, Khammouan, Laos/Vietnam 485
Phong Nha Cave, Vietnam 483, 485, 624
Phong Nha system, Vietnam 102
phosphate paleokarst 560
phosphates 329, 513
phosphorites 514
phosphorus and soil pollution 659
photographing caves **578–581**
    basic tenets of 578
    conservation and 579
    digital cameras 581
    electricity 579
    electronic flashguns 579
    flashbulb 579, 581
    flashpowder 579
    guide number formula 579
    lighting details of photographs 580
    limelight 579
    magnesium lighting 578
    motivations behind 578–579
    panchromatic film 579
    slave units 579, 581
photoreceptors 3, 4–5, 10, 63
photoresponse, behavioral adaptation **3**

phreatic cave passage 521, 522
phreatic zone 454
Phreatoicidea 256, 257
phreatomorphology 454
Phroso (Mercanton) 359
Phyllostomidae 225, 226
physiological adaptations *See* adaptations, physiological.
phytokarst
    features of **581–582**
    formation **582**
Piaggia Bella, Italy 326
Picos de Europa National Park, Spain
    Asopladeru de la Texa 583, 584
    Castil 583, 585
    caves of 582, 583
    closed depressions of 583
    Conjurtao 585
    Cueva de Culiembro 583, 584
    Cueva del Agua 583, 584
    F20 585
    geomorphology 582–583
    hyrdrology 584–585
    Jorcada Blanca 585
    M2 cave (Pozo de Cueltabo) 583, 584
    pitch-ramp systems 583
    Porru la Capilla 585
    Pozo de la Garita Cimera 583
    Pozo del Llastral 583, 584
    Pozo Jultayu 583, 584, 585
    Pozu Cabeza Muxa 584
    Pozu les Cuerries 584
    Pozu Madejuno (Omega 45) 584
    Sil de Oliseda 583, 585
    Sima 56 584
    Sima del Jou de la Canal Parda 584, 585
    Sistema del Xitu 584
    Sistema Trave 583, 584, 585
    Sistema Verdulluenga 585
    speleogenesis 583
    Torca Castil (PC–15) 584, 585
    Torca de los Rebecos (T27) 584
    Torca del Cerro 585
    Torca del Certo del Cuevón (T33) 584
    Torca del Jou de Cerredo 584
    Torca Idoúbeda 583, 584
    Torca Tortorios (CT–1) 584
    Torcas Urriellu 583, 584, 585
Picos d'Europa, Spain 327, **582–583**
pictographs in caves
    Brazil 93
    Caribbean 92
    North America 44, 92
Pierre Saint-Martin, France/Spain 327, 371, 525, **587–588**
    Aranzadi passage 587
    Arresteliako Ziloa 586
    BG63 586
    BU56 cave 585, 586
    caves of 587–588
    Chipi Joseteko 587
    geomorphology 585, 586

Gouffre AN8 586, 587
Gouffre de la Pierre Saint-Martin 586
Gouffre des Bourrugues (B3) 586
Gouffre des Partages (M413) cave 585, 586
Gouffre du Couey Lotdge 586
Gouffre Lonné Peyret 586
Grotte d'Arphidia 586, 587
history of exploration **585**
Las Puertas de Illamina 586
La Verna chamber 585
Réseau de Sondet (BT6) 586
Romy cave 586
SC60 cave 587
Sima de Ukerdi Abajo (UK4) 586, 587
speleogenesis 586–587
Pigeon Grotto, Lebanon 703
Piglaith Cave, France 370
Pignostygus galapagoensis 448
pigotite
    formation of 636
Pikhåg Caves, Norway 705, 706
Pimelodella kroneii 57
Pinargözü Cave, Taurus Mountains, Turkey 741
Pindaya Cave, Thailand 103, 113, 624
pineal organ and photoresponses 3, 10
    in *Proteus anguinus* 63
Pinega gypsum caves, Russia **588–589**
    dendritic cave pattern of 588–589
    geomorphology 588
    Golubinskaya 589
    ice displays 589
    karst conduits 589
    Kulogorskaya 589
    Lomonovskaya 588, 589
    Olympyskaya 588
    permafrost zone 589
    polygonal karst of 588
    Symphoniskaya 589
    Zheleznye Vorota (Iron Gates) 588
Pinega National Park, Russia 589
Pine Point, Northwest Territories, Canada 176, 714
Pin Hole, England 162
pinnacled rockheads and construction 247
Pinnacles, The, Sarawak, Malaysia 532
Pinnacles National Monument, United States 53, 721
Pionerskaja Cave, Georgia 202
Piper's Cave, Scotland 534
piping caves 204, **589–592**
    defined 590
    extraterrestrial 592–593
    geomorphology 591
    humid lands and 591
    overview 589–590
    significance of 590
    speleogenesis 590, 591
    types of 590
piping 252
Pippikin Pot, England 441
Pirin National Park, Bulgaria 330
Pisces **593–597**

Africa 25
Amblyopsidae **595–597**
    adaptations 12
    Amblyopsis spelea 593
    North America, distribution of 46
    Speoplatyrhinus poulsoni 46
    Typhlichthys subterraneus 244
anchialine habitats 65
blind cave fish 98, 148, 151, 152, 182, 183, 594
Characidae
    Astyanax fasciatus 5, 38, 594, 708
Clarias cavernicola 21
Cyprinidae
    Caecobarus geertsii 594
distribution of 593
Gobiidae
    Milyeringa veritas 182, 183, 594, 594
Ictaluridae
    Satan eurystomus 11, 315, 594
    Trogloglanis pattersoni 11, 315, 348, 594
Ostariophysi 593
Pimelodella kroneii 57
Pimelodidae
    Pimelodella kronei 594
Poeciliidae
    Poecilia mexicana 594
Tilapia guinasana 21
troglomorphic features of 594
Pitaka 365
pit caves 189
Pit H Cave, Hawaii, United States 249, 250
Pit of the Jinns Cave, Turkey 208
pitons 242
Pivka Jama, Postojna-Planina Cave System, Slovenia 601
placer deposits in karst **524–515**
plague 294
planar table technique and cave surveying 715–716
Planinsko polje, Slovenia 288
Platé, Désert de, France 326
Plato
    The Allegory of the Cave 206
    and karst hydrology 478
Platycopioida 259, 260
Platyhelmithes 457
    Turbellaria
        anchialine habitats 65
        morphology 457
        Temnocephalida 458
            Scutariellidae 458
        Tricladida 457
Pleocyemata 261
Plesianthropus transvaalensis 17
Plitvice Lakes, Croatia **597–598**
    fauna of 598
    geomorphology 597–598
    hyrdrology 598
    Kozjak Lake 598
    Prošćansko Lake 598
    tufa barriers of 597
Plummer, Wigley and Parkhurst law 187

Plutonium Cave, Turkey 208
PN77 cave, France 373
Pobednaja Cave, Turansky Plain Central Asia 95
Podol'sko-Bukovinsky region, Ukraine 628
point recharge depressions 305
Poland
    Bandzioch Kominiarski Cave 331
    crevice caves 249
    Czatowice Quarry, paleokarst 560
    history of exploration 335–336
    Krakow-Wielun Upland 331
    Niedzwiedzia Cave 331
    Sniezna Studnia 331
    and speleotherapy 698
Smocza Jama 335
Sniezna Studnia 331
    Szachownica Cave 331
    Wielka-Snizena Cave System 331
    Wysoka-Za-Siedmiu Progami Cave 331
Pol-an-Ionain, Burren glaciokarst, Ireland 171
Polar Caves, New Hampshire, United States 721, 722
poljes **599–600**
    anthropogenic influences 599
    classification 599
    and construction 599–600
    defined 599
    Dinaride poljes **291–293**
    distribution of 599
    formation of 599
    hydrology 599
    origin of term 288
    water storage and 599
Polldubh cave, Burren glaciokarst, Ireland 171
pollen, study of *See* palynology.
Poll na gCeim, Burren glaciokarst, Ireland 171
Pollnagollum cave system, Burren glaciokarst, Ireland 171
Poll na Mona, Ireland 591
pollution
    and agricultural activities 244
    America, Central 37
    artificial light 245
    bauxite mining 136
    cave diseases 293
    dispersed **403–404**
        herbicides/pesticides 403
        management of 404
        pathogens 403
        regulation of 403–404
        sources of 403
    effects on cave biota 243, 244, 245
    flue gas desulfurization (FGD) and limestone 490
    human impact on cave biota 244, 245, 246
    indicators of 183
    limestone, use in mitigating 490
    nutrient input, excess 244
    phosphorus and soil 659
    point sources **404–405**
        Clean Water Act (CWA) 405
        Comprehensive Environmental Response, Compensation, and Liability Act (CERCLA) 405
        contaminants, persistence of 405
        National Pollution Discharge Elimination System (NPDES) 405
        non-aqueous phase liquids (NAPL's) 404–405
        pathogens 405
        pesticides 404
        polychlorinated biphenyls (PCB's) 404
        regulation of 405
        Resource Conservation and Recovery Act (RCRA) 405
        Safe Drinking Water Act (SDWA) 405
        sources of 404
        Superfund Act 405
        Underground Injection Control (UIC) Program 405
        volatile organic compounds 404
    remediation **406–407**
        aquifer protection techniques 406
        contaminants, movement of 406
        hydrocarbon fuels 407
        industrial/sewage releases 406
        key points of 406
        non-aqueous phase liquids (NAPL's) 407
        pathogens 407
        polychlorinated biphenyls (PCB's) 407
        strategies for 406
    soils, carbonate karst 659
Polychaeta
    Australia 182
    distribution 66–67
    Lanazarote 179
    morphology 66
    Nereididae 66
    Nerillidae 66
    Serpulidae
        Marifuga cavatica 67, 289, 290
polychlorinated biphenyls (PCB's)
    pollution of groundwater, point sources **404**
    pollution of groundwater, remediation **407**
Polydesmida
    morphology 535
polygonal karst 233, 363
Poneretka Cave, Russia 627
ponors **600–601**
    defined 277, 600
    and flood prevention 600–601
    morphology 600
    origin of term 288
    ponor swallow capacity 600
    pre-ponor retention 600
    role of 600
    sea (marine) ponors 600
    as storage basins 601
    water storage and 600
Pontnewydd Cave, Wales 160
Pope Calixtus III and caves 85
Popov, I. 662
Popovo Polje, Herzegovina 291–292
Porifera 457
    anchialine habitats 65
    Cladorhizidae
        Asbestopluma sp. 457

Dermospongiae: Spongilidae
   Eunapius subterraneus 457
  distribution 457
  morphology 457
Porru la Capilla, Picos de Europa National Park, Spain 585
Port Miou, France 338
postcards **703–705**
Postojna-Planina Cave System, Slovenia 289, **601–602**
  biospeleology **603–604**
    ecology of caves 603
    fauna, aquatic 603–604
    fauna, terrestrial 604
    history of biospeleology 148
    history of research 603
    pollution and 604
    troglomorphism 603
  and cave surveying 715
  Črna Jama 601
  geomorphology 601
  humans and 601
  hydrology 602
  military use of caves 510
  Otoška Jama 601
  speleogenesis 601–602
Postojnska Jama 288, 336, 533, 601, 725, 726
  Valvasor, Johann, documentation of 87, 601
Postojnsko polje, Slovenia 288
Postumia, Grotta di, Italy 336
Posuško Polje, Bosnia/Herzegovina 291
Potential Alpha Energy Concentration (PAEC) 617
Poucha, Pavel 99
Powell, Penelope 166
Pozo de la Garita Cimera, Picos de Europa National Park, Spain 583
Pozo del Llastral, Picos de Europa National Park, Spain 583, 584
Pozo del Madejuno, Spain 525
Pozo Jultayu, Picos de Europa National Park, Spain 583, 584, 585
Pozu Cabeza Muxa, Picos de Europa National Park, Spain 584
Pozu les Cuerries, Picos de Europa National Park, Spain 584
Pozu Madejuno (Omega 45), Picos de Europa National Park, Spain 584
Pozzo A, Italy 345
pre-adaptations 348
Precambrian Shield, Canada 175
precipitation of calcite, rates of 187, 188
Predjama Cave, Slovenia 510
Presle Cave, Belgium 139
pressure springs 699
Priapulida 459
Proceedings of the Univeristy of Bristol Spelaeological Society 464, 465
Proždrikoza ponor, Bosnia/Herzegovina 291
Prolagus sardus 339
Promeszutochnaya *see* Cupp-Coutunn Cave.
Prošćansko Lake, Plitvice Lakes National Park, Croatia 598
Proshchal'naya, Siberia 646
Prosobranchia
  Cyclophoridae
    Pholeoteras euthris 516
    Pholeoteras zilchi 516
  Hydrocenidae
    Hydrocena cattaroensis 516
Prospero's Cave, Bermuda 534
Prostigmata
  Spelaeorhychidae 74
  Trombiculidae 74
  Wenhoeckiidae 74
protection of karst **245–246**
  biodiversity 150, 244
  Ramsar sites 619
  speleothems 246
  U.S. Federal Cave Resource Protection Act 774
  Vistor Experience and Resource Protection 519
protection of groundwater **408**
  COST action 65, groundwater 408
  COST action 620, groundwater 408
  DRASTIC method, groundwater 408
  EPIK system, groundwater 408
  GOD method, groundwater 408
  groundwater 408
  groundwater ecosystems 246
  groundwater remediation, aquifer protection techniques 406
  REKS system, groundwater 408
Protelsonia hungarica 724
Proteus anguinus **61–62**, 279, 289, 366, 370
  adaptations in 6, 11
  and biospeleological study 63
  discovery 148
  distribution 61, 62
  embryonic development 62–63
  in folklore 61, 148, 365
  hearing 63
  lateral-line sensory system 63
  magnetic field sensitivity 63
  morphology 62
  neoteny 62
  nutritional requirements of 63
  threats to 63
  visual system 63
Proteus anguinus parkelj **62, 63**
Protista 366, 507
Protocyon troglodytes 157
Protopithecus brasiliensis 157
protozoa 508
Proval-Krasnaya system, Ukraine 253
prusiking and cave exploration 354
pseudokarst 204, 475, **605–608**
  badlands 591, 606
  classification of 605–606
  consequent 607
  controversies surrounding 605
  crevice 249, 606
  defined 604, 653
  glacial 385–387, 606
  history of research 605
  Mars 608
  multiprocess/interface caves 607
  permafrost 606–607
  piping caves 589–590
  quartzite caves 607–608, 650–651

pseudokarst *(Continued)*
   rheogenic 606
   significance of 604–605
   talus 606, 721–723
   types of 604
   volcanic **760–764**
Pseudoscorpionida 46
   biodiversity 143
   Bochicidae 75
   Canary Islands 182
   Cape Range, Australia 182
   Chthoniidae 75
   in Dinaric karst 289
   France 372
   Neobisiidae 75
   Neobisium spelaeum 603
   North America 46
   Syarinidae 75
   Vachoniidae 75
pseudoscorpions *See* Pseudoscorpionida.
psilocybin mushrooms 425
Psocoptera
   distribution 143, 144, 452
   morphology 452
psychrometer, measurement using 228
Pterygota **451–453**
   Blattodea 25, 110, 143, 180, 451
   Dermaptera 25, 143, 180, 452
   Diptera 143, 453
   Hermiptera 452–453
   Lepidoptera 25, 143, 453
   Orthoptera 451–452
   Psocoptera 143, 452
   Trichoptera 453
Puertas de Illamina, Spain 525
Puerto Princesa Subterranean River National Park, Philippines 105
Puerto Rico
   caves of 192
   composition of 190
   karst 242
   Karst Belt 190
   mogotes of 190
   Mona, Isla de **517–518**
   Sistema del Rio Encantado 190
Pulmonata
   Sciocochlea collasi 516
   Sciocochlea nordsiecki 516
Pulo di Molfetta, Italy 410
purgatory caves 722
Putnam, Frederick Ward 81, 153
PWP equation 187, 188
   dissolution and 295, 296
Pyin-Oo-Lwin, Myanmar 102
Pyramid of the Sun, Teotihuacan, Mexico 40, 42
pyrite
   and breccias 660
   formation of 714
pyrolusite and biomineralization 636
pyrolysis, underground and cave formation 204
pyrothine 530

Qafzeh Cave, Israel 196
Qatar 562
Qattine Azar, Lebanon 114, 463
Qikeng Dong, China 220
Qixing Dong (Seven Star Cave), Yangshuo China 783
Quanaf Cave, Oman 115
quarrying of limestone **609–610**
   blackpowder 609
   blast fracture cones 610
   environmental impacts of 610
   explosives 609–610
   extraction techniques 609
   face-dressing 609
   groundwater pollution 610
   history of 609
   human impacts of 610
   piking 609
   site restoration 611
quartz
   Cheju-Do Lava Caves, South Korea 212
   crystallization rates 300
   formation of 513
   and karst dating 282
   Valea Rea Cave, Romania 329
quartzite, pinnacle fields in 650
quartzite caves 252, 607–608, **650–651**
   America, South **611–613**
      classification 612
      cuesta-type, relief cave 613
      fissure vertical caves 612
      history of research 611–612
      Inficionado karst 612
      South Africa 651–652
   speleogenesis 300
quartzite-dolerite caves 651, 652
Quashies River Cave, Jamaica 189–190
Quasi-National Park of Japan
   Akiyoshi-dai, Japan 29
Queen Charlottes karst, Canada 178
Queenslander Cave, Australia 216
quifeng karst 218
Quintana Roo, Mexico 37
Qumran Caves, Israel 208

Rabanel, France 371
rabies 294
rabies epidemic, 1930s 191
Rachinsky Massif 203
Racoviţă, Emil G.
   biography of **153, 154**
   Biospéologica 154
   Essai sur les problèmes biospéogiques 149, 154
radioactive isotopes and dye tracing 772–773
radioactive waste poisoning 294
radio frequency modulation and cave communication 238
radiocarbon dating of speleothems 557–558
radiolocation 615–616
   applications 615
   and cave surveying 717
   depth, determining, formula 615

electroconductivity and 616
history of 615
methods 615
skin depth 616
radios in caves 238
Radofilao, Jean 493
radon in caves **617–618**
exhalation, factors influencing 618
exposure and health 618
Kusnetz method 617
medicinal/therapeutic uses of 618, 697
Peak District, England 577
radon progeny 617
Radon Sniffer, Thompson and Neilson 618
track etch detectors 617
units, standard 617
Radon Sniffer, Thompson and Neilson 618
Rafinesque, Constantine 49
Rahr Amalou, Algeria 15
Rahr Es Skhoun, Algeria 15
Rahr Medjraba, Algeria 15
Railton, Lewis 166
Rainbow spring, Florida, United States 699
Rakhiot Peak Cave, India 442
Rákóczi Cave, Hungary/Slovakia 26
ramiform mazes, morphology 573
Ramsar Convention 243, 246, 619
Ramsar sites **619–620**
conservation values 619
criteria for inclusion 619
list of 619–620
Ramsar Classification System 619
Ramsar List 619
wetlands *vs.* karst, commonalities **619**
Rana, Buso de la, Italy 326
Rand gold mining, South Africa 22–23
Rankin Ponor, Serbia 329, 330
rappel rack and cave exploration 353–354
Ras el ain, Syria 699
Rassegna Speleologica Italiana 464
rate laws
dissolutional
carbonate rocks **295–298**
evaporite rocks **298–300**
gypsum
dissolution, nonlinear 678
limestones
dissolution, nonlinear 678
quartz dissolution 300
Rat's Nest Cave, Canada 177
Raucherkarhöhle, Austria 174, 525
Raytchova Doupka, Bulgaria 329, 330
reaction, rates of 187
reaction norm 348
Rebel Cave, Philippines 109
recreational caving **621–622**
artificial caves 622
habits, changes in 621
history of 621
human impacts of 621–622

and karst management 622
restrictions on 622
Recreational Opportunity Spectrum 519
Reddan, William 128
Red del Silencio, Spain 525
Reddington Caves, Pennsylvania, United States 510
red imported fire ant *See* Solenopsis invicta.
Red Lady of Paviland 80, 161
redox reactions, microbially mediated 194
Redwall caves, Arizona 392, 393
Reed's Cave-Baker's Pit cave system, England 166
reefs and dolomitization 185
refugium model of colonization 235, 237, 268
regression model of colonization 237
Reka River, Kras, Slovenia 486, 653
REKS system, groundwater protection 408
relative humidity in caves **228–229**
relict
caves
Castleguard Cave, Canada 200
Dobšina Ice Cave, Hungary/Slovakia 25, 27
Khammouan, Laos/Vietnam **483–485**
Mendip Hills, England **503–505**
preservation of 559
Stoke Lane Slocker system, England 503
fauna
Canary Islands 180
Cape Range, Australia 181
Southeast Asia 111
See also individual taxa.
karst 475, 559
See also paleokarst.
religious sites **622–624**
karst landforms and 622–623
management of 624
sites worldwide 623–624
traditional views on 622
Reliquae Diluvianae (Buckland) 164, 208
Remane, A. 149
Remipedia 65, 254
Australia 126, 183
Bahamas 156
Canary Islands 180
Remouchamps Cave, Belgium 139
Renaissance Classicism and cave art 86
Renejevo brezno, Slovenia 468, 469
Reotier travertine cascade, France 737
reproductive behaviour and behavioural adaptation **3**
rescue in caves
bad air 2
critical factors in 1
floods and 2
organizations 1
radios 239
requirements of 1
rock fall/movement and 2
self-rescue 2
Réseau de l'Alpe, France 525
Réseau de Sondet (BT6), Pierre Saint-Martin, France 586
Réseau Jean-Bernard, France 525

Réseau Siegenhengste-Hohgant cave system, Siebenhengste, Switzerland 647
Réseau Verneau, France 326
Reserve Naturelle Integrale, Madagascar 494
reservoirs, artificial underground 739
reservoirs, construction of **277–279**
   classification of 277
   common defects 277–278
   environmental impacts of 278–279
   history 277
   problems with 277
   sealing treatments for 277
   underground fauna, effects on 279
   unfavorable conditions for 277
   water losses, reducing 278
resistance thermometers, measurement using 228
resistivity tomography 379
Resource Conservation and Recovery Act (RCRA) 405
resources, karst **479–481**
   agriculture 479
   biodiversity 480
   burial sites, use as 167, 480
   cultural resource management 481
   dwelling places 337–338, 480
   environment, natural 480
   evolutionary change, study of 480
   forest resources 368–369, 479–480
   hydroelectric power, source of 479
   irrigation 479
   limestone, source of 480
   research, academic 480
   sacred sites, use as 480, 622–624
   speleotherapy 480, 687–698
   tourism 480–481, 725–726
   water supply, source of 479
respiratory quotient 184
Ressell, Emergence Du, France 302–303
restoration of caves **625–627**
   advantages of 626
   conservation and 627
   footprints and 626
   and karst terrains 625
   minimum-impact guidelines 625–626
   speleothems, repair of 626–627
   tools for 626
   volunteers 625
resurgence springs 701
Reynolds numbers, equation 429
Rhadine reyesi 47
Rhar Ain et Tsab, Tunisia 15
Rhar Bou Maza, Algeria 15
Rhar Chara, Morocco 13
Rhar Chiker, Morocco 13
rheogenic pseudokarst 606
Rhinolophidae 225
   Rhinolophus blasii 227
   Rhinolophus ferrumequinum 226
   Rhinolophus hipposideros 227
Rhipidogamarus mivariae 180
Rhoades, R. 671, 673

Rhodamine WT dye (CI Acid Red 388 dye) and dye tracing 769, 772
rhyolite and 762
ribbon worms *See* Nemertea.
Ribeiro, Miranda 60
Ricinulei 76, 143, 144
rift caves *See* caves, crevice.
rifting caves *See* caves, crevice.
rillenkarren 216, 470, 472
rinnenkarren
   defined 216, 470
   erosion rates 323
Rio Chontalcoatlan, Mexico 37
Rio Convention 244
Rio Frio Cave, Belize 735
Rio Stella-Rio Basino, Italy 345
River Cave, Australia 652
Riverhead Cave, Jamaica 191
Riversleigh, Australia 124
Rivière, Emile 85
Rivière de Kverkfjöll, Iceland 342
Rivista Italiana di Speologia 464
Roadcut Cave, Hawaii, United States 251
Robert the Bruce and caves 208
Robin Hood's Cave, England 162, 364
rock bolts and cave exploration 354
Rockhampton, Australia 129
rockhead, construction problems and 247
Rockhouse Cave, South Carolina, United States 722
rock salt, dissolution rates 300
rockslide/rockpile caves 722
rock tablets and erosion rates 322
rock topple caves *See* crevice caves.
Rocky Cape, Australia 123
Rocky Cape North Cave, Australia 127, 128
Rodentia 225
   Hystricidae 25
Roger Rain's House, England 533
Romania
   Avenul din Grind 329
   Avenul din Stanu Foncii 329
   Avenul V5 329
   bauxite production in 135
   Ciclovina Cave 329
   Ciur Ponor 329
   Ghețarul de la Scărișoara 329
   Huempla Poienita system 329
   Movile Cave 329, 365, 507, 516, **528–530**
   Peștera Hodbana 329
   Peștera Românești, musical performances in 533
   Peștera-Românești Timis 704
   Peștera 6S de la Mănzălești 329
   Peștera Vântului 329
   Sistemul Humpleu-Poienița 329
   and speleotherapy 698
   Tăușoare Cave 329
   Topolnița 329
   Șura Mare cave 329
   Valea Rea Cave 329, 693
Romans 86, 87, 623

Romy cave, Pierre Saint-Martin, France 586
Ronan's Well, Cape Peninsula South Africa 651
ropes 353, 354
ropewalking and cave exploration 354
Rose Cottage Cave, South Africa 19
Rosenstingel, Sebastian 87
Rosia Bay, Gibraltar 384
Rossi, G. 494
Ross Ice Shelf, Antarctica 606
Rostovskaya Massif, Georgia 201
Rotifera 458
Rouffignac caves, France 167
roundworms See Nematoda.
Roung Dei-Ho/ Roung Thom Ken system, Cambodia 102
Rousettas aegyptiacus 227
rove-beetles See Coleoptera.
Royal Arch Tower, Australia 216
Ruakuri Cave, North Island New Zealand 542
rubidium 514
rubies
   mining of 514
   Southeast Asia karst 103
Rucheynaya, Siberia 645
Rudaux, Lucien
   Aven Armand 88
   Brambiau 88
   Padrirac 88
Ruined City, Australia 123, 651
Runaway Bay Caves, Jamaica 191
Russell, Israel 53
Russell Cave, Alabama, United States
   archaeology of 43
   national monument 52
Russia
   AB 6 345
   European sector 627–628
   Iron Gates Nature Reserve 588, 589
   Konstitutzionnaja 345
   Kulogorskaja-Troja 345
   Kungur Ice Cave 628, 662
   Kungurskaja Ledjanaja 345
   Olimpijskaja-Lomonosovskaya 345
   Pinega gypsum caves **588–589**
   Poneretka Cave 627
   Siberia/Far East 629, **645–646**
   and speleotherapy 698
   Urals 628–629
   Zolotoj Kljuchik 345
   See also Georgia; Siberia; Soviet Union; Ukraine; Urals.
Russian Speleological Union 663
Rutog, Tibet 220
Rwanda
   Ubuvomo bwa Musanze 22
   Ubuvomo bwa Nyrabadogo 22
Ryadovaya Cave, Siberia 646
Ryrie, Stewart 129

Sabatagi Düdeni, Turkey 741, 742
Sac Actun, Yucatán, Mexico 787
Safe Drinking Water Act (SDWA) 405
Saguo Dong, China 218
Sai Yok, Thailand 107
Sakhalin, Siberia 646
salamander 61–62 See Amphibia.
salamander poisoning, acute 295
Salem cave fish See *Cambarus hubrichti*.
Salman Farsi Dam, Iran 278
Salsman, Gary 52
salt
   breccia pipes, formation of 661
   Canada 176, 178
   caves
      Algeria 15
      Iran 461
      Israel 208, 638
      Romania 329
   and dye tracing 772
   karst 343, 344, 345, 346
   paleokarst 559
   principle karst rock 474
Salt Mine, Wieliczka, Poland 333
saltpetre
   and biomineralization 636
   caves and 410–411
   earth and nitrocalcite 513
   and microorganisms 507
   mining 59, 410–411, 547
   Southeast Asia karst 103
   use of 410
   See also gunpowder.
Salt Pond Cave, Bahamas 676
Salts Cave, Kentucky, United States 44, 81, 92, 168
Salukkan Kallang, Indonesia **631–632**
   fauna, adaptations of 631
   fauna of 631–632
   geomorphology 631
   guano 631
   human impacts on 632
Samaria, karst canyons of, Greece 338
Sam's Point Ice Cave, New York, United States 250
San Andrés de Cuterbo National Park, Peru 55
Sand Cave, Kentucky, United States 206, 360, 534
Sand Hill Bluff Cave, California, United States 491
Sandside Head Cave #2, Scotland 491
San Luis Potosí, Mexico 92
San Miguel, hoya de, Mexico 37
Sano-ana Cave, Japan 30
San Qiao, China 220
San Salvador Island 189
Santa Catalina, Cueva Grande de, Cuba 272
Santang Dong, China 218
Santanna, Hispaniola, caves of 192
Santo Tomás, Gran Caverna de, Cuba 272
São Desidério, Bambuí karst, Brazil 133
São Domingos, Bambuí karst, Brazil 133, 134
sapphires
   mining of 514
   Southeast Asia karst 103
Sarawak Chamber, Malaysia 204, 308, 526
Sardinia
   Corbeddu Cave 339

Sarma Cave, Georgia 201, 487, 525, 663
Sartory, József 715
Satan eurostymous 11, 315, 594
Sato, Denzo 99
sauconite 273
Saudi Arabia
    Ain Hit 115
    Cave of the Falling Star 115
    geomorphology 115
Sauta Cave, Alabama, United States 411
Savarensky, F.P. 605
Savory, James Henry 579
Sawicki, L. 113
Scallops (rock relief) 522, 523
Scandinavia *See individual countries by name.*
scanning spectroflourophotometer and water tracing 772
Schaffenrath, Alois 87
Schiner, Ignaz Rudolph **152**
Schiner, J.R. 149
Schiödt, J.C. 148
Schiödte, Jørgen Mathias Christian
    biography of **152**
    Specimen faunæ subterreaneae 152
Schizomida **76**
    Cape Range, Australia 182
    diversity 143–144
Schloss Hellbrun (Solari) 86
schlottenkarren 177
Schmerling, Phillipe Charles **77, 78**, 81
Schmidl, Adolph 655
    biography **687, 688**
    and cave exploration 335
    Die Grotten und Höhlen von Adelsberg, Leug, Planina und Laas **688**
Schmidt, F. 148
Schreibers, Karl Anton von **152**
Schwäbischer Alb, Germany 341
Schwäbischer Höhlenverein, Germany 336
Schwartzmooskogeleishöhle Gebirge, Austria 174
Schwersystem, Austria 525
Scientific American and history of speleology 51
Sciocochlea collasi 516
Sciocochlea nordsiecki 516
Scladine Cave, Belgium 139
Scolopendromorpha
    Cryptopidae 535
    Lithobiomorpha 535
Scopoli, Giovanni Antonio **152**
Scorpionida 74
    Belisarius xambuei 74
    Chactidae 74
    Chaerilidae 74
    Diplocemtridae 74
    diversity 143
    Ischnuridae 74
    Nepa anopthalma 529
    Vaejovidae 74
Scotland
    archaeological/paleontological caves **160–162**
    Bonnie Prince Charlie and caves 364

Cave of Gold 534
Creag nan Uamh caves 160
Huntley's Cave 534
Oban Bay caves 160
Piper's Cave 534
Sandside Head Cave Q2 491
St Columbia's Cave 427
St Francis Cave 427
scotophilia, behavioural adaptation **3**
Scott, Raymond (*Blue Grotto in Capri*) 534
Scott, Thomas 129
Scott Hollow Cave, West Virginia, United States 70
SC60 cave, Pierre Saint-Martin, France 587
SCUBA 302
Scydmaenus aelleni 449
Scytonema julianum and calcium deposition 508
sea anemones *See* Cnidaria.
sea caves *See* littoral caves.
sea cucumbers *See* Echinodermata.
Seal Canyon Cave, California, United States 491
Sea Maze, California, United States 491
sea squirts *See* Tunicata.
Secret Cave, The (Durst) 360
Sedgwick, Adam 164
sediments
    allonchthonous **633–634**
        basin sedimentation models and 633
        chemical alteration of 634
        defined 633
        as record of history 633
        sorting of 633–634
        water flow and 633
    autochthonous **634–635**
        applications of 635
        breakdown 634–635
        critical thickness, formula 634
        defined 634
        detritous (weathering earth) 635
    biogenic **636–637**
        biomineralization 636
        detrital 636
        internal 636
        matrix mineralization 637
        nucleation biomineralization 637
        secondary (biogenic speleothems) 636
        types of 636
    clastic **553–544**
        allochthonous 553
        analysis of 553–554
        autochthonous 553
        dating of 554–555
        dissolution processes, protection from 553
        paleomagnetism 554
        sliding bed facies 553
        speleothem dating 554
        superposition, laws of 553
        terrestial cosmogenic isotope dating 555
        transport pathways of 553
Sedom, Mount Israel 300, 512, **637–638**
    caves of 638

geomorphology 637
hydraulic limit of water flow 638
hydrology 637
inlet caves 638–639
perennial brine pods 638
Sedom Cave 638
speleogenesis 637
seismic reflection surveys 379
seismic tomography, three-dimensional 379
Sekelak, Škocjanske jama, Slovenia 655
selenite 759
Seler, Edward 40
self-potential methods 379
Selma Plateau Caves, Oman
geomorphology 639
speleogenesis 639–640
Selminum Tem Cave, Papua New Guinea 105, 634
sensory apparatus adaptations, Coleoptera 450
Serbia
Bogovinska Pećina 330
Buronov Ponor 330
Cerjanska Pećina 329, 330
Dubašnica pothole 330
hypogean fauna in 142
Jama du Dubašnici 329
Kucaj mountains 330
Rakin Ponor 329, 330
Ušački Pićinski Sistem 329, 330
Velika Klisura 329, 330
Vrelo Krupac 330
Serendipity Cave, Rocky Mountains Canada 437
Serie Espeologica 464
Serra do Ramalho, Bambuí karst, Brazil 133
Serres, Marcel de 80
Sesser, W.F. 579
Seton Rockshelter, Australia 124
17 Mile Crack, Hawaii, United States 251
Severgin, V. 662
Severn Tunnel, England 301–302
Sewu **641–642** *See* Gunung Sewu cone karst, Java.
shakeholes 306, 309
Shaler, Nathaniel 49, 51
Shalmaneser III 207
Shamanskaya Cave, Siberia 646
Shamshir Ghar, Afghanistan 443
Shandon caves, Ireland 162
Shanidar Cave, Iraq 115 **642–643**
burials in 167, 643
palynology 567, 643
religious views of 623
Shanidar Neanderthals 642–643
Shapour Cave, Iran 461
Shelley, Mary
Frankenstein 206
The Last Man 206
shelly pahoehoe tubes 761
Shelta Cave, Alabama, United States 47
Shen Nung Pên Tshao Ching 98
Sheppard, Jack 166, 302
Shilin Stone Forest, China 218, **643–644**

geomorphology 643–644
Major Stone Forest 644
Minor Stone Forest 644
pinnacle karren of 473, 643
speleogenesis 644
Shivaji Cave, Nepal 623
Shopov cave system, Bulgaria 330
Short, Harry (*A Sea Cave Near Lisbon*) 359
Short Cave, Kentucky, United States 168
Shotoku Taishi 624
show caves *See* tourist caves.
shrimps **262** *See* Decapoda.
Shuzen Lake, Jiuzahaigou, China 424
Shwe Ohm-min Cave, Thailand 624
Siam Society's Journal 113
Siberia **645–646**
Abagy-Dzhie 646
Altajskaja 629, 645
Argarakanskaya Cave 645
Aya Cave 646
Aydashinskaya Cave 646
Badzhevskaja 629, 645
Bol'shaja Oreshnaja Cave 629, 645
Bol'shaya Baydinskaya Cave 646
Bol'shaya Ludarskaya Cave 646
Boro-Khukhan 646
Botovskaya Cave 645, 646, 663
Bukhtarminskaya 646
Chërtovy Vorota 646
Denisova 646
Dolganskaya Yama 646
Dvuglazka Cave 646
Dyuktaiskaya Cave 646
Eleneva Cave 646
Fantaziya 645
Geomorphology 629, 645–646
Grot Proskuryakova Cave 646
Jaschik Pandorry Cave 629
Kal'tsitovaya Cave 646
Karbonatitovaya Cave 645
Kaskadnaya 646
Këktash 629, 645
Klyuch karst shaft 646
Korolëva 645
Kubinskaja 629
Ledovaya Cave 646
Letuchaya Mysh 646
Lurgikanskaya 646
Lysanskaja Cave 629
Mechta Cave 646
Monasytuyskaya 646
Obukheikha 646
Oktyabr'skaya Cave 646
Oyusutskaya-9 Cave 645
paleolithic sites **646**
Proshchal'naya 646
Rucheynaya 645
Ryadovaya Cave 646
Sakhalin 646
Shamanskaya Cave 646

Siberia *(Continued)*
  Skriper 646
  Soantekhnicheskaya 645
  Soktuyskaya 646
  Solyanik 646
  Spasskaya 646
  Stariy Zamok 646
  Strashnaya 646
  Tëmnaya 645
  Tonta 646
  Torgashynskaja 629
  Transbaikalian 646
  Tutkushskaja 629, 645
  Ust'-Kanskaya 646
  Uzurskaya 646
  Vaydinskaya 646
  Vereshchagina 646
  Yashchik Pandory 645
  Zhenevskaja 629
Sichuan, China 218–219
Siderides, N.A. 340
Siebenhengste, Switzerland 326, 525, **647–648**
  A2 647
  Bärenschacht 647
  F1 647
  geomorphology 647, 648
  hydrology 647
  K2 647
  Réseau Siegenhengste-Hohgant cave system 647
  speleogenesis 649
  St Beatus Cave 647
Sierra Madre Oriental, troglobitic fauna 38–39
Siettitia balsetensis 447
Sigurd the Volsung 365
Siju Cave, India 443
Silberberg Grotto, South Africa 17
Silberen System, Hölloch, Switzerland 420
Sil de Oliseda, Picos de Europa National Park, Spain 583, 585
silica and quartz dissolution 300
silicate caves 300, 607–608, 651–652
silicate karst **649–653**
silicates
  dissolution
    quartz 300
    silicate minerals 301
    minerals in caves **513**
silicic acid
  complexation of 300
  dissolution of 300
  production of 301
  solubility and pH 301
silicon 273
Silliman, Benjamin 690
silo filler's disease 293
Silver Spring, Florida, United States 699
silverfish *See* Zygentoma.
Sima 56, Picos de Europa National Park, Spain 584
Sima Acopán 1, Venezuela 611
Sima Aiyán-tepui Noroesti, Venezuela 55, 611, 612
Sima Aonda, Venezuela 55, 611, 612
Sima Aonda 2, Venezuela 55, 612
Sima Aonda 3, Venezuela 55, 612
Sima Aonda Este 2, Venezuela 612
Sima Aonda Superior, Venezuela 611
Sima Auyán-tepui Norte, Venezuela 612
Sima Auyán-tepui Norte 2, Venezuela 612
Sima de la Lluvia, Venezuela 611
Sima de los Huesos, Spain 119, 562
Sima del Elefante, Spain 120
Sima del Jou de la Canal Parda, Picos de Europa National Park, Spain 584, 585
Sima de Ukerdi Abajo (UK4), Pierre Saint-Martin, Spain 586, 587
Sima GESM, Spain 327
Sima Mayor, Venezuela 612
Sima Menor de Sarisariñama, Venezuela 611, 612
Sima Pumacocha, Peru 55
Simon, Eugene 109
Simpson, Eli 166
Sinacori, N.M. 671, 673
Single Rope Techniques (SRT) 353, 372
single-sideband (SSB) operation and cave communication 238
Sinjac (Piva), Yugoslavia 699
Sinjski ponor, Bosnia/Herzegovina 291
Sinkhole Plain caves, Kentucky, United States 184
sinkholes 304, 306
  See also dolines.
Sinocyclocheilus hyalinus 98, 148, 151
Sinoto, Yoshihiko 765
Sipapu Cavern, Arizona, United States 251
Siphonaptera 25, 143, 144
siphon of Lobos, Patagonia, Chile 573
Sipuncula 459
Sirgenstein caves, Germany 333
Sistema Aak Kimin, Yucatán, Mexico 787
Sistema Abejas, Yucatán, Mexico 787
Sistema Actun Ha, Yucatán, Mexico 787
Sistema Aerolito, Yucatán, Mexico 787
Sistema Badalona 327
Sistema Cheve, Mexico 36, 525
Sistema Chikin Ha, Yucatán, Mexico 787
Sistema Cuchillo Cura, Argentina 56
Sistema Cuetzalan, Mexico 36
Sistema del Rio Candaleria, Guatemala 37
Sistema del Rio Encantado, Puerto Rico **317–318**
Sistema del Trave, Spain 525
Sistema del Xitu, Picos de Europa National Park, Spain 584
Sistema Esmerelda, Yucatán Mexico 787
Sistema Hermosura, Colombia 54
Sistema Huautla, Mexico 36, 525
Sistema Naranjal, Yucatán, Mexico 787
Sistema Nohoch Kiin, Yucatán, Mexico 787
Sistema Nohoch Nah Chich, Yucatán, Mexico 786–787, 788
Sistema Ponderosa, Yucatán, Mexico 787
Sistema Purificacion, Mexico 36, 525
Sistema Trave, Picos de Europa National Park, Spain 583, 584, 585
Sistema Verdulluenga, Picos de Europa National Park, Spain 585
Sistema Yaxchen, Yucatán, Mexico 787
Sistemul Humpleu-Poienița, Romania 329
Sivas gypsum karst, Turkey 742
Skadarsko Jezero, Montenegro/Albania 338

Skalarjevo Brezno, Slovenia 468
Škocjanske jama, Slovenia **653–655**
   archaeology 655
   geomorphology 440, 486, 620, 653
   Hankajev Canal 654
   hydrology 653
   Kačna Jama 184, 350, 486
   Sekelak 655
   speleogenesis 655
   Tiha Jama (Silent Cave) 654, 655
Skriper, Siberia 646
Slano Polje, Montenegro 293
Slato polje, Herzegovina 293
Slavka, Ukraine 345, 746
Sleeping Pools, Jamaica 190
slepe dolina 304
sliding fracture caves *See* crevice caves.
slime molds 508
Slivlje Polje, Montenegro 293
Sloane, Hans 191
Sloan's Valley Cave, Kentucky, United States 70, 244
Sloup Cave, Czechoslovakia 335, 688, 715
Slovacka jama, Croatia 525
Slovakia
   Bystra Cave and speleotherapy 698
   Demanovsky Cave System 331
   Stary-hradz 331
   Stratenska-Psie diery Cave System 331
Slovak Karst Protected Landscape Area and Biosphere Reserve, Slovakia 25
Slovak Paradise National Park 333
Slovenia
   Abisso Trebeciano, Kras 486
   Brezno pod Velbom 469, 526
   Brezno pri Risniku, Kras 486
   Čehi 2 468, 469, 525
   Cerknica Lake 211
   Cerknica Polje, history of study 210–211
   Črnelsko brezno 468, 469, 470
   fauna, thermal waters 724–725
   Grotta Gigante, Kras 486
   Hankajev Canal, Škocjanske jama 654
   history of exploration 210–211, 335
   hypogean fauna in 141, 142
   Kačna Jama 184, 350, 486
   Kanin, Mt. Italy/Slovenia 468
   Kanin massif **468–470**
   Kaninski podi 469
   Kras **485–486**
   Lazaro Jerko, Kras 486
   Ljubljanica 699
   Mala Karlovica Cave 210
   Pečina na Hudem Ietu, Kras 486
   Postojna-Planina Cave System 289, 601–602
      and cave surveying 715
      Črna Jama 601
      humans and 601
      Magdalenska Jama 601
      Pivka Jama 601
      speleogenesis 601–602
   Reka River, Kras 486
   Renejevo brezno 468, 469
   Sekelak, Škocjanske jama 655
   Skalarjevo Brezno 468
   Škocjanske jama **653–655**
      biodiversity 143
      geomorphology 440, 486, 620, 653
      Hankajev Canal 654
      hydrology 653
      Kačna Jama 184, 350, 486
      Martel's Hall 486, 653
      Sekelak 655
      speleogenesis 655
      Tiha Jama (Silent Cave) 654, 655
   and speleotherapy 698
   Strabo and Cerknicka Polje 210
   Timavo Springs, Kras 486
   Vandima 468, 469
   Velika Cave 210
   Vilenica Cave, Kras 486, 725
   Vritiglavica Shaft 204, 468, 526
Slovenskej speleologičkej spoločnosti, Slovakia 336
Sluka, Martin 716
Small River Glacier, BC Canada 388
SMAPS software and cave surveying 716
Smart, Dean 113
smectite 587
Smocza Jama, Poland 335
Snead's Cave, Florida, United States 227
Sneznaja-Mezhennogo System, Georgia 202, 487, 525
Sniezna Studnia, Poland 331
snottites, Cueva de Villa Luz, Mexico 146, 758–759
Soantekhnicheskaya, Siberia 645
So-cheon Cave, South Korea 212
Sochinsky artesian basin, Georgia 201
Social and Environmental Monitoring (SEM) programme 519
Sociedad Brasiliera de Espeleologia 60
Sociedade Excursionista e Espeleológica (SEE) 60
Sociedad Peruana de Espeleologia 60
Sociedad Venezolana de Ciencias Naturales, Speleology Section 60
Sociedad Venezolana de Espeleologia 60
Societá Speleologica Italiana, Italy 336
Société de Spéléogie, France 351, 371, 688
Société Internationale de Biospéologie, France 371, 465
Société Québecois de Spéléologie 52
Société Suisse de Spéléologic/Schweizerische Gesellschaft für Höhlenforschung, Switzerland 336
sodium flourescin dye (CI Acid Yellow 73 dye) 253, 769, 771, 772
sodium hypochlorite, use of 734
Sof Omar, Ethiopia 20, **655–656**, 699
   geomorphology 655–656
   hydrology 655
   speleogenesis 655
software and cave surveying 716
Sohoton caves, Phillipines 510
soils on carbonate karst **656–658**
   agriculture and 657
   Alfisol, order 656, 657
   allogenic 656
   autogenic 656, 657

soils on carbonate karst *(Continued)*
    characteristics, determination of 657
    colluvial 656
    conservation of 658
    and construction 657–658
    formation, rate of 657
    Inceptisol, order 656
    Mollisol, order 656
    and pollution 659
    pollution and 657
    tephra 657
    Ultisol, order 657
    water storage and 657
  grouting of, in construction 247
  sources of, forests on karst 368
soil erosion **658–659**
  erosion rates 324, 658–659
Sokolov, D. 662
Sokolov, N. 662
Soktuyskaya, Siberia 646
Solar Activity Index (SA) 695
Solari, Santino (Schloss Hellbrun) 86
Soldatskaya Cave, Ukraine 253
Soldatskaya-Karasubashi system, Ukraine 253
Solenopsis invicta 47, 245
Solomon's Prison, Iran 737
Solue Dong, China 218
solutional denudation rate and karst evolution 475
solution breccia **660–661**
solution caves **203–204**, 205, 667
solution dolines
  diameters of, determining factors 305
  distribution of 305
  glaciation and 307
  and karst evolution 477
  nomenclature 304, 306
  speleogenesis 304–307
solution fluting *See* rillenkarren.
solution pans *See* kamenitza.
solution pipes *See* karst, syngenetic.
solution processes and speleogenesis theories 669
solution runnels *See* rinnenkarren.
solution trough 306
Solyanik, Siberia 646
Somalia
  Ail Afwein 23
  karst of 23
  Las Anod 23
Somiedo, Spain 562
Sophocles and karst hydrology 478
Sotano Akemati, Mexico 525
Sótano de San Augustín, Huautla Cave System, Mexico 425
South Africa
  Apocalypse Pothole 20
  Bushmangat 20, 21
  Cango Cave 20
  Cango Caves, musical performances in 533
Southern cavefish *See Typhlichthys subterraneus.*
South Ice Cave, Oregon, United States 437
South Korea
  Cheju-Do Lava Caves **212–213**

South Nahanni karst, Canada 178, **537–538**
  geomorphology 537, 538
  hydrology 537
  speleothems 538
South Wales Caving Club 166
Soviet Union
  history of speleology **661–663**
  See also Russia; Siberia; Ukraine; Urals.
Sozina railway tunnel, Yugoslavia 738
Spain
  Agua, cueva de 345
  Aguila, sima del 345
  Altamira Cave
    archaeology **33–34**
    art in **34**, 84, 85, 90
    musical performances in 534
    as tourist cave 428
  Atapuerca Caves **119–120**
  Badalona, Sistema 327
  Campamento, sima del 345
  Coral, sima del 345
  Cova de Salnitre 607
  Covadura 345
  Cueva de Nerja 327, 338, 533
  Cueva Mayor 119
  Drac, Coves del 338
  Espluga de Francoli 607
  Estremera Cave 683
  Gouffre AN8, Pierre Saint-Martin 586, 587
  Gran Dolina 119, 564
  Grotte Casteret 327
  Matienzo depression 327
  Ojo Guarena 327, 525
  Ordesa national park 327
  Pareis, Torrent de 338
  Perdido, Monte 327
  Picos de Europa 327, **582–583**
    Asopladeru de la Texa 583, 584
    Castil 583, 585
    caves of **582**, 583
    closed depressions of **583**
    Conjurtao 585
    Cueva de Culiembro 583, 584
    Cueva del Agua 583, 584
    F20 585
    geomorphology **582–583**
    hyrdrology **584–585**
    Jorcada Blanca 585
    M2 cave (Pozo de Cueltabo) 583, 584
    pitch-ramp systems **583**
    Porru la Capilla 585
    Pozo de la Garita Cimera 583
    Pozo del Llastral 583, 584
    Pozo Jultayu 583, 584, 585
    Pozu Cabeza Muxa 584
    Pozu les Cuerries 584
    Pozu Madejuno (Omega 45) 584
    Sil de Oliseda 583, 585
    Sima 56 584
    Sima del Jou de la Canal Parda 584, 585

Sistema del Xitu 584
Sistema Trave 583, 584, 585
Sistema Verdulluenga 585
speleogenesis **583**
Torca Castil (PC–15) 584, 585
Torca de los Rebecos (T27) 525, 584
Torca del Cerro 525
Torca del Certo del Cuevón (T33) 584
Torca del Jou de Cerredo 584
Torca Idoúbeda 583, 584
Torca Tortorios (CT–1) 584
Torcas Urriellu 583, 584, 585
Pierre Saint-Martin **586–587**
Gouffre de la Pierre Saint-Martin 586
Las Puertas de Illamina 586
Sima de Ukerdi Abajo (UK4) 586, 587
Pozo del Madejuno, depth of 525
Puertas de Illamina 525
Red del Silencio 525
Sima de los Huesos 119, 562
Sima del Elefante 120
Sima GESM 327
Sistema del Trave 525
Somiedo, taphonomy 562
Tunel dels Sumidor 345
Spaluga di Lusiana, Italy 117
Spannagel, Karl Hermann 697
Spannagel Cave, Austria **663–664**
geomorphology 663
hydrology 664
speleogenesis 663–664
speleothems 664
sparite 185, 186
Spasskaya, Siberia 646
speciation **665–666**
adaptive shift theory 665
allopatric model 665, 666
aquatic cave communities 665–666
biological species model 665
climatic relict model 665
defined 665
gene interchange and 665
hypogean fauna 348
local habitat model 665
parapatric 665
peripatric 665
rates of 237
species segments, defining 665
Wrightian adaptive shift model 665
Specimen faunæ subterreaneae (Schiödte) 152
Speedwell Cavern, England 576, 714
Spelaeacritus anophthalmus 449
Spelaeogriphacea 126, 256
Spelaeomysis bottazzi 365, 366
Spelälogiches Jahrbuch 464
SpeleoArt 88
speleobiology *See* biospeleology.
Speleo Digest 466
speleogenesis **666–668**
artesian 628, 745, 748
carbon dioxide and 669
coastal/oceanic setting 667, **674–675**
computer models **677–681**
conduit initiation/early development model 668
confined settings 667, 668, **681–683**
decay/obliteration stage of caves 668
dissolutional mechanisms 251, 667
erosion and 669
etymology 666
evaporitic successions and inception of caves 439–440
extraterrestrial caves 355
Four State Model 685–686
groundwater and 667–668
hydrothermal 668, 682
hypogenic 192, 668
ice caves 435–437
inception of caves **437–438, 667**
mixture corrosion theory 671, 673
patterns of caves **573–574**, 667–668, 684–685
phreatic 252
piping caves **590, 591**
quartzite 300, 650
stripe karst **705**
sulfuric acid caves 194, 574, 668
theories, early **669–670**
theories, post-1890 **670–674**
unconfined settings **667, 668**
See also individual cave or region by name.
Speleogenesis: Evolution of Karst Aquifers (Klimchouk, Ford, Dreybrodt and Palmer) 52
Speleological Abstracts 465, 466
Speleological Society of Japan 99
Speleological Society of Korea 99, 212
Speleon 464
Speleophilately International 704
speleothems
and air quality 730, 731
allophane, precipitation of 301
biogenic 636
breakage of 565
carbonate 512, **690–692**
classification of **690**
and CO2 enriched cave air 183
condensation corrosion **240–241**
condensation water types **691**
dating 279, 281, 555–558, 692
erosion rates, measuring **323**
evaporate **692–696**
flowing water types **691**
glacial caves 385
halides 511–512
history of research **689–690**
luminescence of **695–696**
Maya art and 92
as medicine 97, 98, 690
mucolites, formation of 637
paleoclimate studies and 214
paleotectonics **565–566**, 696
pool water types **691**
precipitation of 187

speleothems (Continued)
    protection of 246
    religious views of 622
    repair of 244, **626–627**
    rootsicles 637
    salt 511–512, 692
    seeping water types 691
    snottites 146, 759
    sulfate 513
    volcanic caves 762–763
Speleotheraphy Commission, International Spelelogical Union (UIS) 697
speleotherapy **697–702**
    applications 697, 698
    history of 697
    and karst resources 480
    methods 697–698
    sites 698
Spelunca 464
Spelunka 372
Speonomus spp. 369, 447
Speoplatyrhinus poulsoni 46, 595, 596, 597
Sphaeromatidea 256
sphalerite
    and breccias 660
    formation of 714
    paleokarst, economic resources of 560
Sphinx and limestone 551
Spider Cave, Texas, United States, and biofilms 145, 146
spiders *See* Arachnida.
Spielberg, Steven (*Indiana Jones and the Last Crusade*) 359
spinels 103
Spipola-Aquafredda, Italy 327, 345
Spirit Cave (Tham Pi), Thailand 107
Spitzbergen
    speleogenesis 342
    Trollosen 699
Spluga della Preta, Italy 326
spoil heaps, creation of 244
sponges *See* Porifera.
spongework mazes
    morphology 573
    speleogenesis 574
spring cavefish *See* Forbesichthys agassizi.
Spring Creek, Florida, United States 699
springs, karst **699–702**
    classification of 700
    controls on 699–702
    formation of 699
    gravity springs 699
    hydrology 699
    pressure springs 699
    quality of 702
    resources of 702–703
    vauclusian springs 699
springtails *See* Collembola.
Sprinkle, Leland 534
Spy Cave, Belgium 138–139
Sri Lanka
    geology **445**
    Istripura caves 445
    Vava Pena (Bat Cave) 445
SRT *See* Single Rope Techniques (SRT).
Ssang-ryong Cave, South Korea 212
stalactites 511
    See also speleothems.
stalagmites
    dating 224
    earthquakes and 566
    growth axis of 565
    growth rates 187
    See also speleothems.
stamps and caves **703–705**
Stanton, Rick 303
Stara Mlinica ponor, Bosnia/Herzegovina 291
Stariy Zamok, Siberia 646
Staromuradymovskaja Cave, Urals 629
Star Trek VI: The Undiscovered Country (Meyer) 360
Star Wars (Lucas) 359
Stary-hrad, Slovakia 331
St Beatus Cave, Siebenhengste, Switzerland 647
St Bernadette, Lourdes, France 428, 622, 623, 624
St-Cézaire, Grottes de, France 359
St Columbia's Cave, Scotland 427
St Cuthbert's Swallet, England 504
Steam Caves, Banff, Canada 51
Steatornis caripensis 57, 59, **130–131**, 190, 410
Steffansan, Arni 764
Steinberg, F.A. von 211
Steinernes Meer **173**
Stella, Italy 699
Sterkfontein, South Africa 17, 18
sternal gland 10
Stevens, John Lloyd 40
St Francis Cave, Scotland 427, 510
Sthenerus spp., Australia 124
Stirling, James 127, 129
St John, Issac M 411
St Lawrence platform, composition of 175–176
St Lukas Guild, Austria 88
St Michael's Cave, Gibraltar
    military use of caves 510
    musical performances in 533
Stoke Lane Slocker system, England 503
Stone, Andrea 85
Stone, Ralph (*Caves of Pennsylvania*) 52
stone tools, dating of 279
stoping 308
St Paul's Cathedral and limestone 551
St Paul's Cave, Philippines 113
St Paul's Grotto, Malta 208
St Paul's Church, Rabat, Malta 623
St Paul's Underground River, Philippines 105, 106
St Peter's Square and tufa 551
Strabo
    Cerknicka Polje, Slovenia 210
    karst hydrology 478
Stradchanska, Ukraine 628
Strashnaya, Siberia 646
Stratenska Cave System, Hungary/Slovakia 27

Stratenska-Psie diery Cave System, Slovakia 331
stream-cut caves *See* caves, erosion.
Stringfield, Victor and karst hydrology 479
stripe karst **705–706**
stromatolites 185
Strongs Cave, Australia 120
strontium 514
Studencheskaja, Ukraine 628
Stufe di Neroni (Nero's Oven), Italy 714
Stupishin, A. 662
stygobitic funa
    adaptations of 264
    Africa 24
    anchialine habitats 65
        Asia, Southeast 110
        Australia 125, 126
        Belize 38
        Britain 163
        Canada 200
        Caribbean Islands 39
        Castleguard Cave, Canada 200
    and colonization 236
    defined 258
    Dinaric karst 289
    evolution of 255, 259, 263
    fish, distribution of 593
    hypogean environments 141–142
        Mammoth Cave, Kentucky, United States 499–500
        Panama 38
    Postojna-Planina Cave System, Slovenia 141, 289, 603
    representative species of 256
    Salukkan Kallang, Indonesia 631–632
    Southeast Asia 110, 112
    thermal water habitats 724–725
    as threatened species 262, 315
    Vjetrenica, Bosnia-Herzegovina 141, 289, 759
    Yucatán Phreas, Mexico 39
Stygobromus canadensis 200
Stygobromus pecki 315
Stygocaridacea 255
Stygofauna Mundi (Botosaneanu) 149
Stygoparnus comalensis 315
styolites 185
subjacent collapse dolines 308
subjacent evaporite karst 346
sublimation and glacier caves 385, 387
subsidence 306, 309
    See also suffosion dolines.
subsidence depression 306
subterranean habitats **709–712**
    anchialine **64–65**, 712
        Amphipoda 65, 258, 260
        Belize 38
        Cape Range, Australia 182
        continental 182
        Crustacea 65, 254, 256, 257
        defined 64, 712
        distribution 64, 65
        fauna of 65
        Hawaii, United States 417

        Remipedia 65, 126, 156, 183
        stratification in 64–65
        thalassostygobites 549
        thermal gradient in 707
    aquatic hypogean **711–712**
        euhaline habitats 712
        hyporheal zone 711
        interstitial waters 711
        mixohaline/freshwater habitats 712
        percolation waters 712
        phreatic zone 711
        thermal hypogean waters 712
    aquatic interstitial **454**, 711
        adaptations 454, 455
        evolution, regressive 454, 455
        factors influencing environment 454
        fauna morphology 454–455
    classification **709–712**
    dark cave/crevicular 710–711
    entrance habitats **318–319**
    hypogean **709–712**
        biodiversity in 141–142
        classification of 709–712
        ecological factors influencing 706
        interrelations between 710
    phreatic 258, 711
    terrestrial interstitial 179, 709
    thermal water 712, **724–725**
Subway, The Castleguard Caves, Canada 200
Suctoria, Dinaric karst 289
suffosion caves *See* piping caves
suffosion dolines
    described 305
    nomenclature 304, 308
    speleogenesis 306–307
Sulawesi, Indonesia 105
sulfate deposition, mechanisms of 693
sulfates
    evaporite karst 344
    minerals in caves **513**
sulfides
    minerals in caves **511**
    minerals in karst **713–714**
sulfobacteria 366
sulfur
    and biomineralization 636
    formation of 511
    and sulfuric acid speleogenesis 194
    Ukraine 748
sulfur cycle
    and biomineralization 636
sulfur folia
    Cueva de Villa Luz, Mexico 759
sulfuric acid caves 192, 574
    Boa Vista, Toca da, Brazil **156–157**
    speleogenesis 192, 574
sulfuric acid speleogenesis 668, 682
sulphorhodamine B (CI Acid Red 52) 772
Sumgan-Koutuk Cave, Urals 629
Sumidero Yochib, Mexico 37

sump 301
Sunbeam-by-the-Sea Cave, Baja, Mexico 491
Sunday at Sunset Pot (Perou) 360
Sun Quan and military cave use 509
surface isotope concentration and dating 282
Surtshellir, Iceland 764, 765
Survex software and cave surveying 716
surveying caves **714–717**
   history of 714–715
   radiolocation and 717
   techniques 715–717
   underwater surveying 717
Su-san Cave I, South Korea 212
Suswa, Mount, Kenya 22, 561, 562, 763
Sütlük Düüdeni, Turkey 741, 742
Sutris, Frederich 86
Suyuru-Kaya Cave, Ukraine 253
Svartisen, Norway 522
Svetina, J. 655
swamp fish *See Chologaster cornuta*.
Swartkrans, South Africa 17, 18
Sweden
   Boddagrottorna 722
   Lummelundagrottorna 342
Sweeting, Marjorie Mary 235, 605
   biography of **382**
   Tibet, exploration in 99
Swildon's Hole, England 164, 302, 504, 562, 564, 571
Swinnerton, A.C. 671
Switzerland
   Beatushöhle 87
   Hölloch **525**
   Mammutmünster, Hölloch 422
   Siebenhengste **647–649**
     A2 647
     Bärenschacht 647
     F1 647
     geomorphology 647, 648
     hydrology 647
     K2 647
     Réseau Siegenhengste-Hohgant cave system 647
     speleogenesis 649
     St Beatus Cave, 647
   Silberen System, Hölloch 420
Syjkyrdu Cave, Pamir Central Asia 94, 96
Symphoniskaya, Pinega gypsum caves, Russia 589
Symphyla 144
Syncarida 255, **269–270**
   adaptations of 269
   Anaspididae 269
   Bathynellacea
     Bathynellidae 269, 370
     Parabathynellidae 269
   description of 268–269
   distribution of 268, 269–270, 370
   Koonungidae 269
   Psammaspididae 269
   reproduction of 269
   Stygiocarididae 269
syngenetic karst **717–719**
   defined 717
   evaporite 344, 676
   mixing corrosion and 718
   morphology 718
   variations in 718–719
syngenite minerals
   Tăuşoare Cave 329
Synocyclocheilus hyalinus 594
Syria
   Cater Magara 345
   Peggau Cave 704
   Ras el ain 699
Systema Cheva, Huautla Cave System, Mexico 425, 426
Systema Puficicatión, Huautla Cave System, Mexico 425
Szabadsag Cave, Hungary/Slovakia 26
Szachownica Cave, Poland 331

Tabasco, Mexico, troglobitic fauna 38
Table Rock Cave, Afghanistan 443
Tabon Cave, Philippines 109, 168
Tabun Cave, Israel 167, 195
   Tabun 1 skeleton **195, 197**
Tadarida brasiliensis 226, 227, 409
Tadyaw Cave, Phillipines 168
Tafna underground river, Algeria 15
Tainter Cave, Wisconsin, United States 92
Taisho-da Cave, Japan 30
Taiwan
   geomorphology 96
   history of research 100
Taj Mahal and limestone 551
Takaga-ana Cave, Japan 30
talc 373
Talgua Cave (Cave of the Glowing Skulls), Honduras 42
Talitridae 258
talus caves 606 **721–724**
   classification 721–722
   distribution 721
   glacières 723
   interfaces/multiprocess caves 722–723
   Mars and 721, 723–724
   purgatory caves 722
   rockslide/rockpile caves 722
   significance of 721
talus pseudokarst 606
Tanaidacea 65, 256
Tandarida brasiliensis mexicana 314, 393, 751
tandem accelerator mass spectrometry (TAMS) dating 281
Tangshan, China 223
Tantalhöhle, Austria 173
tantalum 514
Tantulocarida
   anchialine habitats 65
   Stygotantulus stocki, Lanazarote 179
Tanzania
   Hades Cave 22
Taoism and ritual use of caves 622
tapes, measuring and cave surveying 716
taphonomy 17–18
   accumulation of bones, methods **561**

animals entering caves by accident **561–562**
animals living in caves **561**
bone modification **562–564**
bone modification via predation **564**
bone transport via other method **562, 564**
bone transport via predator **562**
cave fauna, analysis of **564**
taranakite 329
Tardigrada 459
Tarkhankutskaya Cave, Ukraine 253
Tasmania, Australia
  cave art in 89
  Exit Cave 122
  history of exploration 129
  Newdegate Cave 123
  Tasman's Arch 704
  troglobitic fauna of 125–126
  Westmoreland Cave 633
Tasmanian devil 124
Tasman's Arch, Tasmania Australia 704
Tassili N'Ajjer, Algeria 14
Tatra Mountains National Park, Europe, Central 333
Tauleve, Honduras Mexico 42
Taung skull 16–17
Taurus Mountains, Turkey
  Aladaglar 742
  Altinbesik Cave 740–741
  Çukurptnar Cave 741
  geomorphology 740
  Kizilelma Mağarasi 741
  Kocain Cave 741
  Kuyukale 741
  Manavgat Gorge 740–741
  Martal Cave 741
  Peynirlikönü Cave (Evren Gunay Cave) 741
  Pinargözü 741
  Pinargözü Cave 741
  Sabatagi Düdeni 741, 742
  Sütlük Düdeni 741, 742
  Tilkiler Düdendi 741
Tăuşoare Cave, Romania 329
Tawi Atayr, Oman 115
Tchernov, Einat 196
tectonic influences on karst evolution 477
Telema tenella 370
telephone, single-wire (SWT) and cave communication 238
Tellkampf, August Otto Theodor 148, 152
Tëmnaya, Siberia 645
Temnocephalida
  in Dinaric karst 289
  Scutariellidae 458
temperature in caves 228
Tempurung Cave, Malaysia 102
Temucin Aygen 740
Tenerife, Canary Islands
  caves of 179
  geomorphology 179
  troglobitic fauna 179, 180
Tenglong, caves of 220
Tennengebirge, Austria 173, 174
Tenochcas 364
Te Reinga Cave System, New Zealand 542
terraces, travertine 218
Terra Ronca State Park, Brazil 134
terrestrial interstitial environment (terrestrial MSS) 709
Teshik-Tash Cave, Uzbekistan 167
Tethyan track 183
Te Wahipounamu-South West New Zealand **543**
Texas Blind Salamander *See Eurycea rathbuni; Typhlomolge rathbuni.*
Thailand
  archaeological caves 107–108
  art in caves 107
  Bingyi Caves 624
  geomorphology 100, 101
  history of exploration 113
  Kanchanaburi 167
  Koanoi 624
  Kogun Cave 624
  Kra Isthmus 110
  Long Rong Rien cave 107–108
  Mae Hong Son 167
  Mimehtu Cave 624
  Nang Nuan oil field 102, 104
  Pindaya Cave 103, 113, 624
  Sai Yok 107
  Shwe Ohm-min Cave 624
  Spirit Cave (Tham Pi) 107
  Tham Chang Dao 102
  Tham Kao Luang 103
  Tham Kubio 102
  Tham Lod 107
  Tham Phra Wang Daeng 102
  Tham Putiharn 103
  Tham Tab Tao 113
  Yadana gas field 102, 104
Tham Chang Dao, Thailand 102
Tham En, Khammouan, Laos/Vietnam 102, 485
Tham Kao Luang, Thailand 103
Tham Koun Dôn, Khammouan, Laos/Vietnam 102, 485
Tham Kubio, Thailand 102
Tham Lod, Thailand 107
Tham Nam Thieng, Khammouan, Laos/Vietnam 102
Thampanna Cave, Nullarbor Plain, Australia 546
Tham Phi Seau, Laos 102
Tham Phi Seua, Khammouan, Laos/Vietnam 485
Tham Phiu, Laos 624
Tham Phra Wang Daeng, Thailand 102
Tham Pong, Laos 167
Tham Putiharn, Thailand 103
Tham Sao Hin, Laos 102
Tham Tab Tao, Thailand 113
Tham Thing, Laos 113
Tham Toutche, Laos 102
Thanksgiving Cave, Canada **178**
The Hole, West Virginia, United States 70
thenardite
  deposition of 694
  formation of 513
theodolite and cave surveying 715, 716
Theopetra Cave, Thrace 340

Theoretical and Applied Karstology 465
The Origin of Maze Caves (Davis) 673
Thérèse shaft, France 284
thermal water habitats, fauna of **724–725**
thermal waves in caves **228, 229**
Thermobathynella adami 724
Thermosbaenacea 255, 256
Thermosbaena mirabilis 24, 724
The Role of Caves in Maya Culture (Thompson) 40
The Spider (Gordon) 359
The Stone Age of Mount Carmel (Garrod) 195
The World About Us (Perou) 360
Thinès, Georges (*L'Évolution régressive des poissons cavernicoles et abyssaux*) 593
thiobacteria 366
Third Unnamed Cave, Tennessee, United States 44
Thirria, C.E. 669
Thomas, Steve (*More People Have Been To The Moon*) 534
Thomas Avenue, Mammoth Cave, Kentucky, United States 570
Thompson, Edward (*Cave of Loltun*) 40, 41
Thompson, J. Eric (*The Role of Caves in Maya Culture*) 40
Thomson, William and Jeita Cave 463
thorium 280, 514
Thrace
    Maronia Cave 340
    Peristeri Cave 340
    Theopetra Cave 340
Thrailkill, J.V. 673
thread-legged bugs *See* Hemiptera.
threatened species
    Cape Range, Australia 183
    Chiroptera 227, 500
    Cockpit Country, Jamaica 235
    Edwards Aquifer, United States 315
    North America 48–49, 500
    Proteus anguinus 63
    stygobites 262
three-step model of colonization 235
Three Window Cave, Selma Plateau, Oman 639
Thrinukagigur, Iceland 762, 764
Thunder Canyon Cave, California, United States 721, 722
Thurston, Lorrin 765
Thurston Lava Tube, Hawaii United States 415–416
Thyphlotriton spelaeus 62
Thysanura 24, 143, 144
tiankengs *See* dolines.
Tibet 220
    Chomolungma (Mt. Everest, Sagarmatha) 622
    history of research **99–100**
tidal flexure vulcanism and lava tube caves 356
Tien Jing Gorge, China 285, 286
Tien Shan, Central Asia **94–95**
    Bolojuk 94
    Bojbulak 95
    Cupp-Coutunn Cave system 95
    Fersman's Cave 95
    Festival'naja-Ledopadnaja system 95
    Geofizicheskaya Cave 95
    geomorphology 94–95
    Karljuksky Cave 94
    karst areas of 95
    Khashim-Ojuk Cave 95
    Khodzharustam Cave 95
    Kievskaja Cave 95, 663
    Kjaptarkhana Cave 95
    Kun-Ee-Gout 95
    Kyrktau Plateau, karst of 95
    Kysyl Dzhar Mountains, karst of 95
    Pobednaja Cave 95
    Sarykamyshskaya 94
    Uluchurskaja Cave 95
    Zajdmana Cave 95
Tiglath Pileser and karst hydrology 478
Tiha Jama (Silent Cave), Škocjanske jama, Slovenia 654
Tikal, Guatemala 37
Tilapia guinasana 21
Timavo, Italy 699
Timavo Springs, Kras, Slovenia 486
Timpanogos Cave, Utah, United States 52
tin 514
Tindoul de la Vayssière, France 371
tinopal CBS-X dye and water tracing 772
Tisu, China 699
Titan Shaft, Peak District, England 238, 576, 577
Tjoarvkrajgge, Norway 342
Tlamaya Cave, Mexico 184
TM 71, France 373
Tobago 190
Tobio, Papua New Guinea 699
Toca da Barriguda, Brazil 55
Toca das Confuspes, Brazil 57
Tolkien, J.R.R. (*Hobbit, The; Lord of the Rings, The*) 206
Tommy Grahams Cave, Australia 183, 545
Tongass National Forest, United States 368
Tonta, Siberia 646
toothless blindcat *See Trogloglanis pattersoni*
topofil
    and cave surveying 716
Topolnița, Romania 329
Toporobot software and cave surveying 716
toppling caves *See* crevice caves.
Tora Bora caves, Afghanistan 443, 510
Torca Castil (PC–15), Picos de Europa National Park, Spain 584, 585
Torca del Cerro, Picos de Europa National Park, Spain 525, 585
Torca del Certo del Cuevón (T33), Picos de Europa National Park, Spain 584
Torca del Jou de Cerredo, Picos de Europa National Park, Spain 584
Torca de los Rebecos, Spain 525, 584
Torca Idoúbeda, Picos de Europa National Park, Spain 583, 584
Torcas Urriellu, Picos de Europa National Park, Spain 583, 584, 585
Torca Tortorios (CT–1), Picos de Europa National Park, Spain 584
Torgashynskaja, Siberia 629
Torghatten sea cave, Norway 342
Torotoro National Park, Brazil 55
torpor, hibernation 226
Torres, Trinidad 119
torricellian airbell and diving 302

Totes Gebirge, Austria 173, 174
Touloumdjian, Claude 302
tourism
    Cape Range, Australia 183
    Carribean Islands 191
    history **725–726**
    and karst 480
    resources, karst 480–481
tourist caves
    Actun Kan (Jobitzina), Guatemala 42
    airborne debris **731–732**
    air quality in **730–732**
    Akiyoshi-do Cave, Japan 30
    algae and lampenflora **733–734**
    Altamira, Spain **33–34**, 84, 85, 428, 534
    Amaterska-Punkva Cave System, Czechia 331
    Anjohibe, Madagascar 494
    Aven d'Orgnac, France 338
    Bear Gulch Cave, California, United States 721
    Belize 140
    Carlsbad Caverns National Park, New Mexico, United States 192
    Castellana, Grotta di, Italy 338
    Castleton, England 577
    Cave of Bolanchen, Mexico 42
    Chillagoe, Australia 89, 192, **215–216**
    Clamouse, Grotte de, France 338, 373
    climate monitoring networks in caves 230
    conferences for guides/workers **728**
    conservation and **728–729**
    Cuertana Casas Archaeological Zone, Mexico 42
    Cueva de las Pinturas, Guatemala 42
    Cuevas de Bellamar, Cuba 191, 272
    Cuevas de Nerja, Spain 327, 338, 533
    Daji Dong, China 218
    defined **726**
    Demoiselles, Grotte des, France 338
    Drac, Coves del, Spain 338
    films, use in 359
    Frasassi caves, Italy 327, **374–375**
    Fuerteventura Island 179, 181
    Ghar Alisadar, Iran 461
    Giant Hole, the, France 373
    Gouffre d'Esparros, France 373
    Gough's Cave, Englandss 504
    Grand Canyon National Park, Arizona, United States **391–393**
    Grotte de Limousis, France 373
    grutus de Juxtlahuaca, Mexico 37, 92
    guide books for **728**
    Ha Long Bay, Vietnam **413–414**
    Harrison's Cave, Barbados 190
    human impacts on **728–729**
    Jeita Cave, Lebanon **463–464**
    Jenolan Caves, New South Wales, Australia 726, 731, 732
    Katelahkor, Iran 461
    Kau Eleku cave, Hawaii, United States 415
    Kula Kai cavern system, Hawaii, United States 416
    Kuzuga-ana Cave, Japan 30
    Lamprechtstofen, Austria 173
    Lanquin Cave, Guatemala 42
    Lascaux, France 428
    Liujiang Cave, China 224
    Loltun Cave, Mexico 42, 84
    Longuin Cave, China 623–624
    Lost River Caves, New Hampshire, United States 721
    Lummelundagrottorna, Sweden 342
    Mammoth Caves, Kentucky, United States 725
    management of 726
    Mark Twain Cave, Missouri, United States 206
    Naracoorte Caves, Australia 124, 633, 726
    Newdegate Cave, Tasmania 123
    Niaux Cave, France 428
    Ochtina Aragonite Cave, Slovakia 26
    Pech Merle, France 428
    Pečina na Hudem Ietu, Kras, Slovenia 486
    physical infrastructure of **728**
    Polar Caves, NH United States 721, 722
    Postojnska Jama, Slovenia 601, 725, 726
    problems of **727**
    Pyramid of the Sun, Mexico 40, 42
    Rákóczi Cave, Hungary/Slovakia 26
    Sam's Point Ice Cave, New York, United States 250
    Santa Catalina, Cueva Grande de Cuba 272
    Škocjanske jama, Slovenia **653–655**
    Sof Omar, Ethiopia 20, **655–656**, 699
    Spannagel Cave, Austria 663–664
    Talgua Cave (Cave of the Glowing Skulls), Honduras 42
    Tauleve, Honduras 42
    Utatlan, Guatemala 42
    Vilenica Cave, Kras, Slovenia 486, 725
    visitor experiences, enhancing **727–728**
    visitor experiences and **726–727**
    Wookey Hole, England 725
    Xochicalco, Mexico 42
    See also individual entries for each cave.
Tournal, Paul 80
Tournefort, J.P. 208, 689
tower karst 217, 241, **734–735**
    Asia, Southeast Islands 104
    cave systems of 735
    Central America 37
    Chillagoe, Australia 216
    China 217, 241, 242
    and cone karst 242
    genetic types of 735
    and karst evolution 477
    morphology 734–735
    Sulawesi, Indonesia 105
Trachysphaera spp. **535**
tracing studies and karst hydrology 479
Transactions of the British Cave Research Association 465
Transbaikalian, Siberia 646
transit and cave surveying 716
Travaux de L'Institute de Spéléologie Émile Racovitza 464
travertine 332, **736–737**
    Afghanistan 443
    Antalya travertine terraces, Turkey 737, 740
    conservation of 737
    deposition of 736
    distribution 737

travertine *(Continued)*
   Eyrie, Lake Basin, Australia 737
   formation of 736
   Grand Canyon National Park, Arizona, United States 393
   Great Close Mine, England 737
   Gutersteiner Falls, Germany 737
   Huangguoshu, China 736
   Huanglong, China 424
   Jiuzahaigou, China 424
   Mammoth Hot Springs, Wyoming, United States 737
   morphology 736–737
   Pamukkale, Turkey 187, 568–569, 737, 740
   Plitvice Lakes, Croatia **597–598**, 737
   Pyin-Oo-Lwin, Myanmar 102
   Reotier travertine cascade, France 737
   Solomon's Prison, Iran 737
   study of 737
   Turner's Falls, Oklahoma, United States 737
Treak Cliff, Peak District, England 576
Trebišnjica, Herzegovina 699
Trebišnjica Hydrosystem, Herzegovina 278, 292
Trebišnjica River, Herzegovina 291–292
Trebistoho Polje, Bosnia/Herzegovina 291
Trechinae
   in Dinaric karst 289
   in France 369
Trichoptera
   Hydrophyshidae 453
      Wormaldia subterranea 453
   terrestrial cave habitats, biodiversity in 144
Trickett, Oliver 128, 352
Tri Nahaci Cave, Iran 461
Trinidad 190, 191
Triplophysa gejieunsis 594
Trissino, Giovanni Giorgio 151
Tritons speleological club 284
Trnsko Polje, Bosnia/Herzegovina 291
troglobites **548**
   troglobitic fauna
      Asia, Southeast 110
      Australia 125–126
      Belize 38
      Britain 163
      Caribbean Islands 39
      Collembola 24, 446
      and conservation 245
      diet of 516
      Guatemala 38
      Hawaii lava tube caves, United States 417
      Ireland 163
      Mexico 38
      and the MSS 457
      Panama 38
      Postojna-Planina cave system, Slovenia 289
      Southeast Asia 110, 112
      and speciation 666
      Vjetrenica, Bosnia-Herzegovina 289, 760
      Yucatán Phreas, Mexico 39
   Troglocaris anophthalmus 603
Trogloglanis pattersoni 11, 315, 348, **594**

troglomorphy **7–9**, 548
Troglomysis vjetrenicensis 760
troglophiles **548**
Troglophilus cavicola 603
trogloxenes **548–549**
Trogulus torosus 759
Trollosen, Spitzbergen 699
trona 212
Trou Al'Wesse Cave, Belgium 138, 139
Trou Da Somme Cave, Belgium 138, 139
Trou du Glaz, France 371
Trouin Sene cave, Haiti **189**
Trou Madame, France 302
Trou Magrite Cave, Belgium 138, 139
Trou qui Souffle, France 755, 756
Trou Sandron Cave, Belgium 138, 139
Troy-Windsor Trail, Jamaica 235
trypanosomiasis 294
TSI Aerodynamic Particle Sizer 732
tsingy 100
   Khammoun karst, Laos 483
   Madagascar 494–495
Tsingy Bemaraha, Madagascar 473, **494–495**
TSOD Cave, NY United States 722
tuberculosis 294
tufa 187, 736
   Basilica of the National Shrine of the Immaculate Conception 551
   Caerwys tufa deposits, Wales 341
   Daleib Wanei, Namibia 23
   formation of 147, 701, 702
   Hammamat Ma'in, Jordan 115
   Huanglong, China 424
   Isha Baidoa, Namibia 23
   Krka river, Croatia 338
   Morocco 14
   ornamental uses of 551
   Pamukkale, Turkey 568
   Plitvice Lakes National Park, Croatia 597
   South Africa waterfall tufas 23
   St Peter's Square 551
tularemia 294
Tulear Region, Madagascar 495
Tully, Montgomery (*Battle Beneath the Earth*) 360
Tumbling Creek Cave, MO United States 557
Túnel de la Atlantida, Lanzarote Canary Islands 179, 302, 762
Tunel dels Sumidor, Spain 345
tungsten 514
Tunicata
   Ascidia 459
Tunisia
   geomorphology 15–16
   Mine cave (Rhar Djebel Serdj) 15–16
   Rhar Ain et Tsab 15
tunnel boring machines (TBMs) 739
Tunnel Cave, Australia 127, 129
tunnelling in karst **738–739**
   artificial underground reservoirs 739
   groundwater and 738–739
   history of 738

planning 738
and water storage 739–740
Tupper, Mount system, Canada 178
Turansky Plain, Central Asia **94–95**
Turbellaria
 anchialine habitats 65
 Sphalloplana mohri 315
 Temnocephalida 289
  Scutariellidae 458
Turkey **740–742**
 Altinbesik Cave 740–741
 Antalya terraces 737, 740
 bauxite production in 135
 Birkilin Caves 207, 742
 Black Sea Coast 742
 Cave of the Seven Sleepers 208
 Çukurptnar Cave 741
 Dog Cave 208
 Dumanli spring 699, 741
 Evren Gunay düdeni (Peynirlikönü) 525, 741
 geomorphology 114, 740–742
 gypsum karst of 742
 Iskender-i-Birkilin cave 207
 Kirkgöz-1 741
 Kirkgözler Springs 699, 741
 Kirkgöz-Suluin 741
 Kizilelma Mağarasi 741
 Kocain Cave 741
 Kurukopru Cave 741
 Kuyukale 741
 Manavgat Gorge 740
 Martal Cave 741
 Oluk Köprü spring 699, 741
 Pamukkale 187, **568–569**, 737
 Peynirlikönü Cave (Evren Gunay Cave) 741
 Pinargözü Cave 741
 Pit of the Jinns Cave 208
 Plutonium Cave 208
 Sivas gypsum karst 742
 Sabatagi Düdeni 741, 742
 Sütlük Düdeni 741, 742
 Taurus Mountains 740–742
 Tilkiler Düdendi **741**
 Tuz Gölü 742
 Yedi Miyarlar 699
Turkish Speleological Society 740
Turkmenistan 95
 Cupp-Coutunn cave 95, **273–275**, 714
Turner's Falls, Oklahoma, United States 737
Turtle Rock lava ball, South Korea 212
Tutkushskaja, Siberia 629, 645
Tu Wan (*Yün Lin Shih Pu*) 98
Tuz Gölü, Turkey 742
Twain, Mark (*The Adventures of Tom Sawyer*) 206, 360
two-cycle model, speleogenesis theory 671
two-step model of colonization 235, 237, 263
Typhlichthys subterraneus 244, 595–596
Typhlocaris salentina 365, 366
Typhlomolge rathbuni 62
Typhleotris madagascariensis, Madagascar 495
tyuyamunite 514

Uai Bobo 1, Indonesia 109
Uai Bobo 2, Indonesia 109
Ubajara National Park, Brazil 473
Ubuvomo bwa Musanze, Rwanda 22
Ubuvomo bwa Nyrabadogo, Rwanda 22
Uchkopisjo, Peru 55
Ugryn, Ukraine 746
UIS *See* International Speleological Union (UIS)
UIS Bulletin 465
UIS Documentation Centres, addresses of 465
Ukraine
 Atlantida 746
 Bukovinka 746
 Crimea **253–254**
 geomorphology 629, 745
 Gostry Govdy 746
 Gypsum caves **746**
 hydrology 745–746
 Kristal'na 345, 746
 maze caves of 746–748
 minerals in caves 748
 Mlynki 345, 746
 Novorossijskaja 628
 Optimistychna 345, 525, 663, 746, 747
 Ozerna 345
 Ozernaja 525, 746, 747
 Podol'sko-Bukovinsky region 628
 Slavka 345, 746
 speleogenesis 745
 and speleotherapy 698
 Stradchanska 628
 Studencheskaja 628
 Ugryn 746
 Vertebra 345, 746
 Zoloushka 329, 330, 345, 525, 746, 748
Ukrainian Speleological Association 663
Ulmer, Edgar G. (*Cavern, The*) 360
Uluchurskaja Cave, Turansky Plain Central Asia 95
Ulu Leang 2, Indonesia 109
Ulu Leang I, Indonesia 109
Umm al Masabih cave, Libya 16
underclothing, thermal and cave exploration 353
Underground Injection Control (UIC) Program **405**
Undersea Research Group 302
United Kingdom **342**
 Agen Allwedd, Ogof 166, 341
 Alyn River Caves 166
 archaeological/paleontological caves **160–162**
 Buckfastleigh cave system 166
 Caerwys tufa deposits 341
 Church Hole 162
 Craig ar Ffynnon, Ogof 166
 Creswell Caves 162
 Dan yr Ogof 166, 341, 580
 Daren Cilau, Ogof 166, 239, 341
 Draenen, Ogof 166, **310–312**, 341, 525, 616
 Draycott, fossils 561
 Dydd Byraf cave system 166
 Eglwys Faen 166
 Ffynnon Ddu, Ogof 166, 341

United Kingdom *(Continued)*
  Forest of Dean 439
  Great Close Mine 737
  Higher Kiln Quarry cave system 166
  history of exploration **164–166**
  Kent's Cavern 77, 78, 82, 341
  Llangynidr Plateau 341
  Llyn Du cave system 166
  Llyn Parc cave system 166
  Mynddyd Llangattock 166, 341
  Mendip Hills 341, **503–505**
    cave development 503
    and cave photography 579
    Charterhouse Caves 504
    dolines of 503
    Eastwater Cave, taphonomy 561
    Fairy Cave Quarry 503
    Ford-Ewers model of speleogenesis 673
    fossils 561, 562, 563, 564
    GB Cavern 504, 562
    Gough's Cave 504
    Lamb Leer Cavern 238
    Long Hole 714
    morphology 503, 504
    Pen Park Hole 164, 714–715
    quarrying and 504
    St Cuthbert's Swallet 504
    Stoke Lane Slocker system 503
    Swildon's Hole 302, 504, 562, 564, 571
    taphonomy 561, 562, 563, 564
    Wookey Hole 79, 302, 503, 504, 562
  Mother Grundy's Parlour 162
  Paviland Cave 80, 281
  Peak District 342, 363, 561, **576–577**
    bedding discontinuities 575
    Blue John Caverns 578
    Buxton sump 302
    drainage 576
    Elderbrush Cave 576
    Eldon Hole 164, 576
    geomorphology 575, 657
    Giants Hole 577
    Giants-Oxlow system 576
    and limestone quarrying 610
    mineral mining in 576, 713
    Peak Cavern 166, 238, 301, 302, 576
      morphology 521
      musical performances in 533
      ropemaking in 426
    Peak-Speedwell system 166, 576
    radon and 577, 618
    Speedwell Cavern 576, 714
    speleogenesis 575
    Treak Cliff 576
    White Peak 575, 576
  Pin Hole 162
  Reed's Cave-Baker's Pit cave system 166
  Robin Hood's Cave 162, 364
  Virgin's Spring, England 491
  Weathercote Cave 87

  Yorkshire Dales, England 342, **783–785**
    Ease Gill Cave System 166, 441, 784, 786
    Gaping Gill 166, 522, 616, 783
    geomorphology 783–785
    inception horizons of 439
    Ingleborough Hill 783–784
    Ingleborough Cave 715, 784
    Ireby Fell 441
    Jordas (Yordas) Cave 87
    Keld Head 302, 438
    King Pot 786
    Kingsdale Cave System 785, 786
    Knock Fell Caverns 682–683
    Lancaster-Ease Gill cave system 166
    Langcliffe Cave 786
    Leck Fell 441
    Lost John's Master Cave 786
    Malham Cove 438, 784
    Malham Cove Rising 303
    Meregill Hole 2
    Mossdale Cave 2, 786
    Notts Pot 441, 786
    Pippikin Pot 441, 784
    Swinsto Hole 785, 786
    Victoria Cave 162
United Nations Building and limestone 551
United States **749–750**
  Anvil Points Claystone Cave, CO 591, 606
  Anza-Borrego, CA 53
  Ape Cave, WA 761
  Appalachia, hypogean fauna in 142
  Arrow Grotto, Feather Cave, New Mexico 92
  Arroyo Tapiado, California 591
  Ash Cave, Kentucky 168
  Badlands National Park, South Dakota 590
  Bangor Cave, Alabama 534
  Bat Cave, North Carolina 722
  Bat Cave, New Mexico
    archaeology of 43
    and cave photography 579
  bauxite production in 135
  Bear Gulch Cave, California 721
  Big Bone Cave, Tennessee 411
  Bigfoot Cave, California 52, 752
  Bluefish Caves, Yukon 43
  Blue Lake, Washington 762
  Blue Spring Cave, Tennessee 70, 750
  Bottomless Pits 251
  Boulder Creek Cave, Washington 722
  Boundless Cave, Virginia 731
  Bracken Cave, Texas 409, 751
  Burnsville Cove, Virginia 69
  Butler-Sinking Creek Cave System, Appalachian Mountains 69
  Carlsbad Caverns National Park, New Mexico 51, 52, **192–194**
    Big Room 193, 194, 195, 308, 580
    and cave photography 579, 580
    guano mining 194, 409
    history of exploration 51
    lampenflora and 734

Main Corridor 193, 194
  minerals in 194
Carver's Cave, Minnesota 50
Catacombs Cave, California 491
Catarina-Confusion Cave System, Texas 591
Cathedral Cave, California 491
Cave in the Rock, Ohio River 50
Caverns of Sonora, Texas 626, 751
Chestnut Ridge Cave System, Appalachian Mountains 69
Christmas Canyon Cave, Washington 591, 592, 606, 608
Chuckanut Mountain Caves, Washington 722
Coldwater Cave, Iowa 751
Columbine Crawl, Wyoming 751
Cotal Cave System, Kentucky 70
Cueva Tres Pisos Cave, California 491
Culverson Creek Cave, West Virginia 70
Cumberland Caverns, Tennessee 70
Danger Cave, Utah 43
Devil's Hole, Nevada 52, 557
Dust Cave, Alabama 43
Eagle Creek Cave, Arizona 409
Edwards Aquifer, Texas 143
El Capitan Pit, Alaska 752
Enchanted Rock Cave, Texas 722
Endless Cavern, Virginia, history of exploration 51
Feather Cave, New Mexico 44, 92
Floyd Collins Crystal Cave, Kentucky 159
Forbidden Fissures, California 491
Friar's Hole Cave, West Virginia 525, 750
Friar's Hole Cave System, West Virginia 70
Frio Cave, Texas 751
Graham Cave, Missouri, archaeology of 43, 44
Grand Canyon National Park, Arizona **391–393**
Grand Caverns (Weyer's Cave), Virginia 50, 534
Grand Meeander Cave, Colorado 722, 723
Grassy Cove, Tennessee 599
Great Expectations (Great X) Cave, Wyoming 751
Great Saltpetre Cave, Kentucky
  and cave surveying 715
  and gunpowder 411
Greenhorn Cave, California 53
Greenhorn Cave System, California 722
Green River, Mammoth Cave, Kentucky 497
Guadalupe Mountains 192
guano mining, Carlsbad Caverns 194
Hawaii lava tube caves 417–418, 765–766
The Hole, West Virginia 70
Honey Creek Cave, Texas 751
Howe's Cave, New York 360, 579
Indian Cave, Tennessee 50
Jester Cave, Oklahoma 345, 751
Jewel Cave, South Dakota
  dating of 57
  geomorphology 775, 776
  history of exploration 52
  length/depth of 525, 750
  minerals in 776
  speleogenesis 775
John Day country, Oregon 53
Jumbo Gumbo, California 491

Kartchner State Caverns, Arizona 229
  cave detections methods 379
  minerals in 513
Kau Eleku cave, Hawaii 415
Kauhako Crater, Hawaii 762
Kaumana cave, Hawaii 416
Kazumura Cave, Hawaii 204, 415, 416, 525, 750, 760–761
Key Cave, Alabama 597
Kula Kai cavern system, Hawaii 416
Lady's Harbor Cave, California 491
Laniakea Cave, Hawaii 765
Lechuguilla Caves, New Mexico 192, 193, 508
  history of speleology 52
  length/depth of 525, 750
  management of 774
  minerals in 511, 513, 751
  speleothems 194
Leon Sinks Cave, Florida 751
Lilburn Cave, California 52, 752
Longhorn Caverns, Texas 534
Lost Creek Pseudokarst, Colorado 606, 721, 722–723
Lost River Cave, Kentucky 534
Lost River Caves, New Hampshire 721
Lower Kane Cave, Wyoming 184
Luray Caverns, Virginia
  accessibility of 750
  and cave photography 579
  history of exploration 51
  musical performances in 534
Madison Cave, Virginia 50, 715
Malaspina Glacier, Alaska 606
Mammoth Cave, Kentucky 44, 191, 204, 206, 214, **495–499**
  art in 92
  bats 500
  biodiversity in 143
  biospeleology **499–501**
  and borehole analysis 398
  burials in 168
  and cave photography 578
  and cave surveying 715
  conservation 500
  dating of 554, 555
  depth of 525
  ecosystems 499
  Fisher Ridge Cave 52, 495, 525, 750
  and gunpowder 411, 499
  Hidden River Cave System 498–499
  history of speleology 50, 52
  length/depth of 525, 750
  Martin Ridge Cave System 495, 750
  mineral deposits in 44, 498
  paragenesis 569, 570
  Rotunda Room 635
  sediments, allochthonous 633
  soils of 657
  speleogenesis 497
  surface drainage 498
Mammoth Hot Springs, Wyoming 737
Mark Twain Cave, Missouri 206
McDougal's Cave, Missouri 206

United States (Continued)
　MDBATHS Cave, New Hampshire 722
　Miller's Cave, Missouri 168
　Model T Cave, Texas 591
　Mount St Helens, Washington 591, 592, 762
　Mud Glyph Cave, Tennessee 82, 83–84, 85
　Mystery Cave, Minnesota 751
　Mystery Cave, Wisconisn 626
　Na One crater, Hawaii 762
　Natural Trap Cave, Wyoming, taphonomy 562
　Ney Cave, Texas 751
　Officer's Cave, Oregon 591, 606
　Oregon Cave National Monument, Oregon 52, 733, 734
　Organ Cave System, West Virginia 51, 70, 525, 750
　Owl Cave, Washington, art in 92
　Painted Cave, California 491
　Paiute Cave, Arizona 251
　Palo Duro canyon, Texas 53
　Paradise Ice Caves, Washington 53, 204, 387, 605, 606
　Park's Ranch 345
　Petrified Forest National Park, Arizona 590, 591
　Pit H cave, Hawaii 249, 250
　Pinnacles National Monument 53, 721
　Polar Caves, New Hampshire 721, 722
　Rainbow spring, Florida 699
　Rockhouse Cave, South Carolina 722
　Russell Cave, Alabama 43, 52
　Salts Cave, Kentucky 44, 81, 92, 168
　Sand Cave, Kentucky 206, 360, 534
　Sand Hill Bluff Cave, California 491
　Sauta Cave, Alabama 411
　Scott Hollow Cave, West Virginia 70
　Seal Canyon Cave, California 491
　Sea Maze, California 491
　Shelta Cave, Alabama 47
　Short Cave, Kentucky 168
　Silver spring, Florida 699
　Sinkhole Plain caves, Kentucky 184
　Sloan's Valley Cave, Kentucky 70
　South Ice Cave, Oregon 437
　Spider Cave, Texas 145, 146
　Spring Creek, Florida 699
　Tainter Cave, Wisconsin, art in 92
　Third Unnamed Cave, Tennessee 44
　Thomas Avenue, Mammoth Cave, Kentucky 570
　Thunder Canyon Cave, California 721, 722
　Timpanogos Cave National Monument, Utah 52
　Triple Eagle Pit 345
　TSOD Cave, New York 722
　Tumbling Creek Cave, Missouri 557
　Turner's Falls, Oklahoma 736
　Valhalla Pit, Alabama 635
　Ventana Cave, Arizona 43
　Waiahuakua Cave, Hawaii 491
　Waialoha Cave, Hawaii 491
　Waiwaipuhi Cave, Hawaii 491
　Wakulla Springs, Florida 52, 302, 303
　Wasden/Owl Cave, Idaho 43
　Weyer's Cave Station (Grand Caverns), Virginia 50, 534, 704
　Wind Cave National Park, South Dakota
　　dating of 57
　　geomorphology 775, 776
　　history of exploration 52
　　length/depth of 525, 750
　　minerals in 776
　　speleogenesis 775
　Wonderland Cave, Arkansas 534
　Woodville Karst Plain system, Florida 52, 302
　Wyandotte Cave, Indiana 44, 81
　Xanadu Cave System, Tennessee 70
　Yosemite Falls Indian Cave, California 722
United States Wilderness Act of 1964 773
University of Bristol Speleological Society 166
Upper Nikšićko Polje, Montenegro 293
Urals 628–629
　Goluboje Ozero spring 629
　Ignatjevskaja Cave 629
　Ishcheevskaja Cave 628
　Kapova Cave 629
　Kinderlinskaja Cave 629
　Kizelovskaja-Viasherskaja System 629
　Krasny Kljuch Spring 629
　Staromuradymovskaja Cave 629
　Sumgan-Koutuk Cave 629
　Şura Mare cave, Romania 329
uranine dye 769, 771
uranium
　and dating methods 280
　mining of 514
　paleokarst, economic resources of 560
uranium-lead dating 556
Ursus spelaearctos 646 *See also* cave bears.
Uruguay
　Cueva de Arequita 56
Ušački Pićinski Sistem, Serbia 329, 330
U-series dating methods
　limitations 555–556
　method 555–556, 557
U-series gamma ray spectrometry 196, 200, 224, 279
U.S. Federal Cave Resource Protection Act 774
U.S. National Monument
　Russell Cave, Alabama, United States 43
Ust'-Kanskaya, Siberia 646
Utatlan, Guatemala Mexico 42
uvala
　defined 305
　origin of term 288
Uzbekistan
　Boj Bulok, depth of 525
　burials in caves 167
　Teshik-Tash Cave, burials in 167
Uzurskaya, Siberia 646

vadose cave morphology **521–523**
Valea Rea Cave, Romania 329, 693
Valhalla Pit, Alabama, United States 635
Valle de Viñales Cuba 271
Vallejo, Diego de 59
valleys in karst **753–754**

blind/semi-blind valleys 753, 754
  classification of 753
  dry valleys 754
  pocket valleys 753–754
  through valleys 753
Vallier cave system, France 755, 756
Valvasor, Johann Weichhard Freiharr von 210, 211
  biography of **152, 687**
  and biospeleology 148
  Die Ehre des Herzogthums Krain 158, 335, 687
  Glory of the Duchy of Carniola 152
  Postojnska Jama, documentation of 87, 601
Vamos Cave, Crete 340
vanadates in caves 513–514
Vancouver Island, Canada 368–369
Vancouver Island karst 178
Vandel, Albert
  biography of **153, 154**
  Biospéologie: La Biologie des Animaux Cavernicoles 154
Vandima, Slovenia 468, 469
Vanguard Caves, Gibraltar 383, 384
vasques 232
Vass Imre, Hungary/Slovakia 26, 27
Vastitas Borealis, Mars 355
vaterite
  formation of 636
  Huanglong, China 424
Vaucluse, Fontaine de, France 301, 302, 303, 338
Vaucluse, France 699
Vaucluse Plateau, France 326
vauclusian resurgence 301
vauclusian springs 699
Vava Pena (Bat Cave), Sri Lanka 445
Vaydinskaya, Siberia 646
Veeder, A. 579
Velika Cave, Slovenia 210
Velika Klisura, Serbia 329, 330
Veliki ponor, Bosnia/Herzegovina 291
Velkovrhia enigmatica 289, 458, 603
Venezuela 251, 252
  Alejandro de Humboldt National Monument 53
  Angel Falls, National Park of Cainama 612
  Cainama National Park 612, 778
  Cueva Alfredo Jahn 53
  Cueva del Cerro Autana 652
  Cueva del Guácharo 53, 55, 59, 131, 409
  Cueva del Samán 53, 55
  Cueva los Encantos 53
  Cueva los Laureles 53
  Cueva Sumidero La Retirada 53
  Haitón del Guarataro 53
  Sima Acopán 1 611
  Sima Aiyán-tepui Noroesti 55, 611, 612
  Sima Aonda 55, 611, 612
  Sima Aonda 2 55, 612
  Sima Aonda 3 55, 612
  Sima Aonda Este 2 612
  Sima Aonda Superior 611
  Sima Auyán-tepui Norte 612
  Sima Auyán-tepui Norte 2 612
  Sima de la Lluvia 611
  Sima Mayor 612
  Sima Menor de Sarisariñama 611, 612
Ventana Cave, Arizona, United States 43
Venus Grotto, Germany 87
Veracruz, troglobitic fauna 38
Verband der deutschen Höhlen-und Karstforscher, Germany 336
Verband österreichischer Höhlensforcher, Austria 336
Vercors, France **754–755**
  geomorphology 754–755
  sediments of 756
  speleogenesis 755–756
Verdon, Gorges du, France 326, 338
Verein für Höhlenkunde, Austria 336, 350
Vereshchagina, Siberia 646
vertical slots *See* grikes
Verne, Jules (*A Journey To The Center of the Earth*) 205–206, 360
Versey, Harold 235
Vertebra, Ukraine 345, 746
vertical caves 204
vertical transition colonization 236
Veshtucoc system, Mexico 36–37
Vespertilionidae 225
Veterani Cave, Romania 510
Veteranische Höhle, Romania 335
Vézère caves, France
  archaeology 757
  art in 90
  history of exploration 427
  religious views of 623
Victoria Cave, Naracoorte, Australia 144
Victoria Cave, England 162
Victoria Fossil Cave, Australia 121
Viento, Cueva del Canary Islands 179
Vietnam
  archaeological caves 108
  art in caves 108
  Cong Nuoc 102
  geomorphology 100
  Ha Long Bay **413–414**
  Hang Khe Ry 102
  history of exploration 113
  Khammouan **483–485**
  Khéo-Phay 167
  Lang Cuom Cave 108, 167
  Marble Caves 103
  Pho Binh Gia, burials in 167
  Phong Nha Cave 483, 485, 624
  Phong Nha system 102
View of the United States of America (Winterbotham) 51
Vigo Cave, Burren glaciokarst, Ireland 171
Vilenica Cave, Kras, Slovenia 486, 725
  612, 778
Villa Luz, Cueva de, Mexico **758–759**
  and biofilms 145, 146
  cave fauna 758
  chemoautotrophic bacteria 758
  minerals in 759
  speleogenesis **758–759**
Vinci, Leonardo da 228
Viré, Armand 149, 154

Virgil (*Aeneid, The*) 205
Virgin's Spring, England 491
Vir Polje, Bosnia/Herzegovina 291
Visitor Activities Management Process 519
Visitor Impact Management (VIM) 519
Vistor Experience and Resource Protection 519
Vjacheslava Pantjukhina, Georgia 525
Vjetrenica, Bosnia-Herzegovina 141, 289
   biospeleology **759–760**
   conservation 760
   fauna, distribution 760
   geomorphology 759
   hydrology 759
Vodyanaya cave, Ukraine 253
Vogelherd caves, Germany 333
void, creation of 357
volatile organic compounds 404
Volcanic Caribees 190
volcanic caves 204, 282–283, **760–764**
volcanic islands, origin of species 236
Volcano Room, Q5, Vancouver Island, Canada 436
voltaite 513
Voronja Cave, Georgai 201, **487–488**
Vorontsovsky Massif, Georgia 201
Vrelo Krupac, Serbia 330
Vrgorac Lake, Dalmatia 291
Vritiglavica cave, Slovenia 204, 468, 526
vulcanospeleology, history of **764–766**

Wad, El cave, Israel 195
Wad, El terrace, Israel 196
Wadi el-Mughara, Israel 195, 196
Wadley, Lyn 19
Wa-hul Cave, South Korea 212
Waiahuakua Cave, Hawaii, United States 491
Waialoha Cave, Hawaii, United States 491
Waianapanapa Caves, Hawaii, United States 415
Waikoropupu Spring, New Zealand 542
Waipu, North Island New Zealand 541
Waitomo Glowworm Cave, New Zealand 541–542, 620, 731
Waitomo Museum of Caves, Australia 728
Waiwaipuhi Cave, Hawaii, United States 491
Wakulla Springs, Florida, United States 52, 302, 303
Waldack, Charles 578
Wallace's line 110
Walsh River, Australia 216
Walsingham Cave System, Bermuda **767–768**
   anchialine habitat 767
   caves of 767–768
   fauna of 141, 767, 768
   geomorphology 767
   impact of human activity on 768
   speleogenesis 767
   threats to 768
Waltham, A.C. 673
Wandjina cult, Australia 364
Wanhuyan Cave **220**
Wankel, Heinreich 336
Wargata Mina, Australia 122, 124, 129, 623
Warm Spring Hill *See* Tangshan, China.
warm traps in caves **229**
Warren, Robert Penn (*The Cave*) 206

Wasden/Owl Cave, Idaho United States 43
Washington Convention on International Trade in Endangered Species of Wild Flora and Fauna 243–244
water
   abstraction, control of 248
   accessibility of 481, 482
   aquifers, paleokarst 560
   allogenic recharge 476
   autogenic recharge 476, 477
   chemistry 213–215, 476–477
   doline hydrology 309
   drinking water standards 403, 404
   flow
      aquifers, principle pathways of 399, 400
      laminar 187
      modelling 671, 673
      regulation of 362
      turbulent 187
   groundwater *See* groundwater
   hardness, testing 214
   resources **481–483**
   supply, for human use 481–482
   table 395, 680–681
   tracing **769–773**
      history of 478
water bears *See* Tardigrada.
water fleas *See* Cladocera.
Watling's Blue Hole, San Salvador Island, Bahamas 155
Watson, Patty Jo 81
wave action/quarrying as coastal process 231
Weathercote Cave, England 87
Weebubbie Cave, Nullarbor Plain, Australia 120, 545
Weidenreich, Frans 222
Weiner, Steve 222
Weller, J.M. 51
Wellington Cave, New South Wales, Australia 124, 127, 128, 547
wells, drilled and karst water supply 482
Westall, William 87
Westbury Quarry, England 162, 562
Western Cordillera, Canada 175, 177–178
Western Interior platform, composition of 175–176
Western Tasmanian Wilderness World Heritage Area, Australia 122, 124
West Indies *See* Caribbean Islands.
Westmoreland Cave, Tasmania 633
Wet Caves, Australia 129
wetlands, conservation 620
Wetlands of International Importance 619
wetting and drying as coastal process 231
Weyer's Cave Station (grand Caverns), Virginia, United States 50, 534, 704
Weyl, P.K. 673
Whalan, Charles 128
Whalan, James 127
Whalemouth Cave, Australia 123, 652
Wheeler, A.O. 51
White, William B. 52, 605, 715
White Desert, Egypt 16
White Peak, Peak District, England 575, 576
widemouth blindcat *See* Satan eurostymous.
Wielka-Snizena Cave System, Poland 331
Wilder, Billy (*Ace In The Hole*) 360
wilderness **773–774**

Wildlife and Countryside Act, United Kingdom and limestone 550
William Pengelly Cave Studies Centre and Trust Fund, England 166
Williams, P.W. 526
Wilson, Jeremiah 128, 726
Wind Cave National Park, South Dakota, United States
   dating of 57
   geomorphology 775, 776
   history of exploration 52
   length/depth of 525, 750
   minerals in 776
   speleogenesis 775
Windsor Great Cave, Jamaica 234, 235
Windward Islands 190
windypits *See* crevice caves.
Wiri Cave, New Zealand 606
Wit Tamdoun, Morocco 13, 14
WKPP *See* Woodville Karst Plain Project (WKPP).
Wolf, Caspar 87
wolframite 514
Wolpoff, Milford 19
Wonambi Fossil Centre, Australia 728
Wonderhole, Cinhoyi Caves, Zimbabwe 21
Wonderland Cave, Arizona, United States 534
Wonderwerk Cave, South Africa 19
Woo, John (*Broken Arrow*) 360
Wood Buffalo National Park, Canada 52, 176
woodlice *See* Oniscidea
Wood Valley Pit Crater, Hawaii, United States 251
Woodville Karst Plain Project (WKPP) 302
Woodville Karst Plain system, Florida, United States 52
Wookey Hole, England
   archaeology in 79, 162, 562
   diving in 166, 302
   exploration 166
   speleogenesis 503, 504
World Conservation Union Red List of Threatened Species 63
world depth record 528
World Heritage Convention 246
World Heritage Sites **777–778**
   Aggtelek and Slovak Karst 25
   Alejandro de Humbolt National Park, Cuba 778
   Altamira Cave, Spain **33–34**
   Atapuerca caves, Spain 119
   Australian Fossil Mammal Sites World Heritage Area, Australia 124
   Baradla-Domica Cave System, Hungary/Slovakia **26–27**, 29, 331, 620, 698
   Canaima National Park, Venezuela 612, 778
   Carlsbad Caverns National Park, New Mexico, United States 51, 192
   Chichen-Itza, Mexico 37, 41
   Cockpit Country, Jamaica **233–235**
   criteria for inclusion **777–778**
   Desembarco del Granma National Park, Cuba 271
   Dobšina Ice Cave, Hungary/Slovakia 25, 27
   Draenen, Ogof, Wales **310–312**
   Durmitor National Park, Yugoslavia 778
   East Rennell, Solomon Islands 778
   Grand Canyon National Park, AZ United States **391–393**
   Greater Blue Mountains, Australia 121
   Gunung Mulu National Park, Sarawak, Malaysia 531
   Ha Long Bay, Vietnam **413–414**
   Hawaii Volcanoes National Park, United States 415
   Henderson Island, England 778
   history of **777**
   Huanglong, China **423–424**
   karst sites, list of **778**
   Jiuzhaigou Valley, China **423–424**
   Kromdraai, South Africa 17, 19
   Lord Howe Island, Australia 778
   Lorentz National Park, Indonesia 778
   Luang Prabang, Lao PDR 778
   Mammoth Cave, Kentucky, United States 495
   Nahanni National Park, Canada 778
   National Park of Cainama, Venezuela 612
   Pamukkale, Turkey 569
   Pirin National Park, Bulgaria 330
   Plitvice Lakes National Park, Croatia **597–598**
   process for inclusion **778**
   Puerto Princesa Subterranean River, Philippines 105
   Pyrennees-Mont Perdu, France 778
   Riversleigh and Naracoorte fossil mammal sites, Australia 778
   Rocky Mountain Parks, Canada 778
   Sian Ka'an, Mexico 778
   Škocjanske jama, Slovenia **653–655**
   Södra Ölands Odlingslandskap, Sweden 778
   Sterkfontein, Swartkrans, Kromdraai fossil homind sites, South Africa 17, 18
   Swartkrans, South Africa 17, 18
   Tasmanian Wilderness 122, 778
   Te Wahipounamu, New Zealand **543**
   Thungyai-Huai Kha Khaeng, Thailand 778
   Tsingy de Bemaraha, Madagascar **494–495**
   Vézère, France **757**
   Victoria Fossil Cave, Australia 121
   Viñales Valley, Cuba 271, 778
   Western Caucasus, Russia 778
   Wulingyuan, China 778
   Zhoukoudian, China **221–222**
   See also individual sites by name.
Worm, Olaus 690
Wormaldia subterranea 453
Wulingyuan, China 778
Wright, Charles 49
Wrightian adaptive shift model of speciation 665
Wujia Dong, China 218
Wujiangdu dam, China 278
Wu Rukang **222**
Wyandotte Cave, Indiana, United States 44, 81
Wyman, Jeffries **152**
Wysoka-Za-Siedmiu Progami Cave, Poland 331
Xanadu Cave System, Tennessee, United States 70
Xé Bang Fai Cave, Laos 102, 483, 485
Xeniaria esau 452
Xiaochai Tiankeng doline, China 286–287
Xiaoyanwan doline, China 219
Xingwen dolines, China 219
Xibalba, Guatemala 37, 364
Xochicalco, Mexico 42
Xu Xiake 98–99

Yadana gas field, Thailand 102, 104
Yamata, Bulgaria 329
Yangshuo karst, China **782–783**
Yanzi Dong, China 218
Yashchik Pandory, Siberia 645

Yedi Miyarlar, Turkey 699
Yellowstone National Park 245
Yemen
    geomorphology 115
    Giniba Cave 115
    Hoq Cave 115
Yi-Jing Jie **151**
Yorkshire Dales, England 342
    Ease Gill Cave System 166, 784, 786
    Gaping Gill 166, 522, 616, 783
    geomorphology **783–785**
    inception horizons of 439
    Ingleborough Hill 783–784
    Ingleborough Cave 715, 784
    Ireby Fell 441
    Jordas (Yordas) Cave 87
    Leck Fell 441
    Keld Head 302, 438
    King Pot 786
    Kingsdale Cave System 785, 786
    Knock Fell Caverns 682–683
    Lancaster Pot 166
    Langcliffe Cave 786
    Leck Fell 441
    Lost John's Master Cave 786
    Malham Cove 438, 784
    Malham Cove Rising 303
    Meregill Hole 2
    Mossdale Cave 2, 786
    Notts Pot 441, 786
    Pippikin Pot 441, 784
    Swinsto Hole 785, 786
    Victoria Cave 162
Yorkshire Dales National Park, England 786
Yorkshire Pot, Canada 52, 177
Yorkshire Rambler's Club, England 166, 351
Yorkshire Speleological Association 166
Yosemite Falls Indian Cave, California, United States 722
Young, Bennett 81
yttrium 514
Yucatán Phreas, Mexico **786–788**
    caves of 787–788
    cenotes 786
    Chichen-Itza 37, 41
    diving 303
    geomorphology 786–787
    human impacts on 788
    Maya and 788
    Nohoch Nah Chich system 303
    Ox Bel Ha system 303
    Ramsar sites 620–621
    Sac Actun 787
    Sistema Aak Kimin 787
    Sistema Abejas 787
    Sistema Actun Ha 787
    Sistema Aerolito 787
    Sistema Chikin Ha 787
    Sistema Esmerelda 787
    Sistema Naranjal 787
    Sistema Nohoch Kiin 787
    Sistema Nohoch Nah Chich 786–787, 788
    Sistema Ponderosa 787
    Sistema Yaxchen 787
    speleogenesis 787
    stygobitic fauna 39
    troglobitic fauna 39
Yugoslavia
    bauxite production in 135
    Bileca reservoir 277
    Buško Blato reservoir 277
    Sinjac (Piva) 699
    Sozina railway tunnel 738
Yukon Territory, Canada 178
Yün Lin Shih Pu (Tu) 98
Yunnan, China 218
Yvert et Tellier's Catalogues 704

Zabrašovké Aragonite Caves, Europe, Central 332, 560
Zacatón, Mexico 303, 391
Zagedanskaya Massif, Georgia 201
Zagedon Massif, Georgia 201
Zajdmana Cave, Turansky Plain Central Asia 95
Zalomka River, Herzegovina 293
Zambia
    Lusaka 22
Zans, V.A. 235
Zebu Well Cave, Madagascar 494
Zhang Dian 99
Zheleznye Vorota (Iron Gates), Pinega gypsum caves, Russia 588
Zhenevskaja, Siberia 629
Zhoukoudian caves, China 79, 220
Zhoukoudian International Paleoanthropological Research Center **222**
Zhoukoudian World Heritage site **221–222**
Zhucaojing Cave, China 219
Zimbabwe
    Big End Chasm 21
    Bounding Pot 21
    Jungle Pot 21
    Mwenga Mwena 20, 21
    Wonderhole, Cinhoyi Caves 21
zinc-alumino-silicates 273
zinc-lead ore 560
zirconium 514
Zitta, Mariolina 534
Zlaté Hory, Czechoslovakia 698
Zmeinaya Cave, Ukraine 253
Zolotoj Kljuchik, Russia 345
Zoloushka, Moldovia/Ukraine 329, 330, 345, 525, 746, 748
zonation model of colonization 235
Zospeum spp. 516
Zygentoma
    distribution 446–447
    morphology 446
    taxonomy 445
Zygomycetes 508